MyMathLab® = Your Resource for Success

In the lab, at home,...

- Access videos, PowerPoint® slides, and animations.
- Complete assigned homework and quizzes.
- Learn from your own personalized Study Plan.
- Print out MyMathGuide for additional practice.
- Explore even more tools for success.

...and on the go.

 Download the free MyDashBoard App to see instructor announcements and check your results on your Apple or Android™ device. MyMathLab log-in is required.

 Download the free Pearson eText App to access the full eText on your Apple® or Android™ device. MyMathLab log-in is required.

 Use your Chapter Test as a study tool! Chapter Test Prep Videos show step-by-step solutions to all Chapter Test exercises. Access these videos in MyMathLab or by scanning the code.

Don't Miss Out! Log In Today.

MyMathLab delivers proven results in helping individual students succeed. It provides engaging experiences that personalize, stimulate, and measure learning for each student. And, it comes from a trusted partner with educational expertise and an eye on the future.

To learn more about how MyMathLab combines proven learning applications with powerful assessment, visit **www.mymathlab.com**

VIDEOS • POWERPOINT SLIDES • ANIMATIONS • HOMEWORK AND QUIZZES • PERSONALIZED STUDY PLAN • TOOLS FOR SUCCESS

NEW! Bittinger Integrated Video and *MyMathGuide* Program

The Bittinger video program and *MyMathGuide* are designed to work hand in hand. Each objective-based video corresponds exactly to the *MyMathGuide* workbook, and vice versa, so students can use both to achieve the conceptual understanding they need to succeed.

Video Program

- **NEW! To-the-Point Objective Videos**
  - **Concise:** Videos get right to the core of each concept.
  - **Objective based:** Simple navigation lets users view a whole section, choose an objective, or go straight to an example.
  - **Interactive: Your Turn Video Checks** let students pause to work problems and check answers.
  - **Integrated:** Designed for use with *MyMathGuide: Notes, Practice, and Video Path.*

- **Chapter Test Prep Videos**
  - **Step-by-step solutions** for every problem in the Chapter Tests
  - Available in MyMathLab,® on YouTube,™ and by scanning the QR code on the Chapter Test pages in the textbook

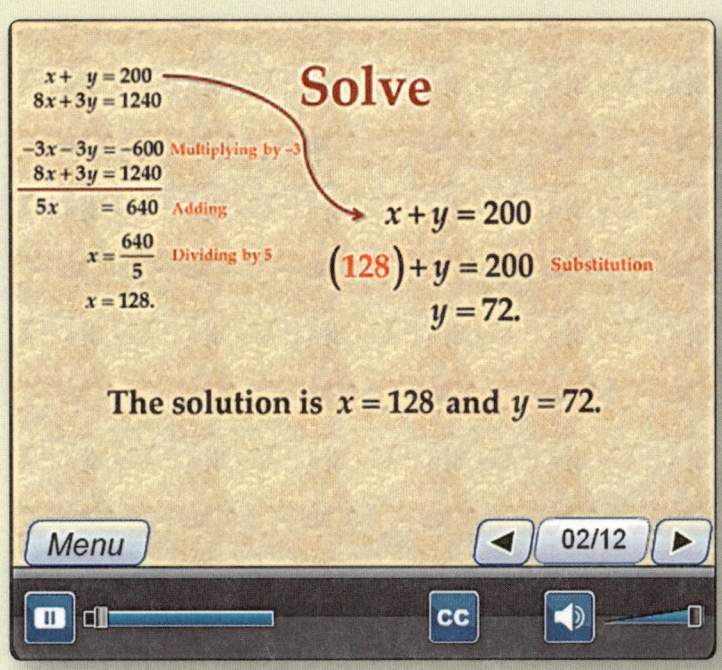

Solve

$$x + y = 200$$
$$8x + 3y = 1240$$

$$-3x - 3y = -600 \quad \text{Multiplying by } -3$$
$$8x + 3y = 1240$$

$$5x \quad = 640 \quad \text{Adding}$$

$$x = \frac{640}{5} \quad \text{Dividing by 5}$$

$$x = 128.$$

$$x + y = 200$$
$$(128) + y = 200 \quad \text{Substitution}$$
$$y = 72.$$

The solution is $x = 128$ and $y = 72$.

Menu ◀ 02/12 ▶

⏸ CC 🔊

NEW! *MyMathGuide: Notes, Practice, and Video Path*

- **Objective-based, hands-on, guided learning:** students follow a textbook, instructor, or video path.
- Notes on **key concepts, skills, and definitions** for each learning objective
- **Vocabulary** practice and review
- **Examples** with guided solutions and Your Turn practice exercises
- **Space** to write questions and notes and to show work
- **Additional Practice Exercises** with Readiness Checks
- Designed to correlate exactly with the new **To-the-Point Objective Videos**

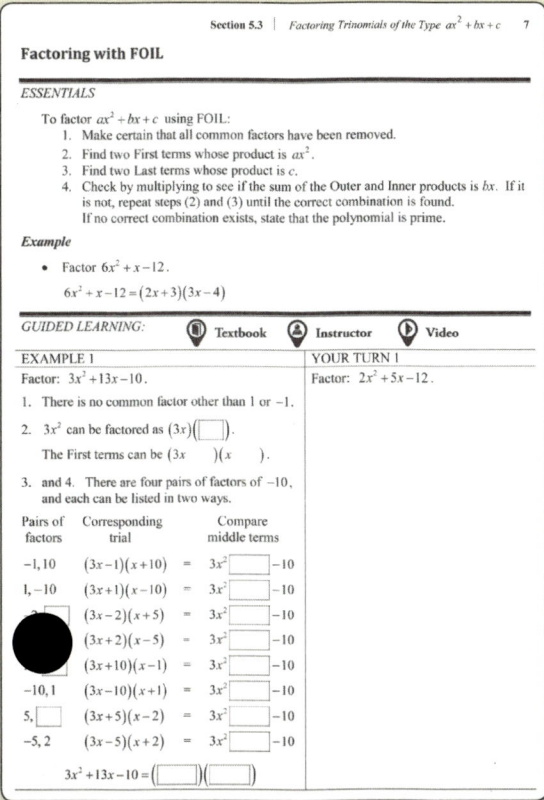

EDITION 6

Elementary and Intermediate Algebra
Concepts and Applications

Marvin L. Bittinger
Indiana University Purdue University Indianapolis

David J. Ellenbogen
Community College of Vermont

Barbara L. Johnson
Indiana University Purdue University Indianapolis

PEARSON

Boston Columbus Indianapolis New York San Francisco Upper Saddle River
Amsterdam Cape Town Dubai London Madrid Milan Munich Paris Montréal Toronto
Delhi Mexico City São Paulo Sydney Hong Kong Seoul Singapore Taipei Tokyo

Editorial Director	Christine Hoag
Editor in Chief	Maureen O'Connor
Executive Editor	Cathy Cantin
Executive Content Editor	Kari Heen
Content Editor	Katherine Minton
Editorial Assistant	Kerianne Okie
Senior Managing Editor	Karen Wernholm
Senior Production Supervisor	Ron Hampton
Senior Author Support/ Technology Specialist	Joe Vetere
Composition	PreMediaGlobal
Production and Editorial Services	Martha K. Morong/Quadrata, Inc.
Art Editor and Photo Researcher	Geri Davis/The Davis Group, Inc.
Rights and Permissions Advisor	Sarah Smith/Creative Compliance
Associate Media Producer	Jonathan Wooding
Content Development Manager	Rebecca Williams (MXL)
Senior Content Developer	Mary Durnwald (TestGen)
Marketing Manager	Rachel Ross
Procurement Manager	Evelyn Beaton
Procurement Specialist	Debbie Rossi
Associate Design Director, USHE North and West	Andrea Nix
Senior Media Procurement Specialist	Ginny Michaud
Text Designer	Geri Davis/The Davis Group, Inc.
Cover Designer	Barbara T. Atkinson
Cover Photograph	Alaska. Denali. Mt. Carpe and Mt. McKinley (North Peak) reflecting in mountain tarn at McGonagall Pass. © Matthias Breiter/AStock/Corbis

Photo Credits
Photo credits appear on page 996.

Library of Congress Cataloging-in-Publication Data
Bittinger, Marvin L.
Elementary and intermediate algebra : concepts and applications.
  Marvin L. Bittinger, David J. Ellenbogen, Barbara L. Johnson—6th ed.
    p. cm.
  Includes indexes.

1. Algebra—Textbooks.   I. Ellenbogen, David II. Johnson, Barbara L. (Barbara Loreen) , 1962- III. Title.

  QA152.3.B546 2014

  512.9--dc23   2012046374

6   16

PEARSON

www.pearsonhighered.com

ISBN-13: 978-0-321-84874-1
ISBN-10: 0-321-84874-8

Contents

A KEY TO THE ICONS IN THE EXERCISE SETS

↳ Concept reinforcement exercises, indicated by blue exercise numbers, provide basic practice with the new concepts and vocabulary.

Aha! Exercises labeled *Aha!* indicate the first time that a new insight can greatly simplify a problem and help students be alert to using that insight on following exercises. They are not more difficult.

▦ Calculator exercises are designed to be worked using either a scientific calculator or a graphing calculator.

▱ Graphing calculator exercises are designed to be worked using a graphing calculator and often provide practice for concepts discussed in the Technology Connections.

▤ Writing exercises are designed to be answered using one or more complete sentences.

✔ A check mark in the annotated instructor's edition indicates Synthesis exercises that the authors consider particularly beneficial for students.

Preface

Welcome to the sixth edition of *Elementary and Intermediate Algebra: Concepts and Applications!* As always, our goal is to present content that is easy to understand yet of enough depth to allow success in future courses. You will recognize many proven features, applications, and explanations; you will also find new material developed as a result of our experience in the classroom as well as from insights from faculty and students throughout North America.

This text is intended for students with a firm background in *arithmetic.* It is one of three texts in an algebra series that also includes *Elementary Algebra: Concepts and Applications,* Ninth Edition, and *Intermediate Algebra: Concepts and Applications,* Ninth Edition.

Approach

Our goal is to help today's students learn and retain mathematical concepts. To achieve this, we feel that we must prepare developmental-mathematics students for the transition from "skills-oriented" elementary algebra courses to more "concept-oriented" college-level mathematics courses. This requires the development of critical thinking skills: to reason mathematically, to communicate mathematically, and to identify and solve mathematical problems. Following are aspects of our approach that we use to help meet the challenges we all face when teaching developmental mathematics.

Problem Solving

We use problem solving and applications to motivate the students wherever possible, and we include real-life applications and problem-solving techniques throughout the text. Problem solving encourages students to think about how mathematics can be used, and it helps to prepare them for more advanced material in future courses.

In Chapter 2, we introduce our five-step process for solving problems: (1) Familiarize, (2) Translate, (3) Carry out, (4) Check, and (5) State the answer. Repeated use of this problem-solving strategy throughout the text provides students with a starting point for any type of problem they encounter, and frees them to focus on the unique aspects of the particular problem. We often use estimation and carefully checked guesses to help with the *Familiarize* and *Check* steps (see pp. 113 and 422).

What's New in the Ninth Edition?

In addition to the following new features and other changes in this edition, including a new design, we have rewritten many key topics in response to user and reviewer feedback and have made significant improvements in design, art, pedagogy, and an expanded supplements package. Detailed information about the content changes outlined on p. ix is available in the form of a conversion guide. Please ask your Pearson representative for more information.

Applications

Interesting applications of mathematics help motivate students and instructors. In the new edition, we have updated real-world data problems and examples to include subjects such as renewable energy (p. 109), graduate-school acceptance (pp. 140 and 611), and zipline rides (p. 676). For a complete list of applications and the page numbers on which they can be found, please refer to the Index of Applications at the back of the book.

Pedagogy

New!

Your Turn Exercises. After every example, students are directed to work a similar exercise. This provides immediate reinforcement of concepts and skills. Answers to these exercises appear at the end of each exercise set. (See pp. 89 and 343.)

New!

Exploring the Concept. Appearing once in almost every chapter, this feature encourages students to think about or visualize a concept. These activities lead into Active Learning Figures available in MyMathLab. Students can manipulate ALFs to further explore and solidify their understanding of the concepts. (See pp. 194, 349, and 674.)

Revised!

Connecting the Concepts. Appearing at least once in every chapter, this feature summarizes concepts from several sections or chapters and illustrates connections between them. It includes a set of mixed exercises to help students make these connections. (See pp. 217, 350, and 658.)

Algebraic–Graphical Connections.   Appearing occasionally throughout the text, this feature draws attention to visualizations of algebraic concepts. (See pp. 506 and 698.)

Technology Connections.   These optional features in each chapter help students use a graphing calculator or a graphing calculator app to visualize concepts. Exercises are included with many of these features, and additional exercises in many exercise sets are marked with a graphing calculator icon to indicate more practice with this optional use of technology. (See pp. 176, 180, and 666.)

Student Notes.   Comments in the margin are addressed directly to students in a conversational tone. Ranging from suggestions for avoiding common mistakes to how to best read new notation, they give students extra explanation of the mathematics appearing on that page. (See pp. 106, 204, and 322.)

Study Skills.   We offer one study tip in every section. Ranging from time management to test preparation, these suggestions for successful study habits apply to any college course and any level of student. (See pp. 154, 287, and 312.)

New!   **Mid-Chapter Review.**   In the middle of every chapter, we offer a brief summary of the concepts covered in the first part of the chapter, two guided solutions to help students work step-by-step through solutions, and a set of mixed review exercises. (See pp. 103, 202, and 335.)

Exercise Sets

New!   **Vocabulary and Reading Check.**   These exercises begin every exercise set and are designed to encourage the student to read the section. Students who can complete these exercises should be prepared to work the remaining exercises in the exercise set. (See pp. 108, 259, and 351.)

Concept Reinforcement Exercises.   These true/false, matching, and fill-in-the-blank exercises appearing near the beginning of many exercise sets build students' confidence and comprehension. Answers to all concept reinforcement exercises appear in the answer section at the back of the book. (See pp. 139, 159, and 208.)

Aha! Exercises.   These exercises are not more difficult than their neighboring exercises; in fact, they can be solved more quickly, without lengthy computation, if the student has the proper insight. Designed to reward students who "look before they leap," the icon indicates the first time a new insight applies, and then it is up to the student to determine when to use that insight on subsequent exercises. (See pp. 236, 401, and 647.)

Revised!   **Skill Review Exercises.**   These exercises appear in every section beginning with Section 1.2. Taken together, each chapter's Skill Review exercises review all the major concepts covered in previous chapters in the text. Often these exercises focus on a single topic, such as solving equations, from multiple perspectives. (See pp. 386, 451, and 509.)

Synthesis Exercises.   Synthesis exercises appear in each exercise set following the Skill Review exercises. Students will need to use skills and concepts from earlier sections to solve these problems, and this will help them develop deeper insights into the current topic. The synthesis exercises are a real strength of the text, and in the annotated instructor's edition, the authors have placed a ✔ next to selected synthesis exercises that they suggest instructors "check out" and consider assigning. These exercises are not more difficult than the others, but they do use previously learned concepts and skills in ways that may be especially beneficial to students as they prepare for future topics. (See pp. 162, 201, and 394.)

Writing Exercises.   Two basic writing exercises appear just before the Skill Review exercises, and at least two more challenging exercises appear in the Synthesis exercises. Writing exercises aid student comprehension by requiring students to use critical thinking to explain concepts in one or more complete sentences. Because correct answers may vary, the only writing exercises for which answers appear at the back of the text are those in the chapter Review Exercises. (See pp. 189, 245, and 756.)

New!   **Quick Quiz.**   Beginning with the second section in each chapter, a five-question Quick Quiz appears near the end of each exercise set. Containing questions from sections already covered in the chapter, these quizzes provide a short but consistent review of the material in the chapter and help students prepare for a chapter test. (See pp. 363 and 531.)

New!   **Prepare to Move On.**   Beginning with Chapter 2, this short set of exercises appears at the end of every exercise set, and reviews concepts and skills previously covered in the text that will be used in the next section of the text. (See pp. 112 and 318.)

End of Chapter

Chapter Resources.   These learning resources appear at the end of each chapter, making them easy to integrate into lessons at the most appropriate time. The mathematics necessary to use the resource has been presented by the end of the section indicated with each resource.

- **Translating for Success** and **Visualizing for Success.** These matching exercises help students learn to translate word problems to mathematical language and to graph equations and inequalities. (See pp. 221 and 364.)
- **Collaborative Activity.** Students who work in groups generally outperform those who do not, so these optional activities direct them to explore mathematics together. Additional collaborative activities and suggestions for directing collaborative learning appear in the *Instructor's Resources Manual with Tests and Mini Lectures.* (See pp. 222, 365, and 566.)

New!
- **Decision Making: Connection.** Although many applications throughout the text involve decision-making situations, this feature specifically applies the math of each chapter to a context in which students may be involved in decision making. (See pp. 294, 433, and 492.)

Revised! **Study Summary.**   At the end of each chapter, this synopsis gives students a fast and effective review of key chapter terms and concepts. Concepts are paired with worked-out examples and practice exercises. (See pp. 146 and 567.)

Chapter Review and Test.   A thorough chapter review and a practice test help prepare students for a test covering the concepts presented in each chapter. (See pp. 369 and 693.)

New! **Quick Response (QR) Codes.**   These have been added to each chapter test, allowing students to link directly to chapter test prep videos with step-by-step solutions to all chapter test exercises. This effective tool allows students to receive help studying exactly when they need it at point of use.

Cumulative Review.   This review appears after every chapter beginning with Chapter 2 to help students retain and apply their knowledge from previous chapters. (See pp. 228 and 440.)

New Design

While incorporating a new layout, a fresh palette of colors, and new features, we retain an open look and a typeface that is easy to read. In addition, we continue to pay close attention to the pedagogical use of color to ensure that it is used to present concepts in the clearest possible manner.

Content Changes

The exposition, examples, and exercises have been carefully reviewed and, as appropriate, revised or replaced.

An increased focus on connections and concepts is made in Exploring the Concept, Connecting the Concepts, and Decision Making: Connection features. This focus is invaluable for student comprehension. In addition, concept and skill review has been expanded in Quick Quiz, Mid-Chapter Review, Study Summary, Skill Review, and Prepare to Move On features, which are described in more detail in the preceding Pedagogy section.

Other content changes include the following.

- Examples and exercises that use real data are updated or replaced with current applications.
- Chapter 1 now discusses the use of prime factorizations to find the LCM of two numbers. Learning this method will prepare students to find the LCM of two polynomials.
- Chapter 2 includes an introduction to motion problems as well as an explanation of percent increase and percent decrease.
- Section 3.7 now provides a comprehensive discussion of finding equations of lines.
- The explanation of simplifying powers of i in Section 10.8 is rewritten to provide students with a better understanding of the concept.
- Composition of functions in Section 12.1 is explained in greater detail to help students move from a concrete example to function notation.
- The optional connection to technology has been revised to include new calculator operating systems and graphing calculator applications for mobile devices.

ANCILLARIES

The following ancillaries are available to help both instructors and students use this text more effectively.

Supplements

NEW! Integrated Bittinger Video Program and MyMathGuide

Bittinger Video Program, available in MyMathLab includes closed captioning and the following video types:

NEW! To-the-Point Objective Videos
- Concise, interactive, and objective based.
- View a whole section, choose an objective, or go straight to an example.
- Interactive Your Turn Video Check pauses for students to work problems and check answers.
- Seamlessly integrated with *MyMathGuide: Notes, Practice, and Video Path*.

Chapter Test Prep Videos
- Step-by-step solution for every problem in the Chapter Tests.
- Also available on YouTube, and by scanning the QR code on the Chapter Test pages in the text.

NEW! MyMathGuide: Notes, Practice, and Video Path
- Guided, hands-on learning for traditional, lab-based, hybrid, and online courses.
- Designed to correlate with *To-the-Point Objective Videos*.
- Highlights key concepts, skills, and definitions for each learning objective.
- Quick Review of key vocabulary terms and vocabulary practice problems.
- Examples with guided solutions and similar Your Turn exercises for practice.
- Space to write questions and notes and to show work.
- Additional Practice Exercises with Readiness Checks.
- Available in MyMathLab and printed.

ISBNs: 0-321-84876-4/978-0-321-84876-5

Student's Solutions Manual
by Christine Verity
- Contains completely worked-out solutions with step-by-step annotations for all the odd-numbered exercises in the text, with the exception of the writing exercises. Also contains all solutions to Chapter Review, Chapter Test, and Connecting the Concepts exercises.
- Available in MyMathLab and printed.

ISBNs: 0-321-84877-2/978-0-321-84877-2

Annotated Instructor's Edition
- Provides answers to all text exercises in color next to the corresponding problems.
- Includes Teaching Tips.
- Icons identify writing and graphing calculator exercises.

ISBNs: 0-321-84881-0/978-0-321-84881-9

Instructor's Solutions Manual (download only)
by Christine Verity
- Contains fully worked-out solutions to the odd-numbered exercises and brief solutions to the even-numbered exercises in the exercise sets.
- Available for download at www.pearsonhighered.com or in MyMathLab.

Instructor's Resource Manual (download only) with Tests and Mini Lectures
by Laurie Hurley
- Includes resources designed to help both new and adjunct faculty with course preparation and classroom management.
- Offers helpful teaching tips correlated to the sections of the text.
- Contains two multiple-choice tests per chapter, six free-response tests per chapter, and eight final exams.
- Available for download at www.pearsonhighered.com or in MyMathLab.

PowerPoint® Lecture Slides (download only)
- Present key concepts and definitions from the text.
- Available for download at www.pearsonhighered.com or in MyMathLab.

TestGen®

TestGen® (www.pearsoned.com/testgen) enables instructors to build, edit, print, and administer tests using a computerized bank of questions developed to cover all the objectives of the text. TestGen is algorithmically based, allowing instructors to create multiple equivalent versions of the same question or test with the click of a button. Instructors can also modify test bank questions or add new questions. The software and testbank are available for download from Pearson Education's online catalog.

AVAILABLE FOR STUDENTS AND INSTRUCTORS

MyMathLab® Online Course (access code required)

The Bittinger courses include all of MyMathLab's robust features, plus these additional highlights:

New! Two MyMathLab course options are now available:

- **Standard MyMathLab courses** allow instructors to build their course their way, offering maximum flexibility and control over all aspects of assignment creation.

New!
- **Ready-to-Go courses** provide students with all the same great MyMathLab features, but make it easier for instructors to get started.

Both Bittinger course options include one pre-made pre-test and post-test for every chapter, section-lecture homework, and a chapter review quiz linked to personalized homework.

New! **Increased coverage** of skill-building, conceptual, and applications exercises, **plus new problem types,** provide even more flexibility when creating homework.

New! **Active Learning Figures**, available for key concepts, foster conceptual understanding for visual and tactile learners. Exploring the Concept in the text leads into ALFs available in MML. Students manipulate ALFs to explore concepts and solidify their understanding.

New! **Bittinger Video Program** includes all new To-the-Point Objective Videos and Chapter Test Prep Videos.

New! *MyMathGuide: Notes, Practice, and Video Path* can be viewed and printed.

MathXL® Online Course (access code required)

MathXL® is the homework and assessment engine that runs MyMathLab. (MyMathLab is MathXL plus a learning management system.)

With MathXL, instructors can

- Create, edit, and assign online homework and tests using algorithmically generated exercises correlated at the objective level to the textbook.
- Create and assign their own online exercises and import TestGen tests for added flexibility.
- Maintain records of all student work tracked in MathXL's online gradebook.

With MathXL, students can

- Take chapter tests in MathXL and receive personalized study plans and/or personalized homework assignments based on their test results.
- Use the study plan and/or the homework to link directly to tutorial exercises for the objectives they need to study.
- Access supplemental animations and video clips directly from selected exercises.

MathXL is available to qualified adopters. For more information, visit our website at www.mathxl.com, or contact your Pearson representative.

Acknowledgments

An outstanding team of professionals was involved in the production of this text. Judy Henn, Laurie Hurley, Helen Medley, Joanne Koratich, Monroe Street, and Holly Martinez carefully checked the book for accuracy and offered thoughtful suggestions. Michelle Lanosga and Daniel Johnson provided exceptional research support, and Christine Verity, Laurie Hurley, and Lisa Collette did remarkable work in preparing supplements. Special thanks are due Nelson Carter and Katherine Carter for their outstanding work on videos.

Martha Morong, of Quadrata, Inc., provided editorial and production services of the highest quality, and Geri Davis, of the Davis Group, Inc., performed superb work as designer, art editor, and photo researcher. Bill Melvin and Network Graphics provided the accurate and creative illustrations and graphs.

The team at Pearson deserves special thanks. Executive Editor Cathy Cantin, Executive Content Editor Kari Heen, Senior Production Supervisor Ron Hampton, Content Editor Katherine Minton, and Editorial Assistant Kerianne Okie provided many fine suggestions, coordinated tasks and schedules, and remained involved and accessible throughout the project. Marketing Manager Rachel Ross skillfully kept in touch with the needs of faculty. Editor in Chief Maureen O'Connor and Editorial Director Chris Hoag deserve credit for assembling this fine team.

We also thank the students at Indiana University Purdue University Indianapolis and the Community College of Vermont and the following professors for their thoughtful reviews and insightful comments.

Darla Aguilar, *Pima Community College–Desert Vista Campus*; Michael Anzzolin, *Waubonsee Community College*; Jan Archibald, *Ventura College*; Ruben Arenas, *East Los Angeles College*; Don Brown, *Macon State College*; Manuel Caramés, *Miami-Dade College–North Campus*; Gary Carpenter, *Pima Community College, Northwest Campus*; Tim Chappell, *Penn Valley Community College*; Phong Chau, *Glendale Community College*; Krista Cohlmia, *Odessa College*; Ola Disu, *Tarrant County College*; Theresa Evans, *Odessa College*; Anissa Florence, *University of Louisville*; Sandy Gordon, *Central Carolina Technical College*; Sharon Hamsa, *Longview Community College*; Doug Harley, *Del Mar College*; Cynthia Harris, *Triton College*; Geoffrey Hirsch, *Ohlone College*; Pat Horacek, *Pensacola Junior College*; Glenn Jablonski, *Triton College*; Joseph Kazimir, *East Los Angeles College*; Sally Keely, *Clark College*; Jennifer Kumi Burkett, *Triton College*; Ana Leon, *Louisville Community College*; Linda Lohman, *Jefferson Community and Technical College*; Bob Martin, *Tarrant County College*; Caroline Martinson, *Jefferson Community and Technical College*; Eric Moller, *Del Mar College*; Agashi Nwogbaga, *Wesley College*; Miriam Pack, *Glendale Community College*; Michelle Parsons, *San Diego Mesa College*; Amy Petty, *South Suburban College*; Anthony Precella, *Del Mar College*; Timothy Precella, *Del Mar College*; Thomas Pulver, *Waubonsee Community College*; Angela Redmon, *Wenatchee Valley College*; Richard Rupp, *Del Mar College*; Mehdi Sadatmousavi, *Pima Community College*; Ahmad Shafiee, *Del Mar College*; Jane Thompson, *Waubonsee Community College*; Ann Thrower, *Kilgore College*

Finally, a special thank-you to all those who so generously agreed to discuss their professional use of mathematics in our chapter openers. These dedicated people all share a desire to make math more meaningful to students. We cannot imagine a finer set of role models.

M.L.B.
D.J.E.
B.L.J.

Introduction to Algebraic Expressions

NUMBER OF SONGS PURCHASED	AMOUNT PAID
2	$1.98
3	2.97
5	4.95
10	9.90

The Media May Change, But the Music Remains.

Thomas Edison's phonograph, patented in 1877, was the first successful method of recording music. At first, music was recorded on cylinders. During the next century, the media evolved to records, then to tapes, and then to CDs. Today, most of the 100,000 albums released each year are available electronically, and online music sales are soaring, due in part to the ability to purchase single tracks. Can you use the data in the table shown to write an equation that *models* the cost of single-track downloads?
(See Example 7 in Section 1.1.)

It's true—even as a musician, I am not exempt from using math, because music is math.

Myra Flynn, a singer/songwriter from Randolph, Vermont, uses math in harmonies, time signatures, tuning systems, and all music theory. Putting an album out requires the use of even more math: calculating the number of hours worked in the studio, payments for producers and musicians, hard-copy and digital distribution regionally, and ticket and concert sales.

P roblem solving is the focus of this text. In Chapter 1, we lay the foundation for the problem-solving approach that is developed in Chapter 2 and used in all chapters that follow. This foundation includes a review of arithmetic, a discussion of real numbers and their properties, and an examination of how real numbers are added, subtracted, multiplied, divided, and raised to powers.

1.1   Introduction to Algebra

Evaluating Algebraic Expressions   •   Translating to Algebraic Expressions   •   Translating to Equations

This section introduces some basic concepts and expressions used in algebra. Our focus is on the wordings and expressions that often arise in real-world problems.

Evaluating Algebraic Expressions

Probably the greatest difference between arithmetic and algebra is the use of *variables*. Suppose that n represents the number of tickets sold in one day for a U2 concert and that each ticket costs $60. Then a total of 60 times n, or $60 \cdot n$, dollars will be collected for tickets.

The letter n is a **variable** because it can represent any one of a set of numbers.

The number 60 is a **constant** because it does not change.

The multiplication sign $\cdot$ is an **operation sign** because it indicates the **operation** of multiplication.

The expression $60 \cdot n$ is a **variable expression** because it contains a variable.

An **algebraic expression** consists of variables and/or numerals, often with operation signs and grouping symbols. Other examples of algebraic expressions are:

$t - 37$;   This contains the variable t, the constant 37, and the operation of subtraction.

$(s + t) \div 2$.   This contains the variables s and t, the constant 2, grouping symbols, and the operations of addition and division.

Multiplication can be written in several ways. For example, "60 times n" can be written as $60 \cdot n, 60 \times n, 60(n), 60 * n$, or simply (and usually) $60n$. Division can also be represented by a fraction bar: $\frac{9}{7}$, or $9/7$, means $9 \div 7$.

To **evaluate** an algebraic expression, we **substitute** a number for each variable in the expression and calculate the result. This result is called the **value** of the expression. The table below lists several values of the expression $60 \cdot n$.

Student Notes

Notation like "$60 \times n$" is not often used in algebra because the "$\times$" symbol can be misread as a variable.

Cost per Ticket (in dollars), 60	Number of Tickets Sold, n	Total Collected (in dollars), $60n$
60	150	9,000
60	200	12,000
60	250	15,000

Student Notes

At the end of each example in this text, you will see ↩ YOUR TURN. This directs you to try an exercise similar to the example. The answers to these exercises appear at the end of each exercise set.

1. Evaluate $m - n$ for $m = 100$ and $n = 64$.

2. Using the formula given in Example 2, find the area of a triangle when b is 30 in. (inches) and h is 10 in.

EXAMPLE 1   Evaluate each expression for the given values.

a) $x + y$ for $x = 37$ and $y = 28$

b) $5ab$ for $a = 2$ and $b = 3$

SOLUTION

a) We substitute 37 for x and 28 for y and carry out the addition:

$$x + y = (37) + (28) = 65.$$

Using parentheses when substituting is not always necessary but is never incorrect.

The value of the expression is 65.

b) We substitute 2 for a and 3 for b and multiply:

$$5ab = 5 \cdot 2 \cdot 3 = 10 \cdot 3 = 30.$$ $5ab$ means 5 times a times b.

↩ YOUR TURN

EXAMPLE 2   The area of a triangle with a base of length b and a height of length h is given by the formula $A = \frac{1}{2}bh$. Find the area when b is 8 m (meters) and h is 6.4 m.

SOLUTION   We substitute 8 m for b and 6.4 m for h and then multiply:

$$A = \frac{1}{2}bh$$
$$= \frac{1}{2}(8\text{ m})(6.4\text{ m})$$
$$= \frac{1}{2}(8)(6.4)(\text{m})(\text{m})$$
$$= 4(6.4)\text{ m}^2$$
$$= 25.6\text{ m}^2, \text{ or } 25.6 \text{ square meters}.$$

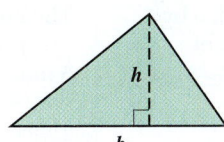

Note that we use square units for area and $(\text{m})(\text{m}) = \text{m}^2$.

↩ YOUR TURN

Translating to Algebraic Expressions

Before attempting to translate problems to equations, we need to be able to translate certain phrases to algebraic expressions. Any variable can be used to represent an unknown quantity; however, it is helpful to choose a descriptive letter. For example, w suggests weight and p suggests population or price. It is important to write down what each variable represents, as well as the unit in which it is measured.

Important Words	Sample Phrase or Sentence	Variable Definition	Translation
Addition (+)			
added to	700 lb was added to the car's weight.	Let w represent the car's weight, in pounds.	$w + 700$
sum of	The sum of a number and 12	Let n represent the number.	$n + 12$
plus	53 plus some number	Let x represent "some number."	$53 + x$
more than	8000 more than Detroit's population	Let p represent Detroit's population.	$p + 8000$
increased by	Alex's original guess, increased by 4	Let n represent Alex's original guess.	$n + 4$
Subtraction (−)			
subtracted from	2 g was subtracted from the weight.	Let w represent the weight, in grams.	$w - 2$
difference of	The difference of two scores	Let m represent the larger score and n represent the smaller score.	$m - n$
minus	A team of size s, minus 2 players	Let s represent the number of players.	$s - 2$
less than	50 lb less than the weight of the lumber	Let w represent the weight of the lumber, in pounds.	$w - 50$
decreased by	The car's speed, decreased by 8 mph	Let s represent the car's speed, in miles per hour.	$s - 8$
Multiplication (·)			
multiplied by	The number of guests, multiplied by 3	Let g represent the number of guests.	$g \cdot 3$
product of	The product of two numbers	Let m and n represent the numbers.	$m \cdot n$
times	5 times the dog's weight	Let w represent the dog's weight, in pounds.	$5w$
twice	Twice the wholesale cost	Let c represent the wholesale cost.	$2c$
of	$\frac{1}{2}$ of Rita's salary	Let s represent Rita's salary.	$\frac{1}{2}s$
Division (÷)			
divided by	A 2-lb coffee cake, divided by 3	*No variables are required for translation.*	$2 \div 3$
quotient of	The quotient of 14 and 7	*No variables are required for translation.*	$14 \div 7$
divided into	4 divided into the delivery fee	Let f represent the delivery fee.	$f \div 4$
ratio of	The ratio of $500 to the price of a new car	Let p represent the price of a new car, in dollars.	$500/p$
per	There were 18 students per teacher.	*No variables are required for translation.*	$18/1$

EXAMPLE 3   Translate each phrase to an algebraic expression.

a) Four less than Ava's height, in inches

b) Eighteen more than a number

c) A day's pay, in dollars, divided by eight

SOLUTION   To help think through a translation, we sometimes begin with a specific number in place of a variable.

a) If the height were 60, then 4 less than 60 would mean $60 - 4$. If we use h to represent "Ava's height, in inches," the translation of "Four less than Ava's height, in inches" is $h - 4$.

b) If we knew the number to be 10, the translation would be $10 + 18$, or $18 + 10$. If we use t to represent "a number," the translation of "Eighteen more than a number" is $t + 18$, or $18 + t$.

3. Translate to an algebraic expression: Twenty less than the number of students registered for the course.

c) We let d represent "a day's pay, in dollars." If the pay were $78, the translation would be $78 \div 8$, or $\frac{78}{8}$. Thus our translation of "A day's pay, in dollars, divided by eight" is $d \div 8$, or $d/8$.

YOUR TURN

> **CAUTION!** The order in which we subtract and divide affects the answer! Answering $4 - h$ or $8 \div d$ in Examples 3(a) and 3(c) is incorrect.

Student Notes

Try looking for "than" or "from" in a phrase and writing what follows it first. Then add or subtract the necessary quantity. (See Examples 4b and 4c.)

4. Translate to an algebraic expression: Half of the sum of two numbers.

EXAMPLE 4 Translate each phrase to an algebraic expression.

a) Half of some number

b) Seven pounds more than twice Owen's weight

c) Six less than the product of two numbers

d) Nine times the difference of a number and 10

e) Eighty-two percent of last year's enrollment

SOLUTION

Phrase	Variable(s)	Algebraic Expression
a) Half of some number	Let n represent the number.	$\frac{1}{2}n$, or $\frac{n}{2}$, or $n \div 2$
b) Seven pounds more than twice Owen's weight	Let w represent Owen's weight, in pounds.	$2w + 7$, or $7 + 2w$
c) Six less than the product of two numbers	Let m and n represent the numbers.	$mn - 6$
d) Nine times the difference of a number and 10	Let a represent the number.	$9(a - 10)$
e) Eighty-two percent of last year's enrollment	Let r represent last year's enrollment.	82% of r, or $0.82r$

YOUR TURN

Translating to Equations

The **equal** symbol, $=$, indicates that the expressions on either side of the equal sign represent the same number. An **equation** is a number sentence with the verb $=$.

It is important to be able to distinguish between expressions and equations. Compare the descriptions in the table below.

Expression	Equation
No $=$ sign appears.	An $=$ sign appears.
Compare to an English *phrase*, like "The interesting book."	Compare to an English *sentence*, like "The book was interesting."
May be of any length: x $3(x - 5) + 4y - 17(3 - y).$	May be of any length: $x = 7$ $3(x - 5) + 4y = 17(3 - y).$

Although we do not study equations until Chapter 2, we can translate certain problem situations to equations now. The words "is the same as," "equal," "is," "are," "was," and "were" often translate to "$=$."

> **Words indicating equality ($=$):** "is the same as," "equal," "is," "are," "was," "were," "represents"

When translating a problem to an equation, we first translate phrases to algebraic expressions, and then the entire statement to an equation containing those expressions.

EXAMPLE 5 Translate the following problem to an equation.

What number plus 478 is 1019?

SOLUTION We let y represent the unknown number. The translation then comes almost directly from the English sentence.

What number plus 478 is 1019?
$$y \qquad + \quad 478 \;=\; 1019$$

Note that "what number plus 478" translates to "$y + 478$" and "is" translates to "$=$."

YOUR TURN

5. Translate to an equation: What number times 12 is 672?

Sometimes it helps to reword a problem before translating.

EXAMPLE 6 Translate the following problem to an equation.

The Siduhe River Bridge in China is the world's highest bridge. At 1550 ft, it is 270 ft higher than the Baluarte Bridge in Mexico. How high is the Baluarte Bridge?

Source: www.highestbridges.com

SOLUTION We let h represent the height, in feet, of the Baluarte Bridge. A rewording and a translation are as follows:

Rewording: The height of the Siduhe River Bridge is 270 ft more than the height of the Baluarte Bridge.

Translating: $1550 \;=\; h + 270$

Baluarte Bridge, Mexico

6. Translate to an equation: Valley College has 13 science instructors. There are 5 more science instructors than math instructors. How many math instructors are there?

YOUR TURN

When we translate a problem into mathematical language, we say that we **model** the problem. A **mathematical model** is a mathematical representation of a real-world situation. Note that the word *model* can be used as either a verb or a noun.

Information about a problem is often given as a set of numbers, called **data.** Sometimes data follow a pattern that can be modeled using an equation.

EXAMPLE 7 *Music.* The table below lists the amount charged for several purchases from an online music store. We let a represent the amount charged, in dollars, and n the number of songs. Find an equation giving a in terms of n.

Number of Songs Purchased, n	Amount Charged, a
2	$1.98
3	2.97
5	4.95
10	9.90

SOLUTION To write an equation for a **in terms of** n means that a will be on one side of the equal sign and an expression involving n will be on the other side.

We look for a pattern in the data. Since the amount charged increases as the number of songs increases, we can try dividing the amount by the number of songs:

$$1.98/2 = 0.99; \qquad 4.95/5 = 0.99;$$
$$2.97/3 = 0.99; \qquad 9.90/10 = 0.99.$$

7. Suppose that an online music store charges $2.58 for 2 songs, $3.87 for 3 songs, and $12.90 for 10 songs. Using the same variables as in Example 7, find an equation giving a in terms of n.

The quotient is the same, 0.99, for each pair of numbers. Thus each song costs $0.99. We reword and translate as follows:

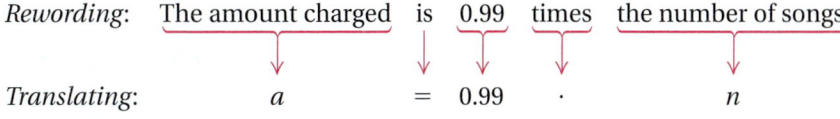

Rewording: The amount charged is 0.99 times the number of songs.

Translating: a = 0.99 · n

YOUR TURN

Technology Connection

Technology Connections are activities that make use of features that are common to most graphing calculators. In some cases, students may find the user's manual for their particular calculator helpful for exact keystrokes.

Although all graphing calculators are not the same, most share the following characteristics.

Screen. The large screen can show graphs and tables as well as the expressions entered. The screen has a different layout for different functions. Computations are performed in the **home screen**. On many calculators, the home screen is accessed by pressing 2ND QUIT. The **cursor** shows location on the screen, and the **contrast** (set by 2ND ⌃ or 2ND ⌄) determines how dark the characters appear.

Keypad. There are options written above the keys as well as on them. To access those above the keys, we press

2ND or ALPHA and then the key. Expressions are usually entered as they would appear in print. For example, to evaluate $3xy + x$ for $x = 65$ and $y = 92$, we press 3 × 65 × 92 + 65 and then ENTER. The value of the expression, 18005, will appear at the right of the screen.

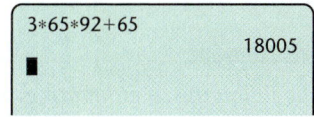

```
3*65*92+65
                      18005
▮
```

Evaluate each of the following.

1. $27a - 18b$, for $a = 136$ and $b = 13$
2. $19xy - 9x + 13y$, for $x = 87$ and $y = 29$

Study Skills

Get the Facts

Throughout this textbook, you will find a feature called Study Skills. These tips are intended to help improve your math study skills. On the first day of class, we recommend that you complete this chart.

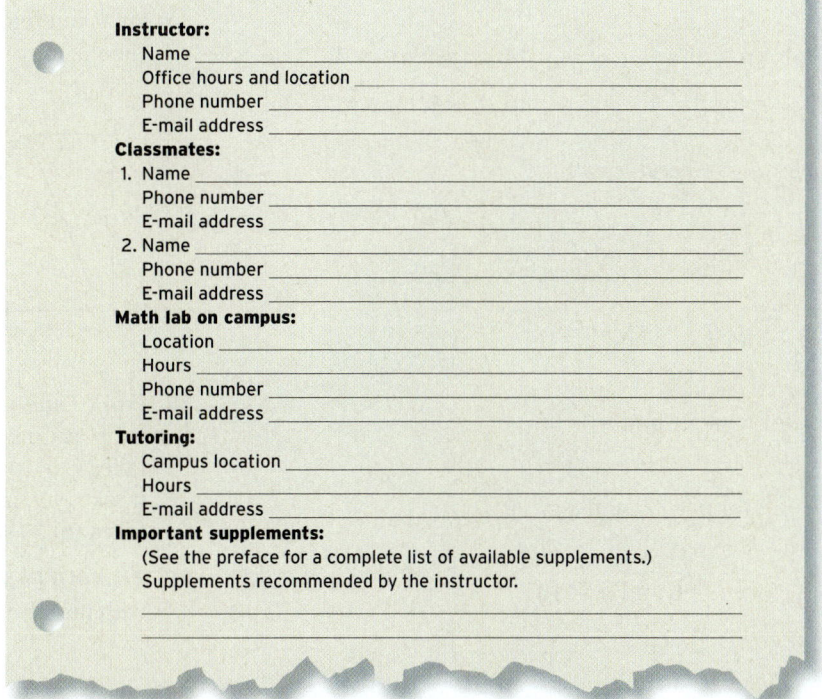

Instructor:
  Name _____
  Office hours and location _____
  Phone number _____
  E-mail address _____
Classmates:
  1. Name _____
    Phone number _____
    E-mail address _____
  2. Name _____
    Phone number _____
    E-mail address _____
Math lab on campus:
  Location _____
  Hours _____
  Phone number _____
  E-mail address _____
Tutoring:
  Campus location _____
  Hours _____
  E-mail address _____
Important supplements:
  (See the preface for a complete list of available supplements.)
  Supplements recommended by the instructor.
  _____
  _____

Chapter Resources:
Translating for Success, p. 69;
Collaborative Activity
(Teamwork), p. 70

1.1 EXERCISE SET

FOR EXTRA HELP

Vocabulary and Reading Check

Choose from the following list of words to complete each statement. Not every word will be used.

constant   model
equation   operation
evaluate   variable
expression

1. In the expression $4 + x$, the number 4 is a(n) _____.

2. To _____ an algebraic expression, we substitute a number for each variable and carry out the operations.

3. When we translate a problem into mathematical language, we say that we _____ the problem.

4. A(n) _____ contains an equal sign.

Concept Reinforcement

Classify each of the following as either an expression or an equation.

5. $10n - 1$

6. $3x = 21$

7. $2x - 5 = 9$

8. $5(x + 2)$

9. $45 = a - 1$

10. $4a - 5b$

11. $2x - 3y = 8$

12. $r(t + 7) + 5$

Evaluating Algebraic Expressions

Evaluate.

13. $5a$, for $a = 9$

14. $11y$, for $y = 7$

15. $12 - r$, for $r = 4$

16. $t + 8$, for $t = 2$

17. $\dfrac{a}{b}$, for $a = 45$ and $b = 9$

18. $\dfrac{c + d}{3}$, for $c = 14$ and $d = 13$

19. $\dfrac{x + y}{4}$, for $x = 2$ and $y = 14$

20. $\dfrac{m}{n}$, for $m = 54$ and $n = 9$

21. $\dfrac{p - q}{7}$, for $p = 55$ and $q = 20$

22. $\dfrac{9m}{q}$, for $m = 6$ and $q = 18$

23. $\dfrac{5z}{y}$, for $z = 9$ and $y = 15$

24. $\dfrac{m - n}{2}$, for $m = 20$ and $n = 8$

Substitute to find the value of each expression.

25. *Hockey.* The area of a rectangle with base b and height h is bh. A regulation hockey goal is 6 ft wide and 4 ft high. Find the area of the opening.

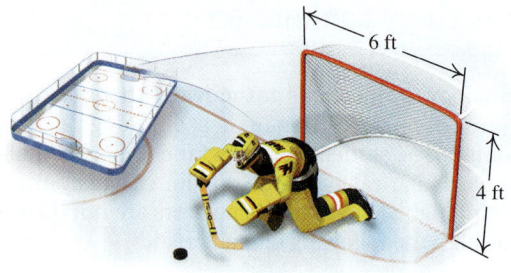

26. *Orbit Time.* A communications satellite orbiting 300 mi above the earth travels about 27,000 mi in one orbit. The time, in hours, for an orbit is

$$\frac{27,000}{v},$$

where v is the velocity, in miles per hour. How long will an orbit take at a velocity of 1125 mph?

27. *Zoology.* A great white shark has triangular teeth. Each tooth measures about 5 cm across the base and has a height of 6 cm. Find the surface area of the front side of one such tooth. (See Example 2.)

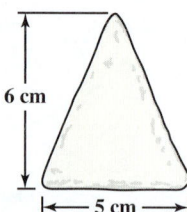 

28. *Work Time.* Justin takes three times as long to do a job as Carl does. Suppose t represents the time it takes Carl to do the job. Then $3t$ represents the time it takes Justin. How long does it take Justin if Carl takes **(a)** 30 sec? **(b)** 90 sec? **(c)** 2 min?

29. *Area of a Parallelogram.* The area of a parallelogram with base b and height h is bh. Edward Tufte's

sculpture *Spring Arcs* is in the shape of a parallelogram with base 67 ft and height 12 ft. What is the area of the parallelogram?

Source: edwardtufte.com

Spring Arcs (2004), Edward Tufte. Solid stainless steel, 12' × 67'.

30. *Women's Softball.* A softball player's batting average is h/a, where h is the number of hits and a is the number of "at bats." In the 2011 Women's College World Series, Michelle Moultrie of the Florida Gators had 13 hits in 24 at bats. What was her batting average? Round to the nearest thousandth.

Translating to Algebraic Expressions

Translate to an algebraic expression.

31. 5 more than Ron's age

32. The product of 4 and a

33. 6 times b

34. 7 more than Patti's weight

35. 9 less than c

36. 4 less than d

37. 6 increased by q

38. 11 increased by z

39. The difference of p and t

40. m subtracted from n

41. x less than y

42. 2 less than Kurt's age

43. x divided by w

44. The quotient of two numbers

45. The sum of the box's length and height

46. The sum of d and f

47. The product of 9 and twice m

48. Abby's speed minus twice the wind speed

49. Thirteen less than one quarter of some number

50. Four less than ten times a number

51. Five times the difference of two numbers

52. One third of the sum of two numbers

53. 64% of the women attending

54. 38% of a number

Translating to Equations

Translate each problem to an equation. Do not solve.

55. What number added to 73 is 201?

56. Seven times what number is 1596?

57. When 42 is multiplied by a number, the result is 2352. Find the number.

58. When 345 is added to a number, the result is 987. Find the number.

59. *Chess.* A chess board has 64 squares. If pieces occupy 19 squares, how many squares are unoccupied?

60. *Hours Worked.* A carpenter charges $35 per hour. How many hours did she work if she billed a total of $3640?

61. *Recycling.* Currently, Americans recycle 34% of all municipal solid waste. This is the same as recycling 82 million tons per year. What is the total amount of waste generated per year?

Source: U.S. EPA, Municipal Solid Waste Department

62. *Travel to Work.* In 2009, the average commuting time to work in New York was 31.4 min. The average commuting time in North Dakota was 15.4 min shorter. How long was the average commute in North Dakota?

Source: American Community Survey

63. *Nutrition.* The number of grams f of dietary fiber recommended daily for children depends on the age a of the child, as shown in the table below. Find an equation for f in terms of a.

Age of Child, a (in years)	Grams of Dietary Fiber Recommended Daily, f
3	8
4	9
5	10
6	11
7	12
8	13

Source: The American Health Foundation

64. *Tuition.*   The table below lists the tuition costs for students taking various numbers of hours of classes. Find an equation for the cost c of tuition for a student taking h hours of classes.

Number of Class Hours, h	Tuition, c
12	$1200
15	1500
18	1800
21	2100

65. *Postage Rates.*   The U.S. Postal Service charges extra for packages that must be processed by hand. The table below lists machinable and nonmachinable costs for certain packages. Find an equation for the nonmachinable cost n in terms of the machinable cost m.

Weight (in pounds)	Machinable Cost, m	Nonmachinable Cost, n
1	$2.74	$4.95
2	3.08	5.29
3	3.42	5.63

Source: pe.usps.gov

66. *Foreign Currency.*   On Emily's trip to Italy, she used her debit card to withdraw money. The table below lists the amounts r that she received and the amounts s that were subtracted from her account. Find an equation for r in terms of s.

Amount Received, r (in U.S. dollars)	Amount Subtracted, s (in U.S. dollars)
$150	$153
75	78
120	123

67. *Number of Drivers.*   The table below lists the number of vehicle miles v traveled annually per household by the number of drivers d in the household. Find an equation for v in terms of d.

Number of Drivers, d	Number of Vehicle Miles Traveled, v
1	10,000
2	20,000
3	30,000
4	40,000

Source: Energy Information Administration

68. *Meteorology.*   The table below lists the number of centimeters of water w to which various amounts of snow s will melt under certain conditions. Find an equation for w in terms of s.

Depth of Snow, s (in centimeters)	Depth of Water, w (in centimeters)
120	12
135	13.5
160	16
90	9

Translating to Algebraic Expressions and Equations

In each of Exercises 69–76, match the phrase or sentence with the appropriate expression or equation from the column on the right.

69. ____   Twice the sum of two numbers

70. ____   Five less than a number is twelve.

71. ____   Twelve more than a number is five.

72. ____   Half of the product of two numbers

73. ____   Three times the sum of a number and five

74. ____   Twice the sum of two numbers is 48.

75. ____   One less than the product of two numbers is 48.

76. ____   Six more than the quotient of two numbers

a) $\dfrac{x}{y} + 6$

b) $2(x + y) = 48$

c) $\dfrac{1}{2} \cdot a \cdot b$

d) $t + 12 = 5$

e) $ab - 1 = 48$

f) $2(m + n)$

g) $3(t + 5)$

h) $x - 5 = 12$

To the student and the instructor: Writing exercises, denoted by , should be answered using one or more English sentences. Because answers to many writing exercises will vary, solutions are not listed in the answers at the back of the book.

 77. What is the difference between a variable, a variable expression, and an equation?

78. What does it mean to evaluate an algebraic expression?

Synthesis

To the student and the instructor: Synthesis exercises are designed to challenge students to extend the concepts or skills studied in each section. Many synthesis exercises will require the assimilation of skills and concepts from several sections.

79. If the lengths of the sides of a square are doubled, is the area doubled? Why or why not?

80. Write a problem that translates to
    $$2006 + t = 2014.$$

81. Signs of Distinction charges $120 per square foot for handpainted signs. The town of Belmar commissioned a triangular sign with a base of 3 ft and a height of 2.5 ft. How much will the sign cost?

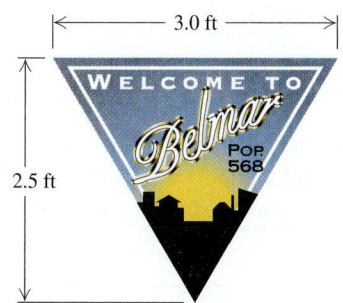

82. Find the area that is shaded.

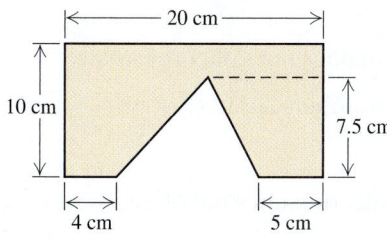

83. Evaluate $\dfrac{x - y}{3}$ when x is twice y and $x = 12$.

84. Evaluate $\dfrac{x + y}{2}$ when y is twice x and $x = 6$.

85. Evaluate $\dfrac{a + b}{4}$ when a is twice b and $a = 16$.

86. Evaluate $\dfrac{a - b}{3}$ when a is three times b and $a = 18$.

Answer each question with an algebraic expression.

87. If $w + 3$ is a whole number, what is the next whole number after it?

88. If $d + 2$ is an odd number, what is the preceding odd number?

Translate to an algebraic expression.

89. The perimeter of a rectangle with length l and width w (perimeter means distance around)

90. The perimeter of a square with side s (perimeter means distance around)

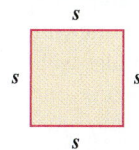

91. Ellie's race time, assuming she took 5 sec longer than Dion and Dion took 3 sec longer than Molly. Assume that Molly's time was t seconds.

92. Kade's age 7 years from now if he is 2 years older than Monique and Monique is a years old

93. If the height of a triangle is doubled, is its area also doubled? Why or why not?

↪ YOUR TURN ANSWERS: SECTION 1.1

1. 36   **2.** 150 in² **3.** Let n represent the number of students registered for the course; $n - 20$ **4.** Let x and y represent the numbers; $\frac{1}{2}(x + y)$ **5.** Let x represent the number; $x \cdot 12 = 672$.   **6.** Let m represent the number of math instructors; $13 = m + 5$ **7.** $a = 1.29n$

1.2   The Commutative, Associative, and Distributive Laws

The Commutative Laws ▪ The Associative Laws ▪ The Distributive Law ▪ The Distributive Law and Factoring

Study Skills

Learn by Example

The examples in each section are designed to prepare you for success with the exercise set. Study the step-by-step solutions of the examples, noting that color is used to indicate substitutions and to call attention to the new steps in multistep examples. The time you spend studying the examples will save you valuable time when you do your assignment.

The commutative, associative, and distributive laws discussed in this section enable us to write *equivalent expressions* that will simplify our work.

The expressions $4 + 4 + 4$, $3 \cdot 4$, and $4 \cdot 3$ all represent the same number, 12. Expressions that represent the same number are said to be **equivalent**. The equivalent expressions $t + 18$ and $18 + t$ are both translations of "eighteen more than a number." These expressions are equivalent because they represent the same number for any value of t. We can illustrate this by making some choices for t.

When $t = 3$, $t + 18 = 3 + 18 = 21$
and $18 + t = 18 + 3 = 21$.

When $t = 40$, $t + 18 = 40 + 18 = 58$
and $18 + t = 18 + 40 = 58$.

The Commutative Laws

Recall that changing the order in addition or multiplication does not change the result. Equations like $3 + 78 = 78 + 3$ and $5 \cdot 14 = 14 \cdot 5$ illustrate this idea and show that addition and multiplication are **commutative**.

THE COMMUTATIVE LAWS

For Addition.   For any numbers a and b,

$a + b = b + a$.

(Changing the order of addition does not affect the answer.)

For Multiplication.   For any numbers a and b,

$ab = ba$.

(Changing the order of multiplication does not affect the answer.)

EXAMPLE 1   Use the commutative laws to write an expression equivalent to each of the following: **(a)** $y + 5$; **(b)** $9x$; **(c)** $7 + ab$.

SOLUTION

a) $y + 5$ is equivalent to $5 + y$ by the commutative law of addition.

b) $9x$ is equivalent to $x \cdot 9$ by the commutative law of multiplication.

c) $7 + ab$ is equivalent to $ab + 7$ by the commutative law of *addition*.

$7 + ab$ is also equivalent to $7 + ba$ by the commutative law of *multiplication*.

$7 + ab$ is also equivalent to $ba + 7$ by the two commutative laws, used together.

1. Use the commutative law of addition to write an expression equivalent to $7a + 3$.

 YOUR TURN

The Associative Laws

Parentheses can be used to indicate groupings. We generally simplify within the parentheses first. For example,

$$3 + (8 + 4) = 3 + 12 = 15$$

and

$$(3 + 8) + 4 = 11 + 4 = 15.$$

Similarly,

$$4 \cdot (2 \cdot 3) = 4 \cdot 6 = 24$$

and

$$(4 \cdot 2) \cdot 3 = 8 \cdot 3 = 24.$$

Note that, so long as only addition or only multiplication appears in an expression, changing the grouping does not change the result. Equations such as $3 + (7 + 5) = (3 + 7) + 5$ and $4(5 \cdot 3) = (4 \cdot 5)3$ illustrate that addition and multiplication are **associative**.

THE ASSOCIATIVE LAWS

For Addition.   For any numbers a, b, and c,

$$a + (b + c) = (a + b) + c.$$

(Numbers can be grouped in any manner for addition.)

For Multiplication.   For any numbers a, b, and c,

$$a \cdot (b \cdot c) = (a \cdot b) \cdot c.$$

(Numbers can be grouped in any manner for multiplication.)

EXAMPLE 2   Use an associative law to write an expression equivalent to each of the following: **(a)** $y + (z + 3)$; **(b)** $(8x)y$.

SOLUTION

a) $y + (z + 3)$ is equivalent to $(y + z) + 3$ by the associative law of addition.

b) $(8x)y$ is equivalent to $8(xy)$ by the associative law of multiplication.

YOUR TURN

2. Use the associative law of multiplication to write an expression equivalent to $37(mp)$.

When only addition or only multiplication is involved, parentheses do not change the result. For that reason, we sometimes omit them altogether. Thus,

$$x + (y + 7) = x + y + 7 \quad \text{and} \quad l(wh) = lwh.$$

A sum such as $(5 + 1) + (3 + 5) + 9$ can be simplified by pairing numbers that add to 10. The associative and commutative laws allow us to do this:

$$(5 + 1) + (3 + 5) + 9 = 5 + 5 + 9 + 1 + 3$$
$$= 10 + 10 + 3 = 23.$$

EXAMPLE 3   Use the commutative and/or associative laws of addition to write two expressions equivalent to $(7 + x) + 3$. Then simplify.

SOLUTION

$$(7 + x) + 3 = (x + 7) + 3 \qquad \text{Using the commutative law; } (x + 7) + 3 \text{ is one equivalent expression.}$$

$$= x + (7 + 3) \qquad \text{Using the associative law; } x + (7 + 3) \text{ is another equivalent expression.}$$

$$= x + 10 \qquad \text{Simplifying}$$

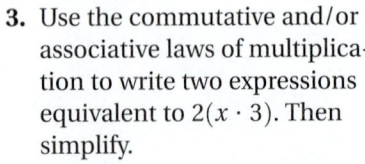 YOUR TURN

3. Use the commutative and/or associative laws of multiplication to write two expressions equivalent to $2(x \cdot 3)$. Then simplify.

The Distributive Law

The *distributive law* is probably the single most important law for manipulating algebraic expressions. Unlike the commutative and associative laws, the distributive law uses multiplication together with addition.

The distributive law relates expressions like $5(x + 2)$ and $5x + 10$, which involve both multiplication and addition. When two numbers are multiplied, the result is a **product**. The parts of the product are called **factors**. When two numbers are added, the result is a **sum**. The parts of the sum are called **terms**.

$5(x + 2)$ is a product. The factors are 5 and $(x + 2)$.

This factor is a sum. Its terms are x and 2.

$5x + 10$ is a sum. The terms are $5x$ and 10.

This term is a product. Its factors are 5 and x.

In general, a term is a number, a variable, or a product or a quotient of numbers and/or variables. Terms are separated by plus signs.

EXAMPLE 4   List the terms in the expression $3s + st + \dfrac{2s}{t}$.

SOLUTION   Terms are separated by plus signs, so the terms in $3s + st + \dfrac{2s}{t}$ are $3s$, st, and $\dfrac{2s}{t}$.

4. List the terms in the expression $5x + 3y$.

 YOUR TURN

EXAMPLE 5   List the factors in the expression $x(3 + y)$.

SOLUTION   Factors are parts of products, so the factors in $x(3 + y)$ are x and $(3 + y)$.

5. List the factors in the expression $4 \cdot x \cdot y$.

YOUR TURN

You have already used the distributive law although you may not have realized it at the time. To illustrate, try to multiply $3 \cdot 21$ mentally. Many people find the product, 63, by thinking of 21 as $20 + 1$ and then multiplying 20 by 3 and 1 by 3. The sum of the two products, $60 + 3$, is 63. Note that if the 3 does not multiply *both* 20 and 1, the result will not be correct.

We can compute $4(7 + 2)$ in two ways. As in the discussion of $3(20 + 1)$ above, to compute $4(7 + 2)$, we can multiply both 7 and 2 by 4 and add the results:

$$4(7 + 2) = 4 \cdot 7 + 4 \cdot 2 = 28 + 8 = 36.$$

By first adding inside the parentheses, we get the same result in a different way:

$$4(7 + 2) = 4(9) = 36.$$

Student Notes

The distributive law involves both addition *and* multiplication. Do not try to "distribute" when only multiplication is involved. For example, $5(3 \cdot x) = (5 \cdot 3) \cdot x = 15x$ (by the associative law of multiplication). *It is incorrect to write* $5(3 \cdot x) = (5 \cdot 3)(5 \cdot x)$.

> ### THE DISTRIBUTIVE LAW
>
> For any numbers a, b, and c,
>
> $$a(b + c) = ab + ac.$$
>
> (The product of a number and a sum can be written as the sum of two products.)

EXAMPLE 6   Multiply: $3(x + 2)$.

SOLUTION   Since $x + 2$ cannot be simplified unless a value for x is given, we use the distributive law:

$$3(x + 2) = 3 \cdot x + 3 \cdot 2 \qquad \text{Using the distributive law}$$
$$= 3x + 6. \qquad \text{Note that } 3 \cdot x \text{ is the same as } 3x.$$

6. Multiply: $5(3 + y)$.

YOUR TURN

The distributive law can also be used when more than two terms are inside the parentheses.

EXAMPLE 7   Multiply: $6(s + 2 + 5w)$.

SOLUTION

$$6(s + 2 + 5w) = 6 \cdot s + 6 \cdot 2 + 6 \cdot 5w \qquad \text{Using the distributive law}$$
$$= 6s + 12 + (6 \cdot 5)w \qquad \text{Using the associative law for multiplication}$$
$$= 6s + 12 + 30w$$

7. Multiply: $4(5a + 6m + 1)$.

YOUR TURN

Because of the commutative law of multiplication, the distributive law can be used on the "right": $(b + c)a = ba + ca$.

EXAMPLE 8   Multiply: $(c + 4)5$.

SOLUTION

$$(c + 4)5 = c \cdot 5 + 4 \cdot 5 \qquad \text{Using the distributive law on the right}$$
$$= 5c + 20 \qquad \text{Using the commutative law; } c \cdot 5 = 5c$$

8. Multiply: $(11 + y)2$.

YOUR TURN

> **CAUTION!**   To use the distributive law for removing parentheses, be sure to multiply *each* term inside the parentheses by the multiplier outside. Thus,
>
> $$a(b + c) \neq ab + c \quad \text{but} \quad a(b + c) = ab + ac.$$

The Distributive Law and Factoring

If we use the distributive law in reverse, we have the basis of a process called **factoring**: $ab + ac = a(b + c)$. To **factor** an expression means to write an equivalent expression that is a product. Recall that the parts of the product are called **factors**. Note that "factor" can be used as either a verb or a noun. A **common factor** is a factor that appears in every term in an expression.

EXAMPLE 9   Use the distributive law to factor each of the following.

a) $3x + 3y$

b) $7x + 21y + 7$

SOLUTION

a) By the distributive law,

$$3x + 3y = 3(x + y). \qquad \text{The common factor for } 3x \text{ and } 3y \text{ is } 3.$$

b) $7x + 21y + 7 = 7 \cdot x + 7 \cdot 3y + 7 \cdot 1$ The common factor is 7.

$$= 7(x + 3y + 1) \qquad \text{Using the distributive law. Be sure to include both the 1 and the common factor, 7.}$$

9. Use the distributive law to factor $15x + 5$.

⟳ YOUR TURN

To check our factoring, we multiply to see if the original expression is obtained. For example, to check the **factorization** in Example 9(b), note that

$$7(x + 3y + 1) = 7 \cdot x + 7 \cdot 3y + 7 \cdot 1$$
$$= 7x + 21y + 7.$$

Since $7x + 21y + 7$ is what we started with in Example 9(b), we have a check.

1.2   EXERCISE SET

FOR EXTRA HELP

 MyMathLab®  Math XL
PRACTICE   WATCH   READ   REVIEW

◥ Vocabulary and Reading Check

Choose from the following list of words to complete each statement. Not every word will be used.

associative	factors
commutative	product
distributive	sum
equivalent	terms

1. _____ expressions represent the same number.

2. Changing the order of multiplication does not affect the answer. This is an example of a(n) _____ law.

3. The result of addition is called a(n) _____.

4. The numbers in a product are called _____.

◥ Concept Reinforcement

Determine whether each statement illustrates a commutative law, an associative law, or a distributive law.

5. $8 + x = x + 8$

6. $5b(c) = 5(bc)$

7. $x(y + z) = xy + xz$

8. $3(t + 4) = 3(4 + t)$

9. $5(x + 2) = (x + 2)5$

10. $2a + 2b = 2(a + b)$

The Commutative Laws

Use the commutative law of addition to write an equivalent expression.

11. $11 + t$

12. $a + 2$

13. $4 + 8x$

14. $ab + c$

15. $9x + 3y$

16. $3a + 7b$

17. $5(a + 1)$

18. $9(x + 5)$

Use the commutative law of multiplication to write an equivalent expression.

19. $7x$

20. xy

21. st

22. $13m$

23. $5 + ab$

24. $x + 3y$

25. $5(a + 1)$

26. $9(x + 5)$

The Associative Laws

Use the associative law of addition to write an equivalent expression.

27. $(x + 8) + y$

28. $(5 + m) + r$

29. $u + (v + 7)$

30. $x + (2 + y)$

31. $(ab + c) + d$

32. $(m + np) + r$

Use the associative law of multiplication to write an equivalent expression.

33. $(10x)y$

34. $(4u)v$

35. $2(ab)$

36. $9(7r)$

37. $3[2(a + b)]$

38. $5[x(2 + y)]$

The Commutative and Associative Laws

Use the commutative and/or associative laws to write two equivalent expressions. Answers may vary.

39. $s + (t + 6)$

40. $7 + (v + w)$

41. $(17a)b$

42. $x(3y)$

Use the commutative and/or associative laws to show why the expression on the left is equivalent to the expression on the right. Write a series of steps with labels, as in Example 3.

43. $(1 + x) + 2$ is equivalent to $x + 3$

44. $(2a)4$ is equivalent to $8a$

45. $(m \cdot 3)7$ is equivalent to $21m$

46. $4 + (9 + x)$ is equivalent to $x + 13$

The Distributive Law

List the terms in each expression.

47. $x + xyz + 1$

48. $9 + 17a + abc$

49. $2a + \dfrac{a}{3b} + 5b$

50. $3xy + 20 + \dfrac{4a}{b}$

51. $4x + 4y$

52. $14 + 2y$

List the factors in each expression.

53. $5n$

54. uv

55. $3(x + y)$

56. $(a + b)12$

57. $7 \cdot a \cdot b$

58. $m \cdot n \cdot 2$

59. $(a - b)(x - y)$

60. $(3 - a)(b + c)$

Multiply.

61. $2(x + 15)$

62. $3(x + 5)$

63. $4(1 + a)$

64. $7(1 + y)$

65. $10(9x + 6)$

66. $9(6m + 7)$

67. $5(r + 2 + 3t)$

68. $4(5x + 8 + 3p)$

69. $(a + b)2$

70. $(x + 2)7$

71. $(x + y + 2)5$

72. $(2 + a + b)6$

The Distributive Law and Factoring

Use the distributive law to factor each of the following. Check by multiplying.

73. $2a + 2b$

74. $5y + 5z$

75. $7 + 7y$

76. $13 + 13x$

77. $32x + 2$

78. $20a + 5$

79. $5x + 10 + 15y$

80. $3 + 27b + 6c$

81. $7a + 35b$

82. $3x + 24y$

83. $44x + 11y + 22z$

84. $14a + 56b + 7$

The Commutative, Associative, and Distributive Laws

Fill in each blank with the law that justifies that step.

85. $3(2 + x)$
$= 3(x + 2)$ _____
$= 3 \cdot x + 3 \cdot 2$ _____
$= 3x + 6$ Multiplying

86. $(y + 4)5$
$= 5(y + 4)$ _____
$= 5 \cdot y + 5 \cdot 4$ _____
$= 5y + 20$ Multiplying

87. $7(2x + 3y)$
$= 7(2x) + 7(3y)$ _____
$= (7 \cdot 2)x + (7 \cdot 3)y$ _____
$= 14x + 21y$ Multiplying

88. $(4a + 2)8$
$= 8(4a + 2)$ _____
$= 8(4a) + 8(2)$ _____
$= (8 \cdot 4)a + 8(2)$ _____
$= 32a + 16$ Multiplying

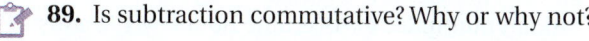 **89.** Is subtraction commutative? Why or why not?

90. Is division associative? Why or why not?

Synthesis

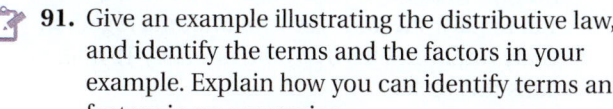 **91.** Give an example illustrating the distributive law, and identify the terms and the factors in your example. Explain how you can identify terms and factors in an expression.

92. Explain how the distributive, commutative, and associative laws can be used to show that $2(3x + 4y)$ is equivalent to $6x + 8y$.

Fill in each blank with the law that justifies that step.

93. $[2(x + 1)] + 3x$
$= [2 \cdot x + 2 \cdot 1] + 3x$ _____
$= [2x + 2] + 3x$ Multiplying
$= 2x + [2 + 3x]$ _____
$= 2x + [3x + 2]$ _____
$= [2x + 3x] + 2$ _____
$= [(2 + 3)x] + 2$ _____
$= [5x] + 2$ Adding
$= 5x + 2$

94. $12a + 4(b + 5)$
$= (4 \cdot 3)a + 4(b + 5)$ Writing 12 as a product
$= 4(3a) + 4(b + 5)$ _____
$= 4(3a) + 4(b) + 4(5)$ _____
$= 4(3a + b + 5)$ _____

Tell whether the expressions in each pairing are equivalent. Then explain why or why not.

95. $8 + 4(a + b)$ and $4(2 + a + b)$

96. $5(a \cdot b)$ and $5 \cdot a \cdot 5 \cdot b$

97. $7 \div 3m$ and $m \cdot 3 \div 7$

98. $(rt + st)5$ and $5t(r + s)$

99. $30y + x \cdot 15$ and $5[2(x + 3y)]$

100. $[c(2 + 3b)]5$ and $10c + 15bc$

101. Evaluate the expressions $3(2 + x)$ and $6 + x$ for $x = 0$. Do your results indicate that $3(2 + x)$ and $6 + x$ are equivalent? Why or why not?

102. Factor $15x + 40$. Then evaluate both $15x + 40$ and the factorization for $x = 4$. Do your results *guarantee* that the factorization is correct? Why or why not? (*Hint:* See Exercise 101.)

103. Aidan, Beth, and Cody consistently work more than 40 hours every week. The first 40 hours are paid at a regular pay rate of $10 per hour. The number of hours over 40 are paid at an overtime pay rate that is one and one-half times the regular pay rate. Each of the employees calculates the week's wages using a different formula.

- Aidan multiplies his overtime hours by 1.5, adds this to his regular hours, and then multiplies the sum by 10.
- Beth multiplies her regular hours by 10, then multiplies her overtime hours by 10 and then by 1.5. Finally, she adds the two amounts together.
- Cody multiplies his overtime hours by 15 and adds this to 400.

Let x represent the number of overtime hours worked in one week.

a) Write an algebraic expression for each method of calculating wages.

b) Use the commutative, associative, and distributive laws to show that all three methods of calculating wages yield the same total.

↪ YOUR TURN ANSWERS: SECTION 1.2

1. $3 + 7a$ **2.** $(37m)p$ **3.** $2(3x)$; $(2 \cdot 3)x$; $6x$;
answers may vary   **4.** $5x, 3y$ **5.** $4, x, y$ **6.** $15 + 5y$
7. $20a + 24m + 4$ **8.** $22 + 2y$ **9.** $5(3x + 1)$

QUICK QUIZ: SECTIONS 1.1 AND 1.2

To the student and the instructor: Beginning in the second section of each chapter, every exercise set contains a short quiz reviewing content already taught in the chapter. The numbers in brackets immediately following the directions or exercise indicate the section in which the skill was introduced. The answers to all quiz exercises appear at the back of the book. Continuous review of chapter content is excellent preparation for a chapter test.

1. Evaluate $x - y$ for $x = 17$ and $y = 8$.   [1.1]

2. Translate to an algebraic expression: Twice the sum of m and 3.   [1.1]

3. Translate to an equation. Do not solve. One-third of what number is 18?   [1.1]

4. Multiply: $3(x + 5y + 7)$.   [1.2]

5. Factor: $14a + 7t + 7$.   [1.2]

1.3 Fraction Notation

Factors and Prime Factorizations   ■   Multiplication, Division, and Simplification of Fractions   ■   Addition and Subtraction of Fractions

This section covers multiplication, addition, subtraction, and division with fractions, including fraction expressions that contain variables.

An example of **fraction notation** for a number is

$$\frac{2}{3}. \quad \begin{array}{l} \leftarrow \text{Numerator} \\ \leftarrow \text{Denominator} \end{array}$$

The top number is called the **numerator,** and the bottom number is called the **denominator**.

Factors and Prime Factorizations

We first review how *natural numbers* are factored. **Natural numbers** can be thought of as the counting numbers:

$$1, 2, 3, 4, 5, \ldots.$$

(The dots indicate that the established pattern continues without ending.)

Since factors are parts of products, to factor a number, we express it as a product of two or more numbers.

Several factorizations of 12 are

$$1 \cdot 12, \quad 2 \cdot 6, \quad 3 \cdot 4, \quad 2 \cdot 2 \cdot 3.$$

It is easy to overlook a factor of a number if the factorizations are not written methodically.

EXAMPLE 1   List all factors of 18.

SOLUTION   Beginning at 1, we check all natural numbers to see if they are factors of 18. If they are, we write the factorization. We stop when we have already included the next natural number in a factorization.

1 is a factor of every number.	$1 \cdot 18$
2 is a factor of 18.	$2 \cdot 9$
3 is a factor of 18.	$3 \cdot 6$

4 is *not* a factor of 18.

5 is *not* a factor of 18.

6 is the next natural number, but we have already listed 6 as a factor in the product $3 \cdot 6$.

We stop at 6 because any natural number greater than 6 would be paired with a factor less than 6. (Remember that multiplication is commutative.)

We now write the factors of 18, going down the above list of factorizations writing the first factors and then up the list writing the second factors:

$$1, \quad 2, \quad 3, \quad 6, \quad 9, \quad 18.$$

 YOUR TURN

Student Notes

If you are asked to "find the factors" of a number, your answer will be a list of numbers. If you are asked to "find a factorization" of a number, your answer will be a product. For example:

The factors of 12 are 1, 2, 3, 4, 6, 12.

A factorization of 12 is $2 \cdot 2 \cdot 3$.

1. List all factors of 54.

Some numbers have only two different factors, the number itself and 1. Such numbers are called **prime**.

PRIME NUMBER

A *prime number* is a natural number that has exactly two different factors: the number itself and 1. The first several primes are 2, 3, 5, 7, 11, 13, 17, 19, and 23.

If a natural number other than 1 is not prime, we call it **composite**.

EXAMPLE 2   Label each number as prime, composite, or neither: 29, 4, 1.

SOLUTION

29 is prime. It has exactly two different factors, 29 and 1.

4 is not prime. It has three different factors, 1, 2, and 4. It is composite.

1 is not prime. It does not have two *different* factors. The number 1 is not considered composite. It is neither prime nor composite.

2. Label 21 as prime, composite, or neither.

 YOUR TURN

Every composite number can be factored into a product of prime numbers. Such a factorization is called the **prime factorization** of that composite number.

EXAMPLE 3   Find the prime factorization of 36.

SOLUTION   We first factor 36 in any way that we can, such as

$$36 = 4 \cdot 9.$$

Since 4 and 9 are not prime, we factor them:

$$36 = 4 \cdot 9$$
$$= 2 \cdot 2 \cdot 3 \cdot 3. \qquad \text{2 and 3 are both prime.}$$

3. Find the prime factorization of 100.

The prime factorization of 36 is $2 \cdot 2 \cdot 3 \cdot 3$.

 YOUR TURN

Student Notes

When writing a factorization, you are writing an equivalent expression for the original number. Some students do this with a tree diagram:

$$36$$
$$36 = 4 \quad \cdot \quad 9$$
$$36 = 2 \cdot 2 \cdot 3 \cdot 3$$
All prime

Multiplication, Division, and Simplification of Fractions

Recall from arithmetic that fractions are multiplied as follows.

MULTIPLICATION OF FRACTIONS

For any two fractions a/b and c/d,

$$\frac{a}{b} \cdot \frac{c}{d} = \frac{ac}{bd}.$$

(The numerator of the product is the product of the two numerators. The denominator of the product is the product of the two denominators.)

EXAMPLE 4 Multiply: (a) $\dfrac{2}{3} \cdot \dfrac{5}{7}$; (b) $\dfrac{4}{x} \cdot \dfrac{8}{y}$; (c) $9 \cdot \dfrac{7}{n}$.

SOLUTION We multiply numerators as well as denominators.

a) $\dfrac{2}{3} \cdot \dfrac{5}{7} = \dfrac{2 \cdot 5}{3 \cdot 7} = \dfrac{10}{21}$ b) $\dfrac{4}{x} \cdot \dfrac{8}{y} = \dfrac{4 \cdot 8}{x \cdot y} = \dfrac{32}{xy}$

4. Multiply: $\dfrac{2}{x} \cdot \dfrac{3}{7}$.

c) $9 \cdot \dfrac{7}{n} = \dfrac{9}{1} \cdot \dfrac{7}{n} = \dfrac{9 \cdot 7}{1 \cdot n} = \dfrac{63}{n}$ 9 can be written $\dfrac{9}{1}$

YOUR TURN

Any nonzero number divided by itself is 1.

> **FRACTION NOTATION FOR 1**
>
> For any number a, except 0,
>
> $$\dfrac{a}{a} = 1.$$

Two numbers whose product is 1 are **reciprocals**, or **multiplicative inverses**, of each other. All numbers, except zero, have reciprocals. For example,

the reciprocal of $\dfrac{2}{3}$ is $\dfrac{3}{2}$ because $\dfrac{2}{3} \cdot \dfrac{3}{2} = \dfrac{6}{6} = 1$;

the reciprocal of 9 is $\dfrac{1}{9}$ because $9 \cdot \dfrac{1}{9} = \dfrac{9}{1} \cdot \dfrac{1}{9} = \dfrac{9}{9} = 1$;

the reciprocal of $\dfrac{1}{4}$ is 4 because $\dfrac{1}{4} \cdot 4 = \dfrac{1}{4} \cdot \dfrac{4}{1} = \dfrac{4}{4} = 1$.

Reciprocals are used to rewrite division in an equivalent form that uses multiplication.

> **DIVISION OF FRACTIONS**
>
> To divide two fractions, multiply by the reciprocal of the divisor:
>
> $$\dfrac{a}{b} \div \dfrac{c}{d} = \dfrac{a}{b} \cdot \dfrac{d}{c}.$$

EXAMPLE 5 Divide: $\dfrac{1}{2} \div \dfrac{3}{5}$.

SOLUTION

$$\dfrac{1}{2} \div \dfrac{3}{5} = \dfrac{1}{2} \cdot \dfrac{5}{3} \qquad \dfrac{5}{3} \text{ is the reciprocal of } \dfrac{3}{5}.$$

$$= \dfrac{5}{6}$$

5. Divide: $\dfrac{3}{4} \div \dfrac{5}{7}$.

YOUR TURN

Multiplying a number by 1 gives that same number because of the *identity property of* 1. A similar property can be stated for division.

Study Skills

Do the Exercises

- When you complete an odd-numbered exercise, you can check your answers at the back of the book. If you miss any, closely examine your work, and if necessary, consult your instructor for guidance.

- Do some even-numbered exercises, even if none are assigned. Because there are no answers given for them, you will gain practice doing exercises in a context similar to taking a quiz or a test. Check your answers later with a friend or your instructor.

> **THE IDENTITY PROPERTY OF 1**
>
> For any number a,
>
> $$a \cdot 1 = 1 \cdot a = a.$$
>
> (Multiplying a number by 1 gives that same number.) The number 1 is called the *multiplicative identity*.

For example, we can multiply $\frac{4}{5} \cdot \frac{6}{6}$ to find an expression equivalent to $\frac{4}{5}$. Since $\frac{6}{6} = 1$, the expression $\frac{4}{5} \cdot \frac{6}{6}$ is equivalent to $\frac{4}{5} \cdot 1$, or simply $\frac{4}{5}$. We have

$$\frac{4}{5} \cdot \frac{6}{6} = \frac{4 \cdot 6}{5 \cdot 6} = \frac{24}{30}.$$

Thus, $\frac{24}{30}$ is equivalent to $\frac{4}{5}$.

We reverse these steps by "removing a factor equal to 1"—in this case, $\frac{6}{6}$. By removing a factor that equals 1, we can *simplify* an expression like $\frac{24}{30}$ to an equivalent expression like $\frac{4}{5}$.

To simplify, we factor the numerator and the denominator, looking for the largest factor common to both. This is sometimes made easier by writing prime factorizations. After identifying common factors, we can express the fraction as a product of two fractions, one of which is in the form a/a.

EXAMPLE 6 Simplify: (a) $\dfrac{15}{40}$; (b) $\dfrac{36}{24}$.

SOLUTION

a) Note that 5 is a factor of both 15 and 40:

$$\frac{15}{40} = \frac{3 \cdot 5}{8 \cdot 5} \qquad \text{Factoring the numerator and the denominator, using the common factor, 5}$$

$$= \frac{3}{8} \cdot \frac{5}{5} \qquad \text{Rewriting as a product of two fractions; } \frac{5}{5} = 1$$

$$= \frac{3}{8} \cdot 1 = \frac{3}{8}. \qquad \text{Using the identity property of 1 (removing a factor equal to 1)}$$

b) $\dfrac{36}{24} = \dfrac{2 \cdot 2 \cdot 3 \cdot 3}{2 \cdot 2 \cdot 2 \cdot 3}$ Writing the prime factorizations and identifying common factors; 12/12 could also be used.

$$= \frac{3}{2} \cdot \frac{2 \cdot 2 \cdot 3}{2 \cdot 2 \cdot 3} \qquad \text{Rewriting as a product of two fractions; } \frac{2 \cdot 2 \cdot 3}{2 \cdot 2 \cdot 3} = 1$$

$$= \frac{3}{2} \cdot 1 = \frac{3}{2} \qquad \text{Using the identity property of 1}$$

6. Simplify: $\dfrac{35}{30}$.

↩ YOUR TURN

It is always wise to check your result to see if any common factors of the numerator and the denominator remain. (This will not occur if prime factorizations are used correctly.) If common factors remain, repeat the process by removing another factor equal to 1 to simplify your result.

"Canceling" is a shortcut that you may have used for removing a factor equal to 1 when working with fraction notation. With *great* concern, we mention it as a possible way to speed up your work. Canceling can be used only when removing common factors in numerators and denominators. Canceling *cannot* be used in sums or differences. Our concern is that "canceling" be used with understanding. Example 6(b) might have been done faster as follows:

$$\frac{36}{24} = \frac{2 \cdot 2 \cdot 3 \cdot \cancel{3}}{2 \cdot 2 \cdot 2 \cdot \cancel{3}} = \frac{3}{2}, \quad \text{or} \quad \frac{36}{24} = \frac{3 \cdot \cancel{12}}{2 \cdot \cancel{12}} = \frac{3}{2}, \quad \text{or} \quad \frac{\overset{\overset{3}{\cancel{18}}}{\cancel{36}}}{\underset{\underset{2}{\cancel{12}}}{\cancel{24}}} = \frac{3}{2}.$$

CAUTION!   Unfortunately, canceling is often performed incorrectly:

$$\frac{\cancel{2} + 3}{\cancel{2}} \neq 3, \qquad \frac{\cancel{4} - 1}{\cancel{4} - 2} \neq \frac{1}{2}, \qquad \frac{1\cancel{5}}{\cancel{5}4} \neq \frac{1}{4}.$$

The above cancellations are incorrect because the expressions canceled are *not* factors. For example, in $2 + 3$, the 2 and the 3 are not factors. Correct simplifications are as follows:

$$\frac{2 + 3}{2} = \frac{5}{2}, \qquad \frac{4 - 1}{4 - 2} = \frac{3}{2}, \qquad \frac{15}{54} = \frac{5 \cdot \cancel{3}}{18 \cdot \cancel{3}} = \frac{5}{18}.$$

Remember: If you can't factor, you can't cancel! If in doubt, don't cancel!

Sometimes it is helpful to use 1 as a factor in the numerator or the denominator when simplifying.

EXAMPLE 7   Simplify: $\dfrac{9}{72}$.

SOLUTION

$$\frac{9}{72} = \frac{1 \cdot 9}{8 \cdot 9}$$ Factoring and using the identity property of 1 to write 9 as $1 \cdot 9$

$$= \frac{1 \cdot \cancel{9}}{8 \cdot \cancel{9}} = \frac{1}{8}$$ Simplifying by removing a factor equal to 1: $\frac{9}{9} = 1$

7. Simplify: $\dfrac{6}{18}$.

YOUR TURN

Addition and Subtraction of Fractions

When denominators are the same, fractions are added or subtracted by adding or subtracting numerators and keeping the same denominator.

ADDITION AND SUBTRACTION OF FRACTIONS

For any two fractions a/d and b/d,

$$\frac{a}{d} + \frac{b}{d} = \frac{a + b}{d} \quad \text{and} \quad \frac{a}{d} - \frac{b}{d} = \frac{a - b}{d}.$$ Note that the denominators are the same.

CAUTION!   When adding or subtracting fractions with the same denominator, add or subtract only the numerators. The denominator does not change.

EXAMPLE 8   Add and simplify: $\dfrac{4}{8} + \dfrac{5}{8}$.

SOLUTION   The common denominator is 8. We add the numerators and keep the common denominator:

$$\frac{4}{8} + \frac{5}{8} = \frac{4+5}{8} = \frac{9}{8}.$$ You can think of this as
$4 \cdot \frac{1}{8} + 5 \cdot \frac{1}{8} = 9 \cdot \frac{1}{8}$, or $\frac{9}{8}$.

8. Add and simplify: $\dfrac{2}{5} + \dfrac{4}{5}$.

 YOUR TURN

In arithmetic, we often write $1\frac{1}{8}$ rather than the "improper" fraction $\frac{9}{8}$. In algebra, $\frac{9}{8}$ is generally more useful and is quite "proper" for our purposes.

When denominators are different, we use the identity property of 1 and multiply to find a common denominator. Then we add, as in Example 8.

A common denominator is a number that is divisible by each fraction's denominator. One method of finding a common denominator uses prime factorizations.

FINDING A COMMON DENOMINATOR USING PRIME FACTORIZATIONS

1. Find the prime factorization of each denominator.
2. Choose one factorization.
3. Multiply that factorization by any factors of the other denominator that it lacks.

EXAMPLE 9   Add: $\dfrac{7}{8} + \dfrac{5}{12}$.

SOLUTION   We follow the steps listed above to find a common denominator.

1. $8 = 2 \cdot 2 \cdot 2$ Find the prime factorization of each denominator.
    $12 = 2 \cdot 2 \cdot 3$

2. $2 \cdot 2 \cdot 2$ Choose one factorization.

3. $2 \cdot 2 \cdot 2 \cdot 3$ The factorization of 8 is "missing" the factor 3 from the factorization of 12.

The common denominator is thus $2 \cdot 2 \cdot 2 \cdot 3$, or 24. It is divisible by both 8 and 12. We multiply both $\frac{7}{8}$ and $\frac{5}{12}$ by suitable forms of 1 to obtain two fractions with denominators of 24:

$$\frac{7}{8} + \frac{5}{12} = \frac{7}{8} \cdot \frac{3}{3} + \frac{5}{12} \cdot \frac{2}{2}$$ Multiplying by 1. Since $8 \cdot 3 = 24$, we multiply $\frac{7}{8}$ by $\frac{3}{3}$. Since $12 \cdot 2 = 24$, we multiply $\frac{5}{12}$ by $\frac{2}{2}$.

$$= \frac{21}{24} + \frac{10}{24}$$ Performing the multiplication

$$= \frac{31}{24}.$$ Adding fractions

9. Add: $\dfrac{9}{8} + \dfrac{4}{5}$.

 YOUR TURN

After adding, subtracting, multiplying, or dividing, we may still need to simplify the answer.

EXAMPLE 10 Perform the indicated operation and, if possible, simplify.

a) $\dfrac{7}{10} - \dfrac{1}{5}$ **b)** $8 \cdot \dfrac{5}{12}$ **c)** $\dfrac{\frac{5}{6}}{\frac{25}{9}}$

SOLUTION

a) $\dfrac{7}{10} - \dfrac{1}{5} = \dfrac{7}{10} - \dfrac{1}{5} \cdot \dfrac{2}{2}$ Using 10 as the common denominator

$= \dfrac{7}{10} - \dfrac{2}{10}$

$= \dfrac{5}{10} = \dfrac{1 \cdot \cancel{5}}{2 \cdot \cancel{5}} = \dfrac{1}{2}$ Removing a factor equal to 1: $\dfrac{5}{5} = 1$

b) $8 \cdot \dfrac{5}{12} = \dfrac{8 \cdot 5}{12}$ Multiplying numerators and denominators. Think of 8 as $\frac{8}{1}$.

$= \dfrac{2 \cdot 2 \cdot 2 \cdot 5}{2 \cdot 2 \cdot 3}$ Factoring; $\dfrac{4 \cdot 2 \cdot 5}{4 \cdot 3}$ can also be used.

$= \dfrac{\cancel{2} \cdot \cancel{2} \cdot 2 \cdot 5}{\cancel{2} \cdot \cancel{2} \cdot 3}$ Removing a factor equal to 1: $\dfrac{2 \cdot 2}{2 \cdot 2} = 1$

$= \dfrac{10}{3}$ Simplifying

c) $\dfrac{\frac{5}{6}}{\frac{25}{9}} = \dfrac{5}{6} \div \dfrac{25}{9}$ Rewriting horizontally. Remember that a fraction bar indicates division.

$= \dfrac{5}{6} \cdot \dfrac{9}{25}$ Multiplying by the reciprocal of $\frac{25}{9}$

$= \dfrac{5 \cdot 3 \cdot 3}{2 \cdot 3 \cdot 5 \cdot 5}$ Writing as one fraction and factoring

$= \dfrac{\cancel{5} \cdot \cancel{3} \cdot 3}{2 \cdot \cancel{3} \cdot \cancel{5} \cdot 5}$ Removing a factor equal to 1: $\dfrac{5 \cdot 3}{3 \cdot 5} = 1$

$= \dfrac{3}{10}$ Simplifying

10. Subtract and, if possible, simplify:

$\dfrac{9}{20} - \dfrac{1}{4}$.

 YOUR TURN

Technology Connection

Many calculators can perform operations using fraction notation. You may find an option n/d on a key or in a **menu** of options that appears when a key is pressed. To select an item from a menu, we highlight its number and press **ENTER** or simply press the number of the item.

For example, to add $\frac{2}{15} + \frac{7}{12}$, we first press **MATH** ▷ ⌃ **ENTER** to choose the n/d option in the MATH NUM submenu. An empty fraction template will appear on the screen. We then press ② ⌄ ① ⑤ to enter the numerator and the denominator of the first fraction. Next, we press ▷ to move outside of the fraction and then press ⊕. To enter the second fraction, we press **MATH** ▷ ⌃ **ENTER** again and then enter the numerator and the denominator of the second fraction. We press **ENTER** to calculate the result.

```
 2    7
─── + ───
 15   12

              43
             ───
              60
```

Since a fraction bar indicates division, we can enter $\frac{2}{15} + \frac{7}{12}$ as 2/15 + 7/12. The answer is typically given in decimal notation. To convert this to fraction notation, we press **MATH** and select the FRAC option. The notation ANS ▶ FRAC shows that the calculator will convert .7166666667 to fraction notation.

```
2/15+7/12
                  .7166666667
Ans▶Frac
                        43/60
```

We see that $\frac{2}{15} + \frac{7}{12} = \frac{43}{60}$.

1.3 EXERCISE SET

FOR EXTRA HELP

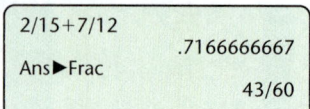 MyMathLab® Math XL PRACTICE WATCH READ REVIEW

🍃 Vocabulary and Reading Check

Choose from the following list of words to complete each statement. Not every word will be used.

add numerator
composite opposite
denominator prime
multiply reciprocal

1. The top number in a fraction is called the
   _____.

2. A(n) _____ number has exactly two different factors.

3. To divide two fractions, multiply by the _____ of the divisor.

4. We need a common denominator in order to _____ fractions.

🍃 Concept Reinforcement

In each of Exercises 5–8, match the description with a number from the list on the right.

5. ____ A factor of 35   **a)** 2

6. ____ A number that has 3 as a factor   **b)** 7

7. ____ An odd composite number   **c)** 60

8. ____ The only even prime number   **d)** 65

To the student and the instructor: *Beginning in this section, selected exercises are marked with the symbol* Aha! *. Students who pause to inspect an Aha! exercise should find the answer more readily than those who proceed mechanically. This is done to discourage rote memorization. Some later "Aha!" exercises in this exercise set are unmarked, to encourage students to always pause before working a problem.*

Factors and Prime Factorizations

Label each of the following numbers as prime, composite, or neither.

9. 9   **10.** 15   **11.** 41   **12.** 49

13. 77   **14.** 37   **15.** 2   **16.** 1

17. 0   **18.** 16

List all the factors of each number.

19. 50   **20.** 70   **21.** 42   **22.** 60

Find the prime factorization of each number. If the number is prime, state this.

23. 39   **24.** 34   **25.** 30

26. 98   **27.** 27   **28.** 54

29. 150   **30.** 56   **31.** 31

32. 180   **33.** 210   **34.** 79

35. 115   **36.** 143

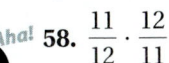

Simplification of Fractions

Simplify.

37. $\dfrac{21}{35}$ **38.** $\dfrac{20}{26}$ **39.** $\dfrac{16}{56}$

40. $\dfrac{72}{27}$ **41.** $\dfrac{12}{48}$ **42.** $\dfrac{18}{84}$

43. $\dfrac{52}{13}$ **44.** $\dfrac{132}{11}$ **45.** $\dfrac{19}{76}$

46. $\dfrac{17}{51}$ **47.** $\dfrac{150}{25}$ **48.** $\dfrac{180}{36}$

49. $\dfrac{42}{50}$ **50.** $\dfrac{75}{80}$ **51.** $\dfrac{120}{82}$

52. $\dfrac{75}{45}$ **53.** $\dfrac{210}{98}$ **54.** $\dfrac{140}{350}$

Multiplication, Division, Addition, and Subtraction of Fractions

Perform the indicated operation and, if possible, simplify.

55. $\dfrac{1}{2} \cdot \dfrac{3}{5}$ **56.** $\dfrac{11}{10} \cdot \dfrac{8}{5}$ **57.** $\dfrac{9}{2} \cdot \dfrac{4}{3}$

Aha! **58.** $\dfrac{11}{12} \cdot \dfrac{12}{11}$ **59.** $\dfrac{1}{8} + \dfrac{3}{8}$ **60.** $\dfrac{1}{2} + \dfrac{1}{8}$

61. $\dfrac{4}{9} + \dfrac{13}{18}$ **62.** $\dfrac{4}{5} + \dfrac{8}{15}$ **63.** $\dfrac{3}{a} \cdot \dfrac{b}{7}$

64. $\dfrac{x}{5} \cdot \dfrac{y}{z}$ **65.** $\dfrac{4}{n} + \dfrac{6}{n}$ **66.** $\dfrac{9}{x} - \dfrac{5}{x}$

67. $\dfrac{3}{10} + \dfrac{8}{15}$ **68.** $\dfrac{7}{8} + \dfrac{5}{12}$ **69.** $\dfrac{11}{7} - \dfrac{4}{7}$

70. $\dfrac{12}{5} - \dfrac{2}{5}$ **71.** $\dfrac{13}{18} - \dfrac{4}{9}$ **72.** $\dfrac{13}{15} - \dfrac{11}{45}$

Aha! **73.** $\dfrac{20}{30} - \dfrac{2}{3}$ **74.** $\dfrac{5}{7} - \dfrac{5}{21}$ **75.** $\dfrac{7}{6} \div \dfrac{3}{5}$

76. $\dfrac{7}{5} \div \dfrac{10}{3}$ **77.** $12 \div \dfrac{4}{9}$ **78.** $\dfrac{9}{4} \div 9$

Aha! **79.** $\dfrac{7}{13} \div \dfrac{7}{13}$ **80.** $\dfrac{1}{10} \div \dfrac{1}{5}$ **81.** $\dfrac{\frac{2}{7}}{\frac{5}{3}}$

82. $\dfrac{\frac{3}{8}}{\frac{1}{5}}$ **83.** $\dfrac{9}{\frac{1}{2}}$ **84.** $\dfrac{\frac{3}{7}}{6}$

85. Under what circumstances would the sum of two fractions be easier to compute than the product of the same two fractions?

 86. Under what circumstances would the product of two fractions be easier to compute than the sum of the same two fractions?

Synthesis

 87. Bryce insists that $(2 + x)/8$ is equivalent to $(1 + x)/4$. What mistake do you think is being made and how could you demonstrate to Bryce that the two expressions are not equivalent?

 88. *Research.* Mathematicians use computers to determine whether very large numbers are prime. Find the largest prime number currently known, and describe how it was found.

89. In the following table, the top number can be factored in such a way that the sum of the factors is the bottom number. For example, in the first column, 56 is factored as $7 \cdot 8$, since $7 + 8 = 15$, the bottom number. Find the missing numbers in each column.

Product	56	63	36	72	140	96	168
Factor	7						
Factor	8						
Sum	15	16	20	38	24	20	29

90. *Packaging.* Tritan Candies uses two sizes of boxes, 6 in. long and 8 in. long. These are packed end to end in larger cartons to be shipped. What is the shortest-length carton that will accommodate boxes of either size without any room left over? (Each carton must contain boxes of only one size; no mixing is allowed.)

Simplify.

91. $\dfrac{16 \cdot 9 \cdot 4}{15 \cdot 8 \cdot 12}$ **92.** $\dfrac{9 \cdot 8xy}{2xy \cdot 36}$

93. $\dfrac{45pqrs}{9prst}$ **94.** $\dfrac{247}{323}$

95. $\dfrac{15 \cdot 4xy \cdot 9}{6 \cdot 25x \cdot 15y}$

96. $\dfrac{10x \cdot 12 \cdot 25y}{2z \cdot 30x \cdot 20y}$

97. $\dfrac{\dfrac{27ab}{15mn}}{\dfrac{18bc}{25np}}$

98. $\dfrac{\dfrac{45xyz}{24ab}}{\dfrac{30xz}{32ac}}$

99. $\dfrac{5\frac{3}{4}rs}{4\frac{1}{2}st}$

100. $\dfrac{3\frac{5}{7}mn}{2\frac{4}{5}np}$

Find the area of each figure.

101.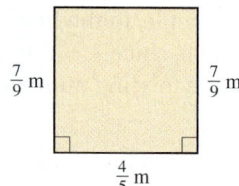

102.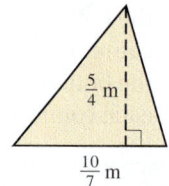

103. Find the perimeter of a square with sides of length $3\frac{5}{9}$ m.

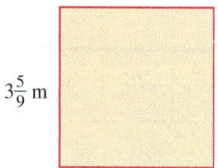

104. Find the perimeter of the rectangle in Exercise 101.

105. Find the total length of the edges of a cube with sides of length $2\frac{3}{10}$ cm.

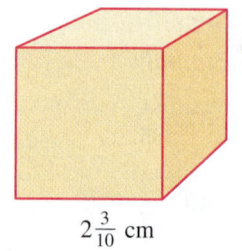

$2\frac{3}{10}$ cm

YOUR TURN ANSWERS: SECTION 1.3

1. 1, 2, 3, 6, 9, 18, 27, 54 **2.** Composite **3.** $2 \cdot 2 \cdot 5 \cdot 5$
4. $\dfrac{6}{7x}$ **5.** $\dfrac{21}{20}$ **6.** $\dfrac{7}{6}$ **7.** $\dfrac{1}{3}$ **8.** $\dfrac{6}{5}$ **9.** $\dfrac{77}{40}$ **10.** $\dfrac{1}{5}$

QUICK QUIZ: SECTIONS 1.1–1.3

1. Translate to an algebraic expression: 4 less than the width of a box. [1.1]

2. Use a commutative law to write an expression equivalent to $5(x + 3)$. There is more than one correct answer. [1.2]

3. Find the prime factorization of 40. [1.3]

4. Simplify: $\dfrac{54}{66}$. [1.3]

5. Add and, if possible, simplify: $\dfrac{4}{x} + \dfrac{7}{x}$. [1.3]

1.4   Positive and Negative Real Numbers

Whole Numbers and Integers ▪ The Rational Numbers ▪ Real Numbers and Order ▪ Absolute Value

A **set** is a collection of objects. The set containing 1, 3, and 7 is usually written $\{1, 3, 7\}$. In this section, we examine some important sets of numbers.

Whole Numbers and Integers

Whole numbers = {0, 1, 2, 3, ...}

The set of **whole numbers** can be written as $\{0, 1, 2, 3, \ldots\}$. We represent this set using dots on the number line, as shown at left.

To create the set of integers, we include all whole numbers, along with their *opposites*. To find the opposite of a number, we locate the number that is the same distance from 0 but on the other side of 0 on the number line. For example,

the opposite of 1 is negative 1, written -1;

and

the opposite of 3 is negative 3, written -3.

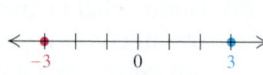

The **integers** consist of all whole numbers and their opposites. Note that, except for 0, opposites occur in pairs. Thus, 5 is the opposite of −5, just as −5 is the opposite of 5. The number 0 acts as its own opposite.

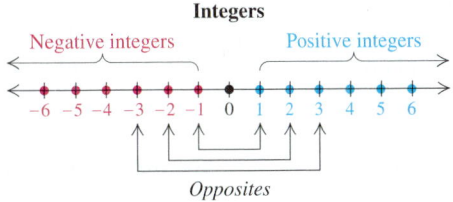

SET OF INTEGERS

The set of integers = {. . . , −4, −3, −2, −1, 0, 1, 2, 3, 4, . . .}.

Integers are associated with many real-world problems and situations.

EXAMPLE 1 State which integer(s) corresponds to each situation.

a) In 2011, there was $13.7 trillion in outstanding mortgage debt in the United States.

   Source: Board of Governors of the Federal Reserve System

b) Badwater Basin in Death Valley, California, is 282 ft below sea level.

SOLUTION

a) The integer −13,700,000,000,000 corresponds to a debt of $13.7 trillion.

b) The integer −282 corresponds to 282 ft below sea level. The elevation is −282 ft.

1. State what integers correspond to this situation: The highest temperature ever recorded in a desert was 136 degrees Fahrenheit (°F) at Al-Aziziyah in the Sahara, Libya. The coldest temperature ever recorded in a desert was 4°F below zero in the McMurdo Dry Valleys, Antarctica.

   Source: Guinness World Records

 YOUR TURN

The Rational Numbers

A number like $\frac{5}{9}$, although built out of integers, is not itself an integer. Another set of numbers, the **rational numbers**, contains integers, fractions, and decimals. Some examples of rational numbers are

$$\frac{5}{9}, \quad -\frac{4}{7}, \quad 95, \quad -16, \quad 0, \quad \frac{-35}{8}, \quad 2.4, \quad -0.31.$$

The number $-\frac{4}{7}$ can be written as $\frac{-4}{7}$ or $\frac{4}{-7}$. Indeed, every number listed above can be written as an integer over an integer. For example, 95 can be written as $\frac{95}{1}$ and 2.4 can be written as $\frac{24}{10}$. In this manner, any *rational* number can be expressed as the *ratio* of two integers. Rather than attempt to list all rational numbers, we use this idea of ratio to describe the set as follows.

> **SET OF RATIONAL NUMBERS**
>
> The set of rational numbers = $\left\{\dfrac{a}{b} \,\middle|\, a \text{ and } b \text{ are integers and } b \neq 0\right\}$.
>
> This is read "the set of all numbers a over b, such that a and b are integers and b does not equal zero."

To *graph* a number is to mark its location on the number line.

EXAMPLE 2   Graph each of the following rational numbers: **(a)** $\frac{5}{2}$; **(b)** -3.2; **(c)** $\frac{11}{8}$.

SOLUTION

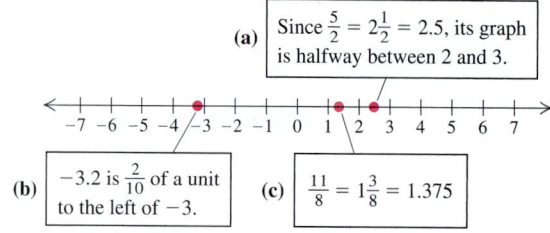

(a) Since $\frac{5}{2} = 2\frac{1}{2} = 2.5$, its graph is halfway between 2 and 3.

(b) -3.2 is $\frac{2}{10}$ of a unit to the left of -3.

(c) $\frac{11}{8} = 1\frac{3}{8} = 1.375$

2. Graph -1.1 on the number line.

YOUR TURN

Every rational number can be written using fraction notation or decimal notation.

EXAMPLE 3   Convert to decimal notation: $-\frac{5}{8}$.

SOLUTION   We first find decimal notation for $\frac{5}{8}$. Since $\frac{5}{8}$ means $5 \div 8$, we divide.

$$
\begin{array}{r}
0.6\,2\,5 \\
8{\overline{)5.0\,0\,0}} \\
\underline{4\,8} \\
2\,0 \\
\underline{1\,6} \\
4\,0 \\
\underline{4\,0} \\
0
\end{array}
$$

 ← The remainder is 0.

3. Convert $-\dfrac{7}{4}$ to decimal notation.

Thus, $\frac{5}{8} = 0.625$, so $-\frac{5}{8} = -0.625$.

YOUR TURN

Because the division in Example 3 ends with the remainder 0, we consider -0.625 a **terminating decimal**. If we are "bringing down" zeros and a remainder reappears, we have a **repeating decimal**, as shown in the next example.

EXAMPLE 4   Convert to decimal notation: $\frac{7}{11}$.

SOLUTION   We divide:

$$
\begin{array}{r}
0.6\,3\,6\,3\ldots \\
11\overline{)7.0\,0\,0\,0} \\
\underline{6\,6} \\
4\,0 \\
\underline{3\,3} \\
7\,0 \\
\underline{6\,6} \\
4\,0
\end{array}
$$

4 reappears as a remainder, so the pattern of 6's and 3's in the quotient will repeat.

4. Convert $-\dfrac{1}{6}$ to decimal notation.

We abbreviate repeating decimals by writing a bar over the repeating part—in this case, $0.\overline{63}$. Thus, $\frac{7}{11} = 0.\overline{63}$.

YOUR TURN

Although we do not prove it here, every rational number can be expressed as either a terminating decimal or a repeating decimal, and every terminating decimal or repeating decimal is a rational number.

Real Numbers and Order

Some numbers, when written in decimal form, neither terminate nor repeat. Such numbers are called **irrational numbers**. These numbers cannot be expressed as a ratio of two integers.

What sort of numbers are irrational? One example is π (the Greek letter *pi*, read "pie"), which is used to find the area and the circumference of a circle: $A = \pi r^2$ and $C = 2\pi r$.

Another irrational number, $\sqrt{2}$ (read "the square root of 2"), is the length of the diagonal of a square with sides of length 1. It is also the number that, when multiplied by itself, gives 2. No rational number can be multiplied by itself to get 2, although some approximations come close:

1.4 is an *approximation* of $\sqrt{2}$ because $(1.4)(1.4) = 1.96$;

1.41 is a better approximation because $(1.41)(1.41) = 1.9881$;

1.4142 is an even better approximation because $(1.4142)(1.4142) = 1.99996164$.

To approximate $\sqrt{2}$ on some calculators, we simply press ② and then ✓. With other calculators, we press ✓ ② **ENTER**. If needed, consult your manual.

EXAMPLE 5   Graph the real number $\sqrt{3}$ on the number line.

SOLUTION   We use a calculator and approximate: $\sqrt{3} \approx 1.732$ ("$\approx$" means "approximately equals"). Then we locate this number on the number line.

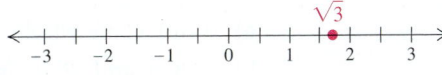

5. Graph $\sqrt{5}$ on the number line.

YOUR TURN

Not every square root is irrational. For example, $\sqrt{9} = 3$, so $\sqrt{9}$ is a rational number.

The rational numbers and the irrational numbers together correspond to all the points on the number line and make up what is called the **real-number system**.

> **SET OF REAL NUMBERS**
>
> The set of real numbers = The set of all numbers corresponding to points on the number line.

The following figure shows the relationships among various kinds of numbers.

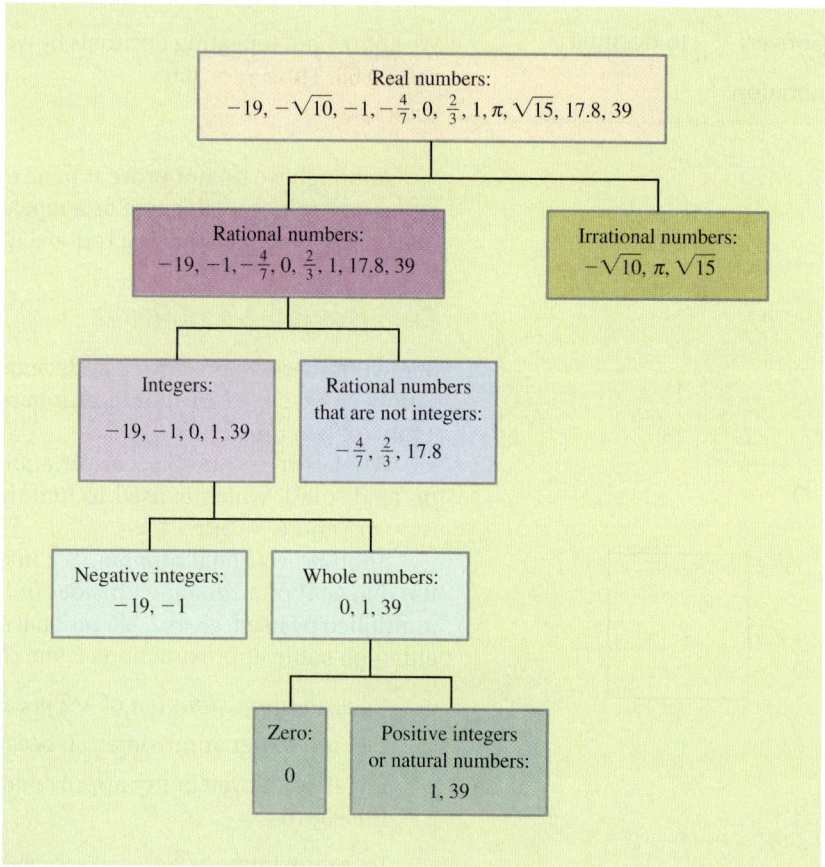

EXAMPLE 6   Which numbers in the following group are **(a)** whole numbers? **(b)** integers? **(c)** rational numbers? **(d)** irrational numbers? **(e)** real numbers?

$$-38, \quad -\frac{8}{5}, \quad 0, \quad 0.\overline{3}, \quad 4.5, \quad \sqrt{30}, \quad 52$$

SOLUTION

a) 0 and 52 are whole numbers.

b) -38, 0, and 52 are integers.

c) -38, $-\frac{8}{5}$, 0, $0.\overline{3}$, 4.5, and 52 are rational numbers.

d) $\sqrt{30}$ is an irrational number.

e) -38, $-\frac{8}{5}$, 0, $0.\overline{3}$, 4.5, $\sqrt{30}$, and 52 are real numbers.

6. Which numbers in the following group are integers?

$$5, \quad -15, \quad 0, \quad -3.6, \quad \frac{2}{3}$$

YOUR TURN

EXPLORING 🔍 THE CONCEPT

Use the number line shown at right to answer the following questions.

1. Graph the numbers 2 and 5.
   a) Which number is farther to the right on the number line?
   b) Which number is greater?
2. Graph the numbers 2 and −5.
   a) Which number is farther to the right on the number line?
   b) Which number is greater?
3. Graph the numbers −2 and −5.
   a) Which number is farther to the right on the number line?
   b) Which number is greater?
4. Why is 4 > 1, but −4 < −1?

ANSWERS

1. (a) 5; (b) 5   2. (a) 2; (b) 2   3. (a) −2; (b) −2
4. 4 is to the right of 1 on the number line, and −1 is to the right of −4 on the number line.

Real numbers are named in order on the number line, with larger numbers farther to the right. For any two numbers, the one to the left is less than the one to the right. We use the symbol **<** to mean "**is less than.**" The sentence −8 < 6 means "−8 is less than 6." The symbol **>** means "**is greater than.**" The sentence −3 > −7 means "−3 is greater than −7."

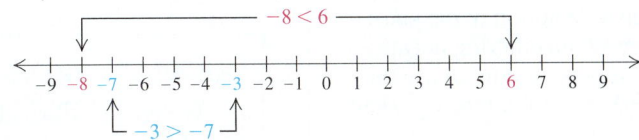

EXAMPLE 7   Use either < or > for ▨ to write a true sentence.

a) 2 ▨ 9

b) −3.45 ▨ 1.32

c) 6 ▨ −12

d) −18 ▨ −5

e) $\frac{7}{11}$ ▨ $\frac{5}{8}$

SOLUTION

a) Since 2 is to the left of 9 on the number line, we know that 2 is less than 9, so 2 < 9.

b) Since −3.45 is to the left of 1.32, we have −3.45 < 1.32.

c) Since 6 is to the right of −12, we have 6 > −12.

d) Since −18 is to the left of −5, we have −18 < −5.

e) We convert to decimal notation: $\frac{7}{11} = 0.\overline{63}$ and $\frac{5}{8} = 0.625$. Thus, $\frac{7}{11} > \frac{5}{8}$.

We also could have used a common denominator: $\frac{7}{11} = \frac{56}{88} > \frac{55}{88} = \frac{5}{8}$.

7. Use either < or > for ▨ to write a true sentence: −8 ▨ −9.

↩ YOUR TURN

Student Notes

It is important to remember that just because an equation or inequality is written or printed, it is not necessarily *true*. For instance, 6 = 7 is an equation and 2 > 5 is an inequality. Of course, both statements are *false*.

Sentences like "$a < -5$" and "$-3 > -8$" are **inequalities**. It may be helpful to think of an inequality sign as "opening" toward the larger number, or as being widest next to the larger number.

Note that $a > 0$ means that a represents a positive real number and $a < 0$ means that a represents a negative real number.

Statements like $a \le b$ and $b \ge a$ are also inequalities. We read $a \le b$ as "a **is less than or equal to** b" and $a \ge b$ as "a **is greater than or equal to** b."

EXAMPLE 8   Write a second inequality with the same meaning as $-11 \le -3$.

SOLUTION   Every inequality can be written in two ways:

$$-11 \le -3 \quad \text{has the same meaning as} \quad -3 \ge -11.$$

If we read both inequalities, we say that

−11 is less than or equal to −3,   and

−3 is greater than or equal to −11.

These statements have the same meaning.

8. Write a second inequality with the same meaning as 5 > *y*.

↩ YOUR TURN

Absolute Value

The distance that a number is from 0 on the number line is called the **absolute value** of the number.

Distance is never negative, so numbers that are opposites have the same absolute value. If a number is nonnegative, its absolute value is the number itself. If a number is negative, its absolute value is its opposite.

> **ABSOLUTE VALUE**
>
> We write $|a|$, read "the absolute value of a," to represent the number of units that a is from zero.

EXAMPLE 9 Find each absolute value: **(a)** $|-3|$; **(b)** $|7.2|$; **(c)** $|0|$.

SOLUTION

a) $|-3| = 3$ since -3 is 3 units from 0.

b) $|7.2| = 7.2$ since 7.2 is 7.2 units from 0.

c) $|0| = 0$ since 0 is 0 units from itself.

9. Find the absolute value: $|-18|$.
_____

↪ YOUR TURN

1.4   EXERCISE SET   FOR EXTRA HELP

MyMathLab® MathXL
PRACTICE   WATCH   READ   REVIEW

↪ Vocabulary and Reading Check

Choose from the following list of words to complete each statement. Not every word will be used.

absolute value   rational number
integer   repeating
irrational number   terminating
opposite   whole number

1. Since $\frac{3}{20} = 0.15$, we can write $\frac{3}{20}$ as a(n) _____ decimal.

2. If a number is a(n) _____, it is either a whole number or the opposite of a whole number.

3. 0 is the only _____ that is not a natural number.

4. A number like $\sqrt{5}$, which cannot be written precisely in fraction notation or decimal notation, is an example of a(n) _____.

5. The _____ of 1 is -1.

6. When two numbers are opposites, they have the same _____.

↪ Concept Reinforcement

Translate each phrase to an algebraic expression.

7. The opposite of n

8. The absolute value of x

9. -10 is less than x

10. 6 is greater than or equal to y

Whole Numbers and Integers

State which real numbers correspond to each situation.

11. *Student Loans and Grants.* The maximum amount that an independent undergraduate first-year student can borrow with a Direct Stafford Loan is $9500. The maximum annual award for the Landscape Architecture Foundation Scholarship is $5000.

Sources: www.studentaid.ed.gov and www.collegescholarships.org

12. Using a NordicTrack exercise machine, Kylie burned 150 calories. She then drank an isotonic drink containing 65 calories.

13. *Record Temperature.* The highest temperature recorded in Alaska is 100 degrees Fahrenheit (°F) at Fort Yukon. The lowest temperature recorded in Alaska is 80°F below zero at Prospect Creek Camp.

Source: www.netstate.com

14. The Dead Sea is 1349 ft below sea level, whereas Mt. Everest is 29,035 ft above sea level.

Source: National Geographic

15. *Stock Market.* The Dow Jones Industrial Average is an indicator of the stock market. On September 29, 2008, the Dow Jones fell a record 777.68 points. On October 13, 2008, the Dow Jones gained a record 936.42 points.

Sources: CME Group Index Services, LLC; Dow Jones Indexes

16. Ignition occurs 10 sec before liftoff. A spent fuel tank is detached 235 sec after liftoff.

17. Kittling County Volunteers received a technology grant of $10,000 and spent $4500 to update its website.

18. Mina deposited $650 in a savings account. Two weeks later, she withdrew $180.

19. The New York Giants gained 8 yd on the first play. They lost 5 yd on the next play.

20. In the 2011 British Open Championship, golfer Rory McIlroy finished 7 over par. In the U.S. Open Championship, he finished 16 under par.

Source: PGA Tour Inc.

The Rational Numbers

Graph each rational number on the number line.

21. -2 **22.** 5

23. -4.3 **24.** 3.87

25. $\frac{10}{3}$ **26.** $-\frac{17}{5}$

Write decimal notation for each number.

27. $\frac{7}{8}$ **28.** $-\frac{1}{8}$ **29.** $-\frac{3}{4}$

30. $\frac{11}{6}$ **31.** $-\frac{7}{6}$ **32.** $-\frac{5}{12}$

33. $\frac{2}{3}$ **34.** $\frac{1}{4}$ **35.** $-\frac{1}{2}$

36. $-\frac{1}{9}$ *Aha!* **37.** $\frac{13}{100}$ **38.** $-\frac{9}{20}$

Real Numbers and Order

Graph each irrational number on the number line.

39. $\sqrt{5}$ **40.** $\sqrt{92}$

41. $-\sqrt{22}$ **42.** $-\sqrt{54}$

Write a true sentence using either $<$ or $>$.

43. $5 \ \square \ 0$ **44.** $8 \ \square \ -8$

45. $-9 \ \square \ 9$ **46.** $0 \ \square \ -7$

47. $-8 \ \square \ -5$ **48.** $-4 \ \square \ -3$

49. $-5 \ \square \ -11$ **50.** $-3 \ \square \ -4$

51. $-12.5 \ \square \ -10.2$ **52.** $-10.3 \ \square \ -14.5$

53. $\frac{5}{12} \ \square \ \frac{11}{25}$ **54.** $-\frac{14}{17} \ \square \ -\frac{27}{35}$

For each of the following, write a second inequality with the same meaning.

55. $-2 > x$ **56.** $a > 9$

57. $10 \le y$ **58.** $-12 \ge t$

For Exercises 59–64, consider the following list:

$$-83, \quad -4.7, \quad 0, \quad \tfrac{5}{9}, \quad 2.\overline{16}, \quad \pi, \quad \sqrt{17}, \quad 62.$$

59. List all rational numbers.

60. List all natural numbers.

61. List all integers.

62. List all irrational numbers.

63. List all real numbers.

64. List all nonnegative integers.

Absolute Value

Find each absolute value.

65. $|-58|$

66. $|-47|$

67. $|-12.2|$

68. $|4.3|$

69. $|\sqrt{2}|$

70. $|-456|$

71. $\left|-\frac{9}{7}\right|$

72. $|-\sqrt{3}|$

73. $|0|$

74. $\left|-\frac{3}{4}\right|$

75. $|x|$, for $x = -8$

76. $|a|$, for $a = -5$

77. Is every integer a rational number? Why or why not?

78. Is every integer a natural number? Why or why not?

Synthesis

79. Is the absolute value of a number always positive? Why or why not?

80. How many rational numbers are there between 0 and 1? Justify your answer.

81. Does "nonnegative" mean the same thing as "positive"? Why or why not?

List in order from least to greatest.

82. $13, -12, 5, -17$

83. $-23, 4, 0, -17$

84. $-\frac{2}{3}, \frac{1}{2}, -\frac{3}{4}, -\frac{5}{6}, \frac{3}{8}, \frac{1}{6}$

85. $\frac{4}{5}, \frac{4}{3}, \frac{4}{8}, \frac{4}{6}, \frac{4}{9}, \frac{4}{2}, -\frac{4}{3}$

Write a true sentence using either $<$, $>$, or $=$.

86. $|-5| \; \blacksquare \; |-2|$

87. $|4| \; \blacksquare \; |-7|$

88. $|-8| \; \blacksquare \; |8|$

89. $|23| \; \blacksquare \; |-23|$

Solve. Consider only integer replacements.

Aha! **90.** $|x| = 19$

91. $|x| < 3$

92. $2 < |x| < 5$

Given that $0.3\overline{3} = \frac{1}{3}$ and $0.6\overline{6} = \frac{2}{3}$, express each of the following as a ratio of two integers.

93. $0.1\overline{1}$

94. $0.9\overline{9}$

95. $5.5\overline{5}$

96. $7.7\overline{7}$

Translate to an inequality.

97. A number a is negative.

98. A number x is nonpositive.

99. The distance from x to 0 is no more than 10.

100. The distance from t to 0 is at least 20.

To the student and the instructor: *The calculator icon,* , *indicates those exercises designed to be solved with a calculator.*

101. When Helga's calculator gives a decimal value for $\sqrt{2}$ and that value is promptly squared, the result is 2. Yet when that same decimal approximation is entered by hand and then squared, the result is not exactly 2. Why do you suppose this is?

102. Is the following statement true? Why or why not?
$$\sqrt{a^2} = |a| \quad \text{for any real number } a.$$

YOUR TURN ANSWERS: SECTION 1.4

1. $136; -4$ **2.**
$$\begin{array}{c} -1.1 \\ \xleftrightarrow{\;\;\mid\;\;\mid\;\bullet\;\mid\;\;\mid\;\;\mid\;\;\mid\;\;\mid\;} \\ {\scriptstyle -4 \;\; -2 \;\;\; 0 \;\;\;\; 2 \;\;\;\; 4} \end{array}$$
3. -1.75

4. $-0.1\overline{6}$ **5.**
$$\begin{array}{c} \sqrt{5} \\ \xleftrightarrow{\;\;\mid\;\;\mid\;\;\mid\;\;\mid\;\;\mid\;\bullet\;\mid\;\;\mid\;} \\ {\scriptstyle -4 \;\; -2 \;\;\; 0 \;\;\;\; 2 \;\;\;\; 4} \end{array}$$
6. $5, -15, 0$

7. $>$ **8.** $y < 5$ **9.** 18

QUICK QUIZ: SECTIONS 1.1–1.4

1. Evaluate $12 - c$ for $c = 11$.   [1.1]

2. Factor: $5x + 5y + 15$.   [1.2]

3. Subtract and, if possible, simplify: $\frac{1}{2} - \frac{1}{3}$.   [1.3]

4. Divide and, if possible, simplify: $\frac{4}{5} \div \frac{2}{5}$.   [1.3]

5. Write a true sentence using either $<$ or $>$: $0 \; \square \; -0.5$.   [1.4]

Mid-Chapter Review

An introduction to algebra involves learning some basic laws and terms.

Commutative Laws: $a + b = b + a;\ ab = ba$

Associative Laws: $a + (b + c) = (a + b) + c;\ a(bc) = (ab)c$

Distributive Law: $a(b + c) = ab + ac$

GUIDED SOLUTIONS

1. Evaluate $\dfrac{x - y}{3}$ for $x = 22$ and $y = 10$.   [1.1]*

Solution

$$\dfrac{x - y}{3} = \dfrac{\boxed{} - \boxed{}}{3} \quad \text{Substituting}$$

$$= \dfrac{\boxed{}}{3} \quad \text{Subtracting}$$

$$= \boxed{} \quad \text{Dividing}$$

2. Factor: $14x + 7$.   [1.2]

Solution

$$14x + 7 = \boxed{} \cdot 2x + \boxed{} \cdot 1$$

Factoring each term using a common factor

$$= \boxed{}(2x + 1) \quad \begin{array}{l}\text{Factoring out the}\\ \text{common factor}\end{array}$$

MIXED REVIEW

Evaluate.   [1.1]

3. $x + y$, for $x = 3$ and $y = 12$

4. $\dfrac{2a}{5}$, for $a = 10$

Translate to an algebraic expression.   [1.1]

5. 10 less than d

6. The product of 8 and the number of hours worked

7. Translate to an equation. Do not solve.   [1.1]

Janine's class has 27 students. This is 5 fewer than the number originally enrolled. How many students originally enrolled in the class?

8. Determine whether 8 is a solution of $13t = 94$. [1.1]

9. Use the commutative law of addition to write an expression equivalent to $7 + 10x$.   [1.2]

10. Use the associative law of multiplication to write an expression equivalent to $3(ab)$.   [1.2]

Multiply.   [1.2]

11. $4(2x + 8)$

12. $3(2m + 5n + 10)$

Factor.   [1.2]

13. $18x + 15$

14. $9c + 12d + 3$

15. Find the prime factorization of 84.   [1.3]

16. Simplify: $\frac{135}{315}$.

Perform the indicated operation and, if possible, simplify.   [1.3]

17. $\dfrac{11}{12} - \dfrac{3}{8}$

18. $\dfrac{8}{15} \div \dfrac{6}{11}$

19. Graph -2.5 on the number line.   [1.4]

20. Write decimal notation for $-\frac{3}{20}$.   [1.4]

Write a true sentence using either < or >.   [1.4]

21. $-16\ \square\ -24$

22. $-\frac{3}{22}\ \square\ -\frac{2}{15}$

23. Write a second inequality with the same meaning as $x \geq 9$.   [1.4]

Find the absolute value.   [1.4]

24. $|-5.6|$

25. $|0|$

---

*The *section reference* [1.1] refers to Chapter 1, Section 1. The concept reviewed in Guided Solution 1 was developed in this section.

1.5 Addition of Real Numbers

Adding with the Number Line ▪ Adding Without the Number Line ▪ Problem Solving ▪ Combining Like Terms

We now consider addition of real numbers. To gain understanding, we will use the number line first and then develop rules that allow us to add without the number line.

Adding with the Number Line

To add $a + b$ on the number line, we start at a and move according to b.

a) If b is positive, we move to the right (the positive direction).

b) If b is negative, we move to the left (the negative direction).

c) If b is 0, we stay at a.

EXAMPLE 1   Add: $-4 + 9$.

SOLUTION   To add on the number line, we locate the first number, -4, and then move 9 units to the right. Note that it requires 4 units to reach 0. The difference between 9 and 4 is where we finish.

$$-4 + 9 = 5$$

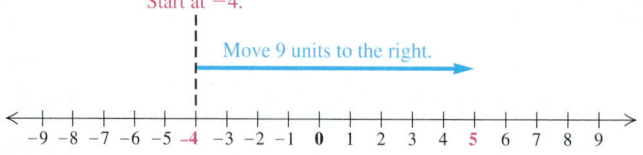

1. Use the number line to add $-3 + 5$.

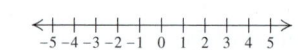

YOUR TURN

EXAMPLE 2   Add: $3 + (-5)$.

SOLUTION   We locate the first number, 3, and then move 5 units to the left. Note that it requires 3 units to reach 0. The difference between 5 and 3 is 2, so we finish 2 units to the left of 0, at -2.

$$3 + (-5) = -2$$

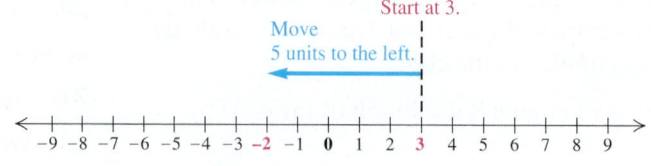

2. Use the number line to add $1 + (-4)$.

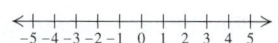

YOUR TURN

EXAMPLE 3   Add: $-4 + (-3)$.

SOLUTION   After locating -4, we move 3 units to the left. We finish a total of 7 units to the left of 0, at -7.

$$-4 + (-3) = -7$$

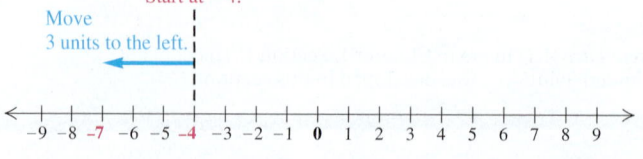

3. Use the number line to add $-2 + (-1)$.

YOUR TURN

EXAMPLE 4   Add: $-5.2 + 0$.

SOLUTION   We locate -5.2 and move 0 units. Thus we finish where we started, at -5.2.

$$-5.2 + 0 = -5.2$$

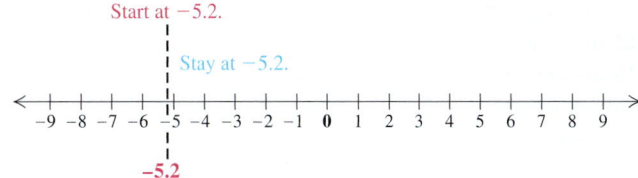

4. Use the number line to add $-4 + 0$.

_____

 YOUR TURN

From Examples 1–4, we can develop the following rules.

RULES FOR ADDITION OF REAL NUMBERS

1. *Positive numbers*: Add as usual. The answer is positive.
2. *Negative numbers*: Add absolute values and make the answer negative (see Example 3).
3. *A positive number and a negative number*: Subtract the smaller absolute value from the greater absolute value. Then:

   a) If the positive number has the greater absolute value, the answer is positive (see Example 1).
   b) If the negative number has the greater absolute value, the answer is negative (see Example 2).
   c) If the numbers have the same absolute value, the answer is 0.

4. *One number is zero*: The sum is the other number (see Example 4).

Rule 4 is known as the **identity property of 0**.

IDENTITY PROPERTY OF 0

For any real number a,

$$a + 0 = 0 + a = a.$$

(Adding 0 to a number gives that same number.) The number 0 is called the *additive identity*.

Adding Without the Number Line

The rules listed above can be used without drawing the number line.

EXAMPLE 5   Add without using the number line.

a) $-12 + (-7)$ **b)** $-1.4 + 8.5$

c) $-36 + 21$ **d)** $1.5 + (-1.5)$

e) $-\frac{7}{8} + 0$ **f)** $\frac{2}{3} + \left(-\frac{5}{8}\right)$

Student Notes

Parentheses are essential when a negative sign follows an operation. Just as we would never write $8 \div \times 2$, it is improper to write $-12 + -7$.

5. Add without using the number line:

$$2.3 + (-9.1).$$

SOLUTION

a) $-12 + (-7) = -19$

Two negatives. *Think:* Add the absolute values, 12 and 7, to get 19. Make the answer *negative*, -19.

b) $-1.4 + 8.5 = 7.1$

A negative and a positive. *Think:* The difference of absolute values is $8.5 - 1.4$, or 7.1. The positive number has the greater absolute value, so the answer is *positive*, 7.1.

c) $-36 + 21 = -15$

A negative and a positive. *Think:* The difference of absolute values is $36 - 21$, or 15. The negative number has the greater absolute value, so the answer is *negative*, -15.

d) $1.5 + (-1.5) = 0$

A negative and a positive. *Think:* Since the numbers are opposites, they have the same absolute value and the answer is 0.

e) $-\dfrac{7}{8} + 0 = -\dfrac{7}{8}$

One number is zero. The sum is the other number, $-\frac{7}{8}$.

f) $\dfrac{2}{3} + \left(-\dfrac{5}{8}\right) = \dfrac{16}{24} + \left(-\dfrac{15}{24}\right)$

$$= \dfrac{1}{24}$$

This is similar to part (b) above. We find a common denominator and then add.

 YOUR TURN

If we are adding several numbers, some positive and some negative, the commutative and associative laws allow us to add all the positives, then add all the negatives, and then add the results. Of course, we can also add from left to right, if we prefer.

EXAMPLE 6   Add: $15 + (-2) + 7 + 14 + (-5) + (-12)$.

SOLUTION

$$15 + (-2) + 7 + 14 + (-5) + (-12)$$

$$= 15 + 7 + 14 + (-2) + (-5) + (-12)$$ Using the commutative law of addition

$$= (15 + 7 + 14) + [(-2) + (-5) + (-12)]$$ Using the associative law of addition

$$= 36 + (-19)$$ Adding the positives; adding the negatives

$$= 17$$ Adding a positive and a negative

6. Add:

$$-3 + 10 + (-11) + (-1) + 4.$$

YOUR TURN

Problem Solving

EXAMPLE 7   *Credit-Card Interest Rates.*   During one four-week period in 2011, the national average credit-card interest rate dropped 0.09%, then rose 0.16%, stayed the same, and then dropped 0.02%. By how much did the national average credit-card interest rate change?

Source: CreditCards.com Weekly Credit Card Rate Report

7. Refer to Example 7. In a different four-week period, the interest rate rose 0.01%, dropped 0.08%, dropped 0.01%, and then did not change. By how much did the interest rate change?

SOLUTION The problem translates to a sum:

Rewording: The 1st change plus the 2nd change plus the 3rd change plus the 4th change is the total change.

Translating: $-0.09 + 0.16 + 0 + (-0.02) = $ Total change

Adding from left to right, we have

$-0.09 + 0.16 + 0 + (-0.02) = 0.07 + 0 + (-0.02) = 0.07 + (-0.02) = 0.05.$

The national average credit-card interest rate rose 0.05% over the four-week period.

YOUR TURN

Combining Like Terms

When two terms have variable factors that are exactly the same, like $5a$ and $-7a$, the terms are called **like**, or **similar**, **terms**. Constants like 6 and 2 are also considered to be like terms. The distributive law enables us to **combine**, or **collect**, **like terms** in order to form equivalent expressions.

EXAMPLE 8 Combine like terms to form equivalent expressions.

a) $-7x + 9x$ **b)** $2a + (-3b) + (-5a) + 9b$

c) $6 + y + (-3.5y) + 2$

SOLUTION

a) $-7x + 9x = (-7 + 9)x$ Using the distributive law
$ = 2x$ Adding -7 and 9

b) $2a + (-3b) + (-5a) + 9b$
$ = 2a + (-5a) + (-3b) + 9b$ Using the commutative law of addition
$ = (2 + (-5))a + (-3 + 9)b$ Using the distributive law
$ = -3a + 6b$ Adding

8. Combine like terms to form an equivalent expression:

$-9x + 5 + 13x + (-11).$

c) $6 + y + (-3.5y) + 2 = y + (-3.5y) + 6 + 2$ Using the commutative law of addition
$ = (1 + (-3.5))y + 6 + 2$ Using the distributive law
$ = -2.5y + 8$ Adding

YOUR TURN

1.5 EXERCISE SET

FOR EXTRA HELP

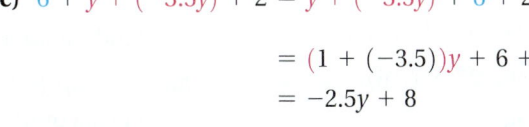 MyMathLab® Math XL
PRACTICE   WATCH   READ   REVIEW

Vocabulary and Reading Check

Choose from the following list of words to complete each statement. Words may be used more than once or not at all.

add   identity   like   negative
positive   subtract   zero

1. To add $-3 + (-6)$, _____ 3 and 6 and make the answer _____.

2. To add $-1 + 8$, _____ 1 from 8 and make the answer _____.

3. To add $-11 + 5$, _____ 5 from 11 and make the answer _____.

4. The number 0 is called the additive _____.

5. The addition $-7 + 0 = -7$ illustrates the _____ property of 0.

6. The expressions $5x$ and $-9x$ are examples of _____ terms.

↳ **Concept Reinforcement**

In each of Exercises 7–12, match the term with a like term from the column on the right.

7. ____ $8n$

8. ____ $7m$

9. ____ 43

10. ____ $28z$

11. ____ $-2x$

12. ____ $-9t$

a) $-3z$

b) $5x$

c) $2t$

d) $-4m$

e) 9

f) $-3n$

Adding with the Number Line

Add using the number line.

13. $5 + (-8)$

14. $2 + (-5)$

15. $-6 + 10$

16. $-3 + 8$

17. $-7 + 0$

18. $-6 + 0$

19. $-3 + (-5)$

20. $-4 + (-6)$

Adding Without the Number Line

Add. Do not use the number line except as a check.

21. $-6 + (-5)$

22. $-8 + (-12)$

23. $10 + (-15)$

24. $12 + (-22)$

25. $12 + (-12)$

26. $17 + (-17)$

27. $-24 + (-17)$

28. $-17 + (-25)$

29. $-13 + 13$

30. $-31 + 31$

31. $20 + (-11)$

32. $8 + (-5)$

33. $-36 + 0$

34. $0 + (-74)$

35. $-3 + 14$

36. $25 + (-6)$

37. $-24 + (-19)$

38. $11 + (-9)$

39. $19 + (-19)$

40. $-20 + (-6)$

41. $23 + (-5)$

42. $-15 + (-7)$

43. $69 + (-85)$

44. $-63 + 13$

45. $-3.6 + 2.8$

46. $-6.5 + 4.7$

47. $-5.4 + (-3.7)$

48. $-3.8 + (-9.4)$

49. $\frac{4}{5} + \left(\frac{-1}{5}\right)$

50. $\frac{-2}{7} + \frac{3}{7}$

51. $\frac{-4}{7} + \frac{-2}{7}$

52. $\frac{-5}{9} + \frac{-2}{9}$

53. $-\frac{2}{5} + \frac{1}{3}$

54. $-\frac{4}{13} + \frac{1}{2}$

55. $\frac{-4}{9} + \frac{2}{3}$

56. $\frac{1}{9} + \left(\frac{-1}{3}\right)$

57. $35 + (-14) + (-19) + (-5)$

58. $-28 + (-44) + 17 + 31 + (-94)$

Aha! 59. $-4.9 + 8.5 + 4.9 + (-8.5)$

60. $24 + 3.1 + (-44) + (-8.2) + 63$

Problem Solving

Solve. Write your answer as a complete sentence.

61. *Gasoline Prices.* During one month, the price of a gallon of 87-octane gasoline dropped 15¢, then dropped 3¢, and then rose 17¢. By how much did the price change during that period?

62. *Natural Gas Prices.* During one winter, the price of a gallon of natural gas dropped 2¢, then rose 25¢, and then dropped 43¢. By how much did the price change during that period?

63. *Telephone Bills.* Chloe's cell-phone bill for July was $82. She sent a check for $50 and then ran up $63 in charges for August. What was her new balance?

64. *Profits and Losses.* The following table lists the profits and losses of Premium Sales over a 3-year period. Find the profit or loss after this period of time.

Year	Profit or loss
2010	−$26,500
2011	−$10,200
2012	+$32,400

65. *Yardage Gained.* In an intramural football game, the quarterback attempted passes with the following results.

First try	13-yd loss
Second try	12-yd gain
Third try	21-yd gain

Find the total gain (or loss).

66. *Account Balance.* Omari has $450 in a checking account. He writes a check for $530, makes a deposit of $75, and then writes a check for $90. What is the balance in the account?

67. *Lake Level.* Between September 2009 and September 2010, the elevation of the Great Salt Lake dropped $\frac{2}{5}$ ft, rose $\frac{9}{10}$ ft, and dropped $\frac{6}{5}$ ft. By how much did the level change?

Source: U.S. Geological Survey

68. *Credit-Card Bills.* Logan's credit-card bill indicates that he owes $470. He sends a check to the credit-card company for $45, charges another $160 in merchandise, and then pays off another $500 of his bill. What is Logan's new balance?

69. *Peak Elevation.* The tallest mountain in the world, as measured from base to peak, is Mauna Kea in Hawaii. From a base 19,684 ft below sea

level, it rises 33,480 ft. What is the elevation of its peak?

Source: Guinness World Records 2007

70. *Class Size.* During the first two weeks of the semester, 5 students withdrew from Hailey's algebra class, 8 students were added to the class, and 4 students were dropped as "no-shows." By how many students did the original class size change?

Combining Like Terms

Combine like terms to form an equivalent expression.

71. $7a + 10a$

72. $3x + 8x$

73. $-3x + 12x$

74. $-2m + (-7m)$

75. $7m + (-9m)$

76. $-4x + 4x$

77. $-8y + (-2y)$

78. $10n + (-17n)$

79. $-3 + 8x + 4 + (-10x)$

80. $8a + 5 + (-a) + (-3)$

81. $6m + 9n + (-9n) + (-10m)$

82. $-11s + (-8t) + (-3s) + 8t$

83. $-4x + 6.3 + (-x) + (-10.2)$

84. $-7 + 10.5y + 13 + (-11.5y)$

Find the perimeter of each figure.

85.
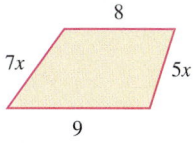

86.
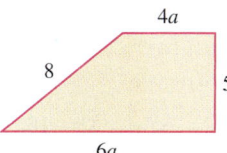

87.
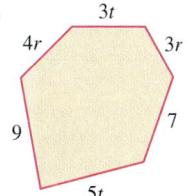

88.
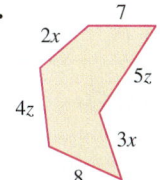

89.
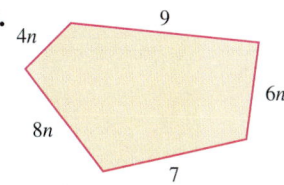

90.
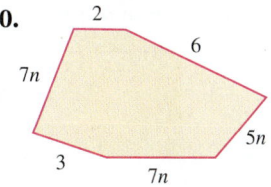

91. Explain in your own words why the sum of two negative numbers is negative.

92. Without performing the actual addition, explain why the sum of all integers from -10 to 10 is 0.

Synthesis

93. Under what circumstances will the sum of one positive number and several negative numbers be positive?

94. Is it possible to add real numbers without knowing how to calculate $a - b$ with a and b both nonnegative and $a \geq b$? Why or why not?

95. *Banking.* Travis had $257.33 in his checking account. After depositing $152 in the account and writing a check, his account was overdrawn by $42.37. What was the amount of the check?

96. *Sports-Card Values.* The value of a sports card dropped $12 and then rose $17.50 before settling at $61. What was the original value of the card?

Find the missing term or terms.

97. $4x + \underline{} + (-9x) + (-2y) = -5x - 7y$

98. $-3a + 9b + \underline{} + 5a = 2a - 6b$

99. $3m + 2n + \underline{} + (-2m) = 2n + (-6m)$

100. $\underline{} + 9x + (-4y) + x = 10x - 7y$

Aha! 101. $7t + 23 + \underline{} + \underline{} = 0$

102. *Geometry.* The perimeter of a rectangle is $7x + 10$. If the length of the rectangle is 5, express the width in terms of x.

103. *Golfing.* After five rounds of golf, a golf pro was 3 under par twice, 2 over par once, 2 under par once, and 1 over par once. On average, how far above or below par was the golfer?

↪ **YOUR TURN ANSWERS: SECTION 1.5**

1. 2   **2.** -3 **3.** -3 **4.** -4 **5.** -6.8 **6.** -1
7. The interest rate dropped 0.08%.   **8.** $4x - 6$

QUICK QUIZ: SECTIONS 1.1–1.5

1. Multiply: $3(6a + 4c + 1)$.   [1.2]

2. Add and, if possible, simplify: $\frac{2}{10} + \frac{6}{5}$.   [1.3]

3. Multiply and, if possible, simplify: $\frac{2}{9} \cdot \frac{3}{8}$.   [1.3]

4. Find the absolute value: $|505|$.   [1.4]

5. Combine like terms: $-12 + 10 + 13x + (-20x)$.   [1.5]

1.6   Subtraction of Real Numbers

Opposites and Additive Inverses   ▪   Subtraction   ▪   Problem Solving

In arithmetic, when a number b is subtracted from another number a, the **difference**, $a - b$, is the number that when added to b gives a. For example, $8 - 5 = 3$ because $3 + 5 = 8$. We will use this approach to develop an efficient way of finding the value of $a - b$ for any real numbers a and b.

Opposites and Additive Inverses

Numbers such as 6 and -6 are *opposites*, or *additive inverses*, of each other. Whenever opposites are added, the result is 0; and whenever two numbers add to 0, those numbers are opposites.

EXAMPLE 1   Find the opposite of each number: **(a)** 34; **(b)** -8.3; **(c)** 0.

SOLUTION

a) The opposite of 34 is -34: $34 + (-34) = 0$.

b) The opposite of -8.3 is 8.3: $-8.3 + 8.3 = 0$.

c) The opposite of 0 is 0: $0 + 0 = 0$.

1. Find the opposite of -15.

YOUR TURN

To write the opposite of a number, we use the symbol $-$, as follows.

> **OPPOSITE**
>
> The *opposite*, or *additive inverse*, of a number a is written $-a$ (read "the opposite of a" or "the additive inverse of a").

Note that if we take a number, say 8, and find its opposite, -8, and then find the opposite of the result, we will have the original number, 8, again.

EXAMPLE 2   Find $-x$ and $-(-x)$ when $x = 16$.

SOLUTION

If $x = 16$, then $-x = -16$.   The opposite of 16 is -16.

If $x = 16$, then $-(-x) = -(-16) = 16$.   The opposite of the opposite of 16 is 16.

2. Find $-x$ and $-(-x)$ when $x = 3.5$.

YOUR TURN

> **THE OPPOSITE OF AN OPPOSITE**
>
> For any real number a,
>
> $$-(-a) = a.$$
>
> (The opposite of the opposite of a is a.)

EXAMPLE 3 Find $-x$ and $-(-x)$ when $x = -3$.

SOLUTION

If $x = -3$, then $-x = -(-3) = 3$.   The opposite of -3 is 3.

Since $-(-x) = x$, it follows that $-(-(-3)) = -3$.   Finding the opposite of an opposite

3. Find $-x$ and $-(-x)$ when $x = -9$.

 YOUR TURN

Note in Example 3 that an extra set of parentheses is used to show that we are substituting the negative number -3 for x. The notation $- -x$ is not used.

A symbol such as -8 is usually read "negative 8." It could be read "the additive inverse of 8," because the additive inverse of 8 is negative 8. It could also be read "the opposite of 8," because the opposite of 8 is -8.

A symbol like $-x$, which has a variable, should be read "the opposite of x" or "the additive inverse of x" and *not* "negative x," since to do so suggests that $-x$ represents a negative number. As we have seen, $-x$ is not always negative.

The symbol "$-$" is read differently depending on where it appears. For example, $-5 - (-x)$ should be read "negative five minus the opposite of x."

Student Notes

As you read mathematics, it is important to verbalize correctly the words and symbols to yourself. Consistently reading the expression $-x$ as "the opposite of x" is a good step in this direction.

EXAMPLE 4 Write each of the following in words.

a) $2 - 8$ **b)** $t - (-4)$ **c)** $-6 - (-x)$

SOLUTION

a) $2 - 8$ is read "two minus eight."

b) $t - (-4)$ is read "t minus negative four."

c) $-6 - (-x)$ is read "negative six minus the opposite of x."

4. Write $-y - 7$ in words.

YOUR TURN

As we saw in Example 3, $-x$ can represent a positive number. This notation can be used to state the *law of opposites*.

THE LAW OF OPPOSITES

For any two numbers a and $-a$,

$$a + (-a) = 0.$$

(When opposites are added, their sum is 0.)

A negative number is said to have a "negative *sign*." A positive number is said to have a "positive *sign*." If we change a number to its opposite, or additive inverse, we say that we have "changed or reversed its sign."

EXAMPLE 5 Change the sign (find the opposite) of each number: **(a)** -3; **(b)** -10; **(c)** 14.

SOLUTION

a) When we change the sign of -3, we obtain 3.

b) When we change the sign of -10, we obtain 10.

c) When we change the sign of 14, we obtain -14.

5. Change the sign of 12.

YOUR TURN

Study Skills

If You Must Miss a Class

Occasionally you may know that you will miss a class. It is usually best to alert your instructor to this as soon as possible. He or she may permit you to make up a missed quiz or test *if you provide enough advance notice*. Ask what assignment will be given in your absence and try your best to learn the material on your own. It may be possible to attend another section of the course for the class that you must miss.

Subtraction

Opposites are helpful when subtraction involves negative numbers. To see why, look for a pattern in the following:

Subtracting	*Reasoning*	*Adding the Opposite*
$9 - 5 = 4$	since $4 + 5 = 9$	$9 + (-5) = 4$
$5 - 8 = -3$	since $-3 + 8 = 5$	$5 + (-8) = -3$
$-6 - 4 = -10$	since $-10 + 4 = -6$	$-6 + (-4) = -10$
$-7 - (-10) = 3$	since $3 + (-10) = -7$	$-7 + 10 = 3$
$-7 - (-2) = -5$	since $-5 + (-2) = -7$	$-7 + 2 = -5$

The matching results suggest that we can subtract by adding the opposite of the number being subtracted. This can always be done and often provides the easiest way to subtract real numbers.

SUBTRACTION OF REAL NUMBERS

For any real numbers a and b,

$$a - b = a + (-b).$$

(To subtract, add the opposite, or additive inverse, of the number being subtracted.)

EXAMPLE 6   Subtract each of the following and then check with addition.

a) $2 - 6$ **b)** $4 - (-9)$ **c)** $-4.2 - (-3.6)$

d) $-1.8 - (-7.5)$ **e)** $\frac{1}{5} - \left(-\frac{3}{5}\right)$

SOLUTION

a) $2 - 6 = 2 + (-6) = -4$ The opposite of 6 is -6. We change the subtraction to addition of the opposite. *Check:* $-4 + 6 = 2$.

b) $4 - (-9) = 4 + 9 = 13$ The opposite of -9 is 9. We change the subtraction to addition of the opposite. *Check:* $13 + (-9) = 4$.

c) $-4.2 - (-3.6) = -4.2 + 3.6$ Adding the opposite of -3.6
$\qquad\qquad\qquad = -0.6$ *Check:* $-0.6 + (-3.6) = -4.2$.

d) $-1.8 - (-7.5) = -1.8 + 7.5$ Adding the opposite
$\qquad\qquad\qquad = 5.7$ *Check:* $5.7 + (-7.5) = -1.8$.

e) $\dfrac{1}{5} - \left(-\dfrac{3}{5}\right) = \dfrac{1}{5} + \dfrac{3}{5}$ Adding the opposite

$\qquad\qquad = \dfrac{1 + 3}{5}$ A common denominator exists so we add in the numerator.

$\qquad\qquad = \dfrac{4}{5}$

Check: $\dfrac{4}{5} + \left(-\dfrac{3}{5}\right) = \dfrac{4}{5} + \dfrac{-3}{5} = \dfrac{4 + (-3)}{5} = \dfrac{1}{5}$.

Technology Connection

On nearly all graphing calculators, it is essential to distinguish between the key for negation and the key for subtraction. To enter a negative number, we use $\boxed{(-)}$; to subtract, we use $\boxed{-}$.

6. Subtract: $-16 - (-20)$.

 YOUR TURN

EXAMPLE 7 Simplify: $8 - (-4) - 2 - (-5) + 3$.

SOLUTION

$$8 - (-4) - 2 - (-5) + 3 = 8 + 4 + (-2) + 5 + 3 \qquad \text{To subtract, we add the opposite.}$$

$$= 18$$

7. Simplify:

$$-3 - (-1) + (-10) - 7.$$

_____

↪ YOUR TURN

The terms of an algebraic expression are separated by plus signs. This means that the terms of $5x - 7y - 9$ are $5x$, $-7y$, and -9, since $5x - 7y - 9 = 5x + (-7y) + (-9)$.

EXAMPLE 8 Identify the terms of $4 - 2ab + 7a - 9$.

SOLUTION We have

$$4 - 2ab + 7a - 9 = 4 + (-2ab) + 7a + (-9), \qquad \text{Rewriting as addition}$$

8. Identify the terms of
$5a - 7x - 10 + y$.

_____

so the terms are 4, $-2ab$, $7a$, and -9.

↪ YOUR TURN

EXAMPLE 9 Combine like terms.

a) $1 + 3x - 7x$ **b)** $-5a - 7b - 4a + 10b$

c) $4 - 3m - 9 + 2m$

SOLUTION

a) $1 + 3x - 7x = 1 + 3x + (-7x)$ Adding the opposite

$$= 1 + (3 + (-7))x \Bigg\} \quad \begin{array}{l}\text{Using the distributive law.}\\ \text{Try to do this mentally.}\end{array}$$

$$= 1 + (-4)x$$

$$= 1 - 4x \qquad \begin{array}{l}\text{Rewriting as subtraction to be more}\\ \text{concise}\end{array}$$

b) $-5a - 7b - 4a + 10b = -5a + (-7b) + (-4a) + 10b$ Adding the opposite

$$= -5a + (-4a) + (-7b) + 10b \qquad \begin{array}{l}\text{Using the com-}\\\text{mutative law of}\\\text{addition}\end{array}$$

$$= -9a + 3b \qquad \begin{array}{l}\text{Combining like}\\\text{terms mentally}\end{array}$$

c) $4 - 3m - 9 + 2m = 4 + (-3m) + (-9) + 2m$ Rewriting as addition

$$= 4 + (-9) + (-3m) + 2m \qquad \begin{array}{l}\text{Using the commutative}\\\text{law of addition}\end{array}$$

$$= -5 + (-1m) \qquad \begin{array}{l}\text{We can write } -1m\\\text{as } -m.\end{array}$$

9. Combine like terms:

$$x - 3y + 7y - 2x.$$

_____

$$= -5 - m$$

↪ YOUR TURN

Problem Solving

The words "difference," "minus," and "less than" are examples of key words that translate to subtraction. Since subtraction is not commutative, the order in which we write a translation is important. Unless the context indicates otherwise, "the difference between 5 and 8" translates to $5 - 8$, and "subtract 10 from 3" means to calculate $3 - 10$.

We use subtraction to solve problems involving differences. These include problems that ask "How much more?" or "How much higher?"

EXAMPLE 10 *Record Elevations.* World records for the highest parachute jump and the lowest manned vessel ocean dive were both set in 1960. On August 16 of that year, Captain Joseph Kittinger jumped from a height of 102,800 ft above sea level. Earlier, on January 23, Jacques Piccard and Navy Lieutenant Donald Walsh descended in a bathyscaphe 35,797 ft below sea level. What was the difference in elevation between the highest parachute jump and the lowest ocean dive?

Sources: www.firstflight.org and www.seasky.org

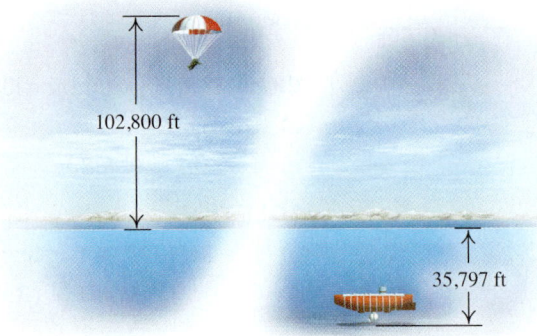

102,800 ft

35,797 ft

10. The lowest elevation in Asia, the Dead Sea, is 1349 ft below sea level. The highest elevation in Asia, Mount Everest, is 29,035 ft. Find the difference in elevation.

Source: National Geographic

SOLUTION To find the difference between two elevations, we always subtract the lower elevation from the higher elevation:

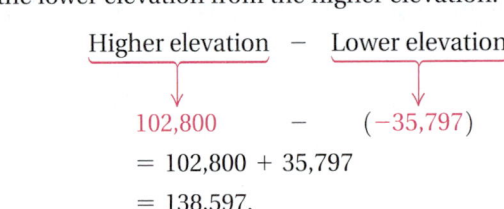

Higher elevation $-$ Lower elevation

$$102,800 \quad - \quad (-35,797)$$
$$= 102,800 + 35,797$$
$$= 138,597.$$

The parachute jump began 138,597 ft higher than the ocean dive ended.

YOUR TURN

1.6 EXERCISE SET

FOR EXTRA HELP

 MyMathLab® MathXL
PRACTICE WATCH READ REVIEW

Vocabulary and Reading Check

Two words appear under each blank in the following sentences. Choose the correct word for each blank.

1. The numbers 5 and −5 are _____
opposites/reciprocals
of each other.

2. The number −100 has a negative _____.
opposite/sign

3. We subtract by adding the _____
opposite/reciprocal
of the number being subtracted.

4. The word _____ usually translates to
difference/quotient
subtraction.

↳ **Concept Reinforcement**

In each of Exercises 5–12, match the expression with the appropriate expression from the column on the right.

5. ___ $-x$

6. ___ $12 - x$

7. ___ $12 - (-x)$

8. ___ $x - 12$

9. ___ $x - (-12)$

10. ___ $-x - 12$

11. ___ $-x - x$

12. ___ $-x - (-12)$

a) x minus negative twelve

b) The opposite of x minus x

c) The opposite of x minus twelve

d) The opposite of x

e) The opposite of x minus negative twelve

f) Twelve minus the opposite of x

g) Twelve minus x

h) x minus twelve

Opposites and Additive Inverses

Write each of the following in words.

13. $6 - 10$

14. $5 - 13$

15. $2 - (-12)$

16. $4 - (-1)$

17. $-x - y$

18. $-a - b$

19. $-3 - (-n)$

20. $-7 - (-m)$

Find the opposite, or additive inverse of each number.

21. 51

22. -17

23. $-\frac{11}{3}$

24. $\frac{7}{2}$

25. -3.14

26. 48.2

Find $-x$ when x is each of the following.

27. -45

28. 26

29. $-\frac{14}{3}$

30. $\frac{1}{328}$

31. 0.101

32. 0

Find $-(-x)$ when x is each of the following.

33. 37

34. 29

35. $-\frac{2}{5}$

36. -9.1

Change the sign. (Find the opposite.)

37. -1

38. -7

39. 15

40. 10

Subtraction

Subtract.

41. $7 - 10$

42. $4 - 13$

43. $0 - 6$

44. $0 - 8$

45. $-4 - 3$

46. $-5 - 6$

47. $-9 - (-3)$

48. $-9 - (-5)$

Aha! 49. $-8 - (-8)$

50. $-10 - (-10)$

51. $14 - 19$

52. $12 - 16$

53. $30 - 40$

54. $20 - 27$

55. $-9 - (-9)$

56. $-40 - (-40)$

57. $5 - 5$

58. $7 - 7$

59. $4 - (-4)$

60. $6 - (-6)$

61. $-7 - 4$

62. $-6 - 8$

63. $6 - (-10)$

64. $3 - (-12)$

65. $-4 - 15$

66. $-14 - 2$

67. $-6 - (-5)$

68. $-4 - (-1)$

69. $5 - (-12)$

70. $5 - (-6)$

71. $0 - (-3)$

72. $0 - (-5)$

73. $-5 - (-2)$

74. $-3 - (-1)$

75. $-7 - 14$

76. $-9 - 16$

77. $0 - 11$

78. $0 - 31$

79. $-8 - 0$

80. $-9 - 0$

81. $-52 - 8$

82. $-63 - 11$

83. $2 - 25$

84. $18 - 63$

85. $-4.2 - 3.1$

86. $-10.1 - 2.6$

87. $-1.3 - (-2.4)$

88. $-5.8 - (-7.3)$

89. $3.2 - 8.7$

90. $1.5 - 9.4$

91. $0.072 - 1$

92. $0.825 - 1$

93. $\frac{2}{11} - \frac{9}{11}$

94. $\frac{3}{7} - \frac{5}{7}$

95. $\frac{-1}{5} - \frac{3}{5}$

96. $\frac{-2}{9} - \frac{5}{9}$

97. $-\frac{4}{17} - \left(-\frac{9}{17}\right)$

98. $-\frac{2}{13} - \left(-\frac{5}{13}\right)$

Simplify.

99. $16 - (-12) - 1 - (-2) + 3$

100. $22 - (-18) + 7 + (-42) - 27$

101. $-31 + (-28) - (-14) - 17$

102. $-43 - (-19) - (-21) + 25$

103. $-34 - 28 + (-33) - 44$

104. $39 + (-88) - 29 - (-83)$

Aha! 105. $-93 + (-84) - (-93) - (-84)$

106. $84 + (-99) + 44 - (-18) - 43$

Identify the terms in each expression.

107. $-3y - 8x$

108. $7a - 9b$

109. $9 - 5t - 3st$

110. $-4 - 3x + 2xy$

Combine like terms.

111. $10x - 13x$

112. $3a - 14a$

113. $7a - 12a + 4$

114. $-9x - 13x + 7$

115. $-8n - 9 + 7n$ **116.** $-7 + 9n - 8n$

117. $5 - 3x - 11$ **118.** $2 + 3a - 7$

119. $2 - 6t - 9 - 2t$

120. $-5 + 4b - 7 - 5b$

121. $5y + (-3x) - 9x + 1 - 2y + 8$

122. $14 - (-5x) + 2z - (-32) + 4z - 2x$

123. $13x - (-2x) + 45 - (-21) - 7x$

124. $8t - (-2t) - 14 - (-5t) + 53 - 9t$

Problem Solving

125. Subtract 32 from -8.

126. Subtract 19 from -7.

127. Subtract -25 from 18.

128. Subtract -31 from -5.

In each of Exercises 129–132, translate the phrase to mathematical language and simplify.

129. The difference between 3.8 and -5.2

130. The difference between -2.1 and -5.9

131. The difference between 114 and -79

132. The difference between 23 and -17

133. *Elevation.* The Jordan River begins in Lebanon at an elevation of 550 m above sea level and empties into the Dead Sea at an elevation of 400 m below sea level. During its 360-km length, by how many meters does it fall?

Source: Brittanica Online

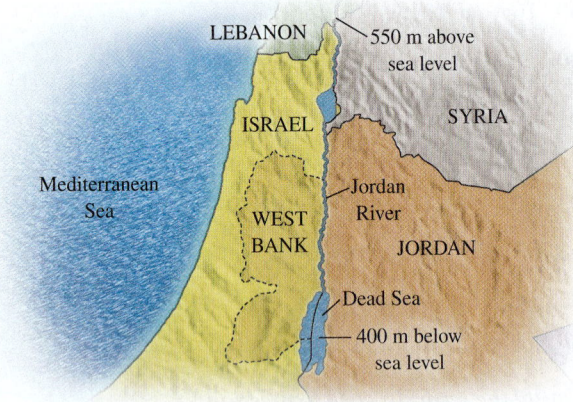

134. *Elevation.* The elevation of Mount Whitney, the highest peak in California, is 14,497 ft. Death Valley, California, is 282 ft below sea level. What is the difference in elevation?

Sources: mount-whitney.com and infoplease.com

135. *Temperature Extremes.* The highest temperature ever recorded in the United States is 134°F in Greenland Ranch, California, on July 10, 1913. The lowest temperature ever recorded is -79.8°F in Prospect Creek, Alaska, on January 23, 1971. How much higher was the temperature in Greenland Ranch than that in Prospect Creek?

Source: infoplease.com

136. *Temperature Change.* In just 12 hr on February 21, 1918, the temperature in Granville, North Dakota, rose from -33°F to 50°F. By how much did the temperature change?

Source: weatherexplained.com

137. *Basketball.* A team's points differential is the difference between points scored and points allowed. The Chicago Bulls improved their points-per-game differential from -1.6 in 2009–2010 to $+7.3$ in 2010–2011. By how many points did they improve?

Source: espn.go.com

138. *Underwater Elevation.* The deepest point in the Pacific Ocean is the Marianas Trench, with a depth of 10,911 m. The deepest point in the Atlantic Ocean is the Puerto Rico Trench, with a depth of 8648 m. What is the difference in elevation of the two trenches?

Source: Guinness World Records 2007

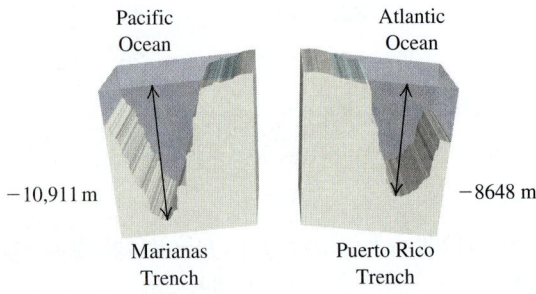

139. Lowell insists that if you can *add* real numbers, then you can also *subtract* real numbers. Do you agree? Why or why not?

140. Are the expressions $-a + b$ and $a + (-b)$ opposites of each other? Why or why not?

Synthesis

141. Explain the different uses of the symbol "$-$". Give examples of each and how they should be read.

142. If a and b are both negative, under what circumstances will $a - b$ be negative?

143. *Power Outages.* During the Northeast's electrical blackout of August 14, 2003, residents of Bloomfield, New Jersey, lost power at 4:00 P.M. One resident returned from vacation at 3:00 P.M. the following day to find the clocks in her apartment reading 8:00 A.M. At what time, and on what day, was power restored?

Tell whether each statement is true or false for all real numbers m and n. Use various replacements for m and n to support your answer.

144. If $m > n$, then $m - n > 0$.

145. If $m > n$, then $m + n > 0$.

146. If m and n are opposites, then $m - n = 0$.

147. If $m = -n$, then $m + n = 0$.

148. A gambler loses a wager and then loses "double or nothing" (meaning the gambler owes twice as much) twice more. After the three losses, the gambler's assets are $-\$20$. Explain how much the gambler originally bet and how the \$20 debt occurred.

149. List the keystrokes needed to compute $-9 - (-7)$.

150. If n is positive and m is negative, what is the sign of $n + (-m)$? Why?

> **YOUR TURN ANSWERS: SECTION 1.6**
>
> **1.** 15 **2.** $-3.5; 3.5$ **3.** $9; -9$ **4.** The opposite of y minus 7 **5.** -12 **6.** 4 **7.** -19 **8.** $5a, -7x, -10, y$ **9.** $-x + 4y$ **10.** 30,384 ft

> **QUICK QUIZ: SECTIONS 1.1–1.6**
>
> **1.** Translate to an equation. Do not solve.
>
>   In the second quarter of 2011, a total of 9.25 million Apple iPad tablets were sold. This was 11% of the total PC sales. How many PCs were sold in all? [1.1]
>   *Sources:* Apple; *Business Insider*
>
> **2.** List all the factors of 52. [1.3]
>
> **3.** Find the prime factorization of 52. [1.3]
>
> **4.** Add: $-10 + (-19)$. [1.5]
>
> **5.** Subtract: $-6 - (-11)$. [1.6]

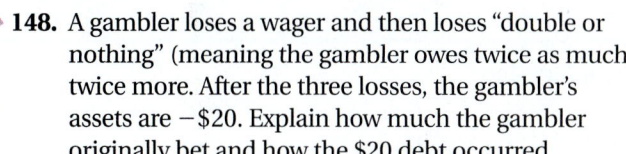

1.7   Multiplication and Division of Real Numbers

Multiplication ■ Division

We now develop rules for multiplication and division of real numbers. Because multiplication and division are closely related, the rules are quite similar.

Multiplication

Student Notes

The multiplication $2 \cdot (-5)$ can be thought of as starting at 0 and *adding* -5 twice:

$$0 + (-5) + (-5) = -10.$$

We already know how to multiply two nonnegative numbers. To see how to multiply a positive number and a negative number, consider the following pattern in which multiplication is regarded as repeated addition:

This number → $4(-5) = (-5) + (-5) + (-5) + (-5) = -20$ ← This number
decreases by $3(-5) = \qquad\quad (-5) + (-5) + (-5) = -15$ increases by
1 each time. $2(-5) = \qquad\qquad\qquad (-5) + (-5) = -10$ 5 each time.
              $1(-5) = \qquad\qquad\qquad\qquad\quad (-5) = \;\; -5$
              $0(-5) = \qquad\qquad\qquad\qquad\qquad\;\; 0 = \;\;\;\; 0$

This pattern illustrates that the product of a negative number and a positive number is negative.

THE PRODUCT OF A NEGATIVE NUMBER AND A POSITIVE NUMBER

To multiply a positive number and a negative number, multiply their absolute values. The answer is negative.

EXAMPLE 1   Multiply: **(a)** $8(-5)$; **(b)** $-\frac{1}{3} \cdot \frac{5}{7}$.

SOLUTION   The product of a negative number and a positive number is negative.

a) $8(-5) = -40$ *Think:* $8 \cdot 5 = 40$; make the answer negative.

b) $-\frac{1}{3} \cdot \frac{5}{7} = -\frac{5}{21}$ *Think:* $\frac{1}{3} \cdot \frac{5}{7} = \frac{5}{21}$; make the answer negative.

1. Multiply: $(-3)(10)$.

YOUR TURN

The pattern developed above also illustrates that the product of 0 and any real number is 0.

THE MULTIPLICATIVE PROPERTY OF ZERO

For any real number a,

$$0 \cdot a = a \cdot 0 = 0.$$

(The product of 0 and any real number is 0.)

EXAMPLE 2   Multiply: $173(-452)0$.

SOLUTION   We have

$$173(-452)0 = 173[(-452)0] \qquad \text{Because of the associative law of multiplication, we can multiply the last two factors first.}$$

$$= 173[0] \qquad \text{Using the multiplicative property of zero}$$

$$= 0. \qquad \text{Using the multiplicative property of zero again}$$

Note that whenever 0 appears as a factor, the product is 0.

2. Multiply: $(-13)(0)(19)$.

YOUR TURN

Student Notes

The multiplication $(-2) \cdot (-5)$ can be thought of as starting at 0 and *subtracting* -5 twice:

$$0 - (-5) - (-5) = 0 + 5 + 5$$
$$= 10.$$

An alternative explanation is to write

$$(-2)(-5) = [(-1)(2)](-5)$$
$$= (-1)[(2)(-5)]$$
$$= -[-10]$$
$$= 10.$$

We can extend the above pattern still further to examine the product of two negative numbers.

This number decreases by 1 each time. $\rightarrow$

$2(-5) =$	$(-5) + (-5) =$	-10	$\leftarrow$ This number increases by 5 each time.
$1(-5) =$	$(-5) =$	-5	
$0(-5) =$	$0 =$	0	
$-1(-5) =$	$-(-5) =$	5	
$-2(-5) =$	$-(-5) - (-5) =$	10	

According to the pattern, the product of two negative numbers is positive.

THE PRODUCT OF TWO NEGATIVE NUMBERS

To multiply two negative numbers, multiply their absolute values. The answer is positive.

EXAMPLE 3   Multiply: **(a)** $(-6)(-8)$; **(b)** $(-1.2)(-3)$.

SOLUTION   The product of two negative numbers is positive.

a) The absolute value of -6 is 6 and the absolute value of -8 is 8. Thus,

$$(-6)(-8) = 6 \cdot 8 \qquad \text{Multiplying absolute values. The answer is positive.}$$
$$= 48.$$

3. Multiply: $\left(-\dfrac{1}{2}\right)\left(-\dfrac{1}{3}\right)$.

b) $(-1.2)(-3) = (1.2)(3) \qquad$ Multiplying absolute values. The answer is positive.
$$= 3.6 \qquad\qquad\quad \text{Try to go directly to this step.}$$

YOUR TURN

When three or more numbers are multiplied, we can order and group the numbers as we please, because of the commutative and associative laws.

EXAMPLE 4   Multiply: **(a)** $-3(-2)(-5)$; **(b)** $-4(-6)(-1)(-2)$.

SOLUTION

a) $-3(-2)(-5) = 6(-5) \qquad$ Multiplying the first two numbers. The product of two negatives is positive.
$$= -30 \qquad\quad \text{The product of a positive and a negative is negative.}$$

b) $-4(-6)(-1)(-2) = 24 \cdot 2 \qquad$ Multiplying the first two numbers and the last two numbers

4. Multiply: $(-1)(-10)(-5)$.

$$= 48$$

YOUR TURN

We can see the following pattern in the results of Example 4.

The product of an even number of negative numbers is positive.

The product of an odd number of negative numbers is negative.

When a number is multiplied by -1, the result is the opposite of that number. For example, $-1(7) = -7$ and $-1(-5) = 5$.

> **THE PROPERTY OF -1**
>
> For any real number a,
>
> $$-1 \cdot a = -a.$$
>
> (Negative one times a is the opposite of a.)

Division

Note that $a \div b$, or $\dfrac{a}{b}$, is the number, if one exists, that when multiplied by b gives a. For example, to show that $10 \div 2$ is 5, we need only note that $5 \cdot 2 = 10$. Thus division can always be checked with multiplication.

The rules for signs for division are the same as those for multiplication: The quotient of a positive number and a negative number is negative; the quotient of two negative numbers is positive.

> **RULES FOR MULTIPLICATION AND DIVISION**
> To multiply or divide two nonzero real numbers:
> 1. Using the absolute values, multiply or divide, as indicated.
> 2. If the signs are the same, the answer is positive.
> 3. If the signs are different, the answer is negative.

EXAMPLE 5 Divide, if possible, and check your answer.

a) $14 \div (-7)$

b) $\dfrac{-32}{-4}$

c) $\dfrac{-10}{2}$

d) $\dfrac{-17}{0}$

SOLUTION

a) $14 \div (-7) = -2$ *Think:* $14 \div 7 = 2$; the answer is negative.
Check: $(-2)(-7) = 14$.

b) $\dfrac{-32}{-4} = 8$ *Think:* $32 \div 4 = 8$; the answer is positive.
Check: $8(-4) = -32$.

c) $\dfrac{-10}{2} = -5$ *Think:* $10 \div 2 = 5$; the answer is negative.
Check: $-5(2) = -10$.

d) $\dfrac{-17}{0}$ is **undefined**.   We look for a number that when multiplied by 0 gives -17. There is no such number because if 0 is a factor, the product is 0, not -17.

 YOUR TURN

Had Example 5(a) been written as $-14 \div 7$ or $-\frac{14}{7}$, rather than $14 \div (-7)$, the result would still have been -2. Thus from Examples 5(a)–5(c), we have the following:

$$\frac{-a}{b} = \frac{a}{-b} = -\frac{a}{b} \quad \text{and} \quad \frac{-a}{-b} = \frac{a}{b}.$$

EXAMPLE 6 Rewrite each of the following in two equivalent forms.

a) $\dfrac{5}{-2}$

b) $-\dfrac{3}{10}$

SOLUTION   We use one of the properties just listed.

a) $\dfrac{5}{-2} = \dfrac{-5}{2}$ and $\dfrac{5}{-2} = -\dfrac{5}{2}$

b) $-\dfrac{3}{10} = \dfrac{-3}{10}$ and $-\dfrac{3}{10} = \dfrac{3}{-10}$

Since $\dfrac{-a}{b} = \dfrac{a}{-b} = -\dfrac{a}{b}$

YOUR TURN

When a fraction contains a negative sign, it can be helpful to rewrite (or simply visualize) the fraction in an equivalent form.

Student Notes

Try to regard "undefined" as a mathematical way of saying "we do not give any meaning to this expression."

5. Divide, if possible, and check your answer: $(-100) \div 25$.

6. Rewrite in two equivalent forms:

$\dfrac{-5}{6}$.

EXAMPLE 7   Perform the indicated operation: **(a)** $\left(-\frac{4}{5}\right)\left(\frac{-7}{3}\right)$; **(b)** $-\frac{2}{7} + \frac{9}{-7}$.

SOLUTION

a) $\left(-\dfrac{4}{5}\right)\left(\dfrac{-7}{3}\right) = \left(-\dfrac{4}{5}\right)\left(-\dfrac{7}{3}\right)$ Rewriting $\dfrac{-7}{3}$ as $-\dfrac{7}{3}$

$\qquad\qquad\qquad = \dfrac{28}{15}$ Try to go directly to this step.

b) Given a choice, we generally choose a positive denominator:

$-\dfrac{2}{7} + \dfrac{9}{-7} = \dfrac{-2}{7} + \dfrac{-9}{7}$ Rewriting both fractions with a common denominator of 7

$\qquad\qquad\qquad = \dfrac{-11}{7}$, or $-\dfrac{11}{7}$.

YOUR TURN

7. Perform the indicated operation:

$\dfrac{-3}{4} - \left(\dfrac{7}{-4}\right)$.

To divide with fraction notation, it is usually easiest to find a reciprocal and then multiply.

EXAMPLE 8   Find the reciprocal of each number, if it exists.

a) -27 **b)** $\dfrac{-3}{4}$ **c)** $-\dfrac{1}{5}$ **d)** 0

SOLUTION   Two numbers are reciprocals of each other if their product is 1.

a) The reciprocal of -27 is $\frac{1}{-27}$. More often, this number is written as $-\frac{1}{27}$. *Check:* $(-27)\left(-\frac{1}{27}\right) = \frac{27}{27} = 1$.

b) The reciprocal of $\frac{-3}{4}$ is $\frac{4}{-3}$, or, equivalently, $-\frac{4}{3}$.   *Check:* $\frac{-3}{4} \cdot \frac{4}{-3} = \frac{-12}{-12} = 1$.

c) The reciprocal of $-\frac{1}{5}$ is -5.   *Check:* $-\frac{1}{5}(-5) = \frac{5}{5} = 1$.

d) The reciprocal of 0 does not exist. To see this, recall that there is no number r for which $0 \cdot r = 1$.

8. Find the reciprocal of $-\frac{10}{9}$, if it exists.

YOUR TURN

EXAMPLE 9   Divide: **(a)** $-\frac{2}{3} \div \left(-\frac{5}{4}\right)$; **(b)** $-\frac{3}{4} \div \frac{3}{10}$.

SOLUTION   We divide by multiplying by the reciprocal of the divisor.

a) $-\dfrac{2}{3} \div \left(-\dfrac{5}{4}\right) = -\dfrac{2}{3} \cdot \left(-\dfrac{4}{5}\right) = \dfrac{8}{15}$ Multiplying by the reciprocal

Be careful not to change the sign when taking a reciprocal!

9. Divide: $-\dfrac{5}{9} \div \dfrac{1}{15}$.

b) $-\dfrac{3}{4} \div \dfrac{3}{10} = -\dfrac{3}{4} \cdot \left(\dfrac{10}{3}\right) = -\dfrac{30}{12} = -\dfrac{5}{2} \cdot \dfrac{6}{6} = -\dfrac{5}{2}$ Removing a factor equal to 1: $\frac{6}{6} = 1$

YOUR TURN

To divide negative numbers with decimal notation, it is usually easiest to carry out the division and then focus on the sign.

EXAMPLE 10 Divide: $27.9 \div (-3)$.

SOLUTION

$$27.9 \div (-3) = \frac{27.9}{-3} = -9.3 \qquad \begin{array}{l} \text{Dividing: } 3\overline{)27.9}\,. \\ \text{The answer is negative.} \end{array}$$

10. Divide: $-96 \div (-0.6)$.

 YOUR TURN

In Example 5(d), we explained why we cannot divide -17 by 0. To see why *no* nonzero number b can be divided by 0, remember that $b \div 0$ would have to be the number that when multiplied by 0 gives b. But since the product of 0 and any number is 0, not b, we say that $b \div 0$ is **undefined** for $b \neq 0$. In the special case of $0 \div 0$, we look for a number r such that $0 \div 0 = r$ and $r \cdot 0 = 0$. But, $r \cdot 0 = 0$ for *any* number r. For this reason, we say that $b \div 0$ is undefined for any choice of b.*

Finally, note that $0 \div 7 = 0$ since $0 \cdot 7 = 0$. This can be written $0/7 = 0$. It is important not to confuse division *by* 0 with division *into* 0.

EXAMPLE 11 Divide, if possible: **(a)** $\frac{0}{-2}$; **(b)** $\frac{5}{0}$.

SOLUTION

a) $\dfrac{0}{-2} = 0$ We can divide 0 by a nonzero number. *Check:* $0(-2) = 0$.

11. Divide, if possible:

$0 \div (-12)$.

b) $\dfrac{5}{0}$ is undefined.   We cannot divide by 0.

 YOUR TURN

DIVISION INVOLVING ZERO

For any real number a, $\dfrac{a}{0}$ is undefined, and for $a \neq 0$, $\dfrac{0}{a} = 0$.

Study Skills

Organize Your Work

When doing homework, consider using a spiral notebook or collecting your work in a three-ring binder. Because your course will probably include graphing, consider purchasing a notebook filled with graph paper. Write legibly, labeling each section and each exercise and showing all steps. Legible, well-organized work will make it easier for those who read your work to give you constructive feedback and for you to use your work to review for a test.

It is important *not* to confuse *opposite* with *reciprocal.* Keep in mind that the opposite, or additive inverse, of a number is what we add to the number in order to get 0. The reciprocal, or multiplicative inverse, is what we multiply the number by in order to get 1.

Compare the following.

Number	Opposite (Change the sign.)	Reciprocal (Invert but do not change the sign.)
$-\dfrac{3}{8}$	$\dfrac{3}{8}$	$-\dfrac{8}{3}$ $\qquad \left(-\dfrac{3}{8}\right)\left(-\dfrac{8}{3}\right) = 1$
19	-19	$\dfrac{1}{19}$ $\qquad -\dfrac{3}{8} + \dfrac{3}{8} = 0$
$\dfrac{18}{7}$	$-\dfrac{18}{7}$	$\dfrac{7}{18}$
-7.9	7.9	$-\dfrac{1}{7.9}$, or $-\dfrac{10}{79}$
0	0	Undefined

---

*Sometimes we say that $0 \div 0$ is *indeterminate*.

CONNECTING THE CONCEPTS

The rules for multiplication and division of real numbers differ significantly from the rules for addition and subtraction. When simplifying an expression, look at the operation first to determine which set of rules to follow.

Addition

If the signs are the same, add absolute values. *The answer has the same sign as the numbers.*

If the signs are different, subtract absolute values. *The answer has the same sign as the number with the larger absolute value. If the absolute values are the same, the answer is 0.*

Subtraction

Add the opposite of the number being subtracted.

Multiplication

If the signs are the same, multiply absolute values. *The answer is positive.*

If the signs are different, multiply absolute values. *The answer is negative.*

Division

Multiply by the reciprocal of the divisor.

EXERCISES

Perform the indicated operation and, if possible, simplify.

1. $-8 + (-2)$
2. $-8 \cdot (-2)$
3. $-8 \div (-2)$
4. $-8 - (-2)$
5. $\dfrac{3}{5} - \dfrac{8}{5}$
6. $\dfrac{12}{5} + \left(\dfrac{-7}{5}\right)$
7. $(1.3)(-2.9)$
8. $-44.1 \div 6.3$
9. $-38 - (-38)$
10. $-46 - 46$

1.7 EXERCISE SET

FOR EXTRA HELP

MyMathLab® Math XL
PRACTICE   WATCH   READ   REVIEW

🢖 Vocabulary and Reading Check

Choose from the following list of words to complete each statement. Words may be used more than once or not at all.

even   positive
negative   reciprocal
odd   undefined
opposite   zero

1. The product of two negative numbers is
_____.

2. The product of a(n) _____ number of negative numbers is negative.

3. Division by zero is _____.

4. To divide by a fraction, multiply by its
_____.

5. The _____ of a negative number is positive.

6. The _____ of a negative number is negative.

🢖 Concept Reinforcement

In each of Exercises 7–16, replace the blank with either 0 or 1 to match the description given.

7. The product of two reciprocals ____

8. The sum of a pair of opposites ____

9. The sum of a pair of additive inverses ____

10. The product of two multiplicative inverses ____

11. This number has no reciprocal. ____

12. This number is its own reciprocal. ____

13. This number is the multiplicative identity. ____

14. This number is the additive identity. ____

15. A nonzero number divided by itself ____

16. Division by this number is undefined. ____

Multiplication

Multiply.

17. $-4 \cdot 10$

18. $-5 \cdot 6$

19. $-8 \cdot 7$

20. $-9 \cdot 2$

21. $4 \cdot (-10)$

22. $9 \cdot (-5)$

23. $-9 \cdot (-8)$

24. $-10 \cdot (-11)$

25. $-19 \cdot (-10)$

26. $-12 \cdot (-10)$

27. $11 \cdot (-12)$

28. $15 \cdot (-43)$

29. $4.5 \cdot (-28)$

30. $-49 \cdot (-2.1)$

31. $-5 \cdot (-2.3)$

32. $-6 \cdot 4.8$

33. $(-25) \cdot 0$

34. $0 \cdot (-4.7)$

35. $\frac{2}{5} \cdot \left(-\frac{5}{7}\right)$

36. $\frac{5}{7} \cdot \left(-\frac{2}{3}\right)$

37. $-\frac{3}{8} \cdot \left(-\frac{2}{9}\right)$

38. $-\frac{5}{8} \cdot \left(-\frac{2}{5}\right)$

39. $(-5.3)(2.1)$

40. $(9.5)(-3.7)$

41. $-\frac{5}{9} \cdot \frac{3}{4}$

42. $-\frac{8}{3} \cdot \frac{9}{4}$

43. $3 \cdot (-7) \cdot (-2) \cdot 6$

44. $9 \cdot (-2) \cdot (-6) \cdot 7$

Aha! **45.** $27 \cdot (-34) \cdot 0$

46. $-43 \cdot (-74) \cdot 0$

47. $-\frac{1}{3} \cdot \frac{1}{4} \cdot \left(-\frac{3}{7}\right)$

48. $-\frac{1}{2} \cdot \frac{3}{5} \cdot \left(-\frac{2}{7}\right)$

49. $-2 \cdot (-5) \cdot (-3) \cdot (-5)$

50. $-3 \cdot (-5) \cdot (-2) \cdot (-1)$

51. $(-31) \cdot (-27) \cdot 0 \cdot (-13)$

52. $7 \cdot (-6) \cdot 5 \cdot (-4) \cdot 3 \cdot (-2) \cdot 1 \cdot 0$

53. $(-8)(-9)(-10)$

54. $(-7)(-8)(-9)(-10)$

55. $(-6)(-7)(-8)(-9)(-10)$

56. $(-5)(-6)(-7)(-8)(-9)(-10)$

Division

Divide, if possible, and check. If a quotient is undefined, state this.

57. $18 \div (-2)$

58. $\frac{24}{-3}$

59. $\frac{36}{-9}$

60. $26 \div (-13)$

61. $\frac{-56}{8}$

62. $\frac{-35}{-7}$

63. $\frac{-48}{-12}$

64. $-63 \div (-9)$

65. $-72 \div 8$

66. $\frac{-50}{25}$

67. $-10.2 \div (-2)$

68. $-2 \div 0.8$

69. $-100 \div (-11)$

70. $\frac{-64}{-7}$

71. $\frac{400}{-50}$

72. $-300 \div (-13)$

73. $\frac{48}{0}$

74. $\frac{0}{-5}$

75. $-4.8 \div 1.2$

76. $-3.9 \div 1.3$

77. $\frac{0}{-9}$

78. $0 \div 18$

Aha! **79.** $\frac{9.7(-2.8)0}{4.3}$

80. $\frac{(-4.9)(7.2)}{0}$

Write each expression in two equivalent forms, as in Example 6.

81. $\frac{-8}{3}$

82. $\frac{-10}{3}$

83. $\frac{29}{-35}$

84. $\frac{18}{-7}$

85. $-\frac{7}{3}$

86. $-\frac{4}{15}$

87. $\frac{-x}{2}$

88. $\frac{9}{-a}$

Find the reciprocal of each number, if it exists.

89. $-\frac{4}{5}$

90. $-\frac{13}{11}$

91. $\frac{51}{-10}$

92. $\frac{43}{-24}$

93. -10

94. 34

95. 4.3

96. -1.7

97. $\frac{-1}{4}$

98. $\frac{-1}{11}$

99. 0

100. -1

Addition, Subtraction, Multiplication, and Division

Perform the indicated operation and, if possible, simplify. If a quotient is undefined, state this.

101. $\left(\frac{-7}{4}\right)\left(\frac{-3}{5}\right)$

102. $\left(-\frac{5}{6}\right)\left(\frac{-1}{3}\right)$

103. $\frac{-3}{8} + \frac{-5}{8}$

104. $\frac{-4}{5} + \frac{7}{5}$

Aha! **105.** $\left(\frac{-9}{5}\right)\left(\frac{5}{-9}\right)$

106. $\left(-\frac{2}{7}\right)\left(\frac{5}{-8}\right)$

107. $\left(-\frac{3}{11}\right) - \left(-\frac{6}{11}\right)$

108. $\left(-\frac{4}{7}\right) - \left(-\frac{2}{7}\right)$

109. $\frac{7}{8} \div \left(-\frac{1}{2}\right)$

110. $\frac{3}{4} \div \left(-\frac{2}{3}\right)$

111. $\frac{9}{5} \cdot \frac{-20}{3}$

112. $\frac{-5}{12} \cdot \frac{7}{15}$

113. $\left(-\frac{18}{7}\right) + \left(\frac{3}{-7}\right)$

114. $\left(\frac{12}{-5}\right) + \left(-\frac{3}{5}\right)$

Aha! **115.** $-\frac{5}{9} \div \left(-\frac{5}{9}\right)$

116. $\frac{-5}{12} \div \frac{15}{7}$

117. $\frac{-3}{10} + \frac{2}{5}$

118. $\frac{-5}{9} + \frac{2}{3}$

119. $\frac{7}{10} \div \left(\frac{-3}{5}\right)$

120. $\left(\frac{-3}{5}\right) \div \frac{6}{15}$

121. $\frac{14}{-9} \div \frac{0}{3}$

122. $\frac{0}{-10} \div \frac{-3}{8}$

123. $\frac{-4}{15} + \frac{2}{-3}$

124. $\frac{3}{-10} + \frac{-1}{5}$

125. Most calculators have a key, often appearing as **1/x**, for finding reciprocals. To use this key, we enter a number and then press **1/x** to find its reciprocal. What should happen if we enter a number and then press the reciprocal key twice? Why?

126. Multiplication can be regarded as repeated addition. Using this idea and the number line, explain why $3 \cdot (-5) = -15$.

Synthesis

127. If two nonzero numbers are opposites of each other, are their reciprocals opposites of each other? Why or why not?

128. If two numbers are reciprocals of each other, are their opposites reciprocals of each other? Why or why not?

Translate to an algebraic expression or equation.

129. The reciprocal of a sum

130. The sum of two reciprocals

131. The opposite of a sum

132. The sum of two opposites

133. A real number is its own opposite.

134. A real number is its own reciprocal.

135. Show that the reciprocal of a sum is *not* the sum of the two reciprocals.

136. Which real numbers are their own reciprocals?

137. Jenna is a meteorologist. On December 10, she notes that the temperature is $-3°F$ at 6:00 A.M. She predicts that the temperature will rise at a rate of $2°$ per hour for 3 hr, and then rise at a rate of $3°$ per hour for 6 hr. She also predicts that the temperature will then fall at a rate of $2°$ per hour for 3 hr, and then fall at a rate of $5°$ per hour for 2 hr. What is Jenna's temperature forecast for 8:00 P.M.?

Tell whether each expression represents a positive number or a negative number when m and n are negative.

138. $\frac{m}{-n}$

139. $\frac{-n}{-m}$

140. $-m \cdot \left(\frac{-n}{m}\right)$

141. $-\left(\frac{n}{-m}\right)$

142. $(m + n) \cdot \frac{m}{n}$

143. $(-n - m)\frac{n}{m}$

144. What must be true of m and n if $-mn$ is to be **(a)** positive? **(b)** zero? **(c)** negative?

145. The following is a proof that a positive number times a negative number is negative. Provide a reason for each step. Assume that $a > 0$ and $b > 0$.

$$a(-b) + ab = a[-b + b]$$
$$= a(0)$$
$$= 0$$

Therefore, $a(-b)$ is the opposite of ab.

146. Is it true that for any numbers a and b, if a is larger than b, then the reciprocal of a is smaller than the reciprocal of b? Why or why not?

147. *Research.* As her airplane was descending, Judy noticed that the onboard monitors indicated an altitude of 20,000 ft and an air temperature of $-15°F$. Find how many degrees air temperature drops for every 1000 ft of altitude. Then predict the temperature on the ground after Judy lands.

↳ YOUR TURN ANSWERS: SECTION 1.7

1. -30 **2.** 0 **3.** $\frac{1}{6}$ **4.** -50 **5.** -4 **6.** $\frac{5}{-6}; -\frac{5}{6}$

7. 1 **8.** $-\frac{9}{10}$ **9.** $-\frac{25}{3}$ **10.** 160 **11.** 0

QUICK QUIZ: SECTIONS 1.1–1.7

1. Factor: $22x + 11 + 33y$.   [1.2]

2. Graph on the number line: -3.5.   [1.4]

3. Simplify: $-3.1 + 1.5 + (-2.8) + (-1.7)$.   [1.5]

4. Combine like terms: $3x - 7m - m - 4x$.   [1.6]

5. Divide: $\frac{3}{8} \div \left(-\frac{3}{16}\right)$.   [1.7]

1.8   Exponential Notation and Order of Operations

Exponential Notation ▪ Order of Operations ▪ Simplifying and the Distributive Law ▪ The Opposite of a Sum

Algebraic expressions often contain *exponential notation*. In this section, we learn how to use exponential notation as well as rules for the *order of operations* in performing certain algebraic manipulations.

Exponential Notation

A product like $3 \cdot 3 \cdot 3 \cdot 3$, in which the factors are the same, is called a **power.** Powers are often written in **exponential notation**. For

$$\underbrace{3 \cdot 3 \cdot 3 \cdot 3}_{4 \text{ factors}}, \quad \text{we write} \quad 3^4.$$

This is read "three to the fourth power," or simply, "three to the fourth." The number 4 is called an **exponent** and the number 3 a **base**. Because $3^4 = 81$, we say that 81 is a power of 3.

Expressions like s^2 and s^3 are usually read "s squared" and "s cubed," respectively. This comes from the fact that a square with sides of length s has an area A given by $A = s^2$ and a cube with sides of length s has a volume V given by $V = s^3$.

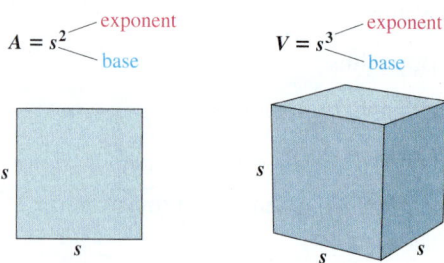

EXAMPLE 1   Write exponential notation for $10 \cdot 10 \cdot 10 \cdot 10 \cdot 10$.

SOLUTION

Exponential notation is 10^5.   5 is the exponent.
10 is the base.

1. Write exponential notation for $(-9)(-9)(-9)$.

YOUR TURN

EXAMPLE 2   Simplify: **(a)** 5^2; **(b)** $(-5)^3$; **(c)** $(2n)^3$.

SOLUTION

a) $5^2 = 5 \cdot 5 = 25$ The exponent 2 indicates two factors of 5.

b) $(-5)^3 = (-5)(-5)(-5)$ The exponent 3 indicates three factors of -5.
$= 25(-5)$ Using the associative law of multiplication
$= -125$

c) $(2n)^3 = (2n)(2n)(2n)$ The exponent 3 indicates three factors of $2n$.
$= 2 \cdot 2 \cdot 2 \cdot n \cdot n \cdot n$ Using the associative and commutative laws of multiplication
$= 8n^3$

2. Simplify: $(-2)^4$.

YOUR TURN

To enter an exponential expression on a graphing calculator, we press ⌃ and then enter the exponent. The x^2 key can be used to enter an exponent of 2.

The following screen shows the calculation

$$\frac{12(9 - 7) + 4 \cdot 5}{2^4 + 3^2}$$

from Example 8 using a calculator with an n/d option.

```
12(9−7)+4*5
─────────
 2⁴+3²
                44
                ──
                25
```

On calculators without an n/d option, we use parentheses to enclose the numerator and the denominator. The following screen shows the above calculation using parentheses. The number has also been converted to fraction notation.

```
(12(9−7)+4*5)/(2^4+3²)
                1.76
Ans▶Frac
                44/25
■
```

Student Notes

Although most scientific and graphing calculators follow the rules for order of operations when evaluating expressions, some do not. Try calculating $4 + 2 \times 5$ on your calculator. If the result shown is 30, your calculator does not follow the rules for order of operations. In this case, you will need to multiply 2×5 first and then add 4.

3. Simplify: $1 - 30 \div 3 + 7$.

To determine what the exponent 1 will mean, look for a pattern in the following:

$$7 \cdot 7 \cdot 7 \cdot 7 = 7^4$$
$$7 \cdot 7 \cdot 7 = 7^3$$
$$7 \cdot 7 = 7^2$$
$$? = 7^1$$

The exponent decreases by 1 each time.

The number of factors decreases by 1 each time. To extend the pattern, we say that

$$7 = 7^1.$$

EXPONENTIAL NOTATION

For any natural number n,

$$b^n \quad \text{means} \quad \overbrace{b \cdot b \cdot b \cdot b \cdots b}^{n \text{ factors}}.$$

Order of Operations

How should $4 + 2 \times 5$ be computed? If we multiply 2 by 5 and then add 4, the result is 14. If we add 2 and 4 first and then multiply by 5, the result is 30. Since these results differ, the order in which we perform operations matters. If grouping symbols such as parentheses ( ), brackets [ ], braces { }, absolute-value symbols | |, or fraction bars appear, they tell us what to do first. For example,

$$(4 + 2) \times 5 \quad \text{indicates} \quad 6 \times 5, \quad \text{resulting in 30,}$$

and

$$4 + (2 \times 5) \quad \text{indicates} \quad 4 + 10, \quad \text{resulting in 14.}$$

Besides grouping symbols, the following conventions exist for determining the order in which operations should be performed.

RULES FOR ORDER OF OPERATIONS

1. Simplify, if possible, within the innermost grouping symbols, ( ), [ ], { }, | |, and above or below any fraction bars.
2. Simplify all exponential expressions.
3. Perform all multiplications and divisions, working from left to right.
4. Perform all additions and subtractions, working from left to right.

Thus the correct way to compute $4 + 2 \times 5$ is to first multiply 2 by 5 and then add 4. The result is 14.

EXAMPLE 3   Simplify: $15 - 2 \cdot 5 + 3$.

SOLUTION   When no groupings or exponents appear, we *always* multiply or divide before adding or subtracting:

$$15 - 2 \cdot 5 + 3 = 15 - 10 + 3 \qquad \text{Multiplying}$$
$$= 5 + 3$$
$$= 8. \qquad \text{Subtracting and adding from left to right}$$

 YOUR TURN

Always calculate within parentheses first. When there are exponents and no parentheses, simplify powers before multiplying or dividing.

EXAMPLE 4   Simplify: **(a)** $(3 \cdot 4)^2$; **(b)** $3 \cdot 4^2$.

SOLUTION

a) $(3 \cdot 4)^2 = (12)^2$ Working within parentheses first
                $= 144$

b) $3 \cdot 4^2 = 3 \cdot 16$ Simplifying the power
              $= 48$ Multiplying

4. Simplify: $(-5) \cdot 3^2$.

Note that $(3 \cdot 4)^2 \neq 3 \cdot 4^2$.

 YOUR TURN

> **CAUTION!**   Example 4 illustrates that, in general, $(ab)^2 \neq ab^2$.

To simplify $-x^2$, it may help to write

$$-x^2 = (-1)x^2.$$

EXAMPLE 5   Evaluate for $x = 5$: **(a)** $(-x)^2$; **(b)** $-x^2$.

SOLUTION

a) $(-x)^2 = (-5)^2 = (-5)(-5) = 25$ We square the opposite of 5.

b) $-x^2 = (-1)x^2 = (-1)(5)^2 = (-1)(25) = -25$ We square 5 and then multiply by -1 (or find the opposite).

5. Evaluate $-x^2$ for $x = 10$.

 YOUR TURN

> **CAUTION!**   Example 5 illustrates that, in general, $(-x)^2 \neq -x^2$.

Student Notes

When simplifying an expression, it is important to copy the entire expression on each line, not just the parts that have been simplified in a given step. As shown in Examples 6 and 7, each line should be equivalent to the line above it.

6. Evaluate $-2 + 100 \div 2y$ for $y = 10$.

EXAMPLE 6   Evaluate $-15 \div 3(6 - a)^3$ for $a = 4$.

SOLUTION

$$-15 \div 3(6 - a)^3 = -15 \div 3(6 - 4)^3 \quad \text{Substituting 4 for } a$$
$$= -15 \div 3(2)^3 \quad \text{Working within parentheses first}$$
$$= -15 \div 3 \cdot 8 \quad \text{Simplifying the exponential expression}$$
$$\left.\begin{array}{l} = -5 \cdot 8 \\ = -40 \end{array}\right\} \quad \text{Dividing and multiplying from left to right}$$

 YOUR TURN

The symbols ( ), [ ], and { } are all used in the same way. Used inside or next to each other, they make it easier to locate the left and right sides of a grouping. When combinations of grouping symbols are used, we begin with the innermost grouping symbols and work to the outside.

EXAMPLE 7 Simplify: $8 \div 4 + 3[9 + 2(3 - 5)^3]$.

SOLUTION

$$8 \div 4 + 3[9 + 2(3 - 5)^3] = 8 \div 4 + 3[9 + 2(-2)^3]$$

Doing the calculations in the innermost grouping symbols first
$(-2)^3 = (-2)(-2)(-2) = -8$

$$= 8 \div 4 + 3[9 + 2(-8)]$$

$$= 8 \div 4 + 3[9 + (-16)]$$

$$= 8 \div 4 + 3[-7]$$

Completing the calculations within the brackets

$$= 2 + (-21)$$

Multiplying and dividing from left to right

$$= -19$$

7. Simplify:

$10 + 5[12 \div (2 - 8)]^2$.

_____

↪ YOUR TURN

EXAMPLE 8 Calculate: $\dfrac{12(9 - 7) + 4 \cdot 5}{2^4 + 3^2}$.

SOLUTION An equivalent expression with brackets is

$$[12(9 - 7) + 4 \cdot 5] \div [2^4 + 3^2].$$

Here the grouping symbols are necessary.

In effect, we need to simplify the numerator, simplify the denominator, and then divide the results:

$$\frac{12(9 - 7) + 4 \cdot 5}{2^4 + 3^2} = \frac{12(2) + 4 \cdot 5}{16 + 9}$$

$$= \frac{24 + 20}{25} = \frac{44}{25}.$$

8. Calculate:

$\dfrac{28 \div 14 \cdot 2 - (6 - 1)}{2^2 + 6^2 \div (-3)}$.

_____

↪ YOUR TURN

Simplifying and the Distributive Law

Sometimes we cannot simplify within grouping symbols. When a sum or a difference is being grouped, we can use the distributive law to remove the grouping symbols.

EXAMPLE 9 Simplify: $5x - 9 + 2(4x + 5)$.

SOLUTION

$$5x - 9 + 2(4x + 5) = 5x - 9 + 8x + 10$$

Using the distributive law

$$= 13x + 1$$

Combining like terms

9. Simplify:

$2x + 5(3x - 7) - 20$.

_____

↪ YOUR TURN

Now that exponents have been introduced, we can make our definition of *like*, or *similar*, *terms* more precise. **Like**, or **similar**, **terms** are either constant terms or terms containing the same variable(s) raised to the same power(s). Thus, 5 and -7, $19xy$ and $2yx$, as well as $4a^3b$ and a^3b are all pairs of like terms.

EXAMPLE 10   Simplify: $7x^2 + 3[x^2 + 2x] - 5x$.

SOLUTION

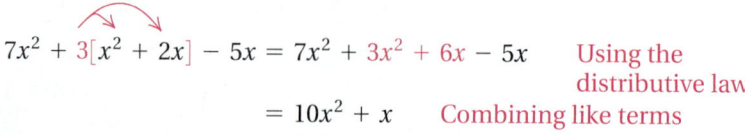

$$7x^2 + 3[x^2 + 2x] - 5x = 7x^2 + 3x^2 + 6x - 5x \qquad \text{Using the}$$
$$\text{distributive law}$$
$$= 10x^2 + x \qquad \text{Combining like terms}$$

10. Simplify:

$\qquad 2a^2 - ab + 10(a^2 + 6ab)$.

 YOUR TURN

The Opposite of a Sum

An expression such as $-(x + y)$ indicates the *opposite*, or *additive inverse*, of the sum of x and y. When a sum within grouping symbols is preceded by a "$-$" symbol, we can multiply the sum by -1 and use the distributive law. In this manner, we can find an equivalent expression for the opposite of a sum.

EXAMPLE 11   Write an expression equivalent to $-(3x + 2y + 4)$ without using parentheses.

SOLUTION

$$-(3x + 2y + 4) = -1(3x + 2y + 4) \qquad \text{Using the property of } -1$$
$$= -1(3x) + (-1)(2y) + (-1)4 \qquad \text{Using the}$$
$$\text{distributive law}$$
$$= -3x - 2y - 4 \qquad \text{Using the property of } -1$$

11. Write an expression equivalent to $-(2 + 5m + 10n)$ without using parentheses.

YOUR TURN

> **THE OPPOSITE OF A SUM**
>
> For any real numbers a and b,
>
> $$-(a + b) = (-a) + (-b) = -a - b.$$
>
> (The opposite of a sum is the sum of the opposites.)

To remove parentheses from an expression like $-(x - 7y + 5)$, we can first rewrite the subtraction as addition:

$$-(x - 7y + 5) = -(x + (-7y) + 5) \qquad \text{Rewriting as addition}$$
$$= -x + 7y - 5. \qquad \text{Taking the opposite of a sum}$$

This procedure is generally streamlined to one step in which we find the opposite by "removing parentheses and changing the sign of every term":

$$-(x - 7y + 5) = -x + 7y - 5.$$

EXAMPLE 12   Simplify: $3x - (4x + 2)$.

SOLUTION

$$3x - (4x + 2) = 3x + [-(4x + 2)] \qquad \text{Adding the opposite of } 4x + 2$$
$$= 3x + [-4x - 2] \qquad \text{Taking the opposite of } 4x + 2$$
$$= 3x + (-4x) + (-2)$$
$$= 3x - 4x - 2 \qquad \text{Try to go directly to this step.}$$
$$= -x - 2 \qquad \text{Combining like terms}$$

 YOUR TURN

Student Notes

If you prefer, you can simplify expressions like the one in Example 12 using multiplication by -1:

$$3x - (4x + 2) = 3x + [-(4x + 2)]$$
$$= 3x + [-1(4x + 2)]$$
$$= 3x + [-4x - 2].$$

The remaining steps are the same.

12. Simplify: $9 - (5x - 11)$.

In practice, the first three steps of Example 12 are generally skipped.

EXAMPLE 13 Simplify: $5t^2 - 2t - (-4t^2 + 9t)$.

SOLUTION

$$5t^2 - 2t - (-4t^2 + 9t) = 5t^2 - 2t + 4t^2 - 9t \qquad \text{Removing parentheses and changing the sign of each term inside}$$

$$= 9t^2 - 11t \qquad \text{Combining like terms}$$

13. Simplify:

$10x - x^2 - (x^2 + 12x)$.

 YOUR TURN

Expressions such as $7 - 3(x + 2)$ can be simplified as follows:

$$7 - 3(x + 2) = 7 + [-3(x + 2)] \qquad \text{Adding the opposite of } 3(x + 2)$$

$$= 7 + [-3x - 6] \qquad \text{Multiplying } x + 2 \text{ by } -3$$

$$= 7 - 3x - 6 \qquad \text{Try to go directly to this step.}$$

$$= 1 - 3x. \qquad \text{Combining like terms}$$

EXAMPLE 14 Simplify: **(a)** $3n - 2(4n - 5)$; **(b)** $7x^3 + 2 - [5(x^3 - 1) + 8]$.

SOLUTION

a) $3n - 2(4n - 5) = 3n - 8n + 10$ Multiplying each term inside the parentheses by -2; $(-2)(4n) = -8n$, and $(-2)(-5) = 10$

$$= -5n + 10 \qquad \text{Combining like terms}$$

Chapter Resources:
Collaborative Activity (Select the Symbols), p. 70; Decision Making: Connection, p. 71

b) $7x^3 + 2 - [5(x^3 - 1) + 8] = 7x^3 + 2 - [5x^3 - 5 + 8]$ Removing parentheses

$$= 7x^3 + 2 - [5x^3 + 3] \qquad \text{Combining like terms}$$

$$= 7x^3 + 2 - 5x^3 - 3 \qquad \text{Removing brackets}$$

$$= 2x^3 - 1 \qquad \text{Combining like terms}$$

14. Simplify:

$6y^2 - 3(2y^2 - 6y - 7) + 10$.

 YOUR TURN

As we progress through our study of algebra, it is important that we be able to distinguish between the two tasks of **simplifying an expression** and **solving an equation**. In this chapter, we have not solved equations, but we have simplified expressions. This enabled us to write equivalent expressions that were simpler than the given expression.

1.8 EXERCISE SET

FOR EXTRA HELP

MyMathLab® MathXL
PRACTICE   WATCH   READ   REVIEW

Vocabulary and Reading Check

In each of Exercises 1–6, match the expression with the best illustration of that expression from the column on the right.

1. An exponent ____
2. A base ____
3. The square of a number ____
4. The cube of a number ____
5. Like terms ____
6. The opposite of a sum ____

a) 10^2
b) The 4 in 4^3
c) The 6 in x^6
d) $-(x + 3) = -x - 3$
e) $3y^4$ and $-y^4$
f) 8^3

Concept Reinforcement

In each of Exercises 7–12, name the operation that should be performed first. Do not perform the calculations.

7. $4 + 8 \div 2 \cdot 2$
8. $7 - 9 + 15$
9. $5 - 2(3 + 4)$
10. $6 + 7 \cdot 3$
11. $18 - 2[4 + (3 - 2)]$
12. $\dfrac{5 - 6 \cdot 7}{2}$

Exponential Notation

Write exponential notation.

13. $x \cdot x \cdot x \cdot x \cdot x \cdot x$
14. $y \cdot y \cdot y \cdot y \cdot y \cdot y$
15. $(-5)(-5)(-5)$
16. $(-7)(-7)(-7)(-7)$
17. $3t \cdot 3t \cdot 3t \cdot 3t \cdot 3t$
18. $5m \cdot 5m \cdot 5m \cdot 5m \cdot 5m$
19. $2 \cdot n \cdot n \cdot n \cdot n$
20. $8 \cdot a \cdot a \cdot a$

Simplify.

21. 4^2
22. 5^3
23. $(-3)^2$
24. $(-7)^2$
25. -3^2
26. -7^2
27. 4^3
28. 9^1
29. $(-5)^4$
30. 5^4
31. 7^1
32. $(-1)^7$
33. $(-2)^5$
34. -2^5
35. $(3t)^4$
36. $(5t)^2$
37. $(-7x)^3$
38. $(-5x)^4$

Order of Operations

Simplify.

39. $5 + 3 \cdot 7$
40. $3 - 4 \cdot 2$
41. $10 \cdot 5 + 1 \cdot 1$
42. $19 - 5 \cdot 4 + 3$
43. $5 - 50 \div 5 \cdot 2$
44. $12 \div 3 + 18 \div 2$
Aha! 45. $14 \cdot 19 \div (19 \cdot 14)$
46. $18 - 6 \div 3 \cdot 2 + 7$
47. $3(-10)^2 - 8 \div 2^2$
48. $9 - 3^2 \div 9(-1)$
49. $8 - (2 \cdot 3 - 9)$
50. $(8 - 2 \cdot 3) - 9$
51. $(8 - 2)(3 - 9)$
52. $32 \div (-2)^2 \cdot 4$
53. $13(-10)^2 + 45 \div (-5)$
54. $2^4 + 2^3 - 10 \div (-1)^4$
55. $5 + 3(2 - 9)^2$
56. $9 - (3 - 5)^3 - 4$
57. $[2 \cdot (5 - 8)]^2 - 12$
58. $2^3 + 2^4 - 5[8 - 4(9 - 10)^2]$
59. $\dfrac{7 + 2}{5^2 - 4^2}$
60. $\dfrac{(5^2 - 3^2)^2}{2 \cdot 6 - 4}$
61. $8(-7) + |3(-4)|$
62. $|10(-5)| + 1(-1)$
63. $36 \div (-2)^2 + 4[5 - 3(8 - 9)^5]$
64. $-48 \div (7 - 9)^3 - 2[1 - 5(2 - 6) + 3^2]$
65. $\dfrac{7^2 - (-1)^7}{5 \cdot 7 - 4 \cdot 3^2 - 2^2}$
66. $\dfrac{(-2)^3 + 4^2}{2 \cdot 3 - 5^2 + 3 \cdot 7}$
67. $\dfrac{-3^3 - 2 \cdot 3^2}{8 \div 2^2 - (6 - |2 - 15|)}$
68. $\dfrac{(-5)^2 - 3 \cdot 5}{3^2 + 4 \cdot |6 - 7| \cdot (-1)^5}$

Evaluate.

69. $9 - 4x$, for $x = 7$
70. $1 + x^3$, for $x = -2$
71. $24 \div t^3$, for $t = -2$
72. $-100 \div a^2$, for $a = -5$
73. $45 \div a \cdot 5$, for $a = -3$
74. $50 \div 2 \cdot t$, for $t = 5$
75. $5x \div 15x^2$, for $x = 3$
76. $6a \div 12a^3$, for $a = 2$

77. $45 \div 3^2 x(x - 1)$, for $x = 3$

78. $-30 \div t(t + 4)^2$, for $t = -6$

79. $-x^2 - 5x$, for $x = -3$

80. $(-x)^2 - 5x$, for $x = -3$

81. $\dfrac{3a - 4a^2}{a^2 - 20}$, for $a = 5$

82. $\dfrac{a^3 - 4a}{a(a - 3)}$, for $a = -2$

Simplifying and the Distributive Law

Simplify.

83. $3x - 2 + 5(2x + 7)$

84. $8n + 3(5n + 2) + 1$

85. $2x^2 + 5(3x^2 - x) - 12x$

86. $6x^2 - x + 4(7x^2 + 3x)$

87. $9t - 7r + 2(3r + 6t)$

88. $4m - 9n + 3(2m - n)$

89. $5t^3 + t + 3(t - 2t^3)$

90. $8n^2 - 3n + 2(n - 4n^2)$

The Opposite of a Sum

Write an equivalent expression without using grouping symbols.

91. $-(9x + 1)$

92. $-(3x + 5)$

93. $-[-7n + 8]$

94. $-(6x - 7)$

95. $-(4a - 3b + 7c)$

96. $-[5n - m - 2p]$

97. $-(3x^2 + 5x - 1)$

98. $-(-9x^3 + 8x + 10)$

Simplify.

99. $8x - (6x + 7)$

100. $2a - (5a - 9)$

101. $2x - 7x - (4x - 6)$

102. $2a + 5a - (6a + 8)$

103. $15x - y - 5(3x - 2y + 5z)$

104. $4a - b - 4(5a - 7b + 8c)$

105. $3x^2 + 11 - (2x^2 + 5)$

106. $5x^4 + 3x - (5x^4 + 3x)$

107. $12a^2 - 3ab + 5b^2 - 5(-5a^2 + 4ab - 6b^2)$

108. $-8a^2 + 5ab - 12b^2 - 6(2a^2 - 4ab - 10b^2)$

109. $-7t^3 - t^2 - 3(5t^3 - 3t)$

110. $9t^4 + 7t - 5(9t^3 - 2t)$

111. $5(2x - 7) - [4(2x - 3) + 2]$

112. $3(6x - 5) - [3(1 - 8x) + 5]$

113. Some students use the mnemonic device PEMDAS to remember the rules for the order of operations. Explain what each letter represents and how the order of the letters in PEMDAS could lead a student to a wrong conclusion about the order of some operations.

114. Jake keys $18/2 \cdot 3$ into his calculator and expects the result to be 3. What mistake is he probably making?

Synthesis

115. Write the sentence $(-x)^2 \neq -x^2$ in words. Explain why $(-x)^2$ and $-x^2$ are not equivalent.

116. Write the sentence $-|x| \neq -x$ in words. Explain why $-|x|$ and $-x$ are not equivalent.

Simplify.

117. $5t - \{7t - [4r - 3(t - 7)] + 6r\} - 4r$

118. $z - \{2z - [3z - (4z - 5z) - 6z] - 7z\} - 8z$

119. $\{x - [f - (f - x)] + [x - f]\} - 3x$

120. Is it true that for all real numbers a and b,
$$ab = (-a)(-b)?$$
Why or why not?

121. Is it true that for all real numbers a, b, and c,
$$a|b - c| = ab - ac?$$
Why or why not?

If $n > 0$, $m > 0$, and $n \neq m$, classify each of the following as either true or false.

122. $-n + m = -(n + m)$

123. $m - n = -(n - m)$

124. $n(-n - m) = -n^2 + nm$

125. $-m(n - m) = -(mn + m^2)$

126. $-n(-n - m) = n(n + m)$

Evaluate.

Aha! **127.** $[x + 3(2 - 5x) \div 7 + x](x - 3)$, for $x = 3$

Aha! **128.** $[x + 2 \div 3x] \div [x + 2 \div 3x]$, for $x = -7$

129. $\dfrac{x^2 + 2^x}{x^2 - 2^x}$, for $x = 3$

130. $\dfrac{x^2 + 2^x}{x^2 - 2^x}$, for $x = 2$

131. In Mexico, between 500 B.C. and 600 A.D., the Mayans represented numbers using powers of 20 and certain symbols. For example, the symbols

represent $4 \cdot 20^3 + 17 \cdot 20^2 + 10 \cdot 20^1 + 0 \cdot 20^0$. Evaluate this number.

Source: National Council of Teachers of Mathematics, 1906 Association Drive, Reston, VA 22091

132. Examine the Mayan symbols and the numbers in Exercise 131. What number does

, , and

each represent?

133. Calculate the volume of the tower shown below.

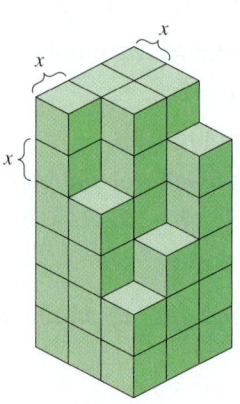

↳ **YOUR TURN ANSWERS: SECTION 1.8**

1. $(-9)^3$ **2.** 16   **3.** -2 **4.** -45 **5.** -100 **6.** 498

7. 30   **8.** $\dfrac{1}{8}$ **9.** $17x - 55$ **10.** $12a^2 + 59ab$

11. $-2 - 5m - 10n$ **12.** $-5x + 20$

13. $-2x^2 - 2x$ **14.** $18y + 31$

QUICK QUIZ: SECTIONS 1.1–1.8

1. Translate to an algebraic expression: Half of the sum of two numbers.   [1.1]

2. Write an inequality with the same meaning as $x > -10$.   [1.4]

3. Find decimal notation: $-\dfrac{11}{9}$.   [1.4]

4. Multiply: $(-2.1)(-1.4)(-1)$.   [1.7]

5. Remove parentheses and simplify:
$$5x - (2y - 3x) - 7y.$$ [1.8]

1. Twice the difference of a number and 11

2. The product of a number and 11 is 2.

3. Twice the difference of two numbers is 11.

4. The quotient of twice a number and 11

5. The quotient of 11 and the product of two numbers

Translating for Success

Use after Section 1.1.

Translate each phrase or sentence to an expression or an equation and match that translation with one of the choices A–O below. Do not solve.

A. $x = 0.2(11)$

B. $\dfrac{2x}{11}$

C. $2x + 2 = 11$

D. $2(11x + 2)$

E. $11x = 2$

F. $0.2x = 11$

G. $11(2x - y)$

H. $2(x - 11)$

I. $11 + 2x = 2$

J. $2x + y = 11$

K. $2(x - y) = 11$

L. $11(x + 2x)$

M. $2(x + y) = 11$

N. $2 + \dfrac{x}{11}$

O. $\dfrac{11}{xy}$

Answers on page A-3

An additional, animated version of this activity appears in MyMathLab. To use MyMathLab, you need a course ID and a student access code. Contact your instructor for more information.

6. Eleven times the sum of a number and twice the number

7. Twice the sum of two numbers is 11.

8. Two more than twice a number is 11.

9. Twice the sum of 11 times a number and 2

10. Twenty percent of some number is 11.

Collaborative Activity   *Teamwork*

Focus: Group problem solving; working collaboratively
Use after: Section 1.1
Time: 15 minutes
Group size: 2

Working and studying as a team often enables students to solve problems that are difficult to solve alone.

Activity

1. The left-hand column below lists the names of 12 colleges. A scrambled list of the names of their sports teams is on the right. As a group, match the names of the colleges to the teams.

   1. University of Texas
   2. Western State College of Colorado
   3. University of North Carolina
   4. University of Massachusetts
   5. Hawaii Pacific University
   6. University of Nebraska
   7. University of California, Santa Cruz
   8. University of Louisiana at Lafayette
   9. Grand Canyon University
   10. Palm Beach Atlantic University
   11. University of Alaska, Anchorage
   12. University of Florida

   a) Antelopes
   b) Banana Slugs
   c) Sea Warriors
   d) Gators
   e) Mountaineers
   f) Sailfish
   g) Longhorns
   h) Tar Heels
   i) Seawolves
   j) Ragin' Cajuns
   k) Cornhuskers
   l) Minutemen

2. After working for 5 min, confer with another group and reach mutual agreement.

3. Does the class agree on all 12 pairs?

4. Do you agree that group collaboration enhances our ability to solve problems?

Collaborative Activity   *Select the Symbols*

Focus: Order of operations
Use after: Section 1.8
Time: 15 minutes
Group size: 2

One way to master the rules for the order of operations is to insert symbols within a display of numbers in order to obtain a predetermined result. For example, the display

$$1 \quad 2 \quad 3 \quad 4 \quad 5$$

can be used to obtain the result 21 as follows:

$$(1 + 2) \div 3 + 4 \cdot 5.$$

Note that without an understanding of the rules for the order of operations, solving a problem of this sort is impossible.

Activity

1. Each group should prepare an exercise similar to the example shown here. (Exponents are not allowed.) To do so, first select five single-digit numbers for display. Then insert operations and grouping symbols and calculate the result.

2. Pair with another group. Each group should give the other its result along with its five-number display, and challenge the other group to insert symbols that will make the display equal the result given.

3. Share with the entire class the various mathematical statements developed by each group.

4. Are the answers always unique for any given set of numbers? Why or why not?

Decision Making *&* Connection   (*Use after Section 1.8.*)

Calorie Differential.   In order to lose weight, a person must burn more calories than he or she consumes. The difference between one's calorie consumption and one's calorie usage is called the *calorie differential.*

	Calorie Consumption	Calorie Usage	Calorie Differential
Monday	1900	2000	
Tuesday	2600	2100	
Wednesday	2400	2500	

1. The table at left shows Porter's calorie consumption and calorie usage for several days. Calculate the calorie differential for each day. If he is on a diet to lose weight, on which day(s) was he successful? If he is on a diet to gain weight, on which day(s) was he successful?

2. The Basal Metabolic Rate (BMR) is the amount of calories a person uses at rest. One formula uses a person's weight in pounds w, height in inches h, and age in years a. Because of different body composition, there is one formula for women and another for men.

   Women: $\text{BMR} = 655 + 4.35w + 4.7h - 4.7a$

   Men: $\text{BMR} = 66 + 6.23w + 12.7h - 6.8a$

   Calorie usage also depends on one's level of activity. To determine your total daily calorie needs to maintain your current weight, you can use the Harris Benedict Formula to multiply your BMR by the appropriate activity factor, as follows.

Activity Level	Daily Calorie Needs to Maintain Weight
Sedentary (little or no exercise)	BMR × 1.2
Lightly active (light exercise/sports 1–3 days/week)	BMR × 1.375
Moderately active (moderate exercise/sports 3–5 days/week)	BMR × 1.55
Very active (hard exercise/sports 6–7 days/week)	BMR × 1.725
Extra active (very hard exercise/sports and physical job)	BMR × 1.9

Source: bmi-calculator.net

   If Porter is lightly active, weighs 180 lb, is 70 in. tall, and is 22 years old, what are his daily calorie needs in order to maintain his weight?

 3. *Research.*   The number of calories burned by *extra* exercise can be added to a person's daily calorie needs in order to calculate total calorie usage. How many calories do you burn during exercise? Choose a favorite exercise, and find the number of calories burned per minute during that exercise.

4. Suppose you want to lose one pound. Follow the steps below to estimate how many calories you can consume each day if you exercise 30 min per day.

   1) Use one of the formulas for BMR given above to estimate your Basal Metabolic Rate. Then use the Harris Benedict Formula to estimate your daily calorie needs to maintain your weight.

   2) Calculate how many calories you would burn during 30 min of your favorite extra exercise.

   3) Add the numbers found in steps (1) and (2) to calculate your daily calorie usage.

   4) Divide −3500 by 7 to calculate the calorie differential needed each day in order to lose one pound in one week.

5. Add the number found in step (4) to the number found in step (3) to estimate the daily calorie consumption needed to lose one pound per week, assuming extra exercise.

Study Summary

KEY TERMS AND CONCEPTS	EXAMPLES	PRACTICE EXERCISES

SECTION 1.1: *Introduction to Algebra*

To **evaluate** an algebraic expression, substitute a number for each variable and carry out the operations. The result is a **value** of that expression.	Evaluate $\dfrac{x + y}{8}$ for $x = 15$ and $y = 9$. $\dfrac{x + y}{8} = \dfrac{15 + 9}{8} = \dfrac{24}{8} = 3$	1. Evaluate $3 + 5c - d$ for $c = 3$ and $d = 10$.
To find the area of a rectangle, a triangle, or a parallelogram, evaluate the appropriate formula for the given values.	Find the area of a triangle with base 3.1 m and height 6 m. $A = \frac{1}{2}bh = \frac{1}{2}(3.1\text{ m})(6\text{ m})$ $\qquad = \frac{1}{2}(3.1)(6)(\text{m} \cdot \text{m}) = 9.3\text{ m}^2$	2. Find the area of a rectangle with length 8 ft and width $\frac{1}{2}$ ft.
Many problems can be solved by **translating** phrases to algebraic expressions and then forming an equation.	Translate to an equation. Do not solve. When 34 is subtracted from a number, the result is 13. What is the number? Let n represent the number. *Rewording:* 34 subtracted from a number is 13 *Translating:* $n - 34$ $= 13$	3. Translate to an equation. Do not solve. 78 is 92 less than some number. What is the number?

SECTION 1.2: *The Commutative, Associative, and Distributive Laws*

The Commutative Laws $a + b = b + a$; $ab = ba$	$3 + (-5) = -5 + 3$; $8(10) = 10(8)$	4. Use the commutative law of addition to write an expression equivalent to $6 + 10n$.
The Associative Laws $a + (b + c) = (a + b) + c$; $a \cdot (b \cdot c) = (a \cdot b) \cdot c$	$-5 + (5 + 6) = (-5 + 5) + 6$; $2 \cdot (5 \cdot 9) = (2 \cdot 5) \cdot 9$	5. Use the associative law of multiplication to write an expression equivalent to $3(ab)$.
The Distributive Law $a(b + c) = ab + ac$	Multiply: $3(2x + 5y)$. $\qquad 3(2x + 5y) = 3 \cdot 2x + 3 \cdot 5y = 6x + 15y$ Factor: $16x + 24y + 8$. $\qquad 16x + 24y + 8 = 8(2x + 3y + 1)$	6. Multiply: $\qquad 10(5m + 9n + 1)$. 7. Factor: $26x + 13$.

SECTION 1.3: *Fraction Notation*

A **prime** number has exactly two different factors, the number itself and 1. Natural numbers that have factors other than 1 and the number itself are **composite** numbers.	2, 3, 5, 7, 11, and 13 are the first six prime numbers. 4, 6, 8, 24, and 100 are examples of composite numbers.	8. Is 15 prime or composite?

The **prime factorization** of a composite number expresses that number as a product of prime numbers.

The prime factorization of 136 is $2 \cdot 2 \cdot 2 \cdot 17$.

9. Find the prime factorization of 84.

For any nonzero number a,
$$\frac{a}{a} = 1.$$

$\frac{15}{15} = 1$ and $\frac{2x}{2x} = 1$.   We assume $x \neq 0$.

10. Simplify: $\frac{t}{t}$. Assume $t \neq 0$.

The Identity Property of 1
$$a \cdot 1 = 1 \cdot a = a$$
The number 1 is called the **multiplicative identity.**

$\frac{2}{3} = \frac{2}{3} \cdot \frac{5}{5}$ since $\frac{5}{5} = 1$.

11. Simplify: $\frac{9}{10} \cdot \frac{13}{13}$.

$$\frac{a}{d} + \frac{b}{d} = \frac{a+b}{d}$$
$$\frac{a}{d} - \frac{b}{d} = \frac{a-b}{d}$$
$$\frac{a}{b} \cdot \frac{c}{d} = \frac{a \cdot c}{b \cdot d}$$
$$\frac{a}{b} \div \frac{c}{d} = \frac{a}{b} \cdot \frac{d}{c}$$

$\frac{1}{6} + \frac{3}{8} = \frac{4}{24} + \frac{9}{24} = \frac{13}{24}$

$\frac{5}{12} - \frac{1}{6} = \frac{5}{12} - \frac{2}{12} = \frac{3}{12} = \frac{1 \cdot 3}{4 \cdot 3} = \frac{1}{4} \cdot \frac{3}{3} = \frac{1}{4} \cdot 1 = \frac{1}{4}$

$\frac{2}{5} \cdot \frac{7}{8} = \frac{2 \cdot 7}{5 \cdot 2 \cdot 4} = \frac{7}{20}$ Removing a factor equal to 1: $\frac{2}{2} = 1$

$\frac{10}{9} \div \frac{4}{15} = \frac{10}{9} \cdot \frac{15}{4} = \frac{2 \cdot 5 \cdot 3 \cdot 5}{3 \cdot 3 \cdot 2 \cdot 2} = \frac{25}{6}$ Removing a factor equal to 1: $\frac{2 \cdot 3}{2 \cdot 3} = 1$

Perform the indicated operation and, if possible, simplify.

12. $\frac{2}{3} + \frac{5}{6}$

13. $\frac{3}{4} - \frac{3}{10}$

14. $\frac{15}{14} \cdot \frac{35}{9}$

15. $15 \div \frac{3}{5}$

SECTION 1.4: *Positive and Negative Real Numbers*

Natural numbers:
$$\{1, 2, 3, 4, \dots\}$$
Whole numbers:
$$\{0, 1, 2, 3, 4, \dots\}$$
Integers:
$$\{\dots, -3, -2, -1, 0, 1, 2, 3, \dots\}$$
Rational numbers:
$$\left\{ \frac{a}{b} \middle| a \text{ and } b \text{ are integers and } b \neq 0 \right\}$$

The rational numbers and the **irrational numbers** make up the set of **real numbers.**

1, 50, and 685 are examples of natural numbers.

0, 37, and 14,615 are examples of whole numbers.

−25, −2, 0, 1, and 2000 are examples of integers.

$\frac{1}{6}, \frac{-3}{7}, 0, 17, 0.758$, and $9.\overline{608}$ are examples of rational numbers.

$\sqrt{7}$ and π are examples of irrational numbers.

16. Which of the following are integers?
$$\frac{9}{10}, \ 0, \ -15, \ \sqrt{2}, \ \frac{30}{3}$$

Every rational number can be written using fraction notation or decimal notation. When written in decimal notation, a rational number either **repeats** or **terminates.**

$-\frac{1}{16} = -0.0625$ This is a terminating decimal.

$\frac{5}{6} = 0.8333\dots = 0.8\overline{3}$ This is a repeating decimal.

17. Find decimal notation: $-\frac{10}{9}$.

Every real number corresponds to a point on the number line. For any two numbers, the one to the left on the number line is less than the one to the right. The symbol $<$ means "**is less than**" and the symbol $>$ means "**is greater than**."

$$4 > -3.1 \qquad -\frac{1}{2} < \sqrt{2}$$

18. Write a true sentence using either $<$ or $>$:
$$-3 \ \ \boxed{}\ \ -4.$$

The **absolute value** of a number is the number of units that number is from 0 on the number line.

$|3| = 3$ since 3 is 3 units from 0.

$|-3| = 3$ since -3 is 3 units from 0.

19. Find the absolute value:
$$|-1.5|.$$

SECTION 1.5: *Addition of Real Numbers*

To **add** two real numbers, use the rules given in Section 1.5.

$-8 + (-3) = -11;$

$-8 + 3 = -5;$

$8 + (-3) = 5;$

$-8 + 8 = 0$

20. Add:
$$-15 + (-10) + 20.$$

The Identity Property of 0

$a + 0 = 0 + a = a$

The number 0 is called the **additive identity.**

$-35 + 0 = -35;$

$0 + \dfrac{2}{9} = \dfrac{2}{9}$

21. Add: $-2.9 + 0.$

SECTION 1.6: *Subtraction of Real Numbers*

The **opposite,** or **additive inverse,** of a number a is written $-a$. The opposite of the opposite of a is a.
$$-(-a) = a$$

Find $-x$ and $-(-x)$ when $x = -11$.
$$-x = -(-11) = 11;$$
$$-(-x) = -(-(-11)) = -11 \qquad \textcolor{red}{-(-x) = x}$$

22. Find $-(-x)$ when $x = -12.$

To **subtract** two real numbers, add the opposite of the number being subtracted.

$-10 - 12 = -10 + (-12) = -22;$

$-10 - (-12) = -10 + 12 = 2$

23. Subtract: $6 - (-9).$

The **terms** of an expression are separated by plus signs. **Like terms** either are constants or have the same variable factors. Like terms can be **combined** using the distributive law.

In the expression $-2x + 3y + 5x - 7y$:

The terms are $-2x, 3y, 5x$, and $-7y$.

The like terms are $-2x$ and $5x$, and $3y$ and $-7y$.

Combining like terms gives
$$-2x + 3y + 5x - 7y = -2x + 5x + 3y - 7y$$
$$= (-2 + 5)x + (3 - 7)y$$
$$= 3x - 4y.$$

24. Combine like terms:
$$3c + d - 10c - 2 + 8d.$$

SECTION 1.7: *Multiplication and Division of Real Numbers*

To **multiply** or **divide** two real numbers, use the rules given in Section 1.7.

Division by 0 is **undefined**.

$(-5)(-2) = 10;$

$30 \div (-6) = -5;$

$0 \div (-3) = 0;$

$-3 \div 0$ is undefined.

25. Multiply: $-3(-7).$

26. Divide: $10 \div (-2.5).$

SECTION 1.8: *Exponential Notation and Order of Operations*

Exponential Notation Exponent $\overbrace{\qquad}^{n \text{ factors}}$ $b^n = b \cdot b \cdot b \cdots b$ Base	$6^2 = 6 \cdot 6 = 36;$ $(-6)^2 = (-6) \cdot (-6) = 36;$ $-6^2 = -(6 \cdot 6) = -36;$ $(6x)^2 = (6x) \cdot (6x) = 36x^2$	**27.** Evaluate: -10^2.
To perform multiple operations, use the rules for **order of operations** given in Section 1.8.	$-3 + (3 - 5)^3 \div 4(-1) = -3 + (-2)^3 \div 4(-1)$ $\qquad\qquad = -3 + (-8) \div 4(-1)$ $\qquad\qquad = -3 + (-2)(-1)$ $\qquad\qquad = -3 + 2$ $\qquad\qquad = -1$	**28.** Simplify: $120 \div (-10) \cdot 2$ $- 3(4 - 5).$
The Opposite of a Sum For any real numbers a and b, $-(a + b) = -a - b.$	$-(2x - 3y) = -(2x) - (-3y) = -2x + 3y$	**29.** Write an equivalent expression without using grouping symbols: $-(-a + 2b - 3c).$
Expressions containing parentheses can be simplified by removing parentheses using the distributive law.	Simplify: $3x^2 - 5(x^2 - 4xy + 2y^2) - 7y^2.$ $3x^2 - 5(x^2 - 4xy + 2y^2) - 7y^2$ $= 3x^2 - 5x^2 + 20xy - 10y^2 - 7y^2$ $= -2x^2 + 20xy - 17y^2$	**30.** Simplify: $2m + n - 3(5 - m - 2n).$

Review Exercises: Chapter 1

 Concept Reinforcement

In each of Exercises 1–10, classify the statement as either true or false.

1. $4x - 5y$ and $12 - 7a$ are both algebraic expressions containing two terms.   [1.2]*

2. $3t + 1 = 7$ and $8 - 2 = 9$ are both equations.   [1.1]

3. The fact that $2 + x$ is equivalent to $x + 2$ is an illustration of the associative law for addition.   [1.2]

4. The statement $4(a + 3) = 4 \cdot a + 4 \cdot 3$ illustrates the distributive law.   [1.2]

5. The number 2 is neither prime nor composite.   [1.3]

6. Every irrational number can be written as a repeating decimal or a terminating decimal.   [1.4]

7. Every natural number is a whole number and every whole number is an integer.   [1.4]

8. The expressions $9r^2s$ and $5rs^2$ are like terms.   [1.8]

9. The opposite of x, written $-x$, never represents a positive number.   [1.6]

10. The number 0 has no reciprocal.   [1.7]

Evaluate.

11. $8t$, for $t = 3$ [1.1]

12. $9 - y^2$, for $y = -5$ [1.8]

13. $-10 + a^2 \div (b + 1)$, for $a = 5$ and $b = -6$ [1.8]

Translate to an algebraic expression.   [1.1]

14. 7 less than y

15. 10 more than the product of x and z

16. 15 times the difference of Brandt's speed and the wind speed

*The notation [1.2] refers to Chapter 1, Section 2.

17. Translate to an equation. Do not solve.   [1.1]

Backpacking burns twice as many calories per hour as housecleaning. If Katie burns 237 calories per hour housecleaning, how many calories per hour would she burn backpacking?

Source: www.myoptumhealth.com

18. The following table lists the number of calories that Kim burns when bowling for various lengths of time. Find an equation for the number of calories burned c when Kim bowls for t hours.   [1.1]

Number of Hours Spent Bowling, t	Number of Calories Burned, c
$\frac{1}{2}$	100
2	400
$2\frac{1}{2}$	500

19. Use the commutative law of multiplication to write an expression equivalent to $3t + 5$.   [1.2]

20. Use the associative law of addition to write an expression equivalent to $(2x + y) + z$.   [1.2]

21. Use the commutative and associative laws to write three expressions equivalent to $4(xy)$.   [1.2]

Multiply.   [1.2]

22. $6(3x + 5y)$ **23.** $8(5x + 3y + 2)$

Factor.   [1.2]

24. $21x + 15y$ **25.** $22a + 99b + 11$

26. Find the prime factorization of 56.   [1.3]

Simplify.   [1.3]

27. $\dfrac{20}{48}$ **28.** $\dfrac{18}{8}$

Perform the indicated operation and, if possible, simplify.   [1.3]

29. $\dfrac{5}{12} + \dfrac{3}{8}$ **30.** $\dfrac{9}{16} \div 3$

31. $\dfrac{2}{3} - \dfrac{1}{15}$ **32.** $\dfrac{9}{10} \cdot \dfrac{6}{5}$

33. Tell which integers correspond to this situation.   [1.4]

Becky borrowed \$3600 to buy a used car. Clayton has \$1350 in his savings account.

34. Graph on a number line: $\frac{-1}{3}$.   [1.4]

35. Write an inequality with the same meaning as $-3 < x$.   [1.4]

36. Write a true sentence using either $<$ or $>$: $-10 \,\square\, 0$.   [1.4]

37. Find decimal notation: $-\frac{4}{9}$.   [1.4]

38. Find the absolute value: $|-1|$.   [1.4]

39. Find $-(-x)$ when x is -12.   [1.6]

Simplify.

40. $-3 + (-7)$ [1.5]

41. $-\frac{2}{3} + \frac{1}{12}$ [1.5]

42. $-3.8 + 5.1 + (-12) + (-4.3) + 10$ [1.5]

43. $-2 - (-10)$ [1.6]

44. $-\frac{9}{10} - \frac{1}{2}$ [1.6]

45. $-2.7(3.4)$ [1.7]

46. $\frac{2}{3} \cdot \left(-\frac{3}{7}\right)$ [1.7]

47. $2 \cdot (-7) \cdot (-2) \cdot (-5)$ [1.7]

48. $35 \div (-5)$ [1.7]

49. $-5.1 \div 1.7$ [1.7]

50. $-\frac{3}{5} \div \left(-\frac{4}{15}\right)$ [1.7]

51. $120 - 6^2 \div 4 \cdot 8$ [1.8]

52. $(120 - 6^2) \div 4 \cdot 8$ [1.8]

53. $(120 - 6^2) \div (4 \cdot 8)$ [1.8]

54. $16 \div (-2)^3 - 5[3 - 1 + 2(4 - 7)]$ [1.8]

55. $|-3 \cdot 5 - 4 \cdot 8| - 3(-2)$ [1.8]

56. $\dfrac{4(18 - 8) + 7 \cdot 9}{9^2 - 8^2}$ [1.8]

Combine like terms.

57. $11a + 2b + (-4a) + (-3b)$ [1.5]

58. $7x - 3y - 11x + 8y$ [1.6]

59. Find the opposite of -7.   [1.6]

60. Find the reciprocal of -7.   [1.7]

61. Write exponential notation for $2x \cdot 2x \cdot 2x \cdot 2x$.   [1.8]

62. Simplify: $(-5x)^3$.   [1.8]

Remove parentheses and simplify.   [1.8]

63. $2a - (5a - 9)$

64. $11x^4 + 2x + 8(x - x^4)$

65. $2n^2 - 5(-3n^2 + m^2 - 4mn) + 6m^2$

66. $8(x + 4) - 6 - [3(x - 2) + 4]$

Synthesis

67. Explain the difference between a constant and a variable.   [1.1]

68. Explain the difference between a term and a factor.   [1.2]

69. Describe at least three ways in which the distributive law was used in this chapter.   [1.2]

70. Devise a rule for determining the sign of a negative number raised to a power.   [1.8]

71. Evaluate $a^{50} - 20a^{25}b^4 + 100b^8$ for $a = 1$ and $b = 2$.   [1.8]

72. If $0.090909\ldots = \frac{1}{11}$ and $0.181818\ldots = \frac{2}{11}$, what rational number is named by each of the following?

    a) $0.272727\ldots$ [1.4]

    b) $0.909090\ldots$ [1.4]

Simplify.   [1.8]

73. $-\left|\frac{7}{8} - \left(-\frac{1}{2}\right) - \frac{3}{4}\right|$

74. $(|2.7 - 3| + 3^2 - |-3|) \div (-3)$

Match each phrase in the left column with the most appropriate choice from the right column.

75. ____ A number is nonnegative.   [1.4]

76. ____ The product of a number and its reciprocal is 1.   [1.7]

77. ____ A number squared   [1.8]

78. ____ A sum of squares   [1.8]

79. ____ The opposite of an opposite is the original number.   [1.6]

80. ____ The order in which numbers are added does not change the result.   [1.2]

81. ____ A number is positive.   [1.4]

82. ____ The absolute value of a product   [1.4]

83. ____ A sum of a number and its reciprocal   [1.7]

84. ____ The square of a sum   [1.8]

85. ____ The absolute value of one number is less than the absolute value of another number.   [1.4]

a) a^2

b) $a + b = b + a$

c) $a > 0$

d) $a + \dfrac{1}{a}$

e) $|ab|$

f) $(a + b)^2$

g) $|a| < |b|$

h) $a^2 + b^2$

i) $a \geq 0$

j) $a \cdot \dfrac{1}{a} = 1$

k) $-(-a) = a$

Test: Chapter 1

For step-by-step test solutions, access the Chapter Test Prep Videos in MyMathLab®, on YouTube (search "Bittinger Combo Alg CA" and click on "Channels"), or by scanning the code.

1. Evaluate $\dfrac{2x}{y}$ for $x = 10$ and $y = 5$.

2. Write an algebraic expression: Nine less than the product of two numbers.

3. Find the area of a triangle when the height h is 30 ft and the base b is 16 ft.

4. Use the commutative law of addition to write an expression equivalent to $3p + q$.

5. Use the associative law of multiplication to write an expression equivalent to $x \cdot (4 \cdot y)$.

6. Translate to an equation. Do not solve.

About 1500 golden lion tamarins, an endangered species of monkey, live in the wild. This is 1050 more than live in zoos worldwide. How many golden lion tamarins live in zoos?

Source: nationalzoo.si.edu

Multiply.

7. $7(5 + x)$

8. $-5(y - 2)$

Factor.

9. $11 + 44x$

10. $7x + 7 + 49y$

11. Find the prime factorization of 300.

12. Simplify: $\dfrac{24}{56}$.

Write a true sentence using either $<$ or $>$.

13. $-4\ \square\ 0$ **14.** $-3\ \square\ -8$

Find the absolute value.

15. $\left|\dfrac{9}{4}\right|$ **16.** $|-3.8|$

17. Find the opposite of $-\dfrac{2}{3}$.

18. Find the reciprocal of $-\dfrac{4}{7}$.

19. Find $-x$ when x is -10.

20. Write an inequality with the same meaning as $x \le -5$.

Perform the indicated operations and, if possible, simplify.

21. $3.1 - (-4.7)$

22. $-8 + 4 + (-7) + 3$

23. $-\dfrac{1}{8} - \dfrac{3}{4}$

24. $4 \cdot (-12)$

25. $-\dfrac{1}{2} \cdot \left(-\dfrac{4}{9}\right)$

26. $-\dfrac{3}{5} \div \left(-\dfrac{4}{5}\right)$

27. $4.864 \div (-0.5)$

28. $10 - 2(-16) \div 4^2 + |2 - 10|$

29. $9 + 7 - 4 - (-3)$

30. $256 \div (-16) \cdot 4$

31. $2^3 - 10[4 - 3(-2 + 18)]$

32. Combine like terms: $18y + 30a - 9a + 4y$.

33. Simplify: $(-2x)^4$.

Remove parentheses and simplify.

34. $4x - (3x - 7)$

35. $4(2a - 3b) + a - 7$

36. $3[5(y - 3) + 9] - 2(8y - 1)$

Synthesis

37. Find $\dfrac{5y - x}{2}$ when $x = 20$ and y is 4 less than half of x.

38. Insert one pair of parentheses to make the following a true statement:

$9 - 3 - 4 + 5 = 15$.

39. Translate to an inequality: A number n is nonnegative.

40. Simplify: $a - \{3a - [4a - (2a - 4a)]\}$.

41. Classify the following as either true or false:

$a|b - c| = |ab| - |ac|$.

Equations, Inequalities, and Problem Solving

NUMBER OF IMAGES USED	FEE FOR IMAGES USED
2	$ 65
20	650
24	780

They Say That a Picture Is Worth a Thousand Words.

When you are using a photographer for an event, the number of images you can afford to buy may be limited. If a photographer charges a $125 session fee plus image fees as shown in the table above, how many images could you purchase for $500? (*See Example 3 in Section 2.5.*)

Whether it is the first photograph or the last print, math is used extensively throughout my photography.

Carlton Riffel, owner of Carlton Riffel Photography in Ocala, Florida, uses math when editing images and calculating payments. He is also constantly adjusting camera settings using math to get the right exposure for his photographs.

Solving equations and inequalities is a recurring theme in much of mathematics. In this chapter, we will study some of the principles used to solve equations and inequalities. We will then use equations and inequalities to solve applied problems.

2.1 Solving Equations

Equations and Solutions   ▪   The Addition Principle   ▪   The Multiplication Principle   ▪   Selecting the Correct Approach

Solving equations is an essential part of problem solving in algebra. In this section, we study two of the most important principles used for this task.

Equations and Solutions

An equation is a number sentence stating that the expressions on either side of the equal sign represent the same number. Equations may be true, false, or neither true nor false.

The equation $8 \cdot 4 = 32$ is *true*.

The equation $7 - 2 = 4$ is *false*.

The equation $x + 6 = 13$ is *neither* true nor false, because we do not know what number x represents.

Equations like $x + 6 = 13$ are true for some replacements for x and are false for others.

> ### SOLUTION OF AN EQUATION
>
> Any replacement for a variable that makes an equation true is called a *solution* of the equation. To *solve* an equation means to find all of its solutions.

Student Notes

At the end of each example in this text, you will see ⟲ YOUR TURN. This directs you to try an exercise similar to the corresponding example. The answers to those exercises appear at the end of each exercise set.

1. Determine whether 8 is a solution of $12 = 20 - t$.

To determine whether a number is a solution of an equation, we substitute that number for the variable. If the values on both sides of the equal sign are the same, then the number that was substituted is a solution.

EXAMPLE 1   Determine whether 7 is a solution of $x + 6 = 13$.

SOLUTION   We have

$$
\begin{array}{ll}
x + 6 = 13 & \text{Writing the equation} \\
\overline{7 + 6 \;\big|\; 13} & \text{Substituting 7 for } x \\
13 \overset{?}{=} 13 \quad \text{TRUE} & 13 = 13 \text{ is a true statement.}
\end{array}
$$

Since the left-hand side and the right-hand side are the same, 7 is a solution.

⟲ YOUR TURN

> **CAUTION!**   Note that in Example 1, the solution is 7, not 13.

EXAMPLE 2 Determine whether -1 is a solution of $7x - 2 = 4x + 5$.

SOLUTION We have

$$
\begin{array}{c|c}
\multicolumn{2}{c}{7x - 2 = 4x + 5} \\
\hline
7(-1) - 2 & 4(-1) + 5 \\
-7 - 2 & -4 + 5 \\
-9 \overset{?}{=} 1 &
\end{array}
$$

Writing the equation
Substituting -1 for x
Carrying out calculations on both sides

FALSE   The statement $-9 = 1$ is false.

Since the left-hand and the right-hand sides differ, -1 is not a solution.

2. Determine whether -1 is a solution of $5 + 2n = 4 - n$.

YOUR TURN

The Addition Principle

Consider the equation

$x = 7.$

We can easily see that the solution of this equation is 7. Replacing x with 7, we get

$7 = 7$, which is true.

In Example 1, we found that the solution of $x + 6 = 13$ is also 7. Although the solution of $x = 7$ may seem more obvious, because $x + 6 = 13$ and $x = 7$ have identical solutions, the equations are said to be **equivalent**.

Student Notes

Be sure to remember the difference between an expression and an equation. For example, $5a - 10$ and $5(a - 2)$ are *equivalent expressions* because they represent the same value for all replacements for a. The *equations* $5a = 10$ and $a = 2$ are *equivalent equations* because they have the same solution, 2.

> **EQUIVALENT EQUATIONS**
>
> Equations with the same solutions are called *equivalent equations*.

There are principles that enable us to begin with one equation and end up with an equivalent equation, like $x = 7$, for which the solution is obvious. One such principle concerns addition. The equation $a = b$ says that a and b stand for the same number. Suppose this is true, and some number c is added to a. We get the same result if we add c to b, because a and b are the same number.

> **THE ADDITION PRINCIPLE**
>
> For any real numbers a, b, and c,
>
> $a = b$ is equivalent to $a + c = b + c$.

To visualize the addition principle, consider a balance similar to one that a jeweler might use. When the two sides of the balance hold equal weight, the balance is level. If weight is then added or removed, equally, on both sides, the balance will remain level.

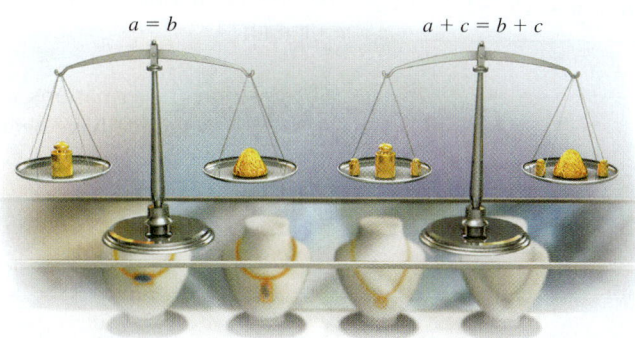

$a = b$ $a + c = b + c$

When using the addition principle, we often say that we "add the same number to both sides of an equation." We can also "subtract the same number from both sides," since subtraction can be regarded as the addition of an opposite.

EXAMPLE 3   Solve: $x + 5 = -7$.

SOLUTION   We can add any number we like to both sides. Since -5 is the opposite, or additive inverse, of 5, we decide to add -5 to each side:

$$x + 5 = -7$$

$$x + 5 - 5 = -7 - 5 \qquad \text{Using the addition principle: adding } -5 \text{ to both sides or subtracting 5 from both sides}$$

$$x + 0 = -12 \qquad \text{Simplifying; } x + 5 - 5 = x + 5 + (-5)$$
$$= x + 0$$

$$x = -12. \qquad \text{Using the identity property of 0}$$

The equation $x = -12$ is equivalent to the equation $x + 5 = -7$ by the addition principle, so the solution of $x = -12$ is the solution of $x + 5 = -7$.

It is obvious that the solution of $x = -12$ is the number -12. To check the answer in the original equation, we substitute.

Check:
$$\begin{array}{c|c} x + 5 = -7 \\ \hline -12 + 5 & -7 \\ -7 \overset{?}{=} -7 & \text{TRUE} \qquad -7 = -7 \text{ is true.} \end{array}$$

The solution of the original equation is -12.

3. Solve: $x + 3 = -10$.
_____

YOUR TURN

In Example 3, note that because we added the *opposite*, or *additive inverse*, of 5, the left-hand side of the equation simplified to x plus the *additive identity*, 0, or simply x. To solve $x + a = b$ for x, we add $-a$ to (or subtract a from) both sides.

EXAMPLE 4   Solve: $-6.5 = y - 8.4$.

SOLUTION   The variable is on the right-hand side this time. We can isolate y by adding 8.4 to each side:

$$-6.5 = y - 8.4 \qquad y - 8.4 \text{ can be regarded as } y + (-8.4).$$

$$-6.5 + 8.4 = y - 8.4 + 8.4 \qquad \text{Using the addition principle: Adding 8.4 to both sides "eliminates" } -8.4 \text{ on the right-hand side.}$$

$$1.9 = y. \qquad y - 8.4 + 8.4 = y + (-8.4) + 8.4$$
$$= y + 0 = y$$

Student Notes

We can also think of "undoing" operations in order to isolate a variable. In Example 4, we began with $y - 8.4$ on the right side. To undo the subtraction, we *add* 8.4.

Check:
$$\begin{array}{c|c} -6.5 = y - 8.4 \\ \hline -6.5 & 1.9 - 8.4 \\ -6.5 \overset{?}{=} -6.5 & \text{TRUE} \qquad -6.5 = -6.5 \text{ is true.} \end{array}$$

The solution is 1.9.

4. Solve: $-8 = y - 19$.
_____

YOUR TURN

Note that the equations $a = b$ and $b = a$ have the same meaning. Thus, $-6.5 = y - 8.4$ could have been rewritten as $y - 8.4 = -6.5$.

The Multiplication Principle

A second principle for solving equations concerns multiplying. Suppose a and b are equal. If a and b are multiplied by some number c, then ac and bc will also be equal.

> **THE MULTIPLICATION PRINCIPLE**
>
> For any real numbers a, b, and c, with $c \neq 0$,
>
> $$a = b \text{ is equivalent to } a \cdot c = b \cdot c.$$

EXAMPLE 5   Solve: $\frac{5}{4}x = 10$.

SOLUTION   We can multiply both sides by any nonzero number we like. Since $\frac{4}{5}$ is the reciprocal of $\frac{5}{4}$, we decide to multiply both sides by $\frac{4}{5}$:

$$\frac{5}{4}x = 10$$
$$\frac{4}{5} \cdot \frac{5}{4}x = \frac{4}{5} \cdot 10 \quad \text{Using the multiplication principle: Multiplying both sides by } \frac{4}{5} \text{ "eliminates" the } \frac{5}{4} \text{ on the left.}$$
$$1 \cdot x = 8 \quad \text{Simplifying}$$
$$x = 8. \quad \text{Using the identity property of 1}$$

Check:
$$\frac{5}{4}x = 10$$

$$\frac{5}{4} \cdot 8 \,\Big|\, 10$$
$$\frac{40}{4} \qquad \text{Think of 8 as } \frac{8}{1}.$$
$$10 \overset{?}{=} 10 \quad \text{TRUE} \qquad 10 = 10 \text{ is true.}$$

The solution is 8.

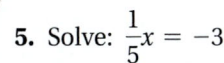

5. Solve: $\dfrac{1}{5}x = -3$.

YOUR TURN

In Example 5, to get x alone on one side, we multiplied by the *reciprocal*, or *multiplicative inverse* of $\frac{5}{4}$. The simplified left-hand side is x times the *multiplicative identity*, 1, or simply x.

Because division is the same as multiplying by a reciprocal, the multiplication principle also tells us that we can "divide both sides by the same nonzero number." That is,

$$\text{if } a = b, \text{ then } \frac{1}{c} \cdot a = \frac{1}{c} \cdot b \text{ and } \frac{a}{c} = \frac{b}{c} \quad (\text{provided } c \neq 0).$$

In a product like $3x$, the multiplier 3 is called the **coefficient**. Examining the coefficient of the variable helps us to decide whether to multiply or to divide in order to solve an equation.

- When the coefficient is an integer or a decimal, it is usually easiest to divide on both sides.
- When the coefficient is a fraction, it is usually easiest to multiply by the reciprocal on both sides.

EXAMPLE 6   Solve: **(a)** $-4x = 9$; **(b)** $-x = 5$.

SOLUTION

Student Notes

In Example 6(a), we can think of undoing the multiplication $-4 \cdot x$ by *dividing* by -4.

a) In $-4x = 9$, the coefficient of x is an integer, so we *divide* on both sides:

$$\frac{-4x}{-4} = \frac{9}{-4}$$ Using the multiplication principle: Dividing both sides by -4 is the same as multiplying by $-\frac{1}{4}$.

$$1 \cdot x = -\frac{9}{4}$$ Simplifying

$$x = -\frac{9}{4}.$$ Using the identity property of 1

Check:
$$\frac{-4x = 9}{-4\left(-\frac{9}{4}\right) \mid 9}$$
$$9 \overset{?}{=} 9 \quad \text{TRUE} \qquad 9 = 9 \text{ is true.}$$

The solution is $-\frac{9}{4}$.

b) To solve an equation like $-x = 5$, remember that when an expression is multiplied or divided by -1, its sign is changed. Here we divide both sides by -1 to change the sign of $-x$:

$$-x = 5$$ Note that $-x = -1 \cdot x$.

$$\frac{-x}{-1} = \frac{5}{-1}$$ Dividing both sides by -1. (Multiplying by -1 would also work. Note that the reciprocal of -1 is -1.)

$$x = -5.$$ Note that $\frac{-x}{-1}$ is the same as $\frac{x}{1}$.

Check:
$$\frac{-x = 5}{-(-5) \mid 5}$$
$$5 \overset{?}{=} 5 \quad \text{TRUE} \qquad 5 = 5 \text{ is true.}$$

6. Solve: $-x = \dfrac{1}{3}$.

The solution is -5.

➷ **YOUR TURN**

MULTIPLYING AND DIVIDING BY -1

Multiplying a number by -1 changes its sign:

$$(-1)(x) = -x \quad \text{and} \quad (-1)(-x) = x.$$

Dividing a number by -1 changes its sign:

$$\frac{x}{-1} = -x \quad \text{and} \quad \frac{-x}{-1} = x.$$

Study Skills

Seeking Help?

A variety of resources are available to help make studying easier and more enjoyable.

- **Textbook supplements.** See the preface for a description of the supplements for this textbook.

- **Your college or university.** Your own school probably has many resources: a learning lab or tutoring center, study skills workshops or group tutoring sessions tailored for the course you are taking, or a bulletin board or network where you can locate a private tutor.

- **Your instructor.** Find out your instructor's office hours and visit when you need additional help. Many instructors also welcome student e-mail.

7. Solve: $\dfrac{4t}{7} = 6$.

EXAMPLE 7   Solve: $\dfrac{2y}{9} = \dfrac{8}{3}$.

SOLUTION   To solve an equation like $\dfrac{2y}{9} = \dfrac{8}{3}$, we can rewrite the left-hand side as $\dfrac{2}{9} \cdot y$ and then use the multiplication principle, multiplying by the reciprocal of $\dfrac{2}{9}$:

$$\frac{2y}{9} = \frac{8}{3}$$

$$\frac{2}{9} \cdot y = \frac{8}{3} \qquad \text{Rewriting } \frac{2y}{9} \text{ as } \frac{2}{9} \cdot y$$

$$\frac{9}{2} \cdot \frac{2}{9} \cdot y = \frac{9}{2} \cdot \frac{8}{3} \qquad \text{Multiplying both sides by } \frac{9}{2}$$

$$1y = \frac{3 \cdot 3 \cdot 2 \cdot 4}{2 \cdot 3} \qquad \text{Removing a factor equal to 1: } \frac{3 \cdot 2}{2 \cdot 3} = 1$$

$$y = 12.$$

Check:
$$\frac{2y}{9} = \frac{8}{3}$$

$$\begin{array}{c|c} \dfrac{2 \cdot 12}{9} & \dfrac{8}{3} \\ \dfrac{24}{9} & \\ \dfrac{8}{3} \stackrel{?}{=} \dfrac{8}{3} \text{ TRUE} \end{array} \qquad \frac{8}{3} = \frac{8}{3} \text{ is true.}$$

The solution is 12.

YOUR TURN

Selecting the Correct Approach

It is important that you be able to determine which principle should be used to solve a particular equation.

EXAMPLE 8   Solve: **(a)** $-8 + x = -3$; **(b)** $18 = 0.3t$.

SOLUTION

a) To undo addition of -8, we subtract -8 from (or add 8 to) both sides. Note that the opposite of *negative* 8 is *positive* 8.

$$-8 + x = -3$$

$$-8 + x + 8 = -3 + 8 \qquad \text{Using the addition principle}$$

$$x = 5$$

Check:
$$-8 + x = -3$$

$$\begin{array}{c|c} -8 + 5 & -3 \\ -3 \stackrel{?}{=} -3 \text{ TRUE} \end{array} \qquad -3 = -3 \text{ is true.}$$

The solution is 5.

b) To undo the multiplication by 0.3, we either divide both sides by 0.3 or multiply both sides by $\frac{1}{0.3}$. Note that the reciprocal of *positive* 0.3 is *positive* $\frac{1}{0.3}$.

$$18 = 0.3t$$

$$\frac{18}{0.3} = \frac{0.3t}{0.3} \qquad \text{Using the multiplication principle}$$

$$60 = t \qquad \text{Simplifying}$$

Check:
$$\begin{array}{c|c} 18 = 0.3t \\ \hline 18 & 0.3(60) \\ 18 \stackrel{?}{=} 18 & \text{TRUE} \qquad 18 = 18 \text{ is true.} \end{array}$$

The solution is 60.

8. Solve: $2.6 + a = -0.46$.

YOUR TURN

2.1 EXERCISE SET

FOR EXTRA HELP

MyMathLab® MathXL
PRACTICE WATCH READ REVIEW

Vocabulary and Reading Check

For each of Exercises 1–6, match the statement with the most appropriate choice from the column on the right.

1. _____ The equations $x + 3 = 7$ and $6x = 24$

2. _____ The expressions $3(x - 2)$ and $3x - 6$

3. _____ A replacement that makes an equation true

4. _____ The role of 9 in $9ab$

5. _____ The principle used to solve $\frac{2}{3} \cdot x = -4$

6. _____ The principle used to solve $\frac{2}{3} + x = -4$

a) A coefficient

b) Equivalent expressions

c) Equivalent equations

d) The multiplication principle

e) The addition principle

f) A solution

Concept Reinforcement

For each of Exercises 7–10, match the equation with the appropriate step from the column on the right that would be used to most readily solve the equation.

7. _____ $6x = 30$

8. _____ $x + 6 = 30$

9. _____ $\frac{1}{6}x = 30$

10. _____ $x - 6 = 30$

a) Add 6 to both sides.

b) Subtract 6 from both sides.

c) Multiply both sides by 6.

d) Divide both sides by 6.

Equations and Solutions

Determine whether the given number is a solution of the given equation.

11. $6 - x = -2$; 4

12. $6 - x = -2$; 8

13. $\frac{2}{3}t = 12$; 18

14. $\frac{2}{3}t = 12$; 8

15. $x + 7 = 3 - x$; -2

16. $-4 + x = 5x$; -1

17. $4 - \frac{1}{5}n = 8$; -20

18. $-3 = 5 - \frac{n}{2}$; 4

The Addition Principle

Solve. Don't forget to check!

19. $x + 10 = 21$

20. $t + 9 = 47$

21. $y + 7 = -18$

22. $x + 12 = -7$

23. $-6 = y + 25$

24. $-5 = x + 8$

25. $x - 18 = 23$

26. $x - 19 = 16$

27. $12 = -7 + y$

28. $15 = -8 + z$

29. $-5 + t = -11$

30. $-6 + y = -21$

31. $r + \frac{1}{3} = \frac{8}{3}$

32. $t + \frac{3}{8} = \frac{5}{8}$

33. $x - \frac{3}{5} = -\frac{7}{10}$

34. $x - \frac{2}{3} = -\frac{5}{6}$

35. $x - \frac{5}{6} = \frac{7}{8}$

36. $y - \frac{3}{4} = \frac{5}{6}$

37. $-\frac{1}{5} + z = -\frac{1}{4}$

38. $-\frac{2}{3} + y = -\frac{3}{4}$

39. $m - 2.8 = 6.3$

40. $y - 5.3 = 8.7$

41. $-9.7 = -4.7 + y$

42. $-7.8 = 2.8 + x$

The Multiplication Principle

Solve. Don't forget to check!

43. $8a = 56$

44. $6x = 72$

45. $84 = 7x$

46. $45 = 9t$

47. $-x = 38$

48. $100 = -x$

Aha! **49.** $-t = -8$

50. $-68 = -r$

51. $-7x = 49$

52. $-4x = 36$

53. $-1.3a = -10.4$

54. $-3.4t = -20.4$

55. $\dfrac{y}{8} = 11$

56. $\dfrac{a}{4} = 13$

57. $\dfrac{4}{5}x = 16$

58. $\dfrac{3}{4}x = 27$

59. $\dfrac{-x}{6} = 9$

60. $\dfrac{-t}{4} = 8$

61. $\dfrac{1}{9} = \dfrac{z}{-5}$

62. $\dfrac{2}{7} = \dfrac{x}{-3}$

Aha! **63.** $-\dfrac{3}{5}r = -\dfrac{3}{5}$

64. $-\dfrac{2}{5}y = -\dfrac{4}{15}$

65. $\dfrac{-3r}{2} = -\dfrac{27}{4}$

66. $\dfrac{5x}{7} = -\dfrac{10}{14}$

Selecting the Correct Approach

Solve. The icon 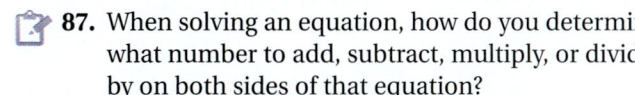 *indicates an exercise designed to provide practice using a calculator.*

67. $4.5 + t = -3.1$

68. $\frac{3}{4}x = 18$

69. $-8.2x = 20.5$

70. $t - 7.4 = -12.9$

71. $x - 4 = -19$

72. $y - 6 = -14$

73. $t - 3 = -8$

74. $t - 9 = -8$

75. $-12x = 14$

76. $-15x = 20$

77. $48 = -\frac{3}{8}y$

78. $14 = t + 27$

79. $a - \dfrac{1}{6} = -\dfrac{2}{3}$

80. $-\dfrac{x}{6} = \dfrac{2}{9}$

81. $-24 = \dfrac{8x}{5}$

82. $\dfrac{1}{5} + y = -\dfrac{3}{10}$

83. $-\frac{4}{3}t = -12$

84. $\frac{17}{35} = -x$

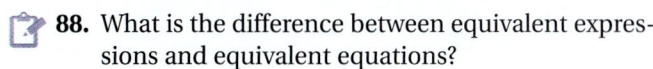 **85.** $-483.297 = -794.053 + t$

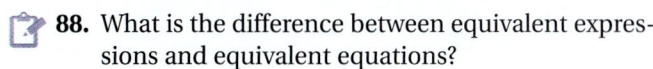 **86.** $-0.2344x = 2028.732$

87. When solving an equation, how do you determine what number to add, subtract, multiply, or divide by on both sides of that equation?

88. What is the difference between equivalent expressions and equivalent equations?

Skill Review

To the student and the instructor: Skill Review exercises *review skills previously studied in the text. The numbers in brackets immediately following the directions or exercise indicate the section in which the skill was introduced. The answer to all Skill Review exercises appear at the back of the book.*

89. Translate to an algebraic expression:

7 less than one-third of y.   [1.1]

90. Multiply: $6(2x + 11)$.   [1.2]

91. Factor: $35a + 55c + 5$.   [1.2]

92. Graph on the number line: $-\frac{11}{5}$.   [1.4]

Synthesis

93. To solve $-3.5 = 14t$, Anika adds 3.5 to both sides. Will this form an equivalent equation? Will it help solve the equation? Explain.

94. Explain why it is not necessary to state a subtraction principle: For any real numbers a, b, and c, $a = b$ is equivalent to $a - c = b - c$.

Solve for x. Assume a, c, m $\neq$ 0.

95. $mx = 11.6m$

96. $x - 4 + a = a$

97. $cx + 5c = 7c$

98. $c \cdot \dfrac{21}{a} = \dfrac{7cx}{2a}$

99. $7 + |x| = 30$

100. $ax - 3a = 5a$

101. If $t - 3590 = 1820$, find $t + 3590$.

102. If $n + 268 = 124$, find $n - 268$.

103. La'Toya estimated her monthly business taxes to be \$225. As her last step, she multiplied by 0.3 when she should have divided by 0.3. What should the correct answer be?

104. Are the equations $x = 5$ and $x^2 = 25$ equivalent? Why or why not?

YOUR TURN ANSWERS: SECTION 2.1

1. Yes **2.** No **3.** -13 **4.** 11 **5.** -15 **6.** $-\frac{1}{3}$
7. $\frac{21}{2}$ **8.** -3.06

PREPARE TO MOVE ON

Simplify. [1.8]

1. $3 \cdot 4 - 18$

2. $14 - 2(7 - 1)$

3. $4x + 10 - 5x$

4. $x - 2(3x - 7) + 14$

2.2   Using the Principles Together

Applying Both Principles ■ Combining Like Terms ■ Clearing Fractions and Decimals ■ Types of Equations

The addition and multiplication principles, along with the properties and laws concerning real numbers, are our tools for solving equations.

Applying Both Principles

EXAMPLE 1   Solve: $5 + 3x = 17$.

SOLUTION   Were we to evaluate $5 + 3x$, the rules for the order of operations direct us to *first* multiply by 3 and *then* add 5. Because of this, we can isolate $3x$ and then x by reversing these operations: We first subtract 5 from both sides and then divide both sides by 3. Our goal is an equivalent equation of the form $x = a$.

$$5 + 3x = 17$$

$$5 + 3x - 5 = 17 - 5 \qquad \text{Using the addition principle: subtracting 5 from both sides (adding } -5)$$

$$5 + (-5) + 3x = 12 \qquad \text{Using a commutative law. Try to perform this step mentally.}$$

Isolate the x-term.
$$3x = 12 \qquad \text{Simplifying}$$

$$\frac{3x}{3} = \frac{12}{3} \qquad \text{Using the multiplication principle: dividing both sides by 3 (multiplying by } \frac{1}{3})$$

Isolate x.
$$x = 4 \qquad \text{Simplifying. This is of the form } x = a.$$

Check:
$$\begin{array}{c|c} 5 + 3x = 17 \\ \hline 5 + 3 \cdot 4 & 17 \\ 5 + 12 & \\ 17 \stackrel{?}{=} 17 & \text{TRUE} \end{array}$$

We use the rules for order of operations: Find the product, $3 \cdot 4$, and then add.

The solution is 4.

1. Solve: $2x - 5 = 7$.

YOUR TURN

Study Skills

Use the Answer Section Carefully

When using the answers listed at the back of this book, try not to "work backward" from the answer. If you frequently require two or more attempts to answer an exercise correctly, you probably need to work more carefully and/or reread the section preceding the exercise set. Remember that on quizzes and tests you have only one attempt per problem and no answer section to consult.

EXAMPLE 2   Solve: $\frac{4}{3}x - 7 = 1$.

SOLUTION   In $\frac{4}{3}x - 7$, we multiply first and then subtract. To reverse these steps, we first add 7 and then either divide by $\frac{4}{3}$ or multiply by $\frac{3}{4}$.

$$\frac{4}{3}x - 7 = 1$$

$$\frac{4}{3}x - 7 + 7 = 1 + 7 \qquad \text{Adding 7 to both sides}$$

$$\frac{4}{3}x = 8$$

$$\frac{3}{4} \cdot \frac{4}{3}x = \frac{3}{4} \cdot 8 \qquad \text{Multiplying both sides by } \frac{3}{4}$$

$$\left. \begin{array}{c} 1 \cdot x = \dfrac{3 \cdot 4 \cdot 2}{4} \\[2mm] x = 6 \end{array} \right\} \quad \text{Simplifying}$$

Check:

$$\frac{\frac{4}{3}x - 7 = 1}{\begin{array}{c|c} \frac{4}{3} \cdot 6 - 7 & 1 \\ 8 - 7 & \\ 1 \stackrel{?}{=} 1 & \text{TRUE} \end{array}}$$

The solution is 6.

2. Solve: $4 + \frac{2}{3}x = 2$.

⟲ YOUR TURN

EXAMPLE 3   Solve: $45 - t = 13$.

SOLUTION   We have

$$45 - t = 13$$
$$45 - t - 45 = 13 - 45 \qquad \text{Subtracting 45 from both sides}$$
$$\left.\begin{array}{l} 45 + (-t) + (-45) = 13 - 45 \\ 45 + (-45) + (-t) = 13 - 45 \end{array}\right\} \quad \text{Try to do these steps mentally.}$$
$$-t = -32 \qquad \text{Try to go directly to this step.}$$
$$(-1)(-t) = (-1)(-32) \qquad \begin{array}{l}\text{Multiplying both sides by } -1. \\ \text{(Dividing by } -1 \text{ would also} \\ \text{work.)}\end{array}$$
$$t = 32.$$

Check:

$$\frac{45 - t = 13}{\begin{array}{c|c} 45 - 32 & 13 \\ 13 \stackrel{?}{=} 13 & \text{TRUE} \end{array}}$$

The solution is 32.

3. Solve: $-3 - y = 8$.

⟲ YOUR TURN

As our skills improve, certain steps can be streamlined.

EXAMPLE 4   Solve: $16.3 - 7.2y = -8.18$.

SOLUTION   We have

$$16.3 - 7.2y = -8.18$$
$$16.3 - 7.2y - 16.3 = -8.18 - 16.3 \qquad \text{Subtracting 16.3 from both sides}$$
$$-7.2y = -24.48 \qquad \text{Simplifying}$$
$$\frac{-7.2y}{-7.2} = \frac{-24.48}{-7.2} \qquad \text{Dividing both sides by } -7.2$$
$$y = 3.4. \qquad \text{Simplifying}$$

Check:

$$\frac{16.3 - 7.2y = -8.18}{\begin{array}{c|c} 16.3 - 7.2(3.4) & -8.18 \\ 16.3 - 24.48 & \\ -8.18 \stackrel{?}{=} -8.18 & \text{TRUE} \end{array}}$$

The solution is 3.4.

4. Solve: $0.8 = 2.8x + 5$.

⟲ YOUR TURN

Combining Like Terms

Technology Connection

A TABLE feature lists the value of a variable expression for different choices of x. For example, to evaluate $6x + 5 - 7x$ for $x = 0, 1, 2, \ldots$, we first use Y= to enter $6x + 5 - 7x$ as y_1. We then use 2ND TBLSET to specify TblStart $= 0$, ΔTbl $= 1$, and select AUTO twice. By pressing 2ND TABLE, we can generate a table in which the value of $6x + 5 - 7x$ is listed for values of x starting at 0 and increasing by one's.

X	Y1	
0	5	
1	4	
2	3	
3	2	
4	1	
5	0	
6	−1	
X = 0		

1. Create the above table on your graphing calculator. Scroll up and down to extend the table.
2. Enter $10 - 4x + 7$ as y_2. Your table should now have three columns.
3. For what x-value is y_1 the same as y_2? Compare this with the solution of Example 5(c). Is this a reliable way to solve equations? Why or why not?

If like terms appear on the same side of an equation, we combine them and then solve. Should like terms appear on both sides of an equation, we can use the addition principle to rewrite all like terms on one side.

EXAMPLE 5   Solve.

a) $3x + 4x = -14$

b) $-x + 5 = -8x + 6$

c) $6x + 5 - 7x = 10 - 4x + 7$

d) $2 - 5(x + 5) = 3(x - 2) - 1$

SOLUTION

a) $3x + 4x = -14$

$\qquad 7x = -14$ Combining like terms

$\qquad \dfrac{7x}{7} = \dfrac{-14}{7}$ Dividing both sides by 7

$\qquad x = -2$ Simplifying

The check is left to the student. The solution is -2.

b) To solve $-x + 5 = -8x + 6$, we must first write only variable terms on one side and only constant terms on the other. This can be done by subtracting 5 from both sides, to get all constant terms on the right, and adding $8x$ to both sides, to get all variable terms on the left. These steps can be performed in either order.

> Isolate variable terms on one side and constant terms on the other side.

$\qquad\qquad -x + 5 = -8x + 6$

$\qquad -x + 8x + 5 = -8x + 8x + 6$ Adding $8x$ to both sides

$\qquad\qquad 7x + 5 = 6$ Simplifying

$\qquad 7x + 5 - 5 = 6 - 5$ Subtracting 5 from both sides

$\qquad\qquad 7x = 1$ Combining like terms

$\qquad\qquad \dfrac{7x}{7} = \dfrac{1}{7}$ Dividing both sides by 7

$\qquad\qquad x = \dfrac{1}{7}$

The check is left to the student. The solution is $\frac{1}{7}$.

c) $6x + 5 - 7x = 10 - 4x + 7$

$\qquad -x + 5 = 17 - 4x$ Combining like terms on each side

$\qquad -x + 5 + 4x = 17 - 4x + 4x$ Adding $4x$ to both sides

$\qquad 5 + 3x = 17$ Simplifying. This is identical to Example 1.

$\qquad 5 + 3x - 5 = 17 - 5$ Subtracting 5 from both sides

$\qquad 3x = 12$ Simplifying

$\qquad \dfrac{3x}{3} = \dfrac{12}{3}$ Dividing both sides by 3

$\qquad x = 4$

Check:

$$\begin{array}{c|c} \hline 6x + 5 - 7x &= 10 - 4x + 7 \\ \hline 6 \cdot 4 + 5 - 7 \cdot 4 & 10 - 4 \cdot 4 + 7 \\ 24 + 5 - 28 & 10 - 16 + 7 \\ 1 & \overset{?}{=} 1 \qquad \text{TRUE} \end{array}$$

d) $2 - 5(x + 5) = 3(x - 2) - 1$

$2 - 5x - 25 = 3x - 6 - 1$	Using the distributive law. This is now similar to part (c) above.
$-5x - 23 = 3x - 7$	Combining like terms on each side
$-5x - 23 + 7 = 3x - 7 + 7$	Adding 7 to both sides
$-5x - 16 = 3x$	Simplifying
$-5x - 16 + 5x = 3x + 5x$	Adding $5x$ to both sides
$-16 = 8x$	
$\dfrac{-16}{8} = \dfrac{8x}{8}$	Dividing both sides by 8
$-2 = x$	This is equivalent to $x = -2$.

The student can confirm that -2 checks and is the solution.

5. Solve:

$$2x - (3 - x) = 7x - 1.$$

YOUR TURN

Clearing Fractions and Decimals

Equations are generally easier to solve when they do not contain fractions or decimals. The multiplication principle can be used to "clear" fractions or decimals, as shown here.

Clearing Fractions	Clearing Decimals
$\frac{1}{2}x + 5 = \frac{3}{4}$	$2.3x + 7 = 5.4$
$4\left(\frac{1}{2}x + 5\right) = 4 \cdot \frac{3}{4}$	$10(2.3x + 7) = 10 \cdot 5.4$
$2x + 20 = 3$	$23x + 70 = 54$

In each case, the resulting equation is equivalent to the original equation, but easier to solve.

AN EQUATION-SOLVING PROCEDURE

1. Use the multiplication principle to clear any fractions or decimals. (This is optional, but can ease computations. See Examples 6 and 7.)
2. If necessary, use the distributive law to remove parentheses. Then combine like terms on each side. (See Example 5.)
3. Use the addition principle, as needed, to isolate all variable terms on one side. Then combine like terms. (See Examples 1–7.)
4. Multiply or divide to solve for the variable, using the multiplication principle. (See Examples 1–7.)
5. Check all possible solutions in the original equation. (See Examples 1–4.)

The easiest way to clear an equation of fractions is to multiply *both sides* of the equation by the smallest, or *least*, common denominator of the fractions in the equation.

EXAMPLE 6   Solve: **(a)** $\frac{2}{3}x - \frac{1}{6} = 2x$; **(b)** $\frac{2}{5}(3x + 2) = 8$.

SOLUTION

a) We multiply both sides by 6, the least common denominator of $\frac{2}{3}$ and $\frac{1}{6}$.

$$6\left(\frac{2}{3}x - \frac{1}{6}\right) = 6 \cdot 2x \qquad \text{Multiplying both sides by 6}$$

$$6 \cdot \frac{2}{3}x - 6 \cdot \frac{1}{6} = 6 \cdot 2x$$

> **CAUTION!** Be sure the distributive law is used to multiply *all* the terms by 6.

$$4x - 1 = 12x \qquad \text{Simplifying. Note that the fractions are cleared: } 6 \cdot \frac{2}{3} = 4, 6 \cdot \frac{1}{6} = 1, \text{ and } 6 \cdot 2 = 12.$$

$$4x - 1 - 4x = 12x - 4x \qquad \text{Subtracting } 4x \text{ from both sides}$$

$$-1 = 8x$$

$$\frac{-1}{8} = \frac{8x}{8} \qquad \text{Dividing both sides by 8}$$

$$-\frac{1}{8} = x \qquad\qquad \frac{-1}{8} = -\frac{1}{8}$$

The student can confirm that $-\frac{1}{8}$ checks and is the solution.

b) To solve $\frac{2}{5}(3x + 2) = 8$, we can multiply both sides by $\frac{5}{2}$ (or divide by $\frac{2}{5}$) to "undo" the multiplication by $\frac{2}{5}$ on the left side.

$$\frac{5}{2} \cdot \frac{2}{5}(3x + 2) = \frac{5}{2} \cdot 8 \qquad \text{Multiplying both sides by } \frac{5}{2}$$

$$3x + 2 = 20 \qquad \text{Simplifying; } \frac{5}{2} \cdot \frac{2}{5} = 1 \text{ and } \frac{5}{2} \cdot \frac{8}{1} = 20$$

$$3x + 2 - 2 = 20 - 2 \qquad \text{Subtracting 2 from both sides}$$

$$3x = 18$$

$$\frac{3x}{3} = \frac{18}{3} \qquad \text{Dividing both sides by 3}$$

$$x = 6$$

The student can confirm that 6 checks and is the solution.

6. Solve: $\frac{1}{4} - 5x = \frac{2}{3}x$.

↩ YOUR TURN

Student Notes

Compare the steps of Examples 4 and 7. Note that the two different approaches yield the same solution. Whenever you can use two approaches to solve a problem, try to do so, both as a check and as a valuable learning experience.

To clear an equation of decimals, we count the greatest number of decimal places in any one number. If the greatest number of decimal places is 1, we multiply both sides by 10; if it is 2, we multiply by 100; and so on.

EXAMPLE 7   Solve: $16.3 - 7.2y = -8.18$.

SOLUTION   The greatest number of decimal places in any one number is *two*. Multiplying by 100 will clear all decimals.

$$100(16.3 - 7.2y) = 100(-8.18) \qquad \text{Multiplying both sides by 100}$$

$$100(16.3) - 100(7.2y) = 100(-8.18) \qquad \text{Using the distributive law}$$

$$1630 - 720y = -818 \qquad \text{Simplifying}$$

$$1630 - 720y - 1630 = -818 - 1630 \qquad \text{Subtracting 1630 from both sides}$$

$$-720y = -2448 \qquad \text{Combining like terms}$$

$$\frac{-720y}{-720} = \frac{-2448}{-720} \qquad \text{Dividing both sides by } -720$$

$$y = 3.4$$

In Example 4, the same solution was found without clearing decimals. Finding the same answer in two ways is a good check. The solution is 3.4.

7. Solve: $0.8 = 2.8x + 5$.

↩ YOUR TURN

Types of Equations

A **linear equation** in one variable—say, x—is an equation equivalent to one of the form $ax = b$ with a and b constants and $a \neq 0$.

We will sometimes refer to the set of solutions, or the **solution set**, of a particular equation. Thus the solution set for Example 7 is $\{3.4\}$. If an equation is true for all replacements, the solution set is $\mathbb{R}$, the set of all real numbers. If an equation is never true, the solution set is the **empty set**, denoted $\varnothing$, or $\{\ \}$. As its name suggests, the empty set is the set containing no elements.

Every equation is either an **identity**, a **contradiction**, or a **conditional equation**.

Type of Equation	Definition	Example	Solution Set
Identity	An equation that is true for all replacements	$x + 5 = 3 + x + 2$	$\mathbb{R}$
Contradiction	An equation that is never true	$n = n + 1$	$\varnothing$, or $\{\ \}$
Conditional equation	An equation that is true for some replacements and false for others	$2x + 5 = 17$	$\{6\}$

EXAMPLE 8   Solve each equation. Then classify the equation as an identity, a contradiction, or a conditional equation.

a) $2x + 7 = 7(x + 1) - 5x$ **b)** $3x - 5 = 3(x - 2)$

c) $3 - 8x = 5 - 7x$

SOLUTION

a) $2x + 7 = 7(x + 1) - 5x$

$\quad 2x + 7 = 7x + 7 - 5x$ Using the distributive law

$\quad 2x + 7 = 2x + 7$ Combining like terms

The equation $2x + 7 = 2x + 7$ is true regardless of what x is replaced with, so all real numbers are solutions. Note that $2x + 7 = 2x + 7$ is equivalent to $2x = 2x$, $7 = 7$, or $0 = 0$. The solution set is $\mathbb{R}$, and the equation is an identity.

b) $3x - 5 = 3(x - 2)$

$\quad\quad 3x - 5 = 3x - 6$ Using the distributive law

$-3x + 3x - 5 = -3x + 3x - 6$ Using the addition principle

$\quad\quad\quad -5 = -6$

Since the original equation is equivalent to $-5 = -6$, which is false regardless of the choice of x, the original equation has no solution. The solution set is $\varnothing$, and the equation is a contradiction.

c) $3 - 8x = 5 - 7x$

$\quad 3 - 8x + 7x = 5 - 7x + 7x$ Using the addition principle

$\quad\quad\quad 3 - x = 5$ Simplifying

$\quad -3 + 3 - x = -3 + 5$ Using the addition principle

$\quad\quad\quad\quad -x = 2$ Simplifying

$\quad\quad\quad\quad x = \dfrac{2}{-1}$, or -2 Dividing both sides by -1 or multiplying both sides by $\frac{1}{-1}$, or -1

There is one solution, -2. For other choices of x, the equation is false. The solution set is $\{-2\}$. This equation is conditional since it can be true or false, depending on the replacement for x.

8. Solve

$y - (2 - y) = 2(y - 1)$.

Then classify the equation as an identity, a contradiction, or a conditional equation.

 YOUR TURN

2.2 EXERCISE SET

Vocabulary and Reading Check

Choose the principle or law from the following list that best completes each sentence. Choices will be used more than once.

Addition principle   Distributive law
Multiplication principle

1. To isolate x in $x - 4 = 7$, we would use the

   _____.

2. To isolate x in $5x = 8$, we would use the

   _____.

3. To clear fractions or decimals, we use the

   _____.

4. To remove parentheses, we use the

   _____.

5. To solve $3x - 1 = 8$, we use the _____

   first.

6. To solve $5(x - 1) + 3(x + 7) = 2$, we use the

   _____ first.

Concept Reinforcement

In each of Exercises 7–12, match the equation with an equivalent equation from the column on the right that could be the next step in finding a solution.

7. ____ $3x - 1 = 7$
8. ____ $4x + 5x = 12$
9. ____ $6(x - 1) = 2$
10. ____ $7x = 9$
11. ____ $4x = 3 - 2x$
12. ____ $8x - 5 = 6 - 2x$

a) $6x - 6 = 2$
b) $4x + 2x = 3$
c) $3x = 7 + 1$
d) $8x + 2x = 6 + 5$
e) $9x = 12$
f) $x = \frac{9}{7}$

Applying Both Principles

Solve and check.

13. $2x + 9 = 25$
14. $5z + 2 = 57$
15. $7t - 8 = 27$
16. $6x - 5 = 2$
17. $3x - 9 = 1$
18. $5x - 9 = 41$
19. $8z + 2 = -54$
20. $4x + 3 = -21$
21. $-37 = 9t + 8$
22. $-39 = 1 + 5t$
23. $12 - t = 16$
24. $9 - t = 21$
25. $-6z - 18 = -132$
26. $-7x - 24 = -129$
27. $5.3 + 1.2n = 1.94$
28. $6.4 - 2.5n = 2.2$
29. $32 - 7x = 11$
30. $27 - 6x = 99$

31. $\frac{3}{5}t - 1 = 8$
32. $\frac{2}{3}t - 1 = 5$
33. $6 + \frac{7}{2}x = -15$
34. $6 + \frac{5}{4}x = -4$
35. $-\dfrac{4a}{5} - 8 = 2$
36. $-\dfrac{8a}{7} - 2 = 4$

Combining Like Terms

Solve and check.

37. $6x + 10x = 18$
38. $-3z + 8z = 45$
39. $4x - 6 = 6x$
40. $7n = 2n + 4$
41. $2 - 5y = 26 - y$
42. $6x - 5 = 7 + 2x$
43. $6x + 3 = 2x + 3$
44. $5y + 3 = 2y + 15$
45. $5 - 2x = 3x - 7x + 25$
46. $10 - 3x = x - 2x + 40$
47. $7 + 3x - 6 = 3x + 5 - x$
48. $5 + 4x - 7 = 4x - 2 - x$

Clearing Fractions and Decimals

Clear fractions or decimals, solve, and check.

49. $\frac{2}{3} + \frac{1}{4}t = 2$
50. $-\frac{5}{6} + x = -\frac{1}{2} - \frac{2}{3}$
51. $\frac{2}{3} + 4t = 6t - \frac{2}{15}$
52. $\frac{1}{2} + 4m = 3m - \frac{5}{2}$
53. $\frac{1}{3}x + \frac{2}{5} = \frac{4}{5} + \frac{3}{5}x - \frac{2}{3}$
54. $1 - \frac{2}{3}y = \frac{9}{5} - \frac{1}{5}y + \frac{3}{5}$
55. $2.1x + 45.2 = 3.2 - 8.4x$
56. $0.91 - 0.2z = 1.23 - 0.6z$
57. $0.76 + 0.21t = 0.96t - 0.49$
58. $1.7t + 8 - 1.62t = 0.4t - 0.32 + 8$
59. $\frac{2}{5}x - \frac{3}{2}x = \frac{3}{4}x + 3$
60. $\frac{5}{16}y + \frac{3}{8}y = 2 + \frac{1}{4}y$
61. $\frac{1}{3}(2x - 1) = 7$
62. $\frac{1}{5}(4x - 1) = 7$

Solving Linear Equations

Solve and check. Label any contradictions or identities.

63. $7(2a - 1) = 21$
64. $5(3 - 3t) = 30$
Aha! 65. $11 = 11(x + 1)$
66. $9 = 3(5x - 2)$
67. $2(3 + 4m) - 6 = 48$
68. $3(5 + 3m) - 8 = 7$
69. $2x = x + x$
70. $5x - x = x + 3x$
71. $2r + 8 = 6r + 10$
72. $3b - 2 = 7b + 4$

73. $4y - 4 + y + 24 = 6y + 20 - 4y$

74. $5y - 10 + y = 7y + 18 - 5y$

75. $3(x + 4) = 3(x - 1)$

76. $5(x - 7) = 3(x - 2) + 2x$

77. $19 - 3(2x - 1) = 7$

78. $5(d + 4) = 7(d - 2)$

79. $2(3t + 1) - 5 = t - (t + 2)$

80. $4x - (x + 6) = 5(3x - 1) + 8$

81. $19 - (2x + 3) = 2(x + 3) + x$

82. $13 - (2c + 2) = 2(c + 2) + 3c$

83. $4 + 7x = 7(x + 1)$

84. $3(t + 2) + t = 2(3 + 2t)$

85. $\frac{3}{4}(3t - 4) = 15$

86. $\frac{3}{2}(2x + 5) = -\frac{15}{2}$

87. $\frac{1}{6}\left(\frac{3}{4}x - 2\right) = -\frac{1}{5}$

88. $\frac{2}{3}\left(\frac{7}{8} - 4x\right) - \frac{5}{8} = \frac{3}{8}$

89. $0.7(3x + 6) = 1.1 - (x - 3)$

90. $0.9(2x - 8) = 4 - (x + 5)$

91. $2(7 - x) - 20 = 7x - 3(2 + 3x)$

92. $5(x - 7) = 3(x - 2) + 2x$

93. $a + (a - 3) = (a + 2) - (a + 1)$

94. $0.8 - 4(b - 1) = 0.2 + 3(4 - b)$

95. Maggie solves $45 - t = 13$ (Example 3) by adding $t - 13$ to both sides. Is this approach preferable to the one used in Example 3? Why or why not?

96. Why must the rules for the order of operations be understood before solving the equations in this section?

Skill Review

97. Add: $\frac{2}{9} + \frac{1}{6}$. [1.3]

98. Multiply: $\frac{2}{7} \cdot \frac{7}{2}$. [1.3]

99. Find decimal notation: $-\frac{1}{9}$. [1.4]

100. Find the absolute value: $|-16|$. [1.4]

Synthesis

101. What procedure would you use to solve an equation like $0.23x + \frac{17}{3} = -0.8 + \frac{3}{4}x$? Could your procedure be streamlined? If so, how?

102. Ethan is determined to solve $3x + 4 = -11$ by first using the multiplication principle to "eliminate" the 3. How should he proceed and why?

Solve. Label any contradictions or identities.

103. $8.43x - 2.5(3.2 - 0.7x) = -3.455x + 9.04$

104. $0.008 + 9.62x - 42.8 = 0.944x + 0.0083 - x$

105. $-2[3(x - 2) + 4] = 4(5 - x) - 2x$

106. $0 = t - (-6) - (-7t)$

107. $3(x + 5) = 3(5 + x)$

108. $5(x - 7) = 3(x - 2) + 2x$

109. $2x(x + 5) - 3(x^2 + 2x - 1) = 9 - 5x - x^2$

110. $9 - 3x = 2(5 - 2x) - (1 - 5x)$

Aha! 111. $[7 - 2(8 \div (-2))]x = 0$

112. $\dfrac{5x + 3}{4} + \dfrac{25}{12} = \dfrac{5 + 2x}{3}$

113. Kyle estimates his long-distance driving time, in hours, by multiplying the number of miles he will drive by $\frac{3}{200}$ and then adding $\frac{1}{4}$ for every major city he will drive through. If he estimates that a trip through two major cities will take 8 hr, how many miles will he be driving?

QUICK QUIZ: SECTIONS 2.1–2.2

1. Determine whether −5 is a solution of $7 - 3x = 22$. [2.1]

Solve and check.

2. $y + 7.5 = 2.1$ [2.1]   3. $\dfrac{2x}{5} = 10$ [2.1]

4. $2 - (x - 7) = 13$ [2.2]   5. $\dfrac{1}{10}r + \dfrac{1}{5} = \dfrac{1}{2}r$ [2.2]

PREPARE TO MOVE ON

Evaluate. [1.8]

1. $3 - 5a$, for $a = 2$ 2. $12 \div 4 \cdot t$, for $t = 5$

3. $7x - 2x$, for $x = -3$ 4. $t(8 - 3t)$, for $t = -2$

2.3 Formulas

Evaluating Formulas   ▪   Solving for a Variable

An equation that shows a relationship between variable quantities will use letters and is known as a **formula**. Most of the letters in this book are variables, but some are constants. For example, c in $E = mc^2$ represents the speed of light.

Evaluating Formulas

EXAMPLE 1 *Event Promotion.*   Event promoters use the formula

$$p = \frac{1.2x}{s}$$

to determine a ticket price p for an event with x dollars of expenses and anticipated ticket sales of s tickets. Grand Events expects expenses for a concert to be $80,000 and anticipates selling 4000 tickets. What should the ticket price be?

Source: The Indianapolis Star, 2/27/03

SOLUTION   We substitute 80,000 for x and 4000 for s and calculate p:

$$p = \frac{1.2x}{s} = \frac{1.2(80,000)}{4000} = 24.$$

The ticket price should be $24.

1. Using the formula in Example 1, determine the ticket price when concert expenses are $40,000 and anticipated sales are 2500 tickets.

YOUR TURN

Technology Connection

Suppose that after calculating $1.2 \cdot 80,000 \div 4000$, we want to find $1.2 \cdot 80,000 \div 5000$. Pressing **2ND** **ENTRY** gives the following.

```
1.2*80000÷4000
                  24
1.2*80000÷4000 ▉
```

Moving the cursor left, we can change 4000 to 5000 and press **ENTER**.

```
1.2*80000÷4000
                  24
1.2*80000÷5000
                19.2
```

1. Verify the work above and then use **2ND** **ENTRY** to find $1.2 \cdot 60000 \div 5000$.

2. Solve for w: $V = lwh$.

YOUR TURN

Solving for a Variable

The **circumference** of a circle is the distance around the circle. The formula $C = 2\pi r$ gives the circumference C of a circle with radius r. For some circles (such as a circular pond), it is easier to measure the circumference than to measure the radius. In order to find the radius of a circle when we know its circumference, we can *solve* the formula for r.

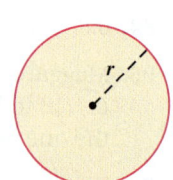

To **solve** for a variable means to write an equivalent equation with that variable alone on one side of the equation. The letter should not appear at all on the other side of the equation.

EXAMPLE 2 *Circumference of a Circle.*   Solve for r: $C = 2\pi r$.

SOLUTION   We want r to appear alone on one side of the equation and not to appear at all on the other side.

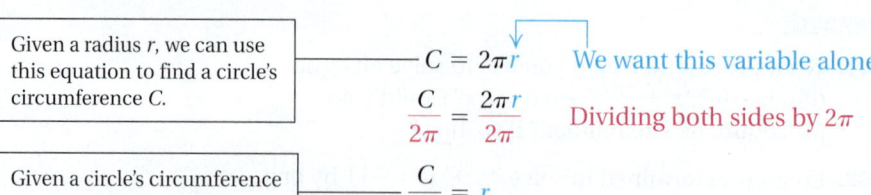

Given a radius r, we can use this equation to find a circle's circumference C.	$C = 2\pi r$	We want this variable alone.
	$\dfrac{C}{2\pi} = \dfrac{2\pi r}{2\pi}$	Dividing both sides by 2π
Given a circle's circumference C, we can use this equation to find the radius r.	$\dfrac{C}{2\pi} = r$	

To see how solving a formula is just like solving an equation, compare the following. In (A), we solve as usual; in (B), we show steps but do not simplify; and in (C), we *cannot* simplify because a, b, and c are unknown.

A. $5x + 2 = 12$

$5x = 12 - 2$

$5x = 10$

$x = \dfrac{10}{5} = 2$

B. $5x + 2 = 12$

$5x = 12 - 2$

$x = \dfrac{12 - 2}{5}$

C. $ax + b = c$

$ax = c - b$

$x = \dfrac{c - b}{a}$

EXAMPLE 3 *Motion.*   The rate r at which an object moves is found by dividing distance d traveled by time t, or

$$r = \frac{d}{t}.$$

Solve for t.

SOLUTION   We use the multiplication principle to clear fractions and then solve for t:

$r = \dfrac{d}{t}$ ⟵ —— We want this variable alone.

$r \cdot t = \dfrac{d}{t} \cdot t$ Multiplying both sides by t

$rt = \dfrac{dt}{t}$ $\dfrac{d}{t} \cdot t = \dfrac{d}{t} \cdot \dfrac{t}{1} = \dfrac{dt}{t}$

$rt = d$ Removing a factor equal to 1: $t/t = 1$.
The equation is cleared of fractions.

$\dfrac{rt}{r} = \dfrac{d}{r}$ Dividing both sides by r

$t = \dfrac{d}{r}.$

This formula can be used to determine the time spent traveling when the distance and the rate are known.

 YOUR TURN

EXAMPLE 4   Solve for y: $3x - 4y = 10$.

SOLUTION   There is one term that contains y, so we begin by isolating that term on one side of the equation.

$3x - 4y = 10$ We want this variable alone.

$-4y = 10 - 3x$ Subtracting $3x$ from both sides

$-\frac{1}{4}(-4y) = -\frac{1}{4}(10 - 3x)$ Multiplying both sides by $-\frac{1}{4}$

$y = -\frac{10}{4} + \frac{3}{4}x$ Multiplying using the distributive law

$y = -\frac{5}{2} + \frac{3}{4}x$ Simplifying the fraction

Student Notes

When working with formulas, a lowercase letter should not be interchanged with its associated uppercase letter; these letters may represent different quantities. For example, a formula for gravity uses both M and m.

3. Solve for x: $y = \dfrac{k}{x}$.

4. Solve for y: $x + 2y = 3$.

5. Solve for p:

$$T = 20(a + p) - 15.$$

EXAMPLE 5 *Nutrition.* The number of calories K needed each day by a moderately active woman who weighs w pounds, is h inches tall, and is a years old, can be estimated using the formula

$$K = 917 + 6(w + h - a).^*$$

Solve for w.

SOLUTION   We undo the operations in the reverse order in which they would be performed on the right side:

$$K = 917 + 6(w + h - a) \qquad \text{We want } w \text{ alone.}$$

$$K - 917 = 6(w + h - a) \qquad \text{Subtracting 917 from both sides}$$

$$\frac{K - 917}{6} = w + h - a \qquad \text{Dividing both sides by 6}$$

$$\frac{K - 917}{6} + a - h = w. \qquad \begin{array}{l}\text{Adding } a \text{ and subtracting } h \\ \text{on both sides}\end{array}$$

This formula can be used to estimate a woman's weight, if we know her age, height, and caloric needs.

YOUR TURN

The above steps are similar to those used in Section 2.2 to solve equations. We use the addition and multiplication principles just as before. An important difference that we will see in the next example is that we will sometimes need to factor.

TO SOLVE A FORMULA FOR A GIVEN VARIABLE

1. If the variable for which you are solving appears in a fraction, use the multiplication principle to clear fractions.
2. Isolate the term(s), with the variable for which you are solving on one side of the equation.
3. If two or more terms contain the variable for which you are solving, factor the variable out.
4. Multiply or divide to solve for the variable in question.

We can also solve for a letter that represents a constant.

EXAMPLE 6 *Surface Area of a Right Circular Cylinder.* The formula $A = 2\pi rh + 2\pi r^2$ gives the surface area A of a right circular cylinder of height h and radius r. Solve for π.

SOLUTION   We have

$$A = 2\pi rh + 2\pi r^2 \qquad \text{We want this letter alone.}$$

$$A = \pi(2rh + 2r^2) \qquad \text{Factoring}$$

$$\frac{A}{2rh + 2r^2} = \pi. \qquad \begin{array}{l}\text{Dividing both sides by } 2rh + 2r^2, \text{ or} \\ \text{multiplying both sides by } 1/(2rh + 2r^2)\end{array}$$

We can also write this as

$$\pi = \frac{A}{2rh + 2r^2}.$$

6. Solve for c: $D = 3c + 6cy$.

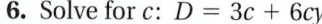

YOUR TURN

*Based on information from M. Parker (ed.), *She Does Math!* (Washington, D.C.: Mathematical Association of America, 1995), p. 96.

CAUTION! Had we performed the following steps in Example 6, we would *not* have solved for π:

$$A = 2\pi rh + 2\pi r^2 \qquad \text{We want } \pi \text{ alone.}$$

$$A - 2\pi r^2 = 2\pi rh \qquad \text{Subtracting } 2\pi r^2 \text{ from both sides}$$

Two occurrences of π

$$\frac{A - 2\pi r^2}{2rh} = \pi. \qquad \text{Dividing both sides by } 2rh$$

The mathematics of each step is correct, but because π occurs on both sides of the formula, *we have not solved the formula for π*. Remember that the letter being solved for should be alone on one side of the equation, with no occurrence of that letter on the other side!

2.3 EXERCISE SET

FOR EXTRA HELP

PRACTICE   WATCH   READ   REVIEW

⤷ Vocabulary and Reading Check

Classify each of the following statements as either true or false.

1. All letters used in algebra represent variables.

2. If a letter appears on both sides of a formula, that formula is not solved for that letter.

Choose the word that best completes each sentence from the choices listed below the blank.

3. The distance around a circle is its

_____.
  area/circumference/volume

4. An equation that uses two or more letters to represent a relationship among quantities is a(n)

_____.
  constant/formula/variable

Evaluating Formulas

Solve.

5. *Outdoor Concerts.* The formula $d = 344t$ can be used to determine how far d, in meters, sound travels through room-temperature air in t seconds. Fans near the back of a large crowd at a concert will experience a time lag between the time they see a note played on stage and the time they hear that note. If the time lag is 0.9 sec, how far are the fans from the stage?

6. *Furnace Output.* Contractors in the Northeast use the formula $B = 30a$ to determine the minimum furnace output B, in British thermal units (Btu's), for a well-insulated house with a square feet of flooring.

Determine the minimum furnace output for an 1800-ft^2 house that is well insulated.

Source: U.S. Department of Energy

7. *College Enrollment.* At many colleges, the number of "full-time-equivalent" students F is given by

$$F = \frac{n}{15},$$

where n is the total number of credits for which students have enrolled in a given semester. Determine the number of full-time-equivalent students on a campus in which students registered for a total of 21,345 credits.

8. *Distance from a Storm.* The formula $M = \frac{1}{5}t$ can be used to determine how far M, in miles, you are from lightning when its thunder takes t seconds to reach your ears. If it takes 10 sec for the sound of thunder to reach you after you have seen the lightning, how far away is the storm?

9. *Federal Funds Rate.* The Federal Reserve Board sets a target f for the federal funds rate, that is, the interest rate that banks charge each other for over-night borrowing of Federal funds. This target rate can be estimated by

$$f = 8.5 + 1.4(I - U),$$

where I is the core inflation rate over the previous 12 months and U is the seasonally adjusted unemployment rate. If core inflation is 0.025 and unemployment is 0.044, what should the federal funds rate be?

Source: Greg Mankiw, Harvard University, www.gregmankiw.blogspot.com/2006/06/what-would-alan-do.html

10. *Calorie Density.* The calorie density D, in calories per ounce, of a food that contains c calories and weighs w ounces is given by

$$D = \frac{c}{w}.^*$$

Eight ounces of fat-free milk contains 84 calories. Find the calorie density of fat-free milk.

11. *Absorption of Ibuprofen.* When 400 mg of the painkiller ibuprofen is swallowed, the number of milligrams n in the bloodstream t hours later (for $0 \le t \le 6$) is estimated by

$$n = 0.5t^4 + 3.45t^3 - 96.65t^2 + 347.7t.$$

How many milligrams of ibuprofen remain in the blood 1 hr after 400 mg has been swallowed?

12. *Size of a League Schedule.* When all n teams in a league play every other team twice, a total of N games are played, where

$$N = n^2 - n.$$

If a soccer league has 7 teams and all teams play each other twice, how many games are played?

Solving for a Variable

In Exercises 13–52, solve each formula for the indicated letter.

13. $A = bh$, for b
(Area of parallelogram with base b and height h)

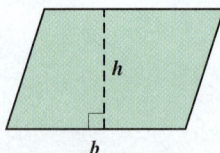

14. $A = bh$, for h

Source: *Nutrition Action Healthletter*, March 2000, p. 9. Center for Science in the Public Interest, Suite 300; 1875 Connecticut Ave NW, Washington, D.C. 20008.

15. $I = Prt$, for P
(Simple-interest formula, where I is interest, P is principal, r is interest rate, and t is time)

16. $I = Prt$, for t

17. $H = 65 - m$, for m
(To determine the number of heating degree days H for a day with m degrees Fahrenheit as the average temperature)

18. $d = h - 64$, for h
(To determine how many inches d above average an h-inch-tall woman is)

19. $P = 2l + 2w$, for l
(Perimeter of a rectangle of length l and width w)

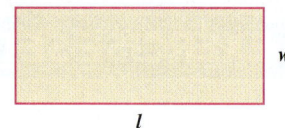

20. $P = 2l + 2w$, for w

21. $A = \pi r^2$, for π
(Area of a circle with radius r)

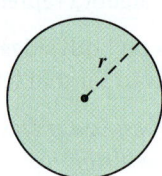

22. $A = \pi r^2$, for r^2

23. $A = \frac{1}{2}bh$, for h
(Area of a triangle with base b and height h)

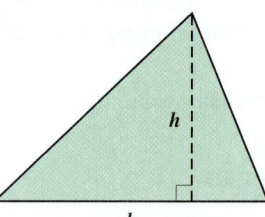

24. $A = \frac{1}{2}bh$, for b

25. $E = mc^2$, for c^2
(A relativity formula from physics)

26. $E = mc^2$, for m

27. $Q = \dfrac{c + d}{2}$, for d

28. $A = \dfrac{a + b + c}{3}$, for b

29. $p - q + r = 2$, for q
(Euler's formula from graph theory)

30. $p = \dfrac{r - q}{2}$, for q

31. $w = \dfrac{r}{f}$, for r

(To compute the wavelength w of a musical note with frequency f and speed of sound r)

32. $M = \dfrac{A}{s}$, for A

(To compute the Mach number M for speed A and speed of sound s)

33. $H = \dfrac{TV}{550}$, for T

(To determine the horsepower H of an airplane propeller with thrust T and airplane velocity V)

34. $P = \dfrac{ab}{c}$, for b

35. $F = \frac{9}{5}C + 32$, for C

(To convert the Celsius temperature C to the Fahrenheit temperature F)

36. $M = \frac{5}{9}n + 18$, for n

37. $2x - y = 1$, for y **38.** $3x - y = 7$, for y

39. $2x + 5y = 10$, for y **40.** $3x + 2y = 12$, for y

41. $4x - 3y = 6$, for y **42.** $5x - 4y = 8$, for y

43. $9x + 8y = 4$, for y **44.** $x + 10y = 2$, for y

45. $3x - 5y = 8$, for y **46.** $7x - 6y = 7$, for y

47. $z = 13 + 2(x + y)$, for x

48. $A = 115 + \frac{1}{2}(p + s)$, for s

49. $t = 27 - \frac{1}{4}(w - l)$, for l

50. $m = 19 - 5(x - n)$, for n

51. $A = at + bt$, for t

52. $S = rx + sx$, for x

53. *Area of a Trapezoid.*   The formula

$$A = \tfrac{1}{2}ah + \tfrac{1}{2}bh$$

can be used to find the area A of a trapezoid with bases a and b and height h. Solve for h. (*Hint*: First clear fractions.)

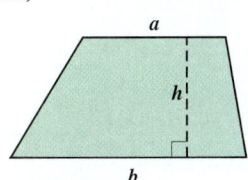

54. *Compounding Interest.*   The formula

$$A = P + Prt$$

is used to find the amount A in an account when simple interest is added to an investment of P dollars (see Exercise 15). Solve for P.

55. *Chess Rating.*   The formula

$$R = r + \frac{400(W - L)}{N}$$

is used to establish a chess player's rating R after that player has played N games, won W of them, and lost L of them. Here r is the average rating of the opponents. Solve for L.

Source: The U.S. Chess Federation

56. *Angle Measure.*   The angle measure S of a sector of a circle is given by

$$S = \frac{360A}{\pi r^2},$$

where r is the radius, A is the area of the sector, and S is in degrees. Solve for r^2.

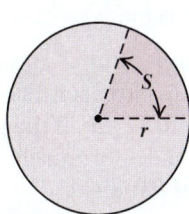

57. Julia has a formula that allows her to convert Celsius temperatures to Fahrenheit temperatures. She needs a formula for converting Fahrenheit temperatures to Celsius temperatures. What advice can you give her?

58. Under what circumstances would it be useful to solve $I = Prt$ for P? (See Exercise 15.)

Skill Review

Simplify.

59. $-2 + 5 - (-4) - 17$ [1.6]

60. $-98 \div \frac{1}{2}$ [1.7]

Aha! **61.** $4.2(-11.75)(0)$ [1.7]

62. $(-2)^5$ [1.8]

63. $20 \div (-4) \cdot 2 - 3$ [1.8]

64. $5|8 - (2 - 7)|$ [1.8]

Synthesis

65. The equations

$$P = 2l + 2w \quad \text{and} \quad w = \frac{P}{2} - l$$

are equivalent formulas involving the perimeter P, length l, and width w of a rectangle. Write a problem for which the second of the two formulas would be more useful.

66. While solving $2A = ah + bh$ for h, Eva writes

$$\frac{2A - ah}{b} = h.$$

What is her mistake?

67. The Harris–Benedict formula gives the number of calories K needed each day by a moderately active man who weighs w kilograms, is h centimeters tall, and is a years old as

$$K = 21.235w + 7.75h - 10.54a + 102.3.$$

If Janos is moderately active, weighs 80 kg, is 190 cm tall, and needs to consume 2852 calories per day, how old is he?

68. *Altitude and Temperature.* Air temperature drops about 1° Celsius (C) for each 100-m rise above ground level, up to 12 km. If the ground level temperature is t°C, find a formula for the temperature T at an elevation of h meters.

Source: A Sourcebook of School Mathematics, Mathematical Association of America, 1980

69. *Surface Area of a Cube.* The surface area A of a cube with side s is given by

$$A = 6s^2.$$

If a cube's surface area is 54 in², find the volume of the cube.

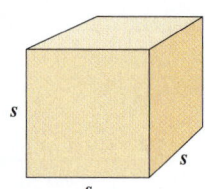

70. *Weight of a Fish.* An ancient fisherman's formula for estimating the weight of a fish is

$$w = \frac{lg^2}{800},$$

where w is the weight, in pounds, l is the length, in inches, and g is the girth (distance around the midsection), in inches. Estimate the girth of a 700-lb yellowfin tuna that is 8 ft long.

71. *Dosage Size.* Clark's rule for determining the size of a particular child's medicine dosage c is

$$c = \frac{w}{a} \cdot d,$$

where w is the child's weight, in pounds, and d is the usual adult dosage for an adult weighing a pounds. Solve for a.

Source: Olsen, June Looby, et al., *Medical Dosage Calculations.* Redwood City, CA: Addison-Wesley, 1995

Solve each formula for the given letter.

72. $\dfrac{y}{z} \div \dfrac{z}{t} = 1$, for y

73. $ac = bc + d$, for c

74. $qt = r(s + t)$, for t

75. $3a = c - a(b + d)$, for a

76. *Furnace Output.* The formula

$$B = 50a$$

is used in New England to estimate the minimum furnace output B, in Btu's, for an old, poorly insulated house with a square feet of flooring. Find an equation for determining the number of Btu's saved by insulating an old house. (*Hint:* See Exercise 6.)

77. Revise the formula in Exercise 67 so that a man's weight in pounds (2.2046 lb = 1 kg) and his height in inches (0.3937 in. = 1 cm) are used.

YOUR TURN ANSWERS: SECTION 2.3

1. $19.20   **2.** $w = \dfrac{V}{lh}$ **3.** $x = \dfrac{k}{y}$ **4.** $y = -\frac{1}{2}x + \frac{3}{2}$

5. $p = \dfrac{T + 15}{20} - a$ **6.** $c = \dfrac{D}{3 + 6y}$

QUICK QUIZ: SECTIONS 2.1–2.3

Solve and check.

1. $-x = -7$ [2.1]

2. $4(n - 7) - n = 36$ [2.2]

3. $1.2t - 0.05 = 3.2t$ [2.2]

4. $3(2x - 4) = 2 - 5(7 - x)$ [2.2]

5. Solve for y: $6x + 2y = 1$.   [2.3]

PREPARE TO MOVE ON

Convert to decimal notation.   [1.4]

1. $\dfrac{1}{4}$ **2.** $\dfrac{9}{8}$ **3.** $\dfrac{2}{3}$ **4.** $\dfrac{5}{6}$

Mid-Chapter Review

We solve equations using the addition and multiplication principles.

For any real numbers a, b, and c:

a) $a = b$ is equivalent to $a + c = b + c$;

b) $a = b$ is equivalent to $ac = bc$, provided $c \neq 0$.

GUIDED SOLUTIONS

Solve.   [2.2]*

1. $2x + 3 = 10$

Solution

$2x + 3 - 3 = 10 - \boxed{}$ Using the addition principle

$\qquad 2x = \boxed{}$ Simplifying

$\dfrac{1}{2} \cdot 2x = \boxed{} \cdot 7$ Using the multiplication principle

$\qquad x = \boxed{}$ Simplifying

2. $\frac{1}{2}(x - 3) = \frac{1}{3}(x - 4)$

Solution

$6 \cdot \frac{1}{2}(x - 3) = \boxed{} \cdot \frac{1}{3}(x - 4)$ Multiplying to clear fractions

$\boxed{}(x - 3) = \boxed{}(x - 4)$ The fractions are cleared.

$3x - \boxed{} = 2x - \boxed{}$ Multiplying

$3x - 9 + 9 = 2x - 8 + \boxed{}$ Using the addition principle

$3x = 2x + \boxed{}$ Simplifying

$3x - \boxed{} = 2x + 1 - 2x$ Using the addition principle

$x = \boxed{}$ Simplifying

MIXED REVIEW

Solve.

3. $x - 2 = -1$ [2.1]

4. $2 - x = -1$ [2.1]

5. $3t = 5$ [2.1]

6. $-\frac{3}{2}x = 12$ [2.1]

7. $\dfrac{y}{8} = 6$ [2.1]

8. $0.06x = 0.03$ [2.1]

9. $3x - 7x = 20$ [2.2]

10. $9x - 7 = 17$ [2.2]

11. $4(t - 3) - t = 6$ [2.2]

12. $8n - (3n - 5) = 5 - n$ [2.2]

13. $\frac{9}{10}y - \frac{7}{10} = \frac{21}{5}$ [2.2]

14. $2(t - 5) - 3(2t - 7) = 12 - 5(3t + 1)$ [2.2]

15. $\frac{2}{3}(x - 2) - 1 = -\frac{1}{2}(x - 3)$ [2.2]

Solve for the indicated variable.   [2.3]

16. $E = wA$, for A

17. $Ax + By = C$, for y

18. $at + ap = m$, for a

19. $m = \dfrac{F}{a}$, for a

20. $v = \dfrac{b - f}{t}$, for b

---

*The notation [2.2] refers to Chapter 2, Section 2.

2.4 Applications with Percent

Converting Between Percent Notation and Decimal Notation   ▪   Solving Percent Problems

Percent problems arise so frequently in everyday life that often we are not even aware of them. In this section, we will solve some real-world percent problems. Before doing so, however, we need to review a few basics.

Converting Between Percent Notation and Decimal Notation

Oceans cover 70% of the earth's surface. This means that of every 100 square miles on the surface of the earth, 70 square miles is ocean. Thus, 70% is a ratio of 70 to 100.

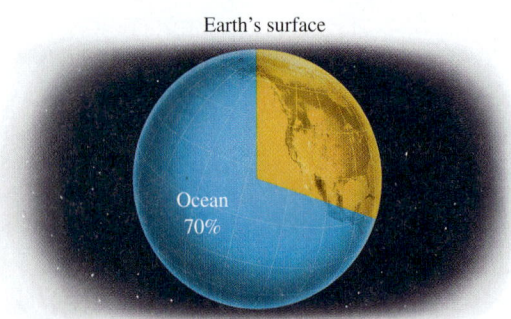

Earth's surface

Ocean 70%

The percent symbol % means "per hundred." We can regard the percent symbol as part of a name for a number. For example,

$$70\% \quad \text{is defined to mean} \quad \frac{70}{100}, \quad \text{or} \quad 70 \times \frac{1}{100}, \quad \text{or} \quad 70 \times 0.01.$$

> **PERCENT NOTATION**
>
> $$n\% \quad \text{means} \quad \frac{n}{100}, \quad \text{or} \quad n \times \frac{1}{100}, \quad \text{or} \quad n \times 0.01.$$

EXAMPLE 1   Convert to decimal notation: **(a)** 78%; **(b)** 1.3%.

SOLUTION

a) $78\% = 78 \times 0.01$ Replacing % with $\times 0.01$

   $= 0.78$

b) $1.3\% = 1.3 \times 0.01$ Replacing % with $\times 0.01$

   $= 0.013$

1. Convert to decimal notation: 120%.

▶ YOUR TURN

As shown above, multiplication by 0.01 simply moves the decimal point two places to the left.

To convert from percent notation to decimal notation, move the decimal point two places to the left and drop the percent symbol.

EXAMPLE 2 Convert the percent notation in the following sentence to decimal notation: Only 20% of teenagers get 8 hr of sleep per night.

Source: National Sleep Foundation

SOLUTION

$$20\% = 20.0\% \qquad 0.20.0 \qquad 20\% = 0.20, \text{ or simply } 0.2$$

Move the decimal point two places to the left.

2. Convert to decimal notation: 4%.

 YOUR TURN

The procedure used in Examples 1 and 2 can be reversed:

$$0.38 = 38 \times 0.01$$
$$= 38\%. \qquad \text{Replacing } \times 0.01 \text{ with } \%$$

To convert from decimal notation to percent notation, move the decimal point two places to the right and write a percent symbol.

EXAMPLE 3 Convert to percent notation: **(a)** 1.27; **(b)** $\frac{1}{4}$; **(c)** 0.3.

SOLUTION

a) We first move the decimal point two places to the right: 1.27.
and then write a % symbol: 127%   This is the same as multiplying 1.27 by 100 and writing %.

b) Note that $\frac{1}{4} = 0.25$. We move the decimal point two places to the right: 0.25.
and then write a % symbol: 25%   Multiplying by 100 and writing %

c) We first move the decimal point two places to the right (recall that $0.3 = 0.30$): 0.30.
and then write a % symbol: 30%   Multiplying by 100 and writing %

3. Convert to percent notation: 0.37.

 YOUR TURN

Solving Percent Problems

In solving percent problems, we first *translate* the problem to an equation. Then we *solve* the equation. The key words in the translation are as follows.

KEY WORDS IN PERCENT TRANSLATIONS

"**Of**" translates to " · " or " × ".   "**Is**" or "**Was**" translates to " = ".

"**What**" translates to a variable.   "**%**" translates to "$\times \frac{1}{100}$" or "$\times 0.01$".

Student Notes

A way of checking answers is by estimating as follows:

$11\% \times 49 \approx 10\% \times 50 = 5.$

Since 5 is close to 5.39, our answer is reasonable.

4. What is 6% of 15?

---

5. 5.2 is 8 percent of what?

---

6. What percent of 80 is 16?

EXAMPLE 4   What is 11% of 49?

SOLUTION

Translate:

$$
\begin{array}{ccccc}
\text{What} & \text{is} & 11\% & \text{of} & 49? \\
\downarrow & \downarrow & \downarrow & \downarrow & \downarrow \\
a & = & 0.11 & \cdot & 49
\end{array}
$$

   "of" means multiply;
   $11\% = 0.11$

$a = 5.39$

Thus, 5.39 is 11% of 49. The answer is 5.39.

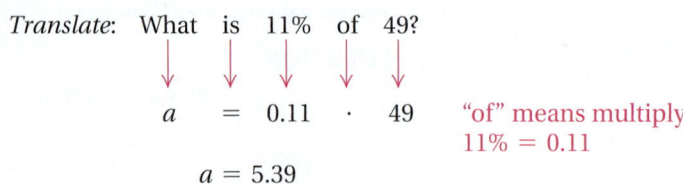 YOUR TURN

EXAMPLE 5   3 is 16 percent of what?

SOLUTION

Translate:

$$
\begin{array}{ccccc}
3 & \text{is} & 16\ \text{percent} & \text{of} & \text{what?} \\
\downarrow & \downarrow & \downarrow & \downarrow & \downarrow \\
3 & = & 0.16 & \cdot & y
\end{array}
$$

$\dfrac{3}{0.16} = y$ Dividing both sides by 0.16

$18.75 = y$

Thus, 3 is 16 percent of 18.75. The answer is 18.75.

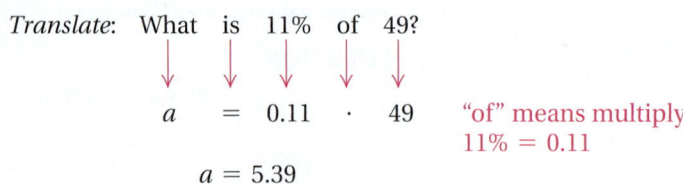 YOUR TURN

EXAMPLE 6   What percent of $50 is $34?

SOLUTION

Translate:

$$
\begin{array}{ccccc}
\text{What percent} & \text{of} & \$50 & \text{is} & \$34? \\
\downarrow & & \downarrow & \downarrow & \downarrow \\
n & \cdot & 50 & = & 34
\end{array}
$$

$n = \dfrac{34}{50}$ Dividing both sides by 50

$n = 0.68 = 68\%$ Converting to percent notation

Thus, $34 is 68% of $50. The answer is 68%.

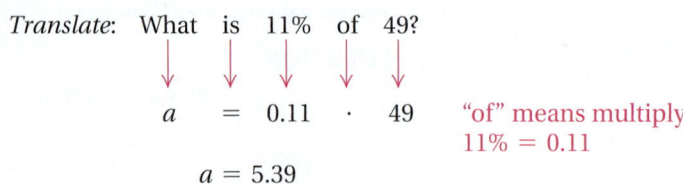 YOUR TURN

Examples 4–6 represent the three basic types of percent problems. Note that in all the problems, the following quantities are present:

- a percent, expressed in decimal notation in the translation,
- a base amount, referred to by the word "of" in the problem, and
- a percentage of the base, found by multiplying the base times the percent.

Student Notes

Always look for connections between examples. Here you should look for similarities between Examples 4 and 7 as well as between Examples 5 and 8 and between Examples 6 and 9.

---

7. Alicia presented her line of jewelry to 250 stores, and 25.6% of those stores placed an order. How many stores placed an order?

EXAMPLE 7 *Carpooling.* In 2009, there were 153 million Americans in the work force, and 10.5% of them carpooled to work. How many carpooled to work?

Source: U.S. Census Bureau

SOLUTION We first reword and then translate. We let $a =$ the number of American workers, in millions, who carpooled to work.

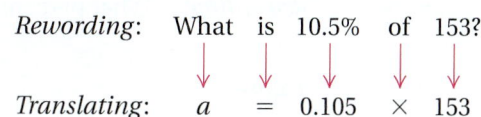

Rewording: What is 10.5% of 153?

Translating: $a = 0.105 \times 153$

The letter is by itself on one side of the equation. To solve the equation, we need only multiply:

$$a = 0.105 \times 153 = 16.065.$$

Since 16.065 million is 10.5% of 153 million, we have found that in 2009, about 16.065 million American workers carpooled to work.

YOUR TURN

---

8. At Valley Heights Community College, 252 freshmen enrolled in a Spanish class. This was 30% of the freshman class. How large was the freshman class?

EXAMPLE 8 *Social Networking.* About 110 million American adults used at least one social-networking website in 2010. This was 47% of all American adults. How many American adults were there in 2010?

Sources: Pew Research Center; U.S. Census Bureau

SOLUTION Before translating the problem to mathematics, we reword and let A represent the total number of American adults, in millions, in 2010.

Rewording: 110 is 47% of A

Translating: $110 = 0.47 \cdot A$

$$\frac{110}{0.47} = A \qquad \text{Dividing both sides by 0.47}$$

$$234 \approx A \qquad \text{The symbol } \approx \text{ means } \textit{is approximately equal to.}$$

There were about 234 million American adults in 2010.

YOUR TURN

EXAMPLE 9 *e-Reader Prices.* During a 2011 promotion, Amazon.com reduced the price of a Kindle e-reader from $139 to $114.

Source: Amazon.com

a) What percent of the regular price does the sale price represent?

b) What is the percent of discount?

SOLUTION

a) We reword and translate, using n for the unknown percent.

Rewording: What percent of 139 is 114?

Translating: $n \cdot 139 = 114$

$$n = \frac{114}{139} \qquad \text{Dividing both sides by 139}$$

$$n \approx 0.82 = 82\% \qquad \text{Converting to percent notation}$$

The sale price is about 82% of the regular price.

b) Since the original price of $139 represents 100% of the regular price, the sale price represents a discount of $(100 - 82)\%$, or 18%.

Alternatively, or as a check, we can find the amount of discount and then calculate the percent of discount, using x for the unknown percent.

Amount of discount: $\$139 - \$114 = \$25$

Rewording: What percent   of   139   is   25?

Translating: x ·   139   =   25

$$x = \frac{25}{139}$$ Dividing both sides by 139

$$x \approx 0.18 = 18\%$$ Converting to percent notation

We have a check. The percent of discount is 18%.

 Chapter Resources:
Collaborative Activity, p. 145;
Decision Making: Connection, p. 145

9. Recently, Elm Cycles reduced the price of a BMW F 800 Gs motorcycle from $15,000 to $13,500. What is the percent of discount?

 YOUR TURN

2.4 EXERCISE SET

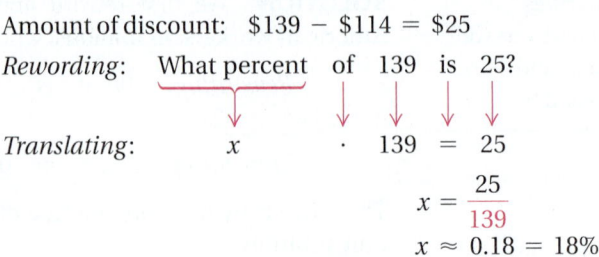

Vocabulary and Reading Check

Choose from the following list the word that best completes each statement. Not every word will be used.

approximately   not
base   percent
decimal   retail
hundred   right
left   sale

1. To convert from percent notation to decimal notation, move the decimal point two places to the _____ and drop the percent symbol.

2. The percent symbol, %, means "per _____."

3. The expression 1.3% is written in _____ notation.

4. The word "of" in a percent problem generally refers to the _____ amount.

5. The _____ price is the original price minus the discount.

6. The symbol ≈ means "is _____ equal to."

Concept Reinforcement

In each of Exercises 7–16, match the question with the most appropriate translation from the following list. Some choices are used more than once.

a) $a = (0.57)23$ **b)** $57 = 0.23y$
c) $n \cdot 23 = 57$ **d)** $n \cdot 57 = 23$
e) $23 = 0.57y$ **f)** $a = (0.23)57$

7. ____ What percent of 57 is 23?

8. ____ What percent of 23 is 57?

9. ____ 23 is 57% of what number?

10. ____ 57 is 23% of what number?

11. ____ 57 is what percent of 23?

12. ____ 23 is what percent of 57?

13. ____ What is 23% of 57?

14. ____ What is 57% of 23?

15. ____ 23% of what number is 57?

16. ____ 57% of what number is 23?

Converting Between Percent Notation and Decimal Notation

Convert the percent notation in each sentence to decimal notation.

17. *Musical Instruments.* Of those who play Guitar Hero and Rock Band but do not currently play an actual musical instrument, 67% indicated that they are likely to begin playing an actual instrument.

Source: Guitar Center Survey, gonintendo.com

18. *Volunteering.* Of all Americans, 55% do volunteer work.

Source: The Nonprofit Almanac in Brief

19. Among American workers, 5% rely on public transportation for their commute.

Source: U.S. Census Bureau

20. If a NASCAR driver loses more than 3% body weight in sweat and does not replace those fluids, the driver's focus and reflexes start declining.
Source: CNN.com

21. Approximately 3.2% of U.S. adults are vegetarians.
Source: Vegetarian Times

22. *Gold.* Gold that is marked 10K is 41.6% gold.

23. *Energy Use.* Using landscape to shade a home's outdoor air-conditioning unit can reduce cooling costs by 10%.
Source: U.S. Department of Energy

24. Semisweet dark chocolate contains about 60% chocolate liquor.

Convert to decimal notation.

25. 6.25%   **26.** 8.375%

27. 0.2%   **28.** 0.8%

29. 175%   **30.** 250%

Convert the decimal notation in each sentence to percent notation.

31. By the end of 2010, about 0.21 of American Internet users had paid to download apps for their cell phones or tablet computers.
Source: Pew Research Center, *Internet and American Life Project*

32. *Baseball Fans.* Baseball is the second most popular sport in the United States, with 0.17 of the adult population saying it is their favorite sport.
Source: Sports Business Daily, 1/25/11

33. In-state tuition at Virginia public colleges and universities cost 0.079 more in 2011–2012 than in 2010–2011.
Source: The State Council of Higher Education for Virginia

34. *Food Security.* The USDA defines food security as access to enough nutritious food for a healthy life. In 2010, approximately 0.057 of U.S. households had very low food security.
Source: USDA

35. *Composition of the Sun.* The sun is 0.7 hydrogen.

36. *Jupiter's Atmosphere.* The atmosphere of Jupiter is 0.1 helium.

Convert to percent notation.

37. 0.0009   **38.** 0.0056

39. 1.06   **40.** 1.08

41. $\frac{3}{5}$ **42.** $\frac{3}{2}$

43. $\frac{8}{25}$ **44.** $\frac{5}{8}$

Solving Percent Problems

Solve.

45. What percent of 76 is 19?

46. What percent of 125 is 30?

47. 14 is 30% of what number?

48. 54 is 24% of what number?

49. 0.3 is 12% of what number?

50. 7 is 175% of what number?

51. What number is 1% of one million?

52. What number is 35% of 240?

53. What percent of 60 is 75?

Aha! **54.** What percent of 70 is 70?

55. What is 2% of 40?

56. What is 40% of 2?

Aha! **57.** 25 is what percent of 50?

58. 0.8 is 2% of what number?

59. What percent of 69 is 23?

60. What percent of 40 is 9?

Renewable Energy. In 2010, renewable energy sources provided 8 quadrillion Btu's of the U.S. energy consumption. The following circle graph shows the percent provided by each renewable source. In each of Exercises 61–64, determine the amount of Btu's contributed by each source.

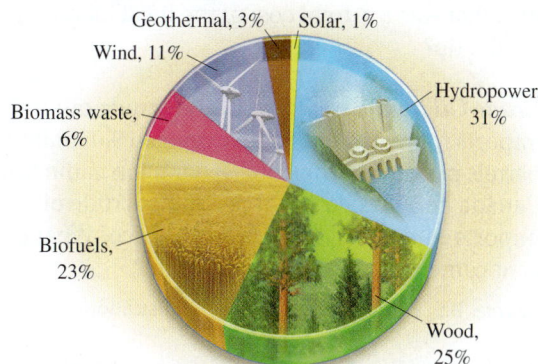

U.S. Renewable Energy

Geothermal, 3%   Solar, 1%
Wind, 11%
Biomass waste, 6%
Hydropower, 31%
Biofuels, 23%
Wood, 25%

Source: U.S. Energy Information Administration

61. Hydropower   **62.** Solar

63. Wind   **64.** Biofuels

65. *College Graduation.* To obtain his bachelor's degree in nursing, Cody must complete 125 credit hours of instruction. If he has completed 60% of his requirement, how many credits did Cody complete?

66. *College Graduation.* To obtain her bachelor's degree in journalism, Addie must complete 125 credit hours of instruction. If 20% of Addie's credit hours remain to be completed, how many credits does she still need to take?

67. *Batting Average.* In the 2011 season, Miguel Cabrera of the Detroit Tigers had 197 hits. His batting average was 0.344, the highest in major league baseball for that season. This means that of the total number of at-bats, 34.4% were hits. How many at-bats did he have?

Source: ESPN

68. *Pass Completions.* In the 2011 season, Drew Brees of the New Orleans Saints completed 468 passes. This was 71.2% of his attempts. How many attempts did he make?

Source: National Football League

69. *Tipping.* Trent left a $4 tip for a meal that cost $25.
   a) What percent of the cost of the meal was the tip?
   b) What was the total cost of the meal including the tip?

70. *Tipping.* Selena left a $12.76 tip for a meal that cost $58.
   a) What percent of the cost of the meal was the tip?
   b) What was the total cost of the meal including the tip?

71. *Crude Oil Imports.* In April 2011, crude oil imports to the United States averaged 9.0 million barrels per day. Of this total, 3.0 million came from Canada and Mexico. What percent of crude oil imports came from Canada and Mexico? What percent came from the rest of the world?

Source: U.S. Energy Information Administration

72. *Alternative-Fuel Vehicles.* Of the 1,500,000 alternative-fuel vehicles produced in the United States in 2008, 300,000 were sport utility vehicles. What percent of alternative-fuel vehicles were sport utility vehicles? What percent were other types of vehicles?

Source: U.S. Energy Information Administration

73. *Student Loans.* Glenn takes out a student loan for $2400. After a year, Glenn decides to pay off the interest, which is 6.80% of $2400. How much will he pay?

74. *Student Loans.* To finance her community college education, LaTonya takes out a student loan for $3500. After a year, LaTonya decides to pay off the interest, which is 4.50% of $3500. How much will she pay?

75. *Infant Health.* In a study of 300 pregnant women with "good-to-excellent" diets, 95% had babies in good or excellent health. How many women in this group had babies in good or excellent health?

76. *Infant Health.* In a study of 300 pregnant women with "poor" diets, 8% had babies in good or excellent health. How many women in this group had babies in good or excellent health?

77. *Cost of Self-Employment.* Because of additional taxes and fewer benefits, it has been estimated that a self-employed person must earn 20% more than a non–self-employed person performing the same task(s). If Tia earns $16 per hour working for Village Copy, how much would she need to earn on her own for a comparable income?

78. Refer to Exercise 77. Rik earns $18 per hour working for Round Edge stairbuilders. How much would Rik need to earn on his own for a comparable income?

79. *Budget Overruns.* The "Big Dig" in Boston, also known as the Central Artery/Tunnel Project, has been labeled the most expensive highway project in U.S. history. The original cost estimate for the project was $2.6 billion, and by the time the last major portion of the project was opened to vehicles, the cost was $12 billion over the original estimate. By what percent did the actual cost exceed the initial estimate?

Source: Associated Press, Dec. 20, 2003, cited by msnbc.msn.com: *Boston's "Big Dig" Opens to Public;* Construction Management Schools

80. *Swimming Records.* On December 18, 2009, Cesar Cielo Filho of Brazil set a world record by swimming 50 m on a long course (50-m pool) in 20.91 sec. This record is 0.61 sec longer than the record on the 50-m short course (25-m pool), which was set by Roland Schoeman of South Africa on August 8, 2009. By what percent is the record for a short course faster than the record for a long course?

Source: Guinness Book of World Records 2011

81. *Body Fat.* One author of this text exercises regularly at a local YMCA that recently offered a body-fat percentage test to its members. The device used measures the passage of a very low voltage of electricity through the body. The author's body-fat percentage was found to be 16.5%, and he weighs 191 lb. What part of his body weight, in pounds, is fat?

82. *Areas of Alaska and Arizona.* The area of Arizona is 19% of the area of Alaska. The area of Alaska is 586,400 mi². What is the area of Arizona?

83. *Spam e-Mail.* About 265 billion of the 294 billion e-mails sent each day are spam and viruses. What percent of e-mails are spam and viruses?

Source: Based on information from Radicati Group

84. *Kissing and Colds.* In a medical study, it was determined that if 800 people kiss someone else who has a cold, only 56 will actually catch the cold. What percent is this?

85. *Calorie Content.* An 8-oz serving of Ocean Spray® Cranberry Juice Cocktail contains 120 calories. This is 240% of the number of calories in an 8-oz serving of Ocean Spray® Cranberry Juice Drink. How many calories are in an 8-oz serving of the Cranberry Juice Drink?

86. *Sodium Content.* Each serving of Planters® Lightly Salted Peanuts contains 95 mg of sodium. This is 50% of the sodium content in each serving of Planters® Dry Roasted Peanuts. How many milligrams of sodium are in a serving of the Dry Roasted Peanuts?

87. How is the use of statistics in the following examples misleading?

a) A business explaining new restrictions on sick leave cited a recent survey indicating that 40% of all sick days were taken on Monday or Friday.

b) An advertisement urging summer installation of a security system quoted FBI statistics stating that over 26% of home burglaries occur between Memorial Day and Labor Day.

88. If Julian leaves a $12 tip for a $90 dinner, is he being generous, stingy, or neither? Explain.

Skill Review

89. Find the opposite of $-\frac{1}{3}$. [1.6]

90. Find the reciprocal of $-\frac{1}{3}$. [1.7]

91. Find $-(-x)$ for $x = -12$. [1.6]

92. Simplify: $(-3x)^2$. [1.8]

Synthesis

93. Campus Bookbuyers pays $30 for a book and sells it for $60. Is this a 100% markup or a 50% markup? Explain.

94. Erin is returning a tent that she bought during a 25%-off storewide sale that has ended. She is offered store credit for 125% of what she paid (not to be used on sale items). Is this fair to Erin? Why or why not?

95. The community of Bardville has 1332 left-handed females. If 48% of the community is female and 15% of all females are left-handed, how many people are in the community?

96. It has been determined that at the age of 10, a girl has reached 84.4% of her final adult height. Dana is 4 ft 8 in. at the age of 10. Predict her final adult height.

97. It has been determined that at the age of 15, a boy has reached 96.1% of his final adult height. Jaraan is 6 ft 4 in. at the age of 15. Predict his final adult height.

98. *Dropout Rate.* Between 2007 and 2009, the high school dropout rate in the United States decreased from 87 per thousand to 81 per thousand. Calculate the percent by which the dropout rate decreased and use that percentage to estimate dropout rates for the United States in 2010 and in 2011.

Source: www.childtrendsdatabank.org

99. *Photography.* A 6-in. by 8-in. photo is framed using a mat meant for a 5-in. by 7-in. photo. What percentage of the photo will be hidden by the mat?

100. Would it be better to receive a 5% raise and then, a year later, an 8% raise or the other way around? Why?

Aha!

101. Jorge is in the 30% tax bracket. This means that 30¢ of each dollar earned goes to taxes. Which would cost him the least: contributing $50 that is tax-deductible or contributing $40 that is not tax-deductible? Explain.

102. *Research.* If you have student loans, find the interest rates on those loans, whether the rates are fixed or variable, and how the interest is calculated. How much interest will you owe when the first payment is due?

 YOUR TURN ANSWERS: SECTION 2.4

1. 1.2   **2.** 0.04   **3.** 37%   **4.** 0.9   **5.** 65   **6.** 20%
7. 64 stores   **8.** 840 students   **9.** 10%

QUICK QUIZ: SECTIONS 2.1–2.4

Solve and check.

1. $6x - (5 - x) = 2x - 5$ [2.2]

2. $\dfrac{x}{2} - 10 = \dfrac{1}{3}$ [2.2]

3. Solve for p: $4p + mp = T$.   [2.3]

4. Convert to decimal notation: 1.2%.   [2.4]

5. 5 is 20% of what number?   [2.4]

PREPARE TO MOVE ON

Translate to an algebraic expression or equation.   [1.1]

1. Twice the length plus twice the width

2. 5% of $180

3. The product of 10 and half of a

4. 10 more than three times a number

5. A board's width is 2 in. less than its length.

6. A number is four times as large as a second number.

2.5   Problem Solving

The Five Steps for Problem Solving   •   Percent Increase and Percent Decrease

Probably the most important use of algebra is as a tool for problem solving. In this section, we develop a five-step problem-solving approach that is used throughout the remainder of the text.

The Five Steps for Problem Solving

FIVE STEPS FOR PROBLEM SOLVING IN ALGEBRA

1. *Familiarize* yourself with the problem.
2. *Translate* to mathematical language. (This often means writing an equation.)
3. *Carry out* some mathematical manipulation. (This often means *solving* an equation.)
4. *Check* your possible answer in the original problem.
5. *State* the answer clearly, using a complete English sentence.

Of the five steps, the most important is probably the first one: becoming familiar with the problem. Here are some hints for familiarization.

TO BECOME FAMILIAR WITH A PROBLEM

1. Read the problem carefully. Try to visualize the problem.
2. Reread the problem, perhaps aloud. Make sure you understand all words as well as any symbols or abbreviations.
3. List the information given and the question(s) to be answered. Choose a variable (or variables) to represent the unknown and specify exactly what the variable represents. For example, let L = length in centimeters, d = distance in miles, and so on.
4. Look for similarities between the problem and other problems you have already solved. Ask yourself what type of problem this is.
5. Find more information. Look up a formula in a book, at a library, or online. Consult a reference librarian or an expert in the field.
6. Make a table that uses the information you have available. Look for patterns that may help in the translation.
7. Make a drawing and label it with known and unknown information, using specific units if given.
8. Think of a possible answer and check your guess. Note the manner in which your guess is checked.

EXAMPLE 1 *Hiking.* In 1957 at the age of 69, Emma "Grandma" Gatewood became the first woman to hike solo all 2100 mi of the Appalachian trail—from Springer Mountain, Georgia, to Mount Katahdin, Maine. Gatewood repeated the feat in 1960 and again in 1963, becoming the first person to hike the trail three times. When Gatewood stood atop Big Walker Mountain, Virginia, she was three times as far from the northern end of the trail as from the southern end. At that point, how far was she from each end of the trail?

SOLUTION

1. **Familiarize.** It may be helpful to make a drawing.

To gain some familiarity, let's suppose that Gatewood stood 600 mi from Springer Mountain. Three times 600 mi is 1800 mi. Since 600 mi + 1800 mi = 2400 mi and 2400 mi > 2100 mi, we see that our guess is too large. Rather than guess again, we let

s = the distance, in miles, to the southern end

and

$3s$ = the distance, in miles, to the northern end.

(We could also let n = the distance to the northern end and $\frac{1}{3}n$ = the distance to the southern end.)

1. In 2005, Ken Looi of New Zealand set a record by covering 235.3 mi in 24 hr on his unicycle. After 8 hr, he was approximately twice as far from the finish line as he was from the start. How far had he traveled?

Source: Guinness World Records

2. **Translate.** From the drawing, we see that the lengths of the two parts of the trail must add up to 2100 mi. This leads to our translation.

Rewording: $\underbrace{\text{Distance to southern end}}$ plus $\underbrace{\text{distance to northern end}}$ is 2100 mi.

Translating: s $+$ $3s$ $=$ 2100

3. **Carry out.** We solve the equation:

$$s + 3s = 2100$$
$$4s = 2100 \qquad \text{Combining like terms}$$
$$s = 525. \qquad \text{Dividing both sides by 4}$$

4. **Check.** As predicted in the *Familiarize* step, s is less than 600 mi. If $s = 525$ mi, then $3s = 1575$ mi. Since 525 mi + 1575 mi = 2100 mi, we have a check.

5. **State.** Atop Big Walker Mountain, Gatewood stood 525 mi from Springer Mountain and 1575 mi from Mount Katahdin.

↩ YOUR TURN

We can represent several *consecutive integers* using one variable.

	Examples	Algebraic Representation
Consecutive Integers	16, 17, 18, 19, 20; −31, −30, −29, −28	x, $x + 1$, $x + 2$, and so on
Consecutive Even Integers	16, 18, 20, 22, 24; −52, −50, −48, −46	x, $x + 2$, $x + 4$, and so on
Consecutive Odd Integers	21, 23, 25, 27, 29; −71, −69, −67, −65	x, $x + 2$, $x + 4$, and so on

EXAMPLE 2 *Interstate Mile Markers.* U.S. interstate highways post numbered markers at every mile to indicate location in case of an emergency. The sum of two consecutive mile markers on I-70 in Kansas is 559. Find the numbers on the markers.

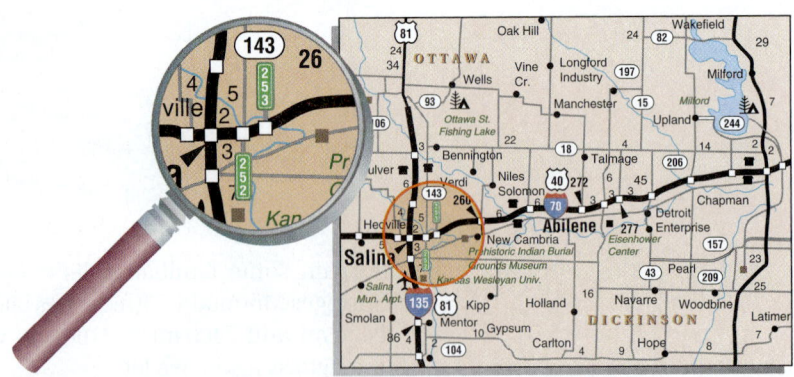

x	$x + 1$	Sum of x and $x + 1$
114	115	229
252	253	505
302	303	605

SOLUTION

1. **Familiarize.** The numbers on the mile markers are consecutive positive integers. Thus if we let $x =$ the smaller number, then $x + 1 =$ the larger number.

   To become familiar with the problem, we can make a table, as shown at left. First, we guess a value for x; then we find $x + 1$. Finally, we add the two numbers and check the sum.

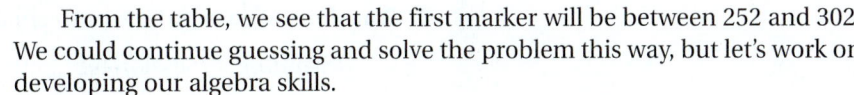

Study Skills

Set Reasonable Expectations

Do not be surprised if your success rate drops some as you work through the exercises in this section. *This is normal.* Your success rate will increase as you gain experience with these types of problems and use some of the study skills already listed.

---

2. The sum of two consecutive mile markers on I-70 in Indiana is 337. (See Example 2.) Find the numbers on the markers.

From the table, we see that the first marker will be between 252 and 302. We could continue guessing and solve the problem this way, but let's work on developing our algebra skills.

2. **Translate.** We reword the problem and translate as follows.

Rewording: First integer   plus   next integer   is   559.

Translating: x $+$ $(x + 1)$ $=$ 559

3. **Carry out.** We solve the equation:

$$x + (x + 1) = 559$$
$$2x + 1 = 559 \qquad \text{Using an associative law and combining like terms}$$
$$2x = 558 \qquad \text{Subtracting 1 from both sides}$$
$$x = 279. \qquad \text{Dividing both sides by 2}$$

If x is 279, then $x + 1$ is 280.

4. **Check.** Our possible answers are 279 and 280. These are consecutive positive integers and $279 + 280 = 559$, so the answers check.

5. **State.** The mile markers are 279 and 280.

YOUR TURN

EXAMPLE 3 *Photography.* Marissa has a budget of $500 for photographs for her company's new website. Several friends have recommended Fine Taste Photography for the job. Marissa has learned that Fine Taste charges $125 for a day session plus a fee for each image used. From her friends, she has gathered the following data.

Number of Images Used	Fee for Images Used
2	$ 65
20	650
24	780

From the information given, how many images can she buy without exceeding her budget?

SOLUTION

1. **Familiarize.** We can determine how much each image will cost. If we divide each fee by the number of images, we find that the cost is $32.50 per image:

$$\$65 \div 2 = \$32.50,$$
$$\$650 \div 20 = \$32.50,$$
$$\$780 \div 24 = \$32.50.$$

Suppose that Marissa buys 14 images. Then the total cost is the fee for the session plus the fee for the images, or

Session fee   plus   cost per image   times   number of images,

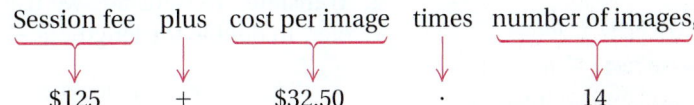

$125 $+$ $32.50 $\cdot$ 14

which is $580. Since this amount is greater than the $500 budgeted, we know that our guess of 14 images was too large. However, the process we used to check our guess can be used to translate the problem to mathematical language. We let $n =$ the number of images that Marissa can buy for a total cost of $500.

2. **Translate.** We reword the problem and translate as follows.

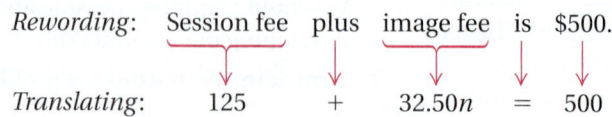

Rewording: Session fee   plus   image fee   is $500.

Translating: 125   +   32.50n   =   500

3. **Carry out.** We solve the equation:

$$125 + 32.50n = 500$$
$$32.50n = 375 \qquad \text{Subtracting 125 from both sides}$$
$$n \approx 11.53. \qquad \text{Dividing both sides by 32.50}$$

We need to round the answer to a whole number. In this case, we must round *down* to avoid going over the budget. We have a possible solution of 11 images.

4. **Check.** The fee for 11 images is $11(\$32.50) = \$357.50$. The total cost is then $\$357.50 + \$125 = \$482.50$. If Marissa spends this amount, she will have $17.50 left, which is not enough to buy another image. Also note that our answer is fewer than 14 images, as we expected from the *Familiarize* step.

5. **State.** Marissa can buy 11 images from Fine Taste Photography without exceeding her budget.

3. Refer to Example 3. If Marissa's budget is increased to $1000, how many images can she buy without exceeding her budget?

YOUR TURN

EXAMPLE 4 *Perimeter of an NBA Court.* The perimeter of an NBA basketball court is 288 ft. The length is 44 ft longer than the width. Find the dimensions of the court.

Source: National Basketball Association

SOLUTION

1. **Familiarize.** Recall that the perimeter of a rectangle is twice the length plus twice the width. Suppose the court were 30 ft wide. The length would then be $30 + 44$, or 74 ft, and the perimeter would be $2 \cdot 30$ ft $+ 2 \cdot 74$ ft, or 208 ft. This shows that in order for the perimeter to be 288 ft, the width must exceed 30 ft. Instead of guessing again, we let $w =$ the width of the court, in feet. Since the court is "44 ft longer than it is wide," we let $w + 44 =$ the length of the court, in feet.

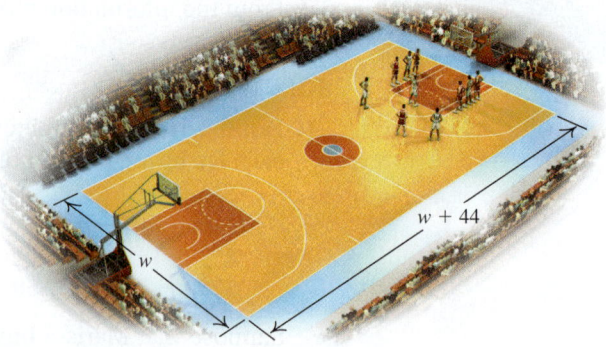

Student Notes

Get in the habit of writing what each variable represents before writing an equation. In Example 4, you might write

width (in feet) $= w$,

length (in feet) $= w + 44$

before translating the problem to an equation. This step becomes more important as problems become more complex.

2. **Translate.** To translate, we use $w + 44$ as the length and 288 as the perimeter. To double the length, $w + 44$, parentheses are essential.

Rewording: Twice the length   plus   twice the width   is   288 ft.

Translating: $2(w + 44)$ + $2w$ =   288

3. Carry out. We solve the equation:

$$2(w + 44) + 2w = 288$$
$$2w + 88 + 2w = 288 \qquad \text{\color{red}Using the distributive law}$$
$$4w + 88 = 288 \qquad \text{\color{red}Combining like terms}$$
$$4w = 200$$
$$w = 50.$$

4. The perimeter of a standard high school basketball court is 268 ft. The length is 34 ft longer than the width. Find the dimensions of the court.

Source: Indiana High School Athletic Association

The dimensions appear to be $w = 50$ ft, and $l = w + 44 = 94$ ft.

4. Check. If the width is 50 ft and the length is 94 ft, then the court is 44 ft longer than it is wide. The perimeter is $2(\color{red}50\text{ ft}\color{black}) + 2(\color{red}94\text{ ft}\color{black}) = 100$ ft $+ 188$ ft, or 288 ft, as specified. We have a check.

5. State. An NBA court is 50 ft wide and 94 ft long.

YOUR TURN

CAUTION!   Always be sure to answer the original problem completely. For instance, in Example 1 we needed to find *two* numbers: the distances from the hiker to *each* end of the trail. Similarly, in Example 4 we needed to find two dimensions, not just the width. Be sure to label each answer with the proper unit.

EXAMPLE 5   *Cross Section of a Roof.*   In a triangular gable end of a roof, the angle of the peak is twice as large as the angle on the back side of the house. The measure of the angle on the front side is 20° greater than the angle on the back side. How large are the angles?

SOLUTION

1. Familiarize. We make a drawing. We label the measure of the back angle x, the measure of the front angle $x + 20$, and the measure of the peak angle $2x$.

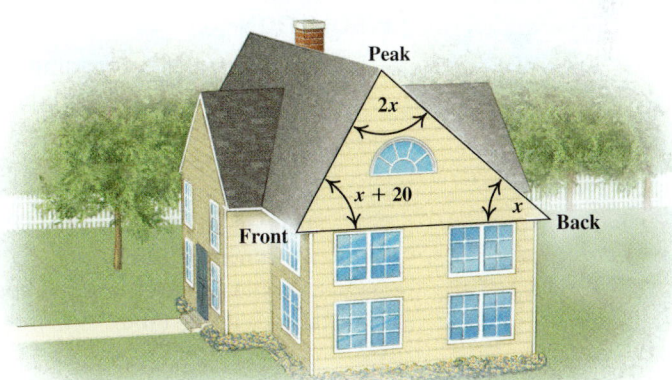

Student Notes

You may be expected to recall material that you have learned in an earlier course. If you have forgotten a formula, refresh your memory by consulting a textbook or a website.

2. Translate. To translate, we need to recall that the sum of the measures of the angles in any triangle is always 180°.

Rewording:	Measure of back angle	+	measure of front angle	+	measure of peak angle	is	180°
Translating:	x	+	$(x + 20)$	+	$2x$	=	180

3. **Carry out.** We solve:

$$x + (x + 20) + 2x = 180$$

$$4x + 20 = 180 \qquad \text{Combining like terms}$$

$$4x = 160 \qquad \text{Subtracting 20 from both sides}$$

$$x = 40. \qquad \text{Dividing both sides by 4}$$

The measures for the angles appear to be:

Back angle: $x = 40°$,

Front angle: $x + 20 = 40 + 20 = 60°$,

Peak angle: $2x = 2(40) = 80°$.

5. The second angle of a triangle is 30° more than the first. The third angle is half as large as the first. Find the measures of the angles.

4. **Check.** Consider 40°, 60°, and 80°, as listed above. The measure of the front angle is 20° greater than the measure of the back angle, the measure of the peak angle is twice the measure of the back angle, and the sum is 180°. These numbers check.

5. **State.** The measures of the angles are 40°, 60°, and 80°.

YOUR TURN

When working with motion problems, we often use tables, as well as the following motion formula.

MOTION FORMULA

$$d = r \cdot t$$

(Distance = Rate × Time)

EXAMPLE 6 *Motion.*   Sharon drove for 3 hr on a highway and then for 1 hr on a side road. Her speed on the highway was 20 mph faster than her speed on the side road. If she traveled a total of 220 mi, how fast did she travel on the side road?

SOLUTION

1. **Familiarize.** After reading the problem carefully, we see that we are asked to find the rate of travel on the side road. We let $x =$ Sharon's speed on the side road, in miles per hour.

   Since we are dealing with motion, we will use the motion formula $d = r \cdot t$. We use the variables in this formula as headings for three columns in a table. The rows of the table correspond to the highway and the side road. We know the times traveled for each type of driving, and we have defined a variable representing the speed on the side road. Thus we can fill in those entries in the table.

Road Type	$d =$	r	$\cdot$	t
Highway				3 hr
Side Road		x mph		1 hr

We want to fill in the remaining entries using the variable x. We know that Sharon's speed on the highway was 20 mph faster than her speed on the side road, or

Highway speed $= (x + 20)$ mph.

Then, since $d = r \cdot t$, we multiply to find each distance:

Highway distance $= (x + 20)(3)$ mi;

Side-road distance $= x(1)$ mi.

Road Type	d	$=$	r	$\cdot$	t
Highway	$(x + 20)(3)$ mi		$(x + 20)$ mph		3 hr
Side Road	$x(1)$ mi		x mph		1 hr

2. **Translate.** We know that Sharon traveled a total of 220 mi. This gives us the translation to an equation.

Rewording: Highway distance   plus   side-road distance   is   total distance.

Translating: $(x + 20)(3)$ $+$ $x(1)$ $=$ 220

3. **Carry out.** We now have an equation to solve:

$$(x + 20)(3) + x(1) = 220$$
$$3x + 60 + x = 220 \qquad \text{Using the distributive law}$$
$$4x + 60 = 220 \qquad \text{Combining like terms}$$
$$4x = 160 \qquad \text{Subtracting 60 from both sides}$$
$$x = 40. \qquad \text{Dividing both sides by 4}$$

4. **Check**. Since x represents Sharon's speed on the side road, we have the following.

	Speed (Rate)	Time	Distance
Highway	$40 + 20 = 60$ mph	3 hr	$60(3) = 180$ mi
Side Road	40 mph	1 hr	$40(1) = 40$ mi

The total distance traveled is 40 mi + 180 mi, or 220 mi. The answer checks.

5. **State.** Sharon's speed on the side road was 40 mph.

 YOUR TURN

6. Erik drove to his first sales appointment at 60 mph. From there, he drove to his second appointment at 40 mph. It took him twice as long to drive to his first appointment as it did to his second. If he traveled a total of 80 mi, how long did it take him to drive from his first appointment to his second?

PROBLEM-SOLVING TIPS

1. The more problems you solve, the more your skills will improve.
2. Look for patterns when solving problems.
3. Clearly define variables before translating to an equation.
4. Consider the dimensions of the variables and constants in the equation. The variables that represent length should all be in the same unit, those that represent money should all be in dollars or all in cents, and so on.
5. Make sure that you have completely answered the original problem using the appropriate units.

Percent Increase and Percent Decrease

Whenever a percent of a quantity is calculated and then added to or subtracted from that quantity, the result is a percent increase or a percent decrease. Some examples follow.

Percent Increase	Percent Decrease
Sales tax	Reduction in crimes
Salary increase	Lowered sale price
Population growth	Body weight loss

For each example of percent increase or decrease, there are three amounts:

The original amount,

The amount of increase or decrease,

The new amount.

Identifying each amount is important in solving percent increase or decrease problems. The following relationships exist among these amounts.

> The amount of increase is added to the original amount to get the new amount.
>
> The amount of decrease is subtracted from the original amount to get the new amount.
>
> The percent increase or percent decrease is calculated on the basis of the original amount.

EXAMPLE 7 *Fitness.* Through diet and exercise, Gabrielle's weight decreased from 150 lb to 145 lb. What was the percent decrease in her body weight?

SOLUTION We begin by identifying the three amounts. Then we recall that the percent of decrease is calculated on the basis of the original amount.

Original Weight	150 lb
Amount of Weight Lost	5 lb
New Weight	145 lb

Original amount − new amount
= 150 lb − 145 lb

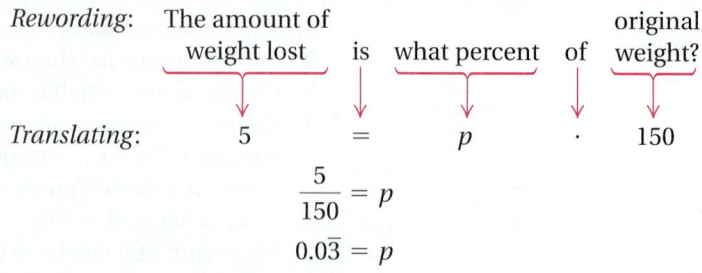

Rewording: The amount of weight lost   is   what percent   of   original weight?

Translating: 5   = p ·   150

$$\frac{5}{150} = p$$

$$0.0\overline{3} = p$$

The percent of decrease is $3.\overline{3}\%$, or $3\frac{1}{3}\%$.

↪ YOUR TURN

7. After he had recovered from a prolonged illness, Victor's weight increased from 140 lb to 150 lb. What was the percent increase in his body weight?

EXAMPLE 8 *Charitable Organizations.* The tax-exempt Hope Food Pantry received a bill of $242.65 for food storage bags. The bill incorrectly included sales tax of 5.5%. How much does the food bank actually owe?

SOLUTION

1. Familiarize. Sales tax is a percent increase and is calculated on the original price of an item. In this example, we are told the amount after the sales tax has been added, so we let $x =$ the original price of the food storage bags. We can organize the information in a table.

Original Price (in dollars)	x
Amount of Sales Tax	5.5% of x, or $0.055x$
Price Including Sales Tax	$242.65

2. Translate. We reword the problem and translate as follows.

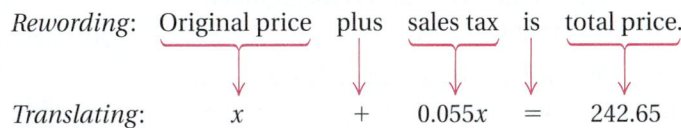

Rewording: Original price plus sales tax is total price.

Translating: x + $0.055x$ =   242.65

3. Carry out. We solve the equation:

$$x + 0.055x = 242.65$$
$$1x + 0.055x = 242.65$$
$$1.055x = 242.65 \quad \text{Combining like terms}$$
$$x = \frac{242.65}{1.055} \quad \text{Dividing both sides by 1.055}$$
$$x = 230.$$

4. Check. To check, we first find 5.5% of $230:

5.5% of $230 = 0.055(\$230) = \$12.65$.   This is the amount of the sales tax.

Next, we add the sales tax to find the total amount billed:

$$\$230 + \$12.65 = \$242.65.$$

Since this is the amount that the food bank was billed, our answer checks.

5. State. The food pantry owes $230.

8. Casey is selling his collection of Transformers at an auction. He wants to be left with $1150 after paying a seller's premium of 8% on the final bid (hammer price) for the collection. What must the hammer price be in order for him to clear $1150?

 YOUR TURN

2.5 EXERCISE SET

Vocabulary and Reading Check

1. List the steps in the five-step problem-solving approach in the correct order.

1) _____ Carry out.

2) _____ Check.

3) _____ Familiarize.

4) _____ State.

5) _____ Translate.

Each of the following was part of a step in at least one example's solution in this section. In the blank, write the name of the step during which each was done. Select from the five steps listed above.

2. _____ Solve an equation.

3. _____ Write the answer clearly.

4. _____ Make and check a guess.

5. _____ Reword the problem.

6. _____ Make a table.

7. _____ Recall a formula.

8. _____ Compare the answer with a prediction from an earlier step.

The Five Steps for Problem Solving

Solve. Even though you might find the answer quickly in another way, use the five-step process in order to build the skill of problem solving.

9. Three less than twice a number is 19. What is the number?

10. Two fewer than ten times a number is 78. What is the number?

11. Five times the sum of 3 and twice some number is 70. What is the number?

12. Twice the sum of 4 and three times some number is 34. What is the number?

13. *Kayaking.* On May 8, 2010, a team of kayakers set a record time of 5 hr 19 min 17 sec to kayak across the length of Loch Ness, Scotland, UK, a distance of 20.5 mi. If the kayakers were four times as far from the starting point as from the finish, how many miles had they traveled?

Source: Guinness World Records 2011

14. *Sled-Dog Racing.* The Iditarod sled-dog race extends for 1049 mi from Anchorage to Nome. If a musher is twice as far from Anchorage as from Nome, how many miles has the musher traveled?

The 1049-mile Iditarod race route

Nome

ALASKA

Anchorage

0   200
miles

15. *NASCAR Racing.* In February 2011, Trevor Bayne became the youngest winner of the Daytona 500, with a time of 03:59:24 for the 500-mi race. At one point, Bayne was 80 mi closer to the finish than to the start. How far had Bayne traveled at that point?

16. *Indy Car Racing.* In October 2011, Ed Carpenter won the Kentucky Indy 300 with a time of 01:42:02 for the 300-mi race. At one point, Carpenter was 20 mi closer to the finish than to the start. How far had Carpenter traveled at that point?

17. *Apartment Numbers.* The apartments in Erica's apartment house are consecutively numbered on each floor. The sum of her number and her

next-door neighbor's number is 2409. What are the two numbers?

18. *Apartment Numbers.* The apartments in Brian's apartment house are numbered consecutively on each floor. The sum of his number and his next-door neighbor's number is 1419. What are the two numbers?

19. *Street Addresses.* The houses on the west side of Lincoln Avenue are consecutive odd numbers. Sam and Colleen are next-door neighbors and the sum of their house numbers is 572. Find their house numbers.

20. *Street Addresses.* The houses on the south side of Elm Street are consecutive even numbers. Wanda and Larry are next-door neighbors and the sum of their house numbers is 794. Find their house numbers.

21. The sum of three consecutive page numbers is 99. Find the numbers.

22. The sum of three consecutive page numbers is 60. Find the numbers.

23. *Longest Marriage.* As half of the world's longest-married couple, the woman was 2 years younger than her husband. Together, their ages totaled 206 years. How old were the man and the woman?

Source: Guinness World Records 2011

24. *Oldest Bride.* The world's oldest bride was 19 years older than her groom. Together, their ages totaled 185 years. How old were the bride and the groom?

Source: Guinness World Records 2010

25. *e-Mail.* In 2010, approximately 294 billion e-mail messages were sent each day. The number of spam messages was about nine times the number of nonspam messages. How many of each type of message were sent each day in 2010?

Source: Ferris Research

26. *Home Remodeling.* In 2009, Americans spent a total of $286 billion to remodel homes. They spent $178 billion more on owner-occupied homes than on rental units. How much was spent on each?

Source: Joint Center for Housing Studies, Harvard University

27. *Page Numbers.* The sum of the page numbers on the facing pages of a book is 281. What are the page numbers?

28. *Perimeter of a Triangle.* The perimeter of a triangle is 195 mm. If the lengths of the sides are consecutive odd integers, find the length of each side.

29. A rectangular community garden is to be enclosed with 92 m of fencing. In order to allow for compost storage, the garden must be 4 m longer than it is wide. Determine the dimensions of the garden.

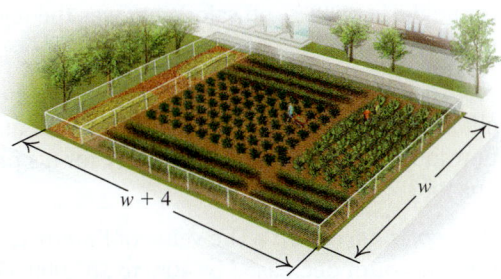

$w + 4$ w

30. *Hancock Building Dimensions.* The top of the John Hancock Building in Chicago is a rectangle whose length is 60 ft more than the width. The perimeter is 520 ft. Find the width and the length of the rectangle. Find the area of the rectangle.

31. *Two-by-four.* The perimeter of a cross section of a "two-by-four" piece of lumber is 10 in. The length is 2 in. longer than the width. Find the actual dimensions of the cross section of a two-by-four.

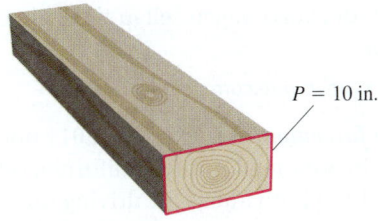

$P = 10$ in.

32. *Standard Billboard Sign.* A standard rectangular highway billboard sign has a perimeter of 124 ft. The length is 6 ft more than three times the width. Find the dimensions of the sign.

33. *Angles of a Triangle.* The second angle of an architect's triangle is three times as large as the first. The third angle is 30° more than the first. Find the measure of each angle.

34. *Angles of a Triangle.* The second angle of a triangular garden is four times as large as the first. The third angle is 45° less than the sum of the other two angles. Find the measure of each angle.

35. *Angles of a Triangle.* The second angle of a triangular kite is four times as large as the first. The third angle is 5° more than the sum of the other two angles. Find the measure of the second angle.

36. *Angles of a Triangle.* The second angle of a triangular building lot is three times as large as the first. The third angle is 10° more than the sum of the other two angles. Find the measure of the third angle.

37. *Rocket Sections.* A rocket is divided into three sections: the payload and navigation section in the top, the fuel section in the middle, and the rocket engine section in the bottom. The top section is one-sixth the length of the bottom section. The middle section is one-half the length of the bottom section. The total length is 240 ft. Find the length of each section.

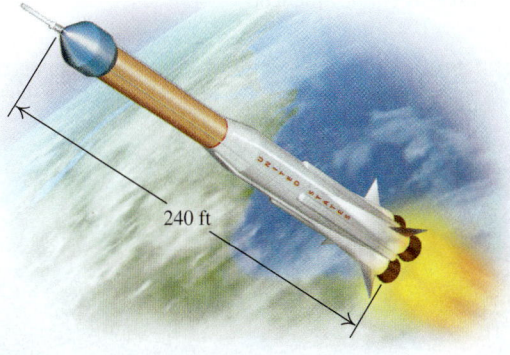

240 ft

38. *Gourmet Sandwiches.* Jenny, Demi, and Joel buy an 18-in. long gourmet sandwich and take it back to their apartment. Since they have different appetites, Jenny cuts the sandwich so that Demi gets half of what Jenny gets and Joel gets three-fourths of what Jenny gets. Find the length of each person's sandwich.

39. *Boating.* Gaston paddled for 3 hr upstream and then for 2 hr downstream. His speed upstream was 10 mph slower than his speed downstream. If he traveled a total of 30 mi, how fast did he travel downstream?

40. *Commuting.* Bjorn rode for $\frac{1}{2}$ hr on the train and then for $\frac{1}{3}$ hr on the bus. The speed of the train was 50 km/h faster than the speed of the bus. If he traveled a total of 37.5 km, what was the speed of the bus?

Aha! **41.** *Long-Distance Running.* Phoebe runs at 12 km/h and walks at 5 km/h. One afternoon, she ran and walked a total of 17 km. If she ran for the same length of time as she walked, for how long did she run?

42. *Driving.* Theodora drove on the Blue Ridge Parkway at 40 mph and then on an interstate highway at 70 mph. She drove three times as long on the Blue Ridge Parkway as she did on the interstate. If she drove a total of 285 mi, for how long did she drive on the interstate?

Percent Increase and Percent Decrease

43. *Town Budget.* The town of Maxwell increased its annual budget from $1,600,000 to $1,800,000. Find the percent increase in the budget.

44. *Employment.* The number of jobs in a state increased from 1,800,000 to 1,816,200. Find the percent increase in the number of jobs.

45. *Endangered Species.* The number of Yangtze finless porpoises dropped from 2000 in the year 2000 to 1000 in the year 2011. Find the percent decrease in the number of porpoises.

Source: Scientific American

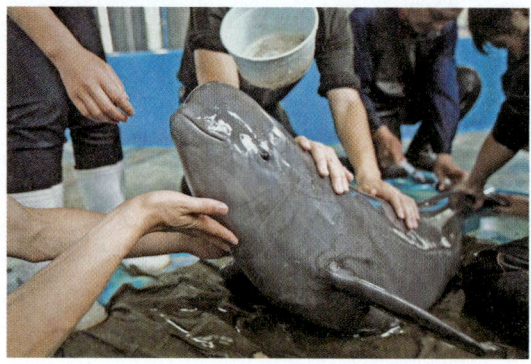

46. *Insurance Rates.* In 2011, the average annual premium for GEICO insurance policy owners in California dropped from $1187 to $1060. Find the percent decrease in the annual premium.

Source: California Department of Insurance

47. *Tax-Exempt Organizations.* A tax-exempt hospital received a bill of $1310.75 for linens. The bill incorrectly included sales tax of 7%. How much does the hospital owe?

48. *Automobile Tax.* Miles paid $5824, including 4% sales tax, for an out-of-state purchase of a car. In order to calculate the amount of sales tax he owes in his state, he must first determine the price of the car without sales tax. How much did the car cost before sales tax?

49. *Income Tax Deductions.* Jinney wants to deduct the sales tax that she paid this year on her state income tax return. She spent $4960.80 during the year, including 6% sales tax. How much sales tax did she pay?

50. *State Sales Tax.* The Tea Chest needs to send the sales tax that it has collected to the state. The total sales, including 4% sales tax, are $7115.68. How much in sales tax does the store need to send to the state?

51. *Discount.* Raena paid $224 for a camera during a 30%-off sale. What was the regular price?

52. *Discount.* Raquel paid $68 for a digital picture frame during a 20%-off sale. What was the regular price?

53. *Job Change.* Bradley took a 15% pay cut when he changed jobs. If he makes $30,600 per year at his new job, what was the annual salary for his previous job?

54. *Retirement Account.* The value of Karen's retirement account decreased by 40% to $87,000. How much was her retirement account worth before the decrease?

Aha! **55.** *Couponing.* Through "extreme couponing," Marie saved 85% of her grocery bill. If she paid $15, what was the original amount of the bill?

56. *Couponing.* Elliot had a coupon for 12% off his first meal at a new restaurant. If he was charged $11 for the meal, what was the original price of the meal?

57. *Automobile Sales.* Volkswagen sold 324,402 vehicles in the United States in 2011. This was a 26.3% increase over sales in 2010. How many vehicles did Volkswagen sell in the United States in 2010?

Source: prnewswire.com

58. *Law Enforcement.* During the 2011 holiday season in Sonoma County, California, officials arrested 163 individuals for driving under the influence of alcohol. This was a 29% increase from the number arrested during the same time in 2010. How many were arrested during the holidays in 2010?

Source: Bay City News Service

59. *Selling a Home.* The Brannons are planning to sell their home. If they want to be left with $117,500 after paying 6% of the selling price to a realtor as a commission, for how much must they sell their home?

60. *Law Enforcement.* In Charlotte, North Carolina, the number of annual crashes at targeted intersections fell 43.6% to 2591 crashes after red-light cameras were installed. How many crashes occurred annually before the cameras were installed?

Source: The National Campaign to Stop Red Light Running

Problem Solving

61. *Taxi Fares.* In New York City, taxis charge $3.00 plus $0.40 per one-fifth mile for peak fares. How far can Ralph travel for $17.50 (assuming a peak fare)?

Source: New York City Taxi and Limousine Commission

62. *Taxi Rates.* In Chicago, a taxi ride costs $2.25 plus $1.80 for each mile traveled. Debbie has budgeted $18 for a taxi ride (excluding tip). How far can she travel on her $18 budget?

Source: City of Chicago

63. *Truck Rentals.* Truck-Rite Rentals rents trucks at a daily rate of $49.95 plus 39¢ per mile. Concert Productions has budgeted $100 for renting a truck to haul equipment to an upcoming concert. How far can they travel in one day and stay within their budget?

64. *Truck Rentals.* Fine Line Trucks rents an 18-ft truck for $42 plus 35¢ per mile. Judy needs a truck for one day to deliver a shipment of plants. How far can she drive and stay within a budget of $70?

65. *Complementary Angles.* The sum of the measures of two *complementary* angles is 90°. If one angle measures 15° more than twice the measure of its complement, find the measure of each angle.

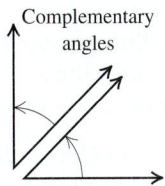

Complementary angles

66. *Complementary Angles.* Two angles are complementary. (See Exercise 65.) The measure of one angle is $1\frac{1}{2}$ times the measure of the other. Find the measure of each angle.

67. *Supplementary Angles.* The sum of the measures of two *supplementary* angles is 180°. If the measure of one angle is $3\frac{1}{2}$ times the measure of the other, find the measure of each angle.

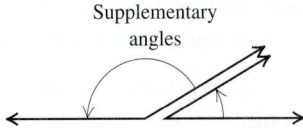

Supplementary angles

68. *Supplementary Angles.* Two angles are supplementary. (See Exercise 67.) If one angle measures 45° less than twice the measure of its supplement, find the measure of each angle.

69. *Copier Paper.* The perimeter of standard-size copier paper is 99 cm. The width is 6.3 cm less than the length. Find the length and the width.

70. *Stock Prices.* Sarah's investment in Jet Blue stock grew 28% to $448. How much did she originally invest?

71. *Savings Interest.* Janeka invested money in a savings account at a rate of 6% simple interest. After 1 year, she has $6996 in the account. How much did Janeka originally invest?

72. *Credit Cards.* The balance in Will's Mastercard® account grew 2%, to $870, in one month. What was his balance at the beginning of the month?

73. *Scrabble®.* In a single game on October 12, 2006, Michael Cresta and Wayne Yorra set three North American Scrabble records: the most points in one game by one player, the most total points in the game, and the most points on a single turn. Cresta scored 340 points more than Yorra, and together they scored 1320 points. What was the winning score?

Source: www.slate.com

74. *Bridge Construction.* The San Francisco–Oakland Bay Bridge consists of two spans connected by a tunnel. The East span is 556 ft longer than the West span, and the total length of the two spans is 19,796 ft. How long is the East span?

Source: historicbridges.org

75. *Cost of Food.* The equation $c = 1.2x + 32.94$ can be used to estimate the cost c of a Thanksgiving dinner for 10 people x years after 2000. Determine in what year the cost of a dinner for 10 people will be $50.94.

Source: Based on information from the American Farm Bureau Federation

76. *Teacher Salaries.* The equation $s = 1352x + 44{,}609$ can be used to estimate the average salary s of teachers in public schools x years after 2000. Determine in what year the average salary will be $63,537.

Source: Based on data from the National Center for Education Statistics

77. *Cricket Chirps and Temperature.* The equation $T = \frac{1}{4}N + 40$ can be used to determine the temperature T, in degrees Fahrenheit, given the number of times N a cricket chirps per minute. Determine the number of chirps per minute for a temperature of 80°F.

78. *Race Time.* The equation $R = -0.028t + 20.8$ can be used to predict the world record in the 200-m dash, where R is the record in seconds and t is the number of years since 1920. In what year will the record be 18.0 sec?

79. Marcus claims that he can solve most of the problems in this section by guessing. Is there anything wrong with this approach? Why or why not?

80. When solving Exercise 24, Jamie used a to represent the bride's age and Ben used a to represent the groom's age. Is one of these approaches preferable to the other? Why or why not?

Skill Review

81. Multiply: $4(2n + 8t + 1)$. [1.8]

82. Factor: $12 + 18x + 21y$. [1.2]

83. Simplify: $x - 3[2x - 4(x - 1) + 2]$. [1.8]

84. Find the absolute value: $|0|$. [1.4]

Synthesis

85. Write a problem for a classmate to solve. Devise it so that the problem can be translated to the equation $x + (x + 2) + (x + 4) = 375$.

86. Write a problem for a classmate to solve. Devise it so that the solution is "Audrey can drive the rental truck for 50 mi without exceeding her budget."

87. *Discounted Dinners.* Kate's "Dining Card" entitles her to $10 off the price of a meal after a 15% tip has been added to the cost of the meal. If, after the discount, the bill is $32.55, how much did the meal originally cost?

88. *Test Scores.* Pam scored 78 on a test that had 4 fill-in questions worth 7 points each and 24 multiple-choice questions worth 3 points each. She had one fill-in question wrong. How many multiple-choice questions did Pam get right?

89. *Gettysburg Address.* Abraham Lincoln's 1863 Gettysburg Address refers to the year 1776 as "four *score* and seven years ago." Determine what a score is.

90. One number is 25% of another. The larger number is 12 more than the smaller. What are the numbers?

91. A storekeeper goes to the bank to get $10 worth of change. She requests twice as many quarters as half dollars, twice as many dimes as quarters, three times as many nickels as dimes, and no pennies or dollars. How many of each coin did the storekeeper get?

92. *Perimeter of a Rectangle.* The width of a rectangle is three-fourths of the length. The perimeter of the rectangle becomes 50 cm when the length and the width are each increased by 2 cm. Find the length and the width.

93. *Discounts.* In exchange for opening a new credit account, Macy's Department Stores® subtracts 10% from all purchases made the day the account is established. Julio is opening an account and has a coupon for which he receives 10% off the first day's reduced price of a camera. If Julio's final price is $77.75, what was the price of the camera before the two discounts?

94. *Sharing Fruit.* Apples are collected in a basket for six people. One-third, one-fourth, one-eighth, and one-fifth of the apples are given to four people, respectively. The fifth person gets ten apples, and one apple remains for the sixth person. Find the original number of apples in the basket.

95. *eBay Purchases.* An eBay seller charges $9.99 for the first DVD purchased and $6.99 for all others. For shipping and handling, he charges the full shipping fee of $3 for the first DVD, one-half of the shipping charge for the second item, and one-third of the shipping charge per item for all remaining items. The total cost of a shipment (excluding tax) was $45.45. How many DVDs were in the shipment?

96. *Winning Percentage.* In a basketball league, the Falcons won 15 of their first 20 games. In order to win 60% of the total number of games, how many more games will they have to play, assuming they win only half of the remaining games?

97. *Taxi Fares.* In New York City, an off-peak taxi ride costs $2.50 plus 40¢ per $\frac{1}{5}$ mile and 40¢ per minute stopped in traffic. Due to traffic, Glenda's taxi took 20 min to complete what is usually a 10-min drive. If she is charged $18.50 for the ride, how far did Glenda travel?

Source: New York City Taxi and Limousine Commission

98. Test Scores. Ella has an average score of 82 on three tests. Her average score on the first two tests is 85. What was the score on the third test?

99. A school purchases a piano and must choose between paying $2000 at the time of purchase or $2150 at the end of one year. Which option should the school select and why?

100. Annette claims the following problem has no solution: "The sum of the page numbers on facing pages is 191. Find the page numbers." Is she correct? Why or why not?

Aha!

101. The perimeter of a rectangle is 101.74 cm. If the length is 4.25 cm longer than the width, find the dimensions of the rectangle.

102. The second side of a triangle is 3.25 cm longer than the first side. The third side is 4.35 cm longer than the second side. If the perimeter of the triangle is 26.87 cm, find the length of each side.

 YOUR TURN ANSWERS: SECTION 2.5

1. About 78.4 mi   **2.** 168 and 169   **3.** 26 images
4. Width: 50 ft; length: 84 ft   **5.** $60°, 90°, 30°$ **6.** $\frac{1}{2}$ hr
7. Approximately 7.1%, or $7\frac{1}{7}\%$ **8.** $1250

QUICK QUIZ: SECTIONS 2.1–2.5

Solve and check. Label any contradictions or identities.

1. $-\frac{2}{5}x = -\frac{5}{3}$ [2.1]

2. $3t - 8 - (9 - t) = 2(5t + 4) - 7$ [2.2]

3. Solve for v: $B = \dfrac{3}{v}$.   [2.3]

4. Caitlan has $500 in a savings account that is paying 2.1% interest per year. How much will she earn in a year?   [2.4]

5. One day Jordan spent a total of 4 hr studying American Literature, College Algebra, and Non-Verbal Communication. He spent twice as much time studying algebra as he did studying literature, and he spent 30 min longer studying communication than he did studying literature. For how long did he study literature?   [2.5]

PREPARE TO MOVE ON

Write a true sentence using either $<$ or $>$.   [1.4]

1. $-8 \,\square\, 1$ **2.** $-2 \,\square\, -5$

Write a second inequality with the same meaning.   [1.4]

3. $x \geq -4$ **4.** $5 > y$

2.6   Solving Inequalities

Solutions of Inequalities ▪ Graphs of Inequalities ▪ Set-Builder Notation and Interval Notation ▪ Solving Inequalities Using the Addition Principle ▪ Solving Inequalities Using the Multiplication Principle ▪ Using the Principles Together

Many real-world situations translate to *inequalities*. For example, a student might need to register for *at least* 12 credits; an elevator might be designed to hold *at most* 2000 pounds; a tax credit might be allowable for families with incomes of *less than* $25,000; and so on. Before solving applications, we must adapt our equation-solving principles to the solving of inequalities.

Solutions of Inequalities

An inequality is a number sentence containing $>$ (is greater than), $<$ (is less than), $\geq$ (is greater than or equal to), $\leq$ (is less than or equal to), or $\neq$ (is not equal to). Inequalities may be true or false. For example,

$-3 \leq 5$ is *true* because $-3 < 5$ is true,

$-3 \leq -3$ is *true* because $-3 = -3$ is true, and

$-5 \geq 4$ is *false* because neither $-5 > 4$ nor $-5 = 4$ is true.

Inequalities like

$$-7 > x, \quad t < 5, \quad 5x - 2 \geq 9, \quad \text{and} \quad -3y + 8 \leq -7$$

are true for some replacements of the variable and false for others.

Any value for the variable that makes an inequality true is called a **solution**. The set of all solutions is called the **solution set**. When all solutions of an inequality have been found, we say that we have **solved** the inequality.

EXAMPLE 1 Determine whether the given number is a solution of $y \geq 6$: **(a)** 6; **(b)** −4.

SOLUTION

a) Since $6 \geq 6$ is true, 6 is a solution.

b) Since $-4 \geq 6$ is false, −4 is not a solution.

YOUR TURN

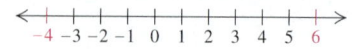

1. Determine whether −1 is a solution of $x > -3$.

Graphs of Inequalities

Because the solutions of inequalities like $x < 2$ are too numerous to list, it is helpful to make a drawing, or **graph**, that represents all the solutions. Graphs of inequalities in one variable can be drawn on the number line by shading all points that are solutions. Parentheses are used to indicate endpoints that *are not* solutions and brackets to indicate endpoints that *are* solutions.*

EXAMPLE 2 Graph each inequality: **(a)** $x < 2$; **(b)** $y \geq -3$; **(c)** $-2 < x \leq 3$.

SOLUTION

a) The solutions of $x < 2$ are those numbers less than 2. They are shown on the graph by shading all points to the left of 2. The parenthesis at 2 and the shading to its left indicate that 2 *is not* part of the graph, but numbers like 1.2 and 1.99 are.

b) The solutions of $y \geq -3$ are shown on the number line by shading the point for −3 and all points to the right of −3. The bracket at −3 indicates that −3 *is* part of the graph.

Student Notes

Note that $-2 < x < 3$ means $-2 < x$ *and* $x < 3$. Because of this, statements like $2 < x < 1$ make no sense—no number is both greater than 2 and less than 1.

c) The inequality $-2 < x \leq 3$ is read "−2 is less than x *and* x is less than or equal to 3," or "x is greater than −2 *and* less than or equal to 3." To be a solution of $-2 < x \leq 3$, a number must be a solution of both $-2 < x$ *and* $x \leq 3$. The number 1 is a solution, as are −0.5, 1.9, and 3. The parenthesis indicates that −2 *is not* a solution, whereas the bracket indicates that 3 *is* a solution. The other solutions are shaded.

2. Graph the inequality:
$0 \leq t < 5$.

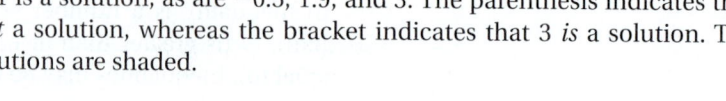

YOUR TURN

---

*An alternative notation uses open dots to indicate endpoints that are not solutions and closed dots to indicate endpoints that are solutions. Using this notation, the solutions of $x < 2$ are graphed as  and the solutions of $y \geq -3$ are graphed as .

Consider the inequality and the graph in Example 2(a). Match each change in the inequality described in the left-hand column with the corresponding change in the graph described in the right-hand column.

Inequality   *Graph*

$x < 2$

←————————————)———→
         0   2

1. Replace $<$ with $\leq$.
2. Replace $<$ with $>$.
3. Replace 2 with -1.

a) The) changes to a (and the shading moves to the right of the (.
b) The) changes to a].
c) The) moves to the left.

ANSWERS
1. (b)   **2.** (a)   **3.** (c)

Set-Builder Notation and Interval Notation

To write the solution set of $x < 3$, we can use **set-builder notation:**

$$\{x \mid x < 3\}.$$

This is read "The set of all x such that x is less than 3."

Another way to write solutions of an inequality in one variable is to use **interval notation.** Interval notation uses parentheses, ( ), and brackets, [ ].

If a and b are real numbers with $a < b$, we define the **open interval (a, b)** as the set of all numbers x for which $a < x < b$. This means that x can be any number between a and b, but it cannot be either a or b.

The **closed interval $[a, b]$** is defined as the set of all numbers x for which $a \leq x \leq b$. **Half-open intervals $(a, b]$** and **$[a, b)$** contain one endpoint and not the other.

We use the symbols ∞ and $-\infty$ to represent positive infinity and negative infinity, respectively. Thus the notation (a, ∞) represents the set of all real numbers greater than a, and $(-\infty, a)$ represents the set of all real numbers less than a.

Interval notation for a set of numbers corresponds to its graph.

> **CAUTION!** Do not confuse the *interval (a, b)* with the *ordered pair (a, b)*. The context in which the notation appears should make the meaning clear.

Student Notes

You may have noticed which inequality signs in set-builder notation correspond to brackets and which correspond to parentheses. The relationship could be written informally as

$$\geq \quad \leq \quad [\,]$$
$$> \quad < \quad (\,).$$

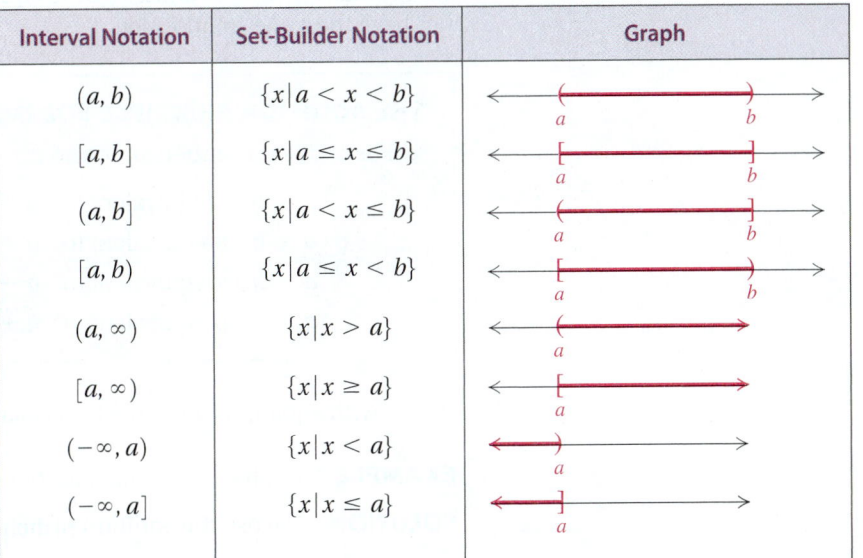

Interval Notation	Set-Builder Notation	Graph
(a, b)	$\{x \mid a < x < b\}$	
$[a, b]$	$\{x \mid a \leq x \leq b\}$	
$(a, b]$	$\{x \mid a < x \leq b\}$	
$[a, b)$	$\{x \mid a \leq x < b\}$	
(a, ∞)	$\{x \mid x > a\}$	
$[a, \infty)$	$\{x \mid x \geq a\}$	
$(-\infty, a)$	$\{x \mid x < a\}$	
$(-\infty, a]$	$\{x \mid x \leq a\}$	

EXAMPLE 3   Graph $t \geq -2$ on the number line and write the solution set using both set-builder notation and interval notation.

SOLUTION   Using set-builder notation, we write the solution set as $\{t \mid t \geq -2\}$.
Using interval notation, we write $[-2, \infty)$.

To graph the solution, we shade all numbers to the right of -2 and use a bracket to indicate that -2 is also a solution.

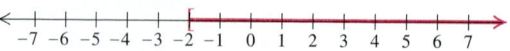

 YOUR TURN

3. Graph $y < 1$ on the number line and write the solution set using both set-builder notation and interval notation.

Solving Inequalities Using the Addition Principle

Consider a balance similar to one that appears in Section 2.1. When one side of the balance holds more weight than the other, the balance tips in that direction. If equal amounts of weight are then added to or subtracted from both sides of the balance, the balance remains tipped in the same direction.

The balance illustrates the idea that when a number, such as 2, is added to (or subtracted from) both sides of a true inequality, such as $3 < 7$, we get another true inequality:

$$3 + 2 < 7 + 2, \quad \text{or} \quad 5 < 9.$$

Similarly, if we add -4 to both sides of $x + 4 < 10$, we get an *equivalent* inequality:

$$x + 4 + (-4) < 10 + (-4), \quad \text{or} \quad x < 6.$$

We say that $x + 4 < 10$ and $x < 6$ are **equivalent**, which means that both inequalities have the same solution set.

THE ADDITION PRINCIPLE FOR INEQUALITIES

For any real numbers a, b, and c:

$a < b$ is equivalent to $a + c < b + c$;
$a \leq b$ is equivalent to $a + c \leq b + c$;
$a > b$ is equivalent to $a + c > b + c$;
$a \geq b$ is equivalent to $a + c \geq b + c$.

As with equations, our goal is to isolate the variable on one side.

EXAMPLE 4   Solve $x + 2 > 8$ and then graph the solution.

SOLUTION   We use the addition principle, subtracting 2 from both sides:

$$x + 2 - 2 > 8 - 2 \qquad \text{Subtracting 2 from, or adding } -2 \text{ to, both sides}$$

$$x > 6.$$

Any number greater than 6 makes $x > 6$ true and is a solution of both that inequality and $x + 2 > 8$. Using set-builder notation, we write the solution set as $\{x \mid x > 6\}$. Using interval notation, we write the solution set as $(6, \infty)$. The graph is as follows:

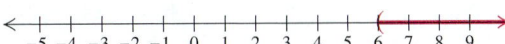

Because most inequalities have an infinite number of solutions, we cannot possibly check them all. A partial check can be made using one of the possible solutions. For this example, we can substitute any number greater than 6—say, 6.1—into the original inequality:

$$
\begin{array}{c|c}
x + 2 > 8 \\
\hline
6.1 + 2 & 8 \\
 & \overset{?}{} \\
8.1 > 8 & \text{TRUE} \qquad 8.1 > 8 \text{ is a true statement.}
\end{array}
$$

Since $8.1 > 8$ is true, 6.1 is a solution. Any number greater than 6 is a solution.

4. Solve $t - 4 < 1$ and then graph the solution.

↩ YOUR TURN

Technology Connection

As a partial check of Example 5, we can let $y_1 = 3x - 1$ and $y_2 = 2x - 5$. We set TblStart $= -5$ and ΔTbl $= 1$ in the TBLSET menu to get the following table. By scrolling up or down, you can note that for $x \le -4$, we have $y_1 \le y_2$.

X	Y₁	Y₂
−5	−16	−15
−4	−13	−13
−3	−10	−11
−2	−7	−9
−1	−4	−7
0	−1	−5
1	2	−3

X = −5

5. Solve $4 + 5n \ge 4n - 1$ and then graph the solution.

↩ YOUR TURN

EXAMPLE 5 Solve $3x - 1 \le 2x - 5$ and then graph the solution.

SOLUTION We have

$$
\begin{aligned}
3x - 1 &\le 2x - 5 \\
3x - 1 + 1 &\le 2x - 5 + 1 && \text{Adding 1 to both sides} \\
3x &\le 2x - 4 && \text{Simplifying} \\
3x - 2x &\le 2x - 4 - 2x && \text{Subtracting } 2x \text{ from both sides} \\
x &\le -4. && \text{Simplifying}
\end{aligned}
$$

The graph is as follows:

The student should check that any number less than or equal to −4 is a solution. The solution set is $\{x \mid x \le -4\}$, or $(-\infty, -4]$.

Solving Inequalities Using the Multiplication Principle

There is a multiplication principle for inequalities similar to that for equations, but it must be modified when multiplying both sides by a negative number. Consider the true inequality

$$3 < 7.$$

If we multiply both sides by a *positive* number—say, 2—we get another true inequality:

$$3 \cdot 2 < 7 \cdot 2, \quad \text{or} \quad 6 < 14. \qquad \text{TRUE}$$

If we multiply both sides by a negative number—say, −2—we get a *false* inequality:

$$3 \cdot (-2) < 7 \cdot (-2), \quad \text{or} \quad -6 < -14. \qquad \text{FALSE}$$

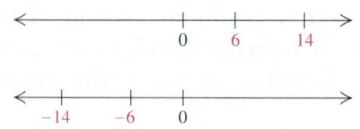

The fact that $6 < 14$ is true, but $-6 < -14$ is false, stems from the fact that the negative numbers, in a sense, *mirror* the positive numbers. Whereas 14 is to the *right* of 6 on the number line, the number -14 is to the *left* of -6. Thus if we reverse the inequality symbol in $-6 < -14$, we get a true inequality:

$$-6 > -14. \quad \text{TRUE}$$

THE MULTIPLICATION PRINCIPLE FOR INEQUALITIES

For any real numbers a and b,
when c is a *positive* number,

$$a < b \quad \text{is equivalent to} \quad ac < bc, \quad \text{and}$$
$$a > b \quad \text{is equivalent to} \quad ac > bc;$$

when c is a *negative* number,

$$a < b \quad \text{is equivalent to} \quad ac > bc, \quad \text{and}$$
$$a > b \quad \text{is equivalent to} \quad ac < bc.$$

Similar statements hold for $\leq$ and $\geq$.

CAUTION! When multiplying or dividing both sides of an inequality by a negative number, don't forget to reverse the inequality symbol!

EXAMPLE 6   Solve and graph each inequality: **(a)** $\frac{1}{4}x < 7$; **(b)** $-2y \leq 18$.

SOLUTION

a) $\frac{1}{4}x < 7$

$4 \cdot \frac{1}{4}x < 4 \cdot 7$ Multiplying both sides by 4, the reciprocal of $\frac{1}{4}$

    The symbol stays the same, since 4 is positive.

$x < 28$ Simplifying

The solution set is $\{x \mid x < 28\}$, or $(-\infty, 28)$. The graph is shown at left.

b) $-2y \leq 18$

$\dfrac{-2y}{-2} \geq \dfrac{18}{-2}$ Multiplying both sides by $-\frac{1}{2}$, or dividing both sides by -2

    At this step, we reverse the inequality, because $-\frac{1}{2}$ is negative.

$y \geq -9$ Simplifying

As a partial check, we substitute a number greater than -9, say -8, into the original inequality:

$$\begin{array}{c} -2y \leq 18 \\ \hline -2(-8) \; \vert \; 18 \\ \overset{?}{} \\ 16 \overset{?}{\leq} 18 \quad \text{TRUE} \qquad 16 \leq 18 \text{ is a true statement.} \end{array}$$

The solution set is $\{y \mid y \geq -9\}$, or $[-9, \infty)$. The graph is shown at left.

6. Solve and graph: $10 > -5x$.

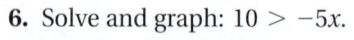

YOUR TURN

Using the Principles Together

We use the addition and multiplication principles together to solve inequalities much as we did when solving equations.

EXAMPLE 7 Solve: $6 - 5y > 7$.

SOLUTION We have

$$6 - 5y > 7$$

$$-6 + 6 - 5y > -6 + 7 \qquad \text{Adding } -6 \text{ to both sides}$$

$$-5y > 1 \qquad \text{Simplifying}$$

$$-\tfrac{1}{5} \cdot (-5y) < -\tfrac{1}{5} \cdot 1 \qquad \text{Multiplying both sides by } -\tfrac{1}{5}, \text{ or dividing} \\ \text{both sides by } -5$$

Remember to reverse the inequality symbol!

$$y < -\tfrac{1}{5}. \qquad \text{Simplifying}$$

As a partial check, we substitute a number smaller than $-\tfrac{1}{5}$, say -1, into the original inequality:

$$\begin{array}{c|c} 6 - 5y > 7 \\ \hline 6 - 5(-1) & 7 \\ 6 - (-5) & \\ 11 \overset{?}{>} 7 & \text{TRUE} \qquad 11 > 7 \text{ is a true statement.} \end{array}$$

The solution set is $\left\{y \,|\, y < -\tfrac{1}{5}\right\}$, or $\left(-\infty, -\tfrac{1}{5}\right)$. The graph is shown at left.

7. Solve: $12 - y \le 3$.

YOUR TURN

EXAMPLE 8 Solve: **(a)** $16.3 - 7.2p \le -8.18$; **(b)** $1 - 3(x - 9) < 5(x + 6) - 2$.

SOLUTION

a) The greatest number of decimal places in any one number is *two*. Multiplying both sides by 100 will clear decimals. Then we proceed as before.

$$16.3 - 7.2p \le -8.18$$

$$100(16.3 - 7.2p) \le 100(-8.18) \qquad \text{Multiplying both sides by 100}$$

$$100(16.3) - 100(7.2p) \le 100(-8.18) \qquad \text{Using the distributive law}$$

$$1630 - 720p \le -818 \qquad \text{Simplifying}$$

$$1630 - 720p - 1630 \le -818 - 1630 \qquad \text{Subtracting 1630 from both sides}$$

$$-720p \le -2448 \qquad \text{Simplifying;} \\ -818 - 1630 = -2448$$

$$\frac{-720p}{-720} \ge \frac{-2448}{-720} \qquad \text{Dividing both sides by } -720$$

Remember to reverse the symbol!

$$p \ge 3.4$$

The solution set is $\{p \,|\, p \ge 3.4\}$, or $[3.4, \infty)$.

b)

$$1 - 3(x - 9) < 5(x + 6) - 2$$

$$1 - 3x + 27 < 5x + 30 - 2 \qquad \text{Using the distributive law to remove} \\ \text{parentheses}$$

$$-3x + 28 < 5x + 28 \qquad \text{Simplifying}$$

$$-3x + 28 - 28 < 5x + 28 - 28 \qquad \text{Subtracting 28 from both sides}$$

$$-3x < 5x$$

$$-3x + 3x < 5x + 3x \qquad \text{Adding } 3x \text{ to both sides}$$

$$0 < 8x$$

$$0 < x \qquad \text{Dividing both sides by 8}$$

The solution set is $\{x \,|\, 0 < x\}$, or $\{x \,|\, x > 0\}$, or $(0, \infty)$.

8. Solve: $\tfrac{1}{4} - \tfrac{2}{3}n \ge -\tfrac{1}{2}$.

YOUR TURN

CONNECTING 🔗 THE CONCEPTS

The procedure for solving inequalities is very similar to that used to solve equations. There are, however, two important differences.

- The multiplication principle for inequalities differs from the multiplication principle for equations: When we multiply or divide on both sides of an inequality by a *negative* number, we must *reverse* the direction of the inequality.

- The solution set of an equation like those solved in this chapter typically consists of one number. The solution set of an inequality typically consists of a set of numbers and is written using either set-builder notation or interval notation.

Compare the following solutions.

Solve: $2 - 3x = x + 10$.

SOLUTION

$$2 - 3x = x + 10$$
$$-3x = x + 8 \quad \text{Subtracting 2 from both sides}$$
$$-4x = 8 \quad \text{Subtracting } x \text{ from both sides}$$
$$x = -2 \quad \text{Dividing both sides by } -4$$

The solution is -2.

Solve: $2 - 3x > x + 10$.

SOLUTION

$$2 - 3x > x + 10$$
$$-3x > x + 8 \quad \text{Subtracting 2 from both sides}$$
$$-4x > 8 \quad \text{Subtracting } x \text{ from both sides}$$
$$x < -2 \quad \text{Dividing both sides by } -4 \text{ and reversing the direction of the inequality symbol}$$

The solution is $\{x | x < -2\}$, or $(-\infty, -2)$.

EXERCISES

Solve.

1. $x - 6 = 15$

2. $x - 6 \leq 15$

3. $3x = -18$

4. $3x > -18$

5. $7 - 3x \geq 8$

6. $7 - 3x = 8$

7. $\dfrac{n}{6} - 6 = 5$

8. $\dfrac{n}{6} - 6 < 5$

9. $10 \geq -2(a - 5)$

10. $10 = -2(a - 5)$

2.6 EXERCISE SET

FOR EXTRA HELP

MyMathLab® Math XL

PRACTICE   WATCH   READ   REVIEW

🐦 Vocabulary and Reading Check

Choose the word from the following list that best completes each sentence. Not every word will be used.

bracket   infinity   parenthesis
closed   interval   set-builder
half-open   open   solution

1. The number -2 is one _____ of the inequality $x < 0$.

2. The solution set $\{x | x \geq 10\}$ is an example of _____ notation.

3. The interval $[6, 10]$ is an example of a(n) _____ interval.

4. When graphing the solution of the inequality $-3 \leq x \leq 2$, place a(n) _____ at both ends of the interval.

🐦 Concept Reinforcement

Insert the symbol $<$, $>$, $\leq$, or $\geq$ to make each pair of inequalities equivalent.

5. $-5x \leq 30$; $x \; \blacksquare \; -6$

6. $-7t \geq 56$; $t \; \blacksquare \; -8$

7. $-2t > -14$; $t \; \blacksquare \; 7$

8. $-3x < -15$; $x \; \blacksquare \; 5$

Classify each pair of inequalities as "equivalent" or "not equivalent."

9. $x < -2$; $-2 > x$

10. $t > -1$; $-1 < t$

11. $-4x - 1 \leq 15$;
    $-4x \leq 16$

12. $-2t + 3 \geq 11$;
    $-2t \geq 14$

Solutions of Inequalities

Determine whether each number is a solution of the given inequality.

13. $x > -4$

    a) 4   b) -6 c) -4

14. $t < 3$

    a) -3 b) 3   c) $2\frac{19}{20}$

15. $y \leq 19$

    a) 18.99   b) 19.01   c) 19

16. $n \geq -4$

    a) 0   b) -4.1 c) -3.9

17. $c \geq -7$

    a) 0   b) $-5\frac{4}{5}$ c) $1\frac{1}{3}$

18. $m \leq -2$

    a) $-1\frac{9}{10}$ b) 0   c) $-2\frac{1}{3}$

Graphs of Inequalities

Graph on the number line.

19. $y < 2$

20. $x \leq 7$

21. $x \geq -1$

22. $t > -2$

23. $0 \leq t$

24. $1 \leq m$

25. $-5 \leq x < 2$

26. $-3 < x \leq 5$

27. $-4 < x < 0$

28. $0 \leq x \leq 5$

Set-Builder Notation and Interval Notation

Graph each inequality, and write the solution set using both set-builder notation and interval notation.

29. $y < 6$

30. $x > 4$

31. $x \geq -4$

32. $t \leq 6$

33. $t > -3$

34. $y < -3$

35. $x \leq -7$

36. $x \geq -6$

Describe each graph using both set-builder notation and interval notation.

37.

38.

39.

40.

41.

42.

43.

44.

Solving Inequalities Using the Addition Principle

Solve. Graph and write both set-builder notation and interval notation for each answer.

45. $y + 6 > 9$

46. $x + 8 \leq -10$

47. $n - 6 < 11$

48. $n - 4 > -3$

49. $2x \leq x - 9$

50. $3x \leq 2x + 7$

51. $5 \geq t + 8$

52. $4 < t + 9$

53. $t - \frac{1}{8} > \frac{1}{2}$

54. $y - \frac{1}{3} > \frac{1}{4}$

55. $-9x + 17 > 17 - 8x$

56. $-8n + 12 > 12 - 7n$

Aha! 57. $-23 < -t$

58. $19 < -x$

Solving Inequalities Using the Multiplication Principle

Solve. Graph and write both set-builder notation and interval notation for each answer.

59. $4x < 28$

60. $3x \geq 24$

61. $-24 > 8t$

62. $-16x < -64$

63. $1.8 \geq -1.2n$

64. $9 \leq -2.5a$

65. $-2y \leq \frac{1}{5}$

66. $-2x \geq \frac{1}{5}$

67. $-\frac{8}{5} > 2x$

68. $-\frac{5}{8} < -10y$

Using the Principles Together

Solve.

69. $2 + 3x < 20$

70. $7 + 4y < 31$

71. $4t - 5 \leq 23$

72. $15x - 7 \leq -7$

73. $39 > 3 - 9x$

74. $5 > 5 - 7y$

75. $5 - 6y > 25$

76. $8 - 2y > 9$

77. $-3 < 8x + 7 - 7x$

78. $-5 < 9x + 8 - 8x$

79. $6 - 4y > 6 - 3y$

80. $7 - 8y > 5 - 7y$

81. $2.1x + 43.2 > 1.2 - 8.4x$

82. $0.96y - 0.79 \leq 0.21y + 0.46$

83. $1.7t + 8 - 1.62t < 0.4t - 0.32 + 8$

84. $0.7n - 15 + n \geq 2n - 8 - 0.4n$

85. $\frac{x}{3} + 4 \leq 1$

86. $\frac{2}{3} - \frac{x}{5} < \frac{4}{15}$

87. $3 < 5 - \dfrac{t}{7}$

88. $2 > 9 - \dfrac{x}{5}$

89. $4(2y - 3) \le -44$

90. $3(2y - 3) > 21$

91. $8(2t + 1) > 4(7t + 7)$

92. $3(t - 2) \ge 9(t + 2)$

93. $3(r - 6) + 2 < 4(r + 2) - 21$

94. $5(t + 3) + 9 \ge 3(t - 2) - 10$

95. $\frac{4}{5}(3x + 4) \le 20$

96. $\frac{2}{3}(2x - 1) \ge 10$

97. $\frac{2}{3}\left(\frac{7}{8} - 4x\right) - \frac{5}{8} < \frac{3}{8}$

98. $\frac{3}{4}\left(3x - \frac{1}{2}\right) - \frac{2}{3} < \frac{1}{3}$

99. Are the inequalities $x > -3$ and $x \ge -2$ equivalent? Why or why not?

100. Are the inequalities $t < -7$ and $t \le -8$ equivalent? Why or why not?

Skill Review

Simplify.   [1.8]

101. $5x - 2(3 - 6x)$

102. $8m - n - 3(2m + 5n)$

103. $x - 2[4y + 3(8 - x) - 1]$

104. $9x - 2\{4 - 5[6 - 2(x + 1) - x]\}$

Synthesis

105. Explain how it is possible for the graph of an inequality to consist of just one number. (*Hint*: See Example 2c.)

106. The statements of the addition and multiplication principles begin with *conditions* described for the variables. Explain the conditions given for each principle.

Solve.

Aha! 107. $x < x + 1$

108. $6[4 - 2(6 + 3t)] > 5[3(7 - t) - 4(8 + 2t)] - 20$

109. $27 - 4[2(4x - 3) + 7] \ge 2[4 - 2(3 - x)] - 3$

Solve for x.

110. $\frac{1}{2}(2x + 2b) > \frac{1}{3}(21 + 3b)$

111. $-(x + 5) \ge 4a - 5$

112. $y < ax + b$ (Assume $a < 0$.)

113. $y < ax + b$ (Assume $a > 0$.)

114. Graph the solutions of $|x| < 3$ on the number line.

Aha! 115. Determine the solution set of $|x| > -3$.

116. Determine the solution set of $|x| < 0$.

117. In order for a meal to be labeled "lowfat," it must have fewer than 3 g of fat per serving *and* the number of calories from fat must be 30% or fewer of the food's total calories.

  a) Each Chik Patties Breaded Veggie Pattie contains 150 calories per serving, and 54 of those calories come from fat. Can these patties be labeled "lowfat"?

  b) Cabot's 50% Reduced Fat Cheddar contains, as the name implies, 50% less fat than Cabot's regular cheddar cheese, but still cannot be labeled lowfat. What can you conclude about the fat content of a serving of Cabot's regular cheddar cheese?

↪ YOUR TURN ANSWERS: SECTION 2.6

1. Yes   **2.**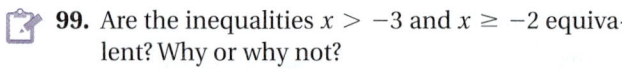

3. 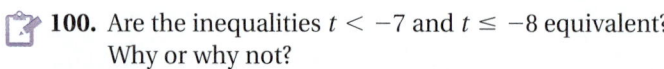 $\{y \mid y < 1\}$, or $(-\infty, 1)$

4. $\{t \mid t < 5\}$, or $(-\infty, 5)$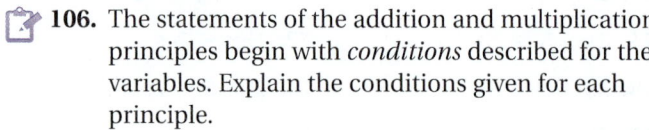

5. $\{n \mid n \ge -5\}$, or $[-5, \infty)$

6. $\{x \mid x > -2\}$, or $(-2, \infty)$

7. $\{y \mid y \ge 9\}$, or $[9, \infty)$ **8.** $\left\{n \mid n \le \frac{9}{8}\right\}$, or $\left(-\infty, \frac{9}{8}\right]$.

QUICK QUIZ: SECTIONS 2.1–2.6

Solve.

1. $1 - (d - 7) = 6d - 2(5d + 10)$ [2.2]

2. $3x - 7x > 10 - x$ [2.6]

3. Pernell left an 18% tip for his lunch service. If the tip was $2.70, what was the price of the lunch before the tip?   [2.4]

4. Melissa's dinner cost $20.89, including 8% sales tax. What was the price of the dinner before the tax was added?   [2.5]

5. Solve for c: $X = 12 - 5(d - c)$.   [2.3]

PREPARE TO MOVE ON

Translate to an equation. Do not solve.   [1.1]

1. The area of a triangle is 5 m^2. The base is 3 m long. What is the height?

2. The perimeter of a triangle is 12 ft. One side is 1 ft longer than the shortest side. The third side is $\frac{1}{2}$ ft longer than the shortest side. How long is the shortest side of the triangle?

2.7   Solving Applications with Inequalities

Translating to Inequalities   ▪   Solving Problems

The five steps for problem solving can be used for problems involving inequalities.

Translating to Inequalities

Before solving problems that involve inequalities, we list some important phrases to look for. Sample translations are listed as well.

Important Words	Sample Sentence	Definition of Variables	Translation
is at least	Kelby walks at least 1.5 mi a day.	Let k represent the length of Kelby's walk, in miles.	$k \geq 1.5$
is at most	At most 5 students dropped the course.	Let n represent the number of students who dropped the course.	$n \leq 5$
cannot exceed	The cost cannot exceed $12,000.	Let c represent the cost, in dollars.	$c \leq 12{,}000$
must exceed	The speed must exceed 40 mph.	Let s represent the speed, in miles per hour.	$s > 40$
is less than	Hamid's weight is less than 130 lb.	Let w represent Hamid's weight, in pounds.	$w < 130$
is more than	Boston is more than 200 mi away.	Let d represent the distance to Boston, in miles.	$d > 200$
is between	The film is between 90 min and 100 min long.	Let t represent the length of the film, in minutes.	$90 < t < 100$
minimum	Ned drank a minimum of 5 glasses of water a day.	Let w represent the number of glasses of water Ned drank.	$w \geq 5$
maximum	The maximum penalty is $100.	Let p represent the penalty, in dollars.	$p \leq 100$
no more than	Alan consumes no more than 1500 calories.	Let c represent the number of calories Alan consumes.	$c \leq 1500$
no less than	Patty scored no less than 80.	Let s represent Patty's score.	$s \geq 80$

The following phrases deserve special attention.

> **TRANSLATING "AT LEAST" AND "AT MOST"**
>
> The quantity x is at least some amount q: $x \geq q$.
>
> (If x is *at least* q, it cannot be less than q.)
>
> The quantity x is at most some amount q: $x \leq q$.
>
> (If x is *at most* q, it cannot be more than q.)

Solving Problems

EXAMPLE 1   *Catering Costs.*   To cater a party, Papa Roux charges a $50 setup fee plus $15 per person. The cost of Hotel Pharmacy's end-of-season softball party cannot exceed $450. How many people can attend the party?

SOLUTION

1. **Familiarize.** Suppose that 20 people were to attend the party. The cost would then be $50 + $15 · 20, or $350. This shows that more than 20 people could attend without exceeding $450. Instead of making another guess, we let $n =$ the number of people in attendance.

2. **Translate.** The cost of the party will be $50 for the setup fee plus $15 times the number of people attending. We can reword as follows:

Rewording: The setup fee plus the cost of the meals cannot exceed $450.

Translating: 50   + $15 \cdot n$ ≤   450

3. **Carry out.** We solve for n:

$$50 + 15n \leq 450$$

$$15n \leq 400 \qquad \text{Subtracting 50 from both sides}$$

$$n \leq \frac{400}{15} \qquad \text{Dividing both sides by 15}$$

$$n \leq 26\frac{2}{3}. \qquad \text{Simplifying}$$

4. **Check.** The solution set is all numbers less than or equal to $26\frac{2}{3}$. Since n represents the number of people in attendance, we round to a whole number. Since the nearest whole number, 27, is not part of the solution set, we round *down* to 26. If 26 people attend, the cost will be $50 + $15 \cdot 26$, or $440, and if 27 attend, the cost will exceed $450.

5. **State.** At most 26 people can attend the party.

1. Refer to Example 1. Suppose that Hotel Pharmacy decides to use a different caterer. This caterer charges a $100 setup fee plus $12 per person. How many people can attend the party?

 YOUR TURN

CAUTION! Solutions of problems should always be checked using the original wording of the problem. In some cases, answers might need to be whole numbers or integers or rounded off in a particular direction.

Some applications with inequalities involve *averages*, or *means*.

AVERAGE, OR MEAN

To find the **average**, or **mean**, of a set of numbers, add the numbers and then divide by the number of addends.

EXAMPLE 2 *Financial Aid.* Full-time students in a health-care education program can receive financial aid and employee benefits from Covenant Health System by working at Covenant while attending school and also agreeing to work there after graduation. Students who work an average of at least 16 hr per week receive extra pay and part-time employee benefits. For the first three weeks of September, Dina worked 20 hr, 12 hr, and 14 hr. How many hours must she work during the fourth week in order to average at least 16 hr per week for the month?

Source: Covenant Health Systems

SOLUTION

1. **Familiarize.** Suppose Dina works 10 hr during the fourth week. Her average for the month would be

$$\frac{20\,\text{hr} + 12\,\text{hr} + 14\,\text{hr} + 10\,\text{hr}}{4} = 14\,\text{hr}. \qquad \text{There are 4 addends, so we divide by 4.}$$

This shows that Dina must work more than 10 hr during the fourth week, if she is to average at least 16 hr of work per week. We let $x =$ the number of hours Dina works during the fourth week.

2. **Translate.** We reword the problem and translate as follows:

Rewording: The average number   should be
                of hours Dina worked   at least   16 hr.

Translating: $\dfrac{20 + 12 + 14 + x}{4}$ $\geq$ 16

3. **Carry out.** Because of the fraction, it is convenient to use the multiplication principle first:

$$\frac{20 + 12 + 14 + x}{4} \geq 16$$

$$4\left(\frac{20 + 12 + 14 + x}{4}\right) \geq 4 \cdot 16 \qquad \text{Multiplying both sides by 4}$$

$$20 + 12 + 14 + x \geq 64$$

$$46 + x \geq 64 \qquad \text{Simplifying}$$

$$x \geq 18. \qquad \text{Subtracting 46 from both sides}$$

Chapter Resource:
Translating for Success, p. 144

4. **Check.** As a partial check, we show that if Dina works 18 hr, she will average at least 16 hr per week:

$$\frac{20 + 12 + 14 + 18}{4} = \frac{64}{4} = 16. \qquad \text{Note that 16 is at least 16.}$$

5. **State.** Dina will average at least 16 hr of work per week for September if she works at least 18 hr during the fourth week.

2. Refer to Example 2. Suppose that for the first three weeks of October, Dina worked 12 hr, 14 hr, and 18 hr. How many hours must she work during the fourth week in order to average at least 16 hr per week for the month?

YOUR TURN

2.7   EXERCISE SET

FOR EXTRA HELP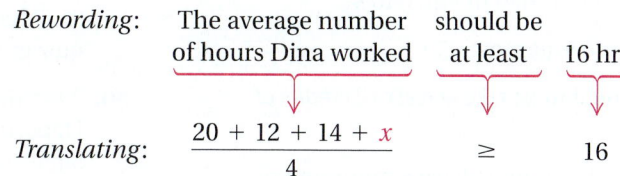

Vocabulary and Reading Check

Choose the phrase that best completes the sentence from the phrases listed below each blank.

1. If Matt's income is always $500 per week or more, then his income _____ $500.
                is at least/is at most

2. If Amy's entertainment budget for each month is $250, then her entertainment expenses _____ $250.
                   must exceed/cannot exceed

3. If Lori works out 30 min or more every day, then her exercise time is _____ 30 min.
                no less than/no more than

4. If Marco must pay $20 or more toward his credit-card loan each month, then $20 is his _____ monthly payment.
   maximum/minimum

Concept Reinforcement

In each of Exercises 5–12, match the sentence with one of the following:

$$a < b; \quad a \leq b; \quad b < a; \quad b \leq a.$$

5. a is at least b.

6. a exceeds b.

7. a is at most b.

8. a is exceeded by b.

9. b is no more than a.

10. b is no less than a.

11. b is less than a.

12. b is more than a.

Translating to Inequalities

Translate to an inequality.

13. A number is less than 10.

14. A number is greater than or equal to 4.

15. The temperature is at most $-3°C$.

16. A full-time student must take at least 12 credits of classes.

17. To rent a car, a driver must have a minimum of 5 years driving experience.

18. Normandale Community College is no more than 15 mi away.

19. The age of the Mayan altar exceeds 1200 years.

20. The maximum safe exposure limit of formaldehyde is 2 parts per million.

21. Bianca earns no less than $12 per hour.

22. The cost of production of software cannot exceed $12,500.

23. Ireland gets between 1100 and 1600 hours of sunshine per year.

24. The cost of gasoline was at most $4 per gallon.

Solving Problems

Use an inequality and the five-step process to solve each problem.

25. *Furnace Repairs.* RJ's Plumbing and Heating charges $55 plus $40 per hour for emergency service. Gary remembers being billed over $150 for an emergency call. How long was RJ's there?

26. *College Tuition.* Vanessa's financial aid stipulates that her tuition not exceed $2500. If her local community college charges a $95 registration fee plus $675 per course, what is the greatest number of courses for which Vanessa can register?

27. *Graduate School.* An unconditional acceptance into the Master of Business Administration (MBA) program at Arkansas State University will be given to students whose GMAT score plus 200 times the undergraduate grade point average is at least 950. Robbin's GMAT score was 500. What must her grade point average be in order to be unconditionally accepted into the program?

Source: graduateschool.astate.edu

28. *Car Payments.* As a rule of thumb, debt payments (other than mortgages) should be less than 8% of a consumer's monthly gross income. Oliver makes $54,000 per year and has a $100 student-loan payment every month. What size car payment can he afford?

Source: money.cnn.com

29. *Quiz Average.* Rod's quiz grades are 73, 75, 89, and 91. What scores on a fifth quiz will make his average quiz grade at least 85?

30. *Nutrition.* Following the guidelines of the U.S. Department of Agriculture, Dale tries to eat at least 5 half-cup servings of vegetables each day. For the first six days of one week, she had 4, 6, 7, 4, 6, and 4 servings. How many servings of vegetables should Dale eat on Saturday, in order to average at least 5 servings per day for the week?

31. *College Course Load.* To remain on financial aid, Millie needs to complete an average of at least 7 credits per quarter each year. In the first three quarters of 2012, Millie completed 5, 7, and 8 credits. How many credits of course work must Millie complete in the fourth quarter if she is to remain on financial aid?

32. *Music Lessons.* Band members at Colchester Middle School are expected to average at least 20 min of practice time per day. One week Monroe practiced 15 min, 28 min, 30 min, 0 min, 15 min, and 25 min. How long must he practice on the seventh day if he is to meet expectations?

33. *Baseball.* In order to qualify for a batting title, a major league baseball player must average at least 3.1 plate appearances per game. For the first nine games of the season, a player had 5, 1, 4, 2, 3, 4, 4, 3, and 2 plate appearances. How many plate appearances must the player have in the tenth game in order to average at least 3.1 per game?

Source: Major League Baseball

34. *Education.* The Mecklenberg County Public Schools stipulate that a standard school day will average at least $5\frac{1}{2}$ hr, excluding meal breaks. For the first four days of one school week, bad weather resulted in school days of 4 hr, $6\frac{1}{2}$ hr, $3\frac{1}{2}$ hr, and $6\frac{1}{2}$ hr. How long must the Friday school day be in order to average at least $5\frac{1}{2}$ hr for the week?

Source: www.meck.k12.va.us

35. *Perimeter of a Triangle.* One side of a triangle is 2 cm shorter than the base. The other side is 3 cm longer than the base. What lengths of the base will allow the perimeter to be greater than 19 cm?

36. *Perimeter of a Sign.* The perimeter of a rectangular sign is not to exceed 50 ft. The length is to be twice the width. What widths will meet these conditions?

37. *Well Drilling.* All Seasons Well Drilling offers two plans. Under the "pay-as-you-go" plan, they charge $500 plus $8 per foot for a well of any depth. Under their "guaranteed-water" plan, they charge a flat fee of $4000 for a well that is guaranteed to provide adequate water for a household. For what depths would it save a customer money to use the pay-as-you-go plan?

38. *Cost of Road Service.* Rick's Automotive charges $50 plus $15 for each quarter hour when making a road call. Twin City Repair charges $70 plus $10 for each quarter hour. Under what circumstances would it be more economical for a motorist to call Rick's?

39. *Insurance-Covered Repairs.* Most insurance companies will replace a vehicle if an estimated repair exceeds 80% of the "blue-book" value of the vehicle. Michele's insurance company paid $8500 for repairs to her Subaru after an accident. What can be concluded about the blue-book value of the car?

40. *Insurance-Covered Repairs.* Following an accident, Jeff's Ford pickup was replaced by his insurance company because the damage was so extensive. Before the damage, the blue-book value of the truck was $21,000. How much would it have cost to repair the truck? (See Exercise 39.)

41. *Sizes of Packages.* The U.S. Postal Service defines a "package" as a parcel for which the sum of the length and the girth is less than 84 in. (Length is the longest side of a package and girth is the distance around the other two sides of the package.) A box has a fixed girth of 29 in. Determine (in terms of an inequality) those lengths for which the box is considered a "package."

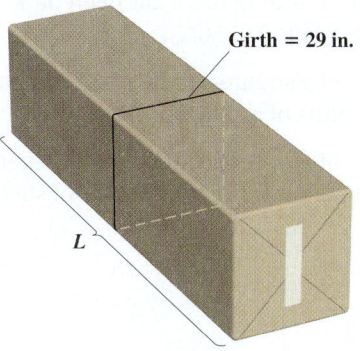

Girth = 29 in.

L

42. *Sizes of Envelopes.* Rhetoric Advertising is a direct-mail company. It determines that for a particular campaign, it can use any envelope with a fixed width of $3\frac{1}{2}$ in. and an area of at least $17\frac{1}{2}$ in^2. Determine (in terms of an inequality) those lengths that will satisfy the company constraints.

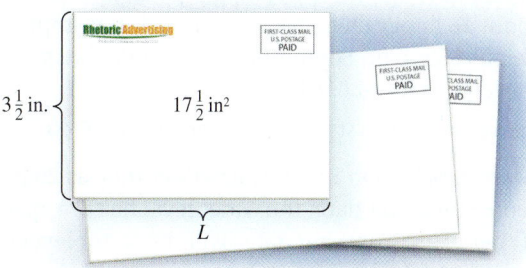

$3\frac{1}{2}$ in. $17\frac{1}{2}$ in^2

L

43. *Body Temperature.* A person is considered to be feverish when his or her temperature is higher than 98.6°F. The formula $F = \frac{9}{5}C + 32$ can be used to convert Celsius temperatures C to Fahrenheit temperatures F. For which Celsius temperatures is a person considered feverish?

44. *Gold Temperatures.* Gold stays solid at Fahrenheit temperatures below 1945.4°. Determine (in terms of an inequality) those Celsius temperatures for which gold stays solid. Use the formula given in Exercise 43.

45. *Area of a Triangular Sign.* Zoning laws in Harrington prohibit displaying signs with areas exceeding 12 ft^2. If Flo's Marina is ordering a triangular sign with an 8-ft base, how tall can the sign be?

8 ft

FLO'S MARINA

?

46. *Area of a Triangular Flag.* As part of an outdoor education course, Trisha needs to make a bright-colored triangular flag with an area of at least 3 ft^2. What lengths can the triangle be if the base is $1\frac{1}{2}$ ft?

47. *Fat Content in Foods.* Reduced Fat Sargento colby cheese contains 5 g of fat per serving. In order for a food to be labeled "reduced fat," it must have at least 25% less fat than the regular item. What can you conclude about the number of grams of fat in a serving of the regular Sargento colby cheese?

Source: Nutrition facts label

48. *Fat Content in Foods.*   Reduced Fat Wheat Thins crackers contain 3.5 g of fat per serving. What can you conclude about the number of grams of fat in a serving of regular Wheat Thins crackers? (See Exercise 47.)

Source: Nutrition facts label

49. *Weight Gain.*   In the last weeks before the yearly Topsfield Weigh In, heavyweight pumpkins gain about 26 lb per day. Charlotte's heaviest pumpkin weighs 532 lb on September 5. For what dates will its weight exceed 818 lb?

Source: Based on a story in the *Burlington Free Press*

50. *Pond Depth.*   On July 1, Garrett's Pond was 25 ft deep. Since that date, the water level has dropped $\frac{2}{3}$ ft per week. For what dates will the water level not exceed 21 ft?

51. *Cell-Phone Budget.*   Liam has budgeted $60 per month for his cell phone. For his service, he pays a monthly fee of $39.95, plus taxes of $6.65, plus 10¢ for each text message sent or received. How many text messages can he send or receive and not exceed his budget?

52. *Banquet Costs.*   The women's volleyball team can spend at most $700 for its awards banquet at a local restaurant. If the restaurant charges a $100 setup fee plus $24 per person, at most how many can attend?

53. *World Records in the Mile Run.*   The formula

$$R = -0.0065t + 4.3259$$

can be used to predict the world record, in minutes, for the 1-mi run t years after 1900. Determine (in terms of an inequality) those years for which the world record will be less than 3.6 min.

Source: Based on information from Information Please Database, Pearson Education, Inc.

54. *World Records in the Women's 1500-m Run.* The formula

$$R = -0.0026t + 4.0807$$

can be used to predict the world record, in minutes, for the 1500-m run t years after 1900. Determine (in terms of an inequality) those years for which the world record will be less than 3.7 min.

Source: Based on information from *Track and Field*

55. *Toll Charges.*   The equation

$$y = 0.078x + 3.13$$

can be used to estimate the cost y, in dollars, of driving x miles on the Pennsylvania Turnpike. For what mileages x will the cost be at most $14?

56. *Price of a Movie Ticket.*   The average price of a movie ticket can be estimated by the equation

$$P = 0.237Y - 468.87,$$

where Y is the year and P is the average price, in dollars. For what years will the average price of a movie ticket be at least $9? (Include the year in which the $9 ticket first occurs.)

Source: Based on data from National Association of Theatre Owners

57. If f represents Fran's age and t represents Todd's age, write a sentence that would translate to $t + 3 < f$.

58. Explain how the meanings of "Five more than a number" and "Five is more than a number" differ.

Skill Review

59. Use the commutative law of addition to write an expression equivalent to $xy + 7$.   [1.2]

60. Find the prime factorization of 225.   [1.3]

61. Change the sign of 18.   [1.6]

62. Find $-(-x)$ for $x = -5$.   [1.6]

Synthesis

63. Write a problem for a classmate to solve. Devise the problem so the answer is "At most 18 passengers can go on the boat." Design the problem so that at least one number in the solution must be rounded down.

64. Write a problem for a classmate to solve. Devise the problem so the answer is "The Rothmans can drive 90 mi without exceeding their truck rental budget."

65. *Ski Wax.*   Green ski wax works best between 5° and 15° Fahrenheit. Determine those Celsius temperatures for which green ski wax works best. (See Exercise 43.)

66. *Parking Fees.*   Mack's Parking Garage charges $4.00 for the first hour and $2.50 for each additional hour. For how long has a car been parked when the charge exceeds $16.50?

67. The area of a square can be no more than 64 cm². What lengths of a side will allow this?

68. The sum of two consecutive odd integers is less than 100. What is the largest pair of such integers?

69. *Frequent Buyer Bonus.*   Alice's Books allows customers to select one free book for every 10 books purchased. The price of that book cannot exceed the average cost of the 10 books. Neoma has bought 9 books whose average cost is $12 per book. How much should her tenth book cost if she wants to select a $15 book for free?

70. *Parking Fees.*   When asked how much the parking charge is for a certain car (see Exercise 66), Mack replies, "between 14 and 24 dollars." For how long has the car been parked?

71. *Grading.*   After 9 quizzes, Blythe's average is 84. Is it possible for Blythe to improve her average by two points with the next quiz? Why or why not?

72. *Discount Card.*   Barnes & Noble offers a member card for $25 per year. This card entitles a customer to a 40% discount off list price on hardcover bestsellers, a 10% discount on other eligible purchases, and $15 off the price of a Nook™ device. Describe two sets of circumstances for which an individual would save money by paying for a membership.

Source: Barnes & Noble

73. *Research.*   Find a formula for the maximum mortgage payment, sometimes called front-end ratio, based on annual income. Using this formula, determine what annual salaries qualify for a $1000 monthly mortgage payment.

QUICK QUIZ: SECTIONS 2.1–2.7
Solve.

1. $3 - x \geq 1 - 2(5 - x)$ [2.6]

2. $0.03t - 2.1 = 10$ [2.2]

3. Kenneth saved $13.75 on a pair of running shoes that originally were priced at $68.75. What percent of the original price did he save?   [2.4]

4. Two hours into a 72-km bicycle race, Jeannette realized that she had twice as far to ride as she had already ridden. At this point, how far was Jeannette from the finish line?   [2.5]

5. On her first three history papers, Nara received scores of 88, 86, and 74. What scores on the fourth paper will make her average grade at least 80?   [2.7]

PREPARE TO MOVE ON
Graph on the number line.   [1.4]

1. -12 **2.** 26

Simplify.   [1.8]

3. $\dfrac{286 - 127}{6 - 4}$ **4.** $\dfrac{1.3 - 9.8}{1 - 2}$

↪ YOUR TURN ANSWERS: SECTION 2.7
1. At most 29 people   **2.** At least 20 hr

1. Consecutive Integers. The sum of two consecutive even integers is 102. Find the integers.

2. Salary Increase. After Susanna earned a 5% raise, her new salary was $25,750. What was her former salary?

3. Dimensions of a Rectangle. The length of a rectangle is 6 in. more than the width. The perimeter of the rectangle is 102 in. Find the length and the width.

4. Population. The population of Kelling Point is decreasing at a rate of 5% per year. The current population is 25,750. What was the population the previous year?

5. Reading Assignment. Quinn has 6 days to complete a 150-page reading assignment. How many pages must he read the first day so that he has no more than 102 pages left to read on the 5 remaining days?

Translating for Success

Use after Section 2.7.

Translate each word problem to an equation or an inequality and select a correct translation from A–O.

A. $0.05(25{,}750) = x$

B. $x + 2x = 102$

C. $2x + 2(x + 6) = 102$

D. $150 - x \leq 102$

E. $x - 0.05x = 25{,}750$

F. $x + (x + 2) = 102$

G. $x + (x + 6) > 102$

H. $x + 5x = 150$

I. $x + 0.05x = 25{,}750$

J. $x + (2x + 6) = 102$

K. $x + (x + 1) = 102$

L. $102 + x > 150$

M. $0.05x = 25{,}750$

N. $102 + 5x > 150$

O. $x + (x + 6) = 102$

Answers on page A-6

An additional, animated version of this activity appears in MyMathLab. *To use MyMathLab, you need a course ID and a student access code. Contact your instructor for more information.*

6. Numerical Relationship. One number is 6 more than twice another. The sum of the numbers is 102. Find the numbers.

7. DVD Collections. Together Mindy and Ken have 102 DVDs. If Ken has 6 more than Mindy, how many does each have?

8. Sales Commissions. Kirk earns a commission of 5% on his sales. One year, he earned commissions totaling $25,750. What were his total sales for the year?

9. Fencing. Jess has 102 ft of fencing that he plans to use to enclose dog runs at two houses. The perimeter of one run is to be twice the perimeter of the other. Into what lengths should the fencing be cut?

10. Quiz Scores. Lupe has a total of 102 points on the first 6 quizzes in her sociology class. How many total points must she earn on the 5 remaining quizzes in order to have more than 150 points for the semester?

Collaborative Activity   *Sales and Discounts*

Focus: Applications and models using percent
Time: 15 minutes
Use after: Section 2.4
Group size: 3
Materials: Calculators are optional.

Often a store will reduce the price of an item by a fixed percentage. When the sale ends, the items are returned to their original prices. Suppose a department store reduces all sporting goods 20%, all clothing 25%, and all electronics 10%.

Activity

1. Each group member should select one of the following items: a $50 basketball, an $80 jacket, or a $200 television. Fill in the first three columns of the first three rows of the chart below.

2. Apply the appropriate discount and determine the sale price of your item. Fill in the fourth column of the chart.

3. Next, find a multiplier that can be used to convert the sale price back to the original price and fill in the remaining column of the chart. Does this multiplier depend on the price of the item?

4. Working as a group, compare the results of part (3) for all three items. Then develop a formula for a multiplier that will restore a sale price to its original price, p, after a discount r has been applied. Complete the fourth row of the table and check that your formula will duplicate the results of part (3).

5. Use the formula from part (4) to find the multiplier that a store would use to return an item to its original price after a "30% off" sale expires. Fill in the last line on the chart.

6. Inspect the last column of your chart. How can these multipliers be used to determine the percentage by which a sale price is increased when a sale ends?

Original Price, p	Discount, r	$1 - r$	Sale Price	Multiplier to convert back to p
p	r	$1 - r$		
	0.30			

Decision Making & Connection   (*Use after Section 2.4.*)

Cost of Self-Employment.   Because of additional taxes and fewer benefits, it has been estimated that a self-employed person must earn 20% more than a non–self-employed person performing the same work.

1. Barry is considering starting his own lawn-care business. He currently earns $16 per hour as a crew manager working for a large lawn-care company. He estimates that he can earn $18 per hour if he starts his own business, but he will lose his benefits. On the basis of these hourly rates alone, which option should Barry choose?

2. Develop a formula that gives the minimum amount A that a self-employed person must earn in order to make a salary equivalent to that of a non–self-employed person making x dollars.

 3. *Research.*   Determine the extra costs that a self-employed person has. Consider self-employment taxes, insurance and vacation benefits, retirement benefits, and other costs. Based on your research, does the 20% rule seem accurate?

Study Summary

KEY TERMS AND CONCEPTS	EXAMPLES	PRACTICE EXERCISES

SECTION 2.1: *Solving Equations*

The Addition Principle for Equations

$a = b$ is equivalent to $a + c = b + c$.

Solve: $x + 5 = -2$.
$$x + 5 = -2$$
$$x + 5 + (-5) = -2 + (-5) \quad \text{Adding } -5 \text{ to both sides}$$
$$x = -7$$

1. Solve: $x - 8 = -3$.

The Multiplication Principle for Equations

$a = b$ is equivalent to $ac = bc$, for $c \neq 0$.

Solve: $-\frac{1}{3}x = 7$.
$$-\frac{1}{3}x = 7$$
$$(-3)\left(-\frac{1}{3}x\right) = (-3)(7) \quad \text{Multiplying both sides by } -3$$
$$x = -21$$

2. Solve: $\frac{1}{4}x = 1.2$.

SECTION 2.2: *Using the Principles Together*

When solving equations, we usually work in the reverse order of the order of operations.

Solve: $-3x - 7 = -8$.
$$-3x - 7 + 7 = -8 + 7 \quad \text{Adding 7 to both sides}$$
$$-3x = -1$$
$$\frac{-3x}{-3} = \frac{-1}{-3} \quad \text{Dividing both sides by } -3$$
$$x = \frac{1}{3}$$

3. Solve: $4 - 3x = 7$.

We can **clear fractions** by multiplying both sides of an equation by the least common multiple of the denominators in the equation.

We can **clear decimals** by multiplying both sides by a power of 10. If there is at most one decimal place in any one number, we multiply by 10. If there are at most two decimal places, we multiply by 100, and so on.

Solve: $\frac{1}{2}x - \frac{1}{3} = \frac{1}{6}x + \frac{2}{3}$.
$$6\left(\frac{1}{2}x - \frac{1}{3}\right) = 6\left(\frac{1}{6}x + \frac{2}{3}\right) \quad \text{Multiplying by 6, the least common denominator}$$
$$6 \cdot \frac{1}{2}x - 6 \cdot \frac{1}{3} = 6 \cdot \frac{1}{6}x + 6 \cdot \frac{2}{3} \quad \text{Using the distributive law}$$
$$3x - 2 = x + 4 \quad \text{Simplifying}$$
$$2x = 6 \quad \text{Subtracting } x \text{ from and adding 2 to both sides}$$
$$x = 3$$

4. Solve: $\frac{1}{6}t - \frac{3}{4} = t - \frac{2}{3}$.

SECTION 2.3: *Formulas*

A **formula** uses letters to show a relationship among two or more quantities. Formulas can be solved for a given letter using the addition and multiplication principles.

Solve for y: $x = \frac{2}{5}y + 7$.
$$x = \frac{2}{5}y + 7 \quad \text{We are solving for } y.$$
$$x - 7 = \frac{2}{5}y \quad \text{Isolating the term containing } y$$
$$\frac{5}{2}(x - 7) = \frac{5}{2} \cdot \frac{2}{5}y \quad \text{Multiplying both sides by } \frac{5}{2}$$
$$\frac{5}{2}x - \frac{5}{2} \cdot 7 = 1 \cdot y \quad \text{Using the distributive law}$$
$$\frac{5}{2}x - \frac{35}{2} = y \quad \text{We have solved for } y.$$

5. Solve for c: $ac - bc = d$.

SECTION 2.4: *Applications with Percent*

Key Words in Percent Translations

"Of" translates to " · " or "×"

"What" translates to a variable

"Is" or "Was" translates to "="

"%" translates to "×$\frac{1}{100}$" or "× 0.01"

What percent of 60 is 7.2?

$$n \cdot 60 = 7.2$$

$$n = \frac{7.2}{60}$$

$$n = 0.12$$

Thus, 7.2 is 12% of 60.

6. 12 is 15% of what number?

SECTION 2.5: *Problem Solving*

Five Steps for Problem Solving in Algebra

1. *Familiarize* yourself with the problem.
2. *Translate* to mathematical language. (This often means writing an equation or an inequality.)
3. *Carry out* some mathematical manipulation. (This often means *solving* an equation or an inequality.)
4. *Check* your possible answer in the original problem.
5. *State* the answer clearly.

The perimeter of a rectangle is 70 cm. The width is 5 cm longer than half the length. Find the length and the width.

1. **Familiarize.** The formula for the perimeter of a rectangle is $P = 2l + 2w$. We can describe the width in terms of the length: $w = \frac{1}{2}l + 5$.

2. **Translate.**

Rewording: Twice the length plus twice the width is the perimeter.

Translating: $2l + 2\left(\frac{1}{2}l + 5\right) = 70$

3. **Carry out.** Solve the equation:

$$2l + 2\left(\tfrac{1}{2}l + 5\right) = 70$$
$$2l + l + 10 = 70 \quad \text{Using the distributive law}$$
$$3l + 10 = 70 \quad \text{Combining like terms}$$
$$3l = 60 \quad \text{Subtracting 10 from both sides}$$
$$l = 20. \quad \text{Dividing both sides by 3}$$

If $l = 20$, then $w = \frac{1}{2}l + 5 = \frac{1}{2} \cdot 20 + 5 = 10 + 5 = 15$.

4. **Check.** The width should be 5 cm longer than half the length. Since half the length is 10 cm, and 15 cm is 5 cm longer, this statement checks. The perimeter should be 70 cm. Since $2l + 2w = 2(20) + 2(15) = 40 + 30 = 70$, this statement also checks.

5. **State.** The length is 20 cm and the width is 15 cm.

7. Deborah rode a total of 120 mi in two bicycle fundraisers. One ride was 25 mi longer than the other. How long was each fundraiser?

50. $(2, -79), (4, -25), (-4, 12)$

51. $(-10, -4), (-16, 7), (3, 15)$

52. $(5, -16), (-7, -4), (12, 3)$

53. $(-100, -5), (350, 20), (800, 37)$

54. $(750, -8), (-150, 17), (400, 32)$

55. $(-83, 491), (-124, -95), (54, -238)$

56. $(738, -89), (-49, -6), (-165, 53)$

57. The following graph was included in a mailing sent by Agway® to their oil customers in 2000. What information is missing from the graph and why is the graph misleading?

Residential Fuel Oil and Natural Gas Prices

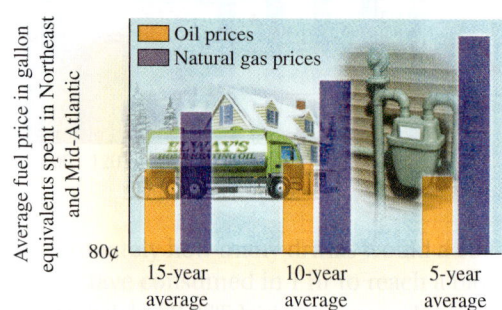

Average fuel price in gallon equivalents spent in Northeast and Mid-Atlantic

☐ Oil prices
☐ Natural gas prices

80¢

15-year average   10-year average   5-year average

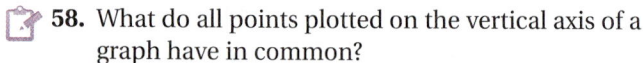

Source: Energy Research Center Inc. *3/1/99–2/29/00

58. What do all points plotted on the vertical axis of a graph have in common?

Skill Review

Simplify.

59. $-\frac{1}{2} + \frac{1}{3}$ [1.5]

60. $-2.6 - 9.1$ [1.6]

61. $3(-40)$ [1.7]

62. $-100 \div (-20)$ [1.7]

63. $(-1)(-2)(-3)(-4)$ [1.7]

64. $3 + (-4) + (-6) + 10$ [1.5]

Synthesis

65. Describe what the result would be if the first and second coordinates of every point in the following graph of an arrow were interchanged.

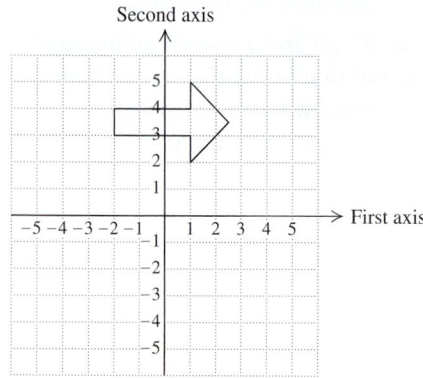

66. The graph accompanying Example 2 flattens out. Why do you think this occurs?

67. In which quadrant(s) could a point be located if its coordinates are opposites of each other?

68. In which quadrant(s) could a point be located if its coordinates are reciprocals of each other?

69. The points $(-1, 1), (4, 1),$ and $(4, -5)$ are three vertices of a rectangle. Find the coordinates of the fourth vertex.

70. The pairs $(-2, -3), (-1, 2),$ and $(4, -3)$ can serve as three (of four) vertices for three different parallelograms. Find the fourth vertex of each parallelogram.

71. Graph eight points such that the sum of the coordinates in each pair is 7. Answers may vary.

72. Find the perimeter of a rectangle if three of its vertices are $(5, -2), (-3, -2),$ and $(-3, 3)$.

73. Find the area of a triangle whose vertices have coordinates $(0, 9), (0, -4),$ and $(5, -4)$.

Sorting Solid Waste. Circle graphs, or pie charts, are often used to show what percent of the whole each item in a group represents. Use the following pie chart to answer Exercises 74–77.

Sorting Solid Waste

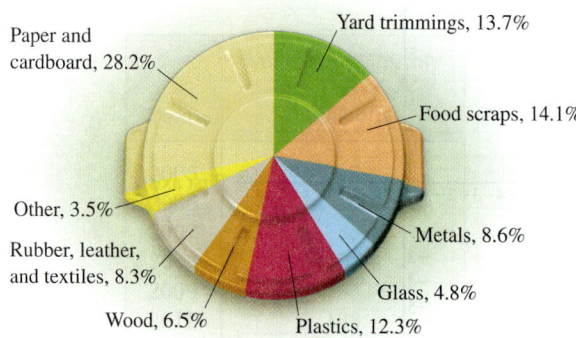

Paper and cardboard, 28.2%
Yard trimmings, 13.7%
Food scraps, 14.1%
Other, 3.5%
Rubber, leather, and textiles, 8.3%
Metals, 8.6%
Glass, 4.8%
Wood, 6.5%
Plastics, 12.3%

Source: U.S. Environmental Protection Agency

74. In 2009, Americans generated 243 million tons of waste. How much of the waste was plastic?

75. In 2009, the average American generated 4.3 lb of waste per day. How much of that was paper and cardboard?

76. Americans are recycling about 25.5% of all glass that is in the waste stream. How much glass did Americans recycle in 2009? (See Exercise 74.)

77. Americans are recycling about 59.9% of all yard trimmings. What amount of yard trimmings did the average American recycle per day in 2009? (Use the information in Exercise 75.)

Coordinates on the Globe. *Coordinates can also be used to describe the location on a sphere:* 0° *latitude is the equator and* 0° *longitude is a line from the North Pole to the South Pole through France and Algeria. In the following figure, hurricane Clara is at a point about 260 mi northwest of Bermuda near latitude 36.0° North, longitude 69.0° West.*

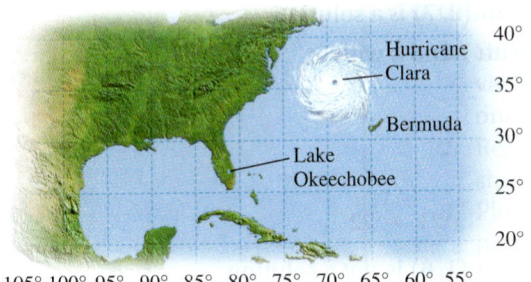

78. Approximate the latitude and the longitude of Bermuda.

79. Approximate the latitude and the longitude of Lake Okeechobee.

 80. In the *Star Trek* science-fiction series, a three-dimensional coordinate system is used to locate objects in space. If the center of a planet is used as the origin, how many "quadrants" will exist? Why? If possible, sketch a three-dimensional coordinate system and label each "quadrant."

 81. *Research.* Find the average monthly wireless phone bill in the United States for the most recent year. (See Exercise 26.) Extend the line graph in Exercise 26 to include this information. How would you describe the trends shown by this graph?

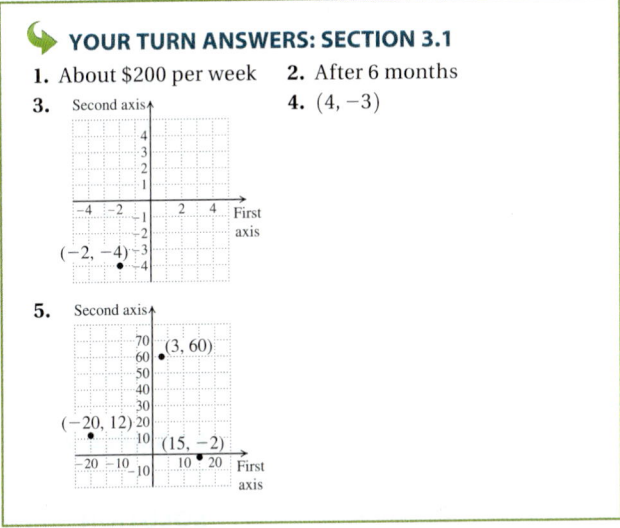

YOUR TURN ANSWERS: SECTION 3.1

1. About $200 per week   **2.** After 6 months
3.   **4.** $(4, -3)$
5.

PREPARE TO MOVE ON

Solve for y.   [2.3]

1. $5y = 2x$ **2.** $2y = -3x$

3. $x - y = 8$ **4.** $2x + 5y = 10$

5. $5x - 8y = 1$

3.2   Graphing Linear Equations

Solutions of Equations ▪ Graphing Linear Equations ▪ Applications

Not only do graphs help us to visualize data, but they can also be used to represent solutions of equations.

Solutions of Equations

When an equation contains two variables, solutions are ordered pairs. Each number in the pair replaces a letter in the equation. Unless stated otherwise, the first number in each pair replaces the variable that occurs first alphabetically.

EXAMPLE 1   Determine whether each of the following pairs is a solution of $4b - 3a = 22$: **(a)** $(2, 7)$; **(b)** $(1, 6)$.

In order for us to graph all three points, the y-axis of our graph must go down to at least -30 and the x-axis must go up to at least 20. Using a scale of 5 units per square allows us to display both intercepts and $(4, -24)$, as well as the origin.

The point $(4, -24)$ appears to line up with the intercepts, so we draw and label the line, as shown below.

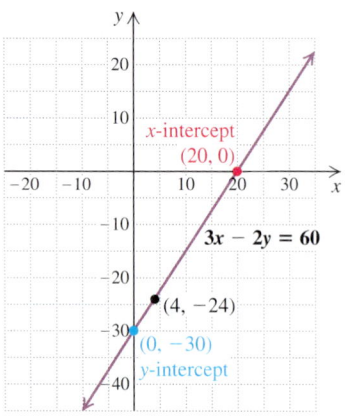

4. Graph $2x - 5y = 30$ using intercepts.

_____

YOUR TURN

Technology Connection

Using Intercepts to Graph

When an equation is graphed using a graphing calculator, we may not always see both intercepts. For example, if $y = -0.8x + 17$ is graphed in the window $[-10, 10, -10, 10]$, neither intercept is visible, as shown in the graph on the left below.

To better view intercepts, we can change the window dimensions or we can zoom out. On many graphing calculator apps, we can zoom in or out using the touch screen. On a graphing calculator without a touch screen, the ZOOM feature allows us to reduce or magnify a graph or a portion of a graph. Before zooming, we must set the ZOOM *factors* in the memory of the ZOOM key. If we zoom out with factors set at 5, both intercepts are visible but the axes are heavily drawn, as shown in the graph in the middle below.

This suggests that the *scales* of the axes should be changed. To do this, we use the WINDOW menu and set Xscl to 5 and Yscl to 5. The resulting graph has tick marks 5 units apart and clearly shows both intercepts, as shown in the graph on the right below. Other choices for Xscl and Yscl can also be made.

Graph each equation so that both intercepts can be easily viewed. Zoom or adjust the window settings so that tick marks can be clearly seen on both axes.

1. $y = -0.72x - 15$ **2.** $y - 2.13x = 27$
3. $5x + 6y = 84$ **4.** $2x - 7y = 150$
5. $19x - 17y = 200$ **6.** $6x + 5y = 159$

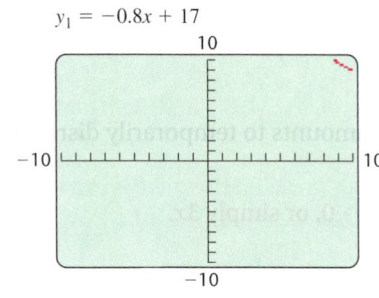

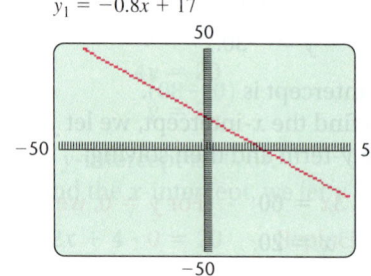

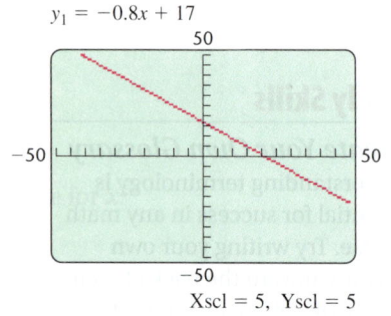

Xscl = 5, Yscl = 5

Graphing Horizontal Lines or Vertical Lines

The equations graphed in Examples 3 and 4 are in the form $Ax + By = C$. We have already stated that any equation in the form $Ax + By = C$ is linear, provided A and B are not both zero. What if A or B (but not both) is zero? We will find that when A is zero, there is no x-term and the graph is a horizontal line. We will also find that when B is zero, there is no y-term and the graph is a vertical line.

EXAMPLE 5 Graph: $y = 3$.

SOLUTION We can regard the equation $y = 3$ as $0 \cdot x + y = 3$. No matter what number we choose for x, we find that y must be 3 if the equation is to be solved. Consider the following table.

$$y = 3$$

Choose any number for x. →

x	y	(x, y)
-2	3	$(-2, 3)$
0	3	$(0, 3)$
4	3	$(4, 3)$

y must be 3.

All pairs will have 3 as the y-coordinate.

When we plot $(-2, 3)$, $(0, 3)$, and $(4, 3)$ and connect the points, we obtain a horizontal line. Any pair of the form $(x, 3)$ is a solution, so the line is parallel to the x-axis with y-intercept $(0, 3)$. Note that the graph of $y = 3$ has no x-intercept.

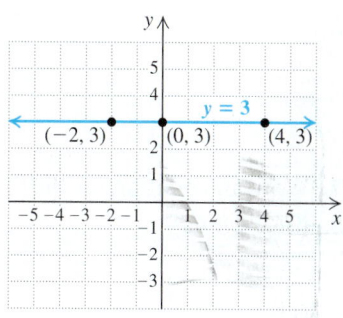

5. Graph: $y = 2$.

_____

YOUR TURN

EXAMPLE 6 Graph: $x = -4$.

SOLUTION We can regard the equation $x = -4$ as $x + 0 \cdot y = -4$. We create a table with all -4's in the x-column.

Student Notes

Sometimes students draw horizontal lines when they should be drawing vertical lines and vice versa. To avoid this mistake, first locate the correct number on the axis whose label is given. Thus, to graph $x = 2$, we locate 2 on the x-axis and then draw a line perpendicular to that axis at that point. Note that the graph of $x = 2$ on a plane is a line, whereas the graph of $x = 2$ on a number line is a point.

$$x = -4$$

x must be -4. →

x	y	(x, y)
-4	-5	$(-4, -5)$
-4	1	$(-4, 1)$
-4	3	$(-4, 3)$

Any number can be used for y.

All pairs will have -4 as the x-coordinate.

When we plot $(-4, -5)$, $(-4, 1)$, and $(-4, 3)$ and connect the points, we obtain a vertical line. Any ordered pair of the form $(-4, y)$ is a solution. The line is parallel to the y-axis with x-intercept $(-4, 0)$. Note that the graph of $x = -4$ has no y-intercept.

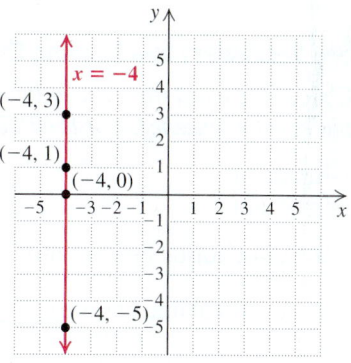

6. Graph: $x = -2$.

_____

YOUR TURN

LINEAR EQUATIONS IN ONE VARIABLE

The graph of $y = b$ is a horizontal line, with y-intercept $(0, b)$.

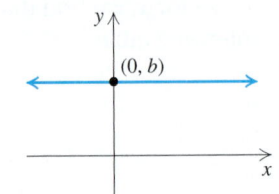

The graph of $x = a$ is a vertical line, with x-intercept $(a, 0)$.

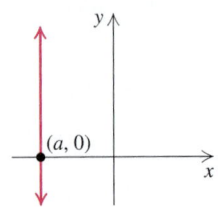

EXAMPLE 7   Write an equation for each graph.

a)

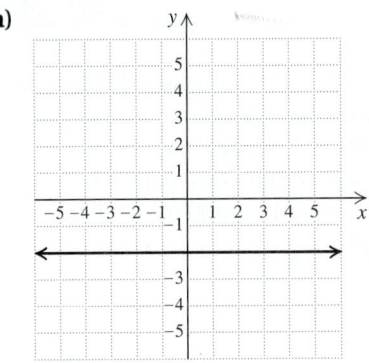

b)

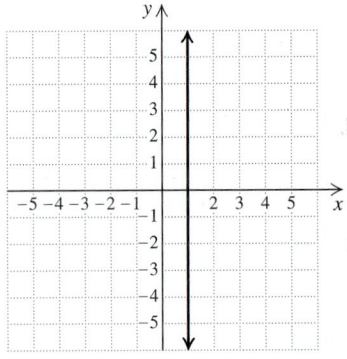

7. Write an equation for the following graph.

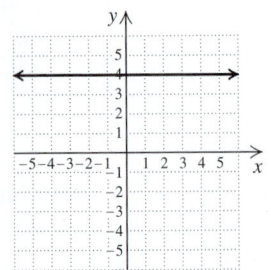

SOLUTION

a) Note that every point on the horizontal line passing through $(0, -2)$ has -2 as the y-coordinate. Thus the equation of the line is $y = -2$.

b) Note that every point on the vertical line passing through $(1, 0)$ has 1 as the x-coordinate. Thus the equation of the line is $x = 1$.

 YOUR TURN

3.3   EXERCISE SET

FOR EXTRA HELP

 MyMathLab® Math XL
PRACTICE   WATCH   READ   REVIEW

↳ Vocabulary and Reading Check

Choose the word from the following list that best completes each sentence. Words will be used more than once.

x-intercept
y-intercept

1. A horizontal line has a(n) _____.

2. A vertical line has a(n) _____.

3. To find a(n) _____, replace x with 0 and solve for y.

4. To find a(n) _____, replace y with 0 and solve for x.

5. The point $(-3, 0)$ could be a(n) _____ of a graph.

6. The point $(0, 7)$ could be a(n) _____ of a graph.

⤷ Concept Reinforcement

In each of Exercises 7–12, match the phrase with the most appropriate choice from the column on the right.

7. _____ A vertical line   **a)** $2x + 5y = 100$

8. _____ A horizontal line   **b)** $(3, -2)$

9. _____ A y-intercept   **c)** $(1, 0)$

10. _____ An x-intercept   **d)** $(0, 2)$

11. _____ A third point as   **e)** $y = 3$
          a check

                                    f) $x = -4$

12. _____ Use a scale of
          10 units per square.

Intercepts

For Exercises 13–20, list **(a)** *the coordinates of the y-intercept and* **(b)** *the coordinates of all x-intercepts.*

13.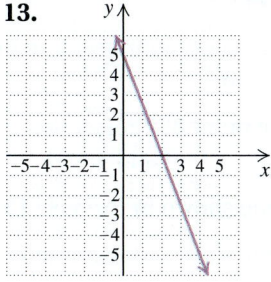

14.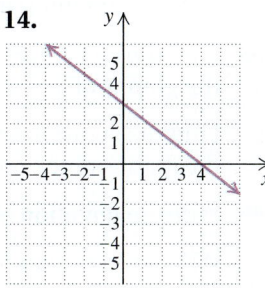

15.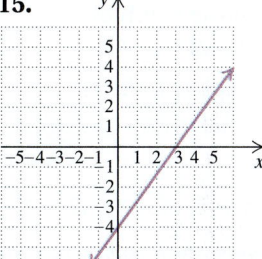

16.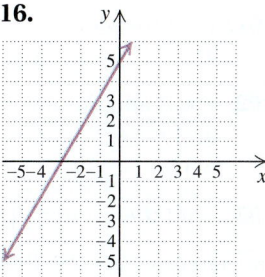

17.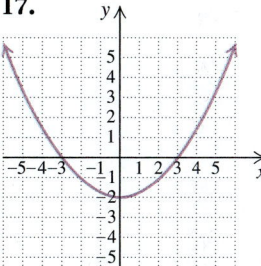

18.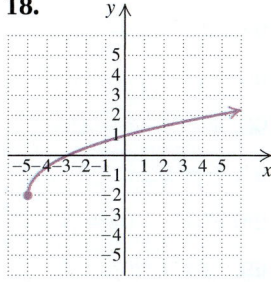

19.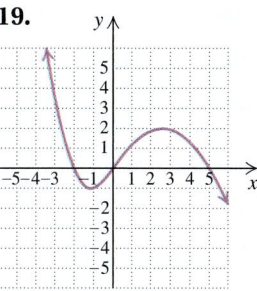

20.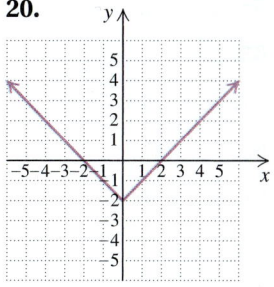

For Exercises 21–30, list **(a)** *the coordinates of any y-intercept and* **(b)** *the coordinates of any x-intercept. Do not graph.*

21. $3x + 5y = 15$ **22.** $2x + 7y = 14$

23. $9x - 2y = 36$ **24.** $10x - 3y = 60$

25. $-4x + 5y = 80$ **26.** $-5x + 6y = 100$

Aha! **27.** $x = 12$ **28.** $y = 10$

29. $y = -9$ **30.** $x = -5$

Using Intercepts to Graph

Find the intercepts. Then graph.

31. $3x + 5y = 15$ **32.** $2x + y = 6$

33. $x + 2y = 4$ **34.** $2x + 5y = 10$

35. $-x + 2y = 8$ **36.** $-x + 3y = 9$

37. $3x + y = 9$ **38.** $2x - y = 8$

39. $y = 2x - 6$ **40.** $y = -3x + 6$

41. $5x - 10 = 5y$ **42.** $3x - 9 = 3y$

43. $2x - 5y = 10$ **44.** $2x - 3y = 6$

45. $6x + 2y = 12$ **46.** $4x + 5y = 20$

47. $4x + 3y = 16$ **48.** $3x + 2y = 8$

49. $2x + 4y = 1$ **50.** $3x - 6y = 1$

51. $5x - 3y = 180$ **52.** $10x + 7y = 210$

53. $y = -30 + 3x$ **54.** $y = -40 + 5x$

55. $-4x = 20y + 80$ **56.** $60 = 20x - 3y$

57. $y - 3x = 0$ **58.** $x + 2y = 0$

Graphing Horizontal Lines or Vertical Lines

Graph.

59. $y = 1$ **60.** $y = 4$

61. $x = 3$ **62.** $x = 6$

63. $y = -2$ **64.** $y = -4$

65. $x = -1$ **66.** $x = -6$

67. $y = -15$ **68.** $x = 20$

69. $y = 0$ **70.** $y = \frac{3}{2}$

71. $x = -\frac{5}{2}$ **72.** $x = 0$

73. $-4x = -100$ **74.** $12y = -360$

75. $35 + 7y = 0$ **76.** $-3x - 24 = 0$

Write an equation for each graph.

77.
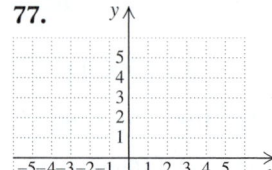

78.
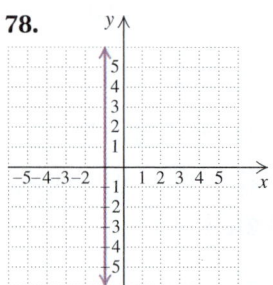

79.
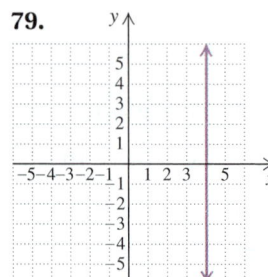

80.
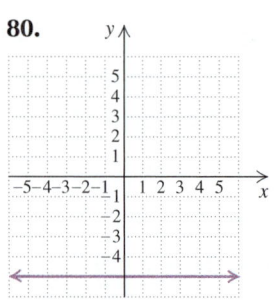

81.
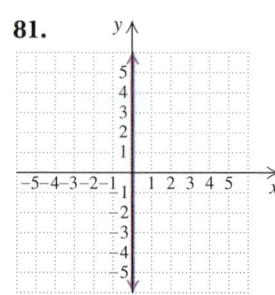

82.
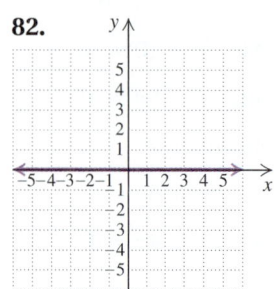

83. Explain in your own words why the graph of $y = 8$ is a horizontal line.

84. Explain in your own words why the graph of $x = -4$ is a vertical line.

Skill Review

Translate to an algebraic expression. [1.1]

85. 7 less than d

86. 5 more than w

87. The sum of 7 and four times a number

88. The product of 3 and a number

89. Twice the sum of two numbers

90. Half of the sum of two numbers

Synthesis

91. Describe what the graph of $x + y = C$ will look like for any choice of C.

92. If the graph of a linear equation has one point that is both the x-intercept and the y-intercept, what is that point? Why?

93. Write an equation for the x-axis.

94. Write an equation of the line parallel to the x-axis and passing through $(3, 5)$.

95. Write an equation of the line parallel to the y-axis and passing through $(-2, 7)$.

96. Find the coordinates of the point of intersection of the graphs of $y = x$ and $y = 6$.

97. Find the coordinates of the point of intersection of the graphs of the equations $x = -3$ and $y = 4$.

98. Write an equation of the line shown in Exercise 13.

99. Write an equation of the line shown in Exercise 16.

100. Find the value of C such that the graph of $3x + C = 5y$ has an x-intercept of $(-4, 0)$.

101. Find the value of C such that the graph of $4x = C - 3y$ has a y-intercept of $(0, -8)$.

102. For A and B nonzero, the graphs of $Ax + D = C$ and $By + D = C$ will be parallel to an axis. Explain why.

103. Find the x-intercept of the graph of $Ax + D = C$.

In Exercises 104–109, find the intercepts of each equation algebraically. Then adjust the window and scale so that the intercepts can be checked graphically with no further window adjustments.

104. $3x + 2y = 50$

105. $2x - 7y = 80$

106. $y = 1.3x - 15$

107. $y = 0.2x - 9$

108. $25x - 20y = 1$

109. $50x + 25y = 1$

110. Draw a graph illustrating the following data. What portion of the graph is horizontal? How is this reflected in the data?

Month	President's Approval Rating
February	60%
March	50
April	40
May	40
June	40
July	40
August	60

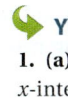 **YOUR TURN ANSWERS: SECTION 3.3**

1. (a) $(-2, 0)$; **(b)** $(0, 3)$ **2.** y-intercept: $(0, 2)$; x-intercept: $(6, 0)$

3.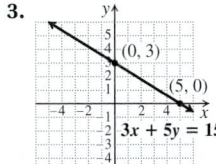

4.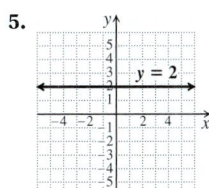

5. **6.**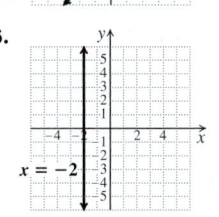

7. $y = 4$

QUICK QUIZ: SECTIONS 3.1–3.3

Graph.

1. $y = 2x - 5$ [3.2] **2.** $y = 4$ [3.3]

3. $x = -1$ [3.3] **4.** $3x - 4y = 12$ [3.3]

5. *Ozone Layer.* Listed below are estimates of the ozone level. Make a line graph of the data. [3.1]

Year	Minimum Ozone Level (in Dobson units)
2005	113.8
2006	98.2
2007	116.2
2008	114.1
2009	107.5
2010	128.0

Source: NASA Ozone Watch

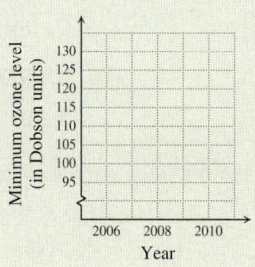

PREPARE TO MOVE ON

Simplify. [1.8]

1. $\dfrac{870 - 630}{12 - 6}$ **2.** $\dfrac{7360 - 2180}{21 - 11}$

3. $\dfrac{25,360 - 18,480}{722 - 310}$ **4.** $\dfrac{3.67 - 1.8376}{6 - 3}$

3.4 Rates

Rates of Change ▪ Visualizing Rates

Rates of Change

Because graphs make use of two axes, they allow us to visualize how two quantities change with respect to each other. A number accompanied by a ratio of two units is used to represent this type of change and is referred to as a *rate*.

> **RATE**
>
> A *rate* is a ratio that indicates how two quantities change with respect to each other.

Study Skills

Find a Study Buddy

It is not always a simple matter to find a study partner. Friendships can become strained and tension can develop if one of the study partners feels that he or she is working too much or too little. Try to find a partner with whom you feel compatible and don't take it personally if your first partner is not a good match. With some effort you will be able to locate a suitable partner. Often tutor centers or learning labs are good places to look for one.

Rates occur often in everyday life:

A website that grows by 10,000 visitors over a period of 2 months has an average *growth rate* of $\frac{10,000}{2}$, or 5000, visitors per month.

A vehicle traveling 150 mi in 3 hr is moving at a *rate* of $\frac{150}{3}$, or 50, mph (miles per hour).

A class of 25 students pays a total of $93.75 to visit a museum. The *rate* is $\frac{\$93.75}{25}$, or $3.75, per student.

> **CAUTION!**   To calculate a rate, it is important to keep track of the units being used.

EXAMPLE 1   *Car Rental.*   On January 3, Alisha rented a Chevrolet Cruze with a full tank of gas and 9312 mi on the odometer. On January 7, she returned the car with 9630 mi on the odometer.* If the rental agency charged Alisha $108 for the rental and needed 12 gal of gas to fill up the gas tank, find the following rates.

a) The car's rate of gas mileage, in miles per gallon

b) The average cost of the rental, in dollars per day

c) The car's rate of travel, in miles per day

SOLUTION

a) The rate of gas mileage, in miles per gallon, is found by dividing the number of miles traveled by the number of gallons used for that amount of driving:

$$\text{Rate, in miles per gallon} = \frac{9630 \text{ mi} - 9312 \text{ mi}}{12 \text{ gal}}$$

| The word "per" indicates division. |

$$= \frac{318 \text{ mi}}{12 \text{ gal}}$$

$$= 26.5 \text{ mi/gal} \qquad \text{Dividing}$$

$$= 26.5 \text{ miles per gallon.}$$

b) The average cost of the rental, in dollars per day, is found by dividing the cost of the rental by the number of days:

$$\text{Rate, in dollars per day} = \frac{108 \text{ dollars}}{4 \text{ days}} \qquad \begin{array}{l}\text{From January 3 to}\\ \text{January 7 is}\\ 7 - 3 = 4 \text{ days.}\end{array}$$

$$= 27 \text{ dollars/day}$$

$$= \$27 \text{ per day.}$$

c) The car's rate of travel, in miles per day, is found by dividing the number of miles traveled by the number of days:

$$\text{Rate, in miles per day} = \frac{318 \text{ mi}}{4 \text{ days}} \qquad \begin{array}{l}9630 \text{ mi} - 9312 \text{ mi} = 318 \text{ mi};\\ \text{From January 3 to January 7}\\ \text{is } 7 - 3 = 4 \text{ days.}\end{array}$$

$$= 79.5 \text{ mi/day}$$

$$= 79.5 \text{ mi per day.}$$

1. Refer to Exercise 1. Find the average daily rate of gas consumption, in gallons per day.

YOUR TURN

---

*For all problems concerning rentals, assume that the pickup time was later in the day than the return time so that no late fees were applied.

> **CAUTION!** Units are a vital part of real-world problems. They must be considered in the translation of a problem and included in the answer to a problem.

Many problems involve a rate of travel, or *speed*. The **speed** of an object is found by dividing the distance traveled by the time required to travel that distance.

EXAMPLE 2 *Transportation.* An Atlantic City Express bus makes regular trips between Paramus and Atlantic City, New Jersey. At 6:00 P.M., the bus is at mileage marker 40 on the Garden State Parkway, and at 8:00 P.M. it is at marker 170. Find the average speed of the bus.

SOLUTION Speed is the distance traveled divided by the time spent traveling:

$$\text{Bus speed} = \frac{\text{Distance traveled}}{\text{Time spent traveling}}$$

$$= \frac{\text{Change in mileage}}{\text{Change in time}}$$

$$= \frac{130\ \text{mi}}{2\ \text{hr}} \qquad \begin{array}{l} 170\ \text{mi} - 40\ \text{mi} = 130\ \text{mi}; \\ 8\text{:00 P.M.} - 6\text{:00 P.M.} = 2\ \text{hr} \end{array}$$

$$= 65\frac{\text{mi}}{\text{hr}}$$

$$= 65\ \text{miles per hour.} \qquad \begin{array}{l} \text{This } \textit{average} \text{ speed does not} \\ \text{indicate by how much the bus} \\ \text{speed may vary along the route.} \end{array}$$

2. A hummingbird flew across Shannon's yard, a distance of 250 ft, in 5 sec. Find the average speed of the hummingbird.

 YOUR TURN

Visualizing Rates

Graphs allow us to visualize a rate of change. As a rule, the quantity listed in the numerator appears on the vertical axis and the quantity listed in the denominator appears on the horizontal axis.

EXAMPLE 3 *Rain Forests.* Between 2004 and 2010, the additional yearly amount of Brazilian rain forest that was deforested decreased by 1250 mi² per year. During 2004, approximately 10,000 mi² of rain forest was deforested. Draw a graph to represent this information.

Source: Based on information from National Institute of Space Research

SOLUTION To label the axes, note that the rate is given as 1250 mi² per year, or

$$1250\frac{\text{mi}^2}{\text{year}}. \qquad \begin{array}{l} \leftarrow \text{Numerator: vertical axis} \\ \leftarrow \text{Denominator: horizontal axis} \end{array}$$

We list *square miles of rain forest deforested* on the vertical axis and *Year* on the horizontal axis. (See the figure on the left at the top of the following page.)

Next, we select a scale for each axis that allows us to plot the information. If we count by increments of 2000 on the vertical axis, we can show 10,000 mi² for 2004 and decreasing amounts for later years. On the horizontal axis, we can count by increments of 1 year. (See the figure in the middle at the top of the following page.)

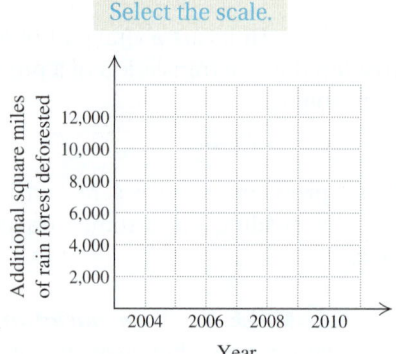

Label the axes.   Select the scale.   Draw the graph.

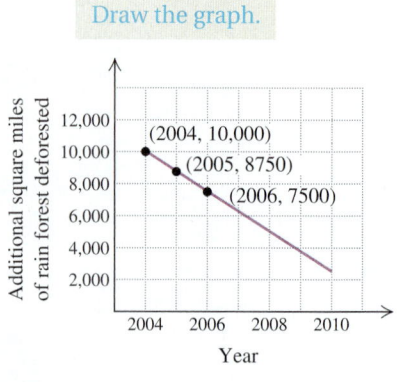

We are given that 10,000 mi^2 of rain forest was deforested during 2004. This gives us one point on the graph: (2004, 10,000).

We use the rate of change to find more points on the graph. Since the amount of yearly deforestation is decreasing, the rate of change is negative:

Deforestation during 2004:   10,000 mi^2;

Deforestation during 2005: $10,000 - 1250 = 8750$ mi^2;

Deforestation during 2006: $8750 - 1250 = 7500$ mi^2.

3. Refer to Example 3. Between 1997 and 2003, the amount of Brazilian rain forest that was deforested each year *increased* at a rate of approximately 700 mi^2 per year. In 1997, approximately 5100 mi^2 of rain forest was deforested. Draw a graph to represent this information.

_____

These calculations give us two more points on the graph: (2005, 8750) and (2006, 7500). After plotting the three points, we draw a line through them, as shown in the figure on the right above. This gives us the graph.

↪ YOUR TURN

EXAMPLE 4 *Banking.* Nadia prepared the following graph from data collected on a recent day at a local bank.

a) What rate can be determined from the graph?

b) What is that rate?

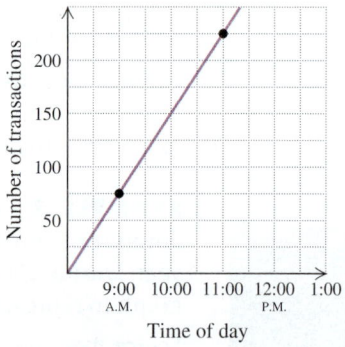

4. Refer to Example 4. Nadia prepared the following graph for data at a different bank location. What is the rate?

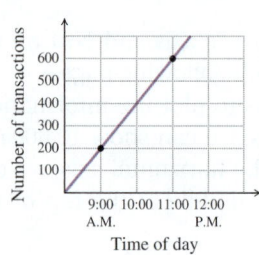

SOLUTION

a) Because the vertical axis shows the number of transactions and the horizontal axis lists the time in hour-long increments, we can find the rate *Number of transactions per hour.*

b) The points (9:00, 75) and (11:00, 225) are both on the graph. This tells us that in the 2 hours between 9:00 and 11:00, there were $225 - 75 = 150$ transactions. Thus the average rate is

$$\frac{225 \text{ transactions} - 75 \text{ transactions}}{11:00 - 9:00} = \frac{150 \text{ transactions}}{2 \text{ hours}}$$

$$= 75 \text{ transactions per hour.}$$

_____   ↪ YOUR TURN

3.4 EXERCISE SET

Vocabulary and Reading Check

Classify each of the following statements as either true or false.

1. A rate is a ratio.

2. In the phrase "meters per second," the word "per" indicates division.

3. To find the speed of an object, divide the time by the distance traveled.

4. A quantity appearing on the vertical axis of a graph should be used in the numerator of the associated rate of change.

Concept Reinforcement

For each of Exercises 5–8, fill in the missing units for each rate.

5. If Eva biked 100 miles in 5 hours, her average rate was 20 _____.

6. If it took Lauren 18 hours to read 6 chapters, her average rate was 3 _____.

7. If Denny's ticket cost $300 for a 150-mile flight, his average rate was 2 _____.

8. If Marc made 8 cakes using 20 cups of flour, his average rate was $2\frac{1}{2}$ _____.

Rates of Change

Solve. For Exercises 9–16, round answers to the nearest cent, where appropriate.

9. *Car Rentals.* Late on June 5, Gaya rented a Ford Focus with a full tank of gas and 13,741 mi on the odometer. On June 8, she returned the car with 14,131 mi on the odometer. The rental agency charged Gaya $118 for the rental and needed 13 gal of gas to fill up the tank.
   a) Find the car's rate of gas consumption, in miles per gallon.
   b) Find the average cost of the rental, in dollars per day.
   c) Find the average rate of travel, in miles per day.
   d) Find the rental rate, in cents per mile.

10. *SUV Rentals.* On February 10, Oscar rented a Chevy Trailblazer with a full tank of gas and 13,091 mi on the odometer. On February 12, he returned the vehicle with 13,322 mi on the odometer. The rental agency charged $92 for the rental and needed 14 gal of gas to fill the tank.
    a) Find the SUV's rate of gas consumption, in miles per gallon.
    b) Find the average cost of the rental, in dollars per day.
    c) Find the average rate of travel, in miles per day.
    d) Find the rental rate, in cents per mile.

11. *Bicycle Rentals.* At 9:00, Jodi rented a mountain bike from The Bike Rack. She returned the bicycle at 11:00, after cycling 14 mi. Jodi paid $15 for the rental.
    a) Find Jodi's average speed, in miles per hour.
    b) Find the rental rate, in dollars per hour.
    c) Find the rental rate, in dollars per mile.

12. *Bicycle Rentals.* At 2:00, Braden rented a mountain bike from Slickrock Cycles. He returned the bike at 5:00, after cycling 18 mi. Braden paid $24 for the rental.
    a) Find Braden's average speed, in miles per hour.
    b) Find the rental rate, in dollars per hour.
    c) Find the rental rate, in dollars per mile.

13. *Proofreading.* Sergei began proofreading at 9:00 A.M., starting at the top of page 93. He worked until 2:00 P.M. that day and finished page 195. He billed the publishers $110 for the day's work.
    a) Find the rate of pay, in dollars per hour.
    b) Find the average proofreading rate, in number of pages per hour.
    c) Find the rate of pay, in dollars per page.

14. *Temporary Help.* A typist for Kelly Services reports to 3E's Properties for work at 10:00 A.M. and leaves at 6:00 P.M. after having typed from the end of page 8 to the end of page 50 of a proposal. 3E's pays $120 for the typist's services.
    a) Find the rate of pay, in dollars per hour.
    b) Find the average typing rate, in number of pages per hour.
    c) Find the rate of pay, in dollars per page.

15. *National Debt.* The U.S. federal budget debt was $8612 billion in 2006 and $15,041 billion in 2011. Find the average rate at which the debt was increasing.
    Source: U.S. Office of Management and Budget

16. *Four-Year-College Tuition.* The average amount of room, board, and tuition at a public four-year college was $17,447 in 2006 and $21,189 in 2010. Find the average rate at which room, board, and tuition are increasing.

Source: U.S. National Center for Education Statistics

17. *Elevators.* At 2:38, Lara entered an elevator on the 34th floor of the Regency Hotel. At 2:40, she stepped off at the 5th floor.
a) Find the elevator's average rate of travel, in number of floors per minute.
b) Find the elevator's average rate of travel, in seconds per floor.

18. *Snow Removal.* By 1:00 P.M., Shani had already shoveled 2 driveways, and by 6:00 P.M. that day, the number was up to 7.
a) Find Shani's average shoveling rate, in number of driveways per hour.
b) Find Shani's average shoveling rate, in hours per driveway.

19. *Mountaineering.* The fastest ascent of Mt. Everest was accomplished by the Sherpa guide Pemba Dorje of Nepal in 2004. Pemba Dorje climbed from base camp, elevation 17,700 ft, to the summit, elevation 29,029 ft, in 8 hr 10 min.

Source: Guinness Book of World Records 2012

a) Find Pemba Dorje's average rate of ascent, in feet per minute.
b) Find Pemba Dorje's average rate of ascent, in minutes per foot.

20. *Pottery.* Master potter Mark Byles holds the record for the most clay pots thrown in 1 hr. He made 150 clay flower pots during the allotted hour.

Source: Guinness Book of World Records 2012

a) Find Mark's average pot-throwing rate, in pots per minute.
b) Find Mark's average pot-throwing rate, in minutes per pot.

Visualizing Rates

In Exercises 21–30, draw a linear graph to represent the given information. Be sure to label and number the axes appropriately (see Example 3).

21. *Recycling.* Between 2001 and 2010, the amount of steel recycled in the United States increased at an average rate of approximately 125,000 tons per year. In 2001, approximately 4,600,000 tons of steel was recycled. Draw a graph to represent this information.

Source: U.S. Environmental Protection Agency

22. *Health Insurance.* In 2008, the average cost for health insurance for a family was about $12,700, and the figure was rising at a rate of about $700 per year.

Source: Based on data from Kaiser/HRET Survey of Health Benefits

23. *Prescription Drug Sales.* In 2009, there were sales of approximately $4 billion of anti-Alzheimer's prescription drug products in the United States, and the figure was increasing at a rate of about $0.5 billion per year.

Source: IMS National Sales Perspectives

24. *Violent Crimes.* In 2009, there were approximately 17.1 violent crimes per 1000 Americans, and the figure was dropping at a rate of about 1.9 violent crimes per 1000 per year.

Source: U.S. Bureau of Justice Statistics

25. *Train Travel.* At 3:00 P.M., the Boston–Washington Metroliner had traveled 230 mi and was cruising at a rate of 90 miles per hour.

26. *Plane Travel.* At 4:00 P.M., the Seattle–Los Angeles shuttle had traveled 400 mi and was cruising at a rate of 300 miles per hour.

27. *Wages.* By 2:00 P.M., Diane had earned $50. She continued earning money at a rate of $15 per hour.

28. *Wages.* By 3:00 P.M., Arnie had earned $70. He continued earning money at a rate of $12 per hour.

29. *Telephone Bills.* Roberta's phone bill was already $7.50 when she made a call for which she was charged at a rate of $0.10 per minute.

30. *Telephone Bills.* At 3:00 P.M., Theo's phone bill was $6.50 and increasing at a rate of 7¢ per minute.

In Exercises 31–40, use the graph provided to calculate a rate of change in which the units of the horizontal axis are used in the denominator.

31. *Call Center.* The following graph shows data from a technical assistance call center. At what rate are calls being handled?

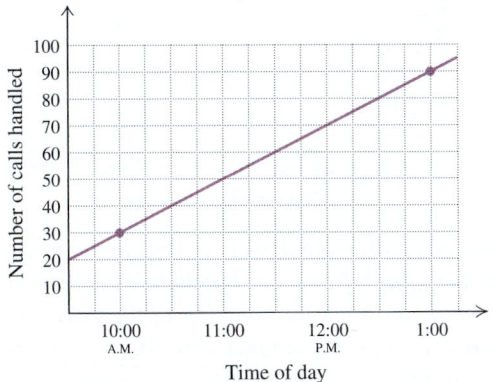

32. *Hairdresser.* Eve's Custom Cuts has a graph displaying data from a recent day of work. At what rate does Eve work?

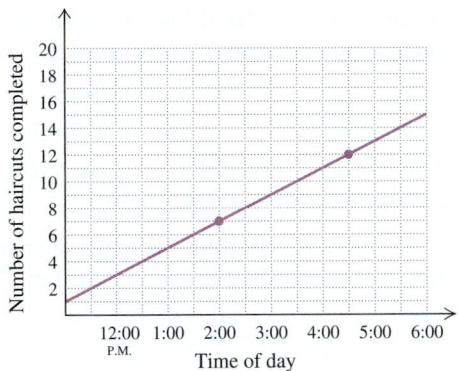

33. *Train Travel.* The following graph shows data from a recent train ride from Chicago to St. Louis. At what rate did the train travel?

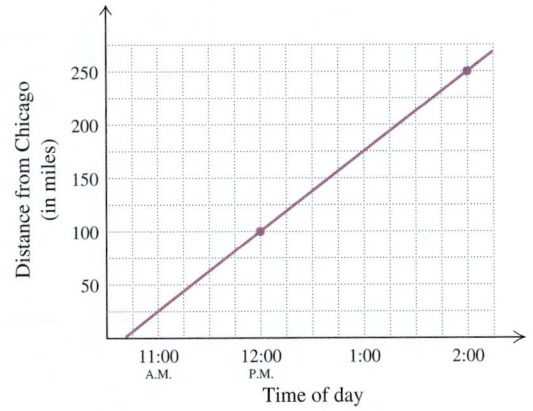

34. *Train Travel.* The following graph shows data from a recent train ride from Denver to Kansas City. At what rate did the train travel?

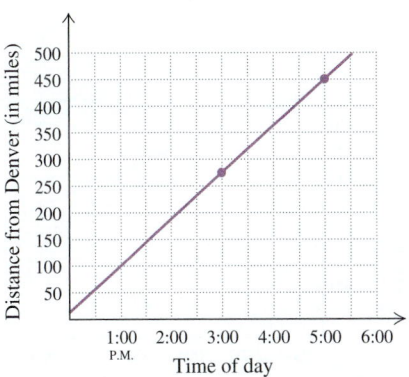

35. *Cost of a Telephone Call.* The following graph shows data from a recent phone call between the United States and Uruguay. At what rate was the customer being billed?

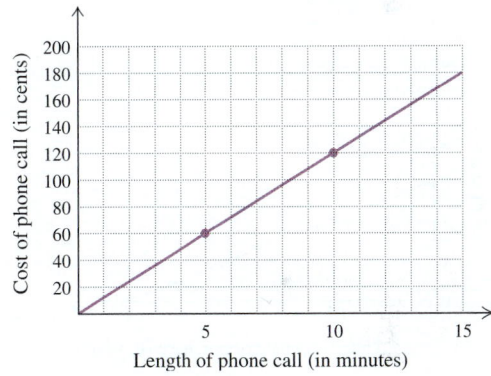

36. *Cost of a Telephone Call.* The following graph shows data from a recent phone call between the United States and China. At what rate was the customer being billed?

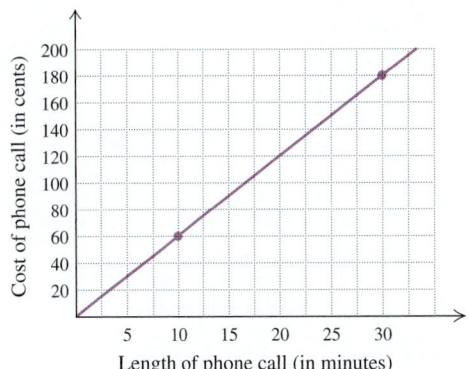

37. *Population.*   The following graph shows data regarding the population of Cleveland, Ohio. At what rate was the population changing?

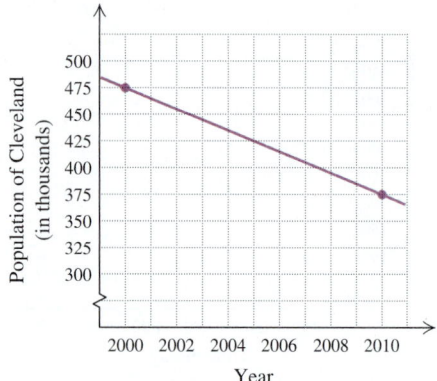

38. *Depreciation of an Office Machine.*   Data regarding the value of a particular color copier is represented in the following graph. At what rate is the value changing?

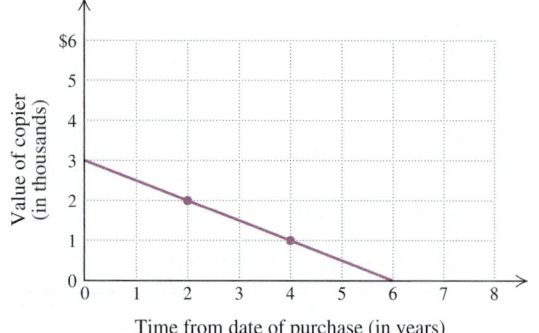

39. *Gas Mileage.*   The following graph shows data for a 2012 Honda Civic Hybrid driven on interstate highways. At what rate was the vehicle consuming gas?

Source: www.fueleconomy.gov

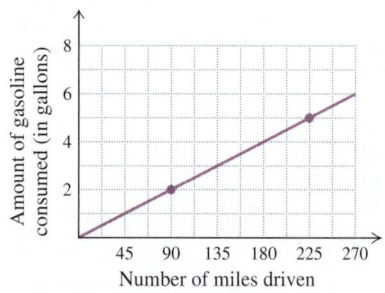

40. *Gas Mileage.*   The following graph shows data for a 2012 Ford Escape Hybrid AWD driven on city streets. At what rate was the vehicle consuming gas?

Source: www.fueleconomy.gov

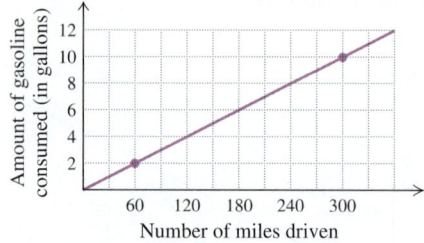

In Exercises 41–46, match each description with the most appropriate graph from the choices below. Scales are intentionally omitted. Assume that of the three sports listed, swimming is the slowest and biking is the fastest.

41. _____ Robin trains for triathlons by running, biking, and then swimming every Saturday.

42. _____ Gene trains for triathlons by biking, running, and then swimming every Sunday.

43. _____ Shirley trains for triathlons by swimming, biking, and then running every Sunday.

44. _____ Evan trains for triathlons by swimming, running, and then biking every Saturday.

45. _____ Angie trains for triathlons by biking, swimming, and then running every Sunday.

46. _____ Mick trains for triathlons by running, swimming, and then biking every Saturday.

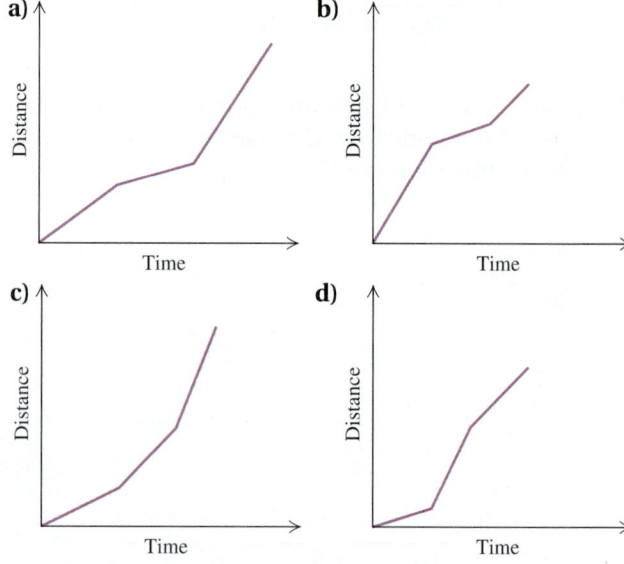

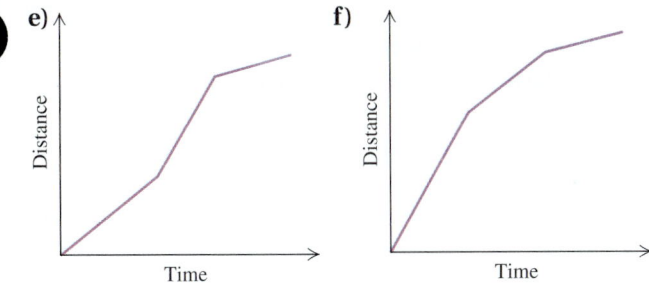

e) Distance / Time

f) Distance / Time

47. What does a negative rate of travel indicate? Explain.

48. Explain how to convert from kilometers per hour to meters per second.

Skill Review

49. Find the prime factorization of 150.   [1.3]

50. Graph on the number line: -3.5.   [1.4]

51. Find decimal notation: $-\dfrac{11}{8}$.   [1.4]

52. Find the absolute value: $\left|-\dfrac{1}{3}\right|$.   [1.4]

53. Find the opposite of $\dfrac{3}{2}$.   [1.6]

54. Find the reciprocal of $\dfrac{3}{2}$.   [1.7]

Synthesis

55. How would the graphs of Jon's and Jenny's total earnings compare in each of the following situations?

 a) Jon earns twice as much per hour as Jenny.
 b) Jon and Jenny earn the same hourly rate, but Jenny received a bonus for a cost-saving suggestion.
 c) Jon is paid by the hour, and Jenny is paid a weekly salary.

56. Write an exercise similar to those in Exercises 9–20 for a classmate to solve. Design the problem so that the solution is "The motorcycle's rate of gas consumption was 65 miles per gallon."

57. *Aviation.* A Boeing 737 airplane climbs from sea level to a cruising altitude of 31,500 ft at a rate of 6300 ft/min. After cruising for 3 min, the jet is forced to land, descending at a rate of 3500 ft/min. Represent the flight with a graph in which altitude is measured on the vertical axis and time on the horizontal axis.

58. *Wages with Commissions.* Each salesperson at Mike's Bikes is paid $140 per week plus 13% of all sales up to $2000, and then 20% on any sales in excess of $2000. Draw a graph in which sales are measured on the horizontal axis and wages on the vertical axis. Then use the graph to estimate the wages paid when a salesperson sells $2700 in merchandise in one week.

59. *Taxi Fares.* The driver of a New York City Yellow Cab recently charged $3.00 plus 40¢ for each fifth of a mile traveled for night service. Draw a graph that could be used to determine the cost of a fare.

60. *Gas Mileage.* A particular Kawasaki motorcycle travels three times as far as a Ford Escape Hybrid on the same amount of gas (see Exercise 40). Draw a graph that reflects this information.

61. *Aviation.* Tim's F-16 jet is moving forward at a deck speed of 95 mph aboard an aircraft carrier that is traveling 39 mph in the same direction. How fast is the jet traveling, in minutes per mile, with respect to the sea?

62. *Navigation.* In 3 sec, Penny walks 24 ft to the bow (front) of a tugboat. The boat is cruising at a rate of 5 ft/sec. What is Penny's rate of travel with respect to land?

63. *Running.* Anne ran from the 4-km mark to the 7-km mark of a 10-km race in 15.5 min. At this rate, how long would it take Anne to run a 5-mi race?

64. *Running.* Jerod ran from the 2-mi marker to the finish line of a 5-mi race in 25 min. At this rate, how long would it take Jerod to run a 10-km race?

65. Trevor picks apples twice as fast as Doug. By 4:30, Doug had already picked 4 bushels of apples. Fifty minutes later, his total reached $5\frac{1}{2}$ bushels. Find Trevor's picking rate. Give your answer in bushels per hour.

66. At 3:00 P.M., Carrie and Chad had already made 46 candles. By 5:00 P.M., the total had reached 100 candles. Assuming a constant production rate, at what time did they make their 82nd candle?

↪ YOUR TURN ANSWERS: SECTION 3.4

1. 3 gal per day   **2.** 50 ft per sec

3.

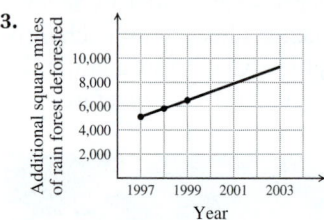

4. 200 transactions per hour

QUICK QUIZ: SECTIONS 3.1–3.4

1. In which quadrant or on which axis is $(0, -15)$ located?   [3.1]

2. Determine whether the ordered pair $(-3, -1)$ is a solution of the equation $2a - 3b = -9$.   [3.2]

3. Graph: $y = -3$.   [3.3]

4. Graph: $y = -3x$.   [3.2]

5. In 2008, there were about 39,200 commercial pilots in the United States. The number of pilots is expected to increase to 46,500 by 2018. Find the rate of increase, in number of pilots per year.   [3.4]

Source: U.S. Department of Labor, Bureau of Labor Statistics

PREPARE TO MOVE ON

Simplify.

1. $-2 - (-7)$ [1.6]

2. $-9 - (-3)$ [1.6]

3. $\dfrac{5 - (-4)}{-2 - 7}$ [1.8]

4. $\dfrac{8 - (-4)}{2 - 11}$ [1.8]

5. $\dfrac{-4 - 8}{11 - 2}$ [1.8]

6. $\dfrac{-5 - (-3)}{4 - 6}$ [1.8]

7. $\dfrac{-6 - (-6)}{-2 - 7}$ [1.8]

8. $\dfrac{-3 - 5}{-1 - (-1)}$ [1.8]

3.5   Slope

Rate and Slope ▪ Horizontal Lines and Vertical Lines ▪ Applications

Study Skills

Add Your Voice to the Author's

If you own your text, consider using it as a notebook. Since many instructors' work closely parallels the book, it is often useful to make notes on the appropriate page as he or she is lecturing.

A *rate* is a measure of how two quantities change with respect to each other. In this section, we will discuss how rate is related to the slope of a line.

Rate and Slope

Gary currently replaces his employees' laptop computers every two years. For a new program, he is considering changing this policy to replacing them every three years with a more durable model. The following tables list the costs of replacing the computers using both programs.

Two-Year Replacement	
Years Since Start of Program	Cost of Computers
0	$ 0
2	3,000
4	6,000
6	9,000
8	12,000

Three-Year Replacement	
Years Since Start of Program	Cost of Computers
0	$ 0
3	4,000
6	8,000
9	12,000
12	16,000

We now graph the pairs of numbers listed in the tables, using the horizontal axis for the number of years since the start of the program and the vertical axis for the cost of the computers.

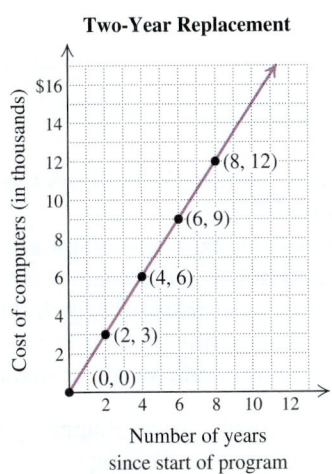

Two-Year Replacement

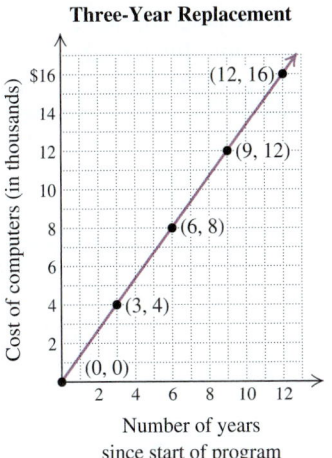

Three-Year Replacement

By comparing the cost of the computers over a specified period of time, we can compare the two rates. Note that the rate of the two-year replacement program is greater so its graph is steeper.

Replacement Program	Cost of Computers	Rate
Two-year	$3000 every 2 years	$\dfrac{3 \text{ thousand dollars}}{2 \text{ years}} = \dfrac{3}{2}$ thousand dollars per year, or $1500/year
Three-year	$4000 every 3 years	$\dfrac{4 \text{ thousand dollars}}{3 \text{ years}} = \dfrac{4}{3}$ thousand dollars per year, or $1333.33/year

The rates $\frac{3}{2}$ and $\frac{4}{3}$ can also be found using the coordinates of any two points that are on each line.

EXAMPLE 1 Use the graph of the Two-Year Replacement program above to find the cost of computers per year.

SOLUTION We can use the points $(6, 9)$ and $(8, 12)$ to find the rate for the two-year program. To do so, remember that these coordinates tell us that after 6 years, the computer cost is $9 thousand, and after 8 years, the computer cost is $12 thousand. In the 2 years between the 6-year and 8-year points, $12 − $9, or $3 thousand, was spent. Thus we have

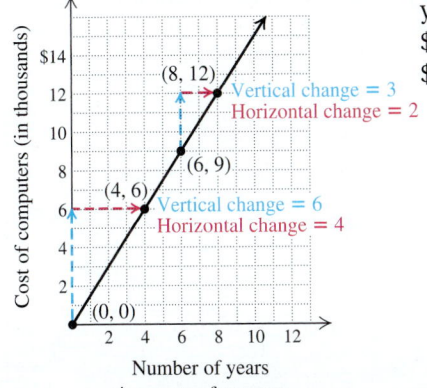

$$\text{Two-year rate} = \frac{\text{change in cost of computers}}{\text{corresponding change in time}}$$
$$= \frac{12 - 9 \text{ thousand dollars}}{8 - 6 \text{ years}}$$
$$= \frac{3 \text{ thousand dollars}}{2 \text{ years}}$$
$$= \frac{3}{2} \text{ thousand dollars per year, or } \$1500/\text{year.}$$

Because the line is straight, the same rate is found using *any* pair of points on the line. For example, using $(0,0)$ and $(4,6)$, we have

$$\text{Two-year rate} = \frac{6-0 \text{ thousand dollars}}{4-0 \text{ years}} = \frac{6 \text{ thousand dollars}}{4 \text{ years}}$$

$$= \frac{3}{2} \text{ thousand dollars per year, or } \$1500/\text{year}.$$

1. Use the graph of the Three-Year Replacement program to find the cost of computers per year.

 YOUR TURN

Note that the rate is always the vertical change divided by the corresponding horizontal change.

When the axes of a graph are simply labeled x and y, the ratio of vertical change to horizontal change is the rate at which y is changing with respect to x. This ratio is a measure of a line's slant, or **slope**.

Consider a line passing through $(2,3)$ and $(6,5)$, as shown below. We find the ratio of vertical change, or *rise*, to horizontal change, or *run*, as follows:

$$\frac{\text{Ratio of vertical change}}{\text{to horizontal change}} = \frac{\text{change in } y}{\text{change in } x} = \frac{\text{rise}}{\text{run}}$$

$$= \frac{5-3}{6-2}$$

$$= \frac{2}{4}, \text{ or } \frac{1}{2}.$$

Note that these calculations can be performed without viewing a graph.

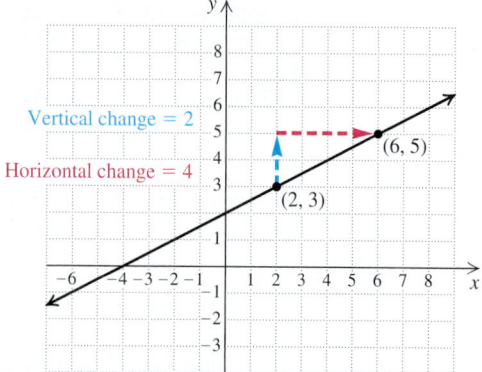

Thus the y-coordinates of points on this line increase at a rate of 2 units for every 4-unit increase in x, which is 1 unit for every 2-unit increase in x, or $\frac{1}{2}$ unit for every 1-unit increase in x. The slope of the line is $\frac{1}{2}$.

In the definition of *slope* below, the *subscripts* 1 and 2 are used to distinguish point 1 and point 2 from each other. The slightly lowered 1's and 2's are not exponents but are used to denote x-values (and y-values) for different points.

Student Notes

The notation x_1 is read "x sub one."

SLOPE

The *slope* of the line containing points (x_1, y_1) and (x_2, y_2) is given by

$$m = \frac{\text{change in } y}{\text{change in } x} = \frac{\text{rise}}{\text{run}} = \frac{y_2 - y_1}{x_2 - x_1}.$$

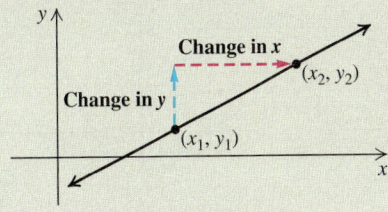

EXAMPLE 2 Find the slope of the line containing the points $(-4, 3)$ and $(2, -6)$.

SOLUTION From $(-4, 3)$ to $(2, -6)$, the change in y, or rise, is $-6 - 3$, or -9. The change in x, or run, is $2 - (-4)$, or 6. Thus,

$$\text{Slope} = \frac{\text{change in } y}{\text{change in } x}$$

$$= \frac{\text{rise}}{\text{run}}$$

$$= \frac{-6 - 3}{2 - (-4)}$$

$$= \frac{-9}{6}$$

$$= -\frac{9}{6}, \text{ or } -\frac{3}{2}.$$

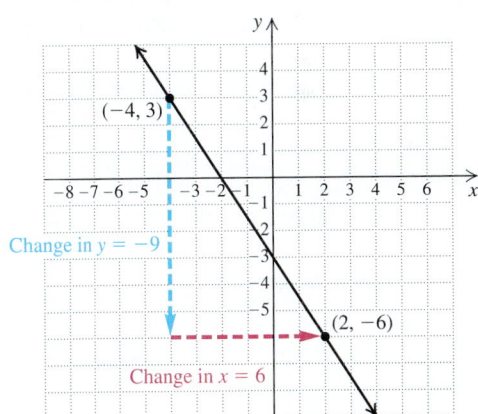

2. Find the slope of the line containing the points $(2, 7)$ and $(-6, 0)$.

The graph of the line is shown above for reference.

YOUR TURN

Student Notes

You may wonder which point should be regarded as (x_1, y_1) and which should be (x_2, y_2). To see that the math works out the same either way, perform both calculations on your own.

CAUTION! When we use the formula

$$m = \frac{y_2 - y_1}{x_2 - x_1},$$

it makes no difference which point is considered (x_1, y_1). What matters is that we subtract the y-coordinates in the same order that we subtract the x-coordinates.

To illustrate, we reverse *both* of the subtractions in Example 2. The slope is still $-\frac{3}{2}$:

$$\text{Slope} = \frac{\text{change in } y}{\text{change in } x} = \frac{3 - (-6)}{-4 - 2} = \frac{9}{-6} = -\frac{3}{2}.$$

═══ EXPLORING 🔍 THE CONCEPT ═══

The *sign* of the slope of a line indicates whether the line slants up or down from left to right. The *absolute value* of the slope indicates the steepness of the slope.

A.
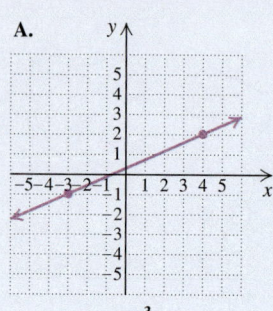
$m = \frac{3}{7}$

B.
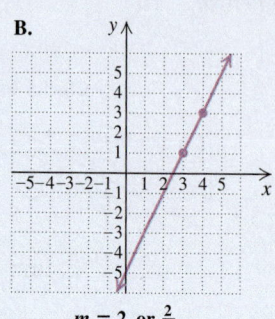
$m = 2$, or $\frac{2}{1}$

C.
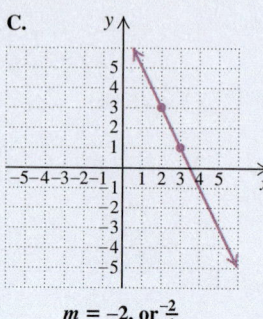
$m = -2$, or $\frac{-2}{1}$

D.
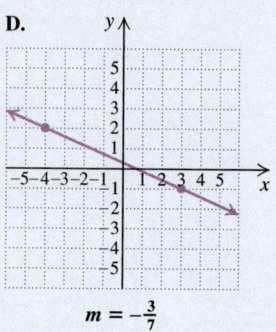
$m = -\frac{3}{7}$

1. Which of the graphs shown have a negative slope?
2. Which of the graphs shown have a positive slope?
3. On the basis of the graphs shown, choose the word that makes each statement true:
   a) A line with a positive slope slants _____ from left to right.
      up/down
   b) A line with a negative slope slants _____ from left to right.
      up/down
4. Which is larger: $\left|\frac{3}{7}\right|$ or $|2|$?
5. Which is larger: $\left|-\frac{3}{7}\right|$ or $|-2|$?
6. On the basis of the graphs shown, choose the word that makes the statement true:
   The _____ the absolute value of the slope, the steeper the line.
      larger/smaller

ANSWERS
1. C, D   **2.** A, B   **3. (a)** Up; **(b)** down   **4.** $|2|$ **5.** $|-2|$ **6.** Larger

A line with positive slope slants up from left to right, and a line with negative slope slants down from left to right. The larger the absolute value of the slope, the steeper the line.

Horizontal Lines and Vertical Lines

What about the slope of a horizontal line or a vertical line?

EXAMPLE 3   Find the slope of the line $y = 4$.

SOLUTION   Consider the points $(2, 4)$ and $(-3, 4)$, which are on the line. The change in y, or the rise, is $4 - 4$, or 0. The change in x, or the run, is $-3 - 2$, or -5. Thus,

$$m = \frac{4 - 4}{-3 - 2}$$

$$= \frac{0}{-5}$$

$$= 0.$$

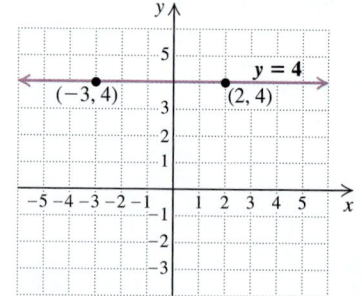

3. Find the slope of the line
$y = 2$.

_____

4. Find the slope of the line
$x = -1$.

_____

Any two points on a horizontal line have the same y-coordinate. Thus the change in y is 0, so the slope is 0.

YOUR TURN

> **A horizontal line has slope 0.**

EXAMPLE 4   Find the slope of the line $x = -3$.

SOLUTION   Consider the points $(-3, 4)$ and $(-3, -2)$, which are on the line. The change in y, or the rise, is $-2 - 4$, or -6. The change in x, or the run, is $-3 - (-3)$, or 0. Thus,

$$m = \frac{-2 - 4}{-3 - (-3)}$$

$$= \frac{-6}{0}. \quad \text{(undefined)}$$

Since division by 0 is not defined, the slope of this line is not defined. The answer to a problem of this type is "The slope of this line is undefined."

YOUR TURN

> **The slope of a vertical line is undefined.**

Applications

Slope has many real-world applications, ranging from car speed to production rate. Slope can also measure steepness. For example, numbers like 2%, 3%, and 6% are often used to represent the **grade** of a road, a measure of a road's steepness. That is, since $3\% = \frac{3}{100}$, a 3% grade means that for every horizontal distance of 100 ft, the road rises or drops 3 ft. The concept of grade also occurs in skiing or snowboarding, where a 7% grade is considered very tame, but a 70% grade is considered steep.

EXAMPLE 5   _Skiing._ Among the steepest skiable terrain in North America, the Headwall on Mount Washington, in New Hampshire, drops 720 ft over a horizontal distance of 900 ft. Find the grade of the Headwall.

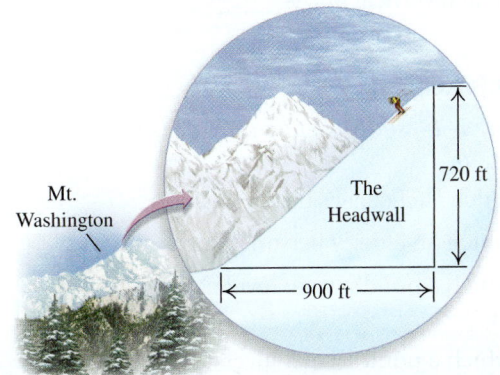

5. A mountain road rises 250 ft over a horizontal distance of 4000 ft. Find the grade of the road.

SOLUTION   The grade of the Headwall is its slope, expressed as a percent:

$$m = \frac{720}{900} = \frac{8}{10} = 80\%.$$ Grade is slope expressed as a percent.

 YOUR TURN

Carpenters use slope when designing stairs, ramps, or roof pitches. Another application occurs in the engineering of a dam—the force or strength of a river depends on how much the river drops over a specified distance.

3.5   EXERCISE SET

FOR EXTRA HELP

MyMathLab® Math XL
PRACTICE WATCH READ REVIEW

✦ Vocabulary and Reading Check

Choose the expression or word from the list below that best completes each statement.

x   negative
y   positive
$x_2 - x_1$ rise
$y_2 - y_1$ run
change in x   undefined
change in y   zero

1. Slope is the rate at which ____ is changing with respect to ____.

2. The slope of a line can be expressed in terms of change as follows: _____.

3. The slope of a line can be expressed using *rise* and *run* as follows: _____.

4. The slope of the line containing (x_1, y_1) and (x_2, y_2) is given by _____.

5. If a line slants up from left to right, the sign of its slope is _____, and if a line slants down from left to right, the sign of its slope is _____.

6. The slope of a horizontal line is _____, and the slope of a vertical line is _____.

✦ Concept Reinforcement

In each of Exercises 7–14, state whether the rate is positive, negative, or zero.

7. The rate at which a teenager's height changes

8. The rate at which an elderly person's height changes

9. The rate at which a pond's water level changes during a drought

10. The rate at which a pond's water level changes during the rainy season

11. The rate at which a person's I.Q. changes during his or her sleep

12. The rate at which the number of people in attendance at a basketball game changes in the moments before the opening tipoff

13. The rate at which the number of people in attendance at a basketball game changes in the moments after the final buzzer sounds

14. The rate at which the number of U.S. Senators changes

Rate and Slope

15. *Blogging.*   Find the rate at which a professional blogger is paid.

Source: Based on information from readwriteweb.com

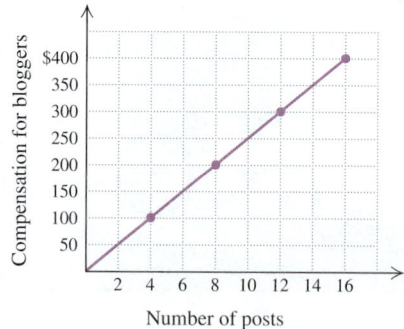

16. *Fitness.*   Find the rate at which a runner burns calories.

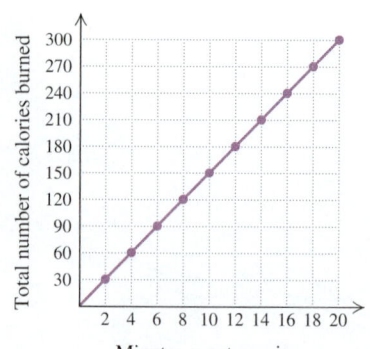

17. *Laptop Computer Prices.* Find the rate of change in the average price of a new laptop computer.

Source: Based on information from PriceGrabber.com Market Reporter Database

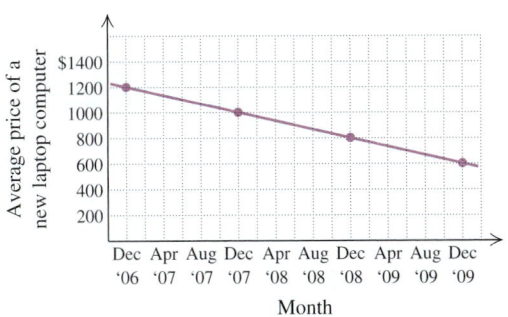

18. *Employment.* Find the rate of change in the number of news reporters and correspondents employed in the United States.

Source: Based on information from the U.S. Department of Labor, Bureau of Labor Statistics

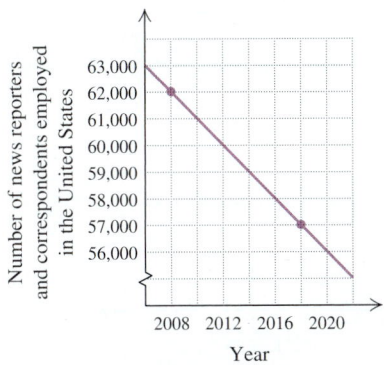

19. *College Admission Tests.* Find the rate of change in SAT critical reading scores with respect to family income.

Source: The College Board, College-Bound Seniors 2011 Total Group Profile Report

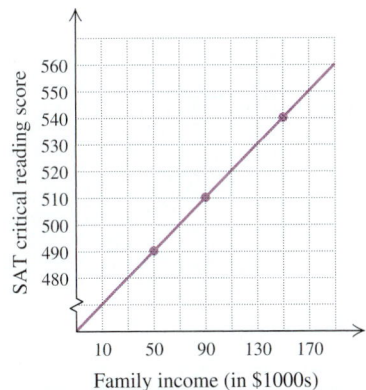

20. *Long-Term Care.* Find the rate of change in the daily cost of a private room in a nursing home.

Source: Based on data from MetLife Market Survey of Nursing Home, Assisted Living, Adult Day Services, and Home Care Costs, 2009–2011

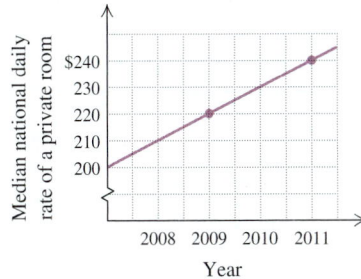

21. *Meteorology.* Find the rate of change in the temperature in Spearfish, Montana, on January 22, 1943, as shown below.

Source: National Oceanic Atmospheric Administration

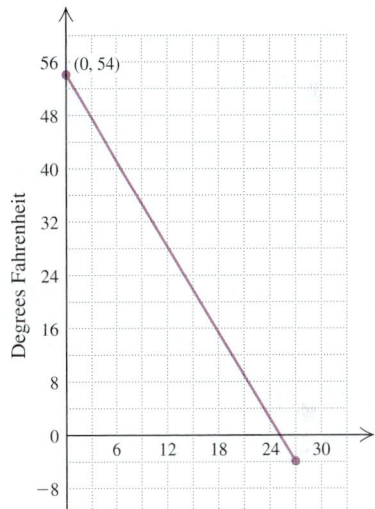

22. *Credit Unions.* Find the rate of change in the number of federal credit unions.

Source: National Credit Union Association

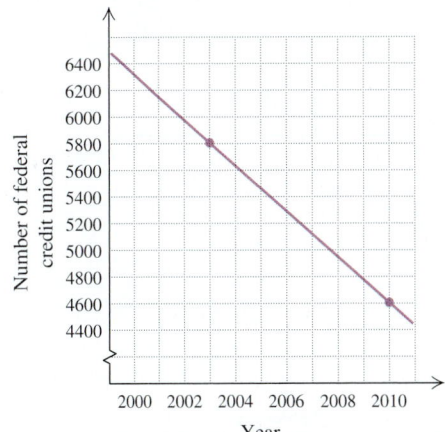

Find the slope, if it is defined, of each line. If the slope is undefined, state this.

23.

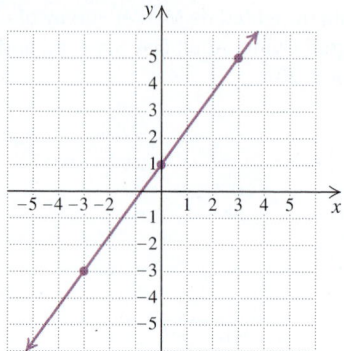

24.

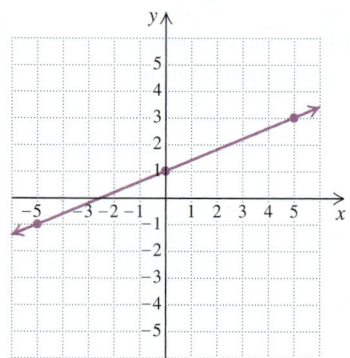

25.

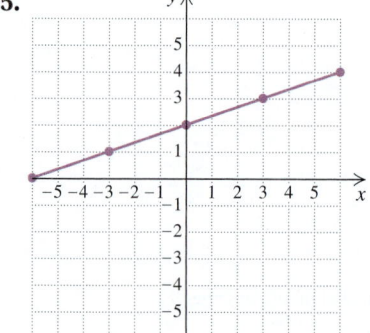

26.

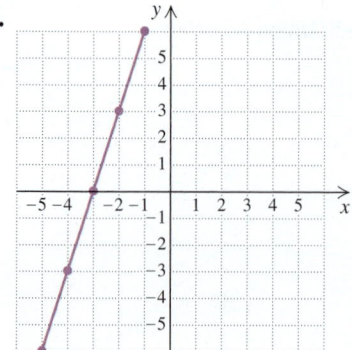

27.

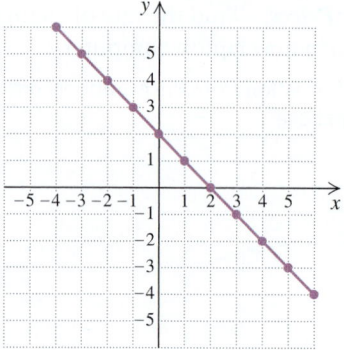

28.

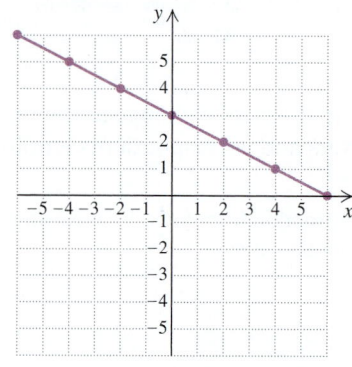

29.

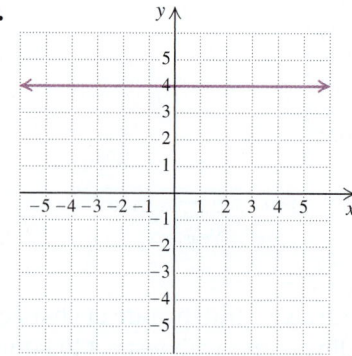

30.

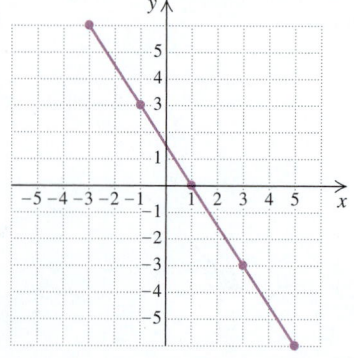

31.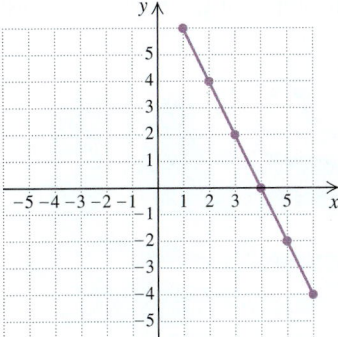

32.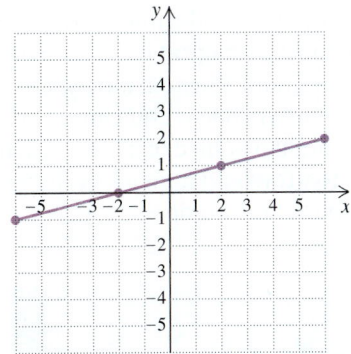

33.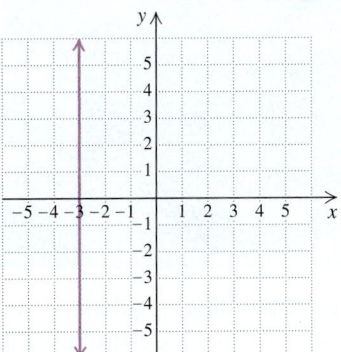

34.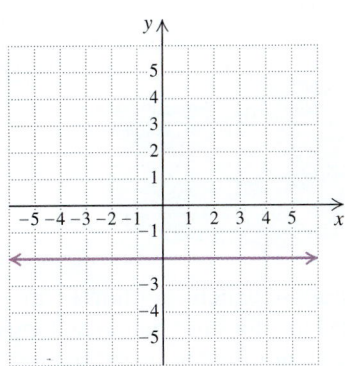

35.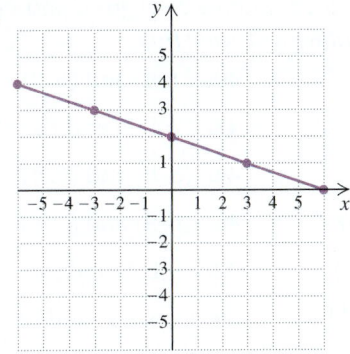

36.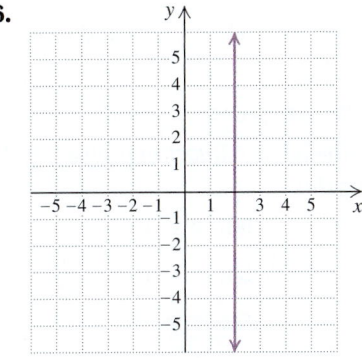

37.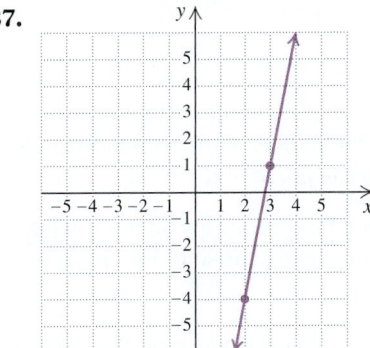

38.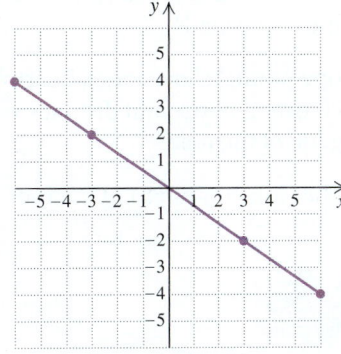

Find the slope of the line containing each given pair of points. If the slope is undefined, state this.

39. $(1, 3)$ and $(5, 8)$

40. $(1, 8)$ and $(6, 9)$

41. $(-2, 4)$ and $(3, 0)$

42. $(-4, 2)$ and $(2, -3)$

43. $(-4, 0)$ and $(5, 6)$

44. $(3, 0)$ and $(6, 9)$

45. $(0, 7)$ and $(-3, 10)$

46. $(0, 9)$ and $(-5, 0)$

47. $(-2, 3)$ and $(-6, 5)$

48. $(-1, 4)$ and $(5, -8)$

Aha! **49.** $\left(-2, \frac{1}{2}\right)$ and $\left(-5, \frac{1}{2}\right)$

50. $(-5, -1)$ and $(2, -3)$

51. $(5, -4)$ and $(2, -7)$

52. $(-10, 3)$ and $(-10, 4)$

53. $(6, -4)$ and $(6, 5)$

54. $(5, -2)$ and $(-4, -2)$

Horizontal Lines and Vertical Lines

Find the slope of each line whose equation is given. If the slope is undefined, state this.

55. $y = 5$

56. $y = 13$

57. $x = -8$

58. $x = 18$

59. $x = 9$

60. $x = -7$

61. $y = -10$

62. $y = -4$

Applications

63. *Surveying.* Lick Skillet Road, near Boulder, Colorado, climbs 230 m over a horizontal distance of 1600 m. What is the grade of the road?

64. *Navigation.* Capital Rapids drops 54 ft vertically over a horizontal distance of 1080 ft. What is the slope of the rapids?

65. *Construction.* Part of New Valley rises 28 ft over a horizontal distance of 80 ft, and is too steep to build on. What is the slope of the land?

66. *Engineering.* At one point, Yellowstone's Beartooth Highway rises 315 ft over a horizontal distance of 4500 ft. Find the grade of the road.

67. *Carpentry.* Find the slope (or pitch) of the roof.

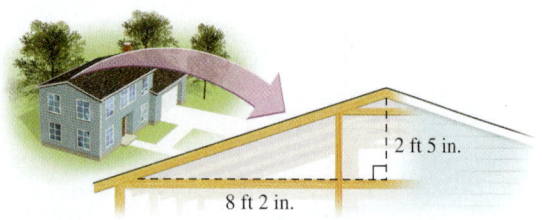

2 ft 5 in.

8 ft 2 in.

68. *Exercise.* Find the slope (or grade) of the treadmill.

0.4 ft

5 ft

69. *Bicycling.* To qualify as a rated climb on the Tour de France, a grade must average at least 4%. The ascent of Dooley Mountain, Oregon, part of the Elkhorn Classic, begins at 3500 ft and climbs to 5400 ft over a horizontal distance of 37,000 ft. What is the grade of the road? Would it qualify as a rated climb if it were part of the Tour de France?

Source: barkercityherald.com

70. *Construction.* Public buildings regularly include steps with 7-in. risers and 11-in. treads. Find the grade of such a stairway.

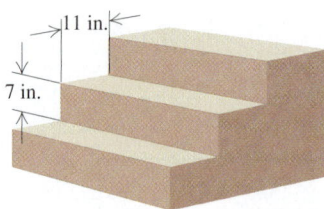

11 in.

7 in.

71. Explain why the order in which coordinates are subtracted to find slope does not matter so long as y-coordinates and x-coordinates are subtracted in the same order.

72. If one line has a slope of -3 and another has a slope of 2, which line is steeper? Why?

Skill Review

73. Multiply: $3(4 + a)$.   [1.2]

74. Factor: $14 + 35x$.   [1.2]

75. Find $-x$ when x is -15.   [1.6]

76. Write another inequality with the same meaning as $x \geq 3$.   [1.4]

77. Write exponential notation for $5t \cdot 5t \cdot 5t$.   [1.8]

78. Simplify: $(-2y)^4$.   [1.8]

Synthesis

79. The points $(-4, -3)$, $(1, 4)$, $(4, 2)$, and $(-1, -5)$ are vertices of a quadrilateral. Use slopes to explain why the quadrilateral is a parallelogram.

80. Which is steeper and why: a ski slope that is $50°$ or one with a grade of 100%?

81. The plans below are for a skateboard "Fun Box." For the ramps labeled A, find the slope or grade.

Source: www.heckler.com

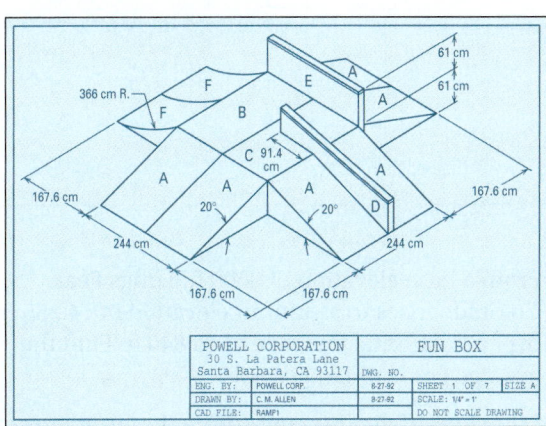

82. A line passes through $(4, -7)$ and never enters the first quadrant. What numbers could the line have for its slope?

83. A line passes through $(2, 5)$ and never enters the second quadrant. What numbers could the line have for its slope?

84. *Architecture.* Architects often use the equation $x + y = 18$ to determine the height y, in inches, of the riser of a step when the tread is x inches wide. Express the slope of stairs designed with this equation without using the variable y.

In Exercises 85 and 86, the slope of the line is $-\frac{2}{3}$, but the numbering on one axis is missing. How many units should each tick mark on that unnumbered axis represent?

85.

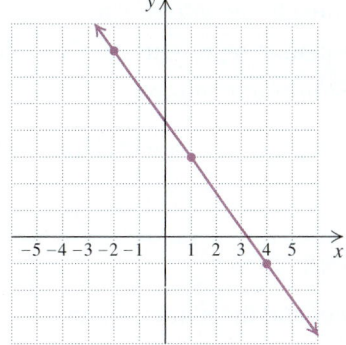

86.

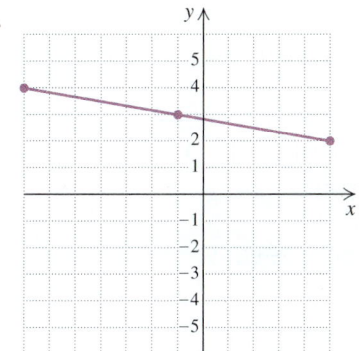

↘ YOUR TURN ANSWERS: SECTION 3.5

1. $\frac{4}{3}$ thousand dollars per year, or $1333.33 per year   **2.** $\frac{7}{8}$
3. 0 **4.** Undefined   **5.** 6.25%

QUICK QUIZ: SECTIONS 3.1– 3.5

Graph.

1. $x = -2$ [3.3]   **2.** $3x - y = 3$ [3.3]

3. $y = 2x - 4$ [3.2]

4. Find the slope of the line containing $(9, -4)$ and $(10, -5)$.   [3.5]

5. Find the slope of the line given by $y = 21$. If the slope is undefined, state this.   [3.5]

PREPARE TO MOVE ON

Solve for y.   [2.3]

1. $2x + 3y = 7$ **2.** $3x - 4y = 8$

3. $ax + by = c$ **4.** $ax - by = c$

Mid-Chapter Review

We can plot points and graph equations on a *Cartesian coordinate plane*.

- A point is represented by an *ordered pair*.
- The graph of an equation represents all of its solutions.
- Equations can be *linear* or *nonlinear*.
- The *slope* of a line represents a rate of change: Slope $= m = \dfrac{\text{change in } y}{\text{change in } x} = \dfrac{y_2 - y_1}{x_2 - x_1}$.

GUIDED SOLUTIONS

1. Find the y-intercept and the x-intercept of the graph of $y - 3x = 6$.   [3.3]

Solution

y-intercept: $y - 3 \cdot \boxed{} = 6$

$\phantom{y\text{-intercept: } y - 3 \cdot } y = \boxed{}$

The y-intercept is $(\boxed{}, \boxed{})$.

x-intercept: $\boxed{} - 3x = 6$

$\phantom{x\text{-intercept: }} -3x = 6$

$\phantom{x\text{-intercept: } -3} x = \boxed{}$

The x-intercept is $(\boxed{}, \boxed{})$.

2. Find the slope of the line containing the points $(1, 5)$ and $(3, -1)$.   [3.5]

Solution

$$m = \frac{y_2 - y_1}{x_2 - x_1} = \frac{-1 - \boxed{}}{3 - \boxed{}}$$

$$= \frac{\boxed{}}{2}$$

$$= \boxed{}$$

MIXED REVIEW

3. Plot the point $(0, -3)$.   [3.1]

4. In which quadrant is the point $(4, -15)$ located? [3.1]

5. Determine whether the ordered pair $(-2, -3)$ is a solution of the equation $y = 5 - x$.   [3.2]

Graph by hand.

6. $y = x - 3$ [3.2]

7. $y = -3x$ [3.2]

8. $3x - y = 2$ [3.2]

9. $4x - 5y = 20$ [3.3]

10. $y = -2$ [3.3]

11. $x = 1$ [3.3]

12. By the end of June, Construction Builders had winterized 10 homes. By the end of August, they had winterized a total of 38 homes. Find the rate at which the company was winterizing homes.   [3.4]

13. From a base elevation of 9600 ft, Longs Peak, Colorado, rises to a summit elevation of 14,255 ft over a horizontal distance of 15,840 ft. Find the average grade of Longs Peak.   [3.5]

Find the slope of the line containing the given pair of points. If the slope is undefined, state this.   [3.5]

14. $(-5, -2)$ and $(1, 8)$

15. $(1, 2)$ and $(4, -7)$

16. $(0, 0)$ and $(0, -2)$

17. $(6, -3)$ and $(2, -3)$

18. What is the slope of the line $y = 4$?   [3.5]

19. What is the slope of the line $x = -7$?   [3.5]

20. Find the x-intercept and the y-intercept of the line given by $2y - 3x = 12$.   [3.3]

3.6 Slope–Intercept Form

Using the *y*-intercept and the Slope to Graph a Line ▪ Equations in Slope–Intercept Form ▪
Graphing and Slope–Intercept Form ▪ Parallel Lines and Perpendicular Lines

If we know the slope and the *y*-intercept of a line, it is possible to graph the line. In this section, we will discover that a line's slope and its *y*-intercept can be determined directly from the line's equation, provided the equation is written in a certain form.

Using the *y*-intercept and the Slope to Graph a Line

Last year, Gary spent $4000 upgrading the local area network (LAN) for his company's computer system. Now he plans to spend $3000 every two years to update his employees' laptop computers. We can make a table and draw a graph showing his costs.

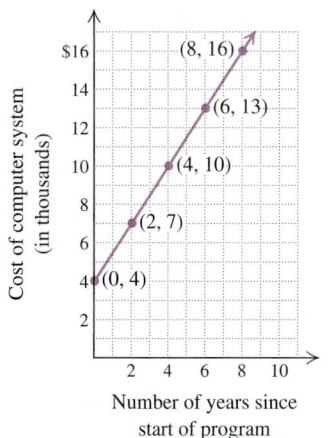

Years Since Start of Program	Cost of Computers (in thousands)
0	$ 4
2	7
4	10
6	13
8	16

The slope of the line is the rate of change in computer cost per year;

$$\text{Slope} = \frac{\text{change in } y}{\text{change in } x} = \frac{\text{rise}}{\text{run}} = \frac{y_2 - y_1}{x_2 - x_1},$$

where (x_1, y_1) and (x_2, y_2) are any two points on the graphed line. Here we select $(0, 4)$ and $(2, 7)$:

$$\text{Slope} = \frac{\text{change in } y}{\text{change in } x} = \frac{7 - 4}{2 - 0} = \frac{3}{2}.$$ The rate of change is $\frac{3}{2}$ thousand dollars per year, or $1500/year

Knowing that the slope is $\frac{3}{2}$, we could have drawn the graph by plotting $(0, 4)$ and from there moving 3 units *up* and 2 units *to the right*. This would have located the point $(2, 7)$. Using $(0, 4)$ and $(2, 7)$, we can then draw the line. This is the method used in the next example.

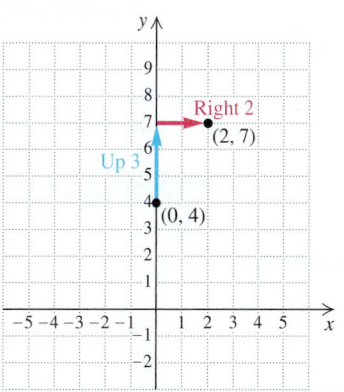

EXAMPLE 1 Draw a line that has slope $\frac{1}{4}$ and y-intercept $(0, 2)$.

SOLUTION We plot $(0, 2)$ and from there move 1 unit *up* and 4 units *to the right*. This locates the point $(4, 3)$. We plot $(4, 3)$ and draw a line passing through $(0, 2)$ and $(4, 3)$, as shown on the right below.

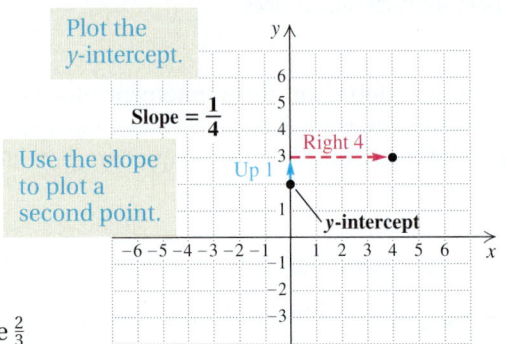

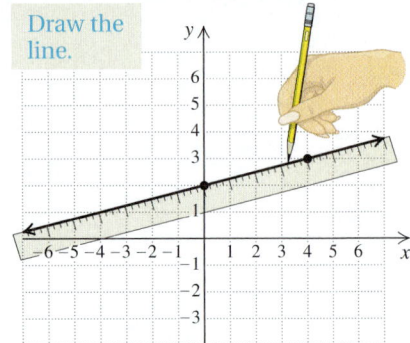

1. Draw a line that has slope $\frac{2}{3}$ and y-intercept $(0, -1)$.

YOUR TURN

Equations in Slope–Intercept Form

It is possible to read the slope and the y-intercept of a line directly from its equation. To find the y-intercept of an equation's graph, we replace x with 0 and solve the resulting equation for y. Compare the following.

$$y = 2x + 3 \qquad\qquad y = mx + b$$
$$= 2 \cdot 0 + 3 = 0 + 3 = 3 \qquad = m \cdot 0 + b = 0 + b = b$$

The y-intercept of the graph of $y = 2x + 3$ is $(0, 3)$.

The y-intercept of the graph of $y = mx + b$ is $(0, b)$.

To calculate the slope of the graph of $y = 2x + 3$, we need two ordered pairs that are solutions of the equation. The y-intercept $(0, 3)$ is one pair; a second pair, $(1, 5)$, can be found by substituting 1 for x. We then have

$$\text{Slope} = \frac{\text{change in } y}{\text{change in } x} = \frac{5 - 3}{1 - 0} = \frac{2}{1} = 2.$$

Note that the slope, 2, is also the x-coefficient in $y = 2x + 3$. It can be similarly shown that the graph of any equation of the form $y = mx + b$ has slope m (see Exercise 83).

> ### THE SLOPE–INTERCEPT EQUATION
>
> The equation $y = mx + b$ is called the *slope–intercept equation*. The graph of $y = mx + b$ has slope m and y-intercept $(0, b)$.

The equation of any nonvertical line can be written in this form. The use of the letter m for slope is derived from the French verb *monter*, to climb.

EXAMPLE 2 Find the slope and the y-intercept of each line whose equation is given.

a) $y = \frac{4}{5}x - 8$ **b)** $2x + y = 5$ **c)** $3x - 4y = 7$

Student Notes

To write an equation "in the form $y = mx + b$" means to solve the equation for y, so that "$y = $" is followed by the term containing x and then the constant term. It may help to write your equation below the general form and align the y, the $=$, and the x. For Example 2(b), you could write

$$y = \quad mx + b$$
$$y = -2x + 5.$$

From this form, you can see that $m = -2$ and $b = 5$.

SOLUTION

a) We rewrite $y = \frac{4}{5}x - 8$ as $y = \frac{4}{5}x + (-8)$. Now we simply read the slope and the y-intercept from the equation:

$$y = \frac{4}{5}x + (-8).$$

The slope is $\frac{4}{5}$. The y-intercept is $(0, -8)$.

b) We first solve for y to find an equivalent equation in the form $y = mx + b$:

$$2x + y = 5$$
$$y = -2x + 5. \qquad \text{Adding } -2x \text{ to both sides}$$

The slope is -2. The y-intercept is $(0, 5)$.

c) We rewrite the equation in the form $y = mx + b$:

$$3x - 4y = 7$$
$$-4y = -3x + 7 \qquad \text{Adding } -3x \text{ to both sides}$$
$$y = -\tfrac{1}{4}(-3x + 7) \qquad \text{Multiplying both sides by } -\tfrac{1}{4}$$
$$y = \tfrac{3}{4}x - \tfrac{7}{4}. \qquad \text{Using the distributive law}$$

The slope is $\frac{3}{4}$. The y-intercept is $\left(0, -\frac{7}{4}\right)$.

2. Find the slope and the y-intercept of the line given by $y = -\frac{1}{3}x - 7$.

YOUR TURN

EXAMPLE 3 A line has slope $-\frac{12}{5}$ and y-intercept $(0, 11)$. Find an equation of the line.

SOLUTION We use the slope–intercept equation, substituting $-\frac{12}{5}$ for m and 11 for b:

$$y = mx + b = -\tfrac{12}{5}x + 11.$$

The desired equation is $y = -\frac{12}{5}x + 11$.

3. A line has slope 4 and y-intercept $(0, -1)$. Find an equation of the line.

YOUR TURN

EXAMPLE 4 *Fast-Food Menus.* The following graph shows the number of items on a McDonald's® lunch menu for various years. Determine an equation for the graph.

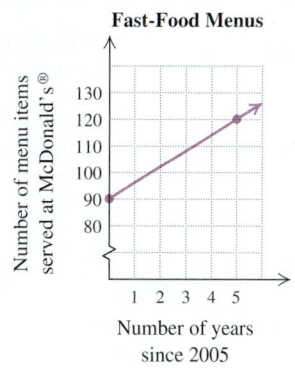

Fast-Food Menus

Number of menu items served at McDonald's®

Number of years since 2005

Source: Based on information from Technomic MenuMonitor

4. Determine an equation for the following graph.

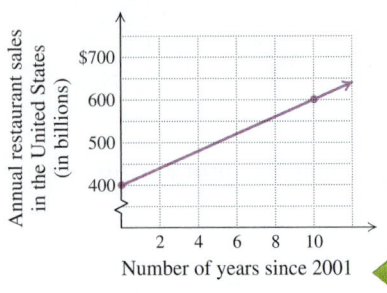

Source: Based on information from the National Restaurant Association

SOLUTION If we know the y-intercept and the slope, we can use slope–intercept form to write an equation for the line. From the graph, we see that $(0, 90)$ is the y-intercept and that the line passes through $(5, 120)$. We calculate the slope:

$$m = \frac{\text{change in } y}{\text{change in } x} = \frac{120 - 90}{5 - 0} = \frac{30}{5} = 6.$$

The desired equation is

$$y = 6x + 90, \qquad \text{\color{red}Using 6 for } m \text{ and 90 for } b$$

where y is the number of menu items x years after 2005.

YOUR TURN

Graphing and Slope–Intercept Form

In Example 1, we drew a graph, knowing only the slope and the y-intercept. In Example 2, we determined the slope and the y-intercept of a line by examining its equation. We now combine the two procedures to develop a quick way to graph a linear equation.

EXAMPLE 5 Graph: **(a)** $y = \frac{3}{4}x + 5$; **(b)** $2x + 3y = 3$.

SOLUTION To graph each equation, we plot the y-intercept and find additional points using the slope.

> Determine the slope and the y-intercept.

a) We can read the slope and the y-intercept from the equation $y = \frac{3}{4}x + 5$:

$$\text{Slope: } \tfrac{3}{4}; \qquad y\text{-intercept: } (0, 5).$$

> Plot the y-intercept.

We plot the y-intercept $(0, 5)$. This gives us one point on the line.
Starting at $(0, 5)$, we use the slope $\frac{3}{4}$ to find another point.

> Use the slope to find a second point.

We move 3 units *up* since the numerator (change in y) is *positive*.

We move 4 units *to the right* since the denominator (change in x) is *positive*.

This gives us a second point on the line, $(4, 8)$.
We can find a third point on the line by rewriting the slope $\frac{3}{4}$ as $\frac{-3}{-4}$, since these fractions are equivalent. Now, starting again at $(0, 5)$, we use the slope $\frac{-3}{-4}$ to find another point.

We move 3 units *down* since the numerator (change in y) is *negative*.

> Use the slope to find a third point.

We move 4 units *to the left* since the denominator (change in x) is *negative*.

This gives us a third point on the line, $(-4, 2)$.
Finally, we draw the line.

> Draw the line.

Student Notes

Recall the following:

$$\frac{3}{4} = \frac{-3}{-4};$$

$$-\frac{3}{4} = \frac{-3}{4} = \frac{3}{-4};$$

$$2 = \frac{2}{1} \quad \text{and} \quad -2 = \frac{-2}{1}.$$

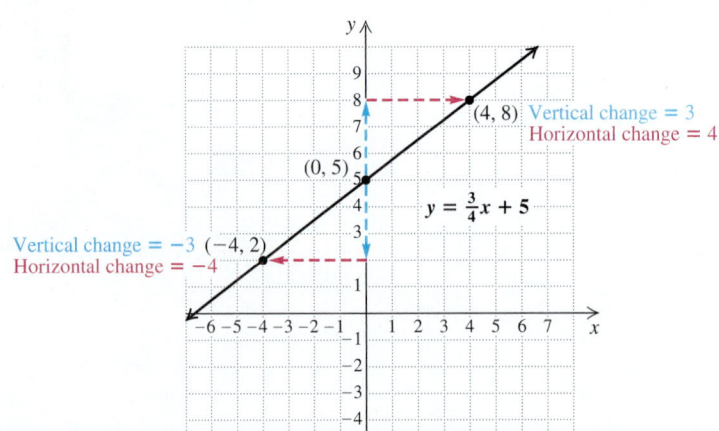

b) To graph $2x + 3y = 3$, we first rewrite it to find the slope and the y-intercept:

$$2x + 3y = 3$$
$$3y = -2x + 3 \qquad \text{Adding } -2x \text{ to both sides}$$
$$y = \tfrac{1}{3}(-2x + 3) \qquad \text{Multiplying both sides by } \tfrac{1}{3}$$
$$y = -\tfrac{2}{3}x + 1. \qquad \text{Using the distributive law}$$

We plot the y-intercept, $(0, 1)$.

The slope is $-\tfrac{2}{3}$. For graphing, we think of this slope as $\tfrac{-2}{3}$ or $\tfrac{2}{-3}$. Starting at $(0, 1)$, we use the slope $\tfrac{-2}{3}$ to find a second point.

We move 2 units *down* since the numerator is *negative*.

We move 3 units *to the right* since the denominator is *positive*.

We plot the new point, $(3, -1)$.

Now, starting at $(3, -1)$ and again using the slope $\tfrac{-2}{3}$, we move to a third point, $(6, -3)$.

Alternatively, we can start at $(0, 1)$ and use the slope $\tfrac{2}{-3}$.

We move 2 units *up* since the numerator is *positive*.

We move 3 units *to the left* since the denominator is *negative*.

This leads to another point on the graph, $(-3, 3)$.

Student Notes

The signs of the numerator and the denominator of the slope indicate whether to move up, down, left, or right. Compare the following slopes.

$\dfrac{1}{2}$ ← 1 unit up
   ← 2 units right

$\dfrac{-1}{-2}$ ← 1 unit down
   ← 2 units left

$\dfrac{-1}{2}$ ← 1 unit down
   ← 2 units right

$\dfrac{1}{-2}$ ← 1 unit up
   ← 2 units left

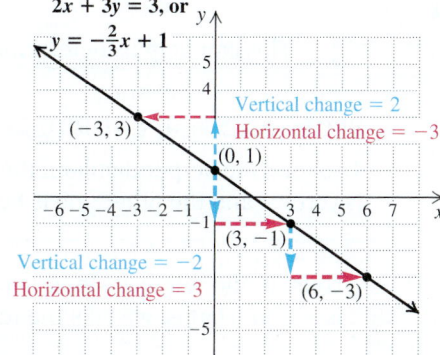

It is important to be able to use both $\tfrac{2}{-3}$ and $\tfrac{-2}{3}$ to draw the graph.

5. Graph: $y = -\tfrac{1}{2}x - 2.$

YOUR TURN

Parallel Lines and Perpendicular Lines

Two lines are parallel if they lie in the same plane and do not intersect no matter how far they are extended. If two lines are vertical, they are parallel. How can we tell if nonvertical lines are parallel? The answer is simple: We look at their slopes.

> **SLOPE AND PARALLEL LINES**
>
> Two lines are parallel if they have the same slope or if both lines are vertical.

EXAMPLE 6   Determine whether the line given by $y = -3x + 4.2$ is parallel to the line given by $6x + 2y = 1$.

SOLUTION   If the slopes of the lines are the same, the lines are parallel.

The slope of $f(x) = -3x + 4.2$ is -3.

To find the slope of $6x + 2y = 1$, we write the equation in slope–intercept form:

$$6x + 2y = 1$$
$$2y = -6x + 1 \qquad \text{Subtracting } 6x \text{ from both sides}$$
$$y = -3x + \tfrac{1}{2}. \qquad \text{Dividing both sides by 2}$$

6. Determine whether the line given by $8x + y = 2$ is parallel to the line given by $y = 8x + 7$.

The slope of the second line is -3. Since the slopes are equal, the lines are parallel.

YOUR TURN

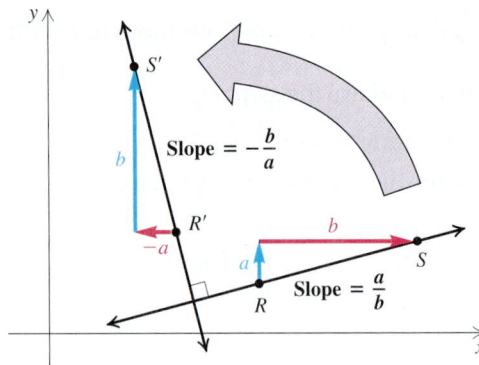

Two lines are perpendicular if they intersect at a right angle. If one line is vertical and another is horizontal, they are perpendicular. There are other instances in which two lines are perpendicular.

Consider a line $\overleftrightarrow{RS}$ as shown at left, with slope a/b. Then think of rotating the figure 90° to get a line $\overleftrightarrow{R'S'}$ perpendicular to $\overleftrightarrow{RS}$. For the new line, the rise and the run are interchanged, but the run is now negative. Thus the slope of the new line is $-b/a$. Let's multiply the slopes:

$$\frac{a}{b}\left(-\frac{b}{a}\right) = -1.$$

This can help us determine which lines are perpendicular.

> **SLOPE AND PERPENDICULAR LINES**
>
> Two lines are perpendicular if the product of their slopes is -1 or if one line is vertical and the other line is horizontal.

Thus, if one line has slope $m\ (m \neq 0)$, the slope of any line perpendicular to it is $-1/m$. That is, we take the reciprocal of $m\ (m \neq 0)$ and change the sign.

EXAMPLE 7   Determine whether the graphs of $2x + y = 8$ and $y = \tfrac{1}{2}x + 7$ are perpendicular.

SOLUTION   The second equation is given in slope–intercept form:

$$y = \tfrac{1}{2}x + 7. \qquad \text{The slope is } \tfrac{1}{2}.$$

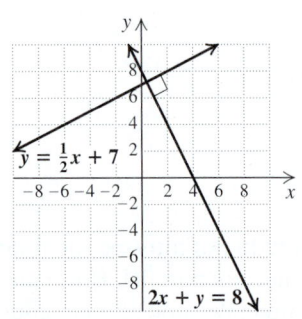

To find the slope of the other line, we solve for y:

$$2x + y = 8$$
$$y = -2x + 8. \qquad \text{Adding } -2x \text{ to both sides}$$
$$\qquad\qquad\qquad \text{The slope is } -2.$$

The lines are perpendicular if the product of their slopes is -1. Since

$$\tfrac{1}{2}(-2) = -1,$$

the graphs are perpendicular. The graphs of both equations are shown at left, and they do appear to be perpendicular.

7. Determine whether the graphs of $x + y = 3$ and $x - y = 8$ are perpendicular.

YOUR TURN

3.6 EXERCISE SET

FOR EXTRA HELP   MyMathLab® Math XL
PRACTICE   WATCH

Vocabulary and Reading Check

Classify each of the following as either "slope" or "y-intercept."

1. m for $y = mx + b$

2. $(0, b)$ for $y = mx + b$

3. The point at which a graph crosses the y-axis

4. A value that is the same for parallel lines

5. $(0, 5)$ for $y = x + 5$

6. 1 for $y = x + 5$

Concept Reinforcement

In each of Exercises 7–12, match the phrase with the most appropriate choice from the following list.

a) $\left(0, \frac{3}{4}\right)$ b) 2

c) $(0, -3)$ d) $\frac{2}{3}$

e) $(0, -2)$ f) 3

7. ____ The slope of the graph of $y = 3x - 2$

8. ____ The slope of the graph of $y = 2x - 3$

9. ____ The slope of the graph of $y = \frac{2}{3}x + 3$

10. ____ The y-intercept of the graph of $y = 2x - 3$

11. ____ The y-intercept of the graph of $y = 3x - 2$

12. ____ The y-intercept of the graph of $y = \frac{2}{3}x + \frac{3}{4}$

Using the y-intercept and the Slope to Graph a Line

Draw a line that has the given slope and y-intercept.

13. Slope $\frac{2}{3}$; y-intercept $(0, 1)$

14. Slope $\frac{3}{5}$; y-intercept $(0, -1)$

15. Slope $\frac{5}{3}$; y-intercept $(0, -2)$

16. Slope $\frac{1}{2}$; y-intercept $(0, 0)$

17. Slope $-\frac{1}{3}$; y-intercept $(0, 5)$

18. Slope $-\frac{4}{5}$; y-intercept $(0, 6)$

19. Slope 2; y-intercept $(0, 0)$

20. Slope -2; y-intercept $(0, -3)$

21. Slope -3; y-intercept $(0, 2)$

22. Slope 3; y-intercept $(0, 4)$

Aha! 23. Slope 0; y-intercept $(0, -5)$

24. Slope 0; y-intercept $(0, 1)$

Equations in Slope–Intercept Form

Find the slope and the y-intercept of each line from the given equation.

25. $y = -\frac{2}{7}x + 5$ 26. $y = -\frac{3}{8}x + 4$

27. $y = \frac{1}{3}x + 7$ 28. $y = \frac{4}{5}x + 1$

29. $y = \frac{9}{5}x - 4$ 30. $y = -\frac{9}{10}x - 5$

31. $-3x + y = 7$ 32. $-4x + y = 7$

33. $4x + 2y = 8$ 34. $3x + 4y = 12$

Aha! 35. $y = 3$ 36. $y - 3 = 5$

37. $2x - 5y = -8$ 38. $12x - 6y = 9$

39. $9x - 8y = 0$ 40. $7x = 5y$

Find the slope–intercept equation of the line with the indicated slope and y-intercept.

41. Slope 5; y-intercept $(0, 7)$

42. Slope -4; y-intercept $\left(0, -\frac{3}{5}\right)$

43. Slope $\frac{7}{8}$; y-intercept $(0, -1)$

44. Slope $\frac{5}{7}$; y-intercept $(0, 4)$

45. Slope $-\frac{5}{3}$; y-intercept $(0, -8)$

46. Slope $\frac{3}{4}$; y-intercept $(0, -35)$

Aha! 47. Slope 0; y-intercept $\left(0, \frac{1}{3}\right)$

48. Slope 7; y-intercept $(0, 0)$

Determine an equation for each graph shown.

49.

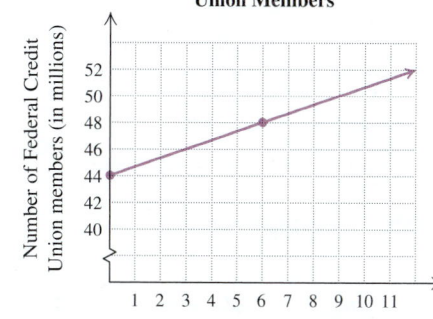

Federal Credit Union Members

Number of Federal Credit Union members (in millions)

Number of years since 2000

Source: Based on information from the National Credit Union Association

50.

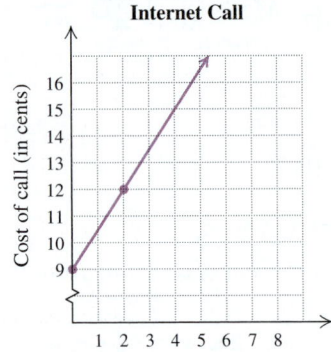

Cost of an Overseas Internet Call

51.

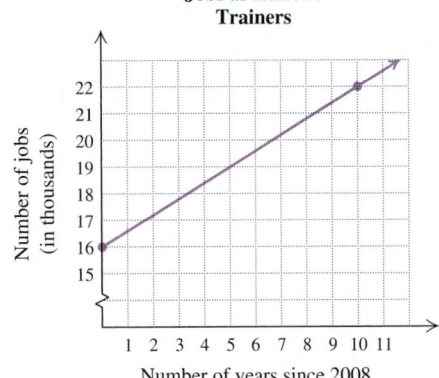

Jobs as Athletic Trainers

Source: Based on data from the U.S. Bureau of Labor Statistics

52.

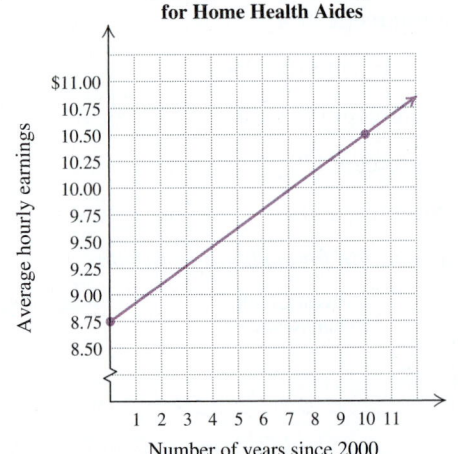

Average Hourly Earnings for Home Health Aides

Source: Based on data from the U.S. Bureau of Labor Statistics

Graphing and Slope–Intercept Form

Graph.

53. $y = \frac{2}{3}x + 2$

54. $y = -\frac{2}{3}x - 3$

55. $y = -\frac{2}{3}x + 3$

56. $y = \frac{2}{3}x - 2$

57. $y = \frac{3}{2}x + 3$

58. $y = \frac{3}{2}x - 2$

59. $y = -\frac{4}{3}x + 3$

60. $y = -\frac{3}{2}x - 2$

61. $2x + y = 1$

62. $3x + y = 2$

63. $3x + y = 0$

64. $2x + y = 0$

65. $4x + 5y = 15$

66. $2x + 3y = 9$

67. $x - 4y = 12$

68. $x + 5y = 20$

Parallel Lines and Perpendicular Lines

Without graphing, tell whether the graphs of each pair of equations are parallel.

69. $x + 2 = y$,
$y - x = -2$

70. $2x - 1 = y$,
$2y - 4x = 7$

71. $y + 9 = 3x$,
$3x - y = -2$

72. $y + 8 = -6x$,
$-2x + y = 5$

73. $y = 3x + 9$,
$2y = 8x - 2$

74. $y = -7x - 9$,
$-3y = 21x + 7$

Without graphing, tell whether the graphs of each pair of equations are perpendicular.

75. $x - 2y = 3$,
$4x + 2y = 1$

76. $2x - 5y = -3$,
$2x + 5y = 4$

77. $y = 3x + 1$,
$6x + 2y = 5$

78. $y = -x + 7$,
$y = x + 3$

79. Can a horizontal line be graphed using the method of Example 5? Why or why not?

80. Can a vertical line be graphed using the method of Example 5? Why or why not?

Skill Review

Solve.

81. Mr. and Mrs. Sturgis left a $3 tip for a meal that cost $25. What percent of the cost of the meal was the tip? [2.4]

82. Irniq is writing a 1000-word essay. He has three times as many words to write as he has already written. How many words has he written? [2.5]

83. In December, the balance in Dalila's college savings account grew 15%, to $2760. What was her balance at the beginning of the month? [2.5]

84. The perimeter of a hospital's rectangular flower garden is 140 ft. The width is 30 ft less than the length. Find the width and the length. [2.5]

Synthesis

85. Explain how it is possible for an incorrect graph to be drawn, even after plotting three points that line up.

86. Which would you prefer, and why: graphing an equation of the form $y = mx + b$ or graphing an equation of the form $Ax + By = C$?

87. Show that the slope of the line given by $y = mx + b$ is m. (*Hint*: Substitute both 0 and 1 for x to find two pairs of coordinates. Then use the formula, Slope = change in y/change in x.)

88. Write an equation of the line with the same slope as the line given by $5x + 2y = 8$ and the same y-intercept as the line given by $3x - 7y = 10$.

89. Write an equation of the line parallel to the line given by $-4x + 8y = 5$ and having the same y-intercept as the line given by $4x - 3y = 0$.

90. Write an equation of the line parallel to the line given by $3x - 2y = 8$ and having the same y-intercept as the line given by $2y + 3x = -4$.

91. Write an equation of the line parallel to the line given by $4x + 5y = 9$ and having the same y-intercept as the line given by $2x + 3y = 12$.

92. Write an equation of the line perpendicular to the line given by $2x + 3y = 7$ and having the same y-intercept as the line given by $5x + 2y = 10$.

93. Write an equation of the line perpendicular to the line given by $3x - 5y = 8$ and having the same y-intercept as the line given by $2x + 4y = 12$.

94. Write an equation of the line perpendicular to the line given by $3x - 2y = 9$ and having the same y-intercept as the line given by $2x + 5y = 0$.

95. Write an equation of the line perpendicular to the line given by $2x + 5y = 6$ that passes through $(2, 6)$. (*Hint*: Draw a graph.)

96. Graph $y_1 = -\frac{3}{4}x - 2$, $y_2 = -\frac{1}{5}x - 2$, $y_3 = -\frac{3}{4}x - 5$, and $y_4 = -\frac{1}{5}x - 5$ using the SIMULTANEOUS mode. Then match each line with the corresponding equation. Check using TRACE.

YOUR TURN ANSWERS: SECTION 3.6

1.
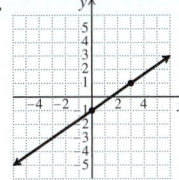

2. Slope: $-\frac{1}{3}$; y-intercept: $(0, -7)$

3. $y = 4x - 1$ **4.** $y = 20x + 400$, where y is the amount of U.S. restaurant sales, in billions of dollars, and x is the number of years since 2001

5.

$y = -\frac{1}{2}x - 2$

6. No   **7.** Yes

QUICK QUIZ: SECTIONS 3.1–3.6

Graph.

1. $y = 3$ [3.3]   **2.** $y = \frac{1}{2}x - 4$ [3.6]

3. In which quadrant or on which axis is $(-12, -0.02)$ located?   [3.1]

4. List the coordinates of the x- and y-intercepts of the graph of $y - x = 7$.   [3.3]

5. Find the slope of the graph of $y - x = 7$.   [3.6]

PREPARE TO MOVE ON

Solve.   [2.3]

1. $y - k = m(x - h)$, for y

2. $y - 9 = -2(x + 4)$, for y

Simplify.   [1.6]

3. $-10 - (-3)$ **4.** $8 - (-5)$

5. $-4 - 5$

3.7   Point–Slope Form and Equations of Lines

Point–Slope Form ◾ Finding the Equation of a Line ◾ Estimations and Predictions Using Two Points

If we know the slope of a line and a point through which the line passes, then we can draw the line. With this information, we can also write an equation of the line.

Point–Slope Form

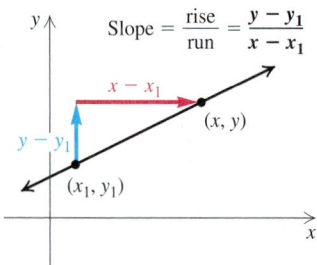

Suppose that a line of slope m passes through the point (x_1, y_1). For any other point (x, y) to lie on this line, we must have

$$\frac{y - y_1}{x - x_1} = m.$$

Note that if (x_1, y_1) itself replaces (x, y), the denominator is 0. To address this concern, we multiply both sides by $x - x_1$:

$$(x - x_1)\frac{y - y_1}{x - x_1} = m(x - x_1)$$

$$y - y_1 = m(x - x_1).$$ This equation *is* true for $(x, y) = (x_1, y_1)$.

Every point on the line is a solution of this equation. This is the **point–slope form** of a linear equation.

> **POINT–SLOPE FORM**
>
> Any equation of the form $y - y_1 = m(x - x_1)$ is said to be written in *point–slope* form and has a graph that is a straight line.
>
> The slope of the line is m.
>
> The line passes through (x_1, y_1).

EXAMPLE 1   Graph: $(y + 4) = -\frac{1}{2}(x - 3)$.

SOLUTION   We first write the equation in point–slope form:

$$y - y_1 = m(x - x_1)$$
$$y - (-4) = -\frac{1}{2}(x - 3).$$ $y + 4 = y - (-4)$

From the equation, we see that $m = -\frac{1}{2}$, $x_1 = 3$, and $y_1 = -4$.
We plot $(3, -4)$, count off a slope of $-\frac{1}{2}$, and draw the line.

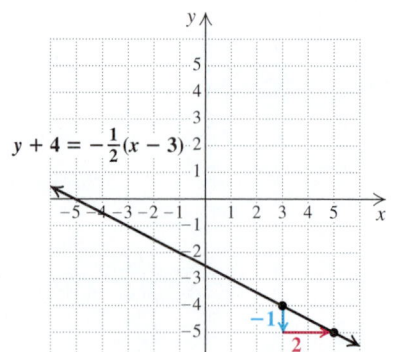

1. Graph: $y - 1 = 3(x + 2)$.

 YOUR TURN

We can use point–slope form to find an equation of the line.

EXAMPLE 2   Use point–slope form to find an equation of the line with slope 3 that passes through $(-7, 8)$.

SOLUTION   We substitute into the point–slope form:

$$y - y_1 = m(x - x_1)$$
$$y - 8 = 3(x - (-7)).$$ Substituting 3 for m, -7 for x_1, and 8 for y_1

This is an equation of the line. If desired, we can solve for y to write it in slope–intercept form.

YOUR TURN

2. Use point–slope form to find an equation of the line with slope -5 that passes through $(1, -6)$.

The point–slope form can be used to find the equation of any line given the slope and a point. Other forms of linear equations can also be used and may be more convenient in some situations.

Finding the Equation of a Line

GIVEN THE SLOPE AND THE y-INTERCEPT

If we know the slope m and the y-intercept $(0, b)$ of a line, we can find an equation of the line by substituting into slope–intercept form, $y = mx + b$.

EXAMPLE 3   Find an equation for the line parallel to $8y = 7x - 24$ with y-intercept $(0, -6)$.

SOLUTION   We first find slope–intercept form of the given line:

$$8y = 7x - 24$$
$$y = \tfrac{7}{8}x - 3.$$ Multiplying both sides by $\tfrac{1}{8}$
The slope is $\tfrac{7}{8}$.

The slope of any line parallel to the line given by $8y = 7x - 24$ is $\tfrac{7}{8}$. For a y-intercept of $(0, -6)$, we must have

$$y = mx + b$$
$$y = \tfrac{7}{8}x - 6.$$ Substituting $\tfrac{7}{8}$ for m and -6 for b

YOUR TURN

3. Find an equation for the line parallel to $3y = 3x + 12$ with y-intercept $(0, 5)$.

Study Skills

Understand Your Mistakes
When your instructor returns a graded quiz, test, or assignment, it is important that you review and understand what your mistakes were. Take advantage of the opportunity to learn from your mistakes.

GIVEN THE SLOPE AND A POINT OR GIVEN TWO POINTS

When we know the slope m of a line and any point on the line, we can find the equation of the line either by using slope–intercept form, $y = mx + b$, and solving for b or by substituting directly into point–slope form, $y - y_1 = m(x - x_1)$.

EXAMPLE 4   Find an equation for the line perpendicular to $2x + y = 5$ that passes through $(1, -3)$.

SOLUTION   We first find slope–intercept form of the given line:

$$2x + y = 5$$
$$y = -2x + 5.$$ Subtracting $2x$ from both sides
The slope is -2.

4. Find an equation in point–slope form for the line perpendicular to $3x - 4y = 7$ that passes through $(8, 2)$.

_____

🔄 YOUR TURN

The slope of any line perpendicular to the line given by $2x + y = 5$ is the opposite of the reciprocal of -2, or $\frac{1}{2}$. Substituting into the point–slope form, we have

$$y - y_1 = m(x - x_1)$$
$$y - (-3) = \tfrac{1}{2}(x - 1). \quad \text{Substituting } \tfrac{1}{2} \text{ for } m, 1 \text{ for } x_1, \text{ and } -3 \text{ for } y_1$$

EXAMPLE 5   Use slope–intercept form to find an equation of the line with slope 4 that passes through $(6, -5)$.

SOLUTION   Since the slope of the line is 4, we have

$$y = mx + b$$
$$y = 4x + b. \quad \text{Substituting 4 for } m$$

To find b, we use the fact that if $(6, -5)$ is a point on the line, then that ordered pair is a solution of the equation of the line.

$$y = 4x + b \qquad \text{We know that } m \text{ is 4.}$$
$$-5 = 4(6) + b \qquad \text{Substituting 6 for } x \text{ and } -5 \text{ for } y$$
$$-5 = 24 + b$$
$$-29 = b \qquad \text{Solving for } b$$

5. Use slope–intercept form to find an equation of the line with slope $\frac{1}{2}$ that passes through $(8, -3)$.

_____

🔄 YOUR TURN

Now we know that $b = -29$, so the equation of the line is

$$y = 4x - 29. \qquad m = 4 \text{ and } b = -29$$

We can also find the equation of a line if we know two points on the line.

EXAMPLE 6   Find the slope–intercept equation of the line passing through $(-1, -5)$ and $(3, -2)$.

SOLUTION   We first determine the slope of the line and then write an equation in point–slope form. (We could also use slope–intercept form as in Example 5.) Note that

> Find the slope.

$$m = \frac{-5 - (-2)}{-1 - 3} = \frac{-3}{-4} = \frac{3}{4}.$$

Since the line passes through $(3, -2)$, we have

> Substitute the point and the slope in the point–slope form.

$$y - (-2) = \tfrac{3}{4}(x - 3) \qquad \text{Substituting into } y - y_1 = m(x - x_1)$$

> Write in slope–intercept form.

$$y + 2 = \tfrac{3}{4}x - \tfrac{9}{4} \qquad \text{Using the distributive law}$$
$$y = \tfrac{3}{4}x - \tfrac{9}{4} - 2 \qquad \text{Subtracting 2 from both sides}$$
$$y = \tfrac{3}{4}x - \tfrac{17}{4}. \qquad -\tfrac{9}{4} - \tfrac{8}{4} = -\tfrac{17}{4}$$

6. Find the slope–intercept equation of the line passing through $(6, -1)$ and $(-2, -3)$.

_____

🔄 YOUR TURN

You can check that using $(-1, -5)$ as (x_1, y_1) in $y - y_1 = \tfrac{3}{4}(x - x_1)$ yields the same slope–intercept equation of the line.

HORIZONTAL LINES OR VERTICAL LINES

If we know that a line is horizontal or vertical and we know one point on the line, we can find an equation for the line.

EXAMPLE 7 Find **(a)** the equation of the horizontal line that passes through $(1, -4)$, and **(b)** the equation of the vertical line that passes through $(1, -4)$.

SOLUTION

a) An equation of a horizontal line is of the form $y = b$. In order for $(1, -4)$ to be a solution of $y = b$, we must have $b = -4$. Thus the equation of the line is $y = -4$.

b) An equation of a vertical line is of the form $x = a$. In order for $(1, -4)$ to be a solution of $x = a$, we must have $a = 1$. Thus the equation of the line is $x = 1$.

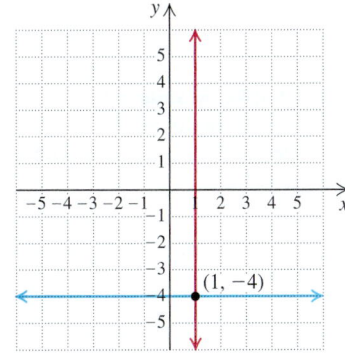

7. Find the equation of the vertical line that passes through $(2, 8)$.

 YOUR TURN

Estimations and Predictions Using Two Points

We can estimate real-life quantities that are not already known by using two points with known coordinates. When the unknown point is located *between* the two points, this process is called **interpolation**. If a graph passing through the known points is *extended* to predict future values, the process is called **extrapolation**. In statistics, methods exist for using a set of several points to interpolate or extrapolate values using lines and other curves.

EXAMPLE 8 *Apps.* The number of apps downloaded to mobile devices has increased rapidly in recent years. This increase is not due entirely to the fact that more mobile devices have been purchased. As the data in the table at left indicate, the number of apps downloaded per iOS device (iPhone, iPad, and iPod) has increased.

a) Graph the data from October 2009 and January 2011. Let x represent the number of months since October 2008 and y the number of apps downloaded per device. Then use the two points to determine an equation for the line.

b) Estimate the number of apps downloaded per device in July 2010 and the number of apps downloaded per device in January 2012.

SOLUTION

a) We first draw and label a horizontal axis to display the month and a vertical axis to display the number of apps downloaded per device. Next, we number the axes, choosing scales that include both the given values and the values to be estimated.

Since $x =$ the number of months since October 2008, we plot $(12, 33)$ and $(27, 63)$ and draw a line passing through both points.

Month	Total Number of Apps Downloaded from the iTunes Store per iOS Device
October 2008	10
April 2009	25
October 2009	33
April 2010	45
October 2010	55
January 2011	63

Source: Based on data from www.asymco.com

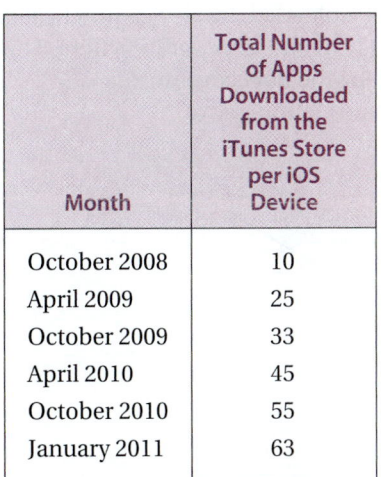

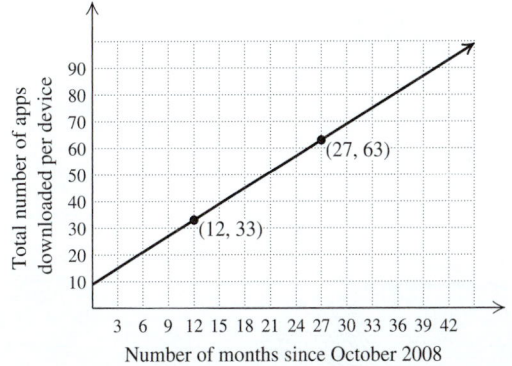

Chapter Resources:
Visualizing for Success, p. 221;
Decision Making: Connection,
p. 222

To find an equation for the line, we first calculate its slope:

$$m = \frac{\text{change in } y}{\text{change in } x} = \frac{63 - 33}{27 - 12} = \frac{30}{15} = 2.$$

The number of total apps downloaded per device purchased increased at a rate of 2 apps per month. We can use either of the given points to write a point–slope equation for the line. Choosing $(12, 33)$, we write an equivalent equation in slope–intercept form:

$y - y_1 = m(x - x_1)$ Writing the general point–slope equation

$y - 33 = 2(x - 12)$ Substituting. This is a point–slope equation.

$y - 33 = 2x - 24$ Using the distributive law

$y = 2x + 9.$ Adding 33 to both sides. This is slope–intercept form.

b) To estimate the total number of apps downloaded per device in July 2010, we substitute 21 for x in the slope–intercept equation:

$y = 2 \cdot 21 + 9 = 51.$ July 2010 is 21 months after October 2008.

In July 2010, there were about 51 apps downloaded per device purchased. Because July 2010 is *between* October 2009 and January 2011, we are *interpolating* here.

To estimate the total number of apps downloaded per device purchased in January 2012, we substitute 39 for x in the slope–intercept equation:

$y = 2 \cdot 39 + 9 = 87.$ January 2012 is 39 months after October 2008.

In January 2012, there were about 87 apps downloaded per device purchased. Because January 2012 is *beyond* January 2011, we are *extrapolating* here. This prediction assumes that the trend indicated by the data continues.

The following graph confirms the estimates found above.

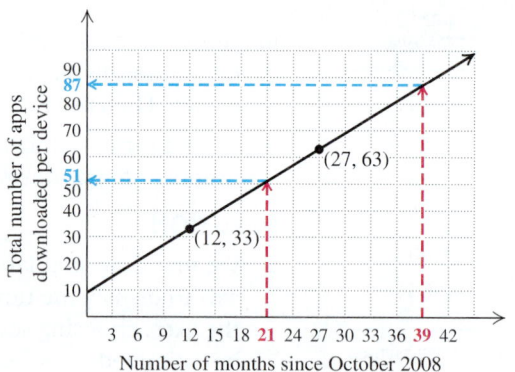

8. The data used for Example 8 do not all lie on a straight line. Thus a different pair of points may yield a different equation for the line used to model the application.

   a) The data from April 2009 and October 2010 correspond to the points $(6, 25)$ and $(24, 55)$. Find an equation for the line containing these points and then rewrite it in slope–intercept form.

   b) Use the equation found in part (a) to estimate the number of apps downloaded per device in July 2010 and to estimate the number of apps downloaded per device in January 2012.

 YOUR TURN

CONNECTING THE CONCEPTS

Any line can be described by a number of equivalent equations. We write the equation in the form that is most useful for us. For example, all four of the equations shown at right describe the same line.

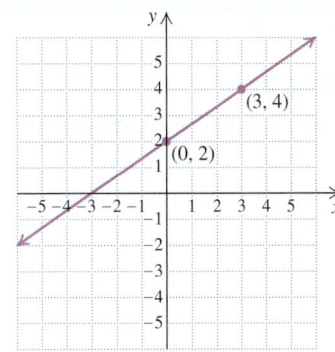

$2x - 3y = -6;$

$y = \frac{2}{3}x + 2;$

$y - 4 = \frac{2}{3}(x - 3);$

$2x + 6 = 3y$

Form of a Linear Equation	Example	Uses
Standard form: $Ax + By = C$	$2x - 3y = -6$	Finding x- and y-intercepts; Graphing using intercepts
Slope–intercept form: $y = mx + b$	$y = \frac{2}{3}x + 2$	Finding slope and y-intercept; Graphing using slope and y-intercept; Writing an equation given slope and y-intercept
Point–slope form: $y - y_1 = m(x - x_1)$	$y - 4 = \frac{2}{3}(x - 3)$	Finding slope and a point on the line; Graphing using slope and a point on the line; Writing an equation given slope and a point on the line

EXERCISES

Tell whether each equation is in standard form, slope–intercept form, point–slope form, or none of these.

1. $y = -\frac{1}{2}x - 7$

2. $5x - 8y = 10$

3. $x = y + 2$

4. $\frac{1}{2}x + \frac{1}{3}y = 5$

5. $y - 2 = 5(x - (-1))$

6. $3y + 7 = x$

Write each equation in standard form.

7. $2x = 5y + 10$

8. $y = 2x + 7$

Write each equation in slope–intercept form.

9. $2x - 7y = 8$

10. $y + 5 = -(x + 3)$

3.7 EXERCISE SET

FOR EXTRA HELP

MyMathLab® Math XL
PRACTICE   WATCH   READ   REVIEW

⮞ Vocabulary and Reading Check

Classify each of the following statements as either true or false.

1. The equation $y = -3x - 1$ is written in point–slope form.

2. The equation $y - 4 = -3(x - 1)$ is written in point–slope form.

3. Knowing the coordinates of just one point on a line is enough to write an equation of the line.

4. Knowing coordinates of just one point on a line and the slope of the line is enough to write an equation of the line.

5. Knowing the coordinates of just two points on a line is enough to write an equation of the line.

6. Point–slope form can be used with any point that is used to calculate the slope of that line.

Point–Slope Form

For each point–slope equation listed, state the slope and a point on the graph.

7. $y - 3 = \frac{1}{4}(x - 5)$

8. $y - 5 = 6(x - 1)$

9. $y + 1 = -7(x - 2)$

10. $y - 4 = -\frac{2}{3}(x + 8)$

11. $y - 6 = -\frac{10}{3}(x + 4)$

12. $y + 1 = -9(x - 7)$

Aha! 13. $y = 5x$

14. $y = \frac{4}{5}x$

Graph.

15. $y - 2 = 3(x - 5)$

16. $y - 4 = 2(x - 3)$

17. $y - 2 = -4(x - 1)$

18. $y - 4 = -5(x - 1)$

19. $y + 4 = \frac{1}{2}(x + 2)$

20. $y + 7 = \frac{1}{3}(x + 5)$

21. $y = -(x - 8)$

22. $y = -3(x + 2)$

Finding the Equation of a Line

Find an equation for each line. Write your final answer in slope–intercept form.

23. Parallel to $y = 3x - 7$; y-intercept $(0, 4)$

24. Parallel to $y = \frac{1}{2}x + 6$; y-intercept $(0, -1)$

25. Perpendicular to $y = -\frac{3}{4}x + 1$; y-intercept $(0, -12)$

26. Perpendicular to $y = \frac{5}{8}x - 2$; y-intercept $(0, 9)$

27. Parallel to $2x - 3y = 4$; y-intercept $\left(0, \frac{1}{2}\right)$

28. Perpendicular to $4x + 7y = 1$; y-intercept $(0, -4.2)$

29. Perpendicular to $x + y = 18$; y-intercept $(0, -32)$

30. Parallel to $x - y = 6$; y-intercept $(0, 27)$

Find an equation in point–slope form for the line having the specified slope and containing the point indicated.

31. $m = 6$, $(7, 1)$

32. $m = 4$, $(3, 8)$

33. $m = -5$, $(3, 4)$

34. $m = -7$, $(1, 2)$

35. $m = \frac{1}{2}$, $(-2, -5)$

36. $m = 1$, $(-4, -6)$

37. $m = -1$, $(9, 0)$

38. $m = -\frac{2}{3}$, $(5, 0)$

Find an equation of the line having the specified slope and containing the indicated point. Write your final answer in slope–intercept form.

39. $m = 2$, $(1, -4)$

40. $m = -4$, $(-1, 5)$

41. $m = -\frac{3}{5}$, $(-4, 8)$

42. $m = -\frac{1}{5}$, $(-2, 1)$

43. $m = -0.6$, $(-3, -4)$

44. $m = 2.3$, $(4, -5)$

Aha! 45. $m = \frac{2}{7}$, $(0, -6)$

46. $m = \frac{1}{4}$, $(0, 3)$

47. $m = \frac{3}{5}$, $(-4, 6)$

48. $m = -\frac{2}{7}$, $(6, -5)$

Write an equation of the line containing the specified point and parallel to the indicated line.

49. $(2, 5)$, $x - 2y = 3$

50. $(1, 4)$, $3x + y = 5$

51. $(-3, 2)$, $x + y = 7$

52. $(-1, -6)$, $x - 5y = 1$

53. $(-2, -3)$, $2x + 3y = -7$

54. $(3, -4)$, $5x - 6y = 4$

Aha! 55. $(5, -4)$, $x = 2$

56. $(-3, 6)$, $y = 7$

Write an equation of the line containing the specified point and perpendicular to the indicated line.

57. $(3, 1)$, $2x - 3y = 4$

58. $(6, 0)$, $5x + 4y = 1$

59. $(-4, 2)$, $x + y = 6$

60. $(-2, -5)$, $x - 2y = 3$

61. $(1, -3)$, $3x - y = 2$

62. $(-5, 6)$, $4x - y = 3$

63. $(-4, -7)$, $3x - 5y = 6$

64. $(-4, 5)$, $7x - 2y = 1$

65. $(-3, 7)$, $y = 5$

66. $(4, -2)$, $x = 1$

Find an equation of the line containing each pair of points. Write your final answer in slope–intercept form.

67. $(2, 3)$ and $(3, 7)$

68. $(3, 8)$ and $(1, 4)$

69. $(1.2, -4)$ and $(3.2, 5)$

70. $(-1, -2.5)$ and $(4, 8.5)$

Aha! **71.** $(2, -5)$ and $(0, -1)$

72. $(-2, 0)$ and $(0, -7)$

73. $(-6, -10)$ and $(-3, -5)$

74. $(-1, -3)$ and $(-4, -9)$

Find an equation of each line.

75. Horizontal line through $(2, -6)$

76. Horizontal line through $(-1, 8)$

77. Vertical line through $(-10, -9)$

78. Vertical line through $(4, 12)$

Estimations and Predictions Using Two Points

In Exercises 79–84, assume the data is linear.

79. *Volunteering.*   The number of volunteers from the Millennial generation (those born in 1982 or later) has increased from 6.1 million in 2003 to 11.6 million in 2010.

Source: www.volunteeringinamerica.gov

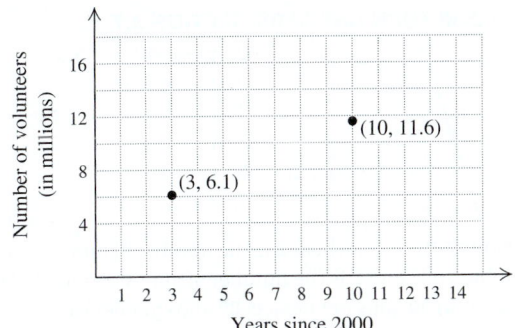

a) Find an equation for the line containing the given points. Let x represent the number of years after 2000 and y the number of volunteers, in millions.

b) Estimate the number of volunteers from the Millennial generation in 2008 and in 2014.

80. *Volunteering.*   The average annual number of volunteer hours per United States resident has decreased from 37.9 hr in 2004 to 33.9 hr in 2010.

Source: www.volunteeringinamerica.gov

a) Find an equation for the line representing the given data. Let x represent the number of years after 2000 and y the average annual number of volunteer hours.

b) Estimate the average annual number of volunteer hours in 2008 and in 2014.

81. *College Enrollment.*   The number of students who completed high school and enrolled in college that same year increased from 1.8 million in 2005 to 2.1 million in 2009.

Source: Digest of Education Statistics, National Center for Education Statistics

a) Find an equation for the line including the given data points. Let x represent the number of years after 2000 and y the number of recent high school graduates (or the equivalent) who enrolled in college that year.

b) Estimate the number of high school graduates (or the equivalent) who enrolled in college in 2007 and in 2015.

82. *Nursery School Enrollment.*   The percent of children ages 3 and 4 who are enrolled in nursery school has decreased from 54.5% in 2007 to 52.4% in 2009.

Source: Digest of Education Statistics, National Center for Education Statistics

a) Find an equation for the line containing the given data points. Let x represent the number of years after 2000 and y the percent of children ages 3 and 4 who are enrolled in nursery school.

b) Estimate the percent of children ages 3 and 4 enrolled in nursery school in 2008 and in 2012.

83. *Urban Population.*   The percent of the U.S. population that lives in metropolitan areas was approximately 79% in 2000 and 76.5% in 2009.

Sources: Census 2000 and 2009 American Community Survey

a) Define variables x and y and find an equation for the line containing the given data points. Answers may vary.

b) Estimate the percent of the U.S. population that lived in metropolitan areas in 2005, and predict the percent living in metropolitan areas in 2012.

84. *Aging Population.*   The number of U.S. residents over the age of 65 was approximately 36.3 million in 2004 and 40.1 million in 2010.

Source: U.S. Census Bureau

a) Define variables x and y and find an equation for the line containing the given data points. Answers may vary.

b) Estimate the number of U.S. residents over the age of 65 in 2006 and in 2016.

 85. Can equations for horizontal or vertical lines be written in point–slope form? Why or why not?

86. Describe a situation in which it is easier to graph the equation of a line in point–slope form than in slope–intercept form.

Skill Review

Solve.

87. $\frac{3}{8}x = -24$ [2.1]

88. $6 - x = -3$ [2.2]

89. $\frac{t}{3} = 6$ [2.1]

90. $\frac{1}{2}n - \frac{1}{3} = \frac{1}{6}n + \frac{3}{2}$ [2.2]

91. $2(x - 7) > 5x + 3$ [2.6]   **92.** $10 - x \le 12$ [2.6]

Synthesis

 93. Describe a procedure that can be used to write the slope–intercept equation for any nonvertical line passing through two given points.

 94. Any nonvertical line has many equations in point–slope form, but only one in slope–intercept form. Why is this?

Graph.

Aha! **95.** $y - 3 = 0(x - 52)$ **96.** $y + 4 = 0(x + 93)$

Write the slope–intercept equation for each line shown.

97.

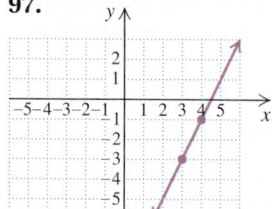

98.

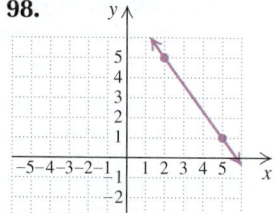

99. Write the slope–intercept equation of the line that has the same y-intercept as the line $x - 3y = 6$ and contains the point $(5, -1)$.

100. Write the slope–intercept equation of the line that contains the point $(-1, 5)$ and is parallel to the line passing through $(2, 7)$ and $(-1, -3)$.

101. Write the slope–intercept equation of the line that has x-intercept $(-2, 0)$ and is parallel to $4x - 8y = 12$.

102. Find k so that the graph of $5y - kx = 7$ and the line containing $(7, -3)$ and $(-2, 5)$ are parallel.

103. Find k so that the graph of $7y - kx = 9$ and the line containing the points $(2, -1)$ and $(-4, 5)$ are perpendicular.

Another form of a linear equation is the **double-intercept form:** $\frac{x}{a} + \frac{y}{b} = 1$. *From this form, we can read the x-intercept $(a, 0)$ and the y-intercept $(0, b)$ directly.*

104. Find the x-intercept and the y-intercept of the graph of $\frac{x}{2} + \frac{y}{5} = 1$.

105. Find the x-intercept and the y-intercept of the graph of $\frac{x}{10} - \frac{y}{3} = 1$.

106. Write the equation $6x + 5y = 30$ in double-intercept form and find the intercepts.

107. If data do not lie exactly on a straight line, more than one equation can be used to model the data. This situation is illustrated in Example 7 and Your Turn Exercise 7. A line of "best fit" can be found using methods not discussed in this section. The most common method used is *linear regression*. To use the linear regression feature of a graphing calculator, use the **STAT** key. Under the STAT EDIT menu, enter x-coordinates in L_1 and y-coordinates in L_2. Then in the STAT CALC menu, choose the Lin Reg option. Use linear regression to find an equation that fits the data in the table in Example 8.

YOUR TURN ANSWERS: SECTION 3.7

1.

2. $y - (-6) = -5(x - 1)$
3. $y = x + 5$
4. $y - 2 = -\frac{4}{3}(x - 8)$
5. $y = \frac{1}{2}x - 7$
6. $y = \frac{1}{4}x - \frac{5}{2}$
7. $x = 2$

$y - 1 = 3(x + 2)$

8. (a) $y = \frac{5}{3}x + 15$, where x is the number of months since October 2008 and y is the number of apps downloaded per device; (b) 50 apps per device; 80 apps per device

---

QUICK QUIZ: SECTIONS 3.1–3.7

Graph.

1. $y - 3 = 2x$ [3.2]   **2.** $x + 7 = 8$ [3.3]

3. $y = -3x + 2$ [3.6]   **4.** $3x - 6y = 6$ [3.3]

5. $y - 1 = 2(x + 1)$ [3.7]

PREPARE TO MOVE ON

Simplify.   [1.8]

1. $(-5)^3$ **2.** $(-2)^6$

3. -2^6 **4.** $3 \cdot 2^4 - 5 \cdot 2^3$

5. $(5 - 7)^2(3 - 2 \cdot 2)$

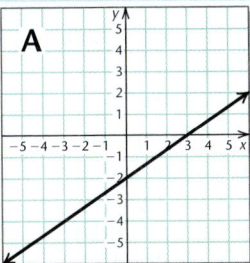

A

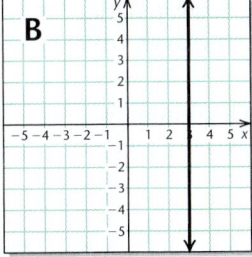

B

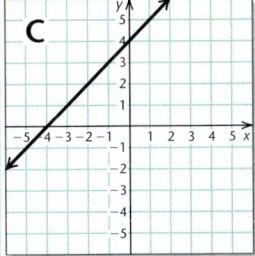

C

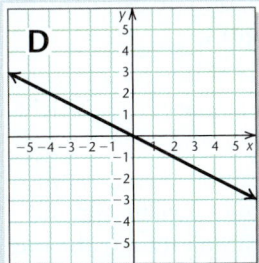

D

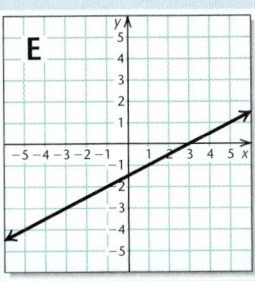

E

Visualizing for Success

Use after Section 3.7.

Match each equation with its graph.

1. $y = x + 4$

2. $y = 2x$

3. $y = 3$

4. $x = 3$

5. $y = -\frac{1}{2}x$

6. $2x - 3y = 6$

7. $y = -3x - 2$

8. $3x + 2y = 6$

9. $y - 3 = 2(x - 1)$

10. $y + 2 = \frac{1}{2}(x + 1)$

Answers on page A-15

An additional, animated version of this activity appears in MyMathLab. *To use MyMathLab, you need a course ID and a student access code. Contact your instructor for more information.*

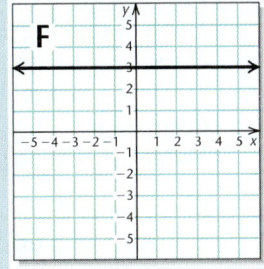

F

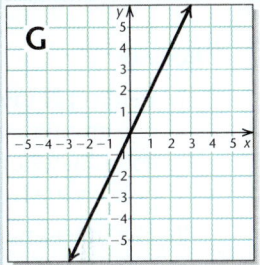

G

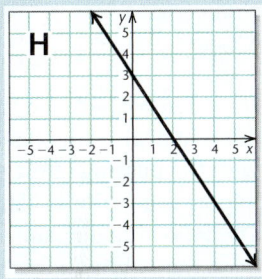

H

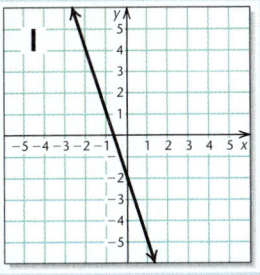

I

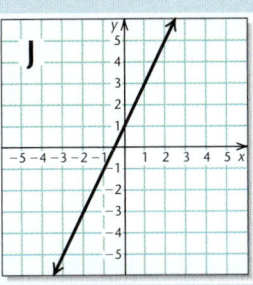

J

Collaborative Activity   *You Sank My Battleship!*

Focus: Graphing points; logical questioning
Use after: Section 3.1
Time: 15–25 minutes
Group size: 3–5
Materials: Graph paper

In the game Battleship®, a player places a miniature ship on a grid that only that player can see. An opponent guesses at coordinates that might "hit" the "hidden" ship. The following activity is similar to this game.

Activity

1. Using only integers from −10 to 10 (inclusive), one group member should secretly record the coordinates of a point on a slip of paper. (This point is the hidden "battleship.")

2. The other group members can then ask up to 10 "yes/no" questions in an effort to determine the coordinates of the secret point. Be sure to phrase each question mathematically (for example, "Is the *x*-coordinate negative?").

3. The group member who selected the point should answer each question. On the basis of the answer given, another group member should cross out the points no longer under consideration. All group members should check that this is done correctly.

4. If the hidden point has not been determined after 10 questions have been answered, the secret coordinates should be revealed to all group members.

5. Repeat parts (1)–(4) until each group member has had the opportunity to select the hidden point and answer questions.

Decision Making & Connection

(Use after Section 3.7.)

Depreciation.   From the minute a new car is driven out of a dealership, it *depreciates*, or drops in value with the passing of time. The Kelley Blue Book Market Report is a periodic listing of the values of used cars. The following data are taken from two such reports from the 2011 and the 2012 Central Editions.

Car	Lending Value, November 2011	Lending Value, March 2012
2006 Saturn Ion	$4725	$4650
2006 Mitsubishi Lancer	5800	5525
2006 Kia Rio	5125	4500

1. Assuming that the values are dropping linearly, draw a line representing the value of each car. Draw all three lines on the same graph, using different colored pens or pencils if possible. Let the horizontal axis represent the time, in years, since January 2011, and the vertical axis the value of each car.

2. At what rate is each car depreciating and how are the different rates illustrated in the graph of part (1)?

3. Find a linear equation that models the value of each car. Let *x* represent the number of years since January 2011, and *y* the value.

4. Use the equations found in part (3) to estimate the value of each car in January 2013.

 5. If you had planned to buy one of the three cars in November 2011 and sell it in January 2013, which one would you have purchased, and why?

 6. *Research.*   Find the trade-in value of your car or of that of a friend. Use the trade-in value and the price originally paid for the car to estimate the rate of depreciation. Then find a linear equation that models the value of the car and use it to estimate its value one year from now.

Study Summary

KEY TERMS AND CONCEPTS	EXAMPLES	PRACTICE EXERCISES

SECTION 3.1: *Reading Graphs, Plotting Points, and Scaling Graphs*

Ordered pairs can be **plotted** or **graphed** using a **coordinate system** that uses two **axes**, which are most often labeled x and y. The axes intersect at the **origin,** $(0,0)$, and divide a plane into four **quadrants**.

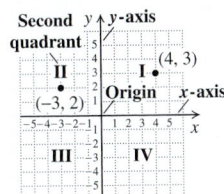

1. Plot the points $(0, -5)$ and $(2, -3)$.

2. In which quadrant is the point $(-10, -20)$ located?

SECTION 3.2: *Graphing Linear Equations*

To **graph** an equation means to make a drawing that represents all of its solutions.

A **linear equation**, such as $y = 2x - 7$ or $2x + 3y = 12$, has a graph that is a straight line.

$$3x = y + 1$$

x	y	(x, y)
1	2	$(1, 2)$
0	-1	$(0, -1)$
-1	-4	$(-1, -4)$

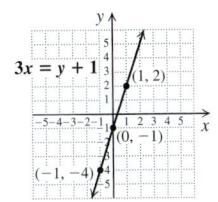

3. Graph: $y = 2x + 1$.

SECTION 3.3: *Graphing and Intercepts*

To find a y-intercept $(0, b)$, let $x = 0$ and solve for y.

To find an x-intercept $(a, 0)$, let $y = 0$ and solve for x.

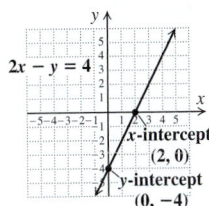

4. Find the x-intercept and the y-intercept of the line given by $10x - y = 10$.

Horizontal Lines
The graph of $y = b$ is a horizontal line, with y-intercept $(0, b)$.

Vertical Lines
The graph of $x = a$ is a vertical line, with x-intercept $(a, 0)$.

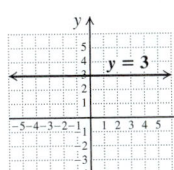 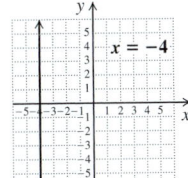

5. Graph: $y = -2$.

6. Graph: $x = 3$.

SECTION 3.4: *Rates*

A **rate** is a ratio that indicates how two quantities change with respect to each other.

Tara had $1500 in her savings account at the beginning of February, and $2400 at the beginning of May. Find the rate at which Tara is saving.

$$\text{Savings rate} = \frac{\text{Amount saved}}{\text{Number of months}} = \frac{\$2400 - \$1500}{3 \text{ months}}$$

$$= \frac{\$900}{3 \text{ months}} = \$300 \text{ per month}$$

7. At 8:30 A.M., a high school had served 47 people at their pancake breakfast. By 9:15 A.M., the total served had reached 67. Find the serving rate, in number of meals per minute.

SECTION 3.5: *Slope*

Slope

$$\text{Slope} = m = \frac{\text{change in } y}{\text{change in } x}$$

$$= \frac{\text{rise}}{\text{run}} = \frac{y_2 - y_1}{x_2 - x_1}$$

The slope of the line containing the points $(-1, -4)$ and $(2, -6)$ is

$$m = \frac{-6 - (-4)}{2 - (-1)} = \frac{-2}{3} = -\frac{2}{3}.$$

8. Find the slope of the line containing the points $(1, 4)$ and $(-9, 3)$.

The slope of a horizontal line is 0.

The slope of a vertical line is undefined.

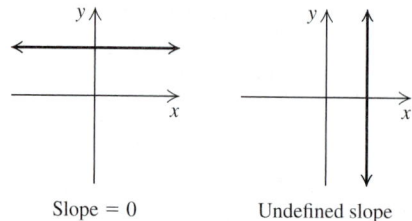

Slope = 0   Undefined slope

9. Find the slope of the line given by $y = 10$.

SECTION 3.6: *Slope–Intercept Form*

Slope–Intercept Form

$$y = mx + b$$

The slope of the line is m.
The y-intercept of the line is $(0, b)$.

For the line given by $y = \frac{2}{3}x - 8$:

The slope is $\frac{2}{3}$ and the y-intercept is $(0, -8)$.

10. Find the slope and the y-intercept of the line given by $y = -4x + \frac{2}{5}$.

To graph a line written in slope–intercept form, plot the y-intercept, count off the slope to locate a second point, and draw the line.

11. Graph: $y = \frac{1}{2}x + 2$.

Parallel Lines
Two lines are parallel if they have the same slope or if both are vertical.

Determine whether the graphs of $y = \frac{2}{3}x - 5$ and $3y - 2x = 7$ are parallel.

$$y = \frac{2}{3}x - 5 \qquad 3y - 2x = 7$$

The slope is $\frac{2}{3}$. $\qquad 3y = 2x + 7$

$$y = \frac{2}{3}x + \frac{7}{3}$$

The slope is $\frac{2}{3}$.

Since the slopes are the same, the graphs are parallel.

12. Determine whether the graphs of $y = 4x - 12$ and $4y = x - 9$ are parallel.

SECTION 3.7: *Point–Slope Form and Equations of Lines*

Point–Slope Form

$$y - y_1 = m(x - x_1)$$

The slope of the line is m.
The line passes through (x_1, y_1).

Write a point–slope equation for the line with slope -2 that contains the point $(3, -5)$.

$$y - y_1 = m(x - x_1)$$
$$y - (-5) = -2(x - 3)$$

13. Write a point–slope equation for the line with slope $\frac{1}{4}$ and containing the point $(-1, 6)$.

Review Exercises: Chapter 3

➤ Concept Reinforcement

Classify each of the following statements as either true or false.

1. Not every ordered pair lies in one of the four quadrants.   [3.1]

2. The equation of a vertical line cannot be written in slope–intercept form.   [3.6]

3. Equations for lines written in slope–intercept form appear in the form $Ax + By = C$.   [3.6]

4. Every horizontal line has an x-intercept.   [3.3]

5. A line's slope is a measure of rate.   [3.5]

6. A positive rate of ascent means that an airplane is flying increasingly higher above the earth.   [3.4]

7. Any two points on a line can be used to determine the slope of a nonvertical line.   [3.5]

8. Knowing a line's slope is enough to write the equation of the line.   [3.6]

9. Knowing two points on a line is enough to write the equation of the line.   [3.7]

10. Parallel lines that are not vertical have the same slope.   [3.6]

The following bar graph shows the number of volunteers in the United States in 2010 by age group.   [3.1]

Source: U.S. Department of Labor, Bureau of Labor Statistics

Volunteering in America

11. Approximately how many more people ages 35 to 44 volunteered than those ages 25 to 34?

12. About 2% of the volunteers ages 16 to 19 volunteered in environmental organizations. How many people ages 16 to 19 volunteered in environmental organizations?

Plot each point.   [3.1]

13. $(5, -1)$ 14. $(2, 3)$ 15. $(-4, 0)$

In which quadrant is each point located?   [3.1]

16. $(-8, -7)$ 17. $(15.3, -13.8)$ 18. $\left(-\frac{1}{2}, \frac{1}{10}\right)$

Find the coordinates of each point in the figure.   [3.1]

19. A 20. B 21. C

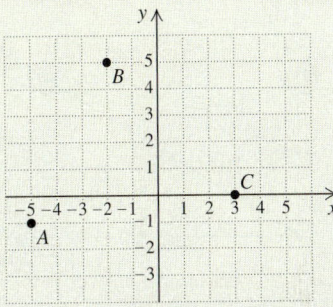

22. Use a grid 10 squares wide and 10 squares high to plot $(-65, -2)$, $(-10, 6)$, and $(25, 7)$. Choose the scale carefully.   [3.1]

23. Determine whether the equation $y = 2x + 7$ has the given ordered pair as a solution: **(a)** $(3, 1)$; **(b)** $(-3, 1)$.   [3.2]

24. Show that the ordered pairs $(0, -3)$ and $(2, 1)$ are solutions of the equation $2x - y = 3$. Then use the graph of the two points to determine another solution. Answers may vary.   [3.2]

Graph.

25. $y = x - 5$ [3.2]   26. $y = -\frac{1}{4}x$ [3.2]

27. $y = -x + 4$ [3.2]   28. $4x + y = 3$ [3.2]

29. $4x + 5 = 3$ [3.3]   30. $5x - 2y = 10$ [3.3]

31. $y = 6$ [3.3]   32. $y = \frac{2}{3}x - 5$ [3.6]

33. $2x + y = 4$ [3.3]

34. $y + 2 = -\frac{1}{2}(x - 3)$ [3.7]

35. *Automobiles.* The average age of vehicles currently driven in the United States is given by $a = 0.2t + 8.8$, where t is the number of years since 2000. Graph the equation with t on the horizontal axis and use the graph to estimate the average age of an American vehicle in 2012.   [3.2]

Source: Based on data from R. L. Polk & Co.

36. At 4:00 P.M., Jesse's Honda Civic was at mile marker 17 of Interstate 290 in Chicago. At 4:45 P.M., Jesse was at mile marker 23.   [3.4]

a) Find Jesse's driving rate, in number of miles per minute.

b) Find Jesse's driving rate, in number of minutes per mile.

37. *Gas Mileage.* The following graph shows data for the gas consumption of a 4-cylinder Ford Explorer driven on city streets. At what rate was the vehicle consuming gas?   [3.4]

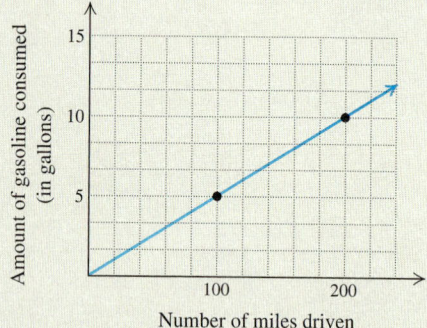

Find the slope of each line.   [3.5]

38.
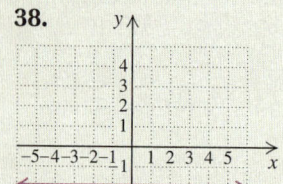

39.
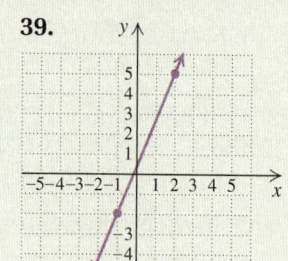

40.
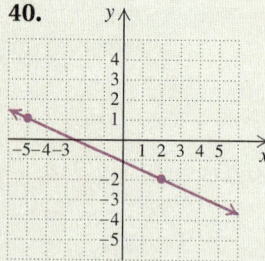

Find the slope of the line containing the given pair of points. If it is undefined, state this.   [3.5]

41. $(-2, 5)$ and $(3, -1)$

42. $(6, 5)$ and $(-2, 5)$

43. $(-3, 0)$ and $(-3, 5)$

44. $(-8.3, 4.6)$ and $(-9.9, 1.4)$

45. Find the x-intercept and the y-intercept of the line given by $5x - 8y = 80$.   [3.3]

46. Find the slope and the y-intercept of the line given by $3x + 5y = 45$.   [3.6]

47. *Architecture.* To meet federal standards, a wheelchair ramp cannot rise more than 1 ft over a horizontal distance of 12 ft. Express this slope as a grade.   [3.5]

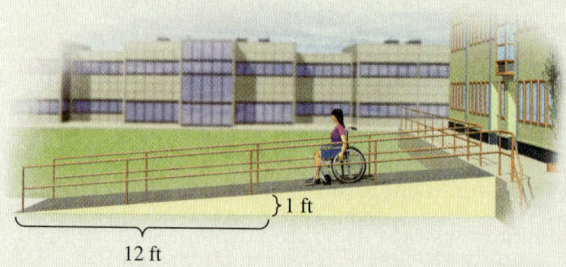

Determine whether each pair of lines is parallel, perpendicular, or neither.   [3.6]

48. $y + 5 = -x,$
$x - y = 2$

49. $3x - 5 = 7y,$
$7y - 3x = 7$

50. Write the slope–intercept equation of the line with slope $\frac{3}{8}$ and y-intercept $(0, 7)$.   [3.6]

51. Write a point–slope equation for the line with slope $-\frac{1}{3}$ that contains the point $(-2, 9)$.   [3.7]

52. Write the slope–intercept equation for the line containing the points $(-2, 5)$ and $(3, 10)$.   [3.7]

53. Write the slope–intercept equation for the line that is perpendicular to the line $3x - 5y = 9$ and that contains the point $(2, -5)$.   [3.7]

54. The average in-state tuition at a public two-year college was \$1847 in 2005 and \$2713 in 2011.

a) Graph the data points and determine an equation for the related line.

b) Calculate the average in-state tuition at a public two-year college in 2010 and estimate the average in-state tuition at a public two-year college in 2013.   [3.7]

Sources: U.S. National Center for Education Statistics, College Board, Trends in College Pricing 2010

Synthesis

55. Can two perpendicular lines share the same y-intercept? Why or why not?   [3.3]

56. Is it possible for a graph to have only one intercept? Why or why not?   [3.3]

57. Find the value of m in $y = mx + 3$ such that $(-2, 5)$ is on the graph.   [3.2]

58. Find the area and the perimeter of a rectangle for which $(-2, 2)$, $(7, 2)$, and $(7, -3)$ are three of the vertices.   [3.1]

59. Find three solutions of $y = 4 - |x|$.   [3.2]

Test: Chapter 3

For step-by-step test solutions, access the Chapter Test Prep Videos in MyMathLab®, on YouTube▶ (search "Bittinger Combo Alg CA" and click on "Channels"), or by scanning the code.

Income.   *The following line graph shows data for median household income in the United States, adjusted for inflation.*

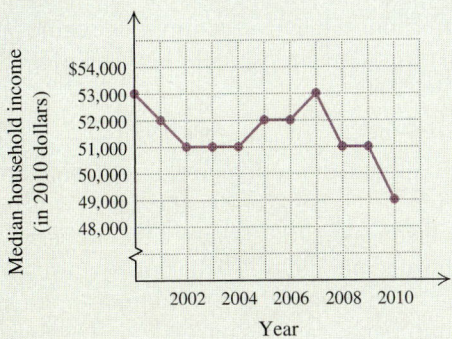

Source: U.S. Census Bureau

1. By how much did median household income decrease from 2007 to 2010?

2. In which year or years shown was median household income the highest?

In which quadrant is each point located?

3. $(-2, -10)$ 4. $(-1.6, 2.3)$

Find the coordinates of each point in the figure.

5. *A*

6. *B*

7. *C*

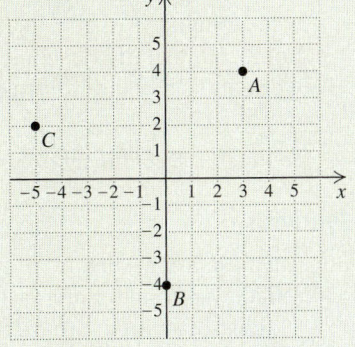

Graph.

8. $y = 2x - 1$ 9. $2x - 4y = -8$

10. $y + 1 = 6$ 11. $y = \frac{3}{4}x$

12. $2x - y = 3$ 13. $x = -1$

14. $y + 4 = -\frac{1}{2}(x - 3)$

Find the slope of the line containing each pair of points. If it is undefined, state this.

15. $(3, -2)$ and $(4, 3)$

16. $(-5, 6)$ and $(-1, -3)$

17. $(4, 7)$ and $(4, -8)$

18. *Running.*   Jon reached the 3-km mark of a race at 2:15 P.M. and the 6-km mark at 2:24 P.M. What was his running rate?

19. At one point Filbert Street, the steepest street in San Francisco, drops 63 ft over a horizontal distance of 200 ft. Find the road grade.

20. Find the x-intercept and the y-intercept of the line given by $5x - y = 30$.

21. Find the slope and the y-intercept of the line given by $y - 8x = 10$.

22. Write the slope–intercept equation of the line with slope $-\frac{1}{3}$ and y-intercept $(0, -11)$.

Determine without graphing whether each pair of lines is parallel, perpendicular, or neither.

23. $4y + 2 = 3x,$ 24. $y = -2x + 5,$
    $-3x + 4y = -12$ $2y - x = 6$

25. Write the slope–intercept equation of the line that is perpendicular to the line $2x - 5y = 8$ and that contains the point $(-3, 2)$.

26. *Aerobic Exercise.*   A person's target heart rate is the number of beats per minute that brings the most aerobic benefit to his or her heart. The target heart rate for a 20-year-old is 150 beats per minute; for a 60-year-old, it is 120 beats per minute.

    a) Graph the data and determine an equation for the related line. Let a = age and r = target heart rate, in number of beats per minute.
    b) Calculate the target heart rate for a 36-year-old.

Synthesis

27. Write an equation of the line that is parallel to the graph of $2x - 5y = 6$ and has the same y-intercept as the graph of $3x + y = 9$.

28. A diagonal of a square connects the points $(-3, -1)$ and $(2, 4)$. Find the area and the perimeter of the square.

29. List the coordinates of three other points that are on the same line as $(-2, 14)$ and $(17, -5)$. Answers may vary.

Cumulative Review: Chapters 1–3

1. Evaluate $\dfrac{x}{5y}$ for $x = 70$ and $y = 2$.   [1.1]

2. Multiply: $6(2a - b + 3)$.   [1.2]

3. Factor: $8x - 4y + 2$.   [1.2]

4. Find the prime factorization of 54.   [1.3]

5. Find decimal notation: $-\frac{3}{20}$.   [1.4]

6. Find decimal notation: 36.7%.   [2.4]

Simplify.

7. $\frac{3}{5} - \frac{5}{12}$ [1.3]

8. $3.4 + (-0.8)$ [1.5]

9. $(-2)(-1.4)(2.6)$ [1.7]

10. $\frac{3}{8} \div \left(-\frac{9}{10}\right)$ [1.7]

11. $1 - [32 \div (4 + 2^2)]$ [1.8]

12. $3(x - 1) - 2[x - (2x + 7)]$ [1.8]

Solve.

13. $\frac{5}{3}x = -45$ [2.1]

14. $3x - 7 = 41$ [2.2]

15. $\dfrac{3}{4} = \dfrac{-n}{8}$ [2.1]

16. $14 - 5x = 2x$ [2.2]

17. $3(5 - x) = 2(3x + 4)$ [2.2]

18. $\frac{1}{4}x - \frac{2}{3} = \frac{3}{4} + \frac{1}{3}x$ [2.2]

19. $x - 28 < 20 - 2x$ [2.6]

20. Solve $A = 2\pi rh + \pi r^2$ for h.   [2.3]

21. In which quadrant is the point $(3, -1)$ located?   [3.1]

22. Graph on the number line: $-1 < x \le 2$.   [2.6]

23. Use a grid 10 squares wide and 10 squares high to plot $(-150, -40)$, $(40, -7)$, and $(0, 6)$. Choose the scale carefully.   [3.1]

Graph.

24. $x = 3$ [3.3]

25. $2x - 5y = 10$ [3.3]

26. $y = -2x + 1$ [3.2]

27. $y = \frac{2}{3}x$ [3.2]

28. $2y - 5 = 3$ [3.3]

29. Find the slope and the y-intercept of the line given by $3x - y = 2$.   [3.6]

30. Find the slope of the line containing the points $(-4, 1)$ and $(2, -1)$.   [3.5]

31. Write an equation of the line with slope $\frac{2}{7}$ and y-intercept $(0, -4)$.   [3.6]

32. Write a point–slope equation of the line with slope $-\frac{3}{8}$ that contains the point $(-6, 4)$.   [3.7]

33. A 150-lb person will burn 240 calories per hour when riding a bicycle at 6 mph. The same person will burn 410 calories per hour when cycling at 12 mph.   [3.7]

    Source: American Heart Association

    a) Graph the data and determine an equation for the related line. Let $r =$ the rate at which the person is cycling and $c =$ the number of calories burned per hour. Use the horizontal axis for r.

    b) Use the equation of part (a) to estimate the number of calories burned per hour by a 150-lb person cycling at 10 mph.

34. In 2010, the mean earnings of individuals with a high school diploma was \$31,003. This was about 54% of the mean earnings of those with a bachelor's degree. What were the mean earnings of individuals with a bachelor's degree in 2010?   [2.4]

    Source: U.S. Census Bureau

35. In order to qualify for availability pay, a criminal investigator must average at least 2 hr of unscheduled duty per workday. For the first four days of one week, Alayna worked 1, 0, 3, and 2 unscheduled hours. How many unscheduled hours must she work on Friday in order to qualify for availability pay?   [2.7]

    Source: U.S. Department of Justice

Synthesis

36. Anya's salary at the end of a year is \$26,780. This reflects a 4% salary increase in February and then a 3% cost-of-living increase in June. What was her salary at the beginning of the year?   [2.5]

Solve. If no solution exists, state this.

37. $4|x| - 13 = 3$ [1.4], [2.2]

38. $\dfrac{2 + 5x}{4} = \dfrac{11}{28} + \dfrac{8x + 3}{7}$ [2.2]

39. $5(7 + x) = (x + 6)5$ [2.2]

Polynomials

ELEPHANT	GIRTH (in centimeters)	LENGTH (in centimeters)	FOOTPAD CIRCUMFERENCE (in centimeters)
Male, age 3 years	244	140	86
Female, age 3 years	231	135	86
Male, age 25 years	404	229	130
Female, age 25 years	366	226	117

It's Not Easy to Weigh an Elephant.

That's why wildlife experts have developed formulas for estimating the weight of an elephant using measurements that are much easier to obtain. One such formula uses the girth of the elephant, measured at the heart, the length of the elephant, and the circumference of the footpad of the elephant to estimate the elephant's weight. We can use that formula to estimate the weight of the elephants described in the table above. *(See Exercises 15 and 16 in Exercise Set 4.7.)*

Source: "How Much Does That Elephant Weigh?" by Mark MacAllister on fieldtripearth.org

As a field biologist, having mathematical skill is incredibly important in order for me to translate the ordinary behavior of animals into a scientific story.

Shermin de Silva, Director of Uda Walawe Elephant Research Project, uses a variety of aspects of math, including algebra, geometry, and calculus, to study how elephants behave and move in their environment, as well as to keep track of changes in their population size for conservation.

Algebraic expressions such as $16t^2, 5a^2 - 3ab$, and $3x^2 - 7x + 5$ are called polynomials. Polynomials occur frequently in applications and appear in most branches of mathematics. Thus learning to add, subtract, multiply, and divide polynomials is an important part of nearly every course in elementary algebra. The focus of this chapter is finding equivalent expressions, not solving equations.

4.1 Exponents and Their Properties

Multiplying Powers with Like Bases ▪ Dividing Powers with Like Bases ▪ Zero as an Exponent ▪
Raising a Power to a Power ▪ Raising a Product or a Quotient to a Power

Before beginning our study of polynomials, we must develop some rules for working with exponents.

Multiplying Powers with Like Bases

An expression like a^3 means $a \cdot a \cdot a$. We can use this fact to find the product of two expressions that have the same base:

$a^3 \cdot a^2 = (a \cdot a \cdot a)(a \cdot a)$ There are three factors in a^3 and two factors in a^2.

$a^3 \cdot a^2 = a \cdot a \cdot a \cdot a \cdot a$ Using an associative law

$a^3 \cdot a^2 = a^5$.

Note that the exponent in a^5 is the sum of the exponents in $a^3 \cdot a^2$. That is, $3 + 2 = 5$. Similarly,

$b^4 \cdot b^3 = (b \cdot b \cdot b \cdot b)(b \cdot b \cdot b)$

$b^4 \cdot b^3 = b^7$,   where $4 + 3 = 7$.

Adding the exponents gives the correct result.

THE PRODUCT RULE

For any number a and any positive integers m and n,

$$a^m \cdot a^n = a^{m+n}.$$

(To multiply powers with the same base, keep the base and add the exponents.)

Study Skills

Helping Yourself by Helping Others

When you feel confident in your command of a topic, don't hesitate to help classmates experiencing trouble. Your understanding and retention of a concept will deepen when you explain it to someone else and your classmate will appreciate your help.

EXAMPLE 1 Multiply and simplify each of the following. (Here "simplify" means express the product as one base to a power whenever possible.)

a) $2^3 \cdot 2^8$

b) $5^3 \cdot 5^8 \cdot 5$

c) $(r + s)^7 (r + s)^6$

d) $(a^3 b^2)(a^3 b^5)$

SOLUTION

a) $2^3 \cdot 2^8 = 2^{3+8}$ Adding exponents: $a^m \cdot a^n = a^{m+n}$

$= 2^{11}$

> **CAUTION!** The base is unchanged: $2^3 \cdot 2^8 \neq 4^{11}$.

b) $5^3 \cdot 5^8 \cdot 5 = 5^3 \cdot 5^8 \cdot 5^1$ Recall that $x^1 = x$ for any number x.

$= 5^{3+8+1}$ Adding exponents

$= 5^{12}$

> **CAUTION!** $5^{12} \neq 5 \cdot 12$.

c) $(r+s)^7(r+s)^6 = (r+s)^{7+6}$ The base here is $r+s$.

$= (r+s)^{13}$

> **CAUTION!** $(r+s)^{13} \neq r^{13} + s^{13}$.

d) $(a^3b^2)(a^3b^5) = a^3b^2a^3b^5$ Using an associative law

$= a^3a^3b^2b^5$ Using a commutative law

$= a^6b^7$ Adding exponents

1. Multiply and simplify:

$x^5 \cdot x \cdot x^4$.

YOUR TURN

Dividing Powers with Like Bases

Recall that any expression that is divided or multiplied by 1 is unchanged. This, together with the fact that any nonzero number divided by itself is 1, leads to a rule for division:

$$\frac{a^5}{a^2} = \frac{a \cdot a \cdot a \cdot a \cdot a}{a \cdot a} = \frac{a \cdot a \cdot a}{1} \cdot \frac{a \cdot a}{a \cdot a}$$

$$\frac{a^5}{a^2} = \frac{a \cdot a \cdot a}{1} \cdot 1$$

$$\frac{a^5}{a^2} = a \cdot a \cdot a = a^3.$$

Note that the exponent in a^3 is the difference of the exponents in a^5/a^2. Similarly,

$$\frac{x^4}{x^3} = \frac{x \cdot x \cdot x \cdot x}{x \cdot x \cdot x} = \frac{x}{1} \cdot \frac{x \cdot x \cdot x}{x \cdot x \cdot x} = \frac{x}{1} \cdot 1 = x = x^1.$$

Subtracting the exponents gives the correct result.

> **THE QUOTIENT RULE**
>
> For any nonzero number a and any positive integers m and n for which $m > n$,
>
> $$\frac{a^m}{a^n} = a^{m-n}.$$
>
> (To divide powers with the same base, subtract the exponent of the denominator from the exponent of the numerator.)

EXAMPLE 2 Divide and simplify. (Here "simplify" means express the quotient as one base to a power whenever possible.)

a) $\dfrac{7^9}{7^4}$ **b)** $\dfrac{(5a)^{12}}{(5a)^4}$ **c)** $\dfrac{4p^5q^7}{6p^2q}$

SOLUTION

a) $\dfrac{7^9}{7^4} = 7^{9-4}$ Subtracting exponents: $\dfrac{a^m}{a^n} = a^{m-n}$

$= 7^5$

> **CAUTION!** The base is unchanged: $\dfrac{7^9}{7^4} \neq 1^5$.

2. Divide and simplify:

$\dfrac{(x+p)^6}{(x+p)}.$

b) $\dfrac{(5a)^{12}}{(5a)^4} = (5a)^{12-4} = (5a)^8$

> **CAUTION!** The base is $5a$, so the parentheses are essential.

c) $\dfrac{4p^5q^7}{6p^2q} = \dfrac{4}{6} \cdot \dfrac{p^5}{p^2} \cdot \dfrac{q^7}{q^1}$

Note that the 4 and the 6 are factors, not exponents!

$= \dfrac{2}{3} \cdot p^{5-2} \cdot q^{7-1} = \dfrac{2}{3}p^3q^6$

Using the quotient rule twice; simplifying

YOUR TURN

Zero as an Exponent

The quotient rule can be used to help determine what 0 means when it appears as an exponent. Consider a^4/a^4, where a is nonzero. Since the numerator and the denominator are the same,

$\dfrac{a^4}{a^4} = 1.$ We assume $a \neq 0$.

On the other hand, using the quotient rule would give us

$\dfrac{a^4}{a^4} = a^{4-4} = a^0.$ Subtracting exponents

Since $a^0 = a^4/a^4 = 1$, this suggests that $a^0 = 1$ for any nonzero value of a.

> **THE EXPONENT ZERO**
>
> For any real number a, with $a \neq 0$,
>
> $a^0 = 1.$
>
> (Any nonzero number raised to the 0 power is 1.)

Note that in the above box, 0^0 is not defined. For this text, we will assume that expressions like a^m do not represent 0^0.

EXAMPLE 3 Simplify: **(a)** 1948^0; **(b)** $(-9)^0$; **(c)** $(3x)^0$; **(d)** $3x^0$; **(e)** $(-1)9^0$; **(f)** -9^0.

SOLUTION

a) $1948^0 = 1$ Any nonzero number raised to the 0 power is 1.

b) $(-9)^0 = 1$ Any nonzero number raised to the 0 power is 1. The base here is -9.

c) $(3x)^0 = 1$, for any $x \neq 0$. The parentheses indicate that the base is $3x$.

d) Since $3x^0$ means $3 \cdot x^0$, the base is x. Recall that simplifying exponential expressions is done before multiplication in the rules for order of operations:

$3x^0 = 3 \cdot 1 = 3,$ for any $x \neq 0$.

e) $(-1)9^0 = (-1)1 = -1$ The base here is 9.

f) -9^0 is read "the opposite of 9^0" and is equivalent to $(-1)9^0$:

$$-9^0 = (-1)9^0 = (-1)1 = -1.$$

3. Simplify: 99^0.

Note from parts (b), (e), and (f) that $-9^0 = (-1)9^0$ and $-9^0 \neq (-9)^0$.

YOUR TURN

Raising a Power to a Power

Consider an expression like $(7^2)^4$:

$$(7^2)^4 = (7^2)(7^2)(7^2)(7^2)$$ There are four factors of 7^2.
$$(7^2)^4 = (7 \cdot 7)(7 \cdot 7)(7 \cdot 7)(7 \cdot 7)$$ We could also use the product rule.
$$(7^2)^4 = 7 \cdot 7 \cdot 7 \cdot 7 \cdot 7 \cdot 7 \cdot 7 \cdot 7$$ Using an associative law
$$(7^2)^4 = 7^8.$$

Note that the exponent in 7^8 is the product of the exponents in $(7^2)^4$. Similarly,

$$(y^5)^3 = y^5 \cdot y^5 \cdot y^5$$ There are three factors of y^5.
$$(y^5)^3 = (y \cdot y \cdot y \cdot y \cdot y)(y \cdot y \cdot y \cdot y \cdot y)(y \cdot y \cdot y \cdot y \cdot y)$$
$$(y^5)^3 = y^{15}.$$

Once again, we get the same result if we multiply exponents:

$$(y^5)^3 = y^{5 \cdot 3} = y^{15}.$$

THE POWER RULE

For any number a and any whole numbers m and n,

$$(a^m)^n = a^{mn}.$$

(To raise a power to a power, multiply the exponents and leave the base unchanged.)

Remember that for this text we assume that 0^0 is not considered.

EXAMPLE 4   Simplify: $(m^2)^5$.

SOLUTION

$$(m^2)^5 = m^{2 \cdot 5}$$ Multiplying exponents: $(a^m)^n = a^{mn}$
$$= m^{10}$$

4. Simplify: $(3^8)^{10}$.

YOUR TURN

Raising a Product or a Quotient to a Power

When an expression inside parentheses is raised to a power, the inside expression is the base. Let's compare $2a^3$ and $(2a)^3$:

$2a^3 = 2 \cdot a \cdot a \cdot a$;   The base is a.

$(2a)^3 = (2a)(2a)(2a)$ The base is $2a$.
$(2a)^3 = (2 \cdot 2 \cdot 2)(a \cdot a \cdot a)$
$(2a)^3 = 2^3 a^3$
$(2a)^3 = 8a^3.$

We see that $2a^3$ and $(2a)^3$ are *not* equivalent. Note too that $(2a)^3$ can be simplified by cubing each factor in $2a$. This leads to the following rule for raising a product to a power.

> **RAISING A PRODUCT TO A POWER**
>
> For any numbers a and b and any whole number n,
>
> $$(ab)^n = a^n b^n.$$
>
> (To raise a product to a power, raise each factor to that power.)

EXAMPLE 5   Simplify: **(a)** $(4a)^3$; **(b)** $(-5x^4)^2$; **(c)** $(a^7 b)^2 (a^3 b^4)$.

SOLUTION

a) $(4a)^3 = 4^3 a^3 = 64a^3$ Raising each factor to the third power and simplifying

b) $(-5x^4)^2 = (-5)^2 (x^4)^2$ Raising each factor to the second power. Parentheses are important here.

$\quad\quad\quad = 25x^8$ Simplifying $(-5)^2$ and using the power rule

c) $(a^7 b)^2 (a^3 b^4) = (a^7)^2 b^2 a^3 b^4$ Raising a product to a power

$\quad\quad\quad\quad\quad = a^{14} b^2 a^3 b^4$ Multiplying exponents

$\quad\quad\quad\quad\quad = a^{17} b^6$ Adding exponents

5. Simplify: $(-2y^4)^3$.

 YOUR TURN

> **CAUTION!**   The rule $(ab)^n = a^n b^n$ applies only to *products* raised to a power, not to sums or differences. For example, $(3 + 4)^2 \neq 3^2 + 4^2$ since $49 \neq 9 + 16$. Similarly, $(5 + x)^2 \neq 5^2 + x^2$.

There is a similar rule for raising a quotient to a power.

> **RAISING A QUOTIENT TO A POWER**
>
> For any numbers a and b, $b \neq 0$, and any whole number n,
>
> $$\left(\frac{a}{b}\right)^n = \frac{a^n}{b^n}.$$
>
> (To raise a quotient to a power, raise the numerator to the power and divide by the denominator to the power.)

EXAMPLE 6   Simplify: **(a)** $\left(\dfrac{x}{5}\right)^2$; **(b)** $\left(\dfrac{5}{a^4}\right)^3$; **(c)** $\left(\dfrac{3a^4}{b^3}\right)^2$.

SOLUTION

a) $\left(\dfrac{x}{5}\right)^2 = \dfrac{x^2}{5^2} = \dfrac{x^2}{25}$ Squaring the numerator and the denominator

b) $\left(\dfrac{5}{a^4}\right)^3 = \dfrac{5^3}{(a^4)^3}$ Raising a quotient to a power

$\quad\quad\quad = \dfrac{125}{a^{4\cdot3}} = \dfrac{125}{a^{12}}$ Using the power rule and simplifying

c) $\left(\dfrac{3a^4}{b^3}\right)^2 = \dfrac{(3a^4)^2}{(b^3)^2}$ Raising a quotient to a power

$\quad\quad\quad = \dfrac{3^2(a^4)^2}{b^{3\cdot2}} = \dfrac{9a^8}{b^6}$ Raising a product to a power and using the power rule

6. Simplify: $\left(\dfrac{-3}{a^3}\right)^2$.

 YOUR TURN

In the following summary of definitions and rules, we assume that no denominators are 0 and that 0^0 is not considered.

DEFINITIONS AND PROPERTIES OF EXPONENTS

For any whole numbers m and n,

1 as an exponent:	$a^1 = a$
0 as an exponent:	$a^0 = 1$
The Product Rule:	$a^m \cdot a^n = a^{m+n}$
The Quotient Rule:	$\dfrac{a^m}{a^n} = a^{m-n}$
The Power Rule:	$(a^m)^n = a^{mn}$
Raising a product to a power:	$(ab)^n = a^n b^n$
Raising a quotient to a power:	$\left(\dfrac{a}{b}\right)^n = \dfrac{a^n}{b^n}$

4.1   EXERCISE SET   FOR EXTRA HELP

PRACTICE   WATCH   READ   REVIEW

➦ Vocabulary and Reading Check

Complete the sentence using the most appropriate phrase from the column on the right. Each phrase is used once.

1. To raise a product to a power, ____
2. To raise a quotient to a power, ____
3. To raise a power to a power, ____
4. To divide powers with the same base, ____
5. Any nonzero number raised to the 0 power ____
6. To multiply powers with the same base, ____
7. To square a fraction, ____
8. To square a product, ____

a) keep the base and add the exponents.
b) multiply the exponents and leave the base unchanged.
c) square the numerator and square the denominator.
d) square each factor.
e) raise each factor to that power.
f) raise the numerator to the power and divide by the denominator to the power.
g) is one.
h) subtract the exponent of the denominator from the exponent of the numerator.

➦ Concept Reinforcement

Identify the base and the exponent in each expression.

9. $(2x)^5$ 10. $(x + 1)^0$ 11. $2x^3$

12. $-y^6$ 13. $\left(\dfrac{4}{y}\right)^7$ 14. $(-5x)^4$

Multiplying Powers with Like Bases

Simplify.

15. $d^3 \cdot d^{10}$ 16. $8^4 \cdot 8^3$

17. $a^6 \cdot a$ 18. $y^7 \cdot y^9$

19. $6^5 \cdot 6^{10}$ 20. $t^0 \cdot t^{16}$

21. $(3y)^4(3y)^8$ 22. $(2t)^8(2t)^{17}$

23. $(5p)^0(5p)^1$ 24. $(8n)(8n)^9$

25. $(x + 3)^5(x + 3)^8$ 26. $(m - 3)^4(m - 3)^5$

27. $(a^2b^7)(a^3b^2)$ 28. $(a^8b^3)(a^4b)$

29. $r^3 \cdot r^7 \cdot r^0$ 30. $s^4 \cdot s^5 \cdot s^2$

31. $(mn^5)(m^3n^4)$ 32. $(a^3b)(ab)^4$

Dividing Powers with Like Bases

Simplify. Assume that no denominator is 0 and that 0^0 is not considered.

33. $\dfrac{7^5}{7^2}$ 34. $\dfrac{4^7}{4^3}$

35. $\dfrac{t^8}{t}$

36. $\dfrac{x^7}{x}$

37. $\dfrac{(5a)^7}{(5a)^6}$

38. $\dfrac{(3m)^9}{(3m)^8}$

Aha! **39.** $\dfrac{(x+y)^8}{(x+y)^8}$

40. $\dfrac{(9x)^{10}}{(9x)^2}$

41. $\dfrac{(r+s)^{12}}{(r+s)^4}$

42. $\dfrac{(a-b)^4}{(a-b)^3}$

43. $\dfrac{12d^9}{15d^2}$

44. $\dfrac{10n^7}{15n^3}$

45. $\dfrac{8a^9b^7}{2a^2b}$

46. $\dfrac{12r^{10}s^7}{4r^2s}$

47. $\dfrac{x^{12}y^9}{x^0y^2}$

48. $\dfrac{a^{10}b^{12}}{a^2b^0}$

Zero as an Exponent

Simplify.

49. t^0 when $t = 15$

50. y^0 when $y = 38$

51. $5x^0$ when $x = -22$

52. $7m^0$ when $m = 1.7$

53. $7^0 + 4^0$

54. $(8 + 5)^0$

55. $(-3)^1 - (-3)^0$

56. $(-4)^0 - (-4)^1$

Raising a Power to a Power

Simplify.

57. $(x^3)^{11}$

58. $(a^5)^8$

59. $(5^8)^4$

60. $(2^5)^2$

61. $(t^{20})^4$

62. $(x^{25})^6$

Raising a Product or a Quotient to a Power

Simplify. Assume that no denominator is 0 and that 0^0 is not considered.

63. $(10x)^2$

64. $(5a)^2$

65. $(-2a)^3$

66. $(-3x)^3$

67. $(-5n^7)^2$

68. $(-4m^4)^2$

69. $(a^2b)^7$

70. $(xy^4)^9$

71. $(r^5t)^3(r^2t^8)$

72. $(a^4b^6)(a^2b)^5$

73. $(2x^5)^3(3x^4)$

74. $(5x^3)^2(2x^7)$

75. $\left(\dfrac{x}{5}\right)^3$

76. $\left(\dfrac{2}{a}\right)^4$

77. $\left(\dfrac{7}{6n}\right)^2$

78. $\left(\dfrac{4x}{3}\right)^3$

79. $\left(\dfrac{a^3}{b^8}\right)^6$

80. $\left(\dfrac{x^5}{y^2}\right)^7$

81. $\left(\dfrac{x^2y}{z^3}\right)^4$

82. $\left(\dfrac{a^4}{b^2c}\right)^5$

83. $\left(\dfrac{a^3}{-2b^5}\right)^4$

84. $\left(\dfrac{x^5}{-3y^3}\right)^4$

85. $\left(\dfrac{5x^7y}{-2z^4}\right)^3$

86. $\left(\dfrac{-4p^5}{3m^2n^3}\right)^3$

Aha! **87.** $\left(\dfrac{4x^3y^5}{3z^7}\right)^0$

88. $\left(\dfrac{5a^7}{2b^5c}\right)^0$

89. Explain in your own words why $-5^2 \neq (-5)^2$.

90. Under what circumstances should exponents be added?

Skill Review

Solve.

91. $-\dfrac{x}{7} = 3$ [2.1]

92. $3x - 2 \leq 5 - x$ [2.6]

93. $\frac{1}{2}x + \frac{1}{3} = \frac{1}{6}x$ [2.2]

94. $6(x - 10) = 3[4(x - 5)]$ [2.2]

95. $8 - 2(n - 7) > 9 - (3 - n)$ [2.6]

96. $8(5 - y) = 32$ [2.2]

Synthesis

97. Under what conditions does a^n represent a negative number? Why?

98. Using the quotient rule, explain why 9^0 is 1.

99. Suppose that the width of a square is three times the width of a second square. How do the areas of the squares compare? Why?

100. Suppose that the width of a cube is twice the width of a second cube. How do the volumes of the cubes compare? Why?

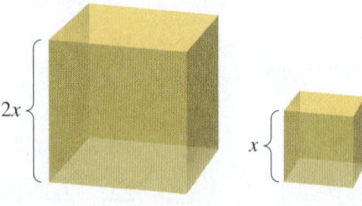

Find a value of the variable that shows that the two expressions are not equivalent. Answers may vary.

101. $3x^2$; $(3x)^2$

102. $(a + 5)^2$; $a^2 + 5^2$

103. $\dfrac{t^6}{t^2}$; t^3

104. $\dfrac{a + 7}{7}$; a

Simplify.

105. $y^{4x} \cdot y^{2x}$

106. $a^{10k} \div a^{2k}$

107. $\dfrac{x^{5t}(x^t)^2}{(x^{3t})^2}$

108. $\dfrac{\left(\frac{1}{2}\right)^3 \left(\frac{2}{3}\right)^4}{\left(\frac{5}{6}\right)^3}$

109. Solve for x:

$$\frac{t^{26}}{t^x} = t^x.$$

Replace ▨ with $>$, $<$, or $=$ to write a true sentence.

110. 3^5 ▨ 3^4

111. 4^2 ▨ 4^3

112. 4^3 ▨ 5^3

113. 4^3 ▨ 3^4

114. 9^7 ▨ 3^{13}

115. 25^8 ▨ 125^5

▦ *Use the fact that $10^3 \approx 2^{10}$ to estimate each of the following powers of 2. Then compute the power of 2 with a calculator and find the difference between the exact value and the approximation.*

116. 2^{14}

117. 2^{22}

118. 2^{26}

119. 2^{31}

▦ *In computer science, 1 KB of memory refers to 1 kilobyte, or 1×10^3 bytes, of memory. This is really an approximation of 1×2^{10} bytes (since computer memory actually uses powers of 2).*

120. The TI-84 Plus graphing calculator has 480 KB of "FLASH ROM." How many bytes is this?

121. The TI-84 Plus Silver Edition graphing calculator has 1.5 MB (megabytes) of FLASH ROM, where 1 MB is 1000 KB (see Exercise 120). How many bytes of FLASH ROM does this calculator have?

↪ **YOUR TURN ANSWERS: SECTION 4.1**

1. x^{10} **2.** $(x + p)^5$ **3.** 1   **4.** 3^{80} **5.** $-8y^{12}$ **6.** $\dfrac{9}{a^6}$

PREPARE TO MOVE ON

Perform the indicated operations.

1. $-10 - 14$ [1.6]

2. $-3(5)$ [1.7]

3. $-16 + 5$ [1.5]

4. $12 - (-4)$ [1.6]

5. $-3 + (-11)$ [1.5]

6. $-8 - (-12)$ [1.6]

4.2   Negative Exponents and Scientific Notation

Negative Integers as Exponents   ▪   Scientific Notation   ▪   Multiplying and Dividing Using Scientific Notation

We now attach a meaning to negative exponents. Once we understand both positive exponents and negative exponents, we can study a method for writing numbers known as *scientific notation*.

Negative Integers as Exponents

Let's define negative exponents so that the rules that apply to whole-number exponents will hold for all integer exponents. To do so, consider a^{-5} and the rule for adding exponents:

$$a^{-5} = a^{-5} \cdot 1 \qquad \text{Using the identity property of 1}$$

$$= \frac{a^{-5}}{1} \cdot \frac{a^5}{a^5} \qquad \text{Writing 1 as } \frac{a^5}{a^5} \text{ and } a^{-5} \text{ as } \frac{a^{-5}}{1}$$

$$= \frac{a^{-5+5}}{a^5} \qquad \text{Adding exponents}$$

$$= \frac{1}{a^5}. \qquad -5 + 5 = 0 \text{ and } a^0 = 1$$

This leads to our definition of negative exponents.

> **NEGATIVE EXPONENTS**
>
> For any real number a that is nonzero and any integer n,
>
> $$a^{-n} = \frac{1}{a^n}.$$
>
> (The numbers a^{-n} and a^n are reciprocals of each other.)

Study Skills

Connect the Dots

Whenever possible, look for connections between concepts covered in different sections or chapters. For example, two sections may both discuss exponents, or two chapters may both cover polynomials.

1. Express using positive exponents and, if possible, simplify: $7x^{-2}$.

EXAMPLE 1   Express using positive exponents and, if possible, simplify.

a) m^{-3} **b)** 4^{-2} **c)** $(-3)^{-2}$ **d)** ab^{-1}

SOLUTION

a) $m^{-3} = \dfrac{1}{m^3}$ m^{-3} is the reciprocal of m^3.

b) $4^{-2} = \dfrac{1}{4^2} = \dfrac{1}{16}$ 4^{-2} is the reciprocal of 4^2. Note that $4^{-2} \neq 4(-2)$.

c) $(-3)^{-2} = \dfrac{1}{(-3)^2} = \dfrac{1}{(-3)(-3)} = \dfrac{1}{9}$ $(-3)^{-2}$ is the reciprocal of $(-3)^2$.

Note that $(-3)^{-2} \neq -\dfrac{1}{3^2}$.

d) $ab^{-1} = a\left(\dfrac{1}{b^1}\right) = a\left(\dfrac{1}{b}\right) = \dfrac{a}{b}$ b^{-1} is the reciprocal of b^1. Note that the base is b, not ab.

 YOUR TURN

> **CAUTION!**   A negative exponent does not, in itself, indicate that an expression is negative. As shown in Example 1,
>
> $$4^{-2} \neq 4(-2) \quad \text{and} \quad (-3)^{-2} \neq -\frac{1}{3^2}.$$

The following is another way to illustrate why negative exponents are defined as they are.

On this side, we divide by 5 at each step.

$$125 = 5^3$$
$$25 = 5^2$$
$$5 = 5^1$$
$$1 = 5^0$$
$$\frac{1}{5} = 5^?$$
$$\frac{1}{25} = 5^?$$

On this side, the exponents decrease by 1.

To continue the pattern, it follows that

$$\frac{1}{5} = \frac{1}{5^1} = 5^{-1}, \quad \frac{1}{25} = \frac{1}{5^2} = 5^{-2}, \quad \text{and, in general,} \quad \frac{1}{a^n} = a^{-n}.$$

EXAMPLE 2   Express $\dfrac{1}{x^7}$ using negative exponents.

2. Express $\dfrac{1}{10^7}$ using negative exponents.

SOLUTION   We know that $\dfrac{1}{a^n} = a^{-n}$. Thus, $\dfrac{1}{x^7} = x^{-7}$.

 YOUR TURN

The rules for exponents still hold when exponents are negative.

EXAMPLE 3 Simplify. Do not use negative exponents in the answer.

a) $t^5 \cdot t^{-2}$

b) $(5x^2y^{-3})^4$

c) $\dfrac{x^{-4}}{x^{-5}}$

d) $\dfrac{1}{t^{-5}}$

e) $\dfrac{s^{-3}}{t^{-5}}$

f) $\dfrac{-10x^{-3}y}{5x^2y^5}$

SOLUTION

a) $t^5 \cdot t^{-2} = t^{5+(-2)} = t^3$ Adding exponents

b) $(5x^2y^{-3})^4 = 5^4(x^2)^4(y^{-3})^4$ Raising each factor to the fourth power

$$= 625x^8y^{-12} = \frac{625x^8}{y^{12}}$$

c) $\dfrac{x^{-4}}{x^{-5}} = x^{-4-(-5)} = x^1 = x$ We subtract exponents even if the exponent in the denominator is negative.

d) Since $\dfrac{1}{a^n} = a^{-n}$, we have $\dfrac{1}{t^{-5}} = t^{-(-5)} = t^5$.

e) $\dfrac{s^{-3}}{t^{-5}} = s^{-3} \cdot \dfrac{1}{t^{-5}}$

$$= \frac{1}{s^3} \cdot t^5 = \frac{t^5}{s^3}$$ Using the result from part (d) above

f) $\dfrac{-10x^{-3}y}{5x^2y^5} = \dfrac{-10}{5} \cdot \dfrac{x^{-3}}{x^2} \cdot \dfrac{y^1}{y^5}$ Note that the -10 and 5 are factors.

$$= -2 \cdot x^{-3-2} \cdot y^{1-5}$$ Using the quotient rule twice; simplifying

$$= -2x^{-5}y^{-4} = \frac{-2}{x^5y^4}$$

3. Simplify:

$$\frac{3x^{-2}}{x^{10}}.$$

Do not use negative exponents in the answer.

YOUR TURN

The result from Example 3(e) can be generalized.

> ### FACTORS AND NEGATIVE EXPONENTS
>
> For any nonzero real numbers a and b and any integers m and n,
>
> $$\frac{a^{-n}}{b^{-m}} = \frac{b^m}{a^n}.$$
>
> (A factor can be moved to the other side of the fraction bar if the sign of the exponent is changed.)

EXAMPLE 4 Simplify: $\dfrac{-15x^{-7}}{5y^2z^{-4}}$.

SOLUTION We can move the factors x^{-7} and z^{-4} to the other side of the fraction bar if we change the sign of each exponent:

$$\frac{-15x^{-7}}{5y^2z^{-4}} = \frac{-15}{5} \cdot \frac{x^{-7}}{y^2z^{-4}}$$ We can simply divide the constant factors.

$$= -3 \cdot \frac{z^4}{y^2x^7}$$

$$= \frac{-3z^4}{x^7y^2}.$$

4. Simplify: $\dfrac{12a^{-1}}{4bc^{-3}}$.

 YOUR TURN

Another way to change the sign of the exponent is to take the reciprocal of the base. To understand why this is true, note that

$$\left(\frac{s}{t}\right)^{-5} = \frac{s^{-5}}{t^{-5}} = \frac{t^5}{s^5} = \left(\frac{t}{s}\right)^5.$$

This often provides the easiest way to simplify an expression containing a negative exponent.

RECIPROCALS AND NEGATIVE EXPONENTS

For any nonzero real numbers a and b and any integer n,

$$\left(\frac{a}{b}\right)^{-n} = \left(\frac{b}{a}\right)^n.$$

(A base to a power is equal to the reciprocal of the base raised to the opposite power.)

EXAMPLE 5 Simplify: $\left(\dfrac{x^4}{2y}\right)^{-3}$.

SOLUTION

$$\left(\frac{x^4}{2y}\right)^{-3} = \left(\frac{2y}{x^4}\right)^3$$
Taking the reciprocal of the base and changing the sign of the exponent

$$= \frac{(2y)^3}{(x^4)^3}$$
Raising a quotient to a power by raising both the numerator and the denominator to the power

$$= \frac{2^3 y^3}{x^{12}}$$
Raising a product to a power; using the power rule in the denominator

$$= \frac{8y^3}{x^{12}}$$
Cubing 2

5. Simplify: $\left(\dfrac{3x}{y^5}\right)^{-2}$.

YOUR TURN

Scientific Notation

Scientific notation provides a useful way of writing the very large or very small numbers that occur in science. The following are examples of scientific notation.

The mass of the earth:

6.0×10^{24} kilograms (kg) = 6,000,000,000,000,000,000,000,000 kg

The mass of a hydrogen atom:

1.7×10^{-24} g = 0.0000000000000000000000017 g

Student Notes

Definitions are usually written as concisely as possible, so that every phrase included is important. The definition for scientific notation states that $1 \le N < 10$. Thus, 2.68×10^5 is written in scientific notation, but 26.8×10^5 and 0.268×10^5 are *not* written in scientific notation.

SCIENTIFIC NOTATION

Scientific notation for a number is an expression of the type

$$N \times 10^m,$$

where N is at least 1 but less than 10 (that is, $1 \le N < 10$), N is expressed in decimal notation, and m is an integer.

Converting from scientific notation to decimal notation involves multiplying by a power of 10. Consider the following.

Scientific Notation	Multiplication	Decimal Notation
4.52×10^2	4.52×100	452.
4.52×10^1	4.52×10	45.2
4.52×10^0	4.52×1	4.52
4.52×10^{-1}	4.52×0.1	0.452
4.52×10^{-2}	4.52×0.01	0.0452

We generally perform this multiplication mentally. Thus to convert $N \times 10^m$ to decimal notation, we move the decimal point.

- When m is positive, we move the decimal point m places to the right.
- When m is negative, we move the decimal point $|m|$ places to the left.

EXAMPLE 6   Convert to decimal notation: **(a)** 7.893×10^5; **(b)** 4.7×10^{-8}.

SOLUTION

a) Since the exponent is positive, the decimal point moves to the right:

$$7.89300. \qquad 7.893 \times 10^5 = 789{,}300 \qquad \text{The decimal point moves}$$
$$\underset{\text{5 places}}{\curvearrowright} \qquad\qquad\qquad\qquad\qquad\qquad \text{5 places to the right.}$$

b) Since the exponent is negative, the decimal point moves to the left:

$$0.00000004.7 \qquad 4.7 \times 10^{-8} = 0.000000047 \qquad \text{The decimal point moves}$$
$$\underset{\text{8 places}}{\curvearrowleft} \qquad\qquad\qquad\qquad\qquad\qquad\qquad \text{8 places to the left.}$$

6. Convert to decimal notation:

  8.04×10^{-3}.

_____

 YOUR TURN

To convert from decimal notation to scientific notation, this procedure is reversed.

EXAMPLE 7   Write in scientific notation: **(a)** 83,000; **(b)** 0.0327.

SOLUTION

a) We need to find m such that $83{,}000 = 8.3 \times 10^m$. To change 8.3 to 83,000 requires moving the decimal point 4 places to the right. This can be accomplished by multiplying by 10^4. Thus,

$$83{,}000 = 8.3 \times 10^4. \qquad \text{This is scientific notation.}$$

b) We need to find m such that $0.0327 = 3.27 \times 10^m$. To change 3.27 to 0.0327 requires moving the decimal point 2 places to the left. This can be accomplished by multiplying by 10^{-2}. Thus,

7. Write in scientific notation:

  600,000.

$$0.0327 = 3.27 \times 10^{-2}. \qquad \text{This is scientific notation.}$$

_____

YOUR TURN

In scientific notation, positive exponents are used to represent large numbers and negative exponents are used to represent small numbers between 0 and 1.

Multiplying and Dividing Using Scientific Notation

Products and quotients of numbers written in scientific notation are found using the rules for exponents.

EXAMPLE 8 Simplify.

a) $(1.8 \times 10^9) \cdot (2.3 \times 10^{-4})$

b) $(3.41 \times 10^5) \div (1.1 \times 10^{-3})$

SOLUTION

a) $(1.8 \times 10^9) \cdot (2.3 \times 10^{-4})$

$= 1.8 \times 2.3 \times 10^9 \times 10^{-4}$ Using the associative and commutative laws

$= 4.14 \times 10^{9+(-4)}$ Adding exponents

$= 4.14 \times 10^5$

b) $(3.41 \times 10^5) \div (1.1 \times 10^{-3})$

$= \dfrac{3.41 \times 10^5}{1.1 \times 10^{-3}}$

$= \dfrac{3.41}{1.1} \times \dfrac{10^5}{10^{-3}}$

$= 3.1 \times 10^{5-(-3)}$ Subtracting exponents

$= 3.1 \times 10^8$

 Chapter Resource:
Decision Making: Connection, p. 294

8. Simplify:

$(2.2 \times 10^{-8}) \times (2.5 \times 10^{-10})$.

 YOUR TURN

When a problem is stated using scientific notation, we generally use scientific notation for the answer. This often requires an additional conversion.

EXAMPLE 9 Simplify.

a) $(3.1 \times 10^5) \cdot (4.5 \times 10^{-3})$

b) $(7.2 \times 10^{-7}) \div (8.0 \times 10^6)$

SOLUTION

a) We have

$$(3.1 \times 10^5) \cdot (4.5 \times 10^{-3}) = 3.1 \times 4.5 \times 10^5 \times 10^{-3}$$
$$= 13.95 \times 10^2.$$

Our answer is not yet in scientific notation because 13.95 is not between 1 and 10. We convert to scientific notation as follows:

$13.95 \times 10^2 = 1.395 \times 10^1 \times 10^2$ Substituting 1.395×10^1 for 13.95

$= 1.395 \times 10^3.$ Adding exponents

Technology Connection

A key labeled $\boxed{10^x}$, $\boxed{\wedge}$, or $\boxed{\text{EE}}$ is used to enter scientific notation into a calculator. Sometimes this is a secondary function, meaning that another key—often labeled SHIFT or $\boxed{\text{2ND}}$ —must be pressed first.

To check Example 9(a), we press

3.1 $\boxed{\text{EE}}$ 5 $\boxed{\times}$ 4.5 $\boxed{\text{EE}}$ $\boxed{(-)}$ 3 $\boxed{\text{ENTER}}$.

The result that appears represents 1.395×10^3. On some calculators, the MODE SCI must be selected in order to display scientific notation.

```
3.1E5*4.5E−3
                        1.395E3
```

Calculate each of the following.

1. $(3.8 \times 10^9) \cdot (4.5 \times 10^7)$
2. $(2.9 \times 10^{-8}) \div (5.4 \times 10^6)$
3. $(9.2 \times 10^7) \div (2.5 \times 10^{-9})$

b) $(7.2 \times 10^{-7}) \div (8.0 \times 10^6) = \dfrac{7.2 \times 10^{-7}}{8.0 \times 10^6} = \dfrac{7.2}{8.0} \times \dfrac{10^{-7}}{10^6}$

$= 0.9 \times 10^{-13}$

$= 9.0 \times 10^{-1} \times 10^{-13}$ Substituting 9.0×10^{-1} for 0.9

$= 9.0 \times 10^{-14}$ Adding exponents

9. Simplify: $\dfrac{1.2 \times 10^6}{3.0 \times 10^{-11}}$.

YOUR TURN

CONNECTING 🔗 THE CONCEPTS

DEFINITIONS AND PROPERTIES OF EXPONENTS

The following summary assumes that no denominators are 0 and that 0^0 is not considered. For any integers m and n,

1 as an exponent: $a^1 = a$

0 as an exponent: $a^0 = 1$

Negative exponents: $a^{-n} = \dfrac{1}{a^n}$,

$\dfrac{a^{-n}}{b^{-m}} = \dfrac{b^m}{a^n}$,

$\left(\dfrac{a}{b}\right)^{-n} = \left(\dfrac{b}{a}\right)^n$

The Product Rule: $a^m \cdot a^n = a^{m+n}$

The Quotient Rule: $\dfrac{a^m}{a^n} = a^{m-n}$

The Power Rule: $(a^m)^n = a^{mn}$

Raising a product to a power: $(ab)^n = a^n b^n$

Raising a quotient to a power: $\left(\dfrac{a}{b}\right)^n = \dfrac{a^n}{b^n}$

EXERCISES

Simplify. Do not use negative exponents in the answer.

1. $x^4 x^{10}$ **2.** $x^{-4} x^{-10}$

3. $\dfrac{x^{-4}}{x^{10}}$ **4.** $\dfrac{x^4}{x^{-10}}$

5. $(x^{-4})^{-10}$ **6.** $(x^4)^{10}$

7. $\dfrac{1}{c^{-8}}$ **8.** c^{-8}

9. $\left(\dfrac{a^3}{b^4}\right)^5$ **10.** $\left(\dfrac{a^3}{b^4}\right)^{-5}$

4.2   EXERCISE SET

FOR EXTRA HELP

MyMathLab®   Math XL
PRACTICE   WATCH   READ   REVIEW

🔖 Vocabulary and Reading Check

Classify each of the following statements as either true or false.

1. A negative exponent indicates a reciprocal.

2. The expressions 3^{-2} and -3^2 are equivalent.

3. A positive exponent of the base 10 in scientific notation indicates a number greater than or equal to 10.

4. The number 18.68×10^{12} is written in scientific notation.

EXAMPLE 3   Identify the coefficient of each term in the polynomial

$$4x^3 - 7x^2 + x - 8.$$

SOLUTION

The coefficient of $4x^3$ is 4.

The coefficient of $-7x^2$ is -7.

The coefficient of the third term is 1, since $x = 1x$.

The coefficient of -8 is simply -8.

3. Identify the coefficient of each term in the polynomial $y^5 - y^2 + 12y - 9$.

YOUR TURN

The **leading term** of a polynomial is the term of highest degree. Its coefficient is called the **leading coefficient**, and its degree is referred to as the **degree of the polynomial**. To see how this terminology is used, consider the polynomial

$$3x^2 - 8x^3 + 5x^4 + 7x - 6.$$

The *terms* are $3x^2$, $-8x^3$, $5x^4$, $7x$,   and -6.

The *coefficients* are 3, -8, 5, 7,   and -6.

The *degree of each term* is 2, 3, 4, 1,   and 0.

The *leading term* is $5x^4$ and the *leading coefficient* is 5.

The *degree of the polynomial* is 4.

Combining Like Terms

Like, or *similar*, *terms* are either constant terms or terms containing the same variable(s) raised to the same power(s). For example, the like terms in

$$4x^3 + 5x - 7x^2 + 2x^3 + x^2$$

are $4x^3$ and $2x^3$ as well as $-7x^2$ and x^2.

Often we can simplify polynomials by *combining*, or *collecting*, like terms.

EXAMPLE 4   Write an equivalent expression by combining like terms.

a) $2x^3 + 6x^3$

b) $5x^2 + 7 + 2x^4 - 6x^2 - 11 - 2x^4$

c) $7a^3 - 5a^2 + 9a^3 + a^2$

d) $\frac{2}{3}x^4 - x^3 - \frac{1}{6}x^4 + \frac{2}{5}x^3 - \frac{3}{10}x^3$

Student Notes

Remember that when we combine like terms, we are not solving equations, but are forming equivalent expressions.

SOLUTION

a) $2x^3 + 6x^3 = (2 + 6)x^3$ Using the distributive law

$\qquad\qquad = 8x^3$

b) $5x^2 + 7 + 2x^4 - 6x^2 - 11 - 2x^4$

$\qquad = 5x^2 - 6x^2 + 2x^4 - 2x^4 + 7 - 11$

$\qquad = (5 - 6)x^2 + (2 - 2)x^4 + (7 - 11)$

$\qquad = -1x^2 + 0x^4 + (-4)$

$\qquad = -x^2 - 4$

These steps are often done mentally.

c) $7a^3 - 5a^2 + 9a^3 + a^2 = 7a^3 - 5a^2 + 9a^3 + 1a^2$ $a^2 = 1 \cdot a^2 = 1a^2$

$\qquad\qquad = 16a^3 - 4a^2$

d) $\frac{2}{3}x^4 - x^3 - \frac{1}{6}x^4 + \frac{2}{5}x^3 - \frac{3}{10}x^3 = \left(\frac{2}{3} - \frac{1}{6}\right)x^4 + \left(-1 + \frac{2}{5} - \frac{3}{10}\right)x^3$

$\qquad\qquad = \left(\frac{4}{6} - \frac{1}{6}\right)x^4 + \left(-\frac{10}{10} + \frac{4}{10} - \frac{3}{10}\right)x^3$

$\qquad\qquad = \frac{3}{6}x^4 - \frac{9}{10}x^3$

$\qquad\qquad = \frac{1}{2}x^4 - \frac{9}{10}x^3$ There are no like terms, so we are done.

4. Write an equivalent expression by combining like terms:

$$2n^2 - 6n + n^2 + 6n.$$

YOUR TURN

Note in Example 4 that the solutions are written so that the term of highest degree appears first, followed by the term of next highest degree, and so on. This is known as **descending order** and is the form in which answers will normally appear.

Evaluating Polynomials and Applications

When each variable in a polynomial is replaced with a number, the polynomial then represents a number, or *value,* that can be calculated using the rules for order of operations.

EXAMPLE 5 Evaluate $-x^2 + 3x + 9$ for $x = -2$.

SOLUTION For $x = -2$, we have

Substitute.

$$-x^2 + 3x + 9 = -(-2)^2 + 3(-2) + 9 \qquad \text{The negative sign in front of } x^2 \text{ remains.}$$
$$= -(4) + 3(-2) + 9 \qquad \text{Evaluating the exponential expression}$$

Simplify.

$$= -4 + (-6) + 9 \qquad \text{Multiplying}$$
$$= -10 + 9 = -1. \qquad \text{Adding}$$

5. Evaluate $t^3 - 5t^2 - 10t + 1$ for $t = -1$.

 YOUR TURN

EXAMPLE 6 *Renewable Energy Sources.* The number of watts of power P generated by a particular wind turbine at a wind speed of x miles per hour can be approximated by the polynomial

$$P = 0.0157x^3 + 0.1163x^2 - 1.3396x + 3.7063.$$

Estimate the power generated by a 10-mph wind.

Source: Based on data from QST, November 2006

SOLUTION To find the power generated by a 10-mph wind, we evaluate the polynomial for $x = 10$:

$$P = 0.0157x^3 + 0.1163x^2 - 1.3396x + 3.7063$$
$$= 0.0157(10)^3 + 0.1163(10)^2 - 1.3396(10) + 3.7063$$
$$= 15.7 + 11.63 - 13.396 + 3.7063$$
$$= 17.6403.$$

6. Use the polynomial given in Example 6 to estimate the power generated by a 20-mph wind.

Since this value is approximate, we round the answer. A 10-mph wind will generate about 17.6 watts.

YOUR TURN

Sometimes, a graph can be used to estimate the value of a polynomial visually.

EXAMPLE 7 *Renewable Energy Sources.* In the following figure, the polynomial from Example 6 has been graphed by evaluating it for several choices of x. Use the graph to estimate the power generated by a 30-mph wind.

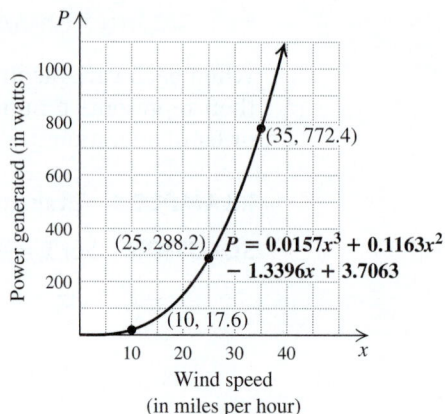

SOLUTION To estimate the power generated by a 30-mph wind, we locate 30 on the horizontal axis of the graph. From there, we move vertically until we meet the curve. From that point, we move horizontally to the P-axis.

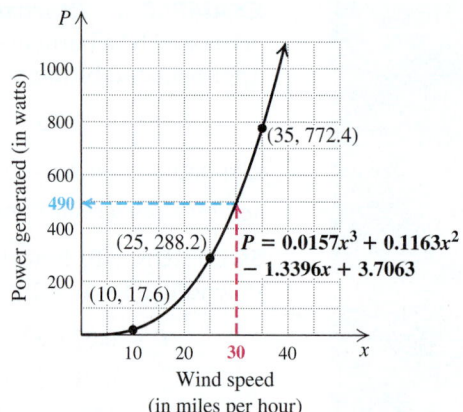

7. Use the graph given in Example 7 to estimate the power generated by a 15-mph wind.

We see that the power generated by a 30-mph wind is approximately 490 watts. (For $x = 30$, the value of $0.0157x^3 + 0.1163x^2 - 1.3396x + 3.7063$ is approximately 490.)

 YOUR TURN

Technology Connection

One way to evaluate a polynomial is to use the TRACE key. For example, to evaluate $0.0157x^3 + 0.1163x^2 - 1.3396x + 3.7063$ in Example 7 for $x = 30$, we can graph the polynomial $y = 0.0157x^3 + 0.1163x^2 - 1.3396x + 3.7063$. We then use TRACE and enter an x-value of 30.

The value of the polynomial appears as y, and the cursor appears at $(30, 492.0883)$. The VALUE option of the CALC menu works in a similar way.

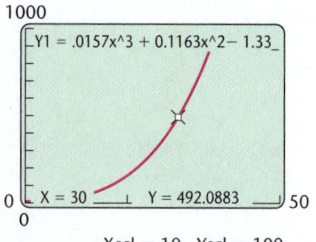

$Xscl = 10, \; Yscl = 100$

1. Use TRACE or CALC VALUE to find the value of $0.0157x^3 + 0.1163x^2 - 1.3396x + 3.7063$ for $x = 40$.

4.3 EXERCISE SET

FOR EXTRA HELP

MyMathLab® Math XL
PRACTICE   WATCH   READ   REVIEW

Vocabulary and Reading Check

Match the description in the left-hand column with the most appropriate algebraic expression from the column on the right.

1. _____ A polynomial with four terms

2. _____ A polynomial with 7 as its leading coefficient

3. _____ A trinomial written in descending order

4. _____ A polynomial with degree 5

5. _____ A binomial with degree 7

6. _____ A monomial of degree 0

7. _____ An expression with two terms that is not a binomial

8. _____ An expression with three terms that is not a trinomial

a) $8x^3 + \dfrac{2}{x^2}$

b) $5x^4 + 3x^3 - 4x + 7$

c) $\dfrac{3}{x} - 6x^2 + 9$

d) $8t - 4t^5$

e) 5

f) $6x^2 + 7x^4 - 2x^3$

g) $4t - 2t^7$

h) $3t^2 + 4t + 7$

Concept Reinforcement

Determine whether each expression is a polynomial.

9. $3x - 7$

10. $-2x^5 + 9 - 7x^2$

11. $\dfrac{x^2 + x + 1}{x^3 - 7}$

12. -10

13. $\frac{1}{4}x^{10} - 8.6$

14. $\dfrac{3}{x^4} - \dfrac{1}{x} + 13$

Terms

Identify the terms of each polynomial.

15. $8x^3 - 11x^2 + 6x + 1$

16. $5a^3 + 4a^2 - a - 7$

17. $-t^6 - 3t^3 + 9t - 4$

18. $n^5 - 4n^3 + 2n - 8$

Types of Polynomials

Classify each polynomial as a monomial, a binomial, a trinomial, or a polynomial with no special name.

19. $x^2 - 23x + 17$

20. $-9x^2$

21. $x^3 - 7x + 2x^2 - 4$

22. $t^3 + 4$

23. $y + 8$

24. $3x^8 + 12x^3 - 9$

25. 17

26. $2x^4 - 7x^3 + x^2 + x - 6$

Degree and Coefficients

27. Complete the following table for the polynomial
$$7x^2 + 8x^5 - 4x^3 + 6 - \tfrac{1}{2}x^4.$$

Term	Coefficient	Degree of the Term	Degree of the Polynomial
		5	
$-\frac{1}{2}x^4$			
	-4		
		2	
	6		

28. Complete the following table for the polynomial
$$-3x^4 + 6x^3 - 2x^2 + 8x + 7.$$

Term	Coefficient	Degree of the Term	Degree of the Polynomial
	-3		
$6x^3$			
		2	
		1	
	7		

Determine the coefficient and the degree of each term in each polynomial.

29. $8x^4 + 2x$

30. $9a^3 - 4a^2$

31. $9t^2 - 3t + 4$

32. $7x^4 + 5x - 3$

33. $x^4 - x^3 + 4x - 3$

34. $2a^5 + a^2 + 8a + 10$

For each of the following polynomials, (a) list the degree of each term; (b) determine the leading term and the leading coefficient; and (c) determine the degree of the polynomial.

35. $5t + t^3 + 8t^4$

36. $1 + 6n + 4n^2$

37. $3a^2 - 7 + 2a^4$

38. $9x^4 + x^2 + x^7 - 12$

39. $8 + 6x^2 - 3x - x^5$

40. $9a - a^4 + 3 + 2a^3$

Combining Like Terms

Combine like terms. Write all answers in descending order.

41. $5n^2 + n + 6n^2$ **42.** $5a + 7a^2 + 3a$

43. $3a^4 - 2a + 2a + a^4$

44. $9b^5 + 3b^2 - 2b^5 - 3b^2$

45. $4b^3 + 5b + 7b^3 + b^2 - 6b$

46. $6x^2 + 2x^4 - 2x^2 - x^4 - 4x^2 + x$

47. $10x^2 + 2x^3 - 3x^3 - 4x^2 - 6x^2 - x^4$

48. $12t^6 - t^3 + 8t^6 + 4t^3 - t^7 - 3t^3$

49. $\frac{1}{5}x^4 + 7 - 2x^2 + 3 - \frac{2}{15}x^4 + 2x^2$

50. $\frac{1}{6}x^3 + 3x^2 - \frac{1}{3}x^3 + 7 + x^2 - 10$

51. $8.3a^2 + 3.7a - 8 - 9.4a^2 + 1.6a + 0.5$

52. $1.4y^3 + 2.9 - 7.7y - 1.3y - 4.1 + 9.6y^3$

Evaluating Polynomials and Applications

Evaluate each polynomial for $x = 3$ and for $x = -3$.

53. $-4x + 9$ **54.** $-6x + 5$

55. $2x^2 - 3x + 7$ **56.** $4x^2 - 6x + 9$

57. $-3x^3 + 7x^2 - 4x - 8$

58. $-2x^3 - 3x^2 + 4x + 2$

59. $2x^4 - \frac{1}{9}x^3$ **60.** $\frac{1}{3}x^4 - 2x^3$

61. *Skydiving.* During the first 13 sec of a jump, the number of feet that a skydiver falls in t seconds is approximated by the polynomial

$$11.12t^2.$$

In 2009, 108 U.S. skydivers fell headfirst in formation from a height of 18,000 ft. How far had they fallen 10 sec after having jumped from the plane?

Source: www.telegraph.co.uk

62. *Skydiving.* For jumps that exceed 13 sec, the polynomial $173t - 369$ can be used to approximate the distance, in feet, that a skydiver has fallen in t seconds. Approximately how far has a skydiver fallen 20 sec after having jumped from a plane?

Circumference. *The circumference of a circle of radius r is given by the polynomial $2\pi r$, where π is an irrational number. For an approximation of π, use 3.14.*

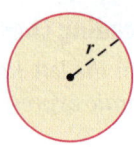

63. Find the circumference of a circle with radius 10 cm.

64. Find the circumference of a circle with radius 5 ft.

Area of a Circle. *The area of a circle of radius r is given by the polynomial πr^2. Use 3.14 for π.*

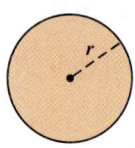

65. Find the area of a circle with radius 7 m.

66. Find the area of a circle with radius 6 ft.

67. *Kayaking.* The distance $s(t)$, in feet, traveled by a body falling freely from rest in t seconds is approximated by $s(t) = 16t^2$. On March 4, 2009, Brazilian kayaker Pedro Olivia set a world record waterfall descent on the Rio Sacre in Brazil. He was airborne for 2.9 sec. How far did he drop?

Source: www.telegraph.co.uk

68. *SCAD Diving.* The SCAD thrill ride is a 2.5-sec free fall into a net. How far does the diver fall? (See Exercise 67.)

Source: "What is SCAD?", www.scadfreefall.co.uk

69. *Stacking Spheres.* In 2004, the journal *Annals of Mathematics* accepted a proof of the so-called Kepler Conjecture: that the most efficient way to pack spheres is in the shape of a square pyramid. The number N of balls in the stack is given by the polynomial function

$$N(x) = \tfrac{1}{3}x^3 + \tfrac{1}{2}x^2 + \tfrac{1}{6}x,$$

where x is the number of layers. Use both the function and the figure to find $N(3)$. Then calculate the number of oranges in a pyramid with 5 layers.

Source: The New York Times 4/6/04

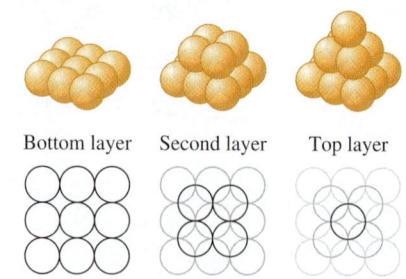

Bottom layer   Second layer   Top layer

70. *Stacking Cannonballs.* The function in Exercise 69 was discovered by Thomas Harriot, assistant to Sir Walter Raleigh, when preparing for an expedition at sea. How many cannonballs did they pack if there were 10 layers to their pyramid?

Source: The New York Times 4/7/04

Veterinary Science. *Gentamicin is an antibiotic frequently used by veterinarians. The concentration, in micrograms per milliliter (mcg/mL), of Gentamicin in a horse's bloodstream t hours after injection can be approximated by the polynomial*

$$C = -0.005t^4 + 0.003t^3 + 0.35t^2 + 0.5t.$$

Use the following graph for Exercises 71 and 72.

Source: Michele Tulis, DVM, telephone interview

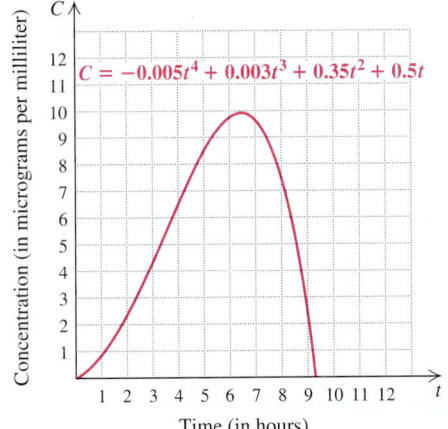

71. Estimate the concentration, in mcg/mL, of Gentamicin in the bloodstream 2 hr after injection.

72. Estimate the concentration, in mcg/mL, of Gentamicin in the bloodstream 4 hr after injection.

73. Explain how it is possible for a term to not be a monomial.

74. Why is it important to understand the rules for order of operations when evaluating polynomials?

Skill Review

Graph.

75. $y = 3x$ [3.2]

76. $y = \frac{1}{2}x - 5$ [3.6]

77. $3x - y = 3$ [3.3]

78. $y = -1$ [3.3]

79. $x = 2$ [3.3]

80. $y + 2 = -2(x - 1)$ [3.7]

Synthesis

81. Suppose that the coefficients of a polynomial are all integers and the polynomial is evaluated for some integer. Must the value of the polynomial then also be an integer? Why or why not?

82. Is it easier to evaluate a polynomial before or after like terms have been combined? Why?

83. Construct a polynomial in x (meaning that x is the variable) of degree 5 with four terms, with coefficients that are consecutive even integers. Write in descending order.

Revenue, Cost, and Profit. *Gigabytes Electronics is selling a new type of computer monitor.* Total revenue *is the total amount of money taken in and* total cost *is the total amount paid for producing the items. The firm estimates that for the monitor's first year, revenue from the sale of x monitors is*

$$250x - 0.5x^2 \text{ dollars,}$$

and the total cost is given by

$$4000 + 0.6x^2 \text{ dollars.}$$

Profit *is the difference between revenue and cost.*

84. Find the profit when 20 monitors are produced and sold.

85. Find the profit when 30 monitors are produced and sold.

Simplify.

86. $\frac{9}{2}x^8 + \frac{1}{9}x^2 + \frac{1}{2}x^9 + \frac{9}{2}x + \frac{9}{2}x^9 + \frac{8}{9}x^2 + \frac{1}{2}x - \frac{1}{2}x^8$

87. $(3x^2)^3 + 4x^2 \cdot 4x^4 - x^4(2x)^2 + ((2x)^2)^3 - 100x^2(x^2)^2$

88. A polynomial in x has degree 3. The coefficient of x^2 is 3 less than the coefficient of x^3. The coefficient of x is three times the coefficient of x^2. The remaining constant is 2 more than the coefficient of x^3. The sum of the coefficients is -4. Find the polynomial.

89. Use the graph for Exercises 71 and 72 to determine the times for which the concentration of Gentamicin is 5 mcg/mL.

90. *Path of the Olympic Arrow.* The Olympic flame at the 1992 Summer Olympics in Spain was lit by a flaming arrow. As the arrow moved d meters horizontally from the archer, its height h, in meters, was approximated by the polynomial

$$-0.0064d^2 + 0.8d + 2.$$

Complete the table for the choices of d given. Then plot the points and draw a graph representing the path of the arrow.

d	$-0.0064d^2 + 0.8d + 2$
0	
30	
60	
90	
120	

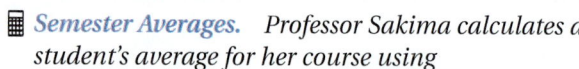**Semester Averages.**   *Professor Sakima calculates a student's average for her course using*

$$A = 0.3q + 0.4t + 0.2f + 0.1h,$$

with q, t, f, and h representing a student's quiz average, test average, final exam score, and homework average, respectively. In Exercises 91 and 92, find the given student's course average rounded to the nearest tenth.

91. Beth: quizzes: 60, 85, 72, 91; final exam: 84; tests: 89, 93, 90; homework: 88

92. Cameron: quizzes: 95, 99, 72, 79; final exam: 91; tests: 68, 76, 92; homework: 86

93. *Research.*   Find out how grades are calculated in one or more of your classes. Write a polynomial that will determine a student's final course average. See Exercises 91 and 92 for examples.

In Exercises 94 and 95, complete the table for the given choices of t. Then plot the points and connect them with a smooth curve representing the graph of the polynomial.

94.

t	$-t^2 + 6t - 4$
1	
2	
3	
4	
5	

95.

t	$-t^2 + 10t - 18$
3	
4	
5	
6	
7	

YOUR TURN ANSWERS: SECTION 4.3

1. $3x, -6, 5x^4$ **2.** 1   **3.** $1, -1, 12, -9$ **4.** $3n^2$
5. 5   **6.** Approximately 149 watts   **7.** Approximately 60 watts

QUICK QUIZ: SECTIONS 4.1–4.3

Simplify.

1. $(-3x^5y^4)^2$ [4.1]   **2.** $\dfrac{48a^5b}{-3ab}$ [4.1]

3. List the coefficients of the polynomial $x^6 - 6x + \frac{1}{2}$. [4.3]

4. Evaluate $-x - 3x^2 + x^3$ for $x = -10$.   [4.3]

5. Evaluate t^0 for $t = -12.5$.   [4.1]

PREPARE TO MOVE ON

Simplify.   [1.8]

1. $2x + 5 - (x + 8)$ **2.** $3x - 7 - (5x - 1)$

3. $4a + 3 - (-2a + 6)$ **4.** $\frac{1}{2}t - \frac{1}{4} - \left(\frac{3}{2}t + \frac{3}{4}\right)$

5. $4t^4 + 8t - (5t^4 - 9t)$

6. $0.1a^2 + 5 - (-0.3a^2 + a - 6)$

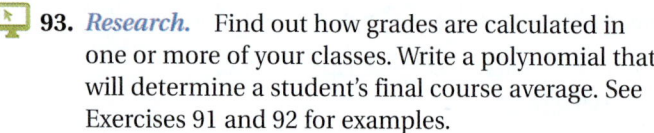

4.4   Addition and Subtraction of Polynomials

Addition of Polynomials   ▪   Opposites of Polynomials   ▪   Subtraction of Polynomials   ▪   Problem Solving

Addition of Polynomials

To add two polynomials, we write a plus sign between them and combine like terms.

EXAMPLE 1   Write an equivalent expression by adding.

a) $(-5x^3 + 6x - 1) + (4x^3 + 3x^2 + 2)$

b) $\left(\frac{2}{3}x^4 + 3x^2 - 7x + \frac{1}{2}\right) + \left(-\frac{1}{3}x^4 + 5x^3 - 3x^2 + 3x - \frac{1}{2}\right)$

CAUTION! Note that equations like those in Examples 1 and 2 are written to show how one expression can be rewritten in an equivalent form. This is very different from solving an equation.

SOLUTION

a) $(-5x^3 + 6x - 1) + (4x^3 + 3x^2 + 2)$

$\quad = -5x^3 + 6x - 1 + 4x^3 + 3x^2 + 2$ Writing without parentheses

$\quad = -5x^3 + 4x^3 + 3x^2 + 6x - 1 + 2$ Using the commutative and associative laws to write like terms together

$\quad = (-5 + 4)x^3 + 3x^2 + 6x + (-1 + 2)$ Combining like terms; using the distributive law

$\quad = -x^3 + 3x^2 + 6x + 1$ Note that $-1x^3 = -x^3$.

b) $\left(\frac{2}{3}x^4 + 3x^2 - 7x + \frac{1}{2}\right) + \left(-\frac{1}{3}x^4 + 5x^3 - 3x^2 + 3x - \frac{1}{2}\right)$

$\quad = \left(\frac{2}{3} - \frac{1}{3}\right)x^4 + 5x^3 + (3 - 3)x^2 + (-7 + 3)x + \left(\frac{1}{2} - \frac{1}{2}\right)$ Combining like terms

$\quad = \frac{1}{3}x^4 + 5x^3 - 4x$

1. Write an equivalent expression by adding:

$(-y^3 + 6y^2 - 4y) + (y^2 + 2y)$.

YOUR TURN

After some practice, polynomial addition is often performed mentally.

EXAMPLE 2   Add: $(2 - 3x + x^2) + (-5 + 7x - 3x^2 + x^3)$.

SOLUTION   We have

$(2 - 3x + x^2) + (-5 + 7x - 3x^2 + x^3)$

$\quad = (2 - 5) + (-3 + 7)x + (1 - 3)x^2 + x^3$ You might do this step mentally.

$\quad = -3 + 4x - 2x^2 + x^3.$ Then you would write only this.

2. Add:

$(-1 - 4n + n^2) +$
$(7 + 10n + 3n^2)$.

YOUR TURN

In the polynomials of the last example, the terms are arranged according to degree, from least to greatest. Such an arrangement is called *ascending order.* As a rule, answers are written in ascending order when the polynomials in the original problem are given in ascending order. If the polynomials in the original problem are given in descending order, the answer is usually written in descending order.

To add using columns, we write the polynomials one under the other, listing like terms under one another and leaving space for any missing terms.

EXAMPLE 3   Add:

$\quad 9x^5 - 2x^3 + 6x^2 + 3$ and $5x^4 - 7x^2 + 6$ and $3x^6 - 5x^5 + x^2 + 5$.

SOLUTION   We arrange the polynomials with like terms in columns.

$$
\begin{array}{r}
9x^5 \quad\quad - 2x^3 + 6x^2 + \ \ 3 \\
5x^4 \quad\quad\quad - 7x^2 + \ \ 6 \\
3x^6 - 5x^5 \quad\quad\quad\quad + 1x^2 + \ \ 5 \\
\hline
3x^6 + 4x^5 + 5x^4 - 2x^3 \quad\quad + 14
\end{array}
$$

We leave spaces for missing terms.

Writing x^2 as $1x^2$

Adding

3. Add using columns:

$12x^3 + 6x - 7$ and

$x^3 - 5x^2 + 6x - 1$.

The answer is $3x^6 + 4x^5 + 5x^4 - 2x^3 + 14$.

YOUR TURN

Opposites of Polynomials

The opposite of the polynomial $x^2 - 3x + 5$ is written $-(x^2 - 3x + 5)$. We can remove the parentheses and the negative sign by changing the sign of every term within the parentheses.

> ### THE OPPOSITE OF A POLYNOMIAL
>
> To find an equivalent polynomial for the *opposite*, or *additive inverse*, of a polynomial, change the sign of every term. This is the same as multiplying the polynomial by -1.

EXAMPLE 4   Write two equivalent expressions for the opposite of $4x^5 - 7x^3 - 8x + \frac{5}{6}$.

SOLUTION

i) $-\left(4x^5 - 7x^3 - 8x + \frac{5}{6}\right)$ This is one way to write the opposite of $4x^5 - 7x^3 - 8x + \frac{5}{6}$.

ii) $-4x^5 + 7x^3 + 8x - \frac{5}{6}$ Changing the sign of every term

4. Write two equivalent expressions for the opposite of $-a^4 - 5a^2 + 10$.

Thus, $-\left(4x^5 - 7x^3 - 8x + \frac{5}{6}\right)$ and $-4x^5 + 7x^3 + 8x - \frac{5}{6}$ are equivalent. Both expressions represent the opposite of $4x^5 - 7x^3 - 8x + \frac{5}{6}$.

YOUR TURN

EXAMPLE 5   Simplify: $-\left(-7x^4 - \frac{5}{9}x^3 + 8x^2 - x + 67\right)$.

SOLUTION   We have

$$-\left(-7x^4 - \tfrac{5}{9}x^3 + 8x^2 - x + 67\right) = 7x^4 + \tfrac{5}{9}x^3 - 8x^2 + x - 67.$$

The same result can be found by multiplying by -1:

5. Simplify:

$-\left(3.5a^4 - \frac{2}{3}a^3 - 7a + 9\right).$

$$-(-7x^4 - \tfrac{5}{9}x^3 + 8x^2 - x + 67)$$
$$= -1(-7x^4) + (-1)\left(-\tfrac{5}{9}x^3\right) + (-1)(8x^2) + (-1)(-x) + (-1)67$$
$$= 7x^4 + \tfrac{5}{9}x^3 - 8x^2 + x - 67.$$

YOUR TURN

Subtraction of Polynomials

We can now subtract one polynomial from another by adding the opposite of the polynomial being subtracted. Recall that $5 - 3 = 5 + (-3)$ and, as a rule, $a - b = a + (-b)$.

EXAMPLE 6   Write an equivalent expression by subtracting.

a) $(9x^5 + x^3 - 2x^2 + 4) - (-2x^5 + x^4 - 4x^3 - 3x^2)$

b) $(7x^5 + x^3 - 9x) - (3x^5 - 4x^3 + 5)$

SOLUTION

a) $(9x^5 + x^3 - 2x^2 + 4) - (-2x^5 + x^4 - 4x^3 - 3x^2)$

$\qquad = 9x^5 + x^3 - 2x^2 + 4 + 2x^5 - x^4 + 4x^3 + 3x^2$ Adding the opposite

$\qquad = 11x^5 - x^4 + 5x^3 + x^2 + 4$ Combining like terms

b) $(7x^5 + x^3 - 9x) - (3x^5 - 4x^3 + 5)$

6. Write an equivalent expression by subtracting:

$(y^3 - 7y^2 - 9) - (-3y^3 - y^2 + 11).$

$\qquad = 7x^5 + x^3 - 9x + (-3x^5) + 4x^3 - 5$ Adding the opposite

$\qquad = 7x^5 + x^3 - 9x - 3x^5 + 4x^3 - 5$ Try to go directly to this step.

$\qquad = 4x^5 + 5x^3 - 9x - 5$ Combining like terms

YOUR TURN

To subtract using columns, we first replace the coefficients in the polynomial being subtracted with their opposites. We then add as before.

EXAMPLE 7 Write in columns and subtract:

$$(5x^2 - 3x + 6) - (9x^2 - 5x - 3).$$

SOLUTION

i) $5x^2 - 3x + 6$ Writing like terms in columns
$-(9x^2 - 5x - 3)$

ii) $5x^2 - 3x + 6$ Changing signs and removing parentheses
$-9x^2 + 5x + 3$ You might start with this step.

7. Write in columns and subtract:

$(12x^2 + x - 1) -$
$(4x^2 + 10x - 8).$

iii) $5x^2 - 3x + 6$
$-9x^2 + 5x + 3$
$\overline{-4x^2 + 2x + 9}$ Adding

YOUR TURN

If you can do so without error, you can mentally find the opposite of each term being subtracted, and write the answer. Lining up like terms is important and may require leaving some blank space.

EXAMPLE 8 Write in columns and subtract:

$$(x^3 + x^2 - 12) - (-2x^3 + x^2 - 3x + 6).$$

SOLUTION We have

8. Write in columns and subtract:

$(5x^4 - 3x^3 - x^2 + 6) -$
$(-4x^3 + 9x - 2).$

$x^3 + x^2 \qquad - 12$ Leaving a blank space for the missing term
$-(-2x^3 + x^2 - 3x + 6)$
$\overline{3x^3 \qquad\qquad + 3x - 18}$

YOUR TURN

Problem Solving

EXAMPLE 9 Find a polynomial for the sum of the areas of rectangles A, B, C, and D.

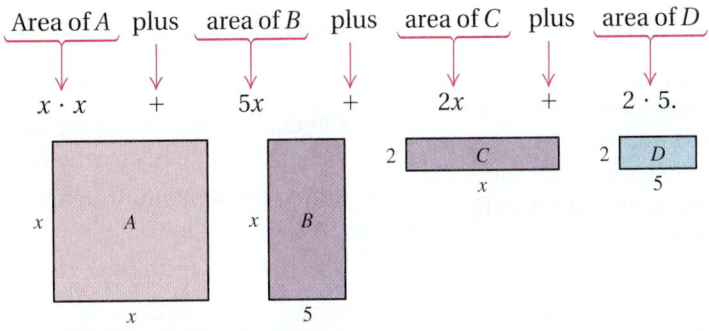

SOLUTION

1. **Familiarize.** Recall that the area of a rectangle is the product of its length and width.

2. **Translate.** We translate the problem to mathematical language. The sum of the areas is a sum of products. We find each product and then add:

Area of A plus   area of B plus   area of C plus   area of D

$x \cdot x \qquad + \qquad 5x \qquad + \qquad 2x \qquad + \qquad 2 \cdot 5.$

3. **Carry out.** We simplify $x \cdot x$ and $2 \cdot 5$ and combine like terms:

$$x^2 + 5x + 2x + 10 = x^2 + 7x + 10.$$

4. **Check.** A partial check is to replace x with a number—say, 3. Then we evaluate $x^2 + 7x + 10$ and compare that result with an alternative calculation:

$$3^2 + 7 \cdot 3 + 10 = 9 + 21 + 10 = 40.$$

9. Find a polynomial for the sum of the areas of rectangles A, B, C, and D.

	C	D
1		
x	A	B

x   3

When we substitute 3 for x and calculate the total area by regarding the figure as one large rectangle, we should also get 40:

$$\text{Total area} = (x + 5)(x + 2) = (3 + 5)(3 + 2) = 8 \cdot 5 = 40.$$

Our check is only partial, since it is possible for an incorrect answer to equal 40 when evaluated for $x = 3$. This would be unlikely, especially if a second choice of x—say, $x = 5$—also checks. We leave that check to the student.

5. **State.** A polynomial for the sum of the areas is $x^2 + 7x + 10$.

YOUR TURN

EXAMPLE 10   A 16-ft wide round fountain is built in a square city park that measures x ft by x ft. Find a polynomial for the remaining area of the park.

SOLUTION

1. **Familiarize.** We make a drawing of the square park and the circular fountain, and let x represent the length of a side of the park.

x ft   ← 16 ft →

x ft

The area of a square of side s is given by $A = s^2$, and the area of a circle of radius r is given by $A = \pi r^2$. Note that a circle with a diameter of 16 ft has a radius of 8 ft.

2. **Translate.** We reword the problem and translate as follows.

Rewording: Area of park   minus   area of fountain   is   area left over.

Translating: x ft $\cdot$ x ft $-$ $\pi \cdot 8$ ft $\cdot 8$ ft $=$ Area left over

3. **Carry out.** We carry out the multiplication:

$$x^2 \text{ ft}^2 - 64\pi \text{ ft}^2 = (x^2 - 64\pi) \text{ ft}^2 = \text{Area left over.}$$

10. Consider the fountain and park described in Example 10. If the fountain is 20 ft wide, find a polynomial for the remaining area of the park.

4. **Check.** As a partial check, note that the units in the answer are square feet (ft^2), a measure of area, as expected.

5. **State.** The remaining area of the park is $(x^2 - 64\pi) \text{ ft}^2$.

 **YOUR TURN**

Technology Connection

To check polynomial addition or subtraction, we can let y_1 = the expression before the addition or subtraction has been performed and y_2 = the simplified sum or difference. If the addition or subtraction is correct, y_1 will equal y_2 and $y_2 - y_1$ will be 0. We enter $y_2 - y_1$ as y_3, using **VARS**. Below is a check of Example 6(b) in which

$$y_1 = (7x^5 + x^3 - 9x) - (3x^5 - 4x^3 + 5),$$
$$y_2 = 4x^5 + 5x^3 - 9x - 5,$$

and

$$y_3 = y_2 - y_1.$$

We graph only y_3. If indeed y_1 and y_2 are equivalent, then y_3 should equal 0. This means its graph should coincide with the x-axis. The TRACE or TABLE features can confirm

that y_3 is always 0, or we can select y_3 to be drawn bold at the ⟨ Y= ⟩ window.

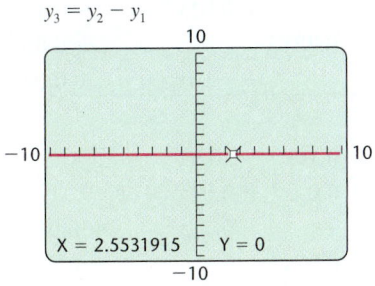

$y_3 = y_2 - y_1$

1. Use a graphing calculator to check Examples 1, 2, and 6.

4.4 EXERCISE SET

FOR EXTRA HELP

MyMathLab® Math*XL*
PRACTICE WATCH READ REVIEW

◆ Vocabulary and Reading Check

Choose the word or words from the following list to complete each sentence. Words may be used more than once or not at all.

ascending   missing
descending   opposite
like   sign

1. To subtract a polynomial, add the _____ of the polynomial.

2. When the terms of a polynomial are arranged according to degree, from least to greatest, the polynomial is written in _____ order.

3. To write an equivalent expression for the _____ of a polynomial, we change the _____ of every term.

4. To add or subtract polynomials using columns, we write _____ terms in columns and leave spaces for _____ terms.

◆ Concept Reinforcement

For each of Exercises 5–8, replace ▮ with the correct expression or operation sign.

5. $(3x^2 + 2) + (6x^2 + 7) = (3 + 6)▮ + (2 + 7)$

6. $(5t - 6) + (4t + 3) = (5 + 4)t + (▮ + 3)$

7. $(9x^3 - x^2) - (3x^3 + x^2) = 9x^3 - x^2 - 3x^3 ▮ x^2$

8. $(-2n^3 + 5) - (n^2 - 2) = -2n^3 + 5 - n^2 ▮ 2$

Addition of Polynomials

Add.

9. $(3x + 2) + (x + 7)$

10. $(x + 1) + (12x + 10)$

11. $(2t + 7) + (-8t + 1)$

12. $(4t - 3) + (-11t + 2)$

13. $(x^2 + 6x + 3) + (-4x^2 - 5)$

14. $(x^2 - 5x + 4) + (8x - 9)$

15. $(7t^2 - 3t - 6) + (2t^2 + 4t + 9)$

16. $(8a^2 + 4a - 7) + (6a^2 - 3a - 1)$

17. $(4m^3 - 7m^2 + m - 5) + (4m^3 + 7m^2 - 4m - 2)$

18. $(5n^3 - n^2 + 4n + 11) + (2n^3 - 4n^2 + n - 11)$

19. $(3 + 6a + 7a^2 + a^3) + (4 + 7a - 8a^2 + 6a^3)$

20. $(7 + 4t - 5t^2 + 6t^3) + (2 + t + 6t^2 - 4t^3)$

21. $(3x^6 + 2x^4 - x^3 + 5x) + (-x^6 + 3x^3 - 4x^2 + 7x^4)$

22. $(4x^5 - 6x^3 - 9x + 1) + (3x^4 + 6x^3 + 9x^2 + x)$

23. $\left(\frac{3}{5}x^4 + \frac{1}{2}x^3 - \frac{2}{3}x + 3\right) + \left(\frac{2}{5}x^4 - \frac{1}{4}x^3 - \frac{3}{4}x^2 - \frac{1}{6}x\right)$

24. $\left(\frac{1}{3}x^9 + \frac{1}{5}x^5 - \frac{1}{2}x^2 + 7\right) + \left(-\frac{1}{5}x^9 + \frac{1}{4}x^4 - \frac{3}{5}x^5\right)$

25. $(5.3t^2 - 6.4t - 9.1) + (4.2t^3 - 1.8t^2 + 7.3)$

26. $(4.9a^3 + 3.2a^2 - 5.1a) + (2.1a^2 - 3.7a + 4.6)$

27. $-4x^3 + 8x^2 + 3x - 2$
$ - 4x^2 + 3x + 2$

28. $-3x^4 + 6x^2 + 2x - 4$
$ - 3x^2 + 2x + 4$

29. $0.05x^4 + 0.12x^3 - 0.5x^2$
$ - 0.02x^3 + 0.02x^2 + 2x$
$1.5x^4 + 0.01x^2 + 0.15$
$0.25x^3 + 0.85$
$-0.25x^4 + 10x^2 - 0.04$

30. $0.15x^4 + 0.10x^3 - 0.9x^2$
$ - 0.01x^3 + 0.01x^2 + x$
$1.25x^4 + 0.11x^2 + 0.01$
$0.27x^3 + 0.99$
$-0.35x^4 + 15x^2 - 0.03$

Opposites of Polynomials

Write two equivalent expressions for the opposite of each polynomial, as in Example 4.

31. $-3t^3 + 4t^2 - 7$

32. $-x^3 - 5x^2 + 2x$

33. $x^4 - 8x^3 + 6x$

34. $5a^3 + 2a - 17$

Simplify.

35. $-(3a^4 - 5a^2 + 1.2)$

36. $-(-6a^3 + 0.2a^2 - 7)$

37. $-(-4x^4 + 6x^2 + \frac{3}{4}x - 8)$

38. $-\left(3x^5 - 2x^3 - \frac{3}{5}x^2 + 16\right)$

Subtraction of Polynomials

Subtract.

39. $(3x + 1) - (5x + 8)$

40. $(7x + 3) - (3x + 2)$

41. $(-9t + 12) - (t^2 + 3t - 1)$

42. $(a^2 - 3a - 2) - (2a^2 - 6a - 2)$

43. $(4a^2 + a - 7) - (3 - 8a^3 - 4a^2)$

44. $(-4x^2 + 2x) - (-5x^2 + 2x^3 + 3)$

Aha! 45. $(7x^3 - 2x^2 + 6) - (6 - 2x^2 + 7x^3)$

46. $(8x^5 + 3x^4 + x - 1) - (8x^5 + 3x^4 - 1)$

47. $(3 + 5a + 3a^2 - a^3) - (2 + 4a - 9a^2 + 2a^3)$

48. $(7 + t - 5t^2 + 2t^3) - (1 + 2t - 4t^2 + 5t^3)$

49. $\left(\frac{5}{8}x^3 - \frac{1}{4}x - \frac{1}{3}\right) - \left(-\frac{1}{2}x^3 + \frac{1}{4}x - \frac{1}{3}\right)$

50. $\left(\frac{1}{5}x^3 + 2x^2 - \frac{3}{10}\right) - \left(-\frac{2}{5}x^3 + 2x^2 + \frac{7}{1000}\right)$

51. $(0.07t^3 - 0.03t^2 + 0.01t) - (0.02t^3 + 0.04t^2 - 1)$

52. $(0.9a^3 + 0.2a - 5) - (0.7a^4 - 0.3a - 0.1)$

53. $x^3 + 3x^2 + 1$
$-(x^3 + x^2 - 5)$

54. $x^2 + 5x + 6$
$-(x^2 + 2x + 1)$

55. $4x^4 - 2x^3$
$-(7x^4 + 6x^3 + 7x^2)$

56. $5x^4 + 6x^3 - 9x^2$
$-(-6x^4 + x^2)$

Problem Solving

57. Solve.
  a) Find a polynomial for the sum of the areas of the rectangles shown in the figure.
  b) Find the sum of the areas when $x = 5$ and $x = 7$.

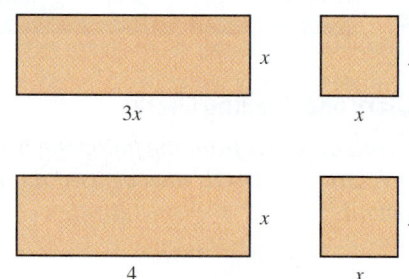

58. Solve. Leave the answers in terms of π.
  a) Find a polynomial for the sum of the areas of the circles shown in the figure.
  b) Find the sum of the areas when $r = 5$ and $r = 11.3$.

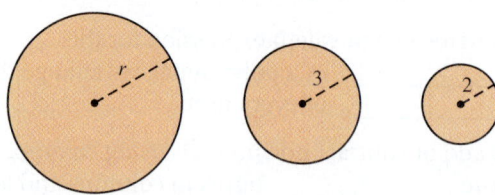

In each of Exercises 59 and 60, find a polynomial for the perimeter of the figure.

59.

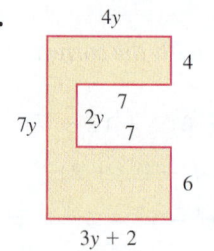

60.

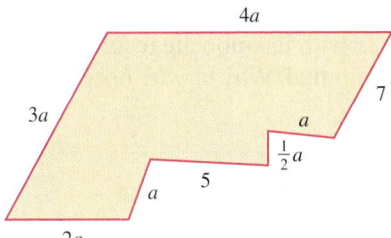

Find two algebraic expressions for the area of each figure. First, regard the figure as one large rectangle, and then regard the figure as a sum of four smaller rectangles.

61.

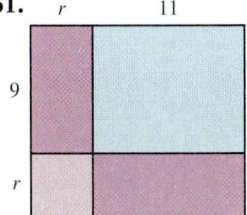

62.

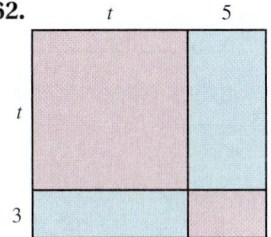

63.

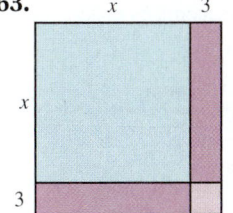

64.

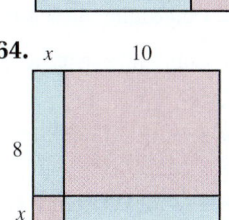

In each of Exercises 65–68, find a polynomial for the shaded area of the figure.

65.

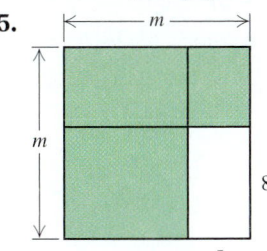

66.

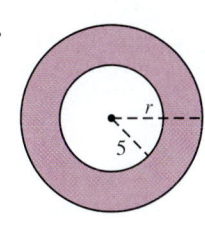

67.

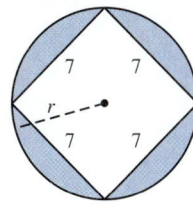

68.

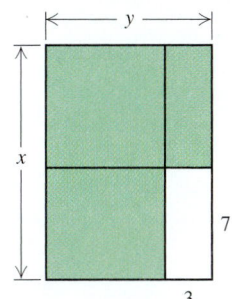

69. A 2-ft by 6-ft bath enclosure is installed in a new bathroom measuring x ft by x ft. Find a polynomial for the remaining floor area.

70. A 5-ft by 7-ft Jacuzzi™ is installed on an outdoor deck measuring y ft by y ft. Find a polynomial for the remaining area of the deck.

71. A 12-ft wide round patio is laid in a garden measuring z ft by z ft. Find a polynomial for the remaining area of the garden.

72. A 10-ft wide round water trampoline is floating in a pool measuring x ft by x ft. Find a polynomial for the remaining surface area of the pool.

73. A 12-m by 12-m mat includes a circle of diameter d meters for wrestling. Find a polynomial for the area of the mat outside the wrestling circle.

74. A 2-m by 3-m rug is spread inside a tepee that has a diameter of x meters. Find a polynomial for the area of the tepee's floor that is not covered.

75. Explain why parentheses are used in the statement of the solution of Example 10: $(x^2 - 64\pi)$ ft^2.

76. Is the sum of two trinomials always a trinomial? Why or why not?

Skill Review

Find the slope–intercept equation for each line described.

77. Slope $\frac{1}{3}$; y-intercept $(0, 2)$ [3.6]

78. Slope 4; contains the point $(-6, 0)$ [3.7]

79. Slope 0; contains the point $(4, 10)$ [3.7]

80. Slope -3; contains the point $(1, 5)$ [3.7]

81. Slope $-\frac{4}{7}$; contains the point $(0, -4)$ [3.6]

82. Contains the points $(-2, 0)$ and $(4, -8)$ [3.7]

Synthesis

83. What can be concluded about two polynomials whose sum is zero?

84. Which, if any, of the commutative, associative, and distributive laws are needed for adding polynomials? Why?

Simplify.

85. $(6t^2 - 7t) + (3t^2 - 4t + 5) - (9t - 6)$

86. $(3x^2 - 4x + 6) - (-2x^2 + 4) + (-5x - 3)$

87. $4(x^2 - x + 3) - 2(2x^2 + x - 1)$

88. $3(2y^2 - y - 1) - (6y^2 - 3y - 3)$

89. $(345.099x^3 - 6.178x) - (94.508x^3 - 8.99x)$

In each of Exercises 90–93, find a polynomial for the surface area of the right rectangular solid.

90.

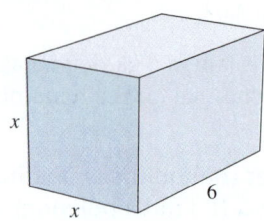

91.

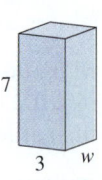

92.

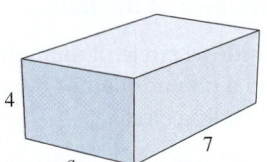

93.

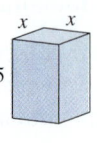

94. Find a polynomial for the total length of all edges in the figure appearing in Exercise 93.

95. *Total Profit.*   Hadley Electronics is marketing a new digital camera. Total revenue is the total amount of money taken in. The firm determines that when it sells x cameras, its total revenue is given by

$$R = 175x - 0.4x^2.$$

Total cost is the total cost of producing x cameras. Hadley Electronics determines that the total cost of producing x cameras is given by

$$C = 5000 + 0.6x^2.$$

The total profit P is

$$(\text{Total Revenue}) - (\text{Total Cost}) = R - C.$$

a) Find a polynomial for total profit.
b) What is the total profit on the production and sale of 75 cameras?
c) What is the total profit on the production and sale of 120 cameras?

96. Does replacing each occurrence of the variable x in $4x^7 - 6x^3 + 2x$ with its opposite result in the opposite of the polynomial? Why or why not?

YOUR TURN ANSWERS: SECTION 4.4

1. $-y^3 + 7y^2 - 2y$ **2.** $6 + 6n + 4n^2$
3. $13x^3 - 5x^2 + 12x - 8$
4. $-(-a^4 - 5a^2 + 10); a^4 + 5a^2 - 10$
5. $-3.5a^4 + \frac{2}{3}a^3 + 7a - 9$ **6.** $4y^3 - 6y^2 - 20$
7. $8x^2 - 9x + 7$ **8.** $5x^4 + x^3 - x^2 - 9x + 8$
9. $x^2 + 4x + 3$ **10.** $(x^2 - 100\pi)\,\text{ft}^2$

QUICK QUIZ: SECTIONS 4.1–4.4

Simplify.

1. $\left(\dfrac{-2x^3}{3y^5}\right)^{-2}$ [4.1]   **2.** $(y + 3)^5(y + 3)^7$ [4.1]

3. Determine the degree of $5n - n^4 + 7 + 6n^2$.   [4.3]

4. Combine like terms and write in descending order:
$2y - 7y^2 - y^3 + y^2$.   [4.3]

5. Subtract: $(5y^2 - y + 7) - (-y - 3)$.   [4.4]

PREPARE TO MOVE ON

Simplify.

1. $2(x^2 - x + 3)$ [1.8]

2. $-5(3x^2 - 2x - 7)$ [1.8]

3. $x^2 \cdot x^6$ [4.1]   **4.** $y^6 \cdot y$ [4.1]

5. $2n \cdot n^2$ [4.1]   **6.** $-6n^4 \cdot n^8$ [4.1]

Mid-Chapter Review

Properties of exponents allow us to simplify exponential expressions and to work with polynomials.

$$a^1 = a \qquad a^m \cdot a^n = a^{m+n}$$

$$a^0 = 1 \qquad (a^m)^n = a^{mn}$$

$$a^{-n} = \frac{1}{a^n} \qquad \frac{a^m}{a^n} = a^{m-n}$$

GUIDED SOLUTIONS

1. Simplify: $\dfrac{x^{-3}y}{x^{-4}y^7}$.   [4.2]

   Solution

   $$\frac{x^{-3}y}{x^{-4}y^7} = x^{-3-\left(\boxed{}\right)}y^{\boxed{}\,-7} = x^{\boxed{}}y^{\boxed{}} = \frac{\boxed{}}{\boxed{}}$$

2. Subtract: $(x^2 + 7x - 12) - (3x^2 - 6x - 1)$.   [4.4]

   Solution

   $$(x^2 + 7x - 12) - (3x^2 - 6x - 1) = x^2 + 7x - 12\,\boxed{}\,3x^2\,\boxed{}\,6x\,\boxed{}\,1 \qquad \text{\textcolor{red}{Adding the opposite}}$$

   $$= \boxed{}\,x^2 + \boxed{}\,x - \boxed{} \qquad \text{\textcolor{red}{Combining like terms}}$$

MIXED REVIEW

Simplify. Do not use negative exponents in the answer.

3. $(x^2y^5)^8$ [4.1]

4. $(4x)^0$ [4.1]

5. $\dfrac{3a^{11}}{12a}$ [4.1]

6. $\dfrac{-48ab^7}{18ab^6}$ [4.1]

7. $5x^{-2}$ [4.2]

8. $(a^{-2}bc^3)^{-1}$ [4.2]

9. $\left(\dfrac{2a^2}{3}\right)^{-3}$ [4.2]

10. $\dfrac{8m^{-2}n^3}{12m^4n^7}$ [4.2]

11. Convert to decimal notation: 1.89×10^{-6}.
    [4.2]

12. Convert to scientific notation: 27,000,000,000.
    [4.2]

13. Determine the degree of $8t^2 - t^3 + 5$.   [4.3]

14. Combine like terms:
    $$3a^2 - 6a - a^2 + 7 + a - 10. \quad [4.3]$$

Perform the indicated operation and simplify.

15. $(3x^2 - 2x + 6) + (5x - 3)$ [4.4]

16. $(9x + 6) - (2x - 1)$ [4.4]

17. $(4x^2 - x - 7) - (10x^2 - 3x + 5)$ [4.4]

18. $(t^9 + 3t^6 - 8t^2) + (5t^7 - 3t^6 + 8t^2)$ [4.4]

19. $(3a^4 - 9a^3 - 7) - (4a^3 + 13a^2 - 3)$ [4.4]

20. $(x^4 - 2x^2 - \frac{1}{2}x) - (x^5 - x^4 + \frac{1}{2}x)$ [4.4]

4.5   Multiplication of Polynomials

Multiplying Monomials   ▪   Multiplying a Monomial and a Polynomial   ▪   Multiplying Any Two Polynomials

We now multiply polynomials using techniques based on the distributive, associative, and commutative laws and the rules for exponents.

Multiplying Monomials

Consider $(3x)(4x)$. We multiply as follows:

$$(3x)(4x) = 3 \cdot x \cdot 4 \cdot x \qquad \text{Using an associative law}$$
$$= 3 \cdot 4 \cdot x \cdot x \qquad \text{Using a commutative law}$$
$$= (3 \cdot 4) \cdot x \cdot x \qquad \text{Using an associative law}$$
$$= 12x^2.$$

Student Notes

Remember that when we compute $(3 \cdot 5)(2 \cdot 4)$, each factor is used only once, even if we change the order:

$$(3 \cdot 5)(2 \cdot 4) = (5 \cdot 2)(3 \cdot 4)$$
$$= 10 \cdot 12 = 120.$$

In the same way,

$$(3 \cdot x)(2 \cdot x) = (3 \cdot 2)(x \cdot x)$$
$$= 6x^2.$$

Some students mistakenly "reuse" a factor.

1. Multiply to form an equivalent expression: $(-6t^3)(9t)$.

TO MULTIPLY MONOMIALS

To find an equivalent expression for the product of two monomials, multiply the coefficients and then multiply any variables using the product rule for exponents.

EXAMPLE 1   Multiply to form an equivalent expression.

a) $(5x)(6x)$ **b)** $(-a)(3a)$ **c)** $(7x^5)(-4x^3)$

SOLUTION

a) $(5x)(6x) = (5 \cdot 6)(x \cdot x)$ Multiplying the coefficients; multiplying the variables
$$= 30x^2 \qquad \text{Simplifying}$$

b) $(-a)(3a) = (-1a)(3a)$ Writing $-a$ as $-1a$ can ease calculations.
$$= (-1 \cdot 3)(a \cdot a) \qquad \text{Using an associative law and a commutative law}$$
$$= -3a^2$$

c) $(7x^5)(-4x^3) = 7(-4)(x^5 \cdot x^3)$
$$= -28x^{5+3}$$
$$= -28x^8 \bigg\} \quad \text{Using the product rule for exponents}$$

↩ YOUR TURN

Multiplying a Monomial and a Polynomial

To find an equivalent expression for the product of a monomial, such as $5x$, and a polynomial, such as $2x^2 - 3x + 4$, we use the distributive law.

EXAMPLE 2   Multiply: **(a)** $x(x + 3)$; **(b)** $5x(2x^2 - 3x + 4)$.

SOLUTION

a) $x(x + 3) = x \cdot x + x \cdot 3$ Using the distributive law
$$= x^2 + 3x$$

b) $5x(2x^2 - 3x + 4) = (5x)(2x^2) - (5x)(3x) + (5x)(4)$ Using the distributive law

$$= 10x^3 - 15x^2 + 20x$$ Performing the three multiplications

2. Multiply: $3a(5a^3 - a + 11)$.

_____

YOUR TURN

The product in Example 2(a) can be visualized as the area of a rectangle with width x and length $x + 3$. Note that the total area can be expressed as $x(x + 3)$ or, by adding the two smaller areas, $x^2 + 3x$.

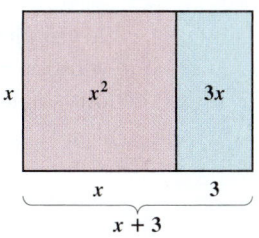

> **THE PRODUCT OF A MONOMIAL AND A POLYNOMIAL**
>
> To multiply a monomial and a polynomial, multiply each term of the polynomial by the monomial.

Try to do this mentally, when possible. Remember that we multiply coefficients and, when the bases match, add exponents.

EXAMPLE 3   Multiply: $2x^2(x^3 - 7x^2 + 10x - 4)$.

SOLUTION

$$\textit{Think:} \quad \underbrace{2x^2 \cdot x^3} - \underbrace{2x^2 \cdot 7x^2} + \underbrace{2x^2 \cdot 10x} - \underbrace{2x^2 \cdot 4}$$

3. Multiply:

$$5y^3(2y^3 - 8y^2 + y - 5).$$

$$2x^2(x^3 - 7x^2 + 10x - 4) = 2x^5 \quad - \quad 14x^4 \quad + \quad 20x^3 \quad - \quad 8x^2$$

_____

YOUR TURN

Multiplying Any Two Polynomials

Before considering the product of *any* two polynomials, let's look at products when both polynomials are binomials. To multiply, we again begin by using the distributive law. This time, however, it is a *binomial* rather than a monomial that is being distributed.

EXAMPLE 4   Multiply each pair of binomials.

a) $x + 5$ and $x + 4$ **b)** $4x - 3$ and $x - 2$

SOLUTION

a) $(x + 5)\,(x + 4) = (x + 5)\,x + (x + 5)\,4$ Using the distributive law

$$= x(x + 5) + 4(x + 5)$$ Using the commutative law for multiplication

$$= x \cdot x + x \cdot 5 + 4 \cdot x + 4 \cdot 5$$ Using the distributive law (twice)

$$= x^2 + 5x + 4x + 20$$ Multiplying the monomials

$$= x^2 + 9x + 20$$ Combining like terms

b) $(4x - 3)(x - 2) = (4x - 3)x - (4x - 3)2$ Using the distributive law

$$= x(4x - 3) - 2(4x - 3)$$ Using the commutative law for multiplication. This step is often omitted.

$$= x \cdot 4x - x \cdot 3 - 2 \cdot 4x - 2(-3)$$ Using the distributive law (twice)

$$= 4x^2 - 3x - 8x + 6$$ Multiplying the monomials

$$= 4x^2 - 11x + 6$$ Combining like terms

4. Multiply $x + 3$ and $5x - 1$.

YOUR TURN

To visualize the product in Example 4(a), consider a rectangle of length $x + 5$ and width $x + 4$. The total area can be expressed as $(x + 5)(x + 4)$ or, by adding the four smaller areas, $x^2 + 5x + 4x + 20$.

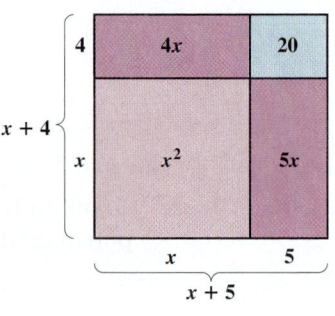

Study Skills

Take a Peek Ahead
Try to at least glance at the next section of material that will be covered in class. This will make it easier to concentrate on your instructor's lecture instead of trying to write to everything down.

Let's consider the product of a binomial and a trinomial. Again we make repeated use of the distributive law.

EXAMPLE 5   Multiply: $(x^2 + 2x - 3)(x + 4)$.

SOLUTION

$(x^2 + 2x - 3)(x + 4)$

$$= (x^2 + 2x - 3)x + (x^2 + 2x - 3)4$$ Using the distributive law

$$= x(x^2 + 2x - 3) + 4(x^2 + 2x - 3)$$ Using the commutative law

$$= x \cdot x^2 + x \cdot 2x - x \cdot 3 + 4 \cdot x^2 + 4 \cdot 2x - 4 \cdot 3$$ Using the distributive law (twice)

$$= x^3 + 2x^2 - 3x + 4x^2 + 8x - 12$$ Multiplying the monomials

$$= x^3 + 6x^2 + 5x - 12$$ Combining like terms

5. Multiply:

$(2t^2 - t + 5)(t + 3)$.

YOUR TURN

To use columns for long multiplication, multiply each term in the top row by every term in the bottom row. We write like terms in columns, and then add the results. Such multiplication is similar to multiplying whole numbers.

321	$300 + 20 + 1$
$\times\ 12$	$\times \qquad\qquad 10 + 2$
642	$600 + 40 + 2$ Multiplying the top row by 2
321	$3000 + 200 + 10$ Multiplying the top row by 10
3852	$3000 + 800 + 50 + 2$ Adding

> ### THE PRODUCT OF TWO POLYNOMIALS
> To multiply two polynomials P and Q, select one of the polynomials—say, P. Then multiply each term of P by every term of Q and combine like terms.

EXAMPLE 6   Multiply: $(5x^4 - 2x^2 + 3x)(x^2 + 2x)$.

SOLUTION

$$
\begin{array}{r}
5x^4 - 2x^2 + 3x \\
x^2 + 2x \\
\hline
10x^5 \qquad - 4x^3 + 6x^2 \\
5x^6 \qquad - 2x^4 + 3x^3 \\
\hline
5x^6 + 10x^5 - 2x^4 - x^3 + 6x^2
\end{array}
$$

Note that each polynomial is written in descending order.

Multiplying the top row by $2x$

Multiplying the top row by x^2

Combining like terms

Line up like terms in columns.

6. Multiply using columns:

$(3x^3 - x^2 + 5)(x - 4)$.

YOUR TURN

Sometimes we multiply horizontally, while still aligning like terms as we write the product.

EXAMPLE 7   Multiply: $(2x^3 + 3x^2 - 4x + 6)(3x + 5)$.

SOLUTION

$$
\begin{aligned}
(2x^3 + 3x^2 - 4x + 6)(3x + 5) &= 6x^4 + 9x^3 - 12x^2 + 18x \\
&\quad + 10x^3 + 15x^2 - 20x + 30 \\
&= 6x^4 + 19x^3 + 3x^2 - 2x + 30
\end{aligned}
$$

Multiplying by $3x$

Multiplying by 5

7. Multiply:

$(5y^3 - 4y^2 + y + 2)(3y + 1)$.

YOUR TURN

CHECKING BY EVALUATING

How can we be certain that our multiplication (or addition or subtraction) of polynomials is correct? One check is to simply review our calculations. A different type of check makes use of the fact that equivalent expressions have the same value when evaluated for the same replacement. Thus a quick, partial, check of Example 7 can be made by selecting a convenient replacement for x (say, 1) and comparing the values of the expressions $(2x^3 + 3x^2 - 4x + 6)(3x + 5)$ and $6x^4 + 19x^3 + 3x^2 - 2x + 30$:

$$
\begin{aligned}
(2x^3 + 3x^2 - 4x + 6)(3x + 5) &= (2 \cdot 1^3 + 3 \cdot 1^2 - 4 \cdot 1 + 6)(3 \cdot 1 + 5) \\
&= (2 + 3 - 4 + 6)(3 + 5) \\
&= 7 \cdot 8 = 56;
\end{aligned}
$$

$$
\begin{aligned}
6x^4 + 19x^3 + 3x^2 - 2x + 30 &= 6 \cdot 1^4 + 19 \cdot 1^3 + 3 \cdot 1^2 - 2 \cdot 1 + 30 \\
&= 6 + 19 + 3 - 2 + 30 \\
&= 28 - 2 + 30 = 56.
\end{aligned}
$$

Since the value of both expressions is 56, the multiplication in Example 7 is very likely correct.

It is possible, by chance, for two expressions that are not equivalent to share the same value when evaluated. For this reason, checking by evaluating is only a partial check. Consult your instructor for the checking approach that he or she prefers.

Technology Connection

Tables can also be used to check polynomial multiplication. To illustrate, we can check Example 7 by entering $y_1 = (2x^3 + 3x^2 - 4x + 6)(3x + 5)$ and $y_2 = 6x^4 + 19x^3 + 3x^2 - 2x + 30$.

When ⟨ TABLE ⟩ is then pressed, we are shown two columns of values—one for y_1 and one for y_2. If our multiplication is correct, the columns of values will match.

X	Y₁	Y₂
−3	36	36
−2	−10	−10
−1	22	22
0	30	30
1	56	56
2	286	286
3	1050	1050

X = −3

1. Form a table and scroll up and down to check Example 6.
2. Check Example 7 by letting

$$y_1 = (2x^3 + 3x^2 - 4x + 6)(3x + 5),$$
$$y_2 = 6x^4 + 19x^3 + 3x^2 - 2x + 30,$$

and

$$y_3 = y_2 - y_1.$$

Then check that y_3 is always 0.

4.5 EXERCISE SET

Vocabulary and Reading Check

For each step, write the letter that gives an explanation of what was done on that step.

a) Combining like terms
b) Multiplying monomials
c) Using the commutative law for multiplication
d) Using the distributive law

$(2x + 3)(5x - 4)$
$= (2x + 3)(5x) - (2x + 3)(4)$ _____(d)
$= (5x)(2x + 3) - 4(2x + 3)$ **1.** ____
$= (5x)(2x) + (5x)(3) - 4(2x) - 4(3)$ **2.** ____
$= 10x^2 + 15x - 8x - 12$ **3.** ____
$= 10x^2 + 7x - 12$ **4.** ____

Concept Reinforcement

In each of Exercises 5–10, match the expression with the correct result from the column on the right. Choices may be used more than once.

5. ____ $3x^2 \cdot 2x^4$ **a)** $6x^8$

6. ____ $3x^8 + 5x^8$ **b)** $8x^6$

7. ____ $4x^3 \cdot 2x^5$ **c)** $6x^6$

8. ____ $3x^5 \cdot 2x^3$ **d)** $8x^8$

9. ____ $4x^6 + 2x^6$

10. ____ $4x^4 \cdot 2x^2$

Multiplying Monomials

Multiply.

11. $(3x^5)7$ **12.** $2x^3 \cdot 11$

13. $(-x^3)(x^4)$ **14.** $(-x^2)(-x)$

15. $(-x^6)(-x^2)$ **16.** $(-x^5)(x^3)$

17. $4t^2(9t^2)$ **18.** $(6a^8)(3a^2)$

19. $(0.3x^3)(-0.4x^6)$ **20.** $(-0.1x^6)(0.2x^4)$

21. $\left(-\frac{1}{4}x^4\right)\left(\frac{1}{5}x^8\right)$ **22.** $\left(-\frac{1}{5}x^3\right)\left(-\frac{1}{3}x\right)$

23. $(-5n^3)(-1)$ **24.** $19t^2 \cdot 0$

25. $(-4y^5)(6y^2)(-3y^3)$ **26.** $7x^2(-2x^3)(2x^6)$

Multiplying a Monomial and a Polynomial

Multiply.

27. $5x(4x + 1)$ **28.** $3x(2x - 7)$

29. $(a - 9)3a$ **30.** $(a - 7)4a$

31. $x^2(x^3 + 1)$ **32.** $-2x^3(x^2 - 1)$

33. $-3n(2n^2 - 8n + 1)$

34. $4n(3n^3 - 4n^2 - 5n + 10)$

35. $-5t^2(3t + 6)$ **36.** $7t^2(2t + 1)$

37. $\frac{2}{3}a^4\left(6a^5 - 12a^3 - \frac{5}{8}\right)$ **38.** $\frac{3}{4}t^5\left(8t^6 - 12t^4 + \frac{12}{7}\right)$

Multiplying Any Two Polynomials

Multiply.

39. $(x + 3)(x + 4)$

40. $(x + 7)(x + 3)$

41. $(t + 7)(t - 3)$

42. $(t - 4)(t + 3)$

43. $(a - 0.6)(a - 0.7)$

44. $(a - 0.4)(a - 0.8)$

45. $(x + 3)(x - 3)$

46. $(x + 6)(x - 6)$

47. $(4 - x)(7 - 2x)$

48. $(5 + x)(5 + 2x)$

49. $\left(t + \frac{3}{2}\right)\left(t + \frac{4}{3}\right)$

50. $\left(a - \frac{2}{5}\right)\left(a + \frac{5}{2}\right)$

51. $\left(\frac{1}{4}a + 2\right)\left(\frac{3}{4}a - 1\right)$

52. $\left(\frac{2}{5}t - 1\right)\left(\frac{3}{5}t + 1\right)$

Draw and label rectangles similar to those following Examples 2 and 4 to illustrate each product.

53. $x(x + 5)$ **54.** $x(x + 2)$

55. $(x + 1)(x + 2)$ **56.** $(x + 3)(x + 1)$

57. $(x + 5)(x + 3)$ **58.** $(x + 4)(x + 6)$

Multiply and check. Write the answer in descending order.

59. $(x^2 - x + 3)(x + 1)$

60. $(x^2 + x - 7)(x + 2)$

61. $(2a + 5)(a^2 - 3a + 2)$

62. $(3t - 4)(t^2 - 5t + 1)$

63. $(y^2 - 7)(3y^4 + y + 2)$

64. $(a^2 + 4)(5a^3 - 3a - 1)$

Aha! **65.** $(3x + 2)(7x + 4x + 1)$

66. $(4x - 5x - 3)(1 + 2x^2)$

67. $(x^2 + 5x - 1)(x^2 - x + 3)$

68. $(x^2 - 3x + 2)(x^2 + x + 1)$

69. $\left(5t^2 - t + \frac{1}{2}\right)(2t^2 + t - 4)$

70. $(2t^2 - 5t - 4)\left(3t^2 - t + \frac{1}{2}\right)$

71. $(x + 1)(x^3 + 7x^2 + 5x + 4)$

72. $(x + 2)(x^3 + 5x^2 + 9x + 3)$

73. Is it possible to understand polynomial multiplication without understanding the distributive law? Why or why not?

74. The polynomials
$$(a + b + c + d) \quad \text{and} \quad (r + s + m + p)$$
are multiplied. Without performing the multiplication, determine how many terms the product will contain. Provide a justification for your answer.

Skill Review

75. Solve $A = \dfrac{b + c}{2}$ for c.   [2.3]

76. What percent of 30 is 24?   [2.4]

77. Graph $t \geq -3$ on the number line.   [2.6]

78. In which quadrant or on what axis is the point $\left(-\frac{1}{2}, 0\right)$ located?   [3.1]

79. Find the slope of the line containing the points $(2, 3)$ and $(5, -1)$.   [3.5]

80. Find the slope and the y-intercept of the line given by $x - 3y = 5$.   [3.6]

Synthesis

81. Under what conditions will the product of two binomials be a trinomial?

82. Explain how the following figure can be used to show that $(x + 3)^2 \neq x^2 + 9$.

Find a polynomial for the shaded area of each figure.

83.

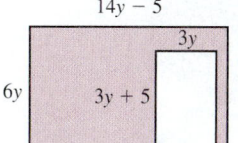

84.

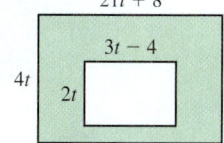

For each figure, determine what the missing number must be in order for the figure to have the given area.

85. Area is $x^2 + 8x + 15$ **86.** Area is $x^2 + 7x + 10$

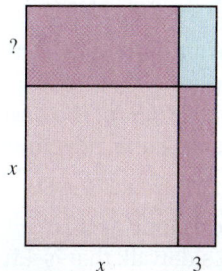

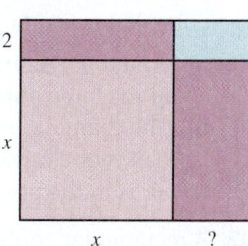

87. A box with a square bottom and no top is to be made from a 12-in.–square piece of cardboard. Squares with side x are cut out of the corners and the sides are folded up. Find the polynomials for the volume and the outside surface area of the box.

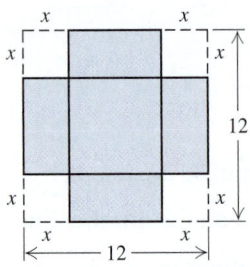

88. Find a polynomial for the volume of the solid shown below.

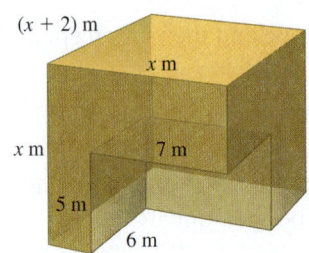

89. An open wooden box is a cube with side x cm. The box, including its bottom, is made of wood that is 1 cm thick. Find a polynomial for the interior volume of the cube.

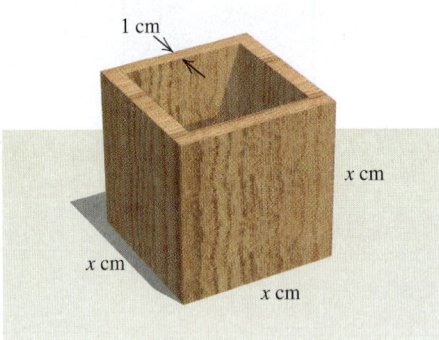

90. A side of a cube is $(x + 2)$ cm long. Find a polynomial for the volume of the cube.

91. A sphere of radius x is enclosed in a plastic cube that is just large enough for the sphere.
 a) Find a polynomial for the amount of air that is in the cube outside the sphere.
 b) The diameter of a baseball is 2.9 in. If a baseball is enclosed in the plastic cube, how much air is in the box?

92. A rectangular garden is twice as long as it is wide and is surrounded by a sidewalk that is 4 ft

wide. The area of the sidewalk is 256 ft². Find the dimensions of the garden.

Compute and simplify.

93. $(x + 3)(x + 6) + (x + 3)(x + 6)$

Aha! **94.** $(x - 2)(x - 7) - (x - 7)(x - 2)$

95. $(x + 5)^2 - (x - 3)^2$

96. $(x + 2)(x + 4)(x - 5)$

97. $(x - 3)^3$

Aha! **98.** Extend the pattern and simplify
$$(x - a)(x - b)(x - c)(x - d) \cdots (x - z).$$

99. Use a graphing calculator to check your answers to Exercises 27, 47, and 59. Use graphs, tables, or both, as directed by your instructor.

↪ YOUR TURN ANSWERS: SECTION 4.5

1. $-54t^4$ **2.** $15a^4 - 3a^2 + 33a$
3. $10y^6 - 40y^5 + 5y^4 - 25y^3$ **4.** $5x^2 + 14x - 3$
5. $2t^3 + 5t^2 + 2t + 15$ **6.** $3x^4 - 13x^3 + 4x^2 + 5x - 20$
7. $15y^4 - 7y^3 - y^2 + 7y + 2$

QUICK QUIZ: SECTIONS 4.1–4.5

1. Multiply: $-2x^4(3x^5 - 8x + 12)$. [4.5]

2. Add: $(2x^4 + 7x^2 - 10x) + (8x^4 - 7x^2 - 5x + 1)$. [4.4]

3. Subtract: $(6a^2 - 8a - 3) - (-a^2 + 2a - 1)$. [4.4]

4. Simplify: $(x^2y)(5xy)^3$. [4.1]

5. Classify $x^7 + x - 3$ as a monomial, a binomial, a trinomial, or a polynomial with no special name. [4.3]

PREPARE TO MOVE ON

Simplify.

1. 0.7^2 [1.8] **2.** $(7x^3)^2$ [4.1]

3. $\left(-\dfrac{3}{2}\right)^2$ [1.8] **4.** $2(3x)\left(\dfrac{1}{4}\right)$ [1.8]

5. $2 \cdot 5a \cdot 0.2$ [1.8] **6.** $\left(\dfrac{1}{3}t^4\right)^2$ [4.1]

4.6   Special Products

Products of Two Binomials   ◼   Multiplying Sums and Differences of Two Terms   ◼   Squaring Binomials   ◼
Multiplications of Various Types

Patterns that we observe in certain polynomial products allow us to compute such products quickly.

Products of Two Binomials

We can find the product $(x + 5)(x + 4)$ by using the distributive law a total of three times. Note that each term in $x + 5$ is multiplied by each term in $x + 4$:

$$(x + 5)(x + 4) = x \cdot x + x \cdot 4 + 5 \cdot x + 5 \cdot 4$$
$$= x^2 + 4x + 5x + 20$$
$$= x^2 + 9x + 20.$$

Note that $x \cdot x$ is found by multiplying the _First_ terms of each binomial, $x \cdot 4$ is found by multiplying the _Outer_ terms of the two binomials, $5 \cdot x$ is the product of the _Inner_ terms of the two binomials, and $5 \cdot 4$ is the product of the _Last_ terms of each binomial:

$$
\begin{array}{cccc}
\text{First} & \text{Outer} & \text{Inner} & \text{Last} \\
\text{terms} & \text{terms} & \text{terms} & \text{terms}
\end{array}
$$

$$(x + 5)(x + 4) = x \cdot x + 4 \cdot x + 5 \cdot x + 5 \cdot 4.$$

To remember this shortcut for multiplying, we use the initials **FOIL**.

THE FOIL METHOD

To multiply two binomials, $A + B$ and $C + D$, multiply the First terms AC, the Outer terms AD, the Inner terms BC, and then the Last terms BD. Then combine like terms, if possible.

$$(A + B)(C + D) = AC + AD + BC + BD$$

1. Multiply **F**irst terms: AC.
2. Multiply **O**uter terms: AD.
3. Multiply **I**nner terms: BC.
4. Multiply **L**ast terms: BD.

   $\downarrow$
   FOIL

$$(A + B)(C + D)$$

Because addition is commutative, the individual multiplications can be performed in any order. Both FLOI and FIOL yield the same result as FOIL, but FOIL is most easily remembered and most widely used.

EXAMPLE 1   Form an equivalent expression by multiplying: $(x + 8)(x^2 + 5)$.

SOLUTION

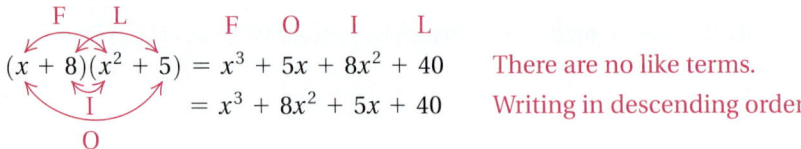

$$(x + 8)(x^2 + 5) = x^3 + 5x + 8x^2 + 40 \qquad \text{There are no like terms.}$$
$$= x^3 + 8x^2 + 5x + 40 \qquad \text{Writing in descending order}$$

1. Form an equivalent expression by multiplying:

$(t + 3)(t^2 - 5)$.

YOUR TURN

After multiplying, remember to combine any like terms.

EXAMPLE 2   Multiply to form an equivalent expression.

a) $(x + 7)(x + 4)$ **b)** $(y + 3)(y - 2)$

c) $(4t^3 + 5t)(3t^2 - 2)$ **d)** $(3 - 4x)(7 - 5x^3)$

SOLUTION

a) $(x + 7)(x + 4) = x^2 + 4x + 7x + 28 \qquad$ Using FOIL
$$\qquad\qquad\qquad = x^2 + 11x + 28 \qquad\qquad \text{Combining like terms}$$

b) $(y + 3)(y - 2) = y^2 - 2y + 3y - 6 \qquad$ Using FOIL
$$\qquad\qquad\qquad = y^2 + y - 6$$

c) $(4t^3 + 5t)(3t^2 - 2) = 12t^5 - 8t^3 + 15t^3 - 10t \qquad$ Using FOIL and the rules for exponents
$$\qquad\qquad\qquad\qquad = 12t^5 + 7t^3 - 10t$$

d) $(3 - 4x)(7 - 5x^3) = 21 - 15x^3 - 28x + 20x^4$
$$\qquad\qquad\qquad\qquad = 21 - 28x - 15x^3 + 20x^4$$

2. Multiply to form an equivalent expression: $(x - 1)(x - 9)$.

In general, if the original binomials are written in *ascending* order, the answer is also written that way.

YOUR TURN

Multiplying Sums and Differences of Two Terms

Consider the product of the sum and the difference of the same two terms, such as

$$(x + 5)(x - 5).$$

Since this is the product of two binomials, we can use FOIL. In doing so, we find that the "outer" and "inner" products are opposites:

$$(x + 5)(x - 5) = x^2 - 5x + 5x - 25 \qquad \text{The "outer" and "inner" terms "drop out." Their sum is zero.}$$
$$= x^2 - 25.$$

Because opposites always add to zero, for products like $(x + 5)(x - 5)$ we can use a shortcut that is faster than FOIL.

> **THE PRODUCT OF A SUM AND A DIFFERENCE**
>
> The product of the sum and the difference of the same two terms is the square of the first term minus the square of the second term:
>
> $$(A + B)(A - B) = A^2 - B^2.$$
>
> This is called a *difference of squares*.

EXAMPLE 3   Multiply.

a) $(x + 4)(x - 4)$ **b)** $(5 + 2w)(5 - 2w)$ **c)** $(3a^4 - 5)(3a^4 + 5)$

SOLUTION

$$(A + B)(A - B) = A^2 - B^2$$

a) $(x + 4)(x - 4) = x^2 - 4^2$ Saying the words can help: "The square of the first term, x^2, minus the square of the second, 4^2"

$$= x^2 - 16$$ Simplifying

b) $(5 + 2w)(5 - 2w) = 5^2 - (2w)^2$

$$= 25 - 4w^2$$ Squaring both 5 and $2w$

c) $(3a^4 - 5)(3a^4 + 5) = (3a^4)^2 - 5^2$

$$= 9a^8 - 25$$ $(3a^4)^2 = 3^2 \cdot (a^4)^2 = 9 \cdot a^{4 \cdot 2} = 9a^8$

3. Multiply: $(y - 10)(y + 10)$.

YOUR TURN

Squaring Binomials

Consider the square of a binomial, such as $(x + 3)^2$. This can be expressed as $(x + 3)(x + 3)$. Since this is the product of two binomials, we can use FOIL. But again, this type of product occurs so often that a faster method has been developed. Look for a pattern in the following:

a) $(x + 3)^2 = (x + 3)(x + 3)$

$$= x^2 + 3x + 3x + 9$$

$$= x^2 + 6x + 9;$$

b) $(a - 7)^2 = (a - 7)(a - 7)$

$$= a^2 - 7a - 7a + 49$$

$$= a^2 - 14a + 49.$$

Perhaps you noticed that in each product the "outer" product and the "inner" product are identical. The other two terms, the "first" product and the "last" product, are squares.

THE SQUARE OF A BINOMIAL

The square of a binomial is the square of the first term, plus twice the product of the two terms, plus the square of the last term:

$$(A + B)^2 = A^2 + 2AB + B^2;$$

$$(A - B)^2 = A^2 - 2AB + B^2.$$

These are called *perfect-square trinomials.**

EXAMPLE 4   Write an equivalent expression for each square of a binomial.

a) $(x + 7)^2$ **b)** $(t - 5)^2$

c) $(3a + 0.4)^2$ **d)** $(5x - 3x^4)^2$

---

*Another name for these is *trinomial squares*.

SOLUTION

$$(A + B)^2 = A^2 + 2 \cdot A \cdot B + B^2$$

a) $(x + 7)^2 = x^2 + 2 \cdot x \cdot 7 + 7^2$ Saying the words can help: "The square of the first term, x^2, plus twice the product of the terms, $2 \cdot 7x$, plus the square of the second term, 7^2"

$$= x^2 + 14x + 49$$

b) $(t - 5)^2 = t^2 - 2 \cdot t \cdot 5 + 5^2$
$$= t^2 - 10t + 25$$

c) $(3a + 0.4)^2 = (3a)^2 + 2 \cdot 3a \cdot 0.4 + 0.4^2$
$$= 9a^2 + 2.4a + 0.16$$

d) $(5x - 3x^4)^2 = (5x)^2 - 2 \cdot 5x \cdot 3x^4 + (3x^4)^2$
$$= 25x^2 - 30x^5 + 9x^8$$ Using the rules for exponents

4. Multiply: $(x^2 - 3)^2$.

YOUR TURN

> **CAUTION!**   Although the square of a product is the product of the squares, the square of a sum is *not* the sum of the squares. That is, $(AB)^2 = A^2B^2$, but
>
> The term $2AB$ is missing.
>
> $$(A + B)^2 \neq A^2 + B^2.$$
>
> To confirm this inequality, note that
>
> $$(7 + 5)^2 = 12^2 = 144,$$
>
> whereas
>
> $$7^2 + 5^2 = 49 + 25 = 74, \quad \text{and} \quad 74 \neq 144.$$

Geometrically, $(A + B)^2$ can be viewed as the area of a square with sides of length $A + B$:

$$(A + B)(A + B) = (A + B)^2.$$

This is equal to the sum of the areas of the four smaller regions:

$$A^2 + AB + AB + B^2 = A^2 + 2AB + B^2.$$

Thus,

$$(A + B)^2 = A^2 + 2AB + B^2.$$

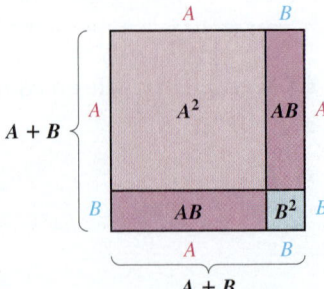

Note that the areas A^2 and B^2 do not fill the area $(A + B)^2$. Two additional areas, AB and AB, are needed.

━━━━━━━━ EXPLORING 🔍 THE CONCEPT ━━━━━━━━

Match each of the following squares with the appropriate square of a binomial and perfect-square trinomial.

1.
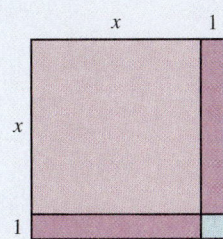

2.
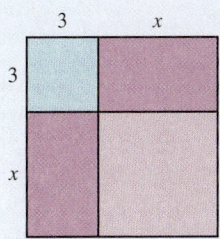

3.
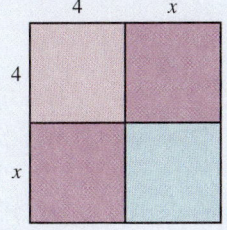

4.

Squared binomial

a) $(x + 2)^2$
b) $(x + 1)^2$
c) $(4 + x)^2$
d) $(3 + x)^2$

Perfect-square trinomial

A) $x^2 + 2x + 1$
B) $x^2 + 4x + 4$
C) $x^2 + 6x + 9$
D) $x^2 + 8x + 16$

ANSWERS
1. (b), (A)   2. (d), (C)
3. (a), (B)   4. (c), (D)

Multiplications of Various Types

Recognizing patterns often helps when new problems are encountered. To simplify a new multiplication problem, always examine what type of product it is so that the best method for finding that product can be used. To do this, ask yourself questions similar to the following.

MULTIPLYING TWO POLYNOMIALS

1. Is the multiplication the product of a monomial and a polynomial? If so, multiply each term of the polynomial by the monomial.
2. Is the multiplication the product of two binomials? If so:

   a) Is it the product of the sum and the difference of the *same* two terms? If so, use the pattern

   $$(A + B)(A - B) = A^2 - B^2.$$

   b) Is the product the square of a binomial? If so, use the pattern

   $$(A + B)(A + B) = (A + B)^2 = A^2 + 2AB + B^2$$

   or

   $$(A - B)(A - B) = (A - B)^2 = A^2 - 2AB + B^2.$$

   c) If neither (a) nor (b) applies, use FOIL.

3. Is the multiplication the product of two polynomials other than those above? If so, multiply each term of one by every term of the other. Use columns if you wish.

EXAMPLE 5 Multiply.

a) $(x + 3)(x - 3)$ b) $(t + 7)(t - 5)$

c) $(x + 7)(x + 7)$ d) $2x^3(9x^2 + x - 7)$

e) $(p + 3)(p^2 + 2p - 1)$ f) $\left(3x - \frac{1}{4}\right)^2$

SOLUTION

a) $(x + 3)(x - 3) = x^2 - 9$ This is the product of the sum and the difference of the same two terms.

b) $(t + 7)(t - 5) = t^2 - 5t + 7t - 35$ Using FOIL
$$= t^2 + 2t - 35$$

c) $(x + 7)(x + 7) = x^2 + 14x + 49$ This is the square of a binomial, $(x + 7)^2$.

d) $2x^3(9x^2 + x - 7) = 18x^5 + 2x^4 - 14x^3$ Multiplying each term of the trinomial by the monomial

Chapter Resource:
Visualizing for Success, p. 293

e) We multiply each term of $p^2 + 2p - 1$ by every term of $p + 3$:

$$(p + 3)(p^2 + 2p - 1) = p^3 + 2p^2 - p \qquad \text{Multiplying by } p$$
$$+ 3p^2 + 6p - 3 \qquad \text{Multiplying by } 3$$
$$= p^3 + 5p^2 + 5p - 3. \qquad \text{Combining like terms}$$

f) $\left(3x - \frac{1}{4}\right)^2 = 9x^2 - 2(3x)\left(\frac{1}{4}\right) + \frac{1}{16}$ Squaring a binomial
$$= 9x^2 - \frac{3}{2}x + \frac{1}{16}$$

5. Multiply: $(4x + 7)(4x - 7)$.

YOUR TURN

4.6   EXERCISE SET

FOR EXTRA HELP   MyMathLab®   MathXL

PRACTICE   WATCH   READ   REVIEW

Vocabulary and Reading Check

Classify each of the following statements as either true or false.

1. FOIL is simply a memory device for finding the product of two binomials.

2. The polynomial $x^2 + 49$ is an example of a perfect-square trinomial.

3. Once FOIL is used, it is always possible to combine like terms.

4. The square of $A + B$ is not the sum of the squares of A and B.

Products of Two Binomials

Multiply.

5. $(x^2 + 2)(x + 3)$ 6. $(x - 5)(x^2 - 6)$

7. $(t^4 - 2)(t + 7)$ 8. $(n^3 + 8)(n - 4)$

9. $(y + 2)(y - 3)$ 10. $(a + 2)(a + 2)$

11. $(3x + 2)(3x + 5)$ 12. $(4x + 1)(2x + 7)$

13. $(5x - 3)(x + 4)$ 14. $(4x - 5)(4x + 5)$

15. $(3 - 2t)(5 - t)$ 16. $(7 - a)(4 - 3a)$

17. $(x^2 + 3)(x^2 - 7)$ 18. $(x^2 + 2)(x^2 - 8)$

Multiplying Sums and Differences of Two Terms

Multiply.

19. $\left(p - \frac{1}{4}\right)\left(p + \frac{1}{4}\right)$ 20. $\left(q + \frac{3}{4}\right)\left(q + \frac{3}{4}\right)$

21. $(x + 0.3)(x - 0.3)$ 22. $(x - 0.1)(x + 0.1)$

23. $(10x^2 + 3)(10x^2 - 3)$ 24. $(5 - 4x^5)(5 + 4x^5)$

25. $(1 - 5t^3)(1 + 5t^3)$ 26. $(x^{10} + 3)(x^{10} - 3)$

Squaring Binomials

Find an equivalent expression for the following squares of binomials.

27. $(t - 2)^2$ 28. $(a - 10)^2$

29. $(x + 10)(x + 10)$ 30. $(x + 12)(x + 12)$

31. $(3x + 2)^2$ 32. $(5x + 1)^2$

33. $(1 - 10a)^2$ 34. $(7 - 6p)^2$

35. $(x^3 + 12)^2$ 36. $(x^5 - 4)^2$

Multiplications of Various Types

Multiply. Try to recognize the type of product before multiplying.

37. $(x^2 + 3)(x^3 - 1)$ 38. $(x^4 - 3)(2x + 1)$

39. $(x + 8)(x - 8)$ 40. $(x + 1)(x - 1)$

41. $(-3n + 2)(n + 7)$

42. $(-m + 5)(2m - 9)$

43. $(x + 3)^2$

44. $(2x - 1)^2$

45. $(7x^3 - 1)^2$

46. $(5x^3 + 2)^2$

47. $(9a^3 + 1)(9a^3 - 1)$

48. $(t^2 - 0.2)(t^2 + 0.2)$

49. $(x^4 + 0.1)(x^4 - 0.1)$

50. $(a^3 + 5)(a^3 - 5)$

51. $\left(t - \frac{3}{4}\right)\left(t + \frac{3}{4}\right)$

52. $\left(m - \frac{2}{3}\right)\left(m + \frac{2}{3}\right)$

53. $(1 - 3t)(1 + 5t^2)$

54. $(1 + 2t)(1 - 3t^2)$

55. $\left(a - \frac{2}{5}\right)^2$

56. $\left(t - \frac{1}{5}\right)^2$

57. $(t^4 + 3)^2$

58. $(a^3 + 6)^2$

59. $(5x - 9)(9x + 5)$

60. $(7x - 2)(2x - 7)$

61. $7n^3(2n^2 - 1)$

62. $5m^3(4 - 3m^2)$

63. $(a - 3)(a^2 + 2a - 4)$

64. $(x^2 - 5)(x^2 + x - 1)$

65. $(7 - 3x^4)(7 - 3x^4)$

66. $(x - 4x^3)^2$

67. $(2 - 3x^4)^2$

68. $(5 - 2t^3)^2$

69. $(5t + 6t^2)^2$

70. $(3p^2 - p)^2$

71. $5x(x^2 + 6x - 2)$

72. $6x(-x^5 + 6x^2 + 9)$

73. $(q^5 + 1)(q^5 - 1)$

74. $(p^4 + 2)(p^4 - 2)$

75. $3t^2(5t^3 - t^2 + t)$

76. $-5x^3(x^2 + 8x - 9)$

77. $(6x^4 - 3x)^2$

78. $(8a^3 + 5)(8a^3 - 5)$

79. $(9a + 0.4)(2a^3 + 0.5)$

80. $(2a - 0.7)(8a^3 - 0.5)$

81. $\left(\frac{1}{5} - 6x^4\right)\left(\frac{1}{5} + 6x^4\right)$

82. $\left(3 + \frac{1}{2}t^5\right)\left(3 + \frac{1}{2}t^5\right)$

83. $(a + 1)(a^2 - a + 1)$

84. $(x - 5)(x^2 + 5x + 25)$

Find the total area of all shaded rectangles.

85.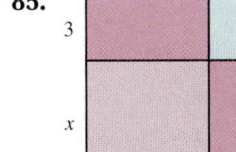

86.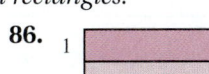

87.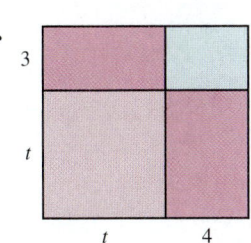

88.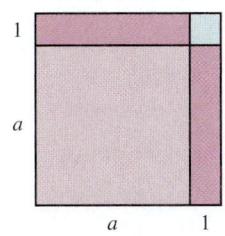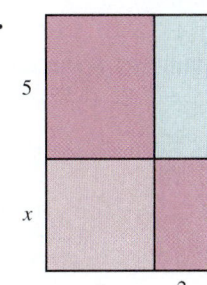

89.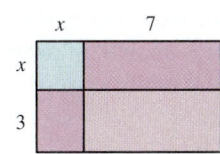

90.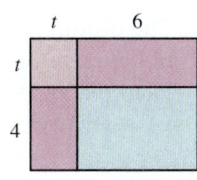

91.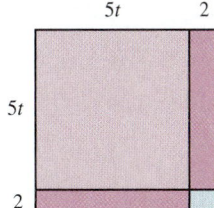

92.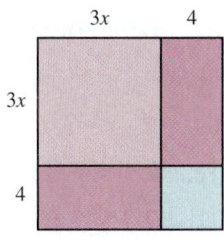

Draw and label rectangles similar to those in Exercises 85–92 to illustrate each of the following.

93. $(x + 5)^2$

94. $(x + 8)^2$

95. $(3 + x)^2$

96. $(7 + t)^2$

97. $(2t + 1)^2$

98. $(3a + 2)^2$

99. Kristi feels that since she can find the product of any two binomials using FOIL, she needn't study the other special products. What advice would you give her?

100. Under what conditions is the product of two binomials a binomial?

Skill Review

Solve.

101. *Energy Use.*   Bailey's refrigerator, upright freezer, and washing machine together use 1200 kilowatt-hours per year (kWh/year) of electricity. Her refrigerator uses three times as much energy as her washing machine, and her freezer uses six times as much energy as her washing machine. How much energy is used by each appliance?   [2.5]

Source: Based on data from ftc.gov/appliancedata

102. *Energy Credit.*   During some years, U.S. tax law allowed a tax credit of 30% of the price of a geo-thermal heat pump, with a maximum credit of $2000. For what prices of a heat pump could a 30% credit be taken?   [2.7]

Solve. [2.3]

103. $3ab = c$, for a

104. $ax - by = c$, for x

Synthesis

105. By writing $19 \cdot 21$ as $(20 - 1)(20 + 1)$, Blair can find the product mentally. How do you think he does this?

106. The product $(A + B)^2$ can be regarded as the sum of the areas of four regions (as shown following Example 4). How might one visually represent $(A + B)^3$? Why?

Multiply.

Aha! **107.** $(4x^2 + 9)(2x + 3)(2x - 3)$

108. $(9a^2 + 1)(3a - 1)(3a + 1)$

Aha! **109.** $(3t - 2)^2(3t + 2)^2$

110. $(5a + 1)^2(5a - 1)^2$

111. $(t^3 - 1)^4(t^3 + 1)^4$

112. $(32.41x + 5.37)^2$

Calculate as the difference of squares.

113. 18×22 [*Hint*: $(20 - 2)(20 + 2)$.]

114. 93×107

Solve.

115. $(x + 2)(x - 5) = (x + 1)(x - 3)$

116. $(2x + 5)(x - 4) = (x + 5)(2x - 4)$

Find a polynomial for the total shaded area in each figure.

117.
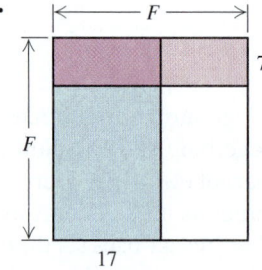

118.
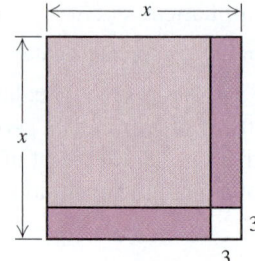

119.
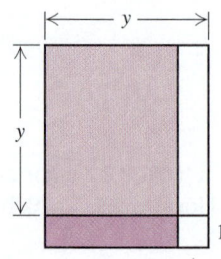

120. Find $(10 - 2x)^2$ by subtracting the white areas from 10^2.

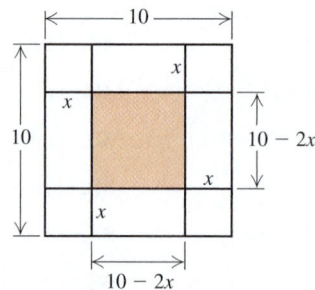

121. Find $(y - 2)^2$ by subtracting the white areas from y^2.

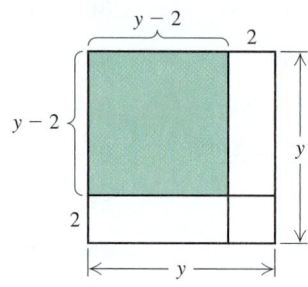

122. Find three consecutive integers for which the sum of the squares is 65 more than three times the square of the smallest integer.

123. Use a graphing calculator and the method developed in Section 4.5 to check your answers to Exercises 22, 47, and 83.

↪ YOUR TURN ANSWERS: SECTION 4.6

1. $t^3 + 3t^2 - 5t - 15$ **2.** $x^2 - 10x + 9$ **3.** $y^2 - 100$
4. $x^4 - 6x^2 + 9$ **5.** $16x^2 - 49$

QUICK QUIZ: SECTIONS 4.1–4.6

Multiply.

1. $(10a^3 b^{10})(-2ab^2)$ [4.1] **2.** $n \cdot n^{13} \cdot n^0$ [4.1]

3. $(x - 3)(4x^2 - x + 2)$ [4.5]

4. $(10x^5 + 1)(10x^5 - 1)$ [4.6]

5. $(3a + 4)^2$ [4.6]

PREPARE TO MOVE ON

Evaluate.   [1.1]

1. $\dfrac{2m}{n}$, for $m = 10$ and $n = 5$

2. $x - y$, for $x = 37$ and $y = 18$

Combine like terms.   [1.6]

3. $2a - 6c - c + 10a$ **4.** $4y + 7y + 2y - w$

4.7 Polynomials in Several Variables

Evaluating Polynomials ▪ Like Terms and Degree ▪ Addition and Subtraction ▪ Multiplication

Thus far, the polynomials that we have studied have had only one variable. Polynomials such as

$$5x + x^2y - 3y + 7, \qquad 9ab^2c - 2a^3b^2 + 8a^2b^3, \quad \text{and} \quad 4m^2 - 9n^2$$

contain two or more variables. In this section, we will add, subtract, multiply, and evaluate such **polynomials in several variables**.

Evaluating Polynomials

To evaluate a polynomial in two or more variables, we substitute numbers for the variables. Then we compute, using the rules for order of operations.

EXAMPLE 1 Evaluate the polynomial $4 + 3x + xy^2 + 8x^3y^3$ for $x = -2$ and $y = 5$.

SOLUTION We substitute -2 for x and 5 for y:

$$4 + 3x + xy^2 + 8x^3y^3 = 4 + 3(-2) + (-2) \cdot 5^2 + 8(-2)^3 \cdot 5^3$$
$$= 4 - 6 - 50 - 8000 = -8052.$$

1. Evaluate the polynomial $x^2y - 3y^2 - 2xy$ for $x = 4$ and $y = -3$.

YOUR TURN

EXAMPLE 2 *Surface Area of a Right Circular Cylinder.* The surface area of a right circular cylinder is given by the polynomial

$$2\pi rh + 2\pi r^2,$$

where h is the height and r is the radius of the base. A 12-oz can has a height of 4.7 in. and a radius of 1.2 in. Approximate its surface area to the nearest tenth of a square inch.

SOLUTION We evaluate the polynomial for $h = 4.7$ in. and $r = 1.2$ in. If 3.14 is used to approximate π, we have

$$2\pi rh + 2\pi r^2 \approx 2(3.14)(1.2 \text{ in.})(4.7 \text{ in.}) + 2(3.14)(1.2 \text{ in.})^2$$
$$= 2(3.14)(1.2 \text{ in.})(4.7 \text{ in.}) + 2(3.14)(1.44 \text{ in}^2)$$
$$= 35.4192 \text{ in}^2 + 9.0432 \text{ in}^2 = 44.4624 \text{ in}^2.$$

If the π key of a calculator is used, we have

$$2\pi rh + 2\pi r^2 \approx 2(3.141592654)(1.2 \text{ in.})(4.7 \text{ in.})$$
$$+ 2(3.141592654)(1.2 \text{ in.})^2$$
$$\approx 44.48495197 \text{ in}^2.$$

2. Use the formula given in Example 2 to find the surface area of a right circular cylinder with a height of 10 cm and a radius of 3 cm. Round to the nearest tenth of a square centimeter.

Note that the unit in the answer (square inches) is a unit of area. The surface area is about 44.5 in^2 (square inches).

YOUR TURN

Like Terms and Degree

Recall that the degree of a monomial is the number of variable factors in the term. For example, the degree of $5x^2$ is 2 because there are two variable factors in $5 \cdot x \cdot x$. Similarly, the degree of $5a^2b^4$ is 6 because there are 6 variable factors in $5 \cdot a \cdot a \cdot b \cdot b \cdot b \cdot b$. Note that 6 can be found by adding the exponents 2 and 4. The degree of a polynomial is the degree of the term of highest degree.

EXAMPLE 3   Identify the degree of each term and the degree of the polynomial

$$9x^2y^3 - 14xy^2z^3 + xy + 4y + 5x^2 + 7.$$

SOLUTION

Term	Degree	Degree of the Polynomial
$9x^2y^3$	5	
$-14xy^2z^3$	6	
xy	2	
$4y$	1	6
$5x^2$	2	
7	0	

 YOUR TURN

3. Identify the degree of each term and the degree of the polynomial $2x^2y + 9y - 6 + 4x^3y^5$.

Note in Example 3 that although both xy and $5x^2$ have degree 2, they are *not* like terms. *Like*, or *similar, terms* either have exactly the same variables with exactly the same exponents or are constants. For example,

$8a^4b^7$ and $5b^7a^4$ are like terms

and

-17 and 3 are like terms,

but

$-2x^2y$ and $9xy^2$ are *not* like terms.

As always, combining like terms is based on the distributive law.

EXAMPLE 4   Combine like terms to form equivalent expressions.

a) $9x^2y + 3xy^2 - 5x^2y - xy^2$

b) $7ab - 5ab^2 + 3ab^2 + 6a^3 + 9ab - 11a^3 + b - 1$

SOLUTION

a) $9x^2y + 3xy^2 - 5x^2y - xy^2 = (9 - 5)x^2y + (3 - 1)xy^2$
$$= 4x^2y + 2xy^2 \qquad \text{Try to go directly to this step.}$$

b) $7ab - 5ab^2 + 3ab^2 + 6a^3 + 9ab - 11a^3 + b - 1$
$$= -5a^3 - 2ab^2 + 16ab + b - 1 \qquad \text{We choose to write descending powers of } a. \text{ Other, equivalent, forms can also be used.}$$

4. Combine like terms:

$$2p^2x - x^2 + 9px^2 - 5p^2x.$$

YOUR TURN

Student Notes

Always read the problem carefully. The difference between

$$(-5x^3 - 3y) + (8x^3 + 4x^2)$$

and

$$(-5x^3 - 3y)(8x^3 + 4x^2)$$

is enormous. To avoid wasting time working on an incorrectly copied exercise, be sure to double-check that you have written the correct problem in your notebook.

5. Add: $(3a^2b - b^2 + 7ab) + (4ab - 7a^2b)$.

6. Subtract:

$(12x^3 - 3xy^2 - xy) - (-4xy^2 + xy - 5x^3)$.

7. Multiply:

$(ab - 2a)(5a^2 + 3ab^2 + b^3)$.

Addition and Subtraction

The same procedures used for adding or subtracting polynomials in one variable are used to add or subtract polynomials in several variables.

EXAMPLE 5   Add.

a) $(-5x^3 + 3y - 5y^2) + (8x^3 + 4x^2 + 7y^2)$

b) $(5ab^2 - 4a^2b + 5a^3 + 2) + (3ab^2 - 2a^2b + 3a^3b - 5)$

SOLUTION

a) $(-5x^3 + 3y - 5y^2) + (8x^3 + 4x^2 + 7y^2)$

$= (-5 + 8)x^3 + 4x^2 + 3y + (-5 + 7)y^2$ Try to do this step mentally.

$= 3x^3 + 4x^2 + 3y + 2y^2$

b) $(5ab^2 - 4a^2b + 5a^3 + 2) + (3ab^2 - 2a^2b + 3a^3b - 5)$

$= 8ab^2 - 6a^2b + 5a^3 + 3a^3b - 3$

YOUR TURN

EXAMPLE 6   Subtract:

$$(4x^2y + x^3y^2 + 3x^2y^3 + 6y) - (4x^2y - 6x^3y^2 + x^2y^2 - 5y).$$

SOLUTION   We find the opposite of $4x^2y - 6x^3y^2 + x^2y^2 - 5y$ and then add.

$(4x^2y + x^3y^2 + 3x^2y^3 + 6y) - (4x^2y - 6x^3y^2 + x^2y^2 - 5y)$

$= 4x^2y + x^3y^2 + 3x^2y^3 + 6y - 4x^2y + 6x^3y^2 - x^2y^2 + 5y$

$= 7x^3y^2 + 3x^2y^3 - x^2y^2 + 11y$ Combining like terms

YOUR TURN

Multiplication

To multiply polynomials in several variables, multiply each term of one polynomial by every term of the other.

EXAMPLE 7   Multiply: $(3x^2y - 2xy + 3y)(xy + 2y)$.

SOLUTION

$$\begin{array}{r} 3x^2y - 2xy + 3y \\ xy + 2y \\ \hline 6x^2y^2 - 4xy^2 + 6y^2 \\ 3x^3y^2 - 2x^2y^2 + 3xy^2 \\ \hline 3x^3y^2 + 4x^2y^2 - xy^2 + 6y^2 \end{array}$$

Multiplying by $2y$

Multiplying by xy

Adding

YOUR TURN

Using patterns for special products, such as the square of a binomial, can speed up our work.

EXAMPLE 8 Multiply.

a) $(p + 5w)(2p - 3w)$ **b)** $(3x + 2y)^2$

c) $(a^3 - 7a^2b)^2$ **d)** $(3x^2y + 2y)(3x^2y - 2y)$

e) $(-2x^3y^2 + 5t)(2x^3y^2 + 5t)$ **f)** $(2x + 3 - 2y)(2x + 3 + 2y)$

SOLUTION

$$\qquad\qquad\qquad \text{F}\qquad\text{O}\qquad\text{I}\qquad\text{L}$$

a) $(p + 5w)(2p - 3w) = 2p^2 - 3pw + 10pw - 15w^2$

$$\qquad\qquad\qquad\qquad = 2p^2 + 7pw - 15w^2 \qquad \text{Combining like terms}$$

$$(A + B)^2 = A^2 + 2 \cdot A \cdot B + B^2$$

b) $(3x + 2y)^2 = (3x)^2 + 2(3x)(2y) + (2y)^2 \qquad$ Squaring a binomial

$$\qquad\qquad = 9x^2 + 12xy + 4y^2$$

$$(A - B)^2 = A^2 - 2 \cdot A \cdot B + B^2$$

c) $(a^3 - 7a^2b)^2 = (a^3)^2 - 2(a^3)(7a^2b) + (7a^2b)^2 \qquad$ Squaring a binomial

$$\qquad\qquad = a^6 - 14a^5b + 49a^4b^2 \qquad \text{Using the rules for exponents}$$

$$(A + B)(A - B) = A^2 - B^2$$

d) $(3x^2y + 2y)(3x^2y - 2y) = (3x^2y)^2 - (2y)^2 \qquad$ Multiplying the sum and the
    difference of two terms

$$\qquad\qquad\qquad = 9x^4y^2 - 4y^2 \qquad \text{Using the rules for exponents}$$

e) $(-2x^3y^2 + 5t)(2x^3y^2 + 5t) = (5t - 2x^3y^2)(5t + 2x^3y^2) \qquad$ Using the commutative law for addition twice

$$\qquad\qquad\qquad = (5t)^2 - (2x^3y^2)^2 \qquad \text{Multiplying the sum and the difference of two terms}$$

$$\qquad\qquad\qquad = 25t^2 - 4x^6y^4$$

$$(A - B)(A + B) = A^2 - B^2$$

f) $(2x + 3 - 2y)(2x + 3 + 2y) = (2x + 3)^2 - (2y)^2 \qquad$ Multiplying a sum and a difference

$$\qquad\qquad\qquad = 4x^2 + 12x + 9 - 4y^2 \qquad \text{Squaring a binomial}$$

CAUTION!

$(3x + 2y)^2 \neq (3x)^2 + (2y)^2$

Technology Connection

One way to evaluate the polynomial in Example 1 for $x = -2$ and $y = 5$ is to store -2 to X and 5 to Y and enter the polynomial.

```
-2 → X
                    -2
5 → Y
                     5
4+3X+XY²+8X³Y³
                 -8052
■
```

Evaluate.

1. $3x^2 - 2y^2 + 4xy + x$, for $x = -6$ and $y = 2.3$

2. $a^2b^2 - 8c^2 + 4abc + 9a$, for $a = 11, b = 15,$ and $c = -7$

---

8. Multiply:

$(5x - 2xy)(x^2 - 2y)$.

 YOUR TURN

 Chapter Resource:
Collaborative Activity, p. 294

In Example 8, we recognized patterns that might not be obvious, particularly in parts (e) and (f). In part (e), we *can* use FOIL, and in part (f), we *can* use long multiplication, but doing so is slower. By carefully inspecting a problem before "jumping in," we can save ourselves considerable work. At least one instructor refers to this as "working smart" instead of "working hard."*

---

*Thanks to Pauline Kirkpatrick of Wharton County Junior College for this language.

4.7 EXERCISE SET

FOR EXTRA HELP

MyMathLab® Math XL
PRACTICE WATCH READ REVIEW

Vocabulary and Reading Check

Consider the polynomial $2x^2y + 7xy^3 + 5x^2y$. *Choose the phrase from the list on the right that describes the expression on the left as it relates to the polynomial.*

1. ____ $2x^2y$ and $5x^2y$
2. ____ x, y
3. ____ 4
4. ____ 3

a) The degree of the polynomial
b) Like terms
c) The number of terms in the polynomial
d) The variables in the polynomial

Concept Reinforcement

Each of the expressions in Exercises 5–10 can be regarded as either **(a)** *the square of a binomial,* **(b)** *the product of the sum and the difference of the same two terms, or* **(c)** *neither of the above. Select the appropriate choice for each expression.*

5. $(3x + 5y)^2$
6. $(4x - 9y)(4x + 9y)$
7. $(5a + 6b)(-6b + 5a)$
8. $(4a - 3b)(4a - 3b)$
9. $(r - 3s)(5r + 3s)$
10. $(2x - 7y)(7y - 2x)$

Evaluating Polynomials

Evaluate each polynomial for $x = 5$ *and* $y = -2$.

11. $x^2 - 2y^2 + 3xy$

12. $x^2 + 5y^2 - 4xy$

Evaluate each polynomial for $x = 2, y = -3$, *and* $z = -4$.

13. $xy^2z - z$

14. $xy - x^2z + yz^2$

Zoology. *The polynomial* $11.5g + 7.55l + 12.5c - 4016$ *can be used to estimate the weight of an elephant, in kilograms, given the girth g of the elephant at the heart, the length l of the elephant, and the circumference c of the elephant's footpad. Use the formula and the data from the following table for Exercises 15 and 16.*

Elephant	Girth, g (in centimeters)	Length, l (in centimeters)	Footpad Circumference, c (in centimeters)
Male, age 3 yr	244	140	86
Female, age 3 yr	231	135	86
Male, age 25 yr	404	229	130
Female, age 25 yr	366	226	117

Source: "How Much Does That Elephant Weigh?" by Mark MacAllister on fieldtripearth.org

15. Estimate the weight of the 3-year-old female elephant described in the table.

16. Estimate the weight of the 25-year-old male elephant described in the table.

Surface Area of a Silo. *A silo is a structure that is shaped like a right circular cylinder with a half sphere on top. The surface area of a silo of height h and radius r (including the area of the base) is given by the polynomial* $2\pi rh + \pi r^2$.

17. A coffee grinder is shaped like a silo, with a height of 7 in. and a radius of $1\frac{1}{2}$ in. Find the surface area of the coffee grinder. Use 3.14 for π.

18. A $1\frac{1}{2}$-oz bottle of roll-on deodorant has a height of 4 in. and a radius of $\frac{3}{4}$ in. Find the surface area of the bottle if the bottle is shaped like a silo. Use 3.14 for π.

Altitude of a Launched Object. *The altitude of an object, in meters, is given by the polynomial*

$$h + vt - 4.9t^2,$$

where h is the height, in meters, at which the launch occurs, v is the initial upward speed (or velocity), in meters per second, and t is the number of seconds for which the object is airborne.

19. A bocce ball is thrown upward with an initial speed of 18 m/sec by a person atop the Leaning Tower of Pisa, which is 50 m above the ground. How high will the ball be 2 sec after it has been thrown?

50 m

20. A golf ball is launched upward with an initial speed of 30 m/sec by a golfer on the third level of The Golf Club at Chelsea Piers, Manhattan, which is 10 m above the ground. How high above the ground will the ball be after 3 sec?

Like Terms and Degree

Identify the degree of each term of each polynomial. Then find the degree of the polynomial.

21. $3x^2y - 5xy + 2y^2 - 11$

22. $xy^3 + 7x^3y^2 - 6xy^4 + 2$

23. $7 - abc + a^2b + 9ab^2$

24. $3p - pq - 7p^2q^3 - 8pq^6$

Combine like terms.

25. $3r + s - r - 7s$

26. $9a + b - 8a - 5b$

27. $5xy^2 - 2x^2y + x + 3x^2$

28. $m^3 + 2m^2n - 3m^2 + 3mn^2$

29. $6u^2v - 9uv^2 + 3vu^2 - 2v^2u + 11u^2$

30. $3x^2 + 6xy + 3y^2 - 5x^2 - 10xy$

31. $5a^2c - 2ab^2 + a^2b - 3ab^2 + a^2c - 2ab^2$

32. $3s^2t + r^2t - 9ts^2 - st^2 + 5t^2s - 7tr^2$

Addition and Subtraction

Add or subtract, as indicated.

33. $(6x^2 - 2xy + y^2) + (5x^2 - 8xy - 2y^2)$

34. $(7r^3 + rs - 5r^2) - (2r^3 - 3rs + r^2)$

35. $(3a^4 - 5ab + 6ab^2) - (9a^4 + 3ab - ab^2)$

36. $(2r^2t - 5rt + rt^2) - (7r^2t + rt - 5rt^2)$

Aha! **37.** $(5r^2 - 4rt + t^2) + (-6r^2 - 5rt - t^2) + (-5r^2 + 4rt - t^2)$

38. $(2x^2 - 3xy + y^2) + (-4x^2 - 6xy - y^2) + (4x^2 + 6xy + y^2)$

39. $(x^3 - y^3) - (-2x^3 + x^2y - xy^2 + 2y^3)$

40. $(a^3 + b^3) - (-5a^3 + 2a^2b - ab^2 + 3b^3)$

41. $(2y^4x^3 - 3y^3x) + (5y^4x^3 - y^3x) - (9y^4x^3 - y^3x)$

42. $(5a^2b - 7ab^2) - (3a^2b + ab^2) + (a^2b - 2ab^2)$

43. Subtract $7x + 3y$ from the sum of $4x + 5y$ and $-5x + 6y$.

44. Subtract $5a + 2b$ from the sum of $2a + b$ and $3a - 4b$.

Multiplication

Multiply.

45. $(4c - d)(3c + 2d)$

46. $(5x + y)(2x - 3y)$

47. $(xy - 1)(xy + 5)$

48. $(ab + 3)(ab - 5)$

49. $(2a - b)(2a + b)$

50. $(a - 3b)(a + 3b)$

51. $(5rt - 2)(4rt - 3)$

52. $(3xy - 1)(4xy + 2)$

53. $(m^3n + 8)(m^3n - 6)$

54. $(9 - u^2v^2)(2 - u^2v^2)$

55. $(6x - 2y)(5x - 3y)$

56. $(7a - 6b)(5a + 4b)$

57. $(pq + 0.1)(-pq + 0.1)$

58. $(rt + 0.2)(-rt + 0.2)$

59. $(x + h)^2$

60. $(a - r)^2$

61. $(4a - 5b)^2$

62. $(2x + 5y)^2$

63. $(ab + cd^2)(ab - cd^2)$

64. $(p^3 - 5q)(p^3 + 5q)$

65. $(2xy + x^2y + 3)(xy + y^2)$

66. $(5cd - c^2 - d^2)(2c - c^2d)$

Aha! **67.** $(a + b - c)(a + b + c)$

68. $(x + y + 2z)(x + y - 2z)$

69. $[a + b + c][a - (b + c)]$

70. $(a + b + c)(a - b - c)$

Find the total area of each figure.

71.

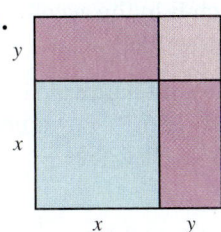

72.

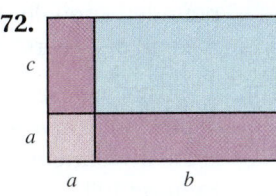

73.

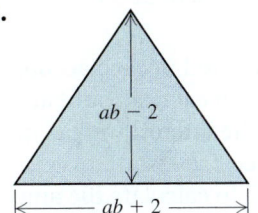

74.

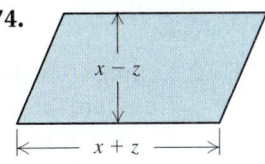

75.

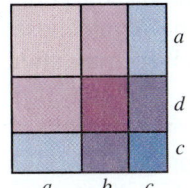

76.

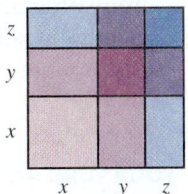

77.

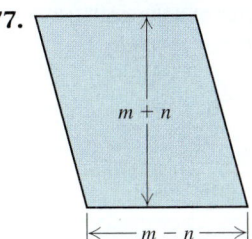

78.

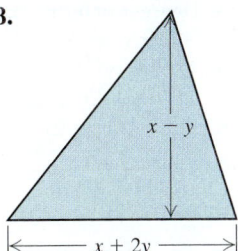

Draw and label rectangles similar to those in Exercises 71, 72, 75, and 76 to illustrate each product.

79. $(r + s)(u + v)$

80. $(m + r)(n + v)$

81. $(a + b + c)(a + d + f)$

82. $(r + s + t)^2$

83. Is it possible for a polynomial in 4 variables to have a degree less than 4? Why or why not?

84. A fourth-degree monomial is multiplied by a third-degree monomial. What is the degree of the product? Explain your reasoning.

Skill Review

Simplify. [1.8]

85. $-16 + 20 \div 2^2 \cdot 5 - 3$

86. $2 - (3 - 8)^2 \div (-5)$

87. $2[3 - 4(5 - 6)^2 - 1] + 10$

88. $-6|-2 - (-1)| + 3(-5)$

89. $2a - 3(-x - 5a + 7)$

90. $4(y + 7) - (y - 2) + (2y - 9)$

Synthesis

91. The concept of "leading term" was intentionally not discussed in this section. Why not?

92. Explain how it is possible for the sum of two trinomials in several variables to be a binomial in one variable.

Find a polynomial for the shaded area in each figure. (Leave results in terms of π where appropriate.)

93.

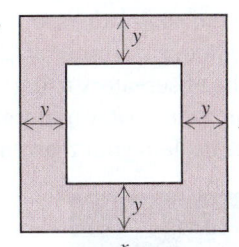

94.

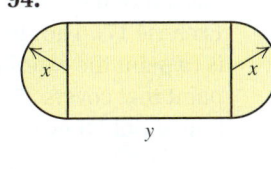

95.

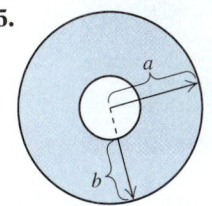

96.

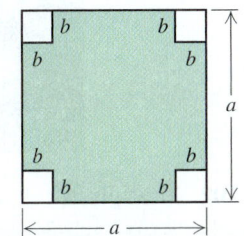

97. Find a polynomial for the total volume of the following figure.

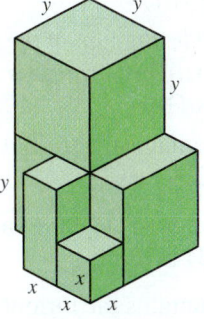

98. Find the shaded area in the following figure using each of the approaches given below. Then check that both answers match.

a) Find the shaded area by subtracting the area of the unshaded square from the total area of the figure.

b) Find the shaded area by adding the areas of the three shaded rectangles.

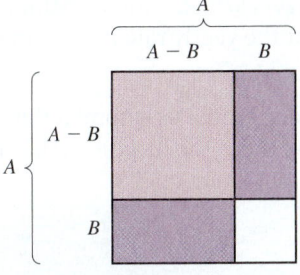

Find a polynomial for the surface area of each solid object shown. (Leave results in terms of π.)

99.

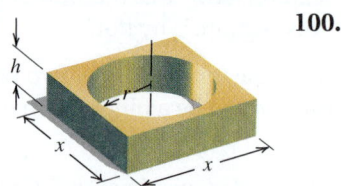

100.

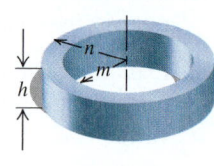

101. The observatory at Danville University is shaped like a silo that is 40 ft high and 30 ft wide (see Exercise 17). The Heavenly Bodies Astronomy Club is to paint the exterior of the observatory using paint that covers 250 ft² per gallon. How many gallons should they purchase? Explain your reasoning.

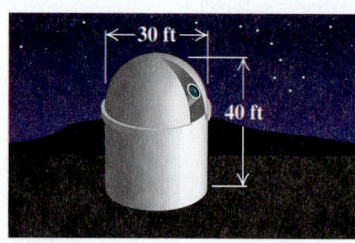

102. Multiply: $(x + a)(x - b)(x - a)(x + b)$.

Spreadsheet applications allow values for cells to be calculated from values in other cells. For example, if the cell C1 contains the formula

$$= A1 + 2 * B1,$$

the value in C1 *will be the sum of the value in* A1 *and twice the value in* B1. *This formula is a polynomial in the two variables* A1 *and* B1.

103. The cell D4 contains the formula

$$= 2 * A4 + 3 * B4.$$

What is the value in D4 if the value in A4 is 5 and the value in B4 is 10?

104. The cell D6 contains the formula

$$= A1 - 0.2 * B1 + 0.3 * C1.$$

What is the value in D6 if the value in A1 is 10, the value in B1 is −3, and the value in C1 is 30?

105. *Interest Compounded Annually.* An amount of money P that is invested at the yearly interest rate r grows to the amount $P(1 + r)^t$ after t years. Find a polynomial that can be used to determine the amount to which P will grow after 2 years.

106. *Yearly Depreciation.* An investment P that drops in value at the yearly rate r drops in value to

$$P(1 - r)^t$$

after t years. Find a polynomial that can be used to determine the value to which P has dropped after 2 years.

107. Suppose that $10,400 is invested at 4.5%, compounded annually. How much is in the account at the end of 5 years? (See Exercise 105.)

108. A $90,000 investment in computer hardware is depreciating at a yearly rate of 12.5%. How much is the investment worth after 4 years? (See Exercise 106.)

109. *Research.* Find the interest rates on several accounts offered by your bank or another local bank.

a) Suppose $10,000 is invested in each account. Compare the amounts in each account at the end of 10 years if the interest were compounded annually. (See Exercise 105.)

b) Find the formula used to calculate the amount in an account if the interest is compounded more often than once a year. Use the compounding period from the bank you chose to compare amounts in each account after 10 years.

YOUR TURN ANSWERS: SECTION 4.7

1. −51 **2.** 245.0 cm²
3. Degrees of terms: 3, 1, 0, 8; degree of polynomial: 8
4. $-3p^2x - x^2 + 9px^2$ **5.** $-4a^2b - b^2 + 11ab$
6. $17x^3 + xy^2 - 2xy$
7. $5a^3b + 3a^2b^3 + ab^4 - 10a^3 - 6a^2b^2 - 2ab^3$
8. $5x^3 - 10xy - 2x^3y + 4xy^2$

QUICK QUIZ: SECTIONS 4.1–4.7

1. Evaluate $-2x^3 - 3x^2 - 5x + 17$ for $x = -2$. [4.3]

2. Subtract:

$(3.1x^2 + 2.7x - 1.1) - (-x^2 + 5.6x - 7.2)$. [4.4]

3. Multiply: $(x + 3)(5x^3 - x + 4)$. [4.5]

4. Multiply: $(a^2 + 10)(a^2 - 10)$. [4.6]

5. Evaluate $xz - z^2 - 2y$ for $x = -2, y = 3$, and $z = -5$. [4.7]

PREPARE TO MOVE ON

Subtract. [4.4]

1.
$$\begin{array}{r} x^2 - 3x - 7 \\ -(5x - 3) \\ \hline \end{array}$$

2.
$$\begin{array}{r} 2x^3 - x + 3 \\ -(x^2 - 1) \\ \hline \end{array}$$

3.
$$\begin{array}{r} 3x^2 + x + 5 \\ -(3x^2 + 3x) \\ \hline \end{array}$$

4.
$$\begin{array}{r} 4x^3 - 3x^2 + x \\ -(4x^3 - 8x^2) \\ \hline \end{array}$$

4.8 Division of Polynomials

Dividing by a Monomial ▪ Dividing by a Binomial

In this section, we study division of polynomials. We will find that polynomial division is similar to division in arithmetic.

Dividing by a Monomial

We first consider division by a monomial. When dividing a monomial by a monomial, we subtract exponents when bases are the same. For example,

$$\frac{42a^2b^5}{-3ab^2} = \frac{42}{-3}a^{2-1}b^{5-2} \qquad \text{Recall that } a^m/a^n = a^{m-n}.$$

$$= -14ab^3.$$

CAUTION! The coefficients are divided but the exponents are subtracted.

To divide a polynomial by a monomial, we note that since

$$\frac{A}{C} + \frac{B}{C} = \frac{A + B}{C},$$

it follows that

$$\frac{A + B}{C} = \frac{A}{C} + \frac{B}{C}. \qquad \text{Switching the left side and the right side of the equation}$$

This is actually how we perform divisions like $86 \div 2$:

$$\frac{86}{2} = \frac{80 + 6}{2} = \frac{80}{2} + \frac{6}{2} = 40 + 3.$$

Similarly, to divide a polynomial by a monomial, we divide each term by the monomial:

$$\frac{80x^5 + 6x^7}{2x^3} = \frac{80x^5}{2x^3} + \frac{6x^7}{2x^3}$$

$$= \frac{80}{2}x^{5-3} + \frac{6}{2}x^{7-3} \qquad \text{Dividing coefficients and subtracting exponents}$$

$$= 40x^2 + 3x^4.$$

EXAMPLE 1 Divide $x^4 + 15x^3 - 6x^2$ by $3x$.

SOLUTION We divide each term of $x^4 + 15x^3 - 6x^2$ by $3x$:

$$\frac{x^4 + 15x^3 - 6x^2}{3x} = \frac{x^4}{3x} + \frac{15x^3}{3x} - \frac{6x^2}{3x}$$

$$= \frac{1}{3}x^{4-1} + \frac{15}{3}x^{3-1} - \frac{6}{3}x^{2-1} \qquad \text{Dividing coefficients and subtracting exponents}$$

$$= \frac{1}{3}x^3 + 5x^2 - 2x. \qquad \text{This is the quotient.}$$

To check, we multiply our answer by $3x$, using the distributive law:

$$3x\left(\frac{1}{3}x^3 + 5x^2 - 2x\right) = 3x \cdot \frac{1}{3}x^3 + 3x \cdot 5x^2 - 3x \cdot 2x$$

$$= x^4 + 15x^3 - 6x^2.$$

This is the polynomial that was being divided, so our answer, $\frac{1}{3}x^3 + 5x^2 - 2x$, checks.

1. Divide $t^3 - 4t^2 + 12t$ by $4t$.

YOUR TURN

EXAMPLE 2   Divide and check: $(10a^5b^4 - 2a^3b^2 + 6a^2b) \div (-2a^2b)$.

SOLUTION   We have

$$\frac{10a^5b^4 - 2a^3b^2 + 6a^2b}{-2a^2b} = \frac{10a^5b^4}{-2a^2b} - \frac{2a^3b^2}{-2a^2b} + \frac{6a^2b}{-2a^2b}$$

We divide coefficients and subtract exponents.

$$= -\frac{10}{2}a^{5-2}b^{4-1} - \left(-\frac{2}{2}\right)a^{3-2}b^{2-1} + \left(-\frac{6}{2}\right)$$

$$= -5a^3b^3 + ab - 3.$$

Check: $-2a^2b(-5a^3b^3 + ab - 3)$

$$= -2a^2b(-5a^3b^3) + (-2a^2b)(ab) + (-2a^2b)(-3)$$

$$= 10a^5b^4 - 2a^3b^2 + 6a^2b$$

2. Divide and check:

$(-3x^2y + 9x^3y^2 - 6x^2y^2)$
$\div (3x^2y).$

Our answer, $-5a^3b^3 + ab - 3$, checks.

YOUR TURN

Dividing by a Binomial

The divisors in Examples 1 and 2 have just one term. For divisors with more than one term, we use long division, much as we do in arithmetic.

EXAMPLE 3   Divide $x^2 + 5x + 6$ by $x + 3$.

SOLUTION   We begin by dividing x^2 by x:

Divide the first term, x^2, by the first term in the divisor: $x^2/x = x$. Ignore the term 3 for the moment.

$$
\begin{array}{r}
x \\
x + 3 \overline{)\, x^2 + 5x + 6} \\
-(x^2 + 3x) \\
\hline
2x
\end{array}
$$

Multiply: $x(x + 3) = x^2 + 3x.$

Subtract both x^2 and $3x$: $x^2 + 5x - (x^2 + 3x) = 2x.$

Now we "bring down" the next term—in this case, 6. The current remainder, $2x + 6$, now becomes the focus of our division. We divide $2x$ by x.

Student Notes

Since long division of polynomials requires many steps, we recommend that you double-check each step of your work as you move forward.

$$
\begin{array}{r}
x + 2 \\
x + 3 \overline{)\, x^2 + 5x + 6} \\
-(x^2 + 3x) \\
\hline
2x + 6 \\
-(2x + 6) \\
\hline
0
\end{array}
$$

Divide $2x$ by x: $2x/x = 2.$

Multiply: $2(x + 3) = 2x + 6.$

Subtract: $(2x + 6) - (2x + 6) = 0.$

The quotient is $x + 2$. The notation R 0 indicates a remainder of 0, although a remainder of 0 is generally not listed in an answer.

Check: To check, we multiply the quotient by the divisor and add any remainder to see if we get the dividend:

$$\underbrace{(x + 3)}_{\text{Divisor}} \; \underbrace{(x + 2)}_{\text{Quotient}} + \underbrace{0}_{\text{Remainder}} = \underbrace{x^2 + 5x + 6.}_{\text{Dividend}}$$

Our answer, $x + 2$, checks.

3. Divide $x^2 - 3x - 10$ by $x - 5$.

YOUR TURN

EXAMPLE 4 Divide: $(2x^2 + 5x - 1) \div (2x - 1)$.

SOLUTION We begin by dividing $2x^2$ by $2x$:

CAUTION! Write the parentheses around the polynomial being subtracted to remind you to subtract all its terms.

$$\begin{array}{r} x \\ 2x - 1 \overline{)2x^2 + 5x - 1} \\ -(2x^2 - x) \\ \hline 6x \end{array}$$

Divide the first term by the first term: $2x^2/(2x) = x$.

Multiply: $x(2x - 1) = 2x^2 - x$.

Subtract by changing signs and adding: $(2x^2 + 5x) - (2x^2 - x) = 6x$.

Now, we bring down the -1 and divide $6x - 1$ by $2x - 1$, focusing on the $6x$ and the $2x$.

$$\begin{array}{r} x + 3 \\ 2x - 1 \overline{)2x^2 + 5x - 1} \\ -(2x^2 - x) \\ \hline 6x - 1 \\ -(6x - 3) \\ \hline 2 \end{array}$$

Divide $6x$ by $2x$: $6x/(2x) = 3$.

Multiply: $3(2x - 1) = 6x - 3$.

Subtract. Note that $-1 - (-3) = -1 + 3 = 2$.

The answer is $x + 3$ with R 2.
Another way to write $x + 3$ R 2 is as

$$\underbrace{x + 3}_{\text{Quotient}} + \underbrace{\dfrac{2}{2x - 1}}_{}.$$ ←Remainder

←Divisor

(This is the way answers will be given at the back of the book.)

Check: To check, we multiply the divisor by the quotient and add the remainder:

$$(2x - 1)(x + 3) + 2 = 2x^2 + 5x - 3 + 2$$
$$= 2x^2 + 5x - 1. \quad \text{Our answer checks.}$$

4. Divide:

$(2x^2 - 3x + 5) \div (2x + 3)$.

YOUR TURN

Our division procedure ends when the degree of the remainder is less than that of the divisor. Check that this was indeed the case in Example 4.

When we are dividing polynomials, it is important that both the divisor and the dividend are written in descending order. Any missing terms in the dividend should be written in, using 0 for the coefficients.

EXAMPLE 5   Divide each of the following.

a) $(x^3 + 1) \div (x + 1)$

b) $(9a^2 + a^3 - 5) \div (a^2 - 1)$

SOLUTION

a)
$$
\begin{array}{r}
x^2 - x + 1 \\
x + 1 \overline{)x^3 + 0x^2 + 0x + 1} \\
-(x^3 + x^2) \\
\hline
-x^2 + 0x \\
-(-x^2 - x) \\
\hline
x + 1 \\
-(x + 1) \\
\hline
0
\end{array}
$$

$\leftarrow$ Writing in the missing terms

$\leftarrow$ Subtracting $x^3 + x^2$ from $x^3 + 0x^2$ and bringing down the $0x$

$\leftarrow$ Subtracting $-x^2 - x$ from $-x^2 + 0x$ and bringing down the 1

The answer is $x^2 - x + 1$.

Check: $(x + 1)(x^2 - x + 1) = x^3 - x^2 + x + x^2 - x + 1$
$$= x^3 + 1.$$

b) We rewrite the problem in descending order.

$$(a^3 + 9a^2 - 5) \div (a^2 - 1).$$

Thus,

$$
\begin{array}{r}
a + 9 \\
a^2 - 1 \overline{)a^3 + 9a^2 + 0a - 5} \\
-(a^3 \quad - a) \\
\hline
9a^2 + a - 5 \\
-(9a^2 \quad - 9) \\
\hline
a + 4
\end{array}
$$

$\leftarrow$ Writing in the missing term

$\leftarrow$ Subtracting $a^3 - a$ from $a^3 + 9a^2 + 0a$ and bringing down the -5

The degree of the remainder is less than the degree of the divisor, so we are finished.

The answer is $a + 9 + \dfrac{a + 4}{a^2 - 1}$.

Check: $(a^2 - 1)(a + 9) + a + 4 = a^3 + 9a^2 - a - 9 + a + 4$
$$= a^3 + 9a^2 - 5.$$

5. Divide: $(2x^2 - 5) \div (x - 1)$.

 YOUR TURN

4.8 EXERCISE SET

FOR EXTRA HELP

MyMathLab® MathXL

PRACTICE   WATCH   READ   REVIEW

Vocabulary and Reading Check

Use the words from the following list to label the numbered expressions from the division shown.

dividend, divisor, quotient, remainder

$$
\begin{array}{r}
②\, x + 2 \\
①x - 3 \overline{)x^2 - x + 9}\, ③ \\
-(x^2 - 3x) \\
\hline
2x + 9 \\
-(2x - 6) \\
\hline
15 \; ④
\end{array}
$$

1. _____

2. _____

3. _____

4. _____

Dividing by a Monomial

Divide and check.

5. $\dfrac{40x^6 - 25x^3}{5}$

6. $\dfrac{16a^5 - 24a^2}{8}$

7. $\dfrac{u - 2u^2 + u^7}{u}$

8. $\dfrac{50x^5 - 7x^4 + 2x}{x}$

9. $(18t^3 - 24t^2 + 6t) \div (3t)$

10. $(20t^3 - 15t^2 + 30t) \div (5t)$

11. $(42x^5 - 36x^3 + 9x^2) \div (6x^2)$

12. $(24x^6 + 18x^4 + 8x^3) \div (4x^3)$

13. $(32t^5 + 16t^4 - 8t^3) \div (-8t^3)$

14. $(36t^6 - 27t^5 - 9t^2) \div (-9t^2)$

15. $\dfrac{8x^2 - 10x + 1}{2x}$

16. $\dfrac{9x^2 + 3x - 2}{3x}$

17. $\dfrac{5x^3y + 10x^5y^2 + 15x^2y}{5x^2y}$

18. $\dfrac{12a^3b^2 + 4a^4b^5 + 16ab^2}{4ab^2}$

19. $\dfrac{9r^2s^2 + 3r^2s - 6rs^2}{-3rs}$

20. $\dfrac{4x^4y - 8x^6y^2 + 12x^8y^6}{4x^4y}$

Dividing by a Binomial

Divide.

21. $(x^2 - 8x + 12) \div (x - 2)$

22. $(x^2 + 2x - 15) \div (x + 5)$

23. $(t^2 - 10t - 20) \div (t - 5)$

24. $(t^2 + 8t - 15) \div (t + 4)$

25. $(2x^2 + 11x - 5) \div (x + 6)$

26. $(3x^2 - 2x - 13) \div (x - 2)$

27. $\dfrac{t^3 + 27}{t + 3}$ **28.** $\dfrac{a^3 + 8}{a + 2}$

29. $\dfrac{a^2 - 21}{a - 5}$ **30.** $\dfrac{t^2 - 13}{t - 4}$

31. $(5x^2 - 16x) \div (5x - 1)$

32. $(3x^2 - 7x + 1) \div (3x - 1)$

33. $(6a^2 + 17a + 8) \div (2a + 5)$

34. $(10a^2 + 19a + 9) \div (2a + 3)$

35. $\dfrac{2t^3 - 9t^2 + 11t - 3}{2t - 3}$

36. $\dfrac{8t^3 - 22t^2 - 5t + 12}{4t + 3}$

37. $(x^3 - x^2 + x - 1) \div (x - 1)$

38. $(t^3 - t^2 + t - 1) \div (t + 1)$

39. $(t^4 + 4t^2 + 3t - 6) \div (t^2 + 5)$

40. $(t^4 - 2t^2 + 4t - 5) \div (t^2 - 3)$

41. $(4x^4 - 3 - x - 4x^2) \div (2x^2 - 3)$

42. $(x + 6x^4 - 4 - 3x^2) \div (1 + 2x^2)$

43. How is the distributive law used when dividing a polynomial by a binomial?

44. On an assignment, Emmy Lou *incorrectly* writes
$$\dfrac{12x^3 - 6x}{3x} = 4x^2 - 6x.$$
What mistake do you think she is making and how might you convince her that a mistake has been made?

Skill Review

Graph.

45. $y = -\frac{2}{3}x + 4$ [3.6]   **46.** $8x = 4y$ [3.2]

47. Find the slope and the y-intercept of the line given by $2y = 8x + 7$.   [3.6]

48. Find the slope–intercept form of the line containing the points $(6, 3)$ and $(-2, -7)$.   [3.7]

Synthesis

49. Explain how to form trinomials for which division by $x - 5$ results in a remainder of 3.

50. Under what circumstances will the quotient of two binomials have more than two terms?

Divide.

51. $(10x^{9k} - 32x^{6k} + 28x^{3k}) \div (2x^{3k})$

52. $(45a^{8k} + 30a^{6k} - 60a^{4k}) \div (3a^{2k})$

53. $(6t^{3h} + 13t^{2h} - 4t^h - 15) \div (2t^h + 3)$

54. $(x^4 + a^2) \div (x + a)$

55. $(5a^3 + 8a^2 - 23a - 1) \div (5a^2 - 7a - 2)$

56. $(15y^3 - 30y + 7 - 19y^2) \div (3y^2 - 2 - 5y)$

57. Divide the sum of $4x^5 - 14x^3 - x^2 + 3$ and $2x^5 + 3x^4 + x^3 - 3x^2 + 5x$ by $3x^3 - 2x - 1$.

58. Divide $5x^7 - 3x^4 + 2x^2 - 10x + 2$ by the sum of $(x - 3)^2$ and $5x - 8$.

If the remainder is 0 when one polynomial is divided by another, the divisor is a factor *of the dividend. Find the value(s) of c for which x − 1 is a factor of each polynomial.*

59. $x^2 - 4x + c$

60. $2x^2 - 3cx - 8$

61. $c^2x^2 + 2cx + 1$

62. *Business.* A company's **revenue** R from an item is defined as the price paid per item times the quantity of items sold.

a) Easy on the Eyes sells high-quality reproductions of original watercolors. Find an expression for the price paid per reproduction if the revenue from the sale of q reproductions is $(80q - q^2)$ dollars.

b) Find an expression for the price paid per reproduction if one more reproduction is sold but the revenue remains $(80q - q^2)$ dollars.

63. The volume of a cube is $(a^3 + 3a^2 + 3a + 1)$ cm^3.

a) Find the length of one side of the cube.
b) Find the area of one side of the cube.

↩ YOUR TURN ANSWERS: SECTION 4.8

1. $\frac{1}{4}t^2 - t + 3$ **2.** $-1 + 3xy - 2y$ **3.** $x + 2$

4. $x - 3 + \dfrac{14}{2x + 3}$ **5.** $2x + 2 + \dfrac{-3}{x - 1}$

QUICK QUIZ: SECTIONS 4.1–4.8

1. Determine the leading term of the polynomial $3t^2 - t^5 + 10 + 2t^4$.   [4.3]

2. Multiply: $3x^4(-5x^2 - 6x + 1)$.   [4.5]

3. Multiply: $\left(\frac{1}{3}a^2 + 6\right)^2$.   [4.6]

4. Add: $(a^2b - 2ab + 5ab^2) + (3ab^2 - a^2b - 8ab)$.   [4.7]

5. Divide: $(16n^4 - 8n^3 + 4n) \div (4n)$.   [4.8]

PREPARE TO MOVE ON

List the factors in each expression.   [1.2]

1. $3(x + 7)$ **2.** $(x - 3)(x + 10)$

3. $x(x + 1)(2x - 5)$

Factor.   [1.2]

4. $9x + 15y$ **5.** $14a + 7c + 7$

1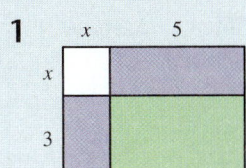

2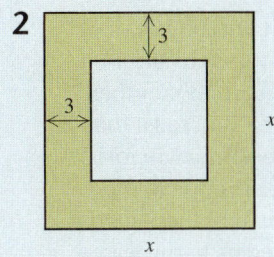

3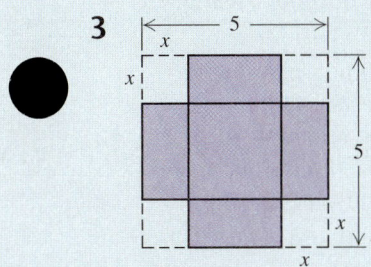

4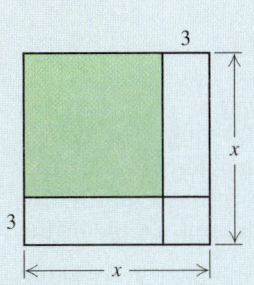

5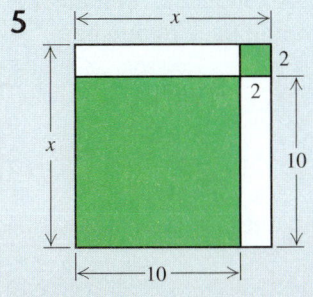

Visualizing for Success

Use after Section 4.6.

In each of Exercises 1–10, find two algebraic expressions for the shaded area of the figure from the list below.

A. $9 - 4x^2$

B. $x^2 - (x - 6)^2$

C. $(x + 3)(x - 3)$

D. $10^2 + 2^2$

E. $8x + 15$

F. $(x + 5)(x + 3) - x^2$

G. $x^2 - 6x + 9$

H. $(3 - 2x)^2 + 4x(3 - 2x)$

I. $(x + 3)^2$

J. $(5x + 3)^2 - 25x^2$

K. $(5 - 2x)^2 + 4x(5 - 2x)$

L. $x^2 - 9$

M. 104

N. $x^2 - 15$

O. $12x - 36$

P. $30x + 9$

Q. $(x - 5)(x - 3) + 3(x - 5)$
  $+ 5(x - 3)$

R. $(x - 3)^2$

S. $25 - 4x^2$

T. $x^2 + 6x + 9$

Answers on page A-22

An additional, animated version of this activity appears in MyMathLab. To use MyMathLab, you need a course ID and a student access code. Contact your instructor for more information.

6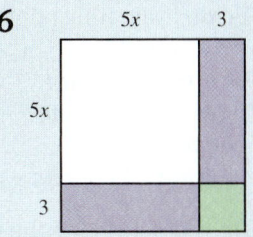

7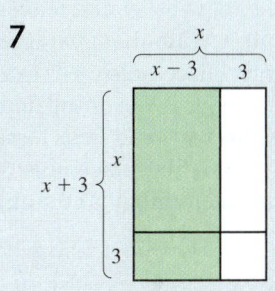

8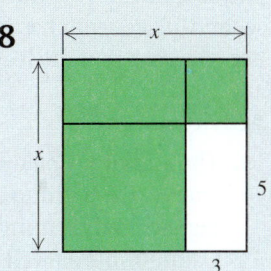

9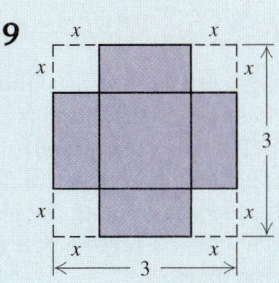

10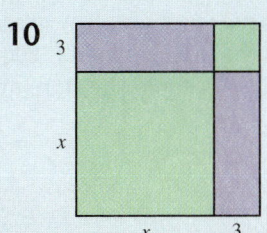

293

Collaborative Activity   *Finding the Magic Number*

Focus: Evaluating polynomials in several
         variables
Use after: Section 4.7
Time: 15–25 minutes
Group size: 3
Materials: A coin for each person

Can you determine the requirements for your
baseball team to clinch first place?

A team's *magic number* is the combined number of wins by that team and losses by the second-
place team that guarantee the leading team a
first-place finish. For example, if the Cubs' magic
number is 3 over the Reds, any combination of
Cubs wins and Reds losses that totals 3 will guarantee a first-place finish for the Cubs. A team's
magic number is computed using the polynomial

$$G - P - L + 1,$$

where G is the length of the season, in games,
P is the number of games that the leading team has
played, and L is the total number of games that the
second-place team has lost minus the total number of games that the leading team has lost.

Activity

1. The standings below are from a fictitious
   league. Each group should calculate the
   Jaguars' magic number with respect to the
   Catamounts as well as the Jaguars' magic
   number with respect to the Wildcats. (Assume
   that the schedule is 162 games long.)

	W	L
Jaguars	92	64
Catamounts	90	66
Wildcats	89	66

2. Each group member should play the role of
   one of the teams, using coin tosses to simu-
   late the remaining games. If a group member
   correctly predicts the side (heads or tails) that
   comes up, the coin toss represents a win for
   that team. Should the other side appear, the
   toss represents a loss. Assume that these games
   are against other (unlisted) teams in the league.
   Each group member should perform three coin
   tosses and then update the standings.

3. Recalculate the two magic numbers, using
   the updated standings from part (2).

4. Slowly—one coin toss at a time—play out the
   remainder of the season. Record all wins and
   losses, update the standings, and recalculate the
   magic numbers each time all three group members have completed a round of coin tosses.

5. Examine the work in part (4) and explain why
   a magic number of 0 indicates that a team
   has been eliminated from contention.

Decision Making & Connection   (*Use after Section 4.2.*)

Buying a Smartphone.   One decision to make when
purchasing a smartphone is the amount of memory
you need. In order to determine that, you need to
know what kind of files you plan to store on the phone.
For these exercises, use the following equivalents:
1 megabyte = 1 million bytes and 1 gigabyte (GB) =
1 billion bytes.

1. Anna is considering a smartphone with 64 GB of
   memory. She plans to use the phone primarily for
   viewing movies. The average size of the movies she
   wants to download is 800 megabytes. How many
   movies could she store on the device?

2. Nate has a collection of 2000 songs. The average
   size of each song is 10 megabytes. The smartphone
   he is considering can be purchased with 16 GB,
   32 GB, or 64 GB. What size phone should he buy?

3. *Research.* Determine the average size of a game
   file. If Nate instead had a collection of 2000 games,
   what size phone should he buy?

4. On the basis of your own use of media, estimate the
   amount of memory you would need to purchase.

Study Summary

KEY TERMS AND CONCEPTS	EXAMPLES	PRACTICE EXERCISES

SECTION 4.1: *Exponents and Their Properties*

1 as an exponent: $\quad a^1 = a$ 0 as an exponent: $\quad a^0 = 1$ The Product Rule: $\quad a^m \cdot a^n = a^{m+n}$ The Quotient Rule: $\quad \dfrac{a^m}{a^n} = a^{m-n}$ The Power Rule: $\quad (a^m)^n = a^{mn}$ Raising a product to a power: $\quad (ab)^n = a^n b^n$ Raising a quotient to a power: $\quad \left(\dfrac{a}{b}\right)^n = \dfrac{a^n}{b^n}$	$3^1 = 3$ $3^0 = 1$ $3^5 \cdot 3^9 = 3^{5+9} = 3^{14}$ $\dfrac{3^7}{3} = 3^{7-1} = 3^6$ $(3^4)^2 = 3^{4 \cdot 2} = 3^8$ $(3x^5)^4 = 3^4(x^5)^4 = 81x^{20}$ $\left(\dfrac{3}{x}\right)^6 = \dfrac{3^6}{x^6}$	*Simplify.* **1.** 6^1 **2.** $(-5)^0$ **3.** $x^5 x^{11}$ **4.** $\dfrac{8^9}{8^2}$ **5.** $(y^5)^3$ **6.** $(x^3 y)^{10}$ **7.** $\left(\dfrac{x^2}{7}\right)^5$

SECTION 4.2: *Negative Exponents and Scientific Notation*

$a^{-n} = \dfrac{1}{a^n}$; $\dfrac{a^{-n}}{b^{-m}} = \dfrac{b^m}{a^n}$ **Scientific notation:** $N \times 10^m, 1 \le N < 10.$	$3^{-2} = \dfrac{1}{3^2} = \dfrac{1}{9}$; $\dfrac{3^{-7}}{x^{-5}} = \dfrac{x^5}{3^7}$ $4100 = 4.1 \times 10^3$; $5 \times 10^{-3} = 0.005$	*Write without negative* *exponents.* **8.** 10^{-1} $\qquad$ **9.** $\dfrac{x^{-1}}{y^{-3}}$ **10.** Convert to scientific notation: 0.000904. **11.** Convert to decimal notation: 6.9×10^5.

SECTION 4.3: *Polynomials*

A **polynomial** is a monomial or a sum of monomials. When a polynomial is written as a sum of monomials, each monomial is a **term** of the polynomial. The **degree of a term** of a polynomial is the number of variable factors in that term. The **coefficient** of a term is the part of the term that is a constant factor. The **leading term** of a polynomial is the term of highest degree. The **leading coefficient** is the coefficient of the leading term. The **degree of the polynomial** is the degree of the leading term.	Polynomial: $10x - x^3 + 4x^5 + 7$ (see table below)	*For Exercises 12–17, consider the polynomial* $x^2 - 10 + 5x - 8x^6$. **12.** List the terms of the polynomial. **13.** What is the degree of the term $5x$? **14.** What is the coefficient of the term x^2? **15.** What is the leading term of the polynomial? **16.** What is the leading coefficient of the polynomial? **17.** What is the degree of the polynomial?

Polynomial: $10x - x^3 + 4x^5 + 7$

Term	$10x$	$-x^3$	$4x^5$	7
Degree of Term	1	3	5	0
Coefficient of Term	10	-1	4	7
Leading Term		$4x^5$		
Leading Coefficient		4		
Degree of Polynomial		5		

A **monomial** has one term.

A **binomial** has two terms.

A **trinomial** has three terms.

Monomial: $4x^3$

Binomial: $x^2 - 5$

Trinomial: $3t^3 + 2t - 10$

18. Classify the polynomial
$$8x - 3 - x^4$$
as a monomial, a binomial, or a trinomial.

Like terms, or **similar terms,** are either constant terms or terms containing the same variable(s) raised to the same power(s). These can be **combined** within a polynomial.

Combine like terms:
$$3y^4 + 6y^2 - 7 - y^4 - 6y^2 + 8.$$
$3y^4 + 6y^2 - 7 - y^4 - 6y^2 + 8$
$= \underline{3y^4 - y^4} + \underline{6y^2 - 6y^2} - \underline{7 + 8}$
$= \quad 2y^4 \quad + \quad 0 \quad + \quad 1$
$= 2y^4 + 1$

19. Combine like terms:
$$3x^2 + 5x - 10x + x.$$

To **evaluate** a polynomial, replace the variable with a number. The **value** is calculated using the rules for order of operations.

Evaluate $t^3 - 2t^2 - 5t + 1$ for $t = -2$.
$t^3 - 2t^2 - 5t + 1$
$= (-2)^3 - 2(-2)^2 - 5(-2) + 1$
$= -8 - 2(4) - (-10) + 1$
$= -8 - 8 + 10 + 1$
$= -5$

20. Evaluate $2 - 3x - x^2$ for $x = -1$.

SECTION 4.4: *Addition and Subtraction of Polynomials*

Add polynomials by combining like terms.

$(2x^2 - 3x + 7) + (5x^3 + 3x - 9)$
$= 2x^2 + (-3x) + 7 + 5x^3 + 3x + (-9)$
$= 5x^3 + 2x^2 - 2$

21. Add: $(9x^2 - 3x) + (4x - x^2)$.

Subtract polynomials by adding the opposite of the polynomial being subtracted.

$(2x^2 - 3x + 7) - (5x^3 + 3x - 9)$
$= 2x^2 - 3x + 7 + (-5x^3 - 3x + 9)$
$= 2x^2 - 3x + 7 - 5x^3 - 3x + 9$
$= -5x^3 + 2x^2 - 6x + 16$

22. Subtract: $(9x^2 - 3x) - (4x - x^2)$.

SECTION 4.5: *Multiplication of Polynomials*

Multiply polynomials by multiplying each term of one polynomial by each term of the other.

$(x + 2)(x^2 - x - 1)$
$= x \cdot x^2 - x \cdot x - x \cdot 1 + 2 \cdot x^2$
$\quad - 2 \cdot x - 2 \cdot 1$
$= x^3 - x^2 - x + 2x^2 - 2x - 2$
$= x^3 + x^2 - 3x - 2$

23. Multiply:
$$(x - 1)(x^2 - x - 2).$$

SECTION 4.6: *Special Products*

FOIL (First, Outer, Inner, Last):

$(A + B)(C + D) = AC + AD + BC + BD$

F   L

$(A + B)(C + D)$

I

O

$(x + 3)(x - 2) = x^2 - 2x + 3x - 6$
$\quad\quad\quad\quad\quad = x^2 + x - 6$

24. Multiply:
$$(x + 4)(2x + 3).$$

The product of a sum and a difference:

$$(A + B)(A - B) = A^2 - B^2$$

$A^2 - B^2$ is called a **difference of squares.**

$$(t^3 + 5)(t^3 - 5) = (t^3)^2 - 5^2$$
$$= t^6 - 25$$

25. Multiply:

$$(5 + 3x)(5 - 3x).$$

The square of a binomial:

$$(A + B)^2 = A^2 + 2AB + B^2;$$

$$(A - B)^2 = A^2 - 2AB + B^2$$

$A^2 + 2AB + B^2$ and $A^2 - 2AB + B^2$ are called **perfect-square trinomials.**

$$(5x + 3)^2 = (5x)^2 + 2(5x)(3) + 3^2$$
$$= 25x^2 + 30x + 9;$$
$$(5x - 3)^2 = (5x)^2 - 2(5x)(3) + 3^2$$
$$= 25x^2 - 30x + 9$$

Multiply.

26. $(x + 9)^2$

27. $(8x - 1)^2$

SECTION 4.7: *Polynomials in Several Variables*

To **evaluate** a polynomial, replace each variable with a number and simplify.

Evaluate $4 - 3xy + x^2y$ for $x = 5$ and $y = -1$.

$$4 - 3xy + x^2y = 4 - 3(5)(-1)$$
$$+ (5)^2(-1)$$
$$= 4 - (-15) + (-25)$$
$$= -6$$

28. Evaluate $xy - y^2 - 4x$ for $x = -2$ and $y = 3$.

The **degree** of a term is the number of variable factors in the term or the sum of the exponents of the variables.

The degree of $-19x^3yz^2$ is 6.

29. What is the degree of $4mn^5$?

Add, subtract, and multiply polynomials in several variables in the same way as polynomials in one variable.

$$(3xy^2 - 4x^2y + 5xy) + (xy - 6x^2y)$$
$$= 3xy^2 - 10x^2y + 6xy;$$
$$(3xy^2 - 4x^2y + 5xy) - (xy - 6x^2y)$$
$$= 3xy^2 + 2x^2y + 4xy;$$
$$(2a^2b + 3a)(5a^2b - a)$$
$$= 10a^4b^2 + 13a^3b - 3a^2$$

30. Add:

$$(3cd^2 + 2c) + (4cd - 9c).$$

31. Subtract:

$$(8pw - p^2w) - (p^2w + 8pw).$$

32. Multiply: $(7xy - x^2)^2$.

SECTION 4.8: *Division of Polynomials*

To divide a polynomial by a monomial, divide each term by the monomial. Divide coefficients and subtract exponents.

$$\frac{3t^5 - 6t^4 + 4t^2 + 9t}{3t} = \frac{3t^5}{3t} - \frac{6t^4}{3t} + \frac{4t^2}{3t} + \frac{9t}{3t}$$
$$= t^4 - 2t^3 + \frac{4}{3}t + 3$$

33. Divide:

$$\frac{4y^5 - 8y^3 + 16y^2}{4y^2}.$$

To divide a polynomial by a binomial, use long division.

Divide: $(x^2 + 5x - 2) \div (x - 3).$

$$\begin{array}{r} x + 8 \\ x - 3{\overline{)}}\,x^2 + 5x - 2 \\ -(x^2 - 3x) \\ \hline 8x - 2 \\ -(8x - 24) \\ \hline 22 \end{array}$$

$$(x^2 + 5x - 2) \div (x - 3)$$
$$= x + 8 + \frac{22}{x - 3}$$

34. Divide:

$$(x^2 - x + 4) \div (x + 1).$$

Test: Chapter 4

For step-by-step test solutions, access the Chapter Test Prep Videos in MyMathLab®, on YouTube® (search "Bittinger Combo Alg CA" and click on "Channels"), or by scanning the code.

Simplify.

1. $x^7 \cdot x \cdot x^5$

2. $\dfrac{3^8}{3^7}$

3. $\dfrac{(3m)^4}{(3m)^4}$

4. $(t^5)^9$

5. $(-3y^2)^3$

6. $(5x^4y)(-2x^5y)^3$

7. $\dfrac{24a^7b^4}{20a^2b}$

8. $\left(\dfrac{4p}{5q^3}\right)^2$

9. Express using a positive exponent: y^{-7}.

10. Express using a negative exponent: $\dfrac{1}{5^6}$.

Simplify.

11. $t^{-4} \cdot t^{-5}$

12. $\dfrac{9x^3y^2}{3x^8y^{-3}}$

13. $(2a^3b^{-1})^{-4}$

14. $\left(\dfrac{ab}{c}\right)^{-3}$

15. Convert to scientific notation: $3{,}060{,}000{,}000$.

16. Convert to decimal notation: 5×10^{-8}.

Multiply or divide, as indicated, and write scientific notation for the result.

17. $\dfrac{5.6 \times 10^6}{3.2 \times 10^{-11}}$

18. $(2.4 \times 10^5)(5.4 \times 10^{16})$

19. Classify $4x^2y - 7y^3$ as a monomial, a binomial, a trinomial, or a polynomial with no special name.

20. Identify the coefficient of each term of the polynomial:
$$3x^5 - x + \tfrac{1}{9}.$$

21. Determine the degree of each term, the leading term and the leading coefficient, and the degree of the polynomial:
$$2t^3 - t + 7t^5 + 4.$$

22. Evaluate $x^2 + 5x - 1$ for $x = -3$.

Combine like terms and write in descending order.

23. $y^2 - 3y - y + \tfrac{3}{4}y^2$

24. $3 - x^2 + 8x + 5x^2 - 6x - 2x + 4x^3$

Add or subtract, as indicated.

25. $(3x^5 + 5x^3 - 5x^2 - 3) + (x^5 + x^4 - 3x^2 + 2x - 4)$

26. $\left(x^4 + \tfrac{2}{3}x + 5\right) + \left(4x^4 + 5x^2 + \tfrac{1}{3}x\right)$

27. $(5a^4 + 3a^3 - a^2 - 2a - 1) - (7a^4 - a^2 - a + 6)$

28. $(t^3 - 0.3t^2 - 20) - (t^4 - 1.5t^3 + 0.3t^2 - 11)$

Multiply.

29. $-2x^2(3x^2 - 3x - 5)$

30. $\left(x - \tfrac{1}{3}\right)^2$

31. $(5t - 7)(5t + 7)$

32. $(3b + 5)(2b - 1)$

33. $(x^6 - 4)(x^8 + 4)$

34. $(8 - y)(6 + 5y)$

35. $(2x + 1)(3x^2 - 5x - 3)$

36. $(8a^3 + 3)^2$

37. Evaluate $2x^2y - 3y^2$ for $x = -3$ and $y = 2$.

38. Combine like terms:
$$2x^3y - y^3 + xy^3 + 8 - 6x^3y - x^2y^2 + 11.$$

39. Subtract:
$$(8a^2b^2 - ab + b^3) - (-6ab^2 - 7ab - ab^3 + 5b^3).$$

40. Multiply: $(3x^5 - y)(3x^5 + y)$.

Divide.

41. $(12x^4 + 9x^3 - 15x^2) \div (3x^2)$

42. $(6x^3 - 8x^2 - 14x + 13) \div (x + 2)$

Synthesis

43. The height of a box is 1 less than its length, and the length is 2 more than its width. Express the volume in terms of the length.

44. Simplify: $2^{-1} - 4^{-1}$.

45. Every day about 265 billion spam e-mails are sent. If each spam e-mail wastes 4 sec of the recipient's time, how many hours are wasted each day due to spam?

Source: Based on data from Radicati Group

Cumulative Review: Chapters 1–4

1. Evaluate $\dfrac{2x + y}{5}$ for $x = 12$ and $y = 6$.   [1.1]

2. Evaluate $5x^2y - xy + y^2$ for $x = -1$ and $y = -2$.   [4.7]

Simplify.

3. $\frac{1}{15} - \frac{2}{9}$ [1.6]

4. $2 - [10 - (5 + 12 \div 2^2 \cdot 3)]$ [1.8]

5. $2y - (y - 7) + 3$ [1.8]

6. $t^4 \cdot t^7 \cdot t$ [4.1]

7. $\dfrac{-100x^6y^8}{25xy^5}$ [4.1]

8. $(2a^2b)(5ab^3)^2$ [4.1]

9. Factor: $10a - 6b + 12$.   [1.2]

10. Determine the degree of the polynomial $-x^4 + 5x^3 + 3x^6 - 1$.   [4.3]

11. In which quadrant is $(-2, 5)$ located?   [3.1]

Graph.

12. $3y + 2x = 0$ [3.2]

13. $3y - 2x = 12$ [3.3]

14. $3y = 2$ [3.3]

15. $3y = 2x + 9$ [3.6]

16. Find the slope and the y-intercept of the line given by $y = \frac{1}{10}x + \frac{3}{8}$.   [3.6]

17. Find the slope of the line containing the points $(2, 3)$ and $(-6, 8)$.   [3.5]

18. Write an equation of the line with slope $-\frac{2}{3}$ and y-intercept $(0, -10)$.   [3.6]

Solve.

19. $\frac{1}{6}n = -\frac{2}{3}$ [2.1]

20. $5y + 7 = 8y - 1$ [2.2]

21. $2 - (x - 7) = 8 - 4(x + 5)$ [2.2]

22. $-\frac{1}{2}t \le 4$ [2.6]

23. $3x - 5 > 9x - 8$ [2.6]

24. Solve $c = \dfrac{5pq}{2t}$ for t.   [2.3]

Add or subtract, as indicated.

25. $(2u^2v - uv^2 + uv) + (3u^2 - v^2u + 5vu^2)$ [4.7]

26. $(2x^5 - x^4 - x) - (x^5 - x^4 + x)$ [4.4]

Multiply.

27. $8x^3(-2x^2 - 6x + 7)$ [4.5]

28. $(x - 2)(x^2 + x - 5)$ [4.5]

29. $(4t^2 + 3)^2$ [4.6]

30. $\left(\frac{1}{2}x + 1\right)\left(\frac{1}{2}x - 1\right)$ [4.6]

31. $(2r^2 + s)(3r^2 - 4s)$ [4.7]

32. Divide: $(x^2 - x + 3) \div (x - 1)$.   [4.8]

Simplify. Do not use negative exponents in the answer.   [4.2]

33. 7^{-10}

34. $(3x^{-7}y^{-2})^{-1}$

35. In 2011, China and the United States together had installed wind turbines capable of producing about 95 thousand megawatts of electricity. China's wind-turbine capacity was 11 thousand megawatts greater than that of the United States. What was China's wind-turbine capacity?   [2.5]

Source: World Wind Energy Association

36. In 2010, U.S. electric utilities used coal to generate 1.8 trillion kilowatt hours (kWh) of electricity. This was 45% of the total amount of electricity generated. How much electricity was generated in 2010?   [2.4]

Source: U.S. Energy Information Administration

37. While studying in the United States under a student visa, Ana can work on campus no more than 20 hr per week. For the first 4 days of one week, she worked 3, 2, 5, and 6 hr. How many hours can she work on the fifth day without violating this restriction?   [2.7]

38. U.S. retail losses due to employee theft have increased from \$15.9 billion in 2008 to \$16.2 billion in 2011. Find the rate of change of retail losses due to employee theft.   [3.4]

Source: National Retail Federation

Synthesis

Solve. If no solution exists, state this.

39. $3x - 2(x + 6) = 4(x - 3)$ [2.2]

40. $x - (2x - 1) = 3x - 4(x + 1) + 10$ [2.2]

Simplify.

41. $7^{-1} + 8^0$ [4.2]

42. $-2x^5(x^7) + (x^3)^4 - (4x^5)^2(-x^2)$ [4.1], [4.3]

Polynomials and Factoring

NUMBER OF MINUTES *t* AFTER INJECTION	NUMBER OF MICROGRAMS *N* OF EPINEPHRINE IN THE BLOODSTREAM
2	160
5	250
8	160

An Epinephrine Shot Could Save Your Life.

An injection of the drug epinephrine (adrenaline) can slow down or stop an extreme allergic reaction. Because the drug is injected directly into the bloodstream, it is carried quickly throughout the body. The table above lists the number of micrograms of epinephrine in an adult's bloodstream *t* minutes after 250 micrograms have been injected. As the data in the table seem to indicate, the effect wears off quickly as well. We can use a *quadratic equation* to model these data and predict the concentration of epinephrine in the bloodstream for other values of *t*.

(See Example 4 in Section 5.8.)

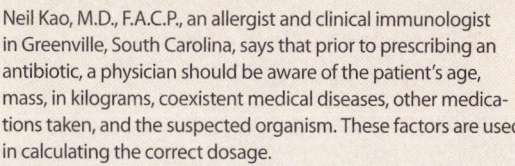

Math is critical when calculating dosages of medication.

Neil Kao, M.D., F.A.C.P., an allergist and clinical immunologist in Greenville, South Carolina, says that prior to prescribing an antibiotic, a physician should be aware of the patient's age, mass, in kilograms, coexistent medical diseases, other medications taken, and the suspected organism. These factors are used in calculating the correct dosage.

F actoring is multiplying reversed. Thus factoring polynomials requires a solid understanding of how to multiply polynomials. Factoring is an important skill that will be used to solve equations and simplify other types of expressions found later in the study of algebra.

5.1   Introduction to Factoring

Factoring Monomials   ▪   Factoring When Terms Have a Common Factor   ▪   Factoring by Grouping   ▪   Checking by Evaluating

Just as a number like 15 can be factored as $3 \cdot 5$, a polynomial like $x^2 + 7x$ can be factored as $x(x + 7)$. In both cases, we ask ourselves, "What was multiplied to obtain the given result?" The situation is much like a popular television game show in which an "answer" is given and participants must find the "question" to which the answer corresponds.

Study Skills

You've Got Mail

Many students overlook an excellent opportunity to get questions answered—e-mail. If your instructor makes his or her e-mail address available, consider using it to get help. Often, just the act of writing out your question brings some clarity. If you do use e-mail, allow some time for your instructor to reply.

> **FACTORING**
>
> To *factor* a polynomial is to find an equivalent expression that is a product. An equivalent expression of this type is called a *factorization* of the polynomial.

Factoring Monomials

To factor a monomial, we find monomials whose product is equivalent to the original monomial. For example, $20x^2$ can be factored as $2 \cdot 10x^2$, $4x \cdot 5x$, or $10x \cdot 2x$, as well as several other ways. To check, we multiply.

EXAMPLE 1   Find three factorizations of $15x^3$.

SOLUTION

a) $15x^3 = (3 \cdot 5)(x \cdot x^2)$ Thinking of how 15 and x^3 can each be factored

$= (3x)(5x^2)$ The factors are $3x$ and $5x^2$. *Check*: $3x \cdot 5x^2 = 15x^3$.

b) $15x^3 = (3 \cdot 5)(x^2 \cdot x)$

$= (3x^2)(5x)$ The factors are $3x^2$ and $5x$. *Check*: $3x^2 \cdot 5x = 15x^3$.

c) $15x^3 = ((-5)(-3))x^3$

$= (-5)(-3x^3)$ The factors are -5 and $-3x^3$.
Check: $(-5)(-3x^3) = 15x^3$.

We see that $(3x)(5x^2)$, $(3x^2)(5x)$, and $(-5)(-3x^3)$ are all factorizations of $15x^3$. Other factorizations exist as well.

1. Find three factorizations of $30a^3$.

↩ YOUR TURN

Factoring When Terms Have a Common Factor

To factor a polynomial with two or more terms, we use the distributive law with the sides of the equation switched.

Multiply: $3(x + 2y - z) = 3 \cdot x + 3 \cdot 2y - 3 \cdot z$

$= 3x + 6y - 3z$

Factor: $3x + 6y - 3z = 3 \cdot x + 3 \cdot 2y - 3 \cdot z$

$= 3(x + 2y - z)$

In the factorization above, note that since 3 appears as a factor of $3x$, $6y$, and $-3z$, it is a *common factor* for all the terms of $3x + 6y - 3z$.

We generally factor out the *largest* common factor.

EXAMPLE 2 Factor to form an equivalent expression: $8a - 12$.

SOLUTION We write the prime factorization of both terms, lining up common factors in columns:

The prime factorization of $8a$ is $2 \cdot 2 \cdot 2 \cdot a$;

The prime factorization of 12 is $2 \cdot 2 \cdot \qquad 3.$

Since both factorizations include two factors of 2, the largest common factor is $2 \cdot 2$, or 4:

$$8a - 12 = 4 \cdot 2a - 4 \cdot 3 \qquad \text{4 is a factor of } 8a \text{ and of 12.}$$
$$8a - 12 = 4(2a - 3). \qquad \text{Try to go directly to this step.}$$

Check: $4(2a - 3) = 4 \cdot 2a - 4 \cdot 3 = 8a - 12$, as expected.

2. Factor to form an equivalent expression: $12x - 30$.

The factorization of $8a - 12$ is $4(2a - 3)$.

YOUR TURN

> **CAUTION!** $2 \cdot 2 \cdot 2 \cdot a - 2 \cdot 2 \cdot 3$ is a factorization of the *terms* of $8a - 12$ but not of the polynomial itself. The factorization of $8a - 12$ is $4(2a - 3)$.

A common factor may contain a variable.

EXAMPLE 3 Factor: $24x^5 + 30x^2$.

SOLUTION

The prime factorization of $24x^5$ is $2 \cdot 2 \cdot 2 \cdot 3 \cdot \qquad x \cdot x \cdot x \cdot x \cdot x.$
The prime factorization of $30x^2$ is $2 \cdot \qquad 3 \cdot 5 \cdot x \cdot x.$

The largest common factor is $2 \cdot 3 \cdot x \cdot x$, or $6x^2$. ←

$$24x^5 + 30x^2 = 6x^2 \cdot 4x^3 + 6x^2 \cdot 5 \qquad \text{Factoring each term}$$
$$= 6x^2(4x^3 + 5) \qquad \text{Factoring out } 6x^2$$

Check: $6x^2(4x^3 + 5) = 6x^2 \cdot 4x^3 + 6x^2 \cdot 5 = 24x^5 + 30x^2$, as expected.

3. Factor: $30x^4 + 40x^3$.

The factorization of $24x^5 + 30x^2$ is $6x^2(4x^3 + 5)$.

YOUR TURN

The largest common factor of a polynomial is the largest common factor of the coefficients times the largest common factor of the variable(s) in all the terms. Suppose in Example 3 that you did not recognize the *largest* common factor, and removed only part of it, as follows:

$$24x^5 + 30x^2 = 2x^2 \cdot 12x^3 + 2x^2 \cdot 15 \qquad 2x^2 \text{ is a common factor.}$$
$$= 2x^2(12x^3 + 15). \qquad 12x^3 + 15 \text{ itself contains a common factor, 3.}$$

Note that $12x^3 + 15$ still has a common factor, 3. To find the largest common factor, we extend the above factoring, as follows, until no more common factors exist:

$$24x^5 + 30x^2 = 2x^2[3(4x^3 + 5)] \qquad \text{Factoring } 12x^3 + 15. \text{ Remember to rewrite the first common factor, } 2x^2.$$
$$= 6x^2(4x^3 + 5). \qquad \text{Using an associative law;}$$
$$2x^2 \cdot 3 = 6x^2$$

Since $4x^3 + 5$ cannot be factored any further, we say that we have factored *completely*. **When we are directed simply to factor, it is understood that we should always factor completely.**

EXAMPLE 4   Factor: $12x^5 - 15x^4 + 27x^3$.

SOLUTION

The prime factorization of $12x^5$ is $2 \cdot 2 \cdot 3 \cdot$ $x \cdot x \cdot x \cdot x \cdot x.$
The prime factorization of $15x^4$ is $3 \cdot$ $5 \cdot x \cdot x \cdot x \cdot x.$
The prime factorization of $27x^3$ is $3 \cdot 3 \cdot 3 \cdot$ $x \cdot x \cdot x.$

The largest common factor is $3 \cdot x \cdot x \cdot x$, or $3x^3$.

$$12x^5 - 15x^4 + 27x^3 = 3x^3 \cdot 4x^2 - 3x^3 \cdot 5x + 3x^3 \cdot 9$$
$$= 3x^3(4x^2 - 5x + 9)$$

Since $4x^2 - 5x + 9$ has no common factor, we are done, except for a check:

$$3x^3(4x^2 - 5x + 9) = 3x^3 \cdot 4x^2 - 3x^3 \cdot 5x + 3x^3 \cdot 9$$
$$= 12x^5 - 15x^4 + 27x^3,$$

as expected. The factorization of $12x^5 - 15x^4 + 27x^3$ is $3x^3(4x^2 - 5x + 9)$.

4. Factor: $8a^5 + 12a^3 - 6a^2$.

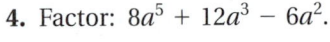 YOUR TURN

Note in Examples 3 and 4 that the *largest* common variable factor is the *smallest* power of x in the original polynomial.

With practice, we can determine the largest common factor without writing the prime factorization of each term. Then, to factor, we write the largest common factor and parentheses and then fill in the parentheses. It is customary for the leading coefficient of the polynomial inside the parentheses to be positive.

EXAMPLE 5   Factor: **(a)** $8r^5s^2 + 16rs^5$; **(b)** $-3xy + 6xz - 3x$.

SOLUTION

a) $8r^5s^2 + 16rs^5 = 8rs^2(r^4 + 2s^3)$ Try to go directly to this step.

The largest common factor is $8rs^2$. $\leftarrow \begin{cases} 8r^5s^2 = 2 \cdot 2 \cdot 2 \cdot & r \cdot r^4 \cdot s^2 \\ 16rs^5 = 2 \cdot 2 \cdot 2 \cdot 2 \cdot r \cdot & s^2 \cdot s^3 \end{cases}$

Check: $8rs^2(r^4 + 2s^3) = 8r^5s^2 + 16rs^5.$

Student Notes

The 1 in $(y - 2z + 1)$ plays an important role in Example 5(b). Without the 1, the term $-3x$ would not appear in the check.

b) $-3xy + 6xz - 3x = -3x(y - 2z + 1)$ Note that either $-3x$ or $3x$ can be the largest common factor.

We generally factor out a negative when the first coefficient is negative. The way we factor can depend on the situation in which we are working. We might also factor as follows:

$$-3xy + 6xz - 3x = 3x(-y + 2z - 1).$$

The checks are left to the student.

5. Factor: $12a^2b^4 - 8a^3b^3$.

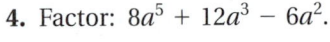 YOUR TURN

In some texts, the largest common factor is referred to as the *greatest* common factor. We have avoided this language because, as shown in Example 5(b), the largest common factor may represent a negative value that is actually *less* than other common factors.

> **TIPS FOR FACTORING**
> 1. Factor out the largest common factor, if one exists.
> 2. The common factor multiplies a polynomial with the same number of terms as the original polynomial.
> 3. Factoring can always be checked by multiplying. Multiplication should yield the original polynomial.

Factoring by Grouping

Sometimes algebraic expressions contain a common factor with two or more terms.

EXAMPLE 6   Factor: $x^2(x + 1) + 2(x + 1)$.

SOLUTION   The binomial $x + 1$ is a factor of both $x^2(x + 1)$ and $2(x + 1)$. Thus, $x + 1$ is a common factor:

$$x^2(x + 1) + 2(x + 1) = (x + 1)x^2 + (x + 1)2 \qquad \text{Using a commutative law twice. Try to do this step mentally.}$$

$$= (x + 1)(x^2 + 2). \qquad \text{Factoring out the common factor, } x + 1$$

To check, we could simply reverse the above steps.
    The factorization is $(x + 1)(x^2 + 2)$.

6. Factor: $x(x - 5) + 3(x - 5)$.

 YOUR TURN

Some polynomials with four terms have a common binomial factor. In order to identify that factor, we regroup into two groups of two terms each.

EXAMPLE 7   Factor: $5x^3 - x^2 + 15x - 3$.

SOLUTION

$$5x^3 - x^2 + 15x - 3 = (5x^3 - x^2) + (15x - 3) \qquad \text{Grouping as two binomials}$$

$$= x^2(5x - 1) + 3(5x - 1) \qquad \text{Factoring each binomial}$$

$$= (5x - 1)(x^2 + 3) \qquad \text{Factoring out the common factor, } 5x - 1$$

Check: $(5x - 1)(x^2 + 3) = 5x \cdot x^2 + 5x \cdot 3 - 1 \cdot x^2 - 1 \cdot 3 \qquad \text{Using FOIL}$
$= 5x^3 - x^2 + 15x - 3.$

7. Factor: $x^3 + 3x^2 + 4x + 12$.

YOUR TURN

If a polynomial can be split into groups of terms and the groups share a common factor, then the original polynomial can be factored. This method, known as **factoring by grouping**, can be tried on any polynomial with four or more terms.

EXAMPLE 8   Factor by grouping.

a) $2x^3 + 8x^2 + x + 4$ **b)** $4t^3 - 15 + 20t^2 - 3t$

SOLUTION

a) $2x^3 + 8x^2 + x + 4 = (2x^3 + 8x^2) + (x + 4)$

$\qquad\qquad\qquad\qquad = 2x^2(x + 4) + 1(x + 4)$ Factoring $2x^3 + 8x^2$ to find a common binomial factor. Writing the 1 helps with the next step.

> **CAUTION!** Be sure to include the term 1.

$\qquad\qquad\qquad\qquad = (x + 4)(2x^2 + 1)$ Factoring out the common factor, $x + 4$. The 1 is essential in the factor $2x^2 + 1$.

     Check: $(x + 4)(2x^2 + 1) = x \cdot 2x^2 + x \cdot 1 + 4 \cdot 2x^2 + 4 \cdot 1$ Using FOIL

$\qquad\qquad\qquad\qquad\qquad = 2x^3 + x + 8x^2 + 4$

$\qquad\qquad\qquad\qquad\qquad = 2x^3 + 8x^2 + x + 4.$ Using a commutative law

   The factorization is $(x + 4)(2x^2 + 1)$.

b) When we try grouping $4t^3 - 15 + 20t^2 - 3t$ as

$\qquad (4t^3 - 15) + (20t^2 - 3t),$

we are unable to factor $4t^3 - 15$. When this happens, we can rearrange the polynomial and try a different grouping:

$\qquad 4t^3 - 15 + 20t^2 - 3t = 4t^3 + 20t^2 - 3t - 15$ Using the commutative law to rearrange the terms

$\qquad\qquad\qquad\qquad = 4t^2(t + 5) - 3(t + 5)$ By factoring out -3, we see that $t + 5$ is a common factor.

$\qquad\qquad\qquad\qquad = (t + 5)(4t^2 - 3).$ To check, we use FOIL.

8. Factor by grouping:

$\qquad 3x^3 - 3x^2 + 2x - 2.$

 YOUR TURN

To reverse the order of subtraction, we can factor out -1.

> **FACTORING OUT −1 TO REVERSE SUBTRACTION**
>
> $b - a = -1(a - b) = -(a - b)$

EXAMPLE 9   Factor: $t^5 - 5t^4 + 10 - 2t$.

SOLUTION

$\qquad t^5 - 5t^4 + 10 - 2t = (t^5 - 5t^4) + (10 - 2t)$ Grouping

$\qquad\qquad\qquad\qquad = t^4(t - 5) + 2(5 - t)$ Factoring each binomial

$\qquad\qquad\qquad\qquad = t^4(t - 5) + 2(-1)(t - 5)$ Factoring out -1 to reverse $5 - t$

$\qquad\qquad\qquad\qquad = t^4(t - 5) - 2(t - 5)$ Simplifying

$\qquad\qquad\qquad\qquad = (t - 5)(t^4 - 2)$ Factoring out $t - 5$

9. Factor: $2x^3 - 16x^2 + 8 - x$.

The check is left to the student.

YOUR TURN

Checking by Evaluating

One way to check a factorization is to multiply. A second type of check uses the fact that equivalent expressions have the same value when evaluated for the same replacement. Thus a quick, partial check of Example 8(a) can be made by using a convenient replacement for x (say, 1) and evaluating both $2x^3 + 8x^2 + x + 4$ and $(x + 4)(2x^2 + 1)$:

$$2 \cdot 1^3 + 8 \cdot 1^2 + 1 + 4 = 2 + 8 + 1 + 4 = 15;$$
$$(1 + 4)(2 \cdot 1^2 + 1) = 5 \cdot 3 = 15.$$

Since the value of both expressions is the same, the factorization is probably correct. Evaluating for several values will make the check more certain.

Chapter Resource:
Decision Making: Connection, p. 366

Technology Connection

A partial check of a factorization can be performed using a table or a graph. To check Example 8(a), we let

$$y_1 = 2x^3 + 8x^2 + x + 4 \quad \text{and} \quad y_2 = (x + 4)(2x^2 + 1).$$

Then we set up a table in AUTO mode. If the factorization is correct, the values of y_1 and y_2 will be the same regardless of the table settings used.

ΔTBL = 1

X	Y₁	Y₂
0	4	4
1	15	15
2	54	54
3	133	133
4	264	264
5	459	459
6	730	730

X = 0

We can also graph $y_1 = 2x^3 + 8x^2 + x + 4$ and $y_2 = (x + 4)(2x^2 + 1)$. If the graphs appear to coincide, the factorization is probably correct. The TRACE feature can be used to confirm this. This approach is illustrated using a graphing calculator app.

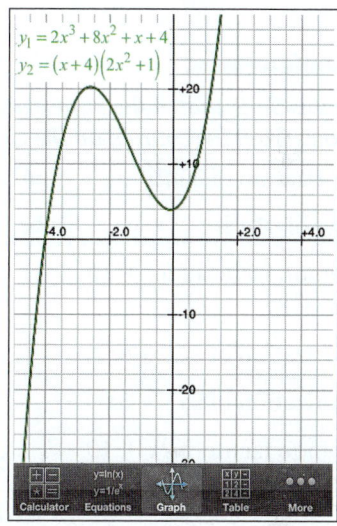

Use a table or a graph to determine whether each of the following factorizations is correct.

1. $x^2 - 7x - 8 = (x - 8)(x + 1)$
2. $4x^2 - 5x - 6 = (4x + 3)(x - 2)$
3. $5x^2 + 17x - 12 = (5x + 3)(x - 4)$
4. $10x^2 + 37x + 7 = (5x - 1)(2x + 7)$
5. $12x^2 - 17x - 5 = (6x + 1)(2x - 5)$
6. $12x^2 - 17x - 5 = (4x + 1)(3x - 5)$
7. $x^2 - 4 = (x - 2)(x - 2)$
8. $x^2 - 4 = (x + 2)(x - 2)$

5.1 EXERCISE SET

FOR EXTRA HELP

MyMathLab®
PRACTICE   WATCH   READ   REVIEW

Vocabulary and Reading Check

Classify each of the following statements as either true or false.

1. The largest common factor of a polynomial always has the same degree as the polynomial itself.

2. When the leading coefficient of a polynomial is negative, we generally factor out a common factor with a negative coefficient.

3. A polynomial is not prime if it contains a common factor other than 1 or −1.

4. All polynomials with four terms can be factored by grouping.

Concept Reinforcement

In each of Exercises 5–12, match the phrase with the most appropriate choice from the column on the right.

5. ____ A factorization of $35a^2b$ a) $3a(1 + 2a)$

6. ____ A factor of $35a^2b$ b) $x + 2$

7. ____ A common factor of $5x + 10$ and $4x + 8$ c) $3x^2(3x^2 - 1)$

8. ____ A factorization of $3x^4 - 9x^2$ d) $1 + 2a$

9. ____ A factorization of $9x^4 - 3x^2$ e) $3x^2(x^2 - 3)$

10. ____ A common factor of $2x + 10$ and $4x + 8$ f) $5a^2$

11. ____ A factor of $3a + 6a^2$ g) 2

12. ____ A factorization of $3a + 6a^2$ h) $7a \cdot 5ab$

Factoring Monomials

Find three factorizations for each monomial. Answers may vary.

13. $14x^3$ 14. $22x^3$ 15. $-15a^4$

16. $-8t^5$ 17. $25t^5$ 18. $9a^4$

Factoring When Terms Have a Common Factor

Factor. Remember to use the largest common factor and to check by multiplying. Factor out a negative factor if the first coefficient is negative.

19. $8x + 24$ 20. $10x + 50$

21. $2x^2 + 2x - 8$ 22. $6x^2 + 3x - 15$

23. $3t^2 + t$ 24. $2t^2 + t$

25. $-5y^2 - 10y$ 26. $-4y^2 - 12y$

27. $x^3 + 6x^2$ 28. $5x^4 - x^2$

29. $16a^4 - 24a^2$ 30. $25a^5 + 10a^3$

31. $-6t^6 + 9t^4 - 4t^2$ 32. $-10t^5 + 15t^4 + 9t^3$

33. $6x^8 + 12x^6 - 24x^4 + 30x^2$

34. $10x^4 - 30x^3 - 50x - 20$

35. $x^5y^5 + x^4y^3 + x^3y^3 - x^2y^2$

36. $x^9y^6 - x^7y^5 + x^4y^4 + x^3y^3$

37. $-35a^3b^4 + 10a^2b^3 - 15a^3b^2$

38. $-21r^5t^4 - 14r^4t^6 + 21r^3t^6$

Factoring by Grouping

Factor.

39. $n(n - 6) + 3(n - 6)$

40. $b(b + 5) + 3(b + 5)$

41. $x^2(x + 3) - 7(x + 3)$

42. $3z^2(2z + 9) + (2z + 9)$

43. $y^2(2y - 9) + (2y - 9)$

44. $x^2(x - 7) - 3(x - 7)$

Factor by grouping, if possible, and check.

45. $x^3 + 2x^2 + 5x + 10$

46. $z^3 + 3z^2 + 7z + 21$

47. $9n^3 - 6n^2 + 3n - 2$

48. $10x^3 - 25x^2 + 2x - 5$

49. $4t^3 - 20t^2 + 3t - 15$

50. $8a^3 - 2a^2 + 12a - 3$

51. $7x^3 + 5x^2 - 21x - 15$

52. $5x^3 + 4x^2 - 10x - 8$

53. $6a^3 + 7a^2 + 6a + 7$

54. $7t^3 - 5t^2 + 7t - 5$

55. $2x^3 - 12x^2 - x + 6$

56. $x^3 - x^2 - 2x + 5$

57. $p^3 + p^2 - 3p + 10$

58. $a^3 - 3a^2 + 6 - 2a$

59. $y^3 + 8y^2 - 2y - 16$

60. $3x^3 + 18x^2 - 5x - 25$

61. $2x^3 + 36 - 8x^2 - 9x$

62. $20g^3 + 5 - 4g^2 - 25g$

 63. In answering a factoring problem, Taylor says the largest common factor is $-5x^2$ and Kimber says the largest common factor is $5x^2$. Can they both be correct? Why or why not?

64. Write a two-sentence paragraph in which the word "factor" is used at least once as a noun and once as a verb.

Skill Review

Simplify.

65. $\frac{2}{5} \div \frac{10}{3}$ [1.3]

66. $-1 + 20 \div 2^2 \cdot 5$ [1.8]

67. $(2xy^{-4})^{-1}$ [4.2]

68. $(a^2b^3c)(a^5b^4c^2)$ [4.1]

69. $(3x^2 - x - 3) - (-x^2 - 6x + 10)$ [4.4]

70. $(w^2y + yz - w^2) + (w^2 - 2w^2y + wz)$ [4.7]

Synthesis

 71. Suresh factors $12x^2y - 18xy^2$ as $6xy \cdot 2x - 6xy \cdot 3y$. Is this the factorization of the polynomial? Why or why not?

72. Azrah recognizes that evaluating usually provides only a partial check of her factoring. Because of this, she often performs a second check with a different replacement value. Is this a good idea? Why or why not?

Factor, if possible.

73. $4x^5 + 6x^2 + 6x^3 + 9$

74. $x^6 + x^2 + x^4 + 1$

75. $2x^4 + 2x^3 - 4x^2 - 4x$

76. $x^3 + x^2 - 2x + 2$

 Aha! **77.** $5x^5 - 5x^4 + x^3 - x^2 + 3x - 3$

Aha! **78.** $ax^2 + 2ax + 3a + x^2 + 2x + 3$

79. Write a trinomial of degree 7 for which $8x^2y^3$ is the largest common factor. Answers may vary.

 80. Kris and Tina are each calculating the amount of sheet metal needed to form a circular canister with a radius of 5 in. and a height of 20 in. Kris uses the formula $A = 2\pi rh + 2\pi r^2$. Tina adds $5 + 20$, and then multiplies that sum by 10 and then by π. Do the methods give the same result? Which seems easier?

↳ **YOUR TURN ANSWERS: SECTION 5.1**

1. $(3a)(10a^2), (6a^2)(5a), (-3)(-10a^3)$; answers may vary
2. $6(2x - 5)$ **3.** $10x^3(3x + 4)$ **4.** $2a^2(4a^3 + 6a - 3)$
5. $4a^2b^3(3b - 2a)$ **6.** $(x - 5)(x + 3)$ **7.** $(x + 3)(x^2 + 4)$
8. $(x - 1)(3x^2 + 2)$ **9.** $(x - 8)(2x^2 - 1)$

PREPARE TO MOVE ON

Multiply.   [4.6]

1. $(x + 2)(x + 7)$

2. $(x - 2)(x - 7)$

3. $(x + 2)(x - 7)$

4. $(x - 2)(x + 7)$

List all factors of each number.   [1.3]

5. 60

6. 18

5.2   Factoring Trinomials of the Type $x^2 + bx + c$

When the Constant Term Is Positive ▪ When the Constant Term Is Negative ▪ Prime Polynomials ▪ Factoring Completely

We now learn how to factor trinomials like

$$x^2 + 5x + 4 \quad \text{or} \quad x^2 + 3x - 10,$$

for which no common factor exists and the leading coefficient is 1. Recall that when factoring, we are writing an equivalent expression that is a product. For these trinomials, the factors will be binomials.

Study Skills

Compare the following multiplications:

$$
\begin{array}{cccc}
\text{F} & \text{O} & \text{I} & \text{L} \\
\downarrow & \downarrow & \downarrow & \downarrow
\end{array}
$$

$$(x + 2)(x + 5) = x^2 + 5x + 2x + 2 \cdot 5$$
$$= x^2 + \quad 7x \quad + \quad 10;$$

$$(x - 2)(x - 5) = x^2 - 5x - 2x + (-2)(-5)$$
$$= x^2 - \quad 7x \quad + \quad 10;$$

$$(x + 3)(x - 7) = x^2 - 7x + 3x + 3(-7)$$
$$= x^2 - \quad 4x \quad - \quad 21;$$

$$(x - 3)(x + 7) = x^2 + 7x - 3x + (-3)7$$
$$= x^2 + \quad 4x \quad - \quad 21.$$

Note that for all four products:

- The product of the two binomials is a trinomial.
- The coefficient of x in the trinomial is the sum of the constant terms in the binomials.
- The constant term in the trinomial is the product of the constant terms in the binomials.

A GEOMETRIC APPROACH TO EXAMPLE 1

The product of two binomials can be regarded as the sum of the areas of four rectangles. Thus we can regard the factoring of $x^2 + 5x + 6$ as a search for p and q so that the sum of areas A, B, C, and D is $x^2 + 5x + 6$.

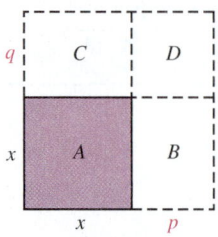

Note that area D is the product of p and q. In order for area D to be 6, p and q must be either 1 and 6 or 2 and 3. We illustrate both below.

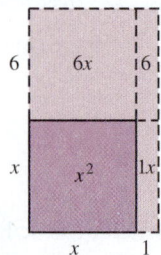 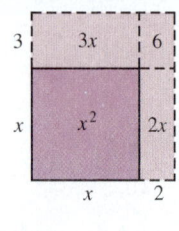

When p and q are 1 and 6, the total area is $x^2 + 7x + 6$, but when p and q are 2 and 3, as shown on the right, the total area is $x^2 + 5x + 6$, as desired. Thus the factorization of $x^2 + 5x + 6$ is $(x + 2)(x + 3)$.

These observations lead to a method for factoring certain trinomials. We first consider those with a positive constant term, as in the first two products above.

When the Constant Term Is Positive

To factor a polynomial like $x^2 + 7x + 10$, we think of FOIL in reverse. The x^2 suggests that the first term of each binomial factor is x. Next, we look for numbers p and q such that

$$x^2 + 7x + 10 = (x + p)(x + q).$$

To get the middle term and the last term of the trinomial, we need two numbers, p and q, whose product is 10 and whose sum is 7. Those numbers are 2 and 5. Thus the factorization is

$$(x + 2)(x + 5).$$

Check: $(x + 2)(x + 5) = x^2 + 5x + 2x + 10$
$$= x^2 + 7x + 10.$$

EXAMPLE 1   Factor to form an equivalent expression:

$$x^2 + 5x + 6.$$

SOLUTION   Think of FOIL in reverse. The first term of each factor is x:

$$(x + \quad)(x + \quad).$$

To complete the factorization, we need a constant term for each binomial. The constants must have a product of 6 and a sum of 5. We list some pairs of numbers that multiply to 6 and then check the sum of each pair of factors.

Pairs of Factors of 6	Sums of Factors
1, 6	7
2, 3	5 ←
−1, −6	−7
−2, −3	−5

The numbers we seek are 2 and 3.

Every pair has a product of 6.   *One* pair has a sum of 5.

Since

$$2 \cdot 3 = 6 \quad \text{and} \quad 2 + 3 = 5,$$

the factorization of $x^2 + 5x + 6$ is $(x + 2)(x + 3)$.

Check: $(x + 2)(x + 3) = x^2 + 3x + 2x + 6 = x^2 + 5x + 6.$

Thus, $(x + 2)(x + 3)$ is a product that is equivalent to $x^2 + 5x + 6$.

Note that since 5 and 6 are both positive, when factoring $x^2 + 5x + 6$ we need not have listed negative factors of 6. Note too that changing the signs of the factors changes only the sign of the sum (see the table above).

1. Factor to form an equivalent expression: $x^2 + 6x + 8$.

YOUR TURN

At the beginning of this section, we considered the multiplication $(x - 2)(x - 5)$. For this product, the resulting trinomial, $x^2 - 7x + 10$, has a positive constant term but a negative coefficient of x. This is because the *product* of two negative numbers is always positive, whereas the *sum* of two negative numbers is always negative.

TO FACTOR $x^2 + bx + c$ WHEN c IS POSITIVE

When the constant term c of a trinomial is positive, look for two numbers with the same sign. Select pairs of numbers with the sign of b, the coefficient of the middle term.

$$x^2 - 7x + 10 = (x - 2)(x - 5); \qquad b \text{ is negative; } c \text{ is positive.}$$

$$x^2 + 7x + 10 = (x + 2)(x + 5) \qquad b \text{ is positive; } c \text{ is positive.}$$

EXAMPLE 2   Factor: $y^2 - 8y + 12$.

SOLUTION   Since the constant term is positive and the coefficient of the middle term is negative, we look for a factorization of 12 in which both factors are negative. Their sum must be -8.

Pairs of Factors of 12	Sums of Factors
−1, −12	−13
−2, −6	−8 ←
−3, −4	−7

We need a sum of −8. The numbers we need are −2 and −6.

2. Factor: $a^2 - 8a + 15$.

The factorization of $y^2 - 8y + 12$ is $(y - 2)(y - 6)$. The check is left to the student.

YOUR TURN

When the Constant Term Is Negative

As we saw in two of the multiplications at the beginning of this section, the product of two binomials can have a negative constant term:

$$(x + 3)(x - 7) = x^2 - 4x - 21$$

and

$$(x - 3)(x + 7) = x^2 + 4x - 21.$$

It is important to note that when the signs of the constants in the binomials are reversed, only the sign of the middle term of the trinomial changes.

EXAMPLE 3   Factor: $x^2 - 8x - 20$.

SOLUTION   The constant term, -20, must be expressed as the product of a negative number and a positive number. Since the sum of these two numbers must be negative (specifically, -8), the negative number must have the greater absolute value.

Pairs of Factors of -20	Sums of Factors
1, −20	−19
2, −10	−8 ←—— The numbers we need are 2 and −10.
4, −5	−1

The numbers that we are looking for are 2 and -10.

Check: $(x + 2)(x - 10) = x^2 - 10x + 2x - 20$
$$= x^2 - 8x - 20.$$

The factorization of $x^2 - 8x - 20$ is $(x + 2)(x - 10)$.

3. Factor: $t^2 - 2t - 15$.

YOUR TURN

TO FACTOR $x^2 + bx + c$ WHEN c IS NEGATIVE

When the constant term c is negative, look for a positive number and a negative number that multiply to c. Consider only pairs of numbers for which the number with the larger absolute value has the sign of b.

$$x^2 - 4x - 21 = (x + 3)(x - 7); \qquad b \text{ is negative}; c \text{ is negative.}$$

$$x^2 + 4x - 21 = (x - 3)(x + 7) \qquad b \text{ is positive}; c \text{ is negative.}$$

Student Notes

Writing a trinomial in descending order will help you to identify b and c.

EXAMPLE 4 Factor: $t^2 - 24 + 5t$.

SOLUTION We first write the trinomial in descending order: $t^2 + 5t - 24$. The factorization of the constant term, -24, must have one positive factor and one negative factor. The sum must be 5, so the positive factor must have the larger absolute value. Thus we consider only pairs of factors in which the positive factor has the larger absolute value.

Pairs of Factors of -24	Sums of Factors	
$-1, 24$	23	
$-2, 12$	10	
$-3,\ 8$	5	← The numbers we need
$-4,\ 6$	2	are -3 and 8.

4. Factor: $t^2 - 28 + 3t$.

The factorization is $(t - 3)(t + 8)$. The check is left to the student.

YOUR TURN

Polynomials in two or more variables, such as $a^2 + 4ab - 21b^2$, are factored in a similar manner.

EXAMPLE 5 Factor: $a^2 + 4ab - 21b^2$.

SOLUTION We look for numbers p and q such that

$$a^2 + 4ab - 21b^2 = (a + pb)(a + qb).$$

Our thinking is much the same as if we were factoring $x^2 + 4x - 21$. We look for factors of -21 whose sum is 4. Those factors are -3 and 7. Thus,

$$a^2 + 4ab - 21b^2 = (a - 3b)(a + 7b).$$

Check: $(a - 3b)(a + 7b) = a^2 + 7ab - 3ba - 21b^2$
$= a^2 + 4ab - 21b^2.$

5. Factor: $x^2 - 3xy - 18y^2$.

The factorization of $a^2 + 4ab - 21b^2$ is $(a - 3b)(a + 7b)$.

YOUR TURN

Prime Polynomials

EXAMPLE 6 Factor: $x^2 - x + 5$.

SOLUTION Since 5 has very few factors, we can easily check all possibilities.

Pairs of Factors of 5	Sums of Factors
$5,\ 1$	6
$-5, -1$	-6

Since there are no factors whose sum is -1, the polynomial is *not* factorable into binomials.

6. Factor: $x^2 + x + 1$.

YOUR TURN

In this text, a polynomial like $x^2 - x + 5$ that cannot be factored using rational numbers is said to be **prime**. In more advanced courses, other types of numbers are considered. There, polynomials like $x^2 - x + 5$ can be factored and are not considered prime.

Factoring Completely

Often factoring requires two or more steps. Remember, when told to factor, we should *factor completely*. This means that the final factorization should contain only prime polynomials.

EXAMPLE 7   Factor: $-2x^3 + 20x^2 - 50x$.

SOLUTION   *Always* look first for a common factor. This time there is one. Since the leading coefficient is negative, we begin by factoring out $-2x$:

$$-2x^3 + 20x^2 - 50x = -2x(x^2 - 10x + 25).$$

Now consider $x^2 - 10x + 25$. Since the constant term is positive and the coefficient of the middle term is negative, we look for a factorization of 25 in which both factors are negative. Their sum must be -10.

Pairs of Factors of 25	Sums of Factors
$-25, -1$	-26
$-5, -5$	-10 ← ── The numbers we need are -5 and -5.

The factorization of $x^2 - 10x + 25$ is $(x - 5)(x - 5)$, or $(x - 5)^2$. The factorization of the original polynomial also includes the factor $-2x$.

> **CAUTION!**   When factoring involves more than one step, be careful to write out the *entire* factorization.

Check: $-2x(x - 5)(x - 5) = -2x[x^2 - 10x + 25]$ Multiplying binomials
$$= -2x^3 + 20x^2 - 50x.$$ Using the distributive law

The factorization of $-2x^3 + 20x^2 - 50x$ is $-2x(x - 5)(x - 5)$, or $-2x(x - 5)^2$.

YOUR TURN

Once any common factors have been factored out, the following summary can be used to factor $x^2 + bx + c$.

> ### TO FACTOR $x^2 + bx + c$
>
> 1. Find a pair of factors that have c as their product and b as their sum.
>   a) When c is positive, both factors will have the same sign as b.
>   b) When c is negative, one factor will be positive and the other will be negative. Select the factors such that the factor with the larger absolute value has the same sign as b.
> 2. Check by multiplying.

Note that each polynomial has a unique factorization (except for the order in which the factors are written).

Student Notes

Whenever a new set of parentheses is created while factoring, check the expression inside the parentheses to see if it can be factored further.

---

7. Factor: $3x^3 + 30x^2 + 27x$.

 Chapter Resource:
Collaborative Activity (Visualizing Factoring), p. 365

5.2 EXERCISE SET

Vocabulary and Reading Check

Classify each of the following statements as either true or false.

1. Whenever the product of a pair of factors is negative, the factors have the same sign.

2. Anytime the sum of a negative number and a positive number is negative, the negative number has the greater absolute value.

3. When factoring any polynomial, it is always best to look first for a common factor.

4. When we factor a trinomial such as $x^2 + 6x + 5$, we try to write it as the product of two binomials.

Concept Reinforcement

For Exercises 5–10, assume that $x^2 + bx + c$ can be factored as $(x + p)(x + q)$. Complete each sentence by replacing each blank with either "positive" or "negative."

5. If b is positive and c is positive, then p will be _____ and q will be _____.

6. If b is negative and c is positive, then p will be _____ and q will be _____.

7. If p is negative and q is negative, then b must be _____ and c must be _____.

8. If p is positive and q is positive, then b must be _____ and c must be _____.

9. If b, c, and p are all negative, then q must be _____.

10. If b and c are negative and p is positive, then q must be _____.

Factoring $x^2 + bx + c$

Factor completely. Remember to look first for a common factor. Check by multiplying. If a polynomial is prime, state this.

11. $x^2 + 8x + 16$

12. $x^2 + 9x + 20$

13. $x^2 + 11x + 10$

14. $y^2 + 8y + 7$

15. $t^2 - 9t + 14$

16. $a^2 - 9a + 20$

17. $b^2 - 5b + 4$

18. $z^2 - 8z + 7$

19. $d^2 - 7d + 10$

20. $x^2 - 8x + 15$

21. $x^2 - 2x - 15$

22. $x^2 - x - 42$

23. $x^2 + 2x - 15$

24. $x^2 + x - 42$

25. $2x^2 - 14x - 36$

26. $3y^2 - 9y - 84$

27. $-x^3 + 6x^2 + 16x$

28. $-x^3 + x^2 + 42x$

29. $4y - 45 + y^2$

30. $7x - 60 + x^2$

31. $x^2 - 72 + 6x$

32. $-2x - 99 + x^2$

33. $-5b^2 - 35b + 150$

34. $-c^4 - c^3 + 56c^2$

35. $x^5 - x^4 - 2x^3$

36. $2a^2 - 4a - 70$

37. $x^2 + 5x + 10$

38. $x^2 + 11x + 18$

39. $32 + 12t + t^2$

40. $y^2 - y + 1$

41. $x^2 + 20x + 99$

42. $x^2 + 20x + 100$

43. $3x^3 - 63x^2 - 300x$

44. $2x^3 - 40x^2 + 192x$

45. $-4x^2 - 40x - 100$

46. $-2x^2 + 42x + 144$

47. $y^2 - 20y + 96$

48. $144 - 25t + t^2$

49. $-a^6 - 9a^5 + 90a^4$

50. $-a^4 - a^3 + 132a^2$

51. $t^2 + \frac{2}{3}t + \frac{1}{9}$

52. $x^2 - \frac{2}{5}x + \frac{1}{25}$

53. $11 + w^2 - 4w$

54. $6 + p^2 + 2p$

55. $p^2 - 7pq + 10q^2$

56. $a^2 - 2ab - 3b^2$

57. $m^2 + 5mn + 5n^2$

58. $x^2 - 11xy + 24y^2$

59. $s^2 - 4st - 12t^2$

60. $b^2 + 8bc - 20c^2$

61. $6a^{10} + 30a^9 - 84a^8$

62. $5a^8 - 20a^7 - 25a^6$

63. Without multiplying $(x - 17)(x - 18)$, explain why it cannot possibly be a factorization of $x^2 + 35x + 306$.

64. Shari factors $x^3 - 8x^2 + 15x$ as $(x^2 - 5x)(x - 3)$. Is she wrong? Why or why not? What advice would you offer?

Skill Review

Solve.

65. $\frac{y}{2} = -5$ [2.1]

66. $x - 0.05 = 1.08$ [2.1]

67. $3x + 7 = 12$ [2.2]

68. $9 - 5x = 4 - 3x$ [2.2]

69. $x - (2x - 5) = 16$ [2.2]   70. $\frac{1}{3}x - \frac{1}{6} = \frac{1}{2}$ [2.2]

Synthesis

71. When searching for a factorization, why do we list pairs of numbers with the correct *product* instead of pairs of numbers with the correct *sum*?

72. When factoring $x^2 + bx + c$ with a large value of c, Riley begins by writing out the prime factorization of c. What is the advantage of doing this?

73. Find all integers b for which $a^2 + ba - 50$ can be factored.

74. Find all integers m for which $y^2 + my + 50$ can be factored.

Factor completely.

75. $y^2 - 0.2y - 0.08$ **76.** $x^2 + \frac{1}{2}x - \frac{3}{16}$

77. $-\frac{1}{3}a^3 + \frac{1}{3}a^2 + 2a$ **78.** $-a^7 + \frac{25}{7}a^5 + \frac{30}{7}a^6$

79. $x^{2m} + 11x^m + 28$ **80.** $-t^{2n} + 7t^n - 10$

Aha! 81. $(a + 1)x^2 + (a + 1)3x + (a + 1)2$

82. $ax^2 - 5x^2 + 8ax - 40x - (a - 5)9$
(*Hint*: See Exercise 81.)

83. Find the volume of a cube if its surface area is $(6x^2 + 36x + 54)$ square meters.

Find a polynomial in factored form for the shaded area in each figure. (Use π in your answers where appropriate.)

84.

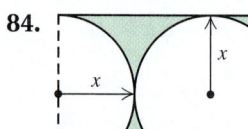

85.

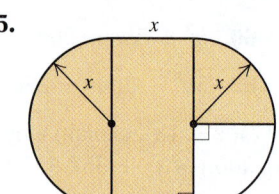

86.

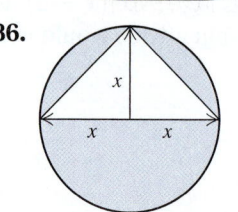

87.

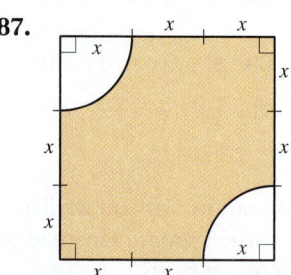

88.

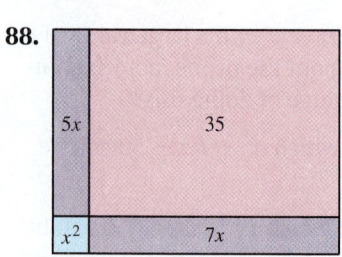

89.

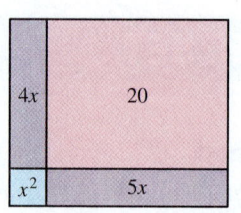

90. A census taker asks a woman, "How many children do you have?"

"Three," she answers.

"What are their ages?"

She responds, "The product of their ages is 36. The sum of their ages is the house number next door."

The math-savvy census taker walks next door, reads the house number, appears puzzled, and returns to the woman, asking, "Is there something you forgot to tell me?"

"Oh yes," says the woman. "I'm sorry. The oldest child is playing a video game."

The census taker records the three ages, thanks the woman for her time, and leaves.

How old is each child? Explain how you reached this conclusion. (*Hint*: Consider factorizations.)

↪ **YOUR TURN ANSWERS: SECTION 5.2**

1. $(x + 2)(x + 4)$ **2.** $(a - 3)(a - 5)$ **3.** $(t + 3)(t - 5)$
4. $(t + 7)(t - 4)$ **5.** $(x - 6y)(x + 3y)$ **6.** Not factorable, or prime   **7.** $3x(x + 9)(x + 1)$

QUICK QUIZ: SECTIONS 5.1–5.2

Factor.

1. $12x^4 - 30x^3 + 6x^2$ [5.1]

2. $3x^3 - 12x^2 + 5x - 20$ [5.1]

3. $x^2 - 3x + 2$ [5.2]

4. $a^2 + 2ab - 8b^2$ [5.2]

5. $3x^2 + 15x + 18$ [5.2]

PREPARE TO MOVE ON

Multiply.   [4.6]

1. $(2x + 3)(3x + 4)$ **2.** $(2x + 3)(3x - 4)$

3. $(2x - 3)(3x + 4)$ **4.** $(2x - 3)(3x - 4)$

5. $(5x - 1)(x - 7)$ **6.** $(x + 6)(3x - 5)$

5.3   Factoring Trinomials of the Type $ax^2 + bx + c$

Factoring with FOIL   ●   The Grouping Method

In this section, we learn to factor trinomials in which the leading, or x^2, coefficient is not 1. First, we will use another FOIL-based method and then we will use an alternative method that involves factoring by grouping. Use the method that you prefer or the one recommended by your instructor.

Factoring with FOIL

Before factoring trinomials of the type $ax^2 + bx + c$, consider the following:

$$
\begin{array}{cccc}
\text{F} & \text{O} & \text{I} & \text{L} \\
\end{array}
$$
$$(2x + 5)(3x + 4) = 6x^2 + 8x + 15x + 20$$
$$= 6x^2 + 23x + 20.$$

To factor $6x^2 + 23x + 20$, we could reverse the multiplication and look for two binomials whose product is this trinomial. We see from above that:

• the product of the First terms must be $6x^2$;
• the product of the Outer terms plus the product of the Inner terms must be $23x$; and
• the product of the Last terms must be 20.

How can such a factorization be found without first seeing the corresponding multiplication? Our first approach relies on trial and error and FOIL.

TO FACTOR $ax^2 + bx + c$ USING FOIL

1. Make certain that all common factors have been removed. If any remain, factor out the largest common factor.
2. Find two First terms whose product is ax^2:

$$(\blacksquare x + \quad)(\blacksquare x + \quad) = ax^2 + bx + c.$$
$$\underline{\qquad\qquad}\text{FOIL}$$

3. Find two Last terms whose product is c:

$$(\blacksquare x + \blacksquare)(\blacksquare x + \blacksquare) = ax^2 + bx + c.$$
$$\underline{\qquad\qquad}\text{FOIL}$$

4. Check by multiplying to see if the sum of the Outer and Inner products is bx. If it is not, repeat steps (2) and (3) until the correct combination is found.

$$(\blacksquare x + \blacksquare)(\blacksquare x + \blacksquare) = ax^2 + bx + c.$$
$$\begin{array}{c} \text{I} \\ \text{O} \end{array} \qquad \text{FOIL}$$

If no correct combination exists, state that the polynomial is prime.

EXAMPLE 1   Factor: $3x^2 - 10x - 8$.

SOLUTION

1. First, check for a common factor. In this case, there is none (other than 1 or -1).

2. Find two **First** terms whose product is $3x^2$. The only possibilities for the **First** terms are $3x$ and x:

$$(3x + \quad)(x + \quad).$$

3. Find two **Last** terms whose product is -8. There are four pairs of factors of -8 and each can be listed in two ways:

$$\begin{array}{ll} -1,\ \ 8 & 8, -1 \\ 1, -8 & -8,\ \ 1 \\ -2,\ \ 4 \quad \text{and} & 4, -2 \\ 2, -4 & -4,\ \ 2. \end{array}$$

Important! Since the First terms are not identical, changing the order of the factors of -8 in the binomial factors results in a different product.

4. Knowing that all **First** and **Last** products will check, systematically inspect the **O**uter and **I**nner products resulting from steps (2) and (3). Look for the combination in which the sum of the products is the middle term, $-10x$.

Pair of Factors	*Corresponding Trial*	*Product*	
$-1,\ \ 8$	$(3x - 1)(x + 8)$	$3x^2 + 24x - x - 8$	
		$= 3x^2 + 23x - 8$	Wrong middle term
$1, -8$	$(3x + 1)(x - 8)$	$3x^2 - 24x + x - 8$	
		$= 3x^2 - 23x - 8$	Wrong middle term
$-2,\ \ 4$	$(3x - 2)(x + 4)$	$3x^2 + 12x - 2x - 8$	
		$= 3x^2 + 10x - 8$	Wrong middle term
$2, -4$	$(3x + 2)(x - 4)$	$3x^2 - 12x + 2x - 8$	
		$= 3x^2 - 10x - 8$	Correct middle term!
$8, -1$	$(3x + 8)(x - 1)$	$3x^2 - 3x + 8x - 8$	
		$= 3x^2 + 5x - 8$	Wrong middle term
$-8,\ \ 1$	$(3x - 8)(x + 1)$	$3x^2 + 3x - 8x - 8$	
		$= 3x^2 - 5x - 8$	Wrong middle term
$4, -2$	$(3x + 4)(x - 2)$	$3x^2 - 6x + 4x - 8$	
		$= 3x^2 - 2x - 8$	Wrong middle term
$-4,\ \ 2$	$(3x - 4)(x + 2)$	$3x^2 + 6x - 4x - 8$	
		$= 3x^2 + 2x - 8$	Wrong middle term

1. Factor: $3x^2 - 13x - 10$.

The correct factorization is $(3x + 2)(x - 4)$. ←

YOUR TURN

Two observations can be made from Example 1. First, we listed all possible trials even though we generally stop after finding the correct factorization. We did this to show that **each trial differs only in the middle term of the product**. Second, note that **only the sign of the middle term changes when the signs in the binomials are reversed**.

A GEOMETRIC APPROACH TO EXAMPLE 2

The factoring of $10x^2 + 37x + 7$ can be regarded as a search for r and s so that the sum of areas A, B, C, and D is $10x^2 + 37x + 7$.

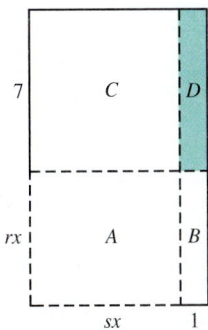

Because A must be $10x^2$, the product rs must be 10. Only when r is 2 and s is 5 will the sum of areas B and C be $37x$ (see below).

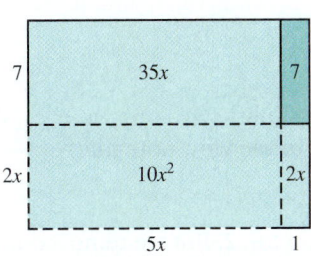

2. Factor: $14x^2 + 13x + 3$.

 **YOUR TURN**

EXAMPLE 2   Factor: $10x^2 + 37x + 7$.

SOLUTION

1. There is no factor (other than 1 or -1) common to all three terms.

2. Because $10x^2$ factors as $10x \cdot x$ or $5x \cdot 2x$, we have two possibilities:

$$(10x +\ \ )(x +\ \ ) \quad \text{or} \quad (5x +\ \ )(2x +\ \ ).$$

3. There are two pairs of factors of 7 and each can be listed in two ways:

$$\begin{array}{cc} 1,\ 7 & 7,\ 1 \\ -1, -7 & -7, -1. \end{array} \quad \text{and}$$

4. Look for **O**uter and **I**nner products for which the sum is the middle term. Because all coefficients in $10x^2 + 37x + 7$ are positive, we need consider only those combinations involving positive factors of 7.

Trial	*Product*	
$(10x + 1)(x + 7)$	$10x^2 + 70x + 1x + 7$	
	$= 10x^2 + 71x + 7$	Wrong middle term
$(10x + 7)(x + 1)$	$10x^2 + 10x + 7x + 7$	
	$= 10x^2 + 17x + 7$	Wrong middle term
$(5x + 7)(2x + 1)$	$10x^2 + 5x + 14x + 7$	
	$= 10x^2 + 19x + 7$	Wrong middle term
$(5x + 1)(2x + 7)$	$10x^2 + 35x + 2x + 7$	
	$= 10x^2 + 37x + 7$	Correct middle term!

The correct factorization is $(5x + 1)(2x + 7)$.

EXAMPLE 3   Factor: $24x^3 - 76x^2 + 40x$.

SOLUTION

1. First, we factor out the largest common factor, $4x$:

$$4x(6x^2 - 19x + 10).$$

2. Next, we factor $6x^2 - 19x + 10$. Since $6x^2$ can be factored as $3x \cdot 2x$ or $6x \cdot x$, we have two possibilities:

$$(3x +\ \ )(2x +\ \ ) \quad \text{or} \quad (6x +\ \ )(x +\ \ ).$$

3. The constant term, 10, can be factored as $1 \cdot 10, 2 \cdot 5, (-1)(-10)$, and $(-2)(-5)$. Since the middle term is negative, we need consider only the factorizations with negative factors:

$$\begin{array}{cc} -1, -10 & -10, -1 \\ -2,\ -5 & -5, -2. \end{array} \quad \text{and}$$

4. The two possibilities from step (2) and the four possibilities from step (3) give $2 \cdot 4$, or 8, possible factorizations to consider.

We first try these factors with $(3x + \quad)(2x + \quad)$, looking for **O**uter and **I**nner products for which the sum is $-19x$. If none gives the correct factorization, then we will consider $(6x + \quad)(x + \quad)$.

Trial	*Product*	
$(3x - 1)(2x - 10)$	$6x^2 - 30x - 2x + 10$	
	$= 6x^2 - 32x + 10$	Wrong middle term
$(3x - 10)(2x - 1)$	$6x^2 - 3x - 20x + 10$	
	$= 6x^2 - 23x + 10$	Wrong middle term
$(3x - 2)(2x - 5)$	$6x^2 - 15x - 4x + 10$	
	$= 6x^2 - 19x + 10$	Correct middle term!
$(3x - 5)(2x - 2)$	$6x^2 - 6x - 10x + 10$	
	$= 6x^2 - 16x + 10$	Wrong middle term

The factorization of $6x^2 - 19x + 10$ is $(3x - 2)(2x - 5)$, but do not forget the common factor! The factorization of $24x^3 - 76x^2 + 40x$ is

$$4x(3x - 2)(2x - 5).$$

3. Factor: $12n^4 + 10n^3 - 12n^2$.

YOUR TURN

In Example 3, look again at the possibility $(3x - 5)(2x - 2)$. Without multiplying, we can reject such a possibility. To see why, note that

$$(3x - 5)(2x - 2) = (3x - 5)2(x - 1).$$

The expression $2x - 2$ has a common factor, 2. But we removed the *largest* common factor in step (1). If $2x - 2$ were one of the factors, then 2 would be *another* common factor in addition to the original, $4x$. Thus, $(2x - 2)$ cannot be part of the factorization of $6x^2 - 19x + 10$. Similar reasoning can be used to reject $(3x - 1)(2x - 10)$ as a possible factorization.

Once the largest common factor is factored out, none of the remaining factors can have a common factor.

Student Notes

Keep your work organized so that you can see what you have already considered. For example, when factoring $6x^2 - 19x + 10$, we can list all possibilities and cross out those in which a common factor appears:

$~~~~\cancel{(3x - 1)(2x - 10)}$
$~~~~(3x - 10)(2x - 1)$
$~~~~(3x - 2)(2x - 5)$
$~~~~\cancel{(3x - 5)(2x - 2)}$
$~~~~(6x - 1)(x - 10)$
$~~~~\cancel{(6x - 10)(x - 1)}$
$~~~~\cancel{(6x - 2)(x - 5)}$
$~~~~(6x - 5)(x - 2)$

By being organized and not erasing, we can see that there are only four possible factorizations.

TIPS FOR FACTORING $ax^2 + bx + c$

To factor $ax^2 + bx + c$ ($a > 0$):

- Make sure that any common factor has been factored out.
- Once the largest common factor has been factored out of the original trinomial, no binomial factor can contain a common factor (other than 1 or -1).
- If c is positive, then the signs in both binomial factors must match the sign of b.
- Reversing the signs in the binomials reverses the sign of the middle term of their product.
- Organize your work so that you can keep track of which possibilities you have checked.
- Remember to include the largest common factor—if there is one—in the final factorization.
- Check by multiplying.

EXAMPLE 4   Factor: $10x + 8 - 3x^2$.

SOLUTION   An important problem-solving strategy is to find a way to make new problems look like problems we already know how to solve. The factoring tips above apply only to trinomials of the form $ax^2 + bx + c$, with $a > 0$. This leads us to rewrite $10x + 8 - 3x^2$ in descending order:

$$10x + 8 - 3x^2 = -3x^2 + 10x + 8. \qquad \text{Using the commutative law to write descending order}$$

Although $-3x^2 + 10x + 8$ looks similar to the trinomials we have factored, the tips above require a positive leading coefficient. This can be found by factoring out -1:

$$-3x^2 + 10x + 8 = -1(3x^2 - 10x - 8) \qquad \text{Factoring out } -1 \text{ changes the signs of the coefficients.}$$
$$= -1(3x + 2)(x - 4). \qquad \text{Using the result from Example 1}$$

The factorization of $10x + 8 - 3x^2$ is $-1(3x + 2)(x - 4)$.

4. Factor: $9y - 10 - 2y^2$.

YOUR TURN

EXAMPLE 5   Factor: $6r^2 - 13rs - 28s^2$.

SOLUTION   In order for the product of the first terms to be $6r^2$ and the product of the last terms to be $-28s^2$, the binomial factors will be of the form

$$(\;\blacksquare r + \blacksquare s)(\blacksquare r + \blacksquare s).$$

We verify that no common factor exists and then examine the first term, $6r^2$. There are two possibilities:

$$(2r + \quad)(3r + \quad) \quad \text{or} \quad (6r + \quad)(r + \quad).$$

The last term, $-28s^2$, has the following pairs of factors:

$$\begin{array}{cc}
s, -28s & -28s, \quad s \\
-s, \quad 28s & 28s, \quad -s \\
2s, -14s & -14s, \quad 2s \\
-2s, \quad 14s & 14s, -2s \\
4s, \quad -7s & -7s, \quad 4s \\
-4s, \quad 7s & 7s, -4s.
\end{array}$$

and

Note that listing the pairs of factors of $-28s^2$ is just like listing the pairs of factors of -28, except that each factor also contains a factor of s.

Some trials, like $(2r + 28s)(3r - s)$ and $(2r + 14s)(3r - 2s)$, cannot be correct because both $(2r + 28s)$ and $(2r + 14s)$ contain a common factor, 2. We try $(2r + 7s)(3r - 4s)$:

$$(2r + 7s)(3r - 4s) = 6r^2 - 8rs + 21rs - 28s^2$$
$$= 6r^2 + 13rs - 28s^2.$$

Our trial is incorrect, but only because of the sign of the middle term. To correctly factor $6r^2 - 13rs - 28s^2$, we simply change the signs in the binomials:

$$(2r - 7s)(3r + 4s) = 6r^2 + 8rs - 21rs - 28s^2$$
$$= 6r^2 - 13rs - 28s^2.$$

The correct factorization of $6r^2 - 13rs - 28s^2$ is $(2r - 7s)(3r + 4s)$.

5. Factor: $12x^2 - 8xy - 15y^2$.

YOUR TURN

The Grouping Method

Another method of factoring trinomials of the type $ax^2 + bx + c$ is known as the *grouping method*. The grouping method relies on rewriting $ax^2 + bx + c$ in the form $ax^2 + px + qx + c$ and then factoring by grouping. To develop this method, consider the following*:

$$(2x + 5)(3x + 4) = 2x \cdot 3x + 2x \cdot 4 + 5 \cdot 3x + 5 \cdot 4 \qquad \text{Using FOIL}$$
$$= 2 \cdot 3 \cdot x^2 + 2 \cdot 4x + 5 \cdot 3x + 5 \cdot 4$$
$$= \underset{\underset{a}{\uparrow\downarrow}}{2 \cdot 3 \cdot x^2} + \underset{\underset{b}{\uparrow\downarrow}}{(2 \cdot 4 + 5 \cdot 3)x} + \underset{\underset{c}{\uparrow\downarrow}}{5 \cdot 4}$$
$$= \quad 6x^2 \quad + \quad 23x \quad + \quad 20.$$

Note that reversing these steps shows that $6x^2 + 23x + 20$ can be rewritten as $6x^2 + 8x + 15x + 20$ and then factored by grouping. Note that the numbers that add to b (in this case, $2 \cdot 4$ and $5 \cdot 3$) also multiply to ac (in this case, $2 \cdot 3 \cdot 5 \cdot 4$).

TO FACTOR $ax^2 + bx + c$ USING THE GROUPING METHOD

1. Factor out the largest common factor, if one exists.
2. Multiply the leading coefficient a and the constant c.
3. Find a pair of factors of ac whose sum is b.
4. Rewrite the middle term, bx, as a sum or a difference using the factors found in step (3).
5. Factor by grouping.
6. Include any common factor from step (1) and check by multiplying.

EXAMPLE 6 Factor: $3x^2 - 10x - 8$.

SOLUTION

1. First, we note that there is no common factor (other than 1 or -1).

2. We multiply the leading coefficient, 3, and the constant, -8:

   $$3(-8) = -24.$$

3. We next look for a factorization of -24 in which the sum of the factors is the coefficient of the middle term, -10.

Pairs of Factors of -24	Sums of Factors
1, -24	-23
-1, 24	23
2, -12	-10 ⟵ — $2 + (-12) = -10$
-2, 12	10
3, -8	-5
-3, 8	5
4, -6	-2
-4, 6	2

We normally stop listing pairs of factors once we have found the one we need.

---

*This discussion was inspired by a lecture given by Irene Doo at Austin Community College.

4. Next, we express the middle term as a sum or a difference using the factors found in step (3):

$$-10x = 2x - 12x.$$

5. We now factor by grouping as follows:

$$3x^2 - 10x - 8 = 3x^2 + 2x - 12x - 8$$

Substituting $2x - 12x$ for $-10x$. We could also use $-12x + 2x$.

$$= x(3x + 2) - 4(3x + 2)$$
$$= (3x + 2)(x - 4).$$

Factoring by grouping

6. *Check:* $(3x + 2)(x - 4) = 3x^2 - 12x + 2x - 8 = 3x^2 - 10x - 8.$

The factorization of $3x^2 - 10x - 8$ is $(3x + 2)(x - 4)$.

6. Factor: $2x^2 - 3x - 20$.

_____

 YOUR TURN

EXAMPLE 7   Factor: $8x^3 + 22x^2 - 6x$.

SOLUTION

1. We factor out the largest common factor, $2x$:

$$8x^3 + 22x^2 - 6x = 2x(4x^2 + 11x - 3).$$

2. To factor $4x^2 + 11x - 3$ by grouping, we multiply the leading coefficient, 4, and the constant term, -3:

$$4(-3) = -12.$$

3. We next look for factors of -12 that add to 11.

Pairs of Factors of -12	Sums of Factors
1, -12	-11
-1, 12	11 ←
.	.
.	.
.	.

Since $-1 + 12 = 11$, there is no need to list other pairs of factors.

4. We then rewrite the $11x$ in $4x^2 + 11x - 3$ using the results of step (3):

$$11x = -1x + 12x, \quad \text{or} \quad 11x = 12x - 1x.$$

5. Next, we factor by grouping:

$$4x^2 + 11x - 3 = 4x^2 - 1x + 12x - 3$$

Rewriting the middle term; $12x - 1x$ could also be used.

$$= x(4x - 1) + 3(4x - 1)$$
$$= (4x - 1)(x + 3).$$

Factoring by grouping

6. The factorization of $4x^2 + 11x - 3$ is $(4x - 1)(x + 3)$. But don't forget the common factor, $2x$. The factorization of the original trinomial is

$$2x(4x - 1)(x + 3).$$

7. Factor: $30n^3 + 5n^2 - 60n$.

_____

YOUR TURN

5.3 EXERCISE SET

FOR EXTRA HELP

MyMathLab® Math XL

PRACTICE   WATCH   READ   REVIEW

Vocabulary and Reading Check

Match each polynomial in Exercises 1–4 with the statement that is true when factoring that polynomial.

a) Factor out a common factor first.

b) The only possibilities for the first terms of the binomial factors are $5x$ and x.

c) Both of the constants in the binomial factors must be negative.

d) The polynomial is prime.

1. ____ $11x^2 + 3x + 1$

2. ____ $12x^2 + 18x + 21$

3. ____ $5x^2 - 14x - 3$

4. ____ $2x^2 - 7x + 5$

Factoring Trinomials of the Type $ax^2 + bx + c$

Factor completely. If a polynomial is prime, state this.

5. $2x^2 + 7x - 4$ 6. $3x^2 + x - 4$

7. $3x^2 - 17x - 6$ 8. $5x^2 - 19x - 4$

9. $4t^2 + 12t + 5$ 10. $6t^2 + 17t + 7$

11. $15a^2 - 14a + 3$ 12. $10a^2 - 11a + 3$

13. $6x^2 + 17x + 12$ 14. $6x^2 + 19x + 10$

15. $6x^2 - 10x - 4$ 16. $5t^3 - 21t^2 + 18t$

17. $7t^3 + 15t^2 + 2t$ 18. $15t^2 + 20t - 75$

19. $10 - 23x + 12x^2$ 20. $-20 + 31x - 12x^2$

21. $-35x^2 - 34x - 8$ 22. $28x^2 + 38x - 6$

23. $4 + 6t^2 - 13t$ 24. $9 + 8t^2 - 18t$

25. $25x^2 + 40x + 16$ 26. $49t^2 + 42t + 9$

27. $20y^2 + 59y - 3$ 28. $25a^2 - 23a - 2$

29. $14x^2 + 73x + 45$ 30. $35x^2 - 57x - 44$

31. $-2x^2 + 15 + x$ 32. $2t^2 - 19 - 6t$

33. $-6x^2 - 33x - 15$ 34. $-12x^2 - 28x + 24$

35. $10a^2 - 8a - 18$ 36. $20y^2 - 25y + 5$

37. $12x^2 + 68x - 24$ 38. $6x^2 + 21x + 15$

39. $4x + 1 + 3x^2$ 40. $-9 + 18x^2 + 21x$

Factor. Use factoring by grouping even though it may seem reasonable to first combine like terms.

41. $x^2 + 3x - 2x - 6$ 42. $x^2 + 4x - 2x - 8$

43. $8t^2 - 6t - 28t + 21$

44. $35t^2 - 40t + 21t - 24$

45. $6x^2 + 4x + 15x + 10$ 46. $3x^2 - 2x + 3x - 2$

47. $2y^2 + 8y - y - 4$ 48. $7n^2 + 35n - n - 5$

49. $6a^2 - 8a - 3a + 4$ 50. $10a^2 - 4a - 5a + 2$

Factor completely. If a polynomial is prime, state this.

51. $16t^2 + 23t + 7$ 52. $9t^2 + 14t + 5$

53. $-9x^2 - 18x - 5$ 54. $-16x^2 - 32x - 7$

55. $10x^2 + 30x - 70$ 56. $10a^2 + 25a - 15$

57. $18x^3 + 21x^2 - 9x$ 58. $6x^3 - 4x^2 - 10x$

59. $89x + 64 + 25x^2$ 60. $47 - 42y + 9y^2$

61. $168x^3 + 45x^2 + 3x$

62. $144x^5 - 168x^4 + 48x^3$

63. $-14t^4 + 19t^3 + 3t^2$

64. $-70a^4 + 68a^3 - 16a^2$

65. $132y + 32y^2 - 54$ 66. $220y + 60y^2 - 225$

67. $2a^2 - 5ab + 2b^2$ 68. $3p^2 - 16pq - 12q^2$

69. $8s^2 + 22st + 14t^2$ 70. $10s^2 + 4st - 6t^2$

71. $27x^2 - 72xy + 48y^2$

72. $-30a^2 - 87ab - 30b^2$

73. $-24a^2 + 34ab - 12b^2$ 74. $15a^2 - 5ab - 20b^2$

75. $19x^3 - 3x^2 + 14x^4$ 76. $10x^5 - 2x^4 + 22x^3$

77. $18a^7 + 8a^6 + 9a^8$ 78. $40a^8 + 16a^7 + 25a^9$

79. Asked to factor $2x^2 - 18x + 36$, Juan *incorrectly* answers

$$2x^2 - 18x + 36 = 2(x^2 + 9x + 18)$$
$$= 2(x + 3)(x + 6).$$

If this were a 10-point quiz question, how many points would you take off? Why?

80. Asked to factor $4x^2 + 28x + 48$, Therese *incorrectly* answers

$$4x^2 + 28x + 48 = (2x + 6)(2x + 8)$$
$$= 2(x + 3)(x + 4).$$

If this were a 10-point quiz question, how many points would you take off? Why?

Skill Review

Graph.

81. $2x - 5y = 10$ [3.3]

82. $-5x = 10$ [3.3]

83. $y = \frac{2}{3}x - 1$ [3.6]

84. $y - 2 = -2(x + 4)$ [3.7]

85. $\frac{1}{2}y = 1$ [3.3]

86. $x = -y$ [3.2]

Synthesis

87. Explain how you would prove to a fellow student that a given trinomial is prime.

88. For the trinomial $ax^2 + bx + c$, suppose that a is the product of three different prime factors and c is the product of another two prime factors. How many possible factorizations (like those in Example 1) exist? Explain your reasoning.

Factor. If a polynomial is prime, state this.

89. $18x^2y^2 - 3xy - 10$

90. $8x^2y^3 + 10xy^2 + 2y$

91. $9a^2b^3 + 25ab^2 + 16$

92. $-9t^{10} - 12t^5 - 4$

93. $16t^{10} - 8t^5 + 1$

94. $9a^2b^2 - 15ab - 2$

95. $-15x^{2m} + 26x^m - 8$

96. $-20x^{2n} - 16x^n - 3$

97. $3a^{6n} - 2a^{3n} - 1$

98. $a^{2n+1} - 2a^{n+1} + a$

99. $7(t - 3)^{2n} + 5(t - 3)^n - 2$

100. $3(a + 1)^{n+1}(a + 3)^2 - 5(a + 1)^n(a + 3)^3$

101. Kara bought a number of sections of decorative fencing, all of equal length, to fence in a rectangular 1500-ft^2 yard. For each of the shorter sides of the rectangle, she used 4 sections plus 2 ft cut from another section. For each of the longer sides, she used 7 sections plus 1 ft cut from another section. How long was each section of fencing?

 YOUR TURN ANSWERS: SECTION 5.3

1. $(3x + 2)(x - 5)$ **2.** $(2x + 1)(7x + 3)$
3. $2n^2(2n + 3)(3n - 2)$ **4.** $-1(2y - 5)(y - 2)$
5. $(2x - 3y)(6x + 5y)$ **6.** $(2x + 5)(x - 4)$
7. $5n(2n + 3)(3n - 4)$

QUICK QUIZ: SECTIONS 5.1–5.3

Factor completely. If a polynomial is prime, state this.

1. $6a^2b^3 - 9ab^4 + 15a^3b^5$ [5.1]

2. $x^3 - 7x^2 + 12x$ [5.2]

3. $p^2 + p - 7$ [5.2]

4. $4a^2 + 16a + 15$ [5.3]

5. $6x^2 - x - 5$ [5.3]

PREPARE TO MOVE ON

Multiply.   [4.6]

1. $(x - 2)^2$ **2.** $(x + 2)^2$

3. $(x + 2)(x - 2)$ **4.** $(5t - 3)^2$

5. $(4a + 1)^2$ **6.** $(2n + 7)(2n - 7)$

5.4 Factoring Perfect-Square Trinomials and Differences of Squares

Recognizing Perfect-Square Trinomials ▪ Factoring Perfect-Square Trinomials ▪
Recognizing Differences of Squares ▪ Factoring Differences of Squares ▪ Factoring Completely

Reversing the rules for special products provides us with shortcuts for factoring certain polynomials.

Recognizing Perfect-Square Trinomials

Some trinomials are squares of binomials. For example, $x^2 + 10x + 25$ is the square of the binomial $x + 5$, because

$$(x + 5)^2 = x^2 + 2 \cdot x \cdot 5 + 5^2 = x^2 + 10x + 25.$$

A trinomial that is the square of a binomial is called a **perfect-square trinomial**.
We can square binomials using the following special-product rule:

$$(A + B)^2 = A^2 + 2AB + B^2;$$
$$(A - B)^2 = A^2 - 2AB + B^2.$$

Reading the right-hand sides first, we can use these equations to factor perfect-square trinomials. In order for a trinomial to be the square of a binomial, it must have the following:

1. Two expressions, A^2 and B^2, must be squares, such as

   $$4, \quad x^2, \quad 81m^2, \quad 16t^2.$$

2. Neither A^2 nor B^2 is being subtracted.

3. The remaining term is either $2 \cdot A \cdot B$ or $-2 \cdot A \cdot B$, where A and B are the **square roots** of A^2 and B^2.

Because $3^2 = 9$, we say that 3 is the *square root* of 9. Also, 2 is the square root of 4, 4 is the square root of 16, and so on.
Note that in order for an expression to be a square, its coefficient must be a perfect square and the power(s) of the variable(s) must be even.

EXAMPLE 1   Determine whether each of the following is a perfect-square trinomial.

a) $x^2 + 6x + 9$ **b)** $t^2 - 8t - 9$ **c)** $16x^2 + 49 - 56x$

SOLUTION

a) **1.** Two expressions, x^2 and 9, are squares.

   2. Neither x^2 nor 9 is being subtracted.

   3. The remaining term, $6x$, is $2 \cdot x \cdot 3$, where x and 3 are the square roots of x^2 and 9.

   Thus, $x^2 + 6x + 9$ *is* a perfect-square trinomial.

b) Both t^2 and 9 are squares. However, since 9 is being subtracted, $t^2 - 8t - 9$ *is not* a perfect-square trinomial.

c) To see if $16x^2 + 49 - 56x$ is a perfect-square trinomial, it helps to first write it in descending order:

   $$16x^2 - 56x + 49.$$

Student Notes

If you're not already quick to recognize the squares that represent $1^2, 2^2, 3^2, \ldots, 12^2$, this would be a good time to memorize these numbers.

Study Skills

Fill in Your Blanks

Don't hesitate to write out any missing steps that you'd like to see included. For instance, in Example 1(c), we state (in red) that $16x^2$ is a square. To solidify your understanding, you may want to write $4x \cdot 4x = 16x^2$ in the margin of your text.

Next, note that:

1. Two expressions, $16x^2$ and 49, are squares.

2. Neither $16x^2$ nor 49 is being subtracted.

3. Twice the product of the square roots, $2 \cdot 4x \cdot 7$, is $56x$. The remaining term, $-56x$, is the opposite of this product.

Thus, $16x^2 + 49 - 56x$ *is* a perfect-square trinomial.

1. Determine whether $100 - 20x + x^2$ is a perfect-square trinomial.

 YOUR TURN

Factoring Perfect-Square Trinomials

To factor perfect-square trinomials, we recognize the following patterns.

> **FACTORING A PERFECT-SQUARE TRINOMIAL**
> $A^2 + 2AB + B^2 = (A + B)^2;$
> $A^2 - 2AB + B^2 = (A - B)^2$

Each factorization uses the square roots of the squared terms and the sign of the remaining term. To verify these equations, you should compute $(A + B)(A + B)$ and $(A - B)(A - B)$.

EXAMPLE 2   Factor.

a) $x^2 + 6x + 9$ **b)** $t^2 + 49 - 14t$ **c)** $100x^2 - 180x + 81$

SOLUTION

a) $x^2 + 6x + 9 = x^2 + 2 \cdot x \cdot 3 + 3^2 = (x + 3)^2$ The sign of the middle term is positive.

$$A^2 + 2 \ A \ B + B^2 = (A + B)^2$$

b) $t^2 + 49 - 14t = t^2 - 14t + 49$ Using a commutative law to write in descending order

$$= t^2 - 2 \cdot t \cdot 7 + 7^2 = (t - 7)^2$$

$$A^2 - 2 \ A \ B + B^2 = (A - B)^2$$

c) $100x^2 - 180x + 81 = (10x)^2 - 2 \cdot 10x \cdot 9 + 9^2 = (10x - 9)^2$ Recall that $(10x)^2 = 100x^2$.

$$A^2 \quad - 2 \ A \ B + B^2 = (A - B)^2$$

2. Factor: $x^2 + 2x + 1$.

The checks are left to the student.

 YOUR TURN

Polynomials in more than one variable can also be perfect-square trinomials.

EXAMPLE 3   Factor: $4p^2 - 12pt + 9t^2$.

SOLUTION   We have

$$4p^2 - 12pt + 9t^2 = (2p)^2 - 2(2p)(3t) + (3t)^2$$ $4p^2$ and $9t^2$ are squares.

$$= (2p - 3t)^2.$$ The sign of the middle term is negative.

Check: $(2p - 3t)(2p - 3t) = 4p^2 - 12pt + 9t^2.$

3. Factor: $100a^2 + 20ab + b^2$.

The factorization is $(2p - 3t)^2$.

 YOUR TURN

EXAMPLE 4   Factor: $-75m^3 - 60m^2 - 12m$.

SOLUTION   *Always* look first for a common factor. This time there is one. We factor out $-3m$ so that the leading coefficient of the polynomial inside the parentheses is positive:

Factor out the common factor.

Factor the perfect-square trinomial.

$$-75m^3 - 60m^2 - 12m = -3m[25m^2 + 20m + 4] \qquad 25m^2 = (5m)^2$$
$$= -3m[(5m)^2 + 2(5m)(2) + 2^2]$$
$$= -3m(5m + 2)^2.$$

Check: $-3m(5m + 2)^2 = -3m(5m + 2)(5m + 2)$
$$= -3m(25m^2 + 20m + 4)$$
$$= -75m^3 - 60m^2 - 12m.$$

The factorization is $-3m(5m + 2)^2$.

4. Factor: $-4n^3 + 12n^2 - 9n$.

YOUR TURN

Recognizing Differences of Squares

Any expression that can be written in the form $A^2 - B^2$, like $x^2 - 9$, is called a **difference of squares**. In order for a binomial to be a difference of squares, it must have the following.

1. There must be two expressions, both squares, such as

$$25, \quad t^2, \quad 4x^2, \quad 1, \quad x^6, \quad 49y^8, \quad 100x^2y^2.$$

2. The terms in the binomial must have different signs.

EXAMPLE 5   Determine whether each of the following is a difference of squares.

a) $9x^2 - 64$ **b)** $25 - t^3$ **c)** $-4x^{10} + 36$

SOLUTION

a) **1.** The first expression is a square: $9x^2 = (3x)^2$.
   The second expression is a square: $64 = 8^2$.

   2. The terms have different signs.

   Thus, $9x^2 - 64$ is a difference of squares, $(3x)^2 - 8^2$.

b) **1.** The expression t^3 is not a square.

   Thus, $25 - t^3$ is not a difference of squares.

c) **1.** The expressions $4x^{10}$ and 36 are squares: $4x^{10} = (2x^5)^2$ and $36 = 6^2$.

   2. The terms have different signs.

5. Determine whether $-x^2 + 49$ is a difference of squares.

Thus, $-4x^{10} + 36$ is a difference of squares, $6^2 - (2x^5)^2$. It is helpful to rewrite $-4x^{10} + 36$ in the equivalent form $36 - 4x^{10}$ before factoring.

YOUR TURN

Factoring Differences of Squares

To factor a difference of squares, we reverse the special product

$$(A + B)(A - B) = A^2 - AB + AB - B^2 = A^2 - B^2.$$

FACTORING A DIFFERENCE OF SQUARES

$$A^2 - B^2 = (A + B)(A - B)$$

Once we have identified the expressions that are playing the roles of A and B, the factorization can be written directly.

EXAMPLE 6 Factor: **(a)** $x^2 - 4$; **(b)** $1 - 9p^2$; **(c)** $s^6 - 16t^{10}$; **(d)** $50x^2 - 8x^8$.

SOLUTION

a) $\underset{\substack{\uparrow \qquad \uparrow \\ A^2 \; - \; B^2}}{x^2 - 4} = \underset{\substack{\;\; \uparrow \quad\;\; \uparrow \quad \uparrow \;\; \uparrow \\ =\;(A+B)(A-B)}}{x^2 - 2^2 = (x + 2)(x - 2)}$

b) $\underset{\substack{\;\;\uparrow \qquad\quad \uparrow \\ A^2 \; - \;\;\; B^2}}{1 - 9p^2 = 1^2 - (3p)^2} = \underset{\substack{\;\; \uparrow \quad\;\;\;\; \uparrow \quad\;\; \uparrow \;\;\;\; \uparrow \\ =\;(A \;+\; B)\;(A \;-\; B)}}{(1 + 3p)(1 - 3p)}$

c) $\underset{\substack{\;\;\;\uparrow \qquad\;\; \uparrow \\ A^2 \;\;\; - \;\;\; B^2}}{s^6 - 16t^{10} = (s^3)^2 - (4t^5)^2}$ Using the rules for powers

$\phantom{s^6 - 16t^{10}} = \underset{\substack{\;\;\uparrow \quad\;\; \uparrow \quad\;\; \uparrow \quad\;\; \uparrow \\ (A \;+\; B) \;\; (A \;-\; B)}}{(s^3 + 4t^5)(s^3 - 4t^5)}$ Try to go directly to this step.

d) *Always* look first for a common factor. This time there is one, $2x^2$:

Factor out the common factor.

$50x^2 - 8x^8 = 2x^2(25 - 4x^6)$

Factor the difference of squares.

$ = 2x^2[5^2 - (2x^3)^2]$ Recognizing $A^2 - B^2$.
Try to do this mentally.

$ = 2x^2(5 + 2x^3)(5 - 2x^3).$

Check: $2x^2(5 + 2x^3)(5 - 2x^3) = 2x^2(25 - 4x^6)$
$ = 50x^2 - 8x^8.$

The factorization of $50x^2 - 8x^8$ is $2x^2(5 + 2x^3)(5 - 2x^3)$.

6. Factor: $a^4 - 9b^2$.

 YOUR TURN

> **CAUTION!** A difference of squares is *not* the square of the difference; that is,
>
> $$A^2 - B^2 \neq (A - B)^2. \quad \text{To see this, note that}$$
> $$(A - B)^2 = A^2 - 2AB + B^2.$$

Factoring Completely

Sometimes, as in Examples 4 and 6(d), a *complete* factorization requires two or more steps. Factoring is complete only when no factor can be factored further.

EXAMPLE 7 Factor: $y^4 - 16$.

SOLUTION We have

Factor a difference of squares.

$y^4 - 16 = (y^2)^2 - 4^2$ Recognizing $A^2 - B^2$

$ = (y^2 + 4)(y^2 - 4)$ Note that $y^2 - 4$ is not prime.

Factor another difference of squares.

$ = (y^2 + 4)(y + 2)(y - 2).$ Note that $y^2 - 4$ is itself a difference of squares.

Check: $(y^2 + 4)(y + 2)(y - 2) = (y^2 + 4)(y^2 - 4)$
$ = y^4 - 16.$

7. Factor: $t^4 - 81$.

The factorization is $(y^2 + 4)(y + 2)(y - 2)$.

YOUR TURN

Note in Example 7 that the factor $y^2 + 4$ is a *sum* of squares that cannot be factored further.

CAUTION!   There is no general formula for factoring a sum of squares. In particular,

$$A^2 + B^2 \neq (A + B)^2 \quad \text{and} \quad A^2 + B^2 \neq (A + B)(A - B).$$

As you proceed through the exercises, these suggestions may prove helpful.

TIPS FOR FACTORING

1. *Always* look first for a common factor! If there is one, factor it out.
2. Be alert for perfect-square trinomials and for binomials that are differences of squares. Once recognized, they can be factored without trial and error.
3. Always factor completely.
4. Check by multiplying.

5.4   EXERCISE SET

FOR EXTRA HELP

MyMathLab® Math XL
PRACTICE   WATCH   READ   REVIEW

Vocabulary and Reading Check

Identify each of the following as a perfect-square trinomial, a difference of squares, a prime polynomial, or none of these.

1. $4x^2 + 49$
2. $x^2 - 64$
3. $t^2 - 100$
4. $x^2 - 5x + 4$
5. $9x^2 + 6x + 1$
6. $a^2 - 8a + 16$
7. $2t^2 + 10t + 6$
8. $-25x^2 - 9$
9. $16t^2 - 25$
10. $4r^2 + 20r + 25$

Recognizing Perfect-Square Trinomials

Determine whether each of the following is a perfect-square trinomial.

11. $x^2 + 18x + 81$
12. $x^2 - 16x + 64$
13. $x^2 - 10x - 25$
14. $x^2 - 14x - 49$
15. $x^2 - 3x + 9$
16. $x^2 + 4x + 4$
17. $9x^2 + 25 - 30x$
18. $36x^2 + 16 - 24x$

Factoring Perfect-Square Trinomials

Factor completely. Remember to look first for a common factor and to check by multiplying. If a polynomial is prime, state this.

19. $x^2 + 16x + 64$
20. $x^2 + 10x + 25$
21. $x^2 - 10x + 25$
22. $x^2 - 16x + 64$
23. $5p^2 + 20p + 20$
24. $3p^2 - 12p + 12$
25. $1 - 2t + t^2$
26. $1 + t^2 + 2t$
27. $18x^2 + 12x + 2$
28. $25x^2 + 10x + 1$
29. $49 - 56y + 16y^2$
30. $75 - 60m + 12m^2$
31. $-x^5 + 18x^4 - 81x^3$
32. $-2x^2 + 40x - 200$
33. $2n^3 + 40n^2 + 200n$
34. $x^3 + 24x^2 + 144x$
35. $20x^2 + 100x + 125$
36. $27m^2 - 36m + 12$
37. $49 - 42x + 9x^2$
38. $64 - 112x + 49x^2$
39. $16x^2 + 24x + 9$
40. $2a^2 + 28a + 98$
41. $2 + 20x + 50x^2$
42. $9x^2 + 30x + 25$

43. $9p^2 + 12px + 4x^2$

44. $x^2 - 3xy + 9y^2$

45. $a^2 - 12ab + 49b^2$

46. $25m^2 - 20mn + 4n^2$

47. $-64m^2 - 16mn - n^2$

48. $-81p^2 + 18pw - w^2$

49. $-32s^2 + 80st - 50t^2$

50. $-36a^2 - 96ab - 64b^2$

Recognizing Differences of Squares

Determine whether each of the following is a difference of squares.

51. $x^2 - 100$

52. $x^2 + 49$

53. $n^4 + 1$

54. $n^4 - 81$

55. $-1 + 64t^2$

56. $-12 + 25t^2$

Factoring Differences of Squares

Factor completely. Remember to look first for a common factor. If a polynomial is prime, state this.

57. $x^2 - 25$

58. $x^2 - 36$

59. $p^2 - 9$

60. $q^2 + 1$

61. $-49 + t^2$

62. $-64 + m^2$

63. $6a^2 - 24$

64. $x^2 - 8x + 16$

65. $49x^2 - 14x + 1$

66. $3t^2 - 3$

67. $200 - 2t^2$

68. $98 - 8w^2$

69. $-80a^2 + 45$

70. $25x^2 - 4$

71. $5t^2 - 80$

72. $-4t^2 + 64$

73. $8x^2 - 162$

74. $24x^2 - 54$

75. $36x - 49x^3$

76. $16x - 81x^3$

77. $49a^4 - 20$

78. $25a^4 - 9$

Factoring Completely

Factor completely.

79. $t^4 - 1$

80. $x^4 - 16$

81. $-3x^3 + 24x^2 - 48x$

82. $-2a^4 + 36a^3 - 162a^2$

83. $75t^3 - 27t$

84. $80s^4 - 45s^2$

85. $a^8 - 2a^7 + a^6$

86. $x^8 - 8x^7 + 16x^6$

87. $10a^2 - 10b^2$

88. $6p^2 - 6q^2$

89. $16x^4 - y^4$

90. $98x^2 - 32y^2$

91. $18t^2 - 8s^2$

92. $a^4 - 81b^4$

 93. Explain in your own words how to determine whether a polynomial is a perfect-square trinomial.

 94. Explain in your own words how to determine whether a polynomial is a difference of squares.

Skill Review

Perform the indicated operation.

95. $(2x^3 - x + 3) + (x^2 + x - 5)$ [4.4]

96. $(3t^2 - 2t - 5) - (t^2 - 8t + 6)$ [4.4]

97. $(2x^2 + y)(3x^2 - y)$ [4.7]

98. $-5x(-x^2 + 3x - 7)$ [4.5]

99. $(21x^3 - 3x^2 + 9x) \div (3x)$ [4.8]

100. $(x^2 - x - 10) \div (x - 5)$ [4.8]

Synthesis

 101. Leon concludes that since $x^2 - 9 = (x - 3)(x + 3)$, it must follow that $x^2 + 9 = (x + 3)(x - 3)$. What mistake(s) is he making?

 102. Write directions that would enable someone to construct a polynomial that contains a perfect-square trinomial, a difference of squares, and a common factor.

Factor completely. If a polynomial is prime, state this.

103. $x^8 - 2^8$

104. $3x^2 - \frac{1}{3}$

105. $18x^3 - \frac{8}{25}x$

106. $0.81t - t^3$

107. $(y - 5)^4 - z^8$

108. $x^2 - \left(\dfrac{1}{x}\right)^2$

109. $-x^4 + 8x^2 + 9$

110. $-16x^4 + 96x^2 - 144$

Aha! **111.** $(y + 3)^2 + 2(y + 3) + 1$

112. $49(x + 1)^2 - 42(x + 1) + 9$

113. $27p^3 - 45p^2 - 75p + 125$

114. $a^{2n} - 49b^{2n}$

115. $81 - b^{4k}$

116. $9b^{2n} + 12b^n + 4$

117. Subtract $(x^2 + 1)^2$ from $x^2(x + 1)^2$.

Factor by grouping. Look for a grouping of three terms that is a perfect-square trinomial.

118. $t^2 + 4t + 4 - 25$

119. $y^2 + 6y + 9 - x^2 - 8x - 16$

Find c such that each polynomial is the square of a binomial.

120. $cy^2 + 6y + 1$

121. $cy^2 - 24y + 9$

122. Find the only positive value of a for which $x^2 + a^2x + a^2$ factors into $(x + a)^2$.

123. Show that the difference of the squares of two consecutive integers is the sum of the integers. (*Hint*: Use x for the smaller number.)

 124. Use a geometric approach, similar to that shown on p. 312, to show that $x^2 + 6x + 9$ is a perfect square.

 YOUR TURN ANSWERS: SECTION 5.4

1. Yes   **2.** $(x + 1)^2$ **3.** $(10a + b)^2$ **4.** $-n(2n - 3)^2$
5. Yes   **6.** $(a^2 + 3b)(a^2 - 3b)$
7. $(t^2 + 9)(t + 3)(t - 3)$

QUICK QUIZ: SECTIONS 5.1–5.4

Factor completely. If a polynomial is prime, state this.

1. $2x^3 + 6x^2 - x - 3$ [5.1]

2. $x^2 - 24x + 80$ [5.2]

3. $9a^3 - 6a^2 - 8a$ [5.3]

4. $y^2 + 16y + 64$ [5.4]

5. $3z^4 - 3$ [5.4]

PREPARE TO MOVE ON

Simplify.   [4.1]

1. $(2x^2y^4)^3$ **2.** $(-5x^2y)^3$

Multiply.   [4.5]

3. $(x - 1)^3$ **4.** $(p + t)^3$

Mid-Chapter Review

The following is a good strategy to follow when you encounter a mixed set of factoring problems.

1. Factor out any common factor.
2. If there are *four* terms, try factoring by grouping.
3. If there are *three* terms, determine whether it is a perfect-square trinomial. If so, factor using the pattern

$$A^2 + 2AB + B^2 = (A + B)^2 \quad \text{or} \quad A^2 - 2AB + B^2 = (A - B)^2.$$

4. If there are *three* terms and it is not a perfect-square trinomial, try factoring using FOIL or by grouping.
5. If there are *two* terms, determine whether it is a difference of squares. If so, factor using the pattern

$$A^2 - B^2 = (A + B)(A - B).$$

GUIDED SOLUTIONS

Factor completely.

1. $12x^3y - 8xy^2 + 24x^2y = \boxed{} \, (3x^2 - 2y + 6x)$ [5.1]   Factoring out the largest common factor. No further factorization is possible.

2. $3a^3 - 3a^2 - 90a = 3a\left(\boxed{} - \boxed{} - \boxed{}\right)$ Factoring out the largest common factor

$$= 3a\left(a - \boxed{}\right)\left(a + \boxed{}\right) \quad [5.2]$$ Factoring the trinomial

MIXED REVIEW

Factor completely. If a polynomial is prime, state this.

3. $6x^5 - 18x^2$ [5.1]

4. $x^2 + 10x + 16$ [5.2]

5. $2x^2 + 13x - 7$ [5.3]

6. $x^3 + 3x^2 + 2x + 6$ [5.1]

7. $64n^2 - 9$ [5.4]

8. $x^2 - 2x - 5$ [5.2]

9. $6p^2 - 6t^2$ [5.4]

10. $b^2 - 14b + 49$ [5.4]

11. $12x^2 - x - 1$ [5.3]

12. $a - 10a^2 + 25a^3$ [5.4]

13. $10x^4 - 10$ [5.4]

14. $t^2 + t - 10$ [5.2]

15. $15d^2 - 30d + 75$ [5.1]

16. $15p^2 + 16px + 4x^2$ [5.3]

17. $-2t^3 - 10t^2 - 12t$ [5.2]

18. $10c^2 + 20c + 10$ [5.4]

19. $5 + 3x - 2x^2$ [5.3]

20. $2m^3n - 10m^2n - 6mn + 30n$ [5.1]

5.5 Factoring Sums or Differences of Cubes

Factoring Sums or Differences of Cubes

Factoring Sums or Differences of Cubes

We have seen that a difference of two squares can be factored but a *sum* of two squares is usually prime. The situation is different with cubes: The difference *or sum* of two cubes can be factored. To see this, consider the following products:

$$(A + B)(A^2 - AB + B^2) = A(A^2 - AB + B^2) + B(A^2 - AB + B^2)$$
$$= A^3 - A^2B + AB^2 + A^2B - AB^2 + B^3$$
$$= A^3 + B^3 \qquad \text{Combining like terms}$$

and

$$(A - B)(A^2 + AB + B^2) = A(A^2 + AB + B^2) - B(A^2 + AB + B^2)$$
$$= A^3 + A^2B + AB^2 - A^2B - AB^2 - B^3$$
$$= A^3 - B^3. \qquad \text{Combining like terms}$$

These products allow us to factor a sum or a difference of two cubes. Note how the location of the $+$ and $-$ signs changes.

N	N^3
0.1	0.001
0.2	0.008
1	1
2	8
3	27
4	64
5	125
6	216

> **FACTORING A SUM OR A DIFFERENCE OF TWO CUBES**
>
> $A^3 + B^3 = (A + B)(A^2 - AB + B^2);$
> $A^3 - B^3 = (A - B)(A^2 + AB + B^2)$

Remembering the list of cubes shown at left may prove helpful when factoring. Since 2 cubed is 8 and 3 cubed is 27, we say that 2 is the *cube root* of 8, that 3 is the cube root of 27, and so on.

EXAMPLE 1 Write an equivalent expression by factoring: $x^3 + 27$.

SOLUTION We first note that

$$x^3 + 27 = x^3 + 3^3. \qquad \text{This is a sum of cubes.}$$

Next, in one set of parentheses, we write the first cube root, x, plus the second cube root, 3:

$$(x + 3)(\qquad).$$

To get the other factor, we think of $x + 3$ and do the following:

Square the first term: x^2.
Multiply the terms and then change the sign: $-3x$.
Square the second term: 3^2, or 9.

$$(x + 3)(x^2 - 3x + 9). \qquad \text{This is factored completely.}$$

Check: $(x + 3)(x^2 - 3x + 9) = x^3 - 3x^2 + 9x + 3x^2 - 9x + 27$
$$= x^3 + 27. \qquad \text{Combining like terms}$$

Thus, $x^3 + 27 = (x + 3)(x^2 - 3x + 9)$.

1. Write an equivalent expression by factoring: $t^3 + 125$.

 YOUR TURN

EXAMPLE 2 Factor.

a) $125x^3 - y^3$

b) $m^6 + 64$

c) $128y^7 - 250x^6y$

d) $r^6 - s^6$

SOLUTION

a) We have

$$125x^3 - y^3 = (5x)^3 - y^3. \qquad \text{Recognizing this as a difference of cubes}$$

In one set of parentheses, we write the cube root of the first term, $5x$, minus the cube root of the second term, y:

$$(5x - y)(\quad). \qquad \text{This can be regarded as } 5x \text{ plus the cube root of } (-y)^3, \text{ since } -y^3 = (-y)^3.$$

To get the other factor, we think of $5x + y$ and do the following:

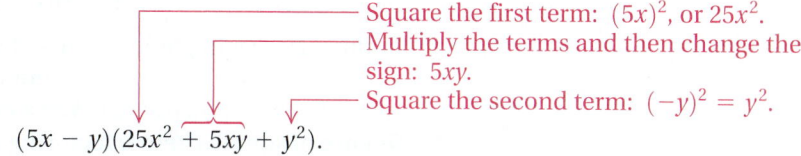

Square the first term: $(5x)^2$, or $25x^2$.
Multiply the terms and then change the sign: $5xy$.
Square the second term: $(-y)^2 = y^2$.

$$(5x - y)(25x^2 + 5xy + y^2).$$

Check:

$$(5x - y)(25x^2 + 5xy + y^2) = 125x^3 + 25x^2y + 5xy^2 - 25x^2y - 5xy^2 - y^3$$

$$= 125x^3 - y^3. \qquad \text{Combining like terms}$$

Thus, $125x^3 - y^3 = (5x - y)(25x^2 + 5xy + y^2)$.

b) We have

$$m^6 + 64 = (m^2)^3 + 4^3. \qquad \text{Rewriting as a sum of quantities cubed}$$

Next, we use the pattern for a sum of cubes:

$$A^3 + B^3 = (A + B)(A^2 - A \cdot B + B^2)$$

$$(m^2)^3 + 4^3 = (m^2 + 4)((m^2)^2 - m^2 \cdot 4 + 4^2) \qquad \text{Regarding } m^2 \text{ as } A \text{ and 4 as } B$$

$$= (m^2 + 4)(m^4 - 4m^2 + 16).$$

The check is left to the student. We have

$$m^6 + 64 = (m^2 + 4)(m^4 - 4m^2 + 16).$$

c) We have

$$128y^7 - 250x^6y = 2y(64y^6 - 125x^6) \qquad \text{Remember: } \textit{Always} \text{ look for a common factor.}$$

$$= 2y[(4y^2)^3 - (5x^2)^3]. \qquad \text{Rewriting as a difference of quantities cubed}$$

To factor $(4y^2)^3 - (5x^2)^3$, we use the pattern for a difference of cubes:

$$A^3 - B^3 = (A - B)(A^2 + A \cdot B + B^2)$$

$$(4y^2)^3 - (5x^2)^3 = (4y^2 - 5x^2)((4y^2)^2 + 4y^2 \cdot 5x^2 + (5x^2)^2)$$

$$= (4y^2 - 5x^2)(16y^4 + 20x^2y^2 + 25x^4).$$

The check is left to the student. We have

$$128y^7 - 250x^6y = 2y(4y^2 - 5x^2)(16y^4 + 20x^2y^2 + 25x^4).$$

Student Notes

If you think of $A^3 - B^3$ as $A^3 + (-B)^3$, you need remember only the pattern for factoring a sum of two cubes. Be sure to simplify your result if you do this.

d) We have

$$r^6 - s^6 = (r^3)^2 - (s^3)^2$$
$$= (r^3 + s^3)(r^3 - s^3) \quad \text{Factoring a difference of two } squares$$
$$= (r + s)(r^2 - rs + s^2)(r - s)(r^2 + rs + s^2). \quad \text{Factoring the sum and the difference of two cubes}$$

2. Factor: $y^7 - y$.

To check, read the steps in reverse order and inspect the multiplication.

YOUR TURN

In Example 2(d), suppose we first factored $r^6 - s^6$ as a difference of two cubes:

$$(r^2)^3 - (s^2)^3 = (r^2 - s^2)(r^4 + r^2s^2 + s^4)$$
$$= (r + s)(r - s)(r^4 + r^2s^2 + s^4).$$

In this case, we might have missed some factors; $r^4 + r^2s^2 + s^4$ can be factored as $(r^2 - rs + s^2)(r^2 + rs + s^2)$, but we probably would never have suspected that such a factorization exists. **Given a choice, it is generally better to factor as a difference of squares before factoring as a sum or a difference of cubes.**

USEFUL FACTORING FACTS

Sum of cubes: $A^3 + B^3 = (A + B)(A^2 - AB + B^2)$

Difference of cubes: $A^3 - B^3 = (A - B)(A^2 + AB + B^2)$

Difference of squares: $A^2 - B^2 = (A + B)(A - B)$

In general, a sum of two squares cannot be factored.

5.5 EXERCISE SET

FOR EXTRA HELP

MyMathLab® MathXL
PRACTICE   WATCH   READ   REVIEW

Vocabulary and Reading Check

Classify each binomial as either a sum of cubes, a difference of cubes, a difference of squares, or none of these.

1. $x^3 - 1$

2. $8 + t^3$

3. $9x^4 - 25$

4. $9x^2 + 25$

5. $1000t^3 + 1$

6. $x^3y^3 - 27z^3$

7. $25x^2 + 8x$

8. $100y^8 - 25x^4$

9. $s^{21} - t^{15}$

10. $14x^3 - 2x$

Factoring Sums or Differences of Cubes

Factor completely.

11. $x^3 - 64$

12. $t^3 - 27$

13. $z^3 + 1$

14. $x^3 + 8$

15. $t^3 - 1000$

16. $m^3 + 125$

17. $27x^3 + 1$

18. $8a^3 + 1$

19. $64 - 125x^3$

20. $27 - 8t^3$

21. $x^3 - y^3$

22. $y^3 - z^3$

23. $a^3 + \frac{1}{8}$

24. $x^3 + \frac{1}{27}$

25. $8t^3 - 8$

26. $2y^3 - 128$

27. $54x^3 + 2$

28. $8a^3 + 1000$

29. $rs^4 + 64rs$

30. $ab^5 + 1000ab^2$

31. $5x^3 - 40z^3$

32. $2y^3 - 54z^3$

33. $y^3 - \frac{1}{1000}$

34. $x^3 - \frac{1}{8}$

35. $x^3 + 0.001$

36. $y^3 + 0.125$

37. $64x^6 - 8t^6$

38. $125c^6 - 8d^6$

39. $54y^4 - 128y$

40. $3z^5 - 3z^2$

41. $z^6 - 1$

42. $t^6 + 1$

43. $t^6 + 64y^6$

44. $p^6 - q^6$

45. $x^{12} - y^3z^{12}$

46. $a^9 + b^{12}c^{15}$

47. How could you use factoring to convince someone that $x^3 + y^3 \neq (x + y)^3$?

48. Explain how to use the pattern for factoring $A^3 + B^3$ to factor $A^3 - B^3$.

Skill Review

Solve.

49. A serving of crackers contains 120 mg of sodium. According to nutrition information on the box, this is 5% of the daily recommended value of sodium. What is the daily recommended value of sodium?   [2.4]

50. Blake stopped for gasoline twice on his drive home from college. He bought a total of 19.2 gal of gasoline. If he bought twice as much on his first stop as on his second stop, how many gallons of gasoline did he buy on his second stop?   [2.5]

51. A tax-exempt charity received a bill of $551.20 for kitchen supplies. The bill incorrectly included sales tax of 6%. How much does the charity actually owe?   [2.5]

52. Katie paid $34.50 for a jacket that was on sale at 25% off. What was the original price of the jacket?   [2.5]

Synthesis

53. Explain how the geometric model below can be used to verify the formula for factoring $a^3 - b^3$.

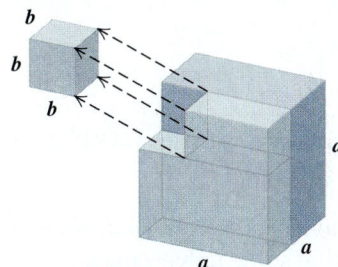

54. Explain how someone could construct a binomial that is both a difference of two cubes and a difference of two squares.

Factor.

55. $x^{6a} - y^{3b}$

56. $2x^{3a} + 16y^{3b}$

Aha! 57. $(x + 5)^3 + (x - 5)^3$

58. $\frac{1}{16}x^{3a} + \frac{1}{2}y^{6a}z^{9b}$

59. $5x^3y^6 - \frac{5}{8}$

60. $x^3 - (x + y)^3$

61. $x^{6a} - (x^{2a} + 1)^3$

62. $(x^{2a} - 1)^3 - x^{6a}$

63. $t^4 - 8t^3 - t + 8$

◆ YOUR TURN ANSWERS: SECTION 5.5

1. $(t + 5)(t^2 - 5t + 25)$
2. $y(y + 1)(y^2 - y + 1)(y - 1)(y^2 + y + 1)$

QUICK QUIZ: SECTIONS 5.1–5.5

Factor.

1. $6x^2p^3 + xp^2 - 3x^3p^4$ [5.1]

2. $x^2 - 11x - 12$ [5.2]

3. $4a^3 - 19a^2 - 5a$ [5.3]

4. $p^2 - w^2$ [5.4]

5. $p^3 - w^3$ [5.5]

PREPARE TO MOVE ON

Complete each statement.

1. When factoring, always check first for a(n) _____ factor.   [5.3]

2. To factor a trinomial of the form $ax^2 + bx + c$, we can use FOIL or the _____ method.   [5.4]

3. The formula for factoring a difference of squares is $A^2 - B^2 =$ _____.   [5.4]

4. A formula for factoring a perfect-square trinomial is $A^2 + 2AB + B^2 =$ _____.   [5.4]

5. The formula for factoring a sum of cubes is $A^3 + B^3 =$ _____.   [5.5]

5.6   Factoring: A General Strategy

Choosing the Right Method

Thus far, each section in this chapter has examined one or two different methods for factoring polynomials. In practice, when the need for factoring a polynomial arises, we must decide on our own which method to use. The following guidelines provide an approach for this.

TO FACTOR A POLYNOMIAL

  A. Always look for a common factor. If there is one, factor out the largest common factor. Be sure to include it in your final answer.

  B. Then look at the number of terms.

    Two terms: If you have a difference of squares, factor accordingly:

$$A^2 - B^2 = (A + B)(A - B).$$

    If you have a sum or a difference of cubes, factor accordingly:

$$A^3 + B^3 = (A + B)(A^2 - AB + B^2);$$
$$A^3 - B^3 = (A - B)(A^2 + AB + B^2).$$

    Three terms: If the trinomial is a perfect-square trinomial, factor accordingly:

$$A^2 + 2AB + B^2 = (A + B)^2;$$
$$A^2 - 2AB + B^2 = (A - B)^2.$$

    If it is not a perfect-square trinomial, try using FOIL or grouping.

    Four terms: Try factoring by grouping.

  C. Always *factor completely*. When a factor can itself be factored, be sure to factor it. Remember that some polynomials, like $x^2 + 9$, are prime.

  D. Check.

We can always check a factorization by multiplying. Another way to check a factorization is to evaluate both the original polynomial and the factorization for one or more choices of the variable(s). Since the polynomial and its factorization are equivalent expressions, they have the same value for any replacements of the variable(s). This check becomes more certain if it is done for more than one replacement. In fact, for a polynomial in one variable of degree n, a check of $n + 1$ replacements is sufficient to show that the factorization is correct.

Student Notes

Quickly checking both the leading term and the constant term of a trinomial to see if they are squares can save you time. If they aren't both squares and there is no common factor, the trinomial can't possibly be a perfect-square trinomial.

Choosing the Right Method

EXAMPLE 1   Factor: $x^2 - 20x + 100$.

SOLUTION

A. We look first for a common factor. There is none.

B. This polynomial is a perfect-square trinomial. We factor it accordingly:

$$x^2 - 20x + 100 = x^2 - 2 \cdot x \cdot 10 + 10^2 \qquad \text{Try to do this step mentally.}$$
$$= (x - 10)^2.$$

C. Nothing can be factored further, so we have factored completely.

D. We check by multiplying: $(x - 10)(x - 10) = x^2 - 20x + 100$.

The factorization is $(x - 10)(x - 10)$, or $(x - 10)^2$.

1. Factor: $a^2 + 2a + 1$.

YOUR TURN

Study Skills

Keep Your Focus

When you are studying with someone else, it can be very tempting to talk about topics not related to what you need to study. If you see that this may happen, explain to your partner(s) that you enjoy the conversation, but would enjoy it more later—after the work has been completed.

EXAMPLE 2 Factor: $2x^3 + 10x^2 + x + 5$.

SOLUTION

A. We look for a common factor. There is none.

B. Because there are four terms, we try factoring by grouping:

$2x^3 + 10x^2 + x + 5$

$= (2x^3 + 10x^2) + (x + 5)$ Separating into two binomials

$= 2x^2(x + 5) + 1(x + 5)$ Factoring out the largest common factor from each binomial. The 1 serves as an aid.

$= (x + 5)(2x^2 + 1)$.   Factoring out the common factor, $x + 5$

C. Nothing can be factored further, so we have factored completely.

D. As a partial check, we evaluate the original polynomial and the factorization for $x = 2$:

$2x^3 + 10x^2 + x + 5 = 2(2)^3 + 10(2)^2 + 2 + 5 = 2 \cdot 8 + 10 \cdot 4 + 2 + 5 = 63$;

$(x + 5)(2x^2 + 1) = (2 + 5)(2(2)^2 + 1) = 7 \cdot 9 = 63$.

The values are the same for $x = 2$. If we were to evaluate both expressions for three additional values of x, the check would be complete.

The factorization is $(x + 5)(2x^2 + 1)$.

2. Factor: $5x^3 + 15x - x^2 - 3$.

YOUR TURN

EXAMPLE 3 Factor: $-10n^5 - 80n^2$.

SOLUTION

A. We note that there is a common factor, $-10n^2$:

$-10n^5 - 80n^2 = -10n^2(n^3 + 8)$.

B. The factor $n^3 + 8$ has two terms and is a sum of cubes. We factor:

$-10n^5 - 80n^2 = -10n^2(n^3 + 8)$

$= -10n^2(n + 2)(n^2 - 2n + 4)$.

C. No factor can be factored further, so we have factored completely.

D. We check by multiplying:

$-10n^2(n + 2)(n^2 - 2n + 4) = -10n^2(n^3 + 8)$

$= -10n^5 - 80n^2$.

The factorization is $-10n^2(n + 2)(n^2 - 2n + 4)$.

3. Factor: $2000x^3 - 2$.

YOUR TURN

EXAMPLE 4   Factor: $6x^2y^4 - 21x^3y^5 + 3x^2y^6$.

SOLUTION

A. We first factor out the largest common factor, $3x^2y^4$:

$$6x^2y^4 - 21x^3y^5 + 3x^2y^6 = 3x^2y^4(2 - 7xy + y^2).$$

B. The constant term in $2 - 7xy + y^2$ is not a square, so we do not have a perfect-square trinomial. Note that x appears only in $-7xy$. The product of a form like $(1 - y)(2 - y)$ has no x in the middle term. Thus, $2 - 7xy + y^2$ cannot be factored.

C. Nothing can be factored further, so we have factored completely.

D. We check by multiplying:

$$3x^2y^4(2 - 7xy + y^2) = 6x^2y^4 - 21x^3y^5 + 3x^2y^6.$$

4. Factor:

$$12p^3w - 18p^2w^2 + 6p^2w.$$

The factorization is $3x^2y^4(2 - 7xy + y^2)$.

YOUR TURN

EXAMPLE 5   Factor: $98x^3 + 280x^2 + 200x$.

SOLUTION

A. We first factor out the largest common factor, $2x$:

$$98x^3 + 280x^2 + 200x = 2x(49x^2 + 140x + 100).$$

B. The trinomial in the parentheses is a perfect-square trinomial: $49x^2 = (7x)^2$, $100 = (10)^2$, and $140x = 2 \cdot 7x \cdot 10$. We factor it accordingly:

$$98x^3 + 280x^2 + 200x = 2x(49x^2 + 140x + 100)$$
$$= 2x(7x + 10)^2.$$

C. Nothing can be factored further, so we have factored completely.

D. As a partial check, we evaluate the original polynomial and the factorization for $x = 1$:

$$98x^3 + 280x^2 + 200x = 98(1)^3 + 280(1)^2 + 200(1) = 98 + 280 + 200 = 578;$$
$$2x(7x + 10)^2 = 2(1)[(7(1) + 10)^2] = 2(17)^2 = 2(289) = 578.$$

The values are the same for $x = 1$. If we were to evaluate both expressions for three additional values of x, the check would be complete.

5. Factor:

$$75a^4 + 30a^3 + 3a^2.$$

The factorization is $2x(7x + 10)^2$.

YOUR TURN

EXAMPLE 6   Factor: $-10m^2 - mn + 3n^2$.

SOLUTION

A. We first look for a common factor. Since the first term is negative, we factor out -1:

$$-1(10m^2 + mn - 3n^2).$$

B. There are three terms in the parentheses. Since none is a square, the trinomial is not a perfect square. We factor using FOIL. (We could also use grouping.) There are two possibilities for the first terms in the binomial factors:

$$(2m + \quad)(5m + \quad) \quad \text{or} \quad (m + \quad)(10m + \quad).$$

The possibilities for the second terms in the binomials are

$$n, -3n \qquad -3n, \quad n$$
$$\text{and}$$
$$-n, \quad 3n \qquad 3n, -n.$$

We check the possible combinations and find that the factorization of $10m^2 + mn - 3n^2$ is $(2m - n)(5m + 3n)$. Thus,

$$-10m^2 - mn + 3n^2 = -1(2m - n)(5m + 3n).$$

C. Nothing can be factored further, so we have factored completely.

D. We check by multiplying:

$$-1(2m - n)(5m + 3n) = -1(10m^2 + mn - 3n^2)$$
$$= -10m^2 - mn + 3n^2.$$

The factorization is $-1(2m - n)(5m + 3n)$, or $-(2m - n)(5m + 3n)$.

6. Factor: $-9x^2 - 37xy - 4y^2$.

YOUR TURN

EXAMPLE 7   Factor: $x^2y^2 + 7xy + 12$.

SOLUTION

A. We first look for a common factor. There is none.

B. Since only one term is a square, we do not have a perfect-square trinomial. We use trial and error, treating the product xy as a single variable:

$$(xy + \quad)(xy + \quad).$$

We factor the last term, 12. All the signs are positive, so we consider only positive factors. Possibilities are 1, 12 and 2, 6 and 3, 4. The pair 3, 4 gives a sum of 7 for the coefficient of the middle term. Thus,

$$x^2y^2 + 7xy + 12 = (xy + 3)(xy + 4).$$

C. Nothing can be factored further, so we have factored completely.

D. *Check:* $(xy + 3)(xy + 4) = x^2y^2 + 7xy + 12$.

The factorization is $(xy + 3)(xy + 4)$.

7. Factor: $c^2t^2 - 9ct + 20$.

YOUR TURN

Compare the variables appearing in Example 6 with those in Example 7. Note that if the leading term contains one variable and a different variable is in the last term, as in Example 6, each factor contains two variable terms. When two variables appear in the leading term and no variables appear in the last term, as in Example 7, each factor contains just one variable term.

EXAMPLE 8   Factor: $a^4 - 16b^4$.

SOLUTION

A. We look first for a common factor. There is none.

B. There are two terms. Since $a^4 = (a^2)^2$ and $16b^4 = (4b^2)^2$, we see that we have a difference of squares. Thus,

$$a^4 - 16b^4 = (a^2 + 4b^2)(a^2 - 4b^2).$$

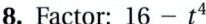 **Chapter Resource:**
Collaborative Activity (Matching Factorizations), p. 366

C. The factor $(a^2 - 4b^2)$ is itself a difference of squares. Thus,

$$a^4 - 16b^4 = (a^2 + 4b^2)(a + 2b)(a - 2b). \qquad \text{Factoring } a^2 - 4b^2$$

D. *Check:* $(a^2 + 4b^2)(a + 2b)(a - 2b) = (a^2 + 4b^2)(a^2 - 4b^2)$
$$= a^4 - 16b^4.$$

The factorization is $(a^2 + 4b^2)(a + 2b)(a - 2b)$.

8. Factor: $16 - t^4$.

YOUR TURN

5.6 EXERCISE SET

FOR EXTRA HELP

MyMathLab® Math XL
PRACTICE   WATCH   READ   REVIEW

↳ Vocabulary and Reading Check

Choose the term from the following list that best completes each sentence. Not every term will be used.

common factor   multiplying
dividing   perfect-square trinomial
grouping   prime polynomial

1. As a first step when factoring polynomials, always check for a(n) _____.

2. When factoring a trinomial, if two terms are not squares, it cannot be a(n) _____.

3. If a polynomial has four terms and no common factor, it may be possible to factor by _____.

4. It is always possible to check a factorization by _____.

Factoring Polynomials

Factor completely. If a polynomial is prime, state this.

5. $5a^2 - 125$

6. $10c^2 - 810$

7. $y^2 + 49 - 14y$

8. $a^2 + 25 + 10a$

9. $3t^2 + 16t + 21$

10. $8t^2 + 31t - 4$

11. $x^3 + 18x^2 + 81x$

12. $x^3 - 24x^2 + 144x$

13. $x^3 - 5x^2 - 25x + 125$

14. $x^3 + 3x^2 - 4x - 12$

15. $27t^3 - 3t$

16. $98t^2 - 18$

17. $9x^3 + 12x^2 - 45x$

18. $20x^3 - 4x^2 - 72x$

19. $t^2 + 25$

20. $4x^2 + 20x - 144$

21. $6y^2 + 18y - 240$

22. $4n^2 + 81$

23. $-2a^6 + 8a^5 - 8a^4$

24. $-x^5 - 14x^4 - 49x^3$

25. $5x^5 - 80x$

26. $4x^4 - 64$

27. $t^4 - 9$

28. $9 + t^8$

29. $-x^6 + 2x^5 - 7x^4$

30. $-x^5 + 4x^4 - 3x^3$

31. $x^3 - y^3$

32. $8t^3 + 1$

33. $ax^2 + ay^2$

34. $12n^2 + 24n^3$

35. $2\pi rh + 2\pi r^2$

36. $4\pi r^2 + 2\pi r$

Aha! 37. $(a + b)5a + (a + b)3b$

38. $5c(a^3 + b) - (a^3 + b)$

39. $x^2 + x + xy + y$

40. $n^2 + 2n + np + 2p$

41. $160a^2m^4 - 10a^2$

42. $32t^4 - 162y^4$

43. $a^2 - 2a - ay + 2y$

44. $2x^2 - 4x + xz - 2z$

45. $3x^2 + 13xy - 10y^2$

46. $-x^2 - y^2 - 2xy$

47. $8m^3n - 32m^2n^2 + 24mn$

48. $a^2 - 7a - 6$

49. $\frac{9}{16} - y^2$

50. $\frac{1}{36}a^2 - m^2$

51. $4b^2 + a^2 - 4ab$

52. $7p^4 - 7q^4$

53. $16x^2 + 24xy + 9y^2$

54. $6a^2b^3 + 12a^3b^2 - 3a^4b^2$

55. $m^2 - 5m + 8$

56. $25z^2 + 10zy + y^2$

57. $10x^2 - 11x - 6$

58. $24x^2 - 47x - 21$

59. $a^4b^4 - 16$

60. $a^5 - 4a^4b - 5a^3b^2$

61. $80cd^2 - 36c^2d + 4c^3$

62. $2p^2 + pq + q^2$

63. $64t^6 - 1$

64. $m^6 - 1$

65. $-12 - x^2y^2 - 8xy$

66. $m^2n^2 - 4mn - 32$

67. $14t + 8t^2 - 15$

68. $8y - 15 + 12y^2$

69. $5p^2t^2 + 25pt - 30$

70. $a^4b^3 + 2a^3b^2 - 15a^2b$

71. $54a^4 + 16ab^3$

72. $54x^3y - 250y^4$

73. $x^6 + x^5y - 2x^4y^2$

74. $2s^6t^2 + 10s^3t^3 + 12t^4$

75. $36a^2 - 15a + \frac{25}{16}$

76. $a^2 + 2a^2bc + a^2b^2c^2$

77. $\frac{1}{81}x^2 - \frac{8}{27}x + \frac{16}{9}$

78. $\frac{1}{4}a^2 + \frac{1}{3}ab + \frac{1}{9}b^2$

79. $1 - 16x^{12}y^{12}$

80. $b^4a - 81a^5$

81. $4a^2b^2 + 12ab + 9$

82. $9c^2 + 6cd + d^2$

83. $z^4 + 6z^3 - 6z^2 - 36z$

84. $t^5 - 2t^4 + 5t^3 - 10t^2$

85. $x^3 + 5x^2 - x - 5$

86. $x^3 + 3x^2 - 16x - 48$

87. Kelly factored $16 - 8x + x^2$ as $(x - 4)^2$, while Tony factored it as $(4 - x)^2$. Are they both correct? Why or why not?

88. Describe in your own words or draw a diagram representing a strategy for factoring polynomials.

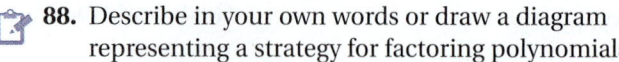

Skill Review

Solve.

89. In 2011, the 1.3 billion people of China owned 900 million mobile phones. Assuming that no one owned more than one mobile phone, what percent of the Chinese population owned a mobile phone? [2.4]

Sources: "China's mobile subscribers rise 1.1 pct to 916.53 mln in July," Reuters, 8/23/2011; *The World Factbook*

90. The number of billionaires in China increased from 120 at the end of 2009 to 189 at the end of 2010. What was the monthly average rate of increase? [3.4]

Source: sify.com

91. In 2011, U.S. shoppers spent $27.4 billion on gifts for Mother's Day and for Father's Day combined. They spent $5.2 billion more for Mother's Day than for Father's Day. How much did shoppers spend for each holiday? [2.5]

Source: National Retail Federation

92. Shelby plans to spend no more than $50 on roses for a Mother's Day gift. If the shipping for the flowers is $15 and each rose costs $6.50, how many roses can he purchase? [2.7]

Synthesis

93. There are third-degree polynomials in x that we are not yet able to factor, despite the fact that they are not prime. Explain how such a polynomial could be created.

94. Describe a method that could be used to find a binomial of degree 16 that can be expressed as the product of prime binomial factors.

Factor.

95. $36 - 12x + x^2 - a^2$

96. $100 + 20t + t^2 - 4y^2$

97. $-(x^5 + 7x^3 - 18x)$

98. $18 + a^3 - 9a - 2a^2$

99. $-x^4 + 7x^2 + 18$

100. $-3a^4 + 15a^2 - 12$

Aha! **101.** $y^2(y + 1) - 4y(y + 1) - 21(y + 1)$

102. $y^2(y - 1) - 2y(y - 1) + (y - 1)$

103. $(y + 4)^2 + 2x(y + 4) + x^2$

104. $6(x - 1)^2 + 7y(x - 1) - 3y^2$

105. $2(a + 3)^4 - (a + 3)^3(b - 2) - (a + 3)^2(b - 2)^2$

106. $5(t - 1)^5 - 6(t - 1)^4(s - 1) + (t - 1)^3(s - 1)^2$

107. $49x^4 + 14x^2 + 1 - 25x^6$

↪ YOUR TURN ANSWERS: SECTION 5.6

1. $(a + 1)^2$ **2.** $(5x - 1)(x^2 + 3)$
3. $2(10x - 1)(100x^2 + 10x + 1)$ **4.** $6p^2w(2p - 3w + 1)$
5. $3a^2(5a + 1)^2$ **6.** $-1(9x + y)(x + 4y)$, or
$-(9x + y)(x + 4y)$ **7.** $(ct - 5)(ct - 4)$
8. $(4 + t^2)(2 + t)(2 - t)$

QUICK QUIZ: SECTIONS 5.1–5.6

Factor.

1. $10m^2 - 90$ [5.4]   **2.** $a + a^2 - 6$ [5.2]

3. $12x^3 - 6x^2 - 10x + 5$ [5.1]

4. $8m^2n^2 - 16mn + 8$ [5.4]

5. $6c^3 - 5c^2 - 6c$ [5.3]

PREPARE TO MOVE ON

Solve. [2.2]

1. $8x - 9 = 0$ **2.** $2x + 7 = 0$

3. $3 - x = 0$ **4.** $3x + 5 = 0$

5. $4x - 1 = 0$ **6.** $22 - 2x = 0$

5.7   Solving Quadratic Equations by Factoring

The Principle of Zero Products   ▪   Factoring to Solve Equations

When we factor a polynomial, we are forming an *equivalent expression*. We now use our factoring skills to *solve equations*. We already know how to solve linear equations like $x + 2 = 7$ and $2x = 9$. The equations we will learn to solve in this section contain a variable raised to a power greater than 1 and will usually have more than one solution.

Second-degree equations like $4t^2 - 7 = 2$ and $x^2 + 6x + 5 = 0$ are called **quadratic equations**.

> ### QUADRATIC EQUATION
>
> A *quadratic equation* is an equation equivalent to one of the form
>
> $$ax^2 + bx + c = 0,$$
>
> where a, b, and c are constants, with $a \neq 0$.

In order to solve quadratic equations, we need to develop a new principle.

The Principle of Zero Products

Suppose we are told that the product of two numbers is 6. From this alone, it is impossible to know the value of either number—the product could be $2 \cdot 3$, $6 \cdot 1$, $12 \cdot \frac{1}{2}$, and so on. However, if we are told that the product of two numbers is 0, we know that at least one of the two factors must itself be 0. For example, if $(x + 3)(x - 2) = 0$, we can conclude that either $x + 3$ is 0 or $x - 2$ is 0.

> ### THE PRINCIPLE OF ZERO PRODUCTS
>
> An equation $AB = 0$ is true if and only if $A = 0$ or $B = 0$, or both. (A product is 0 if and only if at least one factor is 0.)

EXAMPLE 1 Solve: $(x + 3)(x - 2) = 0$.

SOLUTION We look for all values of x that make the equation true. The equation tells us that the product of $x + 3$ and $x - 2$ is 0. In order for any product to be 0, at least one factor must be 0. Thus we look for any value of x for which $x + 3 = 0$, as well as any value of x for which $x - 2 = 0$, that is, either

$$x + 3 = 0 \quad or \quad x - 2 = 0. \qquad \text{Using the principle of zero products. There are two equations to solve.}$$

We solve each equation:

$$\begin{aligned} x + 3 &= 0 \quad or \quad x - 2 = 0 \\ x &= -3 \quad or \qquad x = 2. \end{aligned}$$

Both -3 and 2 should be checked in the original equation.

Check: For -3:

$$\begin{array}{c|c} (x + 3)(x - 2) = 0 \\ \hline (-3 + 3)(-3 - 2) & 0 \\ \end{array}$$

The factor $x + 3$ is 0 when $x = -3$. $\rightarrow 0(-5)$

$$0 \overset{?}{=} 0 \quad \text{TRUE}$$

For 2:

$$\begin{array}{c|c} (x + 3)(x - 2) = 0 \\ \hline (2 + 3)(2 - 2) & 0 \\ \end{array}$$

The factor $x - 2$ is 0 when $x = 2$. $\leftarrow 5(0)$

$$0 \overset{?}{=} 0 \quad \text{TRUE}$$

The solutions are -3 and 2.

YOUR TURN

1. Solve: $(x - 1)(x + 5) = 0$.

When we are using the principle of zero products, the word "or" is meant to emphasize that any one of the factors could be the one that represents 0.

EXAMPLE 2 Solve: $3(5x + 1)(x - 7) = 0$.

SOLUTION The factors in this equation are 3, $5x + 1$, and $x - 7$. Since the factor 3 is constant, the only way in which $3(5x + 1)(x - 7)$ can be 0 is for one of the other factors to be 0, that is,

$5x + 1 = 0$ *or* $x - 7 = 0$ Using the principle of zero products

$5x = -1$ *or* $x = 7$ Solving the two equations separately

$x = -\frac{1}{5}$ *or* $x = 7$. $5x + 1 = 0$ when $x = -\frac{1}{5}$; $x - 7 = 0$ when $x = 7$

Check: For $-\frac{1}{5}$:

$$\frac{3(5x + 1)(x - 7) = 0}{3\left(5\left(-\frac{1}{5}\right) + 1\right)\left(-\frac{1}{5} - 7\right) \mid 0}$$
$$3(-1 + 1)\left(-7\frac{1}{5}\right)$$
$$3(0)\left(-7\frac{1}{5}\right)$$
$$0 \overset{?}{=} 0 \quad \text{TRUE}$$

For 7:

$$\frac{3(5x + 1)(x - 7) = 0}{3(5(7) + 1)(7 - 7) \mid 0}$$
$$3(35 + 1)0$$
$$0 \overset{?}{=} 0 \quad \text{TRUE}$$

2. Solve:

   $6(x + 2)(3x + 5) = 0$.

The solutions are $-\frac{1}{5}$ and 7.

YOUR TURN

The constant factor 3 in Example 2 is never 0 and is not a solution of the equation. However, a variable factor such as x or t *can* equal 0, and must be considered when using the principle of zero products.

EXAMPLE 3 Solve: $7t(t - 5) = 0$.

SOLUTION We have

$7 \cdot t(t - 5) = 0$ The factors are 7, t, and $t - 5$.

$t = 0$ *or* $t - 5 = 0$ Using the principle of zero products

$t = 0$ *or* $t = 5$.   Solving. Note that the constant factor, 7, is never 0.

3. Solve: $10a(a + 9) = 0$.

The solutions are 0 and 5. The check is left to the student.

YOUR TURN

Factoring to Solve Equations

By factoring and using the principle of zero products, we can now solve a variety of quadratic equations.

EXAMPLE 4 Solve: $x^2 + 5x + 6 = 0$.

SOLUTION We first factor the polynomial, and then use the principle of zero products:

$x^2 + 5x + 6 = 0$

$(x + 2)(x + 3) = 0$ Factoring

$x + 2 = 0$ *or* $x + 3 = 0$ Using the principle of zero products

$x = -2$ *or* $x = -3$.

Check: For -2:

$$\begin{array}{c|c}
x^2 + 5x + 6 = 0 \\
\hline
(-2)^2 + 5(-2) + 6 & 0 \\
4 - 10 + 6 & \\
-6 + 6 & \\
& 0 \stackrel{?}{=} 0 \quad \text{TRUE}
\end{array}$$

For -3:

$$\begin{array}{c|c}
x^2 + 5x + 6 = 0 \\
\hline
(-3)^2 + 5(-3) + 6 & 0 \\
9 - 15 + 6 & \\
-6 + 6 & \\
& 0 \stackrel{?}{=} 0 \quad \text{TRUE}
\end{array}$$

The solutions are -2 and -3.

4. Solve: $t^2 - 8t + 15 = 0$.

↩ YOUR TURN

Student Notes

Checking for a common factor is an important step that is often overlooked. In Example 5, the equation must be factored. If we "divide both sides by x," we will not find the solution 0.

EXAMPLE 5   Solve: $x^2 + 7x = 0$.

SOLUTION   Although there is no constant term, because of the x^2-term, the equation is still quadratic. We factor and use the principle of zero products:

$$\begin{aligned}
x^2 + 7x &= 0 \\
x(x + 7) &= 0 \qquad && \text{Factoring out the largest common factor, } x \\
x = 0 \quad &\textit{or} \quad x + 7 = 0 && \text{Using the principle of zero products} \\
x = 0 \quad &\textit{or} \qquad\quad x = -7.
\end{aligned}$$

The solutions are 0 and -7. The check is left to the student.

5. Solve: $n^2 - 8n = 0$.

↩ YOUR TURN

> **CAUTION!**   We *must* have 0 on one side of the equation before the principle of zero products can be used. Get all nonzero terms on one side and 0 on the other.

EXAMPLE 6   Solve: **(a)** $x^2 - 8x = -16$; **(b)** $4t^2 = 25$.

SOLUTION

a)   We first add 16 to get 0 on one side:

$$\begin{aligned}
x^2 - 8x &= -16 \\
x^2 - 8x + 16 &= 0 && \text{Adding 16 to both sides to get 0 on one side} \\
(x - 4)(x - 4) &= 0 && \text{Factoring} \\
x - 4 = 0 \quad &\textit{or} \quad x - 4 = 0 && \text{Using the principle of zero products} \\
x = 4 \quad &\textit{or} \qquad\quad x = 4.
\end{aligned}$$

There is only one solution, 4. The check is left to the student.

b)   We have

Get 0 on one side.

Factor.

Use the principle of zero products.

Solve each equation separately.

$$\begin{aligned}
4t^2 &= 25 \\
4t^2 - 25 &= 0 && \text{Subtracting 25 from both sides to get 0 on one side} \\
(2t - 5)(2t + 5) &= 0 && \text{Factoring a difference of squares} \\
2t - 5 = 0 \quad &\textit{or} \quad 2t + 5 = 0 \\
2t = 5 \quad &\textit{or} \qquad 2t = -5 && \text{Solving the two equations} \\
t = \tfrac{5}{2} \quad &\textit{or} \qquad\; t = -\tfrac{5}{2}. && \text{separately}
\end{aligned}$$

6. Solve: $a^2 = -10a - 25$.

↩ YOUR TURN

The solutions are $\frac{5}{2}$ and $-\frac{5}{2}$. The check is left to the student.

EXAMPLE 7   Solve: $(x + 3)(2x - 1) = 9$.

SOLUTION   Be careful with an equation like this! Since we need 0 on one side, we first multiply the product on the left and then subtract 9 from both sides.

$(x + 3)(2x - 1) = 9$	This is not a product equal to 0.
$2x^2 + 5x - 3 = 9$	Multiplying on the left
$2x^2 + 5x - 3 - 9 = 9 - 9$	Subtracting 9 from both sides to get 0 on one side
$2x^2 + 5x - 12 = 0$	Combining like terms
$(2x - 3)(x + 4) = 0$	Factoring. Now we have a product equal to 0.
$2x - 3 = 0$ $\ \ or\ \ $ $x + 4 = 0$	Using the principle of zero products
$2x = 3$ $\ \ or\ \ $ $x = -4$	
$x = \frac{3}{2}$ $\ \ or\ \ $ $x = -4$	

Check: For $\frac{3}{2}$:

$$
\begin{array}{c|c}
(x + 3)(2x - 1) = 9 & \\
\hline
\left(\frac{3}{2} + 3\right)\left(2 \cdot \frac{3}{2} - 1\right) & 9 \\
\left(\frac{9}{2}\right)(2) & \\
9 \overset{?}{=} 9 & \text{TRUE}
\end{array}
$$

For -4:

$$
\begin{array}{c|c}
(x + 3)(2x - 1) = 9 & \\
\hline
(-4 + 3)(2(-4) - 1) & 9 \\
(-1)(-9) & \\
9 \overset{?}{=} 9 & \text{TRUE}
\end{array}
$$

The solutions are $\frac{3}{2}$ and -4.

7. Solve:

$(3x + 4)(x - 1) = -2$.

YOUR TURN

EXPLORING 🔍 THE CONCEPT

Since an *x*-intercept is on the *x*-axis, the *y*-coordinate of an *x*-intercept will be 0. In order to find any *x*-intercepts of the graph of an equation, we can replace *y* with 0 and solve for *x*.

Graphs of quadratic equations of the form $y = ax^2 + bx + c$ are shaped as shown below. Each *x*-intercept represents a solution of $ax^2 + bx + c = 0$. For example, since 3 is a solution of $x^2 - 2x - 3 = 0$, then $(3, 0)$ is an *x*-intercept of the graph of $y = x^2 - 2x - 3$.

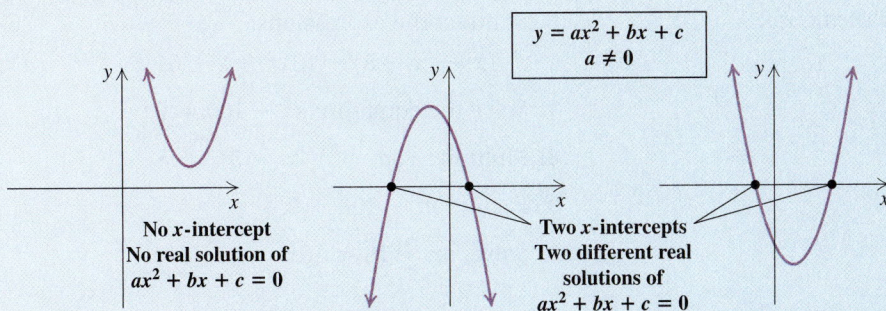

$$y = ax^2 + bx + c$$
$$a \neq 0$$

No *x*-intercept
No real solution of
$ax^2 + bx + c = 0$

Two *x*-intercepts
Two different real
solutions of
$ax^2 + bx + c = 0$

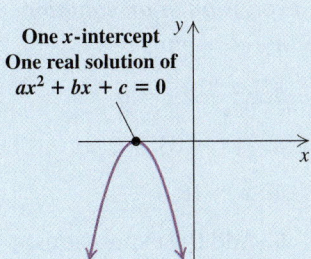

One *x*-intercept
One real solution of
$ax^2 + bx + c = 0$

1. The *x*-intercepts of the graph of $y = x^2 + x - 6$ are shown on the graph at right. What are the solutions of $x^2 + x - 6 = 0$?

2. The solutions of $x^2 + 4x - 5 = 0$ are -5 and 1. What are the *x*-intercepts of the graph of $y = x^2 + 4x - 5$?

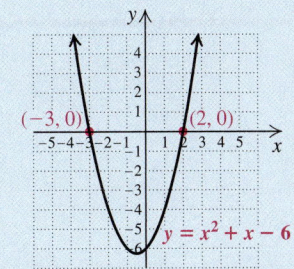

ANSWERS

1. $-3, 2$ **2.** $(-5, 0), (1, 0)$

EXAMPLE 8   Find the *x*-intercepts of the graph of the equation shown. (The grid is intentionally not included.)

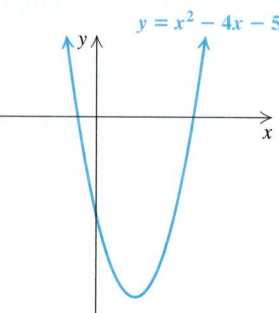

$$y = x^2 - 4x - 5$$

8. Find the *x*-intercepts of the graph of the equation shown.

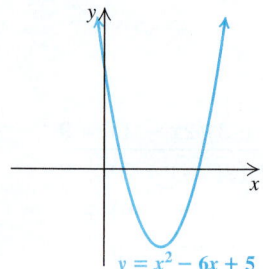

$$y = x^2 - 6x + 5$$

SOLUTION   To find the *x*-intercepts, we let $y = 0$ and solve for *x*:

$$0 = x^2 - 4x - 5 \qquad \text{Substituting 0 for } y$$
$$0 = (x - 5)(x + 1) \qquad \text{Factoring}$$
$$x - 5 = 0 \quad or \quad x + 1 = 0 \qquad \text{Using the principle of zero products}$$
$$x = 5 \quad or \qquad x = -1. \qquad \text{Solving for } x$$

The *x*-intercepts are $(5, 0)$ and $(-1, 0)$. The check is left to the student.

YOUR TURN

CONNECTING 🔗 THE CONCEPTS

Recall that an *equation* is a statement that two *expressions* are equal. When we simplify expressions, combine expressions, and form equivalent expressions, each result is an expression. When we are asked to solve an equation, the result is one or more numbers. Remember to read the directions to an exercise carefully so you do not attempt to "solve" an expression.

EXERCISES

For Exercises 1–4, tell whether each is an expression or an equation.

1. $x^2 - 25$

2. $x^2 - 25 = 0$

3. $x(x + 3) - 2(2x - 7) - (x - 5)$

4. $x = 10$

5. Add the expressions:
$$(2x^3 - 5x + 1) + (x^2 - 3x - 1).$$

6. Subtract the expressions:
$$(x^2 - x - 5) - (3x^2 - x + 6).$$

7. Solve the equation: $t^2 - 100 = 0$.

8. Multiply: $(3a - 2)(2a - 5)$.

9. Factor: $n^2 - 10n + 9$.

10. Solve: $x^2 + 16 = 10x$.

Technology Connection

A graphing calculator allows us to solve polynomial equations even when an equation cannot be solved by factoring. For example, to solve $x^2 - 3x - 5 = 0$, we can let $y_1 = x^2 - 3x - 5$ and $y_2 = 0$. Selecting a bold line type to the left of y_2 in the $\boxed{Y=}$ window makes the line easier to see. Using the INTERSECT option of the CALC menu, we select the two graphs in which we are interested, along with a guess. The graphing calculator displays the nearest point of intersection.

An alternative method uses only y_1 and the ZERO option of the CALC menu. This option requires you to enter an x-value to the left of each x-intercept as a LEFT BOUND. An x-value to the right of the x-intercept is then entered as a RIGHT BOUND. Finally, a GUESS value between the two bounds is entered and the x-intercept, or ZERO, is displayed.

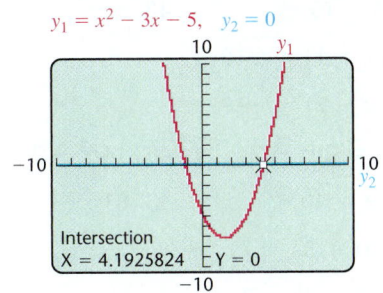

Use a graphing calculator to find the solutions, if they exist, accurate to two decimal places.

1. $x^2 + 4x - 3 = 0$
2. $x^2 - 5x - 2 = 0$
3. $x^2 + 13.54x + 40.95 = 0$
4. $x^2 - 4.43x + 6.32 = 0$
5. $1.235x^2 - 3.409x = 0$

5.7 EXERCISE SET

FOR EXTRA HELP

PRACTICE   WATCH   READ   REVIEW

Vocabulary and Reading Check

For each of Exercises 1–4, match the phrase with the most appropriate choice from the column on the right.

1. ____ The name of equations of the type $ax^2 + bx + c = 0$, with $a \neq 0$

2. ____ The maximum number of solutions of quadratic equations

3. ____ The idea that $A \cdot B = 0$ if and only if $A = 0$ or $B = 0$

4. ____ The number that a product must equal before the principle of zero products is used

   a) 2
   b) 0
   c) Quadratic
   d) The principle of zero products

The Principle of Zero Products

Solve using the principle of zero products.

5. $(x + 2)(x + 9) = 0$

6. $(x + 3)(x + 10) = 0$

7. $(x + 1)(x - 8) = 0$

8. $(x + 5)(x - 4) = 0$

9. $(2t - 3)(t + 6) = 0$

10. $(5t - 8)(t - 1) = 0$

11. $4(7x - 1)(10x - 3) = 0$

12. $6(4x - 3)(2x + 9) = 0$

13. $x(x - 7) = 0$

14. $x(x + 2) = 0$

15. $\left(\frac{2}{3}x - \frac{12}{11}\right)\left(\frac{7}{4}x - \frac{1}{12}\right) = 0$

16. $\left(\frac{1}{9} - 3x\right)\left(\frac{1}{5} + 2x\right) = 0$

17. $6n(3n + 8) = 0$

18. $10n(4n - 5) = 0$

19. $(20 - 0.4x)(7 - 0.1x) = 0$

20. $(1 - 0.05x)(1 - 0.3x) = 0$

Factoring to Solve Equations

Solve by factoring and using the principle of zero products.

21. $x^2 - 7x + 6 = 0$
22. $x^2 - 6x + 5 = 0$

23. $x^2 + 4x - 21 = 0$
24. $x^2 - 7x - 18 = 0$

25. $n^2 + 11n + 18 = 0$
26. $n^2 + 8n + 15 = 0$

27. $x^2 - 10x = 0$
28. $x^2 + 8x = 0$

29. $6t + t^2 = 0$
30. $3t - t^2 = 0$

31. $x^2 - 36 = 0$
32. $x^2 - 100 = 0$

33. $4t^2 = 49$
34. $9t^2 = 25$

35. $0 = 25 + x^2 + 10x$
36. $0 = 6x + x^2 + 9$

37. $64 + x^2 = 16x$ **38.** $x^2 + 1 = 2x$

39. $4t^2 = 8t$ **40.** $12t = 3t^2$

41. $4y^2 = 7y + 15$ **42.** $12y^2 - 5y = 2$

43. $(x - 7)(x + 1) = -16$

44. $(x + 2)(x - 7) = -18$

45. $15z^2 + 7 = 20z + 7$ **46.** $14z^2 - 3 = 21z - 3$

47. $36m^2 - 9 = 40$ **48.** $81x^2 - 5 = 20$

49. $(x + 3)(3x + 5) = 7$

50. $(x - 1)(5x + 4) = 2$

51. $3x^2 - 2x = 9 - 8x$

52. $x^2 - 2x = 18 + 5x$

53. $(6a + 1)(a + 1) = 21$

54. $(2t + 1)(4t - 1) = 14$

55. Use this graph to solve $x^2 - 3x - 4 = 0$.

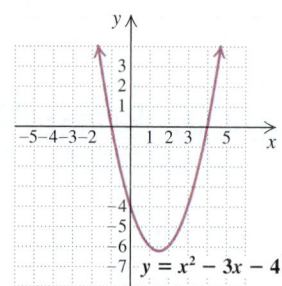

56. Use this graph to solve $x^2 + x - 6 = 0$.

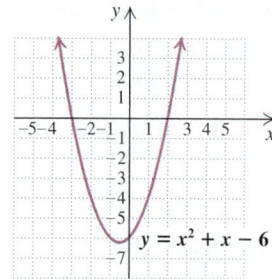

57. Use this graph to solve $-x^2 - x + 6 = 0$.

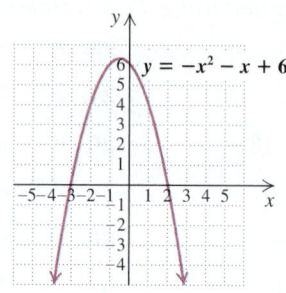

58. Use this graph to solve $-x^2 + 2x + 3 = 0$.

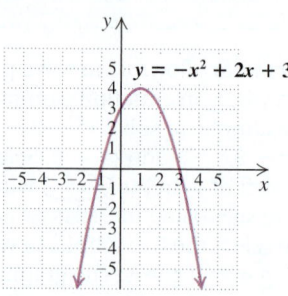

Find the x-intercepts for the graph of each equation. Grids are intentionally not included.

59. $y = x^2 - x - 6$ **60.** $y = x^2 + 3x - 4$

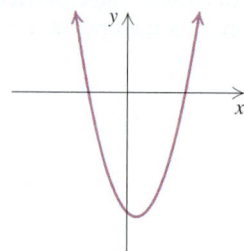   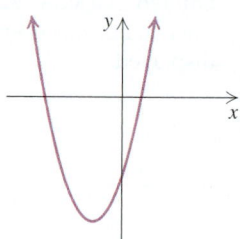

61. $y = x^2 + 2x - 8$ **62.** $y = x^2 - 2x - 15$

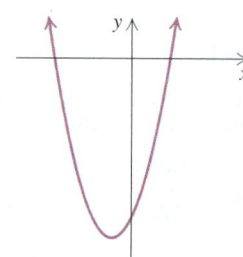   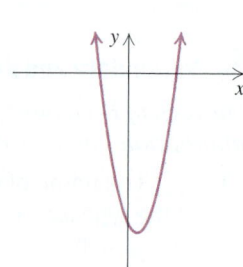

63. $y = 2x^2 + 3x - 9$ **64.** $y = 2x^2 + x - 10$

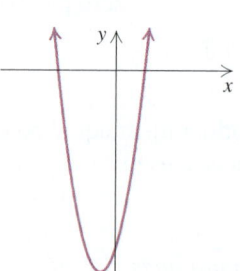   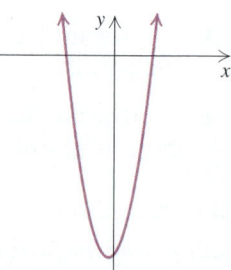

65. The equation $x^2 + 1 = 0$ has no real-number solutions. What implications does this have for the graph of $y = x^2 + 1$?

66. What is the difference between a quadratic polynomial and a quadratic equation?

Skill Review

67. Find the opposite of -65.   [1.6]

68. Find the reciprocal of $\frac{3}{4}$.   [1.7]

69. Find the absolute value of -1.65.   [1.4]

70. Write 0.00068 in scientific notation.   [4.2]

71. Write $-\frac{2}{3}$ in decimal notation.   [1.4]

72. Write $\frac{5}{4}$ in percent notation.   [2.4]

Synthesis

73. What is wrong with solving $x^2 = 3x$ by dividing both sides of the equation by x?

74. When the principle of zero products is used to solve a quadratic equation, will there always be two different solutions? Why or why not?

Solve.

75. $(2x - 11)(3x^2 + 29x + 56) = 0$

76. $(4x + 1)(15x^2 - 7x - 2) = 0$

77. Find an equation with integer coefficients that has the given numbers as solutions. For example, 3 and -2 are solutions of $x^2 - x - 6 = 0$.

a) $-4, 5$ b) $-1, 7$ c) $\frac{1}{4}, 3$
d) $\frac{1}{2}, \frac{1}{3}$ e) $\frac{2}{3}, \frac{3}{4}$ f) $-1, 2, 3$

Solve.

78. $16(x - 1) = x(x + 8)$

79. $a(9 + a) = 4(2a + 5)$

80. $(t - 5)^2 = 2(5 - t)$

81. $-x^2 + \frac{9}{25} = 0$

82. $a^2 = \frac{49}{100}$

Aha! **83.** $(t + 1)^2 = 9$

84. $\frac{27}{25}x^2 = \frac{1}{3}$

85. $x^3 - 2x^2 - x + 2 = 0$

86. $x^3 + 4x^2 - 9x - 36 = 0$

87. For each equation on the left, find an equivalent equation on the right.

a) $x^2 + 10x - 2 = 0$ $4x^2 + 8x + 36 = 0$
b) $(x - 6)(x + 3) = 0$ $(2x + 8)(2x - 5) = 0$
c) $5x^2 - 5 = 0$ $9x^2 - 12x + 24 = 0$
d) $(2x - 5)(x + 4) = 0$ $(x + 1)(5x - 5) = 0$
e) $x^2 + 2x + 9 = 0$ $x^2 - 3x - 18 = 0$
f) $3x^2 - 4x + 8 = 0$ $2x^2 + 20x - 4 = 0$

88. Explain how to construct an equation that has seven solutions.

89. Explain how the graph in Exercise 57 can be used to visualize the solutions of
$$-x^2 - x + 6 = 4.$$

Use a graphing calculator to find the solutions of each equation. Round solutions to the nearest hundredth.

90. $-x^2 + 0.63x + 0.22 = 0$

91. $x^2 - 9.10x + 15.77 = 0$

92. $6.4x^2 - 8.45x - 94.06 = 0$

93. $x^2 + 13.74x + 42.00 = 0$

94. $0.84x^2 - 2.30x = 0$

95. $1.23x^2 + 4.63x = 0$

96. $x^2 + 1.80x - 5.69 = 0$

97. The square of the sum of two consecutive integers is 225. What are the integers?

98. The sum of the squares of two consecutive integers is 313. What are the integers?

↪ YOUR TURN ANSWERS: SECTION 5.7

1. $-5, 1$ **2.** $-2, -\frac{5}{3}$ **3.** $-9, 0$ **4.** $3, 5$ **5.** $0, 8$
6. -5 **7.** $-1, \frac{2}{3}$ **8.** $(1, 0)$ and $(5, 0)$

QUICK QUIZ: SECTIONS 5.1–5.7

1. Factor: $x^2 - 2x - 8$.   [5.2]

2. Solve: $x^2 - 2x - 8 = 0$.   [5.7]

3. Solve: $a^2 = 100$.   [5.7]

4. Factor: $a^2 - 100$.   [5.4]

5. Factor: $2a^2 + 40a + 200$.   [5.4]

PREPARE TO MOVE ON

Translate to an algebraic expression.   [1.1]

1. The square of the sum of two consecutive integers

2. The product of two consecutive integers

Solve.   [2.5]

3. The first angle of a triangle is four times as large as the second. The measure of the third angle is 30° less than that of the second. How large are the angles?

4. A rectangular table top is twice as long as it is wide. The perimeter of the table is 192 in. What are the dimensions of the table?

5.8   Solving Applications

Applications   ◾   The Pythagorean Theorem

Applications

We can use the five-step problem-solving process and our methods of solving quadratic equations to solve new types of problems.

EXAMPLE 1 *Race Numbers.* Terry and Jody registered their boats in the Lakeport Race at the same time. The racing numbers assigned to their boats were consecutive numbers, the product of which was 156. Find the numbers.

SOLUTION

1. **Familiarize.** Consecutive numbers are one apart, like 49 and 50. Let $x =$ the first boat number; then $x + 1 =$ the next boat number.

2. **Translate.** We reword the problem before translating:

   Rewording: The first boat number   times   the next boat number   is   156.

   Translating: x · $(x + 1)$ =   156

3. **Carry out.** We solve the equation as follows:

$$x(x + 1) = 156$$
$$x^2 + x = 156 \qquad \text{Multiplying}$$
$$x^2 + x - 156 = 0 \qquad \text{Subtracting 156 to get 0 on one side}$$
$$(x - 12)(x + 13) = 0 \qquad \text{Factoring}$$
$$x - 12 = 0 \quad or \quad x + 13 = 0 \qquad \text{Using the principle of zero products}$$
$$x = 12 \quad or \qquad x = -13. \qquad \text{Solving each equation}$$

4. **Check.** The solutions of the equation are 12 and -13. Since race numbers are not negative, -13 must be rejected. On the other hand, if x is 12, then $x + 1$ is 13 and $12 \cdot 13 = 156$. Thus the solution 12 checks.

5. **State.** The boat numbers for Terry and Jody were 12 and 13.

   YOUR TURN

1. Refer to Example 1. Suppose that the product of the racing numbers of the boats was 132. Find the numbers.

EXAMPLE 2 *Manufacturing.* Wooden Work, Ltd., builds cutting boards that are twice as long as they are wide. The most popular board that Wooden Work makes has an area of 800 cm². What are the dimensions of the board?

SOLUTION

1. **Familiarize.** We first make a drawing. Recall that the area of any rectangle is Length · Width. We let $x =$ the width of the board, in centimeters. The length is then $2x$, since the board is twice as long as it is wide.

2. **Translate.** We reword and translate as follows:

Rewording: The area of the rectangle is 800 cm^2.

Translating: $2x \cdot x$ $=$ 800

3. **Carry out.** We solve the equation as follows:

$$2x \cdot x = 800$$
$$2x^2 = 800$$

$2x^2 - 800 = 0$ Subtracting 800 to get 0 on one side of the equation

$2(x^2 - 400) = 0$ Factoring out a common factor of 2

$2(x - 20)(x + 20) = 0$ Factoring a difference of squares

$x - 20 = 0$ *or* $x + 20 = 0$ Using the principle of zero products

$x = 20$ *or* $x = -20$.   Solving each equation

4. **Check.** The solutions of the equation are 20 and -20. Since the width must be positive, -20 cannot be a solution. To check 20 cm, we note that if the width is 20 cm, then the length is $2 \cdot 20$ cm $= 40$ cm and the area is 20 cm $\cdot$ 40 cm $= 800\text{ cm}^2$. Thus the solution 20 checks.

5. **State.** The cutting board is 20 cm wide and 40 cm long.

2. Refer to Example 2. Wooden Work also builds a cutting board that has an area of 450 cm^2. This board is also twice as long as it is wide. What are the dimensions of this cutting board?

↪ YOUR TURN

EXAMPLE 3 *Dimensions of a Leaf.* Each leaf of one particular *Philodendron* species is approximately a triangle. A typical leaf has an area of 320 in^2. If the leaf is 12 in. longer than it is wide, find the length and the width of the leaf.

SOLUTION

1. **Familiarize.** The formula for the area of a triangle is Area $= \frac{1}{2} \cdot$ (base) $\cdot$ (height). We let $b =$ the width, in inches, of the triangle's base and $b + 12 =$ the height, in inches.

2. **Translate.** We reword and translate as follows:

Rewording: The area of the leaf is 320 in^2.

Translating: $\frac{1}{2} \cdot b(b + 12)$ $=$ 320

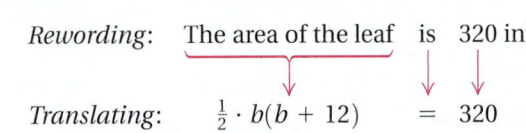

3. **Carry out.** We solve the equation as follows:

$$\tfrac{1}{2} \cdot b \cdot (b + 12) = 320$$

$$\tfrac{1}{2}(b^2 + 12b) = 320 \qquad \text{Multiplying}$$

$$b^2 + 12b = 640 \qquad \text{Multiplying by 2 to clear fractions}$$

$$b^2 + 12b - 640 = 0 \qquad \text{Subtracting 640 to get 0 on one side}$$

$$(b + 32)(b - 20) = 0 \qquad \text{Factoring}$$

$$b + 32 = 0 \quad or \quad b - 20 = 0 \qquad \text{Using the principle of zero products}$$

$$b = -32 \quad or \qquad b = 20.$$

3. Like the *Philodendron* in Example 3, each leaf of the *Triangular Leaf Senecio* approximates a triangle. One leaf of this plant is 6 in. longer than it is wide and has an area of 8 in². Find the length and the width of the leaf.

4. **Check.** The width must be positive, so -32 cannot be a solution. Suppose the base is 20 in. The height would be $20 + 12$, or 32 in., and the area $\tfrac{1}{2}(20)(32)$, or 320 in². These numbers check in the original problem.

5. **State.** The leaf is 32 in. long and 20 in. wide.

 YOUR TURN

Number of Minutes *t* After Injection	Number of Micrograms *N* of Epinephrine in the Bloodstream
2	160
5	250
8	160

EXAMPLE 4 *Medicine.* For certain people suffering an extreme allergic reaction, the drug epinephrine (adrenaline) is sometimes prescribed. The number of micrograms N of epinephrine in an adult's bloodstream t minutes after 250 micrograms have been injected is shown in the table at left for several values of t and can be approximated by

$$-10t^2 + 100t = N.$$

How long after an injection will there be about 210 micrograms of epinephrine in the bloodstream?

Source: Based on information in Chohan, Naina, Rita M. Doyle, and Patricia Nayle (eds.), *Nursing Handbook*, 21st ed. Springhouse, PA: Springhouse Corporation, 2001

SOLUTION

1. **Familiarize.** To familiarize ourselves with this problem, we could calculate N for different choices of t. We leave this to the student. Note that there may be two solutions, one on each side of the time at which the drug's effect peaks.

2. **Translate.** To find the length of time after injection when 210 micrograms are in the bloodstream, we replace N with 210 in the formula above:

$$-10t^2 + 100t = 210. \qquad \text{Substituting 210 for } N. \text{ This is now an equation in one variable.}$$

3. **Carry out.** We solve the equation as follows:

$$-10t^2 + 100t = 210$$

$$-10t^2 + 100t - 210 = 0 \qquad \text{Subtracting 210 from both sides to get 0 on one side}$$

$$-10(t^2 - 10t + 21) = 0 \qquad \text{Factoring out the largest common factor, } -10$$

$$-10(t - 3)(t - 7) = 0 \qquad \text{Factoring}$$

$$t - 3 = 0 \quad or \quad t - 7 = 0 \qquad \text{Using the principle of zero products}$$

$$t = 3 \quad or \qquad t = 7.$$

4. Refer to Example 4. How long after an injection will there be about 160 micrograms of epinephrine in the bloodstream?

4. **Check.** Since $-10 \cdot 3^2 + 100 \cdot 3 = -90 + 300 = 210$, the number 3 checks. Since $-10 \cdot 7^2 + 100 \cdot 7 = -490 + 700 = 210$, the number 7 also checks.

5. **State.** There will be 210 micrograms of epinephrine in the bloodstream approximately 3 minutes and 7 minutes after injection.

 YOUR TURN

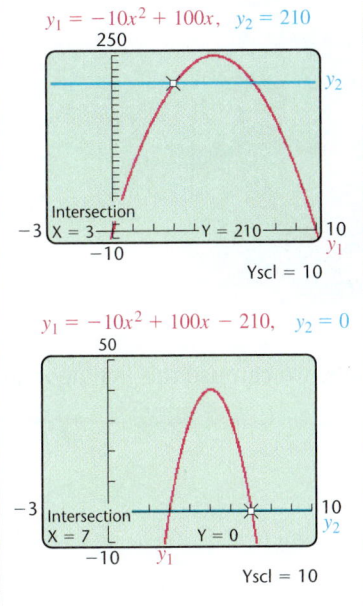
The Pythagorean Theorem

The following problems involve the Pythagorean theorem, which relates the lengths of the sides of a *right* triangle. A triangle is a **right triangle** if it has a 90°, or *right*, angle. The side opposite the 90° angle is called the **hypotenuse**. The other sides are called **legs**.

> ### THE PYTHAGOREAN THEOREM
>
> In any right triangle, if a and b are the lengths of the legs and c is the length of the hypotenuse, then
>
> $$a^2 + b^2 = c^2, \quad \text{or}$$
> $$(\text{Leg})^2 + (\text{Other leg})^2 = (\text{Hypotenuse})^2.$$
>
>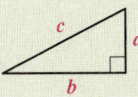
>
> The equation $a^2 + b^2 = c^2$ is called the **Pythagorean equation.***

The Pythagorean theorem is named for the Greek mathematician Pythagoras (569?–500? B.C.). We can think of this relationship as adding areas.

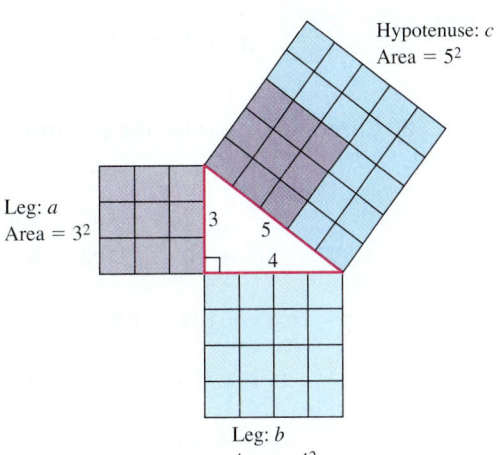

$$a^2 + b^2 = c^2$$
$$3^2 + 4^2 = 5^2$$
$$9 + 16 = 25$$

If we know the lengths of any two sides of a right triangle, we can use the Pythagorean equation to determine the length of the third side.

---

*The *converse* of the Pythagorean theorem is also true. That is, if $a^2 + b^2 = c^2$, then the triangle is a right triangle.

EXAMPLE 5 *Bridge Design.* A 50-ft diagonal brace on a bridge connects a support under the road surface at the center of the bridge to a side support on the bridge. The horizontal distance that it spans is 10 ft longer than the height that it reaches at the center of the bridge. Find both distances.

SOLUTION

1. **Familiarize.** We first make a drawing. The diagonal brace and the missing distances form the hypotenuse and the legs of a right triangle. We let x = the length of the vertical leg. Then $x + 10$ = the length of the horizontal leg. The hypotenuse has length 50 ft.

2. **Translate.** Since the triangle is a right triangle, we can use the Pythagorean theorem:

$$a^2 + b^2 = c^2$$
$$x^2 + (x + 10)^2 = 50^2. \qquad \text{Substituting}$$

3. **Carry out.** We solve the equation as follows:

$$x^2 + (x^2 + 20x + 100) = 2500 \qquad \text{Squaring}$$
$$2x^2 + 20x + 100 = 2500 \qquad \text{Combining like terms}$$
$$2x^2 + 20x - 2400 = 0 \qquad \text{Subtracting 2500 to get 0 on one side}$$
$$2(x^2 + 10x - 1200) = 0 \qquad \text{Factoring out a common factor}$$
$$2(x + 40)(x - 30) = 0 \qquad \text{Factoring. A calculator could be helpful here.}$$
$$x + 40 = 0 \quad or \quad x - 30 = 0 \qquad \text{Using the principle of zero products}$$
$$x = -40 \quad or \qquad x = 30.$$

Chapter Resource:
Translating for Success, p. 364

4. **Check.** The integer -40 cannot be a length of a side because it is negative. If the length is 30 ft, then $x + 10 = 40$, and $30^2 + 40^2 = 900 + 1600 = 2500$, which is 50^2. So the solution 30 checks.

5. Refer to Example 5. Another brace on the bridge is also 50 ft long, but the horizontal distance that it spans is 34 ft longer than the height that it reaches at the center of the bridge. Find both distances.

5. **State.** The height that the brace reaches at the center of the bridge is 30 ft, and the distance that it reaches to the middle of the bridge is 40 ft.

 YOUR TURN

5.8 EXERCISE SET

FOR EXTRA HELP

MyMathLab® MathXL PRACTICE WATCH READ REVIEW

◥ Vocabulary and Reading Check

Complete each statement with the correct variable expression or number.

1. If x is an integer, then the next consecutive integer is _____.

2. If the length of a rectangle is twice its width and its width is w, then its length is _____.

3. A right triangle contains a(n) _____ angle.

4. If the legs of a right triangle have lengths a and b and the hypotenuse is length c, then _____.

Applications

Solve. Use the five-step problem-solving approach.

5. A number is 6 less than its square. Find all such numbers.

6. A number is 30 less than its square. Find all such numbers.

7. *Parking-Space Numbers.* The product of two consecutive parking space numbers is 132. Find the numbers.

8. *Page Numbers.* The product of the page numbers on two facing pages of a book is 420. Find the page numbers.

9. The product of two consecutive even integers is 168. Find the integers.

10. The product of two consecutive odd integers is 195. Find the integers.

11. *Construction.* The front porch on Trent's new home is five times as long as it is wide. If the area of the porch is 320 ft², find the dimensions.

12. *Furnishings.* The work surface of Anita's desk is a rectangle that is twice as long as it is wide. If the area of the desktop is 18 ft², find the length and the width of the desk.

13. A photo is 5 cm longer than it is wide. Find the length and the width if the area is 84 cm².

14. An envelope is 4 cm longer than it is wide. The area is 96 cm². Find the length and the width.

15. *Dimensions of a Sail.* The height of the jib sail on a Lightning sailboat is 5 ft greater than the length of its "foot." If the area of the sail is 42 ft², find the length of the foot and the height of the sail.

16. *Dimensions of a Sail.* The height of a triangular mainsail on a sailboat is 4 ft longer than the base of the sail. If the area of the sail is 38.5 ft², find the height and the base of the sail.

17. *Road Design.* A triangular traffic island has a base that is half as long as its height. Find the base and the height if the island has an area of 64 ft².

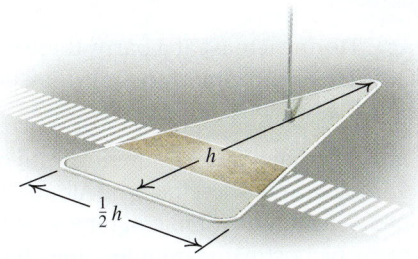

18. *Tent Design.* The triangular entrance to a tent is two-thirds as wide as it is tall. The area of the entrance is 12 ft². Find the height and the base.

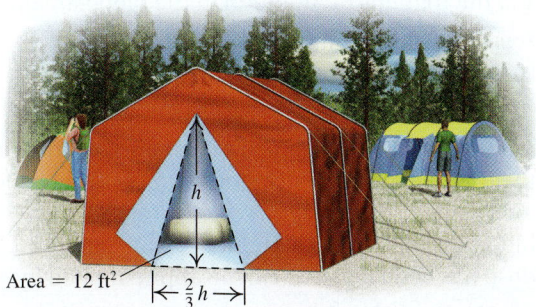

Area = 12 ft²

19. *Cabin Design.* The front of an A-frame cabin in a national park is in the shape of a triangle, with an area of 60 ft². If the height is 1 ft less than twice the base, find the base and the height of the front of the cabin.

20. *Flower Garden Design.* The demonstration flower garden at a state fair is in the shape of a triangle. The base of the triangle is 6 m longer than twice the height. If the area is 28 m², find the base and the height of the flower garden.

Games in a League's Schedule. *In a sports league of x teams in which all teams play each other twice, the total number N of games played is given by*

$$x^2 - x = N.$$

Use this formula for Exercises 21 and 22.

21. The Colchester Youth Soccer League plays a total of 56 games, with all teams playing each other twice. How many teams are in the league?

22. The teams in a women's softball league play each other twice, for a total of 132 games. How many teams are in the league?

Number of Handshakes. *The number of possible handshakes H within a group of n people is given by $H = \frac{1}{2}(n^2 - n)$. Use this formula for Exercises 23–26.*

23. At a meeting, there are 12 people. How many handshakes are possible?

24. At a party, there are 25 people. How many handshakes are possible?

25. *High-Fives.* After winning the championship, all Dallas Maverick teammates exchanged hugs. Altogether there were 66 hugs. How many players were there?

26. *Toasting.* During a toast at a party, there were 105 "clicks" of glasses. How many people took part in the toast?

27. *Medicine.* For many people suffering from constricted bronchial muscles, the drug Albuterol is prescribed. The number of micrograms *A* of Albuterol in a person's bloodstream *t* minutes after 200 micrograms have been inhaled can be approximated by

$$A = -50t^2 + 200t.$$

How long after an inhalation will there be about 150 micrograms of Albuterol in the bloodstream?

Source: Based on information in Chohan, Naina, Rita M. Doyle, and Patricia Nayle (eds.), *Nursing Handbook*, 21st ed. Springhouse, PA: Springhouse Corporation, 2001

28. *Medicine.* For adults with certain heart conditions, the drug Primacor (milrinone lactate) is prescribed. The number of milligrams *M* of Primacor in the bloodstream of a 132-lb patient *t* hours after a 3-mg dose has been injected can be approximated by

$$M = -\frac{1}{2}t^2 + \frac{5}{2}t.$$

How long after an injection will there be about 2 mg in the bloodstream?

Source: Based on information in Chohan, Naina, Rita M. Doyle, and Patricia Nayle (eds.), *Nursing Handbook*, 21st ed. Springhouse, PA: Springhouse Corporation, 2001

29. *Wave Height.* The height of waves in a storm depends on the speed of the wind. Assuming the wind has no obstructions for a long distance, the maximum wave height *H* for a wind speed *x* can be approximated by

$$H = 0.03x^2 + 0.6x - 6,$$

where *H* is in feet and *x* is in knots (nautical miles per hour). For what wind speed would the maximum wave height be 3 ft?

Source: Based on information from Smith, Craig B., *Extreme Waves*, Joseph Henry Press, 2006

30. *Cabinet Making.* Dovetail Woodworking determines that the revenue *R*, in thousands of dollars, from the sale of *x* sets of cabinets is given by $R(x) = 2x^2 + x$. If the cost *C*, in thousands of dollars, of producing *x* sets of cabinets is given by $C(x) = x^2 - 2x + 10$, how many sets must be produced and sold in order for the company to break even?

The Pythagorean Theorem

31. *Construction.* The diagonal braces in a fire tower are 15 ft long and span a horizontal distance of 12 ft. How high does each brace reach vertically?

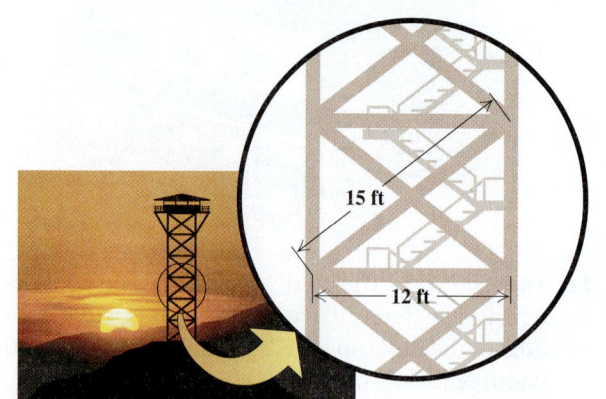

32. *Reach of a Ladder.* Twyla has a 26-ft ladder leaning against her house. If the bottom of the ladder is 10 ft from the base of the house, how high does the ladder reach?

33. *Zipline.* On one zipline in a canopy tour in Costa Rica, riders drop 58 ft while covering a distance of 840 ft along the ground. How long is the zipline?

34. *Roadway Design.* Elliott Street is 24 ft wide when it ends at Main Street in Brattleboro, Vermont. A 40-ft long diagonal crosswalk allows pedestrians to cross Main Street to or from either corner of Elliott Street (see the figure). Determine the width of Main Street.

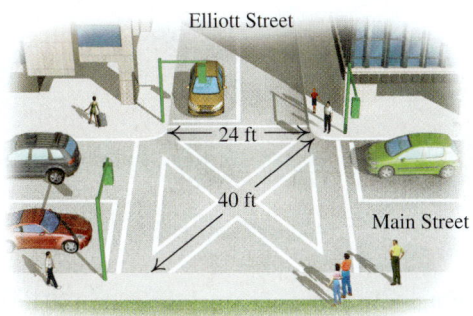

Elliott Street
24 ft
40 ft
Main Street

35. *Archaeology.* Archaeologists have discovered that the 18th-century garden of the Charles Carroll House in Annapolis, Maryland, was a right triangle. One leg of the triangle was formed by a 400-ft long sea wall. The hypotenuse of the triangle was 200 ft longer than the other leg. What were the dimensions of the garden?

Source: www.bsos.umd.edu

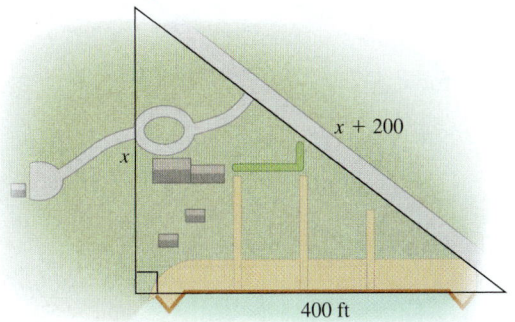

$x + 200$
x
400 ft

36. *Guy Wire.* The height of a wind power assessment tower is 5 m shorter than the guy wire that supports it.

If the guy wire is anchored 15 m from the foot of the antenna, how tall is the antenna?

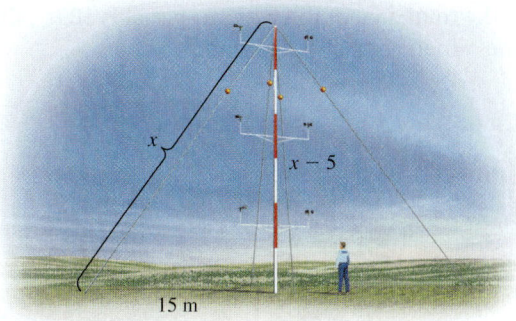

x
$x - 5$
15 m

37. *Right Triangle.* The shortest side of a right triangle measures 7 m. The lengths of the other two sides are consecutive integers. Find the lengths of the other two sides.

38. *Right Triangle.* The shortest side of a right triangle measures 8 cm. The lengths of the other two sides are consecutive odd integers. Find the lengths of the other two sides.

39. *Right Triangle.* The longest side of a right triangle is 1 ft longer than three times the length of the shortest side. The other side of the triangle is 1 ft shorter than three times the length of the shortest side. Find the lengths of the sides of the triangle.

40. *Right Triangle.* The longest side of a right triangle is 5 yd shorter than six times the length of the shortest side. The other side of the triangle is 5 yd longer than five times the length of the shortest side. Find the lengths of the sides of the triangle.

Applications

41. *Architecture.* An architect has allocated a rectangular space of 264 ft² for a square dining room and a 10-ft wide kitchen, as shown in the figure. Find the dimensions of each room.

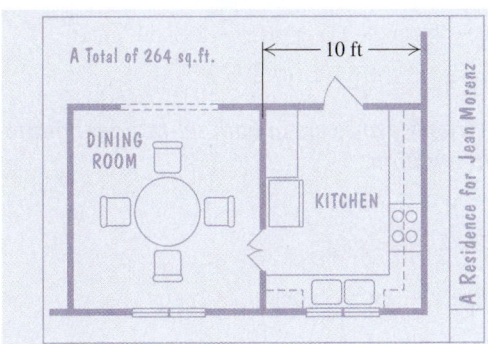

A Total of 264 sq.ft.   ← 10 ft →
DINING ROOM
KITCHEN
A Residence for Jean Morenz

42. *Design.* A window panel for a sun porch consists of a 7-ft tall rectangular window stacked above a square window. The windows have the same width. If the total area of the window panel is 18 ft^2, find the dimensions of each window.

7 ft

Height of a Rocket. For Exercises 43–46, assume that a water rocket is launched upward with an initial velocity of 48 ft/sec. Its height h, in feet, after t seconds, is given by h = 48t − 16t^2.

43. Determine the height of the rocket $\frac{1}{2}$ sec after it has been launched.

44. Determine the height of the rocket 2.5 sec after it has been launched.

45. When will the rocket be exactly 32 ft above the ground?

46. When will the rocket crash into the ground?

47. Do we now have the ability to solve *any* problem that translates to a quadratic equation? Why or why not?

48. Write a problem for a classmate to solve such that only one of two solutions of a quadratic equation can be used as an answer.

Skill Review

Solve. Label any contradictions or identities.

49. $3(x - 2) = 5x + 2(3 - x)$ [2.2]

50. $9x - 2[4(x + 1)] = 4(x - 2 + x)$ [2.2]

51. $10x + 3 = 5(2x + 1) - 2$ [2.2]

Solve. Write the answers in both set-builder notation and interval notation.

52. $3 - y \geq 7$ [2.6]

53. $\frac{1}{3}x + \frac{1}{2} < \frac{1}{6}$ [2.6]

54. Solve $x + 2y = 7$ for y.   [2.3]

Synthesis

The converse of the Pythagorean theorem is also true. That is, if $a^2 + b^2 = c^2$, then the triangle is a right triangle (where a and b are the lengths of the legs and

c is the length of the hypotenuse). Use this result to answer Exercises 55 and 56.

55. An archaeologist has straight rods of 3 ft, 4 ft, and 5 ft. Explain how she could draw a 7-ft by 9-ft rectangle on a piece of land being excavated.

56. Explain how straight rods of 5 cm, 12 cm, and 13 cm can be used to draw a right triangle that has two 45° angles.

57. *Sailing.* The mainsail of a Lightning sailboat is a right triangle in which the hypotenuse is called the leech. If a 24-ft tall mainsail has a leech length of 26 ft and if Dacron® sailcloth costs $1.50 per square foot, find the cost of the fabric for a new mainsail.

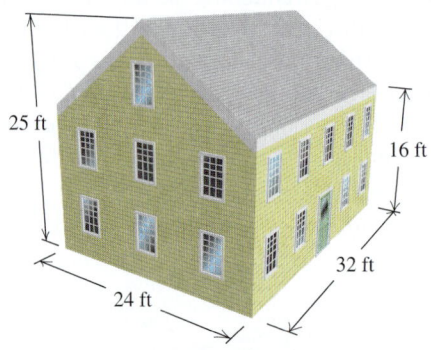

26 ft

24 ft

58. *Roofing.* A *square* of shingles covers 100 ft^2 of surface area. How many squares will be needed to reshingle the house shown?

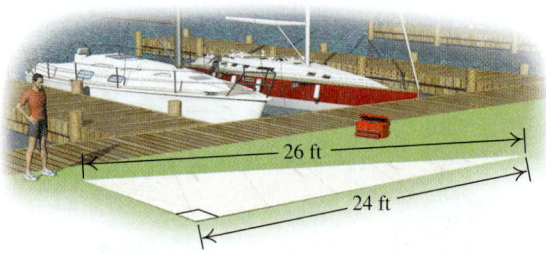

25 ft

16 ft

32 ft

24 ft

59. Solve for x.

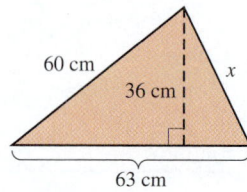

60 cm

36 cm

x

63 cm

60. *Pool Sidewalk.* A cement walk of uniform width is built around a 20-ft by 40-ft rectangular pool. The total area of the pool and the walk is 1500 ft^2. Find the width of the walk.

61. *Folding Sheet Metal.* An open rectangular gutter is made by turning up the sides of a piece of metal 20 in. wide, as shown. The area of the cross section of the gutter is 48 in². Find the possible depths of the gutter.

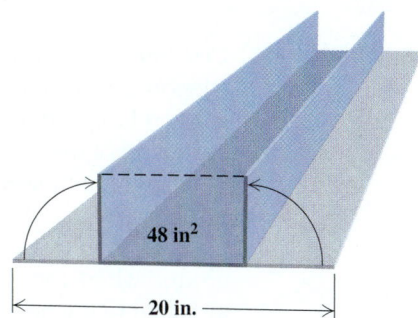

48 in²

20 in.

62. Find a polynomial for the shaded area surrounding the square in the figure below.

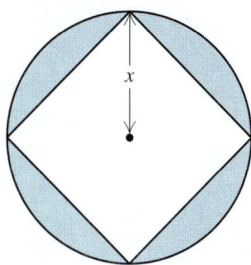

x

63. *Telephone Service.* Use the information in the figure below to determine the height of the telephone pole.

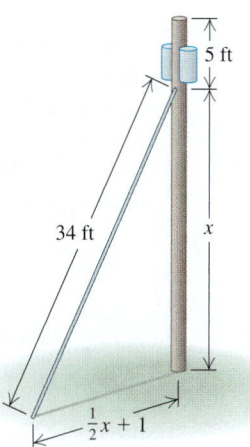

5 ft

34 ft x

$\frac{1}{2}x + 1$

64. *Dimensions of a Closed Box.* The total surface area of a closed box is 350 m². The box is 9 m high and has a square base and lid. Find the length of a side of the base.

65. The maximum length of a postcard that can be mailed with one postcard stamp is $\frac{7}{4}$ in. longer than the maximum width. The maximum area is $\frac{51}{2}$ in². Find the maximum length and the maximum width of a postcard.

Source: USPS

 Medicine. For certain people with acid reflux, the drug Pepcid (famotidine) is used. The number of milligrams N of Pepcid in an adult's bloodstream t hours after a 20-mg tablet has been swallowed can be approximated by

$$N = -0.009t\,(t - 12)^3.$$

Use a graphing calculator with the window $[-1, 13, -1, 25]$ *and the* TRACE *feature to answer Exercises 66–68.*

Source: Based on information in Chohan, Naina, Rita M. Doyle, and Patricia Nayle (eds.), *Nursing Handbook*, 21st ed. Springhouse, PA: Springhouse Corporation, 2001

 66. Approximately how long after a tablet has been swallowed will there be 18 mg in the bloodstream?

67. Approximately how long after a tablet has been swallowed will there be 10 mg in the bloodstream?

68. Approximately how long after a tablet has been swallowed will the peak dosage in the bloodstream occur?

 69. *Research.* Find an online calculator that asks for the width, length, and height of a roof and calculates the roof area.

   a) The roof shown in Exercise 58 is a gable roof. Enter the dimensions of this roof in the online calculator and compare your answer with the one calculated for you.

   b) Develop a formula that could be used by the online calculator to find the roof area.

 YOUR TURN ANSWERS: SECTION 5.8

1. 11 and 12   **2.** 15 cm wide and 30 cm long
3. 2 in. wide and 8 in. long   **4.** Approximately 2 min and 8 min after injection   **5.** Horizontal distance spanned: 48 ft; height at center of bridge: 14 ft

---

QUICK QUIZ: SECTIONS 5.1–5.8

1. Factor: $3x^3 - 15x^2 - 7x + 35$.   [5.1]

2. Solve: $20x^2 = 11x + 3$.   [5.7]

3. Factor: $10x^2 - 17x + 3$.   [5.3]

4. Factor: $6x^8 - 6$.   [5.4]

5. The hypotenuse of a right triangle is 1 ft longer than the length of the longer leg. The length of the shorter leg is 9 ft. Find the length of each side.   [5.8]

PREPARE TO MOVE ON

Simplify.   [1.3]

1. $\frac{24}{28}$ **2.** $\frac{90}{88}$ **3.** $\frac{124}{155}$

Divide, if possible.   [1.7]

4. $\frac{0}{7}$ **5.** $\frac{13}{0}$

1. Angle Measures. The degree measures of the angles of a triangle are three consecutive integers. Find the measures of the angles.

2. Rectangle Dimensions. The area of a rectangle is 3604 ft². The length is 15 ft longer than the width. Find the dimensions of the rectangle.

3. Sales Tax. Claire paid $3604 for a used pickup truck. This included 6% for sales tax. How much did the truck cost before tax?

4. Wire Cutting. A 180-m wire is cut into three pieces. The third piece is 2 m longer than the first. The second is two-thirds as long as the first. How long is each piece?

5. Perimeter. The perimeter of a rectangle is 240 ft. The length is 2 ft greater than the width. Find the length and the width.

Translating for Success

Use after Section 5.8.

Translate each word problem to an equation and select a correct translation from equations A–O.

A. $2x \cdot x = 288$

B. $x(x + 60) = 7021$

C. $59 = x \cdot 60$

D. $x^2 + (x + 15)^2 = 3604$

E. $x^2 + (x + 70)^2 = 130^2$

F. $0.06x = 3604$

G. $2(x + 2) + 2x = 240$

H. $\frac{1}{2}x(x - 1) = 1770$

I. $x + \frac{2}{3}x + (x + 2) = 180$

J. $0.59x = 60$

K. $x + 0.06x = 3604$

L. $2x^2 + x = 288$

M. $x(x + 15) = 3604$

N. $x^2 + 60 = 7021$

O. $x + (x + 1) + (x + 2) = 180$

Answers on page A-26

An additional, animated version of this activity appears in MyMathLab. To use MyMathLab, you need a course ID and a student access code. Contact your instructor for more information.

6. Cell-Phone Tower. A guy wire on a cell-phone tower is 130 ft long and is attached to the top of the tower. The height of the tower is 70 ft longer than the distance from the point on the ground where the wire is attached to the bottom of the tower. Find the height of the tower.

7. Sales Meeting Attendance. PTQ Corporation holds a sales meeting in Tucson. Of the 60 employees, 59 of them attend the meeting. What percent attend the meeting?

8. Dimensions of a Pool. A rectangular swimming pool is twice as long as it is wide. The area of the surface is 288 ft². Find the dimensions of the pool.

9. Dimensions of a Triangle. The height of a triangle is 1 cm less than the length of the base. The area of the triangle is 1770 cm². Find the height and the length of the base.

10. Width of a Rectangle. The length of a rectangle is 60 ft longer than the width. Find the width if the area of the rectangle is 7021 ft².

Collaborative Activity   *Visualizing Factoring*

Focus: Visualizing factoring
Use after: Section 5.2
Time: 20–30 minutes
Group size: 3
Materials: Graph paper and scissors

Factoring a polynomial like $x^2 + 5x + 6$ can be thought of as determining the length and the width of a rectangle that has area $x^2 + 5x + 6$.

Activity

1. **a)** To factor $x^2 + 11x + 10$ geometrically, the group needs to cut out shapes like those below to represent x^2, $11x$, and 10. This can be done by either tracing the figures below or by selecting a value for x, say 4, and using the squares on the graph paper to cut out the following:

   x^2:   Using the value selected for x, cut out a square that is x units on each side.

   $11x$:   Using the value selected for x, cut out a rectangle that is 1 unit wide and x units long. Repeat this to form 11 such strips.

   10:   Cut out two rectangles with whole-number dimensions and an area of 10. One should be 2 units by 5 units and the other 1 unit by 10 units.

   b) The group, working together, should then attempt to use one of the two rectangles with area 10, along with all of the other shapes, to piece together one large rectangle. Only one of the rectangles with area 10 will work.

   c) From the large rectangle formed in part (b), use the length and the width to determine the factorization of $x^2 + 11x + 10$. Where do the dimensions of the rectangle representing 10 appear in the factorization?

2. Repeat step (1) above, but this time use the other rectangle with area 10, and use only 7 of the 11 strips, along with the x^2-shape. Piece together the shapes to form one large rectangle. What factorization do the dimensions of this rectangle suggest?

3. Cut out rectangles with area 12 and use the above approach to factor $x^2 + 8x + 12$. What dimensions should be used for the rectangle with area 12?

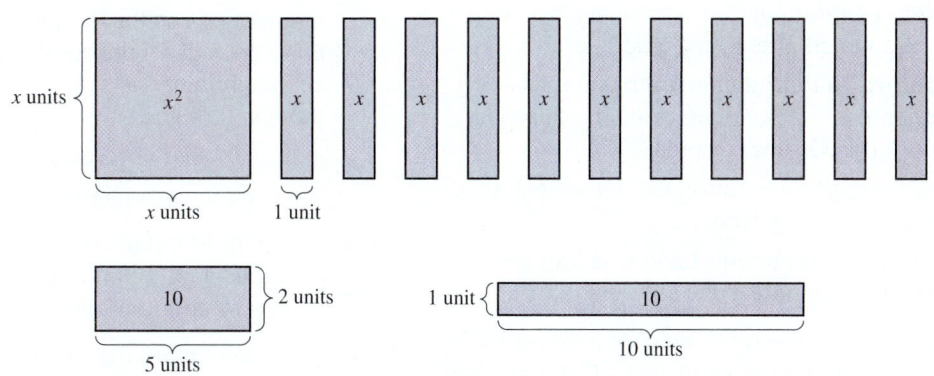

Collaborative Activity   *Matching Factorizations**

Focus: Factoring
Use after: Section 5.6
Time: 20 minutes
Group size: Begin with the entire class. If there is an odd number of students, the instructor should participate.
Materials: Prepared sheets of paper, pins or tape. On half of the sheets, the instructor writes a polynomial. On the remaining sheets, the instructor writes the factorization of those polynomials. The polynomials and factorizations are similar; for example,

$$x^2 - 2x - 8, \quad (x - 2)(x - 4),$$
$$x^2 - 6x + 8, \quad (x - 1)(x - 8),$$
$$x^2 - 9x + 8, \quad (x + 2)(x - 4)$$

Activity

1. As class members enter the room, the instructor pins or tapes either a polynomial or a factorization to the back of each student. Class members are told only whether their sheet of paper contains a polynomial or a factorization.

2. After all students are wearing a sheet of paper, they should mingle with one another, attempting to match up their factorization with the appropriate polynomial or vice versa. They may ask questions of one another that relate to factoring and polynomials. Answers to the questions should be yes or no. For example, a legitimate question might be "Is my last term negative?" or "Do my factors have opposite signs?"

3. The game is over when all factorization/polynomial pairs have "found" one another.

---

*Thanks to Jann MacInnes of Florida Community College at Jacksonville–Kent Campus for suggesting this activity.

Decision Making & Connection   (*Use after Section 5.1.*)

Designing a Tournament. The number of games in a tournament varies widely depending on the type of tournament being played.

In a *single elimination* tournament, a player or team is eliminated after one loss.

In a *double elimination* tournament, a player or team is eliminated after losing twice.

In a *round-robin* tournament, each team plays every other team once, and a winner is chosen on the basis of the win–loss records.

In a *double round-robin* tournament, each team plays every other team twice.

Other types of tournaments include ladder and pyramid tournaments.

1. The number of games G required for a round-robin tournament with n teams is given by $G = \frac{1}{2}n^2 - \frac{1}{2}n$. If 15 basketball teams enter a round-robin tournament, how many games must be played?

2. The number of games G required for a double round-robin tournament with n teams is given by $G = n^2 - n$. If 15 basketball teams enter a double round-robin tournament, how many games must be played?

3. Ideally, in a round-robin tournament, all the teams play at the same time. If there is an even number of teams, then there will be $n/2$ games occurring simultaneously during each round. If there is an odd number of teams, then $(n - 1)/2$ games will occur simultaneously, with one team resting for each round. Use this information and the formula given in Exercise 1 to find a formula for the number of rounds needed for an even number of teams and the number for an odd number of teams.

4. *Research.* Find the number of games required in a single elimination tournament with n teams.

5. *Research.* Find the number of games required in a double elimination tournament with n teams. Is there a minimum number or a maximum number?

6. If you were planning a chess tournament, what type of tournament would you choose? How many players would you invite to participate?

Study Summary

KEY TERMS AND CONCEPTS	EXAMPLES	PRACTICE EXERCISES

SECTION 5.1: *Introduction to Factoring*

To **factor** a polynomial means to write it as a product of polynomials. Always begin by factoring out the **largest common factor**.	$12x^4 - 30x^3 = 6x^3(2x - 5)$	**1.** Factor: $12x^4 - 18x^3 + 30x.$
Some polynomials with four terms can be **factored by grouping**.	$3x^3 - x^2 - 6x + 2 = x^2(3x - 1) - 2(3x - 1)$ $\qquad\qquad\qquad\quad = (3x - 1)(x^2 - 2)$	**2.** Factor: $2x^3 - 6x^2 - x + 3.$

SECTION 5.2: *Factoring Trinomials of the Type $x^2 + bx + c$*

Some trinomials of the type $x^2 + bx + c$ can be factored by reversing the steps of FOIL.	Factor: $x^2 - 11x + 18$. | Pairs of Factors of 18	Sums of Factors | |---|---| | $-1, -18$ | -19 | | $-2, \ -9$ | -11 | The factorization is $(x - 2)(x - 9)$.	**3.** Factor: $x^2 - 7x - 18.$

SECTION 5.3: *Factoring Trinomials of the Type $ax^2 + bx + c$*

One method for factoring trinomials of the type $ax^2 + bx + c$ is a FOIL-based method.	Factor: $6x^2 - 5x - 6$. The factors will be in the form $(3x + \ )(2x + \ )$ or $(6x + \ )(x + \ )$. We list all pairs of factors of -6, and check possible products by multiplying those possibilities that do not contain a common factor. $(3x - 2)(2x + 3) = 6x^2 + 5x - 6$, $(3x + 2)(2x - 3) = 6x^2 - 5x - 6$ ⟵ This is the correct product. The factorization is $(3x + 2)(2x - 3)$.	**4.** Factor: $6x^2 + x - 2$.

Another method for factoring trinomials of the type $ax^2 + bx + c$ involves factoring by grouping.

Factor: $6x^2 - 5x - 6$.

Multiply the leading coefficient and the constant term: $6(-6) = -36$. Look for factors of -36 that add to -5.

Pairs of Factors of -36	Sums of Factors
$1, -36$	-35
$2, -18$	-16
$3, -12$	-9
$4, \ -9$	-5

Rewrite $-5x$ as $4x - 9x$ and factor by grouping:
$$6x^2 - 5x - 6 = 6x^2 + 4x - 9x - 6$$
$$= 2x(3x + 2) - 3(3x + 2)$$
$$= (3x + 2)(2x - 3).$$

5. Factor:
$8x^2 - 22x + 15$.

SECTION 5.4: *Factoring Perfect-Square Trinomials and Differences of Squares*

Factoring a Perfect-Square Trinomial

$A^2 + 2AB + B^2 = (A + B)^2$;
$A^2 - 2AB + B^2 = (A - B)^2$

Factor: $y^2 + 100 - 20y$.

$$A^2 - 2AB + B^2 = (A - B)^2$$

$$y^2 + 100 - 20y = y^2 - 20y + 100 = (y - 10)^2$$

6. Factor:
$100n^2 + 81 + 180n$.

Factoring a Difference of Squares

$A^2 - B^2 = (A + B)(A - B)$

Factor: $9t^2 - 1$.

$$A^2 - B^2 = (A + B)(A - B)$$

$$9t^2 - 1 = (3t + 1)(3t - 1)$$

7. Factor:
$144t^2 - 25$.

SECTION 5.5: *Factoring Sums or Differences of Cubes*

Factoring a Sum or a Difference of Cubes

$A^3 + B^3$
$= (A + B)(A^2 - AB + B^2)$
$A^3 - B^3$
$= (A - B)(A^2 + AB + B^2)$

$x^3 + 1000 = (x + 10)(x^2 - 10x + 100)$

$z^6 - 8w^3 = (z^2 - 2w)(z^4 + 2wz^2 + 4w^2)$

8. Factor: $a^3 - 1$.

SECTION 5.6: *Factoring: A General Strategy*

A general strategy for factoring polynomials can be found in Section 5.6.

Factor: $5x^5 - 80x$.

$5x^5 - 80x = 5x(x^4 - 16)$ $5x$ is the largest common factor.

$= 5x(x^2 + 4)(x^2 - 4)$ $x^4 - 16$ is a difference of squares.

$= 5x(x^2 + 4)(x + 2)(x - 2)$ $x^2 - 4$ is also a difference of squares.

Check:

$$5x(x^2 + 4)(x + 2)(x - 2) = 5x(x^2 + 4)(x^2 - 4)$$
$$= 5x(x^4 - 16)$$
$$= 5x^5 - 80x.$$

9. Factor:
$3x^3 - 36x^2 + 108x$.

| **SECTION 5.7:** *Solving Quadratic Equations by Factoring* | | |

| **The Principle of Zero Products**
An equation $AB = 0$ is true if and only if $A = 0$ or $B = 0$, or both A and B are 0. | Solve: $2x^2 = 8x$.

$\quad 2x^2 - 8x = 0 \qquad$ Getting 0 on one side
$\quad 2x(x - 4) = 0 \qquad$ Factoring
$\quad 2x = 0 \ \ or \ \ x - 4 = 0 \qquad$ Using the principle of zero products

$\quad\ \ x = 0 \ \ or \qquad\quad x = 4$
The solutions are 0 and 4. | **10.** Solve:
$\quad x^2 + 7x = 30.$ |

| **SECTION 5.8:** *Solving Applications* | | |

| **The Pythagorean Theorem**
In any right triangle, if a and b are the lengths of the legs and c is the length of the hypotenuse, then $a^2 + b^2 = c^2$. | Find the lengths of the legs in this triangle.
$\quad x^2 + (x + 1)^2 = 5^2$
$\quad x^2 + x^2 + 2x + 1 = 25$
$\quad 2x^2 + 2x - 24 = 0$
$\quad x^2 + x - 12 = 0$
$\quad (x + 4)(x - 3) = 0$
$\quad x + 4 = 0 \ \ or \ \ x - 3 = 0$
$\quad x = -4 \ \ or \qquad x = 3$
Since lengths are not negative, -4 is not a solution. The lengths of the legs are 3 and 4. 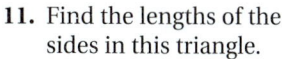 | **11.** Find the lengths of the sides in this triangle.
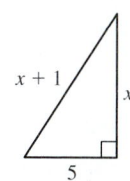 |

Review Exercises: Chapter 5

⤷ Concept Reinforcement

Classify each of the following statements as either true or false.

1. The largest common variable factor is the largest power of the variable in the polynomial.   [5.1]

2. A prime polynomial has no common factor other than 1 or −1.   [5.2]

3. Every perfect-square trinomial can be expressed as a binomial squared.   [5.4]

4. Every binomial can be regarded as a difference of squares.   [5.4]

5. Every quadratic equation has two different solutions.   [5.7]

6. The principle of zero products can be applied whenever a product equals 0.   [5.7]

7. In a right triangle, the hypotenuse is always longer than either leg.   [5.8]

8. The Pythagorean theorem can be applied to any triangle that has an angle measuring at least 90°.   [5.8]

Find three factorizations of each monomial.   [5.1]

9. $20x^3$

10. $-18x^5$

Factor completely. If a polynomial is prime, state this.

11. $12x^4 - 18x^3$ [5.1]

12. $100t^2 - 1$ [5.4]

13. $x^2 + x - 12$ [5.2]

14. $x^2 + 14x + 49$ [5.4]

15. $12x^3 + 12x^2 + 3x$ [5.4]

16. $6x^3 + 9x^2 + 2x + 3$ [5.1]

17. $6a^2 + a - 5$ [5.3]

18. $25t^2 + 9 - 30t$ [5.4]

19. $81a^4 - 1$ [5.4]

20. $9x^3 + 12x^2 - 45x$ [5.3]

21. $2x^3 - 250$ [5.5]

22. $x^4 + 4x^3 - 2x - 8$ [5.1]

23. $a^2b^4 - 64$ [5.4]

24. $-8x^6 + 32x^5 - 4x^4$ [5.1]

25. $75 + 12x^2 - 60x$ [5.4]

26. $y^2 + 9$ [5.4]

27. $-t^3 + t^2 + 42t$ [5.2]

28. $4x^2 - 25$ [5.4]

29. $n^2 - 60 - 4n$ [5.2]

30. $5z^2 - 30z + 10$ [5.1]

31. $8y^3 + 27x^6$ [5.5]

32. $2t^2 - 7t - 4$ [5.3]

33. $7x^3 + 35x^2 + 28x$ [5.2]

34. $-6x^3 + 150x$ [5.4]

35. $15 - 8x + x^2$ [5.2]

36. $3x + x^2 + 5$ [5.2]

37. $x^2y^2 + 6xy - 16$ [5.2]

38. $12a^2 + 84ab + 147b^2$ [5.4]

39. $m^2 + 5m + mt + 5t$ [5.1]

40. $6r^2 + rs - 15s^2$ [5.3]

Solve.   [5.7]

41. $(x - 9)(x + 11) = 0$

42. $x^2 + 2x - 35 = 0$

43. $16x^2 = 9$

44. $3x^2 + 2 = 5x$

45. $(x + 1)(x - 2) = 4$

46. $9t - 15t^2 = 0$

47. $3x^2 + 3 = 6x$

48. The square of a number is 12 more than the number. Find all such numbers.   [5.8]

49. A stone is tossed down from a cliff at an initial velocity of 4 ft/sec. The distance s that the stone falls after t seconds is given by $s = 4t + 16t^2$. If the cliff is 420 ft tall, in how many seconds will the stone reach the bottom of the cliff?   [5.8]

50. Find the x-intercepts of the graph of $y = 2x^2 - 3x - 5$.   [5.7]

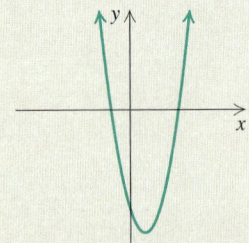

51. The front of a house is a triangle that is as wide as it is tall. Its area is 98 ft². Find the height and the base.   [5.8]

52. Josh needs to add a diagonal brace to his LEGO® robot. The brace must span a height of 8 holes and a width of 6 holes. How long should the brace be?   [5.8]

Synthesis

53. On a quiz, Celia writes the factorization of $4x^2 - 100$ as $(2x - 10)(2x + 10)$. If this were a 10-point question, how many points would you give Celia? Why?   [5.4]

54. How does solving quadratic equations differ from solving linear equations?   [5.7]

Solve.

55. The pages of a book measure 15 cm by 20 cm. Margins of equal width surround the printing on each page and constitute one-half of the area of the page. Find the width of the margins.   [5.8]

15 cm

20 cm

When in the Course of human events, it becomes necessary for one people to dissolve the political bands which have connected them with another, and to assume among the powers of the earth, the separate and equal station to which the Laws of Nature and of

Nature's God entitle them, a decent respect to the opinions of mankind requires that they should declare the causes which impel them to the separation.
We hold these truths to be self-evident, that all men are created equal, that they are endowed by their Creator with certain unalienable

56. The cube of a number is the same as twice the square of the number. Find the number.   [5.8]

57. The length of a rectangle is two times its width. When the length is increased by 20 cm and the width is decreased by 1 cm, the area is 160 cm². Find the original length and width.   [5.8]

Solve.   [5.7]

58. $(x - 2)2x^2 + x(x - 2) - (x - 2)15 = 0$

Aha! **59.** $x^2 + 25 = 0$

Test: Chapter 5

For step-by-step test solutions, access the Chapter Test Prep Videos in MyMathLab®, on YouTube® (search "Bittinger Combo AlgCA" and click on "Channels"), or by scanning the code.

Factor completely. If a polynomial is prime, state this.

1. $x^2 - 13x + 36$

2. $x^2 + 25 - 10x$

3. $6y^2 - 8y^3 + 4y^4$

4. $x^3 + x^2 + 2x + 2$

5. $t^7 - 3t^5$

6. $a^3 + 3a^2 - 4a$

7. $28x - 48 + 10x^2$

8. $4t^2 - 25$

9. $-6m^3 - 9m^2 - 3m$

10. $3r^3 - 3$

11. $45r^2 + 60r + 20$

12. $3x^4 - 48$

13. $49t^2 + 36 + 84t$

14. $x^4 + 2x^3 - 3x - 6$

15. $x^2 + 3x + 6$

16. $6t^3 + 9t^2 - 15t$

17. $3m^2 - 9mn - 30n^2$

Solve.

18. $x^2 - 6x + 5 = 0$

19. $2x^2 - 7x = 15$

20. $4t - 10t^2 = 0$

21. $25t^2 = 1$

22. $x(x - 1) = 20$

23. Find the x-intercepts of the graph of $y = 3x^2 - 5x - 8$.

24. The length of a rectangle is 6 m more than the width. The area of the rectangle is 40 m². Find the length and the width.

25. The number of possible handshakes H within a group of n people is given by $H = \frac{1}{2}(n^2 - n)$. At a meeting, everyone shook hands once with everyone else. If there were 45 handshakes, how many people were at the meeting?

26. A mason wants to be sure she has a right corner in a building's foundation. She marks a point 3 ft from the corner along one wall and another point 4 ft from the corner along the other wall. If the corner is a right angle, what should the distance be between the two marked points?

Synthesis

27. *Dimensions of an Open Box.* A rectangular piece of cardboard is twice as long as it is wide. A 4-cm square is cut out of each corner, and the sides are turned up to make a box with an open top. The volume of the box is 616 cm³. Find the original dimensions of the cardboard.

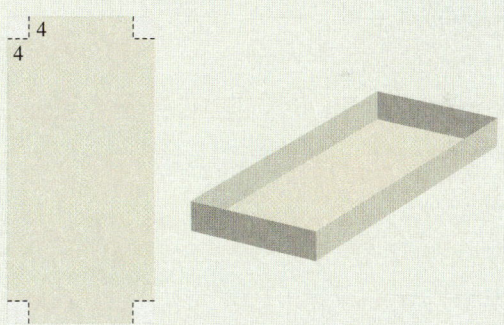

28. Factor: $(a + 3)^2 - 2(a + 3) - 35$.

29. Solve: $20x(x + 2)(x - 1) = 5x^3 - 24x - 14x^2$.

Cumulative Review: Chapters 1–5

Simplify. Do not use negative exponents in the answer.

1. $\frac{3}{8} \div \frac{3}{4}$ [1.3]

2. $\frac{3}{8} + \frac{3}{4}$ [1.3]

3. $-2 + (20 \div 4)^2 - 6 \cdot (-1)^3$ [1.8]

4. $(3x^2y^3)^{-2}$ [4.2]

5. $(t^2)^3 \cdot t^4$ [4.1]

6. $(3x^4 - 2x^2 + x - 7) + (5x^3 + 2x^2 - 3)$ [4.4]

7. $\dfrac{3t^3s^{-1}}{12t^{-5}s}$ [4.2]

8. $\left(\dfrac{-2x^2y}{3z^4}\right)^3$ [4.1]

9. Evaluate $-x$ for $x = -8$. [1.6]

10. Determine the leading term of the polynomial
$$4x^3 - 6x^2 - x^4 + 7. \quad [4.3]$$

11. Divide: $(8x^4 - 20x^3 + 2x^2 - 4x) \div (4x)$. [4.8]

Multiply.

12. $-4t^8(t^3 - 2t - 5)$ [4.5]

13. $(3x - 5)^2$ [4.6]

14. $(10x^5 + y)(10x^5 - y)$ [4.7]

Factor completely.

15. $c^2 - 1$ [5.4]

16. $5x + 5y + 10x^2 + 10xy$ [5.1]

17. $4r^2 - 4rt + t^2$ [5.4]

18. $10y^2 + 40$ [5.1]

19. $x^2y - 3xy + 2y$ [5.2]

20. $12x^2 - 5xy - 2y^2$ [5.3]

Solve.

21. $\frac{1}{3} + 2x = \frac{1}{2}$ [2.2]

22. $8y - 6(y - 2) = 3(2y + 7)$ [2.2]

23. $3x - 7 \geq 4 - 8x$ [2.6]

24. $x^2 + x = 12$ [5.7]

25. $3x^2 = 12$ [5.7]

26. $3x^2 = 12x$ [5.7]

27. Solve $a = bc + dc$ for c. [2.3]

28. Find the slope of the line containing the points $(6, 7)$ and $(-2, 7)$. [3.5]

29. Write the slope–intercept equation for the line with slope 5 that contains the point $\left(-\frac{1}{3}, 0\right)$. [3.7]

Graph.

30. $4(x + 1) = 8$ [3.3]

31. $x + y = 5$ [3.3]

32. $y = \frac{3}{2}x - 2$ [3.6]

33. $3x + 5y = 10$ [3.6]

Solve.

34. On average, men talk 97 min more per month on cell phones than do women. The sum of men's average minutes and women's average minutes is 647 min. What is the average number of minutes per month that men talk on cell phones? [2.5]

Source: International Communications Research for Cingular Wireless

35. A 13-ft ladder is placed against a building in such a way that the distance from the top of the ladder to the ground is 7 ft more than the distance from the bottom of the ladder to the building. Find both distances. [5.8]

36. Donna's quiz grades are 8, 3, 5, and 10. What scores on the fifth quiz will make her average quiz grade at least 7? [2.7]

37. The average amount of sodium in a serving of Chef Boyardee foods dropped from 1100 mg in 2003 to 750 mg in 2011. [3.7]

Source: The Indianapolis Star, 11/25/07; Chef Boyardee

a) Graph the data and determine an equation for the related line. Let s represent the average amount of sodium per serving and t the number of years after 2000.

b) Use the equation of part (a) to estimate the average amount of sodium in a serving of Chef Boyardee foods in 2007.

Synthesis

38. Solve $x = \dfrac{abx}{2 - b}$ for b. [2.3]

39. Solve: $6x^3 + 4x^2 = 2x$. [5.7]

Rational Expressions and Equations

BIRD OR AIRPLANE	WING SPAN	WING WIDTH
Grey heron	180 cm	24 cm
House sparrow	10 in.	1.8 in.
Albatross	9 ft	0.6 ft
Concorde (retired in 2003)	84 ft	47 ft
Boeing 747-400	64 m	8 m

Different Jobs Require Different Wings.

The shape of a wing determines whether a bird or an airplane excels at speed, hovering, or tight maneuvering. One description of a wing's shape is the *wing aspect ratio*: the ratio of the wing span to the wing width. Birds and planes that soar or glide have wings of high aspect ratio; wings of low aspect ratio allow more maneuverability and produce less drag at high speeds. Which of the birds in the table above do you think has a wing aspect ratio the same as that of a small single-engine airplane? (*See Exercises 53 and 54 in Exercise Set 6.7.*)

As a research wildlife biologist, I use math in many ways in every one of my projects.

Cheryl Dykstra, Ph.D., a wildlife biologist in West Chester, Ohio, has used math to calculate the surface area of the wings of nestling bald eagles in order to determine eaglets' heat exchange and energy needs. She also uses math, especially statistics, to understand and to explain her data to other researchers.

Rational expressions are similar to fractions in arithmetic. In this chapter, we learn how to simplify, add, subtract, multiply, and divide rational expressions. These skills are then used to solve equations that arise from real-life problems like the one on the preceding page.

6.1 Rational Expressions

Restricting Replacement Values ▪ Simplifying Rational Expressions ▪ Factors That Are Opposites

Just as a rational number is any number that can be written as a quotient of two integers, a **rational expression** is any expression that can be written as a quotient of two polynomials. The following are examples of rational expressions:

$$\frac{7}{3}, \quad \frac{5}{x+6}, \quad \frac{t^2 - 5t + 6}{4t^2 - 7}.$$

Rational expressions are examples of *algebraic fractions*. They are also examples of *fraction expressions*.

Restricting Replacement Values

Because rational expressions may contain variables in a denominator, and because division by 0 is undefined, we must avoid certain replacement values. For example, in the expression

$$\frac{x+5}{x-7},$$

when x is replaced with 7, the denominator is 0, and the expression is undefined:

$$\frac{x+5}{x-7} = \frac{7+5}{7-7} = \frac{12}{0}. \qquad \text{Division by 0 is undefined.}$$

When x is replaced with a number other than 7, such as 6, the expression *is* defined because the denominator is not 0:

$$\frac{x+5}{x-7} = \frac{6+5}{6-7} = \frac{11}{-1} = -11.$$

The expression is also defined when $x = -5$:

$$\frac{x+5}{x-7} = \frac{-5+5}{-5-7} = \frac{0}{-12} = 0. \qquad \text{0 divided by a nonzero number is 0.}$$

Any replacement for the variable that makes the *denominator* 0 will cause an expression to be undefined.

EXAMPLE 1   Find all numbers for which the rational expression

$$\frac{x+4}{x^2 - 3x - 10}$$

is undefined.

Technology Connection

To check Example 1 with a graphing calculator, let $y_1 = x^2 - 3x - 10$ and $y_2 = (x + 4)/y_1$ and use the TABLE feature. Since $x^2 - 3x - 10$ is 0 for $x = 5$, it is impossible to evaluate y_2 for $x = 5$.

TBLSTART = 0   ΔTBL = 1

X	Y₁	Y₂
0	−10	−.4
1	−12	−.4167
2	−12	−.5
3	−10	−.7
4	−6	−1.333
5	0	ERROR
6	8	1.25

X = 5

1. Find all numbers for which the rational expression $\dfrac{x^2 - x - 2}{x^2 - 25}$ is undefined.

Student Notes

When using a graphing calculator or tutorial software, you may need to use parentheses around the numerator and around the denominator of rational expressions. For example,

$5/3x$ means $\dfrac{5}{3}x$

and

$5/(3x)$ means $\dfrac{5}{3x}$.

SOLUTION To determine which numbers make the rational expression undefined, we set the *denominator* equal to 0 and solve:

$$x^2 - 3x - 10 = 0 \qquad \text{We set the denominator equal to 0.}$$
$$(x - 5)(x + 2) = 0 \qquad \text{Factoring}$$
$$x - 5 = 0 \quad or \quad x + 2 = 0 \qquad \text{Using the principle of zero products}$$
$$x = 5 \quad or \qquad x = -2. \qquad \text{Solving each equation}$$

Check:

For $x = 5$:

$$\dfrac{x + 4}{x^2 - 3x - 10} = \dfrac{5 + 4}{5^2 - 3 \cdot 5 - 10}$$
$$= \dfrac{9}{25 - 15 - 10} = \dfrac{9}{0}. \qquad \text{This expression is undefined, as expected.}$$

For $x = -2$:

$$\dfrac{x + 4}{x^2 - 3x - 10} = \dfrac{-2 + 4}{(-2)^2 - 3(-2) - 10}$$
$$= \dfrac{2}{4 + 6 - 10} = \dfrac{2}{0}. \qquad \text{This expression is undefined, as expected.}$$

Thus, $\dfrac{x + 4}{x^2 - 3x - 10}$ is undefined for $x = 5$ and $x = -2$.

YOUR TURN

Simplifying Rational Expressions

A rational expression is said to be *simplified* when the numerator and the denominator have no factors (other than 1) in common. To simplify a rational expression that contains variables, we use the same process we would use to simplify $\frac{15}{40}$:

$$\dfrac{15}{40} = \dfrac{3 \cdot 5}{8 \cdot 5} \qquad \text{Factoring the numerator and the denominator. Note the common factor, 5.}$$
$$= \dfrac{3}{8} \cdot \dfrac{5}{5} \qquad \text{Rewriting as a product of two fractions}$$
$$= \dfrac{3}{8} \cdot 1 \qquad \dfrac{5}{5} = 1$$
$$= \dfrac{3}{8}. \qquad \text{Using the identity property of 1 to remove the factor 1}$$

Similar steps are followed when simplifying rational expressions: We factor and remove a factor equal to 1, using the fact that

$$\dfrac{ab}{cb} = \dfrac{a}{c} \cdot \dfrac{b}{b} \quad \text{and} \quad \dfrac{a}{c} \cdot \dfrac{b}{b} = \dfrac{a}{c}.$$

EXAMPLE 2   Simplify: $\dfrac{8x^2}{24x}$.

SOLUTION

$$\dfrac{8x^2}{24x} = \dfrac{8 \cdot x \cdot x}{3 \cdot 8 \cdot x}$$ Factoring the numerator and the denominator. Note that the greatest common factor is $8 \cdot x$.

$$= \dfrac{x}{3} \cdot \dfrac{8x}{8x}$$ Rewriting as a product of two rational expressions

$$= \dfrac{x}{3} \cdot 1 \qquad \dfrac{8x}{8x} = 1$$

2. Simplify: $\dfrac{15ab^2}{20b}$.

$$= \dfrac{x}{3}$$ Removing the factor 1

YOUR TURN

We say that $\dfrac{8x^2}{24x}$ *simplifies* to $\dfrac{x}{3}$. In more advanced courses, we would say that $8x^2/(24x)$ simplifies to $x/3$, *with the restriction that* $x \neq 0$. In the work that follows, we assume that all denominators are nonzero.

EXAMPLE 3   Simplify: $\dfrac{5a + 15}{10}$.

SOLUTION

$$\dfrac{5a + 15}{10} = \dfrac{5(a + 3)}{5 \cdot 2}$$ Factoring the numerator and the denominator. The greatest common factor is 5.

$$= \dfrac{5}{5} \cdot \dfrac{a + 3}{2}$$ Rewriting as a product of two rational expressions

$$= 1 \cdot \dfrac{a + 3}{2} \qquad \dfrac{5}{5} = 1$$

3. Simplify: $\dfrac{16}{2x + 10}$.

$$= \dfrac{a + 3}{2}$$ Removing the factor 1

YOUR TURN

The result in Example 3 can be partially checked using a replacement for a—say, $a = 2$.

Original expression:

$$\dfrac{5a + 15}{10} = \dfrac{5 \cdot 2 + 15}{10}$$

$$= \dfrac{25}{10} = \dfrac{5}{2}$$ The results are the same.

Simplified expression:

$$\dfrac{a + 3}{2} = \dfrac{2 + 3}{2}$$

$$= \dfrac{5}{2}$$

If we do not get the same result when evaluating both expressions, we know that a mistake has been made. For example, if $(5a + 15)/10$ is *incorrectly* simplified as $(a + 15)/2$ and we evaluate using $a = 2$, we have the following.

Original expression:

$$\dfrac{5a + 15}{10} = \dfrac{5 \cdot 2 + 15}{10}$$

$$= \dfrac{5}{2}$$ The results are different.

Incorrectly simplified expression:

$$\dfrac{a + 15}{2} = \dfrac{2 + 15}{2}$$

$$= \dfrac{17}{2}$$

This demonstrates that a mistake has been made.

4. Simplify: $\dfrac{x^2 - 5x}{x^2 - 3x - 10}$.

Sometimes the common factor has two or more terms.

EXAMPLE 4 Simplify.

a) $\dfrac{6x - 12}{7x - 14}$

b) $\dfrac{18t^2 + 6t}{6t^2 + 15t}$

c) $\dfrac{x^2 + 3x + 2}{x^2 - 1}$

SOLUTION

a) $\dfrac{6x - 12}{7x - 14} = \dfrac{6(x - 2)}{7(x - 2)}$ Factoring the numerator and the denominator. The greatest common factor is $x - 2$.

$= \dfrac{6}{7} \cdot \dfrac{x - 2}{x - 2}$ Rewriting as a product of two rational expressions

$= \dfrac{6}{7} \cdot 1 \qquad \dfrac{x - 2}{x - 2} = 1$

$= \dfrac{6}{7}$ Removing the factor 1

b) $\dfrac{18t^2 + 6t}{6t^2 + 15t} = \dfrac{3t \cdot 2(3t + 1)}{3t(2t + 5)}$ Factoring the numerator and the denominator. The greatest common factor is $3t$.

$= \dfrac{3t}{3t} \cdot \dfrac{2(3t + 1)}{2t + 5}$ Rewriting as a product of two rational expressions

$= 1 \cdot \dfrac{2(3t + 1)}{2t + 5} \qquad \dfrac{3t}{3t} = 1$

$= \dfrac{2(3t + 1)}{2t + 5}$ Removing the factor 1. The numerator and the denominator have no common factor so the simplification is complete.

c) $\dfrac{x^2 + 3x + 2}{x^2 - 1} = \dfrac{(x + 1)(x + 2)}{(x + 1)(x - 1)}$ Factoring; $x + 1$ is the greatest common factor.

$= \dfrac{x + 1}{x + 1} \cdot \dfrac{x + 2}{x - 1}$ Rewriting as a product of two rational expressions

$= 1 \cdot \dfrac{x + 2}{x - 1} \qquad \dfrac{x + 1}{x + 1} = 1$

$= \dfrac{x + 2}{x - 1}$ Removing the factor 1

YOUR TURN

Canceling is a shortcut that can be used—and easily *misused*—to simplify rational expressions. If done with care and understanding, canceling streamlines the process of removing a factor equal to 1. Example 4(c) could have been streamlined as follows:

$$\dfrac{x^2 + 3x + 2}{x^2 - 1} = \dfrac{(x + 1)(x + 2)}{(x + 1)(x - 1)}$$ When a factor equal to 1 is noted, it is "canceled": $\dfrac{x + 1}{x + 1} = 1$.

$$= \dfrac{x + 2}{x - 1}.$$ Simplifying

> **CAUTION!** Canceling is often used incorrectly:
>
> $$\frac{\cancel{x} + 7}{\cancel{x} + 3}; \qquad \frac{a^2 - \cancel{5}}{\cancel{5}}; \qquad \frac{6\cancel{x}^2 + 5\cancel{x} + 1}{4\cancel{x}^2 - 3\cancel{x}}.$$
>
> Incorrect! Incorrect!   Incorrect!
>
> None of the above cancellations removes a factor equal to 1. Factors are parts of products. For example, in $x \cdot 7$, x and 7 are factors, but in $x + 7$, x and 7 are terms, *not* factors. *Only factors can be canceled.*

EXAMPLE 5   Simplify: $\dfrac{3x^2 - 2x - 1}{x^2 - 3x + 2}$.

SOLUTION   We factor the numerator and the denominator and look for common factors:

$$\frac{3x^2 - 2x - 1}{x^2 - 3x + 2} = \frac{(3x + 1)\cancel{(x - 1)}}{(x - 2)\cancel{(x - 1)}} \qquad \begin{array}{l}\text{Try to visualize this as}\\[4pt] \dfrac{3x + 1}{x - 2} \cdot \dfrac{x - 1}{x - 1}.\end{array}$$

$$= \frac{3x + 1}{x - 2}. \qquad \begin{array}{l}\text{Removing a factor equal to 1:}\\[4pt] \dfrac{x - 1}{x - 1} = 1\end{array}$$

5. Simplify: $\dfrac{n^2 - 16}{2n^2 + 7n - 4}$.

 YOUR TURN

Factors That Are Opposites

Consider

$$\frac{x - 4}{8 - 2x}, \quad \text{or, equivalently,} \quad \frac{x - 4}{2(4 - x)}.$$

At first glance, the numerator and the denominator do not appear to have any common factors. But $x - 4$ and $4 - x$ are opposites, or additive inverses, of each other. Thus we can find a common factor by factoring out -1 in either expression.

EXAMPLE 6   Simplify $\dfrac{x - 4}{8 - 2x}$ and check by evaluating.

SOLUTION   We have

$$\frac{x - 4}{8 - 2x} = \frac{x - 4}{2(4 - x)} \qquad \text{Factoring}$$

$$= \frac{x - 4}{2(-1)(x - 4)} \qquad \text{Note that } 4 - x = -x + 4 = -1(x - 4).$$

$$= \frac{x - 4}{-2(x - 4)} \qquad \begin{array}{l}\text{Had we originally factored out } -2, \text{ we}\\ \text{could have gone directly to this step.}\end{array}$$

$$= \frac{1}{-2} \cdot \frac{x - 4}{x - 4} \qquad \begin{array}{l}\text{Rewriting as a product. It is important}\\ \text{to write the 1 in the numerator.}\end{array}$$

$$= -\frac{1}{2}. \qquad \begin{array}{l}\text{Removing a factor equal to 1:}\\ (x - 4)/(x - 4) = 1\end{array}$$

As a partial check, note that for any choice of x other than 4, the value of the rational expression is $-\frac{1}{2}$. For example, if $x = 5$, then

$$\frac{x - 4}{8 - 2x} = \frac{5 - 4}{8 - 2 \cdot 5}$$

$$= \frac{1}{-2} = -\frac{1}{2}.$$

6. Simplify: $\dfrac{x - 2}{2 - x}$.

 YOUR TURN

6.1 EXERCISE SET

FOR EXTRA HELP

 MyMathLab® MathXL
PRACTICE   WATCH   READ   REVIEW

Vocabulary and Reading Check

Choose the word written under the blank that best completes the statement.

1. A rational expression can be written as a _____ of two polynomials.
   difference/quotient

2. A rational expression is undefined when the _____ is zero.
   denominator/numerator

3. A rational expression is simplified when the numerator and the denominator have no _____ (other than 1) in common.
   factors/terms

4. When we cancel, we remove _____.
   a factor equal to 1/a restricted value

Concept Reinforcement

In each of Exercises 5–8, match the rational expression with the list of numbers in the column on the right for which the rational expression is undefined.

5. ____ $\dfrac{3t}{(t + 1)(t - 4)}$

6. ____ $\dfrac{2t + 7}{(2t - 1)(3t + 4)}$

7. ____ $\dfrac{a + 7}{a^2 - a - 12}$

8. ____ $\dfrac{m - 3}{m^2 - 2m - 15}$

a) $-1, 4$

b) $-3, 5$

c) $-\dfrac{4}{3}, \dfrac{1}{2}$

d) $-3, 4$

Restricting Replacement Values

List all numbers for which each rational expression is undefined.

9. $\dfrac{18}{-11x}$

10. $\dfrac{13}{-5t}$

11. $\dfrac{y - 3}{y + 5}$

12. $\dfrac{a + 6}{a - 10}$

13. $\dfrac{t - 5}{3t - 15}$

14. $\dfrac{x^2 - 4}{5x + 10}$

15. $\dfrac{x^2 - 25}{x^2 - 3x - 28}$

16. $\dfrac{p^2 - 9}{p^2 - 7p + 10}$

17. $\dfrac{t^2 + t - 20}{2t^2 + 11t - 6}$

18. $\dfrac{x^2 + 2x + 1}{3x^2 - x - 14}$

Simplifying Rational Expressions

Simplify by removing a factor equal to 1. Show all steps.

19. $\dfrac{50a^2b}{40ab^3}$

20. $\dfrac{-24x^4y^3}{6x^7y}$

21. $\dfrac{6t + 12}{6t - 18}$

22. $\dfrac{5n - 30}{5n + 5}$

23. $\dfrac{21t - 7}{24t - 8}$

24. $\dfrac{10n + 25}{8n + 20}$

25. $\dfrac{a^2 - 9}{a^2 + 4a + 3}$

26. $\dfrac{a^2 + 5a + 6}{a^2 - 9}$

Simplify, if possible. Then check by evaluating, as in Example 6.

27. $\dfrac{-36x^8}{54x^5}$

28. $\dfrac{45a^4}{30a^6}$

29. $\dfrac{-2y + 6}{-8y}$

30. $\dfrac{4x - 12}{6x}$

31. $\dfrac{6a^2 + 3a}{7a^2 + 7a}$

32. $\dfrac{-4m^2 + 4m}{-8m^2 + 12m}$

33. $\dfrac{t^2 - 16}{t^2 - t - 20}$

34. $\dfrac{a^2 - 4}{a^2 + 5a + 6}$

35. $\dfrac{3a^2 + 9a - 12}{6a^2 - 30a + 24}$

36. $\dfrac{2t^2 - 6t + 4}{4t^2 + 12t - 16}$

37. $\dfrac{x^2 - 8x + 16}{x^2 - 16}$

38. $\dfrac{x^2 - 25}{x^2 + 10x + 25}$

39. $\dfrac{n - 2}{n^3 - 8}$

40. $\dfrac{n^6 + 27}{n^2 + 3}$

41. $\dfrac{t^2 - 1}{t + 1}$

42. $\dfrac{a^2 - 1}{a - 1}$

43. $\dfrac{y^2 + 4}{y + 2}$

44. $\dfrac{m^2 + 9}{m + 3}$

45. $\dfrac{5x^2 + 20}{10x^2 + 40}$

46. $\dfrac{6x^2 + 54}{4x^2 + 36}$

47. $\dfrac{y^2 + 6y}{2y^2 + 13y + 6}$

48. $\dfrac{t^2 + 2t}{2t^2 + t - 6}$

49. $\dfrac{4x^2 - 12x + 9}{10x^2 - 11x - 6}$

50. $\dfrac{4x^2 - 4x + 1}{6x^2 + 5x - 4}$

Factors That Are Opposites

Simplify.

51. $\dfrac{10 - x}{x - 10}$

52. $\dfrac{x - 8}{8 - x}$

53. $\dfrac{7t - 14}{2 - t}$

54. $\dfrac{3 - n}{5n - 15}$

55. $\dfrac{a - b}{4b - 4a}$

56. $\dfrac{2p - 2q}{q - p}$

57. $\dfrac{3x^2 - 3y^2}{2y^2 - 2x^2}$

58. $\dfrac{7a^2 - 7b^2}{3b^2 - 3a^2}$

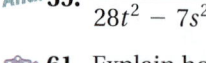 **59.** $\dfrac{7s^2 - 28t^2}{28t^2 - 7s^2}$

60. $\dfrac{9m^2 - 4n^2}{4n^2 - 9m^2}$

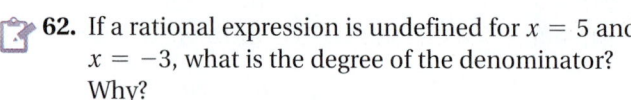 **61.** Explain how simplifying is related to the identity property of 1.

62. If a rational expression is undefined for $x = 5$ and $x = -3$, what is the degree of the denominator? Why?

Skill Review

Factor.

63. $3x^3 + 15x^2 + x + 5$ [5.1]

64. $3x^2 + 16x + 5$ [5.3]

65. $18y^4 - 27y^3 + 3y^2$ [5.1]

66. $25a^2 - 16b^2$ [5.4]

67. $m^3 - 8m^2 + 16m$ [5.4]

68. $5x^2 - 35x + 60$ [5.2]

Synthesis

 69. Luke *incorrectly* simplifies

$$\frac{x^2 + x - 2}{x^2 + 3x + 2} \quad \text{as} \quad \frac{x - 1}{x + 2}.$$

He then checks his simplification by evaluating both expressions for $x = 1$. Use this situation to explain why evaluating is not a foolproof check.

 70. How could you convince someone that $a - b$ and $b - a$ are opposites of each other?

Simplify.

71. $\dfrac{16y^4 - x^4}{(x^2 + 4y^2)(x - 2y)}$

72. $\dfrac{(x - 1)(x^4 - 1)(x^2 - 1)}{(x^2 + 1)(x - 1)^2(x^4 - 2x^2 + 1)}$

73. $\dfrac{x^5 - 2x^3 + 4x^2 - 8}{x^7 + 2x^4 - 4x^3 - 8}$

74. $\dfrac{10t^4 - 8t^3 + 15t - 12}{8 - 10t + 12t^2 - 15t^3}$

75. $\dfrac{(t^4 - 1)(t^2 - 9)(t - 9)^2}{(t^4 - 81)(t^2 + 1)(t + 1)^2}$

76. $\dfrac{(t + 2)^3(t^2 + 2t + 1)(t + 1)}{(t + 1)^3(t^2 + 4t + 4)(t + 2)}$

77. $\dfrac{x^3 + 6x^2 - 4x - 24}{x^2 + 4x - 12}$

78. $\dfrac{10x^2 - 10}{5x^3 - 30x^2 - 5x + 30}$

79. $\dfrac{(x^2 - y^2)(x^2 - 2xy + y^2)}{(x + y)^2(x^2 - 4xy - 5y^2)}$

80. $\dfrac{x^4 - y^4}{(y - x)^4}$

 81. Select any number x, multiply by 2, add 5, multiply by 5, subtract 25, and divide by 10. What do you get? Explain how this procedure can be used for a number trick.

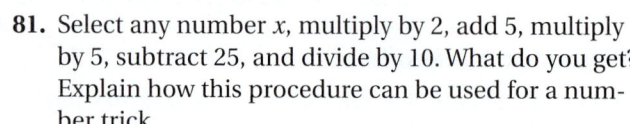

> **YOUR TURN ANSWERS: SECTION 6.1**
>
> **1.** $-5, 5$ **2.** $\dfrac{3ab}{4}$ **3.** $\dfrac{8}{x + 5}$ **4.** $\dfrac{x}{x + 2}$
>
> **5.** $\dfrac{n - 4}{2n - 1}$ **6.** -1

> **PREPARE TO MOVE ON**
>
> *Simplify.*   [1.7]
>
> **1.** $\dfrac{2}{15} \cdot \dfrac{10}{7}$ **2.** $\left(\dfrac{3}{4}\right)\left(\dfrac{-20}{9}\right)$
>
> **3.** $\dfrac{5}{8} \div \dfrac{1}{6}$ **4.** $\dfrac{7}{10} \div \left(-\dfrac{8}{15}\right)$

6.2 Multiplication and Division

Multiplication ▪ Division

Multiplication and division of rational expressions are similar to multiplication and division with fractions. In this section, we again assume that all denominators are nonzero.

Multiplication

Recall that to multiply fractions, we multiply numerator times numerator and denominator times denominator. Rational expressions are multiplied in a similar way.

> **THE PRODUCT OF TWO RATIONAL EXPRESSIONS**
>
> To multiply rational expressions, form the product of the numerators and the product of the denominators:
>
> $$\frac{A}{B} \cdot \frac{C}{D} = \frac{AC}{BD}.$$
>
> Then factor and, if possible, simplify the result.

For example,

$$\frac{3}{5} \cdot \frac{8}{11} = \frac{3 \cdot 8}{5 \cdot 11} \quad \text{and} \quad \frac{x}{3} \cdot \frac{x+2}{y} = \frac{x(x+2)}{3y}.$$

Fraction bars are grouping symbols, so parentheses are needed when writing some products.

EXAMPLE 1 Multiply. (Write the product as a single rational expression.) Then simplify, if possible, by removing a factor equal to 1.

a) $\dfrac{5a^3}{4} \cdot \dfrac{2}{5a}$

b) $(x^2 - 3x - 10) \cdot \dfrac{x+4}{x^2 - 10x + 25}$

c) $\dfrac{10x + 20}{2x^2 - 3x + 1} \cdot \dfrac{x^2 - 1}{5x + 10}$

SOLUTION

a) $\dfrac{5a^3}{4} \cdot \dfrac{2}{5a} = \dfrac{5a^3(2)}{4(5a)}$ Forming the product of the numerators and the product of the denominators

$= \dfrac{5 \cdot a \cdot a \cdot a \cdot 2}{2 \cdot 2 \cdot 5 \cdot a}$ Factoring the numerator and the denominator

$= \dfrac{\cancel{5} \cdot a \cdot \cancel{a} \cdot a \cdot \cancel{2}}{\cancel{2} \cdot 2 \cdot \cancel{5} \cdot \cancel{a}}$

$= \dfrac{a^2}{2}$ Removing a factor equal to 1: $\dfrac{2 \cdot 5 \cdot a}{2 \cdot 5 \cdot a} = 1$

6.2 EXERCISE SET

FOR EXTRA HELP

MyMathLab® MathXL
PRACTICE   WATCH   READ   REVIEW

Vocabulary and Reading Check

Choose from the following list the correct ending for each sentence.

a) interchange the numerator and the denominator.
b) multiply by its reciprocal.
c) multiply numerators and multiply denominators.
d) remove a factor equal to 1.

   1. To simplify rational expressions, _____.

   2. To multiply rational expressions, _____.

   3. To find a reciprocal, _____.

   4. To divide by a rational expression, _____.

Concept Reinforcement

In each of Exercises 5–10, match the product or quotient with an equivalent expression from the column on the right.

5. ____ $\dfrac{x}{2} \cdot \dfrac{5}{y}$ a) $\dfrac{5x}{y}$

6. ____ $\dfrac{x}{2} \div \dfrac{5}{y}$ b) $\dfrac{xy}{5}$

7. ____ $x \cdot \dfrac{5}{y}$ c) $\dfrac{xy}{10}$

8. ____ $\dfrac{x}{2} \div y$ d) $\dfrac{5x}{2y}$

9. ____ $x \div \dfrac{5}{y}$ e) $\dfrac{x}{2y}$

10. ____ $\dfrac{5}{y} \div \dfrac{x}{2}$ f) $\dfrac{10}{xy}$

Multiplication

Multiply. Leave each answer in factored form.

11. $\dfrac{3x}{8} \cdot \dfrac{x + 2}{5x - 1}$

12. $\dfrac{2x}{7} \cdot \dfrac{3x + 5}{x - 1}$

13. $\dfrac{a - 4}{a + 6} \cdot \dfrac{a + 2}{a + 6}$

14. $\dfrac{a + 3}{a + 6} \cdot \dfrac{a + 3}{a - 1}$

15. $\dfrac{n - 4}{n^2 + 4} \cdot \dfrac{n + 4}{n^2 - 4}$

16. $\dfrac{t + 3}{t^2 - 2} \cdot \dfrac{t + 3}{t^2 - 4}$

Multiply and, if possible, simplify.

17. $\dfrac{8t^3}{5t} \cdot \dfrac{3}{4t}$

18. $\dfrac{18}{a^5} \cdot \dfrac{2a^2}{3a}$

19. $\dfrac{3c}{d^2} \cdot \dfrac{8d}{6c^3}$

20. $\dfrac{3x^2y}{2} \cdot \dfrac{4}{xy^3}$

21. $\dfrac{y^2 - 16}{4y + 12} \cdot \dfrac{y + 3}{y - 4}$

22. $\dfrac{m^2 - n^2}{4m + 4n} \cdot \dfrac{m + n}{m - n}$

23. $\dfrac{x^2 - 3x - 10}{(x - 2)^2} \cdot \dfrac{x - 2}{x - 5}$

24. $\dfrac{t + 2}{t - 2} \cdot \dfrac{t^2 - 5t + 6}{(t + 2)^2}$

25. $\dfrac{n^2 - 6n + 5}{n + 6} \cdot \dfrac{n - 6}{n^2 + 36}$

26. $\dfrac{a + 2}{a - 2} \cdot \dfrac{a^2 + 4}{a^2 + 5a + 4}$

27. $\dfrac{a^2 - 9}{a^2} \cdot \dfrac{7a}{a^2 + a - 12}$

28. $\dfrac{x^2 + 10x - 11}{9x} \cdot \dfrac{x^3}{x + 11}$

29. $\dfrac{y^2 - y}{y^2 + 5y + 4} \cdot (y + 4)$

30. $(n - 3) \cdot \dfrac{n^2 + 4n}{n^2 - 5n + 6}$

31. $\dfrac{4v - 8}{5v} \cdot \dfrac{15v^2}{4v^2 - 16v + 16}$

32. $\dfrac{4a^2}{3a^2 - 12a + 12} \cdot \dfrac{3a - 6}{2a}$

33. $\dfrac{t^2 + 2t - 3}{t^2 + 4t - 5} \cdot \dfrac{t^2 - 3t - 10}{t^2 + 5t + 6}$

34. $\dfrac{x^2 + 5x + 4}{x^2 - 6x + 8} \cdot \dfrac{x^2 + 5x - 14}{x^2 + 8x + 7}$

35. $\dfrac{12y + 12}{5y + 25} \cdot \dfrac{3y^2 - 75}{8y^2 - 8}$

36. $\dfrac{9t^2 - 900}{5t^2 - 20} \cdot \dfrac{5t + 10}{3t - 30}$

Aha! 37. $\dfrac{x^2 + 4x + 4}{(x - 1)^2} \cdot \dfrac{x^2 - 2x + 1}{(x + 2)^2}$

38. $\dfrac{x^2 + 7x + 12}{x^2 + 6x + 8} \cdot \dfrac{4 - x^2}{x^2 + x - 6}$

39. $\dfrac{t^2 - 4t + 4}{2t^2 - 7t + 6} \cdot \dfrac{2t^2 + 7t - 15}{t^2 - 10t + 25}$

40. $\dfrac{5y^2 - 4y - 1}{3y^2 + 5y - 12} \cdot \dfrac{y^2 + 6y + 9}{y^2 - 2y + 1}$

41. $(10x^2 - x - 2) \cdot \dfrac{4x^2 - 8x + 3}{10x^2 - 11x - 6}$

42. $\dfrac{2x^2 - 5x + 3}{6x^2 - 5x - 1} \cdot (6x^2 + 13x + 2)$

43. $\dfrac{49x^2 - 25}{4x - 14} \cdot \dfrac{6x^2 - 13x - 28}{28x - 20}$

44. $\dfrac{9t^2 - 4}{8t^2 - 10t + 3} \cdot \dfrac{10t - 5}{3t - 2}$

45. $\dfrac{8x^2 + 14xy - 15y^2}{3x^3 - x^2y} \cdot \dfrac{3x - y}{4xy - 3y^2}$

46. $\dfrac{2x^2 - xy}{6x^2 + 7xy + 2y^2} \cdot \dfrac{9x^2 - 6xy - 8y^2}{3xy - 4y^2}$

47. $\dfrac{c^3 + 8}{c^5 - 4c^3} \cdot \dfrac{c^6 - 4c^5 + 4c^4}{c^2 - 2c + 4}$

48. $\dfrac{t^3 - 27}{t^4 - 9t^2} \cdot \dfrac{t^5 - 6t^4 + 9t}{t^2 + 3t + 9}$

Division

Find the reciprocal of each expression.

49. $\dfrac{2x}{9}$

50. $\dfrac{3 - x}{x^2 + 4}$

51. $a^4 + 3a$

52. $\dfrac{1}{a^2 - b^2}$

Divide and, if possible, simplify.

53. $\dfrac{x}{4} \div \dfrac{5}{x}$

54. $\dfrac{5}{x} \div \dfrac{x}{12}$

55. $\dfrac{a^5}{b^4} \div \dfrac{a^2}{b}$

56. $\dfrac{x^5}{y^2} \div \dfrac{x^2}{y}$

57. $\dfrac{t - 3}{6} \div \dfrac{t + 1}{8}$

58. $\dfrac{10}{a + 3} \div \dfrac{15}{a}$

59. $\dfrac{4y - 8}{y + 2} \div \dfrac{y - 2}{y^2 - 4}$

60. $\dfrac{x^2 - 1}{x} \div \dfrac{x + 1}{2x - 2}$

61. $\dfrac{a}{a - b} \div \dfrac{b}{b - a}$

62. $\dfrac{x - y}{6} \div \dfrac{y - x}{3}$

63. $(n^2 + 5n + 6) \div \dfrac{n^2 - 4}{n + 3}$

64. $(v^2 - 1) \div \dfrac{(v + 1)(v - 3)}{v^2 + 9}$

65. $\dfrac{a + 2}{a - 1} \div \dfrac{3a + 6}{a - 5}$

66. $\dfrac{t - 3}{t + 2} \div \dfrac{4t - 12}{t + 1}$

67. $(2x - 1) \div \dfrac{2x^2 - 11x + 5}{4x^2 - 1}$

68. $(a + 7) \div \dfrac{3a^2 + 14a - 49}{a^2 + 8a + 7}$

69. $\dfrac{w^2 - 14w + 49}{2w^2 - 3w - 14} \div \dfrac{3w^2 - 20w - 7}{w^2 - 6w - 16}$

70. $\dfrac{2m^2 + 59m - 30}{m^2 - 10m + 25} \div \dfrac{2m^2 - 21m + 10}{m^2 + m - 30}$

71. $\dfrac{c^2 + 10c + 21}{c^2 - 2c - 15} \div (5c^2 + 32c - 21)$

72. $\dfrac{z^2 - 2z + 1}{z^2 - 1} \div (4z^2 - z - 3)$

73. $\dfrac{-3 + 3x}{16} \div \dfrac{x - 1}{5}$

74. $\dfrac{-4 + 2x}{15} \div \dfrac{x - 2}{3}$

75. $\dfrac{x - 1}{x + 2} \div \dfrac{1 - x}{4 + x^2}$

76. $\dfrac{-12 + 4x}{12} \div \dfrac{6 - 2x}{6}$

77. $\dfrac{x - y}{x^2 + 2xy + y^2} \div \dfrac{x^2 - y^2}{x^2 - 5xy + 4y^2}$

78. $\dfrac{a^2 - b^2}{a^2 - 4ab + 4b^2} \div \dfrac{a^2 - 3ab + 2b^2}{a - 2b}$

79. $\dfrac{x^3 - 64}{x^3 + 64} \div \dfrac{x^2 - 16}{x^2 - 4x + 16}$

80. $\dfrac{8y^3 - 27}{64y^3 - 1} \div \dfrac{4y^2 - 9}{16y^2 + 4y + 1}$

81. $\dfrac{8a^3 + b^3}{2a^2 + 3ab + b^2} \div \dfrac{8a^2 - 4ab + 2b^2}{4a^2 + 4ab + b^2}$

82. $\dfrac{x^3 + 8y^3}{2x^2 + 5xy + 2y^2} \div \dfrac{x^3 - 2x^2y + 4xy^2}{8x^2 - 2y^2}$

83. Why is it important to insert parentheses when multiplying rational expressions such as

$$\dfrac{x + 2}{5x - 7} \cdot \dfrac{3x - 1}{x + 4}?$$

84. As a first step in dividing $\dfrac{x}{3}$ by $\dfrac{7}{x}$, Jan canceled the x's.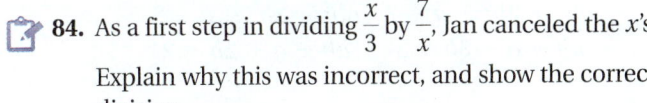

Explain why this was incorrect, and show the correct division.

Skill Review

Graph.

85. $y = \frac{1}{2}x - 5$ [3.5]

86. $3x - 2y = 12$ [3.3]

87. $3(x - 1) = 4$ [3.6]

88. $y - 2 = -(x + 4)$ [3.7]

89. $3y = 5x$ [3.2]

90. $\frac{1}{2}y = 2$ [3.6]

Synthesis

91. Is the reciprocal of a product the product of the two reciprocals? Why or why not?

92. Explain why the quotient

$$\frac{x + 3}{x - 5} \div \frac{x - 7}{x + 1}$$

is undefined for $x = 5$, $x = -1$, and $x = 7$, but *is* defined for $x = -3$.

93. Find the reciprocal of $2\frac{1}{3}x$.

94. Find the reciprocal of $7.25x$.

Simplify.

95. $(x - 2a) \div \dfrac{a^2x^2 - 4a^4}{a^2x + 2a^3}$

96. $\dfrac{3x^2 - 2xy - y^2}{x^2 - y^2} \div (3x^2 + 4xy + y^2)^2$

97. $\dfrac{3a^2 - 5ab - 12b^2}{3ab + 4b^2} \div (3b^2 - ab)^2$

Aha! 98. $\dfrac{a^2 - 3b}{a^2 + 2b} \cdot \dfrac{a^2 - 2b}{a^2 + 3b} \cdot \dfrac{a^2 + 2b}{a^2 - 3b}$

99. $\dfrac{z^2 - 8z + 16}{z^2 + 8z + 16} \div \dfrac{(z - 4)^5}{(z + 4)^5} \div \dfrac{3z + 12}{z^2 - 16}$

100. $\dfrac{(t + 2)^3}{(t + 1)^3} \div \dfrac{t^2 + 4t + 4}{t^2 + 2t + 1} \cdot \dfrac{t + 1}{t + 2}$

101. $\dfrac{a^4 - 81b^4}{a^2c - 6abc + 9b^2c} \cdot \dfrac{a + 3b}{a^2 + 9b^2} \div \dfrac{a^2 + 6ab + 9b^2}{(a - 3b)^2}$

102. $\dfrac{3y^3 + 6y^2}{y^2 - y - 12} \div \dfrac{y^2 - y}{y^2 - 2y - 8} \cdot \dfrac{y^2 + 5y + 6}{y^2}$

103. $\dfrac{xy - 2x + y - 2}{xy + 4x - y - 4} \cdot \dfrac{xy + y + 4x + 4}{xy - y - 2x + 2}$

104. $\dfrac{ab^2 + 2b^2 + a + 2}{ab - a - 3b + 3} \cdot \dfrac{ab^2 - 3b^2 - a + 3}{ab + a + 2b + 2}$

105. $\dfrac{3x^2 - 12x + bx - 4b}{4x^2 - 16x - bx + 4b} \div \dfrac{3bx + b^2 + 6x + 2b}{4bx - b^2 + 8x - 2b}$

106. $\dfrac{2x^2y - xy^2 + 6x^2 - 3xy}{3y^2 - xy + 9y - 3x} \div \dfrac{2x^3 - x^2y + 8x^2 - 4xy}{3xy - x^2 + 6y - 2x}$

107. $\dfrac{8n^2 - 10n + 3}{4n^2 - 4n - 3} \cdot \dfrac{6n^2 - 5n - 6}{6n^2 + 7n - 5} \div \dfrac{12n^2 - 17n + 6}{6n^2 + 7n - 5}$

108. $\dfrac{2p^2 - p - 6}{16p^2 - 25} \cdot \dfrac{12p^2 + 13p - 35}{4p^2 + 12p + 9} \div \dfrac{12p^2 + 43p + 35}{2p^2 - p - 6}$

109. Use a graphing calculator to check that

$$\frac{x - 1}{x^2 + 2x + 1} \div \frac{x^2 - 1}{x^2 - 5x + 4}$$

is equivalent to

$$\frac{x^2 - 5x + 4}{(x + 1)^3}.$$

YOUR TURN ANSWERS: SECTION 6.2

1. $\dfrac{5t}{t - 1}$ **2.** $\dfrac{(x + 2)(x + 1)}{(x - 2)(x - 3)}$ **3.** $\dfrac{x - 1}{x + 1}$

QUICK QUIZ: SECTIONS 6.1–6.2

1. Simplify: $\dfrac{6x^2 - 54}{x^2 - 7x + 12}$.   [6.1]

Perform the indicated operation and, if possible, simplify.   [6.2]

2. $\dfrac{y + 1}{y + 2} \cdot \dfrac{y^2 - y - 6}{2y^2 + 3y + 1}$

3. $\dfrac{5x - 5}{2x + 2} \div \dfrac{10x}{x^2 - 1}$

4. $\dfrac{m - 2}{3m + 9} \cdot \dfrac{m^2 + 6m + 9}{2m^2 - 8}$

5. $\dfrac{x^2 - 16}{x^2 - 25} \div \dfrac{x^2 - 8x + 16}{x^2 - x - 20}$

PREPARE TO MOVE ON

Simplify.

1. $\dfrac{7}{8} + \dfrac{5}{6}$ [1.3]   **2.** $\dfrac{7}{15} - \dfrac{3}{10}$ [1.3]

3. $2x^2 - x + 1 - (x^2 - x - 2)$ [4.4]

4. $3x^2 + x - 7 - (5x^2 + 5x - 8)$ [4.4]

6.3   Addition, Subtraction, and Least Common Denominators

Addition When Denominators Are the Same   ▪   Subtraction When Denominators Are the Same   ▪   Least Common Multiples and Denominators

Addition When Denominators Are the Same

Recall that to add fractions having the same denominator, like $\frac{2}{7}$ and $\frac{3}{7}$, we add the numerators and keep the common denominator: $\frac{2}{7} + \frac{3}{7} = \frac{5}{7}$. The same procedure is used for all rational expressions that share a common denominator. In this section, we assume that all denominators are nonzero.

> **THE SUM OF TWO RATIONAL EXPRESSIONS**
>
> To add when the denominators are the same, add the numerators and keep the common denominator:
>
> $$\frac{A}{B} + \frac{C}{B} = \frac{A + C}{B}.$$

EXAMPLE 1   Add. Simplify the result, if possible.

a) $\dfrac{4}{a} + \dfrac{3 + a}{a}$

b) $\dfrac{2x^2 + 3x - 7}{2x + 1} + \dfrac{x^2 + x - 8}{2x + 1}$

c) $\dfrac{x - 5}{x^2 - 9} + \dfrac{2}{x^2 - 9}$

SOLUTION

a) $\dfrac{4}{a} + \dfrac{3 + a}{a} = \dfrac{7 + a}{a}$ When the denominators are alike, add the numerators and keep the common denominator.

b) $\dfrac{2x^2 + 3x - 7}{2x + 1} + \dfrac{x^2 + x - 8}{2x + 1} = \dfrac{(2x^2 + 3x - 7) + (x^2 + x - 8)}{2x + 1}$

$= \dfrac{3x^2 + 4x - 15}{2x + 1}$ Combining like terms

$= \dfrac{(3x - 5)(x + 3)}{2x + 1}$ Factoring. There are no common factors, so we cannot simplify further.

c) $\dfrac{x - 5}{x^2 - 9} + \dfrac{2}{x^2 - 9} = \dfrac{x - 3}{x^2 - 9}$ Combining like terms in the numerator: $x - 5 + 2 = x - 3$

$= \dfrac{x - 3}{(x - 3)(x + 3)}$ Factoring

$= \dfrac{1 \cdot (x - 3)}{(x - 3)(x + 3)}$ Removing a factor equal to 1:
$\dfrac{x - 3}{x - 3} = 1$

$= \dfrac{1}{x + 3}$

1. Add. Simplify the result, if possible.

$$\dfrac{3x + 1}{x + 2} + \dfrac{x + 7}{x + 2}$$

 YOUR TURN

Subtraction When Denominators Are the Same

When two fractions have the same denominator, we subtract one numerator from the other and keep the common denominator, as in $\frac{5}{7} - \frac{2}{7} = \frac{3}{7}$. The same procedure is used with rational expressions.

THE DIFFERENCE OF TWO RATIONAL EXPRESSIONS

To subtract when the denominators are the same, subtract the second numerator from the first numerator and keep the common denominator:

$$\frac{A}{B} - \frac{C}{B} = \frac{A - C}{B}.$$

CAUTION! The fraction bar under a numerator is a grouping symbol, just like parentheses. Thus, when a numerator is subtracted, it is important to subtract *every* term in that numerator.

EXAMPLE 2 Subtract and, if possible, simplify.

a) $\dfrac{3x}{x + 2} - \dfrac{x - 5}{x + 2}$

b) $\dfrac{x^2}{x - 4} - \dfrac{x + 12}{x - 4}$

SOLUTION

a) $\dfrac{3x}{x + 2} - \dfrac{x - 5}{x + 2} = \dfrac{3x - (x - 5)}{x + 2}$ The parentheses are needed to make sure that we subtract both terms.

$= \dfrac{3x - x + 5}{x + 2}$ Removing the parentheses and changing signs (using the distributive law)

$= \dfrac{2x + 5}{x + 2}$ Combining like terms

b) $\dfrac{x^2}{x - 4} - \dfrac{x + 12}{x - 4} = \dfrac{x^2 - (x + 12)}{x - 4}$ Remember the parentheses!

$= \dfrac{x^2 - x - 12}{x - 4}$ Removing parentheses (using the distributive law)

$= \dfrac{(x - 4)(x + 3)}{x - 4}$ Factoring, in hopes of simplifying

$= \dfrac{(x - 4)(x + 3)}{x - 4}$ Removing a factor equal to 1: $\dfrac{x - 4}{x - 4} = 1$

$= x + 3$

2. Subtract and, if possible, simplify:

$$\frac{3a}{2a - 1} - \frac{a - 1}{2a - 1}.$$

YOUR TURN

Least Common Multiples and Denominators

Thus far, every pair of rational expressions that we have added or subtracted shared a common denominator. To add or subtract rational expressions that have different denominators, we must first find equivalent rational expressions that *do* have a common denominator.

In algebra, we find a common denominator much as we do in arithmetic. Recall that to add $\frac{1}{12}$ and $\frac{7}{30}$, we first identify the smallest number that contains both 12 and 30 as factors. Such a number, the **least common multiple (LCM)** of the denominators, is then used as the **least common denominator (LCD)**.

Let's find the LCM of 12 and 30 using a method that can also be used with polynomials. We begin by writing the prime factorizations of 12 and 30:

$$12 = 2 \cdot 2 \cdot 3;$$
$$30 = 2 \cdot 3 \cdot 5.$$

The LCM must include the factors of each number, so it must include each prime factor the greatest number of times that it appears in either of the factorizations. To find the LCM for 12 and 30, we select one factorization, say

$$2 \cdot 2 \cdot 3,$$

and note that because it lacks a factor of 5, it does not contain the entire factorization of 30. If we multiply $2 \cdot 2 \cdot 3$ by 5, every prime factor occurs just often enough to contain both 12 and 30 as factors.

```
                          ┌───── 12 is a factor of the LCM.
                      ┌─┬─┐
        LCM = 2 · 2 · 3 · 5
                      └─┴─┴─
                          └───── 30 is a factor of the LCM.
```

Note that each prime factor—2, 3, and 5—is used the greatest number of times that it appears in either of the individual factorizations. The factor 2 occurs twice and the factors 3 and 5 once each.

TO FIND THE LEAST COMMON DENOMINATOR (LCD)

1. Write the prime factorization of each denominator.
2. Select one of the factorizations and inspect it to see if it completely contains the other factorization.

   a) If it does, it represents the LCM of the denominators.
   b) If it does not, multiply that factorization by any factors of the other denominator that it lacks. The final product is the LCM of the denominators.

The LCD is the LCM of the denominators. It should contain each factor the greatest number of times that it occurs in any of the individual factorizations.

Student Notes

The terms *least common multiple* and *least common denominator* are similar enough that they may be confusing. We find the LCM of polynomials, and we find the LCD of rational expressions.

EXAMPLE 3   Find the LCD of $\dfrac{5}{36x^2}$ and $\dfrac{7}{24x}$.

SOLUTION

1. We begin by writing the prime factorizations of $36x^2$ and $24x$:

$$36x^2 = 2 \cdot 2 \cdot 3 \cdot 3 \cdot x \cdot x;$$
$$24x = 2 \cdot 2 \cdot 2 \cdot 3 \cdot x.$$

The procedure above can be used to find the LCM of three or more polynomials as well. We factor each polynomial and then construct the LCM using each factor the greatest number of times that it appears in any one factorization.

EXAMPLE 5   For each group of polynomials, find the LCM.

a) $12x$, $16y$, and $8xyz$

b) $x^2 + 4$, $x + 1$, and 5

SOLUTION

a) $12x = 2 \cdot 2 \cdot 3 \cdot x$

$16y = 2 \cdot 2 \cdot 2 \cdot 2 \cdot y$

$8xyz = 2 \cdot 2 \cdot 2 \cdot x \cdot y \cdot z$

$\text{LCM} = 2 \cdot 2 \cdot 3 \cdot x \cdot 2 \cdot 2 \cdot y \cdot z$

We start with the factorization of $12x$. Any one of the three could be chosen.

We multiply by the factors of $16y$ that are missing.

We then multiply by the factor of $8xyz$ that is missing.

The LCM is $2^4 \cdot 3 \cdot xyz$, or $48xyz$.

b)   Since $x^2 + 4$, $x + 1$, and 5 are not factorable, the LCM is their product: $5(x^2 + 4)(x + 1)$.

 YOUR TURN

To add or subtract rational expressions with different denominators, we first write equivalent expressions that have the LCD. To do this, we multiply each rational expression by a carefully constructed form of 1.

EXAMPLE 6   Find equivalent expressions that have the LCD:

$$\frac{x + 3}{x^2 + 5x - 6}, \quad \frac{x + 7}{x^2 - 1}.$$

SOLUTION   From Example 4(c), we know that the LCD is

$$(x + 6)(x - 1)(x + 1).$$

The factor of the LCD that is missing from $x^2 + 5x - 6$ is $x + 1$. We multiply by 1 using $(x + 1)/(x + 1)$:

$$\frac{x + 3}{x^2 + 5x - 6} = \frac{x + 3}{(x + 6)(x - 1)} \cdot \frac{x + 1}{x + 1} = \frac{(x + 3)(x + 1)}{(x + 6)(x - 1)(x + 1)}.$$

The factor of the LCD that is missing from $x^2 - 1$ is $x + 6$. We multiply by 1 using $(x + 6)/(x + 6)$:

$$\frac{x + 7}{x^2 - 1} = \frac{x + 7}{(x + 1)(x - 1)} \cdot \frac{x + 6}{x + 6} = \frac{(x + 7)(x + 6)}{(x + 1)(x - 1)(x + 6)}.$$

Both rational expressions now have the same denominator. We leave the results in factored form.

 YOUR TURN

5. Find the LCM of $x^2 + 5x$, $x^2 - 25$, and $5x$.

6. Find equivalent expressions that have the LCD:

$$\frac{5}{x^2 + x}, \quad \frac{2}{3x}.$$

6.3 EXERCISE SET

FOR EXTRA HELP

MyMathLab® MathXL
PRACTICE   WATCH   READ   REVIEW

Vocabulary and Reading Check

Use one or more words to complete each of the following statements.

1. To add two rational expressions when the denominators are the same, add _____ and keep the common _____.

2. When a numerator is being subtracted, use parentheses to make sure that every _____ in that numerator is subtracted.

3. The least common multiple of two denominators is usually referred to as the _____ and is abbreviated _____.

4. The least common denominator of two rational expressions must contain every _____ that is in either denominator.

Addition and Subtraction When Denominators Are the Same

Perform the indicated operation. Simplify, if possible.

5. $\dfrac{3}{t} + \dfrac{5}{t}$

6. $\dfrac{8}{y^2} + \dfrac{2}{y^2}$

7. $\dfrac{x}{12} + \dfrac{2x + 5}{12}$

8. $\dfrac{a}{7} + \dfrac{3a - 4}{7}$

9. $\dfrac{4}{a + 3} + \dfrac{5}{a + 3}$

10. $\dfrac{5}{x + 2} + \dfrac{8}{x + 2}$

11. $\dfrac{11}{4x - 7} - \dfrac{3}{4x - 7}$

12. $\dfrac{9}{2x + 3} - \dfrac{5}{2x + 3}$

13. $\dfrac{3y + 8}{2y} - \dfrac{y + 1}{2y}$

14. $\dfrac{5 + 3t}{4t} - \dfrac{2t + 1}{4t}$

15. $\dfrac{5x + 7}{x + 3} + \dfrac{x + 11}{x + 3}$

16. $\dfrac{3x + 4}{x - 1} + \dfrac{2x - 9}{x - 1}$

17. $\dfrac{5x + 7}{x + 3} - \dfrac{x + 11}{x + 3}$

18. $\dfrac{3x + 4}{x - 1} - \dfrac{2x - 9}{x - 1}$

19. $\dfrac{a^2}{a - 4} + \dfrac{a - 20}{a - 4}$

20. $\dfrac{x^2}{x + 5} + \dfrac{7x + 10}{x + 5}$

21. $\dfrac{y^2}{y + 2} - \dfrac{5y + 14}{y + 2}$

22. $\dfrac{t^2}{t - 3} - \dfrac{8t - 15}{t - 3}$

Aha! 23. $\dfrac{t^2 - 5t}{t - 1} + \dfrac{5t - t^2}{t - 1}$

24. $\dfrac{y^2 + 6y}{y + 2} + \dfrac{2y + 12}{y + 2}$

25. $\dfrac{x - 6}{x^2 + 5x + 6} + \dfrac{9}{x^2 + 5x + 6}$

26. $\dfrac{x - 5}{x^2 - 4x + 3} + \dfrac{2}{x^2 - 4x + 3}$

27. $\dfrac{3a^2 + 14}{a^2 + 5a - 6} - \dfrac{13a}{a^2 + 5a - 6}$

28. $\dfrac{2a^2 + 15}{a^2 - 7a + 12} - \dfrac{11a}{a^2 - 7a + 12}$

29. $\dfrac{t^2 - 5t}{t^2 + 6t + 9} + \dfrac{4t - 12}{t^2 + 6t + 9}$

30. $\dfrac{y^2 - 7y}{y^2 + 8y + 16} + \dfrac{6y - 20}{y^2 + 8y + 16}$

31. $\dfrac{2y^2 + 3y}{y^2 - 7y + 12} - \dfrac{y^2 + 4y + 6}{y^2 - 7y + 12}$

32. $\dfrac{3a^2 + 7}{a^2 - 2a - 8} - \dfrac{7 + 3a^2}{a^2 - 2a - 8}$

33. $\dfrac{3 - 2x}{x^2 - 6x + 8} + \dfrac{7 - 3x}{x^2 - 6x + 8}$

34. $\dfrac{1 - 2t}{t^2 - 5t + 4} + \dfrac{4 - 3t}{t^2 - 5t + 4}$

35. $\dfrac{x - 9}{x^2 + 3x - 4} - \dfrac{2x - 5}{x^2 + 3x - 4}$

36. $\dfrac{5 - 3x}{x^2 - 2x + 1} - \dfrac{x + 1}{x^2 - 2x + 1}$

Least Common Multiples and Denominators

Find the LCM.

37. 15, 36   38. 18, 30   39. 8, 9

40. 12, 15   41. 6, 12, 15   42. 8, 32, 50

Find the LCM.

43. $18t^2$, $6t^5$ 44. $8x^5$, $24x^2$

45. $15a^4b^7$, $10a^2b^8$ 46. $6a^2b^7$, $9a^5b^2$

47. $2(y - 3)$, $6(y - 3)$ 48. $4(x - 1)$, $8(x - 1)$

49. $x^2 - 2x - 15$, $x^2 - 9$ 50. $t^2 - 4$, $t^2 + 7t + 10$

51. $t^3 + 4t^2 + 4t$, $t^2 - 4t$ 52. $y^3 - y^2$, $y^4 - y^2$

53. $6xz^2$, $8x^2y$, $15y^3z$ 54. $12s^3t$, $15sv^2$, $6t^4v$

55. $a + 1$, $(a - 1)^2$, $a^2 - 1$

56. $x - 2$, $(x + 2)^2$, $x^2 - 4$

57. $2n^2 + n - 1$, $2n^2 + 3n - 2$

58. $m^2 - 2m - 3$, $2m^2 + 3m + 1$

Aha! 59. $t - 3$, $t + 3$, $t^2 - 9$

60. $a - 5$, $a^2 - 10a + 25$

61. $6x^3 - 24x^2 + 18x,\ 4x^5 - 24x^4 + 20x^3$

62. $9x^3 - 9x^2 - 18x,\ 6x^5 - 24x^4 + 24x^3$

63. $2x^3 - 2,\ x^2 - 1$

64. $3a^3 + 24,\ a^2 - 4$

Find equivalent expressions that have the LCD.

65. $\dfrac{5}{6t^4},\ \dfrac{s}{18t^2}$

66. $\dfrac{7}{10y^2},\ \dfrac{x}{5y^6}$

67. $\dfrac{7}{3x^4y^2},\ \dfrac{4}{9xy^3}$

68. $\dfrac{3}{2a^2b},\ \dfrac{7}{8ab^2}$

69. $\dfrac{2x}{x^2 - 4},\ \dfrac{4x}{x^2 + 5x + 6}$

70. $\dfrac{5x}{x^2 - 9},\ \dfrac{2x}{x^2 + 11x + 24}$

71. Explain why the product of two numbers is not always their least common multiple.

72. If the LCM of two numbers is their product, what can you conclude about the two numbers?

Skill Review

Solve.

73. $2x - 7 = 5x + 3$ [2.2]

74. $\frac{1}{3}x < \frac{2}{5}x - 1$ [2.6]

75. $x^2 - 8x = 20$ [5.7]

76. $-x \geq 16 + x$ [2.6]

77. $2x^2 + 4x + 2 = 0$ [5.7]

78. $2 - 3(x - 7) = 15$ [2.2]

Synthesis

79. If the LCM of two third-degree polynomials is a sixth-degree polynomial, what can be concluded about the two polynomials?

80. If the LCM of a binomial and a trinomial is the trinomial, what relationship exists between the two expressions?

Perform the indicated operations. Simplify, if possible.

81. $\dfrac{6x - 1}{x - 1} + \dfrac{3(2x + 5)}{x - 1} + \dfrac{3(2x - 3)}{x - 1}$

82. $\dfrac{2x + 11}{x - 3} \cdot \dfrac{3}{x + 4} + \dfrac{-1}{4 + x} \cdot \dfrac{6x + 3}{x - 3}$

83. $\dfrac{x^2}{3x^2 - 5x - 2} - \dfrac{2x}{3x + 1} \cdot \dfrac{1}{x - 2}$

84. $\dfrac{x + y}{x^2 - y^2} + \dfrac{x - y}{x^2 - y^2} - \dfrac{2x}{x^2 - y^2}$

African Artistry. *In Southeast Mozambique, the design of every woven handbag, or gipatsi (plural, sipatsi) is created by repeating two or more geometric patterns.*

Each pattern encircles the bag, sharing the strands of fabric with any pattern above or below. The length, or period, of each pattern is the number of strands required to construct the pattern. For a gipatsi to be considered beautiful, each individual pattern must fit a whole number of times around the bag.

Source: Gerdes, Paulus, *Women, Art and Geometry in Southern Africa.* Asmara, Eritrea: Africa World Press, Inc., p. 5

85. A weaver is using two patterns to create a gipatsi. Pattern A is 10 strands long, and pattern B is 3 strands long. What is the smallest number of strands that can be used to complete the gipatsi?

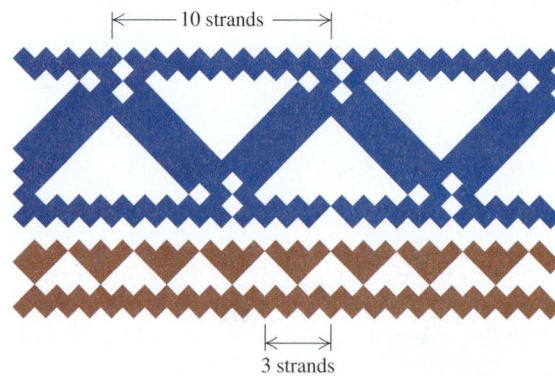

86. A weaver is using a four-strand pattern, a six-strand pattern, and an eight-strand pattern. What is the smallest number of strands that can be used to complete the gipatsi?

87. For technical reasons, the number of strands is generally a multiple of 4. Answer Exercise 83 with this additional requirement in mind.

Find the LCM.

88. 80, 96, 108

89. $4x^2 - 25,\ 6x^2 - 7x - 20,\ (9x^2 + 24x + 16)^2$

90. $9n^2 - 9$, $(5n^2 - 10n + 5)^2$, $15n - 15$

91. *Copiers.*   The Lanier 365C copier can print 60 color pages per minute. The Copystar CS3550ci copier can print 35 color pages per minute. If both machines begin printing at the same instant, how long will it be until they again begin printing a page at exactly the same time?

92. *Running.*   Joni and Trey leave the starting point of a jogging loop at the same time. Joni jogs a lap in 6 min and Trey jogs one in 8 min. Assuming they continue to run at the same pace, when will they next meet at the starting place?

93. *Bus Schedules.*   Beginning at 5:00 A.M., a hotel shuttle bus leaves Salton Airport every 15 min, and the downtown shuttle bus leaves the airport every 25 min. What time will it be when both shuttles again leave at the same time?

94. *Appliances.*   Smoke detectors last an average of 10 years, water heaters an average of 12 years, and refrigerators an average of 15 years. If an apartment house is equipped with new smoke detectors, water heaters, and refrigerators in 2014, in what year will all three appliances need to be replaced at once?

Source: Demesne.info

95. Explain how evaluating can be used to perform a partial check on the result of Example 1(c):

$$\frac{x-5}{x^2-9} + \frac{2}{x^2-9} = \frac{1}{x+3}.$$

96. On p. 389, the second step in finding an LCD is to select one of the factorizations of the denominators. Does it matter which one is selected? Why or why not?

97. *Research.*   Determine the average life of three or more appliances in your apartment or house. If all these appliances were new when your home was built, when can you expect them all to fail in the same year?

QUICK QUIZ: SECTIONS 6.1–6.3

1. Simplify: $\dfrac{x+3}{x^2+10x+21}$.   [6.1]

2. Find the LCM: $x^2 - 6x, x^2 - 7x + 6$.   [6.3]

Perform the indicated operation and, if possible, simplify.

3. $\dfrac{x+1}{x^2+10x+21} \cdot \dfrac{x+7}{x^2-1}$ [6.2]

4. $\dfrac{2t-10}{9t^2} \div \dfrac{t^3-7t^2+10t}{6t}$ [6.2]

5. $\dfrac{x+3}{x-2} - \dfrac{x-5}{x-2}$ [6.3]

PREPARE TO MOVE ON

Using the identity $-\dfrac{a}{b} = \dfrac{-a}{b} = \dfrac{a}{-b}$, *write each number in two equivalent forms.*   [1.7]

1. $-\dfrac{5}{8}$ **2.** $\dfrac{4}{-11}$

Write an equivalent expression without parentheses.   [1.8]

3. $-(x-y)$ **4.** $-(3-a)$

Multiply and simplify.   [1.8]

5. $-1(2x-7)$ **6.** $-1(a-b)$

6.4 Addition and Subtraction with Unlike Denominators

Adding and Subtracting with LCDs   •   When Factors Are Opposites

Adding and Subtracting with LCDs

We now know how to rewrite two rational expressions in equivalent forms that use the LCD. Once rational expressions share a common denominator, they can be added or subtracted. We assume all denominators are nonzero.

TO ADD OR SUBTRACT RATIONAL EXPRESSIONS HAVING DIFFERENT DENOMINATORS

1. Find the LCD.
2. Multiply each rational expression by a form of 1 made up of the factors of the LCD missing from that expression's denominator.
3. Add or subtract the numerators, as indicated. Write the sum or the difference over the LCD.
4. Simplify, if possible.

EXAMPLE 1   Add: $\dfrac{5x^2}{8} + \dfrac{7x}{12}$.

SOLUTION

1. First, we find the LCD:

$$\left.\begin{array}{l} 8 = 2 \cdot 2 \cdot 2 \\ 12 = 2 \cdot 2 \cdot 3 \end{array}\right\} \quad \text{LCD} = 2 \cdot 2 \cdot 2 \cdot 3, \text{ or } 24.$$

2. The denominator 8 must be multiplied by 3 in order to obtain the LCD. The denominator 12 must be multiplied by 2 in order to obtain the LCD. Thus we multiply the first expression by $\frac{3}{3}$ and the second expression by $\frac{2}{2}$ to get the LCD:

$$\frac{5x^2}{8} + \frac{7x}{12} = \frac{5x^2}{2 \cdot 2 \cdot 2} + \frac{7x}{2 \cdot 2 \cdot 3}$$

$$= \frac{5x^2}{2 \cdot 2 \cdot 2} \cdot \frac{3}{3} + \frac{7x}{2 \cdot 2 \cdot 3} \cdot \frac{2}{2} \qquad \text{Multiplying by a form of 1 to get the LCD}$$

$$= \frac{15x^2}{24} + \frac{14x}{24}.$$

3. Next, we add the numerators: $\dfrac{15x^2}{24} + \dfrac{14x}{24} = \dfrac{15x^2 + 14x}{24}$.

4. We factor in hope of simplifying: $\dfrac{x(15x + 14)}{24}$.

   The expression cannot be simplified.

1. Add: $\dfrac{2x}{9} + \dfrac{5}{12}$.

 YOUR TURN

EXAMPLE 2   Subtract: $\dfrac{7}{8x} - \dfrac{5}{12x^2}$.

SOLUTION   We follow the four steps shown above. First, we find the LCD:

$$\left.\begin{array}{l} 8x = 2 \cdot 2 \cdot 2 \cdot x \\ 12x^2 = 2 \cdot 2 \cdot 3 \cdot x \cdot x \end{array}\right\} \quad \text{LCD} = 2 \cdot 2 \cdot 3 \cdot x \cdot x \cdot 2, \text{ or } 24x^2.$$

The denominator $8x$ must be multiplied by $3x$ in order to obtain the LCD. The denominator $12x^2$ must be multiplied by 2 in order to obtain the LCD. Thus we multiply by $\dfrac{3x}{3x}$ and $\dfrac{2}{2}$ to get the LCD. Then we subtract and, if possible, simplify.

$$\frac{7}{8x} - \frac{5}{12x^2} = \frac{7}{8x} \cdot \frac{3x}{3x} - \frac{5}{12x^2} \cdot \frac{2}{2}$$

CAUTION!   Do not simplify *these* rational expressions or you will lose the LCD.

$$= \frac{21x}{24x^2} - \frac{10}{24x^2}$$

$$= \frac{21x - 10}{24x^2} \qquad \text{This cannot be simplified, so we are done.}$$

2. Subtract: $\dfrac{3}{10x^2} - \dfrac{5}{4x}$.

YOUR TURN

EXAMPLE 3   Add: $\dfrac{2a}{a^2 - 1} + \dfrac{1}{a^2 + a}$.

SOLUTION   First, we find the LCD:

Find the LCD.

$$\left.\begin{array}{l} a^2 - 1 = (a - 1)(a + 1) \\ a^2 + a = a(a + 1). \end{array}\right\} \quad \text{LCD} = (a - 1)(a + 1)a$$

We multiply by a form of 1 to get the LCD in each expression:

Write each expression with the LCD.

$$\frac{2a}{a^2 - 1} + \frac{1}{a^2 + a} = \frac{2a}{(a - 1)(a + 1)} \cdot \frac{a}{a} + \frac{1}{a(a + 1)} \cdot \frac{a - 1}{a - 1} \quad \begin{array}{l} \text{Multiplying by} \\ \frac{a}{a} \text{ and } \frac{a - 1}{a - 1} \text{ to} \\ \text{get the LCD} \end{array}$$

$$= \frac{2a^2}{(a - 1)(a + 1)a} + \frac{a - 1}{a(a + 1)(a - 1)}$$

Add numerators.

$$= \frac{2a^2 + a - 1}{a(a - 1)(a + 1)} \qquad \text{Adding numerators}$$

Simplify.

$$\left.\begin{array}{l} = \dfrac{(2a - 1)\cancel{(a + 1)}}{a(a - 1)\cancel{(a + 1)}} \\[2mm] = \dfrac{2a - 1}{a(a - 1)}. \end{array}\right\} \quad \begin{array}{l} \text{Simplifying by factoring and} \\ \text{removing a factor equal to 1:} \\ \dfrac{a + 1}{a + 1} = 1 \end{array}$$

3. Add: $\dfrac{4x}{x^2 - 25} + \dfrac{x}{x + 5}$.

 YOUR TURN

EXAMPLE 4   Subtract.

a) $\dfrac{x + 4}{x - 2} - \dfrac{x - 7}{x + 5}$

b) $\dfrac{x}{x^2 + 5x + 6} - \dfrac{2}{x^2 + 3x + 2}$

SOLUTION

a) First, we find the LCD. It is just the product of the denominators:

$$\text{LCD} = (x - 2)(x + 5).$$

We multiply by a form of 1 to get the LCD in each expression. Then we subtract and try to simplify.

$$\frac{x + 4}{x - 2} - \frac{x - 7}{x + 5} = \frac{x + 4}{x - 2} \cdot \frac{x + 5}{x + 5} - \frac{x - 7}{x + 5} \cdot \frac{x - 2}{x - 2}$$

$$= \frac{x^2 + 9x + 20}{(x - 2)(x + 5)} - \frac{x^2 - 9x + 14}{(x - 2)(x + 5)} \quad \begin{array}{l} \text{Multiplying out} \\ \text{numerators (but} \\ \text{not denominators)} \end{array}$$

$$= \frac{x^2 + 9x + 20 - (x^2 - 9x + 14)}{(x - 2)(x + 5)} \quad \begin{array}{l} \text{Parentheses are} \\ \text{important.} \end{array}$$

$$= \frac{x^2 + 9x + 20 - x^2 + 9x - 14}{(x - 2)(x + 5)} \quad \begin{array}{l} \text{Removing paren-} \\ \text{theses and sub-} \\ \text{tracting every term} \end{array}$$

$$= \frac{18x + 6}{(x - 2)(x + 5)}$$

$$= \frac{6(3x + 1)}{(x - 2)(x + 5)} \quad \text{We cannot simplify.}$$

Student Notes

As you can see, adding or subtracting rational expressions can involve many steps. Therefore, it is wise to double-check each step of your work as you work through each problem. Waiting to inspect your work at the end of each problem is usually a less efficient use of your time.

b) $\dfrac{x}{x^2 + 5x + 6} - \dfrac{2}{x^2 + 3x + 2}$

> **Find the LCD.**

$= \dfrac{x}{(x + 2)(x + 3)} - \dfrac{2}{(x + 2)(x + 1)}$ Factoring denominators. The LCD is $(x + 2)(x + 3)(x + 1)$.

> **Write each expression with the LCD.**

$= \dfrac{x}{(x + 2)(x + 3)} \cdot \dfrac{x + 1}{x + 1} - \dfrac{2}{(x + 2)(x + 1)} \cdot \dfrac{x + 3}{x + 3}$

$= \dfrac{x^2 + x}{(x + 2)(x + 3)(x + 1)} - \dfrac{2x + 6}{(x + 2)(x + 3)(x + 1)}$

> **Subtract numerators.**

$= \dfrac{x^2 + x - (2x + 6)}{(x + 2)(x + 3)(x + 1)}$ Don't forget the parentheses!

$= \dfrac{x^2 + x - 2x - 6}{(x + 2)(x + 3)(x + 1)}$ Remember to subtract each term in $2x + 6$.

$= \dfrac{x^2 - x - 6}{(x + 2)(x + 3)(x + 1)}$ Combining like terms in the numerator

4. Subtract:

> **Simplify.**

$= \dfrac{\cancel{(x + 2)}(x - 3)}{\cancel{(x + 2)}(x + 3)(x + 1)}$ Factoring and simplifying; $\dfrac{x + 2}{x + 2} = 1$

$\dfrac{x}{x^2 + 3x + 2} - \dfrac{3}{x^2 - x - 2}.$

$= \dfrac{x - 3}{(x + 3)(x + 1)}$

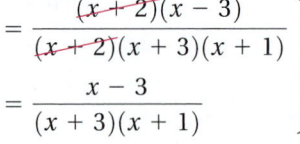 YOUR TURN

When Factors Are Opposites

A special case arises when one denominator is the opposite of the other. When this occurs, we can multiply either expression by 1 using $\dfrac{-1}{-1}$.

EXAMPLE 5   Add: $\dfrac{3}{8a} + \dfrac{1}{-8a}.$

SOLUTION

$\dfrac{3}{8a} + \dfrac{1}{-8a} = \dfrac{3}{8a} + \dfrac{-1}{-1} \cdot \dfrac{1}{-8a}$

$\qquad\qquad = \dfrac{3}{8a} + \dfrac{-1}{8a} = \dfrac{2}{8a}$

> When denominators are opposites, we multiply one rational expression by $-1/-1$ to get the LCD.

5. Add: $\dfrac{t}{2} + \dfrac{3}{-2}.$

$\qquad\qquad = \dfrac{2 \cdot 1}{2 \cdot 4a} = \dfrac{1}{4a}$ Simplifying by removing a factor equal to 1: $\dfrac{2}{2} = 1$

YOUR TURN

Expressions of the form $a - b$ and $b - a$ are opposites of each other. When either of these binomials is multiplied by -1, the result is the other binomial:

$$-1(a - b) = -a + b = b + (-a) = b - a;$$
$$-1(b - a) = -b + a = a + (-b) = a - b.$$

> Multiplication by -1 reverses the order in which subtraction occurs.

EXAMPLE 6 Subtract: $\dfrac{5x}{x-7} - \dfrac{3x}{7-x}$.

SOLUTION

$$\dfrac{5x}{x-7} - \dfrac{3x}{7-x} = \dfrac{5x}{x-7} - \dfrac{-1}{-1} \cdot \dfrac{3x}{7-x} \qquad \text{Note that } x-7 \text{ and } 7-x \text{ are opposites.}$$

$$= \dfrac{5x}{x-7} - \dfrac{-3x}{x-7} \qquad \text{Performing the multiplication. } Note: -1(7-x) = -7 + x = x - 7.$$

$$= \dfrac{5x - (-3x)}{x-7} \qquad \left. \begin{array}{l} \\ \\ \\ \end{array} \right\} \text{ Subtracting. The parentheses are important.}$$

$$= \dfrac{5x + 3x}{x-7}$$

$$= \dfrac{8x}{x-7} \qquad \text{Simplifying}$$

6. Subtract:

$$\dfrac{x}{x-5} - \dfrac{7}{5-x}.$$

↩ YOUR TURN

Sometimes, after factoring to find the LCD, we find a factor in one denominator that is the opposite of a factor in the other denominator. When this happens, multiplication by $-1/-1$ can again be helpful.

EXAMPLE 7 Perform the indicated operations and simplify.

a) $\dfrac{x}{x^2 - 25} + \dfrac{3}{5-x}$

b) $\dfrac{x+9}{x^2-4} + \dfrac{6-x}{4-x^2} - \dfrac{1+x}{x^2-4}$

Student Notes

Your answer may differ slightly from the answer found at the back of the book and still be correct. For example, an equivalent answer to Example 7(a) is $-\dfrac{2x+15}{(x-5)(x+5)}$:

$$\dfrac{-2x-15}{(x-5)(x+5)} = \dfrac{-(2x+15)}{(x-5)(x+5)}$$

$$= -\dfrac{2x+15}{(x-5)(x+5)}.$$

Before reworking an exercise, be sure that your answer is indeed incorrect.

SOLUTION

a) $\dfrac{x}{x^2-25} + \dfrac{3}{5-x} = \dfrac{x}{(x-5)(x+5)} + \dfrac{3}{5-x}$ 　　Factoring

$$= \dfrac{x}{(x-5)(x+5)} + \dfrac{3}{5-x} \cdot \dfrac{-1}{-1} \qquad \begin{array}{l} \text{Multiplication by} \\ -1/-1 \text{ changes} \\ 5-x \text{ to } x-5. \end{array}$$

$$= \dfrac{x}{(x-5)(x+5)} + \dfrac{-3}{x-5} \qquad (5-x)(-1) = x-5$$

$$= \dfrac{x}{(x-5)(x+5)} + \dfrac{-3}{(x-5)} \cdot \dfrac{x+5}{x+5} \qquad \begin{array}{l} \text{The LCD is} \\ (x-5)(x+5). \end{array}$$

$$= \dfrac{x}{(x-5)(x+5)} + \dfrac{-3x-15}{(x-5)(x+5)}$$

$$= \dfrac{-2x-15}{(x-5)(x+5)}$$

Technology Connection

The TABLE feature can be used to check addition or subtraction of rational expressions. Below we check Example 7(a), using

$$y_1 = \dfrac{x}{x^2-25} + \dfrac{3}{5-x}$$

and

$$y_2 = \dfrac{-2x-15}{(x-5)(x+5)}.$$

ΔTBL = 1

X	Y₁	Y₂
1	.70833	.70833
2	.90476	.90476
3	1.3125	1.3125
4	2.5556	2.5556
5	ERROR	ERROR
6	−2.455	−2.455
7	−1.208	−1.208

X = 1

Because the values for y_1 and y_2 match, we have a check.

b) Since $4 - x^2$ is the opposite of $x^2 - 4$, multiplying the second rational expression by $-1/-1$ will lead to a common denominator:

$$\frac{x + 9}{x^2 - 4} + \frac{6 - x}{4 - x^2} - \frac{1 + x}{x^2 - 4} = \frac{x + 9}{x^2 - 4} + \frac{6 - x}{4 - x^2} \cdot \frac{-1}{-1} - \frac{1 + x}{x^2 - 4}$$

$$= \frac{x + 9}{x^2 - 4} + \frac{x - 6}{x^2 - 4} - \frac{1 + x}{x^2 - 4}$$

$$= \frac{x + 9 + x - 6 - 1 - x}{x^2 - 4} \quad \text{Adding and subtracting numerators}$$

$$= \frac{x + 2}{x^2 - 4}$$

$$= \frac{(x + 2) \cdot 1}{(x + 2)(x - 2)} \quad \left.\begin{array}{c} \\ \\ \end{array}\right\} \text{Simplifying}$$

$$= \frac{1}{x - 2}.$$

7. Perform the indicated operations and simplify:

$$\frac{x + 4}{x + 1} - \frac{x}{3 - x} - \frac{4x}{x^2 - 2x - 3}.$$

 YOUR TURN

6.4 EXERCISE SET

FOR EXTRA HELP

MyMathLab® Math XL
PRACTICE   WATCH   READ   REVIEW

Vocabulary and Reading Check

In Exercises 1–4, the four steps for adding rational expressions with different denominators are listed. Fill in the missing word or words for each step.

1. To add or subtract when the denominators are different, first find the _____.

2. Multiply each rational expression by a form of 1 made up of the factors of the LCD that are _____ from that expression's _____.

3. Add or subtract the _____, as indicated. Write the sum or the difference over the _____.

4. _____, if possible.

Adding and Subtracting with LCDs

Perform the indicated operation. Simplify, if possible.

5. $\dfrac{3}{x^2} + \dfrac{5}{x}$

6. $\dfrac{6}{x} + \dfrac{7}{x^2}$

7. $\dfrac{1}{6r} - \dfrac{3}{8r}$

8. $\dfrac{4}{9t} - \dfrac{7}{6t}$

9. $\dfrac{3}{uv^2} + \dfrac{4}{u^3v}$

10. $\dfrac{8}{cd^2} + \dfrac{1}{c^2d}$

11. $\dfrac{-2}{3xy^2} - \dfrac{6}{x^2y^3}$

12. $\dfrac{8}{9t^3} - \dfrac{5}{6t^2}$

13. $\dfrac{x + 3}{8} + \dfrac{x - 2}{6}$

14. $\dfrac{x - 4}{9} + \dfrac{x + 5}{12}$

15. $\dfrac{x - 2}{6} - \dfrac{x + 1}{3}$

16. $\dfrac{a + 2}{2} - \dfrac{a - 4}{4}$

17. $\dfrac{a + 3}{15a} + \dfrac{2a - 1}{3a^2}$

18. $\dfrac{5a + 1}{2a^2} + \dfrac{a + 2}{6a}$

19. $\dfrac{4z - 9}{3z} - \dfrac{3z - 8}{4z}$

20. $\dfrac{x - 1}{4x} - \dfrac{2x + 3}{x}$

21. $\dfrac{3c + d}{cd^2} + \dfrac{c - d}{c^2d}$

22. $\dfrac{u + v}{u^2v} + \dfrac{2u + v}{uv^2}$

23. $\dfrac{4x + 2t}{3xt^2} - \dfrac{5x - 3t}{x^2t}$

24. $\dfrac{5x + 3y}{2x^2y} - \dfrac{3x + 4y}{xy^2}$

25. $\dfrac{3}{x - 2} + \dfrac{3}{x + 2}$

26. $\dfrac{5}{x - 1} + \dfrac{5}{x + 1}$

27. $\dfrac{t}{t + 3} - \dfrac{1}{t - 1}$

28. $\dfrac{y}{y - 3} + \dfrac{12}{y + 4}$

29. $\dfrac{3}{x + 1} + \dfrac{2}{3x}$

30. $\dfrac{2}{x + 5} + \dfrac{3}{4x}$

31. $\dfrac{3}{2t^2 - 2t} - \dfrac{5}{2t - 2}$

32. $\dfrac{8}{3t^2 - 15t} - \dfrac{3}{2t - 10}$

33. $\dfrac{3a}{a^2 - 9} + \dfrac{a}{a + 3}$

34. $\dfrac{5p}{p^2 - 16} + \dfrac{p}{p - 4}$

35. $\dfrac{6}{z + 4} - \dfrac{2}{3z + 12}$

36. $\dfrac{t}{t - 3} - \dfrac{5}{4t - 12}$

37. $\dfrac{5}{q - 1} + \dfrac{2}{(q - 1)^2}$

38. $\dfrac{3}{w + 2} + \dfrac{7}{(w + 2)^2}$

39. $\dfrac{3a}{4a - 20} + \dfrac{9a}{6a - 30}$

40. $\dfrac{4a}{5a - 10} + \dfrac{3a}{10a - 20}$

41. $\dfrac{y}{y - 1} - \dfrac{y - 1}{y}$

42. $\dfrac{x + 4}{x} + \dfrac{x}{x + 4}$

43. $\dfrac{6}{a^2 + a - 2} + \dfrac{4}{a^2 - 4a + 3}$

44. $\dfrac{x}{x^2 + 2x + 1} + \dfrac{1}{x^2 + 5x + 4}$

45. $\dfrac{x}{x^2 + 9x + 20} - \dfrac{4}{x^2 + 7x + 12}$

46. $\dfrac{x}{x^2 + 5x + 6} - \dfrac{2}{x^2 + 3x + 2}$

47. $\dfrac{3z}{z^2 - 4z + 4} + \dfrac{10}{z^2 + z - 6}$

48. $\dfrac{3}{x^2 - 9} + \dfrac{2}{x^2 - x - 6}$

Aha! **49.** $\dfrac{-7}{x^2 + 25x + 24} - \dfrac{0}{x^2 + 11x + 10}$

50. $\dfrac{x}{x^2 + 17x + 72} - \dfrac{1}{x^2 + 15x + 56}$

51. $3 + \dfrac{4}{2x + 1}$

52. $2 + \dfrac{1}{5 - x}$

53. $3 - \dfrac{2}{4 - x}$

54. $4 - \dfrac{3}{3x + 2}$

When Factors Are Opposites

Perform the indicated operation. Simplify, if possible.

55. $\dfrac{5x}{4} - \dfrac{x - 2}{-4}$

56. $\dfrac{x}{6} - \dfrac{2x - 3}{-6}$

Aha! **57.** $\dfrac{x}{x - 5} + \dfrac{x}{5 - x}$

58. $\dfrac{y}{y - 2} - \dfrac{y}{2 - y}$

59. $\dfrac{y^2}{y - 3} + \dfrac{9}{3 - y}$

60. $\dfrac{t^2}{t - 2} + \dfrac{4}{2 - t}$

61. $\dfrac{c - 5}{c^2 - 64} - \dfrac{5 - c}{64 - c^2}$

62. $\dfrac{b - 4}{b^2 - 49} + \dfrac{b - 4}{49 - b^2}$

63. $\dfrac{t - 3}{t^3 - 1} - \dfrac{2}{1 - t^3}$

64. $\dfrac{1 - 6m}{1 - m^3} - \dfrac{5}{m^3 - 1}$

65. $\dfrac{4 - p}{25 - p^2} + \dfrac{p + 1}{p - 5}$

66. $\dfrac{y + 2}{y - 7} + \dfrac{3 - y}{49 - y^2}$

67. $\dfrac{x}{x - 4} - \dfrac{3}{16 - x^2}$

68. $\dfrac{x}{3 - x} - \dfrac{2}{x^2 - 9}$

69. $\dfrac{a}{a^2 - 1} + \dfrac{2a}{a - a^2}$

70. $\dfrac{3x + 2}{3x + 6} + \dfrac{x}{4 - x^2}$

71. $\dfrac{4x}{x^2 - y^2} - \dfrac{6}{y - x}$

72. $\dfrac{4 - a^2}{a^2 - 9} - \dfrac{a - 2}{3 - a}$

Adding and Subtracting

Perform the indicated operations. Simplify, if possible.

73. $\dfrac{x - 3}{2 - x} - \dfrac{x + 3}{x + 2} + \dfrac{x + 6}{4 - x^2}$

74. $\dfrac{t - 5}{1 - t} - \dfrac{t + 4}{t + 1} + \dfrac{t + 2}{t^2 - 1}$

75. $\dfrac{2x + 5}{x + 1} + \dfrac{x + 7}{x + 5} - \dfrac{5x + 17}{(x + 1)(x + 5)}$

76. $\dfrac{x + 5}{x + 3} + \dfrac{x + 7}{x + 2} - \dfrac{7x + 19}{(x + 3)(x + 2)}$

77. $\dfrac{1}{x + y} + \dfrac{1}{x - y} - \dfrac{2x}{x^2 - y^2}$

78. $\dfrac{2r}{r^2 - s^2} + \dfrac{1}{r + s} - \dfrac{1}{r - s}$

79. $\dfrac{1}{x^2 + 7x + 12} - \dfrac{2}{x^2 + 4x + 3} + \dfrac{3}{x^2 + 5x + 4}$

80. $\dfrac{4}{x^2 + x - 2} + \dfrac{2}{x^2 - 4x + 3} - \dfrac{5}{x^2 - x - 6}$

81. What is the advantage of using the *least* common denominator—rather than just *any* common denominator—when adding or subtracting rational expressions?

82. Describe a procedure that can be used to add any two rational expressions.

Skill Review

Simplify.

83. $3 - 12 \div (-4)$ [1.8]

84. $|-6 - 1|$ [1.8]

85. $(1.2 \times 10^8)(2.5 \times 10^6)$ [4.2]

86. $(2a^3b^{-5})(-3ab^4)$ [4.2]

87. $(3a^{-1}b)^{-2}$ [4.2]

88. $-(-12)$ [1.6]

Synthesis

 89. How could you convince someone that

$$\frac{1}{3-x} \quad \text{and} \quad \frac{1}{x-3}$$

are opposites of each other?

 90. Are parentheses as important for adding rational expressions as they are for subtracting rational expressions? Why or why not?

Write expressions for the perimeter and the area of each rectangle.

91.

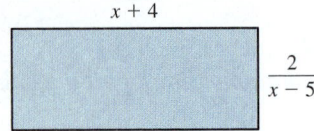

92.

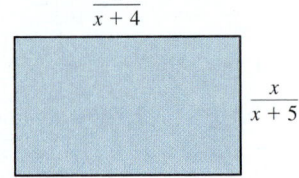

Perform the indicated operations. Simplify, if possible.

93. $\dfrac{x^2}{3x^2 - 5x - 2} - \dfrac{2x}{3x + 1} \cdot \dfrac{1}{x - 2}$

94. $\dfrac{2x + 11}{x - 3} \cdot \dfrac{3}{x + 4} + \dfrac{2x + 1}{4 + x} \cdot \dfrac{3}{3 - x}$

95. $\dfrac{2x - 16}{x^2 - x - 2} - \dfrac{x + 2}{x^2 - 5x + 6} - \dfrac{x + 6}{x^2 - 2x - 3}$

96. $\dfrac{2x + 8}{x^2 + 3x + 2} - \dfrac{x - 2}{x^2 + 5x + 6} - \dfrac{x + 2}{x^2 + 4x + 3}$

Aha! **97.** $\left(\dfrac{x}{x + 7} - \dfrac{3}{x + 2} \right)\left(\dfrac{x}{x + 7} + \dfrac{3}{x + 2} \right)$

98. $\dfrac{1}{ay - 3a + 2xy - 6x} - \dfrac{xy + ay}{a^2 - 4x^2}\left(\dfrac{1}{y - 3} \right)^2$

99. $\left(\dfrac{a}{a - b} + \dfrac{b}{a + b} \right)\left(\dfrac{1}{3a + b} + \dfrac{2a + 6b}{9a^2 - b^2} \right)$

100. $\dfrac{2x^2 + 5x - 3}{2x^2 - 9x + 9} + \dfrac{x + 1}{3 - 2x} + \dfrac{4x^2 + 8x + 3}{x - 3} \cdot \dfrac{x + 3}{9 - 4x^2}$

101. Express

$$\frac{a - 3b}{a - b}$$

as a sum of two rational expressions with denominators that are opposites of each other. Answers may vary.

 102. Use a graphing calculator to check the answer to Exercise 29.

103. Why does the word ERROR appear in the table displayed in the Technology Connection on p. 399?

YOUR TURN ANSWERS: SECTION 6.4

1. $\dfrac{8x + 15}{36}$ **2.** $\dfrac{-25x + 6}{20x^2}$ **3.** $\dfrac{x(x - 1)}{(x - 5)(x + 5)}$

4. $\dfrac{x - 6}{(x - 2)(x + 2)}$ **5.** $\dfrac{t - 3}{2}$ **6.** $\dfrac{x + 7}{x - 5}$

7. $\dfrac{2(x + 2)}{x + 1}$

QUICK QUIZ: SECTIONS 6.1–6.4

Perform the indicated operation and, if possible, simplify.

1. $\dfrac{x^2 - 64}{10x^2} \cdot \dfrac{5x}{2x + 16}$ [6.2]

2. $\dfrac{2x - 3}{x - 4} - \dfrac{x}{x + 2}$ [6.4]

3. $\dfrac{x^2 - 6}{x - 3} - \dfrac{x}{x - 3}$ [6.3]

4. $\dfrac{2t}{2t + 1} + \dfrac{t}{t - 1}$ [6.4]

5. $\dfrac{2x^2 - x - 1}{x^2 - 4x + 4} \div \dfrac{x - 2}{x - 1}$ [6.2]

PREPARE TO MOVE ON

Divide and, if possible, simplify.

1. $\dfrac{\frac{3}{4}}{\frac{5}{6}}$ [1.3] **2.** $\dfrac{\frac{8}{15}}{\frac{9}{10}}$ [1.3]

3. $\dfrac{2x + 6}{x - 1} \div \dfrac{3x + 9}{x - 1}$ [6.2]

4. $\dfrac{x^2 - 9}{x^2 - 4} \div \dfrac{x^2 + 6x + 9}{x^2 + 4x + 4}$ [6.2]

Mid-Chapter Review

The process of adding and subtracting rational expressions is significantly different from multiplying and dividing. The first thing to note when combining rational expressions is the operation sign.

Operation	Need Common Denominator?	Procedure	Tips and Cautions
Addition	Yes	Write equivalent expressions with a common denominator. Add numerators. Keep denominator.	Do not simplify after writing with the LCD. Instead, simplify after adding the numerators.
Subtraction	Yes	Write equivalent expressions with a common denominator. Subtract numerators. Keep denominator.	Use parentheses around the numerator being subtracted. Simplify after subtracting the numerators.
Multiplication	No	Multiply numerators. Multiply denominators.	Do not perform the polynomial multiplication. Instead, factor and try to simplify.
Division	No	Multiply by the reciprocal of the divisor.	Begin by rewriting as a multiplication using the reciprocal of the divisor.

GUIDED SOLUTIONS

1. Divide: $\dfrac{a^2}{a-10} \div \dfrac{a^2+5a}{a^2-100}$. Simplify, if possible.   [6.2]

Solution

$\dfrac{a^2}{a-10} \div \dfrac{a^2+5a}{a^2-100}$

$= \dfrac{a^2}{a-10} \cdot \dfrac{\boxed{}}{\boxed{}}$ — *Multiplying by the reciprocal of the divisor*

$= \dfrac{a \cdot a \cdot (a+10) \cdot \boxed{}}{(a-10) \cdot a \cdot \boxed{}}$ — *Multiplying and factoring*

$= \dfrac{\boxed{}}{\boxed{}} \cdot \dfrac{a(a+10)}{a+5}$ — *Factoring out a factor equal to 1*

$= \dfrac{a(a+10)}{a+5}$ — *Simplifying*

2. Add: $\dfrac{2}{x} + \dfrac{1}{x^2+x}$. Simplify, if possible.   [6.4]

Solution

$\dfrac{2}{x} + \dfrac{1}{x^2+x}$

$= \dfrac{2}{x} + \dfrac{1}{x\boxed{}}$ — *Factoring denominators. The LCD is $x(x+1)$.*

$= \dfrac{2}{x} \cdot \dfrac{\boxed{}}{\boxed{}} + \dfrac{1}{x(x+1)}$ — *Multiplying by 1 to get the LCD in the first denominator*

$= \dfrac{\boxed{}}{x(x+1)} + \dfrac{1}{x(x+1)}$ — *Multiplying*

$= \dfrac{\boxed{}}{x(x+1)}$ — *Adding numerators. We cannot simplify.*

MIXED REVIEW

Perform the indicated operation and, if possible, simplify.

3. $\dfrac{3}{5x} + \dfrac{2}{x^2}$ [6.4]

4. $\dfrac{3}{5x} \cdot \dfrac{2}{x^2}$ [6.2]

5. $\dfrac{3}{5x} \div \dfrac{2}{x^2}$ [6.2]

6. $\dfrac{3}{5x} - \dfrac{2}{x^2}$ [6.4]

7. $\dfrac{2x-6}{5x+10} \cdot \dfrac{x+2}{6x-12}$ [6.2]

8. $\dfrac{2}{x-5} \div \dfrac{6}{x-5}$ [6.2]

9. $\dfrac{x}{x+2} - \dfrac{1}{x-1}$ [6.4]

10. $\dfrac{2}{x+3} + \dfrac{3}{x+4}$ [6.4]

11. $\dfrac{5}{2x-1} + \dfrac{10x}{1-2x}$ [6.4]

12. $\dfrac{3}{x-4} - \dfrac{2}{4-x}$ [6.4]

13. $\dfrac{(x-2)(2x+3)}{(x+1)(x-5)} \div \dfrac{(x-2)(x+1)}{(x-5)(x+3)}$ [6.2]

14. $\dfrac{a}{6a-9b} - \dfrac{b}{4a-6b}$ [6.4]

15. $\dfrac{x^2-16}{x^2-x} \cdot \dfrac{x^2}{x^2-5x+4}$ [6.2]

16. $\dfrac{x+1}{x^2-7x+10} + \dfrac{3}{x^2-x-2}$ [6.4]

17. $\dfrac{3u^2-3}{4} \div \dfrac{4u+4}{3}$ [6.2]

18. $(t^2+t-20) \cdot \dfrac{t+5}{t-4}$ [6.2]

19. $\dfrac{a^2-2a+1}{a^2-4} \div (a^2-3a+2)$ [6.2]

20. $\dfrac{2x-7}{x} - \dfrac{3x-5}{2}$ [6.4]

6.5 Complex Rational Expressions

Using Division to Simplify • Multiplying by the LCD

A **complex rational expression** is a rational expression that has one or more rational expressions within its numerator or denominator. Here are some examples:

$$\dfrac{1+\dfrac{2}{x}}{3}, \quad \dfrac{\dfrac{x+y}{7}}{\dfrac{2x}{x+1}}, \quad \dfrac{\dfrac{4}{3}+\dfrac{1}{5}}{\dfrac{2}{x}-\dfrac{x}{y}}.$$

These are rational expressions within the complex rational expression.

To simplify a complex rational expression is to write an equivalent expression that is no longer complex. We will consider two methods for simplifying complex rational expressions. We assume all denominators are nonzero.

Using Division to Simplify (Method 1)

Our first method for simplifying complex rational expressions involves rewriting the expression as a quotient of two rational expressions.

Study Skills

Multiple Methods

When more than one method is presented, as is the case for simplifying complex rational expressions, it can be helpful to learn both methods. If two different approaches yield the same result, you can be confident that your answer is correct.

TO SIMPLIFY A COMPLEX RATIONAL EXPRESSION BY DIVIDING

1. Add or subtract, as needed, to get a single rational expression in the numerator.
2. Add or subtract, as needed, to get a single rational expression in the denominator.
3. Divide the numerator by the denominator (invert the divisor and multiply).
4. If possible, simplify by removing a factor equal to 1.

The key here is to express a complex rational expression as one rational expression divided by another.

EXAMPLE 1   Simplify: $\dfrac{\dfrac{x}{x-3}}{\dfrac{4}{5x-15}}$.

SOLUTION   Here the numerator and the denominator are already single rational expressions. This allows us to start by dividing (step 3):

$$\dfrac{\dfrac{x}{x-3}}{\dfrac{4}{5x-15}} = \dfrac{x}{x-3} \div \dfrac{4}{5x-15} \qquad \text{Rewriting with a division symbol}$$

$$= \dfrac{x}{x-3} \cdot \dfrac{5x-15}{4} \qquad \text{Multiplying by the reciprocal of the divisor (inverting and multiplying)}$$

$$= \dfrac{x}{x-3} \cdot \dfrac{5(x-3)}{4} \qquad \text{Factoring and removing a factor equal to 1: } \dfrac{x-3}{x-3} = 1$$

$$= \dfrac{5x}{4}.$$

1. Simplify: $\dfrac{\dfrac{1}{x^2+x}}{\dfrac{2}{5x+5}}$.

 YOUR TURN

EXAMPLE 2   Simplify.

a) $\dfrac{\dfrac{5}{2a}+\dfrac{1}{a}}{\dfrac{1}{4a}-\dfrac{5}{6}}$

b) $\dfrac{\dfrac{x^2}{y}-\dfrac{5}{x}}{xz}$

SOLUTION

a)

$$\dfrac{\dfrac{5}{2a}+\dfrac{1}{a}}{\dfrac{1}{4a}-\dfrac{5}{6}} = \left.\dfrac{\dfrac{5}{2a}+\dfrac{1}{a}\cdot\dfrac{2}{2}}{\dfrac{1}{4a}\cdot\dfrac{3}{3}-\dfrac{5}{6}\cdot\dfrac{2a}{2a}}\right\}$$

Multiplying by 1 to get the LCD, $2a$, for the numerator of the complex rational expression

Multiplying by 1 to get the LCD, $12a$, for the denominator of the complex rational expression

1. Form a single rational expression in the numerator.

2. Form a single rational expression in the denominator.

3. Divide the numerator by the denominator.

$$= \dfrac{\dfrac{5}{2a}+\dfrac{2}{2a}}{\dfrac{3}{12a}-\dfrac{10a}{12a}} = \dfrac{\dfrac{7}{2a}}{\dfrac{3-10a}{12a}}$$

← Adding

← Subtracting

$$= \dfrac{7}{2a} \div \dfrac{3-10a}{12a} \qquad \text{Rewriting with a division symbol. This is often done mentally.}$$

$$= \dfrac{7}{2a} \cdot \dfrac{12a}{3-10a} \qquad \text{Multiplying by the reciprocal of the divisor (inverting and multiplying)}$$

4. Simplify.

$$= \dfrac{7}{2a} \cdot \dfrac{2a \cdot 6}{3-10a} \qquad \text{Removing a factor equal to 1: } \dfrac{2a}{2a} = 1$$

$$= \dfrac{42}{3-10a}$$

b) $\dfrac{\dfrac{x^2}{y} - \dfrac{5}{x}}{xz} = \dfrac{\dfrac{x^2}{y}\cdot\dfrac{x}{x} - \dfrac{5}{x}\cdot\dfrac{y}{y}}{xz}$ ⟵ Multiplying by 1 to get the LCD, xy, for the numerator of the complex rational expression

$= \dfrac{\dfrac{x^3}{xy} - \dfrac{5y}{xy}}{xz}$

$= \dfrac{\dfrac{x^3 - 5y}{xy}}{xz}$ ⟵ Subtracting

$\qquad\qquad\qquad$ ⟵ If you prefer, write xz as $\dfrac{xz}{1}$.

$= \dfrac{x^3 - 5y}{xy} \div (xz)$ Rewriting with a division symbol

$= \dfrac{x^3 - 5y}{xy} \cdot \dfrac{1}{xz}$ Multiplying by the reciprocal of the divisor (inverting and multiplying)

$= \dfrac{x^3 - 5y}{x^2yz}$

2. Simplify: $\dfrac{\dfrac{1}{t} - \dfrac{1}{5}}{\dfrac{t-5}{t}}.$

 YOUR TURN

Multiplying by the LCD (Method 2)

A second method for simplifying complex rational expressions relies on multiplying by a carefully chosen expression that is equal to 1. This multiplication by 1 will result in an expression that is no longer complex.

TO SIMPLIFY A COMPLEX RATIONAL EXPRESSION BY MULTIPLYING BY THE LCD

1. Find the LCD of *all* rational expressions within the complex rational expression.
2. Multiply the complex rational expression by an expression equal to 1. Write 1 as the LCD over itself: LCD/LCD.
3. Simplify. No fraction expressions should remain within the complex rational expression.
4. Factor and, if possible, simplify the remaining expression.

EXAMPLE 3 Simplify: $\dfrac{\dfrac{1}{2} + \dfrac{3}{4}}{\dfrac{5}{6} - \dfrac{3}{8}}.$

Student Notes

Pay careful attention to the different ways in which Method 1 and Method 2 use an LCD. In Method 1, one LCD is found in the numerator and one in the denominator. In Method 2, *all* rational expressions are considered in order to find one LCD.

SOLUTION

1. The LCD of $\frac{1}{2}, \frac{3}{4}, \frac{5}{6},$ and $\frac{3}{8}$ is 24.

2. We multiply by an expression equal to 1:

$\dfrac{\dfrac{1}{2} + \dfrac{3}{4}}{\dfrac{5}{6} - \dfrac{3}{8}} = \dfrac{\dfrac{1}{2} + \dfrac{3}{4}}{\dfrac{5}{6} - \dfrac{3}{8}} \cdot \dfrac{24}{24}.$ Multiplying by an expression equal to 1, using the LCD: $\dfrac{24}{24} = 1$

3. Using the distributive law, we perform the multiplication:

$$\frac{\frac{1}{2}+\frac{3}{4}}{\frac{5}{6}-\frac{3}{8}} \cdot \frac{24}{24} = \frac{\left(\frac{1}{2}+\frac{3}{4}\right)24}{\left(\frac{5}{6}-\frac{3}{8}\right)24}$$ ← Multiplying the numerator by 24
Don't forget the parentheses!
← Multiplying the denominator by 24

$$= \frac{\frac{1}{2}(24)+\frac{3}{4}(24)}{\frac{5}{6}(24)-\frac{3}{8}(24)}$$ Using the distributive law

$$= \frac{\frac{24}{2}+\frac{3(24)}{4}}{\frac{5(24)}{6}-\frac{3(24)}{8}}$$

3. Simplify: $\dfrac{\frac{5}{2}-\frac{4}{3}}{\frac{2}{3}+\frac{3}{4}}.$

$$= \frac{12+18}{20-9}, \quad \text{or} \quad \frac{30}{11}.$$ Simplifying

4. The result, $\frac{30}{11}$, cannot be factored or simplified, so we are done.

YOUR TURN

Multiplying like this effectively clears fractions in both the top and the bottom of the complex rational expression.

EXAMPLE 4 Simplify.

a) $\dfrac{\frac{5}{2a}+\frac{1}{a}}{\frac{1}{4a}-\frac{5}{6}}$ **b)** $\dfrac{1-\frac{1}{x}}{1-\frac{1}{x^2}}$

SOLUTION

1. Find the LCD.

a) The denominators within the complex expression are $2a$, a, $4a$, and 6, so the LCD is $12a$. We multiply by 1 using $(12a)/(12a)$:

2. Multiply by LCD/LCD.

$$\frac{\frac{5}{2a}+\frac{1}{a}}{\frac{1}{4a}-\frac{5}{6}} = \frac{\frac{5}{2a}+\frac{1}{a}}{\frac{1}{4a}-\frac{5}{6}} \cdot \frac{12a}{12a} = \frac{\frac{5}{2a}(12a)+\frac{1}{a}(12a)}{\frac{1}{4a}(12a)-\frac{5}{6}(12a)}$$ Using the distributive law

3., 4. Simplify.

$$= \frac{\frac{5(12a)}{2a}+\frac{12a}{a}}{\frac{12a}{4a}-\frac{5(12a)}{6}}$$

$$= \frac{\frac{5\cdot2\cdot6\cdot a}{2\cdot a}+\frac{12\cdot a}{a}}{\frac{4\cdot3\cdot a}{4\cdot a}-\frac{5\cdot2\cdot6\cdot a}{6}}$$ All fractions have been cleared.

$$= \frac{30+12}{3-10a} = \frac{42}{3-10a}.$$

b) $\dfrac{1 - \dfrac{1}{x}}{1 - \dfrac{1}{x^2}} = \dfrac{1 - \dfrac{1}{x}}{1 - \dfrac{1}{x^2}} \cdot \dfrac{x^2}{x^2}$ The LCD is x^2 so we multiply by 1 using x^2/x^2.

$$= \dfrac{1 \cdot x^2 - \dfrac{1}{x} \cdot x^2}{1 \cdot x^2 - \dfrac{1}{x^2} \cdot x^2}$$ Using the distributive law

$$= \dfrac{x^2 - x}{x^2 - 1}$$ All fractions have been cleared within the complex rational expression.

$$= \dfrac{x(x-1)}{(x+1)(x-1)}$$ Factoring and simplifying: $\dfrac{x-1}{x-1} = 1$

$$= \dfrac{x}{x+1}$$

4. Simplify: $\dfrac{1 - \dfrac{2}{3x}}{x - \dfrac{4}{9x}}$.

↩ YOUR TURN

It is important to understand both of the methods studied in this section. Sometimes, as in Example 1, the complex rational expression is either given as—or easily written as—a quotient of two rational expressions. In these cases, Method 1 (using division) is probably the easier method to use. Other times, as in Example 4, it is not difficult to find the LCD of all denominators in the complex rational expression. When this occurs, it is usually easier to use Method 2 (multiplying by the LCD). The more practice you get using both methods, the better you will be at selecting the easier method for any given problem.

6.5 EXERCISE SET

FOR EXTRA HELP

MyMathLab® Math XL

PRACTICE   WATCH   READ   REVIEW

↩ Vocabulary and Reading Check

Consider the expression

$$\dfrac{\dfrac{2}{3} + \dfrac{1}{x}}{\dfrac{5}{x}}.$$

Choose the word from the following list that best completes each statement, with reference to the above expression. Not all words in the list will be used.

complex
denominator
least common denominator
numerator
opposite
reciprocal

1. The expression given above is a(n) _____ rational expression.

2. The expression $\dfrac{5}{x}$ is the _____ of the above expression.

3. The _____ of the rational expressions within the expression above is $3x$.

4. To simplify, we can multiply the numerator by the _____ of $\dfrac{5}{x}$.

↩ Concept Reinforcement

*Each of Exercises 5–8 shows a complex rational expression and the first step taken to simplify that expression. Indicate for each which method is being used: **(a)** using division to simplify (Method 1), or **(b)** multiplying by the LCD (Method 2).*

5. _____ $\dfrac{\dfrac{1}{x} + \dfrac{1}{2}}{\dfrac{1}{3} - \dfrac{1}{x}} = \dfrac{\dfrac{1}{x} + \dfrac{1}{2}}{\dfrac{1}{3} - \dfrac{1}{x}} \cdot \dfrac{6x}{6x}$

6. _____ $\dfrac{\dfrac{1}{x} + \dfrac{1}{2}}{\dfrac{1}{3} - \dfrac{1}{x}} = \dfrac{\dfrac{1}{x} \cdot \dfrac{2}{2} + \dfrac{1}{2} \cdot \dfrac{x}{x}}{\dfrac{1}{3} \cdot \dfrac{x}{x} - \dfrac{1}{x} \cdot \dfrac{3}{3}}$

7. ___ $\dfrac{\dfrac{x-1}{x}}{\dfrac{x^2}{x^2-1}} = \dfrac{x-1}{x} \div \dfrac{x^2}{x^2-1}$

8. ___ $\dfrac{\dfrac{x-1}{x}}{\dfrac{x^2}{x^2-1}} = \dfrac{\dfrac{x-1}{x}}{\dfrac{x^2}{x^2-1}} \cdot \dfrac{x(x+1)(x-1)}{x(x+1)(x-1)}$

Complex Rational Expressions

Simplify. Use either method or the method specified by your instructor.

9. $\dfrac{\dfrac{1}{2}+\dfrac{1}{3}}{\dfrac{1}{4}-\dfrac{1}{6}}$

10. $\dfrac{\dfrac{2}{5}-\dfrac{1}{10}}{\dfrac{7}{20}-\dfrac{4}{15}}$

11. $\dfrac{1+\dfrac{1}{4}}{2+\dfrac{3}{4}}$

12. $\dfrac{3+\dfrac{1}{4}}{1+\dfrac{1}{2}}$

13. $\dfrac{\dfrac{x}{4}+x}{\dfrac{4}{x}+x}$

14. $\dfrac{\dfrac{1}{c}+2}{\dfrac{1}{c}-5}$

15. $\dfrac{\dfrac{x+2}{x-1}}{\dfrac{x+4}{x-3}}$

16. $\dfrac{\dfrac{x-1}{x+3}}{\dfrac{x-6}{x+2}}$

17. $\dfrac{\dfrac{10}{t}}{\dfrac{2}{t^2}-\dfrac{5}{t}}$

18. $\dfrac{\dfrac{5}{x}-\dfrac{2}{x^2}}{\dfrac{2}{x^2}}$

19. $\dfrac{\dfrac{2a-5}{3a}}{\dfrac{a-7}{6a}}$

20. $\dfrac{\dfrac{a+5}{a^2}}{\dfrac{a-2}{3a}}$

21. $\dfrac{\dfrac{x}{6}-\dfrac{3}{x}}{\dfrac{1}{3}+\dfrac{1}{x}}$

22. $\dfrac{\dfrac{2}{x}+\dfrac{x}{4}}{\dfrac{3}{4}-\dfrac{2}{x}}$

23. $\dfrac{\dfrac{1}{s}-\dfrac{1}{5}}{s-5}$

24. $\dfrac{\dfrac{1}{9}-\dfrac{1}{n}}{\dfrac{n+9}{9}}$

25. $\dfrac{\dfrac{1}{t^2}+1}{\dfrac{1}{t}-1}$

26. $\dfrac{2+\dfrac{1}{x}}{2-\dfrac{1}{x^2}}$

27. $\dfrac{\dfrac{x^2}{x^2-y^2}}{\dfrac{x}{x+y}}$

28. $\dfrac{\dfrac{a^2-b^2}{ab}}{\dfrac{a-b}{b}}$

29. $\dfrac{\dfrac{7}{c^2}+\dfrac{4}{c}}{\dfrac{6}{c}-\dfrac{3}{c^3}}$

30. $\dfrac{\dfrac{4}{t^3}-\dfrac{1}{t^2}}{\dfrac{3}{t}+\dfrac{5}{t^2}}$

31. $\dfrac{\dfrac{2}{7a^4}-\dfrac{1}{14a}}{\dfrac{3}{5a^2}+\dfrac{2}{15a}}$

32. $\dfrac{\dfrac{5}{4x^3}-\dfrac{3}{8x}}{\dfrac{3}{2x}+\dfrac{3}{4x^3}}$

Aha! 33. $\dfrac{\dfrac{x}{5y^3}+\dfrac{3}{10y}}{\dfrac{3}{10y}+\dfrac{x}{5y^3}}$

34. $\dfrac{\dfrac{a}{6b^3}+\dfrac{4}{9b^2}}{\dfrac{5}{6b}-\dfrac{1}{9b^3}}$

35. $\dfrac{\dfrac{3}{ab^4}+\dfrac{4}{a^3b}}{\dfrac{5}{a^3b}-\dfrac{3}{ab}}$

36. $\dfrac{\dfrac{2}{x^2y}+\dfrac{3}{xy^2}}{\dfrac{3}{xy^2}+\dfrac{2}{x^2y}}$

37. $\dfrac{t-\dfrac{9}{t}}{t+\dfrac{4}{t}}$

38. $\dfrac{s+\dfrac{2}{s}}{s-\dfrac{3}{s}}$

39. $\dfrac{y+y^{-1}}{y-y^{-1}}$

40. $\dfrac{x-x^{-1}}{x+x^{-1}}$

$\left(\text{Hint: } y^{-1}=\dfrac{1}{y}.\right)$

41. $\dfrac{\dfrac{1}{a-h}-\dfrac{1}{a}}{h}$

42. $\dfrac{\dfrac{1}{x+h}-\dfrac{1}{x}}{h}$

43. $\dfrac{\dfrac{x^{-1}+y^{-1}}{x^2-y^2}}{xy}$

44. $\dfrac{\dfrac{a^{-1}+b^{-1}}{a^2-b^2}}{ab}$

45. $\dfrac{\dfrac{1}{a}+\dfrac{1}{b}}{\dfrac{1}{a^3}+\dfrac{1}{b^3}}$

46. $\dfrac{x-y}{\dfrac{1}{x^3}-\dfrac{1}{y^3}}$

47. $\dfrac{t+5+\dfrac{3}{t}}{t+2+\dfrac{1}{t}}$

48. $\dfrac{a+3+\dfrac{2}{a}}{a+2+\dfrac{5}{a}}$

49. $\dfrac{x-2-\dfrac{1}{x}}{x-5-\dfrac{4}{x}}$

50. $\dfrac{x-3-\dfrac{2}{x}}{x-4-\dfrac{3}{x}}$

51. $\dfrac{\dfrac{a^2-4}{a^2+3a+2}}{\dfrac{a^2-5a-6}{a^2-6a-7}}$

52. $\dfrac{\dfrac{x^2-x-12}{x^2-2x-15}}{\dfrac{x^2+8x+12}{x^2-5x-14}}$

53. $\dfrac{\dfrac{x}{x^2+3x-4}-\dfrac{1}{x^2+3x-4}}{\dfrac{x}{x^2+6x+8}+\dfrac{3}{x^2+6x+8}}$

54. $\dfrac{\dfrac{x}{x^2+5x-6}+\dfrac{6}{x^2+5x-6}}{\dfrac{x}{x^2-5x+4}-\dfrac{2}{x^2-5x+4}}$

55. Is it possible to simplify complex rational expressions without knowing how to divide rational expressions? Why or why not?

56. Why is the distributive law important when simplifying complex rational expressions?

Skill Review

Factor.

57. $6x^3 - 9x^2 - 4x + 6$ [5.1]

58. $12a^2b + 4ab^2 - 8ab$ [5.1]

59. $30n^3 - 3n^2 - 9n$ [5.3]

60. $25a^2 - 40ab + 16b^2$ [5.4]

61. $n^4 - 1$ [5.4]

62. $p^2w^2 - 2pw - 120$ [5.2]

Synthesis

63. Which of the two methods presented would you use to simplify Exercise 36? Why?

64. Which of the two methods presented would you use to simplify Exercise 19? Why?

In each of Exercises 65–68, find all x-values for which the given expression is undefined.

65. $\dfrac{\dfrac{x-5}{x-6}}{\dfrac{x-7}{x-8}}$

66. $\dfrac{\dfrac{x+1}{x+2}}{\dfrac{x+3}{x+4}}$

67. $\dfrac{\dfrac{2x+3}{5x+4}}{\dfrac{3}{7}-\dfrac{x^2}{21}}$

68. $\dfrac{\dfrac{3x-5}{2x-7}}{\dfrac{4x}{5}-\dfrac{8}{15}}$

69. Use multiplication by the LCD (Method 2) to show that

$$\frac{A}{B} \div \frac{C}{D} = \frac{A}{B} \cdot \frac{D}{C}.$$

(*Hint*: Begin by forming a complex rational expression.)

70. The formula

$$\frac{P\left(1 + \dfrac{i}{12}\right)^2}{\dfrac{\left(1 + \dfrac{i}{12}\right)^2 - 1}{\dfrac{i}{12}}},$$

where P is a loan amount and i is an interest rate, arises in certain business situations. Simplify this expression. (*Hint*: Expand the binomials.)

Simplify.

71. $\dfrac{\dfrac{x}{x+5} + \dfrac{3}{x+2}}{\dfrac{2}{x+2} - \dfrac{x}{x+5}}$

72. $\dfrac{\dfrac{z}{1 - \dfrac{z}{2+2z}} - 2z}{\dfrac{2z}{5z-2} - 3}$

Aha! **73.** $\left[\dfrac{\dfrac{x-1}{x-1} - 1}{\dfrac{x+1}{x-1} + 1}\right]^5$

74. $1 + \dfrac{1}{1 + \dfrac{1}{1 + \dfrac{1}{x}}}$

75. $\dfrac{1 - \dfrac{25}{x^2}}{1 + \dfrac{2}{x} - \dfrac{15}{x^2}}$

76. $\dfrac{1 - \dfrac{1}{y} + \dfrac{1}{y^2} - \dfrac{1}{y^3}}{1 - \dfrac{1}{y^4}}$

77. Find the simplified form for the reciprocal of

$$\frac{2}{x-1} - \frac{1}{3x-2}.$$

78. Under what circumstance(s) will there be no restrictions on the variable appearing in a complex rational expression?

  YOUR TURN ANSWERS: SECTION 6.5

1. $\dfrac{5}{2x}$ **2.** $-\dfrac{1}{5}$ **3.** $\dfrac{14}{17}$ **4.** $\dfrac{3}{3x+2}$

QUICK QUIZ: SECTIONS 6.1–6.5

Simplify.

1. $\dfrac{8x^2 - 20x - 12}{4x^2 - 14x + 6}$ [6.1]

2. $\dfrac{\dfrac{2}{x} + \dfrac{2}{3}}{\dfrac{1}{x^2}}$ [6.5]

Perform the indicated operation and, if possible, simplify.

3. $\dfrac{a+2}{a+1} \cdot \dfrac{a+3}{a+2}$ [6.2]

4. $\dfrac{4}{x-3} - \dfrac{5x-2}{x+1}$ [6.4]

5. $\dfrac{x+2}{2x} + \dfrac{x+1}{3x}$ [6.4]

PREPARE TO MOVE ON

Solve.

1. $(x-1)7 - (x+1)9 = 4(x+2)$ [2.2]

2. $\dfrac{5}{9} - \dfrac{2x}{3} = \dfrac{5x}{6} + \dfrac{4}{3}$ [2.2]   **3.** $x^2 - 7x + 12 = 0$ [5.7]

4. $x^2 + 13x - 30 = 0$ [5.7]

6.6   Rational Equations

Solving Rational Equations

Our study of rational expressions allows us to solve a type of equation that we could not have solved prior to this chapter.

Solving Rational Equations

A **rational equation** is an equation containing one or more rational expressions, often with the variable in a denominator. Here are some examples:

$$\frac{2}{3} + \frac{5}{6} = \frac{x}{9}, \qquad t + \frac{7}{t} = -5, \qquad \frac{x^2}{x-1} = \frac{1}{x-1}.$$

TO SOLVE A RATIONAL EQUATION

1. List any restrictions that exist. Numbers that make a denominator equal 0 can never be solutions.
2. Clear the equation of fractions by multiplying both sides by the LCM of the denominators.
3. Solve the resulting equation using the addition principle, the multiplication principle, and the principle of zero products, as needed.
4. Check the possible solution(s) in the original equation.

When clearing an equation of fractions, we use the terminology LCM instead of LCD because the LCM is not used as a denominator in this setting.

EXAMPLE 1   Solve: $\dfrac{x}{6} - \dfrac{x}{8} = \dfrac{1}{12}$.

SOLUTION   Because no variable appears in a denominator, no restrictions exist. The LCM of 6, 8, and 12 is 24, so we multiply both sides by 24:

$$24\left(\frac{x}{6} - \frac{x}{8}\right) = 24 \cdot \frac{1}{12} \qquad \text{Using the multiplication principle to multiply both sides by the LCM. Parentheses are important!}$$

$$24 \cdot \frac{x}{6} - 24 \cdot \frac{x}{8} = 24 \cdot \frac{1}{12} \qquad \text{Using the distributive law}$$

Be sure to multiply *each* term by the LCM.

$$\left.\begin{array}{c}\dfrac{24x}{6} - \dfrac{24x}{8} = \dfrac{24}{12} \\[2mm] 4x - 3x = 2\end{array}\right\} \qquad \text{Simplifying. Note that all fractions have been cleared. If fractions remain, we have either made a mistake or have not used the LCM of the denominators.}$$

$$x = 2.$$

Check:

$$\dfrac{x}{6} - \dfrac{x}{8} = \dfrac{1}{12}$$

$$\begin{array}{c|c} \dfrac{2}{6} - \dfrac{2}{8} & \dfrac{1}{12} \\[6pt] \dfrac{1}{3} - \dfrac{1}{4} & \\[6pt] \dfrac{4}{12} - \dfrac{3}{12} & \\[6pt] \dfrac{1}{12} \stackrel{?}{=} \dfrac{1}{12} & \text{TRUE} \end{array}$$

1. Solve: $\dfrac{t}{3} + \dfrac{t}{5} = 1$.

This checks, so the solution is 2.

▸ YOUR TURN

Recall that the multiplication principle states that $a = b$ is equivalent to $a \cdot c = b \cdot c$, *provided c is not 0.* To clear fractions in rational equations, we often must multiply by a variable expression. For example, to clear the fraction in $1/x = 5$, we multiply both sides of the equation by x. Since x *could* represent 0, the new equation may not be equivalent to the original equation. For this reason, **checking in the original equation is essential**.

EXAMPLE 2   Solve.

a) $\dfrac{2}{3x} + \dfrac{1}{x} = 10$

b) $x + \dfrac{6}{x} = -5$

SOLUTION

List restrictions.

a) Note that in $\dfrac{2}{3x} + \dfrac{1}{x} = 10$, if $x = 0$, both denominators are 0. We list this restriction:

$$x \neq 0.$$

We now clear the equation of fractions and solve, using the LCM of the denominators, $3x$.

Clear fractions.

$$3x\left(\dfrac{2}{3x} + \dfrac{1}{x}\right) = 3x \cdot 10 \qquad \begin{array}{l}\text{Using the multiplication principle to}\\ \text{multiply both sides by the LCM.}\\ \textit{Don't forget the parentheses!}\end{array}$$

$$3x \cdot \dfrac{2}{3x} + 3x \cdot \dfrac{1}{x} = 3x \cdot 10 \qquad \text{Using the distributive law}$$

$$2 + 3 = 30x \qquad \begin{array}{l}\text{Removing factors equal to 1:}\\ (3x)/(3x) = 1 \text{ and } x/x = 1. \text{ This clears}\\ \text{all fractions.}\end{array}$$

Solve.

$$5 = 30x$$

$$\dfrac{5}{30} = x, \quad \text{so } x = \dfrac{1}{6}. \qquad \begin{array}{l}\text{Dividing both sides by 30, or}\\ \text{multiplying both sides by } 1/30\end{array}$$

Since $\frac{1}{6} \neq 0$, and 0 is the only restricted value, $\frac{1}{6}$ *should* check.

Check.

Check:

$$\dfrac{2}{3x} + \dfrac{1}{x} = 10$$

$$\begin{array}{c|c} \dfrac{2}{3 \cdot \frac{1}{6}} + \dfrac{1}{\frac{1}{6}} & 10 \\[10pt] \dfrac{2}{\frac{1}{2}} + \dfrac{1}{\frac{1}{6}} & \\[10pt] 2 \cdot \dfrac{2}{1} + 1 \cdot \dfrac{6}{1} & \\[8pt] 4 + 6 & \\[4pt] 10 \stackrel{?}{=} 10 & \text{TRUE} \end{array}$$

The solution is $\frac{1}{6}$.

b) To solve $x + \dfrac{6}{x} = -5$, note that if $x = 0$, the expression $\dfrac{6}{x}$ is undefined. We list the restriction:

$$x \neq 0.$$

We now clear the equation of fractions and solve:

$$x\left(x + \frac{6}{x} \right) = x(-5) \qquad \text{Multiplying both sides by the LCM, } x.$$
Don't forget the parentheses!

$$x \cdot x + \not{x} \cdot \frac{6}{\not{x}} = -5x \qquad \text{Using the distributive law}$$

$$x^2 + 6 = -5x \qquad \begin{array}{l}\text{Removing a factor equal to 1: } x/x = 1. \text{ We}\\ \text{are left with a quadratic equation.}\end{array}$$

$$x^2 + 5x + 6 = 0 \qquad \begin{array}{l}\text{Using the addition principle to add } 5x \text{ to}\\ \text{both sides}\end{array}$$

$$(x + 3)(x + 2) = 0 \qquad \text{Factoring}$$

$$x + 3 = 0 \quad or \quad x + 2 = 0 \qquad \text{Using the principle of zero products}$$

$$x = -3 \quad or \qquad x = -2. \qquad \begin{array}{l}\text{The only restricted value is 0, so}\\ \text{both answers should check.}\end{array}$$

Student Notes

Not all checking is for finding errors in computation. For these equations, the solution process itself can introduce numbers that do not check.

Check: For -3:

$$x + \frac{6}{x} = -5$$

$$\begin{array}{c|c} -3 + \dfrac{6}{-3} & -5 \\[2mm] -3 - 2 & \\[2mm] -5 \overset{?}{=} -5 & \text{TRUE} \end{array}$$

For -2:

$$x + \frac{6}{x} = -5$$

$$\begin{array}{c|c} -2 + \dfrac{6}{-2} & -5 \\[2mm] -2 - 3 & \\[2mm] -5 \overset{?}{=} -5 & \text{TRUE} \end{array}$$

2. Solve: $\dfrac{2}{3} - \dfrac{1}{n} = \dfrac{7}{3n}$.

Both of these check, so there are two solutions, -3 and -2.

YOUR TURN

EXAMPLE 3 Solve.

a) $1 + \dfrac{3x}{x + 2} = \dfrac{-6}{x + 2}$

b) $\dfrac{3}{x - 5} + \dfrac{1}{x + 5} = \dfrac{2}{x^2 - 25}$

c) $\dfrac{x^2}{x - 1} = \dfrac{1}{x - 1}$

SOLUTION

a) The only denominator in $1 + \dfrac{3x}{x + 2} = \dfrac{-6}{x + 2}$ is $x + 2$. To find restrictions, we set this equal to 0 and solve:

$$x + 2 = 0$$

$$x = -2.$$

If $x = -2$, the rational expressions are undefined. We list the restriction:

$$x \neq -2.$$

Technology Connection

We can use a table to check possible solutions of rational equations. Consider the equation in Example 3(c) and the possible solutions that were found, 1 and -1. To check these solutions, we enter

$$y_1 = \frac{x^2}{x-1} \text{ and } y_2 = \frac{1}{x-1}.$$

After setting Indpnt to Ask and Depend to Auto in the TBLSET menu, we display the table and enter $x = 1$. The ERROR messages indicate that 1 is not a solution because it is not an allowable replacement for x in the equation. Next, we enter $x = -1$. Since y_1 and y_2 have the same value, we know that the equation is true when $x = -1$, and thus -1 is a solution.

X	Y₁	Y₂
1	ERROR	ERROR
-1	-.5	-.5

X =

Use a graphing calculator to check the possible solutions of Examples 3(a) and 3(b).

We clear fractions using the LCM $x + 2$ and solve.

$$(x+2)\left(1 + \frac{3x}{x+2}\right) = (x+2)\frac{-6}{x+2}$$

Multiplying both sides by the LCM. *Don't forget the parentheses!*

$$(x+2) \cdot 1 + (x+2)\frac{3x}{x+2} = (x+2)\frac{-6}{x+2}$$

Using the distributive law; removing a factor equal to 1: $(x+2)/(x+2) = 1$

$$x + 2 + 3x = -6$$
$$4x + 2 = -6$$
$$4x = -8$$
$$x = -2.$$

Above, we stated that $x \neq -2$.

Because of the above restriction, -2 must be rejected as a solution. The check below simply confirms this.

Check:
$$1 + \frac{3x}{x+2} = \frac{-6}{x+2}$$

$$\frac{1 + \dfrac{3(-2)}{-2+2}}{} \,\bigg|\, \frac{\dfrac{-6}{-2+2}}{}$$

$$1 + \frac{-6}{0} \overset{?}{=} \frac{-6}{0} \qquad \text{FALSE}$$

The equation has no solution.

b) The denominators in $\dfrac{3}{x-5} + \dfrac{1}{x+5} = \dfrac{2}{x^2-25}$ are $x-5, x+5,$ and $x^2 - 25$. Setting them equal to 0 and solving, we find that the rational expressions are undefined when $x = 5$ or $x = -5$. We list the restrictions:

$$x \neq 5, \qquad x \neq -5.$$

We clear fractions using the LCM, $(x-5)(x+5)$, and solve.

$$(x-5)(x+5)\left(\frac{3}{x-5} + \frac{1}{x+5}\right) = (x-5)(x+5)\frac{2}{(x-5)(x+5)}$$

$$\frac{(x-5)(x+5)3}{x-5} + \frac{(x-5)(x+5)}{x+5} = \frac{2(x-5)(x+5)}{(x-5)(x+5)}$$

Using the distributive law

$$(x+5)3 + (x-5) = 2$$

Removing factors equal to 1:
$(x-5)/(x-5) = 1,$
$(x+5)/(x+5) = 1,$
and $\dfrac{(x-5)(x+5)}{(x-5)(x+5)} = 1$

$$3x + 15 + x - 5 = 2$$

Using the distributive law

$$4x + 10 = 2$$
$$4x = -8$$
$$x = -2.$$

-2 is not restricted, so it *should* check.

The student can check to confirm that -2 is the solution.

Student Notes

When solving an equation like $\dfrac{x^2}{x-1} = \dfrac{1}{x-1}$ in Example 3(c), we can simply equate numerators since the denominators are equal. We must still note restrictions first.

3. Solve:

$$\frac{6}{x^2-4} = \frac{x+1}{x^2-2x}.$$

c) To solve $\dfrac{x^2}{x-1} = \dfrac{1}{x-1}$, note that if $x = 1$, the denominators are 0. We list the restriction:

$$x \neq 1.$$

We clear fractions using the LCM, $x - 1$, and solve.

$$(x-1) \cdot \frac{x^2}{x-1} = (x-1) \cdot \frac{1}{x-1} \qquad \text{Multiplying both sides by } x - 1, \text{ the LCM}$$

$$x^2 = 1 \qquad \text{Removing a factor equal to 1: } (x-1)/(x-1) = 1$$

$$x^2 - 1 = 0 \qquad \text{Subtracting 1 from both sides}$$

$$(x - 1)(x + 1) = 0 \qquad \text{Factoring}$$

$$x - 1 = 0 \quad or \quad x + 1 = 0 \qquad \text{Using the principle of zero products}$$

$$x = 1 \quad or \qquad x = -1 \qquad \text{Above, we stated that } x \neq 1.$$

Because of the above restriction, 1 must be rejected as a solution. The student should check in the original equation that -1 *does* check. The solution is -1.

↩ YOUR TURN

CONNECTING 🔗 THE CONCEPTS

An equation contains an equal sign; an expression does not. Be careful not to confuse simplifying an expression with solving an equation. When expressions are simplified, the result is an equivalent expression. When equations are solved, the result is a solution. Compare the following.

Simplify: $\dfrac{x-1}{6x} + \dfrac{4}{9}.$

SOLUTION

$$\frac{x-1}{6x} + \frac{4}{9} = \frac{x-1}{6x} \cdot \frac{3}{3} + \frac{4}{9} \cdot \frac{2x}{2x}$$

> The equal signs indicate that all the expressions are equivalent.

$$= \frac{3x-3}{18x} + \frac{8x}{18x} \qquad \text{Writing with the LCD, } 18x$$

$$= \frac{11x-3}{18x} \qquad \text{The result is an expression.}$$

The expressions

$$\frac{11x-3}{18x} \quad \text{and} \quad \frac{x-1}{6x} + \frac{4}{9}$$

are equivalent.

Solve: $\dfrac{x-1}{6x} = \dfrac{4}{9}.$

SOLUTION

$$\frac{x-1}{6x} = \frac{4}{9} \qquad \boxed{\text{Each line is an equivalent equation.}}$$

$$18x \cdot \frac{x-1}{6x} = 18x \cdot \frac{4}{9} \qquad \text{Multiplying by the LCM, } 18x$$

$$3 \cdot 6x \cdot \frac{x-1}{6x} = 2 \cdot 9 \cdot x \cdot \frac{4}{9}$$

$$3(x - 1) = 2x \cdot 4$$

$$3x - 3 = 8x$$

$$-3 = 5x$$

$$-\frac{3}{5} = x \qquad \text{The result is a solution.}$$

The solution is $-\dfrac{3}{5}.$

(continued)

EXERCISES

Tell whether each exercise contains an expression or an equation. Then simplify the expression or solve the equation, as indicated.

1. Add and, if possible, simplify: $\dfrac{2}{5n} + \dfrac{3}{2n-1}$.

2. Solve: $\dfrac{3}{y} - \dfrac{1}{4} = \dfrac{1}{y}$.

3. Solve: $\dfrac{5}{x+3} = \dfrac{3}{x+2}$.

4. Multiply and, if possible, simplify:
$$\frac{8t+8}{2t^2+t-1} \cdot \frac{t^2-1}{t^2-2t+1}.$$

5. Subtract and, if possible, simplify:
$$\frac{2a}{a+1} - \frac{4a}{1-a^2}.$$

6. Solve: $\dfrac{20}{x} = \dfrac{x}{5}$.

6.6   EXERCISE SET

FOR EXTRA HELP

PRACTICE   WATCH   READ   REVIEW

➤ Vocabulary and Reading Check

Classify each of the following statements as either true or false.

1. Every rational equation has at least one solution.

2. It is possible for a rational equation to have more than one solution.

3. When both sides of an equation are multiplied by a variable expression, the result is not always an equivalent equation.

4. All the equation-solving principles studied thus far may be needed when solving a rational equation.

Solving Rational Equations

Solve. If no solution exists, state this.

5. $\dfrac{3}{5} - \dfrac{2}{3} = \dfrac{x}{6}$

6. $\dfrac{5}{8} - \dfrac{3}{5} = \dfrac{x}{10}$

7. $\dfrac{1}{8} + \dfrac{1}{12} = \dfrac{1}{t}$

8. $\dfrac{1}{6} + \dfrac{1}{10} = \dfrac{1}{t}$

9. $\dfrac{x}{6} - \dfrac{6}{x} = 0$

10. $\dfrac{x}{7} - \dfrac{7}{x} = 0$

11. $\dfrac{2}{x} = \dfrac{5}{x} - \dfrac{1}{4}$

12. $\dfrac{3}{t} = \dfrac{4}{t} - \dfrac{1}{5}$

13. $\dfrac{5}{3t} + \dfrac{3}{t} = 1$

14. $\dfrac{3}{4x} + \dfrac{5}{x} = 1$

15. $\dfrac{12}{x} = \dfrac{x}{3}$

16. $\dfrac{x}{2} = \dfrac{18}{x}$

17. $y + \dfrac{4}{y} = -5$

18. $n + \dfrac{3}{n} = -4$

19. $\dfrac{n+2}{n-6} = \dfrac{1}{2}$

20. $\dfrac{a-4}{a+6} = \dfrac{1}{3}$

21. $x + \dfrac{12}{x} = -7$

22. $x + \dfrac{8}{x} = -9$

23. $\dfrac{3}{x-4} = \dfrac{5}{x+1}$

24. $\dfrac{1}{x+3} = \dfrac{4}{x-1}$

25. $\dfrac{a}{6} - \dfrac{a}{10} = \dfrac{1}{6}$

26. $\dfrac{t}{8} - \dfrac{t}{12} = \dfrac{1}{8}$

27. $\dfrac{x+1}{3} - 1 = \dfrac{x-1}{2}$

28. $\dfrac{x+2}{5} - 1 = \dfrac{x-2}{4}$

29. $\dfrac{y+3}{y-3} = \dfrac{6}{y-3}$

30. $\dfrac{3}{a+7} = \dfrac{a+10}{a+7}$

31. $\dfrac{3}{x+4} = \dfrac{5}{x}$

32. $\dfrac{2}{x+3} = \dfrac{7}{x}$

33. $\dfrac{n+1}{n+2} = \dfrac{n-3}{n+1}$

34. $\dfrac{n+2}{n-3} = \dfrac{n+1}{n-2}$

35. $\dfrac{5}{t-2} + \dfrac{3t}{t-2} = \dfrac{4}{t^2-4t+4}$

36. $\dfrac{4}{t-3} + \dfrac{2t}{t-3} = \dfrac{12}{t^2-6t+9}$

37. $\dfrac{x}{x+5} - \dfrac{5}{x-5} = \dfrac{14}{x^2-25}$

38. $\dfrac{5}{x+1} + \dfrac{2x}{x^2-1} = \dfrac{1}{x+1}$

39. $\dfrac{3}{x-3} + \dfrac{5}{x+2} = \dfrac{5x}{x^2-x-6}$

40. $\dfrac{2}{x-2} + \dfrac{1}{x+4} = \dfrac{x}{x^2+2x-8}$

41. $\dfrac{5}{t-3} - \dfrac{30}{t^2-9} = 1$

42. $\dfrac{1}{y+3} + \dfrac{1}{y-3} = \dfrac{1}{y^2-9}$

43. $\dfrac{7}{6-a} = \dfrac{a+1}{a-6}$

44. $\dfrac{t-12}{t-10} = \dfrac{1}{10-t}$

Aha! **45.** $\dfrac{-2}{x+2} = \dfrac{x}{x+2}$

46. $\dfrac{3}{2x-6} = \dfrac{x}{2x-6}$

47. $\dfrac{5}{3x+3} + \dfrac{1}{2x-2} = \dfrac{1}{x^2-1}$

48. $\dfrac{4}{x^2-4} + \dfrac{1}{x+2} + \dfrac{2}{3x-6} = 0$

49. When solving rational equations, why do we multiply each side by the LCM of the denominators?

50. Explain the difference between adding rational expressions and solving rational equations.

Skill Review

51. Find the x-intercept and the y-intercept of the line given by $6x - y = 18$. [3.3]

52. Find the slope of the line containing the points $(6, 0)$ and $(-3, -1)$. [3.5]

53. Find the slope and the y-intercept of the line given by $2x + y = 5$. [3.6]

54. Determine whether the graphs of the following equations are parallel:
$$y = 2 - x,$$
$$3x + 3y = 7. \quad [3.6]$$

55. Write the slope–intercept equation of the line with slope $\frac{1}{3}$ and y-intercept $(0, -2)$. [3.6]

56. Write a point–slope equation for the line with slope -4 that contains the point $(1, -5)$. [3.7]

Synthesis

57. Describe a method that can be used to create rational equations that have no solution.

58. How can a graph be used to determine how many solutions an equation has?

Solve.

59. $1 + \dfrac{x-1}{x-3} = \dfrac{2}{x-3} - x$

60. $\dfrac{4}{y-2} + \dfrac{3}{y^2-4} = \dfrac{5}{y+2} + \dfrac{2y}{y^2-4}$

61. $\dfrac{12-6x}{x^2-4} = \dfrac{3x}{x+2} - \dfrac{3-2x}{2-x}$

62. $\dfrac{x}{x^2+3x-4} + \dfrac{x+1}{x^2+6x+8} = \dfrac{2x}{x^2+x-2}$

63. $7 - \dfrac{a-2}{a+3} = \dfrac{a^2-4}{a+3} + 5$

64. $\dfrac{x^2}{x^2-4} = \dfrac{x}{x+2} - \dfrac{2x}{2-x}$

65. $\dfrac{1}{x-1} + x - 5 = \dfrac{5x-4}{x-1} - 6$

66. $\dfrac{5-3a}{a^2+4a+3} - \dfrac{2a+2}{a+3} = \dfrac{3-a}{a+1}$

67.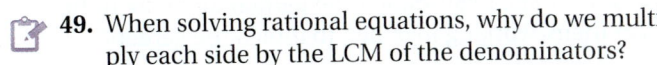

68.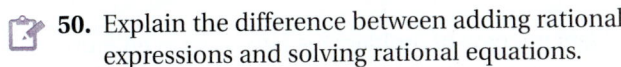

69. Use a graphing calculator to check your answers to Exercises 13, 21, 31, and 59.

70. The reciprocal of a number is the number itself. What is the number?

YOUR TURN ANSWERS: SECTION 6.6

1. $\frac{15}{8}$ **2.** 5 **3.** 1

QUICK QUIZ: SECTIONS 6.1–6.6

1. Multiply and, if possible, simplify:
$$\dfrac{x^2-12x+11}{x+7} \cdot \dfrac{7}{x-1}. \quad [6.2]$$

2. Find the LCM: $x^3 - 4x, x^2 - x - 2$. [6.3]

3. Subtract and, if possible, simplify: $\dfrac{3}{2n} - \dfrac{n-1}{n+2}$. [6.4]

4. Simplify: $\dfrac{\dfrac{x-2}{x+1}}{\dfrac{x+2}{x+1}}$. [6.5]

5. Solve: $\dfrac{1}{x} - \dfrac{2}{x-4} = \dfrac{3}{x^2-4x}$. [6.6]

PREPARE TO MOVE ON

Solve.

1. The sum of two consecutive odd numbers is 276. Find the numbers. [2.5]

2. The product of two consecutive even integers is 48. Find the numbers. [5.8]

3. The height of a triangle is 3 cm longer than its base. If the area of the triangle is 54 cm², find the measurements of the base and the height. [5.8]

4. Between June 9 and June 24, Seth's beard grew 0.9 cm. Find the rate at which Seth's beard grows. [3.4]

6.7   Applications Using Rational Equations and Proportions

Problems Involving Work   ▪   Problems Involving Motion   ▪   Problems Involving Proportions

In many areas of study, applications involving rates, proportions, or reciprocals translate to rational equations.

Problems Involving Work

EXAMPLE 1   Brian and Reba volunteer in a community garden. Brian can mulch the garden alone in 8 hr and Reba can mulch the garden alone in 10 hr. How long will it take the two of them, working together, to mulch the garden?

SOLUTION

1. **Familiarize.** This *work problem* is a type of problem we have not yet encountered. Work problems are often *incorrectly* translated to mathematical language in several ways.

   a) Add the times together: $8 \, \text{hr} + 10 \, \text{hr} = 18 \, \text{hr}.$ ←— Incorrect

   This cannot be the correct approach since it should not take Brian and Reba longer to do the job together than it takes either of them working alone.

   b) Average the times: $(8 \, \text{hr} + 10 \, \text{hr})/2 = 9 \, \text{hr}.$ ←— Incorrect

   Again, this is longer than it would take Brian to do the job alone.

   c) Assume that each person does half the job.   ←— Incorrect

   If each person does half the job, Brian would be finished with his half in 4 hr, and Reba with her half in 5 hr. Since they are working together, Brian would continue to help Reba after completing his half. This does tell us that the job will take between 4 hr and 5 hr when they work together.

   Each incorrect approach began with the time it takes each worker to do the job. The correct approach instead focuses on the *rate* of work, or the amount of the job that each person completes in 1 hr.

   Since it takes Brian 8 hr to mulch the entire garden, in 1 hr he mulches $\frac{1}{8}$ of the garden. Since it takes Reba 10 hr to mulch the entire garden, in 1 hr she mulches $\frac{1}{10}$ of the garden. Together, they mulch $\frac{1}{8} + \frac{1}{10} = \frac{5}{40} + \frac{4}{40} = \frac{9}{40}$ of the garden per hour. The rates are thus

Brian:	$\frac{1}{8}$ of the garden per hour,
Reba:	$\frac{1}{10}$ of the garden per hour,
Together:	$\frac{9}{40}$ of the garden per hour.

   We are looking for the time required to mulch 1 entire garden.

Study Skills

Take Advantage of Free Checking

Always check an answer if it is possible to do so. When an applied problem is being solved, do check an answer in the equation from which it came. However, it is even more important to check the answer with the words of the original problem.

	Fraction of the Garden Mulched		
Time	**By Brian**	**By Reba**	**Together**
1 hr	$\frac{1}{8}$	$\frac{1}{10}$	$\frac{1}{8} + \frac{1}{10}$, or $\frac{9}{40}$
2 hr	$\frac{1}{8} \cdot 2$	$\frac{1}{10} \cdot 2$	$\left(\frac{1}{8} + \frac{1}{10}\right)2$, or $\frac{9}{40} \cdot 2$, or $\frac{9}{20}$
3 hr	$\frac{1}{8} \cdot 3$	$\frac{1}{10} \cdot 3$	$\left(\frac{1}{8} + \frac{1}{10}\right)3$, or $\frac{9}{40} \cdot 3$, or $\frac{27}{40}$
t hr	$\frac{1}{8} \cdot t$	$\frac{1}{10} \cdot t$	$\left(\frac{1}{8} + \frac{1}{10}\right)t$, or $\frac{9}{40} \cdot t$

2. **Translate.** From the table, we see that t must be some number for which

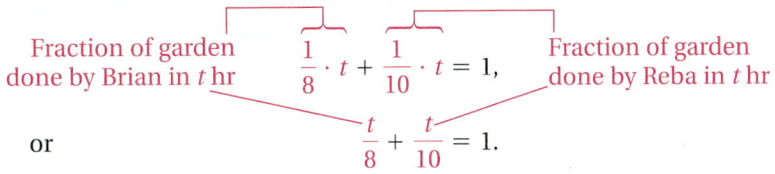

Fraction of garden done by Brian in t hr $\dfrac{1}{8} \cdot t + \dfrac{1}{10} \cdot t = 1,$ Fraction of garden done by Reba in t hr

or $$\frac{t}{8} + \frac{t}{10} = 1.$$

3. **Carry out.** We solve the equation:

$$\frac{t}{8} + \frac{t}{10} = 1 \qquad \text{The LCD is 40.}$$

$$40\left(\frac{t}{8} + \frac{t}{10}\right) = 40 \cdot 1 \qquad \text{Multiplying to clear fractions}$$

$$\frac{40t}{8} + \frac{40t}{10} = 40$$

$$5t + 4t = 40 \qquad \text{Simplifying}$$

$$9t = 40$$

$$t = \tfrac{40}{9}, \text{ or } 4\tfrac{4}{9}.$$

1. Refer to Example 1. Suppose that Brian can pick the produce alone in 60 min and Reba can pick the produce alone in 45 min. How long will it take the two of them, working together, to pick the produce?

4. **Check.** In $\frac{40}{9}$ hr, Brian mulches $\frac{1}{8} \cdot \frac{40}{9}$, or $\frac{5}{9}$, of the garden and Reba mulches $\frac{1}{10} \cdot \frac{40}{9}$, or $\frac{4}{9}$, of the garden. Together, they mulch $\frac{5}{9} + \frac{4}{9}$, or 1 garden. The fact that our solution is between 4 hr and 5 hr (see step 1 above) is also a check.

5. **State.** It will take $4\frac{4}{9}$ hr for Brian and Reba, working together, to mulch the garden.

 YOUR TURN

THE WORK PRINCIPLE

If

 a = the time needed for A to complete the work alone,

 b = the time needed for B to complete the work alone, and

 t = the time needed for A and B to complete the work together,

then

$$\frac{t}{a} + \frac{t}{b} = 1.$$

The following are equivalent equations that can also be used:

$$\left(\frac{1}{a} + \frac{1}{b}\right)t = 1, \qquad \frac{1}{a} \cdot t + \frac{1}{b} \cdot t = 1, \quad \text{and} \quad \frac{1}{a} + \frac{1}{b} = \frac{1}{t}.$$

EXAMPLE 2   It takes Manuel 9 hr longer than it does Zoe to rebuild an engine. Working together, they can do the job in 20 hr. How long would it take each, working alone, to rebuild an engine?

SOLUTION

1. **Familiarize.** Unlike Example 1, this problem does not provide us with the times required by the individuals to do the job alone. We let $z =$ the number of hours it would take Zoe working alone and $z + 9 =$ the number of hours it would take Manuel working alone.

2. **Translate.** Using the same reasoning as in Example 1, we see that Zoe completes $\frac{1}{z}$ of the job in 1 hr and Manuel completes $\frac{1}{z + 9}$ of the job in 1 hr. Thus, in 20 hr, Zoe completes $\frac{1}{z} \cdot 20$ of the job and Manuel completes $\frac{1}{z + 9} \cdot 20$ of the job. Together, Zoe and Manuel can complete the entire job in 20 hr. This gives the following:

Fraction of job done by Zoe in 20 hr $\qquad \dfrac{1}{z} \cdot 20 + \dfrac{1}{z + 9} \cdot 20 = 1,$ $\qquad$ Fraction of job done by Manuel in 20 hr

or $\qquad \dfrac{20}{z} + \dfrac{20}{z + 9} = 1.$

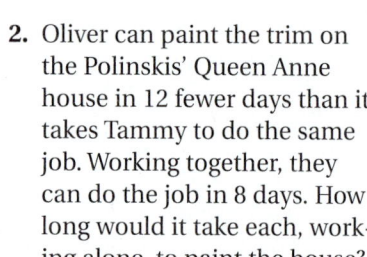

2. Oliver can paint the trim on the Polinskis' Queen Anne house in 12 fewer days than it takes Tammy to do the same job. Working together, they can do the job in 8 days. How long would it take each, working alone, to paint the house?

3. **Carry out.** We solve the equation:

$$\frac{20}{z} + \frac{20}{z + 9} = 1 \qquad \text{The LCM is } z(z + 9).$$

$$z(z + 9)\left(\frac{20}{z} + \frac{20}{z + 9}\right) = z(z + 9)1 \qquad \text{Multiplying to clear fractions}$$

$$(z + 9)20 + z \cdot 20 = z(z + 9) \qquad \text{Distributing and simplifying}$$

$$40z + 180 = z^2 + 9z$$

$$0 = z^2 - 31z - 180 \qquad \text{Getting 0 on one side}$$

$$0 = (z - 36)(z + 5) \qquad \text{Factoring}$$

$$z - 36 = 0 \quad or \quad z + 5 = 0 \qquad \text{Principle of zero products}$$

$$z = 36 \quad or \qquad z = -5.$$

4. **Check.** Since negative time has no meaning in the problem, -5 is not a solution to the original problem. The number 36 checks since, if Zoe takes 36 hr alone and Manuel takes $36 + 9 = 45$ hr alone, in 20 hr they would have finished

$$\tfrac{20}{36} + \tfrac{20}{45} = \tfrac{5}{9} + \tfrac{4}{9} = 1 \text{ complete rebuild.}$$

5. **State.** It would take Zoe 36 hr to rebuild an engine alone, and Manuel 45 hr.

YOUR TURN

Problems Involving Motion

Problems dealing with distance, rate (or speed), and time are called **motion problems**. To translate them, we use either the basic motion formula, $d = rt$, or the formulas $r = d/t$ or $t = d/r$, which can be derived from $d = rt$.

EXAMPLE 3   On her road bike, Olivia bikes 5 km/h faster than Jason does on his mountain bike. In the time it takes Olivia to travel 50 km, Jason travels 40 km. Find the speed of each bicyclist.

SOLUTION

1. **Familiarize.** Let's make a guess and check it.

   Guess: Jason's speed: 10 km/h

   Olivia's speed: 10 + 5, or 15 km/h

   Jason's time: 40/10 = 4 hr ⎱ The times are
   Olivia's time: 50/15 = $3\frac{1}{3}$ hr ⎰ not the same.

   Our guess is wrong, but we can make some observations. If Jason's speed = r, in kilometers/hour, then Olivia's speed = $r + 5$. Jason's travel time is the same as Olivia's travel time.
   We can also make a sketch and label it to help us visualize the situation.

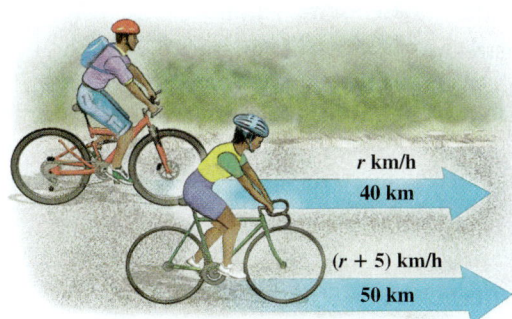

r km/h
40 km

(*r* + 5) km/h
50 km

2. **Translate.** We organize the information in a table. By looking at how we checked our guess, we see that we can fill in the **Time** column of the table using the formula *Time = Distance/Rate*.

	Distance	Speed	Time
Jason's Mountain Bike	40	r	$40/r$
Olivia's Road Bike	50	$r + 5$	$50/(r + 5)$

Since we know that the times are the same, we can write an equation:

$$\frac{40}{r} = \frac{50}{r + 5}.$$

3. **Carry out.** We solve the equation:

$$\frac{40}{r} = \frac{50}{r + 5} \qquad \text{The LCM is } r(r + 5).$$

$$r(r + 5)\frac{40}{r} = r(r + 5)\frac{50}{r + 5} \qquad \text{Multiplying to clear fractions}$$

$$40r + 200 = 50r \qquad \text{Simplifying}$$

$$200 = 10r$$

$$20 = r.$$

3. Peter can drive 25 mph faster on the highway than he can on county roads. In the time it would take Peter to drive 70 mi on county roads, he could drive 120 mi on the highway. How fast can he drive on each type of road?

4. **Check.** If our answer checks, Jason's mountain bike is going 20 km/h and Olivia's road bike is going 20 + 5 = 25 km/h.
   Traveling 50 km at 25 km/h, Olivia is riding for $\frac{50}{25}$ = 2 hr. Traveling 40 km at 20 km/h, Jason is riding for $\frac{40}{20}$ = 2 hr. Our answer checks since the two times are the same.

5. **State.** Olivia's speed is 25 km/h, and Jason's speed is 20 km/h.

 YOUR TURN

EXAMPLE 4   A Hudson River tugboat goes 10 mph in still water. It travels 24 mi upstream and 24 mi back in a total time of 5 hr. What is the speed of the current?

Sources: Based on information from the Department of the Interior, U.S. Geological Survey, and *The Tugboat Captain*, Montgomery County Community College

SOLUTION

1. **Familiarize.** Let's make a guess and check it.

Guess: Speed of current: 4 mph
         Tugboat's speed upstream: $10 - 4 = 6$ mph
         Tugboat's speed downstream: $10 + 4 = 14$ mph
         Travel time upstream: $24/6 = 4$ hr   ⎫ The total time
         Travel time downstream: $24/14 = 1\frac{5}{7}$ hr ⎭ is not 5 hr.

Our guess is wrong, but we can make some observations. If $c = $ the current's rate, in miles per hour, we have the following.

The tugboat's speed upstream is $(10 - c)$ mph.

The tugboat's speed downstream is $(10 + c)$ mph.

The total travel time is 5 hr.

We make a sketch and label it, using the information we know.

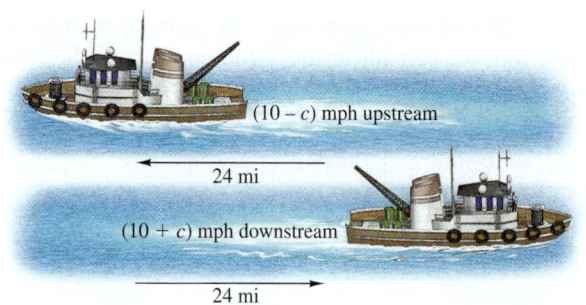

(10 − c) mph upstream

24 mi

(10 + c) mph downstream

24 mi

2. **Translate.** We organize the information in a table. From examining our guess, we see that the time traveled can be represented using the formula *Time = Distance/Rate.*

	Distance	Speed	Time
Upstream	24	$10 - c$	$24/(10 - c)$
Downstream	24	$10 + c$	$24/(10 + c)$

Since the total time upstream and back is 5 hr, we use the last column of the table to form an equation:

$$\frac{24}{10 - c} + \frac{24}{10 + c} = 5.$$

3. **Carry out.** We solve the equation:

$$\frac{24}{10 - c} + \frac{24}{10 + c} = 5 \qquad \text{The LCM is } (10 - c)(10 + c).$$

$$(10 - c)(10 + c)\left[\frac{24}{10 - c} + \frac{24}{10 + c}\right] = (10 - c)(10 + c)5 \qquad \text{Multiplying to clear fractions}$$

$$24(10 + c) + 24(10 - c) = (100 - c^2)5$$

$$480 = 500 - 5c^2 \qquad \text{Simplifying}$$

$$5c^2 - 20 = 0$$

$$5(c^2 - 4) = 0$$

$$5(c - 2)(c + 2) = 0$$

$$c = 2 \quad or \quad c = -2.$$

4. At a length of 4 km, the Comal River in Texas is the shortest navigable river in the United States. Tristan paddled his canoe up and back down the river in $1\frac{1}{3}$ hr. If he paddles 8 km/h in still water, what was the speed of the current?

4. **Check.** Since speed cannot be negative in this problem, -2 cannot be a solution. You should confirm that 2 checks in the original problem.

5. **State.** The speed of the current is 2 mph.

YOUR TURN

EXPLORING 🔍 THE CONCEPT

Motion problems are often much simpler to solve if the information is organized in a table. For each motion problem, fill in the missing entries in the table using the list of options given below.

1. Tara runs 1 km/h faster than Cassie. Tara can run 20 km in the same time that it takes Cassie to run 18 km. Find the speed of each runner.

	Distance	Speed	Time
Tara	20	(a)_____	(b)_____
Cassie	(c)_____	r	(d)_____

Options: 18 $r + 1$ $\dfrac{20}{r+1}$ $\dfrac{18}{r}$

2. Damon rode 50 mi to a state park at a certain speed. Had he been able to ride 3 mph faster, the trip would have been $\frac{1}{4}$ hr shorter. How fast did he ride?

	Distance	Speed	Time
Actual Trip	(a)_____	r	(b)_____
Faster Trip	50	(c)_____	(d)_____

Options: 50 $r + 3$ $\dfrac{50}{r+3}$ $\dfrac{50}{r}$

3. The speed of the Green River current is 4 mph. In the same time that it takes Blair to motor 48 mi downstream, he can travel only 32 mi upstream. What is the speed of the boat in still water?

	Distance	Speed	Time
Downstream	48	(a)_____	$\dfrac{48}{x+4}$
Upstream	(b)_____	(c)_____	(d)_____

Options: 32 $\dfrac{32}{x-4}$ $x + 4$ $x - 4$

ANSWERS

1. (a) $r + 1$; (b) $\dfrac{20}{r+1}$; (c) 18; (d) $\dfrac{18}{r}$

2. (a) 50; (b) $\dfrac{50}{r}$; (c) $r + 3$; (d) $\dfrac{50}{r+3}$

3. (a) $x + 4$; (b) 32; (c) $x - 4$; (d) $\dfrac{32}{x-4}$

Problems Involving Proportions

A **ratio** of two quantities is their quotient. For example, 37% is the ratio of 37 to 100, or $\frac{37}{100}$. A **proportion** is an equation stating that two ratios are equal.

PROPORTION

An equality of ratios,

$$\frac{A}{B} = \frac{C}{D},$$

is called a *proportion*. The fractions within a proportion are said to be *proportional* to each other.

In geometry, if two triangles are **similar**, then their corresponding angles have the same measure and their corresponding sides are *proportional*. To illustrate, if triangle *ABC* is similar to triangle *RST*, then angles *A* and *R* have the same measure, angles *B* and *S* have the same measure, angles *C* and *T* have the same measure, and

$$\frac{a}{r} = \frac{b}{s} = \frac{c}{t}.$$

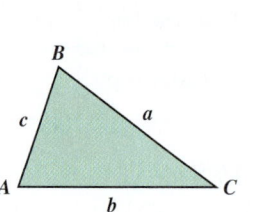

EXAMPLE 5 *Similar Triangles.* Triangles *ABC* and *XYZ* are similar. Solve for *z* if *x* = 10, *a* = 8, and *c* = 5.

SOLUTION We make a drawing, write a proportion, and then solve. Note that side *a* is always opposite angle *A*, side *x* is always opposite angle *X*, and so on.

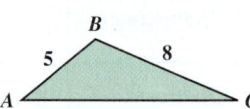
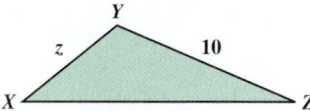

We have

$$\frac{z}{5} = \frac{10}{8} \qquad \text{The proportions } \frac{5}{z} = \frac{8}{10}, \frac{5}{8} = \frac{z}{10}, \text{ or}$$
$$\frac{8}{5} = \frac{10}{z} \text{ could also be used.}$$

$$40 \cdot \frac{z}{5} = 40 \cdot \frac{10}{8} \qquad \text{Multiplying both sides by the LCM, 40}$$

$$8z = 50 \qquad \text{Simplifying}$$

$$z = \frac{50}{8}, \text{ or } 6.25.$$

5. Triangles *QRS* and *MNP* are similar. Solve for *r* if *m* = 5, *n* = 8, and *q* = 4.

YOUR TURN

EXAMPLE 6 *Architecture.* A *blueprint* is a scale drawing of a building representing an architect's plans. Ellia is adding 12 ft to the length of an apartment and needs to indicate the addition on an existing blueprint. If a 10-ft long bedroom is represented by $2\frac{1}{2}$ in. on the blueprint, how much longer should Ellia make the drawing in order to represent the addition?

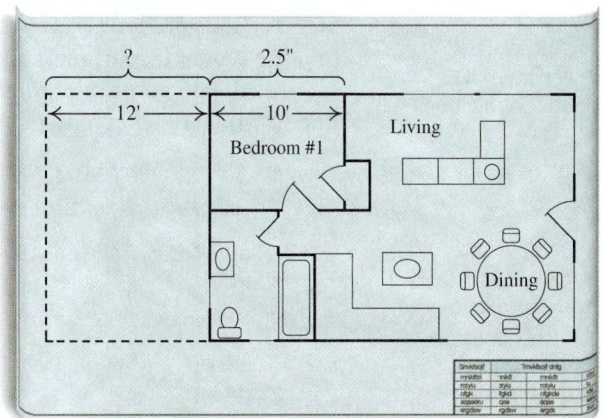

SOLUTION We let w represent the width, in inches, of the addition that Ellia is drawing. Because the drawing must have the correct proportions, we have

$$\text{Inches on drawing} \longrightarrow \frac{w}{12} = \frac{2.5}{10}. \longleftarrow \text{Inches on drawing}$$
$$\text{Feet in real life} \longrightarrow \qquad\qquad \longleftarrow \text{Feet in real life}$$

To solve for w, we multiply both sides by the LCM of the denominators, 60:

$$60 \cdot \frac{w}{12} = 60 \cdot \frac{2.5}{10}$$
$$5w = 6 \cdot 2.5 \qquad \text{Simplifying}$$
$$w = \frac{15}{5}, \text{ or } 3.$$

Ellia should make the blueprint 3 in. longer.

6. Refer to Example 6. Suppose that the width of the apartment is represented on the blueprint by 4 in. How wide is the apartment?

YOUR TURN

EXAMPLE 7 *Text Messaging.* Brent sent or received 384 messages in 8 days. At this rate, how many text messages would he send or receive in 30 days?

SOLUTION We let $x =$ the number of text messages Brent would send or receive in 30 days. We form a proportion in which the ratio of the number of text messages to the number of days is expressed in two ways:

$$\text{Number of text messages} \longrightarrow \frac{384}{8} = \frac{x}{30}. \longleftarrow \text{Number of text messages}$$
$$\text{Number of days} \longrightarrow \qquad\qquad \longleftarrow \text{Number of days}$$

To solve for x, we multiply both sides by the LCM of the denominators, 120:

$$120 \cdot \frac{384}{8} = 120 \cdot \frac{x}{30}$$
$$15 \cdot 8 \cdot \frac{384}{8} = 4 \cdot 30 \cdot \frac{x}{30}$$
$$15 \cdot 384 = 4x$$
$$5760 = 4x$$
$$1440 = x.$$

At this rate, Brent will send or receive 1440 text messages in 30 days.

7. Refer to Example 7. Suppose that Brent sent or received 534 messages in 6 days. At this rate, how many text messages would he send or receive in 30 days?

YOUR TURN

EXAMPLE 8 *Wildlife Population.* To determine the number of brook trout in River Denys, Cape Breton, Nova Scotia, a team of volunteers and professionals caught and marked 1190 brook trout. Later, they captured 915 brook trout, of which 24 were marked. Estimate the number of brook trout in River Denys.

Source: www.gov.ns.ca

SOLUTION We let $T =$ the brook trout population in River Denys. If we assume that the percentage of marked trout in the second group of trout captured is the same as the percentage of marked trout in the entire river, we can form a proportion in which this percentage is expressed in two ways:

Trout originally marked $\longrightarrow \dfrac{1190}{T} = \dfrac{24}{915}.$ $\longleftarrow$ Marked trout in second group
Entire population $\longrightarrow$ $\phantom{\dfrac{1190}{T}}$ $\longleftarrow$ Total trout in second group

To solve for T, we multiply by the LCM, $915T$:

$$915T \cdot \dfrac{1190}{T} = 915T \cdot \dfrac{24}{915}$$ Multiplying both sides by $915T$

$$915 \cdot 1190 = 24T$$ Removing factors equal to 1: $T/T = 1$ and $915/915 = 1$

$$\dfrac{915 \cdot 1190}{24} = T \text{ or } T \approx 45{,}369.$$ Dividing both sides by 24

There are about 45,369 brook trout in the river.

 YOUR TURN

 Chapter Resources:
Translating for Success, p. 432;
Collaborative Activity, p. 433;
Decision Making: Connection, p. 433

8. Refer to Example 8. Suppose that a second team of volunteers in another section of the river captured 875 brook trout, of which 35 were marked. On the basis of these data, estimate the number of brook trout in the River Denys.

6.7 EXERCISE SET

Vocabulary and Reading Check

Classify each of the following statements as either true or false.

1. In order to find the time that it would take two people to complete a job working together, we average the time that it takes each of them to complete the job working separately.

2. To find the rate at which two people work together, we add the rates at which they work separately.

3. Distance equals rate times time.

4. Rate equals distance divided by time.

5. Time equals distance divided by rate.

6. If two triangles are similar, their corresponding sides are of equal length.

Concept Reinforcement

Find each rate.

7. If Sandy can decorate a cake in 2 hr, what is her hourly rate?

8. If Eric can decorate a cake in 3 hr, what is his hourly rate?

9. If Sandy can decorate a cake in 2 hr and Eric can decorate the same cake in 3 hr, what is their hourly rate, working together?

10. If Lisa and Mark can mow a lawn together in 1 hr, what is their hourly rate?

11. If Lisa can mow a lawn by herself in 3 hr, what is her hourly rate?

12. If Lisa and Mark can mow a lawn together in 1 hr, and Lisa can mow the same lawn by herself in 3 hr, what is Mark's hourly rate, working alone?

Problems Involving Work

13. *Volunteerism.* It takes Kelby 10 hr per week to prepare food for delivery to senior citizens. Natalie can do the same job in 15 hr. How long would it take Kelby and Natalie to prepare the food working together?

14. *Pumping Water.* An ABS Robusta 300 TS sump pump can remove water from Martha's flooded basement in 70 min. The Little Giant 1-A sump pump can complete the same job in 30 min. How

long would it take the two pumps together to pump out the basement?

Source: Based on data from www.shoppumps.com

15. *Home Restoration.*   Bryan can refinish the floor of an apartment in 8 hr. Armando can refinish the floor in 6 hr. How long will it take them, working together, to refinish the floor?

16. *Custom Embroidery.*   Chandra can embroider logos on a team's sweatshirts in 6 hr. Traci, a new employee, needs 9 hr to complete the same job. Working together, how long will it take them to do the job?

17. *Filling a Pool.*   The San Paulo community swimming pool can be filled in 12 hr if water enters through a pipe alone or in 30 hr if water enters through a hose alone. If water is entering through both the pipe and the hose, how long will it take to fill the pool?

18. *Filling a Tank.*   A community water tank can be filled in 18 hr by the town office well alone and in 22 hr by the high school well alone. How long will it take to fill the tank if both wells are working?

19. *Scanners.*   The Epson WorkForce G7-1500 takes twice the time required by the Epson WorkForce Pro GT-S50 to scan the manuscript for a book. If, working together, the two machines can complete the job in 5 min, how long would it take each machine, working alone, to scan the manuscript?

Source: epson.com

20. *Cutting Firewood.*   Kent can cut and split a cord of wood twice as fast as Brent can. When they work together, it takes them 4 hr. How long would it take each of them to do the job alone?

21. *Mulching.*   Anita can mulch the college gardens in 3 fewer days than it takes Tori to mulch the same areas. When they work together, it takes them 2 days. How long would it take each of them to do the job alone?

22. *Photo Printing.*   It takes the Canon PIXMA MP495 15 min longer to print a set of photo proofs than it takes the Canon PIXMA MG8220. Together it would take them 10 min to print the photos. How long would it take each machine, working alone, to print the photos?

Source: www.usa.canon.com

23. *Software Development.*   Tristan, an experienced programmer, can write video-game software three times as fast as Sara, who is just learning to program. Working together on one project, it took them 1 month to complete the job. How long would it take each of them to complete the project alone?

24. *Forest Fires.*   The Erickson Air-Crane helicopter can scoop water and douse a certain forest fire four times as fast as an S-58T helicopter. Working together, the two helicopters can douse the fire in 8 hr. How long would it take each helicopter, working alone, to douse the fire?

Sources: Based on information from www.emergency.com and www.arishelicopters.com

25. *Sorting Recyclables.*   Together, it takes Kim and Chris 2 hr 55 min to sort recyclables. Alone, Chris would require 2 fewer hours than Kim. How long would it take Chris to do the job alone? (*Hint:* Convert minutes to hours or hours to minutes.)

26. *Paving.*   Together, Steve and Bill require 4 hr 48 min to pave a driveway. Alone, Steve would require 4 hr more than Bill. How long would it take Bill to do the job alone? (*Hint:* Convert minutes to hours.)

Problems Involving Motion

27. *Train Speeds.*   A CSX freight train is traveling 14 km/h slower than an AMTRAK passenger train. The CSX train travels 330 km in the same time that it takes the AMTRAK train to travel 400 km. Find their speeds. Complete the following table as part of the familiarization.

Distance =	Rate ·	Time
	Distance (in km)	Speed (in km/h)	Time (in hours)
CSX	330		
AMTRAK	400	r	$\dfrac{400}{r}$

28. *Speed of Travel.*   A loaded Roadway truck is moving 40 mph faster than a New York Railways freight train. In the time that it takes the train to travel 150 mi, the truck travels 350 mi. Find their speeds. Complete the following table as part of the familiarization.

Distance =	Rate ·	Time
	Distance (in miles)	Speed (in miles per hour)	Time (in hours)
Truck	350	r	$\dfrac{350}{r}$
Train	150		

29. *Driving Speed.* Sean's Camaro travels 15 mph faster than Rita's Harley. In the same time that Sean travels 156 mi, Rita travels 120 mi. Find their speeds.

30. *Bicycle Speed.* Ada bicycles 5 km/h slower than Elin. In the same time that it takes Ada to ride 48 km, Elin can ride 63 km. How fast does each bicyclist travel?

31. *Kayaking.* The speed of the current in Catamount Creek is 3 mph. Sean can kayak 4 mi upstream in the same time that it takes him to kayak 10 mi downstream. What is the speed of Sean's kayak in still water?

32. *Boating.* The current in the Lazy River moves at a rate of 4 mph. Mickie's dinghy motors 6 mi upstream in the same time that it takes to motor 12 mi downstream. What is the speed of the dinghy in still water?

33. *Moving Sidewalks.* The moving sidewalk at O'Hare Airport in Chicago moves 1.8 ft/sec. Walking on the moving sidewalk, Roslyn travels 105 ft forward in the time that it takes to travel 51 ft in the opposite direction. How fast does Roslyn walk on a nonmoving sidewalk?

34. *Moving Sidewalks.* Newark Airport's moving sidewalk moves at a speed of 1.7 ft/sec. Walking on the moving sidewalk, Drew can travel 120 ft forward in the same time that it takes to travel 52 ft in the opposite direction. What is Drew's walking speed on a nonmoving sidewalk?

Aha! 35. *Tractor Speed.* Manley's tractor is just as fast as Caledonia's. It takes Manley 1 hr more than it takes Caledonia to drive to town. If Manley is 20 mi from town and Caledonia is 15 mi from town, how long does it take Caledonia to drive to town?

36. *Boat Speed.* Tory and Emilio's motorboats travel at the same speed. Tory pilots her boat 40 km before docking. Emilio continues for another 2 hr, traveling a total of 100 km before docking. How long did it take Tory to navigate the 40 km?

37. *Aviation.* A Citation CV jet travels 460 mph in still air and flies 525 mi into the wind and 525 mi with the wind in a total of 2.3 hr. Find the wind speed.

Source: Blue Star Jets, Inc.

38. *Canoeing.* Chad paddles 55 m/min in still water. He paddles 150 m upstream and 150 m downstream in a total time of 5.5 min. What is the speed of the current?

39. *Train Travel.* A freight train covered 120 mi at a certain speed. Had the train been able to travel 10 mph faster, the trip would have been 2 hr shorter. How fast did the train go?

40. *Moped Speed.* Dexter rode 60 mi on his moped at a certain speed. Had he been able to travel 5 mph faster, the trip would have taken 1 hr less time. How fast did Dexter go?

Problems Involving Proportions

Geometry. *For each pair of similar triangles, find the value of the indicated letter.*

41. *b*

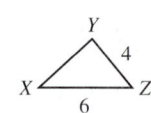

42. *a*

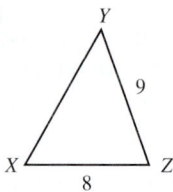
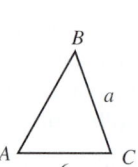

43. *f*

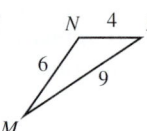
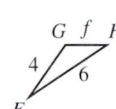

44. *r*

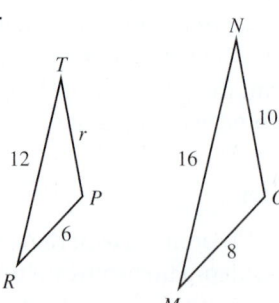

Graphing.   *Find the indicated length.*

45. *r*

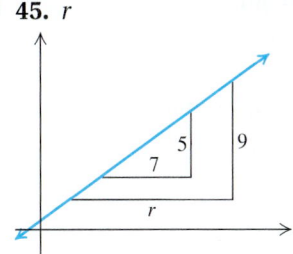

46. *s*

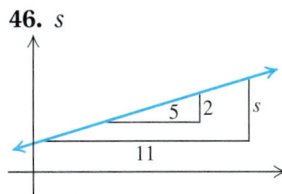

Architecture.   *Use the blueprint below to find the indicated length.*

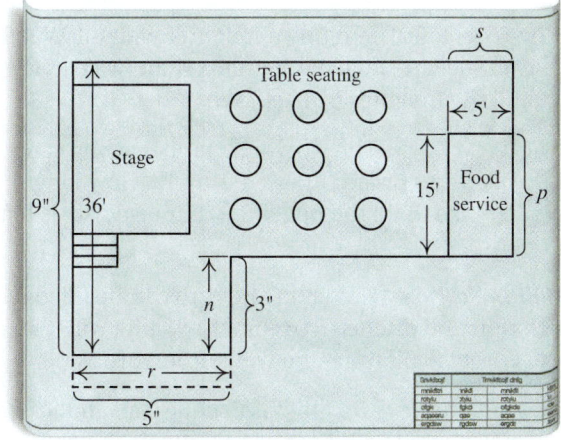

47. *p*, in inches on blueprint

48. *s*, in inches on blueprint

49. *r*, in feet on actual building

50. *n*, in feet on actual building

Find the indicated length.

51. *l*

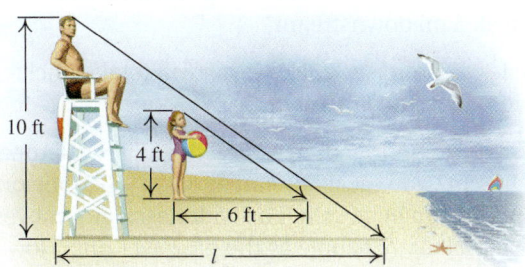

52. *h*

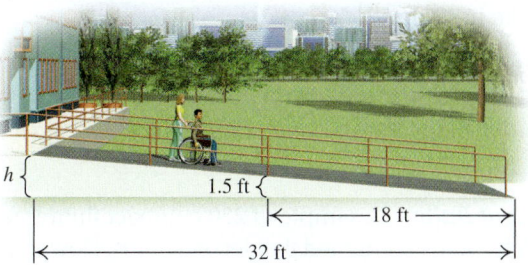

Flight Mechanics.   *The wing aspect ratio for a bird or an airplane is the ratio of the wing span to the wing width. Generally, higher-aspect ratios are more efficient during low-speed flying. Use the table below for Exercises 53 and 54.*

Bird or Airplane	Wing Span	Wing Width
Grey heron	180 cm	24 cm
House sparrow	10 in.	1.8 in.
Albatross	9 ft	0.6 ft
Concorde (retired in 2003)	84 ft	47 ft
Boeing 747-400	64 m	8 m

53. Herons and storks, both waders, have the same wing aspect ratios. A white stork has a wing span of 200 cm. What is the wing width of a white stork?

54. The Piper PA-28 Cherokee airplane, with a wing span of 35 ft, has the same wing aspect ratio as a house sparrow. What is the width of a wing of the Piper airplane?

55. *Spending Habits.*   In the first 8 days of September, Felicia spent $17.40 on coffee. At this rate, how much will she spend in September? (September has 30 days.)

56. *Burning Calories.*   The average 140-lb adult burns about 380 calories bicycling 10 mi at a moderate rate. How far should the average 140-lb adult ride in order to burn 100 calories?

Source: www.livestrong.com

Aha! **57.** *Photography.* Aziza snapped 234 photos over a period of 14 days. At this rate, how many would she take in 42 days?

58. *Mileage.* The Toyota Prius is a hybrid car that travels approximately 204 mi on 4 gal of gas. Find the amount of gas required for a 714-mi trip.

Source: www.toyota.com

59. *Light Bulbs.* A sample of 220 compact fluorescent light bulbs contained 8 defective bulbs. How many defective bulbs would you expect in a batch of 1430 bulbs?

60. *Flash Drives.* A sample of 150 flash drives contained 7 defective drives. How many defective flash drives would you expect in a batch of 2700 flash drives?

61. *Veterinary Science.* The amount of water needed by a small dog depends on its weight. A moderately active 8-lb Shih Tzu needs approximately 12 oz of water per day. How much water does a moderately active 5-lb Bolognese require each day?

Source: www.smalldogsparadise.com

62. *Miles Driven.* Emmanuel is allowed to drive his leased car for 45,000 mi in 4 years without penalty. In the first $1\frac{1}{2}$ years, Emmanuel has driven 16,000 mi. At this rate, will he exceed the mileage allowed for 4 years?

63. *Environmental Science.* To determine the number of humpback whales in a pod, a marine biologist, using tail markings, identifies 27 members of the pod. Several weeks later, 40 whales from the pod are randomly sighted. Of the 40 sighted, 12 are from the 27 originally identified. Estimate the number of whales in the pod.

64. *Fox Population.* To determine the number of foxes in King County, a naturalist catches, tags, and then releases 25 foxes. Later, 36 foxes are caught; 4 of them have tags. Estimate the fox population of the county.

65. Is it correct to assume that two workers will complete a task twice as quickly as one person working alone? Why or why not?

66. If two triangles are exactly the same shape and size, are they similar? Why or why not?

Skill Review

Perform the indicated operation and simplify.

67. $(x^3 - 3x - 7) - (x^2 - 4x + 8)$ [4.4]

68. $(2x^3 - 7)(x + 3)$ [4.5]

69. $(3y^2z - 2yz^2 + y^2) + (4yz^2 + 5y^2 - 6yz)$ [4.7]

70. $(6a^2b^3 + 12ab^2 - 3a^2b) \div (3ab)$ [4.8]

71. $(8n^3 + 3)(8n^3 - 3)$ [4.6]

72. $(x^3 - x + 7) \div (x - 1)$ [4.8]

Synthesis

73. Two steamrollers are paving a parking lot. Working together, will the two steamrollers take less than half as long as the slower steamroller would working alone? Why or why not?

74. Two fuel lines are filling a freighter with oil. Will the faster fuel line take more or less than twice as long to fill the freighter by itself? Why?

75. *Filling a Bog.* The Norwich cranberry bog can be filled in 9 hr and drained in 11 hr. How long will it take to fill the bog if the drainage gate is left open?

76. *Filling a Tub.* Gretchen's hot tub can be filled in 10 min and drained in 8 min. How long will it take to empty a full tub if the water is left on?

77. *Car Cleaning.* Together, Michelle, Sal, and Kristen can clean and wax a car in 1 hr 20 min. To complete the job alone, Michelle needs twice the time that Sal needs and 2 hr more than Kristen. How long would it take each to clean and wax the car working alone?

78. *Grading.* Alma can grade a batch of placement exams in 3 hr. Kevin can grade a batch in 4 hr. If they work together to grade a batch of exams, what percentage of the exams will have been graded by Alma?

79. Refer to Exercise 31. How long will it take Sean to kayak 5 mi downstream?

80. Refer to Exercise 32. How long will it take Mickie to motor 3 mi downstream?

81. *Escalators.* Together, a 100-cm wide escalator and a 60-cm wide escalator can empty a 1575-person auditorium in 14 min. The wider escalator moves twice as many people as the narrower one. How many people per hour does the 60-cm wide escalator move?

Source: McGraw-Hill Encyclopedia of Science and Technology

82. *Aviation.* A Coast Guard plane has enough fuel to fly for 6 hr, and its speed in still air is 240 mph. The plane departs with a 40-mph tailwind and returns to the same airport flying into the same wind. How far can the plane travel under these conditions?

83. *Boating.* Shoreline Travel operates a 3-hr paddleboat cruise on the Missouri River. If the speed of the boat in still water is 12 mph, how far upriver can the pilot travel against a 5-mph current before it is time to turn around?

84. *Travel by Car.* Angenita drives to work at 50 mph and arrives 1 min late. She drives to work at 60 mph and arrives 5 min early. How far does Angenita live from work?

85. According to the U.S. Census Bureau, Population Division, in July 2012, there was one birth every 8 sec, one death every 14 sec, and one new international migrant every 44 sec. How many seconds does it take for a net gain of one person?

86. *Home Maintenance.* Fuel used in many chain saws is made by pouring a 3.2-oz bottle of 2-cycle oil into 160 oz of gasoline. Gus accidentally poured 5.6 oz of 2-cycle oil into 200 oz of gasoline. How much more oil or gasoline should he add in order for the fuel to have the proper ratio of oil to gasoline?

87. At what time after 4:00 will the minute hand and the hour hand of a clock first be in the same position?

88. At what time after 10:30 will the hands of a clock first be perpendicular?

Average speed is defined as total distance divided by total time.

89. Ferdaws drove 200 km. For the first 100 km of the trip, she drove at a speed of 40 km/h. For the second half of the trip, she traveled at a speed of 60 km/h. What was the average speed of the entire trip? (It was *not* 50 km/h.)

90. For the first 50 mi of a 100-mi trip, Garry drove 40 mph. What speed would he have to travel for the last half of the trip so that the average speed for the entire trip would be 45 mph?

91. *Commuting.* To reach an appointment 50 mi away, Dr. Wright allowed 1 hr. After driving 30 mi, she realized that her speed would have to be increased 15 mph for the remainder of the trip. What was her speed for the first 30 mi?

92. Given that
$$\frac{A}{B} = \frac{C}{D},$$
write three other proportions using A, B, C, and D.

93. Show that the four equations in the box labeled "The Work Principle" in this section are equivalent.

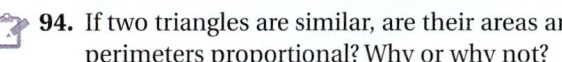 **94.** If two triangles are similar, are their areas and perimeters proportional? Why or why not?

95. Are the equations
$$\frac{A + B}{B} = \frac{C + D}{D} \quad \text{and} \quad \frac{A}{B} = \frac{C}{D}$$
equivalent? Why or why not?

YOUR TURN ANSWERS: SECTION 6.7

1. $25\frac{5}{7}$ min   **2.** Oliver: 12 days; Tammy: 24 days
3. Country roads: 35 mph; highway: 60 mph
4. 4 km/h   **5.** $\frac{32}{5}$, or 6.4   **6.** 16 ft
7. 2670 text messages   **8.** 29,750 brook trout

QUICK QUIZ: SECTIONS 6.1–6.7

1. Simplify: $\dfrac{x^2y^2 - 5x^2y + 6x^2}{7x^3y - 7x^3}$.   [6.1]

2. Divide and, if possible, simplify:
$$\frac{7x + 21}{x - 4} \div \frac{x + 3}{5x - 20}.$$ [6.2]

3. Add and, if possible, simplify:
$$\frac{1}{x^2 + 3x + 2} + \frac{1}{x^2 - x - 2}.$$ [6.4]

4. Solve: $\dfrac{7}{x} = \dfrac{3}{x + 4}$.   [6.6]

5. Scot can prepare a tray of appetizers in 75 min. Steve can prepare the same tray in 60 min. How long would it take them, working together, to prepare the tray of appetizers?   [6.7]

PREPARE TO MOVE ON

Graph each pair of equations on the same set of axes.

1. $y = 2x - 1$,   **2.** $x + y = 3$,
  $y = x + 3$ [3.6] $x - y = 2$ [3.3]

3. $y = \frac{1}{2}x + 2$,   **4.** $x + y = 2$,
  $y = \frac{1}{2}x - 1$ [3.6] $2x + 2y = 4$ [3.3]

CHAPTER 6 RESOURCES

1. *Internet Search-Engine Ads.* In 2010, North American advertisers spent $16.6 billion in marketing through Internet search engines such as Google®. This was a 14% increase over the amount spent in 2009. How much was spent in 2009?

Source: Search Engine Marketing Professional Organization

2. *Bicycling.* The speed of one bicyclist is 2 km/h faster than the speed of another bicyclist. The first bicyclist travels 60 km in the same amount of time that it takes the second to travel 50 km. Find the speed of each bicyclist.

3. *Filling Time.* A swimming pool can be filled in 5 hr by hose A alone and in 6 hr by hose B alone. How long would it take to fill the tank if both hoses were working?

4. *Payroll.* In 2010, the total payroll for Kraftside Productions was $16.6 million. Of this amount, 14% was paid to employees working on an assembly line. How much money was paid to assembly-line workers?

Translating for Success

Use after Section 6.7.

Translate each word problem to an equation and select a correct translation from equations A–O.

A. $2x + 2(x + 1) = 613$

B. $x^2 + (x + 1)^2 = 613$

C. $\dfrac{60}{x + 2} = \dfrac{50}{x}$

D. $x = 14\% \cdot 16.6$

E. $\dfrac{197}{7} = \dfrac{x}{30}$

F. $x + (x + 1) = 613$

G. $\dfrac{7}{197} = \dfrac{x}{30}$

H. $x^2 + (x + 2)^2 = 612$

I. $x^2 + (x + 1)^2 = 612$

J. $\dfrac{50}{x + 2} = \dfrac{60}{x}$

K. $x + 14\% \cdot x = 16.6$

L. $\dfrac{5 + 6}{2} = t$

M. $x^2 + (x + 1)^2 = 452$

N. $\dfrac{1}{5} + \dfrac{1}{6} = \dfrac{1}{t}$

O. $x^2 + (x + 2)^2 = 452$

Answers on page A-31

An additional, animated version of this activity appears in MyMathLab. To use MyMathLab, you need a course ID and a student access code. Contact your instructor for more information.

5. *Cycling Distance.* A bicyclist traveled 197 mi in 7 days. At this rate, how many miles could the cyclist travel in 30 days?

6. *Sides of a Square.* If the sides of a square are increased by 2 ft, the area of the original square plus the area of the enlarged square is 452 ft². Find the length of a side of the original square.

7. *Consecutive Integers.* The sum of two consecutive integers is 613. Find the integers.

8. *Sums of Squares.* The sum of the squares of two consecutive odd integers is 612. Find the integers.

9. *Sums of Squares.* The sum of the squares of two consecutive integers is 613. Find the integers.

10. *Rectangle Dimensions.* The length of a rectangle is 1 ft longer than its width. Find the dimensions of the rectangle if the perimeter is 613 ft.

Collaborative Activity   *Sharing the Workload*

Focus: Modeling, estimation, and work problems
Use after: Section 6.7
Time: 10–15 minutes
Group size: 3
Materials: Paper, pencils, textbooks, and a watch

Many tasks can be done by two people working together. If both people work at the same rate, each does half the task, and the project is completed in half the time. However, when the work rates differ, the faster worker performs more than half of the task.

Activity

1. The project is to write the numbers from 1 to 100 on a sheet of paper. Two of the members in each group should write the numbers, one working slowly and one working quickly. The third group member should record the time required for each to write the numbers.

2. Using the times from step (1), calculate how long it will take the two workers, working together, to complete the task.

3. Next, have the same workers as in step (1)—working at the same speeds as in step (1)—perform the task together. To do this, one person should begin writing with 1, while the other worker, using a separate sheet of paper, begins with 100 and writes the numbers counting backward. The third member is again the timekeeper and should note when the two workers have written all the numbers.

4. Compare the actual experimental time from part (3) with the time predicted by the model in part (2). List reasons that might account for any discrepancy.

5. Let t_1, t_2, and t_3 represent the times required for the first worker, the second worker, and the two workers together, respectively, to complete a task. Then develop a model that can be used to find t_3 when t_1 and t_2 are known.

Decision Making & Connection

Phone Costs.   Unless you use your cell phone strictly for voice calls, your monthly cell bill will include charges for texting and/or data use. In order to save money, many users choose plans that limit the amount of texting or data each month. In order to realize those savings, it is important not to exceed the specified amounts of text messages and data usage.

1. One popular text-messaging plan allows up to 500 text messages (received or sent) each 30-day month. At the end of the 8th day of one month, Xavier checked his account and found that he had sent or received 160 text messages. At that rate, will he exceed the 500-message limit by the end of the month?

2. One popular data package allows up to 250 MB (megabytes) of data (received or sent) each 30-day month. Xavier checked his account at the end of the 16th day of the month and found that he had sent or received 130 MB of data. At that rate, will he exceed the 250-MB limit by the end of the month?

(Use after Section 6.7.)

3. Suppose that Xavier continues to use text messaging and data at the rates described in Exercises 1 and 2. His plan charges $0.20 for every text message sent or received over the limit and $10 for an additional GB (gigabyte; 1 GB = 1000 MB) of data. If his monthly cell bill is $64.99 before text and data charges are added, what will his bill be for the month described?

4. Xavier can change his text messaging to 1000 text messages per month for an additional $5 per month. If the month described in Exercise 1 is a typical month for him, should he change his text messaging plan?

5. *Research.*   Find what phone plans are available in your area. If you already have a phone plan, compare your plan and monthly usage with other plans available. If you do not have a phone plan, estimate your monthly text and data usage and compare the plans using those estimates. Which is the best plan for you?

Study Summary

KEY TERMS AND CONCEPTS	EXAMPLES	PRACTICE EXERCISES

SECTION 6.1: *Rational Expressions*

A **rational expression** can be written as a quotient of two polynomials and is undefined when the denominator is 0. We simplify rational expressions by removing a factor equal to 1.	$$\frac{x^2 - 3x - 4}{x^2 - 1} = \frac{(x + 1)(x - 4)}{(x + 1)(x - 1)}$$ $$= \frac{x - 4}{x - 1} \quad \frac{x + 1}{x + 1} = 1$$	**1.** Simplify: $$\frac{3x^2 - 6x + 3}{x^2 - 4x + 3}.$$

SECTION 6.2: *Multiplication and Division*

The Product of Two Rational Expressions $$\frac{A}{B} \cdot \frac{C}{D} = \frac{AC}{BD}$$	$$\frac{5v + 5}{v - 2} \cdot \frac{2v^2 - 8v + 8}{v^2 - 1}$$ $$= \frac{5(v + 1) \cdot 2(v - 2)(v - 2)}{(v - 2)(v + 1)(v - 1)}$$ $$= \frac{10(v - 2)}{v - 1} \quad \frac{(v + 1)(v - 2)}{(v + 1)(v - 2)} = 1$$	**2.** Multiply and, if possible, simplify: $$\frac{10a + 20}{a^2 - 4} \cdot \frac{a^2 - a - 2}{4a}.$$
The Quotient of Two Rational Expressions $$\frac{A}{B} \div \frac{C}{D} = \frac{A}{B} \cdot \frac{D}{C} = \frac{AD}{BC}$$	$$(x^2 - 5x - 6) \div \frac{x^2 - 1}{x + 6}$$ $$= \frac{x^2 - 5x - 6}{1} \cdot \frac{x + 6}{x^2 - 1}$$ $$= \frac{(x - 6)(x + 1)(x + 6)}{(x + 1)(x - 1)}$$ $$= \frac{(x - 6)(x + 6)}{x - 1} \quad \frac{x + 1}{x + 1} = 1$$	**3.** Divide and, if possible, simplify: $$\frac{t^2 - 100}{t^2 - 2t} \div \frac{t^2 + 8t - 20}{t^3}.$$

SECTION 6.3: *Addition, Subtraction, and Least Common Denominators*

The Sum of Two Rational Expressions $$\frac{A}{B} + \frac{C}{B} = \frac{A + C}{B}$$	$$\frac{7x + 8}{x + 1} + \frac{4x + 3}{x + 1} = \frac{7x + 8 + 4x + 3}{x + 1}$$ $$= \frac{11x + 11}{x + 1}$$ $$= \frac{11(x + 1)}{x + 1}$$ $$= 11 \quad \frac{x + 1}{x + 1} = 1$$	**4.** Add and, if possible, simplify: $$\frac{2x + 4}{x + 5} + \frac{x - 5}{x + 5}.$$
The Difference of Two Rational Expressions $$\frac{A}{B} - \frac{C}{B} = \frac{A - C}{B}$$	$$\frac{7x + 8}{x + 1} - \frac{4x + 3}{x + 1} = \frac{7x + 8 - (4x + 3)}{x + 1}$$ $$= \frac{7x + 8 - 4x - 3}{x + 1}$$ $$= \frac{3x + 5}{x + 1}$$	**5.** Subtract and, if possible, simplify: $$\frac{4x + 1}{3x - 7} - \frac{x + 8}{3x - 7}.$$

To find the **least common multiple, LCM**, of a set of polynomials, write the prime factorizations of the polynomials. The LCM contains each factor the greatest number of times that it occurs in any of the individual factorizations.

Find the LCM of $m^2 - 5m + 6$ and $m^2 - 4m + 4$.

$$m^2 - 5m + 6 = (m - 2)(m - 3) \left.\right\}$$
$$m^2 - 4m + 4 = (m - 2)(m - 2) \left.\right\}$$ Factoring each expression

$$\textbf{LCM} = (m - 2)(m - 2)(m - 3)$$

6. Find the LCM of $a^2 - 25$ and $a^2 - 6a + 5$.

SECTION 6.4: *Addition and Subtraction with Unlike Denominators*

To add or subtract rational expressions with different denominators, first rewrite the expressions as equivalent expressions with a common denominator. The **least common denominator, LCD**, is the LCM of the denominators.

$$\frac{2x}{x^2 - 16} + \frac{x}{x - 4}$$

$$= \frac{2x}{(x + 4)(x - 4)} + \frac{x}{x - 4} \qquad \begin{array}{l}\text{The LCD is}\\(x + 4)(x - 4).\end{array}$$

$$= \frac{2x}{(x + 4)(x - 4)} + \frac{x}{x - 4} \cdot \frac{x + 4}{x + 4}$$

$$= \frac{2x}{(x + 4)(x - 4)} + \frac{x^2 + 4x}{(x + 4)(x - 4)}$$

$$= \frac{x^2 + 6x}{(x + 4)(x - 4)} = \frac{x(x + 6)}{(x + 4)(x - 4)}$$

7. Subtract and, if possible, simplify:

$$\frac{3x - 1}{x - 1} - \frac{x + 1}{x - 2}.$$

SECTION 6.5: *Complex Rational Expressions*

Complex rational expressions contain one or more rational expressions within the numerator and/or the denominator. They can be simplified either by using division or by multiplying by a form of 1 to clear the fractions.

Using division to simplify:

$$\frac{\dfrac{1}{6} - \dfrac{1}{x}}{\dfrac{6 - x}{6}} = \frac{\dfrac{1}{6} \cdot \dfrac{x}{x} - \dfrac{1}{x} \cdot \dfrac{6}{6}}{\dfrac{6 - x}{6}} = \frac{\dfrac{x - 6}{6x}}{\dfrac{6 - x}{6}} \qquad \begin{array}{l}\text{\color{red}Forming a single}\\\text{\color{red}rational expression}\\\text{\color{red}in the numerator}\end{array}$$

$$= \frac{x - 6}{6x} \div \frac{6 - x}{6} = \frac{x - 6}{6x} \cdot \frac{6}{6 - x}$$

$$= \frac{6(x - 6)}{6x(-1)(x - 6)} = \frac{1}{-x} = -\frac{1}{x} \qquad \frac{6(x - 6)}{6(x - 6)} = 1$$

Multiplying by 1 to simplify:

$$\frac{\dfrac{4}{x}}{\dfrac{3}{x} + \dfrac{2}{x^2}} = \frac{\dfrac{4}{x}}{\dfrac{3}{x} + \dfrac{2}{x^2}} \cdot \frac{x^2}{x^2} \qquad \begin{array}{l}\text{\color{red}The LCD of all the}\\\text{\color{red}denominators is }x^2\text{;}\\\text{\color{red}multiplying by }\dfrac{x^2}{x^2}\end{array}$$

$$= \frac{\dfrac{4}{x} \cdot \dfrac{x^2}{1}}{\left(\dfrac{3}{x} + \dfrac{2}{x^2}\right) \cdot \dfrac{x^2}{1}}$$

$$= \frac{\dfrac{4 \cdot x \cdot x}{x}}{\dfrac{3 \cdot x \cdot x}{x} + \dfrac{2 \cdot x^2}{x^2}} = \frac{4x}{3x + 2}$$

8. Simplify:

$$\frac{1 - \dfrac{2}{3x}}{x - \dfrac{4}{9x}}.$$

SECTION 6.6:   *Rational Equations*

To Solve a Rational Equation

1. List any restrictions.
2. Clear the equation of fractions.
3. Solve the resulting equation.
4. Check the possible solution(s) in the original equation.

Solve: $\dfrac{2}{x+1} = \dfrac{1}{x-2}$.   The restrictions are $x \neq -1$ and $x \neq 2$.

$$\frac{2}{x+1} = \frac{1}{x-2}$$

$$(x+1)(x-2) \cdot \frac{2}{x+1} = (x+1)(x-2) \cdot \frac{1}{x-2}$$

$$2(x-2) = x+1$$

$$2x - 4 = x + 1$$

$$x = 5$$

Check: Since $\dfrac{2}{5+1} = \dfrac{1}{5-2}$, the solution is 5.

9. Solve: $\dfrac{1}{3} + \dfrac{1}{6} = \dfrac{1}{t}$.

SECTION 6.7:   *Applications Using Rational Equations and Proportions*

The Work Principle

If a = the time needed for A to complete the work alone,

b = the time needed for B to complete the work alone,

and

t = the time needed for A and B to complete the work together, then:

$$\frac{t}{a} + \frac{t}{b} = 1;$$

$$\left(\frac{1}{a} + \frac{1}{b}\right)t = 1;$$

$$\frac{1}{a} \cdot t + \frac{1}{b} \cdot t = 1;$$

$$\frac{1}{a} + \frac{1}{b} = \frac{1}{t}.$$

Brian and Reba volunteer in a community garden. Brian can mulch the garden alone in 8 hr and Reba can mulch the garden alone in 10 hr. How long would it take them, working together, to mulch the garden?

If t = the time, in hours, that it takes Brian and Reba to do the job working together, then

$$\frac{1}{8} \cdot t + \frac{1}{10} \cdot t = 1 \qquad \text{Using the work principle}$$

$$t = 4\tfrac{4}{9}\,\text{hr}. \qquad \text{Solving the equation}$$

See Example 1 in Section 6.7 for a complete solution of this problem.

10. It takes Kenesha 3 hr to stain a large bookshelf. It takes Fletcher 2 hr to stain the same size bookshelf. How long would it take them, working together, to stain the bookshelf?

The Motion Formula

$$d = r \cdot t,$$

$$r = \frac{d}{t},$$

or

$$t = \frac{d}{r}$$

On her road bike, Olivia bikes 5 km/h faster than Jason does on his mountain bike. In the time it takes Olivia to travel 50 km, Jason travels 40 km. Find the speed of each bicyclist.

If r = Jason's speed, in kilometers/hour, then $r + 5$ = Olivia's speed. Using $t = d/r$ and the fact that the times are equal, we have

$$\frac{40}{r} = \frac{50}{r+5}$$

$$r = 20. \qquad \text{Solving the equation}$$

Olivia's speed is 25 km/h, and Jason's speed is 20 km/h.

See Example 3 in Section 6.7 for a complete solution of this problem.

11. Jerry jogs 5 mph faster than he walks. In the time it would take Jerry to walk 8 mi, he could jog 18 mi. Find how fast Jerry jogs and how fast he walks.

In geometry, proportions arise in the study of **similar triangles**. 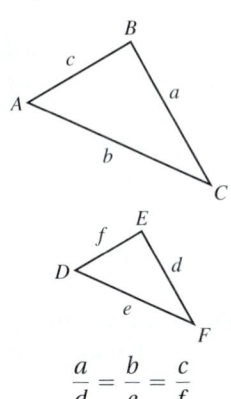 $$\frac{a}{d} = \frac{b}{e} = \frac{c}{f}$$	Triangles *DEF* and *UVW* are similar. Solve for *u*. 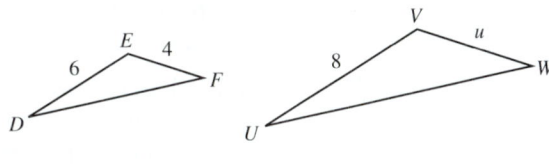 $$\frac{6}{8} = \frac{4}{u}$$ $$u = \frac{32}{6} = \frac{16}{3}$$	**12.** Triangles *ABC* and *QRS* are similar. Solve for *r*. 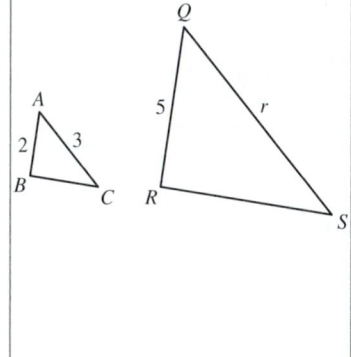

Review Exercises: Chapter 6

➤ Concept Reinforcement

Classify each of the following statements as either true or false.

1. Every rational expression can be simplified.   [6.1]

2. The expression $(t - 3)/(t^2 - 4)$ is undefined for $t = 2$.   [6.1]

3. The expression $(t - 3)/(t^2 - 4)$ is undefined for $t = 3$.   [6.1]

4. To multiply rational expressions, a common denominator is never required.   [6.2]

5. To divide rational expressions, a common denominator is never required.   [6.2]

6. To add rational expressions, a common denominator is never required.   [6.3]

7. To subtract rational expressions, a common denominator is never required.   [6.3]

8. The number 0 can never be a solution of a rational equation.   [6.6]

List all numbers for which each expression is undefined.   [6.1]

9. $\dfrac{17}{-x^2}$

10. $\dfrac{x - 5}{x^2 - 36}$

11. $\dfrac{x^2 + 3x + 2}{x^2 + x - 30}$

12. $\dfrac{-6}{(t + 2)^2}$

Simplify.   [6.1]

13. $\dfrac{3x^2 - 9x}{3x^2 + 15x}$

14. $\dfrac{14x^2 - x - 3}{2x^2 - 7x + 3}$

15. $\dfrac{6y^2 - 36y + 54}{4y^2 - 36}$

16. $\dfrac{5x^2 - 20y^2}{2y - x}$

Multiply or divide and, if possible, simplify.   [6.2]

17. $\dfrac{a^2 - 36}{10a} \cdot \dfrac{2a}{a + 6}$

18. $\dfrac{6y - 12}{2y^2 + 3y - 2} \cdot \dfrac{y^2 - 4}{8y - 8}$

19. $\dfrac{16 - 8t}{3} \div \dfrac{t - 2}{12t}$

20. $\dfrac{4x^4}{x^2 - 1} \div \dfrac{2x^3}{x^2 - 2x + 1}$

21. $\dfrac{x^2 + 1}{x - 2} \cdot \dfrac{2x + 1}{x + 1}$

22. $(t^2 + 3t - 4) \div \dfrac{t^2 - 1}{t + 4}$

Find the LCM.   [6.3]

23. $10a^3b^8,\ 12a^5b$

24. $x^2 - x,\ x^5 - x^3,\ x^4$

25. $y^2 - y - 2,\ y^2 - 4$

Add or subtract and, if possible, simplify.

26. $\dfrac{x + 6}{x + 3} + \dfrac{9 - 4x}{x + 3}$ [6.3]

27. $\dfrac{6x - 3}{x^2 - x - 12} - \dfrac{2x - 15}{x^2 - x - 12}$ [6.3]

28. $\dfrac{3x - 1}{2x} - \dfrac{x - 3}{x}$ [6.4]

29. $\dfrac{2a + 4b}{5ab^2} - \dfrac{5a - 3b}{a^2 b}$ [6.4]

30. $\dfrac{y^2}{y - 2} + \dfrac{6y - 8}{2 - y}$ [6.4]

31. $\dfrac{t}{t + 1} + \dfrac{t}{1 - t^2}$ [6.4]

32. $\dfrac{d^2}{d - 2} + \dfrac{4}{2 - d}$ [6.4]

33. $\dfrac{1}{x^2 - 25} - \dfrac{x - 5}{x^2 - 4x - 5}$ [6.4]

34. $\dfrac{3x}{x + 2} - \dfrac{x}{x - 2} + \dfrac{8}{x^2 - 4}$ [6.4]

35. $\dfrac{3}{4t} + \dfrac{3}{3t + 2}$ [6.4]

Simplify.   [6.5]

36. $\dfrac{\frac{1}{z} + 1}{\frac{1}{z^2} - 1}$

37. $\dfrac{\frac{5}{2x^2}}{\frac{3}{4x} + \frac{4}{x^3}}$

38. $\dfrac{\frac{c}{d} - \frac{d}{c}}{\frac{1}{c} + \frac{1}{d}}$

Solve.   [6.6]

39. $\dfrac{3}{x} - \dfrac{1}{4} = \dfrac{1}{2}$

40. $\dfrac{3}{x + 4} = \dfrac{1}{x - 1}$

41. $x + \dfrac{6}{x} = -7$

42. $1 = \dfrac{2}{x - 1} + \dfrac{2}{x + 2}$

Solve.   [6.7]

43. Jackson can sand the oak floors and stairs in a two-story home in 12 hr. Charis can do the same job in 9 hr. How long would it take if they worked together? (Assume that two sanders are available.)

44. Ben and Jon are working for the summer building trails in a state park. Ben can build one section of the trail in 15 hr less time than Jon. Working together, they can build the section of the trail in 18 hr. How long does it take each to build the section?

45. The Black River's current is 6 mph. A boat travels 50 mi downstream in the same time that it takes to travel 30 mi upstream. What is the speed of the boat in still water?

46. Jennifer's home is 105 mi from her college dorm, and Elizabeth's home is 93 mi away. One Friday afternoon, they left school at the same time and arrived at their homes at the same time. If Jennifer drove 8 mph faster than Elizabeth, how fast did each drive?

47. To estimate the harbor seal population in Bristol Bay, scientists radio-tagged 33 seals. Several days later, they collected a sample of 40 seals, and 24 of them were tagged. Estimate the seal population of the bay.

48. Triangles *ABC* and *XYZ* are similar. Find the value of *x*.

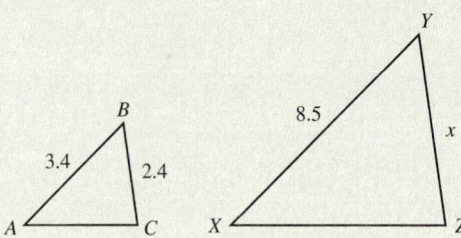

49. A sample of 30 weather-alert radios contained 4 defective ones. How many defective radios would you expect to find in a batch of 540?

Synthesis

 50. For what procedures in this chapter is the LCM of denominators used to clear fractions?   [6.5], [6.6]

51. A student always uses the common denominator found by multiplying the denominators of the expressions being added. How could this approach be improved?   [6.3]

Simplify.

52. $\dfrac{2a^2 + 5a - 3}{a^2} \cdot \dfrac{5a^3 + 30a^2}{2a^2 + 7a - 4} \div \dfrac{a^2 + 6a}{a^2 + 7a + 12}$ [6.2]

53. $\dfrac{12a}{(a - b)(b - c)} - \dfrac{2a}{(b - a)(c - b)}$ [6.4]

Aha! **54.** $\dfrac{5(x - y)}{(x - y)(x + 2y)} - \dfrac{5(x - 3y)}{(x + 2y)(x - 3y)}$ [6.3]

55. The last major-league baseball player to average at least 4 hits in every 10 at bats was Ted Williams in 1941. Suppose that Miguel Cabrera currently has 153 hits after 395 at bats. If he is assured 125 more at bats, what percentage of those must be hits if he is to average 4 hits for every 10 at bats?   [6.7]

Test: Chapter 6

For step-by-step test solutions, access the Chapter Test Prep Videos in MyMathLab®, on YouTube® (search "Bittinger Combo Alg CA" and click on "Channels"), or by scanning the code.

List all numbers for which each expression is undefined.

1. $\dfrac{2 - x}{5x}$

2. $\dfrac{x^2 + x - 30}{x^2 - 3x + 2}$

3. Simplify: $\dfrac{6x^2 + 17x + 7}{2x^2 + 7x + 3}$.

Multiply or divide and, if possible, simplify.

4. $\dfrac{t^2 - 9}{12t} \cdot \dfrac{8t^2}{t^2 - 4t + 3}$

5. $\dfrac{25y^2 - 1}{9y^2 - 6y} \div \dfrac{5y^2 + 9y - 2}{3y^2 + y - 2}$

6. $\dfrac{4a^2 + 1}{4a^2 - 1} \div \dfrac{4a^2}{4a^2 + 4a + 1}$

7. $(x^2 + 6x + 9) \cdot \dfrac{(x - 3)^2}{x^2 - 9}$

8. Find the LCM:

$$y^2 - 9, \quad y^2 + 10y + 21, \quad y^2 + 4y - 21.$$

Add or subtract, and, if possible, simplify.

9. $\dfrac{2 + x}{x^3} + \dfrac{7 - 4x}{x^3}$

10. $\dfrac{5 - t}{t^2 + 1} - \dfrac{t - 3}{t^2 + 1}$

11. $\dfrac{2x - 4}{x - 3} + \dfrac{x - 1}{3 - x}$

12. $\dfrac{2x - 4}{x - 3} - \dfrac{x - 1}{3 - x}$

13. $\dfrac{7}{t - 2} + \dfrac{4}{t}$

14. $\dfrac{y}{y^2 + 6y + 9} + \dfrac{1}{y^2 + 2y - 3}$

15. $\dfrac{1}{x - 1} + \dfrac{4}{x^2 - 1} - \dfrac{2}{x^2 - 2x + 1}$

Simplify.

16. $\dfrac{9 - \dfrac{1}{y^2}}{3 - \dfrac{1}{y}}$

17. $\dfrac{\dfrac{x}{8} - \dfrac{8}{x}}{\dfrac{1}{8} + \dfrac{1}{x}}$

Solve.

18. $\dfrac{1}{t} + \dfrac{1}{3t} = \dfrac{1}{2}$

19. $\dfrac{15}{x} - \dfrac{15}{x - 2} = -2$

20. Kopy Kwik has 2 copiers. One can produce a year-end report in 20 min. The other can produce the same document in 30 min. How long would it take both machines, working together, to produce the report?

21. Katie and Tyler work together to prepare a meal at a soup kitchen in $2\frac{6}{7}$ hr. Working alone, it would take Katie 6 hr more than it would take Tyler. How long would it take each of them to complete the meal, working alone?

22. The average 140-lb adult burns about 320 calories walking 4 mi at a moderate speed. How far should the average 140-lb adult walk in order to burn 100 calories?

Source: www.walking.about.com

23. Ryan drives 20 km/h faster than Alicia. In the same time that Alicia drives 225 km, Ryan drives 325 km. Find the speed of each car.

Synthesis

24. Simplify: $1 - \dfrac{1}{1 - \dfrac{1}{1 - \dfrac{1}{a}}}$.

25. The square of a number is the opposite of the number's reciprocal. Find the number.

Cumulative Review: Chapters 1–6

1. Use the commutative law of multiplication to write an expression equivalent to $a + bc$. [1.2]

2. Evaluate $-x^2$ for $x = 5$. [1.8]

3. Evaluate $(-x)^2$ for $x = 5$. [1.8]

4. Simplify: $-3[2(x - 3) - (x + 5)]$. [1.8]

Solve.

5. $4(y - 5) = -2(y - 2)$ [2.2]

6. $x^2 + 11x + 10 = 0$ [5.7]

7. $49 = x^2$ [5.7]

8. $\frac{4}{9}t + \frac{2}{3} = \frac{1}{3}t - \frac{2}{9}$ [2.2]

9. $\frac{4}{x} + x = 5$ [6.6]

10. $\frac{2}{x - 3} = \frac{5}{3x + 1}$ [6.6]

11. $2x^2 + 7x = 4$ [5.7]

12. $4(x + 7) < 5(x - 3)$ [2.6]

13. $\frac{2}{x^2 - 9} + \frac{5}{x - 3} = \frac{3}{x + 3}$ [6.6]

14. Solve $3a - b + 9 = c$ for b. [2.3]

Graph. [3.2], [3.3], [3.6]

15. $y = \frac{3}{4}x + 5$

16. $x = -3$

17. $4x + 5y = 20$

18. $y = 6$

19. Find the slope of the line containing the points $(1, 5)$ and $(2, 3)$. [3.5]

20. Find the slope and the y-intercept of the line given by $2x - 4y = 1$. [3.6]

21. Write the slope–intercept equation of the line with slope $-\frac{5}{8}$ and y-intercept $(0, -4)$. [3.6]

Simplify.

22. $\frac{x^{-5}}{x^{-3}}$ [4.2]

23. $-(2a^2b^7)^2$ [4.1]

24. Subtract:

$(-8y^2 - y + 2) - (y^3 - 6y^2 + y - 5)$. [4.4]

Multiply.

25. $(2x^2 - 1)(x^3 + x - 3)$ [4.5]

26. $(6x - 5y)^2$ [4.6]

27. $(3n + 2)(n - 5)$ [4.6]

28. $(2x^3 + 1)(2x^3 - 1)$ [4.6]

Factor.

29. $6x - 2x^2 - 24x^4$ [5.1]

30. $16x^2 - 81$ [5.4]

31. $t^2 - 10t + 24$ [5.2]

32. $10t^3 + 10$ [5.5]

33. $6x^2 - 28x + 16$ [5.3]

34. $25t^2 + 40t + 16$ [5.4]

35. $x^2y^2 - xy - 20$ [5.2]

36. $x^4 + 2x^3 - 3x - 6$ [5.1]

Simplify.

37. $\frac{4t - 20}{t^2 - 16} \cdot \frac{t - 4}{t - 5}$ [6.2]

38. $\frac{x^2 - 1}{x^2 - x - 2} \div \frac{x - 1}{x - 2}$ [6.2]

39. $\frac{5ab}{a^2 - b^2} + \frac{a + b}{a - b}$ [6.4]

40. $\frac{x + 2}{4 - x} - \frac{x + 3}{x - 4}$ [6.4]

41. $\dfrac{1 + \dfrac{2}{x}}{1 - \dfrac{4}{x^2}}$ [6.5]

42. Divide:

$(15x^4 - 12x^3 + 6x^2 + 2x + 18) \div (x + 3)$. [4.8]

Solve.

43. For each order, alibris.com charges a shipping fee of $1.80 plus $2.19 per book. The shipping cost for Dae's book order was $34.65. How many books did she order? [2.5]

Source: alibris.com

44. Nikki is laying out two square flower gardens in a client's lawn. Each side of one garden is 2 ft longer than each side of the smaller garden. Together, the area of the gardens is 340 ft^2. Find the length of a side of the smaller garden. [5.8]

45. It takes Wes 25 min to file a week's worth of receipts. Corey, a new employee, takes 75 min to do the same job. How long would it take if they worked together? [6.7]

46. Rachel burned 450 calories in a workout. She burned twice as many in her aerobics session as she did doing calisthenics. How many calories did she burn doing calisthenics? [2.5]

Synthesis

47. Solve: $\frac{1}{3}|n| + 8 = 56$. [1.4], [2.2]

48. Solve: $x(x^2 + 3x - 28) - 12(x^2 + 3x - 28) = 0$. [5.7]

49. Solve: $\frac{2}{x - 3} \cdot \frac{3}{x + 3} - \frac{4}{x^2 - 7x + 12} = 0$. [6.6]

Functions and Graphs

DISTANCE FROM LIGHT SOURCE (in feet)	ILLUMINATION (in foot-candles)
1	400
2	100
4	25

Source: Based on information from Winchip, Susan M., *Fundamentals of Lighting,* Chapter 4, New York: Fairfield Publications, 2008.

Shine a Little Light on Me!

Interior designers consider many details when choosing lighting fixtures, including the amount of illumination that the fixtures provide. Factors such as bulb wattage, type of bulb, the cleanliness of the environment, and the distance from the light source all influence illumination.

As we can see from the table above, illumination decreases as the distance from the light source increases. However, it does not appear to decrease linearly. In this chapter, we will write a formula that relates the distance from a light source and the illumination that source provides. (*See Exercise 79 in Section 7.5.*)

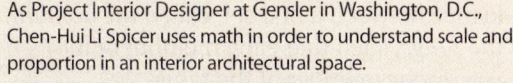

Math becomes an essential tool in all phases of a successful project.

As Project Interior Designer at Gensler in Washington, D.C., Chen-Hui Li Spicer uses math in order to understand scale and proportion in an interior architectural space.

I n this chapter we introduce the concept of a *function. Functions* can often be visualized graphically, as well as added, subtracted, multiplied, and divided. Near the end of the chapter, we use function notation when describing *direct variation* and *inverse variation.*

7.1   Introduction to Functions

Domain and Range ▪ Functions and Graphs ▪ Function Notation and Equations ▪ Applications

We now develop the idea of a *function*—one of the most important concepts in mathematics.

Domain and Range

A function is a special kind of correspondence between two sets. For example,

To each person in a class	there corresponds	a date of birth.
To each bar code in a store	there corresponds	a price.
To each real number	there corresponds	the cube of that number.

In each example, the first set is called the **domain**. The second set is called the **range**. For any member of the domain, there is *exactly one* member of the range to which it corresponds. This kind of correspondence is called a **function**.

Student Notes

Note that not all correspondences are functions.

EXAMPLE 1   Determine whether each correspondence is a function.

a)

Domain	Range
4 ⟶ 2	
1	
−3 ⟶ 5	

b)

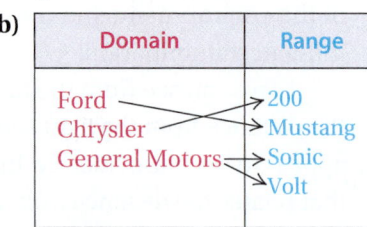

1. Determine whether the correspondence is a function.

Domain	Range
2 ⟶ 4	
⟶ −4	
3 ⟶ 9	
⟶ −9	

SOLUTION

a) The correspondence *is* a function because each member of the domain corresponds to *exactly one* member of the range.

b) The correspondence *is not* a function because a member of the domain (General Motors) corresponds to more than one member of the range.

 YOUR TURN

> **FUNCTION**
>
> A *function* is a correspondence between a first set, called the *domain*, and a second set, called the *range*, such that each member of the domain corresponds to *exactly one* member of the range.

EXAMPLE 2   Determine whether each correspondence is a function. Assume that the item mentioned first is in the domain of the correspondence.

a) The correspondence that assigns to a person his or her weight.

b) The correspondence that assigns to the numbers $-2, 0, 1,$ and 2 each number's square

c) The correspondence that assigns to a best-selling author the titles of books written by that author

SOLUTION

a) For this correspondence, the domain is a set of people and the range is a set of positive numbers (the weights). We ask ourselves, "Does a person have *only one* weight?" Since the answer is Yes, this correspondence *is* a function.

b) The domain is $\{-2, 0, 1, 2\}$, and the range is $\{0, 1, 4\}$. We ask ourselves, "Does each number have *only one* square?" Since the answer is Yes, the correspondence *is* a function.

c) The domain is a set of authors, and the range is a set of book titles. We ask ourselves, "Has each author written *only one* book?" Since many authors have multiple titles published, the answer is No, the correspondence *is not* a function.

YOUR TURN

2. Determine whether the correspondence that assigns to a book the number of pages in the book is a function.
_____

A set of ordered pairs is also a correspondence between two sets. The domain is the set of all first coordinates, and the range is the set of all second coordinates.

EXAMPLE 3   For the correspondence $\{(-6, 7), (1, 4), (-3, 4), (4, -5)\}$, **(a)** write the domain and **(b)** write the range.

SOLUTION

a) The domain is the set of all first coordinates: $\{-6, 1, -3, 4\}$.

b) The range is the set of all second coordinates: $\{7, 4, -5\}$.

YOUR TURN

3. For the correspondence $\{(0, -3), (4, 7), (5, -3)\}$, **(a)** write the domain and **(b)** write the range.
_____

Functions and Graphs

The function in Example 1(a) can be written $\{(-3, 5), (1, 2), (4, 2)\}$ and the function in Example 2(b) $\{(-2, 4), (0, 0), (1, 1), (2, 4)\}$. We graph these functions in black as follows.

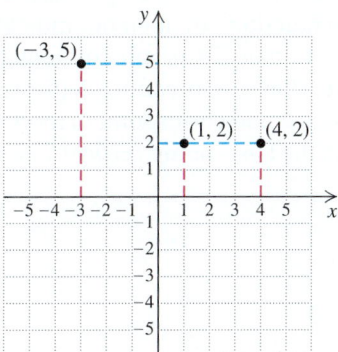

The function $\{(-3, 5), (1, 2), (4, 2)\}$
Domain is $\{-3, 1, 4\}$
Range is $\{5, 2\}$

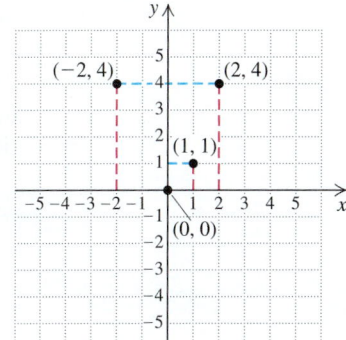

The function $\{(-2, 4), (0, 0), (1, 1), (2, 4)\}$
Domain is $\{-2, 0, 1, 2\}$
Range is $\{4, 0, 1\}$

We can find the domain and the range of a function directly from its graph. Note in the graphs on the preceding page that if we move along the red dashed lines from the points to the horizontal axis, we find the members, or elements, of the domain. Similarly, if we move along the blue dashed lines from the points to the vertical axis, we find the elements of the range.

Functions are generally named using lowercase or uppercase letters. The function in the following example is named *f*.

EXAMPLE 4   For the function *f* represented at left, determine each of the following.

a) The member of the range that is paired with 2

b) The member of the domain that is paired with −3

SOLUTION

a) To determine what member of the range is paired with 2, we first note that we are considering 2 in the domain. Thus we locate 2 on the horizontal axis. Next, we find the point directly above 2 on the graph of *f*. From that point, we can look to the vertical axis to find the corresponding *y*-coordinate, 4. Thus, 4 is the member of the range that is paired with 2.

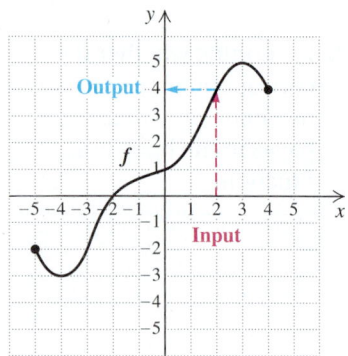

b) To determine what member of the domain is paired with −3, we note that we are considering −3 in the range. Thus we locate −3 on the vertical axis. From there, we look left and right to the graph of *f* to find any points for which −3 is the second coordinate. One such point exists, $(-4, -3)$. We observe that −4 is the only element of the domain paired with −3.

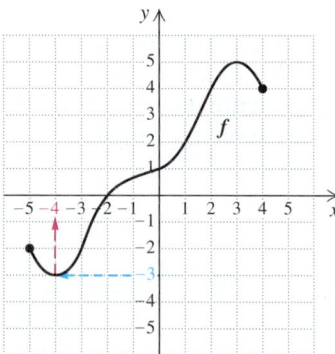

YOUR TURN

A closed dot on a graph, such as in Example 4, indicates that the point is part of the function. An open dot indicates that the point is *not* part of the function.

Note that if a graph contains two or more points with the same first coordinate, that graph cannot represent a function (otherwise one member of the domain would correspond to more than one member of the range). This observation is the basis of the *vertical-line test*.

4. For the function *g* represented below, determine each of the following.

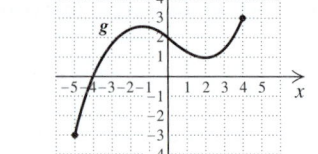

a) The member of the range that is paired with 2

b) The member of the domain that is paired with −3

> **THE VERTICAL-LINE TEST**
>
> If it is possible for a vertical line to cross a graph more than once, then the graph is not the graph of a function.

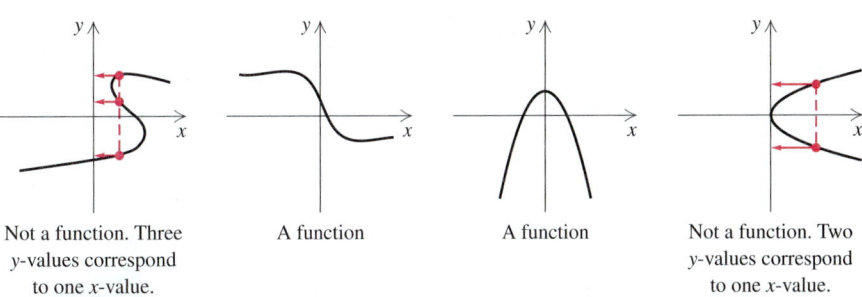

Not a function. Three *y*-values correspond to one *x*-value. A function A function Not a function. Two *y*-values correspond to one *x*-value.

Although not all the graphs above represent functions, they all represent relations.

> **RELATION**
>
> A *relation* is a correspondence between a first set, called the *domain*, and a second set, called the *range*, such that each member of the domain corresponds to *at least one* member of the range.

Relations will appear throughout this book (indeed, every function is a special type of relation), but we will not focus on labeling them as such.

Function Notation and Equations

We often think of an element of the domain of a function as an **input** and its corresponding element of the range as an **output**. For example, consider the function

$$f = \{(-3, 1), (1, -2), (3, 0), (4, 5)\}.$$

Here, for an input of -3, the corresponding output is 1, and for an input of 3, the corresponding output is 0.

We use *function notation* to indicate what output corresponds to a given input. For the function f defined above, we write

$$f(-3) = 1, \qquad f(1) = -2, \qquad f(3) = 0, \quad \text{and} \quad f(4) = 5.$$

The notation $f(x)$ is read "f of x," "f at x," or "the value of f at x." If x is an input, then $f(x)$ is the corresponding output.

> **CAUTION!** $f(x)$ *does not mean* f *times* x.

Most functions are described by equations. For example, $f(x) = 2x + 3$ describes the function that takes an input x, multiplies it by 2, and then adds 3.

$$\underset{\text{Double}}{f(x)} \quad = 2\underset{}{x} + \underset{\text{Add 3}}{3}$$

Input

Student Notes

In mathematics, capitalization makes a difference! The variables r and R, for example, can be used in the same formula to represent two different quantities. Similarly, the function name f is different from the function name F.

To calculate the output $f(4)$, we take the input 4, double it, and add 3 to get 11. That is, we substitute 4 into the formula for $f(x)$:

$$f(x) = 2x \quad + 3$$
$$f(4) = 2 \cdot 4 + 3$$
$$= 11. \longleftarrow \text{Output}$$

To understand function notation, it helps to imagine a "function machine." Think of putting an input into the machine. For $f(x) = 2x + 3$, the machine doubles each input and then adds 3. The result is the output.

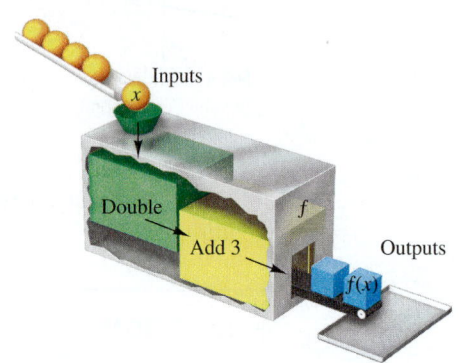

Sometimes, in place of $f(x) = 2x + 3$, we write $y = 2x + 3$, where it is understood that the value of y, the *dependent variable*, depends on our choice of x, the *independent variable*. To understand why $f(x)$ notation is so useful, consider two equivalent statements:

a) Find the member of the range that is paired with 2.

b) Find $f(2)$.

Function notation is not only more concise, it also emphasizes that x is the independent variable.

Note that whether we write $f(x) = 2x + 3$, or $f(t) = 2t + 3$, or $f(\square) = 2 \cdot \square + 3$, we still have $f(4) = 11$. The variable in the parentheses (the independent variable) is the variable used in the algebraic expression.

EXAMPLE 5 Find each indicated function value.

a) $f(5)$, for $f(x) = 3x + 2$ **b)** $g(-2)$, for $g(r) = 5r^2 + 3r$

c) $h(4)$, for $h(x) = 7$ **d)** $F(a + 1)$, for $F(x) = 3x + 2$

e) $F(a) + 1$, for $F(x) = 3x + 2$

SOLUTION Finding a function value is much like evaluating an algebraic expression.

a) $f(x) = 3x + 2$

$f(5) = 3 \cdot 5 + 2 = 17$ 5 is the input; 17 is the output.

b) $g(r) = 5r^2 + 3r$

Substitute. $g(-2) = 5(-2)^2 + 3(-2)$

Evaluate. $= 5 \cdot 4 - 6 = 14$

c) For the function given by $h(x) = 7$, every input has the same output, 7. Therefore, $h(4) = 7$. The function h is an example of a *constant function*.

Student Notes

In Example 5(d), it is important to note that the parentheses on the left are for function notation, whereas those on the right indicate multiplication.

5. Find $f(-1)$ for $f(t) = t^2 - 3$.

d)
$$F(x) = 3x + 2$$
$$F(a + 1) = 3(a + 1) + 2 \qquad \text{The input is } a + 1.$$
$$= 3a + 3 + 2 = 3a + 5$$

e)
$$F(x) = 3x + 2$$
$$F(a) + 1 = [3(a) + 2] + 1 \qquad \text{The input is } a.$$
$$= [3a + 2] + 1 = 3a + 3$$

 YOUR TURN

When we find a function value, we determine an output that corresponds to a given input. To do this, we evaluate the expression for that input value.

Sometimes we want to determine an input that corresponds to a given output. To do this, we may need to solve an equation.

EXAMPLE 6   Let $f(x) = 3x - 7$.

a) What output corresponds to an input of 5?

b) What input corresponds to an output of 5?

SOLUTION

a) We ask ourselves, "$f(5) = \blacksquare$?" Thus we find $f(5)$:

$$f(x) = 3x - 7$$
$$f(5) = 3(5) - 7 \qquad \text{The input is 5. We substitute 5 for } x.$$
$$= 15 - 7 = 8. \qquad \text{Carrying out the calculations}$$

The output 8 corresponds to the input 5; that is, $f(5) = 8$.

b) We ask ourselves, "$f(\blacksquare) = 5$?" Thus we find the value of x for which $f(x) = 5$:

$$f(x) = 3x - 7$$
$$5 = 3x - 7 \qquad \text{The output is 5. We substitute 5 for } f(x).$$
$$12 = 3x$$
$$4 = x. \qquad \text{Solving for } x$$

6. Let $f(x) = \frac{1}{2}x$.
   a) What output corresponds to an input of 10?
   b) What input corresponds to an output of 10?

The input 4 corresponds to the output 5; that is, $f(4) = 5$.

 YOUR TURN

Applications

Function notation is often used in formulas. For example, to emphasize that the area A of a circle is a function of its radius r, instead of

$$A = \pi r^2,$$

we can write

$$A(r) = \pi r^2.$$

EXAMPLE 7   A typical adult dosage of an antihistamine is 24 mg. Young's rule for determining the dosage size $c(a)$ for a typical child of age a is

$$c(a) = \frac{24a}{a + 12}.$$

What should the dosage be for a typical 8-year-old child?

Source: Olsen, June Looby, Leon J. Ablon, and Anthony Patrick Giangrasso, *Medical Dosage Calculations*, 6th ed.

SOLUTION   We find $c(8)$:

7. The area of a square with side s is given by $A(s) = s^2$. What is the area of a square with sides of 9 m?

$$c(8) = \frac{24(8)}{8 + 12} = \frac{192}{20} = 9.6.$$

The dosage for a typical 8-year-old child is 9.6 mg.

YOUR TURN

7.1 EXERCISE SET

FOR EXTRA HELP

 MyMathLab® Math XL

PRACTICE   WATCH   READ   REVIEW

Vocabulary and Reading Check

Choose the word from the following list that best completes each statement. Words may be used more than once.

correspondence   horizontal
domain   range
exactly   vertical
"f of 3"

1. A function is a special kind of _____ between two sets.

2. In any function, each member of the domain is paired with _____ one member of the range.

3. For any function, the set of all inputs, or first values, is called the _____.

4. For any function, the set of all outputs, or second values, is called the _____.

5. When a function is graphed, members of the domain are located on the _____ axis.

6. When a function is graphed, members of the range are located on the _____ axis.

7. The notation $f(3)$ can be read _____.

8. The _____-line test is used to determine whether or not a graph represents a function.

Domain and Range

Determine whether each correspondence is a function.

9. a → 2
        b
        d → 3
        g → 4
        h

10. 2 → a
         → b
         3 → d
         4 → g
         → h

11.

Girl's age (in months)		Average daily weight gain (in grams)
2	——→	21.8
9	——→	11.7
16	——→	8.5
23	——→	7.0

Source: *American Family Physician*, December 1993, p. 1435

12.

Boy's age (in months)		Average daily weight gain (in grams)
2	——→	24.3
9	——→	11.7
16	——→	8.2
23	——→	7.0

Source: *American Family Physician*, December 1993, p. 1435

13. *Actor/Actress*   *Movie*

Julia Roberts → My Best Friend's Wedding
             → Notting Hill
             → Pretty Woman

Denzel Washington → American Gangster
                  → Safe House
                  → Training Day

14. *Celebrity*   *Birthday*

Tom Brady →
P. D. James → August 3
Martha Stewart →
Kim Basinger →
Sinéad O'Connor → December 8
James Galway →

Source: www.leannesbirthdays.com

15. *Predator*   *Prey*

cat → dog
fish → worm
dog → cat
tiger → fish
bat ——→ mosquito

16. *State*   *Neighboring state*

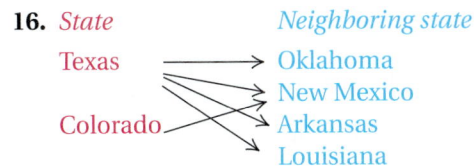

Texas → Oklahoma
New Mexico
Colorado → Arkansas
Louisiana

Determine whether each of the following is a function.

17. The correspondence that assigns to a USB flash drive its storage capacity

18. The correspondence that assigns to a member of a rock band the instrument the person can play

19. The correspondence that assigns to a player on a team that player's uniform number

20. The correspondence that assigns to a triangle its area

For each correspondence, (a) write the domain, (b) write the range, and (c) determine whether the correspondence is a function.

21. $\{(-3, 3), (-2, 5), (0, 9), (4, -10)\}$

22. $\{(0, -1), (1, 3), (2, -1), (5, 3)\}$

23. $\{(1, 1), (2, 1), (3, 1), (4, 1), (5, 1)\}$

24. $\{(1, 1), (1, 2), (1, 3), (1, 4), (1, 5)\}$

25. $\{(4, -2), (-2, 4), (3, -8), (4, 5)\}$

26. $\{(0, 7), (4, 8), (7, 0), (8, 4)\}$

Functions and Graphs

For each graph of a function, determine (a) $f(1)$ and (b) any x-values for which $f(x) = 2$.

27.
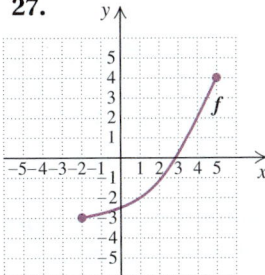

28.
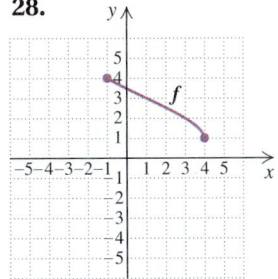

29.
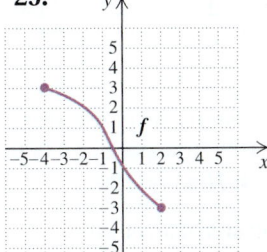

30.
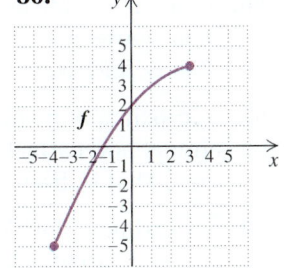

31.
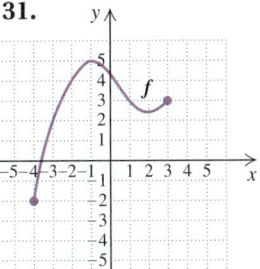

32.
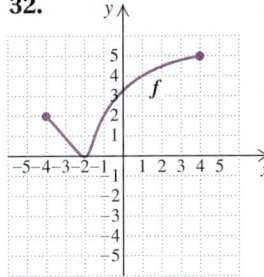

33.
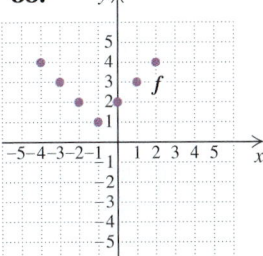

34.
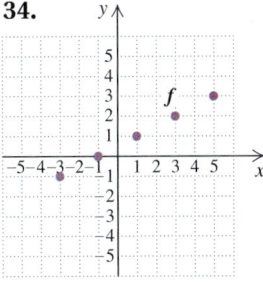

35.
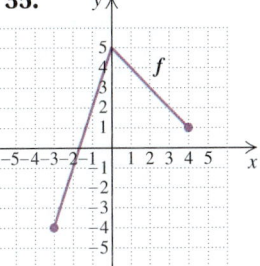

36.
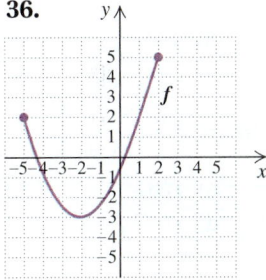

Determine whether each of the following is the graph of a function.

37.
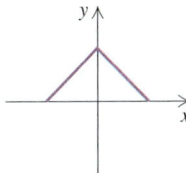

38.
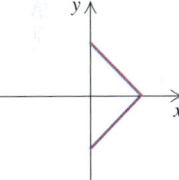

39.
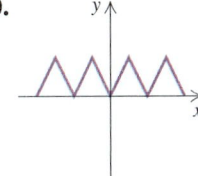

40.
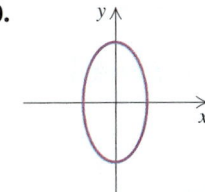

41.
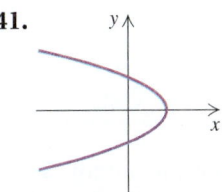

42.
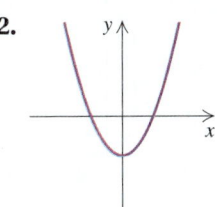

Function Notation and Equations

Find the function values.

43. $g(x) = 2x + 5$

    a) $g(0)$ **b)** $g(-4)$ **c)** $g(-7)$

    d) $g(8)$ **e)** $g(a + 2)$ **f)** $g(a) + 2$

44. $h(x) = 5x - 1$

    a) $h(4)$ **b)** $h(8)$ **c)** $h(-3)$

    d) $h(-4)$ **e)** $h(a - 1)$ **f)** $h(a) + 3$

45. $f(n) = 5n^2 + 4n$

    a) $f(0)$ **b)** $f(-1)$ **c)** $f(3)$

    d) $f(t)$ **e)** $f(2a)$ **f)** $f(3) - 9$

46. $g(n) = 3n^2 - 2n$

    a) $g(0)$ **b)** $g(-1)$ **c)** $g(3)$

    d) $g(t)$ **e)** $g(2a)$ **f)** $g(3) - 4$

47. $f(x) = \dfrac{x - 3}{2x - 5}$

    a) $f(0)$ **b)** $f(4)$ **c)** $f(-1)$

    d) $f(3)$ **e)** $f(x + 2)$ **f)** $f(a + h)$

48. $r(x) = \dfrac{3x - 4}{2x + 5}$

    a) $r(0)$ **b)** $r(2)$ **c)** $r\left(\frac{4}{3}\right)$

    d) $r(-1)$ **e)** $r(x + 3)$ **f)** $r(a + h)$

Fill in the missing values in each table.

$f(x) = 2x - 5$	
x	$f(x)$
49. 8	
50.	13
51.	-5
52. -4	

$f(x) = \frac{1}{3}x + 4$	
x	$f(x)$
53.	$\frac{1}{2}$
54.	$-\frac{1}{3}$
55. $\frac{1}{2}$	
56. $-\frac{1}{3}$	

57. If $f(x) = 4 - x$, for what input is the output 7?

58. If $f(x) = 5x + 1$, for what input is the output $\frac{1}{2}$?

59. If $f(x) = 0.1x - 0.5$, for what input is the output -3?

60. If $f(x) = 2.3 - 1.5x$, for what input is the output 10?

Applications

The function A described by $A(s) = \dfrac{\sqrt{3}}{4}s^2$ gives the area of an equilateral triangle with side s.

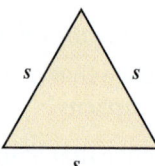

61. Find the area when a side measures 4 cm.

62. Find the area when a side measures 6 in.

The function V described by $V(r) = 4\pi r^2$ gives the surface area of a sphere with radius r.

63. Find the surface area when the radius is 3 in.

64. Find the surface area when the radius is 5 cm.

Heart Attacks and Cholesterol. For Exercises 65–68, use the following graph, which shows the annual heart attack rate per 10,000 men as a function of blood cholesterol level.*

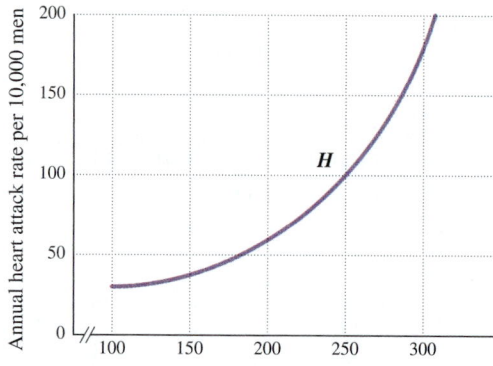

65. Approximate the annual heart attack rate for those men whose blood cholesterol level is 225 mg/dl. That is, find $H(225)$.

66. Approximate the annual heart attack rate for those men whose blood cholesterol level is 275 mg/dl. That is, find $H(275)$.

67. Approximate the blood cholesterol level for an annual heart attack rate of 100 attacks per 10,000 men. That is, find x for which $H(x) = 100$.

68. Approximate the blood cholesterol level for an annual heart attack rate of 50 attacks per 10,000 men. That is, find x for which $H(x) = 50$.

---

*Copyright 1989, CSPI. Adapted from Nutrition Action Health-letter (1875 Connecticut Avenue, N.W., Suite 300, Washington, DC 20009-5728. $20 for 10 issues).

Researchers at Yale University have suggested that the following graphs* may represent three different aspects of love as a function of time.

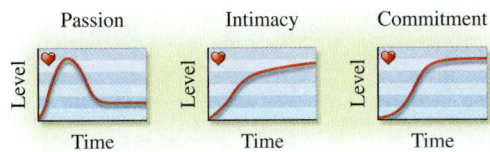

| Passion | Intimacy | Commitment |

69. In what unit would you measure time if the horizontal length of each graph were ten units? Why?

70. Do you agree with the researchers that these graphs should be shaped as they are? Why or why not?

Skill Review

Solve.

71. $2(x - 5) - 3 = 4 - (x - 1)$ [2.2]

72. $(x - 2)^2 = 36$ [5.7]

73. $x^2 = 36$ [5.7]   **74.** $\dfrac{1}{x} = -2$ [6.6]

75. $\dfrac{1}{x} = x$ [6.6]

76. $(x - 2)(x + 3) = 6$ [5.7]

77. $\dfrac{1}{3}x + 2 = \dfrac{5}{4} + 3x$ [2.2]   **78.** $\dfrac{x + 1}{x} = 8$ [6.6]

Synthesis

79. Jaylan is asked to write a function relating the number of fish in an aquarium to the amount of food needed for the fish. Which quantity should he choose as the independent variable? Why?

80. For the function given by $n(z) = ab + wz$, what is the independent variable? How can you tell?

For Exercises 81 and 82, let $f(x) = 3x^2 - 1$ and $g(x) = 2x + 5$.

81. Find $f(g(-4))$ and $g(f(-4))$.

82. Find $f(g(-1))$ and $g(f(-1))$.

83. If f represents the function in Exercise 15, find $f(f(f(f(\text{tiger}))))$.

*From "A Triangular Theory of Love," by R. J. Sternberg, 1986, *Psychological Review*, **93**(2), 119–135. Copyright 1986 by the American Psychological Association, Inc. Reprinted by permission.

84. Suppose that a function g is such that $g(-1) = -7$ and $g(3) = 8$. Find a formula for g if $g(x)$ is of the form $g(x) = mx + b$, where m and b are constants.

For each graph of a function, determine (a) $f(1)$; (b) $f(2)$; and (c) any x-values for which $f(x) = 2$.

85.
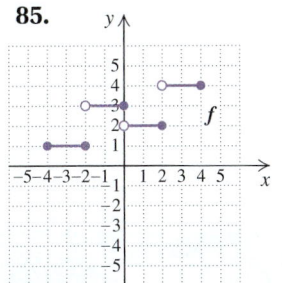

86.
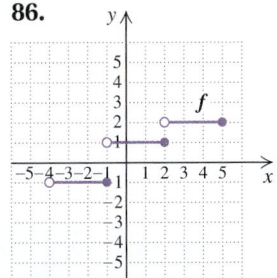

Pregnancy. *For Exercises 87–90, use the following graph of a woman's "stress test." This graph shows the size of a pregnant woman's contractions as a function of time.*

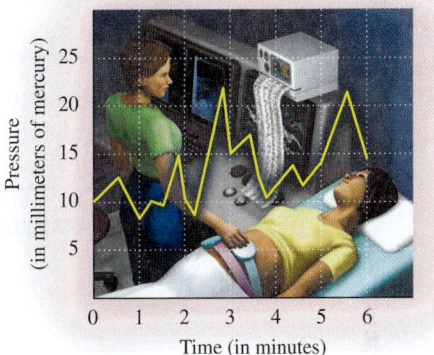

87. How large is the largest contraction that occurred during the test?

88. At what time during the test did the largest contraction occur?

89. What is the frequency of the largest contraction?

90. On the basis of the information provided, how large a contraction would you expect 60 seconds after the end of the test? Why?

91. The *greatest integer function* $f(x) = [x]$ is defined as follows: $[x]$ is the greatest integer that is less than or equal to x. For example, if $x = 3.74$, then $[x] = 3$; and if $x = -0.98$, then $[x] = -1$. Graph the greatest integer function for $-5 \le x \le 5$. (The notation $f(x) = \text{INT}(x)$ is used by many graphing calculators and computers.)

↱ **YOUR TURN ANSWERS: SECTION 7.1**

1. No   **2.** Yes   **3.** (a) $\{0, 4, 5\}$; (b) $\{-3, 7\}$
4. (a) 1; (b) -5 **5.** -2 **6.** (a) 5; (b) 20   **7.** 81 m^2

PREPARE TO MOVE ON

List all numbers for which each rational expression is undefined.   [6.1]

1. $\dfrac{3}{5x}$

2. $\dfrac{n}{3n-1}$

3. $\dfrac{x^2 - 1}{x^2 - x - 12}$

4. $\dfrac{t^2 + 5t + 4}{t^2 - 36}$

7.2   Domain and Range

Determining the Domain and the Range   ▪   Restrictions on Domain   ▪   Piecewise-Defined Functions

A function is a correspondence from a set called the *domain* to a set called the *range*. In this section, we look more closely at the concepts of domain and range.

Determining the Domain and the Range

When a function is given as a set of ordered pairs, the domain is the set of all first coordinates and the range is the set of all second coordinates. We can also determine the domain and the range of a function from its graph.

EXAMPLE 1   Find the domain and the range of the function *f* below.

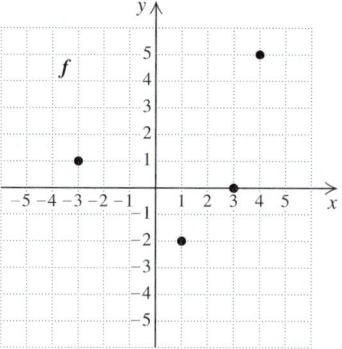

1. Find the domain and the range of the function *g* below.

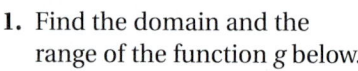
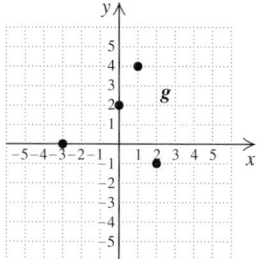

SOLUTION   Here *f* can be written $\{(-3, 1), (1, -2), (3, 0), (4, 5)\}$. The domain is the set of all first coordinates, $\{-3, 1, 3, 4\}$, and the range is the set of all second coordinates, $\{1, -2, 0, 5\}$.

↱ YOUR TURN

In Example 1, we could also have found the domain and the range directly, without first writing *f*, by observing the *x*- and *y*-values used in the graph.

EXAMPLE 2 Find the domain and the range of the function *f* shown here.

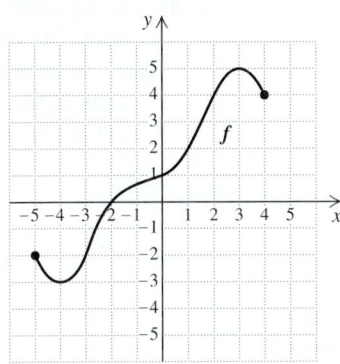

SOLUTION   The domain of the function is the set of all *x*-values that are used in the points on the curve. Because there are no breaks in the graph of *f*, these extend continuously from −5 to 4 and can be viewed as the curve's shadow, or *projection*, on the *x*-axis. Thus the domain is $\{x \mid -5 \le x \le 4\}$, or $[-5, 4]$, shown on the left below.

The range of the function is the set of all *y*-values that are used in the points on the curve. These extend continuously from −3 to 5, and can be viewed as the curve's projection on the *y*-axis. Thus the range is $\{y \mid -3 \le y \le 5\}$, or $[-3, 5]$, shown on the right below.

2. Find the domain and the range of the function *g* shown here.

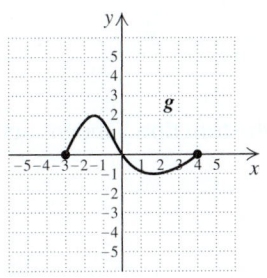

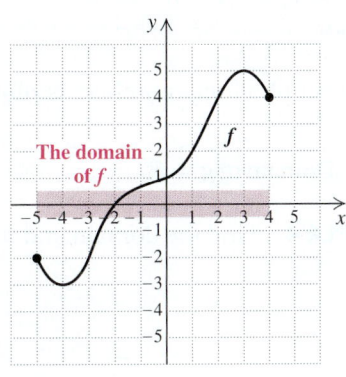

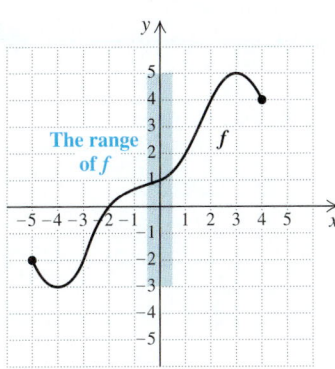

YOUR TURN

In Example 2, the *endpoints* $(-5, -2)$ and $(4, 4)$ emphasize that the function is not defined for values of *x* less than −5 or greater than 4.

The graphs of some functions have no endpoints. Thus a function may have a domain and/or a range that extends without bound toward positive infinity or negative infinity.

EXAMPLE 3   For the function *g* represented below, determine **(a)** the domain of *g* and **(b)** the range of *g*.

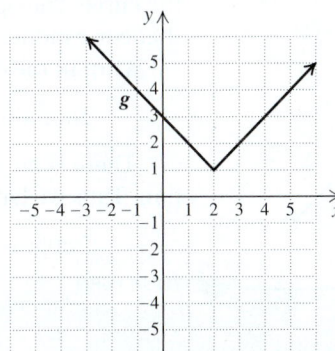

SOLUTION

a) The domain of *g* is the set of all *x*-values that are used in the points on the curve. The arrows on the ends of the graph indicate that it extends both left and right without end. Thus the shadow, or projection, of the graph on the *x*-axis is the entire *x*-axis. (See the graph on the left below.) The domain is $\{x \mid x \text{ is a real number}\}$, or $(-\infty, \infty)$.

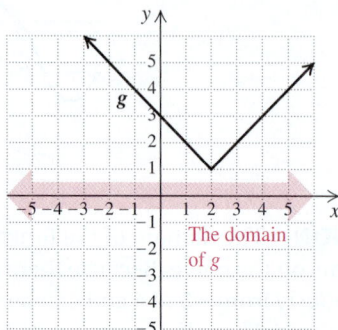 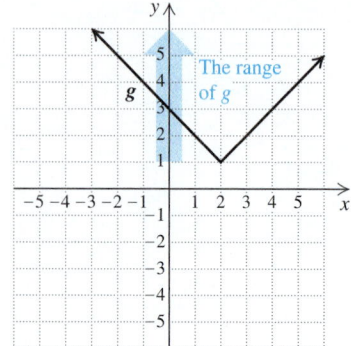

3. For the function *f* represented below, determine **(a)** the domain of *f* and **(b)** the range of *f*.

b) The range of *g* is the set of all *y*-values that are used in the points on the curve. The arrows on the ends of the graph indicate that it extends up without end. Thus the projection of the graph on the *y*-axis is the portion of the *y*-axis greater than or equal to 1. (See the graph on the right above.) The range is $\{y \mid y \geq 1\}$, or $[1, \infty)$.

YOUR TURN

The set of all real numbers is often abbreviated $\mathbb{R}$. Thus, in Example 3, we could write

$$\text{Domain of } g = \mathbb{R}.$$

EXAMPLE 4   Find the domain and the range of the function *f* shown here.

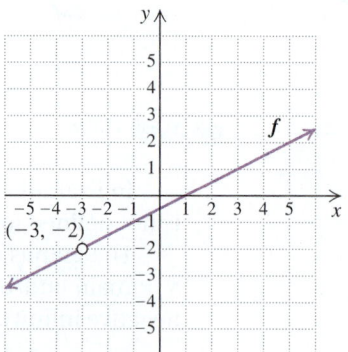

4. Find the domain and the range of the function *g* shown here.

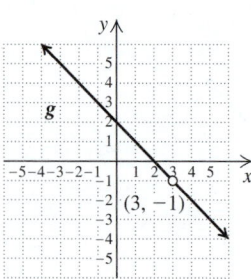

SOLUTION   The domain of *f* is the set of all *x*-values that are used in points on the curve. The open dot in the graph at $(-3, -2)$ indicates that there is no *y*-value that corresponds to $x = -3$; that is, the function is not defined for $x = -3$. Thus, -3 is not in the domain of the function, and

$$\text{Domain of } f = \{x \mid x \text{ is a real number } and \; x \neq -3\}.$$

There is no function value at $(-3, -2)$, so -2 is not in the range of the function. Thus we have

$$\text{Range of } f = \{y \mid y \text{ is a real number } and \; y \neq -2\}.$$

YOUR TURN

When a function is described by an equation, we assume that the domain is the set of all real numbers for which function values can be calculated.

EXAMPLE 5 For each equation, determine the domain of f.

a) $f(x) = |x|$ **b)** $f(x) = \dfrac{x}{2x - 6}$ **c)** $f(t) = \dfrac{t + 1}{t^2 - 4}$

SOLUTION

a) We ask ourselves, "Is there any number x for which we cannot compute $|x|$?" Since we can find the absolute value of *any* number, the answer is no. Thus the domain of f is $\mathbb{R}$, the set of all real numbers.

b) Is there any number x for which $\dfrac{x}{2x - 6}$ cannot be computed? Since $\dfrac{x}{2x - 6}$ cannot be computed when $2x - 6$ is 0, the answer is yes. To determine what x-value causes the denominator to be 0, we solve an equation:

$$2x - 6 = 0 \qquad \text{Setting the denominator equal to 0}$$
$$2x = 6 \qquad \text{Adding 6 to both sides}$$
$$x = 3. \qquad \text{Dividing both sides by 2}$$

Thus, 3 is *not* in the domain of f, whereas all other real numbers are. The domain of f is $\{x \mid x \text{ is a real number } and \ x \neq 3\}$.

c) The expression $\dfrac{t + 1}{t^2 - 4}$ is undefined when $t^2 - 4 = 0$:

$$t^2 - 4 = 0 \qquad \text{Setting the denominator equal to 0}$$
$$(t + 2)(t - 2) = 0 \qquad \text{Factoring}$$
$$t + 2 = 0 \quad or \quad t - 2 = 0 \qquad \text{Using the principle of zero products}$$
$$t = -2 \quad or \qquad t = 2. \qquad \text{Solving; these are the values for which } (t + 1)/(t^2 - 4) \text{ is undefined.}$$

Thus we have

Domain of $f = \{t \mid t \text{ is a real number } and \ t \neq -2 \ and \ t \neq 2\}$.

Note that when the numerator, $t + 1$, is zero, the function value is 0 and *is* defined.

YOUR TURN

Restrictions on Domain

If a function is used as a model for an application, the problem situation may require restrictions on the domain; for example, length and time are generally nonnegative, and a person's age does not increase indefinitely.

EXAMPLE 6 *Prize Tee Shirts.* During intermission at sporting events, it has become common for team mascots to use a powerful slingshot to launch tightly rolled tee shirts into the stands. The height $h(t)$, in feet, of an airborne tee shirt t seconds after being launched can be approximated by

$$h(t) = -15t^2 + 70t + 25.$$

What is the domain of the function?

Technology Connection

To visualize Examples 5(a) and 5(b), note that the graph of $y_1 = |x|$ appears without interruption for any piece of the x-axis that we examine.

In contrast, the graph of $y_2 = \dfrac{x}{2x - 6}$ has a break at $x = 3$.

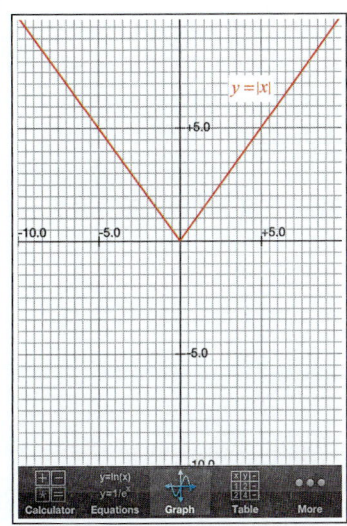

$y_2 = \dfrac{x}{2x - 6}$

5. Find the domain of f, if

$$f(x) = \dfrac{3x}{2x - 5}.$$

SOLUTION   The expression $-15t^2 + 70t + 25$ can be evaluated for any number t, so any restrictions on the domain will come from the problem situation.

First, we note that t cannot be negative, since it represents time from launch, so we have $t \geq 0$. If we make a drawing, we also note that the function will not be defined for values of t that make the height negative.

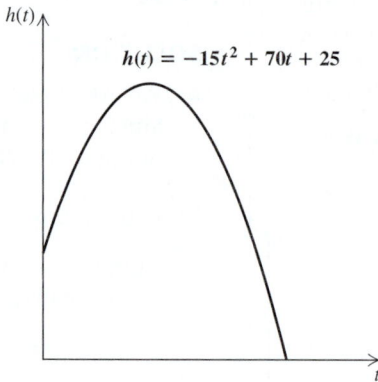

Thus an upper limit for t will be the positive value of t for which $h(t) = 0$. Solving, we obtain

$$h(t) = 0$$
$$-15t^2 + 70t + 25 = 0 \qquad \text{Substituting}$$
$$\left.\begin{array}{l} -5(3t^2 - 14t - 5) = 0 \\ -5(3t + 1)(t - 5) = 0 \end{array}\right\} \qquad \text{Factoring}$$
$$3t + 1 = 0 \quad or \quad t - 5 = 0 \qquad \text{Using the principle of zero products}$$
$$\left.\begin{array}{rcl} 3t = -1 & or & t = 5 \\ t = -\tfrac{1}{3} & or & t = 5. \end{array}\right\} \qquad \text{Solving for } t$$

We already know that $-\tfrac{1}{3}$ is not in the domain of the function because of the restriction $t \geq 0$ above.

The tee shirt will hit the ground after 5 sec, so we have $t \leq 5$. Putting the two restrictions together, we have $t \geq 0$ *and* $t \leq 5$, so the

$$\text{Domain of } h = \{t \,|\, t \text{ is a real number } and\ 0 \leq t \leq 5\}, \text{ or } [0, 5].$$

6. The record R for the 1500-m run t years after 1930 is given by

$$R(t) = 3.85 - 0.0075t.$$

What is the domain of the function?

 YOUR TURN

If the domain of a function is not specifically listed, it can be determined from a table, a graph, an equation, or an application.

DOMAIN OF A FUNCTION

The domain of a function f is the set of all inputs.

- If the correspondence is listed in a table or as a set of ordered pairs, the domain is the set of all first coordinates.

- If the function is described by a graph, the domain is the set of all first coordinates of the points on the graph.

- If the function is described by an equation, the domain is the set of all numbers for which the value of the function can be calculated.

- If the function is used in an application, the domain is the set of all numbers that make sense as inputs in the problem.

Piecewise-Defined Functions

Some functions are defined by different equations for various parts of their domains. Such functions are said to be **piecewise**-defined. For example, the function given by $f(x) = |x|$ can be described by

$$f(x) = \begin{cases} x, & \text{if } x \geq 0, \\ -x, & \text{if } x < 0. \end{cases}$$

To evaluate a piecewise-defined function for an input a, we determine what part of the domain a belongs to and use the appropriate formula for that part of the domain.

EXAMPLE 7 Find each function value for the function f given by

$$f(x) = \begin{cases} 2x, & \text{if } x < 3, \\ x + 1, & \text{if } x \geq 3. \end{cases}$$

a) $f(4)$ **b)** $f(-10)$

SOLUTION

a) The function f is defined using two different equations. To find $f(4)$, we must first determine whether to use the equation $f(x) = 2x$ or the equation $f(x) = x + 1$. To do this, we focus first on the two parts of the domain. It may help to visualize the domain on the number line, as shown below.

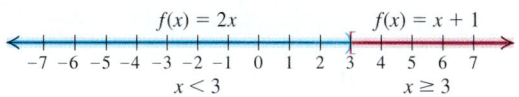

$$f(x) = \begin{cases} 2x, & \text{if } x < 3, \\ x + 1, & \text{if } x \geq 3 \end{cases} \quad \text{4 is in the second part of the domain.}$$

Since $4 \geq 3$, we use $f(x) = x + 1$. Thus, $f(4) = 4 + 1 = 5$.

b) To find $f(-10)$, we first note that $-10 < 3$, so we must use $f(x) = 2x$. Thus, $f(-10) = 2(-10) = -20$.

7. In Example 7, find $f(3)$.

 YOUR TURN

7.2   EXERCISE SET

FOR EXTRA HELP

MyMathLab® Math XL
PRACTICE   WATCH   READ   REVIEW

Vocabulary and Reading Check

Indicate whether each of the following refers to the domain or the range of a function.

1. The set of all first coordinates of ordered pairs giving the function

2. The set of all second coordinates of ordered pairs giving the function

3. The set of all x-values used in points on the graph of the function

4. The set of all y-values used in points on the graph of the function

5. Restricted when a denominator is zero

6. Divided into intervals for piecewise-defined functions

Concept Reinforcement

For Exercises 7–12, use the function f given by

$$f(x) = \begin{cases} x - 5, & \text{if } x < -6, \\ 2x^2, & \text{if } -6 \leq x < -1, \\ |x|, & \text{if } -1 \leq x < 10, \\ 3x + 1, & \text{if } x \geq 10. \end{cases}$$

Write the letter of the equation that should be used to find each function value. Letters may be used more than once or not at all.

7. ___ $f(0)$

8. ___ $f(15)$

9. ___ $f(10)$

10. ___ $f(-6)$

11. ___ $f(-1)$

12. ___ $f(-3)$

a) $f(x) = x - 5$

b) $f(x) = 2x^2$

c) $f(x) = |x|$

d) $f(x) = 3x + 1$

Determining the Domain and the Range

For each graph of a function f, determine the domain and the range of f.

13.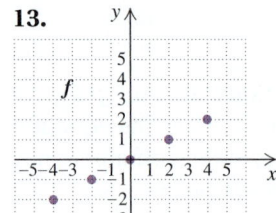

14.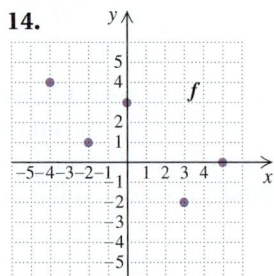

15.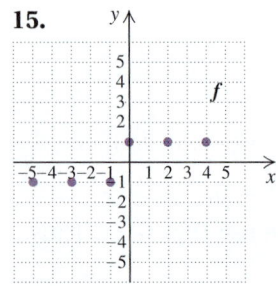

16.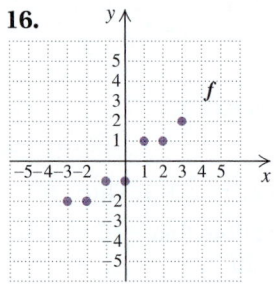

17.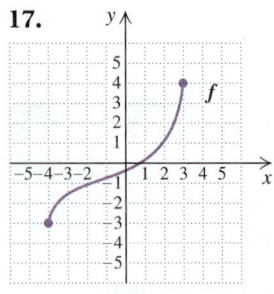

18.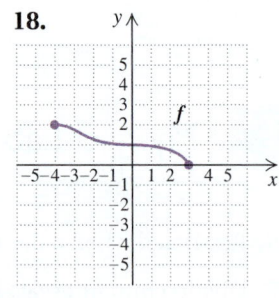

19.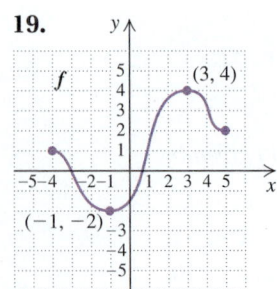

20.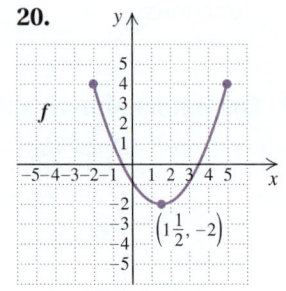

21.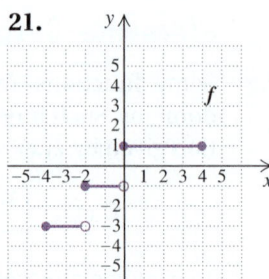

22.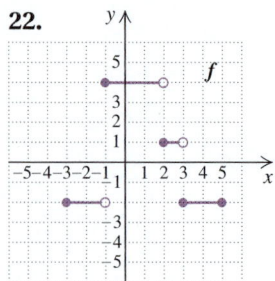

23.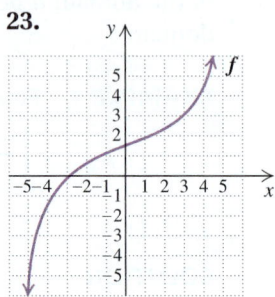

24.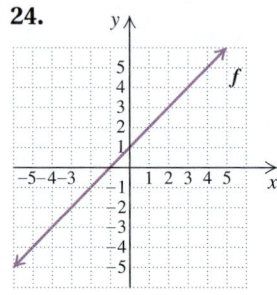

25.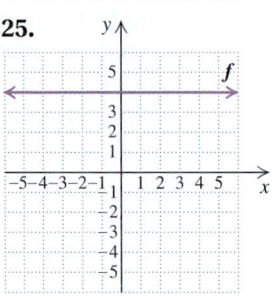

26.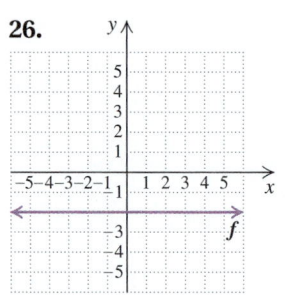

27.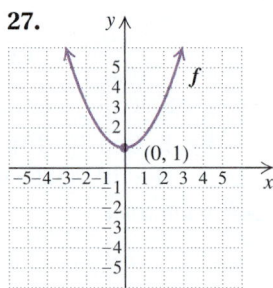

28.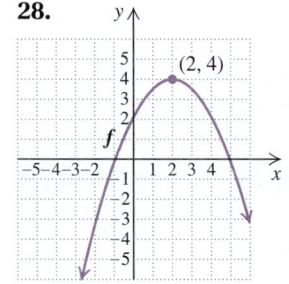

29.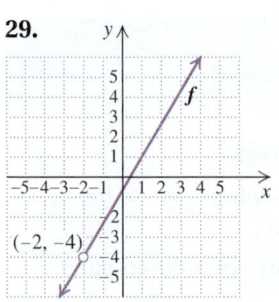

30.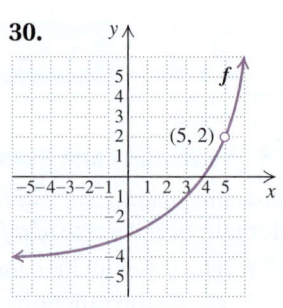

31.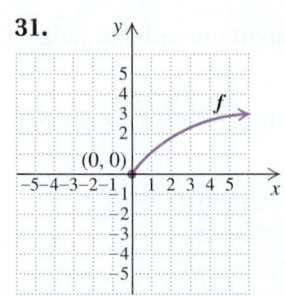

32.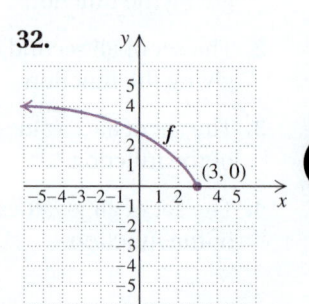

Find the domain of f.

33. $f(x) = \dfrac{5}{x-3}$

34. $f(x) = \dfrac{7}{6-x}$

35. $f(x) = \dfrac{x}{2x-1}$

36. $f(x) = \dfrac{2x}{4x+3}$

37. $f(x) = 2x + 1$

38. $f(x) = x^2 + 3$

39. $f(x) = |5 - x|$

40. $f(x) = |3x - 4|$

41. $f(x) = \dfrac{5}{x^2 - 9}$

42. $f(x) = \dfrac{x}{x^2 - 2x + 1}$

43. $f(x) = x^2 - 9$

44. $f(x) = x^2 - 2x + 1$

45. $f(x) = \dfrac{2x - 7}{x^2 + 8x + 7}$

46. $f(x) = \dfrac{x + 5}{2x^2 - x - 3}$

Restrictions on Domain

47. *Depreciation.* The value $V(t)$ of Anika's phone t years after purchase is given by
$$V(t) = 250 - 50t.$$
What is the domain of the function?

48. *Records in the 400-m Run.* The record R for the 400-m run t years after 1930 is given by
$$R(t) = 46.8 - 0.075t.$$
What is the domain of the function?

49. *Consumer Demand.* The amount A of coffee that consumers are willing to buy at price p is given by
$$A(p) = -2.5p + 26.5.$$
What is the domain of the function?

50. *Seller's Supply.* The amount A of coffee that suppliers are willing to supply at price p is given by
$$A(p) = 2p - 11.$$
What is the domain of the function?

51. *Pressure at Sea Depth.* The pressure P, in atmospheres, at a depth d feet beneath the surface of the ocean is given by
$$P(d) = 0.03d + 1.$$
What is the domain of the function?

52. *Perimeter.* The perimeter P of an equilateral triangle with sides of length s is given by
$$P(s) = 3s.$$
What is the domain of the function?

53. *Fireworks Displays.* The height h, in feet, of a "weeping willow" fireworks display, t seconds after having been launched from an 80-ft high rooftop, is given by
$$h(t) = -16t^2 + 64t + 80.$$
What is the domain of the function?

54. *Safety Flares.* The height h, in feet, of a safety flare, t seconds after having been launched from a height of 224 ft, is given by
$$h(t) = -16t^2 + 80t + 224.$$
What is the domain of the function?

Piecewise-Defined Functions

Find the indicated function values.

55. $f(x) = \begin{cases} x, & \text{if } x < 0, \\ 2x + 1, & \text{if } x \geq 0 \end{cases}$

    a) $f(-5)$ **b)** $f(0)$ **c)** $f(10)$

56. $g(x) = \begin{cases} x - 5, & \text{if } x \leq 5, \\ 3x, & \text{if } x > 5 \end{cases}$

    a) $g(0)$ **b)** $g(5)$ **c)** $g(6)$

57. $G(x) = \begin{cases} x - 5, & \text{if } x \leq -1, \\ x, & \text{if } x > -1 \end{cases}$

    a) $G(-10)$ **b)** $G(0)$ **c)** $G(-1)$

58. $F(x) = \begin{cases} 2x, & \text{if } x < 3, \\ -5x, & \text{if } x \geq 3 \end{cases}$

    a) $F(-1)$ **b)** $F(3)$ **c)** $F(10)$

59. $f(x) = \begin{cases} x^2 - 10, & \text{if } x < -10, \\ x^2, & \text{if } -10 \leq x \leq 10, \\ x^2 + 10, & \text{if } x > 10 \end{cases}$

    a) $f(-10)$ **b)** $f(10)$ **c)** $f(11)$

60. $f(x) = \begin{cases} 2x^2 - 3, & \text{if } x \leq 2, \\ x^2, & \text{if } 2 < x < 4, \\ 5x - 7, & \text{if } x \geq 4 \end{cases}$

    a) $f(0)$ **b)** $f(3)$ **c)** $f(6)$

61. Explain why the domain of the function given by $f(x) = \dfrac{x + 3}{2}$ is $\mathbb{R}$, but the domain of the function given by $g(x) = \dfrac{2}{x + 3}$ is not $\mathbb{R}$.

62. Chloe asserts that for a function described by a set of ordered pairs, the range of the function will always have the same number of elements as there are ordered pairs. Is she correct? Why or why not?

Skill Review

Simplify. Do not use negative exponents in the answer.

63. $3 - 2(1 - 4)^2 \div 6 \cdot 2$ [1.8]

64. $3(x - 7) - 4(2 - 3x)$ [1.8]

65. $(2x^6 y)^2$ [4.1]

66. $\dfrac{24a^{-1}b^{10}}{-14a^{11}b^{-16}}$ [4.2]

Synthesis

67. Ramiro states that $f(x) = \dfrac{x^2}{x}$ and $g(x) = x$ represent the same function. Is he correct? Why or why not?

68. Explain why the domain of a function can be viewed as the projection of its graph on the x-axis.

Sketch the graph of a function for which the domain and the range are as given. Graphs may vary.

69. Domain: $\mathbb{R}$; range: $\mathbb{R}$

70. Domain: $\{3, 1, 4\}$; range: $\{0, 5\}$

71. Domain: $\{x \mid 1 \le x \le 5\}$; range: $\{y \mid 0 \le y \le 2\}$

72. Domain: $\{x \mid x \text{ is a real number } and \ x \ne 1\}$; range: $\{y \mid y \text{ is a real number } and \ y \ne -2\}$

For each graph of a function f, determine the domain and the range of f.

73.

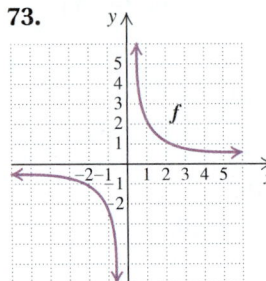

74.

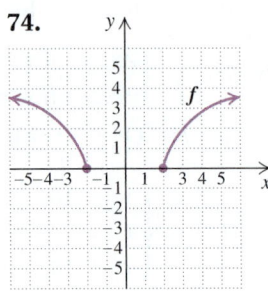

75.

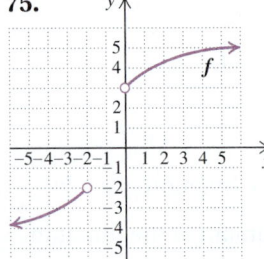

76.

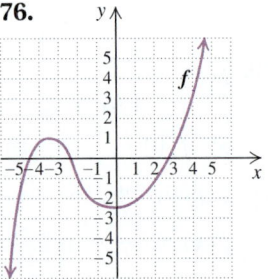

Graph each function on a graphing calculator and estimate its domain and range from the graph.

77. $f(x) = |x - 3|$

78. $f(x) = |x| - 3$

79. $f(x) = \dfrac{3}{x - 2}$

80. $f(x) = \dfrac{-1}{x + 3}$

81. Use a graphing calculator to estimate the range of the function in Exercise 53.

82. Use a graphing calculator to estimate the range of the function in Exercise 54.

83.–86. *For Exercises 83–86, graph the functions given in each of Exercises 55–58, respectively.*

87. A graphing calculator will interpret an expression like $x \ge 1$ as true or false, depending on the value of x. If the expression is true, the graphing calculator assigns a value of 1 to the expression. If the expression is false, the graphing calculator assigns a value of 0. To graph a piecewise-defined function using a graphing calculator, multiply each part of the definition by its domain, using the TEST menu to enter the inequality symbol. Thus the function in Exercise 55 is entered as $y_1 = x(x < 0) + (2x + 1)(x \ge 0)$. Use a graphing calculator in DOT mode to check your answers to Exercises 83–86.

YOUR TURN ANSWERS: SECTION 7.2

1. Domain: $\{-3, 0, 1, 2\}$; range: $\{-1, 0, 2, 4\}$
2. Domain: $\{x \mid -3 \le x \le 4\}$, or $[-3, 4]$; range: $\{y \mid -1 \le y \le 2\}$ or $[-1, 2]$
3. Domain: $\{x \mid x \text{ is a real number}\}$, or $(-\infty, \infty)$; range: $\{y \mid y \ge -2\}$, or $[-2, \infty)$
4. Domain: $\{x \mid x \text{ is a real number } and \ x \ne 3\}$; range: $\{y \mid y \text{ is a real number } and \ y \ne -1\}$
5. $\left\{x \mid x \text{ is a real number } and \ x \ne \frac{5}{2}\right\}$
6. $\left\{t \mid 0 \le t \le 513\frac{1}{3}\right\}$, or $\left[0, 513\frac{1}{3}\right]$ **7.** 4

QUICK QUIZ: SECTIONS 7.1–7.2

1. Find $h(20)$ when $h(x) = \frac{1}{2}x - 10$.   [7.1]

2. If $f(x) = \frac{1}{2}x$, for what input is the output 100?   [7.1]

3. Find the domain of the function given by $f(x) = \dfrac{7}{x}$.

    [7.2]

4. Find the domain of the function given by $g(x) = 3x^2 - 6x + 2$.   [7.2]

5. Find $F(2)$, if $F(x) = \begin{cases} 2x, & \text{if } x \le -1, \\ x - 5, & \text{if } x > -1. \end{cases}$ [7.2]

PREPARE TO MOVE ON

Graph.   [3.6]

1. $y = 2x - 3$ **2.** $y = x + 5$

Find the slope and the y-intercept of each line.   [3.6]

3. $y = \frac{2}{3}x - 4$ **4.** $y = -\frac{1}{4}x + 6$

5. $y = \frac{4}{3}x$ **6.** $y = -5x$

CONNECTING 🔗 THE CONCEPTS

A function is a correspondence. This correspondence can be listed as a set of ordered pairs or described by an equation. The correspondence is between two sets, the domain and the range, and each member of the domain corresponds to exactly one member of the range.

For the function $f: \{(2, 3), (0, -4), (-8, 7)\}$:

    The domain is $\{-8, 0, 2\}$.

    The range is $\{-4, 3, 7\}$.

    The input 2 corresponds to the output 3.

    $f(2) = 3$

For the function given by $g(x) = x^2$:

    The domain is $\mathbb{R}$.

    The range is $[0, \infty)$.

    The input -3 corresponds to the output 9.

    $g(-3) = 9$

    The independent variable is x.

EXERCISES

For Exercises 1–4, let

$$f = \{(3, 5), (-2, 5), (1, -6)\}.$$

1. Find the domain of f. [7.1]

2. Find the range of f. [7.1]

3. Find $f(1)$. [7.1]

4. Find any inputs that correspond to an output of -6. [7.1]

For Exercises 5–8, let

$$g(x) = \frac{x - 1}{2x}.$$

5. Find the domain of g. [7.2]

6. Find $g(1)$. [7.1]

7. Find the output that corresponds to an input of 4. [7.1]

8. Find any inputs that correspond to an output of $\frac{1}{4}$. [7.1]

7.3   Graphs of Functions

Linear Functions   ▪   Nonlinear Functions

Study Skills

Don't Reach for the Sky

When setting personal study goals, make them reasonable. It is better to set several small, reachable goals than one large one that seems unattainable. For example, you might set a goal of memorizing one formula during each study time, instead of a list of formulas at once. Reaching a goal will encourage you to work toward the next one.

A function can be classified both by the type of equation that is used and by the type of graph it represents.

Linear Functions

Any linear equation can be written in *standard form* $Ax + By = C$. If $B \neq 0$, the equation can also be written in *slope–intercept form*. The *point–slope form* is often used to write equations.

EQUATIONS OF LINES

    Standard form: $Ax + By = C$

    Slope–intercept form: $y = mx + b$

    Point–slope form: $y - y_1 = m(x - x_1)$

Two points determine a line. If we know that an equation is linear, we can graph the equation by plotting two points that are on the line and drawing the line that goes through those points.

When an equation is written in slope–intercept form $y = mx + b$, the slope of the line is m and the y-intercept is $(0, b)$. Knowing the y-intercept gives us one point on the line, and we can use the slope to determine another point.

EXAMPLE 1 Graph: $4y = -3x + 8$.

SOLUTION To graph $4y = -3x + 8$, we first rewrite it in slope–intercept form:

$$4y = -3x + 8$$
$$y = \tfrac{1}{4}(-3x + 8) \qquad \text{Multiplying both sides by } \tfrac{1}{4}$$
$$y = -\tfrac{3}{4}x + 2. \qquad \text{Using the distributive law}$$

The slope is $-\tfrac{3}{4}$ and the y-intercept is $(0, 2)$. We plot $(0, 2)$ and think of the slope as either $\tfrac{-3}{4}$ or $\tfrac{3}{-4}$. Using the form $\tfrac{-3}{4}$, we start at $(0, 2)$ and move *down* 3 units (since the numerator is *negative*) and *to the right* 4 units (since the denominator is *positive*). We plot the new point, $(4, -1)$.

Alternatively, we can think of the slope as $\tfrac{3}{-4}$. Starting at $(0, 2)$, we move *up* 3 units (since the numerator is *positive*) and *to the left* 4 units (since the denominator is *negative*). This leads to another point on the graph, $(-4, 5)$. Using the points found, we draw and label the graph.

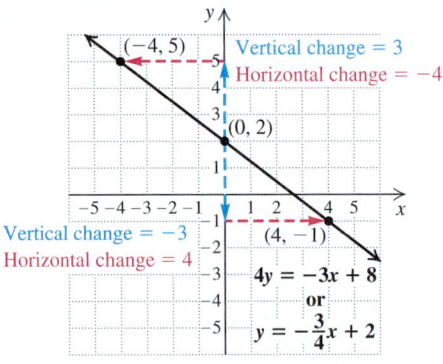

1. Graph: $y = 2x - 3$.

YOUR TURN

Student Notes

Recall that the graphs of equations of the form $y = b$ are horizontal lines, and the graphs of equations of the form $x = a$ are vertical lines.

We can use the vertical-line test to determine which types of linear graphs represent functions. Consider the following graphs.

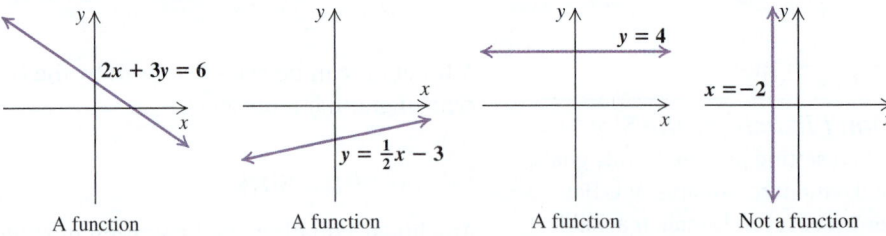

Any vertical line that passes through the graphs of $2x + 3y = 6$, $y = \tfrac{1}{2}x - 3$, and $y = 4$ will cross the graph only once. However, the vertical line through the point $(-2, 0)$ will cross the graph of $x = -2$ at *every* point. In general, *any* straight line that is not vertical is the graph of a function. A **linear function** is a function described by any linear equation whose graph is not vertical. A horizontal line represents a **constant function**.

LINEAR FUNCTION

A function described by an equation of the form $f(x) = mx + b$ is a *linear function*. Its graph is a straight line with slope m and y-intercept $(0, b)$.

When $m = 0$, the function described by $f(x) = b$ is called a *constant function*. Its graph is a horizontal line through $(0, b)$.

EXAMPLE 2   Graph: $f(x) = 3x + 2$.

SOLUTION   The notations

$$f(x) = 3x + 2 \quad \text{and} \quad y = 3x + 2$$

are often used interchangeably. The function notation emphasizes that the second coordinate in each ordered pair is determined by the first coordinate of that pair.

We graph $f(x) = 3x + 2$ in the same way that we would graph $y = 3x + 2$. The vertical axis can be labeled y or $f(x)$. We could use a table of values or, since this is a linear function, use the slope and the y-intercept to graph the function.

Since $f(x) = 3x + 2$ is in the form $f(x) = mx + b$, we can tell from the equation that the slope is 3, or $\frac{3}{1}$, and the y-intercept is $(0, 2)$. We plot $(0, 2)$ and go *up* 3 units and *to the right* 1 unit to determine another point on the line, $(1, 5)$. After we have sketched the line, a third point can be calculated as a check.

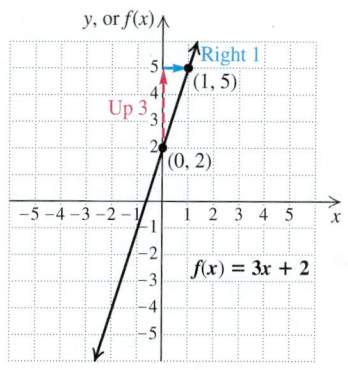

2. Graph: $f(x) = -\frac{1}{2}x + 3$.

YOUR TURN

EXAMPLE 3   Graph: $f(x) = -3$.

SOLUTION   This is a constant function. For every input x, the output is -3. The graph is a horizontal line.

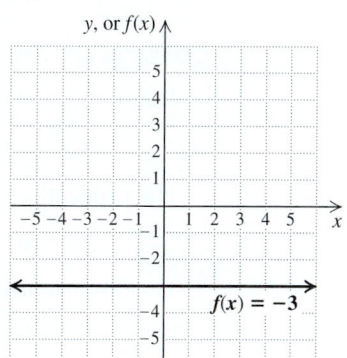

3. Graph: $f(x) = 5$.

YOUR TURN

EXAMPLE 4 *iPad Costs.* In 2012, a 16-GB (gigabyte) Apple iPad cost $630. AT&T offered a plan allowing up to 250 MB (megabytes) of data for $15 per month. Formulate a mathematical model for the total cost of an iPad purchased in 2012 and put into use with a 250-MB data plan. Then use the model to estimate the number of months required for the total cost to reach $750.

Source: www.apple.com

SOLUTION

1. **Familiarize.** For this plan, a monthly fee is charged once the initial purchase is made. After 1 month of service, the total cost is $630 + $15 = $645. After 2 months, the total cost is $630 + $15 · 2 = $660. We can write a general model if we let $C(t)$ represent the total cost, in dollars, for t months of service.

2. **Translate.** We reword and translate as follows:

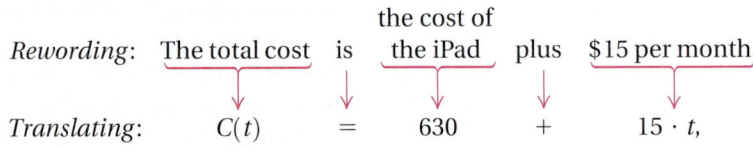

Rewording: The total cost   is   the cost of the iPad   plus $15 per month.

Translating: $C(t)$ $=$ 630 $+$ $15 \cdot t,$

with $t \geq 0$ (since there cannot be a negative number of months).

3. **Carry out.** To determine the time required for the total cost to reach $750, we substitute 750 for $C(t)$ and solve for t:

$C(t) = 630 + 15t$

$750 = 630 + 15t$ Substituting

$120 = 15t$ Subtracting 630 from both sides

$8 = t.$ Dividing both sides by 15

4. **Check.** We evaluate: $C(8) = 15 \cdot 8 + 630 = 120 + 630 = 750.$

5. **State.** It takes 8 months for the total cost to reach $750.

YOUR TURN

4. Use the model in Example 4 to estimate the number of months required for the total cost of the iPad to reach $690.

Since $f(x) = mx + b$ can be evaluated for any choice of x, the domain of all linear functions is $\mathbb{R}$, the set of all real numbers.

The second coordinate of every ordered pair in a constant function $f(x) = b$ is the number b. The range of a constant function thus consists of one number, b. For a nonconstant linear function, the graph extends indefinitely both up and down, so the range is the set of all real numbers, or $\mathbb{R}$.

DOMAIN AND RANGE OF A LINEAR FUNCTION

The domain of any linear function $f(x) = mx + b$ is

$\{x | x$ is a real number$\}$, or $\mathbb{R}$.

The range of any linear function $f(x) = mx + b$, $m \neq 0$, is

$\{y | y$ is a real number$\}$, or $\mathbb{R}$.

The range of any constant function $f(x) = b$ is $\{b\}$.

EXAMPLE 5 Determine the domain and the range of each of the following functions.

a) f, for $f(x) = 2x - 10$ **b)** g, for $g(x) = 4$

SOLUTION

a) Since $f(x) = 2x - 10$ describes a linear function, but not a constant function,

Domain of $f = \mathbb{R}$ and

Range of $f = \mathbb{R}$.

b) The function described by $g(x) = 4$ is a constant function. Thus,

Domain of $g = \mathbb{R}$ and

Range of $g = \{4\}$.

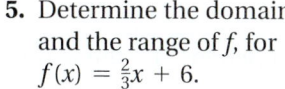

5. Determine the domain and the range of f, for $f(x) = \frac{2}{3}x + 6$.

YOUR TURN

Nonlinear Functions

A function for which the graph is not a straight line is a **nonlinear function**. Some important types of nonlinear functions are listed below.

Type of function	Example
Absolute-value function	$f(x) = |x|$
Polynomial function	$p(x) = x^3 - 4x^2 + 1$
Quadratic function	$q(x) = x^2 + 5x + 2$
Rational function	$r(x) = \dfrac{x + 1}{x - 2}$

Note that linear functions and quadratic functions are special kinds of polynomial functions.

Since the graphs of nonlinear functions are not straight lines, we usually need to calculate more than two or three points to determine the shape of the graph.

EXAMPLE 6 Graph the function given by $f(x) = |x|$, and determine the domain and the range of f.

SOLUTION We calculate function values for several choices of x and list the results in a table.

$f(0) = |0| = 0,$
$f(1) = |1| = 1,$
$f(2) = |2| = 2,$
$f(-1) = |-1| = 1,$
$f(-2) = |-2| = 2$

x	$f(x) = |x|$	$(x, f(x))$
0	0	$(0, 0)$
1	1	$(1, 1)$
2	2	$(2, 2)$
-1	1	$(-1, 1)$
-2	2	$(-2, 2)$

Because we can find the absolute value of any real number, we have

Domain of $f = \mathbb{R}$, or $(-\infty, \infty)$.

Because the absolute value of a number is never negative, we have

Range of $f = \{y \mid y \geq 0\}$, or $[0, \infty)$.

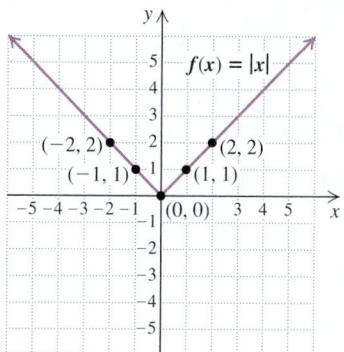

6. Graph the function given by $g(x) = |x| + 2$, and determine the domain and the range of g.

 YOUR TURN

The domain of a polynomial is the set of all real numbers. The domain of a rational function may be restricted.

EXAMPLE 7 Determine the domain of f.

a) $f(x) = x^3 + 5x^2 - 4x + 1$

b) $f(x) = \dfrac{x^2 - 4}{x + 2}$

SOLUTION

 Chapter Resource:
Visualizing for Success, p. 491

a) $f(x) = x^3 + 5x^2 - 4x + 1$ describes a polynomial function. The domain of any polynomial function is $\mathbb{R}$, so the domain of f is $\mathbb{R}$.

b) $f(x) = \dfrac{x^2 - 4}{x + 2}$ describes a rational function. Note that $f(x)$ is undefined for $x + 2 = 0$, or, equivalently, for $x = -2$. Thus the domain of $f = \{x \mid x \text{ is a real number and } x \neq -2\}$.

7. Determine the domain of the function given by $g(x) = 3x^5 - 2x^2 + 7$.

YOUR TURN

7.3 EXERCISE SET

FOR EXTRA HELP

 Vocabulary and Reading Check

Clssify each of the following statements as either true or false.

1. The vertical-line test states that a graph is not that of a function if it contains a vertical line.

2. The graph of a constant function is a horizontal line.

3. The domain of a constant function consists of a single element.

4. The domain of a linear function is the set of all real numbers.

5. Linear functions are typically written in slope–intercept form.

6. Rational functions may have some restrictions on their domains.

Linear Functions

Graph.

7. $y = 2x - 1$

8. $y = \frac{1}{3}x + 2$

9. $y = -\frac{2}{3}x + 3$

10. $y = -4x - 2$

11. $3y = 6 - 4x$

12. $5y = 2x - 15$

13. $x - y = 4$

14. $x + y = 3$

15. $y = -2$

16. $y = 3$

17. $x = 4$

18. $x = -1$

19. $f(x) = x + 3$

20. $f(x) = 2 - x$

21. $f(x) = \frac{3}{4}x + 1$

22. $f(x) = 3x + 2$

23. $g(x) = 4$

24. $g(x) = -5$

25. *Truck Rentals.* Titanium Trucks charges $30 for a one-day truck rental, plus $0.75 per mile. Formulate a linear function to model the cost $C(d)$ of a one-day rental driven d miles, and determine the number of miles driven if the total cost is $75.

26. *Taxis.* A taxi ride in New York City costs $2.50 plus $2.00 per mile.* Formulate a linear function to model the cost $C(d)$ of a d-mile taxi ride, and determine the length of a ride that cost $23.50.

27. *Hair Growth.* Lauren had her hair cut to a length of 5 in. in order to donate the hair to Locks of Love. Her hair then grew at a rate of $\frac{1}{2}$ in. per month. Formulate a linear function to model the length $L(t)$ of Lauren's hair t months after she had the haircut, and determine when her hair will be 15 in. long.

28. *Landscaping.* On Saturday, Shelby Lawncare cut the lawn at Great Harrington Community College to a height of 2 in. Since then, the grass has grown at a rate of $\frac{1}{8}$ in. per day. Formulate a linear function to model the length $L(t)$ of the lawn t days after having been cut, and determine when the grass will be $3\frac{1}{2}$ in. high.

29. *Organic Food.* U.S. sales of organic food was $4.5 billion in 2000 and is increasing at a rate of $2.2 billion each year. Formulate a linear function to model the sales $S(t)$ of organic food, in billions of dollars, t years after 2000, and determine when sales will be $15.5 billion.

Source: Based on data from the Organic Trade Association

---

*Rates are higher between 4 P.M. and 8 P.M. (*Source*: Based on data from New York City Taxi and Limousine Commission, 2012)

30. *Catering.* Chrissie's Catering charges a setup fee of $75 plus $25 per person for catering a party. Formulate a linear function to model the cost $C(x)$ for a party for x people, and determine the number of people at a party if the cost was $775.

In Exercises 31–38, assume that a constant rate of change exists for each model formed.

31. *Recycling.* In 2005, Americans recycled 79.9 million tons of solid waste. In 2010, the figure grew to 85 million tons. Let $N(t)$ represent the number of tons recycled, in millions, and t the number of years since 2000.

Sources: U.S. EPA; Franklin Associates, Ltd.

  a) Find a linear function that fits the data.
  b) Use the function of part (a) to predict the amount recycled in 2016.

32. *National Park Land.* In 1990, the National Park system consisted of about 76.4 million acres. By 2009, the figure had grown to 80.4 million acres. Let $A(t)$ represent the amount of land in the National Park system, in millions of acres, t years after 1990.

Source: U.S. National Park Service

  a) Find a linear function that fits the data.
  b) Use the function of part (a) to predict the amount of land in the National Park system in 2015.

Aha! **33.** *Life Expectancy of Females in the United States.* In 2000, the life expectancy of females born in that year was 79.7 years. In 2010, it was 81.1 years. Let $E(t)$ represent life expectancy and t the number of years since 2000.

Source: National Center for Health Statistics

  a) Find a linear function that fits the data.
  b) Use the function of part (a) to predict the life expectancy of females in 2020.

34. *Life Expectancy of Males in the United States.* In 2000, the life expectancy of males born in that year was 74.3 years. In 2010, it was 76.2 years. Let $E(t)$ represent life expectancy and t the number of years since 2000.

Source: National Center for Health Statistics

  a) Find a linear function that fits the data.
  b) Use the function of part (a) to predict the life expectancy of males in 2020.

35. *History.* During the late 1600s, the capacity of ships in the English, French, and Dutch navies almost doubled, as shown in the graph below. Let $S(t)$ represent the average displacement of a ship, in tons, and t the number of years since 1650.

Naval history

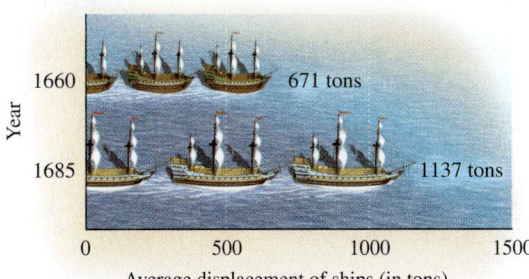

Source: Harding, R., The Evolution of the Sailing Navy, 1509–1815. New York: St. Martin's Press, 1995

  a) Find a linear function that fits the data.
  b) Use the function of part (a) to estimate the average displacement of a ship in 1670.

36. *Consumer Demand.* Suppose that 6.5 million lb of coffee are sold when the price is $12 per pound, and 6.0 million lb are sold when it is $15 per pound.

  a) Find a linear function that expresses the amount of coffee sold as a function of the price per pound.
  b) Use the function of part (a) to predict how much consumers would be willing to buy at a price of $6 per pound.

37. *Pressure at Sea Depth.* The pressure 100 ft beneath the ocean's surface is approximately 4 atm (atmospheres), whereas at a depth of 200 ft, the pressure is about 7 atm.

  a) Find a linear function that expresses pressure as a function of depth.
  b) Use the function of part (a) to determine the pressure at a depth of 690 ft.

38. *Seller's Supply.* Suppose that suppliers are willing to sell 5.0 million lb of coffee at a price of $12 per pound and 7.0 million lb at $15 per pound.

  a) Find a linear function that expresses the amount suppliers are willing to sell as a function of the price per pound.
  b) Use the function of part (a) to predict how much suppliers would be willing to sell at a price of $6 per pound.

Nonlinear Functions

Classify each function as a linear function, an absolute-value function, a quadratic function, another polynomial function, or a rational function, and determine the domain of the function.

39. $f(x) = \dfrac{1}{3}x - 7$

40. $g(x) = \dfrac{x}{x + 1}$

41. $p(x) = x^2 + x + 1$

42. $t(x) = |x - 7|$

43. $f(t) = \dfrac{12}{3t + 4}$

44. $g(n) = 15 - 10n$

45. $f(x) = 0.02x^4 - 0.1x + 1.7$

46. $f(a) = 2|a + 3|$

47. $f(x) = \dfrac{x}{2x - 5}$

48. $g(x) = \dfrac{2x}{3x - 4}$

49. $f(n) = \dfrac{4n - 7}{n^2 + 3n + 2}$

50. $h(x) = \dfrac{x - 5}{2x^2 - 2}$

51. $f(n) = 200 - 0.1n$

52. $g(t) = \dfrac{t^2 - 3t + 7}{8}$

Given the graph of each function, determine the range of f.

53.

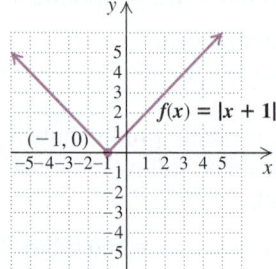

54.

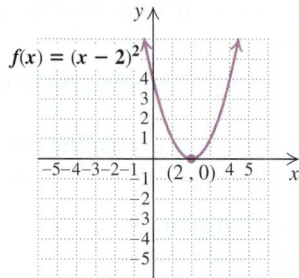

55.

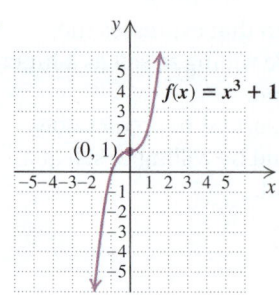

56.

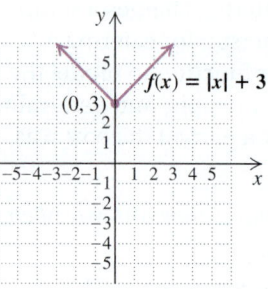

57.

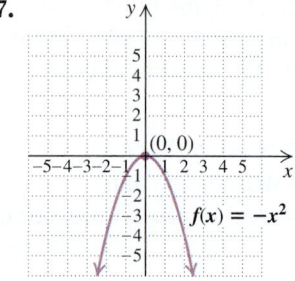

58.

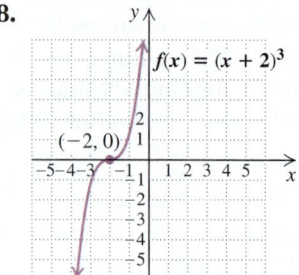

Graph each function and determine its domain and range.

59. $f(x) = x + 3$

60. $f(x) = 2x - 1$

61. $f(x) = -1$

62. $g(x) = 2$

63. $f(x) = |x| + 1$

64. $g(x) = |x - 3|$

65. $g(x) = x^2$

66. $f(x) = x^2 + 2$

 67. Bob believes that the domain and the range of all polynomial functions is $\mathbb{R}$. How could you convince him that he is mistaken?

 68. Explain why linear functions and quadratic functions are special types of polynomial functions.

Skill Review

Factor.

69. $4n^2 - 14n + 49$ [5.4]

70. $4x^2 - 17x - 15$ [5.3]

71. $2x^3 - 14x^2 + 12x$ [5.2]

72. $c^3 - 1000$ [5.5]

73. $d^3 - 64d$ [5.4]

74. $a^3 + 4a^2 + 2a + 8$ [5.1]

 75. Explain why the range of a constant function consists of only one number.

76. On the basis of your answers to Exercises 33 and 34, would you predict that at some point in the future the life expectancy of males will exceed that of females? Why or why not?

Given that $f(x) = mx + b$, *classify each of the following as true or false.*

77. $f(c + d) = f(c) + f(d)$

78. $f(cd) = f(c)f(d)$

79. $f(kx) = kf(x)$

80. $f(c - d) = f(c) - f(d)$

For Exercises 81–84, assume that a linear equation models each situation.

81. *Temperature Conversion.* Water freezes at 32° Fahrenheit and at 0° Celsius. Water boils at 212°F and at 100°C. What Celsius temperature corresponds to a room temperature of 70°F?

82. *Depreciation of a Computer.* After 6 months of use, the value of Don's computer had dropped to $900. After 8 months, the value had gone down to $750. How much did the computer cost originally?

83. *Cell-Phone Charges.* The total cost of Tam's cell phone was $410 after 5 months of service and $690 after 9 months. What costs had Tam already incurred when her service had just begun? Assume that Tam's monthly charge is constant.

84. *Operating Expenses.* The total cost for operating Ming's Wings was $7500 after 4 months and $9250 after 7 months. Predict the total cost after 10 months.

85. For a linear function g, $g(3) = -5$ and $g(7) = -1$.
   a) Find an equation for g.
   b) Find $g(-2)$.
   c) Find a such that $g(a) = 75$.

 86. When several data points are available and they appear to be nearly collinear, a procedure known as *linear regression* can be used to find an equation for the line that most closely fits the data.
   a) Use a graphing calculator with a LINEAR REGRESSION option and the table that follows to find a linear function that predicts the wattage of a CFL (compact fluorescent) lightbulb as a function of the wattage of a standard incandescent bulb of equivalent brightness. Round coefficients to the nearest thousandth.

Energy Conservation

Incandescent Wattage	CFL Equivalent
25 W	5 W
50 W	9 W
60 W	15 W
100 W	25 W
120 W	28 W

Source: U.S. Department of Energy

   b) Use the function from part (a) to estimate the CFL wattage that is equivalent to a 75-watt incandescent bulb.

YOUR TURN ANSWERS: SECTION 7.3

1.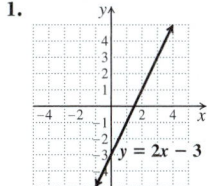

2. 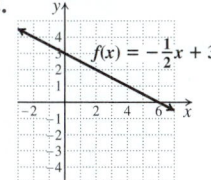 $f(x) = -\frac{1}{2}x + 3$

3. 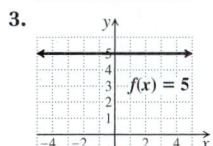 $f(x) = 5$

4. 4 months

5. Domain: $\mathbb{R}$; range: $\mathbb{R}$

6. 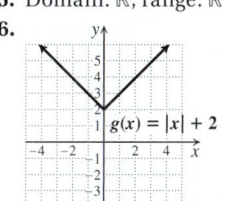 $g(x) = |x| + 2$

Domain: $\mathbb{R}$; range: $\{y \mid y \geq 2\}$, or $[2, \infty)$ **7.** $\mathbb{R}$

QUICK QUIZ: SECTIONS 7.1–7.3

Let $f(x) = \dfrac{x - 1}{x - 2}$.

1. Find $f(6)$. [7.1]   **2.** Find $f(a + 1)$. [7.1]

3. Find the domain of f. [7.2], [7.3]

Let $g(x) = 3x^2 - 6x - 7$.

4. Determine whether g is a linear function, a quadratic function, or a rational function. [7.3]

5. Find the domain of g. [7.2], [7.3]

PREPARE TO MOVE ON

Perform the indicated operations.

1. $(x^2 + 2x + 7) + (3x^2 - 8)$ [4.4]

2. $(3x^3 - x^2 + x) - (x^3 + 2x - 7)$ [4.4]

3. $(2x + 1)(x - 7)$ [4.6]

4. $(x - 3)(x + 4)$ [4.6]

Mid-Chapter Review

- A function is a correspondence between two sets: the domain and the range.
- In a function, each member of the domain corresponds to exactly one member of the range.
- A graph is that of a function if it passes the vertical-line test.
- The domain of a function is the set of all inputs.
- The range of a function is the set of all outputs.
- The notation $f(2) = 7$ means that for the function f, an input of 2 corresponds to an output of 7.

GUIDED SOLUTIONS

1. Find $f(-5)$ for $f(x) = 2x^2 - 3x$.   [7.1]

Solution

$$f(x) = 2x^2 - 3x$$
$$f(-5) = 2(\boxed{})^2 - 3(\boxed{})$$
$$= 2(\boxed{}) - (\boxed{})$$
$$= \boxed{} + \boxed{}$$
$$= \boxed{}$$

2. For $g(x) = \dfrac{3x}{x - 7}$, determine the domain of g.
[7.2]

Solution

$$x - 7 = \boxed{}$$
$$x = \boxed{}$$

Domain $= \{x \mid x \text{ is a real number } and \ x \neq \boxed{}\}$

MIXED REVIEW

Let $f = \{(3, 6), (4, 8), (-1, -2), (0, 0)\}$.

3. Find the domain of f.

4. Find the range of f.

5. Find $f(-1)$.

Let $g(x) = x - 1$ and $h(x) = \dfrac{2}{x}$.

6. Find the domain of g.

7. Find the domain of h.

8. Find $h(10)$.

9. Determine whether g is a linear function, a quadratic function, or a rational function.

10. Determine whether h is a linear function, a quadratic function, or a rational function.

11. Determine the range of g.

12. Find the domain of the function given by

$$f(x) = \frac{x - 3}{x^2 + 6x - 40}.$$

Use the following graph of F for Exercises 13–17.

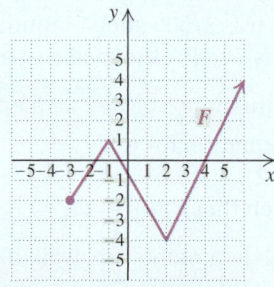

13. Determine from the graph whether F is a function.

14. Find $F(2)$.

15. Find any inputs for which the output is 1.

16. Find the domain of F.

17. Find the range of F.

Let $G(x) = \begin{cases} 1 - x, & \text{if } x < 1, \\ 10, & \text{if } x = 1, \\ x + 1, & \text{if } x > 1. \end{cases}$

18. Find $G(3)$.

19. Find $G(1)$.

20. Find $G(-12)$.

7.4 The Algebra of Functions

The Sum, Difference, Product, or Quotient of Two Functions   •   Domains and Graphs

We now examine four ways in which functions can be combined.

The Sum, Difference, Product, or Quotient of Two Functions

Suppose that a is in the domain of two functions, f and g. The input a is paired with $f(a)$ by f and with $g(a)$ by g. The outputs can then be added to get $f(a) + g(a)$.

EXAMPLE 1 Let $f(x) = x + 4$ and $g(x) = x^2 + 1$. Find $f(2) + g(2)$.

SOLUTION We visualize two function machines, as shown below. Because 2 is in the domain of each function, we can compute $f(2)$ and $g(2)$.

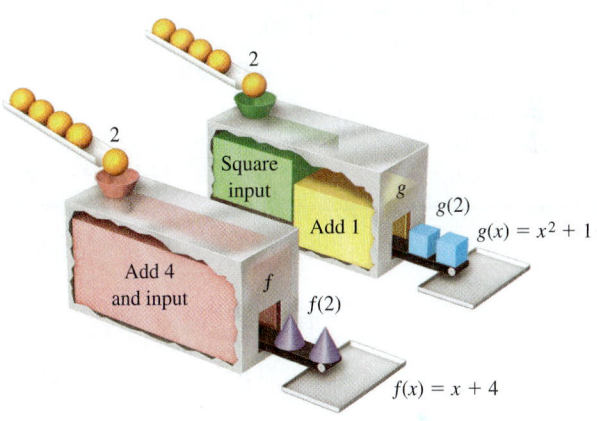

Since

$$f(2) = 2 + 4 = 6 \quad \text{and} \quad g(2) = 2^2 + 1 = 5,$$

we have

$$f(2) + g(2) = 6 + 5 = 11.$$

YOUR TURN

1. Using the functions defined in Example 1, find $f(-5) + g(-5)$.

In Example 1, suppose that we were to write $f(x) + g(x)$ as $(x + 4) + (x^2 + 1)$, or $f(x) + g(x) = x^2 + x + 5$. This could then be regarded as a "new" function and visualized as one function machine, shown at left, combining the two machines from Example 1. The notation $(f + g)(x)$ is generally used to indicate the output of a function formed in this manner. Similar notations exist for subtraction, multiplication, and division of functions.

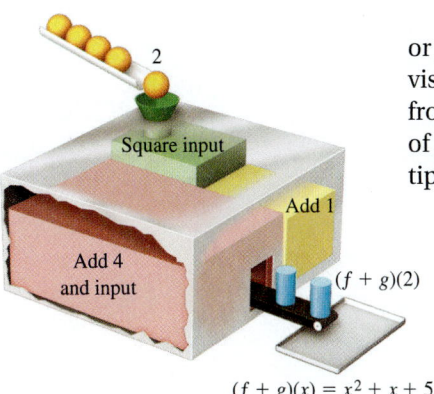

$(f + g)(x) = x^2 + x + 5$

THE ALGEBRA OF FUNCTIONS

If f and g are functions and x is in the domain of both functions, then:

1. $(f + g)(x) = f(x) + g(x);$
2. $(f - g)(x) = f(x) - g(x);$
3. $(f \cdot g)(x) = f(x) \cdot g(x);$
4. $(f/g)(x) = f(x)/g(x),$ provided $g(x) \neq 0.$

EXAMPLE 2 For $f(x) = x^2 - x$ and $g(x) = x + 2$, find the following.

a) $(f + g)(4)$

b) $(f - g)(x)$ and $(f - g)(-1)$

c) $(f/g)(x)$ and $(f/g)(-3)$

d) $(f \cdot g)(4)$

SOLUTION

a) Since $f(4) = 4^2 - 4 = 12$ and $g(4) = 4 + 2 = 6$, we have

$$(f + g)(4) = f(4) + g(4)$$
$$= 12 + 6 \qquad \text{Substituting}$$
$$= 18.$$

Alternatively, we could first find $(f + g)(x)$:

$$(f + g)(x) = f(x) + g(x)$$
$$= x^2 - x + x + 2$$
$$= x^2 + 2. \qquad \text{Combining like terms}$$

Thus,

$$(f + g)(4) = 4^2 + 2 = 18. \qquad \text{Our results match.}$$

b) We have

$$(f - g)(x) = f(x) - g(x)$$
$$= x^2 - x - (x + 2) \qquad \text{Substituting}$$
$$= x^2 - 2x - 2. \qquad \begin{array}{l}\text{Removing parentheses and}\\ \text{combining like terms}\end{array}$$

Then,

$$(f - g)(-1) = (-1)^2 - 2(-1) - 2 \qquad \begin{array}{l}\text{Using } (f - g)(x) \text{ is faster than}\\ \text{using } f(x) - g(x).\end{array}$$
$$= 1. \qquad \text{Simplifying}$$

c) We have

$$(f/g)(x) = f(x)/g(x)$$
$$= \frac{x^2 - x}{x + 2}. \qquad \text{We assume that } x \neq -2.$$

Then,

$$(f/g)(-3) = \frac{(-3)^2 - (-3)}{-3 + 2} \qquad \text{Substituting}$$
$$= \frac{12}{-1} = -12.$$

d) Using our work in part (a), we have

$$(f \cdot g)(4) = f(4) \cdot g(4)$$
$$= 12 \cdot 6$$
$$= 72.$$

2. Using the functions defined in Example 2, find $(g - f)(x)$ and $(g - f)(-1)$.

 YOUR TURN

Domains and Graphs

Applications involving sums or differences of functions often appear in print. For example, the following graphs are similar to those published by the California Department of Education to promote breakfast programs in which students eat a balanced meal of fruit or juice, toast or cereal, and 2% or whole milk. The combination of carbohydrate, protein, and fat gives a sustained release of energy, delaying the onset of hunger for several hours.

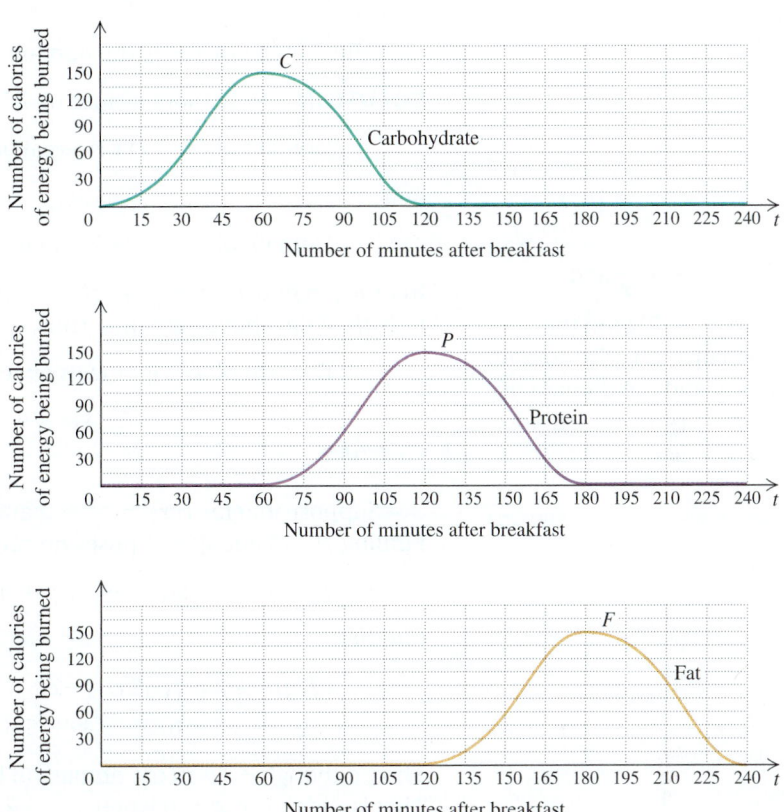

When the three graphs are superimposed, and the calorie expenditures added, it becomes clear that a balanced meal results in a steady, sustained supply of energy.

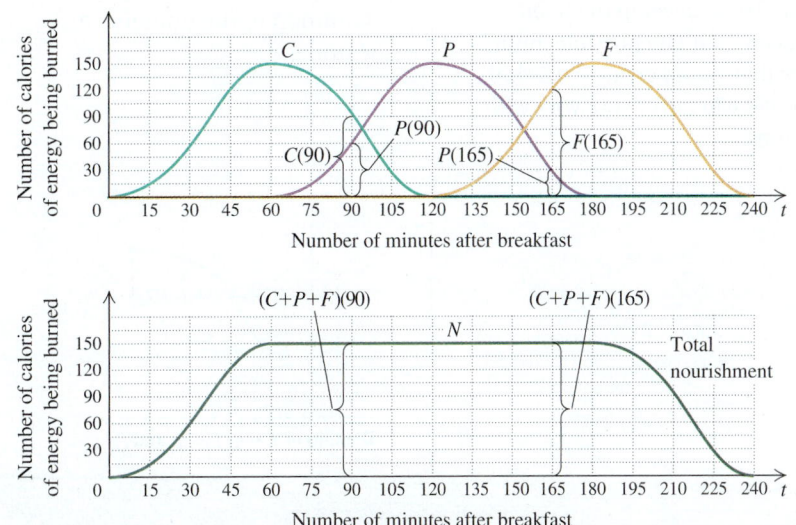

For any point $(t, N(t))$, we have

$$N(t) = (C + P + F)(t) = C(t) + P(t) + F(t).$$

To find $(f + g)(a)$, $(f - g)(a)$, or $(f \cdot g)(a)$, we must know that $f(a)$ and $g(a)$ exist. This means that a must be in the domain of both f and g.

EXAMPLE 3   Let

$$f(x) = \frac{5}{x} \quad \text{and} \quad g(x) = \frac{2x - 6}{x + 1}.$$

Find the domain of $f + g$, the domain of $f - g$, and the domain of $f \cdot g$.

SOLUTION   Note that because division by 0 is undefined, we have

Domain of $f = \{x \,|\, x \text{ is a real number } and \ x \neq 0\}$

and

Domain of $g = \{x \,|\, x \text{ is a real number } and \ x \neq -1\}$.

In order to find $f(a) + g(a)$, $f(a) - g(a)$, or $f(a) \cdot g(a)$, we must know that a is in *both* of the above domains. Thus,

Domain of $f + g = $ Domain of $f - g = $ Domain of $f \cdot g$

$$= \{x \,|\, x \text{ is a real number } and \ x \neq 0 \ and \ x \neq -1\}.$$

> **3.** Let $f(x) = \dfrac{x}{x - 8}$ and
> $g(x) = x^2 - 7$. Find the domain of $f + g$, the domain of $f - g$, and the domain of $f \cdot g$.

YOUR TURN

Suppose that for $f(x) = x^2 - x$ and $g(x) = x + 2$, we want to find $(f/g)(-2)$. Finding $f(-2)$ and $g(-2)$ poses no problem:

$$f(-2) = 6 \quad \text{and} \quad g(-2) = 0;$$

but then

$$(f/g)(-2) = f(-2)/g(-2)$$
$$= 6/0. \qquad \text{Division by 0 is undefined.}$$

Thus, although -2 is in the domain of both f and g, it is not in the domain of f/g. That is, since $x + 2 = 0$ when $x = -2$, the domain of f/g must exclude -2.

Student Notes

The concern over a denominator being 0 arises throughout this course. Try to develop the habit of checking for any possible input values that would create a denominator of 0 whenever you work with functions.

DETERMINING THE DOMAIN

The domain of $f + g$, $f - g$, or $f \cdot g$ is the set of all values common to the domains of f and g.

The domain of f/g is the set of all values common to the domains of f and g, excluding any values for which $g(x)$ is 0.

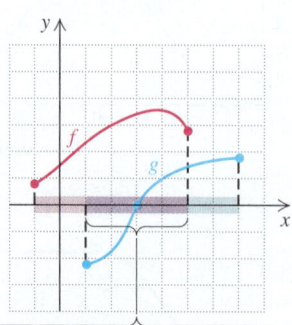

Domain of $f + g$, $f - g$, and $f \cdot g$

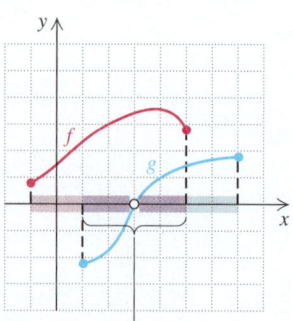

Domain of f/g

EXAMPLE 4 Given

$$f(x) = \frac{1}{x - 3} \quad \text{and} \quad g(x) = 2x - 7,$$

find the domains of $f + g, f - g, f \cdot g$, and f/g.

SOLUTION We first find the domain of f and the domain of g:

> Find the domains of f and g.

The domain of f is $\{x \mid x \text{ is a real number } and \ x \neq 3\}$.

The domain of g is $\mathbb{R}$.

The domains of $f + g$, $f - g$, and $f \cdot g$ are the set of all elements common to the domains of f and g. This consists of all real numbers except 3.

> Find the domains of $f + g, f - g,$ and $f \cdot g.$

The domain of $f + g =$ the domain of $f - g =$ the domain of $f \cdot g$
$$= \{x \mid x \text{ is a real number } and \ x \neq 3\}.$$

Because we cannot divide by 0, the domain of f/g must also exclude any values of x for which $g(x)$ is 0. We determine those values by solving $g(x) = 0$:

> Find any values of x for which $g(x) = 0.$

$$g(x) = 0$$
$$2x - 7 = 0 \qquad \text{Replacing } g(x) \text{ with } 2x - 7$$
$$2x = 7$$
$$x = \tfrac{7}{2}.$$

The domain of f/g is the domain of the sum, the difference, and the product of f and g, found above, excluding $\tfrac{7}{2}$.

> Find the domain of f/g.

The domain of $f/g = \left\{ x \mid x \text{ is a real number } and \ x \neq 3 \ and \ x \neq \tfrac{7}{2} \right\}$.

YOUR TURN

4. Given $f(x) = \dfrac{3}{2x + 1}$ and $g(x) = x - 4$, find the domains of $f + g, f - g, f \cdot g$, and f/g.

Technology Connection

A partial check of Example 4 can be performed by setting up a table so the TBLSTART is 1 and the increment of change (ΔTbl) is 0.5. (Other choices, like 0.1, will also work.) Next, we let $y_1 = \dfrac{1}{x - 3}$ and $y_2 = 2x - 7$. Using Y-VARS to write $y_3 = y_1 + y_2$ and $y_4 = y_1/y_2$, we can create the table of values shown here. Note that when x is 3.5, a value for y_3 can be found, but y_4 is undefined. If we "de-select" y_1 and y_2 as we enter them, the columns for y_3 and y_4 appear without scrolling through the table.

X	Y3	Y4
1	−5.5	.1
1.5	−4.667	.16667
2	−4	.33333
2.5	−4	1
3	ERROR	ERROR
3.5	2	ERROR
4	2	1

X = 3.5

Use a similar approach to partially check Example 3.

Chapter Resource:
Collaborative Activity, p. 492

Division by 0 is not the only condition that can force restrictions on the domain of a function. When we examine functions similar to that given by $f(x) = \sqrt{x}$, the concern is taking the square root of a negative number.

7.4 EXERCISE SET

Vocabulary and Reading Check

Make each of the following statements true by selecting the correct word for each blank.

1. If f and g are functions, then $(f + g)(x)$ is the _____ of the functions.
   sum/difference

2. One way to compute $(f - g)(2)$ is to _____ $g(2)$ from $f(2)$.
   erase/subtract

3. One way to compute $(f - g)(2)$ is to simplify $f(x) - g(x)$ and then _____ the result for $x = 2$.
   evaluate/substitute

4. The domain of $f + g, f - g$, and $f \cdot g$ is the set of all values common to the _____ of f and g.
   domains/ranges

5. The domain of f/g is the set of all values common to the domains of f and g, _____ any values for which $g(x)$ is 0.
   including/excluding

6. The height of $(f + g)(a)$ on a graph is the _____ of the heights of $f(a)$ and $g(a)$.
   product/sum

The Sum, Difference, Product, or Quotient of Two Functions

Let $f(x) = -2x + 3$ and $g(x) = x^2 - 5$. Find each of the following.

7. $f(3) + g(3)$
8. $f(4) + g(4)$
9. $f(1) - g(1)$
10. $f(2) - g(2)$
11. $f(-2) \cdot g(-2)$
12. $f(-1) \cdot g(-1)$
13. $f(-4)/g(-4)$
14. $f(3)/g(3)$
15. $g(1) - f(1)$
16. $g(-3)/f(-3)$
17. $(f + g)(x)$
18. $(f - g)(x)$
19. $(g - f)(x)$
20. $(g/f)(x)$

Let $F(x) = x^2 - 2$ and $G(x) = 5 - x$. Find each of the following.

21. $(F + G)(x)$
22. $(F + G)(a)$
23. $(F - G)(3)$
24. $(F - G)(2)$

25. $(F \cdot G)(a)$
26. $(G \cdot F)(x)$
27. $(F/G)(x)$
28. $(G - F)(x)$
29. $(G/F)(-2)$
30. $(F/G)(-1)$
31. $(F + F)(1)$
32. $(G \cdot G)(6)$

Let
$$r(x) = \frac{5}{x^2} \quad \text{and} \quad t(x) = \frac{3}{2x}.$$

Find each of the following.

33. $(r \cdot t)(x)$
34. $(r/t)(x)$
35. $(r - t)(x)$
36. $(r + t)(x)$
37. $(t/r)(x)$
38. $(t - r)(x)$

Let
$$f(x) = \frac{x - 1}{x^2 - x - 6} \quad \text{and} \quad g(x) = \frac{x + 2}{x^2 - 9}.$$

Find each of the following.

39. $(g - f)(x)$
40. $(f - g)(x)$
41. $(f + g)(x)$
42. $(f \cdot g)(x)$
43. $(f/g)(x)$
44. $(g/f)(x)$

Domains and Graphs

The following graph shows the number of births in the United States, in millions, from 1970–2010. Here, $C(t)$ represents the number of Caesarean section births, $B(t)$ the number of non-Caesarean section births, and $N(t)$ the total number of births in year t.

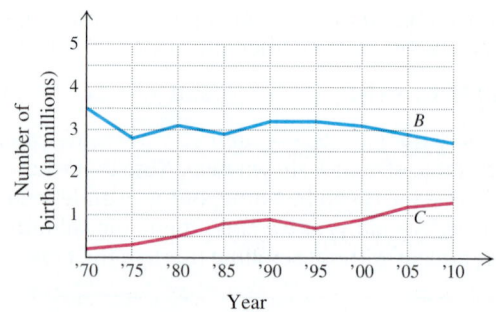

Source: National Center for Health Statistics

45. Use estimates of $C(2005)$ and $B(2005)$ to estimate $N(2005)$.

46. Use estimates of $C(1985)$ and $B(1985)$ to estimate $N(1985)$.

In the following graph, $S(t)$ represents the number of gallons of carbonated soft drinks, $M(t)$ the number of gallons of milk, $J(t)$ the number of gallons of fruit juice, and $W(t)$ the number of gallons of bottled water consumed by the average American in year t.

Beverage consumption

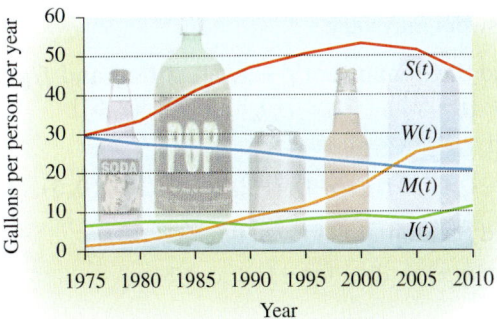

Source: Economic Research Service, U. S. Department of Agriculture

47. Use estimates of $S(2010)$ and $W(2010)$ to estimate $(S - W)(2010)$.

48. Use estimates of $M(2010)$ and $J(2010)$ to estimate $(M - J)(2010)$.

Often function addition is represented by stacking the graphs of individual functions directly on top of each other. The following graph indicates how U.S. municipal solid waste has been managed. The braces indicate the values of the individual functions.

Talking trash

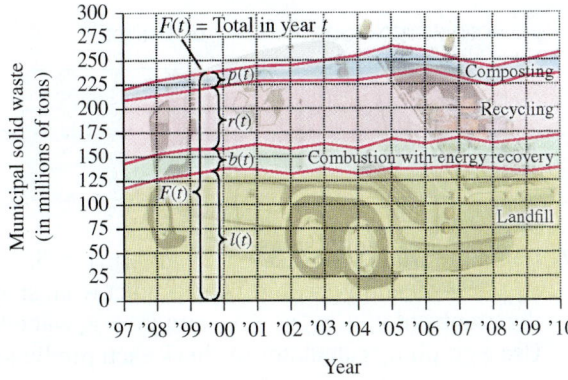

Source: Environmental Protection Agency

49. Estimate $(p + r)('09)$. What does it represent?

50. Estimate $(p + r + b)('09)$. What does it represent?

51. Estimate $F('00)$. What does it represent?

52. Estimate $F('10)$. What does it represent?

53. Estimate $(F - p)('08)$. What does it represent?

54. Estimate $(F - l)('07)$. What does it represent?

For each pair of functions f and g, determine the domain of the sum, the difference, and the product of the two functions.

55. $f(x) = x^2$,
$g(x) = 7x - 4$

56. $f(x) = 5x - 1$,
$g(x) = 2x^2$

57. $f(x) = \dfrac{1}{x + 5}$,
$g(x) = 4x^3$

58. $f(x) = 3x^2$,
$g(x) = \dfrac{1}{x - 9}$

59. $f(x) = \dfrac{2}{x}$,
$g(x) = x^2 - 4$

60. $f(x) = x^3 + 1$,
$g(x) = \dfrac{5}{x}$

61. $f(x) = x + \dfrac{2}{x - 1}$,
$g(x) = 3x^3$

62. $f(x) = 9 - x^2$,
$g(x) = \dfrac{3}{x + 6} + 2x$

63. $f(x) = \dfrac{3}{2x + 9}$,
$g(x) = \dfrac{5}{1 - x}$

64. $f(x) = \dfrac{5}{3 - x}$,
$g(x) = \dfrac{1}{4x - 1}$

For each pair of functions f and g, determine the domain of f/g.

65. $f(x) = x^4$,
$g(x) = x - 3$

66. $f(x) = 2x^3$,
$g(x) = 5 - x$

67. $f(x) = 3x - 2$,
$g(x) = 2x + 8$

68. $f(x) = 5 + x$,
$g(x) = 6 - 2x$

69. $f(x) = \dfrac{3}{x - 4}$,
$g(x) = 5 - x$

70. $f(x) = \dfrac{1}{2 - x}$,
$g(x) = 7 + x$

71. $f(x) = \dfrac{2x}{x + 1}$,
$g(x) = 2x + 5$

72. $f(x) = \dfrac{7x}{x - 2}$,
$g(x) = 3x + 7$

For Exercises 73–80, consider the functions F and G as shown.

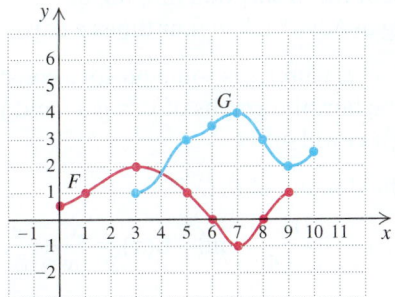

73. Determine $(F + G)(5)$ and $(F + G)(7)$.

74. Determine $(F \cdot G)(6)$ and $(F \cdot G)(9)$.

75. Determine $(G - F)(7)$ and $(G - F)(3)$.

76. Determine $(F/G)(3)$ and $(F/G)(7)$.

77. Find the domains of F, G, $F + G$, and F/G.

78. Find the domains of $F - G$, $F \cdot G$, and G/F.

79. Graph $F + G$.

80. Graph $G - F$.

81. Examine the graph for Exercises 47 and 48. Between what years did the average American drink more soft drinks than juice, bottled water, and milk combined? Explain how you determined this.

82. Examine the graph for Exercises 45 and 46. Did the total number of births increase or decrease from 1970 to 2010? Did the percent of births by Caesarean section increase or decrease from 1970 to 2010? Explain how you determined your answers.

Skill Review

Solve.

83. One angle of a triangle measures twice the second angle. The third angle measures three times the second angle. Find the measures of the angles.   [2.5]

84. In one basketball game, Terrence scored 5 fewer points than Isaiah. Together, they scored 27 points. How many points did Terrence score?   [2.5]

85. A *mole* of a substance contains 6.022×10^{23} molecules. If a mole of water weighs 18.015 g, how much does each molecule weigh?   [4.2]

86. The flower bed in front of a school's administration building is 6 ft longer than five times its width. If the area of the bed is 216 ft^2, how long is the flower bed?   [5.8]

Synthesis

87. Examine the graphs following Example 2 and explain how similar graphs could be drawn to represent the absorption into the bloodstream of 200 mg of Advil® taken four times a day.

88. If $f(x) = c$, where c is some positive constant, describe how the graphs of $y = g(x)$ and $y = (f + g)(x)$ will differ.

89. Find the domain of F/G, if

$$F(x) = \frac{1}{x - 4} \quad \text{and} \quad G(x) = \frac{x^2 - 4}{x - 3}.$$

90. Find the domain of f/g, if

$$f(x) = \frac{3x}{2x + 5} \quad \text{and} \quad g(x) = \frac{x^4 - 1}{3x + 9}.$$

91. Sketch the graph of two functions f and g such that the domain of f/g is

$$\{x \mid -2 \le x \le 3 \text{ and } x \ne 1\}.$$

Answers may vary.

92. Find the domains of $f + g, f - g, f \cdot g$, and f/g, if $f = \{(-2, 1), (-1, 2), (0, 3), (1, 4), (2, 5)\}$

and

$g = \{(-4, 4), (-3, 3), (-2, 4), (-1, 0), (0, 5), (1, 6)\}$.

93. Find the domain of m/n, if

$$m(x) = 3x \quad \text{for } -1 < x < 5$$

and

$$n(x) = 2x - 3.$$

94. For f and g as defined in Exercise 92, find $(f + g)(-2)$, $(f \cdot g)(0)$, and $(f/g)(1)$.

95. Write equations for two functions f and g such that the domain of $f + g$ is

$$\{x \mid x \text{ is a real number } \text{and } x \ne -2 \text{ and } x \ne 5\}.$$

Answers may vary.

96. Let $y_1 = 2.5x + 1.5$, $y_2 = x - 3$, and $y_3 = y_1/y_2$. Depending on whether the CONNECTED or DOT mode is used, the graph of y_3 appears as follows. Use algebra to determine which graph more accurately represents y_3.

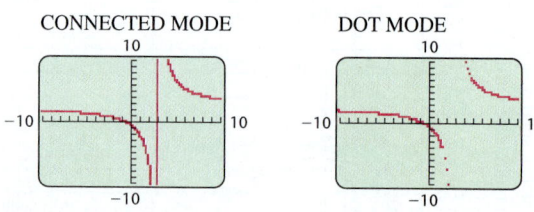

97. Using the window $[-5, 5, -1, 9]$, graph $y_1 = 5$, $y_2 = x + 2$, and $y_3 = \sqrt{x}$. Then predict what shape the graphs of $y_1 + y_2$, $y_1 + y_3$, and $y_2 + y_3$ will take. Use a graphing calculator to check each prediction.

98. Use the TABLE feature on a graphing calculator to check your answers to Exercises 59, 61, 69, and 71.

QUICK QUIZ: SECTIONS 7.1–7.4

Let $f(x) = \begin{cases} x^2 - 1, & \text{if } x \leq 0, \\ x, & \text{if } x > 0. \end{cases}$

1. Find $f(-3)$. [7.1]   **2.** Find $f(0)$. [7.1]

3. Graph $g(x) = 3x - 2$. [7.3]

Let $F(x) = x^2 - x$ and $G(x) = 3 - 2x$.

4. Find $(F + G)(0)$. [7.4]   **5.** Find $(G - F)(x)$. [7.4]

PREPARE TO MOVE ON

Solve. [2.3]

1. $ac = b$, for c **2.** $x - wz = y$, for w

3. $pq - rq = st$, for q **4.** $ab = d - cb$, for b

5. $ab - cd = 3b + d$, for b

7.5 Formulas, Applications, and Variation

Formulas ▪ Direct Variation ▪ Inverse Variation ▪ Joint Variation and Combined Variation

Formulas

Formulas occur frequently as mathematical models. Many formulas contain rational expressions, and to solve such formulas for a specified letter, we proceed as we do when solving rational equations.

EXAMPLE 1 *Electronics.* The formula

$$\frac{1}{R} = \frac{1}{r_1} + \frac{1}{r_2}$$

is used by electricians to determine the resistance R of two resistors r_1 and r_2 connected in parallel. Solve for r_1.

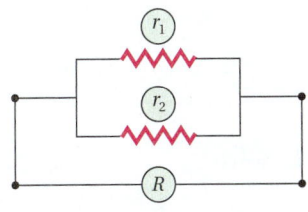

SOLUTION We use the same approach as for solving a rational equation:

$$Rr_1r_2 \cdot \frac{1}{R} = Rr_1r_2 \cdot \left(\frac{1}{r_1} + \frac{1}{r_2} \right) \qquad \text{Multiplying both sides by the LCM of the denominators}$$

$$Rr_1r_2 \cdot \frac{1}{R} = Rr_1r_2 \cdot \frac{1}{r_1} + Rr_1r_2 \cdot \frac{1}{r_2} \qquad \text{Multiplying to remove parentheses}$$

$$r_1r_2 = Rr_2 + Rr_1. \qquad \text{Simplifying by removing factors equal to 1: } \frac{R}{R} = 1; \frac{r_1}{r_1} = 1; \frac{r_2}{r_2} = 1$$

Student Notes

The subscripts 1 and 2 in Example 1 indicate that r_1 and r_2 are different variables representing similar quantities.

At this point it is tempting to multiply by $1/r_2$ to get r_1 alone on the left, *but* note that there is an r_1 on the right. We must get all the terms involving r_1 on the *same side* of the equation.

$$r_1 r_2 - R r_1 = R r_2 \qquad \text{Subtracting } R r_1 \text{ from both sides}$$

$$r_1 (r_2 - R) = R r_2 \qquad \text{Factoring out } r_1 \text{ in order to combine like terms}$$

$$r_1 = \frac{R r_2}{r_2 - R} \qquad \text{Dividing both sides by } r_2 - R \text{ to get } r_1 \text{ alone}$$

This formula can be used to calculate r_1 whenever R and r_2 are known.

1. Solve $\dfrac{1}{x} = \dfrac{1}{y} + \dfrac{1}{z}$ for x.

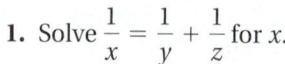 YOUR TURN

EXAMPLE 2 *Astronomy.* The formula

$$\frac{V^2}{R^2} = \frac{2g}{R + h}$$

is used to find a satellite's *escape velocity* V, where R is a planet's radius, h is the satellite's height above the planet, and g is the planet's gravitational constant. Solve for h.

SOLUTION We first multiply by the LCM, $R^2(R + h)$, to clear fractions:

$$\frac{V^2}{R^2} = \frac{2g}{R + h}$$

$$R^2(R + h)\frac{V^2}{R^2} = R^2(R + h)\frac{2g}{R + h} \qquad \text{Multiplying to clear fractions}$$

$$\frac{R^2(R + h)V^2}{R^2} = \frac{R^2(R + h)2g}{R + h}$$

$$(R + h)V^2 = R^2 \cdot 2g. \qquad \begin{array}{l}\text{Removing factors equal to 1:}\\ \dfrac{R^2}{R^2} = 1 \text{ and } \dfrac{R + h}{R + h} = 1\end{array}$$

Remember: We are solving for h. Although we could distribute V^2, since h appears only within the factor $R + h$, it is easier to divide both sides by V^2:

$$\frac{(R + h)V^2}{V^2} = \frac{2R^2 g}{V^2} \qquad \text{Dividing both sides by } V^2$$

$$R + h = \frac{2R^2 g}{V^2} \qquad \text{Removing a factor equal to 1: } \dfrac{V^2}{V^2} = 1$$

2. Solve

$$\frac{a}{x} = \frac{b}{x + c}$$

for c.

$$h = \frac{2R^2 g}{V^2} - R. \qquad \text{Subtracting } R \text{ from both sides}$$

The last equation can be used to determine the height of a satellite above a planet when the planet's radius and gravitational constant, along with the satellite's escape velocity, are known.

 YOUR TURN

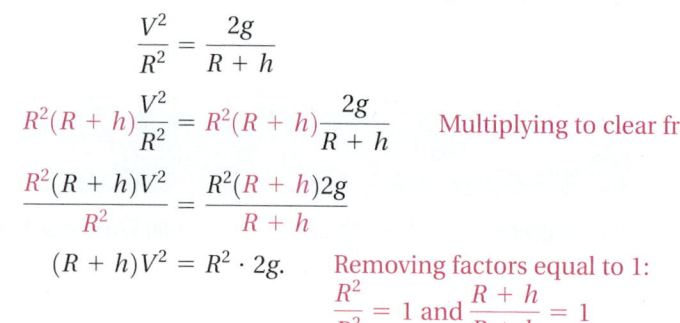

EXAMPLE 3 *Acoustics (the Doppler Effect).* The formula

$$f = \frac{sg}{s + v}$$

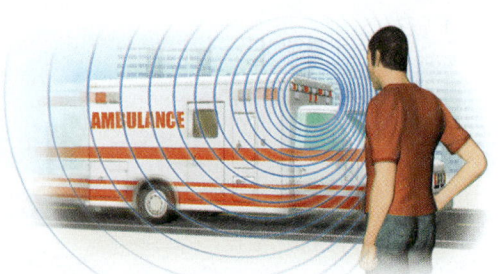

is used to determine the frequency f of a sound that is moving at velocity v toward a listener who hears the sound as frequency g. Here s is the speed of sound in a particular medium. Solve for s.

Student Notes

The steps used to solve equations are precisely the same steps used to solve formulas. If you feel "rusty" in this regard, study the earlier section in which this type of equation first appears. When you can consistently solve those equations, you are ready to work with formulas.

3. Solve

$$w = \frac{tv}{v + x}$$

for v.

SOLUTION We first clear fractions by multiplying by the LCM, $s + v$:

$$f \cdot (s + v) = \frac{sg}{s + v}(s + v)$$

$$fs + fv = sg. \qquad \text{The variable for which we are solving, } s, \text{ appears on both sides, forcing us to distribute on the left side.}$$

Next, we must get all terms containing s on one side:

$$fv = sg - fs \qquad \text{Subtracting } fs \text{ from both sides}$$

$$fv = s(g - f) \qquad \text{Factoring out } s. \text{ This is like combining like terms.}$$

$$\frac{fv}{g - f} = s. \qquad \text{Dividing both sides by } g - f$$

Since s is isolated on one side, we have solved for s. This last equation can be used to determine the speed of sound whenever f, v, and g are known.

YOUR TURN

TO SOLVE A RATIONAL EQUATION FOR A SPECIFIED VARIABLE

1. Multiply both sides by the LCM of the denominators to clear fractions, if necessary.
2. Multiply to remove parentheses, if necessary.
3. Get all terms with the specified variable alone on one side.
4. Factor out the specified variable if it is in more than one term.
5. Multiply or divide on both sides to isolate the specified variable.

Variation

We now examine three real-world situations: direct variation, inverse variation, and combined variation.

DIRECT VARIATION

A fitness trainer earns $22 per hour. In 1 hr, $22 is earned. In 2 hr, $44 is earned. In 3 hr, $66 is earned, and so on. This gives rise to a set of ordered pairs:

$$(1, 22), (2, 44), (3, 66), (4, 88), \quad \text{and so on.}$$

Note that the ratio of earnings E to time t is $\frac{22}{1}$ in every case.

If a situation is modeled by pairs for which the ratio is constant, we say there is **direct variation**. Here earnings *vary directly* as the time:

We have $\dfrac{E}{t} = 22$, so $E = 22t$ or, using function notation, $E(t) = 22t$.

DIRECT VARIATION

When a situation is modeled by a linear function of the form $f(x) = kx$, or $y = kx$, where k is a nonzero constant, we say that there is *direct variation*, that *y varies directly* as *x*, or that *y is proportional to x*. The number k is called the *variation constant*, or the *constant of proportionality*.

Note that for $k > 0$, any equation of the form $y = kx$ indicates that as x increases, y increases as well.

EXAMPLE 4 Find the variation constant and an equation of variation if y varies directly as x, and $y = 32$ when $x = 2$.

SOLUTION We know that $(2, 32)$ is a solution of $y = kx$. Therefore,

$$32 = k \cdot 2 \qquad \text{Substituting}$$

$$\frac{32}{2} = k, \quad \text{or} \quad k = 16. \qquad \text{Solving for } k$$

The variation constant is 16. The equation of variation is $y = 16x$. The notation $y(x) = 16x$ or $f(x) = 16x$ is also used.

4. Find the variation constant and an equation of variation if y varies directly as x, and $y = 5$ when $x = 10$.

 YOUR TURN

EXAMPLE 5 *Ocean Waves.* The speed v of a train of ocean waves varies directly as the swell period t, the time between successive waves. Waves with a swell period of 12 sec are traveling 21 mph. How fast are waves traveling that have a swell period of 20 sec?

Source: www.rodntube.com

SOLUTION

1. **Familiarize.** Because of the phrase "v ... varies directly as ... t," we express the speed of the wave v, in miles per hour, as a function of the swell period t, in seconds. Thus, $v(t) = kt$, where k is the variation constant. Because we are using ratios, we can use the units "seconds" and "miles per hour" without converting sec to hr or hr to sec. Knowing that waves with a swell period of 12 sec are traveling 21 mph, we have $v(12) = 21$.

2. **Translate.** We find the variation constant using the data and then use it to write the equation of variation:

$$v(t) = kt$$

$$v(12) = k \cdot 12 \qquad \text{Replacing } t \text{ with 12}$$

$$21 = k \cdot 12 \qquad \text{Replacing } v(12) \text{ with 21}$$

$$\frac{21}{12} = k \qquad \text{Solving for } k$$

$$1.75 = k. \qquad \text{This is the variation constant.}$$

The equation of variation is $v(t) = 1.75t$. This is the translation.

3. **Carry out.** To find the speed of waves with a swell period of 20 sec, we compute $v(20)$:

$$v(t) = 1.75t$$

$$v(20) = 1.75(20) \qquad \text{Substituting 20 for } t$$

$$= 35.$$

4. **Check.** To check, we could reexamine all our calculations. Note that our answer seems reasonable since the ratios $21/12$ and $35/20$ are both 1.75.

5. **State.** Waves with a swell period of 20 sec are traveling 35 mph.

5. The amount of vegetables produced varies directly as the amount spent on seeds and fertilizer. According to a recent survey, an investment of $25 for seeds and fertilizer can produce vegetables worth $625 at a grocery store. What is the market value of vegetables produced with a $40 investment in seeds and fertilizer?

Source: W. Atlee Burpee & Co.

 YOUR TURN

INVERSE VARIATION

Suppose a bus travels 20 mi. At 20 mph, the trip takes 1 hr. At 40 mph, it takes $\frac{1}{2}$ hr. At 60 mph, it takes $\frac{1}{3}$ hr, and so on. This gives pairs of numbers, all having the same product:

$$(20, 1), \left(40, \tfrac{1}{2}\right), \left(60, \tfrac{1}{3}\right), \left(80, \tfrac{1}{4}\right), \quad \text{and so on.}$$

Note that the product of each pair is 20. When a situation is modeled by pairs for which the product is constant, we say that there is **inverse variation**. Since $r \cdot t = 20$, we have

$$t = \frac{20}{r} \quad \text{or, using function notation,} \quad t(r) = \frac{20}{r}.$$

> ### INVERSE VARIATION
>
> When a situation is modeled by a rational function of the form $f(x) = k/x$, or $y = k/x$, where k is a nonzero constant, we say that there is *inverse variation*, that *y varies inversely as x*, or that *y is inversely proportional to x*. The number k is called the *variation constant*, or the *constant of proportionality*.

Note that for $k > 0$, any equation of the form $y = k/x$ indicates that as x increases, y decreases.

EXAMPLE 6   Find the variation constant and an equation of variation if y varies inversely as x, and $y = 32$ when $x = 0.2$.

SOLUTION   We know that $(0.2, 32)$ is a solution of

$$y = \frac{k}{x}.$$

Therefore,

$$32 = \frac{k}{0.2} \qquad \text{Substituting}$$

$$(0.2)32 = k \qquad \text{Multiplying both sides by 0.2}$$

$$6.4 = k. \qquad \text{Solving for } k$$

The variation constant is 6.4. The equation of variation is

$$y = \frac{6.4}{x}.$$

6. Find the variation constant and an equation of variation if y varies inversely as x, and $y = \frac{1}{2}$ when $x = 10$.

 YOUR TURN

There are many real-life quantities that vary inversely.

EXAMPLE 7   *Fuel Efficiency.*   The number of gallons of fuel that a vehicle uses varies inversely as its fuel efficiency. Maria's 2011 Ford Escape gets 25 mpg (miles per gallon) in combined city and highway driving. Last year, her vehicle used 480 gal of gasoline. How many gallons of gasoline would she have used if she had driven a 2012 Ford Focus, which gets 30 mpg, instead?

Source: www.fueleconomy.gov

99. *Intensity of Light.*   The intensity I of light from a bulb varies directly as the wattage of the bulb and inversely as the square of the distance d from the bulb. If the wattage of a light source and its distance from reading matter are both doubled, how does the intensity change?

100. Describe in words the variation represented by

$W = \dfrac{km_1M_1}{d^2}$. Assume k is a constant.

101. *Tension of a Musical String.*   The tension T on a string in a musical instrument varies jointly as the string's mass per unit length m, the square of its length l, and the square of its fundamental frequency f. A 2-m long string of mass 5 gm/m with a fundamental frequency of 80 has a tension of 100 N (Newtons). How long should the same string be if its tension is going to be changed to 72 N?

102. *Volume and Cost.*   A peanut butter jar in the shape of a right circular cylinder is 4 in. high and 3 in. in diameter and sells for $2.40. If we assume that cost is proportional to volume, how much should a jar 6 in. high and 6 in. in diameter cost?

103. *Golf Distance Finder.*   A device used in golf to estimate the distance d to a hole measures the size s that the 7-ft pin *appears* to be in a view-finder. The viewfinder uses the principle, dia-grammed here, that s gets bigger when d gets smaller. If $s = 0.56$ in. when $d = 50$ yd, find an equation of variation that expresses d as a function of s. What is d when $s = 0.40$ in.?

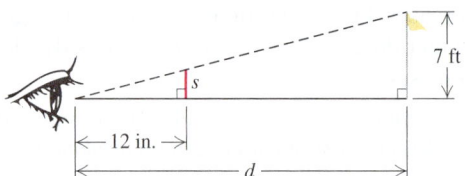

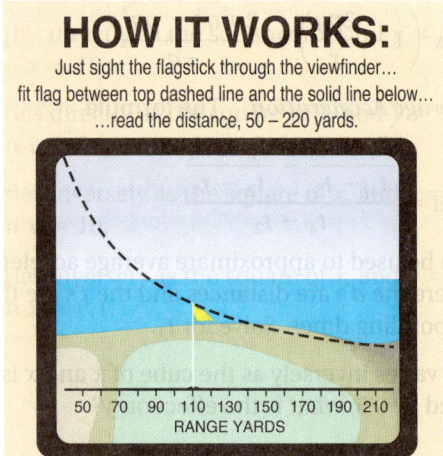

HOW IT WORKS:
Just sight the flagstick through the viewfinder…
fit flag between top dashed line and the solid line below…
…read the distance, 50 – 220 yards.

QUICK QUIZ: SECTIONS 7.1–7.5

1. Determine whether this graph is that of a function.   [7.1]

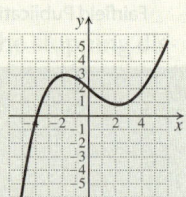

Determine the domain and the range of each function. [7.2]

2.

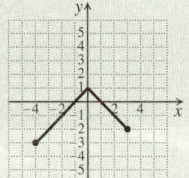

3.

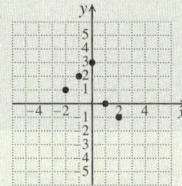

4. Graph: $f(x) = |x| - 1$.   [7.4]

5. The number of kilograms W of water in a human body varies directly as the mass of the body. A 96-kg person contains 64 kg of water. How many kilograms of water are in a 60-kg person?   [7.5]

PREPARE TO MOVE ON
Solve.   [2.3]

1. $x - 6y = 3$, for y

2. $3x - 8y = 5$, for y

3. $5x + 2y = -3$, for y

Translate each of the following. Do not solve.   [1.1]

4. Three less than half of some number is 57.

5. The sum of two consecutive integers is 145.

6. The difference between a number and its opposite is 20.

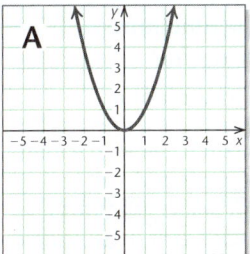

A

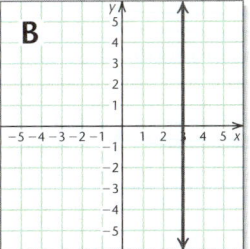

B

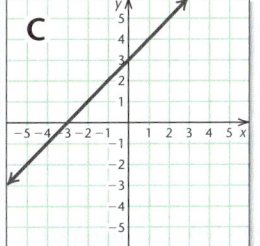

C

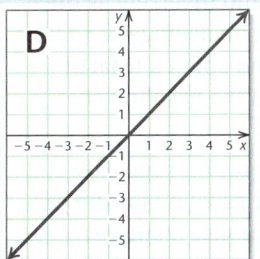

D

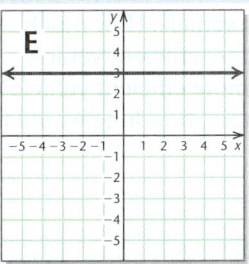

E

Visualizing for Success

Use after Section 7.3.

Match each equation or function with its graph.

1. $f(x) = x$

2. $f(x) = |x|$

3. $f(x) = x^2$

4. $f(x) = 3$

5. $x = 3$

6. $f(x) = x + 3$

7. $f(x) = x - 3$

8. $f(x) = 2x$

9. $f(x) = -2x$

10. $f(x) = \dfrac{1}{x}$

Answers on page A-35

An alternate, animated version of this activity appears in MyMathLab. To use MyMathLab, you need a course ID and a student access code. Contact your instructor for more information.

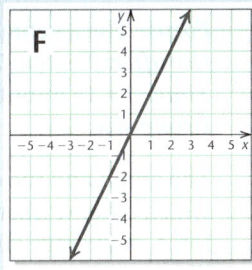

F

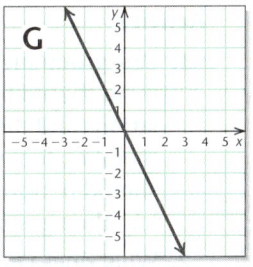

G

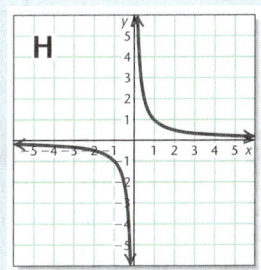

H

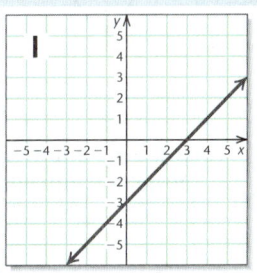

I

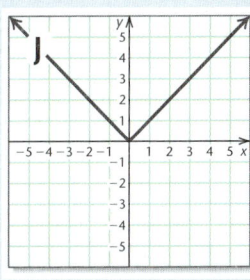

J

For each of the graphs in Exercises 15–18, **(a)** *determine whether the graph represents a function and* **(b)** *if so, determine the domain and the range of the function.*   [7.1], [7.2]

15.

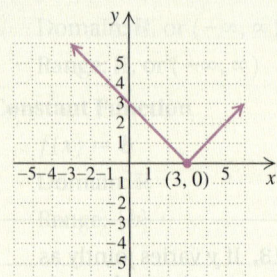

16.

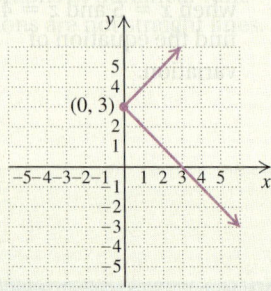

17.

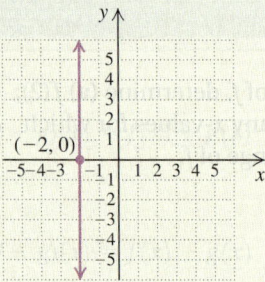

18.

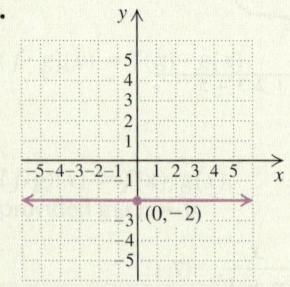

Find the domain of each function.

19. $f(x) = 3x^2 - 7$ [7.2]

20. $g(x) = \dfrac{x^2}{x - 1}$ [7.2]

21. $f(t) = \dfrac{1}{t^2 + 5t + 4}$ [7.2]

22. If a service agreement is cancelled, the amount that Vale Appliances will refund on the agreement is given by the function

$$r(t) = 900 - 15t,$$

where t is the number of weeks since the date of purchase. What is the domain of the function?   [7.2]

23. For the function given by

$$f(x) = \begin{cases} 2 - x, & \text{if } x \le -2, \\ x^2, & \text{if } -2 < x \le 5, \\ x + 10, & \text{if } x > 5, \end{cases}$$

find **(a)** $f(-3)$; **(b)** $f(-2)$; **(c)** $f(4)$; and **(d)** $f(25)$.   [7.2]

24. It costs \$90 plus \$30 per month to join the Family Fitness Center. Formulate a linear function to model the cost $C(t)$ for t months of membership, and determine the time required for the cost to reach \$300.   [7.3]

25. *Records in the 200-meter Run.*   In 1983, the record for the 200-m run was 19.75 sec.* In 2011, it was 19.19 sec. Let $R(t)$ represent the record in the 200-m run t years after 1980.   [7.3]

   Source: International Association of Athletics Federation

   a) Find a linear function that fits the data.
   b) Use the function of part (a) to predict the record in 2015 and in 2020.

Classify each function as a linear function, an absolute-value function, a quadratic function, another polynomial function, or a rational function.   [7.3]

26. $f(x) = |3x - 7|$

27. $g(x) = 4x^5 - 8x^3 + 7$

28. $p(x) = x^2 + x - 10$

29. $h(n) = 4n - 17$ **30.** $s(t) = \dfrac{t + 1}{t + 2}$

---

*Records are for elevations less than 1000 m.

Graph each function and determine its domain and range.   [7.3]

31. $f(x) = 3$

32. $f(x) = 2x + 1$

33. $g(x) = |x + 1|$

Let $g(x) = 3x - 6$ and $h(x) = x^2 + 1$. Find the following.

34. $(g \cdot h)(4)$ [7.4]

35. $(g - h)(-2)$ [7.4]

36. $(g/h)(-1)$ [7.4]

37. The domains of $g + h$ and $g \cdot h$ [7.4]

38. The domain of h/g [7.4]

Solve.   [7.5]

39. $I = \dfrac{2V}{R + 2r}$, for r

40. $S = \dfrac{H}{m(t_1 - t_2)}$, for m

41. $\dfrac{1}{ac} = \dfrac{2}{ab} - \dfrac{3}{bc}$, for c

42. $T = \dfrac{A}{v(t_2 - t_1)}$, for t_1

43. Find an equation of variation in which y varies directly as x, and $y = 30$ when $x = 4$.   [7.5]

44. Find an equation of variation in which y varies inversely as x, and $y = 3$ when $x = \frac{1}{4}$.   [7.5]

45. Find an equation of variation in which y varies jointly as x and the square of w and inversely as z, and $y = 150$ when $x = 6$, $w = 10$, and $z = 2$.   [7.5]

Solve.   [7.5]

46. For a triangle with a fixed area, the base of the triangle varies inversely as the height. If the base of a triangle with area A is 8 cm when the height is 10 cm, what is the base when the height is 4 cm?

47. *Electrical Safety.*   The amount of time t needed for an electrical shock to stop a 150-lb person's heart varies inversely as the square of the current flowing through the body. It is known that a 0.089-amp current is deadly to a 150-lb person after 3.4 sec. How long would it take a 0.096-amp current to be deadly?

Source: Safety Consulting Services

48. The number of centimeters W of water produced from melting snow varies directly as the number of centimeters S of snow. Meterologists know that under certain conditions, 150 cm of snow will melt to 16.8 cm of water. The average annual snowfall in Alta, Utah, is about 500 in. Assuming the above conditions, how much water will replace the 500 in. of snow?

Synthesis

49. If two functions have the same domain and range, are the functions identical? Why or why not?   [7.2]

50. Jenna believes that 0 is never in the domain of a rational function. Is she correct? Why or why not?   [7.2]

51. Treasure Tea charges $7.99 for each package of loose tea. Shipping charges are $2.95 per package plus $20 per order for overnight delivery. Find a linear function for determining the cost of one order of x packages of tea, including shipping and overnight delivery.   [7.3]

52. Determine the domain and the range of the function graphed below.   [7.2]

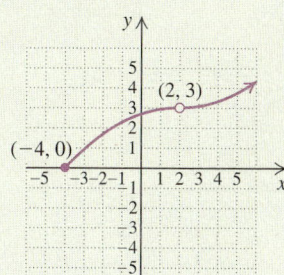

For step-by-step test solutions, access the Chapter Test Prep Videos in MyMathLab®, on YouTube (search "Bittinger Combo Alg CA" and click on "Channels"), or by scanning the code.

Test: Chapter 7

1. For the following graph of f, determine **(a)** $f(-2)$; **(b)** the domain of f; **(c)** any x-value for which $f(x) = \frac{1}{2}$; and **(d)** the range of f.

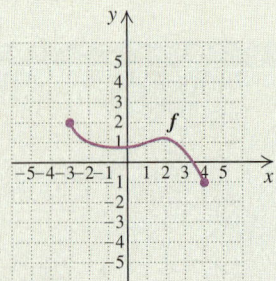

*For each of the following graphs, **(a)** determine whether the graph represents a function and **(b)** if so, determine the domain and the range of the function.*

2.

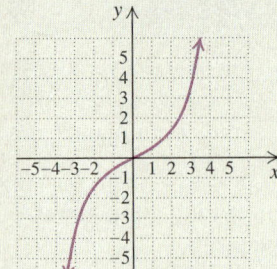

3.

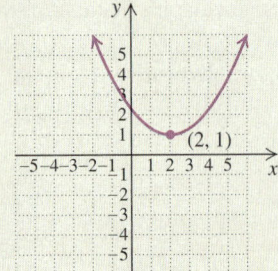

4.

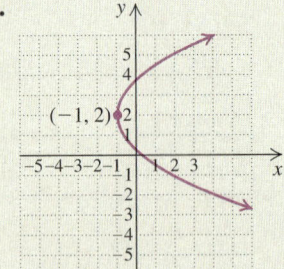

5. The distance d, in miles, that Kerry is from Chicago is given by the function $d(t) = 240 - 60t$, where t is the number of hours since he left Indianapolis. What is the domain of the function?

6. For the function given by
$$f(x) = \begin{cases} x^2, & \text{if } x < 0, \\ 3x - 5, & \text{if } 0 \le x \le 2, \\ x + 7, & \text{if } x > 2, \end{cases}$$
find **(a)** $f(0)$ and **(b)** $f(3)$.

7. Porter paid \$180 for his phone. His monthly service fee is \$55. Formulate a linear function to model the cost $C(t)$ for t months of service, and determine the amount of time required for the total cost to reach \$840.

8. If you rent a truck for one day and drive it 250 mi, the cost is \$100. If you rent it for one day and drive it 300 mi, the cost is \$115. Let $C(m)$ represent the cost, in dollars, of driving m miles.

   a) Find a linear function that fits the data.
   b) Use the function to determine how much it will cost to rent the truck for one day and drive it 500 mi.

Classify each function as a linear function, a quadratic function, another polynomial function, an absolute-value function, or a rational function. Then find the domain of each function.

9. $f(x) = \frac{1}{4}x + 7$

10. $g(x) = \dfrac{3}{x^2 - 16}$

11. $p(x) = 4x^2 + 7$

Graph each function and determine its domain and range.

12. $f(x) = \frac{1}{3}x - 2$

13. $g(x) = x^2 - 1$

14. $h(x) = -\frac{1}{2}$

Find the following, given that $g(x) = \dfrac{1}{x}$ and $h(x) = 2x + 1$.

15. $g(-1)$

16. $h(5a)$

17. $(g + h)(x)$

18. The domain of g

19. The domain of $g + h$

20. The domain of g/h

21. Solve $R = \dfrac{gs}{g + s}$ for s.

22. Find an equation of variation in which y varies directly as x, and $y = 10$ when $x = 20$.

23. The number of workers n needed to clean a stadium after a game varies inversely as the amount of time t allowed for the cleanup. If it takes 25 workers to clean the stadium when there are 6 hr allowed for the job, how many workers are needed if the stadium must be cleaned in 5 hr?

24. The surface area of a balloon varies directly as the square of its radius. The area is 325 in^2 when the radius is 5 in. What is the area when the radius is 7 in.?

Synthesis

25. The function $f(t) = 5 + 15t$ can be used to determine a bicycle racer's location, in miles from the starting line, measured t hours after passing the 5-mi mark.

   a) How far from the start will the racer be 1 hr and 40 min after passing the 5-mi mark?

   b) Assuming a constant rate, how fast is the racer traveling?

26. Given that $f(x) = 5x^2 + 1$ and $g(x) = 4x - 3$, find an expression for $h(x)$ so that the domain of $f/g/h$ is

$$\left\{ x \mid x \text{ is a real number } and \ x \neq \tfrac{3}{4} \ and \ x \neq \tfrac{2}{7} \right\}.$$

Answers may vary.

Cumulative Review: Chapters 1–7

1. Convert to scientific notation: 391,000,000.   [4.2]

2. Determine the slope and the y-intercept for the line given by $7x - 4y = 12$.   [3.6]

3. Find an equation for the line that passes through the points $(-1, 7)$ and $(4, -3)$.   [3.7]

4. If
$$f(x) = \frac{x - 3}{x^2 - 11x + 30},$$
find **(a)** $f(3)$ and **(b)** the domain of f.   [7.1], [7.2]

Graph on a plane.

5. $5x = y$ [3.2]

6. $8y + 2x = 16$ [3.3]

7. $f(x) = -4$ [7.2]

8. $y = \frac{1}{3}x - 2$ [3.6]

Perform the indicated operations and simplify.

9. $(5x^2 - 2x + 1)(3x^2 + x - 2)$ [4.5]

10. $(3x^2 + y)^2$ [4.6]

11. $(2x^2 - 9)(2x^2 + 9)$ [4.6]

12. $(-5m^3n^2 - 3mn^3) + (-4m^2n^2 + 4m^3n^2) - (2mn^3 - 3m^2n^2)$ [4.4]

13. $\dfrac{x^4 - 1}{x^2 - x - 2} \div \dfrac{x^2 + 1}{x - 2}$ [6.2]

14. $\dfrac{5ab}{a^2 - b^2} + \dfrac{a + b}{a - b}$ [6.4]

Factor.

15. $4x^3 + 400x$ [5.1]

16. $x^2 + 8x - 84$ [5.2]

17. $16y^2 - 25$ [5.4]

18. $64x^3 + 8$ [5.5]

19. $t^2 - 16t + 64$ [5.4]

20. $3t^2 + 17t - 28$ [5.3]

Solve.

21. $8x = 1 + 16x^2$ [5.7]

22. $288 = 2y^2$ [5.7]

23. $\frac{1}{3}x - \frac{1}{5} \geq \frac{1}{5}x - \frac{1}{3}$ [2.6]

24. $5(x - 2) - (x - 3) = 7x - 2(5 - x)$ [2.2]

25. $\dfrac{6}{x - 5} = \dfrac{2}{2x}$ [6.6]

26. $P = \dfrac{4a}{a + b}$, for a [7.5]

27. Find the slope of the line containing $(2, 5)$ and $(1, 10)$.   [3.5]

28. Find an equation of the line containing $(5, -2)$ and perpendicular to the line given by $x - y = 5$.   [3.7]

Find the following, given that $f(x) = x + 5$ and $g(x) = x^2 - 1$.

29. $g(-10)$ [7.1]

30. $(g/f)(x)$ [7.4]

31. A rectangular quilted wall hanging is 4 in. longer than it is wide. The area of the quilt is 320 in². Find the perimeter of the quilt.   [5.8]

32. The IQAir HealthPro Plus air purifier can clean the air in a 20-ft by 25-ft meeting room in 5 fewer minutes than it takes the Austin Healthmate HM400 to do the same job. Together the two machines can purify the air in the room in 6 min. How long would it take each machine, working alone, to purify the air in the room?   [6.7]

Source: Manufacturers' and retailers' websites

33. The time t that it takes for Johann to drive to work varies inversely as his speed. On a day when Johann averages 45 mph, it takes him 20 min to drive to work. How long will it take him to drive to work when he averages only 40 mph?   [7.5]

Synthesis

34. Multiply: $(x - 4)^3$.   [4.5]

35. Find all roots for $f(x) = x^4 - 34x^2 + 225$.   [5.6], [7.3]

36. An Everyday Photo Book costs $12 for the first 20 pages plus $0.75 for each additional page. Formulate a piecewise-defined function to model the cost $C(x)$ for a book with x pages.   [7.2]

Systems of Linear Equations and Problem Solving

SERVING SIZE	CALORIES	PROTEIN (in grams)	VITAMIN C (in milligrams)
Roast beef, 3 oz	300	20	0
Baked potato, 1	100	5	20
Broccoli, 156 g	50	5	100
Asparagus, 180 g	50	5	44

What's Good for You Ought to Be Good to You.

Dieticians and nutritionists face the daily challenge of making nutritious meals both attractive and appetizing. Using nutrition information such as that in the table above, they design meal plans that meet individual needs and tastes.

In this chapter, we design two simple meals with equivalent requirements using the foods listed here. You may find that you would prefer eating one of these meals over the other. *(See Exercises 23 and 24 in Section 8.5.)*

In the field of nutrition and dietetics, the use of math is absolutely essential.

Anju Agarwal, a nutritionist in Huntersville, North Carolina, uses math to interpret food labels, to analyze the nutrition composition of foods, and to calculate body mass index (BMI), energy needs, energy expenditure, and nutritional intake.

In fields ranging from business to zoology, problems arise that are most easily solved using a *system of equations*. In this chapter, we solve systems and applications using graphing, substitution, elimination, and matrices.

8.1 Systems of Equations in Two Variables

Translating ▪ Identifying Solutions ▪ Solving Systems Graphically

The whooping crane was one of the first animals on the U.S. endangered list. Conservation efforts have increased the number of whooping cranes from 16 in 1941 to over 500 in 2012.

1. Refer to Example 1. In 2012, there were 25 species of amphibians in the United States that were considered threatened or endangered. The number of species considered endangered was 1.5 times the number considered threatened. Write a system of equations that models the number of amphibian species considered endangered or threatened in 2012.

Source: U.S. Fish and Wildlife Service

Translating

Problems involving two unknown quantities are often translated most easily using two equations in two unknowns. Together, these equations form a **system of equations**. We look for a solution to the problem by attempting to find a pair of numbers for which *both* equations are true.

EXAMPLE 1 *Endangered Species.* In 2012, there were 92 species of birds in the United States that were considered threatened (likely to become endangered) or endangered (in danger of becoming extinct). The number of species considered threatened was 3 less than one-fourth of the number considered endangered. Write a system of equations that models the number of U.S. bird species considered endangered or threatened in 2012.

Source: U.S. Fish and Wildlife Service

SOLUTION

1. **Familiarize.** We let t represent the number of threatened bird species and d the number of endangered bird species in 2012.

2. **Translate.** There are two statements to translate. First, we look at the total number of endangered or threatened species of birds:

Rewording: The number of threatened species plus the number of endangered species was 92.

Translating: t $+$ d $=$ 92

The second statement compares the two amounts, d and t:

Rewording: The number of threatened species was 3 less than one-fourth of the number of endangered species.

Translating: t $=$ $\frac{1}{4}d - 3$

We have now translated the problem to a pair, or **system**, **of equations**:

$$t + d = 92,$$
$$t = \tfrac{1}{4}d - 3.$$

 YOUR TURN

> ## SYSTEM OF EQUATIONS
>
> A *system of equations* is a set of two or more equations, in two or more variables, for which a common solution is sought.

Problems like Example 1 *can* be solved using one variable; however, as problems become complicated, you will find that using more than one variable (and more than one equation) is often the preferable approach.

EXAMPLE 2 *Jewelry Design.* Star Bright Jewelry Design purchased 80 beads for a total of $39 (excluding tax) to make a necklace. Some of the beads were sterling silver beads costing 40¢ each and the rest were gemstone beads costing 65¢ each. How many of each type were bought? Translate to a system of equations.

SOLUTION

1. **Familiarize.** To familiarize ourselves with this problem, let's guess that the designer bought 20 beads at 40¢ each and 60 beads at 65¢ each. The total cost would then be

$$20 \cdot 40¢ + 60 \cdot 65¢ = 800¢ + 3900¢, \quad \text{or} \quad 4700¢.$$

Since 4700¢ = $47 and $47 ≠ $39, our guess is incorrect. Rather than guess again, let's see how algebra can be used to translate the problem.

2. **Translate.** We let s = the number of silver beads and g = the number of gemstone beads. Since the cost of each bead is given in cents and the total cost is in dollars, we must choose one of the units to use throughout the problem. We choose to work in cents, so the total cost is 3900¢. The information can be organized in a table, which will help with the translating.

Type of Bead	Silver	Gemstone	Total	
Number Bought	s	g	80	→ $s + g = 80$
Price	40¢	65¢		
Amount	$40s$¢	$65g$¢	3900¢	→ $40s + 65g = 3900$

The first row of the table and the first sentence of the problem indicate that a total of 80 beads were bought:

$$s + g = 80.$$

Since each silver bead cost 40¢ and s of them were bought, $40s$ cents was paid for the silver beads. Similarly, $65g$ cents was paid for the gemstone beads. This leads to a second equation:

$$40s + 65g = 3900.$$

We now have the following system of equations as the translation:

$$s + g = 80,$$
$$40s + 65g = 3900.$$

2. Refer to Example 2. For another necklace, Star Bright Jewelry Design purchased 60 sterling silver beads and gemstone beads for a total of $30. How many of each type did the designer buy? Translate to a system of equations.

 YOUR TURN

Student Notes

We complete the solutions of Examples 1 and 2 in Section 8.3.

> **CAUTION!** Be sure to check the ordered pair in *both* equations.

3. Determine whether $(6, -1)$ is a solution of the system

$$x + y = 5,$$
$$x - 3y = 3.$$

Identifying Solutions

A *solution* of a system of two equations in two variables is an ordered pair of numbers that makes *both* equations true.

EXAMPLE 3 Determine whether $(-4, 7)$ is a solution of the system

$$x + y = 3,$$
$$5x - y = -27.$$

SOLUTION Unless stated otherwise, we use alphabetical order of the variables. Thus we replace x with -4 and y with 7:

$$\frac{x + y = 3}{-4 + 7 3}$$
$$3 \overset{?}{=} 3 \quad \text{TRUE}$$

$$\frac{5x - y = -27}{5(-4) - 7 -27}$$
$$-20 - 7$$
$$-27 \overset{?}{=} -27 \quad \text{TRUE}$$

The pair $(-4, 7)$ makes both equations true, so it is a solution of the system. We can also describe the solution by writing $x = -4$ and $y = 7$. Set notation can also be used to list the solution set $\{(-4, 7)\}$.

YOUR TURN

Solving Systems Graphically

Recall that the graph of an equation is a drawing that represents its solution set. If we graph the equations in Example 3, we find that $(-4, 7)$ is the only point common to both lines. Thus one way to solve a system of two equations is to graph both equations and identify any points of intersection. **The coordinates of each point of intersection represent a solution of that system.**

$$x + y = 3,$$
$$5x - y = -27$$

The point of intersection of the graphs is $(-4, 7)$.

The solution of the system is $(-4, 7)$.

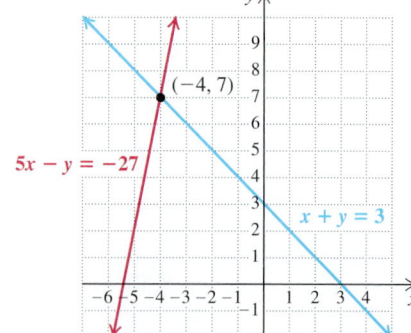

Most pairs of lines have exactly one point in common. We will soon see, however, that this is not always the case.

EXAMPLE 4 Solve each system graphically.

a) $y - x = 1,$
$\, y + x = 3$

b) $y = -3x + 5,$
$\, y = -3x - 2$

c) $3y - 2x = 6,$
$\, -12y + 8x = -24$

SOLUTION

a) We begin by graphing the equations. All ordered pairs from line L_1 are solutions of the first equation. All ordered pairs from line L_2 are solutions of the second equation. The point of intersection has coordinates that make *both* equations true. Apparently, $(1, 2)$ is the solution. Graphs are not always accurate, so solving by graphing may yield approximate answers. Our check below shows that $(1, 2)$ is indeed the solution.

Graph both equations.

Look for any points in common.

$$y - x = 1,$$
$$y + x = 3$$

The solution of the system is $(1, 2)$.

Check.

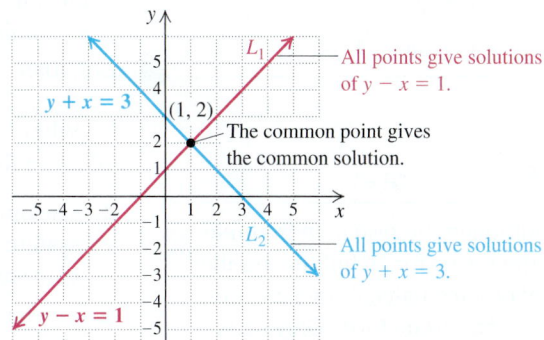

All points give solutions of $y - x = 1$.

The common point gives the common solution.

All points give solutions of $y + x = 3$.

Check:

$$\begin{array}{c|c} y - x = 1 \\ \hline 2 - 1 & 1 \\ 1 \stackrel{?}{=} 1 & \text{TRUE} \end{array}$$

$$\begin{array}{c|c} y + x = 3 \\ \hline 2 + 1 & 3 \\ 3 \stackrel{?}{=} 3 & \text{TRUE} \end{array}$$

b) We graph the equations. The lines have the same slope, -3, and different y-intercepts, so they are parallel. There is no point at which they cross, so the system has no solution.

$$y = -3x + 5,$$
$$y = -3x - 2$$

The system has no solution.

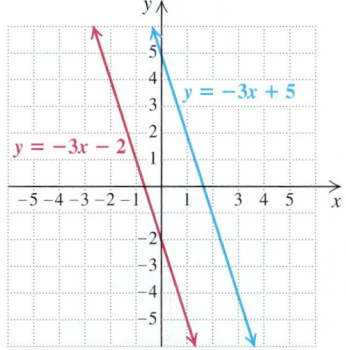

$y = -3x + 5$

$y = -3x - 2$

c) We graph the equations and find that the same line is drawn twice. Thus any solution of one equation is a solution of the other. Each point on the line is a solution of both equations, so the system itself has an infinite number of solutions. We check one solution, $(0, 2)$, which is the y-intercept of each equation.

$$3y - 2x = 6,$$
$$-12y + 8x = -24$$

The solution of the system is $\{(x, y) \mid 3y - 2x = 6\}$.

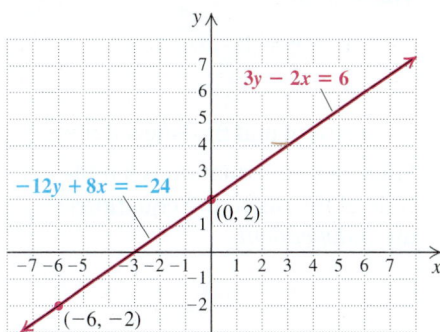

$3y - 2x = 6$

$-12y + 8x = -24$

$(0, 2)$

$(-6, -2)$

Check:

$$\begin{array}{c|c} 3y - 2x = 6 \\ \hline 3(2) - 2(0) & 6 \\ 6 - 0 & \\ 6 \stackrel{?}{=} 6 & \text{TRUE} \end{array}$$

$$\begin{array}{c|c} -12y + 8x = -24 \\ \hline -12(2) + 8(0) & -24 \\ -24 + 0 & \\ -24 \stackrel{?}{=} -24 & \text{TRUE} \end{array}$$

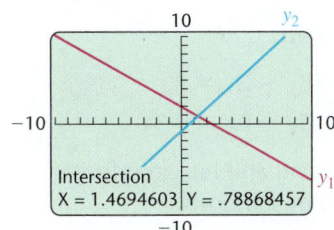

4. Solve graphically:

$$x = y,$$
$$y = 2x - 4.$$

You can check that $(-6, -2)$ is another solution of both equations. In fact, any pair that is a solution of one equation is a solution of the other equation as well. Thus the solution set is $\{(x, y) \mid 3y - 2x = 6\}$ or, in words, "the set of all pairs (x, y) for which $3y - 2x = 6$." Since the two equations are equivalent, we could have written instead $\{(x, y) \mid -12y + 8x = -24\}$.

 YOUR TURN

Student Notes

Although the system in Example 4(c) is true for an infinite number of ordered pairs, those pairs must be of a certain form. Only pairs that are solutions of $3y - 2x = 6$ or $-12y + 8x = -24$ are solutions of the system. It is incorrect to think that *all* ordered pairs are solutions.

When we graph a system of two linear equations in two variables, one of the following three outcomes will occur.

1. The lines have one point in common, and that point is the only solution of the system (see Example 4a). Any system that has *at least one solution* is said to be **consistent**.

2. The lines are parallel, with no point in common, and the system has *no solution* (see Example 4b). This type of system is called **inconsistent**.

3. The lines coincide, sharing the same graph. Because every solution of one equation is a solution of the other, the system has an infinite number of solutions (see Example 4c). Since it has at least one solution, this type of system is also **consistent**.

When one equation in a system can be obtained by multiplying both sides of another equation by a constant, the two equations are said to be **dependent**. Thus the equations in Example 4(c) are dependent, but those in Examples 4(a) and 4(b) are **independent**. For systems of three or more equations, the definitions of dependent and independent will be slightly modified.

Algebraic 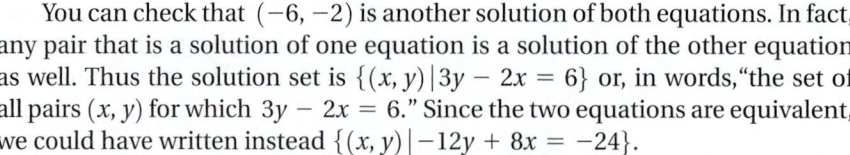 Graphical Connection

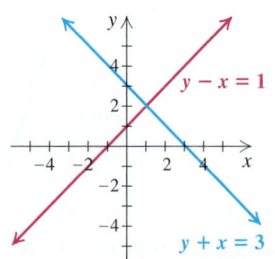

Graphs intersect at one point.

The system

$$y - x = 1,$$
$$y + x = 3$$

is *consistent* and has one solution.

Since neither equation is a multiple of the other, the equations are *independent*.

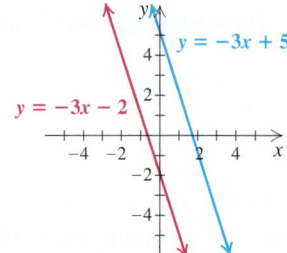

Graphs are parallel.

The system

$$y = -3x - 2,$$
$$y = -3x + 5$$

is *inconsistent* because there is no solution.

Since neither equation is a multiple of the other, the equations are *independent*.

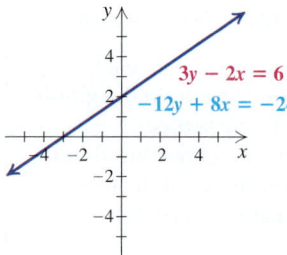

Equations have the same graph.

The system

$$3y - 2x = 6,$$
$$-12y + 8x = -24$$

is *consistent* and has an infinite number of solutions.

Since one equation is a multiple of the other, the equations are *dependent*.

 Chapter Resource:
Visualizing for Success, p. 565

Graphing is helpful when solving systems because it allows us to "see" the solution. It can also be used on systems of nonlinear equations, and in many applications, it provides a satisfactory answer. However, graphing often lacks precision, especially when fraction solutions or decimal solutions are involved.

8.1 EXERCISE SET

FOR EXTRA HELP

MyMathLab® MathⓍL

PRACTICE   WATCH   READ   REVIEW

Vocabulary and Reading Check

Classify each of the following statements as either true or false.

1. Every system of equations has at least one solution.

2. It is possible for a system of equations to have an infinite number of solutions.

3. Every point of intersection of the graphs of the equations in a system corresponds to a solution of the system.

4. The graphs of the equations in a system of two equations may coincide.

5. The graphs of the equations in a system of two equations could be parallel lines.

6. Any system of equations that has at most one solution is said to be consistent.

7. Any system of equations that has more than one solution is said to be inconsistent.

8. The equations $x + y = 5$ and $2(x + y) = 2(5)$ are dependent.

Identifying Solutions

Determine whether the ordered pair is a solution of the given system of equations. Remember to use alphabetical order of variables.

9. $(2, 3)$; $2x - y = 1,$
$\quad 5x - 3y = 1$

10. $(4, 0)$; $2x + 7y = 8,$
$\quad x - 9y = 4$

11. $(-5, 1)$; $x + 5y = 0,$
$\quad y = 2x + 9$

12. $(-1, -2)$; $x + 3y = -7,$
$\quad 3x - 2y = 12$

13. $(0, -5)$; $x - y = 5,$
$\quad y = 3x - 5$

14. $(5, 2)$; $a + b = 7,$
$\quad 2a - 8 = b$

Aha! 15. $(3, -1)$; $3x - 4y = 13,$
$\quad 6x - 8y = 26$

16. $(4, -2)$; $-3x - 2y = -8,$
$\quad 8 = 3x + 2y$

Solving Systems Graphically

Solve each system graphically. Be sure to check your solution. If a system has an infinite number of solutions, use set-builder notation to write the solution set. If a system has no solution, state this.

17. $x - y = 1,$
$\quad x + y = 5$

18. $x + y = 6,$
$\quad x - y = 4$

19. $3x + y = 5,$
$\quad x - 2y = 4$

20. $2x - y = 4,$
$\quad 5x - y = 13$

21. $2y = 3x + 5,$
$\quad x = y - 3$

22. $4x - y = 9,$
$\quad x - 3y = 16$

23. $x = y - 1,$
$\quad 2x = 3y$

24. $a = 1 + b,$
$\quad b = 5 - 2a$

25. $y = -1,$
$\quad x = 3$

26. $y = 2,$
$\quad x = -4$

27. $t + 2s = -1,$
$\quad s = t + 10$

28. $b + 2a = 2,$
$\quad a = -3 - b$

29. $2b + a = 11,$
$\quad a - b = 5$

30. $y = -\frac{1}{3}x - 1,$
$\quad 4x - 3y = 18$

31. $y = -\frac{1}{4}x + 1,$
$\quad 2y = x - 4$

32. $6x - 2y = 2,$
$\quad 9x - 3y = 1$

33. $y - x = 5,$
$\quad 2x - 2y = 10$

34. $y = x + 2,$
$\quad 3y - 2x = 4$

35. $y = 3 - x,$
$\quad 2x + 2y = 6$

36. $2x - 3y = 6,$
$\quad 3y - 2x = -6$

37. For the systems in the odd-numbered exercises 17–35, which are consistent?

38. For the systems in the even-numbered exercises 18–36, which are consistent?

39. For the systems in the odd-numbered exercises 17–35, which contain dependent equations?

40. For the systems in the even-numbered exercises 18–36, which contain dependent equations?

Translating

Translate each problem situation to a system of equations. Do not attempt to solve, but save for later use.

41. The sum of two numbers is 10. The first number is $\frac{2}{3}$ of the second number. What are the numbers?

42. The sum of two numbers is 30. The first number is twice the second number. What are the numbers?

43. *e-Mail Usage.*   In 2012, the average e-mail user sent or received 115 e-mails each day. The number of e-mails received was 1 more than twice the number sent. How many were sent and how many were received each day?

Source: The Radicati Group, Inc.

44. *Nontoxic Furniture Polish.*   A nontoxic wood furniture polish can be made by mixing mineral (or olive) oil with vinegar. To make a 16-oz batch for a squirt bottle, Jazmun uses an amount of mineral oil that is 4 oz more than twice the amount of vinegar. How much of each ingredient is required?

Sources: Based on information from Chittenden Solid Waste District and *Clean House, Clean Planet* by Karen Logan

45. *Geometry.*   Two angles are supplementary.* One angle is 3° less than twice the other. Find the measures of the angles.

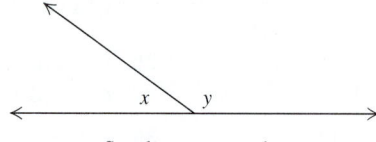

Supplementary angles

46. *Geometry.*   Two angles are complementary.† The sum of the measures of the first angle and half the second angle is 64°. Find the measures of the angles.

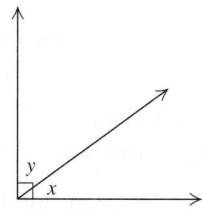

Complementary angles

47. *Basketball Scoring.*   Wilt Chamberlain once scored 100 points, setting a record for points scored in an NBA game. Chamberlain took only two-point shots and (one-point) foul shots and made a total of 64 shots. How many shots of each type did he make?

48. *Basketball Scoring.*   The Fenton College Cougars made 40 field goals in a recent basketball game, some 2-pointers and the rest 3-pointers. Altogether, the 40 baskets counted for 89 points. How many of each type of field goal was made?

---

*The sum of the measures of two supplementary angles is 180°.
†The sum of the measures of two complementary angles is 90°.

49. *Retail Sales.*   Simply Souvenirs sold 45 hats and tee shirts. The hats sold for $14.50 each and the tee shirts for $19.50 each. In all, $697.50 was taken in for the souvenirs. How many of each type of souvenir were sold?

50. *Retail Sales.*   Cool Treats sold 60 ice cream cones. Single-dip cones sold for $2.50 each and double-dip cones for $4.15 each. In all, $179.70 was taken in for the cones. How many of each size cone were sold?

51. *Fitness Equipment.*   In 2012, the Mio® Classic Heart Rate watch cost $39.99 and the Mio® Drive Plus Heart Rate Watch cost $59.99. A nonprofit community health organization purchased 35 heart-rate watches for use at a wellness center. If the organization spent $1639.65 for the watches, how many of each type did they purchase?

52. *Fundraising.*   The Buck Creek Fire Department served 250 dinners. A child's plate cost $5.50 and an adult's plate cost $9.00. A total of $1935 was collected. How many of each type of plate were served?

53. *Lacrosse.*   The perimeter of an NCAA men's lacrosse field is 340 yd. The length is 50 yd longer than the width. Find the dimensions.

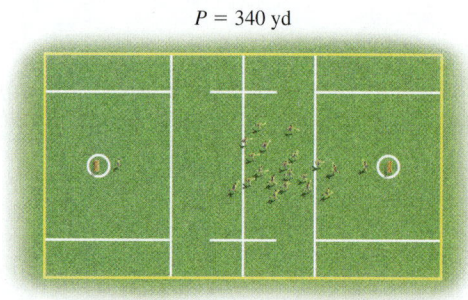

$P = 340$ yd

54. *Tennis.*   The perimeter of a standard tennis court used for doubles is 228 ft. The width is 42 ft less than the length. Find the dimensions.

55. Write a problem for a classmate to solve that requires writing a system of two equations. Devise the problem so that the solution is "The Bucks made 6 three-point baskets and 31 two-point baskets."

56. Write a problem for a classmate to solve that can be translated into a system of two equations. Devise the problem so that the solution is "In 2012, Diana took five 3-credit classes and two 4-credit classes."

Skill Review

Simplify. Do not use negative exponents in the answer.

57. $-\frac{1}{2} - \frac{3}{10}$ [1.6]

58. $(0.05)(-1.2)$ [1.7]

59. $(-3)^2 - 2 - 4 \cdot 6 \div 2 \cdot 3$ [1.8]

60. $(-3x^2y^{-4})(-2x^{-7}y^{12})$ [4.2]

61. -10^{-2} [4.2]

62. $(5a^2b^6)^0$ [4.1]

Synthesis

Advertising Media. For Exercises 63 and 64, consider the following graph showing the U.S. market share for various advertising media.

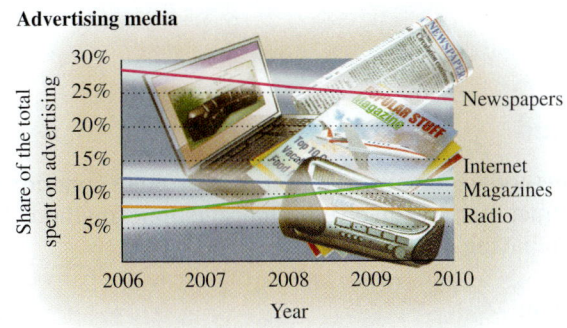

Source: ZenithOptimedia

63. In what year did the Internet and radio have the same advertising market share? Explain.

64. Will Internet advertising ever exceed that of newspapers? If so, when? Explain your answers.

65. For each of the following conditions, write a system of equations. Answers may vary.

   a) $(5, 1)$ is a solution.

   b) There is no solution.

   c) There is an infinite number of solutions.

66. A system of linear equations has $(1, -1)$ and $(-2, 3)$ as solutions. Determine:

   a) a third point that is a solution (answers may vary), and

   b) how many solutions there are.

67. The solution of the following system is $(4, -5)$. Find A and B.

$$Ax - 6y = 13,$$
$$x - By = -8.$$

Translate to a system of equations. Do not solve.

68. *Ages.* Tyler is twice as old as his son. Ten years ago, Tyler was three times as old as his son. How old are they now?

69. *Work Experience.* Dell and Juanita are mathematics professors at a state university. Together, they have 46 years of service. Two years ago, Dell had taught 2.5 times as many years as Juanita. How long has each taught at the university?

70. *Design.* A piece of posterboard has a perimeter of 156 in. If you cut 6 in. off the width, the length becomes four times the width. What are the dimensions of the original piece of posterboard?

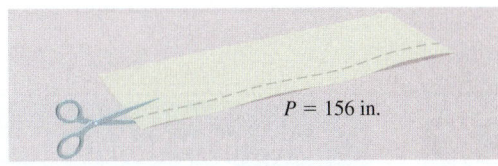

$P = 156$ in.

71. *Nontoxic Scouring Powder.* A nontoxic scouring powder is made up of 4 parts baking soda and 1 part vinegar. How much of each ingredient is needed for a 16-oz mixture?

72. Solve Exercise 41 graphically.

73. Solve Exercise 44 graphically.

Solve graphically.

74. $y = |x|,$
$3y - x = 8$

75. $x - y = 0,$
$y = x^2$

In Exercises 76–79, use a graphing calculator to solve each system of linear equations for x and y. Round all coordinates to the nearest hundredth.

76. $y = 8.23x + 2.11,$
$y = -9.11x - 4.66$

77. $y = -3.44x - 7.72,$
$y = 4.19x - 8.22$

78. $14.12x + 7.32y = 2.98,$
$21.88x - 6.45y = -7.22$

79. $5.22x - 8.21y = -10.21,$
$-12.67x + 10.34y = 12.84$

80. Solve graphically using the grid below:

$$2x - 3y = 0,$$
$$-4x + 3y = -1.$$

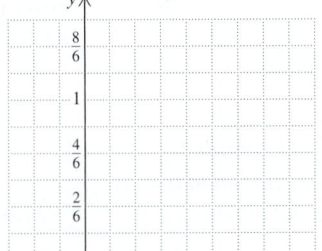

81. *Research.* Refer to Exercise 63. Find advertising market share for a recent year and extend the graph before Exercise 63 to include the new data. Do the 2006–2010 trends continue? If any new trends emerge, try to find reasons for the change.

PREPARE TO MOVE ON

Solve. [2.2]

1. $3x + 2(5x - 1) = 6$

2. $4(3y + 2) - 7y = 3$

3. $2x - (x - 7) = 18$

Solve. [2.3]

4. $3x - y = 4$, for y

5. $5y - 2x = 7$, for x

YOUR TURN ANSWERS: SECTION 8.1

1. Let t represent the number of threatened amphibian species and d the number of endangered amphibian species; $t + d = 25, d = 1.5t$
2. $s + g = 60, 40s + 65g = 3000$ **3.** No **4.** $(4, 4)$

8.2 Solving by Substitution or Elimination

The Substitution Method ▪ The Elimination Method

Study Skills

Learn from Your Mistakes

Immediately after each quiz or test, write out a step-by-step solution to any questions you missed. Visit your professor during office hours or consult a tutor for help with problems that still give you trouble. Misconceptions tend to persist if not corrected as soon as possible.

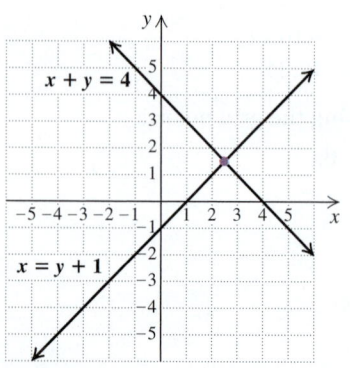

A visualization of Example 1. Note that the coordinates of the intersection are not obvious.

The Substitution Method

Algebraic (nongraphical) methods for solving systems are often superior to graphing, especially when fractions are involved. One algebraic method, the *substitution method*, relies on having a variable isolated.

EXAMPLE 1 Solve the system

$$x + y = 4, \quad (1)$$
$$x = y + 1. \quad (2)$$

For easy reference, we have numbered the equations.

SOLUTION Equation (2) says that x and $y + 1$ name the same number. Thus we can substitute $y + 1$ for x in equation (1):

$$x + y = 4 \qquad \text{Equation (1)}$$
$$(y + 1) + y = 4. \qquad \text{Substituting } y + 1 \text{ for } x$$

We solve this last equation, using methods learned earlier:

$$(y + 1) + y = 4$$
$$2y + 1 = 4 \qquad \text{Removing parentheses and combining like terms}$$
$$2y = 3 \qquad \text{Subtracting 1 from both sides}$$
$$y = \tfrac{3}{2}. \qquad \text{Dividing both sides by 2}$$

We now return to the original pair of equations and substitute $\tfrac{3}{2}$ for y in either equation so that we can solve for x. For this problem, calculations are slightly easier if we use equation (2):

$$x = y + 1 \qquad \text{Equation (2)}$$
$$= \tfrac{3}{2} + 1 \qquad \text{Substituting } \tfrac{3}{2} \text{ for } y$$
$$= \tfrac{3}{2} + \tfrac{2}{2} = \tfrac{5}{2}.$$

We obtain the ordered pair $\left(\tfrac{5}{2}, \tfrac{3}{2}\right)$. A check ensures that it is a solution.

1. Solve the system

$$x - y = 3,$$
$$x = 2 - y.$$

Check:

$x + y = 4$	
$\frac{5}{2} + \frac{3}{2}$	4
	$\frac{8}{2}$
4 $\overset{?}{=}$ 4	TRUE

$x = y + 1$	
$\frac{5}{2}$	$\frac{3}{2} + 1$
	$\frac{3}{2} + \frac{2}{2}$
$\frac{5}{2} \overset{?}{=} \frac{5}{2}$	TRUE

Since $\left(\frac{5}{2}, \frac{3}{2}\right)$ checks, it is the solution.

YOUR TURN

The exact solution to Example 1 is difficult to find graphically because it involves fractions. The graph shown serves as a partial check and provides a visualization of the problem.

If neither equation in a system has a variable alone on one side, we first isolate a variable in one equation and then substitute.

EXAMPLE 2   Solve the system

$$2x + y = 6, \qquad (1)$$
$$3x + 4y = 4. \qquad (2)$$

SOLUTION   First, we select an equation and solve for one variable. We can isolate y by subtracting $2x$ from both sides of equation (1):

$$2x + y = 6 \qquad (1)$$
$$y = 6 - 2x. \qquad (3) \qquad \text{Subtracting } 2x \text{ from both sides}$$

Next, we proceed as in Example 1, by substituting:

$$3x + 4(6 - 2x) = 4 \qquad \text{Substituting } 6 - 2x \text{ for } y \text{ in equation (2).}$$
$$\qquad\qquad\qquad\qquad\quad \text{Use parentheses!}$$
$$3x + 24 - 8x = 4 \qquad \text{Distributing to remove parentheses}$$
$$3x - 8x = 4 - 24 \qquad \text{Subtracting 24 from both sides}$$
$$-5x = -20$$
$$x = 4. \qquad \text{Dividing both sides by } -5$$

Next, we substitute 4 for x in equation (1), (2), or (3). It is easiest to use equation (3) because it has already been solved for y:

$$y = 6 - 2x$$
$$= 6 - 2(4)$$
$$= 6 - 8 = -2.$$

A visualization of Example 2

The pair $(4, -2)$ appears to be the solution. We check in equations (1) and (2).

Check:

$2x + y = 6$	
$2(4) + (-2)$	6
$8 - 2$	
6 $\overset{?}{=}$ 6	TRUE

$3x + 4y = 4$	
$3(4) + 4(-2)$	4
$12 - 8$	
4 $\overset{?}{=}$ 4	TRUE

2. Solve the system

$$x + 2y = 4,$$
$$2x + 3y = 1.$$

Since $(4, -2)$ checks, it is the solution.

YOUR TURN

For a system with no solution, the graphs of the equations do not intersect. How do we recognize such systems when solving by an algebraic method?

A visualization of Example 3

3. Solve the system

$$x + y = 1,$$
$$x + y = 2.$$

EXAMPLE 3 Solve the system

$$y = -3x + 5 \quad (1)$$
$$y = -3x - 2. \quad (2)$$

SOLUTION If we solve this system graphically, we see that the lines are parallel and the system has no solution. Let's now try to solve the system by substitution. Proceeding as in Example 1, we substitute $-3x - 2$ for y in the first equation:

$$-3x - 2 = -3x + 5 \qquad \text{Substituting } -3x - 2 \text{ for } y \text{ in equation (1)}$$
$$-2 = 5. \qquad \text{Adding } 3x \text{ to both sides; } -2 = 5 \text{ is a contradiction. The equation is always false.}$$

Since there is no solution of $-2 = 5$, there is no solution of the system. We state that there is no solution.

YOUR TURN

When solving a system algebraically yields a contradiction, the system has no solution. As we will see in Example 7, when solving a system of two equations algebraically yields an equation that is always true, the system has an infinite number of solutions.

The Elimination Method

The *elimination method* for solving systems of equations makes use of the *addition principle*: If $a = b$, then $a + c = b + c$.

EXAMPLE 4 Solve the system

$$2x - 3y = 0, \quad (1)$$
$$-4x + 3y = -1. \quad (2)$$

SOLUTION According to equation (2), $-4x + 3y$ and -1 are the same number. Thus we can use the addition principle and add $-4x + 3y$ to the left side of equation (1) and -1 to the right side:

$$2x - 3y = 0 \qquad (1)$$
$$\underline{-4x + 3y = -1} \qquad (2)$$
$$-2x + 0y = -1. \quad \text{Adding. Note that } y \text{ has been "eliminated."}$$

The resulting equation has just one variable, x, for which we solve:

$$-2x = -1$$
$$x = \tfrac{1}{2}.$$

A visualization of Example 4

4. Solve the system

$$-2x - 7y = 3,$$
$$6x + 7y = -2.$$

Next, we substitute $\tfrac{1}{2}$ for x in equation (1) and solve for y:

$$2 \cdot \tfrac{1}{2} - 3y = 0 \qquad \text{Substituting. We also could have used equation (2).}$$
$$1 - 3y = 0$$
$$-3y = -1, \text{ so } y = \tfrac{1}{3}.$$

Check:

$$\begin{array}{c|c}
\underline{2x - 3y = 0} & \\
2\left(\tfrac{1}{2}\right) - 3\left(\tfrac{1}{3}\right) \,\big|\, 0 & \\
1 - 1 \,\big| & \\
0 \stackrel{?}{=} 0 \quad \text{TRUE}
\end{array}
\qquad
\begin{array}{c|c}
\underline{-4x + 3y = -1} & \\
-4\left(\tfrac{1}{2}\right) + 3\left(\tfrac{1}{3}\right) \,\big|\, -1 & \\
-2 + 1 \,\big| & \\
-1 \stackrel{?}{=} -1 \quad \text{TRUE}
\end{array}$$

Since $\left(\tfrac{1}{2}, \tfrac{1}{3}\right)$ checks, it is the solution. See also the graph at left.

YOUR TURN

Adding in Example 4 eliminated the variable y because two terms, $-3y$ in equation (1) and $3y$ in equation (2), are opposites. For some systems, we must multiply before adding.

Student Notes

It is wise to double-check each step of your work as you go, rather than checking all steps at the end of a problem. One common error is to forget to multiply *both* sides of an equation when using the multiplication principle.

EXAMPLE 5 Solve the system

$$5x + 4y = 22, \quad (1)$$
$$-3x + 8y = 18. \quad (2)$$

SOLUTION If we add the left sides of the two equations, we will not eliminate a variable. However, if the $4y$ in equation (1) were changed to $-8y$, we would. To accomplish this change, we multiply both sides of equation (1) by -2:

$$-10x - 8y = -44 \qquad \text{Multiplying both sides of equation (1) by } -2$$
$$\underline{-3x + 8y = 18}$$
$$-13x + 0 = -26 \qquad \text{Adding}$$
$$x = 2. \qquad \text{Solving for } x$$

Then
$$-3 \cdot 2 + 8y = 18 \qquad \text{Substituting 2 for } x \text{ in equation (2)}$$
$$-6 + 8y = 18$$
$$\left.\begin{array}{r} 8y = 24 \\ y = 3. \end{array}\right\} \text{Solving for } y$$

We obtain $(2, 3)$, or $x = 2, y = 3$. We leave it to the student to confirm that this checks and is the solution.

5. Solve the system

$$2x - 3y = 8,$$
$$6x + 5y = 4.$$

🔁 YOUR TURN

Sometimes we must multiply twice in order to make two terms become opposites.

EXAMPLE 6 Solve the system

$$2x + 3y = 17, \quad (1)$$
$$5x + 7y = 29. \quad (2)$$

SOLUTION We multiply so that the x-terms will be eliminated when we add.

Multiply to get terms that are opposites.

$$2x + 3y = 17, \xrightarrow[\text{sides by 5}]{\text{Multiplying both}} 10x + 15y = 85$$

Solve for one variable.

$$5x + 7y = 29 \xrightarrow[\text{sides by } -2]{\text{Multiplying both}} \underline{-10x - 14y = -58}$$
$$0 + y = 27 \qquad \text{Adding}$$
$$y = 27$$

Next, we substitute to find x:

Substitute.

$$2x + 3 \cdot 27 = 17 \qquad \text{Substituting 27 for } y \text{ in equation (1)}$$

Solve for the other variable.

$$2x + 81 = 17$$

Check in both equations.

$$\left.\begin{array}{r} 2x = -64 \\ x = -32. \end{array}\right\} \text{Solving for } x$$

State the solution as an ordered pair.

Check:
$$\begin{array}{c|c} 2x + 3y = 17 \\ \hline 2(-32) + 3(27) & 17 \\ -64 + 81 & \\ 17 \overset{?}{=} 17 & \text{TRUE} \end{array} \qquad \begin{array}{c|c} 5x + 7y = 29 \\ \hline 5(-32) + 7(27) & 29 \\ -160 + 189 & \\ 29 \overset{?}{=} 29 & \text{TRUE} \end{array}$$

6. Solve the system

$$4x + 3y = 11,$$
$$3x + 2y = 7.$$

We obtain $(-32, 27)$, or $x = -32, y = 27$, as the solution.

🔁 YOUR TURN

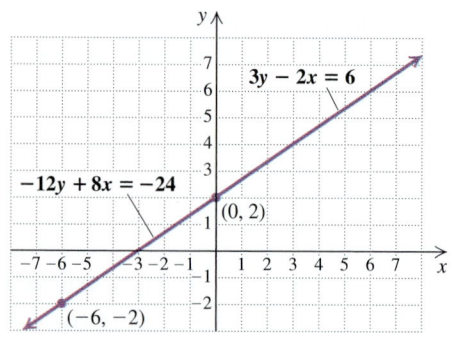

A visualization of Example 7

7. Solve the system

$$x - 3y = 2,$$
$$-5x + 15y = -10.$$

YOUR TURN

EXAMPLE 7 Solve the system

$$3y - 2x = 6, \qquad (1)$$
$$-12y + 8x = -24. \qquad (2)$$

SOLUTION If we solve this system graphically, as shown at left, we find that the lines coincide and the system has an infinite number of solutions. Suppose we were to solve this system using the elimination method:

$$12y - 8x = 24 \qquad \text{Multiplying both sides of equation (1) by 4}$$
$$\underline{-12y + 8x = -24}$$
$$0 = 0. \qquad \text{We obtain an identity; } 0 = 0 \text{ is always true.}$$

Note that both variables have been eliminated and what remains is an identity—that is, an equation that is always true. Any pair that is a solution of equation (1) is also a solution of equation (2). The equations are dependent and the solution set is infinite:

$$\{(x, y) \mid 3y - 2x = 6\}, \quad \text{or equivalently,} \quad \{(x, y) \mid -12y + 8x = -24\}.$$

Example 3 and Example 7 illustrate how to tell algebraically whether a system of two equations is inconsistent or whether the equations are dependent.

RULES FOR SPECIAL CASES

When solving a system of two linear equations in two variables:

1. If we obtain an identity such as $0 = 0$, then the system has an infinite number of solutions. The equations are dependent and, since a solution exists, the system is consistent.*
2. If we obtain a contradiction such as $0 = 7$, then the system has no solution. The system is inconsistent.

When decimals or fractions appear, it often helps to *clear* before solving.

EXAMPLE 8 Solve the system

$$0.2x + 0.3y = 1.7,$$
$$\tfrac{1}{7}x + \tfrac{1}{5}y = \tfrac{29}{35}.$$

SOLUTION We have

$$0.2x + 0.3y = 1.7, \;\rightarrow\; \text{Multiplying both sides by 10} \;\rightarrow\; 2x + 3y = 17$$
$$\tfrac{1}{7}x + \tfrac{1}{5}y = \tfrac{29}{35} \;\rightarrow\; \text{Multiplying both sides by 35} \;\rightarrow\; 5x + 7y = 29.$$

8. Solve the system

$$\tfrac{1}{2}x - \tfrac{1}{3}y = \tfrac{1}{6},$$
$$0.3x + 0.4y = 1.2.$$

We multiplied both sides of the first equation by 10 to clear the decimals. Multiplication by 35, the least common denominator, clears the fractions in the second equation. The problem now happens to be identical to Example 6. The solution is $(-32, 27)$, or $x = -32$, $y = 27$.

YOUR TURN

---

*Consistent systems and dependent equations are discussed in greater detail in Section 8.4.

The steps in each algebraic method for solving systems of two equations are given below. Note that in both methods, we find the value of one variable and then substitute to find the corresponding value of the other variable.

TO SOLVE A SYSTEM USING SUBSTITUTION

1. Isolate a variable in one of the equations (unless one is already isolated).
2. Substitute for that variable in the other equation, using parentheses.
3. Solve the equation in which the substitution was made.
4. Substitute the solution from step (3) in any of the equations, and solve for the other variable.
5. Form an ordered pair and check in the original equations.

Chapter Resource:
Collaborative Activity, p. 566

TO SOLVE A SYSTEM USING ELIMINATION

1. Write both equations in standard form.
2. Multiply both sides of one or both equations by a constant, if necessary, so that the coefficients of one of the variables are opposites.
3. Add the left sides and the right sides of the resulting equations. One variable should be eliminated in the sum.
4. Solve for the remaining variable.
5. Substitute the value of the second variable in any of the equations, and solve for the other variable.
6. Form an ordered pair and check in the original equations.

CONNECTING 🔗 THE CONCEPTS

We now have three different methods for solving systems of two linear equations. Each method has certain strengths and weaknesses, as outlined below.

Method	Strengths	Weaknesses
Graphical	Solutions are displayed graphically. Can be used with any system that can be graphed.	For some systems, only approximate solutions can be found graphically. The graph drawn may not be large enough to show the solution.
Substitution	Yields exact solutions. Easy to use when a variable has a coefficient of 1.	Introduces extensive computations with fractions when solving more complicated systems. Solutions are not displayed graphically.
Elimination	Yields exact solutions. Easy to use when fractions or decimals appear in the system.	Solutions are not displayed graphically.

(continued)

EXERCISES

Solve using an appropriate method.

1. $x = y$,
$x + y = 2$

2. $x + y = 10$,
$x - y = 8$

3. $y = \frac{1}{2}x + 1$,
$y = 2x - 5$

4. $y = 2x - 3$,
$x + y = 12$

5. $12x - 19y = 13$,
$8x + 19y = 7$

6. $2x - 5y = 1$,
$3x + 2y = 11$

7. $y = \frac{5}{3}x + 7$,
$y = \frac{5}{3}x - 8$

8. $x = 2 - y$,
$3x + 3y = 6$

8.2 EXERCISE SET

FOR EXTRA HELP

PRACTICE   WATCH   READ   REVIEW

↳ Vocabulary and Reading Check

Choose the word from the following list that best completes each statement. Not all words will be used.

consistent   opposite
elimination   substitution
inconsistent

1. To use the _____ method, a variable must be isolated.

2. The _____ method makes use of the addition principle.

3. To eliminate a variable by adding, two terms must be _____.

4. A(n) _____ system has no solution.

↳ Concept Reinforcement

In each of Exercises 5–10, match the system listed with the choice from the column on the right that would be a subsequent step in solving the system.

5. ____ $3x - 4y = 6$,
$5x + 4y = 1$

6. ____ $2x - y = 8$,
$y = 5x + 3$

7. ____ $x - 2y = 3$,
$5x + 3y = 4$

8. ____ $8x + 6y = -15$,
$5x - 3y = 8$

9. ____ $y = 4x - 7$,
$6x + 3y = 19$

10. ____ $y = 4x - 1$,
$y = -\frac{2}{3}x - 1$

a) $-5x + 10y = -15$,
$5x + 3y = 4$

b) The lines intersect at $(0, -1)$.

c) $6x + 3(4x - 7) = 19$

d) $8x = 7$

e) $2x - (5x + 3) = 8$

f) $8x + 6y = -15$,
$10x - 6y = 16$

For Exercises 11–58, if a system has an infinite number of solutions, use set-builder notation to write the solution set. If a system has no solution, state this.

The Substitution Method

Solve using the substitution method.

11. $y = 3 - 2x$,
$3x + y = 5$

12. $3y + x = 4$,
$x = 2y - 1$

13. $3x + 5y = 3$,
$x = 8 - 4y$

14. $9x - 2y = 3$,
$3x - 6 = y$

15. $3s - 4t = 14$,
$5s + t = 8$

16. $m - 2n = 16$,
$4m + n = 1$

17. $4x - 2y = 6$,
$2x - 3 = y$

18. $t = 4 - 2s$,
$t + 2s = 6$

19. $-5s + t = 11$,
$4s + 12t = 4$

20. $5x + 6y = 14$,
$-3y + x = 7$

21. $2x + 2y = 2$,
$3x - y = 1$

22. $4p - 2q = 16$,
$5p + 7q = 1$

23. $2a + 6b = 4$,
$3a - b = 6$

24. $3x - 4y = 5$,
$2x - y = 1$

25. $2x - 3 = y$,
$y - 2x = 1$

26. $a - 2b = 3$,
$3a = 6b + 9$

The Elimination Method

Solve using the elimination method.

27. $x + 3y = 7$,
$-x + 4y = 7$

28. $2x + y = 6$,
$x - y = 3$

29. $x - 2y = 11$,
$3x + 2y = 17$

30. $5x - 3y = 8$,
$-5x + y = 4$

31. $9x + 3y = -3$,
$2x - 3y = -8$

32. $6x - 3y = 18$,
$6x + 3y = -12$

33. $5x + 3y = 19,$
$x - 6y = 11$

34. $3x + 2y = 3,$
$9x - 8y = -2$

35. $5r - 3s = 24,$
$3r + 5s = 28$

36. $5x - 7y = -16,$
$2x + 8y = 26$

37. $6s + 9t = 12,$
$4s + 6t = 5$

38. $10a + 6b = 8,$
$5a + 3b = 2$

39. $\frac{1}{2}x - \frac{1}{6}y = 10,$
$\frac{2}{5}x + \frac{1}{2}y = 8$

40. $\frac{1}{3}x + \frac{1}{5}y = 7,$
$\frac{1}{6}x - \frac{2}{5}y = -4$

41. $\frac{x}{2} + \frac{y}{3} = \frac{7}{6},$
$\frac{2x}{3} + \frac{3y}{4} = \frac{5}{4}$

42. $\frac{2x}{3} + \frac{3y}{4} = \frac{11}{12},$
$\frac{x}{3} + \frac{7y}{18} = \frac{1}{2}$

Aha! **43.** $12x - 6y = -15,$
$-4x + 2y = 5$

44. $8s + 12t = 16,$
$6s + 9t = 12$

45. $0.3x + 0.2y = 0.3,$
$0.5x + 0.4y = 0.4$

46. $0.3x + 0.2y = 5,$
$0.5x + 0.4y = 11$

The Substitution and Elimination Methods

Solve using any algebraic method.

47. $a - 2b = 16,$
$b + 3 = 3a$

48. $5x - 9y = 7,$
$7y - 3x = -5$

49. $10x + y = 306,$
$10y + x = 90$

50. $3(a - b) = 15,$
$4a = b + 1$

51. $6x - 3y = 3,$
$4x - 2y = 2$

52. $x + 2y = 8,$
$x = 4 - 2y$

53. $3s - 7t = 5,$
$7t - 3s = 8$

54. $2s - 13t = 120,$
$-14s + 91t = -840$

55. $0.05x + 0.25y = 22,$
$0.15x + 0.05y = 24$

56. $2.1x - 0.9y = 15,$
$-1.4x + 0.6y = 10$

57. $13a - 7b = 9,$
$2a - 8b = 6$

58. $3a - 12b = 9,$
$4a - 5b = 3$

59. Describe a procedure that can be used to write an inconsistent system of equations.

60. Describe a procedure that can be used to write a system that has an infinite number of solutions.

Skill Review

61. Use an associative law to write an equation equivalent to $(4 + m) + n$.   [1.2]

62. Combine like terms: $a^2 - 4a - 3a^2 + 4a + 7$.   [1.8]

63. Simplify: $8x - 3[5x + 2(6 - 9x)]$.   [1.8]

64. Evaluate $-p^2$ for $p = -1$.   [1.8]

65. Convert 30,050,000 to scientific notation.   [4.2]

66. Convert 6.1×10^{-4} to decimal notation.   [4.2]

Synthesis

67. Some systems are more easily solved by substitution and some are more easily solved by elimination. What guidelines could be used to help someone determine which method to use?

68. Explain how it is possible to solve Exercise 43 mentally.

69. If $(1, 2)$ and $(-3, 4)$ are two solutions of $f(x) = mx + b$, find m and b.

70. If $(0, -3)$ and $\left(-\frac{3}{2}, 6\right)$ are two solutions of $px - qy = -1$, find p and q.

71. Determine a and b for which $(-4, -3)$ is a solution of the system
$$ax + by = -26,$$
$$bx - ay = 7.$$

72. Solve for x and y in terms of a and b:
$$5x + 2y = a,$$
$$x - y = b.$$

Solve.

73. $\frac{x + y}{2} - \frac{x - y}{5} = 1,$
$\frac{x - y}{2} + \frac{x + y}{6} = -2$

74. $3.5x - 2.1y = 106.2,$
$4.1x + 16.7y = -106.28$

Each of the following is a system of nonlinear equations. However, each is reducible to linear, *since an appropriate substitution (say, u for $1/x$ and v for $1/y$) yields a linear system. Make such a substitution, solve for the new variables, and then solve for the original variables.*

75. $\frac{2}{x} + \frac{1}{y} = 0,$
$\frac{5}{x} + \frac{2}{y} = -5$

76. $\frac{1}{x} - \frac{3}{y} = 2,$
$\frac{6}{x} + \frac{5}{y} = -34$

Solve using a system of equations.

77. *Energy Consumption.*   With average use, a toaster oven and a convection oven together consume 15 kilowatt hours (kWh) of electricity each month. A convection oven uses four times as much electricity as a toaster oven. How much does each use per month?

Source: Lee County Electric Cooperative

78. *Communication.*   Terri has two monthly bills: one for her cell phone and one for the data package for her tablet. The total of the two bills is $69.98 per month. Her cell-phone bill is $40 more per month than the bill for her tablet's data package. How much is each bill?

79. To solve the system

$$17x + 19y = 102,$$
$$136x + 152y = 826$$

Preston graphs both equations on a graphing calculator and gets the following screen. He then (incorrectly) concludes that the equations are dependent and the solution set is infinite. How can algebra be used to convince Preston that a mistake has been made?

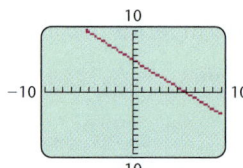

YOUR TURN ANSWERS: SECTION 8.2

1. $\left(\frac{5}{2}, -\frac{1}{2}\right)$ **2.** $(-10, 7)$ **3.** No solution   **4.** $\left(\frac{1}{4}, -\frac{1}{2}\right)$
5. $\left(\frac{13}{7}, -\frac{10}{7}\right)$ **6.** $(-1, 5)$ **7.** $\{(x, y) \mid x - 3y = 2\}$
8. $\left(\frac{14}{9}, \frac{11}{6}\right)$

QUICK QUIZ: SECTIONS 8.1–8.2

1. Determine whether $(4, -1)$ is a solution of

$$x + y = 3,$$
$$x - y = 5. \quad [8.1]$$

2. Solve graphically:

$$x + y = 4,$$
$$y = 2x - 5. \quad [8.1]$$

3. Solve using substitution:

$$3x - y = 1,$$
$$y = 2x - 4. \quad [8.2]$$

4. Solve using elimination:

$$x - y = 2,$$
$$2x + 3y = 1. \quad [8.2]$$

5. Solve using any appropriate method:

$$2x = 1 - 3y,$$
$$y - 3x = 0. \quad [8.1], [8.2]$$

PREPARE TO MOVE ON

Solve.   [2.5]

1. After her condominium had been on the market for 6 months, Gilena reduced the price to $94,500. This was $\frac{9}{10}$ of the original asking price. How much did Gilena originally ask for her condominium?

2. Ellia needs to average 80 on her tests in order to earn a B in her math class. Her average after four tests is 77.5. What score is needed on the fifth test in order to raise the average to 80?

3. North American Truck and Trailer rents vans for $53 per day plus 5¢ per mile. Anazi rented a van for 2 days. The bill was $120.50. How far did Anazi drive the van?

Source: www.northamericantrucktrailer.com

8.3 Solving Applications: Systems of Two Equations

Total-Value Problems and Mixture Problems • Motion Problems

You are in a much better position to solve problems now that you know how systems of equations can be used. Using systems often makes the translating step easier.

EXAMPLE 1 *Endangered Species.* In 2012, there were 92 species of birds in the United States that were considered threatened or endangered. The number considered threatened was 3 less than one-fourth of the number considered endangered. How many U.S. bird species were considered endangered or threatened in 2012?

Source: U.S. Fish and Wildlife Service

SOLUTION The *Familiarize* and *Translate* steps were completed in Example 1 of Section 8.1. The resulting system of equations is

$$t + d = 92,$$
$$t = \tfrac{1}{4}d - 3,$$

where d is the number of endangered bird species and t is the number of threatened bird species in the United States in 2012.

3. **Carry out.** We solve the system of equations. Since one equation already has a variable isolated, let's use the substitution method:

$$t + d = 92$$
$$\tfrac{1}{4}d - 3 + d = 92 \qquad \text{Substituting } \tfrac{1}{4}d - 3 \text{ for } t$$
$$\tfrac{5}{4}d - 3 = 92 \qquad \text{Combining like terms}$$
$$\tfrac{5}{4}d = 95 \qquad \text{Adding 3 to both sides}$$
$$d = \tfrac{4}{5} \cdot 95 \qquad \text{Multiplying both sides by } \tfrac{4}{5}: \tfrac{4}{5} \cdot \tfrac{5}{4} = 1$$
$$d = 76. \qquad \text{Simplifying}$$

Next, using either of the original equations, we substitute and solve for t:

$$t = \tfrac{1}{4} \cdot 76 - 3 = 19 - 3 = 16.$$

4. **Check.** The sum of 76 and 16 is 92, so the total number of species is correct. Since 3 less than one-fourth of 76 is $19 - 3$, or 16, the numbers check.

5. **State.** In 2012, there were 76 bird species considered endangered and 16 considered threatened.

 YOUR TURN

1. Refer to Example 1. In 2012, there were 25 species of amphibians in the United States that were considered threatened or endangered. The number of species considered endangered was 1.5 times the number considered threatened. How many U.S. amphibian species were considered endangered and how many were considered threatened in 2012?

Source: U.S. Fish and Wildlife Service

Total-Value Problems and Mixture Problems

EXAMPLE 2 *Jewelry Design.* In order to make a necklace, Star Bright Jewelry Design purchased 80 beads for a total of $39 (excluding tax). Some of the beads were sterling silver beads costing 40¢ each and the rest were gemstone beads costing 65¢ each. How many of each type were bought?

Student Notes

It is very important that you clearly label precisely what each variable represents. Not only will this help with writing equations, but it will help you identify and state solutions.

2. Refer to Example 2. For another necklace, the jewelry designer purchased 60 sterling silver beads and gemstone beads for a total of $30. How many of each type did the designer buy?

SOLUTION   The *Familiarize* and *Translate* steps were completed in Example 2 of Section 8.1.

3. Carry out. We are to solve the system of equations

$$s + g = 80, \qquad (1)$$
$$40s + 65g = 3900, \qquad (2) \qquad \text{Working in cents rather than dollars}$$

where s is the number of sterling silver beads bought and g is the number of gemstone beads bought. Because both equations are in the form $Ax + By = C$, let's use the elimination method to solve the system. We can eliminate s by multiplying both sides of equation (1) by -40 and adding them to the corresponding sides of equation (2):

$$
\begin{array}{rl}
-40s - 40g = -3200 & \text{Multiplying both sides of equation (1) by } -40 \\
\underline{40s + 65g = 3900} & \\
25g = 700 & \text{Adding} \\
g = 28. & \text{Solving for } g
\end{array}
$$

To find s, we substitute 28 for g in equation (1) and solve for s:

$$
\begin{array}{rl}
s + g = 80 & \text{Equation (1)} \\
s + 28 = 80 & \text{Substituting 28 for } g \\
s = 52. & \text{Solving for } s
\end{array}
$$

We obtain $(28, 52)$, or $g = 28$ and $s = 52$.

4. Check. We check in the original problem. Recall that g is the number of gemstone beads and s is the number of silver beads.

Number of beads: $g + s = 28 + 52 = 80$
Cost of gemstone beads: $65g = 65 \times 28 = 1820¢$
Cost of silver beads: $40s = 40 \times 52 = \underline{2080¢}$
                    $\text{Total} = 3900¢$

The numbers check.

5. State. The designer bought 28 gemstone beads and 52 sterling silver beads.

🔄 YOUR TURN

Example 2 involved two types of items (sterling silver beads and gemstone beads), the quantity of each type bought, and the total value of the items. We refer to this type of problem as a *total-value problem*.

EXAMPLE 3   *Blending Teas.* TeaPots n Treasures sells loose Oolong tea for $2.15 per ounce. Donna mixed Oolong tea with shaved almonds that sell for $0.95 per ounce to create the Market Street Oolong blend that sells for $1.85 per ounce. One week, she made 300 oz of Market Street Oolong. How much tea and how much shaved almonds did Donna use?

SOLUTION

1. **Familiarize.** Since we know the price per ounce of the blend, we can find the total value of the blend by multiplying 300 ounces times $1.85 per ounce. We let $l =$ the number of ounces of Oolong tea and $a =$ the number of ounces of shaved almonds.

2. **Translate.** Since a 300-oz batch was made, we must have

$$l + a = 300.$$

To find a second equation, note that the total value of the 300-oz blend must match the combined value of the separate ingredients:

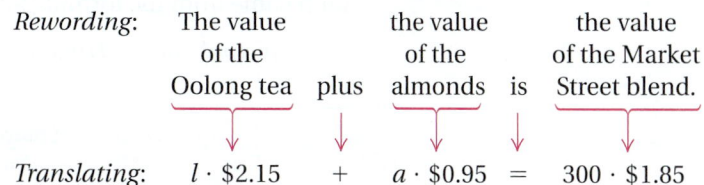

These equations can also be obtained from a table.

	Oolong Tea	Almonds	Market Street Blend	
Number of Ounces	l	a	300	→ $l + a = 300$
Price per Ounce	$2.15	$0.95	$1.85	
Value of Tea	$2.15l$	$0.95a$	$300 \cdot \$1.85$, or $555	→ $2.15l + 0.95a = 555$

Study Skills

Expect to be Challenged
Do not be surprised if your success rate drops some as you work on real-world problems. *This is normal.* Your success rate will increase as you gain experience with these types of problems and use some of the study skills already listed.

Clearing decimals in the second equation, we have $215l + 95a = 55{,}500.$ We have translated to a system of equations:

$$l + a = 300, \quad (1)$$
$$215l + 95a = 55{,}500. \quad (2)$$

3. Carry out. We can solve using substitution. When equation (1) is solved for l, we have $l = 300 - a$. Substituting $300 - a$ for l in equation (2), we find a:

$215(300 - a) + 95a = 55{,}500$	Substituting
$64{,}500 - 215a + 95a = 55{,}500$	Using the distributive law
$-120a = -9000$	Combining like terms; subtracting 64,500 from both sides
$a = 75.$	Dividing both sides by -120

We have $a = 75$ and, from equation (1) above, $l + a = 300$. Thus, $l = 225$.

3. Refer to Example 3. TeaPots n Treasures also sells loose rooibos tea for $1.50 per ounce. Donna mixed rooibos tea with shaved almonds to create the State Street Rooibos blend that sells for $1.39 per ounce. One week, she made 200 oz of State Street Rooibos. How much tea and how much shaved almonds did Donna use?

4. Check. Combining 225 oz of Oolong tea and 75 oz of almonds will give a 300-oz blend. The value of 225 oz of Oolong tea is $225(\$2.15)$, or $483.75. The value of 75 oz of almonds is $75(\$0.95)$, or $71.25. Thus the combined value of the blend is $483.75 + \$71.25$, or $555. A 300-oz blend priced at $1.85 per ounce would also be worth $555, so our answer checks.

5. State. The Market Street blend was made by combining 225 oz of Oolong tea and 75 oz of almonds.

 YOUR TURN

EXAMPLE 4 *Student Loans.* Rani's student loans totaled $9600. Part was a PLUS loan made at 7.9% interest, and the rest was a Stafford loan made at 6.8% interest. After one year, Rani's loans accumulated $702.30 in interest. What was the original amount of each loan?

SOLUTION

1. Familiarize. We begin with a guess. If $3000 was borrowed at 7.9% and $6600 was borrowed at 6.8%, the two loans would total $9600. The interest would then be $0.079(\$3000)$, or $237, and $0.068(\$6600)$, or $448.80, for a total of only $685.80 in interest. Our guess was wrong, but checking the guess familiarized us with the problem. More than $3000 was borrowed at the higher rate.

2. Translate. We let p = the amount of the PLUS loan and s = the amount of the Stafford loan. Next, we organize a table in which the entries in each column come from the formula for simple interest:

Principal · Rate · Time = Interest.

	PLUS Loan	Stafford Loan	Total
Principal	p	s	$9600	→ $p + s = 9600$
Rate of Interest	7.9%	6.8%	
Time	1 year	1 year	
Interest	$0.079p$	$0.068s$	$702.30	→ $0.079p + 0.068s = 702.30$

The total amount borrowed is found in the first row of the table:

$p + s = 9600.$

A second equation, representing accumulated interest, is found in the last row:

$0.079p + 0.068s = 702.30,$ or $79p + 68s = 702,300.$ Clearing decimals

3. Carry out. The system can be solved by elimination:

$p + s = 9600,$ → Multiplying both → $-79p - 79s = -758,400$
$79p + 68s = 702,300$ sides by -79

$\underline{ 79p + 68s = 702,300}$
$-11s = -56,100$

$p + s = 9600$ ←——— $s = 5100$
$p + 5100 = 9600$
$p = 4500.$

We find that $p = 4500$ and $s = 5100.$

4. Check. The total amount borrowed is $4500 + $5100, or $9600. The interest on $4500 at 7.9% for 1 year is 0.079($4500), or $355.50. The interest on $5100 at 6.8% for 1 year is 0.068($5100), or $346.80. The total amount of interest is $355.50 + $346.80, or $702.30, so the numbers check.

5. State. The PLUS loan was for $4500, and the Stafford loan was for $5100.

4. Refer to Example 4. Chin-Sun's student loans totaled $8400. Part was a PLUS loan at 7.9% interest, and the rest was a Stafford loan at 6.8% interest. After one year, Chin-Sun's loans accumulated $620.70 in interest. What was the original amount of each loan?

YOUR TURN

Before proceeding to Example 5, briefly scan Examples 2–4 for similarities. Note that in each case, one of the equations in the system is a simple sum while the other equation represents a sum of products. Example 5 continues this pattern with what is commonly called a *mixture problem.*

PROBLEM-SOLVING TIP

When solving a problem, see if it is patterned or modeled after a problem that you have already solved.

EXAMPLE 5 *Mixing Fertilizers.* Nature's Green Gardening, Inc., carries two brands of fertilizer containing nitrogen and water. "Gentle Grow" is 3% nitrogen and "Sun Saver" is 8% nitrogen. Nature's Green needs to combine the two types of solution into a 90-L mixture that is 6% nitrogen. How much of each brand should be used?

SOLUTION

1. **Familiarize.** We must consider not only the size of the mixture, but also its strength.

EXPLORING 🔍 THE CONCEPT

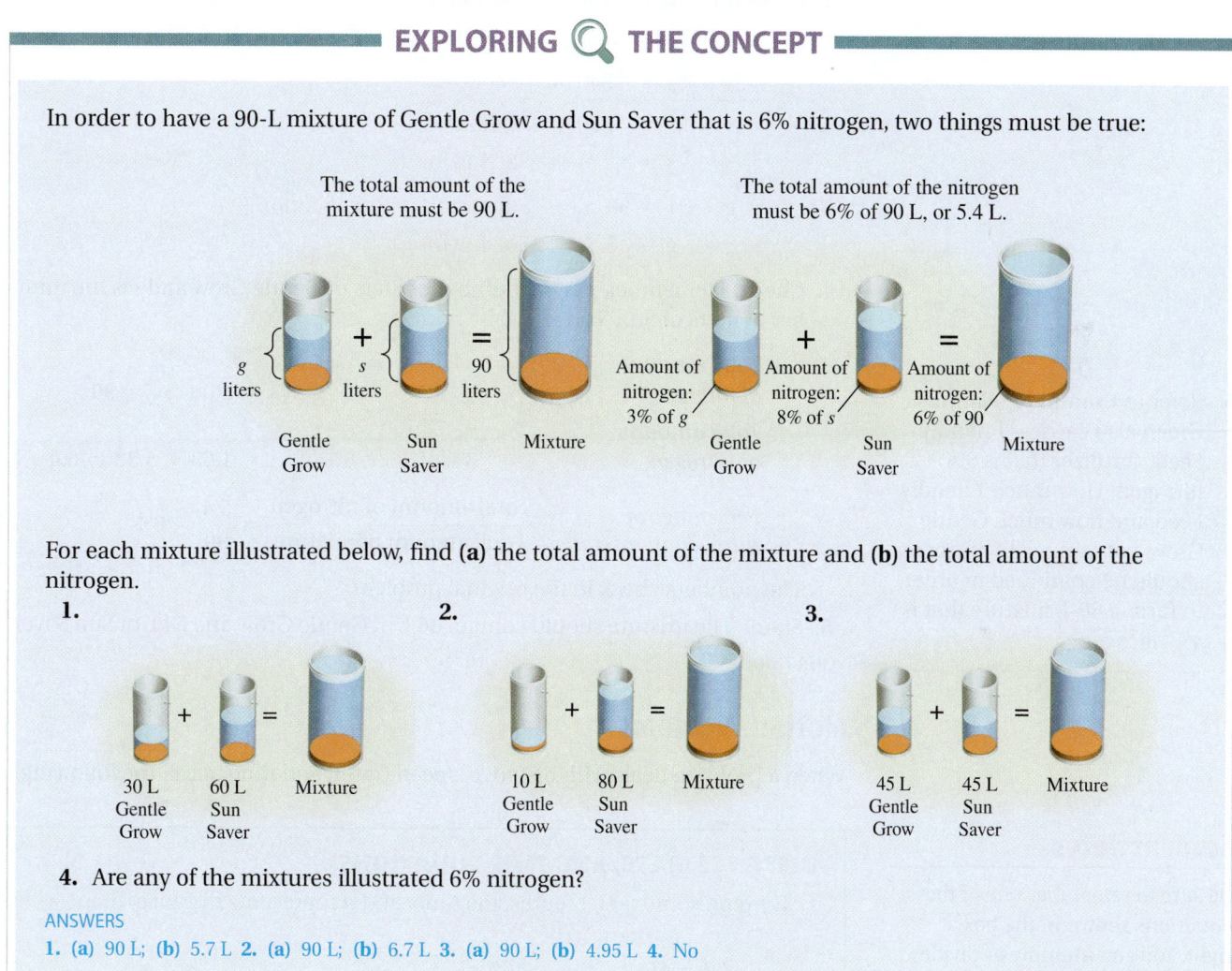

In order to have a 90-L mixture of Gentle Grow and Sun Saver that is 6% nitrogen, two things must be true:

The total amount of the mixture must be 90 L.

The total amount of the nitrogen must be 6% of 90 L, or 5.4 L.

For each mixture illustrated below, find **(a)** the total amount of the mixture and **(b)** the total amount of the nitrogen.

1.   **2.**   **3.**

30 L Gentle Grow   60 L Sun Saver   Mixture

10 L Gentle Grow   80 L Sun Saver   Mixture

45 L Gentle Grow   45 L Sun Saver   Mixture

4. Are any of the mixtures illustrated 6% nitrogen?

ANSWERS
1. (a) 90 L; **(b)** 5.7 L **2. (a)** 90 L; **(b)** 6.7 L **3. (a)** 90 L; **(b)** 4.95 L **4.** No

2. **Translate.** We let g = the number of liters of Gentle Grow and s = the number of liters of Sun Saver. The information can be organized in a table.

	Gentle Grow	Sun Saver	Mixture	
Number of Liters	g	s	90	→ $g + s = 90$
Percent of Nitrogen	3%	8%	6%	
Amount of Nitrogen	$0.03g$	$0.08s$	0.06×90, or 5.4 liters	→ $0.03g + 0.08s = 5.4$

If we add g and s in the first row, we get one equation. It represents the total amount of mixture: $g + s = 90$.

If we add the amounts of nitrogen listed in the third row, we get a second equation. This equation represents the amount of nitrogen in the mixture: $0.03g + 0.08s = 5.4$.

After clearing decimals, we have translated the problem to the system

$$g + \ s = 90, \qquad (1)$$
$$3g + 8s = 540. \qquad (2)$$

3. Carry out. We use the elimination method to solve the system:

$$
\begin{array}{ll}
-3g - 3s = -270 & \text{Multiplying both sides of equation (1) by } -3 \\
\underline{3g + 8s = 540} & \\
5s = 270 & \text{Adding} \\
s = 54; & \text{Solving for } s \\
g + 54 = 90 & \text{Substituting into equation (1)} \\
g = 36. & \text{Solving for } g
\end{array}
$$

4. Check. Remember, g is the number of liters of Gentle Grow and s is the number of liters of Sun Saver.

Total amount of mixture:	$g + s = 36 + 54 = 90$
Total amount of nitrogen:	3% of 36 + 8% of 54 = $1.08 + 4.32 = 5.4$
Percentage of nitrogen in mixture:	$\dfrac{\text{Total amount of nitrogen}}{\text{Total amount of mixture}} = \dfrac{5.4}{90} = 6\%$

The numbers check in the original problem.

5. State. The mixture should contain 36 L of Gentle Grow and 54 L of Sun Saver.

 YOUR TURN

5. Refer to Example 5. Nature's Green also carries "Friendly Feed" fertilizer that is 9% nitrogen. How much Friendly Feed and how much Gentle Grow, containing 3% nitrogen, should be combined in order to form a 60-L mixture that is 4% nitrogen?

Motion Problems

When a problem deals with distance, speed (rate), and time, recall the following.

> **DISTANCE, RATE, AND TIME EQUATIONS**
>
> If r represents rate, t represents time, and d represents distance, then:
>
> $$d = rt, \quad r = \frac{d}{t}, \quad \text{and} \quad t = \frac{d}{r}.$$

Student Notes

Be sure to remember one of the equations shown in the box at right. You can multiply or divide on both sides, as needed, to obtain the others.

EXAMPLE 6 *Train Travel.* A Vermont Railways freight train, loaded with logs, leaves Boston, heading to Washington D.C., at a speed of 60 km/h. Two hours later, an Amtrak® Metroliner leaves Boston, bound for Washington D.C., on a parallel track at 90 km/h. At what point will the Metroliner catch up to the freight train?

SOLUTION

1. Familiarize. Let's make a guess and check to see if it is correct. Suppose the trains meet after traveling 180 km. We can calculate the time for each train.

	Distance	Rate	Time
Freight Train	180 km	60 km/h	$\frac{180}{60} = 3$ hr
Metroliner	180 km	90 km/h	$\frac{180}{90} = 2$ hr

We see that the distance cannot be 180 km, since the difference in travel times for the trains is *not* 2 hr. Although our guess is wrong, we can use a similar chart to organize the information in this problem.

The distance at which the trains meet is unknown, but we do know that the trains will have traveled the same distance when they meet. We let d = this distance.

The time that the trains are running is also unknown, but we do know that the freight train has a 2-hr head start. Thus if we let t = the number of hours that the freight train is running before they meet, then $t - 2$ is the number of hours that the Metroliner runs before catching up to the freight train.

60 km/h
d kilometers
t hours

90 km/h
d kilometers
t − 2 hours

Trains meet here

2. **Translate.** We can organize the information in a chart. The formula *Distance = Rate · Time* guides our choice of rows and columns.

	Distance	Rate	Time
Freight Train	d	60	t	→ $d = 60t$
Metroliner	d	90	$t - 2$	→ $d = 90(t - 2)$

Using *Distance = Rate · Time* twice, we get two equations:

$$d = 60t, \qquad (1)$$
$$d = 90(t - 2). \qquad (2)$$

3. **Carry out.** We solve the system using substitution:

$\quad 60t = 90(t - 2)$ — Substituting $60t$ for d in equation (2)

$\quad 60t = 90t - 180$

$\quad -30t = -180$

$\quad t = 6.$

Student Notes

Always be careful to answer the question asked in the problem. In Example 6, the problem asks for distance, not time. Answering "6 hr" would be incorrect.

The time for the freight train is 6 hr, which means that the time for the Metroliner is $6 - 2$, or 4 hr. Remember that it is distance, not time, that the problem asked for. Thus for $t = 6$, we have $d = 60 \cdot 6 = 360$ km.

4. **Check.** At 60 km/h, the freight train travels $60 \cdot 6$, or 360 km, in 6 hr. At 90 km/h, the Metroliner travels $90 \cdot (6 - 2) = 360$ km in 4 hr. The numbers check.

5. **State.** The Metroliner catches up to the freight train 360 km from Boston.

6. An Amtrak® Metroliner traveling 90 mph leaves Washington, D.C., 3 hr after a freight train traveling 60 mph. If they travel on parallel tracks, at what point will the Metroliner catch up to the freight train?

 YOUR TURN

EXAMPLE 7 *Jet Travel.* A Boeing 747-400 jet flies 4 hr west with a 60-mph tailwind. Returning *against* the wind takes 5 hr. Find the speed of the jet with no wind.

SOLUTION

1. **Familiarize.** We imagine the situation and make a drawing. Note that the wind *speeds up* the outbound flight but *slows down* the return flight.

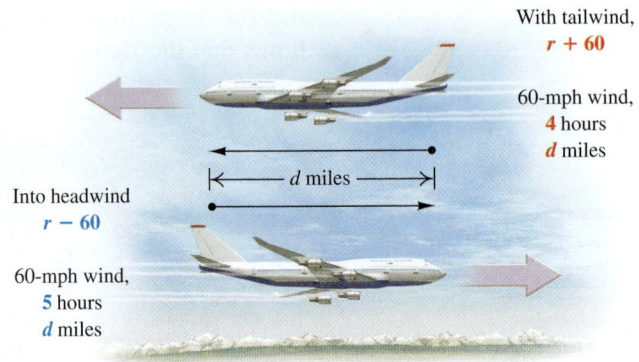

Let's make a guess of 400 mph for the jet's speed if there were no wind. Note that the distances traveled each way must be the same.

Speed with no wind:	400 mph
Speed with the wind:	400 + 60 = 460 mph
Speed against the wind:	400 − 60 = 340 mph
Distance with the wind:	460 · 4 = 1840 mi
Distance against the wind:	340 · 5 = 1700 mi

These must match.

Since the distances are not the same, our guess of 400 mph is incorrect.

We let r = the speed, in miles per hour, of the jet in still air. Then $r + 60$ = the jet's speed with the wind and $r - 60$ = the jet's speed against the wind. We also let d = the distance traveled, in miles.

2. **Translate.** The information can be organized in a chart. The distances traveled are the same, so we use *Distance = Rate* (or *Speed*) *· Time*. Each row of the chart gives an equation.

	Distance	Rate	Time	
With Wind	d	$r + 60$	4	→ $d = (r + 60)4$
Against Wind	d	$r - 60$	5	→ $d = (r - 60)5$

We now have a system of equations:

$$d = (r + 60)4, \quad (1)$$
$$d = (r - 60)5. \quad (2)$$

3. **Carry out.** We solve the system using substitution:

$(r - 60)5 = (r + 60)4$	Substituting $(r - 60)5$ for d in equation (1)
$5r - 300 = 4r + 240$	Using the distributive law
$r = 540.$	Solving for r

7. A motorboat travels 30 min upstream against a 4-mph current. Returning with the current takes 18 min. Find the speed of the motorboat in still water.

4. Check. When $r = 540$, the speed with the wind is $540 + 60 = 600$ mph, and the speed against the wind is $540 - 60 = 480$ mph. The distance with the wind, $600 \cdot 4 = 2400$ mi, matches the distance into the wind, $480 \cdot 5 = 2400$ mi, so we have a check.

5. State. The speed of the jet with no wind is 540 mph.

YOUR TURN

TIPS FOR SOLVING MOTION PROBLEMS

1. Draw a diagram using an arrow or arrows to represent distance and the direction of each object in motion.
2. Organize the information in a chart.
3. Look for times, distances, or rates that are the same. These often can lead to an equation.
4. Translating to a system of equations allows for the use of two variables. Label the variables carefully.
5. Always make sure that you have answered the question asked.

8.3 EXERCISE SET

FOR EXTRA HELP

MyMathLab® MathXL   PRACTICE   WATCH   READ   REVIEW

Vocabulary and Reading Check

Choose the word from the following list that best completes each statement. Not every word will be used.

difference   distance   mixture
principal   sum   total value

1. If 10 coffee mugs are sold for $8 each, the _____ of the mugs is $80.

2. To find simple interest, multiply the _____ by the rate and the time.

3. To solve a motion problem, we often use the fact that _____ divided by rate equals time.

4. When a boat travels downstream, its speed is the _____ of the speed of the current and the speed of the boat in still water.

Applications

5.–18. *For Exercises 5–18, solve Exercises 41–54, respectively, from Section 8.1.*

19. *Renewable Energy.* In 2010, solar and wind energy generation totaled 1033 trillion Btu's. Wind generated 52 trillion Btu's more than eight times that generated by solar energy. How much was generated by each source?

Source: U.S. Energy Information Administration

20. *Snowmen.* The tallest snowman ever recorded—really a snow *woman* named Olympia—was built by residents of Bethel, Maine, and surrounding towns. Her body and head together made up her total record height of 122 ft. The body was 2 ft longer than 14 times the height of the head. What were the separate heights of Olympia's head and body?

Source: Based on information from *Guinness World Records*

Total-Value Problems and Mixture Problems

21. *College Credits.* Each course at Mt. Regis College is worth either 3 or 4 credits. The members of the men's swim team are taking a total of 48 courses that are worth a total of 155 credits. How many 3-credit courses and how many 4-credit courses are being taken?

22. *College Credits.* Each course at Pease County Community College is worth either 3 or 4 credits. The members of the women's golf team are taking a total of 27 courses that are worth a total of 89 credits. How many 3-credit courses and how many 4-credit courses are being taken?

23. *Recycled Paper.* Staples® recently charged $38.39 per case of regular paper and $44.79 per case of paper made of recycled fibers. Last semester, Valley College Copy Center spent $1106.93 for 27 cases of paper. How many of each type were purchased?

24. *Photocopying.* Quick Copy recently charged 49¢ per page for color copies and 9¢ per page for black-and-white copies. If Shirlee's bill for 90 copies was $12.90, how many copies of each type were made?

25. *Lighting.* The Home Depot® recently sold 23-watt CFL bulbs for $2.50 each and 42-watt CFL bulbs for $8.25 each. If River Memorial Hospital purchased 200 such bulbs for a total of $1305, how many of each type did they purchase?

26. *Office Supplies.* Hancock County Social Services is preparing materials for a seminar. They purchase a combination of 80 large binders and small binders. The large binders cost $8.49 each and the small ones cost $5.99 each. If the total cost of the binders was $544.20, how many of each size were purchased?

27. *Museum Admission.* The Indianapolis Children's Museum charges $12.50 for a one-day youth admission and $17.50 for a one-day adult admission. One Friday, the museum collected $1535 from a total of 110 youths and adults. How many admissions of each type were sold?

28. *Amusement Park Admission.* Cedar Point Amusement Park charges $51.99 for an adult admission and $26.99 for a junior admission. One Thursday, the park collected $15,591.40 from a total of 360 adults and juniors. How many admissions of each type were sold?

Aha! **29.** *Blending Coffees.* The Roasted Bean charges $17.00 per pound for Fair Trade Organic Mexican coffee and $15.00 per pound for Fair Trade Organic Peruvian coffee. How much of each type should be used to make a 28-lb blend that sells for $16.00 per pound?

30. *Mixed Nuts.* Oh Nuts! sells pistachio kernels for $10.00 per pound and almonds for $8.00 per pound. How much of each type should be used to make a 50-lb mixture that sells for $8.80 per pound?

31. *Event Planning.* As part of the refreshments for Yvette's 25th birthday party, Kim plans to provide a bowl of M&M candies. She wants to mix custom-printed M&Ms costing 60¢ per ounce with bulk M&Ms costing 25¢ per ounce to create 20 lb of a mixture costing 32¢ per ounce. How much of each type of M&M should she use?

Source: www.mymms.com

32. *Blending Spices.* Spice of Life sells ground sumac for $1.35 per ounce and ground thyme for $1.85 per ounce. Aman wants to make a 20-oz Zahtar seasoning blend using the two spices that sells for $1.65 per ounce. How much of each spice should Aman use?

33. *Catering.* Cati's Catering is planning an office reception. The office administrator has requested a candy mixture that is 25% chocolate. Cati has available mixtures that are either 50% chocolate or 10% chocolate. How much of each type should be mixed to get a 20-lb mixture that is 25% chocolate?

34. *Ink Remover.* Etch Clean Graphics uses one cleanser that is 25% acid and a second that is 50% acid. How many liters of each should be mixed to get 30 L of a solution that is 40% acid?

35. *Blending Granola.* Deep Thought Granola is 25% nuts and dried fruit. Oat Dream Granola is 10% nuts and dried fruit. How much of Deep Thought and how much of Oat Dream should be mixed to form a 20-lb batch of granola that is 19% nuts and dried fruit?

36. *Livestock Feed.* Soybean meal is 16% protein and corn meal is 9% protein. How many pounds of each should be mixed to get a 350-lb mixture that is 12% protein?

37. *Student Loans.* Stacey's two student loans totaled $12,000. One of her loans was at 6.5% simple interest and the other at 7.2%. After one year, Stacey owed $811.50 in interest. What was the amount of each loan?

38. *Investments.* A self-employed contractor nearing retirement made two investments totaling $15,000. In one year, these investments yielded $573 in simple interest. Part of the money was invested at 3% and the rest at 4.5%. How much was invested at each rate?

39. *Automotive Maintenance.* "Steady State" antifreeze is 18% alcohol and "Even Flow" is 10% alcohol. How many liters of each should be mixed to get 20 L of a mixture that is 15% alcohol?

40. *Chemistry.* E-Chem Testing has a solution that is 80% base and another that is 30% base. A technician needs 150 L of a solution that is 62% base. The 150 L will be prepared by mixing the two solutions on hand. How much of each should be used?

41. *Octane Ratings.* The octane rating of a gasoline is a measure of the amount of isooctane in a gallon of gas. Manufacturers recommend using 93-octane gasoline on retuned motors. How much 87-octane gas and how much 95-octane gas should Yousef mix in order to make 10 gal of 93-octane gas for his retuned Ford F-150?

Source: Champlain Electric and Petroleum Equipment

42. *Octane Ratings.* The octane rating of a gasoline is a measure of the amount of isooctane in a gallon of gas. Subaru recommends 91-octane gasoline for the 2008 Legacy 3.0 R. How much 87-octane gas and how much 93-octane gas should Kelsey mix in order to make 12 gal of 91-octane gas for her Legacy?

Sources: Champlain Electric and Petroleum Equipment; Dean Team Ballwin

43. *Food Science.* The following bar graph shows the milk fat percentages in three dairy products.

How many pounds each of whole milk and cream should be mixed to form 200 lb of milk for cream cheese?

Milk fat

44. *Food Science.* How much lowfat milk (1% fat) and how much whole milk (4% fat) should be mixed to make 5 gal of reduced fat milk (2% fat)?

Motion Problems

45. *Train Travel.* A train leaves Danville Union and travels north at 75 km/h. Two hours later, an express train leaves on a parallel track and travels north at 125 km/h. How far from the station will they meet?

46. *Car Travel.* Two cars leave Salt Lake City, traveling in opposite directions. One car travels at a speed of 80 km/h and the other at 96 km/h. In how many hours will they be 528 km apart?

47. *Canoeing.* Kahla paddled for 4 hr with a 6-km/h current to reach a campsite. The return trip against the same current took 10 hr. Find the speed of Kahla's canoe in still water.

48. *Boating.* Chen's motorboat took 3 hr to make a trip downstream with a 6-mph current. The return trip against the same current took 5 hr. Find the speed of the boat in still water.

49. *Point of No Return.* A plane flying the 3458-mi trip from New York City to London has a 50-mph tailwind. The flight's *point of no return* is the point at which the flight time required to return to New York is the same as the time required to continue to London. If the speed of the plane in still air is 360 mph, how far is New York from the point of no return?

50. *Point of No Return.* A plane is flying the 2553-mi trip from Los Angeles to Honolulu into a 60-mph headwind. If the speed of the plane in still air is 310 mph, how far from Los Angeles is the plane's point of no return? (See Exercise 49.)

Applications

51. *Architecture.* The rectangular ground floor of the John Hancock building has a perimeter of 860 ft. The length is 100 ft more than the width. Find the length and the width.

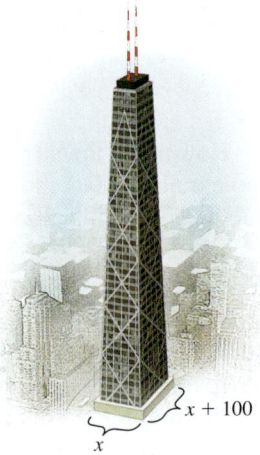

$x + 100$
x

52. *Real Estate.* The perimeter of a rectangular ocean-front lot is 190 m. The width is one-fourth of the length. Find the dimensions.

53. By February 2012, Nintendo had sold one and one-half as many Wii game machines worldwide as Sony had sold PlayStation 3 consoles. Together, they had sold 154 million game machines. How many of each were sold?

Source: vgchartz.com

54. *Hockey Rankings.* Hockey teams receive 2 points for a win and 1 point for a tie. The Wildcats once won a championship with 60 points. They won 9 more games than they tied. How many wins and how many ties did the Wildcats have?

55. At one time, Netflix offered members an unlimited number of movies and television episodes streamed to their TVs and computers for $7.99 per month. For $7.99 more per month, Netflix offered an unlimited one DVD-at-a-time by mail rental plan in addition to the unlimited streaming. During one week, 280 new subscribers paid a total of $2956.30 for their plans. How many $7.99 plans and how many $15.98 plans were purchased?

56. *Radio Airplay.* Akio must play 12 commercials during his 1-hr radio show. Each commercial is either 30 sec or 60 sec long. If the total commercial time during that hour is 10 min, how many commercials of each type does Akio play?

57. *Making Change.* Monica buys a $9.25 book using a $20 bill. The store has no bills and gives change in quarters and fifty-cent pieces. There are 30 coins in all. How many of each kind are there?

58. *Teller Work.* Sabina goes to a bank and changes a $50 bill for $5 bills and $1 bills. There are 22 bills in all. How many of each kind are there?

59. In what ways are Examples 3 and 4 similar? In what sense are their systems of equations similar?

60. Write at least three study tips of your own for someone beginning this exercise set.

Skill Review

Let $h(x) = x - 7$ and $f(x) = x^2 + 2$. Find the following.

61. $h(0)$ [7.1]

62. $f(-10)$ [7.1]

63. $(h \cdot f)(7)$ [7.4]

64. $(h + f)(x)$ [7.4]

65. The domain of $h + f$ [7.4]

66. The domain of f/h [7.4]

Synthesis

67. Suppose that in Example 3 you are asked only for the amount of almonds needed for the Market Street blend. Would the method of solving the problem change? Why or why not?

68. Write a problem similar to Example 2 for a classmate to solve. Design the problem so that the solution is "The bakery sold 24 loaves of bread and 18 packages of sandwich rolls."

69. *Recycled Paper.* Unable to purchase 60 reams of paper that contains 20% post-consumer fiber, the Naylor School bought paper that was either 0% post-consumer fiber or 30% post-consumer fiber. How many reams of each should be purchased in order to use the same amount of post-consumer fiber as if the 20% post-consumer fiber paper were available?

70. *Automotive Maintenance.* The radiator in Natalie's car contains 6.3 L of antifreeze and water. This mixture is 30% antifreeze. How much of this mixture should she drain and replace with pure antifreeze so that there will be a mixture of 50% antifreeze?

71. *Metal Alloys.* For a metal to be labeled "sterling silver," the silver alloy must contain at least 92.5% pure silver. Nicole has 32 oz of coin silver, which is 90% pure silver. How much pure silver must she add to the coin silver in order to have a sterling-silver alloy?

Source: The Jewelry Repair Manual, R. Allen Hardy, Courier Dover Publications, 1996, p. 271.

72. *Exercise.* Huan jogs and walks to school each day. She averages 4 km/h walking and 8 km/h jogging. From home to school is 6 km and Huan makes the trip in 1 hr. How far does she jog in a trip?

73. *Bakery.* Gigi's Cupcakes offers a gift box with six cupcakes for $15.99. Gigi's also sells cupcakes individually for $3 each. Gigi's sold a total of 256 cupcakes one Saturday for a total of $701.67 in sales (excluding tax). How many six-cupcake gift boxes were included in that day's sales total?

74. The tens digit of a two-digit positive integer is 2 more than three times the units digit. If the digits are interchanged, the new number is 13 less than half the given number. Find the given integer. (*Hint*: Let x = the tens-place digit and y = the units-place digit; then $10x + y$ is the number.)

75. *Wood Stains.* Williams' Custom Flooring has 0.5 gal of stain that is 20% brown and 80% neutral. A customer orders 1.5 gal of a stain that is 60% brown and 40% neutral. How much pure brown stain and how much neutral stain should be added to the original 0.5 gal in order to make up the order?*

76. *Train Travel.* A train leaves Union Station for Central Station, 216 km away, at 9 A.M. One hour later, a train leaves Central Station for Union Station. They meet at noon. If the second train had started at 9 A.M. and the first train at 10:30 A.M., they would still have met at noon. Find the speed of each train.

77. *Fuel Economy.* Grady's station wagon gets 18 miles per gallon (mpg) in city driving and 24 mpg in highway driving. The car is driven 465 mi on 23 gal of gasoline. How many miles were driven in the city and how many were driven on the highway?

78. *Biochemistry.* Industrial biochemists routinely use a machine to mix a buffer of 10% acetone by adding 100% acetone to water. One day, instead of adding 5 L of acetone to a vat of water to create the buffer, a machine added 10 L. How much additional water was needed to bring the concentration down to 10%?

 79. See Exercise 75 above. Let x = the amount of pure brown stain added to the original 0.5 gal. Find a function $P(x)$ that can be used to determine the percentage of brown stain in the 1.5-gal mixture. On a graphing calculator, draw the graph of P and use INTERSECT to confirm the answer to Exercise 75.

80. *Siblings.* Fred and Phyllis are twins. Phyllis has twice as many brothers as she has sisters. Fred has the same number of brothers as sisters. How many girls and how many boys are in the family?

↳ YOUR TURN ANSWERS: SECTION 8.3

1. Endangered species: 15; threatened species: 10
2. Silver beads: 36; gemstone beads: 24 **3.** Rooibos: 160 oz; almonds: 40 oz **4.** PLUS loan: $4500; Stafford loan: $3900 **5.** Friendly Feed: 10 L; Gentle Grow: 50 L
6. 540 mi from Washington D.C. **7.** 16 mph

QUICK QUIZ: SECTIONS 8.1–8.3

Solve. If a system has an infinite number of solutions, use set-builder notation to write the solution set. If a system has no solution, state this. [8.1], [8.2]

1. $x - 2y = 7,$
$x = 2y - 5$

2. $3x - 4y = 11,$
$x + 4y = 12$

3. $y = 2x - 4,$
$y = \frac{1}{2}x + 2$

4. $x + 3y = 3,$
$2x + 5y = 6$

5. In order to raise funds for a concert tour, Arie's choir sold rolls of trash bags. Large trash bags sold for $17 per roll and small trash bags sold for $12 per roll. If Arie sold 28 rolls and collected $441, how many rolls of each type of trash bag did he sell? [8.3]

PREPARE TO MOVE ON

Evaluate. [4.7]

1. $2x - 3y - z$, for $x = 5$, $y = 2$, and $z = 3$

2. $4x + y - 6z$, for $x = \frac{1}{2}$, $y = \frac{1}{2}$, and $z = \frac{1}{3}$

3. $3a - b + 2c$, for $a = 1$, $b = -6$, and $c = 4$

4. $a - 2b - 3c$, for $a = -2$, $b = 3$, and $c = -5$

---

*This problem was suggested by Professor Chris Burditt of Yountville, California, and is based on a real-world situation.

8.4 | Systems of Equations in Three Variables

Identifying Solutions ▪ Solving Systems in Three Variables ▪ Dependency, Inconsistency, and Geometric Considerations

Some problems naturally call for a translation to three or more equations. In this section, we learn how to solve systems of three linear equations. Later, we will use such systems in problem-solving situations.

Identifying Solutions

A **linear equation in three variables** is an equation equivalent to one of the form $Ax + By + Cz = D$, where A, B, C, and D are real numbers. We refer to the form $Ax + By + Cz = D$ as *standard form* for a linear equation in three variables.

A solution of a system of three equations in three variables is an ordered triple (x, y, z) that makes *all three* equations true. The numbers in an ordered triple correspond to the variables in alphabetical order unless otherwise indicated.

EXAMPLE 1 Determine whether $\left(\frac{3}{2}, -4, 3\right)$ is a solution of the system

$$4x - 2y - 3z = 5,$$
$$-8x - y + z = -5,$$
$$2x + y + 2z = 5.$$

SOLUTION We substitute $\left(\frac{3}{2}, -4, 3\right)$ into all three equations, using alphabetical order:

$$\begin{array}{c} 4x - 2y - 3z = 5 \\ \hline 4 \cdot \frac{3}{2} - 2(-4) - 3 \cdot 3 \ \big|\ 5 \\ 6 + 8 - 9 \ \big| \\ \hline 5 \overset{?}{=} 5 \quad \text{TRUE} \end{array}$$

$$\begin{array}{c} -8x - y + z = -5 \\ \hline -8 \cdot \frac{3}{2} - (-4) + 3 \ \big|\ -5 \\ -12 + 4 + 3 \ \big| \\ \hline -5 \overset{?}{=} -5 \quad \text{TRUE} \end{array}$$

$$\begin{array}{c} 2x + y + 2z = 5 \\ \hline 2 \cdot \frac{3}{2} + (-4) + 2 \cdot 3 \ \big|\ 5 \\ 3 - 4 + 6 \ \big| \\ \hline 5 \overset{?}{=} 5 \quad \text{TRUE} \end{array}$$

The triple makes all three equations true, so it is a solution.

1. Determine whether $\left(-2, \frac{1}{2}, 5\right)$ is a solution of the system

$$x - 2y + z = 2,$$
$$3x - 4y + 2z = 3,$$
$$x + 6y - z = -10.$$

YOUR TURN

Solving Systems in Three Variables

The graph of a linear equation in three variables is a plane. Because a three-dimensional coordinate system is required, solving systems in three variables graphically is difficult. The substitution method *can* be used but is generally cumbersome. Fortunately, the elimination method works well for any system of three equations in three variables.

EXAMPLE 2 Solve the following system of equations:

$$x + y + z = 4, \qquad (1)$$
$$x - 2y - z = 1, \qquad (2)$$
$$2x - y - 2z = -1. \qquad (3)$$

SOLUTION We select *any* two of the three equations and work to get an equation in two variables. Let's add equations (1) and (2):

$$\begin{array}{ll} x + y + z = 4 & (1) \\ \underline{x - 2y - z = 1} & (2) \\ 2x - y \phantom{{}+{}} = 5. & (4) \qquad \text{Adding to eliminate } z \end{array}$$

Study Skills

Helping Others Will Help You Too

When you thoroughly understand a topic, don't hesitate to help class-mates experiencing trouble. Your understanding and retention of the material will deepen and your classmate will appreciate your help.

Next, we select a different pair of equations and eliminate the *same variable* that we did above. Let's use equations (1) and (3) to again eliminate z. Be careful! A common error is to eliminate a different variable in this step.

$$x + y + z = 4,$$
$$2x - y - 2z = -1$$

Multiplying both sides of equation (1) by 2

$$2x + 2y + 2z = 8$$
$$2x - y - 2z = -1$$
$$4x + y = 7 \quad (5)$$

Now we solve the resulting system of equations (4) and (5). That solution will give us two of the numbers in the solution of the original system.

$$2x - y = 5 \quad (4)$$
$$4x + y = 7 \quad (5)$$
$$6x = 12 \quad \text{Adding}$$
$$x = 2$$

Note that we now have two equations in two variables. Had we not eliminated the *same* variable in both of the above steps, this would not be the case.

We can use either equation (4) or (5) to find y. We choose equation (5):

$$4x + y = 7 \quad (5)$$
$$4 \cdot 2 + y = 7 \quad \text{Substituting 2 for } x \text{ in equation (5)}$$
$$8 + y = 7$$
$$y = -1.$$

We now have $x = 2$ and $y = -1$. To find the value for z, we use any of the original three equations and substitute to find the third number, z. Let's use equation (1) and substitute our two numbers in it:

$$x + y + z = 4 \quad (1)$$
$$2 + (-1) + z = 4 \quad \text{Substituting 2 for } x \text{ and } -1 \text{ for } y$$
$$1 + z = 4$$
$$z = 3.$$

We have obtained the triple $(2, -1, 3)$. It should check in *all three* equations:

$$x + y + z = 4$$
$$2 + (-1) + 3 \,|\, 4$$
$$4 \overset{?}{=} 4 \quad \text{TRUE}$$

$$x - 2y - z = 1$$
$$2 - 2(-1) - 3 \,|\, 1$$
$$1 \overset{?}{=} 1 \quad \text{TRUE}$$

$$2x - y - 2z = -1$$
$$2 \cdot 2 - (-1) - 2 \cdot 3 \,|\, -1$$
$$-1 \overset{?}{=} -1 \quad \text{TRUE}$$

The solution is $(2, -1, 3)$.

YOUR TURN

2. Solve the following system of equations:

$$x + y + z = 6,$$
$$2x - y - z = 3,$$
$$x - 2y + 2z = 13.$$

SOLVING SYSTEMS OF THREE LINEAR EQUATIONS

To use the elimination method to solve systems of three linear equations:

1. Write all equations in standard form $Ax + By + Cz = D$.
2. Clear any decimals or fractions.
3. Choose a variable to eliminate. Then select two of the three equations and multiply and add, as needed, to produce one equation in which the selected variable is eliminated.
4. Next, use a different pair of equations and eliminate the same variable as in step (3).
5. Solve the system of equations resulting from steps (3) and (4).
6. Substitute the solution from step (5) into one of the original three equations and solve for the third variable. Then check.

EXAMPLE 3   Solve the system

$$4x - 2y - 3z = 5, \qquad (1)$$
$$-8x - y + z = -5, \qquad (2)$$
$$2x + y + 2z = 5. \qquad (3)$$

SOLUTION

Write in standard form.

1., 2. The equations are already in standard form with no fractions or decimals.

3. Select a variable to eliminate. We decide on y because the y-terms are opposites of each other in equations (2) and (3). We add:

Eliminate a variable. (We choose y).

$$-8x - y + z = -5 \qquad (2)$$
$$\underline{2x + y + 2z = 5} \qquad (3)$$
$$-6x + 3z = 0. \qquad (4) \qquad \text{Adding}$$

4. We use another pair of equations to create a second equation in x and z. That is, we again eliminate y. To do so, we use equations (1) and (3):

Eliminate the same variable using a different pair of equations.

$$4x - 2y - 3z = 5,$$
$$2x + y + 2z = 5 \xrightarrow[\text{of equation (3) by 2}]{\text{Multiplying both sides}}$$

$$4x - 2y - 3z = 5$$
$$\underline{4x + 2y + 4z = 10}$$
$$8x + z = 15. \quad (5)$$

5. Now we solve the resulting system of equations (4) and (5). That allows us to find two of the three variables.

Solve the system of two equations in two variables.

$$-6x + 3z = 0,$$
$$8x + z = 15 \xrightarrow[\text{of equation (5) by } -3]{\text{Multiplying both sides}}$$

$$-6x + 3z = 0$$
$$\underline{-24x - 3z = -45}$$
$$-30x = -45$$
$$x = \frac{-45}{-30} = \frac{3}{2}$$

We use equation (5) to find z:

$$8x + z = 15$$
$$8 \cdot \tfrac{3}{2} + z = 15 \qquad \text{Substituting } \tfrac{3}{2} \text{ for } x$$
$$12 + z = 15$$
$$z = 3.$$

Solve for the remaining variable and check.

6. Finally, we use any of the original equations and substitute to find the third number, y. To do so, we choose equation (3):

$$2x + y + 2z = 5 \qquad (3)$$
$$2 \cdot \tfrac{3}{2} + y + 2 \cdot 3 = 5 \qquad \text{Substituting } \tfrac{3}{2} \text{ for } x \text{ and } 3 \text{ for } z$$
$$3 + y + 6 = 5$$
$$y + 9 = 5$$
$$y = -4.$$

The solution is $\left(\tfrac{3}{2}, -4, 3\right)$. The check was performed as Example 1.

3. Solve the system

$$x - 3y + z = 13,$$
$$2x + 3y + 2z = 20,$$
$$-3x - 6y + z = 3.$$

YOUR TURN

Sometimes, certain variables are missing at the outset.

EXAMPLE 4   Solve the system

$$x + y + z = 180, \qquad (1)$$
$$x - z = -70, \qquad (2)$$
$$2y - z = 0. \qquad (3)$$

SOLUTION

1., 2. The equations appear in standard form with no fractions or decimals.

3., 4. Note that there is no y in equation (2). Thus, at the outset, we already have y eliminated from one equation. We need another equation with no y-term, so we work with equations (1) and (3):

$$x + y + z = 180, \xrightarrow[\text{of equation (1) by } -2]{\text{Multiplying both sides}} \begin{array}{r} -2x - 2y - 2z = -360 \\ 2y - z = 0 \\ \hline -2x \quad\quad - 3z = -360. \end{array} \quad (4)$$

5., 6. Now we solve the resulting system of equations (2) and (4):

$$\begin{array}{r} x - z = -70, \\ -2x - 3z = -360 \end{array} \xrightarrow[\text{of equation (2) by } 2]{\text{Multiplying both sides}} \begin{array}{r} 2x - 2z = -140 \\ -2x - 3z = -360 \\ \hline -5z = -500 \\ z = 100. \end{array}$$

Continuing as in Examples 2 and 3, we get the solution $(30, 50, 100)$. The check is left to the student.

4. Solve the system

$$\begin{aligned} x - y \quad\quad &= 8, \\ x + y + z &= 17, \\ x \quad\quad + z &= 5. \end{aligned}$$

🔄 YOUR TURN

Dependency, Inconsistency, and Geometric Considerations

Each equation in Examples 2, 3, and 4 has a graph that is a plane in three dimensions. The solutions are points common to the planes of each system. Since three planes can have an infinite number of points in common or no points at all in common, we need to generalize the concept of *consistency*.

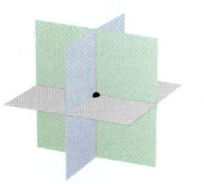

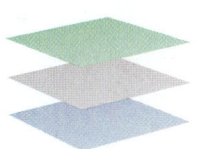

Planes intersect at one point. System is *consistent* and has one solution.

Planes intersect along a common line. System is *consistent* and has an infinite number of solutions.

Three parallel planes. System is *inconsistent;* it has no solution.

Planes intersect two at a time, with no point common to all three. System is *inconsistent;* it has no solution.

CONSISTENCY

A system of equations that has at least one solution is said to be **consistent**.

A system of equations that has no solution is said to be **inconsistent**.

EXAMPLE 5 Solve:

$$\begin{aligned} y + 3z &= 4, &&(1) \\ -x - y + 2z &= 0, &&(2) \\ x + 2y + z &= 1. &&(3) \end{aligned}$$

SOLUTION   There is no x-term in equation (1). By adding equations (2) and (3), we can find a second equation in which x is again absent:

$$
\begin{array}{rl}
-x - \ y + 2z = 0 & (2) \\
\underline{x + 2y + \ z = 1} & (3) \\
y + 3z = 1. & (4) \qquad \text{Adding}
\end{array}
$$

Equations (1) and (4) form a system in y and z. We solve as before:

$$
\begin{array}{l}
y + 3z = 4, \\
y + 3z = 1
\end{array}
\quad
\xrightarrow[\text{of equation (1) by } -1]{\text{Multiplying both sides}}
\quad
\begin{array}{r}
-y - 3z = -4 \\
\underline{y + 3z = \quad 1}
\end{array}
$$

This is a contradiction. $\longrightarrow 0 = -3.$ Adding

Since we end up with a *false* equation, or contradiction, we state that the system has no solution. It is *inconsistent*.

5. Solve:

$$
\begin{aligned}
x - 2y + 2z &= 6, \\
2x + 3y\quad\ &= 1, \\
-3x - 8y + 2z &= 0.
\end{aligned}
$$

YOUR TURN

The notion of *dependency* can also be extended to systems of three equations.

EXAMPLE 6   Solve:

$$
\begin{array}{rl}
2x + \ y + \ z = 3, & (1) \\
x - 2y - \ z = 1, & (2) \\
3x + 4y + 3z = 5. & (3)
\end{array}
$$

SOLUTION   Our plan is to first use equations (1) and (2) to eliminate z. Then we will select another pair of equations and again eliminate z:

$$
\begin{array}{r}
2x + \ y + z = 3 \\
\underline{x - 2y - z = 1} \\
3x - \ y \quad\ = 4. \qquad (4)
\end{array}
$$

Next, we use equations (2) and (3) to eliminate z again:

$$
\begin{array}{l}
x - 2y - \ z = 1, \\
3x + 4y + 3z = 5
\end{array}
\quad
\xrightarrow[\text{of equation (2) by } 3]{\text{Multiplying both sides}}
\quad
\begin{array}{r}
3x - 6y - 3z = 3 \\
\underline{3x + 4y + 3z = 5} \\
6x - 2y \quad\ = 8. \qquad (5)
\end{array}
$$

We now solve the resulting system of equations (4) and (5):

$$
\begin{array}{l}
3x - \ y = 4, \\
6x - 2y = 8
\end{array}
\quad
\xrightarrow[\text{of equation (4) by } -2]{\text{Multiplying both sides}}
\quad
\begin{array}{r}
-6x + 2y = \ -8 \\
\underline{6x - 2y = \quad 8} \\
0 = \quad 0. \qquad (6)
\end{array}
$$

6. Solve:

$$
\begin{aligned}
x - y + \ z &= -1, \\
2x + y + 2z &= 5, \\
4x - y + 4z &= 3.
\end{aligned}
$$

Equation (6), which is an identity, indicates that equations (1), (2), and (3) are *dependent*. This means that the original system of three equations is equivalent to a system of two equations. One way to see this is to note that two times equation (1), minus equation (2), is equation (3). Thus removing equation (3) from the system does not affect the solution of the system.* In writing an answer to this problem, we simply state that "the equations are dependent."

YOUR TURN

---

* A set of equations is dependent if at least one equation can be expressed as a sum of multiples of other equations in that set.

In a system of two equations, when equations are dependent the system is consistent. This is not always the case for systems of three or more equations. The following figures illustrate some possibilities geometrically.

The planes intersect along a common line. The equations are *dependent* and the system is *consistent*. There is an infinite number of solutions.

The planes coincide. The equations are *dependent* and the system is *consistent*. There is an infinite number of solutions.

Two planes coincide. The third plane is parallel. The equations are *dependent* and the system is *inconsistent*. There is no solution.

8.4 EXERCISE SET

FOR EXTRA HELP

PRACTICE WATCH READ REVIEW

Vocabulary and Reading Check

Classify each of the following statements as either true or false.

1. $3x + 5y + 4z = 7$ is a linear equation in three variables.

2. Every system of three equations in three unknowns has at least one solution.

3. It is not difficult to solve a system of three equations in three unknowns by graphing.

4. If, when we are solving a system of three equations, a false equation results from adding a multiple of one equation to another, the system is inconsistent.

5. If, when we are solving a system of three equations, an identity results from adding a multiple of one equation to another, the equations are dependent.

6. Whenever a system of three equations contains dependent equations, there is an infinite number of solutions.

Identifying Solutions

7. Determine whether $(2, -1, -2)$ is a solution of the system

$$x + y - 2z = 5,$$
$$2x - y - z = 7,$$
$$-x - 2y - 3z = 6.$$

8. Determine whether $(-1, -3, 2)$ is a solution of the system

$$x - y + z = 4,$$
$$x - 2y - z = 3,$$
$$3x + 2y - z = 1.$$

Solving Systems in Three Variables

Solve each system. If a system's equations are dependent or if there is no solution, state this.

9. $x - y - z = 0,$
   $2x - 3y + 2z = 7,$
   $-x + 2y + z = 1$

10. $x + y - z = 0,$
    $2x - y + z = 3,$
    $-x + 5y - 3z = 2$

11. $x - y - z = 1,$
    $2x + y + 2z = 4,$
    $x + y + 3z = 5$

12. $x + y - 3z = 4,$
    $2x + 3y + z = 6,$
    $2x - y + z = -14$

13. $3x + 4y - 3z = 4,$
    $5x - y + 2z = 3,$
    $x + 2y - z = -2$

14. $2x - 3y + z = 5,$
    $x + 3y + 8z = 22,$
    $3x - y + 2z = 12$

15. $x + y + z = 0,$
    $2x + 3y + 2z = -3,$
    $-x - 2y - z = 1$

16. $3a - 2b + 7c = 13,$
    $a + 8b - 6c = -47,$
    $7a - 9b - 9c = -3$

17. $2x - 3y - z = -9,$
    $2x + 5y + z = 1,$
    $x - y + z = 3$

18. $4x + y + z = 17,$
    $x - 3y + 2z = -8,$
    $5x - 2y + 3z = 5$

Aha! **19.** $a + b + c = 5$,
$2a + 3b - c = 2$,
$2a + 3b - 2c = 4$

20. $u - v + 6w = 8$,
$3u - v + 6w = 14$,
$-u - 2v - 3w = 7$

21. $-2x + 8y + 2z = 4$,
$x + 6y + 3z = 4$,
$3x - 2y + z = 0$

22. $x - y + z = 4$,
$5x + 2y - 3z = 2$,
$4x + 3y - 4z = -2$

23. $2u - 4v - w = 8$,
$3u + 2v + w = 6$,
$5u - 2v + 3w = 2$

24. $4p + q + r = 3$,
$2p - q + r = 6$,
$2p + 2q - r = -9$

25. $r + \frac{3}{2}s + 6t = 2$,
$2r - 3s + 3t = 0.5$,
$r + s + t = 1$

26. $5x + 3y + \frac{1}{2}z = \frac{7}{2}$,
$0.5x - 0.9y - 0.2z = 0.3$,
$3x - 2.4y + 0.4z = -1$

27. $4a + 9b = 8$,
$8a + 6c = -1$,
$6b + 6c = -1$

28. $3p + 2r = 11$,
$q - 7r = 4$,
$p - 6q = 1$

29. $x + y + z = 57$,
$-2x + y = 3$,
$x - z = 6$

30. $x + y + z = 105$,
$10y - z = 11$,
$2x - 3y = 7$

31. $a - 3c = 6$,
$b + 2c = 2$,
$7a - 3b - 5c = 14$

32. $2a - 3b = 2$,
$7a + 4c = \frac{3}{4}$,
$2c - 3b = 1$

Aha! **33.** $x + y + z = 83$,
$y = 2x + 3$,
$z = 40 + x$

34. $l + m = 7$,
$3m + 2n = 9$,
$4l + n = 5$

35. $x + z = 0$,
$x + y + 2z = 3$,
$y + z = 2$

36. $x + y = 0$,
$x + z = 1$,
$2x + y + z = 2$

37. $x + y + z = 1$,
$-x + 2y + z = 2$,
$2x - y = -1$

38. $y + z = 1$,
$x + y + z = 1$,
$x + 2y + 2z = 2$

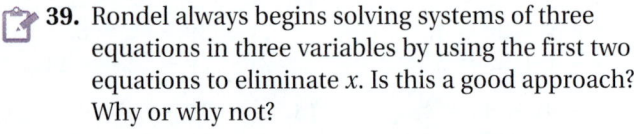 **39.** Rondel always begins solving systems of three equations in three variables by using the first two equations to eliminate x. Is this a good approach? Why or why not?

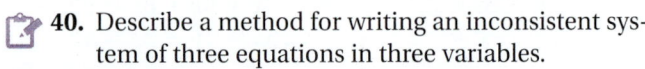 **40.** Describe a method for writing an inconsistent system of three equations in three variables.

Skill Review

41. Find the slope and the y-intercept of the graph of $x - 3y = 7$. [3.6]

42. Find the slope of the graph of $f(x) = 8$. If the slope is undefined, state this. [7.3]

43. Find the intercepts of the graph of $2x - 5y = 20$. [3.3]

44. Find the slope of the line containing $(6, 9)$ and $(-2, 4)$. [3.5]

Determine whether each pair of lines is parallel, perpendicular, or neither. [3.6]

45. $3x - y = 12$,
$y = 3x + 7$

46. $2x - 5y = 6$,
$2x + 5y = 1$

Synthesis

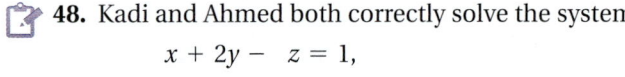 **47.** Is it possible for a system of three linear equations to have exactly two ordered triples in its solution set? Why or why not?

48. Kadi and Ahmed both correctly solve the system
$$x + 2y - z = 1,$$
$$-x - 2y + z = 3,$$
$$2x + 4y - 2z = 2.$$
Kadi states "the equations are dependent" while Ahmed states "there is no solution." How is this possible?

Solve.

49. $\dfrac{x + 2}{3} - \dfrac{y + 4}{2} + \dfrac{z + 1}{6} = 0$,
$\dfrac{x - 4}{3} + \dfrac{y + 1}{4} - \dfrac{z - 2}{2} = -1$,
$\dfrac{x + 1}{2} + \dfrac{y}{2} + \dfrac{z - 1}{4} = \dfrac{3}{4}$

50. $w + x - y + z = 0$,
$w - 2x - 2y - z = -5$,
$w - 3x - y + z = 4$,
$2w - x - y + 3z = 7$

51. $w + x + y + z = 2$,
$w + 2x + 2y + 4z = 1$,
$w - x + y + z = 6$,
$w - 3x - y + z = 2$

For Exercises 52 and 53, let u represent $1/x$, v represent $1/y$, and w represent $1/z$. Solve for u, v, and w, and then solve for x, y, and z.

52. $\dfrac{2}{x} + \dfrac{2}{y} - \dfrac{3}{z} = 3$,
$\dfrac{1}{x} - \dfrac{2}{y} - \dfrac{3}{z} = 9$,
$\dfrac{7}{x} - \dfrac{2}{y} + \dfrac{9}{z} = -39$

53. $\dfrac{2}{x} - \dfrac{1}{y} - \dfrac{3}{z} = -1$,
$\dfrac{2}{x} - \dfrac{1}{y} + \dfrac{1}{z} = -9$,
$\dfrac{1}{x} + \dfrac{2}{y} - \dfrac{4}{z} = 17$

Determine k so that each system is dependent.

54. $x - 3y + 2z = 1,$
$2x + y - z = 3,$
$9x - 6y + 3z = k$

55. $5x - 6y + kz = -5,$
$x + 3y - 2z = 2,$
$2x - y + 4z = -1$

In each case, three solutions of an equation in x, y, and z are given. Find the equation.

56. $Ax + By + Cz = 12;$
$\left(1, \frac{3}{4}, 3\right), \left(\frac{4}{3}, 1, 2\right),$ and $(2, 1, 1)$

57. $z = b - mx - ny;$
$(1, 1, 2), (3, 2, -6),$ and $\left(\frac{3}{2}, 1, 1\right)$

58. Write an inconsistent system of equations that contains dependent equations.

↪ **YOUR TURN ANSWERS: SECTION 8.4**

1. No   **2.** $(3, -1, 4)$ **3.** $\left(3, -\frac{2}{3}, 8\right)$ **4.** $(20, 12, -15)$

5. No solution   **6.** The equations are dependent.

QUICK QUIZ: SECTIONS 8.1–8.4

Solve. If a system has an infinite number of solutions, use set-builder notation to write the solution set. If a system has no solution, state this. [8.1], [8.2]

1. $3x + 2y = 9,$
$x - 6y = 1$

2. $2x - y = 4,$
$3y = 6x - 12$

3. Solve: $x + y + z = 8,$
$2x + y - z = -5,$
$x - 2y - z = 8.$ [8.4]

4. Jared's motorboat took 2 hr to make a trip downstream with a 4-mph current. The return trip against the same current took 3 hr. Find the speed of the boat in still water. [8.3]

5. Julia has paint with 15% red pigment and paint with 10% red pigment. How much of each should she use to form 3 gal of paint with 12% red pigment? [8.3]

PREPARE TO MOVE ON

Translate each statement to an equation. [1.1]

1. The sum of three consecutive numbers is 100.

2. The sum of three numbers is 100.

3. The product of two numbers is five times a third number.

4. The product of two numbers is twice their sum.

Mid-Chapter Review

Systems of two equations can be solved using graphical or algebraic methods. Since graphing in three dimensions is difficult, algebraic methods are used to solve systems of three equations. Both substitution and elimination work well for systems of two equations, but elimination is usually the preferred method for systems of three equations.

GUIDED SOLUTIONS

Solve. [8.2]

1. $2x - 3y = 5,$
$y = x - 1$

Solution

$2x - 3\left(\boxed{}\right) = 5$ Substituting $x - 1$ for y

$2x - \boxed{} + \boxed{} = 5$ Using the distributive law

$\boxed{} + 3 = 5$ Combining like terms

$-x = \boxed{}$ Subtracting 3 from both sides

$x = \boxed{}$ Dividing both sides by -1

$y = x - 1$

$y = \boxed{} - 1$ Substituting

$y = \boxed{}$

The solution is $\left(\boxed{}, \boxed{}\right)$.

2. $2x - 5y = 1,$
$x + 5y = 8$

Solution

$2x - 5y = 1$

$\underline{x + 5y = 8}$

$\boxed{} = \boxed{}$

$x = \boxed{}$

$x + 5y = 8$

$\boxed{} + 5y = 8$ Substituting

$5y = \boxed{}$

$y = \boxed{}$

The solution is $\left(\boxed{}, \boxed{}\right)$.

MIXED REVIEW

Solve using any appropriate method. [8.1], [8.2], [8.4]

3. $x = y,$
$x + y = 2$

4. $x + y = 10,$
$x - y = 8$

5. $y = \frac{1}{2}x + 1,$
$y = 2x - 5$

6. $y = 2x - 3,$
$x + y = 12$

7. $x = 5,$
$y = 10$

8. $3x + 5y = 8,$
$3x - 5y = 4$

9. $2x - y = 1,$
$2y - 4x = 3$

10. $x = 2 - y,$
$3x + 3y = 6$

11. $1.1x - 0.3y = 0.8,$
$2.3x + 0.3y = 2.6$

12. $\frac{1}{4}x = \frac{1}{3}y,$
$\frac{1}{2}x - \frac{1}{15}y = 2$

13. $3x + y - z = -1,$
$2x - y + 4z = 2,$
$x - y + 3z = 3$

14. $2x + y - 3z = -4,$
$4x + y + 3z = -1,$
$2x - y + 6z = 7$

15. $3x + 5y - z = 8,$
$x + 6y = 4,$
$x - 7y - z = 3$

16. $x - y = 4,$
$2x + y - z = 5,$
$3x - z = 9$

Solve. [8.3]

17. In 2010, 12.3% of the amount spent on Internet advertising went to Yahoo! and Microsoft. Yahoo! received 0.3% more of the amount spent on Internet advertising than twice the amount Microsoft received. What percent of Internet advertising went to each company?

Source: ZenithOptimedia

18. As part of a fundraiser, the Cobblefield Daycare collected 430 returnable bottles and cans, some worth 5 cents each and the rest worth 10 cents each. If the total value of the cans and bottles was $26.20, how many 5-cent bottles or cans and how many 10-cent bottles or cans were collected?

19. Pecan Morning granola is 25% nuts and dried fruit. Oat Dream granola is 10% nuts and dried fruit. How much of Pecan Morning and how much of Oat Dream should be mixed in order to form a 20-lb batch of granola that is 19% nuts and dried fruit?

20. The Grand Royale cruise ship takes 3 hr to make a trip up the Amazon River against a 6-mph current. The return trip with the same current takes 1.5 hr. Find the speed of the ship in still water.

8.5 Solving Applications: Systems of Three Equations

Applications of Three Equations in Three Unknowns

Applications of Three Equations in Three Unknowns

Systems of three or more equations arise in the natural and social sciences, business, and engineering. To begin, let's first look at a purely numerical application.

EXAMPLE 1 The sum of three numbers is 4. The first number minus twice the second, minus the third is 1. Twice the first number minus the second, minus twice the third is −1. Find the numbers.

SOLUTION

1. **Familiarize.** There are three statements involving the same three numbers. Let's label these numbers x, y, and z.

2. **Translate.** We can translate directly as follows.

The sum of the three numbers is 4.
$$x + y + z = 4$$

The first number minus twice the second minus the third is 1.
$$x - 2y - z = 1$$

Twice the first number minus the second minus twice the third is −1.
$$2x - y - 2z = -1$$

We now have a system of three equations:

$$x + y + z = 4,$$
$$x - 2y - z = 1,$$
$$2x - y - 2z = -1.$$

3. **Carry out.** The solution of this system is $(2, -1, 3)$, as we found in Example 2 of Section 8.4.

4. **Check.** The first statement of the problem says that the sum of the three numbers is 4. That checks, because $2 + (-1) + 3 = 4$. The second statement says that the first number minus twice the second, minus the third is 1: $2 - 2(-1) - 3 = 1$. That checks. The check of the third statement is left to the student.

5. **State.** The three numbers are 2, −1, and 3.

 YOUR TURN

1. The sum of three numbers is 10. Twice the first number plus the second equals the third. Half the first number plus the second plus the third is 6. Find the numbers.

EXAMPLE 2 *Architecture.* In a triangular cross section of a roof, the largest angle is 70° greater than the smallest angle. The largest angle is twice as large as the remaining angle. Find the measure of each angle.

SOLUTION

1. **Familiarize.** The first thing we do is make a drawing, or a sketch.

Student Notes

It is quite likely that you are expected to remember that the sum of the measures of the angles in any triangle is 180°. You may want to ask your instructor which other formulas from geometry and elsewhere you are expected to know.

Since we don't know the size of any angle, we use x, y, and z to represent the three measures, from smallest to largest. Recall that the measures of the angles in any triangle add up to 180°.

2. **Translate.** This geometric fact about triangles gives us one equation:

$x + y + z = 180$.

Two of the statements can be translated almost directly.

The largest angle   is   70° greater than the smallest angle.

z = $x + 70$

The largest angle   is   twice as large as the remaining angle.

z = $2y$

We now have a system of three equations:

$$x + y + z = 180, \qquad x + y + z = 180,$$
$$x + 70 = z, \qquad \text{or} \quad x - z = -70, \qquad \text{Rewriting in}$$
$$2y = z; \qquad\qquad 2y - z = 0. \qquad \text{standard form}$$

3. **Carry out.** This system was solved in Example 4 of Section 8.4. The solution is $(30, 50, 100)$.

4. **Check.** The sum of the numbers is 180, so that checks. The measure of the largest angle, 100°, is 70° greater than the measure of the smallest angle, 30°, so that checks. The measure of the largest angle is also twice the measure of the remaining angle, 50°. Thus we have a check.

5. **State.** The angles in the triangle measure 30°, 50°, and 100°.

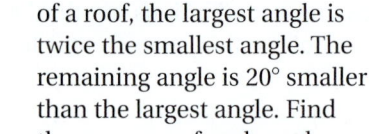

2. In a triangular cross section of a roof, the largest angle is twice the smallest angle. The remaining angle is 20° smaller than the largest angle. Find the measure of each angle.

 YOUR TURN

EXAMPLE 3 *Downloads.* Kaya frequently downloads music, TV shows, and movies. In January, she downloaded 5 songs, 10 TV shows, and 3 movies for a total of $55. In February, she spent $195 on 25 songs, 25 TV shows, and 12 movies. In March, she spent $81 on 15 songs, 8 TV shows, and 5 movies. Assuming each song is the same price, each TV show is the same price, and each movie is the same price, how much does each type of download cost?

SOLUTION

1. **Familiarize.** We let s = the cost, in dollars, per song, t = the cost, in dollars, per TV show, and m = the cost, in dollars, per movie. The total cost is the sum of the cost per item times the number of items purchased.

2. **Translate.** In January, Kaya spent $5 \cdot s$ for songs, $10 \cdot t$ for TV shows, and $3 \cdot m$ for movies. The total of these amounts was $55. Each month's downloads will translate to an equation. We can organize the information in a table.

	Cost of Songs	Cost of TV Shows	Cost of Movies	Total Cost	
January	$5s$	$10t$	$3m$	55	$\longrightarrow 5s + 10t + 3m = 55$
February	$25s$	$25t$	$12m$	195	$\longrightarrow 25s + 25t + 12m = 195$
March	$15s$	$8t$	$5m$	81	$\longrightarrow 15s + 8t + 5m = 81$

We now have a system of three equations:

$$5s + 10t + 3m = 55, \quad (1)$$
$$25s + 25t + 12m = 195, \quad (2)$$
$$15s + 8t + 5m = 81. \quad (3)$$

3. **Carry out.** We begin by using equations (1) and (2) to eliminate s.

$$5s + 10t + 3m = 55,$$
$$25s + 25t + 12m = 195$$

Multiplying both sides of equation (1) by -5 $\longrightarrow$

$$-25s - 50t - 15m = -275$$
$$\underline{25s + 25t + 12m = \quad 195}$$
$$-25t - 3m = \quad -80 \quad (4)$$

We then use equations (1) and (3) to again eliminate s.

$$5s + 10t + 3m = 55,$$
$$15s + 8t + 5m = 81$$

Multiplying both sides of equation (1) by -3 $\longrightarrow$

$$-15s - 30t - 9m = -165$$
$$\underline{15s + 8t + 5m = \quad 81}$$
$$-22t - 4m = \quad -84 \quad (5)$$

Now we solve the resulting system of equations (4) and (5).

$$-25t - 3m = -80,$$

Multiplying both sides of equation (4) by -4 $\longrightarrow$

$$100t + 12m = \quad 320$$

$$-22t - 4m = -84$$

Multiplying both sides of equation (5) by 3 $\longrightarrow$

$$\underline{-66t - 12m = -252}$$
$$34t \quad = \quad 68$$
$$t = 2$$

To find m, we use equation (4):

$$-25t - 3m = -80$$
$$-25 \cdot 2 - 3m = -80 \quad \text{Substituting 2 for } t$$
$$-50 - 3m = -80$$
$$-3m = -30$$
$$m = 10.$$

3. Eli frequently downloads music, HDTV shows, and games. In April, he downloaded 3 albums, 10 HDTV shows, and 8 games for a total of $74. In May, he spent $100 for 5 albums, 12 HDTV shows, and 4 games. In June, he spent $79 for 2 albums, 15 HDTV shows, and 10 games. Assuming each album is the same price, each HDTV show is the same price, and each game is the same price, how much does each type of download cost?

Finally, we use equation (1) to find s:

$$5s + 10t + 3m = 55$$
$$5s + 10 \cdot 2 + 3 \cdot 10 = 55 \qquad \text{Substituting 2 for } t \text{ and 10 for } m$$
$$5s + 20 + 30 = 55$$
$$5s + 50 = 55$$
$$5s = 5$$
$$s = 1.$$

4. Check. If a song costs $1, a TV show costs $2, and a movie costs $10, then the total cost for each month's downloads is as follows:

January: $5 \cdot \$1 + 10 \cdot \$2 + 3 \cdot \$10 = \$5 + \$20 + \$30 = \$55;$
February: $25 \cdot \$1 + 25 \cdot \$2 + 12 \cdot \$10 = \$25 + \$50 + \$120 = \$195;$
March: $15 \cdot \$1 + 8 \cdot \$2 + 5 \cdot \$10 = \$15 + \$16 + \$50 = \$81.$

This checks with the information given in the problem.

5. State. A song costs $1, a TV show costs $2, and a movie costs $10.

 YOUR TURN

8.5 EXERCISE SET

FOR EXTRA HELP

PRACTICE   WATCH   READ   REVIEW

Vocabulary and Reading Check

Match each statement with a translation from the list below.

a) $x + y + z = 50$
b) $x - y - z = -50$
c) $x - y + z = 50$
d) $x - y - z = 50$

1. ____ The sum of three numbers is 50.

2. ____ The first number minus the second plus the third is 50.

3. ____ The first number is 50 more than the sum of the other two numbers.

4. ____ The first number is 50 less than the sum of the other two numbers.

Applications of Three Equations in Three Unknowns

Solve.

5. The sum of three numbers is 85. The second is 7 more than the first. The third is 2 more than four times the second. Find the numbers.

6. The sum of three numbers is 5. The first number minus the second plus the third is 1. The first minus the third is 3 more than the second. Find the numbers.

7. The sum of three numbers is 26. Twice the first minus the second is 2 less than the third. The third is the second minus three times the first. Find the numbers.

8. The sum of three numbers is 105. The third is 11 less than ten times the second. Twice the first is 7 more than three times the second. Find the numbers.

9. *Geometry.* In triangle *ABC*, the measure of angle *B* is three times that of angle *A*. The measure of angle *C* is 20° more than that of angle *A*. Find the angle measures.

10. *Geometry.* In triangle *ABC*, the measure of angle *B* is twice the measure of angle *A*. The measure of angle *C* is 80° more than that of angle *A*. Find the angle measures.

11. *Scholastic Aptitude Test.*   Many high-school students take the Scholastic Aptitude Test (SAT). Those taking the SAT receive three scores: a critical reading score, a mathematics score, and a writing score. In 2011, the average total score of high-school seniors was 1500. The average mathematics score exceeded the reading score by 17 points, and the average writing score was 8 points less than the reading score. What was the average score for each category?

Source: College Entrance Examination Board

12. *Advertising.*   In 2011, U.S. companies spent a total of $111 billion on newspaper, television, and Internet ads. The amount spent on television ads was $11 billion more than the amount spent on newspaper and Internet ads combined. The amount spent on newspaper ads was $8 billion less than the amount spent on Internet ads. How much was spent on each form of advertising?

Source: eMarketer

13. *Nutrition.*   Most nutritionists now agree that a healthy adult diet includes 25–35 g of fiber each day. A breakfast of 2 bran muffins, 1 banana, and a 1-cup serving of Wheaties® contains 9 g of fiber; a breakfast of 1 bran muffin, 2 bananas, and a 1-cup serving of Wheaties® contains 10.5 g of fiber; and a breakfast of 2 bran muffins and a 1-cup serving of Wheaties® contains 6 g of fiber. How much fiber is in each of these foods?

Sources: usda.gov and InteliHealth.com

14. *Nutrition.*   Refer to Exercise 13. A breakfast consisting of 2 pancakes and a 1-cup serving of strawberries contains 4.5 g of fiber, whereas a breakfast of 2 pancakes and a 1-cup serving of Cheerios® contains 4 g of fiber. When a meal consists of 1 pancake, a 1-cup serving of Cheerios®, and a 1-cup serving of strawberries, it contains 7 g of fiber. How much fiber is in each of these foods?

Source: InteliHealth.com

Aha! 15. *Automobile Pricing.*   The base model of a 2012 Jeep Grand Cherokee Laredo (2WD) with a tow package cost $29,515. When equipped with a tow package and a sunroof, the vehicle's price rose to $30,365. The cost of the base model with a sunroof was $29,770. Find the base price, the cost of a tow package, and the cost of a sunroof.

Source: www.jeep.com

16. *Telemarketing.*   Sven, Tina, and Laurie can process 740 telephone orders per day. Sven and Tina together can process 470 orders, while Tina and Laurie together can process 520 orders per day. How many orders can each person process alone?

17. *Coffee Prices.*   Reba works at a Starbucks® coffee shop where a 12-oz cup of coffee costs $1.75, a 16-oz cup costs $1.95, and a 20-oz cup costs $2.25. During one busy period, Reba served 55 cups of coffee, emptying six 144-oz "brewers" while collecting a total of $107.75. How many cups of each size did Reba fill?

18. *Restaurant Management.*   Chick-fil-A® recently sold 14-oz lemonades for $1.49 each, 20-oz lemonades for $1.69 each, and 32-oz lemonades for $2.05 each. During a lunchtime rush, Chris sold 40 lemonades, using $6\frac{1}{4}$ gal of lemonade while collecting a total of $67.40. How many drinks of each size were sold? (*Hint:* 1 gal = 128 oz.)

19. *Small-Business Loans.*   Chelsea took out three loans for a total of $120,000 to start an organic orchard. Her business-equipment loan was at an interest rate of 11%, the small-business loan was at an interest rate of 5%, and her home-equity loan was at an interest rate of 5.5%. The total simple interest due on the loans in one year was $7250. The annual simple interest on the home-equity loan was $2200 more than the interest on the business-equipment loan. How much did she borrow from each source?

20. *Investments.*   A business class divided an imaginary investment of $80,000 among three mutual funds. The first fund grew by 4%, the second by 6%, and the third by 8%. Total earnings were $4400. The earnings from the third fund were $200 more than the earnings from the first. How much was invested in each fund?

21. *Gold Alloys.*   Gold used to make jewelry is often a blend of gold, silver, and copper. The relative amounts of the metals determine the color of the alloy. Red gold is 75% gold, 5% silver, and 20% copper. Yellow gold is 75% gold, 12.5% silver, and 12.5%

copper. White gold is 37.5% gold and 62.5% silver. If 100 g of red gold costs $4177.15, 100 g of yellow gold costs $4185.25, and 100 g of white gold costs $2153.875, how much do gold, silver, and copper cost per gram?

Source: World Gold Council

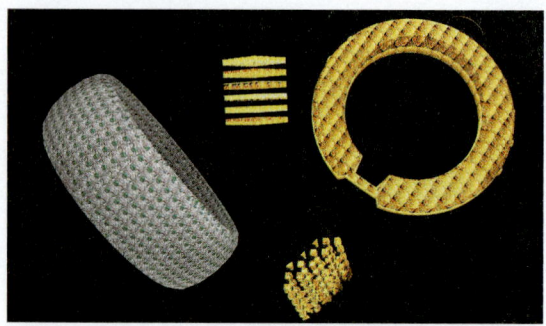

22. *Gardening.* Dana is designing three large perennial flower beds for her yard and is planning to use combinations of three types of flowers. In her traditional cottage-style garden, Dana will include 7 purple coneflower plants, 6 yellow foxglove plants, and 8 white lupine plants at a total cost of $93.31. The flower bed around her deck will be planted with 12 yellow foxglove plants and 12 white lupine plants at a total cost of $126.00. A third garden area in a corner of Dana's yard will contain 4 purple coneflower plants, 5 yellow foxglove plants, and 7 white lupine plants at a total cost of $72.82. What is the price per plant for the coneflowers, the foxgloves, and the lupines?

23. *Nutrition.* A dietician in a hospital prepares meals under the guidance of a physician. Suppose that for a particular patient a physician prescribes a meal to have 800 calories, 55 g of protein, and 220 mg of vitamin C. The dietician prepares a meal of roast beef, baked potatoes, and broccoli according to the data in the following table.

Serving Size	Calories	Protein (in grams)	Vitamin C (in milligrams)
Roast Beef, 3 oz	300	20	0
Baked Potato, 1	100	5	20
Broccoli, 156 g	50	5	100

How many servings of each food are needed in order to satisfy the doctor's orders?

24. *Nutrition.*   Repeat Exercise 23 but replace the broccoli with asparagus, for which a 180-g serving contains 50 calories, 5 g of protein, and 44 mg of vitamin C. Which meal would you prefer eating?

25. Students in a Listening Responses class bought 40 tickets for a piano concert. The number of tickets purchased for seats in either the first mezzanine or the main floor was the same as the number purchased for seats in the second mezzanine. First mezzanine seats cost $52 each, main floor seats cost $38 each, and second mezzanine seats cost $28 each. The total cost of the tickets was $1432. How many of each type of ticket were purchased?

26. *Basketball Scoring.*   The New York Knicks recently scored a total of 92 points on a combination of 2-point field goals, 3-point field goals, and 1-point foul shots. Altogether, the Knicks made 50 baskets and 19 more 2-pointers than foul shots. How many shots of each kind were made?

27. *World Population Growth.*   The world population is projected to be 9.4 billion in 2050. At that time, there is expected to be approximately 2.9 billion more people in Asia than in Africa. The population for the rest of the world will be approximately 0.6 billion more than one-fourth of the population of Asia. Find the projected populations of Asia, Africa, and the rest of the world in 2050.

Source: U.S. Census Bureau

28. *History.* Find the year in which the first U.S. transcontinental railroad was completed. The sum of the digits in the year is 24. The ones digit is 1 more than the hundreds digit. Both the tens and the ones digits are multiples of 3.

29. Jaci knows that one angle in a triangle is twice as large as another. Does she have enough information to find the measures of the angles in the triangle? Why or why not?

30. Write a problem for a classmate to solve. Design the problem so that it translates to a system of three equations in three variables.

Skill Review

Graph.

31. $y = 4$ [3.3]

32. $y = -\frac{2}{5}x + 3$ [3.6]

33. $y - 3x = 3$ [3.3]

34. $2x = -8$ [3.3]

35. $f(x) = 2x - 1$ [7.3]

36. $3x - y = 2$ [3.2]

Synthesis

37. Consider Exercise 26. Suppose there were no foul shots made. Would there still be a solution? Why or why not?

38. Consider Exercise 17. Suppose Reba collected $50. Could the problem still be solved? Why or why not?

39. *Health Insurance.* In 2012, UnitedHealthOne Plan 100 health insurance for a 35-year-old and his or her spouse cost $304 per month. That rate increased to $420 per month if a child were included and $537 per month if two children were included. The rate dropped to $266 per month for just the applicant and one child. Find the separate costs for insuring the applicant, the spouse, the first child, and the second child.

    Source: UNICARE Life and Health Insurance Company® through www.ehealth.com

40. Find a three-digit number such that the sum of the digits is 14, the tens digit is 2 more than the ones digit, and if the digits are reversed, the number is unchanged.

41. *Ages.* Tammy's age is the sum of the ages of Carmen and Dennis. Carmen's age is 2 more than the sum of the ages of Dennis and Mark. Dennis's age is four times Mark's age. The sum of all four ages is 42. How old is Tammy?

42. *Ticket Revenue.* A magic show's audience of 100 people consists of adults, students, and children. The ticket prices are $10 each for adults, $3 each for students, and 50¢ each for children. A total of $100 is taken in. How many adults, students, and children are in attendance? Does there seem to be some information missing? Do some more careful reasoning.

43. *Sharing Raffle Tickets.* Hal gives Tom as many raffle tickets as Tom first had and Gary as many as Gary first had. In like manner, Tom then gives Hal and Gary as many tickets as each then has. Similarly, Gary gives Hal and Tom as many tickets as each then has. If each finally has 40 tickets, with how many tickets does Tom begin?

44. Find the sum of the angle measures at the tips of the star in this figure.

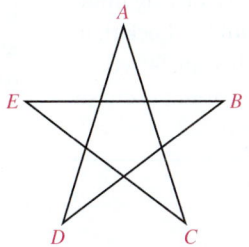

QUICK QUIZ: SECTIONS 8.1–8.5

Solve. [8.1], [8.2], [8.4]

1. $y = 2x - 5,$
   $y = \frac{1}{2}x + 1$

2. $x + 2y = 3,$
   $3x = 4 - y$

3. $10x + 20y = 40,$
   $x - \quad y = 7$

4. $9x + 8y = 0,$
   $11x - 7y = 0$

5. $2x + \ y + \ z = 3,$
   $x + \ y - 4z = 13,$
   $4x + 3y + 2z = 11$

PREPARE TO MOVE ON

Simplify. [1.8]

1. $-2(2x - 3y)$

2. $-6(x - 2y) + (6x - 5y)$

3. $-(2a - b - 6c)$

4. $-2(3x - y + z) + 3(-2x + y - 2z)$

5. $(8x - 10y + 7z) + 5(3x + 2y - 4z)$

8.6 Elimination Using Matrices

Matrices and Systems ▪ Row-Equivalent Operations

In solving systems of equations, we perform computations with the constants. If we agree to keep all like terms in the same column, we can simplify writing a system by omitting the variables. For example, if we do not write the variables, the operation of addition, and the equal signs, the system

$$\begin{aligned} 3x + 4y &= 5, \\ x - 2y &= 1 \end{aligned} \quad \text{simplifies to} \quad \begin{array}{ccc} 3 & 4 & 5 \\ 1 & -2 & 1 \end{array}.$$

Matrices and Systems

In the example above, we have written a rectangular array of numbers. Such an array is called a **matrix** (plural, **matrices**). We ordinarily write brackets around matrices. The following are matrices:

$$\begin{bmatrix} -3 & 1 \\ 0 & 5 \end{bmatrix}, \quad \begin{bmatrix} 2 & 0 & -1 & 3 \\ -5 & 2 & 7 & -1 \\ 4 & 5 & 3 & 0 \end{bmatrix}, \quad \begin{bmatrix} 2 & 3 \\ 7 & 15 \\ -2 & 23 \\ 4 & 1 \end{bmatrix}.$$

The individual numbers are called *elements*, or *entries*.

The **rows** of a matrix are horizontal, and the **columns** are vertical.

$$\begin{bmatrix} 5 & -2 & 2 \\ 1 & 0 & 1 \\ 0 & 1 & 2 \end{bmatrix} \begin{array}{l} \longrightarrow \text{row 1} \\ \longrightarrow \text{row 2} \\ \longrightarrow \text{row 3} \end{array}$$

column 1   column 2   column 3

Let's see how matrices can be used to solve a system.

EXAMPLE 1   Solve the system

$$\begin{aligned} 5x - 4y &= -1, \\ -2x + 3y &= 2. \end{aligned}$$

To better explain each step, we list the corresponding system in the margin.

$$\begin{aligned} 5x - 4y &= -1, \\ -2x + 3y &= 2 \end{aligned}$$

SOLUTION   We write a matrix using only coefficients and constants, listing x-coefficients in the first column and y-coefficients in the second. A dashed line separates the coefficients from the constants:

$$\begin{bmatrix} 5 & -4 & \vdots & -1 \\ -2 & 3 & \vdots & 2 \end{bmatrix}. \qquad \text{Refer to the notes in the margin for further information.}$$

Our goal is to transform

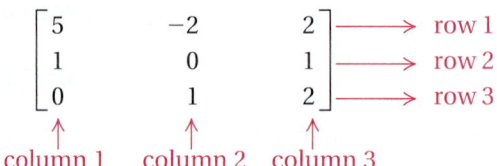

We can then reinsert the variables x and y, form equations, and complete the solution.

Our calculations are similar to those done if we wrote the entire equations. The first step is to multiply and/or interchange the rows so that each number in the first column below the first number is a multiple of that number. Here that means multiplying Row 2 by 5. This corresponds to multiplying both sides of the second equation by 5.

$$5x - 4y = -1,$$
$$-10x + 15y = 10$$

$$\begin{bmatrix} 5 & -4 & \vdots & -1 \\ -10 & 15 & \vdots & 10 \end{bmatrix}$$

New Row 2 = 5(Row 2 from the step above)
$$= 5(-2 \quad 3 \quad \vdots \quad 2) = (-10 \quad 15 \quad \vdots \quad 10)$$

Next, we multiply the first row by 2, add this to Row 2, and write that result as the "new" Row 2. This corresponds to multiplying the first equation by 2 and adding the result to the second equation in order to eliminate a variable. Write out these computations as necessary.

$$5x - 4y = -1,$$
$$7y = 8$$

$$\begin{bmatrix} 5 & -4 & \vdots & -1 \\ 0 & 7 & \vdots & 8 \end{bmatrix}$$

2(Row 1) $= 2(5 \quad -4 \quad \vdots \quad -1) = (10 \quad -8 \quad \vdots \quad -2)$
New Row 2 $= (10 \quad -8 \quad \vdots \quad -2) + (-10 \quad 15 \quad \vdots \quad 10)$
$= (0 \quad 7 \quad \vdots \quad 8)$

If we now reinsert the variables, we have

$$5x - 4y = -1, \quad (1) \qquad \text{From Row 1}$$
$$7y = 8. \quad (2) \qquad \text{From Row 2}$$

Solving equation (2) for y gives us

$$7y = 8 \quad (2)$$
$$y = \tfrac{8}{7}.$$

Next, we substitute $\tfrac{8}{7}$ for y in equation (1):

$$5x - 4y = -1 \quad (1)$$
$$5x - 4 \cdot \tfrac{8}{7} = -1 \qquad \text{Substituting } \tfrac{8}{7} \text{ for } y \text{ in equation (1)}$$
$$x = \tfrac{5}{7}. \qquad \text{Solving for } x$$

The solution is $\left(\tfrac{5}{7}, \tfrac{8}{7}\right)$. The check is left to the student.

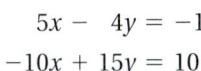 YOUR TURN

1. Solve using matrices:

$$-x + 5y = 4,$$
$$3x - y = 6.$$

EXAMPLE 2 Solve the system

$$2x - y + 4z = -3,$$
$$x \qquad - 4z = 5,$$
$$6x - y + 2z = 10.$$

SOLUTION We first write a matrix, using only the constants. Where there are missing terms, we must write 0's:

$$2x - y + 4z = -3,$$
$$x \qquad - 4z = 5,$$
$$6x - y + 2z = 10$$

$$\begin{bmatrix} 2 & -1 & 4 & \vdots & -3 \\ 1 & 0 & -4 & \vdots & 5 \\ 6 & -1 & 2 & \vdots & 10 \end{bmatrix}.$$

Our goal is to transform the matrix to one of the form

$$ax + by + cz = d,$$
$$ey + fz = g,$$
$$hz = i$$

$$\begin{bmatrix} a & b & c & \vdots & d \\ 0 & e & f & \vdots & g \\ 0 & 0 & h & \vdots & i \end{bmatrix}.$$

This matrix is in a form called *row-echelon form.*

A matrix of this form can be rewritten as a system of equations that is equivalent to the original system, and from which a solution can be easily found.

The first step is to multiply and/or interchange the rows so that each number in the first column is a multiple of the first number in the first row. In this case, we begin by interchanging Rows 1 and 2:

$$x \quad\; - 4z = 5,$$
$$2x - y + 4z = -3,$$
$$6x - y + 2z = 10$$

$$\begin{bmatrix} 1 & 0 & -4 & \vdots & 5 \\ 2 & -1 & 4 & \vdots & -3 \\ 6 & -1 & 2 & \vdots & 10 \end{bmatrix}.$$

This corresponds to interchanging the first two equations.

Next, we multiply the first row by -2, add it to the second row, and replace Row 2 with the result:

$$x \quad\;\; - 4z = 5,$$
$$\quad\; - y + 12z = -13,$$
$$6x - y + \; 2z = 10$$

$$\begin{bmatrix} 1 & 0 & -4 & \vdots & 5 \\ 0 & -1 & 12 & \vdots & -13 \\ 6 & -1 & 2 & \vdots & 10 \end{bmatrix}.$$

$-2(1\;\; 0\;\; -4\; \vdots\; 5) = (-2\;\; 0\;\; 8\; \vdots\; -10)$ and
$(-2\;\; 0\;\; 8\; \vdots\; -10) + (2\;\; -1\;\; 4\; \vdots\; -3) =$
$(0\;\; -1\;\; 12\; \vdots\; -13)$

Now we multiply the first row by -6, add it to the third row, and replace Row 3 with the result:

$$x \quad\;\; - 4z = 5,$$
$$-y + 12z = -13,$$
$$-y + 26z = -20$$

$$\begin{bmatrix} 1 & 0 & -4 & \vdots & 5 \\ 0 & -1 & 12 & \vdots & -13 \\ 0 & -1 & 26 & \vdots & -20 \end{bmatrix}.$$

$-6(1\;\; 0\;\; -4\; \vdots\; 5) = (-6\;\; 0\;\; 24\; \vdots\; -30)$ and
$(-6\;\; 0\;\; 24\; \vdots\; -30) + (6\;\; -1\;\; 2\; \vdots\; 10) =$
$(0\;\; -1\;\; 26\; \vdots\; -20)$

Next, we multiply Row 2 by -1, add it to the third row, and replace Row 3 with the result:

$$x \quad\;\; - 4z = 5,$$
$$-y + 12z = -13,$$
$$14z = -7$$

$$\begin{bmatrix} 1 & 0 & -4 & \vdots & 5 \\ 0 & -1 & 12 & \vdots & -13 \\ 0 & 0 & 14 & \vdots & -7 \end{bmatrix}.$$

$-1(0\;\; -1\;\; 12\; \vdots\; -13) = (0\;\; 1\;\; -12\; \vdots\; 13)$ and
$(0\;\; 1\;\; -12\; \vdots\; 13) + (0\;\; -1\;\; 26\; \vdots\; -20) =$
$(0\;\; 0\;\; 14\; \vdots\; -7)$

Reinserting the variables gives us

$$x \quad\;\; - \; 4z = \;\;\; 5,$$
$$-y + 12z = -13,$$
$$14z = \; -7.$$

2. Solve using matrices:

$$2x + \; y + 3z = 1,$$
$$x + 2y + 4z = 6,$$
$$-2x \qquad\; - \; z = 7.$$

Solving this last equation for z, we get $z = -\frac{1}{2}$. Next, we substitute $-\frac{1}{2}$ for z in the preceding equation and solve for y: $-y + 12\left(-\frac{1}{2}\right) = -13$, so $y = 7$. Since there is no y-term in the first equation of this last system, we need only substitute $-\frac{1}{2}$ for z to solve for x: $x - 4\left(-\frac{1}{2}\right) = 5$, so $x = 3$. The solution is $\left(3, 7, -\frac{1}{2}\right)$. The check is left to the student.

 YOUR TURN

The operations used in the preceding example correspond to those used to produce equivalent systems of equations, that is, systems of equations that have the same solution. We call the matrices **row-equivalent** and the operations that produce them **row-equivalent operations.**

Row-Equivalent Operations

ROW-EQUIVALENT OPERATIONS

Each of the following row-equivalent operations produces a row-equivalent matrix:

a) Interchanging any two rows.
b) Multiplying all elements of a row by a nonzero constant.
c) Replacing a row with the sum of that row and a multiple of another row.

Student Notes

Note that row-equivalent matrices are not *equal*. It is the solutions of the corresponding systems that are the same.

Computers solve systems of equations using row-equivalent matrices. Matrices are part of a branch of mathematics known as linear algebra. They are also studied in many courses in finite mathematics.

Technology Connection

Row-equivalent operations can be performed on a graphing calculator. For example, to interchange the first and second rows of the matrix, as in step (1) of Example 2 above, we enter the matrix as matrix **A** and select "rowSwap" from the MATRIX MATH menu.

To store the result of the operation as **B**, we use **STO**, as shown in the window at right.

```
rowSwap([A],1,2) → [B]
        [1   0  -4   5]
        [2  -1   4  -3]
        [6  -1   2  10]
```

1. Use a graphing calculator to proceed through all the steps in Example 2.

8.6 EXERCISE SET

FOR EXTRA HELP

MyMathLab® Math XL
PRACTICE   WATCH   READ   REVIEW

Vocabulary and Reading Check

Complete each of the following statements.

1. A(n) _____ is a rectangular array of numbers.

2. The _____ of a matrix are horizontal, and the columns are _____.

3. Each number in a matrix is called a(n) _____ or element.

4. The plural of the word matrix is _____.

5. As part of solving a system using matrices, we can interchange any two _____.

6. When a matrix is in row-echelon form, the leftmost column in the matrix has zeros in all rows except the _____ one.

Matrices and Systems

Solve using matrices.

7. $x + 2y = 11,$
$3x - y = 5$

8. $x + 3y = 16,$
$6x + y = 11$

9. $3x + y = -1,$
$6x + 5y = 13$

10. $2x - y = 6,$
$8x + 2y = 0$

11. $6x - 2y = 4,$
$7x + y = 13$

12. $3x + 4y = 7,$
$-5x + 2y = 10$

13. $3x + 2y + 2z = 3,$
$x + 2y - z = 5,$
$2x - 4y + z = 0$

14. $4x - y - 3z = 19,$
$8x + y - z = 11,$
$2x + y + 2z = -7$

15. $p - 2q - 3r = 3,$
$2p - q - 2r = 4,$
$4p + 5q + 6r = 4$

16. $x + 2y - 3z = 9,$
$2x - y + 2z = -8,$
$3x - y - 4z = 3$

17. $3p + 2r = 11,$
$q - 7r = 4,$
$p - 6q = 1$

18. $4a + 9b = 8,$
$8a + 6c = -1,$
$6b + 6c = -1$

19. $2x + 2y - 2z - 2w = -10,$
$w + y + z + x = -5,$
$x - y + 4z + 3w = -2,$
$w - 2y + 2z + 3x = -6$

20. $-w - 3y + z + 2x = -8,$
$x + y - z - w = -4,$
$w + y + z + x = 22,$
$x - y - z - w = -14$

Solve using matrices.

21. *Coin Value.* A collection of 42 coins consists of dimes and nickels. The total value is $3.00. How many dimes and how many nickels are there?

22. *Coin Value.* A collection of 43 coins consists of dimes and quarters. The total value is $7.60. How many dimes and how many quarters are there?

23. *Snack Mix.* Bree sells a dried-fruit mixture for $5.80 per pound and Hawaiian macadamia nuts for $14.75 per pound. She wants to blend the two to get a 15-lb mixture that she will sell for $9.38 per pound. How much of each should she use?

24. *Mixing Paint.* Higher quality paint typically contains more solids. Alex has available paint that contains 45% solids and paint that contains 25% solids. How much of each should he use to create 20 gal of paint that contains 39% solids?

25. *Investments.* Elena receives $112 per year in simple interest from three investments totaling $2500. Part is invested at 3%, part at 4%, and part at 5%. There is $1100 more invested at 5% than at 4%. Find the amount invested at each rate.

26. *Investments.* Miguel receives $160 per year in simple interest from three investments totaling $3200. Part is invested at 2%, part at 3%, and part at 6%. There is $1900 more invested at 6% than at 3%. Find the amount invested at each rate.

27. Explain how you can recognize dependent equations when solving with matrices.

28. Explain how you can recognize an inconsistent system when solving with matrices.

Skill Review

Simplify.

29. $\dfrac{x^3 - x}{x + 2} \cdot \dfrac{x^2 + 3x + 2}{x - 1}$ [6.2]

30. $\dfrac{12x^2y^4}{5z} \div \dfrac{10xz}{3y^2}$ [6.2]

31. $\dfrac{3}{3a^2 + 2a - 1} - \dfrac{1}{a^2 + 2a + 1}$ [6.4]

32. $\dfrac{\dfrac{1}{t} + \dfrac{2}{t^2}}{\dfrac{3}{5t}}$ [6.5]

Synthesis

33. If the matrices

$$\begin{bmatrix} a_1 & b_1 & c_1 \\ d_1 & e_1 & f_1 \end{bmatrix} \text{ and } \begin{bmatrix} a_2 & b_2 & c_2 \\ d_2 & e_2 & f_2 \end{bmatrix}$$

share the same solution, does it follow that the corresponding entries are all equal to each other ($a_1 = a_2$, $b_1 = b_2$, etc.)? Why or why not?

34. Explain how the row-equivalent operations make use of the addition, multiplication, and distributive properties.

35. The sum of the digits in a four-digit number is 10. Twice the sum of the thousands digit and the tens digit is 1 less than the sum of the other two digits.

The tens digit is twice the thousands digit. The ones digit equals the sum of the thousands digit and the hundreds digit. Find the four-digit number.

36. Solve for x and y:

$$ax + by = c,$$
$$dx + ey = f.$$

 YOUR TURN ANSWERS: SECTION 8.6

1. $\left(\frac{17}{7}, \frac{9}{7}\right)$ **2.** $(-10, -18, 13)$

QUICK QUIZ: SECTIONS 8.1–8.6

Solve. If a system has an infinite number of solutions, use set-builder notation to write the solution set. If a system has no solution, state this.

1. $2x + y = 3,$
    $6x + 2y = 4$ [8.2]

2. $y = \dfrac{5}{3}x + 7,$
    $y = \dfrac{5}{3}x - 8$ [8.1], [8.2]

3. Solve: $x + 5y = 6,$
        $x + 2z = 3,$
        $5y + 2z = 8.$ [8.4]

4. Network Community College bought 42 packages of dry-erase markers. Some packages contained 4 markers and some contained 6 markers. If they purchased a total of 200 markers, how many of each size package did they buy?   [8.3]

5. Drink Fresh contains 30% juice, and Summer Light contains 5% juice. How much of each should be mixed to obtain 6 L of a beverage that contains 10% juice?   [8.3]

PREPARE TO MOVE ON

Simplify.   [1.8]

1. $3(-1) - (-4)(5)$

2. $7(-5) - 2(-8)$

3. $-2(5 \cdot 3 - 4 \cdot 6) - 3(2 \cdot 7 - 15) + 4(3 \cdot 8 - 5 \cdot 4)$

4. $6(2 \cdot 7 - 3(-4)) - 4(3(-8) - 10) + 5(4 \cdot 3 - (-2)7)$

8.7   Determinants and Cramer's Rule

Determinants of 2 × 2 Matrices ▪ Cramer's Rule: 2 × 2 Systems ▪ Cramer's Rule: 3 × 3 Systems

Determinants of 2 × 2 Matrices

When a matrix has m rows and n columns, it is called an "m by n" matrix, and its *dimensions* are $m \times n$. If a matrix has the same number of rows and columns, it is called a **square matrix**. Associated with every square matrix is a number called its **determinant**, defined as follows for 2 × 2 matrices.

> **2 × 2 DETERMINANTS**
>
> The determinant of a two-by-two matrix $\begin{bmatrix} a & c \\ b & d \end{bmatrix}$ is denoted $\begin{vmatrix} a & c \\ b & d \end{vmatrix}$
>
> and is defined as follows:
>
> $$\begin{vmatrix} a & c \\ b & d \end{vmatrix} = ad - bc.$$

EXAMPLE 1   Evaluate: $\begin{vmatrix} 2 & -5 \\ 6 & 7 \end{vmatrix}$.

SOLUTION   We multiply and subtract as follows:

$$\begin{vmatrix} 2 & -5 \\ 6 & 7 \end{vmatrix} = 2 \cdot 7 - 6 \cdot (-5) = 14 + 30 = 44.$$

YOUR TURN

1. Evaluate: $\begin{vmatrix} -3 & -1 \\ 5 & 7 \end{vmatrix}$.

Cramer's Rule: 2 × 2 Systems

One of the many uses for determinants is in solving systems of linear equations in which the number of variables is the same as the number of equations and the constants are not all 0. Let's consider a system of two equations:

$$a_1 x + b_1 y = c_1,$$
$$a_2 x + b_2 y = c_2.$$

If we use the elimination method, a series of steps can show that

$$x = \frac{c_1 b_2 - c_2 b_1}{a_1 b_2 - a_2 b_1} \quad \text{and} \quad y = \frac{a_1 c_2 - a_2 c_1}{a_1 b_2 - a_2 b_1}.$$

These fractions can be rewritten using determinants.

CRAMER'S RULE: 2 × 2 SYSTEMS

The solution of the system

$$a_1 x + b_1 y = c_1,$$
$$a_2 x + b_2 y = c_2,$$

if it is unique, is given by

$$x = \frac{\begin{vmatrix} c_1 & b_1 \\ c_2 & b_2 \end{vmatrix}}{\begin{vmatrix} a_1 & b_1 \\ a_2 & b_2 \end{vmatrix}}, \qquad y = \frac{\begin{vmatrix} a_1 & c_1 \\ a_2 & c_2 \end{vmatrix}}{\begin{vmatrix} a_1 & b_1 \\ a_2 & b_2 \end{vmatrix}}.$$

These formulas apply only if the denominator is not 0.

To use Cramer's rule, we find the determinants and compute x and y as shown above. Note that in the denominators, which are identical, the coefficients of x and y appear in the same position as in the original equations. In the numerator of x, the constants c_1 and c_2 replace a_1 and a_2. In the numerator of y, the constants c_1 and c_2 replace b_1 and b_2.

EXAMPLE 2   Solve using Cramer's rule:

$$2x + 5y = 7,$$
$$5x - 2y = -3.$$

SOLUTION   We have

$$x = \frac{\begin{vmatrix} 7 & 5 \\ -3 & -2 \end{vmatrix}}{\begin{vmatrix} 2 & 5 \\ 5 & -2 \end{vmatrix}} \quad \begin{array}{l} \leftarrow \text{The constants } \genfrac{}{}{0pt}{}{7}{-3} \text{ replace } \genfrac{}{}{0pt}{}{a_1}{a_2} \text{ in the first column.} \\ \\ \leftarrow \text{The columns are the coefficients, } \genfrac{}{}{0pt}{}{a_1 \ \ b_1}{a_2 \ \ b_2}. \end{array}$$

$$= \frac{7(-2) - (-3)5}{2(-2) - 5 \cdot 5} = \frac{1}{-29} = -\frac{1}{29}$$

and

$$y = \frac{\begin{vmatrix} 2 & 7 \\ 5 & -3 \end{vmatrix}}{\begin{vmatrix} 2 & 5 \\ 5 & -2 \end{vmatrix}} \quad \begin{array}{l} \leftarrow \text{The constants } \genfrac{}{}{0pt}{}{7}{-3} \text{ replace } \genfrac{}{}{0pt}{}{b_1}{b_2} \text{ in the second column.} \\ \\ \leftarrow \text{The denominator is the same as in the expression for } x. \end{array}$$

$$= \frac{2(-3) - 5 \cdot 7}{-29} = \frac{-41}{-29} = \frac{41}{29}.$$

2. Solve using Cramer's rule:

$$-2x - y = 7,$$
$$3x + 4y = 1.$$

The solution is $\left(-\frac{1}{29}, \frac{41}{29}\right)$. The check is left to the student.

↩ YOUR TURN

Cramer's Rule: 3 × 3 Systems

Cramer's rule can be extended for systems of three linear equations. However, before doing so, we must define what a 3 × 3 determinant is.

3 × 3 DETERMINANTS

The determinant of a three-by-three matrix can be defined as follows:

$$\begin{vmatrix} a_1 & b_1 & c_1 \\ a_2 & b_2 & c_2 \\ a_3 & b_3 & c_3 \end{vmatrix} = a_1 \begin{vmatrix} b_2 & c_2 \\ b_3 & c_3 \end{vmatrix} \overset{\text{Subtract.}}{-} a_2 \begin{vmatrix} b_1 & c_1 \\ b_3 & c_3 \end{vmatrix} \overset{\text{Add.}}{+} a_3 \begin{vmatrix} b_1 & c_1 \\ b_2 & c_2 \end{vmatrix}.$$

Student Notes

Cramer's rule and the evaluation of determinants rely on patterns. Recognizing and remembering the patterns will help you understand and use the definitions.

Note that the a's come from the first column. Note too that the 2×2 determinants above can be obtained by crossing out the row and the column in which the a occurs.

For a_1:

$$\begin{vmatrix} \cancel{a_1} & \cancel{b_1} & \cancel{c_1} \\ a_2 & b_2 & c_2 \\ a_3 & b_3 & c_3 \end{vmatrix}$$

For a_2:

$$\begin{vmatrix} a_1 & b_1 & c_1 \\ \cancel{a_2} & \cancel{b_2} & \cancel{c_2} \\ a_3 & b_3 & c_3 \end{vmatrix}$$

For a_3:

$$\begin{vmatrix} a_1 & b_1 & c_1 \\ a_2 & b_2 & c_2 \\ \cancel{a_3} & \cancel{b_3} & \cancel{c_3} \end{vmatrix}$$

EXAMPLE 3 Evaluate:

$$\begin{vmatrix} -1 & 0 & 1 \\ -5 & 1 & -1 \\ 4 & 8 & 1 \end{vmatrix}.$$

SOLUTION We have

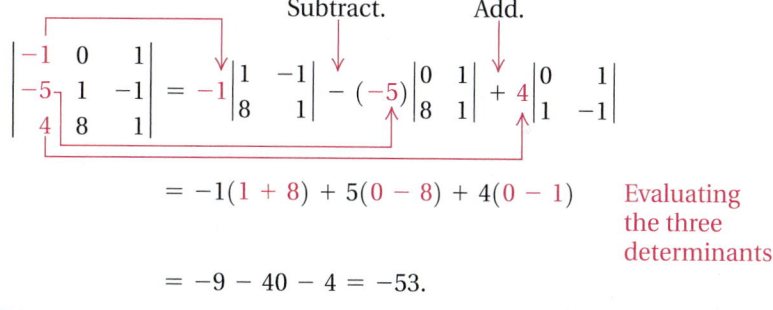

$$= -1(1 + 8) + 5(0 - 8) + 4(0 - 1) \qquad \text{Evaluating the three determinants}$$

$$= -9 - 40 - 4 = -53.$$

3. Evaluate: $\begin{vmatrix} 2 & -1 & 0 \\ 1 & 0 & -3 \\ 6 & 2 & 4 \end{vmatrix}.$

 YOUR TURN

Technology Connection

Determinants can be evaluated on most graphing calculators using **2ND** (MATRIX). After entering a matrix, we select the determinant operation from the MATRIX MATH menu and enter the name of the matrix. The graphing calculator will return the value of the determinant of the matrix. For example, if

$$\mathbf{A} = \begin{bmatrix} 1 & 6 & -1 \\ -3 & -5 & 3 \\ 0 & 4 & 2 \end{bmatrix},$$

we have

```
det([A])
                26
```

1. Confirm the calculations in Example 4.

CRAMER'S RULE: 3 × 3 SYSTEMS

The solution of the system

$$a_1x + b_1y + c_1z = d_1,$$
$$a_2x + b_2y + c_2z = d_2,$$
$$a_3x + b_3y + c_3z = d_3$$

can be found using the following determinants:

$$D = \begin{vmatrix} a_1 & b_1 & c_1 \\ a_2 & b_2 & c_2 \\ a_3 & b_3 & c_3 \end{vmatrix}, \qquad D_x = \begin{vmatrix} d_1 & b_1 & c_1 \\ d_2 & b_2 & c_2 \\ d_3 & b_3 & c_3 \end{vmatrix},$$

$$D_y = \begin{vmatrix} a_1 & d_1 & c_1 \\ a_2 & d_2 & c_2 \\ a_3 & d_3 & c_3 \end{vmatrix}, \qquad D_z = \begin{vmatrix} a_1 & b_1 & d_1 \\ a_2 & b_2 & d_2 \\ a_3 & b_3 & d_3 \end{vmatrix}.$$

D contains only coefficients.
In D_x the *d*'s replace the *a*'s.
In D_y, the *d*'s replace the *b*'s.
In D_z, the *d*'s replace the *c*'s.

If a unique solution exists, it is given by

$$x = \frac{D_x}{D}, \qquad y = \frac{D_y}{D}, \qquad z = \frac{D_z}{D}.$$

These formulas apply only if $D \neq 0$.

EXAMPLE 4 Solve using Cramer's rule:

$$x - 3y + 7z = 13,$$
$$x + y + z = 1,$$
$$x - 2y + 3z = 4.$$

SOLUTION We compute D, D_x, D_y, and D_z:

$$D = \begin{vmatrix} 1 & -3 & 7 \\ 1 & 1 & 1 \\ 1 & -2 & 3 \end{vmatrix} = -10; \qquad D_x = \begin{vmatrix} 13 & -3 & 7 \\ 1 & 1 & 1 \\ 4 & -2 & 3 \end{vmatrix} = 20;$$

$$D_y = \begin{vmatrix} 1 & 13 & 7 \\ 1 & 1 & 1 \\ 1 & 4 & 3 \end{vmatrix} = -6; \qquad D_z = \begin{vmatrix} 1 & -3 & 13 \\ 1 & 1 & 1 \\ 1 & -2 & 4 \end{vmatrix} = -24.$$

4. Solve using Cramer's rule:

$$x - y + 2z = 5,$$
$$2x + y - z = 6,$$
$$-x + 2y - 2z = 3.$$

Then

$$x = \frac{D_x}{D} = \frac{20}{-10} = -2; \qquad y = \frac{D_y}{D} = \frac{-6}{-10} = \frac{3}{5}; \qquad z = \frac{D_z}{D} = \frac{-24}{-10} = \frac{12}{5}.$$

The solution is $\left(-2, \frac{3}{5}, \frac{12}{5}\right)$. The check is left to the student.

YOUR TURN

In Example 4, we need not have evaluated D_z. Once x and y were found, we could have substituted them into one of the equations to find z.

When we are using Cramer's rule, if the denominator is 0 and at least one of the other determinants is not 0, the system is inconsistent. If *all* the determinants are 0, then the equations in the system are dependent.

8.7   EXERCISE SET

FOR EXTRA HELP

 MyMathLab® Math XL
PRACTICE   WATCH   READ   REVIEW

◆ Vocabulary and Reading Check

Classify each of the following statements as either true or false.

1. A square matrix has the same number of rows and columns.

2. A 3 × 4 matrix has 3 rows and 4 columns.

3. A determinant is a number.

4. Cramer's rule exists only for 2 × 2 systems.

5. Whenever Cramer's rule yields a denominator that is 0, the system has no solution.

6. Whenever Cramer's rule yields a numerator that is 0, the equations are dependent.

Determinants

Evaluate.

7. $\begin{vmatrix} 3 & 5 \\ 4 & 8 \end{vmatrix}$

8. $\begin{vmatrix} 3 & 2 \\ 2 & -3 \end{vmatrix}$

9. $\begin{vmatrix} 10 & 8 \\ -5 & -9 \end{vmatrix}$

10. $\begin{vmatrix} 3 & 2 \\ -7 & 11 \end{vmatrix}$

11. $\begin{vmatrix} 1 & 4 & 0 \\ 0 & -1 & 2 \\ 3 & -2 & 1 \end{vmatrix}$

12. $\begin{vmatrix} 2 & 4 & -2 \\ 1 & 0 & 2 \\ 0 & 1 & 3 \end{vmatrix}$

13. $\begin{vmatrix} -1 & -2 & -3 \\ 3 & 4 & 2 \\ 0 & 1 & 2 \end{vmatrix}$

14. $\begin{vmatrix} 5 & 2 & 2 \\ 0 & 1 & -1 \\ 3 & 3 & 1 \end{vmatrix}$

15. $\begin{vmatrix} -4 & -2 & 3 \\ -3 & 1 & 2 \\ 3 & 4 & -2 \end{vmatrix}$

16. $\begin{vmatrix} 2 & -1 & 1 \\ 1 & 2 & -1 \\ 3 & 4 & -3 \end{vmatrix}$

Cramer's Rule

Solve using Cramer's rule.

17. $5x + 8y = 1,$
    $3x + 7y = 5$

18. $3x - 4y = 6,$
    $5x + 9y = 10$

19. $5x - 4y = -3,$
    $7x + 2y = 6$

20. $-2x + 4y = 3,$
    $3x - 7y = 1$

21. $3x - y + 2z = 1,$
    $x - y + 2z = 3,$
    $-2x + 3y + z = 1$

22. $3x + 2y - z = 4,$
    $3x - 2y + z = 5,$
    $4x - 5y - z = -1$

23. $2x - 3y + 5z = 27,$
    $x + 2y - z = -4,$
    $5x - y + 4z = 27$

24. $x - y + 2z = -3,$
    $x + 2y + 3z = 4,$
    $2x + y + z = -3$

25. $r - 2s + 3t = 6,$
    $2r - s - t = -3,$
    $r + s + t = 6$

26. $a - 3c = 6,$
    $b + 2c = 2,$
    $7a - 3b - 5c = 14$

27. Describe at least one of the patterns that you see in Cramer's rule.

28. Which version of Cramer's rule do you find more useful: the version for 2 × 2 systems or the version for 3 × 3 systems? Why?

Skill Review

Factor.

29. $6x^2 + 23x - 4$ [5.3]

30. $5a^4 - 5a^3 + 15a^2 - 15$ [5.1]

31. $25y^8 - 49z^2$ [5.4]

32. $8n^3 + 8$ [5.5]

Synthesis

33. Cramer's rule states that if $a_1x + b_1y = c_1$ and $a_2x + b_2y = c_2$ are dependent, then

$$\begin{vmatrix} a_1 & b_1 \\ a_2 & b_2 \end{vmatrix} = 0.$$

Explain why this will always happen.

34. Under what conditions can a 3 × 3 system of linear equations be consistent but unable to be solved using Cramer's rule?

Solve.

35. $\begin{vmatrix} y & -2 \\ 4 & 3 \end{vmatrix} = 44$

36. $\begin{vmatrix} 2 & x & -1 \\ -1 & 3 & 2 \\ -2 & 1 & 1 \end{vmatrix} = -12$

37. $\begin{vmatrix} m+1 & -2 \\ m-2 & 1 \end{vmatrix} = 27$

38. Show that an equation of the line through (x_1, y_1) and (x_2, y_2) can be written

$$\begin{vmatrix} x & y & 1 \\ x_1 & y_1 & 1 \\ x_2 & y_2 & 1 \end{vmatrix} = 0.$$

 YOUR TURN ANSWERS: SECTION 8.7

1. -16 **2.** $\left(-\frac{29}{5}, \frac{23}{5}\right)$ **3.** 34   **4.** $\left(\frac{9}{5}, 8, \frac{28}{5}\right)$

QUICK QUIZ: SECTIONS 8.1–8.7

1. Solve graphically:

$$y = 2x - 1,$$
$$y = \tfrac{1}{3}x + 4. \quad [8.1]$$

2. Solve using the substitution method:

$$2x - y = 7,$$
$$y = 3x + 1. \quad [8.2]$$

3. Solve using the elimination method:

$$4x + 3y = 1,$$
$$2x + 3y = 5. \quad [8.2]$$

4. Solve using matrices:

$$5x + 3y = 5,$$
$$x + 2y = 1. \quad [8.6]$$

5. Solve using Cramer's rule:

$$8x + 5y = 4,$$
$$9x - 6y = 1. \quad [8.7]$$

PREPARE TO MOVE ON

For $f(x) = 80x + 2500$ and $g(x) = 150x$, find the following.

1. $(g - f)(x)$ [7.4]

2. $(g - f)(100)$ [7.4]

3. All values of x for which $f(x) = g(x)$ [2.2], [7.1]

4. All values of x for which $(g - f)(x) = 0$ [2.2], [7.4]

8.8   Business and Economics Applications

Break-Even Analysis ▪ Supply and Demand

Study Skills

Try to Look Ahead

If you are able to at least skim through an upcoming section before your instructor covers that lesson, you will be better able to focus on what is being emphasized in class. Similarly, if you can begin studying for a quiz or test a day or two before you really must, you will reap great rewards for doing so.

Break-Even Analysis

The money that a business spends to manufacture a product is its *cost*. The **total cost** of production can be thought of as a function C, where $C(x)$ is the cost of producing x units. When a company sells its product, it takes in money. This is *revenue* and can be thought of as a function R, where $R(x)$ is the **total revenue** from the sale of x units. **Total profit** is the money taken in less the money spent, or total revenue minus total cost. Total profit from the production and sale of x units is a function P given by

$$\textbf{Profit = Revenue − Cost,} \quad \text{or} \quad P(x) = R(x) - C(x).$$

If $R(x)$ is greater than $C(x)$, there is a gain and $P(x)$ is positive. If $C(x)$ is greater than $R(x)$, there is a loss and $P(x)$ is negative. When $R(x) = C(x)$, the company breaks even.

There are two kinds of costs. First, there are costs like rent, insurance, machinery, and so on. These costs, which must be paid regardless of how many items are produced, are called *fixed costs*. Second, costs for labor, materials, marketing, and so on are called *variable costs*, because they vary according to the amount being produced. The sum of the fixed cost and the variable cost gives the **total cost**.

> **CAUTION!** Do not confuse "cost" with "price." When we discuss the *cost* of an item, we are referring to what it costs to produce the item. The *price* of an item is what a consumer pays to purchase the item and is used when calculating revenue.

EXAMPLE 1 *Manufacturing Chairs.* Renewable Designs is planning to make a new chair. Fixed costs will be $90,000, and it will cost $150 to produce each chair. Each chair sells for $400.

a) Find the total cost $C(x)$ of producing x chairs.

b) Find the total revenue $R(x)$ from the sale of x chairs.

c) Find the total profit $P(x)$ from the production and sale of x chairs.

d) What profit will the company realize from the production and sale of 300 chairs? of 800 chairs?

e) Graph the total-cost, total-revenue, and total-profit functions using the same set of axes. Determine the break-even point.

SOLUTION

a) Total cost, in dollars, is given by

$$C(x) = \text{(Fixed costs) plus (Variable costs)},$$

or $\quad C(x) = \quad 90{,}000 \quad + \quad 150x,$

where x is the number of chairs produced.

b) Total revenue, in dollars, is given by

$$R(x) = 400x. \qquad \text{\$400 times the number of chairs sold. We assume that every chair produced is sold.}$$

c) Total profit, in dollars, is given by

$$P(x) = R(x) - C(x) \qquad \text{Profit is revenue minus cost.}$$
$$= 400x - (90{,}000 + 150x) \qquad \text{Parentheses are important.}$$
$$= 250x - 90{,}000.$$

d) Profits are

$$P(300) = 250 \cdot 300 - 90,000 = -\$15,000$$

when 300 chairs are produced and sold, and

$$P(800) = 250 \cdot 800 - 90,000 = \$110,000$$

when 800 chairs are produced and sold. Thus the company loses money if only 300 chairs are sold, but makes money if 800 are sold.

e) The graphs of each of the three functions are shown below:

$$C(x) = 90,000 + 150x,$$ This represents the cost function.
$$R(x) = 400x,$$ This represents the revenue function.
$$P(x) = 250x - 90,000.$$ This represents the profit function.

$C(x)$, $R(x)$, and $P(x)$ are all in dollars.

The revenue function has a graph that goes through the origin and has a slope of 400. The cost function has an intercept on the $-axis of 90,000 and has a slope of 150. The profit function has an intercept on the $-axis of $-90,000$ and has a slope of 250. It is shown by the red and black dashed line. The red portion of the dashed line shows a "negative" profit, which is a loss. (That is what is known as "being in the red.") The black portion of the dashed line shows a "positive" profit, or gain. (That is what is known as "being in the black.")

Student Notes

If you plan to study business or economics, you may want to consult the material in this section when these topics arise in your other courses.

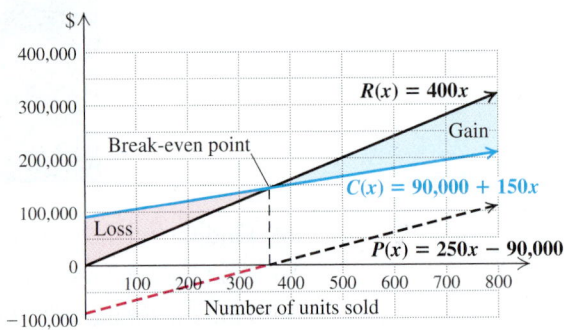

Gains occur when revenue exceeds cost. Losses occur when revenue is less than cost. The **break-even point** occurs where the graphs of R and C cross. Thus to find the break-even point, we solve a system:

$$R(x) = 400x,$$
$$C(x) = 90,000 + 150x.$$

Since revenue and cost are equal at the break-even point, the system can be rewritten as

$$d = 400x, \qquad (1)$$
$$d = 90,000 + 150x, \qquad (2)$$

where d is the dollar figure at the break-even point. We solve using substitution:

$$400x = 90,000 + 150x \qquad \text{Substituting } 400x \text{ for } d \text{ in equation (2)}$$
$$250x = 90,000$$
$$x = 360.$$

Renewable Designs breaks even if it produces and sells 360 chairs and takes in a total of $R(360) = 400(360) = \$144,000$ in revenue. Note that the x-coordinate of the break-even point can also be found by solving $P(x) = 0$. The break-even point is (360 chairs, $144,000).

1. Refer to Example 1. Renewable Designs is also planning to make a new table. Fixed costs will be $70,000, and it will cost $250 to produce each table. Each table sells for $600.

a) Find the total cost $C(x)$ of producing x tables.

b) Find the total revenue $R(x)$ from the sale of x tables.

c) Find the total profit $P(x)$ from the production and sale of x tables.

d) What profit will the company realize from the production and sale of 500 tables?

e) Determine the break-even point.

 YOUR TURN

Supply and Demand

As the price of coffee varies, so too does the amount sold. The table and the graph below show that *consumers will demand less as the price goes up.*

Demand Function, D

Price, p, per Kilogram	Quantity, D(p) (in millions of kilograms)
$16.00	25
18.00	20
20.00	15
22.00	10
24.00	5

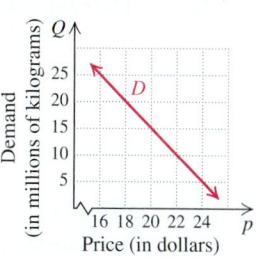

As the price of coffee varies, the amount made available varies as well. The table and the graph below show that *sellers will supply more as the price goes up.*

Supply Function, S

Price, p, per Kilogram	Quantity, S(p) (in millions of kilograms)
$18.00	5
19.00	10
20.00	15
21.00	20
22.00	25

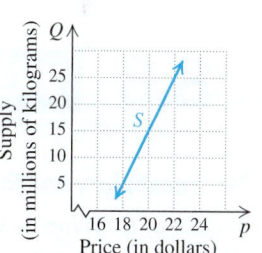

Considering demand and supply together, we see that as price increases, demand decreases. As price increases, supply increases. The point of intersection is called the **equilibrium point**. At that price, the amount that the seller will supply is the same amount that the consumer will buy. The situation is similar to a buyer and a seller negotiating the price of an item. The equilibrium point is the price and quantity that they finally agree on.

Any ordered pair of coordinates from the graph is (price, quantity), because the horizontal axis is the price axis and the vertical axis is the quantity axis. If D is a demand function and S is a supply function, then the equilibrium point is where demand equals supply:

$$D(p) = S(p).$$

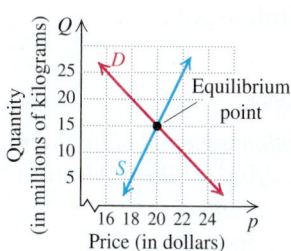

EXAMPLE 2   Find the equilibrium point for the demand and supply functions given:

$$D(p) = 1000 - 60p, \quad (1)$$
$$S(p) = 200 + 4p. \quad (2)$$

SOLUTION   Since both demand and supply are *quantities* and they are equal at the equilibrium point, we rewrite the system as

$$q = 1000 - 60p, \quad (1)$$
$$q = 200 + 4p. \quad (2)$$

We substitute $200 + 4p$ for q in equation (1) and solve:

$$200 + 4p = 1000 - 60p \qquad \text{Substituting } 200 + 4p \text{ for } q \text{ in equation (1)}$$
$$200 + 64p = 1000 \qquad\qquad \text{Adding } 60p \text{ to both sides}$$
$$64p = 800 \qquad\qquad\quad \text{Adding } -200 \text{ to both sides}$$
$$p = \tfrac{800}{64} = 12.5.$$

Chapter Resource:
Decision Making: Connection,
p. 566

Thus the equilibrium price is \$12.50 per unit.
    To find the equilibrium quantity, we substitute \$12.50 into either $D(p)$ or $S(p)$. We use $S(p)$:

$$S(12.5) = 200 + 4(12.5) = 200 + 50 = 250.$$

The equilibrium quantity is 250 units, and the equilibrium point is ($\$12.50, 250$).

2. Find the equilibrium point for the demand and supply functions given:

$$D(p) = 850 - 30p,$$
$$S(p) = 550 + 10p.$$

 YOUR TURN

8.8   EXERCISE SET

FOR EXTRA HELP

PRACTICE   WATCH   READ   REVIEW

Vocabulary and Reading Check

In each of Exercises 1–8, match the word or phrase with the most appropriate choice from the list below.

1. ____   Total cost

2. ____   Fixed costs

3. ____   Variable costs

4. ____   Total revenue

5. ____   Total profit

6. ____   Price

7. ____   Break-even point

8. ____   Equilibrium point

a) The amount of money that a company takes in

b) The sum of fixed costs and variable costs

c) The point at which total revenue equals total cost

d) What consumers pay per item

e) The difference between total revenue and total cost

f) What companies spend whether or not a product is produced

g) The point at which supply equals demand

h) The costs that vary according to the number of items produced

Break-Even Analysis

For each of the following pairs of total-cost and total-revenue functions, find **(a)** *the total-profit function and* **(b)** *the break-even point.*

9. $C(x) = 35x + 200{,}000,$
$R(x) = 55x$

10. $C(x) = 20x + 500{,}000,$
$R(x) = 70x$

11. $C(x) = 15x + 3100,$
$R(x) = 40x$

12. $C(x) = 30x + 49{,}500,$
$R(x) = 85x$

13. $C(x) = 40x + 22{,}500,$
$R(x) = 85x$

14. $C(x) = 20x + 10{,}000$,
$R(x) = 100x$

15. $C(x) = 24x + 50{,}000$,
$R(x) = 40x$

16. $C(x) = 40x + 8010$,
$R(x) = 58x$

Aha! **17.** $C(x) = 75x + 100{,}000$,
$R(x) = 125x$

18. $C(x) = 20x + 120{,}000$,
$R(x) = 50x$

Supply and Demand

Find the equilibrium point for each of the following pairs of demand and supply functions.

19. $D(p) = 2000 - 15p$,
$S(p) = 740 + 6p$

20. $D(p) = 1000 - 8p$,
$S(p) = 350 + 5p$

21. $D(p) = 760 - 13p$,
$S(p) = 430 + 2p$

22. $D(p) = 800 - 43p$,
$S(p) = 210 + 16p$

23. $D(p) = 7500 - 25p$,
$S(p) = 6000 + 5p$

24. $D(p) = 8800 - 30p$,
$S(p) = 7000 + 15p$

25. $D(p) = 1600 - 53p$,
$S(p) = 320 + 75p$

26. $D(p) = 5500 - 40p$,
$S(p) = 1000 + 85p$

Solve.

27. *Manufacturing.* SoundGen, Inc., plans to manufacture a new type of cell phone. The fixed costs are $45,000, and the variable costs are estimated to be $40 per unit. The revenue from each cell phone is to be $130. Find the following.

a) The total cost $C(x)$ of producing x cell phones
b) The total revenue $R(x)$ from the sale of x cell phones
c) The total profit $P(x)$ from the production and sale of x cell phones
d) The profit or loss from the production and sale of 3000 cell phones; of 400 cell phones
e) The break-even point

28. *Computer Manufacturing.* Current Electronics plans to introduce a new laptop computer. The fixed costs are $125,300, and the variable costs are $450 per unit. The revenue from each computer is $800. Find the following.

a) The total cost $C(x)$ of producing x computers
b) The total revenue $R(x)$ from the sale of x computers
c) The total profit $P(x)$ from the production and sale of x computers
d) The profit or loss from the production and sale of 100 computers; of 400 computers
e) The break-even point

29. *Pet Safety.* Christine designed and is now producing a pet car seat. The fixed costs for setting up production are $10,000, and the variable costs are $30 per unit. The revenue from each seat is to be $80. Find the following.

a) The total cost $C(x)$ of producing x seats
b) The total revenue $R(x)$ from the sale of x seats
c) The total profit $P(x)$ from the production and sale of x seats
d) The profit or loss from the production and sale of 2000 seats; of 50 seats
e) The break-even point

30. *Manufacturing Caps.* Martina's Custom Printing is adding painter's caps to its product line. For the first year, the fixed costs for setting up production are $16,404. The variable costs for producing a dozen caps are $6.00. The revenue on each dozen caps will be $18.00. Find the following.

a) The total cost $C(x)$ of producing x dozen caps
b) The total revenue $R(x)$ from the sale of x dozen caps
c) The total profit $P(x)$ from the production and sale of x dozen caps
d) The profit or loss from the production and sale of 3000 dozen caps; of 1000 dozen caps
e) The break-even point

31. In Example 1, the slope of the line representing Revenue is the sum of the slopes of the other two lines. This is not a coincidence. Explain why.

32. Variable costs and fixed costs are often compared to the slope and the y-intercept, respectively, of an equation for a line. Explain why you feel this analogy is or is not valid.

Skill Review

Solve.

33. $25x^2 = 1$ [5.7]

34. $10x^2 = 20x - 10$ [5.7]

35. $\dfrac{1}{x} = \dfrac{2}{5}$ [6.6]

36. $x - 3 = \dfrac{18}{x}$ [6.6]

Synthesis

37. Bernadette claims that since her fixed costs are $3000, she need sell only 10 custom birdbaths at $300 each in order to break even. Is her reasoning valid? Why or why not?

38. In this section, we examined supply and demand functions for coffee. Does it seem realistic to you for the graph of D to have a constant slope? Why or why not?

39. *Yo-yo Production.* Bing Boing Hobbies is willing to produce 100 yo-yo's at $2.00 each and 500 yo-yo's at $8.00 each. Research indicates that the public will buy 500 yo-yo's at $1.00 each and 100 yo-yo's at $9.00 each. Find the equilibrium point.

40. *Loudspeaker Production.* Sonority Speakers, Inc., has fixed costs of $15,400 and variable costs of $100 for each pair of speakers produced. If the speakers sell for $250 per pair, how many pairs of speakers must be produced (and sold) in order to have enough profit to cover the fixed costs of two additional facilities? Assume that all fixed costs are identical.

Use a graphing calculator to solve.

41. *Dog Food Production.* Puppy Love, Inc., is producing a new line of puppy food. The marketing department predicts that the demand function will be $D(p) = -14.97p + 987.35$ and the supply function will be $S(p) = 98.55p - 5.13$.

a) To the nearest cent, what price per unit should be charged in order to have equilibrium between supply and demand?

b) The production of the puppy food involves $87,985 in fixed costs and $5.15 per unit in variable costs. If the price per unit is the value you found in part (a), how many units must be sold in order to break even?

42. *Computer Production.* Brushstroke Computers, Inc., is planning a new line of computers, each of which will sell for $970. The fixed costs in setting up production are $1,235,580, and the variable costs for each computer are $697.

a) What is the break-even point?

b) The marketing department at Brushstroke is not sure that $970 is the best price. Their demand function for the new computers is given by $D(p) = -304.5p + 374,580$ and their supply function is given by $S(p) = 788.7p - 576,504$. To the nearest dollar, what price p would result in equilibrium between supply and demand?

c) If the computers are sold for the equilibrium price found in part (b), what is the break-even point?

YOUR TURN ANSWERS: SECTION 8.8

1. (a) $C(x) = 70,000 + 250x$; **(b)** $R(x) = 600x$; **(c)** $P(x) = 350x - 70,000$; **(d)** $105,000; **(e)** (200 units, $120,000)   **2.** ($7.50, 625)

QUICK QUIZ: SECTIONS 8.1–8.8

1. The perimeter of a rectangular classroom is 140 ft. The width is 10 ft shorter than the length. Find the dimensions.   [8.3]

2. Joanna has in her refrigerator low-fat milk, containing 1% fat, and whole milk, containing 3.5% fat. How much of each should she mix to obtain 16 oz of milk containing 2% fat?   [8.3]

3. The measure of the largest angle in a triangle is equal to the sum of the measures of the other two angles. The smallest angle is one-third the size of the middle angle. Find the measures of the angles.   [8.5]

Evaluate.   [8.7]

4. $\begin{vmatrix} -5 & -2 \\ 3 & -4 \end{vmatrix}$

5. $\begin{vmatrix} 2 & 1 & 0 \\ 3 & -1 & 5 \\ 0 & 2 & 1 \end{vmatrix}$

PREPARE TO MOVE ON

Solve.   [2.2]

1. $4x - 3 = 21$

2. $5 - x = 7$

3. $x - 4 = 9x - 10$

4. $3 - (x + 2) = 7$

5. $1 - 3(2x + 1) = 3 - 5x$

A
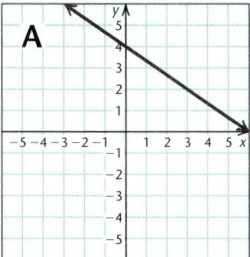

B
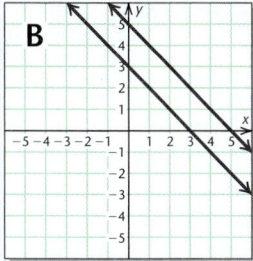

C
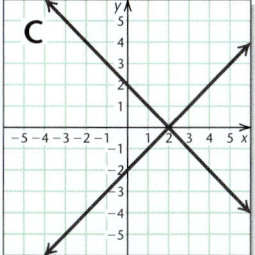

D
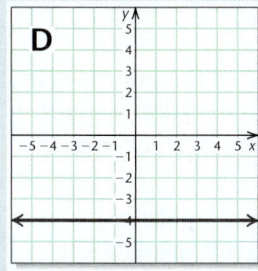

E
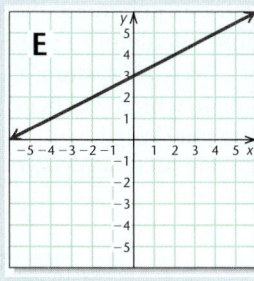

Visualizing for Success

Use after Section 8.1.

Match each equation or system of equations with its graph.

1. $x + y = 2,$
   $x - y = 2$

2. $y = \frac{1}{3}x - 5$

3. $4x - 2y = -8$

4. $2x + y = 1,$
   $x + 2y = 1$

5. $8y + 32 = 0$

6. $f(x) = -x + 4$

7. $\frac{2}{3}x + y = 4$

8. $x = 4,$
   $y = 3$

9. $y = \frac{1}{2}x + 3,$
   $2y - x = 6$

10. $y = -x + 5,$
    $y = 3 - x$

Answers on page A-39

An additional, animated version of this activity appears in MyMathLab. To use MyMathLab, you need a course ID and a student access code. Contact your instructor for more information.

F
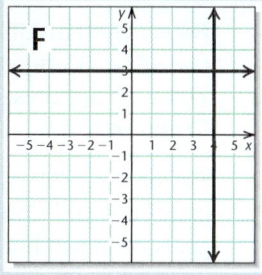

G
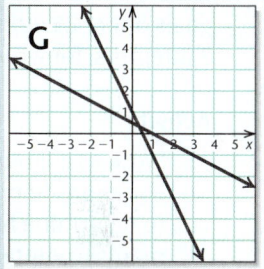

H
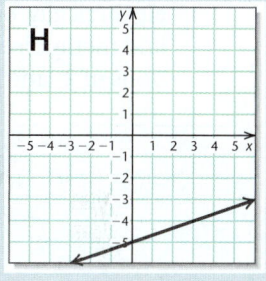

I
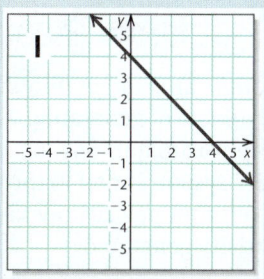

J
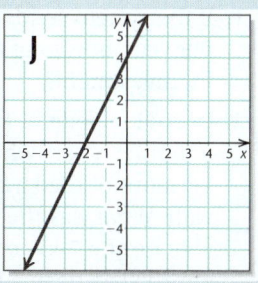

Collaborative Activity   *How Many Two's? How Many Three's?*

Focus: Systems of linear equations
Use after: Section 8.2
Time: 20 minutes
Group size: 3

The box score at right, from the 2012 NBA All-Star game, contains information on how many field goals (worth either 2 or 3 points) and free throws (worth 1 point) each player attempted and made. For example, the line "Durant 14–25 5–7 36" means that the West's Kevin Durant made 14 field goals out of 25 attempts and 5 free throws out of 7 attempts, for a total of 36 points.

Activity

1. Work as a group to develop a system of two equations in two unknowns that can be used to determine how many 2-pointers and how many 3-pointers were made by the West.

2. Each group member should solve the system from part (1) in a different way: one person algebraically, one person by making a table and methodically checking all combinations of 2- and 3-pointers, and one person by

guesswork. Compare answers when this has been completed.

3. Determine, as a group, how many 2-pointers and how many 3-pointers the East made.

West (152)
Bryant 9–17 7–8 27, Paul 3–7 0–0 8, Bynum 0–3 0–1 1, Durant 14–25 5–7 36, Griffin 9–12 3–6 22, Westbrook 10–17 0–2 21, Love 7–12 1–3 17, Nowitzki 3–8 0–0 7, Gasol 2–5 0–0 4, Parker 3–5 0–0 6, Aldridge 2–5 0–0 4, Nash 0–0 0–0 0
Totals 62–116 16–26 152

East (149)
Wade 11–15 2–2 24, Rose 6–8 0–0 14, Howard 4–9 1–2 9, James 15–23 0–0 36, Anthony 7–15 5–7 19, Bosh 3–9 0–0 7, Williams 8–11 0–0 20, Rondo 1–3 0–0 2, Iguodala 6–7 0–0 12, Pierce 1–8 0–0 3, Hibbert 1–3 1–1 3, Deng 0–2 0–0 0
Totals 63–113 9–12 149

West 39 49 36 28 — 152
East 28 41 43 37 — 149

Decision Making & Connection   (*Use after Section 8.8.*)

Solar Energy. A photovoltaic (PV) electric system capable of generating 7 kW per hour of electricity cost approximately $45,000 in Vermont in 2011. Because the PV system is tied to the electric grid, local utilities in Vermont pay 20¢ per kilowatt-hour (kWh) for the electricity generated.

Source: Based on data from Bill Heigis, Habie Guion, and David Ellenbogen

1. If the system sends to the grid, on average, 8000 kW per year, how long will it take to break even on the PV investment?

2. In 2011, a Federal Tax Credit of 30% was available to homeowners who installed a PV system. Also, Vermont offered a state rebate of $0.75 per system watt. What was the total tax credit and rebate available for the 7-kW system described? (*Hint*: There are 1000 watts in a kilowatt.)

3. Given the incentives described in Exercise 2, what is the final cost to the owner of the 7-kW system? How long will it take to break even on the investment?

4. Assuming no maintenance costs and a lifespan of 20 years, how much will the system generate in profit?

5. *Research.* Determine the rate that your local utility, or a nearby utility, pays for electricity generated by a home PV system. Use an online calculator to estimate the amount of electricity a 7-kW PV system will generate per year in the area in which you live. Use this information to estimate how long it will take you to break even on a $45,000 PV investment.

Study Summary

KEY TERMS AND CONCEPTS	EXAMPLES	PRACTICE EXERCISES

SECTION 8.1: *Systems of Equations in Two Variables*

A solution of a **system of two equations** is an ordered pair that makes both equations true. The intersection of the graphs of the equations represents the solution of the system.

A system is **consistent** if it has at least one solution. Otherwise it is **inconsistent.**

The equations in a system are **dependent** if one of them can be written as a multiple and/or a sum of the other equation(s). Otherwise, they are **independent.**

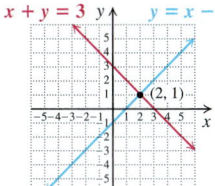

$x + y = 3,$
$y = x - 1$

The graphs intersect at $(2, 1)$.
The solution is $(2, 1)$.
The system is consistent.
The equations are independent.

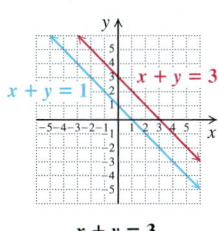

$x + y = 3,$
$x + y = 1$

The graphs do not intersect.
There is no solution.
The system is inconsistent.
The equations are independent.

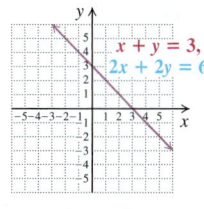

$x + y = 3,$
$2x + 2y = 6$

The graphs are the same.
The solution set is
$\{(x, y) \mid x + y = 3\}$.
The system is consistent.
The equations are dependent.

1. Solve by graphing:
$$x - y = 3,$$
$$y = 2x - 5.$$

SECTION 8.2: *Solving by Substitution or Elimination*

To use the **substitution method,** we solve one equation for a variable and substitute the expression for that variable in the other equation.

Solve:
$$2x + 3y = 8,$$
$$x = y + 1.$$

Substitute and solve for y: Substitute and solve for x:

$$2(y + 1) + 3y = 8 \qquad x = y + 1$$
$$2y + 2 + 3y = 8 \qquad\quad = \tfrac{6}{5} + 1$$
$$y = \tfrac{6}{5}. \qquad\qquad = \tfrac{11}{5}.$$

The solution is $\left(\tfrac{11}{5}, \tfrac{6}{5}\right)$.

2. Solve by substitution:
$$x = 3y - 2,$$
$$y - x = 1.$$

To use the **elimination method,** we add to eliminate a variable.

Solve:
$$4x - 2y = 6,$$
$$3x + y = 7.$$

Eliminate y and solve for x: Substitute and solve for y:

$$4x - 2y = 6 \qquad\qquad 3x + y = 7$$
$$\underline{6x + 2y = 14} \qquad\quad 3 \cdot 2 + y = 7$$
$$10x \quad\;\; = 20 \qquad\qquad\quad y = 1.$$
$$x = 2.$$

The solution is $(2, 1)$.

3. Solve by elimination:
$$2x - y = 5,$$
$$x + 3y = 1.$$

567

SECTION 8.3:   *Solving Applications: Systems of Two Equations*

Total-value, mixture, and **motion problems** often translate directly to systems of equations.

Motion problems use one of the following relationships:

$$d = rt, \quad r = \frac{d}{t}, \quad t = \frac{d}{r}.$$

Simple-interest problems use the formula

Principal · Rate · Time = Interest.

Total Value
In order to make a necklace, Star Bright Jewelry Design purchased 80 beads for a total of $39 (excluding tax). Some of the beads were sterling silver beads costing 40¢ each and the rest were gemstone beads costing 65¢ each. How many of each type were bought? (See Example 2 in Section 8.3 for a solution.)

Mixture
Nature's Green Gardening, Inc., carries two brands of fertilizer containing nitrogen and water. "Gentle Grow" is 3% nitrogen and "Sun Saver" is 8% nitrogen. Nature's Green needs to combine the two types of solutions into a 90-L mixture that is 6% nitrogen. How much of each brand should be used? (See Example 5 in Section 8.3 for a solution.)

Motion
A Boeing 747-400 jet flies 4 hr west with a 60-mph tailwind. Returning against the wind takes 5 hr. Find the speed of the jet with no wind. (See Example 7 in Section 8.3 for a solution.)

4. Sure Supply charges $17.49 for a box of gel pens and $16.49 for a box of mechanical pencils. If Valley College purchased 120 such boxes for $2010.80, how many boxes of each type did they purchase?

5. A cleaning solution that is 40% nitric acid is being mixed with a solution that is 15% nitric acid in order to create 2 L of a solution that is 25% nitric acid. How much 40%-acid and how much 15%-acid should be used?

6. Ruth paddled for $1\frac{1}{2}$ hr with a 2-mph current. The return trip against the same current took $2\frac{1}{2}$ hr. Find the speed of Ruth's canoe in still water.

SECTION 8.4:   *Systems of Equations in Three Variables*

Systems of three equations in three variables are usually most easily solved using elimination.

Solve:

$$
\begin{aligned}
x + y - z &= 3, &(1)\\
-x + y + 2z &= -5, &(2)\\
2x - y - 3z &= 9. &(3)
\end{aligned}
$$

Eliminate x using two equations:

$$
\begin{array}{rl}
x + y - z = 3 & (1)\\
\underline{-x + y + 2z = -5} & (2)\\
2y + z = -2 &
\end{array}
$$

Solve the system of two equations for y and z:

$$
\begin{array}{rl}
2y + z = -2 &\\
\underline{-3y - z = 3} &\\
-y = 1 &\\
y = -1 &
\end{array}
$$

$$
\begin{aligned}
2(-1) + z &= -2\\
z &= 0.
\end{aligned}
$$

The solution is $(4, -1, 0)$.

Eliminate x again using two different equations:

$$
\begin{array}{rl}
-2x - 2y + 2z = -6 & (1)\\
\underline{2x - y - 3z = 9} & (3)\\
-3y - z = 3 &
\end{array}
$$

Substitute and solve for x:

$$
\begin{aligned}
x + y - z &= 3\\
x + (-1) - 0 &= 3\\
x &= 4.
\end{aligned}
$$

7. Solve:

$$
\begin{aligned}
x - 2y - z &= 8,\\
2x + 2y - z &= 8,\\
x - 8y + z &= 1.
\end{aligned}
$$

SECTION 8.5:   *Solving Applications: Systems of Three Equations*

Many problems with three unknowns can be solved after translating to a system of three equations.	In a triangular cross section of a roof, the largest angle is 70° greater than the smallest angle. The largest angle is twice as large as the remaining angle. Find the measure of each angle. The angles in the triangle measure 30°, 50°, and 100°. (See Example 2 in Section 8.5 for a complete solution.)	**8.** The sum of three numbers is 9. The third number is half the sum of the first and second numbers. The second number is 2 less than the sum of the first and third numbers. Find the numbers.

SECTION 8.6:   *Elimination Using Matrices*

A **matrix** (plural, **matrices**) is a rectangular array of numbers. The individual numbers are called **entries**, or **elements**. By using **row-equivalent** operations, we can solve systems of equations using matrices.	Solve: $x + 4y = 1,$ $2x - y = 3.$ Write as a matrix in row-echelon form: $$\begin{bmatrix} 1 & 4 & \vdots & 1 \\ 2 & -1 & \vdots & 3 \end{bmatrix} \longrightarrow \begin{bmatrix} 1 & 4 & \vdots & 1 \\ 0 & -9 & \vdots & 1 \end{bmatrix}.$$ Rewrite as equations and solve: $-9y = 1 \longrightarrow x + 4\left(-\frac{1}{9}\right) = 1$ $y = -\frac{1}{9} \qquad\qquad x = \frac{13}{9}.$ The solution is $\left(\frac{13}{9}, -\frac{1}{9}\right)$.	**9.** Solve using matrices: $3x - 2y = 10,$ $x + y = 5.$

SECTION 8.7:   *Determinants and Cramer's Rule*

Determinant of a 2 × 2 Matrix $$\begin{vmatrix} a & c \\ b & d \end{vmatrix} = ad - bc$$ **Determinant of a 3 × 3 Matrix** $$\begin{vmatrix} a_1 & b_1 & c_1 \\ a_2 & b_2 & c_2 \\ a_3 & b_3 & c_3 \end{vmatrix} =$$ $a_1 \begin{vmatrix} b_2 & c_2 \\ b_3 & c_3 \end{vmatrix} - a_2 \begin{vmatrix} b_1 & c_1 \\ b_3 & c_3 \end{vmatrix}$ $+ a_3 \begin{vmatrix} b_1 & c_1 \\ b_2 & c_2 \end{vmatrix}$	$$\begin{vmatrix} 2 & 3 \\ -1 & 5 \end{vmatrix} = 2 \cdot 5 - (-1)(3) = 13$$ $$\begin{vmatrix} 2 & 3 & 2 \\ 0 & 1 & 0 \\ -1 & 5 & -4 \end{vmatrix}$$ $= 2 \begin{vmatrix} 1 & 0 \\ 5 & -4 \end{vmatrix} - 0 \begin{vmatrix} 3 & 2 \\ 5 & -4 \end{vmatrix} + (-1)\begin{vmatrix} 3 & 2 \\ 1 & 0 \end{vmatrix}$ $= 2(-4 - 0) - 0 - 1(0 - 2)$ $= -8 + 2 = -6$	*Evaluate.* **10.** $\begin{vmatrix} 3 & -5 \\ 2 & 6 \end{vmatrix}$ **11.** $\begin{vmatrix} 1 & 2 & -1 \\ 2 & 0 & 3 \\ 0 & 1 & 5 \end{vmatrix}$
We can use determinants and **Cramer's rule** to solve systems of equations. Cramer's rule for 2 × 2 matrices and for 3 × 3 matrices can be found in Section 8.7.	Solve: $x - 3y = 7,$ $2x + 5y = 4.$ $$x = \dfrac{\begin{vmatrix} 7 & -3 \\ 4 & 5 \end{vmatrix}}{\begin{vmatrix} 1 & -3 \\ 2 & 5 \end{vmatrix}}; \qquad y = \dfrac{\begin{vmatrix} 1 & 7 \\ 2 & 4 \end{vmatrix}}{\begin{vmatrix} 1 & -3 \\ 2 & 5 \end{vmatrix}}$$ $x = \frac{47}{11} \qquad\qquad y = \frac{-10}{11}$ The solution is $\left(\frac{47}{11}, -\frac{10}{11}\right)$.	**12.** Solve using Cramer's rule: $3x - 5y = 12,$ $2x + 6y = 1.$

SECTION 8.8:   *Business and Economics Applications*

The **break-even point** occurs where the **revenue** equals the **cost,** or where **profit** is 0.

Find **(a)** the total-profit function and **(b)** the break-even point for the total-cost and total-revenue functions

$$C(x) = 38x + 4320 \quad \text{and} \quad R(x) = 62x.$$

a) Profit = Revenue − Cost

$$P(x) = R(x) - C(x)$$
$$P(x) = 62x - (38x + 4320)$$
$$P(x) = 24x - 4320$$

b) $C(x) = R(x)$ At the break-even point, revenue = cost.

$$38x + 4320 = 62x$$

$$180 = x$$ Solving for x

$$R(180) = 11{,}160$$ Finding the revenue (or cost) at the break-even point

The break-even point is $(180, \$11{,}160)$.

13. Find **(a)** the total-profit function and **(b)** the break-even point for the total-cost and total-revenue functions

$$C(x) = 15x + 9000,$$
$$R(x) = 90x.$$

An **equilibrium point** occurs where **supply** equals **demand**.

Find the equilibrium point for the demand and supply functions

$$S(p) = 60 + 7p \quad \text{and} \quad D(p) = 90 - 13p.$$

$$S(p) = D(p)$$ At the equilibrium point, supply = demand.

$$60 + 7p = 90 - 13p$$

$$20p = 30$$

$$p = 1.5$$ Solving for p

$$S(1.5) = 70.5$$ Finding the supply (or demand) at the equilibrium point

The equilibrium point is $(\$1.50, 70.5)$.

14. Find the equilibrium point for the supply and demand functions

$$S(p) = 60 + 9p,$$
$$D(p) = 195 - 6p.$$

Review Exercises: Chapter 8

 Concept Reinforcement

Choose the word from the list below that best completes each statement.

contradiction
dependent
determinant
elimination
graphical

inconsistent
parallel
square
substitution
zero

1. The system

$$5x + 3y = 7,$$
$$y = 2x + 1$$

is most easily solved using the _____ method.   [8.2]

2. The system

$$-2x + 3y = 8,$$
$$2x + 2y = 7$$

is most easily solved using the _____ method.   [8.2]

3. Of the methods used to solve systems of equations, the _____ method may yield only approximate solutions.   [8.1], [8.2]

4. When one equation in a system is a multiple of another equation in that system, the equations are said to be _____.   [8.1]

5. A system for which there is no solution is said to be _____.   [8.1]

6. When we are using an algebraic method to solve a system of equations, obtaining a(n) _____ tells us that the system is inconsistent. [8.2]

7. When we are graphing to solve a system of two equations, if there is no solution, the lines will be _____ . [8.1]

8. When a matrix has the same number of rows and columns, it is said to be _____. [8.7]

9. Cramer's rule is a formula in which the numerator and the denominator of each fraction is a(n) _____. [8.7]

10. At the break-even point, the value of the profit function is _____. [8.8]

For Exercises 11–20, if a system has an infinite number of solutions, use set-builder notation to write the solution set. If a system has no solution, state this.

Solve graphically. [8.1]

11. $y = x - 3,$
$y = \frac{1}{4}x$

12. $2x - 3y = 12,$
$4x + y = 10$

Solve using the substitution method. [8.2]

13. $5x - 2y = 4,$
$x = y - 2$

14. $y = x + 2,$
$y - x = 8$

Solve using the elimination method. [8.2]

15. $2x + 5y = 8,$
$6x - 5y = 10$

16. $3x - 5y = 9,$
$5x - 3y = -1$

Solve using any appropriate method. [8.1], [8.2]

17. $x - 3y = -2,$
$7y - 4x = 6$

18. $4x - 7y = 18,$
$9x + 14y = 40$

19. $1.5x - 3 = -2y,$
$3x + 4y = 6$

20. $y = 2x - 5,$
$y = \frac{1}{2}x + 1$

Solve. [8.3]

21. Jillian charges $25 for a private guitar lesson and $18 for a group guitar lesson. One day in August, Jillian earned $265 from 12 students. How many students of each type did Jillian teach?

22. A freight train leaves Houston at midnight traveling north at 44 mph. One hour later, a passenger train, going 55 mph, travels north from Houston on a parallel track. How long will it take the passenger train to overtake the freight train?

23. D'Andre wants 14 L of fruit punch that is 10% juice. At the store, he finds only punch that is 15% juice or punch that is 8% juice. How much of each should he purchase?

Solve. If a system's equations are dependent or if there is no solution, state this. [8.4]

24. $x + 4y + 3z = 2,$
$2x + y + z = 10,$
$-x + y + 2z = 8$

25. $4x + 2y - 6z = 34,$
$2x + y + 3z = 3,$
$6x + 3y - 3z = 37$

26. $2x - 5y - 2z = -4,$
$7x + 2y - 5z = -6,$
$-2x + 3y + 2z = 4$

27. $3x + y = 2,$
$x + 3y + z = 0,$
$x + z = 2$

28. $2x - 3y + z = 1,$
$x - y + 2z = 5,$
$3x - 4y + 3z = -2$

Solve. [8.5]

29. In triangle ABC, the measure of angle A is four times the measure of angle C, and the measure of angle B is 45° more than the measure of angle C. What are the measures of the angles of the triangle?

30. The sum of the average number of times a man, a woman, and a one-year-old child cry each month is 56.7. A woman cries 3.9 more times than a man. The average number of times a one-year-old cries per month is 43.3 more than the average number of times combined that a man and a woman cry. What is the average number of times per month that each cries?

Solve using matrices. Show your work. [8.6]

31. $3x + 4y = -13,$
$5x + 6y = 8$

32. $3x - y + z = -1,$
$2x + 3y + z = 4,$
$5x + 4y + 2z = 5$

Evaluate. [8.7]

33. $\begin{vmatrix} -2 & -5 \\ 3 & 10 \end{vmatrix}$

34. $\begin{vmatrix} 2 & 3 & 0 \\ 1 & 4 & -2 \\ 2 & -1 & 5 \end{vmatrix}$

Solve using Cramer's rule. Show your work. [8.7]

35. $2x + 3y = 6,$
$x - 4y = 14$

36. $2x + y + z = -2,$
$2x - y + 3z = 6,$
$3x - 5y + 4z = 7$

37. Find **(a)** the total-profit function and **(b)** the break-even point for the total-cost and total-revenue functions

$$C(x) = 30x + 15{,}800,$$
$$R(x) = 50x. \quad [8.8]$$

38. Find the equilibrium point for the demand and supply functions

$$S(p) = 60 + 7p$$

and

$$D(p) = 120 - 13p. \quad [8.8]$$

39. Danae is beginning to produce organic honey. For the first year, the fixed costs for setting up production are $54,000. The variable costs for producing each pint of honey are $4.75. The revenue from each pint of honey is $9.25. Find the following. [8.8]

  a) The total cost $C(x)$ of producing x pints of honey

  b) The total revenue $R(x)$ from the sale of x pints of honey

  c) The total profit $P(x)$ from the production and sale of x pints of honey

  d) The profit or loss from the production and sale of 5000 pints of honey; of 15,000 pints of honey

  e) The break-even point

Synthesis

 40. How would you go about solving a problem that involves four variables? [8.5]

 41. Explain how a system of equations can be both dependent and inconsistent. [8.4]

42. Danae is leaving a job that pays $36,000 per year to make honey (see Exercise 39). How many pints of honey must she produce and sell in order to make as much money as she earned at her previous job? [8.8]

43. Solve graphically:

$$y = x + 2,$$
$$y = x^2 + 2. \quad [8.1]$$

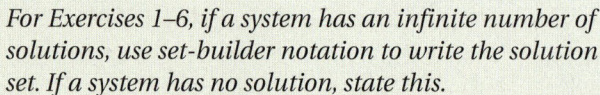
Test: Chapter 8

For step-by-step test solutions, access the Chapter Test Prep Videos in MyMathLab, on YouTube (search "Bittinger Combo Alg CA" and click on "Channels"), or by scanning the code.

For Exercises 1–6, if a system has an infinite number of solutions, use set-builder notation to write the solution set. If a system has no solution, state this.

1. Solve graphically:

$$2x + y = 8,$$
$$y - x = 2.$$

2. Solve using the substitution method:

$$x + 3y = -8,$$
$$4x - 3y = 23.$$

Solve using the elimination method.

3. $3x - y = 7,$
  $x + y = 1$

4. $4y + 2x = 18,$
  $3x + 6y = 26$

Solve using any appropriate method.

5. $2x - 4y = -6,$
  $x = 2y - 3$

6. $4x - 6y = 3,$
  $6x - 4y = -3$

7. The perimeter of a standard basketball court is 288 ft. The length is 44 ft longer than the width. Find the dimensions.

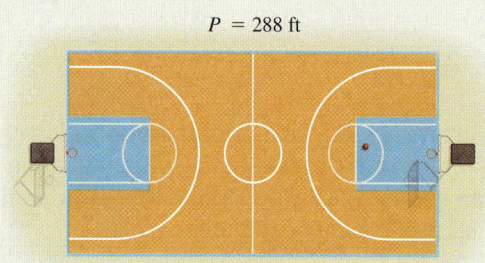

$P = 288$ ft

8. Pepperidge Farm® Goldfish is a snack food for which 40% of its calories come from fat. Rold Gold® Pretzels receive 9% of their calories from fat. How many grams of each would be needed to make 620 g of a snack mix for which 15% of the calories are from fat?

9. Kylie's motorboat took 3 hr to make a trip downstream on a river flowing at 5 mph. The return trip against the same current took 5 hr. Find the speed of the boat in still water.

Solve. If a system's equations are dependent or if there is no solution, state this.

10. $-3x + y - 2z = 8,$
$-x + 2y - z = 5,$
$2x + y + z = -3$

11. $6x + 2y - 4z = 15,$
$-3x - 4y + 2z = -6,$
$4x - 6y + 3z = 8$

12. $2x + 2y = 0,$
$4x + 4z = 4,$
$2x + y + z = 2$

13. $3x + 3z = 0,$
$2x + 2y = 2,$
$3y + 3z = 3$

Solve using matrices.

14. $4x + y = 12,$
$3x + 2y = 2$

15. $x + 3y - 3z = 12,$
$3x - y + 4z = 0,$
$-x + 2y - z = 1$

Evaluate.

16. $\begin{vmatrix} 4 & -2 \\ 3 & -5 \end{vmatrix}$

17. $\begin{vmatrix} 3 & 4 & 2 \\ -2 & -5 & 4 \\ 0 & 5 & -3 \end{vmatrix}$

18. Solve using Cramer's rule:
$3x + 4y = -1,$
$5x - 2y = 4.$

19. An electrician, a carpenter, and a plumber are hired to work on a house. The electrician earns $30 per hour, the carpenter $28.50 per hour, and the plumber $34 per hour. The first day on the job, they worked a total of 21.5 hr and earned a total of $673.00. If the plumber worked 2 more hours than the carpenter, how many hours did each work?

20. Find the equilibrium point for the demand and supply functions
$$D(p) = 79 - 8p \text{ and } S(p) = 37 + 6p,$$
where p is the price, in dollars, $D(p)$ is the number of units demanded, and $S(p)$ is the number of units supplied.

21. Kick Back, Inc., is producing a new hammock. For the first year, the fixed costs for setting up production are $44,000. The variable costs for producing each hammock are $25. The revenue from each hammock is $80. Find the following.

a) The total cost $C(x)$ of producing x hammocks

b) The total revenue $R(x)$ from the sale of x hammocks

c) The total profit $P(x)$ from the production and sale of x hammocks

d) The profit or loss from the production and sale of 300 hammocks; of 900 hammocks

e) The break-even point

Synthesis

22. The graph of the function $f(x) = mx + b$ contains the points $(-1, 3)$ and $(-2, -4)$. Find m and b.

23. Some of the world's best and most expensive coffee is Hawaii's Kona coffee. In order for coffee to be labeled "Kona Blend," it must contain at least 30% Kona beans. Bean Town Roasters has 40 lb of Mexican coffee. How much Kona coffee must they add if they wish to market it as Kona Blend?

Cumulative Review: Chapters 1–8

Simplify. Do not leave negative exponents in your answers.

1. $x^4 \cdot x^{-6} \cdot x^{13}$ [4.2]

2. $\dfrac{-10a^7b^{-11}}{25a^{-4}b^{22}}$ [4.2]

3. $\left(\dfrac{3x^4y^{-2}}{4x^{-5}}\right)^4$ [4.2]

4. $\dfrac{2.42 \times 10^5}{6.05 \times 10^{-2}}$ [4.2]

5. Solve $A = \frac{1}{2}h(b + t)$ for b. [2.3]

6. Solve $B = \dfrac{r + s}{r}$ for r. [7.5]

Solve.

7. $3n - (4n - 2) = 7$ [2.2]

8. $9c - [3 - 4(2 - c)] = 10$ [2.2]

9. $x^2 + 5x + 6 = 0$ [5.7]

10. $t + \dfrac{6}{t} = 5$ [6.6]

11. $2y + 9 \le 5y + 11$ [2.6]

12. $3x + y = 4,$
$y = 6x - 5$ [8.2]

13. $6x - 10y = -22,$
$-11x - 15y = 27$
[8.2]

14. $x + y + z = -5,$
$2x + 3y - 2z = 8,$
$x - y + 4z = -21$
[8.4]

Graph.

15. $f(x) = -2x + 8$ [7.3]

16. $x = 2y$ [3.2]

17. $4x + 16 = 0$ [3.3]

18. $-3x + 2y = 6$ [3.3]

19. Find the slope and the y-intercept of the line with equation $-4y + 9x = 12$. [3.6]

20. Find an equation in slope–intercept form of the line containing the points $(-6, 3)$ and $(4, 2)$. [3.6]

21. Determine whether the lines given by the following equations are parallel, perpendicular, or neither:
$2x = 4y + 7,$
$x - 2y = 5.$ [3.6]

22. Find an equation of the line containing the point $(2, 1)$ and perpendicular to the line $x - 2y = 5$.
[3.7]

23. Determine the domain of the function given by
$$f(x) = \dfrac{7}{x + 10}.$$ [7.2]

Given $g(x) = 4x - 3$ and $h(x) = -2x^2 + 1$, find the following.

24. $h(4)$ [7.1]

25. $(g - h)(a)$ [7.4]

Factor.

26. $5x^2 - 20y^2$ [5.4]

27. $6x^2 + 13x + 6$ [5.3]

Simplify.

28. $(5a + 7)^2$ [4.6]

29. $(2x^2 - 5x - 3) \div (x - 3)$ [4.8]

30. $\dfrac{x + 5}{x - 2} - \dfrac{x - 1}{2 - x}$ [6.4]

31. $\dfrac{x^2 - x}{4x^2 + 8x} \div \dfrac{x^2 - 1}{2x}$ [6.2]

Solve.

32. A snowmobile is traveling 40 mph faster than a dog sled. In the same time that the dog sled travels 24 mi, the snowmobile travels 104 mi. Find the speeds of the sled and the snowmobile. [6.7]

33. *Professional Memberships.* A one-year professional membership in Investigative Reporters and Editors, Inc. (IRE), costs $70. A one-year student membership costs $25. If, in May, 150 members joined IRE for a total of $6810 in membership dues, how many professionals and how many students joined that month? [8.3]

Source: Investigative Reporters and Editors, Inc.

34. *Saline Solutions.* "Sea Spray" is 25% salt and the rest water. "Ocean Mist" is 5% salt and the rest water. How many ounces of each would be needed to obtain 120 oz of a mixture that is 20% salt? [8.3]

35. *Test Scores.* Franco's scores on four tests are 93, 85, 100, and 86. What must the score be on the fifth test for his average to be 90? [2.5]

Synthesis

36. Simplify: $(6x^{a+2}y^{b+2})(-2x^{a-2}y^{y+1})$. [4.2]

37. Given that $f(x) = mx + b$ and that $f(5) = -3$ and $f(-4) = 2$, find m and b. [7.1], [8.3]

Inequalities and Problem Solving

YEAR	MUNICIPAL SOLID WASTE GENERATED (in pounds per person per day)
2000	4.72
2005	4.67
2007	4.64
2008	4.53
2009	4.35
2010	4.43

Reduce, Reuse, Recycle!

All three of these strategies are effective in reducing the amount of waste that ends up in landfills. Although it is difficult to measure how many items are reused, the Environmental Protection Agency (EPA) does measure how much waste is produced and how much is recycled. As the data in the table above indicate, the amount of solid waste *generated* per person per day has slowly been decreasing. Although not shown here, the EPA has evidence that the amount of solid waste *recovered* per person per day has been increasing. In this chapter, we will use functions that model waste recovery and recycling. *(See Example 7 and Exercise 93 in Section 9.1.)*

I use math all the time both to do my work correctly and to provide accurate accounting for reimbursement.

Gonzalo Calderone, a supervisor for a recycling center in California, uses math to calculate percentages of mandatory diversion rates for all recyclable materials. He must have accurate and detailed accounting for all products that can be recycled.

Inequalities are mathematical sentences containing symbols such as $<$ (is less than). We solve inequalities using principles similar to those used to solve equations. In this chapter, we solve a variety of inequalities, systems of inequalities, and real-world applications.

9.1   Inequalities and Applications

Solving Inequalities   •   Applications

Solving Inequalities

An **inequality** is any sentence containing $<, >, \leq, \geq,$ or $\neq$. Some examples of inequalities are

$$-2 < a, \qquad x > 4, \qquad x + 3 \leq 6, \qquad 6 - 7y \geq 10y - 4, \quad \text{and} \quad 5x \neq 10.$$

Any replacement for the variable that makes an inequality true is called a **solution**. The set of all solutions is called the **solution set**. When all solutions of an inequality are found, we say that we have **solved** the inequality.

We can use two principles to solve inequalities.

THE ADDITION PRINCIPLE FOR INEQUALITIES

For any real numbers a, b, and c:

$$a < b \text{ is equivalent to } a + c < b + c;$$
$$a > b \text{ is equivalent to } a + c > b + c.$$

Similar statements hold for $\leq$ and $\geq$.

THE MULTIPLICATION PRINCIPLE FOR INEQUALITIES

For any real numbers a and b, and for any *positive* number c,

$$a < b \text{ is equivalent to } ac < bc;$$
$$a > b \text{ is equivalent to } ac > bc.$$

For any real numbers a and b, and for any *negative* number c,

$$a < b \text{ is equivalent to } ac > bc;$$
$$a > b \text{ is equivalent to } ac < bc.$$

Similar statements hold for $\leq$ and $\geq$.

EXAMPLE 1 Solve: $2x + 4 < -x + 1$.

SOLUTION

$$2x + 4 < -x + 1$$
$$2x + 4 - 4 < -x + 1 - 4 \qquad \text{Subtracting 4 from both sides}$$
$$2x < -x - 3$$
$$2x + x < -x - 3 + x \qquad \text{Adding } x \text{ to both sides}$$
$$3x < -3$$
$$\frac{3x}{3} < \frac{-3}{3} \qquad \text{Dividing both sides by 3. The } < \text{ symbol stays the same since 3 is positive.}$$
$$x < -1$$

The solution set is $\{x \mid x < -1\}$, or $(-\infty, -1)$.

1. Solve: $3n - 6 < 7n + 4$.

YOUR TURN

We now look at a graphical method for solving inequalities.

To solve the inequality in Example 1, $2x + 4 < -x + 1$, we let $f(x) = 2x + 4$ and $g(x) = -x + 1$. Consider the graphs of the functions $f(x) = 2x + 4$ and $g(x) = -x + 1$.

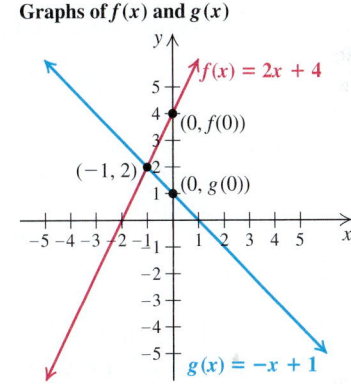

Graphs of $f(x)$ and $g(x)$

The graphs intersect at the point $(-1, 2)$. Thus, when $x = -1$, $f(x) = g(x)$. At all x-values except -1, either $f(x) > g(x)$ or $f(x) < g(x)$. Note from the graphs that $f(x) > g(x)$ when the graph of f lies above the graph of g. Also, $f(x) < g(x)$ when the graph of f lies below the graph of g.

Compare $f(0)$ and $g(0)$. Note from the graphs that $f(0)$ lies above $g(0)$. In fact, for all values of x greater than -1, $f(x) > g(x)$. For all values of x less than -1, $f(x) < g(x)$. In this way, the point of intersection of the graphs marks the endpoint of the solution set of an inequality.

Note that using the graphs of $f(x)$ and $g(x)$ to solve an inequality is not the same as graphing the solutions of the inequality.

For $f(x) = 2x + 4$ and $g(x) = -x + 1$, compare the following.

Equation/Inequality	*Solution Set*	*Graph of Solution Set*
$f(x) = g(x)$	$\{-1\}$	
$f(x) < g(x)$	$(-\infty, -1)$	
$f(x) > g(x)$	$(-1, \infty)$	

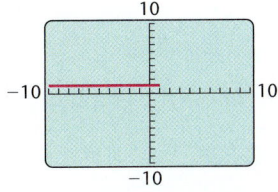
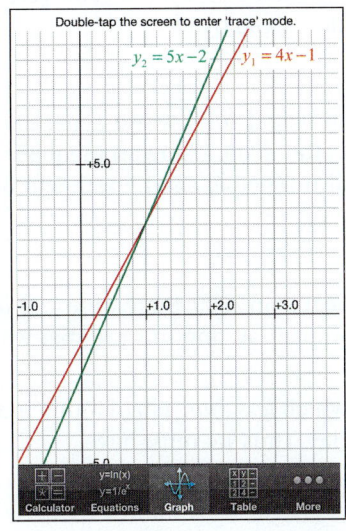

EXAMPLE 2   Solve graphically: $-3x + 1 \geq x - 7$.

SOLUTION   We let $f(x) = -3x + 1$ and $g(x) = x - 7$, and graph both functions. The solution set will consist of the interval for which the graph of f lies on or above the graph of g.

To find the point of intersection, we solve the system of equations

$$y = -3x + 1,$$
$$y = x - 7.$$

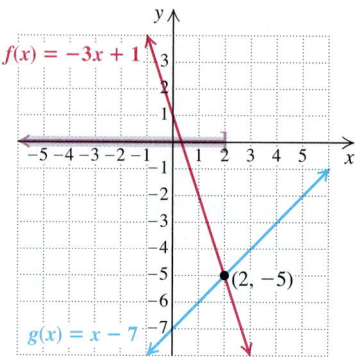

Using substitution, we have

$$-3x + 1 = x - 7$$
$$-4x = -8$$
$$x = 2.$$

Then $y = -3(2) + 1 = -5$.

Thus the point of intersection is

$$(2, -5).$$

The graph of f lies *on* the graph of g when $x = 2$. It lies *above* the graph of g when $x < 2$. Thus the solution of $-3x + 1 \geq x - 7$ is

$$\{x | x \leq 2\}, \quad \text{or} \quad (-\infty, 2].$$

This set is indicated by the purple shading on the x-axis.

2. Solve graphically:

$$6 - x < 4 + x.$$

YOUR TURN

Applications

Year	Municipal Solid Waste Generated (in pounds per person per day)
2000	4.72
2005	4.67
2007	4.64
2008	4.53
2009	4.35
2010	4.43

Source: U.S. Environmental Protection Agency

EXAMPLE 3   *Solid Waste.*   The table at left shows the amount of municipal solid waste generated per person per day in the United States for several years. Although the data are not exactly linear, the function given by

$$m(t) = -0.03t + 4.72$$

is a good model. Here, $m(t)$ is the amount of municipal solid waste generated, in number of pounds per person per day, t years after 2000. Using an inequality, determine those years for which less than 4 lb of municipal solid waste will be generated per person per day.

SOLUTION

1. **Familiarize.** By examining the formula, we see that in 2000, there were 4.72 lb of municipal solid waste generated per person per day, and this number was changing at a rate of -0.03 lb per year.

2. **Translate.** We are asked to find the years for which *less than* 4 lb of municipal solid waste will be generated per person per day. Thus we must have

$$m(t) < 4$$
$$-0.03t + 4.72 < 4.$$

3. **Carry out.** We solve the inequality:

$$-0.03t + 4.72 < 4$$
$$-0.03t < -0.72$$
$$t > 24.$$

Note that this corresponds to years after 2024.

4. **Check.** We can partially check our answer by finding $m(t)$ for a value of t greater than 24. For example,

$$m(25) = -0.03(25) + 4.72 = 3.97, \quad \text{and} \quad 3.97 < 4.$$

3. Refer to Example 3. Determine the years for which 4.3 lb or less of municipal solid waste will be generated per person per day.

5. **State.** Less than 4 lb of municipal solid waste will be generated per person per day for years after 2024.

 YOUR TURN

EXAMPLE 4 *Job Offers.* After graduation, Jessica had two job offers in sales:

Uptown Fashions: A salary of $1500 per month, plus a commission of 4% of sales;

Ergo Designs: A salary of $1700 per month, plus a commission of 6% of sales in excess of $10,000.

If sales always exceed $10,000, for what amount of sales would Uptown Fashions provide higher pay?

SOLUTION

1. **Familiarize.** Suppose that Jessica sold a certain amount—say, $12,000—in one month. Which plan would be better? Working for Uptown, she would earn $1500 plus 4% of $12,000, or $1500 + 0.04($12,000) = $1980.

Since with Ergo Designs commissions are paid only on sales in excess of $10,000, Jessica would earn $1700 plus 6% of ($12,000 − $10,000), or $1700 + 0.06($2000) = $1820.

Thus for monthly sales of $12,000, Uptown pays more. Similar calculations show that for sales of $30,000 per month, Ergo pays more. To determine *all* values for which Uptown pays more, we solve an inequality based on the above calculations.

Let S = the amount of monthly sales, in dollars, and assume $S > 10,000$ as stated above. We list the given information in a table.

Uptown Fashions Monthly Income	Ergo Designs Monthly Income
$1500 salary 4% of sales = $0.04S$ *Total*: $1500 + 0.04S$	$1700 salary 6% of sales over $10,000 = $0.06(S − 10,000)$ *Total*: $1700 + 0.06(S − 10,000)$

Chapter Resources:
Collaborative Activity, p. 620;
Decision Making: Connection, p. 620

2. **Translate.** We want to find all values of S for which

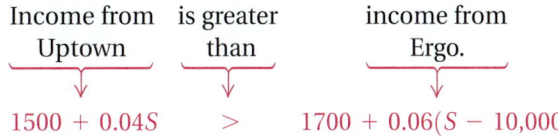

Income from Uptown   is greater than   income from Ergo.

$$1500 + 0.04S \quad > \quad 1700 + 0.06(S − 10,000)$$

3. Carry out. We solve the inequality:

$$1500 + 0.04S > 1700 + 0.06(S - 10{,}000)$$

$$1500 + 0.04S > 1700 + 0.06S - 600 \qquad \text{Using the distributive law}$$

$$1500 + 0.04S > 1100 + 0.06S \qquad \text{Combining like terms}$$

$$400 > 0.02S \qquad \begin{array}{l}\text{Subtracting 1100 and } 0.04S \\ \text{from both sides}\end{array}$$

$$20{,}000 > S, \text{ or } S < 20{,}000. \qquad \text{Dividing both sides by 0.02}$$

4. Refer to Example 4. Suppose that after salary negotiations, Uptown Fashions offers a salary of $1800 per month, plus a commission of 3% of sales, and Ergo Designs offers a salary of $1900 per month, plus a commission of 5% of sales in excess of $15,000. If sales always exceed $15,000, for what amount of sales would Ergo Designs provide higher pay?

4. Check. The above steps indicate that income from Uptown Fashions is higher than income from Ergo Designs for sales less than $20,000. In the *Familiarize* step, we saw that for sales of $12,000, Uptown pays more. Since $12{,}000 < 20{,}000$, this is a partial check.

5. State. When monthly sales are less than $20,000, Uptown Fashions provides the higher pay (assuming sales are greater than $10,000).

 YOUR TURN

9.1 EXERCISE SET

PRACTICE　WATCH　READ　REVIEW

◆ Vocabulary and Reading Check

Choose the word from the following list that best completes each statement. Not every word will be used.

above	negative
below	on
equation	positive
inequality	solution

1. A(n) _____ is a sentence containing $<, >, \le, \ge,$ or $\ne$.

2. Because $-8 < -1$ is true, -8 is a(n) _____ of $x < -1$.

3. We reverse the direction of the inequality symbol when we multiply both sides of an inequality by a(n) _____ number.

4. When $f(x) > g(x)$, the graph of f lies _____ the graph of g.

◆ Concept Reinforcement

Classify each of the following as equivalent inequalities, equivalent equations, equivalent expressions, or not equivalent.

5. $5x + 7 = 6 - 3x, \ 8x + 7 = 6$

6. $2(4x + 1), \ 8x + 2$

7. $x - 7 > -2, \ x > 5$

8. $-4t \le 12, \ t \le -3$

9. $\frac{3}{5}a + \frac{1}{5} = 2, \ 3a + 1 = 10$

10. $-\frac{1}{3}t \le -5, \ t \ge 15$

Solving Inequalities

Solve algebraically.

11. $3x + 1 < 7$

12. $2x - 5 \ge 9$

12. $3 - x \ge 12$

13. $8 - x < 15$

15. $\dfrac{2x + 7}{5} < -9$

16. $\dfrac{5y + 13}{4} > -2$

17. $\dfrac{3t - 7}{-4} \le 5$

18. $\dfrac{2t - 9}{-3} \ge 7$

19. $3 - 8y \ge 9 - 4y$

20. $4m + 7 \ge 9m - 3$

21. $5(t - 3) + 4t < 2(7 + 2t)$

22. $2(4 + 2x) > 2x + 3(2 - 5x)$

23. $5[3m - (m + 4)] > -2(m - 4)$

24. $8x - 3(3x + 2) - 5 \ge 3(x + 4) - 2x$

Solve each inequality using the given graph.

25. $f(x) \geq g(x)$

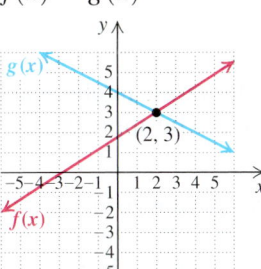

26. $f(x) < g(x)$

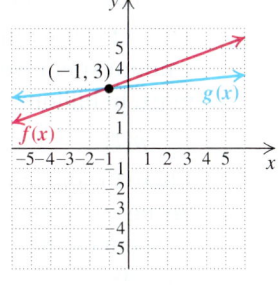

27. $f(x) < g(x)$

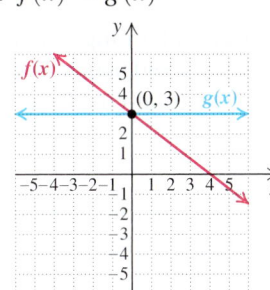

28. $f(x) \geq g(x)$

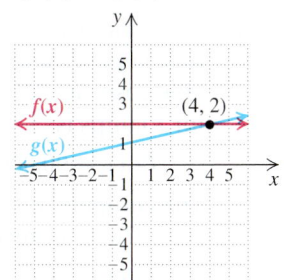

Solve graphically.

29. $x - 3 < 4$

30. $x + 4 \geq 6$

31. $2x - 3 \geq 1$

32. $3x + 1 < 1$

33. $x + 3 > 2x - 5$

34. $3x - 5 \leq 3 - x$

35. $\frac{1}{2}x - 2 \leq 1 - x$

36. $x + 5 > \frac{1}{3}x - 1$

37. Let $f(x) = 7 - 3x$ and $g(x) = 2x - 3$. Find all values of x for which $f(x) \leq g(x)$.

38. Let $f(x) = 2x + 1$ and $g(x) = -\frac{1}{2}x + 6$. Find all values of x for which $f(x) < g(x)$.

Applications

Solve.

39. *Photography.* Eli will photograph a wedding for a flat fee of $900 or for an hourly rate of $120. For what lengths of time would the hourly rate be less expensive?

40. *Truck Rentals.* Jenn can rent a moving truck for either $99 with unlimited mileage or $49 plus 80¢ per mile. For what mileages would the unlimited mileage plan save money?

41. *Exam Scores.* There are 80 questions on a college entrance examination. Two points are awarded for

each correct answer, and one-half point is deducted for each incorrect answer. How many questions does Tami need to answer correctly in order to score at least 100 on the test? Assume that Tami answers every question.

42. *Insurance Claims.* After a serious automobile accident, most insurance companies will replace the damaged car with a new one if repair costs exceed 80% of the NADA, or "blue-book," value of the car. Lorenzo's car recently sustained $9200 worth of damage but was not replaced. What was the blue-book value of his car?

43. *Wages.* Toni can be paid in one of two ways:

    Plan A: A salary of $400 per month, plus a commission of 8% of gross sales;

    Plan B: A salary of $610 per month, plus a commission of 5% of gross sales.

For what amount of gross sales should Toni select plan A?

44. *Wages.* Eric can be paid for his masonry work in one of two ways:

    Plan A: $300 plus $15.00 per hour;

    Plan B: Straight $17.50 per hour.

Suppose that the job takes n hours. For what values of n is plan B better for Eric?

45. *ATM Rates.* The Intercity Bank offers two account plans. Their Local plan charges a $5 monthly service fee plus $3.00 for each ATM (automated teller machine) transaction beyond 4 transactions per month. Their Anywhere plan charges a $15 monthly service fee plus $1.75 per ATM transaction. For what number of ATM transactions per month will the Anywhere plan cost less?

46. *Checking Accounts.* North Bank charges $10 per month for a student checking account. The first 8 checks are free, and each additional check costs $0.75. South Bank offers a student checking account with no monthly charge. The first 8 checks are free, and each additional check costs $3. For what numbers of checks is the South Bank plan more expensive? (Assume that the student will always write more than 8 checks.)

47. *Digital Music.* The amount of money spent worldwide for streaming music services t years after 2010 can be approximated by

$$m(t) = 0.42t + 0.532,$$

where $m(t)$ is in billions of dollars. Determine (using an inequality) those years for which more than $3 billion will be spent for streaming music services.

Source: Based on data from Gartner

9.2   Intersections, Unions, and Compound Inequalities

Intersections of Sets and Conjunctions of Sentences   ▪   Unions of Sets and Disjunctions of Sentences   ▪
Interval Notation and Domains

Two inequalities joined by the word "and" or the word "or" are called **compound inequalities**. To solve compound inequalities, we must know how sets can be combined.

Intersections of Sets and Conjunctions of Sentences

$A \cap B$

The **intersection** of two sets A and B is the set of all elements that are common to both A and B. We denote the intersection of sets A and B as

$$A \cap B.$$

The intersection of two sets is represented by the purple region shown in the figure at left. For example, if $A = \{$all students who are taking a math class$\}$ and $B = \{$all students who are taking a history class$\}$, then $A \cap B = \{$all students who are taking a math class *and* a history class$\}$.

EXAMPLE 1   Find the intersection: $\{1, 2, 3, 4, 5\} \cap \{-2, -1, 0, 1, 2, 3\}$.

SOLUTION   The numbers 1, 2, and 3 are common to both sets, so the intersection is $\{1, 2, 3\}$.

1. Find the intersection:

$$\{4, 5, 6, 7\} \cap \{2, 3, 5, 7\}.$$

YOUR TURN

When two or more sentences are joined by the word *and* to make a compound sentence, the new sentence is called a **conjunction** of the sentences. The following is a conjunction of inequalities:

$$-2 < x \quad and \quad x < 1.$$

A number is a solution of a conjunction if it is a solution of *both* of the separate parts. For example, -1 is a solution because it is a solution of $-2 < x$ as well as $x < 1$; that is, -1 is *both* greater than -2 *and* less than 1.

The solution set of a conjunction is the intersection of the solution sets of the individual sentences.

EXAMPLE 2   Graph and write interval notation for the conjunction

$$-2 < x \quad and \quad x < 1.$$

SOLUTION   We first graph $-2 < x$, then $x < 1$, and finally the conjunction $-2 < x$ and $x < 1$.

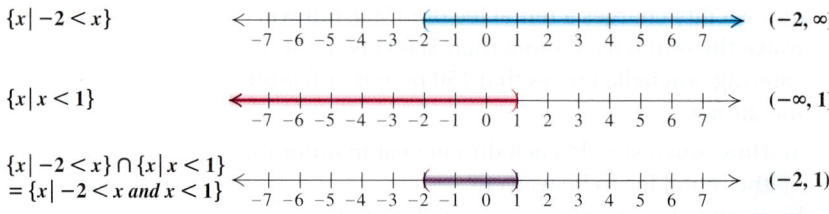

2. Graph and write interval notation for the conjunction

$$1 < x \quad and \quad x \le 4.$$

The solution set of the conjunction $-2 < x$ and $x < 1$ is the interval $(-2, 1)$. In set-builder notation, this is written $\{x \mid -2 < x < 1\}$, the set of all numbers that are *simultaneously* greater than -2 *and* less than 1.

YOUR TURN

Study Skills

Guess What Comes Next

If you have at least skimmed over the day's material before you go to class, you will be better able to follow the instructor. As you listen, pay attention to the direction the lecture is taking, and try to predict what topic or idea the instructor will present next. As you take a more active role in listening, you will grasp more of the material taught.

For $a < b$,

$$a < x \quad and \quad x < b \quad \text{can be abbreviated} \quad a < x < b;$$

and, equivalently,

$$b > x \quad and \quad x > a \quad \text{can be abbreviated} \quad b > x > a.$$

MATHEMATICAL USE OF THE WORD "AND"

The word "and" corresponds to "intersection" and to the symbol "$\cap$". Any solution of a conjunction must make each part of the conjunction true.

EXAMPLE 3 Solve and graph: $-1 \le 2x + 5 < 13$.

SOLUTION This inequality is an abbreviation for the conjunction

$$-1 \le 2x + 5 \quad and \quad 2x + 5 < 13.$$

The word *and* corresponds to set *intersection*. To solve the conjunction, we solve each inequality separately and then find the intersection of the solution sets:

$-1 \le 2x + 5$	and	$2x + 5 < 13$	
$-6 \le 2x$	and	$2x < 8$	Subtracting 5 from both sides of each inequality
$-3 \le x$	and	$x < 4.$	Dividing both sides of each inequality by 2

The solution of the conjunction is the intersection of the separate solution sets.

$\{x \mid -3 \le x\}$ $[-3, \infty)$

$\{x \mid x < 4\}$ $(-\infty, 4)$

$\{x \mid -3 \le x\} \cap \{x \mid x < 4\}$ $= \{x \mid -3 \le x < 4\}$ $[-3, 4)$

3. Solve and graph:

$$-5 < 3x - 1 < 0.$$

We can now abbreviate the answer as $-3 \le x < 4$. The solution set is $\{x \mid -3 \le x < 4\}$, or, in interval notation, $[-3, 4)$.

YOUR TURN

The steps in Example 3 are often combined as follows:

$$-1 \le 2x + 5 < 13$$
$$-1 - 5 \le 2x + 5 - 5 < 13 - 5 \qquad \text{Subtracting 5 from all three regions}$$
$$-6 \le 2x < 8$$
$$-3 \le x < 4. \qquad \text{Dividing by 2 in all three regions}$$

CAUTION! The abbreviated form of a conjunction, like $-3 \le x < 4$, can be written only if both inequality symbols point in the same direction.

EXAMPLE 4 Solve and graph: $2x - 5 \geq -3$ *and* $5x + 2 \geq 17$.

SOLUTION We first solve each inequality, retaining the word *and*:

$$2x - 5 \geq -3 \quad and \quad 5x + 2 \geq 17$$
$$2x \geq 2 \quad and \quad 5x \geq 15$$
$$x \geq 1 \quad and \quad x \geq 3.$$

————— Keep the word "and."

Next, we find the intersection of the two separate solution sets.

$\{x \mid x \geq 1\}$ [1, ∞)

$\{x \mid x \geq 3\}$ [3, ∞)

$\{x \mid x \geq 1\} \cap \{x \mid x \geq 3\}$
$= \{x \mid x \geq 3\}$ [3, ∞)

The numbers common to both sets are those greater than or equal to 3. Thus the solution set is $\{x \mid x \geq 3\}$, or, in interval notation, $[3, \infty)$. You should check that any number in $[3, \infty)$ satisfies the conjunction whereas numbers outside $[3, \infty)$ do not.

4. Solve and graph:

$$5x < 10 \quad and \quad x + 3 \leq 1.$$

↩ YOUR TURN

Sometimes there is no way to solve both parts of a conjunction at once.

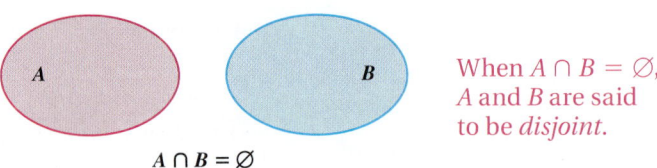

$A \cap B = \varnothing$

When $A \cap B = \varnothing$, A and B are said to be *disjoint*.

EXAMPLE 5 Solve and graph: $2x - 3 > 1$ *and* $3x - 1 < 2$.

SOLUTION We solve each inequality separately:

$$2x - 3 > 1 \quad and \quad 3x - 1 < 2$$
$$2x > 4 \quad and \quad 3x < 3$$
$$x > 2 \quad and \quad x < 1.$$

The solution set is the intersection of the individual inequalities.

$\{x \mid x > 2\}$ (2, ∞)

$\{x \mid x < 1\}$ (−∞, 1)

$\{x \mid x > 2\} \cap \{x \mid x < 1\}$
$= \{x \mid x > 2 \ and \ x < 1\} = \varnothing$ ∅

5. Solve and graph:

$$x + 6 < 5 \quad and \quad 3x + 1 > 7.$$

Since no number is both greater than 2 and less than 1, the solution set is the empty set, $\varnothing$.

↩ YOUR TURN

Unions of Sets and Disjunctions of Sentences

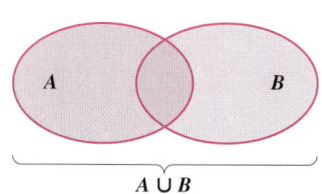

$A \cup B$

The **union** of two sets A and B is the collection of elements belonging to A and/or B. We denote the union of A and B by

$$A \cup B.$$

The union of two sets is often pictured as shown at left. For example, if $A = \{$all students who are taking a math class$\}$ and $B = \{$all students who are taking a history class$\}$, then $A \cup B = \{$all students who are taking a math class *or* a history class$\}$. Note that this set includes students who are taking a math class *and* a history class. Mathematically, the word "or" can be regarded as "and/or."

EXAMPLE 6   Find the union: $\{2, 3, 4\} \cup \{3, 5, 7\}$.

6. Find the union:

$\{4, 6, 8\} \cup \{7, 8, 9\}$.

SOLUTION   The numbers in either or both sets are 2, 3, 4, 5, and 7, so the union is $\{2, 3, 4, 5, 7\}$.

YOUR TURN

When two or more sentences are joined by the word *or* to make a compound sentence, the new sentence is called a **disjunction** of the sentences. Here is an example:

$$x < -3 \quad or \quad x > 3.$$

A number is a solution of a disjunction if it is a solution of at least one of the separate parts. For example, -5 is a solution of this disjunction since -5 is a solution of $x < -3$.

Student Notes

Remember that the union or the intersection of two sets is itself a set and is written using set notation.

The solution set of a disjunction is the union of the solution sets of the individual sentences.

EXAMPLE 7   Graph and write interval notation for the disjunction

$$x < -3 \quad or \quad x > 3.$$

SOLUTION   We graph $x < -3$, then $x > 3$, and finally $x < -3$ or $x > 3$.

$\{x | x < -3\}$ — number line — $(-\infty, -3)$

$\{x | x > 3\}$ — number line — $(3, \infty)$

$\{x | x < -3\} \cup \{x | x > 3\}$
$= \{x | x < -3 \text{ or } x > 3\}$ — number line — $(-\infty, -3) \cup (3, \infty)$

7. Graph and write interval notation for the disjunction

$x < -2 \quad or \quad x > 3.$

The solution set of $x < -3$ or $x > 3$ is $\{x | x < -3 \text{ or } x > 3\}$, or, in interval notation, $(-\infty, -3) \cup (3, \infty)$. There is no simpler way to write the solution.

YOUR TURN

> **MATHEMATICAL USE OF THE WORD "OR"**
>
> The word "or" corresponds to "union" and to the symbol "$\cup$". For a number to be a solution of a disjunction, it must be in *at least one* of the solution sets of the individual sentences.

EXAMPLE 8   Solve and graph: $7 + 3x < 3 \; or \; 13 - 5x \leq 3$.

SOLUTION   We solve each inequality separately, retaining the word *or*:

$$7 + 3x < 3 \quad or \quad 13 - 5x \leq 3$$
$$3x < -4 \quad or \qquad -5x \leq -10$$

Keep the word "or." ——————————————

Dividing by a negative number and reversing the symbol

$$x < -\tfrac{4}{3} \quad or \qquad x \geq 2.$$

To find the solution set of the disjunction, we consider the individual graphs. We graph $x < -\tfrac{4}{3}$ and then $x \geq 2$. Then we take the union of these graphs.

$\left\{ x \,|\, x < -\tfrac{4}{3} \right\}$

number line from −6 to 6, shaded left of −4/3 $\left(-\infty, -\tfrac{4}{3}\right)$

$\{ x \,|\, x \geq 2 \}$

number line from −6 to 6, shaded right of 2 $[2, \infty)$

$\left\{ x \,|\, x < -\tfrac{4}{3} \right\} \cup \{ x \,|\, x \geq 2 \}$
$= \left\{ x \,|\, x < -\tfrac{4}{3} \; or \; x \geq 2 \right\}$ number line $\left(-\infty, -\tfrac{4}{3}\right) \cup [2, \infty)$

8. Solve and graph:

$2 - x > 1 \quad or \quad 4x - 9 > 7.$

The solution set is $\left\{ x \,|\, x < -\tfrac{4}{3} \; or \; x \geq 2 \right\}$, or $\left(-\infty, -\tfrac{4}{3}\right) \cup [2, \infty)$.

YOUR TURN

> **CAUTION!**   A compound inequality like
>
> $$x < -4 \quad or \quad x \geq 2$$
>
> *cannot* be expressed as $2 \leq x < -4$. Doing so would be to say that x is *simultaneously* less than −4 and greater than or equal to 2. No number is both less than −4 *and* greater than 2, but many are less than −4 *or* greater than 2.

EXAMPLE 9   Solve: $3x - 11 < 4 \; or \; 4x + 9 \geq 1$.

SOLUTION   We solve the individual inequalities separately, retaining the word *or*:

$$3x - 11 < 4 \quad or \quad 4x + 9 \geq 1$$
$$3x < 15 \quad or \qquad 4x \geq -8$$
$$x < 5 \quad or \qquad x \geq -2.$$

——————— Keep the word "or."

To find the solution set, we first look at the individual graphs.

$\{ x \,|\, x < 5 \}$ number line shaded left of 5 $(-\infty, 5)$

$\{ x \,|\, x \geq -2 \}$ number line shaded right of −2 $[-2, \infty)$

$\{ x \,|\, x < 5 \} \cup \{ x \,|\, x \geq -2 \}$
$= \{ x \,|\, x < 5 \; or \; x \geq -2 \}$ number line fully shaded $(-\infty, \infty) = \mathbb{R}$

9. Solve:

$6x - 2 > 4 \quad or \quad 3x - 5 < 1.$

Since *all* numbers are less than 5 or greater than or equal to −2, the two sets fill the entire number line. Thus the solution set is $\mathbb{R}$, the set of all real numbers.

YOUR TURN

Interval Notation and Domains

If $g(x) = \dfrac{5x - 2}{x - 3}$, then the number 3 is not in the domain of g. We can represent the domain of g using set-builder notation or interval notation.

EXAMPLE 10   Use interval notation to write the domain of g if $g(x) = \dfrac{5x - 2}{x - 3}$.

SOLUTION   The expression $\dfrac{5x - 2}{x - 3}$ is not defined when the denominator is 0. We set $x - 3$ equal to 0 and solve:

$$x - 3 = 0$$
$$x = 3. \qquad \text{The number 3 is } not \text{ in the domain.}$$

We have the domain of $g = \{x \mid x$ is a real number $and\ x \neq 3\}$. If we graph this set, we see that the domain can be written as a union of two intervals.

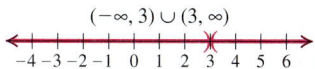
$$(-\infty, 3) \cup (3, \infty)$$

10. Use interval notation to write the domain of f if

$f(x) = \dfrac{x}{2x + 1}.$

Thus the domain of $g = (-\infty, 3) \cup (3, \infty)$.

↩ YOUR TURN

Only nonnegative numbers have square roots that are real numbers. Thus finding the domain of a radical function often involves solving an inequality.

EXAMPLE 11   Find the domain of f if $f(x) = \sqrt{7 - x}$.

SOLUTION   In order for $\sqrt{7 - x}$ to exist as a real number, $7 - x$ must be non-negative. Thus we solve $7 - x \geq 0$:

$$7 - x \geq 0 \qquad 7 - x \text{ must be nonnegative.}$$
$$-x \geq -7 \qquad \text{Subtracting 7 from both sides}$$
$$\qquad\qquad\qquad \text{The symbol must be reversed.}$$
$$x \leq 7. \qquad \text{Multiplying both sides by } -1$$

11. Find the domain of g if $g(x) = \sqrt{x + 3}$.

For $x \leq 7$, we have $7 - x \geq 0$. Thus the domain of f is $\{x \mid x \leq 7\}$, or $(-\infty, 7]$.

↩ YOUR TURN

9.2   EXERCISE SET

FOR EXTRA HELP

MyMathLab® Math XL
PRACTICE   WATCH   READ   REVIEW

↪ **Vocabulary and Reading Check**

Complete each statement using the word intersection *or the word* union.

1. The _____ of two sets is the set of all elements that are in both sets.

2. The symbol ∪ indicates _____.

3. The word "and" corresponds to _____.

4. The symbol ∩ indicates _____.

5. The solution of a disjunction is the _____ of the solution sets of the individual sentences.

6. The _____ of two sets is the set of all elements that are in either set or in both sets.

↳ Concept Reinforcement

In each of Exercises 7–16, match the set with the most appropriate choice below.

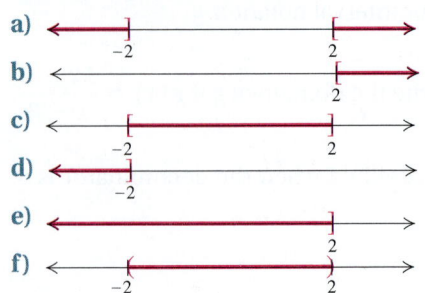

a)
b)
c)
d)
e)
f)
g)
h)

i) $\mathbb{R}$

j) $\varnothing$

7. ____ $\{x \mid x < -2 \text{ or } x > 2\}$

8. ____ $\{x \mid x < -2 \text{ and } x > 2\}$

9. ____ $\{x \mid x > -2\} \cap \{x \mid x < 2\}$

10. ____ $\{x \mid x \le -2\} \cup \{x \mid x \ge 2\}$

11. ____ $\{x \mid x \le -2\} \cup \{x \mid x \le 2\}$

12. ____ $\{x \mid x \le -2\} \cap \{x \mid x \le 2\}$

13. ____ $\{x \mid x \ge -2\} \cap \{x \mid x \ge 2\}$

14. ____ $\{x \mid x \ge -2\} \cup \{x \mid x \ge 2\}$

15. ____ $\{x \mid x \le 2\} \text{ and } \{x \mid x \ge -2\}$

16. ____ $\{x \mid x \le 2\} \text{ or } \{x \mid x \ge -2\}$

Intersections of Sets and Unions of Sets

Find each indicated intersection or union.

17. $\{2, 4, 16\} \cap \{4, 16, 256\}$

18. $\{1, 2, 4\} \cup \{4, 6, 8\}$

19. $\{0, 5, 10, 15\} \cup \{5, 15, 20\}$

20. $\{2, 5, 9, 13\} \cap \{5, 8, 10\}$

21. $\{a, b, c, d, e, f\} \cap \{b, d, f\}$

22. $\{u, v, w\} \cup \{u, w\}$

23. $\{x, y, z\} \cup \{u, v, x, y, z\}$

24. $\{m, n, o, p\} \cap \{m, o, p\}$

25. $\{3, 6, 9, 12\} \cap \{5, 10, 15\}$

26. $\{1, 5, 9\} \cup \{4, 6, 8\}$

27. $\{1, 3, 5\} \cup \varnothing$

28. $\{1, 3, 5\} \cap \varnothing$

Conjunctions of Sentences and Disjunctions of Sentences

Graph and write interval notation for each compound inequality.

29. $1 < x < 3$

30. $0 \le y \le 5$

31. $-6 \le y \le 0$

32. $-8 < x \le -2$

33. $x < -1 \text{ or } x > 4$

34. $x < -5 \text{ or } x > 1$

35. $x \le -2 \text{ or } x > 1$

36. $x \le -5 \text{ or } x > 2$

37. $-4 \le -x < 2$

38. $x > -7 \text{ and } x < -2$

39. $x > -2 \text{ and } x < 4$

40. $3 > -x \ge -1$

41. $5 > a \text{ or } a > 7$

42. $t \ge 2 \text{ or } -3 > t$

43. $x \ge 5 \text{ or } -x \ge 4$

44. $-x < 3 \text{ or } x < -6$

45. $7 > y \text{ and } y \ge -3$

46. $6 > -x \ge 0$

47. $-x < 7 \text{ and } -x \ge 0$

48. $x \ge -3 \text{ and } x < 3$

Aha! 49. $t < 2 \text{ or } t < 5$

50. $t > 4 \text{ or } t > -1$

Solve and graph each solution set.

51. $-3 \le x + 2 < 9$

52. $-1 < x - 3 < 5$

53. $0 < t - 4 \text{ and } t - 1 \le 7$

54. $-6 \le t + 1 \text{ and } t + 8 < 2$

55. $-7 \le 2a - 3 \text{ and } 3a + 1 < 7$

Aha! 56. $-4 \le 3n + 5 \text{ and } 2n - 3 \le 7$

57. $x + 3 \le -1 \text{ or } x + 3 > -2$

58. $x + 5 < -3 \text{ or } x + 5 \ge 4$

59. $-10 \le 3x - 1 \le 5$

60. $-18 \le 4x + 2 \le 30$

61. $5 > \dfrac{x - 3}{4} > 1$

62. $3 \ge \dfrac{x - 1}{2} \ge -4$

63. $-2 \le \dfrac{x + 2}{-5} \le 6$

64. $-10 \le \dfrac{x + 6}{-3} \le -8$

65. $2 \le f(x) \le 8$, where $f(x) = 3x - 1$

66. $7 \ge g(x) \ge -2$, where $g(x) = 3x - 5$

67. $-21 \le f(x) < 0$, where $f(x) = -2x - 7$

68. $4 > g(t) \ge 2$, where $g(t) = -3t - 8$

69. $f(t) < 3$ *or* $f(t) > 8$, where $f(t) = 5t + 3$

70. $g(x) \leq -2$ *or* $g(x) \geq 10$, where $g(x) = 3x - 5$

71. $6 > 2a - 1$ *or* $-4 \leq -3a + 2$

72. $3a - 7 > -10$ *or* $5a + 2 \leq 22$

73. $a + 3 < -2$ *and* $3a - 4 < 8$

74. $1 - a < -2$ *and* $2a + 1 > 9$

75. $3x + 2 < 2$ *and* $3 - x < 1$

76. $2x - 1 > 5$ *and* $2 - 3x > 11$

77. $2t - 7 \leq 5$ *or* $5 - 2t > 3$

78. $5 - 3a \leq 8$ *or* $2a + 1 > 7$

Interval Notation and Domains

For $f(x)$ as given, use interval notation to write the domain of f.

79. $f(x) = \dfrac{9}{x + 6}$

80. $f(x) = \dfrac{2}{x - 5}$

81. $f(x) = \dfrac{1}{x}$

82. $f(x) = -\dfrac{6}{x}$

83. $f(x) = \dfrac{x + 3}{2x - 8}$

84. $f(x) = \dfrac{x - 1}{3x + 6}$

85. $f(x) = \sqrt{x - 10}$

86. $f(x) = \sqrt{x + 2}$

87. $f(x) = \sqrt{3 - x}$

88. $f(x) = \sqrt{11 - x}$

89. $f(x) = \sqrt{2x + 7}$

90. $f(x) = \sqrt{8 - 5x}$

91. $f(x) = \sqrt{8 - 2x}$

92. $f(x) = \sqrt{2x - 10}$

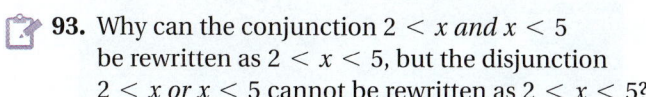 **93.** Why can the conjunction $2 < x$ *and* $x < 5$ be rewritten as $2 < x < 5$, but the disjunction $2 < x$ *or* $x < 5$ cannot be rewritten as $2 < x < 5$?

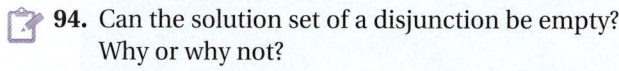 **94.** Can the solution set of a disjunction be empty? Why or why not?

Skill Review

Factor.

95. $x^2 - 12x + 20$ [5.2]

96. $50c^6 - 18d^2$ [5.4]

97. $10c^6 - 10$ [5.5]

98. $48x^3z^8 - 90x^4z^5 + 16x^2z^5$ [5.1]

Synthesis

99. What can you conclude about a, b, c, and d, if $[a, b] \cup [c, d] = [a, d]$? Why?

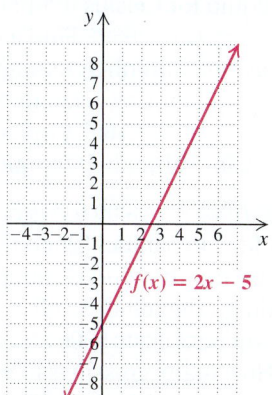 **100.** What can you conclude about a, b, c, and d, if $[a, b] \cap [c, d] = [a, b]$? Why?

101. Use the following graph of $f(x) = 2x - 5$ to solve $-7 < 2x - 5 < 7$.

$f(x) = 2x - 5$

102. Use the following graph of $g(x) = 4 - x$ to solve $4 - x < -2$ *or* $4 - x > 7$.

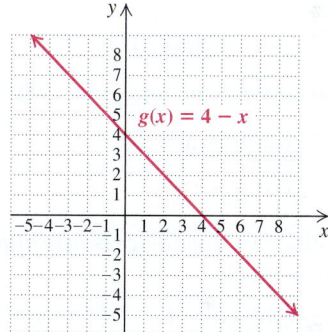

$g(x) = 4 - x$

103. *Pressure at Sea Depth.* The function given by

$$P(d) = 1 + \frac{d}{33}$$

gives the pressure, in atmospheres (atm), at a depth of d feet in the sea. For what depths d is the pressure at least 1 atm and at most 7 atm?

104. *Converting Dress Sizes.* The function given by

$$f(x) = 2(x + 10)$$

can be used to convert dress sizes x in the United States to dress sizes $f(x)$ in Italy. For what dress sizes in the United States will dress sizes in Italy be between 32 and 46?

105. *Body Fat Percentage.* The function given by

$$F(d) = (4.95/d - 4.50) \times 100$$

can be used to estimate the body fat percentage $F(d)$ of a person with an average body density d, in kilograms per liter. A woman's body fat percentage is considered healthy if $25 \leq F(d) \leq 31$. What body densities are considered healthy for a woman?

106. *Temperatures of Liquids.* The formula

$$C = \tfrac{5}{9}(F - 32)$$

is used to convert Fahrenheit temperatures F to Celsius temperatures C.

a) Gold is liquid for Celsius temperatures C such that $1063° \le C < 2660°$. Find a comparable inequality for Fahrenheit temperatures.

b) Silver is liquid for Celsius temperatures C such that $960.8° \le C < 2180°$. Find a comparable inequality for Fahrenheit temperatures.

107. *Minimizing Tolls.* A $6.00 toll is charged to cross the bridge to Sanibel Island from mainland Florida. A six-month reduced-fare pass costs $50 and reduces the toll to $2.00. A six-month unlimited-trip pass costs $300 and allows for free crossings. How many crossings in six months does it take for the reduced-fare pass to be the more economical choice?

Source: www.leewayinfo.com

Solve and graph.

108. $4a - 2 \le a + 1 \le 3a + 4$

109. $4m - 8 > 6m + 5$ or $5m - 8 < -2$

110. $x - 10 < 5x + 6 \le x + 10$

111. $3x < 4 - 5x < 5 + 3x$

Determine whether each sentence is true or false for all real numbers a, b, and c.

112. If $-b < -a$, then $a < b$.

113. If $a \le c$ and $c \le b$, then $b > a$.

114. If $a < c$ and $b < c$, then $a < b$.

115. If $-a < c$ and $-c > b$, then $a > b$.

For $f(x)$ as given, use interval notation to write the domain of f.

116. $f(x) = \dfrac{\sqrt{5 + 2x}}{x - 1}$

117. $f(x) = \dfrac{\sqrt{3 - 4x}}{x + 7}$

118. For $f(x) = \sqrt{x - 5}$ and $g(x) = \sqrt{9 - x}$, use interval notation to write the domain of $f + g$.

119. Let $y_1 = -1$, $y_2 = 2x + 5$, and $y_3 = 13$. Then use the graphs of y_1, y_2, and y_3 to check the solution to Example 3.

120. Let $y_1 = 3x - 11$, $y_2 = 4$, $y_3 = 4x + 9$, and $y_4 = 1$. Then use the graphs of y_1, y_2, y_3, and y_4 to check the solution to Example 9.

Readability. The reading difficulty of a textbook can be estimated by the Flesch Reading Ease Formula

$$r = 206.835 - 1.015n - 84.6s,$$

where r is the reading ease, n is the average number of words in a sentence, and s is the average number of syllables in a word. Sample reading-level scores are shown in the following table. Use this information for Exercises 121 and 122.

Score	Reading Ease
$90 \le r \le 100$	5th grade
$60 \le r \le 70$	8th and 9th grades
$0 \le r \le 30$	College graduates

Source: readabilityformulas.com

121. Bryan is writing a book for 5th-graders using an average of 1.2 syllables per word. How long should his average sentence length be?

122. The reading score for Alexa's new book for young adults indicates that it should be read with ease by 8th- and 9th-graders. If she averages 8 words per sentence, what is the average number of syllables per word?

123. A machine filling water bottles pours 16 oz of water into each bottle, with a margin of error of 0.1 oz. Write an inequality and interval notation for the amount of water that the machine pours into a bottle.

124. At one point during the 2012 presidential campaign, a Gallup poll indicated that Barack Obama had an approval rating of 46%, with a margin of error of 1%. Write an inequality and interval notation for Obama's approval rating.

Source: www.upi.com

125. Use a graphing calculator to check your answers to Exercises 51–54 and Exercises 69–72.

126. On many graphing calculators, the TEST key provides access to inequality symbols, while the LOGIC option of that same key accesses the conjunction *and* and the disjunction *or*. Thus, if $y_1 = x > -2$ and $y_2 = x < 4$, Exercise 39 can be checked by forming the expression $y_3 = y_1$ *and* y_2. The interval(s) in the solution set appears as a horizontal line 1 unit above the *x*-axis. (Be careful to "deselect" y_1 and y_2 so that only y_3 is drawn.) Use the TEST key to check Exercises 41, 45, 47, and 49.

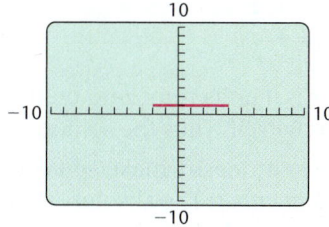

 127. Use a graphing calculator to confirm the domains of the functions in Exercises 85, 87, and 91.

128. *Research.* Find a formula for body mass index (BMI), and find the range for which your BMI would be considered healthy. For your height, what weights will result in an acceptable BMI?

129. *Research.* Find what a "95% confidence interval" means, and explain it in writing or to your class.

 YOUR TURN ANSWERS: SECTION 9.2

1. $\{5, 7\}$ **2.** $\{x \mid 1 < x \le 4\}$, or $(1, 4]$

3. $\{x \mid -\frac{4}{3} < x < \frac{1}{3}\}$, or $\left(-\frac{4}{3}, \frac{1}{3}\right)$

4. $\{x \mid x \le -2\}$, or $(-\infty, -2]$

5. $\varnothing$ **6.** $\{4, 6, 7, 8, 9\}$

7. $\{x \mid x < -2 \text{ or } x > 3\}$, or $(-\infty, -2) \cup (3, \infty)$

8. $\{x \mid x < 1 \text{ or } x > 4\}$, or $(-\infty, 1) \cup (4, \infty)$

9. $\mathbb{R}$, or $(-\infty, \infty)$ **10.** $\{x \mid x \text{ is a real number } and\, x \ne -\frac{1}{2}\}$, or $\left(-\infty, -\frac{1}{2}\right) \cup \left(-\frac{1}{2}, \infty\right)$ **11.** $\{x \mid x \ge -3\}$, or $[-3, \infty)$

QUICK QUIZ: SECTIONS 9.1–9.2

Solve. Write the solution set using both set-builder notation and interval notation.

1. $5 - 6x < x + 3$ [9.1]

2. $x - (9 - x) \ge 3(7 - x)$ [9.1]

3. $-\frac{2}{3}m - 5 > 7$ [9.1]

4. $3 > 7 - 2y$ or $6y < y$ [9.2]

5. $-1 < 7 - x < 4$ [9.2]

PREPARE TO MOVE ON

Find the absolute value. [1.4]

1. $\left|\frac{2}{3}\right|$ **2.** $|-16|$ **3.** $|0|$ **4.** $|8 - 15|$

5. Given that $f(x) = 3x - 10$, find all x for which $f(x) = 8$. [7.1]

9.3 Absolute-Value Equations and Inequalities

Equations with Absolute Value • Inequalities with Absolute Value

Equations with Absolute Value

The following is a formal definition of absolute value.

Study Skills

What Was That All About?

Start your notes or homework by writing the date, the course name or number, and the topic being discussed. Include as well the section number in the text where appropriate.

ABSOLUTE VALUE

The absolute value of x, denoted $|x|$, is defined as

$$|x| = \begin{cases} x, & \text{if } x \ge 0, \\ -x, & \text{if } x < 0. \end{cases}$$

(When x is nonnegative, the absolute value of x is x. When x is negative, the absolute value of x is the opposite of x.)

To better understand this definition, suppose x is -5. Then $|x| = |-5| = 5$, and 5 is the opposite of -5. This shows that when x represents a negative number, the absolute value of x is the opposite of x (which is positive).

Since distance is always nonnegative, we can think of a number's absolute value as its distance from zero on the number line.

EXAMPLE 1   Find each solution set: **(a)** $|x| = 4$; **(b)** $|x| = 0$; **(c)** $|x| = -7$.

SOLUTION

a) We interpret $|x| = 4$ to mean that the number x is 4 units from zero on the number line. There are two such numbers, 4 and -4. Thus the solution set is $\{-4, 4\}$.

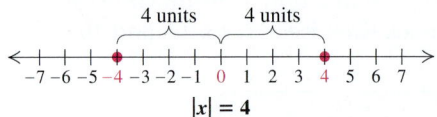

$$|x| = 4$$

b) We interpret $|x| = 0$ to mean that x is 0 units from zero on the number line. The only number that satisfies this is 0 itself. Thus the solution set is $\{0\}$.

c) Since distance is always nonnegative, it doesn't make sense to talk about a number that is -7 units from zero. *Remember*: The absolute value of a number is never negative. Thus, $|x| = -7$ has no solution; the solution set is $\varnothing$.

1. Find the solution set: $|x| = 6$.

 YOUR TURN

Example 1 leads us to the following principle for solving equations.

> **THE ABSOLUTE-VALUE PRINCIPLE FOR EQUATIONS**
>
> For any positive number p and any algebraic expression X:
>
> **a)** The solutions of $|X| = p$ are those numbers that satisfy
>
> $$X = -p \quad or \quad X = p.$$
>
> **b)** The equation $|X| = 0$ is equivalent to the equation $X = 0$.
> **c)** The equation $|X| = -p$ has no solution.

EXAMPLE 2   Find each solution set: **(a)** $|2x + 5| = 13$; **(b)** $|4 - 7x| = -8$.

SOLUTION

a) We use the absolute-value principle, knowing that $2x + 5$ is either 13 or -13:

$$|X| = p$$
$$|2x + 5| = 13 \qquad \text{Substituting}$$
$$2x + 5 = -13 \quad or \quad 2x + 5 = 13$$
$$2x = -18 \quad or \qquad 2x = 8$$
$$x = -9 \quad or \qquad x = 4.$$

Check: For -9:

$$\frac{|2x + 5| = 13}{\begin{array}{c|c} |2(-9) + 5| & 13 \\ |-18 + 5| & \\ |-13| & \\ & 13 \overset{?}{=} 13 \quad \text{TRUE} \end{array}}$$

For 4:

$$\frac{|2x + 5| = 13}{\begin{array}{c|c} |2 \cdot 4 + 5| & 13 \\ |8 + 5| & \\ |13| & \\ & 13 \overset{?}{=} 13 \quad \text{TRUE} \end{array}}$$

The number $2x + 5$ is 13 units from zero if x is replaced with -9 or 4. The solution set is $\{-9, 4\}$.

2. Find the solution set:

$$|3x - 5| = 7.$$

b) The absolute-value principle reminds us that absolute value is never negative. The equation $|4 - 7x| = -8$ has no solution. The solution set is $\varnothing$.

 YOUR TURN

To use the absolute-value principle, we must be sure that the absolute-value expression is alone on one side of the equation.

EXAMPLE 3   Given that $f(x) = 2|x + 3| + 1$, find all x for which $f(x) = 15$.

SOLUTION   Since we are looking for $f(x) = 15$, we substitute:

$$f(x) = 15$$
$$2|x + 3| + 1 = 15 \qquad \text{Replacing } f(x) \text{ with } 2|x + 3| + 1$$
$$2|x + 3| = 14 \qquad \text{Subtracting 1 from both sides}$$
$$|x + 3| = 7 \qquad \text{Dividing both sides by 2}$$
$$x + 3 = -7 \quad or \quad x + 3 = 7 \qquad \text{Using the absolute-value principle for equations}$$
$$x = -10 \quad or \qquad x = 4.$$

3. Given that $g(x) = 2|5x| - 4$, find all x for which $g(x) = 10$.

We leave it to the student to check that $f(-10) = f(4) = 15$. The solution set is $\{-10, 4\}$.

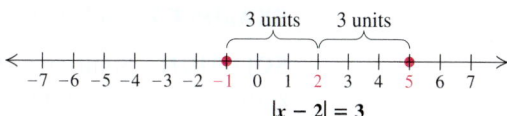 YOUR TURN

EXAMPLE 4   Solve: $|x - 2| = 3$.

SOLUTION   Because this is of the form $|a - b| = c$, it can be solved in two ways.

Method 1.   We interpret $|x - 2| = 3$ as stating that the number $x - 2$ is 3 units from zero. Using the absolute-value principle, we replace X with $x - 2$ and p with 3:

$$|X| = p$$
$$|x - 2| = 3 \qquad \text{We use this approach in Examples 1–3.}$$
$$x - 2 = -3 \quad or \quad x - 2 = 3 \qquad \text{Using the absolute-value principle}$$
$$x = -1 \quad or \qquad x = 5.$$

> **CAUTION!**   There are two solutions of $|x - 2| = 3$. Simply solving $x - 2 = 3$ will yield only one of those solutions.

Method 2.   The expressions $|a - b|$ and $|b - a|$ both represent the *distance between a and b* on the number line. For example, the distance between 7 and 8 is given by $|8 - 7|$ or $|7 - 8|$. From this viewpoint, the equation $|x - 2| = 3$ states that the distance between x and 2 is 3 units. We draw the number line and locate all numbers that are 3 units from 2.

3 units   3 units

‹—+—+—+—+—+—●—+—+—+—+—●—+—+—›
　−7 −6 −5 −4 −3 −2 −1　0　1　2　3　4　5　6　7

$|x - 2| = 3$

The solutions of $|x - 2| = 3$ are -1 and 5.

Check: The check consists of observing that both methods give the same solutions. The solution set is $\{-1, 5\}$.

4. Solve: $|x - 5| = 1$.

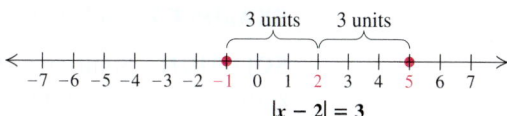 YOUR TURN

Some equations contain two absolute-value expressions. Consider $|a| = |b|$. This means that a and b are the same distance from zero. If a and b are the same distance from zero, they are either the same number or opposites.

For any algebraic expressions X and Y:

$$\text{If } |X| = |Y|, \quad \text{then } X = Y \quad \text{or} \quad X = -Y.$$

It turns out that *any* point on the same side of $y = x$ as $(4, 2)$ is also a solution. Thus, if one point in a half-plane is a solution, then *all* points in that half-plane are solutions.

We finish drawing the solution set by shading the half-plane below $y = x$. The solution set consists of the shaded half-plane as well as the boundary line itself.

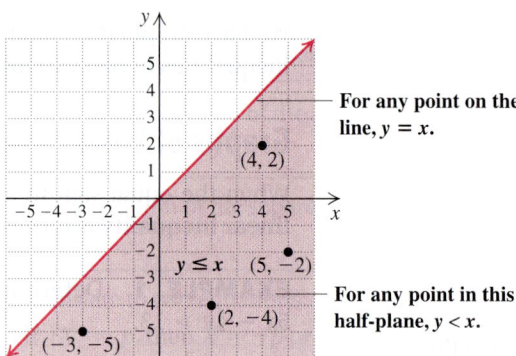

2. Graph: $y \leq x + 1$.

YOUR TURN

From Example 2, we see that for any inequality of the form $y \leq f(x)$ or $y < f(x)$, we shade *below* the graph of $y = f(x)$.

EXAMPLE 3 Graph: $8x + 3y > 24$.

SOLUTION First, we sketch the graph of $8x + 3y = 24$. A convenient way to graph this equation is to use the x-intercept, $(3, 0)$, and the y-intercept, $(0, 8)$. Since the inequality sign is $>$, points on this line do not represent solutions of the inequality, and the line is drawn dashed. Points representing solutions of $8x + 3y > 24$ are in either the half-plane above the line or the half-plane below the line. To determine which, we select a point that is not on the line and check whether it is a solution of $8x + 3y > 24$. Let's use $(0, 0)$ as this *test point*:

$$\begin{array}{c|c} \multicolumn{2}{c}{8x + 3y > 24} \\ \hline 8(0) + 3(0) & 24 \\ 0 \overset{?}{>} 24 & \text{FALSE} \end{array}$$

Since $0 > 24$ is *false*, $(0, 0)$ is not a solution. Thus no point in the half-plane containing $(0, 0)$ is a solution. The points in the other half-plane *are* solutions, so we shade that half-plane and obtain the graph shown below.

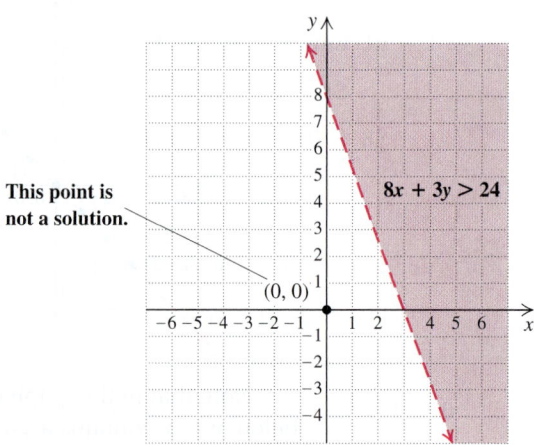

3. Graph: $2x + y < 6$.

YOUR TURN

> **STEPS FOR GRAPHING LINEAR INEQUALITIES**
>
> 1. Replace the inequality sign with an equal sign and graph this line as the boundary. If the inequality symbol is $<$ or $>$, draw the line dashed. If the symbol is $\leq$ or $\geq$, draw the line solid.
> 2. The graph of the inequality consists of a half-plane on one side of the line and, if the line is solid, the line as well.
>
>   a) For an inequality of the form $y < mx + b$ or $y \leq mx + b$, shade *below* the line.
>
>   For an inequality of the form $y > mx + b$ or $y \geq mx + b$, shade *above* the line.
>
>   b) If y is not isolated, use a test point not on the line as in Example 3. If the test point *is* a solution, shade the half-plane containing the point. If it is *not* a solution, shade the other half-plane. Additional test points can also be used as a check.

EXAMPLE 4 Graph: $6x - 2y < 12$.

SOLUTION We could graph $6x - 2y = 12$ and use a test point, as in Example 3. Instead, let's solve $6x - 2y < 12$ for y:

$$6x - 2y < 12$$
$$-2y < -6x + 12 \qquad \text{Adding } -6x \text{ to both sides}$$
$$y > 3x - 6. \qquad \text{Dividing both sides by } -2 \text{ and reversing the } < \text{ symbol}$$

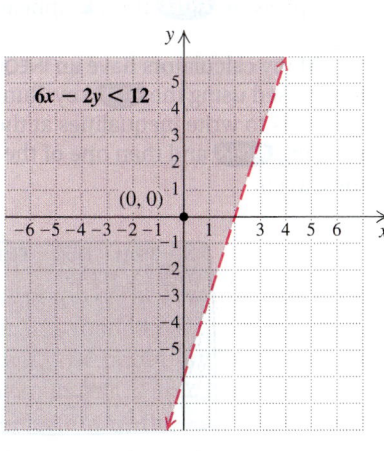

The graph consists of the half-plane above the dashed boundary line $y = 3x - 6$ (see the graph at right). As a check, note that the test point $(0, 0)$ is a solution of the inequality and is in the half-plane that we shaded.

4. Graph: $x > 6y - 6$.

YOUR TURN

EXAMPLE 5 Graph $x > -3$ on a plane.

SOLUTION There is only one variable in this inequality. If we graph the inequality on a line, its graph is as follows:

However, we can also write this inequality as $x + 0y > -3$ and graph it on a plane. Using the same technique as in the examples above, we graph the boundary $x = -3$ in the plane, using a dashed line. Then we test some point, say, $(2, 5)$:

$$\frac{x + 0y > -3}{2 + 0 \cdot 5 \mid -3}$$
$$2 \overset{?}{>} -3 \quad \text{TRUE}$$

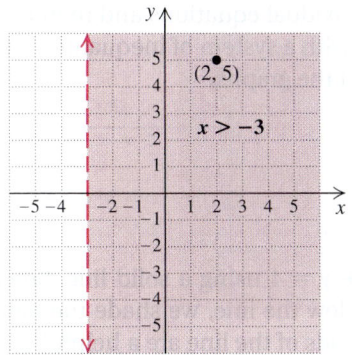

Since $(2, 5)$ is a solution, all points in the half-plane containing $(2, 5)$ are solutions. We shade that half-plane. We can also simply note that solutions of $x > -3$ are pairs with first coordinates greater than -3.

5. Graph $x \leq 2$ on a plane.

YOUR TURN

EXERCISES

Graph each solution on the number line.

1. $x + 2 = 7$

2. $x + 2 > 7$

3. $x + 2 \leq 7$

Graph on a plane.

4. $x + y = 2$

5. $x + y < 2$

6. $x + y \geq 2$

7. $x + 2 \leq 7$

8. $y = x - 1,$
   $y = -x + 1$

9. $y \geq 1 - x,$
   $y \leq x - 3,$
   $y \leq 2$

9.4 EXERCISE SET

FOR EXTRA HELP

MyMathLab® MathXL
PRACTICE WATCH READ REVIEW

Vocabulary and Reading Check

In each of Exercises 1–6, match the phrase with the most appropriate choice from the column on the right.

1. _____ The solution set of a linear inequality

2. _____ The graph of a linear inequality

3. _____ The graph of a system of linear inequalities

4. _____ Often a convenient test point

5. _____ The name for the corners of a graph of a system of linear inequalities

6. _____ A dashed line

a) $(0, 0)$

b) Vertices

c) A half-plane

d) The intersection of two or more half-planes

e) All ordered pairs that satisfy the inequality

f) Indicates the line is not part of the solution

Graphs of Linear Inequalities

Determine whether each ordered pair is a solution of the given inequality.

7. $(-2, 3);\ 2x - y > -4$

8. $(1, -6);\ 3x + y \geq -3$

9. $(5, 8);\ 3y - 5x \leq 0$

10. $(6, 20);\ 5y - 8x < 40$

Graph on a plane.

11. $y \geq \frac{1}{2}x$

12. $y \leq 3x$

13. $y > x - 3$

14. $y < x + 3$

15. $y \leq x + 2$

16. $y \geq x - 5$

17. $x - y \leq 4$

18. $x + y < 4$

19. $2x + 3y < 6$

20. $3x + 4y \leq 12$

21. $2y - x \leq 4$

22. $2y - 3x > 6$

23. $2x - 2y \geq 8 + 2y$

24. $3x - 2 \leq 5x + y$

25. $x > -2$

26. $x \geq 3$

27. $y \leq 6$

28. $y < -1$

Systems of Linear Inequalities

Graph.

29. $-2 < y < 7$

30. $-4 < y < -1$

31. $-5 \leq x < 4$

32. $-2 < y \leq 1$

33. $0 \leq y \leq 3$

34. $0 \leq x \leq 6$

35. $y > x,$
    $y < -x + 3$

36. $y < x,$
    $y > -x + 1$

37. $y \leq x,$
    $y \leq 2x - 5$

38. $y \geq x,$
    $y \leq -x + 4$

39. $y \leq -3,$
    $x \geq -1$

40. $y \geq -3,$
    $x \geq 1$

41. $x > -4,$
    $y < -2x + 3$

42. $x < 3,$
    $y > -3x + 2$

43. $y \leq 5,$
    $y \geq -x + 4$

44. $y \geq -2,$
    $y \geq x + 3$

45. $x + y \leq 6,$
    $x - y \leq 4$

46. $x + y < 1,$
    $x - y < 2$

47. $y + 3x > 0,$
    $y + 3x < 2$

48. $y - 2x \geq 1,$
    $y - 2x \leq 3$

Graph each system of inequalities. Find the coordinates of any vertices formed.

49. $y \leq 2x - 3,$
    $y \geq -2x + 1,$
    $x \leq 5$

50. $2y - x \leq 2,$
    $y - 3x \geq -4,$
    $y \geq -1$

51. $x + 2y \leq 12,$
    $2x + y \leq 12,$
    $x \geq 0,$
    $y \geq 0$

52. $x - y \leq 2,$
    $x + 2y \geq 8,$
    $y \leq 4$

53. $8x + 5y \leq 40,$
$x + 2y \leq 8,$
$x \geq 0,$
$y \geq 0$

54. $4y - 3x \geq -12,$
$4y + 3x \geq -36,$
$y \leq 0,$
$x \leq 0$

55. $y - x \geq 2,$
$y - x \leq 4,$
$2 \leq x \leq 5$

56. $3x + 4y \geq 12,$
$5x + 6y \leq 30,$
$1 \leq x \leq 3$

57. Explain in your own words why the boundary line is drawn dashed for the symbols $<$ and $>$ and why it is drawn solid for the symbols $\leq$ and $\geq$.

58. When graphing linear inequalities, Ron makes a habit of always shading above the line when the symbol $\geq$ is used. Is this wise? Why or why not?

Skill Review

Simplify.

59. $(2p^3 - 3w^4)^2$ [4.7]

60. $(5x^2 + 7)(5x^2 - 7)$ [4.6]

61. $(3t^4 + 2t^2 + 1) - (6t^4 - t^2 + 2)$ [4.4]

62. $(x^2 - x - 10) \div (x + 1)$ [4.8]

63. $\dfrac{15x^2 - 65x - 50}{2x^2 - x - 3} \cdot \dfrac{x^2 + 2x + 1}{10x^2 - 250}$ [6.2]

64. $\dfrac{2x}{x^2 + 6x + 9} - \dfrac{x + 2}{x^2 - x - 12}$ [6.4]

Synthesis

65. Explain how a system of linear inequalities could have a solution set consisting of one ordered pair.

66. In Example 7 of this section, is the point $(4, 0)$ part of the solution set? Why or why not?

Graph.

67. $x + y > 8,$
$x + y \leq -2$

68. $x + y \geq 1,$
$-x + y \geq 2,$
$x \geq -2,$
$y \geq 2,$
$y \leq 4,$
$x \leq 2$

69. $x - 2y \leq 0,$
$-2x + y \leq 2,$
$x \leq 2,$
$y \leq 2,$
$x + y \leq 4$

70. Write four systems of four inequalities that describe a 2-unit by 2-unit square that has $(0, 0)$ as one of the vertices.

71. *Luggage Size.* Unless an additional fee is paid, most major airlines will not check any luggage

for which the sum of the item's length, width, and height exceeds 62 in. The U.S. Postal Service will ship a package only if the sum of the package's length and girth (distance around its midsection) does not exceed 130 in. Video Promotions is ordering several 30-in. long cases that will be both mailed and checked as luggage. Using w and h for width and height (in inches), respectively, write and graph an inequality that represents all acceptable combinations of width and height.

Sources: U.S. Postal Service; www.case2go.com

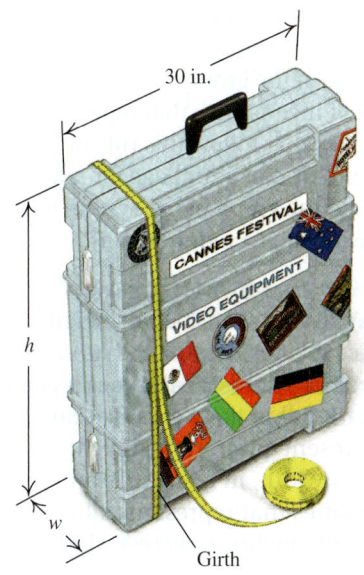

30 in.

h

w

Girth

72. *Hockey Wins and Losses.* The Skating Stars believe they need at least 60 points for the season in order to make the playoffs. A win is worth 2 points, a tie is worth 1 point, and a loss is worth 0 points. The team plays no more than 50 games. Graph a system of inequalities that describes the situation. Let w represent the number of wins and t the number of ties.

73. *Graduate-School Admissions.* Students entering a master's degree program at the University of Louisiana at Lafayette must meet minimum score requirements on the Graduate Records Examination (GRE). The GRE Verbal score must be at least 145 and the sum of the GRE Quantitative and Verbal scores must be at least 287. Each score has a maximum of 170. Using q for the quantitative score and v for the verbal score, write and graph a system of inequalities that represents all combinations that meet the requirements for entrance into the program.

Source: University of Louisiana at Lafayette

74. *Widths of a Basketball Floor.* Sizes of basketball floors vary due to building sizes and other constraints such as cost. The length L is to be at most 94 ft and the width W is to be at most 50 ft. Graph

We find the coordinates of each vertex by solving a system of two linear equations. The coordinates of point A are obviously $(0, 0)$. To find the coordinates of point C, we solve the system

$$b + r = 12, \quad (1)$$
$$9b + 15r = 120. \quad (2)$$

We multiply both sides of equation (1) by -9 and add:

$$
\begin{array}{r}
-9b - 9r = -108 \\
\underline{9b + 15r = 120} \\
6r = 12 \\
r = 2.
\end{array}
$$

Vertex (b, r)	Total Number of Points $T = 70b + 80r$
$A\ (0, 0)$	0
$B\ (12, 0)$	840
$C\ (10, 2)$	860
$D\ (0, 8)$	640

Substituting, we find that $b = 10$. Thus the coordinates of C are $(10, 2)$. Point B is the intersection of $b + r = 12$ and $r = 0$, so B is $(12, 0)$. Point D is the intersection of $9b + 15r = 120$ and $b = 0$, so D is $(0, 8)$. Computing the score for each ordered pair, we obtain the table at left. The greatest value in the table is 860, obtained when $b = 10$ and $r = 2$.

2. Refer to Example 2. Suppose Cy may turn in no more than 10 summaries and/or projects. How many of each should he submit in order to receive the greatest number of points?

4. Check. We can check that $T \le 860$ for several other points in the shaded region. This is left to the student.

5. State. In order to maximize his points, Cy should submit 10 book summaries and 2 research projects.

 YOUR TURN

9.5   EXERCISE SET

Vocabulary and Reading Check

Complete each of the following statements.

1. In linear programming, the quantity we wish to maximize or minimize is represented by the _____ function.

2. In linear programming, the demands arising from the given situation are known as _____.

3. To solve a linear programming problem, we make use of the _____ principle.

4. The shaded portion of a graph that represents all points that satisfy a problem's constraints is known as the _____ region.

5. In linear programming, the corners of the shaded portion of the graph are referred to as _____.

6. If it exists, the maximum value of an objective function occurs at a(n) _____ of the feasible region.

Linear Programming

Find the maximum and the minimum values of each objective function and the values of x and y at which they occur.

7. $F = 2x + 14y$,
subject to
$5x + 3y \le 34$,
$3x + 5y \le 30$,
$x \ge 0$,
$y \ge 0$

8. $G = 7x + 8y$,
subject to
$3x + 2y \le 12$,
$2y - x \le 4$,
$x \ge 0$,
$y \ge 0$

9. $P = 8x - y + 20$,
subject to
$6x + 8y \le 48$,
$0 \le y \le 4$,
$0 \le x \le 7$

10. $Q = 24x - 3y + 52$,
subject to
$5x + 4y \le 20$,
$0 \le y \le 4$,
$0 \le x \le 3$

11. $F = 2y - 3x$,
subject to
$y \le 2x + 1$,
$y \ge -2x + 3$,
$x \le 3$

12. $G = 5x + 2y + 4$,
subject to
$y \le 2x + 1$,
$y \ge -x + 3$,
$x \le 5$

13. *Transportation Cost.*   It takes Caroline 1 hr to ride the train to work and 1.5 hr to ride the bus. Every week, she must make at least 5 trips to work, and she plans to spend no more than 6 hr in travel time. If a train trip costs $5 and a bus trip costs $4, how many times per week should she ride each in order to minimize her cost?

14. *Food Service.*   Chad sells shrimp gumbo and shrimp sandwiches. He uses 3 oz of shrimp in each bowl of gumbo and 5 oz of shrimp in each sandwich. One Saturday morning, he realizes that he has only 120 oz of shrimp and that he must make a total of at least 30 shrimp meals. If his profit is $2 per gumbo order and $3 per sandwich, how many of each item should Chad make in order to maximize profit? (Assume that he sells everything that he makes.)

15. *Photo Albums.*   Photo Perfect prints pages of photographs for albums. A page containing 4 photos costs $3 and a page containing 6 photos costs $5. Ann can spend no more than $90 for photo pages of her recent vacation, and she can use no more than 20 pages in her album. What combination of 4-photo pages and 6-photo pages will maximize the number of photos she can display? What is the maximum number of photos that she can display?

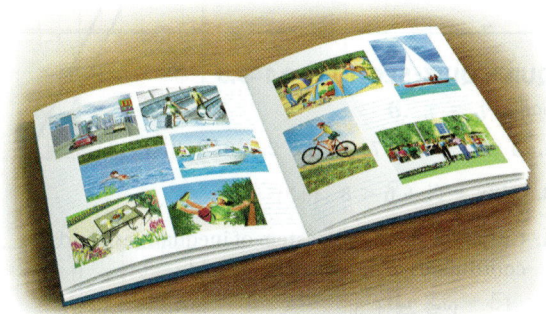

16. *Recycling.*   Mack collects bottles and cans from trash cans to turn in at the recycling center. It takes him 1.5 min to prepare a large container for return and 0.5 min to prepare a small container. He has at most 30 min per day to spend cleaning containers, and he is allowed to return no more than 30 containers per day. If he receives 10¢ for every large container and 5¢ for every small container, how many of each should he return in order to maximize his daily income? What is the maximum amount that he can make each day?

Aha! **17.** *Investing.*   Rosa is planning to invest up to $40,000 in corporate or municipal bonds, or both. She must invest from $6000 to $22,000 in corporate bonds, and she won't invest more than $30,000 in municipal bonds. The interest on corporate bonds is 4% and on municipal bonds is $3\frac{1}{2}$ %. This is simple interest for one year. How much should Rosa invest in each type of bond in order to earn the most interest? What is the maximum interest?

18. *Investing.*   Jamaal is planning to invest up to $22,000 in City Bank or the Southwick Credit Union, or both. He wants to invest at least $2000 but no more than $14,000 in City Bank. He will invest no more than $15,000 in the Southwick Credit Union. Interest is 6% at City Bank and is $6\frac{1}{2}$% at the Credit Union. This is simple interest for one year. How much should Jamaal invest in each bank in order to earn the most interest? What is the maximum interest?

19. *Test Scores.*   Corinna is taking a test in which short-answer questions are worth 10 points each and essay questions are worth 15 points each. She estimates that it takes 3 min to answer each short-answer question and 6 min to answer each essay question. The total time allowed is 60 min, and no more than 16 questions can be answered. Assuming that all her answers are correct, how many questions of each type should Corinna answer in order to get the best score?

20. *Test Scores.*   Edy is about to take a test that contains short-answer questions worth 4 points each and word problems worth 7 points each. Edy must do at least 5 short-answer questions, but time restricts doing more than 10. She must do at least 3 word problems, but time restricts doing more than 10. Edy can do no more than 18 questions in total. How many of each type of question should Edy do in order to maximize her score? What is this maximum score?

21. *Grape Growing.*   Auggie's vineyard consists of 240 acres upon which he wishes to plant Merlot grapes and Cabernet grapes. Profit per acre of Merlot is $400, and profit per acre of Cabernet is $300. The number of hours of labor available is 3200. Each acre of Merlot requires 20 hr of labor, and each acre of Cabernet requires 10 hr of labor. Determine how the land should be divided between Merlot and Cabernet in order to maximize profit.

10

Exponents and Radicals

HEIGHT (in centimeters)	MASS (in kilograms)			
	55	60	65	70
100	1.11	1.15	1.19	1.23
110	1.19	1.24	1.28	1.32
120	1.27	1.32	1.36	1.41

Your Largest Organ System Renews Itself Every Month.

Your skin is your largest organ system, but just how big is it? The total surface area of the human body is called the body surface area (BSA) and is used by professionals such as doctors and pharmacologists to determine drug dosage, how much fluid to administer intravenously, and thermal energy loss. BSA cannot be measured directly but is calculated from a person's mass and height using one of several formulas. In this chapter, we will calculate the BSA of a child using the DuBois formula. (*See Exercise 123 in Exercise Set 10.2.*)

I use math every day to calculate important factors related to patient care.

Beth Bennett Lopez, MPAS, PA-C, a dermatology physician assistant in Phoenix, Arizona, uses math when calculating pediatric drug doses, measuring patient's moles, and calculating body surface area for a psoriasis patient.

10.1 Radical Expressions and Functions

10.2 Rational Numbers as Exponents

10.3 Multiplying Radical Expressions

10.4 Dividing Radical Expressions

10.5 Expressions Containing Several Radical Terms

CONNECTING THE CONCEPTS

MID-CHAPTER REVIEW

10.6 Solving Radical Equations

10.7 The Distance Formula, the Midpoint Formula, and Other Applications

10.8 The Complex Numbers

CHAPTER RESOURCES

Visualizing for Success
Collaborative Activity
Decision Making: Connection

STUDY SUMMARY

REVIEW EXERCISES
CHAPTER TEST
CUMULATIVE REVIEW

In this chapter, we learn about square roots, cube roots, fourth roots, and so on. These roots can be expressed in radical notation or in exponential notation using exponents that are fractions. The chapter closes with an introduction to the complex-number system.

10.1   Radical Expressions and Functions

Square Roots and Square-Root Functions   ▪   Expressions of the Form $\sqrt{a^2}$ ▪   Cube Roots   ▪   Odd and Even nth Roots

In this section, we consider roots, such as square roots and cube roots, and the *radical* expressions involving such roots.

Study Skills

Advance Planning Pays Off

The best way to prepare for a final exam is to do so over a period of at least two weeks. First, review each chapter, studying the terminology, formulas, problems, properties, and procedures in the Study Summaries. Then retake your quizzes and tests. If you miss any questions, spend extra time reviewing the corresponding topics. Also consider participating in a study group or attending a tutoring or review session.

Square Roots and Square-Root Functions

When a number is multiplied by itself, we say that the number is squared. If we can find a number that was squared in order to produce some value a, we call that first number a *square root* of a.

> **SQUARE ROOT**
>
> The number c is a *square root* of a if $c^2 = a$.

For example,

   9 has -3 and 3 as square roots because $(-3)^2 = 9$ and $3^2 = 9$.

   25 has -5 and 5 as square roots because $(-5)^2 = 25$ and $5^2 = 25$.

   -4 does not have a real-number square root because there is no real number c for which $c^2 = -4$.

Every positive number has two square roots, and 0 has only itself as a square root. Negative numbers do not have real-number square roots; however, they do have nonreal square roots. Later in this chapter we introduce the *complex-number* system in which such square roots exist.

EXAMPLE 1   Find the two square roots of 36.

SOLUTION   The square roots are 6 and -6, because $6^2 = 36$ and $(-6)^2 = 36$.

YOUR TURN

1. Find the two square roots of 49.

Whenever we refer to *the* square root of a number, we mean the nonnegative square root of that number. This is also called the *principal square root* of the number.

Student Notes

It is important to remember the difference between *the* square root of 9 and *a* square root of 9. *A* square root of 9 means either 3 or -3, but *the* square root of 9, denoted $\sqrt{9}$, means the principal square root of 9, or 3.

> **PRINCIPAL SQUARE ROOT**
>
> The *principal square root* of a nonnegative number is its nonnegative square root. The symbol $\sqrt{}$ is called a *radical sign* and is used to indicate the principal square root of the number over which it appears.

EXAMPLE 2   Simplify each of the following.

a) $\sqrt{25}$ **b)** $\sqrt{\dfrac{25}{64}}$ **c)** $-\sqrt{64}$ **d)** $\sqrt{0.0049}$

SOLUTION

a) $\sqrt{25} = 5$ $\sqrt{}$ indicates the principal square root. Note that $\sqrt{25} \neq -5$.

b) $\sqrt{\dfrac{25}{64}} = \dfrac{5}{8}$ Since $\left(\dfrac{5}{8}\right)^2 = \dfrac{25}{64}$

c) $-\sqrt{64} = -8$ Since $\sqrt{64} = 8, -\sqrt{64} = -8.$

d) $\sqrt{0.0049} = 0.07$ $(0.07)(0.07) = 0.0049.$ Note too that
$$\sqrt{0.0049} = \sqrt{\dfrac{49}{10{,}000}} = \dfrac{7}{100}.$$

2. Simplify: $-\sqrt{\dfrac{1}{9}}$.

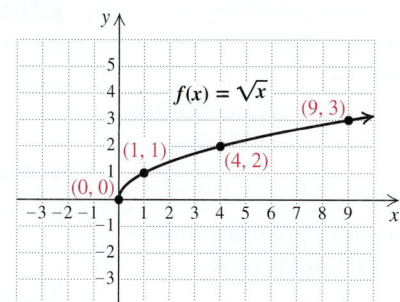 YOUR TURN

In addition to being read as "the principal square root of a," $\sqrt{a}$ is also read as "the square root of a," "root a," or "radical a." Any expression in which a radical sign appears is called a *radical expression*. The following are radical expressions:

$$\sqrt{5}, \quad \sqrt{a}, \quad -\sqrt{3x}, \quad \sqrt{\dfrac{y^2 + 7}{y}}, \quad \sqrt{x} + 8.$$

The expression under the radical sign is called the **radicand**. In the expressions above, the radicands are 5, a, $3x$, $(y^2 + 7)/y$, and x, respectively.

Values for square roots found on calculators are, for the most part, approximations. For example, a calculator will show a number like

   2.23606798

for $\sqrt{5}$. The exact value of $\sqrt{5}$ is not given by any repeating or terminating decimal. In general, for any number a that is not a perfect square, $\sqrt{a}$ is a nonterminating, nonrepeating decimal or an *irrational number*.

The square-root function, given by

$$f(x) = \sqrt{x},$$

has $[0, \infty)$ as its domain and $[0, \infty)$ as its range. We can draw its graph by selecting convenient values for x and calculating the corresponding outputs. Once these ordered pairs have been graphed, a smooth curve can be drawn.

$$f(x) = \sqrt{x}$$

x	$\sqrt{x}$	$(x, f(x))$
0	0	$(0, 0)$
1	1	$(1, 1)$
4	2	$(4, 2)$
9	3	$(9, 3)$

EXAMPLE 3   For each function, find the indicated function value.

a) $f(x) = \sqrt{3x - 2}; \; f(1)$ **b)** $g(z) = -\sqrt{6z + 4}; \; g(3)$

SOLUTION

a) $f(1) = \sqrt{3 \cdot 1 - 2}$ Substituting
     $= \sqrt{1} = 1$ Simplifying

3. If $f(x) = \sqrt{1 - x}$, find $f(-3)$.

b) $g(3) = -\sqrt{6 \cdot 3 + 4}$ Substituting

$= -\sqrt{22}$ Simplifying. This answer is exact.

≈ -4.69041576 Using a calculator to write an approximation

YOUR TURN

Expressions of the Form $\sqrt{a^2}$

As the next example shows, $\sqrt{a^2}$ does not always simplify to a.

EXAMPLE 4   Evaluate $\sqrt{a^2}$ for each of the following values: **(a)** 5; **(b)** 0; **(c)** −5.

SOLUTION

a) $\sqrt{5^2} = \sqrt{25} = 5$

         Same

b) $\sqrt{0^2} = \sqrt{0} = 0$

         Same

4. Evaluate $\sqrt{t^2}$ for $t = -3$ and for $t = 3$.

c) $\sqrt{(-5)^2} = \sqrt{25} = 5$

         Opposites   Note that $\sqrt{(-5)^2} \neq -5$.

YOUR TURN

You may have noticed that evaluating $\sqrt{a^2}$ is just like evaluating $|a|$.

Technology Connection

To see the necessity of absolute-value signs, let y_1 represent the left side and y_2 the right side of each of the following equations. Then use a graph or a table to determine whether these equations are true.

1. $\sqrt{x^2} \overset{?}{=} x$

2. $\sqrt{x^2} \overset{?}{=} |x|$

3. $x \overset{?}{=} |x|$

> **SIMPLIFYING $\sqrt{a^2}$**
>
> For any real number a,
>
> $$\sqrt{a^2} = |a|.$$
>
> (The principal square root of a^2 is the absolute value of a.)

When a radicand is the square of a variable expression, like $(x + 1)^2$ or $36t^2$, absolute-value signs are needed when simplifying. **We use absolute-value signs unless we know that the expression being squared is nonnegative.** This ensures that our result is never negative.

Student Notes

Some absolute-value notation can be simplified.

- $|ab| = |a| \cdot |b|$, so an expression like $|3x|$ can be written

 $|3x| = |3| \cdot |x| = 3|x|$.

- Even powers of real numbers are never negative, so

 $|x^2| = x^2$,

 $|x^4| = x^4$,

 $|x^6| = x^6$, and so on.

EXAMPLE 5   Simplify each expression. Assume that the variable can represent any real number.

a) $\sqrt{36t^2}$ **b)** $\sqrt{(x + 1)^2}$ **c)** $\sqrt{x^2 - 8x + 16}$

d) $\sqrt{t^6}$ **e)** $\sqrt{a^8}$

SOLUTION

a) $\sqrt{36t^2} = \sqrt{(6t)^2} = |6t|$, or $6|t|$ Since t can be negative, absolute-value notation is necessary.

b) $\sqrt{(x + 1)^2} = |x + 1|$ Since $x + 1$ can be negative (for example, if $x = -3$), absolute-value notation is necessary.

c) $\sqrt{x^2 - 8x + 16} = \sqrt{(x - 4)^2} = |x - 4|$ Since $x - 4$ can be negative, absolute-value notation is necessary.

5. Simplify $\sqrt{(3x - 7)^2}$. Assume that x can represent any real number.

d) Recall that $(a^m)^n = a^{mn}$, and note that $(t^3)^2 = t^6$. Thus,

$$\sqrt{t^6} = |t^3|.$$ Since t^3 can be negative, absolute-value notation is necessary.

e) $\sqrt{a^8} = \sqrt{(a^4)^2} = |a^4| = a^4$ Since a^4 is never negative, $|a^4| = a^4$.

 YOUR TURN

If we assume that the expression being squared is nonnegative, then absolute-value notation is not necessary.

EXAMPLE 6   Simplify each expression. Assume that the expressions being squared are nonnegative. Thus absolute-value notation is not necessary.

a) $\sqrt{y^2}$ **b)** $\sqrt{a^{10}}$ **c)** $\sqrt{9x^2 - 6x + 1}$

SOLUTION

a) $\sqrt{y^2} = y$ We assume that y is nonnegative, so no absolute-value notation is necessary. When y *is* negative, $\sqrt{y^2} \neq y$.

b) $\sqrt{a^{10}} = a^5$ Assuming that a^5 is nonnegative. Note that $(a^5)^2 = a^{10}$.

6. Simplify $\sqrt{n^{14}}$. Assume that n is nonnegative.

c) $\sqrt{9x^2 - 6x + 1} = \sqrt{(3x - 1)^2} = 3x - 1$ Assuming that $3x - 1$ is nonnegative

 YOUR TURN

Cube Roots

We often need to know what number cubed produces a certain value. When such a number is found, we say that we have found a *cube root*. For example,

2 is the cube root of 8 because $2^3 = 2 \cdot 2 \cdot 2 = 8$;

-4 is the cube root of -64 because $(-4)^3 = (-4)(-4)(-4) = -64$.

> **CUBE ROOT**
>
> The number c is the *cube root* of a if $c^3 = a$. In symbols, we write $\sqrt[3]{a}$ to denote the cube root of a.

Each real number has only one real-number cube root. The cube-root function, given by

$$f(x) = \sqrt[3]{x},$$

has $\mathbb{R}$ as its domain and $\mathbb{R}$ as its range. To draw its graph, we select convenient values for x and calculate the corresponding outputs. Once these ordered pairs have been graphed, a smooth curve is drawn. Note that the cube root of a positive number is positive, and the cube root of a negative number is negative.

$f(x) = \sqrt[3]{x}$

x	$\sqrt[3]{x}$	$(x, f(x))$
0	0	$(0, 0)$
1	1	$(1, 1)$
8	2	$(8, 2)$
-1	-1	$(-1, -1)$
-8	-2	$(-8, -2)$

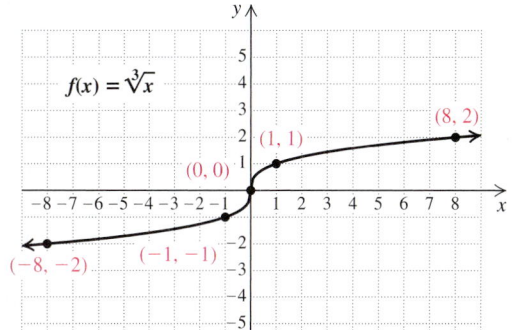

5. If a is a whole number that is not a perfect square, then $\sqrt{a}$ is a(n) _____ number.
            irrational/rational

6. The domain of the function f given by $f(x) = \sqrt[3]{x}$ is the set of all _____ numbers.
            whole/real/positive

7. If $\sqrt[4]{x}$ is a real number, then x must be _____ .
            negative/positive/nonnegative.

8. If $\sqrt[3]{x}$ is negative, then x must be _____ .
            negative/positive

Square Roots and Square-Root Functions

For each number, find all of its square roots.

9. 64

10. 81

11. 100

12. 121

13. 400

14. 2500

15. 625

16. 225

Simplify.

17. $\sqrt{49}$

18. $\sqrt{144}$

19. $-\sqrt{16}$

20. $-\sqrt{100}$

21. $\sqrt{\dfrac{36}{49}}$

22. $\sqrt{\dfrac{4}{9}}$

23. $-\sqrt{\dfrac{16}{81}}$

24. $-\sqrt{\dfrac{81}{144}}$

25. $\sqrt{0.04}$

26. $\sqrt{0.36}$

27. $\sqrt{0.0081}$

28. $\sqrt{0.0016}$

For each function, find the specified function value, if it exists.

29. $f(t) = \sqrt{5t - 10};\ f(3), f(2), f(1), f(-1)$

30. $g(x) = \sqrt{x^2 - 25};\ g(-6), g(3), g(6), g(13)$

31. $t(x) = -\sqrt{2x^2 - 1};\ t(5), t(0), t(-1), t\left(-\tfrac{1}{2}\right)$

32. $p(z) = \sqrt{2z - 20};\ p(4), p(10), p(12), p(0)$

33. $f(t) = \sqrt{t^2 + 1};\ f(0), f(-1), f(-10)$

34. $g(x) = -\sqrt{(x + 1)^2};\ g(-3), g(4), g(-5)$

Expressions of the Form $\sqrt{a^2}$

Simplify. Variables may represent any real number, so remember to use absolute-value notation when necessary. If a root cannot be simplified, state this.

35. $\sqrt{100x^2}$

36. $\sqrt{16t^2}$

37. $\sqrt{(-4b)^2}$

38. $\sqrt{(-7c)^2}$

39. $\sqrt{(8 - t)^2}$

40. $\sqrt{(a + 3)^2}$

41. $\sqrt{y^2 + 16y + 64}$

42. $\sqrt{x^2 - 4x + 4}$

43. $\sqrt{4x^2 + 28x + 49}$

44. $\sqrt{9x^2 - 30x + 25}$

45. $\sqrt{a^{22}}$

46. $\sqrt{x^{10}}$

47. $\sqrt{-25}$

48. $\sqrt{-16}$

Cube Roots

Simplify.

49. $\sqrt[3]{-1}$

50. $-\sqrt[3]{-1000}$

51. $-\sqrt[3]{64}$

52. $\sqrt[3]{27}$

53. $-\sqrt[3]{-125y^3}$

54. $\sqrt[3]{-64x^3}$

Odd and Even nth Roots

Identify the radicand and the index for each expression.

55. $5\sqrt{p^2 + 4}$

56. $-7\sqrt{y^2 - 8}$

57. $x^2y^3 \sqrt[5]{\dfrac{x}{y + 4}}$

58. $\dfrac{a^2}{b}\sqrt[6]{a(a + b)}$

Simplify. Use absolute-value notation when necessary.

59. $-\sqrt[4]{256}$

60. $-\sqrt[4]{625}$

61. $-\sqrt[5]{-\dfrac{32}{243}}$

62. $\sqrt[5]{-\dfrac{1}{32}}$

63. $\sqrt[6]{x^6}$

64. $\sqrt[8]{y^8}$

65. $\sqrt[9]{t^9}$

66. $\sqrt[5]{a^5}$

67. $\sqrt[4]{(6a)^4}$

68. $\sqrt[4]{(7b)^4}$

69. $\sqrt[10]{(-6)^{10}}$

70. $\sqrt[12]{(-10)^{12}}$

71. $\sqrt[414]{(a + b)^{414}}$

72. $\sqrt[1976]{(2a + b)^{1976}}$

Simplify. Assume that no radicands were formed by raising negative quantities to even powers. Thus absolute-value notation is not necessary.

73. $\sqrt{16x^2}$

74. $\sqrt{25t^2}$

75. $-\sqrt{(3t)^2}$

76. $-\sqrt{(7c)^2}$

77. $\sqrt{(-5b)^2}$

78. $\sqrt{(-10a)^2}$

79. $\sqrt{a^2 + 2a + 1}$

80. $\sqrt{9 - 6y + y^2}$

81. $\sqrt[4]{16x^4}$

82. $\sqrt[4]{81x^4}$

83. $\sqrt[3]{(x - 1)^3}$

84. $-\sqrt[3]{(7y)^3}$

85. $\sqrt{t^{18}}$

86. $\sqrt{a^{14}}$

87. $\sqrt{(x - 2)^8}$

88. $\sqrt{(x + 3)^{10}}$

For each function, find the specified function value, if it exists.

89. $f(x) = \sqrt[3]{x + 1};\ f(7), f(26), f(-9), f(-65)$

90. $g(x) = -\sqrt[3]{2x - 1};\ g(0), g(-62), g(-13), g(63)$

91. $g(t) = \sqrt[4]{t - 3};\ g(19), g(-13), g(1), g(84)$

92. $f(t) = \sqrt[4]{t + 1};\ f(0), f(15), f(-82), f(80)$

Determine the domain of each function described.

93. $f(x) = \sqrt{x - 6}$

94. $g(x) = \sqrt{x + 8}$

95. $g(t) = \sqrt[4]{t + 8}$

96. $f(x) = \sqrt[4]{x - 9}$

97. $g(x) = \sqrt[4]{10 - 2x}$

98. $g(t) = \sqrt[3]{2t - 6}$

99. $f(t) = \sqrt[5]{2t + 7}$

100. $f(t) = \sqrt[6]{4 + 3t}$

101. $h(z) = -\sqrt[6]{5z + 2}$

102. $d(x) = -\sqrt[4]{5 - 7x}$

Aha! **103.** $f(t) = 7 + \sqrt[8]{t^8}$

104. $g(t) = 9 + \sqrt[6]{t^6}$

105. Explain how to write the negative square root of a number using radical notation.

106. Does the square root of a number's absolute value always exist? Why or why not?

Skill Review

Let $f(x) = 3x - 1$ and $g(x) = \dfrac{1}{x}$.

107. Find $f\left(\frac{1}{3}\right)$. [7.1]

108. Find the domain of f. [7.2]

109. Find the domain of g. [7.2]

110. Find $(f + g)(x)$. [7.4]

111. Find $(fg)(x)$. [7.4]

112. Which function, f or g, is a linear function? [7.3]

Synthesis

113. Under what conditions does the *n*th root of x^3 exist as a real number? Explain your reasoning.

114. Under what conditions does the *n*th root of x^2 exist? Explain your reasoning.

Firefighting. *The number of gallons per minute discharged from a fire hose depends on the diameter of the hose and the nozzle pressure. For a 2-in. diameter solid bore nozzle, the discharge can be modeled by*

$$f(p) = 118.8\sqrt{p},$$

where p is the nozzle pressure, in pounds per square inch (psi), and $f(p)$ is the water flow, in gallons per minute (GPM). Use this function for Exercises 115 and 116.

115. Estimate the water flow when the nozzle pressure is 50 psi.

116. Estimate the water flow when the nozzle pressure is 175 psi.

117. *Biology.* The number of species S of plants in Guyana in an area of A hectares can be estimated using the formula

$$S = 88.63\sqrt[4]{A}.$$

The Kaieteur National Park in Guyana has an area of 63,000 hectares. How many species of plants are in the park?

Source: Hans ter Steege, "A Perspective on Guyana and its Plant Richness," as found on www.bio.uu.nl

118. *Spaces in a Parking Lot.* A parking lot has attendants to park the cars. The number of spaces N needed for waiting cars before attendants can get to them is given by the formula $N = 2.5\sqrt{A}$, where A is the number of arrivals in peak hours. Find the number of spaces needed for the given number of arrivals in peak hours: **(a)** 25; **(b)** 36; **(c)** 49; **(d)** 64.

Determine the domain of each function described. Then draw the graph of each function.

119. $f(x) = \sqrt{x} + 5$

120. $g(x) = \sqrt{x} + 5$

121. $g(x) = \sqrt{x} - 2$

122. $f(x) = \sqrt{x - 2}$

123. Find the domain of f if

$$f(x) = \frac{\sqrt{x + 3}}{\sqrt[4]{2 - x}}.$$

124. Find the domain of g if

$$g(x) = \frac{\sqrt[4]{5 - x}}{\sqrt[6]{x + 4}}.$$

125. Find the domain of F if

$$F(x) = \frac{x}{\sqrt{x^2 - 5x - 6}}.$$

126. Use a graphing calculator to check your answers to Exercises 39, 45, and 67. On some graphing calculators, a MATH key is needed to enter higher roots.

127. *Length of Marriage.* John Tierney and Garth Sundem have developed an equation that can be used to predict the probability that a celebrity marriage will last. The probability P that a couple will still be married in T years is given by

$$P = 50\sqrt[15]{\frac{\text{NYT(Ah + Aw)}}{\text{ENQ(Sc + 5)}} \cdot \text{Md} \cdot \left[\frac{\text{Md}}{(\text{Md} + 2)}\right]^{T^2}}$$

where NYT = the number of times since 1990 that the wife's name appeared in the *New York Times*, ENQ = the number of times since 1990 that the wife's name appeared in the *National Enquirer*, Ah = the husband's age, in years, Aw = the wife's age, in years, Md = the number of months that the couple dated before marriage, and Sc = the number of the top five photos returned by a Google images search for the wife's name in which she was "scantily clad."

Kate Middleton and Prince William were both 29 when they were married and had dated for 120 months. Kate appeared 258 times in the *New York Times* and 44 times in the *National Enquirer*, and her Sc value was 0. What is the probability that the marriage will last 5 years?

Source: "Refining the Formula That Predicts Celebrity Marriages' Doom," www.nytimes.com, March 12, 2012

PREPARE TO MOVE ON

Simplify. Do not use negative exponents in your answer.
[4.1], [4.2]

1. $(3xy^8)(5x^2y)$

2. $(2a^{-1}b^2c)^{-3}$

3. $\left(\dfrac{10x^{-1}y^5}{5x^2y^{-1}}\right)^{-1}$

4. $\left(\dfrac{8x^3y^{-2}}{2xz^4}\right)^{-2}$

YOUR TURN ANSWERS: SECTION 10.1

1. $-7, 7$ **2.** $-\frac{1}{3}$ **3.** 2 **4.** $3; 3$ **5.** $|3x - 7|$
6. n^7 **7.** -2 **8.** $10t$ **9.** $3n$ **10.** $|2x|$, or $2|x|$
11. $\{x | x \geq -\frac{3}{2}\}$, or $\left[-\frac{3}{2}, \infty\right)$

10.2 Rational Numbers as Exponents

Rational Exponents ■ Negative Rational Exponents ■ Laws of Exponents ■ Simplifying Radical Expressions

We have already considered natural-number exponents and integer exponents. We now expand the study of exponents further to include all rational numbers. This will give meaning to expressions like $7^{1/3}$ and $(2x)^{-4/5}$.

Study Skills

This Looks Familiar

Sometimes a topic seems familiar and students are tempted to assume that they already know the material. Try to avoid this tendency. Often new extensions or applications are included when a topic reappears.

Rational Exponents

When defining rational exponents, we want the rules for exponents to hold for them just as they do for integer exponents. For example, we want to continue to add exponents when multiplying expressions with the same base. Then $a^{1/2} \cdot a^{1/2} = a^{1/2+1/2} = a^1$ suggests that $a^{1/2}$ means $\sqrt{a}$, and $a^{1/3} \cdot a^{1/3} \cdot a^{1/3} = a^{1/3+1/3+1/3} = a^1$ suggests that $a^{1/3}$ means $\sqrt[3]{a}$.

$a^{1/n} = \sqrt[n]{a}$

$a^{1/n}$ means $\sqrt[n]{a}$. When a is nonnegative, n can be any natural number greater than 1. When a is negative, n can be any odd natural number greater than 1.

Thus, $a^{1/5} = \sqrt[5]{a}$ and $a^{1/10} = \sqrt[10]{a}$.

The denominator of the exponent is the index of the radical expression.

EXAMPLE 1   Write an equivalent expression using radical notation and, if possible, simplify.

a) $16^{1/2}$ **b)** $(-8)^{1/3}$ **c)** $(abc)^{1/5}$ **d)** $(25x^{16})^{1/2}$

1. Write an expression equivalent to $49^{1/2}$ using radical notation and, if possible, simplify.

CAUTION! When we are converting from radical notation to exponential notation, parentheses are often needed to indicate the base.

$$\sqrt{5x} = (5x)^{1/2}$$

---

2. Write an expression equivalent to $\sqrt[4]{2ac}$ using exponential notation.

SOLUTION

a) $16^{1/2} = \sqrt{16} = 4$

b) $(-8)^{1/3} = \sqrt[3]{-8} = -2$

c) $(abc)^{1/5} = \sqrt[5]{abc}$

> The denominator of the exponent becomes the index. The base becomes the radicand. Recall that for square roots, the index 2 is understood without being written.

d) $(25x^{16})^{1/2} = 25^{1/2}x^8 = \sqrt{25} \cdot x^8 = 5x^8$

YOUR TURN

EXAMPLE 2 Write an equivalent expression using exponential notation.

a) $\sqrt[5]{7ab}$ b) $\sqrt[7]{\dfrac{x^3y}{4}}$ c) $\sqrt{5x}$

SOLUTION Parentheses are required to indicate the base.

a) $\sqrt[5]{7ab} = (7ab)^{1/5}$

b) $\sqrt[7]{\dfrac{x^3y}{4}} = \left(\dfrac{x^3y}{4}\right)^{1/7}$

> The index becomes the denominator of the exponent. The radicand becomes the base.

c) $\sqrt{5x} = (5x)^{1/2}$

> The index 2 is understood without being written. We assume $x \geq 0$.

YOUR TURN

How shall we define $a^{2/3}$? If the property for multiplying exponents is to hold, we must have $a^{2/3} = (a^{1/3})^2$, as well as $a^{2/3} = (a^2)^{1/3}$. This would suggest that $a^{2/3} = (\sqrt[3]{a})^2$ and $a^{2/3} = \sqrt[3]{a^2}$. We make our definition accordingly.

Student Notes

It is important to remember both meanings of $a^{m/n}$. When the root of the base a is known, $(\sqrt[n]{a})^m$ is generally easier to work with. When $\sqrt[n]{a}$ is not known, $\sqrt[n]{a^m}$ is often more convenient.

POSITIVE RATIONAL EXPONENTS

For any natural numbers m and n ($n \neq 1$) and any real number a for which $\sqrt[n]{a}$ exists,

$$a^{m/n} \quad \text{means} \quad (\sqrt[n]{a})^m, \quad \text{or} \quad \sqrt[n]{a^m}.$$

EXAMPLE 3 Write an equivalent expression using radical notation and simplify.

a) $27^{2/3}$ b) $25^{3/2}$

SOLUTION

a) $27^{2/3}$ means $(\sqrt[3]{27})^2$ or, equivalently, $\sqrt[3]{27^2}$. Let's see which is easier to simplify:

$$(\sqrt[3]{27})^2 = 3^2 \qquad \sqrt[3]{27^2} = \sqrt[3]{729}$$
$$= 9; \qquad\qquad = 9.$$

The simplification on the left is probably easier for most people.

b) $25^{3/2}$ means $(\sqrt[2]{25})^3$ or, equivalently, $\sqrt[2]{25^3}$ (the index 2 is normally omitted). Since $\sqrt{25}$ is more commonly known than $\sqrt{25^3}$, we use that form:

$$25^{3/2} = (\sqrt{25})^3 = 5^3 = 125.$$

3. Write an expression equivalent to $1000^{2/3}$ using radical notation and simplify.

YOUR TURN

EXAMPLE 4   Write an equivalent expression using exponential notation.

a) $\sqrt[3]{9^4}$ **b)** $\left(\sqrt[4]{7xy}\right)^5$

SOLUTION

4. Write an expression equivalent to $\left(\sqrt[5]{4x}\right)^3$ using exponential notation.

a) $\sqrt[3]{9^4} = 9^{4/3}$
b) $\left(\sqrt[4]{7xy}\right)^5 = (7xy)^{5/4}$ The index becomes the denominator of the fraction that is the exponent.

YOUR TURN

Negative Rational Exponents

Recall that $x^{-2} = 1/x^2$. Negative rational exponents behave similarly.

> **NEGATIVE RATIONAL EXPONENTS**
>
> For any rational number m/n and any nonzero real number a for which $a^{m/n}$ exists,
>
> $$a^{-m/n} \quad \text{means} \quad \frac{1}{a^{m/n}}.$$

> **CAUTION!**   A negative exponent does not indicate that the expression in which it appears is negative: $a^{-1} \neq -a$.

EXAMPLE 5   Write an equivalent expression with positive exponents and, if possible, simplify.

a) $9^{-1/2}$ **b)** $(5xy)^{-4/5}$ **c)** $64^{-2/3}$

d) $4x^{-2/3}y^{1/5}$ **e)** $\left(\dfrac{3r}{7s}\right)^{-5/2}$

SOLUTION

a) $9^{-1/2} = \dfrac{1}{9^{1/2}}$ $9^{-1/2}$ is the reciprocal of $9^{1/2}$.

Since $9^{1/2} = \sqrt{9} = 3$, the answer simplifies to $\dfrac{1}{3}$.

b) $(5xy)^{-4/5} = \dfrac{1}{(5xy)^{4/5}}$ $(5xy)^{-4/5}$ is the reciprocal of $(5xy)^{4/5}$.

c) $64^{-2/3} = \dfrac{1}{64^{2/3}}$ $64^{-2/3}$ is the reciprocal of $64^{2/3}$.

Since $64^{2/3} = \left(\sqrt[3]{64}\right)^2 = 4^2 = 16$, the answer simplifies to $\dfrac{1}{16}$.

d) $4x^{-2/3}y^{1/5} = 4 \cdot \dfrac{1}{x^{2/3}} \cdot y^{1/5} = \dfrac{4y^{1/5}}{x^{2/3}}$

5. Write an expression equivalent to $16^{-1/2}p^{2/3}w^{-3/5}$ with positive exponents and, if possible, simplify.

e) Recall that $(a/b)^{-n} = (b/a)^n$. This property holds for *any* negative exponent:
$$\left(\dfrac{3r}{7s}\right)^{-5/2} = \left(\dfrac{7s}{3r}\right)^{5/2}.$$ Writing the reciprocal of the base and changing the sign of the exponent

YOUR TURN

Laws of Exponents

The same laws hold for rational exponents as for integer exponents.

LAWS OF EXPONENTS

For any real numbers a and b and any rational exponents m and n for which a^m, a^n, and b^m are defined:

1. $a^m \cdot a^n = a^{m+n}$ When multiplying, add exponents if the bases are the same.

2. $\dfrac{a^m}{a^n} = a^{m-n}$ When dividing, subtract exponents if the bases are the same. (Assume $a \neq 0$.)

3. $(a^m)^n = a^{m \cdot n}$ To raise a power to a power, multiply the exponents.

4. $(ab)^m = a^m b^m$ To raise a product to a power, raise each factor to the power and multiply.

EXAMPLE 6 Use the laws of exponents to simplify.

a) $3^{1/5} \cdot 3^{3/5}$

b) $\dfrac{a^{1/4}}{a^{1/2}}$

c) $(7.2^{2/3})^{3/4}$

d) $(a^{-1/3} b^{2/5})^{1/2}$

SOLUTION

a) $3^{1/5} \cdot 3^{3/5} = 3^{1/5+3/5} = 3^{4/5}$ Adding exponents

b) $\dfrac{a^{1/4}}{a^{1/2}} = a^{1/4-1/2} = a^{1/4-2/4}$ Subtracting exponents after finding a common denominator

$= a^{-1/4}$, or $\dfrac{1}{a^{1/4}}$ $a^{-1/4}$ is the reciprocal of $a^{1/4}$.

c) $(7.2^{2/3})^{3/4} = 7.2^{(2/3)(3/4)} = 7.2^{6/12}$ Multiplying exponents

$= 7.2^{1/2}$ Using arithmetic to simplify the exponent

d) $(a^{-1/3} b^{2/5})^{1/2} = a^{(-1/3)(1/2)} \cdot b^{(2/5)(1/2)}$ Raising a product to a power and multiplying exponents

$= a^{-1/6} b^{1/5}$, or $\dfrac{b^{1/5}}{a^{1/6}}$

6. Use the laws of exponents to simplify $\dfrac{8^{3/4}}{8^{-1/6}}$.

YOUR TURN

Simplifying Radical Expressions

Many radical expressions can be simplified by using rational exponents.

TO SIMPLIFY RADICAL EXPRESSIONS

1. Convert radical expressions to exponential expressions.
2. Use arithmetic and the laws of exponents to simplify.
3. Convert back to radical notation as needed.

Technology Connection

One way to check Example 7(a) is to let $y_1 = (5x)^{3/6}$ and $y_2 = \sqrt{5x}$. Then we use GRAPH or TABLE to see if $y_1 = y_2$. An alternative check is to let $y_3 = y_2 - y_1$ and see if $y_3 = 0$. Check Example 7(a) using one of these two methods.

1. Why are rational exponents especially useful when working on a graphing calculator?

7. Use rational exponents to simplify $\left(\sqrt[3]{ab}\right)^{15}$. Do not use fraction exponents in the final answer.

EXAMPLE 7 Use rational exponents to simplify. Do not use exponents that are fractions in the final answer.

a) $\sqrt[6]{(5x)^3}$ **b)** $\sqrt[5]{t^{20}}$

c) $\left(\sqrt[3]{ab^2c}\right)^{12}$ **d)** $\sqrt{\sqrt[3]{x}}$

SOLUTION

a) $\sqrt[6]{(5x)^3} = (5x)^{3/6}$ Converting to exponential notation
$\qquad\qquad = (5x)^{1/2}$ Simplifying the exponent
$\qquad\qquad = \sqrt{5x}$ Returning to radical notation

b) $\sqrt[5]{t^{20}} = t^{20/5}$ Converting to exponential notation
$\qquad\quad = t^4$ Simplifying the exponent

c) $\left(\sqrt[3]{ab^2c}\right)^{12} = (ab^2c)^{12/3}$ Converting to exponential notation
$\qquad\qquad\quad = (ab^2c)^4$ Simplifying the exponent
$\qquad\qquad\quad = a^4b^8c^4$ Using the laws of exponents

d) $\sqrt{\sqrt[3]{x}} = \sqrt{x^{1/3}}$ Converting the radicand to exponential notation
$\qquad\quad = (x^{1/3})^{1/2}$ Try to go directly to this step.
$\qquad\quad = x^{1/6}$ Using the laws of exponents
$\qquad\quad = \sqrt[6]{x}$ Returning to radical notation

YOUR TURN

10.2 EXERCISE SET

FOR EXTRA HELP

PRACTICE   WATCH   READ   REVIEW

Vocabulary and Reading Check

Choose the word from the list below that best completes each statement. Not every word will be used.

add   radical
equivalent   rational
multiply   subtract

1. The expression $\sqrt{3x}$ is an example of a(n) _____ expression.

2. When dividing one exponential expression by another with the same base, we _____ exponents.

3. The expressions $\sqrt[3]{5mn}$ and $(5mn)^{1/3}$ are _____.

4. The exponent in the expression $7^{2/3}$ is a(n) _____ exponent.

Concept Reinforcement

In Exercises 5–12, match each expression with the equivalent expression from the column on the right.

5. ____ $x^{2/5}$ **a)** $x^{3/5}$

6. ____ $x^{5/2}$ **b)** $\left(\sqrt[5]{x}\right)^4$

7. ____ $x^{-5/2}$ **c)** $\sqrt{x^5}$

8. ____ $x^{-2/5}$ **d)** $x^{1/2}$

9. ____ $x^{1/5} \cdot x^{2/5}$ **e)** $\dfrac{1}{(\sqrt{x})^5}$

10. ____ $(x^{1/5})^{5/2}$ **f)** $\sqrt[4]{x^5}$

11. ____ $\sqrt[5]{x^4}$ **g)** $\sqrt[5]{x^2}$

12. ____ $\left(\sqrt[4]{x}\right)^5$ **h)** $\dfrac{1}{(\sqrt[5]{x})^2}$

Note: Assume for all exercises that all variables are non-negative and that all denominators are nonzero.

Rational Exponents

Write an equivalent expression using radical notation and, if possible, simplify.

13. $y^{1/3}$ **14.** $t^{1/4}$

15. $36^{1/2}$ **16.** $125^{1/3}$

17. $32^{1/5}$

18. $81^{1/4}$

19. $64^{1/2}$

20. $100^{1/2}$

21. $(xyz)^{1/2}$

22. $(ab)^{1/4}$

23. $(a^2b^2)^{1/5}$

24. $(x^3y^3)^{1/4}$

25. $t^{5/6}$

26. $a^{3/2}$

27. $16^{3/4}$

28. $4^{7/2}$

29. $125^{4/3}$

30. $9^{5/2}$

31. $(81x)^{3/4}$

32. $(125a)^{2/3}$

33. $(25x^4)^{3/2}$

34. $(9y^6)^{3/2}$

Write an equivalent expression using exponential notation.

35. $\sqrt[3]{18}$

36. $\sqrt[4]{10}$

37. $\sqrt{30}$

38. $\sqrt{22}$

39. $\sqrt{x^7}$

40. $\sqrt{a^3}$

41. $\sqrt[5]{m^2}$

42. $\sqrt[5]{n^4}$

43. $\sqrt[4]{xy}$

44. $\sqrt[3]{cd}$

45. $\sqrt[5]{xy^2z}$

46. $\sqrt{x^3y^2z^2}$

47. $(\sqrt{3mn})^3$

48. $(\sqrt[3]{7xy})^4$

49. $(\sqrt[7]{8x^2y})^5$

50. $(\sqrt[6]{2a^5b})^7$

51. $\dfrac{2x}{\sqrt[3]{z^2}}$

52. $\dfrac{3a}{\sqrt[5]{c^2}}$

Negative Rational Exponents

Write an equivalent expression with positive exponents and, if possible, simplify.

53. $8^{-1/3}$

54. $10{,}000^{-1/4}$

55. $(2rs)^{-3/4}$

56. $(5xy)^{-5/6}$

57. $\left(\dfrac{1}{16}\right)^{-3/4}$

58. $\left(\dfrac{1}{8}\right)^{-2/3}$

59. $\dfrac{8c}{a^{-3/5}}$

60. $\dfrac{3b}{a^{-5/7}}$

61. $2a^{3/4}b^{-1/2}c^{2/3}$

62. $5x^{-2/3}y^{4/5}z$

63. $3^{-5/2}a^3b^{-7/3}$

64. $2^{-1/3}x^4y^{-2/7}$

65. $\left(\dfrac{2ab}{3c}\right)^{-5/6}$

66. $\left(\dfrac{7x}{8yz}\right)^{-3/5}$

67. $xy^{-1/4}$

68. $aw^{-1/3}$

Laws of Exponents

Use the laws of exponents to simplify. Do not use negative exponents in any answers.

69. $11^{1/2} \cdot 11^{1/3}$

70. $5^{1/4} \cdot 5^{1/8}$

71. $\dfrac{3^{5/8}}{3^{-1/8}}$

72. $\dfrac{8^{7/11}}{8^{-2/11}}$

73. $\dfrac{4.3^{-1/5}}{4.3^{-7/10}}$

74. $\dfrac{2.7^{-11/12}}{2.7^{-1/6}}$

75. $(10^{3/5})^{2/5}$

76. $(5^{5/4})^{3/7}$

77. $a^{2/3} \cdot a^{5/4}$

78. $x^{3/4} \cdot x^{1/3}$

Aha! **79.** $(64^{3/4})^{4/3}$

80. $(27^{-2/3})^{3/2}$

81. $(m^{2/3}n^{-1/4})^{1/2}$

82. $(x^{-1/3}y^{2/5})^{1/4}$

Simplifying Radical Expressions

Use rational exponents to simplify. Do not use fraction exponents in the final answer. Write answers using radical notation.

83. $\sqrt[9]{x^3}$

84. $\sqrt[12]{a^3}$

85. $\sqrt[3]{y^{15}}$

86. $\sqrt[4]{y^{40}}$

87. $\sqrt[12]{a^6}$

88. $\sqrt[30]{x^5}$

89. $(\sqrt[7]{xy})^{14}$

90. $(\sqrt[3]{ab})^{15}$

91. $\sqrt[4]{(7a)^2}$

92. $\sqrt[8]{(3x)^2}$

93. $(\sqrt[8]{2x})^6$

94. $(\sqrt[10]{3a})^5$

95. $\sqrt{\sqrt[5]{m}}$

96. $\sqrt[6]{\sqrt{n}}$

97. $\sqrt[4]{(xy)^{12}}$

98. $\sqrt{(ab)^6}$

99. $(\sqrt[5]{a^2b^4})^{15}$

100. $(\sqrt[3]{x^2y^5})^{12}$

101. $\sqrt[3]{\sqrt[4]{xy}}$

102. $\sqrt[5]{\sqrt[3]{2a}}$

 103. If $f(x) = (x + 5)^{1/2}(x + 7)^{-1/2}$, find the domain of f. Explain how you found your answer.

 104. Let $f(x) = 5x^{-1/3}$. Under what condition will we have $f(x) > 0$? Why?

Skill Review

Solve.

105. $2(t + 3) - 5 = 1 - (6 - t)$ [2.2]

106. $10 - 5y > 4$ [2.6]

107. $-3 \le 5x + 7 \le 10$ [9.2]

108. $x^2 = x + 6$ [5.7]

109. $\dfrac{15}{x} - \dfrac{15}{x + 2} = 2$ [6.6]

110. $2x - y = 3,$
$x = 1 - y$ [8.2]

Synthesis

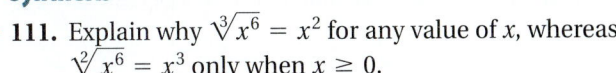 **111.** Explain why $\sqrt[3]{x^6} = x^2$ for any value of x, whereas $\sqrt{x^6} = x^3$ only when $x \ge 0$.

 112. If $g(x) = x^{3/n}$, why does the domain of g depend on whether n is odd or even?

Use rational exponents to simplify.

113. $\sqrt{x\sqrt[3]{x^2}}$

114. $\sqrt[4]{\sqrt[3]{8x^3y^6}}$

115. $\sqrt[14]{c^2 - 2cd + d^2}$

Music.   The function given by $f(x) = k2^{x/12}$ can be used to determine the frequency, in cycles per second, of a musical note that is x half-steps above a note with frequency k.

116. The frequency of concert A for a trumpet is 440 cycles per second. Find the frequency of the A that is two octaves (24 half-steps) above concert A. (Skilled trumpeters can reach this note.)

117. Show that the G that is 7 half-steps (a "perfect fifth") above middle C (262 cycles per second) has a frequency that is about 1.5 times that of middle C.

118. Show that the C sharp that is 4 half-steps (a "major third") above concert A (see Exercise 116) has a frequency that is about 25% greater than that of concert A.

119. *Road Pavement Messages.*   In a psychological study, it was determined that the proper length L of the letters of a word printed on pavement is given by

$$L = \frac{0.000169d^{2.27}}{h},$$

where d is the distance of a car from the lettering and h is the height of the eye above the surface of the road. All units are in meters. This formula says that from a vantage point h meters above the surface of the road, if a driver is to be able to recognize a message d meters away, that message will be the most recognizable if the length of the letters is L. Find L to the nearest tenth of a meter, given d and h.

*This application was inspired by information provided by Dr. Homer B. Tilton of Pima Community College East.

a) $h = 1$ m, $d = 60$ m
b) $h = 0.9906$ m, $d = 75$ m
c) $h = 2.4$ m, $d = 80$ m
d) $h = 1.1$ m, $d = 100$ m

120. *Baseball.*   The statistician Bill James has found that a baseball team's winning percentage P can be approximated by

$$P = \frac{r^{1.83}}{r^{1.83} + \sigma^{1.83}},$$

where r is the total number of runs scored by that team and σ (sigma) is the total number of runs scored by their opponents. During a recent season, the San Francisco Giants scored 799 runs and their opponents scored 749 runs. Use James's formula to predict the Giants' winning percentage. (The team actually won 55.6% of their games.)

Source: Bittinger, M., *One Man's Journey Through Mathematics.* Boston: Addison-Wesley, 2004

121. *Forestry.*   The total wood volume T, in cubic feet, in a California black oak can be estimated using the formula

$$T = 0.936\,d^{1.97}h^{0.85},$$

where d is the diameter of the tree at breast height and h is the total height of the tree. How much wood is in a California black oak that is 3 ft in diameter at breast height and 80 ft high?

Source: Norman H. Pillsbury and Michael L. Kirkley, 1984. Equations for total, wood, and saw-log volume for thirteen California hardwoods, USDA Forest Service PNW Research Note No. 414: 52 p.

122. *Physics.*   The equation $m = m_0(1 - v^2c^{-2})^{-1/2}$, developed by Albert Einstein, is used to determine the mass m of an object that is moving v meters per second and has mass m_0 before the motion begins. The constant c is the speed of light, approximately 3×10^8 m/sec. Suppose that a particle with mass 8 mg is accelerated to a speed of $\frac{9}{5} \times 10^8$ m/sec. Without using a calculator, find the new mass of the particle.

123. The total surface area of the human body is called the body surface area (BSA). Values of the BSA can be read from tables such as the following or approximated using a formula. A person's body surface area (BSA) can be approximated by the DuBois formula

$$\text{BSA} = 0.007184w^{0.425}h^{0.725},$$

where w is mass, in kilograms, h is height, in centimeters, and BSA is in square meters. What is the

BSA of a child who is 122 cm tall and has a mass of 29.5 kg?

Source: www.halls.md

Height (in centimeters)	Mass (in kilograms)			
	55	60	65	70
100	1.11	1.15	1.19	1.23
110	1.19	1.24	1.28	1.32
120	1.27	1.32	1.36	1.41

124. Using a graphing calculator, select **MODE** SIMUL and **FORMAT** EXPROFF. Then graph

$$y_1 = x^{1/2}, \qquad y_2 = 3x^{2/5},$$
$$y_3 = x^{4/7}, \quad \text{and} \quad y_4 = \tfrac{1}{5}x^{3/4}.$$

Looking only at coordinates, match each graph with its equation.

 125. a) Graph $f(x) = x^{1/3}$, $g(x) = (x^{1/6})^2$, and $h(x) = (x^2)^{1/6}$. How do the graphs and the domains differ?

b) Study the definition of $a^{m/n}$ carefully and then predict which of the graphs in part (a), if any, would best represent the graph of $k(x) = x^{2/6}$. Explain why.

c) Check using a graphing calculator.

> **YOUR TURN ANSWERS: SECTION 10.2**
> **1.** $\sqrt{49}$, or 7   **2.** $(2ac)^{1/4}$ **3.** $(\sqrt[3]{1000})^2$, or 100
> **4.** $(4x)^{3/5}$ **5.** $\dfrac{p^{2/3}}{16^{1/2}w^{3/5}}$, or $\dfrac{p^{2/3}}{4w^{3/5}}$ **6.** $8^{11/12}$ **7.** a^5b^5

QUICK QUIZ: SECTIONS 10.1–10.2
Simplify. Assume that a variable can represent any real number.

1. $-\sqrt{81}$ [10.1]   **2.** $\sqrt[6]{x^6}$ [10.1]

3. $16^{1/4}$ [10.2]   **4.** $(\sqrt[4]{2})^8$ [10.2]

5. Find $f(-5)$ if $f(t) = \sqrt{10 - 3t}$.   [10.1]

PREPARE TO MOVE ON
Multiply.

1. $(x + 5)(x - 5)$ [4.6]   **2.** $(x - 2)(x^2 + 2x + 4)$ [4.5]

Factor.   [5.4]

3. $9a^2 - 24a + 16$ **4.** $3n^2 + 12n + 12$

10.3 Multiplying Radical Expressions

Multiplying Radical Expressions ▪ Simplifying by Factoring ▪ Multiplying and Simplifying

Multiplying Radical Expressions

Note that $\sqrt{4}\sqrt{25} = 2 \cdot 5 = 10$ and $\sqrt{4 \cdot 25} = \sqrt{100} = 10$. Likewise,

$$\sqrt[3]{27}\,\sqrt[3]{8} = 3 \cdot 2 = 6 \quad \text{and} \quad \sqrt[3]{27 \cdot 8} = \sqrt[3]{216} = 6.$$

These examples suggest the following.

> **THE PRODUCT RULE FOR RADICALS**
> For any real numbers $\sqrt[n]{a}$ and $\sqrt[n]{b}$,
>
> $$\sqrt[n]{a} \cdot \sqrt[n]{b} = \sqrt[n]{a \cdot b}.$$
>
> (The product of two nth roots is the nth root of the product of the two radicands.)

> **CAUTION!** The product rule for radicals applies only when radicals have the same index.

Rational exponents can be used to derive this rule:

$$\sqrt[n]{a} \cdot \sqrt[n]{b} = a^{1/n} \cdot b^{1/n} = (a \cdot b)^{1/n} = \sqrt[n]{a \cdot b}.$$

EXAMPLE 1 Multiply.

a) $\sqrt{2} \cdot \sqrt{7}$ **b)** $\sqrt{x+3}\,\sqrt{x-3}$ **c)** $\sqrt[4]{\dfrac{y}{5}} \cdot \sqrt[4]{\dfrac{7}{x}}$

SOLUTION

a) When no index is written, roots are understood to be square roots with an unwritten index of two. We apply the product rule:

$$\sqrt{2} \cdot \sqrt{7} = \sqrt{2 \cdot 7} = \sqrt{14}.$$

> **CAUTION!**
> $\sqrt{x^2 - 9} \neq \sqrt{x^2} - \sqrt{9}.$

b) $\sqrt{x+3}\sqrt{x-3} = \sqrt{(x+3)(x-3)}$ The product of two square roots is the square root of the product.

$$= \sqrt{x^2 - 9}$$

c) The index in each radical expression is 4, so in order to multiply we can use the product rule:

$$\sqrt[4]{\frac{y}{5}} \cdot \sqrt[4]{\frac{7}{x}} = \sqrt[4]{\frac{y}{5} \cdot \frac{7}{x}} = \sqrt[4]{\frac{7y}{5x}}.$$

1. Multiply: $\sqrt[3]{4} \cdot \sqrt[3]{5}$.

↩ YOUR TURN

Simplifying by Factoring

The number p is a *perfect square* if there exists a rational number q for which $q^2 = p$. We say that p is a *perfect cube* if $q^3 = p$ for some rational number q. In general, p is a *perfect nth power* if $q^n = p$ for some rational number q. The product rule allows us to simplify $\sqrt[n]{ab}$ when a or b is a perfect nth power.

> **USING THE PRODUCT RULE TO SIMPLIFY**
> $$\sqrt[n]{ab} = \sqrt[n]{a} \cdot \sqrt[n]{b}.$$
> $\left(\sqrt[n]{a} \text{ and } \sqrt[n]{b} \text{ must both be real numbers.}\right)$

To illustrate, suppose we wish to simplify $\sqrt{20}$. Since this is a *square* root, we check to see if there is a factor of 20 that is a perfect *square*. There is one, 4, so we express 20 as $4 \cdot 5$ and use the product rule:

$$\sqrt{20} = \sqrt{4 \cdot 5} \quad \text{Factoring the radicand (4 is a perfect square)}$$
$$= \sqrt{4} \cdot \sqrt{5} \quad \text{Factoring into two radicals}$$
$$= 2\sqrt{5}. \quad \text{Finding the square root of 4}$$

> **TO SIMPLIFY A RADICAL EXPRESSION WITH INDEX *n* BY FACTORING**
>
> 1. If possible, express the radicand as a product in which one or more factors are perfect *n*th powers.
> 2. Rewrite the expression as the *n*th root of each factor.
> 3. Simplify any expressions containing perfect *n*th powers.
> 4. Simplification is complete when no radicand has a factor that is a perfect *n*th power. (All factors in the radicand can be written with exponents less than the index.)

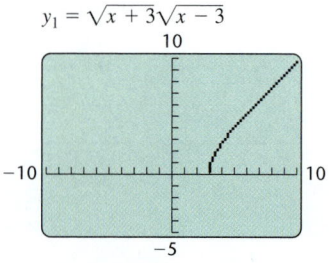
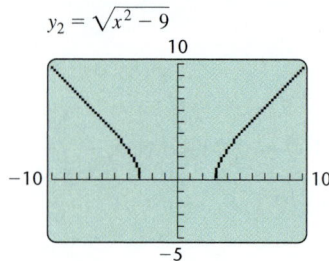

It is often safe to assume that a radicand does not represent a negative number raised to an even power. We will henceforth make this assumption—unless functions are involved—and discontinue use of absolute-value notation when taking even roots.

EXAMPLE 2 Simplify by factoring: **(a)** $\sqrt{200}$; **(b)** $\sqrt{50x^2y}$; **(c)** $\sqrt[3]{-72}$; **(d)** $\sqrt[4]{162x^6}$.

SOLUTION

a) $\sqrt{200} = \sqrt{100 \cdot 2}$ 100 is the largest perfect-square factor of 200.
$= \sqrt{100} \cdot \sqrt{2} = 10\sqrt{2}$ $\sqrt{100} = 10$

Express the radicand as a product.

Rewrite as the *n*th root of each factor.

Simplify.

b) $\sqrt{50x^2y} = \sqrt{25 \cdot 2 \cdot x^2 \cdot y}$ $25x^2$ is the largest perfect-square factor of $50x^2y$.
$= \sqrt{25x^2} \cdot \sqrt{2y}$ Factoring into two radicals
$= 5x\sqrt{2y}$ $\sqrt{25x^2} = 5x$. We have assumed that $x \geq 0$.

c) $\sqrt[3]{-72} = \sqrt[3]{-8 \cdot 9}$ -8 is a perfect-cube (third-power) factor of -72.
$= \sqrt[3]{-8} \cdot \sqrt[3]{9} = -2\sqrt[3]{9}$ Taking the cube root of -8

d) $\sqrt[4]{162x^6} = \sqrt[4]{81 \cdot 2 \cdot x^4 \cdot x^2}$ $81 \cdot x^4$ is the largest perfect fourth-power factor of $162x^6$.
$= \sqrt[4]{81x^4} \cdot \sqrt[4]{2x^2}$ Factoring into two radicals
$= 3x\sqrt[4]{2x^2}$ $\sqrt[4]{81x^4} = 3x$. We have assumed that $x \geq 0$.

Let's look at this example another way. We write a complete factorization and look for quadruples of factors. Each quadruple makes a perfect fourth power:

$$\sqrt[4]{162x^6} = \sqrt[4]{3 \cdot 3 \cdot 3 \cdot 3 \cdot 2 \cdot x \cdot x \cdot x \cdot x \cdot x \cdot x} \qquad 3 \cdot 3 \cdot 3 \cdot 3 = 3^4 \text{ and}$$
$$x \cdot x \cdot x \cdot x = x^4$$
$$= 3 \cdot x \cdot \sqrt[4]{2 \cdot x \cdot x} \qquad \sqrt[4]{3^4} = 3 \text{ and } \sqrt[4]{3^4} = x$$
$$= 3x\sqrt[4]{2x^2}.$$

2. Simplify by factoring:
$\sqrt{18ab^2}$.

YOUR TURN

EXAMPLE 3 If $f(x) = \sqrt{3x^2 - 6x + 3}$, find a simplified form for $f(x)$. Because we are working with a function, assume that x can be any real number.

SOLUTION

$$f(x) = \sqrt{3x^2 - 6x + 3}$$
$$= \sqrt{3(x^2 - 2x + 1)}$$ Factoring the radicand; $x^2 - 2x + 1$ is a perfect square.
$$= \sqrt{(x - 1)^2 \cdot 3}$$
$$= \sqrt{(x - 1)^2} \cdot \sqrt{3}$$ Factoring into two radicals
$$= |x - 1|\sqrt{3}$$ Finding the square root of $(x - 1)^2$

3. If $f(x) = \sqrt{10x^2 + 60x + 90}$, find a simplified form for $f(x)$. Assume that x can be any real number.

YOUR TURN

Technology Connection

To check Example 3, let $y_1 = \sqrt{3x^2 - 6x + 3}$, $y_2 = |x - 1|\sqrt{3}$, and $y_3 = (x - 1)\sqrt{3}$. Do the graphs all coincide? Why or why not?

EXAMPLE 4 Simplify: **(a)** $\sqrt{x^7y^{11}z^9}$; **(b)** $\sqrt[3]{16a^7b^{14}}$.

SOLUTION

a) There are many ways to factor $x^7y^{11}z^9$. Because of the square root (index of 2), we identify the largest exponents that are multiples of 2:

$$\sqrt{x^7y^{11}z^9} = \sqrt{x^6 \cdot x \cdot y^{10} \cdot y \cdot z^8 \cdot z}$$ The largest perfect-square factor is $x^6y^{10}z^8$.
$$= \sqrt{x^6}\sqrt{y^{10}}\sqrt{z^8}\sqrt{xyz}$$ Factoring into several radicals
$$= x^{6/2}y^{10/2}z^{8/2}\sqrt{xyz}$$ Converting to rational exponents. Try to do this mentally.
$$= x^3y^5z^4\sqrt{xyz}.$$ Simplifying

Check: $\left(x^3y^5z^4\sqrt{xyz}\right)^2 = (x^3)^2(y^5)^2(z^4)^2\left(\sqrt{xyz}\right)^2$

$$= x^6 \cdot y^{10} \cdot z^8 \cdot xyz = x^7y^{11}z^9.$$

Our check shows that $x^3y^5z^4\sqrt{xyz}$ is the square root of $x^7y^{11}z^9$.

b) There are many ways to factor $16a^7b^{14}$. Because of the cube root (index of 3), we identify factors with the largest exponents that are multiples of 3:

$$\sqrt[3]{16a^7b^{14}} = \sqrt[3]{8 \cdot 2 \cdot a^6 \cdot a \cdot b^{12} \cdot b^2} \qquad \text{The largest perfect-cube factor is } 8a^6b^{12}.$$

$$= \sqrt[3]{8}\sqrt[3]{a^6}\sqrt[3]{b^{12}}\sqrt[3]{2ab^2} \qquad \text{Rewriting as a product of cube roots}$$

$$= 2a^2b^4\sqrt[3]{2ab^2}. \qquad \text{Simplifying the expressions containing perfect cubes}$$

As a check, let's redo the problem using a complete factorization of the radicand:

$$\sqrt[3]{16a^7b^{14}} = \sqrt[3]{2 \cdot 2 \cdot 2 \cdot 2 \cdot a \cdot a \cdot a \cdot a \cdot a \cdot a \cdot a \cdot b \cdot b \cdot b \cdot b \cdot b \cdot b \cdot b \cdot b \cdot b \cdot b \cdot b \cdot b \cdot b \cdot b}$$

Each triple of factors makes a cube.

$$= 2 \cdot a \cdot a \cdot b \cdot b \cdot b \cdot b \cdot \sqrt[3]{2 \cdot a \cdot b \cdot b} = 2a^2b^4\sqrt[3]{2ab^2}. \qquad \text{Our answer checks.}$$

4. Simplify: $\sqrt[3]{5000x^{12}y^{13}z^2}$.  YOUR TURN

> To simplify an nth root, identify factors in the radicand with exponents that are multiples of n.

Multiplying and Simplifying

We have used the product rule for radicals to find products and also to simplify radical expressions. For some radical expressions, it is possible to do both: First find a product and then simplify.

EXAMPLE 5 Multiply and simplify.

a) $\sqrt{15}\sqrt{6}$ **b)** $3\sqrt[3]{25} \cdot 2\sqrt[3]{5}$ **c)** $\sqrt[4]{8x^3y^5}\sqrt[4]{4x^2y^3}$

SOLUTION

a) $\sqrt{15}\sqrt{6} = \sqrt{15 \cdot 6}$ Multiplying radicands

$\qquad\qquad = \sqrt{90} = \sqrt{9}\sqrt{10}$ 9 is a perfect square.

$\qquad\qquad = 3\sqrt{10}$

b) $3\sqrt[3]{25} \cdot 2\sqrt[3]{5} = 3 \cdot 2 \cdot \sqrt[3]{25 \cdot 5}$ Using a commutative law; multiplying radicands

$\qquad\qquad = 6 \cdot \sqrt[3]{125}$ 125 is a perfect cube.

$\qquad\qquad = 6 \cdot 5$, or 30

Student Notes

To multiply $\sqrt{x} \cdot \sqrt{x}$, remember what $\sqrt{x}$ represents and go directly to the product, x. For $x \geq 0$,

$\sqrt{x} \cdot \sqrt{x} = x,$

$\left(\sqrt{x}\right)^2 = x,$ and

$\sqrt{x^2} = x.$

c) $\sqrt[4]{8x^3y^5}\sqrt[4]{4x^2y^3} = \sqrt[4]{32x^5y^8}$ Multiplying radicands

$\qquad\qquad = \sqrt[4]{16x^4y^8 \cdot 2x}$ Identifying the largest perfect fourth-power factor

$\qquad\qquad = \sqrt[4]{16x^4y^8}\sqrt[4]{2x}$ Factoring into radicals

$\qquad\qquad = 2xy^2\sqrt[4]{2x}$ Finding the fourth root. We assume that $x \geq 0$.

To check, we can use complete factorizations of the radicands. Checking part (c), we have

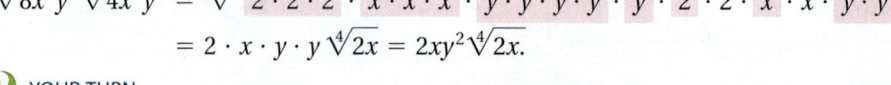

$$\sqrt[4]{8x^3y^5}\sqrt[4]{4x^2y^3} = \sqrt[4]{2 \cdot 2 \cdot 2 \cdot x \cdot x \cdot x \cdot y \cdot y \cdot y \cdot y \cdot y \cdot 2 \cdot 2 \cdot x \cdot x \cdot y \cdot y \cdot y}$$

$$= 2 \cdot x \cdot y \cdot y \sqrt[4]{2x} = 2xy^2\sqrt[4]{2x}.$$

5. Multiply and simplify:

$\sqrt{10x}\sqrt{15x}$.

 YOUR TURN

10.3 EXERCISE SET

Vocabulary and Reading Check

Classify each of the following statements as either true or false.

1. For any real numbers $\sqrt[n]{a}$ and $\sqrt[n]{b}$, $\sqrt[n]{a} \cdot \sqrt[n]{b} = \sqrt[n]{ab}$.

2. For any real numbers $\sqrt[n]{a}$ and $\sqrt[n]{b}$, $\sqrt[n]{a} + \sqrt[n]{b} = \sqrt[n]{a + b}$.

3. For $x > 0$, $\sqrt{x^2 - 9} = x - 3$.

4. Since $(-10)^3 = -1000$, the number -1000 is a perfect cube.

5. The expression $\sqrt[3]{X}$ is not simplified if X contains a factor that is a perfect cube.

6. When no index is written, as in $\sqrt{5}$, the root is understood to be a square root.

Multiplying Radical Expressions

Multiply.

7. $\sqrt{3}\,\sqrt{10}$
8. $\sqrt{6}\,\sqrt{5}$
9. $\sqrt[3]{7}\,\sqrt[3]{5}$
10. $\sqrt[3]{2}\,\sqrt[3]{3}$
11. $\sqrt[4]{6}\,\sqrt[4]{9}$
12. $\sqrt[4]{4}\,\sqrt[4]{10}$
13. $\sqrt{2x}\,\sqrt{13y}$
14. $\sqrt{5a}\,\sqrt{6b}$
15. $\sqrt[5]{8y^3}\,\sqrt[5]{10y}$
16. $\sqrt[5]{9t^2}\,\sqrt[5]{2t}$
17. $\sqrt{y-b}\,\sqrt{y+b}$
18. $\sqrt{x-a}\,\sqrt{x+a}$
19. $\sqrt[3]{0.7y}\,\sqrt[3]{0.3y}$
20. $\sqrt[3]{0.5x}\,\sqrt[3]{0.2x}$
21. $\sqrt[5]{x-2}\,\sqrt[5]{(x-2)^2}$
22. $\sqrt[4]{x-1}\,\sqrt[4]{x^2+x+1}$
23. $\sqrt{\dfrac{2}{t}}\,\sqrt{\dfrac{3s}{11}}$
24. $\sqrt{\dfrac{7p}{6}}\,\sqrt{\dfrac{5}{q}}$
25. $\sqrt[7]{\dfrac{x-3}{4}}\,\sqrt[7]{\dfrac{5}{x+2}}$
26. $\sqrt[6]{\dfrac{a}{b-2}}\,\sqrt[6]{\dfrac{3}{b+2}}$

Simplifying by Factoring

Simplify. Assume that no radicands were formed by raising negative numbers to even powers.

27. $\sqrt{12}$
28. $\sqrt{300}$
29. $\sqrt{45}$
30. $\sqrt{27}$
31. $\sqrt{8x^9}$
32. $\sqrt{75y^5}$
33. $\sqrt{120}$
34. $\sqrt{350}$
35. $\sqrt{36a^4b}$
36. $\sqrt{175y^8}$
37. $\sqrt[3]{8x^3y^2}$
38. $\sqrt[3]{27ab^6}$
39. $\sqrt[3]{-16x^6}$
40. $\sqrt[3]{-32a^6}$

Find a simplified form of $f(x)$. Assume that x can be any real number.

41. $f(x) = \sqrt[3]{40x^6}$
42. $f(x) = \sqrt[3]{27x^5}$
43. $f(x) = \sqrt{49(x-3)^2}$
44. $f(x) = \sqrt{81(x-1)^2}$
45. $f(x) = \sqrt{5x^2 - 10x + 5}$
46. $f(x) = \sqrt{2x^2 + 8x + 8}$

Simplify. Assume that no radicands were formed by raising negative numbers to even powers.

47. $\sqrt{a^{10}b^{11}}$
48. $\sqrt{x^8y^7}$
49. $\sqrt[3]{x^5y^6z^{10}}$
50. $\sqrt[3]{a^6b^7c^{13}}$
51. $\sqrt[4]{16x^5y^{11}}$
52. $\sqrt[5]{-32a^7b^{11}}$
53. $\sqrt[5]{x^{13}y^8z^{17}}$
54. $\sqrt[5]{a^6b^8c^9}$
55. $\sqrt[3]{-80a^{14}}$
56. $\sqrt[4]{810x^9}$

Multiplying and Simplifying

Multiply and simplify. Assume that no radicands were formed by raising negative numbers to even powers.

57. $\sqrt{5}\,\sqrt{10}$
58. $\sqrt{2}\,\sqrt{6}$
59. $\sqrt{6}\,\sqrt{33}$
60. $\sqrt{10}\,\sqrt{35}$
61. $\sqrt[3]{9}\,\sqrt[3]{3}$
62. $\sqrt[3]{2}\,\sqrt[3]{4}$
Aha! 63. $\sqrt{24y^5}\,\sqrt{24y^5}$
64. $\sqrt{120t^9}\,\sqrt{120t^9}$
65. $\sqrt[3]{5a^2}\,\sqrt[3]{2a}$
66. $\sqrt[3]{7x}\,\sqrt[3]{3x^2}$
67. $3\sqrt{2x^5} \cdot 4\sqrt{10x^2}$
68. $3\sqrt{5a^7} \cdot 2\sqrt{15a^3}$
69. $\sqrt[3]{s^2t^4}\,\sqrt[3]{s^4t^6}$
70. $\sqrt[3]{x^2y^4}\,\sqrt[3]{x^2y^6}$
71. $\sqrt[3]{(x-y)^2}\,\sqrt[3]{(x-y)^{10}}$
72. $\sqrt[3]{(t+4)^5}\,\sqrt[3]{(t+4)}$
73. $\sqrt[4]{20a^3b^7}\,\sqrt[4]{4a^2b^5}$
74. $\sqrt[4]{9x^7y^2}\,\sqrt[4]{9x^2y^9}$
75. $\sqrt[5]{x^3(y+z)^6}\,\sqrt[5]{x^3(y+z)^4}$
76. $\sqrt[5]{a^3(b-c)^4}\,\sqrt[5]{a^7(b-c)^4}$

77. Explain how you could convince a friend that $\sqrt{x^2 - 16} \neq \sqrt{x^2} - \sqrt{16}$.

78. Why is it incorrect to say that, in general, $\sqrt{x^2} = x$?

Skill Review

Perform the indicated operation and, if possible, simplify.

79. $\dfrac{15a^2x}{8b} \cdot \dfrac{24b^2x}{5a}$ [6.2]

80. $\dfrac{x^2 - 1}{x^2 - 4} \div \dfrac{x^2 - x - 2}{x^2 + x - 2}$ [6.2]

81. $\dfrac{x - 3}{2x - 10} - \dfrac{3x - 5}{x^2 - 25}$ [6.4]

82. $\dfrac{6x}{25y^2} + \dfrac{3y}{10x}$ [6.4]

83. $\dfrac{a^{-1} + b^{-1}}{ab}$ [6.5]

84. $\dfrac{\dfrac{1}{x + 1} - \dfrac{2}{x}}{\dfrac{3}{x} + \dfrac{1}{x + 1}}$ [6.5]

Synthesis

85. Explain why it is true that $\sqrt[n]{ab} = \sqrt[n]{a} \cdot \sqrt[n]{b}$ for any real numbers $\sqrt[n]{a}$ and $\sqrt[n]{b}$.

86. Is $\sqrt{(2x + 3)^8} = (2x + 3)^4$ always, sometimes, or never true? Why?

87. *Radar Range.* The function given by
$$R(x) = \frac{1}{2} \sqrt[4]{\frac{x \cdot 3.0 \times 10^6}{\pi^2}}$$
can be used to determine the maximum range $R(x)$, in miles, of an ARSR-3 surveillance radar with a peak power of x watts. Determine the maximum radar range when the peak power is 5×10^4 watts.

Source: Introduction to RADAR Techniques, Federal Aviation Administration, 1988

88. *Speed of a Skidding Car.* Police can estimate the speed at which a car was traveling by measuring its skid marks. The function given by
$$r(L) = 2\sqrt{5L}$$
can be used, where L is the length of a skid mark, in feet, and $r(L)$ is the speed, in miles per hour. Find the exact speed and an estimate (to the nearest tenth mile per hour) for the speed of a car that left skid marks **(a)** 20 ft long; **(b)** 70 ft long; **(c)** 90 ft long. See also Exercise 102.

89. *Wind Chill Temperature.* When the temperature is T degrees Celsius and the wind speed is v meters per second, the *wind chill temperature*, T_w, is the temperature (with no wind) that it feels like. Here is a formula for finding wind chill temperature:
$$T_w = 33 - \frac{(10.45 + 10\sqrt{v} - v)(33 - T)}{22}.$$
Estimate the wind chill temperature (to the nearest tenth of a degree) for the given actual temperatures and wind speeds.

a) $T = 7°C$, $v = 8$ m/sec
b) $T = 0°C$, $v = 12$ m/sec
c) $T = -5°C$, $v = 14$ m/sec
d) $T = -23°C$, $v = 15$ m/sec

Simplify. Assume that all variables are nonnegative.

90. $\left(\sqrt{r^3 t}\right)^7$

91. $\left(\sqrt[3]{25x^4}\right)^4$

92. $\left(\sqrt[3]{a^2 b^4}\right)^5$

93. $\left(\sqrt{a^3 b^5}\right)^7$

Draw and compare the graphs of each group of equations.

94. $f(x) = \sqrt{x^2 - 2x + 1}$,
$g(x) = x - 1$,
$h(x) = |x - 1|$

95. $f(x) = \sqrt{x^2 + 2x + 1}$,
$g(x) = x + 1$,
$h(x) = |x + 1|$

96. If $f(t) = \sqrt{t^2 - 3t - 4}$, what is the domain of f?

97. What is the domain of g, if $g(x) = \sqrt{x^2 - 6x + 8}$?

Solve.

98. $\sqrt[3]{5x^{k+1}} \, \sqrt[3]{25x^k} = 5x^7$, for k

99. $\sqrt[5]{4a^{3k+2}} \, \sqrt[5]{8a^{6-k}} = 2a^4$, for k

100. Use a graphing calculator to check your answers to Exercises 21 and 41.

101. Antonio is puzzled. When he uses a graphing calculator to graph $y = \sqrt{x} \cdot \sqrt{x}$, he gets the following screen. Explain why Antonio did not get the complete line $y = x$.

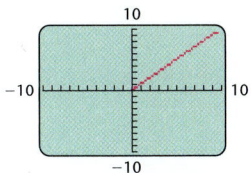

102. Does a car traveling twice as fast as another car leave a skid mark that is twice as long? (See Exercise 88.) Why or why not?

↪ **YOUR TURN ANSWERS: SECTION 10.3**

1. $\sqrt[3]{20}$ **2.** $3b\sqrt{2a}$ **3.** $f(x) = |x + 3|\sqrt{10}$
4. $10x^4y^4\sqrt[3]{5yz^2}$ **5.** $5x\sqrt{6}$

QUICK QUIZ: SECTIONS 10.1–10.3
Simplify. Assume that no radicands were formed by raising negative quantities to even powers.

1. $\sqrt{(5 - y)^{10}}$ [10.1]

2. $\sqrt[3]{40x^5y^6}$ [10.3]

3. Use rational exponents to simplify: $\sqrt[20]{c^4}$. [10.2]

4. Use the laws of exponents to simplify: $3^{2/5} \cdot 3^{1/2}$. [10.2]

5. Multiply and simplify: $\sqrt[3]{4x^2y}\,\sqrt[3]{2xy^2}$. [10.3]

PREPARE TO MOVE ON
Simplify. [4.1]

1. $\dfrac{82ab}{2}$ 2. $\dfrac{120m^2}{64n^3}$

3. $\dfrac{34xy^5}{2y}$ 4. $\dfrac{45x^5y^2}{3xy^2}$

10.4 Dividing Radical Expressions

Dividing and Simplifying ▪ Rationalizing Denominators or Numerators with One Term

Dividing and Simplifying

Study Skills

Professors Are Human
Even the best professors sometimes make mistakes. If, as you review your notes, you find that something doesn't make sense, it may be due to your instructor having made a mistake. If, after double-checking, you still perceive a mistake, politely ask him or her about it. Your instructor will welcome the opportunity to correct any errors.

Just as the root of a product can be expressed as the product of two roots, the root of a quotient can be expressed as the quotient of two roots. For example,

$$\sqrt[3]{\frac{27}{8}} = \frac{3}{2} \quad \text{and} \quad \frac{\sqrt[3]{27}}{\sqrt[3]{8}} = \frac{3}{2}.$$

This example suggests the following.

THE QUOTIENT RULE FOR RADICALS
For any real numbers $\sqrt[n]{a}$ and $\sqrt[n]{b}$, $b \neq 0$,

$$\sqrt[n]{\frac{a}{b}} = \frac{\sqrt[n]{a}}{\sqrt[n]{b}}.$$

Remember that an nth root is simplified when its radicand has no factors that are perfect nth powers. **Unless functions are involved, we assume that no radicands represent negative quantities raised to an even power.**

EXAMPLE 1 Simplify by taking the roots of the numerator and the denominator.

a) $\sqrt[3]{\dfrac{27}{125}}$

b) $\sqrt{\dfrac{25}{y^2}}$

SOLUTION

a) $\sqrt[3]{\dfrac{27}{125}} = \dfrac{\sqrt[3]{27}}{\sqrt[3]{125}} = \dfrac{3}{5}$ Taking the cube roots of the numerator and the denominator

1. Simplify by taking the roots of the numerator and the denominator: $\sqrt{\dfrac{25}{49}}$.

b) $\sqrt{\dfrac{25}{y^2}} = \dfrac{\sqrt{25}}{\sqrt{y^2}} = \dfrac{5}{y}$ Taking the square roots of the numerator and the denominator. Assume $y > 0$.

YOUR TURN

Any radical expressions appearing in the answers should be simplified as much as possible.

EXAMPLE 2   Simplify: **(a)** $\sqrt{\dfrac{16x^3}{y^8}}$; **(b)** $\sqrt[3]{\dfrac{27y^{14}}{8x^3}}$.

SOLUTION

a) $\sqrt{\dfrac{16x^3}{y^8}} = \dfrac{\sqrt{16x^3}}{\sqrt{y^8}}$

$\qquad = \dfrac{\sqrt{16x^2 \cdot x}}{\sqrt{y^8}} = \dfrac{4x\sqrt{x}}{y^4}$ Simplifying the numerator and the denominator

b) $\sqrt[3]{\dfrac{27y^{14}}{8x^3}} = \dfrac{\sqrt[3]{27y^{14}}}{\sqrt[3]{8x^3}}$

$\qquad = \dfrac{\sqrt[3]{27y^{12}y^2}}{\sqrt[3]{8x^3}}$ y^{12} is the largest perfect-cube factor of y^{14}.

$\qquad = \dfrac{\sqrt[3]{27y^{12}}\,\sqrt[3]{y^2}}{\sqrt[3]{8x^3}}$

2. Simplify: $\sqrt[3]{\dfrac{16x^5}{27y^6}}$.

$\qquad = \dfrac{3y^4\sqrt[3]{y^2}}{2x}$ Simplifying the numerator and the denominator

YOUR TURN

If we read from right to left, the quotient rule tells us that to divide two radical expressions that have the same index, we can divide the radicands.

EXAMPLE 3   Divide and, if possible, simplify.

a) $\dfrac{\sqrt{80}}{\sqrt{5}}$ **b)** $\dfrac{5\sqrt[3]{32}}{\sqrt[3]{2}}$

c) $\dfrac{\sqrt{72xy}}{2\sqrt{2}}$ **d)** $\dfrac{\sqrt[4]{18a^9b^5}}{\sqrt[4]{3b}}$

Student Notes

When writing radical signs, be careful what you include in the radicand. The following represent *different* numbers:

$$\sqrt{\dfrac{5 \cdot 2}{3}}, \quad \dfrac{\sqrt{5 \cdot 2}}{3}, \quad \dfrac{\sqrt{5} \cdot 2}{3}.$$

SOLUTION

a) $\dfrac{\sqrt{80}}{\sqrt{5}} = \sqrt{\dfrac{80}{5}} = \sqrt{16} = 4$

> Because the indices match, we can divide the radicands.

b) $\dfrac{5\sqrt[3]{32}}{\sqrt[3]{2}} = 5\sqrt[3]{\dfrac{32}{2}} = 5\sqrt[3]{16}$

$\qquad = 5\sqrt[3]{8 \cdot 2}$ 8 is the largest perfect-cube factor of 16.

$\qquad = 5\sqrt[3]{8}\,\sqrt[3]{2} = 5 \cdot 2\sqrt[3]{2}$

$\qquad = 10\sqrt[3]{2}$

c) $\dfrac{\sqrt{72xy}}{2\sqrt{2}} = \dfrac{1}{2}\sqrt{\dfrac{72xy}{2}}$

$= \dfrac{1}{2}\sqrt{36xy} = \dfrac{1}{2} \cdot 6\sqrt{xy}$

$= 3\sqrt{xy}$

> Because the indices match, we can divide the radicands.

3. Divide and, if possible, simplify:

$\dfrac{3\sqrt{8x}}{\sqrt{2}}.$

d) $\dfrac{\sqrt[4]{18a^9b^5}}{\sqrt[4]{3b}} = \sqrt[4]{\dfrac{18a^9b^5}{3b}}$

$= \sqrt[4]{6a^9b^4} = \sqrt[4]{a^8b^4}\,\sqrt[4]{6a}$

$= a^2b\sqrt[4]{6a}$

Note that 8 is the largest power less than 9 that is a multiple of the index 4.

 YOUR TURN

Rationalizing Denominators or Numerators with One Term*

The expressions

$$\frac{1}{\sqrt{2}} \quad \text{and} \quad \frac{\sqrt{2}}{2}$$

are equivalent, but the second expression does not have a radical expression in the denominator.[†] We can **rationalize the denominator** of a radical expression if we multiply by 1 in either of two ways.

One way is to multiply by 1 *under* the radical to make the denominator of the radicand a perfect power.

EXAMPLE 4   Rationalize each denominator.

a) $\sqrt{\dfrac{7}{3}}$
     b) $\sqrt[3]{\dfrac{5}{16}}$

SOLUTION

a) We multiply by 1 under the radical, using $\tfrac{3}{3}$. We do this so that the denominator of the radicand will be a perfect square:

$$\sqrt{\frac{7}{3}} = \sqrt{\frac{7}{3} \cdot \frac{3}{3}} \qquad \text{\color{red}Multiplying by 1 under the radical}$$

$$= \sqrt{\frac{21}{9}} \qquad \text{\color{red}The denominator, 9, is now a perfect square.}$$

$$= \frac{\sqrt{21}}{\sqrt{9}} = \frac{\sqrt{21}}{3}.$$

Student Notes

When we rationalize a denominator, the resulting expression is equivalent to the original expression. You can check this in Example 4(a) by approximating $\sqrt{7/3}$ and $\sqrt{21}/3$ using a calculator.

b) Note that the index is 3, so we want the denominator to be a perfect cube. Since $16 = 4^2$, we multiply under the radical by $\tfrac{4}{4}$:

$$\sqrt[3]{\frac{5}{16}} = \sqrt[3]{\frac{5}{4 \cdot 4} \cdot \frac{4}{4}} \qquad \text{\color{red}Since the index is 3, we need 3 identical factors in the denominator.}$$

$$= \sqrt[3]{\frac{20}{4^3}} \qquad \text{\color{red}The denominator is now a perfect cube.}$$

$$= \frac{\sqrt[3]{20}}{\sqrt[3]{4^3}} = \frac{\sqrt[3]{20}}{4}.$$

4. Rationalize the denominator:

$\sqrt[3]{\dfrac{5x^2}{y}}.$

 YOUR TURN

*Denominators and numerators with two terms are rationalized in Section 10.5.
[†]See Exercise 75 in Exercise Set 10.4.

Another way to rationalize a denominator is to multiply by 1 *outside* the radical.

EXAMPLE 5   Rationalize each denominator.

a) $\dfrac{\sqrt{4}}{5\sqrt{x}}$

b) $\dfrac{3b\sqrt[3]{a}}{\sqrt[3]{25bc^5}}$

SOLUTION

a)
$$\dfrac{\sqrt{4}}{5\sqrt{x}} = \dfrac{2}{5\sqrt{x}}$$ Simplifying. We assume $x > 0$.

$$= \dfrac{2}{5\sqrt{x}} \cdot \dfrac{\sqrt{x}}{\sqrt{x}}$$ Multiplying by 1. Since the factor 5 in the denominator is rational, we do not need to include it in the form of 1.

$$= \dfrac{2\sqrt{x}}{5(\sqrt{x})^2}$$ Try to do this step mentally.

$$= \dfrac{2\sqrt{x}}{5x}$$

b) Note that the radicand $25bc^5$ is $5 \cdot 5 \cdot b \cdot c \cdot c \cdot c \cdot c \cdot c$. In order for this to be a cube, we need another factor of 5, two more factors of b, and one more factor of c. Thus we multiply by 1, using $\sqrt[3]{5b^2c}/\sqrt[3]{5b^2c}$:

$$\dfrac{3b\sqrt[3]{a}}{\sqrt[3]{25bc^5}} = \dfrac{3b\sqrt[3]{a}}{\sqrt[3]{25bc^5}} \cdot \dfrac{\sqrt[3]{5b^2c}}{\sqrt[3]{5b^2c}}$$ Multiplying by 1

$$= \dfrac{3b\sqrt[3]{5ab^2c}}{\sqrt[3]{125b^3c^6}}$$ $\longleftarrow$ This radicand is now a perfect cube.

$$= \dfrac{3b\sqrt[3]{5ab^2c}}{5bc^2}$$

$$= \dfrac{3\sqrt[3]{5ab^2c}}{5c^2}.$$ Always simplify if possible.

5. Rationalize the denominator:

$$\dfrac{\sqrt{6a}}{2\sqrt{10ab}}.$$

 YOUR TURN

Sometimes it is necessary to rationalize a numerator. To do so, we multiply by 1 to make the radicand in the *numerator* a perfect power.

EXAMPLE 6   Rationalize the numerator: $\dfrac{\sqrt[3]{4a^2}}{\sqrt[3]{5b}}.$

SOLUTION

$$\dfrac{\sqrt[3]{4a^2}}{\sqrt[3]{5b}} = \dfrac{\sqrt[3]{4a^2}}{\sqrt[3]{5b}} \cdot \dfrac{\sqrt[3]{2a}}{\sqrt[3]{2a}}$$ Multiplying by 1

$$= \dfrac{\sqrt[3]{8a^3}}{\sqrt[3]{10ba}}$$ $\longleftarrow$ This radicand is now a perfect cube.

$$= \dfrac{2a}{\sqrt[3]{10ab}}$$

6. Rationalize the numerator:

$$\sqrt{\dfrac{7}{20}}.$$

 YOUR TURN

10.4 EXERCISE SET

FOR EXTRA HELP

MyMathLab® MathXL
PRACTICE   WATCH   READ   REVIEW

↪ Vocabulary and Reading Check

Give a justification for each equation by indicating either the *quotient rule for radicals or multiplying by 1.*

1. $\sqrt{\dfrac{x}{100}} = \dfrac{\sqrt{x}}{\sqrt{100}}$ _____

2. $\sqrt{\dfrac{5}{2}} = \sqrt{\dfrac{5}{2} \cdot \dfrac{2}{2}}$ _____

3. $\dfrac{\sqrt[3]{y^4}}{\sqrt[3]{x}} = \dfrac{\sqrt[3]{y^4}}{\sqrt[3]{x}} \cdot \dfrac{\sqrt[3]{x^2}}{\sqrt[3]{x^2}}$ _____

4. $\dfrac{\sqrt{50a^3}}{\sqrt{2a}} = \sqrt{\dfrac{50a^3}{2a}}$ _____

↪ Concept Reinforcement

In Exercises 5–10, match each expression with an equivalent expression from the column on the right. Assume $a, b > 0$.

5. ____ $\sqrt[4]{\dfrac{16a^6}{a^2}}$ a) $\dfrac{a^2}{b^3}$

6. ____ $\dfrac{\sqrt[3]{a^6}}{\sqrt[3]{b^9}}$ b) $\sqrt{\dfrac{a \cdot b}{b^3 \cdot b}}$

7. ____ $\sqrt[5]{\dfrac{a^6}{b^4}}$ c) $\sqrt{a}$

                              d) $\dfrac{\sqrt[3]{a^2}}{b^2}$

8. ____ $\sqrt{\dfrac{a}{b^3}}$

                              e) $\sqrt[5]{\dfrac{a^6 b}{b^4 \cdot b}}$

9. ____ $\dfrac{\sqrt{5a^4}}{\sqrt{5a^3}}$ f) $2a$

10. ____ $\sqrt[3]{\dfrac{a^2}{b^6}}$

Dividing and Simplifying

Simplify by taking the roots of the numerator and the denominator. Assume that all variables represent positive numbers.

11. $\sqrt{\dfrac{49}{100}}$ 12. $\sqrt{\dfrac{81}{25}}$ 13. $\sqrt[3]{\dfrac{125}{8}}$

14. $\sqrt[3]{\dfrac{1000}{27}}$ 15. $\sqrt{\dfrac{121}{t^2}}$ 16. $\sqrt{\dfrac{144}{p^2}}$

17. $\sqrt{\dfrac{36y^3}{x^4}}$ 18. $\sqrt{\dfrac{25a^5}{b^6}}$ 19. $\sqrt[3]{\dfrac{27a^4}{8b^3}}$

20. $\sqrt[3]{\dfrac{64x^7}{216y^6}}$ 21. $\sqrt[4]{\dfrac{32a^4}{2b^4c^8}}$ 22. $\sqrt[4]{\dfrac{81x^4}{y^8z^4}}$

23. $\sqrt[4]{\dfrac{a^5b^8}{c^{10}}}$ 24. $\sqrt[4]{\dfrac{x^9y^{12}}{z^6}}$ 25. $\sqrt[5]{\dfrac{32x^6}{y^{11}}}$

26. $\sqrt[5]{\dfrac{243a^9}{b^{13}}}$ 27. $\sqrt[6]{\dfrac{x^6y^8}{z^{15}}}$ 28. $\sqrt[6]{\dfrac{a^9b^{12}}{c^{13}}}$

Divide and, if possible, simplify. Assume that all variables represent positive numbers.

29. $\dfrac{\sqrt{18y}}{\sqrt{2y}}$ 30. $\dfrac{\sqrt{700x}}{\sqrt{7x}}$

31. $\dfrac{\sqrt[3]{26}}{\sqrt[3]{13}}$ 32. $\dfrac{\sqrt[3]{35}}{\sqrt[3]{5}}$

33. $\dfrac{\sqrt{40xy^3}}{\sqrt{8x}}$ 34. $\dfrac{\sqrt{56ab^3}}{\sqrt{7a}}$

35. $\dfrac{\sqrt[3]{96a^4b^2}}{\sqrt[3]{12a^2b}}$ 36. $\dfrac{\sqrt[3]{189x^5y^7}}{\sqrt[3]{7x^2y^2}}$

37. $\dfrac{\sqrt{100ab}}{5\sqrt{2}}$ 38. $\dfrac{\sqrt{75ab}}{3\sqrt{3}}$

39. $\dfrac{\sqrt[4]{48x^9y^{13}}}{\sqrt[4]{3xy^{-2}}}$ 40. $\dfrac{\sqrt[5]{64a^{11}b^{28}}}{\sqrt[5]{2ab^{-2}}}$

41. $\dfrac{\sqrt[3]{x^3 - y^3}}{\sqrt[3]{x - y}}$ ←   → 42. $\dfrac{\sqrt[3]{r^3 + s^3}}{\sqrt[3]{r + s}}$

Hint: Factor and then simplify.

Rationalizing Denominators or Numerators with One Term

Rationalize each denominator. Assume that all variables represent positive numbers.

43. $\sqrt{\dfrac{2}{5}}$ 44. $\sqrt{\dfrac{7}{2}}$ 45. $\dfrac{2\sqrt{5}}{7\sqrt{3}}$

46. $\dfrac{3\sqrt{5}}{2\sqrt{7}}$ 47. $\sqrt[3]{\dfrac{5}{4}}$ 48. $\sqrt[3]{\dfrac{2}{9}}$

49. $\dfrac{\sqrt[3]{3a}}{\sqrt[3]{5c}}$ 50. $\dfrac{\sqrt[3]{7x}}{\sqrt[3]{3y}}$ 51. $\dfrac{\sqrt[4]{5y^6}}{\sqrt[4]{9x}}$

52. $\dfrac{\sqrt[5]{3a^4}}{\sqrt[5]{2b^7}}$ 53. $\sqrt[3]{\dfrac{2}{x^2y}}$ 54. $\sqrt[3]{\dfrac{5}{ab^2}}$

55. $\sqrt{\dfrac{7a}{18}}$ 56. $\sqrt{\dfrac{3x}{20}}$ 57. $\sqrt[5]{\dfrac{9}{32x^5y}}$

58. $\sqrt[4]{\dfrac{7}{64a^2b^4}}$ *Aha!* 59. $\sqrt{\dfrac{10ab^2}{72a^3b}}$ 60. $\sqrt{\dfrac{21x^2y}{75xy^5}}$

Rationalize each numerator. Assume that all variables represent positive numbers.

61. $\sqrt{\dfrac{5}{11}}$

62. $\sqrt{\dfrac{2}{3}}$

63. $\dfrac{2\sqrt{6}}{5\sqrt{7}}$

64. $\dfrac{3\sqrt{10}}{2\sqrt{3}}$

65. $\dfrac{\sqrt{8}}{2\sqrt{3x}}$

66. $\dfrac{\sqrt{12}}{\sqrt{5y}}$

67. $\dfrac{\sqrt[3]{7}}{\sqrt[3]{2}}$

68. $\dfrac{\sqrt[3]{5}}{\sqrt[3]{4}}$

69. $\sqrt{\dfrac{7x}{3y}}$

70. $\sqrt{\dfrac{7a}{6b}}$

71. $\sqrt[3]{\dfrac{2a^5}{5b}}$

72. $\sqrt[3]{\dfrac{2a^4}{7b}}$

73. $\sqrt{\dfrac{x^3y}{2}}$

74. $\sqrt{\dfrac{ab^5}{3}}$

 75. Explain why it is easier to approximate
$$\dfrac{\sqrt{2}}{2} \quad \text{than} \quad \dfrac{1}{\sqrt{2}}$$
if no calculator is available and $\sqrt{2} \approx 1.414213562$.

 76. A student *incorrectly* claims that
$$\dfrac{5 + \sqrt{2}}{\sqrt{18}} = \dfrac{5 + \sqrt{1}}{\sqrt{9}} = \dfrac{5 + 1}{3}.$$
How could you convince the student that a mistake has been made? How would you explain the correct way of rationalizing the denominator?

Skill Review

Perform the indicated operations and, if possible, simplify.

77. $-\dfrac{2}{9} \div \dfrac{4}{6}$ [1.2]

78. $-\dfrac{2}{9}\left(-\dfrac{4}{6}\right)$ [1.2]

79. $12 - 100 \div 5 \cdot (-2)^2 - 3(6 - 7)$ [1.2]

80. $\left(9x^3 - 3x - \dfrac{1}{2}\right) - \left(x^2 - 12x - \dfrac{1}{2}\right)$ [5.1]

81. $(12x^3 - 6x - 8) \div (x + 1)$ [6.6], [6.7]

82. $(7m - 2n)^2$ [5.5]

Synthesis

83. Is the quotient of two irrational numbers always an irrational number? Why or why not?

84. Is it possible to understand how to rationalize a denominator without knowing how to multiply rational expressions? Why or why not?

85. *Pendulums.* The *period* of a pendulum is the time it takes the pendulum to complete one cycle, swinging to and fro. For a pendulum that is L centimeters long, the period T is given by the formula
$$T = 2\pi\sqrt{\dfrac{L}{980}},$$
where T is in seconds. Find, to the nearest hundredth of a second, the period of a pendulum of length **(a)** 65 cm; **(b)** 98 cm; **(c)** 120 cm. Use a calculator's $\boxed{\pi}$ key if possible.

Perform the indicated operations.

86. $\dfrac{7\sqrt{a^2b}\,\sqrt{25xy}}{5\sqrt{a^{-4}b^{-1}}\,\sqrt{49x^{-1}y^{-3}}}$

87. $\dfrac{\left(\sqrt[3]{81mn^2}\right)^2}{\left(\sqrt[3]{mn}\right)^2}$

88. $\dfrac{\sqrt{44x^2y^9z}\,\sqrt{22y^9z^6}}{\left(\sqrt{11xy^8z^2}\right)^2}$

89. $\sqrt{a^2 - 3} - \dfrac{a^2}{\sqrt{a^2 - 3}}$

90. $5\sqrt{\dfrac{x}{y}} + 4\sqrt{\dfrac{y}{x}} - \dfrac{3}{\sqrt{xy}}$

91. Provide a reason for each step in the following derivation of the quotient rule:
$$\sqrt[n]{\dfrac{a}{b}} = \left(\dfrac{a}{b}\right)^{1/n} \quad \underline{\hspace{2cm}}$$
$$= \dfrac{a^{1/n}}{b^{1/n}} \quad \underline{\hspace{2cm}}$$
$$= \dfrac{\sqrt[n]{a}}{\sqrt[n]{b}} \quad \underline{\hspace{2cm}}$$

92. Show that $\dfrac{\sqrt[n]{a}}{\sqrt[n]{b}}$ is the nth root of $\dfrac{a}{b}$ by raising it to the nth power and simplifying.

93. Let $f(x) = \sqrt{18x^3}$ and $g(x) = \sqrt{2x}$. Find $(f/g)(x)$ and specify the domain of f/g.

94. Let $f(t) = \sqrt{2t}$ and $g(t) = \sqrt{50t^3}$. Find $(f/g)(t)$ and specify the domain of f/g.

95. Let $f(x) = \sqrt{x^2 - 9}$ and $g(x) = \sqrt{x - 3}$. Find $(f/g)(x)$ and specify the domain of f/g.

96. *Research.* A *Foucault pendulum* is designed to demonstrate the earth's rotation.

a) Find the lengths of several Foucault pendulums, typically found in museums or universities. Calculate the period of each pendulum.

b) Explain how a Foucault pendulum demonstrates the earth's rotation using words, pictures, or a model.

10.5 Expressions Containing Several Radical Terms

Adding and Subtracting Radical Expressions ▪ Products of Two or More Radical Terms ▪
Rationalizing Denominators or Numerators With Two Terms ▪ Terms with Differing Indices

Radical expressions like $6\sqrt{7} + 4\sqrt{7}$ or $(\sqrt{a} + \sqrt{b})(\sqrt{a} - \sqrt{b})$ contain more than one *radical term* and can sometimes be simplified.

Student Notes

Combining like radicals is similar to combining like terms. Recall the following:

$$3x + 8x = (3+8)x = 11x$$

and

$$6x^2 - 7x^2 = (6-7)x^2 = -x^2.$$

Adding and Subtracting Radical Expressions

When two radical expressions have the same indices and radicands, they are said to be **like radicals**. Like radicals can be combined (added or subtracted) in much the same way that we combine like terms.

EXAMPLE 1 Simplify by combining like radical terms.

a) $6\sqrt{7} + 4\sqrt{7}$

b) $6\sqrt[5]{4x} + 3\sqrt[5]{4x} - \sqrt[3]{4x}$

SOLUTION

a) $6\sqrt{7} + 4\sqrt{7} = (6+4)\sqrt{7}$ Using the distributive law (factoring out $\sqrt{7}$)

$\qquad\qquad\qquad = 10\sqrt{7}$ *Think*: 6 square roots of 7 plus 4 square roots of 7 is 10 square roots of 7.

b) $6\sqrt[5]{4x} + 3\sqrt[5]{4x} - \sqrt[3]{4x} = (6+3)\sqrt[5]{4x} - \sqrt[3]{4x}$ Try to do this step mentally.

$\qquad\qquad\qquad\qquad\qquad = 9\sqrt[5]{4x} - \sqrt[3]{4x}$ The indices are different. We cannot combine these terms.

1. Simplify by combining like radical terms:

$$2\sqrt{5} - 7\sqrt{5} + 4\sqrt{5}.$$

 YOUR TURN

Our ability to simplify radical expressions can help us to find like radicals even when, at first, it may appear that there are none.

EXAMPLE 2 Simplify by combining like radical terms, if possible.

a) $3\sqrt{8} - 5\sqrt{2}$ **b)** $9\sqrt{5} - 4\sqrt{3}$

SOLUTION

a) $3\sqrt{8} - 5\sqrt{2} = 3\sqrt{4 \cdot 2} - 5\sqrt{2}$

$\qquad = 3\sqrt{4} \cdot \sqrt{2} - 5\sqrt{2}$ } Simplifying $\sqrt{8}$

$\qquad = 3 \cdot 2 \cdot \sqrt{2} - 5\sqrt{2}$

$\qquad = 6\sqrt{2} - 5\sqrt{2}$

$\qquad = \sqrt{2}$ Combining like radicals

2. Simplify by combining like radical terms, if possible:

$5\sqrt{2} + 3\sqrt{8} + \sqrt{18}.$

b) $9\sqrt{5} - 4\sqrt{3}$ cannot be simplified.   The radicands are different.

YOUR TURN

If terms contain the same radical factor, we can factor out that radical expression in order to combine like terms, if possible.

EXAMPLE 3 Simplify by combining like terms, if possible.

a) $\sqrt[3]{2} - 7x\sqrt[3]{2} + 5\sqrt[3]{2}$ **b)** $\sqrt[3]{2x^6y^4} + 7\sqrt[3]{2y}$

SOLUTION

a) $\sqrt[3]{2} - 7x\sqrt[3]{2} + 5\sqrt[3]{2} = (1 - 7x + 5)\sqrt[3]{2}$ Factoring out $\sqrt[3]{2}$

$\qquad\qquad\qquad\qquad\quad = (6 - 7x)\sqrt[3]{2}$ These parentheses are important!

b) $\sqrt[3]{2x^6y^4} + 7\sqrt[3]{2y} = \sqrt[3]{x^6y^3 \cdot 2y} + 7\sqrt[3]{2y}$

$\qquad\qquad\qquad\quad = \sqrt[3]{x^6y^3} \cdot \sqrt[3]{2y} + 7\sqrt[3]{2y}$ } Simplifying $\sqrt[3]{2x^6y^4}$. $\sqrt[3]{2y}$ is a factor of both terms.

$\qquad\qquad\qquad\quad = x^2y \cdot \sqrt[3]{2y} + 7\sqrt[3]{2y}$

$\qquad\qquad\qquad\quad = (x^2y + 7)\sqrt[3]{2y}$ Factoring to combine like radical terms

3. Simplify by combining like radical terms, if possible:

$3\sqrt{x} + \sqrt{25x^3}.$

YOUR TURN

Products of Two or More Radical Terms

Radical expressions often contain products with factors that have more than one term. Multiplying such products is similar to multiplying polynomials. Some products will yield like radical terms, which we can now combine.

EXAMPLE 4 Multiply.

a) $\sqrt{3}(x - \sqrt{5})$ **b)** $\sqrt[3]{y}(\sqrt[3]{y^2} + \sqrt[3]{2})$ **c)** $(4 - \sqrt{7})^2$

d) $(4\sqrt{3} + \sqrt{2})(\sqrt{3} - 5\sqrt{2})$ **e)** $(\sqrt{a} + \sqrt{b})(\sqrt{a} - \sqrt{b})$

SOLUTION

a) $\sqrt{3}(x - \sqrt{5}) = \sqrt{3} \cdot x - \sqrt{3} \cdot \sqrt{5}$ Using the distributive law

$\qquad\qquad\quad = x\sqrt{3} - \sqrt{15}$ Multiplying radicals

b) $\sqrt[3]{y}(\sqrt[3]{y^2} + \sqrt[3]{2}) = \sqrt[3]{y} \cdot \sqrt[3]{y^2} + \sqrt[3]{y} \cdot \sqrt[3]{2}$ Using the distributive law

$\qquad\qquad\qquad\quad = \sqrt[3]{y^3} + \sqrt[3]{2y}$ Multiplying radicals

$\qquad\qquad\qquad\quad = y + \sqrt[3]{2y}$ Simplifying $\sqrt[3]{y^3}$

c) $(4 - \sqrt{7})^2 = (4 - \sqrt{7})(4 - \sqrt{7})$ We could also use the pattern $(A - B)^2 = A^2 - 2AB + B^2$.

$\qquad\qquad\qquad\quad\; \overset{\text{F}\qquad\text{O}\qquad\text{I}\qquad\text{L}}{}$

$\qquad\qquad = 4^2 - 4\sqrt{7} - 4\sqrt{7} + (\sqrt{7})^2$

$\qquad\qquad = 16 - 8\sqrt{7} + 7$ Squaring and combining like terms

$\qquad\qquad = 23 - 8\sqrt{7}$ Adding 16 and 7

$$\overset{\text{F}\qquad\qquad\text{O}\qquad\qquad\text{I}\qquad\qquad\text{L}}{}$$

d) $(4\sqrt{3} + \sqrt{2})(\sqrt{3} - 5\sqrt{2}) = 4(\sqrt{3})^2 - 20\sqrt{3} \cdot \sqrt{2} + \sqrt{2} \cdot \sqrt{3} - 5(\sqrt{2})^2$

$= 4 \cdot 3 - 20\sqrt{6} + \sqrt{6} - 5 \cdot 2$ Multiplying radicals

$= 12 - 20\sqrt{6} + \sqrt{6} - 10$

$= 2 - 19\sqrt{6}$ Combining like terms

e) $(\sqrt{a} + \sqrt{b})(\sqrt{a} - \sqrt{b}) = (\sqrt{a})^2 - \sqrt{a}\sqrt{b} + \sqrt{a}\sqrt{b} - (\sqrt{b})^2$ Using FOIL

$= a - b$ Combining like terms

4. Multiply: $(7 + \sqrt{x})^2$.

 YOUR TURN

In Example 4(e) above, note that the outer and inner products in FOIL are opposites, so that $a - b$ is not itself a radical expression. Pairs of radical expressions like $\sqrt{a} + \sqrt{b}$ and $\sqrt{a} - \sqrt{b}$ are called **conjugates**.

Rationalizing Denominators or Numerators with Two Terms

The use of conjugates allows us to rationalize denominators or numerators that contain two terms.

EXAMPLE 5   Write an equivalent expression with a rationalized denominator.

a) $\dfrac{4}{7 + \sqrt{3}}$

b) $\dfrac{4 + \sqrt{2}}{\sqrt{5} - \sqrt{2}}$

SOLUTION

a) $\dfrac{4}{7 + \sqrt{3}} = \dfrac{4}{7 + \sqrt{3}} \cdot \dfrac{7 - \sqrt{3}}{7 - \sqrt{3}}$ Multiplying by 1, using the conjugate of $7 + \sqrt{3}$, which is $7 - \sqrt{3}$

$= \dfrac{4(7 - \sqrt{3})}{(7 + \sqrt{3})(7 - \sqrt{3})}$ Multiplying numerators and denominators

$= \dfrac{4(7 - \sqrt{3})}{7^2 - (\sqrt{3})^2}$ Using $(A + B)(A - B) = A^2 - B^2$

$= \dfrac{28 - 4\sqrt{3}}{49 - 3}$ No radicals remain in the denominator.

$\left.\begin{array}{l} = \dfrac{28 - 4\sqrt{3}}{46} \\[2ex] = \dfrac{2(14 - 2\sqrt{3})}{2 \cdot 23} \\[2ex] = \dfrac{14 - 2\sqrt{3}}{23} \end{array}\right\}$ Simplifying

b) $\dfrac{4 + \sqrt{2}}{\sqrt{5} - \sqrt{2}} = \dfrac{4 + \sqrt{2}}{\sqrt{5} - \sqrt{2}} \cdot \dfrac{\sqrt{5} + \sqrt{2}}{\sqrt{5} + \sqrt{2}}$ Multiplying by 1, using the conjugate of $\sqrt{5} - \sqrt{2}$, which is $\sqrt{5} + \sqrt{2}$

$= \dfrac{(4 + \sqrt{2})(\sqrt{5} + \sqrt{2})}{(\sqrt{5} - \sqrt{2})(\sqrt{5} + \sqrt{2})}$ Multiplying numerators and denominators

$= \dfrac{4\sqrt{5} + 4\sqrt{2} + \sqrt{2}\sqrt{5} + (\sqrt{2})^2}{(\sqrt{5})^2 - (\sqrt{2})^2}$ Multiplying

$= \dfrac{4\sqrt{5} + 4\sqrt{2} + \sqrt{10} + 2}{5 - 2}$ Squaring in the denominator and the numerator

$= \dfrac{4\sqrt{5} + 4\sqrt{2} + \sqrt{10} + 2}{3}$ No radicals remain in the denominator.

5. Rationalize the denominator:

$\dfrac{\sqrt{2}}{\sqrt{3} - \sqrt{2}}.$

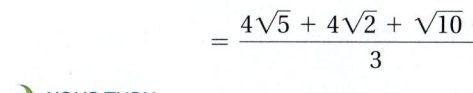 YOUR TURN

To rationalize a numerator with two terms, we use the conjugate of the numerator.

EXAMPLE 6 Rationalize the numerator: $\dfrac{4 + \sqrt{2}}{\sqrt{5} - \sqrt{2}}$.

SOLUTION

$$\frac{4 + \sqrt{2}}{\sqrt{5} - \sqrt{2}} = \frac{4 + \sqrt{2}}{\sqrt{5} - \sqrt{2}} \cdot \frac{4 - \sqrt{2}}{4 - \sqrt{2}} \qquad \text{Multiplying by 1, using the conjugate of } 4 + \sqrt{2}, \text{ which is } 4 - \sqrt{2}$$

$$= \frac{16 - (\sqrt{2})^2}{4\sqrt{5} - \sqrt{5}\sqrt{2} - 4\sqrt{2} + (\sqrt{2})^2}$$

$$= \frac{14}{4\sqrt{5} - \sqrt{10} - 4\sqrt{2} + 2}$$

6. Rationalize the numerator:

$$\frac{5 - \sqrt{y}}{\sqrt{7}}.$$

 YOUR TURN

CONNECTING 🔗 THE CONCEPTS

To rationalize denominators with one term or those with two terms, we multiply by 1.

One Term

Multiply by 1, using the factor(s) needed to make the denominator a perfect *n*th power.

$$\frac{3}{\sqrt{5}} = \frac{3}{\sqrt{5}} \cdot \frac{\sqrt{5}}{\sqrt{5}} = \frac{3\sqrt{5}}{5}$$

Two Terms

Multiply by 1, using the conjugate of the denominator.

$$\frac{3}{7 + \sqrt{5}} = \frac{3}{7 + \sqrt{5}} \cdot \frac{7 - \sqrt{5}}{7 - \sqrt{5}} = \frac{21 - 3\sqrt{5}}{44}$$

EXERCISES

Write an equivalent expression with a rationalized denominator.

1. $\dfrac{6}{\sqrt{7}}$

2. $\dfrac{1}{3 - \sqrt{2}}$

3. $\dfrac{2}{\sqrt{xy}}$

4. $\dfrac{5}{\sqrt{8}}$

5. $\dfrac{\sqrt{2}}{\sqrt{5} + \sqrt{3}}$

6. $\dfrac{2}{1 - \sqrt{5}}$

7. $\dfrac{1}{\sqrt[3]{x^2y}}$

8. $\dfrac{a}{\sqrt[4]{a^3b^2}}$

Terms with Differing Indices

If radical terms have different indices, we cannot multiply or divide the terms by simply multiplying or dividing radicands. However, we can often convert to exponential notation, use the rules for exponents, and then convert back to radical notation.

EXAMPLE 7 Divide and, if possible, simplify: $\dfrac{\sqrt[4]{(x+y)^3}}{\sqrt{x+y}}$.

SOLUTION

$$\dfrac{\sqrt[4]{(x+y)^3}}{\sqrt{x+y}} = \dfrac{(x+y)^{3/4}}{(x+y)^{1/2}}$$ Converting to exponential notation

$$= (x+y)^{3/4-1/2}$$ Since the bases are identical, we can subtract exponents: $\frac{3}{4} - \frac{1}{2} = \frac{3}{4} - \frac{2}{4} = \frac{1}{4}.$

$$\left.\begin{array}{l} = (x+y)^{1/4} \\ = \sqrt[4]{x+y} \end{array}\right\}$$ Converting back to radical notation

7. Divide and, if possible, simplify:

$$\dfrac{\sqrt{3x}}{\sqrt[5]{(3x)^2}}.$$

 YOUR TURN

TO SIMPLIFY PRODUCTS OR QUOTIENTS WITH DIFFERING INDICES

1. Convert all radical expressions to exponential notation.
2. When the bases are identical, subtract exponents to divide and add exponents to multiply. This may require finding a common denominator for the exponents.
3. Convert back to radical notation and, if possible, simplify.

EXAMPLE 8 Multiply and simplify: $\sqrt{x^3}\sqrt[3]{x}$.

SOLUTION

$$\sqrt{x^3}\sqrt[3]{x} = x^{3/2} \cdot x^{1/3}$$ Converting to exponential notation

$$= x^{11/6}$$ Adding exponents: $\frac{3}{2} + \frac{1}{3} = \frac{9}{6} + \frac{2}{6}$

$$= \sqrt[6]{x^{11}}$$ Converting back to radical notation

$$\left.\begin{array}{l} = \sqrt[6]{x^6}\sqrt[6]{x^5} \\ = x\sqrt[6]{x^5} \end{array}\right\}$$ Simplifying

8. Multiply and simplify:

$$\sqrt{x}\sqrt[3]{x^4}.$$

 YOUR TURN

EXAMPLE 9 If $f(x) = \sqrt[3]{x^2}$ and $g(x) = \sqrt{x} + \sqrt[4]{x}$, find $(f \cdot g)(x)$.

SOLUTION Recall that $(f \cdot g)(x) = f(x) \cdot g(x)$. Thus,

$$(f \cdot g)(x) = \sqrt[3]{x^2}\left(\sqrt{x} + \sqrt[4]{x}\right)$$ x is assumed to be nonnegative.

$$= x^{2/3}(x^{1/2} + x^{1/4})$$ Converting to exponential notation

$$= x^{2/3} \cdot x^{1/2} + x^{2/3} \cdot x^{1/4}$$ Using the distributive law

$$= x^{2/3+1/2} + x^{2/3+1/4}$$ Adding exponents:

$$= x^{7/6} + x^{11/12}$$ $\frac{2}{3} + \frac{1}{2} = \frac{4}{6} + \frac{3}{6}; \frac{2}{3} + \frac{1}{4} = \frac{8}{12} + \frac{3}{12}$

$$= \sqrt[6]{x^7} + \sqrt[12]{x^{11}}$$ Converting back to radical notation

$$\left.\begin{array}{l} = \sqrt[6]{x^6}\sqrt[6]{x} + \sqrt[12]{x^{11}} \\ = x\sqrt[6]{x} + \sqrt[12]{x^{11}}. \end{array}\right\}$$ Simplifying

9. If $f(x) = 2\sqrt{x}$ and $g(x) = \sqrt[3]{x} - \sqrt[5]{x^2}$, find $(f \cdot g)(x)$.

YOUR TURN

We often can write the final result as a single radical expression by finding a common denominator in the exponents.

EXAMPLE 10 Divide and, if possible, simplify: $\dfrac{\sqrt[3]{a^2b^4}}{\sqrt{ab}}$.

SOLUTION

$$\frac{\sqrt[3]{a^2b^4}}{\sqrt{ab}} = \frac{(a^2b^4)^{1/3}}{(ab)^{1/2}} \qquad \text{Converting to exponential notation}$$

$$= \frac{a^{2/3}b^{4/3}}{a^{1/2}b^{1/2}} \qquad \text{Using the product and power rules}$$

$$= a^{2/3-1/2}b^{4/3-1/2} \qquad \text{Subtracting exponents}$$

$$= a^{1/6}b^{5/6}$$

$$= \sqrt[6]{a}\,\sqrt[6]{b^5} \qquad \text{Converting to radical notation}$$

$$= \sqrt[6]{ab^5} \qquad \text{Using the product rule for radicals}$$

10. Divide and, if possible, simplify:

$$\dfrac{\sqrt[5]{x^{15}y^7}}{\sqrt{x^2y}}.$$

YOUR TURN

10.5 EXERCISE SET

FOR EXTRA HELP

MyMathLab® MathXL

PRACTICE WATCH READ REVIEW

Vocabulary and Reading Check

For Exercises 1–6, fill in each blank by selecting from the following words (which may be used more than once).

base(s)   indice(s)
conjugate(s)   numerator(s)
denominator(s)   radicand(s)

1. To add radical expressions, both the _____ and the _____ must be the same.

2. To multiply radicands, the _____ must be the same.

3. To find a product by adding exponents, the _____ must be the same.

4. To add the numerators of rational expressions, the _____ must be the same.

5. To rationalize the _____ of $\dfrac{\sqrt{c} - \sqrt{a}}{5}$, we multiply by a form of 1, using the _____ of $\sqrt{c} - \sqrt{a}$, which is $\sqrt{c} + \sqrt{a}$, to write 1.

6. To find a quotient by subtracting exponents, the _____ must be the same.

Adding and Subtracting Radical Expressions

Add or subtract. Simplify by combining like radical terms, if possible. Assume that all variables and radicands represent positive real numbers.

7. $4\sqrt{3} + 7\sqrt{3}$

8. $6\sqrt{5} + 2\sqrt{5}$

9. $7\sqrt[3]{4} - 5\sqrt[3]{4}$

10. $14\sqrt[5]{2} - 8\sqrt[5]{2}$

11. $\sqrt[3]{y} + 9\sqrt[3]{y}$

12. $4\sqrt[4]{t} - \sqrt[4]{t}$

13. $8\sqrt{2} - \sqrt{2} + 5\sqrt{2}$

14. $\sqrt{6} + 3\sqrt{6} - 8\sqrt{6}$

15. $9\sqrt[3]{7} - \sqrt{3} + 4\sqrt[3]{7} + 2\sqrt{3}$

16. $5\sqrt{7} - 8\sqrt[4]{11} + \sqrt{7} + 9\sqrt[4]{11}$

17. $4\sqrt{27} - 3\sqrt{3}$

18. $9\sqrt{50} - 4\sqrt{2}$

19. $3\sqrt{45} - 8\sqrt{20}$

20. $5\sqrt{12} + 16\sqrt{27}$

21. $3\sqrt[3]{16} + \sqrt[3]{54}$

22. $\sqrt[3]{27} - 5\sqrt[3]{8}$

23. $\sqrt{a} + 3\sqrt{16a^3}$

24. $2\sqrt{9x^3} - \sqrt{x}$

25. $\sqrt[3]{6x^4} - \sqrt[3]{48x}$

26. $\sqrt[3]{54x} - \sqrt[3]{2x^4}$

27. $\sqrt{4a - 4} + \sqrt{a - 1}$

28. $\sqrt{9y + 27} + \sqrt{y + 3}$

29. $\sqrt{x^3 - x^2} + \sqrt{9x - 9}$

30. $\sqrt{4x - 4} - \sqrt{x^3 - x^2}$

Products of Two or More Radical Terms

Multiply. Assume that all variables represent nonnegative real numbers.

31. $\sqrt{2}(5 + \sqrt{2})$

32. $\sqrt{3}(6 - \sqrt{3})$

33. $3\sqrt{5}(\sqrt{6} - \sqrt{7})$

34. $4\sqrt{2}(\sqrt{3} + \sqrt{5})$

35. $\sqrt{2}(3\sqrt{10} - \sqrt{8})$

36. $\sqrt{3}(2\sqrt{15} - 3\sqrt{4})$

37. $\sqrt[3]{3}(\sqrt[3]{9} - 4\sqrt[3]{21})$

38. $\sqrt[3]{2}(\sqrt[3]{4} - 2\sqrt[3]{32})$

39. $\sqrt[3]{a}\left(\sqrt[3]{a^2} + \sqrt[3]{24a^2}\right)$

40. $\sqrt[3]{x}\left(\sqrt[3]{3x^2} - \sqrt[3]{81x^2}\right)$

41. $(2 + \sqrt{6})(5 - \sqrt{6})$

42. $(4 - \sqrt{5})(2 + \sqrt{5})$

43. $(\sqrt{2} + \sqrt{7})(\sqrt{3} - \sqrt{7})$

44. $(\sqrt{7} - \sqrt{2})(\sqrt{5} + \sqrt{2})$

45. $(2 - \sqrt{3})(2 + \sqrt{3})$

46. $(3 + \sqrt{11})(3 - \sqrt{11})$

47. $(\sqrt{10} - \sqrt{15})(\sqrt{10} + \sqrt{15})$

48. $(\sqrt{12} + \sqrt{5})(\sqrt{12} - \sqrt{5})$

49. $(3\sqrt{7} + 2\sqrt{5})(2\sqrt{7} - 4\sqrt{5})$

50. $(4\sqrt{5} - 3\sqrt{2})(2\sqrt{5} + 4\sqrt{2})$

51. $(4 + \sqrt{7})^2$ **52.** $(3 + \sqrt{10})^2$

53. $(\sqrt{3} - \sqrt{2})^2$ **54.** $(\sqrt{5} - \sqrt{3})^2$

55. $(\sqrt{2t} + \sqrt{5})^2$ **56.** $(\sqrt{3x} - \sqrt{2})^2$

57. $(3 - \sqrt{x + 5})^2$ **58.** $(4 + \sqrt{x - 3})^2$

59. $\left(2\sqrt[4]{7} - \sqrt[4]{6}\right)\left(3\sqrt[4]{9} + 2\sqrt[4]{5}\right)$

60. $\left(4\sqrt[3]{3} + \sqrt[3]{10}\right)\left(2\sqrt[3]{7} + 5\sqrt[3]{6}\right)$

Rationalizing Denominators or Numerators with Two Terms

Rationalize each denominator. If possible, simplify your result.

61. $\dfrac{6}{3 - \sqrt{2}}$ **62.** $\dfrac{5}{4 - \sqrt{5}}$

63. $\dfrac{2 + \sqrt{5}}{6 + \sqrt{3}}$ **64.** $\dfrac{1 + \sqrt{2}}{3 + \sqrt{5}}$

65. $\dfrac{\sqrt{a}}{\sqrt{a} + \sqrt{b}}$ **66.** $\dfrac{\sqrt{z}}{\sqrt{x} - \sqrt{z}}$

Aha! 67. $\dfrac{\sqrt{7} - \sqrt{3}}{\sqrt{3} - \sqrt{7}}$ **68.** $\dfrac{\sqrt{7} + \sqrt{5}}{\sqrt{5} + \sqrt{2}}$

69. $\dfrac{3\sqrt{2} - \sqrt{7}}{4\sqrt{2} + 2\sqrt{5}}$ **70.** $\dfrac{5\sqrt{3} - \sqrt{11}}{2\sqrt{3} - 5\sqrt{2}}$

Rationalize each numerator. If possible, simplify your result.

71. $\dfrac{\sqrt{5} + 1}{4}$ **72.** $\dfrac{\sqrt{15} - 3}{6}$

73. $\dfrac{\sqrt{6} - 2}{\sqrt{3} + 7}$ **74.** $\dfrac{\sqrt{10} + 4}{\sqrt{2} - 3}$

75. $\dfrac{\sqrt{x} - \sqrt{y}}{\sqrt{x} + \sqrt{y}}$ **76.** $\dfrac{\sqrt{a} + \sqrt{b}}{\sqrt{a} - \sqrt{b}}$

77. $\dfrac{\sqrt{a + h} - \sqrt{a}}{h}$ **78.** $\dfrac{\sqrt{x - h} - \sqrt{x}}{h}$

Terms with Differing Indices

Perform the indicated operation and simplify. Assume that all variables represent positive real numbers.

79. $\sqrt[3]{a}\sqrt[6]{a}$ **80.** $\sqrt[10]{a}\sqrt[5]{a^2}$

81. $\sqrt{b^3}\sqrt[5]{b^4}$ **82.** $\sqrt[3]{b^4}\sqrt[4]{b^3}$

83. $\sqrt{xy^3}\,\sqrt[3]{x^2y}$ **84.** $\sqrt[5]{a^3b}\,\sqrt{ab}$

85. $\sqrt[4]{9ab^3}\sqrt{3a^4b}$ **86.** $\sqrt{2x^3y^3}\,\sqrt[3]{4xy^2}$

87. $\sqrt{a^4b^3c^4}\,\sqrt[3]{ab^2c}$ **88.** $\sqrt[3]{xy^2z}\sqrt{x^3yz^2}$

89. $\dfrac{\sqrt[3]{a^2}}{\sqrt[4]{a}}$ **90.** $\dfrac{\sqrt[3]{x^2}}{\sqrt[5]{x}}$

91. $\dfrac{\sqrt[4]{x^2y^3}}{\sqrt[3]{xy}}$ **92.** $\dfrac{\sqrt[5]{a^4b}}{\sqrt[3]{ab}}$

93. $\dfrac{\sqrt{ab^3}}{\sqrt[5]{a^2b^3}}$ **94.** $\dfrac{\sqrt[5]{x^3y^4}}{\sqrt{xy}}$

95. $\dfrac{\sqrt{(7 - y)^3}}{\sqrt[3]{(7 - y)^2}}$ **96.** $\dfrac{\sqrt[5]{(y - 9)^3}}{\sqrt{y - 9}}$

97. $\dfrac{\sqrt[4]{(5 + 3x)^3}}{\sqrt[3]{(5 + 3x)^2}}$ **98.** $\dfrac{\sqrt[3]{(2x + 1)^2}}{\sqrt[5]{(2x + 1)^2}}$

99. $\sqrt[3]{x^2y}\left(\sqrt{xy} - \sqrt[5]{xy^3}\right)$

100. $\sqrt[4]{a^2b}\left(\sqrt[3]{a^2b} - \sqrt[5]{a^2b^2}\right)$

101. $\left(m + \sqrt[3]{n^2}\right)\left(2m + \sqrt[4]{n}\right)$

102. $\left(r - \sqrt[4]{s^3}\right)\left(3r - \sqrt[5]{s}\right)$

Products of Two or More Radical Terms

In Exercises 103–106, $f(x)$ and $g(x)$ are as given. Find $(f \cdot g)(x)$. Assume that all variables represent nonnegative real numbers.

103. $f(x) = \sqrt[4]{x}$, $g(x) = 2\sqrt{x} - \sqrt[3]{x^2}$

104. $f(x) = \sqrt[4]{2x} + 5\sqrt{2x}$, $g(x) = \sqrt[3]{2x}$

105. $f(x) = x + \sqrt{7}$, $g(x) = x - \sqrt{7}$

106. $f(x) = x - \sqrt{2}$, $g(x) = x + \sqrt{6}$

Let $f(x) = x^2$. Find each of the following.

107. $f(3 - \sqrt{2})$

108. $f(5 - \sqrt{3})$

109. $f(\sqrt{6} + \sqrt{21})$

110. $f(\sqrt{2} + \sqrt{10})$

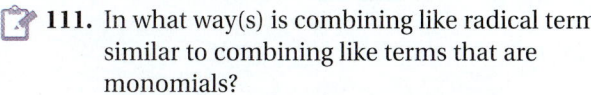 **111.** In what way(s) is combining like radical terms similar to combining like terms that are monomials?

112. Why do we need to know how to multiply radical expressions before learning how to add them?

Skill Review

113. In which quadrant is the point $\left(6, -\frac{1}{2}\right)$ located? [2.1]

114. Find the slope of the line containing the points $(9, 10)$ and $(6, 7)$. [2.3]

115. Find the x-intercept and the y-intercept of the line given by $x - y = 10$. [2.4]

116. Find the slope and the y-intercept of the line given by $3y + 5x = 1$. [2.3]

117. Write the slope–intercept equation of the line that is perpendicular to the line $y = \frac{1}{2}x - 7$ and has a y-intercept of $(0, 12)$. [2.5]

118. Write the slope–intercept equation of the line that contains the points $(-1, -6)$ and $(-4, 0)$. [2.5]

Synthesis

 119. Nadif *incorrectly* writes

$$\sqrt[5]{x^2} \cdot \sqrt{x^3} = x^{2/5} \cdot x^{3/2} = \sqrt[5]{x^3}.$$

What mistake do you suspect he is making?

 120. After examining the expression $\sqrt[4]{25xy^3}\,\sqrt{5x^4y}$, Sofia (correctly) concludes that x and y are both nonnegative. Explain how she could reach this conclusion.

Find a simplified form for $f(x)$. Assume $x \geq 0$.

121. $f(x) = \sqrt{x^3 - x^2} + \sqrt{9x^3 - 9x^2} - \sqrt{4x^3 - 4x^2}$

122. $f(x) = \sqrt{20x^2 + 4x^3} - 3x\sqrt{45 + 9x} + \sqrt{5x^2 + x^3}$

123. $f(x) = \sqrt[4]{x^5 - x^4} + 3\sqrt[4]{x^9 - x^8}$

124. $f(x) = \sqrt[4]{16x^4 + 16x^5} - 2\sqrt[4]{x^8 + x^9}$

Simplify.

125. $7x\sqrt{(x+y)^3} - 5xy\sqrt{x+y} - 2y\sqrt{(x+y)^3}$

126. $\sqrt{27a^5(b+1)}\,\sqrt[3]{81a(b+1)^4}$

127. $\sqrt{8x(y+z)^5}\,\sqrt[3]{4x^2(y+z)^2}$

128. $\frac{1}{2}\sqrt{36a^5bc^4} - \frac{1}{2}\sqrt[3]{64a^4bc^6} + \frac{1}{6}\sqrt{144a^3bc^6}$

129. $\dfrac{\dfrac{1}{\sqrt{w}} - \sqrt{w}}{\dfrac{\sqrt{w} + 1}{\sqrt{w}}}$

130. $\dfrac{1}{4 + \sqrt{3}} + \dfrac{1}{\sqrt{3}} + \dfrac{1}{\sqrt{3} - 4}$

Express each of the following as the product of two radical expressions.

131. $x - 5$

132. $y - 7$

133. $x - a$

Multiply.

134. $\sqrt{9 + 3\sqrt{5}}\,\sqrt{9 - 3\sqrt{5}}$

135. $\left(\sqrt{x+2} - \sqrt{x-2}\right)^2$

 136. Use a graphing calculator to check your answers to Exercises 25, 39, and 81.

137. A formula for factoring a sum of squares is

$$A^2 + B^2 = (A + B + \sqrt{2AB})(A + B - \sqrt{2AB}).$$

a) Show that this is an identity.

b) What must be true of A and B in order for $A^2 + B^2$ to be written as the product of two binomials with rational coefficients?

↪ YOUR TURN ANSWERS: SECTION 10.5

1. $-\sqrt{5}$ **2.** $14\sqrt{2}$ **3.** $(3 + 5x)\sqrt{x}$ **4.** $49 + 14\sqrt{x} + x$

5. $\sqrt{6} + 2$ **6.** $\dfrac{25 - y}{5\sqrt{7} + \sqrt{7y}}$ **7.** $\sqrt[10]{3x}$ **8.** $x\sqrt[6]{x^5}$

9. $2\sqrt[6]{x^5} - 2\sqrt[10]{x^9}$ **10.** $x^2\sqrt[10]{y^9}$

QUICK QUIZ: SECTIONS 10.1–10.5

1. Simplify $\sqrt{100t^2 + 20t + 1}$. Assume that t can represent any real number. [10.1]

2. Write an equivalent expression using radical notation: $(3xy)^{7/8}$. [10.2]

3. Write an equivalent expression using exponential notation: $\sqrt{17ab}$. [10.2]

4. Rationalize the denominator: $\sqrt[3]{\dfrac{5}{4x^5y}}$. [10.4]

5. Rationalize the numerator: $\dfrac{2 + \sqrt{3}}{\sqrt{10}}$. [10.5]

PREPARE TO MOVE ON

Solve.

1. $3x - 1 = 125$ [2.2]

2. $x^2 + 2x + 1 = 2(11 - x)$ [5.7]

3. $-6 = 5 - x$ [2.2]

4. $x^2 - 10x + 25 = 4(x - 2)$ [5.7]

Mid-Chapter Review

Many radical expressions can be simplified. It is important to know under which conditions radical expressions can be multiplied and divided and radical terms can be combined.

Multiplication and division: The indices must be the same.

$$\frac{\sqrt{50t^5}}{\sqrt{2t^{11}}} = \sqrt{\frac{50t^5}{2t^{11}}} = \sqrt{\frac{25}{t^6}} = \frac{5}{t^3}; \qquad \sqrt[4]{8x^3} \cdot \sqrt[4]{2x} = \sqrt[4]{16x^4} = 2x$$

Combining like radical terms: The indices and the radicands must both be the same.

$$\sqrt{75x} + \sqrt{12x} - \sqrt{3x} = 5\sqrt{3x} + 2\sqrt{3x} - \sqrt{3x} = 6\sqrt{3x}$$

Radical expressions with differing indices can sometimes be simplified using rational exponents.

$$\sqrt[3]{x^2}\sqrt{x} = x^{2/3}x^{1/2} = x^{4/6}x^{3/6} = x^{7/6} = \sqrt[6]{x^7} = x\sqrt[6]{x}$$

GUIDED SOLUTIONS

1. Multiply and simplify: $\sqrt{6x^9} \cdot \sqrt{2xy}$. [10.3]

Solution

$$\sqrt{6x^9} \cdot \sqrt{2xy} = \sqrt{6x^9 \cdot \boxed{}}$$

$$= \sqrt{12x^{10}y}$$

$$= \sqrt{\boxed{} \cdot 3y}$$

$$= \sqrt{\boxed{}} \cdot \sqrt{3y}$$

$$= \boxed{}\sqrt{3y} \qquad \text{Taking the square root}$$

2. Combine like radical terms:

$$\sqrt{12} - 3\sqrt{75} + \sqrt{8}. \quad [10.5]$$

Solution

$$\sqrt{12} - 3\sqrt{75} + \sqrt{8}$$

$$= \boxed{} - 3 \cdot 5\sqrt{3} + \boxed{} \qquad \text{Simplifying each term}$$

$$= 2\sqrt{3} - \boxed{} + 2\sqrt{2} \qquad \text{Multiplying}$$

$$= \boxed{} + 2\sqrt{2} \qquad \text{Combining like radical terms}$$

MIXED REVIEW

Simplify. Assume that variables can represent any real number.

3. $\sqrt{81}$ [10.1]

4. $-\sqrt{\dfrac{9}{100}}$ [10.1]

5. $\sqrt{64t^2}$ [10.1]

6. $\sqrt[5]{x^5}$ [10.1]

7. Find $f(-5)$ if $f(x) = \sqrt[3]{12x - 4}$. [10.1]

8. Determine the domain of g if $g(x) = \sqrt[4]{10 - x}$. [10.1]

9. Write an equivalent expression using radical notation and simplify: $8^{2/3}$. [10.2]

Simplify. Assume for Exercises 10–25 that no radicands were formed by raising negative numbers to even powers.

10. $\sqrt[6]{\sqrt{a}}$ [10.2]

11. $\sqrt[3]{y^{24}}$ [10.2]

12. $\sqrt{(t + 5)^2}$ [10.1]

13. $\sqrt[3]{-27a^{12}}$ [10.1]

14. $\sqrt{6x}\,\sqrt{15x}$ [10.3]

15. $\dfrac{\sqrt{20y}}{\sqrt{45y}}$ [10.4]

16. $\sqrt{6}(\sqrt{10} - \sqrt{33})$ [10.5]

17. $\dfrac{\sqrt{t}}{\sqrt[8]{t^3}}$ [10.5]

18. $\sqrt[5]{\dfrac{3a^{12}}{96a^2}}$ [10.4]

19. $2\sqrt{3} - 5\sqrt{12}$ [10.5]

20. $(\sqrt{5} + 3)(\sqrt{5} - 3)$ [10.5]

21. $(\sqrt{15} + \sqrt{10})^2$ [10.5]

22. $\sqrt{25x - 25} - \sqrt{9x - 9}$ [10.5]

23. $\sqrt{x^3y}\,\sqrt[3]{xy^4}$ [10.5]

24. $\sqrt[3]{5000} + \sqrt[3]{625}$ [10.5]

25. $\sqrt[3]{12x^2y^5}\,\sqrt[3]{18x^7y}$ [10.3]

10.6 Solving Radical Equations

The Principle of Powers ■ Equations with Two Radical Terms

Now that we know how to work with radicals and rational exponents, we can learn how to solve a new type of equation.

The Principle of Powers

A **radical equation** is an equation in which the variable appears in a radicand. Examples are

$$\sqrt[3]{2x} + 1 = 5, \quad \sqrt{a - 2} = 7, \quad \text{and} \quad 4 - \sqrt{3x + 1} = \sqrt{6 - x}.$$

To solve such equations, we need a new principle. Suppose $a = b$ is true. If we square both sides, we get another true equation: $a^2 = b^2$. This can be generalized.

> **THE PRINCIPLE OF POWERS**
>
> If $a = b$, then $a^n = b^n$ for any exponent n.

Note that the principle of powers is an "if–then" statement. If we interchange the two parts of the sentence, then we have the statement "If $a^n = b^n$ for some exponent n, then $a = b$." **This statement is not always true**. For example, "if $x = 3$, then $x^2 = 9$" is true, but the statement "if $x^2 = 9$, then $x = 3$" is *not* true when x is replaced with -3.

When n is even, every solution of $x = a$ is a solution of $x^n = a^n$, but not every solution of $x^n = a^n$ is a solution of $x = a$.

> When we raise both sides of an equation to an even exponent, it is essential that we check the answer in the *original* equation.

EXAMPLE 1 Solve: $\sqrt{x} - 3 = 4$.

SOLUTION Before using the principle of powers, we must isolate the radical term:

$$\sqrt{x} - 3 = 4$$
$$\sqrt{x} = 7 \qquad \text{Isolating the radical by adding 3 to both sides}$$
$$(\sqrt{x})^2 = 7^2 \qquad \text{Using the principle of powers}$$
$$x = 49.$$

Check:
$$\begin{array}{c|c} \sqrt{x} - 3 = 4 & \\ \hline \sqrt{49} - 3 & 4 \\ 7 - 3 & \\ 4 \overset{?}{=} 4 & \text{TRUE} \end{array}$$

The solution is 49.

1. Solve: $\sqrt{x + 1} - 3 = 2$.

↵ YOUR TURN

It is important that we isolate a radical term before using the principle of powers. Suppose in Example 1 that both sides of the equation were squared *before* we isolated the radical. We would have had the expression $(\sqrt{x} - 3)^2$ or $x - 6\sqrt{x} + 9$ on the left side, and the radical would have remained in the problem.

EXAMPLE 2   Solve: $\sqrt{x} + 5 = 3$.

SOLUTION

$$\sqrt{x} + 5 = 3$$

$$\sqrt{x} = -2 \qquad \text{Isolating the radical by adding } -5 \text{ to both sides}$$

> The equation $\sqrt{x} = -2$ has no solution because the principal square root of a number is never negative. We continue as in Example 1 for comparison.

$$(\sqrt{x})^2 = (-2)^2 \qquad \text{Using the principle of powers}$$

$$x = 4$$

Check:
$$\sqrt{x} + 5 = 3$$
$$\begin{array}{c|c} \sqrt{4} + 5 & 3 \\ 2 + 5 & \\ 7 \overset{?}{=} 3 & \text{FALSE} \end{array}$$

The number 4 does not check. Thus, $\sqrt{x} + 5 = 3$ has no solution.

2. Solve: $\sqrt{x} + 6 = 1$.

YOUR TURN

Note in Example 2 that $x = 4$ has the solution 4, but $\sqrt{x} + 5 = 3$ has *no* solution. Thus, $x = 4$ and $\sqrt{x} + 5 = 3$ are *not* equivalent equations.

TO SOLVE AN EQUATION WITH A RADICAL TERM

1. Isolate the radical term on one side of the equation.
2. Use the principle of powers and solve the resulting equation.
3. Check any possible solution in the original equation.

EXAMPLE 3   Solve: $x = \sqrt{x + 7} + 5$.

SOLUTION

$$x = \sqrt{x + 7} + 5$$

$$x - 5 = \sqrt{x + 7} \qquad \text{Isolating the radical by subtracting 5 from both sides}$$

$$\left. \begin{array}{l} (x - 5)^2 = (\sqrt{x + 7})^2 \\ x^2 - 10x + 25 = x + 7 \end{array} \right\} \qquad \text{Using the principle of powers; squaring both sides}$$

$$x^2 - 11x + 18 = 0 \qquad \text{Adding } -x - 7 \text{ to both sides to form a quadratic equation in standard form}$$

$$(x - 9)(x - 2) = 0 \qquad \text{Factoring}$$

$$x = 9 \quad or \quad x = 2 \qquad \text{Using the principle of zero products}$$

The possible solutions are 9 and 2. Let's check.

Check:

For 9:

$$x = \sqrt{x + 7} + 5$$

$$\frac{}{9 \mid \sqrt{9 + 7} + 5}$$

$$9 \overset{?}{=} 9 \qquad \text{TRUE}$$

For 2:

$$x = \sqrt{x + 7} + 5$$

$$\frac{}{2 \mid \sqrt{2 + 7} + 5}$$

$$2 \overset{?}{=} 8 \qquad \text{FALSE}$$

3. Solve: $x = \sqrt{x + 10} + 10$.

Since 9 checks but 2 does not, the solution is 9.

YOUR TURN

EXAMPLE 4 Solve: $(2x + 1)^{1/3} + 5 = 0$.

SOLUTION We can use exponential notation to solve:

$$(2x + 1)^{1/3} + 5 = 0$$

$$(2x + 1)^{1/3} = -5 \qquad \text{Subtracting 5 from both sides}$$

$$[(2x + 1)^{1/3}]^3 = (-5)^3 \qquad \text{Cubing both sides}$$

$$(2x + 1)^1 = (-5)^3 \qquad \text{Multiplying exponents}$$

$$2x + 1 = -125$$

$$2x = -126 \qquad \text{Subtracting 1 from both sides}$$

$$x = -63.$$

Student Notes

In Example 4, $(2x + 1)^{1/3}$ can also be written $\sqrt[3]{2x + 1}$. Then cubing both sides would show

$$\left(\sqrt[3]{2x + 1}\right)^3 = (-5)^3$$

$$2x + 1 = -125.$$

4. Solve: $(3x - 5)^{1/5} = 2$.

Because both sides were raised to an *odd* power, a check is not *essential*. It is wise, however, for the student to confirm that -63 checks and is the solution.

YOUR TURN

Equations with Two Radical Terms

A strategy for solving equations with two or more radical terms is as follows.

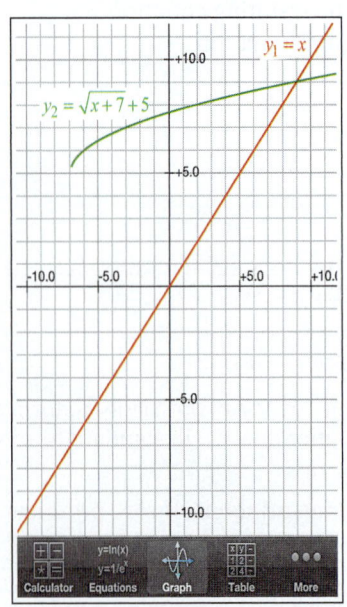

Technology Connection

To solve Example 3, we can graph $y_1 = x$ and $y_2 = \sqrt{x + 7} + 5$ and then find any point(s) of intersection. The intersection occurs at $x = 9$. Note that there is no intersection when $x = 2$, as predicted in the check of Example 3.

1. Use a graphing calculator to solve Examples 1, 2, 4, 5, and 6.

> ### TO SOLVE AN EQUATION WITH TWO OR MORE RADICAL TERMS
>
> **1.** Isolate one of the radical terms.
> **2.** Use the principle of powers.
> **3.** If a radical remains, perform steps (1) and (2) again.
> **4.** Solve the resulting equation.
> **5.** Check possible solutions in the original equation.

EXAMPLE 5 Solve: $\sqrt{2x - 5} = 1 + \sqrt{x - 3}$.

SOLUTION

$$\sqrt{2x - 5} = 1 + \sqrt{x - 3}$$

$$\left(\sqrt{2x - 5}\right)^2 = \left(1 + \sqrt{x - 3}\right)^2 \qquad \text{One radical is already isolated. We square both sides.}$$

This is similar to squaring a binomial. We square 1, then find twice the product of 1 and $\sqrt{x - 3}$, and finally square $\sqrt{x - 3}$. Study this carefully.

$$2x - 5 = 1 + 2\sqrt{x - 3} + \left(\sqrt{x - 3}\right)^2$$

$$2x - 5 = 1 + 2\sqrt{x - 3} + (x - 3)$$

$$x - 3 = 2\sqrt{x - 3} \qquad \text{Isolating the remaining radical term}$$

$$(x - 3)^2 = (2\sqrt{x - 3})^2 \qquad \text{Squaring both sides}$$

$$x^2 - 6x + 9 = 4(x - 3) \qquad \text{Remember to square both the 2 and the } \sqrt{x - 3} \text{ on the right side.}$$

$$x^2 - 6x + 9 = 4x - 12$$
$$x^2 - 10x + 21 = 0$$
$$(x - 7)(x - 3) = 0 \qquad \text{Factoring}$$
$$x = 7 \quad or \quad x = 3 \qquad \text{Using the principle of zero products}$$

5. Solve:

$$\sqrt{x - 2} + 2 = \sqrt{2x + 3}.$$

We leave it to the student to show that 7 and 3 both check and are the solutions.

YOUR TURN

Chapter Resources:
Collaborative Activity, p. 689;
Decision Making: Connection,
p. 689

CAUTION! A common error in solving equations like

$$\sqrt{2x - 5} = 1 + \sqrt{x - 3}$$

is to obtain $1 + (x - 3)$ as the square of the right side. This is wrong because $(A + B)^2 \neq A^2 + B^2$. Placing parentheses around each side when squaring serves as a reminder to square the entire expression.

EXAMPLE 6 Let

$$f(x) = \sqrt{x + 5} - \sqrt{x - 7}.$$

Find all x-values for which $f(x) = 2$.

SOLUTION   We must have $f(x) = 2$, or

$$\sqrt{x + 5} - \sqrt{x - 7} = 2. \qquad \text{Substituting for } f(x)$$

To solve, we isolate one radical term and square both sides:

Isolate a radical term.

$$\sqrt{x + 5} = 2 + \sqrt{x - 7} \qquad \begin{array}{l}\text{Adding } \sqrt{x - 7} \text{ to both sides}\\ \text{to isolate a radical term}\end{array}$$

Raise both sides to the same power.

$$(\sqrt{x + 5})^2 = (2 + \sqrt{x - 7})^2 \qquad \begin{array}{l}\text{Using the principle of powers}\\ \text{(squaring both sides)}\end{array}$$

$$x + 5 = 4 + 4\sqrt{x - 7} + (x - 7) \qquad \begin{array}{l}\text{Using}\\ (A + B)^2 = A^2 + 2AB + B^2\end{array}$$

$$5 = 4\sqrt{x - 7} - 3 \qquad \begin{array}{l}\text{Adding } -x \text{ to both sides and}\\ \text{combining like terms}\end{array}$$

Isolate a radical term.

$$8 = 4\sqrt{x - 7} \qquad \begin{array}{l}\text{Isolating the remaining}\\ \text{radical term}\end{array}$$

$$2 = \sqrt{x - 7}$$

Raise both sides to the same power.

$$2^2 = (\sqrt{x - 7})^2 \qquad \text{Squaring both sides}$$

$$4 = x - 7$$

Solve.

$$11 = x.$$

Check.

Check: $f(11) = \sqrt{11 + 5} - \sqrt{11 - 7}$
$$= \sqrt{16} - \sqrt{4}$$

6. Let $f(x) = \sqrt{x} + \sqrt{x + 1}$.
Find all x-values for which
$f(x) = 2$.

$$= 4 - 2 = 2.$$

We have $f(x) = 2$ when $x = 11$.

YOUR TURN

10.6 EXERCISE SET

Vocabulary and Reading Check

Choose the word from the following list that best completes each statement. Not every word will be used.

even	radical
isolate	raise
odd	rational
powers	square roots

1. When we "square both sides" of an equation, we are using the principle of _____.

2. The equation $\sqrt{2x - 5} = 7$ is a(n) _____ equation.

3. To solve an equation with a radical term, we first _____ the radical term on one side of the equation.

4. A check is essential when we raise both sides of an equation to a(n) _____ power.

Concept Reinforcement

Classify each of the following statements as either true or false.

5. If $t = 7$, then $t^2 = 49$.

6. If $\sqrt{x} = 3$, then $(\sqrt{x})^2 = 3^2$.

7. If $x^2 = 36$, then $x = 6$.

8. $\sqrt{x} - 8 = 7$ is equivalent to $\sqrt{x} = 15$.

The Principle of Powers

Solve.

9. $\sqrt{5x + 1} = 4$

10. $\sqrt{7x - 3} = 5$

11. $\sqrt{3x} + 1 = 5$

12. $\sqrt{2x} - 1 = 2$

13. $\sqrt{y + 5} - 4 = 1$

14. $\sqrt{x - 2} - 7 = -4$

15. $\sqrt{8 - x} + 7 = 10$

16. $\sqrt{y + 4} + 6 = 7$

17. $\sqrt[3]{y + 3} = 2$

18. $\sqrt[3]{x - 2} = 3$

19. $\sqrt[4]{t - 10} = 3$

20. $\sqrt[4]{t + 5} = 2$

21. $6\sqrt{x} = x$

22. $7\sqrt{y} = y$

23. $2y^{1/2} - 13 = 7$

24. $3x^{1/2} + 12 = 9$

25. $\sqrt[3]{x} = -5$

26. $\sqrt[3]{y} = -4$

27. $z^{1/4} + 8 = 10$

28. $x^{1/4} - 2 = 1$

 29. $\sqrt{n} = -2$

30. $\sqrt{a} = -1$

31. $\sqrt[3]{3x + 1} - 4 = -1$

32. $\sqrt[4]{2x + 3} - 5 = -2$

33. $(21x + 55)^{1/3} = 10$

34. $(5y + 31)^{1/4} = 2$

35. $\sqrt[3]{3y + 6} + 7 = 8$

36. $\sqrt[3]{6x + 9} + 5 = 2$

37. $3 + \sqrt{5 - x} = x$

38. $x = \sqrt{x - 1} + 3$

Equations with Two Radical Terms

Solve.

39. $\sqrt{3t + 4} = \sqrt{4t + 3}$

40. $\sqrt{2t - 7} = \sqrt{3t - 12}$

41. $3(4 - t)^{1/4} = 6^{1/4}$

42. $2(1 - x)^{1/3} = 4^{1/3}$

43. $\sqrt{4x - 3} = 2 + \sqrt{2x - 5}$

44. $3 + \sqrt{z - 6} = \sqrt{z + 9}$

45. $\sqrt{20 - x} + 8 = \sqrt{9 - x} + 11$

46. $4 + \sqrt{10 - x} = 6 + \sqrt{4 - x}$

47. $\sqrt{x + 2} + \sqrt{3x + 4} = 2$

48. $\sqrt{6x + 7} - \sqrt{3x + 3} = 1$

49. If $f(x) = \sqrt{x} + \sqrt{x - 9}$, find any x for which $f(x) = 1$.

50. If $g(x) = \sqrt{x} + \sqrt{x - 5}$, find any x for which $g(x) = 5$.

51. If $f(t) = \sqrt{t - 2} - \sqrt{4t + 1}$, find any t for which $f(t) = -3$.

52. If $g(t) = \sqrt{2t + 7} - \sqrt{t + 15}$, find any t for which $g(t) = -1$.

53. If $f(x) = \sqrt{2x - 3}$ and $g(x) = \sqrt{x + 7} - 2$, find any x for which $f(x) = g(x)$.

54. If $f(x) = 2\sqrt{3x + 6}$ and $g(x) = 5 + \sqrt{4x + 9}$, find any x for which $f(x) = g(x)$.

55. If $f(t) = 4 - \sqrt{t - 3}$ and $g(t) = (t + 5)^{1/2}$, find any t for which $f(t) = g(t)$.

56. If $f(t) = 7 + \sqrt{2t - 5}$ and $g(t) = 3(t + 1)^{1/2}$, find any t for which $f(t) = g(t)$.

57. Explain in your own words why it is important to check your answers when using the principle of powers.

58. The principle of powers is an "if–then" statement that becomes false when the sentence parts are interchanged. Give an example of another such if–then statement from everyday life (answers will vary).

Skill Review

Solve.

59. In order to earn at least a B in her Economics class, Taylor must average at least 80% on the five tests. Her grades on the first four tests are 74%, 88%, 76%, and 78%. What must she score on the last test in order to earn at least a B in the course?   [2.7]

60. The number of building permits issued for single-family homes in the United States decreased from 1,378,000 in 2006 to 419,000 in 2011. What was the rate of change?   [3.4]

Source: U.S. Census Bureau

61. A flood rescue team uses a boat that travels 10 mph in still water. To reach a stranded family, they travel 7 mi against the current and return 7 mi with the current in a total time of $1\frac{2}{3}$ hr. What is the speed of the current?   [6.7]

62. Melted Goodness mixes Swiss chocolate and whipping cream to make a dessert fondue. Swiss chocolate costs $1.20 per ounce and whipping cream costs $0.30 per ounce. How much of each does Melted Goodness use in order to make 65 oz of fondue at a cost of $60.00?   [8.3]

Synthesis

63. Describe a procedure for creating radical equations that have no solution.

64. Is checking essential when the principle of powers is used with an odd power n? Why or why not?

65. *Firefighting.*   The velocity of water flow, in feet per second, from a nozzle is given by

$$v(p) = 12.1\sqrt{p},$$

where p is the nozzle pressure, in pounds per square inch (psi). Find the nozzle pressure if the water flow is 100 feet per second.

Source: Houston Fire Department Continuing Education

66. *Firefighting.*   The velocity of water flow, in feet per second, from a water tank that is h feet high is given by

$$v(h) = 8\sqrt{h}.$$

Find the height of a water tank that provides a water flow of 60 feet per second.

Source: Houston Fire Department Continuing Education

67. *Music.*   The frequency of a violin string varies directly with the square root of the tension on the string. A violin string vibrates with a frequency of 260 Hz when the tension on the string is 28 N. What is the frequency when the tension is 32 N?

68. *Music.*   The frequency of a violin string varies inversely with the square root of the density of the string. A nylon violin string with a density of 1200 kg/m^3 vibrates with a frequency of 250 Hz. What is the frequency of a silk and steel-core violin string with a density of 1300 kg/m^3?

Source: www.speech.kth.se

Steel Manufacturing.   *In the production of steel and other metals, the temperature of the molten metal is so great that conventional thermometers melt. Instead, sound is transmitted across the surface of the metal to a receiver on the far side and the speed of the sound is measured. The formula*

$$S(t) = 1087.7\sqrt{\frac{9t + 2617}{2457}}$$

gives the speed of sound $S(t)$, in feet per second, at a temperature of t degrees Celsius.

69. Find the temperature of a blast furnace where sound travels 1880 ft/sec.

70. Find the temperature of a blast furnace where sound travels 1502.3 ft/sec.

71. Solve the above equation for t.

Escape Velocity.   *A formula for the escape velocity v of a satellite is*

$$v = \sqrt{2gr}\sqrt{\frac{h}{r + h}},$$

where g is the force of gravity, r is the planet or star's radius, and h is the height of the satellite above the planet or star's surface.

72. Solve for h.

73. Solve for r.

Solve.

74. $\left(\dfrac{z}{4} - 5\right)^{2/3} = \dfrac{1}{25}$

75. $\dfrac{x + \sqrt{x + 1}}{x - \sqrt{x + 1}} = \dfrac{5}{11}$

76. $\sqrt{\sqrt{y} + 49} = 7$

77. $(z^2 + 17)^{3/4} = 27$

78. $x^2 - 5x - \sqrt{x^2 - 5x - 2} = 4$
(*Hint:* Let $u = x^2 - 5x - 2$.)

79. $\sqrt{8 - b} = b\sqrt{8 - b}$

Without graphing, determine the x-intercepts of the graphs given by each of the following.

80. $f(x) = \sqrt{x - 2} - \sqrt{x + 2} + 2$

81. $g(x) = 6x^{1/2} + 6x^{-1/2} - 37$

82. $f(x) = (x^2 + 30x)^{1/2} - x - (5x)^{1/2}$

 83. Use a graphing calculator to check your answers to Exercises 9, 17, and 33.

 84. Use a graphing calculator to check your answers to Exercises 29, 39, and 43.

↪ YOUR TURN ANSWERS: SECTION 10.6

1. 24   **2.** No solution   **3.** 15   **4.** $\frac{37}{3}$ **5.** 3, 11   **6.** $\frac{9}{16}$

QUICK QUIZ: SECTIONS 10.1–10.6
Simplify. Write all answers using radical notation. Assume that all variables represent positive numbers.

1. $\sqrt{121n^2}$ [10.1]   **2.** $\sqrt[3]{\sqrt{a}}$ [10.2]

3. $2\sqrt{12} - \sqrt{75}$ [10.5]   **4.** $\sqrt{x}\sqrt[5]{x^3}$ [10.3]

5. Solve: $7 - \sqrt{3x + 1} = 5$.   [10.6]

PREPARE TO MOVE ON
Solve.

1. The largest sign in the United States is a rectangle with a perimeter of 430 ft. The length of the rectangle is 5 ft longer than thirteen times the width. Find the dimensions of the sign.   [2.5]

   Source: Florida Center for Instructional Technology

2. The base of a triangular sign is 4 in. longer than twice the height. The area of the sign is 255 in². Find the dimensions of the sign.   [5.8]

3. The length of a rectangular lawn between classroom buildings is 2 yd less than twice the width of the lawn. A path that is 34 yd long stretches diagonally across the area. What are the dimensions of the lawn?   [5.8]

4. One leg of a right triangle is 5 cm long. The hypotenuse is 1 cm longer than the other leg. Find the length of the hypotenuse.   [5.8]

10.7   The Distance Formula, the Midpoint Formula, and Other Applications

Using the Pythagorean Theorem   •   Two Special Triangles   •   The Distance Formula and the Midpoint Formula

Study Skills

Making Sketches

One need not be an artist to make highly useful mathematical sketches. That said, it is important to make sure that your sketches are drawn accurately enough to represent the relative sizes within each shape. For example, if one side of a triangle is clearly the longest, make sure your drawing reflects this.

Using the Pythagorean Theorem

There are many kinds of problems that involve powers and roots. Many also involve right triangles and the Pythagorean theorem.

THE PYTHAGOREAN THEOREM*

In any right triangle, if a and b are the lengths of the legs and c is the length of the hypotenuse, then

$$a^2 + b^2 = c^2.$$

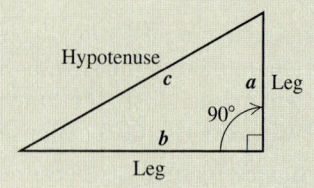

---

*The converse of the Pythagorean theorem also holds. That is, if a, b, and c are the lengths of the sides of a triangle and $a^2 + b^2 = c^2$, then the triangle is a right triangle.

In applying the Pythagorean theorem, we often make use of the following principle.

THE PRINCIPLE OF SQUARE ROOTS

For any nonnegative real number n,

$$\text{If}\quad x^2 = n,\quad \text{then}\quad x = \sqrt{n}\quad \text{or}\quad x = -\sqrt{n}.$$

For most real-world applications involving length or distance, $-\sqrt{n}$ is not needed.

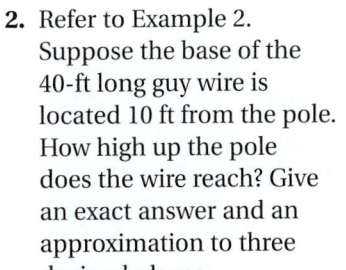

EXAMPLE 1 *Baseball.* A baseball diamond is a square with sides of length 90 ft. Suppose a catcher fields a ball while standing on the third-base line 10 ft from home plate, as shown in the figure. How far is the catcher's throw to first base? Give an exact answer and an approximation to three decimal places.

SOLUTION We make a drawing and let $d =$ the distance, in feet, to first base. Note that a right triangle is formed in which the leg from home plate to first base measures 90 ft and the leg from home plate to where the catcher fields the ball measures 10 ft.

We substitute these values into the Pythagorean theorem to find d:

$$d^2 = 90^2 + 10^2$$
$$= 8100 + 100$$
$$= 8200.$$

We now use the principle of square roots: If $d^2 = 8200$, then $d = \sqrt{8200}$ or $d = -\sqrt{8200}$. Since d represents length, it must be the *positive* square root of 8200:

$$d = \sqrt{8200}\ \text{ft} = 10\sqrt{82}\ \text{ft}\qquad \text{This is an exact answer.}$$
$$\approx 90.554\ \text{ft}.\qquad \text{Using a calculator for an approximation}$$

1. Refer to Example 1. How far would the catcher's throw to first base be from a point on the third-base line 20 ft from home plate? Give an exact answer and an approximation to three decimal places.

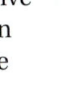 YOUR TURN

EXAMPLE 2 *Guy Wires.* The base of a 40-ft long guy wire is 15 ft from the telephone pole that it anchors. How high up the pole does the guy wire reach? Give an exact answer and an approximation to three decimal places.

SOLUTION We make a drawing and let $h =$ the height, in feet, to which the guy wire reaches. A right triangle is formed in which one leg measures 15 ft and the hypotenuse measures 40 ft. Using the Pythagorean theorem, we have

$$h^2 + 15^2 = 40^2$$
$$h^2 + 225 = 1600$$
$$h^2 = 1375$$
$$h = \sqrt{1375}.$$

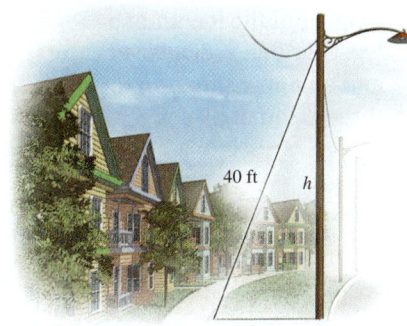

Exact answer:

$$h = \sqrt{1375}\ \text{ft}\qquad \text{Using the positive square root}$$
$$= 5\sqrt{55}\ \text{ft}$$

Approximation:

$$h \approx 37.081\ \text{ft}\qquad \text{Using a calculator}$$

2. Refer to Example 2. Suppose the base of the 40-ft long guy wire is located 10 ft from the pole. How high up the pole does the wire reach? Give an exact answer and an approximation to three decimal places.

 YOUR TURN

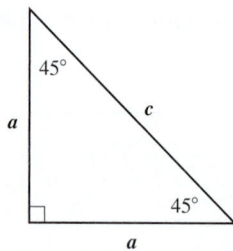

Two Special Triangles

When both legs of a right triangle are the same size, as shown at left, we call the triangle an *isosceles right triangle*, or a 45°–45°–90° triangle. If one leg of an isosceles right triangle has length a, we can find a formula for the length of the hypotenuse as follows:

$$c^2 = a^2 + b^2$$

$$= a^2 + a^2 \qquad \text{Because the triangle is isosceles, both legs are the same size: } a = b.$$

$$= 2a^2. \qquad \text{Combining like terms}$$

Next, we use the principle of square roots. Because a, b, and c are lengths, there is no need to consider negative square roots or absolute values. Thus,

$$c = \sqrt{2a^2} \qquad \text{Using the principle of square roots}$$

$$= \sqrt{a^2 \cdot 2} = a\sqrt{2}. \qquad \text{The equation } c = a\sqrt{2} \text{ is worth remembering.}$$

EXAMPLE 3   One leg of an isosceles right triangle measures 7 cm. Find the length of the hypotenuse. Give an exact answer and an approximation to three decimal places.

3. One leg of an isosceles right triangle measures 5 m. Find the length of the hypotenuse. Give an exact answer and an approximation to three decimal places.

SOLUTION   We substitute:

$$c = a\sqrt{2}$$

$$= 7\sqrt{2}.$$

Exact answer: $c = 7\sqrt{2}$ cm

Approximation: $c \approx 9.899$ cm

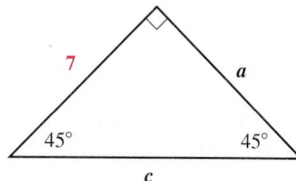

YOUR TURN

When the hypotenuse of an isosceles right triangle is known, the length of the legs can be found.

EXAMPLE 4   The hypotenuse of an isosceles right triangle is 5 ft long. Find the length of a leg. Give an exact answer and an approximation to three decimal places.

SOLUTION   We replace c with 5 and solve for a:

$$5 = a\sqrt{2} \qquad \text{Substituting 5 for } c \text{ in } c = a\sqrt{2}$$

$$\frac{5}{\sqrt{2}} = a \qquad \text{Dividing both sides by } \sqrt{2}$$

$$\frac{5\sqrt{2}}{2} = a. \qquad \text{Rationalize the denominator if desired.}$$

4. The hypotenuse of an isosceles right triangle is 12 ft long. Find the length of a leg. Give an exact answer and an approximation to three decimal places.

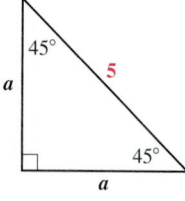

Exact answer: $a = \dfrac{5}{\sqrt{2}}$ ft, or $\dfrac{5\sqrt{2}}{2}$ ft

Approximation: $a \approx 3.536$ ft   Using a calculator

YOUR TURN

LENGTHS WITHIN ISOSCELES RIGHT TRIANGLES

The length of the hypotenuse in an isosceles right triangle, or a 45°–45°–90° triangle, is the length of a leg times $\sqrt{2}$.

$$c = a\sqrt{2}$$

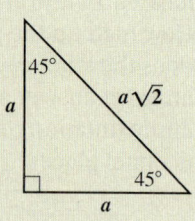

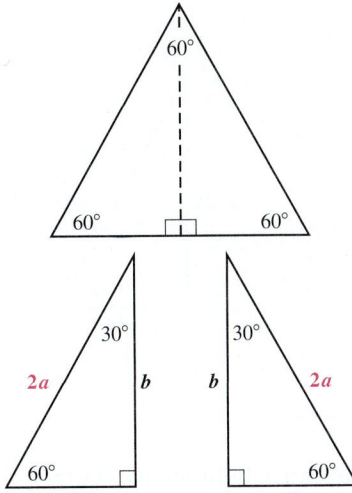

A second special triangle is known as a 30°–60°–90° triangle, so named because of the measures of its angles. Note that in an equilateral triangle, all sides have the same length and all angles are 60°. An altitude, drawn dashed in the figure, bisects, or splits in half, one angle and one side. Two 30°–60°–90° right triangles are thus formed.

If we let a represent the length of the shorter leg in a 30°–60°–90° triangle, then $2a$ represents the length of the hypotenuse. We have

$$a^2 + b^2 = (2a)^2 \qquad \text{Using the Pythagorean theorem}$$
$$a^2 + b^2 = 4a^2$$
$$b^2 = 3a^2 \qquad \text{Subtracting } a^2 \text{ from both sides}$$
$$b = \sqrt{3a^2} \qquad \text{Considering only the positive square root}$$
$$= \sqrt{a^2 \cdot 3}$$
$$= a\sqrt{3}. \qquad \text{This relationship is worth remembering.}$$

EXAMPLE 5   The shorter leg of a 30°–60°–90° triangle measures 8 in. Find the lengths of the other sides. Give exact answers and, where appropriate, an approximation to three decimal places.

SOLUTION   The hypotenuse is twice as long as the shorter leg, so we have

$$c = 2a \qquad\qquad \text{This relationship is worth remembering.}$$
$$= 2 \cdot 8 = 16 \text{ in.} \quad \text{This is the length of the hypotenuse.}$$

The length of the longer leg is the length of the shorter leg times $\sqrt{3}$. This gives us

$$b = a\sqrt{3} \qquad \text{This holds for all 30°–60°–90° triangles.}$$
$$= 8\sqrt{3} \text{ in.} \quad \text{This is the length of the longer leg.}$$

Exact answer: $c = 16$ in., $b = 8\sqrt{3}$ in.

Approximation: $b \approx 13.856$ in.

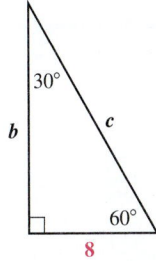

5. The shorter leg of a 30°–60°–90° triangle measures 7 in. Find the lengths of the other sides. Give exact answers and, where appropriate, an approximation to three decimal places.

YOUR TURN

EXAMPLE 6   The length of the longer leg of a 30°–60°–90° triangle is 14 cm. Find the length of the hypotenuse. Give an exact answer and an approximation to three decimal places.

SOLUTION   The length of the hypotenuse is twice the length of the shorter leg. We first find a, the length of the shorter leg, by using the length of the longer leg:

$$14 = a\sqrt{3} \qquad \text{Substituting 14 for } b \text{ in } b = a\sqrt{3}$$
$$\frac{14}{\sqrt{3}} = a. \qquad \text{Dividing by } \sqrt{3}$$

Since the hypotenuse is twice as long as the shorter leg, we have

$$c = 2a$$
$$= 2 \cdot \frac{14}{\sqrt{3}} \qquad \text{Substituting}$$
$$= \frac{28}{\sqrt{3}} \text{ cm.}$$

Exact answer: $c = \dfrac{28}{\sqrt{3}}$ cm, or $\dfrac{28\sqrt{3}}{3}$ cm

Approximation: $c \approx 16.166$ cm

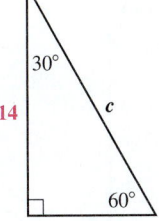

6. The length of the longer leg of a 30°–60°–90° triangle is 6 cm. Find the length of the hypotenuse. Give an exact answer and an approximation to three decimal places.

YOUR TURN

Student Notes

Perhaps the easiest way to remember the important results listed for these special triangles is to write out, on your own, the derivations shown in this section.

LENGTHS WITHIN 30°–60°–90° TRIANGLES

The length of the longer leg in a 30°–60°–90° triangle is the length of the shorter leg times $\sqrt{3}$. The hypotenuse is twice as long as the shorter leg.

$$b = a\sqrt{3},$$
$$c = 2a$$

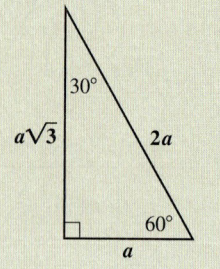

The Distance Formula and the Midpoint Formula

We can use the Pythagorean theorem to find the distance between two points. To find the distance between two points on the number line, we subtract. Depending on the order in which we subtract, the difference may be positive or negative. However, if we take the absolute value of the difference, we obtain the same positive value for the distance regardless of the order in which we subtract:

7 units

$|4 - (-3)| = |7| = 7;$
$|-3 - 4| = |-7| = 7.$

In a plane, if two points are on a horizontal line, they have the same second coordinate. We can find the distance between them by subtracting their first coordinates and taking the absolute value of that difference.

EXPLORING THE CONCEPT

1. Find the coordinates of point A.
2. Find the distance from $(4, -4)$ to A.
3. Find the distance from $(1, 2)$ to A.
4. Use the Pythagorean theorem to find the distance from $(1, 2)$ to $(4, -4)$.

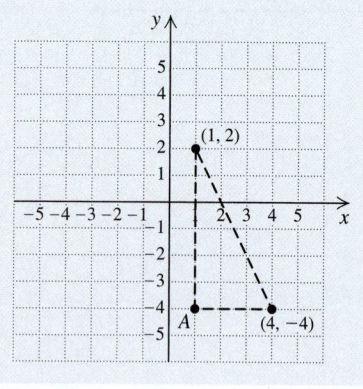

ANSWERS

1. $(1, -4)$ 2. 3 units   3. 6 units
4. $\sqrt{45}$ units $= 3\sqrt{5}$ units

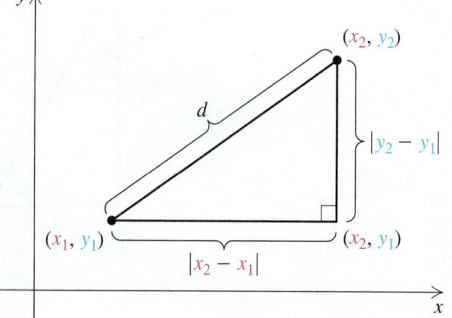

Generalizing, we have that the distance between the points (x_1, y_1) and (x_2, y_1) on a horizontal line is $|x_2 - x_1|$. Similarly, the distance between (x_2, y_1) and (x_2, y_2) on a vertical line is $|y_2 - y_1|$.

So long as $x_1 \neq x_2$ and $y_1 \neq y_2$, the points (x_1, y_1) and (x_2, y_2), along with the point (x_2, y_1), describe a right triangle. The lengths of the legs are $|x_2 - x_1|$ and $|y_2 - y_1|$. We find d, the length of the hypotenuse, by using the Pythagorean theorem:

$$d^2 = |x_2 - x_1|^2 + |y_2 - y_1|^2.$$

Since the square of a number is the same as the square of its opposite, we can replace the absolute-value signs with parentheses:

$$d^2 = (x_2 - x_1)^2 + (y_2 - y_1)^2.$$

Taking the principal square root, we have a formula for distance.

THE DISTANCE FORMULA

The distance d between any two points (x_1, y_1) and (x_2, y_2) is given by

$$d = \sqrt{(x_2 - x_1)^2 + (y_2 - y_1)^2}.$$

EXAMPLE 7   Find the distance between $(5, -1)$ and $(-4, 6)$. Find an exact answer and an approximation to three decimal places.

SOLUTION   We substitute into the distance formula:

$$d = \sqrt{(-4 - 5)^2 + [6 - (-1)]^2} \qquad \text{Substituting. A drawing is optional.}$$
$$= \sqrt{(-9)^2 + 7^2}$$
$$= \sqrt{130} \qquad \text{This is exact.}$$
$$\approx 11.402. \qquad \text{Using a calculator for an approximation}$$

 YOUR TURN

7. Find the distance between $(4, 7)$ and $(-8, 2)$.

The distance formula can be used to verify a formula for the coordinates of the *midpoint* of a segment connecting two points. We state the midpoint formula and leave its proof to the exercises.

THE MIDPOINT FORMULA

If the endpoints of a segment are (x_1, y_1) and (x_2, y_2), then the coordinates of the midpoint are

$$\left(\frac{x_1 + x_2}{2}, \frac{y_1 + y_2}{2} \right).$$

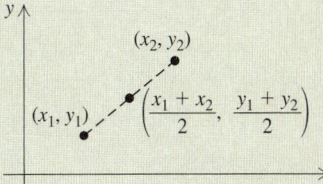

(To locate the midpoint, average the x-coordinates and average the y-coordinates.)

Student Notes

To help remember the formulas correctly, note that the distance formula is a variation of the Pythagorean theorem and the result is a number. The midpoint formula involves averages and the result is an ordered pair.

EXAMPLE 8   Find the midpoint of the segment with endpoints $(-2, 3)$ and $(4, -6)$.

SOLUTION   Using the midpoint formula, we obtain

$$\left(\frac{-2 + 4}{2}, \frac{3 + (-6)}{2} \right), \quad \text{or} \quad \left(\frac{2}{2}, \frac{-3}{2} \right), \quad \text{or} \quad \left(1, -\frac{3}{2} \right).$$

8. Find the midpoint of the segment with endpoints $(-3, -4)$ and $(0, -10)$.

 YOUR TURN

10.7 EXERCISE SET

FOR EXTRA HELP

MyMathLab® Math
PRACTICE WATCH READ REVIEW

Vocabulary and Reading Check

Complete each statement with the best choice from the following list.

1. In any ____ triangle, the square of the length of the hypotenuse is the sum of the squares of the lengths of the legs.

2. The shortest side of a right triangle is always one of the two ____.

3. The principle of ____ states that if $x^2 = n$, then $x = \sqrt{n}$ or $x = -\sqrt{n}$.

4. In a(n) ____ right triangle, both legs have the same length.

5. In a(n) ____ right triangle, the hypotenuse is twice as long as the shorter leg.

6. If both legs in a right triangle have measure a, then the ____ measures $a\sqrt{2}$.

a) Hypotenuse

b) Isosceles

c) Legs

d) Right

e) Square roots

f) 30°–60°–90°

Using the Pythagorean Theorem

In a right triangle, find the length of the side not given. Give an exact answer and, where appropriate, an approximation to three decimal places.

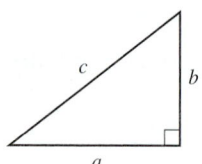

7. $a = 5$, $b = 3$

8. $a = 8$, $b = 10$

Aha! 9. $a = 9$, $b = 9$

10. $a = 10$, $b = 10$

11. $b = 15$, $c = 17$

12. $a = 7$, $c = 25$

In Exercises 13–18, give an exact answer and, where appropriate, an approximation to three decimal places.

13. A right triangle's hypotenuse is 8 m, and one leg is $4\sqrt{3}$ m. Find the length of the other leg.

14. A right triangle's hypotenuse is 6 cm, and one leg is $\sqrt{5}$ cm. Find the length of the other leg.

15. The hypotenuse of a right triangle is $\sqrt{20}$ in., and one leg measures 1 in. Find the length of the other leg.

16. The hypotenuse of a right triangle is $\sqrt{15}$ ft, and one leg measures 2 ft. Find the length of the other leg.

Aha! 17. One leg in a right triangle is 1 m, and the hypotenuse measures $\sqrt{2}$ m. Find the length of the other leg.

18. One leg of a right triangle is 1 yd, and the hypotenuse measures 2 yd. Find the length of the other leg.

In Exercises 19–28, give an exact answer and, where appropriate, an approximation to three decimal places.

19. *Bicycling.* Clare routinely bicycles across a rectangular parking lot on her way to work. If the lot is 200 ft long and 150 ft wide, how far does Clare travel when she rides across the lot diagonally?

20. *Guy Wire.* How long is a guy wire that reaches from the top of a 15-ft pole to a point on the ground 10 ft from the pole?

21. *Zipline.* For Super Bowl XLVI in Indianapolis, a temporary zipline was constructed on Capitol Street. The ride extended 800 ft along the street, and riders dropped 60 ft. How long was the zipline?

22. *Baseball.* Suppose the catcher in Example 1 makes a throw to second base from the same location. How far is that throw?

23. *Television Sets.* What does it mean to refer to a 51-in. TV set? Such units refer to the diagonal of the screen. A 51-in. TV set has a width of 45 in. What is its height?

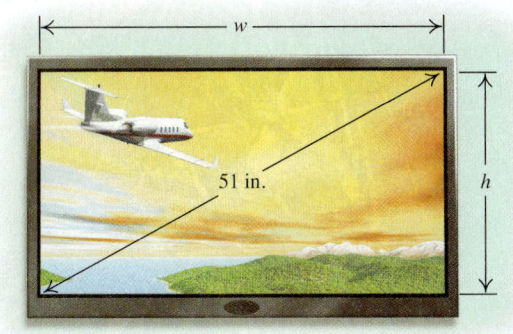

24. *Television Sets.* A 53-in. TV set has a screen with a height of 28 in. What is its width? (See Exercise 23.)

25. *Speaker Placement.* A stereo receiver is in a corner of a 12-ft by 14-ft room. Wire will run under a rug, diagonally, to a subwoofer in the far corner. If 4 ft of slack is required on each end, how long a piece of wire should be purchased?

26. *Distance over Water.* To determine the width of a pond, a surveyor locates two stakes at either end of the pond and uses instrumentation to place a third stake so that the distance across the pond is the length of a hypotenuse. If the third stake is 90 m from one stake and 70 m from the other, what is the distance across the pond?

27. *Walking.* Students at Pohlman Community College have worn a path that cuts diagonally across the campus "quad." If the quad is a rectangle that Marissa measures as 70 paces long and 40 paces wide, how many paces will Marissa save by using the diagonal path?

28. *Crosswalks.* The diagonal crosswalk at the intersection of State St. and Main St. is the hypotenuse of a triangle in which the crosswalks across State St. and Main St. are the legs. If State St. is 28 ft wide and Main St. is 40 ft wide, how much distance is saved by using the diagonal crosswalk rather than crossing both streets?

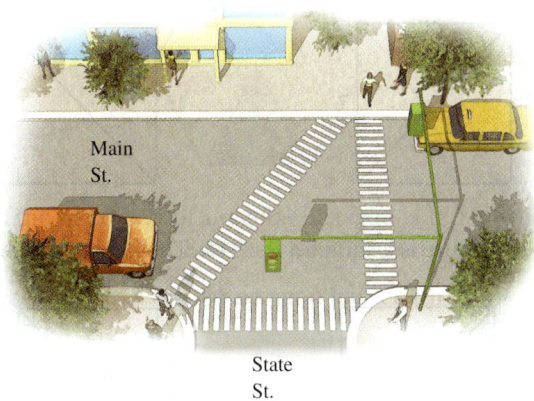

Two Special Triangles

For each triangle, find the missing length(s). Give an exact answer and, where appropriate, an approximation to three decimal places.

29.
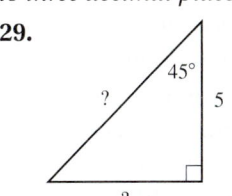

30.
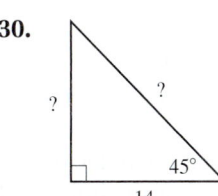

31.
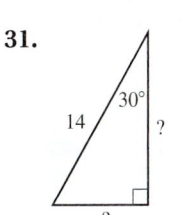

32.
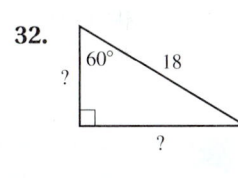

33.
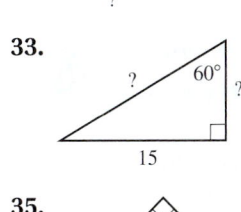

34.
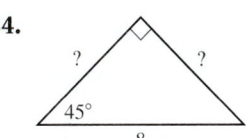

35.
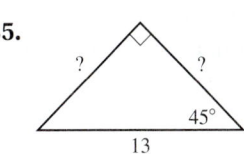

36.
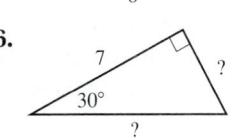

37.
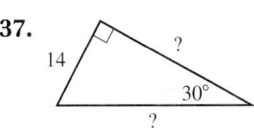

38.
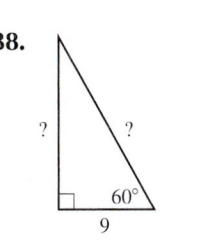

92. The length and the width of a rectangle are given by consecutive integers. The area of the rectangle is 90 cm². Find the length of a diagonal of the rectangle.

93. A cube measures 5 cm on each side. How long is the diagonal that connects two opposite corners of the cube? Give an exact answer.

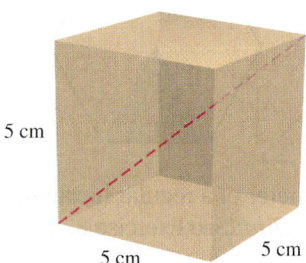

5 cm

5 cm   5 cm

94. Prove the midpoint formula by showing that

i) the distance from (x_1, y_1) to

$$\left(\frac{x_1 + x_2}{2}, \frac{y_1 + y_2}{2} \right)$$

equals the distance from (x_2, y_2) to

$$\left(\frac{x_1 + x_2}{2}, \frac{y_1 + y_2}{2} \right);$$

and

ii) the points

$$(x_1, y_1), \left(\frac{x_1 + x_2}{2}, \frac{y_1 + y_2}{2} \right),$$

and

$$(x_2, y_2)$$

lie on the same line (see Exercise 86).

⤷ **YOUR TURN ANSWERS: SECTION 10.7**

1. $\sqrt{8500}$ ft $= 10\sqrt{85}$ ft; 92.195 ft

2. $\sqrt{1500}$ ft $= 10\sqrt{15}$ ft; 38.730 ft   **3.** $5\sqrt{2}$ m; 7.071 m

4. $\dfrac{12}{\sqrt{2}}$ ft $= 6\sqrt{2}$ ft; 8.485 ft   **5.** Leg: $7\sqrt{3}$ in.; 12.124 in.;

hypotenuse: 14 in.   **6.** $\dfrac{12}{\sqrt{3}}$ cm $= 4\sqrt{3}$ cm; 6.928 cm

7. 13   **8.** $\left(-\dfrac{3}{2}, -7 \right)$

QUICK QUIZ: SECTIONS 10.1–10.7

Simplify. Assume that all variables represent positive numbers.

1. $\sqrt[3]{\dfrac{54x^4}{125t^6}}$ [10.4]   **2.** $\sqrt[4]{32x^5} - \sqrt[4]{2x^{13}}$ [10.5]

3. $\sqrt{40x^3} \sqrt{45x}$ [10.3]

4. Solve: $5 - \sqrt{x+1} = \sqrt{x}$.   [10.6]

5. Find the distance between $(4, -1)$ and $(-3, -7)$. Give an exact answer and an approximation to three decimal places.   [10.7]

PREPARE TO MOVE ON

Find the conjugate of each number.   [10.5]

1. $2 - \sqrt{3}$ **2.** $\sqrt{7} + 6$

Multiply.   [4.5], [4.6]

3. $(3x - 2y)(3x + 2y)$ **4.** $(5w - 2x)^2$

5. $-4c(2a - 7c)$ **6.** $(4a + p)(6a - 5p)$

10.8   The Complex Numbers

Imaginary Numbers and Complex Numbers • Addition and Subtraction • Multiplication •
Conjugates and Division • Powers of *i*

Imaginary Numbers and Complex Numbers

Negative numbers do not have square roots in the real-number system. However, a larger number system that contains the real-number system is designed so that negative numbers *do* have square roots. That system is called the **complex-number system**, and it will allow us to solve equations like $x^2 + 1 = 0$. The complex-number system makes use of *i*, a number that is, by definition, a square root of -1.

Study Skills

Studying Together by Phone

Working with a classmate over the telephone can be an effective way to receive or give help. Because you cannot point to figures on paper, you must verbalize the mathematics. This tends to improve understanding of the material. In some cases, it may be more effective to study with a classmate over the phone than in person.

1. Express in terms of i: $\sqrt{-36}$.

THE NUMBER i

i is the unique number for which $i = \sqrt{-1}$ and $i^2 = -1$.

We can now define the square root of a negative number as follows:

$$\sqrt{-p} = \sqrt{-1}\sqrt{p} = i\sqrt{p} \text{ or } \sqrt{p}\,i, \text{ for any positive number } p.$$

EXAMPLE 1 Express in terms of i: **(a)** $\sqrt{-7}$; **(b)** $\sqrt{-16}$; **(c)** $-\sqrt{-50}$.

SOLUTION

a) $\sqrt{-7} = \sqrt{-1 \cdot 7} = \sqrt{-1} \cdot \sqrt{7} = i\sqrt{7}$, or $\sqrt{7}\,i$ $\boxed{i \text{ is } not \text{ under the radical.}}$

b) $\sqrt{-16} = \sqrt{-1 \cdot 16} = \sqrt{-1} \cdot \sqrt{16} = i \cdot 4 = 4i$

c) $-\sqrt{-50} = -\sqrt{-1} \cdot \sqrt{25} \cdot \sqrt{2} = -i \cdot 5 \cdot \sqrt{2} = -5i\sqrt{2}$, or $-5\sqrt{2}\,i$

YOUR TURN

IMAGINARY NUMBERS

An *imaginary number* is any number that can be written in the form $a + bi$, where a and b are real numbers and $b \neq 0$.

Don't let the name "imaginary" fool you. Imaginary numbers appear in fields such as engineering and the physical sciences. The following are examples of imaginary numbers:

$$5 + 4i, \qquad \text{Here } a = 5, b = 4.$$
$$\sqrt{3} - \pi i, \qquad \text{Here } a = \sqrt{3}, b = -\pi.$$
$$\sqrt{7}\,i. \qquad \text{Here } a = 0, b = \sqrt{7}.$$

The union of the set of all imaginary numbers and the set of all real numbers is the set of all **complex numbers**.

COMPLEX NUMBERS

A *complex number* is any number that can be written in the form $a + bi$, where a and b are real numbers. (Note that a and b both can be 0.)

The following are examples of complex numbers:

$7 + 3i$ (here $a \neq 0, b \neq 0$); $4i$ (here $a = 0, b \neq 0$);
8 (here $a \neq 0, b = 0$); 0 (here $a = 0, b = 0$).

Complex numbers like $17i$ or $4i$, in which $a = 0$ and $b \neq 0$, are called *pure imaginary numbers*.

SOLUTION

4. Find the conjugate of $2 + 9i$.

a) $-3 - 7i$ The conjugate is $-3 + 7i$.

b) $4i$ The conjugate is $-4i$. Note that $4i = 0 + 4i$.

YOUR TURN

The product of a complex number and its conjugate is a real number.

EXAMPLE 5   Multiply: $(5 + 7i)(5 - 7i)$.

SOLUTION

5. Multiply:

$(-3 + 2i)(-3 - 2i)$.

$$(5 + 7i)(5 - 7i) = 5^2 - (7i)^2 \quad\quad \text{Using } (A + B)(A - B) = A^2 - B^2$$
$$= 25 - 49i^2$$
$$= 25 - 49(-1) \quad i^2 = -1$$
$$= 25 + 49 = 74$$

YOUR TURN

Conjugates are used when dividing by an imaginary number. The procedure is much like that used to rationalize denominators with two terms.

EXAMPLE 6   Divide and simplify to the form $a + bi$.

a) $\dfrac{-2 + 9i}{1 - 3i}$

b) $\dfrac{7 + 4i}{5i}$

SOLUTION

a) To divide and simplify $(-2 + 9i)/(1 - 3i)$, we multiply by 1, using the conjugate of the denominator to form 1:

$$\frac{-2 + 9i}{1 - 3i} = \frac{-2 + 9i}{1 - 3i} \cdot \frac{1 + 3i}{1 + 3i}$$ Multiplying by 1 using the conjugate of the denominator in the symbol for 1

$$= \frac{(-2 + 9i)(1 + 3i)}{(1 - 3i)(1 + 3i)}$$ Multiplying numerators; multiplying denominators

$$= \frac{-2 - 6i + 9i + 27i^2}{1^2 - 9i^2}$$ Using FOIL

$$= \frac{-2 + 3i + (-27)}{1 - (-9)}$$ $i^2 = -1$

$$= \frac{-29 + 3i}{10}$$ Writing in the form $a + bi$

$$= -\frac{29}{10} + \frac{3}{10}i.$$ Recall that $\dfrac{M + N}{D} = \dfrac{M}{D} + \dfrac{N}{D}$.

b) The conjugate of $5i$ is $-5i$, so we *could* multiply by $-5i/(-5i)$. However, when the denominator is a pure imaginary number, it is easiest if we multiply by i/i:

$$\frac{7 + 4i}{5i} = \frac{7 + 4i}{5i} \cdot \frac{i}{i}$$ Multiplying by 1 using i/i. We can also use the conjugate of $5i$ to write 1 as $-5i/(-5i)$.

$$= \frac{7i + 4i^2}{5i^2}$$ Multiplying

6. Divide and simplify to the form $a + bi$:

$$\frac{2 + 4i}{1 - 2i}.$$

$$= \frac{7i + 4(-1)}{5(-1)}$$ $i^2 = -1$

$$= \frac{7i - 4}{-5} = \frac{-4}{-5} + \frac{7}{-5}i, \text{ or } \frac{4}{5} - \frac{7}{5}i.$$ Writing in the form $a + bi$

YOUR TURN

Powers of i

In the following discussion, we show why there is no need to use powers of i (other than 1) when writing answers.

We use the following to simplify powers of i.

- $i^2 = -1$
- $i^n = i \cdot i^{n-1}$
- $(-1)^n = 1$ when n is even
- $(-1)^n = -1$ when n is odd

To simplify i^n when n is even, we rewrite i^n as a power of -1. Even powers of i are 1 or -1.

EXAMPLE 7   Simplify: i^{30}.

SOLUTION

$$
\begin{aligned}
i^{30} &= (i^2)^{15} &&(a^m)^n = a^{mn} \\
&= (-1)^{15} &&i^2 = -1 \\
&= -1 &&(-1)^n = -1 \text{ when } n \text{ is odd.}
\end{aligned}
$$

7. Simplify: i^{26}.

 YOUR TURN

Student Notes

You may notice that the powers of i cycle through $i, -1, -i, 1$:

$$
\begin{aligned}
i^1 &= i, \\
i^2 &= -1, \\
i^3 &= -i, \\
i^4 &= 1, \\
i^5 &= i, \text{ and so on.}
\end{aligned}
$$

8. Simplify: i^{33}.

To simplify i^n when n is odd, we rewrite i^n as $i \cdot i^{n-1}$ and simplify i^{n-1}. Odd powers of i are i or $-i$.

EXAMPLE 8   Simplify: i^{49}.

SOLUTION

$$
\begin{aligned}
i^{49} &= i \cdot i^{48} &&i^n = i \cdot i^{n-1} \\
&= i(i^2)^{24} &&(a^m)^n = a^{mn} \\
&= i(-1)^{24} &&i^2 = -1 \\
&= i(1) &&(-1)^n = 1 \text{ when } n \text{ is even.} \\
&= i
\end{aligned}
$$

YOUR TURN

EXAMPLE 9   Simplify: **(a)** i^{24}; **(b)** i^{75}.

SOLUTION

a) $\begin{aligned}[t] i^{24} &= (i^2)^{12} \\ &= (-1)^{12} = 1 \end{aligned}$

b) $\begin{aligned}[t] i^{75} &= i \cdot i^{74} &&\text{Writing } i^n \text{ as } i \cdot i^{n-1} \\ &= i(i^2)^{37} \\ &= i(-1)^{37} \\ &= i(-1) \\ &= -i \end{aligned}$

9. Simplify: i^{60}.

YOUR TURN

10.8 **EXERCISE SET**   FOR EXTRA HELP   MyMathLab® Math

↪ Vocabulary and Reading Check

Classify each of the following statements as either true or false.

1. Imaginary numbers are so named because they have no real-world applications.

2. Every real number is imaginary, but not every imaginary number is real.

3. Every imaginary number is a complex number, but not every complex number is imaginary.

4. Every real number is a complex number, but not every complex number is real.

5. We add complex numbers by combining real parts and combining imaginary parts.

6. The product of a complex number and its conjugate is always a real number.

7. The square of a complex number is always a real number.

8. The quotient of two complex numbers is always a complex number.

Imaginary Numbers and Complex Numbers

Express in terms of i.

9. $\sqrt{-100}$

10. $\sqrt{-9}$

11. $\sqrt{-5}$

12. $\sqrt{-7}$

13. $\sqrt{-8}$

14. $\sqrt{-12}$

15. $-\sqrt{-11}$

16. $-\sqrt{-17}$

17. $-\sqrt{-49}$

18. $-\sqrt{-81}$

19. $-\sqrt{-300}$

20. $-\sqrt{-75}$

21. $6 - \sqrt{-84}$

22. $4 - \sqrt{-60}$

23. $-\sqrt{-76} + \sqrt{-125}$

24. $\sqrt{-4} + \sqrt{-12}$

25. $\sqrt{-18} - \sqrt{-64}$

26. $\sqrt{-72} - \sqrt{-25}$

Addition and Subtraction

Add or subtract and simplify. Write each answer in the form a + bi.

27. $(3 + 4i) + (2 - 7i)$

28. $(5 - 6i) + (8 + 9i)$

29. $(9 + 5i) - (2 + 3i)$

30. $(8 + 7i) - (2 + 4i)$

31. $(7 - 4i) - (5 - 3i)$

32. $(5 - 3i) - (9 + 2i)$

33. $(-5 - i) - (7 + 4i)$

34. $(-2 + 6i) - (-7 + i)$

Multiplication

Multiply and simplify. Write each answer in the form a + bi.

35. $5i \cdot 8i$

36. $3i \cdot 9i$

37. $(-4i)(-6i)$

38. $7i \cdot (-8i)$

39. $\sqrt{-36}\sqrt{-9}$

40. $\sqrt{-49}\sqrt{-16}$

41. $\sqrt{-3}\sqrt{-10}$

42. $\sqrt{-6}\sqrt{-7}$

43. $\sqrt{-6}\sqrt{-21}$

44. $\sqrt{-15}\sqrt{-10}$

45. $5i(2 + 6i)$

46. $2i(7 + 3i)$

47. $-7i(3 + 4i)$

48. $-4i(6 - 5i)$

49. $(1 + i)(3 + 2i)$

50. $(4 + i)(2 + 3i)$

51. $(6 - 5i)(3 + 4i)$

52. $(5 - 6i)(2 + 5i)$

53. $(7 - 2i)(2 - 6i)$

54. $(-4 + 5i)(3 - 4i)$

55. $(3 + 8i)(3 - 8i)$

56. $(1 + 2i)(1 - 2i)$

57. $(-7 + i)(-7 - i)$

58. $(-4 + 5i)(-4 - 5i)$

59. $(4 - 2i)^2$

60. $(1 - 2i)^2$

61. $(2 + 3i)^2$

62. $(3 + 2i)^2$

63. $(-2 + 3i)^2$

64. $(-5 - 2i)^2$

Conjugates and Division

Divide and simplify. Write each answer in the form a + bi.

65. $\dfrac{10}{3 + i}$

66. $\dfrac{26}{5 + i}$

67. $\dfrac{2}{3 - 2i}$

68. $\dfrac{4}{2 - 3i}$

69. $\dfrac{2i}{5 + 3i}$

70. $\dfrac{3i}{4 + 2i}$

71. $\dfrac{5}{6i}$

72. $\dfrac{4}{7i}$

73. $\dfrac{5 - 3i}{4i}$

74. $\dfrac{2 + 7i}{5i}$

Aha! 75. $\dfrac{7i + 14}{7i}$

76. $\dfrac{6i + 3}{3i}$

77. $\dfrac{4 + 5i}{3 - 7i}$

78. $\dfrac{5 + 3i}{7 - 4i}$

79. $\dfrac{2 + 3i}{2 + 5i}$

80. $\dfrac{3 + 2i}{4 + 3i}$

81. $\dfrac{3 - 2i}{4 + 3i}$

82. $\dfrac{5 - 2i}{3 + 6i}$

Powers of i

Simplify.

83. i^{32}

84. i^{19}

85. i^{15}

86. i^{38}

87. i^{42}

88. i^{64}

89. i^9

90. i^{17}

91. $(-i)^6$

92. $(-i)^4$

93. $(5i)^3$

94. $(-3i)^5$

95. $i^2 + i^4$

96. $5i^5 + 4i^3$

97. Is the product of two imaginary numbers always an imaginary number? Why or why not?

98. In what way(s) are conjugates of complex numbers similar to the conjugates used in Section 7.5?

Skill Review

Factor completely.

99. $x^2 - 100$ [5.4]

100. $t^3 + 1000$ [5.5]

101. $2x - 63 + x^2$ [5.2]

102. $12a^3 - 5a^2 - 3a$ [5.3]

103. $w^3 - 4w + 3w^2 - 12$ [5.1], [5.4]

104. $24x^3y^2 - 60x^2y^4 - 12x^2y^2$ [5.1]

Synthesis

105. Is the set of real numbers a subset of the set of complex numbers? Why or why not?

106. Explain why there is no need to use powers of i (other than 1) when writing complex numbers.

Complex numbers are often graphed on a plane. The horizontal axis is the real axis and the vertical axis is the imaginary axis. A complex number such as $5 - 2i$ then corresponds to 5 on the real axis and -2 on the imaginary axis.

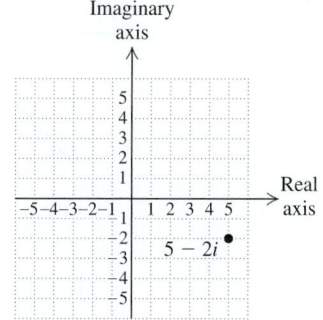

107. Graph each of the following.

   a) $3 + 2i$

   b) $-1 + 4i$

   c) $3 - i$

   d) $-5i$

108. Graph each of the following.

   a) $1 - 4i$

   b) $-2 - 3i$

   c) i

   d) 4

The absolute value of a complex number $a + bi$ is its distance from the origin. (See the graph above.) Using the distance formula, we have $|a + bi| = \sqrt{a^2 + b^2}$. Find the absolute value of each complex number.

109. $|3 + 4i|$

110. $|8 - 6i|$

111. $|-1 + i|$

112. $|-3 - i|$

Consider the function g given by

$$g(z) = \frac{z^4 - z^2}{z - 1}.$$

113. Find $g(3i)$.

114. Find $g(1 + i)$.

115. Find $g(5i - 1)$.

116. Find $g(2 - 3i)$.

117. Evaluate

$$\frac{1}{w - w^2} \quad \text{for} \quad w = \frac{1 - i}{10}.$$

Simplify.

118. $\dfrac{i^5 + i^6 + i^7 + i^8}{(1 - i)^4}$

119. $(1 - i)^3(1 + i)^3$

120. $\dfrac{5 - \sqrt{5}\,i}{\sqrt{5}\,i}$

121. $\dfrac{6}{1 + \dfrac{3}{i}}$

122. $\left(\dfrac{1}{2} - \dfrac{1}{3}i\right)^2 - \left(\dfrac{1}{2} + \dfrac{1}{3}i\right)^2$

123. $\dfrac{i - i^{38}}{1 + i}$

↳ **YOUR TURN ANSWERS: SECTION 10.8**

1. $6i$ **2.** $-4 - 3i$ **3.** $22 - 3i$ **4.** $2 - 9i$ **5.** 13

6. $-\frac{6}{5} + \frac{8}{5}i$ **7.** -1 **8.** i **9.** 1

QUICK QUIZ: SECTIONS 10.1–10.8

Let $f(x) = \sqrt{2x - 1}$.

1. Find $f(5)$.   [10.1]

2. Find the domain of f.   [10.1]

3. Find all a such that $f(a) = 7$.   [10.6]

4. Simplify: $(5 - \sqrt{2})(1 - 3\sqrt{6})$.   [10.5]

5. Simplify: $(3 - i)(5 + 2i)$.   [10.8]

PREPARE TO MOVE ON

Solve.   [5.7]

1. $x^2 - x - 6 = 0$

2. $(x - 5)^2 = 0$

3. $2t^2 - 50 = 0$

4. $15x^2 = 14x + 8$

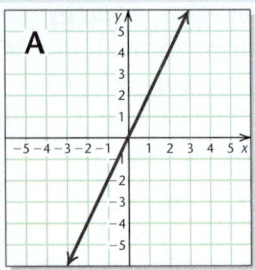

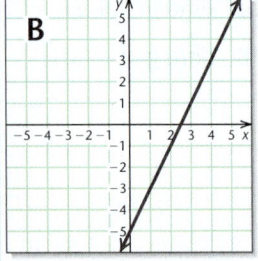

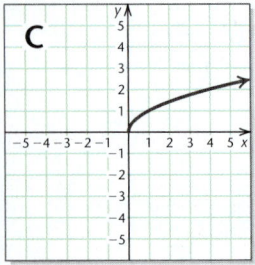

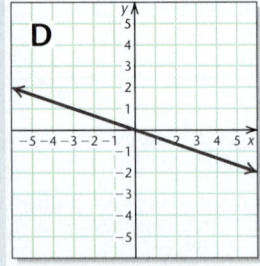

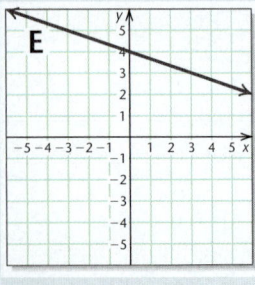

Visualizing for Success

Use after Section 10.1.

Match each function with its graph.

1. $f(x) = 2x - 5$

2. $f(x) = x^2 - 1$

3. $f(x) = \sqrt{x}$

4. $f(x) = x - 2$

5. $f(x) = -\frac{1}{3}x$

6. $f(x) = 2x$

7. $f(x) = 4 - x$

8. $f(x) = |2x - 5|$

9. $f(x) = -2$

10. $f(x) = -\frac{1}{3}x + 4$

Answers on page A-51

An additional, animated version of this activity appears in MyMathLab. *To use MyMathLab, you need a course ID and a student access code. Consult your instructor for more information.*

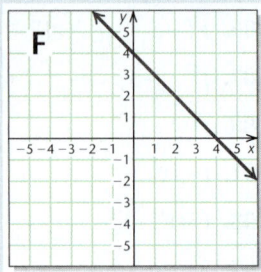

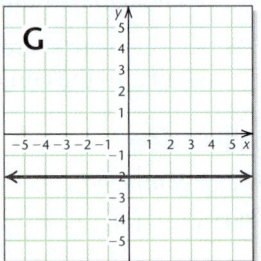

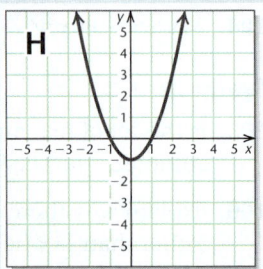

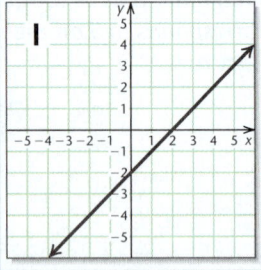

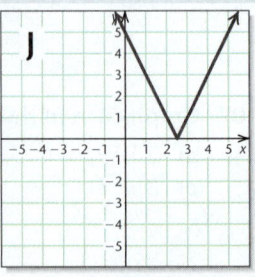

Collaborative Activity   *Tailgater Alert*

Focus: Radical equations and problem solving
Use after: Section 10.6
Time: 15–25 minutes
Group size: 2–3
Materials: Calculators

The faster a car travels, the more distance it needs to stop. Police recommend that for each 10 mph of speed, a driver allow 1 car length between vehicles. Thus a driver traveling at 30 mph should have at least 3 car lengths between his or her vehicle and the one in front.
     The function $r(L) = 2\sqrt{5L}$ can be used to find the speed, in miles per hour, that a car was traveling when it left skid marks L feet long.

Activity

1. Each group member should estimate the length of a car in which he or she frequently travels. (Each should use a different length, if possible.)

2. Using a calculator as needed, each group member should complete the table below.

Column 1 gives a car's speed s, and column 2 lists the minimum amount of space between cars traveling s miles per hour, as recommended by police. Column 3 is the speed that a vehicle *could* travel were it forced to stop in the distance listed in column 2, using the above function.

Column 1 s (in miles per hour)	Column 2 $L(s)$ (in feet)	Column 3 $r(L)$ (in miles per hour)
20		
30		
40		
50		
60		

3. Compare tables to determine whether there are any speeds at which the "1 car length per 10 mph" guideline might not suffice. What recommendations would your group make to a new driver?

Decision Making & Connection   (*Use after Section 10.6.*)

Distance.   The distance D, in miles, that one can see to the horizon from a height of h feet can be approximated by $D = \sqrt{1.5h}$.

1. On a clear day, Max stands at the top of an observation tower on a mountain overlooking the ocean. According to his GPS, he is at an elevation of 1500 ft. How far can he see to the horizon?

2. On Judy's first flight in a hot air balloon, she noticed that she could detect, on the horizon, a landmark that she knew was 30 mi away from the ground under her balloon. How high was the balloon?

3. Cole plans to erect two radio antennas, each of which is 100 ft tall. Assume for this situation that all terrain is at sea level.

   a) How far could one see to the horizon from the top of one antenna?
   b) How far apart should he place the antennas if he wants the top of one antenna to be just visible from the top of the other over the horizon? (This distance is called the visual line-of-sight (VLOS).)
   c) Radio waves can "bend" slightly around the horizon. The radio line-of-sight (RLOS) is

the maximum distance between two antennas in radio communication. Under standard atmospheric conditions, the RLOS is approximately $\frac{4}{3}$ of the VLOS. How far apart can Cole erect the antennas and still maintain radio communication?

 4. *Research.*   Because the surface of the earth is spherical, distance between points described by latitude and longitude coordinates is found using trigonometry.

   a) Find the latitude and longitude coordinates of two locations that are between 100 and 1000 mi apart. Use a GPS or an online tool to determine the distance between those points.
   b) If points on the earth's surface are mapped on a plane, they can be described using UTM (Universal Transverse Mercator) coordinates, and the distance between the points can be found using the Pythagorean theorem. Use a GPS or an online tool to convert the coordinates in part (a) to UTM coordinates and to determine the distance between the points.
   c) Explain why the distances found in parts (a) and (b) are not exactly the same.

Study Summary

KEY TERMS AND CONCEPTS	EXAMPLES	PRACTICE EXERCISES

SECTION 10.1: Radical Expressions and Functions

c is a **square root** of a if $c^2 = a$.	The square roots of 25 are -5 and 5.	*Simplify.* **1.** $-\sqrt{81}$
c is a **cube root** of a if $c^3 = a$.	The cube root of -8 is -2.	**2.** $\sqrt[3]{-1}$
$\sqrt{a}$ indicates the **principal** square root of a.	$\sqrt{25} = 5$	
$\sqrt[n]{a}$ indicates the **nth root** of a.	$\sqrt[3]{-8} = -2$	
index $\sqrt[n]{a}$ —— **radicand**	The index of $\sqrt[3]{-8}$ is 3. The radicand of $\sqrt[3]{-8}$ is -8.	
For all a, $\sqrt[n]{a^n} = \lvert a \rvert$ when n is even; $\sqrt[n]{a^n} = a$ when n is odd. If a represents a nonnegative number, $\sqrt[n]{a^n} = a$.	Assume that x can represent any real number. $$\sqrt{(3+x)^2} = \lvert 3 + x \rvert$$ Assume that x represents a nonnegative number. $$\sqrt{(7x)^2} = 7x$$	**3.** Simplify $\sqrt{36x^2}$. Assume that x can represent any real number. **4.** Simplify $\sqrt[4]{x^4}$. Assume that x represents a nonnegative number.

SECTION 10.2: Rational Numbers as Exponents

$a^{1/n}$ means $\sqrt[n]{a}$. $a^{m/n}$ means $(\sqrt[n]{a})^m$ or $\sqrt[n]{a^m}$. $a^{-m/n}$ means $\dfrac{1}{a^{m/n}}$.	$64^{1/2} = \sqrt{64} = 8$ $125^{2/3} = \left(\sqrt[3]{125}\right)^2 = 5^2 = 25$ $8^{-1/3} = \dfrac{1}{8^{1/3}} = \dfrac{1}{2}$	**5.** Simplify: $100^{-1/2}$.

SECTION 10.3: Multiplying Radical Expressions

The Product Rule for Radicals For any real numbers $\sqrt[n]{a}$ and $\sqrt[n]{b}$, $\sqrt[n]{a} \cdot \sqrt[n]{b} = \sqrt[n]{a \cdot b}$.	$\sqrt[3]{4x} \cdot \sqrt[3]{5y} = \sqrt[3]{20xy}$ $\sqrt{75x^8y^{11}} = \sqrt{25 \cdot x^8 \cdot y^{10} \cdot 3 \cdot y}$ $\phantom{\sqrt{75x^8y^{11}}} = \sqrt{25} \cdot \sqrt{x^8} \cdot \sqrt{y^{10}} \cdot \sqrt{3y}$ $\phantom{\sqrt{75x^8y^{11}}} = 5x^4y^5\sqrt{3y}$	**6.** Multiply: $\sqrt{7x} \cdot \sqrt{3y}$. **7.** Simplify: $\sqrt{200x^5y^{18}}$.

SECTION 10.4: Dividing Radical Expressions

The Quotient Rule for Radicals For any real numbers $\sqrt[n]{a}$ and $\sqrt[n]{b}$, $b \neq 0$, $\sqrt[n]{\dfrac{a}{b}} = \dfrac{\sqrt[n]{a}}{\sqrt[n]{b}}$.	$\sqrt[3]{\dfrac{8y^4}{125}} = \dfrac{\sqrt[3]{8y^4}}{\sqrt[3]{125}} = \dfrac{2y\sqrt[3]{y}}{5}$ $\dfrac{\sqrt{18a^9}}{\sqrt{2a^3}} = \sqrt{\dfrac{18a^9}{2a^3}} = \sqrt{9a^6} = 3a^3$ Assuming a is positive	**8.** Simplify: $\sqrt{\dfrac{12x^3}{25}}$.

| We can **rationalize a denominator** by multiplying by 1. | $\dfrac{1}{\sqrt{2}} = \dfrac{1}{\sqrt{2}} \cdot \dfrac{\sqrt{2}}{\sqrt{2}} = \dfrac{\sqrt{2}}{2}$ | **9.** Rationalize the denominator: $\sqrt{\dfrac{2x}{3y}}.$ |

SECTION 10.5:   *Expressions Containing Several Radical Terms*

Like radicals have both the same indices and radicands.	$\sqrt{12} + 5\sqrt{3} = \sqrt{4 \cdot 3} + 5\sqrt{3} = 2\sqrt{3} + 5\sqrt{3} = 7\sqrt{3}$	**10.** Simplify: $5\sqrt{8} - 3\sqrt{50}.$
Radical expressions are multiplied in much the same way that polynomials are multiplied.	$(1 + 5\sqrt{6})(4 - \sqrt{6})$ $\quad = 1 \cdot 4 - 1\sqrt{6} + 4 \cdot 5\sqrt{6} - 5\sqrt{6} \cdot \sqrt{6}$ $\quad = 4 - \sqrt{6} + 20\sqrt{6} - 5 \cdot 6$ $\quad = -26 + 19\sqrt{6}$	**11.** Simplify: $(2 - \sqrt{3})(5 - 7\sqrt{3}).$
To rationalize a denominator containing two terms, we use the **conjugate** of the denominator to write a form of 1.	$\dfrac{2}{1 - \sqrt{3}} = \dfrac{2}{1 - \sqrt{3}} \cdot \dfrac{1 + \sqrt{3}}{1 + \sqrt{3}} \quad \begin{array}{l}1 + \sqrt{3} \text{ is the} \\ \text{conjugate of } 1 - \sqrt{3}.\end{array}$ $\qquad = \dfrac{2(1 + \sqrt{3})}{-2} = -1 - \sqrt{3}$	**12.** Rationalize the denominator: $\dfrac{\sqrt{15}}{3 + \sqrt{5}}.$
When terms have different indices, we can often use rational exponents to simplify.	$\sqrt[3]{p} \cdot \sqrt[4]{q^3} = p^{1/3} \cdot q^{3/4}$ $\qquad = p^{4/12} \cdot q^{9/12} \quad \begin{array}{l}\text{Finding a common} \\ \text{denominator}\end{array}$ $\qquad = \sqrt[12]{p^4 q^9}$	**13.** Simplify: $\dfrac{\sqrt{x^5}}{\sqrt[3]{x}}.$

SECTION 10.6:   *Solving Radical Equations*

| **The Principle of Powers** If $a = b$, then $a^n = b^n$.

 Solutions found using the principle of powers must be checked in the original equation. | $\begin{aligned} x - 7 &= \sqrt{x - 5} \\ (x - 7)^2 &= (\sqrt{x - 5})^2 \\ x^2 - 14x + 49 &= x - 5 \\ x^2 - 15x + 54 &= 0 \\ (x - 6)(x - 9) &= 0 \\ x = 6 \quad &or \quad x = 9 \quad \begin{array}{l}\text{Only 9 checks and} \\ \text{is the solution.}\end{array} \end{aligned}$ | **14.** Solve: $\sqrt{2x + 3} = x.$ |

SECTION 10.7:   *The Distance Formula, the Midpoint Formula, and Other Applications*

| **The Pythagorean Theorem** In any right triangle, if a and b are the lengths of the legs and c is the length of the hypotenuse, then $a^2 + b^2 = c^2.$ | Find the length of the hypotenuse of a right triangle with legs of lengths 4 and 7. Give an exact answer in radical notation, as well as a decimal approximation to three decimal places.
 $\begin{aligned} a^2 + b^2 &= c^2 \\ 4^2 + 7^2 &= c^2 \quad \text{Substituting} \\ 16 + 49 &= c^2 \\ 65 &= c^2 \\ \sqrt{65} &= c \quad \text{This is exact.} \\ 8.062 &\approx c \quad \text{This is approximate.} \end{aligned}$ | **15.** The hypotenuse of a right triangle is 10 m long, and one leg is 7 m long. Find the length of the other leg. Give an exact answer in radical notation, as well as a decimal approximation to three decimal places. |

Special Triangles
The length of the hypotenuse in an isosceles right triangle (45°–45°–90° triangle) is the length of a leg times $\sqrt{2}$.

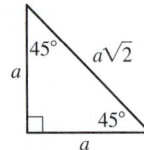

The length of the longer leg in a 30°–60°–90° triangle is the length of the shorter leg times $\sqrt{3}$. The hypotenuse is twice as long as the shorter leg.

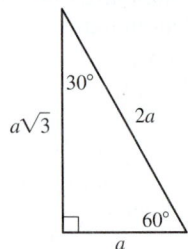

Find the missing lengths. Give an exact answer and, where appropriate, an approximation to three decimal places.

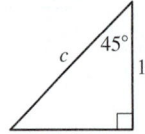

$$a = 10; \quad c = a\sqrt{2}$$
$$c = 10\sqrt{2}$$
$$c \approx 14.142$$

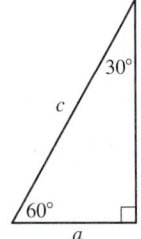

$$b = a\sqrt{3} \qquad c = 2a$$
$$18 = a\sqrt{3} \qquad c = 2\left(\frac{18}{\sqrt{3}}\right)$$
$$\frac{18}{\sqrt{3}} = a \qquad c = \frac{36}{\sqrt{3}} = 12\sqrt{3}$$
$$10.392 \approx a; \qquad c \approx 20.785$$

Find the missing lengths. Give an exact answer and, where appropriate, an approximation to three decimal places.

16.

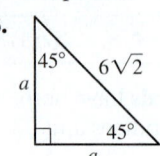

17.

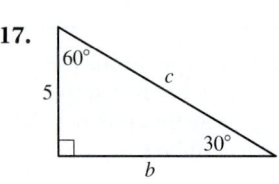

The Distance Formula
The distance d between any two points (x_1, y_1) and (x_2, y_2) is given by $d = \sqrt{(x_2 - x_1)^2 + (y_2 - y_1)^2}$.

Find the distance between $(3, -5)$ and $(-1, -2)$.
$$d = \sqrt{(-1 - 3)^2 + (-2 - (-5))^2}$$
$$= \sqrt{(-4)^2 + (3)^2}$$
$$= \sqrt{16 + 9} = \sqrt{25} = 5$$

18. Find the distance between $(-2, 1)$ and $(6, -10)$. Give an exact answer and an approximation to three decimal places.

The Midpoint Formula
If the endpoints of a segment are (x_1, y_1) and (x_2, y_2), then the coordinates of the midpoint are $\left(\dfrac{x_1 + x_2}{2}, \dfrac{y_1 + y_2}{2}\right)$.

Find the midpoint of the segment with endpoints $(3, -5)$ and $(-1, -2)$.
$$\left(\frac{3 + (-1)}{2}, \frac{-5 + (-2)}{2}\right), \quad \text{or} \quad \left(1, -\frac{7}{2}\right)$$

19. Find the midpoint of the segment with endpoints $(-2, 1)$ and $(6, -10)$.

SECTION 10.8: *The Complex Numbers*

A **complex number** is any number that can be written in the form $a + bi$, where a and b are real numbers, $i = \sqrt{-1}$, and $i^2 = -1$.

$(3 + 2i) + (4 - 7i) = 7 - 5i;$
$(8 + 6i) - (5 + 2i) = 3 + 4i;$
$(2 + 3i)(4 - i) = 8 - 2i + 12i - 3i^2$
$$= 8 + 10i - 3(-1) = 11 + 10i;$$
$$\frac{1 - 4i}{3 - 2i} = \frac{1 - 4i}{3 - 2i} \cdot \frac{3 + 2i}{3 + 2i} = \frac{3 + 2i - 12i - 8i^2}{9 + 6i - 6i - 4i^2}$$
$$= \frac{3 - 10i - 8(-1)}{9 - 4(-1)} = \frac{11 - 10i}{13} = \frac{11}{13} - \frac{10}{13}i$$
$$i^{38} = (i^2)^{19} = (-1)^{19} = -1$$

20. Add:
$(5 - 3i) + (-8 - 9i).$

21. Subtract:
$(2 - i) - (-1 + i).$

22. Multiply:
$(1 - 7i)(3 - 5i).$

23. Divide: $\dfrac{1 + i}{1 - i}.$

24. Simplify: $i^{39}.$

Review Exercises: Chapter 10

⤵ Concept Reinforcement

Classify each of the following statements as either true or false.

1. $\sqrt{ab} = \sqrt{a} \cdot \sqrt{b}$ for any real numbers $\sqrt{a}$ and $\sqrt{b}$. [10.3]

2. $\sqrt{a + b} = \sqrt{a} + \sqrt{b}$ for any real numbers $\sqrt{a}$ and $\sqrt{b}$. [10.5]

3. $\sqrt{a^2} = a$, for any real number a. [10.1]

4. $\sqrt[3]{a^3} = a$, for any real number a. [10.1]

5. $x^{2/5}$ means $\sqrt[5]{x^2}$ and $\left(\sqrt[5]{x}\right)^2$. [10.2]

6. The hypotenuse of a right triangle is never shorter than either leg. [10.7]

7. Some radical equations have no solution. [10.6]

8. If $f(x) = \sqrt{x - 5}$, then the domain of f is the set of all nonnegative real numbers. [10.1]

Simplify. [10.1]

9. $\sqrt{\dfrac{100}{121}}$

10. $-\sqrt{0.36}$

Let $f(x) = \sqrt{x + 10}$. Find the following. [10.1]

11. $f(15)$

12. The domain of f

Simplify. Assume that each variable can represent any real number. [10.1]

13. $\sqrt{64t^2}$

14. $\sqrt{(c + 7)^2}$

15. $\sqrt{4x^2 + 4x + 1}$

16. $\sqrt[5]{-32}$

17. Write an equivalent expression using exponential notation: $\left(\sqrt[3]{5ab}\right)^4$. [10.2]

18. Write an equivalent expression using radical notation: $(3a^4)^{1/5}$. [10.2]

Use rational exponents to simplify. Assume $x, y \geq 0$. [10.2]

19. $\sqrt{x^6 y^{10}}$

20. $\left(\sqrt[6]{x^2 y}\right)^2$

Simplify. Do not use negative exponents in the answers. [10.2]

21. $(x^{-2/3})^{3/5}$

22. $\dfrac{7^{-1/3}}{7^{-1/2}}$

23. If $f(x) = \sqrt{25(x - 6)^2}$, find a simplified form for $f(x)$. [10.3]

Simplify. Write all answers using radical notation. Assume that all variables represent positive numbers.

24. $\sqrt[4]{16x^{20}y^8}$ [10.3]

25. $\sqrt{250x^3y^2}$ [10.3]

26. $\sqrt{5a}\sqrt{7b}$ [10.3]

27. $\sqrt[3]{3x^4b}\sqrt[3]{9xb^2}$ [10.3]

28. $\sqrt[3]{-24x^{10}y^8}\,\sqrt[3]{18x^7y^4}$ [10.3]

29. $\dfrac{\sqrt[3]{60xy^3}}{\sqrt[3]{10x}}$ [10.4]

30. $\dfrac{\sqrt{75x}}{2\sqrt{3}}$ [10.4]

31. $\sqrt[4]{\dfrac{48a^{11}}{c^8}}$ [10.4]

32. $5\sqrt[3]{4y} + 2\sqrt[3]{4y}$ [10.5]

33. $2\sqrt{75} - 9\sqrt{3}$ [10.5]

34. $\sqrt{50} + 2\sqrt{18} + \sqrt{32}$ [10.5]

35. $(3 + \sqrt{10})(3 - \sqrt{10})$ [10.5]

36. $(\sqrt{3} - 3\sqrt{8})(\sqrt{5} + 2\sqrt{8})$ [10.5]

37. $\sqrt[4]{x}\,\sqrt{x}$ [10.5]

38. $\dfrac{\sqrt[3]{x^2}}{\sqrt[4]{x}}$ [10.5]

39. If $f(x) = x^2$, find $f(2 - \sqrt{a})$. [10.5]

40. Rationalize the denominator:
$$\sqrt{\dfrac{x}{8y}}.$$ [10.4]

41. Rationalize the denominator:
$$\dfrac{4\sqrt{5}}{\sqrt{2} + \sqrt{3}}.$$ [10.5]

42. Rationalize the numerator of the expression in Exercise 41. [10.5]

Solve. [10.6]

43. $\sqrt{y + 6} - 2 = 3$

44. $(x + 1)^{1/3} = -5$

45. $1 + \sqrt{x} = \sqrt{3x - 3}$

46. If $f(x) = \sqrt{x + 2} + x$, find a such that $f(a) = 4$. [10.6]

Solve. Give an exact answer and, where appropriate, an approximation to three decimal places.   [10.7]

47. The diagonal of a square has length 10 cm. Find the length of a side of the square.

48. A skate-park jump has a ramp that is 6 ft long and is 2 ft high. How long is its base?

6 ft   2 ft
?

49. Find the missing lengths. Give exact answers and, where appropriate, an approximation to three decimal places.

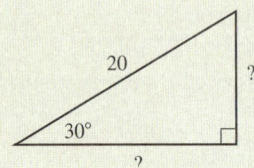

20   ?
30°
?

50. Find the distance between $(-6, 4)$ and $(-1, 5)$. Give an exact answer and an approximation to three decimal places.   [10.7]

51. Find the midpoint of the segment with endpoints $(-7, -2)$ and $(3, -1)$.   [10.7]

52. Express in terms of i and simplify: $\sqrt{-45}$.   [10.8]

53. Add: $(-4 + 3i) + (2 - 12i)$.   [10.8]

54. Subtract: $(9 - 7i) - (3 - 8i)$.   [10.8]

Simplify.   [10.8]

55. $(2 + 5i)(2 - 5i)$

56. i^{34}

57. $(6 - 3i)(2 - i)$

58. Divide. Write the answer in the form $a + bi$.

$$\frac{7 - 2i}{3 + 4i}$$ [10.8]

Synthesis

59. What makes some complex numbers real and others imaginary?   [10.8]

60. Explain why $\sqrt[n]{x^n} = |x|$ when n is even, but $\sqrt[n]{x^n} = x$ when n is odd.   [10.1]

61. Write a quotient of two imaginary numbers that is a real number (answers may vary).   [10.8]

62. Solve: $\sqrt{11x} + \sqrt{6 + x} = 6$.   [10.6]

63. Simplify:

$$\frac{2}{1 - 3i} - \frac{3}{4 + 2i}.$$ [10.8]

64. Don's Discount Shoes has two locations. The sign at the original location is shaped like an isosceles right triangle. The sign at the newer location is shaped like a 30°–60°–90° triangle. The hypotenuse of each sign measures 6 ft. Which sign has the greater area and by how much? (Round to three decimal places.)   [10.7]

Test: Chapter 10

For step-by-step test solutions, access the Chapter Test Prep Videos in MyMathLab®, on You Tube® (search "Bittinger Combo Alg CA" and click on "Channels"), or by scanning the code.

Simplify. Assume that variables can represent any real number.

1. $\sqrt{50}$

2. $\sqrt[3]{-\dfrac{8}{x^6}}$

3. $\sqrt{81a^2}$

4. $\sqrt{x^2 - 8x + 16}$

5. Write an equivalent expression using exponential notation: $\sqrt{7xy}$.

6. Write an equivalent expression using radical notation: $(4a^3b)^{5/6}$.

7. If $f(x) = \sqrt{2x - 10}$, determine the domain of f.

8. If $f(x) = x^2$, find $f(5 + \sqrt{2})$.

Simplify. Write all answers using radical notation. Assume that all variables represent positive numbers.

9. $\sqrt[5]{32x^{16}y^{10}}$

10. $\sqrt[3]{4w}\,\sqrt[3]{4v^2}$

11. $\sqrt{\dfrac{100a^4}{9b^6}}$

12. $\dfrac{\sqrt[5]{48x^6y^{10}}}{\sqrt[5]{16x^2y^9}}$

13. $\sqrt[4]{x^3}\,\sqrt{x}$

14. $\dfrac{\sqrt{y}}{\sqrt[10]{y}}$

15. $8\sqrt{2} - 2\sqrt{2}$

16. $\sqrt{50xy} + \sqrt{72xy} - \sqrt{8xy}$

17. $(7 + \sqrt{x})(2 - 3\sqrt{x})$

18. Rationalize the denominator:
$$\dfrac{\sqrt[3]{x}}{\sqrt[3]{4y}}.$$

Solve.

19. $6 = \sqrt{x - 3} + 5$

20. $x = \sqrt{3x + 3} - 1$

21. $\sqrt{2x} = \sqrt{x + 1} + 1$

Solve. For Exercises 22–24, give exact answers and approximations to three decimal places.

22. A referee jogs diagonally from one corner of a 50-ft by 90-ft basketball court to the far corner. How far does she jog?

23. The hypotenuse of a 30°–60°–90° triangle is 10 cm long. Find the lengths of the legs.

24. Find the distance between the points $(3, 7)$ and $(-1, 8)$.

25. Find the midpoint of the segment with endpoints $(2, -5)$ and $(1, -7)$.

26. Express in terms of i and simplify: $\sqrt{-50}$.

27. Subtract: $(9 + 8i) - (-3 + 6i)$.

28. Multiply. Write the answer in the form $a + bi$.
$$(4 - i)^2$$

29. Divide. Write the answer in the form $a + bi$.
$$\dfrac{-2 + i}{3 - 5i}$$

30. Simplify: i^{37}.

Synthesis

31. Solve: $\sqrt{2x - 2} + \sqrt{7x + 4} = \sqrt{13x + 10}$.

32. Simplify:
$$\dfrac{1 - 4i}{4i(1 + 4i)^{-1}}.$$

33. The function $D(h) = 1.2\sqrt{h}$ can be used to approximate the distance D, in miles, that a person can see to the horizon from a height h, in feet. How far above sea level must a pilot fly in order to see a horizon that is 180 mi away?

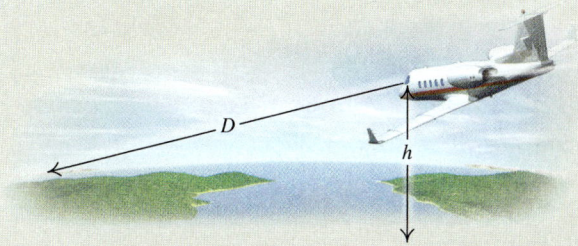

Cumulative Review: Chapters 1–10

Solve.

1. $x(x + 2) = 35$ [5.7]

2. $\dfrac{1}{x} = \dfrac{2}{5}$ [6.6]

3. $\sqrt[3]{t} = -1$ [10.6]

4. $25x^2 - 10x + 1 = 0$ [5.7]

5. $|x - 2| \le 5$ [9.3]

6. $2x + 5 > 6 \ or \ x - 3 \le 9$ [9.2]

7. $\dfrac{2x}{x - 1} + \dfrac{x}{x - 3} = 2$ [6.6]

8. $x = \sqrt{2x - 5} + 4$ [10.6]

9. $2x - \ y + \ z = 1,$
$\quad x + 2y + \ z = -3,$
$\quad 5x - \ y + 3z = 0$ [8.4]

Graph on a plane.

10. $3y = -6$ [3.3]

11. $y = -x + 5$ [3.6]

12. $x + y \le 2$ [9.4]

13. $2x = y$ [3.2]

14. Find an equation for the line parallel to the line given by $y = 7x$ and passing through the point $(0, -11)$.   [3.7]

Perform the indicated operations and, if possible, simplify.

15. $18 \div 3 \cdot 2 - 6^2 \div (2 + 4)$ [1.8]

16. $(2a - 5b)^2$ [4.6]

17. $(c^2 - 3d)(c^2 + 3d)$ [4.6]

18. $\dfrac{x + 3}{x - 2} - \dfrac{x + 5}{x + 1}$ [6.4]

19. $\dfrac{a^2 - a - 6}{a^2 - 1} \div \dfrac{a^2 - 6a + 9}{2a^2 + 3a + 1}$ [6.2]

20. $\dfrac{\dfrac{1}{x} + \dfrac{1}{x + 1}}{\dfrac{x}{x + 1}}$ [6.5]

21. $\sqrt{200} - 5\sqrt{8}$ [10.5]

22. $(1 + \sqrt{5})(4 - \sqrt{5})$ [10.5]

23. $\sqrt[3]{y}\sqrt[5]{y}$ [10.5]

Factor.

24. $x^2 - 5x - 14$ [5.2]

25. $4y^8 - 4y^5$ [5.5]

26. $3t^2 - 5t - 8$ [5.3]

27. $yt - xt - yz^2 + xz^2$ [5.1]

Find the domain of each function.

28. $f(x) = \dfrac{2x - 3}{x^2 - 6x + 9}$ [7.2]

29. $f(x) = \sqrt{2x - 11}$ [10.1]

Find each of the following, if $f(x) = \sqrt{2x - 3}$ and $g(x) = x^2$.

30. $g(1 - \sqrt{5})$ [10.5]

31. $(f + g)(x)$ [7.2], [10.5]

32. *Emergency Shelter.* The entrance to a tent used by a rescue team is the shape of an equilateral triangle. If the base of the tent is 4 ft wide, how tall is the tent? Give an exact answer and an approximation to three decimal places.   [10.7]

33. *Age at Marriage.* The median age at first marriage for U.S. men has grown from 25.1 in 2001 to 25.5 in 2006. Let $m(t)$ represent the median age of men at first marriage t years after 2000.   [7.3]

    Source: U.S. Census Bureau

    a) Find a linear function that fits the data.

    b) Use the function from part (a) to predict the median age of men at first marriage in 2020.

    c) In what year will the median age of men at first marriage reach 28 for the first time?

34. *Salary.* Nell's annual salary is \$38,849. This includes a 6% superior performance raise. What would Nell's salary have been without the performance raise?   [2.4]

35. *Landscaping.* A rectangular parking lot is 80 ft by 100 ft. Part of the asphalt is removed in order to install a landscaped border of uniform width around it. The area of the new parking lot is 6300 ft². How wide is the landscaped border?   [5.8]

Synthesis

36. Give an equation in standard form for the line whose x-intercept is $(-3, 0)$ and whose y-intercept is $(0, 5)$.   [3.7]

Solve.

37. $\dfrac{\dfrac{1}{x} + \dfrac{1}{x + 1}}{\dfrac{1}{x} - 1} = 1$ [6.5], [6.6]

38. $2\sqrt{3x - 2} = 2 + \sqrt{7x + 1}$ [10.6]

Quadratic Functions and Equations

YEAR	NUMBER OF SELF-PUBLISHED BOOKS
2006	51,000
2007	63,000
2008	74,000
2009	95,000
2010	133,000

Source: Based on data from Bowker's Books in Print database, in "Secret of Self-Publishing: Success," *The Wall Street Journal*, October 31, 2011

You Too Can Write a Book!

Although publication carries no guarantee of sales, many new and established authors are turning to self-publishing to see their works appear, either in print or as e-books. As the table above indicates, the number of self-published books is increasing dramatically and can be modeled using a *quadratic function*. (*See Example 3 in Section 11.8.*)

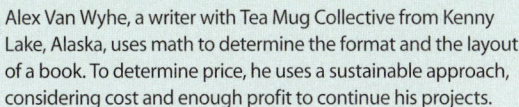

A self-published author uses math in some way in almost every stage of production.

Alex Van Wyhe, a writer with Tea Mug Collective from Kenny Lake, Alaska, uses math to determine the format and the layout of a book. To determine price, he uses a sustainable approach, considering cost and enough profit to continue his projects.

T he mathematical translation of a problem is often a function or an equation containing a second-degree polynomial in one variable. Such functions or equations are said to be *quadratic*. In this chapter, we examine a variety of ways to solve quadratic equations and look at graphs and applications of quadratic functions.

11.1   Quadratic Equations

The Principle of Square Roots   ▪   Completing the Square   ▪   Problem Solving

The general form of a quadratic function is

$$f(x) = ax^2 + bx + c, \quad \text{with } a \neq 0,$$

and its graph is a *parabola*. Such graphs open up or down and can have 0, 1, or 2 x-intercepts. We learn to graph quadratic functions later in this chapter.

Algebraic 🔗 Graphical Connection

The graphs of the quadratic function $f(x) = x^2 + 6x + 8$ and the linear function $g(x) = 0$ are shown below.

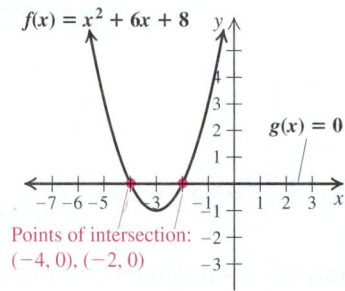

Note that $(-4, 0)$ and $(-2, 0)$ are the points of intersection of the graphs of $f(x) = x^2 + 6x + 8$ and $g(x) = 0$ (the x-axis).

As we saw in Section 5.8, we can solve equations like $x^2 + 6x + 8 = 0$ by factoring:

$$x^2 + 6x + 8 = 0$$

$$(x + 4)(x + 2) = 0 \qquad\qquad \text{Factoring}$$

$$x + 4 = 0 \quad or \quad x + 2 = 0 \qquad \text{Using the principle of zero products}$$

$$x = -4 \quad or \qquad x = -2.$$

Note that -4 and -2 are the first coordinates of the points of intersection (or the x-intercepts) of the graph of $f(x)$ above.

To solve a quadratic equation by factoring, we write the equation in the *standard form* $ax^2 + bx + c = 0$, factor, and use the principle of zero products.

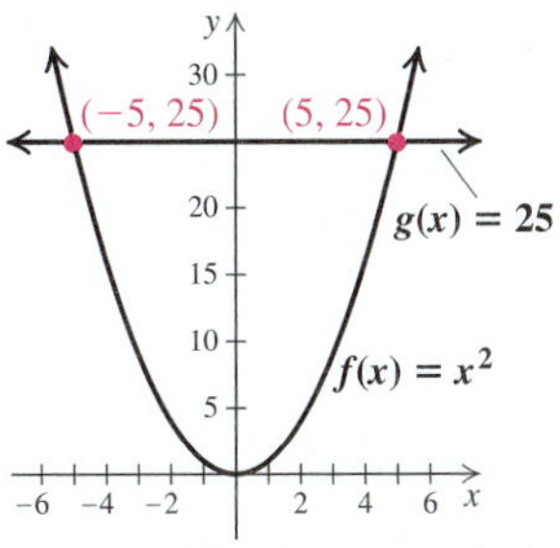

A visualization of Example 1

1. Solve: $64 = y^2$.

EXAMPLE 1 Solve: $x^2 = 25$.

SOLUTION We have

$$x^2 = 25$$
$$x^2 - 25 = 0 \qquad \text{Writing in standard form}$$
$$(x - 5)(x + 5) = 0 \qquad \text{Factoring}$$
$$x - 5 = 0 \quad or \quad x + 5 = 0 \qquad \text{Using the principle of zero products}$$
$$x = 5 \quad or \qquad x = -5.$$

The solutions are 5 and -5. A graph in which $f(x) = x^2$ represents the left side of the equation and $g(x) = 25$ represents the right side provides a check (see the figure at left). We can also check by substituting 5 and -5 into the original equation.

YOUR TURN

In this section and the next, we develop algebraic methods for solving *any* quadratic equation, including those that cannot be solved by factoring.

The Principle of Square Roots

Let's reconsider $x^2 = 25$. The number 25 has two square roots, 5 and -5, the solutions of the equation. Square roots provide quick solutions for equations of the type $x^2 = k$.

THE PRINCIPLE OF SQUARE ROOTS

For any real number k, if $x^2 = k$, then

$$x = \sqrt{k} \quad or \quad x = -\sqrt{k}.$$

EXAMPLE 2 Solve: $3x^2 = 6$. Give exact solutions and approximations to three decimal places.

SOLUTION We have

$$3x^2 = 6$$
$$x^2 = 2 \qquad \text{Isolating } x^2$$
$$x = \sqrt{2} \quad or \quad x = -\sqrt{2}. \qquad \text{Using the principle of square roots}$$

We can use the symbol $\pm \sqrt{2}$ to represent both of the solutions.

CAUTION! There are *two* solutions: $\sqrt{2}$ and $-\sqrt{2}$. Don't forget the second solution.

A visualization of Example 2

2. Solve: $t^2 = 10$. Give exact solutions and approximations to three decimal places.

Check: For $\sqrt{2}$:

$$\begin{array}{c|c} 3x^2 = 6 \\ \hline 3(\sqrt{2})^2 & 6 \\ 3 \cdot 2 & \\ 6 \stackrel{?}{=} 6 & \text{TRUE} \end{array}$$

For $-\sqrt{2}$:

$$\begin{array}{c|c} 3x^2 = 6 \\ \hline 3(-\sqrt{2})^2 & 6 \\ 3 \cdot 2 & \\ 6 \stackrel{?}{=} 6 & \text{TRUE} \end{array}$$

The solutions are $\sqrt{2}$ and $-\sqrt{2}$, or $\pm \sqrt{2}$, which round to 1.414 and -1.414.

YOUR TURN

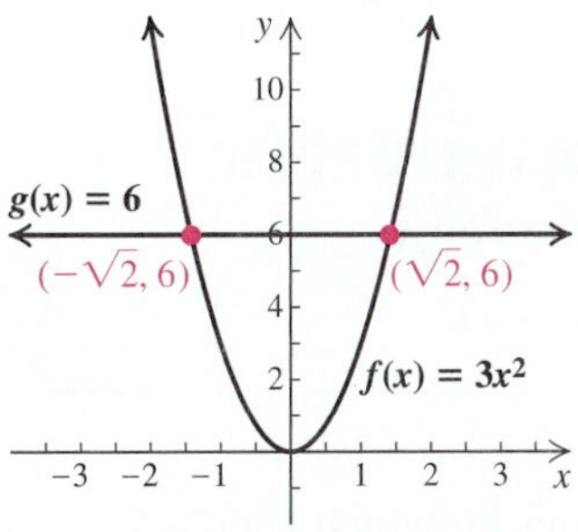

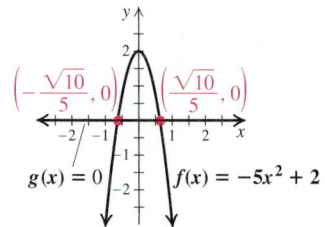

A visualization of Example 3

3. Solve: $3x^2 = 1$.

EXAMPLE 3 Solve: $-5x^2 + 2 = 0$.

SOLUTION We have

$$-5x^2 + 2 = 0$$
$$x^2 = \tfrac{2}{5} \qquad \text{Isolating } x^2$$
$$x = \sqrt{\tfrac{2}{5}} \quad \text{or} \quad x = -\sqrt{\tfrac{2}{5}}. \qquad \text{Using the principle of square roots}$$

The solutions are $\sqrt{\tfrac{2}{5}}$ and $-\sqrt{\tfrac{2}{5}}$, or simply $\pm\sqrt{\tfrac{2}{5}}$. If we rationalize the denominator, the solutions are written $\pm\frac{\sqrt{10}}{5}$. The checks are left to the student.

YOUR TURN

We can now find imaginary-number solutions. Unless the context dictates otherwise, you should find all complex-number solutions of equations.

EXAMPLE 4 Solve: $4x^2 + 9 = 0$. (Find all solutions in the complex-number system.)

SOLUTION We have

$$4x^2 + 9 = 0$$
$$x^2 = -\tfrac{9}{4} \qquad \text{Isolating } x^2$$
$$x = \sqrt{-\tfrac{9}{4}} \quad \text{or} \quad x = -\sqrt{-\tfrac{9}{4}} \qquad \text{Using the principle of square roots}$$
$$x = \sqrt{\tfrac{9}{4}}\sqrt{-1} \quad \text{or} \quad x = -\sqrt{\tfrac{9}{4}}\sqrt{-1}$$
$$x = \tfrac{3}{2}i \quad \text{or} \quad x = -\tfrac{3}{2}i. \qquad \text{Recall that } \sqrt{-1} = i.$$

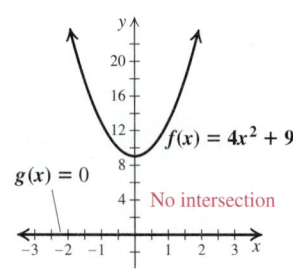

A visualization of Example 4

4. Solve: $2t^2 + 200 = 0$.

Check:

For $\tfrac{3}{2}i$:

$$\begin{array}{c|c} 4x^2 + 9 = 0 & \\ \hline 4\left(\tfrac{3}{2}i\right)^2 + 9 & 0 \\ 4 \cdot \tfrac{9}{4} \cdot i^2 + 9 & \\ 9(-1) + 9 & \\ & 0 \overset{?}{=} 0 \quad \text{TRUE} \end{array}$$

For $-\tfrac{3}{2}i$:

$$\begin{array}{c|c} 4x^2 + 9 = 0 & \\ \hline 4\left(-\tfrac{3}{2}i\right)^2 + 9 & 0 \\ 4 \cdot \tfrac{9}{4} \cdot i^2 + 9 & \\ 9(-1) + 9 & \\ & 0 \overset{?}{=} 0 \quad \text{TRUE} \end{array}$$

The solutions are $\tfrac{3}{2}i$ and $-\tfrac{3}{2}i$, or $\pm\tfrac{3}{2}i$. The graph at left confirms that there are no real-number solutions.

YOUR TURN

The principle of square roots can be restated in a more general form.

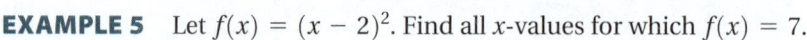

THE PRINCIPLE OF SQUARE ROOTS (GENERALIZED FORM)

For any real number k and any algebraic expression X:

If $X^2 = k$, then $X = \sqrt{k}$ or $X = -\sqrt{k}$.

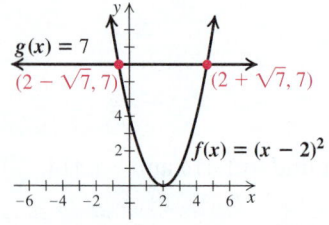

A visualization of Example 5

EXAMPLE 5 Let $f(x) = (x - 2)^2$. Find all x-values for which $f(x) = 7$.

SOLUTION We are asked to find all x-values for which

$$f(x) = 7,$$

or

$$(x - 2)^2 = 7. \qquad \text{Substituting } (x - 2)^2 \text{ for } f(x)$$

The generalized principle of square roots gives us

$$x - 2 = \sqrt{7} \qquad \text{or} \quad x - 2 = -\sqrt{7}$$
$$x = 2 + \sqrt{7} \quad \text{or} \qquad x = 2 - \sqrt{7}.$$

Check: $f(2 + \sqrt{7}) = (2 + \sqrt{7} - 2)^2 = (\sqrt{7})^2 = 7.$

Similarly,

$$f(2 - \sqrt{7}) = (2 - \sqrt{7} - 2)^2 = (-\sqrt{7})^2 = 7.$$

The solutions are $2 + \sqrt{7}$ and $2 - \sqrt{7}$, or simply $2 \pm \sqrt{7}$.

5. Let $f(x) = (x + 5)^2$. Find all x-values for which $f(x) = 3$.

YOUR TURN

Example 5 is of the form $(x - a)^2 = c$, where a and c are constants. Sometimes we must factor in order to obtain this form.

EXAMPLE 6 Solve: $x^2 + 6x + 9 = 2$.

SOLUTION We have

$$x^2 + 6x + 9 = 2 \qquad \text{The left side is the square of a binomial.}$$
$$(x + 3)^2 = 2 \qquad \text{Factoring}$$
$$x + 3 = \sqrt{2} \qquad or \quad x + 3 = -\sqrt{2} \qquad \begin{array}{l}\text{Using the principle}\\ \text{of square roots}\end{array}$$
$$x = -3 + \sqrt{2} \quad or \qquad x = -3 - \sqrt{2}. \qquad \begin{array}{l}\text{Adding } -3 \text{ to both}\\ \text{sides}\end{array}$$

$f(x) = x^2 + 6x + 9$

$(-3 - \sqrt{2}, 2)$ $g(x) = 2$

$(-3 + \sqrt{2}, 2)$

A visualization of Example 6

The solutions are $-3 + \sqrt{2}$ and $-3 - \sqrt{2}$, or $-3 \pm \sqrt{2}$. The checks are left to the student.

YOUR TURN

6. Solve: $t^2 - 10t + 25 = 3$.

Completing the Square

By using a method called *completing the square*, we can use the principle of square roots to solve *any* quadratic equation. To see how this is done, consider

$$x^2 + 6x + 4 = 0.$$

The trinomial $x^2 + 6x + 4$ is not a perfect square. We can, however, create an equivalent equation with a perfect-square trinomial on one side:

$$x^2 + 6x + 4 = 0$$
$$x^2 + 6x = -4 \qquad \text{Only variable terms are on the left side.}$$
$$x^2 + 6x + 9 = -4 + 9 \qquad \begin{array}{l}\text{Adding 9 to both sides. We explain this}\\ \text{decision shortly.}\end{array}$$
$$(x + 3)^2 = 5. \qquad \begin{array}{l}\text{Factoring the perfect-square trinomial. We}\\ \text{could continue to solve as in Example 6.}\end{array}$$

We chose to add 9 to both sides because it creates a perfect-square trinomial on the left side. The 9 was found by taking half of the coefficient of x and squaring it.

To understand why this procedure works, examine the following drawings.

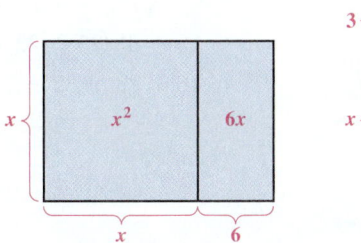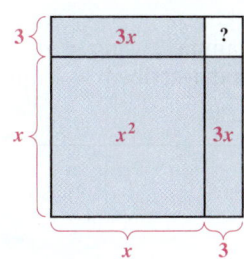

Note that the shaded areas in both figures represent the same area, $x^2 + 6x$. However, only the figure on the right, in which the $6x$ is halved, can be converted into a square with the addition of a constant term. The constant 9 is the "missing" piece that *completes* the square.

To complete the square for $x^2 + bx$, we add $(b/2)^2$.

Example 7, which follows, provides practice in finding numbers that complete the square. We will then use this skill to solve equations.

EXAMPLE 7 Replace the blanks in each equation with constants to form a true equation.

a) $x^2 + 14x + \underline{\hspace{1cm}} = (x + \underline{\hspace{1cm}})^2$

b) $x^2 - 5x + \underline{\hspace{1cm}} = (x - \underline{\hspace{1cm}})^2$

c) $x^2 + \frac{3}{4}x + \underline{\hspace{1cm}} = (x + \underline{\hspace{1cm}})^2$

Student Notes

In problems like Examples 7(b) and (c), it is best to avoid decimal notation. Most students have an easier time recognizing $\frac{9}{64}$ as $\left(\frac{3}{8}\right)^2$ than seeing 0.140625 as 0.375^2.

7. Replace the blanks with constants to form a true equation:
$x^2 + 7x + \underline{\hspace{1cm}} = (x + \underline{\hspace{1cm}})^2$.

SOLUTION

a) Take half of the coefficient of x: Half of 14 is 7.

Square this number: $7^2 = 49$.

Add 49 to complete the square: $x^2 + 14x + 49 = (x + 7)^2$.

b) Take half of the coefficient of x: Half of -5 is $-\frac{5}{2}$.

Square this number: $\left(-\frac{5}{2}\right)^2 = \frac{25}{4}$.

Add $\frac{25}{4}$ to complete the square: $x^2 - 5x + \frac{25}{4} = \left(x - \frac{5}{2}\right)^2$.

c) Take half of the coefficient of x: Half of $\frac{3}{4}$ is $\frac{3}{8}$.

Square this number: $\left(\frac{3}{8}\right)^2 = \frac{9}{64}$.

Add $\frac{9}{64}$ to complete the square: $x^2 + \frac{3}{4}x + \frac{9}{64} = \left(x + \frac{3}{8}\right)^2$.

YOUR TURN

We can now use the method of completing the square to solve equations.

EXAMPLE 8 Solve: $x^2 - 8x - 7 = 0$.

SOLUTION We begin by adding 7 to both sides:

$$x^2 - 8x - 7 = 0$$

$$x^2 - 8x \qquad = 7 \qquad \text{Adding 7 to both sides. We can now complete the square on the left side.}$$

$$x^2 - 8x + 16 = 7 + 16 \qquad \text{Adding 16 to both sides to complete the square: } \frac{1}{2}(-8) = -4, \text{ and } (-4)^2 = 16$$

$$(x - 4)^2 = 23 \qquad \text{Factoring and simplifying}$$

$$x - 4 = \pm\sqrt{23} \qquad \text{Using the principle of square roots}$$

$$x = 4 \pm \sqrt{23}. \qquad \text{Adding 4 to both sides}$$

Check: For $4 + \sqrt{23}$:

$$
\begin{array}{c|c}
x^2 - 8x - 7 = 0 & \\
\hline
(4 + \sqrt{23})^2 - 8(4 + \sqrt{23}) - 7 & 0 \\
16 + 8\sqrt{23} + 23 - 32 - 8\sqrt{23} - 7 & \\
16 + 23 - 32 - 7 + 8\sqrt{23} - 8\sqrt{23} & \\
0 \overset{?}{=} 0 & \text{TRUE}
\end{array}
$$

Technology Connection

One way to check Example 8 is to store $4 + \sqrt{23}$ as x using the STO▸ key. We can then evaluate $x^2 - 8x - 7$ by entering $x^2 - 8x - 7$ and pressing ENTER.

1. Check Example 8 using the method described above.

For $4 - \sqrt{23}$:

$$\begin{array}{c|c}
x^2 - 8x - 7 = 0 & \\
\hline
(4 - \sqrt{23})^2 - 8(4 - \sqrt{23}) - 7 & 0 \\
16 - 8\sqrt{23} + 23 - 32 + 8\sqrt{23} - 7 & \\
16 + 23 - 32 - 7 - 8\sqrt{23} + 8\sqrt{23} & \\
0 \overset{?}{=} 0 \quad \text{TRUE} &
\end{array}$$

The solutions are $4 + \sqrt{23}$ and $4 - \sqrt{23}$, or $4 \pm \sqrt{23}$.

8. Solve: $x^2 + 6x - 2 = 0$.

YOUR TURN

Recall that the value of $f(x)$ must be 0 at any x-intercept of the graph of f. If $f(a) = 0$, then $(a, 0)$ is an x-intercept of the graph.

EXAMPLE 9   Find the x-intercepts of the graph of $f(x) = x^2 + 5x - 3$.

SOLUTION   We set $f(x)$ equal to 0 and solve:

$$f(x) = 0$$
$$x^2 + 5x - 3 = 0 \qquad \text{Substituting}$$
$$x^2 + 5x = 3 \qquad \text{Adding 3 to both sides}$$
$$x^2 + 5x + \frac{25}{4} = 3 + \frac{25}{4} \qquad \begin{array}{l}\text{Completing the square:} \\ \frac{1}{2} \cdot 5 = \frac{5}{2}, \text{ and } \left(\frac{5}{2}\right)^2 = \frac{25}{4}\end{array}$$
$$\left(x + \frac{5}{2}\right)^2 = \frac{37}{4} \qquad \text{Factoring and simplifying}$$
$$x + \frac{5}{2} = \pm \frac{\sqrt{37}}{2} \qquad \begin{array}{l}\text{Using the principle of square roots} \\ \text{and the quotient rule for radicals}\end{array}$$
$$x = -\frac{5}{2} \pm \frac{\sqrt{37}}{2}, \text{ or } \frac{-5 \pm \sqrt{37}}{2}. \qquad \text{Adding } -\frac{5}{2} \text{ to both sides}$$

The x-intercepts are

$$\left(-\frac{5}{2} - \frac{\sqrt{37}}{2}, 0\right) \text{ and } \left(-\frac{5}{2} + \frac{\sqrt{37}}{2}, 0\right).$$

The checks are left to the student.

YOUR TURN

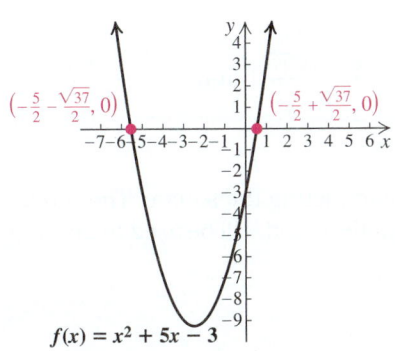

$\left(-\frac{5}{2} - \frac{\sqrt{37}}{2}, 0\right)$ $\left(-\frac{5}{2} + \frac{\sqrt{37}}{2}, 0\right)$

$f(x) = x^2 + 5x - 3$

A visualization of Example 9

9. Find the x-intercepts of the graph of $f(x) = x^2 + 4x + 1$.

Before we complete the square in a quadratic equation, the leading coefficient must be 1. When it is not 1, we divide both sides of the equation by whatever that coefficient may be.

> **TO SOLVE A QUADRATIC EQUATION IN x BY COMPLETING THE SQUARE**
>
> 1. Isolate the terms with variables on one side of the equation, and arrange them in descending order.
> 2. Divide both sides by the coefficient of x^2 if that coefficient is not 1.
> 3. Complete the square by taking half of the coefficient of x and adding its square to both sides.
> 4. Express the trinomial as the square of a binomial (factor the trinomial) and simplify the other side.
> 5. Use the principle of square roots (find the square roots of both sides).
> 6. Solve for x by adding or subtracting on both sides.

EXAMPLE 10 Solve: $3x^2 + 7x - 2 = 0$.

SOLUTION We follow the steps listed above:

$$3x^2 + 7x - 2 = 0$$

Isolate the variable terms.

$$3x^2 + 7x = 2 \qquad \text{Adding 2 to both sides}$$

Divide both sides by the coefficient of x^2.

$$x^2 + \frac{7}{3}x = \frac{2}{3} \qquad \text{Dividing both sides by 3}$$

Complete the square.

$$x^2 + \frac{7}{3}x + \frac{49}{36} = \frac{2}{3} + \frac{49}{36} \qquad \begin{array}{l} \text{Completing the square:} \\ \left(\frac{1}{2} \cdot \frac{7}{3}\right)^2 = \frac{49}{36} \end{array}$$

Factor the trinomial.

$$\left(x + \frac{7}{6}\right)^2 = \frac{73}{36} \qquad \text{Factoring and simplifying}$$

Use the principle of square roots.

$$x + \frac{7}{6} = \pm \frac{\sqrt{73}}{6} \qquad \begin{array}{l} \text{Using the principle of square roots} \\ \text{and the quotient rule for radicals} \end{array}$$

Solve for x.

$$x = -\frac{7}{6} \pm \frac{\sqrt{73}}{6}, \quad \text{or} \quad \frac{-7 \pm \sqrt{73}}{6}. \qquad \text{Adding } -\frac{7}{6} \text{ to both sides}$$

The checks are left to the student. The solutions are $-\frac{7}{6} \pm \frac{\sqrt{73}}{6}$, or $\frac{-7 \pm \sqrt{73}}{6}$.

This can be written as

$$-\frac{7}{6} + \frac{\sqrt{73}}{6} \quad \text{and} \quad -\frac{7}{6} - \frac{\sqrt{73}}{6}, \quad \text{or} \quad \frac{-7 + \sqrt{73}}{6} \quad \text{and} \quad \frac{-7 - \sqrt{73}}{6}.$$

10. Solve: $2x^2 - 8x - 3 = 0$.

YOUR TURN

Any quadratic equation can be solved by completing the square. The procedure is also useful when graphing quadratic equations and will be used to develop a formula for solving quadratic equations.

Problem Solving

After one year, an amount of money P, invested at 4% per year, is worth 104% of P, or $P(1.04)$. If that amount continues to earn 4% interest per year, after the second year the investment will be worth 104% of $P(1.04)$, or $P(1.04)^2$. This is called **compounding interest** since after the first time period, interest is earned on both the initial investment *and* the interest from the first time period. Continuing the above pattern, we see that after the third year, the investment will be worth 104% of $P(1.04)^2$, or $P(1.04)^3$. Generalizing, we have the following.

> **THE COMPOUND-INTEREST FORMULA**
>
> If an amount of money P is invested at interest rate r, compounded annually, then in t years, it will grow to the amount A given by
>
> $$A = P(1 + r)^t. \qquad (r \text{ is written in decimal notation.})$$

EXAMPLE 11 *Investment Growth.* Katia invested $4000 at interest rate r, compounded annually. In 2 years, it grew to $4410. What was the interest rate?

SOLUTION

1. **Familiarize.** The compound-interest formula is given above.

2. **Translate.** The translation consists of substituting into the formula:

$$A = P(1 + r)^t$$

$$4410 = 4000(1 + r)^2. \qquad \text{Substituting}$$

3. Carry out. We solve for r:

$$4410 = 4000(1 + r)^2$$

$$\frac{4410}{4000} = (1 + r)^2 \qquad \text{Dividing both sides by 4000}$$

$$\frac{441}{400} = (1 + r)^2 \qquad \text{Simplifying}$$

$$\pm\sqrt{\frac{441}{400}} = 1 + r \qquad \text{Using the principle of square roots}$$

$$\pm\frac{21}{20} = 1 + r \qquad \text{Simplifying}$$

$$-\frac{20}{20} \pm \frac{21}{20} = r \qquad \text{Adding } -1, \text{ or } -\frac{20}{20}, \text{ to both sides}$$

$$\frac{1}{20} = r \quad or \quad -\frac{41}{20} = r.$$

11. Max invested $1600 at interest rate r, compounded annually. In 2 years, it grew to $1936. What was the interest rate?

4. Check. Since the interest rate cannot be negative, we need check only $\frac{1}{20}$, or 5%. If $4000 were invested at 5% interest, compounded annually, then in 2 years it would grow to $4000(1.05)^2$, or $4410. The rate 5% checks.

5. State. The interest rate was 5%.

 YOUR TURN

EXAMPLE 12 *Free-Falling Objects.* The formula $s = 16t^2$ is used to approximate the distance s, in feet, that an object falls freely from rest in t seconds. The Grand Canyon Skywalk is 4000 ft above the Colorado River. How long will it take a stone to fall from the Skywalk to the river? Round to the nearest tenth of a second.

Source: www.grandcanyonskywalk.com

SOLUTION

1. Familiarize. We agree to disregard air resistance and use the given formula.

2. Translate. We substitute into the formula:

$$s = 16t^2$$
$$4000 = 16t^2.$$

3. Carry out. We solve for t:

$$4000 = 16t^2$$

$$250 = t^2$$

$$\sqrt{250} = t \qquad \text{Using the principle of square roots; rejecting the negative square root since } t \text{ cannot be negative in this problem}$$

$$15.8 \approx t. \qquad \text{Using a calculator and rounding to the nearest tenth}$$

4. Check. Since $16(15.8)^2 = 3994.24 \approx 4000$, our answer checks.

12. The Willis Tower in Chicago is 1454 ft tall. How long would it take an object to fall freely from the top? Round to the nearest tenth of a second.

5. State. It takes about 15.8 sec for a stone to fall freely from the Grand Canyon Skywalk to the river.

 YOUR TURN

Technology Connection

To check Example 10, we graph $y = 3x^2 + 7x - 2$ and use the ZERO or ROOT option of the CALC menu. We enter a Left Bound, a Right Bound, and a Guess, and a value for the root then appears. Since $-7/6 - \sqrt{73}/6 \approx -2.590667$, the answer checks.

1. Use a graphing calculator to confirm the solutions in Example 9.
2. Use a graphing calculator to confirm that there are no real-number solutions of $x^2 - 6x + 11 = 0$.

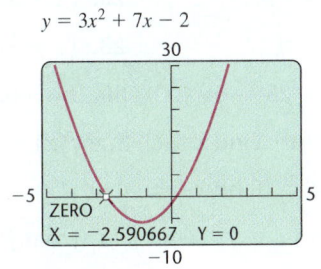

11.1 EXERCISE SET

FOR EXTRA HELP

MyMathLab® MathXL
PRACTICE   WATCH   READ   REVIEW

Vocabulary and Reading Check

Choose the word or phrase from the following list that best completes each statement.

complete the square   square roots
parabola   standard form
quadratic function   zero products

1. The general form of a(n) _____ is $f(x) = ax^2 + bx + c$.

2. The graph of a quadratic function is a(n)

   _____.

3. The quadratic equation $ax^2 + bx + c = 0$ is written in _____.

4. If $(x - 2)(x + 3) = 0$, we know that $x - 2 = 0$ or $x + 3 = 0$ because of the principle of

   _____.

5. If $x^2 = 7$, we know that $x = \sqrt{7}$ or $x = -\sqrt{7}$ because of the principle of _____.

6. We add 25 to $x^2 + 10x$ in order to

   _____.

The Principle of Square Roots

Solve. (Find all complex-number solutions.)

7. $x^2 = 100$

8. $t^2 = 144$

9. $p^2 - 50 = 0$

10. $c^2 - 8 = 0$

11. $5y^2 = 30$

12. $4y^2 = 12$

13. $9x^2 - 49 = 0$

14. $36a^2 - 25 = 0$

15. $6t^2 - 5 = 0$

16. $7x^2 - 5 = 0$

17. $a^2 + 1 = 0$

18. $t^2 + 4 = 0$

19. $4d^2 + 81 = 0$

20. $25y^2 + 16 = 0$

21. $(x - 3)^2 = 16$

22. $(x + 1)^2 = 100$

23. $(t + 5)^2 = 12$

24. $(y - 4)^2 = 18$

25. $(x + 1)^2 = -9$

26. $(x - 1)^2 = -49$

27. $\left(y + \frac{3}{4}\right)^2 = \frac{17}{16}$

28. $\left(t + \frac{3}{2}\right)^2 = \frac{7}{2}$

29. $x^2 - 10x + 25 = 64$

30. $x^2 - 6x + 9 = 100$

31. Let $f(x) = x^2$. Find x such that $f(x) = 19$.

32. Let $f(x) = x^2$. Find x such that $f(x) = 11$.

33. Let $f(x) = (x - 5)^2$. Find x such that $f(x) = 16$.

34. Let $g(x) = (x - 2)^2$. Find x such that $g(x) = 25$.

35. Let $F(t) = (t + 4)^2$. Find t such that $F(t) = 13$.

36. Let $f(t) = (t + 6)^2$. Find t such that $f(t) = 15$.

Aha! **37.** Let $g(x) = x^2 + 14x + 49$. Find x such that $g(x) = 49$.

38. Let $F(x) = x^2 + 8x + 16$. Find x such that $F(x) = 9$.

Completing the Square

Replace the blanks in each equation with constants to complete the square and form a true equation.

39. $x^2 + 16x + \underline{\quad} = (x + \underline{\quad})^2$

40. $x^2 + 12x + \underline{\quad} = (x + \underline{\quad})^2$

41. $t^2 - 10t + \underline{\quad} = (t - \underline{\quad})^2$

42. $t^2 - 6t + \underline{\quad} = (t - \underline{\quad})^2$

43. $t^2 - 2t + \underline{\quad} = (t - \underline{\quad})^2$

44. $x^2 + 2x + \underline{\quad} = (x + \underline{\quad})^2$

45. $x^2 + 3x + \underline{\quad} = (x + \underline{\quad})^2$

46. $t^2 - 9t + \underline{\quad} = (t - \underline{\quad})^2$

47. $x^2 + \frac{2}{5}x + \underline{\quad} = (x + \underline{\quad})^2$

48. $x^2 + \frac{2}{3}x + \underline{\quad} = (x + \underline{\quad})^2$

49. $t^2 - \frac{5}{6}t + \underline{\quad} = (t - \underline{\quad})^2$

50. $t^2 - \frac{5}{3}t - \underline{\quad} = (t - \underline{\quad})^2$

Solve by completing the square. Show your work.

51. $x^2 + 6x = 7$

52. $x^2 + 8x = 9$

53. $t^2 - 10t = -23$

54. $t^2 - 4t = -1$

55. $x^2 + 12x + 32 = 0$

56. $x^2 + 16x + 15 = 0$

57. $t^2 + 8t - 3 = 0$

58. $t^2 + 6t - 5 = 0$

Complete the square to find the x-intercepts of each function given by the equation listed.

59. $f(x) = x^2 + 6x + 7$

60. $f(x) = x^2 + 10x - 2$

61. $g(x) = x^2 + 9x - 25$

62. $g(x) = x^2 + 5x + 2$

63. $f(x) = x^2 - 10x - 22$

64. $f(x) = x^2 - 8x - 10$

Solve by completing the square. Remember to first divide, as in Example 10, to make sure that the coefficient of x^2 is 1.

65. $9x^2 + 18x = -8$

66. $4x^2 + 8x = -3$

67. $3x^2 - 5x - 2 = 0$

68. $2x^2 - 5x - 3 = 0$

69. $5x^2 + 4x - 3 = 0$

70. $4x^2 + 3x - 5 = 0$

71. Find the *x*-intercepts of the graph of
$f(x) = 4x^2 + 2x - 3$.

72. Find the *x*-intercepts of the graph of
$f(x) = 3x^2 + x - 5$.

73. Find the *x*-intercepts of the graph of
$g(x) = 2x^2 - 3x - 1$.

74. Find the *x*-intercepts of the graph of
$g(x) = 3x^2 - 5x - 1$.

Problem Solving

Interest.   Use $A = P(1 + r)^t$ to find the interest rate in Exercises 75–78. Refer to Example 11.

75. $2000 grows to $2205 in 2 years

76. $1000 grows to $1060.90 in 2 years

77. $6250 grows to $6760 in 2 years

78. $6250 grows to $7290 in 2 years

Free-Falling Objects.   Use $s = 16t^2$ for Exercises 79–82. Refer to Example 12 and neglect air resistance. Round answers to the nearest tenth of a second.

79. At 121 ft, Excalibur Tower in Groningen, the Netherlands, is the world's tallest man-made climbing wall. How long would it take an object to fall freely from the top?

Source: http://enpundit.com/2012/excalibur-worlds-highest-climbing-wall

80. At a height of approximately 1200 ft, Tushuk Tash in Xinjiang Uyghur Autonomous Region, China, is the world's highest natural arch. How long would it take an object to fall freely from the top of the arch?

Source: www.naturalarches.org

81. El Capitan in Yosemite National Park is 3593 ft high. How long would it take a carabiner to fall freely from the top?

Source: *Guinness World Records* 2008

82. At 2063 ft, the KVLY-TV tower in North Dakota is the world's tallest supported tower. How long would it take an object to fall freely from the top?

Source: North Dakota Tourism Division

83. Explain in your own words a sequence of steps that can be used to solve any quadratic equation in the quickest way.

84. Write an interest-rate problem for a classmate to solve. Devise the problem so that the solution is "The loan was made at 7% interest."

Skill Review

Factor completely.

85. $3y^3 - 300y$ [5.4]

86. $12t + 36 + t^2$ [5.4]

87. $6x^2 + 6x + 6$ [5.1]

88. $10a^5 - 10a^4 - 60a^3$ [5.2]

89. $20x^2 + 7x - 6$ [5.3]

90. $n^6 - 1$ [5.5]

Synthesis

91. What would be better: to receive 3% interest every 6 months, or to receive 6% interest every 12 months? Why?

92. Write a problem involving a free-falling object for a classmate to solve (see Example 12). Devise the problem so that the solution is "The object takes about 4.5 sec to fall freely from the top of the structure."

Find b such that each trinomial is a square.

93. $x^2 + bx + 81$

94. $x^2 + bx + 49$

95. If $f(x) = 2x^5 - 9x^4 - 66x^3 + 45x^2 + 280x$ and $x^2 - 5$ is a factor of $f(x)$, find all five values of a for which $f(a) = 0$.

96. If $f(x) = \left(x - \frac{1}{3}\right)(x^2 + 6)$ and $g(x) = \left(x - \frac{1}{3}\right)\left(x^2 - \frac{2}{3}\right)$, find all a for which $(f + g)(a) = 0$.

97. *Boating.* A barge and a fishing boat leave a dock at the same time, traveling at a right angle to each other. The barge travels 7 km/h slower than the fishing boat. After 4 hr, the boats are 68 km apart. Find the speed of each boat.

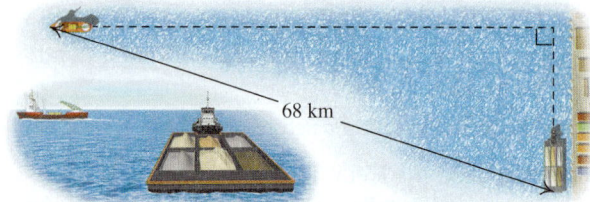

98. Find three consecutive integers such that the square of the first plus the product of the other two is 67.

99. Exercises 29, 33, and 53 can be solved on a graphing calculator without first rewriting in standard form. Simply let y_1 represent the left side of the equation and y_2 the right side. Then use a graphing calculator to determine the x-coordinate of any point of intersection. Use a graphing calculator to solve Exercises 29, 33, and 53 in this manner.

100. Use a graphing calculator to check your answers to Exercises 7, 13, 71, and 73.

 101. Example 11 can be solved with a graphing calculator by graphing each side of

$$4410 = 4000(1 + r)^2.$$

How could you determine, from a reading of the problem, a suitable viewing window? What might that window be?

↪ **YOUR TURN ANSWERS: SECTION 11.1**

1. $8, -8$ **2.** $\sqrt{10}, -\sqrt{10}$, or $3.162, -3.162$

3. $\sqrt{\dfrac{1}{3}}, -\sqrt{\dfrac{1}{3}}$, or $\dfrac{\sqrt{3}}{3}, -\dfrac{\sqrt{3}}{3}$ **4.** $10i, -10i$

5. $-5 + \sqrt{3}, -5 - \sqrt{3}$ **6.** $5 + \sqrt{3}, 5 - \sqrt{3}$

7. $x^2 + 7x + \dfrac{49}{4} = \left(x + \dfrac{7}{2}\right)^2$ **8.** $-3 + \sqrt{11}, -3 - \sqrt{11}$

9. $(-2 - \sqrt{3}, 0), (-2 + \sqrt{3}, 0)$ **10.** $2 + \dfrac{\sqrt{22}}{2}, 2 - \dfrac{\sqrt{22}}{2}$,

or $\dfrac{4 + \sqrt{22}}{2}, \dfrac{4 - \sqrt{22}}{2}$ **11.** 10%   **12.** About 9.5 sec

---

PREPARE TO MOVE ON

Evaluate.   [1.8]

1. $b^2 - 4ac$, for $a = 3$, $b = 2$, and $c = -5$

2. $b^2 - 4ac$, for $a = 1$, $b = -1$, and $c = 4$

Simplify.   [10.3], [10.8]

3. $\sqrt{200}$ **4.** $\sqrt{-4}$

5. $\sqrt{-8}$

11.2 The Quadratic Formula

Solving Using the Quadratic Formula ■ Approximating Solutions

Study Skills

Know It "By Heart"

When memorizing something like the quadratic formula, try to first understand and write out the derivation. Doing this two or three times will help you remember this important formula.

We can use the process of completing the square to develop a general formula for solving quadratic equations.

Solving Using the Quadratic Formula

Each time we solve by completing the square, the procedure is the same. Here we develop a formula that condenses this work.

We begin with a quadratic equation in standard form,

$$ax^2 + bx + c = 0,$$

with $a > 0$. For $a < 0$, a slightly different derivation is needed (see Exercise 60), but the result is the same. Let's solve by completing the square. As the steps are performed, compare them with Example 10 in Section 11.1.

$$ax^2 + bx = -c \qquad \text{Adding } -c \text{ to both sides}$$

$$x^2 + \frac{b}{a}x = -\frac{c}{a} \qquad \text{Dividing both sides by } a$$

Half of $\frac{b}{a}$ is $\frac{b}{2a}$ and $\left(\frac{b}{2a}\right)^2$ is $\frac{b^2}{4a^2}$. We add $\frac{b^2}{4a^2}$ to both sides.

$$x^2 + \frac{b}{a}x + \frac{b^2}{4a^2} = -\frac{c}{a} + \frac{b^2}{4a^2}$$

Adding $\frac{b^2}{4a^2}$ to complete the square

$$\left(x + \frac{b}{2a}\right)^2 = -\frac{4ac}{4a^2} + \frac{b^2}{4a^2}$$

Factoring on the left side; finding a common denominator on the right side

$$\left(x + \frac{b}{2a}\right)^2 = \frac{b^2 - 4ac}{4a^2}$$

$$x + \frac{b}{2a} = \pm\frac{\sqrt{b^2 - 4ac}}{2a}$$

Using the principle of square roots and the quotient rule for radicals. Since $a > 0$, $\sqrt{4a^2} = 2a$.

$$x = \frac{-b \pm \sqrt{b^2 - 4ac}}{2a}$$

Adding $-\frac{b}{2a}$ to both sides

It is important to remember the quadratic formula and know how to use it.

THE QUADRATIC FORMULA

The solutions of $ax^2 + bx + c = 0$, $a \neq 0$, are given by

$$x = \frac{-b \pm \sqrt{b^2 - 4ac}}{2a}.$$

EXAMPLE 1   Solve $5x^2 + 8x = -3$ using the quadratic formula.

SOLUTION   We first find standard form and determine a, b, and c:

$5x^2 + 8x + 3 = 0$;   Adding 3 to both sides to get 0 on one side

$a = 5$, $b = 8$, $c = 3$.

Next, we use the quadratic formula:

$$x = \frac{-b \pm \sqrt{b^2 - 4ac}}{2a}$$

It is important to remember this formula.

$$x = \frac{-8 \pm \sqrt{8^2 - 4 \cdot 5 \cdot 3}}{2 \cdot 5}$$

Substituting

$$x = \frac{-8 \pm \sqrt{64 - 60}}{10}$$

Be sure to write the fraction bar all the way across.

$$x = \frac{-8 \pm \sqrt{4}}{10}$$

$$= \frac{-8 \pm 2}{10}$$

$$x = \frac{-8 + 2}{10} \quad or \quad x = \frac{-8 - 2}{10}$$

The symbol $\pm$ indicates two solutions.

$$x = \frac{-6}{10} \quad or \quad x = \frac{-10}{10}$$

$$x = -\frac{3}{5} \quad or \quad x = -1.$$

1. Solve $12x^2 - 8x - 15$ using the quadratic formula.

The solutions are $-\frac{3}{5}$ and -1. The checks are left to the student.

YOUR TURN

Because $5x^2 + 8x + 3$ can be factored, the quadratic formula may not have been the fastest way of solving Example 1. However, because the quadratic formula works for *any* quadratic equation, we need not spend too much time struggling to solve a quadratic equation by factoring.

TO SOLVE A QUADRATIC EQUATION

1. If the equation can be easily written in the form $ax^2 = p$ or $(x + k)^2 = d$, use the principle of square roots.
2. If step (1) does not apply, write the equation in the form $ax^2 + bx + c = 0$.
3. Try factoring and using the principle of zero products.
4. If factoring seems difficult or impossible, use the quadratic formula. Completing the square can also be used.

The solutions of a quadratic equation can always be found using the quadratic formula. They cannot always be found by factoring.

A second-degree polynomial in one variable is said to be quadratic. Similarly, a second-degree polynomial function in one variable is said to be a **quadratic function**.

EXAMPLE 2   For the quadratic function given by $f(x) = 3x^2 - 6x - 4$, find all x for which $f(x) = 0$.

SOLUTION   We substitute and solve for x:

$$f(x) = 0$$
$$3x^2 - 6x - 4 = 0. \qquad \text{Substituting}$$

Since $3x^2 - 6x - 4$ does not factor, we use the quadratic formula with $a = 3$, $b = -6$, and $c = -4$:

$$x = \frac{-(-6) \pm \sqrt{(-6)^2 - 4 \cdot 3 \cdot (-4)}}{2 \cdot 3}$$

$$= \frac{6 \pm \sqrt{36 + 48}}{6} \qquad {\color{red}(-6)^2 - 4 \cdot 3 \cdot (-4) = 36 - (-48) = 36 + 48}$$

$$= \frac{6 \pm \sqrt{84}}{6} \qquad {\color{red}\text{Note that 4 is a perfect-square factor of 84.}}$$

$$= \frac{6}{6} \pm \frac{\sqrt{84}}{6} \qquad {\color{red}\text{Writing as two fractions to simplify each separately}}$$

$$= 1 \pm \frac{\sqrt{4}\sqrt{21}}{6} \qquad {\color{red}84 = 4 \cdot 21}$$

$$= 1 \pm \frac{2\sqrt{21}}{2 \cdot 3}$$

$$= 1 \pm \frac{\sqrt{21}}{3}. \qquad {\color{red}\text{Removing a factor of 1: } \frac{2}{2} = 1}$$

The solutions are $1 - \dfrac{\sqrt{21}}{3}$ and $1 + \dfrac{\sqrt{21}}{3}$. The checks are left to the student.

 YOUR TURN

When we use the quadratic formula to solve equations, we will be able to find imaginary-number solutions.

Technology Connection

To check Example 2, graph $y_1 = 3x^2 - 6x - 4$, press TRACE, and enter $1 + \sqrt{21}/3$. A rational approximation and the y-value 0 should appear.

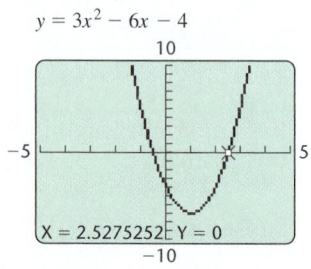

$y = 3x^2 - 6x - 4$

X = 2.5275252   Y = 0

Use this approach to check the other solution of Example 2.

2. For the quadratic function given by

$$f(x) = 2x^2 - 2x - 3,$$

find all x for which $f(x) = 0$.

EXAMPLE 3   Solve: $x(x + 5) = 2(2x - 1)$. (Find all complex-number solutions.)

SOLUTION   We first find standard form:

$$x^2 + 5x = 4x - 2 \quad \text{Multiplying}$$
$$x^2 + x + 2 = 0. \quad \text{Subtracting } 4x \text{ and adding 2 to both sides}$$

Since we cannot factor $x^2 + x + 2$, we use the quadratic formula with $a = 1$, $b = 1$, and $c = 2$:

$$x = \frac{-1 \pm \sqrt{1^2 - 4 \cdot 1 \cdot 2}}{2 \cdot 1} \quad \text{Substituting}$$

$$= \frac{-1 \pm \sqrt{1 - 8}}{2}$$

$$= \frac{-1 \pm \sqrt{-7}}{2}$$

$$= \frac{-1 \pm i\sqrt{7}}{2}, \text{ or } -\frac{1}{2} \pm \frac{\sqrt{7}}{2}i.$$

The solutions are $-\frac{1}{2} - \frac{\sqrt{7}}{2}i$ and $-\frac{1}{2} + \frac{\sqrt{7}}{2}i$. The checks are left to the student.

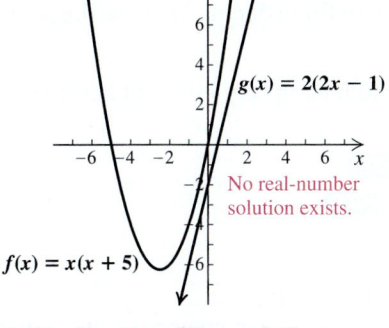

A visualization of Example 3

3. Solve: $x(x + 2) = -5$.

YOUR TURN

EXAMPLE 4   If $f(t) = 2 + \frac{7}{t}$ and $g(t) = \frac{4}{t^2}$, find all t for which $f(t) = g(t)$.

SOLUTION   We set $f(t)$ equal to $g(t)$ and solve:

$$f(t) = g(t)$$
$$2 + \frac{7}{t} = \frac{4}{t^2}. \quad \text{Substituting. Note that } t \neq 0.$$

This is a rational equation. To solve, we multiply both sides by the LCM of the denominators, t^2:

$$t^2\left(2 + \frac{7}{t}\right) = t^2 \cdot \frac{4}{t^2}$$
$$2t^2 + 7t = 4 \quad \text{Simplifying}$$
$$2t^2 + 7t - 4 = 0. \quad \text{Subtracting 4 from both sides}$$

We use the quadratic formula with $a = 2$, $b = 7$, and $c = -4$:

$$t = \frac{-7 \pm \sqrt{7^2 - 4 \cdot 2 \cdot (-4)}}{2 \cdot 2}$$

$$= \frac{-7 \pm \sqrt{49 + 32}}{4} \quad \begin{array}{l} 7^2 - 4 \cdot 2 \cdot (-4) = 49 - (-32) \\ = 49 + 32 \end{array}$$

$$= \frac{-7 \pm \sqrt{81}}{4}$$

$$= \frac{-7 \pm 9}{4}$$

$$t = \frac{-7 + 9}{4} \quad \text{or} \quad t = \frac{-7 - 9}{4}$$

$$t = \frac{2}{4} = \frac{1}{2} \quad \text{or} \quad t = \frac{-16}{4} = -4. \quad \begin{array}{l}\text{Both answers should check since} \\ t \neq 0.\end{array}$$

You can confirm that $f\left(\frac{1}{2}\right) = g\left(\frac{1}{2}\right)$ and $f(-4) = g(-4)$. The solutions are $\frac{1}{2}$ and -4.

YOUR TURN

Technology Connection

To determine whether quadratic equations are solved more quickly on a graphing calculator or by using the quadratic formula, solve Examples 2 and 4 both ways. Which method is faster? Which method is more precise? Why?

4. If $f(x) = 2 + \frac{1}{x}$ and $g(x) = \frac{3}{x^2}$, find all x for which $f(x) = g(x)$.

Approximating Solutions

When the solution of an equation is irrational, a rational-number approximation is often useful in real-world applications.

EXAMPLE 5 Use a calculator to approximate, to three decimal places, the solutions of Example 2.

SOLUTION On most calculators, one of the following sequences of keystrokes can be used to approximate $1 + \sqrt{21}/3$:

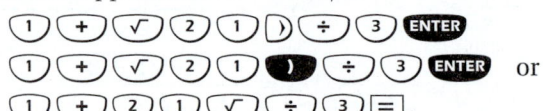

Similar keystrokes can be used to approximate $1 - \sqrt{21}/3$.

The solutions are approximately 2.527525232 and −0.5275252317. Rounded to three decimal places, the solutions are approximately 2.528 and −0.528.

 YOUR TURN

Student Notes

It is important that you understand both the rules for order of operations *and* the manner in which your calculator applies those rules.

5. Use a calculator to approximate, to three decimal places, the solutions of Your Turn Exercise 2.

CONNECTING 🔗 THE CONCEPTS

We have studied four different ways of solving quadratic equations. Each method has advantages and disadvantages, as outlined below. Note that although the quadratic formula can be used to solve *any* quadratic equation, sometimes other methods are faster and easier to use.

Method	Advantages	Disadvantages
Factoring	Can be very fast.	Can be used only on certain equations. Many equations are difficult or impossible to solve by factoring.
The principle of square roots	Fastest way to solve equations of the form $X^2 = k$. Can be used to solve *any* quadratic equation.	Can be slow when original equation is not written in the form $X^2 = k$.
Completing the square	Works well on equations of the form $x^2 + bx = -c$, when b is even. Can be used to solve *any* quadratic equation.	Can be complicated when $a \neq 1$ or when b in $x^2 + bx = -c$ is not even.
The quadratic formula	Can be used to solve *any* quadratic equation.	Can be slower than factoring or using the principle of square roots for certain equations.

EXERCISES

Solve. Examine each exercise carefully, and solve using the easiest method.

1. $x^2 - 3x - 10 = 0$

2. $x^2 = 121$

3. $x^2 + 6x = 10$

4. $x^2 + x - 3 = 0$

5. $(x + 1)^2 = 2$

6. $x^2 - 10x + 25 = 0$

7. $x^2 - 2x = 6$

8. $4t^2 = 11$

11.2 EXERCISE SET

FOR EXTRA HELP

MyMathLab® MathXL

PRACTICE   WATCH   READ   REVIEW

Vocabulary and Reading Check

Classify each of the following statements as either true or false.

1. The quadratic formula can be used to solve *any* quadratic equation.

2. The steps used to derive the quadratic formula are the same as those used when solving by completing the square.

3. The quadratic formula does not work if solutions are imaginary numbers.

4. Solving by factoring is always slower than using the quadratic formula.

5. A quadratic equation can have as many as four solutions.

6. It is possible for a quadratic equation to have no real-number solutions.

Solving Using the Quadratic Formula

Solve. (Find all complex-number solutions.)

7. $2x^2 + 3x - 5 = 0$

8. $3x^2 - 7x + 2 = 0$

9. $u^2 + 2u - 4 = 0$

10. $u^2 - 2u - 2 = 0$

11. $t^2 + 3 = 6t$

12. $t^2 + 4t = 1$

13. $x^2 = 3x + 5$

14. $x^2 + 5x + 3 = 0$

15. $3t(t + 2) = 1$

16. $2t(t + 2) = 1$

17. $\dfrac{1}{x^2} - 3 = \dfrac{8}{x}$

18. $\dfrac{9}{x} - 2 = \dfrac{5}{x^2}$

19. $t^2 + 10 = 6t$

20. $t^2 + 10t + 26 = 0$

21. $p^2 - p + 1 = 0$

22. $p^2 + p + 4 = 0$

23. $x^2 + 4x + 6 = 0$

24. $x^2 + 11 = 6x$

25. $12t^2 + 17t = 40$

26. $15t^2 + 7t = 2$

27. $25x^2 - 20x + 4 = 0$

28. $36x^2 + 84x + 49 = 0$

29. $7x(x + 2) + 5 = 3x(x + 1)$

30. $5x(x - 1) - 7 = 4x(x - 2)$

31. $14(x - 4) - (x + 2) = (x + 2)(x - 4)$

32. $11(x - 2) + (x - 5) = (x + 2)(x - 6)$

33. $51p = 2p^2 + 72$

34. $72 = 3p^2 + 50p$

35. $x(x - 3) = x - 9$

36. $x(x - 1) = 2x - 7$

37. $x^3 - 8 = 0$ (*Hint:* Factor the difference of cubes. Then use the quadratic formula.)

38. $x^3 + 1 = 0$

39. Let $f(x) = 6x^2 - 7x - 20$. Find x such that $f(x) = 0$.

40. Let $g(x) = 4x^2 - 2x - 3$. Find x such that $g(x) = 0$.

41. Let
$$f(x) = \frac{7}{x} + \frac{7}{x + 4}.$$
Find all x for which $f(x) = 1$.

42. Let
$$g(x) = \frac{2}{x} + \frac{2}{x + 3}.$$
Find all x for which $g(x) = 1$.

43. Let
$$F(x) = \frac{3 - x}{4} \quad \text{and} \quad G(x) = \frac{1}{4x}.$$
Find all x for which $F(x) = G(x)$.

44. Let
$$f(x) = x + 5 \quad \text{and} \quad g(x) = \frac{3}{x - 5}.$$
Find all x for which $f(x) = g(x)$.

Approximating Solutions

Solve using the quadratic formula. Then use a calculator to approximate, to three decimal places, the solutions as rational numbers.

45. $x^2 + 6x + 4 = 0$

46. $x^2 + 4x - 7 = 0$

Aha! 47. $x^2 - 6x + 4 = 0$

48. $x^2 - 4x + 1 = 0$

49. $2x^2 - 3x - 7 = 0$

50. $3x^2 - 3x - 2 = 0$

51. Are there any equations that can be solved by the quadratic formula but not by completing the square? Why or why not?

52. Suppose you are solving a quadratic equation with no constant term ($c = 0$). Would you use factoring or the quadratic formula to solve? Why?

Skill Review

Simplify.

53. $(-3x^2 y^6)^0$ [4.1]

54. $100^{3/2}$ [10.2]

55. $x^{1/4} \cdot x^{2/3}$ [10.2]

56. $(27^{-2})^{1/3}$ [10.2]

57. $\dfrac{18a^5 bc^{10}}{24a^{-5} bc^3}$ [4.2]

58. $\left(\dfrac{2xw^{-3}}{3x^{-4}w}\right)^{-2}$ [4.2]

Synthesis

 59. Explain how you could use the quadratic formula to help factor a quadratic polynomial.

 60. If $a < 0$ and $ax^2 + bx + c = 0$, then $-a$ is positive and the equivalent equation, $-ax^2 - bx - c = 0$, can be solved using the quadratic formula.

    a) Find this solution, replacing a, b, and c in the formula with $-a$, $-b$, and $-c$ from the equation.

    b) Why does the result of part (a) indicate that the quadratic formula "works" regardless of the sign of a?

For Exercises 61–63, let

$$f(x) = \frac{x^2}{x-2} + 1 \quad and \quad g(x) = \frac{4x-2}{x-2} + \frac{x+4}{2}.$$

61. Find the x-intercepts of the graph of f.

62. Find the x-intercepts of the graph of g.

63. Find all x for which $f(x) = g(x)$.

Solve. Approximate the solutions to three decimal places.

64. $x^2 - 0.75x - 0.5 = 0$

65. $z^2 + 0.84z - 0.4 = 0$

Solve.

66. $(1 + \sqrt{3})x^2 - (3 + 2\sqrt{3})x + 3 = 0$

67. $\sqrt{2}x^2 + 5x + \sqrt{2} = 0$

68. $ix^2 - 2x + 1 = 0$

69. One solution of $kx^2 + 3x - k = 0$ is -2. Find the other.

 70. Use a graphing calculator to solve Exercises 9, 27, and 43.

 71. Use a graphing calculator to solve Exercises 11, 33, and 41. Use the method of graphing each side of the equation.

 72. Can a graphing calculator be used to solve *any* quadratic equation? Why or why not?

↘ YOUR TURN ANSWERS: SECTION 11.2

1. $-\dfrac{5}{6}, \dfrac{3}{2}$ **2.** $\dfrac{1}{2} + \dfrac{\sqrt{7}}{2}, \dfrac{1}{2} - \dfrac{\sqrt{7}}{2}$, or $\dfrac{1}{2} \pm \dfrac{\sqrt{7}}{2}$

3. $-1 + 2i, -1 - 2i$, or $-1 \pm 2i$ **4.** $-\dfrac{3}{2}, 1$

5. $1.823, -0.823$

QUICK QUIZ: SECTIONS 11.1–11.2

1. Solve using the principle of zero products:

$$(z + 3)(2z - 5) = 0. \quad [11.1]$$

2. Solve using the principle of square roots:

$$(x - 2)^2 = 3. \quad [11.1]$$

3. Solve by completing the square:

$$t^2 - 6t + 4 = 0. \quad [11.1]$$

4. Solve using the quadratic formula:

$$x^2 - 3x - 1 = 0. \quad [11.2]$$

5. Solve using any appropriate method:

$$2x^2 + x + 5 = 0. \quad [11.2]$$

PREPARE TO MOVE ON

Multiply and simplify.

1. $(x - 2i)(x + 2i)$ [10.8]

2. $(x - 6\sqrt{5})(x + 6\sqrt{5})$ [10.5]

3. $(x - (2 - \sqrt{7}))(x - (2 + \sqrt{7}))$ [10.5]

4. $(x - (-3 + 5i))(x - (-3 - 5i))$ [10.8]

11.3 Studying Solutions of Quadratic Equations

The Discriminant ● Writing Equations from Solutions

Study Skills

Sharpen Your Skills

Every so often, you may encounter a lesson that you remember from a previous math course. When this occurs, make sure that you understand *all* of that lesson. Take time to review any new concepts and to sharpen any old skills.

The Discriminant

It is sometimes enough to know what *type* of number a solution will be, without actually solving the equation. Suppose we want to know if $2x^2 + 7x - 15 = 0$ has rational solutions (and thus can be solved by factoring). Using the quadratic formula, we would have

$$x = \frac{-b \pm \sqrt{b^2 - 4ac}}{2a}$$

$$= \frac{-(7) \pm \sqrt{(7)^2 - 4 \cdot 2(-15)}}{2 \cdot 2} = \frac{-7 \pm \sqrt{169}}{4}.$$

Since 169 is a perfect square ($\sqrt{169} = 13$), we know that the solutions of the equation are rational numbers. This means that $2x^2 + 7x - 15 = 0$ *can* be solved by factoring. Note that the radicand, 169, determines what type of number the solutions will be.

The radicand $b^2 - 4ac$ is known as the **discriminant**. If a, b, and c are rational, then we can make the following observations.

Discriminant	Observation	Example
$b^2 - 4ac = 0$	We get the same solution twice. There is one *repeated* solution and it is rational.	$9x^2 + 6x + 1 = 0$ $b^2 - 4ac = 6^2 - 4 \cdot 9 \cdot 1 = 0$ Solving, we have $x = \dfrac{-6 \pm \sqrt{0}}{2 \cdot 9}$. The (repeated) solution is $-\frac{1}{3}$.
$b^2 - 4ac$ is positive. 1. $b^2 - 4ac$ is a perfect square. 2. $b^2 - 4ac$ is *not* a perfect square.	There are two different real-number solutions. 1. The solutions are rational numbers. 2. The solutions are irrational conjugates.	**1.** $6x^2 + 5x + 1 = 0$ $b^2 - 4ac = 5^2 - 4 \cdot 6 \cdot 1 = 1$ Solving, we have $x = \dfrac{-5 \pm \sqrt{1}}{2 \cdot 6}$. The solutions are $-\frac{1}{3}$ and $-\frac{1}{2}$. **2.** $x^2 + 4x + 2 = 0$ $b^2 - 4ac = 4^2 - 4 \cdot 1 \cdot 2 = 8$ Solving, we have $x = \dfrac{-4 \pm \sqrt{8}}{2 \cdot 1}$. The solutions are $-2 + \sqrt{2}$ and $-2 - \sqrt{2}$.
$b^2 - 4ac$ is negative.	There are two different imaginary-number solutions. They are complex conjugates.	$x^2 + 4x + 5 = 0$ $b^2 - 4ac = 4^2 - 4 \cdot 1 \cdot 5 = -4$ Solving, we have $x = \dfrac{-4 \pm \sqrt{-4}}{2 \cdot 1}$. The solutions are $-2 + i$ and $-2 - i$.

Note that all quadratic equations have either one or two solutions. These solutions can always be found algebraically; only real-number solutions can be found graphically. Also, note that any equation for which $b^2 - 4ac$ is a perfect square can be solved by factoring.

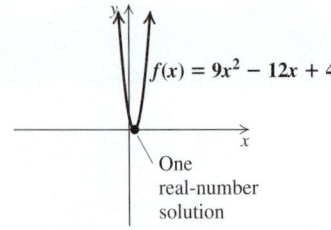

A visualization of part (a)

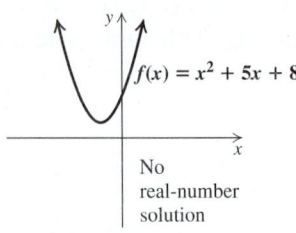

A visualization of part (b)

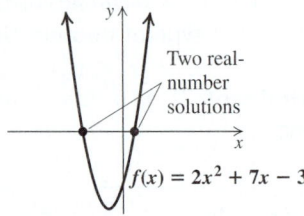

A visualization of part (c)

1. Determine what type of number the solutions are and how many solutions exist:

$$4x^2 - 9x + 2 = 0.$$

EXAMPLE 1   For each equation, determine what type of number the solutions are and how many solutions exist.

a) $9x^2 - 12x + 4 = 0$ **b)** $x^2 + 5x + 8 = 0$ **c)** $2x^2 + 7x - 3 = 0$

SOLUTION

a) For $9x^2 - 12x + 4 = 0$, we have

$$a = 9, \quad b = -12, \quad c = 4.$$

We substitute and compute the discriminant:

$$b^2 - 4ac = (-12)^2 - 4 \cdot 9 \cdot 4$$
$$= 144 - 144 = 0.$$

There is exactly one solution (it is repeated), and it is rational. Thus,

$$9x^2 - 12x + 4 = 0$$

can be solved by factoring.

b) For $x^2 + 5x + 8 = 0$, we have

$$a = 1, \quad b = 5, \quad c = 8.$$

We substitute and compute the discriminant:

$$b^2 - 4ac = 5^2 - 4 \cdot 1 \cdot 8$$
$$= 25 - 32 = -7.$$

Since the discriminant is negative, there are two different imaginary-number solutions that are complex conjugates of each other.

c) For $2x^2 + 7x - 3 = 0$, we have

$$a = 2, \quad b = 7, \quad c = -3.$$

We substitute and compute the discriminant:

$$b^2 - 4ac = 7^2 - 4 \cdot 2(-3)$$
$$= 49 - (-24) = 73.$$

The discriminant is a positive number that is not a perfect square. Thus there are two different irrational solutions that are conjugates of each other.

YOUR TURN

Discriminants can also be used to determine the number of *x*-intercepts of the graph of a quadratic function.

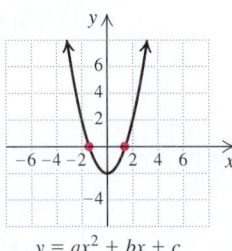

$y = ax^2 + bx + c$
$b^2 - 4ac > 0$
Two real solutions
of $ax^2 + bx + c = 0$
Two *x*-intercepts

$y = ax^2 + bx + c$
$b^2 - 4ac = 0$
One real solution
of $ax^2 + bx + c = 0$
One *x*-intercept

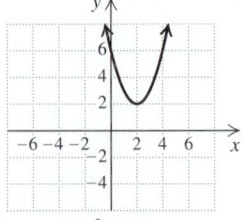

$y = ax^2 + bx + c$
$b^2 - 4ac < 0$
No real solutions
of $ax^2 + bx + c = 0$
No *x*-intercept

Writing Equations from Solutions

We know by the principle of zero products that $(x - 2)(x + 3) = 0$ has solutions 2 and -3. If we wish for two given numbers to be solutions of an equation, we can create such an equation, using the principle in reverse.

EXAMPLE 2   Find an equation for which the given numbers are solutions.

a) 3 and $-\frac{2}{5}$

b) $2i$ and $-2i$

c) $5\sqrt{7}$ and $-5\sqrt{7}$

d) $-4, 0,$ and 1

SOLUTION

a)
$$x = 3 \quad or \quad x = -\tfrac{2}{5}$$
$$x - 3 = 0 \quad or \quad x + \tfrac{2}{5} = 0 \qquad \text{Getting 0's on one side}$$
$$(x - 3)\left(x + \tfrac{2}{5}\right) = 0 \qquad \text{Using the principle of zero products (multiplying)}$$
$$x^2 + \tfrac{2}{5}x - 3x - 3 \cdot \tfrac{2}{5} = 0 \qquad \text{Multiplying}$$
$$x^2 - \tfrac{13}{5}x - \tfrac{6}{5} = 0 \qquad \text{Combining like terms}$$
$$5x^2 - 13x - 6 = 0 \qquad \text{Multiplying both sides by 5 to clear fractions}$$

Note that multiplying both sides by 5 clears the equation of fractions. Had we preferred, we could have multiplied $x + \frac{2}{5} = 0$ by 5, thus clearing fractions *before* using the principle of zero products.

b)
$$x = 2i \quad or \quad x = -2i$$
$$x - 2i = 0 \quad or \quad x + 2i = 0 \qquad \text{Getting 0's on one side}$$
$$(x - 2i)(x + 2i) = 0 \qquad \text{Using the principle of zero products (multiplying)}$$
$$x^2 - (2i)^2 = 0 \qquad \text{Finding the product of a sum and a difference}$$
$$x^2 - 4i^2 = 0$$
$$x^2 + 4 = 0 \qquad i^2 = -1$$

c)
$$x = 5\sqrt{7} \quad or \quad x = -5\sqrt{7}$$
$$x - 5\sqrt{7} = 0 \quad or \quad x + 5\sqrt{7} = 0 \qquad \text{Getting 0's on one side}$$
$$(x - 5\sqrt{7})(x + 5\sqrt{7}) = 0 \qquad \text{Using the principle of zero products}$$
$$x^2 - (5\sqrt{7})^2 = 0 \qquad \text{Finding the product of a sum and a difference}$$
$$x^2 - 25 \cdot 7 = 0$$
$$x^2 - 175 = 0$$

d)
$$x = -4 \quad or \quad x = 0 \quad or \quad x = 1$$
$$x + 4 = 0 \quad or \quad x = 0 \quad or \quad x - 1 = 0 \qquad \text{Getting 0's on one side}$$
$$(x + 4)x(x - 1) = 0 \qquad \text{Using the principle of zero products}$$
$$x(x^2 + 3x - 4) = 0 \qquad \text{Multiplying}$$
$$x^3 + 3x^2 - 4x = 0$$

2. Find an equation for which 0 and $\frac{1}{5}$ are solutions.

 YOUR TURN

To check any of these equations, we can simply substitute one or more of the given solutions. For example, in Example 2(d) above,

$$(-4)^3 + 3(-4)^2 - 4(-4) = -64 + 3 \cdot 16 + 16$$
$$= -64 + 48 + 16 = 0.$$

11.3 EXERCISE SET

FOR EXTRA HELP

MyMathLab® Math XL

PRACTICE   WATCH   READ   REVIEW

Vocabulary and Reading Check

Match the description of the solution(s) with each discriminant. Answers may be used more than once.

1. ____ $b^2 - 4ac = 9$ **a)** One rational solution

2. ____ $b^2 - 4ac = 0$ **b)** Two different rational solutions

3. ____ $b^2 - 4ac = -1$

4. ____ $b^2 - 4ac = 1$ **c)** Two different irrational solutions

5. ____ $b^2 - 4ac = 8$ **d)** Two different imaginary-number solutions

6. ____ $b^2 - 4ac = 12$

The Discriminant

For each equation, determine what type of number the solutions are and how many solutions exist.

7. $x^2 - 7x + 5 = 0$ 8. $x^2 - 5x + 3 = 0$

9. $x^2 + 11 = 0$ 10. $x^2 + 7 = 0$

11. $x^2 - 11 = 0$ 12. $x^2 - 7 = 0$

13. $4x^2 + 8x - 5 = 0$ 14. $4x^2 - 12x + 9 = 0$

15. $x^2 + 4x + 6 = 0$ 16. $x^2 - 2x + 4 = 0$

17. $9t^2 - 48t + 64 = 0$ 18. $10t^2 - t - 2 = 0$

Aha! 19. $9t^2 + 3t = 0$ 20. $4m^2 + 7m = 0$

21. $x^2 + 4x = 8$ 22. $x^2 + 5x = 9$

23. $2a^2 - 3a = -5$ 24. $3a^2 + 5 = -7a$

25. $7x^2 = 19x$ 26. $5x^2 = 48x$

27. $y^2 + \frac{9}{4} = 4y$ 28. $x^2 = \frac{1}{2}x - \frac{3}{5}$

Writing Equations from Solutions

Write a quadratic equation having the given numbers as solutions.

29. $-5, 4$ 30. $-2, 8$

31. 3, only solution (*Hint*: It must be a repeated solution.)

32. -5, only solution

33. $-1, -3$ 34. $-2, -5$

35. $5, \frac{3}{4}$ 36. $4, \frac{2}{3}$

37. $-\frac{1}{4}, -\frac{1}{2}$ 38. $\frac{1}{2}, \frac{1}{3}$

39. $2.4, -0.4$ 40. $-0.6, 1.4$

41. $-\sqrt{3}, \sqrt{3}$ 42. $-\sqrt{7}, \sqrt{7}$

43. $2\sqrt{5}, -2\sqrt{5}$ 44. $3\sqrt{2}, -3\sqrt{2}$

45. $4i, -4i$ 46. $3i, -3i$

47. $2 - 7i, 2 + 7i$ 48. $5 - 2i, 5 + 2i$

49. $3 - \sqrt{14}, 3 + \sqrt{14}$ 50. $2 - \sqrt{10}, 2 + \sqrt{10}$

51. $1 - \dfrac{\sqrt{21}}{3}, 1 + \dfrac{\sqrt{21}}{3}$ 52. $\dfrac{5}{4} - \dfrac{\sqrt{33}}{4}, \dfrac{5}{4} + \dfrac{\sqrt{33}}{4}$

Write a third-degree equation having the given numbers as solutions.

53. $-2, 1, 5$ 54. $-5, 0, 2$

55. $-1, 0, 3$ 56. $-2, 2, 3$

57. Explain why there are not two different solutions when the discriminant is 0.

58. Describe a procedure that could be used to write an equation having the first 7 natural numbers as solutions.

Skill Review

Simplify.

59. $\sqrt{270a^7b^{12}}$ [10.3]   60. $\sqrt[4]{8w^3}\,\sqrt[4]{4w^7}$ [10.3]

61. $\sqrt[3]{x}\,\sqrt{x}$ [10.5]   62. $\sqrt{-3}\,\sqrt{-2}$ [10.8]

63. $(2 - i)(3 + i)$ [10.8]   64. i^{18} [10.8]

Synthesis

65. If we assume that a quadratic equation has integers for coefficients, will the product of the solutions always be a real number? Why or why not?

66. Can a fourth-degree equation with rational coefficients have exactly three irrational solutions? Why or why not?

67. The graph of an equation of the form
$$y = ax^2 + bx + c$$
is shown below. Determine *a*, *b*, and *c* from the identified points.

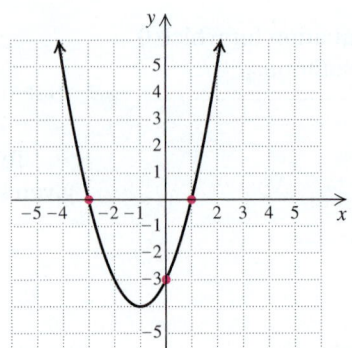

68. Show that the product of the solutions of $ax^2 + bx + c = 0$ is c/a.

For each equation under the given condition, (a) find k and (b) find the other solution.

69. $kx^2 - 2x + k = 0$; one solution is -3

70. $x^2 - kx + 2 = 0$; one solution is $1 + i$

71. $x^2 - (6 + 3i)x + k = 0$; one solution is 3

72. Show that the sum of the solutions of $ax^2 + bx + c = 0$ is $-b/a$.

73. Show that whenever there is just one solution of $ax^2 + bx + c = 0$, that solution is of the form $-b/(2a)$.

74. Find h and k such that $3x^2 - hx + 4k = 0$, the sum of the solutions is -12, and the product of the solutions is 20. (*Hint:* See Exercises 68 and 72.)

75. Suppose that $f(x) = ax^2 + bx + c$, with $f(-3) = 0$, $f\left(\frac{1}{2}\right) = 0$, and $f(0) = -12$. Find a, b, and c.

76. Find an equation for which $2 - \sqrt{3}$, $2 + \sqrt{3}$, $5 - 2i$, and $5 + 2i$ are solutions.

Aha! **77.** Write a quadratic equation with integer coefficients for which $-\sqrt{2}$ is one solution.

78. Write a quadratic equation with integer coefficients for which $10i$ is one solution.

79. Find an equation with integer coefficients for which $1 - \sqrt{5}$ and $3 + 2i$ are two of the solutions.

80. A discriminant that is a perfect square indicates that factoring can be used to solve the quadratic equation. Why?

81. While solving an equation of the form $ax^2 + bx + c = 0$ with a graphing calculator, Keisha gets the following screen. How could the sign of the discriminant help her check the graph?

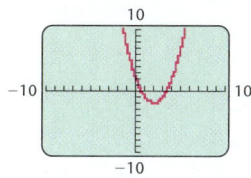

 YOUR TURN ANSWERS: SECTION 11.3

1. Two rational   **2.** $5x^2 - x = 0$; answers may vary

QUICK QUIZ: SECTIONS 11.1–11.3

Solve.

1. $x^2 + 16x + 64 = 3$ [11.1]

2. $3x^2 - 1 = x$ [11.2]

3. Find the x-intercepts of the graph of $f(x) = x^2 - 8$.   [11.1]

4. Solve $x^2 + 2x - 1 = 0$. Use a calculator to approximate the solutions with rational numbers. Round to three decimal places.   [11.2]

5. Determine how many different solutions there are of $3x^2 - 5x - 7 = 0$ and whether they are real or imaginary. If they are real, specify whether they are rational or irrational.   [11.3]

PREPARE TO MOVE ON

Solve each formula for the specified variable.   [7.5]

1. $\dfrac{c}{d} = c + d$, for c

2. $x = \dfrac{3}{1 - y}$, for y

Solve.

3. Kiara's motorboat took 4 hr to make a trip downstream with a 2-mph current. The return trip against the same current took 6 hr. Find the speed of the boat in still water.   [8.3]

4. Homer walks 1.5 mph faster than Gladys. In the time it takes Homer to walk 7 mi, Gladys walks 4 mi. Find the speed of each person.   [6.7]

11.4 Applications Involving Quadratic Equations

Solving Formulas ▪ Solving Problems

1. Water flow F from a hose, in number of gallons per minute, is given by $F = 118.8\sqrt{x}$, where x is the nozzle pressure, in pounds per square inch. Solve for x.

   Source: Based on data from www.firetactics.com

Solving Formulas

To solve a formula for a certain letter, we use the principles for solving equations to get that letter alone on one side.

EXAMPLE 1 *Period of a Pendulum.* The time T required for a pendulum of length l to swing back and forth (complete one period) is given by $T = 2\pi\sqrt{l/g}$, where g is the earth's gravitational constant. Solve for l.

SOLUTION We have

$$T = 2\pi\sqrt{\frac{l}{g}} \qquad \text{This is a radical equation.}$$

$$T^2 = \left(2\pi\sqrt{\frac{l}{g}}\right)^2 \qquad \text{Squaring both sides}$$

$$T^2 = 2^2\pi^2\frac{l}{g}$$

$$gT^2 = 4\pi^2 l \qquad \text{Multiplying both sides by } g \text{ to clear fractions}$$

$$\frac{gT^2}{4\pi^2} = l. \qquad \text{Dividing both sides by } 4\pi^2$$

We now have l alone on one side and l does not appear on the other side, so the formula is solved for l.

🔄 YOUR TURN

In formulas for which variables represent only nonnegative numbers, there is no need for absolute-value signs when taking square roots.

EXAMPLE 2 *Hang Time.** An athlete's *hang time* is the amount of time that the athlete can remain airborne when jumping. A formula relating an athlete's vertical leap V, in inches, to hang time T, in seconds, is $V = 48T^2$. Solve for T.

SOLUTION We have

$$48T^2 = V$$

$$T^2 = \frac{V}{48} \qquad \text{Dividing by 48 to isolate } T^2$$

$$T = \frac{\sqrt{V}}{\sqrt{48}} \qquad \begin{array}{l}\text{Using the principle of square roots}\\ \text{and the quotient rule for radicals.}\\ \text{We assume } V, T \geq 0.\end{array}$$

$$= \frac{\sqrt{V}}{\sqrt{16}\sqrt{3}}$$

*This formula is taken from an article by Peter Brancazio, "The Mechanics of a Slam Dunk," *Popular Mechanics,* November 1991. Courtesy of Professor Peter Brancazio, Brooklyn College.

$$= \frac{\sqrt{V}}{4\sqrt{3}}$$

$$= \frac{\sqrt{V}}{4\sqrt{3}} \cdot \frac{\sqrt{3}}{\sqrt{3}}$$

$$= \frac{\sqrt{3V}}{12} \qquad \text{Rationalizing the denominator}$$

2. Solve $y = (x - 2)^2$ for x.

↩ YOUR TURN

EXAMPLE 3 *Falling Distance.*   An object tossed downward with an initial speed (velocity) of v_0 will travel a distance of s meters, where $s = 4.9t^2 + v_0 t$ and t is measured in seconds. Solve for t.

SOLUTION   Since t is squared in one term and raised to the first power in the other term, the equation is quadratic in t.

$$4.9t^2 + v_0 t = s$$

$$4.9t^2 + v_0 t - s = 0 \qquad \text{Writing standard form}$$

$$a = 4.9, \quad b = v_0, \quad c = -s$$

$$t = \frac{-v_0 \pm \sqrt{(v_0)^2 - 4(4.9)(-s)}}{2(4.9)} \qquad \text{Using the quadratic formula}$$

Since the negative square root would yield a negative value for t, we use only the positive root:

$$t = \frac{-v_0 + \sqrt{(v_0)^2 + 19.6s}}{9.8}.$$

3. Solve $d = -16h^2 + 64h$ for h.

↩ YOUR TURN

The following list of steps should help you when solving formulas for a given letter. Remember that solving a formula requires the same approach as solving an equation.

TO SOLVE A FORMULA FOR A LETTER—SAY, h

1. Clear fractions and use the principle of powers, as needed. Perform these steps until any radicals containing h are gone and h is not in any denominator.
2. Combine all like terms.
3. If the only power of h is h^1, the equation can be solved as a linear equation or a rational equation. (See Example 1.)
4. If h^2 appears but h does not, solve for h^2 and use the principle of square roots to then solve for h. (See Example 2.)
5. If there are terms containing both h and h^2, put the equation in standard form and use the quadratic formula. (See Example 3.)

Solving Problems

Some problems translate to rational equations. The solution of such rational equations can involve quadratic equations.

EXAMPLE 4 *Motorcycle Travel.*   Fiona rode her motorcycle 300 mi at a certain average speed. Had she traveled 10 mph faster, the trip would have taken 1 hr less. Find Fiona's average speed.

SOLUTION

1. **Familiarize.** We make a drawing, labeling it with the information provided, and create a table. To do so, we let r represent the rate, in miles per hour, and t the time, in hours, for Fiona's trip.

300 miles

Time t Speed r

300 miles

Time $t - 1$ Speed $r + 10$

Recall that the definition of speed, $r = d/t$, relates rate, time, and distance.

Distance	Speed	Time
300	r	t
300	$r + 10$	$t - 1$

$\longrightarrow r = \dfrac{300}{t}$

$\longrightarrow r + 10 = \dfrac{300}{t - 1}$

2. **Translate.** From the table, we obtain

$$r = \frac{300}{t} \quad \text{and} \quad r + 10 = \frac{300}{t - 1}.$$

3. **Carry out.** A system of equations has been formed. We solve using substitution:

$$\frac{300}{t} + 10 = \frac{300}{t - 1}$$
Substituting $300/t$ for r in the second equation

$$t(t - 1) \cdot \left[\frac{300}{t} + 10\right] = t(t - 1) \cdot \frac{300}{t - 1}$$
Multiplying by the LCM to clear fractions

$$t(t - 1) \cdot \frac{300}{t} + t(t - 1) \cdot 10 = t(t - 1) \cdot \frac{300}{t - 1}$$
Using the distributive law

$$\frac{t(t - 1)}{1} \cdot \frac{300}{t} + t(t - 1) \cdot 10 = \frac{t(t - 1)}{1} \cdot \frac{300}{t - 1}$$
Removing factors that equal 1: $t/t = 1$ and $(t - 1)/(t - 1) = 1$

$$300(t - 1) + 10(t^2 - t) = 300t$$
$$300t - 300 + 10t^2 - 10t = 300t$$
$$10t^2 - 10t - 300 = 0$$
Rewriting in standard form

$$t^2 - t - 30 = 0$$
Dividing by 10

$$(t - 6)(t + 5) = 0$$
Factoring

$$t = 6 \quad or \quad t = -5.$$
Using the principle of zero products

4. Check. Note that we have solved for t, not r as required. Since negative time has no meaning here, we disregard the -5 and use 6 hr to find r:

$$r = \frac{300 \text{ mi}}{6 \text{ hr}} = 50 \text{ mph}.$$

> **CAUTION!** Always make sure that you find the quantity asked for in the problem.

To see if 50 mph checks, we increase the speed 10 mph to 60 mph and see how long the trip would have taken at that speed:

$$t = \frac{d}{r} = \frac{300 \text{ mi}}{60 \text{ mph}} = 5 \text{ hr}.$$ Note that mi/mph $= \text{mi} \div \frac{\text{mi}}{\text{hr}} =$

$$\cancel{\text{mi}} \cdot \frac{\text{hr}}{\cancel{\text{mi}}} = \text{hr}.$$

4. Taryn's Cessna travels 120 mph in still air. She flies 140 mi into the wind and 140 mi with the wind in a total of 2.4 hr. Find the wind speed.

This is 1 hr less than the trip actually took, so the answer checks.

5. State. Fiona traveled at an average speed of 50 mph.

⮑ YOUR TURN

11.4 EXERCISE SET

FOR EXTRA HELP

PRACTICE   WATCH   READ   REVIEW

⮑ Vocabulary and Reading Check

Match each formula with its description from the column on the right.

1. ___ $T = 2\pi\sqrt{\dfrac{l}{g}}$

2. ___ $V = 48T^2$

3. ___ $s = 4.9t^2 + v_0 t$

4. ___ $t = \dfrac{d}{r}$

a) Falling distance

b) Motion formula

c) Period of a pendulum

d) Vertical leap

Solving Formulas

Solve each formula for the indicated letter. Assume that all variables represent positive numbers.

5. $A = 4\pi r^2$, for r
(Surface area of a sphere)

6. $A = 6s^2$, for s
(Surface area of a cube)

7. $A = 2\pi r^2 + 2\pi rh$, for r
(Surface area of a right cylindrical solid)

8. $N = \dfrac{k^2 - 3k}{2}$, for k
(Number of diagonals of a polygon)

9. $F = \dfrac{Gm_1 m_2}{r^2}$, for r
(Law of gravity)

10. $N = \dfrac{kQ_1 Q_2}{s^2}$, for s
(Number of phone calls between two cities)

11. $c = \sqrt{gH}$, for H
(Velocity of ocean wave)

12. $r = 2\sqrt{5L}$, for L
(Speed of car based on length of skid marks)

13. $a^2 + b^2 = c^2$, for b
(Pythagorean formula in two dimensions)

14. $a^2 + b^2 + c^2 = d^2$, for c
(Pythagorean formula in three dimensions)

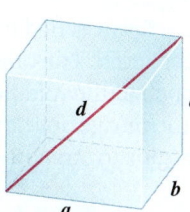

15. $s = v_0 t + \dfrac{gt^2}{2}$, for t
(A motion formula)

16. $A = \pi r^2 + \pi rs$, for r
(Surface area of a cone)

17. $N = \frac{1}{2}(n^2 - n)$, for n
(Number of games if n teams play each other once)

18. $A = A_0(1 - r)^2$, for r
(A business formula)

19. $T = I\sqrt{\dfrac{s}{d}}$, for d
(True airspeed)

20. $W = \sqrt{\dfrac{1}{LC}}$, for L
(An electricity formula)

Aha! **21.** $at^2 + bt + c = 0$, for t
(An algebraic formula)

22. $A = P_1(1 + r)^2 + P_2(1 + r)$, for r
(Amount in an account)

Solve.

23. *Falling Distance.*   (Use $4.9t^2 + v_0t = s$.)
   a) A bolt falls off an airplane at an altitude of 500 m. Approximately how long does it take the bolt to reach the ground?
   b) A ball is thrown downward at a speed of 30 m/sec from an altitude of 500 m. Approximately how long does it take the ball to reach the ground?
   c) Approximately how far will an object fall in 5 sec, when thrown downward at an initial velocity of 30 m/sec from a plane?

24. *Falling Distance.*   (Use $4.9t^2 + v_0t = s$.)
   a) A life preserver is dropped from a helicopter at an altitude of 75 m. Approximately how long does it take the life preserver to reach the water?
   b) A coin is tossed downward with an initial velocity of 30 m/sec from an altitude of 75 m. Approximately how long does it take the coin to reach the ground?
   c) Approximately how far will an object fall in 2 sec, if thrown downward at an initial velocity of 20 m/sec from a helicopter?

25. *Bungee Jumping.*   Wyatt is tied to one end of a 40-m elasticized (bungee) cord. The other end of the cord is secured to a winch at the middle of a bridge. If Wyatt jumps off the bridge, for how long will he fall before the cord begins to stretch? (Use $4.9t^2 = s$.)

40 m

26. *Bungee Jumping.*   Chika is tied to a bungee cord (see Exercise 25) and falls for 2.5 sec before her cord begins to stretch. How long is the bungee cord?

27. *Hang Time.*   The NBA's Russell Westbrook has a vertical leap of 36.5 in. What is his hang time? (Use $V = 48T^2$.)
Source: www.nbadraft.net

28. *League Schedules.*   In a bowling league, each team plays each of the other teams once. If a total of 66 games is played, how many teams are in the league? (See Exercise 17.)

For Exercises 29 and 30, use $4.9t^2 + v_0t = s$.

29. *Downward Speed.*   A stone thrown downward from a 100-m cliff travels 51.6 m in 3 sec. What was the initial velocity of the object?

30. *Downward Speed.*   A pebble thrown downward from a 200-m cliff travels 91.2 m in 4 sec. What was the initial velocity of the object?

For Exercises 31 and 32, use $A = P_1(1 + r)^2 + P_2(1 + r)$. (See Exercise 22.)

31. *Compound Interest.*   A firm invests $3200 in a savings account for 2 years. At the beginning of the second year, an additional $1800 is invested. If a total of $5375.48 is in the account at the end of the second year, what is the annual interest rate?

32. *Compound Interest.*   A business invests $10,000 in a savings account for 2 years. At the beginning of the second year, an additional $3500 is invested. If a total of $14,822.75 is in the account at the end of the second year, what is the annual interest rate?

Solving Problems

Solve.

33. *Car Trips.*   During the first part of a trip, Tara drove 120 mi at a certain speed. Tara then drove another 100 mi at a speed that was 10 mph slower. If the total time of Tara's trip was 4 hr, what was her speed on each part of the trip?

34. *Canoeing.*   During the first part of a canoe trip, Ken covered 60 km at a certain speed. He then traveled 24 km at a speed that was 4 km/h slower. If the total time for the trip was 8 hr, what was the speed on each part of the trip?

35. *Car Trips.*   Diane's Dodge travels 200 mi averaging a certain speed. If the car had gone 10 mph faster, the trip would have taken 1 hr less. Find Diane's average speed.

36. *Car Trips.*   Stan's Subaru travels 280 mi averaging a certain speed. If the car had gone 5 mph faster, the trip would have taken 1 hr less. Find Stan's average speed.

37. *Air Travel.* A Cessna flies 600 mi at a certain speed. A Beechcraft flies 1000 mi at a speed that is 50 mph faster, but takes 1 hr longer. Find the speed of each plane.

38. *Air Travel.* A turbo-jet flies 50 mph faster than a super-prop plane. If a turbo-jet goes 2000 mi in 3 hr less time than it takes the super-prop to go 2800 mi, find the speed of each plane.

39. *Bicycling.* Naoki bikes the 36 mi to Hillsboro averaging a certain speed. The return trip is made at a speed that is 3 mph slower. Total time for the round trip is 7 hr. Find Naoki's average speed on each part of the trip.

40. *Car Speed.* On a sales trip, Samir drives the 600 mi to Richmond averaging a certain speed. The return trip is made at an average speed that is 10 mph slower. Total time for the round trip is 22 hr. Find Samir's average speed on each part of the trip.

41. *Navigation.* The Hudson River flows at a rate of 3 mph. A patrol boat travels 60 mi upriver and returns in a total time of 9 hr. What is the speed of the boat in still water?

42. *Navigation.* The current in a typical Mississippi River shipping route flows at a rate of 4 mph. In order for a barge to travel 24 mi upriver and then return in a total of 5 hr, approximately how fast must the barge be able to travel in still water?

43. *Filling a Pool.* A well and a spring are filling a swimming pool. Together, they can fill the pool in 3 hr. The well, working alone, can fill the pool in 8 hr less time than it would take the spring. How long would the spring take, working alone, to fill the pool?

44. *Filling a Tank.* Two pipes are connected to the same tank. Working together, they can fill the tank in 4 hr. The larger pipe, working alone, can fill the tank in 6 hr less time than it would take the smaller one. How long would the smaller one take, working alone, to fill the tank?

45. *Paddleboats.* Kofi paddles 1 mi upstream and 1 mi back in a total time of 1 hr. The speed of the river is 2 mph. Find the speed of Kofi's paddleboat in still water.

46. *Rowing.* Abby rows 10 km upstream and 10 km back in a total time of 3 hr. The speed of the river is 5 km/h. Find Abby's speed in still water.

47. *Reforestation.* Working together, Katherine and Julianna can plant new trees on their recently reforested land in 6 days. Working alone, it would take Julianna 2 days longer than it would take Katherine to plant the trees. How long would it take Katherine, working alone, to plant the trees?

48. *Team Teaching.* Working together, Tanner and Joel can grade their students' projects in 2 hr. Working alone, it would take Tanner 2 hr longer than it would take Joel to grade the projects. How long would it take Joel, working alone, to grade the projects?

49. Marti is tied to a bungee cord that is twice as long as the cord tied to Rafe. Will Marti's fall take twice as long as Rafe's before their cords begin to stretch? Why or why not? (See Exercises 25 and 26.)

50. Under what circumstances would a negative value for *t*, time, have meaning?

Skill Review

Solve.

51. $\frac{1}{2}(x - 7) = \frac{1}{3}x + 4$ [2.2]

52. $\frac{1}{x} = -8$ [6.6]

53. $6|3x + 2| = 12$ [9.2]

54. $2x - y = 4,$
$x + 2y = 3$ [8.2]

55. $\sqrt{2x - 8} = 15$ [10.6]

56. $2 + \sqrt{t} = \sqrt{t + 5}$ [10.6]

Synthesis

57. Write a problem for a classmate to solve. Devise the problem so that **(a)** the solution is found after solving a rational equation and **(b)** the solution is "The express train travels 90 mph."

58. Sophia has no difficulty creating a table for motion problems but cannot write equations from the table. What suggestion can you offer Sophia?

59. Biochemistry. The equation

$$A = 6.5 - \frac{20.4t}{t^2 + 36}$$

is used to calculate the acid level A in a person's blood t minutes after sugar is consumed. Solve for t.

60. Special Relativity. Einstein found that an object with initial mass m_0 and traveling velocity v has mass

$$m = \frac{m_0}{\sqrt{1 - \dfrac{v^2}{c^2}}},$$

where c is the speed of light. Solve the formula for c.

61. Find a number for which the reciprocal of 1 less than the number is the same as 1 more than the number.

62. Purchasing. A discount store bought a quantity of potted plants for $250 and sold all but 15 at a profit of $3.50 per plant. With the total amount received, the manager could buy 4 more than twice as many as were bought before. Find the cost per plant.

63. Art and Aesthetics. For over 2000 years, artists, sculptors, and architects have regarded the proportions of a "golden" rectangle as visually appealing. A rectangle of width w and length l is considered "golden" if

$$\frac{w}{l} = \frac{l}{w + l}.$$

Solve for l.

64. Diagonal of a Cube. Find a formula that expresses the length of the three-dimensional diagonal of a cube as a function of the cube's surface area.

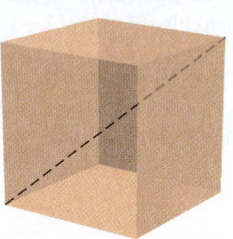

65. Solve for n:

$$mn^4 - r^2pm^3 - r^2n^2 + p = 0.$$

66. Surface Area. Find a formula that expresses the diameter of a right cylindrical solid as a function of its surface area and its height. (See Exercise 7.)

67. A sphere is inscribed in a cube as shown in the figure below. Express the surface area of the sphere as a function of the surface area S of the cube. (See Exercise 5.)

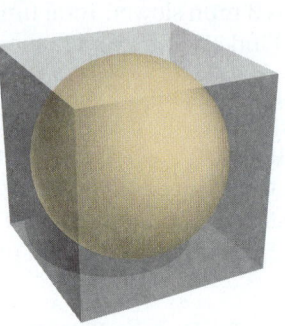

↪ YOUR TURN ANSWERS: SECTION 11.4

1. $x = \dfrac{F^2}{14{,}113.44}$ **2.** $x = 2 \pm \sqrt{y}$

3. $h = 2 \pm \dfrac{\sqrt{64 - d}}{4}$ **4.** 20 mph

QUICK QUIZ: SECTIONS 11.1–11.4

Solve. [11.1], [11.2]

1. $x^2 - 20x = 15$

2. $x(x - 2) = 25$

3. $2500 grows to $2601 in 2 years. Use the formula $A = P(1 + r)^t$ to determine the interest rate.   [11.1]

4. Write a quadratic equation having the solutions $2\sqrt{3}$ and $-2\sqrt{3}$.   [11.3]

5. Solve $n = d^2 + 2d$ for d.   [11.4]

PREPARE TO MOVE ON

Simplify.

1. $(m^{-1})^2$ [4.2]   2. $(y^{1/6})^2$ [10.2]

Solve.

3. $t^{-1} = \dfrac{1}{2}$ [6.6]   4. $x^{1/4} = 3$ [10.6]

11.5 Equations Reducible to Quadratic

Equations in Quadratic Form ▪ Radical Equations and Rational Equations

Student Notes

To identify an equation in quadratic form, look for two variable expressions in the equation. In order for an equation to be in quadratic form, the exponent in one expression must be twice the exponent in the other expression.

Equations in Quadratic Form

Certain equations that are not really quadratic can be regarded in a manner that allows us to use the methods developed for quadratic equations. For example, because the square of x^2 is x^4, the equation $x^4 - 9x^2 + 8 = 0$ is said to be "quadratic in x^2":

$$x^4 - 9x^2 + 8 = 0$$

$$(x^2)^2 - 9(x^2) + 8 = 0 \qquad \text{Thinking of } x^4 \text{ as } (x^2)^2$$

$$u^2 - 9u + 8 = 0. \qquad \text{To make this clearer, write } u \text{ instead of } x^2.$$

The equation $u^2 - 9u + 8 = 0$ can be solved for u by factoring or by the quadratic formula. Then, remembering that $u = x^2$, we can solve for x. Equations that can be solved like this are *reducible to quadratic* and are said to be *in quadratic form*.

EXAMPLE 1 Solve: $x^4 - 9x^2 + 8 = 0$.

SOLUTION We begin by letting $u = x^2$ and finding u^2.

$$\text{If we let} \quad u = x^2,$$
$$\text{then} \qquad u^2 = (x^2)^2 = x^4.$$

Next, we substitute u^2 for x^4 and u for x^2:

$$u^2 - 9u + 8 = 0$$
$$(u - 8)(u - 1) = 0 \qquad \text{Factoring}$$
$$u - 8 = 0 \quad or \quad u - 1 = 0 \qquad \text{Principle of zero products}$$
$$u = 8 \quad or \qquad u = 1. \qquad \text{We have solved for } u.$$

We now replace u with x^2 and solve these equations:

$$x^2 = 8 \qquad or \quad x^2 = 1$$
$$x = \pm\sqrt{8} \quad or \quad x = \pm 1$$
$$x = \pm 2\sqrt{2} \quad or \quad x = \pm 1. \qquad \text{We have solved for } x.$$

To check, note that for both $x = 2\sqrt{2}$ and $-2\sqrt{2}$, we have $x^2 = 8$ and $x^4 = 64$. Similarly, for both $x = 1$ and -1, we have $x^2 = 1$ and $x^4 = 1$. Thus instead of making four checks, we need make only two.

> **CAUTION!** A common error when working on problems like Example 1 is to solve for u but forget to solve for x. Remember to solve for the *original* variable!

Check: For $\pm 2\sqrt{2}$:

$$\frac{x^4 - 9x^2 + 8 = 0}{(\pm 2\sqrt{2})^4 - 9(\pm 2\sqrt{2})^2 + 8 \mid 0}$$
$$64 - 9 \cdot 8 + 8$$
$$0 \overset{?}{=} 0 \quad \text{TRUE}$$

For ± 1:

$$\frac{x^4 - 9x^2 + 8 = 0}{(\pm 1)^4 - 9(\pm 1)^2 + 8 \mid 0}$$
$$1 - 9 + 8$$
$$0 \overset{?}{=} 0 \quad \text{TRUE}$$

The solutions are $1, -1, 2\sqrt{2}$, and $-2\sqrt{2}$.

1. Solve: $x^4 - 8x^2 - 9 = 0$.

 YOUR TURN

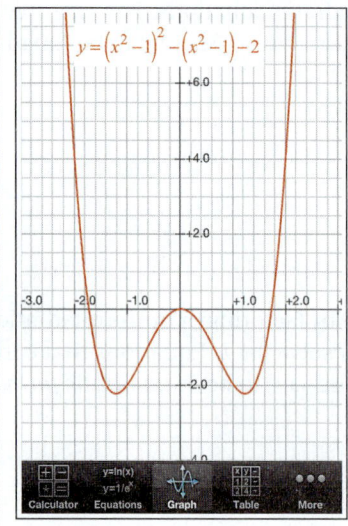
2. Find the x-intercepts of the graph of $f(x) = (x^2 - 2)^2 - 3(x^2 - 2) - 4 = 0$.

Equations like those in Example 1 can be solved by factoring:

$$x^4 - 9x^2 + 8 = 0$$
$$(x^2 - 1)(x^2 - 8) = 0$$
$$x^2 - 1 = 0 \quad or \quad x^2 - 8 = 0$$
$$x^2 = 1 \quad or \quad x^2 = 8$$
$$x = \pm 1 \quad or \quad x = \pm 2\sqrt{2}.$$

However, it can become difficult to solve an equation without first making a substitution.

EXAMPLE 2   Find the x-intercepts of the graph of
$$f(x) = (x^2 - 1)^2 - (x^2 - 1) - 2.$$

SOLUTION   The x-intercepts occur where $f(x) = 0$ so we must have

$$(x^2 - 1)^2 - (x^2 - 1) - 2 = 0. \qquad \text{Setting } f(x) \text{ equal to 0}$$

If we let $u = x^2 - 1,$

then $u^2 = (x^2 - 1)^2.$

Substituting, we have

$$u^2 - u - 2 = 0 \qquad \text{Substituting in } (x^2 - 1)^2 - (x^2 - 1) - 2 = 0$$
$$(u - 2)(u + 1) = 0$$
$$u = 2 \quad or \quad u = -1. \quad \text{Using the principle of zero products}$$

Next, we replace u with $x^2 - 1$ and solve these equations:

$$x^2 - 1 = 2 \quad or \quad x^2 - 1 = -1$$
$$x^2 = 3 \quad or \quad x^2 = 0 \qquad \text{Adding 1 to both sides}$$
$$x = \pm\sqrt{3} \quad or \quad x = 0. \qquad \text{Using the principle of square roots}$$

The x-intercepts occur at $(-\sqrt{3}, 0), (0, 0),$ and $(\sqrt{3}, 0)$.

YOUR TURN

The following tips may prove useful.

TO SOLVE AN EQUATION THAT IS REDUCIBLE TO QUADRATIC

1. Look for two variable expressions in the equation. One expression should be the square of the other. (The exponent in one expression will be twice the exponent in the other expression.)
2. Write down any substitutions that you are making.
3. Remember to solve for the variable that is used in the original equation.
4. Check possible answers in the original equation.

Radical Equations and Rational Equations

Sometimes rational equations, radical equations, or equations containing exponents that are fractions are reducible to quadratic. It is especially important that answers to these equations be checked in the original equation.

EXAMPLE 3   Solve: $x - 3\sqrt{x} - 4 = 0$.

SOLUTION   This radical equation could be solved using the principle of powers. However, if we note that the square of $\sqrt{x}$ is x, we can regard the equation as "quadratic in $\sqrt{x}$."

If we let $u = \sqrt{x}$,

then $u^2 = x$.

Substituting, we have

$$x - 3\sqrt{x} - 4 = 0$$
$$u^2 - 3u - 4 = 0$$
$$(u - 4)(u + 1) = 0$$
$$u = 4 \quad or \quad u = -1. \qquad \text{Using the principle of zero products}$$

Next, we replace u with $\sqrt{x}$ and solve these equations:

$$\sqrt{x} = 4 \quad or \quad \sqrt{x} = -1.$$

Squaring gives us $x = 16$ or $x = 1$ and also makes checking essential.

Check: For 16:

$$\begin{array}{c|c} x - 3\sqrt{x} - 4 = 0 \\ \hline 16 - 3\sqrt{16} - 4 & 0 \\ 16 - 3 \cdot 4 - 4 & \\ & 0 \stackrel{?}{=} 0 \quad \text{TRUE} \end{array}$$

For 1:

$$\begin{array}{c|c} x - 3\sqrt{x} - 4 = 0 \\ \hline 1 - 3\sqrt{1} - 4 & 0 \\ 1 - 3 \cdot 1 - 4 & \\ & -6 \stackrel{?}{=} 0 \quad \text{FALSE} \end{array}$$

The number 16 checks, but 1 does not. Had we noticed that $\sqrt{x} = -1$ has no solution (since principal square roots are never negative), we could have focused only on $\sqrt{x} = 4$. The solution is 16.

YOUR TURN

EXAMPLE 4   Solve: $2m^{-2} + m^{-1} - 15 = 0$.

SOLUTION   Note that the square of m^{-1} is $(m^{-1})^2$, or m^{-2}.

If we let $u = m^{-1}$,

then $u^2 = m^{-2}$.

Substituting, we have

$$2u^2 + u - 15 = 0 \qquad \text{Substituting in } 2m^{-2} + m^{-1} - 15 = 0$$
$$(2u - 5)(u + 3) = 0$$
$$2u - 5 = 0 \quad or \quad u + 3 = 0 \qquad \text{Using the principle of zero products}$$
$$2u = 5 \quad or \qquad u = -3$$
$$u = \tfrac{5}{2} \quad or \qquad u = -3.$$

Now, we replace u with m^{-1} and solve:

$$m^{-1} = \frac{5}{2} \quad or \quad m^{-1} = -3$$

$$\frac{1}{m} = \frac{5}{2} \quad or \quad \frac{1}{m} = -3 \qquad \text{Recall that } m^{-1} = \frac{1}{m}.$$

$$1 = \frac{5}{2}m \quad or \quad 1 = -3m \qquad \text{Multiplying both sides by } m$$

$$\frac{2}{5} = m \quad or \quad -\frac{1}{3} = m. \qquad \text{Solving for } m$$

Student Notes

Note that $x^2 = -1$ has complex solutions but $\sqrt{x} = -1$ does not.

Technology Connection

Check Example 3 with a graphing calculator. Use the ZERO, ROOT, or INTERSECT option, if possible.

3. Solve: $x - 5\sqrt{x} - 14 = 0$.

Determine u and u^2.

Substitute.

Solve for u.

Solve for the original variable.

Check.

Check:

For $\frac{2}{5}$:

$$2m^{-2} + m^{-1} - 15 = 0$$

$$\begin{array}{c|c} 2\left(\frac{2}{5}\right)^{-2} + \left(\frac{2}{5}\right)^{-1} - 15 & 0 \\ 2\left(\frac{5}{2}\right)^2 + \left(\frac{5}{2}\right) - 15 & \\ 2\left(\frac{25}{4}\right) + \frac{5}{2} - 15 & \\ \frac{25}{2} + \frac{5}{2} - 15 & \\ \frac{30}{2} - 15 & \\ & 0 \overset{?}{=} 0 \quad \text{TRUE} \end{array}$$

For $-\frac{1}{3}$:

$$2m^{-2} + m^{-1} - 15 = 0$$

$$\begin{array}{c|c} 2\left(-\frac{1}{3}\right)^{-2} + \left(-\frac{1}{3}\right)^{-1} - 15 & 0 \\ 2\left(-\frac{3}{1}\right)^2 + \left(-\frac{3}{1}\right) - 15 & \\ 2(9) + (-3) - 15 & \\ 18 - 3 - 15 & \\ & 0 \overset{?}{=} 0 \quad \text{TRUE} \end{array}$$

4. Solve: $m^{-4} - 5m^{-2} + 6 = 0$.

Both numbers check. The solutions are $-\frac{1}{3}$ and $\frac{2}{5}$.

YOUR TURN

Note that Example 4 can also be written $2/m^2 + 1/m - 15 = 0$. It can then be solved by letting $u = 1/m$ and $u^2 = 1/m^2$ or by clearing fractions.

EXAMPLE 5   Solve: $t^{2/5} - t^{1/5} - 2 = 0$.

SOLUTION   Note that the square of $t^{1/5}$ is $(t^{1/5})^2$, or $t^{2/5}$. The equation is therefore quadratic in $t^{1/5}$.

$$\text{If we let} \quad u = t^{1/5},$$
$$\text{then} \quad u^2 = t^{2/5}.$$

Substituting, we have

$$u^2 - u - 2 = 0 \qquad \text{Substituting in } t^{2/5} - t^{1/5} - 2 = 0$$
$$(u - 2)(u + 1) = 0$$
$$u = 2 \quad or \quad u = -1. \qquad \text{Using the principle of zero products}$$

Now, we replace u with $t^{1/5}$ and solve:

$$t^{1/5} = 2 \quad or \quad t^{1/5} = -1$$
$$t = 32 \quad or \quad t = -1. \qquad \text{Principle of powers; raising both sides to the 5th power}$$

Check:

For 32:

$$t^{2/5} - t^{1/5} - 2 = 0$$

$$\begin{array}{c|c} 32^{2/5} - 32^{1/5} - 2 & 0 \\ (32^{1/5})^2 - 32^{1/5} - 2 & \\ 2^2 - 2 - 2 & \\ & 0 \overset{?}{=} 0 \quad \text{TRUE} \end{array}$$

For -1:

$$t^{2/5} - t^{1/5} - 2 = 0$$

$$\begin{array}{c|c} (-1)^{2/5} - (-1)^{1/5} - 2 & 0 \\ [(-1)^{1/5}]^2 - (-1)^{1/5} - 2 & \\ (-1)^2 - (-1) - 2 & \\ & 0 \overset{?}{=} 0 \quad \text{TRUE} \end{array}$$

5. Solve: $2t^{2/3} - t^{1/3} - 3 = 0$.

Both numbers check. The solutions are 32 and -1.

YOUR TURN

EXAMPLE 6   Solve: $(5 + \sqrt{r})^2 + 6(5 + \sqrt{r}) + 2 = 0$.

SOLUTION

$$\text{If we let} \quad u = 5 + \sqrt{r},$$
$$\text{then} \quad u^2 = (5 + \sqrt{r})^2.$$

Substituting, we have

$$u^2 + 6u + 2 = 0$$

$$u = \frac{-6 \pm \sqrt{6^2 - 4 \cdot 1 \cdot 2}}{2 \cdot 1} \qquad \text{Using the quadratic formula}$$

$$\left. \begin{aligned} &= \frac{-6 \pm \sqrt{28}}{2} \\ &= \frac{-6}{2} \pm \frac{2\sqrt{7}}{2} \\ &= -3 \pm \sqrt{7}. \end{aligned} \right\} \qquad \text{Simplifying; } \sqrt{28} = \sqrt{4}\sqrt{7}$$

Now, we replace u with $5 + \sqrt{r}$ and solve for r:

$$\begin{aligned} u &= -3 + \sqrt{7} \quad or \qquad u = -3 - \sqrt{7} \\ 5 + \sqrt{r} &= -3 + \sqrt{7} \quad or \quad 5 + \sqrt{r} = -3 - \sqrt{7} \\ \sqrt{r} &= -8 + \sqrt{7} \quad or \qquad \sqrt{r} = -8 - \sqrt{7}. \end{aligned}$$

6. Solve:

$(3 - \sqrt{x})^2 - 2(3 - \sqrt{x}) - 8 = 0.$

Both $-8 + \sqrt{7}$ and $-8 - \sqrt{7}$ are negative. Since $\sqrt{r}$ is never negative, both values of $\sqrt{r}$ must be rejected. The equation has no solution.

YOUR TURN

11.5 EXERCISE SET

FOR EXTRA HELP

MyMathLab® Math XL
PRACTICE   WATCH   READ   REVIEW

Vocabulary and Reading Check

Classify each of the following statements as either true or false.

1. Some equations that are not really quadratic are quadratic in form.

2. Some radical equations and rational equations are reducible to quadratic.

3. We have not completed solving an equation that is quadratic in form until we have solved for the original variable.

4. When solving an equation that is quadratic in form, we should check any possible solutions in the original equation.

Concept Reinforcement

Write the substitution that could be used to make each equation quadratic in u.

5. For $3p - 4\sqrt{p} + 6 = 0$, use $u = $ _____.

6. For $x^{1/2} - x^{1/4} - 2 = 0$, use $u = $ _____.

7. For $(x^2 + 3)^2 + (x^2 + 3) - 7 = 0$, use $u = $

_____.

8. For $t^{-6} + 5t^{-3} - 6 = 0$, use $u = $ _____.

9. For $(1 + t)^4 + (1 + t)^2 + 4 = 0$, use $u = $

_____.

10. For $w^{1/3} - 3w^{1/6} + 8 = 0$, use $u = $ _____.

Equations in Quadratic Form

Solve.

11. $x^4 - 13x^2 + 36 = 0$

12. $x^4 - 17x^2 + 16 = 0$

13. $t^4 - 7t^2 + 12 = 0$

14. $t^4 - 11t^2 + 18 = 0$

15. $4x^4 - 9x^2 + 5 = 0$

16. $9x^4 - 38x^2 + 8 = 0$

17. $w + 4\sqrt{w} - 12 = 0$

18. $s + 3\sqrt{s} - 40 = 0$

19. $(x^2 - 7)^2 - 3(x^2 - 7) + 2 = 0$

20. $(x^2 - 2)^2 - 12(x^2 - 2) + 20 = 0$

21. $x^4 + 5x^2 - 36 = 0$

22. $x^4 + 5x^2 + 4 = 0$

23. $(n^2 + 6)^2 - 7(n^2 + 6) + 10 = 0$

24. $(m^2 + 7)^2 - 6(m^2 + 7) - 16 = 0$

Radical Equations and Rational Equations

Solve.

25. $r - 2\sqrt{r} - 6 = 0$

26. $s - 4\sqrt{s} - 1 = 0$

27. $(1 + \sqrt{x})^2 + 5(1 + \sqrt{x}) + 6 = 0$

28. $(3 + \sqrt{x})^2 + 3(3 + \sqrt{x}) - 10 = 0$

29. $x^{-2} - x^{-1} - 6 = 0$

11.6   Quadratic Functions and Their Graphs

The Graph of $f(x) = ax^2$ •   The Graph of $f(x) = a(x - h)^2$ •   The Graph of $f(x) = a(x - h)^2 + k$

The graph of any *linear* function $f(x) = mx + b$ is a straight line. In this section and the next, we will see that the graph of any *quadratic* function $f(x) = ax^2 + bx + c$ is a *parabola*. We examine the shape of such graphs by first looking at quadratic functions with $b = 0$ and $c = 0$.

The Graph of $f(x) = ax^2$

The most basic quadratic function is $f(x) = x^2$.

EXAMPLE 1   Graph: $f(x) = x^2$.

SOLUTION   We choose some values for x and compute $f(x)$ for each. Then we plot the ordered pairs and connect them with a smooth curve.

x	$f(x) = x^2$	$(x, f(x))$
-3	9	$(-3, 9)$
-2	4	$(-2, 4)$
-1	1	$(-1, 1)$
0	0	$(0, 0)$
1	1	$(1, 1)$
2	4	$(2, 4)$
3	9	$(3, 9)$

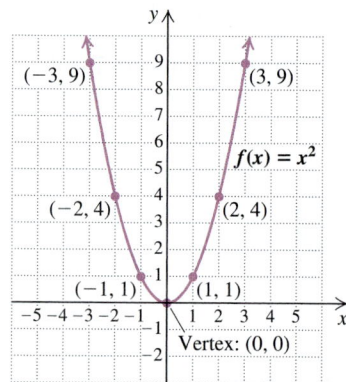

1. Graph: $f(x) = 2x^2$.

 YOUR TURN

All quadratic functions have graphs similar to the one in Example 1. Such curves are called **parabolas**. They are U-shaped and can open upward, as in Example 1, or downward. The "turning point" of the graph is called the **vertex** of the parabola. The vertex of the graph in Example 1 is (0, 0).

A parabola is symmetric with respect to a line that goes through the center of the parabola and the vertex. This line is known as the parabola's **axis of symmetry**. In Example 1, the y-axis (the vertical line $x = 0$) is the axis of symmetry. Were the paper folded on this line, the two halves of the curve would match.

Student Notes

By paying attention to the symmetry of each parabola and the location of the vertex, you save yourself considerable work. Note that the x-values 1 unit to the right or to the left of the vertex are paired with the y-value a units above the vertex. Thus the graph of $y = \frac{3}{2}x^2$ includes the points $\left(-1, \frac{3}{2}\right)$ and $\left(1, \frac{3}{2}\right)$.

The graph of any function of the form $y = ax^2$ has the vertex $(0, 0)$ and the axis of symmetry $x = 0$. By plotting points, we can compare the graphs of $g(x) = \frac{1}{2}x^2$ and $h(x) = 2x^2$ with the graph of $f(x) = x^2$.

x	$g(x) = \frac{1}{2}x^2$
-3	$\frac{9}{2}$
-2	2
-1	$\frac{1}{2}$
0	0
1	$\frac{1}{2}$
2	2
3	$\frac{9}{2}$

x	$h(x) = 2x^2$
-3	18
-2	8
-1	2
0	0
1	2
2	8
3	18

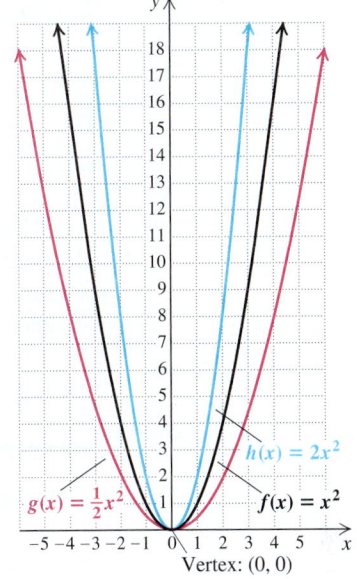

Technology Connection

To explore the effect of a on the graph of $y = ax^2$, let $y_1 = x^2$, $y_2 = 3x^2$, and $y_3 = \frac{1}{3}x^2$. Graph the equations and use ⟨TRACE⟩ to see how the y-values compare, using ⌃ or ⌄ to hop the cursor from one curve to the next.

Many graphing calculators include a Transfrm application. If you run that application and let $y_1 = Ax^2$, the graph becomes interactive. The value of A can be entered while viewing the graph, or the values can be stepped up or down by pressing ⟨ or ⟩.

1. Compare the graphs of $y_1 = \frac{1}{5}x^2$, $y_2 = x^2$, $y_3 = \frac{5}{2}x^2$, $y_4 = -\frac{1}{5}x^2$, $y_5 = -x^2$, and $y_6 = -\frac{5}{2}x^2$.
2. Describe the effect that A has on each graph.

Note that the graph of $g(x) = \frac{1}{2}x^2$ is "wider" than the graph of $f(x) = x^2$, and the graph of $h(x) = 2x^2$ is "narrower." The vertex and the axis of symmetry, however, remain $(0, 0)$ and the line $x = 0$, respectively.

When we consider the graph of $k(x) = -\frac{1}{2}x^2$, we see that the parabola is the same shape as the graph of $g(x) = \frac{1}{2}x^2$, but opens downward. We say that the graphs of k and g are *reflections* of each other across the x-axis.

x	$k(x) = -\frac{1}{2}x^2$
-3	$-\frac{9}{2}$
-2	-2
-1	$-\frac{1}{2}$
0	0
1	$-\frac{1}{2}$
2	-2
3	$-\frac{9}{2}$

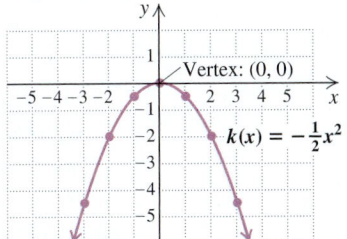

GRAPHING $f(x) = ax^2$

The graph of $f(x) = ax^2$ is a parabola with $x = 0$ as its axis of symmetry. Its vertex is the origin.

For $a > 0$, the parabola opens upward. For $a < 0$, the parabola opens downward.

If $|a|$ is greater than 1, the parabola is narrower than $y = x^2$.

If $|a|$ is between 0 and 1, the parabola is wider than $y = x^2$.

The width of a parabola and whether it opens upward or downward are determined by the coefficient a in $f(x) = ax^2 + bx + c$. In the remainder of this section, we graph quadratic functions that are written in a form from which the vertex can be read directly.

The Graph of $f(x) = a(x - h)^2$

EXAMPLE 2   Graph: $f(x) = (x - 3)^2$.

SOLUTION   We choose some values for x and compute $f(x)$. Since $(x - 3)^2 = 1 \cdot (x - 3)^2$, $a = 1$, and the graph opens upward. It is important to note that for any input that is 3 more than an input for Example 1, the outputs match. We plot the points and draw the curve.

x	$f(x) = (x - 3)^2$
-1	16
0	9
1	4
2	1
3	0
4	1
5	4
6	9

$\longleftarrow$ Vertex

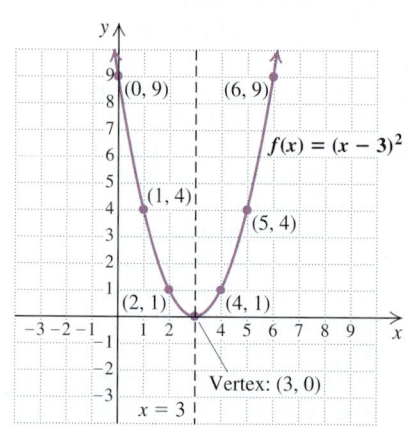

The line $x = 3$ is the axis of symmetry and the point $(3, 0)$ is the vertex. Had we recognized earlier that $x = 3$ is the axis of symmetry, we could have computed some values on one side, such as $(4, 1)$, $(5, 4)$, and $(6, 9)$, and then used symmetry to get their mirror images $(2, 1)$, $(1, 4)$, and $(0, 9)$ without further computation.

2. Graph: $f(x) = (x + 2)^2$.

YOUR TURN

The result of Example 2 can be generalized:

The vertex of the graph of $f(x) = a(x - h)^2$ is $(h, 0)$.

EXAMPLE 3   Graph: $g(x) = -2(x + 4)^2$.

SOLUTION   We choose some values for x and compute $g(x)$. Since $a = -2$, the graph will open downward. If we rewrite the equation as $g(x) = -2(x - (-4))^2$, we see that $(-4, 0)$ is the vertex. The axis of symmetry is then $x = -4$. We plot some points and draw the curve.

x	$g(x) = -2(x + 4)^2$
-6	-8
-5	-2
-4	0
-3	-2
-2	-8

←Vertex

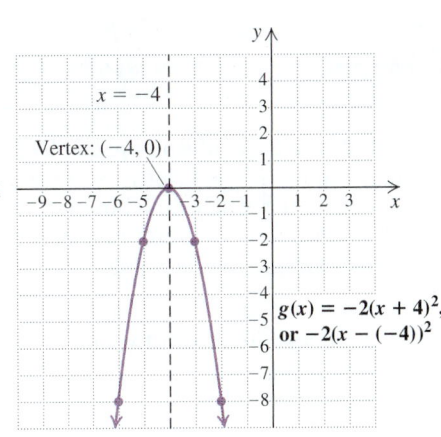

$x = -4$

Vertex: $(-4, 0)$

$g(x) = -2(x + 4)^2$, or $-2(x - (-4))^2$

3. Graph: $g(x) = -(x - 3)^2$.

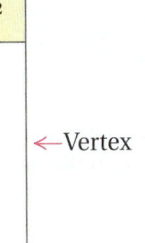 YOUR TURN

In Example 2, the graph of $f(x) = (x - 3)^2$ looks just like the graph of $y = x^2$, except that it is moved, or *translated*, 3 units to the right. In Example 3, the graph of $g(x) = -2(x + 4)^2$ looks like the graph of $y = -2x^2$, except that it is shifted 4 units to the left. These results are generalized as follows.

Technology Connection

To explore the effect of h on the graph of $f(x) = a(x - h)^2$, let $y_1 = 7x^2$ and $y_2 = 7(x - 1)^2$. Graph both y_1 and y_2 and compare y-values, beginning at $x = 1$ and increasing x by one unit at a time. The G-T or HORIZ **MODE** can be used to view a split screen showing both the graph and a table.

Next, let $y_3 = 7(x - 2)^2$ and compare its graph and y-values with those of y_1 and y_2. Then let $y_4 = 7(x + 1)^2$ and $y_5 = 7(x + 2)^2$.

1. Compare graphs and y-values and describe the effect of h on the graph of $f(x) = a(x - h)^2$.
2. If the Transfrm application is available, let $y_1 = A(x - B)^2$ and describe the effect that A and B have on each graph.

> **GRAPHING $f(x) = a(x - h)^2$**
>
> The graph of $f(x) = a(x - h)^2$ has the same shape as the graph of $y = ax^2$.
>
> - If h is positive, the graph of $y = ax^2$ is shifted h units to the right.
> - If h is negative, the graph of $y = ax^2$ is shifted $|h|$ units to the left.
> - The vertex is $(h, 0)$, and the axis of symmetry is $x = h$.

The Graph of $f(x) = a(x - h)^2 + k$

If we add a positive constant k to the right side of $f(x) = a(x - h)^2$, the graph of $f(x)$ is moved up. If we add a negative constant k, the curve is moved down. The axis of symmetry for the parabola remains $x = h$, but the vertex will be at (h, k), or, equivalently, $(h, f(h))$.

Because of the shape of their graphs, quadratic functions have either a *minimum* value or a *maximum* value. Many real-world applications involve finding that value. For example, a business owner is concerned with minimizing cost and maximizing profit. If a parabola opens upward ($a > 0$), the function value, or y-value, at the vertex is a least, or minimum, value. That is, it is less than the y-value at any other point on the graph. If the parabola opens downward ($a < 0$), the function value at the vertex is a greatest, or maximum, value.

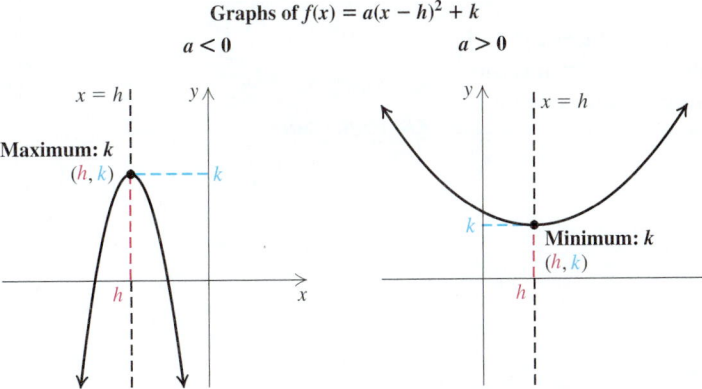

Graphs of $f(x) = a(x - h)^2 + k$

$a < 0$ $a > 0$

$x = h$ Maximum: k (h, k) k

$x = h$ k Minimum: k (h, k)

Technology Connection

To study the effect of k on the graph of $f(x) = a(x - h)^2 + k$, let $y_1 = 7(x - 1)^2$ and $y_2 = 7(x - 1)^2 + 2$. Graph both y_1 and y_2 in the window $[-5, 5, -5, 5]$ and use TRACE or a TABLE to compare the y-values for any given x-value.

1. Let $y_3 = 7(x - 1)^2 - 4$ and compare its graph and y-values with those of y_1 and y_2.

2. Try other values of k, including decimals and fractions. Describe the effect of k on the graph of $f(x) = a(x - h)^2$.
3. If the Transfrm application is available, let $y_1 = A(x - B)^2 + C$ and describe the effect that A, B, and C have on each graph.

GRAPHING $f(x) = a(x - h)^2 + k$

The graph of $f(x) = a(x - h)^2 + k$ has the same shape as the graph of $y = a(x - h)^2$.

- If k is positive, the graph of $y = a(x - h)^2$ is shifted k units up.
- If k is negative, the graph of $y = a(x - h)^2$ is shifted $|k|$ units down.
- The vertex is (h, k), and the axis of symmetry is $x = h$.
- For $a > 0$, the minimum function value is k. For $a < 0$, the maximum function value is k.

EXAMPLE 4 Graph $g(x) = (x - 3)^2 - 5$, and find the vertex and the minimum function value.

SOLUTION The graph will look like that of $f(x) = (x - 3)^2$ (see Example 2) but shifted 5 units down. You can confirm this by plotting some points.

The vertex is now $(3, -5)$, and the minimum function value is -5.

x	$g(x) = (x - 3)^2 - 5$
0	4
1	-1
2	-4
3	-5	← Vertex
4	-4
5	-1
6	4

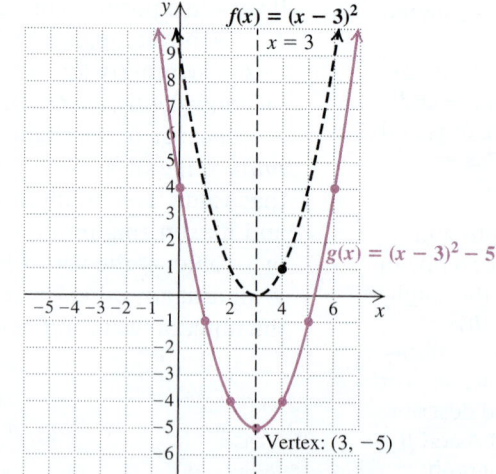

4. Graph $g(x) = (x + 2)^2 - 1$, and find the vertex and the minimum function value.

 YOUR TURN

EXAMPLE 5   Graph $h(x) = \frac{1}{2}(x - 3)^2 + 6$, and find the vertex and the minimum function value.

SOLUTION   The graph looks just like that of $f(x) = \frac{1}{2}x^2$ but is shifted 3 units to the right and 6 units up. The vertex is $(3, 6)$, and the axis of symmetry is $x = 3$. We draw $f(x) = \frac{1}{2}x^2$ and then shift the curve over and up. The minimum function value is 6. By plotting some points, we have a check.

x	$h(x) = \frac{1}{2}(x - 3)^2 + 6$
0	$10\frac{1}{2}$
1	8
3	6
5	8
6	$10\frac{1}{2}$

← Vertex

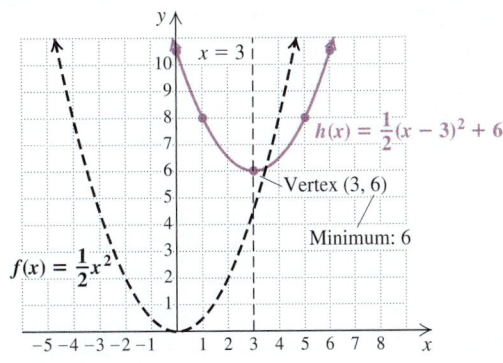

5. Graph
$f(x) = 2(x - 1)^2 + 4$, and find the vertex and the minimum function value.

YOUR TURN

EXAMPLE 6   Graph $y = -2(x + 3)^2 + 5$. Find the vertex, the axis of symmetry, and the maximum or minimum value.

SOLUTION   We first express the equation in the equivalent form

$$y = -2[x - (-3)]^2 + 5.$$ This is in the form $y = a(x - h)^2 + k$.

The graph looks like that of $y = -2x^2$ translated 3 units to the left and 5 units up. The vertex is $(-3, 5)$, and the axis of symmetry is $x = -3$. Since -2 is negative, the graph opens downward, and we know that 5, the second coordinate of the vertex, is the maximum y-value.

   We compute a few points as needed, selecting convenient x-values on either side of the vertex. The graph is shown here.

Chapter Resource:
Collaborative Activity, p. 769

x	$y = -2(x + 3)^2 + 5$
-4	3
-3	5
-2	3

← Vertex

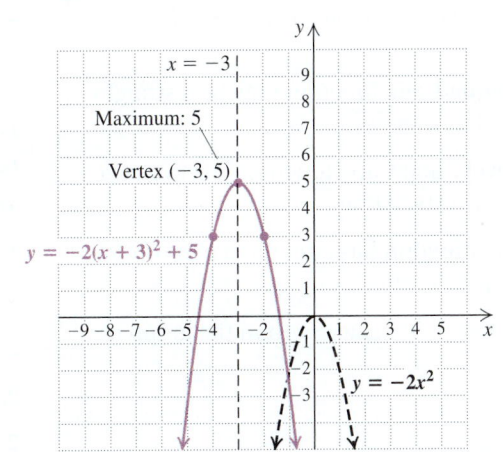

6. Graph $y = -\frac{1}{2}(x - 2)^2 - 1$. Find the vertex, the axis of symmetry, and the maximum or minimum value.

YOUR TURN

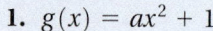

EXPLORING THE CONCEPT

The graph shown at right is that of a quadratic function $f(x) = ax^2$.
Match each of the following functions with the appropriate transformation
of the graph of f.

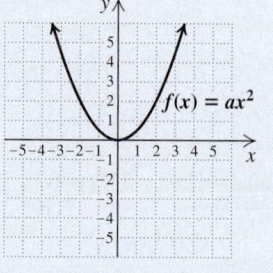

1. $g(x) = ax^2 + 1$
2. $p(x) = a(x + 1)^2$
3. $h(x) = -ax^2$
4. $q(x) = 2ax^2$

a)

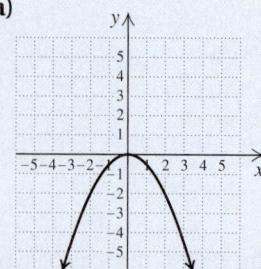

b)

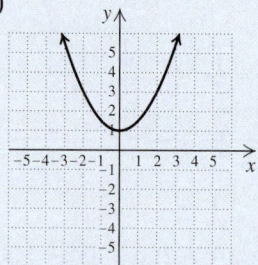

c)

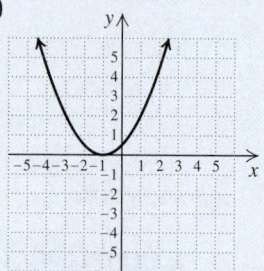

d)

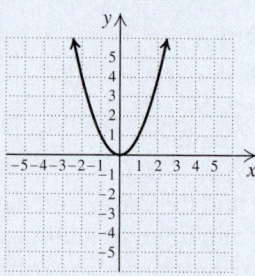

ANSWERS

1. (b)   2. (c)   3. (a)   4. (d)

11.6   EXERCISE SET

FOR EXTRA HELP

MyMathLab® Math XL

PRACTICE   WATCH   READ   REVIEW

Vocabulary and Reading Check

Classify each of the following statements as either true or false.

1. The graph of a quadratic function may be a straight line or a parabola.

2. The graph of every quadratic function is symmetric with respect to a vertical line.

3. Every quadratic function has either a maximum value or a minimum value.

4. The graph of $f(x) = 5x^2$ is wider than the graph of $f(x) = 3x^2$.

The Graph of $f(x) = ax^2$

Graph.

5. $f(x) = x^2$

6. $f(x) = -x^2$

7. $f(x) = -2x^2$

8. $f(x) = -3x^2$

9. $g(x) = \frac{1}{3}x^2$

10. $g(x) = \frac{1}{4}x^2$

Aha! 11. $h(x) = -\frac{1}{3}x^2$

12. $h(x) = -\frac{1}{4}x^2$

13. $f(x) = \frac{5}{2}x^2$

14. $f(x) = \frac{3}{2}x^2$

The Graph of $f(x) = a(x - h)^2$

For each of the following, graph the function, label the vertex, and draw the axis of symmetry.

15. $g(x) = (x + 1)^2$

16. $g(x) = (x + 4)^2$

17. $f(x) = (x - 2)^2$

18. $f(x) = (x - 1)^2$

19. $f(x) = -(x + 1)^2$

20. $f(x) = -(x - 1)^2$

21. $g(x) = -(x - 2)^2$

22. $g(x) = -(x + 4)^2$

23. $f(x) = 2(x + 1)^2$

24. $f(x) = 2(x + 4)^2$

25. $g(x) = 3(x - 4)^2$

26. $g(x) = 3(x - 5)^2$

27. $h(x) = -\frac{1}{2}(x - 4)^2$

28. $h(x) = -\frac{3}{2}(x - 2)^2$

29. $f(x) = \frac{1}{2}(x - 1)^2$

30. $f(x) = \frac{1}{3}(x + 2)^2$

31. $f(x) = -2(x + 5)^2$

32. $f(x) = -3(x + 7)^2$

33. $h(x) = -3\left(x - \frac{1}{2}\right)^2$

34. $h(x) = -2\left(x + \frac{1}{2}\right)^2$

The Graph of $f(x) = a(x - h)^2 + k$

For each of the following, graph the function and find the vertex, the axis of symmetry, and the maximum value or the minimum value.

35. $f(x) = (x - 5)^2 + 2$

36. $f(x) = (x + 3)^2 - 2$

37. $f(x) = (x + 1)^2 - 3$

38. $f(x) = (x - 1)^2 + 2$

39. $g(x) = \frac{1}{2}(x + 4)^2 + 1$

40. $g(x) = -(x - 2)^2 - 4$

41. $h(x) = -2(x - 1)^2 - 3$

42. $h(x) = -2(x + 1)^2 + 4$

43. $f(x) = 2(x + 3)^2 + 1$

44. $f(x) = 2(x - 5)^2 - 3$

45. $g(x) = -\frac{3}{2}(x - 2)^2 + 4$

46. $g(x) = \frac{3}{2}(x + 2)^2 - 1$

Without graphing, find the vertex, the axis of symmetry, and the maximum value or the minimum value.

47. $f(x) = 5(x - 3)^2 + 9$

48. $f(x) = 2(x - 1)^2 - 10$

49. $f(x) = -\frac{3}{7}(x + 8)^2 + 2$

50. $f(x) = -\frac{1}{4}(x + 4)^2 - 12$

51. $f(x) = \left(x - \frac{7}{2}\right)^2 - \frac{29}{4}$

52. $f(x) = -\left(x + \frac{3}{4}\right)^2 + \frac{17}{16}$

53. $f(x) = -\sqrt{2}(x + 2.25)^2 - \pi$

54. $f(x) = 2\pi(x - 0.01)^2 + \sqrt{15}$

55. Explain, without plotting points, why the graph of $y = x^2 - 4$ looks like the graph of $y = x^2$ translated 4 units down.

56. Explain, without plotting points, why the graph of $y = (x + 2)^2$ looks like the graph of $y = x^2$ translated 2 units to the left.

Skill Review

Add or subtract, as indicated. Simplify if possible.

57. $\dfrac{3}{x} + \dfrac{x}{x + 2}$ [6.4]

58. $3\sqrt{2x} + \sqrt{50x}$ [10.5]

59. $\sqrt[3]{8t} - \sqrt[3]{27t} + \sqrt{25t}$ [10.5]

60. $(2a^2 - 3a - 7) - (9a^2 - 6a + 1)$ [4.4]

61. $\dfrac{1}{x - 1} - \dfrac{x - 2}{x + 3}$ [6.4]

62. $\dfrac{1}{4 - x} + \dfrac{8}{x^2 - 16} - \dfrac{2}{x + 4}$ [6.4]

Synthesis

63. Before graphing a quadratic function, Martha always plots five points. First, she calculates and plots the coordinates of the vertex. Then she plots *four* more points after calculating *two* more ordered pairs. How is this possible?

64. If the graphs of $f(x) = a_1(x - h_1)^2 + k_1$ and $g(x) = a_2(x - h_2)^2 + k_2$ have the same shape, what, if anything, can you conclude about the a's, the h's, and the k's? Why?

Write an equation for a function having a graph with the same shape as the graph of $f(x) = \frac{3}{5}x^2$, but with the given point as the vertex.

65. $(1, 3)$ **66.** $(2, 8)$

67. $(4, -7)$ **68.** $(9, -6)$

69. $(-2, -5)$ **70.** $(-4, -2)$

For each of the following, write the equation of the parabola that has the shape of $f(x) = 2x^2$ or $g(x) = -2x^2$ and has a maximum value or a minimum value at the specified point.

71. Minimum: $(2, 0)$ **72.** Minimum: $(-4, 0)$

73. Maximum: $(0, -5)$ **74.** Maximum: $(3, 8)$

Use the following graph of $f(x) = a(x - h)^2 + k$ for Exercises 75–78.

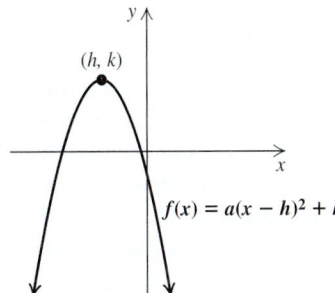

75. Describe what will happen to the graph if h is increased.

76. Describe what will happen to the graph if k is decreased.

77. Describe what will happen to the graph if a is replaced with $-a$.

78. Describe what will happen to the graph if $(x - h)$ is replaced with $(x + h)$.

Find an equation for the quadratic function F that satisfies the following conditions.

79. The graph of F is the same shape as the graph of f, where $f(x) = 3(x + 2)^2 + 7$, and $F(x)$ is a minimum at the same point that $g(x) = -2(x - 5)^2 + 1$ is a maximum.

80. The graph of F is the same shape as the graph of f, where $f(x) = -\frac{1}{3}(x - 2)^2 + 7$, and $F(x)$ is a maximum at the same point that $g(x) = 2(x + 4)^2 - 6$ is a minimum.

Functions other than parabolas can be translated. When calculating $f(x)$, if we replace x with $x - h$, where h is a constant, the graph will be moved horizontally. If we replace $f(x)$ with $f(x) + k$, the graph will be moved vertically. Use the graph below for Exercises 81–86.

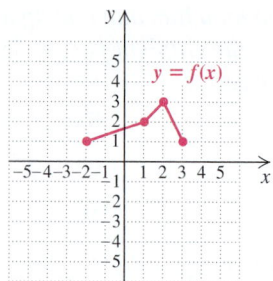

Draw a graph of each of the following.

81. $y = f(x - 1)$

82. $y = f(x + 2)$

83. $y = f(x) + 2$

84. $y = f(x) - 3$

85. $y = f(x + 3) - 2$

86. $y = f(x - 3) + 1$

87. Use the TRACE and/or TABLE features of a graphing calculator to confirm the maximum and minimum values given as answers to Exercises 47, 49, and 51. Be sure to adjust the window appropriately. On many graphing calculators, a maximum or minimum option may be available by using a CALC key.

88. Use a graphing calculator to check your graphs for Exercises 14, 24, and 44.

89. While trying to graph $y = -\frac{1}{2}x^2 + 3x + 1$, Yusef gets the following screen. How can Yusef tell at a glance that a mistake has been made?

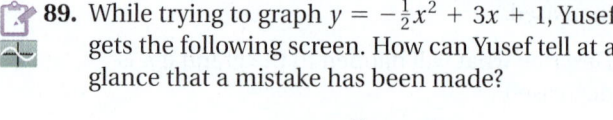

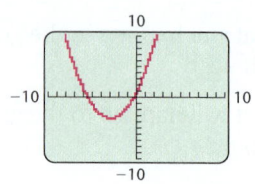

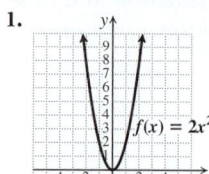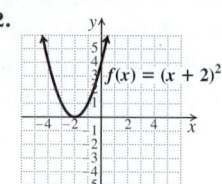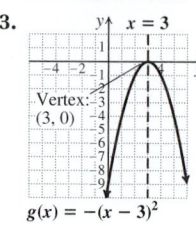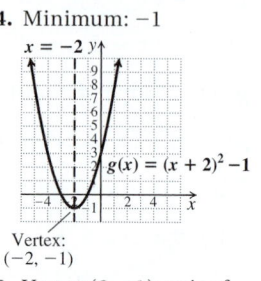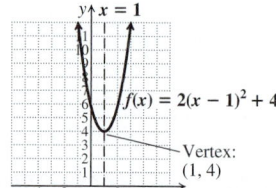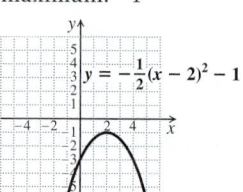
QUICK QUIZ: SECTIONS 11.1–11.6

Solve.

1. $(t + 7)^2 = 13$ [11.1]

2. $x^2 - 3x + 3 = 0$ [11.2]

3. $2m^{-2} - m^{-1} = 15$ [11.5]

4. Solve $t = \dfrac{xy}{3z^2}$ for z.   [11.4]

5. Graph $f(x) = (x - 3)^2 - 2$. Find the vertex, the axis of symmetry, and the maximum or minimum value.   [11.6]

PREPARE TO MOVE ON

Find the x-intercept and the y-intercept.   [3.3]

1. $8x - 6y = 24$

2. $3x + 4y = 8$

Find the x-intercepts of the graph of each equation.   [5.7]

3. $f(x) = x^2 + 8x + 15$

4. $g(x) = 2x^2 - x - 3$

Replace the blanks with constants to form a true equation.   [11.1]

5. $x^2 - 14x + \underline{\quad} = (x - \underline{\quad})^2$

6. $x^2 + 7x + \underline{\quad} = (x + \underline{\quad})^2$

11.7 More About Graphing Quadratic Functions

Graphing $f(x) = ax^2 + bx + c$ ■   Finding Intercepts

By *completing the square*, we can rewrite any polynomial $ax^2 + bx + c$ in the form $a(x - h)^2 + k$. This will allow us to graph any quadratic function.

Graphing $f(x) = ax^2 + bx + c$

EXAMPLE 1   Graph: $g(x) = x^2 - 6x + 4$. Label the vertex and the axis of symmetry.

SOLUTION   We have

$$g(x) = x^2 - 6x + 4$$
$$= (x^2 - 6x) + 4.$$

To complete the square inside the parentheses, we take half the x-coefficient, $\frac{1}{2} \cdot (-6) = -3$, and square it to get $(-3)^2 = 9$. Then we add $9 - 9$ inside the parentheses:

$$g(x) = (x^2 - 6x + 9 - 9) + 4 \qquad \text{The effect is of adding 0.}$$
$$= (x^2 - 6x + 9) + (-9 + 4) \qquad \text{Using an associative law}$$
$$= (x - 3)^2 - 5. \qquad \text{Factoring and simplifying}$$

The graph is that of $f(x) = x^2$ translated 3 units right and 5 units down. The vertex is $(3, -5)$, and the axis of symmetry is $x = 3$. As a simple check, note that $g(0) = 4$ and $(0, 4)$ is on the graph.

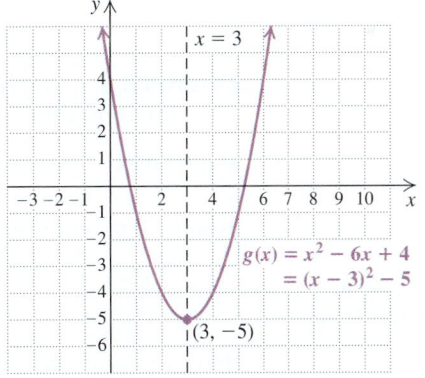

$$g(x) = x^2 - 6x + 4$$
$$= (x - 3)^2 - 5$$

1. Graph: $f(x) = x^2 - 2x + 3$. Label the vertex and the axis of symmetry.

 YOUR TURN

When the leading coefficient is not 1, we factor out that number from the first two terms. Then we complete the square and use the distributive law.

EXAMPLE 2   Graph: $f(x) = 3x^2 + 12x + 13$. Label the vertex and the axis of symmetry.

SOLUTION   Since the coefficient of x^2 is not 1, we need to factor out that number—in this case, 3—from the first two terms. Remember that we want the form $f(x) = a(x - h)^2 + k$:

$$f(x) = 3x^2 + 12x + 13 = 3(x^2 + 4x) + 13.$$

Student Notes

In this section, we add and subtract the same number when completing the square instead of adding the same number to both sides of an equation. The effect is the same with both approaches: An equivalent equation is formed.

Now we complete the square as before. We take half the x-coefficient, $\frac{1}{2} \cdot 4 = 2$, and square it: $2^2 = 4$. Then we add $4 - 4$ inside the parentheses:

$$f(x) = 3(x^2 + 4x + 4 - 4) + 13. \qquad \text{Adding } 4 - 4, \text{ or } 0, \text{ inside the parentheses}$$

The distributive law allows us to separate the -4 from the perfect-square trinomial so long as it is multiplied by 3. *This step is critical*:

$$f(x) = 3(x^2 + 4x + 4 - 4) + 13$$
$$= 3(x^2 + 4x + 4) + 3(-4) + 13 \qquad \text{This leaves a perfect-square trinomial inside the parentheses.}$$
$$= 3(x + 2)^2 + 1. \qquad \text{Factoring and simplifying}$$

The vertex is $(-2, 1)$, and the axis of symmetry is $x = -2$. The coefficient of x^2 is 3, so the graph is narrow and opens upward. We choose x-values on either side of the vertex, compute y-values, and then graph the parabola.

x	$f(x) = 3(x + 2)^2 + 1$
-2	1
-3	4
-1	4

$\longleftarrow$ Vertex

2. Graph:

$$f(x) = 2x^2 + 12x + 16.$$

Label the vertex and the axis of symmetry.

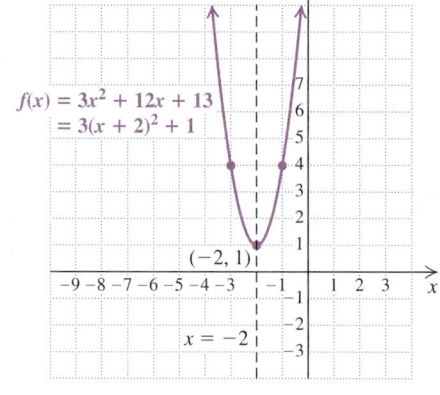

$f(x) = 3x^2 + 12x + 13$
$= 3(x + 2)^2 + 1$

$(-2, 1)$

$x = -2$

 YOUR TURN

EXAMPLE 3 Graph $f(x) = -2x^2 + 10x - 7$, and find the maximum or minimum function value.

SOLUTION We first find the vertex by completing the square. To do so, we factor out -2 from the first two terms of the expression. This makes the coefficient of x^2 inside the parentheses 1:

$$f(x) = -2x^2 + 10x - 7$$
$$= -2(x^2 - 5x) - 7.$$

Factor out a from both variable terms.

Now we complete the square as before. We take half of the x-coefficient and square it to get $\frac{25}{4}$. Then we add $\frac{25}{4} - \frac{25}{4}$ inside the parentheses:

$$f(x) = -2\left(x^2 - 5x + \frac{25}{4} - \frac{25}{4}\right) - 7$$

Complete the square inside the parentheses.

$$= -2\left(x^2 - 5x + \frac{25}{4}\right) + (-2)\left(-\frac{25}{4}\right) - 7 \qquad \text{Multiplying by } -2, \text{ using the distributive law, and regrouping}$$

Regroup.

Factor and simplify.

$$= -2\left(x - \frac{5}{2}\right)^2 + \frac{11}{2}. \qquad \text{Factoring and simplifying}$$

The vertex is $\left(\frac{5}{2}, \frac{11}{2}\right)$, and the axis of symmetry is $x = \frac{5}{2}$. The coefficient of x^2, -2, is negative, so the graph opens downward and the second coordinate of the vertex, $\frac{11}{2}$, is the maximum function value.

We plot a few points on either side of the vertex, including the y-intercept, $f(0)$, and graph the parabola.

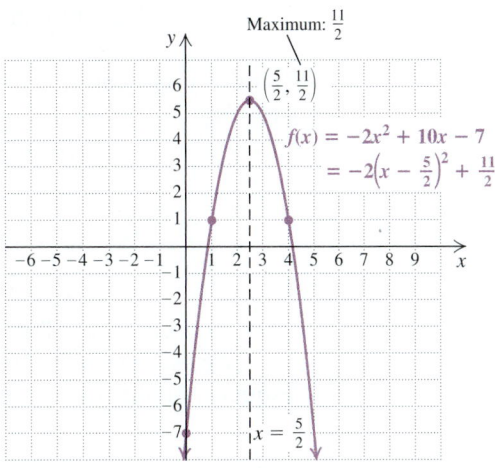

x	$f(x)$	
$\frac{5}{2}$	$\frac{11}{2}$	⟵ Vertex
0	-7	⟵ y-intercept
1	1	
4	1	

3. Graph $f(x) = -2x^2 - 2x - 2$, and find the maximum or minimum function value.

YOUR TURN

The method used in Examples 1–3 can be generalized to find a formula for locating the vertex. We complete the square as follows:

$$f(x) = ax^2 + bx + c$$

$$= a\left(x^2 + \frac{b}{a}x\right) + c. \qquad \text{Factoring } a \text{ out of the first two terms. Check by multiplying.}$$

Half of the x-coefficient, $\frac{b}{a}$, is $\frac{b}{2a}$. We square it to get $\frac{b^2}{4a^2}$ and add $\frac{b^2}{4a^2} - \frac{b^2}{4a^2}$ inside the parentheses. Then we distribute the a and regroup terms:

$$f(x) = a\left(x^2 + \frac{b}{a}x + \frac{b^2}{4a^2} - \frac{b^2}{4a^2}\right) + c$$

$$= a\left(x^2 + \frac{b}{a}x + \frac{b^2}{4a^2}\right) + a\left(-\frac{b^2}{4a^2}\right) + c \qquad \text{Using the distributive law}$$

$$= a\left(x + \frac{b}{2a}\right)^2 + \frac{-b^2}{4a} + \frac{4ac}{4a} \qquad \text{Factoring and finding a common denominator}$$

$$= a\left[x - \left(-\frac{b}{2a}\right)\right]^2 + \frac{4ac - b^2}{4a}.$$

Thus we have the following.

Student Notes

It is easier to remember a formula when you understand its derivation. Check with your instructor to determine what formulas you will be expected to remember.

> **THE VERTEX OF A PARABOLA**
>
> The vertex of the parabola given by $f(x) = ax^2 + bx + c$ is
>
> $$\left(-\frac{b}{2a},\ f\left(-\frac{b}{2a}\right)\right), \quad \text{or} \quad \left(-\frac{b}{2a},\ \frac{4ac - b^2}{4a}\right).$$
>
> - The x-coordinate of the vertex is $-b/(2a)$.
> - The axis of symmetry is $x = -b/(2a)$.
> - The second coordinate of the vertex is most commonly found by computing $f\left(-\frac{b}{2a}\right)$.

Let's reexamine Example 3 to see how we could have found the vertex directly. From the formula above,

$$\text{the } x\text{-coordinate of the vertex is } -\frac{b}{2a} = -\frac{10}{2(-2)} = \frac{5}{2}.$$

Substituting $\frac{5}{2}$ into $f(x) = -2x^2 + 10x - 7$, we find the second coordinate of the vertex:

$$\begin{aligned} f\left(\tfrac{5}{2}\right) &= -2\left(\tfrac{5}{2}\right)^2 + 10\left(\tfrac{5}{2}\right) - 7 \\ &= -2\left(\tfrac{25}{4}\right) + 25 - 7 \\ &= -\tfrac{25}{2} + 18 \\ &= -\tfrac{25}{2} + \tfrac{36}{2} = \tfrac{11}{2}. \end{aligned}$$

The vertex is $\left(\frac{5}{2}, \frac{11}{2}\right)$. The axis of symmetry is $x = \frac{5}{2}$.

We have developed two methods for finding the vertex, one by completing the square and the other using a formula.

Finding Intercepts

All quadratic functions have a y-intercept and 0, 1, or 2 x-intercepts. For $f(x) = ax^2 + bx + c$, the y-intercept is $(0, f(0))$, or $(0, c)$. To find x-intercepts, if any exist, we look for points where $y = 0$ or $f(x) = 0$. Thus, for $f(x) = ax^2 + bx + c$, the x-intercepts occur at those x-values for which $ax^2 + bx + c = 0$.

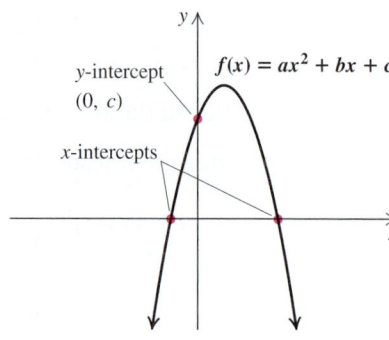

Chapter Resource:
Visualizing for Success, p. 768

EXAMPLE 4   Find any x-intercepts and the y-intercept of the graph of $f(x) = x^2 - 2x - 2$.

SOLUTION   The y-intercept is simply $(0, f(0))$, or $(0, -2)$. To find any x-intercepts, we solve

$$0 = x^2 - 2x - 2.$$

We are unable to factor $x^2 - 2x - 2$, so we use the quadratic formula and get $x = 1 \pm \sqrt{3}$. Thus the x-intercepts are $(1 - \sqrt{3}, 0)$ and $(1 + \sqrt{3}, 0)$.

If graphing, we would approximate, to get $(-0.7, 0)$ and $(2.7, 0)$.

4. Find any x-intercepts and the y-intercept of the graph of $f(x) = 3x^2 + 7x - 20$.

 YOUR TURN

If the solutions of $f(x) = 0$ are imaginary, the graph of f has no x-intercepts.

Algebraic 🔗 Graphical Connection

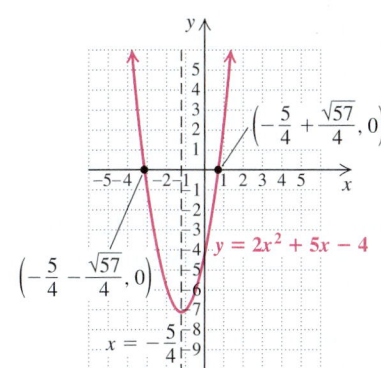

Because the graph of a quadratic equation is symmetric, the x-intercepts of the graph, if they exist, will be symmetric with respect to the axis of symmetry. This symmetry can be seen directly if the x-intercepts are found using the quadratic formula.

For example, the x-intercepts of the graph of $y = 2x^2 + 5x - 4$ are

$$\left(-\frac{5}{4} + \frac{\sqrt{57}}{4}, 0\right) \quad \text{and} \quad \left(-\frac{5}{4} - \frac{\sqrt{57}}{4}, 0\right).$$

For this equation, the axis of symmetry is $x = -\dfrac{5}{4}$ and the x-intercepts are $\dfrac{\sqrt{57}}{4}$ units to the left and right of $-\dfrac{5}{4}$ on the x-axis.

11.7 EXERCISE SET

FOR EXTRA HELP   MyMathLab® Math XL
PRACTICE WATCH READ REVIEW

👉 Vocabulary and Reading Check

Classify each of the following statements as either true or false.

1. The graph of $f(x) = 3x^2 - x + 6$ opens upward.

2. The function given by $g(x) = -x^2 + 3x + 1$ has a minimum value.

3. The graph of $f(x) = -2(x - 3)^2 + 7$ has its vertex at $(3, 7)$.

4. The graph of $g(x) = 4(x + 6)^2 - 2$ has its vertex at $(-6, -2)$.

5. The graph of $g(x) = \frac{1}{2}\left(x - \frac{3}{2}\right)^2 + \frac{1}{4}$ has $x = \frac{1}{4}$ as its axis of symmetry.

6. The function given by $f(x) = (x - 2)^2 - 5$ has a minimum value of -5.

7. The y-intercept of the graph of $f(x) = 2x^2 - 6x + 7$ is $(7, 0)$.

8. If the graph of a quadratic function f opens upward and has a vertex of $(1, 5)$, then the graph has no x-intercepts.

Graphing $f(x) = ax^2 + bx + c$

Complete the square to write each function in the form $f(x) = a(x - h)^2 + k$.

9. $f(x) = x^2 - 8x + 2$

10. $f(x) = x^2 - 6x - 1$

11. $f(x) = x^2 + 3x - 5$

12. $f(x) = x^2 + 5x + 3$

13. $f(x) = 3x^2 + 6x - 2$

14. $f(x) = 2x^2 - 20x - 3$

15. $f(x) = -x^2 - 4x - 7$

16. $f(x) = -2x^2 - 8x + 4$

17. $f(x) = 2x^2 - 5x + 10$

18. $f(x) = 3x^2 + 7x - 3$

*For each quadratic function, **(a)** find the vertex and the axis of symmetry and **(b)** graph the function.*

19. $f(x) = x^2 + 4x + 5$

20. $f(x) = x^2 + 2x - 5$

21. $f(x) = x^2 + 8x + 20$

22. $f(x) = x^2 - 10x + 21$

23. $h(x) = 2x^2 - 16x + 25$

24. $h(x) = 2x^2 + 16x + 23$

25. $f(x) = -x^2 + 2x + 5$

26. $f(x) = -x^2 - 2x + 7$

27. $g(x) = x^2 + 3x - 10$

28. $g(x) = x^2 + 5x + 4$

29. $h(x) = x^2 + 7x$

30. $h(x) = x^2 - 5x$

31. $f(x) = -2x^2 - 4x - 6$

32. $f(x) = -3x^2 + 6x + 2$

For each quadratic function, (a) find the vertex, the axis of symmetry, and the maximum or minimum function value and (b) graph the function.

33. $g(x) = x^2 - 6x + 13$

34. $g(x) = x^2 - 4x + 5$

35. $g(x) = 2x^2 - 8x + 3$

36. $g(x) = 2x^2 + 5x - 1$

37. $f(x) = 3x^2 - 24x + 50$

38. $f(x) = 4x^2 + 16x + 13$

39. $f(x) = -3x^2 + 5x - 2$

40. $f(x) = -3x^2 - 7x + 2$

41. $h(x) = \frac{1}{2}x^2 + 4x + \frac{19}{3}$

42. $h(x) = \frac{1}{2}x^2 - 3x + 2$

Finding Intercepts

Find any x-intercepts and the y-intercept. If no x-intercepts exist, state this.

43. $f(x) = x^2 - 6x + 3$

44. $f(x) = x^2 + 5x + 4$

45. $g(x) = -x^2 + 2x + 3$

46. $g(x) = x^2 - 6x + 9$

Aha! **47.** $f(x) = x^2 - 9x$

48. $f(x) = x^2 - 7x$

49. $h(x) = -x^2 + 4x - 4$

50. $h(x) = -2x^2 - 20x - 50$

51. $g(x) = x^2 + x - 5$

52. $g(x) = 2x^2 + 3x - 1$

53. $f(x) = 2x^2 - 4x + 6$

54. $f(x) = x^2 - x + 2$

55. The graph of a quadratic function f opens downward and has no x-intercepts. In what quadrant(s) must the vertex lie? Explain your reasoning.

56. Is it possible for the graph of a quadratic function to have only one x-intercept if the vertex is off the x-axis? Why or why not?

Skill Review

Multiply or divide, as indicated. Simplify if possible.

57. $(x^2 - 7)(x^2 + 3)$ [4.6]

58. $\dfrac{x^2 - x - 2}{x^2 - 9} \cdot \dfrac{x^2 + 7x + 12}{x^2 + x}$ [6.2]

59. $\sqrt[3]{18x^4y} \cdot \sqrt[3]{6x^2y}$ [10.3]

60. $(2x^3 - x - 3) \div (x - 1)$ [4.8]

61. $\dfrac{4a^2 - b^2}{2ab} \div \dfrac{2a^2 - ab - b^2}{6a^2}$ [6.2]

62. $\dfrac{\sqrt[4]{x^3}}{\sqrt[3]{x^4}}$ [10.5]

Synthesis

63. If the graphs of two quadratic functions have the same x-intercepts, will they also have the same vertex? Why or why not?

64. Suppose that the graph of $f(x) = ax^2 + bx + c$ has $(x_1, 0)$ and $(x_2, 0)$ as x-intercepts. Explain why the graph of $g(x) = -ax^2 - bx - c$ will also have $(x_1, 0)$ and $(x_2, 0)$ as x-intercepts.

For each quadratic function, find (a) the maximum or minimum value and (b) any x-intercepts and the y-intercept.

65. $f(x) = 2.31x^2 - 3.135x - 5.89$

66. $f(x) = -18.8x^2 + 7.92x + 6.18$

67. Graph the function
$$f(x) = x^2 - x - 6.$$
Then use the graph to approximate solutions of the following equations.

   a) $x^2 - x - 6 = 2$

   b) $x^2 - x - 6 = -3$

68. Graph
$$f(x) = \frac{x^2}{2} + x - \frac{3}{2}.$$
Then use the graph to approximate solutions of the following equations.

   a) $\dfrac{x^2}{2} + x - \dfrac{3}{2} = 0$

   b) $\dfrac{x^2}{2} + x - \dfrac{3}{2} = 1$

   c) $\dfrac{x^2}{2} + x - \dfrac{3}{2} = 2$

Find an equivalent equation of the type
$$f(x) = a(x - h)^2 + k.$$

69. $f(x) = mx^2 - nx + p$

70. $f(x) = 3x^2 + mx + m^2$

71. The graph of a quadratic function has $(-1, 0)$ as one intercept and $(3, -5)$ as its vertex. Find an equation for the function.

72. The graph of a quadratic function has $(4, 0)$ as one intercept and $(-1, 7)$ as its vertex. Find an equation for the function.

Graph.

73. $f(x) = |x^2 - 1|$

74. $f(x) = |x^2 - 3x - 4|$

75. $f(x) = |2(x - 3)^2 - 5|$

76. Use a graphing calculator to check your answers to Exercises 25, 41, 53, 65, and 67.

↪ **YOUR TURN ANSWERS: SECTION 11.7**

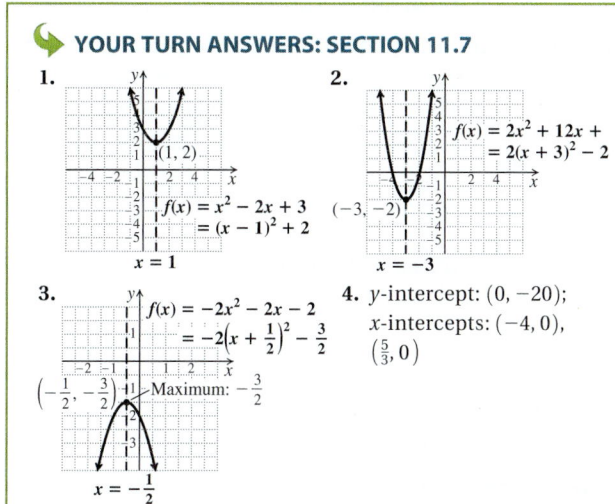

1. $f(x) = x^2 - 2x + 3 = (x - 1)^2 + 2$; $(1, 2)$; $x = 1$

2. $f(x) = 2x^2 + 12x + 16 = 2(x + 3)^2 - 2$; $(-3, -2)$; $x = -3$

3. $f(x) = -2x^2 - 2x - 2 = -2\left(x + \frac{1}{2}\right)^2 - \frac{3}{2}$; $\left(-\frac{1}{2}, -\frac{3}{2}\right)$; Maximum: $-\frac{3}{2}$; $x = -\frac{1}{2}$

4. *y*-intercept: $(0, -20)$; *x*-intercepts: $(-4, 0)$, $\left(\frac{5}{3}, 0\right)$

QUICK QUIZ: SECTIONS 11.1–11.7

Solve.

1. $3t^2 = 5$ [11.1]

2. $2x^2 + 3x = 6$ [11.2]

3. Write a quadratic equation having the solutions $\frac{2}{5}$ and -1. [11.3]

Graph.

4. $f(x) = -3(x + 1)^2$ [11.6]

5. $f(x) = x^2 - 2x + 3$ [11.7]

PREPARE TO MOVE ON

Solve. [8.4]

1. $x - y + z = -6$,
$2x + y + z = 2$,
$3x + y + z = 0$

2. $z = -5$,
$2x - y + 3z = -27$,
$x + 2y + 7z = -26$

3. $\frac{1}{2} = c$,
$5 = 9a + 6b + 2c$,
$29 = 81a + 9b + c$

11.8   Problem Solving and Quadratic Functions

Maximum and Minimum Problems ▪ Fitting Quadratic Functions to Data

Let's look now at some of the many situations in which quadratic functions are used for problem solving.

Maximum and Minimum Problems

We have seen that for any quadratic function f, the value of $f(x)$ at the vertex is either a maximum or a minimum. Thus problems in which a quantity must be maximized or minimized can be solved by finding the coordinates of a vertex, assuming the problem can be modeled with a quadratic function.

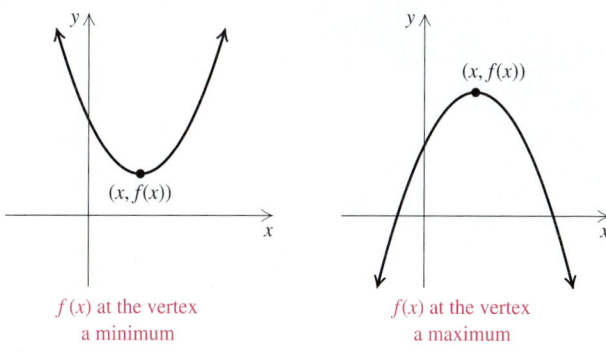

$f(x)$ at the vertex a minimum $f(x)$ at the vertex a maximum

Year	PET Plastic Bottle Recycling Rate
1995	39.7%
2000	22.3
2005	23.1
2010	29.1

Source: National Association for PET Container Resources

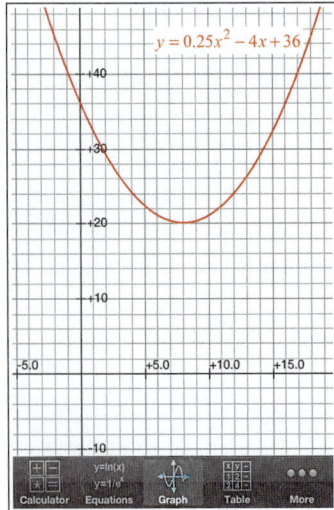

1. The amount of precipitation, in inches, that falls in Seattle, Washington, during the nth month of the year can be approximated by $p(n) = 0.1n^2 - 1.6n + 7$. Here, $n = 1$ represents January, $n = 2$ represents February, and so on. Find the minimum monthly precipitation and the month in which it occurs.

Source: Based on data from city-data.com

EXAMPLE 1 *Recycling.* After dropping for several years, the recyling rate of PET plastic bottles has begun to rise, as indicated by the data in the table at left. The recycling rate, in percent, can be approximated by $r(t) = 0.25t^2 - 4t + 36$, where t is the number of years after 1995. In what year was the recycling rate the lowest, and what percent of PET plastic bottles were recycled that year?

SOLUTION

1., 2. **Familiarize** and **Translate.** The function given is quadratic. Since the coefficient of the t^2-term is positive, the graph opens upward so a minimum value exists. The calculator-generated graph at left confirms this.

3. **Carry out.** We can either complete the square or use the formula for the vertex. Completing the square, we have

$$r(t) = 0.25t^2 - 4t + 36$$
$$= 0.25(t^2 - 16t) + 36$$
$$= 0.25(t^2 - 16t + 64 - 64) + 36 \qquad \text{Completing the square}$$
$$= 0.25(t^2 - 16t + 64) + (0.25)(-64) + 36 \qquad \text{Using the distributive law}$$
$$= 0.25(t - 8)^2 + 20. \qquad \text{Factoring and simplifying}$$

There is a minimum value of 20 when $t = 8$.

4. **Check.** Using the formula, we have $-b/(2a) = -(-4)/(2 \cdot 0.25) = 8$. Then

$$r(8) = 0.25(8)^2 - 4(8) + 36 = 20.$$

Both approaches give the same minimum, and that minimum is also confirmed by the graph. The answer checks.

5. **State.** The minimum recycling rate was 20%. It occurred 8 years after 1995, or in 2003.

YOUR TURN

EXAMPLE 2 *Swimming Area.* A lifeguard has 100 m of linked flotation devices with which to cordon off a rectangular swimming area at North Beach. If the shoreline forms one side of the rectangle, what dimensions will maximize the size of the area for swimming?

SOLUTION

1. **Familiarize.** We make a drawing and label it, letting w = the width of the rectangle, in meters, and l = the length of the rectangle, in meters.

Recall that Area = $l \cdot w$ and Perimeter = $2w + 2l$. Since the beach forms one length of the rectangle, the flotation devices comprise three sides. Thus

$$2w + l = 100.$$

The table below shows some possible dimensions for a rectangular area that can be enclosed with 100 m of flotation devices. All possibilities are chosen so that $2w + l = 100$.

l	w	Rope Length	Area, A
40 m	30 m	100 m	1200 m^2
30 m	35 m	100 m	1050 m^2
20 m	40 m	100 m	800 m^2
⋮	⋮	⋮	⋮

What choice of l and w will maximize A?

Technology Connection

To generate a table of values for Example 2, let x represent the width of the swimming area, in meters. If l represents the length, in meters, we must have $100 = 2x + l$. Next, solve for l and use that expression for y_1. Then let $y_2 = x \cdot y_1$ (to enter y_1, press **VARS** and select Y-VARS and then FUNCTION and then 1) so that y_2 represents the area. Scroll through the resulting table, adjusting the settings as needed, to determine the point at which area is maximized.

2. Refer to Example 2. If the lifeguard has 160 m of flotation devices, what dimensions will maximize the size of the swimming area?

2. **Translate.** We have two equations: One guarantees that 100 m of flotation devices are used; the other expresses area in terms of length and width.

$$2w + l = 100,$$
$$A = l \cdot w$$

3. **Carry out.** We need to express A as a function of either l or w but not both. To do so, we solve for l in the first equation to obtain $l = 100 - 2w$. Substituting for l in the second equation, we get a quadratic function:

$$A = (100 - 2w)w \qquad \text{Substituting for } l$$
$$= 100w - 2w^2 \qquad \text{This represents a parabola opening downward,}$$
$$\qquad\qquad\qquad \text{so a maximum exists.}$$
$$= -2w^2 + 100w.$$

Factoring and completing the square, we get

$$A = -2(w^2 - 50w + 625 - 625) \qquad \text{We could also use the vertex}$$
$$\qquad\qquad\qquad\qquad\qquad\qquad \text{formula.}$$
$$= -2(w - 25)^2 + 1250.$$

There is a maximum value of 1250 when $w = 25$.

4. **Check.** If $w = 25$ m, then $l = 100 - 2 \cdot 25 = 50$ m. These dimensions give an area of 1250 m². Note that 1250 m² is greater than any of the values for A found in the *Familiarize* step. To be more certain, we could check values other than those used in that step. For example, if $w = 26$ m, then $l = 48$ m, and $A = 26 \cdot 48 = 1248$ m². Since 1250 m² is greater than 1248 m², it appears that we have a maximum.

5. **State.** The largest rectangular area for swimming that can be enclosed is 25 m by 50 m. Note that the beach is a long side of this rectangle.

YOUR TURN

Fitting Quadratic Functions to Data

Whenever a certain quadratic function fits a situation, that function can be determined if three inputs and their outputs are known.

EXAMPLE 3 *Publishing.* The number of self-published books has increased from 2006 to 2010. As the table and graph suggest, the number of self-published books can be modeled by a quadratic function if we consider the right half of the graph of the function.

Year	Number of Self-Published Books
2006	51,000
2007	63,000
2008	74,000
2009	95,000
2010	133,000

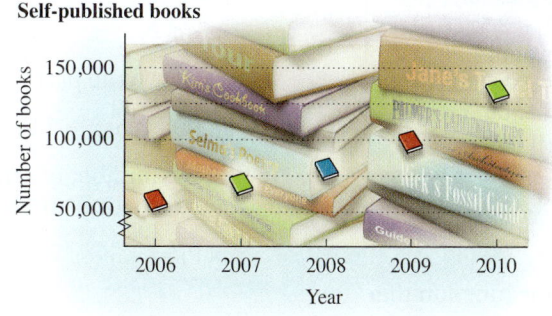

Self-published books

Source: Based on data from Bowker's Books in Print database, in "Secret of Self-Publishing: Success," *The Wall Street Journal*, October 31, 2011.

Chapter Resource:
Decision Making: Connection,
p. 769

a) Let t represent the number of years since 2006 and $p(t)$ the number of self-published books, in thousands. Use the data points $(0, 51)$, $(2, 74)$, and $(4, 133)$ to find a quadratic function that fits the data.

b) Use the function from part (a) to estimate the number of self-published books in 2013.

SOLUTION

a) We are looking for a function of the form $p(t) = at^2 + bt + c$ given that $p(0) = 51$, $p(2) = 74$, and $p(4) = 133$. Thus,

$$51 = a \cdot 0^2 + b \cdot 0 + c, \qquad \text{Using the data point } (0, 51)$$
$$74 = a \cdot 2^2 + b \cdot 2 + c, \qquad \text{Using the data point } (2, 74)$$
$$133 = a \cdot 4^2 + b \cdot 4 + c. \qquad \text{Using the data point } (4, 133)$$

After simplifying, we see that we need to solve the system

$$51 = c, \qquad\qquad (1)$$
$$74 = 4a + 2b + c, \qquad (2)$$
$$133 = 16a + 4b + c. \qquad (3)$$

We know from equation (1) that $c = 51$. Substituting that value into equations (2) and (3), we have

$$74 = 4a + 2b + 51,$$
$$133 = 16a + 4b + 51.$$

Subtracting 51 from both sides of each equation, we have

$$23 = 4a + 2b, \qquad (4)$$
$$82 = 16a + 4b. \qquad (5)$$

To solve, we multiply equation (4) by -2, and then add to eliminate b:

$$-46 = -8a - 4b$$
$$\underline{82 = 16a + 4b}$$
$$36 = 8a$$
$$4.5 = a. \qquad \text{Solving for } a$$

Next, we solve for b, using equation (4) above:

$$23 = 4(4.5) + 2b \qquad \text{Substituting}$$
$$23 = 18 + 2b$$
$$5 = 2b$$
$$2.5 = b. \qquad\qquad \text{Solving for } b$$

We can now write $p(t) = at^2 + bt + c$ as

$$p(t) = 4.5t^2 + 2.5t + 51.$$

b) To find the number of self-published books in 2013, we evaluate the function for $t = 7$, because 2013 is 7 years after 2006:

$$p(7) = 4.5(7)^2 + 2.5(7) + 51$$
$$= 289.$$

We estimate that 289,000 books will be self-published in 2013.

 YOUR TURN

Technology Connection

To use a graphing calculator to fit a quadratic function to the data in Example 3, we first select EDIT in the $\boxed{\text{STAT}}$ menu and enter the given data.

L1	L2	L3 2
0	51	------
2	74	
4	133	
L2(4) =		

We then press $\boxed{\text{STAT}}$ $\boxed{\triangleright}$ $\boxed{5}$ $\boxed{\text{VARS}}$ $\boxed{\triangleright}$ $\boxed{1}$ $\boxed{1}$ $\boxed{\text{ENTER}}$. The first three keystrokes select QuadReg from the STAT CALC menu. The keystrokes $\boxed{\text{VARS}}$ $\boxed{\triangleright}$ $\boxed{1}$ $\boxed{1}$ copy the regression equation to the equation-editor screen as y_1.

```
QuadReg
y=ax2+bx+c
a=4.5
b=2.5
c=51
```

We see that the regression equation is $y = 4.5x^2 + 2.5x + 51$.

To check Example 3(b), we set Indpnt to Ask in the Table Setup and enter $X = 7$ in the table. A Y1-value of 289 confirms our answer.

3. Find a quadratic function that fits the points $(0, 6)$, $(1, 8)$, and $(3, 4)$.

11.8 EXERCISE SET

↳ Vocabulary and Reading Check

In each of Exercises 1–6, match the description with the graph that displays that characteristic.

1. ____ A minimum value of $f(x)$ exists.

2. ____ A maximum value of $f(x)$ exists.

3. ____ No maximum or minimum value of $f(x)$ exists.

4. ____ The data points appear to suggest a linear model for g.

5. ____ The data points appear to suggest that g is a quadratic function with a maximum.

6. ____ The data points appear to suggest that g is a quadratic function with a minimum.

a)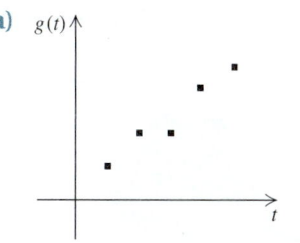

b)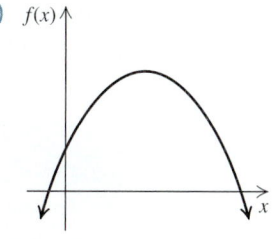

c)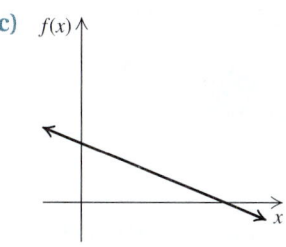

d)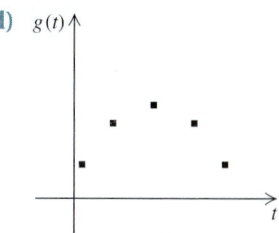

e)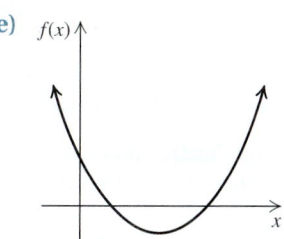

f)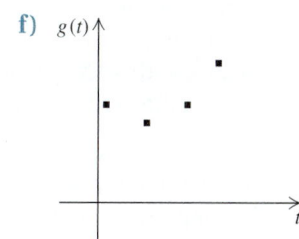

Maximum and Minimum Problems

Solve.

7. *Newborn Calves.* The number of pounds of milk per day recommended for a calf that is x weeks old can be approximated by $p(x)$, where

$$p(x) = -0.2x^2 + 1.3x + 6.2.$$

When is a calf's milk consumption greatest and how much milk does it consume at that time?

Source: C. Chaloux, University of Vermont

8. *Stock Prices.* The value of a share of I. J. Solar can be represented by $V(x) = x^2 - 6x + 13$, where x is the number of months after January 2009. What is the lowest value $V(x)$ will reach, and when did that occur?

9. *Minimizing Cost.* Sweet Harmony Crafts has determined that when x hundred dulcimers are built, the average cost per dulcimer can be estimated by

$$C(x) = 0.1x^2 - 0.7x + 2.425,$$

where $C(x)$ is in hundreds of dollars. What is the minimum average cost per dulcimer and how many dulcimers should be built in order to achieve that minimum?

10. *Maximizing Profit.* Recall that total profit P is the difference between total revenue R and total cost C. Given $R(x) = 1000x - x^2$ and $C(x) = 3000 + 20x$, find the total profit, the maximum value of the total profit, and the value of x at which it occurs.

11. *Architecture.* An architect is designing an atrium for a hotel. The atrium is to be rectangular with a perimeter of 720 ft of brass piping. What dimensions will maximize the area of the atrium?

12. *Furniture Design.* A furniture builder is designing a rectangular end table with a perimeter of 128 in. What dimensions will yield the maximum area?

13. *Patio Design.* A stone mason has enough stones to enclose a rectangular patio with 60 ft of perimeter, assuming that the attached house forms one side of the rectangle. What is the maximum area that the mason can enclose? What should the dimensions of the patio be in order to yield this area?

14. *Garden Design.* Ginger is fencing in a rectangular garden, using the side of her house as one side of the rectangle. What is the maximum area that she can enclose with 40 ft of fence? What should the dimensions of the garden be in order to yield this area?

15. *Molding Plastics.* Economite Plastics plans to produce a one-compartment vertical file by bending the long side of an 8-in. by 14-in. sheet of plastic along two lines to form a U shape. How tall should the file be in order to maximize the volume that the file can hold?

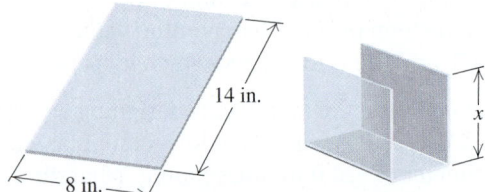

16. *Composting.* A rectangular compost container is to be formed in a corner of a fenced yard, with 8 ft of chicken wire completing the other two sides of the rectangle. If the chicken wire is 3 ft high, what dimensions of the base will maximize the container's volume?

17. What is the maximum product of two numbers that add to 18? What numbers yield this product?

18. What is the maximum product of two numbers that add to 26? What numbers yield this product?

19. What is the minimum product of two numbers that differ by 8? What are the numbers?

20. What is the minimum product of two numbers that differ by 7? What are the numbers?

Aha! **21.** What is the maximum product of two numbers that add to −10? What numbers yield this product?

22. What is the maximum product of two numbers that add to −12? What numbers yield this product?

Fitting Quadratic Functions to Data

Choosing Models. For the scatterplots and graphs in Exercises 23–34, determine which, if any, of the following functions might be used as a model for the data: Linear, with $f(x) = mx + b$; quadratic, with $f(x) = ax^2 + bx + c, a > 0$; quadratic, with $f(x) = ax^2 + bx + c, a < 0$; neither quadratic nor linear.

23. Sonoma Sunshine

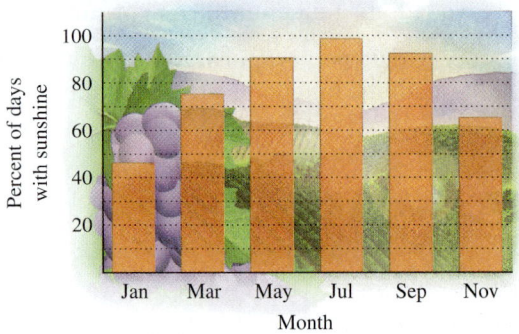

Source: www.city-data.com

24. Sonoma Precipitation

Source: www.city-data.com

25. Safe sight distance to the left

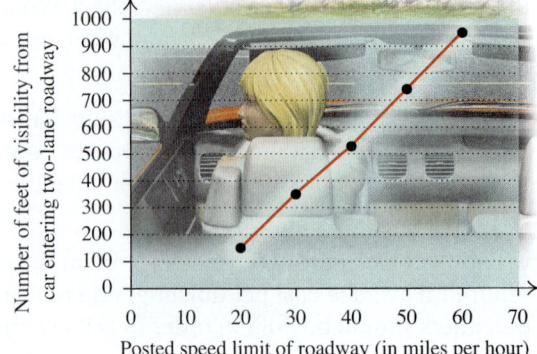

Source: Institute of Traffic Engineers

26. Safe sight distance to the right

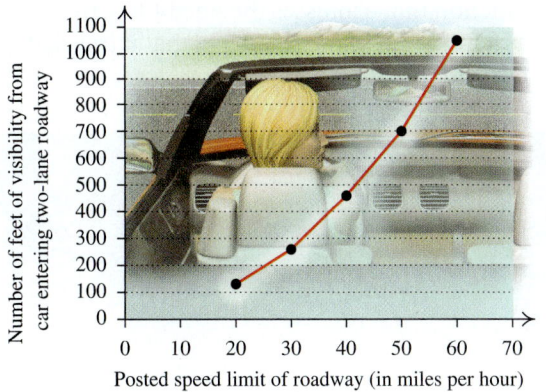

Source: Institute of Traffic Engineers

27. Dow Jones Industrial Average

28. U.S. senior population

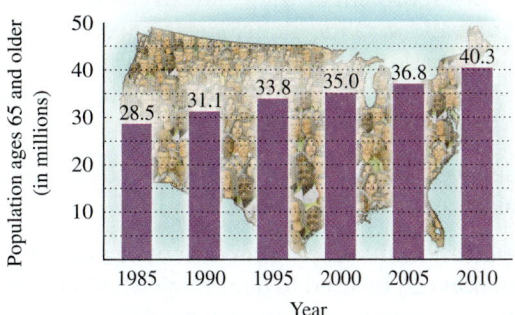

Source: U.S. Bureau of Labor Statistics

29. Working longer

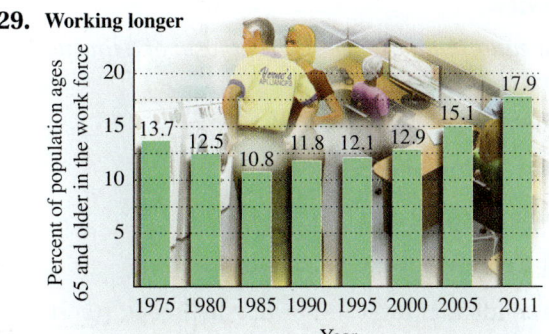

Source: U.S. Department of Labor, Bureau of Labor Statistics

30. Airline bumping rate

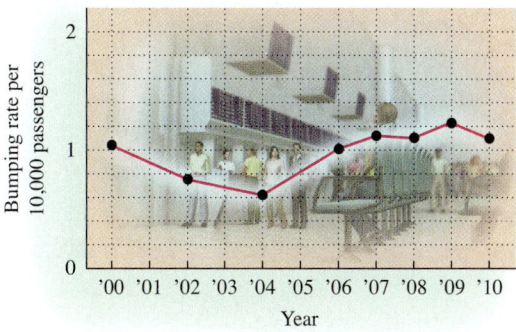

Source: U.S. Department of Transportation

31. Carbon footprint

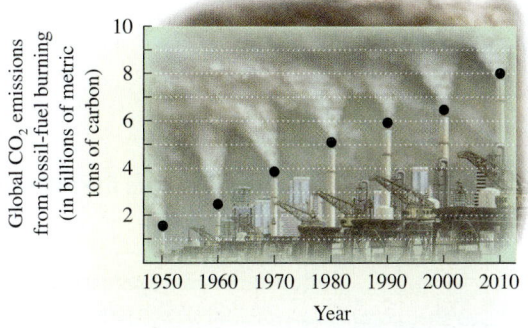

Source: Carbon Dioxide Information Analysis Center, U.S. Department of Energy

32. Angie's List

Source: *The Wall Street Journal*, August 30, 2011

33. The worth of a dollar

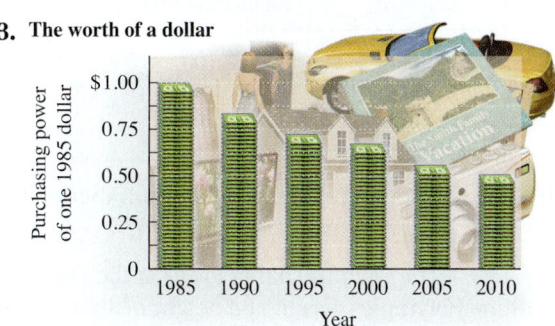

Source: Based on data from U.S. Bureau of Labor Statistics

34. **Average number of live births per 1000 women, 2009**

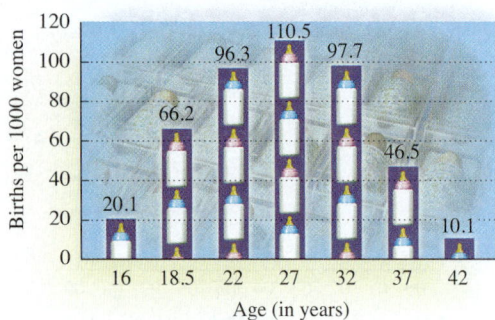

Source: U.S. Centers for Disease Control

Find a quadratic function that fits the set of data points.

35. $(1, 4), (-1, -2), (2, 13)$

36. $(1, 4), (-1, 6), (-2, 16)$

37. $(2, 0), (4, 3), (12, -5)$

38. $(-3, -30), (3, 0), (6, 6)$

39. a) Find a quadratic function that fits the following data.

Travel Speed (in kilometers per hour)	Number of Nighttime Accidents (for every 200 million kilometers driven)
60	400
80	250
100	250

b) Use the function to estimate the number of nighttime accidents that occur at 50 km/h.

40. a) Find a quadratic function that fits the following data.

Travel Speed (in kilometers per hour)	Number of Daytime Accidents (for every 200 million kilometers driven)
60	100
80	130
100	200

b) Use the function to estimate the number of daytime accidents that occur at 50 km/h.

41. *Archery.* The Olympic flame tower at the 1992 Summer Olympics was lit at a height of about 27 m by a flaming arrow that was launched about 63 m from the base of the tower. If the arrow landed

about 63 m beyond the tower, find a quadratic function that expresses the height h of the arrow as a function of the distance d that it traveled horizontally.

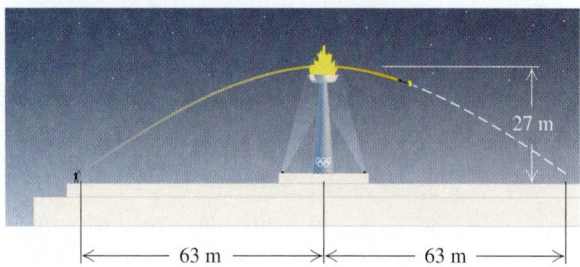

42. *Sump Pump.* The lift distances for a Liberty 250 sump pump moving fluid at various flow rates are shown in the following table.

Gallons per Minute	Lift Distance (in feet)
10	21
20	18
40	8

Source: Liberty Pumps

a) Let x represent the flow rate, in gallons per minute, and $d(x)$ the lift distance, in feet. Find a quadratic function that fits the data.

b) Use the function to find the lift distance for a flow rate of 50 gal per min.

 43. Does every nonlinear function have a minimum or a maximum value? Why or why not?

 44. Explain how the leading coefficient of a quadratic function can be used to determine whether a maximum or a minimum function value exists.

Skill Review

Find an equation in slope–intercept form of a line with the given characteristics.

45. Slope: $-\frac{1}{3}$; y-intercept: $(0, 16)$ [3.6]

46. Slope: 2; contains $(-3, 7)$ [3.7]

47. Contains $(4, 8)$ and $(10, 0)$ [3.7]

48. Parallel to $y = \frac{2}{3}x + 5$; contains $(-2, -9)$ [3.7]

49. Perpendicular to $2x + y = 3$; y-intercept: $(0, -6)$ [3.7]

50. Horizontal line through $(7, -4)$ [3.7]

Synthesis

The following graphs can be used to compare the base-ball statistics of pitcher Roger Clemens with the 31 other pitchers since 1968 who started at least 10 games in at least 15 seasons and pitched at least 3000 innings. Use the graphs to answer questions 51 and 52.

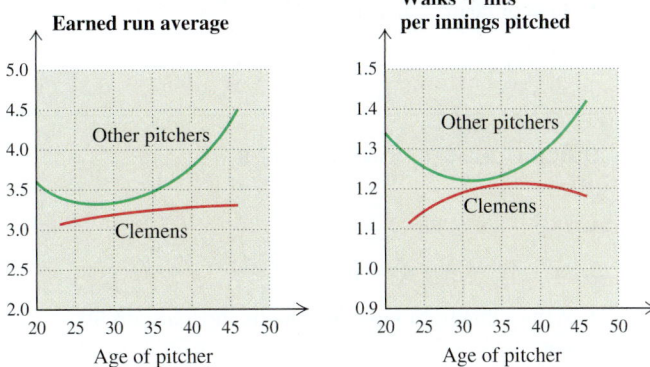

Source: *The New York Times*, February 10, 2008;
Eric Bradlow, Shane Jensen, Justin Wolfers and Adi Wyner

51. The earned run average describes how many runs a pitcher has allowed per game. The lower the earned run average, the better a pitcher. Compare, in terms of maximums or minimums, the earned run average of Roger Clemens with that of other pitchers. Is there any reason to suspect that the aging process was unusual for Clemens? Explain.

52. The statistic "Walks + hits per innings pitched" is related to how often a pitcher allows a batter to reach a base. The lower this statistic, the better. Compare, in terms of maximums or minimums, the "walks + hits" statistic of Roger Clemens with that of other pitchers.

53. *Bridge Design.* The cables supporting a straight-line suspension bridge are nearly parabolic in shape. Suppose that a suspension bridge is being designed with concrete supports 160 ft apart and with vertical cables 30 ft above road level at the midpoint of the bridge and 80 ft above road level at a point 50 ft from the midpoint of the bridge. How long are the longest vertical cables?

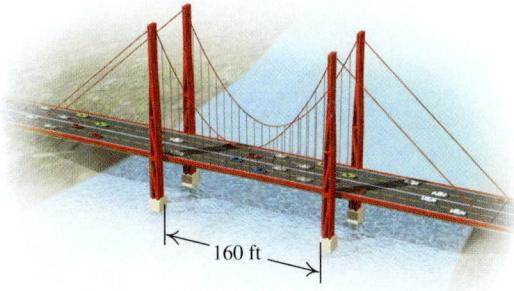

54. *Trajectory of a Launched Object.* The height above the ground of a launched object is a quadratic function of the time that it is in the air. Suppose that a flare is launched from a cliff 64 ft above sea level. If 3 sec after being launched the flare is again level with the cliff, and if 2 sec after that it lands in the sea, what is the maximum height that the flare reaches?

55. *Cover Charges.* When the owner of Sweet Sounds charges a $10 cover charge, an average of 80 people will attend a show. For each 25¢ increase in admission price, the average number attending decreases by 1. What should the owner charge in order to make the most money?

56. *Crop Yield.* An orange grower finds that she gets an average yield of 40 bushels (bu) per tree when she plants 20 trees on an acre of ground. Each time she adds one tree per acre, the yield per tree decreases by 1 bu, due to congestion. How many trees per acre should she plant for maximum yield?

57. *Norman Window.* A *Norman window* is a rectangle with a semicircle on top. Big Sky Windows is designing a Norman window that will require 24 ft of trim. What dimensions will allow the maximum amount of light to enter a house?

58. *Minimizing Area.* A 36-in. piece of string is cut into two pieces. One piece is used to form a circle while the other is used to form a square. How should the string be cut so that the sum of the areas is a minimum?

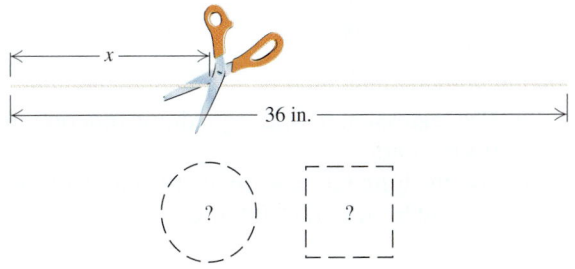

Regression can be used to find the "best"-fitting quadratic function when more than three data points are provided.

59. *Business.* The numbers of paid subscriptions per year to Angie's List, a company that provides business reviews, are shown in the following table.

Year	Number of Paid Subscriptions (in thousands)
2006	152.1
2007	234.9
2008	333.5
2009	411.5
2010	602.9
2011	821.8*

Source: *The Wall Street Journal,* August 30, 2011

*As of June 30, 2011

a) Use regression to find a quadratic function that can be used to estimate the number of paid subscriptions $a(x)$ to Angie's List x years after 2006.

b) Use the function found in part (a) to predict the number of paid subscriptions to Angie's List in 2014.

60. *Hydrology.* The drawing below shows the cross section of a river. Typically rivers are deepest in the middle, with the depth decreasing to 0 at the edges. A hydrologist measures the depths D, in feet, of a river at distances x, in feet, from one bank. The results are listed in the table below.

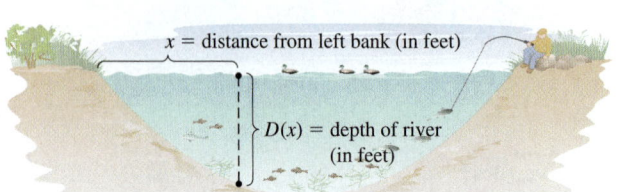

x = distance from left bank (in feet)

$D(x)$ = depth of river (in feet)

Distance x, from the Left Bank (in feet)	Depth D of the River (in feet)
0	0
15	10.2
25	17
50	20
90	7.2
100	0

a) Use regression to find a quadratic function that fits the data.

b) Use the function to estimate the depth of the river 70 ft from the left bank.

61. *Research.* Find the number of self-published books in 2013 and compare it with the estimate in Example 3(b).

⤷ YOUR TURN ANSWERS: SECTION 11.8

1. Minimum precipitation: 0.6 in. in month 8, or August

2. 40 m by 80 m **3.** $f(x) = -\dfrac{4}{3}x^2 + \dfrac{10}{3}x + 6$

QUICK QUIZ: SECTIONS 11.1–11.8

Solve.   [11.1], [11.2], [11.5]

1. $12x^2 + 7x = 10$

2. $(x - 4)^2 - (x - 4) = 6$

3. Write a quadratic equation having the solutions $5i$ and $-5i$. [11.3]

4. Solve $V = 3.5\sqrt{h}$ for h. [11.4]

5. Graph: $f(x) = 2(x - 4)^2 + 3$. [11.6]

PREPARE TO MOVE ON

Solve.

1. $4 - x \le 7$ [2.6]   **2.** $|4x + 1| < 11$ [9.2]

Subtract to find an equivalent expression for $f(x)$ and list all restrictions on the domain. [6.4], [7.2]

3. $f(x) = \dfrac{x - 3}{x + 4} - 5$ **4.** $f(x) = \dfrac{x}{x - 1} - 1$

Solve. [6.6]

5. $\dfrac{x}{x - 1} = 1$ **6.** $\dfrac{(x + 6)(x - 9)}{x + 5} = 0$

11.9   Polynomial Inequalities and Rational Inequalities

Quadratic and Other Polynomial Inequalities   ▪   Rational Inequalities

Quadratic and Other Polynomial Inequalities

Inequalities like the following are called *polynomial inequalities*:

$$x^3 - 5x > x^2 + 7, \qquad 4x - 3 < 9, \qquad 5x^2 - 3x + 2 \geq 0.$$

Second-degree polynomial inequalities in one variable are called *quadratic inequalities*. To solve polynomial inequalities, we can focus attention on where the outputs of a polynomial function are positive and where they are negative.

EXAMPLE 1   Solve: $x^2 + 3x - 10 > 0$.

SOLUTION   Consider the "related" function $f(x) = x^2 + 3x - 10$. We are looking for those x-values for which $f(x) > 0$. Graphically, function values are positive when the graph is above the x-axis.

The graph of f opens upward since the leading coefficient is positive. Thus function values are positive outside the interval formed by the x-intercepts. To find the intercepts, we set the polynomial equal to 0 and solve:

$$x^2 + 3x - 10 = 0$$
$$(x + 5)(x - 2) = 0$$
$$x + 5 = 0 \quad or \quad x - 2 = 0$$
$$x = -5 \quad or \qquad x = 2. \qquad \text{The } x\text{-intercepts are } (-5, 0) \text{ and}$$
$$(2, 0).$$

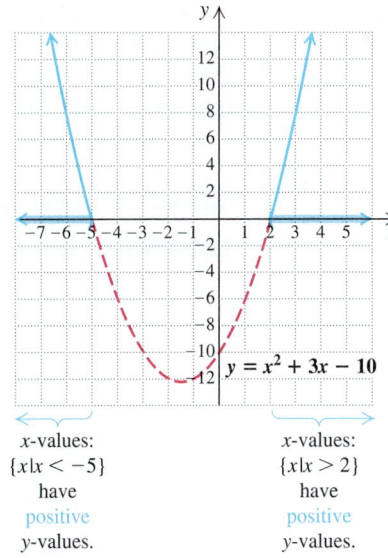

$y = x^2 + 3x - 10$

| x-values: $\{x \mid x < -5\}$ have positive y-values. | x-values: $\{x \mid x > 2\}$ have positive y-values. |

Thus the solution set of the inequality is

$$(-\infty, -5) \cup (2, \infty), \quad \text{or} \quad \{x \mid x < -5 \, or \, x > 2\}.$$

1. Solve: $x^2 - 2x - 8 > 0$.

YOUR TURN

Any inequality with 0 on one side can be solved by considering a graph of the related function and finding intercepts as in Example 1.

EXAMPLE 2   Solve: $x^2 - 2x \leq 2$.

SOLUTION   We first write the quadratic inequality in standard form:

$$x^2 - 2x - 2 \leq 0.$$ This is equivalent to the original inequality.

The graph of $f(x) = x^2 - 2x - 2$ is a parabola opening upward. Values of $f(x)$ are negative for x-values between the x-intercepts. We find the x-intercepts by solving $f(x) = 0$:

$$x = \frac{-b \pm \sqrt{b^2 - 4ac}}{2a}$$

$$= \frac{-(-2) \pm \sqrt{(-2)^2 - 4 \cdot 1(-2)}}{2 \cdot 1}$$

$$= \frac{2 \pm \sqrt{12}}{2} = \frac{2}{2} \pm \frac{2\sqrt{3}}{2} = 1 \pm \sqrt{3}.$$

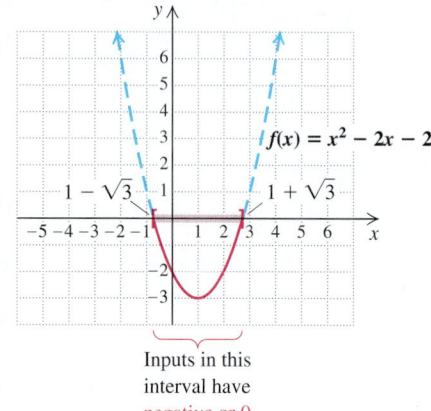

Inputs in this interval have negative or 0 outputs.

At the x-intercepts, $1 - \sqrt{3}$ and $1 + \sqrt{3}$, the value of $f(x)$ is 0. Since the inequality symbol is $\leq$, the solution set will include all values of x for which $f(x)$ is negative *or* $f(x)$ is 0. Thus the solution set of the inequality is

$$[1 - \sqrt{3}, 1 + \sqrt{3}], \quad \text{or} \quad \{x \mid 1 - \sqrt{3} \leq x \leq 1 + \sqrt{3}\}.$$

2. Solve: $x^2 + 4x \leq 2$.

YOUR TURN

In Example 2, it was not essential to draw the graph. The important information came from finding the x-intercepts and the sign of $f(x)$ on each side of those intercepts. We now solve a polynomial inequality, without graphing the related function f, but instead by locating the x-intercepts, or **zeros**, of f and then using *test points* to determine the sign of $f(x)$ over each interval of the x-axis.

EXAMPLE 3   For $f(x) = 5x^3 + 10x^2 - 15x$, find all x-values for which $f(x) > 0$.

SOLUTION   We first solve the related equation:

$$f(x) = 0$$
$5x^3 + 10x^2 - 15x = 0$ Substituting
$\left.\begin{array}{l} 5x(x^2 + 2x - 3) = 0 \\ 5x(x + 3)(x - 1) = 0 \end{array}\right\}$ Since $f(x)$ is third-degree, we expect up to three zeros.
$5x = 0 \quad or \quad x + 3 = 0 \quad or \quad x - 1 = 0$
$x = 0 \quad or \quad x = -3 \quad or \quad x = 1.$

The zeros of f are $-3, 0$, and 1. These zeros divide the number line, or x-axis, into four intervals: A, B, C, and D.

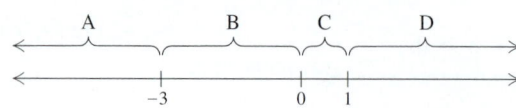

Student Notes

When we are evaluating test values, there is often no need to do lengthy computations since all we need to determine is the sign of the result.

Next, selecting one convenient test value from each interval, we determine the sign of $f(x)$ over that interval. We know that, within each interval, the sign of $f(x)$ cannot change. If it did, there would need to be another zero in that interval. Using the factored form of $f(x)$ eases the computations:

$$f(x) = (5x) \cdot (x + 3) \cdot (x - 1).$$

For interval A,

$$f(-4) = (5(-4)) \cdot (-4 + 3) \cdot (-4 - 1) \qquad -4 \text{ is a convenient value in interval A.}$$

$$= (-20) \cdot (-1) \cdot (-5)$$

$$= -100. \qquad f(-4) \text{ is negative.}$$

For interval B,

$$f(-1) = (5(-1)) \cdot (-1 + 3) \cdot (-1 - 1) \qquad -1 \text{ is a convenient value in interval B.}$$

$$= (-5) \cdot (2) \cdot (-2)$$

$$= 20. \qquad f(-1) \text{ is positive.}$$

For interval C,

$$f\left(\tfrac{1}{2}\right) = \left(5 \cdot \tfrac{1}{2}\right) \cdot \left(\tfrac{1}{2} + 3\right) \cdot \left(\tfrac{1}{2} - 1\right) \qquad \tfrac{1}{2} \text{ is a convenient value in interval C.}$$

$$= \left(\tfrac{5}{2}\right) \cdot \left(\tfrac{7}{2}\right) \cdot \left(-\tfrac{1}{2}\right)$$

$$= -\tfrac{35}{8}. \qquad f\left(\tfrac{1}{2}\right) \text{ is negative.}$$

For interval D,

$$f(2) = (5 \cdot 2) \cdot (2 + 3) \cdot (2 - 1) \qquad 2 \text{ is a convenient value in interval D.}$$

$$= (10) \cdot (5) \cdot (1)$$

$$= 50. \qquad f(2) \text{ is positive.}$$

We indicate on the number line the sign of $f(x)$ in each interval.

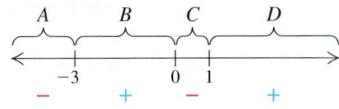

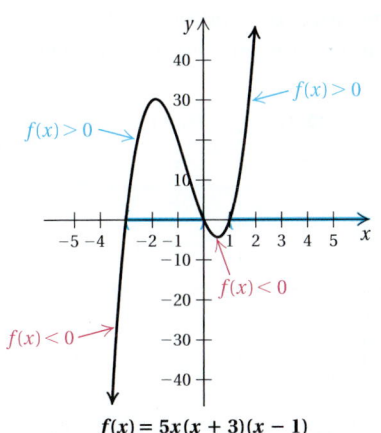

$$f(x) = 5x(x + 3)(x - 1)$$

A computer-generated visualization of Example 3

Recall that we are looking for all x for which $5x^3 + 10x^2 - 15x > 0$. The calculations above indicate that $f(x)$ is positive for any number in intervals B and D. The solution set of the original inequality is

$$(-3, 0) \cup (1, \infty), \quad \text{or} \quad \{x \mid -3 < x < 0 \text{ or } x > 1\}.$$

3. For $f(x) = 3x^3 + 9x^2 + 6x$, find all x-values for which $f(x) < 0$.

YOUR TURN

TO SOLVE A POLYNOMIAL INEQUALITY USING FACTORS

1. Add or subtract to get 0 on one side and solve the related polynomial equation.
2. Use the numbers found in step (1) to divide the number line into intervals.
3. Using a test value from each interval, determine the sign of the function over each interval.
4. Select the interval(s) for which the inequality is satisfied and write interval notation or set-builder notation for the solution set. Include endpoints of intervals when $\leq$ or $\geq$ appear.

We need focus only on the *sign* of $f(x)$. By looking at the number of positive and negative factors, we can determine the sign of the polynomial function.

EXAMPLE 4   For $f(x) = 4x^4 - 4x^2$, find all x-values for which $f(x) \leq 0$.

SOLUTION   We first solve the related equation:

Solve $f(x) = 0$.

$$f(x) = 0$$
$$4x^4 - 4x^2 = 0 \qquad \text{Substituting}$$
$$\left.\begin{array}{l} 4x^2(x^2 - 1) = 0 \\ 4x^2(x + 1)(x - 1) = 0 \end{array}\right\} \quad \begin{array}{l}\text{We expect up to four zeros of a fourth-}\\ \text{degree polynomial function.}\end{array}$$
$$4x^2 = 0 \quad or \quad x + 1 = 0 \quad or \quad x - 1 = 0$$
$$x = 0 \quad or \qquad x = -1 \quad or \qquad x = 1.$$

Divide the number line into intervals.

Since f has zeros at $-1, 0$, and 1, we divide the number line into four intervals:

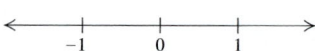

The product $4x^2(x + 1)(x - 1)$ is positive or negative, depending on the signs of $4x^2, x + 1$, and $x - 1$. The sign of the product can be determined by making a chart.

Determine the sign of the function over each interval.

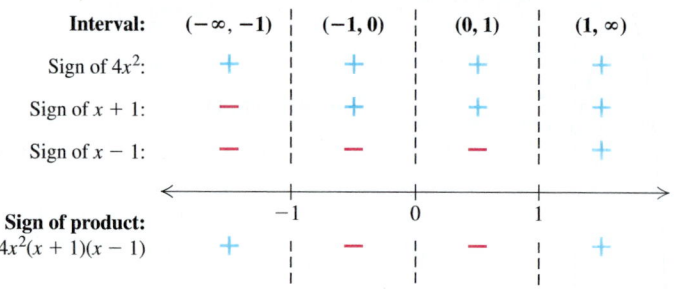

Select the interval(s) for which the inequality is satisfied.

A product is negative when it has an odd number of negative factors. Since the $\leq$ sign allows for equality, the zeros $-1, 0$, and 1 are solutions. From the chart, we see that the solution set is

$$[-1, 0] \cup [0, 1], \quad \text{or simply} \quad [-1, 1], \quad \text{or} \quad \{x | -1 \leq x \leq 1\}.$$

4. For $f(x) = 5x^3 - 5x$, find all x-values for which $f(x) \geq 0$.

YOUR TURN

Technology Connection

To solve $2.3x^2 \leq 9.11 - 2.94x$, we write the inequality in the form $2.3x^2 + 2.94x - 9.11 \leq 0$ and graph the function $f(x) = 2.3x^2 + 2.94x - 9.11$.

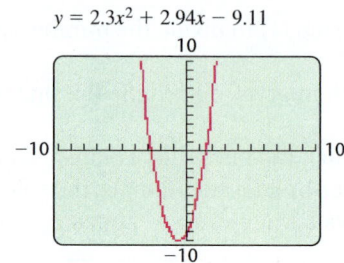

The x-values for which the graph lies *on or below* the x-axis begin somewhere between -3 and -2, and continue to somewhere between 1 and 2. Using the ZERO option of CALC and rounding, we find that these endpoints are -2.73 and 1.45. The solution set is approximately $\{x | -2.73 \leq x \leq 1.45\}$.

Use a graphing calculator to solve each inequality. Round the values of the endpoints to the nearest hundredth.

1. $4.32x^2 - 3.54x - 5.34 \leq 0$
2. $7.34x^2 - 16.55x - 3.89 \geq 0$
3. $5.79x^3 - 5.68x^2 + 10.68x$
       $> 2.11x^3 + 16.90x - 11.69$

Rational Inequalities

Inequalities involving rational expressions are called **rational inequalities**. Like polynomial inequalities, rational inequalities can be solved using test values. Unlike polynomials, however, rational expressions often have values for which the expression is undefined. These values must be used when dividing the number line into intervals.

EXAMPLE 5   Solve: $\dfrac{x-3}{x+4} \geq 2$.

SOLUTION   We write the related equation by changing the $\geq$ symbol to $=$:

$$\frac{x-3}{x+4} = 2. \qquad \textcolor{red}{\text{Note that } x \neq -4.}$$

Next, we solve this related equation:

$$\textcolor{red}{(x+4)} \cdot \frac{x-3}{x+4} = \textcolor{red}{(x+4)} \cdot 2 \qquad \textcolor{red}{\text{Clearing fractions by multiplying both sides by } x+4}$$

$$x - 3 = 2x + 8$$

$$-11 = x. \qquad \textcolor{red}{\text{Solving for } x. \text{ Note that } x \neq -4.}$$

Since -11 is a solution of the related equation, we use -11 when dividing the number line into intervals. Since the rational expression is undefined for $x = -4$, we must also use -4:

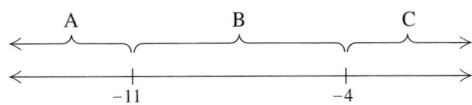

We test a number from each interval to see where the original inequality is satisfied:

$$\frac{x-3}{x+4} \geq 2.$$

For interval A,

$$\text{Test } -15, \quad \frac{-15-3}{-15+4} = \frac{-18}{-11} \qquad \textcolor{red}{\text{Any } x < -11 \text{ could be the test value.}}$$

$$= \frac{18}{11} \not\geq 2. \qquad \textcolor{red}{-15 \text{ is not a solution, so interval A is not part of the solution set.}}$$

For interval B,

$$\text{Test } -8, \quad \frac{-8-3}{-8+4} = \frac{-11}{-4} \qquad \textcolor{red}{\text{Any } x \text{ between } -11 \text{ and } -4 \text{ could be used.}}$$

$$= \frac{11}{4} \geq 2. \qquad \textcolor{red}{-8 \text{ is a solution, so interval B is part of the solution set.}}$$

For interval C,

$$\text{Test } 1, \quad \frac{1-3}{1+4} = \frac{-2}{5} \qquad \textcolor{red}{\text{Any } x > -4 \text{ could be used.}}$$

$$= -\frac{2}{5} \not\geq 2. \qquad \textcolor{red}{1 \text{ is not a solution, so interval C is not part of the solution set.}}$$

5. Solve: $\dfrac{x + 2}{x - 1} \leq 5$.

The solution set includes interval B. The endpoint -11 is included because the inequality symbol is $\geq$ and -11 is a solution of the related equation. The number -4 is *not* included because $(x - 3)/(x + 4)$ is undefined for $x = -4$. Thus the solution set of the original inequality is

$$[-11, -4), \quad \text{or} \quad \{x \mid -11 \leq x < -4\}.$$

↩ YOUR TURN

Algebraic 🔗 Graphical Connection

To compare the algebraic solution of Example 5 with a graphical solution, we graph $f(x) = (x - 3)/(x + 4)$ and the line $y = 2$. The solutions of $(x - 3)/(x + 4) \geq 2$ are found by locating all x-values for which $f(x) \geq 2$.

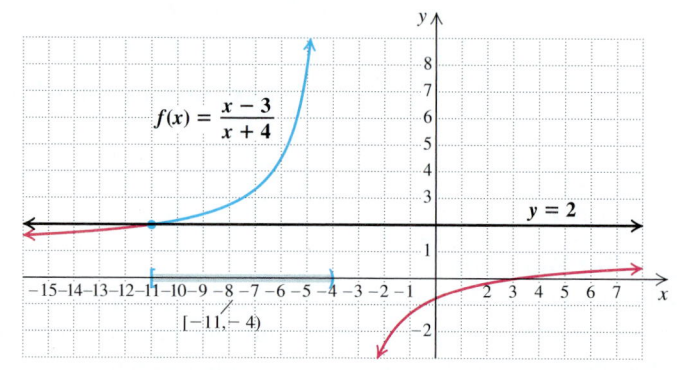

TO SOLVE A RATIONAL INEQUALITY

1. Find any replacements for which the rational expression is undefined.
2. Change the inequality symbol to an equal sign and solve the related equation.
3. Use the numbers found in steps (1) and (2) to divide the number line into intervals.
4. Substitute a test value from each interval into the inequality. If the number is a solution, then the interval to which it belongs is part of the solution set.
5. Select all interval(s) and endpoints for which the inequality is satisfied and use interval notation or set-builder notation to write the solution set. If the inequality symbol is $\leq$ or $\geq$, then solutions from step (2) are included in the solution set. (All numbers found in step (1) are excluded from the solution set.)

11.9 EXERCISE SET

FOR EXTRA HELP

MyMathLab® MathXL
PRACTICE WATCH READ REVIEW

Vocabulary and Reading Check

Classify each of the following statements as either true or false.

1. The solution of $(x - 3)(x + 2) \le 0$ is $[-2, 3]$.

2. The solution of $(x + 5)(x - 4) \ge 0$ is $[-5, 4]$.

3. The solution of $(x - 1)(x - 6) > 0$ is $\{x \mid x < 1 \text{ or } x > 6\}$.

4. The solution of $(x + 4)(x + 2) < 0$ is $(-4, -2)$.

5. To solve $\dfrac{x + 2}{x - 3} < 0$ using intervals, we divide the number line into the intervals $(-\infty, -2)$ and $(-2, \infty)$.

6. To solve $\dfrac{x - 5}{x + 4} \ge 0$ using intervals, we divide the number line into the intervals $(-\infty, -4)$, $(-4, 5)$, and $(5, \infty)$.

Polynomial Inequalities and Rational Inequalities

Solve each inequality using the graph provided.

7. $p(x) \le 0$

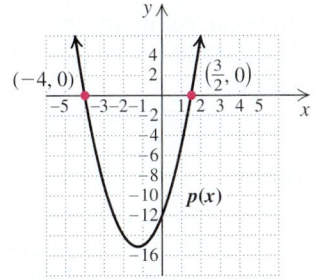

8. $p(x) < 0$

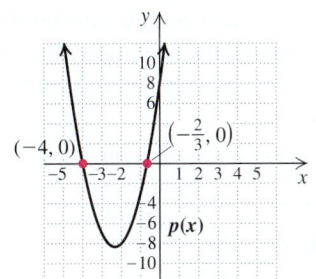

9. $x^4 + 12x > 3x^3 + 4x^2$

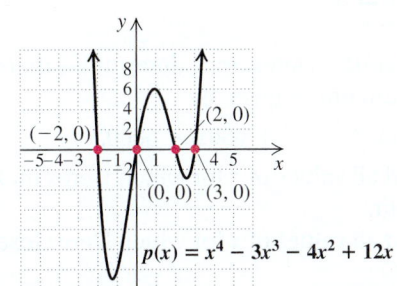

10. $x^4 + x^3 \ge 6x^2$

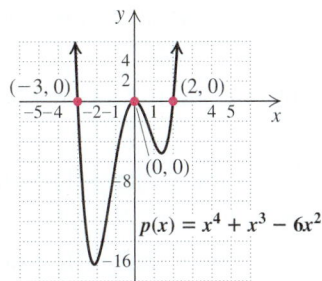

11. $\dfrac{x - 1}{x + 2} < 3$

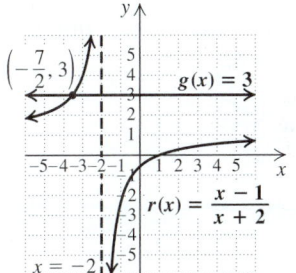

12. $\dfrac{2x - 1}{x - 5} \ge 1$

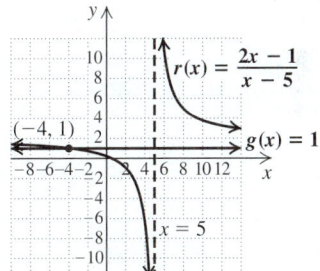

Quadratic and Other Polynomial Inequalities

Solve.

13. $(x - 6)(x - 5) < 0$

14. $(x + 8)(x + 10) > 0$

15. $(x + 7)(x - 2) \ge 0$

16. $(x - 1)(x + 4) \le 0$

17. $x^2 - x - 2 > 0$

18. $x^2 + x - 2 < 0$

Aha! 19. $x^2 + 4x + 4 < 0$

20. $x^2 + 6x + 9 < 0$

21. $x^2 - 4x \le 3$

22. $x^2 + 6x \ge 2$

23. $3x(x + 2)(x - 2) < 0$

24. $5x(x + 1)(x - 1) > 0$

25. $(x - 1)(x + 2)(x - 4) \geq 0$

26. $(x + 3)(x + 2)(x - 1) < 0$

27. For $f(x) = 7 - x^2$, find all x-values for which $f(x) \geq 3$.

28. For $f(x) = 14 - x^2$, find all x-values for which $f(x) > 5$.

29. For $g(x) = (x - 2)(x - 3)(x + 1)$, find all x-values for which $g(x) > 0$.

30. For $g(x) = (x + 3)(x - 2)(x + 1)$, find all x-values for which $g(x) < 0$.

31. For $F(x) = x^3 - 7x^2 + 10x$, find all x-values for which $F(x) \leq 0$.

32. For $G(x) = x^3 - 8x^2 + 12x$, find all x-values for which $G(x) \geq 0$.

Rational Inequalities

Solve.

33. $\dfrac{1}{x - 5} < 0$

34. $\dfrac{1}{x + 4} > 0$

35. $\dfrac{x + 1}{x - 3} \geq 0$

36. $\dfrac{x - 2}{x + 4} \leq 0$

37. $\dfrac{x + 1}{x + 6} \geq 1$

38. $\dfrac{x - 1}{x - 2} \leq 1$

39. $\dfrac{(x - 2)(x + 1)}{x - 5} \leq 0$

40. $\dfrac{(x + 4)(x - 1)}{x + 3} \geq 0$

41. $\dfrac{x}{x + 3} \geq 0$

42. $\dfrac{x - 2}{x} \leq 0$

43. $\dfrac{x - 5}{x} < 1$

44. $\dfrac{x}{x - 1} > 2$

45. $\dfrac{x - 1}{(x - 3)(x + 4)} \leq 0$

46. $\dfrac{x + 2}{(x - 2)(x + 7)} \geq 0$

47. For $f(x) = \dfrac{5 - 2x}{4x + 3}$, find all x-values for which $f(x) \geq 0$.

48. For $g(x) = \dfrac{2 + 3x}{2x - 4}$, find all x-values for which $g(x) \geq 0$.

49. For $G(x) = \dfrac{1}{x - 2}$, find all x-values for which $G(x) \leq 1$.

50. For $F(x) = \dfrac{1}{x - 3}$, find all x-values for which $F(x) \leq 2$.

51. Explain how any quadratic inequality can be solved by examining a parabola.

52. Describe a method for creating a quadratic inequality for which there is no solution.

Skill Review

53. On a typical weekday, the average full-time college student spends a total of 7.1 hr in educational or leisure activities. The student spends 0.7 hr more in leisure activities than in educational activities. On an average weekday, how many hours does the student spend on educational activities?   [2.5]

Source: U.S. Bureau of Labor Statistics

54. Kent paddled for 2 hr with a 5-km/h current to reach a campsite. The return trip against the same current took 7 hr. Find the speed of Kent's canoe in still water.   [8.3]

55. Josh and Lindsay plan to rent a moving truck. The truck costs $70 plus 40¢ per mile. They have budgeted $90 for the truck rental. For what mileages will they not exceed their budget? [2.7]

56. It takes Deanna twice as long to set up a fundraising auction as it takes Donna. Together they can set up for the auction in 4 hr. How long would it take each of them to do the job alone?   [6.7]

Synthesis

57. When solving a rational inequality containing the symbol $\leq$ or $\geq$, endpoints of some intervals may not be part of the solution set. Why?

58. Describe a method that could be used to create a quadratic inequality that has $(-\infty, a] \cup [b, \infty)$ as the solution set. Assume $a < b$.

Find each solution set.

59. $x^2 + 2x < 5$

60. $x^4 + 2x^2 \geq 0$

61. $x^4 + 3x^2 \leq 0$

62. $\left| \dfrac{x + 2}{x - 1} \right| \leq 3$

63. *Total Profit.*   Derex, Inc., determines that its total-profit function is given by
$$P(x) = -3x^2 + 630x - 6000.$$

  a) Find all values of x for which Derex makes a profit.

  b) Find all values of x for which Derex loses money.

64. *Height of a Thrown Object.* The function

$$S(t) = -16t^2 + 32t + 1920$$

gives the height S, in feet, of an object thrown from a cliff that is 1920 ft high. Here t is the time, in seconds, that the object is in the air.

a) For what times does the height exceed 1920 ft?
b) For what times is the height less than 640 ft?

65. *Number of Handshakes.* There are n people in a room. The number N of possible handshakes by the people is given by the function

$$N(n) = \frac{n(n-1)}{2}.$$

For what number of people n is $66 \le N \le 300$?

66. *Number of Diagonals.* A polygon with n sides has D diagonals, where D is given by the function

$$D(n) = \frac{n(n-3)}{2}.$$

Find the number of sides n if

$$27 \le D \le 230.$$

Use a graphing calculator to graph each function and find solutions of $f(x) = 0$. Then solve the inequalities $f(x) < 0$ and $f(x) > 0$.

67. $f(x) = x^3 - 2x^2 - 5x + 6$

68. $f(x) = \frac{1}{3}x^3 - x + \frac{2}{3}$

69. $f(x) = x + \frac{1}{x}$

70. $f(x) = x - \sqrt{x}, x \ge 0$

71. $f(x) = \frac{x^3 - x^2 - 2x}{x^2 + x - 6}$

72. $f(x) = x^4 - 4x^3 - x^2 + 16x - 12$

Find the domain of each function.

73. $f(x) = \sqrt{x^2 - 4x - 45}$

74. $f(x) = \sqrt{9 - x^2}$

75. $f(x) = \sqrt{x^2 + 8x}$

76. $f(x) = \sqrt{x^2 + 2x + 1}$

77. Describe a method that could be used to create a rational inequality that has $(-\infty, a] \cup (b, \infty)$ as the solution set. Assume $a < b$.

78. Use a graphing calculator to solve Exercises 43 and 49 by drawing two curves, one for each side of the inequality.

YOUR TURN ANSWERS: SECTION 11.9

1. $(-\infty, -2) \cup (4, \infty)$, or $\{x \mid x < -2 \text{ or } x > 4\}$
2. $[-2 - \sqrt{6}, -2 + \sqrt{6}]$, or $\{x \mid -2 - \sqrt{6} \le x \le -2 + \sqrt{6}\}$
3. $(-\infty, -2) \cup (-1, 0)$, or $\{x \mid x < -2 \text{ or } -1 < x < 0\}$
4. $[-1, 0] \cup [1, \infty)$, or $\{x \mid -1 \le x \le 0 \text{ or } x \ge 1\}$
5. $(-\infty, 1) \cup [\frac{7}{4}, \infty)$, or $\{x \mid x < 1 \text{ or } x \ge \frac{7}{4}\}$

QUICK QUIZ: SECTIONS 11.1–11.9

Solve.

1. $3x^2 + 1 = 0$ [11.1]

2. $x - 3\sqrt{x} - 4 = 0$ [11.5]

3. $5c^2 - c - 1 = 0$ [11.2]

4. $2x^2 - x - 3 \ge 0$ [11.9]

5. Find any x-intercepts and the y-intercept of the graph of $f(x) = 4x^2 - 3x$. [11.7]

PREPARE TO MOVE ON

Graph each function. [7.3]

1. $f(x) = x^3 - 2$

2. $g(x) = \frac{2}{x}$

3. If $g(x) = x^2 - 3$, find $g(\sqrt{a - 5})$. [7.1], [10.1]

4. If $g(x) = x^2 + 2$, find $g(2a + 5)$. [7.1]

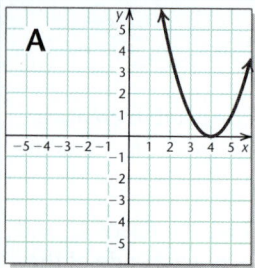

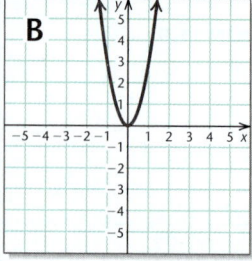

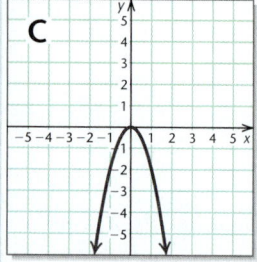

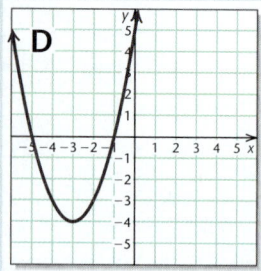

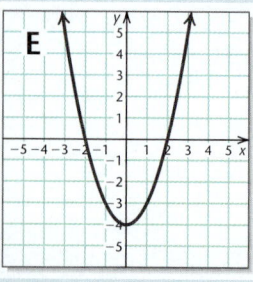

Visualizing for Success

Use after Section 11.7.

Match each function with its graph.

1. $f(x) = 3x^2$

2. $f(x) = x^2 - 4$

3. $f(x) = (x - 4)^2$

4. $f(x) = x - 4$

5. $f(x) = -2x^2$

6. $f(x) = x + 3$

7. $f(x) = |x + 3|$

8. $f(x) = (x + 3)^2$

9. $f(x) = \sqrt{x + 3}$

10. $f(x) = (x + 3)^2 - 4$

Answers on page A-58

An additional, animated version of this activity appears in MyMathLab. To use MyMathLab, you need a course ID and a student access code. Contact your instructor for more information.

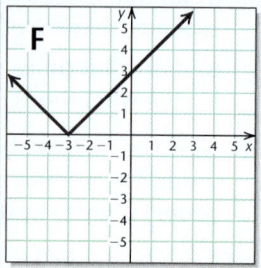

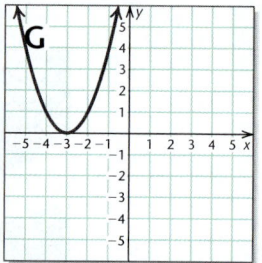

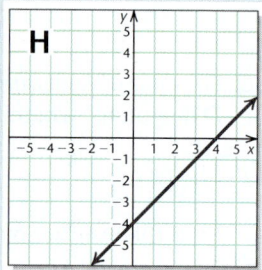

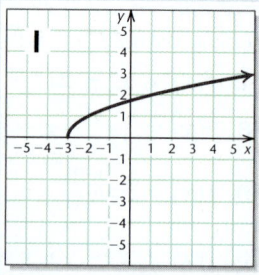

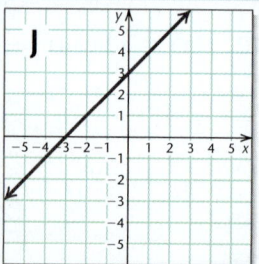

Collaborative Activity   *Match the Graph*

Focus: Graphing quadratic functions
Use after: Section 11.6
Time: 15–20 minutes
Group size: 6
Materials: Index cards

Activity

1. On each of six index cards, write one of the following equations:

$$y = \tfrac{1}{2}(x - 3)^2 + 1; \qquad y = \tfrac{1}{2}(x - 1)^2 + 3;$$
$$y = \tfrac{1}{2}(x + 1)^2 - 3; \qquad y = \tfrac{1}{2}(x + 3)^2 + 1;$$
$$y = \tfrac{1}{2}(x + 3)^2 - 1; \qquad y = \tfrac{1}{2}(x + 1)^2 + 3.$$

2. Fold each index card and mix up the six cards in a hat or bag. Then, one by one, each group member should select one of the equations. Do not let anyone see your equation.

3. Each group member should carefully graph the equation selected. Make the graph large enough so that when it is finished, it can be easily viewed by the rest of the group. Be sure to scale the axes and label the vertex, but **do not label the graph with the equation used**.

4. When all group members have drawn a graph, place the graphs in a pile. The group should then match and agree on the correct equation for each graph *with no help from the person who drew the graph.* If a mistake has been made and a graph has no match, determine what its equation *should* be.

5. Compare your group's labeled graphs with those of other groups to reach consensus within the class on the correct label for each graph.

Decision Making & Connection   *(Use after Section 11.8.)*

Pizza Pricing.   Papa Romeo's Pizza in Chicago, Illinois, sells a 10-in. diameter cheese pizza for $9, a 14-in. diameter pizza for $13, and an 18-in. diameter pizza for $21. Which better models the price of the pizza: a linear function or a quadratic function of the diameter?

Source: mypaparomeospizza.com

1. Graph ordered pairs from the data above using the form (diameter, price). Do the data appear to be quadratic or linear?

2. Fit each of the following models to the data, where $p(x)$ is the price, in dollars, of an x-inch diameter pizza. Using a different color for each, graph the functions on the same graph as the ordered pairs. Then determine visually which model best fits the data.

   a) Linear function $p(x) = mx + b$, using the points $(10, 9)$ and $(14, 13)$
   b) Linear function $p(x) = mx + b$, using the points $(10, 9)$ and $(18, 21)$
   c) Quadratic function $p(x) = ax^2 + bx + c$, using all three points

3. One way to tell whether a function is a good fit is to see how well it predicts another known value. Papa Romeo's also sells a 22-in. diameter cheese pizza for $28. Which function from part (2) comes closest to predicting the actual value?

4. Because the area of a circle is given by $A = \pi r^2$, would you expect the price of a cheese pizza to be quadratic or linear?

5. *Research.*   Find another restaurant that sells at least four sizes of pizza. Listing diameter on the horizontal axis and price on the vertical axis, graph their pizza prices and determine whether a linear model or a quadratic model appears to be the best fit. Use two or three of the prices to find a function that models the data, and test your model by predicting a known price not used to form the model.

Study Summary

KEY TERMS AND CONCEPTS	EXAMPLES	PRACTICE EXERCISES

SECTION 11.1: *Quadratic Equations*

A **quadratic equation in standard form** is written $ax^2 + bx + c = 0$, with a, b, and c constant and $a \neq 0$. Some quadratic equations can be solved by factoring.	$x^2 - 3x - 10 = 0$ $(x + 2)(x - 5) = 0$ $x + 2 = 0 \quad or \quad x - 5 = 0$ $x = -2 \quad or \qquad x = 5$	1. Solve: $x^2 - 12x + 11 = 0.$
The Principle of Square Roots For any real number k, if $X^2 = k$, then $X = \sqrt{k}$ or $X = -\sqrt{k}$.	$x^2 - 8x + 16 = 25$ $(x - 4)^2 = 25$ $x - 4 = -5 \quad or \quad x - 4 = 5$ $x = -1 \quad or \qquad x = 9$	2. Solve: $x^2 - 18x + 81 = 5.$
Any quadratic equation can be solved by **completing the square.**	$x^2 + 6x = 1$ $x^2 + 6x + \left(\frac{6}{2}\right)^2 = 1 + \left(\frac{6}{2}\right)^2$ $x^2 + 6x + 9 = 1 + 9$ $(x + 3)^2 = 10$ $x + 3 = \pm\sqrt{10}$ $x = -3 \pm \sqrt{10}$	3. Solve by completing the square: $x^2 + 20x = 21.$

SECTION 11.2: *The Quadratic Formula*

The Quadratic Formula The solutions of $ax^2 + bx + c = 0$ are given by $x = \dfrac{-b \pm \sqrt{b^2 - 4ac}}{2a}.$	$3x^2 - 2x - 5 = 0 \qquad a = 3, b = -2, c = -5$ $x = \dfrac{-(-2) \pm \sqrt{(-2)^2 - 4 \cdot 3(-5)}}{2 \cdot 3}$ $x = \dfrac{2 \pm \sqrt{4 + 60}}{6}$ $x = \dfrac{2 \pm \sqrt{64}}{6}$ $x = \dfrac{2 \pm 8}{6}$ $x = \dfrac{10}{6} = \dfrac{5}{3} \quad or \quad x = \dfrac{-6}{6} = -1$	4. Solve: $2x^2 - 3x - 9 = 0.$

SECTION 11.3: *Studying Solutions of Quadratic Equations*

The **discriminant** of the quadratic formula is $b^2 - 4ac$. $b^2 - 4ac = 0 \rightarrow$ One solution; a rational number	For $4x^2 - 12x + 9 = 0$, $b^2 - 4ac = (-12)^2 - 4(4)(9)$ $= 144 - 144 = 0.$ Thus, $4x^2 - 12x + 9 = 0$ has one rational solution.	5. Use the discriminant to determine the number and type of solutions of $2x^2 + 5x + 9 = 0.$

$b^2 - 4ac > 0 \rightarrow$ Two real solutions; both are rational if $b^2 - 4ac$ is a perfect square.

$b^2 - 4ac < 0 \rightarrow$ Two imaginary-number solutions

For $x^2 + 6x - 2 = 0$, $b^2 - 4ac = (6)^2 - 4(1)(-2)$
$$= 36 + 8 = 44.$$
Thus, $x^2 + 6x - 2 = 0$ has two irrational real-number solutions.

For $2x^2 - 3x + 5 = 0$, $b^2 - 4ac = (-3)^2 - 4(2)(5)$
$$= 9 - 40 = -31.$$
Thus, $2x^2 - 3x + 5 = 0$ has two imaginary-number solutions.

SECTION 11.4:   *Applications Involving Quadratic Equations*

To solve a formula for a letter, use the same principles used for solving equations.

Solve $y = pn^2 + dn$ for n.
$$pn^2 + dn - y = 0 \qquad a = p, b = d, c = -y$$
$$n = \frac{-d \pm \sqrt{d^2 - 4p(-y)}}{2 \cdot p}$$
$$n = \frac{-d \pm \sqrt{d^2 + 4py}}{2p}$$

6. Solve $a = n^2 + 1$ for n.

SECTION 11.5:   *Equations Reducible to Quadratic*

Equations that are **reducible to quadratic** or in **quadratic form** can be solved by making an appropriate substitution.

$$x^4 - 10x^2 + 9 = 0 \qquad \text{Let } u = x^2. \text{ Then } u^2 = x^4.$$
$$u^2 - 10u + 9 = 0 \qquad \text{Substituting}$$
$$(u - 9)(u - 1) = 0$$
$$u - 9 = 0 \quad or \quad u - 1 = 0$$
$$u = 9 \quad or \quad u = 1 \qquad \text{Solving for } u$$
$$x^2 = 9 \quad or \quad x^2 = 1 \qquad \text{Replacing } u \text{ with } x^2$$
$$x = \pm 3 \quad or \quad x = \pm 1 \qquad \text{Solving for } x$$

7. Solve:
$$x - \sqrt{x} - 30 = 0.$$

SECTION 11.6:   *Quadratic Functions and Their Graphs*
SECTION 11.7:   *More About Graphing Quadratic Functions*

The graph of a quadratic function
$$f(x) = ax^2 + bx + c$$
$$= a(x - h)^2 + k$$
is a **parabola.** The graph opens upward for $a > 0$ and downward for $a < 0$.

The **vertex** is (h, k), and the **axis of symmetry** is $x = h$.

If $a > 0$, the function has a **minimum** value of k, and if $a < 0$, the function has a **maximum** value of k.

The vertex and the axis of symmetry occur where
$$x = -\frac{b}{2a}.$$

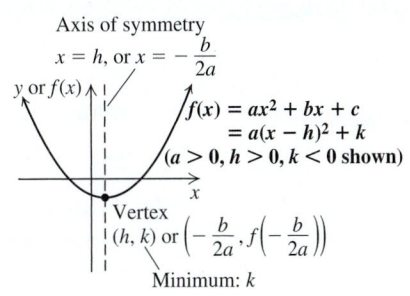

8. Graph
$$f(x) = 2x^2 - 12x + 3.$$
Label the vertex and the axis of symmetry, and identify the minimum or maximum function value.

SECTION 11.8:   *Problem Solving and Quadratic Functions*

Some problem situations can be **modeled** using quadratic functions. For those problems, a quantity can often be **maximized** or **minimized** by finding the coordinates of a vertex.	A lifeguard has 100 m of linked flotation devices with which to cordon off a rectangular swimming area at North Beach. If the shoreline forms one side of the rectangle, what dimensions will maximize the size of the area for swimming? This problem and its solution appear as Example 2 in Section 11.8.	**9.** Loretta is putting fencing around a rectangular vegetable garden. She can afford to buy 120 ft of fencing. What dimensions should she plan for the garden in order to maximize its area?

SECTION 11.9:   *Polynomial Inequalities and Rational Inequalities*

When solving a **polynomial inequality**, use the x-intercepts, or **zeros**, of a function to divide the x-axis into intervals. When solving a **rational inequality**, use the solutions of a rational equation along with any replacements that make a denominator zero to divide the x-axis into intervals.	Solve: $x^2 - 2x - 15 > 0$. $\quad x^2 - 2x - 15 = 0$ Solving the related equation $\quad (x - 5)(x + 3) = 0$ $\qquad\qquad x = 5 \;\; or \;\; x = -3$ −3 and 5 divide the number line into three intervals. <table><tr><td>+</td><td>−</td><td>+</td></tr><tr><td>−3</td><td></td><td>5</td></tr></table> For $f(x) = x^2 - 2x - 15 = (x - 5)(x + 3)$: $\quad f(x)$ is positive for $x < -3$; $\quad f(x)$ is negative for $-3 < x < 5$; $\quad f(x)$ is positive for $x > 5$. Thus, $x^2 - 2x - 15 > 0$ for $(-\infty, -3) \cup (5, \infty)$, or $\{x \,|\, x < -3 \; or \; x > 5\}$.	**10.** Solve: $x^2 - 11x - 12 < 0$.

Review Exercises: Chapter 11

 Concept Reinforcement

Classify each of the following statements as either true or false.

1. Every quadratic equation has two different solutions. [11.3]

2. Every quadratic equation has at least one solution. [11.3]

3. If an equation cannot be solved by completing the square, it cannot be solved by the quadratic formula. [11.2]

4. A negative discriminant indicates two imaginary-number solutions of a quadratic equation. [11.3]

5. The graph of $f(x) = 2(x + 3)^2 - 4$ has its vertex at $(3, -4)$. [11.6]

6. The graph of $g(x) = 5x^2$ has $x = 0$ as its axis of symmetry. [11.6]

7. The graph of $f(x) = -2x^2 + 1$ has no minimum value. [11.6]

8. The zeros of $g(x) = x^2 - 9$ are -3 and 3. [11.6]

9. If a quadratic function has two different imaginary-number zeros, the graph of the function has two x-intercepts. [11.7]

10. To solve a polynomial inequality, we often must solve a polynomial equation. [11.9]

Solve.

11. $9x^2 - 2 = 0$ [11.1]

12. $8x^2 + 6x = 0$ [11.1]

13. $x^2 - 12x + 36 = 9$ [11.1]

14. $x^2 - 4x + 8 = 0$ [11.2]

15. $x(3x + 4) = 4x(x - 1) + 15$ [11.2]

16. $x^2 + 9x = 1$ [11.2]

17. Solve $x^2 - 5x - 2 = 0$, using a calculator to approximate the solutions to three decimal places.   [11.2]

18. Let $f(x) = 4x^2 - 3x - 1$. Find all x such that $f(x) = 0$.   [11.2]

Replace the blanks with constants to form a true equation.   [11.1]

19. $x^2 - 18x +$ ___ $= (x -$ ___$)^2$

20. $x^2 + \frac{3}{5}x +$ ___ $= (x +$ ___$)^2$

21. Solve by completing the square. Show your work.
$$x^2 - 6x + 1 = 0 \quad [11.1]$$

22. \$2500 grows to \$2704 in 2 years. Use the formula $A = P(1 + r)^t$ to find the interest rate.   [11.1]

23. The London Eye observation wheel is 443 ft tall. Use $s = 16t^2$ to approximate how long it would take an object to fall from the top.   [11.1]

For each equation, determine whether the solutions are real or imaginary. If they are real, specify whether they are rational or irrational.   [11.3]

24. $x^2 + 3x - 6 = 0$

25. $x^2 + 2x + 5 = 0$

26. Write a quadratic equation having the solutions $3i$ and $-3i$.   [11.3]

27. Write a quadratic equation having -5 as its only solution.   [11.3]

Solve.   [11.4]

28. Horizons has a manufacturing plant 300 mi from company headquarters. Their corporate pilot must fly from headquarters to the plant and back in 4 hr. If there is a 20-mph headwind going and a 20-mph tailwind returning, how fast must the plane be able to travel in still air?

29. Working together, Dani and Cheri can reply to a day's worth of customer-service e-mails in 4 hr. Working alone, Dani takes 6 hr longer than Cheri. How long would it take Cheri alone to reply to the e-mails?

30. Find all x-intercepts of the graph of $f(x) = x^4 - 13x^2 + 36$.   [11.5]

Solve.   [11.5]

31. $15x^{-2} - 2x^{-1} - 1 = 0$

32. $(x^2 - 4)^2 - (x^2 - 4) - 6 = 0$

33. a) Graph: $f(x) = -3(x + 2)^2 + 4$.   [11.6]
    b) Label the vertex.
    c) Draw the axis of symmetry.
    d) Find the maximum or the minimum value.

34. For the function given by $f(x) = 2x^2 - 12x + 23$:   [11.7]

    a) find the vertex and the axis of symmetry;
    b) graph the function.

35. Find any x-intercepts and the y-intercept of the graph of
$$f(x) = x^2 - 9x + 14. \quad [11.7]$$

36. Solve $N = 3\pi\sqrt{\dfrac{1}{p}}$ for p.   [11.4]

37. Solve $2A + T = 3T^2$ for T.   [11.4]

State whether each graph appears to represent a quadratic function or a linear function.   [11.8]

38.

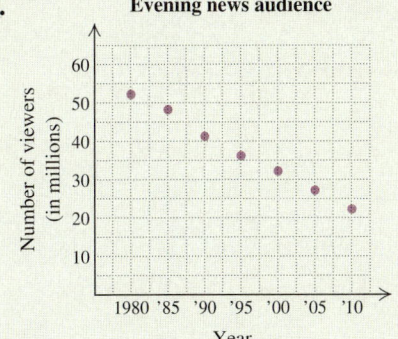

Evening news audience

Source: Nielsen Media Research

39.

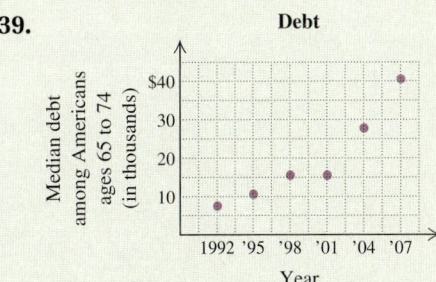

Debt

Source: U.S. Federal Reserve

40. Eastgate Consignments wants to build a rectangular area in a corner for children to play in while their parents shop. They have 30 ft of low fencing. What is the maximum area they can enclose? What dimensions will yield this area?   [11.8]

41. The following table lists the median debt among Americans ages 65 to 74 x years after 1992. (See Exercise 39.)   [11.8]

Years Since 1992	Median Debt Among Americans Ages 65 to 74 (in thousands)
0	$ 7
6	15
12	40

a) Find the quadratic function that fits the data.
b) Use the function to estimate the median debt among Americans ages 65 to 74 in 2013.

Solve.   [11.9]

42. $x^3 - 3x > 2x^2$

43. $\dfrac{x - 5}{x + 3} \le 0$

Synthesis

44. Explain how the x-intercepts of a quadratic function can be used to help find the maximum or minimum value of the function.   [11.7], [11.8]

45. Explain how the x-intercepts of a quadratic function can be used to factor a quadratic polynomial.   [11.4], [11.7]

46. Discuss two ways in which completing the square was used in this chapter.   [11.1], [11.2], [11.7]

47. A quadratic function has x-intercepts at -3 and 5. If the y-intercept is at -7, find an equation for the function.   [11.7]

48. Find h and k if, for $3x^2 - hx + 4k = 0$, the sum of the solutions is 20 and the product of the solutions is 80.   [11.3]

49. The average of two positive integers is 171. One of the numbers is the square root of the other. Find the integers.   [11.5]

Test: Chapter 11

For step-by-step test solutions, access the Chapter Test Prep Videos in MyMathLab®, on YouTube (search "Bittinger Combo AlgCA"), or by scanning the code.

Solve.

1. $25x^2 - 7 = 0$

2. $4x(x - 2) - 3x(x + 1) = -18$

3. $x^2 + 2x + 3 = 0$

4. $2x + 5 = x^2$

5. $x^{-2} - x^{-1} = \frac{3}{4}$

6. Solve $x^2 + 3x = 5$, using a calculator to approximate the solutions to three decimal places.

7. Let $f(x) = 12x^2 - 19x - 21$. Find x such that $f(x) = 0$.

Replace the blanks with constants to form a true equation.

8. $x^2 - 20x + \underline{\hspace{1cm}} = (x - \underline{\hspace{1cm}})^2$

9. $x^2 + \frac{2}{7}x + \underline{\hspace{1cm}} = (x + \underline{\hspace{1cm}})^2$

10. Solve by completing the square. Show your work.
$$x^2 + 10x + 15 = 0$$

11. Determine the type of number that the solutions of $x^2 + 2x + 5 = 0$ will be.

12. Write a quadratic equation having solutions $\sqrt{11}$ and $-\sqrt{11}$.

Solve.

13. The Connecticut River flows at a rate of 4 km/h for the length of a popular scenic route. In order for a cruiser to travel 60 km upriver and then return in a total of 8 hr, how fast must the boat be able to travel in still water?

14. Dal and Kim can assemble a swing set in $1\frac{1}{2}$ hr. Working alone, it takes Kim 4 hr longer than Dal to assemble the swing set. How long would it take Dal, working alone, to assemble the swing set?

15. Find all x-intercepts of the graph of
$$f(x) = x^4 - 15x^2 - 16.$$

16. a) Graph: $f(x) = 4(x - 3)^2 + 5$.
 b) Label the vertex.
 c) Draw the axis of symmetry.
 d) Find the maximum or the minimum function value.

17. For the function $f(x) = 2x^2 + 4x - 6$:
 a) find the vertex and the axis of symmetry;
 b) graph the function.

18. Find the x- and y-intercepts of
$$f(x) = x^2 - x - 6.$$

19. Solve $V = \frac{1}{3}\pi(R^2 + r^2)$ for r. Assume all variables are positive.

20. State whether the graph appears to represent a linear function, a quadratic function, or neither.

Chicago air quality

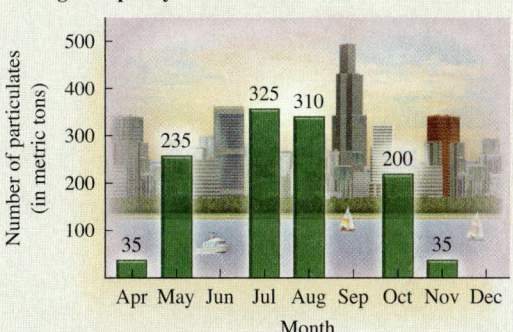

Source: The National Arbor Day Foundation

21. Jay's Metals has determined that when x hundred storage cabinets are built, the average cost per cabinet is given by
$$C(x) = 0.2x^2 - 1.3x + 3.4025,$$
where $C(x)$ is in hundreds of dollars. What is the minimum cost per cabinet and how many cabinets should be built to achieve that minimum?

22. Find the quadratic function that fits the data points $(0, 0)$, $(3, 0)$, and $(5, 2)$.

Solve.

23. $x^2 + 5x < 6$

24. $x - \dfrac{1}{x} \geq 0$

Synthesis

25. One solution of $kx^2 + 3x - k = 0$ is -2. Find the other solution.

26. Find a fourth-degree polynomial equation, with integer coefficients, for which $-\sqrt{3}$ and $2i$ are solutions.

27. Solve: $x^4 - 4x^2 - 1 = 0$.

Cumulative Review: Chapters 1–11

Simplify.

1. $-3 \cdot 8 \div (-2)^3 \cdot 4 - 6(5 - 7)$ [1.8]

2. $(5x^2y - 8xy - 6xy^2) - (2xy - 9x^2y + 3xy^2)$ [4.4]

3. $(9p^2q + 8t)(9p^2q - 8t)$ [4.6]

4. $\dfrac{t^2 - 25}{9t^2 + 24t + 16} \div \dfrac{3t^2 - 11t - 20}{t^2 + t}$ [6.2]

5. $(3\sqrt{2} + i)(2\sqrt{2} - i)$ [10.8]

Factor.

6. $12x^4 - 75y^4$ [5.4]

7. $x^3 - 24x^2 + 80x$ [5.2]

8. $100m^6 - 100$ [5.5]

9. $6t^2 + 35t + 36$ [5.3]

Solve.

10. $2(5x - 3) - 8x = 4 - (3 - x)$ [2.2]

11. $2(5x - 3) - 8x < 4 - (3 - x)$ [2.6]

12. $\begin{aligned} 2x - 6y &= 3, \\ -3x + 8y &= -5 \end{aligned}$ [8.2]

13. $x(x - 5) = 66$ [5.7]

14. $\dfrac{2}{t} + \dfrac{1}{t - 1} = 2$ [6.6]

15. $\sqrt{x} = 1 + \sqrt{2x - 7}$ [10.6]

16. $m^2 + 10m + 25 = 2$ [11.1]

17. $3x^2 + 1 = x$ [11.2]

Graph.

18. $9x - 2y = 18$ [3.3]   **19.** $x < \frac{1}{2}y$ [9.3]

20. $y = 2(x - 3)^2 + 1$ [11.6]

21. $f(x) = x^2 + 4x + 3$ [11.7]

22. Find an equation in slope–intercept form whose graph has slope -5 and y-intercept $\left(0, \frac{1}{2}\right)$. [3.6]

23. Find the slope of the line containing $(8, 3)$ and $(-2, 10)$. [3.5]

Find the domain of f.

24. $f(x) = \sqrt{10 - x}$ [7.2]

25. $f(x) = \dfrac{x + 3}{x - 4}$ [7.2], [9.2]

Solve each formula for the specified letter.

26. $b = \dfrac{a + c}{2a}$, for a [7.5]

27. $p = 2\sqrt{\dfrac{r}{3t}}$, for t [11.4]

Solve.

28. *Gold Prices.* Marisa is selling some of her gold jewelry. She has 4 bracelets and 1 necklace that weigh a total of 3 oz. [2.5]

   a) Marisa's jewelry is 58% gold. How many ounces of gold does her jewelry contain?

   b) A gold dealer offers Marisa $2088 for the jewelry. How much per ounce of gold was she offered?

   c) The retail price of gold at the time of Marisa's sale was $1600 per ounce. What percent of the gold price was she offered for her jewelry?

29. *Television Viewing.* The average number of Americans who watch the evening news has decreased from 52 million in 1980 to 22 million in 2010. [7.3]

   Source: Nielsen Media Research

   a) Let $f(t)$ represent the average number of viewers of the evening news, in millions, t years after 1980. Find a linear function that fits the data.

   b) Use the function from part (a) to predict the number of Americans watching the evening news in 2014.

   c) If the trend continues, in what year will no one watch the evening news?

30. *Education.* Andres ordered number tiles at $9 per set and alphabet tiles at $15 per set for his classroom. He ordered a total of 36 sets for $384. How many sets of each did he order? [8.3]

31. *Minimizing Cost.* Dormitory Furnishings has determined that when x bunk beds are built, the average cost, in dollars, per bunk bed can be estimated by $c(x) = 0.004375x^2 - 3.5x + 825$. What is the minimum average cost per bunk bed and how many bunk beds should be built to achieve that minimum? [11.8]

Synthesis

Solve.

32. $\dfrac{\dfrac{1}{x}}{2 + \dfrac{1}{x - 1}} = 3$ [6.5], [11.2]

33. $x^4 + 5x^2 \leq 0$ [11.9]

34. Find the points of intersection of the graphs of $f(x) = x^2 + 8x + 1$ and $g(x) = 10x + 6$. [11.2]

Exponential Functions and Logarithmic Functions

NOTE	FREQUENCY (in hertz)
A	27.5
A#	29.1352
B	30.8677
C	32.7032

Music Contains Mathematics.

Math and music are closely connected, and much of the beauty and pleasure of music can be described in mathematical terms. From rhythms to harmonies to Bach's fugues, the languages and structure of music and math overlap. One example that we will consider in this chapter deals with frequencies in music. When we hear different musical pitches, we are detecting differences in vibrations of sound, described in oscillations per second, or hertz. The table above lists the frequencies for the lowest four notes on an 88-key piano. The frequencies between notes do not increase at the same rate; instead, they can be modeled using an *exponential function*. (*See Exercise 9 in Exercise Set 12.7.*)

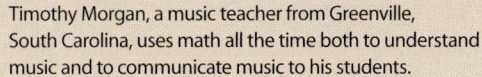

Math is the mechanism that moves music. Music without math becomes noise.

Timothy Morgan, a music teacher from Greenville, South Carolina, uses math all the time both to understand music and to communicate music to his students.

The *exponential* functions and *logarithmic* functions that we consider in this chapter have rich applications in many fields, such as epidemiology (the study of the spread of disease), population growth, and marketing. Exponential functions have variable exponents, and logarithmic functions are their closely related *inverse* functions.

12.1 | Composite Functions and Inverse Functions

Composite Functions ▪ Inverses and One-to-One Functions ▪ Finding Formulas for Inverses

Graphing Functions and Their Inverses ▪ Inverse Functions and Composition

Later in this chapter, we introduce two closely related types of functions: exponential functions and logarithmic functions. In order to properly understand the link between these functions, we must first understand composite functions and inverse functions.

Composite Functions

Functions frequently occur in which some quantity depends on a variable that, in turn, depends on another variable. For instance, a firm's profits may be a function of the number of items the firm produces, which may in turn be a function of the number of employees hired. In this case, the firm's profits may be considered a **composite function**.

Tea Mug Collective's Shane Kimberlin, Alaskan artist

Let's consider an example of a profit function. Tea Mug Collective sells hand-painted tee shirts. The monthly profit p, in dollars, from the sale of m shirts is given by $p = 15m - 1200$. The number of shirts m produced in a month by x employees is given by $m = 40x$.

If Tea Mug Collective employs 10 people, then in one month they can produce $m = 40(10) = 400$ shirts. The profit from selling these 400 shirts would be $p = 15(400) - 1200 = 4800$ dollars. Can we find an equation that would allow us to calculate the monthly profit on the basis of the number of employees? We begin with the profit equation and substitute:

$$p = 15m - 1200$$
$$= 15(40x) - 1200 \qquad \text{Substituting } 40x \text{ for } m$$
$$= 600x - 1200.$$

The equation $p = 600x - 1200$ gives the monthly profit when Tea Mug Collective has x employees.

To find a composition of functions, we follow the same reasoning above using function notation:

$$p(m) = 15m - 1200, \qquad \text{Profit as a function of the number of shirts produced}$$
$$m(x) = 40x; \qquad \text{Number of shirts as a function of the number of employees}$$

$$p(m(x)) = p(40x)$$
$$= 15(40x) - 1200$$
$$= 600x - 1200.$$

If we call this new function P, then $P(x) = 600x - 1200$. This gives profit as a function of the number of employees.

Study Skills

Divide and Conquer

In longer sections of reading, there are almost always subsections. Rather than feel obliged to read an entire section at once, use the subsections as natural resting points. Taking a break between subsections can increase your comprehension and can be an efficient use of your time.

We call P the *composition* of p and m. In general, the composition of f and g is written $f \circ g$ and is read "the composition of f and g," "f composed with g," or "f circle g."

COMPOSITION OF FUNCTIONS

The *composite function* $f \circ g$, the *composition* of f and g, is defined as

$$(f \circ g)(x) = f(g(x)).$$

It is not uncommon to use the same variable to represent the input in more than one function.

> **Throughout this chapter, keep in mind that equations such as $m(x) = 40x$ and $m(t) = 40t$ describe the same function. Both equations tell us to find a function value by multiplying the input by 40.**

We can visualize the composition of functions as follows.

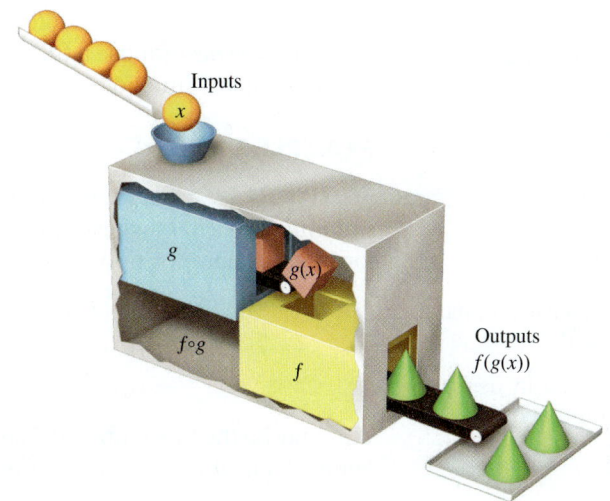

EXAMPLE 1   Given $f(x) = 3x$ and $g(x) = 1 + x^2$:

a) Find $(f \circ g)(5)$ and $(g \circ f)(5)$.   **b)** Find $(f \circ g)(x)$ and $(g \circ f)(x)$.

SOLUTION   Consider each function separately:

$$f(x) = 3x \qquad \text{This function multiplies each input by 3.}$$

and

$$g(x) = 1 + x^2. \qquad \text{This function adds 1 to the square of each input.}$$

a) To find $(f \circ g)(5)$, we find $g(5)$ and then use that as an input for f:

$$(f \circ g)(5) = f(g(5)) = f(1 + 5^2) \qquad \text{Using } g(x) = 1 + x^2$$
$$= f(26) = 3 \cdot 26 = 78. \qquad \text{Using } f(x) = 3x$$

To find $(g \circ f)(5)$, we find $f(5)$ and then use that as an input for g:

$$(g \circ f)(5) = g(f(5)) = g(3 \cdot 5) \qquad \text{Note that } f(5) = 3 \cdot 5 = 15.$$
$$= g(15) = 1 + 15^2 = 1 + 225 = 226.$$

b) We find $(f \circ g)(x)$ by substituting $g(x)$ for x in the equation for $f(x)$:

$$(f \circ g)(x) = f(g(x)) = f(1 + x^2) \qquad \text{Using } g(x) = 1 + x^2$$
$$= 3 \cdot (1 + x^2) = 3 + 3x^2. \qquad \text{Using } f(x) = 3x$$

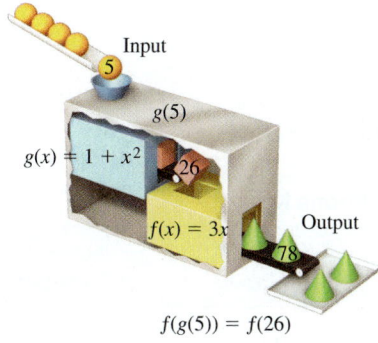

A composition machine for Example 1

To find $(g \circ f)(x)$, we substitute $f(x)$ for x in the equation for $g(x)$:

$$(g \circ f)(x) = g(f(x)) = g(3x) \qquad \text{Substituting } 3x \text{ for } f(x)$$
$$= 1 + (3x)^2 = 1 + 9x^2.$$

We can now find the function values of part (a) using the functions of part (b):

$$(f \circ g)(5) = 3 + 3(5)^2 = 3 + 3 \cdot 25 = 78;$$
$$(g \circ f)(5) = 1 + 9(5)^2 = 1 + 9 \cdot 25 = 226.$$

1. Given $f(x) = x^2 - 2$ and $g(x) = 5x$, find $(f \circ g)(x)$.

YOUR TURN

Example 1 shows that, in general, $(f \circ g)(x) \neq (g \circ f)(x)$.

EXAMPLE 2   Given $f(x) = \sqrt{x}$ and $g(x) = x - 1$, find $(f \circ g)(x)$ and $(g \circ f)(x)$.

SOLUTION

$$(f \circ g)(x) = f(g(x)) = f(x - 1) = \sqrt{x - 1}; \qquad \text{Using } g(x) = x - 1$$
$$(g \circ f)(x) = g(f(x)) = g(\sqrt{x}) = \sqrt{x} - 1 \qquad \text{Using } f(x) = \sqrt{x}$$

2. Given $f(x) = 2x - 7$ and $g(x) = \sqrt[3]{x}$, find $(g \circ f)(x)$.

YOUR TURN

When we *decompose* a function, we think of the function as the composition of two "simpler" functions.

EXAMPLE 3   If $h(x) = (7x + 3)^2$, find f and g such that $h(x) = (f \circ g)(x)$.

SOLUTION   We can think of $h(x)$ as the result of first evaluating $7x + 3$ and then squaring that. This suggests that we let $g(x) = 7x + 3$ and $f(x) = x^2$. We check by forming the composition:

$$(f \circ g)(x) = f(g(x))$$
$$= f(7x + 3)$$
$$= (7x + 3)^2 = h(x), \text{ as desired.}$$

This may be the most "obvious" solution, but there are other less obvious answers. For example, if $f(x) = (x - 1)^2$ and $g(x) = 7x + 4$, then

$$(f \circ g)(x) = f(g(x))$$
$$= f(7x + 4)$$
$$= (7x + 4 - 1)^2 = (7x + 3)^2 = h(x).$$

3. If $h(x) = \sqrt{3 - x}$, find f and g such that $h(x) = (f \circ g)(x)$.
Answers may vary.

YOUR TURN

Technology Connection

In Example 3, we see that if $g(x) = 7x + 3$ and $f(x) = x^2$, then $f(g(x)) = (7x + 3)^2$. One way to show this is to let $y_1 = 7x + 3$ and $y_2 = x^2$. If we let $y_3 = (7x + 3)^2$ and $y_4 = y_2(y_1)$, we can use graphs or a table to show that $y_3 = y_4$.

1. Check Example 2 by using the above approach.

Inverses and One-to-One Functions

Let's view the following two functions as relations, or correspondences.

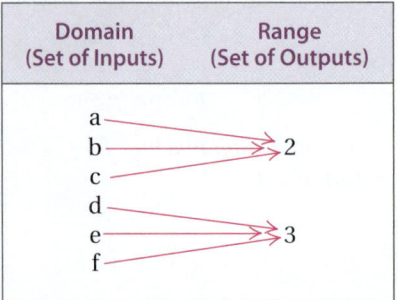

Countries and Their Capitals

Domain (Set of Inputs)	Range (Set of Outputs)
Australia	Canberra
China	Beijing
Germany	Berlin
Madagascar	Antananaviro
Turkey	Ankara
United States	Washington, D.C.

Phone Keys

Domain (Set of Inputs)	Range (Set of Outputs)
a	
b	2
c	
d	
e	3
f	

Suppose we reverse the arrows. We obtain what is called the **inverse relation**. Are these inverse relations functions?

Countries and Their Capitals

Range (Set of Outputs)	Domain (Set of Inputs)
Australia ←	Canberra
China ←	Beijing
Germany ←	Berlin
Madagascar ←	Antananaviro
Turkey ←	Ankara
United States ←	Washington, D.C.

Phone Keys

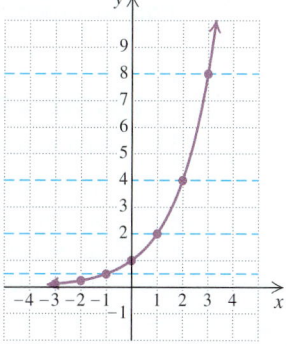

Recall that for a function, each input has exactly one output. In some functions, different inputs correspond to the same output. Only when this possibility is *excluded* will the inverse be a function. For the functions listed above, this means the inverse of the "Capitals" correspondence is a function, but the inverse of the "Phone Keys" correspondence is not.

In the Capitals function, each input has its own output, so it is a **one-to-one function**. In the Phone Keys function, a and b are both paired with 2. Thus the Phone Keys function is not a one-to-one function.

ONE-TO-ONE FUNCTION

A function f is *one-to-one* if different inputs have different outputs. That is, f is one-to-one if for any a and b in the domain of f with $a \neq b$, we have $f(a) \neq f(b)$. If a function is one-to-one, then its inverse correspondence is also a function.

How can we tell graphically whether a function is one-to-one?

EXAMPLE 4   Below is the graph of a function. Determine whether the function is one-to-one and thus has an inverse that is a function.

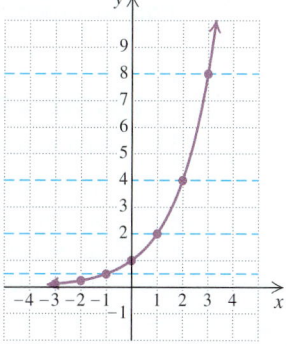

4. Determine whether the function graphed below is one-to-one.

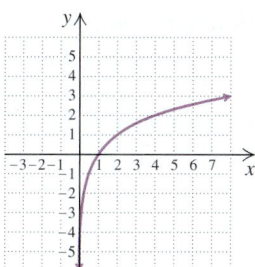

SOLUTION   A function is one-to-one if different inputs have different outputs—that is, if no two x-values share the same y-value. For this function, we cannot find two x-values that have the same y-value. To see this, note that no horizontal line can be drawn so that it crosses the graph more than once. The function is one-to-one so its inverse is a function.

 YOUR TURN

The graph of every function must pass the vertical-line test. In order for a function to have an inverse that is a function, it must pass the *horizontal-line test* as well.

> **THE HORIZONTAL-LINE TEST**
>
> If it is impossible to draw a horizontal line that intersects a function's graph more than once, then the function is one-to-one. For every one-to-one function, an inverse function exists.

EXAMPLE 5   Determine whether the function given by $f(x) = x^2$ is one-to-one and thus has an inverse that is a function.

SOLUTION   The graph of $f(x) = x^2$ is shown here. Many horizontal lines cross the graph more than once. For example, the line $y = 4$ crosses where the first coordinates are -2 and 2. Although these are different inputs, they have the same output. That is, $-2 \neq 2$, but

$$f(-2) = (-2)^2 = 4 = 2^2 = f(2).$$

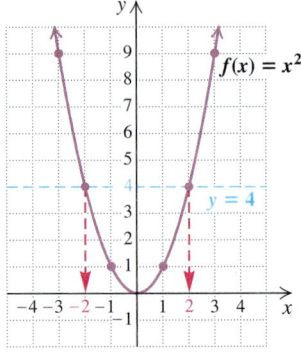

Thus the function is not one-to-one and its inverse is not a function.

5. Determine whether the function given by $f(x) = (x - 2)^2$ is one-to-one.

 YOUR TURN

Finding Formulas for Inverses

When the inverse of f is also a function, it is denoted f^{-1} (read "f-inverse").

> **CAUTION!**   The -1 in f^{-1} is *not* an exponent!

For any equation in two variables, if we interchange the variables, we form an equation of the inverse correspondence. If the inverse correspondence is a function, we proceed as follows to find a formula for f^{-1}.

> **TO FIND A FORMULA FOR f^{-1}**
>
> First make sure that f is one-to-one. Then:
>
> **1.** Replace $f(x)$ with y.
> **2.** Interchange x and y. (This gives the inverse function.)
> **3.** Solve for y.
> **4.** Replace y with $f^{-1}(x)$. (This is inverse function notation.)

EXAMPLE 6   Determine whether each function is one-to-one and if it is, find a formula for $f^{-1}(x)$.

a) $f(x) = 2x - 3$ **b)** $f(x) = \dfrac{3}{x}$

Student Notes

If you are not certain whether a function is one-to-one, you can try to find a formula for its inverse. If it is possible to solve *uniquely* for y after x and y have been interchanged, then the function is one-to-one.

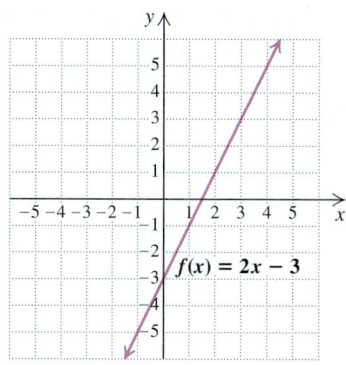

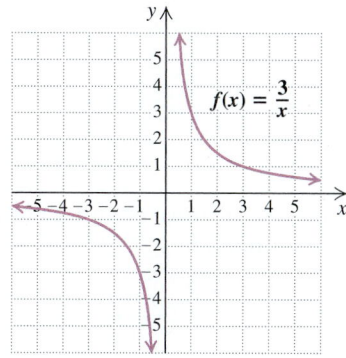

6. Determine whether the function given by $f(x) = x + 2$ is one-to-one. If it is, find a formula for $f^{-1}(x)$.

SOLUTION

a) The graph of $f(x) = 2x - 3$ is shown at left. This function, like any linear function that is not constant, passes the horizontal-line test. Thus, f is one-to-one and we can find a formula for $f^{-1}(x)$.

   1. Replace $f(x)$ with y: $y = 2x - 3$.

   2. Interchange x and y: $x = 2y - 3$.

   3. Solve for y: $x + 3 = 2y$

$$\frac{x + 3}{2} = y.$$

   4. Replace y with $f^{-1}(x)$: $f^{-1}(x) = \dfrac{x + 3}{2}$.

In this case, the function f doubles all inputs and then subtracts 3. Thus, to "undo" f, the function f^{-1} adds 3 to each input and then divides by 2.

b) The graph of $f(x) = 3/x$ is shown at left. The function passes the horizontal-line test. Thus it is one-to-one and its inverse is a function.

   1. Replace $f(x)$ with y: $y = \dfrac{3}{x}$.

   2. Interchange x and y: $x = \dfrac{3}{y}$.

   3. Solve for y: $xy = 3$

$$y = \frac{3}{x}.$$

   4. Replace y with $f^{-1}(x)$: $f^{-1}(x) = \dfrac{3}{x}$.

Note that this function and its inverse are the same function.

YOUR TURN

Graphing Functions and Their Inverses

How do the graphs of a function and its inverse compare?

EXAMPLE 7   Graph $f(x) = 2x - 3$ and $f^{-1}(x) = (x + 3)/2$ on the same set of axes. Then compare.

SOLUTION   The graph of each function follows. Note that the graph of f^{-1} can be drawn by reflecting the graph of f across the line $y = x$. That is, if we graph $f(x) = 2x - 3$ in wet ink and fold the paper along the line $y = x$, the graph of $f^{-1}(x) = (x + 3)/2$ will appear as the impression made by f.

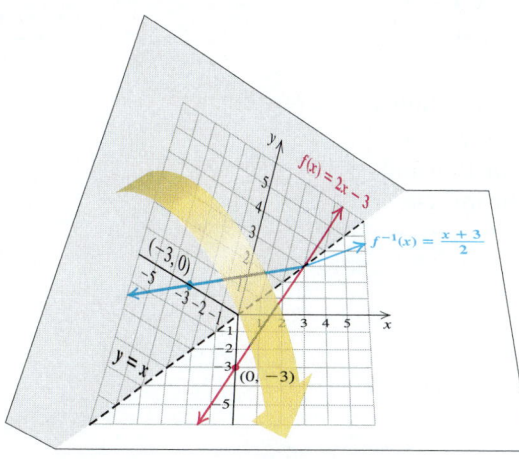

7. Graph $f(x) = \frac{1}{2}x + 1$ and $f^{-1}(x) = 2x - 2$ on the same set of axes.

When x and y are interchanged to find a formula for the inverse, we are, in effect, reflecting or flipping the graph of $f(x) = 2x - 3$ across the line $y = x$. For example, when $(0, -3)$, the coordinates of the y-intercept of the graph of f, are reversed, we get $(-3, 0)$, the x-intercept of the graph of f^{-1}.

YOUR TURN

VISUALIZING INVERSES

The graph of f^{-1} is a reflection of the graph of f across the line $y = x$.

EXAMPLE 8 Consider $g(x) = x^3 + 2$.

a) Determine whether g is one-to-one.

b) If it is one-to-one, find a formula for its inverse.

c) Graph the inverse, if it exists.

SOLUTION

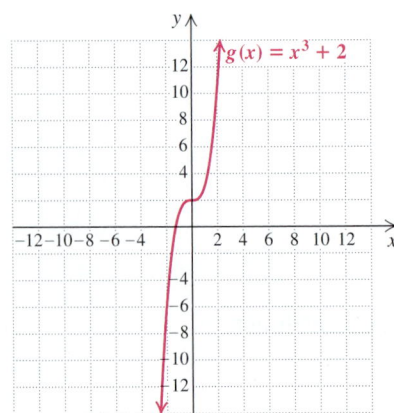

a) The graph of $g(x) = x^3 + 2$ is shown at left. It passes the horizontal-line test and thus is one-to-one and has an inverse that is a function.

b) **1.** Replace $g(x)$ with y: $y = x^3 + 2$.   Using $g(x) = x^3 + 2$

    2. Interchange x and y: $x = y^3 + 2$.   This represents the inverse relation.

    3. Solve for y: $x - 2 = y^3$

                                 $\sqrt[3]{x - 2} = y$.   Each real number has only one cube root, so we can solve uniquely for y.

    4. Replace y with $g^{-1}(x)$: $g^{-1}(x) = \sqrt[3]{x - 2}$.

c) To graph g^{-1}, we can reflect the graph of $g(x) = x^3 + 2$ across the line $y = x$, as we did in Example 7. We also could graph $g^{-1}(x) = \sqrt[3]{x - 2}$ by plotting points. As a check, note that $(2, 10)$ is on the graph of g and $(10, 2)$ is on the graph of g^{-1}.

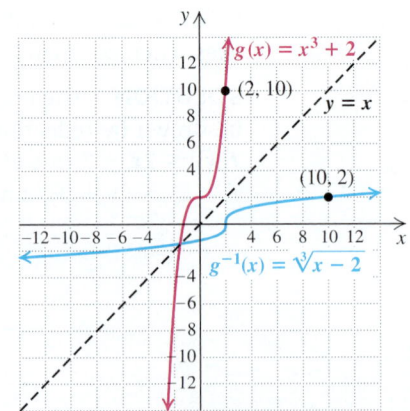

8. Consider $f(x) = \sqrt[3]{x + 1}$.

  a) Determine whether f is one-to-one.

  b) If it is one-to-one, find a formula for its inverse.

  c) Graph the function and, if it exists, the inverse.

YOUR TURN

EXPLORING 🔍 THE CONCEPT

The graph of an inverse function is the reflection of the graph of the function across the line $y = x$.
Match each function below with the graph of its inverse function.

1. 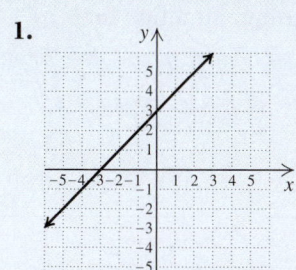 **2.** 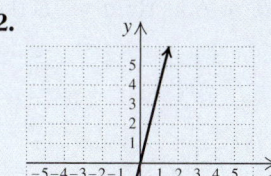 **3.** 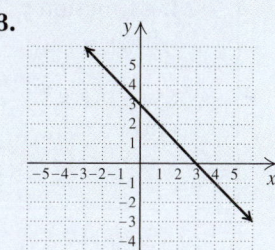 **4.**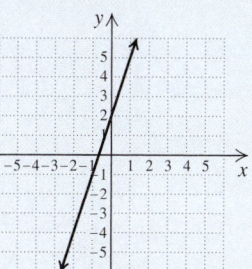

(a) 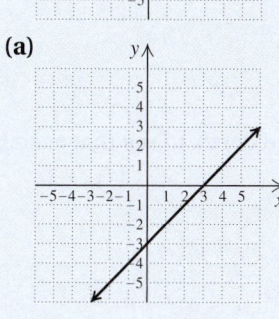 **(b)** 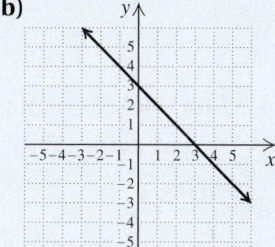 **(c)** 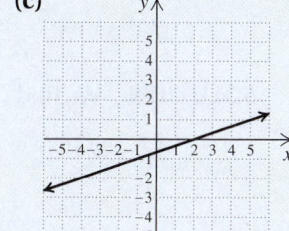 **(d)**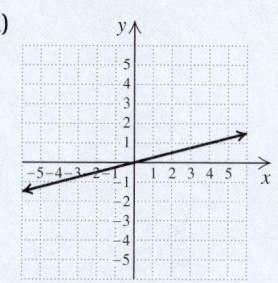

ANSWERS
1. (a) **2.** (d) **3.** (b) **4.** (c)

Inverse Functions and Composition

Let's consider inverses of functions in terms of function machines. Suppose that a one-to-one function f is programmed into a machine. If the machine is run in reverse, it will perform the inverse function f^{-1}. Inputs then enter at the opposite end, and the entire process is reversed.

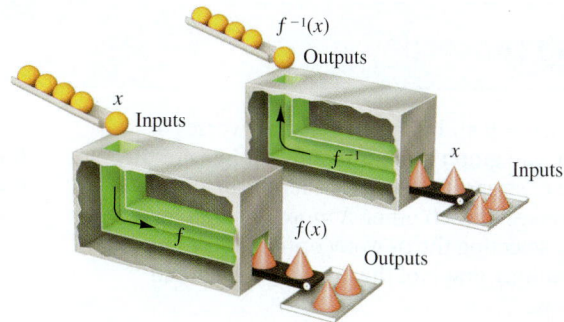

Consider $f(x) = x^3 + 2$ and $f^{-1}(x) = \sqrt[3]{x - 2}$ from Example 8. For the input 3,

$$f(3) = 3^3 + 2 = 27 + 2 = 29.$$

The output, $f(3)$, is 29. Let's now use 29 as an input in the inverse:

$$f^{-1}(29) = \sqrt[3]{29 - 2} = \ = \sqrt[3]{27} = 3.$$

The function f takes 3 to 29. The inverse function f^{-1} takes the number 29 back to 3.

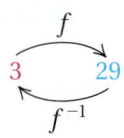

In general, if f is one-to-one, then f^{-1} takes the output $f(x)$ back to x. Similarly, f takes the output $f^{-1}(x)$ back to x.

COMPOSITION AND INVERSES

If a function f is one-to-one, then f^{-1} is the unique function for which

$$(f^{-1} \circ f)(x) = f^{-1}(f(x)) = x$$

and

$$(f \circ f^{-1})(x) = f(f^{-1}(x)) = x.$$

EXAMPLE 9   Let $f(x) = 2x + 1$. Show that

$$f^{-1}(x) = \frac{x - 1}{2}.$$

SOLUTION   We find $(f^{-1} \circ f)(x)$ and $(f \circ f^{-1})(x)$ and check to see that each is x.

$$(f^{-1} \circ f)(x) = f^{-1}(f(x)) = f^{-1}(2x + 1)$$

$$= \frac{(2x + 1) - 1}{2}$$

$$= \frac{2x}{2} = x; \qquad \text{Thus, } (f^{-1} \circ f)(x) = x.$$

$$(f \circ f^{-1})(x) = f(f^{-1}(x)) = f\left(\frac{x - 1}{2}\right)$$

$$= 2 \cdot \frac{x - 1}{2} + 1$$

$$= x - 1 + 1 = x \qquad \text{Thus, } (f \circ f^{-1})(x) = x.$$

9. Let $f(x) = 3x - 5$. Show that

$$f^{-1}(x) = \frac{x + 5}{3}.$$

 YOUR TURN

Technology Connection

To see if $y_1 = 2x + 6$ and $y_2 = \frac{1}{2}x - 3$ are inverses of each other, we can graph both functions, along with the line $y = x$, on a "squared" set of axes. It *appears* that y_1 and y_2 are inverses of each other. A more precise check is achieved by selecting the DRAWINV option of the (DRAW) menu. The resulting graph of the inverse of y_1 should coincide with y_2.

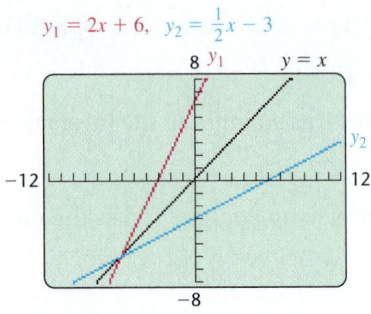

$y_1 = 2x + 6, \quad y_2 = \frac{1}{2}x - 3$

For a more dependable check, examine a TABLE in which $y_1 = 2x + 6$ and $y_2 = \frac{1}{2} \cdot y_1 - 3$. Note that y_2 "undoes" what y_1 does.

TBLSTART $= -3$ ΔTBL $= 1$ $y_2 = \frac{1}{2}y_1 - 3$

X	Y1	Y2
-3	0	-3
-2	2	-2
-1	4	-1
0	6	0
1	8	1
2	10	2
3	12	3

X = 3

1. Use a graphing calculator to check Examples 7, 8, and 9.
2. Will DRAWINV work for *any* choice of y_1? Why or why not?

12.1 EXERCISE SET

FOR EXTRA HELP

MyMathLab® MathXL
PRACTICE WATCH READ REVIEW

Vocabulary and Reading Check

Classify each of the following statements as either true or false.

1. The composition of two functions f and g is written $f \circ g$.

2. The notation $(f \circ g)(x)$ means $f(g(x))$.

3. If $f(x) = x^2$ and $g(x) = x + 3$, then $(g \circ f)(x) = (x + 3)^2$.

4. For any function h, there is only one way to decompose the function as $h = f \circ g$.

5. The function f is one-to-one if $f(1) = 1$.

6. The -1 in f^{-1} is an exponent.

7. The function f is the inverse of f^{-1}.

8. If g and h are inverses of each other, then $(g \circ h)(x) = x$.

Composite Functions

For each pair of functions, find **(a)** $(f \circ g)(1)$; **(b)** $(g \circ f)(1)$; **(c)** $(f \circ g)(x)$; **(d)** $(g \circ f)(x)$.

9. $f(x) = x^2 + 1$; $g(x) = x - 3$

10. $f(x) = x + 4$; $g(x) = x^2 - 5$

11. $f(x) = 5x + 1$; $g(x) = 2x^2 - 7$

12. $f(x) = 3x^2 + 4$; $g(x) = 4x - 1$

13. $f(x) = x + 7$; $g(x) = 1/x^2$

14. $f(x) = 1/x^2$; $g(x) = x + 2$

15. $f(x) = \sqrt{x}$; $g(x) = x + 3$

16. $f(x) = 10 - x$; $g(x) = \sqrt{x}$

17. $f(x) = \sqrt{4x}$; $g(x) = 1/x$

18. $f(x) = \sqrt{x + 3}$; $g(x) = 13/x$

19. $f(x) = x^2 + 4$; $g(x) = \sqrt{x - 1}$

20. $f(x) = x^2 + 8$; $g(x) = \sqrt{x + 17}$

Find $f(x)$ and $g(x)$ such that $h(x) = (f \circ g)(x)$. Answers may vary.

21. $h(x) = (3x - 5)^4$

22. $h(x) = (2x + 7)^3$

23. $h(x) = \sqrt{9x + 1}$

24. $h(x) = \sqrt[3]{4x - 5}$

25. $h(x) = \dfrac{6}{5x - 2}$

26. $h(x) = \dfrac{3}{x} + 4$

Inverses and One-to-One Functions

Determine whether each function is one-to-one.

27. $f(x) = -x$

28. $f(x) = x + 5$

Aha! 29. $f(x) = x^2 + 3$

30. $f(x) = 3 - x^2$

31.

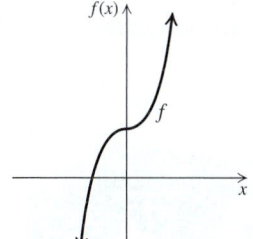

32.

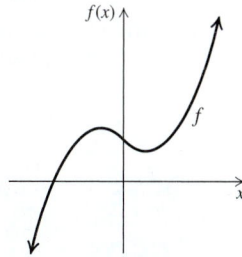

33.

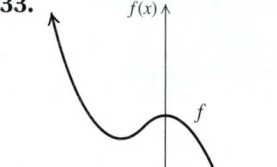

34.

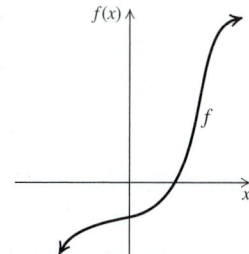

Finding Formulas for Inverses

For each function, **(a)** *determine whether it is one-to-one;* **(b)** *if it is one-to-one, find a formula for the inverse.*

35. $f(x) = x + 3$

36. $f(x) = x + 2$

37. $f(x) = 2x$

38. $f(x) = 3x$

39. $g(x) = 3x - 1$

40. $g(x) = 2x - 3$

41. $f(x) = \frac{1}{2}x + 1$

42. $f(x) = \frac{1}{3}x + 2$

43. $g(x) = x^2 + 5$

44. $g(x) = x^2 - 4$

45. $h(x) = -10 - x$

46. $h(x) = 7 - x$

Aha! 47. $f(x) = \dfrac{1}{x}$

48. $f(x) = \dfrac{4}{x}$

49. $g(x) = 1$

50. $h(x) = 8$

51. $f(x) = \dfrac{2x + 1}{3}$

52. $f(x) = \dfrac{3x + 2}{5}$

53. $f(x) = x^3 + 5$

54. $f(x) = x^3 - 4$

55. $g(x) = (x - 2)^3$

56. $g(x) = (x + 7)^3$

57. $f(x) = \sqrt{x}$

58. $f(x) = \sqrt{x - 1}$

59. *Dress Sizes in the United States and Italy.* A size-6 dress in the United States is size 36 in Italy. A function that converts dress sizes in the United States to those in Italy is

$$f(x) = 2(x + 12).$$

a) Find the dress sizes in Italy that correspond to sizes 8, 10, 14, and 18 in the United States.
b) Does f have an inverse that is a function? If so, find a formula for the inverse.
c) Use the inverse function to find dress sizes in the United States that correspond to sizes 40, 44, 52, and 60 in Italy.

60. *Dress Sizes in the United States and France.* A size-6 dress in the United States is size 38 in France. A function that converts dress sizes in the United States to those in France is

$$f(x) = x + 32.$$

a) Find the dress sizes in France that correspond to sizes 8, 10, 14, and 18 in the United States.
b) Does f have an inverse that is a function? If so, find a formula for the inverse.
c) Use the inverse function to find dress sizes in the United States that correspond to sizes 40, 42, 46, and 50 in France.

Graphing Functions and Their Inverses

Graph each function and its inverse using the same set of axes.

61. $f(x) = \frac{2}{3}x + 4$ **62.** $g(x) = \frac{1}{4}x + 2$

63. $f(x) = x^3 + 1$ **64.** $f(x) = x^3 - 1$

65. $g(x) = \frac{1}{2}x^3$ **66.** $g(x) = \frac{1}{3}x^3$

67. $F(x) = -\sqrt{x}$ **68.** $f(x) = \sqrt{x}$

69. $f(x) = -x^2, x \geq 0$

70. $f(x) = x^2 - 1, x \leq 0$

Inverse Functions and Composition

71. Let $f(x) = \sqrt[3]{x} - 4$. Show that
$$f^{-1}(x) = x^3 + 4.$$

72. Let $f(x) = 3/(x + 2)$. Show that
$$f^{-1}(x) = \frac{3}{x} - 2.$$

73. Let $f(x) = (1 - x)/x$. Show that
$$f^{-1}(x) = \frac{1}{x + 1}.$$

74. Let $f(x) = x^3 - 5$. Show that
$$f^{-1}(x) = \sqrt[3]{x + 5}.$$

75. Is there a one-to-one relationship between items in a store and the price of each of those items? Why or why not?

76. Mathematicians usually try to select "logical" words when forming definitions. Does the term "one-to-one" seem logical? Why or why not?

Skill Review

Simplify.

77. $t^{1/5}t^{2/3}$ [10.2]   **78.** $\sqrt[3]{40a^5b^{12}}$ [10.3]

79. $(-3x^{-6}y^4)^{-2}$ [4.2]   **80.** i^{43} [10.8]

81. $3^3 + 2^2 - (32 \div 4 - 16 \div 8)$ [1.8]

82. $(1.5 \times 10^{-3})(4.2 \times 10^{-12})$ [4.2]

Synthesis

83. The function $V(t) = 750(1.2)^t$ is used to predict the value $V(t)$ of a certain rare stamp t years from 2008. Do not calculate $V^{-1}(t)$, but explain how V^{-1} could be used.

84. An organization determines that the cost per person $C(x)$, in dollars, of chartering a bus with x passengers is given by

$$C(x) = \frac{100 + 5x}{x}.$$

Determine $C^{-1}(x)$ and explain how this inverse function could be used.

For Exercises 85 and 86, graph the inverse of f.

85. 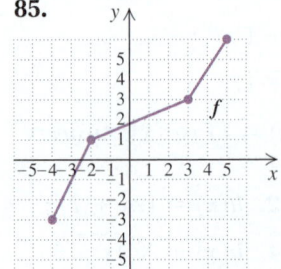   **86.**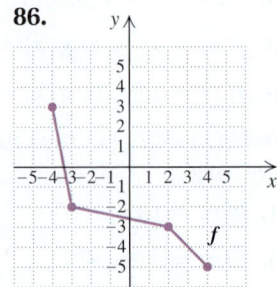

87. *Dress Sizes in France and Italy.* Use the information in Exercises 59 and 60 to find a function for the French dress size that corresponds to a size x dress in Italy.

88. *Dress Sizes in Italy and France.* Use the information in Exercises 59 and 60 to find a function for the Italian dress size that corresponds to a size x dress in France.

89. What relationship exists between the answers to Exercises 87 and 88? Explain how you determined this.

90. Show that function composition is associative by showing that $((f \circ g) \circ h)(x) = (f \circ (g \circ h))(x)$.

91. Show that if $h(x) = (f \circ g)(x)$, then $h^{-1}(x) = (g^{-1} \circ f^{-1})(x)$. (*Hint*: Use Exercise 90.)

Determine whether or not the given pairs of functions are inverses of each other.

92. $f(x) = 0.75x^2 + 2;\ g(x) = \sqrt{\dfrac{4(x-2)}{3}}$

93. $f(x) = 1.4x^3 + 3.2;\ g(x) = \sqrt[3]{\dfrac{x - 3.2}{1.4}}$

94. $f(x) = \sqrt{2.5x + 9.25};$
$g(x) = 0.4x^2 - 3.7, x \geq 0$

95. $f(x) = 0.8x^{1/2} + 5.23;$
$g(x) = 1.25(x^2 - 5.23), x \geq 0$

96. $f(x) = 2.5(x^3 - 7.1);$
$g(x) = \sqrt[3]{0.4x + 7.1}$

97. Match each function in Column A with its inverse from Column B.

Column A

(1) $y = 5x^3 + 10$

(2) $y = (5x + 10)^3$

(3) $y = 5(x + 10)^3$

(4) $y = (5x)^3 + 10$

Column B

A. $y = \dfrac{\sqrt[3]{x} - 10}{5}$

B. $y = \sqrt[3]{\dfrac{x}{5}} - 10$

C. $y = \sqrt[3]{\dfrac{x - 10}{5}}$

D. $y = \dfrac{\sqrt[3]{x - 10}}{5}$

98. Examine the following table. Is it possible that f and g are inverses of each other? Why or why not?

x	$f(x)$	$g(x)$
6	6	6
7	6.5	8
8	7	10
9	7.5	12
10	8	14
11	8.5	16
12	9	18

99. The following window appears on a graphing calculator.

X	Y₁	Y₂
0	1	−2
1	1.5	0
2	2	2
3	2.5	4
4	3	6
5	3.5	8
6	4	10
X = 0		

a) What evidence is there that the functions Y₁ and Y₂ are inverses of each other?

b) Find equations for Y₁ and Y₂, assuming that both are linear functions.

c) On the basis of your answer to part (b), are Y₁ and Y₂ inverses of each other?

YOUR TURN ANSWERS: SECTION 12.1

1. $(f \circ g)(x) = 25x^2 - 2$ **2.** $(g \circ f)(x) = \sqrt[3]{2x - 7}$
3. $f(x) = \sqrt{x}, g(x) = 3 - x$ **4.** Yes **5.** No
6. f is one-to-one; $f^{-1}(x) = x - 2$
7.

8. (a) Yes;
(b) $f^{-1}(x) = x^3 - 1$;
(c)

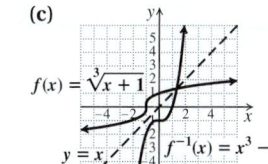

9. $(f^{-1} \circ f)(x) = f^{-1}(f(x)) = f^{-1}(3x - 5)$
$= \dfrac{(3x - 5) + 5}{3} = \dfrac{3x}{3} = x;$
$(f \circ f^{-1})(x) = f(f^{-1}(x)) = f\left(\dfrac{x + 5}{3}\right) = 3\left(\dfrac{x + 5}{3}\right) - 5$
$= (x + 5) - 5 = x$

PREPARE TO MOVE ON

Simplify.

1. 2^{-3} [4.2]

2. $5^{(1-3)}$ [4.2]

3. $4^{5/2}$ [10.2]

4. $3^{7/10}$ [10.2]

Graph. [3.2]

5. $y = x^3$

6. $x = y^3$

12.2   Exponential Functions

Graphing Exponential Functions   ▪   Equations with x and y Interchanged   ▪   Applications of Exponential Functions

In this section, we introduce a new type of function, the *exponential function*. These functions and their inverses, called *logarithmic functions*, have applications in many fields.

Consider the graph below. The rapidly rising curve approximates the graph of an *exponential function*.

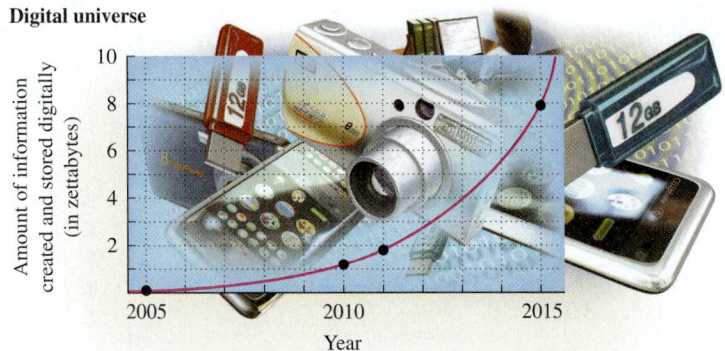

Digital universe

Source: Based on data from IDC and EMC Corporation, in "Digital Universe to Add 1.8 Zettabytes in 2011," Rich Miller, June 28, 2011, datacenterknowledge.com

Graphing Exponential Functions

In Chapter 10, we studied exponential expressions with rational-number exponents. For example, $5^{1.73}$, or $5^{173/100}$, represents the 100th root of 5 raised to the 173rd power. What about expressions with irrational exponents, such as $5^{\sqrt{3}}$ or $7^{-\pi}$? To attach meaning to $5^{\sqrt{3}}$, consider a rational approximation, r, of $\sqrt{3}$. As r gets closer to $\sqrt{3}$, the value of 5^r gets closer to some real number p.

r closes in on $\sqrt{3}$.	5^r closes in on some real number p.
$1.7 < r < 1.8$	$15.426 \approx 5^{1.7} < p < 5^{1.8} \approx 18.119$
$1.73 < r < 1.74$	$16.189 \approx 5^{1.73} < p < 5^{1.74} \approx 16.452$
$1.732 < r < 1.733$	$16.241 \approx 5^{1.732} < p < 5^{1.733} \approx 16.267$

We define $5^{\sqrt{3}}$ to be the number p. To eight decimal places,

$$5^{\sqrt{3}} \approx 16.24245082.$$

Any positive irrational exponent can be interpreted in a similar way. Negative irrational exponents are then defined using reciprocals. Thus, so long as a is positive, a^x has meaning for *any* real number x, and all the laws of exponents still hold. We can now define an *exponential function*.

> **EXPONENTIAL FUNCTION**
>
> The function $f(x) = a^x$, where a is a positive constant, $a \neq 1$, and x is any real number, is called the *exponential function*, base a.

We require the base a to be positive to avoid imaginary numbers that would result from taking even roots of negative numbers. The restriction $a \neq 1$ is made to exclude the constant function $f(x) = 1^x$, or $f(x) = 1$.

The following are examples of exponential functions:

$$f(x) = 2^x, \qquad f(x) = \left(\tfrac{1}{3}\right)^x, \qquad f(x) = 5^{-7x}. \qquad \text{Note that } 5^{-7x} = (5^{-7})^x.$$

Like polynomial functions, the domain of an exponential function is the set of all real numbers. Unlike polynomial functions, exponential functions have a variable exponent. Because of this, graphs of exponential functions either rise or fall dramatically.

EXAMPLE 1   Graph the exponential function given by $y = f(x) = 2^x$.

SOLUTION   We compute some function values, thinking of y as $f(x)$, and list the results in a table. It is helpful to start by letting $x = 0$. This gives us the y-intercept.

x	y, or $f(x)$
0	1
1	2
2	4
3	8
-1	$\frac{1}{2}$
-2	$\frac{1}{4}$
-3	$\frac{1}{8}$

$$f(0) = 2^0 = 1; \qquad f(-1) = 2^{-1} = \frac{1}{2^1} = \frac{1}{2};$$
$$f(1) = 2^1 = 2;$$
$$f(2) = 2^2 = 4; \qquad f(-2) = 2^{-2} = \frac{1}{2^2} = \frac{1}{4};$$
$$f(3) = 2^3 = 8;$$
$$f(-3) = 2^{-3} = \frac{1}{2^3} = \frac{1}{8}$$

Next, we plot these points and connect them with a smooth curve.

The curve comes very close to the x-axis, but does not touch or cross it.

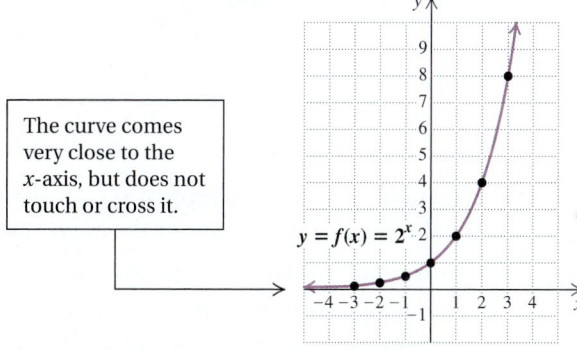

Be sure to plot enough points to determine how steeply the curve rises.

$y = f(x) = 2^x$

Note that as x increases, the function values increase without bound. As x decreases, the function values decrease, getting closer to 0. The x-axis, or the line $y = 0$, is a horizontal *asymptote*, meaning that the curve gets closer and closer to this line the further we move to the left.

1. Graph: $y = f(x) = 10^x$.

YOUR TURN

EXAMPLE 2   Graph: $y = f(x) = \left(\tfrac{1}{2}\right)^x$.

SOLUTION   We compute some function values, thinking of y as $f(x)$, and list the results in a table. Before we do this, note that

$$y = f(x) = \left(\tfrac{1}{2}\right)^x = (2^{-1})^x = 2^{-x}.$$

Then we have

x	y, or $f(x)$
0	1
1	$\frac{1}{2}$
2	$\frac{1}{4}$
3	$\frac{1}{8}$
-1	2
-2	4
-3	8

$$f(0) = 2^{-0} = 1; \qquad f(3) = 2^{-3} = \frac{1}{2^3} = \frac{1}{8};$$
$$f(1) = 2^{-1} = \frac{1}{2^1} = \frac{1}{2}; \qquad f(-1) = 2^{-(-1)} = 2^1 = 2;$$
$$f(-2) = 2^{-(-2)} = 2^2 = 4;$$
$$f(2) = 2^{-2} = \frac{1}{2^2} = \frac{1}{4}; \qquad f(-3) = 2^{-(-3)} = 2^3 = 8.$$

Next, we plot these points and connect them with a smooth curve. This curve is a mirror image, or *reflection*, of the graph of $y = 2^x$ (see Example 1) across the *y*-axis. The line $y = 0$ is again the horizontal asymptote.

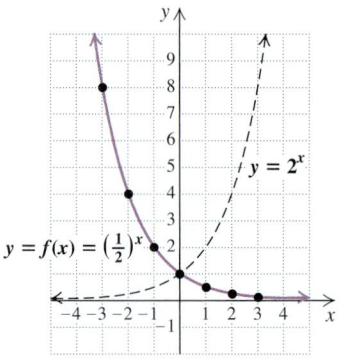

2. Graph: $y = f(x) = \left(\frac{1}{10}\right)^x$.

YOUR TURN

From Examples 1 and 2, we can make the following observations.

- For $a > 1$, the graph of $f(x) = a^x$ increases from left to right. The greater the value of a, the steeper the curve. (See the figure on the left below.)

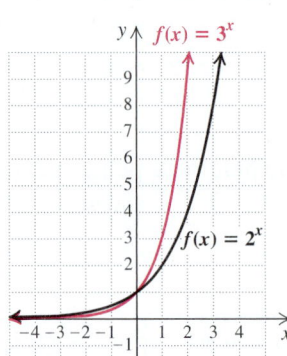
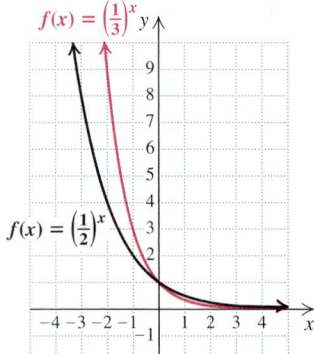

- For $0 < a < 1$, the graph of $f(x) = a^x$ decreases from left to right. The smaller the value of a, the steeper the curve. (See the figure on the right above.)
- All graphs of $f(x) = a^x$ go through the *y*-intercept $(0, 1)$.
- All graphs of $f(x) = a^x$ have the *x*-axis as the horizontal asymptote.
- If $f(x) = a^x$, with $a > 0$ and $a \ne 1$, the domain of f is all real numbers, and the range of f is all positive real numbers.
- For $a > 0$ and $a \ne 1$, the function given by $f(x) = a^x$ is one-to-one. Its graph passes the horizontal-line test.

EXAMPLE 3   Graph: $y = f(x) = 2^{x-2}$.

SOLUTION   We construct a table of values. Then we plot the points and connect them with a smooth curve. Here $x - 2$ is the *exponent*.

$$f(0) = 2^{0-2} = 2^{-2} = \tfrac{1}{4};$$
$$f(1) = 2^{1-2} = 2^{-1} = \tfrac{1}{2};$$
$$f(2) = 2^{2-2} = 2^{0} = 1;$$
$$f(3) = 2^{3-2} = 2^{1} = 2;$$
$$f(4) = 2^{4-2} = 2^{2} = 4;$$
$$f(-1) = 2^{-1-2} = 2^{-3} = \tfrac{1}{8};$$
$$f(-2) = 2^{-2-2} = 2^{-4} = \tfrac{1}{16}$$

x	y, or $f(x)$
0	$\frac{1}{4}$
1	$\frac{1}{2}$
2	1
3	2
4	4
−1	$\frac{1}{8}$
−2	$\frac{1}{16}$

Technology Connection

Graphing calculators are helpful when graphing equations like $y = 5000(1.075)^x$. To set the window, note that *y*-values are positive and increase rapidly. One suitable window is $[-10, 10, 0, 15000]$, with a *y*-scale of 1000.

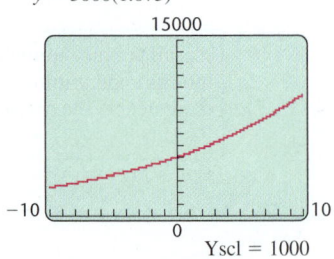

$y = 5000(1.075)^x$

Graph each pair of functions. Select an appropriate window and scale.

1. $y_1 = \left(\frac{5}{2}\right)^x$ and $y_2 = \left(\frac{2}{5}\right)^x$
2. $y_1 = 3.2^x$ and $y_2 = 3.2^{-x}$
3. $y_1 = \left(\frac{3}{7}\right)^x$ and $y_2 = \left(\frac{7}{3}\right)^x$
4. $y_1 = 5000(1.08)^x$ and $y_2 = 5000(1.08)^{x-3}$

Student Notes

When using translations, make sure that you are shifting in the correct direction. When in doubt, substitute a value for *x* and make some calculations.

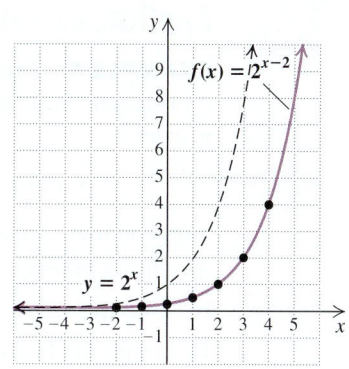

The graph looks just like the graph of $y = 2^x$, but translated 2 units to the right. The y-intercept of $y = 2^x$ is $(0, 1)$. The y-intercept of $y = 2^{x-2}$ is $\left(0, \frac{1}{4}\right)$. The line $y = 0$ is again the horizontal asymptote.

3. Graph: $y = f(x) = 2^{x+2}$.

_____

YOUR TURN

Equations with x and y Interchanged

It will be helpful in later work to be able to graph an equation in which the x and the y in $y = a^x$ are interchanged.

EXAMPLE 4   Graph: $x = 2^y$.

SOLUTION   Note that x is alone on one side of the equation. To find ordered pairs that are solutions, we choose values for y and then compute values for x.

For $y = 0$, $x = 2^0 = 1$.

For $y = 1$, $x = 2^1 = 2$.

For $y = 2$, $x = 2^2 = 4$.

For $y = 3$, $x = 2^3 = 8$.

For $y = -1$, $x = 2^{-1} = \frac{1}{2}$.

For $y = -2$, $x = 2^{-2} = \frac{1}{4}$.

For $y = -3$, $x = 2^{-3} = \frac{1}{8}$.

x	y
1	0
2	1
4	2
8	3
$\frac{1}{2}$	-1
$\frac{1}{4}$	-2
$\frac{1}{8}$	-3

(1) Choose values for y.
(2) Compute values for x.

We plot the points and connect them with a smooth curve.

This curve does not touch or cross the y-axis, which serves as a vertical asymptote.

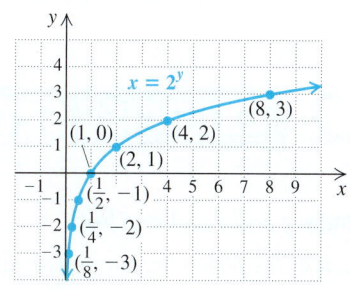

Note too that this curve looks just like the graph of $y = 2^x$, except that it is reflected across the line $y = x$, as shown at left.

Technology Connection

To practice graphing equations that are translations of each other, use **MODE** SIMUL and **FORMAT** EXPROFF to graph $y_1 = 2^x$, $y_2 = 2^{x+1}$, $y_3 = 2^{x-1}$, $y_4 = 2^x + 1$, and $y_5 = 2^x - 1$. Use a bold curve for y_1 and then predict which curve represents which equation. Use **TRACE** to confirm your predictions. Switching **FORMAT** to EXPRON and using **TRACE** provides a definitive check (see also Exercise 75).

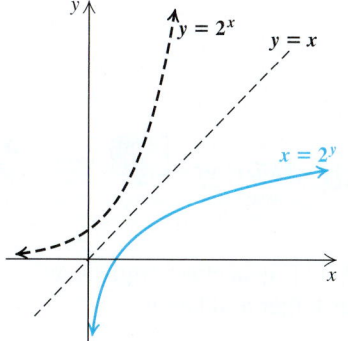

4. Graph: $x = 10^y$.

YOUR TURN

Applications of Exponential Functions

EXAMPLE 5 *Interest Compounded Annually.* The amount of money A that a principal P will be worth after t years at interest rate r, compounded annually, is given by

$$A = P(1 + r)^t.$$

Suppose that \$100,000 is invested at 5% interest, compounded annually.

a) Find a function for the amount in the account after t years.

b) Find the amount of money in the account at $t = 0$, $t = 4$, $t = 8$, and $t = 10$.

c) Graph the function.

SOLUTION

a) If $P = \$100{,}000$ and $r = 5\% = 0.05$, we can substitute these values and form the following function:

$$A(t) = \$100{,}000(1 + 0.05)^t \qquad \text{Using } A = P(1 + r)^t$$
$$= \$100{,}000(1.05)^t.$$

b) To find the function values, a calculator with a power key is helpful.

$$A(0) = \$100{,}000(1.05)^0 \qquad\qquad A(8) = \$100{,}000(1.05)^8$$
$$= \$100{,}000(1) \qquad\qquad\qquad \approx \$100{,}000(1.477455444)$$
$$= \$100{,}000; \qquad\qquad\qquad\quad \approx \$147{,}745.54;$$

$$A(4) = \$100{,}000(1.05)^4 \qquad\qquad A(10) = \$100{,}000(1.05)^{10}$$
$$= \$100{,}000(1.21550625) \qquad\quad \approx \$100{,}000(1.628894627)$$
$$\approx \$121{,}550.63; \qquad\qquad\qquad \approx \$162{,}889.46$$

c) We use the function values computed in part (b), and others if we wish, to draw the graph as follows. Note that the axes are scaled differently because of the large numbers.

$A(t) = \$100,000(1.05)^t$

Technology Connection

Graphing calculators can quickly find many function values. Let $y_1 = 100{,}000(1.05)^x$. Then use the TABLE feature with INDPNT set to ASK to check Example 5(b).

 Chapter Resource:
Collaborative Activity, p. 840

5. Refer to Example 5. Suppose that \$20,000 is invested at 4% interest, compounded annually.

a) Find a function for the amount in the account after t years.

b) Find the amount of money in the account at $t = 0$ and $t = 10$.

YOUR TURN

12.2 EXERCISE SET

FOR EXTRA HELP

MyMathLab® *Math XL*
PRACTICE   WATCH   READ   REVIEW

Vocabulary and Reading Check

Classify each of the following statements as either true or false.

1. The graph of $f(x) = a^x$ always passes through the point $(0, 1)$.

2. The graph of $g(x) = \left(\frac{1}{2}\right)^x$ gets closer and closer to the x-axis as x gets larger and larger.

3. The graph of $f(x) = 2^{x-3}$ looks just like the graph of $y = 2^x$, but it is translated 3 units to the right.

4. The graph of $g(x) = 2^x - 3$ looks just like the graph of $y = 2^x$, but it is translated 3 units up.

5. The graph of $y = 3^x$ gets close to, but never touches, the y-axis.

6. The graph of $x = 3^y$ gets close to, but never touches, the y-axis.

Graphing Exponential Functions

Graph.

7. $y = f(x) = 3^x$ **8.** $y = f(x) = 4^x$

9. $y = 6^x$ **10.** $y = 5^x$

11. $y = 2^x + 1$ **12.** $y = 2^x + 3$

13. $y = 3^x - 2$ **14.** $y = 3^x - 1$

15. $y = 2^x - 5$ **16.** $y = 2^x - 4$

17. $y = 2^{x-3}$ **18.** $y = 2^{x-1}$

19. $y = 2^{x+1}$ **20.** $y = 2^{x+3}$

21. $y = \left(\frac{1}{4}\right)^x$ **22.** $y = \left(\frac{1}{5}\right)^x$

23. $y = \left(\frac{1}{3}\right)^x$ **24.** $y = \left(\frac{1}{6}\right)^x$

25. $y = 2^{x+1} - 3$ **26.** $y = 2^{x-3} - 1$

Equations with x and y Interchanged

Graph.

27. $x = 6^y$ **28.** $x = 3^y$

29. $x = 3^{-y}$ **30.** $x = 2^{-y}$

31. $x = 4^y$ **32.** $x = 5^y$

33. $x = \left(\frac{4}{3}\right)^y$ **34.** $x = \left(\frac{3}{2}\right)^y$

Graph each pair of equations on the same set of axes.

35. $y = 3^x,\ x = 3^y$ **36.** $y = 2^x,\ x = 2^y$

37. $y = \left(\frac{1}{2}\right)^x,\ x = \left(\frac{1}{2}\right)^y$ **38.** $y = \left(\frac{1}{4}\right)^x,\ x = \left(\frac{1}{4}\right)^y$

Applications of Exponential Functions

Solve.

39. *Digital Music.* Digital music sales, in billions of dollars, t years after 2010 can be approximated by
$$M(t) = 3.2(1.23)^t.$$
Source: Forrester Research

a) Estimate digital music sales in 2010, in 2012, and in 2014.

b) Graph the function.

40. *Growth of Bacteria.* The bacteria *Escherichia coli* are commonly found in the human bladder. Suppose that 3000 bacteria are present at time

$t = 0$. Then t minutes later, the number of bacteria present can be approximated by
$$N(t) = 3000(2)^{t/20}.$$

a) How many bacteria will be present after 10 min? 20 min? 30 min? 40 min? 60 min?

b) Graph the function.

41. *Smoking Cessation.* The percentage of smokers P who receive telephone counseling to quit smoking and are still successful t months later can be approximated by
$$P(t) = 21.4(0.914)^t.$$
Sources: New England Journal of Medicine; data from California's Smokers' Hotline

a) Estimate the percentage of smokers receiving telephone counseling who are successful in quitting for 1 month, 3 months, and 1 year.

b) Graph the function.

42. *Smoking Cessation.* The percentage of smokers P who, without telephone counseling, have successfully quit smoking for t months (see Exercise 41) can be approximated by
$$P(t) = 9.02(0.93)^t.$$
Sources: New England Journal of Medicine; data from California's Smokers' Hotline

a) Estimate the percentage of smokers not receiving telephone counseling who are successful in quitting for 1 month, 3 months, and 1 year.

b) Graph the function.

43. *Marine Biology.* Due to excessive whaling prior to the mid-1970s, the humpback whale is considered an endangered species. The worldwide population of humpbacks, $P(t)$, in thousands, t years after 1900 ($t < 70$) can be approximated by
$$P(t) = 150(0.960)^t.$$
Sources: Based on information from the American Cetacean Society and the ASK Archive

a) How many humpback whales were alive in 1930? in 1960?

b) Graph the function.

44. *Salvage Value.* A laser printer is purchased for $1200. Its value each year is about 80% of the value of the preceding year. Its value, in dollars, after t years is given by the exponential function
$$V(t) = 1200(0.8)^t.$$

a) Find the value of the printer after 0 year, 1 year, 2 years, 5 years, and 10 years.

b) Graph the function.

45. *Marine Biology.* As a result of preservation efforts in most countries in which whaling was common, the humpback whale population has grown since the 1970s. The worldwide population of humpbacks, $P(t)$, in thousands, t years after 1982 can be approximated by

$$P(t) = 5.5(1.047)^t.$$

Sources: Based on information from the American Cetacean Society and the ASK Archive

a) How many humpback whales were alive in 1992? in 2004?

b) Graph the function.

46. *Recycling Aluminum Cans.* About three-fifths of all aluminum cans distributed will be recycled each year. A beverage company distributes 250,000 cans. The number still in use after time t, in years, is given by the exponential function

$$N(t) = 250,000\left(\tfrac{3}{5}\right)^t.$$

Source: The Aluminum Association, Inc., 2011

a) The aluminum from how many cans is still in use after 0 year? 1 year? 4 years? 10 years?

b) Graph the function.

47. *Animal Population.* The moose population in New York is growing exponentially. The number of moose in the state t years after 1997 can be approximated by

$$M(t) = 50(1.25)^t.$$

Source: Based on data from the New York State Department of Environmental Conservation

a) Estimate the number of moose in New York in 1997, in 2012, and in 2020.

b) Graph the function.

48. *Vehicle Sales.* The number of cars and light trucks sold in China is increasing exponentially and can be estimated by

$$N(t) = 1.09(1.3)^t,$$

where $N(t)$ is in millions of vehicles and t is the number of years after 2000.

Source: Based on data from Wards Auto

a) Estimate the number of cars and light trucks sold in China in 2000, in 2010, and in 2014.

b) Graph the function.

49. Without using a calculator, explain why 2^π must be greater than 8 but less than 16.

50. Suppose that $1000 is invested for 5 years at 7% interest, compounded annually. In what year will the most interest be earned? Why?

Skill Review

Factor.

51. $3x^2 - 48$ [5.4]

52. $x^2 - 20x + 100$ [5.4]

53. $6x^2 + x - 12$ [5.3]

54. $8x^6 - 64y^6$ [5.5]

55. $t^2 - y^2 + 2y - 1$ [5.4]

56. $5x^4 - 10x^3 - 3x^2 + 6x$ [5.1]

Synthesis

57. Examine Exercise 48. Do you believe that the equation for the number of cars and light trucks sold in China will be accurate 20 years from now? Why or why not?

58. Explain why the graph of $x = 2^y$ is the graph of $y = 2^x$ reflected across the line $y = x$.

Determine which of the two numbers is larger. Do not use a calculator.

59. $\pi^{1.3}$ or $\pi^{2.4}$

60. $\sqrt{8^3}$ or $8^{\sqrt{3}}$

Graph.

61. $f(x) = 2.5^x$

62. $f(x) = 0.5^x$

63. $y = 2^x + 2^{-x}$

64. $y = \left|\left(\tfrac{1}{2}\right)^x - 1\right|$

65. $y = |2^x - 2|$

66. $y = 2^{-(x-1)^2}$

67. $y = |2^{x^2} - 1|$

68. $y = 3^x + 3^{-x}$

Graph both equations using the same set of axes.

69. $y = 3^{-(x-1)}, \ x = 3^{-(y-1)}$

70. $y = 1^x, \ x = 1^y$

71. *Invasive Species.* Ruffe is a species of freshwater fish that is considered invasive where it is not native. In 1984, there were about 100 ruffe in Loch Lomond, Scotland. By 1988, there were about 3000 ruffe in the lake, and by 1992 there were about 14,000 ruffe. After pressing **STAT** and entering the data, use the EXP REG option in the STAT CALC menu to find an exponential function that models the number of ruffe in Loch Lomond t years after 1984. Then use that function to estimate the number of ruffe in the lake in 1990.

Source: Drake, John M., "Risk Analysis for Species Introductions: Forecasting Population Growth of Eurasian Ruffe (*Gymnocephalus cernus*)," 2005. Retrieved from dragonfly.ecology.uga.edu

 72. *Keyboarding Speed.* Trey is studying keyboarding. After he has studied for t hours, Trey's speed, in words per minute, is given by the exponential function

$$S(t) = 200[1 - (0.99)^t].$$

Use a graph and/or table of values to predict Trey's speed after studying for 10 hr, 40 hr, and 80 hr.

73. The following graph shows growth in the height of ocean waves over time, assuming a steady surface wind.

Source: magicseaweed.com

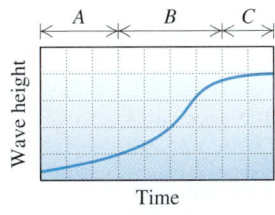

Source: magicseaweed.com

a) Consider the portions of the graph marked A, B, and C. Suppose that each portion can be labeled Exponential Growth, Linear Growth, or Saturation. How would you label each portion?

b) Small vertical movements in wind, surface roughness of water, and gravity are three forces that create waves. How might these forces be related to the shape of the wave-height graph?

 74. Consider any exponential function of the form $f(x) = a^x$ with $a > 1$. Will it always follow that $f(3) - f(2) > f(2) - f(1)$, and, in general, $f(n + 2) - f(n + 1) > f(n + 1) - f(n)$? Why or why not? (*Hint:* Think graphically.)

 75. On many graphing calculators, it is possible to enter and graph $y_1 = A \wedge (X - B) + C$ after first pressing **APPS** Transfrm. Use this application to graph $f(x) = 2.5^{x-3} + 2, g(x) = 2.5^{x+3} + 2,$ $h(x) = 2.5^{x-3} - 2,$ and $k(x) = 2.5^{x+3} - 2.$

 76. *Research.* Ruffe (see Exercise 71) have been introduced into the Great Lakes and several rivers in the United States. What impact does this species now have on fishing and the environment?

 YOUR TURN ANSWERS: SECTION 12.2

1. 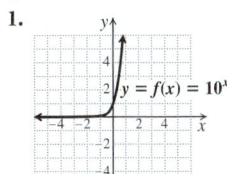   **2.**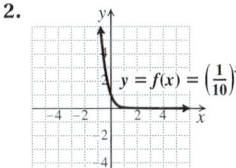

3. 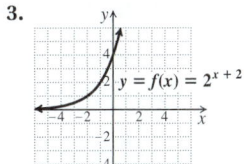   **4.**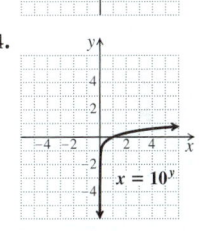

5. **(a)** $A(t) = 20,000(1.04)^t$; **(b)** \$20,000; \$29,604.89

QUICK QUIZ: SECTIONS 12.1–12.2

1. Find $(f \circ g)(x)$ for $f(x) = 5 - x$ and $g(x) = 3x^2 - x - 4.$ [12.1]

2. Find $f(x)$ and $g(x)$ such that $(f \circ g)(x) = h(x)$ and $h(x) = (x - 6)^4.$ Answers may vary. [12.1]

3. Find a formula for the inverse of $g(x) = 2x + 5.$ [12.1]

Graph. [12.2]

4. $y = 2^x - 3$ **5.** $x = 2^y$

PREPARE TO MOVE ON

Graph.

1. $f(x) = \sqrt{x} - 3$ [10.1]   **2.** $g(x) = \sqrt[3]{x} + 1$ [10.1]

3. $g(x) = x^3 + 2$ [7.1]   **4.** $f(x) = 1 - x^2$ [11.7]

12.3   Logarithmic Functions

The Meaning of Logarithms • Graphs of Logarithmic Functions • Equivalent Equations •
Solving Certain Logarithmic Equations

We are now ready to study inverses of exponential functions. These functions have many applications and are called *logarithm*, or *logarithmic, functions.*

The Meaning of Logarithms

Consider the exponential function $f(x) = 2^x$. Like all exponential functions, f is one-to-one. Let's attempt to find a formula for $f^{-1}(x)$:

1. Replace $f(x)$ with y: $y = 2^x$.
2. Interchange x and y: $x = 2^y$.
3. Solve for y: y = the exponent to which we raise 2 to get x.
4. Replace y with $f^{-1}(x)$: $f^{-1}(x)$ = the exponent to which we raise 2 to get x.

We now define a new symbol to replace the words "the exponent to which we raise 2 to get x":

> $\log_2 x$, read "the logarithm, base 2, of x," or "log, base 2, of x," means "the exponent to which we raise 2 to get x."

Thus if $f(x) = 2^x$, then $f^{-1}(x) = \log_2 x$. Note that $f^{-1}(8) = \log_2 8 = 3$, because 3 is *the exponent to which we raise 2 to get* 8.

EXAMPLE 1   Simplify: **(a)** $\log_2 32$; **(b)** $\log_2 1$; **(c)** $\log_2 \frac{1}{8}$.

SOLUTION

a) Think of $\log_2 32$ as the exponent to which we raise 2 to get 32. That exponent is 5. Therefore, $\log_2 32 = 5$.

b) We ask ourselves: "To what exponent do we raise 2 in order to get 1?" That exponent is 0 (recall that $2^0 = 1$). Thus, $\log_2 1 = 0$.

c) To what exponent do we raise 2 in order to get $\frac{1}{8}$? Since $2^{-3} = \frac{1}{8}$, we have $\log_2 \frac{1}{8} = -3$.

1. Simplify: $\log_2 16$.

YOUR TURN

Although numbers like $\log_2 13$ can be only approximated, we must remember that $\log_2 13$ represents *the exponent to which we raise* 2 *to get* 13. That is, $2^{\log_2 13} = 13$. A calculator indicates that $\log_2 13 \approx 3.7$ and $2^{3.7} \approx 13$.

For any exponential function $f(x) = a^x$, the inverse is called a **logarithmic function, base a**, and is written $f^{-1}(x) = \log_a x$. The graph of the inverse can be drawn by reflecting the graph of $f(x) = a^x$ across the line $y = x$.

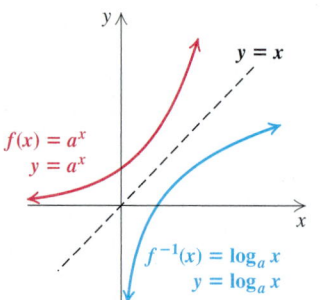

> **THE MEANING OF $\log_a x$**
>
> For $x > 0$ and a a positive constant other than 1, $\log_a x$ is the exponent to which a must be raised in order to get x. Thus,
>
> $$\log_a x = m \quad \text{means} \quad a^m = x$$
>
> or equivalently,
>
> $$\log_a x \text{ is that unique exponent for which } a^{\log_a x} = x.$$

It is important to remember that *a logarithm is an exponent*. It might help to verbalize: "The logarithm, base a, of a number x is the exponent to which a must be raised in order to get x."

EXAMPLE 2 Simplify: $7^{\log_7 85}$.

SOLUTION Remember that $\log_7 85$ is the exponent to which 7 is raised to get 85. Raising 7 to that exponent, we have

$$7^{\log_7 85} = 85.$$

2. Simplify: $5^{\log_5 14}$.

YOUR TURN

Because logarithmic functions and exponential functions are inverses of each other, the result in Example 2 should come as no surprise: If $f(x) = \log_7 x$, then

$$\text{for} \quad f(x) = \log_7 x, \text{ we have } f^{-1}(x) = 7^x$$
$$\text{and} \quad f^{-1}(f(x)) = f^{-1}(\log_7 x) = 7^{\log_7 x} = x.$$

Thus, $f^{-1}(f(85)) = 7^{\log_7 85} = 85$.

The following is a comparison of exponential functions and logarithmic functions.

Exponential Function	Logarithmic Function
$y = a^x$	$x = a^y$
$f(x) = a^x$	$g(x) = \log_a x$
$a > 0, a \neq 1$	$a > 0, a \neq 1$
The domain is $\mathbb{R}$.	The range is $\mathbb{R}$.
$y > 0$ (Outputs are positive.)	$x > 0$ (Inputs are positive.)
$f^{-1}(x) = \log_a x$	$g^{-1}(x) = a^x$

Graphs of Logarithmic Functions

EXAMPLE 3 Graph: $y = f(x) = \log_5 x$.

SOLUTION If $y = \log_5 x$, then $5^y = x$. We can find ordered pairs that are solutions by choosing values for y and computing the x-values.

For $y = 0, x = 5^0 = 1$.

For $y = 1, x = 5^1 = 5$.

For $y = 2, x = 5^2 = 25$.

For $y = -1, x = 5^{-1} = \frac{1}{5}$.

For $y = -2, x = 5^{-2} = \frac{1}{25}$.

(1) Select y.

(2) Compute x.

x, or 5^y	y
1	0
5	1
25	2
$\frac{1}{5}$	-1
$\frac{1}{25}$	-2

This table shows the following:

$\log_5 1 = 0;$

$\log_5 5 = 1;$

$\log_5 25 = 2;$

$\log_5 \frac{1}{5} = -1;$

$\log_5 \frac{1}{25} = -2.$

These can all be checked using the equations above.

Technology Connection

To see that $f(x) = 10^x$ and $g(x) = \log_{10} x$ are inverses of each other, let $y_1 = 10^x$ and $y_2 = \log_{10} x = \log x$. Then, using a squared window, compare both graphs. If possible, select DrawInv from the (DRAW) menu and then press (VARS) (▷) (1) (1) (ENTER) to see another representation of f^{-1}. Finally, let $y_3 = y_1(y_2)$ and $y_4 = y_2(y_1)$ to show, using a table or graphs, that, for $x > 0$, $y_3 = y_4 = x$.

We plot the set of ordered pairs and connect the points with a smooth curve. The graphs of $y = 5^x$ and $y = x$ are shown only for reference.

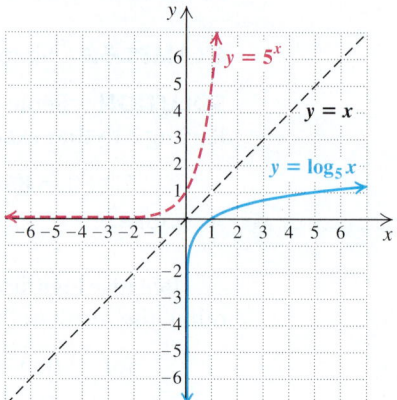

3. Graph: $y = f(x) = \log_8 x$.

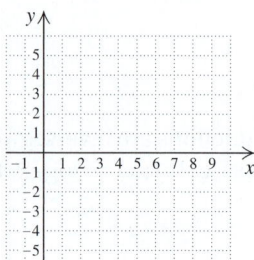

YOUR TURN

Equivalent Equations

We use the definition of logarithm to rewrite a *logarithmic equation* as an equivalent *exponential equation* or the other way around:

$$m = \log_a x \quad \text{is equivalent to} \quad a^m = x.$$

> **CAUTION!**   **Do not forget this relationship!** It may be the most important statement in the chapter. Many times this is used to justify a property we are considering.

EXAMPLE 4   Rewrite each as an equivalent exponential equation.

a) $y = \log_3 5$ **b)** $-2 = \log_a 7$ **c)** $a = \log_b d$

SOLUTION

a) $y = \log_3 5$ is equivalent to $3^y = 5$ The logarithm is the exponent.
                                                   The base remains the base.

4. Rewrite $6 = \log_3 x$ as an equivalent exponential equation.

b) $-2 = \log_a 7$ is equivalent to $a^{-2} = 7$

c) $a = \log_b d$ is equivalent to $b^a = d$

YOUR TURN

EXAMPLE 5   Rewrite each as an equivalent logarithmic equation.

a) $8 = 2^x$ **b)** $y^{-1} = 4$ **c)** $a^b = c$

SOLUTION

a) $8 = 2^x$ is equivalent to $x = \log_2 8$ The exponent is the logarithm.
                                               The base remains the base.

5. Rewrite $t^{-3} = 5$ as an equivalent logarithmic equation.

b) $y^{-1} = 4$ is equivalent to $-1 = \log_y 4$

c) $a^b = c$ is equivalent to $b = \log_a c$

YOUR TURN

Solving Certain Logarithmic Equations

Many logarithmic equations can be solved by rewriting them as equivalent exponential equations.

EXAMPLE 6   Solve: **(a)** $\log_2 x = -3$; **(b)** $\log_x 16 = 2$.

SOLUTION

a) $\log_2 x = -3$

$\qquad 2^{-3} = x$ Rewriting as an exponential equation

$\qquad \frac{1}{8} = x$ Computing 2^{-3}

Check: $\log_2 \frac{1}{8}$ is the exponent to which 2 is raised to get $\frac{1}{8}$. Since that exponent is -3, we have a check. The solution is $\frac{1}{8}$.

b) $\log_x 16 = 2$

$\qquad x^2 = 16$ Rewriting as an exponential equation

$\qquad x = 4 \;\; or \;\; x = -4$ Principle of square roots

Check: $\log_4 16 = 2$ because $4^2 = 16$. Thus, 4 is a solution of $\log_x 16 = 2$. Because all logarithmic bases must be positive, -4 cannot be a solution. Logarithmic bases must be positive because logarithms are defined using exponential functions that require positive bases. The solution is 4.

6. Solve: $\log_3 x = 2$.

⟵ YOUR TURN

One method for solving certain logarithmic and exponential equations relies on the following property, which results from the fact that exponential functions are one-to-one.

THE PRINCIPLE OF EXPONENTIAL EQUALITY

For any real number b, where $b \neq -1, 0,$ or $1,$

$\qquad b^m = b^n$ is equivalent to $m = n.$

(Powers of the same base are equal if and only if the exponents are equal.)

EXAMPLE 7   Solve: **(a)** $\log_{10} 1000 = x$; **(b)** $\log_4 1 = t$.

SOLUTION

a) We rewrite $\log_{10} 1000 = x$ in exponential form and solve:

$\qquad 10^x = 1000$ Rewriting as an exponential equation

$\qquad 10^x = 10^3$ Writing 1000 as a power of 10

$\qquad x = 3.$ Equating exponents

Check: This equation can also be solved directly by determining the exponent to which we raise 10 in order to get 1000. In both cases we find that $\log_{10} 1000 = 3$, so we have a check. The solution is 3.

b) We rewrite $\log_4 1 = t$ in exponential form and solve:

$$4^t = 1 \qquad \text{Rewriting as an exponential equation}$$
$$4^t = 4^0 \qquad \text{Writing 1 as a power of 4}$$
$$t = 0. \qquad \text{Equating exponents}$$

Check: As in part (a), this equation can be solved directly by determining the exponent to which we raise 4 in order to get 1. In both cases we find that $\log_4 1 = 0$, so we have a check. The solution is 0.

7. Solve: $\log_9 9 = x$.

 YOUR TURN

Example 7 illustrates an important property of logarithms.

Study Skills

When a Turn Is Trouble
Occasionally a page turn can interrupt your thoughts as you work through a section. You may find it helpful to rewrite (in pencil) the last equation or sentence appearing on a page at the very top of the next page.

$\log_a 1$

The logarithm, base a, of 1 is 0: $\log_a 1 = 0$.

This follows from the fact that $a^0 = 1$ is equivalent to the logarithmic equation $\log_a 1 = 0$. Thus, $\log_{10} 1 = 0$, $\log_7 1 = 0$, and so on.

Another property results from the fact that $a^1 = a$. This is equivalent to the logarithmic equation $\log_a a = 1$.

$\log_a a$

The logarithm, base a, of a is 1: $\log_a a = 1$.

Thus, $\log_{10} 10 = 1$, $\log_8 8 = 1$, and so on.

12.3 EXERCISE SET

FOR EXTRA HELP
PRACTICE   WATCH   READ   REVIEW

Vocabulary and Reading Check

Choose the word from those listed below the blank that best completes each statement.

1. The inverse of the function given by $f(x) = 3^x$ is a(n) _____ function.
exponential/logarithmic

2. A logarithm is a(n) _____.
base/exponent

3. Logarithm bases are _____.
negative/positive

4. The logarithm, base a, of a is _____.
0/1

Concept Reinforcement

In each of Exercises 5–12, match the expression or equation with an equivalent expression or equation from the column on the right.

5. ____ $\log_5 25$ **a)** 1

6. ____ $2^5 = x$ **b)** x

7. ____ $\log_5 5$ **c)** $x^5 = 2$

8. ____ $\log_2 1$ **d)** $\log_2 x = 5$

9. ____ $\log_5 5^x$ **e)** $\log_2 5 = x$

10. ____ $\log_x 2 = 5$ **f)** $5^2 = x$

11. ____ $5 = 2^x$ **g)** 2

12. ____ $\log_5 x = 2$ **h)** 0

The Meaning of Logarithms

Simplify.

13. $\log_{10} 1000$

14. $\log_{10} 100$

15. $\log_7 49$

16. $\log_3 9$

17. $\log_5 \frac{1}{25}$

18. $\log_5 \frac{1}{5}$

19. $\log_8 \frac{1}{8}$

20. $\log_8 \frac{1}{64}$

21. $\log_5 625$

22. $\log_5 125$

23. $\log_7 7$

24. $\log_9 1$

25. $\log_3 1$

26. $\log_3 3$

Aha! **27.** $\log_6 6^5$

28. $\log_6 6^9$

29. $\log_{10} 0.01$

30. $\log_{10} 0.1$

31. $\log_{16} 4$

32. $\log_{100} 10$

33. $\log_9 27$

34. $\log_4 32$

35. $\log_{1000} 100$

36. $\log_{16} 8$

37. $3^{\log_3 29}$

38. $6^{\log_6 13}$

Graphs of Logarithmic Functions

Graph.

39. $y = \log_{10} x$

40. $y = \log_2 x$

41. $y = \log_3 x$

42. $y = \log_7 x$

43. $f(x) = \log_6 x$

44. $f(x) = \log_4 x$

45. $f(x) = \log_{2.5} x$

46. $f(x) = \log_{1/2} x$

Graph each pair of functions using one set of axes.

47. $f(x) = 3^x, \ f^{-1}(x) = \log_3 x$

48. $f(x) = 4^x, \ f^{-1}(x) = \log_4 x$

Equivalent Equations

Rewrite each of the following as an equivalent exponential equation. Do not solve.

49. $x = \log_{10} 8$

50. $y = \log_8 10$

51. $\log_9 9 = 1$

52. $\log_6 36 = 2$

53. $\log_{10} 0.1 = -1$

54. $\log_{10} 0.01 = -2$

55. $\log_{10} 7 = 0.845$

56. $\log_{10} 3 = 0.4771$

57. $\log_c m = 8$

58. $\log_b n = 23$

59. $\log_r C = t$

60. $\log_m P = a$

61. $\log_e 0.25 = -1.3863$

62. $\log_e 0.989 = -0.0111$

63. $\log_r T = -x$

64. $\log_c M = -w$

Rewrite each of the following as an equivalent logarithmic equation. Do not solve.

65. $10^2 = 100$

66. $10^4 = 10,000$

67. $5^{-3} = \frac{1}{125}$

68. $2^{-5} = \frac{1}{32}$

69. $16^{1/4} = 2$

70. $8^{1/3} = 2$

71. $10^{0.4771} = 3$

72. $10^{0.3010} = 2$

73. $z^m = 6$

74. $m^n = r$

75. $p^t = q$

76. $y^t = x$

77. $e^3 = 20.0855$

78. $e^2 = 7.3891$

Solving Certain Logarithmic Equations

Solve.

79. $\log_6 x = 2$

80. $\log_4 x = 3$

81. $\log_2 32 = x$

82. $\log_5 25 = x$

83. $\log_x 9 = 1$

84. $\log_x 12 = 1$

85. $\log_x 7 = \frac{1}{2}$

86. $\log_x 9 = \frac{1}{2}$

87. $\log_3 x = -2$

88. $\log_2 x = -1$

89. $\log_{32} x = \frac{2}{5}$

90. $\log_8 x = \frac{2}{3}$

 91. Explain why a logarithm base must be positive.

92. Is it easier to find x given $x = \log_9 \frac{1}{3}$ or given $9^x = \frac{1}{3}$? Explain your reasoning.

Skill Review

Simplify.

93. $\sqrt{18a^3 b}\sqrt{50ab^7}$ [10.3]

94. $(2\sqrt{3} + \sqrt{5})(2\sqrt{3} - \sqrt{10})$ [10.5]

95. $\sqrt{192x} - \sqrt{75x}$ [10.5]

96. $\sqrt[4]{\sqrt[3]{x}}$ [10.2]

97. $\dfrac{\sqrt[3]{24xy^8}}{\sqrt[3]{3xy}}$ [10.4]

98. $\dfrac{\sqrt[5]{a^4 y^6}}{\sqrt{ay}}$ [10.5]

Synthesis

99. Would a manufacturer be pleased or unhappy if sales of a product grew logarithmically? Why?

100. Explain why the number $\log_2 13$ must be between 3 and 4.

101. Graph both equations using one set of axes:
$$y = \left(\tfrac{3}{2}\right)^x, \qquad y = \log_{3/2} x.$$

Graph.

102. $y = \log_2(x - 1)$

103. $y = \log_3 |x + 1|$

Solve.

104. $|\log_3 x| = 2$

105. $\log_4(3x - 2) = 2$

106. $\log_8(2x + 1) = -1$

107. $\log_{10}(x^2 + 21x) = 2$

Simplify.

108. $\log_{1/4} \frac{1}{64}$

109. $\log_{1/5} 25$

110. $\log_{81} 3 \cdot \log_3 81$

111. $\log_{10} (\log_4 (\log_3 81))$

112. $\log_2 (\log_2 (\log_4 256))$

113. Show that $b^{x_1} = b^{x_2}$ is *not* equivalent to $x_1 = x_2$ for $b = 0$ or $b = 1$.

 114. If $\log_b a = x$, does it follow that $\log_a b = 1/x$? Why or why not?

 YOUR TURN ANSWERS: SECTION 12.3

1. 4   **2.** 14   **3.**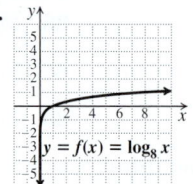
4. $3^6 = x$
5. $\log_t 5 = -3$
6. 9   **7.** 1

QUICK QUIZ: SECTIONS 12.1–12.3

1. Determine whether $f(x) = x^2 + 2$ is one-to-one. [12.1]

2. Graph $y = \left(\frac{1}{3}\right)^x$ and $x = \left(\frac{1}{3}\right)^y$ on the same set of axes. [12.2]

3. Simplify: $\log_3 81$. [12.3]

4. Rewrite as an equivalent exponential equation: $x = \log_3 t$. [12.3]

5. Solve: $\log_4 x = -2$. [12.3]

PREPARE TO MOVE ON

Use the rules for exponents to simplify each expression. [4.1]

1. $c^7 c^9$ **2.** $(x^5)^6$

3. $\dfrac{a^{15}}{a^3}$

Rewrite using rational exponents. [10.2]

4. $\sqrt{3}$ **5.** $\sqrt[3]{t^2}$

12.4   Properties of Logarithmic Functions

Logarithms of Products ▪ Logarithms of Powers ▪ Logarithms of Quotients ▪ Using the Properties Together

Logarithmic functions are important in many applications and in more advanced mathematics. We now establish some basic properties that are useful in manipulating expressions involving logarithms. As their proofs reveal, the properties of logarithms are related to the properties of exponents.

Logarithms of Products

The first property we discuss is related to the product rule for exponents: $a^m \cdot a^n = a^{m+n}$. Its proof appears immediately after Example 2.

> **THE PRODUCT RULE FOR LOGARITHMS**
>
> For any positive numbers M, N, and a ($a \neq 1$),
>
> $$\log_a (MN) = \log_a M + \log_a N.$$
>
> (The logarithm of a product is the sum of the logarithms of the factors.)

EXAMPLE 1   Express as an equivalent expression that is a sum of logarithms: $\log_2 (4 \cdot 16)$.

SOLUTION   We have

$$\log_2 (4 \cdot 16) = \log_2 4 + \log_2 16. \qquad \text{Using the product rule for logarithms}$$

As a check, note that

$$\log_2 (4 \cdot 16) = \log_2 64 = 6, \quad \text{since } 2^6 = 64,$$

and that

$$\log_2 4 + \log_2 16 = 2 + 4 = 6, \quad \text{since } 2^2 = 4 \text{ and } 2^4 = 16.$$

1. Express as an equivalent expression that is a sum of logarithms: $\log_{10} (10 \cdot 1000)$.

YOUR TURN

EXAMPLE 2   Express as an equivalent expression that is a single logarithm: $\log_b 7 + \log_b 5$.

SOLUTION

$$\log_b 7 + \log_b 5 = \log_b (7 \cdot 5) \qquad \text{Using the product rule for logarithms}$$
$$= \log_b 35.$$

2. Express as an equivalent expression that is a single logarithm: $\log_3 x + \log_3 y$.

YOUR TURN

A Proof of the Product Rule.   Let $\log_a M = x$ and $\log_a N = y$. Converting to exponential equations, we have $a^x = M$ and $a^y = N$.

Now we multiply the left side of the first exponential equation by the left side of the second equation and similarly multiply the right sides to obtain

$$MN = a^x \cdot a^y, \quad \text{or} \quad MN = a^{x+y}.$$

Converting back to a logarithmic equation, we get

$$\log_a (MN) = x + y.$$

Recalling what x and y represent, we have

$$\log_a (MN) = \log_a M + \log_a N. \qquad \blacksquare$$

Logarithms of Powers

The second basic property is related to the power rule for exponents: $(a^m)^n = a^{mn}$. Its proof follows Example 3.

THE POWER RULE FOR LOGARITHMS

For any positive numbers M and a ($a \neq 1$), and any real number p,

$$\log_a M^p = p \cdot \log_a M.$$

(The logarithm of a power of M is the exponent times the logarithm of M.)

To better understand the power rule, note that

$$\log_a M^3 = \log_a (M \cdot M \cdot M) = \log_a M + \log_a M + \log_a M = 3 \log_a M.$$

EXAMPLE 3 Use the power rule for logarithms to write an equivalent expression that is a product: **(a)** $\log_a 9^{-5}$; **(b)** $\log_7 \sqrt[3]{x}$.

SOLUTION

3. Use the power rule for logarithms to write an equivalent expression that is a product: $\log_5 25^7$.

a) $\log_a 9^{-5} = -5\log_a 9$ Using the power rule for logarithms

b) $\log_7 \sqrt[3]{x} = \log_7 x^{1/3}$ Writing exponential notation

   $= \frac{1}{3}\log_7 x$ Using the power rule for logarithms

 YOUR TURN

A Proof of the Power Rule. Let $x = \log_a M$. We then write the equivalent exponential equation, $a^x = M$. Raising both sides to the pth power, we get

$$(a^x)^p = M^p, \quad \text{or} \quad a^{xp} = M^p. \quad \text{Multiplying exponents}$$

Converting back to a logarithmic equation gives us

$$\log_a M^p = xp.$$

But $x = \log_a M$, so substituting, we have

$$\log_a M^p = (\log_a M)p = p \cdot \log_a M.$$

Student Notes

Without understanding and remembering the rules of this section, it will be extremely difficult to solve exponential and logarithmic equations.

Logarithms of Quotients

The third property that we study is related to the quotient rule for exponents: $a^m/a^n = a^{m-n}$. Its proof follows Example 5.

> **THE QUOTIENT RULE FOR LOGARITHMS**
>
> For any positive numbers M, N, and a ($a \neq 1$),
>
> $$\log_a \frac{M}{N} = \log_a M - \log_a N.$$
>
> (The logarithm of a quotient is the logarithm of the dividend minus the logarithm of the divisor.)

To better understand the quotient rule, note that

$$\log_2 \tfrac{8}{32} = \log_2 \tfrac{1}{4} = -2$$

and $\log_2 8 - \log_2 32 = 3 - 5 = -2$.

EXAMPLE 4 Express as an equivalent expression that is a difference of logarithms: $\log_t (6/U)$.

4. Express as an equivalent expression that is a difference of logarithms: $\log_4\left(\frac{x}{8}\right)$.

SOLUTION

$$\log_t \frac{6}{U} = \log_t 6 - \log_t U \quad \text{Using the quotient rule for logarithms}$$

 YOUR TURN

EXAMPLE 5 Express as an equivalent expression that is a single logarithm:

$$\log_b 17 - \log_b 27.$$

SOLUTION

5. Express as an equivalent expression that is a single logarithm: $\log_3 a - \log_3 16$.

$$\log_b 17 - \log_b 27 = \log_b \frac{17}{27} \quad \text{Using the quotient rule for logarithms "in reverse"}$$

 YOUR TURN

A Proof of the Quotient Rule.   Our proof uses both the product rule and the power rule:

$$\log_a \frac{M}{N} = \log_a (MN^{-1}) \qquad \text{Rewriting } \frac{M}{N} \text{ as } MN^{-1}$$

$$= \log_a M + \log_a N^{-1} \qquad \text{Using the product rule for logarithms}$$

$$= \log_a M + (-1)\log_a N \qquad \text{Using the power rule for logarithms}$$

$$= \log_a M - \log_a N. \qquad \blacksquare$$

Using the Properties Together

EXAMPLE 6   Express as an equivalent expression, using the logarithms of x, y, and z.

a) $\log_b \dfrac{x^3}{yz}$

b) $\log_a \sqrt[4]{\dfrac{xy}{z^3}}$

SOLUTION

a)

$$\log_b \frac{x^3}{yz} = \log_b x^3 - \log_b yz \qquad \begin{array}{l}\text{Using the quotient rule for}\\\text{logarithms}\end{array}$$

$$= 3\log_b x - \log_b yz \qquad \text{Using the power rule for logarithms}$$

$$= 3\log_b x - (\log_b y + \log_b z) \qquad \begin{array}{l}\text{Using the product rule for}\\\text{logarithms. Because of the}\\\text{subtraction, parentheses are}\\\text{essential.}\end{array}$$

$$= 3\log_b x - \log_b y - \log_b z \qquad \text{Using the distributive law}$$

b)

$$\log_a \sqrt[4]{\frac{xy}{z^3}} = \log_a\left(\frac{xy}{z^3}\right)^{1/4} \qquad \text{Writing exponential notation}$$

$$= \frac{1}{4} \cdot \log_a \frac{xy}{z^3} \qquad \begin{array}{l}\text{Using the power rule for}\\\text{logarithms}\end{array}$$

$$= \frac{1}{4}(\log_a xy - \log_a z^3) \qquad \begin{array}{l}\text{Using the quotient rule for}\\\text{logarithms. Parentheses are}\\\text{important.}\end{array}$$

$$= \frac{1}{4}(\log_a x + \log_a y - 3\log_a z) \qquad \begin{array}{l}\text{Using the product rule and the}\\\text{power rule for logarithms}\end{array}$$

6. Express as an equivalent expression, using the logarithms of x, y, and z:

$$\log_a \frac{x^2 y}{z^4}.$$

 YOUR TURN

> **CAUTION!**   Because the product rule and the quotient rule replace one term with two, it is often essential to apply the rules within parentheses, as in Example 6.

EXAMPLE 7   Express as an equivalent expression that is a single logarithm.

a) $\dfrac{1}{2}\log_a x - 7\log_a y + \log_a z$

b) $\log_a \dfrac{b}{\sqrt{x}} + \log_a \sqrt{bx}$

SOLUTION

a) $\frac{1}{2}\log_a x - 7\log_a y + \log_a z$

$= \log_a x^{1/2} - \log_a y^7 + \log_a z$ — Using the power rule for logarithms

$= (\log_a \sqrt{x} - \log_a y^7) + \log_a z$ — Using parentheses to emphasize the order of operations; $x^{1/2} = \sqrt{x}$

$= \log_a \dfrac{\sqrt{x}}{y^7} + \log_a z$ — Using the quotient rule for logarithms. Note that all terms have the same base.

$= \log_a \dfrac{z\sqrt{x}}{y^7}$ — Using the product rule for logarithms

b) $\log_a \dfrac{b}{\sqrt{x}} + \log_a \sqrt{bx} = \log_a \dfrac{b \cdot \sqrt{bx}}{\sqrt{x}}$ — Using the product rule for logarithms

$= \log_a b\sqrt{b}$ — Removing a factor equal to 1: $\dfrac{\sqrt{x}}{\sqrt{x}} = 1$

7. Express as an equivalent expression that is a single logarithm:

$2\log_a x - 3\log_a y - \log_a z.$

$= \log_a b^{3/2}$, or $\dfrac{3}{2}\log_a b$ — Noting that $b\sqrt{b} = b^1 \cdot b^{1/2}$

 **YOUR TURN**

If we know the logarithms of two different numbers (with the same base), the properties allow us to calculate other logarithms.

EXAMPLE 8   Given $\log_a 2 = 0.431$ and $\log_a 3 = 0.683$, use the properties of logarithms to calculate the value of each of the following. If this is not possible, state so.

a) $\log_a 6$ **b)** $\log_a \frac{2}{3}$ **c)** $\log_a 81$

d) $\log_a \frac{1}{3}$ **e)** $\log_a (2a)$ **f)** $\log_a 5$

SOLUTION

a) $\log_a 6 = \log_a(2 \cdot 3) = \log_a 2 + \log_a 3$ — Using the product rule for logarithms

$= 0.431 + 0.683 = 1.114$

Check: $a^{1.114} = a^{0.431} \cdot a^{0.683} = 2 \cdot 3 = 6$

b) $\log_a \frac{2}{3} = \log_a 2 - \log_a 3$ — Using the quotient rule for logarithms

$= 0.431 - 0.683 = -0.252$

c) $\log_a 81 = \log_a 3^4 = 4\log_a 3$ — Using the power rule for logarithms

$= 4(0.683) = 2.732$

d) $\log_a \frac{1}{3} = \log_a 1 - \log_a 3$ — Using the quotient rule for logarithms

$= 0 - 0.683 = -0.683$

8. Given $\log_a 2 = 0.431$ and $\log_a 3 = 0.683$, use the properties of logarithms to calculate the value of $\log_a \frac{9}{4}$.

e) $\log_a (2a) = \log_a 2 + \log_a a$ — Using the product rule for logarithms

$= 0.431 + 1 = 1.431$

f) $\log_a 5$ *cannot be found using these properties.* $(\log_a 5 \neq \log_a 2 + \log_a 3)$

 **YOUR TURN**

A final property follows from the product rule: Since $\log_a a^k = k\log_a a$, and $\log_a a = 1$, we have $\log_a a^k = k$.

> ### THE LOGARITHM OF THE BASE TO AN EXPONENT
>
> For any base a,
>
> $$\log_a a^k = k.$$
>
> (The logarithm, base a, of a to an exponent is the exponent.)

This property also follows directly from the definition of logarithm: k is the exponent to which you raise a in order to get a^k.

EXAMPLE 9   Simplify: **(a)** $\log_3 3^7$; **(b)** $\log_{10} 10^{-5.2}$.

SOLUTION

a) $\log_3 3^7 = 7$ 7 is the exponent to which you raise 3 in order to get 3^7.

b) $\log_{10} 10^{-5.2} = -5.2$

9. Simplify: $\log_8 8^5$.

_____

 YOUR TURN

We summarize the properties of logarithms as follows.

> For any positive numbers M, N, and a ($a \neq 1$):
>
> $$\log_a (MN) = \log_a M + \log_a N; \qquad \log_a M^p = p \cdot \log_a M;$$
>
> $$\log_a \frac{M}{N} = \log_a M - \log_a N; \qquad \log_a a^k = k.$$

> **CAUTION!**   Keep in mind that, in general,
>
> $$\log_a(M + N) \neq \log_a M + \log_a N, \qquad \log_a (MN) \neq (\log_a M)(\log_a N),$$
>
> $$\log_a(M - N) \neq \log_a M - \log_a N, \qquad \log_a \frac{M}{N} \neq \frac{\log_a M}{\log_a N}.$$

12.4 EXERCISE SET

FOR EXTRA HELP   MyMathLab® MathXL
PRACTICE   WATCH   READ   REVIEW

Vocabulary and Reading Check

Match each expression with an equivalent expression from the column on the right.

1. ____ $\log_7 20$

2. ____ $\log_7 5^4$

3. ____ $\log_7 \frac{5}{4}$

4. ____ $\log_7 7$

5. ____ $\log_7 7^4$

6. ____ $\log_7 5 + \log_7 6$

a) $\log_7 5 - \log_7 4$

b) 1

c) 4

d) $\log_7 30$

e) $\log_7 5 + \log_7 4$

f) $4 \log_7 5$

Logarithms of Products

Express as an equivalent expression that is a sum of logarithms.

7. $\log_3 (81 \cdot 27)$

8. $\log_2 (16 \cdot 32)$

9. $\log_4 (64 \cdot 16)$

10. $\log_5 (25 \cdot 125)$

11. $\log_c (rst)$

12. $\log_t (3ab)$

Express as an equivalent expression that is a single logarithm.

13. $\log_a 2 + \log_a 10$

14. $\log_b 5 + \log_b 9$

15. $\log_c t + \log_c y$

16. $\log_t H + \log_t M$

Mid-Chapter Review

We use the following properties to simplify expressions and to rewrite equivalent logarithmic equations and exponential equations.

$$\log_a x = m \text{ means } a^m = x \qquad \log_a a^k = k$$
$$\log_a (MN) = \log_a M + \log_a N \qquad \log_a a = 1$$
$$\log_a \frac{M}{N} = \log_a M - \log_a N \qquad \log_a 1 = 0$$
$$\log_a M^p = p \cdot \log_a M$$

GUIDED SOLUTIONS

1. Find a formula for the inverse of $f(x) = 2x - 5$. [12.1]

Solution

$$y = 2x - 5$$
$$\boxed{} = 2\,\boxed{} - 5 \qquad \text{Interchanging } x \text{ and } y$$
$$\left.\begin{array}{l} \boxed{} = 2y \\[4pt] \dfrac{\boxed{}}{2} = y \end{array}\right\} \quad \text{Solving for } y$$
$$f^{-1}(x) = \boxed{}$$

2. Solve: $\log_4 x = 1$. [12.3]

Solution

$$\log_4 x = 1$$
$$\boxed{}^{\boxed{}} = x \qquad \text{Rewriting as an exponential equation}$$
$$\boxed{} = x$$

MIXED REVIEW

3. Find $(f \circ g)(x)$ if $f(x) = x^2 + 1$ and $g(x) = x - 5$. [12.1]

4. If $h(x) = \sqrt{5x - 3}$, find $f(x)$ and $g(x)$ such that $h(x) = (f \circ g)(x)$. Answers may vary. [12.1]

5. Find a formula for the inverse of $g(x) = 6 - x$. [12.1]

6. Graph: $f(x) = 2^x + 3$. [12.2]

Simplify.

7. $\log_4 16$ [12.3]

8. $\log_5 \frac{1}{5}$ [12.3]

9. $\log_{100} 10$ [12.3]

10. $\log_b b$ [12.4]

11. $\log_8 8^{19}$ [12.4]

12. $\log_t 1$ [12.4]

Rewrite each of the following as an equivalent exponential equation.

13. $\log_x 3 = m$ [12.3]

14. $\log_2 1024 = 10$ [12.3]

Rewrite each of the following as an equivalent logarithmic equation.

15. $e^t = x$ [12.3]

16. $64^{2/3} = 16$ [12.3]

17. Express as an equivalent expression using $\log x$, $\log y$, and $\log z$:

$$\log \sqrt{\frac{x^2}{yz^3}}.$$ [12.4]

18. Express as an equivalent expression that is a single logarithm:

$$\log a - 2 \log b - \log c.$$ [12.4]

Solve. [12.3]

19. $\log_x 64 = 3$

20. $\log_3 x = -1$

12.5 Common Logarithms and Natural Logarithms

Common Logarithms on a Calculator ● The Base *e* and Natural Logarithms on a Calculator ●
Changing Logarithmic Bases ● Graphs of Exponential Functions and Logarithmic Functions, Base *e*

Study Skills

Is Your Answer Reasonable?
It is always a good idea—especially when using a calculator—to check that your answer is reasonable. It is easy for an incorrect calculation or keystroke to result in an answer that is clearly too big or too small.

Any positive number other than 1 can serve as the base of a logarithmic function. However, there are logarithmic bases that fit into certain applications more naturally than others.

Base-10 logarithms, called **common logarithms**, are useful because they have the same base as our "commonly" used decimal system. Before calculators became widely available, common logarithms helped with tedious calculations. In fact, that is why logarithms were devised.

The logarithmic base most widely used today is an irrational number named *e*. We will consider *e* and base *e*, or *natural*, logarithms later in this section. First we examine common logarithms.

Common Logarithms on a Calculator

Before the advent of scientific calculators, printed tables listed common logarithms. Today we find common logarithms using calculators.

The abbreviation **log**, with no base written, is generally understood to mean logarithm base 10, that is, a common logarithm. Thus,

$$\log 17 \quad \text{means} \quad \log_{10} 17.$$

The key for common logarithms is usually marked **LOG**. To find the common logarithm of a number, we key in that number and press **LOG**. With most graphing calculators, we press **LOG**, the number, and then **ENTER**.

EXAMPLE 1 Use a calculator to approximate each number to four decimal places.

a) $\log 5312$

b) $\dfrac{\log 6500}{\log 0.007}$

SOLUTION

a) We enter 5312 and then press **LOG**. On most graphing calculators, we press **LOG**, followed by 5312 and **ENTER**. We find that

$$\log 5312 \approx 3.7253. \qquad \textcolor{red}{\text{Rounded to four decimal places}}$$

b) We enter 6500 and then press **LOG**. Next, we press (÷), enter 0.007, and then press **LOG** (=). On most graphing calculators, we press **LOG**, key in 6500, press ( )) (÷) **LOG**, key in 0.007, and then press ( )) **ENTER**. Be careful not to round until the end:

$$\frac{\log 6500}{\log 0.007} \approx -1.7694. \qquad \textcolor{red}{\text{Rounded to four decimal places}}$$

YOUR TURN

The inverse of a logarithmic function is an exponential function. Because of this, on many calculators the **LOG** key doubles as the (10ˣ) key after a **2ND** or [SHIFT] key is pressed. Calculators lacking a (10ˣ) key may have a key labeled [xʸ], [aˣ], or .

Technology Connection

To find log 6500/log 0.007 on a graphing calculator, we must use parentheses with care.

1. What keystrokes are needed to create the following?

> log(7)/log(3)
> 1.771243749

---

1. Use a calculator to approximate $\dfrac{\log 5}{\log 2}$ to four decimal places.

EXAMPLE 5   Find $\log_5 8$ using the change-of-base formula.

SOLUTION   We use the change-of-base formula with $a = 10$, $b = 5$, and $M = 8$:

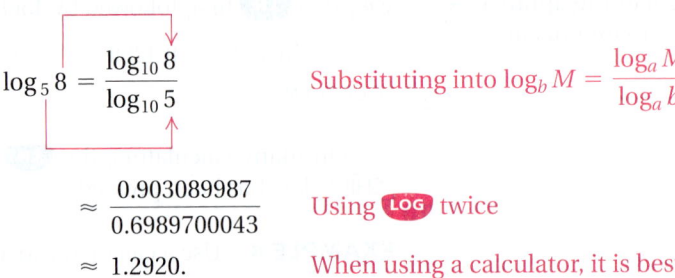

$$\log_5 8 = \frac{\log_{10} 8}{\log_{10} 5} \qquad \text{Substituting into } \log_b M = \frac{\log_a M}{\log_a b}$$

$$\approx \frac{0.903089987}{0.6989700043} \qquad \text{Using } \boxed{\text{LOG}} \text{ twice}$$

$$\approx 1.2920. \qquad \text{When using a calculator, it is best to wait until the end to round.}$$

To check, note that $\ln 8 / \ln 5 \approx 1.2920$. We can also use a calculator to verify that $5^{1.2920} \approx 8$. As a quick partial check, note that since $5^1 = 5$ and $5^2 = 25$, we can expect an answer between 1 and 2.

5. Find $\log_2 6$ using the change-of-base formula.

↩ YOUR TURN

Student Notes

The choice of the logarithm base a in the change-of-base formula should be either 10 or e so that the logarithms can be found using a calculator. Either choice will yield the same end result.

EXAMPLE 6   Find $\log_4 31$.

SOLUTION   As shown in the check of Example 5, base e can also be used.

$$\log_4 31 = \frac{\log_e 31}{\log_e 4} \qquad \text{Substituting into } \log_b M = \frac{\log_a M}{\log_a b}$$

$$= \frac{\ln 31}{\ln 4} \approx \frac{3.433987204}{1.386294361} \qquad \text{Using } \boxed{\text{LN}} \text{ twice}$$

$$\approx 2.4771. \qquad \textit{Check: } 4^{2.4771} \approx 31.$$

6. Find $\log_8 3$.

↩ YOUR TURN

Graphs of Exponential Functions and Logarithmic Functions, Base e

EXAMPLE 7   Graph $f(x) = e^x$ and $g(x) = e^{-x}$ and state the domain and the range of f and g.

SOLUTION   We use a calculator with an $\boxed{e^x}$ key to find approximate values of e^x and e^{-x}. Using these values, we can graph the functions.

x	e^x	e^{-x}
0	1	1
1	2.7	0.4
2	7.4	0.1
−1	0.4	2.7
−2	0.1	7.4

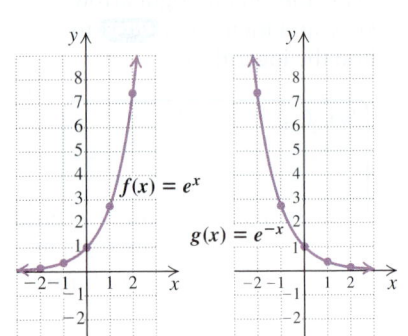

7. Graph $f(x) = 3e^x$ and state the domain and the range of f.

The domain of each function is $\mathbb{R}$, and the range of each function is $(0, \infty)$.

↩ YOUR TURN

EXAMPLE 8   Graph $f(x) = e^{-x} + 2$ and state the domain and the range of f.

SOLUTION   We find some solutions with a calculator, plot them, and then draw the graph. For example, $f(2) = e^{-2} + 2 \approx 0.1 + 2 \approx 2.1$. The graph is exactly like the graph of $g(x) = e^{-x}$, but is translated up 2 units.

x	$e^{-x} + 2$
0	3
1	2.4
2	2.1
−1	4.7
−2	9.4

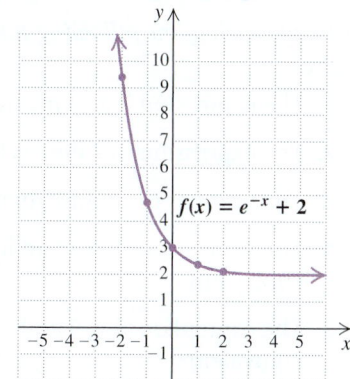

8. Graph $g(x) = e^x + 1$ and state the domain and the range of g.

_____

The domain of f is $\mathbb{R}$, and the range is $(2, \infty)$.

YOUR TURN

Technology Connection

Logarithmic functions with bases other than 10 or e can be drawn using the change-of-base formula. For example, $y = \log_5 x$ can be written $y = \ln x / \ln 5$. If your calculator has a LOGBASE(option, the change-of-base formula is not needed.

1. Graph $y = \log_7 x$.
2. Graph $y = \log_5 (x + 2)$.
3. Graph $y = \log_7 x + 2$.

EXAMPLE 9   Graph and state the domain and the range of each function.

a) $g(x) = \ln x$ **b)** $f(x) = \ln (x + 3)$

SOLUTION

a) We find some solutions with a calculator and then draw the graph. As expected, the graph is a reflection across the line $y = x$ of the graph of $y = e^x$.

x	$\ln x$
1	0
4	1.4
7	1.9
0.5	−0.7

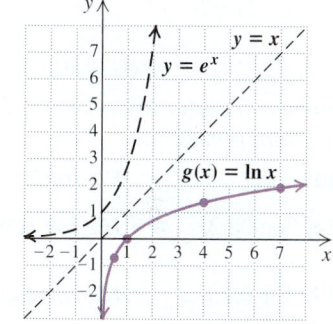

The domain of g is $(0, \infty)$, and the range is $\mathbb{R}$.

b) We find some solutions with a calculator, plot them, and draw the graph.

Chapter Resource:
Visualizing for Success, p. 839

x	$\ln (x + 3)$
0	1.1
1	1.4
2	1.6
3	1.8
4	1.9
−1	0.7
−2	0
−2.5	−0.7

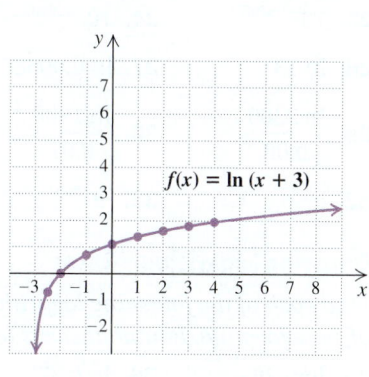

9. Graph $f(x) = \ln x + 2$ and state the domain and the range of the function.

_____

The graph of $y = \ln (x + 3)$ is the graph of $y = \ln x$ translated 3 units to the left. Since $x + 3$ must be positive, the domain is $(-3, \infty)$ and the range is $\mathbb{R}$.

 YOUR TURN

12.5 EXERCISE SET

FOR EXTRA HELP

MyMathLab® MathXL
PRACTICE WATCH READ REVIEW

⮕ Vocabulary and Reading Check

Classify each of the following statements as either true or false.

1. The expression log 23 means $\log_{10} 23$.

2. The expression ln 7 means $\log_e 7$.

3. The number e is approximately 2.7.

4. The expressions log 9 and log 18/log 2 are equivalent.

5. The expressions log 9 and log 18 − log 2 are equivalent.

6. The expressions $\log_2 9$ and ln 9/ln 2 are equivalent.

7. The expressions ln 81 and 2 ln 9 are equivalent.

8. The domain of the function given by $f(x) = \ln(x + 2)$ is $(-2, \infty)$.

9. The range of the function given by $g(x) = e^x$ is $(0, \infty)$.

10. The range of the function given by $f(x) = \ln x$ is $(-\infty, \infty)$.

Logarithms on a Calculator

Use a calculator to find each of the following to four decimal places.

11. log 7

12. log 2

13. log 13.7

14. log 98.3   *Aha!* 15. log 1000

16. log 100

17. log 0.75

18. log 0.25

19. $\dfrac{\log 8200}{\log 2}$

20. $\dfrac{\log 5700}{\log 5}$

21. $10^{1.7}$

22. $10^{0.59}$

23. $10^{-2.9523}$

24. $10^{-3.2046}$

25. ln 9

26. ln 13

27. ln 0.0062

28. ln 0.00073

29. $\dfrac{\ln 2300}{0.08}$

30. $\dfrac{\ln 1900}{0.07}$

31. $e^{2.71}$

32. $e^{3.06}$

33. $e^{-3.49}$

34. $e^{-2.64}$

Changing Logarithmic Bases

Find each of the following logarithms using the change-of-base formula. Round answers to four decimal places.

35. $\log_3 28$

36. $\log_6 37$

37. $\log_2 100$

38. $\log_7 100$

39. $\log_4 5$

40. $\log_8 7$

41. $\log_{0.1} 2$

42. $\log_{0.25} 25$

43. $\log_2 0.1$

44. $\log_{25} 0.25$

45. $\log_\pi 10$

46. $\log_\pi 100$

Graphs of Exponential Functions and Logarithmic Functions, Base e

Graph and state the domain and the range of each function.

47. $f(x) = e^x$

48. $f(x) = e^{-x}$

49. $f(x) = e^x + 3$

50. $f(x) = e^x + 2$

51. $f(x) = e^x - 2$

52. $f(x) = e^x - 3$

53. $f(x) = 0.5e^x$

54. $f(x) = 2e^x$

55. $f(x) = 0.5e^{2x}$

56. $f(x) = 2e^{-0.5x}$

57. $f(x) = e^{x-3}$

58. $f(x) = e^{x-2}$

59. $f(x) = e^{x+2}$

60. $f(x) = e^{x+3}$

61. $f(x) = -e^x$

62. $f(x) = -e^{-x}$

63. $g(x) = \ln x + 1$

64. $g(x) = \ln x + 3$

65. $g(x) = \ln x - 2$

66. $g(x) = \ln x - 1$

67. $g(x) = 2 \ln x$

68. $g(x) = 3 \ln x$

69. $g(x) = -2 \ln x$

70. $g(x) = -\ln x$

71. $g(x) = \ln(x + 2)$

72. $g(x) = \ln(x + 1)$

73. $g(x) = \ln(x - 1)$

74. $g(x) = \ln(x - 3)$

75. Using a calculator, Adan gives an *incorrect* approximation for log 79 that is between 4 and 5. How could you convince him, without using a calculator, that he is mistaken?

76. Examine Exercise 75. What mistake do you believe Adan made?

Skill Review

Find each of the following, given that

$$f(x) = \dfrac{1}{x + 2} \quad and \quad g(x) = 5x - 8.$$

77. $f(-1)$ [7.1]

78. $(f + g)(0)$ [7.4]

79. $(g - f)(x)$ [7.4]

80. The domain of f [7.2]

81. The domain of f/g [7.2]

82. $gg(x)$ [7.4]

Synthesis

83. Explain how the graph of $f(x) = e^x$ could be used to graph the function given by $g(x) = 1 + \ln x$.

84. How would you explain to a classmate why $\log_2 5 = \log 5/\log 2$ *and* $\log_2 5 = \ln 5/\ln 2$?

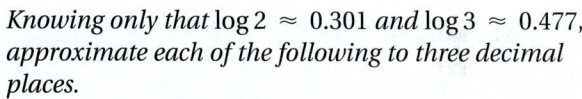

Knowing only that $\log 2 \approx 0.301$ *and* $\log 3 \approx 0.477$, *approximate each of the following to three decimal places.*

85. $\log_6 81$ **86.** $\log_9 16$ **87.** $\log_{12} 36$

88. Find a formula for converting common logarithms to natural logarithms.

89. Find a formula for converting natural logarithms to common logarithms.

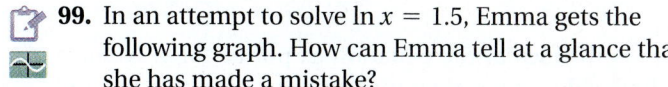 *Solve for x. Give an approximation to four decimal places.*

90. $\log (275x^2) = 38$

91. $\log (492x) = 5.728$

92. 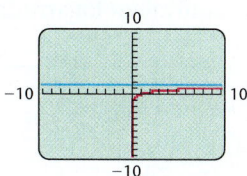 $\dfrac{3.01}{\ln x} = \dfrac{28}{4.31}$

93. $\log 692 + \log x = \log 3450$

For each function given below, **(a)** *determine the domain and the range,* **(b)** *set an appropriate window, and* **(c)** *draw the graph. Graphs may vary, depending on the scale used.*

94. $f(x) = 7.4e^x \ln x$

95. $f(x) = 3.4 \ln x - 0.25e^x$

96. $f(x) = x \ln (x - 2.1)$

97. $f(x) = 2x^3 \ln x$

98. Use a graphing calculator to check your answers to Exercises 49, 57, and 71.

99. In an attempt to solve $\ln x = 1.5$, Emma gets the following graph. How can Emma tell at a glance that she has made a mistake?

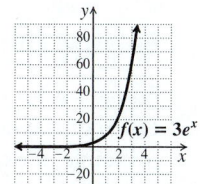 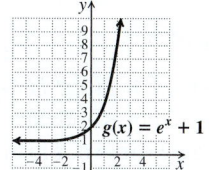
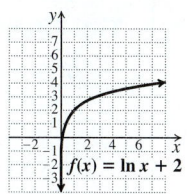
QUICK QUIZ: SECTIONS 12.1–12.5

1. Find $f(x)$ and $g(x)$ such that $h(x) = (f \circ g)(x)$ and $h(x) = \sqrt{3x - 7}$. Answers may vary.   [12.1]

Express as an equivalent expression that is a single logarithm and, if possible, simplify.   [12.4]

2. $2 \log_a x - 3 \log_a y$

3. $\log_a (x + 1) + \log_a (x - 1)$

Graph.

4. $y = 2^{x-5}$ [12.2]   **5.** $y = \ln x + 2$ [12.5]

PREPARE TO MOVE ON

Solve.

1. $x(x - 3) = 28$ [5.7]   **2.** $5x^2 - 7x = 0$ [5.7]

3. $17x - 15 = 0$ [2.2]   **4.** $\dfrac{5}{3} = 2t$ [2.2]

5. $(x - 5) \cdot 9 = 11$ [2.2]   **6.** $\dfrac{x + 3}{x - 3} = 7$ [6.6]

12.6   Solving Exponential Equations and Logarithmic Equations

Solving Exponential Equations ▪ Solving Logarithmic Equations

Solving Exponential Equations

Equations with variables in exponents, such as $5^x = 12$ and $2^{7x} = 64$, are called **exponential equations**. We can solve certain exponential equations by using the principle of exponential equality, first stated in Section 12.3.

> **THE PRINCIPLE OF EXPONENTIAL EQUALITY**
>
> For any real number b, where $b \neq -1, 0,$ or $1,$
>
> $$b^m = b^n \quad \text{is equivalent to} \quad m = n.$$
>
> (Powers of the same base are equal if and only if the exponents are equal.)

EXAMPLE 1   Solve: $4^{3x} = 16$.

SOLUTION   Note that $16 = 4^2$. Thus we can write each side as a power of the same base:

$$4^{3x} = 4^2 \qquad \text{Rewriting 16 as a power of 4}$$
$$3x = 2 \qquad \text{The base on each side is 4, so the exponents must be equal.}$$
$$x = \tfrac{2}{3}. \qquad \text{Solving for } x$$

Since $4^{3x} = 4^{3(2/3)} = 4^2 = 16$, the answer checks. The solution is $\tfrac{2}{3}$.

1. Solve: $3^{4x} = 9$.

↩ YOUR TURN

In Example 1, we wrote both sides of the equation as powers of 4. When it seems impossible to write both sides of an equation as powers of the same base, we use the following principle and write an equivalent logarithmic equation.

> **THE PRINCIPLE OF LOGARITHMIC EQUALITY**
>
> For any logarithmic base a, and for $x, y > 0,$
>
> $$x = y \quad \text{is equivalent to} \quad \log_a x = \log_a y.$$
>
> (Two expressions are equal if and only if the logarithms of those expressions are equal.)

The principle of logarithmic equality, used together with the power rule for logarithms, allows us to solve equations in which the variable is an exponent.

EXAMPLE 2   Solve: $7^{x-2} = 60$.

SOLUTION   We have

$$7^{x-2} = 60$$

Take the logarithm of both sides.

$$\log 7^{x-2} = \log 60$$

Using the principle of logarithmic equality to take the common logarithm on both sides. We could also use a logarithm with another base, such as e.

Use the power rule for logarithms.

$$(x-2)\log 7 = \log 60$$ Using the power rule for logarithms

$$x - 2 = \frac{\log 60}{\log 7}$$ ← **CAUTION!** This is *not* $\log 60 - \log 7$.

$$x = \frac{\log 60}{\log 7} + 2$$ Adding 2 to both sides

Solve for x.

$$\approx 4.1041.$$ Using a calculator and rounding to four decimal places

Check.   Since $7^{4.1041-2} \approx 60.0027$, we have a check. We can also note that since $7^{4-2} = 49$, we expect a solution greater than 4. The solution is $\frac{\log 60}{\log 7} + 2$, or approximately 4.1041.

2. Solve: $5^{x+1} = 12$.

 YOUR TURN

EXAMPLE 3   Solve: $e^{0.06t} = 1500$.

SOLUTION   Since one side is a power of e, it is easiest to take the *natural logarithm* on both sides:

$$\ln e^{0.06t} = \ln 1500$$ Taking the natural logarithm on both sides

$$0.06t = \ln 1500$$ Finding the logarithm of the base to a power: $\log_a a^k = k$. Logarithmic and exponential functions are inverses of each other.

$$t = \frac{\ln 1500}{0.06}$$ Dividing both sides by 0.06

$$\approx 121.8870.$$ Using a calculator and rounding to four decimal places

3. Solve: $e^{-0.03t} = 120$.

 YOUR TURN

> **TO SOLVE AN EQUATION OF THE FORM $a^t = b$ FOR t**
>
> **1.** Take the logarithm (either natural or common) of both sides.
> **2.** Use the power rule for logarithms so that the variable is a factor instead of an exponent.
> **3.** Divide both sides by the coefficient of the variable to isolate the variable.
> **4.** If appropriate, use a calculator to find an approximate solution.

Solving Logarithmic Equations

Certain logarithmic equations can be solved by writing an equivalent exponential equation.

EXAMPLE 4 Solve: **(a)** $\log_4(8x - 6) = 3$; **(b)** $\ln(5x) = 27$.

SOLUTION

a) $\log_4(8x - 6) = 3$

$$4^3 = 8x - 6 \qquad \textcolor{red}{\text{Writing the equivalent exponential equation}}$$

$$64 = 8x - 6$$

$$70 = 8x \qquad \textcolor{red}{\text{Adding 6 to both sides}}$$

$$x = \frac{70}{8}, \text{ or } \frac{35}{4}$$

Check:

$$\begin{array}{c|c} \log_4(8x - 6) = 3 & \\ \hline \log_4\left(8 \cdot \frac{35}{4} - 6\right) & 3 \\ \log_4(2 \cdot 35 - 6) & \\ \log_4 64 & \\ 3 \overset{?}{=} 3 & \textcolor{red}{\text{TRUE}} \end{array}$$

The solution is $\frac{35}{4}$.

b) $\ln(5x) = 27 \qquad \textcolor{red}{\text{Remember: } \ln(5x) \text{ means } \log_e(5x).}$

$$e^{27} = 5x \qquad \textcolor{red}{\text{Writing the equivalent exponential equation}}$$

$$\frac{e^{27}}{5} = x \qquad \textcolor{red}{\text{This is a very large number.}}$$

The solution is $\dfrac{e^{27}}{5}$. The check is left to the student.

4. Solve: $\log(5x - 3) = 2$.

◀ YOUR TURN

Often the properties for logarithms are needed in order to solve a logarithmic equation. The goal is to first write an equivalent equation in which the variable appears in just one logarithmic expression. We then isolate that expression and solve as in Example 4.

Student Notes

It is essential that you remember the properties of logarithms from Section 12.4. Consider reviewing the properties before attempting to solve equations similar to those in Example 5.

EXAMPLE 5 Solve.

a) $\log x + \log(x - 3) = 1$

b) $\log_2(x + 7) - \log_2(x - 7) = 3$

c) $\log_7(x + 1) + \log_7(x - 1) = \log_7 8$

SOLUTION

a) Here, log means $\log_{10}$, so we write in the base, 10, for both logarithmic expressions.

Find a
single logarithm.

Write an equivalent
exponential equation.

Solve.

$$\log_{10} x + \log_{10}(x - 3) = 1$$

$$\log_{10}[x(x - 3)] = 1 \qquad \textcolor{red}{\begin{array}{l}\text{Using the product rule for logarithms} \\ \text{to obtain a single logarithm}\end{array}}$$

$$x(x - 3) = 10^1 \qquad \textcolor{red}{\begin{array}{l}\text{Writing an equivalent exponential} \\ \text{equation}\end{array}}$$

$$x^2 - 3x = 10$$

$$x^2 - 3x - 10 = 0$$

$$(x + 2)(x - 5) = 0 \qquad \textcolor{red}{\text{Factoring}}$$

$$x + 2 = 0 \quad or \quad x - 5 = 0 \qquad \textcolor{red}{\begin{array}{l}\text{Using the principle of zero} \\ \text{products}\end{array}}$$

$$x = -2 \quad or \qquad x = 5$$

Check.

Check:

For -2:

$$\underline{\log x + \log (x - 3) = 1}$$

$$\log (-2) + \log (-2 - 3) \overset{?}{=} 1 \quad \text{FALSE}$$

For 5:

$$\begin{array}{c|c} \log x + \log (x - 3) = 1 & \\ \hline \log 5 + \log (5 - 3) & 1 \\ \log 5 + \log 2 & \\ \log 10 & \\ & 1 \overset{?}{=} 1 \quad \text{TRUE} \end{array}$$

The number -2 *does not check* because the logarithm of a negative number is undefined. The solution is 5.

b) We have

$$\log_2 (x + 7) - \log_2 (x - 7) = 3$$

$$\log_2 \frac{x + 7}{x - 7} = 3 \qquad \begin{array}{l} \text{Using the quotient rule} \\ \text{for logarithms to obtain} \\ \text{a single logarithm} \end{array}$$

$$\frac{x + 7}{x - 7} = 2^3 \qquad \begin{array}{l} \text{Writing an equivalent} \\ \text{exponential equation} \end{array}$$

$$\frac{x + 7}{x - 7} = 8$$

$$x + 7 = 8(x - 7) \qquad \begin{array}{l} \text{Multiplying by } x - 7 \text{ to clear} \\ \text{fractions} \end{array}$$

$$x + 7 = 8x - 56 \qquad \text{Using the distributive law}$$

$$63 = 7x$$

$$9 = x. \qquad \text{Dividing by 7}$$

Check:

$$\begin{array}{c|c} \log_2 (x + 7) - \log_2 (x - 7) = 3 & \\ \hline \log_2 (9 + 7) - \log_2 (9 - 7) & 3 \\ \log_2 16 - \log_2 2 & \\ 4 - 1 & \\ & 3 \overset{?}{=} 3 \quad \text{TRUE} \end{array}$$

The solution is 9.

c) We have

$$\log_7 (x + 1) + \log_7 (x - 1) = \log_7 8$$

$$\log_7 [(x + 1)(x - 1)] = \log_7 8 \qquad \begin{array}{l} \text{Using the product rule for} \\ \text{logarithms} \end{array}$$

$$\log_7 (x^2 - 1) = \log_7 8 \qquad \begin{array}{l} \text{Multiplying. Note that } both \\ \textit{sides are base-7 logarithms.} \end{array}$$

$$x^2 - 1 = 8 \qquad \begin{array}{l} \text{Using the principle of} \\ \text{logarithmic equality} \end{array}$$

$$x^2 - 9 = 0$$

$$(x - 3)(x + 3) = 0 \qquad \text{Solving the quadratic equation}$$

$$x = 3 \quad or \quad x = -3.$$

We leave it to the student to show that 3 checks but -3 does not. The solution is 3.

Technology Connection

To solve exponential equations and logarithmic equations, we can determine the x-coordinate at any point of intersection. For example, to solve $e^{0.5x} - 7 = 2x + 6$, we graph $y_1 = e^{0.5x} - 7$ and $y_2 = 2x + 6$ as shown.

We find that the x-coordinates at the intersections are approximately -6.48 and 6.52.

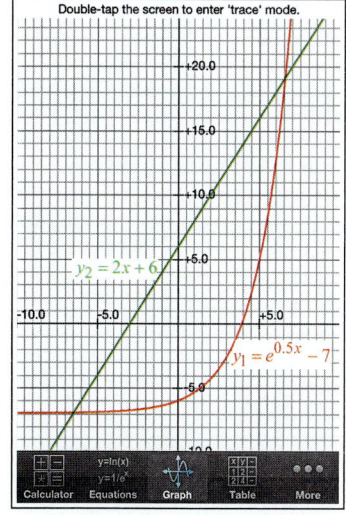

Double-tap the screen to enter 'trace' mode.

Use a graphing calculator to solve each equation to the nearest hundredth.

1. $e^{7x} = 14$
2. $8e^{0.5x} = 3$
3. $xe^{3x-1} = 5$
4. $4 \ln (x + 3.4) = 2.5$
5. $\ln 3x = 0.5x - 1$
6. $\ln x^2 = -x^2$

5. Solve:

$$\log_3 (x + 4) + \log_3 (x + 2) = 1.$$

YOUR TURN

CONNECTING THE CONCEPTS

We have used several procedures for solving exponential equations and logarithmic equations. Carefully inspecting an equation helps us choose the best method to use. Compare the following.

Equation	Description of Equation	Solution
$8^{2x} = 8^5$	Exponential equation Each side is an exponential expression with the same base.	The expressions $2x$ and 5 must be equal. $2x = 5$ $x = \frac{5}{2}$
$2^t = 7$	Exponential equation Expressions have different bases.	Take the logarithm of both sides. $\log 2^t = \log 7$ $t \cdot \log 2 = \log 7$ $t = \dfrac{\log 7}{\log 2}$
$\log_5 x = 3$	Logarithmic equation One logarithmic expression	Rewrite as an equivalent exponential equation. $5^3 = x$ $125 = x$
$\log_3 (x + 1) = \log_3 (2x)$	Logarithmic equation Each side is a logarithmic expression with the same base.	The expressions $x + 1$ and $2x$ must be equal. $x + 1 = 2x$ $1 = x$

EXERCISES

Solve.

1. $\log (2x) = 3$

2. $2^{x+1} = 2^9$

3. $e^t = 5$

4. $\log_3 (x^2 + 1) = \log_3 26$

5. $\ln 2x = \ln 8$

6. $3^{5x} = 4$

7. $\log_2 (x + 1) = -1$

8. $9^{x+1} = 27^{-x}$

12.6 EXERCISE SET

FOR EXTRA HELP

MyMathLab® Math XL
PRACTICE   WATCH   READ   REVIEW

Vocabulary and Reading Check

Classify each of the following statements as either true or false.

1. The solution of a logarithmic equation is never a negative number.

2. To solve an exponential equation, we can take the common logarithm of both sides of the equation.

3. We cannot calculate the logarithm of a negative number.

4. To solve an exponential equation, we can take the natural logarithm of both sides of the equation.

Concept Reinforcement

Match each equation with an equivalent equation from the list below that could be the next step in the solution process.

a) $5^3 = x$

b) $\log_5 (x^2 - 2x) = 3$

c) $\log_5 \dfrac{x}{x-2} = 3$

d) $\log 5^x = \log 3$

5. ____ $5^x = 3$

6. ____ $\log_5 x = 3$

7. ____ $\log_5 x + \log_5 (x - 2) = 3$

8. ____ $\log_5 x - \log_5 (x - 2) = 3$

Solving Exponential Equations

Solve. Where appropriate, include approximations to three decimal places.

9. $3^{2x} = 81$

10. $2^{3x} = 64$

11. $4^x = 32$

12. $9^x = 27$

13. $2^x = 10$

14. $2^x = 24$

15. $2^{x+5} = 16$

16. $2^{x-1} = 8$

17. $8^{x-3} = 19$

18. $5^{x+2} = 15$

19. $e^t = 50$

20. $e^t = 20$

21. $e^{-0.02t} = 8$

22. $e^{-0.01t} = 100$

23. $4.9^x - 87 = 0$

24. $7.2^x - 65 = 0$

25. $19 = 2e^{4x}$

26. $29 = 3e^{2x}$

27. $7 + 3e^{-x} = 13$

28. $4 + 5e^{-x} = 9$

Solving Logarithmic Equations

Solve. Where appropriate, include approximations to three decimal places. If no solution exists, state this.

Aha! 29. $\log_3 x = 4$

30. $\log_2 x = 6$

31. $\log_4 x = -2$

32. $\log_5 x = -3$

33. $\ln x = 5$

34. $\ln x = 4$

35. $\ln (4x) = 3$

36. $\ln (3x) = 2$

37. $\log x = 1.2$

38. $\log x = 0.6$

39. $\ln (2x + 1) = 4$

40. $\ln (4x - 2) = 3$

Aha! 41. $\ln x = 1$

42. $\log x = 1$

43. $5 \ln x = -15$

44. $3 \ln x = -3$

45. $\log_2 (8 - 6x) = 5$

46. $\log_5 (7 - 2x) = 3$

47. $\log (x - 9) + \log x = 1$

48. $\log (x + 9) + \log x = 1$

49. $\log x - \log (x + 3) = 1$

50. $\log x - \log (x + 7) = -1$

Aha! 51. $\log (2x + 1) = \log 5$

52. $\log (x + 1) - \log x = 0$

53. $\log_4 (x + 3) = 2 + \log_4 (x - 5)$

54. $\log_2 (x + 3) = 4 + \log_2 (x - 3)$

55. $\log_7 (x + 1) + \log_7 (x + 2) = \log_7 6$

56. $\log_6 (x + 3) + \log_6 (x + 2) = \log_6 20$

57. $\log_5 (x + 4) + \log_5 (x - 4) = \log_5 20$

58. $\log_4 (x + 2) + \log_4 (x - 7) = \log_4 10$

59. $\ln (x + 5) + \ln (x + 1) = \ln 12$

60. $\ln (x - 6) + \ln (x + 3) = \ln 22$

61. $\log_2 (x - 3) + \log_2 (x + 3) = 4$

62. $\log_3 (x - 4) + \log_3 (x + 4) = 2$

63. $\log_{12} (x + 5) - \log_{12} (x - 4) = \log_{12} 3$

64. $\log_6 (x + 7) - \log_6 (x - 2) = \log_6 5$

65. $\log_2 (x - 2) + \log_2 x = 3$

66. $\log_4 (x + 6) - \log_4 x = 2$

67. Madison finds that the solution of $\log_3 (x + 4) = 1$ is -1, but rejects -1 as an answer. What mistake do you suspect she is making?

EXAMPLE 2 *Chemistry: pH of Liquids.* In chemistry, the pH of a liquid is a measure of its acidity. We calculate pH as follows:

$$pH = -\log[H^+],$$

where $[H^+]$ is the hydrogen ion concentration in moles per liter.

a) The hydrogen ion concentration of human blood is normally about 3.98×10^{-8} moles per liter. Find the pH.

Source: www.merck.com

b) The average pH of seawater is about 8.2. Find the hydrogen ion concentration.

Source: www.seafriends.org.nz

SOLUTION

a) To find the pH of blood, we use the above formula:

$$
\begin{aligned}
pH &= -\log[H^+] \\
&= -\log[3.98 \times 10^{-8}] && \text{Substituting} \\
&\approx -(-7.400117) && \text{Using a calculator} \\
&\approx 7.4.
\end{aligned}
$$

The pH of human blood is normally about 7.4.

b) We substitute and solve for $[H^+]$:

$$
\begin{aligned}
8.2 &= -\log[H^+] && \text{Using } pH = -\log[H^+] \\
-8.2 &= \log[H^+] && \text{Dividing both sides by } -1 \\
10^{-8.2} &= [H^+] && \text{Converting to an exponential equation} \\
6.31 \times 10^{-9} &\approx [H^+]. && \text{Using a calculator; writing scientific notation}
\end{aligned}
$$

2. The pH of the soil in Jeannette's garden is 6.3. Find the hydrogen ion concentration.

The hydrogen ion concentration of seawater is about 6.31×10^{-9} moles per liter.

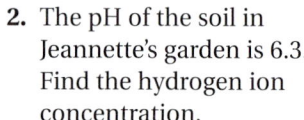  YOUR TURN

Applications of Exponential Functions

EXAMPLE 3 *Interest Compounded Annually.* Suppose that $25,000 is invested at 4% interest, compounded annually. In t years, it will grow to the amount A given by

$$A(t) = 25{,}000(1.04)^t.$$

a) How long will it take to have $80,000 in the account?

b) Find the amount of time it takes for the $25,000 to double itself.

SOLUTION

a) We set $A(t) = 80{,}000$ and solve for t:

$$
\begin{aligned}
80{,}000 &= 25{,}000(1.04)^t \\
3.2 &= 1.04^t && \text{Dividing both sides by } 25{,}000 \\
\log 3.2 &= \log 1.04^t && \text{Taking the common logarithm on both sides} \\
\log 3.2 &= t \log 1.04 && \text{Using the power rule for logarithms} \\
\frac{\log 3.2}{\log 1.04} &= t && \text{Dividing both sides by } \log 1.04 \\
29.7 &\approx t. && \text{Using a calculator}
\end{aligned}
$$

Student Notes

Study the different steps in the solution of Example 3(b). Note that if 50,000 and 25,000 are replaced with 6000 and 3000, the doubling time is unchanged.

3. If $25,000 is invested at 5% interest, compounded annually, in t years it will grow to the amount A given by $A(t) = 25,000(1.05)^t$.

   a) How long will it take to have $80,000 in the account?

   b) Find the amount of time it takes for the $25,000 to double itself.

As always, when doing a calculation like this, it is best to wait until the end to round. At an interest rate of 4% per year, it will take about 29.7 years for $25,000 to grow to $80,000.

b) To find the *doubling time*, we replace $A(t)$ with 50,000 and solve for t:

$$50,000 = 25,000(1.04)^t$$

$$2 = (1.04)^t \qquad \text{Dividing both sides by 25,000}$$

$$\log 2 = \log (1.04)^t \qquad \text{Taking the common logarithm on both sides}$$

$$\log 2 = t \log 1.04 \qquad \text{Using the power rule for logarithms}$$

$$t = \frac{\log 2}{\log 1.04} \approx 17.7. \qquad \text{Dividing both sides by } \log 1.04 \text{ and using a calculator}$$

At an interest rate of 4% per year, the doubling time is about 17.7 years.

YOUR TURN

Like investments, populations often grow exponentially.

EXPONENTIAL GROWTH

An **exponential growth model** is a function of the form

$$P(t) = P_0 e^{kt}, \quad k > 0,$$

where P_0 is the population at time 0, $P(t)$ is the population at time t, and k is the **exponential growth rate** for the situation. The **doubling time** is the amount of time necessary for the population to double in size.

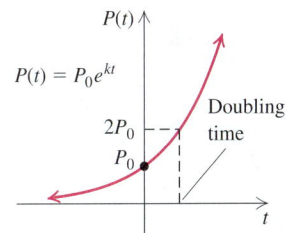

The exponential growth rate is the rate of growth of a population or other quantity at any *instant* in time. Since the change in population is continually growing, the percent of total growth after one year exceeds the exponential growth rate.

EXAMPLE 4   *Invasive Species.*   Beginning in 1988, infestations of zebra mussels started spreading through North American waters. These mussels spread with such speed that water treatment facilities, power plants, and entire ecosystems can become threatened. The area of an infestation of zebra mussels can have an annual exponential growth rate of 350%.

Source: Based on information from Dr. Gerald Mackie, Department of Zoology, University of Guelph in Ontario

a) A zoologist discovers an infestation of zebra mussels covering 10 cm². Find the exponential growth function that models the data. Let t represent the number of years since the discovery of the infestation.

b) Use the function found in part (a) to estimate the size of the infestation after 5 years.

SOLUTION

a) At $t = 0$, the size of the infestation is 10 cm². We substitute 10 for P_0 and 350%, or 3.5, for k. This gives the exponential growth function

$$P(t) = 10e^{3.5t}.$$

4. Refer to Example 4. Suppose that the area of infestation increases exponentially at a rate of 400% per year. To what size would a 10-cm² infestation grow after 5 years?

b) To estimate the size of the infestation after 5 years, we compute $P(5)$:

$$P(5) = 10e^{3.5(5)} \qquad \text{Using } P(t) = 10e^{3.5t} \text{ from part (a)}$$
$$= 10e^{17.5} \approx 398{,}247{,}844. \qquad \text{Using a calculator}$$

After 5 years, the zebra mussels will cover about 398,247,844 cm², or about 39,825 m².

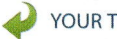 YOUR TURN

EXAMPLE 5 *Social Networking.* The number of unique visitors to the virtual pinboard Pinterest.com increased exponentially in the last six months of 2011. In June 2011, there were 608,000 unique visitors to Pinterest; this number had increased to 11,716,000 by January 2012.

Source: comScore, in "Pinterest Hits 10 Million U.S. Monthly Uniques Faster Than Any Standalone Site Ever," Josh Constine, Feb. 7, 2012, on techcrunch.com

a) Find the exponential growth rate and the exponential growth function.

b) Assuming that the trend continues, estimate the month in which there will be 30 million unique visitors to Pinterest.

SOLUTION

a) We let $P(t) = P_0 e^{kt}$, where t is the number of months since June 2011 and $P(t)$ is the number of unique visitors to Pinterest, in thousands. Next, we substitute 608 for P_0:

$$P(t) = 608e^{kt}.$$

To find the exponential growth rate k, we note that after 7 months, there were 11,716,000, or 11,716 thousand, unique visitors:

$$\left.\begin{array}{l} P(7) = 608e^{k \cdot 7} \\ 11{,}716 = 608e^{7k} \end{array}\right\} \quad \text{Substituting}$$

$$\frac{11{,}716}{608} = e^{7k} \qquad \text{Dividing both sides by 608}$$

$$\ln\,(11{,}716/608) = \ln e^{7k} \qquad \text{Taking the natural logarithm on both sides}$$

$$\ln\,(11{,}716/608) = 7k \qquad \ln e^{7k} = \log_e e^{7k} = 7k$$

$$\frac{\ln\,(11{,}716/608)}{7} = k \qquad \text{Dividing both sides by 7}$$

$$0.423 \approx k. \qquad \text{Using a calculator and rounding}$$

The exponential growth rate is 42.3%, and the exponential growth function is given by $P(t) = 608e^{0.423t}$.

b) To estimate the month in which Pinterest has 30 million unique visitors, we replace $P(t)$ with 30,000 (since 30 million is 30,000 thousand) and solve for t:

$$30{,}000 = 608e^{0.423t}$$

$$\frac{30{,}000}{608} = e^{0.423t} \qquad \text{Dividing both sides by 608}$$

$$\ln\,(30{,}000/608) = \ln e^{0.423t} \qquad \text{Taking the natural logarithm on both sides}$$

$$\ln\,(30{,}000/608) = 0.423t \qquad \ln e^a = a$$

$$\frac{\ln\,(30{,}000/608)}{0.423} = t \qquad \text{Dividing both sides by 0.423}$$

$$9.2 \approx t. \qquad \text{Using a calculator and rounding}$$

According to this model, Pinterest had 30 million unique visitors 9.2 months after June 2011, or sometime during March 2012.

5. *Malware.* The number of malware variants targeted to Android devices increased exponentially from 10 in the first quarter of 2011 to 37 in the first quarter of 2012. Find the exponential growth rate and the exponential growth function.

Source: crn.com

 YOUR TURN

EXAMPLE 6 *Interest Compounded Continuously.* When an amount of money P_0 is invested at interest rate k, compounded *continuously*, interest is computed every "instant" and added to the original amount. The balance $P(t)$, after t years, is given by the exponential growth model

$$P(t) = P_0 e^{kt}.$$

a) Suppose that \$30,000 is invested and grows to \$44,754.75 in 5 years. Find the exponential growth function.

b) What is the doubling time?

SOLUTION

a) We have $P(0) = 30{,}000$. Thus the exponential growth function is

$$P(t) = 30{,}000 e^{kt}, \quad \text{where } k \text{ must still be determined.}$$

Knowing that for $t = 5$ we have $P(5) = 44{,}754.75$, it is possible to solve for k:

$$44{,}754.75 = 30{,}000 e^{k(5)}$$
$$44{,}754.75 = 30{,}000 e^{5k}$$
$$\frac{44{,}754.75}{30{,}000} = e^{5k} \qquad \text{Dividing both sides by 30,000}$$
$$1.491825 = e^{5k}$$
$$\ln 1.491825 = \ln e^{5k} \qquad \text{Taking the natural logarithm on both sides}$$
$$\ln 1.491825 = 5k \qquad \ln e^a = a$$
$$\frac{\ln 1.491825}{5} = k \qquad \text{Dividing both sides by 5}$$
$$0.08 \approx k. \qquad \text{Using a calculator and rounding}$$

The interest rate is about 0.08, or 8%, compounded continuously. Because interest is being compounded continuously, the yearly interest rate is a bit more than 8%. The exponential growth function is

$$P(t) = 30{,}000 e^{0.08t}.$$

b) To find the doubling time T, we replace $P(T)$ with 60,000 and solve for T:

$$60{,}000 = 30{,}000 e^{0.08T}$$
$$2 = e^{0.08T} \qquad \text{Dividing both sides by 30,000}$$
$$\ln 2 = \ln e^{0.08T} \qquad \text{Taking the natural logarithm on both sides}$$
$$\ln 2 = 0.08T \qquad \ln e^a = a$$
$$\frac{\ln 2}{0.08} = T \qquad \text{Dividing both sides by 0.08}$$
$$8.7 \approx T. \qquad \text{Using a calculator and rounding}$$

Thus the original investment of \$30,000 will double in about 8.7 years.

YOUR TURN

For any specified interest rate, continuous compounding gives the highest yield and the shortest doubling time.

In some real-life situations, a quantity or population is *decreasing* or *decaying* exponentially.

$P(t) = 30{,}000 e^{0.08t}$

Doubling time = ?

A visualization of Example 6

6. Refer to Example 6. If \$20,000 is invested and grows to \$22,103.42 in 5 years, find the exponential growth function and the doubling time.

EXPONENTIAL DECAY

An **exponential decay model** is a function of the form

$$P(t) = P_0 e^{-kt}, \quad k > 0,$$

where P_0 is the quantity present at time 0, $P(t)$ is the amount present at time t, and k is the **decay rate**. The **half-life** is the amount of time necessary for half of the quantity to decay.

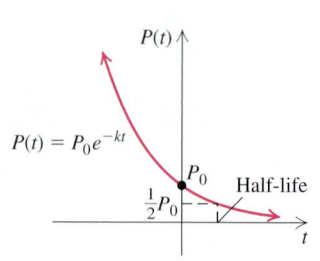

EXAMPLE 7 *Carbon Dating.* The radioactive element carbon-14 has a half-life of 5750 years. The percentage of carbon-14 in the remains of organic matter can be used to determine the age of that material. Recently, near Patuxent River, Maryland, archaeologists discovered charcoal that had lost 8.1% of its carbon-14. The age of this charcoal was evidence that this is the oldest dwelling ever discovered in Maryland. What was the age of the charcoal?

Source: Based on information from *The Baltimore Sun.* "Digging Where Indians Camped Before Columbus," by Frank D. Roylance, July 2, 2009

SOLUTION We first find k. To do so, we use the concept of half-life. When $t = 5750$ (the half-life), $P(t)$ is half of P_0. Then

$$0.5P_0 = P_0 e^{-k(5750)} \qquad \text{Substituting in } P(t) = P_0 e^{-kt}$$

$$0.5 = e^{-5750k} \qquad \text{Dividing both sides by } P_0$$

$$\ln 0.5 = \ln e^{-5750k} \qquad \text{Taking the natural logarithm on both sides}$$

$$\ln 0.5 = -5750k \qquad \ln e^a = a$$

$$\frac{\ln 0.5}{-5750} = k \qquad \text{Dividing both sides by } -5750$$

$$0.00012 \approx k. \qquad \text{Using a calculator and rounding}$$

Now we have a function for the decay of carbon-14:

$$P(t) = P_0 e^{-0.00012t}. \qquad \text{This completes the first part of our solution.}$$

(*Note:* This equation can be used for subsequent carbon-dating problems.) If the charcoal has lost 8.1% of its carbon-14 from an initial amount P_0, then $100\% - 8.1\%$, or 91.9%, of P_0 is still present. To find the age t of the charcoal, we solve this equation for t:

$$0.919P_0 = P_0 e^{-0.00012t} \qquad \text{We want to find } t \text{ for which } P(t) = 0.919P_0.$$

$$0.919 = e^{-0.00012t} \qquad \text{Dividing both sides by } P_0$$

$$\ln 0.919 = \ln e^{-0.00012t} \qquad \text{Taking the natural logarithm on both sides}$$

$$\ln 0.919 = -0.00012t \qquad \ln e^a = a$$

$$\frac{\ln 0.919}{-0.00012} = t \qquad \text{Dividing both sides by } -0.00012$$

$$700 \approx t. \qquad \text{Using a calculator}$$

The charcoal is about 700 years old.

Chapter Resource:
Decision Making: Connection, p. 840

7. In Chaco Canyon, New Mexico, archaeologists found corn pollen that had lost 38.1% of its carbon-14. What was the age of the pollen?

 YOUR TURN

12.7 EXERCISE SET

FOR EXTRA HELP

MyMathLab® MathXL
PRACTICE   WATCH   READ   REVIEW

Vocabulary and Reading Check

For the exponential growth model $P(t) = P_0 e^{kt}$, $k > 0$, match each variable with its description.

1. ____ k
2. ____ $P(t)$
3. ____ P_0
4. ____ T, where $2P_0 = P_0 e^{kT}$

a) Doubling time
b) Exponential growth rate
c) Population at time 0
d) Population at time t

Applications of Exponential Functions and Logarithmic Functions

5. *Digital Storage.* The amount of information created and stored digitally can be estimated by
$$A(t) = 0.08(1.6)^t,$$
where $A(t)$ is in zettabytes (1 zettabyte $= 10^{21}$ bytes) and t is the number of years after 2005.

Source: Based on data from IDC and EMC Corporation, in "Digital Universe to Add 1.8 Zettabytes in 2011." Rich Miller, June 28, 2011, datacenterknowledge.com

a) Determine the year in which the amount of information created and stored digitally reached 1 zettabyte.
b) What is the doubling time for the amount of information created and stored digitally?

6. *Health.* The rate of the number of deaths due to stroke in the United States can be estimated by
$$S(t) = 180(0.97)^t,$$
where $S(t)$ is the number of deaths per 100,000 people and t is the number of years after 1960.

Source: Based on data from Centers for Disease Control and Prevention

a) In what year was the death rate due to stroke 100 per 100,000 people?
b) In what year will the death rate due to stroke be 25 per 100,000 people?

7. *Student Loan Repayment.* A college loan of $29,000 is made at 3% interest, compounded annually. After t years, the amount due, A, is given by the function
$$A(t) = 29{,}000(1.03)^t.$$

a) After what amount of time will the amount due reach $35,000?
b) Find the doubling time.

8. *Spread of a Rumor.* The number of people who have heard a rumor increases exponentially. If each person who hears a rumor repeats it to two people per day, and if 20 people start the rumor, the number of people N who have heard the rumor after t days is given by
$$N(t) = 20(3)^t.$$

a) After what amount of time will 1000 people have heard the rumor?
b) What is the doubling time for the number of people who have heard the rumor?

9. The frequency, in hertz (Hz), of the nth key on an 88-key piano is given by
$$f(n) = 27.5\left(\sqrt[12]{2}\right)^{n-1},$$
where $n = 1$ corresponds to the lowest key on the piano keyboard, an A.

Source: "Piano Key Frequencies," on en.wikipedia.org

a) What number key on the keyboard has a frequency of 440 Hz?
b) How many keys does it take for the frequency to double?

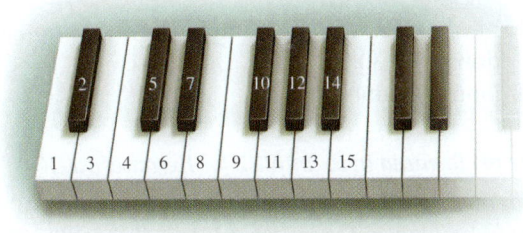

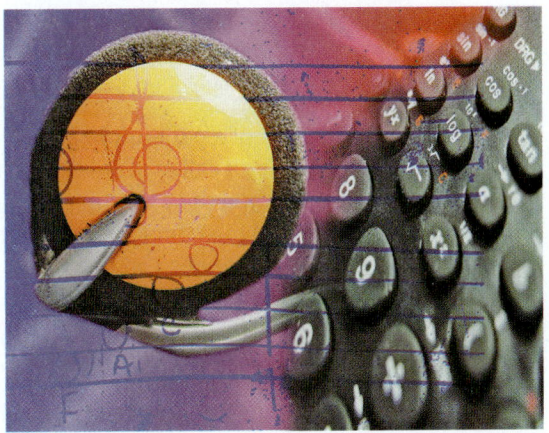

10. *Smoking.* The percentage of smokers who receive telephone counseling and successfully quit smoking for t months is given by

$$P(t) = 21.4(0.914)^t.$$

Sources: *New England Journal of Medicine; data from California's Smoker's Hotline*

a) In what month will 15% of those who quit and used telephone counseling still be smoke-free?

b) In what month will 5% of those who quit and used telephone counseling still be smoke-free?

11. *Internet Traffic.* The number of unique visitors $V(t)$ to tumblr.com, a blogging platform, can be approximated by

$$V(t) = 12(1.044)^t,$$

where $V(t)$ is in millions of visitors and t is the number of months after March 2011.

Source: *Based on data from siteanalytics.compete.com/tumblr.com/*

a) In what month were there 50 million unique visitors to tumblr.com?

b) Find the doubling time.

12. *Marine Biology.* As a result of preservation efforts in countries in which whaling was once common, the humpback whale population has grown since the 1970s. The worldwide population $P(t)$, in thousands, t years after 1982 can be estimated by

$$P(t) = 5.5(1.047)^t.$$

a) In what year will the humpback whale population reach 30,000?

b) Find the doubling time.

Use the pH formula in Example 2 for Exercises 13–16.

13. *Chemistry.* The hydrogen ion concentration of fresh-brewed coffee is about 1.3×10^{-5} moles per liter. Find the pH.

14. *Chemistry.* The hydrogen ion concentration of milk is about 1.6×10^{-7} moles per liter. Find the pH.

15. *Medicine.* When the pH of a patient's blood drops below 7.4, a condition called *acidosis* sets in. Acidosis can be deadly when the pH drops to 7.0. What would the hydrogen ion concentration of the patient's blood be at that point?

16. *Medicine.* When the pH of a patient's blood rises above 7.4, a condition called *alkalosis* sets in. Alkalosis can be deadly when the patient's pH reaches 7.8. What would the hydrogen ion concentration of the patient's blood be at that point?

Use the formula in Example 1 for Exercises 17–20.

17. *Racing.* The intensity of sound from a race car in full throttle is about 10 W/m^2. How loud in decibels is this sound level?

Source: *nascar.about.com*

18. *Audiology.* The intensity of sound in normal conversation is about $3.2 \times 10^{-6} \text{ W/m}^2$. How loud in decibels is this sound level?

19. *Concerts.* The crowd at a Hearsay concert at Wembley Arena in London cheered at a sound level of 128.8 dB. What is the intensity of such a sound?

Source: *www.peterborough.gov.uk*

20. *City Ordinances.* In New York City, the maximum allowable sound level of music from a commercial establishment, as measured inside nearby residences, is 42 dB. What is the intensity of such sound?

Source: *New York City Department of Environmental Protection, Bureau of Environmental Compliance, Nov. 2011*

21. *E-mail Volume.* The SenderBase® Security Network ranks e-mail volume using a logarithmic scale. The magnitude M of a network's daily e-mail volume is given by

$$M = \log \frac{v}{1.34},$$

where v is the number of e-mail messages sent each day. How many e-mail messages are sent each day by a network that has a magnitude of 7.5?

Source: *forum.spamcop.net*

22. *Richter Scale.* The Richter scale, developed in 1935, has been used for years to measure earthquake magnitude. The Richter magnitude m of an earthquake is given by

$$m = \log \frac{A}{A_0},$$

where A is the maximum amplitude of the earthquake and A_0 is a constant. What is the magnitude on the Richter scale of an earthquake with an amplitude that is a million times A_0?

Use the compound-interest formula in Example 6 for Exercises 23 and 24.

23. *Interest Compounded Continuously.* Suppose that P_0 is invested in a savings account where interest is compounded continuously at 2.5% per year.

a) Express $P(t)$ in terms of P_0 and 0.025.

b) Suppose that $5000 is invested. What is the balance after 1 year? after 2 years?

c) When will an investment of $5000 double itself?

24. *Interest Compounded Continuously.* Suppose that P_0 is invested in a savings account where interest is compounded continuously at 3.1% per year.

  a) Express $P(t)$ in terms of P_0 and 0.031.

  b) Suppose that $1000 is invested. What is the balance after 1 year? after 2 years?

  c) When will an investment of $1000 double itself?

25. *Population Growth.* In 2012, the population of the United States was 314 million and the exponential growth rate was 0.963% per year.

  Source: U.S. Census Bureau

  a) Find the exponential growth function.

  b) Predict the U.S. population in 2020.

  c) When will the U.S. population reach 400 million?

26. *World Population Growth.* In 2012, the world population was 7 billion and the exponential growth rate was 1.12% per year.

  Source: U.S. Census Bureau

  a) Find the exponential growth function.

  b) Predict the world population in 2016.

  c) When will the world population be 10 billion?

27. *Population Growth.* The exponential growth rate of the population of United Arab Emirates is 3.3% per year (one of the highest in the world). What is the doubling time?

  Source: CIA World Factbook

28. *Bacteria Growth.* The number of bacteria in a culture grows at an exponential growth rate of 139% per hour. What is the doubling time for these bacteria?

29. *World Population.* The function

$$Y(x) = 89.29 \ln \frac{x}{7}$$

can be used to estimate the number of years $Y(x)$ after 2012 required for the world population to reach x billion people.

  Sources: Based on data from U.S. Census Bureau; International Data Base

  a) In what year will the world population reach 10 billion?

  b) In what year will the world population reach 12 billion?

  c) Graph the function.

30. *Marine Biology.* The function

$$Y(x) = 21.77 \ln \frac{x}{5.5}$$

can be used to estimate the number of years $Y(x)$ after 1982 required for the world's humpback whale population to reach x thousand whales.

  a) In what year will the whale population reach 15,000?

  b) In what year will the whale population reach 25,000?

  c) Graph the function.

31. *Computer Usage.* The number of tablet PC users in the United States can be estimated by

$$c(t) = 26 + 36 \ln t, \quad t \geq 1,$$

where $c(t)$ is in millions of users and t is the number of years after 2010.

  Source: Based on data from Forrester Research Report, "Tablets Will Grow As Fast As MP3 Players"

  a) How many tablet PC users were there in the United States in 2012?

  b) Graph the function.

  c) In what year will there be 100 million tablet PC users in the United States?

32. *Forgetting.* Students in an English class took a final exam. They took equivalent forms of the exam at monthly intervals thereafter. The average score $S(t)$, in percent, after t months was found to be

$$S(t) = 78 - 20 \log (t + 1), \quad t \geq 0.$$

  a) What was the average score when they initially took the test, $t = 0$?

  b) What was the average score after 4 months? after 24 months?

  c) Graph the function.

  d) After what time t was the average score 60%?

33. *Wind Power.* U.S. wind-power capacity has grown exponentially from 4232 megawatts in 2001 to 46,919 megawatts in 2011.

  Source: U.S. Department of Energy.

  a) Find the exponential growth rate k and write an equation for an exponential function that can be used to predict U.S. wind-power capacity, in megawatts, t years after 2001.

  b) Estimate the year in which wind-power capacity will reach 100,000 megawatts.

34. *Spread of a Computer Virus.* The number of computers infected by a virus t days after it first appears usually increases exponentially. In 2009, the "Conflicker" worm spread from about 2.4 million computers on January 12 to about 3.2 million computers on January 13.

  Source: Based on data from PC World

  a) Find the exponential growth rate k and write an equation for an exponential function that can be used to predict the number of computers infected t days after January 12, 2009.

  b) Assuming exponential growth, estimate how long it took the Conflicker worm to infect 10 million computers.

35. *Pharmaceuticals.* The concentration of acetaminophen in the body decreases exponentially after a dosage is given. In one clinical study, adult subjects averaged 11 micrograms/milliliter (mcg/mL) of the drug in their blood plasma 1 hr after a 1000-mg dosage and 2 micrograms/milliliter 6 hr after dosage.

Sources: Based on information from tylenolprofessional.com and Mark Knopp, M.D.

a) Find the value k, and write an equation for an exponential function that can be used to predict the concentration of acetaminophen, in micrograms/milliliter, t hours after a 1000-mg dosage.

b) Estimate the concentration of acetaminophen 3 hr after a 1000-mg dosage.

c) To relieve a fever, the concentration of acetaminophen should go no lower than 4 mcg/mL. After how many hours will a 1000-mg dosage drop to that level?

d) Find the half-life of acetaminophen.

36. *Atmospheric Pressure.* The atmospheric pressure in the lower stratosphere decreases exponentially from 473 lb/ft^2 at 36,152 ft to 51 lb/ft^2 at 82,345 ft.

Source: Based on information from grc.nasa.gov

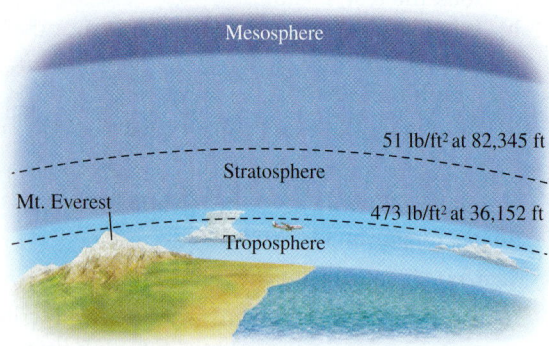

a) Find the exponential decay rate k, and write an equation for an exponential function that can be used to estimate the atmospheric pressure in the stratosphere h feet above 36,152 ft.

b) Estimate the atmospheric pressure at 50,000 ft ($h = 50,000 - 36,152$).

c) At what height is the atmospheric pressure 100 lb/ft^2?

d) What change in altitude will result in atmospheric pressure being halved?

37. *Archaeology.* A date palm seedling is growing in Kibbutz Ketura, Israel, from a seed found in King Herod's palace at Masada. The seed had lost 21% of its carbon-14. How old was the seed? (See Example 7.)

Source: Based on information from www.sfgate.com

38. *Archaeology.* Soil from beneath the Kish Church in Azerbaijan was found to have lost 12% of its carbon-14. How old was the soil? (See Example 7.)

Source: Based on information from www.azer.com

39. *Chemistry.* The exponential decay rate of iodine-131 is 9.6% per day. What is its half-life?

40. *Chemistry.* The decay rate of krypton-85 is 6.3% per year. What is its half-life?

41. *Caffeine.* The half-life of caffeine in the human body for a healthy adult is approximately 5 hr.

a) What is the exponential decay rate?

b) How long will it take 95% of the caffeine consumed to leave the body?

42. *Home Construction.* The chemical urea formaldehyde was used in some insulation in houses built during the mid to late 1960s. Unknown at the time was the fact that urea formaldehyde emitted toxic fumes as it decayed. The half-life of urea formaldehyde is 1 year.

a) What is its decay rate?

b) How long will it take 95% of the urea formaldehyde present to decay?

43. *Art Masterpieces.* As of August 2012, the highest auction price for a sculpture was $104.3 million, paid in 2010 for Alberto Giacometti's bronze sculpture *Walking Man I*. The same sculpture was purchased for about $9 million in 1990.

Source: Based on information from *The New York Times*, 02/02/10

a) Find the exponential growth rate k, and determine the exponential growth function that can be used to estimate the sculpture's value $V(t)$, in millions of dollars, t years after 1990.

b) Estimate the value of the sculpture in 2020.

c) What is the doubling time for the value of the sculpture?

d) How long after 1990 will the value of the sculpture be $1 billion?

44. *Value of a Sports Card.* Legend has it that because he objected to teenagers smoking, and because his first baseball card was issued in cigarette packs, the great shortstop Honus Wagner halted production of his card before many were produced. One of these cards was purchased in 1991 by hockey great Wayne Gretzky (and a partner) for $451,000. The same card was sold in 2007 for $2.8 million. For the following questions, assume that the card's value increases exponentially, as it has for many years.

a) Find the exponential growth rate k, and determine an exponential function that can be used to estimate the dollar value, $V(t)$, of the card t years after 1991.

b) Predict the value of the card in 2015.

c) What is the doubling time for the value of the card?

d) In what year will the value of the card first exceed $8,000,000?

45. Write a problem for a classmate to solve in which information is provided and the classmate is asked to find an exponential growth function. Make the problem as realistic as possible.

46. Examine the restriction on t in Exercise 32.

a) What upper limit might be placed on t?

b) In practice, would this upper limit ever be enforced? Why or why not?

Skill Review

Find a linear function whose graph has the given characteristics.

47. Slope: 18; y-intercept: $\left(0, \frac{1}{2}\right)$ [3.6], [7.3]

48. Contains $(6, 11)$ and $(-6, -11)$ [3.7], [7.3]

49. Parallel to $2x - 3y = 4$; contains $(-3, 7)$ [3.7], [7.3]

50. Perpendicular to $y = \frac{1}{2}x + 3$; y-intercept: $(0, 8)$ [3.7], [7.3]

Synthesis

51. Can the model used in Example 5 to predict the number of unique visitors to Pinterest still be valid? Why or why not?

52. *Atmospheric Pressure.* Atmospheric pressure P at an elevation a feet above sea level can be estimated by

$$P = P_0 e^{-0.00004a},$$

where P_0 is the pressure at sea level, which is approximately 29.9 in. of mercury (Hg). Explain how a barometer, or some other device for measuring atmospheric pressure, can be used to find the height of a skyscraper.

53. *Sports Salaries.* As of August 2012, Alex Rodriguez of the New York Yankees had the largest contract in sports history. The 10-year $275-million deal, signed in 2007, stipulated that he receive $20 million in 2016. How much money would have been invested in 2007, at 4% interest compounded continuously, in order to have $20 million for Rodriguez in 2016? (This is much like determining what $20 million in 2016 is worth in 2007 dollars.)

Source: The San Francisco Chronicle

54. *Supply and Demand.* The supply and demand for the sale of stereos by Sound Ideas are given by

$$S(x) = e^x \quad \text{and} \quad D(x) = 162,755e^{-x},$$

where $S(x)$ is the price at which the company is willing to supply x stereos and $D(x)$ is the demand price for a quantity of x stereos. Find the equilibrium point. (For reference, see Section 8.8.)

55. *Stellar Magnitude.* The apparent stellar magnitude m of a star with received intensity I is given by

$$m(I) = -(19 + 2.5 \cdot \log I),$$

where I is in watts per square meter (W/m^2). The smaller the apparent stellar magnitude, the brighter the star appears.

Source: The Columbus Optical SETI Observatory

a) The intensity of light received from the sun is $1390 \text{ W}/\text{m}^2$. What is the apparent stellar magnitude of the sun?

b) The 5-m diameter Hale telescope on Mt. Palomar can detect a star with magnitude $+23$. What is the received intensity of light from such a star?

56. *Growth of Bacteria.* The bacteria *Escherichia coli* (*E. coli*) are commonly found in the human bladder. Suppose that 3000 of the bacteria are present at time $t = 0$. Then t minutes later, the number of bacteria present is

$$N(t) = 3000(2)^{t/20}.$$

If 100,000,000 bacteria accumulate, a bladder infection can occur. If, at 11:00 A.M., a patient's bladder contains 25,000 *E. coli* bacteria, at what time can infection occur?

57. Show that for exponential growth at rate k, the doubling time T is given by $T = \dfrac{\ln 2}{k}$.

58. Show that for exponential decay at rate k, the half-life T is given by $T = \dfrac{\ln 2}{k}$.

 YOUR TURN ANSWERS: SECTION 12.7

1. $10^6\,\text{W/m}^2$ **2.** 5.01×10^{-7} moles per liter
3. **(a)** About 23.8 years; **(b)** about 14.2 years
4. $4{,}851{,}651{,}954\,\text{cm}^2$, or $485{,}165\,\text{m}^2$
5. $k = 1.308$; $P(t) = 10e^{1.308t}$, where $P(t)$ is the number of malware variants t years after 2011
6. $P(t) = 20{,}000e^{0.02t}$; 34.7 years   **7.** About 4000 years

---

QUICK QUIZ: SECTIONS 12.1–12.7

1. Determine whether $f(x) = 7 - x$ is one-to-one. [12.1]

2. Simplify: $\log_b \sqrt[4]{b^3}$.   [12.4]

3. Use a calculator to find $\dfrac{\log 15}{\log 2}$. Round to four decimal places.   [12.5]

4. Solve: $\log_2 (x - 3) + \log_2 (x + 3) = 4$.   [12.6]

5. Stephanie invests \$10,000 at 3% interest, compounded annually. How long does it take for the investment to double itself?   [12.7]

PREPARE TO MOVE ON

1. Find the distance between $(-3, 7)$ and $(-2, 6)$.   [10.7]

2. Find the coordinates of the midpoint of the segment connecting $(3, -8)$ and $(5, -6)$.   [10.7]

3. Solve by completing the square: $x^2 + 8x = 1$.   [12.1]

4. Graph: $y = x^2 - 5x - 6$.   [12.7]

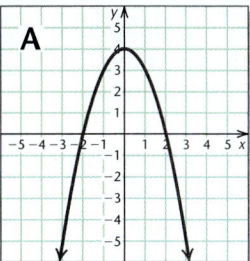

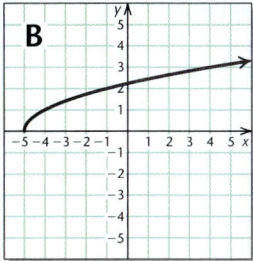

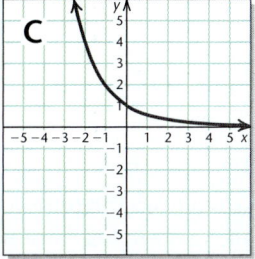

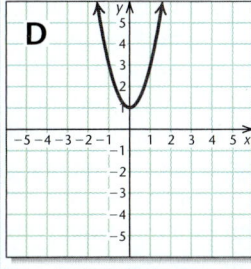

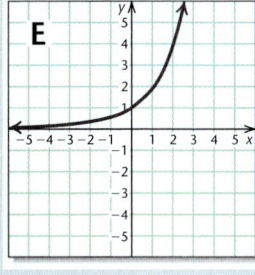

Visualizing for Success

Use after Section 12.5.

Match each function with its graph.

1. $f(x) = 2x - 3$

2. $f(x) = 2x^2 + 1$

3. $f(x) = \sqrt{x + 5}$

4. $f(x) = |x - 4|$

5. $f(x) = \ln x$

6. $f(x) = 2^{-x}$

7. $f(x) = -4$

8. $f(x) = \log x + 3$

9. $f(x) = 2^x$

10. $f(x) = 4 - x^2$

Answers on page A-67

An additional, animated version of this activity appears in MyMathLab. *To use MyMathLab, you need a course ID and a student access code. Contact your instructor for more information.*

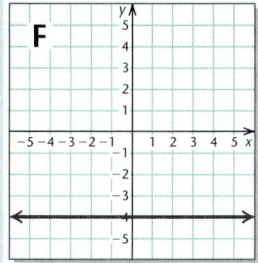

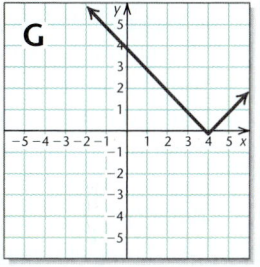

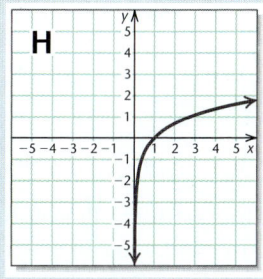

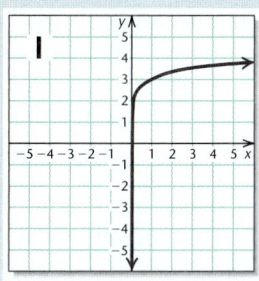

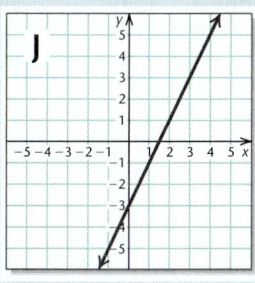

Collaborative Activity   *The True Cost of a New Car*

Focus: Car loans and exponential functions
Use after: Section 12.2
Time: 30 minutes
Group size: 2
Materials: Calculators with exponentiation keys

The formula

$$M = \frac{Pr}{1 - (1 + r)^{-n}}$$

is used to determine the payment size, M, when a loan of P dollars is to be repaid in n equally sized monthly payments. Here, r represents the monthly interest rate. Loans repaid in this fashion are said to be *amortized* (spread out equally) over a period of n months.

Activity

1. Suppose one group member is selling the other a car for $2600, financed at 1% interest per month for 24 months. What should be the size of each monthly payment?

2. Suppose both group members are shopping for the same model new car. To save time, each group member visits a different dealer. One dealer offers the car for $13,000 at 10.5% interest (0.00875 monthly interest) for 60 months (no down payment). The other dealer offers the same car for $12,000, but at 12% interest (0.01 monthly interest) for 48 months (no down payment).

   a) Determine the monthly payment size for each offer. Then determine the total amount paid for the car under each offer. How much of each total is interest?

   b) Work together to find the annual interest rate for which the total cost of 60 monthly payments for the $13,000 car would equal the total amount paid for the $12,000 car (as found in part a above).

Decision Making & Connection   *(Use after Section 12.7.)*

College Costs. It is difficult to plan for future college costs, but you can use historical data of costs from previous years to help predict future costs. Here, we assume that the costs are rising exponentially.

1. The average cost of in-state tuition and fees for a public four-year college, adjusted for inflation, was $4430 in 2000–2001 and $8244 in 2010–2011. Find an exponential function that fits the data.
   Source: collegeboard.com

2. Many students receive financial aid. Although this amount varies widely, the average amount of education tax benefits and grant aid per student, adjusted for inflation, was $2050 in 2000–2001 and $5750 in 2010–2011. Find an exponential function that fits the data.
   Source: collegeboard.com

3. Use the functions found in steps (1) and (2) to estimate the average in-state tuition and fees and the average amount of financial aid awarded for the school year following the one in which you are currently enrolled.

4. Subtract the financial aid awarded from the tuition and fees to estimate the average in-state net tuition cost to a student for the next school year.

5. *Research.* Find the cost of tuition and fees for two years for the college you currently attend. If possible, use the numbers for this year and for a year 5–10 years in the past. Use this information to find an exponential function that fits the data and then estimate the cost of tuition and fees for next year. If you have received financial aid for more than one year, use the data for two years to estimate your financial aid for next year. Then estimate what your net cost for the next school year would be.

Study Summary

KEY TERMS AND CONCEPTS	EXAMPLES	PRACTICE EXERCISES

SECTION 12.1:   *Composite Functions and Inverse Functions*

The **composition** of f and g is defined as
$$(f \circ g)(x) = f(g(x)).$$

If $f(x) = \sqrt{x}$ and $g(x) = 2x - 5$, then
$$(f \circ g)(x) = f(g(x)) = f(2x - 5)$$
$$= \sqrt{2x - 5}.$$

1. Find $(f \circ g)(x)$ if $f(x) = 1 - 6x$ and $g(x) = x^2 - 3$.

A function f is **one-to-one** if different inputs always have different outputs. The graph of a one-to-one function passes the **horizontal-line test.**

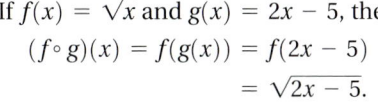

f is *not* one-to-one f is one-to-one

2. Determine whether $f(x) = 5x - 7$ is one-to-one.

If f is one-to-one, it is possible to find its inverse:

1. Replace $f(x)$ with y.
2. Interchange x and y.
3. Solve for y.

4. Replace y with $f^{-1}(x)$.

If $f(x) = 2x - 3$, find $f^{-1}(x)$.

1. $y = 2x - 3$
2. $x = 2y - 3$
3. $x + 3 = 2y$
$$\frac{x + 3}{2} = y$$

4. $\dfrac{x + 3}{2} = f^{-1}(x)$

3. If $f(x) = 5x + 1$, find $f^{-1}(x)$.

SECTION 12.2:   *Exponential Functions*
SECTION 12.3:   *Logarithmic Functions*

For an **exponential function** f:
  $f(x) = a^x,\ a > 0,\ a \neq 1$;
  Domain of f is $\mathbb{R}$;
  $f^{-1}(x) = \log_a x$.
For a **logarithmic function** g:
  $g(x) = \log_a x,\ a > 0,\ a \neq 1$;
  Domain of g is $(0, \infty)$;
  $g^{-1}(x) = a^x$.

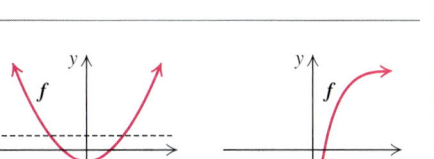

$f(x) = a^x,\ a > 1$

$g(x) = \log_a x,\ a > 1$

4. Graph by hand:
$f(x) = 2^x$.

5. Graph by hand:
$f(x) = \log x$.

$\log_a x = m$ means $a^m = x$.

Solve: $\log_8 x = 2$.
$\qquad 8^2 = x$ *Rewriting as an exponential equation*
$\qquad 64 = x$

6. Rewrite as an equivalent logarithmic equation:
$5^4 = 625$.

841

There are a variety of applications and equations with graphs that are conic sections. A circle is one example of a *conic section*, meaning that it can be regarded as a cross section of a cone.

13.1   Conic Sections: Parabolas and Circles

Parabolas   •   Circles

This section and the next two examine curves formed by cross sections of cones. These curves are all graphs of $Ax^2 + By^2 + Cxy + Dx + Ey + F = 0$. The constants $A, B, C, D, E,$ and F determine which of the following shapes serve as the graph.

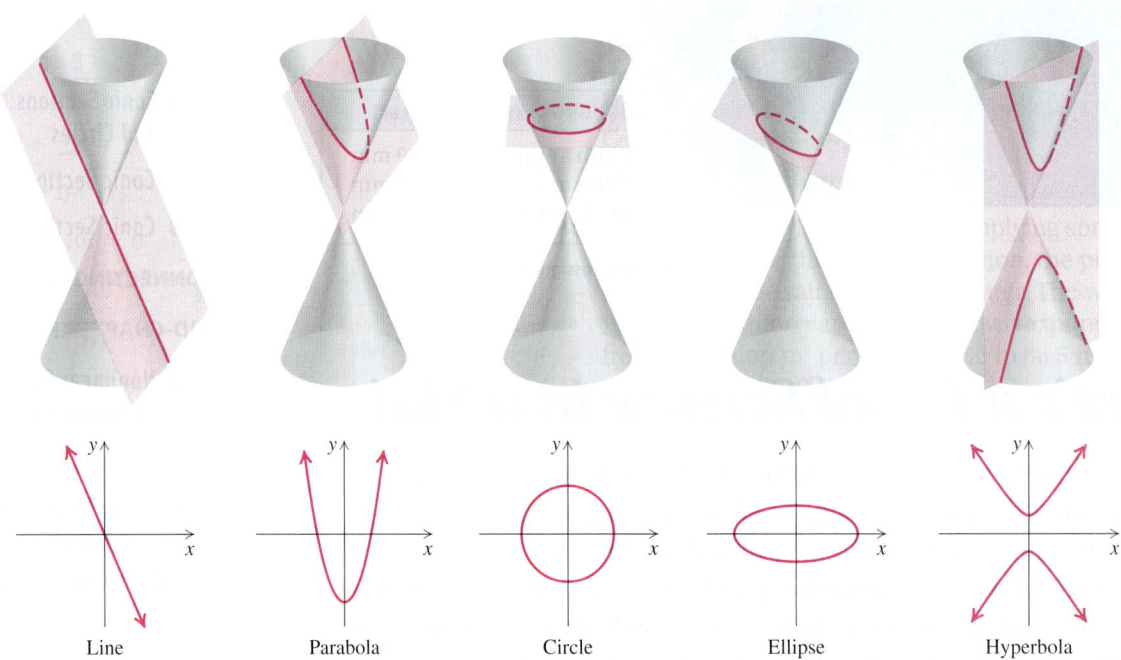

| Line | Parabola | Circle | Ellipse | Hyperbola |

Parabolas

When a cone is sliced as in the second figure above, the conic section formed is a **parabola**. Parabolas have many applications in electricity, mechanics, and optics. A cross section of a contact lens or a satellite dish is a parabola, and arches that support certain bridges are parabolas.

EQUATION OF A PARABOLA

A parabola with a vertical axis of symmetry opens upward or downward and has an equation that can be written in the form

$$y = ax^2 + bx + c.$$

A parabola with a horizontal axis of symmetry opens to the right or to the left and has an equation that can be written in the form

$$x = ay^2 + by + c.$$

Parabolas with equations of the form $f(x) = ax^2 + bx + c$ were graphed in Chapter 11.

EXAMPLE 1   Graph: $y = x^2 - 4x + 9$.

SOLUTION   To locate the vertex, we can use either of two approaches. One way is to complete the square:

$$y = (x^2 - 4x) + 9 \qquad \text{Note that half of } -4 \text{ is } -2, \text{ and } (-2)^2 = 4.$$
$$= (x^2 - 4x + 4 - 4) + 9 \qquad \text{Adding and subtracting 4}$$
$$= (x^2 - 4x + 4) + (-4 + 9) \qquad \text{Regrouping}$$
$$= (x - 2)^2 + 5. \qquad \text{Factoring and simplifying}$$

The vertex is $(2, 5)$.

A second way to find the vertex is to recall that the x-coordinate of the vertex of the parabola given by $y = ax^2 + bx + c$ is $-b/(2a)$:

$$x = -\frac{b}{2a} = -\frac{-4}{2(1)} = 2.$$

To find the y-coordinate of the vertex, we substitute 2 for x:

$$y = x^2 - 4x + 9 = 2^2 - 4(2) + 9 = 5.$$

Either way, the vertex is $(2, 5)$. Next, we calculate and plot some points on each side of the vertex. As expected for a positive coefficient of x^2, the graph opens upward.

x	y	
2	5	←Vertex
0	9	←y-intercept
1	6	
3	6	
4	9	

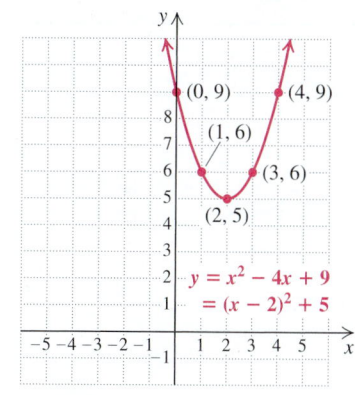

1. Graph: $y = x^2 + 2x - 3$.

 YOUR TURN

TO GRAPH AN EQUATION OF THE FORM $y = ax^2 + bx + c$

1. Find the vertex (h, k) either by completing the square to find an equivalent equation

   $$y = a(x - h)^2 + k,$$

   or by using $-b/(2a)$ to find the x-coordinate and substituting to find the y-coordinate.
2. Choose other values for x on each side of the vertex, and compute the corresponding y-values.
3. The graph opens upward for $a > 0$ and downward for $a < 0$.

If we interchange x and y in the equation in Example 1, we obtain an equation for the *inverse* relation, $x = y^2 - 4y + 9$. The graph of this equation will be the reflection of the graph in Example 1 across $y = x$.

Any equation of the form $x = ay^2 + by + c$ represents a horizontal parabola that opens to the right for $a > 0$, opens to the left for $a < 0$, and has an axis of symmetry parallel to the x-axis.

EXAMPLE 2 Graph: $x = y^2 - 4y + 9$.

SOLUTION This equation is like that in Example 1 but with x and y interchanged. The vertex is $(5, 2)$ instead of $(2, 5)$. To find ordered pairs, we choose values for y on each side of the vertex. Then we compute values for x. Note that the x- and y-values of the table in Example 1 are now switched. You should confirm that, by completing the square, we have $x = (y - 2)^2 + 5$.

x	y	
5	2	←Vertex
9	0	←x-intercept
6	1	
6	3	
9	4	

(1) Choose values for y.
(2) Compute values for x.

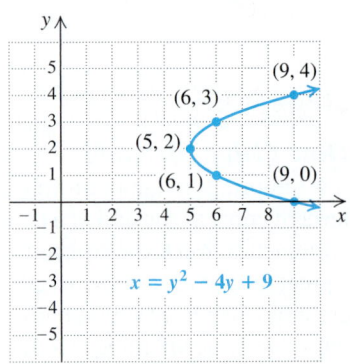

$x = y^2 - 4y + 9$

2. Graph: $x = y^2 + 2y - 3$.

YOUR TURN

TO GRAPH AN EQUATION OF THE FORM $x = ay^2 + by + c$

1. Find the vertex (h, k) either by completing the square to find an equivalent equation

$$x = a(y - k)^2 + h,$$

or by using $-b/(2a)$ to find the y-coordinate and substituting to find the x-coordinate.
2. Choose other values for y that are on either side of k and compute the corresponding x-values.
3. The graph opens to the right if $a > 0$ and to the left if $a < 0$.

EXAMPLE 3 Graph: $x = -2y^2 + 10y - 7$.

SOLUTION We find the vertex by completing the square:

$$x = -2y^2 + 10y - 7$$
$$= -2(y^2 - 5y \qquad) - 7$$
$$= -2\left(y^2 - 5y + \tfrac{25}{4}\right) - 7 - (-2)\tfrac{25}{4} \qquad \tfrac{1}{2}(-5) = \tfrac{-5}{2}; \left(\tfrac{-5}{2}\right)^2 = \tfrac{25}{4}; \text{ we add and subtract } (-2)\tfrac{25}{4}.$$
$$= -2\left(y - \tfrac{5}{2}\right)^2 + \tfrac{11}{2}. \qquad \text{Factoring and simplifying}$$

The vertex is $\left(\tfrac{11}{2}, \tfrac{5}{2}\right)$.

For practice, we also find the vertex by first computing its y-coordinate, $-b/(2a)$, and then substituting to find the x-coordinate:

$$y = -\frac{b}{2a} = -\frac{10}{2(-2)} = \frac{5}{2}$$

$$x = -2y^2 + 10y - 7 = -2\left(\tfrac{5}{2}\right)^2 + 10\left(\tfrac{5}{2}\right) - 7$$
$$= \tfrac{11}{2}.$$

To find ordered pairs, we choose values for y on each side of the vertex and then compute values for x. A table is shown below, together with the graph. The graph opens to the left because the y^2-coefficient, -2, is negative.

x	y	
$\frac{11}{2}$	$\frac{5}{2}$	← Vertex
-7	0	← x-intercept
5	2	
5	3	
1	1	
1	4	
-7	5	

(1) Choose these values for y.

(2) Compute these values for x.

3. Graph: $x = -3y^2 - 6y + 1$.

⟲ YOUR TURN

Circles

Another conic section, the **circle**, is the set of points in a plane that are a fixed distance r, called the **radius** (plural, **radii**), from a fixed point (h, k), called the **center**. Note that the word radius can mean either any segment connecting a point on a circle to the center or the length of such a segment. Using the idea of a fixed distance r and the distance formula,

$$d = \sqrt{(x_2 - x_1)^2 + (y_2 - y_1)^2},$$

we can find the equation of a circle.

If (x, y) is on a circle of radius r, centered at (h, k), then by the definition of a circle and the distance formula, it follows that

$$r = \sqrt{(x - h)^2 + (y - k)^2}.$$

Squaring both sides gives the equation of a circle in standard form.

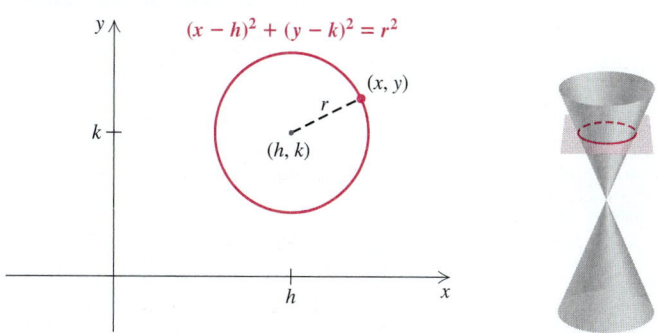

$$(x - h)^2 + (y - k)^2 = r^2$$

> **EQUATION OF A CIRCLE (STANDARD FORM)**
> The equation of a circle, centered at (h, k), with radius r, is given by
> $$(x - h)^2 + (y - k)^2 = r^2.$$

Note that for $h = 0$ and $k = 0$, the circle is centered at the origin. Otherwise, the circle is translated $|h|$ units horizontally and $|k|$ units vertically.

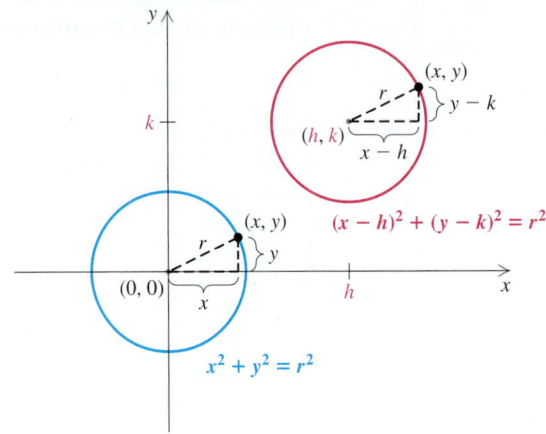

EXAMPLE 4   Find an equation of the circle centered at $(4, -5)$ with radius 6.

SOLUTION   Using the standard form, we obtain

$$(x - 4)^2 + (y - (-5))^2 = 6^2, \quad \text{Using } (x - h)^2 + (y - k)^2 = r^2$$

or $(x - 4)^2 + (y + 5)^2 = 36$.

4. Find an equation of the circle centered at $(-2, 7)$ with radius 9.

↩ YOUR TURN

EXAMPLE 5   Find the center and the radius and then graph each circle.

a) $(x - 2)^2 + (y + 3)^2 = 4^2$

b) $x^2 + y^2 + 8x - 2y + 15 = 0$

SOLUTION

a)   We write standard form:

$$(x - 2)^2 + [y - (-3)]^2 = 4^2.$$

The center is $(2, -3)$ and the radius is 4. To graph, we plot the points $(2, 1)$, $(2, -7)$, $(-2, -3)$, and $(6, -3)$, which are, respectively, 4 units above, below, left, and right of $(2, -3)$. We then either sketch a circle by hand or use a compass.

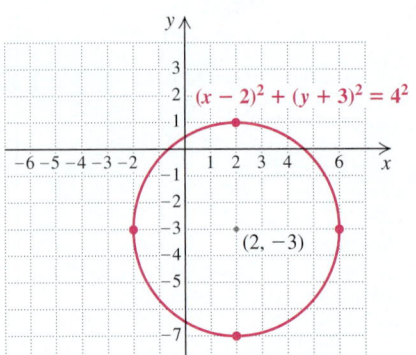

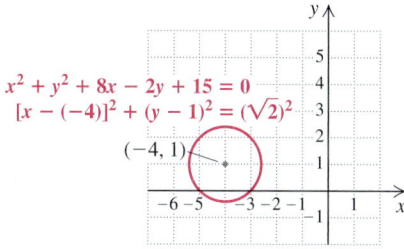

$x^2 + y^2 + 8x - 2y + 15 = 0$
$[x - (-4)]^2 + (y - 1)^2 = (\sqrt{2})^2$

$(-4, 1)$

b) To write the equation $x^2 + y^2 + 8x - 2y + 15 = 0$ in standard form, we complete the square twice, once with $x^2 + 8x$ and once with $y^2 - 2y$:

$$x^2 + y^2 + 8x - 2y + 15 = 0$$

$$x^2 + 8x \qquad + y^2 - 2y \qquad = -15 \qquad \text{Grouping the } x\text{-terms and the } y\text{-terms; subtracting 15 from both sides}$$

$$x^2 + 8x + 16 + y^2 - 2y + 1 = -15 + 16 + 1 \qquad \text{Adding } \left(\tfrac{8}{2}\right)^2, \text{ or 16, and } \left(-\tfrac{2}{2}\right)^2, \text{ or 1, to both sides to get standard form}$$

$$(x + 4)^2 + (y - 1)^2 = 2 \qquad \text{Factoring}$$

$$[x - (-4)]^2 + (y - 1)^2 = (\sqrt{2})^2. \qquad \text{Writing standard form}$$

5. Find the center and the radius and then graph the circle:

$$x^2 + (y - 3)^2 = 5.$$

The center is $(-4, 1)$ and the radius is $\sqrt{2}$.

YOUR TURN

Technology Connection

Graphing the equation of a circle using a graphing calculator usually requires two steps:

1. Solve the equation for y. The result will include a $\pm$ sign in front of a radical.
2. Graph two functions, one for the $+$ sign and the other for the $-$ sign, on the same set of axes.

For example, to graph $(x - 3)^2 + (y + 1)^2 = 16$, solve for $y + 1$ and then y:

$$(y + 1)^2 = 16 - (x - 3)^2$$

$$y + 1 = \pm\sqrt{16 - (x - 3)^2}$$

$$y = -1 \pm \sqrt{16 - (x - 3)^2},$$

or $\qquad y_1 = -1 + \sqrt{16 - (x - 3)^2}$

and $\qquad y_2 = -1 - \sqrt{16 - (x - 3)^2}.$

When both functions are graphed in a "squared" window (to eliminate distortion), the result is as follows.

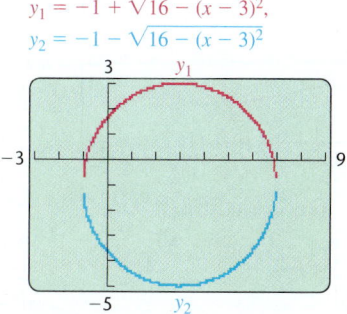

$y_1 = -1 + \sqrt{16 - (x - 3)^2},$
$y_2 = -1 - \sqrt{16 - (x - 3)^2}$

On many calculators, pressing **APPS** and selecting Conics and then Circle accesses a program in which equations in standard form can be graphed directly and then Traced.

Graph each of the following equations.

1. $x^2 + y^2 - 16 = 0$
2. $(x - 1)^2 + (y - 2)^2 = 25$
3. $(x + 3)^2 + (y - 5)^2 = 16$

13.1 **EXERCISE SET**

FOR EXTRA HELP

MyMathLab® MathXL

PRACTICE   WATCH   READ   REVIEW

Vocabulary and Reading Check

Choose the word from the following list that best completes each statement. Words may be used more than once or not at all.

center	horizontal	vertex
circle	parabola	vertical
conic sections	radii	

1. Parabolas and circles are examples of

_____.

2. A(n) _____ is the set of points in a plane that are a fixed distance from its center.

3. A parabola with a(n) _____ axis of symmetry opens to the right or to the left.

4. In the equation of a parabola, the point (h, k) represents the _____ of the parabola.

5. In the equation of a circle, the point (h, k) represents the _____ of the circle.

6. A radius is the distance from a point on a circle to the _____.

 Concept Reinforcement

In each of Exercises 7–12, match the equation with the graph of that equation from those shown.

7. ____ $(x - 2)^2 + (y + 5)^2 = 9$

8. ____ $(x + 2)^2 + (y - 5)^2 = 9$

9. ____ $y = (x - 2)^2 - 5$

10. ____ $y = (x - 5)^2 - 2$

11. ____ $x = (y - 2)^2 - 5$

12. ____ $x = (y - 5)^2 - 2$

a)
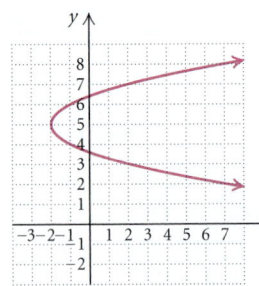

b)
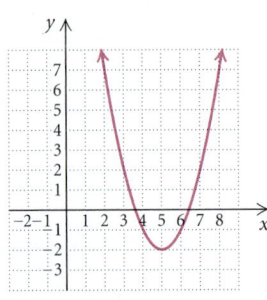

c)
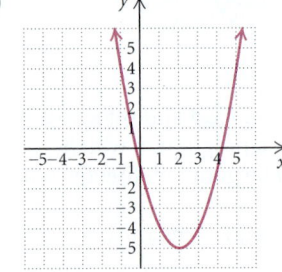

d)
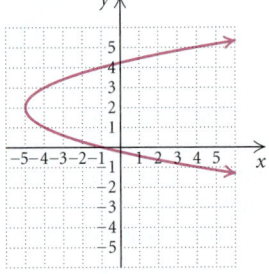

e)
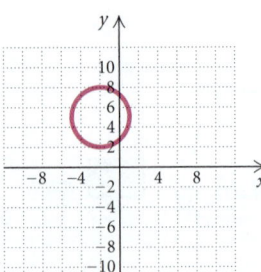

f)
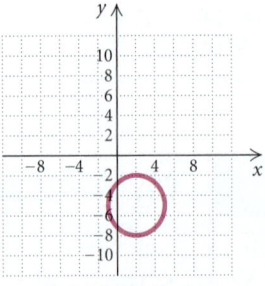

Parabolas

Graph. Be sure to label each vertex.

13. $y = -x^2$

14. $y = 2x^2$

15. $y = -x^2 + 4x - 5$

16. $x = 4 - 3y - y^2$

17. $x = y^2 - 4y + 2$

18. $y = x^2 + 2x + 3$

19. $x = y^2 + 3$

20. $x = -y^2$

21. $x = 2y^2$

22. $x = y^2 - 1$

23. $x = -y^2 - 4y$

24. $x = y^2 + 3y$

25. $y = x^2 - 2x + 1$

26. $y = x^2 + 2x + 1$

27. $x = -\frac{1}{2}y^2$

28. $y = -\frac{1}{2}x^2$

29. $x = -y^2 + 2y - 1$

30. $x = -y^2 - 2y + 3$

31. $x = -2y^2 - 4y + 1$

32. $x = 2y^2 + 4y - 1$

Circles

Find an equation of the circle satisfying the given conditions.

33. Center $(0, 0)$, radius 8

34. Center $(0, 0)$, radius 11

35. Center $(7, 3)$, radius $\sqrt{6}$

36. Center $(5, 6)$, radius $\sqrt{11}$

37. Center $(-4, 3)$, radius $3\sqrt{2}$

38. Center $(-2, 7)$, radius $2\sqrt{5}$

39. Center $(-5, -8)$, radius $10\sqrt{3}$

40. Center $(-7, -2)$, radius $5\sqrt{2}$

Aha! **41.** Center $(0, 0)$, passing through $(-3, 4)$

42. Center $(0, 0)$, passing through $(11, -10)$

43. Center $(-4, 1)$, passing through $(-2, 5)$

44. Center $(-1, -3)$, passing through $(-4, 2)$

Find the center and the radius of each circle. Then graph the circle.

45. $x^2 + y^2 = 1$

46. $x^2 + y^2 = 25$

47. $(x + 1)^2 + (y + 3)^2 = 49$

48. $(x - 2)^2 + (y + 3)^2 = 100$

49. $(x - 4)^2 + (y + 3)^2 = 10$

50. $(x + 5)^2 + (y - 1)^2 = 15$

51. $x^2 + y^2 = 8$

52. $x^2 + y^2 = 20$

53. $(x - 5)^2 + y^2 = \frac{1}{4}$

54. $x^2 + (y - 1)^2 = \frac{1}{25}$

55. $x^2 + y^2 + 8x - 6y - 15 = 0$

56. $x^2 + y^2 + 6x - 4y - 15 = 0$

57. $x^2 + y^2 - 8x + 2y + 13 = 0$

58. $x^2 + y^2 + 6x + 4y + 12 = 0$

59. $x^2 + y^2 + 10y - 75 = 0$

60. $x^2 + y^2 - 8x - 84 = 0$

61. $x^2 + y^2 + 7x - 3y - 10 = 0$

62. $x^2 + y^2 - 21x - 33y + 17 = 0$

63. $36x^2 + 36y^2 = 1$

64. $4x^2 + 4y^2 = 1$

65. Does the graph of an equation of a circle include the point that is the center? Why or why not?

66. Is a point a conic section? Why or why not?

Skill Review

Simplify. Assume all variables represent positive numbers.

67. $\sqrt[4]{48x^7y^{12}}$ [10.3]

68. $\sqrt{y}\sqrt[3]{y^2}$ [10.5]

69. $\dfrac{\sqrt{200x^4w^2}}{\sqrt{2w}}$ [10.4]

70. $\dfrac{\sqrt[3]{t}}{\sqrt[10]{t}}$ [10.5]

71. $\sqrt{8} - 2\sqrt{2} + \sqrt{12}$ [10.5]

72. $\left(3 + \sqrt{2}\right)\left(4\sqrt{3} - \sqrt{2}\right)$ [10.5]

Synthesis

73. On a piece of graph paper, draw a line and a point not on the line. Then plot several points that are equidistant from the point and the line. What shape do the points appear to form? How could you confirm this?

74. If an equation has two variable terms with the same degree, can its graph be a parabola? Why or why not?

Find an equation of a circle satisfying the given conditions.

75. Center $(3, -5)$ and tangent to (touching at one point) the y-axis

76. Center $(-7, -4)$ and tangent to the x-axis

77. The endpoints of a diameter are $(7, 3)$ and $(-1, -3)$.

78. Center $(-3, 5)$ with a circumference of 8π units

79. *Wrestling.* The equation $x^2 + y^2 = \frac{81}{4}$, where x and y represent the number of meters from the center, can be used to draw the outer circle on a wrestling mat used in International, Olympic, and World Championship wrestling. The equation $x^2 + y^2 = 16$ can be used to draw the inner edge of the red zone. Find the area of the red zone.

Source: Based on data from the Government of Western Australia

80. *Snowboarding.* Each side edge of the Burton X8 155 snowboard is an arc of a circle with a "running length" of 1180 mm and a "sidecut depth" of 23 mm (see the figure below).

Source: evogear.com

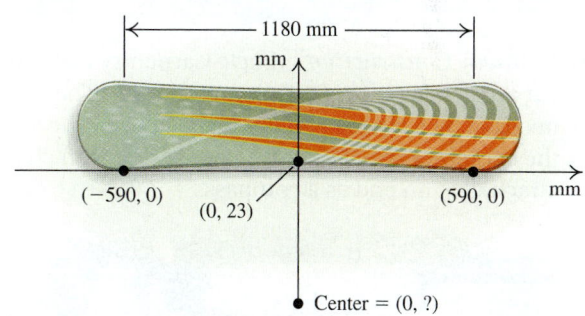

a) Using the coordinates shown, locate the center of the circle. (*Hint*: Equate distances.)

b) What radius is used for the edge of the board?

81. *Snowboarding.* The Never Summer Infinity 149 snowboard has a running length of 1160 mm and a sidecut depth of 23.5 mm (see Exercise 80). What radius is used for the edge of this snowboard?

Source: neversummer.com

82. *Skiing.* The Rossignol Experience 98 ski, when lying flat and viewed from above, has edges that are arcs of a circle. (Actually, each edge is made of two arcs of slightly different radii. The arc for the rear half of the ski edge has a slightly larger radius.)

Source: rossignol.com

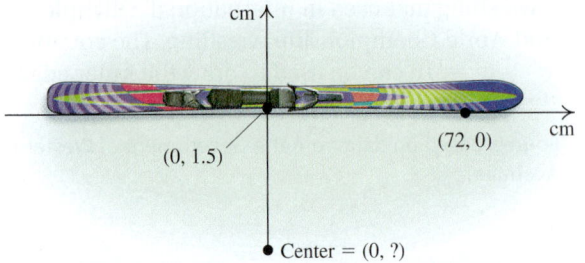

a) Using the coordinates shown, locate the center of the circle. (*Hint*: Equate distances.)
b) What radius is used for the arc passing through $(0, 1.5)$ and $(72, 0)$?

83. *Doorway Construction.* Engle Carpentry needs to cut an arch for the top of an entranceway. The arch needs to be 8 ft wide and 2 ft high. To draw the arch, the carpenters will use a stretched string with chalk attached at an end as a compass.

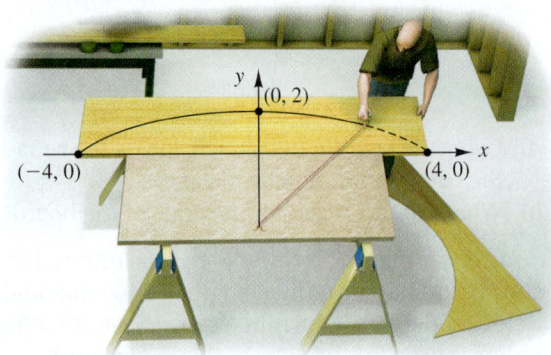

a) Using a coordinate system, locate the center of the circle.
b) What radius should the carpenters use to draw the arch?

84. *Archaeology.* During an archaeological dig, Estella finds the bowl fragment shown below. What was the original diameter of the bowl?

85. *Ferris Wheel Design.* A ferris wheel has a radius of 24.3 ft. Assuming that the center is 30.6 ft above the base of the ferris wheel and that the origin is below the center, as in the following figure, find an equation of the circle.

86. Use a graph of $x = y^2 - y - 6$ to approximate to the nearest tenth the solutions of each of the following.

a) $y^2 - y - 6 = 2$
b) $y^2 - y - 6 = -3$

87. *Power of a Motor.* The horsepower of a certain kind of engine is given by the formula

$$H = \frac{D^2 N}{2.5},$$

where N is the number of cylinders and D is the diameter, in inches, of each piston. Graph this equation, assuming that $N = 6$ (a six-cylinder engine). Let D run from 2.5 to 8. Then use the graph to estimate the diameter of each piston in a six-cylinder 120-horsepower engine.

88. If the equation $x^2 + y^2 - 6x + 2y - 6 = 0$ is written as $y^2 + 2y + (x^2 - 6x - 6) = 0$, it can be regarded as quadratic in y.

a) Use the quadratic formula to solve for y.
b) Show that the graph of your answer to part (a) coincides with the graph in the Technology Connection on p. 853.

89. Why should a graphing calculator's window be "squared" before graphing a circle?

YOUR TURN ANSWERS: SECTION 13.1

1.
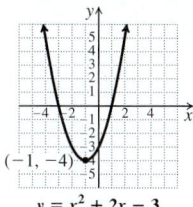

$y = x^2 + 2x - 3$

2.
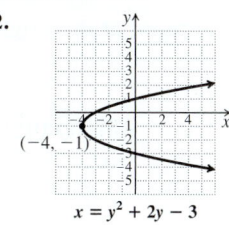

$x = y^2 + 2y - 3$

3.
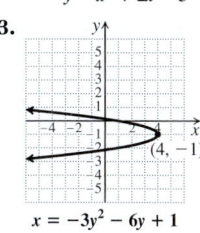

$x = -3y^2 - 6y + 1$

4. $(x - (-2))^2 + (y - 7)^2 = 81,$
or $(x + 2)^2 + (y - 7)^2 = 81$

5. $(0, 3); \sqrt{5}$

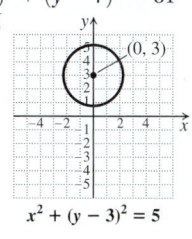

$x^2 + (y - 3)^2 = 5$

PREPARE TO MOVE ON

Solve. [11.1]

1. $\dfrac{y^2}{16} = 1$

2. $\dfrac{x^2}{a^2} = 1$

3. $\dfrac{(x - 1)^2}{25} = 1$

4. $\dfrac{1}{4} + \dfrac{(y + 3)^2}{36} = 1$

13.2 Conic Sections: Ellipses

Ellipses Centered at $(0, 0)$ ▪ Ellipses Centered at (h, k)

Study Skills

Preparing for the Final Exam

It is never too early to begin studying for a final exam. If you have at least three days, consider the following:

- Reviewing the highlighted or boxed information in each chapter;
- Studying the Chapter Tests, Review Exercises, Cumulative Reviews, and Study Summaries;
- Re-taking on your own all quizzes and tests;
- Attending any review sessions being offered;
- Organizing or joining a study group;
- Asking a tutor or a professor about specific trouble spots;
- Asking for previous final exams (and answers) to work for practice.

When a cone is sliced at an angle, as shown below, the conic section formed is an *ellipse*. To draw an ellipse, stick two tacks in a piece of cardboard. Then tie a loose string to the tacks, place a pencil as shown, and draw an oval by moving the pencil while stretching the string tight.

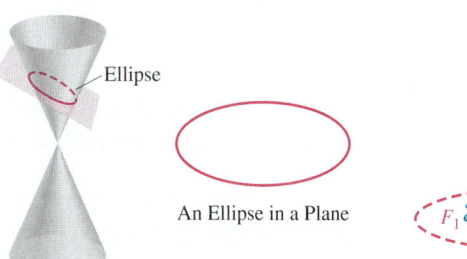

An Ellipse in a Plane

Ellipses Centered at $(0, 0)$

An **ellipse** is defined as the set of all points in a plane for which the sum of the distances from two fixed points F_1 and F_2 is constant. The points F_1 and F_2 are called **foci** (pronounced fō-sī), the plural of focus. In the figure above, the tacks are at the foci and the length of the string is the constant sum of the distances from the tacks to the pencil. The midpoint of the segment F_1F_2 is the **center**. The equation of an ellipse follows. Its derivation is outlined in Exercise 51.

> **EQUATION OF AN ELLIPSE CENTERED AT THE ORIGIN**
>
> The equation of an ellipse centered at the origin and symmetric with respect to both axes is
>
> $$\frac{x^2}{a^2} + \frac{y^2}{b^2} = 1, \quad a, b > 0. \qquad \text{(Standard form)}$$

To graph an ellipse centered at the origin, it helps to first find the intercepts. If we replace x with 0, we can find the y-intercepts:

$$\frac{0^2}{a^2} + \frac{y^2}{b^2} = 1$$

$$\frac{y^2}{b^2} = 1$$

$$y^2 = b^2 \quad \text{or} \quad y = \pm b.$$

Thus the y-intercepts are $(0, b)$ and $(0, -b)$. Similarly, the x-intercepts are $(a, 0)$ and $(-a, 0)$. If $a > b$, the ellipse is said to be horizontal and $(-a, 0)$ and $(a, 0)$ are referred to as the **vertices** (singular, **vertex**). If $b > a$, the ellipse is said to be vertical and $(0, -b)$ and $(0, b)$ are then the vertices.

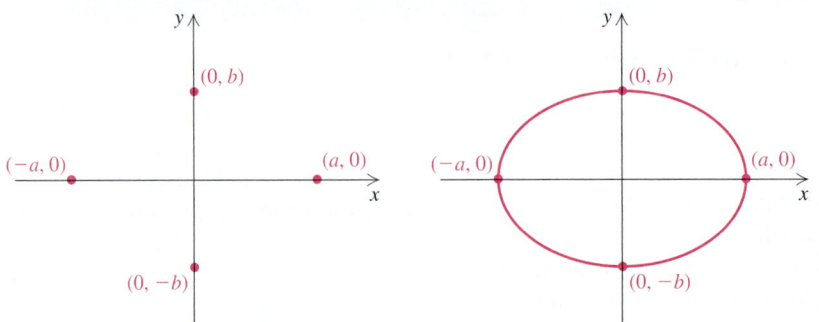

Plotting these four points and drawing an oval-shaped curve, we graph the ellipse. If a more precise graph is desired, we can plot more points.

> **USING a AND b TO GRAPH AN ELLIPSE**
>
> For the ellipse
>
> $$\frac{x^2}{a^2} + \frac{y^2}{b^2} = 1,$$
>
> the x-intercepts are $(-a, 0)$ and $(a, 0)$. The y-intercepts are $(0, -b)$ and $(0, b)$. For $a^2 > b^2$, the ellipse is horizontal. For $b^2 > a^2$, the ellipse is vertical.

EXAMPLE 1   Graph the ellipse

$$\frac{x^2}{4} + \frac{y^2}{9} = 1.$$

SOLUTION   Note that

$$\frac{x^2}{4} + \frac{y^2}{9} = \frac{x^2}{2^2} + \frac{y^2}{3^2}. \qquad \text{Identifying } a \text{ and } b. \text{ Since } b^2 > a^2,\\ \text{the ellipse is vertical.}$$

Since $a = 2$ and $b = 3$, the x-intercepts are $(-2, 0)$ and $(2, 0)$, and the y-intercepts are $(0, -3)$ and $(0, 3)$. We plot these points and connect them with an oval-shaped curve. To plot two other points, we let $x = 1$ and solve for y:

$$\frac{1^2}{4} + \frac{y^2}{9} = 1$$

$$36\left(\frac{1}{4} + \frac{y^2}{9}\right) = 36 \cdot 1$$

$$36 \cdot \frac{1}{4} + 36 \cdot \frac{y^2}{9} = 36$$

$$9 + 4y^2 = 36$$

$$4y^2 = 27$$

$$y^2 = \frac{27}{4}$$

$$y = \pm\sqrt{\frac{27}{4}}$$

$$y \approx \pm 2.6.$$

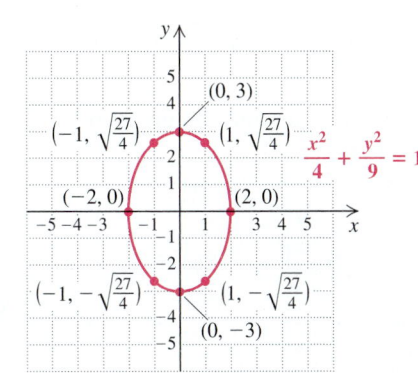

1. Graph the ellipse

$$\frac{x^2}{25} + \frac{y^2}{9} = 1.$$

Thus, $(1, 2.6)$ and $(1, -2.6)$ can also be used to draw the graph. We leave it to the student to confirm that $(-1, 2.6)$ and $(-1, -2.6)$ also appear on the graph.

YOUR TURN

EXAMPLE 2 Graph: $4x^2 + 25y^2 = 100$.

SOLUTION To write the equation in standard form, we divide both sides by 100 to get 1 on the right side:

$$\frac{4x^2 + 25y^2}{100} = \frac{100}{100} \qquad \text{Dividing by 100 to get 1 on the right side}$$

$$\left.\begin{array}{c}\dfrac{4x^2}{100} + \dfrac{25y^2}{100} = 1 \\[2mm] \dfrac{x^2}{25} + \dfrac{y^2}{4} = 1\end{array}\right\} \qquad \text{Simplifying}$$

$$\frac{x^2}{5^2} + \frac{y^2}{2^2} = 1. \qquad a = 5, b = 2$$

Student Notes

Note that any equation of the form $Ax^2 + By^2 = C$ (with $A \neq B$ and $A, B > 0$) can be rewritten as an equivalent equation in standard form. The graph is an ellipse.

The x-intercepts are $(-5, 0)$ and $(5, 0)$, and the y-intercepts are $(0, -2)$ and $(0, 2)$. We plot the intercepts and connect them with an oval-shaped curve. Other points can also be computed and plotted.

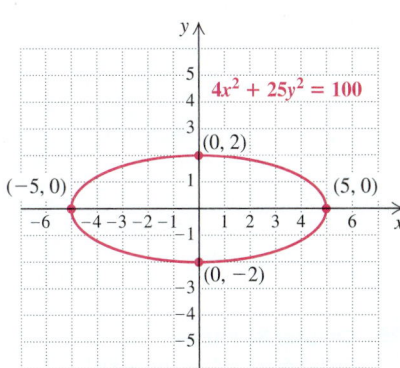

2. Graph: $4x^2 + y^2 = 4$.

YOUR TURN

Ellipses Centered at (h, k)

Horizontal and vertical translations can be used to graph ellipses that are not centered at the origin.

Student Notes

The graph of

$$\frac{(x-h)^2}{a^2} + \frac{(y-k)^2}{b^2} = 1$$

is the same shape as the graph of

$$\frac{x^2}{a^2} + \frac{y^2}{b^2} = 1,$$

with its center moved from $(0,0)$ to (h,k).

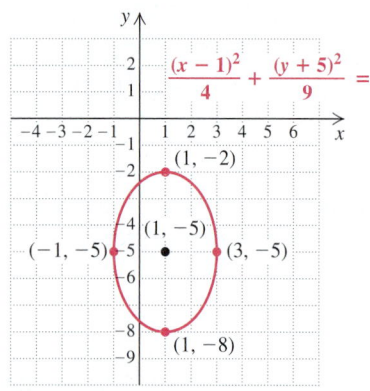

3. Graph the ellipse

$$\frac{(x+2)^2}{9} + \frac{(y-3)^2}{4} = 1.$$

Chapter Resource:
Collaborative Activity, p. 883

EQUATION OF AN ELLIPSE CENTERED AT (h, k)

The standard form of a horizontal or vertical ellipse centered at (h, k) is

$$\frac{(x-h)^2}{a^2} + \frac{(y-k)^2}{b^2} = 1.$$

The vertices are $(h + a, k)$ and $(h - a, k)$ if horizontal; $(h, k + b)$ and $(h, k - b)$ if vertical.

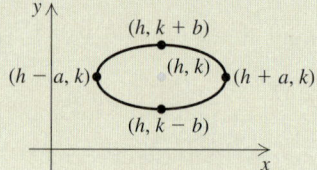

EXAMPLE 3   Graph the ellipse

$$\frac{(x-1)^2}{4} + \frac{(y+5)^2}{9} = 1.$$

SOLUTION   Note that

$$\frac{(x-1)^2}{4} + \frac{(y+5)^2}{9} = \frac{(x-1)^2}{2^2} + \frac{(y+5)^2}{3^2}.$$

Thus, $a = 2$ and $b = 3$. To determine the center of the ellipse, (h, k), note that

$$\frac{(x-1)^2}{2^2} + \frac{(y+5)^2}{3^2} = \frac{(x-1)^2}{2^2} + \frac{(y-(-5))^2}{3^2}.$$

Thus the center is $(1, -5)$. We plot points 2 units to the left and right of center, as well as 3 units above and below center. These are the points $(3, -5)$, $(-1, -5)$, $(1, -2)$, and $(1, -8)$. The graph of the ellipse is shown at left.

Note that this ellipse is the same as the ellipse in Example 1 but translated 1 unit to the right and 5 units down.

YOUR TURN

Ellipses have many applications. Communications satellites move in elliptical orbits with the earth as a focus while the earth itself follows an elliptical path around the sun. A medical instrument, the lithotripter, uses shock waves originating at one focus to crush a kidney stone located at the other focus.

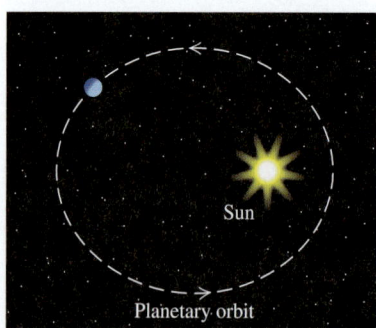

Planetary orbit

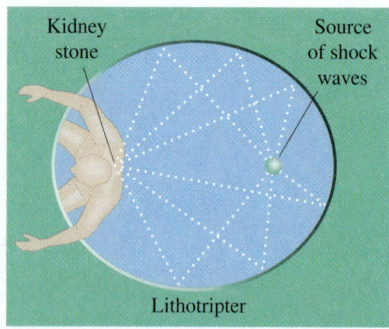

Lithotripter

In some buildings, an ellipsoidal ceiling creates a "whispering gallery" in which a person at one focus can whisper and still be heard clearly at the other focus. This happens because sound waves coming from one focus are all reflected to the other focus. Similarly, light waves bouncing off an ellipsoidal mirror are used in a dentist's or surgeon's reflector light. The light source is located at one focus while the patient's mouth or surgical field is at the other.

Technology Connection

To graph an ellipse on a graphing calculator, we solve for y and graph two functions.

To illustrate, let's check Example 2:

$$4x^2 + 25y^2 = 100$$
$$25y^2 = 100 - 4x^2$$
$$y^2 = 4 - \frac{4}{25}x^2$$
$$y = \pm\sqrt{4 - \frac{4}{25}x^2}.$$

Using a squared window, we have our check:

$$y_1 = -\sqrt{4 - \frac{4}{25}x^2}, \quad y_2 = \sqrt{4 - \frac{4}{25}x^2}$$

On many calculators, pressing **APPS** and selecting Conics and then Ellipse accesses a program in which equations in Standard Form can be graphed directly.

13.2 EXERCISE SET

FOR EXTRA HELP

MyMathLab® MathXL

PRACTICE   WATCH   READ   REVIEW

Vocabulary and Reading Check

For each term, write the letter of the appropriate labeled part of the drawing.

1. ____ Ellipse
2. ____ Focus
3. ____ Center
4. ____ Vertex

Concept Reinforcement

Classify each of the following statements as either true or false.

5. The graph of $\dfrac{x^2}{25} + \dfrac{y^2}{50} = 1$ is a vertical ellipse.

6. The graph of $\dfrac{x^2}{25} - \dfrac{y^2}{9} = 1$ is a horizontal ellipse.

7. The graph of $\dfrac{x^2}{9} + \dfrac{y^2}{25} = 1$ includes the points $(-3, 0)$ and $(3, 0)$.

8. The graph of $\dfrac{(x + 3)^2}{25} + \dfrac{(y - 2)^2}{36} = 1$ is an ellipse centered at $(-3, 2)$.

Ellipses Centered at $(0, 0)$

Graph.

9. $\dfrac{x^2}{1} + \dfrac{y^2}{4} = 1$

10. $\dfrac{x^2}{4} + \dfrac{y^2}{1} = 1$

11. $\dfrac{x^2}{25} + \dfrac{y^2}{9} = 1$

12. $\dfrac{x^2}{16} + \dfrac{y^2}{25} = 1$

13. $4x^2 + 9y^2 = 36$

14. $9x^2 + 4y^2 = 36$

15. $16x^2 + 9y^2 = 144$

16. $9x^2 + 16y^2 = 144$

17. $2x^2 + 3y^2 = 6$

18. $5x^2 + 7y^2 = 35$

Aha! 19. $5x^2 + 5y^2 = 125$

20. $8x^2 + 5y^2 = 80$

21. $3x^2 + 7y^2 - 63 = 0$

22. $3x^2 + 3y^2 - 48 = 0$

23. $16x^2 = 16 - y^2$

24. $9y^2 = 9 - x^2$

25. $16x^2 + 25y^2 = 1$

26. $9x^2 + 4y^2 = 1$

Ellipses Centered at (h, k)

Graph.

27. $\dfrac{(x-3)^2}{9} + \dfrac{(y-2)^2}{25} = 1$

28. $\dfrac{(x-2)^2}{25} + \dfrac{(y-4)^2}{9} = 1$

29. $\dfrac{(x+4)^2}{16} + \dfrac{(y-3)^2}{49} = 1$

30. $\dfrac{(x+5)^2}{4} + \dfrac{(y-2)^2}{36} = 1$

31. $12(x-1)^2 + 3(y+4)^2 = 48$
(*Hint:* Divide both sides by 48.)

32. $4(x-6)^2 + 9(y+2)^2 = 36$

Aha! **33.** $4(x+3)^2 + 4(y+1)^2 - 10 = 90$

34. $9(x+6)^2 + (y+2)^2 - 20 = 61$

35. How can you tell from the equation of an ellipse whether its graph is horizontal or vertical?

36. Can an ellipse ever be the graph of a function? Why or why not?

Skill Review

Solve.

37. $x^2 - 5x + 3 = 0$ [11.2]

38. $\log_x 81 = 4$ [12.6]

39. $\dfrac{4}{x+2} + \dfrac{3}{2x-1} = 2$ [6.6]

40. $3 - \sqrt{2x-1} = 1$ [10.6]

41. $x^2 = 11$ [11.1]

42. $x^2 + 4x = 60$ [5.7]

Synthesis

43. Explain how it is possible to recognize that the graph of $9x^2 + 18x + y^2 - 4y + 4 = 0$ is an ellipse.

44. As the foci get closer to the center of an ellipse, what shape does the graph begin to resemble? Explain why this happens.

Find an equation of an ellipse that contains the following points.

45. $(-9, 0), (9, 0), (0, -11)$, and $(0, 11)$

46. $(-7, 0), (7, 0), (0, -10)$, and $(0, 10)$

47. $(-2, -1), (6, -1), (2, -4)$, and $(2, 2)$

48. $(4, 3), (-6, 3), (-1, -1)$, and $(-1, 7)$

49. *Theatrical Lighting.* A spotlight on a violin soloist casts an ellipse of light on the floor below her that is 6 ft wide and 10 ft long. Find an equation of that ellipse if the performer is in its center, x is the distance from the performer to the nearest side of the ellipse, and y is the distance from the performer to the top of the ellipse.

50. *Astronomy.* The maximum distance of Mars from the sun is 2.48×10^8 mi. Its minimum distance is 3.46×10^7 mi. The sun is one focus of the elliptical orbit. Find the distance from the sun to the other focus.

51. Let $(-c, 0)$ and $(c, 0)$ be the foci of an ellipse. Any point $P(x, y)$ is on the ellipse if the sum of the distances from the foci to P is some constant. Use $2a$ to represent this constant.

    a) Show that an equation for the ellipse is given by

$$\frac{x^2}{a^2} + \frac{y^2}{a^2 - c^2} = 1.$$

    b) Substitute b^2 for $a^2 - c^2$ to get standard form.

52. *President's Office.* The Oval Office of the President of the United States is an ellipse 31 ft wide and 38 ft long. Show in a sketch precisely where the President and an adviser could sit to best hear each other using the room's acoustics. (*Hint:* See Exercise 51(b) and the discussion following Example 3.)

53. *Dentistry.* The light source in some dental lamps shines against a reflector that is shaped like a portion of an ellipse in which the light source is one focus of the ellipse. Reflected light enters a patient's mouth at the other focus of the ellipse. If the ellipse from which the reflector was formed is 2 ft wide and 6 ft long, how far should the patient's mouth be from the light source? (*Hint*: See Exercise 51(b).)

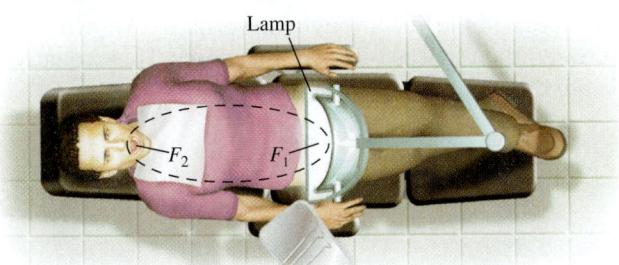

Lamp

54. *Firefighting.* The size and shape of certain forest fires can be approximated as the union of two "half-ellipses." For the blaze modeled below, the equation of the smaller ellipse—the part of the fire moving *into* the wind—is

$$\frac{x^2}{40{,}000} + \frac{y^2}{10{,}000} = 1.$$

The equation of the other ellipse—the part moving *with* the wind—is

$$\frac{x^2}{250{,}000} + \frac{y^2}{10{,}000} = 1.$$

Determine the width and the length of the fire.

Source for figure: "Predicting Wind-Driven Wild Land Fire Size and Shape," Hal E. Anderson, Research Paper INT-305, U.S. Department of Agriculture, Forest Service, February 1983

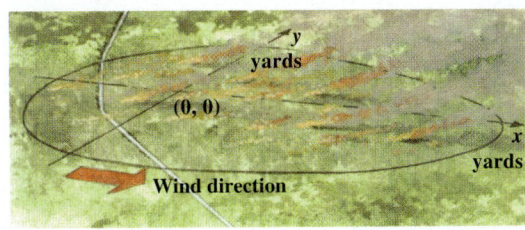

For each of the following equations, complete the square as needed and find an equivalent equation in standard form. Then graph the ellipse.

55. $x^2 - 4x + 4y^2 + 8y - 8 = 0$

56. $4x^2 + 24x + y^2 - 2y - 63 = 0$

57. Use a graphing calculator to check your answers to Exercises 11, 25, 29, and 33.

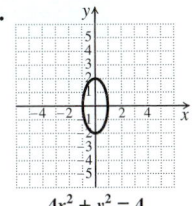

YOUR TURN ANSWERS: SECTION 13.2

1.

$\dfrac{x^2}{25} + \dfrac{y^2}{9} = 1$

2.

$4x^2 + y^2 = 4$

3.

$\dfrac{(x+2)^2}{9} + \dfrac{(y-3)^2}{4} = 1$

QUICK QUIZ: SECTIONS 13.1–13.2

1. Find the center and the radius of the circle:
$$(x+5)^2 + (y+1)^2 = 4. \quad [13.1]$$

2. Find an equation of the circle with center $(9, -23)$ and radius $10\sqrt{2}$. [13.1]

Graph.

3. $(x+3)^2 + (y+1)^2 = 4$ [13.1]

4. $y = x^2 - 2x - 2$ [13.1]

5. $\dfrac{x^2}{1} + \dfrac{y^2}{25} = 1$ [13.2]

PREPARE TO MOVE ON

1. Solve $xy = 4$ for y. [6.6]

2. Write $6x - 3y = 12$ in the form $y = mx + b$. [3.6]

3. Write $5x = x^2 - 7$ in the form $ax^2 + bx + c = 0, a > 0$. [11.2]

<table>
<tr><td>**13.3**</td><td>**Conic Sections: Hyperbolas**</td></tr>
</table>

Hyperbolas ▪ Hyperbolas (Nonstandard Form) ▪ Classifying Graphs of Equations

Study Skills

Listen Up!

Many professors make a point of telling their classes what topics or chapters will (or will not) be covered on the final exam. Take special note of this information and use it to plan your studying.

Hyperbolas

A **hyperbola** looks like a pair of parabolas, but the shapes are not quite parabolic. Every hyperbola has two **vertices** and the line through the vertices is known as the **axis**. The point halfway between the vertices is called the **center**. The two curves that comprise a hyperbola are called **branches**.

Parabola

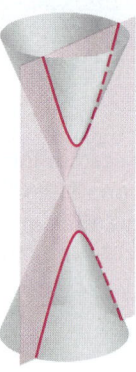

Hyperbola in three dimensions

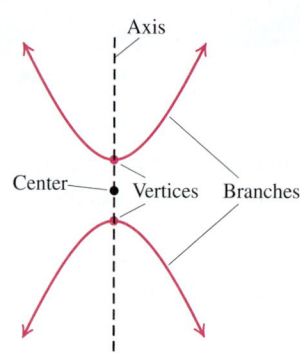

Hyperbola in a plane

> ### EQUATION OF A HYPERBOLA CENTERED AT THE ORIGIN
>
> A hyperbola with its center at the origin* has its equation as follows:
>
> $$\frac{x^2}{a^2} - \frac{y^2}{b^2} = 1 \qquad \text{(Horizontal axis)};$$
>
> $$\frac{y^2}{b^2} - \frac{x^2}{a^2} = 1 \qquad \text{(Vertical axis)}.$$

Note that both equations have a positive term and a negative term on the left-hand side. For the discussion that follows, we assume $a, b > 0$.

To graph a hyperbola, it helps to begin by graphing two lines called **asymptotes**. Although the asymptotes themselves are not part of the graph, they serve as guidelines for an accurate drawing.

As a hyperbola gets farther away from the origin, it gets closer and closer to its asymptotes. That is, the larger $|x|$ gets, the closer the graph gets to an asymptote. The asymptotes act to "constrain" the graph of a hyperbola. Parabolas are *not* constrained by any asymptotes.

---

*Hyperbolas with horizontal or vertical axes and centers *not* at the origin are discussed in Exercises 57–62.

ASYMPTOTES OF A HYPERBOLA

For hyperbolas with equations as shown below, the asymptotes are the lines

$$y = \frac{b}{a}x \quad \text{and} \quad y = -\frac{b}{a}x.$$

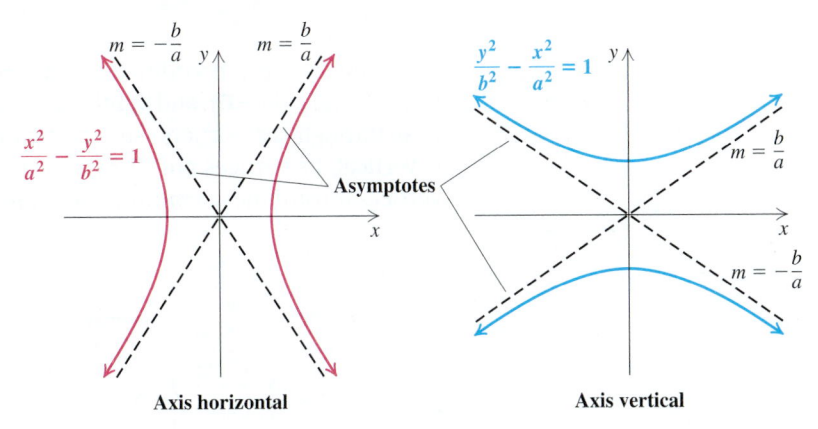

Axis horizontal   Axis vertical

In Section 13.2, we used a and b to determine the width and the length of an ellipse. For hyperbolas, a and b are used to determine the base and the height of a rectangle that can be used as an aid in sketching asymptotes and locating vertices.

EXAMPLE 1   Graph: $\dfrac{x^2}{4} - \dfrac{y^2}{9} = 1$.

SOLUTION   Note that

$$\frac{x^2}{4} - \frac{y^2}{9} = \frac{x^2}{2^2} - \frac{y^2}{3^2}, \qquad \text{Identifying } a \text{ and } b$$

so $a = 2$ and $b = 3$. The asymptotes are thus

$$y = \frac{3}{2}x \quad \text{and} \quad y = -\frac{3}{2}x.$$

To help us sketch asymptotes and locate vertices, we use a and b—in this case, 2 and 3—to form the pairs $(-2, 3)$, $(2, 3)$, $(2, -3)$, and $(-2, -3)$. We plot these pairs and lightly sketch a rectangle. The asymptotes pass through the corners and, since this is a horizontal hyperbola, the vertices $(-2, 0)$ and $(2, 0)$ are where the rectangle intersects the x-axis. Finally, we draw the hyperbola, as shown below.

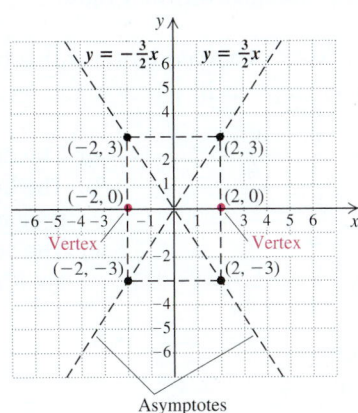

Asymptotes

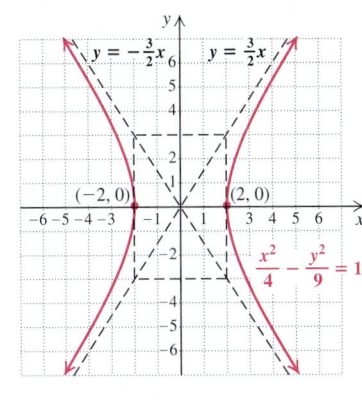

1. Graph: $\dfrac{x^2}{25} - \dfrac{y^2}{4} = 1$.

YOUR TURN

EXAMPLE 2   Graph: $\dfrac{y^2}{36} - \dfrac{x^2}{4} = 1$.

SOLUTION   Note that

$$\dfrac{y^2}{36} - \dfrac{x^2}{4} = \dfrac{y^2}{6^2} - \dfrac{x^2}{2^2} = 1.$$

> Whether the hyperbola is horizontal or vertical is determined by which term is nonnegative. Here the y^2-term is nonnegative, so the hyperbola is vertical.

Using ± 2 as x-coordinates and ± 6 as y-coordinates, we plot $(2, 6)$, $(2, -6)$, $(-2, 6)$, and $(-2, -6)$, and lightly sketch a rectangle through them. The asymptotes pass through the corners (see the figure on the left below). Since the hyperbola is vertical, its vertices are $(0, 6)$ and $(0, -6)$. Finally, we draw curves through the vertices toward the asymptotes, as shown below.

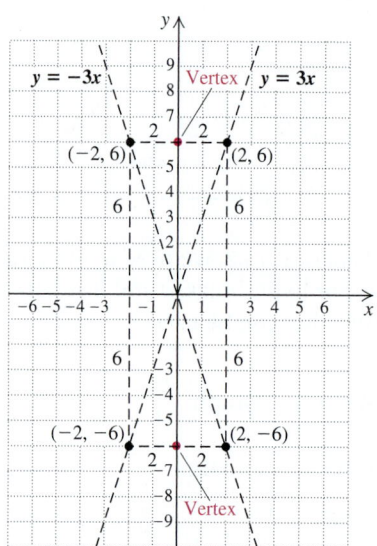

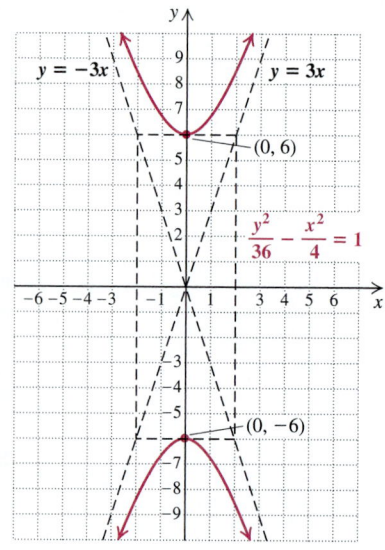

2. Graph: $\dfrac{y^2}{9} - \dfrac{x^2}{16} = 1$.

YOUR TURN

EXPLORING 🔍 THE CONCEPT

The graphs of $\dfrac{x^2}{4} - \dfrac{y^2}{9} = 1$, $\dfrac{y^2}{9} - \dfrac{x^2}{4} = 1$, and $\dfrac{x^2}{4} + \dfrac{y^2}{9} = 1$ are all related to the rectangle shown at right.

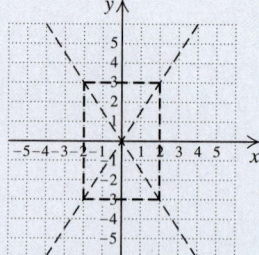

Match each equation with a description of its graph as it relates to the rectangle.

1. $\dfrac{x^2}{4} - \dfrac{y^2}{9} = 1$

2. $\dfrac{y^2}{9} - \dfrac{x^2}{4} = 1$

3. $\dfrac{x^2}{4} + \dfrac{y^2}{9} = 1$

a) The ellipse inscribed in the rectangle
b) The horizontal hyperbola with asymptotes that are diagonals of the rectangle
c) The vertical hyperbola with asymptotes that are diagonals of the rectangle

ANSWERS
1. (b)   **2.** (c)   **3.** (a)

Hyperbolas (Nonstandard Form)

The equations for hyperbolas just examined are the standard ones, but there are other hyperbolas. We consider some of them.

EQUATION OF A HYPERBOLA IN NONSTANDARD FORM

Hyperbolas having the x- and y-axes as asymptotes have equations that can be written in the form

$$xy = c, \quad \text{where } c \text{ is a nonzero constant.}$$

EXAMPLE 3   Graph: $xy = -8$.

SOLUTION   We first solve for y:

$$y = -\frac{8}{x}. \qquad \text{Dividing both sides by } x. \text{ Note that } x \neq 0.$$

Next, we find some solutions and form a table. Note that x cannot be 0 and that for large values of $|x|$, the value of y is close to 0. Thus the x- and y-axes serve as asymptotes. We plot the points and draw two curves.

x	y
2	-4
-2	4
4	-2
-4	2
1	-8
-1	8
8	-1
-8	1

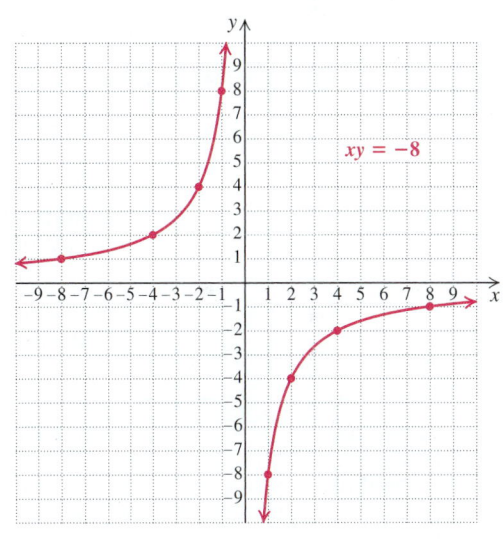

3. Graph: $xy = 6$.

YOUR TURN

Hyperbolas have many applications. A jet breaking the sound barrier creates a sonic boom with a wave front the shape of a cone. The intersection of the cone with the ground is one branch of a hyperbola. Some comets travel in hyperbolic orbits, and a cross section of many lenses is hyperbolic in shape.

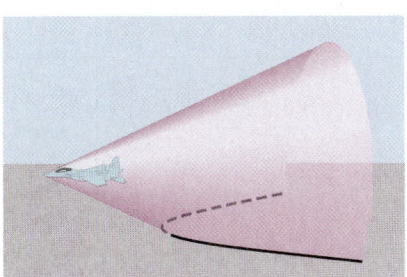

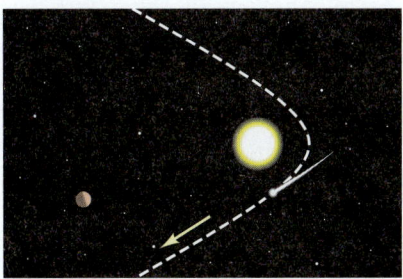

Technology Connection

The procedure used to graph a hyperbola in standard form is similar to that used to draw a circle or an ellipse. Consider the graph of

$$\frac{x^2}{25} - \frac{y^2}{49} = 1.$$

The student should confirm that solving for y yields

$$y_1 = \frac{\sqrt{49x^2 - 1225}}{5} = \frac{7}{5}\sqrt{x^2 - 25}$$

and

$$y_2 = \frac{-\sqrt{49x^2 - 1225}}{5} = -\frac{7}{5}\sqrt{x^2 - 25},$$

or $y_2 = -y_1.$

When the two pieces are drawn on the same squared window, the result is as shown. The gaps occur where the graph is nearly vertical.

$$y_1 = \frac{7}{5}\sqrt{x^2 - 25},$$
$$y_2 = -\frac{7}{5}\sqrt{x^2 - 25}$$

On many calculators, pressing **APPS** and selecting Conics and then Hyperbola accesses a program in which hyperbolas in standard form can be graphed directly.

Graph each of the following.

1. $\dfrac{x^2}{16} - \dfrac{y^2}{60} = 1$ 2. $16x^2 - 3y^2 = 64$

3. $\dfrac{y^2}{20} - \dfrac{x^2}{64} = 1$ 4. $45y^2 - 9x^2 = 441$

Classifying Graphs of Equations

By writing an equation of a conic section in a standard form, we can classify its graph as a parabola, a circle, an ellipse, or a hyperbola. Every conic section can also be represented by an equation of the form

$$Ax^2 + By^2 + Cxy + Dx + Ey + F = 0.$$

We can also classify graphs using values of A and B.

Graph	Standard Form		$Ax^2 + By^2 + Cxy + Dx + Ey + F = 0$
Parabola	$y = ax^2 + bx + c;$	Vertical parabola	Either $A = 0$ or $B = 0$, but not both.
	$x = ay^2 + by + c$	Horizontal parabola	
Circle	$x^2 + y^2 = r^2;$	Center at the origin	$A = B$, and neither A nor B is 0.
	$(x - h)^2 + (y - k)^2 = r^2$	Center at (h, k)	
Ellipse	$\dfrac{x^2}{a^2} + \dfrac{y^2}{b^2} = 1;$	Center at the origin	$A \neq B$, and A and B have the same sign.
	$\dfrac{(x - h)^2}{a^2} + \dfrac{(y - k)^2}{b^2} = 1$	Center at (h, k)	
Hyperbola	$\dfrac{x^2}{a^2} - \dfrac{y^2}{b^2} = 1;$	Horizontal hyperbola	A and B have opposite signs.
	$\dfrac{y^2}{b^2} - \dfrac{x^2}{a^2} = 1$	Vertical hyperbola	
	$xy = c$	Asymptotes are axes	Only C and F are nonzero.

Algebraic manipulations are often needed to express an equation in one of the preceding forms.

EXAMPLE 4   Classify the graph of each equation as a circle, an ellipse, a parabola, or a hyperbola. Refer to the above table as needed.

a) $5x^2 = 20 - 5y^2$ **b)** $x + 3 + 8y = y^2$

c) $x^2 = y^2 + 4$ **d)** $x^2 = 16 - 4y^2$

SOLUTION

a) We get the terms with variables on one side by adding $5y^2$ to both sides:

$$5x^2 + 5y^2 = 20.$$

Since *both* x and y are squared, we do not have a parabola. The fact that the squared terms are *added* tells us that we have an ellipse or a circle. Since the coefficients are the same, we factor 5 out of both terms on the left and then divide by 5:

$$5(x^2 + y^2) = 20 \qquad \text{\color{red}Factoring out 5}$$
$$x^2 + y^2 = 4 \qquad \text{\color{red}Dividing both sides by 5}$$
$$x^2 + y^2 = 2^2. \qquad \text{\color{red}This is an equation for a circle.}$$

We see that the graph is a circle centered at the origin with radius 2.
  We can also write the equation in the form

$$5x^2 + 5y^2 - 20 = 0. \qquad \color{red}A = 5, B = 5$$

Since $A = B$, the graph is a circle.

b) The equation $x + 3 + 8y = y^2$ has only one variable that is squared, so we solve for the other variable:

$$x = y^2 - 8y - 3. \qquad \text{\color{red}This is an equation for a parabola.}$$

The graph is a horizontal parabola that opens to the right.
  We can also write the equation in the form

$$y^2 - x - 8y - 3 = 0. \qquad \color{red}A = 0, B = 1$$

Since $A = 0$ and $B \neq 0$, the graph is a parabola.

c) In $x^2 = y^2 + 4$, both variables are squared, so the graph is not a parabola. We subtract y^2 on both sides and divide by 4 to obtain

$$\frac{x^2}{2^2} - \frac{y^2}{2^2} = 1. \qquad \text{\color{red}This is an equation for a hyperbola.}$$

The subtraction indicates that the graph is a hyperbola. Because it is the x^2-term that is nonnegative, the hyperbola is horizontal.
  We can also write the equation in the form

$$x^2 - y^2 - 4 = 0. \qquad \color{red}A = 1, B = -1$$

Since A and B have opposite signs, the graph is a hyperbola.

d) In $x^2 = 16 - 4y^2$, both variables are squared, so the graph cannot be a parabola. We add $4y^2$ to both sides to obtain the following equivalent equation:

$$x^2 + 4y^2 = 16.$$

If the coefficients of the terms were the same, the graph would be a circle, as in part (a). Since they are not, we divide both sides by the constant term, 16:

$$\frac{x^2}{16} + \frac{y^2}{4} = 1. \qquad \text{\color{red}This is an equation for an ellipse.}$$

The graph of this equation is a horizontal ellipse.

Chapter Resource:
Visualizing for Success, p. 882

4. Classify the graph of $7x^2 = 12 + 7y^2$ as a circle, an ellipse, a parabola, or a hyperbola.

We can also write the equation in the form

$$x^2 + 4y^2 - 16 = 0. \qquad A = 1, B = 4$$

Since $A \neq B$ and both A and B are positive, the graph is an ellipse.

 YOUR TURN

CONNECTING 🔗 THE CONCEPTS

When graphing equations of conic sections, it is usually helpful to first determine what type of graph the equation represents. We then find the coordinates of key points and equations of lines that determine the shape and the location of the graph.

Graph	Equation	Key Points	Equations of Lines
Parabola	$y = a(x - h)^2 + k$ $x = a(y - k)^2 + h$	Vertex: (h, k) Vertex: (h, k)	Axis of symmetry: $x = h$ Axis of symmetry: $y = k$
Circle	$(x - h)^2 + (y - k)^2 = r^2$	Center: (h, k)	
Ellipse	$\dfrac{x^2}{a^2} + \dfrac{y^2}{b^2} = 1$	x-intercepts: $(-a, 0), (a, 0)$; y-intercepts: $(0, -b), (0, b)$	
Hyperbola	$\dfrac{x^2}{a^2} - \dfrac{y^2}{b^2} = 1$ $\dfrac{y^2}{b^2} - \dfrac{x^2}{a^2} = 1$	Vertices: $(-a, 0), (a, 0)$ Vertices: $(0, -b), (0, b)$	Asymptotes (for both equations): $y = \dfrac{b}{a}x, \quad y = -\dfrac{b}{a}x$
	$xy = c$		Asymptotes: $x = 0, y = 0$

EXERCISES

1. Find the vertex and the axis of symmetry of the graph of
$$y = 3(x - 4)^2 + 1.$$

2. Find the vertex and the axis of symmetry of the graph of
$$x = y^2 + 2y + 3.$$

3. Find the center of the graph of
$$(x - 3)^2 + (y - 2)^2 = 5.$$

4. Find the center of the graph of
$$x^2 + 6x + y^2 + 10y = 12.$$

5. Find the x-intercepts and the y-intercepts of the graph of
$$\frac{x^2}{144} + \frac{y^2}{81} = 1.$$

6. Find the vertices of the graph of
$$\frac{x^2}{9} - \frac{y^2}{121} = 1.$$

7. Find the vertices of the graph of
$$4y^2 - x^2 = 4.$$

8. Find the asymptotes of the graph of
$$\frac{y^2}{9} - \frac{x^2}{4} = 1.$$

13.3 EXERCISE SET

FOR EXTRA HELP

MyMathLab® MathXL
PRACTICE   WATCH   READ   REVIEW

Vocabulary and Reading Check

For each term, write the letter of the appropriate labeled part of the drawing.

1. ____ Asymptote

2. ____ Axis

3. ____ Branch

4. ____ Center

5. ____ Hyperbola

6. ____ Vertex

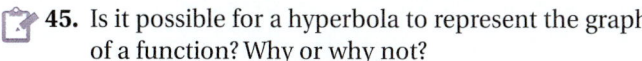

Hyperbolas

Graph each hyperbola. Label all vertices and sketch all asymptotes.

7. $\dfrac{y^2}{16} - \dfrac{x^2}{16} = 1$

8. $\dfrac{x^2}{9} - \dfrac{y^2}{9} = 1$

9. $\dfrac{x^2}{4} - \dfrac{y^2}{25} = 1$

10. $\dfrac{y^2}{16} - \dfrac{x^2}{9} = 1$

11. $\dfrac{y^2}{36} - \dfrac{x^2}{9} = 1$

12. $\dfrac{x^2}{25} - \dfrac{y^2}{36} = 1$

13. $y^2 - x^2 = 25$

14. $x^2 - y^2 = 4$

15. $25x^2 - 16y^2 = 400$

16. $4y^2 - 9x^2 = 36$

Hyperbolas (Nonstandard Form)

Graph.

17. $xy = -6$

18. $xy = 8$

19. $xy = 4$

20. $xy = -9$

21. $xy = -2$

22. $xy = -1$

23. $xy = 1$

24. $xy = 2$

Classifying Graphs of Equations

Classify each of the following as the equation of a circle, an ellipse, a parabola, or a hyperbola.

25. $x^2 + y^2 - 6x + 10y - 40 = 0$

26. $y - 4 = 2x^2$

27. $9x^2 + 4y^2 - 36 = 0$

28. $x + 3y = 2y^2 - 1$

29. $4x^2 - 9y^2 - 72 = 0$

30. $y^2 + x^2 = 8$

31. $y^2 = 20 - x^2$

32. $2y + 13 + x^2 = 8x - y^2$

33. $x - 10 = y^2 - 6y$

34. $y = \dfrac{5}{x}$

35. $x - \dfrac{3}{y} = 0$

36. $9x^2 = 9 - y^2$

37. $y + 6x = x^2 + 5$

38. $x^2 = 49 + y^2$

39. $25y^2 = 100 + 4x^2$

40. $3x^2 + 5y^2 + x^2 = y^2 + 49$

41. $3x^2 + y^2 - x = 2x^2 - 9x + 10y + 40$

42. $4y^2 + 20x^2 + 1 = 8y - 5x^2$

43. $16x^2 + 5y^2 - 12x^2 + 8y^2 - 3x + 4y = 568$

44. $56x^2 - 17y^2 = 234 - 13x^2 - 38y^2$

45. Is it possible for a hyperbola to represent the graph of a function? Why or why not?

46. Explain how the equation of a hyperbola differs from the equation of an ellipse.

Skill Review

Factor completely.

47. $16 - y^4$ [5.4]

48. $9x^2y^2 - 30xy + 25$ [5.4]

49. $10c^3 - 80c^2 + 150c$ [5.2]

50. $x^3 + x^2 + 3x + 3$ [5.1]

51. $8t^4 - 8t$ [5.5]

52. $6a^2 + 11ab - 10b^2$ [5.3]

Synthesis

53. What is it in the equation of a hyperbola that controls how wide open the branches are? Explain your reasoning.

54. If, in

$$\frac{x^2}{a^2} - \frac{y^2}{b^2} = 1,$$

$a = b$, what are the asymptotes of the graph? Why?

Find an equation of a hyperbola satisfying the given conditions.

55. Having intercepts $(0, 6)$ and $(0, -6)$ and asymptotes $y = 3x$ and $y = -3x$

56. Having intercepts $(8, 0)$ and $(-8, 0)$ and asymptotes $y = 4x$ and $y = -4x$

The standard form for equations of horizontal or vertical hyperbolas centered at (h, k) are as follows:

$$\frac{(x - h)^2}{a^2} - \frac{(y - k)^2}{b^2} = 1$$

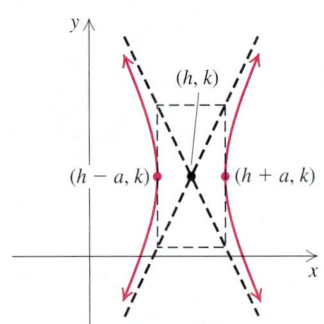

$$\frac{(y - k)^2}{b^2} - \frac{(x - h)^2}{a^2} = 1$$

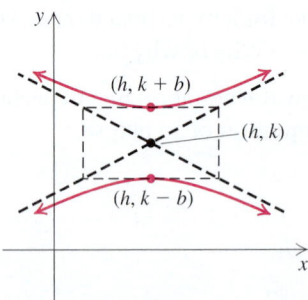

The vertices are as labeled and the asymptotes are

$$y - k = \frac{b}{a}(x - h) \quad and \quad y - k = -\frac{b}{a}(x - h).$$

For each of the following equations of hyperbolas, complete the square, if necessary, and write in standard form. Find the center, the vertices, and the asymptotes. Then graph the hyperbola.

57. $\dfrac{(x - 5)^2}{36} - \dfrac{(y - 2)^2}{25} = 1$

58. $\dfrac{(x - 2)^2}{9} - \dfrac{(y - 1)^2}{4} = 1$

59. $8(y + 3)^2 - 2(x - 4)^2 = 32$

60. $25(x - 4)^2 - 4(y + 5)^2 = 100$

61. $4x^2 - y^2 + 24x + 4y + 28 = 0$

62. $4y^2 - 25x^2 - 8y - 100x - 196 = 0$

63. Use a graphing calculator to check your answers to Exercises 13, 25, 31, and 57.

64. *Research.* What conic sections have been used to model paths of comets? Find what comets have recently passed close to the earth, and describe the path of each.

YOUR TURN ANSWERS: SECTION 13.3

1. 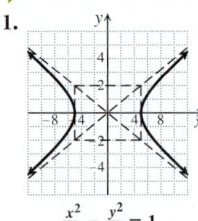 $\frac{x^2}{25} - \frac{y^2}{4} = 1$

2. 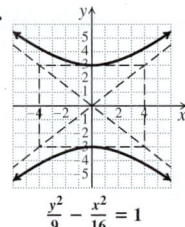 $\frac{y^2}{9} - \frac{x^2}{16} = 1$

3.

$xy = 6$

4. Hyperbola

QUICK QUIZ: SECTIONS 13.1–13.3

Graph.

1. $(x - 2)^2 + y^2 = 16$ [13.1]   **2.** $xy = 6$ [13.3]

3. $\dfrac{x^2}{36} + \dfrac{y^2}{4} = 1$ [13.2]   **4.** $\dfrac{x^2}{36} - \dfrac{y^2}{4} = 1$ [13.3]

5. $x = -y^2 - 4y + 5$ [13.1]

PREPARE TO MOVE ON

Solve.

1. $5x + 2y = -3,$
   $2x + 3y = 12$ [8.2]

2. $3x - y = 2,$
   $y = 2x + 1$ [8.2]

3. $\frac{3}{4}x^2 + x^2 = 7$ [11.2]   **4.** $3x^2 + 10x - 8 = 0$ [11.2]

5. $x^2 - 3x - 1 = 0$ [11.2]

6. $x^2 + \dfrac{25}{x^2} = 26$ [11.5]

Mid-Chapter Review

Parabolas, circles, ellipses, and hyperbolas are all conic sections, that is, curves formed by cross sections of cones. Section 13.3 contains a summary of the characteristics of the graphs of these conic sections.

GUIDED SOLUTIONS

1. Find the center and the radius:

$$x^2 + y^2 - 4x + 2y = 6. \quad [13.1]$$

Solution

$$\left(x^2 - 4x\right) + \left(y^2 + \boxed{}\right) = 6$$

$$\left(x^2 - 4x + \boxed{}\right) + \left(y^2 + 2y + \boxed{}\right) = 6 + \boxed{} + \boxed{}$$

$$\left(x - \boxed{}\right)^2 + \left(y + \boxed{}\right)^2 = 11$$

The center of the circle is $\left(\boxed{}, \boxed{}\right)$.

The radius is $\boxed{}$.

2. Classify the equation as representing a circle, an ellipse, a parabola, or a hyperbola:

$$x^2 - \frac{y^2}{25} = 1. \quad [13.1], [13.2], [13.3]$$

Solution

To answer this, complete the following:

a) Is there both an x^2-term and a y^2-term? $\boxed{}$
b) Do both the x^2-term and the y^2-term have the same sign? $\boxed{}$
c) The graph of the equation is a(n) $\boxed{}$.

MIXED REVIEW

3. Find an equation of the circle with center $(-4, 9)$ and radius $2\sqrt{5}$. [13.1]

4. Find the center and the radius of the graph of $x^2 - 10x + y^2 + 2y = 10$. [13.1]

Classify each of the following as the graph of a parabola, a circle, an ellipse, or a hyperbola. Then graph. [13.1], [13.2], [13.3]

5. $x^2 + y^2 = 36$

6. $y = x^2 - 5$

7. $\dfrac{x^2}{25} + \dfrac{y^2}{49} = 1$

8. $\dfrac{x^2}{25} - \dfrac{y^2}{49} = 1$

9. $x = (y + 3)^2 + 2$

10. $4x^2 + 9y^2 = 36$

11. $xy = -4$

12. $(x + 2)^2 + (y - 3)^2 = 1$

13. $x^2 + y^2 - 8y - 20 = 0$

14. $x = y^2 + 2y$

15. $16y^2 - x^2 = 16$

16. $x = \dfrac{9}{y}$

13.4 Nonlinear Systems of Equations

Systems Involving One Nonlinear Equation ▪ Systems of Two Nonlinear Equations ▪ Problem Solving

We now consider systems of two equations in which at least one equation is nonlinear.

Systems Involving One Nonlinear Equation

Suppose that a system consists of an equation of a circle and an equation of a line. The figures below represent three ways in which the circle and the line can intersect. We see that such a system will have 0, 1, or 2 real solutions.

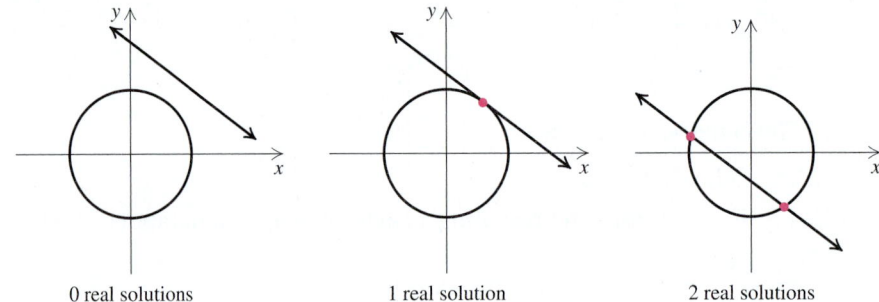

0 real solutions 1 real solution 2 real solutions

Recall that *graphing*, *elimination*, and *substitution* were all used to solve systems of linear equations. To solve systems in which one equation is of first degree and one is of second degree, it is preferable to use the *substitution* method.

EXAMPLE 1 Solve the system

$$x^2 + y^2 = 25, \quad (1) \quad \text{(The graph is a circle.)}$$
$$3x - 4y = 0. \quad (2) \quad \text{(The graph is a line.)}$$

SOLUTION First, we solve the linear equation, (2), for x:

$$x = \tfrac{4}{3}y. \quad (3) \quad \textcolor{red}{\text{We could have solved for } y \text{ instead.}}$$

Then we substitute $\tfrac{4}{3}y$ for x in equation (1) and solve for y:

$$\left(\tfrac{4}{3}y\right)^2 + y^2 = 25$$
$$\tfrac{16}{9}y^2 + y^2 = 25$$
$$\tfrac{25}{9}y^2 = 25$$
$$y^2 = 9 \qquad \textcolor{red}{\text{Multiplying both sides by } \tfrac{9}{25}}$$
$$y = \pm 3. \qquad \textcolor{red}{\text{Using the principle of square roots}}$$

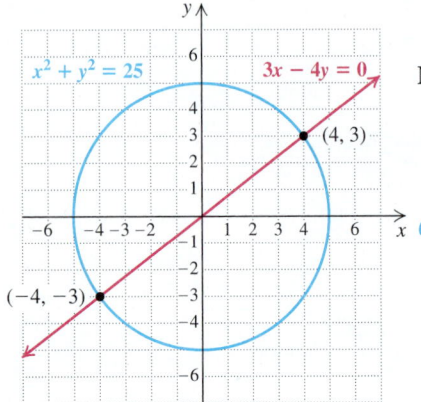

Now we substitute these numbers for y in equation (3) and solve for x:

for $y = 3$, $\quad x = \tfrac{4}{3}(3) = 4;$ $\qquad \textcolor{red}{\text{The ordered pair is } (4, 3).}$
for $y = -3$, $\quad x = \tfrac{4}{3}(-3) = -4.$ $\qquad \textcolor{red}{\text{The ordered pair is } (-4, -3).}$

Check: For $(4, 3)$:

$$\frac{x^2 + y^2 = 25}{\textcolor{red}{4^2 + 3^2} \mid 25}$$
$$16 + 9$$
$$25 \overset{?}{=} 25 \quad \textcolor{red}{\text{TRUE}}$$

$$\frac{3x - 4y = 0}{3(4) - 4(3) \mid 0}$$
$$12 - 12$$
$$0 \overset{?}{=} 0 \quad \textcolor{red}{\text{TRUE}}$$

1. Solve the system

$$x^2 + y^2 = 169,$$
$$x - 2y = 2.$$

We leave it to the student to confirm that $(-4, -3)$ also checks in both equations. The pairs $(4, 3)$ and $(-4, -3)$ check, so they are solutions. The graph on the preceding page serves as a check. Intersections occur at $(4, 3)$ and $(-4, -3)$.

YOUR TURN

EXAMPLE 2   Solve the system

$$y + 3 = 2x, \quad (1) \quad \text{(A first-degree equation)}$$
$$x^2 + 2xy = -1. \quad (2) \quad \text{(A second-degree equation)}$$

SOLUTION   First, we solve the linear equation (1) for y:

$$y = 2x - 3. \quad (3)$$

Then we substitute $2x - 3$ for y in equation (2) and solve for x:

$$x^2 + 2x(2x - 3) = -1$$
$$x^2 + 4x^2 - 6x = -1$$
$$5x^2 - 6x + 1 = 0$$
$$(5x - 1)(x - 1) = 0 \qquad \text{Factoring}$$
$$5x - 1 = 0 \quad or \quad x - 1 = 0 \qquad \text{Using the principle of zero products}$$
$$x = \tfrac{1}{5} \quad or \qquad x = 1.$$

Now we substitute these numbers for x in equation (3) and solve for y:

for $x = \tfrac{1}{5}$, $y = 2(\tfrac{1}{5}) - 3 = -\tfrac{13}{5}$; The ordered pair is $(\tfrac{1}{5}, -\tfrac{13}{5})$.
for $x = 1$, $y = 2(1) - 3 = -1$. The ordered pair is $(1, -1)$.

You can confirm that $(\tfrac{1}{5}, -\tfrac{13}{5})$ and $(1, -1)$ check, so they are both solutions.

YOUR TURN

Student Notes

Be sure to either list each solution of a system as an ordered pair or separately state the value of each variable.

2. Solve the system

$$y^2 = xy + 3,$$
$$y = 2x - 1.$$

EXAMPLE 3   Solve the system

$$x + y = 5, \quad (1) \quad \text{(The graph is a line.)}$$
$$y = 3 - x^2. \quad (2) \quad \text{(The graph is a parabola.)}$$

SOLUTION   We substitute $3 - x^2$ for y in the first equation:

$$x + 3 - x^2 = 5$$
$$-x^2 + x - 2 = 0 \qquad \text{Adding } -5 \text{ to both sides and rearranging}$$
$$x^2 - x + 2 = 0. \qquad \text{Multiplying both sides by } -1$$

Since $x^2 - x + 2$ does not factor, we need the quadratic formula:

$$x = \frac{-b \pm \sqrt{b^2 - 4ac}}{2a}$$
$$= \frac{-(-1) \pm \sqrt{(-1)^2 - 4 \cdot 1 \cdot 2}}{2(1)} \qquad \text{Substituting}$$
$$= \frac{1 \pm \sqrt{1 - 8}}{2} = \frac{1 \pm \sqrt{-7}}{2} = \frac{1}{2} \pm \frac{\sqrt{7}}{2}i.$$

Solving equation (1) for y, we have $y = 5 - x$. Substituting values for x gives

$$y = 5 - \left(\frac{1}{2} + \frac{\sqrt{7}}{2}i\right) = \frac{9}{2} - \frac{\sqrt{7}}{2}i \quad \text{and}$$
$$y = 5 - \left(\frac{1}{2} - \frac{\sqrt{7}}{2}i\right) = \frac{9}{2} + \frac{\sqrt{7}}{2}i.$$

The solutions are

$$\left(\frac{1}{2} + \frac{\sqrt{7}}{2}i, \frac{9}{2} - \frac{\sqrt{7}}{2}i\right) \quad \text{and} \quad \left(\frac{1}{2} - \frac{\sqrt{7}}{2}i, \frac{9}{2} + \frac{\sqrt{7}}{2}i\right).$$

There are no real-number solutions. Note in the figure at right that the graphs do not intersect. Getting only nonreal solutions indicates that the graphs do not intersect.

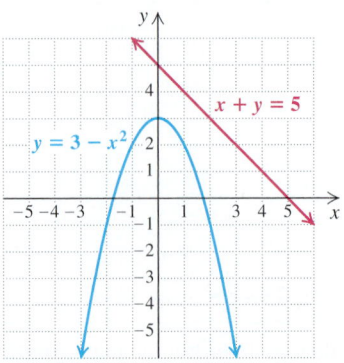

3. Solve the system

$$x = \tfrac{1}{2}y^2,$$
$$x = y - 3.$$

YOUR TURN

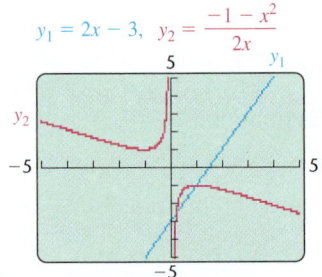
Systems of Two Nonlinear Equations

We now consider systems of two second-degree equations. Graphs of such systems can involve any two nonlinear conic sections. The following figure shows some ways in which a circle and a hyperbola can intersect.

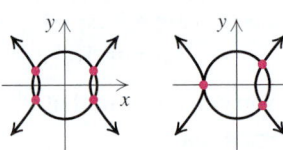

4 real solutions   3 real solutions   2 real solutions

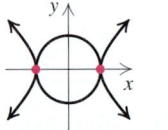 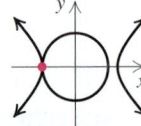 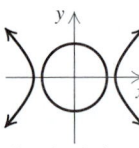

2 real solutions   1 real solution   0 real solutions

To solve systems of two second-degree equations, we either substitute or eliminate. The elimination method is generally better when both equations are of the form $Ax^2 + By^2 = C$. Then we can eliminate an x^2-term or a y^2-term in a manner similar to the procedure used for systems of linear equations.

EXAMPLE 4 Solve the system

$$2x^2 + 5y^2 = 22, \quad (1) \quad \text{(The graph is an ellipse.)}$$
$$3x^2 - y^2 = -1. \quad (2) \quad \text{(The graph is a hyperbola.)}$$

SOLUTION Here we multiply equation (2) by 5 and then add:

$$\begin{aligned} 2x^2 + 5y^2 &= 22 \\ 15x^2 - 5y^2 &= -5 \qquad \text{Multiplying both sides of equation (2) by 5} \\ \hline 17x^2 &= 17 \qquad \text{Adding} \\ x^2 &= 1 \\ x &= \pm 1. \end{aligned}$$

There is no x-term in equation (2), and whether x is -1 or 1, we still have $x^2 = 1$. Thus we can simultaneously substitute 1 and -1 for x in equation (2):

$$\left.\begin{array}{r} 3 \cdot (\pm 1)^2 - y^2 = -1 \\ 3 - y^2 = -1 \\ -y^2 = -4 \end{array}\right\} \quad \begin{array}{l} \text{Since } (-1)^2 = 1^2, \text{ we can evaluate for} \\ x = -1 \text{ and } x = 1 \text{ simultaneously.} \end{array}$$

$$y^2 = 4, \quad \text{or} \quad y = \pm 2.$$

Thus, if $x = 1$, then $y = 2$ or $y = -2$; and if $x = -1$, then $y = 2$ or $y = -2$. The four possible solutions are $(1, 2)$, $(1, -2)$, $(-1, 2)$, and $(-1, -2)$.

Check: Since $(2)^2 = (-2)^2$ and $(1)^2 = (-1)^2$, we can check all four pairs at once.

$$\begin{array}{c|c} \hline 2x^2 + 5y^2 = 22 \\ \hline 2(\pm 1)^2 + 5(\pm 2)^2 & 22 \\ 2 + 20 & \\ 22 \overset{?}{=} 22 & \text{TRUE} \end{array} \qquad \begin{array}{c|c} \hline 3x^2 - y^2 = -1 \\ \hline 3(\pm 1)^2 - (\pm 2)^2 & -1 \\ 3 - 4 & \\ -1 \overset{?}{=} -1 & \text{TRUE} \end{array}$$

4. Solve the system

$$x^2 - 6y^2 = 1,$$
$$2x^2 - 5y^2 = 30.$$

The solutions are $(1, 2)$, $(1, -2)$, $(-1, 2)$, and $(-1, -2)$.

 YOUR TURN

When a product of variables appears in one equation and the other equation is of the form $Ax^2 + By^2 = C$, we often solve for a variable in the equation with the product and then use substitution.

EXAMPLE 5 Solve the system

$$x^2 + 4y^2 = 20, \quad (1) \qquad \text{(The graph is an ellipse.)}$$
$$xy = 4. \qquad (2) \qquad \text{(The graph is a hyperbola.)}$$

SOLUTION First, we solve equation (2) for y:

$$y = \frac{4}{x}. \qquad \text{Dividing both sides by } x. \text{ Note that } x \neq 0.$$

Then we substitute $4/x$ for y in equation (1) and solve for x:

$$x^2 + 4\left(\frac{4}{x}\right)^2 = 20$$

$$x^2 + \frac{64}{x^2} = 20$$

$$x^4 + 64 = 20x^2 \qquad \text{Multiplying by } x^2$$

$$x^4 - 20x^2 + 64 = 0 \qquad \begin{array}{l} \text{Obtaining standard form.} \\ \text{This equation is reducible} \\ \text{to quadratic.} \end{array}$$

$$(x^2 - 4)(x^2 - 16) = 0 \qquad \begin{array}{l} \text{Factoring. If you prefer, let} \\ u = x^2 \text{ and substitute.} \end{array}$$

$$(x - 2)(x + 2)(x - 4)(x + 4) = 0 \qquad \text{Factoring again}$$

$$x = 2 \quad or \quad x = -2 \quad or \quad x = 4 \quad or \quad x = -4. \qquad \begin{array}{l} \text{Using the principle} \\ \text{of zero products} \end{array}$$

5. Solve the system

$$2x^2 + 3y^2 = 21,$$
$$xy = 3.$$

Since $y = 4/x$, for $x = 2$, we have $y = 4/2$, or 2. Thus, $(2, 2)$ is a solution. Similarly, $(-2, -2)$, $(4, 1)$, and $(-4, -1)$ are solutions. You can show that all four pairs check.

 YOUR TURN

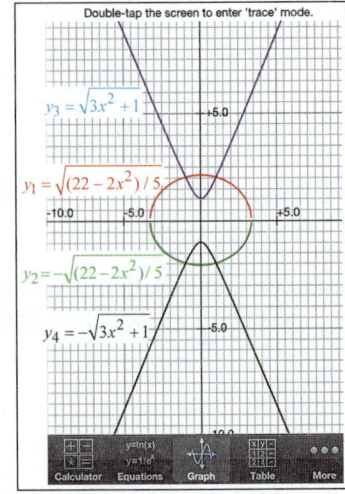
Problem Solving

We now consider applications that can be modeled by a system of equations in which at least one equation is not linear.

EXAMPLE 6 *Architecture.* For a college fitness center, an architect plans to lay out a rectangular piece of land that has a perimeter of 204 m and an area of 2565 m². Find the dimensions of the piece of land.

SOLUTION

1. **Familiarize.** We draw and label a sketch, letting $l =$ the length and $w =$ the width, both in meters.

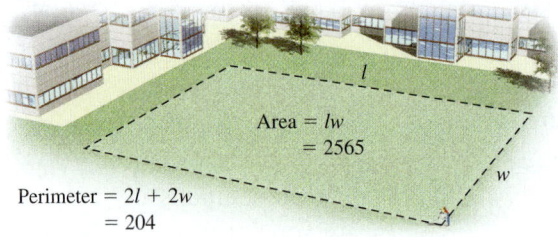

Area = lw
= 2565

Perimeter = $2l + 2w$
= 204

2. **Translate.** We then have the following translation:

Perimeter: $2w + 2l = 204$;

Area: $lw = 2565$.

3. **Carry out.** We solve the system

$2w + 2l = 204$,

$lw = 2565$.

Solving the second equation for l gives us $l = 2565/w$. Then we substitute $2565/w$ for l in the first equation and solve for w:

$$2w + 2\left(\frac{2565}{w}\right) = 204$$

$2w^2 + 2(2565) = 204w$ Multiplying both sides by w

$2w^2 - 204w + 2(2565) = 0$ Standard form

$w^2 - 102w + 2565 = 0$ Multiplying by $\frac{1}{2}$

> Factoring could be used instead of the quadratic formula, but the numbers are quite large.

$$w = \frac{-(-102) \pm \sqrt{(-102)^2 - 4 \cdot 1 \cdot 2565}}{2 \cdot 1}$$

$$w = \frac{102 \pm \sqrt{144}}{2} = \frac{102 \pm 12}{2}$$

$w = 57 \quad or \quad w = 45.$

If $w = 57$, then $l = 2565/w = 2565/57 = 45$. If $w = 45$, then $l = 2565/w = 2565/45 = 57$. Since length is usually considered to be longer than width, we have the solution $l = 57$ and $w = 45$, or $(57, 45)$.

4. **Check.** If $l = 57$ and $w = 45$, the perimeter is $2 \cdot 57 + 2 \cdot 45$, or 204. The area is $57 \cdot 45$, or 2565. The numbers check.

5. **State.** The length is 57 m, and the width is 45 m.

6. Gretchen used 94 ft of fencing to enclose her rectangular garden. If the area of the garden is 550 ft², what are the garden's dimensions?

↩ YOUR TURN

EXAMPLE 7 *HDTV Dimensions.* The Kaplans' new HDTV has a 40-in. diagonal screen with an area of 768 in². Find the width and the length of the screen.

SOLUTION

1. **Familiarize.** We make a drawing and label it. Note the right triangle in the figure. We let l = the length and w = the width, both in inches.

Chapter Resource:
Decision Making: Connection, p. 883

2. **Translate.** We translate to a system of equations:

$$l^2 + w^2 = 40^2,$$ Using the Pythagorean theorem

$$lw = 768.$$ Using the formula for the area of a rectangle

3. **Carry out.** We solve the system

$$\left. \begin{array}{l} l^2 + w^2 = 1600, \\ lw = 768 \end{array} \right\}$$ You should complete the solution of this system.

to get $(32, 24)$, $(24, 32)$, $(-32, -24)$, and $(-24, -32)$.

7. The area of a rectangular stamp is 60 mm², and the length of a diagonal is 13 mm. Find the width and the length of the stamp.

4. **Check.** Measurements must be positive and length is usually greater than width, so we check only $(32, 24)$. In the right triangle, $32^2 + 24^2 = 1024 + 576 = 1600$, or 40^2. The area is $32 \cdot 24 = 768$, so our answer checks.

5. **State.** The length is 32 in., and the width is 24 in.

 YOUR TURN

13.4 EXERCISE SET

FOR EXTRA HELP

MyMathLab® MathXL
PRACTICE   WATCH   READ   REVIEW

Vocabulary and Reading Check

Classify each of the following statements as either true or false.

1. A system of equations that represent a line and an ellipse can have 0, 1, or 2 solutions.

2. A system of equations that represent a parabola and a circle can have up to 4 solutions.

3. A system of equations representing a hyperbola and a circle can have no fewer than 2 solutions.

4. A system of equations representing an ellipse and a line has either 0 or 2 solutions.

5. Systems containing one first-degree equation and one second-degree equation are most easily solved using the substitution method.

6. Systems containing two second-degree equations of the form $Ax^2 + By^2 = C$ are most easily solved using the elimination method.

Systems Involving One Nonlinear Equation

Solve. Remember that graphs can be used to confirm all real solutions.

7. $x^2 + y^2 = 41,$
   $y - x = 1$

8. $x^2 + y^2 = 45,$
   $y - x = 3$

9. $4x^2 + 9y^2 = 36,$
   $3y + 2x = 6$

10. $9x^2 + 4y^2 = 36,$
    $3x + 2y = 6$

11. $y^2 = x + 3,$
    $2y = x + 4$

12. $y = x^2,$
    $3x = y + 2$

13. $x^2 - xy + 3y^2 = 27,$
$\quad x - y = 2$

14. $2y^2 + xy + x^2 = 7,$
$\quad x - 2y = 5$

15. $x^2 + 4y^2 = 25,$
$\quad x + 2y = 7$

16. $x^2 - y^2 = 16,$
$\quad x - 2y = 1$

17. $x^2 - xy + 3y^2 = 5,$
$\quad x - y = 2$

18. $m^2 + 3n^2 = 10,$
$\quad m - n = 2$

19. $3x + y = 7,$
$\quad 4x^2 + 5y = 24$

20. $2y^2 + xy = 5,$
$\quad 4y + x = 7$

21. $a + b = 6,$
$\quad ab = 8$

22. $p + q = -1,$
$\quad pq = -12$

23. $2a + b = 1,$
$\quad b = 4 - a^2$

24. $4x^2 + 9y^2 = 36,$
$\quad x + 3y = 3$

25. $a^2 + b^2 = 89,$
$\quad a - b = 3$

26. $xy = 10,$
$\quad x + y = 7$

Systems of Two Nonlinear Equations

Solve. Remember that graphs can be used to confirm all real solutions.

27. $y = x^2,$
$\quad x = y^2$

28. $x^2 + y^2 = 25,$
$\quad y^2 = x + 5$

Aha! **29.** $x^2 + y^2 = 16,$
$\quad x^2 - y^2 = 16$

30. $y^2 - 4x^2 = 25,$
$\quad 4x^2 + y^2 = 25$

31. $x^2 + y^2 = 25,$
$\quad xy = 12$

32. $x^2 - y^2 = 16,$
$\quad x + y^2 = 4$

33. $x^2 + y^2 = 9,$
$\quad 25x^2 + 16y^2 = 400$

34. $x^2 + y^2 = 4,$
$\quad 9x^2 + 16y^2 = 144$

35. $x^2 + y^2 = 14,$
$\quad x^2 - y^2 = 4$

36. $x^2 + y^2 = 16,$
$\quad y^2 - 2x^2 = 10$

37. $x^2 + y^2 = 10,$
$\quad xy = 3$

38. $x^2 + y^2 = 5,$
$\quad xy = 2$

39. $x^2 + 4y^2 = 20,$
$\quad xy = 4$

40. $x^2 + y^2 = 13,$
$\quad xy = 6$

41. $2xy + 3y^2 = 7,$
$\quad 3xy - 2y^2 = 4$

42. $3xy + x^2 = 34,$
$\quad 2xy - 3x^2 = 8$

43. $4a^2 - 25b^2 = 0,$
$\quad 2a^2 - 10b^2 = 3b + 4$

44. $xy - y^2 = 2,$
$\quad 2xy - 3y^2 = 0$

45. $ab - b^2 = -4,$
$\quad ab - 2b^2 = -6$

46. $x^2 - y = 5,$
$\quad x^2 + y^2 = 25$

Problem Solving

Solve.

47. *Art.*   Elliot is designing a rectangular stained glass miniature that has a perimeter of 28 cm and a diagonal of length 10 cm. What should the dimensions of the glass be?

48. *Geometry.*   A rectangle has an area of 2 yd^2 and a perimeter of 6 yd. Find its dimensions.

49. *Tile Design.*   The Clay Works tile company wants to make a new rectangular tile that has a perimeter of 6 in. and a diagonal of length $\sqrt{5}$ in. What should the dimensions of the tile be?

50. *Geometry.*   A rectangle has an area of 20 in^2 and a perimeter of 18 in. Find its dimensions.

51. *Design of a Van.*   The cargo area of a delivery van must be 60 ft^2, and the length of a diagonal must accommodate a 13-ft board. Find the dimensions of the cargo area.

52. *Dimensions of a Rug.*   The diagonal of a Persian rug is 25 ft, and the area of the rug is 300 ft^2. Find the length and the width of the rug.

53. The product of two numbers is 90. The sum of their squares is 261. Find the numbers.

54. *Investments.*   A certain amount of money saved for 1 year at a certain interest rate yielded $125 in simple interest. If $625 more had been invested and the rate had been 1% less, the interest would have been the same. Find the principal and the rate.

55. *Laptop Dimensions.*   The screen on Ashley's new laptop has an area of 90 in^2 and a $\sqrt{200.25}$-in. diagonal. Find the width and the length of the screen.

56. *Garden Design.* A garden contains two square flower beds. Find the length of each bed if the sum of their areas is 832 ft^2 and the difference of their areas is 320 ft^2.

57. The area of a rectangle is $\sqrt{3}$ m^2, and the length of a diagonal is 2 m. Find the dimensions.

58. The area of a rectangle is $\sqrt{2}$ m^2, and the length of a diagonal is $\sqrt{3}$ m. Find the dimensions.

59. How can an understanding of conic sections be helpful when a system of nonlinear equations is being solved algebraically?

60. Write a problem for a classmate to solve. Devise the problem so that a system of two nonlinear equations with exactly one real solution is solved.

Skill Review

Simplify.

61. $(3a^{-4})^2(2a^{-5})^{-1}$ [1.6]

62. $16^{-1/2}$ [7.2]

63. $\log 10,000$ [9.5]

64. i^{71} [7.8]

65. $-10^2 \div 2 \cdot 5 - 3$ [1.2]

66. $\sqrt{500}$ [7.3]

Synthesis

67. Write a problem that translates to a system of two equations. Design the problem so that at least one equation is nonlinear and so that no real solution exists.

68. Suppose a system of equations is comprised of one linear equation and one nonlinear equation. Is it possible for such a system to have three solutions? Why or why not?

Solve.

69. $p^2 + q^2 = 13,$
$\dfrac{1}{pq} = -\dfrac{1}{6}$

70. $a + b = \dfrac{5}{6},$
$\dfrac{a}{b} + \dfrac{b}{a} = \dfrac{13}{6}$

71. *Fence Design.* A roll of chain-link fencing contains 100 ft of fence. The fencing is bent at a 90° angle to enclose a rectangular work area of 2475 ft^2, as shown. Find the length and the width of the rectangle.

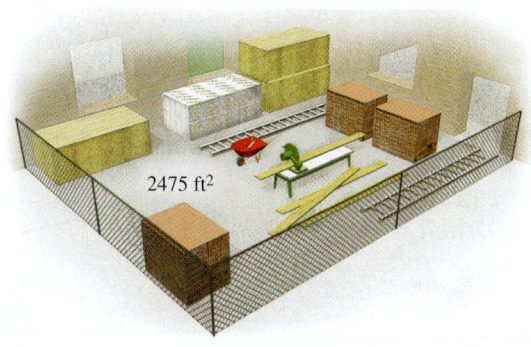

2475 ft^2

72. A piece of wire 100 cm long is to be cut into two pieces and those pieces are each to be bent to make a square. The area of one square is to be 144 cm^2 greater than that of the other. How should the wire be cut?

73. *Box Design.* Four squares with sides 5 in. long are cut from the corners of a rectangular metal sheet that has an area of 340 in^2. The edges are bent up to form an open box with a volume of 350 in^3. Find the dimensions of the box.

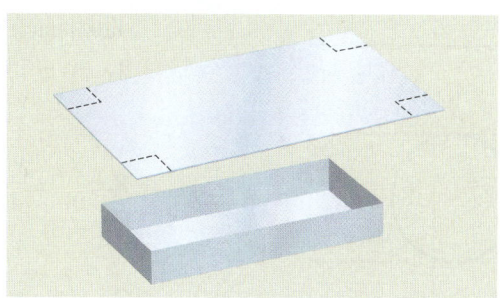

↪ YOUR TURN ANSWERS: SECTION 13.4

1. $(12, 5), \left(-\dfrac{56}{5}, -\dfrac{33}{5}\right)$ **2.** $(2, 3), \left(-\dfrac{1}{2}, -2\right)$

3. $(-2 + \sqrt{5}i, 1 + \sqrt{5}i), (-2 - \sqrt{5}i, 1 - \sqrt{5}i)$

4. $(5, 2), (5, -2), (-5, 2), (-5, -2)$

5. $(3, 1), (-3, -1), \left(\dfrac{\sqrt{6}}{2}, \sqrt{6}\right), \left(-\dfrac{\sqrt{6}}{2}, -\sqrt{6}\right)$

6. Length: 25 ft; width: 22 ft   **7.** Length: 12 mm; width: 5 mm

---

QUICK QUIZ: SECTIONS 13.1–13.4

Classify each of the following as the equation of a circle, an ellipse, a parabola, or a hyperbola. [13.3]

1. $y^2 = 8 - x^2$ **2.** $2x^2 - x - y^2 + 4 = 0$

3. $5x^2 + 10y^2 = 50$ **4.** $4x + y^2 + y = 7$

5. Solve:

$$x - y = 3,$$
$$x^2 + y^2 = 5. \quad [13.4]$$

PREPARE TO MOVE ON

Simplify. [1.8]

1. $(-1)^9(-3)^2$ **2.** $(-1)^{10}(-3)^3$

Evaluate each of the following. [1.8]

3. $\dfrac{(-1)^k}{k - 5}$, for $k = 10$ **4.** $\dfrac{n}{2}(3 + n)$, for $n = 11$

5. $\dfrac{7(1 - r^2)}{1 - r}$, for $r = \dfrac{1}{2}$

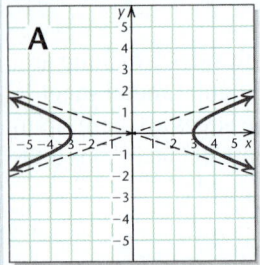

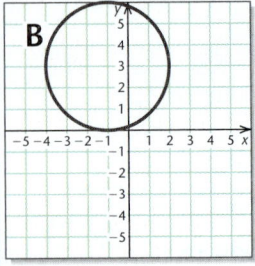

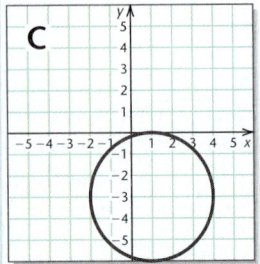

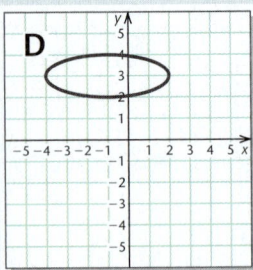

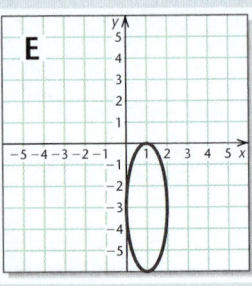

Visualizing for Success

Use after Section 13.3.

Match each equation with its graph.

1. $(x - 1)^2 + (y + 3)^2 = 9$

2. $\dfrac{x^2}{9} - \dfrac{y^2}{1} = 1$

3. $y = (x - 1)^2 - 3$

4. $(x + 1)^2 + (y - 3)^2 = 9$

5. $x = (y - 1)^2 - 3$

6. $\dfrac{(x + 1)^2}{9} + \dfrac{(y - 3)^2}{1} = 1$

7. $xy = 3$

8. $y = -(x + 1)^2 + 3$

9. $\dfrac{y^2}{9} - \dfrac{x^2}{1} = 1$

10. $\dfrac{(x - 1)^2}{1} + \dfrac{(y + 3)^2}{9} = 1$

Answers on page A-74

An additional, animated version of this activity appears in MyMathLab. To use MyMathLab, you need a course ID and a student access code. Contact your instructor for more information.

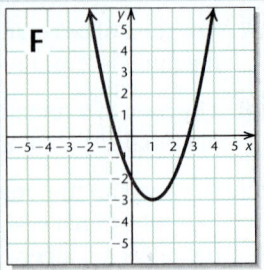

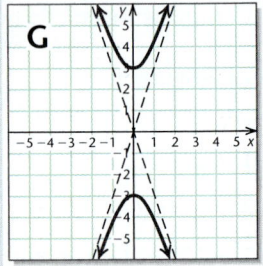

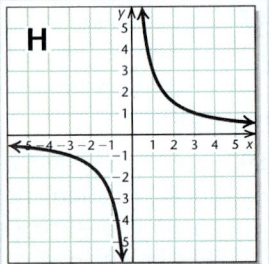

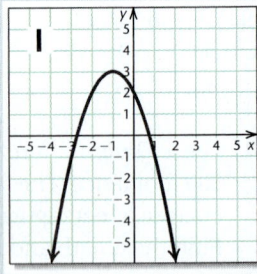

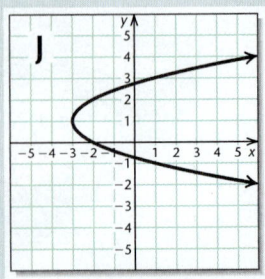

Collaborative Activity   *A Cosmic Path*

Focus: Ellipses
Use after: Section 13.2
Time: 20–30 minutes
Group size: 2
Materials: Scientific calculators

On May 4, 2007, Comet 17P/Holmes was at the point closest to the sun in its orbit. Comet 17P is traveling in an elliptical orbit with the sun as one focus, and one orbit takes about 6.88 years. One astronomical unit (AU) is 93,000,000 mi. One group member should do the following calculations in AU and the other in millions of miles.

Source: Harvard-Smithsonian Center for Astrophysics

Activity

1. At its *perihelion*, a comet with an elliptical orbit is at the point in its orbit closest to the sun. At its *aphelion*, the comet is at the point farthest from the sun. The perihelion distance for Comet 17P is 2.053218 AU, and the aphelion distance is 5.183610 AU. Use these distances to find *a*. (See the following diagram.)

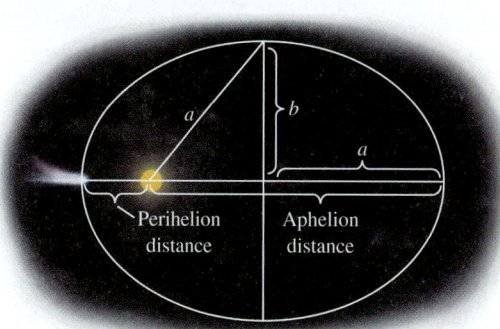

Perihelion distance   Aphelion distance

2. Using the figure above, express b^2 as a function of *a*. Then find *b* using the value found for *a* in part (1).

3. One formula for approximating the perimeter of an ellipse is

$$P = \pi\left(3a + 3b - \sqrt{(3a + b)(a + 3b)}\,\right),$$

developed by the Indian mathematician S. Ramanujan in 1914. How far does Comet 17P travel in one orbit?

4. What is the speed of the comet? Find the answer in AU per year and in miles per hour.

5. Which calculations—AUs or mi—were easier to use? Why?

Decision Making & Connection   *(Use after Section 13.4.)*

Aspect Ratio.   Photographs, television screens, and movie screens are often described in terms of an aspect ratio, as follows.

Area: $A = bh$
Diagonal: $d = \sqrt{b^2 + h^2}$
Aspect ratio: $r = \dfrac{b}{h}$

1. Wide-screen televisions have an aspect ratio of $\frac{16}{9}$. If a manufacturer plans a diagonal of length 40 in., what must the dimensions of the screen be?

2. Older televisions had an aspect ratio of $\frac{4}{3}$. For a diagonal of length 40 in., what would the dimensions of this television be?

3. For a fixed diagonal of 40 in., which has more screen area: a wide-screen TV or an older TV?

4. Find a formula for the height of a rectangle given the diagonal and the aspect ratio.

 5. *Research.*   The aspect ratio of a rectangle that is considered most pleasing aesthetically is called the *golden ratio*. Find the value of the golden ratio. What aspect ratios used today for photography or video are closest to the golden ratio?

Study Summary

SECTION 13.1: *Conic Sections: Parabolas and Circles*

Parabola

$y = ax^2 + bx + c$

$\quad = a(x - h)^2 + k$

    Opens upward ($a > 0$)
    or downward ($a < 0$)
    Vertex: (h, k)

$x = ay^2 + by + c$

$\quad = a(y - k)^2 + h$

    Opens right ($a > 0$)
    or left ($a < 0$)
    Vertex: (h, k)

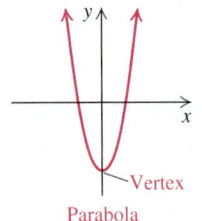

Parabola

$x = -y^2 + 4y - 1$

$\quad = -(y^2 - 4y \quad\quad) - 1$

$\quad = -(y^2 - 4y + 4) - 1 - (-1)(4)$

$\quad = -(y - 2)^2 + 3$ $a = -1$; parabola opens left

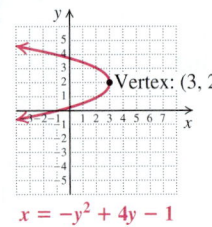

Vertex: (3, 2)

$x = -y^2 + 4y - 1$

---

1. Graph

$x = y^2 + 6y + 7$.

Label the vertex.

Circle

$x^2 + y^2 = r^2$

    Radius: r
    Center: $(0, 0)$

$(x - h)^2 + (y - k)^2 = r^2$

    Radius: r
    Center: (h, k)

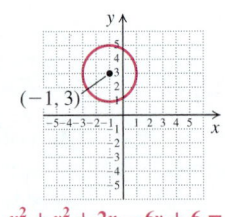

Circle

$x^2 + y^2 + 2x - 6y + 6 = 0$

$x^2 + 2x + \quad\ y^2 - 6y \quad\quad = -6$

$x^2 + 2x + 1 + y^2 - 6y + 9 = -6 + 1 + 9$

$\quad\quad (x + 1)^2 + (y - 3)^2 = 4$

$\quad [x - (-1)]^2 + (y - 3)^2 = 2^2$ Radius: 2; center: $(-1, 3)$

$(-1, 3)$

$x^2 + y^2 + 2x - 6y + 6 = 0$

---

2. Find the center and the radius and then graph

$x^2 + y^2 - 6x + 5 = 0$.

SECTION 13.2: *Conic Sections: Ellipses*

Ellipse

$$\frac{x^2}{a^2} + \frac{y^2}{b^2} = 1 \quad \text{Center: } (0, 0)$$

$$\frac{(x - h)^2}{a^2} + \frac{(y - k)^2}{b^2} = 1$$

Center: (h, k)

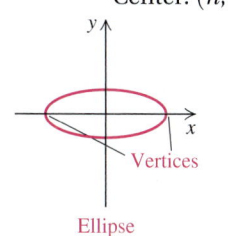

Ellipse

$$\frac{(x - 4)^2}{4} + \frac{(y + 1)^2}{9} = 1$$

$$\frac{(x - 4)^2}{2^2} + \frac{[y - (-1)]^2}{3^2} = 1 \qquad \begin{array}{l} 3 > 2; \text{ ellipse is vertical} \\ \text{with center } (4, -1) \end{array}$$

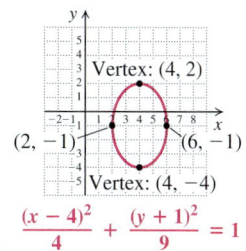

$$\frac{(x - 4)^2}{4} + \frac{(y + 1)^2}{9} = 1$$

3. Graph: $\dfrac{x^2}{9} + y^2 = 1$.

SECTION 13.3: *Conic Sections: Hyperbolas*

Hyperbola

$$\frac{x^2}{a^2} - \frac{y^2}{b^2} = 1$$

Two branches opening right and left

$$\frac{y^2}{b^2} - \frac{x^2}{a^2} = 1$$

Two branches opening upward and downward

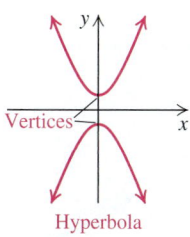

Hyperbola

$$\frac{x^2}{4} - \frac{y^2}{1} = 1$$

$$\frac{x^2}{2^2} - \frac{y^2}{1^2} = 1 \qquad \text{Opens right and left}$$

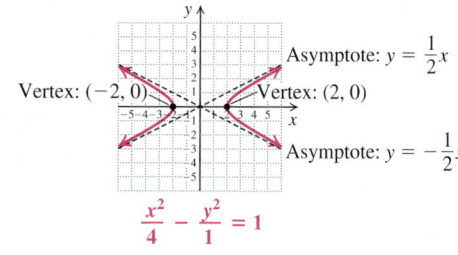

$$\frac{x^2}{4} - \frac{y^2}{1} = 1$$

4. Graph: $\dfrac{y^2}{16} - \dfrac{x^2}{4} = 1$.

SECTION 13.4: *Nonlinear Systems of Equations*

We can solve a system containing at least one nonlinear equation using substitution or elimination.

Solve:

$x^2 - y = -1,$ (1)   (The graph is a parabola.)

$x + 2y = 3.$ (2)   (The graph is a line.)

$x = 3 - 2y$ Solving for x

$(3 - 2y)^2 - y = -1$ Substituting

$9 - 12y + 4y^2 - y = -1$

$4y^2 - 13y + 10 = 0$

$(4y - 5)(y - 2) = 0$

$4y - 5 = 0 \quad or \quad y - 2 = 0$

$y = \frac{5}{4} \quad or \quad y = 2$

If $y = \frac{5}{4}$, then $x = 3 - 2\left(\frac{5}{4}\right) = \frac{1}{2}$. $\left(\frac{1}{2}, \frac{5}{4}\right)$ is a solution.

If $y = 2$, then $x = 3 - 2(2) = -1$. $(-1, 2)$ is a solution.

The solutions are $\left(\frac{1}{2}, \frac{5}{4}\right)$ and $(-1, 2)$.

5. Solve:

$x^2 + y^2 = 41,$

$y - x = 1.$

Review Exercises: Chapter 13

Concept Reinforcement

Classify each of the following statements as either true or false.

1. Any parabola that opens upward or downward represents the graph of a function.   [13.1]

2. The center of a circle is part of the circle itself. [13.1]

3. The foci of an ellipse are part of the ellipse itself. [13.2]

4. It is possible for a hyperbola to represent the graph of a function.   [13.3]

5. If an equation of a conic section has only one term of degree 2, its graph cannot be a circle, an ellipse, or a hyperbola.   [13.3]

6. Two nonlinear graphs can intersect in more than one point.   [13.4]

7. Every system of nonlinear equations has at least one real solution.   [13.4]

8. Both substitution and elimination can be used as methods for solving a system of nonlinear equations.   [13.4]

Find the center and the radius of each circle.   [13.1]

9. $(x + 3)^2 + (y - 2)^2 = 16$

10. $(x - 5)^2 + y^2 = 11$

11. $x^2 + y^2 - 6x - 2y + 1 = 0$

12. $x^2 + y^2 + 8x - 6y = 20$

13. Find an equation of the circle with center $(-4, 3)$ and radius 4.   [13.1]

14. Find an equation of the circle with center $(7, -2)$ and radius $2\sqrt{5}$.   [13.1]

Classify each equation as a circle, an ellipse, a parabola, or a hyperbola. Then graph.

15. $5x^2 + 5y^2 = 80$ [13.1], [13.3]

16. $9x^2 + 2y^2 = 18$ [13.2], [13.3]

17. $y = -x^2 + 2x - 3$ [13.1], [13.3]

18. $\dfrac{y^2}{9} - \dfrac{x^2}{4} = 1$ [13.3]

19. $xy = 9$ [13.3]

20. $x = y^2 + 2y - 2$ [13.1], [13.3]

21. $\dfrac{(x + 1)^2}{3} + (y - 3)^2 = 1$ [13.2], [13.3]

22. $x^2 + y^2 + 6x - 8y - 39 = 0$ [13.1], [13.3]

Solve.   [13.4]

23. $x^2 - y^2 = 21$,
    $x + y = 3$

24. $x^2 - 2x + 2y^2 = 8$,
    $2x + y = 6$

25. $x^2 - y = 5$,
    $2x - y = 5$

26. $x^2 + y^2 = 25$,
    $x^2 - y^2 = 7$

27. $x^2 - y^2 = 3$,
    $y = x^2 - 3$

28. $x^2 + y^2 = 18$,
    $2x + y = 3$

29. $x^2 + y^2 = 100$,
    $2x^2 - 3y^2 = -120$

30. $x^2 + 2y^2 = 12$,
    $xy = 4$

31. A rectangular bandstand has a perimeter of 38 m and an area of 84 m². What are the dimensions of the bandstand?   [13.4]

32. One type of carton used by tableproducts.com exactly fits both a rectangular plate of area 108 in² and chopsticks of length 15 in., laid diagonally on top of the plate. Find the length and the width of the carton.   [13.4]

15 in.

33. The perimeter of a square mounting board is 12 cm more than the perimeter of a square mirror. The board's area exceeds the area of the mirror by 39 cm². Find the perimeter of each object.   [13.4]

34. The sum of the areas of two circles is 130π ft². The difference of the circumferences is 16π ft. Find the radius of each circle.   [13.4]

Synthesis

35. How does the graph of a hyperbola differ from the graph of a parabola?   [13.1], [13.3]

36. Explain why function notation rarely appears in this chapter, and list the types of graphs discussed for which function notation could be used.   [13.1], [13.2], [13.3]

37. Solve:
$$4x^2 - x - 3y^2 = 9,$$
$$-x^2 + x + y^2 = 2. \quad \text{[13.4]}$$

38. Find an equation of the circle that passes through $(-2, -4)$, $(5, -5)$, and $(6, 2)$. [13.1], [13.4]

39. Find an equation of the ellipse with the following intercepts: $(-10, 0)$, $(10, 0)$, $(0, -1)$, and $(0, 1)$. [13.2]

Test: Chapter 13

For step-by-step test solutions, access the Chapter Test Prep Videos in MyMathLab®, on YouTube™ (search "Bittinger Combo Alg CA" and click on Channels"), or by scanning the code.

1. Find an equation of the circle with center $(3, -4)$ and radius $2\sqrt{3}$.

Find the center and the radius of each circle.

2. $(x - 4)^2 + (y + 1)^2 = 5$

3. $x^2 + y^2 + 4x - 6y + 4 = 0$

Classify the equation as a circle, an ellipse, a parabola, or a hyperbola. Then graph.

4. $y = x^2 - 4x - 1$

5. $x^2 + y^2 + 2x + 6y + 6 = 0$

6. $\dfrac{x^2}{16} - \dfrac{y^2}{9} = 1$

7. $16x^2 + 4y^2 = 64$

8. $xy = -5$

9. $x = -y^2 + 4y$

Solve.

10. $x^2 + y^2 = 36,$
$\quad 3x + 4y = 24$

11. $x^2 - y = 3,$
$\quad 2x + y = 5$

12. $x^2 - 2y^2 = 1,$
$\quad xy = 6$

13. $x^2 + y^2 = 10,$
$\quad x^2 = y^2 + 2$

14. A rectangular bookmark with diagonal of length $5\sqrt{5}$ has an area of 22. Find the dimensions of the bookmark.

15. Two squares are such that the sum of their areas is 8 m^2 and the difference of their areas is 2 m^2. Find the length of a side of each square.

16. A rectangular dance floor has a diagonal of length 40 ft and a perimeter of 112 ft. Find the dimensions of the dance floor.

17. Brett invested a certain amount of money for 1 year and earned $72 in interest. Erin invested $240 more than Brett at an interest rate that was $\frac{5}{6}$ of the rate given to Brett, but she earned the same amount of interest. Find the principal and the interest rate for Brett's investment.

Synthesis

18. Find an equation of the ellipse passing through $(6, 0)$ and $(6, 6)$ with vertices at $(1, 3)$ and $(11, 3)$.

19. The sum of two numbers is 36, and the product is 4. Find the sum of the reciprocals of the numbers.

20. *Theatrical Production.* An E.T.C. spotlight for a college's production of *Hamlet* projects an ellipse of light on a stage that is 8 ft wide and 14 ft long. Find an equation of that ellipse if an actor is in its center and x represents the number of feet, horizontally, from the actor to the edge of the ellipse and y represents the number of feet, vertically, from the actor to the edge of the ellipse.

Cumulative Review: Chapters 1–13

Simplify.

1. $(4t^2 - 5s)^2$ [4.6]

2. $\dfrac{1}{3t} + \dfrac{1}{t-3}$ [6.4]

3. $\sqrt{6t}\,\sqrt{15t^3w}$ [10.3]

4. $(81a^{2/3}b^{1/4})^{3/4}$ [10.2]

5. $\log_2 \dfrac{1}{16}$ [12.3]

6. $(4+3i)(4-3i)$ [12.8]

7. -8^{-2} [4.2]

Factor.

8. $100x^2 - 60xy + 9y^2$ [5.4]

9. $3m^6 - 24$ [5.5]

10. $ax + by - ay - bx$ [5.1]

11. $32x^2 - 20x - 3$ [5.3]

Solve. Where appropriate, give an approximation to four decimal places.

12. $3(x-5) - 4x \geq 2(x+5)$ [2.6]

13. $16x^2 - 18x = 0$ [5.7]

14. $\dfrac{2}{x} + \dfrac{1}{x-2} = 1$ [6.6]

15. $5x^2 + 5 = 0$ [11.2]

16. $\log_x 64 = 3$ [12.6]

17. $3^x = 1.5$ [12.6]

18. $x = \sqrt{2x-5} + 4$ [10.6]

19. $x^2 + 2y^2 = 5,$
    $2x^2 + y^2 = 7$ [13.4]

Graph.

20. $3x - y = 9$ [3.3]

21. $y = \log_5 x$ [12.3]

22. $\dfrac{x^2}{25} + \dfrac{y^2}{1} = 1$ [13.2]

23. $f(x) = 2^{x-1}$ [12.2]

24. $x^2 + (y-3)^2 = 4$ [13.1]

25. $x < 2y + 1$ [9.4]

26. Graph: $f(x) = -(x+2)^2 + 3$.   [11.7]
    a) Label the vertex.
    b) Draw the axis of symmetry.
    c) Find the maximum or minimum value.

27. Find the slope–intercept equation of the line containing the points $(-3,6)$ and $(1,2)$.   [3.7]

28. Write a quadratic equation having the solutions $\sqrt{3}$ and $-\sqrt{3}$. Answers may vary.   [11.3]

Solve.

29. *Aviation.* BlueAir owns two types of airplanes. One type flies 60 mph faster than the other. Laura often rents a plane from BlueAir to visit her parents. The flight takes 4 hr with the faster plane and 4 hr 24 min with the slower plane. What distance does she fly?   [8.3]

30. *Retail.* The median size of a grocery store in the United States was 48,750 ft^2 in 2006 and 46,000 ft^2 in 2010.   [7.3]
    Source: Food Marketing Industry
    a) Find a linear function that fits the data. Let t represent the number of years since 2006.
    b) Use the function from part (a) to predict the median size of a grocery store in 2014.
    c) Assuming the trend continues, in what year will the median size of a grocery store be 40,000 ft^2?

31. *Population.* The population of Latvia was 2.19 million in 2012 and was decreasing exponentially at a rate of 0.84% per year.   [12.7]
    Sources: Encyclopedia of the Nations; The CIA World Factbook
    a) Write an exponential function that could be used to find $P(t)$, the population of Latvia, in millions, t years after 2012.
    b) Predict what the population will be in 2025.
    c) What is the half life of the population?

32. *Art.* Elyse is designing a rectangular tray. She wants to put a row of beads around the tray, and has enough beads to make an edge that is 32 in. long. What dimensions of the tray will give it the greatest area?   [11.8]

Synthesis

33. If y varies inversely as the square root of x and x is multiplied by 100, what is the effect on y?   [7.5], [10.1]

34. For $f(x) = x - \dfrac{1}{x^2}$, find all x-values for which $f(x) \leq 0$.   [11.9]

Sequences, Series, and the Binomial Theorem

YEAR	GLOBAL SOLAR PV CAPACITY (in gigawatts)
2012	70
2013	84
2014	100.8

Sources: Based on data from The European Photovoltaic Industry Association; GlobalData

Get Clean Power from the Sun.

As the cost of generating electricity from sunlight decreases, more solar photovoltaic (PV) panels are being installed around the world. The table above shows one estimate of the total global solar PV capacity for several years. Later in this chapter, we will use a *geometric series* to estimate how much total power will be generated by solar energy from 2012 through 2020. (*See Exercise 67 in Exercise Set 14.3.*)

It would be impossible to track utility usage and costs and to determine savings without using math in my daily work.

Wayne Benbow, an energy management consultant from Lakewood, Colorado, uses math to track energy consumption and the associated costs in over 13 million square feet of floor space in 150 schools.

A *sequence* is, quite simply, an ordered list. When the members of a sequence are numbers, we can discuss their sum. Such a sum is called a *series*.

Section 14.4 presents the *binomial theorem*, which is used to expand expressions of the form $(a + b)^n$. Such an expansion is itself a series.

14.1 Sequences and Series

Sequences ▪ Finding the General Term ▪ Sums and Series ▪ Sigma Notation

Sequences

Suppose that $10,000 is invested at 5%, compounded annually. The value of the account at the start of years 1, 2, 3, 4, and so on, is

$10,000, $10,500, $11,025, $11,576.25, ...

We can regard this as a function that pairs 1 with $10,000, 2 with $10,500, 3 with $11,025, and so on. This is an example of a **sequence** (or **progression**). The domain of a sequence is a set of counting numbers beginning with 1.

If we stop after a certain number of years, we obtain a **finite sequence**:

$10,000, $10,500, $11,025, $11,576.25.

If we continue listing the amounts in the account, we obtain an **infinite sequence**:

$10,000, $10,500, $11,025, $11,576.25, $12,155.06, ...

The three dots near the end indicate that the sequence goes on without stopping.

SEQUENCES

An *infinite sequence* is a function having for its domain the set of natural numbers: $\{1, 2, 3, 4, 5, \ldots\}$.

A *finite sequence* is a function having for its domain a set of natural numbers: $\{1, 2, 3, 4, 5, \ldots, n\}$, for some natural number n.

As another example, consider the sequence given by

$$a(n) = 2^n, \quad \text{or} \quad a_n = 2^n.$$

The notation a_n means $a(n)$ but is used more commonly with sequences. Some function values (also called *terms* of the sequence) follow:

$$a_1 = 2^1 = 2,$$
$$a_2 = 2^2 = 4,$$
$$a_3 = 2^3 = 8,$$
$$a_6 = 2^6 = 64.$$

Note that n gives the position in the sequence and a_n defines the number that is in that position. The first term of the sequence is a_1, the fifth term is a_5, and the *n*th term, or **general term**, is a_n. This sequence can also be denoted in the following ways:

$$2, 4, 8, \ldots;$$

or $$2, 4, 8, \ldots, 2^n, \ldots.$$ The 2^n emphasizes that the *n*th term of this sequence is found by raising 2 to the *n*th power.

EXAMPLE 1   Find the first four terms and the 57th term of the sequence for which the general term is $a_n = (-1)^n/(n+1)$.

SOLUTION   We have

$$a_1 = \frac{(-1)^1}{1+1} = -\frac{1}{2},$$ Substituting in $a_n = \frac{(-1)^n}{n+1}$

$$a_2 = \frac{(-1)^2}{2+1} = \frac{1}{3},$$

$$a_3 = \frac{(-1)^3}{3+1} = -\frac{1}{4},$$

$$a_4 = \frac{(-1)^4}{4+1} = \frac{1}{5},$$

$$a_{57} = \frac{(-1)^{57}}{57+1} = -\frac{1}{58}.$$

Note that the factor $(-1)^n$ causes the signs of the terms to alternate between positive and negative, depending on whether *n* is even or odd.

YOUR TURN

Finding the General Term

By looking for a pattern, we can often write an expression for the general term of a sequence. When only a few terms are given, more than one pattern may fit.

EXAMPLE 2   For each sequence, predict the general term.

a) $1, 4, 9, 16, 25, \ldots$ **b)** $2, 4, 8, \ldots$ **c)** $-1, 2, -4, 8, -16, \ldots$

SOLUTION

a) $1, 4, 9, 16, 25 \ldots$

These are squares of consecutive positive integers, so the general term could be n^2.

b) $2, 4, 8, \ldots$

We regard the pattern as powers of 2, in which case 16 would be the next term and 2^n the general term.

c) $-1, 2, -4, 8, -16, \ldots$

These are powers of 2 with alternating signs, so the general term may be

$$(-1)^n[2^{n-1}].$$

Making sure the signs of the terms alternate

Raising 2 to a power that is 1 less than the term's position in the sequence

To check, note that -4 is the third term, and $(-1)^3[2^{3-1}] = -1 \cdot 2^2 = -4$.

YOUR TURN

In Example 2(b), if 2^n is the general term, the sequence can be written with more terms as

$$2, 4, 8, 16, 32, 64, 128, \ldots.$$

Suppose instead that the second term is found by adding 2, the third term by adding 4, the next term by adding 6, and so on. In this case, 14 would be the next term and the sequence would be

$$2, 4, 8, 14, 22, 32, 44, 58, \ldots.$$

This illustrates that the fewer terms we are given, the greater the uncertainty about determining the nth term.

Sums and Series

> **SERIES**
>
> Given the infinite sequence
>
> $$a_1, a_2, a_3, a_4, \ldots, a_n, \ldots,$$
>
> the sum of the terms
>
> $$a_1 + a_2 + a_3 + \cdots + a_n + \cdots$$
>
> is called an *infinite series* and is denoted S_∞. The nth *partial sum* is the sum of the first n terms:
>
> $$a_1 + a_2 + a_3 + \cdots + a_n.$$
>
> A partial sum is also called a *finite series* and is denoted S_n.

EXAMPLE 3   For the sequence $-2, 4, -6, 8, -10, 12, -14$, find: **(a)** S_2; **(b)** S_3; **(c)** S_7.

SOLUTION

3. For the sequence

$$1, -1, 1 -1, \ldots,$$

find S_{12}.

a) $S_2 = -2 + 4 = 2$ This is the sum of the first 2 terms.

b) $S_3 = -2 + 4 + (-6) = -4$ This is the sum of the first 3 terms.

c) $S_7 = -2 + 4 + (-6) + 8 + (-10) + 12 + (-14) = -8$ This is the sum of the first 7 terms.

 YOUR TURN

Sigma Notation

Student Notes

A great deal of information is condensed into sigma notation. Be careful to pay attention to what values the index of summation will take on. Evaluate the general term for each value and then add the results.

When the general term of a sequence is known, the Greek letter Σ (uppercase sigma) can be used to write a series. For example, the sum of the first four terms of the sequence $3, 5, 7, 9, 11, \ldots, 2k + 1, \ldots$ can be named as follows, using *sigma notation*, or *summation notation*:

$$\sum_{k=1}^{4} (2k + 1).$$

This represents $(2 \cdot 1 + 1) + (2 \cdot 2 + 1) + (2 \cdot 3 + 1) + (2 \cdot 4 + 1)$, and is read "the sum as k goes from 1 to 4 of $(2k + 1)$." The letter k is called the *index of summation*. The index need not always start at 1.

EXAMPLE 4 Write out and evaluate each sum.

a) $\displaystyle\sum_{k=1}^{5} k^2$ **b)** $\displaystyle\sum_{k=4}^{6} (-1)^k (2k)$

SOLUTION

a) $\displaystyle\sum_{k=1}^{5} k^2 = 1^2 + 2^2 + 3^2 + 4^2 + 5^2 = 1 + 4 + 9 + 16 + 25 = 55$

Evaluate k^2 for all integers from 1 through 5. Then add.

4. Write out and evaluate

$$\sum_{k=0}^{3} (2^k + 5).$$

b) $\displaystyle\sum_{k=4}^{6} (-1)^k (2k) = (-1)^4 (2 \cdot 4) + (-1)^5 (2 \cdot 5) + (-1)^6 (2 \cdot 6)$
$$= 8 - 10 + 12 = 10$$

YOUR TURN

EXAMPLE 5 Write sigma notation for each sum.

a) $1 + 4 + 9 + 16 + 25$

b) $3 + 9 + 27 + 81 + \cdots$

c) $-1 + 3 - 5 + 7$

SOLUTION

a) $1 + 4 + 9 + 16 + 25$

Note that this is a sum of squares, $1^2 + 2^2 + 3^2 + 4^2 + 5^2$, so the general term is k^2. Sigma notation is

$$\sum_{k=1}^{5} k^2.$$ The sum starts with 1^2 and ends with 5^2.

Answers may vary. For example, another—perhaps less obvious—way of writing $1 + 4 + 9 + 16 + 25$ is

$$\sum_{k=2}^{6} (k - 1)^2.$$

b) $3 + 9 + 27 + 81 + \cdots$

This is a sum of powers of 3, and it is also an infinite series. We use the symbol ∞ for infinity and write the series using sigma notation:

$$\sum_{k=1}^{\infty} 3^k.$$

c) $-1 + 3 - 5 + 7$

Except for the alternating signs, this is the sum of the first four positive odd numbers. It is useful to remember that $2k - 1$ is a formula for the kth positive odd number. It is also important to remember that the factor $(-1)^k$ can be used to create the alternating signs. The general term is thus $(-1)^k (2k - 1)$, beginning with $k = 1$. Sigma notation is

$$\sum_{k=1}^{4} (-1)^k (2k - 1).$$

5. Write sigma notation for

$10 + 20 + 30 + 40 + \cdots$.

To check, we can evaluate $(-1)^k (2k - 1)$ using 1, 2, 3, and 4, and write the sum of the four terms. We leave this to the student.

YOUR TURN

91. Find S_{100} and S_{101} for the sequence in which $a_n = (-1)^n$.

Find the first five terms of each sequence; then find S_5.

92. $a_n = \dfrac{1}{2^n} \log 1000^n$

93. $a_n = i^n, i = \sqrt{-1}$

94. Find all values for x that solve the following:

$$\sum_{k=1}^{x} i^k = -1.$$

95. The nth term of a sequence is given by

$$a_n = n^5 - 14n^4 + 6n^3 + 416n^2 - 655n - 1050.$$

Use a graphing calculator with a TABLE feature to determine which term in the sequence is 6144.

96. To define a sequence recursively on a graphing calculator (see Exercises 87 and 88), we use the SEQ MODE. The general term U_n or V_n can often be expressed in terms of U_{n-1} or V_{n-1} by pressing **2ND** **7** or **2ND** **8**. The starting values of U_n, V_n, and n are set as one of the WINDOW variables.

Use recursion to determine how many different handshakes occur when 50 people shake hands with one another. To develop the recursion formula, begin with a group of 2 and determine how many additional handshakes occur with the arrival of each new person.

YOUR TURN ANSWERS: SECTION 14.1

1. $-1, 4, -9, 16; 2500$ **2.** $a_n = 5n$ **3.** 0
4. $(2^0 + 5) + (2^1 + 5) + (2^2 + 5) + (2^3 + 5) = 35$
5. $\displaystyle\sum_{k=1}^{\infty} 10k$

PREPARE TO MOVE ON

Evaluate. [1.8]

1. $\dfrac{7}{2}(a_1 + a_7)$, for $a_1 = 8$ and $a_7 = 20$

2. $a_1 + (n - 1)d$, for $a_1 = 3$, $n = 10$, and $d = -2$

Simplify. [1.8]

3. $(a_1 + 5d) + (a_n - 5d)$

4. $(a_1 + 8d) - (a_1 + 7d)$

14.2   Arithmetic Sequences and Series

Arithmetic Sequences ▪ Sum of the First *n* Terms of an Arithmetic Sequence ▪ Problem Solving

Study Skills

Rest Before a Test

The final exam is probably your most important test of the semester. Do yourself a favor and see to it that you get a good night's sleep the night before. Being well rested will help you put forth your best work.

In this section, we concentrate on sequences and series that are said to be arithmetic (pronounced ar-ith-MET-ik).

Arithmetic Sequences

In an **arithmetic sequence** (or **progression**), adding the same number to any term gives the next term in the sequence. For example, the sequence 2, 5, 8, 11, 14, 17, . . . is arithmetic because adding 3 to any term produces the next term.

> **ARITHMETIC SEQUENCE**
>
> A sequence is *arithmetic* if there exists a number d, called the *common difference*, such that $a_{n+1} = a_n + d$ for any integer $n \geq 1$.

EXAMPLE 1   For each arithmetic sequence, identify the first term, a_1, and the common difference, d.

a) $4, 9, 14, 19, 24, \ldots$ **b)** $27, 20, 13, 6, -1, -8, \ldots$

SOLUTION   To find a_1, we simply use the first term listed. To find d, we choose any term other than a_1 and subtract the preceding term from it.

Sequence	First Term, a_1	Common Difference, d
a) $4, 9, 14, 19, 24, \ldots$	4	$5 \leftarrow 9 - 4 = 5$
b) $27, 20, 13, 6, -1, -8, \ldots$	27	$-7 \leftarrow 20 - 27 = -7$

To find the common difference, we subtracted a_1 from a_2. Had we subtracted a_2 from a_3 or a_3 from a_4, we would have found the same values for d.

Check: As a check, note that when d is added to each term, the result is the next term in the sequence.

1. Identify the first term, a_1, and the common difference, d, for the arithmetic sequence

$$0, \tfrac{1}{2}, 1, \tfrac{3}{2}, 2, \ldots.$$

YOUR TURN

To develop a formula for the general, nth, term of any arithmetic sequence, we denote the common difference by d and write out the first few terms:

$a_1,$

$a_2 = a_1 + d,$

$a_3 = a_2 + d = (a_1 + d) + d = a_1 + 2d,$ Substituting $a_1 + d$ for a_2

$a_4 = a_3 + d = (a_1 + 2d) + d = a_1 + 3d.$ Substituting $a_1 + 2d$ for a_3

Note that the coefficient of d in each case is 1 less than the subscript.

Generalizing, we obtain the following formula.

TO FIND a_n FOR AN ARITHMETIC SEQUENCE

The nth term of an arithmetic sequence with common difference d is

$$a_n = a_1 + (n - 1)d, \quad \text{for any integer } n \geq 1.$$

EXAMPLE 2   Find the 14th term of the arithmetic sequence $6, 9, 12, 15, \ldots$.

SOLUTION   First we note that $a_1 = 6$, $d = 3$, and $n = 14$. Using the formula for the nth term of an arithmetic sequence, we have

$$a_n = a_1 + (n - 1)d$$
$$a_{14} = 6 + (14 - 1) \cdot 3 = 6 + 13 \cdot 3 = 6 + 39 = 45.$$

The 14th term is 45.

2. Find the 20th term of the arithmetic sequence

$$100, 97, 94, 91, \ldots.$$

YOUR TURN

EXAMPLE 3 For the sequence $6, 9, 12, 15, \ldots$, which term is 300?

SOLUTION Determining which term is 300 is the same as finding n if $a_n = 300$. In Example 2, we found that for this sequence we have $a_1 = 6$ and $d = 3$. Thus,

$$a_n = a_1 + (n - 1)d \qquad \text{Using the formula for the } n\text{th term of an arithmetic sequence}$$
$$300 = 6 + (n - 1) \cdot 3 \qquad \text{Substituting}$$
$$300 = 6 + 3n - 3$$
$$297 = 3n$$
$$99 = n.$$

3. For the sequence

$100, 97, 94, 91, \ldots$,

which term is -8?

The term 300 is the 99th term of the sequence.

YOUR TURN

Given two terms and their places in an arithmetic sequence, we can construct the sequence.

EXAMPLE 4 The 3rd term of an arithmetic sequence is 14, and the 16th term is 79. Find a_1 and d and construct the sequence.

SOLUTION We know that $a_3 = 14$ and $a_{16} = 79$. Thus we would have to add d a total of 13 times to get from 14 to 79. That is,

$$14 + 13d = 79. \qquad a_3 \text{ and } a_{16} \text{ are 13 terms apart; } 16 - 3 = 13$$

Solving $14 + 13d = 79$, we obtain

$$13d = 65 \qquad \text{Subtracting 14 from both sides}$$
$$d = 5. \qquad \text{Dividing both sides by 13}$$

We subtract d twice from a_3 to get to a_1. Thus,

4. The 4th term of an arithmetic sequence is 5, and the 21st term is 175. Find a_1 and d and construct the sequence.

$$a_1 = 14 - 2 \cdot 5 = 4. \qquad a_1 \text{ and } a_3 \text{ are 2 terms apart; } 3 - 1 = 2$$

The sequence is $4, 9, 14, 19, \ldots$. Note that we could have subtracted d a total of 15 times from a_{16} in order to find a_1.

YOUR TURN

In general, d should be subtracted $(n - 1)$ times from a_n in order to find a_1.

Sum of the First *n* Terms of an Arithmetic Sequence

When the terms of an arithmetic sequence are added, an **arithmetic series** is formed. To develop a formula for computing S_n when the series is arithmetic, we list the first n terms of the sequence as follows:

This is the next-to-last term. If you add d to this term, the result is a_n.

$$a_1, (a_1 + d), (a_1 + 2d), \ldots, (a_n - 2d), (a_n - d), a_n$$

This term is two terms back from the end. If you add d to this term, you get the next-to-last term, $a_n - d$.

Thus, S_n is given by

$$S_n = a_1 + (a_1 + d) + (a_1 + 2d) + \cdots + (a_n - 2d) + (a_n - d) + a_n.$$

Using a commutative law, we have a second equation:

$$S_n = a_n + (a_n - d) + (a_n - 2d) + \cdots + (a_1 + 2d) + (a_1 + d) + a_1.$$

Adding corresponding terms on each side of the two equations above, we get

$$2S_n = [a_1 + a_n] + [(a_1 + d) + (a_n - d)] + [(a_1 + 2d) + (a_n - 2d)]$$
$$+ \cdots + [(a_n - 2d) + (a_1 + 2d)] + [(a_n - d) + (a_1 + d)] + [a_n + a_1].$$

This simplifies to

$$2S_n = [a_1 + a_n] + [a_1 + a_n] + [a_1 + a_n]$$
$$+ \cdots + [a_n + a_1] + [a_n + a_1] + [a_n + a_1].$$

There are n bracketed sums.

Since $[a_1 + a_n]$ is being added n times, it follows that

$$2S_n = n[a_1 + a_n].$$

Dividing both sides by 2 leads to the following formula.

Student Notes

The formula for the sum of an arithmetic sequence is very useful, but remember that it does not work for sequences that are not arithmetic.

> ### TO FIND S_n FOR AN ARITHMETIC SEQUENCE
>
> The sum of the first n terms of an arithmetic sequence is given by
>
> $$S_n = \frac{n}{2}(a_1 + a_n).$$

EXAMPLE 5   Find the sum of the first 100 positive even numbers.

SOLUTION   The sum is

$$2 + 4 + 6 + \cdots + 198 + 200.$$

This is the sum of the first 100 terms of the arithmetic sequence for which

$$a_1 = 2, \quad n = 100, \quad \text{and} \quad a_n = 200.$$

We use the formula for S_n for an arithmetic sequence:

$$S_n = \frac{n}{2}(a_1 + a_n),$$

5. Find the sum of the first 100 positive odd numbers.

$$S_{100} = \frac{100}{2}(2 + 200) = 50(202) = 10,100.$$

 YOUR TURN

The above formula is useful when we know the first and last terms, a_1 and a_n. To find S_n when a_n is unknown, but a_1, n, and d are known, we can use $a_n = a_1 + (n - 1)d$ to calculate a_n and then proceed as in Example 5.

EXAMPLE 6   Find the sum of the first 15 terms of the arithmetic sequence $13, 10, 7, 4, \ldots$.

SOLUTION   Note that

$$a_1 = 13, \quad n = 15, \quad \text{and} \quad d = -3.$$

Before using the formula for S_n, we find a_{15}:

$$a_{15} = 13 + (15 - 1)(-3) \qquad \text{Substituting into the formula for } a_n$$
$$= 13 + 14(-3) = -29.$$

6. Find the sum of the first 18 terms of the arithmetic sequence

$$10, 8, 6, 4, \ldots.$$

Knowing that $a_{15} = -29$, we have

$$S_{15} = \tfrac{15}{2}(13 + (-29)) \qquad \text{Using the formula for } S_n$$
$$= \tfrac{15}{2}(-16) = -120.$$

 YOUR TURN

Problem Solving

In problem-solving situations, translation may involve sequences or series. As always, there is often a variety of ways in which a problem can be solved. You should use the approach that is best or easiest for you. In this chapter, however, we will try to emphasize sequences and series and their related formulas.

EXAMPLE 7   *Hourly Wages.*   Chris accepts a job managing a music store, starting with an hourly wage of $14.60, and is promised a raise of 25¢ per hour every 2 months for 5 years. After 5 years of work, what will be Chris's hourly wage?

SOLUTION

1. **Familiarize.** It helps to write down the hourly wage for several two-month time periods.

   Beginning: 14.60,

   After two months: 14.85,

   After four months: 15.10,

   and so on.

   What appears is a sequence of numbers: 14.60, 14.85, 15.10, ... Since the same amount is added each time, the sequence is arithmetic.

   Because we want to know a particular term in the sequence, we will use the formula $a_n = a_1 + (n - 1)d$. To do so, we need a_1, n, and d. From our list above, we have

   $$a_1 = 14.60 \quad \text{and} \quad d = 0.25.$$

   What is n? After 1 year, there have been 6 raises, since Chris gets a raise every 2 months. There are 5 years, so the total number of raises will be $5 \cdot 6$, or 30. Altogether, there will be 31 terms: the original wage and 30 increased rates.

2. **Translate.** We want to find a_n for the arithmetic sequence in which $a_1 = 14.60$, $n = 31$, and $d = 0.25$.

3. **Carry out.** Substituting in the formula for a_n gives us

   $$a_{31} = 14.60 + (31 - 1) \cdot 0.25$$
   $$= 22.10.$$

4. **Check.** We can check by redoing the calculations or we can calculate in a slightly different way for another check. For example, at the end of a year, there will be 6 raises, for a total raise of $1.50. At the end of 5 years, the total raise will be $5 \times \$1.50$, or $7.50. If we add that to the original wage of $14.60, we obtain $22.10. The answer checks.

5. **State.** After 5 years, Chris's hourly wage will be $22.10.

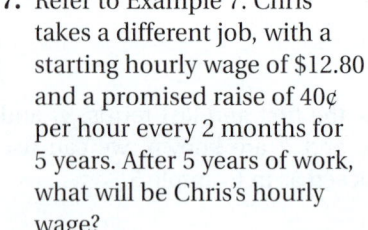

7. Refer to Example 7. Chris takes a different job, with a starting hourly wage of $12.80 and a promised raise of 40¢ per hour every 2 months for 5 years. After 5 years of work, what will be Chris's hourly wage?

 YOUR TURN

EXAMPLE 8 *Telephone Pole Storage.* A stack of telephone poles has 30 poles in the bottom row. There are 29 poles in the second row, 28 in the next row, and so on. How many poles are in the stack if there are 5 poles in the top row?

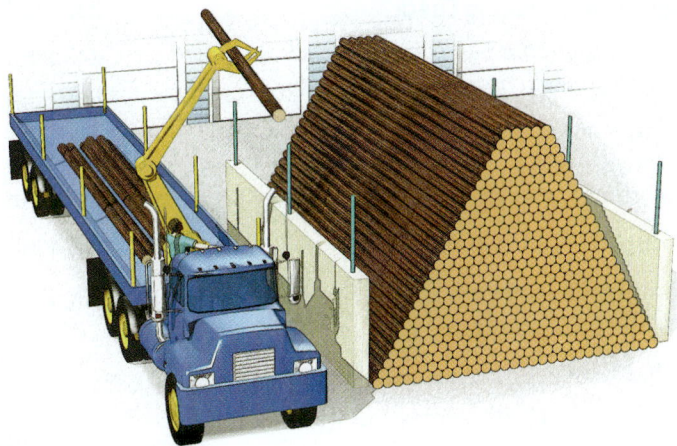

SOLUTION

1. **Familiarize.** The following figure shows the ends of the poles.

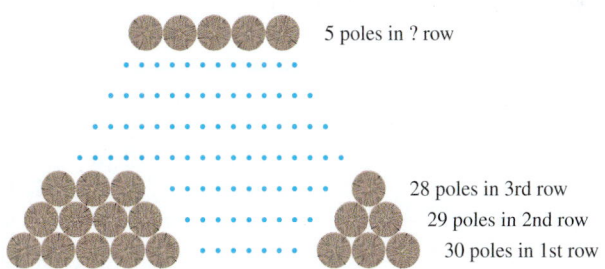

Note that there are $30 - 1 = 29$ poles in the 2nd row, $30 - 2 = 28$ poles in the 3rd row, $30 - 3 = 27$ poles in the 4th row, and so on. The pattern leads to $30 - 25 = 5$ poles in the 26th row.

The situation is represented by the equation

$$30 + 29 + 28 + \cdots + 5. \qquad \text{There are 26 terms in this series.}$$

Thus we have an arithmetic series. We recall the formula

$$S_n = \frac{n}{2}(a_1 + a_n).$$

2. **Translate.** We want to find the sum of the first 26 terms of an arithmetic sequence in which $a_1 = 30$ and $a_{26} = 5$.

3. **Carry out.** Substituting into the above formula gives us

$$S_{26} = \frac{26}{2}(30 + 5)$$
$$= 13 \cdot 35 = 455.$$

4. **Check.** In this case, we can check the calculations by doing them again. A longer, more difficult way would be to do the entire addition:

$$30 + 29 + 28 + \cdots + 5.$$

5. **State.** There are 455 poles in the stack.

8. How many poles will be in a pile of telephone poles if there are 50 in the first layer, 49 in the second, and so on, until there are 6 in the top layer?

 YOUR TURN

14.2 EXERCISE SET

FOR EXTRA HELP

MyMathLab® Math*XL*
PRACTICE   WATCH READ REVIEW

Vocabulary and Reading Check

Choose the expression that best completes each statement. Not every expression will be used.

arithmetic sequence   first term
arithmetic series   sum
common difference

1. $5 + 7 + 9 + 11$ is an example of a(n) _____.

2. In an arithmetic sequence, subtracting a_n from a_{n+1} will give the _____.

3. In $5, 7, 9, 11$, the _____ is 5.

4. For $5 + 7 + 9 + 11$, the expression $\frac{4}{2}(5 + 11)$ gives the _____.

Arithmetic Sequences

Find the first term and the common difference.

5. $8, 13, 18, 23, \ldots$

6. $2.5, 3, 3.5, 4, \ldots$

7. $7, 3, -1, -5, \ldots$

8. $-8, -5, -2, 1, \ldots$

9. $\frac{3}{2}, \frac{9}{4}, 3, \frac{15}{4}, \ldots$

10. $\frac{3}{5}, \frac{1}{10}, -\frac{2}{5}, \ldots$

11. $\$8.16, \$8.46, \$8.76, \$9.06, \ldots$

12. $\$825, \$804, \$783, \$762, \ldots$

13. Find the 19th term of the arithmetic sequence $10, 18, 26, \ldots.$

14. Find the 23rd term of the arithmetic sequence $10, 16, 22, \ldots.$

15. Find the 18th term of the arithmetic sequence $8, 2, -4, \ldots.$

16. Find the 14th term of the arithmetic sequence $3, \frac{7}{3}, \frac{5}{3}, \ldots.$

17. Find the 13th term of the arithmetic sequence $\$1200, \$964.32, \$728.64, \ldots.$

18. Find the 10th term of the arithmetic sequence $\$2345.78, \$2967.54, \$3589.30, \ldots.$

19. In the sequence of Exercise 13, what term is 210?

20. In the sequence of Exercise 14, what term is 208?

21. In the sequence of Exercise 15, what term is -328?

22. In the sequence of Exercise 16, what term is -27?

23. Find a_{18} when $a_1 = 8$ and $d = 10$.

24. Find a_{20} when $a_1 = 12$ and $d = -5$.

25. Find a_1 when $d = 4$ and $a_8 = 33$.

26. Find a_1 when $d = 8$ and $a_{11} = 26$.

27. Find n when $a_1 = 5, d = -3$, and $a_n = -76$.

28. Find n when $a_1 = 25, d = -14$, and $a_n = -507$.

29. For an arithmetic sequence in which $a_{17} = -40$ and $a_{28} = -73$, find a_1 and d. Write the first five terms of the sequence.

30. In an arithmetic sequence, $a_{17} = \frac{25}{3}$ and $a_{32} = \frac{95}{6}$. Find a_1 and d. Write the first five terms of the sequence.

Aha! 31. Find a_1 and d if $a_{13} = 13$ and $a_{54} = 54$.

32. Find a_1 and d if $a_{12} = 24$ and $a_{25} = 50$.

Sum of the First *n* Terms of an Arithmetic Sequence

33. Find the sum of the first 20 terms of the arithmetic series $1 + 5 + 9 + 13 + \cdots.$

34. Find the sum of the first 14 terms of the arithmetic series $11 + 7 + 3 + \cdots.$

35. Find the sum of the first 250 natural numbers.

36. Find the sum of the first 400 natural numbers.

37. Find the sum of the even numbers from 2 to 100, inclusive.

38. Find the sum of the odd numbers from 1 to 99, inclusive.

39. Find the sum of all multiples of 6 from 6 to 102, inclusive.

40. Find the sum of all multiples of 4 that are between 15 and 521.

41. An arithmetic series has $a_1 = 4$ and $d = 5$. Find S_{20}.

42. An arithmetic series has $a_1 = 9$ and $d = -3$. Find S_{32}.

Problem Solving

Solve.

43. *Band Formations.* The South Brighton Drum and Bugle Corps has 7 musicians in the front row, 9 in the second row, 11 in the third row, and so on, for 15 rows. How many musicians are in the last row? How many musicians are there altogether?

44. *Gardening.* A gardener is planting tulip bulbs at the entrance to a college. She puts 50 bulbs in the first row, 46 in the second row, 42 in the third row, and so on, for 13 rows. How many bulbs will be in the last row? How many bulbs will she plant altogether?

45. *Archaeology.* Many ancient Mayan pyramids were constructed over a span of several generations. Each layer of the pyramid has a stone perimeter, enclosing a layer of dirt or debris on which a structure once stood. One drawing of such a pyramid indicates that the perimeter of the bottom layer contains 36 stones, the next level up contains 32 stones, and so on, up to the top row, which contains 4 stones. How many stones are in the pyramid?

46. *Auditorium Design.* Theaters are often built with more seats per row as the rows move toward the back. The Community Theater has 20 seats in the first row, 22 in the second, 24 in the third, and so on, for 16 rows. How many seats are in the theater?

47. *Accumulated Savings.* If 10¢ is saved on October 1, another 20¢ on October 2, another 30¢ on October 3, and so on, how much is saved during October? (October has 31 days.)

48. *Accumulated Savings.* Carrie saves money in an arithmetic sequence: $700 for the first year, another $850 the second, and so on, for 20 years. How much does she save in all (disregarding interest)?

49. It is said that as a young child, the mathematician Karl F. Gauss (1777–1855) was able to compute the sum $1 + 2 + 3 + \cdots + 100$ very quickly in his head. Explain how Gauss might have done this and present a formula for the sum of the first n natural numbers. (*Hint*: $1 + 99 = 100$.)

50. Write a problem for a classmate to solve. Devise the problem so that its solution requires computing S_{17} for an arithmetic sequence.

Skill Review

Find an equation of the line satisfying the given conditions.

51. Slope $\frac{1}{3}$, y-intercept $(0, 10)$ [3.6]

52. Containing the points $(2, 3)$ and $(4, -5)$ [3.7]

53. Containing the point $(5, 0)$ and parallel to the line given by $2x + y = 8$ [3.7]

54. Containing the point $(-1, -4)$ and perpendicular to the line given by $3x - 4y = 7$ [3.7]

Find an equation of the circle satisfying the given conditions. [13.1]

55. Center $(0, 0)$, radius 4

56. Center $(-2, 1)$, radius $2\sqrt{5}$

Synthesis

57. When every term in an arithmetic sequence is an integer, S_n must also be an integer. Given that n, a_1, and a_n may each, at times, be even or odd, explain why $\frac{n}{2}(a_1 + a_n)$ is always an integer.

58. The sum of the first n terms of an arithmetic sequence is also given by

$$S_n = \frac{n}{2}\big[2a_1 + (n - 1)d\big].$$

Use the earlier formulas for a_n and S_n to explain how this equation was developed.

59. A frog is at the bottom of a 100-ft well. With each jump, the frog climbs 4 ft, but then slips back 1 ft. How many jumps does it take for the frog to reach the top of the hole?

60. Find a formula for the sum of the first n consecutive odd numbers starting with 1:

$$1 + 3 + 5 + \cdots + (2n - 1).$$

61. Prove that if p, m, and q are consecutive terms in an arithmetic sequence, then

$$m = \frac{p + q}{2}.$$

62. *Straight-Line Depreciation.*   A company buys a copier for $5200 on January 1 of a given year. The machine is expected to last for 8 years, at the end of which time its *trade-in*, or *salvage*, *value* will be $1100. If the company figures the decline in value to be the same each year, then the trade-in values, after t years, $0 \leq t \leq 8$, form an arithmetic sequence given by

$$a_t = C - t\left(\frac{C - S}{N}\right),$$

where C is the original cost of the item, N the years of expected life, and S the salvage value.

a) Find the formula for a_t for the straight-line depreciation of the copier.

b) Find the trade-in value after 0 year, 1 year, 2 years, 3 years, 4 years, 7 years, and 8 years.

c) Find a formula that expresses a_t recursively. (See Exercises 87 and 88 in Exercise Set 14.1.)

63. Use your answer to Exercise 35 to find the sum of all integers from 501 through 750.

👉 **YOUR TURN ANSWERS: SECTION 14.2**

1. $a_1 = 0; d = \frac{1}{2}$ **2.** 43   **3.** 37th
4. $a_1 = -25; d = 10; -25, -15, -5, 5, 15, \ldots$ **5.** 10,000
6. -126 **7.** $24.80   **8.** 1260 poles

14.3   Geometric Sequences and Series

Geometric Sequences   •   Sum of the First n Terms of a Geometric Sequence   •   Infinite Geometric Series   •
Problem Solving

In an arithmetic sequence, a certain number is added to each term to get the next term. When each term in a sequence is *multiplied* by a certain fixed number to get the next term, the sequence is **geometric**. In this section, we examine both geometric sequences (or progressions) and geometric series.

Geometric Sequences

Consider the sequence

$$2, 6, 18, 54, 162, \ldots$$

If we multiply each term by 3, we obtain the next term. The multiplier is called the *common ratio* because it is found by dividing any term by the preceding term.

> ### GEOMETRIC SEQUENCE
> A sequence is *geometric* if there exists a number r, called the *common ratio*, for which
> $$\frac{a_{n+1}}{a_n} = r, \quad \text{or} \quad a_{n+1} = a_n \cdot r \quad \text{for any integer } n \geq 1.$$

EXAMPLE 1 For each geometric sequence, find the common ratio.

a) $4, 20, 100, 500, 2500, \ldots$

b) $3, -6, 12, -24, 48, -96, \ldots$

c) $\$5200, \$3900, \$2925, \$2193.75, \ldots$

SOLUTION

Sequence		*Common Ratio*
a) $4, 20, 100, 500, 2500, \ldots$	5	$\frac{20}{4} = 5, \frac{100}{20} = 5$, and so on
b) $3, -6, 12, -24, 48, -96, \ldots$	-2	$\frac{-6}{3} = -2, \frac{12}{-6} = -2$, and so on
c) $\$5200, \$3900, \$2925, \$2193.75, \ldots$	0.75	$\frac{\$3900}{\$5200} = 0.75, \frac{\$2925}{\$3900} = 0.75$

1. Find the common ratio for the geometric sequence

$$20, 10, 5, 2\tfrac{1}{2}, \ldots.$$

 YOUR TURN

Note that when the signs of the terms alternate, the common ratio is negative.

To develop a formula for the general, or nth, term of a geometric sequence, let a_1 be the first term and let r be the common ratio. We write out a few terms:

$$a_1,$$
$$a_2 = a_1 r,$$
$$a_3 = a_2 r = (a_1 r)r = a_1 r^2, \qquad \text{Substituting } a_1 r \text{ for } a_2$$
$$a_4 = a_3 r = (a_1 r^2)r = a_1 r^3. \qquad \text{Substituting } a_1 r^2 \text{ for } a_3$$

Note that the exponent is 1 less than the subscript.

Generalizing, we obtain the following.

> ### TO FIND a_n FOR A GEOMETRIC SEQUENCE
> The nth term of a geometric sequence with common ratio r is given by
> $$a_n = a_1 r^{n-1}, \quad \text{for any integer } n \geq 1.$$

EXAMPLE 2 Find the 7th term of the geometric sequence $4, 20, 100, \ldots.$

SOLUTION First, we note that

$$a_1 = 4 \quad \text{and} \quad n = 7.$$

To find the common ratio, we can divide any term (other than the first) by the term preceding it. Since the second term is 20 and the first is 4,

$$r = \frac{20}{4}, \quad \text{or } 5.$$

2. Find the 8th term of the geometric sequence

$$3, 6, 12, \ldots.$$

Substituting in the formula $a_n = a_1 r^{n-1}$, we have

$$a_7 = 4 \cdot 5^{7-1} = 4 \cdot 5^6 = 4 \cdot 15{,}625 = 62{,}500.$$

 YOUR TURN

EXAMPLE 3 Find the 10th term of the geometric sequence

$$64, -32, 16, -8, \ldots.$$

SOLUTION First, we note that

$$a_1 = 64, \quad n = 10, \quad \text{and} \quad r = \frac{-32}{64} = -\frac{1}{2}.$$

Then, using the formula for the nth term of a geometric sequence, we have

$$a_{10} = 64 \cdot \left(-\frac{1}{2}\right)^{10-1} = 64 \cdot \left(-\frac{1}{2}\right)^9 = 2^6 \cdot \left(-\frac{1}{2^9}\right) = -\frac{1}{2^3} = -\frac{1}{8}.$$

3. Find the 9th term of the geometric sequence

$$-5, 10, -20, 40, \ldots.$$

YOUR TURN

Sum of the First n Terms of a Geometric Sequence

We next develop a formula for S_n when a sequence is geometric:

$$a_1, \ a_1r, \ a_1r^2, \ a_1r^3, \ldots, a_1r^{n-1}, \ldots.$$

The **geometric series** S_n is given by

$$S_n = a_1 + a_1r + a_1r^2 + \cdots + a_1r^{n-2} + a_1r^{n-1}. \tag{1}$$

Multiplying both sides by r gives us

$$rS_n = a_1r + a_1r^2 + a_1r^3 + \cdots + a_1r^{n-1} + a_1r^n. \tag{2}$$

Student Notes

The three determining characteristics of a geometric sequence or series are the first term (a_1), the number of terms (n), and the common ratio (r). Be sure you understand how to use these characteristics to write out a sequence or a series.

When we subtract corresponding sides of equation (2) from equation (1), the color terms drop out, leaving

$$S_n - rS_n = a_1 - a_1r^n$$
$$S_n(1 - r) = a_1(1 - r^n), \qquad \text{Factoring}$$

or

$$S_n = \frac{a_1(1 - r^n)}{1 - r}. \qquad \text{Dividing both sides by } 1 - r$$

TO FIND S_n FOR A GEOMETRIC SEQUENCE

The sum of the first n terms of a geometric sequence with common ratio r is given by

$$S_n = \frac{a_1(1 - r^n)}{1 - r}, \quad \text{for any } r \neq 1.$$

EXAMPLE 4 Find the sum of the first 7 terms of the geometric sequence $3, 15, 75, 375, \ldots.$

SOLUTION First, we note that

$$a_1 = 3, \quad n = 7, \quad \text{and} \quad r = \frac{15}{3} = 5.$$

4. Find the sum of the first 9 terms of the geometric sequence

$-5, 10, -20, 40, \ldots.$

Then, substituting in the formula $S_n = \dfrac{a_1(1 - r^n)}{1 - r}$, we have

$$S_7 = \frac{3(1 - 5^7)}{1 - 5} = \frac{3(1 - 78{,}125)}{-4}$$

$$= \frac{3(-78{,}124)}{-4} = 58{,}593.$$

YOUR TURN

Infinite Geometric Series

Suppose we consider the sum of the terms of an infinite geometric sequence, such as $3, 6, 12, 24, 48, \ldots.$ We get what is called an **infinite geometric series**:

$$3 + 6 + 12 + 24 + 48 + \cdots.$$

Here, as n increases, the sum of the first n terms, S_n, increases without bound. There are also infinite series that get closer and closer to some specific number. For example, consider the sequence

$$\frac{1}{2} + \frac{1}{4} + \frac{1}{8} + \frac{1}{16} + \cdots + \frac{1}{2^n} + \cdots,$$

and evaluate S_n for the first four values of n:

$$
\begin{aligned}
S_1 &= \tfrac{1}{2} &&= \tfrac{1}{2} = 0.5, \\
S_2 &= \tfrac{1}{2} + \tfrac{1}{4} &&= \tfrac{3}{4} = 0.75, \\
S_3 &= \tfrac{1}{2} + \tfrac{1}{4} + \tfrac{1}{8} &&= \tfrac{7}{8} = 0.875, \\
S_4 &= \tfrac{1}{2} + \tfrac{1}{4} + \tfrac{1}{8} + \tfrac{1}{16} &&= \tfrac{15}{16} = 0.9375.
\end{aligned}
$$

> The denominator of each sum is 2^n, where n is the subscript of S. The numerator is $2^n - 1$.

Thus, for this particular series, we have

$$S_n = \frac{2^n - 1}{2^n} = \frac{2^n}{2^n} - \frac{1}{2^n} = 1 - \frac{1}{2^n}.$$

Note that the value of S_n is less than 1 for any value of n, but as n gets larger and larger, the value of $1/2^n$ gets closer to 0, so the value of S_n gets closer to 1. We can visualize S_n by considering a square with area 1. For S_1, we shade half the square. For S_2, we shade half the square plus half the remaining part, or $\tfrac{1}{4}$. For S_3, we shade the parts shaded in S_2 plus half the remaining part. Again we see that the values of S_n will continue to get close to 1 (shading the complete square). We say that 1 is the **limit** of S_n and that 1 is the sum of this infinite geometric series.

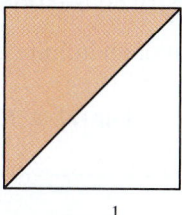 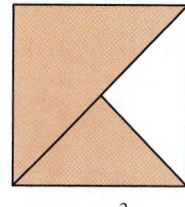 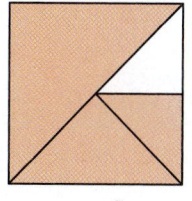 

$S_1 = \dfrac{1}{2}$ $S_2 = \dfrac{3}{4}$ $S_3 = \dfrac{7}{8}$ $S_4 = \dfrac{15}{16}$

An infinite geometric series is denoted S_∞. It can be shown (but we will not do so here) that the sum of the terms of an infinite geometric sequence exists if and only if $|r| < 1$ (that is, the common ratio's absolute value is less than 1).

To find a formula for the sum of an infinite geometric series, we first consider the sum of the first n terms:

$$S_n = \frac{a_1(1 - r^n)}{1 - r} = \frac{a_1 - a_1 r^n}{1 - r}. \qquad \text{\color{red}Using the distributive law}$$

For $|r| < 1$, it follows that the value of r^n gets closer to 0 as n gets larger. (Check this by selecting a number between -1 and 1 and finding larger and larger powers on a calculator.) As r^n gets closer to 0, so too does $a_1 r^n$. Thus, S_n gets closer to $a_1/(1 - r)$.

EXPLORING 🔍 THE CONCEPT

Graphically, a geometric series has a limit if the graph of the sequence gets closer to $n = 0$ as n increases.

1. For which of the following sequences does it appear that the series will have a limit?

A.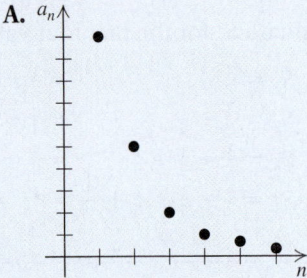

B.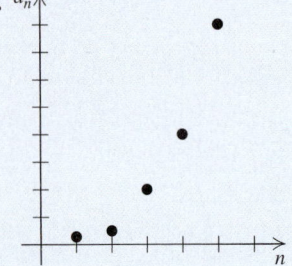

C.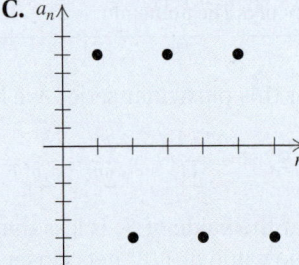

D.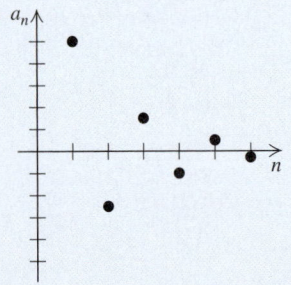

ANSWER

1. A, D

THE LIMIT OF AN INFINITE GEOMETRIC SERIES

For $|r| < 1$, the limit of an infinite geometric series is given by

$$S_\infty = \frac{a_1}{1 - r}. \qquad (\text{For } |r| \geq 1, \text{ no limit exists.})$$

EXAMPLE 5   Determine whether each series has a limit. If a limit exists, find it.

a) $1 + 3 + 9 + 27 + \cdots$ **b)** $-35 + 7 - \frac{7}{5} + \frac{7}{25} + \cdots$

SOLUTION

a) Here $r = 3$, so $|r| = |3| = 3$. Since $|r| \not< 1$, the series *does not* have a limit.

b) Here $r = -\frac{1}{5}$, so $|r| = \left|-\frac{1}{5}\right| = \frac{1}{5}$. Since $|r| < 1$, the series *does* have a limit. We find the limit by substituting into the formula for S_∞:

$$S_\infty = \frac{-35}{1 - \left(-\frac{1}{5}\right)} = \frac{-35}{\frac{6}{5}} = -35 \cdot \frac{5}{6} = \frac{-175}{6} = -29\frac{1}{6}.$$

5. Determine whether $20 + 10 + 5 + \cdots$ has a limit. If a limit exists, find it.

 YOUR TURN

EXAMPLE 6 Find fraction notation for 0.63636363. . . .

SOLUTION We can express this as

$$0.63 + 0.0063 + 0.000063 + \cdots.$$

This is an infinite geometric series, where $a_1 = 0.63$ and $r = 0.01$. Since $|r| < 1$, this series has a limit:

$$S_\infty = \frac{a_1}{1 - r} = \frac{0.63}{1 - 0.01} = \frac{0.63}{0.99} = \frac{63}{99}.$$

6. Find fraction notation for 0.575757. . . .

Thus fraction notation for 0.63636363. . . . is $\frac{63}{99}$, or $\frac{7}{11}$.

YOUR TURN

CONNECTING 🔗 THE CONCEPTS

If a sequence is arithmetic or geometric, the general term and a partial sum can be found using a formula. An infinite sum exists only for geometric series with $|r| < 1$.

Arithmetic Sequences	Geometric Sequences
Common difference: d	Common ratio: r
$a_n = a_1 + (n - 1)d$	$a_n = a_1 r^{n-1}$
$S_n = \dfrac{n}{2}(a_1 + a_n)$	$S_n = \dfrac{a_1(1 - r^n)}{1 - r};$ $S_\infty = \dfrac{a_1}{1 - r},\ \|r\| < 1$

EXERCISES

1. Find the common difference for the arithmetic sequence 115, 112, 109, 106, ...

2. Find the common ratio for the geometric sequence $\frac{1}{3}, -\frac{1}{6}, \frac{1}{12}, -\frac{1}{24}, \ldots$.

3. Find the 21st term of the arithmetic sequence 10, 15, 20, 25, ...

4. Find the 8th term of the geometric sequence 5, 10, 20, 40, ...

5. Find S_{30} for the arithmetic series $2 + 12 + 22 + 32 + \cdots$.

6. Find S_{10} for the geometric series $\$100 + \$100(1.03) + \$100(1.03)^2 + \cdots$.

7. Determine whether the infinite geometric series $0.9 + 0.09 + 0.009 + \cdots$ has a limit. If a limit exists, find it.

8. Determine whether the infinite geometric series $0.9 + 9 + 90 + \cdots$ has a limit. If a limit exists, find it.

Problem Solving

For some problem-solving situations, the translation may involve geometric sequences or series.

EXAMPLE 7 *Loan Repayment.*   Francine's student loan is in the amount of $6000. Interest is 9% compounded annually, and the entire amount is to be paid after 10 years. How much is to be paid back?

SOLUTION

1. **Familiarize.** Suppose we let P represent any principal amount. At the end of one year, the amount owed will be $P + 0.09P$, or $1.09P$. That amount will be the principal for the second year. The amount owed at the end of the second year will be $1.09 \times$ New principal $= 1.09(1.09P)$, or 1.09^2P. Thus the amount owed at the beginning of successive years is as follows:

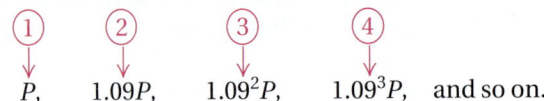

$$P, \quad 1.09P, \quad 1.09^2P, \quad 1.09^3P, \quad \text{and so on.}$$

We have a geometric sequence. The amount owed at the beginning of the 11th year will be the amount owed at the end of the 10th year.

2. **Translate.** This is a geometric sequence with $a_1 = 6000$, $r = 1.09$, and $n = 11$. The appropriate formula for finding the nth term is

$$a_n = a_1 r^{n-1}.$$

3. **Carry out.** We substitute and calculate:

$$a_{11} = \$6000(1.09)^{11-1} = \$6000(1.09)^{10}$$
$$\approx \$14{,}204.18. \qquad \text{Using a calculator and rounding to the nearest hundredth}$$

4. **Check.** A check, by repeating the calculations, is left to the student.

5. **State.** Francine will owe $14,204.18 at the end of 10 years.

7. Refer to Example 7. If Francine's loan amount is $8000, with 5% interest compounded annually, how much is owed after 10 years?

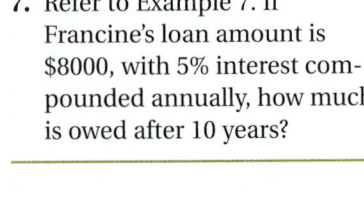 YOUR TURN

EXAMPLE 8 *Bungee Jumping.*   A bungee jumper rebounds 60% of the height jumped. Clyde's bungee jump is made using a cord that stretches to 200 ft.

a) After jumping and then rebounding 9 times, how far has Clyde traveled upward (the total rebound distance)?

b) Theoretically, how far will Clyde travel upward (bounce) before coming to rest?

SOLUTION

1. **Familiarize.** Let's do some calculations and look for a pattern.

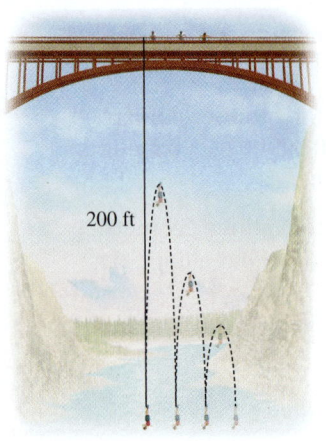

200 ft

First fall:	200 ft
First rebound:	0.6×200, or 120 ft
Second fall:	120 ft, or 0.6×200
Second rebound:	0.6×120, or $0.6(0.6 \times 200)$, which is 72 ft
Third fall:	72 ft, or $0.6(0.6 \times 200)$
Third rebound:	0.6×72, or $0.6(0.6(0.6 \times 200))$, which is 43.2 ft

The rebound distances form a geometric sequence:

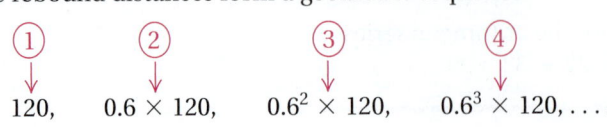

$$120, \quad 0.6 \times 120, \quad 0.6^2 \times 120, \quad 0.6^3 \times 120, \dots.$$

2. Translate.

a) The total rebound distance after 9 bounces is the sum of a geometric sequence. The first term is 120 and the common ratio is 0.6. There will be 9 terms, so we can use the formula

$$S_n = \frac{a_1(1 - r^n)}{1 - r}.$$

b) Theoretically, Clyde will never stop bouncing. Realistically, the bouncing will eventually stop. To approximate the actual distance bounced, we consider an infinite number of bounces and use the formula

$$S_\infty = \frac{a_1}{1 - r}.$$ Since $r = 0.6$ and $|0.6| < 1$, we know that S_∞ exists.

3. Carry out.

a) We substitute into the formula and calculate:

$$S_9 = \frac{120[1 - (0.6)^9]}{1 - 0.6} \approx 297.$$ Using a calculator

b) We substitute and calculate:

$$S_\infty = \frac{120}{1 - 0.6} = 300.$$

4. Check. We can do the calculations again. It makes sense that $S_\infty > S_9$.

5. State.

8. Refer to Example 8. If Clyde's cord stretches to 300 ft, how far will he "bounce" or "travel upward" before coming to rest?

a) In 9 bounces, Clyde will have traveled upward a total distance of about 297 ft.

b) Theoretically, Clyde will travel upward a total of 300 ft before coming to rest.

YOUR TURN

14.3 EXERCISE SET

FOR EXTRA HELP

MyMathLab® Math XL

PRACTICE   WATCH   READ   REVIEW

Vocabulary and Reading Check

Complete each statement by selecting the appropriate word or expression from those listed below each blank.

1. The list 16, 8, 4, 2, 1, . . . is a(n) _____
   finite/infinite

   _____   _____.
   arithmetic/geometric   sequence/series

2. The sum 16 + 8 + 4 + 2 + 1 is a(n) _____
   finite/infinite

   _____   _____.
   arithmetic/geometric   sequence/series

3. For 16 + 8 + 4 + 2 + 1 + · · · , the common _____ is _____ than 1, so the
   difference/ratio   less/more

   limit _____ exist.
   does/does not

4. The number $0.\overline{2}$ is equal to _____
   0.2222/0.22 . . .

   and can be written as the _____
   arithmetic/geometric

   _____ 0.2 + 0.02 + 0.002 + · · · .
   sequence/series

Concept Reinforcement

Classify each of the following as an arithmetic sequence, a geometric sequence, an arithmetic series, a geometric series, or none of these.

5. 3, 6, 12, 24, . . .

6. 10, 7, 4, 1, −2, . . .

7. 4 + 20 + 100 + 500 + 2500 + 12,500

8. 10 + 12 + 14 + 16 + 18 + 20

9. $3 - \frac{3}{2} + \frac{3}{4} - \frac{3}{8} + \frac{3}{16} - \cdots$

10. $1 + \frac{1}{2} + \frac{1}{3} + \frac{1}{4} + \frac{1}{5} + \frac{1}{6} + \cdots$

Geometric Sequences

Find the common ratio for each geometric sequence.

11. $10, 20, 40, 80, \ldots$

12. $5, 20, 80, 320, \ldots$

13. $6, -0.6, 0.06, -0.006, \ldots$

14. $-5, -0.5, -0.05, -0.005, \ldots$

15. $\frac{1}{2}, -\frac{1}{4}, \frac{1}{8}, -\frac{1}{16}, \ldots$

16. $\frac{2}{3}, -\frac{4}{3}, \frac{8}{3}, -\frac{16}{3}, \ldots$

17. $75, 15, 3, \frac{3}{5}, \ldots$

18. $12, -4, \frac{4}{3}, -\frac{4}{9}, \ldots$

19. $\frac{1}{m}, \frac{6}{m^2}, \frac{36}{m^3}, \frac{216}{m^4}, \ldots$

20. $4, \frac{4m}{5}, \frac{4m^2}{25}, \frac{4m^3}{125}, \ldots$

Find the indicated term for each geometric sequence.

21. $2, 6, 18, \ldots$; the 7th term

22. $2, 8, 32, \ldots$; the 9th term

23. $\sqrt{3}, 3, 3\sqrt{3}, \ldots$; the 10th term

24. $2, 2\sqrt{2}, 4, \ldots$; the 8th term

25. $-\frac{8}{243}, \frac{8}{81}, -\frac{8}{27}, \ldots$; the 14th term

26. $\frac{7}{625}, \frac{-7}{125}, \frac{7}{25}, \ldots$; the 13th term

27. $\$1000, \$1040, \$1081.60, \ldots$; the 10th term

28. $\$1000, \$1050, \$1102.50, \ldots$; the 12th term

Find the nth, or general, term for each geometric sequence.

29. $1, 5, 25, 125, \ldots$

30. $2, 4, 8, \ldots$

31. $1, -1, 1, -1, \ldots$

32. $\frac{1}{4}, \frac{1}{16}, \frac{1}{64}, \ldots$

33. $\frac{1}{x}, \frac{1}{x^2}, \frac{1}{x^3}, \ldots$

34. $5, \frac{5m}{2}, \frac{5m^2}{4}, \ldots$

Sum of the First *n* Terms of a Geometric Sequence

For Exercises 35–42, use the formula for S_n to find the indicated sum for each geometric series.

35. S_9 for $6 + 12 + 24 + \cdots$

36. S_6 for $16 - 8 + 4 - \cdots$

37. S_7 for $\frac{1}{18} - \frac{1}{6} + \frac{1}{2} - \cdots$

Aha! **38.** S_5 for $7 + 0.7 + 0.07 + \cdots$

39. S_8 for $1 + x + x^2 + x^3 + \cdots$

40. S_{10} for $1 + x^2 + x^4 + x^6 + \cdots$

41. S_{16} for $\$200 + \$200(1.06) + \$200(1.06)^2 + \cdots$

42. S_{23} for $\$1000 + \$1000(1.08) + \$1000(1.08)^2 + \cdots$

Infinite Geometric Series

Determine whether each infinite geometric series has a limit. If a limit exists, find it.

43. $18 + 6 + 2 + \cdots$

44. $80 + 20 + 5 + \cdots$

45. $7 + 3 + \frac{9}{7} + \cdots$

46. $12 + 9 + \frac{27}{4} + \cdots$

47. $3 + 15 + 75 + \cdots$

48. $2 + 3 + \frac{9}{2} + \cdots$

49. $4 - 6 + 9 - \frac{27}{2} + \cdots$

50. $-6 + 3 - \frac{3}{2} + \frac{3}{4} - \cdots$

51. $0.43 + 0.0043 + 0.000043 + \cdots$

52. $0.37 + 0.0037 + 0.000037 + \cdots$

53. $\$500(1.02)^{-1} + \$500(1.02)^{-2} + \$500(1.02)^{-3} + \cdots$

54. $\$1000(1.08)^{-1} + \$1000(1.08)^{-2} + \$1000(1.08)^{-3} + \cdots$

Find fraction notation for each repeating decimal.

55. $0.5555\ldots$

56. $0.8888\ldots$

57. $3.4646\ldots$

58. $1.2323\ldots$

59. $0.15151515\ldots$

60. $0.12121212\ldots$

Problem Solving

Solve. Use a calculator as needed for evaluating formulas.

61. *Rebound Distance.* A ping-pong ball is dropped from a height of 20 ft and always rebounds one-fourth of the distance fallen. How high does it rebound the 6th time?

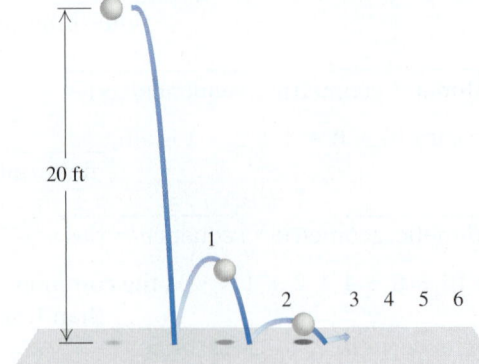

62. *Rebound Distance.* Approximate the total of the rebound heights of the ball in Exercise 61.

63. *Population Growth.* Yorktown has a current population of 100,000 that is increasing by 3% each year. What will the population be in 15 years?

64. *Amount Owed.* Gilberto borrows $15,000. The loan is to be repaid in 13 years at 5.5% interest, compounded annually. How much will he owe at the end of 13 years?

65. *Shrinking Population.* A population of 5000 fruit flies is dying off at a rate of 4% per minute. How many flies will be alive after 15 min?

66. *Shrinking Population.* For the population of fruit flies in Exercise 65, how long will it take for only 1800 fruit flies to remain alive? (*Hint*: Use logarithms.) Round to the nearest minute.

67. *Solar Energy.* The global solar photovoltaic (PV) installed capacity was about 70 GW (gigawatts) in 2012 and was expected to grow by 20% each year, as indicated in the data below. If all installed solar PV cells were operating at capacity, how many gigawatts of power would be produced by solar energy from 2012 through 2020?

Sources: The European Photovoltaic Industry Association; GlobalData

Year	Solar PV Capacity (in gigawatts)
2012	70
2013	84
2014	100.8

68. *Daily Wages.* Suppose you were offered a job for the month of September (30 days) under the following conditions. You will be paid $0.01 for the first day, $0.02 for the second, $0.04 for the third, and so on, doubling your previous day's salary each day. How much would you earn? (Would you take the job? Make a guess before reading further.)

69. *Rebound Distance.* A superball dropped from the top of the Washington Monument (556 ft high) rebounds three-fourths of the distance fallen. How far (up and down) will the ball have traveled when it hits the ground for the 6th time?

70. *Rebound Distance.* Approximate the total distance that the ball of Exercise 69 will have traveled when it comes to rest.

71. *Stacking Paper.* Construction paper is about 0.02 in. thick. Beginning with just one piece, a stack is doubled again and again 10 times. Find the height of the final stack.

72. *Monthly Earnings.* Suppose you accepted a job for the month of February (28 days) under the following conditions. You will be paid $0.01 the first day, $0.02 the second, $0.04 the third, and so on, doubling your previous day's salary each day. How much would you earn?

Aha! **73.** Under what circumstances is it possible for the 5th term of a geometric sequence to be greater than the 4th term but less than the 7th term?

74. When r is negative, a series is said to be *alternating*. Why do you suppose this terminology is used?

Skill Review

Solve.

75. $|x - 3| = 11$ [9.3]

76. $|2x + 5| < 6$ [9.3]

77. $|3x - 7| \geq 1$ [9.3]

78. $-5 < 6 - 3x < 7$ [9.2]

79. $x^2 - 5x - 14 < 0$ [11.9]

80. $x \geq \dfrac{1}{x}$ [11.9]

Synthesis

81. Write a problem for a classmate to solve. Devise the problem so that a geometric series is involved and the solution is "The total amount in the bank is $900(1.08)^{40}$, or about $19,550."

82. The infinite series

$$S_\infty = 2 + \frac{1}{2} + \frac{1}{2 \cdot 3} + \frac{1}{2 \cdot 3 \cdot 4} + \frac{1}{2 \cdot 3 \cdot 4 \cdot 5}$$
$$+ \frac{1}{2 \cdot 3 \cdot 4 \cdot 5 \cdot 6} + \cdots$$

is not geometric, but it does have a sum. Using S_1, S_2, S_3, S_4, S_5, and S_6, predict the value of S_∞ and explain your reasoning.

Calculate each of the following sums.

83. $\displaystyle\sum_{k=1}^{\infty} 6(0.9)^k$

84. $\displaystyle\sum_{k=1}^{\infty} 5(-0.7)^k$

85. Find the sum of the first n terms of
$$x^2 - x^3 + x^4 - x^5 + \cdots.$$

86. Find the sum of the first n terms of
$$1 + x + x^2 + x^3 + \cdots.$$

87. The sides of a square are each 16 cm long. A second square is inscribed by joining the midpoints of the sides, successively. In the second square we repeat the process, inscribing a third square. If this process is continued indefinitely, what is the sum of all of the areas of all the squares? (*Hint*: Use an infinite geometric series.)

88. Show that $0.999\ldots$ is 1.

 89. Using Example 5 and Exercises 43–54, explain how the graph of a geometric sequence can be used to determine whether a geometric series has a limit.

 90. To compare the *graphs* of an arithmetic sequence and a geometric sequence, we plot n on the horizontal axis and a_n on the vertical axis. Graph Example 1(a) of Section 14.2 and Example 1(a) of Section 14.3 on the same set of axes. How do the graphs of geometric sequences differ from the graphs of arithmetic sequences?

 91. *Research.*   How are items such as computers depreciated on an income tax return? Form a sequence listing the value of a computer from the time of its purchase until it is completely depreciated using one of the methods allowed by the IRS. Is the sequence arithmetic, geometric, or neither? Is it realistic?

QUICK QUIZ: SECTIONS 14.1–14.3

1. Find the first 4 terms and a_{20} of the sequence with the general term $a_n = (-1)^n n$.   [14.1]

2. Find S_7 for the sequence 5, 10, 15, 20, ...   [14.1]

3. Find the sum of the first 15 terms of the arithmetic series $50 + 47 + 44 + \cdots$.   [14.2]

4. Find the sum of the first 10 terms of the geometric series $\frac{1}{2} + 1 + 2 + 4 + \cdots$.   [14.3]

5. Find fraction notation for the repeating decimal $1.565656\ldots$.   [14.3]

PREPARE TO MOVE ON

Multiply.   [4.5], [4.6]

1. $(x + y)^2$

2. $(x + y)^3$

3. $(x - y)^3$

4. $(x - y)^4$

5. $(2x + y)^3$

6. $(2x - y)^3$

Mid-Chapter Review

A *sequence* is simply an ordered list. A *series* is a sum of consecutive terms in a sequence. Some sequences of numbers have patterns, and a formula can be found for a general term. When every pair of consecutive terms has a common difference, the sequence is *arithmetic*. When every pair of consecutive terms has a common ratio, the sequence is *geometric*.

GUIDED SOLUTIONS

1. Find the 14th term of the arithmetic sequence $-6, -1, 4, 9, \ldots$. [14.2]

   Solution

   $$a_n = a_1 + (n-1)d$$

   $$n = \boxed{}, \qquad a_1 = \boxed{}, \qquad d = \boxed{}$$

   $$a_{14} = \boxed{} + (\boxed{} - 1)$$

   $$a_{14} = \boxed{}$$

2. Find the 7th term of the geometric sequence $\frac{1}{9}, -\frac{1}{3}, 1, -3, \ldots$. [14.3]

   Solution

   $$a_n = a_1 r^{n-1}$$

   $$n = \boxed{}, \qquad a_1 = \boxed{}, \qquad r = \boxed{}$$

   $$a_7 = \boxed{} \cdot (\boxed{})^{\boxed{} - 1}$$

   $$a_7 = \boxed{}$$

MIXED REVIEW

3. Find a_{20} if $a_n = n^2 - 5n$. [14.1]

4. Write an expression for the general term a_n of the sequence $\frac{1}{2}, \frac{1}{3}, \frac{1}{4}, \frac{1}{5}, \ldots$. [14.1]

5. Find S_{12} for the sequence $1, 2, 3, 4, \ldots$. [14.1]

6. Write out and evaluate the sum

$$\sum_{k=2}^{5} k^2. \quad [14.1]$$

7. Rewrite using sigma notation:

   $1 - 2 + 3 - 4 + 5 - 6$. [14.1]

8. Which term is 22 in the arithmetic sequence $10, 10.2, 10.4, 10.6, \ldots$? [14.2]

9. For an arithmetic sequence, find a_{25} when $a_1 = 9$ and $d = -2$. [14.2]

10. Find the 12th term of the geometric sequence $1000, 100, 10, \ldots$. [14.3]

11. Find the nth, or general, term for the geometric sequence $2, -2, 2, -2, \ldots$. [14.3]

12. Determine whether the infinite geometric series $100 - 20 + 4 - \cdots$ has a limit. If the limit exists, find it. [14.3]

13. Renata earns \$1 on June 1, another \$2 on June 2, another \$3 on June 3, another \$4 on June 4, and so on. How much does she earn during the 30 days of June? [14.2]

14. Dwight earns \$1 on June 1, another \$2 on June 2, another \$4 on June 3, another \$8 on June 4, and so on. How much does he earn during the 30 days of June? [14.3]

17. $\dfrac{9!}{4!\,5!}$

18. $\dfrac{10!}{6!\,4!}$

19. $\dbinom{10}{4}$

20. $\dbinom{8}{5}$

Aha! **21.** $\dbinom{9}{9}$

22. $\dbinom{7}{7}$

23. $\dbinom{30}{2}$

24. $\dbinom{51}{49}$

25. $\dbinom{40}{38}$

26. $\dbinom{35}{2}$

Binomial Expansion

Expand. Use both of the methods shown in this section.

27. $(a - b)^4$

28. $(m + n)^5$

29. $(p + w)^7$

30. $(x - y)^6$

31. $(3c - d)^7$

32. $(x^2 - 3y)^5$

33. $(t^{-2} + 2)^6$

34. $(3c - d)^6$

35. $\left(3s + \dfrac{1}{t}\right)^9$

36. $\left(x + \dfrac{2}{y}\right)^9$

37. $(x^3 - 2y)^5$

38. $(a^2 - b^3)^5$

39. $(\sqrt{5} + t)^6$

40. $(\sqrt{3} - t)^4$

41. $\left(\dfrac{1}{\sqrt{x}} - \sqrt{x}\right)^6$

42. $(x^{-2} + x^2)^4$

Find the indicated term for each binomial expression.

43. 3rd, $(a + b)^6$

44. 6th, $(x + y)^7$

45. 12th, $(a - 3)^{14}$

46. 11th, $(x - 2)^{12}$

47. 5th, $(2x^3 + \sqrt{y})^8$

48. 4th, $\left(\dfrac{1}{b^2} + c\right)^7$

49. Middle, $(2u + 3v^2)^{10}$

50. Middle two, $(\sqrt{x} + \sqrt{3})^5$

Aha! **51.** 9th, $(x - y)^8$

52. 13th, $(a - \sqrt{b})^{12}$

53. Maya claims that she can calculate mentally the first two terms and the last two terms of the expansion of $(a + b)^n$ for any whole number n. How do you think she does this?

54. Without performing any calculations, explain why the expansions of $(x - y)^8$ and $(y - x)^8$ must be equal.

Skill Review

Graph.

55. $y = x^2 - 5$ [11.7]

56. $y = x - 5$ [3.6]

57. $y \geq x - 5$ [9.4]

58. $y = 5^x$ [12.2]

59. $f(x) = \log_5 x$ [12.3]

60. $x^2 + y^2 = 5$ [13.1]

Synthesis

61. Explain how someone can determine the x^2-term of the expansion of $\left(x - \dfrac{3}{x}\right)^{10}$ without calculating any other terms.

62. Devise two problems requiring the use of the binomial theorem. Design the problems so that one is solved more easily using Form 1 and the other is solved more easily using Form 2. Then explain what makes one form easier to use than the other in each case.

63. The notation $\dbinom{n}{r}$ is read "n choose r" because it can be used to calculate the number of ways in which a set of r elements can be chosen from a set containing n elements. Show that there are exactly $\dbinom{5}{3}$ ways of choosing a subset of size 3 from $\{a, b, c, d, e\}$.

64. *Baseball.* During the 2011 season, Michael Young of the Texas Rangers had a batting average of 0.338. In that season, if someone were to randomly select 5 of his "at-bats," the probability of Young's getting exactly 3 hits would be the 3rd term of the binomial expansion of $(0.338 + 0.662)^5$. Find that term and use a calculator to estimate the probability.

Source: www.baseball-reference.com

65. *Widows or Divorcees.* The probability that a woman will be either widowed or divorced is 85%. If 8 women are randomly selected, the probability that exactly 5 of them will be either widowed or divorced is the 6th term of the binomial expansion of $(0.15 + 0.85)^8$. Use a calculator to estimate that probability.

66. *Baseball.* In reference to Exercise 64, the probability that Young will get *at most* 3 hits is found by adding the last 4 terms of the binomial expansion of $(0.338 + 0.662)^5$. Find these terms and use a calculator to estimate the probability.

67. *Widows or Divorcees.* In reference to Exercise 65, the probability that *at least* 6 of the women will be widowed or divorced is found by adding the last three terms of the binomial expansion of $(0.15 + 0.85)^8$. Find these terms and use a calculator to estimate the probability.

68. Find the term of

$$\left(\frac{3x^2}{2} - \frac{1}{3x} \right)^{12}$$

that does not contain x.

69. Prove that

$$\binom{n}{r} = \binom{n}{n-r}.$$

for any whole numbers n and r. Assume $r \le n$.

70. Form 1 of the binomial theorem can be proved using form 2 of the binomial theorem. The key step in that proof is showing that the coefficients inside Pascal's triangle are found by adding the two terms above. Prove this fact by showing that

$$\binom{n}{r} = \binom{n-1}{r-1} + \binom{n-1}{r}.$$

71. Find the middle term of $(x^2 - 6y^{3/2})^6$.

72. Find the ratio of the 4th term of

$$\left(p^2 - \frac{1}{2} p \sqrt[3]{q} \right)^5$$

to the 3rd term.

73. Find the term containing $\dfrac{1}{x^{1/6}}$ of

$$\left(\sqrt[3]{x} - \frac{1}{\sqrt{x}} \right)^7.$$

Aha! **74.** Multiply: $(x^2 + 2xy + y^2)(x^2 + 2xy + y^2)^2(x + y)$.

75. What is the degree of $(x^3 + 2)^4$?

 YOUR TURN ANSWERS: SECTION 14.4

1. $u^4 - 4u^3v + 6u^2v^2 - 4uv^3 + v^4$
2. $x^5 + 20x^3 + 160x + \dfrac{640}{x} + \dfrac{1280}{x^3} + \dfrac{1024}{x^5}$ **3.** 45   **4.** 84
5. $a^5 + 10a^4c + 40a^3c^2 + 80a^2c^3 + 80ac^4 + 32c^5$
6. $w^8 - 4w^6t^2 + 6w^4t^4 - 4w^2t^6 + t^8$ **7.** $2016a^4w^5$

QUICK QUIZ: SECTIONS 14.1–14.4

1. Write out and evaluate $\displaystyle\sum_{k=1}^{4} \frac{k+1}{k}$.   [14.1]

2. Find the common difference for the arithmetic sequence $2.5, 2.1, 1.7, 1.3, \ldots$.   [14.2]

3. Find the common ratio for the geometric sequence $-200, 100, -50, 25, \ldots$.   [14.3]

4. Simplify: $\dbinom{12}{9}$.   [14.4]

5. Expand: $(x + w)^4$.   [14.4]

EXAMPLE 9 *Energy Use.*   Translate the following problem to an equation. Do not solve.

The average energy use of food-service buildings is 51 Btu per square foot. This is three times the energy use of buildings providing lodging. What is the energy use per square foot of buildings providing lodging?

Source: Energy Information Administration

9. Translate the following problem to an equation. Do not solve. Charron paid $285 for season passes to all the home college games. This was five times what she paid for her parking pass. How much did she pay for her parking pass?

SOLUTION   We let c represent the energy use per square foot of a building providing lodging. We then reword the problem to make the translation more direct.

Rewording: 3   times   lodging energy use   is   51.

Translating: 3   · c =   51

↪ YOUR TURN

R.1 EXERCISE SET   FOR EXTRA HELP

PRACTICE   WATCH   READ   REVIEW

The Real Numbers

Use either $<$ or $>$ for ▨ to write a true sentence.

1. 8 ▨ -16

2. -7 ▨ 1

3. -6.1 ▨ -1.3

4. -4.2 ▨ -10.7

5. 0 ▨ -15

6. -98 ▨ 0

Find each absolute value.

7. $|22|$

8. $\left|\frac{11}{4}\right|$

9. $|-1.3|$

10. $|-105|$

Operations on Real Numbers

Simplify.

11. $(-14) + (-11)$

12. $3 - (-2)$

13. $-\frac{1}{3} - \frac{2}{5}$

14. $\frac{3}{8} \div \frac{3}{5}$

15. $4.2 - 10.7$

16. $(-1.3)(2.8)$

17. $-9 + 0$

18. $\left(-\frac{1}{2}\right) + \frac{1}{8}$

19. $0 \div (-10)$

20. $0 - 32$

21. $\left(-\frac{3}{10}\right) + \left(-\frac{1}{5}\right)$

22. $\left(-\frac{4}{7}\right)\left(\frac{7}{4}\right)$

23. $-3.8 + 9.6$

24. $-0.01 + 1$

25. $(-12) \div 4$

26. $(-87)(0)$

27. $32 - (-7)$

28. $-100 + 35$

29. $(-10)(-17.5)$

30. $-10 - 2.68$

31. $(-68) + 36$

32. $175 \div (-25)$

33. $2 + (-3) + 7 + 10$

34. $-5 + (-15) + 13 + (-1)$

35. $3 \cdot (-2) \cdot (-1) \cdot (-1)$

36. $(-6) \cdot (-5) \cdot (-4) \cdot (-3) \cdot (-2) \cdot (-1)$

37. $(-1)^4 + 2^3$

38. $(-1)^5 + 2^4$

39. $2 \times 6 - 3 \times 5$

40. $12 \div 4 + 15 \div 3$

41. $3 - (11 + 2 \cdot 4)$

42. $3 - 11 + 2 \cdot 4$

43. $4 \cdot 5^2$

44. $7 \cdot 2^3$

45. $25 - 8 \times 3 + 1$

46. $12 - 16 \times 5 + 4$

47. $2 - (3^3 + 16 \div (-2)^3)$

48. $-7 - (8 + 10 \cdot 2^2)$

49. $|6(-3)| + |(-2)(-9)|$

50. $3 - |2 - 7 + 4|$

51. $\dfrac{7000 + (-10)^3}{10^2 \times (2 + 4)}$

52. $\dfrac{3 - 2 \times 6 - 5}{2(3 + 7)^2}$

53. $2 + 8 \div 2 \times 2$

54. $2 + 8 \div (2 \times 2)$

Algebraic Expressions

Evaluate.

55. $y - x$, for $x = 10$ and $y = 3$

56. $n - 2m$, for $m = 6$ and $n = 11$

57. $-3 - x^2 + 12x$, for $x = 5$

58. $14 + (y - 5)^2 - 12 \div y$, for $y = -2$

59. The area of a parallelogram with base b and height h is bh. Find the area of the parallelogram when the height is 3.5 cm and the base is 8 cm.

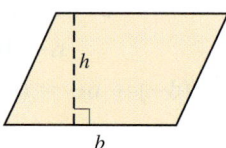

60. The area of a triangle with base b and height h is $\frac{1}{2}bh$. Find the area of the triangle when the height is 2 in. and the base is 6.2 in.

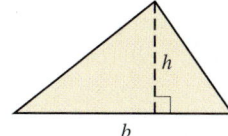

Multiply.

61. $4(2x + 7)$

62. $3(5y + 1)$

63. $-2(15 - 3x)$

64. $-7(3x - 5)$

65. $2(4a + 6b - 3c)$

66. $5(8p + q - 5r)$

67. $-3(2x - y + z)$

68. $-10(-6 - y - z)$

Factor.

69. $8x + 6y$

70. $7p + 14q$

71. $3 + 3w$

72. $4x + 4y$

73. $10x + 50y + 100$

74. $81p + 27q + 9$

Combine like terms.

75. $3p - 2p$

76. $4x + 3x$

77. $4m + 10 - 5m + 12$

78. $3a - 4b - b - 6a$

79. $-6x + 7 + x$

80. $16r + (-7r) + 3s$

Remove parentheses and simplify.

81. $2p - (7 - 4p)$

82. $4r - (3r + 5)$

83. $6x + 5y - 7(x - y)$

84. $14m - 6(2n - 3m) + n$

85. $6[2a + 4(a - 2b)]$

86. $2[2a + 1 - (3a - 6)]$

87. $3 - 2[5(x - 10y) - (3 + 2y)]$

88. $7 - 4[2(3 - 2x) - 5(4x - 3)]$

Translate each problem to an equation. Do not solve.

89. Three times what number is 348?

90. What number added to 256 is 113?

91. *Coca-Cola® Consumption.* The average U.S. citizen consumes 412 servings of Coca-Cola beverages each year. This is 5.15 times the international average. What is the international per capita consumption of Coca-Cola beverages?

Source: thecoca-colacompany.com

92. *Vegetable Production.* It takes 42 gal of water to produce 1 lb of broccoli. This is twice the amount of water used to produce 1 lb of lettuce. How many gallons of water does it take to produce 1 lb of lettuce?

93. *Breakfast Calories.* International House of Pancakes strawberry banana French toast contains 1060 calories. This is 20 more calories than 2 orders of McDonald's hotcakes and sausage provide. How many calories are in an order of McDonald's hotcakes and sausage?

Source: www.calorieking.com

94. *University Revenue.* In 2009, public university revenue was approximately $36,000 per student. This was $4000 more than four times the tuition revenue per student. What was the tuition revenue per student?

Source: www.freeby50.com

YOUR TURN ANSWERS: SECTION R.1

1. $<$ **2.** 287 **3.** -25 **4.** 18 **5.** 14 m

6. $7(1 + 2m + 10x)$ **7.** $-7x - 7$ **8.** $-3x + 24$

9. Let p represent what Charron paid for her parking pass; $285 = 5p$

REFERENCES

Absolute value, p. 34
Addition, Section 1.5
Algebraic expression, p. 2
Combine like terms, p. 41
Constant, p. 2
Division, Section 1.7
Division by 0, p. 56
Equivalent expressions, p. 12
Evaluate, p. 2
Exponential notation, p. 60
Factor, pp. 14, 16
Irrational numbers, p. 31
Multiplication, Section 1.7
Opposite, p. 44
Order, p. 33
Order of operations, p. 61
Real numbers, p. 32
Repeating decimals, p. 30
Sets of real numbers, Section 1.4
Substitute, p. 2
Subtraction, Section 1.6
Terminating decimals, p. 30
Terms, p. 16
Translating to algebraic expressions, p. 3
Translating to equations, p. 5
Variable, p. 2

R.2 Equations, Inequalities, and Problem Solving

Solving Equations and Formulas ▪ Solving Inequalities ▪ Problem Solving

Solving Equations and Formulas

Any replacement for the variable in an equation that makes the equation true is called a *solution* of the equation. To **solve** an equation means to find all of its solutions.

We use the following principles to write equations with the same solutions, called **equivalent equations**.

THE ADDITION AND MULTIPLICATION PRINCIPLES FOR EQUATIONS

The Addition Principle

For any real numbers a, b, and c,

$$a = b \quad \text{is equivalent to} \quad a + c = b + c.$$

The Multiplication Principle

For any real numbers a, b, and c, with $c \neq 0$,

$$a = b \quad \text{is equivalent to} \quad a \cdot c = b \cdot c.$$

To solve $x + a = b$ for x, we add $-a$ to (or subtract a from) both sides. To solve $ax = b$ for x, we multiply both sides by $\frac{1}{a}$ (or divide both sides by a).

To solve an equation like $-3x - 10 = 14$, we first isolate the variable term, $-3x$, using the addition principle. Then we use the multiplication principle to get the variable by itself.

EXAMPLE 1   Solve: $-3x - 10 = 14$.

SOLUTION

$$-3x - 10 = 14$$

$$-3x - 10 + 10 = 14 + 10 \qquad \text{Using the addition principle:}$$
$$\text{Adding 10 to both sides}$$

Isolate the x-term. $-3x = 24$ \qquad Simplifying

$$\frac{-3x}{-3} = \frac{24}{-3} \qquad \text{Dividing both sides by } -3$$

Isolate x. $x = -8$ \qquad Simplifying

Check: $$\begin{array}{c|c} -3x - 10 = 14 \\ \hline -3(-8) - 10 & 14 \\ 24 - 10 & \\ 14 \overset{?}{=} 14 & \text{TRUE} \end{array}$$

The solution is -8.

1. Solve: $5 + 2x = 4$.

YOUR TURN

Equations are generally easier to solve when they do not contain fractions. In general, the easiest way to clear an equation of fractions is to multiply *every term on both sides* of the equation by the least common denominator.

EXAMPLE 2 Solve: $\frac{5}{2} - \frac{1}{6}t = \frac{2}{3}$.

SOLUTION The number 6 is the least common denominator, so we multiply both sides by 6.

$$\frac{5}{2} - \frac{1}{6}t = \frac{2}{3}$$

$$6\left(\frac{5}{2} - \frac{1}{6}t\right) = 6 \cdot \frac{2}{3} \qquad \text{Multiplying both sides by 6}$$

$$6 \cdot \frac{5}{2} - 6 \cdot \frac{1}{6}t = 6 \cdot \frac{2}{3} \qquad \text{Using the distributive law. Be sure to multiply every term by 6.}$$

$$15 - t = 4 \qquad \text{The fractions are cleared.}$$

$$15 - t - 15 = 4 - 15 \qquad \text{Subtracting 15 from both sides}$$

$$-t = -11 \qquad \begin{array}{l}15 - t - 15 = 15 + (-t) + (-15) \\ \qquad = -t + 15 + (-15) = -t\end{array}$$

$$(-1)(-t) = (-1)(-11) \qquad \text{Multiplying both sides by } -1 \text{ to change the sign}$$

$$t = 11$$

Check:

$$\begin{array}{c|c}\frac{5}{2} - \frac{1}{6}t = \frac{2}{3} \\ \hline \frac{5}{2} - \frac{1}{6}(11) & \frac{2}{3} \\ \frac{5}{2} - \frac{11}{6} & \\ \frac{15}{6} - \frac{11}{6} & \\ \frac{2}{3} \stackrel{?}{=} \frac{2}{3} & \text{TRUE}\end{array}$$

The solution is 11.

2. Solve: $\dfrac{x}{4} - 2 = \dfrac{1}{5}$.

YOUR TURN

To solve equations that contain parentheses, we can use the distributive law to first remove the parentheses. If like terms appear in an equation, we combine them and then solve.

EXAMPLE 3 Solve: $1 - 3(4 - x) = 2(x + 5) - 3x$.

SOLUTION

$$1 - 3(4 - x) = 2(x + 5) - 3x$$

$$1 - 12 + 3x = 2x + 10 - 3x \qquad \text{Using the distributive law}$$

$$-11 + 3x = -x + 10 \qquad \begin{array}{l}\text{Combining like terms;} \\ 1 - 12 = -11 \text{ and } 2x - 3x = -x\end{array}$$

$$-11 + 3x + x = 10 \qquad \begin{array}{l}\text{Adding } x \text{ to both sides to get all} \\ x\text{-terms on one side}\end{array}$$

$$-11 + 4x = 10 \qquad \text{Combining like terms}$$

$$4x = 10 + 11 \qquad \begin{array}{l}\text{Adding 11 to both sides to isolate} \\ \text{the } x\text{-term}\end{array}$$

$$4x = 21 \qquad \text{Simplifying}$$

$$x = \tfrac{21}{4} \qquad \text{Dividing both sides by 4}$$

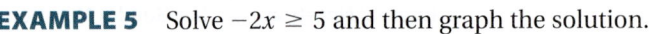

Note that when we multiply both sides of an inequality by a negative number, we must reverse the direction of the inequality symbol in order to have an equivalent inequality.

EXAMPLE 5   Solve $-2x \geq 5$ and then graph the solution.

SOLUTION   We have

$$-2x \geq 5$$

$$\frac{-2x}{-2} \leq \frac{5}{-2} \qquad \text{Multiplying by } -\frac{1}{2} \text{ or dividing by } -2$$

The symbol must be reversed!

$$x \leq -\frac{5}{2}.$$

Any number less than or equal to $-\frac{5}{2}$ is a solution. The graph is as follows:

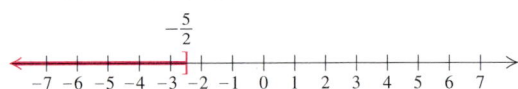

5. Solve $12 > -3x$ and graph the solution.

_____

The solution set is $\left\{x \mid x \leq -\frac{5}{2}\right\}$, or $\left(-\infty, -\frac{5}{2}\right]$.

YOUR TURN

We can use the addition and multiplication principles together to solve inequalities. We can also combine like terms, multiply to remove parentheses, and clear fractions and decimals.

EXAMPLE 6   Solve: $2 - 3(x + 5) > 4 - 6(x - 1)$.

SOLUTION   We have

$$2 - 3(x + 5) > 4 - 6(x - 1)$$

$$2 - 3x - 15 > 4 - 6x + 6 \qquad \begin{array}{l}\text{Using the distributive law to remove} \\ \text{parentheses}\end{array}$$

$$-3x - 13 > -6x + 10 \qquad \text{Simplifying}$$

$$-3x + 6x > 10 + 13 \qquad \begin{array}{l}\text{Adding } 6x \text{ and also } 13, \text{ to get all} \\ x\text{-terms on one side and all constant} \\ \text{terms on the other side}\end{array}$$

$$3x > 23 \qquad \text{Combining like terms}$$

$$x > \frac{23}{3}. \qquad \begin{array}{l}\text{Multiplying by } \frac{1}{3}. \text{ The inequality} \\ \text{symbol stays the same because } \frac{1}{3} \\ \text{is positive.}\end{array}$$

6. Solve:

$10(x - 2) - 3x \leq 5 - (2 - x).$

_____

The solution set is $\left\{x \mid x > \frac{23}{3}\right\}$, or $\left(\frac{23}{3}, \infty\right)$.

YOUR TURN

Problem Solving

One of the most important uses of algebra is as a tool for problem solving. The following five steps can be used to help solve problems of many types.

FIVE STEPS FOR PROBLEM SOLVING IN ALGEBRA

1. *Familiarize* yourself with the problem.
2. *Translate* to mathematical language. (This often means writing an equation.)
3. *Carry out* some mathematical manipulation. (This often means *solving* an equation.)
4. *Check* your possible answer in the original problem.
5. *State* the answer clearly, using a complete English sentence.

EXAMPLE 7 *Kitchen Cabinets.* Cherry kitchen cabinets cost 10% more than oak cabinets. Shelby Custom Cabinets designs a kitchen using $7480 worth of cherry cabinets. How much would the same kitchen cost using oak cabinets?

SOLUTION

1. **Familiarize.** The *Familiarize* step is often the most important of the five steps, and may require a significant amount of time. Sometimes it helps to make a drawing or a table, make a guess and check it, or look up further information. For this problem, we could review percent notation, and note that 10% of the price of the oak cabinets must be added to the price of the oak cabinets in order to get the price of the cherry cabinets. We let c = the cost of the oak cabinets.

2. **Translate.** We rewrite and translate:

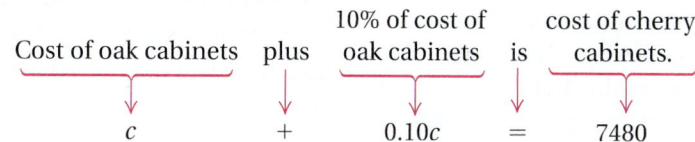

Cost of oak cabinets	plus	10% of cost of oak cabinets	is	cost of cherry cabinets.
c	$+$	$0.10c$	$=$	7480

3. **Carry out.** We solve the equation:

$$c + 0.10c = 7480$$
$$1c + 0.10c = 7480 \quad \text{\color{red}{Writing } }c\text{ \color{red}{as} }1c\text{ \color{red}{before combining terms}}$$
$$1.10c = 7480 \quad \text{\color{red}{Combining like terms}}$$
$$c = \frac{7480}{1.10} \quad \text{\color{red}{Dividing by 1.10}}$$
$$c = 6800.$$

4. **Check.** We check in the wording of the stated problem: Cherry cabinets cost 10% more, so the additional cost is

$$10\% \text{ of } \$6800 = 0.10(\$6800) = \$680.$$

The total cost of the cherry cabinets is then

$$\$6800 + \$680 = \$7480,$$

which is the amount stated in the problem.

5. **State.** The oak cabinets would cost $6800.

7. Refer to Example 7. The cherry cabinets cost 5% more than hickory cabinets. How much would the same kitchen cost using hickory cabinets?

YOUR TURN

Sometimes the translation of a problem is an inequality.

EXAMPLE 8   *Fitness Club.*   Elyse pays $30 per month to be a member of Perfect Fitness. An hour-long class costs an additional $2.50. How many classes can she take per month and not exceed a cost of $50 per month?

SOLUTION

1. **Familiarize.** Suppose that Elyse takes 10 classes in one month. Her cost would then be the monthly membership fee plus the fee for classes, or $30 + $2.50(10) = $55. This exceeds $50, so we know that the number of classes must be less than 10. We let c = the number of classes taken in one month.

2. **Translate.** The *Familiarize* step helps us reword and translate.

	The monthly fee	plus	the cost of classes	cannot exceed	$50.
Rewording:					
Translating:	30	+	2.5c	≤	50

3. **Carry out.** We solve the inequality:

$$30 + 2.5c \le 50$$
$$2.5c \le 20 \qquad \text{Subtracting 30 from both sides}$$
$$m \le 8. \qquad \text{Dividing by 2.5. The inequality symbol stays the same.}$$

4. **Check.** As a partial check, note that the monthly cost including the cost for 8 classes is

$$\$30 + \$2.50(8) = \$50.$$

Since fewer classes will cost even less, our answer checks. We also note that 8 is less than 10 classes, as noted in the *Familiarize* step.

5. **State.** Elyse will not exceed a cost of $50 if she takes no more than 8 classes.

8. Refer to Example 8. Elyse can work out in the Super Athlete room for $2 per hour. For how many hours can she work out in that room and not exceed a cost of $50 for the month?

YOUR TURN

R.2 EXERCISE SET

FOR EXTRA HELP

MyMathLab® MathXL

PRACTICE   WATCH   READ   REVIEW

Solving Equations and Formulas

Solve.

1. $-6 + x = 10$

2. $y + 7 = -3$

3. $t + \frac{1}{3} = \frac{1}{4}$

4. $-\frac{2}{3} + p = \frac{1}{6}$

5. $-1.9 = x - 1.1$

6. $x + 4.6 = 1.7$

7. $-x = \frac{5}{3}$

8. $-y = -\frac{2}{5}$

9. $-\frac{2}{7}x = -12$

10. $-\frac{1}{4}x = 3$

11. $\dfrac{-t}{5} = 1$

12. $\dfrac{2}{3} = -\dfrac{z}{8}$

13. $3y + 10 = 15$

14. $12 = 5y + 18$

15. $4x + 7 = 3 - 5x$

16. $2x = 5 + 7x$

17. $2x - 7 = 5x + 1 - x$

18. $a + 7 - 2a = 14 + 7a - 10$

19. $\frac{2}{5} + \frac{1}{3}t = 5$

20. $-\frac{5}{6} + t = \frac{1}{2}$

21. $x + 0.45 = 2.6x$

22. $1.8x + 0.16 = 4.2 - 0.05x$

23. $8(3 - m) + 7 = 47$

24. $2(5 - m) = 5(6 + m)$

25. $4 - (6 + x) = 13$

26. $18 = 9 - (3 - x)$

27. $2 + 3(4 + c) = 1 - 5(6 - c)$

28. $b + (b + 5) - 2(b - 5) = 18 + b$

29. $0.1(a - 0.2) = 1.2 + 2.4a$

30. $\frac{2}{3}\left(\frac{1}{2} - x\right) + \frac{5}{6} = \frac{3}{2}\left(\frac{2}{3}x + 1\right)$

31. $A = lw$, for l

32. $A = lw$, for w

33. $I = \frac{P}{V}$, for P

34. $b = \frac{A}{h}$, for A

35. $q = \frac{p + r}{2}$, for p

36. $q = \frac{p - r}{2}$, for r

37. $A = \pi r^2 + \pi r^2 h$, for π

38. $ax + by = c$, for a

Solving Inequalities

Solve and graph. Write each answer in set-builder notation and in interval notation.

39. $x + 3 \leq 15$

40. $y + 7 < -10$

41. $m - 17 > -5$

42. $x + 9 \geq -8$

43. $2x \geq -3$

44. $-\frac{1}{2}n \leq 4$

45. $-5t > 15$

46. $3x > 10$

Solve. Write each answer in set-builder notation and in interval notation.

47. $2y - 7 > 13$

48. $2 - 6y \leq 18$

49. $6 - 5a \leq a$

50. $4b + 7 > 2 - b$

51. $2(3 + 5x) \geq 7(10 - x)$

52. $2(x + 5) < 8 - 3x$

53. $\frac{2}{3}(6 - x) < \frac{1}{4}(x + 3)$

54. $\frac{2}{3}t + \frac{8}{9} \geq \frac{4}{6} - \frac{1}{4}t$

55. $0.7(2 + x) \geq 1.1x + 5.75$

56. $0.4x + 5.7 \leq 2.6 - 3(1.2x - 7)$

Problem Solving

Solve. Use the five-step problem-solving process.

57. Three less than the sum of 2 and some number is 6. What is the number?

58. Five times some number is 10 less than the number. What is the number?

59. The sum of two consecutive even integers is 34. Find the numbers.

60. The sum of three consecutive integers is 195. Find the numbers.

61. *Reading.* Leisa is reading a 500-page book. She has twice as many pages to read as she has already finished. How many pages has she already read?

62. *Mowing.* It takes Caleb 50 min to mow his lawn. As he takes a break to answer his phone, he calculates that it will take him three times as many minutes to finish as he has already spent mowing. How long has he already spent mowing?

63. *Perimeter of a Rectangle.* The perimeter of a rectangle is 28 cm. The width is 5 cm less than the length. Find the width and the length.

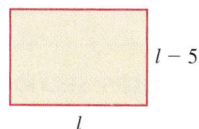

64. *Triangles.* The second angle of a triangle is one-third as large as the first. The third angle is 5° more than the first. Find the measure of the second angle.

65. *Water Usage.* Rural Water Company charges a monthly service fee of $9.70 plus a volume charge of $2.60 for every hundred cubic feet of water used. How much water was used if the monthly bill is $33.10?

66. *Telephone Bills.* Brandon pays $2.95 per month for a long-distance telephone service that offers a flat rate of 5¢ per minute. One month his total long-distance telephone bill was $7.05. How many minutes of long-distance telephone calls were made that month?

67. *Sale Price.* A can of tomatoes is on sale at 20% off for 88¢. What is the normal selling price of the tomatoes?

68. *Plywood.* The price of a piece of plywood rose 5% to $42. What was the original price of the plywood?

69. *Practice.* Dierdre's basketball coach requires each team member to average at least 15 min per day shooting baskets. One week Dierdre spent 10 min, 20 min, 5 min, 0 min, 25 min, and 15 min shooting baskets. How long must she practice shooting baskets on the seventh day if she is to meet the requirement?

70. *Perimeter of a Garden.* The perimeter of Garry's rectangular garden cannot exceed the 100 ft of fencing that he purchased. He wants the length to be twice the width. What widths of the garden will meet these conditions?

71. *Meeting Costs.* Great Space charges a $75 cleaning fee plus $45 per hour for the use of its meeting room. Complete Consultants has budgeted $200 to rent a room for a seminar. For how many hours can they rent the meeting room at Great Space?

72. *Meeting Costs.* Spring Haven charges a $15 setup fee, a $30 cleanup fee, and $50 per hour for the use of its meeting room. For what lengths of time will Spring Haven's room be less expensive than the room at Great Space (see Exercise 71)?

 YOUR TURN ANSWERS: SECTION R.2

1. $-\frac{1}{2}$ **2.** $\frac{44}{5}$ **3.** -15 **4.** $p = 2Q + q$

5. $\{x \mid x > -4\}$, or $(-4, \infty)$;

6. $\{x \mid x \leq \frac{23}{6}\}$, or $\left(-\infty, \frac{23}{6}\right]$ **7.** $7123.81

8. No more than 10 hr

REFERENCES
The addition principle, p. 81
The addition principle for inequalities, p. 130
Clearing fractions, p. 91
Conditional equation, p. 93
Contradiction, p. 93
Equivalent equations, p. 81
Equivalent inequalities, p. 130
Familiarization step, p. 113
Formulas, Section 2.3
Graphs of inequalities, p. 128
Identity, p. 93
Interval notation, p. 129
The multiplication principle, p. 83
The multiplication principle for inequalities, p. 132
Percent, Section 2.4
Problem solving, Section 2.5
Set-builder notation, p. 129
Solutions of inequalities, p. 128
Solving applications with inequalities, Section 2.7

R.3   Introduction to Graphing

Points and Ordered Pairs ▪ Graphs and Slope ▪ Linear Equations

Points and Ordered Pairs

We can represent, or graph, pairs of numbers such as $(2, -5)$ on a plane. To do so, we use two perpendicular number lines called **axes**. The axes cross at a point called the **origin**. Arrows on the axes show the positive directions.

The order of the **coordinates**, or numbers in a pair, is important. The **first coordinate** indicates horizontal position and the **second coordinate** indicates vertical position. Such pairs of numbers are called **ordered pairs**. Thus the ordered pairs $(1, -2)$ and $(-2, 1)$ correspond to different points, as shown in the accompanying figure.

The axes divide the plane into four regions, or **quadrants**, as indicated by Roman numerals in the figure below. Points on the axes are not considered to be in any quadrant. The horizontal axis is often labeled the *x*-axis, and the vertical axis the *y*-axis.

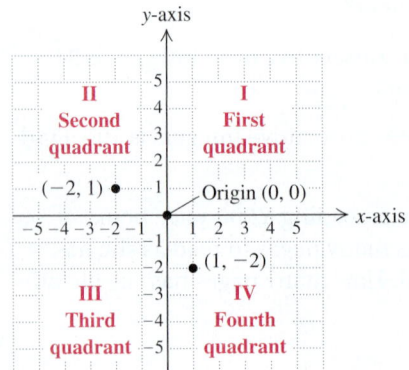

Graphs and Slope

When an equation contains two variables, solutions must be ordered pairs. Unless stated otherwise, the first number in each pair replaces the variable that occurs first alphabetically.

EXAMPLE 1   Determine whether $(1, 4)$ is a solution of $y - x = 3$.

SOLUTION   We substitute 1 for x and 4 for y since x occurs first alphabetically:

$$\begin{array}{c|c} y - x = 3 \\ \hline 4 - 1 & 3 \\ 3 \overset{?}{=} 3 & \text{TRUE} \end{array}$$

Since $3 = 3$ is true, the pair $(1, 4)$ *is* a solution.

1. Determine whether $(-2, 5)$ is a solution of $y - x = 3$.

↩ YOUR TURN

The **graph** of an equation represents all of its solutions.

EXAMPLE 2   Graph: $y = -2x + 1$.

SOLUTION   We select a value for x, calculate the corresponding value of y, and form an ordered pair.

If $x = 0$, then $y = -2 \cdot 0 + 1 = 1$, and $(0, 1)$ is a solution. Repeating this step, we find other ordered pairs and list the results in a table. We then plot the points corresponding to the pairs. They appear to form a straight line, so we draw a line through the points.

$y = -2x + 1$

x	y	(x, y)
0	1	$(0, 1)$
−1	3	$(-1, 3)$
3	−5	$(3, -5)$
1	−1	$(1, -1)$

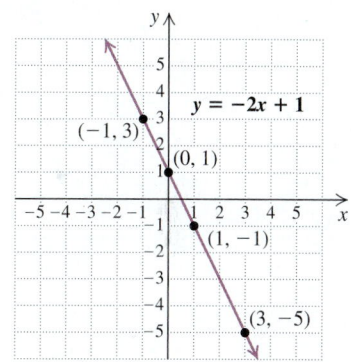

2. Graph: $y = 2x - 1$.

↩ YOUR TURN

The graph in Example 2 is a straight line. An equation whose graph is a straight line is a **linear equation**. The *rate of change* of y with respect to x is called the **slope** of a graph. A linear graph has constant slope. It can be found using any two points on a line.

SLOPE

The *slope* of the line containing points (x_1, y_1) and (x_2, y_2) is given by

$$m = \frac{\text{change in } y}{\text{change in } x} = \frac{\text{rise}}{\text{run}} = \frac{y_2 - y_1}{x_2 - x_1}.$$

EXAMPLE 3   Find the slope of the line containing the points $(-2, 1)$ and $(3, -4)$.

SOLUTION   From $(-2, 1)$ to $(3, -4)$, the change in y, or the rise, is $-4 - 1$, or -5. The change in x, or the run, is $3 - (-2)$, or 5. Thus

3. Find the slope of the line containing the points $(0, -9)$ and $(4, 3)$.

$$\text{Slope} = \frac{\text{change in } y}{\text{change in } x} = \frac{\text{rise}}{\text{run}} = \frac{-4 - 1}{3 - (-2)} = \frac{-5}{5} = -1.$$

YOUR TURN

The slope of a line indicates the direction and the steepness of its slant. The larger the absolute value of the slope, the steeper the line. The direction of the slant is indicated by the sign of the slope, as shown in the figures below.

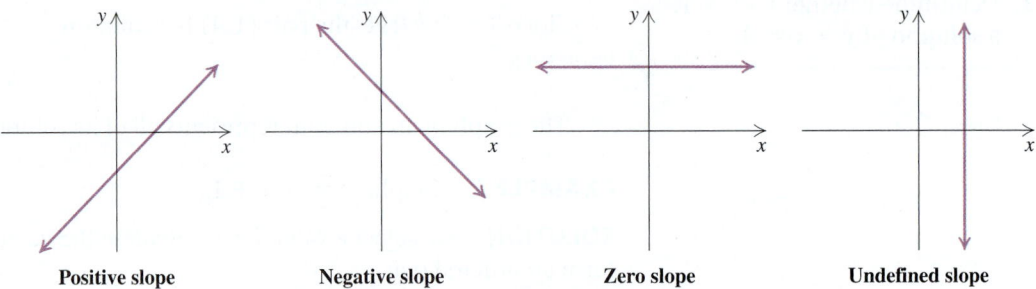

Positive slope   Negative slope   Zero slope   Undefined slope

The **x-intercept** of a line, if it exists, is the point at which the graph crosses the x-axis. To find an x-intercept, we replace y with 0 and solve for x.

The **y-intercept** of a line, if it exists, is the point at which the graph crosses the y-axis. To find a y-intercept, we replace x with 0 and solve for y.

Linear Equations

Any equation that can be written in the **standard form** $Ax + By = C$ is linear. Linear equations can also be written in other forms.

FORMS OF LINEAR EQUATIONS

Standard form: $Ax + By = C$

Slope–intercept form: $y = mx + b$

Point–slope form: $y - y_1 = m(x - x_1)$

The slope and the y-intercept of a line can be read from the slope–intercept form of the line's equation.

SLOPE AND *y*-INTERCEPT

For the graph of any equation $y = mx + b$,

- the slope is m, and

- the y-intercept is $(0, b)$.

EXAMPLE 4   Find the slope and the y-intercept of the line given by the equation $4x - 3y = 9$.

SOLUTION   We write the equation in slope–intercept form $y = mx + b$:

$$4x - 3y = 9 \qquad \text{We must solve for } y.$$
$$-3y = -4x + 9 \qquad \text{Adding } -4x \text{ to both sides}$$
$$y = \tfrac{4}{3}x - 3. \qquad \text{Dividing both sides by } -3$$

The slope is $\frac{4}{3}$ and the y-intercept is $(0, -3)$.

4. Find the slope and the y-intercept of the line given by the equation $y - x = 10$.

YOUR TURN

If we know that an equation is a straight line, we can plot two points on the line and draw the line through those points. The intercepts are often convenient points to use. A third point can be used as a check.

EXAMPLE 5   Graph $2x - 5y = 10$ using intercepts.

SOLUTION   To find the x-intercept, we let $y = 0$ and solve for x:

$$2x - 5 \cdot 0 = 10 \qquad \text{Replacing } y \text{ with } 0$$
$$2x = 10$$
$$x = 5.$$

To find the y-intercept, we let $x = 0$ and solve for y:

$$2 \cdot 0 - 5y = 10 \qquad \text{Replacing } x \text{ with } 0$$
$$-5y = 10$$
$$y = -2.$$

Thus the x-intercept is $(5, 0)$ and the y-intercept is $(0, -2)$. The graph is a line, since $2x - 5y = 10$ is in the form $Ax + By = C$. It passes through these two points. We leave it to the student to calculate a third point as a check.

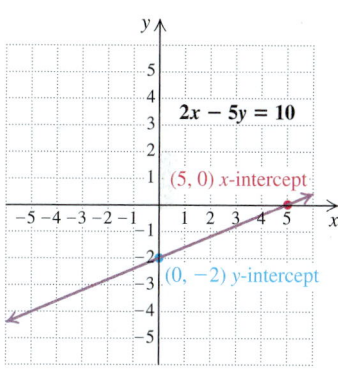

5. Graph $x + 3y = 3$ using intercepts.

YOUR TURN

Alternatively, if we know a point on the line and its slope, we can plot the point and "count off" its slope to locate another point on the line.

EXAMPLE 6   Graph: $y = -\frac{1}{2}x + 3$.

SOLUTION   The equation is in slope–intercept form, so we can read the slope and the y-intercept directly from the equation.

Slope: $-\frac{1}{2}$

y-intercept: $(0, 3)$

We plot the y-intercept and use the slope to find another point.

Another way to write the slope is $\dfrac{-1}{2}$. This means that for a run of 2 units, there is a negative rise, or a drop, of 1 unit. Starting at $(0, 3)$, we move 2 units in the positive horizontal direction and then 1 unit down, to locate the point $(2, 2)$. Then we draw the graph. A third point can be calculated and plotted as a check.

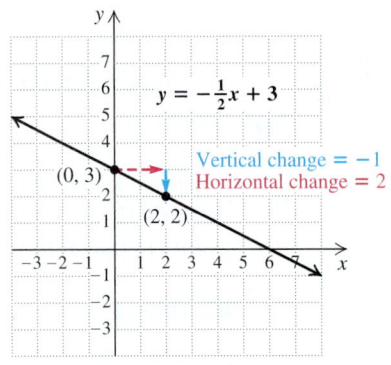

6. Graph: $y = \frac{2}{3}x - 5$.

 YOUR TURN

Horizontal lines and vertical lines intersect only one axis.

HORIZONTAL LINES AND VERTICAL LINES

Horizontal Line	*Vertical Line*
$y = b$	$x = a$
y-intercept $(0, b)$	x-intercept $(a, 0)$
Slope is 0	Undefined slope
Example: $y = -3$	Example: $x = 2$

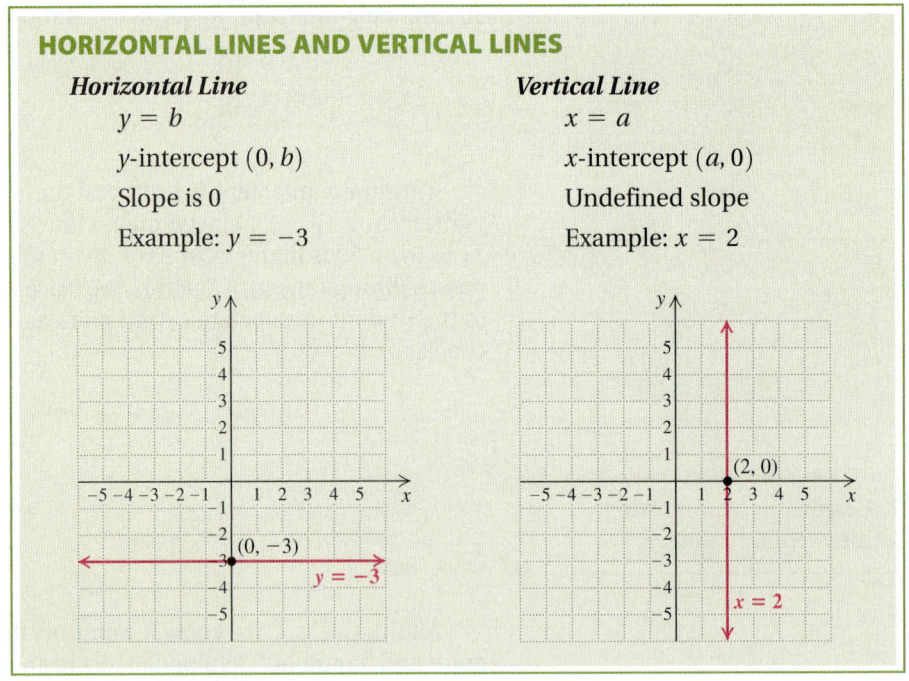

If we know the slope of a line and the coordinates of a point on the line, we can find an equation of the line, using either the slope–intercept equation, $y = mx + b$, or the point–slope equation, $y - y_1 = m(x - x_1)$.

EXAMPLE 7   Find the slope–intercept equation of a line given the following:

a) The slope is 2, and the y-intercept is $(0, -5)$.

b) The graph contains the points $(-2, 1)$ and $(3, -4)$.

SOLUTION

a) Since the slope and the y-intercept are given, we use the slope–intercept equation:

$$y = mx + b$$
$$y = 2x - 5. \qquad \text{Substituting 2 for } m \text{ and } -5 \text{ for } b$$

b) To use the point–slope equation, we need to know a point on the line and the line's slope. The slope can be found from the points given:

Find the slope.

$$m = \frac{1 - (-4)}{-2 - 3} = \frac{5}{-5} = -1.$$

Either point can be used as (x_1, y_1). Using $(-2, 1)$, we have

Find a point–slope equation for the line.

$$y - y_1 = m(x - x_1)$$
$$y - 1 = -1(x - (-2)) \qquad \text{Substituting } -2 \text{ for } x_1, 1 \text{ for } y_1, \text{ and } -1 \text{ for } m$$
$$y - 1 = -(x + 2)$$

Find the slope–intercept equation for the line.

$$y - 1 = -x - 2$$
$$y = -x - 1. \qquad \text{This is in slope–intercept form.}$$

 YOUR TURN

7. Find the slope–intercept equation of the line containing the points $(1, 2)$ and $(5, -3)$.

We can tell from the slopes of two lines whether they are parallel or perpendicular.

PARALLEL LINES AND PERPENDICULAR LINES

Two lines are parallel if they have the same slope.

Two lines are perpendicular if the product of the slopes is -1.

EXAMPLE 8   Determine whether the graphs of $4x - y = 8$ and $3x + 12y = 24$ are parallel, perpendicular, or neither.

SOLUTION   We solve both equations for y in order to determine the slopes of the lines:

$$4x - y = 8$$
$$-y = -4x + 8$$
$$y = 4x - 8.$$

The slope of $4x - y = 8$ is 4.
   For the second line, we have

$$3x + 12y = 24$$
$$12y = -3x + 24$$
$$y = -\tfrac{1}{4}x + 2.$$

8. Determine whether the graphs of the lines are parallel, perpendicular, or neither:

$$2x + 5y = 7,$$
$$2x = 5y + 1.$$

The slope of $3x + 12y = 24$ is $-\tfrac{1}{4}$. Since $4 \cdot \left(-\tfrac{1}{4}\right) = -1$, the lines are perpendicular.

 YOUR TURN

R.3   EXERCISE SET

Points and Ordered Pairs

1. Plot the following points.
$(2, -3), (5, 1), (0, 2), (-1, 0),$
$(0, 0), (-2, -5), (-1, 1), (1, -1)$

2. Plot the following points.
$(0, -4), (-4, 0), (5, -2), (2, 5),$
$(3, 3), (-3, -1), (-1, 4), (0, 1)$

In which quadrant or on what axis is each point located?

3. $(-2, 5)$

4. $(15, 27)$

5. $(-6, 0)$

6. $(-1.7, -5.9)$

7. $(18, -7)$

8. $(0, 3)$

Graphs and Slope

Determine whether each equation has the given ordered pair as a solution.

9. $y = 2x - 5; \ (1, 3)$

10. $4x + 3y = 8; \ (-1, 4)$

11. $a - 5b = -3; \ (2, 1)$

12. $c = d + 1; \ (1, 2)$

Graph.

13. $y = \frac{1}{3}x + 3$

14. $y = -x - 2$

15. $y = -4x$

16. $y = \frac{3}{4}x + 1$

Find the slope of the line containing each given pair of points. If the slope is undefined, state this.

17. $(3, 6)$ and $(2, 7)$

18. $(-1, 7)$ and $(-5, 1)$

19. $(-2, -\frac{1}{2})$ and $(5, -\frac{1}{2})$

20. $(6.8, 7.5)$ and $(6.8, -3.2)$

Linear Equations

Find the slope and the y-intercept of each equation.

21. $y = 2x - 5$

22. $y = 4 - x$

23. $2x + 7y = 1$

24. $x - 2y = 3$

Find the intercepts. Then graph.

25. $3 - y = 2x$

26. $2x + 5y = 10$

27. $y = 3x + 5$

28. $y = -x + 7$

29. $3x - 2y = 6$

30. $2y + 1 = x$

Determine the coordinates of the y-intercept of each equation. Then graph the equation.

31. $y = 2x - 5$

32. $y = -\frac{5}{4}x - 3$

33. $2y + 4x = 6$

34. $3y + x = 4$

Find the slope of each line. Then graph.

35. $y = 4$

36. $x = -5$

37. $x = 3$

38. $y = -1$

Find the slope–intercept equation of a line with the given conditions.

39. The slope is 5 and the y-intercept is $(0, 9)$.

40. The slope is $\frac{2}{3}$ and the y-intercept is $(0, -5)$.

41. The graph contains the points $(0, 3)$ and $(-1, 4)$.

42. The graph contains the points $(5, 1)$ and $(8, 0)$.

Determine whether each pair of lines is parallel, perpendicular, or neither.

43. $x + y = 5,$
$x - y = 1$

44. $2x + y = 3,$
$y = 4 - 2x$

45. $2x + 3y = 1,$
$2x - 3y = 5$

46. $y = \frac{1}{3}x - 7,$
$y + 3x = 1$

 YOUR TURN ANSWERS: SECTION R.3

1. No   2.   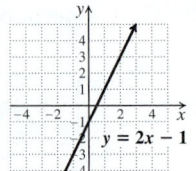   3. 3
4. Slope: 1;
   y-intercept: $(0, 10)$

5.   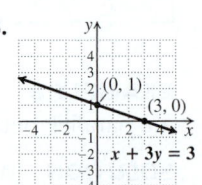   6.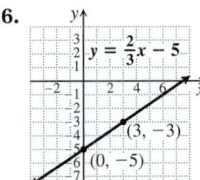

7. $y = -\dfrac{5}{4}x + \dfrac{13}{4}$ 8. Neither

REFERENCES

Coordinates, p. 156
Graphing ordered pairs, p. 156
Horizontal line, p. 178
Parallel lines, p. 207
Perpendicular lines, p. 208
Quadrants, p. 157
Slope, p. 190
Solutions of equations, p. 163
Vertical line, p. 178
x-intercept, p. 173
y-intercept, p. 173

R.4   Polynomials

Exponents ▪ Polynomials ▪ Addition and Subtraction of Polynomials ▪ Multiplication of Polynomials ▪ Division of Polynomials

Exponents

The following properties hold for any integers m and n and any real numbers a and b, provided no denominators are 0 and 0^0 is not considered.

PROPERTIES OF EXPONENTS

The Zero Exponent:	$a^0 = 1$
Negative Exponents:	$a^{-n} = \dfrac{1}{a^n}$
The Product Rule:	$a^m \cdot a^n = a^{m+n}$
The Quotient Rule:	$\dfrac{a^m}{a^n} = a^{m-n}$
The Power Rule:	$(a^m)^n = a^{mn}$
Raising a product to a power:	$(ab)^n = a^n b^n$
Raising a quotient to a power:	$\left(\dfrac{a}{b}\right)^n = \dfrac{a^n}{b^n}$

EXAMPLE 1   Simplify. Do not use negative exponents in your answer.

a) $(97)^0$

b) $7y^{-1}$

c) $\dfrac{1}{x^{-2}}$

SOLUTION

a) $(97)^0 = 1$ Any number (other than 0 itself) raised to the 0 power is 1.

b) $7y^{-1} = 7\left(\dfrac{1}{y^1}\right) = \dfrac{7}{y}$ y^{-1} is the reciprocal of y^1.

c) $\dfrac{1}{x^{-2}} = x^{-(-2)} = x^2$ The reciprocal of x^{-2} is $x^{-(-2)}$, or x^2.

1. Simplify $(2x)^0$ for $x = -10$.

YOUR TURN

EXAMPLE 2   Simplify. If negative exponents appear in the answer, write an equivalent answer using only positive exponents.

a) $(x^2 y^{-1})(xy^{-3})$

b) $\dfrac{(3p)^3}{(3p)^{-2}}$

c) $\left(\dfrac{ab^2}{3c^3}\right)^{-4}$

SOLUTION

a) $(x^2 y^{-1})(xy^{-3}) = x^2 y^{-1} x y^{-3}$ Using an associative law

$\qquad\qquad\qquad = x^2 x^1 y^{-1} y^{-3}$ Using a commutative law; $x = x^1$

$\qquad\qquad\qquad = x^{2+1} y^{-1+(-3)}$ Using the product rule: Adding exponents

$\qquad\qquad\qquad = x^3 y^{-4}$, or $\dfrac{x^3}{y^4}$

b) $\dfrac{(3p)^3}{(3p)^{-2}} = (3p)^{3-(-2)}$ Using the quotient rule: Subtracting exponents

$\qquad\qquad = (3p)^5$

$\qquad\qquad = 3^5 p^5$ Raising each factor to the fifth power

$\qquad\qquad = 243p^5$

c) $\left(\dfrac{ab^2}{3c^3}\right)^{-4} = \dfrac{(ab^2)^{-4}}{(3c^3)^{-4}}$ Raising the numerator and the denominator to the -4 power

$\qquad\qquad = \dfrac{a^{-4}(b^2)^{-4}}{3^{-4}(c^3)^{-4}}$ Raising each factor to the -4 power

$\qquad\qquad = \dfrac{a^{-4}b^{-8}}{3^{-4}c^{-12}}$ Multiplying exponents

$\qquad\qquad = \dfrac{3^4 c^{12}}{a^4 b^8},\ \text{or}\ \dfrac{81c^{12}}{a^4 b^8}$ Rewriting without negative exponents

2. Simplify. If negative exponents appear in the answer, write an equivalent answer using only positive exponents.

$(2x^{-3}y^5)^{-2}$

_____

 YOUR TURN

Polynomials

The algebraic expression $4y^2 - 8y^5 + y^3 - 6y + 7$ is an example of a **polynomial**. A polynomial has *terms, coefficients*, and a *degree*.

Definition	Example, for the polynomial $4y^2 - 8y^5 + y^3 - 6y + 7$
The **terms** of a polynomial are separated by addition signs.	Terms: $4y^2, -8y^5, y^3, -6y, 7$
The part of a term that is a constant factor is the **coefficient** of the term.	Coefficients: $4, -8, 1, -6, 7$
The **degree of a term** is the number of variable factors in that term.	Degree of each term: $2, 5, 3, 1, 0$
The **leading term** of a polynomial is the term of highest degree.	Leading term: $-8y^5$
The **degree of a polynomial** is the degree of the leading term.	Degree of polynomial: 5
The coefficient of the leading term is the **leading coefficient** of the polynomial.	Leading coefficient: -8

A polynomial is written in *descending order* when the leading term appears first, followed by the term of next highest degree, and so on.

Polynomials are classified by the number of terms and by degree.

A **monomial** has one term.   *Example:* $-2x^3 y$

A **binomial** has two terms.   *Example:* $1.4x^2 - 10$

A **trinomial** has three terms.   *Example:* $x^2 - 3x - 6$

A **constant** polynomial has degree 0.   *Example:* 7

A **linear** polynomial has degree 1.   *Example:* $3x + 5$

A **quadratic** polynomial has degree 2.   *Example:* $5x^2 - x$

A **cubic** polynomial has degree 3.   *Example:* $x^3 + 2x^2 - \frac{1}{3}$

A **quartic** polynomial has degree 4.   *Example:* $-6x^4 - 2x^2 + 19$

Like, or *similar*, *terms* are either constant terms or terms containing the same variable(s) raised to the same power(s). Polynomials containing like terms can be simplified by *combining* those terms.

EXAMPLE 3 Combine like terms: $4x^2 + 2x - x^2 + x^3$. Write the answer in descending order.

3. Combine like terms:

$-4 + 6x^3 - x^2 + 1 + x^2$.

Write the answer in descending order.

SOLUTION The like terms are $4x^2$ and $-x^2$. Thus we have

$$4x^2 + 2x - x^2 + x^3 = 4x^2 - x^2 + 2x + x^3$$
$$= 3x^2 + 2x + x^3 \quad \text{Combining like terms}$$
$$= x^3 + 3x^2 + 2x. \quad \text{Writing in descending order}$$

YOUR TURN

A polynomial is evaluated by replacing the variable or variables with a number or numbers.

EXAMPLE 4 Evaluate $-a^2 + 2ab + 5b^2$ for $a = -1$ and $b = 3$.

SOLUTION We replace a with -1 and b with 3 and calculate the value using the rules for order of operations:

$$-a^2 + 2ab + 5b^2 = -(-1)^2 + 2 \cdot (-1) \cdot 3 + 5 \cdot 3^2$$
$$= -1 - 6 + 45 = 38.$$

4. Evaluate $2x^3 - x^2 - 3x + 1$ for $x = -1$.

YOUR TURN

Polynomials can be added, subtracted, multiplied, and divided.

Addition and Subtraction of Polynomials

To add polynomials, we write a plus sign between them and combine like terms.

EXAMPLE 5 Add: $(4x^3 + 3x^2 + 2x - 7) + (-5x^2 + x - 10)$.

SOLUTION

$$(4x^3 + 3x^2 + 2x - 7) + (-5x^2 + x - 10)$$
$$= 4x^3 + (3 - 5)x^2 + (2 + 1)x + (-7 - 10)$$
$$= 4x^3 - 2x^2 + 3x - 17$$

5. Add:

$(3x^2 + x - 12) + (x - 6)$.

YOUR TURN

To find the **opposite of a polynomial**, we replace each term with its opposite. This process is called *changing the sign* of each term. Thus the opposite of

$$3y^4 - 7y^2 - \tfrac{1}{3}y + 17$$

is

$$-\left(3y^4 - 7y^2 - \tfrac{1}{3}y + 17\right) = -3y^4 + 7y^2 + \tfrac{1}{3}y - 17.$$

To subtract polynomials, we add the opposite of the polynomial being subtracted.

EXAMPLE 6   Subtract: $(3a^4 - 2a + 7) - (-a^3 + 5a - 1)$.

SOLUTION

$$(3a^4 - 2a + 7) - (-a^3 + 5a - 1)$$

6. Subtract:

$(5y^3 - 6y - 8) - (y^3 - y^2 + 7).$

$$= 3a^4 - 2a + 7 + a^3 - 5a + 1 \qquad \text{Adding the opposite}$$
$$= 3a^4 + a^3 - 7a + 8 \qquad \text{Combining like terms}$$

YOUR TURN

Multiplication of Polynomials

To multiply two monomials, we multiply coefficients and then multiply variables using the product rule for exponents. To multiply a monomial and a polynomial, we multiply each term of the polynomial by the monomial, using the distributive law.

EXAMPLE 7   Multiply: $4x^3(3x^4 - 2x^3 + 7x - 5)$.

SOLUTION

7. Multiply:

$2xy(3x^2 - 4y^2 + 5).$

$$\text{Think:} \quad 4x^3 \cdot 3x^4 - 4x^3 \cdot 2x^3 + 4x^3 \cdot 7x - 4x^3 \cdot 5$$
$$4x^3(3x^4 - 2x^3 + 7x - 5) = 12x^7 \quad - \quad 8x^6 \quad + \quad 28x^4 \quad - \quad 20x^3$$

YOUR TURN

To multiply any two polynomials P and Q, we select one of the polynomials— say, P. We then multiply each term of P by every term of Q and combine like terms.

EXAMPLE 8   Multiply: $(2a^3 + 3a - 1)(a^2 - 4a)$.

SOLUTION   It is often helpful to use columns for a long multiplication. We multiply each term at the top by every term at the bottom, write like terms in columns, and add the results.

$$
\begin{array}{r}
2a^3 + 3a - 1 \\
a^2 - 4a \\
\hline
-8a^4 - 12a^2 + 4a \\
2a^5 + 3a^3 - a^2 \\
\hline
2a^5 - 8a^4 + 3a^3 - 13a^2 + 4a
\end{array}
$$

Multiplying the top row by $-4a$

Multiplying the top row by a^2

Combining like terms. Be sure that like terms are lined up in columns.

8. Multiply:

$(t + 2)(2t^2 - 3t + 4).$

YOUR TURN

Two binomials can be multiplied efficiently using the FOIL method.

> **THE FOIL METHOD**
>
> To multiply two binomials, $A + B$ and $C + D$, multiply the First terms AC, the Outer terms AD, the Inner terms BC, and then the Last terms BD. Then combine like terms, if possible.
>
> $$(A + B)(C + D) = AC + AD + BC + BD$$
>
> **1.** Multiply First terms: AC.
> **2.** Multiply Outer terms: AD.
> **3.** Multiply Inner terms: BC.
> **4.** Multiply Last terms: BD.
>
>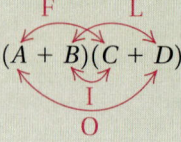
>
> FOIL

EXAMPLE 9   Multiply: $(3x + 4)(x - 2)$.

SOLUTION

$$(3x + 4)(x - 2) = 3x^2 - 6x + 4x - 8$$
$$= 3x^2 - 2x - 8 \quad \text{Combining like terms}$$

9. Multiply: $(3a^2 + 1)(a - 5)$.

YOUR TURN

Special products occur so often that specific formulas or methods for computing them have been developed.

> **SPECIAL PRODUCTS**
>
> The product of a sum and a difference of the same two terms:
> $$(A + B)(A - B) = A^2 - B^2$$
> This is called a *difference of squares*.
> The square of a binomial:
> $$(A + B)^2 = A^2 + 2AB + B^2$$
> $$(A - B)^2 = A^2 - 2AB + B^2$$

EXAMPLE 10   Multiply: **(a)** $(x + 3y)(x - 3y)$; **(b)** $(x^3 + 2)^2$.

SOLUTION

$$(A + B)(A - B) = A^2 - B^2$$

a) $(x + 3y)(x - 3y) = x^2 - (3y)^2 \quad A = x \text{ and } B = 3y$
$$= x^2 - 9y^2$$

$$(A + B)^2 = A^2 + 2 \cdot A \cdot B + B^2$$

b) $(x^3 + 2)^2 = (x^3)^2 + 2 \cdot x^3 \cdot 2 + 2^2 \quad A = x^3 \text{ and } B = 2$
$$= x^6 + 4x^3 + 4$$

10. Multiply: $(y + 1)(y - 1)$.

YOUR TURN

Division of Polynomials

To divide a polynomial by a monomial, we divide each term by the monomial.

EXAMPLE 11   Divide: $(3x^5 + 8x^3 - 12x) \div (4x)$.

SOLUTION   This division can be written

$$\frac{3x^5 + 8x^3 - 12x}{4x} = \frac{3x^5}{4x} + \frac{8x^3}{4x} - \frac{12x}{4x} \quad \text{Dividing each term by } 4x$$

$$= \frac{3}{4}x^{5-1} + \frac{8}{4}x^{3-1} - \frac{12}{4}x^{1-1} \quad \text{Dividing coefficients and subtracting exponents}$$

$$= \frac{3}{4}x^4 + 2x^2 - 3.$$

11. Divide:

$(15w^4 - 35w^2 + 5w) \div (5w)$.

YOUR TURN

To check, we multiply the quotient by $4x$:

$$\left(\tfrac{3}{4}x^4 + 2x^2 - 3\right)4x = 3x^5 + 8x^3 - 12x. \quad \text{The answer checks.}$$

To use long division, we write polynomials in descending order, including terms with 0 coefficients for missing terms. As shown below in Example 12, the procedure ends when the degree of the remainder is less than the degree of the divisor.

EXAMPLE 12   Divide: $(4x^3 - 7x + 1) \div (2x + 1)$.

SOLUTION   The polynomials are already written in descending order, but there is no x^2-term in the dividend. We fill in $0x^2$ for that term.

$$\begin{array}{r} 2x^2 \\ 2x + 1{\overline{\smash{)}4x^3 + 0x^2 - 7x + 1}} \\ \underline{4x^3 + 2x^2} \\ -2x^2 \end{array}$$

Divide the first term of the dividend, $4x^3$, by the first term in the divisor, $2x$: $4x^3/(2x) = 2x^2$.

Multiply $2x^2$ by the divisor, $2x + 1$.

Subtract: $(4x^3 + 0x^2) - (4x^3 + 2x^2) = -2x^2$.

Then we bring down the next term of the dividend, $-7x$.

$$\begin{array}{r} 2x^2 - x \\ 2x + 1{\overline{\smash{)}4x^3 + 0x^2 - 7x + 1}} \\ \underline{4x^3 + 2x^2} \\ -2x^2 - 7x \\ \underline{-2x^2 - x} \\ -6x \end{array}$$

Divide the first term of $-2x^2 - 7x$ by the first term in the divisor: $-2x^2/(2x) = -x$.

The $-7x$ has been "brought down."

Multiply $-x$ by the divisor, $2x + 1$.

Subtract: $(-2x^2 - 7x) - (-2x^2 - x) = -6x$.

Since the degree of the remainder, $-6x$, is *not* less than the degree of the divisor, we must continue dividing.

$$\begin{array}{r} 2x^2 - x - 3 \\ 2x + 1{\overline{\smash{)}4x^3 + 0x^2 - 7x + 1}} \\ \underline{4x^3 + 2x^2} \\ -2x^2 - 7x \\ \underline{-2x^2 - x} \\ -6x + 1 \\ \underline{-6x - 3} \\ 4 \end{array}$$

Divide the first term of $-6x + 1$ by the first term in the divisor: $-6x/(2x) = -3$.

The 1 has been "brought down."

Multiply -3 by $2x + 1$.

Subtract.

The answer is $2x^2 - x - 3$ with R4, or

$$\text{Quotient} \longrightarrow 2x^2 - x - 3 + \frac{4}{2x + 1}. \quad \begin{array}{l} \leftarrow \text{Remainder} \\ \leftarrow \text{Divisor} \end{array}$$

Check: To check, we can multiply by the divisor and add the remainder:

$$(2x + 1)(2x^2 - x - 3) + 4 = 4x^3 - 7x - 3 + 4$$
$$= 4x^3 - 7x + 1.$$

12. Divide:

$(6x^3 - 15x^2 + 2x - 5) \div (2x - 5)$.

YOUR TURN

R.4 EXERCISE SET

Exponents

Simplify.

1. a^0, for $a = -25$

2. y^0, for $y = 6.97$

3. $4^0 - 4^1$

4. $8^1 - 8^0$

Write an equivalent expression using positive exponents. Then, if possible, simplify.

5. 8^{-2}

6. $(-2)^{-3}$

7. $10x^{-5}$

8. $-16y^{-3}$

9. $(ab)^{-2}$

10. ab^{-2}

11. $\dfrac{1}{y^{-10}}$

12. $\dfrac{1}{x^{-t}}$

Simplify.

13. $x^5 \cdot x^{10}$

14. $a^4 \cdot a^{-2}$

15. $\dfrac{a}{a^{-5}}$

16. $\dfrac{p^{-3}}{p^{-8}}$

17. $\dfrac{(4x)^{11}}{(4x)^2}$

18. $\dfrac{a^2 b^9}{a^9 b^2}$

19. $(7^8)^5$

20. $(x^3)^{-7}$

21. $(x^{-2}y^{-3})^{-4}$

22. $(-2a^2)^3$

23. $\left(\dfrac{y^2}{4}\right)^3$

24. $\left(\dfrac{ab^2}{c^3}\right)^4$

25. $\left(\dfrac{2p^3}{3q^4}\right)^{-2}$

26. $\left(\dfrac{2}{x}\right)^{-5}$

Polynomials

Identify the terms of each polynomial.

27. $8x^3 - 6x^2 + x - 7$

28. $-a^2 b + 4a^2 - 8b + 17$

Determine the coefficient and the degree of each term in each polynomial. Then find the degree of each polynomial.

29. $18x^3 + 36x^9 - 7x + 3$

30. $-8y^7 + y + 19$

31. $-x^2 y + 4y^3 - 2xy$

32. $8 - x^2 y^4 + y^7$

Determine the leading term and the leading coefficient of each polynomial.

33. $-p^2 + 5 + 8p^4 - 7p$

34. $13 + 20t - 30t^2 - t^3$

Combine like terms. Write each answer in descending order.

35. $3x^3 - x^2 + x^4 + x^2$

36. $5t - 8t^2 + 4t^2$

37. $3 - 2t^2 + 8t - 3t - 5t^2 + 7$

38. $8x^5 - \frac{1}{3} + \frac{4}{5}x + 1 - \frac{1}{2}x$

Evaluate each polynomial for the given replacements of the variables.

39. $3x^2 - 7x + 10$, for $x = -2$

40. $-y + 3y^2 + 2y^3$, for $y = 3$

41. $a^2 b^3 + 2b^2 - 6a$, for $a = 2$ and $b = -1$

42. $2pq^3 - 5q^2 + 8p$, for $p = -4$ and $q = -2$

Memorizing Words. *Participants in a psychology experiment were able to memorize an average of M words in t minutes, where $M = -0.001t^3 + 0.1t^2$.*

43. Find the average number of words memorized in 10 min.

44. Find the average number of words memorized in 14 min.

Addition and Subtraction of Polynomials

Add or subtract, as indicated.

45. $(3x^3 + 2x^2 + 8x) + (x^3 - 5x^2 + 7)$

46. $(-6x^4 + 3x^2 - 16) + (4x^2 + 4x - 7)$

47. $(8y^2 - 2y - 3) - (9y^2 - 7y - 1)$

48. $(4t^2 + 6t - 7) - (t + 5)$

49. $(-x^2 y + 2y^2 + y) - (3y^2 + 2x^2 y - 7y)$

50. $(ab + x^2 y^2) + (2ab - x^2 y^2)$

Multiplication of Polynomials

Multiply.

51. $4x^2(3x^3 - 7x + 7)$

52. $a^2 b(a^3 + b^2 - ab - 2b)$

53. $(2a + y)(4a + b)$

54. $(x + 7y)(y - 3x)$

55. $(x + 7)(x^2 - 3x + 1)$

56. $(2x - 3)(x^2 - x - 1)$

57. $(x + 7)(x - 7)$

58. $(2x + 1)^2$

59. $(x + y)^2$

60. $(xy + 1)(xy - 1)$

61. $(2x^2 + 7)(3x^2 - 2)$

62. $(1.1x^2 + 5)(0.1x^2 - 2)$

63. $(6a - 5y)(7a + 3y)$

64. $(3p^2 - q^3)^2$

Division of Polynomials

Divide and check.

65. $(3t^5 + 9t^3 - 6t^2 + 15t) \div (-3t)$

66. $(4x^5 + 10x^4 - 16x^2) \div (4x^2)$

67. $(15x^2 - 16x - 15) \div (3x - 5)$

68. $(x^3 - 2x^2 - 14x + 1) \div (x - 5)$

69. $(2x^3 - x^2 + 1) \div (x + 1)$

70. $(2x^3 + 3x^2 - 50) \div (2x - 5)$

71. $(5x^3 + 3x^2 - 5x) \div (x^2 - 1)$

72. $(2x^3 + 3x^2 + 6x + 10) \div (x^2 + 3)$

REFERENCES

Addition of polynomials, Section 4.4
Coefficient, p. 247
Degree of a polynomial, p. 248
Degree of a term, p. 247
Division of polynomials, Section 4.6
Evaluating a polynomial, p. 249
The exponent zero, p. 232
FOIL, p. 271
Leading coefficient, p. 248
Leading term, p. 248
Multiplication of polynomials, Section 4.5
Multiplying sums and differences of
    two terms, p. 272
Negative exponents, p. 238
Opposite of a polynomial, p. 256
Polynomials, Section 4.3
The power rule, p. 233
The product rule, p. 230
The quotient rule, p. 231
Raising a product to a power, p. 234
Raising a quotient to a power, p. 234
Squaring binomials, p. 273
Subtraction of polynomials, Section 4.4
Term, p. 246

R.5 Polynomials and Factoring

Common Factors and Factoring by Grouping • Factoring Trinomials • Factoring Special Forms •
Solving Polynomial Equations by Factoring

Common Factors and Factoring by Grouping

To *factor* a polynomial is to find an equivalent expression that is a product. For example, three factorizations of $50x^6$ are $5 \cdot 10x^6$, $5x^3 \cdot 10x^3$, and $2x \cdot 25x^5$.

If all terms in a polynomial share a common factor, that factor can be "factored out." When factoring a polynomial with two or more terms, we use the largest common factor of the terms, if one exists.

EXAMPLE 1   Factor: $3x^6 + 15x^4 - 9x^3$.

SOLUTION   The largest factor common to 3, 15, and -9 is 3. The largest power of x common to x^6, x^4, and x^3 is x^3. Thus the largest common factor of the terms of the polynomial is $3x^3$. We factor as follows:

$$3x^6 + 15x^4 - 9x^3 = 3x^3 \cdot x^3 + 3x^3 \cdot 5x - 3x^3 \cdot 3 \qquad \text{Factoring each term}$$

$$= 3x^3(x^3 + 5x - 3). \qquad \text{Factoring out } 3x^3$$

Factorizations can always be checked by multiplying:

1. Factor:

$$2a^4d^3 - 15ad^2 - 8a^2d^4.$$

$$3x^3(x^3 + 5x - 3) = 3x^6 + 15x^4 - 9x^3.$$

YOUR TURN

A polynomial with two or more terms can itself be a common factor. For example, $3x^2(x - 2) + 5(x - 2) = (x - 2)(3x^2 + 5)$.

If a polynomial with four terms can be split into two groups of terms, and both groups share a common binomial factor, the polynomial can be factored. This method is known as **factoring by grouping**.

EXAMPLE 2   Factor by grouping: $2x^3 + 6x^2 - x - 3$.

SOLUTION   First, we consider the polynomial as two groups of terms, $2x^3 + 6x^2$ and $-x - 3$. Then we factor each group separately:

$$2x^3 + 6x^2 - x - 3 = 2x^2(x + 3) - 1(x + 3) \qquad \text{Factoring out } 2x^2 \text{ and } -1 \text{ to find the common binomial factor, } x + 3$$

2. Factor by grouping:

$$3x^3 + 12x^2 - x - 4.$$

$$= (x + 3)(2x^2 - 1).$$

The check is left to the student.

YOUR TURN

Not every polynomial with four terms is factorable by grouping. Any polynomial that is not factorable is said to be **prime**.

Student Notes

When factoring $x^2 + bx + c$, we look for a pair of factors that have c as their product and b as their sum. The signs of b and c help us determine the signs of the factors.

Sign of c	Sign of b	Signs of factors
$+$	$+$	$+, +$
$+$	$-$	$-, -$
$-$	$+$	$+, -$
$-$	$-$	$+, -$

Factoring Trinomials

Many trinomials that have no common factor can be written as the product of two binomials. We look first at trinomials of the form $x^2 + bx + c$, for which the leading coefficient is 1.

Factoring trinomials involves a trial-and-error process. In order for the product of two binomials to be $x^2 + bx + c$, the binomials must look like

$$(x + p)(x + q),$$

where p and q are constants that must be determined. We look for two numbers whose product is c and whose sum is b.

EXAMPLE 3   Factor.

a) $x^2 - 2x - 24$

b) $3t^2 - 33st + 84s^2$

SOLUTION

a) The factorization is of the form

$$(x +)(x +).$$

To find the constant terms, we need a pair of factors whose product is -24 and whose sum is -2. Since -24 is negative, one factor of -24 will be negative and one will be positive. Since b is also negative, the negative factor must have the larger absolute value.

Pairs of Factors of −24	Sums of Factors
1, −24	−23
2, −12	−10
3, −8	−5
4, −6	−2 ←

The numbers we need are 4 and −6.

The factorization is $(x + 4)(x - 6)$.

Check: $(x + 4)(x - 6) = x^2 - 6x + 4x - 24 = x^2 - 2x - 24.$

b) Always look first for a common factor. There is a common factor, 3, which we factor out first:

$$3t^2 - 33st + 84s^2 = 3(t^2 - 11st + 28s^2).$$

Now we consider $t^2 - 11st + 28s^2$. Think of $28s^2$ as the "constant" term c and $-11s$ as the "coefficient" b of the middle term. We try to express $28s^2$ as the product of two factors whose sum is $-11s$. These factors are $-4s$ and $-7s$. Thus the factorization of $t^2 - 11st + 28s^2$ is

$(t - 4s)(t - 7s).$ This is not the entire factorization of $3t^2 - 33st + 84s^2$.

Finally, we include the common factor, 3, and write

$$3t^2 - 33st + 84s^2 = 3(t - 4s)(t - 7s).$$ This is the factorization.

Check: $3(t - 4s)(t - 7s) = 3(t^2 - 11st + 28s^2) = 3t^2 - 33st + 84s^2.$

3. Factor: $n^2 + 2n - 24.$

YOUR TURN

When the leading coefficient of a trinomial is not 1, the number of trials needed to find a factorization can increase dramatically. We will consider two methods for factoring trinomials of the type $ax^2 + bx + c$: factoring with FOIL and the grouping method.

TO FACTOR $ax^2 + bx + c$ USING FOIL

1. Make certain that all common factors have been removed. If any remain, factor out the largest common factor.

2. Find two First terms whose product is ax^2:

$(\ x + \ )(\ x + \ ) = ax^2 + bx + c.$
FOIL

3. Find two Last terms whose product is c:

$(\ x + \ )(\ x + \ ) = ax^2 + bx + c.$
FOIL

4. Check by multiplying to see if the sum of the Outer and Inner products is bx. If necessary, repeat steps (2) and (3) until the correct combination is found.

$(\ x + \ )(\ x + \ ) = ax^2 + bx + c.$
I
O   FOIL

If no correct combination exists, state that the polynomial is prime.

EXAMPLE 4   Factor: $20x^3 - 22x^2 - 12x$.

SOLUTION

1. First, we factor out the largest common factor, $2x$:

$$20x^3 - 22x^2 - 12x = 2x(10x^2 - 11x - 6).$$

2. Next, in order to factor the trinomial $10x^2 - 11x - 6$, we search for two terms whose product is $10x^2$. The possibilities are

$$(x + \quad)(10x + \quad) \quad \text{or} \quad (2x + \quad)(5x + \quad).$$

3. There are four pairs of factors of -6. The first terms of the binomials differ, so the order of the factors is important. There are eight possibilities for the last terms:

$$\begin{array}{lll}
1, -6 & & -6, \ 1 \\
-1, \ 6 & & 6, -1 \\
2, -3 & \text{and} & -3, \ 2 \\
-2, \ 3 & & 3, -2.
\end{array}$$

4. Since each of the eight possibilities from step (3) could be used in either of the two possibilities from step (2), there are $2 \cdot 8$, or 16, possible factorizations. We check the possibilities systematically until we find one that gives the correct factorization. Let's first try factors with $(2x + \quad)(5x + \quad)$.

Pair of Factors	*Corresponding Trial*	*Product*	
1, −6	$(2x + 1)(5x - 6)$	$10x^2 - 7x - 6$	← Wrong middle term
−1, 6	$(2x - 1)(5x + 6)$	$10x^2 + 7x - 6$	← Wrong middle term. Changing the signs in the binomials changed only the sign of the middle term in the product.
2, −3	$(2x + 2)(5x - 3)$	$10x^2 + 4x - 6$	← Wrong middle term
−6, 1	$(2x - 6)(5x + 1)$	$10x^2 - 28x - 6$	← Wrong middle term
−3, 2	$(2x - 3)(5x + 2)$	$10x^2 - 11x - 6$	← Correct middle term

We can stop when we find a correct factorization. Including the common factor $2x$, we now have

$$20x^3 - 22x^2 - 12x = 2x(2x - 3)(5x + 2).$$

This can be checked by multiplying.

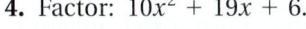 YOUR TURN

With practice, some of the trials can be skipped or performed mentally.

The second method of factoring trinomials of the type $ax^2 + bx + c$ involves factoring by grouping.

Student Notes

Both the third and the fourth trial factorizations in Example 4 contain a binomial with a common factor: $2x + 2 = 2(x + 1)$ and $2x - 6 = 2(x - 3)$. If all common factors are factored out as a first step, you can immediately eliminate any trial factorization containing a binomial with a common factor.

4. Factor: $10x^2 + 19x + 6$.

> ### TO FACTOR $ax^2 + bx + c$ USING THE GROUPING METHOD
> 1. Factor out the largest common factor, if one exists.
> 2. Multiply the leading coefficient a and the constant c.
> 3. Find a pair of factors of ac whose sum is b.
> 4. Rewrite the middle term, bx, as a sum or a difference using the factors found in step (3).
> 5. Factor by grouping.
> 6. Include any common factor from step (1) and check by multiplying.

EXAMPLE 5   Factor: $7x^2 + 31x + 12$.

SOLUTION

1. There is no common factor (other than 1 or -1).
2. We multiply the leading coefficient, 7, and the constant, 12: $7 \cdot 12 = 84$.
3. We look for a pair of factors of 84 whose sum is 31. Since both 84 and 31 are positive, we need consider only positive factors.

Pairs of Factors of 84	Sums of Factors
1, 84	85
2, 42	44
3, 28	31 ← ——— $3 + 28 = 31$

4. Next, we rewrite $31x$ using the factors 3 and 28: $31x = 3x + 28x$.
5. We now factor by grouping:

$$7x^2 + 31x + 12 = 7x^2 + 3x + 28x + 12 \qquad \text{Substituting } 3x + 28x \text{ for } 31x$$
$$= x(7x + 3) + 4(7x + 3)$$
$$= (7x + 3)(x + 4). \qquad \text{Factoring out the common factor, } 7x + 3$$

5. Factor:

$$8n^3 - 34n^2 + 15n.$$

6. *Check:* $(7x + 3)(x + 4) = 7x^2 + 31x + 12$.

 YOUR TURN

Factoring Special Forms

We can factor certain types of polynomials directly, without using trial and error.

> ### FACTORING FORMULAS
> | | |
> |---|---|
> | Perfect-square trinomial: | $A^2 + 2AB + B^2 = (A + B)^2$, |
> | | $A^2 - 2AB + B^2 = (A - B)^2$ |
> | Difference of squares: | $A^2 - B^2 = (A + B)(A - B)$ |
> | Sum of cubes: | $A^3 + B^3 = (A + B)(A^2 - AB + B^2)$ |
> | Difference of cubes: | $A^3 - B^3 = (A - B)(A^2 + AB + B^2)$ |

Before using the factoring formulas, it is important to check carefully that the expression being factored is indeed in one of the forms listed. Note that there is no factoring formula for the sum of two squares.

EXAMPLE 6   Factor.

a) $2x^2 - 2$

b) $x^2y^2 + 20xy + 100$

c) $p^3 - 64$

d) $3y^2 + 27$

SOLUTION

a) We first factor out a common factor, 2:

$$2x^2 - 2 = 2(x^2 - 1).$$

Looking at $x^2 - 1$, we see that it is a difference of squares, $A^2 - B^2$, with $A = x$ and $B = 1$. The factorization is thus

$$2x^2 - 2 = 2(x^2 - 1) = 2(x + 1)(x - 1).$$

$$A^2 - B^2 \qquad (A + B)(A - B)$$

b) First, we check for a common factor; there is none. The polynomial is a perfect-square trinomial, $A^2 + 2AB + B^2$, with $A = xy$ and $B = 10$. The factorization is thus

$$x^2y^2 + 20xy + 100 = (xy)^2 + 2 \cdot xy \cdot 10 + 10^2 = (xy + 10)^2.$$

$$A^2 + 2 \cdot A \cdot B + B^2 = (A + B)^2$$

c) This is a difference of cubes, $A^3 - B^3$, with $A = p$ and $B = 4$:

$$p^3 - 64 = (p)^3 - (4)^3$$
$$= (p - 4)(p^2 + 4p + 16).$$

d) We factor out the common factor, 3:

$$3y^2 + 27 = 3(y^2 + 9).$$

Since $y^2 + 9$ is a sum of squares, no further factorization is possible.

6. Factor: $4y^2 - 20y + 25$.

 YOUR TURN

A polynomial is *factored completely* when no factor can be factored further.

EXAMPLE 7   Factor completely: $x^4 - 1$.

SOLUTION

7. Factor completely:

$x^3 + 2x^2 - x - 2$.

$$x^4 - 1 = (x^2 + 1)(x^2 - 1) \qquad \text{Factoring a difference of squares}$$
$$= (x^2 + 1)(x + 1)(x - 1) \qquad \text{The factor } x^2 - 1 \text{ is itself a difference of squares.}$$

YOUR TURN

Solving Polynomial Equations by Factoring

A **polynomial equation** is formed by setting two polynomials equal to each other. A **quadratic equation** is a polynomial equation equivalent to one of the form $ax^2 + bx + c = 0$, where $a \neq 0$. Polynomial equations that can be factored can be solved using the principle of zero products.

THE PRINCIPLE OF ZERO PRODUCTS

An equation $ab = 0$ is true if and only if $a = 0$ or $b = 0$, or both. (A product is 0 if and only if at least one factor is 0.)

EXAMPLE 8   Solve: **(a)** $x^2 - 11x = 12$;   **(b)** $5x^2 + 10x + 5 = 0$.

SOLUTION

a) We must have 0 on one side of the equation before using the principle of zero products:

Get 0 on one side.

$$x^2 - 11x = 12$$

Factor.

$$x^2 - 11x - 12 = 0 \qquad \text{Subtracting 12 from both sides}$$

Use the principle of zero products.

$$(x - 12)(x + 1) = 0 \qquad \text{Factoring}$$

$$x - 12 = 0 \quad or \quad x + 1 = 0 \qquad \text{Using the principle of zero products}$$

Solve.

$$x = 12 \quad or \qquad x = -1.$$

The solutions are 12 and -1. To check, we substitute 12 and then -1 for x in the original equation and confirm that each makes the equation true.

b) We have

$$5x^2 + 10x + 5 = 0$$

$$5(x^2 + 2x + 1) = 0 \qquad \text{Factoring out a common factor}$$

$$5(x + 1)(x + 1) = 0 \qquad \text{Factoring completely}$$

$$x + 1 = 0 \quad or \quad x + 1 = 0 \qquad \text{Using the principle of zero products}$$

$$x = -1 \quad or \qquad x = -1.$$

There is only one solution, -1. The check is left to the student.

YOUR TURN

Student Notes

In the factorization

$$5(x + 1)(x + 1),$$

5 is a factor, but it does not contain a variable and thus does not yield a solution of $5(x + 1)(x + 1) = 0$.

---

8. Solve: $9x^2 = 1$.

Quadratic equations can be used to solve problems. One important result that uses squared quantities is the Pythagorean theorem.

THE PYTHAGOREAN THEOREM

The sum of the squares of the legs of a right triangle is equal to the square of the hypotenuse:

$$a^2 + b^2 = c^2.$$

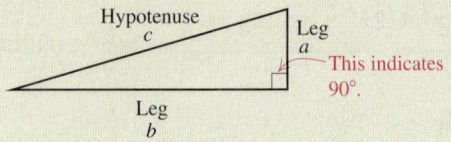

Hypotenuse c

Leg a

This indicates 90°.

Leg b

EXAMPLE 9 *Hiking.* Cheri hiked 500 ft up a steep incline. Her global positioning unit (GPS) indicated that her horizontal position had changed by 100 ft more than her vertical position had changed. What was the change in altitude?

SOLUTION

1. **Familiarize.** We first make a drawing and let x = the change in altitude, in feet. We know then that the horizontal change is $x + 100$, since Cheri's horizontal position has changed by 100 ft more than her vertical position. The hypotenuse has length 500 ft.

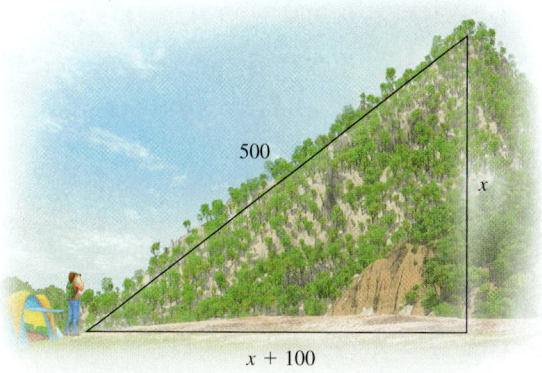

500

x

$x + 100$

2. **Translate.** Applying the Pythagorean theorem gives us

$$a^2 + b^2 = c^2$$
$$x^2 + (x + 100)^2 = 500^2.$$

3. **Carry out.** We solve the equation:

$$x^2 + (x + 100)^2 = 500^2$$

$x^2 + x^2 + 200x + 10{,}000 = 250{,}000$ Squaring $x + 100$; squaring 500

$2x^2 + 200x + 10{,}000 = 250{,}000$ Combining like terms

$2x^2 + 200x - 240{,}000 = 0$ Getting 0 on one side

$2(x^2 + 100x - 120{,}000) = 0$ Factoring out a common factor

$2(x + 400)(x - 300) = 0$ Factoring a trinomial

$x + 400 = 0$ or $x - 300 = 0$ Using the principle of zero products

$x = -400$ or $x = 300$.

4. **Check.** We know that the change in altitude is positive, so -400 cannot be a solution. When $x = 300$, the horizontal change is 400. Since $300^2 + 400^2 = 500^2$, the solution checks.

5. **State.** Cheri's change in altitude was 300 ft.

9. The length of a hypotenuse of a right triangle is 13 ft. One leg of the triangle is 7 ft longer than the other leg. Find the lengths of the legs.

 YOUR TURN

R.5 EXERCISE SET

FOR EXTRA HELP

MyMathLab® Math XL

PRACTICE   WATCH   READ   REVIEW

Factoring Polynomials

Factor completely. If a polynomial is prime, state this.

1. $18t^5 - 12t^4 + 6t^3$ **2.** $x^2y^4 - 2xy^5 + 3x^3y^6$

3. $y^2 - 6y + 9$ **4.** $4z^2 - 25$

5. $2p^3(p + 2) + (p + 2)$ **6.** $6y^2 + y - 1$

7. $x^2 + 100$ **8.** $y^3 - 1$

9. $8t^3 + 27$ **10.** $a^2b^2 + 24ab + 144$

11. $m^2 + 13m + 42$ **12.** $2x^3 - 6x^2 + x - 3$

13. $x^4 - 81$ **14.** $x^2 + x + 3$

15. $8x^2 + 22x + 15$ **16.** $4x^2 - 40x + 100$

17. $x^3 + 2x^2 - x - 2$

18. $(x + 2y)(x - 1) + (x + 2y)(x - 2)$

19. $0.001t^6 - 0.008$

20. $x^2 - 20 - x$

21. $-\frac{1}{16} + x^4$

22. $5x^8 - 5z^{16}$

23. $mn - 2m + 3n - 6$

24. $t^6 - p^6$

25. $5mn + m^2 - 150n^2$

26. $\frac{1}{27} + x^3$

27. $24x^2y - 6y - 10xy$ **28.** $-3y^2 - 12y - 12$

29. $y^2 + 121 - 22y$ **30.** $t^3 - 2t^2 - 5t + 10$

Solving Polynomial Equations by Factoring

Solve.

31. $(x - 1)(x + 3) = 0$

32. $(3x - 5)(7 - 4x) = 0$

33. $8x(11 - x) = 0$

34. $(x - 3)(x + 1)(2x - 9) = 0$

35. $x^2 = 9$ **36.** $8x^2 = 2x$

37. $4x^2 - 18x = 70$ **38.** $x^2 + 6x + 9 = 0$

39. $2x^2 - 10x = 0$ **40.** $100x^2 = 9$

41. $(a + 1)(a - 5) = 7$ **42.** $d(d - 3) = 40$

43. $x^2 + 6x - 55 = 0$ **44.** $x^2 + 7x - 60 = 0$

45. $\frac{1}{2}x^2 + 5x + \frac{25}{2} = 0$ **46.** $3 + 10x^2 = 11x$

47. *Design.* The base of a triangular window is 3 ft longer than its height. The area of the window is 20 ft². What are the dimensions of the window?

48. *Page Numbers.* The product of the page numbers on two facing pages of a book is 156. Find the page numbers.

49. *Swing Sets.* The length of a slide on a swing set is 10 ft. The distance from the base of the ladder to the base of the slide is 2 ft more than the height of the ladder. Find the height of the ladder.

10 ft

h

$h + 2$

50. *Right Triangles.* The hypotenuse of a right triangle is 17 ft. One leg is 1 ft shorter than twice the length of the other leg. Find the lengths of the legs.

YOUR TURN ANSWERS: SECTION R.5

1. $ad^2(2a^3d - 15 - 8ad^2)$ **2.** $(x + 4)(3x^2 - 1)$
3. $(n + 6)(n - 4)$ **4.** $(5x + 2)(2x + 3)$
5. $n(2n - 1)(4n - 15)$ **6.** $(2y - 5)^2$
7. $(x + 2)(x + 1)(x - 1)$ **8.** $-\frac{1}{3}, \frac{1}{3}$ **9.** 5 ft, 12 ft

---

REFERENCES

Common factor, p. 304
Difference of squares, p. 330
Factor by grouping, p. 307
Factoring completely, p. 316
Factoring trinomials of the type
   $ax^2 + bx + c$, Section 5.3
Factoring trinomials of the type
   $x^2 + bx + c$, Section 5.2
Factoring with FOIL, p. 319
The grouping method, p. 324
Perfect-square trinomials, p. 328
Prime polynomial, p. 316
The principle of zero products, p. 346
The Pythagorean theorem, p. 357
Quadratic equation, p. 345
Sum or difference of cubes, Section 5.5

R.6 Rational Expressions and Equations

Rational Expressions ▪ Multiplication and Division of Rational Expressions ▪
Addition and Subtraction of Rational Expressions ▪ Complex Rational Expressions ▪ Solving Rational Equations

Rational Expressions

A **rational expression** is a quotient of two polynomials. Because division by 0 is undefined, a rational expression is undefined for any number that will make its denominator 0.

EXAMPLE 1 Find all numbers for which the rational expression

$$\frac{2x + 5}{x^2 - 9x - 10}$$

is undefined.

SOLUTION We set the denominator equal to 0 and solve:

$$x^2 - 9x - 10 = 0$$
$$(x - 10)(x + 1) = 0 \qquad \text{Factoring}$$
$$x - 10 = 0 \quad or \quad x + 1 = 0 \qquad \text{Using the principle of zero products}$$
$$x = 10 \quad or \qquad x = -1.$$

If x is replaced with 10 or with -1, the denominator is 0. Thus,

$$\frac{2x + 5}{x^2 - 9x - 10} \quad \text{is undefined for } x = 10 \text{ and } x = -1.$$

YOUR TURN

1. Find all numbers for which the rational expression

$$\frac{x - 7}{2x^2 + 4x}$$

is undefined.

Multiplication and Division of Rational Expressions

Multiplication and division of rational expressions is similar to multiplication and division with fractions.

> **THE PRODUCT AND THE QUOTIENT OF TWO RATIONAL EXPRESSIONS**
>
> To multiply two rational expressions, multiply numerators and multiply denominators:
>
> $$\frac{A}{B} \cdot \frac{C}{D} = \frac{AC}{BD}.$$
>
> To divide by a rational expression, multiply by its reciprocal:
>
> $$\frac{A}{B} \div \frac{C}{D} = \frac{A}{B} \cdot \frac{D}{C} = \frac{AD}{BC}.$$

We simplify rational expressions by removing a factor equal to 1.

EXAMPLE 2   Simplify: $\dfrac{9x^2 + 12x}{6x^2 - 3x}$.

SOLUTION

$$\frac{9x^2 + 12x}{6x^2 - 3x} = \frac{3x(3x + 4)}{3x(2x - 1)}$$ Factoring the numerator and the denominator

$$\frac{3x(3x + 4)}{3x(2x - 1)} = \frac{3x}{3x} \cdot \frac{3x + 4}{2x - 1}$$ Rewriting as a product of two rational expressions

$$= 1 \cdot \frac{3x + 4}{2x - 1}$$ $\dfrac{3x}{3x} = 1$

$$= \frac{3x + 4}{2x - 1}$$ Removing a factor equal to 1

2. Simplify: $\dfrac{x^2 - x - 6}{2x^2 - 3x - 9}$.

YOUR TURN

After multiplying or dividing rational expressions, we simplify, if possible.

EXAMPLE 3   Perform each indicated operation and, if possible, simplify.

a) $\dfrac{x^2 - x - 6}{3x} \cdot \dfrac{12x^3}{x + 2}$

b) $\dfrac{x^2 - 1}{x + 5} \div \dfrac{x^2 + 2x + 1}{2x + 10}$

SOLUTION

a) $\dfrac{x^2 - x - 6}{3x} \cdot \dfrac{12x^3}{x + 2} = \dfrac{(x^2 - x - 6)(12x^3)}{3x(x + 2)}$ Multiplying the numerators and the denominators

$$= \frac{(x - 3)(x + 2)(3x)(4x^2)}{3x(x + 2)}$$ Factoring the numerator. Try to go directly to this step.

$$= \frac{(x - 3)(\cancel{x + 2})(\cancel{3x})(4x^2)}{(\cancel{3x})(\cancel{x + 2})}$$ Removing a factor equal to 1: $\dfrac{(x + 2)(3x)}{(x + 2)(3x)} = 1$

$$= 4x^2(x - 3)$$

b) $\dfrac{x^2 - 1}{x + 5} \div \dfrac{x^2 + 2x + 1}{2x + 10} = \dfrac{x^2 - 1}{x + 5} \cdot \dfrac{2x + 10}{x^2 + 2x + 1}$ Multiplying by the reciprocal of the divisor

$$= \frac{(x + 1)(x - 1)(2)(x + 5)}{(x + 5)(x + 1)(x + 1)}$$ Multiplying rational expressions and factoring

$$= \frac{(\cancel{x + 1})(x - 1)(2)(\cancel{x + 5})}{(\cancel{x + 5})(\cancel{x + 1})(x + 1)}$$ Removing a factor equal to 1: $\dfrac{(x + 1)(x + 5)}{(x + 1)(x + 5)} = 1$

3. Divide and, if possible, simplify:

$$\frac{a - b}{b^2} \div \frac{a^2 - b^2}{b^2 - b}.$$

$$= \frac{2(x - 1)}{x + 1}$$ We leave the numerator in factored form.

YOUR TURN

Addition and Subtraction of Rational Expressions

Like multiplication and division, addition and subtraction of rational expressions is similar to addition and subtraction of fractions.

THE SUM AND THE DIFFERENCE OF TWO RATIONAL EXPRESSIONS

To add when denominators are the same, add the numerators and keep the same denominator:

$$\frac{A}{B} + \frac{C}{B} = \frac{A + C}{B}.$$

To subtract when denominators are the same, subtract the second numerator from the first and keep the same denominator:

$$\frac{A}{B} - \frac{C}{B} = \frac{A - C}{B}.$$

EXAMPLE 4 Add and, if possible, simplify:

$$\frac{x - 6}{x^2 - 6x + 5} + \frac{5}{x^2 - 6x + 5}.$$

SOLUTION

$$\frac{x - 6}{x^2 - 6x + 5} + \frac{5}{x^2 - 6x + 5} = \frac{x - 6 + 5}{x^2 - 6x + 5} \qquad \text{Adding numerators}$$

$$= \frac{x - 1}{(x - 5)(x - 1)} \qquad \text{Factoring the denominator}$$

$$= \frac{1(x - 1)}{(x - 5)(x - 1)} \qquad \begin{array}{l}\text{Removing a} \\ \text{factor equal to 1:} \\ \frac{x - 1}{x - 1} = 1\end{array}$$

$$= \frac{1}{x - 5}$$

4. Subtract and, if possible, simplify:

$$\frac{t + 5}{t^2 - 9} - \frac{t}{t^2 - 9}.$$

 YOUR TURN

When two rational expressions have different denominators, we must rewrite them in an equivalent form with a common denominator. We generally rewrite them using their **least common denominator (LCD)**, which is the **least common multiple (LCM)** of the denominators.

TO FIND THE LEAST COMMON DENOMINATOR (LCD)

1. Write the prime factorization of each denominator.
2. Select one factorization and inspect it to see if it completely contains the other.

    a) If it does, it represents the LCM of the denominators.
    b) If it does not, multiply that factorization by any factors of the other denominator that it lacks. The final product is the LCM of the denominators.

EXAMPLE 5 Subtract: $\dfrac{x+2}{x^2-1} - \dfrac{x}{x^2+4x+3}$.

SOLUTION To find the LCD, we factor each denominator and use these factorizations to construct the LCM:

$$x^2 - 1 = (x+1)(x-1);$$
$$x^2 + 4x + 3 = (x+1)(x+3).$$

We select the factorization of $x^2 - 1$. Since it does not contain the factor $(x+3)$, we multiply $(x+1)(x-1)$ by $(x+3)$:

Find the LCD.

$$\text{LCM} = (x+1)(x-1)(x+3).$$

We multiply each rational expression by a form of 1 that is made up of the factors of the LCM that it is missing:

Rewrite each expression with the LCD.

$$\frac{x+2}{x^2-1} - \frac{x}{x^2+4x+3} = \frac{x+2}{(x+1)(x-1)} \cdot \frac{x+3}{x+3} - \frac{x}{(x+1)(x+3)} \cdot \frac{x-1}{x-1}$$

$$= \frac{x^2+5x+6}{(x+1)(x-1)(x+3)} - \frac{x^2-x}{(x+1)(x-1)(x+3)}$$

Subtract numerators. Keep the denominator.

$$= \frac{x^2+5x+6-(x^2-x)}{(x+1)(x-1)(x+3)} \qquad \text{Parentheses are important.}$$

$$= \frac{x^2+5x+6-x^2+x}{(x+1)(x-1)(x+3)} \qquad \text{Removing parentheses by subtracting each term}$$

$$= \frac{6x+6}{(x+1)(x-1)(x+3)}$$

Simplify if possible.

$$= \frac{6\cancel{(x+1)}}{\cancel{(x+1)}(x-1)(x+3)} \qquad \begin{array}{l}\text{Factoring and removing}\\ \text{a factor equal to 1:}\\ (x+1)/(x+1) = 1\end{array}$$

$$= \frac{6}{(x-1)(x+3)}.$$

5. Add and, if possible, simplify:

$$\frac{n+7}{n^2-n-2} + \frac{1}{n^2+n-6}.$$

 YOUR TURN

When denominators are opposites, we can find a common denominator by multiplying either rational expression by $-1/-1$.

EXAMPLE 6 Add: $\dfrac{a}{a-b} + \dfrac{5}{b-a}$.

SOLUTION

$$\frac{a}{a-b} + \frac{5}{b-a} = \frac{a}{a-b} + \frac{5}{b-a} \cdot \frac{-1}{-1} \qquad \begin{array}{l}\text{Writing 1 as } -1/-1 \text{ and}\\ \text{multiplying to obtain a}\\ \text{common denominator}\end{array}$$

$$= \frac{a}{a-b} + \frac{-5}{a-b} \qquad \begin{array}{l}(b-a)(-1) = -b+a\\ \qquad\qquad\quad = a-b\end{array}$$

$$= \frac{a-5}{a-b}$$

6. Add and, if possible, simplify:

$$\frac{2x+3}{x-1} + \frac{x+4}{1-x}.$$

 YOUR TURN

Complex Rational Expressions

A **complex rational expression** is a rational expression that has one or more rational expressions within its numerator or denominator. We examine two methods for simplifying complex rational expressions. The first involves writing the expression as a quotient of two rational expressions.

TO SIMPLIFY A COMPLEX RATIONAL EXPRESSION BY DIVIDING

1. Add or subtract, as needed, to get a single rational expression in the numerator.
2. Add or subtract, as needed, to get a single rational expression in the denominator.
3. Divide the numerator by the denominator.
4. If possible, simplify by removing a factor equal to 1.

EXAMPLE 7 Simplify by dividing: $\dfrac{\dfrac{2}{x+1}}{\dfrac{1}{x+2}+\dfrac{1}{x}}$.

SOLUTION

1. There is already a single rational expression in the numerator.

2. We add to get a single rational expression in the denominator:

$$\frac{\dfrac{2}{x+1}}{\dfrac{1}{x+2}+\dfrac{1}{x}}=\frac{\dfrac{2}{x+1}}{\dfrac{1}{x+2}\cdot\dfrac{x}{x}+\dfrac{1}{x}\cdot\dfrac{x+2}{x+2}}$$

Multiplying by 1 to get the LCD, $x(x+2)$, for the denominator

$$=\frac{\dfrac{2}{x+1}}{\dfrac{x}{x(x+2)}+\dfrac{x+2}{x(x+2)}}=\frac{\dfrac{2}{x+1}}{\dfrac{2x+2}{x(x+2)}}.$$

Adding in the denominator

3. Next, we divide the numerator by the denominator:

$$\frac{\dfrac{2}{x+1}}{\dfrac{2x+2}{x(x+2)}}=\frac{2}{x+1}\div\frac{2x+2}{x(x+2)}=\frac{2}{x+1}\cdot\frac{x(x+2)}{2x+2}.$$

4. Simplifying, we have

$$\frac{2}{(x+1)}\cdot\frac{x(x+2)}{2x+2}=\frac{2\cdot x(x+2)}{2(x+1)(x+1)}$$

Factoring in the denominator and removing a factor equal to 1: $\frac{2}{2}=1$

$$=\frac{x(x+2)}{(x+1)^2}.$$

7. Simplify by dividing: $\dfrac{\dfrac{2x}{x-1}}{\dfrac{1}{x}-3}$.

YOUR TURN

A second method involves multiplying by the LCD.

TO SIMPLIFY A COMPLEX RATIONAL EXPRESSION BY MULTIPLYING BY THE LCD

1. Find the LCD of *all* rational expressions within the complex rational expression.
2. Multiply the complex rational expression by a factor equal to 1. Write 1 as the LCD over itself (LCD/LCD).
3. Simplify. No rational expressions should remain within the complex rational expression.
4. Factor and, if possible, simplify.

EXAMPLE 8   Simplify by multiplying by the LCD: $\dfrac{1 + \dfrac{2}{t}}{\dfrac{4}{t^2} - 1}$.

SOLUTION

1. The denominators *within* the complex rational expression are t and t^2, so the LCD is t^2.

2. We multiply by a form of 1 using t^2/t^2:

$$\frac{1 + \dfrac{2}{t}}{\dfrac{4}{t^2} - 1} = \frac{1 + \dfrac{2}{t}}{\dfrac{4}{t^2} - 1} \cdot \frac{t^2}{t^2}.$$

3. We distribute and simplify:

$$\frac{1 + \dfrac{2}{t}}{\dfrac{4}{t^2} - 1} \cdot \frac{t^2}{t^2} = \frac{1 \cdot t^2 + \dfrac{2}{t} \cdot t^2}{\dfrac{4}{t^2} \cdot t^2 - 1 \cdot t^2}$$

$$= \frac{t^2 + 2t}{4 - t^2}. \qquad \text{\color{red}No rational expression remains within the numerator or the denominator.}$$

8. Simplify by multiplying by the LCD:

$$\frac{\dfrac{1}{a} + \dfrac{1}{b}}{\dfrac{a}{b}}.$$

4. Finally, we simplify:

$$\frac{t^2 + 2t}{4 - t^2} = \frac{t(t + 2)}{(2 + t)(2 - t)} \qquad \text{\color{red}Factoring and simplifying; } \frac{t + 2}{t + 2} = 1$$

$$= \frac{t}{2 - t}.$$

 YOUR TURN

Solving Rational Equations

A **rational equation** is an equation containing one or more rational expressions, often with the variable in a denominator.

TO SOLVE A RATIONAL EQUATION

1. List any restrictions that exist. Numbers that make a denominator equal 0 can never be solutions.
2. Clear the equation of fractions by multiplying both sides by the LCM of the denominators.
3. Solve the resulting equation using the addition principle, the multiplication principle, and the principle of zero products, as needed.
4. Check the possible solution(s) in the original equation.

Because a possible solution in step (3) may make a denominator 0, checking is essential when solving rational equations.

EXAMPLE 9 Solve: $x + \dfrac{10}{x} = 7$.

SOLUTION First, we note that x cannot be 0. The LCD is x, so we multiply both sides by x:

List restrictions.

$$x + \frac{10}{x} = 7 \qquad\qquad \textit{Note: } x \neq 0.$$

Clear fractions.

$$x\left(x + \frac{10}{x}\right) = 7x \qquad\qquad \text{Don't forget the parentheses!}$$

$$x \cdot x + x \cdot \frac{10}{x} = 7x \qquad\qquad \text{Using the distributive law}$$

$$x^2 + 10 = 7x \qquad\qquad \text{We have a quadratic equation.}$$

$$x^2 - 7x + 10 = 0 \qquad\qquad \text{Getting 0 on one side}$$

$$(x - 2)(x - 5) = 0 \qquad\qquad \text{Factoring}$$

Solve the equation.

$$x - 2 = 0 \quad or \quad x - 5 = 0 \qquad \text{Using the principle of zero products}$$

$$x = 2 \quad or \qquad x = 5.$$

Check.

Check: For 2:   For 5:

$$x + \frac{10}{x} = 7 \qquad\qquad\qquad x + \frac{10}{x} = 7$$

$$2 + \frac{10}{2} \;\bigg|\; 7 \qquad\qquad\qquad 5 + \frac{10}{5} \;\bigg|\; 7$$

$$2 + 5 \qquad\qquad\qquad\qquad 5 + 2$$

$$7 \overset{?}{=} 7 \;\;\text{TRUE} \qquad\qquad\qquad 7 \overset{?}{=} 7 \;\;\text{TRUE}$$

Both numbers check, so there are two solutions, 2 and 5.

9. Solve: $\dfrac{1}{x} + \dfrac{1}{x + 2} = \dfrac{7}{24}$.

↩ YOUR TURN

Many problems translate to rational equations. **Work problems**, which involve the time that it takes to complete a task, can often be solved using the work principle.

THE WORK PRINCIPLE

If

$a =$ the time needed for A to complete the work alone,

$b =$ the time needed for B to complete the work alone, and

$t =$ the time needed for A and B to complete the work together,

then $\dfrac{t}{a} + \dfrac{t}{b} = 1.$

EXAMPLE 10 *Painting.* It takes Kerry 30 hr to repaint a mural. It takes Jesse 45 hr to repaint the same mural. How long would it take Kerry and Jesse, working together, to repaint the mural?

SOLUTION

1. Familiarize. We reason that if Kerry and Jesse each repaint half the mural, it would take Kerry 15 hr and Jesse $22\frac{1}{2}$ hr. So the time it takes them working together should be between 15 hr and $22\frac{1}{2}$ hr. We let $t =$ the time that it takes them to repaint the mural, working together.

2. **Translate.** We use the work principle to translate the problem:

$$\frac{t}{a} + \frac{t}{b} = 1 \qquad \text{\textit{a} is the time that it takes Kerry to repaint the mural;}$$
$$\text{\textit{b} is the time that it takes Jesse to repaint the mural.}$$

$$\frac{t}{30} + \frac{t}{45} = 1.$$

3. **Carry out.** We solve the equation:

$$\frac{t}{30} + \frac{t}{45} = 1$$

$$90\left(\frac{t}{30} + \frac{t}{45}\right) = 90 \cdot 1 \qquad \text{The LCD is } 2 \cdot 3 \cdot 3 \cdot 5, \text{ or } 90.$$

$$90 \cdot \frac{t}{30} + 90 \cdot \frac{t}{45} = 90$$

$$3t + 2t = 90$$

$$5t = 90$$

$$t = 18.$$

10. Refer to Example 10. Suppose that it takes Kerry 10 hr to paint a garage, and it takes Jesse 8 hr to paint the same garage. How long would it take them, working together, to paint the garage?

4. **Check.** We note that, as predicted in the *Familiarize* step, the answer is between 15 hr and $22\frac{1}{2}$ hr. Also, if each works 18 hr, Kerry will do $\frac{18}{30}$ of the job and Jesse will do $\frac{18}{45}$ of the job, and

$$\frac{18}{30} + \frac{18}{45} = \frac{3}{5} + \frac{2}{5} = 1. \qquad \text{The entire job will be completed.}$$

5. **State.** Together it will take them 18 hr to repaint the mural.

YOUR TURN

Motion problems deal with distance, speed (or rate), and time, and are often translated using the distance formula $d = rt$.

EXAMPLE 11 *Driving Time.* Karen and Eva are each driving to a sales meeting. Because of road conditions, Karen is able to drive 15 mph faster than Eva. In the same time that it takes Karen to travel 120 mi, Eva travels only 90 mi. Find their speeds.

SOLUTION

1. **Familiarize.** We let $t =$ the time, in hours, that is spent traveling and $r =$ Karen's speed, in miles per hour. Then Eva's speed $= r - 15$. From the distance formula, we have $t = d/r$, so we can organize the information as follows.

	Distance	Speed	Time
Karen	120	r	$120/r$
Eva	90	$r - 15$	$90/(r-15)$

2. **Translate.** Since the times are the same, we have the equation

$$\frac{120}{r} = \frac{90}{r-15}.$$

3. **Carry out.** We solve the equation:

$$\frac{120}{r} = \frac{90}{r-15}$$

$$r(r-15)\frac{120}{r} = r(r-15)\frac{90}{r-15} \qquad \text{The LCD is } r(r-15).$$

$$120(r-15) = 90r \qquad \text{Simplifying}$$

$$120r - 1800 = 90r \qquad \text{Removing parentheses}$$

$$-1800 = -30r \qquad \text{Subtracting } 120r \text{ from both sides}$$

$$60 = r. \qquad \text{Dividing both sides by } -30$$

11. Refer to Example 11. Suppose that Karen is able to drive 25 mph faster than Eva. In the same time that it takes Karen to travel 240 mi, Eva travels only 140 mi. Find their speeds.

4. **Check.** If $r = 60$, then $r - 15 = 45$. If Karen travels 120 mi at 60 mph, she will have traveled for 2 hr. If Eva travels 90 mi at 45 mph, she will also have traveled for 2 hr. Since the times are the same, the speeds check.

5. **State.** Karen is traveling at 60 mph, while Eva is traveling at 45 mph.

YOUR TURN

Another type of problem that translates to a rational equation involves proportions. A **ratio** of two quantities is their quotient. A **proportion** is an equation stating that two ratios are equal.

EXAMPLE 12 *Baking.* Rob discovers there is $2\frac{1}{2}$ cups of pancake mix left in the box. The directions on the mix indicate that $1\frac{1}{3}$ cups of milk should be added to 2 cups of mix. How much milk should Rob add to the $2\frac{1}{2}$ cups of mix?

SOLUTION Since the problem translates directly to a proportion, we will not follow all five steps of the problem-solving process. We write the ratio of mix to milk in two ways:

$$\begin{array}{c}\text{Mix} \longrightarrow \\ \text{Milk} \longrightarrow\end{array} \frac{2}{1\frac{1}{3}} = \frac{2\frac{1}{2}}{x} \begin{array}{c}\longleftarrow \text{Mix} \\ \longleftarrow \text{Milk}\end{array}.$$

The LCD is $x\left(1\frac{1}{3}\right)$. We solve for x:

$$x\left(1\tfrac{1}{3}\right)\frac{2}{1\frac{1}{3}} = x\left(1\tfrac{1}{3}\right)\frac{2\frac{1}{2}}{x} \qquad \text{Multiplying by the LCD}$$

$$2x = \left(1\tfrac{1}{3}\right)\left(2\tfrac{1}{2}\right) \qquad \text{Simplifying}$$

$$2x = \tfrac{10}{3} \qquad \text{Converting to fraction notation and multiplying}$$

$$x = \tfrac{5}{3}. \qquad \text{Multiplying both sides by } \tfrac{1}{2} \text{ and simplifying}$$

12. It takes Keri $3\frac{1}{2}$ hr to grade 50 journal entries. At that rate, how long would it take her to grade 40 entries?

Rob needs to add $\frac{5}{3}$, or $1\frac{2}{3}$, cups of milk.

YOUR TURN

R.6   EXERCISE SET

Rational Expressions

List all numbers for which each rational expression is undefined.

1. $\dfrac{x - 7}{3x + 1}$

2. $\dfrac{10 - y}{-5y}$

3. $\dfrac{p^2 - 1}{p^2 - 4}$

4. $\dfrac{10x}{x^2 + 9x + 8}$

Simplify by removing a factor equal to 1.

5. $\dfrac{16x^2y}{18xy^2}$

6. $\dfrac{2x + 10}{6x + 30}$

7. $\dfrac{t^2 - 2t - 8}{t^2 - 16}$

8. $\dfrac{a^3 + 2a^2 + a}{a^2 + 4a + 3}$

9. $\dfrac{2 - x}{x^2 - 4}$

10. $\dfrac{n - 3}{3 - n}$

Addition, Subtraction, Multiplication, and Division of Rational Expressions

Perform each indicated operation. Then, if possible, simplify.

11. $\dfrac{3x}{x + y} \cdot \dfrac{2x + 2y}{x^2}$

12. $\dfrac{5}{x + 7} \cdot \dfrac{x + 7}{10}$

13. $\dfrac{a^2 + 2a + 1}{a} \div \dfrac{a^2}{a^2 - 1}$

14. $\dfrac{x}{x + 3} + \dfrac{3 - x}{x + 3}$

15. $\dfrac{2x}{x - 7} - \dfrac{x + 7}{x - 7}$

16. $\dfrac{x}{x + y} \div \dfrac{y}{x + y}$

17. $\dfrac{5}{x} + \dfrac{4}{x^2}$

18. $\dfrac{x^2 + 4x + 3}{x^2 + x - 2} \cdot \dfrac{x^2 + 3x + 2}{x^2 + 2x - 3}$

19. $\dfrac{2a + b}{a - b} - \dfrac{4}{3a - 3b}$

20. $(x^2 - 16) \div \dfrac{4x + 16}{3x^2}$

21. $\dfrac{2 - x}{5x^2} \div \dfrac{x^2 - 4}{3x}$

22. $\dfrac{2x}{x - 5} + \dfrac{3}{x + 4}$

23. $\dfrac{x^3 + 2x^2 + x}{x^2 - 4} \cdot \dfrac{x^2 - x - 2}{x^4 + x^3}$

24. $\dfrac{-1}{x^2 + 7x + 10} - \dfrac{3}{x^2 + 8x + 15}$

25. $\dfrac{2}{(x + 1)^2} + \dfrac{1}{x + 1}$

26. $\dfrac{2x}{x^2 - 3x} \div (x - 3)$

27. $\dfrac{x}{x - 2} + \dfrac{2}{2 - x}$

28. $\dfrac{3}{y - 1} - \dfrac{y}{1 - y}$

29. $\dfrac{t}{t^2 - 1} - \dfrac{1}{1 - t}$

30. $\dfrac{1}{5 - x} + \dfrac{x}{2x - 10}$

31. $\dfrac{x - y}{2x} \cdot \dfrac{3x^2}{y - x}$

32. $\dfrac{1}{x + y} + \dfrac{2}{x^2 + y^2}$

33. $\dfrac{x - 2}{x + 5} - \dfrac{x + 3}{x - 4}$

34. $\dfrac{z^2 + 2z + 1}{8z} \div \dfrac{z^2 - z - 2}{4z^2 - 4}$

Complex Rational Expressions

Simplify.

35. $\dfrac{\dfrac{2}{x} - \dfrac{1}{x^2}}{\dfrac{x}{4}}$

36. $\dfrac{\dfrac{x}{3} - \dfrac{3}{x}}{\dfrac{1}{x} + \dfrac{1}{3}}$

37. $\dfrac{\dfrac{3}{x - 7}}{\dfrac{4x + 3}{x + 1}}$

38. $\dfrac{\dfrac{a}{a - b}}{\dfrac{a^2}{a^2 - b^2}}$

39. $\dfrac{x - \dfrac{3}{x - 2}}{x - \dfrac{12}{x + 1}}$

40. $\dfrac{t + \dfrac{1}{t}}{t - \dfrac{2}{t}}$

41. $\dfrac{\dfrac{1}{2} - \dfrac{1}{x}}{\dfrac{2 - x}{2}}$

42. $\dfrac{\dfrac{x}{2y^2} + \dfrac{y}{3x^2}}{\dfrac{1}{6xy} + \dfrac{2}{x^2y}}$

Solving Rational Equations

Solve.

43. $\dfrac{1}{2} + \dfrac{1}{3} = \dfrac{1}{t}$

44. $\dfrac{1}{4} + \dfrac{1}{t} = \dfrac{1}{3}$

45. $x + \dfrac{1}{x} = 2$

46. $\dfrac{x-7}{x+1} = \dfrac{2}{3}$

47. $\dfrac{3}{y+7} = \dfrac{1}{y-8}$

48. $\dfrac{x+1}{x-2} = \dfrac{3}{x-2}$

49. $\dfrac{1}{x-3} - \dfrac{x-4}{x^2-9} = 1$

50. $\dfrac{3}{a+4} = \dfrac{a-1}{4-a}$

51. *Painting.* Quentin can paint the turret on a Queen Anne house in 40 hr. It takes Austin 50 hr to paint the same turret. How long would it take them, working together, to paint the turret?

52. *Building Fences.* Lindsay can build a fence in 6 hr. Laura can do the same job in 5 hr. How long will it take them, working together, to build the fence?

53. *Snowmobiling.* Jessica can ride her snowmobile through the fields 20 km/h faster than Josh can ride his through the woods. In the time that it takes Jessica to ride 18 km, Josh travels 10 km. Find the speed of each snowmobile.

54. *Bicycling.* Ani bicycles 8 mi and Lia bicycles 12 mi to meet at a park for lunch. Because Ani's trip is mostly uphill, she rides 5 mph slower than Lia. Ani and Lia leave their homes at the same time and arrive at the park at the same time. Find the speed of each bicyclist.

55. *Elk Population.* To determine the size of a park's elk population, rangers tag 15 elk and set them free. Months later, 40 elk are caught, of which 12 have tags. Estimate the size of the elk population.

56. *Manufacturing Pegs.* A sample of 136 wooden pegs contained 17 defective pegs. How many defective pegs would you expect in a sample of 840 pegs?

↪ **YOUR TURN ANSWERS: SECTION R.6**

1. $-2, 0$ **2.** $\dfrac{x+2}{2x+3}$ **3.** $\dfrac{b-1}{b(a+b)}$ **4.** $\dfrac{5}{t^2-9}$

5. $\dfrac{n^2+11n+22}{(n-2)(n+1)(n+3)}$ **6.** 1 **7.** $\dfrac{-2x^2}{(x-1)(3x-1)}$

8. $\dfrac{a+b}{a^2}$ **9.** $-\dfrac{8}{7}, 6$ **10.** $\dfrac{40}{9}$ hr, or $4\dfrac{4}{9}$ hr   **11.** Karen: 60 mph; Eva: 35 mph   **12.** 2.8 hr

---

REFERENCES

Addition and subtraction of rational expressions, Sections 6.3–6.4
Complex rational expressions, Section 6.5
Factors that are opposites, p. 378
Least common denominator, p. 389
Least common multiple, p. 389
Motion problems, p. 420
Multiplication and division of rational expressions, Section 6.2
Proportion, p. 424
Ratio, p. 424
Rational expressions, p. 374
Solving rational equations, Section 6.6
Work problems, p. 418

16. *Male Height.* Jason's brothers are 174 cm, 180 cm, 179 cm, and 172 cm tall. The average male is 176.5 cm tall. How tall is Jason if he and his brothers have an average height of 176.5 cm?

> ↳ **YOUR TURN ANSWERS: APPENDIX A**
> **1.** 19   **2.** 6   **3.** 1.6   **4.** 2, 3   **5.** No mode

---

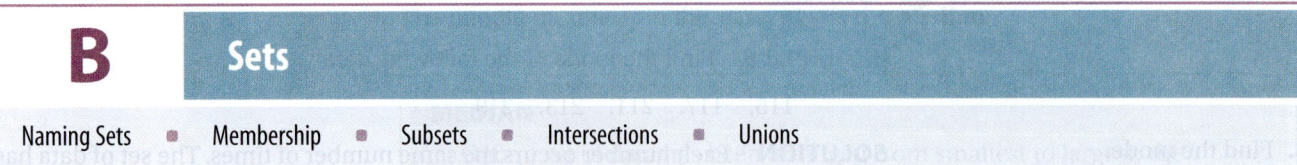

B   Sets

Naming Sets • Membership • Subsets • Intersections • Unions

Naming Sets

A **set** is a collection of objects. In mathematics the objects, or **elements**, of a set are generally numbers.

To name the set of whole numbers less than 6, we can use *roster notation*, as follows:

$$\{0, 1, 2, 3, 4, 5\}.$$

The set of real numbers x for which x is less than 6 cannot be named by listing all its members because there is an infinite number of them. We name such a set using *set-builder notation*, as follows:

$$\{x \,|\, x < 6\}.$$

This is read

"The set of all x such that x is less than 6."

The **empty set** contains no elements and is written $\varnothing$. It can also be written $\{\ \}$. Note that $\{0\}$ contains 0 and is thus *not* empty.

Membership

$x \in A$

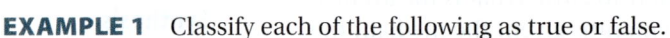

The symbol $\in$ means *is a member of* or *belongs to*, or *is an element of.* Thus, $x \in A$ means

x is a member of A,   or x belongs to A,   or x is an element of A.

EXAMPLE 1 Classify each of the following as true or false.

a) $1 \in \{1, 2, 3\}$ **b)** $1 \in \{2, 3\}$

c) $4 \in \{x \,|\, x$ is an even whole number$\}$

SOLUTION

a) Since 1 is listed as a member of the set, $1 \in \{1, 2, 3\}$ is true.

b) Since 1 is *not* a member of $\{2, 3\}$, the statement $1 \in \{2, 3\}$ is false.

c) Since 4 is an even whole number, $4 \in \{x \,|\, x$ is an even whole number$\}$ is true.

1. Classify as true or false.

$5 \in \{x \,|\, x$ is an even whole number$\}$.

↳ YOUR TURN

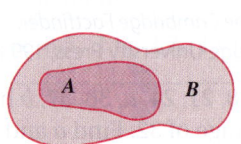

Subsets

If every element of A is also an element of B, then A is a *subset* of B. This is denoted $A \subseteq B$.

For example, the set of whole numbers is a subset of the set of integers, and the set of rational numbers is a subset of the set of real numbers.

EXAMPLE 2 Classify each of the following as true or false.

a) $\{1, 2\} \subseteq \{1, 2, 3, 4\}$ **b)** $\{p, q, r, w\} \subseteq \{a, p, r, z\}$

c) $\{x \mid x < 6\} \subseteq \{x \mid x \le 11\}$

SOLUTION

a) Since every element of $\{1, 2\}$ is in the set $\{1, 2, 3, 4\}$, it follows that $\{1, 2\} \subseteq \{1, 2, 3, 4\}$ is true.

b) Since $q \in \{p, q, r, w\}$, but $q \notin \{a, p, r, z\}$, it follows that $\{p, q, r, w\} \subseteq \{a, p, r, z\}$ is false.

2. Classify as true or false:

$\{8, 9, 10\} \subseteq \{8, 9\}$.

c) Since every number that is less than 6 is also less than 11, the statement $\{x \mid x < 6\} \subseteq \{x \mid x \le 11\}$ is true.

YOUR TURN

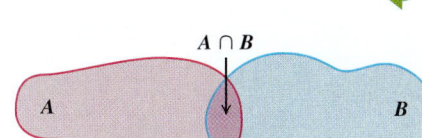

Intersections

The *intersection* of sets A and B, denoted $A \cap B$, is the set of members common to both sets.

EXAMPLE 3 Find each intersection.

a) $\{0, 1, 3, 5, 25\} \cap \{2, 3, 4, 5, 6, 7, 9\}$

b) $\{a, p, q, w\} \cap \{m, n, x, y\}$

SOLUTION

a) $\{0, 1, 3, 5, 25\} \cap \{2, 3, 4, 5, 6, 7, 9\} = \{3, 5\}$

3. Find the intersection:

$\{3, 4, 5\} \cap \{3, 4, 5, 6\}$.

b) $\{a, p, q, w\} \cap \{m, n, x, y\} = \varnothing$ These sets have no members in common. Their intersection is the empty set.

YOUR TURN

$A \cup B$ is shaded.

Unions

Two sets A and B can be combined to form a set that contains the members of both A and B. The new set is called the *union* of A and B, denoted $A \cup B$.

EXAMPLE 4 Find each union.

a) $\{0, 5, 7, 13, 27\} \cup \{0, 2, 3, 4, 5\}$ **b)** $\{a, c, e, g\} \cup \{b, d, f\}$

SOLUTION

a) $\{0, 5, 7, 13, 27\} \cup \{0, 2, 3, 4, 5\} = \{0, 2, 3, 4, 5, 7, 13, 27\}$

Note that the 0 and the 5 are *not* listed twice in the union.

4. Find the union:

$\{1, 3, 5, 7\} \cup \{2, 3, 5, 8\}$.

b) $\{a, c, e, g\} \cup \{b, d, f\} = \{a, b, c, d, e, f, g\}$

YOUR TURN

B **EXERCISE SET** FOR EXTRA HELP MyMathLab® Math
PRACTICE WATCH READ REVIEW

Naming Sets

Name each set using the roster method.

1. The set of whole numbers 8 through 11

2. The set of whole numbers 83 through 89

3. The set of odd numbers between 40 and 50

4. The set of multiples of 5 between 10 and 40

5. $\{x \mid$ the square of x is 9$\}$

6. $\{x \mid x$ is the cube of $\frac{1}{2}\}$

Membership and Subsets

Classify each statement as either true or false.

7. $5 \in \{x \mid x$ is an odd number$\}$

8. $8 \in \{x \mid x$ is an odd number$\}$

9. Skiing ∈ The set of all sports

10. Pharmacist ∈ The set of all professions requiring a college degree

11. $3 \in \{-4, -3, 0, 1\}$

12. $0 \in \{-4, -3, 0, 1\}$

13. $\frac{2}{3} \in \{x \mid x \text{ is a rational number}\}$

14. $\frac{2}{3} \in \{x \mid x \text{ is a real number}\}$

15. $\{-1, 0, 1\} \subseteq \{-3, -2, -1, 1, 2, 3\}$

16. The set of vowels ⊂ The set of consonants

17. The set of integers ⊆ The set of rational numbers

18. $\{2, 4, 6\} \subseteq \{1, 2, 3, 4, 5, 6, 7\}$

Intersections

Find each intersection.

19. $\{a, b, c, d, e\} \cap \{c, d, e, f, g\}$

20. $\{a, e, i, o, u\} \cap \{q, u, i, c, k\}$

21. $\{1, 2, 3, 4, 6, 12\} \cap \{1, 2, 3, 6, 9, 18\}$

22. $\{1, 2, 3, 4, 6, 12\} \cap \{1, 5, 7, 35\}$

23. $\{2, 4, 6, 8\} \cap \{1, 3, 5, 7\}$

24. $\{a, e, i, o, u\} \cap \{m, n, f, g, h\}$

Unions

Find each union.

25. $\{a, e, i, o, u\} \cup \{q, u, i, c, k\}$

26. $\{a, b, c, d, e\} \cup \{c, d, e, f, g\}$

27. $\{1, 2, 3, 4, 6, 12\} \cup \{1, 2, 3, 6, 9, 18\}$

28. $\{1, 2, 3, 4, 6, 12\} \cup \{1, 5, 7, 35\}$

29. $\{2, 4, 6, 8\} \cup \{1, 3, 5, 7\}$

30. $\{a, e, i, o, u\} \cup \{m, n, f, g, h\}$

31. What advantage(s) does set-builder notation have over roster notation?

32. What advantage(s) does roster notation have over set-builder notation?

Synthesis

33. Find the union of the set of integers and the set of whole numbers.

34. Find the intersection of the set of odd integers and the set of even integers.

35. Find the union of the set of rational numbers and the set of irrational numbers.

36. Find the intersection of the set of even integers and the set of positive rational numbers.

37. Find the intersection of the set of rational numbers and the set of irrational numbers.

38. Find the union of the set of negative integers, the set of positive integers, and the set containing 0.

39. For a set A, find each of the following.
    a) $A \cup \varnothing$
    b) $A \cup A$
    c) $A \cap A$
    d) $A \cap \varnothing$

Classify each statement as either true or false.

40. The empty set can be written $\varnothing$, { }, or {0}.

41. For any set A, $\varnothing \subseteq A$.

42. For any set A, $A \subseteq A$.

43. For any sets A and B, $A \cap B \subseteq A$.

44. A set is *closed* under an operation if, when the operation is performed on its members, the result is in the set. For example, the set of real numbers is closed under the operation of addition since the sum of any two real numbers is a real number.
    a) Is the set of even numbers closed under addition?
    b) Is the set of odd numbers closed under addition?
    c) Is the set {0, 1} closed under addition?
    d) Is the set {0, 1} closed under multiplication?
    e) Is the set of real numbers closed under multiplication?
    f) Is the set of integers closed under division?

45. Experiment with sets of various types and determine whether the following distributive law for sets is true:

$$A \cap (B \cup C) = (A \cap B) \cup (A \cap C).$$

 YOUR TURN ANSWERS: APPENDIX B

1. False   2. False   3. $\{3, 4, 5\}$ 4. $\{1, 2, 3, 5, 7, 8\}$

Synthetic Division   ▪   The Remainder Theorem

Synthetic Division

To divide a polynomial by a binomial of the type $x - a$, we can streamline the usual procedure to develop a process called *synthetic division*.

Compare the following. In each stage, we attempt to write less than in the previous stage, while retaining enough essentials to solve the problem.

When a polynomial is written in descending order, the coefficients provide the essential information.

Long Division

$$
\begin{array}{r}
4x^2 + 5x + 11 \\
x - 2 \overline{\smash{\big)}\, 4x^3 - 3x^2 + x + 7} \\
\underline{4x^3 - 8x^2} \\
5x^2 + x \\
\underline{5x^2 - 10x} \\
11x + 7 \\
\underline{11x - 22} \\
29
\end{array}
$$

Stage 1

$$
\begin{array}{r}
4 + 5 + 11 \\
1 - 2 \overline{\smash{\big)}\, 4 - 3 + 1 + 7} \\
\underline{4 - 8} \\
5 + 1 \\
\underline{5 - 10} \\
11 + 7 \\
\underline{11 - 22} \\
29
\end{array}
$$

Because the leading coefficient in $x - 2$ is 1, each time we multiply it by a term in the answer, the leading coefficient of that product is the same as the coefficient in the answer. In stage 2, rather than duplicate these numbers, we focus on where -2 is used. We also drop the 1 from the divisor. To simplify further, in stage 3, we reverse the sign of the -2 in the divisor and, in exchange, *add* at each step in the long division.

Stage 2

$$
\begin{array}{r}
4 + 5 + 11 \\
-2 \overline{\smash{\big)}\, 4 - 3 + 1 + 7} \\
\underline{-8} \\
5 + 1 \\
\underline{-10} \\
11 + 7 \\
\underline{-22} \\
29
\end{array}
$$

Multiply: $-2 \cdot 4 = -8.$
Subtract: $-3 - (-8) = 5.$
Multiply: $-2 \cdot 5 = -10.$
Subtract: $1 - (-10) = 11.$
Multiply: $-2 \cdot 11 = -22.$
Subtract: $7 - (-22) = 29.$

Stage 3

Replace the -2 with 2.

$$
\begin{array}{r}
4 + 5 + 11 \\
2 \overline{\smash{\big)}\, 4 - 3 + 1 + 7} \\
8 \\
5 + 1 \\
10 \\
11 + 7 \\
22 \\
29
\end{array}
$$

Multiply: $2 \cdot 4 = 8.$
Add: $-3 + 8 = 5.$
Multiply: $2 \cdot 5 = 10.$
Add: $1 + 10 = 11.$
Multiply: $2 \cdot 11 = 22.$
Add: $7 + 22 = 29.$

The blue numbers can be eliminated if we look at the red numbers instead, as shown in stage 4 below. Note that the 5 and the 11 preceding the remainder 29 coincide with the 5 and the 11 following the 4 on the top line. By writing a 4 to the left of 5 on the bottom line, we can eliminate the top line in stage 4 and read our answer from the bottom line. This final stage is commonly called **synthetic division**.

Student Notes

You will not need to write out all five stages when performing synthetic division on your own. We show the steps to help you understand the reasoning behind the method.

Stage 4

$$
\begin{array}{r}
4 \quad\ 5 \quad 11 \\
2\overline{)4 \ -3 \quad\ 1 \quad\ 7} \\
\underline{8 \quad 10 \quad 22} \\
5 \quad 11 \quad 29
\end{array}
$$

Stage 5

$$
\begin{array}{r}
\underline{2}\,|\,4 \ -3 \quad\ 1 \quad\ 7 \\
\underline{8 \quad 10 \quad 22} \\
4 \quad\ 5\,|\,11 \quad 29
\end{array}
$$

← This is the remainder.

└── This is the zero-degree coefficient.

└── This is the first-degree coefficient.

└── This is the second-degree coefficient.

The quotient is $4x^2 + 5x + 11$ with a remainder of 29.

> Remember that in order for this method to work, the divisor must be of the form $x - a$, that is, a variable minus a constant. Note that the coefficient of the variable is 1.

Before using synthetic division in Example 1, let's review how stage 5, above, is formed.

This is the constant being subtracted in the divisor. → $\underline{2}\,|\,4 \ -3 \quad 1 \quad 7$ ← These are the coefficients in the dividend.

This line is formed by adding in each column. → $\underline{\ 8 \quad 10 \quad 22}$ $\ 4 \quad 5 \quad 11\,|\,29$ ← These are found by multiplying 2 by the number below and to the left.

This is found first by bringing down the 4.

EXAMPLE 1   Use synthetic division to divide: $(x^3 + 6x^2 - x - 30) \div (x - 2)$.

SOLUTION

$$
\underline{2}\,|\,1 \quad 6 \ -1 \ -30
$$

Write the 2 of $x - 2$ and the coefficients of the dividend.

$$
1
$$

Bring down the first coefficient.

$$
\begin{array}{r}
\underline{2}\,|\,1 \quad 6 \ -1 \ -30 \\
\underline{2} \\
1 \quad 8
\end{array}
$$

Multiply 1 by 2 to get 2.

Add 6 and 2.

$$
\begin{array}{r}
\underline{2}\,|\,1 \quad 6 \ -1 \ -30 \\
\underline{2 \quad 16} \\
1 \quad 8 \quad 15
\end{array}
$$

Multiply 8 by 2.

Add -1 and 16.

$$
\begin{array}{r}
\underline{2}\,|\,1 \quad 6 \ -1 \ -30 \\
\underline{2 \quad 16 \quad 30} \\
1 \quad 8 \quad 15 \quad 0
\end{array}
$$

Multiply 15 by 2 and add.

The answer is $x^2 + 8x + 15$ with R 0, or just $x^2 + 8x + 15$.

↩ YOUR TURN

1. Use synthetic division to divide:

$(x^3 + 2x^2 - 17x + 6) \div (x - 3).$

EXAMPLE 2   Use synthetic division to divide.

a) $(2x^3 + 7x^2 - 5) \div (x + 3)$

b) $(10x^2 - 13x + 3x^3 - 20) \div (4 + x)$

SOLUTION

a) $(2x^3 + 7x^2 - 5) \div (x + 3)$

The dividend has no x-term, so we need to write 0 as the coefficient of x. Note that $x + 3 = x - (-3)$, so we write -3 inside the ⌐.

$$
\begin{array}{r|rrrr}
-3 & 2 & 7 & 0 & -5 \\
   &   & -6 & -3 & 9 \\
\hline
   & 2 & 1 & -3 & \;|\; 4
\end{array}
$$

The answer is $2x^2 + x - 3$, with R 4, or $2x^2 + x - 3 + \dfrac{4}{x + 3}$.

b) We first rewrite $(10x^2 - 13x + 3x^3 - 20) \div (4 + x)$ in descending order:

$$(3x^3 + 10x^2 - 13x - 20) \div (x + 4).$$

Next, we use synthetic division. Note that $x + 4 = x - (-4)$.

$$
\begin{array}{r|rrrr}
-4 & 3 & 10 & -13 & -20 \\
   &   & -12 & 8 & 20 \\
\hline
   & 3 & -2 & -5 & \;|\; 0
\end{array}
$$

The answer is $3x^2 - 2x - 5$.

YOUR TURN

2. Use synthetic division to divide:

$$(3x - 9 + 2x^3) \div (x + 2).$$

The Remainder Theorem

When a polynomial function $f(x)$ is divided by $x - a$, the remainder is related to the function value $f(a)$. Compare the following from Examples 1 and 2.

Polynomial Function	Divisor	Remainder	Function Value
Example 1: $f(x) = x^3 + 6x^2 - x - 30$	$x - 2$	0	$f(2) = 0$
Example 2(a): $g(x) = 2x^3 + 7x^2 - 5$	$x + 3$, or $x - (-3)$	4	$g(-3) = 4$
Example 2(b): $p(x) = 10x^2 - 13x + 3x^3 - 20$	$4 + x$, or $x - (-4)$	0	$p(-4) = 0$

When the remainder is 0, the divisor $x - a$ is a factor of the polynomial being divided. We can write $f(x) = x^3 + 6x^2 - x - 30$ in Example 1 as

$$f(x) = (x - 2)(x^2 + 8x + 15).$$

Then

$$
\left.
\begin{aligned}
f(2) &= (2 - 2)(2^2 + 8 \cdot 2 + 15) \\
     &= 0(2^2 + 8 \cdot 2 + 15) = 0.
\end{aligned}
\right\}
\quad \text{Since } x - 2 \text{ is a factor, } f(2) = 0.
$$

Technology Connection

In Example 1, the division by $x - 2$ gave a remainder of 0. The remainder theorem tells us that this means that when $x = 2$, the value of $x^3 + 6x^2 - x - 30$ is 0. Check this both graphically and algebraically (by substitution). Then perform a similar check for Example 2(b).

Thus, when the remainder is 0 after division by $x - a$, the function value $f(a)$ is also 0. Remarkably, this pattern extends to nonzero remainders as well. The fact that the remainder and the function value coincide is predicted by the remainder theorem.

> ### THE REMAINDER THEOREM
> The remainder obtained by dividing $P(x)$ by $x - r$ is $P(r)$.

A proof of this result is outlined in Exercise 37.

EXAMPLE 3   Let $f(x) = 8x^5 - 6x^3 + x - 8$. Use synthetic division to find $f(2)$.

SOLUTION   The remainder theorem tells us that $f(2)$ is the remainder when $f(x)$ is divided by $x - 2$. We use synthetic division to find that remainder:

$$
\begin{array}{r|rrrrrr}
2 & 8 & 0 & -6 & 0 & 1 & -8 \\
  &   & 16 & 32 & 52 & 104 & 210 \\
\hline
  & 8 & 16 & 26 & 52 & 105 & 202
\end{array}
$$

Although the bottom line can be used to find the quotient for the division $(8x^5 - 6x^3 + x - 8) \div (x - 2)$, what we are really interested in is the remainder. It tells us that $f(2) = 202$. The calculations are easier than the more typical calculation of $f(2)$.

3. Let
$$f(x) = 5x^4 + 3x^2 + 2x - 7.$$
Use synthetic division to find $f(-2)$.

 YOUR TURN

The remainder theorem is often used to check division. Thus Example 2(a) can be checked by computing $g(-3) = 2(-3)^3 + 7(-3)^2 - 5$. Since $g(-3) = 4$ and the remainder in Example 2(a) is also 4, our division was probably correct.

 EXERCISE SET   FOR EXTRA HELP   MyMathLab® Math⬥XL
PRACTICE   WATCH   READ   REVIEW

Vocabulary and Reading Check

Classify each of the following statements as either true or false.

1. If $P(-5) = 39$ and $P(x) = x^3 + 7x^2 + 3x + 4$, then
$$
\begin{array}{r|rrrr}
-5 & 1 & 7 & 3 & 4 \\
   &   & -5 & -10 & 35 \\
\hline
   & 1 & 2 & -7 & 39
\end{array}
$$

2. In order to use synthetic division, we must be sure that the divisor is of the form $x - a$.

3. Synthetic division can be used in problems in which long division could not be used.

4. In order for $f(x)/g(x)$ to exist, $g(x)$ must be 0.

5. If $x - 2$ is a factor of some polynomial $P(x)$, then $P(2) = 0$.

6. If $p(3) = 0$ for some polynomial $p(x)$, then $x - 3$ is a factor of $p(x)$.

Synthetic Division

Use synthetic division to divide.

7. $(x^3 - 4x^2 - 2x + 5) \div (x - 1)$

8. $(x^3 - 4x^2 + 5x - 6) \div (x - 3)$

9. $(a^2 + 8a + 11) \div (a + 3)$

10. $(a^2 + 8a + 11) \div (a + 5)$

11. $(2x^3 - x^2 - 7x + 14) \div (x + 2)$

12. $(3x^3 - 10x^2 - 9x + 15) \div (x - 4)$

13. $(a^3 - 10a + 12) \div (a - 2)$

14. $(a^3 - 14a + 15) \div (a - 3)$

15. $(3y^3 - 7y^2 - 20) \div (y - 3)$

16. $(2x^3 - 3x^2 + 8) \div (x + 2)$

17. $(x^5 - 32) \div (x - 2)$

18. $(y^5 - 1) \div (y - 1)$

19. $(3x^3 + 1 - x + 7x^2) \div \left(x + \frac{1}{3}\right)$

20. $(8x^3 - 1 + 7x - 6x^2) \div \left(x - \frac{1}{2}\right)$

The Remainder Theorem

Use synthetic division to find the indicated function value.

21. $f(x) = 5x^4 + 12x^3 + 28x + 9; f(-3)$

22. $g(x) = 3x^4 - 25x^2 - 18; g(3)$

23. $P(x) = 2x^4 - x^3 - 7x^2 + x + 2; P(-3)$

24. $F(x) = 3x^4 + 8x^3 + 2x^2 - 7x - 4; F(-2)$

25. $f(x) = x^4 - 6x^3 + 11x^2 - 17x + 20; f(4)$

26. $p(x) = x^4 + 7x^3 + 11x^2 - 7x - 12; p(2)$

27. Why is it that we *add* when performing synthetic division, but *subtract* when performing long division?

28. Explain how synthetic division could be useful when attempting to factor a polynomial.

Synthesis

29. Let $Q(x)$ be a polynomial function with $p(x)$ a factor of $Q(x)$. If $p(3) = 0$, does it follow that $Q(3) = 0$? Why or why not? If $Q(3) = 0$, does it follow that $p(3) = 0$? Why or why not?

30. What adjustments must be made if synthetic division is to be used to divide a polynomial by a binomial of the form $ax + b$, with $a > 1$?

31. To prove the remainder theorem, note that any polynomial $P(x)$ can be rewritten as $(x - r) \cdot Q(x) + R$, where $Q(x)$ is the quotient polynomial that arises when $P(x)$ is divided by $x - r$, and R is some constant (the remainder).

a) How do we know that R must be a constant?
b) Show that $P(r) = R$ (this says that $P(r)$ is the remainder when $P(x)$ is divided by $x - r$).

32. Let $f(x) = 6x^3 - 13x^2 - 79x + 140$. Find $f(4)$ and then solve the equation $f(x) = 0$.

33. Let $f(x) = 4x^3 + 16x^2 - 3x - 45$. Find $f(-3)$ and then solve the equation $f(x) = 0$.

34. Use the TRACE feature on a graphing calculator to check your answer to Exercise 32.

35. Use the TRACE feature on a graphing calculator to check your answer to Exercise 33.

Nested Evaluation. One way to evaluate a polynomial function like $P(x) = 3x^4 - 5x^3 + 4x^2 - 1$ is to successively factor out x as shown:

$$P(x) = x(x(x(3x - 5) + 4) + 0) - 1.$$

Computations are then performed using this "nested" form of $P(x)$.

36. Use nested evaluation to find $f(4)$ in Exercise 32. Note the similarities to the calculations performed with synthetic division.

37. Use nested evaluation to find $f(-3)$ in Exercise 33. Note the similarities to the calculations performed with synthetic division.

YOUR TURN ANSWERS: APPENDIX C

1. $x^2 + 5x - 2$ **2.** $2x^2 - 4x + 11 + \dfrac{-31}{x + 2}$ **3.** 81

TABLES

TABLE 1   Fraction and Decimal Equivalents

Fraction Notation	$\frac{1}{10}$	$\frac{1}{8}$	$\frac{1}{6}$	$\frac{1}{5}$	$\frac{1}{4}$	$\frac{3}{10}$	$\frac{1}{3}$	$\frac{3}{8}$	$\frac{2}{5}$	$\frac{1}{2}$
Decimal Notation	0.1	0.125	$0.16\overline{6}$	0.2	0.25	0.3	$0.333\overline{3}$	0.375	0.4	0.5
Percent Notation	10%	12.5%, or $12\frac{1}{2}\%$	$16.6\overline{6}\%$, or $16\frac{2}{3}\%$	20%	25%	30%	$33.3\overline{3}\%$, or $33\frac{1}{3}\%$	37.5%, or $37\frac{1}{2}\%$	40%	50%
Fraction Notation	$\frac{3}{5}$	$\frac{5}{8}$	$\frac{2}{3}$	$\frac{7}{10}$	$\frac{3}{4}$	$\frac{4}{5}$	$\frac{5}{6}$	$\frac{7}{8}$	$\frac{9}{10}$	$\frac{1}{1}$
Decimal Notation	0.6	0.625	$0.666\overline{6}$	0.7	0.75	0.8	$0.83\overline{3}$	0.875	0.9	1
Percent Notation	60%	62.5%, or $62\frac{1}{2}\%$	$66.6\overline{6}\%$, or $66\frac{2}{3}\%$	70%	75%	80%	$83.3\overline{3}\%$, or $83\frac{1}{3}\%$	87.5%, or $87\frac{1}{2}\%$	90%	100%

TABLE 2   Squares and Square Roots with Approximations to Three Decimal Places

N	$\sqrt{N}$	N^2	N	$\sqrt{N}$	N^2	N	$\sqrt{N}$	N^2	N	$\sqrt{N}$	N^2
1	1	1	26	5.099	676	51	7.141	2601	76	8.718	5776
2	1.414	4	27	5.196	729	52	7.211	2704	77	8.775	5929
3	1.732	9	28	5.292	784	53	7.280	2809	78	8.832	6084
4	2	16	29	5.385	841	54	7.348	2916	79	8.888	6241
5	2.236	25	30	5.477	900	55	7.416	3025	80	8.944	6400
6	2.449	36	31	5.568	961	56	7.483	3136	81	9	6561
7	2.646	49	32	5.657	1024	57	7.550	3249	82	9.055	6724
8	2.828	64	33	5.745	1089	58	7.616	3364	83	9.110	6889
9	3	81	34	5.831	1156	59	7.681	3481	84	9.165	7056
10	3.162	100	35	5.916	1225	60	7.746	3600	85	9.220	7225
11	3.317	121	36	6	1296	61	7.810	3721	86	9.274	7396
12	3.464	144	37	6.083	1369	62	7.874	3844	87	9.327	7569
13	3.606	169	38	6.164	1444	63	7.937	3969	88	9.381	7744
14	3.742	196	39	6.245	1521	64	8	4096	89	9.434	7921
15	3.873	225	40	6.325	1600	65	8.062	4225	90	9.487	8100
16	4	256	41	6.403	1681	66	8.124	4356	91	9.539	8281
17	4.123	289	42	6.481	1764	67	8.185	4489	92	9.592	8464
18	4.243	324	43	6.557	1849	68	8.246	4624	93	9.644	8649
19	4.359	361	44	6.633	1936	69	8.307	4761	94	9.695	8836
20	4.472	400	45	6.708	2025	70	8.367	4900	95	9.747	9025
21	4.583	441	46	6.782	2116	71	8.426	5041	96	9.798	9216
22	4.690	484	47	6.856	2209	72	8.485	5184	97	9.849	9409
23	4.796	529	48	6.928	2304	73	8.544	5329	98	9.899	9604
24	4.899	576	49	7	2401	74	8.602	5476	99	9.950	9801
25	5	625	50	7.071	2500	75	8.660	5625	100	10	10,000

PHOTO CREDITS

ANSWERS

CHAPTER 1

Technology Connection, p. 7

1. 3438   **2.** 47,531

Exercise Set 1.1, pp. 8–11

1. Constant   **2.** Evaluate   **3.** Model   **4.** Equation
5. Expression   **6.** Equation   **7.** Equation
8. Expression   **9.** Equation   **10.** Expression
11. Equation   **12.** Expression   **13.** 45   **15.** 8   **17.** 5
19. 4   **21.** 5   **23.** 3   **25.** 24 ft^2 **27.** 15 cm^2
29. 804 ft^2 **31.** Let r represent Ron's age; $r + 5$, or $5 + r$
33. $6b$, or $b \cdot 6$ **35.** $c - 9$ **37.** $6 + q$, or $q + 6$
39. $p - t$ **41.** $y - x$ **43.** $x \div w$, or $\dfrac{x}{w}$
45. Let l represent the length of the box and h the height; $l + h$, or $h + l$ **47.** $9 \cdot 2m$, or $2m \cdot 9$
49. Let y represent "some number"; $\dfrac{1}{4}y - 13$, or $\dfrac{y}{4} - 13$
51. Let a and b represent the two numbers; $5(a - b)$
53. Let w represent the number of women attending; 64% of w, or $0.64w$ **55.** Let x represent the unknown number; $73 + x = 201$ **57.** Let x represent the unknown number; $42x = 2352$ **59.** Let s represent the number of unoccupied squares; $s + 19 = 64$ **61.** Let w represent the amount of solid waste generated, in millions of tons; 34% of $w = 82$, or $0.34w = 82$ **63.** $f = a + 5$ **65.** $n = m + 2.21$
67. $v = 10{,}000d$ **69.** (f)   **71.** (d)   **73.** (g)
75. (e)   **77.** 📋   **79.** 📋   **81.** \$450   **83.** 2
85. 6   **87.** $w + 4$ **89.** $l + w + l + w$, or $2l + 2w$
91. $t + 8$ **93.** 📋

Exercise Set 1.2, pp. 16–18

1. Equivalent   **2.** Commutative   **3.** Sum   **4.** Factors
5. Commutative   **6.** Associative   **7.** Distributive
8. Commutative   **9.** Commutative   **10.** Distributive
11. $t + 11$ **13.** $8x + 4$ **15.** $3y + 9x$ **17.** $5(1 + a)$
19. $x \cdot 7$ **21.** ts **23.** $5 + ba$ **25.** $(a + 1)5$
27. $x + (8 + y)$ **29.** $(u + v) + 7$ **31.** $ab + (c + d)$
33. $10(xy)$ **35.** $(2a)b$ **37.** $(3 \cdot 2)(a + b)$
39. $(s + t) + 6$; $(t + 6) + s$ **41.** $17(ab)$; $b(17a)$
43. $(1 + x) + 2 = (x + 1) + 2$ Commutative law
$\qquad\qquad\qquad = x + (1 + 2)$ Associative law
$\qquad\qquad\qquad = x + 3$ Simplifying
45. $(m \cdot 3)7 = m(3 \cdot 7)$ Associative law
$\qquad\qquad = m \cdot 21$ Simplifying
$\qquad\qquad = 21m$ Commutative law
47. $x, xyz, 1$ **49.** $2a, \dfrac{a}{3b}, 5b$ **51.** $4x, 4y$ **53.** $5, n$
55. $3, (x + y)$ **57.** $7, a, b$ **59.** $(a - b), (x - y)$

61. $2x + 30$ **63.** $4 + 4a$ **65.** $90x + 60$
67. $5r + 10 + 15t$ **69.** $2a + 2b$ **71.** $5x + 5y + 10$
73. $2(a + b)$ **75.** $7(1 + y)$ **77.** $2(16x + 1)$
79. $5(x + 2 + 3y)$ **81.** $7(a + 5b)$ **83.** $11(4x + y + 2z)$
85. Commutative law of addition; distributive law
87. Distributive law; associative law of multiplication
89. 📋   **91.** 📋   **93.** Distributive law; associative law of addition; commutative law of addition; associative law of addition; distributive law   **95.** Yes; distributive law
97. No; for example, let $m = 1$. Then $7 \div 3 \cdot 1 = \frac{7}{3}$ and $1 \cdot 3 \div 7 = \frac{3}{7}$.   **99.** No; for example, let $x = 1$ and $y = 2$. Then $30 \cdot 2 + 1 \cdot 15 = 60 + 15 = 75$ and $5[2(1 + 3 \cdot 2)] = 5[2(7)] = 5 \cdot 14 = 70$.   **101.** 📋
103. Answers may vary. **(a)** Aidan: $10(1.5x + 40)$; Beth: $10 \cdot 40 + 10(1.5)x$; Cody: $15x + 400$; **(b)** all expressions are equivalent to $15x + 400$.

Quick Quiz: Sections 1.1–1.2, p. 18

1. 9   **2.** $2(m + 3)$ **3.** Let n represent the number; $\frac{1}{3} \cdot n = 18$ **4.** $3x + 15y + 21$ **5.** $7(2a + t + 1)$

Exercise Set 1.3, pp. 26–28

1. Numerator   **2.** Prime   **3.** Reciprocal   **4.** Add
5. (b)   **6.** (c)   **7.** (d)   **8.** (a)   **9.** Composite
11. Prime   **13.** Composite   **15.** Prime   **17.** Neither
19. 1, 2, 5, 10, 25, 50   **21.** 1, 2, 3, 6, 7, 14, 21, 42
23. $3 \cdot 13$ **25.** $2 \cdot 3 \cdot 5$ **27.** $3 \cdot 3 \cdot 3$ **29.** $2 \cdot 3 \cdot 5 \cdot 5$
31. Prime   **33.** $2 \cdot 3 \cdot 5 \cdot 7$ **35.** $5 \cdot 23$ **37.** $\frac{3}{5}$
39. $\frac{2}{7}$ **41.** $\frac{1}{4}$ **43.** 4   **45.** $\frac{1}{4}$ **47.** 6   **49.** $\frac{21}{25}$
51. $\frac{60}{41}$ **53.** $\frac{15}{7}$ **55.** $\frac{3}{10}$ **57.** 6   **59.** $\frac{1}{2}$ **61.** $\frac{7}{6}$
63. $\dfrac{3b}{7a}$ **65.** $\dfrac{10}{n}$ **67.** $\frac{5}{6}$ **69.** 1   **71.** $\frac{5}{18}$ **73.** 0
75. $\frac{35}{18}$ **77.** 27   **79.** 1   **81.** $\frac{6}{35}$ **83.** 18   **85.** 📋
87. 📋   **89.** Row 1: 7, 2, 36, 14, 8, 8; row 2: 9, 18, 2, 10, 12, 21
91. $\frac{2}{5}$ **93.** $\dfrac{5q}{t}$ **95.** $\frac{6}{25}$ **97.** $\dfrac{5ap}{2cm}$ **99.** $\dfrac{23r}{18t}$
101. $\dfrac{28}{45}$ m^2 **103.** $14\frac{2}{9}$ m   **105.** $27\frac{3}{5}$ cm

Quick Quiz: Sections 1.1–1.3, p. 28

1. Let w represent the width of the box; $w - 4$
2. $5(3 + x)$, or $(x + 3)5$ **3.** $2 \cdot 2 \cdot 2 \cdot 5$ **4.** $\frac{9}{11}$ **5.** $\dfrac{11}{x}$

Technology Connection, p. 32

1. 2.236067977   **2.** 2.645751311   **3.** 3.605551275
4. 5.196152423

Exercise Set 1.4, pp. 34–36

1. Terminating   **2.** Integer   **3.** Whole number
4. Irrational number   **5.** Opposite   **6.** Absolute value
7. $-n$ **8.** $|x|$ **9.** $-10 < x$ **10.** $6 \geq y$
11. -9500; 5000 **13.** 100; -80 **15.** -777.68; 936.42
17. $10{,}000$; -4500 **19.** 8; -5
21.
23.
25.
27. 0.875 **29.** -0.75 **31.** $-1.1\overline{6}$ **33.** $0.\overline{6}$
35. -0.5 **37.** 0.13 **39.**
41. $-\sqrt{22}$ **43.** $>$ **45.** $<$
47. $<$ **49.** $>$ **51.** $<$ **53.** $<$ **55.** $x < -2$
57. $y \geq 10$ **59.** $-83, -4.7, 0, \frac{5}{9}, 2.\overline{16}, 62$ **61.** $-83, 0, 62$
63. $-83, -4.7, 0, \frac{5}{9}, 2.\overline{16}, \pi, \sqrt{17}, 62$ **65.** 58 **67.** 12.2
69. $\sqrt{2}$ **71.** $\frac{9}{7}$ **73.** 0 **75.** 8 **77.** ✏
79. ✏   **81.** ✏   **83.** $-23, -17, 0, 4$
85. $-\frac{4}{3}, \frac{4}{9}, \frac{4}{8}, \frac{4}{6}, \frac{4}{5}, \frac{4}{3}, \frac{4}{2}$ **87.** $<$ **89.** $=$ **91.** $-2, -1, 0, 1, 2$
93. $\frac{1}{9}$ **95.** $\frac{50}{9}$ **97.** $a < 0$ **99.** $|x| \leq 10$ **101.** ✏

Quick Quiz: Sections 1.1–1.4, p 36

1. 1 **2.** $5(x + y + 3)$ **3.** $\frac{1}{6}$ **4.** 2 **5.** $0 > -0.5$

Mid-Chapter Review: Chapter 1, p. 37

1. $\dfrac{x - y}{3} = \dfrac{22 - 10}{3} = \dfrac{12}{3} = 4$
2. $14x + 7 = 7 \cdot 2x + 7 \cdot 1 = 7(2x + 1)$
3. 15 **4.** 4 **5.** $d - 10$ **6.** Let h represent the
number of hours worked; $8h$ **7.** Let s represent the
number of students originally enrolled; $s - 5 = 27$
8. No   **9.** $10x + 7$ **10.** $(3a)b$ **11.** $8x + 32$
12. $6m + 15n + 30$ **13.** $3(6x + 5)$ **14.** $3(3c + 4d + 1)$
15. $2 \cdot 2 \cdot 3 \cdot 7$ **16.** $\frac{3}{7}$ **17.** $\frac{13}{24}$ **18.** $\frac{44}{45}$
19.
22. $<$ **23.** $9 \leq x$ **24.** 5.6 **25.** 0 **20.** -0.15 **21.** $>$

Exercise Set 1.5, pp. 41–43

1. Add; negative   **2.** Subtract; positive   **3.** Subtract;
negative   **4.** Identity   **5.** Identity   **6.** Like
7. (f)   **8.** (d)   **9.** (e)   **10.** (a)   **11.** (b)   **12.** (c)
13. -3 **15.** 4 **17.** -7 **19.** -8 **21.** -11
23. -5 **25.** 0 **27.** -41 **29.** 0 **31.** 9 **33.** -36
35. 11 **37.** -43 **39.** 0 **41.** 18 **43.** -16
45. -0.8 **47.** -9.1 **49.** $\frac{3}{5}$ **51.** $\frac{-6}{7}$ **53.** $-\frac{1}{15}$
55. $\frac{2}{9}$ **57.** -3 **59.** 0 **61.** The price dropped 1¢.
63. Her new balance was $95.   **65.** The total gain was 20 yd.
67. The lake dropped $\frac{7}{10}$ ft.   **69.** The elevation of the
peak is $13{,}796$ ft.   **71.** $17a$ **73.** $9x$ **75.** $-2m$

77. $-10y$ **79.** $1 - 2x$ **81.** $-4m$ **83.** $-5x - 3.9$
85. $12x + 17$ **87.** $7r + 8t + 16$ **89.** $18n + 16$
91. ✏   **93.** ✏   **95.** $451.70   **97.** $-5y$
99. $-7m$ **101.** $-7t, -23$ **103.** 1 under par

Quick Quiz: Sections 1.1–1.5, p. 43

1. $18a + 12c + 3$ **2.** $\frac{7}{5}$ **3.** $\frac{1}{12}$ **4.** 505 **5.** $-2 - 7x$

Exercise Set 1.6, pp. 48–51

1. Opposites   **2.** Sign   **3.** Opposite   **4.** Difference
5. (d)   **6.** (g)   **7.** (f)   **8.** (h)   **9.** (a)   **10.** (c)
11. (b)   **12.** (e)   **13.** Six minus ten   **15.** Two
minus negative twelve   **17.** The opposite of x minus y
19. Negative three minus the opposite of n **21.** -51
23. $\frac{11}{3}$ **25.** 3.14 **27.** 45 **29.** $\frac{14}{3}$ **31.** -0.101
33. 37 **35.** $-\frac{2}{5}$ **37.** 1 **39.** -15 **41.** -3
43. -6 **45.** -7 **47.** -6 **49.** 0 **51.** -5
53. -10 **55.** 0 **57.** 0 **59.** 8 **61.** -11 **63.** 16
65. -19 **67.** -1 **69.** 17 **71.** 3 **73.** -3
75. -21 **77.** -11 **79.** -8 **81.** -60 **83.** -23
85. -7.3 **87.** 1.1 **89.** -5.5 **91.** -0.928 **93.** $-\frac{7}{11}$
95. $-\frac{4}{5}$ **97.** $\frac{5}{17}$ **99.** 32 **101.** -62 **103.** -139
105. 0 **107.** $-3y, -8x$ **109.** $9, -5t, -3st$ **111.** $-3x$
113. $-5a + 4$ **115.** $-n - 9$ **117.** $-3x - 6$
119. $-8t - 7$ **121.** $-12x + 3y + 9$ **123.** $8x + 66$
125. -40 **127.** 43 **129.** $3.8 - (-5.2)$; 9
131. $114 - (-79)$; 193 **133.** 950 m   **135.** $213.8°$F
137. 8.9 points   **139.** ✏   **141.** ✏
143. 11:00 P.M., on August 14   **145.** False. For example,
let $m = -3$ and $n = -5$. Then $-3 > -5$, but
$-3 + (-5) = -8 \not> 0$.   **147.** True. For example,
for $m = 4$ and $n = -4$, $4 = -(-4)$ and $4 + (-4) = 0$;
for $m = -3$ and $n = 3$, $-3 = -3$ and $-3 + 3 = 0$.
149. (-) 9 − (-) 7 ENTER

Quick Quiz: Sections 1.1–1.6, p. 51

1. Let p represent the total PC sales; $0.11p = 9.25$
2. $1, 2, 4, 13, 26, 52$ **3.** $2 \cdot 2 \cdot 13$ **4.** -29 **5.** 5

Connecting the Concepts, p. 57

1. -10 **2.** 16 **3.** 4 **4.** -6 **5.** -1 **6.** 1
7. -3.77 **8.** -7 **9.** 0 **10.** -92

Exercise Set 1.7, pp. 57–59

1. Positive   **2.** Odd   **3.** Undefined   **4.** Reciprocal
5. Opposite   **6.** Reciprocal   **7.** 1 **8.** 0 **9.** 0
10. 1 **11.** 0 **12.** 1 **13.** 1 **14.** 0 **15.** 1
16. 0 **17.** -40 **19.** -56 **21.** -40 **23.** 72
25. 190 **27.** -132 **29.** -126 **31.** 11.5 **33.** 0
35. $-\frac{2}{7}$ **37.** $\frac{1}{12}$ **39.** -11.13 **41.** $-\frac{5}{12}$ **43.** 252
45. 0 **47.** $\frac{1}{28}$ **49.** 150 **51.** 0 **53.** -720
55. $-30{,}240$ **57.** -9 **59.** -4 **61.** -7 **63.** 4
65. -9 **67.** 5.1 **69.** $\frac{100}{11}$ **71.** -8 **73.** Undefined
75. -4 **77.** 0 **79.** 0 **81.** $-\frac{8}{3}; \frac{8}{-3}$ **83.** $-\frac{29}{35}; \frac{-29}{35}$

85. $\frac{-7}{3}; \frac{7}{-3}$ **87.** $-\frac{x}{2}; \frac{x}{-2}$ **89.** $-\frac{5}{4}$ **91.** $-\frac{10}{51}$ **93.** $-\frac{1}{10}$
95. $\frac{1}{4.3}$, or $\frac{10}{43}$ **97.** -4 **99.** Does not exist   **101.** $\frac{21}{20}$
103. -1 **105.** 1   **107.** $\frac{3}{11}$ **109.** $-\frac{7}{4}$ **111.** -12
113. -3 **115.** 1   **117.** $\frac{1}{10}$ **119.** $-\frac{7}{6}$
121. Undefined   **123.** $-\frac{14}{15}$ **125.** 🖊️   **127.** 🖊️
129. $\dfrac{1}{a+b}$ **131.** $-(a+b)$ **133.** $x = -x$
135. For 2 and 3, the reciprocal of the sum is $1/(2+3)$, or $1/5$. But $1/5 \neq 1/2 + 1/3$.   **137.** 5°F   **139.** Positive
141. Positive   **143.** Positive   **145.** Distributive law; law of opposites; multiplicative property of zero   **147.** 🖥️

Quick Quiz: Sections 1.1–1.7, p. 59

1. $11(2x + 1 + 3y)$ **2.**
$$-3.5$$
(number line from -5 to 5, point at -3.5)
3. -6.1 **4.** $-x - 8m$ **5.** -2

Exercise Set 1.8, pp. 66–68

1. (c)   **2.** (b)   **3.** (a)   **4.** (f)   **5.** (e)   **6.** (d)
7. Division   **8.** Subtraction   **9.** Addition
10. Multiplication   **11.** Subtraction   **12.** Multiplication
13. x^6 **15.** $(-5)^3$ **17.** $(3t)^5$ **19.** $2n^4$ **21.** 16
23. 9   **25.** -9 **27.** 64   **29.** 625   **31.** 7   **33.** -32
35. $81t^4$ **37.** $-343x^3$ **39.** 26   **41.** 51   **43.** -15
45. 1   **47.** 298   **49.** 11   **51.** -36 **53.** 1291
55. 152   **57.** 24   **59.** 1   **61.** -44 **63.** 41
65. -10 **67.** -5 **69.** -19 **71.** -3 **73.** -75
75. 9   **77.** 30   **79.** 6   **81.** -17 **83.** $13x + 33$
85. $17x^2 - 17x$ **87.** $21t - r$ **89.** $-t^3 + 4t$
91. $-9x - 1$ **93.** $7n - 8$ **95.** $-4a + 3b - 7c$
97. $-3x^2 - 5x + 1$ **99.** $2x - 7$ **101.** $-9x + 6$
103. $9y - 25z$ **105.** $x^2 + 6$ **107.** $37a^2 - 23ab + 35b^2$
109. $-22t^3 - t^2 + 9t$ **111.** $2x - 25$ **113.** 🖊️
115. 🖊️   **117.** $-6r - 5t + 21$ **119.** $-2x - f$
121. 🖊️   **123.** True   **125.** False   **127.** 0   **129.** 17
131. 39,000   **133.** $44x^3$

Quick Quiz: Sections 1.1–1.8, p. 68

1. Let m and n represent the numbers; $\frac{1}{2}(m + n)$
2. $-10 < x$ **3.** $-1.\overline{2}$ **4.** -2.94 **5.** $8x - 9y$

Translating for Success, p. 69

1. H   **2.** E   **3.** K   **4.** B   **5.** O   **6.** L   **7.** M
8. C   **9.** D   **10.** F

Decision Making: Connection, p. 71

1. $-100; 500; -100$; weight-loss diet: Monday, Wednesday; weight-gain diet: Tuesday   **2.** 2649 calories

Study Summary: Chapter 1, pp. 72–75

1. 8   **2.** 4 ft^2 **3.** Let n represent some number; $78 = n - 92$ **4.** $10n + 6$ **5.** $(3a)b$
6. $50m + 90n + 10$ **7.** $13(2x + 1)$ **8.** Composite

9. $2 \cdot 2 \cdot 3 \cdot 7$ **10.** 1   **11.** $\frac{9}{10}$ **12.** $\frac{3}{2}$ **13.** $\frac{9}{20}$
14. $\frac{25}{6}$ **15.** 25   **16.** $0, -15, \frac{30}{3}$ **17.** $-1.\overline{1}$ **18.** >
19. 1.5   **20.** -5 **21.** -2.9 **22.** -12 **23.** 15
24. $-7c + 9d - 2$ **25.** 21   **26.** -4 **27.** -100
28. -21 **29.** $a - 2b + 3c$ **30.** $5m + 7n - 15$

Review Exercises: Chapter 1, pp. 75–77

1. True   **2.** True   **3.** False   **4.** True   **5.** False
6. False   **7.** True   **8.** False   **9.** False   **10.** True
11. 24   **12.** -16 **13.** -15 **14.** $y - 7$
15. $xz + 10$, or $10 + xz$ **16.** Let b represent Brandt's speed and w the wind speed; $15(b - w)$ **17.** Let b represent the number of calories per hour that Katie burns while backpacking; $b = 2 \cdot 237$ **18.** $c = 200t$ **19.** $t \cdot 3 + 5$
20. $2x + (y + z)$ **21.** $(4x)y, 4(yx), (4y)x$; answers may vary
22. $18x + 30y$ **23.** $40x + 24y + 16$ **24.** $3(7x + 5y)$
25. $11(2a + 9b + 1)$ **26.** $2 \cdot 2 \cdot 2 \cdot 7$ **27.** $\frac{5}{12}$ **28.** $\frac{9}{4}$
29. $\frac{19}{24}$ **30.** $\frac{3}{16}$ **31.** $\frac{3}{5}$ **32.** $\frac{27}{25}$ **33.** $-3600; 1350$
34.
$$\frac{-1}{3}$$
(number line from -5 to 5, point at $-\frac{1}{3}$)
35. $x > -3$ **36.** <
37. $-0.\overline{4}$ **38.** 1   **39.** -12 **40.** -10 **41.** $-\frac{7}{12}$
42. -5 **43.** 8   **44.** $-\frac{7}{5}$ **45.** -9.18 **46.** $-\frac{2}{7}$
47. -140 **48.** -7 **49.** -3 **50.** $\frac{9}{4}$ **51.** 48
52. 168   **53.** $\frac{21}{8}$ **54.** 18   **55.** 53   **56.** $\frac{103}{17}$
57. $7a - b$ **58.** $-4x + 5y$ **59.** 7   **60.** $-\frac{1}{7}$
61. $(2x)^4$ **62.** $-125x^3$ **63.** $-3a + 9$ **64.** $3x^4 + 10x$
65. $17n^2 + m^2 + 20mn$ **66.** $5x + 28$
67. 🖊️ The value of a constant never varies. A variable can represent a variety of numbers.   **68.** 🖊️ A term is one of the parts of an expression that is separated from the other parts by plus signs. A factor is part of a product.   **69.** 🖊️ The distributive law is used in factoring algebraic expressions, multiplying algebraic expressions, combining like terms, finding the opposite of a sum, and subtracting algebraic expressions.   **70.** 🖊️ A negative number raised to an even power is positive; a negative number raised to an odd power is negative.   **71.** 25,281   **72.** (a) $\frac{3}{11}$; (b) $\frac{10}{11}$
73. $-\frac{5}{8}$ **74.** -2.1 **75.** (i)   **76.** (j)   **77.** (a)
78. (h)   **79.** (k)   **80.** (b)   **81.** (c)   **82.** (e)
83. (d)   **84.** (f)   **85.** (g)

Test: Chapter 1, p. 78

1. [1.1] 4   **2.** [1.1] Let x and y represent the numbers; $xy - 9$ **3.** [1.1] 240 ft^2 **4.** [1.2] $q + 3p$
5. [1.2] $(x \cdot 4) \cdot y$ **6.** [1.1] Let t represent the number of golden lion tamarins living in zoos; $1500 = t + 1050$
7. [1.2] $35 + 7x$ **8.** [1.7] $-5y + 10$ **9.** [1.2] $11(1 + 4x)$
10. [1.2] $7(x + 1 + 7y)$ **11.** [1.3] $2 \cdot 2 \cdot 3 \cdot 5 \cdot 5$
12. [1.3] $\frac{3}{7}$ **13.** [1.4] <   **14.** [1.4] >   **15.** [1.4] $\frac{9}{4}$
16. [1.4] 3.8   **17.** [1.6] $\frac{2}{3}$ **18.** [1.7] $-\frac{7}{4}$ **19.** [1.6] 10
20. [1.4] $-5 \geq x$ **21.** [1.6] 7.8   **22.** [1.5] -8
23. [1.6] $-\frac{7}{8}$ **24.** [1.7] -48 **25.** [1.7] $\frac{2}{9}$ **26.** [1.7] $\frac{3}{4}$
27. [1.7] -9.728 **28.** [1.8] 20   **29.** [1.6] 15
30. [1.8] -64 **31.** [1.8] 448   **32.** [1.6] $21a + 22y$
33. [1.8] $16x^4$ **34.** [1.8] $x + 7$ **35.** [1.8] $9a - 12b - 7$
36. [1.8] $-y - 16$ **37.** [1.1] 5
38. [1.8] $9 - (3 - 4) + 5 = 15$ **39.** [1.4] $n \geq 0$
40. [1.8] $4a$ **41.** [1.8] False

CHAPTER 2

Exercise Set 2.1, pp. 86–87

1. (c)   **2.** (b)   **3.** (f)   **4.** (a)   **5.** (d)   **6.** (e)
7. (d)   **8.** (b)   **9.** (c)   **10.** (a)   **11.** No   **13.** Yes
15. Yes   **17.** Yes   **19.** 11   **21.** -25 **23.** -31
25. 41   **27.** 19   **29.** -6 **31.** $\frac{7}{3}$ **33.** $-\frac{1}{10}$
35. $\frac{41}{24}$ **37.** $-\frac{1}{20}$ **39.** 9.1   **41.** -5 **43.** 7
45. 12   **47.** -38 **49.** 8   **51.** -7 **53.** 8
55. 88   **57.** 20   **59.** -54 **61.** $-\frac{5}{9}$ **63.** 1
65. $\frac{9}{2}$ **67.** -7.6 **69.** -2.5 **71.** -15 **73.** -5
75. $-\frac{7}{6}$ **77.** -128 **79.** $-\frac{1}{2}$ **81.** -15 **83.** 9
85. 310.756   **87.** 🖼   **89.** $\frac{1}{3}y - 7$ **90.** $12x + 66$

91. $5(7a + 11c + 1)$ **92.**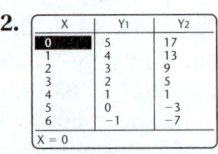
93. 🖼   **95.** 11.6   **97.** 2   **99.** $-23, 23$
101. 9000   **103.** $2500

Prepare to Move On, p. 87

1. -6 **2.** 2   **3.** $-x + 10$ **4.** $-5x + 28$

Technology Connection, p. 90

1.

X	Y1
0	5
1	4
2	3
3	2
4	1
5	0
6	-1

X = 0

2.

X	Y1	Y2
0	5	17
1	4	13
2	3	9
3	2	5
4	1	1
5	0	-3
6	-1	-7

X = 0

3. 4; not reliable because, depending on the choice of ΔTbl, it is easy to scroll past a solution without realizing it.

Exercise Set 2.2, pp. 94–95

1. Addition principle   **2.** Multiplication principle
3. Multiplication principle   **4.** Distributive law
5. Addition principle   **6.** Distributive law
7. (c)   **8.** (e)   **9.** (a)   **10.** (f)   **11.** (b)   **12.** (d)
13. 8   **15.** 5   **17.** $\frac{10}{3}$ **19.** -7 **21.** -5 **23.** -4
25. 19   **27.** -2.8 **29.** 3   **31.** 15   **33.** -6
35. $-\frac{25}{2}$ **37.** $\frac{9}{8}$ **39.** -3 **41.** -6 **43.** 0
45. 10   **47.** 4   **49.** $\frac{16}{3}$ **51.** $\frac{2}{5}$ **53.** 1   **55.** -4
57. $1.\overline{6}$ **59.** $-\frac{60}{37}$ **61.** 11   **63.** 2   **65.** 0   **67.** 6
69. All real numbers; identity   **71.** $-\frac{1}{2}$ **73.** 0
75. No solution; contradiction   **77.** $\frac{5}{2}$ **79.** $\frac{1}{6}$ **81.** 2
83. No solution; contradiction   **85.** 8   **87.** $\frac{16}{15}$
89. $-\frac{1}{31}$ **91.** All real numbers; identity   **93.** 2
95. 🖼   **97.** $\frac{7}{8}$ **98.** 1   **99.** $-0.\overline{1}$ **100.** 16
101. 🖼   **103.** $\frac{1136}{909}$, or $1.\overline{2497}$ **105.** No solution;
contradiction   **107.** All real numbers are solutions;
identity   **109.** $\frac{2}{3}$ **111.** 0   **113.** 500 mi

Quick Quiz: Sections 2.1–2.2, p. 95

1. Yes   **2.** -5.4 **3.** 25   **4.** -4 **5.** $\frac{1}{2}$

Prepare to Move On, p. 95

1. -7 **2.** 15   **3.** -15 **4.** -28

Technology Connection, p. 96

1. 14.4

Exercise Set 2.3, pp. 99–102

1. False   **2.** True   **3.** Circumference   **4.** Formula
5. 309.6 m   **7.** 1423 students   **9.** 8.4734
11. 255 mg   **13.** $b = \dfrac{A}{h}$ **15.** $P = \dfrac{I}{rt}$
17. $m = 65 - H$ **19.** $l = \dfrac{P - 2w}{2}$, or $l = \dfrac{P}{2} - w$
21. $\pi = \dfrac{A}{r^2}$ **23.** $h = \dfrac{2A}{b}$ **25.** $c^2 = \dfrac{E}{m}$
27. $d = 2Q - c$ **29.** $q = p + r - 2$ **31.** $r = wf$
33. $T = \dfrac{550H}{V}$ **35.** $C = \frac{5}{9}(F - 32)$ **37.** $y = 2x - 1$
39. $y = -\frac{2}{5}x + 2$ **41.** $y = \frac{4}{3}x - 2$ **43.** $y = -\frac{9}{8}x + \frac{1}{2}$
45. $y = \frac{3}{5}x - \frac{8}{5}$ **47.** $x = \dfrac{z - 13}{2} - y$, or $x = \dfrac{z - 13 - 2y}{2}$
49. $l = 4(t - 27) + w$ **51.** $t = \dfrac{A}{a + b}$ **53.** $h = \dfrac{2A}{a + b}$
55. $L = W - \dfrac{N(R - r)}{400}$, or $L = \dfrac{400W - NR + Nr}{400}$
57. 🖼   **59.** -10 **60.** -196 **61.** 0
62. -32 **63.** -13 **64.** 65   **65.** 🖼
67. 40 years old   **69.** 27 in³   **71.** $a = \dfrac{w}{c} \cdot d$
73. $c = \dfrac{d}{a - b}$ **75.** $a = \dfrac{c}{3 + b + d}$
77. $K = 9.632w + 19.685h - 10.54a + 102.3$

Quick Quiz: Sections 2.1–2.3, p. 102

1. 7   **2.** $\frac{64}{3}$ **3.** -0.025 **4.** -21 **5.** $y = -3x + \frac{1}{2}$

Prepare to Move On, p. 102

1. 0.25   **2.** 1.125   **3.** $0.\overline{6}$ **4.** $0.8\overline{3}$

Mid-Chapter Review: Chapter 2, p. 103

1. $2x + 3 - 3 = 10 - 3$
$\qquad 2x = 7$
$\qquad \frac{1}{2} \cdot 2x = \frac{1}{2} \cdot 7$
$\qquad x = \frac{7}{2}$
2. $6 \cdot \frac{1}{2}(x - 3) = 6 \cdot \frac{1}{3}(x - 4)$
$\qquad 3(x - 3) = 2(x - 4)$
$\qquad 3x - 9 = 2x - 8$
$\qquad 3x - 9 + 9 = 2x - 8 + 9$
$\qquad 3x = 2x + 1$
$\qquad 3x - 2x = 2x + 1 - 2x$
$\qquad x = 1$
3. 1   **4.** 3   **5.** $\frac{5}{3}$ **6.** -8 **7.** 48   **8.** 0.5
9. -5 **10.** $\frac{8}{3}$ **11.** 6   **12.** 0   **13.** $\frac{49}{9}$
14. $-\frac{4}{11}$ **15.** $\frac{23}{7}$ **16.** $A = \dfrac{E}{w}$ **17.** $y = \dfrac{C - Ax}{B}$
18. $a = \dfrac{m}{t + p}$ **19.** $a = \dfrac{F}{m}$ **20.** $b = vt + f$

Exercise Set 2.4, pp. 108–112

1. Left   **2.** Hundred   **3.** Percent   **4.** Base
5. Sale   **6.** Approximately   **7.** (d)   **8.** (c)   **9.** (e)
10. (b)   **11.** (c)   **12.** (d)   **13.** (f)   **14.** (a)
15. (b)   **16.** (e)   **17.** 0.67   **19.** 0.05   **21.** 0.032
23. 0.1   **25.** 0.0625   **27.** 0.002   **29.** 1.75
31. 21%   **33.** 7.9%   **35.** 70%   **37.** 0.09%
39. 106%   **41.** 60%   **43.** 32%   **45.** 25%
47. $46\frac{2}{3}$, or $\frac{140}{3}$ **49.** 2.5   **51.** 10,000   **53.** 125%
55. 0.8   **57.** 50%   **59.** $33.\overline{3}\%$, or $33\frac{1}{3}\%$
61. 2.48 quadrillion Btu's   **63.** 0.88 quadrillion Btu's
65. 75 credits   **67.** 573 at-bats   **69.** (a) 16%;
(b) \$29   **71.** $33.\overline{3}\%$, or $33\frac{1}{3}\%$; $66.\overline{6}\%$, or $66\frac{2}{3}\%$
73. \$163.20   **75.** 285 women   **77.** \$19.20 per hour
79. About 462%   **81.** About 31.5 lb   **83.** About 90%
85. 50 calories   **87.** ☐   **89.** $\frac{1}{3}$ **90.** -3 **91.** -12
92. $9x^2$ **93.** ☐   **95.** 18,500 people
97. About 6 ft 7 in.   **99.** About 27%   **101.** ☐

Quick Quiz: Sections 2.1–2.4, p. 112

1. 0   **2.** $\frac{62}{3}$ **3.** $p = \dfrac{T}{4+m}$ **4.** 0.012   **5.** 25

Prepare to Move On, p. 112

1. Let l represent the length and w represent the width;
$2l + 2w$ **2.** $0.05 \cdot 180$ **3.** $10\left(\frac{1}{2}a\right)$
4. Let n represent the number; $3n + 10$
5. Let l represent the board's length and w represent the
width; $w = l - 2$ **6.** Let x represent the first number
and y represent the second number; $x = 4y$

Exercise Set 2.5, pp. 121–127

1. (1) Familiarize. (2) Translate. (3) Carry out.
(4) Check. (5) State.   **2.** Carry out.   **3.** State.
4. Familiarize.   **5.** Translate.   **6.** Familiarize.
7. Familiarize.   **8.** Check.   **9.** 11   **11.** $\frac{11}{2}$
13. 16.4 mi   **15.** 290 mi   **17.** 1204 and 1205
19. 285 and 287   **21.** 32, 33, 34   **23.** Man: 104 years;
woman: 102 years   **25.** Spam: 264.6 billion messages;
nonspam: 29.4 billion messages   **27.** 140 and 141
29. Width: 21 m; length: 25 m   **31.** $1\frac{1}{2}$ in. by $3\frac{1}{2}$ in.
33. 30°, 90°, 60°   **35.** 70°   **37.** Bottom: 144 ft; middle:
72 ft; top: 24 ft   **39.** 12 mph   **41.** 1 hr   **43.** 12.5%
45. 50%   **47.** \$1225   **49.** \$280.80   **51.** \$320
53. \$36,000   **55.** \$100   **57.** 256,850 vehicles
59. \$125,000   **61.** $7\frac{1}{4}$ mi   **63.** $128\frac{1}{3}$ mi   **65.** 65°, 25°
67. 140°, 40°   **69.** Length: 27.9 cm; width: 21.6 cm
71. \$6600   **73.** 830 points   **75.** 2015
77. 160 chirps per minute   **79.** ☐   **81.** $8n + 32t + 4$
82. $3(4 + 6x + 7y)$ **83.** $7x - 18$ **84.** 0   **85.** ☐
87. \$37   **89.** 20   **91.** Half-dollars: 5; quarters: 10;
dimes: 20; nickels: 60   **93.** \$95.99   **95.** 5 DVDs
97. 6 mi   **99.** ☐   **101.** Width: 23.31 cm;
length: 27.56 cm

Quick Quiz: Sections 2.1–2.5, p. 127

1. $\frac{25}{6}$ **2.** -3 **3.** $v = \dfrac{3}{B}$ **4.** \$10.50   **5.** $52\frac{1}{2}$ min

Prepare to Move On, p. 127

1. $<$ **2.** $>$ **3.** $-4 \le x$ **4.** $y < 5$

Connecting the Concepts, p. 134

1. 21   **2.** $\{x \mid x \le 21\}$, or $(-\infty, 21]$ **3.** -6
4. $\{x \mid x > -6\}$, or $(-6, \infty)$ **5.** $\{x \mid x \le -\frac{1}{3}\}$, or $\left(-\infty, -\frac{1}{3}\right]$
6. $-\frac{1}{3}$ **7.** 66   **8.** $\{n \mid n < 66\}$, or $(-\infty, 66)$
9. $\{a \mid a \ge 0\}$, or $[0, \infty)$ **10.** 0

Exercise Set 2.6, pp. 134–136

1. Solution   **2.** Set-builder   **3.** Closed   **4.** Bracket
5. $\ge$ **6.** $\le$ **7.** $<$ **8.** $>$ **9.** Equivalent
10. Equivalent   **11.** Equivalent   **12.** Not equivalent
13. (a) Yes; (b) no; (c) no   **15.** (a) Yes; (b) no; (c) yes
17. (a) Yes; (b) yes; (c) yes

19. $y < 2$

21. $x \ge -1$

23. $0 \le t$

25. $-5 \le x < 2$

27. $-4 < x < 0$

29. $\{y \mid y < 6\}$, $(-\infty, 6)$
31. $\{x \mid x \ge -4\}$, $[-4, \infty)$
33. $\{t \mid t > -3\}$, $(-3, \infty)$
35. $\{x \mid x \le -7\}$, $(-\infty, -7]$
37. $\{x \mid x > -4\}$, $(-4, \infty)$ **39.** $\{x \mid x \le 2\}$, $(-\infty, 2]$
41. $\{x \mid x < -1\}$, $(-\infty, -1)$ **43.** $\{x \mid x \ge 0\}$, $[0, \infty)$
45. $\{y \mid y > 3\}$, $(3, \infty)$
47. $\{n \mid n < 17\}$, $(-\infty, 17)$,
49. $\{x \mid x \le -9\}$, $(-\infty, -9]$,
51. $\{t \mid t \le -3\}$, $(-\infty, -3]$,
53. $\{t \mid t > \frac{5}{8}\}$, $\left(\frac{5}{8}, \infty\right)$,
55. $\{x \mid x < 0\}$, $(-\infty, 0)$,
57. $\{t \mid t < 23\}$, $(-\infty, 23)$,
59. $\{x \mid x < 7\}$, $(-\infty, 7)$,
61. $\{t \mid t < -3\}$, $(-\infty, -3)$,
63. $\{n \mid n \ge -1.5\}$, $[-1.5, \infty)$,
65. $\{y \mid y \ge -\frac{1}{10}\}$, $\left[-\frac{1}{10}, \infty\right)$,
67. $\{x \mid x < -\frac{4}{5}\}$, $\left(-\infty, -\frac{4}{5}\right)$,
69. $\{x \mid x < 6\}$, or $(-\infty, 6)$ **71.** $\{t \mid t \le 7\}$, or $(-\infty, 7]$
73. $\{x \mid x > -4\}$, or $(-4, \infty)$ **75.** $\{y \mid y < -\frac{10}{3}\}$, or $\left(-\infty, -\frac{10}{3}\right)$
77. $\{x \mid x > -10\}$, or $(-10, \infty)$ **79.** $\{y \mid y < 0\}$, or $(-\infty, 0)$
81. $\{x \mid x > -4\}$, or $(-4, \infty)$ **83.** $\{t \mid t > 1\}$, or $(1, \infty)$
85. $\{x \mid x \le -9\}$, or $(-\infty, -9]$ **87.** $\{t \mid t < 14\}$, or $(-\infty, 14)$
89. $\{y \mid y \le -4\}$, or $(-\infty, -4]$ **91.** $\{t \mid t < -\frac{5}{3}\}$, or $\left(-\infty, -\frac{5}{3}\right)$
93. $\{r \mid r > -3\}$, or $(-3, \infty)$ **95.** $\{x \mid x \le 7\}$, or $(-\infty, 7]$

97. $\left\{x \mid x > -\frac{5}{32}\right\}$, or $\left(-\frac{5}{32}, \infty\right)$ **99.** 🖉   **101.** $17x - 6$
102. $2m - 16n$ **103.** $7x - 8y - 46$ **104.** $-21x + 32$
105. 🖉   **107.** $\{x \mid x \text{ is a real number}\}$, or $(-\infty, \infty)$
109. $\left\{x \mid x \leq \frac{5}{6}\right\}$, or $\left(-\infty, \frac{5}{6}\right]$ **111.** $\{x \mid x \leq -4a\}$, or
$(-\infty, -4a]$ **113.** $\left\{x \mid x > \dfrac{y - b}{a}\right\}$, or $\left(\dfrac{y - b}{a}, \infty\right)$
115. $\{x \mid x \text{ is a real number}\}$, or $(-\infty, \infty)$ **117. (a)** No;
(b) There is more than 6 g of fat per serving.

Quick Quiz: Sections 2.1–2.6, p. 136

1. $-\frac{28}{3}$ **2.** $\left\{x \mid x < -\frac{10}{3}\right\}$, or $\left(-\infty, -\frac{10}{3}\right)$ **3.** $15.00
4. $19.34   **5.** $c = \dfrac{X - 12 + 5d}{5}$

Prepare to Move On, p. 136

1. Let h represent the height of the triangle; $\frac{1}{2} \cdot 3 \cdot h = 5$
2. Let s represent the shortest side of the triangle;
$s + (s + 1) + \left(s + \frac{1}{2}\right) = 12$

Exercise Set 2.7, pp. 139–143

1. Is at least   **2.** Cannot exceed   **3.** No less than
4. Minimum   **5.** $b \leq a$ **6.** $b < a$ **7.** $a \leq b$
8. $a < b$ **9.** $b \leq a$ **10.** $a \leq b$ **11.** $b < a$
12. $a < b$ **13.** Let n represent the number; $n < 10$
15. Let t represent the temperature; $t \leq -3$
17. Let d represent the number of years of driving
experience; $d \geq 5$ **19.** Let a represent the age of the
altar; $a > 1200$ **21.** Let h represent Bianca's hourly
wage, in dollars; $h \geq 12$ **23.** Let s represent the number
of hours of sunshine; $1100 < s < 1600$ **25.** More than
2.375 hr   **27.** At least 2.25   **29.** Scores greater than or
equal to 97   **31.** 8 credits or more   **33.** At least 3 plate
appearances   **35.** Lengths greater than 6 cm
37. Depths less than 437.5 ft   **39.** Blue-book value is
greater than or equal to $10,625   **41.** Lengths less than
55 in.   **43.** Temperatures greater than 37°C   **45.** No
more than 3 ft tall   **47.** A serving contains at least $6\frac{2}{3}$ g of
fat.   **49.** Dates after September 16   **51.** No more than
134 text messages   **53.** Years after 2012
55. Distances less than or equal to 139 mi   **57.** 🖉
59. $7 + xy$ **60.** $3 \cdot 3 \cdot 5 \cdot 5$ **61.** -18 **62.** -5
63. 🖉   **65.** Temperatures between $-15°C$ and $-9\frac{4}{9}°C$
67. Lengths less than or equal to 8 cm
69. At least $42   **71.** 🖉   **73.** 🖥

Quick Quiz: Sections 2.1–2.7, p. 143

1. $\{x \mid x \leq 4\}$, or $(-\infty, 4]$ **2.** $\frac{1210}{3}$ **3.** 20%   **4.** 48 km
5. 72 or higher

Prepare to Move On, p. 143

1. ←●—┼—→
        -12 0
2. ←—┼——●→
          0 26
3. 79.5   **4.** 8.5

Translating for Success, p. 144

1. F   **2.** I   **3.** C   **4.** E   **5.** D   **6.** J   **7.** O
8. M   **9.** B   **10.** L

Decision Making: Connection, p. 145

1. Barry should remain a crew manager.   **2.** $A = 1.2x$
3. 🖥

Study Summary: Chapter 2, pp. 146–148

1. 5   **2.** 4.8   **3.** -1 **4.** $-\frac{1}{10}$ **5.** $c = \dfrac{d}{a - b}$
6. 80   **7.** $47\frac{1}{2}$ mi; $72\frac{1}{2}$ mi   **8.** $(-\infty, 0]$
9. $\{x \mid x > 7\}$, or $(7, \infty)$ **10.** $\left\{x \mid x \geq -\frac{1}{4}\right\}$, or $\left[-\frac{1}{4}, \infty\right)$
11. Let d represent the distance Luke runs, in miles; $d \geq 3$

Review Exercises: Chapter 2, pp. 149–150

1. True   **2.** False   **3.** True   **4.** True
5. True   **6.** False   **7.** True   **8.** True   **9.** -25
10. 7   **11.** -65 **12.** 1.11   **13.** $\frac{1}{2}$ **14.** $-\frac{3}{2}$
15. -8 **16.** -4 **17.** $-\frac{1}{3}$ **18.** 4   **19.** No
solution; contradiction   **20.** 4   **21.** 16   **22.** 1
23. $-\frac{7}{5}$ **24.** 0   **25.** All real numbers; identity
26. $d = \dfrac{C}{\pi}$ **27.** $B = \dfrac{3V}{h}$ **28.** $y = \frac{5}{2}x - 5$
29. $x = \dfrac{b}{t - a}$ **30.** 0.012   **31.** 44%   **32.** 70%
33. 140   **34.** No   **35.** Yes   **36.** No
37. ←┼┼┼┼┼┼○┼┼┼→
          $5x - 6 < 2x + 3$
   $-5\,-4\,-3\,-2\,-1\ 0\ 1\ 2\ 3\ 4\ 5$
          $-2 < x \leq 5$
38. ←┼┼┼┼○┼┼┼┼┼┼→
   $-5\,-4\,-3\,-2\,-1\ 0\ 1\ 2\ 3\ 4\ 5$
          $t > 0$
39. ←┼┼┼┼┼┼┼○┼┼┼→
   $-5\,-4\,-3\,-2\,-1\ 0\ 1\ 2\ 3\ 4\ 5$
40. $\left\{t \mid t \geq -\frac{1}{2}\right\}$, or $\left[-\frac{1}{2}, \infty\right)$ **41.** $\{y \mid y > 3\}$, or $(3, \infty)$
42. $\{y \mid y \leq -4\}$, or $(-\infty, -4]$ **43.** $\{y \mid y > -7\}$, or $(-7, \infty)$
44. $\{x \mid x > -6\}$, or $(-6, \infty)$ **45.** $\left\{x \mid x > -\frac{9}{11}\right\}$, or $\left(-\frac{9}{11}, \infty\right)$
46. $\{t \mid t \leq -12\}$, or $(-\infty, -12]$ **47.** $\{x \mid x \leq -8\}$,
or $(-\infty, -8]$ **48.** 68.75%   **49.** 8 ft, 10 ft
50. Indian students: 104,000; Chinese students: 158,000
51. About 157,000 students   **52.** 57, 59
53. Width: 11 cm; length: 17 cm   **54.** $160
55. $35°, 85°, 60°$ **56.** $105 or less   **57.** 14 or fewer
copies   **58.** 🖉 Multiplying both sides of an equation
by *any* nonzero number results in an equivalent equation.
When multiplying on both sides of an inequality, the sign of
the number being multiplied by must be considered. If the
number is positive, the direction of the inequality symbol
remains unchanged; if the number is negative, the direction of the inequality symbol must be reversed to produce
an equivalent inequality.   **59.** 🖉 The solutions of an
equation can usually each be checked. The solutions of an
inequality are normally too numerous to check. Checking
a few numbers from the solution set found cannot guarantee that the answer is correct, although if any number
does not check, the answer found is incorrect.

60. About 1 hr 36 min　　**61.** Nile: 4160 mi; Amazon: 4225 mi　　**62.** $-23, 23$　　**63.** $-20, 20$
64. $a = \dfrac{y-3}{2-b}$　　**65.** $F = \dfrac{0.3(12w)}{9}$, or $F = 0.4w$

Test: Chapter 2, p. 151

1. [2.1] 9　　**2.** [2.1] -3　　**3.** [2.1] 49　　**4.** [2.1] -12
5. [2.2] 2　　**6.** [2.1] -8　　**7.** [2.2] $-\frac{23}{67}$　　**8.** [2.2] 7
9. [2.2] $\frac{23}{3}$　　**10.** All real numbers; identity
11. [2.6] $\{x \mid x > -5\}$, or $(-5, \infty)$　　**12.** [2.6] $\{x \mid x > -13\}$, or $(-13, \infty)$　　**13.** [2.6] $\{y \mid y \le -13\}$, or $(-\infty, -13]$
14. [2.6] $\{n \mid n < -5\}$, or $(-\infty, -5)$　　**15.** [2.6] $\{x \mid x < -7\}$, or $(-\infty, -7)$　　**16.** [2.6] $\{t \mid t \ge -1\}$, or $[-1, \infty)$
17. [2.6] $\{x \mid x \le -1\}$, or $(-\infty, -1]$　　**18.** [2.3] $r = \dfrac{A}{2\pi h}$
19. [2.3] $l = 2w - P$　　**20.** [2.4] 2.3　　**21.** [2.4] 0.3%
22. [2.4] 14.8　　**23.** [2.4] 44%

24. [2.6]
$$y < 4$$
$$\overset{}{\xleftarrow{\hspace{1em}}\underset{-10\,-8\,-6\,-4\,-2\ \ 0\ \ 2\ \ 4\ \ 6\ \ 8\ 10}{\rule{0pt}{0pt}}\xrightarrow{\hspace{1em}}}$$
$$-2 \le x \le 2$$

25. [2.6]
$$\overset{}{\xleftarrow{\hspace{1em}}\underset{-5\,-4\,-3\,-2\,-1\ 0\ 1\ 2\ 3\ 4\ 5}{\rule{0pt}{0pt}}\xrightarrow{\hspace{1em}}}$$

26. [2.5] Width: 7 cm; length: 11 cm
27. [2.5] 123.75 mi from Buffalo River and 41.25 mi from Lake Ft. Smith State Park　　**28.** [2.5] 81 mm, 83 mm, 85 mm
29. [2.4] $65　　**30.** [2.7] More than 35 one-way trips per month　　**31.** [2.3] $d = \dfrac{a}{3}$　　**32.** [1.4], [2.2] $-15, 15$
33. [2.7] Let h = the number of hours of sun each day; $4 \le h \le 6$　　**34.** [2.5] 60 tickets

Cumulative Review: Chapters 1–2, p. 152

1. -12　　**2.** $\frac{3}{4}$　　**3.** -4.2　　**4.** 10　　**5.** 134
6. $\frac{1}{2}$　　**7.** $2x + 1$　　**8.** $-21n + 36$

9.
$$\overset{-\frac{5}{2}}{\xleftarrow{\hspace{1em}}\underset{-5\,-4\,-3\,-2\,-1\ 0\ 1\ 2\ 3\ 4\ 5}{\rule{0pt}{0pt}}\xrightarrow{\hspace{1em}}}$$
10. $2(3x + 2y + 4z)$
11. 16　　**12.** 9　　**13.** $\frac{13}{18}$　　**14.** 1　　**15.** $-\frac{7}{2}$
16. $z = \dfrac{x}{4y}$　　**17.** $y = \frac{4}{9}x - \frac{1}{9}$　　**18.** $n = \dfrac{p}{a + r}$
19. 1.83　　**20.** 37.5%　　**21.**
$$\overset{t > -\frac{5}{2}}{\xleftarrow{\hspace{1em}}\underset{-5\,-4\,-3\,-2\,-1\ 0\ 1\ 2\ 3\ 4\ 5}{\rule{0pt}{0pt}}\xrightarrow{\hspace{1em}}}$$
22. $\{t \mid t \le -2\}$, or $(-\infty, -2]$　　**23.** $\{t \mid t < 3\}$, or $(-\infty, 3)$
24. $\{x \mid x < 30\}$, or $(-\infty, 30)$　　**25.** $\{n \mid n \ge 2\}$, or $[2, \infty)$
26. 48 million　　**27.** $14\frac{1}{3}$ m　　**28.** 9 ft, 15 ft
29. No more than $52　　**30.** About 54%　　**31.** 105°
32. For widths greater than 27 cm　　**33.** $4t$　　**34.** $-5, 5$
35. $1025

CHAPTER 3

Exercise Set 3.1, pp. 159–163

1. E　　**2.** C　　**3.** G　　**4.** B　　**5.** H　　**6.** I
7. (a)　　**8.** (c)　　**9.** (b)　　**10.** (d)　　**11.** 2 drinks
13. The person weighs more than 140 lb.
15. $17 billion　　**17.** 2000　　**19.** 2006

21.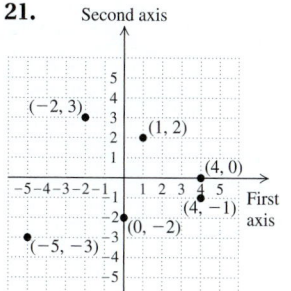

23.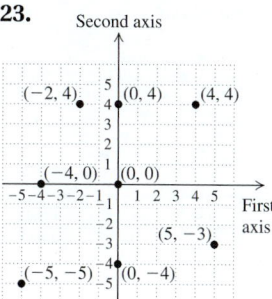

25.
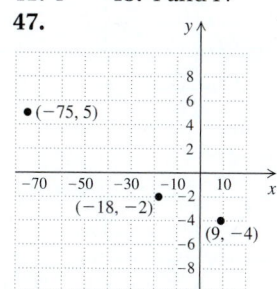

Text Messaging

27. $A(-4, 5)$; $B(-3, -3)$; $C(0, 4)$; $D(3, 4)$; $E(3, -4)$
29. $A(4, 1)$; $B(0, -5)$; $C(-4, 0)$; $D(-3, -2)$; $E(3, 0)$
31. IV　　**33.** III　　**35.** y-axis　　**37.** II　　**39.** x-axis
41. I　　**43.** I and IV　　**45.** I and III

47.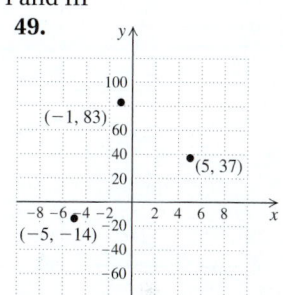

49.

51.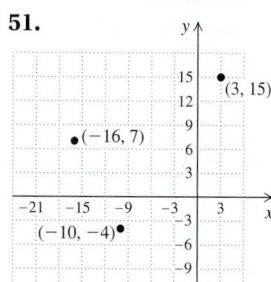

53.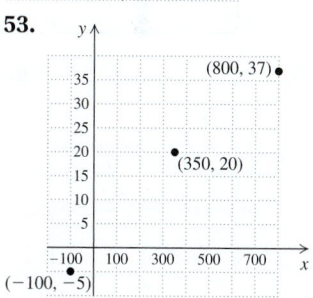

55.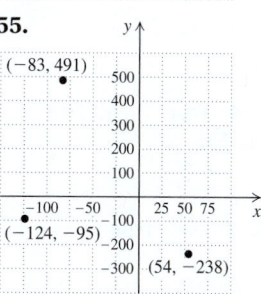

57. 📋　　**59.** $-\frac{1}{6}$　　**60.** -11.7　　**61.** -120　　**62.** 5
63. 24　　**64.** 3　　**65.** 📋　　**67.** II or IV　　**69.** $(-1, -5)$

71.

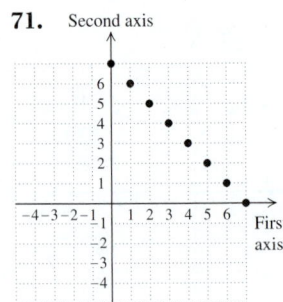

73. $\frac{65}{2}$ sq units   **75.** About 1.2 lb   **77.** About 0.4 lb
79. Latitude 27° North; longitude 81° West   **81.**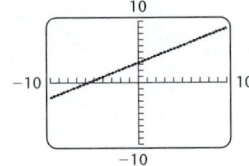

Prepare to Move On, p. 163

1. $y = \frac{2}{5}x$ **2.** $y = -\frac{3}{2}x$ **3.** $y = x - 8$
4. $y = -\frac{2}{5}x + 2$ **5.** $y = \frac{5}{8}x - \frac{1}{8}$

Technology Connection, p. 169

1. $y = -5x + 6.5$ **2.** $y = 3x + 4.5$

3. $7y - 4x = 22$, or   **4.** $5y + 11x = -20$, or
$y = \frac{4}{7}x + \frac{22}{7}$ $y = -\frac{11}{5}x - 4$

5. $2y - x^2 = 0$, or   **6.** $y + x^2 = 8$, or
$y = 0.5x^2$ $y = -x^2 + 8$

Exercise Set 3.2, pp. 170–173

1. False   **2.** True   **3.** True   **4.** True   **5.** True
6. False   **7.** Yes   **9.** No   **11.** No
13. $y = x + 3$ $y = x + 3$
$2 \mid -1 + 3$ $7 \mid 4 + 3$
$2 \stackrel{?}{=} 2$ True $7 \stackrel{?}{=} 7$ True
$(2, 5)$; answers may vary

15. $y = \frac{1}{2}x + 3$ $y = \frac{1}{2}x + 3$
$5 \mid \frac{1}{2} \cdot 4 + 3$ $2 \mid \frac{1}{2}(-2) + 3$
$ 2 + 3$ $ -1 + 3$
$5 \stackrel{?}{=} 5$ True $2 \stackrel{?}{=} 2$ True
$(0, 3)$; answers may vary

17. $y + 3x = 7$ $y + 3x = 7$
$1 + 3 \cdot 2 \mid 7$ $-5 + 3 \cdot 4 \mid 7$
$ 1 + 6$ $ -5 + 12$
$7 \stackrel{?}{=} 7$ True $7 \stackrel{?}{=} 7$ True
$(1, 4)$; answers may vary

19. $4x - 2y = 10$ $4x - 2y = 10$
$4 \cdot 0 - 2(-5) \mid 10$ $4 \cdot 4 - 2 \cdot 3 \mid 10$
$ 0 + 10$ $ 16 - 6$
$10 \stackrel{?}{=} 10$ True $10 \stackrel{?}{=} 10$ True
$(2, -1)$; answers may vary

21.
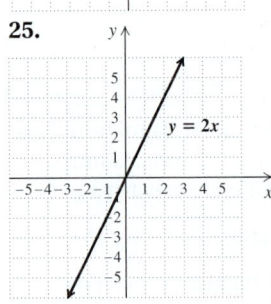

23.
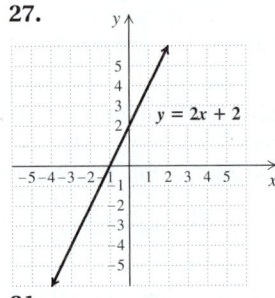

25.
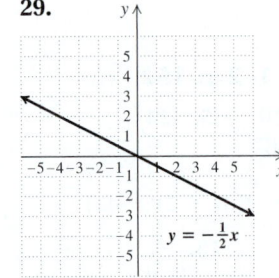

27.
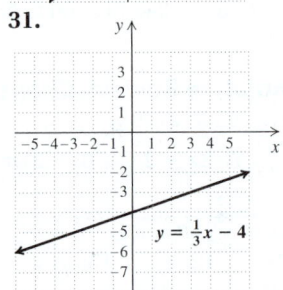

29.
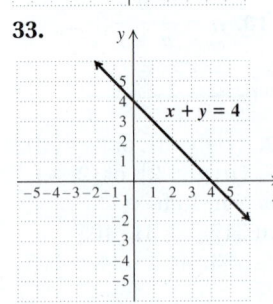

31.
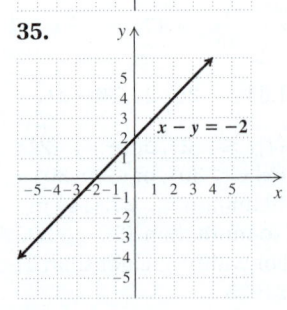

33.
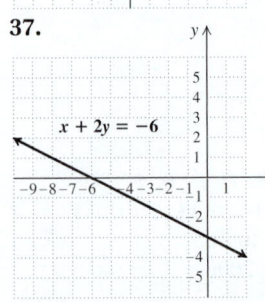

35.
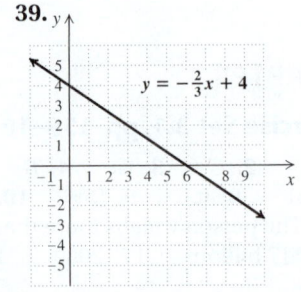

37.

39.

41.
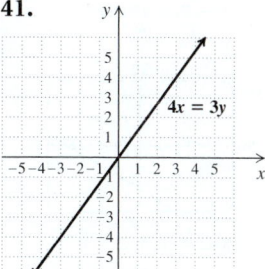
$4x = 3y$

43.
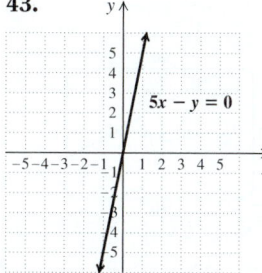
$5x - y = 0$

45.
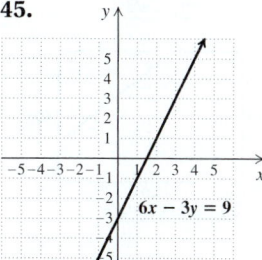
$6x - 3y = 9$

47.
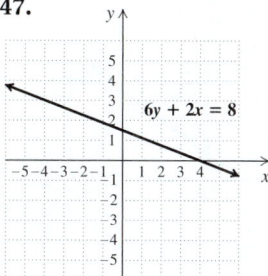
$6y + 2x = 8$

49. About $6300
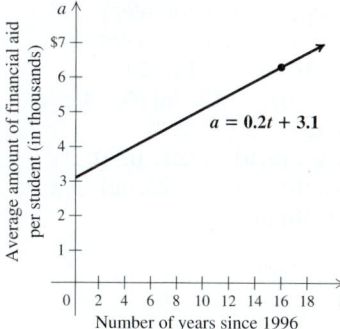
$a = 0.2t + 3.1$
Number of years since 1996

51. About 6.5 mpg
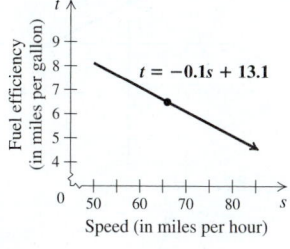
$t = -0.1s + 13.1$
Speed (in miles per hour)

53. About $34
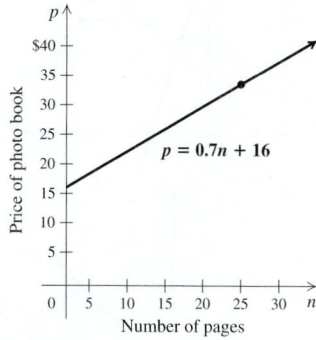
$p = 0.7n + 16$
Number of pages

55. About 85,000 apps
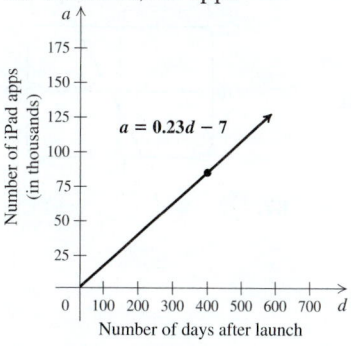
$a = 0.23d - 7$
Number of days after launch

57. About $1700
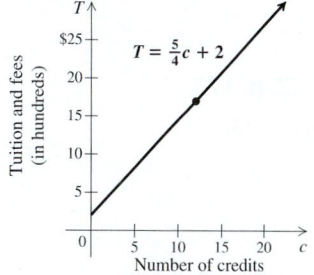
$T = \frac{5}{4}c + 2$
Number of credits

59. 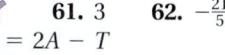   **61.** 3   **62.** $-\frac{21}{5}$

63. $Q = 2A - T$

64. $p = \dfrac{w}{q + 1}$

65. $y = \dfrac{C - Ax}{B}$

66. $y = m(x - h) + k$

67.

69.
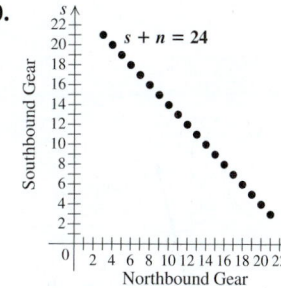
$s + n = 24$
Northbound Gear

71. $x + y = 5$, or $y = -x + 5$

73. $y = x + 2$

75.
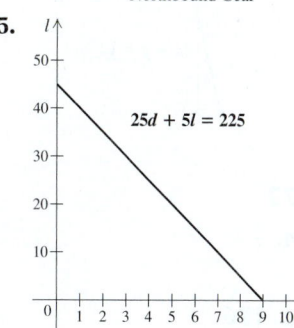
$25d + 5l = 225$

Answers may vary.
1 dinner, 40 lunches;
5 dinners, 20 lunches;
8 dinners, 5 lunches

77.
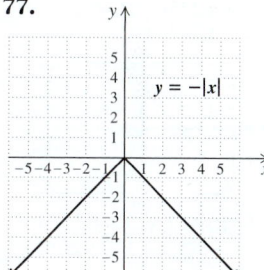
$y = -|x|$

79.
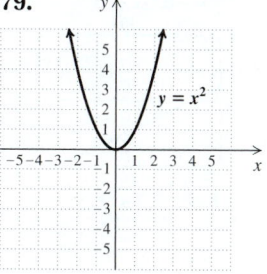
$y = x^2$

81.
$y = -2.8x + 3.5$
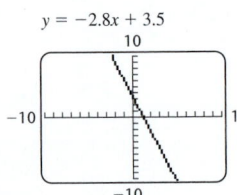

83.
$y = 2.8x - 3.5$
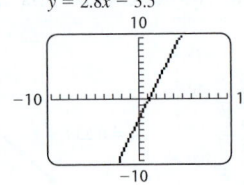

85.
$y = x^2 + 4x + 1$
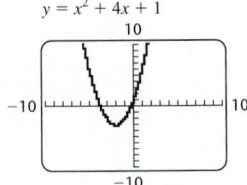

87. 🗒

3.
$5x + 6y = 84$, or
$y = -\frac{5}{6}x + 14$
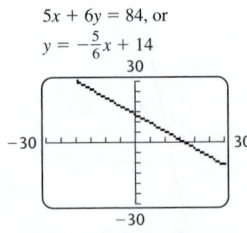
Xscl = 5, Yscl = 5

4.
$2x - 7y = 150$, or
$y = \frac{2}{7}x - \frac{150}{7}$
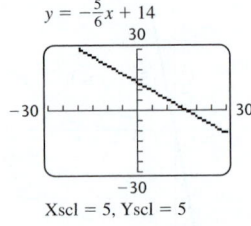
Xscl = 10, Yscl = 5

5.
$19x - 17y = 200$, or
$y = \frac{19}{17}x - \frac{200}{17}$
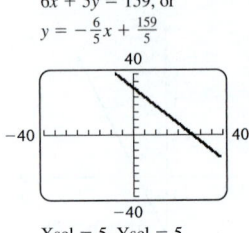

6.
$6x + 5y = 159$, or
$y = -\frac{6}{5}x + \frac{159}{5}$
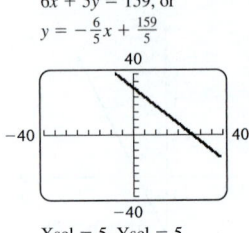
Xscl = 5, Yscl = 5

Quick Quiz: Sections 3.1–3.2, p. 173

1.
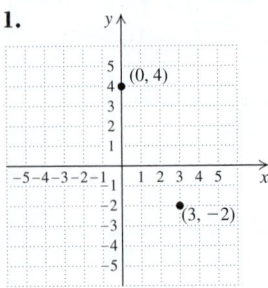

2. IV　　**3.** No

4.
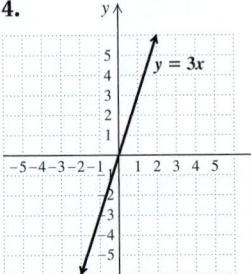

5.
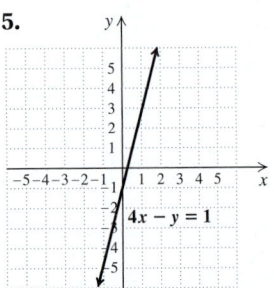

Prepare to Move On, p. 173

1. $\frac{12}{5}$　　**2.** $-\frac{3}{4}$　　**3.** $-\frac{7}{8}$　　**4.** $\frac{7}{6}$

Technology Connection, p. 176

1.
$y = -0.72x - 15$
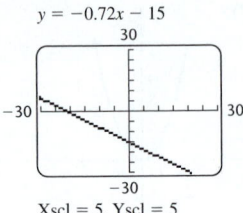
Xscl = 5, Yscl = 5

2.
$y - 2.13x = 27$, or
$y = 2.13x + 27$
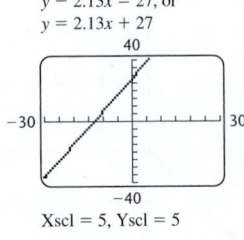
Xscl = 5, Yscl = 5

Exercise Set 3.3, pp. 178–181

1. y-intercept　　**2.** x-intercept　　**3.** y-intercept
4. x-intercept　　**5.** x-intercept　　**6.** y-intercept
7. (f)　　**8.** (e)　　**9.** (d)　　**10.** (c)　　**11.** (b)
12. (a)　　**13.** (a) $(0, 5)$; (b) $(2, 0)$　　**15.** (a) $(0, -4)$;
(b) $(3, 0)$　　**17.** (a) $(0, -2)$; (b) $(-3, 0), (3, 0)$
19. (a) $(0, 0)$; (b) $(-2, 0), (0, 0), (5, 0)$　　**21.** (a) $(0, 3)$;
(b) $(5, 0)$　　**23.** (a) $(0, -18)$; (b) $(4, 0)$　　**25.** (a) $(0, 16)$;
(b) $(-20, 0)$　　**27.** (a) None; (b) $(12, 0)$
29. (a) $(0, -9)$; (b) none

31.
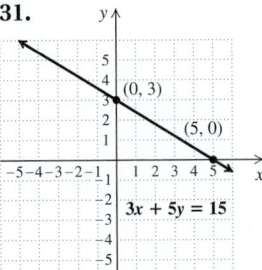
$3x + 5y = 15$

33.
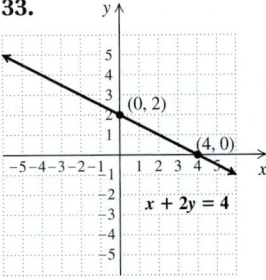
$x + 2y = 4$

35.
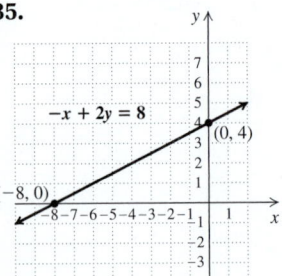
$-x + 2y = 8$

37.
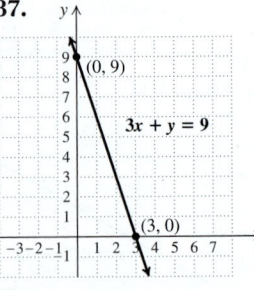
$3x + y = 9$

39.
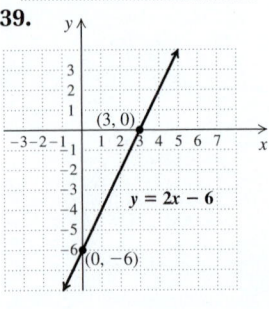
$y = 2x - 6$

41.
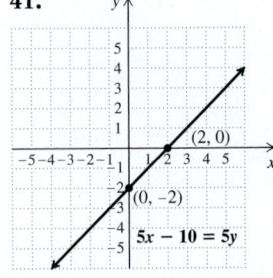
$5x - 10 = 5y$

43.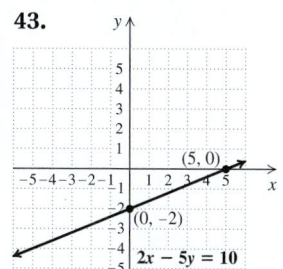

45.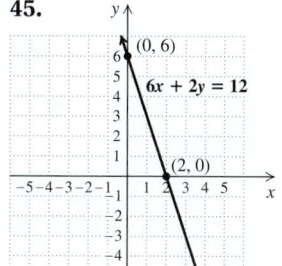

47.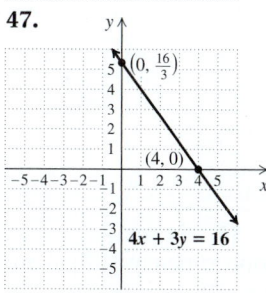

49.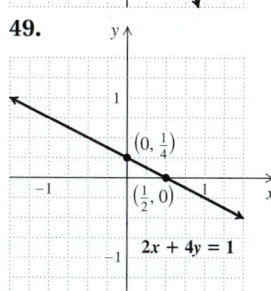

51.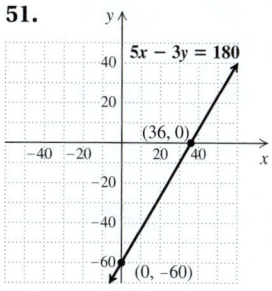

53.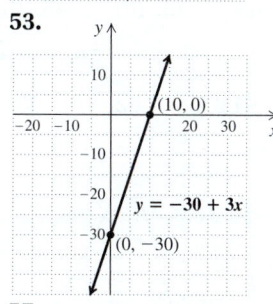

55.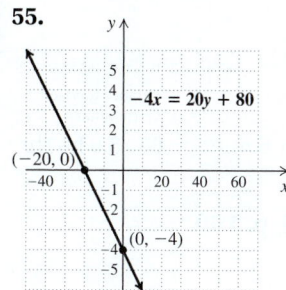

57.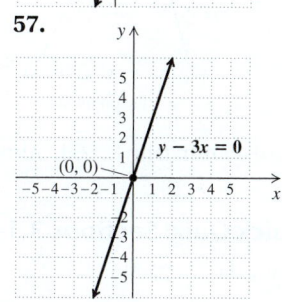

59.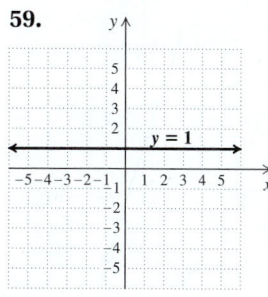

61.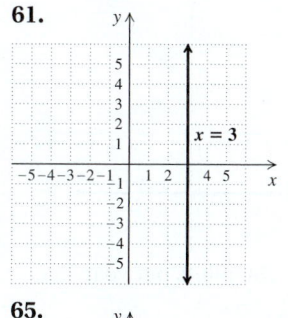

63.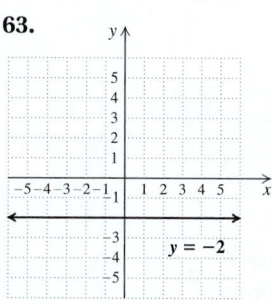

65.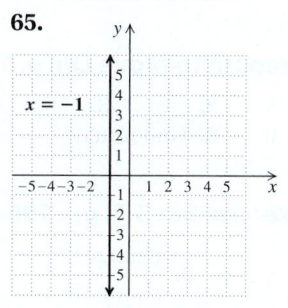

67.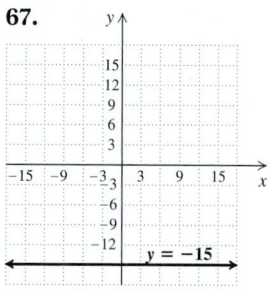

69.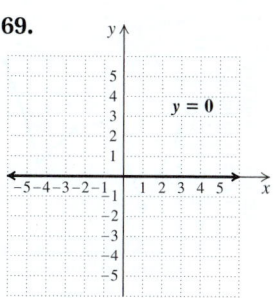

71.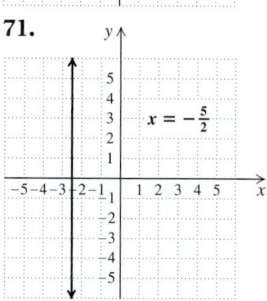

73.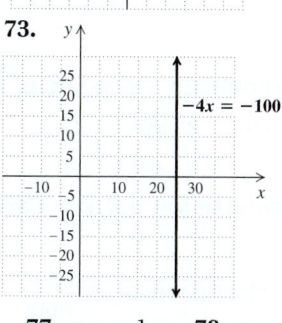

75.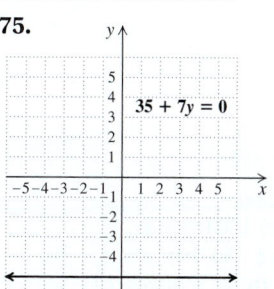

77. $y = -1$ **79.** $x = 4$
81. $x = 0$ **83.** ☑
85. $d - 7$ **86.** $w + 5$,
or $5 + w$ **87.** Let n
represent the number;
$7 + 4n$ **88.** Let n
represent the number; $3n$

89. Let x and y represent the numbers; $2(x + y)$
90. Let a and b represent the numbers; $\frac{1}{2}(a + b)$
91. ☑   **93.** $y = 0$ **95.** $x = -2$
97. $(-3, 4)$ **99.** $-5x + 3y = 15$, or $y = \frac{5}{3}x + 5$
101. -24 **103.** $\left(\dfrac{C - D}{A}, 0\right)$ **105.** $\left(0, -\dfrac{80}{7}\right)$, or
$(0, -11.\overline{428571})$; $(40, 0)$ **107.** $(0, -9)$; $(45, 0)$
109. $\left(0, \frac{1}{25}\right)$, or $(0, 0.04)$; $\left(\frac{1}{50}, 0\right)$, or $(0.02, 0)$

Quick Quiz: Sections 3.1–3.3, p. 181

1.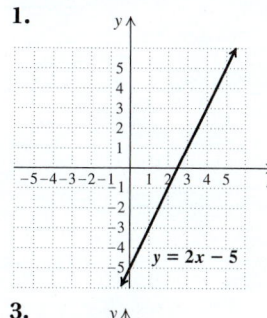

2.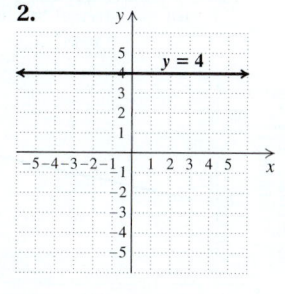

3.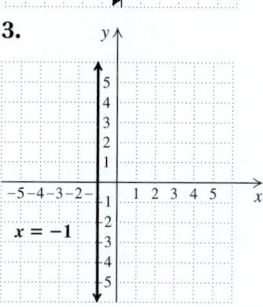

4.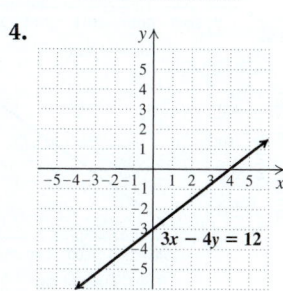

5.

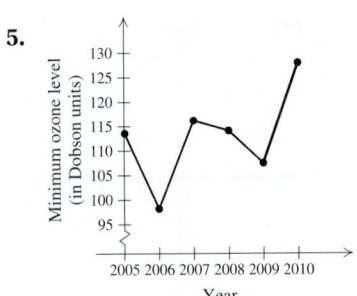

Prepare to Move On, p. 181

1. 40   **2.** 518   **3.** $\frac{1720}{103}$ **4.** 0.6108

Exercise Set 3.4, pp. 185–190

1. True   **2.** True   **3.** False   **4.** True

5. Miles per hour, or $\dfrac{\text{miles}}{\text{hour}}$

6. Hours per chapter, or $\dfrac{\text{hours}}{\text{chapter}}$

7. Dollars per mile, or $\dfrac{\text{dollars}}{\text{mile}}$

8. Cups of flour per cake, or $\dfrac{\text{cups of flour}}{\text{cake}}$

9. (a) 30 mpg; **(b)** \$39.33/day; **(c)** 130 mi/day;
(d) 30¢/mi   **11. (a)** 7 mph; **(b)** \$7.50/hr;
(c) \$1.07/mi   **13. (a)** \$22/hr; **(b)** 20.6 pages/hr;
(c) \$1.07/page   **15.** \$1285.8 billion/year
17. (a) 14.5 floors/min; **(b)** 4.14 sec/floor
19. (a) 23.12 ft/min; **(b)** 0.04 min/ft
21.

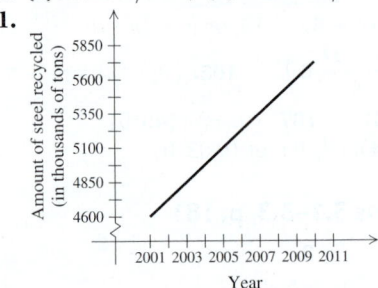

23.

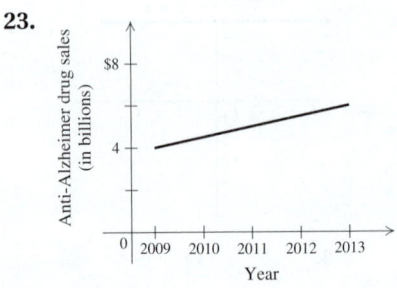

25.

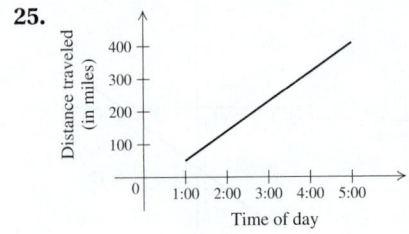

27.

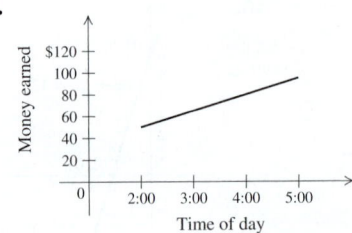

29.

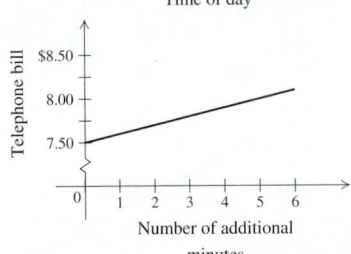

31. 20 calls/hr   **33.** 75 mi/hr   **35.** 12¢/min
37. −10,000 people/year   **39.** Approximately 0.02 gal/mi
41. (e)   **43.** (d)   **45.** (b)   **47.** 📝
49. $2 \cdot 3 \cdot 5 \cdot 5$ **50.**

-3.5 marked on a number line from −5 to 5

51. −1.375   **52.** $\frac{1}{3}$ **53.** $-\frac{3}{2}$ **54.** $\frac{2}{3}$ **55.** 📝
57.   **59.**

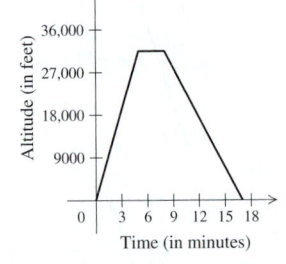 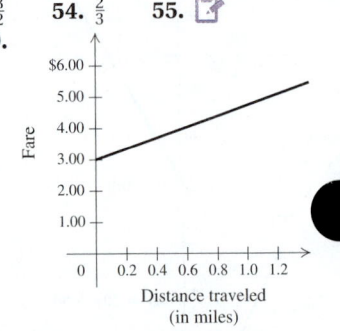

61. 0.45 min/mi   **63.** About 41.6 min   **65.** 3.6 bu/hr

Quick Quiz: Sections 3.1–3.4, p. 190

1. y-axis   **2.** No
3.   **4.**

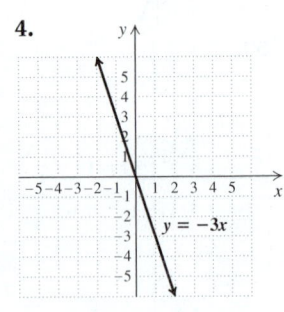

5. 730 pilots per year

Prepare to Move On, p. 190

1. 5   **2.** −6   **3.** −1   **4.** $-\frac{4}{3}$ **5.** $-\frac{4}{3}$ **6.** 1
7. 0   **8.** Undefined

Exercise Set 3.5, pp. 196–201

1. y, x **2.** $\dfrac{\text{change in } y}{\text{change in } x}$ **3.** $\dfrac{\text{rise}}{\text{run}}$ **4.** $\dfrac{y_2 - y_1}{x_2 - x_1}$
5. Positive, negative   **6.** Zero, undefined   **7.** Positive

8. Negative   **9.** Negative   **10.** Positive
11. Zero   **12.** Positive   **13.** Negative   **14.** Zero
15. $25/post   **17.** -$200/year   **19.** $\frac{1}{2}$ point/$1000 income
21. About $-2.1°$/min   **23.** $\frac{4}{3}$ **25.** $\frac{1}{3}$ **27.** -1
29. 0   **31.** -2 **33.** Undefined   **35.** $-\frac{1}{3}$
37. 5   **39.** $\frac{5}{4}$ **41.** $-\frac{4}{5}$ **43.** $\frac{2}{3}$ **45.** -1
47. $-\frac{1}{2}$ **49.** 0   **51.** 1   **53.** Undefined   **55.** 0
57. Undefined   **59.** Undefined   **61.** 0
63. 14.375%, or $14\frac{3}{8}\%$ **65.** 35%   **67.** $\frac{29}{98}$, or about 30%
69. About 5.1%; yes   **71.** 🔲   **73.** $12 + 3a$
74. $7(2 + 5x)$ **75.** 15   **76.** $3 \le x$ **77.** $(5t)^3$
78. $16y^4$ **79.** 🔲   **81.** 0.364, or 36.4%
83. $\{m\,|\,m \ge \frac{5}{2}\}$ **85.** $\frac{1}{2}$

Quick Quiz: Sections 3.1–3.5, p. 201

1. 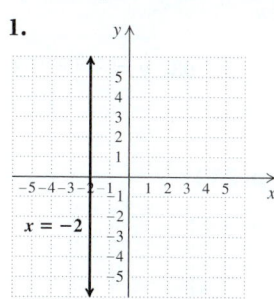   **2.**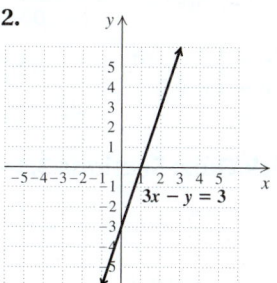

3. 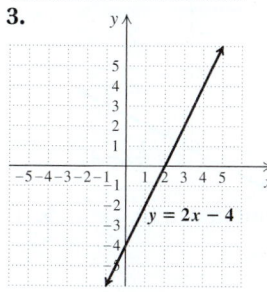   **4.** -1 **5.** 0

Prepare to Move On, p. 201

1. $y = -\frac{2}{3}x + \frac{7}{3}$ **2.** $y = \frac{3}{4}x - 2$ **3.** $y = \dfrac{c - ax}{b}$

4. $y = \dfrac{ax - c}{b}$

Mid-Chapter Review: Chapter 3, p. 202

1. *y-intercept*: $y - 3 \cdot 0 = 6$
$\qquad\qquad\qquad y = 6$
The *y*-intercept is $(0, 6)$.
x-intercept: $0 - 3x = 6$
$\qquad\qquad -3x = 6$
$\qquad\qquad\quad x = -2$
The *x*-intercept is $(-2, 0)$.

2. $m = \dfrac{y_2 - y_1}{x_2 - x_1} = \dfrac{-1 - 5}{3 - 1}$
$\qquad = \dfrac{-6}{2}$
$\qquad = -3$

3. 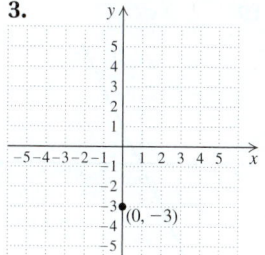   **4.** IV   **5.** No

6. 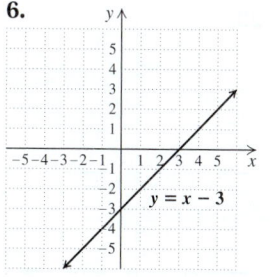   **7.**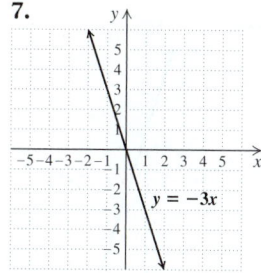

8. 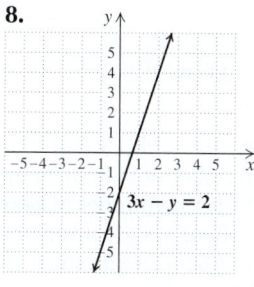   **9.**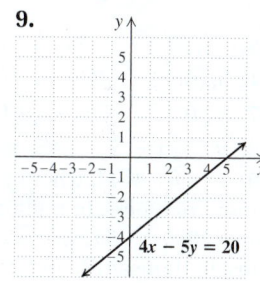

10. 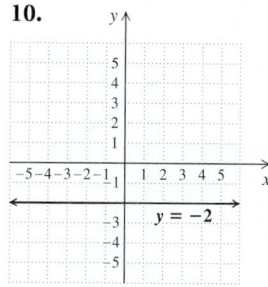   **11.**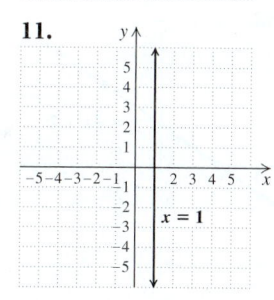

12. 14 homes/month   **13.** About 29%   **14.** $\frac{5}{3}$
15. -3 **16.** Undefined   **17.** 0   **18.** 0
19. Undefined   **20.** $(-4, 0), (0, 6)$

Exercise Set 3.6, pp. 209–211

1. Slope   **2.** *y*-intercept   **3.** *y*-intercept   **4.** Slope
5. *y*-intercept   **6.** Slope   **7.** (f)   **8.** (b)
9. (d)   **10.** (c)   **11.** (e)   **12.** (a)
13. 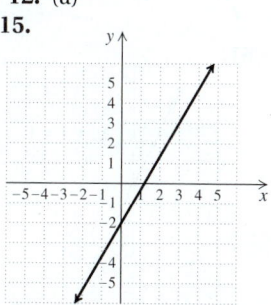   **15.**

17.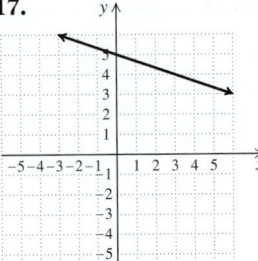

19.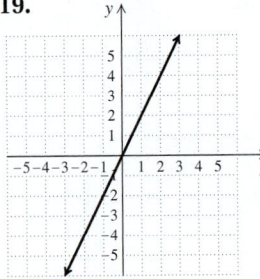

21.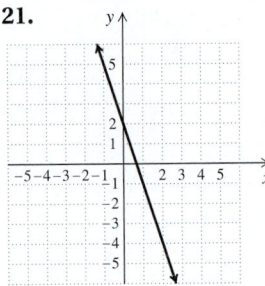

23.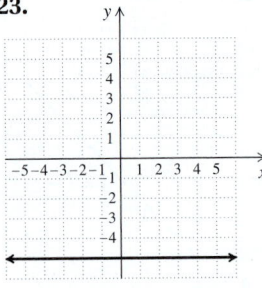

25. $-\frac{2}{7}$; $(0, 5)$ **27.** $\frac{1}{3}$; $(0, 7)$ **29.** $\frac{9}{5}$; $(0, -4)$

31. 3; $(0, 7)$ **33.** -2; $(0, 4)$ **35.** 0; $(0, 3)$

37. $\frac{2}{5}$; $\left(0, \frac{8}{5}\right)$ **39.** $\frac{9}{8}$; $(0, 0)$ **41.** $y = 5x + 7$

43. $y = \frac{7}{8}x - 1$ **45.** $y = -\frac{5}{3}x - 8$ **47.** $y = \frac{1}{3}$

49. $y = \frac{2}{3}x + 44$, where y is the number of federal credit union members, in millions, and x is the number of years since 2000   **51.** $y = \frac{3}{5}x + 16$, where y is the number of jobs, in thousands, and x is the number of years since 2008

53.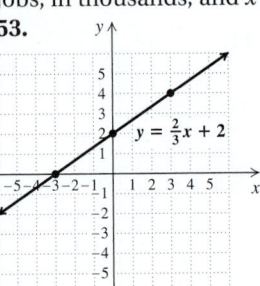

55.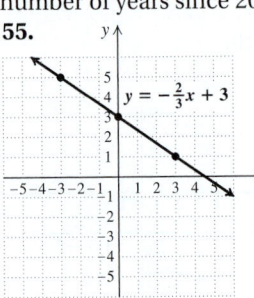

57.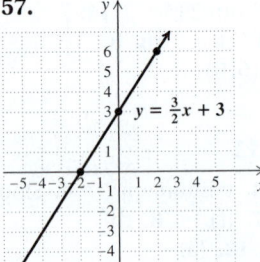

59.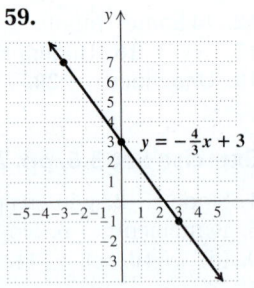

61.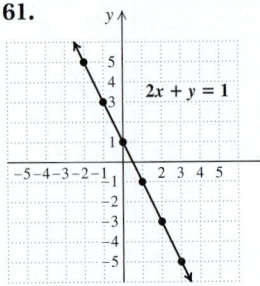

63.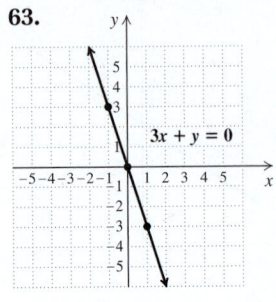

65.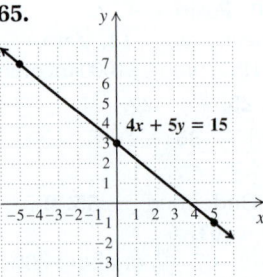

67.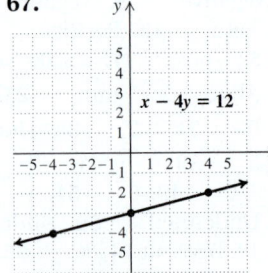

69. Yes   **71.** Yes   **73.** No   **75.** Yes   **77.** No
79. 🗒   **81.** 12%   **82.** 250 words   **83.** $2400
84. Width: 20 ft; length: 50 ft   **85.** 🗒
87. When $x = 0$, $y = b$, so $(0, b)$ is on the line. When $x = 1$, $y = m + b$, so $(1, m + b)$ is on the line. Then

$$\text{slope} = \frac{(m + b) - b}{1 - 0} = m.$$

89. $y = \frac{1}{2}x$ **91.** $y = -\frac{4}{5}x + 4$ **93.** $y = -\frac{5}{3}x + 3$
95. $y = \frac{5}{2}x + 1$

Quick Quiz: Sections 3.1–3.6, p. 211

1.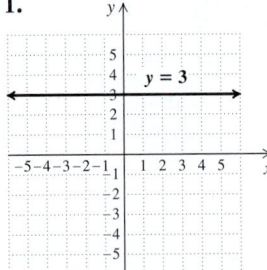

2.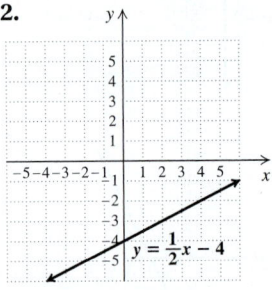

3. III   **4.** x-intercept: $(-7, 0)$; y-intercept: $(0, 7)$ **5.** 1

Prepare to Move On, p. 211

1. $y = m(x - h) + k$ **2.** $y = -2(x + 4) + 9$
3. -7 **4.** 13   **5.** -9

Connecting the Concepts, p. 217

1. Slope–intercept form   **2.** Standard form
3. None of these   **4.** Standard form   **5.** Point–slope form   **6.** None of these   **7.** $2x - 5y = 10$
8. $-2x + y = 7$, or $2x - y = -7$ **9.** $y = \frac{2}{7}x - \frac{8}{7}$
10. $y = -x - 8$

Exercise Set 3.7, pp. 218–220

1. False   **2.** True   **3.** False   **4.** True
5. True   **6.** True   **7.** $\frac{1}{4}$; $(5, 3)$ **9.** -7; $(2, -1)$
11. $-\frac{10}{3}$; $(-4, 6)$ **13.** 5; $(0, 0)$

15.

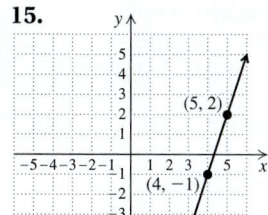

$y - 2 = 3(x - 5)$

17.

$y - 2 = -4(x - 1)$

19.

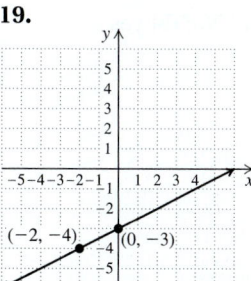

$y - (-4) = \frac{1}{2}(x - (-2))$, or

$y + 4 = \frac{1}{2}(x + 2)$

21.

$y - 0 = -1(x - 8)$, or
$y = -(x - 8)$

23. $y = 3x + 4$ **25.** $y = \frac{4}{3}x - 12$
27. $y = \frac{2}{3}x + \frac{1}{2}$ **29.** $y = x - 32$
31. $y - 1 = 6(x - 7)$ **33.** $y - 4 = -5(x - 3)$
35. $y - (-5) = \frac{1}{2}(x - (-2))$ **37.** $y - 0 = -1(x - 9)$
39. $y = 2x - 6$ **41.** $y = -\frac{3}{5}x + \frac{28}{5}$
43. $y = -0.6x - 5.8$ **45.** $y = \frac{2}{7}x - 6$
47. $y = \frac{3}{5}x + \frac{42}{5}$ **49.** $y = \frac{1}{2}x + 4$ **51.** $y = -x - 1$
53. $y = -\frac{2}{3}x - \frac{13}{3}$ **55.** $x = 5$ **57.** $y = -\frac{3}{2}x + \frac{11}{2}$
59. $y = x + 6$ **61.** $y = -\frac{1}{3}x - \frac{8}{3}$ **63.** $y = -\frac{5}{3}x - \frac{41}{3}$
65. $x = -3$ **67.** $y = 4x - 5$ **69.** $y = 4.5x - 9.4$
71. $y = -2x - 1$ **73.** $y = \frac{5}{3}x$ **75.** $y = -6$
77. $x = -10$ **79. (a)** $y = \frac{11}{14}x + \frac{131}{25}$; **(b)** 10.0 million
volunteers; 14.7 million volunteers   **81. (a)** $y = 0.075x +$
1.425; **(b)** 1.95 million graduates; 2.55 million
graduates   **83. (a)** Let x represent the number of
years after 2000 and y the percent of the U.S. population
living in metropolitan areas; $y = -\frac{5}{18}x + 79$;
(b) 77.6%; 75.7%   **85.** 📋   **87.** -64 **88.** 9
89. 18   **90.** $\frac{11}{2}$ **91.** $\left\{x \mid x < -\frac{17}{3}\right\}$, or $\left(-\infty, -\frac{17}{3}\right)$
92. $\{x \mid x \geq -2\}$, or $[-2, \infty)$ **93.** 📋
95.

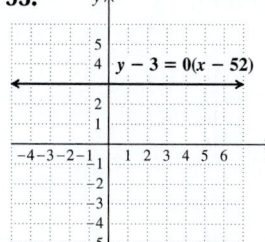

97. $y = 2x - 9$
99. $y = \frac{1}{5}x - 2$
101. $y = \frac{1}{2}x + 1$
103. 7
105. $(10, 0), (0, -3)$
107. $y = 1.876712329x +$
11.28767123

Quick Quiz: Sections 3.1–3.7, p. 220

1.

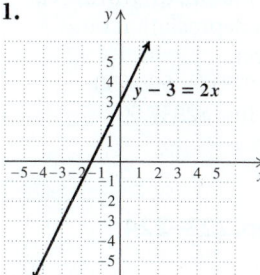

2.

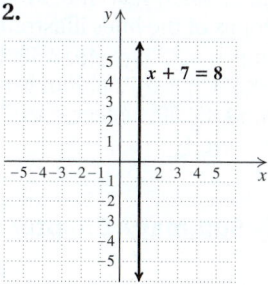

3.

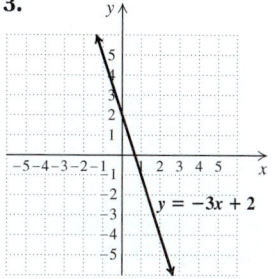

4.

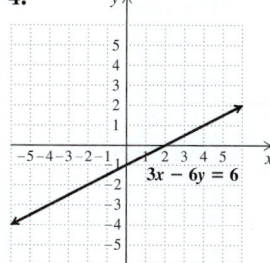

5.

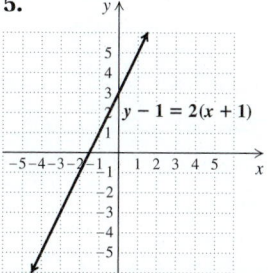

Prepare to Move On, p. 220

1. -125 **2.** 64   **3.** -64 **4.** 8   **5.** -4

Visualizing for Success, p. 221

1. C   **2.** G   **3.** F   **4.** B   **5.** D   **6.** A
7. I   **8.** H   **9.** J   **10.** E

Decision Making: Connection, p. 222

1.

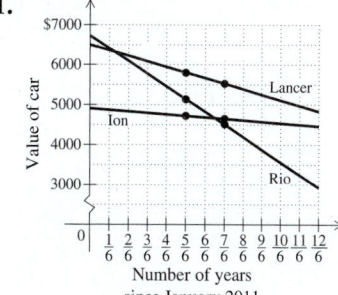

2. Ion: \$18.75/month, or \$225/year; Lancer: \$68.75/month, or \$825/year; Rio: \$156.25/month, or \$1875/year. The slopes of the lines illustrate the depreciation rates.
3. Ion: $y = -225x + 4912.50$; Lancer: $y = -825x + 6487.50$; Rio: $y = -1875x + 6687.50$
4. Ion: \$4462.50; Lancer: \$4837.50; Rio: \$2937.50
5. 📝   **6.** 🖥

Study Summary: Chapter 3, pp. 223–224

1.
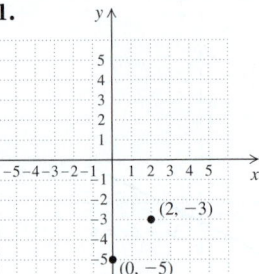

2. III

3.
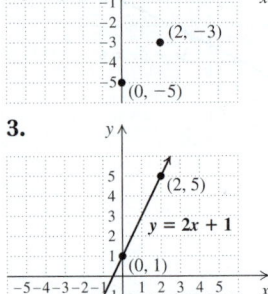

4. x-intercept: $(1, 0)$; y-intercept: $(0, -10)$

5.
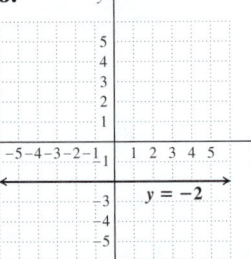

6.
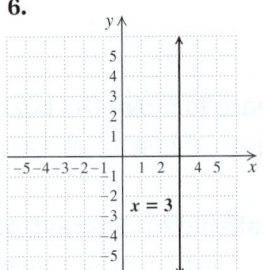

7. $\frac{4}{9}$ meal/min   **8.** $\frac{1}{10}$ **9.** 0
10. Slope: -4; y-intercept: $\left(0, \frac{2}{5}\right)$
11.
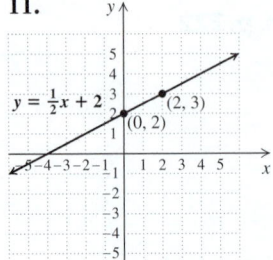

12. No

13. $y - 6 = \frac{1}{4}(x - (-1))$

Review Exercises: Chapter 3, pp. 225–226

1. True   **2.** True   **3.** False   **4.** False   **5.** True
6. True   **7.** True   **8.** False   **9.** True   **10.** True

11. 4 million people   **12.** About 80,000 people
13.–15.
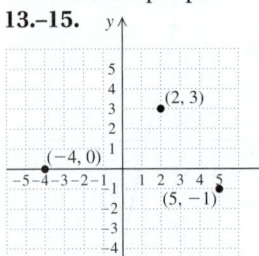

16. III   **17.** IV
18. II   **19.** $(-5, -1)$
20. $(-2, 5)$
21. $(3, 0)$

22.
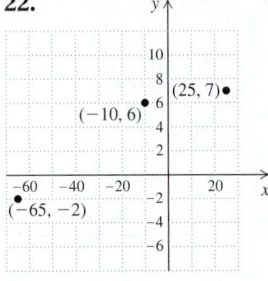

23. (a) No; **(b)** yes

24.
$$\frac{2x - y = 3}{\begin{array}{c|c} 2 \cdot 0 - (-3) & 3 \\ 0 + 3 & \\ & 3 \overset{?}{=} 3 \quad \text{True} \end{array}}$$
$$\frac{2x - y = 3}{\begin{array}{c|c} 2 \cdot 2 - 1 & 3 \\ 4 - 1 & \\ & 3 \overset{?}{=} 3 \quad \text{True} \end{array}}$$
$(-1, -5)$; answers may vary

25.
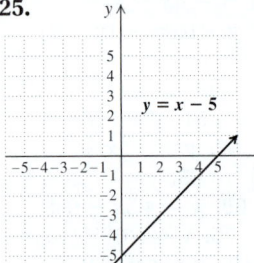

26.
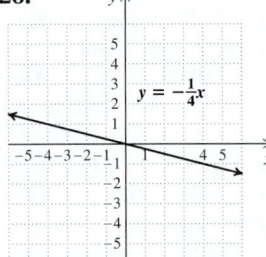

27.
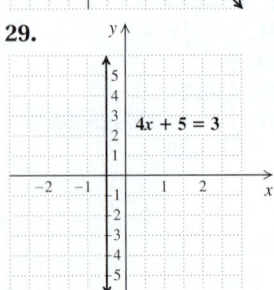

28.
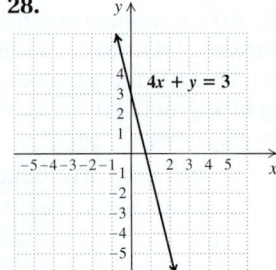

29.
(graph with $4x + 5 = 3$)

30.
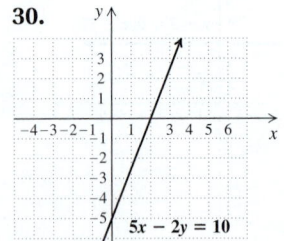

31.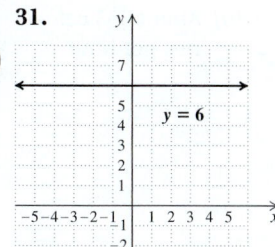

32.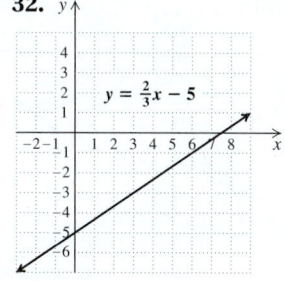

33.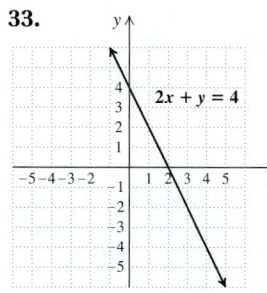

34.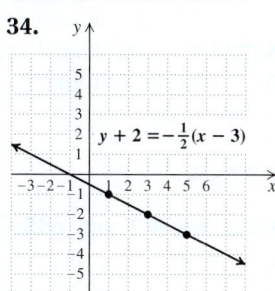

35. About 11.2 years

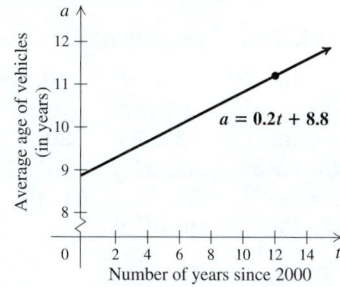

36. (a) $\frac{2}{15}$ mi/min; **(b)** 7.5 min/mi   **37.** $\frac{1}{20}$ gal/mi
38. 0   **39.** $\frac{7}{3}$ **40.** $-\frac{3}{7}$ **41.** $-\frac{6}{5}$ **42.** 0
43. Undefined   **44.** 2   **45.** $(16, 0), (0, -10)$
46. $-\frac{3}{5}; (0, 9)$ **47.** $8.\overline{3}\%$ **48.** Perpendicular
49. Parallel   **50.** $y = \frac{3}{8}x + 7$
51. $y - 9 = -\frac{1}{3}(x - (-2))$ **52.** $y = x + 7$
53. $y = -\frac{5}{3}x - \frac{5}{3}$ **54. (a)** Let x represent the number of years after 2000 and y the average in-state tuition at a public two-year college; $y = 144.\overline{3}x + 1125.\overline{3}$; Answers may vary. **(b)** \$2568.67; \$3001.67   **55.** 📋 Two perpendicular lines share the same y-intercept if their point of intersection is on the y-axis.   **56.** 📋 The graph of a vertical line has only an x-intercept. The graph of a horizontal line has only a y-intercept. The graph of a nonvertical, nonhorizontal line will have only one intercept if it passes through the origin: $(0, 0)$ is both the x-intercept and the y-intercept.   **57.** -1
58. Area: 45 sq units; perimeter: 28 units
59. $(0, 4), (1, 3), (-1, 3)$; answers may vary

Test: Chapter 3, p. 227

1. [3.1] \$4000   **2.** [3.1] 2000 and 2007   **3.** [3.1] III
4. [3.1] II   **5.** [3.1] $(3, 4)$ **6.** [3.1] $(0, -4)$
7. [3.1] $(-5, 2)$

8. [3.2]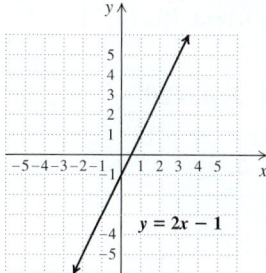

9. [3.3]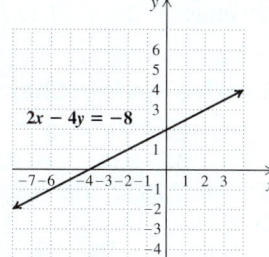

10. [3.3]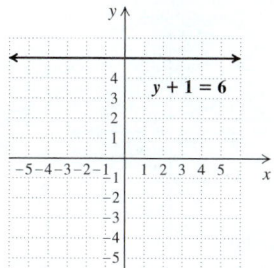

11. [3.2]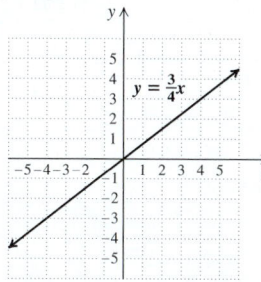

12. [3.2]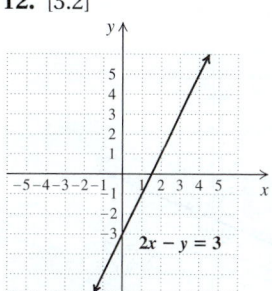

13. [3.3]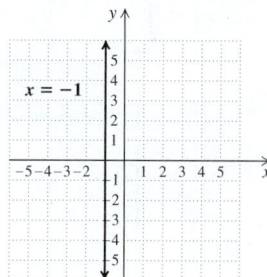

14. [3.7]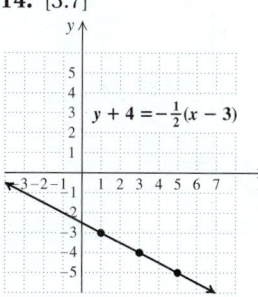

15. [3.5] 5   **16.** [3.5] $-\frac{9}{4}$ **17.** [3.5] Undefined
18. [3.4] $\frac{1}{3}$ km/min   **19.** [3.5] 31.5%
20. [3.3] $(6, 0), (0, -30)$ **21.** [3.6] 8; $(0, 10)$
22. [3.6] $y = -\frac{1}{3}x - 11$
23. [3.6] Parallel   **24.** [3.6] Perpendicular
25. [3.7] $y = -\frac{5}{2}x - \frac{11}{2}$
26. [3.7] **(a)**   **(b)** 138 beats per minute

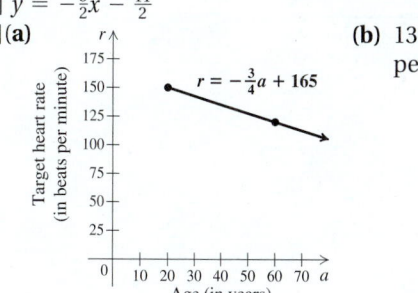

27. [3.6] $y = \frac{2}{5}x + 9$ **28.** [3.1] Area: 25 sq units; perimeter: 20 units   **29.** [3.2], [3.7] $(0, 12), (-3, 15), (5, 7)$

Cumulative Review: Chapters 1–3, p. 228

1. 7   **2.** $12a - 6b + 18$ **3.** $2(4x - 2y + 1)$
4. $2 \cdot 3^3$ **5.** -0.15 **6.** 0.367 **7.** $\frac{11}{60}$
8. 2.6   **9.** 7.28   **10.** $-\frac{5}{12}$ **11.** -3
12. $5x + 11$ **13.** -27 **14.** 16   **15.** -6
16. 2   **17.** $\frac{7}{9}$ **18.** -17 **19.** $\{x \mid x < 16\}$, or

$(-\infty, 16)$ **20.** $h = \dfrac{A - \pi r^2}{2\pi r}$ **21.** IV

22. $-1 < x \le 2$
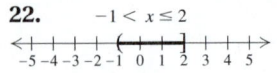

23.
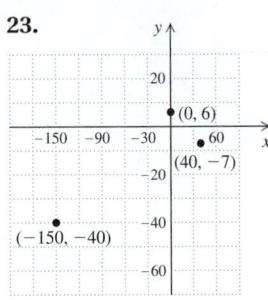

24.
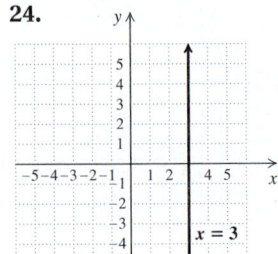

25.
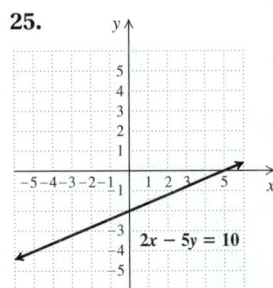

26.
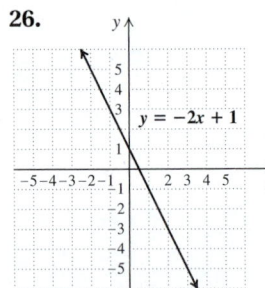

27.
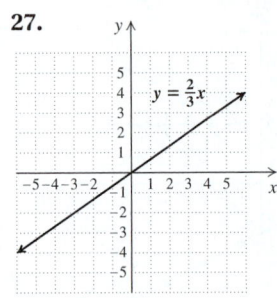

28.
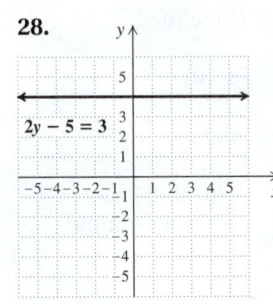

29. $3; (0, -2)$
30. $-\frac{1}{3}$ **31.** $y = \frac{2}{7}x - 4$
32. $y - 4 = -\frac{3}{8}(x - (-6))$

33. (a)
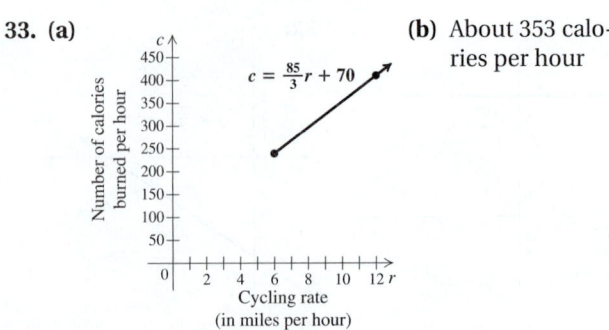
(b) About 353 calories per hour

34. \$57,413   **35.** 4 hr   **36.** \$25,000   **37.** $-4, 4$
38. 3   **39.** No solution

CHAPTER 4

Exercise Set 4.1, pp. 235–237

1. (e)   **2.** (f)   **3.** (b)   **4.** (h)   **5.** (g)   **6.** (a)
7. (c)   **8.** (d)   **9.** Base: $2x$; exponent: 5
10. Base: $x + 1$; exponent: 0   **11.** Base: x; exponent: 3
12. Base: y; exponent: 6   **13.** Base: $\dfrac{4}{y}$; exponent: 7
14. Base: $-5x$; exponent: 4   **15.** d^{13} **17.** a^7 **19.** 6^{15}
21. $(3y)^{12}$ **23.** $5p$ **25.** $(x + 3)^{13}$ **27.** a^5b^9 **29.** r^{10}
31. m^4n^9 **33.** 7^3, or 343   **35.** t^7 **37.** $5a$ **39.** 1
41. $(r + s)^8$ **43.** $\frac{4}{5}d^7$ **45.** $4a^7b^6$ **47.** $x^{12}y^7$ **49.** 1
51. 5   **53.** 2   **55.** -4 **57.** x^{33} **59.** 5^{32} **61.** t^{80}
63. $100x^2$ **65.** $-8a^3$ **67.** $25n^{14}$ **69.** $a^{14}b^7$
71. $r^{17}t^{11}$ **73.** $24x^{19}$ **75.** $\dfrac{x^3}{125}$ **77.** $\dfrac{49}{36n^2}$
79. $\dfrac{a^{18}}{b^{48}}$ **81.** $\dfrac{x^8y^4}{z^{12}}$ **83.** $\dfrac{a^{12}}{16b^{20}}$ **85.** $-\dfrac{125x^{21}y^3}{8z^{12}}$
87. 1   **89.** 📋   **91.** -21 **92.** $\{x \mid x \le \frac{7}{4}\}$, or $\left(-\infty, \frac{7}{4}\right]$
93. -1 **94.** 0   **95.** $\{n \mid n < \frac{16}{3}\}$, or $\left(-\infty, \frac{16}{3}\right)$
96. 1   **97.** 📋   **99.** 📋   **101.** Let $x = 1$;
then $3x^2 = 3$, but $(3x)^2 = 9$.   **103.** Let $t = -1$;
then $\dfrac{t^6}{t^2} = 1$, but $t^3 = -1$.   **105.** y^{6x} **107.** x^t
109. 13   **111.** $<$ **113.** $<$ **115.** $>$
117. 4,000,000; 4,194,304; 194,304   **119.** 2,000,000,000;
2,147,483,648; 147,483,648   **121.** 1,536,000 bytes,
or approximately 1,500,000 bytes

Prepare to Move On, p. 237

1. -24 **2.** -15 **3.** -11 **4.** 16   **5.** -14 **6.** 4

Technology Connection, p. 242

1. 1.71×10^{17} **2.** $5.\overline{370} \times 10^{-15}$ **3.** 3.68×10^{16}

Connecting the Concepts, p. 243

1. x^{14} **2.** $\dfrac{1}{x^{14}}$ **3.** $\dfrac{1}{x^{14}}$ **4.** x^{14} **5.** x^{40} **6.** x^{40}

7. c^8 **8.** $\dfrac{1}{c^8}$ **9.** $\dfrac{a^{15}}{b^{20}}$ **10.** $\dfrac{b^{20}}{a^{15}}$

Exercise Set 4.2, pp. 243–246

1. True   **2.** False   **3.** True   **4.** False   **5.** (c)
6. (d)   **7.** (a)   **8.** (b)   **9.** Positive power of 10
10. Negative power of 10   **11.** Negative power of 10
12. Positive power of 10   **13.** Positive power of 10

14. Negative power of 10   **15.** $\dfrac{1}{x^3} = \dfrac{1}{8}$

17. $\dfrac{1}{(-2)^6} = \dfrac{1}{64}$ **19.** $\dfrac{1}{t^9}$ **21.** $\dfrac{8}{x^3}$ **23.** a^8

25. $\frac{1}{7}$ **27.** $\dfrac{3a^8}{b^6}$ **29.** $\left(\dfrac{2}{x}\right)^5 = \dfrac{32}{x^5}$ **31.** $\dfrac{1}{3x^5z^4}$

33. 9^{-2} **35.** y^{-3} **37.** t^{-1} **39.** 2^3, or 8

41. $\dfrac{1}{x^{12}}$ **43.** $\dfrac{1}{t^2}$ **45.** $\dfrac{10}{a^6b^2}$ **47.** $\dfrac{1}{n^{15}}$ **49.** t^{18}

51. $\dfrac{1}{m^7n^7}$ **53.** $\dfrac{9}{x^5}$ **55.** $\dfrac{25t^6}{r^8}$ **57.** t^{14}

59. $\dfrac{1}{y^4}$ **61.** $5y^3$ **63.** $2x^5$ **65.** $\dfrac{-3b^9}{2a^7}$ **67.** 1

69. $3s^2t^4u^4$ **71.** $\dfrac{1}{x^{12}y^{15}}$ **73.** $\dfrac{m^{10}n^6}{9}$ **75.** $\dfrac{b^5c^4}{a^8}$

77. $\dfrac{9}{a^8}$ **79.** $\dfrac{n^{12}}{m^3}$ **81.** $\dfrac{27b^{12}}{8a^6}$ **83.** 1   **85.** $\dfrac{2b^3}{a^4}$

87. $\dfrac{5y^4z^{10}}{4x^{11}}$ **89.** 4920   **91.** 0.00892

93. 904,000,000   **95.** 0.000003497   **97.** 3.6×10^7
99. 5.83×10^{-3} **101.** 7.8×10^{10} **103.** 1.032×10^{-6}
105. 6×10^{13} **107.** 2.47×10^8 **109.** 3.915×10^{-16}
111. 2.5×10^{13} **113.** 5.0×10^{-6} **115.** 3×10^{-21}
117. 📋   **119.** $-\frac{1}{6}$ **120.** 7   **121.** $-32a^5$
122. 1   **123.** 5   **124.** -15 **125.** 📋   **127.** 📋
129. 2^{-12} **131.** 5   **133.** 5^6 **135.** $\frac{1}{3} + \frac{1}{4} = \frac{7}{12}$
137. 3×10^8 mi   **139.** 2.31×10^8
141. 2×10^4 strands

Quick Quiz: Sections 4.1–4.2, p. 246

1. $200n^8m^{11}$ **2.** $\dfrac{1}{6^{11}}$ **3.** $\dfrac{y^3}{8x^6}$ **4.** 1   **5.** 3.007×10^7

Prepare to Move On, p. 246

1. $8x$ **2.** $-3a - 6b$ **3.** $-2x - 7$ **4.** $-4t - r - 5$
5. 1004   **6.** 9

Technology Connection, p. 250

1. 1141.0023

Exercise Set 4.3, pp. 251–254

1. (b)   **2.** (f)   **3.** (h)   **4.** (d)   **5.** (g)   **6.** (e)
7. (a)   **8.** (c)   **9.** Yes   **10.** Yes   **11.** No   **12.** Yes
13. Yes   **14.** No   **15.** $8x^3, -11x^2, 6x, 1$ **17.** $-t^6, -3t^3,$
$9t, -4$ **19.** Trinomial   **21.** Polynomial with no
special name   **23.** Binomial   **25.** Monomial
27.

Term	Coefficient	Degree of the Term	Degree of the Polynomial
$8x^5$	8	5	
$-\frac{1}{2}x^4$	$-\frac{1}{2}$	4	
$-4x^3$	-4	3	5
$7x^2$	7	2	
6	6	0	

29. Coefficients: 8, 2; degrees: 4, 1
31. Coefficients: 9, -3, 4; degrees: 2, 1, 0
33. Coefficients: 1, -1, 4, -3; degrees: 4, 3, 1, 0
35. (a) 1, 3, 4; (b) $8t^4$, 8; (c) 4
37. (a) 2, 0, 4; (b) $2a^4$, 2; (c) 4
39. (a) 0, 2, 1, 5; (b) $-x^5$, -1; (c) 5
41. $11n^2 + n$ **43.** $4a^4$ **45.** $11b^3 + b^2 - b$
47. $-x^4 - x^3$ **49.** $\frac{1}{15}x^4 + 10$ **51.** $-1.1a^2 + 5.3a - 7.5$
53. $-3; 21$ **55.** 16; 34   **57.** $-38; 148$ **59.** 159; 165
61. 1112 ft   **63.** 62.8 cm   **65.** 153.86 m²
67. About 135 ft   **69.** 14; 55 oranges
71. About 2.3 mcg/mL   **73.** 📋
75.

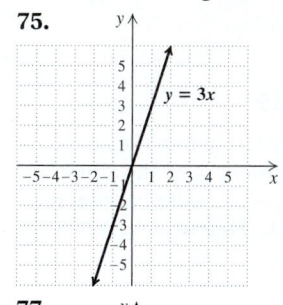

76.

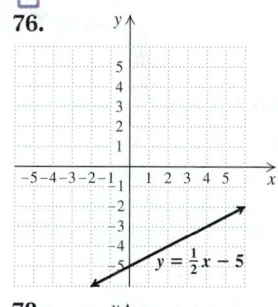

77.

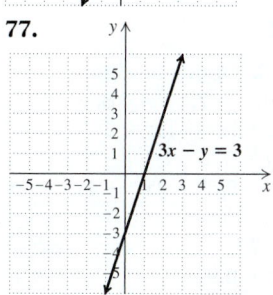

78.

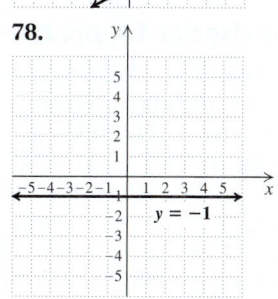

79.

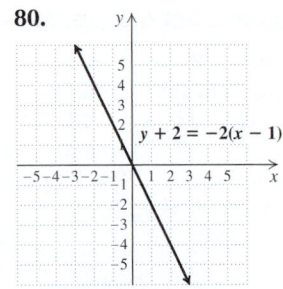

$x = 2$

80.

$y + 2 = -2(x - 1)$

81. 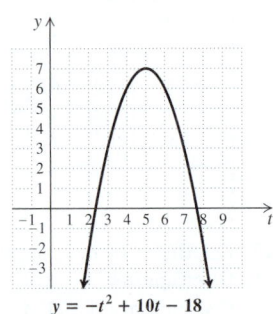   **83.** $2x^5 + 4x^4 + 6x^3 + 8$; answers may vary
85. \$2510   **87.** $3x^6$ **89.** After about 3.4 hr and 8.5 hr
91. 85.0   **93.**

95.

t	$-t^2 + 10t - 18$
3	3
4	6
5	7
6	6
7	3

$y = -t^2 + 10t - 18$

Quick Quiz: Sections 4.1–4.3, p. 254

1. $9x^{10}y^8$ **2.** $-16a^4$ **3.** $1, -6, \frac{1}{2}$ **4.** -1290
5. 1

Prepare to Move On, p. 254

1. $x - 3$ **2.** $-2x - 6$ **3.** $6a - 3$ **4.** $-t - 1$
5. $-t^4 + 17t$ **6.** $0.4a^2 - a + 11$

Technology Connection, p. 259

1. In each case, let y_1 = the expression before the addition or subtraction has been performed, y_2 = the simplified sum or difference, and $y_3 = y_2 - y_1$; and note that the graph of y_3 coincides with the x-axis. That is, $y_3 = 0$.

Exercise Set 4.4, pp. 259–262

1. Opposite   **2.** Ascending   **3.** Opposite; sign
4. Like; missing   **5.** x^2 **6.** -6 **7.** $-$ **8.** $+$
9. $4x + 9$ **11.** $-6t + 8$ **13.** $-3x^2 + 6x - 2$
15. $9t^2 + t + 3$ **17.** $8m^3 - 3m - 7$
19. $7 + 13a - a^2 + 7a^3$ **21.** $2x^6 + 9x^4 + 2x^3 - 4x^2 + 5x$
23. $x^4 + \frac{1}{4}x^3 - \frac{3}{4}x^2 - \frac{5}{6}x + 3$
25. $4.2t^3 + 3.5t^2 - 6.4t - 1.8$ **27.** $-4x^3 + 4x^2 + 6x$
29. $1.3x^4 + 0.35x^3 + 9.53x^2 + 2x + 0.96$
31. $-(-3t^3 + 4t^2 - 7); 3t^3 - 4t^2 + 7$
33. $-(x^4 - 8x^3 + 6x); -x^4 + 8x^3 - 6x$
35. $-3a^4 + 5a^2 - 1.2$ **37.** $4x^4 - 6x^2 - \frac{3}{4}x + 8$
39. $-2x - 7$ **41.** $-t^2 - 12t + 13$
43. $8a^3 + 8a^2 + a - 10$ **45.** 0
47. $1 + a + 12a^2 - 3a^3$ **49.** $\frac{9}{8}x^3 - \frac{1}{2}x$
51. $0.05t^3 - 0.07t^2 + 0.01t + 1$
53. $2x^2 + 6$ **55.** $-3x^4 - 8x^3 - 7x^2$
57. (a) $5x^2 + 4x$; **(b)** 145; 273   **59.** $16y + 26$

61. $(r + 11)(r + 9); 9r + 99 + 11r + r^2$
63. $(x + 3)^2; x^2 + 3x + 9 + 3x$ **65.** $m^2 - 40$
67. $\pi r^2 - 49$ **69.** $(x^2 - 12)\,\text{ft}^2$ **71.** $(z^2 - 36\pi)\,\text{ft}^2$
73. $\left(144 - \dfrac{d^2}{4}\pi\right)\text{m}^2$ **75.**   **77.** $y = \frac{1}{3}x + 2$
78. $y = 4x + 24$ **79.** $y = 10$ **80.** $y = -3x + 8$
81. $y = -\frac{4}{7}x - 4$ **82.** $y = -\frac{4}{3}x - \frac{8}{3}$
83.   **85.** $9t^2 - 20t + 11$ **87.** $-6x + 14$
89. $250.591x^3 + 2.812x$ **91.** $20w + 42$ **93.** $2x^2 + 20x$
95. (a) $P = -x^2 + 175x - 5000$; **(b)** \$2500; **(c)** \$1600

Quick Quiz: Sections 4.1–4.4, p. 262

1. $\dfrac{9y^{10}}{4x^6}$ **2.** $(y + 3)^{12}$ **3.** 4   **4.** $-y^3 - 6y^2 + 2y$
5. $5y^2 + 10$

Prepare to Move On, p. 262

1. $2x^2 - 2x + 6$ **2.** $-15x^2 + 10x + 35$ **3.** x^8
4. y^7 **5.** $2n^3$ **6.** $-6n^{12}$

Mid-Chapter Review, p. 263

1. $\dfrac{x^{-3}y}{x^{-4}y^7} = x^{-3-(-4)}y^{1-7} = x^1y^{-6} = \dfrac{x}{y^6}$

2. $(x^2 + 7x - 12) - (3x^2 - 6x - 1) =$
$x^2 + 7x - 12 - 3x^2 + 6x + 1 = -2x^2 + 13x - 11$

3. $x^{16}y^{40}$ **4.** 1   **5.** $\frac{1}{4}a^{10}$ **6.** $-\frac{8}{3}b$ **7.** $\dfrac{5}{x^2}$
8. $\dfrac{a^2}{bc^3}$ **9.** $\dfrac{27}{8a^6}$ **10.** $\dfrac{2}{3m^6n^4}$ **11.** 0.00000189
12. 2.7×10^{10} **13.** 3   **14.** $2a^2 - 5a - 3$
15. $3x^2 + 3x + 3$ **16.** $7x + 7$
17. $-6x^2 + 2x - 12$ **18.** $t^9 + 5t^7$
19. $3a^4 - 13a^3 - 13a^2 - 4$ **20.** $-x^5 + 2x^4 - 2x^2$

Technology Connection, p. 268

1. Let $y_1 = (5x^4 - 2x^2 + 3x)(x^2 + 2x)$ and
$y_2 = 5x^6 + 10x^5 - 2x^4 - x^3 + 6x^2$. With the table set in
AUTO mode, note that the value in the Y_1- and Y_2-columns
match, regardless of how far we scroll up or down.
2. Use TRACE, a table, or a boldly drawn graph to confirm
that y_3 is always 0.

Exercise Set 4.5, pp. 268–270

1. (c)   **2.** (d)   **3.** (b)   **4.** (a)   **5.** (c)
6. (d)   **7.** (d)   **8.** (a)   **9.** (c)   **10.** (b)
11. $21x^5$ **13.** $-x^7$ **15.** x^8 **17.** $36t^4$
19. $-0.12x^9$ **21.** $-\frac{1}{20}x^{12}$ **23.** $5n^3$ **25.** $72y^{10}$
27. $20x^2 + 5x$ **29.** $3a^2 - 27a$ **31.** $x^5 + x^2$
33. $-6n^3 + 24n^2 - 3n$ **35.** $-15t^3 - 30t^2$
37. $4a^9 - 8a^7 - \frac{5}{12}a^4$ **39.** $x^2 + 7x + 12$
41. $t^2 + 4t - 21$ **43.** $a^2 - 1.3a + 0.42$ **45.** $x^2 - 9$
47. $28 - 15x + 2x^2$ **49.** $t^2 + \frac{17}{6}t + 2$
51. $\frac{3}{16}a^2 + \frac{5}{4}a - 2$

53.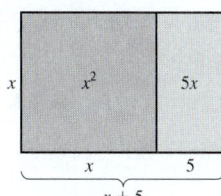

55.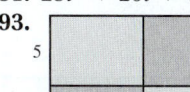

57.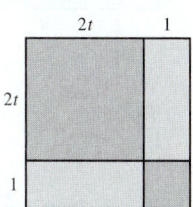

59. $x^3 + 2x + 3$ **61.** $2a^3 - a^2 - 11a + 10$
63. $3y^6 - 21y^4 + y^3 + 2y^2 - 7y - 14$
65. $33x^2 + 25x + 2$ **67.** $x^4 + 4x^3 - 3x^2 + 16x - 3$
69. $10t^4 + 3t^3 - 20t^2 + \frac{9}{2}t - 2$
71. $x^4 + 8x^3 + 12x^2 + 9x + 4$
73. ⌨   **75.** $c = 2A - b$ **76.** 80%
77. ⟵⊦⊦⟶
⁻³ 0
78. x-axis
79. $-\frac{4}{3}$ **80.** Slope: $\frac{1}{3}$; y-intercept: $\left(0, -\frac{5}{3}\right)$
81. ⌨   **83.** $75y^2 - 45y$ **85.** 5
87. $V = (4x^3 - 48x^2 + 144x)$ in^3; $S = (-4x^2 + 144)$ in^2
89. $(x^3 - 5x^2 + 8x - 4)$ cm^3
91. (a) $\left(8 - \frac{4}{3}\pi\right)x^3$; (b) 11.6 in^3
93. $2x^2 + 18x + 36$ **95.** $16x + 16$
97. $x^3 - 9x^2 + 27x - 27$ **99.** 〰

Quick Quiz: Sections 4.1–4.5, p. 270

1. $-6x^9 + 16x^5 - 24x^4$ **2.** $10x^4 - 15x + 1$
3. $7a^2 - 10a - 2$ **4.** $125x^5y^4$ **5.** Trinomial

Prepare to Move On, p. 270

1. 0.49   **2.** $49x^6$ **3.** $\frac{9}{4}$ **4.** $\frac{3}{2}x$ **5.** $2a$ **6.** $\frac{1}{9}t^8$

Exercise Set 4.6, pp. 276–278

1. True   **2.** False   **3.** False   **4.** True
5. $x^3 + 3x^2 + 2x + 6$ **7.** $t^5 + 7t^4 - 2t - 14$
9. $y^2 - y - 6$ **11.** $9x^2 + 21x + 10$
13. $5x^2 + 17x - 12$ **15.** $15 - 13t + 2t^2$
17. $x^4 - 4x^2 - 21$ **19.** $p^2 - \frac{1}{16}$
21. $x^2 - 0.09$ **23.** $100x^4 - 9$ **25.** $1 - 25t^6$
27. $t^2 - 4t + 4$ **29.** $x^2 + 20x + 100$
31. $9x^2 + 12x + 4$ **33.** $1 - 20a + 100a^2$
35. $x^6 + 24x^3 + 144$ **37.** $x^5 + 3x^3 - x^2 - 3$
39. $x^2 - 64$ **41.** $-3n^2 - 19n + 14$
43. $x^2 + 6x + 9$ **45.** $49x^6 - 14x^3 + 1$
47. $81a^6 - 1$ **49.** $x^8 - 0.01$ **51.** $t^2 - \frac{9}{16}$
53. $1 - 3t + 5t^2 - 15t^3$ **55.** $a^2 - \frac{4}{5}a + \frac{4}{25}$
57. $t^8 + 6t^4 + 9$ **59.** $45x^2 - 56x - 45$
61. $14n^5 - 7n^3$ **63.** $a^3 - a^2 - 10a + 12$
65. $49 - 42x^4 + 9x^8$ **67.** $4 - 12x^4 + 9x^8$
69. $25t^2 + 60t^3 + 36t^4$ **71.** $5x^3 + 30x^2 - 10x$

73. $q^{10} - 1$ **75.** $15t^5 - 3t^4 + 3t^3$
77. $36x^8 - 36x^5 + 9x^2$ **79.** $18a^4 + 0.8a^3 + 4.5a + 0.2$
81. $\frac{1}{25} - 36x^8$ **83.** $a^3 + 1$ **85.** $x^2 + 6x + 9$
87. $t^2 + 7t + 12$ **89.** $x^2 + 10x + 21$
91. $25t^2 + 20t + 4$
93. 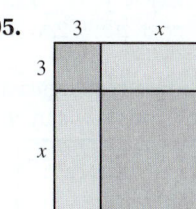   **95.**

97.   **99.** 📋

101. Refrigerator: 360 kWh/year; freezer: 720 kWh/year; washing machine: 120 kWh/year
102. For heat pumps costing \$6666.67 or less
103. $a = \dfrac{c}{3b}$ **104.** $x = \dfrac{by + c}{a}$
105. 📋   **107.** $16x^4 - 81$ **109.** $81t^4 - 72t^2 + 16$
111. $t^{24} - 4t^{18} + 6t^{12} - 4t^6 + 1$ **113.** 396   **115.** -7
117. $17F + 7(F - 17)$, $F^2 - (F - 17)(F - 7)$; other equivalent expressions are possible.
119. $(y + 1)(y - 1)$, $y(y + 1) - y - 1$; other equivalent expressions are possible.   **121.** $y^2 - 4y + 4$ **123.** 〰

Quick Quiz: Sections 4.1–4.6, p. 278

1. $-20a^4b^{12}$ **2.** n^{14} **3.** $4x^3 - 13x^2 + 5x - 6$
4. $100x^{10} - 1$ **5.** $9a^2 + 24a + 16$

Prepare to Move On, p. 278

1. 4   **2.** 19   **3.** $12a - 7c$ **4.** $13y - w$

Technology Connection, p. 282

1. 36.22   **2.** 22,312

Exercise Set 4.7, pp. 283–286

1. (b)   **2.** (d)   **3.** (a)   **4.** (c)   **5.** (a)   **6.** (b)
7. (b)   **8.** (a)   **9.** (c)   **10.** (c)   **11.** -13
13. -68 **15.** About 735 kg   **17.** 73.005 in^2
19. 66.4 m   **21.** 3, 2, 2, 0; 3   **23.** 0, 3, 3, 3; 3
25. $2r - 6s$ **27.** $5xy^2 - 2x^2y + x + 3x^2$
29. $9u^2v - 11uv^2 + 11u^2$ **31.** $6a^2c - 7ab^2 + a^2b$
33. $11x^2 - 10xy - y^2$ **35.** $-6a^4 - 8ab + 7ab^2$
37. $-6r^2 - 5rt - t^2$ **39.** $3x^3 - x^2y + xy^2 - 3y^3$
41. $-2y^4x^3 - 3y^3x$ **43.** $-8x + 8y$
45. $12c^2 + 5cd - 2d^2$ **47.** $x^2y^2 + 4xy - 5$
49. $4a^2 - b^2$ **51.** $20r^2t^2 - 23rt + 6$
53. $m^6n^2 + 2m^3n - 48$ **55.** $30x^2 - 28xy + 6y^2$

57. $0.01 - p^2q^2$ **59.** $x^2 + 2xh + h^2$
61. $16a^2 - 40ab + 25b^2$ **63.** $a^2b^2 - c^2d^4$
65. $x^3y^2 + x^2y^3 + 2x^2y^2 + 2xy^3 + 3xy + 3y^2$
67. $a^2 + 2ab + b^2 - c^2$ **69.** $a^2 - b^2 - 2bc - c^2$
71. $x^2 + 2xy + y^2$ **73.** $\frac{1}{2}a^2b^2 - 2$
75. $a^2 + c^2 + ab + 2ac + ad + bc + bd + cd$
77. $m^2 - n^2$
79. We draw a rectangle with   **81.**
dimensions $r + s$ by $u + v$.

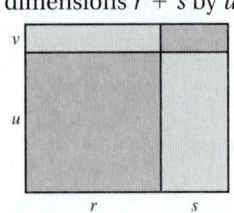
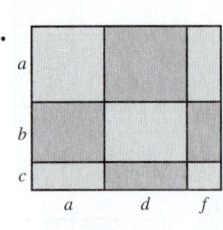

83. 🗒   **85.** 6   **86.** 7   **87.** 6   **88.** -21
89. $17a + 3x - 21$ **90.** $5y + 21$ **91.** 🗒
93. $4xy - 4y^2$ **95.** $2\pi ab - \pi b^2$
97. $x^3 + 2y^3 + x^2y + xy^2$ **99.** $2x^2 - 2\pi r^2 + 4xh + 2\pi rh$
101. 🗒   **103.** 40   **105.** $P + 2Pr + Pr^2$
107. \$12,960.29

Quick Quiz: Sections 4.1–4.7, p. 286

1. 31   **2.** $4.1x^2 - 2.9x + 6.1$
3. $5x^4 + 15x^3 - x^2 + x + 12$ **4.** $a^4 - 100$ **5.** -21

Prepare to Move On, p. 286

1. $x^2 - 8x - 4$ **2.** $2x^3 - x^2 - x + 4$
3. $-2x + 5$ **4.** $5x^2 + x$

Exercise Set 4.8, pp. 290–292

1. Divisor   **2.** Quotient   **3.** Dividend
4. Remainder   **5.** $8x^6 - 5x^3$ **7.** $1 - 2u + u^6$
9. $6t^2 - 8t + 2$ **11.** $7x^3 - 6x + \frac{3}{2}$ **13.** $-4t^2 - 2t + 1$
15. $4x - 5 + \dfrac{1}{2x}$ **17.** $x + 2x^3y + 3$
19. $-3rs - r + 2s$ **21.** $x - 6$ **23.** $t - 5 + \dfrac{-45}{t - 5}$
25. $2x - 1 + \dfrac{1}{x + 6}$ **27.** $t^2 - 3t + 9$
29. $a + 5 + \dfrac{4}{a - 5}$ **31.** $x - 3 - \dfrac{3}{5x - 1}$
33. $3a + 1 + \dfrac{3}{2a + 5}$ **35.** $t^2 - 3t + 1$ **37.** $x^2 + 1$
39. $t^2 - 1 + \dfrac{3t - 1}{t^2 + 5}$ **41.** $2x^2 + 1 + \dfrac{-x}{2x^2 - 3}$
43. 🗒
45.

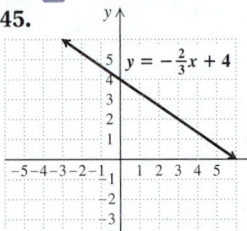

46.

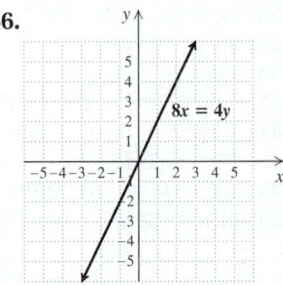

47. Slope: 4; y-intercept: $\left(0, \frac{7}{2}\right)$ **48.** $y = \frac{5}{4}x - \frac{9}{2}$
49. 🗒   **51.** $5x^{6k} - 16x^{3k} + 14$ **53.** $3t^{2h} + 2t^h - 5$
55. $a + 3 + \dfrac{5}{5a^2 - 7a - 2}$ **57.** $2x^2 + x - 3$ **59.** 3
61. -1 **63. (a)** $(a + 1)$ cm; **(b)** $(a^2 + 2a + 1)$ cm^2

Quick Quiz: Sections 4.1–4.8, p. 292

1. $-t^5$ **2.** $-15x^6 - 18x^5 + 3x^4$ **3.** $\frac{1}{9}a^4 + 4a^2 + 36$
4. $8ab^2 - 10ab$ **5.** $4n^3 - 2n^2 + 1$

Prepare to Move On, p. 292

1. $3, x + 7$ **2.** $x - 3, x + 10$ **3.** $x, x + 1, 2x - 5$
4. $3(3x + 5y)$ **5.** $7(2a + c + 1)$

Visualizing for Success, p. 293

1. E, F   **2.** B, O   **3.** S, K   **4.** R, G   **5.** D, M
6. J, P   **7.** C, L   **8.** N, Q   **9.** A, H   **10.** I, T

Decision Making: Connection, p. 294

1. 80 movies   **2.** 32 GB or 64 GB   **3.** 🖥   **4.** 🖱

Study Summary: Chapter 4, pp. 295–297

1. 6   **2.** 1   **3.** x^{16} **4.** 8^7 **5.** y^{15} **6.** $x^{30}y^{10}$
7. $\dfrac{x^{10}}{7^5}$ **8.** $\frac{1}{10}$ **9.** $\dfrac{y^3}{x}$ **10.** 9.04×10^{-4}
11. 690,000   **12.** $x^2, -10, 5x, -8x^6$ **13.** 1   **14.** 1
15. $-8x^6$ **16.** -8 **17.** 6   **18.** Trinomial
19. $3x^2 - 4x$ **20.** 4   **21.** $8x^2 + x$ **22.** $10x^2 - 7x$
23. $x^3 - 2x^2 - x + 2$ **24.** $2x^2 + 11x + 12$
25. $25 - 9x^2$ **26.** $x^2 + 18x + 81$ **27.** $64x^2 - 16x + 1$
28. -7 **29.** 6   **30.** $3cd^2 + 4cd - 7c$ **31.** $-2p^2w$
32. $49x^2y^2 - 14x^3y + x^4$ **33.** $y^3 - 2y + 4$
34. $x - 2 + \dfrac{6}{x + 1}$

Review Exercises: Chapter 4, pp. 298–299

1. True   **2.** True   **3.** True   **4.** False
5. False   **6.** False   **7.** True   **8.** True   **9.** n^{12}
10. $(7x)^{10}$ **11.** t^6 **12.** 4^3, or 64   **13.** 1
14. $-9c^4d^2$ **15.** $-8x^3y^6$ **16.** $18x^5$ **17.** a^7b^6
18. $\dfrac{4t^{10}}{9s^8}$ **19.** $\dfrac{1}{8^6}$ **20.** a^{-9} **21.** $\dfrac{1}{4^2}$, or $\dfrac{1}{16}$ **22.** $\dfrac{2b^9}{a^{13}}$
23. $\dfrac{1}{w^{15}}$ **24.** $\dfrac{x^6}{4y^2}$ **25.** $\dfrac{y^3}{8x^3}$ **26.** 470,000,000
27. 1.09×10^{-5} **28.** 2.09×10^4 **29.** 5.12×10^{-5}
30. $-4y^5, 7y^2, -3y, -2$ **31.** $7, -\frac{5}{6}, -4, 10$
32. (a) 2, 0, 5; **(b)** $15t^5$, 15; **(c)** 5   **33. (a)** 5, 0, 2, 1;
(b) $-2x^5, -2$; **(c)** 5   **34.** Trinomial
35. Polynomial with no special name   **36.** Monomial
37. $t - 1$ **38.** $14a^5 - 2a^2 - a - \frac{2}{3}$ **39.** -24

40. 16   **41.** $x^5 + 8x^4 + 6x^3 - 2x - 9$
42. $6a^5 - a^3 - 12a^2$ **43.** $x^5 - 3x^3 - 2x^2 + 8$
44. $\frac{3}{4}x^4 + \frac{1}{4}x^3 - \frac{1}{3}x^2 - \frac{7}{4}x + \frac{3}{8}$
45. $-x^5 + x^4 - 5x^3 - 2x^2 + 2x$
46. **(a)** $4w + 6$; **(b)** $w^2 + 3w$ **47.** $-30x^5$
48. $49x^2 + 14x + 1$ **49.** $a^2 - 3a - 28$ **50.** $d^2 - 64$
51. $12x^3 - 23x^2 + 13x - 2$ **52.** $15t^5 - 6t^4 + 12t^3$
53. $4a^2 - 81$ **54.** $x^2 - 1.3x + 0.4$
55. $x^7 + x^5 - 3x^4 + 3x^3 - 2x^2 + 5x - 3$
56. $16y^6 - 40y^3 + 25$ **57.** $2t^4 - 11t^2 - 21$
58. $a^2 + \frac{1}{6}a - \frac{1}{3}$ **59.** $-49 + 4n^2$ **60.** 49
61. Coefficients: 1, −7, 9, −8; degrees: 6, 2, 2, 0; 6
62. Coefficients: 1, −1, 1; degrees: 13, 22, 15; 22
63. $-4u + 4v - 7$ **64.** $6m^3 + 4m^2n - mn^2$
65. $2a^2 - 16ab$ **66.** $11x^3y^2 - 8x^2y - 6x^2 - 6x + 6$
67. $2x^2 - xy - 15y^2$ **68.** $25a^2b^2 - 10abcd^2 + c^2d^4$
69. $\frac{1}{2}x^2 - \frac{1}{2}y^2$ **70.** $y^4 - \frac{1}{3}y + 4$
71. $3x^2 - 7x + 4 + \dfrac{1}{2x + 3}$ **72.** $t^3 + 2t - 3$
73. ✎ In the expression $5x^3$, the exponent refers only to the x. In the expression $(5x)^3$, the entire expression $5x$ is the base.
74. ✎ It is possible to determine two possibilities for the binomial that was squared by using the equation $(A - B)^2 = A^2 - 2AB + B^2$ in reverse. Since, in $x^2 - 6x + 9$, $A^2 = x^2$ and $B^2 = 9$, or 3^2, the binomial that was squared was $A - B$, or $x - 3$. If the polynomial is written $9 - 6x + x^2$, then $A^2 = 9$ and $B^2 = x^2$, so the binomial that was squared was $3 - x$. We cannot determine without further information whether the binomial squared was $x - 3$ or $3 - x$.   **75.** **(a)** 9; **(b)** 28   **76.** $64x^{16}$
77. $8x^4 + 4x^3 + 5x - 2$ **78.** $-16x^6 + x^2 - 10x + 25$
79. $\frac{94}{13}$ **80.** 2.28×10^{11} platelets

Test: Chapter 4, p. 300

1. [4.1] x^{13} **2.** [4.1] 3   **3.** [4.1] 1   **4.** [4.1] t^{45}
5. [4.1] $-27y^6$ **6.** [4.1] $-40x^{19}y^4$ **7.** [4.1] $\frac{6}{5}a^5b^3$
8. [4.1] $\dfrac{16p^2}{25q^6}$ **9.** [4.2] $\dfrac{1}{y^7}$ **10.** [4.2] 5^{-6}
11. [4.2] $\dfrac{1}{t^9}$ **12.** [4.2] $\dfrac{3y^5}{x^5}$ **13.** [4.2] $\dfrac{b^4}{16a^{12}}$
14. [4.2] $\dfrac{c^3}{a^3b^3}$ **15.** [4.2] 3.06×10^9
16. [4.2] 0.00000005   **17.** [4.2] 1.75×10^{17}
18. [4.2] 1.296×10^{22} **19.** [4.3] Binomial
20. [4.3] 3, −1, $\frac{1}{9}$ **21.** [4.3] Degrees of terms: 3, 1, 5, 0; leading term: $7t^5$; leading coefficient: 7; degree of polynomial: 5   **22.** [4.3] −7   **23.** [4.3] $\frac{7}{4}y^2 - 4y$
24. [4.3] $4x^3 + 4x^2 + 3$
25. [4.4] $4x^5 + x^4 + 5x^3 - 8x^2 + 2x - 7$
26. [4.4] $5x^4 + 5x^2 + x + 5$
27. [4.4] $-2a^4 + 3a^3 - a - 7$
28. [4.4] $-t^4 + 2.5t^3 - 0.6t^2 - 9$
29. [4.5] $-6x^4 + 6x^3 + 10x^2$ **30.** [4.6] $x^2 - \frac{2}{3}x + \frac{1}{9}$
31. [4.6] $25t^2 - 49$ **32.** [4.6] $6b^2 + 7b - 5$
33. [4.6] $x^{14} - 4x^8 + 4x^6 - 16$ **34.** [4.6] $48 + 34y - 5y^2$
35. [4.5] $6x^3 - 7x^2 - 11x - 3$
36. [4.6] $64a^6 + 48a^3 + 9$ **37.** [4.7] 24
38. [4.7] $-4x^3y - x^2y^2 + xy^3 - y^3 + 19$
39. [4.7] $8a^2b^2 + 6ab + 6ab^2 + ab^3 - 4b^3$

40. [4.7] $9x^{10} - y^2$ **41.** [4.8] $4x^2 + 3x - 5$
42. [4.8] $6x^2 - 20x + 26 + \dfrac{-39}{x + 2}$
43. [4.5], [4.6] $V = l(l - 2)(l - 1) = l^3 - 3l^2 + 2l$
44. [4.2] $\frac{1}{2} - \frac{1}{4} = \frac{1}{4}$ **45.** [4.6] About 2.9×10^8 hr

Cumulative Review: Chapters 1–4, p. 301

1. 6   **2.** −8   **3.** $-\frac{7}{45}$ **4.** 6   **5.** $y + 10$ **6.** t^{12}
7. $-4x^5y^3$ **8.** $50a^4b^7$ **9.** $2(5a - 3b + 6)$ **10.** 6
11. II
12. 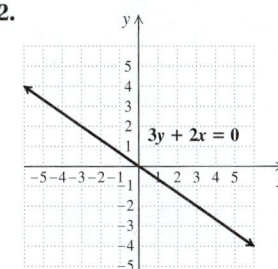   **13.**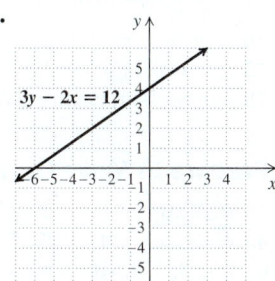

14. 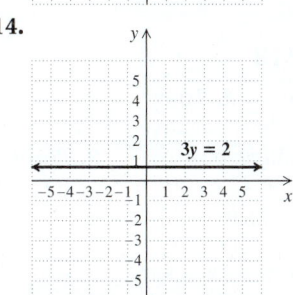   **15.**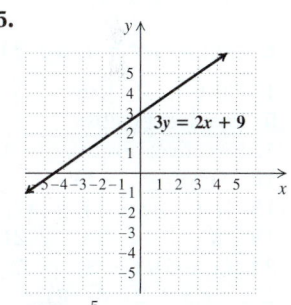

16. Slope: $\frac{1}{10}$; y-intercept: $\left(0, \frac{3}{8}\right)$ **17.** $-\frac{5}{8}$
18. $y = -\frac{2}{3}x - 10$ **19.** −4   **20.** $\frac{8}{3}$ **21.** −7
22. $\{t \mid t \geq -8\}$, or $[-8, \infty)$ **23.** $\left\{x \mid x < \frac{1}{2}\right\}$, or $\left(-\infty, \frac{1}{2}\right)$
24. $t = \dfrac{5pq}{2c}$ **25.** $7u^2v - 2uv^2 + uv + 3u^2$
26. $x^5 - 2x$ **27.** $-16x^5 - 48x^4 + 56x^3$
28. $x^3 - x^2 - 7x + 10$ **29.** $16t^4 + 24t^2 + 9$
30. $\frac{1}{4}x^2 - 1$ **31.** $6r^4 - 5r^2s - 4s^2$ **32.** $x + \dfrac{3}{x - 1}$
33. $\dfrac{1}{7^{10}}$ **34.** $\dfrac{x^7y^2}{3}$ **35.** 53 thousand megawatts
36. 4 trillion kWh   **37.** No more than 4 hr
38. $0.1 billion per year   **39.** 0   **40.** No solution
41. $\frac{1}{7} + 1 = \frac{8}{7}$ **42.** $15x^{12}$

CHAPTER 5

Technology Connection, p. 309

1. Correct   **2.** Correct   **3.** Not correct
4. Not correct   **5.** Not correct   **6.** Correct
7. Not correct   **8.** Correct

Exercise Set 5.1, pp. 310–311

1. False   **2.** True   **3.** True   **4.** False   **5.** (h)
6. (f)   **7.** (b)   **8.** (e)   **9.** (c)   **10.** (g)   **11.** (d)
12. (a)   **13.** Answers may vary.
$(14x)(x^2), (7x^2)(2x), (-2)(-7x^3)$ **15.** Answers may vary.
$(-15)(a^4), (-5a)(3a^3), (-3a^2)(5a^2)$ **17.** Answers may vary. $(5t^2)(5t^3), (25t)(t^4), (-5t)(-5t^4)$ **19.** $8(x + 3)$

21. $2(x^2 + x - 4)$ **23.** $t(3t + 1)$ **25.** $-5y(y + 2)$
27. $x^2(x + 6)$ **29.** $8a^2(2a^2 - 3)$ **31.** $-t^2(6t^4 - 9t^2 + 4)$
33. $6x^2(x^6 + 2x^4 - 4x^2 + 5)$ **35.** $x^2y^2(x^3y^3 + x^2y + xy - 1)$
37. $-5a^2b^2(7ab^2 - 2b + 3a)$ **39.** $(n - 6)(n + 3)$
41. $(x + 3)(x^2 - 7)$ **43.** $(2y - 9)(y^2 + 1)$
45. $(x + 2)(x^2 + 5)$ **47.** $(3n - 2)(3n^2 + 1)$
49. $(t - 5)(4t^2 + 3)$ **51.** $(7x + 5)(x^2 - 3)$
53. $(6a + 7)(a^2 + 1)$ **55.** $(x - 6)(2x^2 - 1)$
57. Not factorable by grouping **59.** $(y + 8)(y^2 - 2)$
61. $(x - 4)(2x^2 - 9)$ **63.** 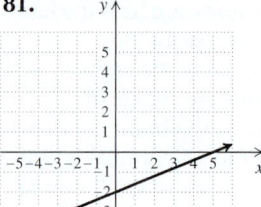 **65.** $\frac{3}{25}$ **66.** 24

67. $\frac{y^4}{2x}$ **68.** $a^7b^7c^3$ **69.** $4x^2 + 5x - 13$

70. $-w^2y + yz + wz$ **71.** 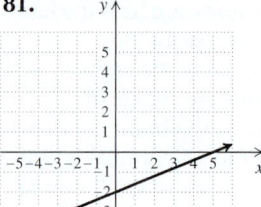 **73.** $(2x^3 + 3)(2x^2 + 3)$
75. $2x(x + 1)(x^2 - 2)$ **77.** $(x - 1)(5x^4 + x^2 + 3)$
79. Answers may vary. $8x^4y^3 - 24x^2y^4 + 16x^3y^4$

Prepare to Move On, p. 311

1. $x^2 + 9x + 14$ **2.** $x^2 - 9x + 14$ **3.** $x^2 - 5x - 14$
4. $x^2 + 5x - 14$ **5.** 1, 2, 3, 4, 5, 6, 10, 12, 15, 20, 30, 60
6. 1, 2, 3, 6, 9, 18

Exercise Set 5.2, pp. 317–318

1. False **2.** True **3.** True **4.** True
5. Positive; positive **6.** Negative; negative
7. Negative; positive **8.** Positive; positive **9.** Positive
10. Negative **11.** $(x + 4)(x + 4)$ **13.** $(x + 1)(x + 10)$
15. $(t - 2)(t - 7)$ **17.** $(b - 4)(b - 1)$
19. $(d - 2)(d - 5)$ **21.** $(x - 5)(x + 3)$
23. $(x + 5)(x - 3)$ **25.** $2(x + 2)(x - 9)$
27. $-x(x + 2)(x - 8)$ **29.** $(y - 5)(y + 9)$
31. $(x - 6)(x + 12)$ **33.** $-5(b - 3)(b + 10)$
35. $x^3(x - 2)(x + 1)$ **37.** Prime **39.** $(t + 4)(t + 8)$
41. $(x + 9)(x + 11)$ **43.** $3x(x - 25)(x + 4)$
45. $-4(x + 5)(x + 5)$ **47.** $(y - 12)(y - 8)$
49. $-a^4(a - 6)(a + 15)$ **51.** $\left(t + \frac{1}{3}\right)\left(t + \frac{1}{3}\right)$
53. Prime **55.** $(p - 5q)(p - 2q)$ **57.** Prime
59. $(s - 6t)(s + 2t)$ **61.** $6a^8(a - 2)(a + 7)$
63. 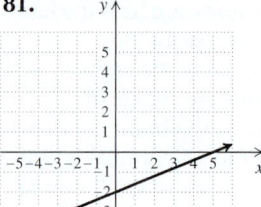 **65.** -10 **66.** 1.13 **67.** $\frac{5}{3}$ **68.** $\frac{5}{2}$
69. -11 **70.** 2 **71.** 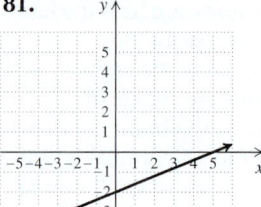 **73.** $-5, 5, -23, 23, -49, 49$
75. $(y + 0.2)(y - 0.4)$ **77.** $-\frac{1}{3}a(a - 3)(a + 2)$
79. $(x^m + 4)(x^m + 7)$ **81.** $(a + 1)(x + 2)(x + 1)$
83. $(x + 3)^3$, or $(x^3 + 9x^2 + 27x + 27)$ cubic meters
85. $x^2\left(\frac{3}{4}\pi + 2\right)$, or $\frac{1}{4}x^2(3\pi + 8)$
87. $x^2\left(9 - \frac{1}{2}\pi\right)$ **89.** $(x + 4)(x + 5)$

Quick Quiz: Sections 5.1–5.2, p. 318

1. $6x^2(2x^2 - 5x + 1)$ **2.** $(x - 4)(3x^2 + 5)$
3. $(x - 2)(x - 1)$ **4.** $(a + 4b)(a - 2b)$
5. $3(x + 2)(x + 3)$

Prepare to Move On, p. 318

1. $6x^2 + 17x + 12$ **2.** $6x^2 + x - 12$ **3.** $6x^2 - x - 12$
4. $6x^2 - 17x + 12$ **5.** $5x^2 - 36x + 7$
6. $3x^2 + 13x - 30$

Exercise Set 5.3, pp. 326–327

1. (d) **2.** (a) **3.** (b) **4.** (c) **5.** $(2x - 1)(x + 4)$

7. $(3x + 1)(x - 6)$ **9.** $(2t + 1)(2t + 5)$
11. $(5a - 3)(3a - 1)$ **13.** $(3x + 4)(2x + 3)$
15. $2(3x + 1)(x - 2)$ **17.** $t(7t + 1)(t + 2)$
19. $(4x - 5)(3x - 2)$ **21.** $-1(7x + 4)(5x + 2)$, or
$-(7x + 4)(5x + 2)$ **23.** Prime **25.** $(5x + 4)^2$
27. $(20y - 1)(y + 3)$ **29.** $(7x + 5)(2x + 9)$
31. $-1(x - 3)(2x + 5)$, or $-(x - 3)(2x + 5)$
33. $-3(2x + 1)(x + 5)$ **35.** $2(a + 1)(5a - 9)$
37. $4(3x - 1)(x + 6)$ **39.** $(3x + 1)(x + 1)$
41. $(x + 3)(x - 2)$ **43.** $(4t - 3)(2t - 7)$
45. $(3x + 2)(2x + 5)$ **47.** $(y + 4)(2y - 1)$
49. $(3a - 4)(2a - 1)$ **51.** $(16t + 7)(t + 1)$
53. $-1(3x + 1)(3x + 5)$, or $-(3x + 1)(3x + 5)$
55. $10(x^2 + 3x - 7)$ **57.** $3x(3x - 1)(2x + 3)$
59. $(x + 1)(25x + 64)$ **61.** $3x(7x + 1)(8x + 1)$
63. $-t^2(2t - 3)(7t + 1)$ **65.** $2(2y + 9)(8y - 3)$
67. $(2a - b)(a - 2b)$ **69.** $2(s + t)(4s + 7t)$
71. $3(3x - 4y)^2$ **73.** $-2(3a - 2b)(4a - 3b)$
75. $x^2(2x + 3)(7x - 1)$ **77.** $a^6(3a + 4)(3a + 2)$
79.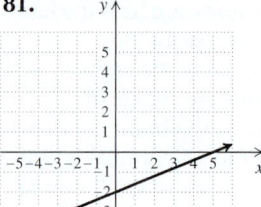
81. 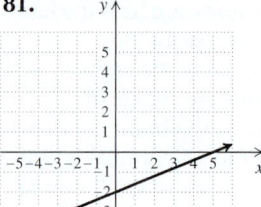 **82.**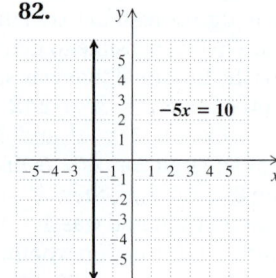

83. 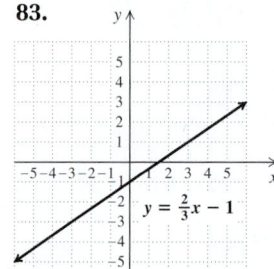 **84.**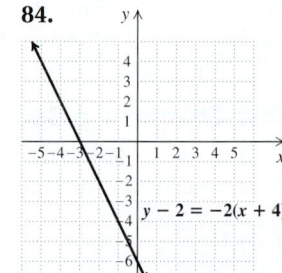

85. 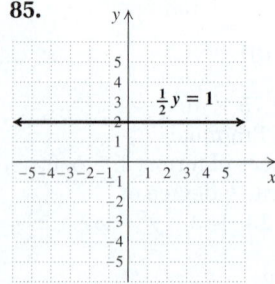 **86.**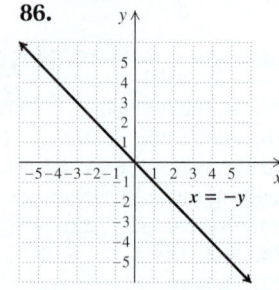

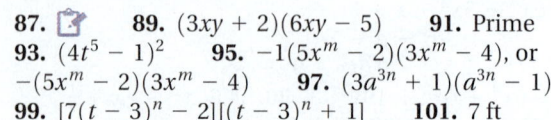
87. 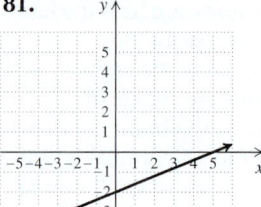 **89.** $(3xy + 2)(6xy - 5)$ **91.** Prime
93. $(4t^5 - 1)^2$ **95.** $-1(5x^m - 2)(3x^m - 4)$, or
$-(5x^m - 2)(3x^m - 4)$ **97.** $(3a^{3n} + 1)(a^{3n} - 1)$
99. $[7(t - 3)^n - 2][(t - 3)^n + 1]$ **101.** 7 ft

Quick Quiz: Sections 5.1–5.3, p. 327

1. $3ab^3(2a - 3b + 5a^2b^2)$ **2.** $x(x - 3)(x - 4)$
3. Prime **4.** $(2a + 3)(2a + 5)$ **5.** $(6x + 5)(x - 1)$

Prepare to Move On, p. 327

1. $x^2 - 4x + 4$ **2.** $x^2 + 4x + 4$ **3.** $x^2 - 4$
4. $25t^2 - 30t + 9$ **5.** $16a^2 + 8a + 1$ **6.** $4n^2 - 49$

Exercise Set 5.4, pp. 332–334

1. Prime polynomial **2.** Difference of squares
3. Difference of squares **4.** None of these
5. Perfect-square trinomial **6.** Perfect-square trinomial
7. None of these **8.** Prime polynomial **9.** Difference
of squares **10.** Perfect-square trinomial **11.** Yes
13. No **15.** No **17.** Yes **19.** $(x + 8)^2$
21. $(x - 5)^2$ **23.** $5(p + 2)^2$ **25.** $(1 - t)^2$, or $(t - 1)^2$
27. $2(3x + 1)^2$ **29.** $(7 - 4y)^2$, or $(4y - 7)^2$
31. $-x^3(x - 9)^2$ **33.** $2n(n + 10)^2$ **35.** $5(2x + 5)^2$
37. $(7 - 3x)^2$, or $(3x - 7)^2$ **39.** $(4x + 3)^2$
41. $2(1 + 5x)^2$, or $2(5x + 1)^2$ **43.** $(3p + 2x)^2$
45. Prime **47.** $-1(8m + n)^2$, or $-(8m + n)^2$
49. $-2(4s - 5t)^2$ **51.** Yes **53.** No **55.** Yes
57. $(x + 5)(x - 5)$ **59.** $(p + 3)(p - 3)$
61. $(7 + t)(-7 + t)$, or $(t + 7)(t - 7)$
63. $6(a + 2)(a - 2)$ **65.** $(7x - 1)^2$
67. $2(10 + t)(10 - t)$ **69.** $-5(4a + 3)(4a - 3)$
71. $5(t + 4)(t - 4)$ **73.** $2(2x + 9)(2x - 9)$
75. $x(6 + 7x)(6 - 7x)$ **77.** Prime
79. $(t^2 + 1)(t + 1)(t - 1)$ **81.** $-3x(x - 4)^2$
83. $3t(5t + 3)(5t - 3)$ **85.** $a^6(a - 1)^2$
87. $10(a + b)(a - b)$ **89.** $(4x^2 + y^2)(2x + y)(2x - y)$
91. $2(3t + 2s)(3t - 2s)$ **93.** 📋 **95.** $2x^3 + x^2 - 2$
96. $2t^2 + 6t - 11$ **97.** $6x^4 + x^2y - y^2$
98. $5x^3 - 15x^2 + 35x$ **99.** $7x^2 - x + 3$
100. $x + 4 + \dfrac{10}{x - 5}$ **101.** 📋
103. $(x^4 + 2^4)(x^2 + 2^2)(x + 2)(x - 2)$, or
$(x^4 + 16)(x^2 + 4)(x + 2)(x - 2)$
105. $2x\left(3x - \frac{2}{5}\right)\left(3x + \frac{2}{5}\right)$
107. $[(y - 5)^2 + z^4][(y - 5) + z^2][(y - 5) - z^2]$, or
$(y^2 - 10y + 25 + z^4)(y - 5 + z^2)(y - 5 - z^2)$
109. $-1(x^2 + 1)(x + 3)(x - 3)$, or
$-(x^2 + 1)(x + 3)(x - 3)$ **111.** $(y + 4)^2$
113. $(3p + 5)(3p - 5)^2$ **115.** $(9 + b^{2k})(3 + b^k)(3 - b^k)$
117. $2x^3 - x^2 - 1$ **119.** $(y + x + 7)(y - x - 1)$
121. 16 **123.** $(x + 1)^2 - x^2 = [(x + 1) + x][(x + 1) - x] =$
$2x + 1 = (x + 1) + x$

Quick Quiz: Sections 5.1–5.4, p. 334

1. $(x + 3)(2x^2 - 1)$ **2.** $(x - 20)(x - 4)$
3. $a(3a - 4)(3a + 2)$ **4.** $(y + 8)^2$
5. $3(z^2 + 1)(z + 1)(z - 1)$

Prepare to Move On, p. 334

1. $8x^6y^{12}$ **2.** $-125x^6y^3$ **3.** $x^3 - 3x^2 + 3x - 1$
4. $p^3 + 3p^2t + 3pt^2 + t^3$

Mid-Chapter Review: Chapter 5, p. 335

1. $12x^3y - 8xy^2 + 24x^2y = 4xy(3x^2 - 2y + 6x)$
2. $3a^3 - 3a^2 - 90a = 3a(a^2 - a - 30) =$
$3a(a - 6)(a + 5)$ **3.** $6x^2(x^3 - 3)$ **4.** $(x + 2)(x + 8)$

5. $(x + 7)(2x - 1)$ **6.** $(x + 3)(x^2 + 2)$
7. $(8n + 3)(8n - 3)$ **8.** Prime **9.** $6(p + t)(p - t)$
10. $(b - 7)^2$ **11.** $(3x - 1)(4x + 1)$ **12.** $a(1 - 5a)^2$
13. $10(x^2 + 1)(x + 1)(x - 1)$ **14.** Prime
15. $15(d^2 - 2d + 5)$ **16.** $(3p + 2x)(5p + 2x)$
17. $-2t(t + 2)(t + 3)$ **18.** $10(c + 1)^2$
19. $-1(2x - 5)(x + 1)$ **20.** $2n(m - 5)(m^2 - 3)$

Exercise Set 5.5, pp. 338–339

1. Difference of cubes **2.** Sum of cubes **3.** Difference
of squares **4.** None of these **5.** Sum of cubes
6. Difference of cubes **7.** None of these
8. Difference of squares **9.** Difference of cubes
10. None of these **11.** $(x - 4)(x^2 + 4x + 16)$
13. $(z + 1)(z^2 - z + 1)$ **15.** $(t - 10)(t^2 + 10t + 100)$
17. $(3x + 1)(9x^2 - 3x + 1)$
19. $(4 - 5x)(16 + 20x + 25x^2)$
21. $(x - y)(x^2 + xy + y^2)$ **23.** $\left(a + \frac{1}{2}\right)\left(a^2 - \frac{1}{2}a + \frac{1}{4}\right)$
25. $8(t - 1)(t^2 + t + 1)$
27. $2(3x + 1)(9x^2 - 3x + 1)$
29. $rs(s + 4)(s^2 - 4s + 16)$
31. $5(x - 2z)(x^2 + 2xz + 4z^2)$
33. $\left(y - \frac{1}{10}\right)\left(y^2 + \frac{1}{10}y + \frac{1}{100}\right)$
35. $(x + 0.1)(x^2 - 0.1x + 0.01)$
37. $8(2x^2 - t^2)(4x^4 + 2x^2t^2 + t^4)$
39. $2y(3y - 4)(9y^2 + 12y + 16)$
41. $(z + 1)(z^2 - z + 1)(z - 1)(z^2 + z + 1)$
43. $(t^2 + 4y^2)(t^4 - 4t^2y^2 + 16y^4)$
45. $(x^4 - yz^4)(x^8 + x^4yz^4 + y^2z^8)$ **47.** 📋
49. 2400 mg **50.** 6.4 gal **51.** \$520 **52.** \$46
53. 📋 **55.** $(x^{2a} - y^b)(x^{4a} + x^{2a}y^b + y^{2b})$
57. $2x(x^2 + 75)$ **59.** $5\left(xy^2 - \frac{1}{2}\right)\left(x^2y^4 + \frac{1}{2}xy^2 + \frac{1}{4}\right)$
61. $-(3x^{4a} + 3x^{2a} + 1)$ **63.** $(t - 8)(t - 1)(t^2 + t + 1)$

Quick Quiz: Sections 5.1–5.5, p. 339

1. $xp^2(6xp + 1 - 3x^2p)$ **2.** $(x - 12)(x + 1)$
3. $a(4a + 1)(a - 5)$ **4.** $(p + w)(p - w)$
5. $(p - w)(p^2 + pw + w^2)$

Prepare to Move On, p. 339

1. Common **2.** Grouping or ac **3.** $(A + B)(A - B)$
4. $(A + B)^2$ **5.** $(A + B)(A^2 - AB + B^2)$

Exercise Set 5.6, pp. 344–345

1. Common factor **2.** Perfect-square trinomial
3. Grouping **4.** Multiplying **5.** $5(a + 5)(a - 5)$
7. $(y - 7)^2$ **9.** $(3t + 7)(t + 3)$ **11.** $x(x + 9)^2$
13. $(x - 5)^2(x + 5)$ **15.** $3t(3t + 1)(3t - 1)$
17. $3x(3x - 5)(x + 3)$ **19.** Prime **21.** $6(y - 5)(y + 8)$
23. $-2a^4(a - 2)^2$ **25.** $5x(x^2 + 4)(x + 2)(x - 2)$
27. $(t^2 + 3)(t^2 - 3)$ **29.** $-x^4(x^2 - 2x + 7)$
31. $(x - y)(x^2 + xy + y^2)$ **33.** $a(x^2 + y^2)$
35. $2\pi r(h + r)$ **37.** $(a + b)(5a + 3b)$
39. $(x + 1)(x + y)$ **41.** $10a^2(4m^2 + 1)(2m + 1)(2m - 1)$
43. $(a - 2)(a - y)$ **45.** $(3x - 2y)(x + 5y)$
47. $8mn(m^2 - 4mn + 3)$ **49.** $\left(\frac{3}{4} + y\right)\left(\frac{3}{4} - y\right)$
51. $(a - 2b)^2$ **53.** $(4x + 3y)^2$ **55.** Prime

57. $(2x - 3)(5x + 2)$ **59.** $(a^2b^2 + 4)(ab + 2)(ab - 2)$
61. $4c(4d - c)(5d - c)$
63. $(2t + 1)(4t^2 - 2t + 1)(2t - 1)(4t^2 + 2t + 1)$
65. $-1(xy + 2)(xy + 6)$, or $-(xy + 2)(xy + 6)$
67. $(2t + 5)(4t - 3)$ **69.** $5(pt + 6)(pt - 1)$
71. $2a(3a + 2b)(9a^2 - 6ab + 4b^2)$
73. $x^4(x + 2y)(x - y)$ **75.** $\left(6a - \frac{5}{4}\right)^2$ **77.** $\left(\frac{1}{9}x - \frac{4}{3}\right)^2$
79. $(1 + 4x^6y^6)(1 + 2x^3y^3)(1 - 2x^3y^3)$ **81.** $(2ab + 3)^2$
83. $z(z + 6)(z^2 - 6)$ **85.** $(x + 5)(x + 1)(x - 1)$
87. ☑   **89.** About 69.2%   **90.** 5.75 billionaires per
month   **91.** Mother's Day: \$16.3 billion; Father's Day:
\$11.1 billion   **92.** 5 or fewer roses   **93.** ☑
95. $(6 - x + a)(6 - x - a)$ **97.** $-x(x^2 + 9)(x^2 - 2)$
99. $-1(x^2 + 2)(x + 3)(x - 3)$, or $-(x^2 + 2)(x + 3)(x - 3)$
101. $(y + 1)(y - 7)(y + 3)$ **103.** $(y + 4 + x)^2$
105. $(a + 3)^2(2a + b + 4)(a - b + 5)$
107. $(7x^2 + 1 + 5x^3)(7x^2 + 1 - 5x^3)$

Quick Quiz: Sections 5.1–5.6, p. 345

1. $10(m + 3)(m - 3)$ **2.** $(a + 3)(a - 2)$
3. $(2x - 1)(6x^2 - 5)$ **4.** $8(mn - 1)^2$
5. $c(3c + 2)(2c - 3)$

Prepare to Move On, p. 345

1. $\frac{9}{8}$ **2.** $-\frac{7}{2}$ **3.** 3   **4.** $-\frac{5}{3}$ **5.** $\frac{1}{4}$ **6.** 11

Connecting the Concepts, p. 350

1. Expression   **2.** Equation   **3.** Expression
4. Equation   **5.** $2x^3 + x^2 - 8x$ **6.** $-2x^2 - 11$
7. $-10, 10$ **8.** $6a^2 - 19a + 10$ **9.** $(n - 1)(n - 9)$
10. 2, 8

Technology Connection, p. 351

1. $-4.65, 0.65$ **2.** $-0.37, 5.37$ **3.** $-8.98, -4.56$
4. No solution   **5.** 0, 2.76

Exercise Set 5.7, pp. 351–353

1. (c)   **2.** (a)   **3.** (d)   **4.** (b)   **5.** $-9, -2$
7. $-1, 8$ **9.** $-6, \frac{3}{2}$ **11.** $\frac{1}{7}, \frac{3}{10}$ **13.** 0, 7   **15.** $\frac{1}{21}, \frac{18}{11}$
17. $-\frac{8}{3}, 0$ **19.** 50, 70   **21.** 1, 6   **23.** $-7, 3$
25. $-9, -2$ **27.** 0, 10   **29.** $-6, 0$ **31.** $-6, 6$
33. $-\frac{7}{2}, \frac{7}{2}$ **35.** -5 **37.** 8   **39.** 0, 2   **41.** $-\frac{5}{4}, 3$
43. 3   **45.** $0, \frac{4}{3}$ **47.** $-\frac{7}{6}, \frac{7}{6}$ **49.** $-4, -\frac{2}{3}$ **51.** $-3, 1$
53. $-\frac{5}{2}, \frac{4}{3}$ **55.** $-1, 4$ **57.** $-3, 2$ **59.** $(-2, 0), (3, 0)$
61. $(-4, 0), (2, 0)$ **63.** $(-3, 0), \left(\frac{3}{2}, 0\right)$ **65.** ☑   **67.** 65
68. $\frac{4}{3}$ **69.** 1.65   **70.** 6.8×10^{-4} **71.** $-0.\overline{6}$
72. 125%   **73.** ☑   **75.** $-7, -\frac{8}{3}, \frac{11}{2}$
77. (a) $x^2 - x - 20 = 0$; (b) $x^2 - 6x - 7 = 0$;
(c) $4x^2 - 13x + 3 = 0$; (d) $6x^2 - 5x + 1 = 0$;
(e) $12x^2 - 17x + 6 = 0$; (f) $x^3 - 4x^2 + x + 6 = 0$
79. $-5, 4$ **81.** $-\frac{3}{5}, \frac{3}{5}$ **83.** $-4, 2$ **85.** $-1, 1, 2$
87. (a) $2x^2 + 20x - 4 = 0$; (b) $x^2 - 3x - 18 = 0$;
(c) $(x + 1)(5x - 5) = 0$; (d) $(2x + 8)(2x - 5) = 0$;
(e) $4x^2 + 8x + 36 = 0$; (f) $9x^2 - 12x + 24 = 0$
89. ☑   **91.** 2.33, 6.77   **93.** $-9.15, -4.59$
95. $-3.76, 0$ **97.** -8 and -7, or 7 and 8

Quick Quiz: Sections 5.1–5.7, p. 353

1. $(x - 4)(x + 2)$ **2.** $-2, 4$ **3.** $-10, 10$
4. $(a + 10)(a - 10)$ **5.** $2(a + 10)^2$

Prepare to Move On, p. 353

1. Let n represent the first integer; $[n + (n + 1)]^2$
2. Let n represent the first integer; $n(n + 1)$
3. $140°, 35°, 5°$ **4.** Length: 64 in.; width: 32 in.

Exercise Set 5.8, pp. 359–363

1. $x + 1$ **2.** $2w$ **3.** $90°$ **4.** $c^2 = a^2 + b^2$
5. $-2, 3$ **7.** 11, 12   **9.** -14 and -12; 12 and 14
11. Length: 40 ft; width: 8 ft   **13.** Length: 12 cm; width:
7 cm   **15.** Foot: 7 ft; height: 12 ft   **17.** Base: 8 ft;
height: 16 ft   **19.** Base: 8 ft; height: 15 ft   **21.** 8 teams
23. 66 handshakes   **25.** 12 players   **27.** 1 min, 3 min
29. 10 knots   **31.** 9 ft   **33.** 842 ft   **35.** 300 ft by
400 ft by 500 ft   **37.** 24 m, 25 m   **39.** 12 ft, 35 ft, 37 ft
41. Dining room: 12 ft by 12 ft; kitchen: 12 ft by 10 ft
43. 20 ft   **45.** 1 sec, 2 sec   **47.** ☑
49. $\varnothing$; contradiction   **50.** 0   **51.** $\mathbb{R}$; identity
52. $\{y \mid y \le -4\}$, or $(-\infty, -4]$
53. $\{x \mid x < -1\}$, or $(-\infty, -1)$ **54.** $y = -\frac{1}{2}x + \frac{7}{2}$
55. ☑   **57.** \$180   **59.** 39 cm   **61.** 4 in., 6 in.
63. 35 ft   **65.** Length: 6 in.; width: $4\frac{1}{4}$ in.
67. 0.8 hr, 6.4 hr   **69.** 🖳

Quick Quiz: Sections 5.1–5.8, p. 363

1. $(x - 5)(3x^2 - 7)$ **2.** $-\frac{1}{5}, \frac{3}{4}$ **3.** $(2x - 3)(5x - 1)$
4. $6(x^4 + 1)(x^2 + 1)(x + 1)(x - 1)$ **5.** Longer leg: 40 ft;
hypotenuse: 41 ft

Prepare to Move On, p. 363

1. $\frac{6}{7}$ **2.** $\frac{45}{44}$ **3.** $\frac{4}{5}$ **4.** 0   **5.** Undefined

Translating for Success, p. 364

1. O   **2.** M   **3.** K   **4.** I   **5.** G   **6.** E   **7.** C
8. A   **9.** H   **10.** B

Decision Making: Connection, p. 366

1. 105 games   **2.** 210 games   **3.** Even number of
teams: $n - 1$ rounds; odd number of teams: n rounds
4. 🖳   **5.** 🖳   **6.** ☑

Study Summary: Chapter 5, pp. 367–369

1. $6x(2x^3 - 3x^2 + 5)$ **2.** $(x - 3)(2x^2 - 1)$
3. $(x - 9)(x + 2)$ **4.** $(3x + 2)(2x - 1)$
5. $(2x - 3)(4x - 5)$ **6.** $(10n + 9)^2$
7. $(12t + 5)(12t - 5)$ **8.** $(a - 1)(a^2 + a + 1)$
9. $3x(x - 6)^2$ **10.** $-10, 3$ **11.** 12, 13

Review Exercises: Chapter 5, pp. 369–370

1. False   **2.** True   **3.** True   **4.** False   **5.** False
6. True   **7.** True   **8.** False   **9.** Answers may vary.

$(4x)(5x^2), (-2x^2)(-10x), (x^3)(20)$ **10.** Answers may vary. $(-3x^2)(6x^3), (2x^4)(-9x), (-18x)(x^4)$
11. $6x^3(2x - 3)$ **12.** $(10t + 1)(10t - 1)$
13. $(x + 4)(x - 3)$ **14.** $(x + 7)^2$ **15.** $3x(2x + 1)^2$
16. $(2x + 3)(3x^2 + 1)$ **17.** $(6a - 5)(a + 1)$
18. $(5t - 3)^2$ **19.** $(9a^2 + 1)(3a + 1)(3a - 1)$
20. $3x(3x - 5)(x + 3)$ **21.** $2(x - 5)(x^2 + 5x + 25)$
22. $(x + 4)(x^3 - 2)$ **23.** $(ab^2 + 8)(ab^2 - 8)$
24. $-4x^4(2x^2 - 8x + 1)$ **25.** $3(2x - 5)^2$ **26.** Prime
27. $-t(t + 6)(t - 7)$ **28.** $(2x + 5)(2x - 5)$
29. $(n + 6)(n - 10)$ **30.** $5(z^2 - 6z + 2)$
31. $(2y + 3x^2)(4y^2 - 6x^2y + 9x^4)$ **32.** $(2t + 1)(t - 4)$
33. $7x(x + 1)(x + 4)$ **34.** $-6x(x + 5)(x - 5)$
35. $(5 - x)(3 - x)$ **36.** Prime   **37.** $(xy + 8)(xy - 2)$
38. $3(2a + 7b)^2$ **39.** $(m + 5)(m + t)$
40. $(3r + 5s)(2r - 3s)$ **41.** $-11, 9$ **42.** $-7, 5$
43. $-\frac{3}{4}, \frac{3}{4}$ **44.** $\frac{2}{3}, 1$ **45.** $-2, 3$ **46.** $0, \frac{3}{5}$
47. 1 **48.** $-3, 4$ **49.** 5 sec   **50.** $(-1, 0), (\frac{5}{2}, 0)$
51. Height: 14 ft; base: 14 ft   **52.** 10 holes
53. 🖊 Answers may vary. Because Celia did not first factor out the largest common factor, 4, her factorization will not be "complete" until she removes a common factor of 2 from each binomial. The answer should be $4(x - 5)(x + 5)$. Awarding 3 to 7 points would seem reasonable.   **54.** 🖊 The principle of zero products is used to solve quadratic equations and is not used to solve linear equations.
55. 2.5 cm   **56.** $0, 2$ **57.** Length: 12 cm; width: 6 cm   **58.** $-3, 2, \frac{5}{2}$ **59.** No real solution

Test: Chapter 5, p. 371

1. [5.2] $(x - 4)(x - 9)$ **2.** [5.4] $(x - 5)^2$
3. [5.1] $2y^2(2y^2 - 4y + 3)$ **4.** [5.1] $(x + 1)(x^2 + 2)$
5. [5.1] $t^5(t^2 - 3)$ **6.** [5.2] $a(a + 4)(a - 1)$
7. [5.3] $2(5x - 6)(x + 4)$ **8.** [5.4] $(2t + 5)(2t - 5)$
9. [5.3] $-3m(2m + 1)(m + 1)$
10. [5.5] $3(r - 1)(r^2 + r + 1)$ **11.** [5.4] $5(3r + 2)^2$
12. [5.4] $3(x^2 + 4)(x + 2)(x - 2)$ **13.** [5.4] $(7t + 6)^2$
14. [5.1] $(x + 2)(x^3 - 3)$ **15.** [5.2] Prime
16. [5.3] $3t(2t + 5)(t - 1)$
17. [5.2] $3(m - 5n)(m + 2n)$ **18.** [5.7] $1, 5$
19. [5.7] $-\frac{3}{2}, 5$ **20.** [5.7] $0, \frac{2}{5}$ **21.** [5.7] $-\frac{1}{5}, \frac{1}{5}$
22. [5.7] $-4, 5$ **23.** [5.7] $(-1, 0), (\frac{8}{3}, 0)$
24. [5.8] Length: 10 m; width: 4 m   **25.** [5.8] 10 people
26. [5.8] 5 ft   **27.** [5.8] 15 cm by 30 cm
28. [5.2] $(a - 4)(a + 8)$ **29.** [5.7] $-\frac{8}{3}, 0, \frac{2}{5}$

Cumulative Review: Chapters 1–5, p. 372

1. $\frac{1}{2}$ **2.** $\frac{9}{8}$ **3.** 29 **4.** $\frac{1}{9x^4y^6}$ **5.** t^{10}
6. $3x^4 + 5x^3 + x - 10$ **7.** $\frac{t^8}{4s^2}$ **8.** $-\frac{8x^6y^3}{27z^{12}}$
9. 8 **10.** $-x^4$ **11.** $2x^3 - 5x^2 + \frac{1}{2}x - 1$
12. $-4t^{11} + 8t^9 + 20t^8$ **13.** $9x^2 - 30x + 25$
14. $100x^{10} - y^2$ **15.** $(c + 1)(c - 1)$
16. $5(x + y)(1 + 2x)$ **17.** $(2r - t)^2$ **18.** $10(y^2 + 4)$
19. $y(x - 1)(x - 2)$ **20.** $(3x - 2y)(4x + y)$
21. $\frac{1}{12}$ **22.** $-\frac{9}{4}$ **23.** $\{x | x \geq 1\}$, or $[1, \infty)$ **24.** $-4, 3$
25. $-2, 2$ **26.** $0, 4$ **27.** $c = \frac{a}{b + d}$ **28.** 0
29. $y = 5x + \frac{5}{3}$

30.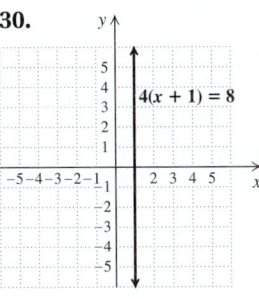

31.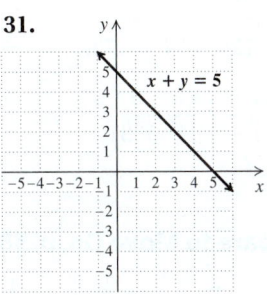

32.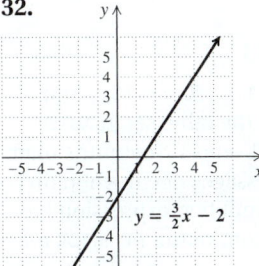

33.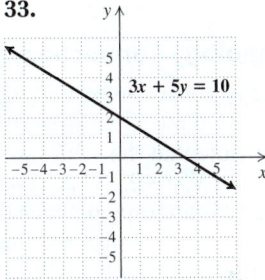

34. 372 min   **35.** Bottom of ladder to building: 5 ft; top of ladder to ground: 12 ft   **36.** Scores that are 9 and higher
37. (a)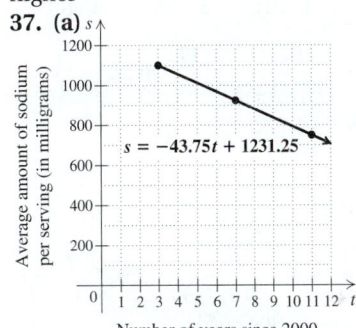

(b) 925 mg per serving   **38.** $b = \frac{2}{a + 1}$ **39.** $-1, 0, \frac{1}{3}$

CHAPTER 6

Technology Connection, p. 377

1. Correct   **2.** Not correct

Exercise Set 6.1, pp. 379–380

1. Quotient   **2.** Denominator   **3.** Factors
4. A factor equal to 1   **5.** (a)   **6.** (c)   **7.** (d)
8. (b)   **9.** 0 **11.** -5 **13.** 5 **15.** $-4, 7$
17. $-6, \frac{1}{2}$ **19.** $\frac{5a}{4b^2}$ **21.** $\frac{t + 2}{t - 3}$ **23.** $\frac{7}{8}$
25. $\frac{a - 3}{a + 1}$ **27.** $-\frac{2x^3}{3}$ **29.** $\frac{y - 3}{4y}$ **31.** $\frac{3(2a + 1)}{7(a + 1)}$
33. $\frac{t - 4}{t - 5}$ **35.** $\frac{a + 4}{2(a - 4)}$ **37.** $\frac{x - 4}{x + 4}$
39. $\frac{1}{n^2 + 2n + 4}$ **41.** $t - 1$ **43.** $\frac{y^2 + 4}{y + 2}$ **45.** $\frac{1}{2}$
47. $\frac{y}{2y + 1}$ **49.** $\frac{2x - 3}{5x + 2}$ **51.** -1 **53.** -7
55. $-\frac{1}{4}$ **57.** $-\frac{3}{2}$ **59.** -1 **61.** 🖊
63. $(x + 5)(3x^2 + 1)$ **64.** $(3x + 1)(x + 5)$
65. $3y^2(6y^2 - 9y + 1)$ **66.** $(5a + 4b)(5a - 4b)$
67. $m(m - 4)^2$ **68.** $5(x - 4)(x - 3)$ **69.** 🖊

71. $-(2y + x)$ **73.** $\dfrac{x^3 + 4}{(x^3 + 2)(x^2 + 2)}$

75. $\dfrac{(t - 1)(t - 9)^2}{(t + 1)(t^2 + 9)}$ **77.** $x + 2$ **79.** $\dfrac{(x - y)^3}{(x + y)^2(x - 5y)}$

81. 📋

Prepare to Move On, p. 380

1. $\frac{4}{21}$ **2.** $-\frac{5}{3}$ **3.** $\frac{15}{4}$ **4.** $-\frac{21}{16}$

Technology Connection, p. 383

1. Let $y_1 = ((x^2 + 3x + 2)/(x^2 + 4))/(5x^2 + 10x)$ and $y_2 = (x + 1)/((x^2 + 4)(5x))$. With the table set in AUTO mode, note that the values in the Y1- and Y2-columns match except for $x = -2$.   **2.** ERROR messages occur when division by 0 is attempted. Since the simplified expression has no factor of $x + 5$ or $x + 1$ in a denominator, no ERROR message occurs in Y2 for $x = -5$ or -1.

Exercise Set 6.2, pp. 384–386

1. (d)   **2.** (c)   **3.** (a)   **4.** (b)   **5.** (d)   **6.** (c)

7. (a)   **8.** (e)   **9.** (b)   **10.** (f)   **11.** $\dfrac{3x(x + 2)}{8(5x - 1)}$

13. $\dfrac{(a - 4)(a + 2)}{(a + 6)^2}$ **15.** $\dfrac{(n - 4)(n + 4)}{(n^2 + 4)(n^2 - 4)}$ **17.** $\dfrac{6t}{5}$

19. $\dfrac{4}{c^2 d}$ **21.** $\dfrac{y + 4}{4}$ **23.** $\dfrac{x + 2}{x - 2}$

25. $\dfrac{(n - 5)(n - 1)(n - 6)}{(n + 6)(n^2 + 36)}$ **27.** $\dfrac{7(a + 3)}{a(a + 4)}$ **29.** $\dfrac{y(y - 1)}{y + 1}$

31. $\dfrac{3v}{v - 2}$ **33.** $\dfrac{t - 5}{t + 5}$ **35.** $\dfrac{9(y - 5)}{10(y - 1)}$ **37.** 1

39. $\dfrac{(t - 2)(t + 5)}{(t - 5)(t - 5)}$ **41.** $(2x - 1)^2$ **43.** $\dfrac{(7x + 5)(3x + 4)}{8}$

45. $\dfrac{2x + 5y}{x^2 y}$ **47.** $c(c - 2)$ **49.** $\dfrac{9}{2x}$ **51.** $\dfrac{1}{a^4 + 3a}$

53. $\dfrac{x^2}{20}$ **55.** $\dfrac{a^3}{b^3}$ **57.** $\dfrac{4(t - 3)}{3(t + 1)}$ **59.** $4(y - 2)$

61. $-\dfrac{a}{b}$ **63.** $\dfrac{(n + 3)(n + 3)}{n - 2}$ **65.** $\dfrac{a - 5}{3(a - 1)}$

67. $\dfrac{(2x - 1)(2x + 1)}{x - 5}$ **69.** $\dfrac{(w - 7)(w - 8)}{(2w - 7)(3w + 1)}$

71. $\dfrac{1}{(c - 5)(5c - 3)}$ **73.** $\frac{15}{16}$ **75.** $\dfrac{-x^2 - 4}{x + 2}$

77. $\dfrac{(x - 4y)(x - y)}{(x + y)^3}$ **79.** $\dfrac{x^2 + 4x + 16}{(x + 4)^2}$

81. $\dfrac{(2a + b)^2}{2(a + b)}$ **83.** 📋

85.

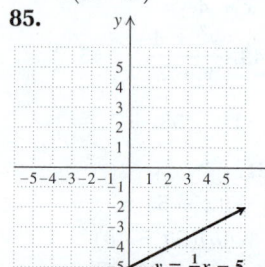

86.

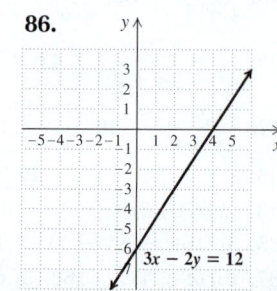

87.

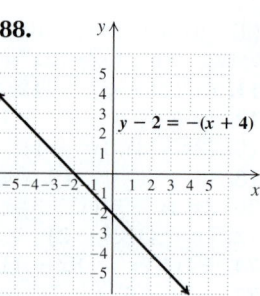

88.

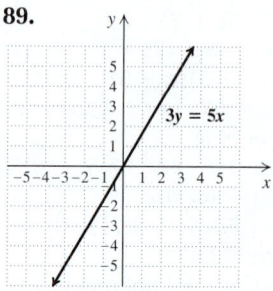

89.

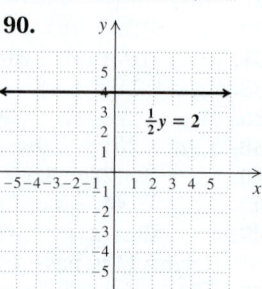

90.

91. 📋   **93.** $\dfrac{3}{7x}$ **95.** 1   **97.** $\dfrac{1}{b^3(a - 3b)}$

99. $\dfrac{(z + 4)^3}{3(z - 4)^2}$ **101.** $\dfrac{a - 3b}{c}$ **103.** $\dfrac{(x + 1)^2}{(x - 1)^2}$

105. 1   **107.** $\dfrac{(2n - 1)(3n + 2)}{(2n + 1)(3n - 2)}$ **109.**

Quick Quiz: Sections 6.1–6.2, p. 386

1. $\dfrac{6(x + 3)}{x - 4}$ **2.** $\dfrac{y - 3}{2y + 1}$ **3.** $\dfrac{(x - 1)^2}{4x}$

4. $\dfrac{m + 3}{6(m + 2)}$ **5.** $\dfrac{(x + 4)^2}{(x + 5)(x - 4)}$

Prepare to Move On, p. 386

1. $\frac{41}{24}$ **2.** $\frac{1}{6}$ **3.** $x^2 + 3$ **4.** $-2x^2 - 4x + 1$

Exercise Set 6.3, pp. 393–395

1. Numerators; denominator   **2.** Term   **3.** Least common denominator; LCD   **4.** Factor   **5.** $\dfrac{8}{t}$

7. $\dfrac{3x + 5}{12}$ **9.** $\dfrac{9}{a + 3}$ **11.** $\dfrac{8}{4x - 7}$ **13.** $\dfrac{2y + 7}{2y}$

15. 6   **17.** $\dfrac{4(x - 1)}{x + 3}$ **19.** $a + 5$ **21.** $y - 7$

23. 0   **25.** $\dfrac{1}{x + 2}$ **27.** $\dfrac{(3a - 7)(a - 2)}{(a + 6)(a - 1)}$

29. $\dfrac{t - 4}{t + 3}$ **31.** $\dfrac{y + 2}{y - 4}$ **33.** $-\dfrac{5}{x - 4}$, or $\dfrac{5}{4 - x}$

35. $-\dfrac{1}{x - 1}$, or $\dfrac{1}{1 - x}$ **37.** 180   **39.** 72   **41.** 60

43. $18t^5$ **45.** $30a^4 b^8$ **47.** $6(y - 3)$

49. $(x - 5)(x + 3)(x - 3)$ **51.** $t(t - 4)(t + 2)^2$

53. $120x^2 y^3 z^2$ **55.** $(a + 1)(a - 1)^2$

57. $(2n - 1)(n + 1)(n + 2)$ **59.** $(t + 3)(t - 3)$

61. $12x^3(x - 5)(x - 3)(x - 1)$

63. $2(x + 1)(x - 1)(x^2 + x + 1)$

65. $\dfrac{15}{18t^4}, \dfrac{st^2}{18t^4}$ **67.** $\dfrac{21y}{9x^4 y^3}, \dfrac{4x^3}{9x^4 y^3}$

69. $\dfrac{2x(x+3)}{(x-2)(x+2)(x+3)}, \dfrac{4x(x-2)}{(x-2)(x+2)(x+3)}$

71. 🔲   **73.** $-\frac{10}{3}$ **74.** $\{x|x>15\}$, or $(15,\infty)$

75. $-2, 10$ **76.** $\{x|x\le -8\}$, or $(-\infty,-8]$ **77.** -1

78. $\frac{8}{3}$ **79.** 🔲   **81.** $\dfrac{18x+5}{x-1}$ **83.** $\dfrac{x}{3x+1}$

85. 30 strands   **87.** 60 strands

89. $(2x+5)(2x-5)(3x+4)^4$ **91.** 12 sec

93. 6:15 A.M.   **95.** 🔲   **97.** 🖥️

Quick Quiz: Sections 6.1–6.3, p. 395

1. $\dfrac{1}{x+7}$ **2.** $x(x-6)(x-1)$ **3.** $\dfrac{1}{(x+3)(x-1)}$

4. $\dfrac{4}{3t^2(t-2)}$ **5.** $\dfrac{8}{x-2}$

Prepare to Move On, p. 395

1. $\frac{-5}{8}, \frac{5}{-8}$ **2.** $\frac{-4}{11}, -\frac{4}{11}$ **3.** $-x+y$, or $y-x$

4. $-3+a$, or $a-3$ **5.** $-2x+7$, or $7-2x$

6. $-a+b$, or $b-a$

Exercise Set 6.4, pp. 400–402

1. LCD   **2.** Missing; denominator   **3.** Numerators;

LCD   **4.** Simplify   **5.** $\dfrac{3+5x}{x^2}$ **7.** $-\dfrac{5}{24r}$

9. $\dfrac{3u^2+4v}{u^3v^2}$ **11.** $\dfrac{-2(xy+9)}{3x^2y^3}$ **13.** $\dfrac{7x+1}{24}$

15. $\dfrac{-x-4}{6}$ **17.** $\dfrac{a^2+13a-5}{15a^2}$ **19.** $\dfrac{7z-12}{12z}$

21. $\dfrac{(3c-d)(c+d)}{c^2d^2}$ **23.** $\dfrac{4x^2-13xt+9t^2}{3x^2t^2}$

25. $\dfrac{6x}{(x+2)(x-2)}$ **27.** $\dfrac{(t-3)(t+1)}{(t-1)(t+3)}$ **29.** $\dfrac{11x+2}{3x(x+1)}$

31. $\dfrac{-5t+3}{2t(t-1)}$ **33.** $\dfrac{a^2}{(a-3)(a+3)}$ **35.** $\dfrac{16}{3(z+4)}$

37. $\dfrac{5q-3}{(q-1)^2}$ **39.** $\dfrac{9a}{4(a-5)}$ **41.** $\dfrac{2y-1}{y(y-1)}$

43. $\dfrac{10}{(a-3)(a+2)}$ **45.** $\dfrac{x-5}{(x+5)(x+3)}$

47. $\dfrac{3z^2+19z-20}{(z-2)^2(z+3)}$ **49.** $\dfrac{-7}{x^2+25x+24}$ **51.** $\dfrac{6x+7}{2x+1}$

53. $\dfrac{10-3x}{4-x}$ **55.** $\dfrac{3x-1}{2}$ **57.** 0   **59.** $y+3$

61. 0   **63.** $\dfrac{1}{t^2+t+1}$ **65.** $\dfrac{p^2+7p+1}{(p-5)(p+5)}$

67. $\dfrac{(x+1)(x+3)}{(x-4)(x+4)}$ **69.** $\dfrac{-a-2}{(a+1)(a-1)}$, or

$\dfrac{a+2}{(1+a)(1-a)}$ **71.** $\dfrac{2(5x+3y)}{(x-y)(x+y)}$ **73.** $\dfrac{2x-3}{2-x}$

75. 3   **77.** 0   **79.** $\dfrac{2}{(x+3)(x+4)}$ **81.** 🔲

83. 6   **84.** 7   **85.** 3×10^{14} **86.** $\dfrac{-6a^4}{b}$ **87.** $\dfrac{a^2}{9b^2}$

88. 12   **89.** 🔲   **91.** Perimeter: $\dfrac{2(5x-7)}{(x-5)(x+4)}$; area:

93. $\dfrac{6}{(x-5)(x+4)}$ **93.** $\dfrac{x}{3x+1}$ **95.** $\dfrac{-29}{(x-3)(x+1)}$

97. $\dfrac{x^4+4x^3-5x^2-126x-441}{(x+2)^2(x+7)^2}$

99. $\dfrac{5(a^2+2ab-b^2)}{(a-b)(3a+b)(3a-b)}$ **101.** $\dfrac{a}{a-b}+\dfrac{3b}{b-a}$;

answers may vary.   **103.** 🔲, 〰️

Quick Quiz: Sections 6.1–6.4, p. 402

1. $\dfrac{x-8}{4x}$ **2.** $\dfrac{(x-1)(x+6)}{(x-4)(x+2)}$ **3.** $x+2$

4. $\dfrac{t(4t-1)}{(2t+1)(t-1)}$ **5.** $\dfrac{(x-1)^2(2x+1)}{(x-2)^3}$

Prepare to Move On, p. 402

1. $\frac{9}{10}$ **2.** $\frac{16}{27}$ **3.** $\frac{2}{3}$ **4.** $\dfrac{(x-3)(x+2)}{(x-2)(x+3)}$

Mid-Chapter Review: Chapter 6, pp. 403–404

1. $\dfrac{a^2}{a-10}\div\dfrac{a^2+5a}{a^2-100}=\dfrac{a^2}{a-10}\cdot\dfrac{a^2-100}{a^2+5a}$

$=\dfrac{a\cdot a\cdot(a+10)\cdot(a-10)}{(a-10)\cdot a\cdot(a+5)}$

$=\dfrac{a(a-10)}{a(a-10)}\cdot\dfrac{a(a+10)}{a+5}$

$=\dfrac{a(a+10)}{a+5}$

2. $\dfrac{2}{x}+\dfrac{1}{x^2+x}=\dfrac{2}{x}+\dfrac{1}{x(x+1)}$

$=\dfrac{2}{x}\cdot\dfrac{x+1}{x+1}+\dfrac{1}{x(x+1)}$

$=\dfrac{2x+2}{x(x+1)}+\dfrac{1}{x(x+1)}$

$=\dfrac{2x+3}{x(x+1)}$

3. $\dfrac{3x+10}{5x^2}$ **4.** $\dfrac{6}{5x^3}$ **5.** $\dfrac{3x}{10}$ **6.** $\dfrac{3x-10}{5x^2}$

7. $\dfrac{x-3}{15(x-2)}$ **8.** $\frac{1}{3}$ **9.** $\dfrac{x^2-2x-2}{(x-1)(x+2)}$

10. $\dfrac{5x+17}{(x+3)(x+4)}$ **11.** -5 **12.** $\dfrac{5}{x-4}$

13. $\dfrac{(2x+3)(x+3)}{(x+1)^2}$ **14.** $\frac{1}{6}$ **15.** $\dfrac{x(x+4)}{(x-1)^2}$

16. $\dfrac{x+7}{(x-5)(x+1)}$ **17.** $\dfrac{9(u-1)}{16}$ **18.** $(t+5)^2$

19. $\dfrac{a-1}{(a+2)(a-2)^2}$ **20.** $\dfrac{-3x^2+9x-14}{2x}$

Exercise Set 6.5, pp. 408–410

1. Complex   **2.** Denominator   **3.** Least common
denominator   **4.** Reciprocal   **5.** (b)   **6.** (a)

7. (a)   **8.** (b)   **9.** 10   **11.** $\frac{5}{11}$ **13.** $\dfrac{5x^2}{4(x^2+4)}$

15. $\dfrac{(x+2)(x-3)}{(x-1)(x+4)}$ **17.** $\dfrac{-10t}{5t-2}$ **19.** $\dfrac{2(2a-5)}{a-7}$

21. $\dfrac{x^2 - 18}{2(x + 3)}$ **23.** $-\dfrac{1}{5}$ **25.** $\dfrac{1 + t^2}{t(1 - t)}$ **27.** $\dfrac{x}{x - y}$

29. $\dfrac{c(4c + 7)}{3(2c^2 - 1)}$ **31.** $\dfrac{15(4 - a^3)}{14a^2(9 + 2a)}$ **33.** 1

35. $\dfrac{3a^2 + 4b^3}{b^3(5 - 3a^2)}$ **37.** $\dfrac{(t - 3)(t + 3)}{t^2 + 4}$ **39.** $\dfrac{y^2 + 1}{(y + 1)(y - 1)}$

41. $\dfrac{1}{a(a - h)}$ **43.** $\dfrac{1}{x - y}$ **45.** $\dfrac{a^2 b^2}{b^2 - ab + a^2}$

47. $\dfrac{t^2 + 5t + 3}{(t + 1)^2}$ **49.** $\dfrac{x^2 - 2x - 1}{x^2 - 5x - 4}$ **51.** $\dfrac{(a - 2)(a - 7)}{(a + 1)(a - 6)}$

53. $\dfrac{x + 2}{x + 3}$ **55.** 🖻   **57.** $(2x - 3)(3x^2 - 2)$

58. $4ab(3a + b - 2)$ **59.** $3n(2n + 1)(5n - 3)$

60. $(5a - 4b)^2$ **61.** $(n^2 + 1)(n + 1)(n - 1)$

62. $(pw + 10)(pw - 12)$ **63.** 🖻   **65.** 6, 7, 8

67. $-3, -\dfrac{4}{5}, 3$

69. $\dfrac{A}{B} \div \dfrac{C}{D} = \dfrac{\dfrac{A}{B}}{\dfrac{C}{D}} = \dfrac{\dfrac{A}{B}}{\dfrac{C}{D}} \cdot \dfrac{BD}{BD} = \dfrac{AD}{BC} = \dfrac{A}{B} \cdot \dfrac{D}{C}$

71. $\dfrac{x^2 + 5x + 15}{-x^2 + 10}$ **73.** 0   **75.** $\dfrac{x - 5}{x - 3}$

77. $\dfrac{(x - 1)(3x - 2)}{5x - 3}$

Quick Quiz: Sections 6.1–6.5, p. 410

1. $\dfrac{2(2x + 1)}{2x - 1}$ **2.** $\dfrac{2x(x + 3)}{3}$ **3.** $\dfrac{a + 3}{a + 1}$

4. $\dfrac{-5x^2 + 21x - 2}{(x - 3)(x + 1)}$ **5.** $\dfrac{5x + 8}{6x}$

Prepare to Move On, p. 410

1. -4 **2.** $-\dfrac{14}{27}$ **3.** 3, 4   **4.** $-15, 2$

Connecting the Concepts, pp. 415–416

1. Expression; $\dfrac{19n - 2}{5n(2n - 1)}$ **2.** Equation; 8

3. Equation; $-\dfrac{1}{2}$ **4.** Expression; $\dfrac{8(t + 1)}{(t - 1)(2t - 1)}$

5. Expression; $\dfrac{2a}{a - 1}$ **6.** Equation; $-10, 10$

Exercise Set 6.6, pp. 416–417

1. False   **2.** True   **3.** True   **4.** True   **5.** $-\dfrac{2}{5}$
7. $\dfrac{24}{5}$ **9.** $-6, 6$ **11.** 12   **13.** $\dfrac{14}{3}$ **15.** $-6, 6$
17. $-4, -1$ **19.** -10 **21.** $-4, -3$ **23.** $\dfrac{23}{2}$
25. $\dfrac{5}{2}$ **27.** -1 **29.** No solution   **31.** -10
33. $-\dfrac{7}{3}$ **35.** $-2, \dfrac{7}{3}$ **37.** $-3, 13$ **39.** No solution
41. 2   **43.** -8 **45.** No solution   **47.** No solution
49. 🖻   **51.** x-intercept: $(3, 0)$; y-intercept: $(0, -18)$
52. $\dfrac{1}{9}$ **53.** Slope: -2; y-intercept: $(0, 5)$ **54.** Yes
55. $y = \dfrac{1}{3}x - 2$ **56.** $y - (-5) = -4(x - 1)$
57. 🖻   **59.** -2 **61.** 3   **63.** 4   **65.** 4
67. -2 **69.** 🖼

Quick Quiz: Sections 6.1–6.6, p. 417

1. $\dfrac{7(x - 11)}{x + 7}$ **2.** $x(x + 2)(x - 2)(x + 1)$

3. $\dfrac{-2n^2 + 5n + 6}{2n(n + 2)}$ **4.** $\dfrac{x - 2}{x + 2}$ **5.** -7

Prepare to Move On, p. 417

1. 137, 139   **2.** $-8, -6; 6, 8$ **3.** Base: 9 cm;
height: 12 cm   **4.** 0.06 cm per day

Exercise Set 6.7, pp. 426–431

1. False   **2.** True   **3.** True   **4.** True   **5.** True
6. False   **7.** $\dfrac{1}{2}$ cake per hour   **8.** $\dfrac{1}{3}$ cake per hour
9. $\dfrac{5}{6}$ cake per hour   **10.** 1 lawn per hour   **11.** $\dfrac{1}{3}$ lawn
per hour   **12.** $\dfrac{2}{3}$ lawn per hour   **13.** 6 hr
15. $3\dfrac{3}{7}$ hr   **17.** $8\dfrac{4}{7}$ hr   **19.** GT-S50: $7\dfrac{1}{2}$ min; GT-1500:
15 min   **21.** Anita: 3 days; Tori: 6 days   **23.** Tristan:
$\dfrac{4}{3}$ months; Sara: 4 months   **25.** 300 min, or 5 hr
27. AMTRAK: 80 km/h; CSX: 66 km/h

	Distance (in km)	Speed (in km/h)	Time (in hours)
CSX	330	$r - 14$	$\dfrac{330}{r - 14}$
AMTRAK	400	r	$\dfrac{400}{r}$

29. Rita: 50 mph; Sean: 65 mph   **31.** 7 mph
33. 5.2 ft/sec   **35.** 3 hr   **37.** 40 mph   **39.** 20 mph
41. 10.5   **43.** $\dfrac{8}{3}$ **45.** 12.6   **47.** $3\dfrac{3}{4}$ in.
49. 20 ft   **51.** 15 ft   **53.** About 26.7 cm   **55.** $65.25
57. 702 photos   **59.** 52 bulbs   **61.** $7\dfrac{1}{2}$ oz
63. 90 whales   **65.** 🖻   **67.** $x^3 - x^2 + x - 15$
68. $2x^4 + 6x^3 - 7x - 21$ **69.** $3y^2z + 2yz^2 + 6y^2 - 6yz$
70. $2ab^2 + 4b - a$ **71.** $64n^6 - 9$ **72.** $x^2 + x + \dfrac{7}{x - 1}$
73. 🖻   **75.** $49\dfrac{1}{2}$ hr   **77.** Michelle: 6 hr; Sal: 3 hr;
Kristen: 4 hr   **79.** 30 min   **81.** 2250 people per hour
83. $14\dfrac{7}{8}$ mi   **85.** About 13 sec   **87.** $21\dfrac{9}{11}$ min after 4:00
89. 48 km/h   **91.** 45 mph

93. Equation 1: $\dfrac{1}{a} \cdot t + \dfrac{1}{b} \cdot t = 1$;

Equation 2: $\left(\dfrac{1}{a} + \dfrac{1}{b}\right)t = 1$;

Equation 3: $\dfrac{t}{a} + \dfrac{t}{b} = 1$;

Equation 4: $\dfrac{1}{a} + \dfrac{1}{b} = \dfrac{1}{t}$

$\dfrac{1}{a} \cdot t + \dfrac{1}{b} \cdot t = 1$ Equation 1

$$t\left(\frac{1}{a} + \frac{1}{b}\right) = 1 \qquad \text{Factoring out } t; \\ \text{equation 1 = equation 2}$$

$$t \cdot \frac{1}{a} + t \cdot \frac{1}{b} = 1 \qquad \text{Using the distributive law}$$

$$\frac{t}{a} + \frac{t}{b} = 1 \qquad \text{Multiplying;} \\ \text{equation 2 = equation 3}$$

$$\frac{1}{t} \cdot \left(\frac{t}{a} + \frac{t}{b}\right) = \frac{1}{t} \cdot 1 \qquad \text{Multiplying both sides} \\ \text{by } \frac{1}{t}$$

$$\frac{1}{t} \cdot \frac{t}{a} + \frac{1}{t} \cdot \frac{t}{b} = \frac{1}{t} \cdot 1 \qquad \text{Using the distributive law}$$

$$\frac{1}{a} + \frac{1}{b} = \frac{1}{t} \qquad \text{Multiplying;} \\ \text{equation 3 = equation 4}$$

95. ☑

Quick Quiz: Sections 6.1–6.7, p. 431

1. $\dfrac{(y-3)(y-2)}{7x(y-1)}$ **2.** 35 **3.** $\dfrac{2x}{(x-2)(x+1)(x+2)}$

4. -7 **5.** $\frac{100}{3}$ min, or $33\frac{1}{3}$ min

Prepare to Move On, p. 431

1.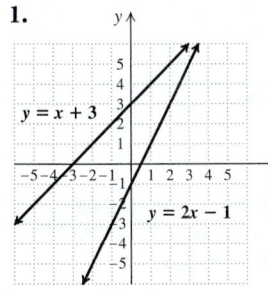

2.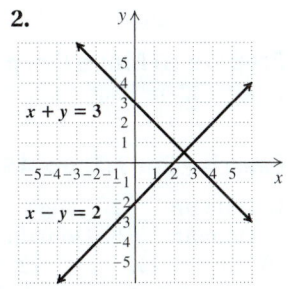

3.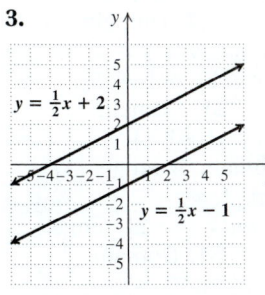

4.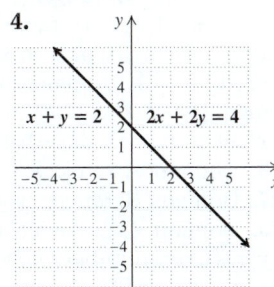

Translating for Success, p. 432

1. K **2.** C **3.** N **4.** D **5.** E **6.** O **7.** F
8. H **9.** B **10.** A

Decision Making: Connection, p. 433

1. Yes **2.** No **3.** $84.99 **4.** Yes

Study Summary: Chapter 6, pp. 434–437

1. $\dfrac{3(x-1)}{(x-3)}$ **2.** $\dfrac{5(a+1)}{2a}$ **3.** $\dfrac{t^2(t-10)}{(t-2)^2}$

4. $\dfrac{3x-1}{x+5}$ **5.** 1 **6.** $(a+5)(a-5)(a-1)$

7. $\dfrac{(x-3)(2x-1)}{(x-1)(x-2)}$ **8.** $\dfrac{3}{3x+2}$ **9.** 2 **10.** $\frac{6}{5}$ hr, or $1\frac{1}{5}$ hr **11.** Jogging: 9 mph; walking: 4 mph **12.** $\frac{15}{2}$

Review Exercises: Chapter 6, pp. 437–438

1. False **2.** True **3.** False **4.** True **5.** True
6. False **7.** False **8.** False **9.** 0 **10.** $-6, 6$

11. $-6, 5$ **12.** -2 **13.** $\dfrac{x-3}{x+5}$ **14.** $\dfrac{7x+3}{x-3}$

15. $\dfrac{3(y-3)}{2(y+3)}$ **16.** $-5(x+2y)$ **17.** $\dfrac{a-6}{5}$

18. $\dfrac{3(y-2)^2}{4(2y-1)(y-1)}$ **19.** $-32t$ **20.** $\dfrac{2x(x-1)}{x+1}$

21. $\dfrac{(x^2+1)(2x+1)}{(x-2)(x+1)}$ **22.** $\dfrac{(t+4)^2}{t+1}$ **23.** $60a^5b^8$

24. $x^4(x-1)(x+1)$ **25.** $(y-2)(y+2)(y+1)$

26. $\dfrac{15-3x}{x+3}$ **27.** $\dfrac{4}{x-4}$ **28.** $\dfrac{x+5}{2x}$

29. $\dfrac{2a^2-21ab+15b^2}{5a^2b^2}$ **30.** $y-4$ **31.** $\dfrac{t(t-2)}{(t-1)(t+1)}$

32. $d+2$ **33.** $\dfrac{-x^2+x+26}{(x+1)(x-5)(x+5)}$ **34.** $\dfrac{2(x-2)}{x+2}$

35. $\dfrac{3(7t+2)}{4t(3t+2)}$ **36.** $\dfrac{z}{1-z}$ **37.** $\dfrac{10x}{3x^2+16}$

38. $c-d$ **39.** 4 **40.** $\frac{7}{2}$ **41.** $-6, -1$
42. $-1, 4$ **43.** $5\frac{1}{7}$ hr **44.** Ben: 30 hr; Jon: 45 hr
45. 24 mph **46.** Jennifer: 70 mph; Elizabeth: 62 mph
47. 55 seals **48.** 6 **49.** 72 radios
50. ☑ The LCM of denominators is used to clear fractions when simplifying a complex rational expression using the method of multiplying by the LCD, and when solving rational equations. **51.** ☑ Although multiplying the denominators of the expressions being added results in a common denominator, it is often not the *least* common denominator. Using a common denominator other than the LCD makes the expressions more complicated, requires additional simplifying after the additon has been performed, and leaves more room for error.
52. $\dfrac{5(a+3)^2}{a}$ **53.** $\dfrac{10a}{(a-b)(b-c)}$ **54.** 0 **55.** 44%

Test: Chapter 6, p. 439

1. [6.1] 0 **2.** [6.1] 1, 2 **3.** [6.1] $\dfrac{3x+7}{x+3}$

4. [6.2] $\dfrac{2t(t+3)}{3(t-1)}$ **5.** [6.2] $\dfrac{(5y+1)(y+1)}{3y(y+2)}$

6. [6.2] $\dfrac{(2a+1)(4a^2+1)}{4a^2(2a-1)}$ **7.** [6.2] $(x+3)(x-3)$

8. [6.3] $(y-3)(y+3)(y+7)$ **9.** [6.3] $\dfrac{-3x+9}{x^3}$

10. [6.3] $\dfrac{-2t+8}{t^2+1}$ **11.** [6.4] 1 **12.** [6.4] $\dfrac{3x-5}{x-3}$

13. [6.4] $\dfrac{11t-8}{t(t-2)}$ **14.** [6.4] $\dfrac{y^2+3}{(y-1)(y+3)^2}$

15. [6.4] $\dfrac{x^2+2x-7}{(x+1)(x-1)^2}$ **16.** [6.5] $\dfrac{3y+1}{y}$

17. [6.5] $x-8$ **18.** [6.6] $\frac{8}{3}$ **19.** [6.6] $-3, 5$

20. [6.7] 12 min   **21.** [6.5] Tyler: 4 hr; Katie: 10 hr
22. [6.7] $1\frac{1}{4}$ mi   **23.** [6.7] Ryan: 65 km/h; Alicia: 45 km/h
24. [6.5] a **25.** [6.7] -1

Cumulative Review: Chapters 1–6, p. 440

1. $a + cb$ **2.** -25 **3.** 25 **4.** $-3x + 33$ **5.** 4
6. $-10, -1$ **7.** $-7, 7$ **8.** -8 **9.** $1, 4$ **10.** -17
11. $-4, \frac{1}{2}$ **12.** $\{x \mid x > 43\}$, or $(43, \infty)$ **13.** -13
14. $b = 3a - c + 9$
15.

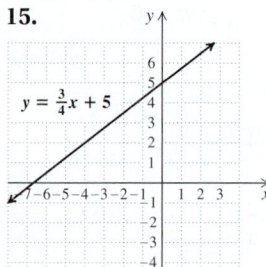

16.

$x = -3$

17.

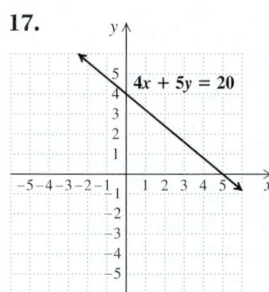

$4x + 5y = 20$

18.

$y = 6$

19. -2 **20.** Slope: $\frac{1}{2}$; y-intercept: $\left(0, -\frac{1}{4}\right)$
21. $y = -\frac{5}{8}x - 4$ **22.** $\frac{1}{x^2}$ **23.** $-4a^4 b^{14}$
24. $-y^3 - 2y^2 - 2y + 7$ **25.** $2x^5 + x^3 - 6x^2 - x + 3$
26. $36x^2 - 60xy + 25y^2$ **27.** $3n^2 - 13n - 10$
28. $4x^6 - 1$ **29.** $2x(3 - x - 12x^3)$ **30.** $(4x + 9)(4x - 9)$
31. $(t - 4)(t - 6)$ **32.** $10(t + 1)(t^2 - t + 1)$
33. $2(3x - 2)(x - 4)$ **34.** $(5t + 4)^2$ **35.** $(xy - 5)(xy + 4)$
36. $(x + 2)(x^3 - 3)$ **37.** $\frac{4}{t + 4}$ **38.** 1
39. $\frac{a^2 + 7ab + b^2}{(a + b)(a - b)}$ **40.** $\frac{2x + 5}{4 - x}$ **41.** $\frac{x}{x - 2}$
42. $15x^3 - 57x^2 + 177x - 529 + \frac{1605}{x + 3}$ **43.** 15 books
44. 12 ft   **45.** $\frac{75}{4}$ min, or $18\frac{3}{4}$ min   **46.** 150 calories
47. $-144, 144$ **48.** $-7, 4, 12$ **49.** 18

CHAPTER 7

Exercise Set 7.1, pp. 448–452

1. Correspondence   **2.** Exactly   **3.** Domain
4. Range   **5.** Horizontal   **6.** Vertical   **7.** "f of 3"
8. Vertical   **9.** Yes   **11.** Yes   **13.** No   **15.** Yes
17. Function   **19.** Function   **21. (a)** $\{-3, -2, 0, 4\}$;
(b) $\{-10, 3, 5, 9\}$; **(c)** yes   **23. (a)** $\{1, 2, 3, 4, 5\}$; **(b)** $\{1\}$;
(c) yes   **25. (a)** $\{-2, 3, 4\}$; **(b)** $\{-8, -2, 4, 5\}$; **(c)** no
27. (a) -2; **(b)** 4   **29. (a)** -2; **(b)** -2 **31. (a)** 3;
(b) -3 **33. (a)** 3; **(b)** $-2, 0$ **35. (a)** 4; **(b)** $-1, 3$

37. Yes   **39.** Yes   **41.** No   **43. (a)** 5; **(b)** -3;
(c) -9; **(d)** 21; **(e)** $2a + 9$; **(f)** $2a + 7$ **45. (a)** 0;
(b) 1; **(c)** 57; **(d)** $5t^2 + 4t$; **(e)** $20a^2 + 8a$; **(f)** 48
47. (a) $\frac{3}{5}$; **(b)** $\frac{1}{3}$; **(c)** $\frac{4}{7}$; **(d)** 0; **(e)** $\frac{x - 1}{2x - 1}$; **(f)** $\frac{a + h - 3}{2a + 2h - 5}$
49. 11   **51.** 0   **53.** $-\frac{21}{2}$ **55.** $\frac{25}{6}$ **57.** -3
59. -25 **61.** $4\sqrt{3}$ cm$^2 \approx 6.93$ cm^2
63. 36π in$^2 \approx 113.10$ in^2 **65.** 75 heart attacks per
10,000 men   **67.** 250 mg/dl   **69.** 📋   **71.** 6
72. $-4, 8$ **73.** $-6, 6$ **74.** $-\frac{1}{2}$ **75.** $-1, 1$
76. $-4, 3$ **77.** $\frac{9}{32}$ **78.** $\frac{1}{7}$ **79.** 📋   **81.** 26; 99
83. Worm   **85. (a)** 2; **(b)** 2; **(c)** $\{x \mid 0 < x \leq 2\}$, or $(0, 2]$
87. About 22 mm   **89.** 1 every 3 min
91.

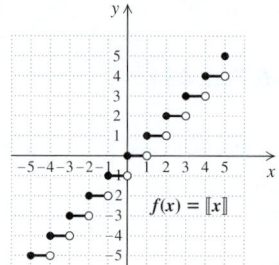

$f(x) = [\![x]\!]$

Prepare to Move On, p. 452

1. 0   **2.** $\frac{1}{3}$ **3.** $-3, 4$ **4.** $-6, 6$

Exercise Set 7.2, pp. 457–460

1. Domain   **2.** Range   **3.** Domain   **4.** Range
5. Domain   **6.** Domain   **7. (c)**   **8. (d)**   **9. (d)**
10. (b)   **11. (c)**   **12. (b)**   **13.** Domain:
$\{-4, -2, 0, 2, 4\}$; range, $\{-2, -1, 0, 1, 2\}$ **15.** Domain:
$\{-5, -3, -1, 0, 2, 4\}$; range: $\{-1, 1\}$ **17.** Domain:
$\{x \mid -4 \leq x \leq 3\}$, or $[-4, 3]$; range: $\{y \mid -3 \leq y \leq 4\}$, or
$[-3, 4]$ **19.** Domain: $\{x \mid -4 \leq x \leq 5\}$, or $[-4, 5]$;
range: $\{y \mid -2 \leq y \leq 4\}$, or $[-2, 4]$ **21.** Domain:
$\{x \mid -4 \leq x \leq 4\}$, or $[-4, 4]$; range: $\{-3, -1, 1\}$
23. Domain: $\mathbb{R}$; range: $\mathbb{R}$ **25.** Domain: $\mathbb{R}$; range: $\{4\}$
27. Domain: $\mathbb{R}$; range: $\{y \mid y \geq 1\}$, or $[1, \infty)$
29. Domain: $\{x \mid x$ is a real number $and\ x \neq -2\}$; range:
$\{y \mid y$ is a real number $and\ y \neq -4\}$ **31.** Domain:
$\{x \mid x \geq 0\}$, or $[0, \infty)$; range: $\{y \mid y \geq 0\}$, or $[0, \infty)$
33. $\{x \mid x$ is a real number $and\ x \neq 3\}$
35. $\{x \mid x$ is a real number $and\ x \neq \frac{1}{2}\}$ **37.** $\mathbb{R}$ **39.** $\mathbb{R}$
41. $\{x \mid x$ is a real number $and\ x \neq 3\ and\ x \neq -3\}$ **43.** $\mathbb{R}$
45. $\{x \mid x$ is a real number $and\ x \neq -1\ and\ x \neq -7\}$
47. $\{t \mid 0 \leq t \leq 5\}$, or $[0, 5]$
49. $\{p \mid \$0 \leq p \leq \$10.60\}$, or $[0, 10.60]$
51. $\{d \mid d \geq 0\}$, or $[0, \infty)$ **53.** $\{t \mid 0 \leq t \leq 5\}$, or $[0, 5]$
55. (a) -5; **(b)** 1; **(c)** 21   **57. (a)** -15; **(b)** 0; **(c)** -6
59. (a) 100; **(b)** 100; **(c)** 131   **61.** 📋   **63.** -3
64. $15x - 29$ **65.** $4x^{12}y^2$ **66.** $-\frac{12b^{26}}{7a^{12}}$ **67.** 📋

69. 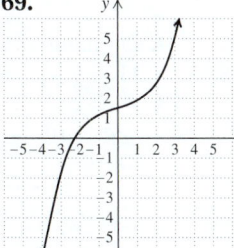   **71.**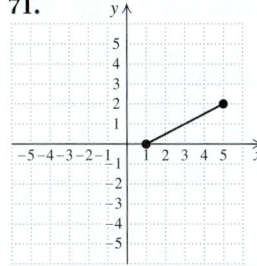

73. Domain: $\{x\,|\,x$ is a real number *and* $x \neq 0\}$; range: $\{y\,|\,y$ is a real number *and* $y \neq 0\}$ **75.** Domain: $\{x\,|\,x < -2$ *or* $x > 0\}$; range: $\{y\,|\,y < -2$ *or* $y > 3\}$
77. Domain: $\mathbb{R}$; range: $\{y\,|\,y \geq 0\}$, or $[0, \infty)$
79. Domain: $\{x\,|\,x$ is a real number *and* $x \neq 2\}$; range: $\{y\,|\,y$ is a real number *and* $y \neq 0\}$
81. $\{h\,|\,0 \leq h \leq 144\}$, or $[0, 144]$
83. 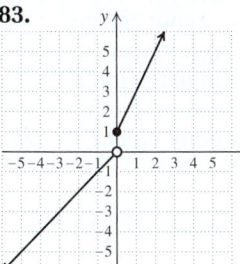   **85.**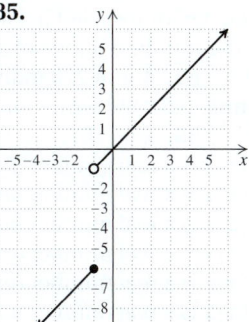

87.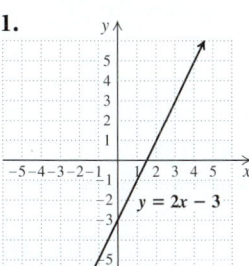

Quick Quiz: Sections 7.1–7.2, p. 460

1. 0   **2.** 200   **3.** $\{x\,|\,x$ is a real number *and* $x \neq 0\}$
4. $\mathbb{R}$ **5.** -3

Prepare to Move On, p. 460

1. 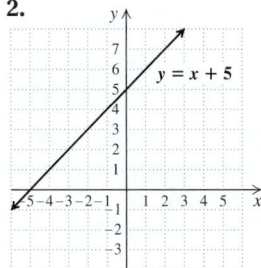   **2.**

3. Slope: $\frac{2}{3}$; y-intercept: $(0, -4)$
4. Slope: $-\frac{1}{4}$; y-intercept: $(0, 6)$
5. Slope: $\frac{4}{3}$; y-intercept: $(0, 0)$
6. Slope: -5; y-intercept: $(0, 0)$

Connecting the Concepts, p. 461

1. $\{-2, 1, 3\}$ **2.** $\{-6, 5\}$ **3.** -6 **4.** 1
5. $\{x\,|\,x$ is a real number *and* $x \neq 0\}$ **6.** 0
7. $\frac{3}{8}$ **8.** 2

Exercise Set 7.3, pp. 466–469

1. False   **2.** True   **3.** False   **4.** True
5. True   **6.** True
7. 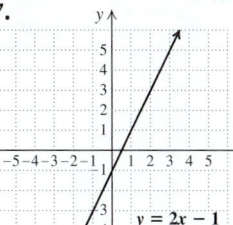   **9.**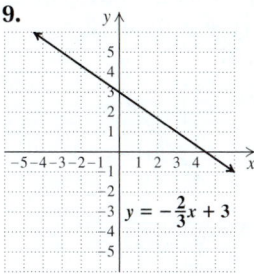
11. 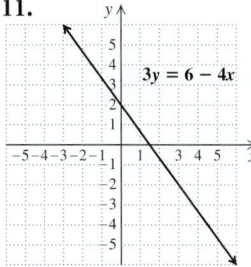   **13.**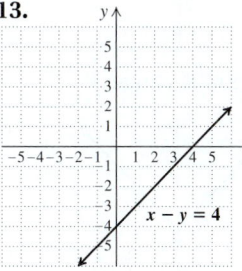
15. 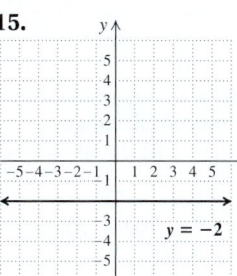   **17.**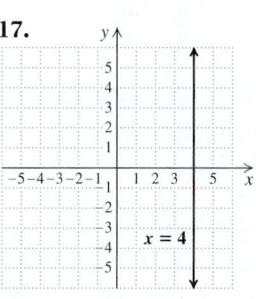
19. 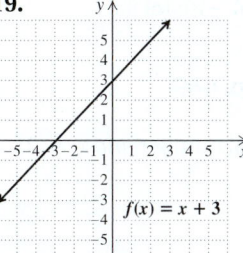   **21.**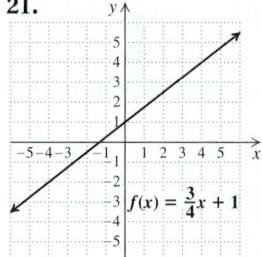
23. 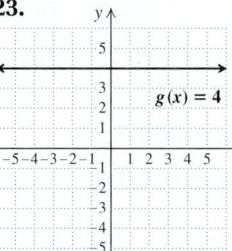   **25.** $C(d) = 0.75d + 30$; 60 miles
27. $L(t) = \frac{1}{2}t + 5$; 20 months after the haircut
29. $S(t) = 2.2t + 4.5$; 2005
31. (a) $N(t) = 1.02t + 74.8$; (b) 91.12 million tons
33. (a) $E(t) = 0.14t + 79.7$; (b) 82.5 years
35. (a) $S(t) = \frac{466}{25}t + \frac{2423}{5}$; (b) 857.4 tons
37. (a) $P(d) = 0.03d + 1$; (b) 21.7 atm   **39.** Linear function; $\mathbb{R}$ **41.** Quadratic function; $\mathbb{R}$ **43.** Rational function; $\{t\,|\,t$ is a real number *and* $t \neq -\frac{4}{3}\}$

45. Polynomial function; $\mathbb{R}$ **47.** Rational function; $\{x \mid x \text{ is a real number } and\ x \neq \frac{5}{2}\}$
49. Rational function; $\{n \mid n \text{ is a real number } and\ n \neq -1 \text{ and } n \neq -2\}$ **51.** Linear function; $\mathbb{R}$
53. $\{y \mid y \geq 0\}$, or $[0, \infty)$ **55.** $\mathbb{R}$ **57.** $\{y \mid y \leq 0\}$, or $(-\infty, 0]$
59.
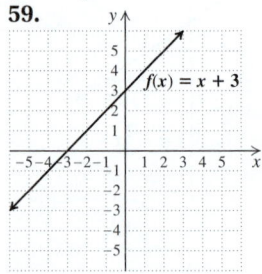
61.
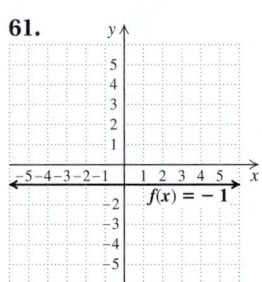

Domain: $\mathbb{R}$; range: $\mathbb{R}$ Domain: $\mathbb{R}$; range: $\{-1\}$
63.
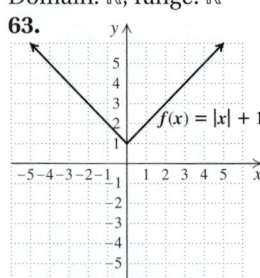
65.
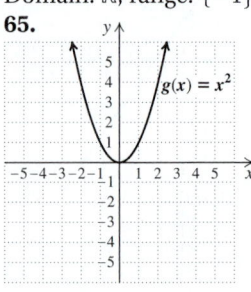

Domain: $\mathbb{R}$; range: $\{y \mid y \geq 1\}$,   Domain: $\mathbb{R}$; range:
or $[1, \infty)$ $\{y \mid y \geq 0\}$, or $[0, \infty)$

67. 🖎   **69.** $(2n - 7)^2$ **70.** $(4x + 3)(x - 5)$
71. $2x(x - 1)(x - 6)$ **72.** $(c - 10)(c^2 + 10c + 100)$
73. $d(d + 8)(d - 8)$ **74.** $(a + 4)(a^2 + 2)$
75. 🖎   **77.** False   **79.** False   **81.** 21.1°C
83. $60   **85. (a)** $g(x) = x - 8$; **(b)** -10; **(c)** 83

Quick Quiz: Sections 7.1–7.3, p. 469

1. $\frac{5}{4}$ **2.** $\frac{a}{a - 1}$ **3.** $\{x \mid x \text{ is a real number } and\ x \neq 2\}$
4. Quadratic   **5.** $\mathbb{R}$

Prepare to Move On, p. 469

1. $4x^2 + 2x - 1$ **2.** $2x^3 - x^2 - x + 7$
3. $2x^2 - 13x - 7$ **4.** $x^2 + x - 12$

Mid-Chapter Review: Chapter 7, p. 470

1. $f(-5) = 2(-5)^2 - 3(-5) = 2(25) - (-15) =$
$50 + 15 = 65$ **2.** $x - 7 = 0$
$x = 7$
Domain $= \{x \mid x \text{ is a real number } and\ x \neq 7\}$
3. $\{-1, 0, 3, 4\}$ **4.** $\{-2, 0, 6, 8\}$ **5.** -2
6. $\mathbb{R}$ **7.** $\{x \mid x \text{ is a real number } and\ x \neq 0\}$
8. $\frac{1}{5}$ **9.** Linear function   **10.** Rational function
11. $\mathbb{R}$ **12.** $\{x \mid x \text{ is a real number } and\ x \neq -10$
$and\ x \neq 4\}$ **13.** Yes   **14.** -4 **15.** $-1, 4\frac{1}{2}$
16. $\{x \mid x \geq -3\}$, or $[-3, \infty)$
17. $\{y \mid y \geq -4\}$, or $[-4, \infty)$ **18.** 4   **19.** 10
20. 13

Exercise Set 7.4, pp. 476–479

1. Sum   **2.** Subtract   **3.** Evaluate   **4.** Domains
5. Excluding   **6.** Sum   **7.** 1   **9.** 5   **11.** -7
13. 1   **15.** -5 **17.** $x^2 - 2x - 2$ **19.** $x^2 + 2x - 8$
21. $x^2 - x + 3$ **23.** 5   **25.** $-a^3 + 5a^2 + 2a - 10$
27. $\frac{x^2 - 2}{5 - x}, x \neq 5$ **29.** $\frac{7}{2}$ **31.** -2
33. $\frac{15}{2x^3}$ **35.** $\frac{10 - 3x}{2x^2}$ **37.** $\frac{3x}{10}$
39. $\frac{2x + 7}{(x + 2)(x - 3)(x + 3)}$ **41.** $\frac{2x^2 + 6x + 1}{(x + 2)(x - 3)(x + 3)}$
43. $\frac{(x - 1)(x + 3)}{(x + 2)^2}$ **45.** $1.2 + 2.9 = 4.1$ million
47. $45 - 28 = 17$ **49.** About 85 million; the number of tons of municipal solid waste that was composted or recycled in 2009   **51.** About 240 million; the number of tons of municipal solid waste in 2000   **53.** About 220 million; the number of tons of municipal solid waste that was not composted in 2008   **55.** $\mathbb{R}$ **57.** $\{x \mid x \text{ is a real number } and\ x \neq -5\}$
59. $\{x \mid x \text{ is a real number } and\ x \neq 0\}$
61. $\{x \mid x \text{ is a real number } and\ x \neq 1\}$
63. $\{x \mid x \text{ is a real number } and\ x \neq -\frac{9}{2} \text{ and } x \neq 1\}$
65. $\{x \mid x \text{ is a real number } and\ x \neq 3\}$
67. $\{x \mid x \text{ is a real number } and\ x \neq -4\}$
69. $\{x \mid x \text{ is a real number } and\ x \neq 4 \text{ and } x \neq 5\}$
71. $\{x \mid x \text{ is a real number } and\ x \neq -1 \text{ and } x \neq -\frac{5}{2}\}$
73. 4; 3   **75.** 5; -1 **77.** $\{x \mid 0 \leq x \leq 9\}$;
$\{x \mid 3 \leq x \leq 10\}$; $\{x \mid 3 \leq x \leq 9\}$; $\{x \mid 3 \leq x \leq 9\}$
79.
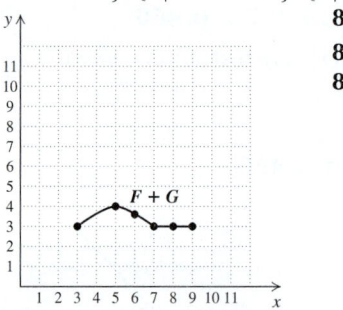
81. 🖎
83. 60°, 30°, 90°
84. 11 points

85. 2.992×10^{-23} g   **86.** 36 ft   **87.** 🖎
89. $\{x \mid x \text{ is a real number } and\ x \neq 4 \text{ and } x \neq 3 \text{ and } x \neq 2$
$and\ x \neq -2\}$
91. Answers may vary.
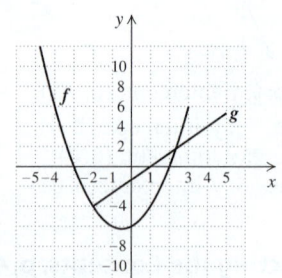
93. $\{x \mid x \text{ is a real number } and\ -1 < x < 5 \text{ and } x \neq \frac{3}{2}\}$
95. Answers may vary. $f(x) = \frac{1}{x + 2}, g(x) = \frac{1}{x - 5}$
97. 🔁

Quick Quiz: Sections 7.1–7.4, p. 479

1. 8   **2.** −1

3.

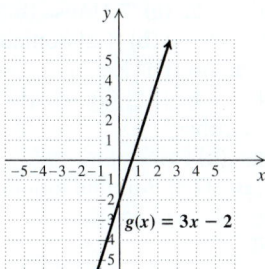

$g(x) = 3x - 2$

4. 3

5. $-x^2 - x + 3$

Prepare to Move On, p. 479

1. $c = \dfrac{b}{a}$ **2.** $w = \dfrac{x - y}{z}$ **3.** $q = \dfrac{st}{p - r}$

4. $b = \dfrac{d}{a + c}$ **5.** $b = \dfrac{cd + d}{a - 3}$

Exercise Set 7.5, pp. 485–490

1. (d)   **2.** (f)   **3.** (e)   **4.** (b)   **5.** (a)   **6.** (c)
7. Inverse   **8.** Direct   **9.** Direct   **10.** Inverse

11. Inverse   **12.** Direct   **13.** $d = \dfrac{L}{f}$

15. $v_1 = \dfrac{2s}{t} - v_2$, or $\dfrac{2s - tv_2}{t}$ **17.** $b = \dfrac{at}{a - t}$

19. $g = \dfrac{Rs}{s - R}$ **21.** $n = \dfrac{IR}{E - Ir}$ **23.** $q = \dfrac{pf}{p - f}$

25. $t_1 = \dfrac{H}{Sm} + t_2$, or $\dfrac{H + Smt_2}{Sm}$ **27.** $r = \dfrac{Re}{E - e}$

29. $r = 1 - \dfrac{a}{S}$, or $\dfrac{S - a}{S}$ **31.** $a + b = \dfrac{f}{c^2}$

33. $r = \dfrac{A}{P} - 1$, or $\dfrac{A - P}{P}$ **35.** $t_1 = t_2 - \dfrac{d_2 - d_1}{v}$, or

$\dfrac{vt_2 - d_2 + d_1}{v}$ **37.** $t = \dfrac{ab}{b + a}$ **39.** $Q = \dfrac{2Tt - 2AT}{A - q}$

41. $w = \dfrac{4.15c - 98.42}{p + 0.082}$ **43.** $k = 6; y = 6x$

45. $k = 1.7; y = 1.7x$ **47.** $k = 10; y = 10x$

49. $k = 100; y = \dfrac{100}{x}$ **51.** $k = 44; y = \dfrac{44}{x}$

53. $k = 9; y = \dfrac{9}{x}$ **55.** $33\frac{1}{3}$ cm   **57.** 3.5 hr

59. 12 people   **61.** 600 tons   **63.** 286 Hz
65. About 21 min   **67.** About 33.06   **69.** $y = \frac{1}{2}x^2$

71. $y = \dfrac{5000}{x^2}$ **73.** $y = 1.5xz$ **75.** $y = \dfrac{4wx^2}{z}$

77. 61.3 ft   **79.** 64 foot-candles   **81.** About 57 mph
83.   **85.** $9x^2$ **86.** $9 + 6x + x^2$ **87.** $\frac{8}{35}$

88. $\dfrac{5x}{9y^5}$ **89.** −6   **90.** $3x - 10$ **91.**

93. 567 mi   **95.** Ratio is $\dfrac{a + 12}{a + 6}$; percent increase is

$\dfrac{6}{a + 6} \cdot 100\%$, or $\dfrac{600}{a + 6}\%$

97. $t_1 = t_2 + \dfrac{(d_2 - d_1)(t_4 - t_3)}{a(t_4 - t_2)(t_4 - t_3) + d_3 - d_4}$
99. The intensity is halved.   **101.** About 1.7 m
103. $d(s) = \dfrac{28}{s}$; 70 yd

Quick Quiz: Sections 7.1–7.5, p. 490

1. Yes   **2.** Domain: $\{x \mid -4 \le x \le 3\}$, or $[-4, 3]$;
range: $\{y \mid -3 \le y \le 1\}$, or $[-3, 1]$ **3.** Domain:
$\{-2, -1, 0, 1, 2\}$; range: $\{-1, 0, 1, 2, 3\}$
4.

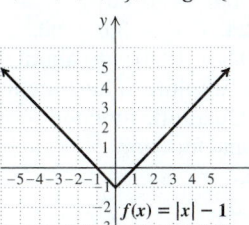

$f(x) = |x| - 1$

5. 40 kg

Prepare to Move On, p. 490

1. $y = \frac{1}{6}x - \frac{1}{2}$ **2.** $y = \frac{3}{8}x - \frac{5}{8}$ **3.** $y = -\frac{5}{2}x - \frac{3}{2}$
4. Let x represent the number; $\frac{1}{2}x - 3 = 57$
5. Let x represent the first integer; $x + (x + 1) = 145$
6. Let n represent the number; $n - (-n) = 20$

Visualizing for Success, p. 491

1. D   **2.** J   **3.** A   **4.** E   **5.** B   **6.** C   **7.** I
8. F   **9.** G   **10.** H

Decision Making: Connection, p. 492

1. 51.2 gal   **2.** $2250   **3.** $320; $720; $960   **4.**

Study Summary: Chapter 7, pp. 493–495

1. 5   **2.** Yes   **3.** Domain: $\mathbb{R}$; range: $\{y \mid y \ge -2\}$
4. $\mathbb{R}$
5.

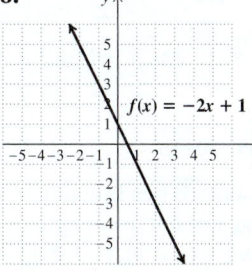

$f(x) = -2x + 1$

6.

$f(x) = x^2 - 2$

Domain: $\mathbb{R}$; range: $\{y \mid y \ge -2\}$,
or $[-2, \infty)$

7. $2x - 9$ **8.** 5   **9.** −6   **10.** $\dfrac{x - 2}{x - 7}, x \ne 7$

11. $y = 50x$ **12.** $y = \dfrac{40}{x}$ **13.** $y = \frac{1}{10}xz$

Review Exercises: Chapter 7, pp. 495–497

1. True   **2.** True   **3.** False   **4.** True   **5.** False
6. True   **7.** True   **8.** True   **9.** True   **10.** False
11. (a) 3; (b) $\{x \mid -2 \le x \le 4\}$, or $[-2, 4]$; (c) -1;
(d) $\{y \mid 1 \le y \le 5\}$, or $[1, 5]$ **12.** $\frac{3}{2}$
13. $4a^2 + 4a - 3$ **14.** 19.5%   **15.** (a) Yes;
(b) domain: $\mathbb{R}$; range: $\{y \mid y \ge 0\}$, or $[0, \infty)$
16. (a) No   **17.** (a) No   **18.** (a) Yes; (b) domain: $\mathbb{R}$;
range: $\{-2\}$ **19.** $\mathbb{R}$ **20.** $\{x \mid x$ is a real number
and $x \ne 1\}$ **21.** $\{t \mid t$ is a real number *and* $t \ne -1$
and $t \ne -4\}$ **22.** $\{t \mid 0 \le t \le 60\}$, or $[0, 60]$
23. (a) 5; (b) 4; (c) 16; (d) 35   **24.** $C(t) = 30t + 90$;
7 months   **25.** (a) $R(t) = -0.02t + 19.81$; (b) about
19.11 sec; about 19.01 sec   **26.** Absolute-value function
27. Polynomial function   **28.** Quadratic function
29. Linear function   **30.** Rational function
31.

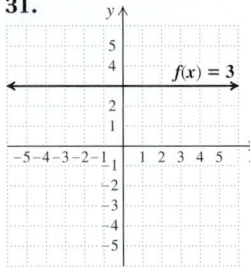

32.

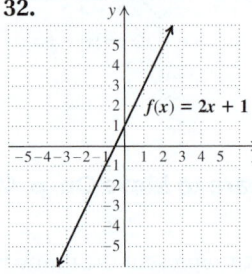

Domain: $\mathbb{R}$; range: $\{3\}$ Domain: $\mathbb{R}$; range: $\mathbb{R}$
33.   **34.** 102
35. -17 **36.** $-\frac{9}{2}$
37. $\mathbb{R}$

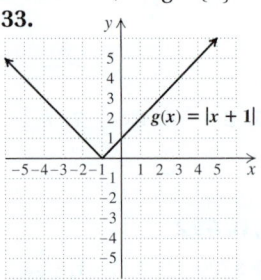

Domain: $\mathbb{R}$; range: $\{y \mid y \ge 0\}$,
or $[0, \infty)$

38. $\{x \mid x$ is a real number and $x \ne 2\}$ **39.** $r = \dfrac{2V - IR}{2I}$,
or $\dfrac{V}{I} - \dfrac{R}{2}$ **40.** $m = \dfrac{H}{S(t_1 - t_2)}$ **41.** $c = \dfrac{b + 3a}{2}$
42. $t_1 = \dfrac{-A}{vT} + t_2$, or $t_1 = \dfrac{-A + vTt_2}{vT}$ **43.** $y = \frac{15}{2}x$

44. $y = \dfrac{\frac{3}{4}}{x}$ **45.** $y = \dfrac{1}{2}\dfrac{xw^2}{z}$ **46.** 20 cm

47. About 2.9 sec   **48.** 56 in.   **49.** Two functions
that have the same domain and range are not necessarily
identical. For example, the functions f: $\{(-2, 1), (-3, 2)\}$
and g: $\{(-2, 2), (-3, 1)\}$ have the same domain and range
but are different functions.   **50.** Jenna is not correct.
Any value of the variable that makes a denominator 0 is
not in the domain; 0 itself may or may not make a denominator 0.   **51.** $f(x) = 10.94x + 20$ **52.** Domain:
$\{x \mid x \ge -4$ *and* $x \ne 2\}$; range: $\{y \mid y \ge 0$ *and* $y \ne 3\}$

Test: Chapter 7, pp. 498–499

1. [7.1], [7.2] (a) 1; (b) $\{x \mid -3 \le x \le 4\}$, or $[-3, 4]$; (c) 3;
(d) $\{y \mid -1 \le y \le 2\}$, or $[-1, 2]$ **2.** (a) [7.1] Yes; (b) [7.2]
domain: $\mathbb{R}$; range: $\mathbb{R}$ **3.** (a) [7.1] Yes; (b) [7.2] domain: $\mathbb{R}$;
range: $\{y \mid y \ge 1\}$, or $[1, \infty)$ **4.** (a) [7.1] No
5. [7.2] $\{t \mid 0 \le t \le 4\}$, or $[0, 4]$ **6.** [7.2] (a) -5; (b) 10
7. [7.3] $C(t) = 55t + 180$; 12 months
8. [7.3] (a) $C(m) = 0.3m + 25$; (b) \$175
9. [7.3] Linear function; $\mathbb{R}$ **10.** [7.3] Rational function;
$\{x \mid x$ is a real number *and* $x \ne -4$ *and* $x \ne 4\}$
11. [7.3] Quadratic function; $\mathbb{R}$
12. [7.3]   **13.** [7.3]

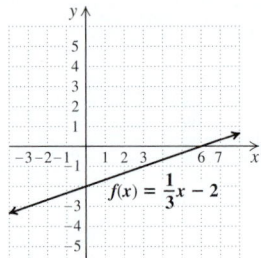

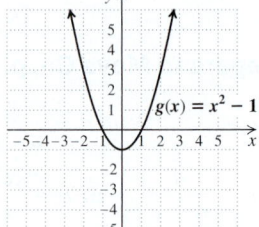

Domain: $\mathbb{R}$; range: $\mathbb{R}$ Domain: $\mathbb{R}$; range:
$\{y \mid y \ge -1\}$, or $[-1, \infty)$
14. [7.3]   **15.** [7.1] -1
16. [7.1] $10a + 1$
17. [7.4] $\dfrac{1}{x} + 2x + 1$

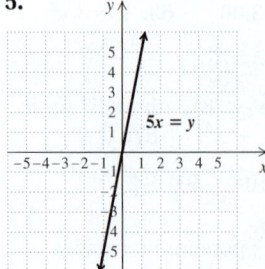

Domain: $\mathbb{R}$; range: $\left\{-\frac{1}{2}\right\}$
18. [7.1] $\{x \mid x$ is a real number *and* $x \ne 0\}$
19. [7.4] $\{x \mid x$ is a real number *and* $x \ne 0\}$
20. [7.4] $\left\{x \mid x$ is a real number *and* $x \ne 0$ *and* $x \ne -\frac{1}{2}\right\}$
21. [7.5] $s = \dfrac{Rg}{g - R}$ **22.** [7.5] $y = \frac{1}{2}x$
23. [7.5] 30 workers   **24.** [7.5] 637 in²   **25.** [7.3]
(a) 30 mi; (b) 15 mph   **26.** [7.4] $h(x) = 7x - 2$

Cumulative Review: Chapters 1–7, p. 500

1. 3.91×10^8 **2.** Slope: $\frac{7}{4}$; y-intercept: $(0, -3)$
3. $y = -2x + 5$ **4.** (a) 0; (b) $\{x \mid x$ is a real number *and*
$x \ne 5$ *and* $x \ne 6\}$
5.   **6.**

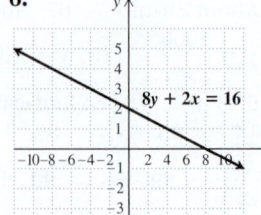

7.

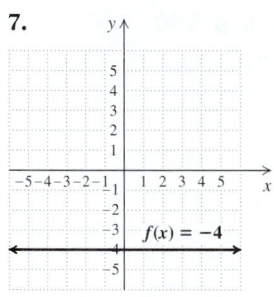

$f(x) = -4$

8.

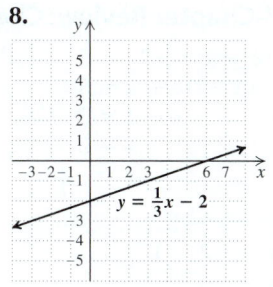

$y = \frac{1}{3}x - 2$

9. $15x^4 - x^3 - 9x^2 + 5x - 2$ **10.** $9x^4 + 6x^2y + y^2$
11. $4x^4 - 81$ **12.** $-m^3n^2 - m^2n^2 - 5mn^3$
13. $x - 1$ **14.** $\dfrac{a^2 + 7ab + b^2}{(a - b)(a + b)}$ **15.** $4x(x^2 + 100)$
16. $(x - 6)(x + 14)$ **17.** $(4y - 5)(4y + 5)$
18. $8(2x + 1)(4x^2 - 2x + 1)$ **19.** $(t - 8)^2$
20. $(3t - 4)(t + 7)$ **21.** $\frac{1}{4}$ **22.** $-12, 12$
23. $\{x \mid x \ge -1\}$, or $[-1, \infty)$ **24.** $\frac{3}{5}$ **25.** -1
26. $a = \dfrac{Pb}{4 - P}$ **27.** -5 **28.** $y = -x + 3$
29. 99 **30.** $\dfrac{x^2 - 1}{x + 5}$ **31.** 72 in.   **32.** IQAir
HealthPro: 10 min; Austin Healthmate: 15 min
33. $22\frac{1}{2}$ min   **34.** $x^3 - 12x^2 + 48x - 64$
35. $-3, 3, -5, 5$
36. $C(x) = \begin{cases} 12, & \text{if } x \le 20, \\ 12 + 0.75(x - 20), & \text{if } x > 20 \end{cases}$

CHAPTER 8

Technology Connection, p. 505

1. $(1.53, 2.58)$ **2.** $(0.87, -0.32)$

Exercise Set 8.1, pp. 507–510

1. False   **2.** True   **3.** True   **4.** True   **5.** True
6. False   **7.** False   **8.** True   **9.** Yes   **11.** No
13. Yes   **15.** Yes   **17.** $(3, 2)$ **19.** $(2, -1)$
21. $(1, 4)$ **23.** $(-3, -2)$ **25.** $(3, -1)$ **27.** $(3, -7)$
29. $(7, 2)$ **31.** $(4, 0)$ **33.** No solution
35. $\{(x, y) \mid y = 3 - x\}$ **37.** All except Exercise 33
39. Exercise 35   **41.** Let x represent the first number
and y the second number; $x + y = 10, x = \frac{2}{3}y$ **43.** Let s
represent the number of e-mails sent and r the number of
e-mails received; $s + r = 115, r = 2s + 1$ **45.** Let x and
y represent the angles; $x + y = 180, x = 2y - 3$ **47.** Let
x represent the number of two-point shots and y the number of foul shots; $x + y = 64, 2x + y = 100$ **49.** Let x
represent the number of hats sold and y the number of tee
shirts sold; $x + y = 45, 14.50x + 19.50y = 697.50$
51. Let c represent the number of Classic Heart Rate
watches purchased and d the number of Drive Plus Heart
Rate watches; $c + d = 35, 39.99c + 59.99d = 1639.65$
53. Let l represent the length, in yards, and w the width, in
yards; $2l + 2w = 340; l = w + 50$ **55.** 📝   **57.** $-\frac{4}{5}$
58. -0.06 **59.** -29 **60.** $\dfrac{6y^8}{x^5}$ **61.** $-\frac{1}{100}$ **62.** 1

63. 📝   **65.** Answers may vary. **(a)** $x + y = 6, x - y = 4$;
(b) $x + y = 1, 2x + 2y = 3$; **(c)** $x + y = 1, 2x + 2y = 2$
67. $A = -\frac{17}{4}, B = -\frac{12}{5}$ **69.** Let x and y represent the
number of years that Dell and Juanita have taught at the
university, respectively; $x + y = 46, x - 2 = 2.5(y - 2)$
71. Let s and v represent the number of ounces of baking
soda and vinegar needed, respectively; $s = 4v, s + v = 16$
73. Mineral oil: 12 oz; vinegar: 4 oz   **75.** $(0, 0), (1, 1)$
77. $(0.07, -7.95)$ **79.** $(0.00, 1.25)$ **81.** 🖥

Prepare to Move On, p. 510

1. $\frac{8}{13}$ **2.** -1 **3.** 11   **4.** $y = 3x - 4$
5. $x = \frac{5}{2}y - \frac{7}{2}$

Connecting the Concepts, pp. 515–516

1. $(1, 1)$ **2.** $(9, 1)$ **3.** $(4, 3)$ **4.** $(5, 7)$
5. $\left(1, -\frac{1}{19}\right)$ **6.** $(3, 1)$ **7.** No solution
8. $\{(x, y) \mid x = 2 - y\}$

Exercise Set 8.2, pp. 516–518

1. Substitution   **2.** Elimination   **3.** Opposites
4. Inconsistent   **5.** (d)   **6.** (e)   **7.** (a)   **8.** (f)
9. (c)   **10.** (b)   **11.** $(2, -1)$ **13.** $(-4, 3)$
15. $(2, -2)$ **17.** $\{(x, y) \mid 2x - 3 = y\}$ **19.** $(-2, 1)$
21. $\left(\frac{1}{2}, \frac{1}{2}\right)$ **23.** $(2, 0)$ **25.** No solution   **27.** $(1, 2)$
29. $(7, -2)$ **31.** $(-1, 2)$ **33.** $\left(\frac{49}{11}, -\frac{12}{11}\right)$ **35.** $(6, 2)$
37. No solution   **39.** $(20, 0)$ **41.** $(3, -1)$
43. $\{(x, y) \mid -4x + 2y = 5\}$ **45.** $\left(2, -\frac{3}{2}\right)$ **47.** $(-2, -9)$
49. $(30, 6)$ **51.** $\{(x, y) \mid 4x - 2y = 2\}$ **53.** No
solution   **55.** $(140, 60)$ **57.** $\left(\frac{1}{3}, -\frac{2}{3}\right)$ **59.** 📝
61. $4 + (m + n)$ **62.** $-2a^2 + 7$ **63.** $47x - 36$
64. -1 **65.** 3.005×10^7 **66.** 0.00061 **67.** 📝
69. $m = -\frac{1}{2}, b = \frac{5}{2}$ **71.** $a = 5, b = 2$ **73.** $\left(-\frac{32}{17}, \frac{38}{17}\right)$
75. $\left(-\frac{1}{5}, \frac{1}{10}\right)$ **77.** Toaster oven: 3 kWh; convection
oven: 12 kWh   **79.** 📝

Quick Quiz: Sections 8.1–8.2, p. 518

1. Yes   **2.** $(3, 1)$ **3.** $(-3, -10)$ **4.** $\left(\frac{7}{5}, -\frac{3}{5}\right)$
5. $\left(\frac{1}{11}, \frac{3}{11}\right)$

Prepare to Move On, p. 518

1. $105,000   **2.** 90   **3.** 290 mi

Exercise Set 8.3, pp. 527–531

1. Total value   **2.** Principal   **3.** Distance   **4.** Sum
5. 4, 6   **7.** e-mails sent: 38; e-mails received: 77
9. $119°, 61°$ **11.** Two-point shots: 36; foul shots: 28
13. Hats: 36; tee shirts: 9   **15.** Classic Heart Rate
watches: 23; Drive Plus Heart Rate watches: 12
17. Length: 110 yd; width: 60 yd   **19.** Wind: 924 trillion Btu's; solar: 109 trillion Btu's   **21.** 3-credit courses:
37; 4-credit courses: 11   **23.** Regular paper: 16 cases;
recycled paper: 11 cases   **25.** 23-watt bulbs: 60; 42-watt
bulbs: 140   **27.** Youths: 78; adults: 32   **29.** Mexican:

14 lb; Peruvian: 14 lb   **31.** Custom-printed M&Ms: 64 oz; bulk M&Ms: 256 oz   **33.** 50%-chocolate: 7.5 lb;10%-chocolate: 12.5 lb   **35.** Deep Thought: 12 lb; Oat Dream: 8 lb   **37.** $7500 at 6.5%; $4500 at 7.2%   **39.** Steady State: 12.5 L; Even Flow: 7.5 L   **41.** 87-octane: 2.5 gal; 95-octane: 7.5 gal   **43.** Whole milk: $169\frac{3}{13}$ lb; cream: $30\frac{10}{13}$ lb   **45.** 375 km   **47.** 14 km/h   **49.** About 1489 mi   **51.** Length: 265 ft; width: 165 ft   **53.** Wii game machines: 92.4 million; PlayStation 3 consoles: 61.6 million   **55.** $7.99 plans: 190; $15.98 plans: 90   **57.** Quarters: 17; fifty-cent pieces: 13   **59.** 📝   **61.** -7 **62.** 102   **63.** 0   **64.** $x^2 + x - 5$ **65.** $\mathbb{R}$ **66.** $\{x\,|\,x \text{ is a real number } and\ x \neq 7\}$ **67.** 📝   **69.** 0%: 20 reams; 30%: 40 reams   **71.** $10\frac{2}{3}$ oz   **73.** 33 boxes   **75.** Brown: 0.8 gal; neutral: 0.2 gal   **77.** City: 261 mi; highway: 204 mi   **79.** $P(x) = \dfrac{0.1 + x}{1.5}$

(This expresses the percent as a decimal quantity.)

Quick Quiz: Sections 8.1–8.3, p. 531

1. No solution   **2.** $\left(\frac{23}{4}, \frac{25}{16}\right)$ **3.** $(4, 4)$
4. $(3, 0)$ **5.** Large trash bags: 21 rolls; small trash bags: 7 rolls

Prepare to Move On, p. 531

1. 1   **2.** $\frac{1}{2}$ **3.** 17   **4.** 7

Exercise Set 8.4, pp. 537–539

1. True   **2.** False   **3.** False   **4.** True   **5.** True
6. False   **7.** Yes   **9.** $(3, 1, 2)$ **11.** $(1, -2, 2)$
13. $(2, -5, -6)$ **15.** No solution   **17.** $(-2, 0, 5)$
19. $(21, -14, -2)$ **21.** The equations are dependent.
23. $\left(3, \frac{1}{2}, -4\right)$ **25.** $\left(\frac{1}{2}, \frac{1}{3}, \frac{1}{6}\right)$ **27.** $\left(\frac{1}{2}, \frac{2}{3}, -\frac{5}{6}\right)$
29. $(15, 33, 9)$ **31.** $(3, 4, -1)$ **33.** $(10, 23, 50)$
35. No solution   **37.** The equations are dependent.
39. 📝   **41.** Slope: $\frac{1}{3}$; y-intercept: $\left(0, -\frac{7}{3}\right)$ **42.** 0
43. x-intercept: $(10, 0)$; y-intercept: $(0, -4)$ **44.** $\frac{5}{8}$
45. Parallel   **46.** Neither   **47.** 📝   **49.** $(1, -1, 2)$
51. $(1, -2, 4, -1)$ **53.** $\left(-1, \frac{1}{5}, -\frac{1}{2}\right)$ **55.** 14
57. $z = 8 - 2x - 4y$

Quick Quiz: Sections 8.1–8.4, p. 539

1. $\left(\frac{14}{5}, \frac{3}{10}\right)$ **2.** $\{(x, y)\,|\,2x - y = 4\}$ **3.** $(5, -6, 9)$
4. 20 mph   **5.** 15% pigment: $1\frac{1}{5}$ gal; 10% pigment: $1\frac{4}{5}$ gal

Prepare to Move On, p. 539

1. Let x represent the first number;
$x + (x + 1) + (x + 2) = 100$ **2.** Let x, y, and z represent the numbers; $x + y + z = 100$ **3.** Let x, y, and z represent the numbers; $xy = 5z$ **4.** Let x and y represent the numbers; $xy = 2(x + y)$

Mid-Chapter Review: Chapter 8, p. 540

1.
$$2x - 3(x - 1) = 5$$
$$2x - 3x + 3 = 5$$
$$-x + 3 = 5$$
$$-x = 2$$
$$x = -2$$

$$y = x - 1$$
$$y = -2 - 1$$
$$y = -3$$

The solution is $(-2, -3)$.

2.
$$2x - 5y = 1$$
$$\underline{x + 5y = 8}$$
$$3x = 9$$
$$x = 3$$

$$x + 5y = 8$$
$$3 + 5y = 8$$
$$5y = 5$$
$$y = 1$$

The solution is $(3, 1)$.

3. $(1, 1)$ **4.** $(9, 1)$ **5.** $(4, 3)$ **6.** $(5, 7)$ **7.** $(5, 10)$
8. $\left(2, \frac{2}{5}\right)$ **9.** No solution   **10.** $\{(x, y)\,|\,x = 2 - y\}$
11. $(1, 1)$ **12.** $\left(\frac{40}{9}, \frac{10}{3}\right)$ **13.** $(2, -10, -3)$
14. $\left(\frac{1}{2}, -4, \frac{1}{3}\right)$ **15.** No solution   **16.** The equations are dependent.   **17.** Microsoft: 4%; Yahoo!: 8.3%
18. 5-cent bottles or cans: 336; 10-cent bottles or cans: 94
19. Pecan Morning: 12 lb; Oat Dream: 8 lb   **20.** 18 mph

Exercise Set 8.5, pp. 544–547

1. (a)   **2.** (c)   **3.** (d)   **4.** (b)   **5.** 8, 15, 62
7. 8, 21, -3 **9.** 32°, 96°, 52°   **11.** Reading: 497; mathematics: 514; writing: 489   **13.** Bran muffin: 1.5 g; banana: 3 g; 1 cup of Wheaties: 3 g   **15.** Base price: $28,920; tow package: $595; sunroof: $850
17. 12-oz cups: 17; 16-oz cups: 25; 20-oz cups: 13
19. Business-equipment loan: $15,000; small-business loan: $35,000; home-equity loan: $70,000   **21.** Gold: $55.62/g; silver: $1.09/g; copper: $0.01/g   **23.** Roast beef: 2 servings; baked potato: 1 serving; broccoli: 2 servings
25. First mezzanine: 8 tickets; main floor: 12 tickets; second mezzanine: 20 tickets   **27.** Asia: 5.2 billion; Africa: 2.3 billion; rest of the world: 1.9 billion   **29.** 📝

31.
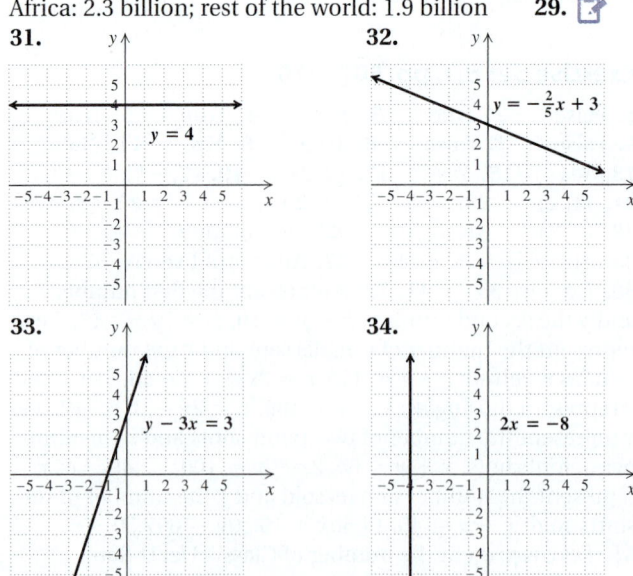

35.

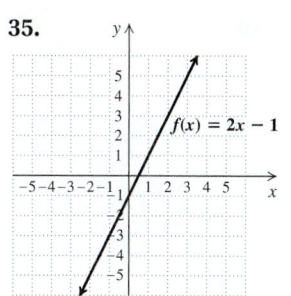

$f(x) = 2x - 1$

36.

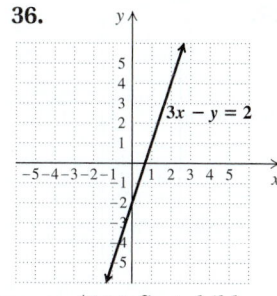

$3x - y = 2$

37. 🖮   **39.** Applicant: $150; spouse: $154; first child: $116; second child: $117   **41.** 20 years   **43.** 35 tickets

Quick Quiz: Sections 8.1–8.5, p. 547

1. $(4, 3)$ **2.** $(1, 1)$ **3.** $(6, -1)$ **4.** $(0, 0)$
5. $(0, 5, -2)$

Prepare to Move On, p. 547

1. $-4x + 6y$ **2.** $7y$ **3.** $-2a + b + 6c$
4. $-12x + 5y - 8z$ **5.** $23x - 13z$

Exercise Set 8.6, pp. 551–552

1. Matrix   **2.** Rows; vertical   **3.** Entry   **4.** Matrices
5. Rows   **6.** First   **7.** $(3, 4)$ **9.** $(-2, 5)$
11. $\left(\frac{3}{2}, \frac{5}{2}\right)$ **13.** $\left(2, \frac{1}{2}, -2\right)$ **15.** $(2, -2, 1)$
17. $\left(4, \frac{1}{2}, -\frac{1}{2}\right)$ **19.** $(1, -3, -2, -1)$ **21.** Dimes: 18; nickels: 24   **23.** Dried fruit: 9 lb; macadamia nuts: 6 lb
25. $400 at 3%; $500 at 4%; $1600 at 5%   **27.** 🖮
29. $x(x + 1)^2$ **30.** $\dfrac{18xy^6}{25z^2}$ **31.** $\dfrac{4}{(3a - 1)(a + 1)^2}$
32. $\dfrac{5t + 10}{3t}$ **33.** 🖮   **35.** 1324

Quick Quiz: Sections 8.1–8.6, p. 552

1. $(-1, 5)$ **2.** No solution   **3.** $\left(\frac{1}{2}, \frac{11}{10}, \frac{5}{4}\right)$
4. 4-marker packages: 26; 6-marker packages: 16
5. Drink Fresh: $1\frac{1}{5}$ L; Summer Light: $4\frac{4}{5}$ L

Prepare to Move On, p. 552

1. 17   **2.** -19 **3.** 37   **4.** 422

Exercise Set 8.7, pp. 557–558

1. True   **2.** True   **3.** True   **4.** False   **5.** False
6. False   **7.** 4   **9.** -50 **11.** 27   **13.** -3
15. -5 **17.** $(-3, 2)$ **19.** $\left(\frac{9}{19}, \frac{51}{38}\right)$ **21.** $\left(-1, -\frac{6}{7}, \frac{11}{7}\right)$
23. $(2, -1, 4)$ **25.** $(1, 2, 3)$ **27.** 🖮
29. $(6x - 1)(x + 4)$ **30.** $5(a - 1)(a^2 + 3)$
31. $(5y^4 + 7z)(5y^4 - 7z)$ **32.** $8(n + 1)(n^2 - n + 1)$
33. 🖮   **35.** 12   **37.** 10

Quick Quiz: Sections 8.1–8.7, p. 558

1. $(3, 5)$ **2.** $(-8, -23)$ **3.** $(-2, 3)$ **4.** $(1, 0)$
5. $\left(\frac{29}{93}, \frac{28}{93}\right)$

Prepare to Move On, p. 558

1. $70x - 2500$ **2.** 4500   **3.** $\frac{250}{7}$ **4.** $\frac{250}{7}$

Exercise Set 8.8, pp. 562–564

1. (b)   **2.** (f)   **3.** (h)   **4.** (a)   **5.** (e)   **6.** (d)
7. (c)   **8.** (g)   **9.** (a) $P(x) = 20x - 200,000$;
(b) (10,000 units, $550,000)   **11.** (a) $P(x) = 25x - 3100$;
(b) (124 units, $4960)   **13.** (a) $P(x) = 45x - 22,500$;
(b) (500 units, $42,500)   **15.** (a) $P(x) = 16x - 50,000$;
(b) (3125 units, $125,000)   **17.** (a) $P(x) = 50x - 100,000$;
(b) (2000 units, $250,000)   **19.** ($60, 1100)
21. ($22, 474)   **23.** ($50, 6250)   **25.** ($10, 1070)
27. (a) $C(x) = 45,000 + 40x$; **(b)** $R(x) = 130x$;
(c) $P(x) = 90x - 45,000$; **(d)** $225,000 profit, $9000 loss
(e) (500 phones, $65,000)   **29.** (a) $C(x) = 10,000 + 30x$;
(b) $R(x) = 80x$; **(c)** $P(x) = 50x - 10,000$; **(d)** $90,000 profit, $7500 loss; **(e)** (200 seats, $16,000)   **31.** 🖮
33. $-\frac{1}{5}, \frac{1}{5}$ **34.** 1   **35.** $\frac{5}{2}$ **36.** $-3, 6$
37. 🖮   **39.** ($5, 300 yo-yo's)
41. (a) $8.74; (b) 24,509 units

Quick Quiz: Sections 8.1–8.8, p. 564

1. Length: 40 ft; width: 30 ft   **2.** Low-fat milk: $9\frac{3}{5}$ oz; whole milk: $6\frac{2}{5}$ oz   **3.** $90°, 67.5°, 22.5°$ **4.** 26
5. -25

Prepare to Move On, p. 564

1. 6   **2.** -2 **3.** $\frac{3}{4}$ **4.** -6 **5.** -5

Visualizing for Success, p. 565

1. C   **2.** H   **3.** J   **4.** G   **5.** D   **6.** I   **7.** A
8. F   **9.** E   **10.** B

Decision Making: Connection, p. 566

1. $28\frac{1}{8}$ years   **2.** $18,750   **3.** $26,250; about 16.4 years
4. $5750   **5.** 🖥

Study Summary: Chapter 8, pp. 567–570

1. $(2, -1)$ **2.** $\left(-\frac{1}{2}, \frac{1}{2}\right)$ **3.** $\left(\frac{16}{7}, -\frac{3}{7}\right)$ **4.** Pens: 32 boxes; pencils: 88 boxes   **5.** 40%-acid: 0.8 L; 15%-acid: 1.2 L   **6.** 8 mph   **7.** $\left(2, -\frac{1}{2}, -5\right)$
8. $(2.5, 3.5, 3)$ **9.** $(4, 1)$ **10.** 28   **11.** -25
12. $\left(\frac{11}{4}, -\frac{3}{4}\right)$ **13.** (a) $P(x) = 75x - 9000$;
(b) (120, $10,800)   **14.** ($9, 141)

Review Exercises: Chapter 8, pp. 570–572

1. Substitution   **2.** Elimination   **3.** Graphical
4. Dependent   **5.** Inconsistent   **6.** Contradiction
7. Parallel   **8.** Square   **9.** Determinant   **10.** Zero
11. $(4, 1)$ **12.** $(3, -2)$ **13.** $\left(\frac{8}{3}, \frac{14}{3}\right)$
14. No solution   **15.** $\left(\frac{9}{4}, \frac{7}{10}\right)$ **16.** $(-2, -3)$
17. $\left(-\frac{4}{5}, \frac{2}{5}\right)$ **18.** $\left(\frac{76}{17}, -\frac{2}{119}\right)$ **19.** $\{(x, y) \mid 3x + 4y = 6\}$
20. $(4, 3)$ **21.** Private lessons: 7 students; group lessons: 5 students   **22.** 4 hr   **23.** 8% juice: 10 L; 15% juice: 4 L

24. $(4, -8, 10)$ **25.** The equations are dependent.
26. $(2, 0, 4)$ **27.** $\left(\frac{8}{9}, -\frac{2}{3}, \frac{10}{9}\right)$ **28.** No solution
29. A: $90°$; B: $67.5°$; C: $22.5°$ **30.** Man: 1.4;
woman: 5.3; one-year-old child: 50   **31.** $\left(55, -\frac{89}{2}\right)$
32. $(-1, 1, 3)$ **33.** -5 **34.** 9   **35.** $(6, -2)$
36. $(-3, 0, 4)$ **37. (a)** $P(x) = 20x - 15{,}800$;
(b) (790 units, $39,500)   **38.** ($3, 81)
39. (a) $C(x) = 4.75x + 54{,}000$; **(b)** $R(x) = 9.25x$;
(c) $P(x) = 4.5x - 54{,}000$; **(d)** $31,500 loss, $13,500
profit; **(e)** (12,000 pints of honey, $111,000)   **40.** 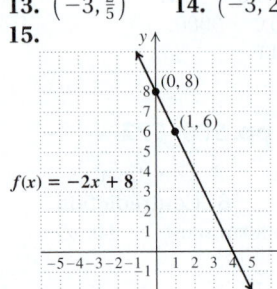 To
solve a problem involving four variables, go through the
Familiarize and *Translate* steps as usual. The resulting
system of equations can be solved using the elimination
method just as for three variables but likely with more
steps.   **41.** 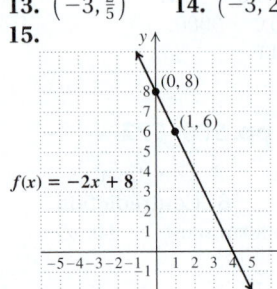 A system of equations can be both
dependent and inconsistent if it is equivalent to a system
with fewer equations that has no solution. An example is a
system of three equations in three unknowns in which two
of the equations represent the same plane, and the third
represents a parallel plane.   **42.** 20,000 pints
43. $(0, 2), (1, 3)$

Test: Chapter 8, pp. 572–573

1. [8.1] $(2, 4)$ **2.** [8.2] $\left(3, -\frac{11}{3}\right)$ **3.** [8.2] $(2, -1)$
4. [8.2] No solution   **5.** [8.2] $\{(x, y) \mid x = 2y - 3\}$
6. [8.2] $\left(-\frac{3}{2}, -\frac{3}{2}\right)$ **7.** [8.3] Length: 94 ft; width: 50 ft
8. [8.3] Pepperidge Farm Goldfish: 120 g; Rold Gold
Pretzels: 500 g   **9.** [8.3] 20 mph   **10.** [8.4] The equations are dependent.   **11.** [8.4] $\left(2, -\frac{1}{2}, -1\right)$
12. [8.4] No solution   **13.** [8.4] $(0, 1, 0)$
14. [8.6] $\left(\frac{22}{5}, -\frac{28}{5}\right)$ **15.** [8.6] $(3, 1, -2)$ **16.** [8.7] -14
17. [8.7] -59 **18.** [8.7] $\left(\frac{7}{13}, -\frac{17}{26}\right)$
19. [8.5] Electrician: 3.5 hr; carpenter: 8 hr; plumber: 10 hr
20. [8.8] ($3, 55)   **21.** [8.8] **(a)** $C(x) = 25x + 44{,}000$;
(b) $R(x) = 80x$; **(c)** $P(x) = 55x - 44{,}000$; **(d)** $27,500 loss,
$5500 profit; **(e)** (800 hammocks, $64,000)
22. [7.3], [8.3] $m = 7, b = 10$ **23.** [8.3] $\frac{120}{7}$ lb

Cumulative Review: Chapters 1–8, p. 574

1. x^{11} **2.** $-\frac{2a^{11}}{5b^{33}}$ **3.** $\frac{81x^{36}}{256y^8}$ **4.** 4.00×10^6
5. $b = \frac{2A}{h} - t$, or $b = \frac{2A - ht}{h}$ **6.** $r = \frac{s}{B - 1}$
7. -5 **8.** 1   **9.** $-3, -2$ **10.** 2, 3
11. $\left\{y \mid y \geq -\frac{2}{3}\right\}$, or $\left[-\frac{2}{3}, \infty\right)$ **12.** $(1, 1)$
13. $\left(-3, \frac{2}{5}\right)$ **14.** $(-3, 2, -4)$
15.

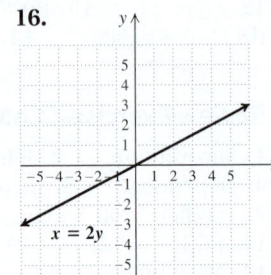

$f(x) = -2x + 8$

16.

$x = 2y$

17.

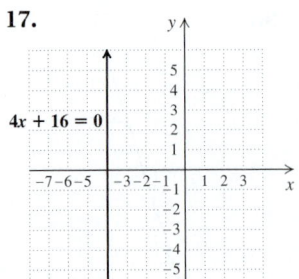

$4x + 16 = 0$

18.

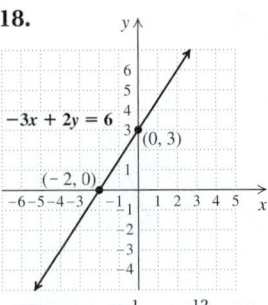

$-3x + 2y = 6$

19. Slope: $\frac{9}{4}$; y-intercept: $(0, -3)$ **20.** $y = -\frac{1}{10}x + \frac{12}{5}$
21. Parallel   **22.** $y = -2x + 5$
23. $\{x \mid x$ is a real number *and* $x \neq -10\}$ **24.** -31
25. $2a^2 + 4a - 4$ **26.** $5(x + 2y)(x - 2y)$
27. $(3x + 2)(2x + 3)$ **28.** $25a^2 + 70a + 49$
29. $2x + 1$ **30.** $\frac{2(x + 2)}{x - 2}$ **31.** $\frac{x}{2(x + 1)(x + 2)}$
32. Dog sled: 12 mph; snowmobile: 52 mph
33. 68 professionals; 82 students   **34.** Sea Spray: 90 oz;
Ocean Mist: 30 oz   **35.** 86   **36.** $-12x^{2a}y^{b+y+3}$
37. $m = -\frac{5}{9}, b = -\frac{2}{9}$

CHAPTER 9

Exercise Set 9.1, pp. 580–583

1. Inequality   **2.** Solution   **3.** Negative
4. Above   **5.** Equivalent equations   **6.** Equivalent
expressions   **7.** Equivalent inequalities
8. Not equivalent   **9.** Equivalent equations
10. Equivalent inequalities   **11.** $\{x \mid x < 2\}$, or $(-\infty, 2)$
13. $\{x \mid x \leq -9\}$, or $(-\infty, -9]$ **15.** $\{x \mid x < -26\}$, or
$(-\infty, -26)$ **17.** $\left\{t \mid t \geq -\frac{13}{3}\right\}$, or $\left[-\frac{13}{3}, \infty\right)$
19. $\left\{y \mid y \leq -\frac{3}{2}\right\}$, or $\left(-\infty, -\frac{3}{2}\right]$ **21.** $\left\{t \mid t < \frac{29}{5}\right\}$, or
$\left(-\infty, \frac{29}{5}\right)$ **23.** $\left\{m \mid m > \frac{7}{3}\right\}$, or $\left(\frac{7}{3}, \infty\right)$
25. $\{x \mid x \geq 2\}$, or $[2, \infty)$ **27.** $\{x \mid x > 0\}$, or $(0, \infty)$
29. $\{x \mid x < 7\}$, or $(-\infty, 7)$ **31.** $\{x \mid x \geq 2\}$, or $[2, \infty)$
33. $\{x \mid x < 8\}$, or $(-\infty, 8)$ **35.** $\{x \mid x \leq 2\}$, or $(-\infty, 2]$
37. $\{x \mid x \geq 2\}$, or $[2, \infty)$
39. Lengths of time less than $7\frac{1}{2}$ hr   **41.** At least 56 questions correct   **43.** Gross sales greater than $7000
45. For more than 17 transactions   **47.** Years after 2015
49. (a) Body densities less than $\frac{99}{95}$ kg/L, or about 1.04 kg/L;
(b) body densities less than $\frac{495}{482}$ kg/L, or about 1.03 kg/L
51. Years after 2032   **53. (a)** $\left\{x \mid x < 3913\frac{1}{23}\right\}$, or
$\{x \mid x \leq 3913\}$; **(b)** $\left\{x \mid x > 3913\frac{1}{23}\right\}$, or $\{x \mid x \geq 3914\}$
55. 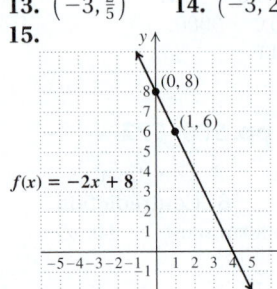   **57.** $-\frac{6}{5}$ **58.** $-5, 3$ **59.** $\left(\frac{4}{3}, -\frac{7}{9}\right)$
60. $\left(\frac{8}{5}, \frac{27}{5}\right)$ **61.** $r = \frac{b}{a + c}$ **62.** $n = \frac{a}{y - b}$
63. 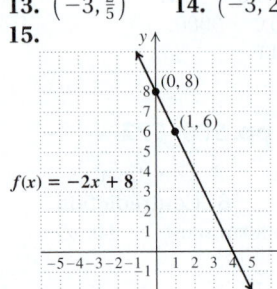   **65.** $\left\{x \mid x \leq \frac{2}{a - 1}\right\}$ **67.** $\left\{y \mid y \geq \frac{2a + 5b}{b(a - 2)}\right\}$
69. False; $2 < 3$ and $4 < 5$, but $2 - 4 = 3 - 5$.
71. About 7.8 gal or less   **73. (a)** More than $40;
(b) costs greater than $30   **75.**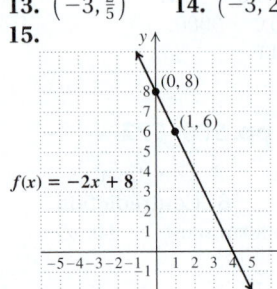

Prepare to Move On, p. 583

1. $\{x \mid x \text{ is a real number } and\ x \neq 0\}$
2. $\{x \mid x \text{ is a real number } and\ x \neq \frac{7}{5}\}$ **3.** $\mathbb{R}$
4. $\{x \mid x \text{ is a real number } and\ x \neq 0\}$

Exercise Set 9.2, pp. 589–593

1. Intersection   **2.** Union   **3.** Intersection
4. Intersection   **5.** Union   **6.** Union   **7.** (h)
8. (j)   **9.** (f)   **10.** (a)   **11.** (e)   **12.** (d)   **13.** (b)
14. (g)   **15.** (c)   **16.** (i)   **17.** $\{4, 16\}$
19. $\{0, 5, 10, 15, 20\}$ **21.** $\{b, d, f\}$ **23.** $\{u, v, x, y, z\}$
25. $\varnothing$ **27.** $\{1, 3, 5\}$
29. $(1, 3)$
31. $[-6, 0]$
33. $(-\infty, -1) \cup (4, \infty)$
35. $(-\infty, -2] \cup (1, \infty)$
37. $(-2, 4]$
39. $(-2, 4)$
41. $(-\infty, 5) \cup (7, \infty)$
43. $(-\infty, -4] \cup [5, \infty)$
45. $[-3, 7)$
47. $(-7, 0]$
49. $(-\infty, 5)$
51. $\{x \mid -5 \leq x < 7\}$, or $[-5, 7)$
53. $\{t \mid 4 < t \leq 8\}$, or $(4, 8]$
55. $\{a \mid -2 \leq a < 2\}$, or $[-2, 2)$
57. $\mathbb{R}$, or $(-\infty, \infty)$
59. $\{x \mid -3 \leq x \leq 2\}$, or $[-3, 2]$
61. $\{x \mid 7 < x < 23\}$, or $(7, 23)$
63. $\{x \mid -32 \leq x \leq 8\}$, or $[-32, 8]$
65. $\{x \mid 1 \leq x \leq 3\}$, or $[1, 3]$
67. $\{x \mid -\frac{7}{2} < x \leq 7\}$, or $(-\frac{7}{2}, 7]$
69. $\{t \mid t < 0 \text{ } or\ t > 1\}$, or $(-\infty, 0) \cup (1, \infty)$
71. $\{a \mid a < \frac{7}{2}\}$, or $(-\infty, \frac{7}{2})$
73. $\{a \mid a < -5\}$, or $(-\infty, -5)$
75. $\varnothing$
77. $\{t \mid t \leq 6\}$, or $(-\infty, 6]$
79. $(-\infty, -6) \cup (-6, \infty)$ **81.** $(-\infty, 0) \cup (0, \infty)$
83. $(-\infty, 4) \cup (4, \infty)$ **85.** $\{x \mid x \geq 10\}$, or $[10, \infty)$
87. $\{x \mid x \leq 3\}$, or $(-\infty, 3]$ **89.** $\{x \mid x \geq -\frac{7}{2}\}$, or
$[-\frac{7}{2}, \infty)$ **91.** $\{x \mid x \leq 4\}$, or $(-\infty, 4]$ **93.**
95. $(x - 2)(x - 10)$ **96.** $2(5c^3 + 3d)(5c^3 - 3d)$
97. $10(c + 1)(c^2 - c + 1)(c - 1)(c^2 + c + 1)$
98. $16x^2z^5(3xz^3 - 5x^2 + 1)$ **99.**

101. $(-1, 6)$ **103.** $0 \text{ ft} \leq d \leq 198 \text{ ft}$ **105.** Densities between 1.03 kg/L and 1.04 kg/L   **107.** More than 12 trips and fewer than 125 trips
109. $\{m \mid m < \frac{6}{5}\}$, or $(-\infty, \frac{6}{5})$
111. $\{x \mid -\frac{1}{8} < x < \frac{1}{2}\}$, or $(-\frac{1}{8}, \frac{1}{2})$
113. False   **115.** True   **117.** $(-\infty, -7) \cup (-7, \frac{3}{4}]$
119.   **121.** Between 5.24 words and 15.09 words per sentence   **123.** Let w represent the number of ounces in a bottle; $15.9 \leq w \leq 16.1$; $[15.9, 16.1]$ **125.**
127.   **129.**

Quick Quiz: Sections 9.1–9.2, p. 593

1. $\{x \mid x > \frac{2}{7}\}$, or $(\frac{2}{7}, \infty)$ **2.** $\{x \mid x \geq 6\}$, or $[6, \infty)$
3. $\{m \mid m < -18\}$, or $(-\infty, -18)$
4. $\{y \mid y < 0 \text{ } or\ y > 2\}$, or $(-\infty, 0) \cup (2, \infty)$
5. $\{x \mid 3 < x < 8\}$, or $(3, 8)$

Prepare to Move On, p. 593

1. $\frac{2}{3}$ **2.** 16   **3.** 0   **4.** 7   **5.** 6

Technology Connection, p. 598

1. The x-values on the graph of $y_1 = |4x + 2|$ that are *below* the line $y = 6$ solve the inequality $|4x + 2| < 6$.
2. The x-values on the graph of $y_1 = |3x - 2|$ that are below the line $y = 4$ are in the interval $(-\frac{2}{3}, 2)$.
3. The graphs of $y_1 = |4x + 2|$ and $y_2 = -6$ do not intersect.

Exercise Set 9.3, pp. 599–601

1. True   **2.** False   **3.** True   **4.** True   **5.** True
6. True   **7.** False   **8.** False   **9.** (g)   **10.** (h)
11. (d)   **12.** (a)   **13.** (a)   **14.** (b)   **15.** $\{-10, 10\}$
17. $\varnothing$ **19.** $\{0\}$ **21.** $\{-\frac{1}{2}, \frac{7}{2}\}$ **23.** $\varnothing$ **25.** $\{-4, 8\}$
27. $\{6, 8\}$ **29.** $\{-5.5, 5.5\}$ **31.** $\{-8, 8\}$ **33.** $\{-1, 1\}$
35. $\{-\frac{11}{2}, \frac{13}{2}\}$ **37.** $\{-2, 12\}$ **39.** $\{-\frac{1}{3}, 3\}$ **41.** $\{-7, 1\}$
43. $\{-8.7, 8.7\}$ **45.** $\{-\frac{9}{2}, \frac{11}{2}\}$ **47.** $\{-8, 2\}$ **49.** $\{-\frac{1}{2}\}$
51. $\{-\frac{3}{5}, 5\}$ **53.** $\mathbb{R}$ **55.** $\{\frac{1}{4}\}$
57. $\{a \mid -3 \leq a \leq 3\}$, or $[-3, 3]$
59. $\{t \mid t < 0 \text{ } or\ t > 0\}$, or $(-\infty, 0) \cup (0, \infty)$
61. $\{x \mid -3 < x < 5\}$, or $(-3, 5)$
63. $\{n \mid -8 \leq n \leq 4\}$, or $[-8, 4]$
65. $\{x \mid x < -2 \text{ } or\ x > 8\}$, or $(-\infty, -2) \cup (8, \infty)$
67. $\mathbb{R}$, or $(-\infty, \infty)$
69. $\{a \mid a \leq -\frac{10}{3} \text{ } or\ a \geq \frac{2}{3}\}$, or $(-\infty, -\frac{10}{3}] \cup [\frac{2}{3}, \infty)$
71. $\{y \mid -9 < y < 15\}$, or $(-9, 15)$

73. $\{x|x \le -8 \text{ or } x \ge 0\}$, or $(-\infty, -8] \cup [0, \infty)$

75. $\{x|x < -\frac{1}{2} \text{ or } x > \frac{7}{2}\}$, or $\left(-\infty, -\frac{1}{2}\right) \cup \left(\frac{7}{2}, \infty\right)$

77. $\varnothing$

79. $\{x|x < -\frac{43}{24} \text{ or } x > \frac{9}{8}\}$, or $\left(-\infty, -\frac{43}{24}\right) \cup \left(\frac{9}{8}, \infty\right)$

81. $\{m|-9 \le m \le 3\}$, or $[-9, 3]$

83. $\{a|-6 < a < 0\}$, or $(-6, 0)$

85. $\{x|-\frac{1}{2} \le x \le \frac{7}{2}\}$, or $\left[-\frac{1}{2}, \frac{7}{2}\right]$

87. $\{x|x \le -\frac{7}{3} \text{ or } x \ge 5\}$, or $\left(-\infty, -\frac{7}{3}\right] \cup [5, \infty)$

89. $\{x|-4 < x < 5\}$, or $(-4, 5)$

91.   **93.** $y = \frac{1}{3}x - 2$ **94.** $y - 7 = -8(x - 3)$
95. $y = 2x + 5$ **96.** $y = -x - 1$ **97.**
99. $\left\{\frac{5}{4}, \frac{5}{2}\right\}$ **101.** $\{x|-4 \le x \le -1 \text{ or } 3 \le x \le 6\}$,
or $[-4, -1] \cup [3, 6]$ **103.** $\{t|t \le \frac{5}{2}\}$, or $\left(-\infty, \frac{5}{2}\right]$
105. $|x| < 3$ **107.** $|x| \ge 6$ **109.** $|x + 3| > 5$
111. $|x - 7| < 2$, or $|7 - x| < 2$ **113.** $|x - 3| \le 4$
115. $|x + 4| < 3$ **117.** Between 80 ft and 100 ft
119.

Quick Quiz: Sections 9.1–9.3, p. 601

1. $\{x|x \ge -\frac{13}{2}\}$, or $\left[-\frac{13}{2}, \infty\right)$ **2.** $\{2, 3, 5, 7\}$
3. $\{1, 2, 3, 4, 6, 8\}$ **4.** $\{-3, 11\}$ **5.** $\{x|-5 < x < 5\}$,
or $(-5, 5)$

Prepare to Move On, p. 601

1.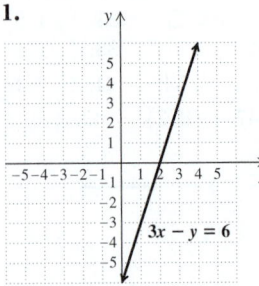

2.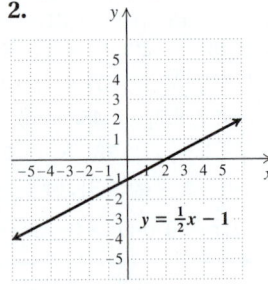

3.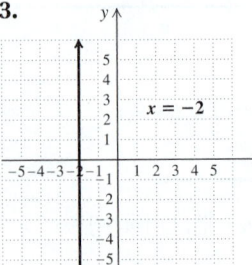

4.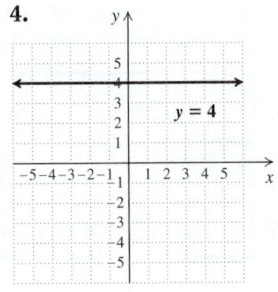

5. $\left(4, -\frac{4}{3}\right)$ **6.** $(5, 1)$

Mid-Chapter Review: Chapter 9, p. 602

1. $2 \le x \le 11$
   The solution is $[2, 11]$.
2. $x - 1 < -9 \quad or \quad 9 < x - 1$
   $\quad\quad x < -8 \quad or \quad 10 < x$
   The solution is $(-\infty, -8) \cup (10, \infty)$.
3. $\{-15, 15\}$ **4.** $\{t|-10 < t < 10\}$, or $(-10, 10)$
5. $\{p|p < -15 \text{ or } p > 15\}$, or $(-\infty, -15) \cup (15, \infty)$
6. $\{-4, 3\}$ **7.** $\{x|2 < x < 11\}$, or $(2, 11)$
8. $\{t|-4 < t < 4\}$, or $(-4, 4)$
9. $\{x|x < -6 \text{ or } x > 13\}$, or $(-\infty, -6) \cup (13, \infty)$
10. $\{x|-7 \le x \le 3\}$, or $[-7, 3]$ **11.** $\left\{-\frac{8}{3}, \frac{8}{3}\right\}$
12. $\{x|x < -\frac{13}{2}\}$, or $\left(-\infty, -\frac{13}{2}\right)$
13. $\{n|-9 < n \le \frac{8}{3}\}$, or $\left(-9, \frac{8}{3}\right]$
14. $\{x|x \le -\frac{17}{2} \text{ or } x \ge \frac{7}{2}\}$, or $\left(-\infty, -\frac{17}{2}\right] \cup \left[\frac{7}{2}, \infty\right)$
15. $\{x|x \ge -2\}$, or $[-2, \infty)$ **16.** $\{-42, 38\}$ **17.** $\varnothing$
18. $\{a|-7 < a < -5\}$, or $(-7, -5)$ **19.** $\mathbb{R}$, or $(-\infty, \infty)$
20. $\mathbb{R}$, or $(-\infty, \infty)$

Technology Connection, p. 606

1.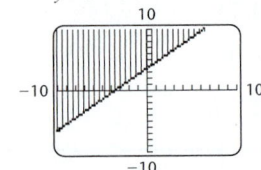

2.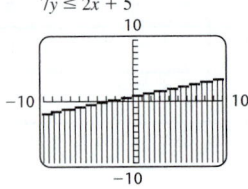

3.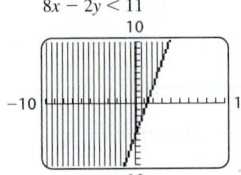

4.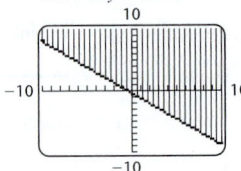

Technology Connection, p. 608

1.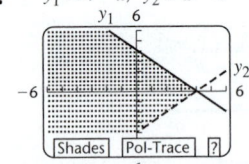

Connecting the Concepts, pp. 609–610

1.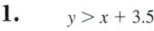

2.

3.

4.
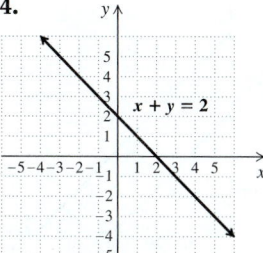
$x + y = 2$

5.
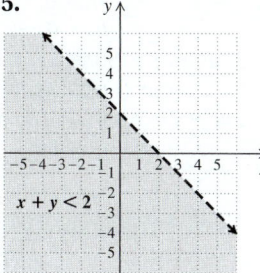
$x + y < 2$

19.
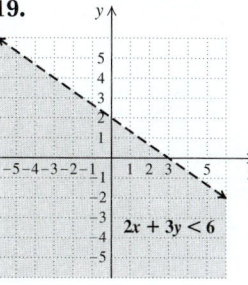
$2x + 3y < 6$

21.
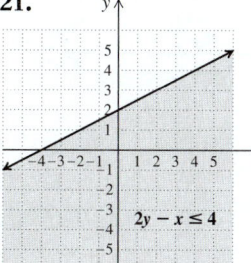
$2y - x \le 4$

6.
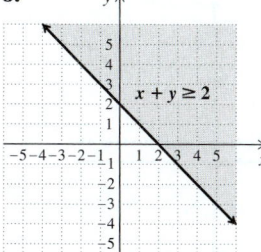
$x + y \ge 2$

7.
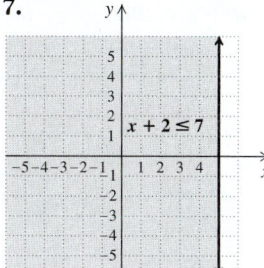
$x + 2 \le 7$

23.
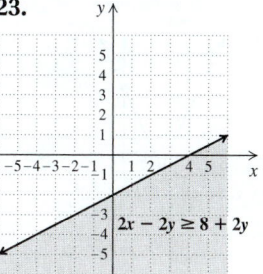
$2x - 2y \ge 8 + 2y$

25.
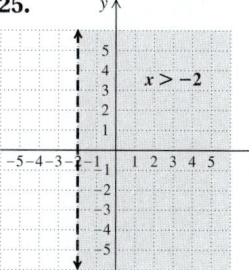
$x > -2$

8.
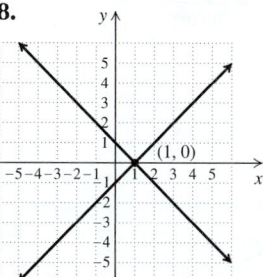
$(1, 0)$

9.
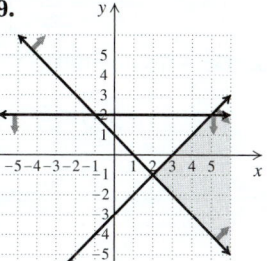

27.
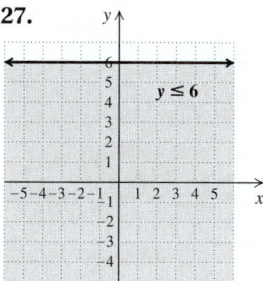
$y \le 6$

29.
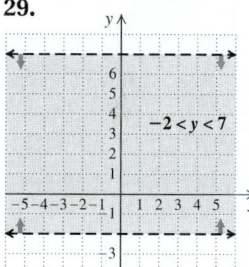
$-2 < y < 7$

Exercise Set 9.4, pp. 610–612

1. (e)   **2.** (c)   **3.** (d)   **4.** (a)   **5.** (b)   **6.** (f)
7. No   **9.** Yes

11.
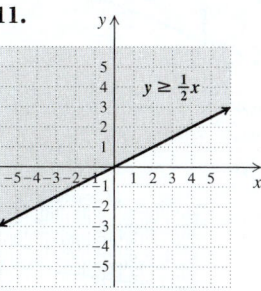
$y \ge \frac{1}{2}x$

13.
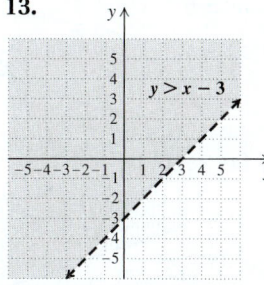
$y > x - 3$

31.
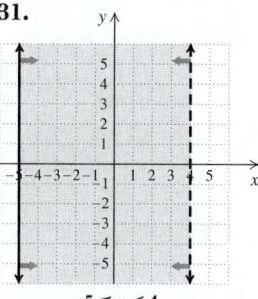
$-5 \le x < 4$

33.
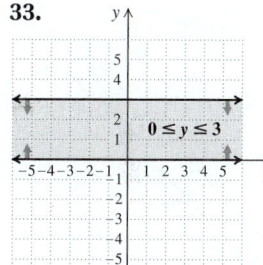
$0 \le y \le 3$

15.
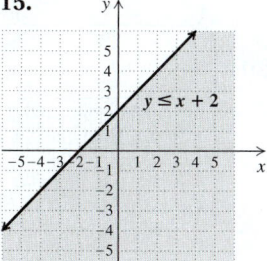
$y \le x + 2$

17.
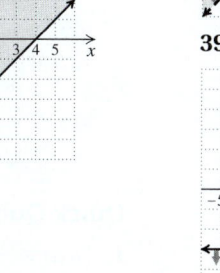
$x - y \le 4$

35.
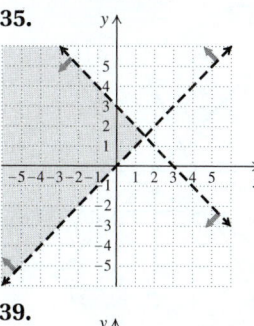

37.
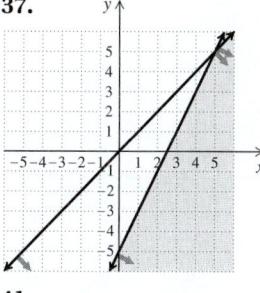

39.
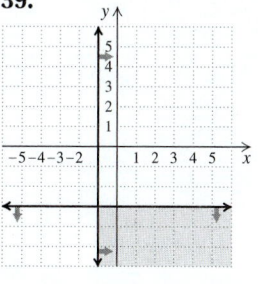

41.
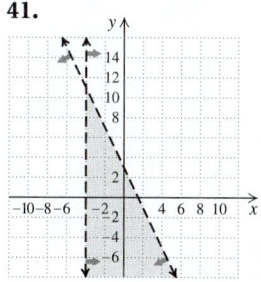

43.
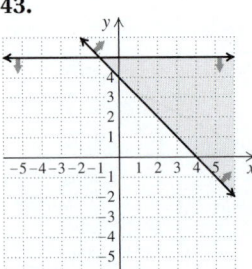

45.
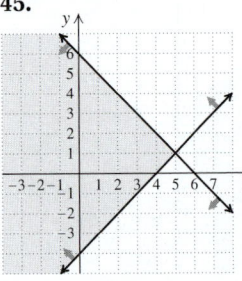

47.
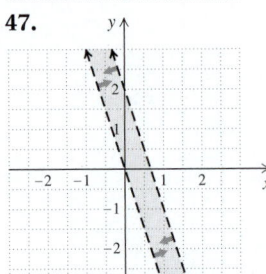

49.
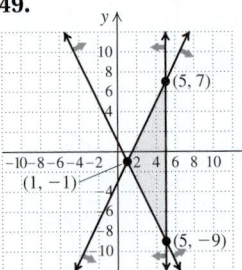

51.
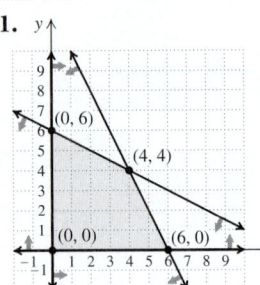

53.
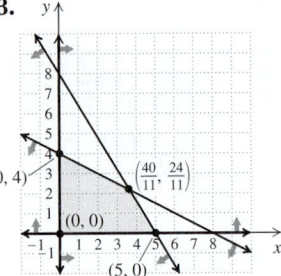

55.
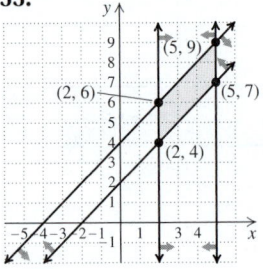

57.
59. $4p^6 - 12p^3w^4 + 9w^8$
60. $25x^4 - 49$
61. $-3t^4 + 3t^2 - 1$
62. $x - 2 + \dfrac{-8}{x + 1}$
63. $\dfrac{(3x + 2)(x + 1)}{2(2x - 3)(x + 5)}$
64. $\dfrac{x^2 - 13x + 6}{(x - 4)(x + 3)^2}$
65.

67.
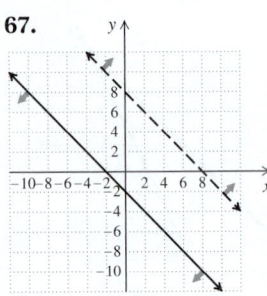

69.
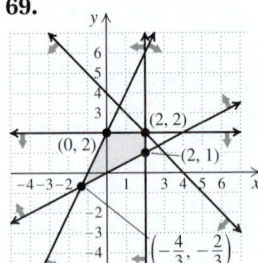

71.
$$w > 0,$$
$$h > 0,$$
$$w + h + 30 \leq 62, \text{ or}$$
$$w + h \leq 32,$$
$$2w + 2h + 30 \leq 130, \text{ or}$$
$$w + h \leq 50$$
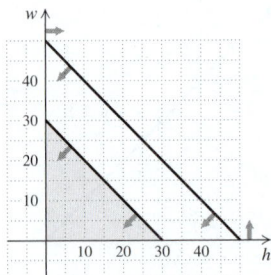

73. $q + v \geq 287,$
$v \geq 145,$
$q \leq 170,$
$v \leq 170$
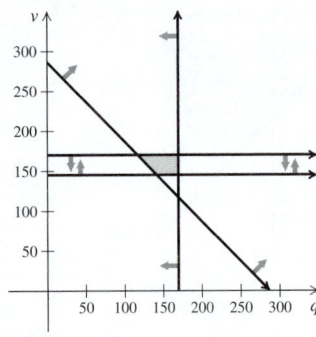

75. $35c + 75a > 1000,$
$c \geq 0,$
$a \geq 0$
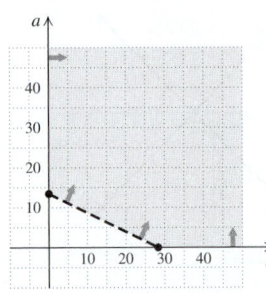

77. $h < 2w,$
$w \leq 1.5h,$
$h \leq 3200,$
$h \geq 0,$
$w \geq 0$
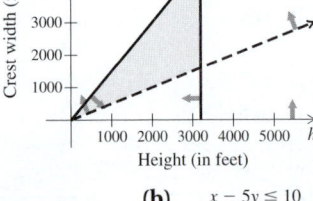

79. (a) $3x + 6y > 2$
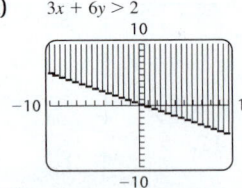
(b) $x - 5y \leq 10$
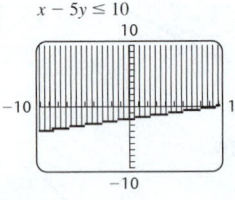
(c) $13x - 25y + 10 \leq 0$
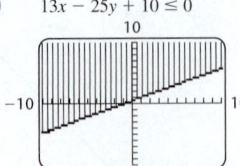
(d) $2x + 5y > 0$
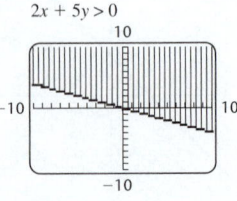

Quick Quiz: Sections 9.1–9.4, p. 612

1. $\{a \mid a \geq -3.6\}$, or $[-3.6, \infty)$
2. $\mathbb{R}$, or $(-\infty, \infty)$

3. $\left\{x \mid x < -\frac{11}{2} \, or \, x > 3\right\}$, or $\left(-\infty, -\frac{11}{2}\right) \cup (3, \infty)$

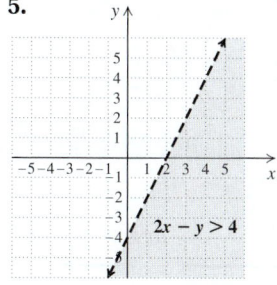

4. $\varnothing$ **5.**

Prepare to Move On, p. 612

1. 3%: $3600; 5%: $6400   **2.** Student tickets: 62; adult tickets: 108   **3.** Corn: 240 acres; soybeans: 160 acres

Exercise Set 9.5, pp. 616–618

1. Objective   **2.** Constraints   **3.** Corner
4. Feasible   **5.** Vertices   **6.** Vertex
7. Maximum 84 when $x = 0$, $y = 6$; minimum 0 when $x = 0$, $y = 0$ **9.** Maximum 76 when $x = 7$, $y = 0$; minimum 16 when $x = 0$, $y = 4$ **11.** Maximum 5 when $x = 3$, $y = 7$; minimum -15 when $x = 3$, $y = -3$
13. Bus: 2 trips; train: 3 trips   **15.** 4-photo pages: 5; 6-photo pages: 15; 110 photos   **17.** Corporate bonds: $22,000; municipal bonds: $18,000; maximum: $1510
19. Short-answer questions: 12; essay questions: 4
21. Merlot: 80 acres; Cabernet: 160 acres   **23.** 2.5 servings of each   **25.** 🖎   **27.** $\frac{1}{100}$ **28.** y^{16} **29.** $-2x^{12}$
30. $-\dfrac{8}{a^9 b^{12}}$ **31.** $\dfrac{3d^3}{2c}$ **32.** 1   **33.** 🖎
35. T3's: 30; S5's: 10

Quick Quiz: Sections 9.1–9.5, p. 618

1. $\left\{x \mid -\frac{4}{5} < x \le -\frac{1}{5}\right\}$, or $\left(-\frac{4}{5}, -\frac{1}{5}\right]$ **2.** $\left\{-\frac{2}{3}, \frac{10}{3}\right\}$
3. $\{x \mid -1 < x < 0\}$, or $(-1, 0)$
4.

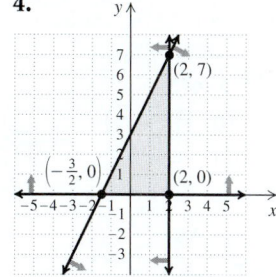

5. Maximum 8 when $x = 2$, $y = 0$; minimum -6 when $x = -\frac{3}{2}$, $y = 0$

Prepare to Move On, p. 618

1. a^{12} **2.** $9t^{10}$ **3.** $x^2 + 4x + 4$ **4.** $9a^2 - 6a + 1$

Visualizing for Success, p. 619

1. B   **2.** F   **3.** J   **4.** A   **5.** E   **6.** G   **7.** C
8. D   **9.** I   **10.** H

Decision Making: Connection, p. 620

1. Bills greater than $9500   **2.** United Health One: $1175.76; SmartSense Plus: $2366.04   **3.** The United Health One plan will save Elisabeth money no matter how great her bills are.

Study Summary: Chapter 9, pp. 621–622

1. $\{x \mid x \ge -4\}$, or $[-4, \infty)$ **2.** $\left\{x \mid -2 < x \le -\frac{3}{4}\right\}$, or $\left(-2, -\frac{3}{4}\right]$ **3.** $\{x \mid x \le 13 \, or \, x > 22\}$, or $(-\infty, 13] \cup (22, \infty)$ **4.** $\left\{-1, \frac{9}{2}\right\}$ **5.** $\{x \mid 11 \le x \le 13\}$, or $[11, 13]$ **6.** $\{x \mid x < -5 \, or \, x > 2\}$, or $(-\infty, -5) \cup (2, \infty)$
7.

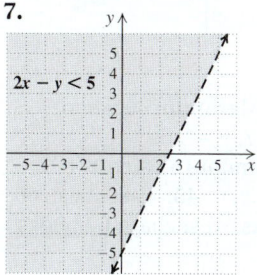

8. Maximum 8 when $x = 4$ and $y = 0$

Review Exercises: Chapter 9, pp. 622–623

1. True   **2.** True   **3.** True   **4.** True   **5.** False
6. False   **7.** $\left\{x \mid x > -\frac{3}{2}\right\}$, or $\left(-\frac{3}{2}, \infty\right)$
8. $\{x \mid x < -3\}$, or $(-\infty, -3)$ **9.** $\left\{x \mid x \le -\frac{5}{2}\right\}$, or $\left(-\infty, -\frac{5}{2}\right]$ **10.** $\{x \mid x < 5\}$, or $(-\infty, 5)$
11. $\{x \mid x < 1\}$, or $(-\infty, 1)$ **12.** $\{x \mid x \ge -4\}$, or $[-4, \infty)$
13. $\{x \mid x \ge 6\}$, or $[6, \infty)$ **14.** $\{x \mid x \le 2\}$, or $(-\infty, 2]$
15. More than 125 hr   **16.** $3000   **17.** $\{a, c\}$
18. $\{a, b, c, d, e, f, g\}$
19. ⟵———(———|———|———⟶ $(-3, 2]$
    -3 0   2
20. ⟵————————|————————⟶ $(-\infty, \infty)$
    0
21. $\{x \mid -8 < x \le 0\}$, or $(-8, 0]$ ⟵——(———|——]⟶
    -8 0
22. $\left\{x \mid -\frac{5}{4} < x < \frac{5}{2}\right\}$, or $\left(-\frac{5}{4}, \frac{5}{2}\right)$ ⟵——(——|——)——⟶
    $-\frac{5}{4}$ 0 $\frac{5}{2}$
23. $\{x \mid x < -3 \, or \, x > 1\}$, or $(-\infty, -3) \cup (1, \infty)$

⟵——)——|——(——⟶
   -3 0 1

24. $\{x \mid x < -11 \, or \, x \ge -6\}$, or $(-\infty, -11) \cup [-6, \infty)$

⟵——)————[————|——⟶
  -11 -6 0

25. $\{x \mid x \le -6 \, or \, x \ge 8\}$, or $(-\infty, -6] \cup [8, \infty)$

⟵——]————|————[——⟶
  -6 0   8

26. $\left\{x \mid x < -\frac{2}{5} \text{ or } x > \frac{8}{5}\right\}$, or $\left(-\infty, -\frac{2}{5}\right) \cup \left(\frac{8}{5}, \infty\right)$

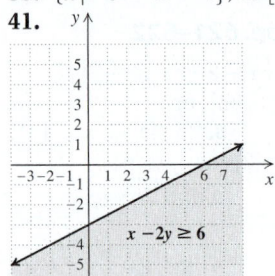

27. $(-\infty, -3) \cup (-3, \infty)$ **28.** $[2, \infty)$ **29.** $\left(-\infty, \frac{1}{4}\right]$
30. $\{-11, 11\}$ **31.** $\{t \mid t \le -21 \text{ or } t \ge 21\}$, or
$(-\infty, -21] \cup [21, \infty)$ **32.** $\{5, 11\}$
33. $\left\{a \mid -\frac{7}{2} < a < 2\right\}$, or $\left(-\frac{7}{2}, 2\right)$
34. $\left\{x \mid x \le -\frac{11}{3} \text{ or } x \ge \frac{19}{3}\right\}$, or $\left(-\infty, -\frac{11}{3}\right] \cup \left[\frac{19}{3}, \infty\right)$
35. $\left\{-14, \frac{4}{3}\right\}$ **36.** $\varnothing$ **37.** $\{x \mid -16 \le x \le 8\}$, or
$[-16, 8]$ **38.** $\{x \mid x < 0 \text{ or } x > 10\}$, or $(-\infty, 0) \cup (10, \infty)$
39. $\{x \mid -6 \le x \ge 4\}$, or $[-6, 4]$ **40.** $\varnothing$
41.

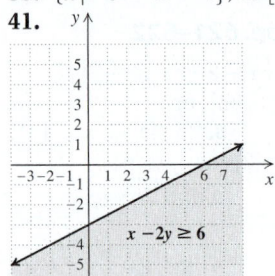

42.

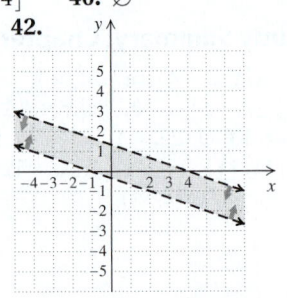

43.

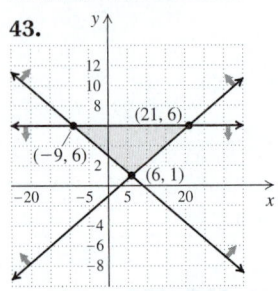

44. Maximum 40 when
$x = 7, y = 15$;
minimum 10 when $x = 1$,
$y = 3$ **45.** East coast: 40
books; West coast: 60 books

46. ☛ The equation $|X| = p$ has two solutions when p is
positive because X can be either p or $-p$. The same equation has no solution when p is negative because no number
has a negative absolute value.   **47.** ☛ The solution set
of a system of inequalities is all ordered pairs that make *all*
the individual inequalities true. This consists of ordered
pairs that are common to all the individual solution sets, or
the intersection of the graphs.   **48.** $\left\{x \mid -\frac{8}{3} \le x \le -2\right\}$,
or $\left[-\frac{8}{3}, -2\right]$

Test: Chapter 9, p. 624

1. [9.1] $\{y \mid y \le -2\}$, or $(-\infty, -2]$ **2.** [9.1] $\left\{x \mid x > \frac{16}{5}\right\}$,
or $\left(\frac{16}{5}, \infty\right)$ **3.** [9.1] $\left\{x \mid x \le \frac{9}{16}\right\}$, or $\left(-\infty, \frac{9}{16}\right]$
4. [9.1] $\{x \mid x > 1\}$, or $(1, \infty)$ **5.** [9.1] $\{x \mid x \ge 4\}$,
or $[4, \infty)$ **6.** [9.1] $\{x \mid x > 1\}$, or $(1, \infty)$ **7.** [9.1] More
than $187\frac{1}{2}$ mi   **8.** [9.1] Less than or equal to 2.5 hr
9. [9.2] $\{a, e\}$ **10.** [9.2] $\{a, b, c, d, e, i, o, u\}$
11. [9.1] $(-\infty, 2]$ **12.** [9.2] $(-\infty, 7) \cup (7, \infty)$
13. [9.2] $\left\{x \mid -\frac{3}{2} < x \le \frac{1}{2}\right\}$, or $\left(-\frac{3}{2}, \frac{1}{2}\right]$

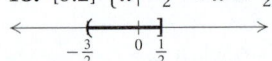

14. [9.2] $\{x \mid x < 3 \text{ or } x > 6\}$, or $(-\infty, 3) \cup (6, \infty)$

15. [9.2] $\left\{x \mid x < -4 \text{ or } x \ge -\frac{5}{2}\right\}$, or $(-\infty, -4) \cup \left[-\frac{5}{2}, \infty\right)$

16. [9.2] $\{x \mid -3 \le x \le 1\}$, or $[-3, 1]$

17. [9.3] $\{-15, 15\}$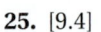
18. [9.3] $\{a \mid a < -5 \text{ or } a > 5\}$, or $(-\infty, -5) \cup (5, \infty)$

19. [9.3] $\left\{x \mid -2 < x < \frac{8}{3}\right\}$, or $\left(-2, \frac{8}{3}\right)$

20. [9.3] $\left\{t \mid t \le -\frac{13}{5} \text{ or } t \ge \frac{7}{5}\right\}$, or $\left(-\infty, -\frac{13}{5}\right] \cup \left[\frac{7}{5}, \infty\right)$

21. [9.3] $\varnothing$ **22.** [9.2] $\left\{x \mid x < \frac{1}{2} \text{ or } x > \frac{7}{2}\right\}$, or
$\left(-\infty, \frac{1}{2}\right) \cup \left(\frac{7}{2}, \infty\right)$
23. [9.3] $\left\{-\frac{3}{2}\right\}$
24. [9.4]

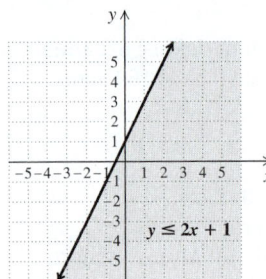

25. [9.4]

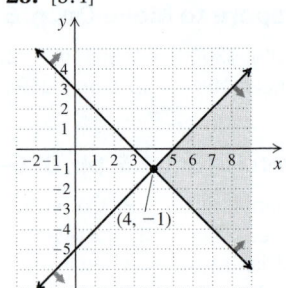

26. [9.4]

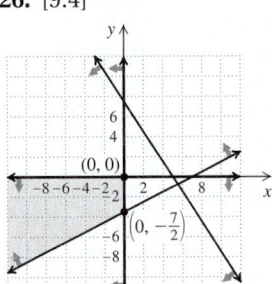

27. [9.5] Maximum 57 when
$x = 6, y = 9$; minimum
5 when $x = 1, y = 0$
28. [9.5] Manicures: 35;
haircuts: 15; maximum:
$690
29. [9.3] $[-1, 0] \cup [4, 6]$
30. [9.2] $\left(\frac{1}{5}, \frac{4}{5}\right)$
31. [9.3] $|x + 3| \le 5$

Cumulative Review: Chapters 1–9, p. 625

1. 22   **2.** $c - 6$ **3.** $-\dfrac{6x^4}{y^3}$ **4.** $\dfrac{9a^6}{4b^4}$

5. $\dfrac{2}{x^2 + 5x + 25}$ **6.** 5   **7.** $\left\{-\frac{7}{2}, \frac{9}{2}\right\}$

8. $\{t \mid t < -3 \text{ or } t > 3\}$, or $(-\infty, -3) \cup (3, \infty)$

9. $\left\{x \mid -2 \le x \le \frac{10}{3}\right\}$, or $\left[-2, \frac{10}{3}\right]$ **10.** $\left(\frac{22}{17}, -\frac{2}{17}\right)$

11. No solution   **12.** $\left\{x \mid x < \frac{13}{2}\right\}$, or $\left(-\infty, \frac{13}{2}\right)$

13. $-2, 4$ **14.** $-4, 3$
15.

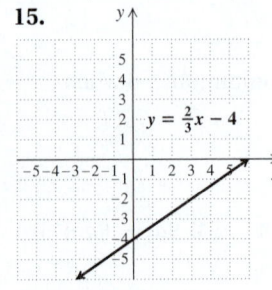

16.

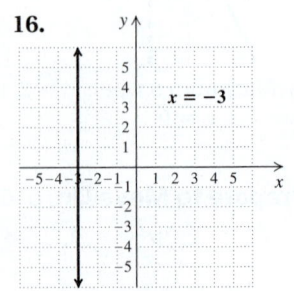

17.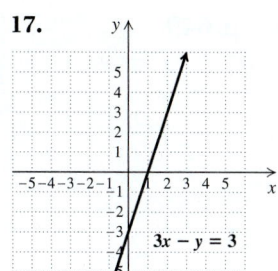

18.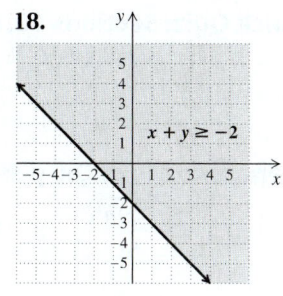

19.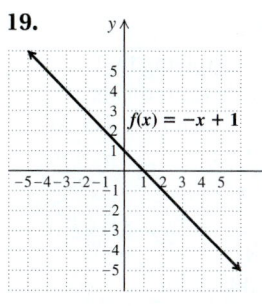

20.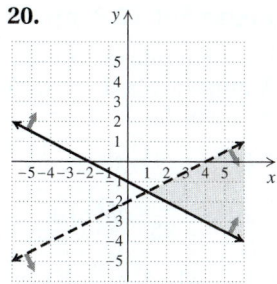

21. Slope: $\frac{4}{9}$; y-intercept: $(0, -2)$ **22.** $y = \frac{2}{3}x + 4$
23. Domain: $\mathbb{R}$; range: $\{y \mid y \geq -2\}$, or $[-2, \infty)$

24. 22   **25.** $x^2 + 6x - 9$ **26.** $t = \dfrac{c}{a - d}$

27. Beef: 9300 gal; wheat: 2300 gal
28. **(a)** $f(t) = 0.7t + 0.56$; **(b)** \$4.76 billion; **(c)** approximately 2018   **29.** \$50 billion   **30.** 183 cannons
31. $m = \frac{1}{3}, b = \frac{16}{3}$ **32.** $[-4, 0) \cup (0, \infty)$

CHAPTER 10

Technology Connection, p. 630

1. False   **2.** True   **3.** False

Exercise Set 10.1, pp. 633–636

1. Two   **2.** Negative   **3.** Positive   **4.** Negative
5. Irrational   **6.** Real   **7.** Nonnegative   **8.** Negative
9. 8, −8   **11.** 10, −10   **13.** 20, −20   **15.** 25, −25
17. 7   **19.** −4   **21.** $\frac{6}{7}$ **23.** $-\frac{4}{9}$ **25.** 0.2
27. 0.09   **29.** $\sqrt{5}$; 0; does not exist; does not exist
31. −7; does not exist; −1; does not exist
33. 1; $\sqrt{2}$; $\sqrt{101}$ **35.** $10|x|$ **37.** $|-4b|$, or $4|b|$
39. $|8 - t|$ **41.** $|y + 8|$ **43.** $|2x + 7|$ **45.** $|a^{11}|$
47. Cannot be simplified   **49.** −1   **51.** −4   **53.** $5y$
55. $p^2 + 4$; 2   **57.** $\dfrac{x}{y + 4}$; 5   **59.** −4   **61.** $\frac{2}{3}$
63. $|x|$ **65.** t **67.** $6|a|$ **69.** 6   **71.** $|a + b|$
73. $4x$ **75.** $-3t$ **77.** $5b$ **79.** $a + 1$ **81.** $2x$
83. $x - 1$ **85.** t^9 **87.** $(x - 2)^4$ **89.** 2; 3; −2; −4
91. 2; does not exist; does not exist; 3   **93.** $\{x \mid x \geq 6\}$,
or $[6, \infty)$ **95.** $\{t \mid t \geq -8\}$, or $[-8, \infty)$ **97.** $\{x \mid x \leq 5\}$,
or $(-\infty, 5]$ **99.** $\mathbb{R}$ **101.** $\{z \mid z \geq -\frac{2}{5}\}$, or
$[-\frac{2}{5}, \infty)$ **103.** $\mathbb{R}$ **105.** 🖊   **107.** 0   **108.** $\mathbb{R}$

109. $\{x \mid x \neq 0\}$, or $(-\infty, 0) \cup (0, \infty)$
110. $(f + g)(x) = 3x - 1 + \dfrac{1}{x}$ **111.** $(fg)(x) = 3 - \dfrac{1}{x}$
112. f **113.** 🖊   **115.** About 840 GPM
117. About 1404 species
119. $\{x \mid x \geq -5\}$, or $[-5, \infty)$ **121.** $\{x \mid x \geq 0\}$, or $[0, \infty)$

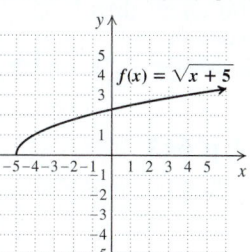   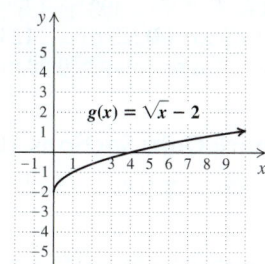

123. $\{x \mid -3 \leq x < 2\}$, or $[-3, 2)$
125. $\{x \mid x < -1 \ or \ x > 6\}$, or $(-\infty, -1) \cup (6, \infty)$ **127.** 89%

Prepare to Move On, p. 636

1. $15x^3y^9$ **2.** $\dfrac{a^3}{8b^6c^3}$ **3.** $\dfrac{x^3}{2y^6}$ **4.** $\dfrac{y^4z^8}{16x^4}$

Technology Connection, p. 638

1. Without parentheses, the expression entered would be $\dfrac{7^2}{3}$.

2. For $x = 0$ or $x = 1$, $y_1 = y_2 = y_3$; on $(0, 1)$, $y_1 > y_2 > y_3$; on $(1, \infty)$, $y_1 < y_2 < y_3$.

Technology Connection, p. 640

1. Many graphing calculators do not have keys for radicals of index 3 or higher. On those graphing calculators that offer $\sqrt[x]{\ }$ in a MATH menu, rational exponents still require fewer keystrokes.

Exercise Set 10.2, pp. 640–643

1. Radical   **2.** Subtract   **3.** Equivalent
4. Rational   **5.** (g)   **6.** (c)   **7.** (e)   **8.** (h)
9. (a)   **10.** (d)   **11.** (b)   **12.** (f)   **13.** $\sqrt[3]{y}$
15. 6   **17.** 2   **19.** 8   **21.** $\sqrt{xyz}$ **23.** $\sqrt[4]{a^2b^2}$
25. $\sqrt[6]{t^5}$ **27.** 8   **29.** 625   **31.** $27\sqrt[4]{x^3}$
33. $125x^6$ **35.** $18^{1/3}$ **37.** $30^{1/2}$ **39.** $x^{7/2}$
41. $m^{2/5}$ **43.** $(xy)^{1/4}$ **45.** $(xy^2z)^{1/5}$ **47.** $(3mn)^{3/2}$
49. $(8x^2y)^{5/7}$ **51.** $\dfrac{2x}{z^{2/3}}$ **53.** $\frac{1}{2}$ **55.** $\dfrac{1}{(2rs)^{3/4}}$
57. 8   **59.** $8a^{3/5}c$ **61.** $\dfrac{2a^{3/4}c^{2/3}}{b^{1/2}}$ **63.** $\dfrac{a^3}{3^{5/2}b^{7/3}}$
65. $\left(\dfrac{3c}{2ab}\right)^{5/6}$ **67.** $\dfrac{x}{y^{1/4}}$ **69.** $11^{5/6}$ **71.** $3^{3/4}$
73. $4.3^{1/2}$ **75.** $10^{6/25}$ **77.** $a^{23/12}$ **79.** 64   **81.** $\dfrac{m^{1/3}}{n^{1/8}}$
83. $\sqrt[3]{x}$ **85.** y^5 **87.** $\sqrt{a}$ **89.** x^2y^2 **91.** $\sqrt{7a}$

93. $\sqrt[4]{8x^3}$ **95.** $\sqrt[10]{m}$ **97.** x^3y^3 **99.** a^6b^{12}
101. $\sqrt[12]{xy}$ **103.** 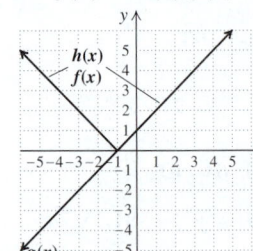   **105.** -6 **106.** $\{y\,|\,y < \frac{6}{5}\}$, or $(-\infty, \frac{6}{5})$ **107.** $\{x\,|\,-2 \le x \le \frac{3}{5}\}$, or $[-2, \frac{3}{5}]$
108. $-2, 3$ **109.** $-5, 3$ **110.** $(\frac{4}{3}, -\frac{1}{3})$ **111.**
113. $\sqrt[6]{x^5}$ **115.** $\sqrt[7]{c - d}, c \ge d$
117. $2^{7/12} \approx 1.498 \approx 1.5$ **119. (a)** 1.8 m; **(b)** 3.1 m;
(c) 1.5 m; **(d)** 5.3 m   **121.** 338 cubic feet
123. Approximately 0.99 m²   **125.**

Quick Quiz: Sections 10.1–10.2, p. 643

1. -9 **2.** $|x|$ **3.** 2 **4.** 4 **5.** 5

Prepare to Move On, p. 643

1. $x^2 - 25$ **2.** $x^3 - 8$ **3.** $(3a - 4)^2$ **4.** $3(n + 2)^2$

Technology Connection, p. 644

1. The graphs differ in appearance because the domain of y_1 is the intersection of $[-3, \infty)$ and $[3, \infty)$, or $[3, \infty)$. The domain of y_2 is $(-\infty, -3] \cup [3, \infty)$.

Exercise Set 10.3, pp. 647–649

1. True   **2.** False   **3.** False   **4.** True   **5.** True
6. True   **7.** $\sqrt{30}$ **9.** $\sqrt[3]{35}$ **11.** $\sqrt[4]{54}$
13. $\sqrt{26xy}$ **15.** $\sqrt[5]{80y^4}$ **17.** $\sqrt{y^2 - b^2}$
19. $\sqrt[3]{0.21y^2}$ **21.** $\sqrt[5]{(x - 2)^3}$ **23.** $\sqrt{\dfrac{6s}{11t}}$
25. $\sqrt[7]{\dfrac{5x - 15}{4x + 8}}$ **27.** $2\sqrt{3}$ **29.** $3\sqrt{5}$ **31.** $2x^4\sqrt{2x}$
33. $2\sqrt{30}$ **35.** $6a^2\sqrt{b}$ **37.** $2x\sqrt[3]{y^2}$ **39.** $-2x^2\sqrt[3]{2}$
41. $f(x) = 2x^2\sqrt[5]{5}$ **43.** $f(x) = |7(x - 3)|$, or $7|x - 3|$
45. $f(x) = |x - 1|\sqrt{5}$ **47.** $a^5b^5\sqrt{b}$ **49.** $xy^2z^3\sqrt[3]{x^2z}$
51. $2xy^2\sqrt[4]{xy^3}$ **53.** $x^2yz^3\sqrt[5]{x^3y^3z^2}$ **55.** $-2a^4\sqrt[3]{10a^2}$
57. $5\sqrt{2}$ **59.** $3\sqrt{22}$ **61.** 3 **63.** $24y^5$ **65.** $a\sqrt[3]{10}$
67. $24x^3\sqrt{5x}$ **69.** $s^2t^3\sqrt[4]{t}$ **71.** $(x - y)^4$
73. $2ab^3\sqrt[4]{5a}$ **75.** $x(y + z)^2\sqrt[5]{x}$ **77.**
79. $9abx^2$ **80.** $\dfrac{(x - 1)^2}{(x - 2)^2}$ **81.** $\dfrac{x + 1}{2(x + 5)}$
82. $\dfrac{3(4x^2 + 5y^3)}{50xy^2}$ **83.** $\dfrac{b + a}{a^2b^2}$ **84.** $\dfrac{-x - 2}{4x + 3}$ **85.**
87. 175.6 mi   **89. (a)** $-3.3°$C; **(b)** $-16.6°$C; **(c)** $-25.5°$C;
(d) $-54.0°$C   **91.** $25x^5\sqrt[3]{25x}$ **93.** $a^{10}b^{17}\sqrt{ab}$
95. $f(x) = h(x); f(x) \ne g(x)$

97. $\{x\,|\,x \le 2 \text{ or } x \ge 4\}$, or $(-\infty, 2] \cup [4, \infty)$ **99.** 6
101.

Quick Quiz: Sections 10.1–10.3, p. 649

1. $(5 - y)^5$ **2.** $2xy^2\sqrt[3]{5x^2}$ **3.** $\sqrt[5]{c}$ **4.** $3^{9/10}$
5. $2xy$

Prepare to Move On, p. 649

1. $41ab$ **2.** $\dfrac{15m^2}{8n^3}$ **3.** $17xy^4$ **4.** $15x^4$

Exercise Set 10.4, pp. 653–655

1. Quotient rule for radicals   **2.** Multiplying by 1
3. Multiplying by 1   **4.** Quotient rule for radicals
5. (f)   **6.** (a)   **7.** (e)   **8.** (b)   **9.** (c)   **10.** (d)
11. $\dfrac{7}{10}$ **13.** $\dfrac{5}{2}$ **15.** $\dfrac{11}{t}$ **17.** $\dfrac{6y\sqrt{y}}{x^2}$ **19.** $\dfrac{3a\sqrt[3]{a}}{2b}$
21. $\dfrac{2a}{bc^2}$ **23.** $\dfrac{ab^2}{c^2}\sqrt[4]{\dfrac{a}{c^2}}$ **25.** $\dfrac{2x}{y^2}\sqrt[5]{\dfrac{x}{y}}$ **27.** $\dfrac{xy}{z^2}\sqrt[6]{\dfrac{y^2}{z^3}}$
29. 3 **31.** $\sqrt[3]{2}$ **33.** $y\sqrt{5y}$ **35.** $2\sqrt[3]{a^2b}$
37. $\sqrt{2ab}$ **39.** $2x^2y^3\sqrt[4]{y^3}$ **41.** $\sqrt[3]{x^2 + xy + y^2}$
43. $\dfrac{\sqrt{10}}{5}$ **45.** $\dfrac{2\sqrt{15}}{21}$ **47.** $\dfrac{\sqrt[3]{10}}{2}$ **49.** $\dfrac{\sqrt[3]{75ac^2}}{5c}$
51. $\dfrac{y\sqrt[4]{45y^2x^3}}{3x}$ **53.** $\dfrac{\sqrt[3]{2xy^2}}{xy}$ **55.** $\dfrac{\sqrt{14a}}{6}$
57. $\dfrac{\sqrt[5]{9y^4}}{2xy}$ **59.** $\dfrac{\sqrt{5b}}{6a}$ **61.** $\dfrac{5}{\sqrt{55}}$ **63.** $\dfrac{12}{5\sqrt{42}}$
65. $\dfrac{2}{\sqrt{6x}}$ **67.** $\dfrac{7}{\sqrt[3]{98}}$ **69.** $\dfrac{7x}{\sqrt{21xy}}$ **71.** $\dfrac{2a^2}{\sqrt[3]{20ab}}$
73. $\dfrac{x^2y}{\sqrt{2xy}}$ **75.**   **77.** $-\dfrac{1}{3}$ **78.** $\dfrac{4}{27}$ **79.** -65
80. $9x^3 - x^2 + 9x$ **81.** $12x^2 - 12x + 6 + \dfrac{-14}{x + 1}$
82. $49m^2 - 28mn + 4n^2$ **83.**   **85. (a)** 1.62 sec;
(b) 1.99 sec; **(c)** 2.20 sec   **87.** $9\sqrt[3]{9n^2}$
89. $\dfrac{-3\sqrt{a^2 - 3}}{a^2 - 3}$, or $\dfrac{-3}{\sqrt{a^2 - 3}}$ **91.** Step 1: $\sqrt[n]{a} = a^{1/n}$,
by definition; Step 2: $\left(\dfrac{a}{b}\right)^n = \dfrac{a^n}{b^n}$, raising a quotient to a
power; Step 3: $a^{1/n} = \sqrt[n]{a}$, by definition
93. $(f/g)(x) = 3x$, where x is a real number and
$x > 0$ **95.** $(f/g)(x) = \sqrt{x + 3}$, where x is a real
number and $x > 3$

Quick Quiz: Sections 10.1–10.4, p. 655

1. $\dfrac{3}{10}$ **2.** $x^{1/3}$ **3.** $6mn^2\sqrt{10mp}$ **4.** $4x\sqrt{x}$
5. $\{x\,|\,x \ge -8\}$, or $[-8, \infty)$

Prepare to Move On, p. 655

1. $-\dfrac{1}{4}$ **2.** $\dfrac{11}{18}$ **3.** $a^2 - b^2$ **4.** $a^4 - 4y^2$
5. $15 + 4x - 4x^2$ **6.** $6x^6 + 6x^5$

Connecting the Concepts, p. 658

1. $\dfrac{6\sqrt{7}}{7}$ 2. $\dfrac{3+\sqrt{2}}{7}$ 3. $\dfrac{2\sqrt{xy}}{xy}$ 4. $\dfrac{5\sqrt{2}}{4}$

5. $\dfrac{\sqrt{10}-\sqrt{6}}{2}$ 6. $\dfrac{-1-\sqrt{5}}{2}$ 7. $\dfrac{\sqrt[3]{xy^2}}{xy}$

8. $\dfrac{\sqrt[4]{ab^2}}{b}$

Exercise Set 10.5, pp. 660–662

1. Radicands; indices   2. Indices   3. Bases
4. Denominators   5. Numerator; conjugate
6. Bases   7. $11\sqrt{3}$ 9. $2\sqrt[3]{4}$ 11. $10\sqrt[3]{y}$
13. $12\sqrt{2}$ 15. $13\sqrt[3]{7}+\sqrt{3}$ 17. $9\sqrt{3}$
19. $-7\sqrt{5}$ 21. $9\sqrt[3]{2}$ 23. $(1+12a)\sqrt{a}$
25. $(x-2)\sqrt[3]{6x}$ 27. $3\sqrt{a-1}$ 29. $(x+3)\sqrt{x-1}$
31. $5\sqrt{2}+2$ 33. $3\sqrt{30}-3\sqrt{35}$ 35. $6\sqrt{5}-4$
37. $3-4\sqrt[3]{63}$ 39. $a+2a\sqrt[3]{3}$ 41. $4+3\sqrt{6}$
43. $\sqrt{6}-\sqrt{14}+\sqrt{21}-7$ 45. 1   47. -5
49. $2-8\sqrt{35}$ 51. $23+8\sqrt{7}$ 53. $5-2\sqrt{6}$
55. $2t+5+2\sqrt{10t}$ 57. $14+x-6\sqrt{x+5}$
59. $6\sqrt[4]{63}+4\sqrt[4]{35}-3\sqrt[4]{54}-2\sqrt[4]{30}$
61. $\dfrac{18+6\sqrt{2}}{7}$ 63. $\dfrac{12-2\sqrt{3}+6\sqrt{5}-\sqrt{15}}{33}$
65. $\dfrac{a-\sqrt{ab}}{a-b}$ 67. -1 69. $\dfrac{12-3\sqrt{10}-2\sqrt{14}+\sqrt{35}}{6}$
71. $\dfrac{1}{\sqrt{5}-1}$ 73. $\dfrac{2}{14+2\sqrt{3}+3\sqrt{2}+7\sqrt{6}}$
75. $\dfrac{x-y}{x+2\sqrt{xy}+y}$ 77. $\dfrac{1}{\sqrt{a+h}+\sqrt{a}}$ 79. $\sqrt{a}$
81. $b^2\sqrt[10]{b^3}$ 83. $xy\sqrt[6]{xy^5}$ 85. $3a^2b\sqrt[4]{ab}$
87. $a^2b^2c^2\sqrt[6]{a^2bc^2}$ 89. $\sqrt[12]{a^5}$ 91. $\sqrt[12]{x^2y^5}$ 93. $\sqrt[10]{ab^9}$
95. $\sqrt[6]{(7-y)^5}$ 97. $\sqrt[12]{5+3x}$ 99. $x\sqrt[6]{xy^5}-\sqrt[15]{x^{13}y^{14}}$
101. $2m^2+m\sqrt[4]{n}+2m\sqrt[3]{n^2}+\sqrt[12]{n^{11}}$
103. $2\sqrt[4]{x^3}-\sqrt[12]{x^{11}}$ 105. x^2-7 107. $11-6\sqrt{2}$
109. $27+6\sqrt{14}$ 111. ✏   113. IV   114. 1
115. x-intercept: $(10,0)$; y-intercept: $(0,-10)$
116. Slope: $-\frac{5}{3}$; y-intercept: $(0,\frac{1}{3})$ 117. $y=-2x+12$
118. $y=-2x-8$ 119. ✏   121. $f(x)=2x\sqrt{x-1}$
123. $f(x)=(x+3x^2)\sqrt[4]{x-1}$ 125. $(7x^2-2y^2)\sqrt{x+y}$
127. $4x(y+z)^3\sqrt[6]{2x(y+z)}$ 129. $1-\sqrt{w}$
131. $(\sqrt{x}+\sqrt{5})(\sqrt{x}-\sqrt{5})$
133. $(\sqrt{x}+\sqrt{a})(\sqrt{x}-\sqrt{a})$ 135. $2x-2\sqrt{x^2-4}$
137. (a) $(A+B+\sqrt{2AB})(A+B-\sqrt{2AB})=$
$(A+B)^2-(\sqrt{2AB})^2=A^2+2AB+B^2-2AB=A^2+B^2$;
(b) $2AB$ must be a perfect square.

Quick Quiz: Sections 10.1–10.5, p. 662

1. $|10t+1|$ 2. $\sqrt[8]{(3xy)^7}$, or $(\sqrt[8]{3xy})^7$ 3. $(17ab)^{1/2}$
4. $\dfrac{\sqrt[3]{10xy^2}}{2x^2y}$ 5. $\dfrac{1}{2\sqrt{10}-\sqrt{30}}$

Prepare to Move On, p. 662

1. 42   2. $-7,3$ 3. 11   4. $3,11$

Mid-Chapter Review: Chapter 10, p. 663

1. $\sqrt{6x^9}\cdot\sqrt{2xy}=\sqrt{6x^9\cdot 2xy}$
$=\sqrt{12x^{10}y}$
$=\sqrt{4x^{10}\cdot 3y}$
$=\sqrt{4x^{10}}\cdot\sqrt{3y}$
$=2x^5\sqrt{3y}$
2. $\sqrt{12}-3\sqrt{75}+\sqrt{8}=2\sqrt{3}-3\cdot 5\sqrt{3}+2\sqrt{2}$
$=2\sqrt{3}-15\sqrt{3}+2\sqrt{2}$
$=-13\sqrt{3}+2\sqrt{2}$
3. 9   4. $-\frac{3}{10}$ 5. $|8t|$, or $8|t|$ 6. x 7. -4
8. $\{x\,|\,x\le 10\}$, or $(-\infty,10]$ 9. 4   10. $\sqrt[12]{a}$ 11. y^8
12. $t+5$ 13. $-3a^4$ 14. $3x\sqrt{10}$ 15. $\frac{2}{3}$
16. $2\sqrt{15}-3\sqrt{22}$ 17. $\sqrt[8]{t}$ 18. $\dfrac{a^2}{2}$ 19. $-8\sqrt{3}$
20. -4 21. $25+10\sqrt{6}$ 22. $2\sqrt{x-1}$
23. $xy\sqrt[10]{x^7y^3}$ 24. $15\sqrt[3]{5}$ 25. $6x^3y^2$

Technology Connection, p. 666

1. The x-coordinates of the points of intersection should approximate the solutions of the examples.

Exercise Set 10.6, pp. 668–670

1. Powers   2. Radical   3. Isolate   4. Even
5. True   6. True   7. False   8. True   9. 3
11. $\frac{16}{3}$ 13. 20   15. -1 17. 5   19. 91
21. $0,36$ 23. 100   25. -125 27. 16
29. No solution   31. $\frac{80}{3}$ 33. 45   35. $-\frac{5}{3}$ 37. 4
39. 1   41. $\frac{106}{27}$ 43. $3,7$ 45. $\frac{80}{9}$ 47. -1
49. No solution   51. $2,6$ 53. 2   55. 4
57. ✏   59. At least 84%   60. $-191{,}800$ permits per year   61. 4 mph   62. Swiss chocolate: 45 oz; whipping cream: 20 oz   63. ✏   65. About 68 psi   67. About
278 Hz   69. 524.8°C   71. $t=\dfrac{1}{9}\left(\dfrac{S^2\cdot 2457}{1087.7^2}-2617\right)$
73. $r=\dfrac{v^2h}{2gh-v^2}$ 75. $-\frac{8}{9}$ 77. $-8,8$ 79. $1,8$
81. $\left(\frac{1}{36},0\right),(36,0)$ 83. ▨

Quick Quiz: Sections 10.1–10.6, p. 670

1. $11n$ 2. $\sqrt[6]{a}$ 3. $-\sqrt{3}$ 4. $x\sqrt[10]{x}$ 5. 1

Prepare to Move On, p. 670

1. Length: 200 ft; width: 15 ft   2. Base: 34 in.; height: 15 in.   3. Length: 30 yd; width: 16 yd   4. 13 cm

Exercise Set 10.7, pp. 676–680

1. (d)   **2.** (c)   **3.** (e)   **4.** (b)   **5.** (f)   **6.** (a)
7. $\sqrt{34}$; 5.831   **9.** $9\sqrt{2}$; 12.728   **11.** 8   **13.** 4 m
15. $\sqrt{19}$ in.; 4.359 in.   **17.** 1 m   **19.** 250 ft
21. $\sqrt{643{,}600}$ ft $= 20\sqrt{1609}$ ft; 802.247 ft   **23.** 24 in.
25. $(\sqrt{340} + 8)$ ft $= (2\sqrt{85} + 8)$ ft; 26.439 ft
27. $(110 - \sqrt{6500})$ paces $= (110 - 10\sqrt{65})$ paces;
29.377 paces   **29.** Leg $= 5$; hypotenuse $= 5\sqrt{2} \approx 7.071$
31. Shorter leg $= 7$; longer leg $= 7\sqrt{3} \approx 12.124$
33. Leg $= 5\sqrt{3} \approx 8.660$; hypotenuse $= 10\sqrt{3} \approx 17.321$
35. Both legs $= \dfrac{13\sqrt{2}}{2} \approx 9.192$ **37.** Leg $= 14\sqrt{3} \approx$
24.249; hypotenuse $= 28$ **39.** $5\sqrt{3} \approx 8.660$
41. $7\sqrt{2} \approx 9.899$ **43.** $\dfrac{15\sqrt{2}}{2} \approx 10.607$
45. $\sqrt{10{,}561}$ ft ≈ 102.767 ft   **47.** $\dfrac{1089}{4}\sqrt{3}$ ft$^2 \approx$
471.551 ft^2 **49.** $(0, -4)$, $(0, 4)$ **51.** 5
53. $\sqrt{10} \approx 3.162$ **55.** $\sqrt{200} = 10\sqrt{2} \approx 14.142$
57. 17.8   **59.** $\dfrac{\sqrt{13}}{6} \approx 0.601$ **61.** $\sqrt{12} \approx 3.464$
63. $\sqrt{101} \approx 10.050$ **65.** $(3, 4)$ **67.** $\left(\frac{7}{2}, \frac{7}{2}\right)$
69. $(-1, -3)$ **71.** $(0.7, 0)$ **73.** $\left(-\frac{1}{12}, \frac{1}{24}\right)$
75. $\left(\dfrac{\sqrt{2} + \sqrt{3}}{2}, \dfrac{3}{2}\right)$ **77.** 📋

79.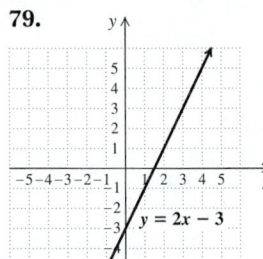
$y = 2x - 3$

80.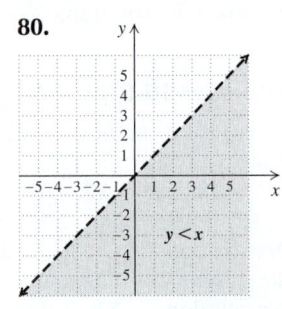
$y < x$

81.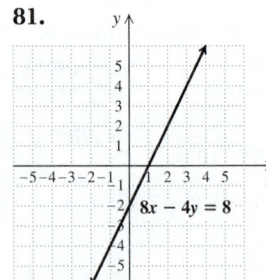
$8x - 4y = 8$

82.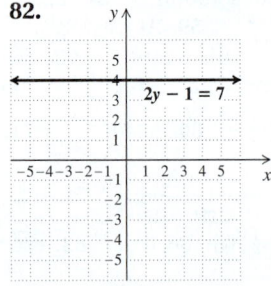
$2y - 1 = 7$

83.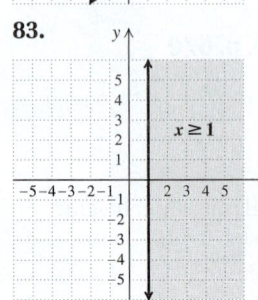
$x \geq 1$

84.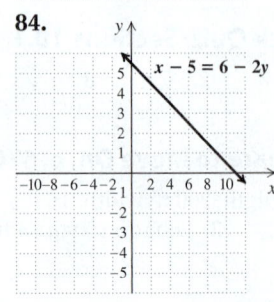
$x - 5 = 6 - 2y$

85. 📋   **87.** $36\sqrt{3}$ cm^2; 62.354 cm^2
89. (a) $d = s + s\sqrt{2}$, or $d = s(1 + \sqrt{2})$;
(b) $A = 2s^2 + 2\sqrt{2}s^2$, or $A = 2s^2(1 + \sqrt{2})$
91. 60.28 ft by 60.28 ft   **93.** $\sqrt{75}$ cm

Quick Quiz: Sections 10.1–10.7, p. 680

1. $\dfrac{3x\sqrt[3]{2x}}{5t^2}$ **2.** $x\sqrt[4]{2x}(2 - x^2)$ **3.** $30x^2\sqrt{2}$
4. $\frac{144}{25}$ **5.** $\sqrt{85} \approx 9.220$

Prepare to Move On, p. 680

1. $2 + \sqrt{3}$ **2.** $\sqrt{7} - 6$ **3.** $9x^2 - 4y^2$
4. $25w^2 - 20wx + 4x^2$ **5.** $-8ac + 28c^2$
6. $24a^2 - 14ap - 5p^2$

Exercise Set 10.8, pp. 686–687

1. False   **2.** False   **3.** True   **4.** True   **5.** True
6. True   **7.** False   **8.** True   **9.** $10i$ **11.** $i\sqrt{5}$, or
$\sqrt{5}i$ **13.** $2i\sqrt{2}$, or $2\sqrt{2}i$ **15.** $-i\sqrt{11}$, or $-\sqrt{11}i$
17. $-7i$ **19.** $-10i\sqrt{3}$, or $-10\sqrt{3}i$ **21.** $6 - 2i\sqrt{21}$, or
$6 - 2\sqrt{21}i$ **23.** $(-2\sqrt{19} + 5\sqrt{5})i$ **25.** $(3\sqrt{2} - 8)i$
27. $5 - 3i$ **29.** $7 + 2i$ **31.** $2 - i$ **33.** $-12 - 5i$
35. -40 **37.** -24 **39.** -18 **41.** $-\sqrt{30}$
43. $-3\sqrt{14}$ **45.** $-30 + 10i$ **47.** $28 - 21i$
49. $1 + 5i$ **51.** $38 + 9i$ **53.** $2 - 46i$ **55.** 73
57. 50   **59.** $12 - 16i$ **61.** $-5 + 12i$ **63.** $-5 - 12i$
65. $3 - i$ **67.** $\frac{6}{13} + \frac{4}{13}i$ **69.** $\frac{3}{17} + \frac{5}{17}i$ **71.** $-\frac{5}{6}i$
73. $-\frac{3}{4} - \frac{5}{4}i$ **75.** $1 - 2i$ **77.** $-\frac{23}{58} + \frac{43}{58}i$
79. $\frac{19}{29} - \frac{4}{29}i$ **81.** $\frac{6}{25} - \frac{17}{25}i$ **83.** 1   **85.** $-i$
87. -1 **89.** i **91.** -1 **93.** $-125i$ **95.** 0
97. 📋   **99.** $(x + 10)(x - 10)$
100. $(t + 10)(t^2 - 10t + 100)$ **101.** $(x + 9)(x - 7)$
102. $a(4a - 3)(3a + 1)$ **103.** $(w + 2)(w - 2)(w + 3)$
104. $12x^2y^2(2x - 5y^2 - 1)$ **105.** 📋
107.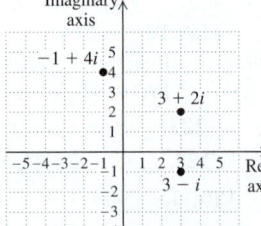
109. 5   **111.** $\sqrt{2}$
113. $-9 - 27i$
115. $50 - 120i$
117. $\frac{250}{41} + \frac{200}{41}i$ **119.** 8
121. $\frac{3}{5} + \frac{9}{5}i$ **123.** 1

Quick Quiz: Sections 10.1–10.8, p. 687

1. 3   **2.** $\left\{x \mid x \geq \frac{1}{2}\right\}$, or $\left[\frac{1}{2}, \infty\right)$ **3.** 25
4. $5 - 15\sqrt{6} - \sqrt{2} + 6\sqrt{3}$ **5.** $17 + i$

Prepare to Move On, p. 687

1. $-2, 3$ **2.** 5   **3.** $-5, 5$ **4.** $-\frac{2}{5}, \frac{4}{3}$

Visualizing for Success, p. 688

1. B **2.** H **3.** C **4.** I **5.** D **6.** A
7. F **8.** J **9.** G **10.** E

Decision Making: Connection, p. 689

1. Approximately 47.4 mi **2.** 600 ft
3. (a) Approximately 12.2 mi; **(b)** approximately 24.4 mi;
(c) approximately 32.5 mi

Study Summary: Chapter 10, pp. 690–692

1. -9 **2.** -1 **3.** $|6x|$, or $6|x|$ **4.** x **5.** $\frac{1}{10}$
6. $\sqrt{21xy}$ **7.** $10x^2y^9\sqrt{2x}$ **8.** $\frac{2x\sqrt{3x}}{5}$ **9.** $\frac{\sqrt{6xy}}{3y}$
10. $-5\sqrt{2}$ **11.** $31 - 19\sqrt{3}$ **12.** $\frac{3\sqrt{15} - 5\sqrt{3}}{4}$
13. $x^2\sqrt[6]{x}$ **14.** 3 **15.** $\sqrt{51}$ m ≈ 7.141 m
16. $a = 6$ **17.** $b = 5\sqrt{3} \approx 8.660; c = 10$
18. $\sqrt{185} \approx 13.601$ **19.** $\left(2, -\frac{9}{2}\right)$ **20.** $-3 - 12i$
21. $3 - 2i$ **22.** $-32 - 26i$ **23.** i **24.** $-i$

Review Exercises: Chapter 10, pp. 693–694

1. True **2.** False **3.** False **4.** True **5.** True
6. True **7.** True **8.** False **9.** $\frac{10}{11}$ **10.** -0.6
11. 5 **12.** $\{x|x \geq -10\}$, or $[-10, \infty)$ **13.** $8|t|$
14. $|c + 7|$ **15.** $|2x + 1|$ **16.** -2 **17.** $(5ab)^{4/3}$
18. $\sqrt[5]{3a^4}$ **19.** x^3y^5 **20.** $\sqrt[3]{x^2y}$ **21.** $\frac{1}{x^{2/5}}$
22. $7^{1/6}$ **23.** $f(x) = 5|x - 6|$ **24.** $2x^5y^2$
25. $5xy\sqrt{10x}$ **26.** $\sqrt{35ab}$ **27.** $3xb\sqrt[3]{x^2}$
28. $-6x^5y^4\sqrt[3]{2x^2}$ **29.** $y\sqrt[3]{6}$ **30.** $\frac{5\sqrt{x}}{2}$ **31.** $\frac{2a^2\sqrt[4]{3a^3}}{c^2}$
32. $7\sqrt[3]{4y}$ **33.** $\sqrt{3}$ **34.** $15\sqrt{2}$ **35.** -1
36. $\sqrt{15} + 4\sqrt{6} - 6\sqrt{10} - 48$ **37.** $\sqrt[4]{x^3}$ **38.** $\sqrt[12]{x^5}$
39. $4 - 4\sqrt{a} + a$ **40.** $\frac{\sqrt{2xy}}{4y}$ **41.** $-4\sqrt{10} + 4\sqrt{15}$
42. $\frac{20}{\sqrt{10} + \sqrt{15}}$ **43.** 19 **44.** -126 **45.** 4
46. 2 **47.** $5\sqrt{2}$ cm; 7.071 cm **48.** $\sqrt{32}$ ft; 5.657 ft
49. Short leg $= 10$; long leg $= 10\sqrt{3} \approx 17.321$
50. $\sqrt{26} \approx 5.099$ **51.** $\left(-2, -\frac{3}{2}\right)$ **52.** $3i\sqrt{5}$, or $3\sqrt{5}i$
53. $-2 - 9i$ **54.** $6 + i$ **55.** 29 **56.** -1
57. $9 - 12i$ **58.** $\frac{13}{25} - \frac{34}{25}i$ **59.** ✍ A complex number
$a + bi$ is real when $b = 0$. It is imaginary when $b \neq 0$.
60. ✍ An absolute-value sign must be used to simplify
$\sqrt[n]{x^n}$ when n is even, since x may be negative. If x is negative while n is even, the radical expression cannot be
simplified to x, since $\sqrt[n]{x^n}$ represents the principal, or nonnegative, root. When n is odd, there is only one root, and
it will be positive or negative depending on the sign of x.
Thus there is no absolute-value sign when n is odd.
61. $\frac{2i}{3i}$; answers may vary **62.** 3 **63.** $-\frac{2}{5} + \frac{9}{10}i$
64. The isosceles right triangle is larger by about 1.206 ft².

Test: Chapter 10, p. 695

1. [10.3] $5\sqrt{2}$ **2.** [10.4] $-\frac{2}{x^2}$ **3.** [10.1] $9|a|$
4. [10.1] $|x - 4|$ **5.** [10.2] $(7xy)^{1/2}$ **6.** [10.2] $\sqrt[6]{(4a^3b)^5}$
7. [10.1] $\{x|x \geq 5\}$, or $[5, \infty)$ **8.** [10.5] $27 + 10\sqrt{2}$
9. [10.3] $2x^3y^2\sqrt[5]{x}$ **10.** [10.3] $2\sqrt[3]{2wv^2}$ **11.** [10.4] $\frac{10a^2}{3b^3}$
12. [10.4] $\sqrt[5]{3x^4y}$ **13.** [10.5] $x\sqrt[4]{x}$ **14.** [10.5] $\sqrt[5]{y^2}$
15. [10.5] $6\sqrt{2}$ **16.** [10.5] $9\sqrt{2xy}$
17. [10.5] $14 - 19\sqrt{x} - 3x$ **18.** [10.4] $\frac{\sqrt[3]{2xy^2}}{2y}$
19. [10.6] 4 **20.** [10.6] $-1, 2$ **21.** [10.6] 8
22. [10.7] $\sqrt{10,600}$ ft ≈ 102.956 ft **23.** [10.7] 5 cm;
$5\sqrt{3}$ cm ≈ 8.660 cm **24.** [10.7] $\sqrt{17} \approx 4.123$
25. [10.7] $\left(\frac{3}{2}, -6\right)$ **26.** [10.8] $5i\sqrt{2}$, or $5\sqrt{2}i$
27. [10.8] $12 + 2i$ **28.** [10.8] $15 - 8i$
29. [10.8] $-\frac{11}{34} - \frac{7}{34}i$ **30.** [10.8] i **31.** [10.6] 3
32. [10.8] $-\frac{17}{4}i$ **33.** [10.6] 22,500 ft

Cumulative Review: Chapters 1–10, p. 696

1. $-7, 5$ **2.** $\frac{5}{2}$ **3.** -1 **4.** $\frac{1}{5}$
5. $\{x|-3 \leq x \leq 7\}$, or $[-3, 7]$ **6.** $\mathbb{R}$, or $(-\infty, \infty)$
7. $-3, 2$ **8.** 7 **9.** $(1, -1, -2)$
10.
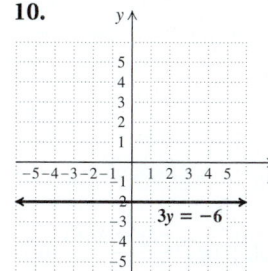
11.
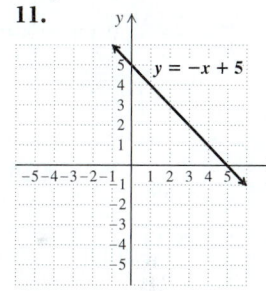
12.
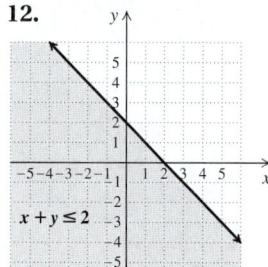
13.
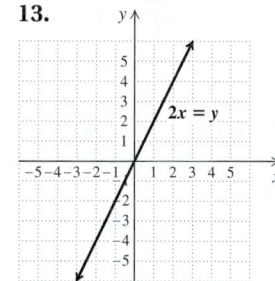
14. $y = 7x - 11$ **15.** 6 **16.** $4a^2 - 20ab + 25b^2$
17. $c^4 - 9d^2$ **18.** $\frac{x + 13}{(x - 2)(x + 1)}$ **19.** $\frac{(a + 2)(2a + 1)}{(a - 3)(a - 1)}$
20. $\frac{2x + 1}{x^2}$ **21.** 0 **22.** $-1 + 3\sqrt{5}$ **23.** $\sqrt[15]{y^8}$
24. $(x - 7)(x + 2)$ **25.** $4y^5(y - 1)(y^2 + y + 1)$
26. $(3t - 8)(t + 1)$ **27.** $(y - x)(t - z^2)$
28. $\{x|x$ is a real number $and\ x \neq 3\}$, or $(-\infty, 3) \cup (3, \infty)$
29. $\{x|x \geq \frac{11}{2}\}$, or $[\frac{11}{2}, \infty)$ **30.** $6 - 2\sqrt{5}$
31. $(f + g)(x) = x^2 + \sqrt{2x - 3}$
32. $2\sqrt{3}$ ft ≈ 3.464 ft **33. (a)** $m(t) = 0.08t + 25.02$;
(b) 26.62; **(c)** 2037 **34.** \$36,650 **35.** 5 ft
36. $5x - 3y = -15$ **37.** -2 **38.** 9

CHAPTER 11

Technology Connection, p. 705

1. x-intercepts should be approximations of $-5/2 + \sqrt{37}/2$ and $-5/2 - \sqrt{37}/2$.
2. The graph of $y = x^2 - 6x + 11$ has no x-intercepts.

Exercise Set 11.1, pp. 706–708

1. Quadratic function **2.** Parabola **3.** Standard form
4. Zero products **5.** Square roots **6.** Complete the
square **7.** ± 10 **9.** $\pm 5\sqrt{2}$ **11.** $\pm\sqrt{6}$
13. $\pm\frac{7}{3}$ **15.** $\pm\sqrt{\frac{5}{6}}$, or $\pm\frac{\sqrt{30}}{6}$ **17.** $\pm i$ **19.** $\pm\frac{9}{2}i$
21. $-1, 7$ **23.** $-5 \pm 2\sqrt{3}$ **25.** $-1 \pm 3i$
27. $-\frac{3}{4} \pm \frac{\sqrt{17}}{4}$, or $\frac{-3 \pm \sqrt{17}}{4}$ **29.** $-3, 13$ **31.** $\pm\sqrt{19}$
33. $1, 9$ **35.** $-4 \pm \sqrt{13}$ **37.** $-14, 0$
39. $x^2 + 16x + 64 = (x + 8)^2$
41. $t^2 - 10t + 25 = (t - 5)^2$
43. $t^2 - 2t + 1 = (t - 1)^2$ **45.** $x^2 + 3x + \frac{9}{4} = \left(x + \frac{3}{2}\right)^2$
47. $x^2 + \frac{2}{5}x + \frac{1}{25} = \left(x + \frac{1}{5}\right)^2$
49. $t^2 - \frac{5}{6}t + \frac{25}{144} = \left(t - \frac{5}{12}\right)^2$ **51.** $-7, 1$ **53.** $5 \pm \sqrt{2}$
55. $-8, -4$ **57.** $-4 \pm \sqrt{19}$
59. $(-3 - \sqrt{2}, 0), (-3 + \sqrt{2}, 0)$
61. $\left(-\frac{9}{2} - \frac{\sqrt{181}}{2}, 0\right), \left(-\frac{9}{2} + \frac{\sqrt{181}}{2}, 0\right)$, or
$\left(\frac{-9 - \sqrt{181}}{2}, 0\right), \left(\frac{-9 + \sqrt{181}}{2}, 0\right)$
63. $(5 - \sqrt{47}, 0), (5 + \sqrt{47}, 0)$ **65.** $-\frac{4}{3}, -\frac{2}{3}$
67. $-\frac{1}{3}, 2$ **69.** $-\frac{2}{5} \pm \frac{\sqrt{19}}{5}$, or $\frac{-2 \pm \sqrt{19}}{5}$
71. $\left(-\frac{1}{4} - \frac{\sqrt{13}}{4}, 0\right), \left(-\frac{1}{4} + \frac{\sqrt{13}}{4}, 0\right)$, or $\left(\frac{-1 - \sqrt{13}}{4}, 0\right)$,
$\left(\frac{-1 + \sqrt{13}}{4}, 0\right)$ **73.** $\left(\frac{3}{4} - \frac{\sqrt{17}}{4}, 0\right), \left(\frac{3}{4} + \frac{\sqrt{17}}{4}, 0\right)$,
or $\left(\frac{3 - \sqrt{17}}{4}, 0\right), \left(\frac{3 + \sqrt{17}}{4}, 0\right)$ **75.** 5% **77.** 4%
79. About 2.8 sec **81.** About 15.0 sec **83.** ☑️
85. $3y(y + 10)(y - 10)$ **86.** $(t + 6)^2$
87. $6(x^2 + x + 1)$ **88.** $10a^3(a + 2)(a - 3)$
89. $(4x + 3)(5x - 2)$
90. $(n + 1)(n^2 - n + 1)(n - 1)(n^2 + n + 1)$
91. ☑️ **93.** ± 18 **95.** $-\frac{7}{2}, -\sqrt{5}, 0, \sqrt{5}, 8$
97. Barge: 8 km/h; fishing boat: 15 km/h **99.** 📉
101. ☑️, 📉

Prepare to Move On, p. 708

1. 64 **2.** -15 **3.** $10\sqrt{2}$ **4.** $2i$
5. $2i\sqrt{2}$, or $2\sqrt{2}i$

Connecting the Concepts, p. 712

1. $-2, 5$ **2.** ± 11 **3.** $-3 \pm \sqrt{19}$ **4.** $-\frac{1}{2} \pm \frac{\sqrt{13}}{2}$
5. $-1 \pm \sqrt{2}$ **6.** 5 **7.** $1 \pm \sqrt{7}$ **8.** $\pm\frac{\sqrt{11}}{2}$

Exercise Set 11.2, pp. 713–714

1. True **2.** True **3.** False **4.** False **5.** False
6. True **7.** $-\frac{5}{2}, 1$ **9.** $-1 \pm \sqrt{5}$ **11.** $3 \pm \sqrt{6}$
13. $\frac{3}{2} \pm \frac{\sqrt{29}}{2}$ **15.** $-1 \pm \frac{2\sqrt{3}}{3}$ **17.** $-\frac{4}{3} \pm \frac{\sqrt{19}}{3}$
19. $3 \pm i$ **21.** $\frac{1}{2} \pm \frac{\sqrt{3}}{2}i$ **23.** $-2 \pm \sqrt{2}i$ **25.** $-\frac{8}{3}, \frac{5}{4}$
27. $\frac{2}{5}$ **29.** $-\frac{11}{8} \pm \frac{\sqrt{41}}{8}$ **31.** $5, 10$ **33.** $\frac{3}{2}, 24$
35. $2 \pm \sqrt{5}i$ **37.** $2, -1 \pm \sqrt{3}i$ **39.** $-\frac{4}{3}, \frac{5}{2}$
41. $5 \pm \sqrt{53}$ **43.** $\frac{3}{2} \pm \frac{\sqrt{5}}{2}$ **45.** $-5.236, -0.764$
47. $0.764, 5.236$ **49.** $-1.266, 2.766$ **51.** ☑️
53. 1 **54.** 1000 **55.** $x^{11/12}$ **56.** $\frac{1}{9}$ **57.** $\frac{3a^{10}c^7}{4}$
58. $\frac{9w^8}{4x^{10}}$ **59.** ☑️ **61.** $(-2, 0), (1, 0)$
63. $4 - 2\sqrt{2}, 4 + 2\sqrt{2}$ **65.** $-1.179, 0.339$
67. $\frac{-5\sqrt{2} \pm \sqrt{34}}{4}$ **69.** $\frac{1}{2}$ **71.** 📉

Quick Quiz: Sections 11.1–11.2, p. 714

1. $-3, \frac{5}{2}$ **2.** $2 \pm \sqrt{3}$ **3.** $3 \pm \sqrt{5}$ **4.** $\frac{3}{2} \pm \frac{\sqrt{13}}{2}$
5. $-\frac{1}{4} \pm \frac{\sqrt{39}}{4}i$

Prepare to Move On, p. 714

1. $x^2 + 4$ **2.** $x^2 - 180$ **3.** $x^2 - 4x - 3$
4. $x^2 + 6x + 34$

Exercise Set 11.3, pp. 718–719

1. (b) **2.** (a) **3.** (d) **4.** (b) **5.** (c) **6.** (c)
7. Two irrational **9.** Two imaginary **11.** Two irrational
13. Two rational **15.** Two imaginary **17.** One rational
19. Two rational **21.** Two irrational
23. Two imaginary **25.** Two rational
27. Two irrational **29.** $x^2 + x - 20 = 0$
31. $x^2 - 6x + 9 = 0$ **33.** $x^2 + 4x + 3 = 0$
35. $4x^2 - 23x + 15 = 0$ **37.** $8x^2 + 6x + 1 = 0$
39. $x^2 - 2x - 0.96 = 0$ **41.** $x^2 - 3 = 0$
43. $x^2 - 20 = 0$ **45.** $x^2 + 16 = 0$
47. $x^2 - 4x + 53 = 0$ **49.** $x^2 - 6x - 5 = 0$
51. $3x^2 - 6x - 4 = 0$ **53.** $x^3 - 4x^2 - 7x + 10 = 0$

55. $x^3 - 2x^2 - 3x = 0$ **57.** 📋   **59.** $3a^3b^6\sqrt{30a}$
60. $2w^2\sqrt[4]{2w^2}$ **61.** $\sqrt[6]{x^5}$ **62.** $-\sqrt{6}$ **63.** $7 - i$
64. -1 **65.** 📋   **67.** $a = 1, b = 2, c = -3$
69. (a) $-\frac{3}{5}$; **(b)** $-\frac{1}{3}$ **71. (a)** $9 + 9i$; **(b)** $3 + 3i$
73. The solutions of $ax^2 + bx + c = 0$ are

$x = \dfrac{-b \pm \sqrt{b^2 - 4ac}}{2a}$. When there is just one solution,

$b^2 - 4ac$ must be 0, so $x = \dfrac{-b \pm 0}{2a} = \dfrac{-b}{2a}$.

75. $a = 8, b = 20, c = -12$ **77.** $x^2 - 2 = 0$
79. $x^4 - 8x^3 + 21x^2 - 2x - 52 = 0$ **81.** 📋, 📉

Quick Quiz: Sections 11.1–11.3, p. 719

1. $-8 \pm \sqrt{3}$ **2.** $\dfrac{1}{6} \pm \dfrac{\sqrt{13}}{6}$ **3.** $(-2\sqrt{2}, 0), (2\sqrt{2}, 0)$
4. $-2.414, 0.414$ **5.** Two irrational solutions

Prepare to Move On, p. 719

1. $c = \dfrac{d^2}{1 - d}$ **2.** $y = \dfrac{x - 3}{x}$, or $1 - \dfrac{3}{x}$
3. 10 mph   **4.** Homer: 3.5 mph; Gladys: 2 mph

Exercise Set 11.4, pp. 723–726

1. (c)   **2.** (d)   **3.** (a)   **4.** (b)
5. $r = \dfrac{1}{2}\sqrt{\dfrac{A}{\pi}}$ **7.** $r = \dfrac{-\pi h + \sqrt{\pi^2 h^2 + 2\pi A}}{2\pi}$
9. $r = \dfrac{\sqrt{Gm_1 m_2}}{F}$ **11.** $H = \dfrac{c^2}{g}$
13. $b = \sqrt{c^2 - a^2}$ **15.** $t = \dfrac{-v_0 + \sqrt{(v_0)^2 + 2gs}}{g}$
17. $n = \dfrac{1 + \sqrt{1 + 8N}}{2}$ **19.** $d = \dfrac{I^2 s}{T^2}$
21. $t = \dfrac{-b \pm \sqrt{b^2 - 4ac}}{2a}$ **23. (a)** 10.1 sec; **(b)** 7.49 sec;
(c) 272.5 m   **25.** 2.9 sec   **27.** 0.872 sec   **29.** 2.5 m/sec
31. 4.5%   **33.** First part: 60 mph; second part: 50 mph
35. 40 mph   **37.** Cessna: 150 mph, Beechcraft: 200 mph;
or Cessna: 200 mph, Beechcraft: 250 mph
39. To Hillsboro: 12 mph; return trip: 9 mph
41. About 14 mph   **43.** 12 hr   **45.** About 3.24 mph
47. About 11 days   **49.** 📋   **51.** 45   **52.** $-\frac{1}{8}$
53. $-\frac{4}{3}, 0$ **54.** $\left(\frac{11}{5}, \frac{2}{5}\right)$ **55.** $\frac{233}{2}$ **56.** $\frac{1}{16}$ **57.** 📋
59. $t = \dfrac{-10.2 + 6\sqrt{-A^2 + 13A - 39.36}}{A - 6.5}$ **61.** $\pm\sqrt{2}$
63. $l = \dfrac{w + w\sqrt{5}}{2}$
65. $n = \pm\sqrt{\dfrac{r^2 \pm \sqrt{r^4 + 4m^4 r^2 p - 4mp}}{2m}}$
67. $A(S) = \dfrac{\pi S}{6}$

Quick Quiz: Sections 11.1–11.4, p. 726

1. $10 \pm \sqrt{115}$ **2.** $1 \pm \sqrt{26}$ **3.** 2%
4. $x^2 - 12 = 0$ **5.** $d = -1 \pm \sqrt{1 + n}$

Prepare to Move On, p. 726

1. m^{-2}, or $\dfrac{1}{m^2}$ **2.** $y^{1/3}$ **3.** 2   **4.** 81

Exercise Set 11.5, pp. 731–732

1. True   **2.** True   **3.** True   **4.** True   **5.** $\sqrt{p}$
6. $x^{1/4}$ **7.** $x^2 + 3$ **8.** t^{-3} **9.** $(1 + t)^2$ **10.** $w^{1/6}$
11. $\pm 2, \pm 3$ **13.** $\pm\sqrt{3}, \pm 2$ **15.** $\pm\dfrac{\sqrt{5}}{2}, \pm 1$
17. 4   **19.** $\pm 2\sqrt{2}, \pm 3$ **21.** $\pm 2, \pm 3i$ **23.** $\pm i, \pm 2i$
25. $8 + 2\sqrt{7}$ **27.** No solution   **29.** $-\frac{1}{2}, \frac{1}{3}$
31. $-4, 1$ **33.** $-27, 8$ **35.** 729   **37.** 1
39. No solution   **41.** $\frac{12}{5}$ **43.** $\left(\frac{4}{25}, 0\right)$
45. $\left(\dfrac{3}{2} + \dfrac{\sqrt{33}}{2}, 0\right), \left(\dfrac{3}{2} - \dfrac{\sqrt{33}}{2}, 0\right), (4, 0), (-1, 0)$
47. $(-243, 0), (32, 0)$ **49.** No x-intercepts   **51.** 📋

53.
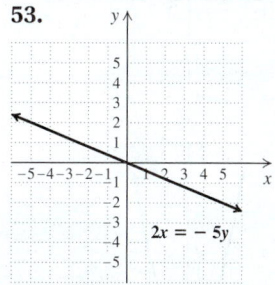
$2x = -5y$

54.
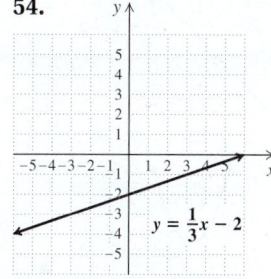
$y = \frac{1}{3}x - 2$

55.
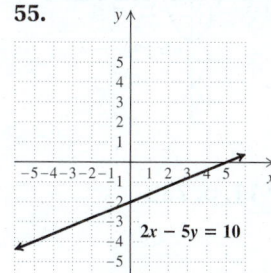
$2x - 5y = 10$

56.
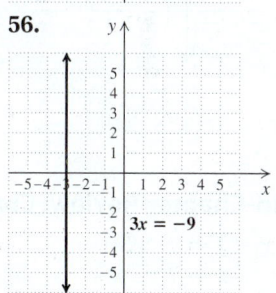
$3x = -9$

57.
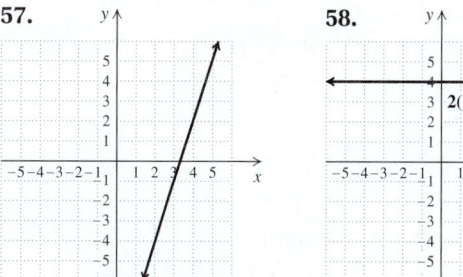

58.
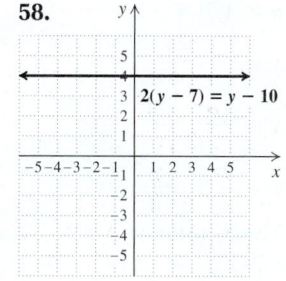
$2(y - 7) = y - 10$

59. 📋   **61.** $\pm\sqrt{\dfrac{-5 \pm \sqrt{37}}{6}}$ **63.** $\frac{100}{99}$
65. 9   **67.** $-2, 1, 1 + \sqrt{3}i, 1 - \sqrt{3}i, -\dfrac{1}{2} + \dfrac{\sqrt{3}}{2}i,$
$-\dfrac{1}{2} - \dfrac{\sqrt{3}}{2}i$ **69.** $-3, -1, 1, 4$

Quick Quiz: Sections 11.1–11.5, p. 732

1. $\frac{1}{2}, 1$ **2.** $-\frac{1}{2} \pm \frac{\sqrt{3}}{2}i$ **3.** $-3, -1, 1, 3$

4. $\frac{7}{4} \pm \frac{\sqrt{33}}{4}$ **5.** $x = \frac{9c^2}{80y}$

Prepare to Move On, p. 732

1. 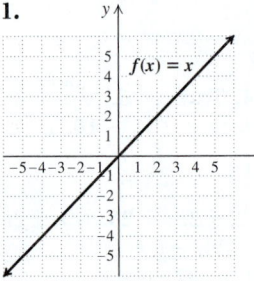   **2.**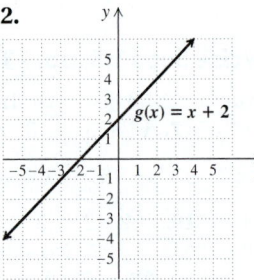

3. 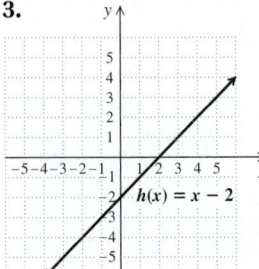   **4.**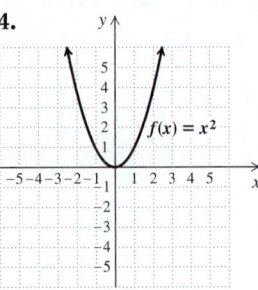

5. 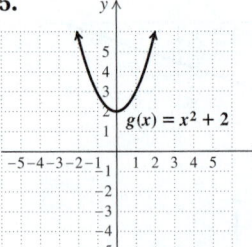   **6.**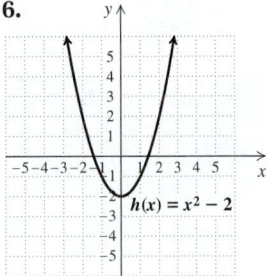

Mid-Chapter Review: Chapter 11, p. 733

1. $x - 7 = \pm\sqrt{5}$
  $x = 7 \pm \sqrt{5}$
  The solutions are $7 + \sqrt{5}$ and $7 - \sqrt{5}$.

2. $a = 1, b = -2, c = -1$
$$x = \frac{-(-2) \pm \sqrt{(-2)^2 - 4 \cdot 1 \cdot (-1)}}{2 \cdot 1}$$
$$x = \frac{2 \pm \sqrt{8}}{2}$$
$$x = \frac{2}{2} \pm \frac{2\sqrt{2}}{2}$$
  The solutions are $1 + \sqrt{2}$ and $1 - \sqrt{2}$.

3. $-7, 3$ **4.** ± 14 **5.** $1 \pm \sqrt{6}$ **6.** $1 \pm 2i$
7. $\pm 2, \pm 2i$ **8.** $-3 \pm \sqrt{7}$ **9.** $-5, 0$ **10.** $-\frac{5}{6}, 2$
11. $0, \frac{7}{16}$ **12.** $-\frac{5}{6} \pm \frac{\sqrt{37}}{6}$ **13.** $-6, 5$ **14.** $\pm\sqrt{2}, \pm 2i$
15. Two irrational   **16.** Two rational   **17.** Two imaginary
18. $v = 20\sqrt{\frac{F}{A}}$, or $\frac{20\sqrt{FA}}{A}$ **19.** $D = d + \sqrt{d^2 + 2hd}$
20. South: 75 mph; north: 45 mph

Technology Connection, p. 735

1. The graphs of y_1, y_2, and y_3 open upward. The graphs of y_4, y_5, and y_6 open downward. The graph of y_1 is wider than the graph of y_2. The graph of y_3 is narrower than the graph of y_2. Similarly, the graph of y_4 is wider than the graph of y_5, and the graph of y_6 is narrower than the graph of y_5.
2. If A is positive, the graph opens upward. If A is negative, the graph opens downward. Compared with the graph of $y = x^2$, the graph of $y = Ax^2$ is wider if $|A| < 1$ and narrower if $|A| > 1$.

Technology Connection, p. 737

1. Compared with the graph of $y = ax^2$, the graph of $y = a(x - h)^2$ is shifted left or right. It is shifted left if h is negative and right if h is positive.   **2.** The value of A makes the graph wider or narrower, and makes the graph open downward if A is negative. The value of B shifts the graph left or right.

Technology Connection, p. 738

1. The graph of y_2 looks like the graph of y_1 shifted up 2 units, and the graph of y_3 looks like the graph of y_1 shifted down 4 units.   **2.** Compared with the graph of $y = a(x - h)^2$, the graph of $y = a(x - h)^2 + k$ is shifted up or down $|k|$ units. It is shifted down if k is negative and up if k is positive.   **3.** The value of A makes the graph wider or narrower, and makes the graph open downward if A is negative. The value of B shifts the graph left or right. The value of C shifts the graph up or down.

Exercise Set 11.6, pp. 740–742

1. False   **2.** True   **3.** True   **4.** False
5. 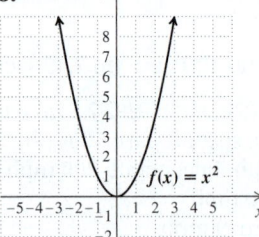   **7.**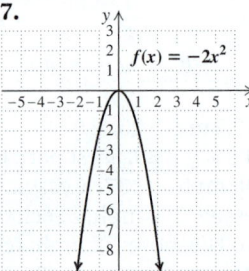
9. 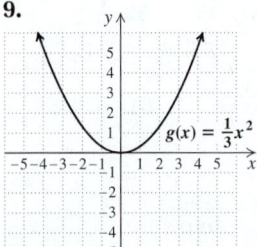   **11.**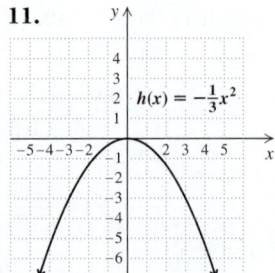

13.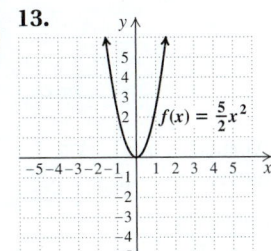

15. Vertex: $(-1, 0)$; axis of symmetry: $x = -1$
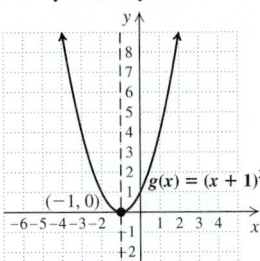

33. Vertex: $\left(\frac{1}{2}, 0\right)$; axis of symmetry: $x = \frac{1}{2}$
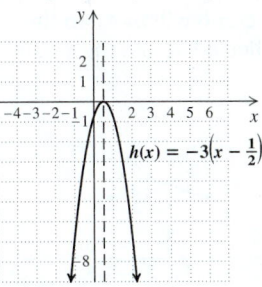

35. Vertex: $(5, 2)$; axis of symmetry: $x = 5$; minimum: 2
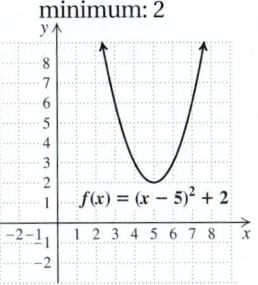

17. Vertex $(2, 0)$; axis of symmetry: $x = 2$
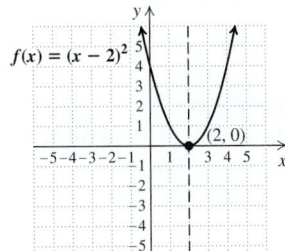

19. Vertex: $(-1, 0)$; axis of symmetry: $x = -1$
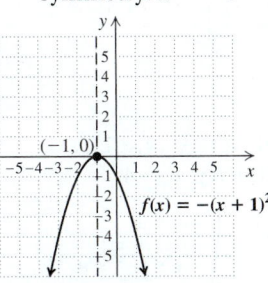

37. Vertex: $(-1, -3)$; axis of symmetry: $x = -1$; minimum: -3
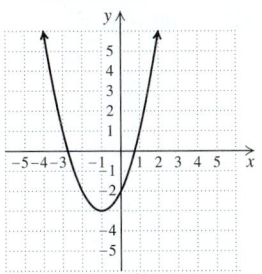

$f(x) = (x + 1)^2 - 3$

39. Vertex: $(-4, 1)$; axis of symmetry: $x = -4$; minimum: 1
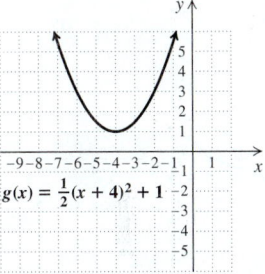

21. Vertex: $(2, 0)$; axis of symmetry: $x = 2$
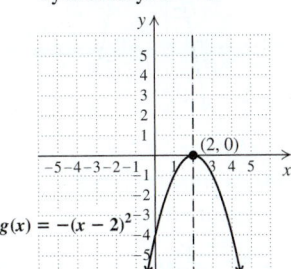

23. Vertex: $(-1, 0)$; axis of symmetry: $x = -1$
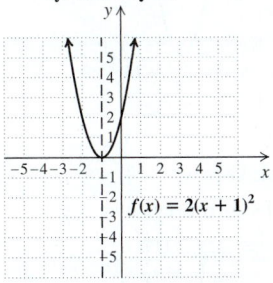

41. Vertex: $(1, -3)$; axis of symmetry: $x = 1$; maximum: -3
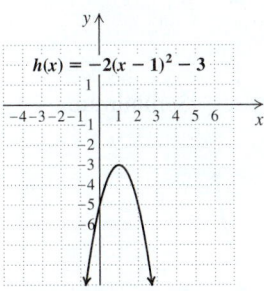

43. Vertex: $(-3, 1)$; axis of symmetry: $x = -3$; minimum: 1
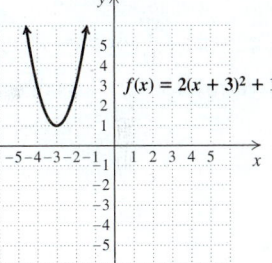

25. Vertex: $(4, 0)$; axis of symmetry: $x = 4$
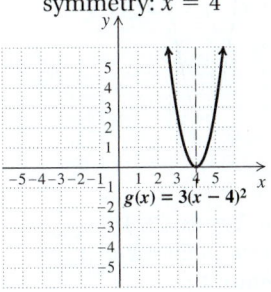

27. Vertex: $(4, 0)$; axis of symmetry: $x = 4$
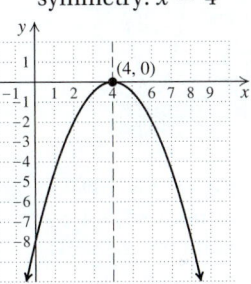

45. Vertex: $(2, 4)$; axis of symmetry: $x = 2$; maximum: 4
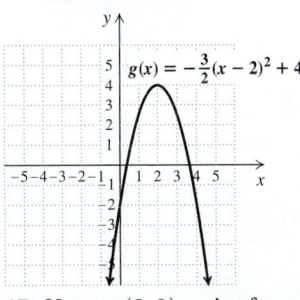

29. Vertex: $(1, 0)$; axis of symmetry: $x = 1$
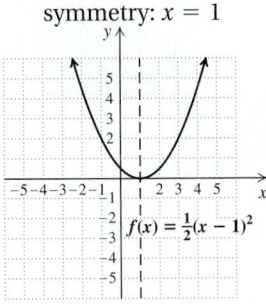

31. Vertex: $(-5, 0)$; axis of symmetry: $x = -5$
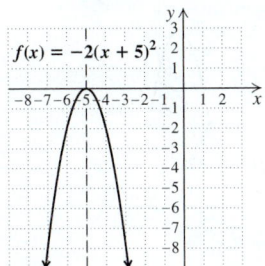

47. Vertex: $(3, 9)$; axis of symmetry: $x = 3$; minimum: 9
49. Vertex: $(-8, 2)$; axis of symmetry: $x = -8$; maximum: 2
51. Vertex: $\left(\frac{7}{2}, -\frac{29}{4}\right)$; axis of symmetry: $x = \frac{7}{2}$; minimum: $-\frac{29}{4}$
53. Vertex: $(-2.25, -\pi)$; axis of symmetry: $x = -2.25$; maximum: $-\pi$ **55.** 🗒   **57.** $\dfrac{x^2 + 3x + 6}{x(x + 2)}$
58. $8\sqrt{2x}$ **59.** $-\sqrt[3]{t} + 5\sqrt{t}$ **60.** $-7a^2 + 3a - 8$
61. $\dfrac{-x^2 + 4x + 1}{(x - 1)(x + 3)}$ **62.** $\dfrac{-3}{x + 4}$ **63.** 🗒

65. $f(x) = \frac{3}{5}(x-1)^2 + 3$ **67.** $f(x) = \frac{3}{5}(x-4)^2 - 7$
69. $f(x) = \frac{3}{5}(x+2)^2 - 5$ **71.** $f(x) = 2(x-2)^2$
73. $g(x) = -2x^2 - 5$ **75.** The graph will move to the
right.   **77.** The graph will be reflected across the x-axis.
79. $F(x) = 3(x-5)^2 + 1$

81.
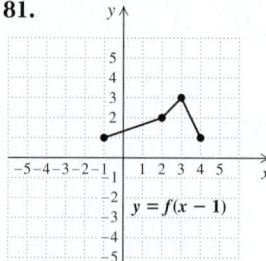
$y = f(x-1)$

83.
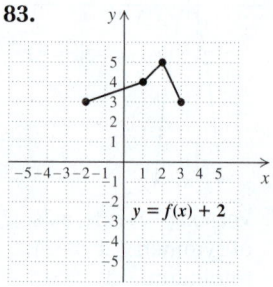
$y = f(x) + 2$

85.
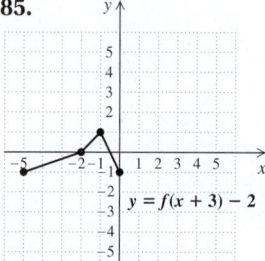
$y = f(x+3) - 2$

87.   **89.**

Quick Quiz: Sections 11.1–11.6, p. 742

1. $-7 \pm \sqrt{13}$ **2.** $\frac{3}{2} \pm \frac{\sqrt{3}}{2}i$ **3.** $-\frac{2}{5}, \frac{1}{3}$

4. $z = \pm\sqrt{\frac{xy}{3t}}$, or $z = \pm\frac{\sqrt{3txy}}{3t}$

5. Vertex: $(3, -2)$; axis of symmetry: $x = 3$; minimum: -2

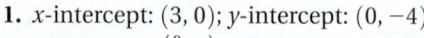
$f(x) = (x-3)^2 - 2$

Prepare to Move On, p. 742

1. x-intercept: $(3, 0)$; y-intercept: $(0, -4)$
2. x-intercept: $\left(\frac{8}{3}, 0\right)$; y-intercept: $(0, 2)$
3. $(-5, 0), (-3, 0)$ **4.** $(-1, 0), \left(\frac{3}{2}, 0\right)$
5. $x^2 - 14x + 49 = (x-7)^2$
6. $x^2 + 7x + \frac{49}{4} = \left(x + \frac{7}{2}\right)^2$

Exercise Set 11.7, pp. 747–749

1. True   **2.** False   **3.** True   **4.** True   **5.** False
6. True   **7.** False   **8.** True
9. $f(x) = (x-4)^2 + (-14)$

11. $f(x) = \left(x - \left(-\frac{3}{2}\right)\right)^2 + \left(-\frac{29}{4}\right)$
13. $f(x) = 3(x - (-1))^2 + (-5)$
15. $f(x) = -(x - (-2))^2 + (-3)$
17. $f(x) = 2\left(x - \frac{5}{4}\right)^2 + \frac{55}{8}$
19. (a) Vertex: $(-2, 1)$; axis of symmetry: $x = -2$;
(b)
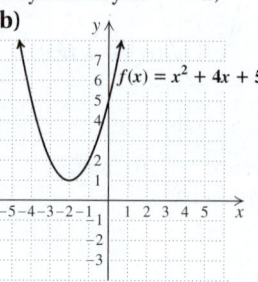
$f(x) = x^2 + 4x + 5$

21. (a) Vertex: $(-4, 4)$; axis of symmetry: $x = -4$;
(b)
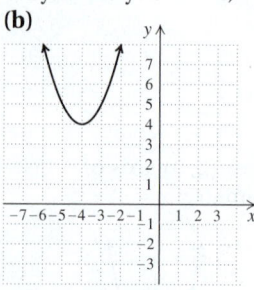
$f(x) = x^2 + 8x + 20$

23. (a) Vertex: $(4, -7)$; axis of symmetry: $x = 4$;
(b)
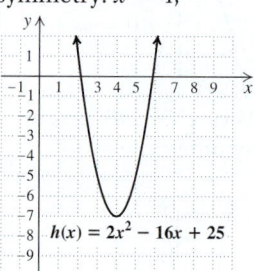
$h(x) = 2x^2 - 16x + 25$

25. (a) Vertex: $(1, 6)$; axis of symmetry: $x = 1$;
(b)
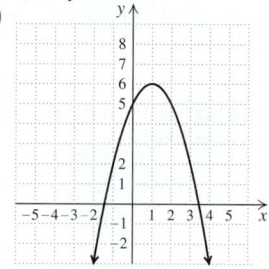
$f(x) = -x^2 + 2x + 5$

27. (a) Vertex: $\left(-\frac{3}{2}, -\frac{49}{4}\right)$; axis of symmetry: $x = -\frac{3}{2}$;
(b)
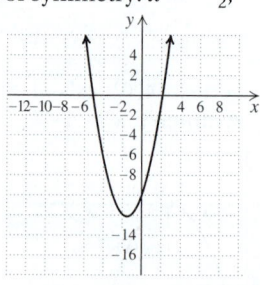
$g(x) = x^2 + 3x - 10$

29. (a) Vertex: $\left(-\frac{7}{2}, -\frac{49}{4}\right)$; axis of symmetry: $x = -\frac{7}{2}$;
(b)
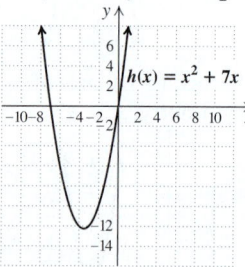
$h(x) = x^2 + 7x$

31. (a) Vertex: $(-1, -4)$; axis of symmetry: $x = -1$;
(b)
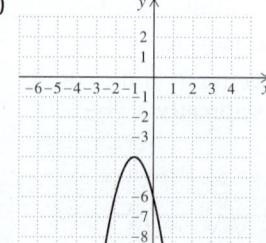
$f(x) = -2x^2 - 4x - 6$

33. (a) Vertex: $(3, 4)$; axis of symmetry: $x = 3$; minimum: 4;
(b)
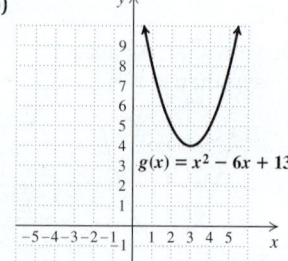
$g(x) = x^2 - 6x + 13$

18.
20.

35. (a) Vertex: $(2, -5)$; axis of symmetry: $x = 2$; minimum: -5;
(b)

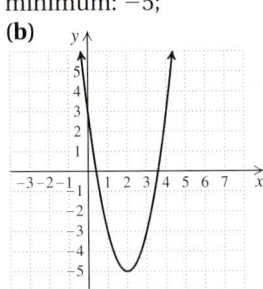

$g(x) = 2x^2 - 8x + 3$

37. (a) Vertex: $(4, 2)$; axis of symmetry: $x = 4$; minimum: 2;
(b)

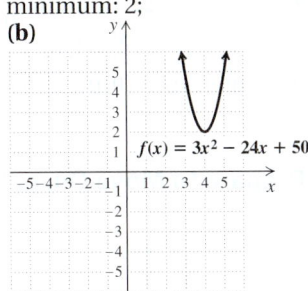

$f(x) = 3x^2 - 24x + 50$

39. (a) Vertex: $\left(\frac{5}{6}, \frac{1}{12}\right)$; axis of symmetry: $x = \frac{5}{6}$; maximum: $\frac{1}{12}$;
(b)

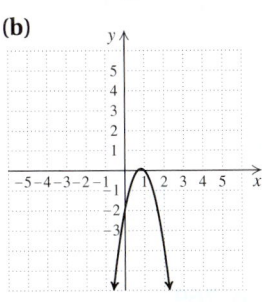

$f(x) = -3x^2 + 5x - 2$

41. (a) Vertex: $\left(-4, -\frac{5}{3}\right)$; axis of symmetry: $x = -4$; minimum: $-\frac{5}{3}$;
(b)

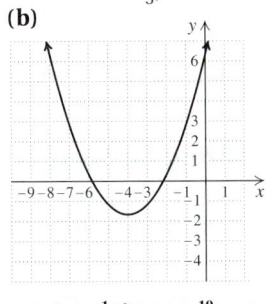

$h(x) = \frac{1}{2}x^2 + 4x + \frac{19}{3}$

43. $(3 - \sqrt{6}, 0), (3 + \sqrt{6}, 0); (0, 3)$ **45.** $(-1, 0), (3, 0)$; $(0, 3)$ **47.** $(0, 0), (9, 0); (0,0)$ **49.** $(2, 0); (0, -4)$

51. $\left(-\frac{1}{2} - \frac{\sqrt{21}}{2}, 0\right), \left(-\frac{1}{2} + \frac{\sqrt{21}}{2}, 0\right); (0, -5)$

53. No x-intercept; $(0, 6)$ **55.** ☑ **57.** $x^4 - 4x^2 - 21$

58. $\dfrac{(x - 2)(x + 4)}{x(x - 3)}$ **59.** $3x^2\sqrt[3]{4y^2}$

60. $2x^2 + 2x + 1 + \dfrac{-2}{x - 1}$ **61.** $\dfrac{3a(2a - b)}{b(a - b)}$ **62.** $\dfrac{1}{\sqrt[12]{x^7}}$

63. ☑ **65. (a)** Minimum: -6.953660714;
(b) $(-1.056433682, 0), (2.413576539, 0); (0, -5.89)$
67. (a) $-2.4, 3.4$; **(b)** $-1.3, 2.3$
69. $f(x) = m\left(x - \dfrac{n}{2m}\right)^2 + \dfrac{4mp - n^2}{4m}$
71. $f(x) = \frac{5}{16}x^2 - \frac{15}{8}x - \frac{35}{16}$, or $f(x) = \frac{5}{16}(x - 3)^2 - 5$
73.

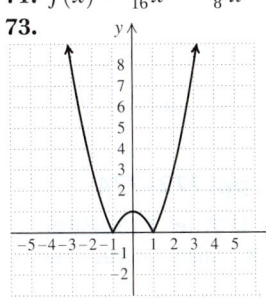

$f(x) = |x^2 - 1|$

75.

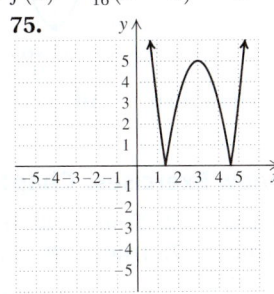

$f(x) = |2(x - 3)^2 - 5|$

4.
6.

Quick Quiz: Sections 11.1–11.7, p. 749

1. $\pm\dfrac{\sqrt{15}}{3}$ **2.** $-\dfrac{3}{4} \pm \dfrac{\sqrt{57}}{4}$ **3.** $5x^2 + 3x - 2 = 0$

4.

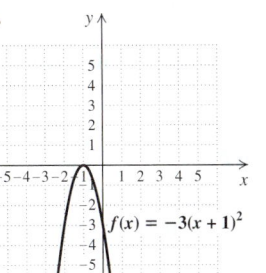

$f(x) = -3(x + 1)^2$

5.

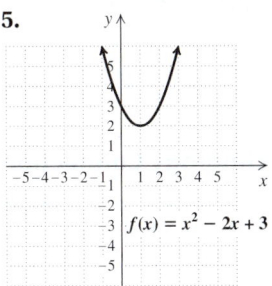

$f(x) = x^2 - 2x + 3$

Prepare to Move On, p. 749

1. $(-2, 5, 1)$ **2.** $(-3, 6, -5)$ **3.** $\left(\frac{1}{3}, \frac{1}{6}, \frac{1}{2}\right)$

Exercise Set 11.8, pp. 753–758

1. (e) **2.** (b) **3.** (c) **4.** (a) **5.** (d) **6.** (f)
7. $3\frac{1}{4}$ lb weeks; 8.3 lb of milk per day **9.** \$120/dulcimer; 350 dulcimers **11.** 180 ft by 180 ft **13.** 450 ft²; 15 ft by 30 ft (The house serves as a 30-ft side.) **15.** 3.5 in.
17. 81; 9 and 9 **19.** -16; 4 and -4 **21.** 25; -5 and -5
23. $f(x) = ax^2 + bx + c, a < 0$ **25.** $f(x) = mx + b$
27. Neither quadratic nor linear
29. $f(x) = ax^2 + bx + c, a > 0$ **31.** $f(x) = mx + b$
33. $f(x) = mx + b$ **35.** $f(x) = 2x^2 + 3x - 1$
37. $f(x) = -\frac{1}{4}x^2 + 3x - 5$
39. (a) $A(s) = \frac{3}{16}s^2 - \frac{135}{4}s + 1750$; **(b)** about 531 accidents for every 200 million km driven
41. $h(d) = -0.0068d^2 + 0.8571d$ **43.** ☑
45. $y = -\frac{1}{3}x + 16$ **46.** $y = 2x + 13$
47. $y = -\frac{4}{3}x + \frac{40}{3}$ **48.** $y = \frac{2}{3}x - \frac{23}{3}$ **49.** $y = \frac{1}{2}x - 6$
50. $y = -4$ **51.** ☑ **53.** 158 ft **55.** \$15
57. The radius of the circular portion of the window and the height of the rectangular portion should each be $\dfrac{24}{\pi + 4}$ ft.
59. (a) $a(x) = 18.78035714x^2 + 35.54107143x + 165.1107143$; **(b)** 1,651,382 subscriptions **61.** 🖥

Quick Quiz: Sections 11.1–11.8, p. 758

1. $-\frac{5}{4}, \frac{2}{3}$ **2.** 2, 7 **3.** $x^2 + 25 = 0$

4. $h = \dfrac{V^2}{12.25}$ **5.**

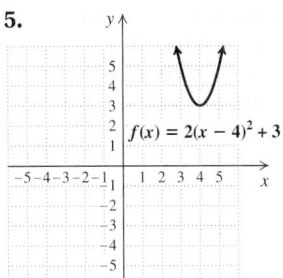

$f(x) = 2(x - 4)^2 + 3$

27.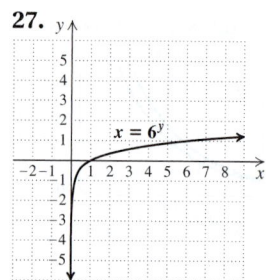

29.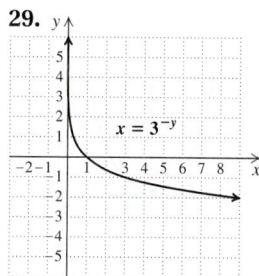

31.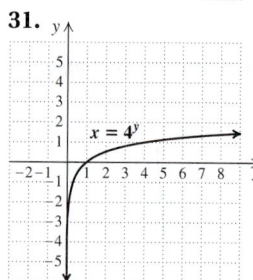

33.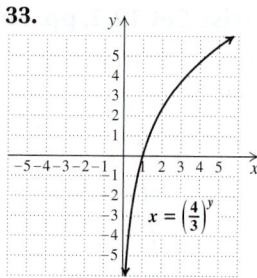

35.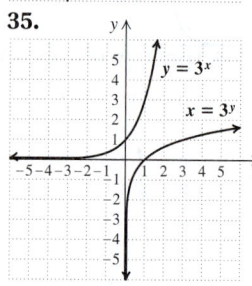

37.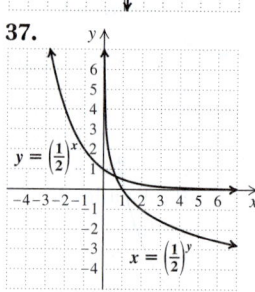

39. (a) \$3.2 billion; \$4.8 billion; \$7.3 billion;

(b)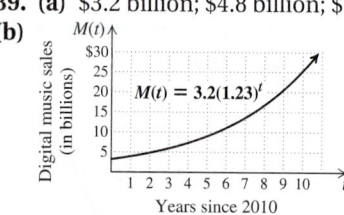

41. (a) 19.6%; 16.3%; 7.3%;

(b)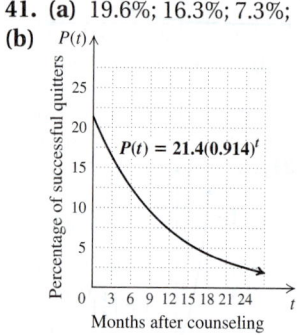

43. (a) About 44,079 whales; about 12,953 whales;

(b)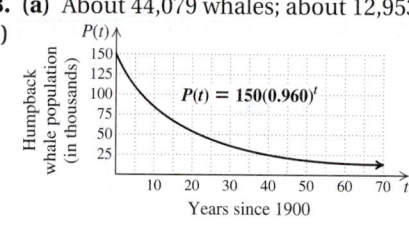

45. (a) About 8706 whales; about 15,107 whales;

(b)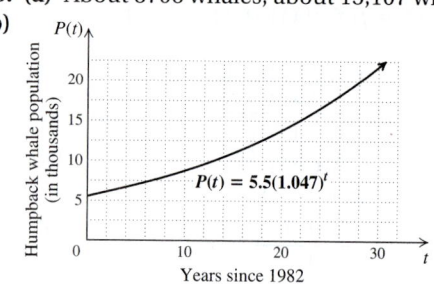

47. (a) About 50 moose; about 1421 moose; about 8470 moose; **(b)**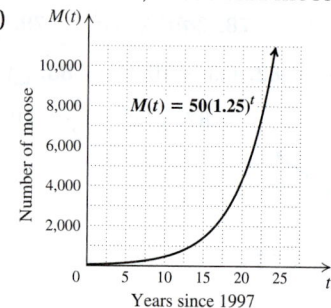

49. 📋   **51.** $3(x + 4)(x - 4)$ **52.** $(x - 10)^2$

53. $(2x + 3)(3x - 4)$

54. $8(x^2 - 2y^2)(x^4 + 2x^2y^2 + 4y^4)$

55. $(t - y + 1)(t + y - 1)$ **56.** $x(x - 2)(5x^2 - 3)$

57. 📋   **59.** $\pi^{2.4}$

61.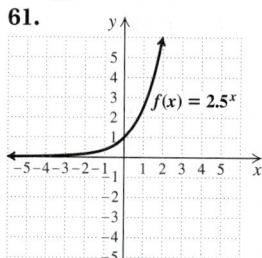

63.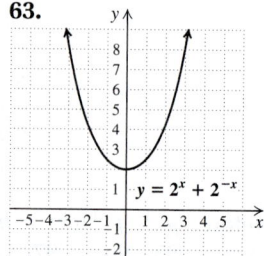

65.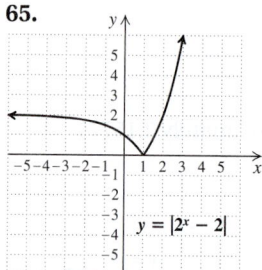

67.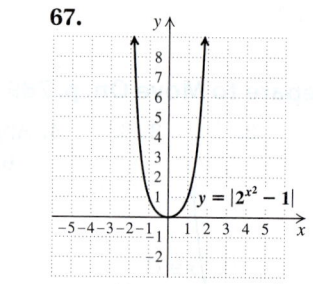

69.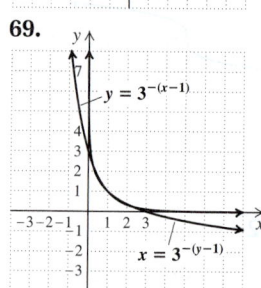

71. $N(t) = 136(1.85)^t$; about 5550 ruffe

73. 📋   **75.**

Quick Quiz: Sections 12.1–12.2, p. 797

1. $(f \circ g)(x) = -3x^2 + x + 9$

2. $f(x) = x^4; g(x) = x - 6$ **3.** $g^{-1}(x) = \dfrac{x-5}{2}$

4.

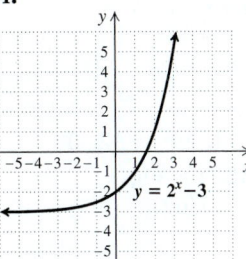

$y = 2^x - 3$

5.

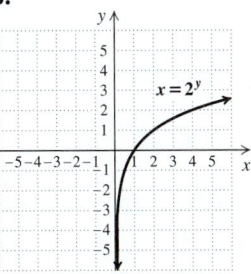

$x = 2^y$

Prepare to Move On, p. 797

1.

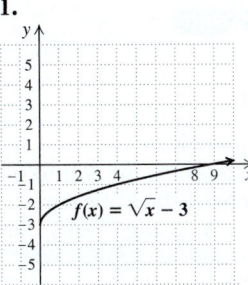

$f(x) = \sqrt{x} - 3$

2.

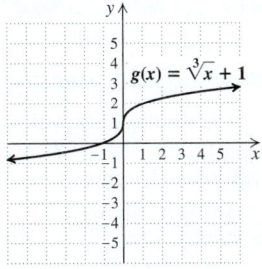

$g(x) = \sqrt[3]{x} + 1$

3.

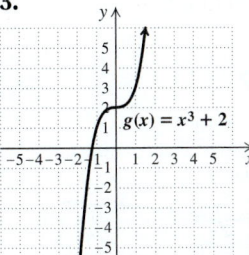

$g(x) = x^3 + 2$

4.

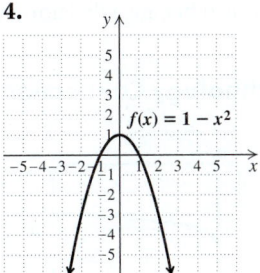

$f(x) = 1 - x^2$

Exercise Set 12.3, pp. 802–804

1. Logarithmic   **2.** Exponent   **3.** Positive   **4.** 1
5. (g)   **6.** (d)   **7.** (a)   **8.** (h)   **9.** (b)   **10.** (c)
11. (e)   **12.** (f)   **13.** 3   **15.** 2   **17.** -2 **19.** -1
21. 4   **23.** 1   **25.** 0   **27.** 5   **29.** -2 **31.** $\frac{1}{2}$
33. $\frac{3}{2}$ **35.** $\frac{2}{3}$ **37.** 29
39.

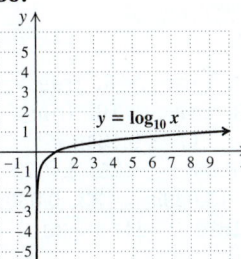

$y = \log_{10} x$

41.

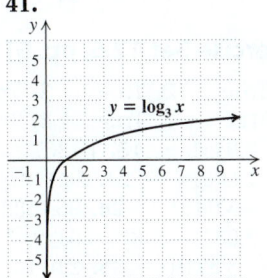

$y = \log_3 x$

43.

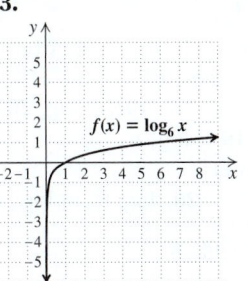

$f(x) = \log_6 x$

45.

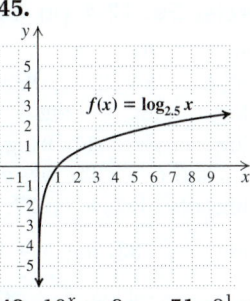

$f(x) = \log_{2.5} x$

47.

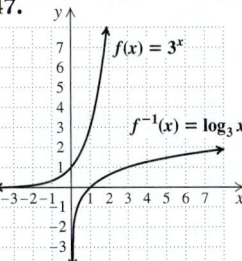

$f(x) = 3^x$

$f^{-1}(x) = \log_3 x$

49. $10^x = 8$ **51.** $9^1 = 9$
53. $10^{-1} = 0.1$
55. $10^{0.845} = 7$
57. $c^8 = m$ **59.** $r^t = C$
61. $e^{-1.3863} = 0.25$
63. $r^{-x} = T$
65. $2 = \log_{10} 100$
67. $-3 = \log_5 \frac{1}{125}$
69. $\frac{1}{4} = \log_{16} 2$
71. $0.4771 = \log_{10} 3$
73. $m = \log_z 6$

75. $t = \log_p q$ **77.** $3 = \log_e 20.0855$ **79.** 36
81. 5   **83.** 9   **85.** 49   **87.** $\frac{1}{9}$ **89.** 4   **91.** 📋
93. $30a^2b^4$ **94.** $12 - 2\sqrt{30} + 2\sqrt{15} - 5\sqrt{2}$
95. $3\sqrt{3x}$ **96.** $\sqrt[12]{x}$ **97.** $2y^2\sqrt[3]{y}$ **98.** $\sqrt[10]{a^3 y^7}$
99. 📋

101.

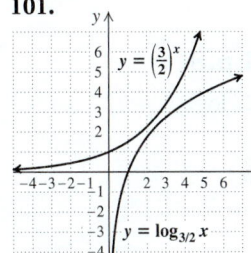

$y = \left(\frac{3}{2}\right)^x$

$y = \log_{3/2} x$

103.

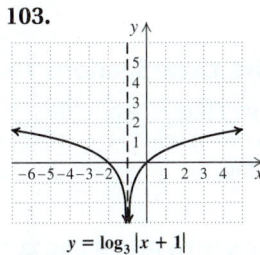

$y = \log_3 |x + 1|$

105. 6   **107.** $-25, 4$ **109.** -2 **111.** 0
113. Let $b = 0$, and suppose that $x_1 = 1$ and $x_2 = 2$. Then $0^1 = 0^2$, but $1 \neq 2$. Then let $b = 1$, and suppose that $x_1 = 1$ and $x_2 = 2$. Then $1^1 = 1^2$, but $1 \neq 2$.

Quick Quiz: Sections 12.1–12.3, p. 804

1. No   **2.**

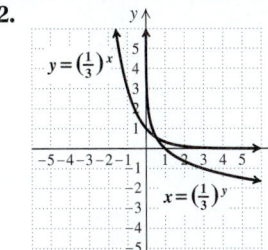

$y = \left(\frac{1}{3}\right)^x$

$x = \left(\frac{1}{3}\right)^y$

3. 4   **4.** $3^x = t$
5. $\frac{1}{16}$

Prepare to Move On, p. 804

1. c^{16} **2.** x^{30} **3.** a^{12} **4.** $3^{1/2}$ **5.** $t^{2/3}$

Exercise Set 12.4, pp. 809–811

1. (e) **2.** (f) **3.** (a) **4.** (b) **5.** (c) **6.** (d)
7. $\log_3 81 + \log_3 27$ **9.** $\log_4 64 + \log_4 16$
11. $\log_c r + \log_c s + \log_c t$ **13.** $\log_a (2 \cdot 10)$, or $\log_a 20$
15. $\log_c (t \cdot y)$ **17.** $8 \log_a r$ **19.** $\frac{1}{3} \log_2 y$
21. $-3 \log_b C$ **23.** $\log_2 5 - \log_2 11$
25. $\log_b m - \log_b n$ **27.** $\log_a \frac{19}{2}$ **29.** $\log_b \frac{36}{4}$, or $\log_b 9$
31. $\log_a \dfrac{x}{y}$ **33.** $\log_a x + \log_a y + \log_a z$
35. $3 \log_a x + 4 \log_a z$ **37.** $2 \log_a w - 2 \log_a x + \log_a y$
39. $5 \log_a x - 3 \log_a y - \log_a z$
41. $\log_b x + 2 \log_b y - \log_b w - 3 \log_b z$
43. $\frac{1}{2}(7 \log_a x - 5 \log_a y - 8 \log_a z)$
45. $\frac{1}{3}(6 \log_a x + 3 \log_a y - 2 - 7 \log_a z)$ **47.** $\log_a 3x$
49. $\log_a (x^8 z^3)$ **51.** $\log_b \dfrac{w^2}{zy^4}$ **53.** $\log_a x$
55. $\log_a \dfrac{y^5}{x^{3/2}}$ **57.** $\log_a (x - 3)$ **59.** 1.953
61. -0.369 **63.** -1.161 **65.** $\frac{3}{2}$ **67.** Cannot be
found **69.** 10 **71.** m **73.**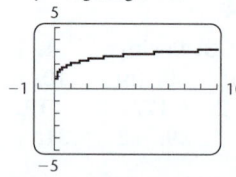
75. $\left\{ x \mid -\frac{11}{3} \le x \le -1 \right\}$, or $\left[-\frac{11}{3}, -1 \right]$ **76.** $-\frac{3}{4}, \frac{5}{6}$
77. $-2 \pm i$ **78.** $\frac{1}{2}$ **79.** 16, 256 **80.** $\frac{1}{4}$, 9
81. ✏️ **83.** $\log_a (x^6 - x^4 y^2 + x^2 y^4 - y^6)$
85. $\frac{1}{2} \log_a (1 - s) + \frac{1}{2} \log_a (1 + s)$ **87.** $\frac{10}{3}$
89. -2 **91.** $\frac{2}{5}$ **93.** True

Quick Quiz: Sections 12.1–12.4, p. 811

1. $(g \circ f)(x) = 2x^2 - 20x + 50$ **2.** $\log_m 5 = 10$
3. 5 **4.** $2 \log_a x + 3 \log_a y - \log_a z$
5. $\frac{1}{2} \log_a x + \frac{1}{2} \log_a y + \log_a z$

Prepare to Move On, p. 811

1. $(-\infty, -7) \cup (-7, \infty)$, or
$\{x \mid x$ is a real number $and\, x \ne -7\}$
2. $(-\infty, -3) \cup (-3, 2) \cup (2, \infty)$, or
$\{x \mid x$ is a real number $and\, x \ne -3$ $and\, x \ne 2\}$
3. $(-\infty, 10]$, or $\{x \mid x \le 10\}$ **4.** $(-\infty, \infty)$, or $\mathbb{R}$

Mid-Chapter Review: Chapter 12, p. 812

1.
$$y = 2x - 5$$
$$x = 2y - 5$$
$$x + 5 = 2y$$
$$\frac{x + 5}{2} = y$$
$$f^{-1}(x) = \frac{x + 5}{2}$$
2. $\log_4 x = 1$
$$4^1 = x$$
$$4 = x$$
3. $x^2 - 10x + 26$ **4.** $f(x) = \sqrt{x}; g(x) = 5x - 3$

5. $g^{-1}(x) = 6 - x$ **6.**

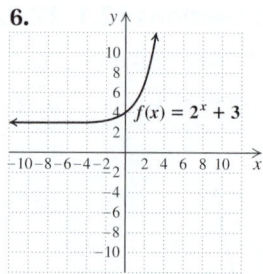

7. 2 **8.** -1 **9.** $\frac{1}{2}$ **10.** 1 **11.** 19 **12.** 0
13. $x^m = 3$ **14.** $2^{10} = 1024$ **15.** $t = \log_e x$
16. $\frac{2}{3} = \log_{64} 16$ **17.** $\log x - \frac{1}{2} \log y - \frac{3}{2} \log z$
18. $\log \dfrac{a}{b^2 c}$ **19.** 4 **20.** $\frac{1}{3}$

Technology Connection, p. 813

1. (LOG) (7) ( ) (÷) (LOG) (3) ( ) (ENTER)

Technology Connection, p. 814

1. As x gets larger, the value of y_1 approaches
2.7182818284.... **2.** For large values of x, the graphs of
y_1 and y_2 will be very close or appear to be the same curve,
depending on the window chosen. **3.** Using (TRACE),
no y-value is given for $x = 0$. Using a table, an error message appears for y_1 when $x = 0$. The domain does not
include 0 because division by 0 is undefined.

Technology Connection, p. 817

1. $y = \log x / \log 7$ **2.** $y = \log (x+2)/\log 5$

3. $y = \log x / \log 7 + 2$

Exercise Set 12.5, pp. 818–819

1. True **2.** True **3.** True **4.** False **5.** True
6. True **7.** True **8.** True **9.** True **10.** True
11. 0.8451 **13.** 1.1367 **15.** 3 **17.** -0.1249
19. 13.0014 **21.** 50.1187 **23.** 0.0011 **25.** 2.1972
27. -5.0832 **29.** 96.7583 **31.** 15.0293 **33.** 0.0305
35. 3.0331 **37.** 6.6439 **39.** 1.1610 **41.** -0.3010
43. -3.3219 **45.** 2.0115

47. Domain: $\mathbb{R}$; range: $(0, \infty)$

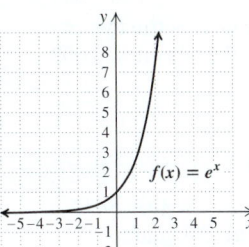

49. Domain: $\mathbb{R}$; range: $(3, \infty)$

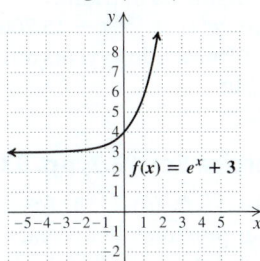

51. Domain: $\mathbb{R}$; range: $(-2, \infty)$

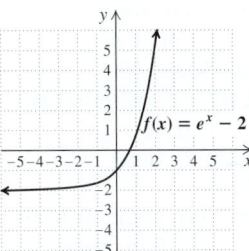

53. Domain: $\mathbb{R}$; range: $(0, \infty)$

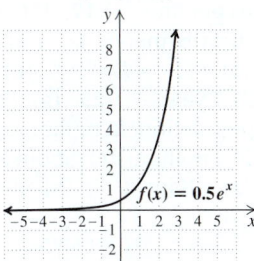

55. Domain: $\mathbb{R}$; range: $(0, \infty)$

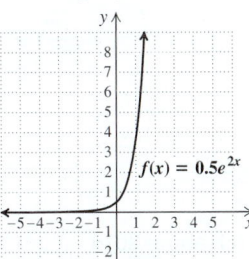

57. Domain: $\mathbb{R}$; range: $(0, \infty)$

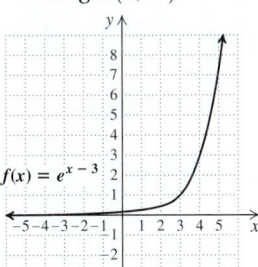

59. Domain: $\mathbb{R}$; range: $(0, \infty)$

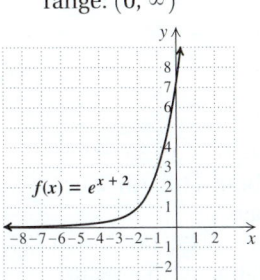

61. Domain: $\mathbb{R}$; range: $(-\infty, 0)$

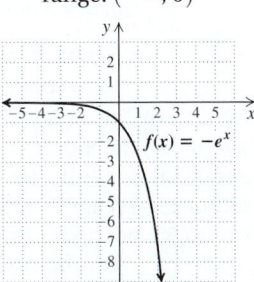

63. Domain: $(0, \infty)$; range: $\mathbb{R}$

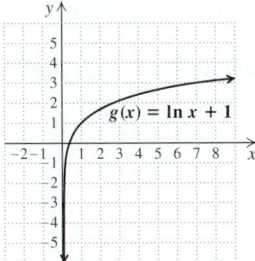

65. Domain: $(0, \infty)$; range: $\mathbb{R}$

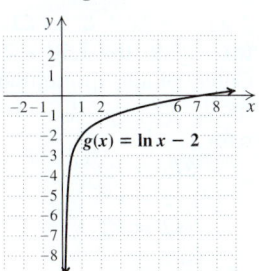

67. Domain: $(0, \infty)$; range: $\mathbb{R}$

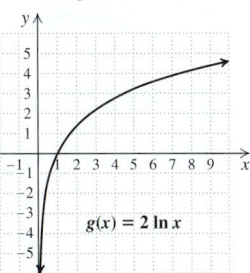

69. Domain: $(0, \infty)$; range: $\mathbb{R}$

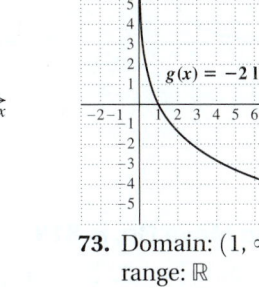

71. Domain: $(-2, \infty)$; range: $\mathbb{R}$

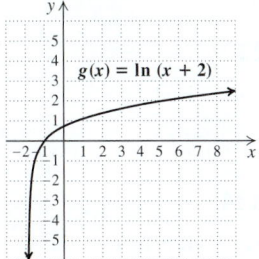

73. Domain: $(1, \infty)$; range: $\mathbb{R}$

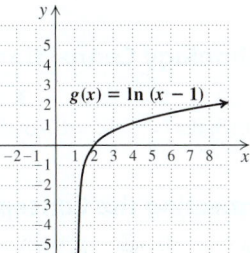

75. 📋   **77.** 1   **78.** $-7\frac{1}{2}$

79. $(g - f)(x) = 5x - 8 - \dfrac{1}{x + 2}$

80. $\{x \mid x \text{ is a real number } and\ x \neq -2\}$, or $(-\infty, -2) \cup (-2, \infty)$

81. $\left\{x \mid x \text{ is a real number } and\ x \neq -2\ and\ x \neq \frac{8}{5}\right\}$, or $(-\infty, -2) \cup \left(-2, \frac{8}{5}\right) \cup \left(\frac{8}{5}, \infty\right)$

82. $gg(x) = 25x^2 - 80x + 64$ **83.** 📋   **85.** 2.452

87. 1.442   **89.** $\log M = \dfrac{\ln M}{\ln 10}$ **91.** 1086.5129

93. 4.9855   **95. (a)** Domain: $\{x \mid x > 0\}$, or $(0, \infty)$; range: $\{y \mid y < 0.5135\}$, or $(-\infty, 0.5135)$; **(b)** $[-1, 5, -10, 5]$; **(c)** $y = 3.4 \ln x - 0.25e^x$

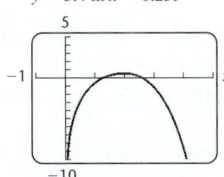

97. (a) Domain: $\{x \mid x > 0\}$, or $(0, \infty)$; range: $\{y \mid y > -0.2453\}$, or $(-0.2453, \infty)$; **(b)** $[-1, 5, -1, 10]$; **(c)** $y = 2x^3 \ln x$

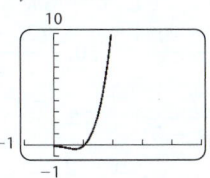

99. 📈, 📋

Quick Quiz: Sections 12.1–12.5, p. 819

1. $f(x) = \sqrt{x}$; $g(x) = 3x - 7$ **2.** $\log_a \dfrac{x^2}{y^3}$

3. $\log_a (x^2 - 1)$

4.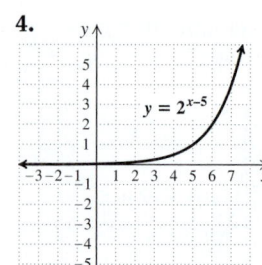

5.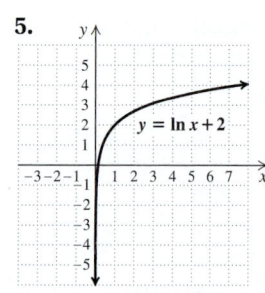

Prepare to Move On, p. 819

1. $-4, 7$ **2.** $0, \frac{7}{5}$ **3.** $\frac{15}{17}$ **4.** $\frac{5}{6}$ **5.** $\frac{56}{9}$ **6.** 4

Technology Connection, p. 823

1. 0.38   **2.** -1.96 **3.** 0.90   **4.** -1.53
5. 0.13, 8.47   **6.** $-0.75, 0.75$

Connecting the Concepts, p. 824

1. 500   **2.** 8   **3.** $\ln 5 \approx 1.6094$ **4.** $-5, 5$ **5.** 4
6. $\dfrac{\ln 4}{5 \ln 3} \approx 0.2524$ **7.** $-\frac{1}{2}$ **8.** $-\frac{2}{5}$

Exercise Set 12.6, pp. 825–826

1. False   **2.** True   **3.** True   **4.** True   **5.** (d)
6. (a)   **7.** (b)   **8.** (c)   **9.** 2   **11.** $\frac{5}{2}$
13. $\dfrac{\log 10}{\log 2} \approx 3.322$ **15.** -1 **17.** $\dfrac{\log 19}{\log 8} + 3 \approx 4.416$
19. $\ln 50 \approx 3.912$ **21.** $\dfrac{\ln 8}{-0.02} \approx -103.972$
23. $\dfrac{\log 87}{\log 4.9} \approx 2.810$ **25.** $\dfrac{\ln\left(\frac{19}{2}\right)}{4} \approx 0.563$
27. $\dfrac{\ln 2}{-1} \approx -0.693$ **29.** 81   **31.** $\frac{1}{16}$
33. $e^5 \approx 148.413$ **35.** $\dfrac{e^3}{4} \approx 5.021$
37. $10^{1.2} \approx 15.849$ **39.** $\dfrac{e^4 - 1}{2} \approx 26.799$
41. $e \approx 2.718$ **43.** $e^{-3} \approx 0.050$ **45.** -4
47. 10   **49.** No solution   **51.** 2   **53.** $\frac{83}{15}$ **55.** 1
57. 6   **59.** 1   **61.** 5   **63.** $\frac{17}{2}$ **65.** 4   **67.** ☑
69. $\dfrac{(a+2)(a-2)^2}{a^4}$ **70.** $\dfrac{t(2t-3)(t-1)}{2t+3}$
71. $\dfrac{5m-7}{(m+1)(m-5)}$ **72.** $\dfrac{-x^2+4x+2}{x(x-2)}$
73. $\dfrac{x(3y-2)}{2y+x}$ **74.** $\dfrac{x+2}{x+1}$ **75.** ☑   **77.** -4
79. 2   **81.** $\pm\sqrt{34}$ **83.** $-3, -1$ **85.** $-625, 625$
87. $\frac{1}{2}, 5000$ **89.** $-3, -1$ **91.** $\frac{1}{100{,}000}, 100{,}000$
93. $-\frac{1}{3}$ **95.** 38   **97.** 1

Quick Quiz: Sections 12.1–12.6, p. 826

1. $g^{-1}(x) = x + 6$ **2.** -1 **3.** 11   **4.** 8
5. $10^{2.7} \approx 501.187$

Prepare to Move On, p. 826

1. Length: 9.5 ft; width: 3.5 ft
2. (a) $w(t) = \frac{5900}{11}t + 15{,}200$, where $w(t)$ is the average cost of a wedding t years after 1990; **(b)** about \$28,609
3. $1\frac{1}{5}$ hr

Exercise Set 12.7, pp. 833–838

1. (b)   **2.** (d)   **3.** (c)   **4.** (a)   **5. (a)** 2010;
(b) approximately 1.5 years   **7. (a)** 6.4 years;
(b) 23.4 years   **9. (a)** The 49th key; **(b)** 12 keys
11. (a) 33 months after March 2011, or in December 2013;
(b) approximately 16 months   **13.** 4.9   **15.** 10^{-7}
moles per liter   **17.** 130 dB   **19.** 7.6 W/m²
21. Approximately 42.4 million messages per day
23. (a) $P(t) = P_0 e^{0.025t}$; **(b)** \$5126.58; \$5256.36;
(c) 27.7 years   **25. (a)** $P(t) = 314e^{0.00963t}$, where $P(t)$ is the population, in millions, t years after 2012;
(b) 339 million; **(c)** about 2037   **27.** 21 years
29. (a) About 2044; **(b)** about 2060;
(c) 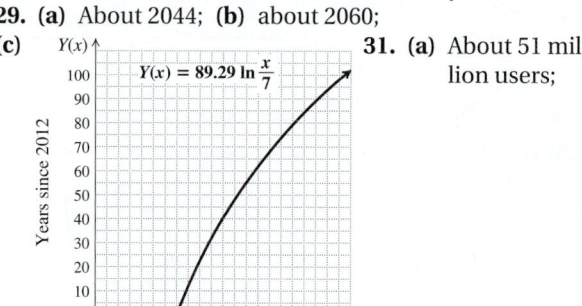   **31. (a)** About 51 million users;

(b) 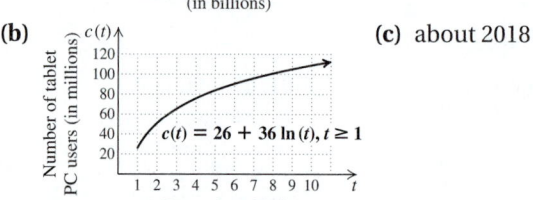   **(c)** about 2018

33. (a) $k \approx 0.241$; $P(t) = 4232e^{0.241t}$;
(b) about 2014 **35. (a)** $k \approx 0.341$; $P(t) = 15.5e^{-0.341t}$;
(b) about 5.6 mcg/mL; **(c)** after about 4 hr; **(d)** 2 hr
37. About 1964 years   **39.** 7.2 days
41. (a) 13.9% per hour; **(b)** 21.6 hr   **43. (a)** $k \approx 0.123$;
$V(t) = 9e^{0.123t}$; **(b)** about \$360 million; **(c)** 5.6 years;
(d) 38.3 years   **45.** ☑   **47.** $f(x) = 18x + \frac{1}{2}$
48. $f(x) = \frac{11}{6}x$ **49.** $f(x) = \frac{2}{3}x + 9$
50. $f(x) = -2x + 8$ **51.** ☑   **53.** \$14.0 million
55. (a) -26.9; **(b)** 1.58×10^{-17} W/m²
57. Consider an exponential growth function $P(t) = P_0 e^{kt}$.
At time T, $P(T) = 2P_0$.
Solve for T:
$$2P_0 = P_0 e^{kT}$$
$$2 = e^{kT}$$
$$\ln 2 = kT$$
$$\frac{\ln 2}{k} = T.$$

Quick Quiz: Sections 12.1–12.7, p. 838

1. Yes **2.** $\frac{3}{4}$ **3.** 3.9069 **4.** 5
5. Approximately 23.4 years

Prepare to Move On, p. 838

1. $\sqrt{2}$ **2.** $(4, -7)$ **3.** $-4 \pm \sqrt{17}$
4.

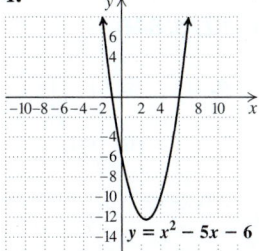

Visualizing for Success, p. 839

1. J **2.** D **3.** B **4.** G **5.** H **6.** C **7.** F
8. I **9.** E **10.** A

Decision Making: Connection, p. 840

1. $F(t) = 4430e^{0.062t}$, where t is the number of school years since 2000–2001 **2.** $G(t) = 2050e^{0.103t}$, where t is the number of school years since 2000–2001
3. Answers will vary. **4.** Answers will vary. **5.** 🖥

Study Summary: Chapter 12, pp. 841–842

1. $(f \circ g)(x) = 19 - 6x^2$ **2.** Yes **3.** $f^{-1}(x) = \dfrac{x - 1}{5}$
4.

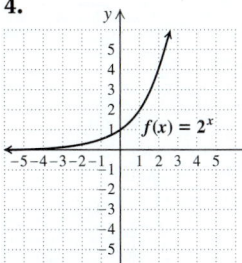

5.

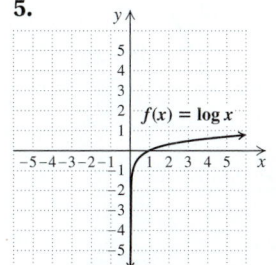

6. $\log_5 625 = 4$ **7.** $\log_9 x + \log_9 y$ **8.** $\log_6 7 - \log_6 10$
9. 0 **10.** 19 **11.** 2.3219 **12.** $\frac{4}{3}$
13. $\dfrac{\ln 10}{0.1} \approx 23.0259$ **14.** (a) $P(t) = 15{,}000e^{0.023t}$;
(b) 30.1 years **15.** 35 days

Review Exercises: Chapter 12, pp. 843–844

1. True **2.** True **3.** True **4.** False **5.** False
6. True **7.** False **8.** False **9.** True **10.** False
11. $(f \circ g)(x) = 4x^2 - 12x + 10$; $(g \circ f)(x) = 2x^2 - 1$
12. $f(x) = \sqrt{x}$; $g(x) = 3 - x$ **13.** No
14. $f^{-1}(x) = x + 10$ **15.** $g^{-1}(x) = \dfrac{2x - 1}{3}$
16. $f^{-1}(x) = \dfrac{\sqrt[3]{x}}{3}$

17.

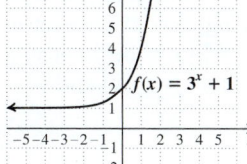

18.

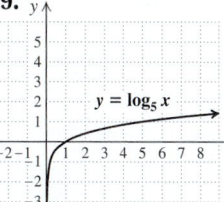

19.

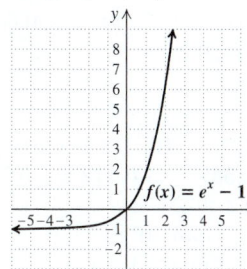

20. 2 **21.** -2 **22.** 11
23. $\frac{1}{2}$ **24.** $\log_2 \frac{1}{8} = -3$
25. $\log_{25} 5 = \frac{1}{2}$
26. $16 = 4^x$ **27.** $1 = 8^0$

28. $4 \log_a x + 2 \log_a y + 3 \log_a z$
29. $5 \log_a x - (\log_a y + 2 \log_a z)$, or
$5 \log_a x - \log_a y - 2 \log_a z$
30. $\frac{1}{4}(2 \log z - 3 \log x - \log y)$ **31.** $\log_a (5 \cdot 8)$, or $\log_a 40$
32. $\log_a \frac{48}{12}$, or $\log_a 4$ **33.** $\log \dfrac{a^{1/2}}{bc^2}$ **34.** $\log_a \sqrt[3]{\dfrac{x}{y^2}}$
35. 1 **36.** 0 **37.** 17 **38.** 6.93 **39.** -3.2698
40. 8.7601 **41.** 3.2698 **42.** 2.54995 **43.** -3.6602
44. 1.8751 **45.** 61.5177 **46.** -1.2040 **47.** 0.3753
48. 2.4307 **49.** 0.8982
50. Domain: $\mathbb{R}$; range: $(-1, \infty)$
51. Domain: $(0, \infty)$; range: $\mathbb{R}$

52. 3 **53.** -1 **54.** $\frac{1}{81}$ **55.** 2 **56.** $\frac{1}{1000}$
57. $e^3 \approx 20.0855$ **58.** $\frac{1}{2}\left(\dfrac{\log 19}{\log 4} + 5\right) \approx 3.5620$
59. $\dfrac{\log 12}{\log 2} \approx 3.5850$ **60.** $\dfrac{\ln 0.03}{-0.1} \approx 35.0656$
61. $e^{-3} \approx 0.0498$ **62.** $\frac{15}{2}$ **63.** 16 **64.** 5
65. (a) 82; (b) 66.8; (c) 35 months
66. (a) 2.3 years; (b) 3.1 years
67. (a) $k \approx 1.565$; $A(t) = 175e^{1.565t}$;
(b) \$91.6 billion; **(c)** about 2014; **(d)** 0.44 year
68. (a) $k \approx 0.060$; $F(t) = 2500e^{-0.060t}$;
(b) approximately 3 fatalities; **(c)** about 2014
69. 11.553% per year **70.** 16.5 years **71.** 3463 years

72. 5.1 **73.** About 114 dB **74.** 📋 Negative numbers do not have logarithms because logarithm bases are positive, and there is no exponent to which a positive number can be raised to yield a negative number.   **75.** ✏️ If $f(x) = e^x$, then to find the inverse function, we let $y = e^x$ and interchange x and y: $x = e^y$. If $x = e^y$, then $\log_e x = y$ by the definition of logarithms. Since $\log_e x = \ln x$, we have $y = \ln x$ or $f^{-1}(x) = \ln x$. Thus, $g(x) = \ln x$ is the inverse of $f(x) = e^x$. Another approach is to find $(f \circ g)(x)$ and $(g \circ f)(x)$:

$$(f \circ g)(x) = e^{\ln x} = x, \text{ and}$$
$$(g \circ f)(x) = \ln e^x = x.$$

Thus, g and f are inverse functions.   **76.** e^{e^3}
77. $-3, -1$ **78.** $\left(\frac{8}{3}, -\frac{2}{3}\right)$

Test: Chapter 12, p. 845

1. [12.1] $(f \circ g)(x) = 2 + 6x + 4x^2$; $(g \circ f)(x) = 2x^2 + 2x + 1$

2. [12.1] $f(x) = \frac{1}{x}$; $g(x) = 2x^2 + 1$ **3.** [12.1] No

4. [12.1] $f^{-1}(x) = \frac{x - 4}{3}$ **5.** [12.1] $g^{-1}(x) = \sqrt[3]{x} - 1$

6. [12.2]   **7.** [12.3]

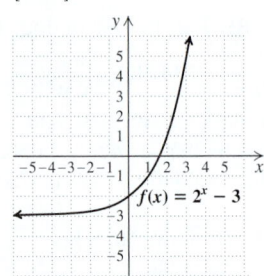

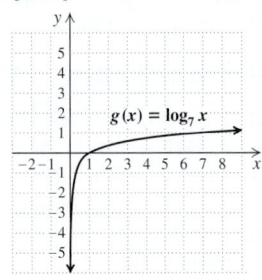

8. [12.3] 3   **9.** [12.3] $\frac{1}{2}$ **10.** [12.4] 1   **11.** [12.4] 0
12. [12.3] $\log_5 \frac{1}{625} = -4$ **13.** [12.3] $2^m = \frac{1}{2}$
14. [12.4] $3 \log a + \frac{1}{2} \log b - 2 \log c$
15. [12.4] $\log_a \left(z^2 \sqrt[3]{x}\right)$
16. [12.4] 1.146   **17.** [12.4] 0.477   **18.** [12.4] 1.204
19. [12.5] 1.3979   **20.** [12.5] 0.1585   **21.** [12.5] −0.9163
22. [12.5] 121.5104   **23.** [12.5] 2.4022
24. [12.5]   Domain: $\mathbb{R}$; range: $(3, \infty)$

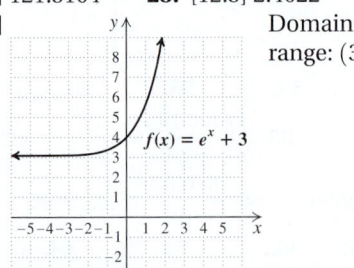

25. [12.5]   Domain: $(4, \infty)$; range: $\mathbb{R}$

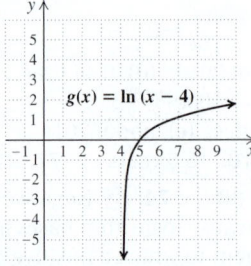

26. [12.6] -5 **27.** [12.6] 2   **28.** [12.6] $\frac{1}{100}$
29. [12.6] $\frac{\log 1.2}{\log 7} \approx 0.0937$ **30.** [12.6] 4

31. [12.7] **(a)** 2.26 ft/sec; **(b)** 1,290,000
32. [12.7] **(a)** $P(t) = 155e^{0.019t}$, where t is the number of years after 2012 and $P(t)$ is in millions; **(b)** 167 million; 319 million; **(c)** about 2025; **(d)** 36.5 years
33. [12.7] **(a)** $k \approx 0.024$; $C(t) = 30{,}360e^{0.024t}$; **(b)** $43,516; **(c)** about 2022   **34.** [12.7] 4.3%   **35.** [12.7] 7.0
36. [12.6] $-309{,}316$ **37.** [12.4] 2

Cumulative Review: Chapters 1–12, p. 846

1. $\frac{y^{12}}{16x^8}$ **2.** $-\frac{y^4}{3z^5}$ **3.** $\frac{20x^6 z^2}{y}$ **4.** 8   **5.** $(3, -1)$

6. $(1, -2, 0)$ **7.** $-7, 10$ **8.** $\frac{9}{2}$ **9.** $\frac{3}{4}$ **10.** $\pm 4i$

11. $\pm 2, \pm 3$ **12.** 9   **13.** $\frac{\log 7}{5 \log 3} \approx 0.3542$

14. $\frac{8e}{e - 1} \approx 12.6558$ **15.** $(-\infty, -5) \cup (1, \infty)$, or $\{x \mid x < -5 \text{ or } x > 1\}$ **16.** $-3 \pm 2\sqrt{5}$
17. $\{x \mid x \le -2 \text{ or } x \ge 5\}$, or $(-\infty, -2] \cup [5, \infty)$
18. $\frac{a + 2}{6}$ **19.** $\frac{7x + 4}{(x + 6)(x - 6)}$ **20.** $\sqrt[10]{(x + 5)^7}$
21. $15 - 4\sqrt{3}i$ **22.** $(3 + 4n)(9 - 12n + 16n^2)$
23. $2(3x - 2y)(x + 2y)$ **24.** $2(m + 3n)^2$
25. $(x - 2y)(x + 2y)(x^2 + 4y^2)$ **26.** $\frac{6 + \sqrt{y} - y}{4 - y}$
27. $f^{-1}(x) = \frac{x - 9}{-2}$, or $f^{-1}(x) = \frac{9 - x}{2}$
28. $f(x) = -10x - 8$
29.   **30.**

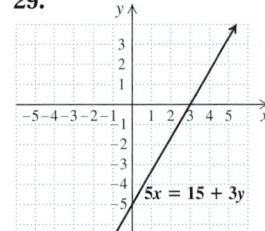

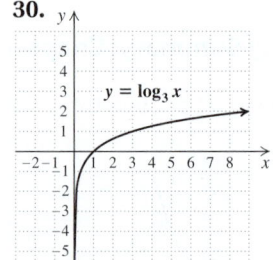

31.   **32.**

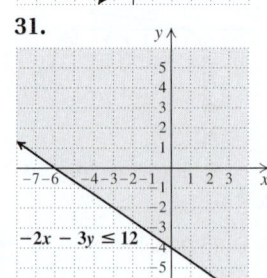

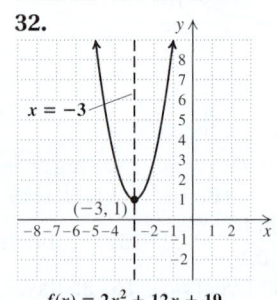

$f(x) = 2x^2 + 12x + 19$
Minimum: 1

33.   Domain: $\mathbb{R}$; range: $(0, \infty)$

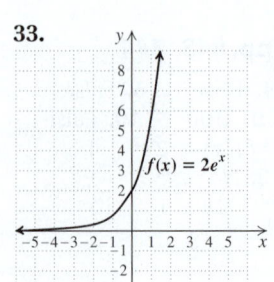

34. 13.5 million acre-feet **35.** **(a)** $k \approx 0.076$; $D(t) = 15e^{0.076t}$; **(b)** 108 million m^3 per day; **(c)** 2020 **36.** $5\frac{5}{11}$ min **37.** Thick and Tasty: 6 oz; Light and Lean: 9 oz **38.** $\frac{1}{3}, \frac{10{,}000}{3}$ **39.** 35 mph

CHAPTER 13

Technology Connection, p. 853

1. $x^2 + y^2 - 16 = 0$

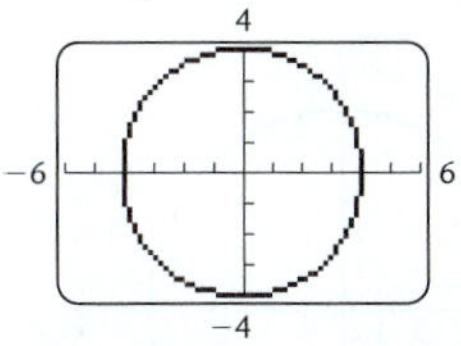

2. $(x - 1)^2 + (y - 2)^2 = 25$

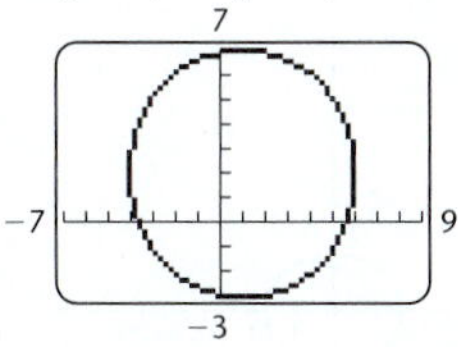

3. $(x + 3)^2 + (y - 5)^2 = 16$

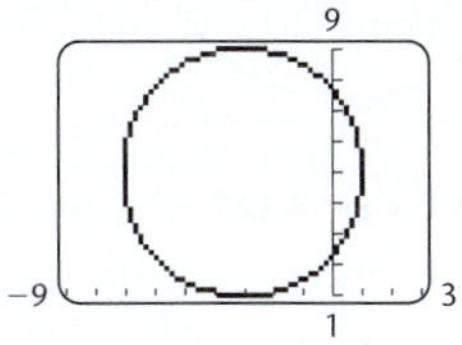

Exercise Set 13.1, pp. 853–857

1. Conic sections **2.** Circle **3.** Horizontal **4.** Vertex **5.** Center **6.** Center **7.** (f) **8.** (e) **9.** (c) **10.** (b) **11.** (d) **12.** (a)

13.

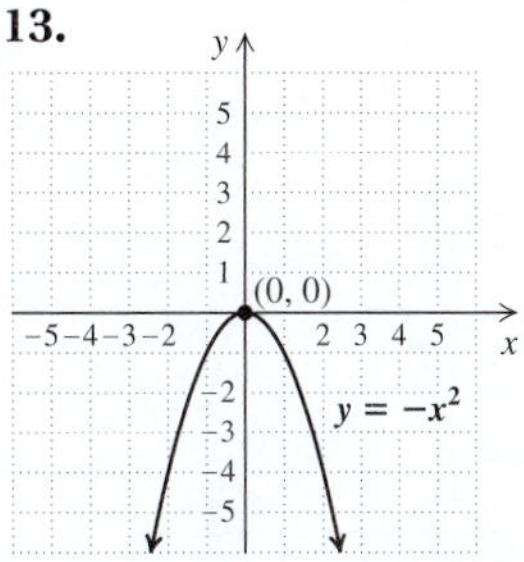

15.

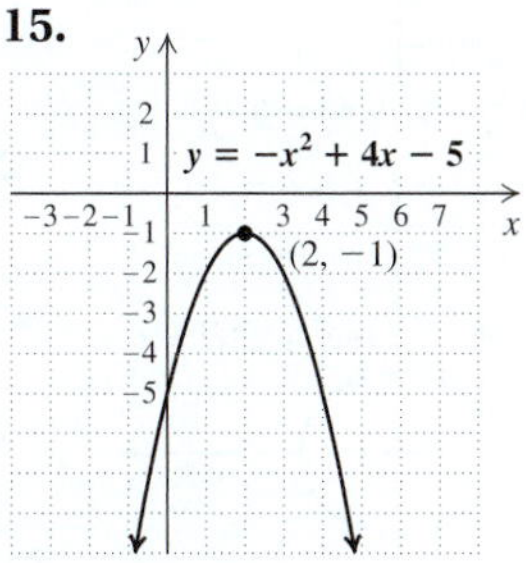

17.

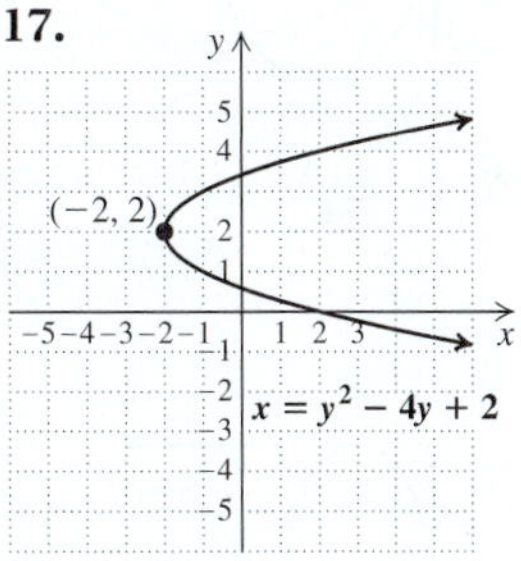

19.

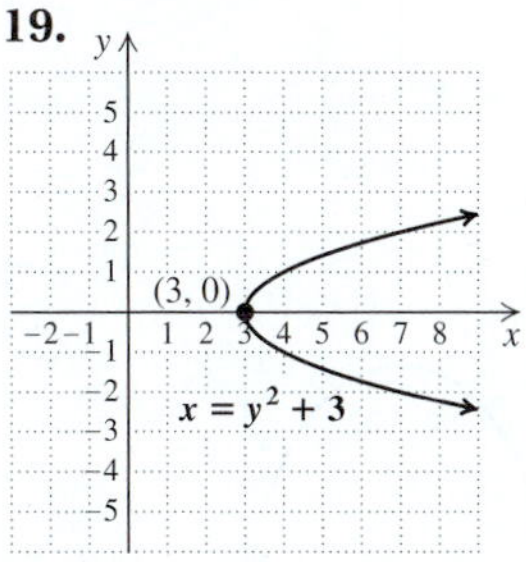

21.

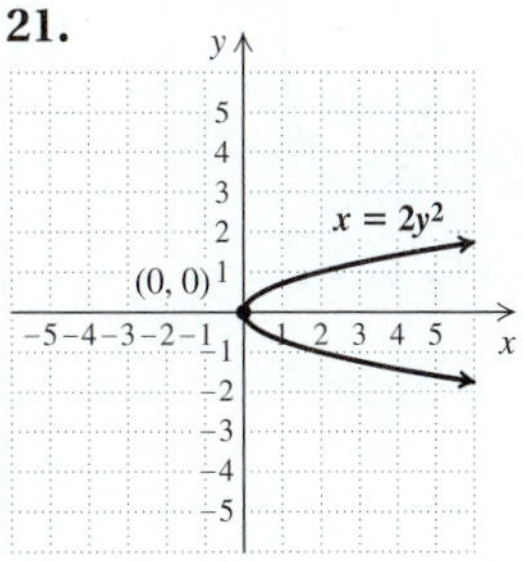

23.

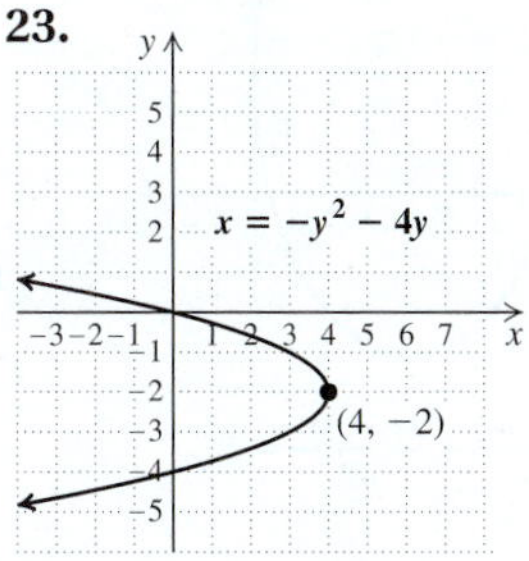

25.

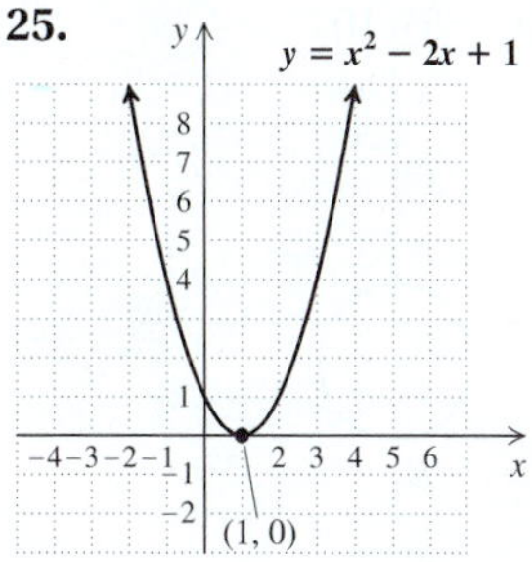

27.

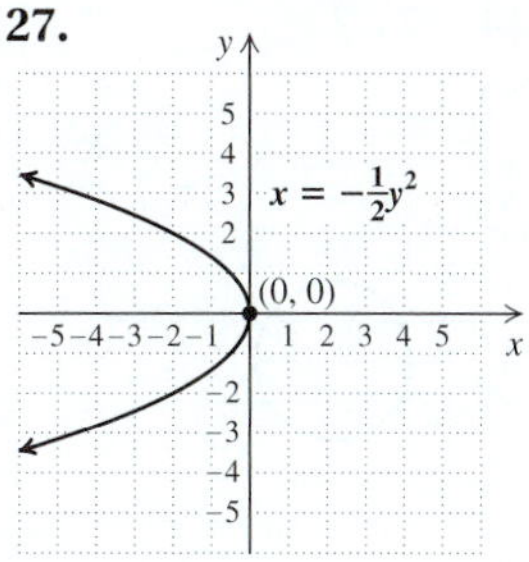

29.

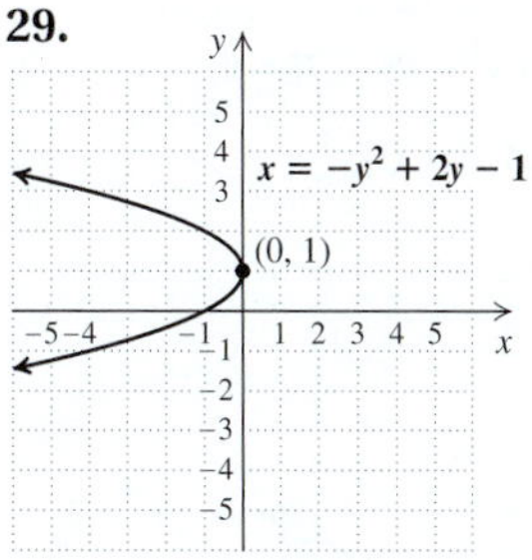

31.

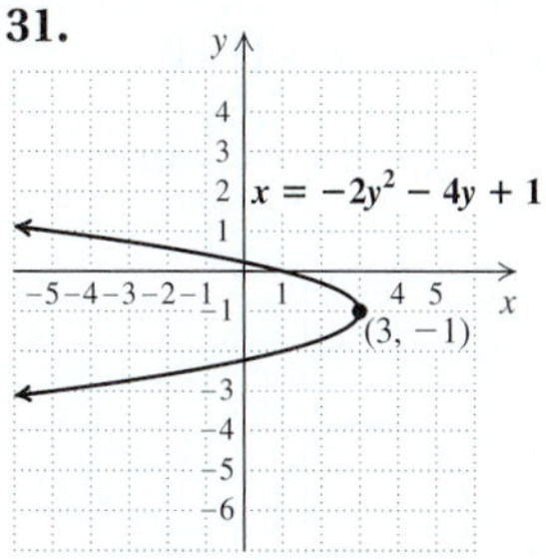

33. $x^2 + y^2 = 64$ **35.** $(x - 7)^2 + (y - 3)^2 = 6$ **37.** $(x + 4)^2 + (y - 3)^2 = 18$ **39.** $(x + 5)^2 + (y + 8)^2 = 300$ **41.** $x^2 + y^2 = 25$ **43.** $(x + 4)^2 + (y - 1)^2 = 20$

45. $(0, 0)$; 1

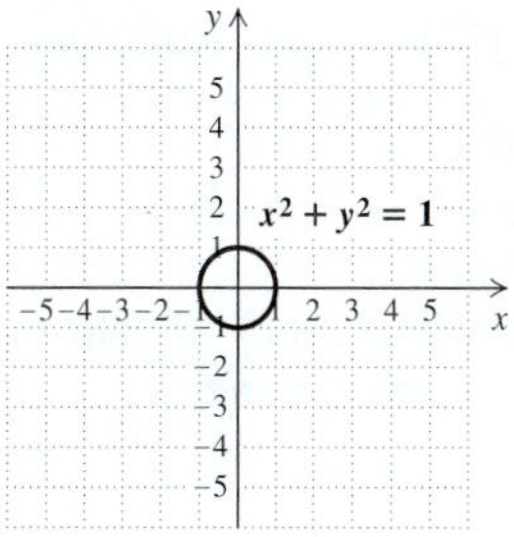

47. $(-1, -3)$; 7

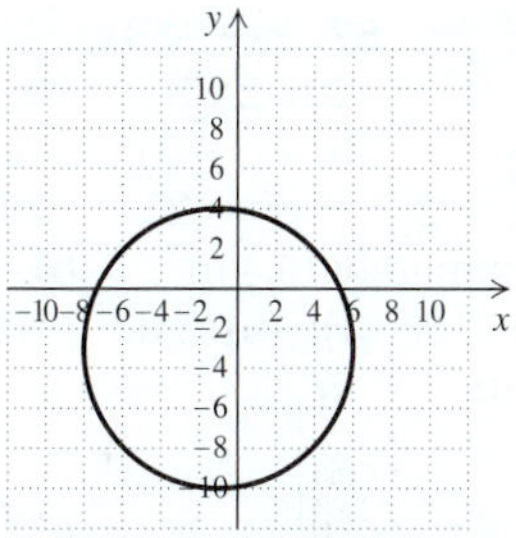

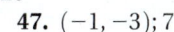

$(x + 1)^2 + (y + 3)^2 = 49$

49. $(4, -3)$; $\sqrt{10}$

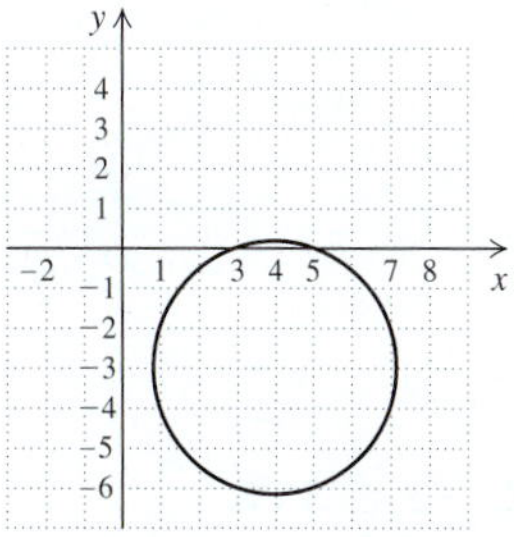

$(x - 4)^2 + (y + 3)^2 = 10$

51. $(0, 0)$; $2\sqrt{2}$

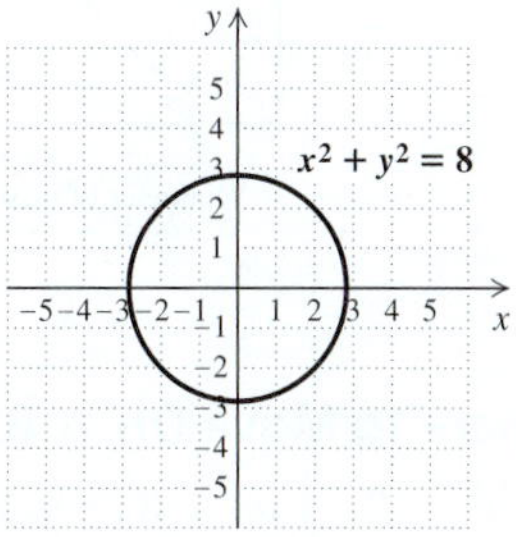

53. $(5, 0)$; $\frac{1}{2}$

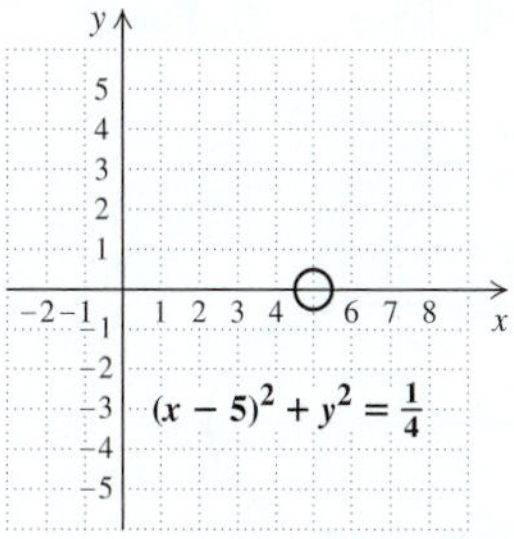

55. $(-4, 3)$; $\sqrt{40}$, or $2\sqrt{10}$

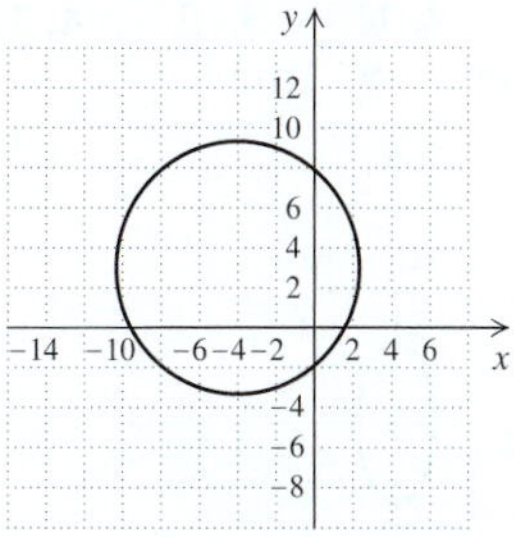

$x^2 + y^2 + 8x - 6y - 15 = 0$

57. $(4, -1); 2$

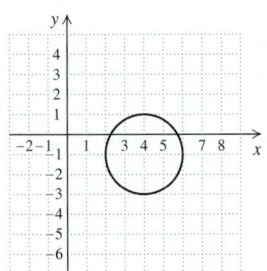

$x^2 + y^2 - 8x + 2y + 13 = 0$

59. $(0, -5); 10$

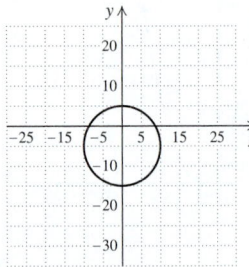

$x^2 + y^2 + 10y - 75 = 0$

61. $\left(-\dfrac{7}{2}, \dfrac{3}{2}\right); \sqrt{\dfrac{98}{4}}$, or $\dfrac{7\sqrt{2}}{2}$

63. $(0, 0); \dfrac{1}{6}$

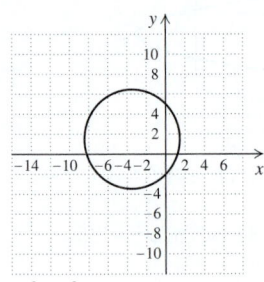

$x^2 + y^2 + 7x - 3y - 10 = 0$

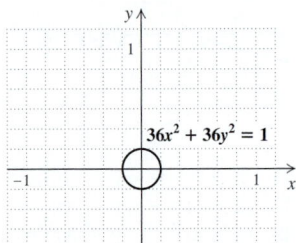

$36x^2 + 36y^2 = 1$

65. 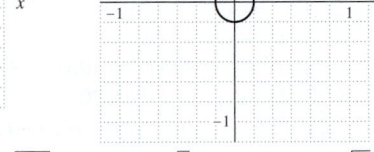 **67.** $2xy^3\sqrt[4]{3x^3}$ **68.** $y\sqrt[6]{y}$ **69.** $10x^2\sqrt{w}$
70. $\sqrt[30]{t^7}$ **71.** $2\sqrt{3}$ **72.** $12\sqrt{3} - 3\sqrt{2} + 4\sqrt{6} - 2$
73. 📋 **75.** $(x - 3)^2 + (y + 5)^2 = 9$
77. $(x - 3)^2 + y^2 = 25$ **79.** $\dfrac{17}{4}\pi$ m², or
approximately 13.4 m² **81.** 7169 mm
83. (a) $(0, -3);$ **(b)** 5 ft **85.** $x^2 + (y - 30.6)^2 = 590.49$
87. 7 in.

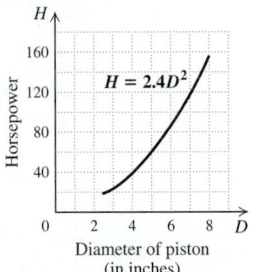

$H = 2.4D^2$

Horsepower (vertical axis)

Diameter of piston (in inches)

Prepare to Move On, p. 857

1. ± 4 **2.** $\pm a$ **3.** $-4, 6$ **4.** $-3 \pm 3\sqrt{3}$

Exercise Set 13.2, pp. 861–863

1. A **2.** C **3.** B **4.** D **5.** True **6.** False
7. True **8.** True
9.

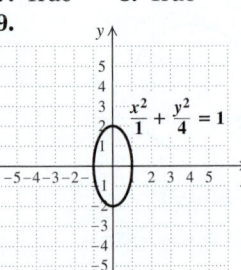

$\dfrac{x^2}{1} + \dfrac{y^2}{4} = 1$

11.

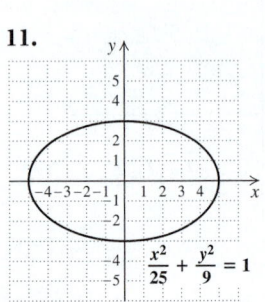

$\dfrac{x^2}{25} + \dfrac{y^2}{9} = 1$

13.

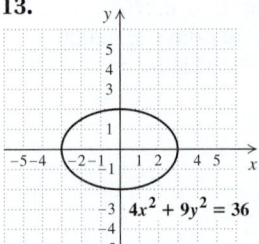

$4x^2 + 9y^2 = 36$

15.

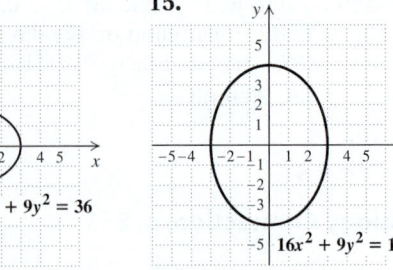

$16x^2 + 9y^2 = 144$

17.

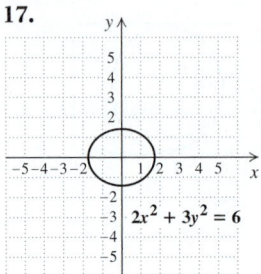

$2x^2 + 3y^2 = 6$

19.

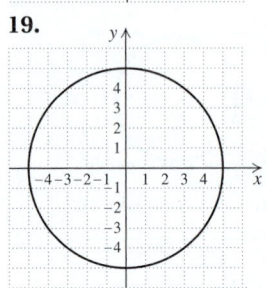

$5x^2 + 5y^2 = 125$

21.

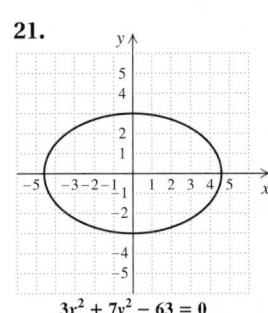

$3x^2 + 7y^2 - 63 = 0$

23.

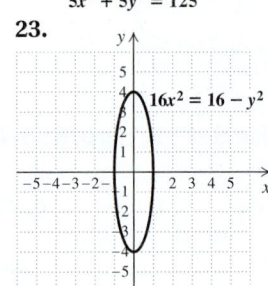

$16x^2 = 16 - y^2$

25.

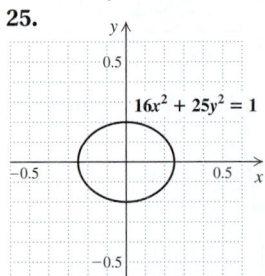

$16x^2 + 25y^2 = 1$

27.

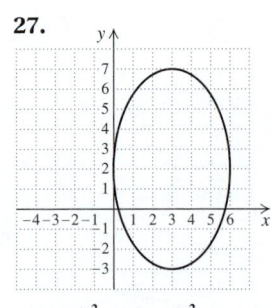

$\dfrac{(x - 3)^2}{9} + \dfrac{(y - 2)^2}{25} = 1$

29.

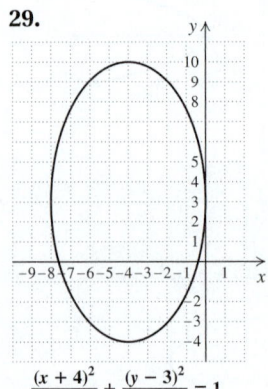

$\dfrac{(x + 4)^2}{16} + \dfrac{(y - 3)^2}{49} = 1$

31.

$12(x - 1)^2 + 3(y + 4)^2 = 48$

33.

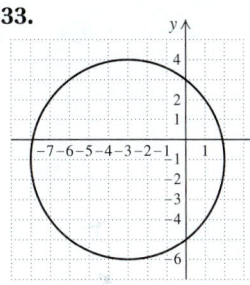

$4(x + 3)^2 + 4(y + 1)^2 - 10 = 90$

49. $\dfrac{x^2}{9} + \dfrac{y^2}{25} = 1$

51. (a) Let $F_1 = (-c, 0)$ and $F_2 = (c, 0)$. Then the sum of the distances from the foci to P is $2a$. By the distance formula,

$$\sqrt{(x + c)^2 + y^2} + \sqrt{(x - c)^2 + y^2} = 2a, \text{ or}$$
$$\sqrt{(x + c)^2 + y^2} = 2a - \sqrt{(x - c)^2 + y^2}.$$

Squaring, we get

$$(x + c)^2 + y^2 = 4a^2 - 4a\sqrt{(x - c)^2 + y^2} + (x - c)^2 + y^2,$$

or

$$x^2 + 2cx + c^2 + y^2 = 4a^2 - 4a\sqrt{(x - c)^2 + y^2} + x^2 - 2cx + c^2 + y^2.$$

Thus,

$$-4a^2 + 4cx = -4a\sqrt{(x - c)^2 + y^2}$$
$$a^2 - cx = a\sqrt{(x - c)^2 + y^2}.$$

Squaring again, we get

$$a^4 - 2a^2cx + c^2x^2 = a^2(x^2 - 2cx + c^2 + y^2)$$
$$a^4 - 2a^2cx + c^2x^2 = a^2x^2 - 2a^2cx + a^2c^2 + a^2y^2,$$

or

$$x^2(a^2 - c^2) + a^2y^2 = a^2(a^2 - c^2)$$
$$\dfrac{x^2}{a^2} + \dfrac{y^2}{a^2 - c^2} = 1.$$

(b) When P is at $(0, b)$, it follows that $b^2 = a^2 - c^2$. Substituting, we have

$$\dfrac{x^2}{a^2} + \dfrac{y^2}{b^2} = 1.$$

53. 5.66 ft   **55.**

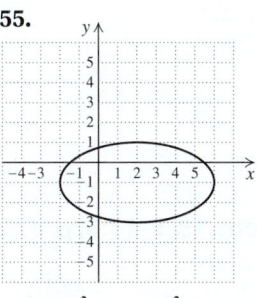

$\dfrac{(x - 2)^2}{16} + \dfrac{(y + 1)^2}{4} = 1$

35. 📋   **37.** $\dfrac{5}{2} \pm \dfrac{\sqrt{13}}{2}$

38. 3   **39.** $-\dfrac{3}{4}, 2$

40. $\dfrac{5}{2}$ **41.** $-\sqrt{11}, \sqrt{11}$

42. $-10, 6$ **43.** 📋

45. $\dfrac{x^2}{81} + \dfrac{y^2}{121} = 1$

47. $\dfrac{(x - 2)^2}{16} + \dfrac{(y + 1)^2}{9} = 1$

57.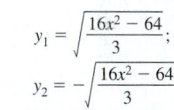

Quick Quiz: Sections 13.1–13.2, p. 863

1. $(-5, -1); 2$ **2.** $(x - 9)^2 + (y - (-23))^2 = 200$, or $(x - 9)^2 + (y + 23)^2 = 200$

3.

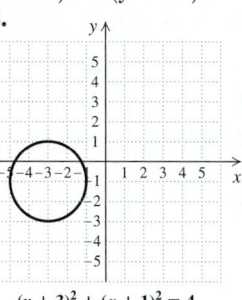

$(x + 3)^2 + (y + 1)^2 = 4$

4.

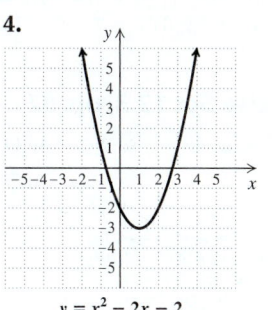

$y = x^2 - 2x - 2$

5.

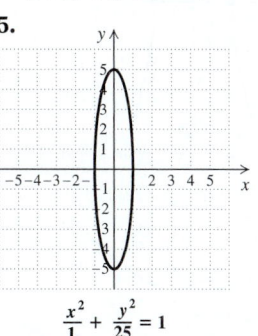

$\dfrac{x^2}{1} + \dfrac{y^2}{25} = 1$

Prepare to Move On, p. 863

1. $y = \dfrac{4}{x}$ **2.** $y = 2x - 4$ **3.** $x^2 - 5x - 7 = 0$

Technology Connection, p. 868

1. $y_1 = \dfrac{\sqrt{15x^2 - 240}}{2}$;

$y_2 = -\dfrac{\sqrt{15x^2 - 240}}{2}$

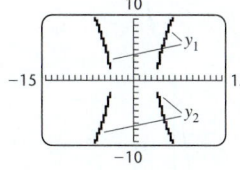

2. $y_1 = \sqrt{\dfrac{16x^2 - 64}{3}}$;

$y_2 = -\sqrt{\dfrac{16x^2 - 64}{3}}$

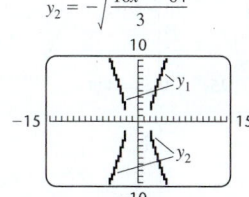

3. $y_1 = \dfrac{\sqrt{5x^2 + 320}}{4}$;

$y_2 = -\dfrac{\sqrt{5x^2 + 320}}{4}$

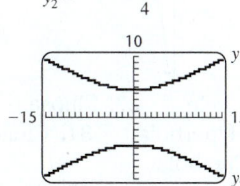

4. $y_1 = \sqrt{\dfrac{9x^2 + 441}{45}}$;

$y_2 = -\sqrt{\dfrac{9x^2 + 441}{45}}$

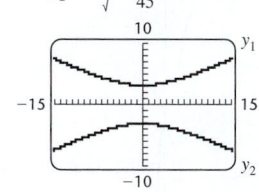

Connecting the Concepts, p. 870

1. $(4, 1); x = 4$ **2.** $(2, -1); y = -1$ **3.** $(3, 2)$
4. $(-3, -5)$ **5.** $(-12, 0), (12, 0), (0, -9), (0, 9)$
6. $(-3, 0), (3, 0)$ **7.** $(0, -1), (0, 1)$ **8.** $y = \frac{3}{2}x, y = -\frac{3}{2}x$

Exercise Set 13.3, pp. 871–872

1. B   **2.** E   **3.** A   **4.** D   **5.** F   **6.** C

7.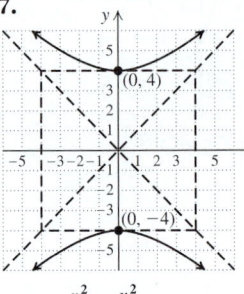

$$\frac{y^2}{16} - \frac{x^2}{16} = 1$$

9.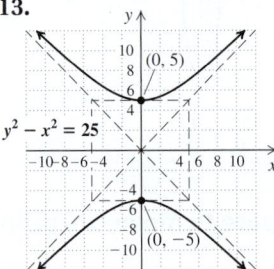

$$\frac{x^2}{4} - \frac{y^2}{25} = 1$$

11.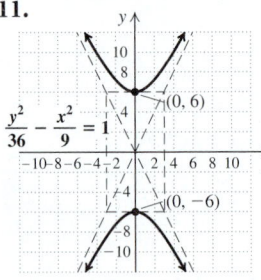

$$\frac{y^2}{36} - \frac{x^2}{9} = 1$$

13.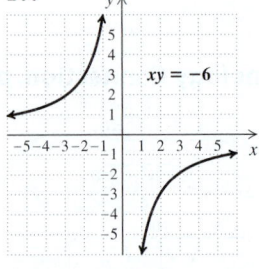

$$y^2 - x^2 = 25$$

15.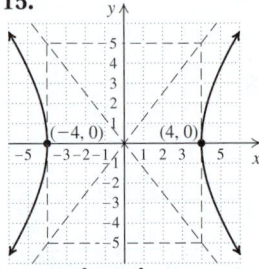

$$25x^2 - 16y^2 = 400$$

17.

$$xy = -6$$

19.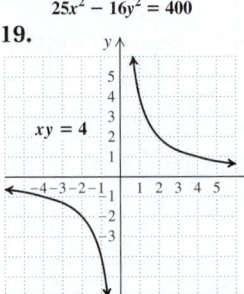

$$xy = 4$$

21.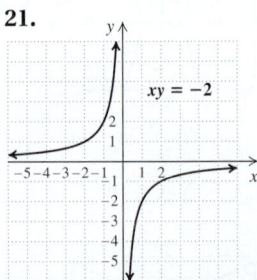

$$xy = -2$$

23.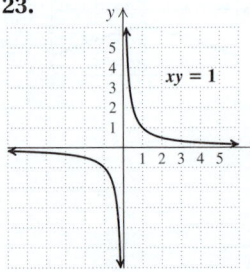

$$xy = 1$$

25. Circle   **27.** Ellipse
29. Hyperbola   **31.** Circle
33. Parabola
35. Hyperbola   **37.** Parabola
39. Hyperbola   **41.** Circle
43. Ellipse   **45.** 📋
47. $(4 + y^2)(2 + y)(2 - y)$
48. $(3xy - 5)^2$

49. $10c(c - 3)(c - 5)$ **50.** $(x + 1)(x^2 + 3)$
51. $8t(t - 1)(t^2 + t + 1)$ **52.** $(3a - 2b)(2a + 5b)$
53. 📋   **55.** $\dfrac{y^2}{36} - \dfrac{x^2}{4} = 1$

57. $C: (5, 2); V: (-1, 2), (11, 2)$; asymptotes:
$y - 2 = \frac{5}{6}(x - 5), y - 2 = -\frac{5}{6}(x - 5)$

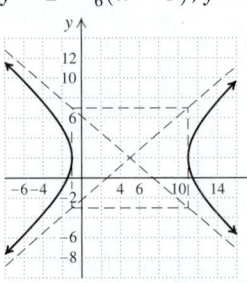

$$\frac{(x - 5)^2}{36} - \frac{(y - 2)^2}{25} = 1$$

59. $\dfrac{(y + 3)^2}{4} - \dfrac{(x - 4)^2}{16} = 1; C: (4, -3); V: (4, -5), (4, -1);$
asymptotes: $y + 3 = \frac{1}{2}(x - 4), y + 3 = -\frac{1}{2}(x - 4)$

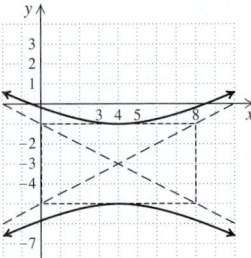

$$8(y + 3)^2 - 2(x - 4)^2 = 32$$

61. $\dfrac{(x + 3)^2}{1} - \dfrac{(y - 2)^2}{4} = 1; C: (-3, 2); V: (-4, 2), (-2, 2);$
asymptotes: $y - 2 = 2(x + 3), y - 2 = -2(x + 3)$

63.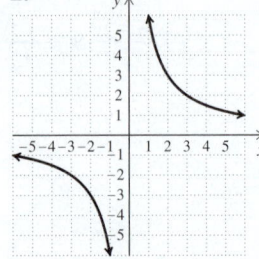

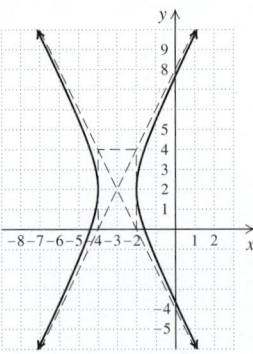

$$4x^2 - y^2 + 24x + 4y + 28 = 0$$

Quick Quiz: Sections 13.1–13.3, p. 872

1.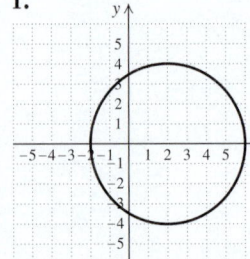

$$(x - 2)^2 + y^2 = 16$$

2.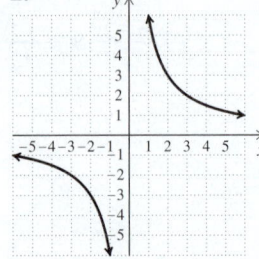

$$xy = 6$$

3.
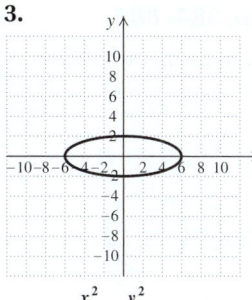
$$\frac{x^2}{36} + \frac{y^2}{4} = 1$$

4.
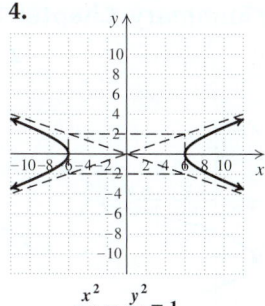
$$\frac{x^2}{36} - \frac{y^2}{4} = 1$$

5.
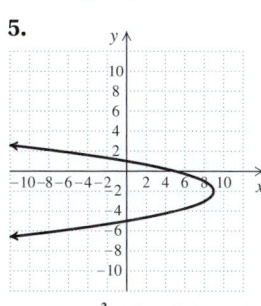
$$x = -y^2 - 4y + 5$$

Prepare to Move On, p. 872

1. $(-3, 6)$ **2.** $(3, 7)$ **3.** $-2, 2$ **4.** $-4, \frac{2}{3}$
5. $\frac{3}{2} \pm \frac{\sqrt{13}}{2}$ **6.** $\pm 1, \pm 5$

Mid-Chapter Review: Chapter 13, p. 873

1.
$$(x^2 - 4x) + (y^2 + 2y) = 6$$
$$(x^2 - 4x + 4) + (y^2 + 2y + 1) = 6 + 4 + 1$$
$$(x - 2)^2 + (y + 1)^2 = 11$$

The center of the circle is $(2, -1)$. The radius is $\sqrt{11}$.
2. (a) Is there both an x^2-term and a y^2-term? Yes;
(b) Do both the x^2-term and the y^2-term have the same sign? No; **(c)** The graph of the equation is a hyperbola.
3. $(x + 4)^2 + (y - 9)^2 = 20$ **4.** Center: $(5, -1)$; radius: 6
5. Circle   **6.** Parabola

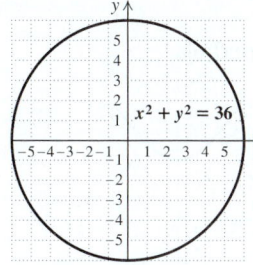
$$x^2 + y^2 = 36$$

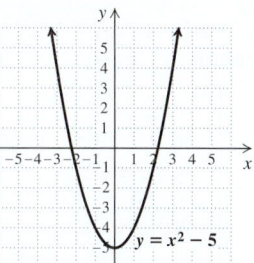
$$y = x^2 - 5$$

7. Ellipse   **8.** Hyperbola

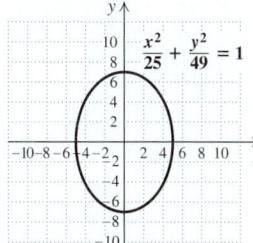
$$\frac{x^2}{25} + \frac{y^2}{49} = 1$$

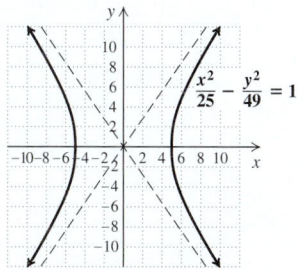
$$\frac{x^2}{25} - \frac{y^2}{49} = 1$$

9. Parabola
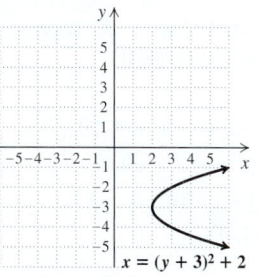
$$x = (y + 3)^2 + 2$$

10. Ellipse
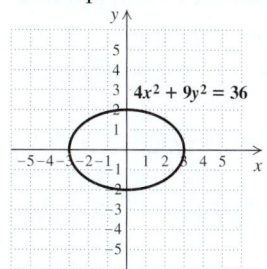
$$4x^2 + 9y^2 = 36$$

11. Hyperbola
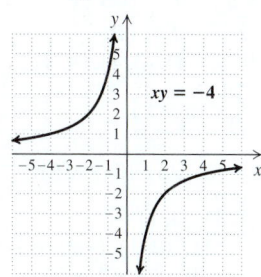
$$xy = -4$$

12. Circle
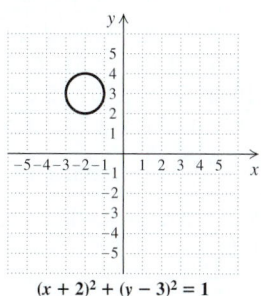
$$(x + 2)^2 + (y - 3)^2 = 1$$

13. Circle
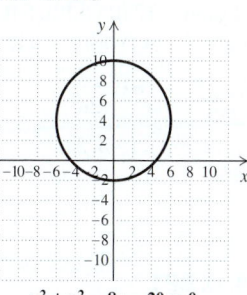
$$x^2 + y^2 - 8y - 20 = 0$$

14. Parabola
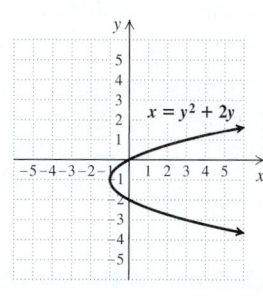
$$x = y^2 + 2y$$

15. Hyperbola
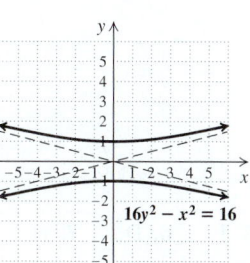
$$16y^2 - x^2 = 16$$

16. Hyperbola
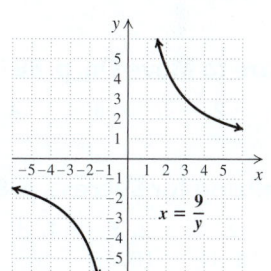
$$x = \frac{9}{y}$$

Technology Connection, p. 876

1. $(-1.50, -1.17); (3.50, 0.50)$
2. $(-2.77, 2.52); (-2.77, -2.52)$

Technology Connection, p. 878

1. $y_1 = \sqrt{(20 - x^2)/4}; \ y_2 = -\sqrt{(20 - x^2)/4}; \ y_3 = 4/x$
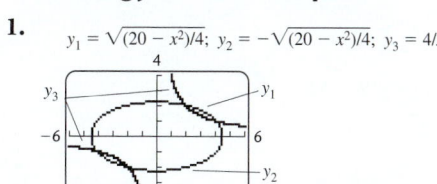

Exercise Set 13.4, pp. 879–881

1. True   **2.** True   **3.** False   **4.** False   **5.** True
6. True   **7.** $(-5, -4), (4, 5)$ **9.** $(0, 2), (3, 0)$
11. $(-2, 1)$

13. $\left(\dfrac{5 + \sqrt{70}}{3}, \dfrac{-1 + \sqrt{70}}{3}\right), \left(\dfrac{5 - \sqrt{70}}{3}, \dfrac{-1 - \sqrt{70}}{30}\right)$

15. $\left(4, \frac{3}{2}\right), (3, 2)$ **17.** $\left(\frac{7}{3}, \frac{1}{3}\right), (1, -1)$ **19.** $\left(\frac{11}{4}, -\frac{5}{4}\right), (1, 4)$
21. $(2, 4), (4, 2)$ **23.** $(3, -5), (-1, 3)$
25. $(-5, -8), (8, 5)$ **27.** $(0, 0), (1, 1),$

$\left(-\dfrac{1}{2} + \dfrac{\sqrt{3}}{2}i, -\dfrac{1}{2} - \dfrac{\sqrt{3}}{2}i\right), \left(-\dfrac{1}{2} - \dfrac{\sqrt{3}}{2}i, -\dfrac{1}{2} + \dfrac{\sqrt{3}}{2}i\right)$

29. $(-4, 0), (4, 0)$ **31.** $(-4, -3), (-3, -4), (3, 4), (4, 3)$
33. $\left(\dfrac{16}{3}, \dfrac{5\sqrt{7}}{3}i\right), \left(\dfrac{16}{3}, -\dfrac{5\sqrt{7}}{3}i\right), \left(-\dfrac{16}{3}, \dfrac{5\sqrt{7}}{3}i\right),$

$\left(-\dfrac{16}{3}, -\dfrac{5\sqrt{7}}{3}i\right)$ **35.** $(-3, -\sqrt{5}), (-3, \sqrt{5}), (3, -\sqrt{5}),$

$(3, \sqrt{5})$ **37.** $(-3, -1), (-1, -3), (1, 3), (3, 1)$
39. $(4, 1), (-4, -1), (2, 2), (-2, -2)$ **41.** $(2, 1), (-2, -1)$
43. $\left(2, -\frac{4}{5}\right), \left(-2, -\frac{4}{5}\right), (5, 2), (-5, 2)$ **45.** $(-\sqrt{2}, \sqrt{2}),$
$(\sqrt{2}, -\sqrt{2})$ **47.** Length: 8 cm; width: 6 cm
49. Length: 2 in.; width: 1 in.   **51.** Length: 12 ft; width: 5 ft
53. 6 and 15; -6 and -15 **55.** Length: 12 in.; width: 7.5 in.

57. Length: $\sqrt{3}$ m; width: 1 m   **59.** 🗒   **61.** $\dfrac{9}{2a^3}$
62. $\frac{1}{4}$ **63.** 4   **64.** $-i$ **65.** -253
66. $10\sqrt{5}$ **67.** 🗒   **69.** $(-2, 3), (2, -3), (-3, 2),$
$(3, -2)$ **71.** Length: 55 ft; width: 45 ft   **73.** 10 in.
by 7 in. by 5 in.

Quick Quiz: Sections 13.1–13.4, p. 881

1. Circle   **2.** Hyperbola   **3.** Ellipse   **4.** Parabola
5. $(2, -1), (1, -2)$

Prepare to Move On, p. 881

1. -9 **2.** -27 **3.** $\frac{1}{5}$ **4.** 77   **5.** $\frac{21}{2}$

Visualizing for Success, p. 882

1. C   **2.** A   **3.** F   **4.** B   **5.** J   **6.** D   **7.** H
8. I   **9.** G   **10.** E

Decision Making: Connection, p. 883

1. $b \approx 34.9$ in.; $h \approx 19.6$ in.   **2.** $b = 32$ in.; $h = 24$ in.

3. An older TV   **4.** $h = \dfrac{d}{\sqrt{1 + r^2}}$ **5.** 🖥

Study Summary: Chapter 13, pp. 884–885

1.
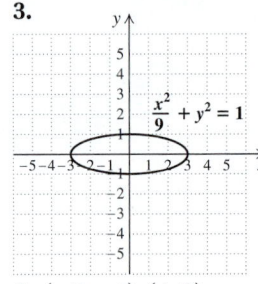

2. Center: $(3, 0)$; radius: 2
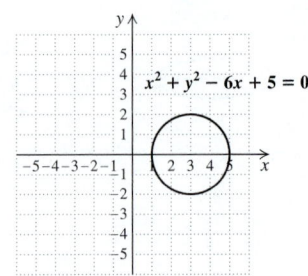

3.
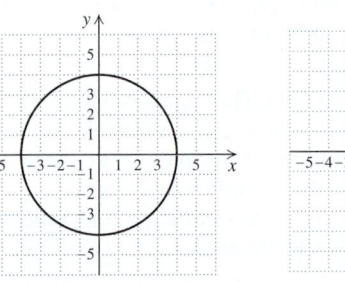

4.
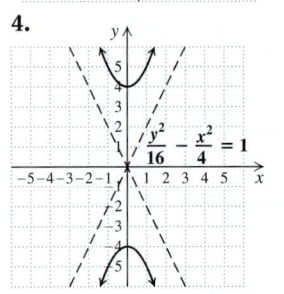

5. $(-5, -4), (4, 5)$

Review Exercises: Chapter 13, pp. 886–887

1. True   **2.** False   **3.** False   **4.** True   **5.** True
6. True   **7.** False   **8.** True   **9.** $(-3, 2), 4$
10. $(5, 0), \sqrt{11}$ **11.** $(3, 1), 3$ **12.** $(-4, 3), 3\sqrt{5}$
13. $(x + 4)^2 + (y - 3)^2 = 16$
14. $(x - 7)^2 + (y + 2)^2 = 20$
15. Circle   **16.** Ellipse

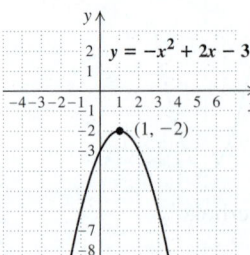

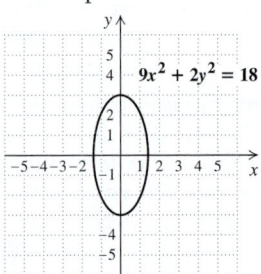

17. Parabola   **18.** Hyperbola

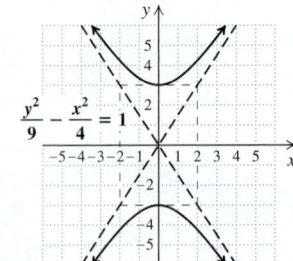

19. Hyperbola

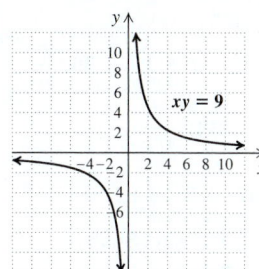

20. Parabola

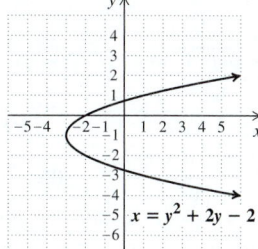

21. Ellipse

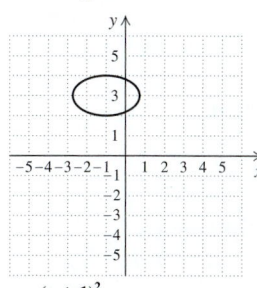

22. Circle

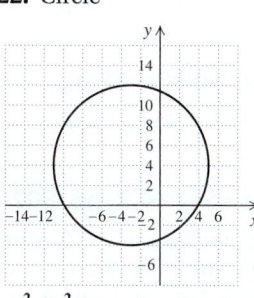

$\frac{(x+1)^2}{3} + (y-3)^2 = 1$ $x^2 + y^2 + 6x - 8y - 39 = 0$

23. $(5, -2)$ **24.** $(2, 2)$, $\left(\frac{32}{9}, -\frac{10}{9}\right)$ **25.** $(0, -5)$, $(2, -1)$
26. $(4, 3)$, $(4, -3)$, $(-4, 3)$, $(-4, -3)$ **27.** $(2, 1)$, $(\sqrt{3}, 0)$,
$(-2, 1)$, $(-\sqrt{3}, 0)$ **28.** $(3, -3)$, $\left(-\frac{3}{5}, \frac{21}{5}\right)$ **29.** $(6, 8)$,
$(6, -8)$, $(-6, 8)$, $(-6, -8)$ **30.** $(2, 2)$, $(-2, -2)$,
$(2\sqrt{2}, \sqrt{2})$, $(-2\sqrt{2}, -\sqrt{2})$ **31.** Length: 12 m; width: 7 m
32. Length: 12 in.; width: 9 in.   **33.** Board: 32 cm; mirror:
20 cm   **34.** 3 ft, 11 ft   **35.** ✒ The graph of a parabola
has one branch whereas the graph of a hyperbola has two
branches. A hyperbola has asymptotes, but a parabola does
not.   **36.** ✒ Function notation rarely appears in this
chapter because many of the relations are not functions.
Function notation could be used for vertical parabolas and
for hyperbolas that have the axes as asymptotes.
37. $(-5, -4\sqrt{2})$, $(-5, 4\sqrt{2})$, $(3, -2\sqrt{2})$, $(3, 2\sqrt{2})$
38. $(x-2)^2 + (y+1)^2 = 25$ **39.** $\frac{x^2}{100} + \frac{y^2}{1} = 1$

Test: Chapter 13, p. 887

1. [13.1] $(x-3)^2 + (y+4)^2 = 12$ **2.** [13.1] $(4, -1)$, $\sqrt{5}$
3. [13.1] $(-2, 3)$, 3
4. [13.1], [13.3] Parabola   **5.** [13.1], [13.3] Circle

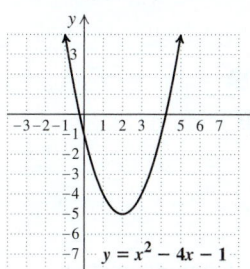

$y = x^2 - 4x - 1$

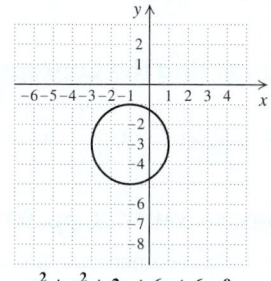

$x^2 + y^2 + 2x + 6y + 6 = 0$

6. [13.3] Hyperbola

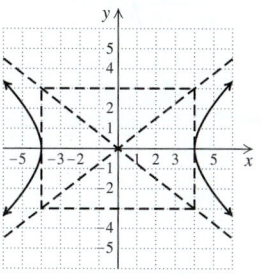

$\frac{x^2}{16} - \frac{y^2}{9} = 1$

7. [13.2], [13.3] Ellipse

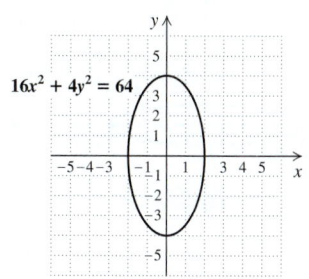

$16x^2 + 4y^2 = 64$

8. [13.3] Hyperbola

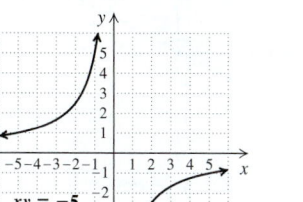

$xy = -5$

9. [13.1], [13.3] Parabola

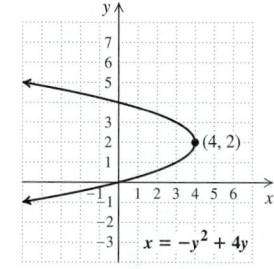

$(4, 2)$ $x = -y^2 + 4y$

10. [13.4] $(0, 6)$, $\left(\frac{144}{25}, \frac{42}{25}\right)$ **11.** [13.4] $(-4, 13)$, $(2, 1)$
12. [13.4] $(3, 2)$, $(-3, -2)$, $\left(-2\sqrt{2}\,i, \frac{3\sqrt{2}}{2}i\right)$,
$\left(2\sqrt{2}\,i, -\frac{3\sqrt{2}}{2}i\right)$ **13.** [13.4] $(\sqrt{6}, 2)$, $(\sqrt{6}, -2)$,
$(-\sqrt{6}, 2)$, $(-\sqrt{6}, -2)$ **14.** [13.4] 2 by 11
15. [13.4] $\sqrt{5}$ m, $\sqrt{3}$ m
16. [13.4] Length: 32 ft; width: 24 ft
17. [13.4] \$1200, 6%   **18.** [13.2] $\frac{(x-6)^2}{25} + \frac{(y-3)^2}{9} = 1$
19. [13.4] 9   **20.** [13.2] $\frac{x^2}{16} + \frac{y^2}{49} = 1$

Cumulative Review: Chapters 1–13, p. 888

1. $16t^4 - 40t^2s + 25s^2$ **2.** $\frac{4t-3}{3t(t-3)}$ **3.** $3t^2\sqrt{10w}$
4. $27a^{1/2}b^{3/16}$ **5.** -4 **6.** 25   **7.** $-\frac{1}{64}$
8. $(10x - 3y)^2$ **9.** $3(m^2 - 2)(m^4 + 2m^2 + 4)$
10. $(x-y)(a-b)$ **11.** $(4x-3)(8x+1)$
12. $\left(-\infty, -\frac{25}{3}\right]$, or $\left\{x \mid x \le -\frac{25}{3}\right\}$ **13.** $0, \frac{9}{8}$ **14.** 1, 4
15. $\pm i$ **16.** 4   **17.** $\frac{\log 1.5}{\log 3} \approx 0.3691$ **18.** 7
19. $(-\sqrt{3}, -1)$, $(-\sqrt{3}, 1)$, $(\sqrt{3}, -1)$, $(\sqrt{3}, 1)$
20.   **21.**

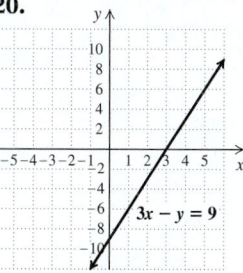

$3x - y = 9$ $y = \log_5 x$

22.

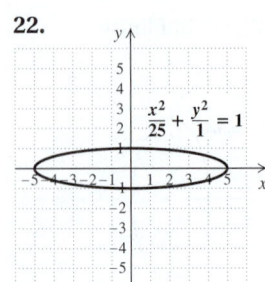

23.

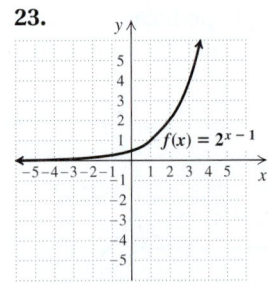

24.

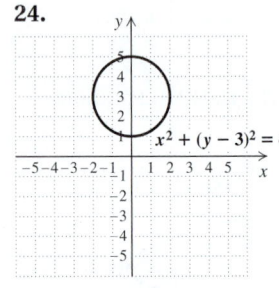

25.

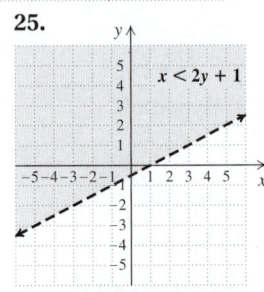

26.

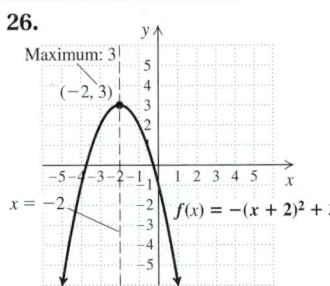

27. $y = -x + 3$
28. $x^2 - 3 = 0$
29. 2640 mi

30. (a) $g(t) = -687.5t + 48,750$;   **(b)** $43,250\ \text{ft}^2$;
(c) about 2019   **31. (a)** $P(t) = 2.19e^{-0.0084t}$;
(b) 1.96 million; **(c)** 83 years   **32.** 8 in. by 8 in.
33. y is divided by 10.   **34.** $(-\infty, 0) \cup (0, 1]$, or
$\{x \mid x < 0\ or\ 0 < x \le 1\}$

CHAPTER 14

Exercise Set 14.1, pp. 894–896

1. B   **2.** A   **3.** B   **4.** A   **5.** A   **6.** B   **7.** (f)
8. (a)   **9.** (d)   **10.** (b)   **11.** (c)   **12.** (e)
13. 43   **15.** 364   **17.** -23.5 **19.** -363 **21.** $\frac{441}{400}$
23. 2, 5, 8, 11; 29; 44   **25.** 3, 6, 11, 18; 102; 227
27. $\frac{1}{2}, \frac{2}{3}, \frac{3}{4}, \frac{4}{5}; \frac{10}{11}; \frac{15}{16}$ **29.** $1, -\frac{1}{2}, \frac{1}{4}, -\frac{1}{8}; -\frac{1}{512}; \frac{1}{16,384}$
31. $-1, \frac{1}{2}, -\frac{1}{3}, \frac{1}{4}; \frac{1}{10}; -\frac{1}{15}$ **33.** $0, 7, -26, 63; 999; -3374$
35. $2n$ **37.** $(-1)^n$ **39.** $(-1)^{n+1} \cdot n$ **41.** $2n + 1$
43. $n^2 - 1$, or $(n+1)(n-1)$ **45.** $\dfrac{n}{n+1}$
47. $(0.1)^n$, or 10^{-n} **49.** $(-1)^n \cdot n^2$ **51.** 5
53. 1.11111, or $1\frac{11,111}{100,000}$ **55.** $\frac{1}{2} + \frac{1}{4} + \frac{1}{6} + \frac{1}{8} + \frac{1}{10} = \frac{137}{120}$
57. $10^0 + 10^1 + 10^2 + 10^3 + 10^4 = 11,111$
59. $2 + \frac{3}{2} + \frac{4}{3} + \frac{5}{4} + \frac{6}{5} + \frac{7}{6} + \frac{8}{7} = \frac{1343}{140}$
61. $(-1)^2 2^1 + (-1)^3 2^2 + (-1)^4 2^3 + (-1)^5 2^4 +$
$(-1)^6 2^5 + (-1)^7 2^6 + (-1)^8 2^7 + (-1)^9 2^8 = -170$
63. $(0^2 - 2 \cdot 0 + 3) + (1^2 - 2 \cdot 1 + 3) +$
$(2^2 - 2 \cdot 2 + 3) + (3^2 - 2 \cdot 3 + 3) +$
$(4^2 - 2 \cdot 4 + 3) + (5^2 - 2 \cdot 5 + 3) = 43$
65. $\dfrac{(-1)^3}{3 \cdot 4} + \dfrac{(-1)^4}{4 \cdot 5} + \dfrac{(-1)^5}{5 \cdot 6} = -\dfrac{1}{15}$ **67.** $\displaystyle\sum_{k=1}^{5} \dfrac{k+1}{k+2}$

69. $\displaystyle\sum_{k=1}^{6} k^2$ **71.** $\displaystyle\sum_{k=2}^{n} (-1)^k k^2$ **73.** $\displaystyle\sum_{k=1}^{\infty} 6k$
75. $\displaystyle\sum_{k=1}^{\infty} \dfrac{1}{k(k+1)}$ **77.** 📋   **79.** $t^2 - t + 1$ **80.** $\dfrac{x}{a}$
81. $\dfrac{5a - 3}{a(a-1)(a+1)}$ **82.** $\dfrac{2t}{t-1}$ **83.** $\dfrac{x-4}{4(x+2)}$
84. $\dfrac{3(y+1)(y-1)}{y}$ **85.** 📋   **87.** 1, 3, 13, 63, 313, 1563
89. $2500, $2000, $1600, $1280, $1024, $819.20, $655.36,
$524.29, $419.43, $335.54   **91.** $S_{100} = 0; S_{101} = -1$
93. $i, -1, -i, 1, i; i$ **95.** 11th term

Prepare to Move On, p. 896

1. 98   **2.** -15 **3.** $a_1 + a_n$ **4.** d

Exercise Set 14.2, pp. 902–904

1. Arithmetic series   **2.** Common difference
3. First term   **4.** Sum   **5.** $a_1 = 8, d = 5$
7. $a_1 = 7, d = -4$ **9.** $a_1 = \frac{3}{2}, d = \frac{3}{4}$
11. $a_1 = $8.16, d = 0.30 **13.** 154   **15.** -94
17. $-$1628.16$ **19.** 26th   **21.** 57th   **23.** 178
25. 5   **27.** 28   **29.** $a_1 = 8; d = -3; 8, 5, 2, -1, -4$
31. $a_1 = 1; d = 1$ **33.** 780   **35.** 31,375   **37.** 2550
39. 918   **41.** 1030   **43.** 35 musicians; 315 musicians
45. 180 stones   **47.** $49.60   **49.** 📋
51. $y = \frac{1}{3}x + 10$ **52.** $y = -4x + 11$ **53.** $y = -2x + 10$
54. $y = -\frac{4}{3}x - \frac{16}{3}$ **55.** $x^2 + y^2 = 16$
56. $(x+2)^2 + (y-1)^2 = 20$ **57.** 📋
59. 33 jumps   **61.** Let $d = $ the common difference.
Since p, m, and q form an arithmetic sequence, $m = p + d$
and
$$q = p + 2d.\ \text{Then } \dfrac{p+q}{2} = \dfrac{p + (p + 2d)}{2} = p + d = m.$$
63. 156,375

Quick Quiz: Sections 14.1–14.2, p. 904

1. 19   **2.** $a_n = (n-1)^2$ **3.** $\displaystyle\sum_{k=1}^{\infty} 2k$ **4.** -0.5 **5.** 1

Prepare to Move On, p. 904

1. 315   **2.** 50   **3.** $\frac{2}{3}$

Connecting the Concepts, p. 909

1. -3 **2.** $-\frac{1}{2}$ **3.** 110   **4.** 640   **5.** 4410
6. $1146.39   **7.** 1   **8.** No

Exercise Set 14.3, pp. 911–914

1. Infinite; geometric; sequence   **2.** Finite; geometric;
series   **3.** Ratio; less; does   **4.** $0.22 \ldots$; geometric;
series   **5.** Geometric sequence   **6.** Arithmetic sequence
7. Geometric series   **8.** Arithmetic series
9. Geometric series   **10.** None of these   **11.** 2
13. -0.1 **15.** $-\frac{1}{2}$ **17.** $\frac{1}{5}$ **19.** $\dfrac{6}{m}$ **21.** 1458

23. 243 **25.** 52,488 **27.** \$1423.31 **29.** $a_n = 5^{n-1}$
31. $a_n = (-1)^{n-1}$, or $a_n = (-1)^{n+1}$
33. $a_n = \dfrac{1}{x^n}$, or $a_n = x^{-n}$ **35.** 3066 **37.** $\frac{547}{18}$
39. $\dfrac{1-x^8}{1-x}$, or $(1+x)(1+x^2)(1+x^4)$ **41.** \$5134.51
43. 27 **45.** $\frac{49}{4}$ **47.** No **49.** No **51.** $\frac{43}{9}$
53. \$25,000 **55.** $\frac{5}{9}$ **57.** $\frac{343}{99}$ **59.** $\frac{5}{33}$ **61.** $\frac{5}{1024}$ ft
63. 155,797 **65.** 2710 flies **67.** About 1456 GW
69. 3100.35 ft **71.** 20.48 in. **73.**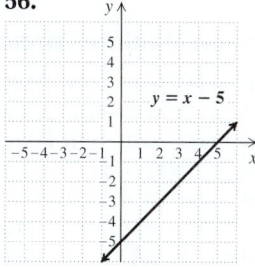
75. $\{-8, 14\}$ **76.** $\left(x\middle|-\frac{11}{2} < x < \frac{1}{2}\right)$, or $\left(-\frac{11}{2}, \frac{1}{2}\right)$
77. $\left\{x\middle|x \le 2\ or\ x \ge \frac{8}{3}\right\}$, or $(-\infty, 2] \cup \left[\frac{8}{3}, \infty\right)$
78. $\left\{x\middle|-\frac{1}{3} < x < \frac{11}{3}\right\}$, or $\left(-\frac{1}{3}, \frac{11}{3}\right)$
79. $\{x\mid -2 < x < 7\}$, or $(-2, 7)$
80. $\{x\mid -1 \le x < 0\ or\ x \ge 1\}$, or $[-1, 0) \cup [1, \infty)$
81. **83.** 54 **85.** $\dfrac{x^2[1 - (-x)^n]}{1+x}$ **87.** 512 cm^2
89. **91.**

Quick Quiz: Sections 14.1–14.3, p. 914

1. $-1, 2, -3, 4; 20$ **2.** 140 **3.** 435 **4.** $511\frac{1}{2}$
5. $\frac{155}{99}$

Prepare to Move On, p. 914

1. $x^2 + 2xy + y^2$ **2.** $x^3 + 3x^2y + 3xy^2 + y^3$
3. $x^3 - 3x^2y + 3xy^2 - y^3$
4. $x^4 - 4x^3y + 6x^2y^2 - 4xy^3 + y^4$
5. $8x^3 + 12x^2y + 6xy^2 + y^3$
6. $8x^3 - 12x^2y + 6xy^2 - y^3$

Mid-Chapter Review: Chapter 14, p. 915

1. $a_n = a_1 + (n-1)d$ **2.** $a_n = a_1 r^{n-1}$
$\quad n = 14, a_1 = -6, d = 5 \qquad n = 7, a_1 = \frac{1}{9}, r = -3$
$\quad a_{14} = -6 + (14-1)5 \qquad a_7 = \frac{1}{9} \cdot (-3)^{7-1}$
$\quad a_{14} = 59 \qquad\qquad\qquad a_7 = 81$
3. 300 **4.** $\dfrac{1}{n+1}$ **5.** 78 **6.** $2^2 + 3^2 + 4^2 + 5^2 = 54$
7. $\displaystyle\sum_{k=1}^{6}(-1)^{k+1}\cdot k$ **8.** 61st **9.** -39 **10.** $\dfrac{1}{10^8}$
11. $2(-1)^{n+1}$ **12.** Yes, a limit exists; $\frac{250}{3}$, or $83\frac{1}{3}$
13. \$465 **14.** \$1,073,741,823

Technology Connection, p. 919

1. 479,001,600 **2.** 56; 792

Exercise Set 14.4, pp. 921–923

1. Binomial **2.** Expansion **3.** First **4.** Third
5. Factorial **6.** Binomial **7.** 2^5, or 32 **8.** 8
9. 9 **10.** 1 **11.** 24 **13.** 3,628,800 **15.** 90
17. 126 **19.** 210 **21.** 1 **23.** 435 **25.** 780
27. $a^4 - 4a^3b + 6a^2b^2 - 4ab^3 + b^4$
29. $p^7 + 7p^6q + 21p^5q^2 + 35p^4q^3 + 35p^3q^4 + 21p^2q^5 + 7pq^6 + q^7$

31. $2187c^7 - 5103c^6d + 5103c^5d^2 - 2835c^4d^3 + 945c^3d^4 - 189c^2d^5 + 21cd^6 - d^7$
33. $t^{-12} + 12t^{-10} + 60t^{-8} + 160t^{-6} + 240t^{-4} + 192t^{-2} + 64$
35. $19{,}683s^9 + \dfrac{59{,}049s^8}{t} + \dfrac{78{,}732s^7}{t^2} + \dfrac{61{,}236s^6}{t^3} + \dfrac{30{,}618s^5}{t^4} + \dfrac{10{,}206s^4}{t^5} + \dfrac{2268s^3}{t^6} + \dfrac{324s^2}{t^7} + \dfrac{27s}{t^8} + \dfrac{1}{t^9}$
37. $x^{15} - 10x^{12}y + 40x^9y^2 - 80x^6y^3 + 80x^3y^4 - 32y^5$
39. $125 + 150\sqrt{5}t + 375t^2 + 100\sqrt{5}t^3 + 75t^4 + 6\sqrt{5}t^5 + t^6$
41. $x^{-3} - 6x^{-2} + 15x^{-1} - 20 + 15x - 6x^2 + x^3$
43. $15a^4b^2$ **45.** $-64{,}481{,}508a^3$ **47.** $1120x^{12}y^2$
49. $1{,}959{,}552u^5v^{10}$ **51.** y^8 **53.**
55. **56.**
57. **58.**
59. **60.**
61. **63.** List all the subsets of size 3: $\{a, b, c\}, \{a, b, d\},$ $\{a, b, e\}, \{a, c, d\}, \{a, c, e\}, \{a, d, e\}, \{b, c, d\}, \{b, c, e\},$ $\{b, d, e\}, \{c, d, e\}.$ There are exactly 10 subsets of size 3 and $\binom{5}{3} = 10$, so there are exactly $\binom{5}{3}$ ways of forming a subset of size 3 from $\{a, b, c, d, e\}$.
65. $\binom{8}{5}(0.15)^3(0.85)^5 \approx 0.084$ **67.** $\binom{8}{6}(0.15)^2(0.85)^6 + \binom{8}{7}(0.15)(0.85)^7 + \binom{8}{8}(0.85)^8 \approx 0.89$
69.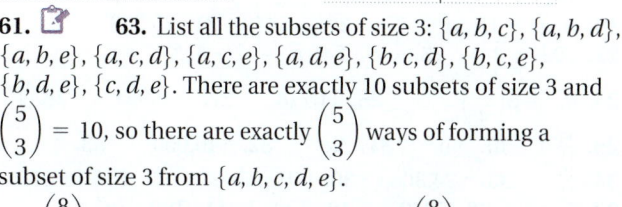

71. $-4320x^6y^{9/2}$ **73.** $-\dfrac{35}{x^{1/6}}$ **75.** 12

Quick Quiz: Sections 14.1–14.4, p. 923

1. $\dfrac{1+1}{1} + \dfrac{2+1}{2} + \dfrac{3+1}{3} + \dfrac{4+1}{4} = \dfrac{73}{12}$ **2.** -0.4
3. $-\frac{1}{2}$ **4.** 220   **5.** $x^4 + 4x^3w + 6x^2w^2 + 4xw^3 + w^5$

Visualizing for Success, p. 924

1. J   **2.** G   **3.** A   **4.** H   **5.** I   **6.** B   **7.** E
8. D   **9.** F   **10.** C

Decision Making: Connection, p. 925

1. (a) $1000, 1040, 1080, 1120, \ldots$;
(b) $a_n = 1000 + 40(n-1)$;   (c) $\$1800$
2. (a) $1000, 1040, 1081.60, 1124.87, \ldots$;
(b) $a_n = 1000(1.04)^{n-1}$;   (c) $\$2191.12$
3. (a) $1000, 1010, 1020, 1030, \ldots$; or
$a_n = 1000 + 10(n-1)$; $\$1800$
(b) $1000, 1010, 1020.10, 1030.30, \ldots$; or $a_n = 1000(1.01)^{n-1}$;
$\$2216.72$ **4.** ⌨

Study Summary: Chapter 14, pp. 926–927

1. 143   **2.** -25 **3.** $5 \cdot 0 + 5 \cdot 1 + 5 \cdot 2 + 5 \cdot 3 = 30$
4. 15.5   **5.** 215   **6.** -640 **7.** $-20{,}475$
8. 16   **9.** $39{,}916{,}800$ **10.** 84
11. $x^{10} - 10x^8 + 40x^6 - 80x^4 + 80x^2 - 32$ **12.** $3240t^7$

Review Exercises: Chapter 14, pp. 927–928

1. False   **2.** True   **3.** True   **4.** False   **5.** False
6. True   **7.** False   **8.** False   **9.** $1, 11, 21, 31$; 71; 111
10. $0, \frac{1}{5}, \frac{1}{5}, \frac{3}{17}, \frac{7}{65}, \frac{11}{145}$ **11.** $a_n = -5n$
12. $a_n = (-1)^n(2n-1)$
13. $-2 + 4 + (-8) + 16 + (-32) = -22$
14. $-3 + (-5) + (-7) + (-9) + (-11) + (-13) = -48$
15. $\displaystyle\sum_{k=1}^{6} 7k$ **16.** $\displaystyle\sum_{k=1}^{5} \dfrac{1}{(-2)^k}$ **17.** -55 **18.** $\frac{1}{5}$
19. $a_1 = -15, d = 5$ **20.** -544 **21.** $25{,}250$
22. $1024\sqrt{2}$ **23.** $\frac{3}{4}$ **24.** $a_n = 2(-1)^n$
25. $a_n = 3\left(\dfrac{x}{4}\right)^{n-1}$ **26.** $11{,}718$ **27.** $-4095x$ **28.** 12
29. $\frac{49}{11}$ **30.** No   **31.** No   **32.** $\$40{,}000$ **33.** $\frac{5}{9}$
34. $\frac{16}{11}$ **35.** $\$24.30$ **36.** 903 poles   **37.** $\$15{,}791.18$
38. 6 m   **39.** 5040   **40.** 120   **41.** $190a^{18}b^2$
42. $x^4 - 8x^3y + 24x^2y^2 - 32xy^3 + 16y^4$
43. ⌨ For a geometric sequence with $|r| < 1$, as n gets larger, the absolute value of the terms gets smaller, since $|r^n|$ gets smaller.   **44.** ⌨ The first form of the binomial theorem draws the coefficients from Pascal's triangle; the second form uses factorial notation. The second form avoids the need to compute all preceding rows of Pascal's triangle, and is generally easier to use when only one term of an expansion is needed. When several terms of an expansion are needed and n is not large (say, $n \le 8$), it is often easier to use Pascal's triangle.   **45.** $\dfrac{1 - (-x)^n}{x + 1}$
46. $x^{-15} + 5x^{-9} + 10x^{-3} + 10x^3 + 5x^9 + x^{15}$

Test: Chapter 14, p. 929

1. [14.1] $\frac{1}{2}, \frac{1}{5}, \frac{1}{10}, \frac{1}{17}, \frac{1}{26}; \frac{1}{145}$ **2.** [14.1] $a_n = 4\left(\frac{1}{3}\right)^n$
3. [14.1] $-3 + (-7) + (-15) + (-31) = -56$
4. [14.1] $\displaystyle\sum_{k=1}^{5} (-1)^{k+1} k^3$ **5.** [14.2] $\frac{13}{2}$
6. [14.2] $a_1 = 31.2$; $d = -3.8$ **7.** [14.2] 2508
8. [14.3] 1536   **9.** [14.3] $\frac{2}{3}$ **10.** [14.3] 3^n
11. [14.3] 5621   **12.** [14.3] 1   **13.** [14.3] No
14. [14.3] $\frac{\$25{,}000}{23} \approx \$1086.96$ **15.** [14.3] $\frac{85}{99}$
16. [14.2] 63 seats   **17.** [14.2] $\$17{,}100$
18. [14.3] $\$5987.37$ **19.** [14.3] 36 m   **20.** [14.4] 220
21. [14.4] $x^5 - 15x^4y + 90x^3y^2 - 270x^2y^3 + 405xy^4 - 243y^5$ **22.** [14.4] $220a^9x^3$
23. [14.2] $n(n+1)$ **24.** [14.3] $\dfrac{1 - \left(\frac{1}{x}\right)^n}{1 - \frac{1}{x}}$, or $\dfrac{x^n - 1}{x^{n-1}(x-1)}$

Cumulative Review/Final Exam: Chapters 1–14, pp. 930–931

1. $\frac{7}{15}$ **2.** $-4y + 17$ **3.** 280   **4.** 8.4×10^{-15}
5. $3a^2 - 8ab - 15b^2$ **6.** $4a^2 - 1$
7. $9a^4 - 30a^2y + 25y^2$ **8.** $\dfrac{4}{x+2}$
9. $\dfrac{(x+y)(x^2 + xy + y^2)}{x^2 + y^2}$ **10.** $x - a$ **11.** $12a^2\sqrt{b}$
12. $-27x^{10}y^{-2}$, or $-\dfrac{27x^{10}}{y^2}$ **13.** $25x^4y^{1/3}$
14. $y\sqrt[12]{x^5y^2}, y \ge 0$ **15.** $14 + 8i$
16. $(2x-3)^2$ **17.** $(3a-2)(9a^2 + 6a + 4)$
18. $12(s^2 + 2t)(s^2 - 2t)$ **19.** $3(y^2 + 3)(5y^2 - 4)$
20. $7x^3 + 9x^2 + 19x + 38 + \dfrac{72}{x-2}$
21. $[4, \infty)$, or $\{x \mid x \ge 4\}$ **22.** $(-\infty, 5) \cup (5, \infty)$,
or $\{x \mid x < 5 \text{ or } x > 5\}$ **23.** $y = 3x - 8$
24. $x^2 - 50 = 0$ **25.** $(2, -3)$; 6
26. $\log_a \dfrac{\sqrt[3]{x^2} \cdot z^5}{\sqrt{y}}$ **27.** $a^5 = c$ **28.** 2.0792
29. 0.6826   **30.** 5   **31.** -121 **32.** 875
33. $16\left(\frac{1}{4}\right)^{n-1}$ **34.** $13{,}440a^4b^6$ **35.** $\frac{3}{5}$ **36.** $-\frac{6}{5}, 4$
37. $\mathbb{R}$, or $(-\infty, \infty)$ **38.** $\left(-1, \frac{1}{2}\right)$ **39.** $(2, -1, 1)$
40. 2   **41.** $\pm 2, \pm 5$ **42.** $(\sqrt{5}, \sqrt{3}), (\sqrt{5}, -\sqrt{3})$,
$(-\sqrt{5}, \sqrt{3}), (-\sqrt{5}, -\sqrt{3})$ **43.** 1.7925   **44.** 1005
45. $-\frac{1}{2}$ **46.** $\{x \mid -2 \le x \le 3\}$, or $[-2, 3]$
47. $\pm i\sqrt{3}$ **48.** $-2 \pm \sqrt{7}$ **49.** $\{y \mid y < -5 \text{ or } y > 2\}$,
or $(-\infty, -5) \cup (2, \infty)$ **50.** $-8, 10$ **51.** $R = \dfrac{Ir}{1 - I}$

52.
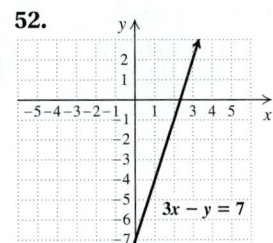

53.
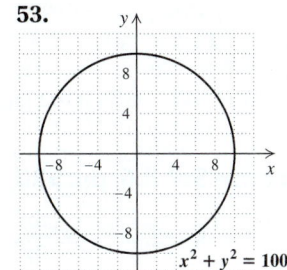

54.
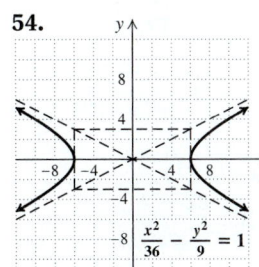

55.
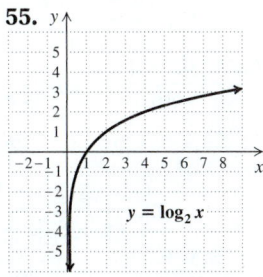

56.
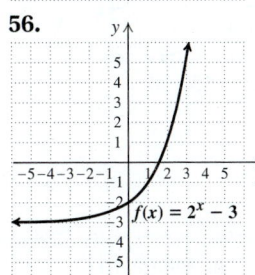

57.
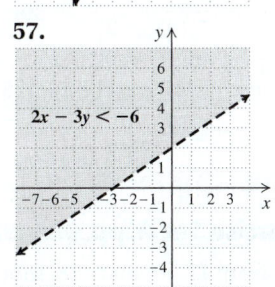

58.
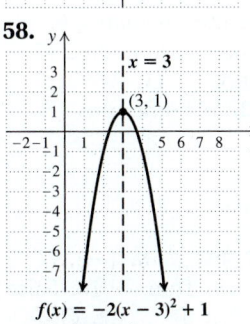

59. 5000 ft² **60.** 5 ft by 12 ft
61. More than 25 downloads
62. $2.68 herb: 10 oz; $4.60 herb: 14 oz **63.** 350 mph
64. $8\frac{2}{5}$ hr, or 8 hr 24 min
65. (a) $1.5 trillion per year;
(b) $f(t) = 1.5t + 9$;
(c) $P(t) = 9e^{0.128t}$;
(d) $27 trillion;
(e) 41.8 trillion;
(f) about 5.4 years
66. All real numbers except 0 and −12 **67.** 81
68. y gets divided by 8 **69.** 84 years

CHAPTER R

Exercise Set R.1, pp. 938–939

1. > **3.** < **5.** > **7.** 22 **9.** 1.3
11. −25 **13.** $-\frac{11}{15}$ **15.** −6.5 **17.** −9 **19.** 0
21. $-\frac{1}{2}$ **23.** 5.8 **25.** −3 **27.** 39 **29.** 175
31. −32 **33.** 16 **35.** −6 **37.** 9 **39.** −3
41. −16 **43.** 100 **45.** 2 **47.** −23 **49.** 36
51. 10 **53.** 10 **55.** −7 **57.** 32 **59.** 28 cm²
61. $8x + 28$ **63.** $-30 + 6x$ **65.** $8a + 12b - 6c$
67. $-6x + 3y - 3z$ **69.** $2(4x + 3y)$ **71.** $3(1 + w)$
73. $10(x + 5y + 10)$ **75.** p **77.** $-m + 22$
79. $-5x + 7$ **81.** $6p - 7$ **83.** $-x + 12y$
85. $36a - 48b$ **87.** $-10x + 104y + 9$ **89.** Let n represent the number; $3n = 348$ **91.** Let c represent the international per capita consumption of Coca-Cola beverages; $412 = 5.15c$ **93.** Let h represent the number of calories in an order of McDonald's hotcakes and sausage; $1060 = 2h + 20$

Exercise Set R.2, pp. 946–948

1. 16 **3.** $-\frac{1}{12}$ **5.** −0.8 **7.** $-\frac{5}{3}$ **9.** 42 **11.** −5
13. $\frac{5}{3}$ **15.** $-\frac{4}{9}$ **17.** −4 **19.** $\frac{69}{5}$ **21.** $\frac{9}{32}$
23. −2 **25.** −15 **27.** $\frac{43}{2}$ **29.** $-\frac{61}{115}$ **31.** $l = \frac{A}{w}$
33. $P = IV$ **35.** $p = 2q - r$ **37.** $\pi = \frac{A}{r^2 + r^2h}$
39. $\{x \mid x \le 12\}$, or $(-\infty, 12]$
41. $\{m \mid m > 12\}$, or $(12, \infty)$
43. $\{x \mid x \ge -\frac{3}{2}\}$, or $[-\frac{3}{2}, \infty)$
45. $\{t \mid t < -3\}$, or $(-\infty, -3)$
47. $\{y \mid y > 10\}$, or $(10, \infty)$ **49.** $\{a \mid a \ge 1\}$, or $[1, \infty)$
51. $\{x \mid x \ge \frac{64}{17}\}$, or $[\frac{64}{17}, \infty)$ **53.** $\{x \mid x > \frac{39}{11}\}$, or $(\frac{39}{11}, \infty)$
55. $\{x \mid x \le -10.875\}$, or $(-\infty, -10.875]$ **57.** 7
59. 16, 18 **61.** $166\frac{2}{3}$ pages **63.** 4.5 cm, 9.5 cm
65. 900 cubic feet **67.** 110¢, or $1.10 **69.** 30 min or more **71.** For $2\frac{7}{9}$ hr or less

Exercise Set R.3, p. 954

1.
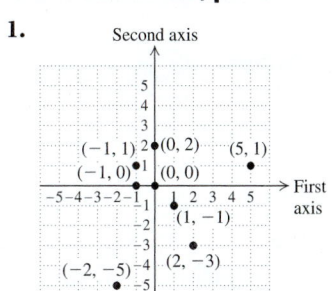
3. II **5.** x-axis
7. IV **9.** No
11. Yes

13. **15.**
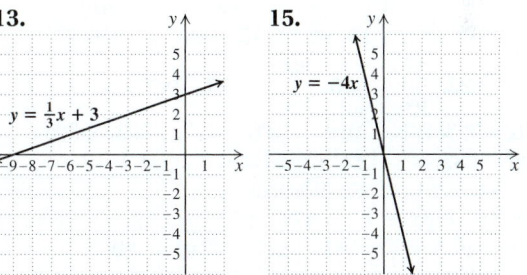

17. −1 **19.** 0 **21.** Slope: 2; y-intercept: $(0, -5)$
23. Slope: $-\frac{2}{7}$; y-intercept: $(0, \frac{1}{7})$
25. **27.**
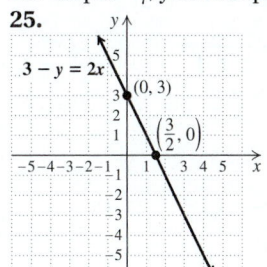 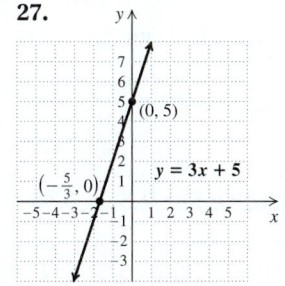

29.

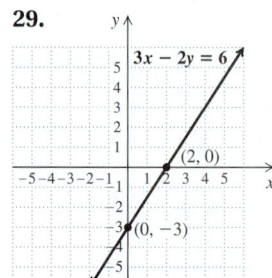

31.

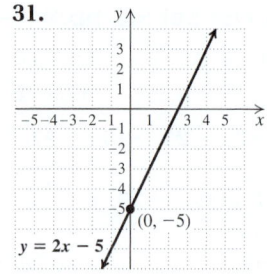

33.

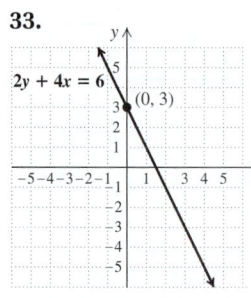

35. 0

37. Undefined

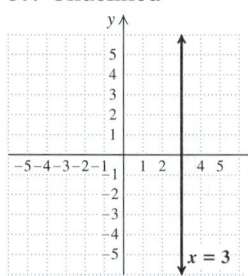

39. $y = 5x + 9$
41. $y = -x + 3$
43. Perpendicular
45. Neither

Exercise Set R.4, pp. 961–962

1. 1   **3.** -3 **5.** $\dfrac{1}{8^2} = \dfrac{1}{64}$ **7.** $\dfrac{10}{x^5}$ **9.** $\dfrac{1}{(ab)^2}$

11. y^{10} **13.** x^{15} **15.** a^6 **17.** $(4x)^9$ **19.** 7^{40}

21. $x^8 y^{12}$ **23.** $\dfrac{y^6}{64}$ **25.** $\dfrac{9q^8}{4p^6}$ **27.** $8x^3, -6x^2, x, -7$

29. $18, 36, -7, 3; 3, 9, 1, 0; 9$ **31.** $-1, 4, -2; 3, 3, 2; 3$
33. $8p^4; 8$ **35.** $x^4 + 3x^3$ **37.** $-7t^2 + 5t + 10$
39. 36   **41.** -14 **43.** 9 words
45. $4x^3 - 3x^2 + 8x + 7$ **47.** $-y^2 + 5y - 2$
49. $-3x^2y - y^2 + 8y$ **51.** $12x^5 - 28x^3 + 28x^2$
53. $8a^2 + 2ab + 4ay + by$ **55.** $x^3 + 4x^2 - 20x + 7$
57. $x^2 - 49$ **59.** $x^2 + 2xy + y^2$ **61.** $6x^4 + 17x^2 - 14$
63. $42a^2 - 17ay - 15y^2$ **65.** $-t^4 - 3t^2 + 2t - 5$

67. $5x + 3$ **69.** $2x^2 - 3x + 3 + \dfrac{-2}{x + 1}$

71. $5x + 3 + \dfrac{3}{x^2 - 1}$

Exercise Set R.5, p. 970

1. $6t^3(3t^2 - 2t + 1)$ **3.** $(y - 3)^2$
5. $(p + 2)(2p^3 + 1)$ **7.** Prime
9. $(2t + 3)(4t^2 - 6t + 9)$ **11.** $(m + 6)(m + 7)$
13. $(x^2 + 9)(x + 3)(x - 3)$ **15.** $(2x + 3)(4x + 5)$
17. $(x + 2)(x + 1)(x - 1)$
19. $(0.1t^2 - 0.2)(0.01t^4 + 0.02t^2 + 0.04)$
21. $\left(x^2 + \frac{1}{4}\right)\left(x + \frac{1}{2}\right)\left(x - \frac{1}{2}\right)$ **23.** $(n - 2)(m + 3)$

25. $(m + 15n)(m - 10n)$ **27.** $2y(3x + 1)(4x - 3)$
29. $(y - 11)^2$ **31.** $-3, 1$ **33.** $0, 11$ **35.** $-3, 3$
37. $-\frac{5}{2}, 7$ **39.** $0, 5$ **41.** $-2, 6$ **43.** $-11, 5$
45. -5 **47.** Base: 8 ft; height: 5 ft   **49.** 6 ft

Exercise Set R.6, pp. 980–981

1. $-\frac{1}{3}$ **3.** $-2, 2$ **5.** $\dfrac{8x}{9y}$ **7.** $\dfrac{t + 2}{t + 4}$ **9.** $\dfrac{-1}{x + 2}$

11. $\dfrac{6}{x}$ **13.** $\dfrac{(a + 1)^3(a - 1)}{a^3}$ **15.** 1   **17.** $\dfrac{5x + 4}{x^2}$

19. $\dfrac{6a + 3b - 4}{3a - 3b}$ **21.** $\dfrac{-3}{5x(x + 2)}$ **23.** $\dfrac{(x + 1)^2}{x^2(x + 2)}$

25. $\dfrac{x + 3}{(x + 1)^2}$ **27.** 1   **29.** $\dfrac{2t + 1}{(t + 1)(t - 1)}$

31. $\dfrac{-3x}{2}$ **33.** $\dfrac{-7(2x + 1)}{(x + 5)(x - 4)}$ **35.** $\dfrac{8x - 4}{x^3}$

37. $\dfrac{3(x + 1)}{(x - 7)(4x + 3)}$ **39.** $\dfrac{(x + 1)^2}{(x - 2)(x + 4)}$ **41.** $\dfrac{-1}{x}$

43. $\frac{6}{5}$ **45.** 1   **47.** $\frac{31}{2}$ **49.** $-4, 4$ **51.** $22\frac{2}{9}$ hr
53. Jessica: 45 km/h; Josh: 25 km/h   **55.** 50

APPENDIXES

Exercise Set A, pp. 985–986

1. Mean: 17; median: 18; mode: 13   **3.** Mean: $10.\overline{6}$; median: 9; modes: 3, 20   **5.** Mean: 4.06; median: 4.6; mode: none   **7.** Mean: $239.\overline{3}$; median: 234; mode: 234
9. Average: 34.875; median: 22; mode: 0   **11.** Average: $218.\overline{3}$; median: 222; mode: 202   **13.** 10 home runs
15. $a = 30, b = 58$

Exercise Set B, pp. 987–988

1. $\{8, 9, 10, 11\}$ **3.** $\{41, 43, 45, 47, 49\}$ **5.** $\{-3, 3\}$
7. True   **9.** True   **11.** False   **13.** True   **15.** False
17. True   **19.** $\{c, d, e\}$ **21.** $\{1, 2, 3, 6\}$ **23.** $\varnothing$
25. $\{a, e, i, o, u, q, c, k\}$ **27.** $\{1, 2, 3, 4, 6, 9, 12, 18\}$
29. $\{1, 2, 3, 4, 5, 6, 7, 8\}$ **31.** 📝   **33.** The set of
integers   **35.** The set of real numbers   **37.** $\varnothing$
39. (a) A; (b) A; (c) A; (d) $\varnothing$ **41.** True   **43.** True
45. True

Exercise Set C, pp. 992–993

1. True   **2.** True   **3.** False   **4.** False   **5.** True
6. True   **7.** $x^2 - 3x - 5$ **9.** $a + 5 + \dfrac{-4}{a + 3}$

11. $2x^2 - 5x + 3 + \dfrac{8}{x + 2}$ **13.** $a^2 + 2a - 6$

15. $3y^2 + 2y + 6 + \dfrac{-2}{y - 3}$ **17.** $x^4 + 2x^3 + 4x^2 + 8x + 16$

19. $3x^2 + 6x - 3 + \dfrac{2}{x + \frac{1}{3}}$ **21.** 6   **23.** 125   **25.** 0

27. 📝   **29.** 📝   **31.** (a) The degree of R must
be less than 1, the degree of $x - r$, (b) Let $x = r$. Then
$P(r) = (r - r) \cdot Q(r) + R = 0 \cdot Q(r) + R = R$.
33. $0; -3; -\frac{5}{2}, \frac{3}{2}$ **35.** 📊   **37.** 0

GLOSSARY

Absolute value [R.1, 1.4]   The distance that a number is from 0 on the number line

Additive identity [1.5]   The number 0

Additive inverse [1.6]   A number's opposite; two numbers are additive inverses of each other if when added the result is zero

Algebraic expression [R.1, 1.1]   An expression consisting of variables and/or numerals, often with operation signs and grouping symbols

Arithmetic sequence (or **Progression**) **[14.2]**   A sequence in which the difference between any two successive terms is constant

Arithmetic series [14.2]   A series for which the associated sequence is arithmetic

Ascending order [4.4]   A polynomial in one variable written with the terms arranged according to degree, from least to greatest

Associative law for addition [R.1, 1.2]   The statement that when three numbers are added, changing the grouping does not change the result

Associative law for multiplication [R.1, 1.2]   The statement that when three numbers are multiplied, changing the grouping does not change the result

Asymptote [13.3]   A line that a graph approaches more and more closely as x increases or as x decreases

Average [2.7, App A]   Most commonly, the *mean* of a set of numbers, found by adding the numbers and then dividing by the number of addends

Axes (singular, **Axis**) **[R.3, 3.1]**   Two perpendicular number lines used to identify points in a plane

Axis of symmetry [11.6]   A line that can be drawn through a graph such that the part of the graph on one side of the line is an exact reflection of the part on the opposite side

Bar graph [3.1]   A graphic display of data using bars proportional in length to the numbers represented

Base [1.8]   In exponential notation, the number being raised to a power

Binomial [R.4, 4.3]   A polynomial with two terms

Branches [13.3]   The two curves that comprise a hyperbola

Break-even point [8.8]   In business, the point of intersection of the revenue function and the cost function

Center point [App A]   A single representative number used to analyze a set of data; also called *measure of central tendency*

Circle [13.1]   A set of points in a plane that are a fixed distance r, called the radius, from a fixed point (h, k), called the center

Circle graph [3.1]   A graphic display of data often used to show what percent of the whole each item in a group represents

Circumference [2.3]   The distance around a circle

Closed interval [a, b] [R.2, 2.6]   The set of all numbers x for which $a \le x \le b$; thus, $[a, b] = \{x \mid a \le x \le b\}$

Coefficient [R.4, 2.1, 4.3]   The numerical multiplier of a variable

Combined variation [7.5]   A mathematical relationship in which a variable varies directly and/or inversely, at the same time, with more than one other variable

Common factor [R.5, 1.2]   A factor that appears in every term in an expression

Common logarithm [12.5]   A logarithm with base 10

Commutative law for addition [R.1, 1.2]   The statement that when two numbers are added, changing the order of addition does not affect the answer

Commutative law for multiplication [R.1, 1.2]   The statement that when two numbers are multiplied, changing the order of multiplication does not affect the answer

Completing the square [11.1]   A method of adding a particular constant to an expression so that the resulting sum is a perfect square

Complex number [10.8]   Any number that can be written in the form $a + bi$, where a and b are real numbers and $i = \sqrt{-1}$

Complex rational expression [R.6, 6.5]   A rational expression that contains rational expressions within its numerator and/or its denominator

Complex-number system [10.8]   A number system that contains the real-number system and is designed so that negative numbers have defined square roots

Composite function [12.1]   A function in which some quantity depends on a variable that, in turn, depends on another variable

Composite number [1.3]   A natural number other than 1 that is not prime

Compound inequality [9.2]   A statement in which two or more inequalities are joined by the word "and" or the word "or"

Compound interest [11.1]   Interest earned on both the initial investment and the interest from previous periods

Conditional equation [R.2, 2.2]   An equation that is true for some replacements and false for others

Conic section [13.1]   A curve formed by the intersection of a plane and a cone

Conjugate of a complex number [10.8]   The *conjugate* of a complex number $a + bi$ is $a - bi$, and the *conjugate* of $a - bi$ is $a + bi$.

Conjugates [10.5]   Pairs of radical expressions like $\sqrt{a} + \sqrt{b}$ and $\sqrt{a} - \sqrt{b}$

Conjunction [9.2]   A sentence in which two or more statements are joined by the word "and"

Consecutive even integers [2.5]   Integers that are even and two units apart

Consecutive integers [2.5]   Integers that are one unit apart

Consecutive odd integers [2.5]   Integers that are odd and two units apart

Consistent system of equations [8.1, 8.4]   A system of equations that has at least one solution

Constant [R.1, 1.1]   A particular number that never changes

Constant of proportionality [7.5]   The constant, k, in an equation of variation; also called *variation constant*

Constant function [7.3]   A function given by an equation of the form $f(x) = b$, where b is a real number

Constant polynomial [R.4]   A polynomial of degree 0

Contradiction [R.2, 2.2]   An equation that is never true

Coordinates [R.3, 3.1]   The numbers in an ordered pair

Counting numbers [1.3]   The set of numbers used for counting: $\{1, 2, 3, 4, 5, \ldots\}$; also called *natural numbers*

Cube root [10.1]   The number c is the *cube root* of a if $c^3 = a$.

Cubic polynomial [R.4]   A polynomial in one variable of degree 3

Data [1.1]   A set of numbers used to represent information

Decay rate [12.7]   The rate of decay of a population or other quantity at any instant in time

Degree of a monomial [R.4, 4.3]   The number of variable factors in the monomial

Degree of a polynomial [R.4, 4.3, 4.7]   The degree of the term of highest degree in a polynomial

Demand function [8.8]   A function modeling the relationship between the price of a good and the quantity of that good demanded

Denominator [1.3]   The number below the fraction bar in a fraction

Dependent equations [8.1, 8.4]   Equations in a system from which one equation can be removed without changing the solution set

Descending order [R.4, 4.4]   A polynomial in one variable written with the terms arranged according to degree, from greatest to least

Determinant [8.7]   A descriptor of a square matrix; the determinant of a two-by-two matrix $\begin{bmatrix} a & c \\ b & d \end{bmatrix}$ is denoted by $\begin{vmatrix} a & c \\ b & d \end{vmatrix}$ and is defined as $ad - bc$.

Difference [1.2]   The result when two numbers are subtracted

Difference of cubes [R.5, 5.5]   An expression that can be written in the form $A^3 - B^3$

Difference of squares [R.4, R.5, 4.6, 5.4]   An expression that can be written in the form $A^2 - B^2$

Direct variation [7.5]   A situation that can be modeled by a linear function of the form $f(x) = kx$, or $y = kx$, where k is a nonzero constant

Discriminant [11.3]   The radicand $b^2 - 4ac$ from the quadratic formula

Disjunction [9.2]   A sentence in which two or more statements are joined by the word "or"

Distance formula [10.7]   The formula $d = \sqrt{(x_2 - x_1)^2 + (y_2 - y_1)^2}$, where d is the distance between any two points (x_1, y_1) and (x_2, y_2)

Distributive law [R.1, 1.2]   The statement that multiplying a factor by the sum of two numbers gives the same result as multiplying the factor by each of the two numbers and then adding

Domain [7.1, 7.2]   The set of all first coordinates of the ordered pairs in a function

Doubling time [12.7]   The amount of time necessary for a population to double in size

Elements [App B]   The objects of a set

Elements of a matrix [8.6]   The individual entries in a matrix

Elimination method [8.2]   An algebraic method that uses the addition principle to solve a system of equations

Ellipse [13.2]   The set of all points in a plane for which the sum of the distances from two fixed points F_1 and F_2 is constant

Empty set [2.2, App B]   The set containing no elements, written $\varnothing$ or $\{\ \}$

Equation [1.1]   A number sentence formed by placing an equals sign between two expressions

Equilibrium point [8.8]   The point of intersection between the demand function and the supply function

Equivalent equations [R.2, 2.1]   Equations that have the same solutions

Equivalent expressions [R.1, 1.2]   Expressions that have the same value for all allowable replacements

Equivalent inequalities [R.2, 2.6]   Inequalities that have the same solution set

Evaluate [R.1, 1.1]   To substitute a number for each variable in the expression and calculate the result

Even root [10.1]   A root with an even index

Exponent [1.8]   In an expression of the form b^n, the number n is an exponent.

Exponential decay [12.7]   A decrease in quantity over time that can be modeled by an exponential function of the form $P(t) = P_0 e^{-kt}$, $k > 0$, where P_0 is the quantity present at time 0, $P(t)$ is the amount present at time t, and k is the exponential decay rate

Exponential equation [12.6]   An equation with a variable in an exponent.

Exponential function [12.2]   A function $f(x) = a^x$, where a is a positive constant, $a \neq 1$, and x is any real number

Exponential growth [12.7]   An increase in quantity over time that can be modeled by an exponential function of the form $P(t) = P_0 e^{kt}$, $k > 0$, where P_0 is the quantity present at time 0, $P(t)$ is the amount present at time t, and k is the exponential growth rate

Exponential notation [R.1, 1.8]   A representation of a number using a base raised to an exponent

Extrapolation [3.7]   The process of estimating a value that goes beyond the given data

Factor [R.1, 1.2, 5.1]   *Verb:* To write an equivalent expression that is a product; *noun:* part of a product

Factoring [1.2]   The process of rewriting an expression as a product

Factoring by grouping [R.5, 5.1]   If a polynomial can be split into groups of terms and the groups share a common factor, then the original polynomial can be factored. This method can be tried on any polynomial with four or more terms.

Finite sequence [14.1]   A function having for its domain a set of natural numbers: $\{1, 2, 3, 4, 5, \ldots, n\}$, for some natural number n

Finite series [14.1]   The sum of the first n terms of a sequence: $a_1 + a_2 + a_3 + \cdots + a_n$; also called *partial sum*

Fixed costs [8.8]   In business, costs that must be paid regardless of how many items are produced

Foci (singular, **Focus**) **[13.2]**   Two fixed points that determine the points of an ellipse

FOIL method [R.4, 4.6]   To multiply two binomials $A + B$ and $C + D$, multiply the First terms AC, the Outside terms AD, the Inner terms BC, and then the Last terms BD. Then add the results.

Formula [R.2, 2.3]   An equation using numbers and/or letters to represent a relationship between two or more quantities

Fraction notation [1.3]   A number written using a numerator and a denominator

Function [7.1]   A correspondence between a first set, called the *domain*, and a second set, called the *range*, such that each member of the domain corresponds to *exactly one* member of the range

General term of a sequence [14.1]   The nth term, denoted a_n

Geometric sequence [14.3]   A sequence in which the ratio of every pair of successive terms is constant

Geometric series [14.3]   A series for which the associated sequence is geometric

Grade [3.5]   The ratio of the vertical distance a road rises over the horizontal distance it runs, expressed as a percent

Graph [R.2, 2.6]   A picture or a diagram of the data in a table; a line, a curve, a plane, a collection of points, etc., that represents all the solutions of an equation or an inequality

Grouping method [5.3]   A method for factoring a trinomial of the type $ax^2 + bx + c$, that uses factoring by grouping

Growth rate [12.7]   The rate of growth of a population or other quantity at any instant in time

Half-life [12.7]   The amount of time necessary for half of a quantity to decay

Half-open intervals $(a, b]$ and $[a, b)$ [R.2, 2.6]   An interval that contains one endpoint and not the other; thus, $(a, b] = \{x \mid a < x \leq b\}$ and $[a, b) = \{x \mid a \leq x < b\}$

Horizontal line [R.3, 3.3, 3.7]   The graph of $y = b$ is a horizontal line, with y-intercept $(0, b)$.

Horizontal-line test [12.1]   If it is impossible to draw a horizontal line that intersects a function's graph more than once, then the function is one-to-one.

Hyperbola [13.3]   The set of all points P in the plane such that the difference of the distance from P to two fixed points is constant

Hypotenuse [R.5, 5.8, 10.7]   In a right triangle, the side opposite the 90° angle

i [10.8]   The square root of -1; that is, $i = \sqrt{-1}$ and $i^2 = -1$

Identity [R.2, 2.2]   An equation that is true for all replacements

Identity property of 0 [R.1, 1.5]   The statement that the sum of 0 and a number is the original number

Identity property of 1 [R.1, 1.3]   The statement that the product of 1 and a number is the original number

Imaginary number [10.8]   A number that can be written in the form $a + bi$, where a and b are real numbers and $b \neq 0$ and $i = \sqrt{-1}$

Inconsistent system of equations [8.1, 8.4]   A system of equations for which there is no solution

Independent equations [8.1, 8.4]   Equations that are not dependent

Index (plural, **Indices**) **[10.1]**   In the radical, $\sqrt[n]{a}$, the number n is called the index.

Inequality [1.4, 9.1]   A mathematical sentence using $<, >, \leq, \geq$, or $\neq$

Infinite geometric series [14.3]   The sum of the terms of an infinite geometric sequence

Infinite sequence [14.1]   A function having for its domain the set of natural numbers: $\{1, 2, 3, 4, 5, \ldots\}$

Infinite series [14.1]   Given the infinite sequence $a_1, a_2, a_3, a_4, \ldots a_n, \ldots$, the sum of the terms $a_1 + a_2 + a_3 + a_4 + \cdots a_n + \cdots$ is called an infinite series.

Input [7.1]   An element of the domain of a function

Integers [R.1, 1.4]   The set of all whole numbers and their opposites: $\{\ldots, -4, -3, -2, -1, 0, 1, 2, 3, 4, \ldots\}$

Interpolation [3.7]   The process of estimating a value between given values

Intersection of sets A and B [9.2, App B]   The set of all elements that are common to both A and B; denoted $A \cap B$

Interval notation [R.2, 2.6]   The use of a pair of numbers inside parentheses and/or brackets to represent the set of all numbers between and sometimes including those two numbers; see also *open, closed,* and *half-open intervals*

Inverse relation [12.1]   The relation formed by interchanging the members of the domain and the range of a relation

Inverse variation [7.5]   A situation that can be modeled by a rational function of the form $f(x) = k/x$, or $y = k/x$, where k is a nonzero constant

Irrational number [R.1, 1.4, 10.1]   A real number that cannot be written as the ratio of two integers; when written in decimal notation, an irrational number neither terminates nor repeats

Isosceles right triangle [10.7]   A right triangle in which both legs have the same length

Joint variation [7.5]   A situation that can be modeled by an equation of the form $y = kxz$, where k is a nonzero constant

Largest common factor [R.5, 5.1]   The largest common factor of a polynomial is the largest common factor of the coefficients times the largest common factor of the variable(s) in all of the terms.

Leading coefficient [R.4, 4.3]   The coefficient of the term of highest degree in a polynomial

Leading term [R.4, 4.3]   The term of highest degree in a polynomial

Least common denominator (LCD) [R.6, 6.3]   The least common multiple of the denominators of two or more fractions

Least common multiple (LCM) [R.6, 6.3]   The smallest number that is a multiple of two or more numbers

Legs [R.5, 5.8, 10.7]   In a right triangle, the two sides that form the 90° angle

Like radicals [10.5]   Radical expressions that have the same indices and radicands

Like terms [R.1, R.4, 1.5, 1.8, 4.7]   Terms containing the same variable(s) raised to the same power(s); also called *similar terms*

Line graph [3.1]   A graph in which quantities are represented as points connected by straight-line segments

Linear equation [R.3, 2.2, 3.2]   In two variables, any equation whose graph is a straight line and can be written in the form $y = mx + b$ or $Ax + By = C$, where x and y are variables and $m, b, A, B,$ and C are constants

Linear equation in three variables [8.4]   An equation equivalent to one of the form $Ax + By + Cz = D$, where $A, B, C,$ and D are constants

Linear function [7.3]   A function whose graph is a straight line and can be described by an equation of the form $f(x) = mx + b$, where m and b are constants

Linear inequality [9.4]   An inequality whose related equation is a linear equation

Linear polynomial [R.4]   A polynomial of degree 1

Linear regression [3.7]   A method for finding an equation for a line that best fits a set of data

Logarithmic equation [12.6]   An equation containing a logarithmic expression

Logarithmic function, base a [12.3]   The inverse of an exponential function $f(x) = a^x$, written $f^{-1}(x) = \log_a x$

Mathematical model [1.1]   A mathematical representation of a real-world situation

Matrix (plural, **Matrices**) **[8.6]**   A rectangular array of numbers

Maximum value [11.6]   The greatest function value (output) achieved by a function

Mean [2.7, App A]   The sum of a set of numbers, divided by the number of addends

Measure of central tendency [App A]   A single representative number used to analyze a set of data; also called *center point*

Median [App A]   After arranging a set of data from smallest to largest, the middle number if there is an odd number of data numbers, or the average of the two middle numbers if there is an even number of data numbers

Midpoint formula [10.7]   The formula $\left(\dfrac{x_1 + x_2}{2}, \dfrac{y_1 + y_2}{2} \right)$, which represents the midpoint of the line segment with endpoints (x_1, y_1) and (x_2, y_2)

Minimum value [11.6]   The least function value (output) achieved by a function

Mode [App A]   In a set of data, the number or numbers that occur most often; if each number occurs the same number of times, then there is *no* mode

Monomial [R.4, 4.3]   A constant, a variable, or a product of a constant and one or more variables

Motion problem [R.6, 2.5, 6.7, 8.3]   A problem dealing with distance, rate (or speed), and time

Multiplicative identity [1.3]   The number 1

Multiplicative inverses [1.3]   *Reciprocals*; two numbers whose product is 1

Multiplicative property of zero [R.1, 1.7]   The statement that the product of 0 and any real number is 0

Natural logarithm [12.5]   A logarithm with base e; also called Napierian logarithm

Natural numbers [R.1, 1.3]   The numbers used for counting: $\{1, 2, 3, 4, 5, \dots\}$; also called *counting numbers*

Nonlinear equation [3.2]   An equation whose graph is not a straight line

Nonlinear function [7.3]   A function for which the graph is not a straight line

nth root [10.1]   A number c is called the nth root of a if $c^n = a$.

Numerator [1.3]   The number above the fraction bar in a fraction

Odd root [10.1]   A root with an odd index

One-to-one function [12.1]   A function for which different inputs have different outputs

Open interval (a, b) [R.2, 2.6]   The set of all numbers x for which $a < x < b$; thus, $(a, b) = \{x \mid a < x < b\}$

Opposite of a polynomial [R.4, 4.4]   To find the *opposite* of a polynomial, change the sign of every term; this is the same as multiplying the polynomial by -1

Opposites [R.1, 1.6]   Two expressions whose sum is 0; *additive inverses*

Ordered pair [R.3, 3.1]   A pair of numbers of the form (x, y) for which the order in which the numbers are listed is important

Origin [R.3, 3.1]   The point on a coordinate plane where the two axes intersect

Output [7.1]   An element of the range of a function

Parabola [11.1, 11.6, 13.1]   A graph of a quadratic function

Parallel lines [R.3, 3.6]   Lines in the same plane that never intersect; two lines are parallel if they have the same slope or if both lines are vertical

Partial sum [14.1]   The sum of the first n terms of a sequence: $a_1 + a_2 + a_3 + \cdots + a_n$; also called *finite series*

Pascal's triangle [14.4]   A triangular array of coefficients of the expansion $(a + b)^n$ for $n = 0, 1, 2, \dots$

Percent notation [2.4]   The percent symbol % means "per hundred."

Perfect-square trinomial [R.5, 5.4]   A trinomial that is the square of a binomial

Perpendicular lines [R.3, 3.6]   Two lines that intersect to form a right angle; two lines are perpendicular if the product of their slopes is -1 or if one line is vertical and the other line is horizontal

Piecewise-defined function [7.2]   A function defined by different equations for various parts of the domain

Point–slope form [R.3, 3.7, 7.3]   Any equation of the form $y - y_1 = m(x - x_1)$, where the slope of the line is m and the line passes through (x_1, y_1)

Polynomial [R.4, 4.3]   A monomial or a sum of monomials

Polynomial equation [R.5]   An equation in which two polynomials are set equal to each other

Polynomial inequality [11.9]   An inequality that is equivalent to an inequality with a polynomial as one side and 0 as the other

Prime factorization [1.3]   The factorization of a composite number into a product of prime numbers

Prime number [1.3]   A natural number that has exactly two different natural number factors: the number itself and 1

Prime polynomial [R.5, 5.2]   A polynomial that cannot be factored using rational numbers

Principal square root [10.1]   The nonnegative square root of a number

Product [1.2]   The result when two numbers are multiplied

Progression [14.1]   A function for which the domain is a set of counting numbers beginning with 1; also called *sequence*

Proportion [R.6, 6.7]   An equation stating that two ratios are equal

Pure imaginary number [10.8]   A complex number of the form $a + bi$, in which $a = 0$ and $b \neq 0$

Pythagorean theorem [R.5, 5.8, 10.7]   In any right triangle, if a and b are the lengths of the legs and c is the length of the hypotenuse, then $a^2 + b^2 = c^2$.

Quadrants [R.3, 3.1]   The four regions into which the horizontal axis and the vertical axis divide a plane

Quadratic equation [R.5, 5.7, 11.1]   An equation equivalent to one of the form $ax^2 + bx + c = 0$, where a, b, and c are constants, with $a \neq 0$

Quadratic formula [11.2]   The formula
$$x = \frac{-b \pm \sqrt{b^2 - 4ac}}{2a},$$ which gives the solutions of
$ax^2 + bx + c = 0, a \neq 0$

Quadratic function [11.2]   A second-degree polynomial function in one variable

Quadratic inequality [11.9]   A second-degree polynomial inequality in one variable

Quadratic polynomial [R.4]   A polynomial in one variable of degree 2

Quartic polynomial [R.4]   A polynomial in one variable of degree 4

Quotient [1.2]   The result when two numbers are divided

Radical equation [10.6]   An equation in which a variable appears in a radicand

Radical expression [10.1]   Any expression in which a radical sign appears

Radical sign [10.1]   The symbol $\sqrt{}$

Radical term [10.5]   A term in which a radical sign appears

Radicand [10.1]   The expression under a radical sign

Radius (plural, Radii) [13.1]   The distance from the center of a circle to a point on the circle; a segment connecting a point on the circle to the center of the circle

Range [7.1, 7.2]   The set of all second coordinates of the ordered pairs in a function

Rate [3.4]   A ratio that indicates how two quantities change with respect to each other

Ratio [R.6, 6.7]   The quotient of two quantities

Rational equation [R.6, 6.6]   An equation that contains one or more rational expressions

Rational expression [R.6, 6.1]   A quotient of two polynomials

Rational inequality [11.9]   An inequality involving a rational expression

Rational numbers [R.1, 1.4]   The set of all numbers a/b, such that a and b are integers and $b \neq 0$

Rationalizing the denominator [10.4]   A procedure for finding an equivalent expression without a radical expression in the denominator

Rationalizing the numerator [10.4]   A procedure for finding an equivalent expression without a radical expression in the numerator

Real numbers [R.1, 1.4]   The set of all numbers corresponding to points on the number line

Reciprocals [1.3]   Two numbers whose product is 1; *multiplicative inverses*

Reflection [12.2]   The mirror image of a graph

Relation [7.1]   A correspondence between a first set, called the *domain*, and a second set, called the *range*, such that each member of the domain corresponds to at least one member of the range

Remainder theorem [App C]   The remainder obtained by dividing the polynomial $P(x)$ by $x - r$ is $P(r)$.

Repeating decimal [R.1, 1.4]   A decimal in which a block of digits repeats indefinitely

Revenue [4.8]   The price per item times the quantity of items sold

Right triangle [5.8]   A triangle that has a 90° angle

Roster notation [App B]   Set notation in which the elements of a set are listed within { }

Row-equivalent operations [8.6]   Operations used to produce equivalent systems of equations

Scientific notation [4.2]   An expression of the type $N \times 10^m$, where N is at least 1 but less than 10 (that is, $1 \leq N < 10$), N is expressed in decimal notation, and m is an integer

Sequence [14.1]   A function for which the domain is a set of consecutive counting numbers beginning with 1; also called *progression*

Series [14.1]   The sum of specified terms in a sequence

Set [1.4, App B]   A collection of objects

Set-builder notation [R.2, 2.6, App B]   The naming of a set by describing basic characteristics of the elements in the set

Sigma notation [14.1]   The naming of a sum using the Greek letter Σ (sigma) as part of an abbreviated form; also called *summation notation*

Similar terms [R.1, R.4, 1.5, 1.8, 4.7]   Terms containing the same variable(s) raised to the same power(s); also called *like terms*

Similar triangles [6.7]   Triangles in which corresponding angles have the same measure and corresponding sides are proportional

Slope [R.3, 3.5]   The ratio of vertical change to horizontal change for any two points on a line

Slope–intercept equation [R.3, 3.6, 7.3]   An equation of the form $y = mx + b$, with slope m and y-intercept $(0, b)$

Solution [R.2, 2.1, 2.6, 9.1]   Any replacement or substitution for a variable that makes an equation or inequality true

Solution of a system [8.1]   A solution of a system of two equations makes *both* equations true.

Solution set [2.2, 2.6, 9.1]   The set of all solutions of an equation, an inequality, or a system of equations or inequalities

Solve [R.2, 2.1, 2.6, 9.1]   To find all solutions of an equation, an inequality, or a system of equations or inequalities

Speed [3.4]   The speed of an object is found by dividing the distance traveled by the time required to travel that distance.

Square matrix [8.7]   A matrix with the same number of rows and columns

Square root [10.1]   The number c is a *square root* of a if $c^2 = a$.

Standard form of a linear equation [R.3, 3.7, 7.3]   Any equation of the form $Ax + By = C$, where A, B, and C are real numbers and A and B are not both 0

Subset [App B]   If every element of A is also an element of B, then A is a *subset* of B; denoted $A \subseteq B$

Substitute [R.1, 1.1]   To replace a variable with a number or an expression

Substitution method [8.2]   An algebraic method for solving a system of equations

Sum [1.2]   The result when two numbers are added

Sum of cubes [R.5, 5.5]   An expression that can be written in the form $A^3 + B^3$

Summation notation [14.1]   The naming of a sum using the Greek letter Σ (sigma) as part of an abbreviated form; also called *sigma notation*

Supply function [8.8]   A function modeling the relationship between the price of a good and the quantity of that good supplied

Synthetic division [App C]   A method used to divide a polynomial by a binomial of the type $x - a$

System of equations [8.1]   A set of two or more equations, in two or more variables, for which a common solution is sought

Term [R.1, 1.2, 4.3]   A number, a variable, a product of numbers and/or variables, or a quotient of numbers and/or variables

Terminating decimal [R.1, 1.4]   A decimal that can be written using a finite number of decimal places

Total cost [8.8]   The amount spent to produce a product

Total profit [8.8]   The money taken in less the money spent, or total revenue minus total cost

Total revenue [8.8]   The amount taken in from the sale of a product

Trinomial [R.4, 4.3]   A polynomial with three terms

Undefined [R.1, 1.7]   An expression that has no meaning attached to it

Union of sets A and B [9.2, App B]   The collection of elements belonging to A and/or B; denoted $A \cup B$

Value [1.1]   The numerical result after a number has been substituted for a variable in an expression and calculations have been carried out

Variable [R.1, 1.1]   A letter that represents an unknown number

Variable costs [8.8]   In business, costs that vary according to the quantity being produced

Variable expression [1.1]   An expression containing a variable

Variation constant [7.5]   The constant k in an equation of direct variation or inverse variation; also called *constant of proportionality*

Vertex [11.6]   The "turning point" of the graph of a quadratic equation

Vertical line [R.3, 3.3, 3.7]   The graph of $x = a$ is a vertical line, with x-intercept $(a, 0)$.

Vertical-line test [7.1]   The statement that if it is possible for a vertical line to cross a graph more than once, then the graph is not the graph of a function

Whole numbers [R.1, 1.4]   The set of natural numbers and 0: $\{0, 1, 2, 3, 4, 5, \ldots\}$

x-axis [R.3, 3.1]   The horizontal axis in a coordinate plane

x-intercept [R.3, 3.3]   A point at which a graph crosses the x-axis

y-axis [R.3, 3.1]   The vertical axis in a coordinate plane

y-intercept [R.3, 3.3]   A point at which a graph crosses the y-axis

Zeros [11.9]   The x-values for which $f(x)$ is 0, for any function f

INDEX

Simplifying radical expressions, 639–640
  by dividing, 649–650
  by factoring, 644–646
  of the form $\sqrt{a^2}$, 630–631
  by multiplying and factoring, 646
  with terms with different indices,
    658–660, 691
Slope, 190–196, 224, 949–950
  applications involving, 195–196
  finding equation of a line and,
    213–214
  of a horizontal line, 194–195, 224
  of parallel lines, 207–208, 224
  of perpendicular lines, 208
  rate and, 190–194
  of a vertical line, 195, 224
Slope-intercept form, 203–208, 224,
    461, 462
  equations in, 204–206, 224
  graphing and, 206–207, 224
Solutions
  of an inequality, 942
  of equations, 80–81, 163–164
  of systems of equations in three
    variables, 532
  of systems of equations in two vari-
    ables, 504
  writing quadratic equations from,
    717, 770–771
Solution sets, 93, 128
Solving equations, 65, 80–86, 940–942
  addition principle for, 81–82, 146
  multiplication principle for, 83–85, 146
  procedure for, 91
  selecting correct approach for, 85–86
  solutions and, 80–81
Solving formulas for a variable, 96–99
Solving inequalities, 127–134, 148,
    942–944
Solving rational equations, 976–979
  for a specific variable, 481
Special products, 330–331, 959
Square(s)
  of a binomial, 273–275, 297
  completing, 701–704, 770
  differences of. See Differences of
    squares
Square matrices, 553
Square roots, 628–630, 690
  principal, 628–629, 690
  principle of, 699–701
Standard form of an equation, 950
  of a circle, 690, 852
  linear, in three variables, 532
  quadratic, 698–699, 770
Study Skills
  abbreviations, 820
  active class participation, 576
  asking questions, 280, 503
  asking questions by e-mail, 304
  avoiding temptation, 138

chart of facts, 7
checking your answers, 418
continual review, 104
doing exercises, 22
doing multiple problems, 203
doing your homework promptly, 381
double-checking your copying, 548
exercise breaks, 374
finishing a chapter, 354
guessing what comes next, 585
helping classmates, 230, 533
highlighting text, 346
improving your study skills, 604
including steps in work, 60
information about final exam, 864
instructor mistakes, 649
introduction of new topics, 443
keeping ahead of instructor, 30
keeping math relevant, 541
keeping your focus, 341
knowing how to use your
  calculator, 790
learning from examples, 12
learning from mistakes, 510
learning multiple methods, 404
learning new vocabulary, 727
looking ahead, 266, 558
looking for connections between
  concepts, 238
maintaining your level of effort, 849
memorizing, 708
missing classes, 46
moving at your own speed, 805
multiple solutions, 411
music while studying, 704
note taking, 593
organizing homework, 56
pacing yourself, 98
planning next semester's courses, 664
preparing for study sessions, 734
preparing for tests, 472, 628, 680
quiz questions that stump you, 396
reading instructions, 319
reading subsections, 778
reading the text, 246
recording important dates, 749
resources, 85
reviewing for exams, 287
reviewing material, 657, 722
reviewing your final, 904
reviewing your mistakes, 213
setting personal study goals, 461
sharpening your skills, 715
sitting near the front of the classroom, 456
sleep, 130, 896
sorting problems by type, 827
study area, 643
study buddies, 182
study groups, 40
studying for finals, 890

studying together by phone, 681
success rate, 521
time management, 763, 916
topics that appear familiar, 636
turning pages, 802
using a second text, 271
using supporting work, 312
value of practice, 613
verbalizing your questions, 167, 553
visualizing the steps in problems
  solving, 387, 483
working to master exercises, 257
working with pencil, 154
writing out missing steps, 328
writing your own glossary, 175
Subsets, 986–987
Substitution, 2
Substitution method
  evaluating algebraic expressions
    using, 936
  solving nonlinear systems of equa-
    tions involving one equation
    using, 874–876
  solving nonlinear systems of equa-
    tions involving two equations
    using, 682–683
  for solving systems of equations,
    510–512, 515, 567
Subtraction. See also Differences
  of complex numbers, 682
  factoring out −1 to reverse, 308
  of fractions, 23–24
  of like radicals, 655–656
  of polynomials, 256–257, 296,
    957–958
  of polynomials in several variables,
    281, 297
  of rational expressions, 972–974
  of rational expressions when
    denominators are the same,
    388, 434
  of rational expressions with LCDs,
    395–398, 435
  of real numbers, 44–48, 46–47, 74, 934
Subtraction principle
  for equations, 940
  for inequalities, 943
Sum(s), 14. See also Addition
  of cubes, 966
  of cubes, factoring of, 336–338, 368
  of first n terms of an arithmetic
    series, 898–899, 926
  opposites of, 64–65, 75, 936
  partial (finite series), 892
  of terms of a sequence, 892
  of two functions, 471–472, 494
  of two terms, multiplication of,
    272–273, 297
Summation, index of, 892, 926
Summation notation, 892–893, 926
Supply and demand, 561–562, 570

INDEX OF APPLICATIONS

Formulas

$$m = \frac{y_2 - y_1}{x_2 - x_1}$$
Slope of a line

$$y = mx + b$$
Slope–intercept form of a linear equation

$$y - y_1 = m(x - x_1)$$
Point–slope form of a linear equation

$$(A + B)(A - B) = A^2 - B^2$$
Product of the sum and difference of the same two terms

$$\left.\begin{array}{l}(A + B)^2 = A^2 + 2AB + B^2, \\ (A - B)^2 = A^2 - 2AB + B^2\end{array}\right\}$$
Square of a binomial

$$d = rt$$
Formula for distance traveled

$$\frac{1}{a} \cdot t + \frac{1}{b} \cdot t = 1$$
Work principle

$$s = 16t^2$$
Free-fall distance

$$y = kx$$
Direct variation

$$y = \frac{k}{x}$$
Inverse variation

$$x = \frac{-b \pm \sqrt{b^2 - 4ac}}{2a}$$
Quadratic formula

Plane Geometry

Rectangle
Area: $A = lw$
Perimeter: $P = 2l + 2w$

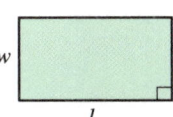

Square
Area: $A = s^2$
Perimeter: $P = 4s$

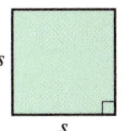

Triangle
Area: $A = \frac{1}{2}bh$

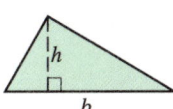

Triangle
Sum of Angle Measures:
$A + B + C = 180°$

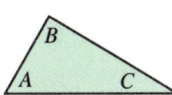

Right Triangle
Pythagorean Theorem (Equation):
$a^2 + b^2 = c^2$

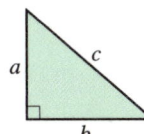

Parallelogram
Area: $A = bh$

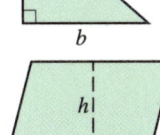

Trapezoid
Area: $A = \frac{1}{2}h(b_1 + b_2)$

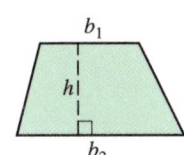

Circle
Area: $A = \pi r^2$
Circumference:
$C = \pi d = 2\pi r$
$\left(\frac{22}{7}\text{ and } 3.14 \text{ are different approximations for } \pi\right)$

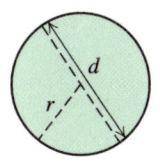

Solid Geometry

Rectangular Solid
Volume: $V = lwh$

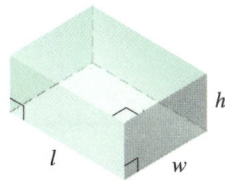

Cube
Volume: $V = s^3$

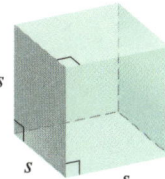

Right Circular Cylinder
Volume: $V = \pi r^2 h$
Total Surface Area:
$S = 2\pi rh + 2\pi r^2$

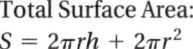
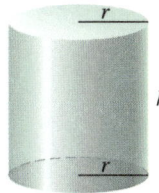

Right Circular Cone
Volume: $V = \frac{1}{3}\pi r^2 h$
Total Surface Area:
$S = \pi r^2 + \pi rs$
Slant Height:
$s = \sqrt{r^2 + h^2}$

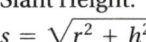
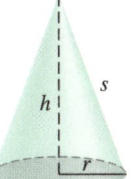

Sphere
Volume: $V = \frac{4}{3}\pi r^3$
Surface Area: $S = 4\pi r^2$

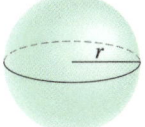

Selected Keys of the Scientific Calculator

This secondary function takes the square root of number displayed.

Squares number displayed.

Activates secondary functions printed above certain keys. Also denoted INV or 2nd.

Used when entering numbers in scientific notation. Also denoted EXP.

Finds reciprocal of number displayed.

Used to raise any base to a power. Also denoted y^x, a^x, or ∧.

Stores number displayed in memory. Also denoted MIN or M.

Recalls number stored in memory. Also denoted MR.

This secondary function raises 10 to any power entered.

Clears all preceding numbers and operations. Also used to turn calculator on.

Used as an approximation for pi.

Used to perform indicated operation.

Used to control order in which certain operations are performed.

Clears last number displayed but not preceding operations.

Used when entering decimal notation.

Used to change sign of number displayed.

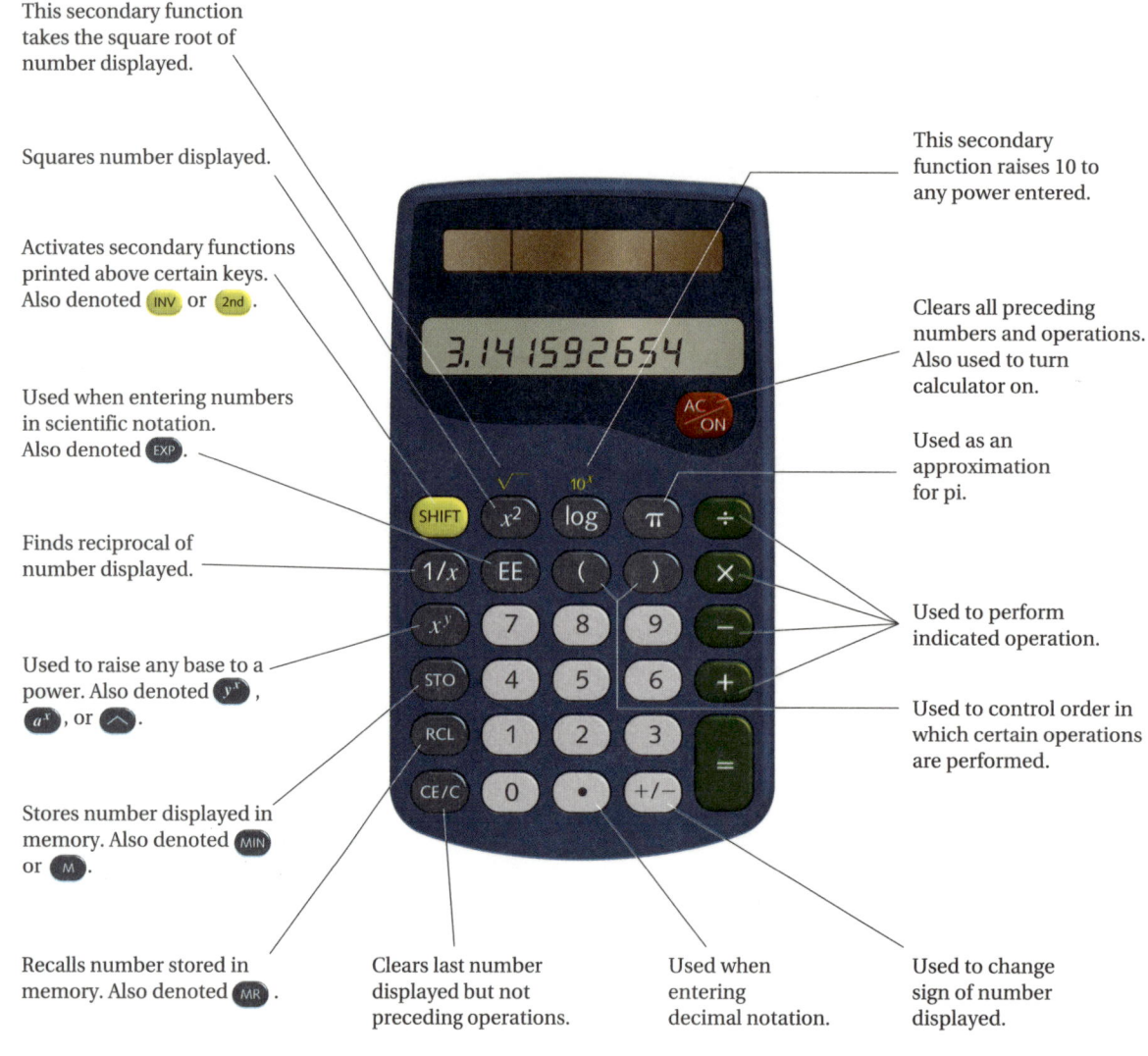

Selected Keys of the Graphing Calculator*

Controls the values that are used when creating a table.

Determines the portion of the curve(s) shown and the scale of the graph.

Used to enter the equation(s) that is to be graphed.

Controls whether graphs are drawn sequentially or simultaneously and if the window is split.

Activates the secondary functions printed above many keys in blue or green.

Used to delete previously entered characters.

Accesses preprogrammed applications and tutorials.

These keys are similar to those found on a scientific calculator.

Magnifies or reduces a portion of the curve being viewed and can "square" the graph to reduce distortion.

Used to determine certain important values associated with a graph.

Used to display the coordinates of points on a curve.

Used to display x- and y-values in a table.

Used to graph equations that were entered using the Y= key.

Used to move the cursor and adjust contrast.

Used to fit curves to data.

Used to access a previously named function or equation.

Used to raise a base to a power.

Used to write the variable, x.

Used as a negative sign.

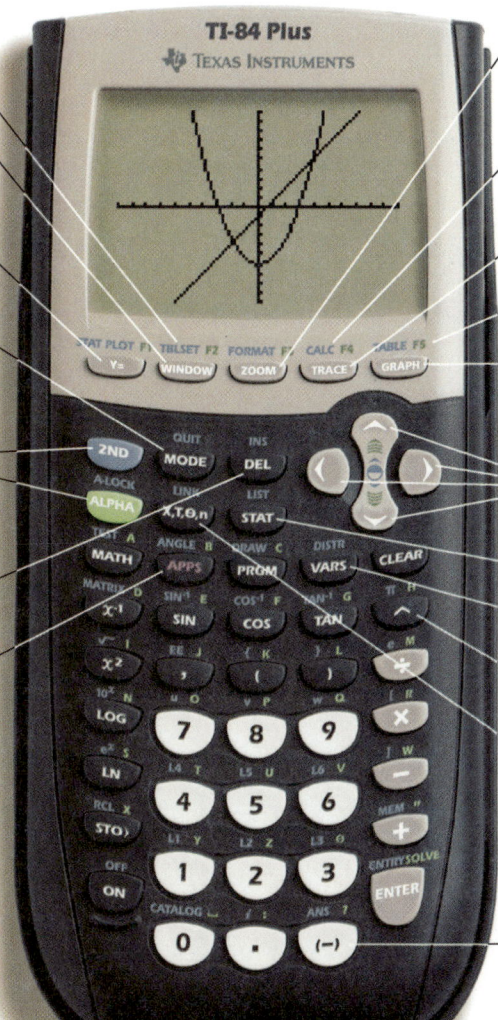

*Key functions and locations are the same for the TI-83 Plus.